Microvascular Research

VOLUME 1

Microvascular Research

VOLUME 1

Editor-in-Chief

David Shepro, Ph.D.
Boston University
Boston, MA, USA

ELSEVIER
ACADEMIC
PRESS

Amsterdam Boston Heidelberg London New York Oxford
Paris San Diego San Francisco Singapore Sydney Tokyo

Elsevier Academic Press
30 Corporate Drive, Suite 400, Burlington, MA 01803, USA
525 B Street, Suite 1900, San Diego, California 92101-4495, USA
84 Theobald's Road, London WC1X 8RR, UK

This book is printed on acid-free paper. ∞

Library of Congress Cataloging-in-Publication Data
Application Submitted.

British Library Cataloguing in Publication Data
A catalogue record for this book is available from the British Library

ISBN 13: 978-0-12-639510-5 (set)
ISBN 10: 0-12-639510-1 (set)
ISBN 10: 0-12-639511-X (Volume 1)
ISBN 13: 978-0-12-639511-2 (Volume 1)
ISBN 10: 0-12-639512-8 (Volume 2)
ISBN 13: 978-0-12-639512-9 (Volume 2)

For all information on all Elsevier Academic Press publications
visit our Web site at www.books.elsevier.com

Printed in the United States of America
05 06 07 08 09 10 9 8 7 6 5 4 3 2 1

Cover image based on an Aboriginal rock painting showing what may be the first depiction of the circulatory system. The original artwork dates from circa 6000 B.C.

Foreword I

Dr. David Shepro has compiled an outstanding book on Microvascular Research that summarizes the state-of-the-art advances in this exciting field and points to the future horizons in this interdisciplinary area of great important in health and disease.

The book starts with a Section on Basic Science that covers the molecular and cellular bases of endothelial cell structure and function; roles of various molecules and cells in regulating vascular development; *in vitro* models ranging from zebrafish to mammalian and human systems; physiological regulation of permeability, vascular tone and hemodynamics, and the process of transendothelial transport in relation to endothelial junctions and adhesion molecules. This Basic Science Section provides an excellent foundation and underpinning for the subsequent Sections.

The second Section presents the current knowledge on Microvascular Adaptation in Organs, including the heart, central nervous system and eye, gastrointestinal tract, hematopoietic system, kidney, liver, lung, lymphatics, pancreas, skeletal muscle, skin, and reproductive systems. The comprehensive presentations in this Section point out the uniqueness of each organ/tissue and at the same time provide insights into the similarity in microvascular adaptation among different organs and tissues. This organ-level treatment of microvascular physiology leverages on the molecular and cellular approaches in the Basic Science Section and set the stage for the Sections on Pathology and Therapy that follow.

The third Section on Pathology covers the microvascular basis of pathological changes seen in various disease states. Some of the clinical conditions are well-documented as stemming from microvascular abnormalities, e.g., thrombosis and hemostasis, diabetes, inflammation, wound healing, etc. With the expansion of knowledge on microvascular research, there is now increasing evidence that the microvasculature also plays a significant role in Alzheimer's disease, arthritis, transplant rejection, tumor growth and metastasis, etc. This book provides an in-depth analysis of the role of microvasculature in a variety of pathological conditions.

The fourth Section on Therapy provides new insights into the great potential of applying the fruits of microvascular research to clinical conditions such as fluid treatment of hemorrhagic shock, the betterment of delayed preconditioning, therapeutic use of statins, therapeutic angiogenesis, and the delivery of molecular and genetic therapeutic agents to vascular endothelium. This Section points to the translational potential of microvascular research to enhance human health and combat disease.

The fifth Section on New Research Modes and Procedures presents a stimulating and inspiring view of the vista of microvascular research. It covers important recent advances such as DNA microarray, tissue engineering, endothelial biomarkers, knockout/transgenic models, mathematical models, and proteomics, as well as stem cells and their therapeutic promises. This Section provides a great ending of the book with the exciting future of microvascular research.

In summary, this book provides a comprehensive and logical progression from basic science through organ adaptation to pathology and therapy, ending with new research approaches. It is the most definitive book ever written on Microvascular Research and it is a must for all researchers and students in this field.

Dr. Shepro's superb background and experience made him uniquely qualified to edit such a state-of-the-art treatise on Microvascular Research. He has made important contributions to microvascular research, publishing over 200 peer-reviewed articles and four books. He is a Founding Editor and the Editor-in-Chief of *MICROVASCULAR RESEARCH: An International Journal*. In this capacity, Dr. Shepro has a marvelous knowledge of the leading scientists working at the frontiers of microvascular research, and hence he was able to gather such a stellar group of authors contributing to this wonderful book. As a member of the microvascular research community, I wish to congratulate Dr. Shepro and the contributors to this book for their outstanding accomplishments and to thank them for disseminating their precious knowledge to everyone working on this subject. This book undoubtedly will have a major impact on the further advancement of this important field.

Shu Chien, M.D., Ph.D.
University Professor of Bioengineering and Medicine
Chair, Department of Bioengineering
University of California, San Diego
La Jolla, California, U.S.A.
March 2005

Foreword II

The present compilation of articles on the topic of Microvascular Research is both the most comprehensive and most current of the works ever written on this subject. Its breadth is dazzling and the assembled cast of contributors is a Who's Who of the field's best minds. There is virtually no area of scientific discovery relevant to the structure, function and pathology of the microvasculature that is not addressed in this text. Together Drs Shepro and D'Amore have logged more that 75 years of research in vascular biology and their knowledge of the field and of the top scientists in this field is unparalleled.

The volume starts with the basic biology of the cells that comprise the microvascular bed and moves from there to vascular development and permeability, *in vivo* models, peri-vascular transport and on to the role of the microvasculature in specific organ beds as well as in human disease. What is noticeable is the extremely large list of human diseases that are caused or influenced by alterations in the microvasculature. One begins to think that the list of diseases that are *not* influenced by microvascular alterations must be considerable shorter. Of course, a large amount of the research on microvessels over the past 35 years has been conducted on the topic of tumor angiogenesis and this topic is well covered in this volume.

It is remarkable to see how much progress has been made in microvascular research over the past few years and how much information is now known. In some cases, the text reveals new progress on topics that have been studied for decades. Among these are the relationship of the endothelium with the extracellular matrix, endothelial metabolism, endothelial heterogeneity, the distinction between developmental vessel formation and pathologic vessel formation, the traditional experimental models such as the chick chorioallantoic membrane and hamster cheek pouch, the maintenance of vascular tone and the regulation of microvascular hemodynamics, a long-term interest of Dr. Shepro.

At the same time, there are a variety of new areas of microvascular research that have burst onto the scene in just the last few years and these, too, are wonderfully represented in this volume. Among these are the role of ephrins as signaling mediators in vascular cells, the ability of zebrafish to function as model system for vascular biology, the effect of mechanical signals on microvascular structure and function, and the extraordinary new work on VEGF and VEGF antagonists that is leading to important new treatments for both cancer and for vascular diseases of the eye.

If there is a message that emanates from this book as a whole, it is the concept that blood vessels have a crucial yet differing role in the function of virtually every tissue and in the pathogenesis of a host of diseases. No longer is it sufficient to consider blood vessels as passive tubes that merely carry nutrients to tissues. Rather the microvessels are a vital part of every tissue that interact mechanically, biochemically and metabolically with the parenchymal cells of the tissues they support. The communication is elaborate and goes in both directions, from the vessels to tissue and vice versa. If you study the vasculature, you must look at it from the context of the tissue in which it occurs. If you study any specific tissue, you must understand its relationship with the microvessels that inhabit it. Reading this volume allows a new appreciation of these interactions.

It should be noted that Dr. D'Amore began her career as a graduate student in Dr. Shepro's laboratory and is now one of the leading investigators in the field of vascular biology. Perhaps they didn't realize at the time that they would one day put together a volume as definitive and thoughtful as this one. One only hopes that there is a graduate student in Dr. D'Amore's laboratory who can be counted on to update this volume in ten or twenty years.

Bruce Zetter, Ph.D.
Charles Nowiszewski Professor of Cancer Biology
Children's Hospital
Harvard Medical School
Boston, Massachusetts

Introduction

Rationale and Reader's Guide

I can safely state that no one truly comprehends the physiology and pathophysiology of the microvasculature. In a large measure, this less than desirable state is the result of an explosion of new scientific data that creates paradoxically [or maybe poetically] a concomitant explosion of ignorance. But challenges enhance curiosity. Regardless how far scientists' understanding may be from the natural laws that regulate microvessel functions, we share an incumbency to accumulate and disseminate, by whatever means, current new concepts based upon the latest tested investigations. Hence, the *Biology and Pathology Microvasculature Research* in printed and electronic formats. As Lord Florey once stated: "I do not think there is any sharp line between physiology and experimental pathology," hence the subtitle.

In 1967, at a Boston conference the *Microcirculation as Related to Shock*, (D. Shepro and G.P. Fulton, Eds. Acad. Press, 1968) Robert Ebert, then dean of Harvard Medical School, summarized the proceedings in his keynote address that: *"in shock all physiological parameters are disturbed but if there is a common denominator it would be the microcirculation."*

Endothelium is metabolically one of the most active tissues in the body . . . probably second only to nervous tissue. Microvessels bridge the subcellular to the cellular to the organ to the system. **The breaching of the microvascular barrier is part and parcel of every disease. An inflammatory response, initially localized by microvessels, is frequently a persistent parameter of every disease.** Hence, whatever the therapeutic intervention, continuous attenuation of this defense mechanism, when awry, is required. As a closing but not a final argument, one important current concept is that there are specific *microvascular diseases* that to a large extent are independent of events in other parts of the cardiovascular system; e.g. angina with patent coronary arteries, cardiac slow flow states, long flight induced edema. Some scientists have speculated that the first stage of all forms of dementia is a breakdown of the blood brain barrier.

In the past, writers and publishers at some point had to agree on an endpoint, regardless of the manuscript's blemishes, and submit what was on hand for publication. As the "*ink dried*," the accumulated responses from critics and readers and new information would form the basis to kick-off a new edition, more frequently than not, years after the initial publication. Informatics technology, in all research endeavors, adds a new dimension to publishing, namely instant editing when the need arises, whether it be daily, weekly, quarterly, biannually. Because the editors and publishers are mindful of the fact that the currency of each scientific publication devaluates rapidly, the electronic version of *Microvascular Research* will be updated frequently. If "break throughs" should occur, an epilogue can be added immediately or reverse, obsolete "facts" and speculations that do not dovetail with new documented data can be immediately deleted.

The readers will quickly note that each presentation is limited to about 10–15 printed pages, unusual especially for review articles, and with a limited annotated (author's) bibliography. This editorial decision is based upon the realization that investigators are swamped with data and with the new tools to share information, speed has become an essential parameter. Global updating is so easily and readily available that, *less can be an advantage*. Given the space constraints, our contributors are challenged to provide the readers with their best objective and subjective take on the state of their sub-discipline. The redeeming fall-out of condensation is that lengthy introductions, redundant citations, speculations beyond the call, are eliminated. The creative sifting and contracting prior to submitting an article serves as a *needle's eye*.

A Perspective on Historical Recordings

In keeping with the editorial consensus that brevity can be a virtue and in light of the speed of scientific discoveries [see Rationale], a classical account that chronologically recapitulates microcirculatory milestones is not included in the volume. Some readers will regard this decision as an intellectual miscalculation; that without a record of discoveries there is blinding ignorance. Yet, it is an accepted fact of academia that recorders of events cannot prevent their imaginations from slipping into their "factual accounts." The noted historian Pieter Geyl states "history is an argument without end". For example: W*ho was the first to posit the existence of capillaries and their connecting role: Erisistratus? Galen? Malpighi? Harvey? Leeuwenhoek? Boerhaave? Henle?* The winner of the paternity title will go to the histiographer of your choice. Fortunately, a wealth of microcirculatory data are readily available in paper and electronic format, which lessens the need for such a chapter in the Encyclopedia. However, to accent the editors' recognition and appreciation of history, an illustration may provide the wherewithal to unravel one scientific puzzle in the history of microvascular research.

Why did it take over a century for scientists to appreciate the heterogeneity and high metabolic activity of the ubiquitous endothelial cell [not to mention the pericyte]? Elie Metchnikoff (1883), at the Pasteur Institute, in describing his new theory on inflammation that laid the foundation for the modern concept of this defense mechanism, stated:

"Movement [contractility] of endothelial cells plays an important role in the formation of stomata during inflammation as Klebs thought and as I imagined and stated in my first paper on inflammation."

Another Nobel laureate, August Krogh, described and illustrated endothelial cell motility in "The Anatomy and Physiology of Capillaries" [1922], the publication of his Sterling Lectures given at Yale University. Although both works are highly quoted, the scientific concepts that they embrace were virtually ignored in endothelial research until a scant 30 years ago. In 1966, the laureate Lord Florey, in his address entitled, "The Endothelial Cell", honoring the noted pathologist Sir Roy Cameron, spoke exclusively on the morphology of these cells. Presciently, he concluded that his was an interim report and "*I should expect to see in the next ten years a rich harvest of new [functional] knowledge. . . .*"

Again the question—why the achievements of three laureates, whose collective contributions significantly codified microvascular biology, were largely ignored? One explanation certainly can be argued on the importance of technology over imagination; that in the life sciences technology has always played a dominant role (review the history of the microscope) in contrast to Newton's and Einstein's science, where imagination was dominant. In other words, the technology for accurately profiling mural cells was unavailable. But in this scientific teaser I believe the answer lies in the incandescence of another stellar scientist, Earnest H. Starling, whose hypothesis on fluid transport still remains a major physiological principal. But for years this monumental work may have put blinders on investigators regarding the true nature of the microvascular wall. The Starling quotation in part supports this opinion:

"We have no sufficient evidence to conclude that endothelial cells of capillary walls take any part in the formation of lymph" [1896]

The elegance of Starling's hypothesis and research on filtration was so convincing that for decades microvascular physiologists devoted their efforts, for the most part, to fine tuning his formula. I hasten to add that this opinion does not diminish Starling's achievements and legacy an iota. Newton's unique contributions were eventually and correctly challenged and even the theories of Einstein, the genius of the 20th century, are now under scrutiny.

I would add one very personal speculation (a chronicler's prerogative) that might explain why Starling ignored other scientific data that could have affected his unilateral view of the microvessel wall as a passive barrier. Notwithstanding their correspondence on science and other matters, Starling did not appear to appreciate Krogh's experimental data. My perception is that he viewed Krogh as a junior level scientist and I "*imagine*" he was a tad jealous of his Danish colleague's talents.

In summary, I respectfully suggest to those who are interested in the historical analysis of microcirculation to select the format that best matches their interest. The chronological approach is the most common and of course should include examples from ancient oriental cultures as well as those from western civilization. Another choice would be to select the superstars of different eras. For example, Andreas Vesalius's "**De Humani Corporis Fabrica**" could be the starting point of modern science since this publication marked the deathknell for skewing scientific data to comply with authority (Aristotle; Galen). Vesalius's work was based on careful observations. Would there have been a Harvey [17 century] without a Versalius [16 century]? Advances in technology, such as the history of the microscope, can also be an artful approach to chronicle advances in microcirculatory knowledge. Whatever your choice, the endpoint will be similar.

References

Florey, H.W. 1966. The endothelial cell. *Brit. Med. J.* 27: 487–490

Krogh, A. Anatomy and Physiology of Capillaries. New Haven: Yale University Press. 1922. pp. 47–69

Metchnikoff, E. Lectures on the Comparative Pathology of Inflammation. New York: Diver Publications, Inc., 1968. pp. 137–156.

Starling. E.H. 1896. On the absorption of fluids from connective tissue spaces. *J. Physiol. Lond.* 19: 312–326

Acknowledgements

I thank all of the contributors, the editors who developed and shaped the contents of these volumes, our Elsevier Editor, Noelle Gracy, who inaugurated the project, and provided often needed encouragement, notwithstanding a transfer to Holland and a *maternity leave*. And to Karen Dempsey, at Elsevier, San Diego, who managed to organize our input, in bits and pieces, into a professional, valuable publication—always positive in her responses and never displays an "ounce of static." To my assistant, Christina Pitcher-Cozzone, for her "can do" handling of correspondence and almost daily revisions of contributors, titles and contents. Lastly, my gratitude to my wife, Marilyn, for her encouragement whenever I whined about the problems of a head editor and who did not complain about lengthy disappearances to my computer during the final stages of production.

Editorial Board Members

Editor-in-Chief

David Shepro, Ph.D.

Dr. Shepro is Professor of Surgery and Professor of Biology at Boston University. His research interests include cellular signaling and gene regulation and was the first to culture successfully large numbers of aortic endothelial cells that provided the wherewithal to research their metabolic activity and responses to inflammatory agonists.

David Shepro is a Fellow of the American Association for the Advancement of Science and a Fellow of the American Physiological Society. He has received the national and international awards from the American Microcirculatory Society, the Landis Award and the Zweifach Gold Medal Award, respectively, and the Taylor Smith Gold Medal Award from the New England Ophthalmological Society, the latter for his research on the biology/pathology of the retinal microcirculation. He was Burroughs Welcome Visiting Professor in Basic Medical Sciences and Distinguished Visiting Professor at the University of Cagliari School of Medicine, Sardinia.

Dr. Shepro is the Chair of the Literature Selection Technical Review Committee for the National Library of Medicine and Chairman of the Scientific User Committee for the Marine Biological Laboratory—Woods Hole Oceanographic Institute Library. He is the founding editor of the journal *Microvascular Research*.

Associate Editor

Patricia A. D'Amore, Ph.D.

Dr. Patricia A. D'Amore is a Senior Scientist and the Associate Director of Research at The Schepens Eye Research Institute. She is Professor of Ophthalmology and Pathology at Harvard Medical School, and Chair of the Angiogenesis, Invasion, and Metastasis Program in Development at the Dana Farber Harvard Cancer Center. She is the author of 95 publications which have appeared in journals such as *Nature Medicine, Journal of Clinical Investigation, Development*, and *Journal of Cell Biology*. She is Co-Editor-in-Chief of the journal *Microvascular Research*.

Editorial Board

Dame Carol M. Black, Ph.D.

Professor Dame Black is currently President of the Royal College of Physicians (UK) and Professor of Rheumatology at the Centre of Rheumatology, the Royal Free and University College Medical School. Her department at the Royal Free Hospital is a major centre for scleroderma in Europe and one of the largest in the world, with an extensive programme of clinical and basic research.

Joe G. N. Garcia, M.D.

Dr. Garcia is the Lowell T. Coggeshall Professor and Chair, Department of Medicine at the University of Chicago. His research program studies endothelial cytoskeletal regulation of the pulmonary microvascular endothelial barrier with a focus on the signaling events regulating barrier function of the endothelial cell monolayers. This work has resulted in over 200 peer-reviewed publications and numerous book chapters. He is Co-Editor-in-Chief of *Microvascular Research*, and Associate Editor of the *American Journal of Respiratory Cell & Molecular Biology*, and the *Journal of Stem Cell Research*.

D. Neil Granger, Ph.D.

Dr D. Neil Granger is Boyd Professor and Head of the Department of Molecular & Cellular Physiology at LSU Health Sciences Center. He has authored over 500 publications dealing with the role of the microcirculation in health and disease, and has served as Editor-in-Chief of Microcirculation, President of the Microcirculatory Society, and President of the American Physiological Society. His areas of research interest include ischemia-reperfusion injury, inflammation, and oxidative stress.

Christian Haudenschild, M.D.

Christian Haudenschild directs the Experimental Pathology Department at the American Red Cross Jerome H. Holland Laboratory for the Biomedical Sciences. His laboratory's main area of research is angiogenesis.

Herbert B. Hechtman, M.D.

Dr. Hechtman is Professor of Surgery at Harvard Medical School and Brigham and Womens Hospital. His laboratory's areas of research include the consequences of ischemia and inflammatory reactions in the lungs.

Rakesh K. Jain, Ph.D.

Rakesh K. Jain is Andrew Werk Cook Professor of Tumor Biology, Department of Radiation Oncology, Harvard Medical School and Director of the Edwin L. Steele Laboratory of Tumor Biology at Massachusetts General Hospital. Dr. Jain is regarded as a leader in the areas of vascular and interstitial biology of tumors and is known for integrating tumor biology with bioengineering. He is a member of the Institute of Medicine and the National Academy of Engineering of the U. S. National Academies.

Joseph A. Madri, Ph.D., M.D.

Joseph A. Madri received his BS and MS in Biology from St. John's University, NY and his Ph.D. in Chemistry and an M.D. from Indiana University. He is a Professor of Pathology at Yale University with appointments in the Medical School and Graduate School of Arts & Sciences. His areas of interest include vascular and cardiac development, angiogenesis, immune cell-endothelial cell interactions, and extracellular matrix biology.

Contributors

David Abraham
Centre for Rheumatology, Royal Free Hospital, London NW3 2QG, UNITED KINGDOM

Anthony P. Adamis
Research Center, Eyetech Pharmaceuticals, Woburn, MA 01801

Alexei V. Agapitov
East Amherst, NY 14051

William Cameron Aird
Department of Medicine, Division of Molecular/Vascular Biology, Beth Israel Deaconess Medical Center, Harvard Medical School, Boston, MA 02215

J. Steven Alexander
Department of Molecular and Cellular Physiology, Louisiana State University Health Sciences Center, Shreveport, LA 71130-3932

Abu-Bakr Al-Mehdi
Department of Pharmacology, College of Medicine, University of South Alabama, Mobile, AL 36688

Arne Andersson
Department of Medical Cell Biology, University of Uppsala, Uppsala, SE-75123, Sweden

Nick Anim-Nyame
Department of Maternal and Fetal Medicine, Faculty of Medicine, Imperial College of Science, Technology and Medicine at Chelsea and Westminster Hospital, London SW10 9NH, UNITED KINGDOM

Peter B. Anning
Department of Medical Biochemistry and Immunology, University of Wales College of Medicine, Cardiff, Wales CF14 4XN, UNITED KINGDOM

John D. Ash
Departments of Ophthalmology and Cell Biology, University of Oklahoma Health Sciences Center, Oklahoma City, OK 73104

James W. Baish
Department of Mechanical Engineering, Bucknell University, Lewisburg, PA 17837

Ann L. Baldwin
Department of Physiology, University of Arizona College of Medicine, Tucson, AZ 85724-5051

William A. Banks
Geriatrics Research, Educational, and Clinical Center, Veterans Affairs Medical Center, St Louis, MO 63106

Margery A. Barrand
Department of Pharmacology, University of Cambridge, Cambridge CB2 1PD, UNITED KINGDOM

Matthias Barton
Medical Policlinic, Departement of Internal Medicine, University Hospital Zurich, Zürich, CH-8091, SWITZERLAND

David O. Bates
Microvascular Research Laboratories, Department of Physiology, Preclinical Veterinary School, University of Bristol, Clifton, Bristol BS2 8EJ, UNITED KINGDOM

John F. Beltrame
Cardiology Unit, The Queen Elizabeth Hospital, Woodville South, SA-5011, SOUTH AUSTRALIA

Monica M. Bertagnolli
Division of Surgical Oncology, Department of Surgery, Brigham & Women's Hospital, Boston, MA 02115

Christer Betsholtz
Department of Medical Biochemistry, Göteborg University, Goteburg, SE-40330, SWEDEN

Jahar Bhattacharya
Lung Biology Laboratory, College of Physicians and Surgeons, St. Luke's/Roosevelt Hospital, Columbia University, New York, NY 10019

Imran Bhutto
Ophthalmology Department, Johns Hopkins Hospital, Baltimore, MD 21287-9115

David G. Binion
Division of Gastroenterology and Hepatology, Medical College of Wisconsin, Milwaukee, WI 53226

Konstantin Birukov
Department of Medicine, Section of Pulmonary Critical Medicine, University of Chicago, Chicago IL 60611

Anna Birukova
Department of Medicine, Section of Pulmonary Critical Medicine, University of Chicago, Chicago IL 60611

Joyce E. Bischoff
Vascular Biology Program, Department of Surgery, Children's Hospital, Harvard Medical School, Boston, MA 02115

Carol M. Black
Centre for Rheumatology, Royal Free Hospital, London NW3 2QG, UNITED KINGDOM

Bruce S. Bochner
Department of Medicine—Allergy and Clinical Immunology, Johns Hopkins Unversity School of Medicine, Baltimore, MD 21224

Claudine S. Bonder
Departments of Physiology and Biophysics, University of Calgary Health Sciences Centre, Calgary, AB T2N 4N1, CANADA

Nick Brindle
Cardiovascular Research Institute and Department of Surgery, University of Leicester, Leicester LE2 7LX, UNITED KINGDOM

Amy Brock
Vascular Biology Program, Children's Hospital, Harvard Medical School, Boston, MA 02115

Peter H. Burri
Institute of Anatomy, Developmental Biology, University of Berne, Bern 9, CH 3000, SWITZERLAND

Pedro Cabrales
Microhemodynamics Laboratory, Department of Bioengineering, University of California—San Diego, La Jolla, CA 92093-0412

Edward C. Carlson
Department of Anatomy & Cell Biology, University of North Dakota, Grand Forks, ND 58203

Per-Ola Carlsson
Department of Medical Cell Biology, University of Uppsala, Uppsala, SE-75123, Sweden

Gediminas Cepinskas
Vascular Biology/Inflammation Program, Lawson Health Research Institute, London, Ontario N6A 465, CANADA

Subrata Chakrabarti
Department of Pathology, Pathologist, London Health Sc. Ctr, The University of Western Ontario, London, Ontario N6A 5A5, CANADA

Rodney K. Chan
Division of Surgery, Brigham and Women's Hospital, Boston, MA 02115

Pierre Charbord
Laboratoire d'Hématopoïèse, Bâtiment Bretonneau, Faculté de Médecine, Université François-Rabelais, Tours Cedex, F-37032, FRANCE

Shi-Yuan Cheng
Research Pavilion at the Hillman Cancer Center, University of Pittsburgh Cancer Institute, Pittsburgh, PA 15213-1863

Francesco Chiarelli
Department of Medicine, Section of Pediatrics, University of Chieti, Chieti Scalo, I-66100, ITALY

Brian W. Christman
Department of Medicine, Division of Allergy, Pulmonary and Critical Care Medicine, Vanderbilt University, Nashville, TN 37232-2650

Martin K. Church
Allergy and Inflammation Sciences, University of Southampton School of Medicine, Southampton SO16 6YD, UNITED KINGDOM

Geraldine Clough
Infection Inflammation and Repair, University of Southampton School of Medicine, Southampton SO16 6YD, UNITED KINGDOM

Cecile Colpaert
Department of Pathology, University Hospital Antwerp, Edegem-Antwerp, B-2650, BELGIUM

Hiran C. Cooray
Department of Pharmacology, University of Cambridge, Cambridge CB2 1PD, UNITED KINGDOM

Patricia A. D'Amore
Schepens Eye Research Institute, Boston, MA 02114

Sergei M. Danilov
Department of Anesthesiology, University of Illinois in Chicago, Chicago, IL 60612

Michael R. Dashwood
Department of Clinical Biochemistry, Royal Free and University College Medical School, London NW3 2QG, UNITED KINGDOM

Marc Dellian
Department of Otorhinolaryngology, Head and Neck Surgery, Klinikum Grosshadern, University of Munich, München, D-81377, GERMANY

Robert H. Demling
Brigham and Women's Hospital, Harvard Medical School, Boston, MA 02115

Michael Detmar
Cutaneous Biology Research Center, Massachusetts General Hospital, Charlestown, MA 02129

Mario Di Gioacchino
Department of Medicine and Science of Aging, "G. D'Annunzio" University School of Medicine, Chieti, I-66013, ITALY

Gabriele Dini
Department of Neurosurgery, The Cleveland Clinic Foundation, Cleveland, OH 44195

Valentin Djonov
Institute of Anatomy, University of Berne, Bern, CH 3012, SWITZERLAND

Jorge Domenech
Laboratoire d'Hématologie, Hôpital Bretonneau, CHU de Tours, Tours Cedex, F-37044, FRANCE

Steven M. Dudek
Department of Medicine, Section of Pulmonary Critical Medicine, University of Chicago, Chicago IL 60611

Michael Dunn
Department of Neuroscience, Institute of Psychiatry, King's College London, London SE5 8AF, UNITED KINGDOM

Danuta Dus
Ludwik Hirszfeld Institute of Immunology and Experimental Therapy, Polish Academy of Sciences, Centre of Excellence "IMMUNE", Wroclaw, PL-53-114

Ann M. Dvorak
Department of Pathology, Beth Israel Deasoness Medical Center, Boston, MA 02215

Harold F. Dvorak
Department of Pathology, Beth Israel Deaconess Medical Center, Harvard Medical School, Boston, MA 02215-5491

Britta Engelhardt
Max-Planck Institute for Vascular Biology, ZMBE-Institute for Cell Biology, Muenster, D-48149, GERMANY

Denis English
Experimental Cell Research Laboratory and Allied Health Sciences, Methodist Research Institute, Indiana University School of Medicine, Indianapolis, IN 46202

Frank M. Faraci
Department of Internal Medicine, Division of Cardiovascular Diseases, University of Iowa College of Medicine, Iowa City, IA 52242-1081

Susan A. Farr
Department of Internal Medicine, Division of Geriatrics, Saint Louis University School of Medicine, St Louis, MO 63104

Dian Feng
Department of Pathology, Beth Israel Deaconess Medical Center, Boston, MA 02215

Shuju Feng
Department of Medicine, Section of Thrombosis Research, Baylor College of Medicine, Michael E. DeBakey VA Medical Center, Houston, TX 77030

Cecilia A. Fernández
Laboratory of Surgical Research, Vascular Biology Program, Children's Hospital, Harvard Medical School, Boston, MA 02115

Iwon Fijolkowska
Department of Cardiopulmonary Pathology, The Johns Hopkins University, Baltimore, MD 21205

Howard A. Fine
Neuro-oncology Branch, National Cancer Institute, National Institutes of Health, Bethesda, MD 20814

James H. Finigan
Department of Medicine, Pulmonary, Johns Hopkins University School of Medicine, Baltimore, MD 21224

Eric B. Finkelstein
Schepens Eye Research Institute, Boston, MA 02114

Susan J. Fisher
Department of Stomatology, University of California, San Francisco, San Francisco, CA 94143-0512

Ingrid Fleming
Institut für Kardiovaskuläre Physiologie, Klinikum der J.W. Goethe-Universität, Frankfurt am Main, D-60596, GERMANY

Michael M. Flessner
Department of Medicine; Division of Nephrology, University of Mississippi School of Medicine, Jackson, MS 39216

Alpha A. Fowler
Division of Pulmonary and Critical Care Medicine, Department of Internal Medicine, Virginia Commonwealth University, Richmond, VA 23298

Dai Fukumura
Department of Radiation Oncology, Massachusetts General Hospital, Harvard Medical School, Boston, MA 02114

John Gamble
Department of Maternal and Fetal Medicine, Faculty of Medicine, Imperial College of Science, Technology and Medicine at Chelsea and Westminster Hospital, London SW10 9NH, UNITED KINGDOM

Joe G.N. Garcia
Department of Medicine, Section of Pulmonary Critical Medicine, University of Chicago, Chicago IL 60611

Guillermo Garcia-Cardena
Center for Excellence in Vascular Biology, Brigham and Women's Hospital, Boston, MA 02115

Giampietro Gasparini
Medical Oncology, San Filippo Neri Hospital, Rome, I-00135, ITALY

Ursula Gehling
Abteilung für Hämatologie und Onkologie, University Hospital Hamburg-Eppendorf, Hamburg, D-20246, GERMANY

Jeffrey M. Gelfand
Pathology Dept., Vascular Research, Brigham and Women's Hospital, Boston, MA 02115

Holger Gerhardt
Department of Medical Biochemistry, Göteborg University, Goteburg, SE-40330, SWEDEN

Berhane Ghebrehiwet
Department of Medicine, State University of New York at Stony Brook, Stony Brook, NY 11794-8161

Derek W. Gilroy
Department of Experimental Pathology, William Harvey Research Institute, Queen Mary University of London, London EC1M 6BQ, UNITED KINGDOM

Reda Girgis
Division of Pulmonary and Critical Care Medicine, The Johns Hopkins Medical Institutions, The Johns Hopkins Hospital, Baltimore, MD 21205

John Glod
Cancer Institute of New Jersey, New Brunswick, NJ 08903

Nobuhito Goda
Department of Biochemistry and Integrative Medical Biology, School of Medicine, Keio University, Shinjuku-ku, Tokyo, JAPAN

Andrew P. Golden
Department of Biomedical Engineering, Boston University, Boston, MA 02215

Norberto C. Gonzalez
Department of Molecular and Integrative Physiology, University of Kansas Medical Center, Kansas City, KS 66160-7401

Andrea Moreira Gonzalez
Department of Plastic and Reconstructive Surgery, The Cleveland Clinic Foundation, Cleveland, OH 44195

D. Neil Granger
Department of Molecular and Cellular Physiology, Louisiana State University Health Sciences Center, Shreveport, LA 71130-3932

Anna T. Grazul-Bilska
Department of Animal and Range Science, North Dakota State University, Fargo, ND 58105-5727

Rita Greco
Department of Medicine, Section of Pediatrics, University of Chieti, Chieti Scalo, I-66013, ITALY

Dmitry Grigoryev
Division of Pulmonary and Critical Care Medicine, Johns Hopkins University School of Medicine, Baltimore, MD 21224

Matthew B. Grisham
Department of Molecular & Cellular Physiology, Louisiana State University Health Sciences Center, Shreveport, LA 71130-3932

Danièle Gulino-DeBrac
Laboratoire d'Ingénierie des Macromolécules, Jean-Pierre Ebel Institute of Structural Biology (CEA/CNRS, UJF), Grenoble Cedex 1, F-38027, FRANCE

J. Silvio Gutkind
Oral and Pharyngeal Cancer Branch, National Institute of Dental and Craniofacial Research, National Institutes of Health, Bethesda, MD 20892-4340

Mark Haas
Department of Pathology, Johns Hopkins University School of Medicine, Baltimore, MD 21287

Kerri L. Hallene
Department of Neurosurgery, The Cleveland Clinic Foundation, Cleveland, OH 44195

Hans-Peter Hammes
Abt. für Nephrologie/Endokrinologie, 5. Medizinische Klinik, Mannheim der Universität Heidelberg, Mannheim, D-68167, GERMANY

Kevin A. Harvey
Experimental Cell Research Laboratory and Allied Health Sciences, Methodist Research Institute, Indiana University School of Medicine, Indianapolis, IN 46202

Paul M. Hassoun
Department of Medicine—Pulmonary, Johns Hopkins University School of Medicine, Baltimore, MD 21224

Ossama A. Hatoum
Division of Gastroenterology and Hepatology, Medical College of Wisconsin, Milwaukee, WI 53226

William G. Haynes
Department of Internal Medicine, University of Iowa, Iowa City, IA 52242

Herbert B. Hechtman
Division of Surgery, Brigham and Women's Hospital, Boston, MA 02115

Donald D. Heistad
Department of Internal Medicine, Division of Cardiovascular Diseases, University of Iowa College of Medicine, Iowa City, IA 52242-1081

Ira M. Herman
Program in Cellular and Molecular Physiology, Tufts University School of Medicine, Boston, MA 02111

Baskin Hermant
Laboratoire d'Ingénierie des Macromolécules, Jean-Pierre Ebel Institute of Structural Biology (CEA/CNRS, UJF), Grenoble Cedex 1, F-38027, FRANCE

Dan D. Hershko
Department of Surgery A, Rambam Medical Center, Haifa, IL-31096, ISRAEL

Robert L. Hester
Department of Physiology and Biophysics, University of Mississippi Medical Center, Jackson, MS 39216-4505

Stéphanie Heyraud
Laboratoire d'Ingénierie des Macromolécules, Jean-Pierre Ebel Institute of Structural Biology (CEA/CNRS, UJF), Grenoble Cedex 1, F-38027, FRANCE

Caryl E. Hill
Division of Neuroscience, John Curtin School of Medical Research, Australian National University, Canberra, ACT 0200, AUSTRALIA

May Ho
Faculty of Medicine, Department of Microbiology and Infectious Diseases, University of Calgary, Calgary, Alberta T2N 4N1, CANADA

Virginia Huxley
Department of Medical Pharmacology and Physiology, University of Missouri—Columbia, Columbia, MO 65211

Richard O. Hynes
Center for Cancer Research, Massachusetts Institute of Technology, Cambridge, MA 02139

Shahrul Izham Ibrahim
Division of Surgery, Brigham and Women's Hospital, Boston, MA 02115

Donald Ingber
Vascular Biology Program, Children's Hospital, Harvard Medical School, Boston, MA 02115

Marcos Intaglietta
Microhemodynamics Laboratory, Department of Bioengineering, University of California—San Diego, La Jolla, CA 92093-0412

Sofia Ioannidou
Eyetech Research Center, Eyetech Pharmaceuticals, Inc., Woburn, MA 01801

Renato V. Iozzo
Department of Pathology, Anatomy, and Cell Biology, Kimmel Cancer Center, Thomas Jefferson University, Philadelphia, PA 19107

Shingo Itou
Laboratory of Food Science, Department of Food Science and Technology College of Bioresource Sciences, The University of Shizuoka, Shizuoka-shi, JAPAN

Yuichi Izumi
Department of Periodontology, Kagoshima University Graduate School of Medical and Dental Sciences, Kagoshima 890-8544, JAPAN

William F. Jackson
Department of Biological Sciences, Western Michigan University, Kalamazoo, MI 49008-5410

Jeffrey R. Jacobson
Department of Medicine, Section of Pulmonary Critical Medicine, University of Chicago, Chicago IL 60611

Sirpa Jalkanen
MediCity Research Laboratory, University of Turku, Turku, FIN-20520, FINLAND

Damir Janigro
Department of Neurological Surgery, Cerebrovascular Research Center, The Cleveland Clinic Foundation, Cleveland, OH 44195

Leif Jansson
Department of Medical Cell Biology, University of Uppsala, Uppsala, SE-75123, Sweden

Michael J. Jarzynka
Research Pavilion at the Hillman Cancer Center, University of Pittsburgh Cancer Institute, Pittsburgh, PA 15213-1863

Jeffrey L. Jasperse
Department of Sports Medicine, Natural Science Division, Pepperdine University, Malibu, CA 90263

Jamie Y. Jeremy
Division of Cardiac, Anaesthetic and Radiological Sciences, University of Bristol, Bristol BS2 8HW, UNITED KINGDOM

Randall S. Johnson
Division of Biological Sciences, Molecular Biology, University of California, San Diego, La Jolla, CA 92093-0366

Michael Jünger
Klinik und Poliklinik fur Hautkrankheiten, der Ernst-Moritz-Arndt-Universitat Greifswald, Griefswald, D-17487, GERMANY

Örjan Källskog
Department of Medical Cell Biology, University of Uppsala, Uppsala, SE-75123, Sweden

Makio Kawakam
Department of Pathology, Jikei University School of Medicine, Minato-ku, Tokyo 105-8471, JAPAN

Usamah S. Kayyali
Pulmonary and Critical Care Division, Tupper Research Institute, Tufts University School of Medicine, New England Medical Center, Boston, MA 02111

Zia Ali Khan
Department of Pathology, Pathologist, London Health Sc. Ctr, The University of Western Ontario, London, Ontario N6A 5A5, CANADA

Kameha R. Kidd
Laboratory of Molecular Genetics, National Institute of Child Health and Human Development, National Institutes of Health, Bethesda, MD 20892

Claudine Kieda
Centre de Biophysique Moléculaire-Glycobiologie, Centre National de la Recherche Scientifique, Orleans, F-45071

Bela Kis
Department of Physiology/Pharmacology, Wake Forest University School of Medicine, Winston-Salem, NC 27157

Michael Klagsbrun
Departments of Surgical Research and Pathology, Vascular Biology Program, Children's Hospital, Harvard Medical School, Boston, MA 02115-5737

Moritz A. Konerding
Anatomisches Institut * Makroskopischer Bereich, Johannes Gutenberg-Universität Mainz, Mainz, D-55099, GERMANY

Robert D. Koos
Department of Physiology, Unviersity of Maryland School of Medicine, Baltimore, MD 21201-1559

Ronald J. Korthuis
Department of Medical Pharmacology and Physiology, University of Missouri-Columbia School of Medicine, Columbia, MO 65212

Michael H. Koval
Department of Physiology and Institute for Environmental Medicine, University of Pennsylvania School of Medicine, Philadelphia, PA 19104

Michael M. Krausz
Department of Surgery A, Rambam Medical Center, Haifa, IL-31096, ISRAEL

Michael M. Krausz
Department of Surgery A, Rambam Medical Center, Haifa, IL-31096, ISRAEL

Agnieszka Krawczenko
Ludwik Hirszfeld Institute of Immunology and Experimental Therapy, Polish Academy of Sciences, Centre of Excellence "IMMUNE", Wroclaw, PL-53-114

Michael H. Kroll
Department of Medicine, Section of Thrombosis Research, Baylor College of Medicine, Michael E. DeBakey VA Medical Center, Houston, TX 77030

Paul Kubes
Departments of Physiology and Biophysics, University of Calgary Health Sciences Centre, Calgary, AB T2N 4N1, CANADA

Rainer Kunstfeld
Cutaneous Biology Research Center, Massachusetts General Hospital, Charlestown, MA 02129

Peter R. Kvietys
Vascular Biology/Inflammation Program, Lawson Health Research Institute, London, Ontario N6A 465, CANADA

Evanthia Lalla
Division of Periodontics, Columbia University School of Dental and Oral Surgery, New York, NY 10032

Ira B. Lamster
Division of Periodontics, Columbia University School of Dental and Oral Surgery, New York, NY 10032

Robert Langley
Department of Cancer Biology, The University of Texas M. D. Anderson Cancer Center, Houston, TX 77030

Stephen Laroux
Department of Molecular & Cellular Physiology, Louisiana State University Health Sciences Center, Shreveport, LA 71130-3932

Maurice H. Laughlin
Department of Biomedical Sciences, College of Veterinary Medicine, University of Missouri-Columbia, Columbia, MO 65211

Andrew Leask
Centre for Rheumatology, Royal Free Hospital, London NW3 2PF, UNITED KINGDOM

Whaseon Lee-Kwon
Division of Medicine and Nephrology, University of Maryland Hospital—North, Baltimore, M.D. 21201-1595

Hans-Anton Lehr
Institute of Pathology, University of Mainz, Medical Center, Mainz, D-55101, GERMANY

Martin Leuwer
University Department of Anaesthesia, University of Liverpool, Liverpool L69 3GA, UNITED KINGDOM

Klaus F. Ley
Cardiovascular Research Center, University of Virginia Health System, Charlottesville, VA 22908-1394

Lina H.K. Lim
Department of Physiology, National University of Singapore, Singapore 117597, Republic of Singapore

Michelle I. Lin
Department of Pharmacology, Yale University School of Medicine, New Haven, CT 06510-0812

Ute Lindauer
Experimental Neurology, Charité, Universitätsmedizin Berlin, Berlin, D-10098

Bengt Lindholm
Baxter Novum and Renal Medicine, Department of Clinical Science, Carolinska Institutet, Stockholm, S-141 86, SWEDEN

Herbert H. Lipowsky
Department of Bioengineering, Pennsylvania State University, University Park, PA 16802

Perry Hsien-Tsung Liu
Department of Plastic Surgery, Brigham and Women's Hospital, Harvard Medical School, Boston, MA 02115

Julian H. Lombard
Department of Physiology, Medical College of Wisconsin, Milwaukee, WI 53226-1408

Raffaele Longo
Medical Oncology, San Filippo Neri Hospital, Rome, I-00135, ITALY

Ming Lu
Department of Ophthamology, Massachusetts Eye & Ear Infirmary, Boston, MA 02114

Gerard Lutty
Ophthalmology Department, Johns Hopkins Hospital, Baltimore, MD 21287-9115

Shwu-Fan Ma
Department of Medicine, Section of Pulmonary Critical Medicine, University of Chicago, Chicago IL 60611

Miho Machigashira
Department of Periodontology, Kagoshima University Graduate School of Medical and Dental Sciences, Kagoshima 890-8544, JAPAN

William MacNee
ELEGI Colt Laboratories, University of Edinburgh Medical School, Edinburgh EH8 9YG, SCOTLAND

Roni Mamluk
Departments of Surgical Research and Pathology, Vascular Biology Program, Children's Hospital, Harvard Medical School, Boston, MA 02115-5737

Philipp C. Manegold
Department of Otorhinolaryngology, Head and Neck Surgery, Klinikum Grosshadern, University of Munich, München, D-81377, GERMANY

Federica M. Marelli-Berg
Department of Immunology, Division of Medicine, Faculty of Medicine, Imperial College at Hammersmith Campus, London W6 8RF, UNITED KINGDOM

Hildegard R. Maricq
Department of Medicine, Medical University of South Carolina, Charleston, SC 29412

Sean P. Marrelli
Department of Anesthesiology, Baylor College of Medicine, Houston, TX 77030

Marie B. Marron
Department of Cardiovascular Sciences, University of Leicester, Leicester LE2 7LX, UNITED KINGDOM

Gernot Marx
University Department of Anaesthesia, University of Liverpool, Liverpool L69 3GA, UNITED KINGDOM

William G. Mayhan
Department of Physiology and Biophysics, University of Nebraska Medical Center, Omaha, NE 68198-4575

Joseph H. McCarty
Center for Cancer Research, Massachusetts Institute of Technology, Cambridge, MA 02139

Robert S. McCuskey
Department of Cell Biology and Anatomy, University of Arizona College of Medicine, Tucson, AZ 85724-5044

Laura Linz McGillem
Department of Pulmonary and Critical Care Medicine, The Johns Hopkins University, Baltimore, MD 21224

Emma McGregor
Proteome Sciences plc, Institute of Psychiatry, King's College London, London SE5 8AF, UNITED KINGDOM

Michael D. Menger
Institute for Clinical and Experimental Surgery, Universität des Saarlandes, Homburg, D-66421, GERMANY

Christopher Charles Michel
Faculty of Medicine, Division of Biomedical Sciences, Imperial College School of Medicine, London SW7 2AZ, UNITED KINGDOM

Ross Milner
Department of Surgery, Emory University School of Medicine, Atlanta, GA 30322

Fred L. Minnear
Robert C. Byrd Health Sciences Center, West Virginia University, Morgantown, NY 26506-9104

Richard D. Minshall
Department of Anesthesiology and Pharmacology, The University of Illinois College of Medicine, Chicago, IL 60612

David R. Mole
Wellcome Trust Center for Human Genetics, University of Oxford, Oxford OX3 7BN, UNITED KINGDOM

Silvia Montaner
Oral and Pharyngeal Cancer Branch, National Institute of Dental and Craniofacial Research, National Institutes of Health, Bethesda, MD 20892-4340

Francis Daniels Moore
Division of General and GI Surgery, Department of Surgery, Brigham & Women's Hospital, Harvard Medical School, Boston, MA

Guido Morgese
Dipartimento di Pediatria, Ostetricia e Medicina della Riproduzione, University of Siena, Siena, I-53100, ITALY

John E. Morley
Department of Internal Medicine, Division of Geriatrics, Saint Louis University School of Medicine, St Louis, MO 63104

Marsha A. Moses
Surgical Research, Room 1009, Children's Hospital, Harvard Medical School, Boston, MA 02115

Lance L. Munn
Department of Radiation Oncology, Massachusetts General Hospital, Harvard University, Boston, MA 02114

Kazunori Murakami
Department of Anesthesiology, University of Texas Medical Branch, Galveston, TX 77555-0591

Vladimir R. Muzykantov
Department of Pharmacology, IFEM, University of Pennsylvania Medical Center, Philadelphia, PA 19104-6068

Keigo Nakatani
Department of Pediatrics, Kobayashi city Hospital, Kobayashi city, Miyazaki 886-8503, JAPAN

Masaaki Nakayama
Division of Kidney and Hypertension, Department of Internal Medicine, Jikei University School of Medicine, Minato-ku, Tokyo 105-8471, JAPAN

Yasuo Nara
Laboratory of Health Science, Department of Biological Pharmacy, School of Pharmacy, Shujitu University, Okayama, JAPAN

Gerard B. Nash
Department of Physiology, The Medical School, The University of Birmingham, Birmingham B15 2TT, UNITED KINGDOM

Craig M. Neville
Department of Surgery, Massachusettes Generl Hospital, Boston, MA 02114

Naoko Nonaka
Department of Internal Medicine, Division of Geriatrics, Saint Louis University School of Medicine, St Louis, MO 63104

Miwako Numata
Division of Kidney and Hypertension, Department of Internal Medicine, Jikei University School of Medicine, Minato-ku, Tokyo 105-8471, JAPAN

Valerie B. O'Donnell
Department of Medical Biochemistry and Immunology, University of Wales College of Medicine, Cardiff, Wales CF14 4XN, UNITED KINGDOM

Kenji Okajima
Department of Diagnostic Medicine, Graduate School of Medical Science, Kumamoto University, Kumamoto 860-0811, JAPAN

Dennis P. Orgill
Department of Plastic Surgery, Brigham and Women's Hospital, Harvard Medical School, Boston, MA 02115

Ewa Paleolog
Kennedy Institute of Rheumatology, Faculty of Medicine, Imperial College, Hammersmith, London W6 8LH, UNITED KINGDOM

Thomas L. Pallone
Division of Nephrology, University of Maryland at Baltimore, Baltimore, M.D. 21201-1595

Carlo Palombo
Department of Internal Medicine, University of Pisa, Pisa, I-56126, ITALY

Panos N. Papapanou
Division of Periodontics, Columbia University School of Dental and Oral Surgery, New York, NY 10032

Mohan Parigi
Division of Pulmonary and Critical Care Medicine, Johns Hopkins University School of Medicine, Baltimore, MD 21224

Jeremy Pearson
Centre for Cardiovascular Biology & Medicine, School of Biomedical Sciences, King's College London, London SE1 1UL, UNITED KINGDOM

Ellinor I. B. Peerschke
Department of Pathology, Weill Medical College of Cornell University, New York, NY 10021

Michael Sean Pepper
Department of Morphology, University Medical Center, l'Université de Genève, Geneva 4, F-1211, SWITZERLAND

Mauro Perretti
Department of Biochemical Pharmacology, William Harvey Research Institute, Queen Mary University of London, London EC1M 6BQ, UNITED KINGDOM

Irina Petrache
Division of Pulmonary and Critical Care Medicine, Johns Hopkins School of Medicine, Baltimore, MD 21224

Emanuel Petricoin
Food & Drug Administration Clinical Proteomics Program, National Cancer Institute, Bethesda, MD 20892

Janos Pittner
Division of Medicine and Nephrology, University of Maryland Medical School, Baltimore, M.D. 21201-1595

Thomas J. Poole
Department of Cell and Developmental Biology, State University of New York, Upstate Medical University, Syracuse, NY 13210

Aleksander S. Popel
Department of Biomedical Engineering, Whitaker Biomedical Engineering Institute, School of Medicine, Johns Hopkins University, Baltimore, MD 21205

Marcus J. Post
Department of Physiology and of Biomedical Technology, Maastricht University, Maastricht, D-6200 MD, GERMANY

Robert G. Presson
Department of Anesthesiology, Indiana University School of Medicine, Indianapolis, IN 46202-5200

Marco Presta
Department of Biomedical Sciences and Biotechnology, The University of Brescia School of Medicine, Brescia, I-25123, ITALY

Donald G. Puro
Department of Ophthalmology and Visual Sciences, University of Michigan, Ann Arbor, MI 48105

Hamid Rabb
Kidney Transplant Program, Johns Hopkins University School of Medicine, Baltimore, MD 21205

Vineeth Rajkumar
Centre for Rheumatology, Royal Free Hospital, London NW3 2QG, UNITED KINGDOM

Peter J. Ratcliffe
Wellcome Trust Center for Human Genetics, University of Oxford, Oxford OX3 7BN, UNITED KINGDOM

Dale A. Redmer
Department of Animal and Range Science, North Dakota State University, Fargo, ND 58105-5727

Lawrence P. Reynolds
Center for Nutrition and Pregnancy, Department of Animal and Range Science, North Dakota State University, Fargo, ND 58105-5727

Domenico Ribatti
Department of Human Anatomy and Histology, University of Bari, Bari, ITALY

Sarah Riley
Centre for Cardiovascular Biology & Medicine, School of Biomedical Sciences, King's College London, London SE1 1UL, UNITED KINGDOM

Stephen M. Robbins
Faculty of Medicine, Department of Microbiology and Infectious Diseases, University of Calgary, Calgary, Alberta T2N 4N1, CANADA

Ivan M. Robbins
Department of Medicine, Division of Allergy, Pulmonary and Critical Care Medicine, Vanderbilt University, Nashville, TN 37232-2650

L. Christie Rockwell
Department of Anthropology, Temple University, Philadelphia, PA 19122-6089

Marco Rossi
Department of Internal Medicine, University of Pisa, Pisa, I-56126, ITALY

Kristofer Rubin
Department of Medical Biochemistry and Microbiology, Uppsala University, Uppsala, SE-75123

Tao Rui
Vascular Biology/Inflammation Program, Lawson Health Research Institute, London, Ontario N6A 465, CANADA

Rolando E. Rumbaut
Department of Medicine, Baylor College of Medicine, Houston, TX 77030

Nicole May Rummery
Division of Neuroscience, John Curtin School of Medical Research, Australian National University, Canberra, ACT 0200, AUSTRALIA

Hiromi Sakai
Department of Polymer Chemistry, Waseda University, Shinjuku-ku, Tokyo 169-8050, JAPAN

Marko Salmi
Director, Cell Trafficking, National Public Health Institute, Turku, FIN-20520, FINLAND

Alexei V. Salnikov
Department of Medical Biochemistry and Microbiology, Uppsala University, Uppsala, SE-75123

Steven J. Samuelsson
Eyetech Research Center, Eyetech Pharmaceuticals, Inc., Woburn, MA 01801

James D. San Antonio
Department of Medicine, Thomas Jefferson University, Philadelphia, PA 19107-5099

David C. Sane
Department of Cardiology, Wake Forest University School of Medicine, Winston-Salem, NC 27157-1045

Ingrid H. Sarelius
Department of Pharmacology and Physiology, University of Rochester School of Medicine and Dentistry, Rochester, NY 14642

Vladimir Savransky
Kidney Transplant Program, Johns Hopkins University School of Medicine, Baltimore, MD 21205

Kane L. Schaphorst
Center for Lung Biology, College of Medicine, University of South Alabama, Mobile, AL 36688
Anja Schlez
Universitats-Hautklinik, Eberhard-Karls Universitat, Tubingen, D-72076, GERMANY
Geert W. Schmid-Schoenbein
Department of Biomedical Engineering, University of California San Diego, La Jolla, CA 92093-0412
Geert W. Schmid-Schoenbein
Deparmtent of Biengineering, The Whitaker Institute for Biomedical Engineering, University of California, San Diego, La Jolla, CA 92093-0412
Ann Marie Schmidt
Department of Surgery, College of Physicians & Surgeons, Columbia University, New York, NY 10032
Tiffany N. Seagroves
Department of Biology, University of California, San Diego, La Jolla, CA 92093
William C. Sessa
Department of Pharmacology, Yale University School of Medicine, New Haven, CT 06510-0812
William L. Shao
Division of Pulmonary and Critical Care Medicine, Gene Expression Profiling Core, Johns Hopkins University School of Medicine, Baltimore, MD 21224
David T. Shima
Eyetech Research Center, Eyetech Pharmaceuticals, Inc., Woburn, MA 01801
Maria Siemionow
Department of Plastic and Reconstructive Surgery, The Cleveland Clinic Foundation, Cleveland, OH 44195
Michael Simons
Section of Cardiology, Dartmouth-Hitchcock Medical Center, Lebanon, NH 03756
Robert B. Smith
Department of Surgery, Emory University School of Medicine, Atlanta, GA 30322
Akrit Sodhi
Oral and Pharyngeal Cancer Branch, National Institute of Dental and Craniofacial Research, National Institutes of Health, Bethesda, MD 20892
Philip J. Steer
Department of Maternal and Fetal Medicine, Faculty of Medicine, Imperial College of Science, Technology and Medicine at Chelsea and Westminster Hospital, London SW10 9NH, UNITED KINGDOM
Troy Stevens
Department of Pharmacology, College of Medicine, University of South Alabama, Mobile, AL 36688
Karen Stokes
Department of Molecular and Cellular Physiology, Louisiana State University Health Sciences Center, Shreveport, LA 71130-3932
John H. Stone
Department of Medicine, The Johns Hopkins Vasculitis Center, Baltimore, MD 21224
Makoto Suematsu
Department of Biochemistry and Integrative Medical Biology, School of Medicine, Keio University, Shinjuku-ku, Tokyo Tokyo, JAPAN
Christian Sundberg
Department of Pediatrics, Uppsala University, Uppsala, SE-751 85
Kanyarat Suthin
Department of Conservative Dentistry, Faculty of Dentistry, Prince of Songkla University, Hadyai, Songkhla 90112, THAILAND
Erik Svensjö
Laboratorio de Imunologia Molecular, Instituto Biofisica Carlos Chagas Filho, Universidade Federal do Rio de Janeiro, Rio de Janeiro, RJ 21944-900, BRAZIL
Motoki Tagami
Department of Internal Medicine, Sanraku Hospital, Tokyo, JAPAN
Seiichiro Takeshita
Faculty of Education, Dept. of School Health Science, Ibaraki University, Mito, Ibaraki 310-8512, JAPAN
Min D. Tang
Department of Biomedical Engineering, Boston University, Boston, MA 02215
Miranda ten Kate
Laboratory of Experimental Surgery and Oncology, Erasmus Medical Center, Rotterdam 3015 GD, THE NETHERLANDS
Joe Tien
Department of Biomedical Engineering, Boston University, Boston, MA 02215
Daniel L. Traber
Department of Anesthesiology, University of Texas Medical Branch, Galveston, TX 77555
Amy G. Tsai
Microhemodynamics Laboratory, Department of Bioengineering, University of California—San Diego, La Jolla, CA 92093-0412
Rubin Tuder
Division of Pulmonary and Critical Care Medicine, Department of Pathology, The Johns Hopkins University School of Medicine, Baltimore, MD 21205
Karel Tyml
Department of Medical Biophysics, University of Western Ontario, London, Ontario N6A 5C1, CANADA
Joseph P. Vacanti
Department of Surgery, Massachusettes Generl Hospital, Boston, MA 02114
C.H.J. Van Eijck
Laboratory of Experimental Surgery and Oncology, Erasmus Medical Center, Rotterdam 3015 GD, THE NETHERLANDS

Alexander D. Verin
Division of Pulmonary and Critical Care Medicine, Johns Hopkins School of Medicine, Baltimore, MD 21224

Peter B. Vermeulen
Lab Pathology, AZ Sint-Augustinus, Wilrijk—Antwerp, B-2610, BELGIUM

Nicola Verna
Department of Medicine and Science of Aging, Section of Allergy, Clinical Immunology, and Occupational Medicine, "G. D'Annunzio" University School of Medicine, Pescara, I-65132, ITALY

Alberto Verrotti
Department of Medicine, Section of Pediatrics, University of Chieti, Chieti 66013, ITALY

Dietmar Vestweber
Institute of Cell Biology, ZMBE, University of Muenster, Muenster, D-48149, GERMANY

Stanley A. Vinores
Wilmer Ophthalmological Institute, Johns Hopkins Hospital, Johns Hopkins University School of Medicine, Baltimore, MD 21287

Stephen M. Vogel
Department of Anesthesiology and Pharmacology, The University of Illinois College of Medicine, Chicago, IL 60612

Brigitte Vollmar
Faculty of Medicine, Department of Experimental Surgery, University of Rostock, Rostock, D-18055, GERMANY

Thorsten Vowinkel
Department of Molecular and Cellular Physiology, Louisiana State University Health Sciences Center, Shreveport, LA 71130-3932

Wiltz W. Wagner
Department of Anesthesia, Indiana University School of Medicine, Indianapolis, IN 46202

Jean-Luc Wautier
Institut National de la Transfusion Sanguine, University Paris VII, Paris cedex 15 75739, FRANCE

Marie-Paule Wautier
Institut National de la Transfusion Sanguine, University Paris VII, Paris cedex 15 75739, FRANCE

Marianne Weidenhaupt
European Molecular Biology Laboratory (EMBL), Grenoble, F-38027, FRANCE

Brant M. Weinstein
Laboratory of Molecular Genetics, National Institute of Child Health and Human Development, National Institutes of Health, Bethesda, MD 20892

William J. Welch
Department of Medicine, Division of Nephrology and Hypertension, Georgetown University Medical Center, Washington, DC 20057

John B. West
Department of Medicine, Physiology, University of California, San Diego, La Jolla, CA 92093

Michael S. Wolin
Department of Physiology, New York Medical College, Valhalla, NY 10595

John G. Wood
Department of Molecular and Integrative Physiology, University of Kansas Medical Center, Kansas City, KS 66160

Mack H. Wu
Department of Surgery, Cardiovascular Research Institute, Texas A & M University Health Sciences Center, Temple, TX 76504

Kazuo Yamagata
College of Bioresource Science, Department of Food Science & Technology, Nihon University, Fujisawa, Kanagawa 252-8510, JAPAN

Yukio Yamori
Collaborating Center for Research on Primary Prevention of Cardiovascular Diseases, World Health Organization, Kyoto, JAPAN

Shui Qing Ye
Division of Pulmonary and Critical Care Medicine, Gene Expression Profiling Core, Johns Hopkins University School of Medicine, Baltimore, MD 21224

Bryan G. Yipp
Faculty of Medicine, Department of Microbiology and Infectious Diseases, University of Calgary, Calgary, Alberta T2N 4N1, CANADA

Hidehiko Yokogoshi
College of Bioresource Science, Department of Food Science & Technology, Nihon University, Fujisawa, Kanagawa 252-8510, JAPAN

Sarah Y. Yuan
Department of Surgery, Cardiovascular Research Institute, Texas A & M University Health Sciences Center, Temple, TX 76504

Hellmuth Zöltzer
Institut für Biologie, Abteilung Zellbiologie, Arbeitsgruppe Lymphologie, Universität Kassel, Kassel, D-34132, GERMANY

Contents

VOLUME 1

PART II
ORGAN MICROVASCULAR ADAPTATIONS

Section A: Cardiology

Section B: Central Nervous System and Eye

Section H: Lymphatics

VOLUME 2

Section I: Pancreas

Section B: Acute Respiratory Distress Syndrome

Section C: Arthritis

Section D: Blood, Thrombosis and Homeostasis

Hemoglobin-Induced Microvascular Dysfunction

Section E: Coronary Slow Flow

Section J: Inflammation

PART IV
THERAPY

PART V
NEW RESEARCH MODES AND PROCEDURES

Section F: Proteomics

Section G: Stem Cells

PART I

Basic Science

SECTION A

Cell Biology

CHAPTER 1

Endothelin in the Microvasculature

Matthias Barton

University of Zurich School of Medicine, Medical Policlinic, Department of Medicine, University Hospital Zurich, Switzerland

Introduction

Vascular activity of a peptide secreted from endothelial cells was discovered in the mid-1980s. In 1988, Yanagisawa and collegues published the sequencences of the gene and the peptide, which belongs to a peptide family structurally related to vasoconstrictor snake venoms. According to their cellular origin, the peptides were named *endothelins*. The predominant isoform, endothelin-1, is the most potent vasoconstrictor known and also promotes cell growth throughout the cardiovascular system. Endothelin is synthesized from its inactive precursor, big-endothelin, by endothelin-converting enzymes and other peptidases, and in mammals signal transduction is mediated through activation of two G-protein-coupled endothelin receptors. The endothelin system is activated in conditions associated with vascular injury and disease. This chapter discusses the role of endothelin in the microvasculature for some of the clinically most prevalent entities, atherogenesis, inflammation, and cancer development.

Endothelins: Vasoactive Peptides

After the discovery of endothelium-derived relaxing factor, later identified as nitric oxide [1], the vascular activity of a peptide secreted from endothelial cells was described in the mid-1980s [1,2]. In 1988, the gene sequencence of a 21-amino acid protein was identified [3] and found to be structurally similar to snake venoms, the sarafatoxins. The protein was named *endothelin* based on its cellular origin, and it soon turned out that three functionally different isoforms exist [4]. In the late 1990s, additional endothelin isoforms consisting of 31 and 32 amino acids were discovered [5, 6]. The predominant isoform of the endothelin peptide family, endothelin-1_{1-21}, is the most potent vasoconstrictor known in terms of duration of action and potency and also stimulates cell growth throughout the cardiovascular system. In the vasculature, endothelin is synthesized from its inactive precursor, big-endothelin, by endothelin-converting enzymes and other peptidases (Figure 1). In mammals, endothelin signaling is mediated through activation of two high-affinity, G-protein-coupled endothelin receptors [5, 6]. Endothelin ET_A receptors predominantly promote vasoconstriction and growth, whereas activation of the ET_B receptor—which is highly expressed in endothelial cells—is mainly coupled to the release of the vasodilators and growth inhibitors nitric oxide and prostacyclin [5]. ET_A receptors preferably bind endothelin-1, whereas the ET_B receptor binds all three isoforms with equal affinity.

Regulatory Role of Endothelin for Microvascular Function

Expression of endothelin receptors in the microvasculature differs from the expressional pattern observed in larger vessels. Specifically, the endothelin ET_A receptor—which is rarely detected in macrovascular endothelial cells—is expressed in microvascular endothelium. Microvascular endothelial cells are also structurally different from endothelial cells in other parts of the vascular tree. They differ in terms of abundency of cytoskeleton stress fibers, metabolic activity, and number of tight junctions [7]. Also, microvascular endothelial cells contain more vesicles. These cell organelles play a central role for endothelin synthesis: Endothelin-converting enzyme-2 protein expression is localized to endothelial cell vesicles, structures that store

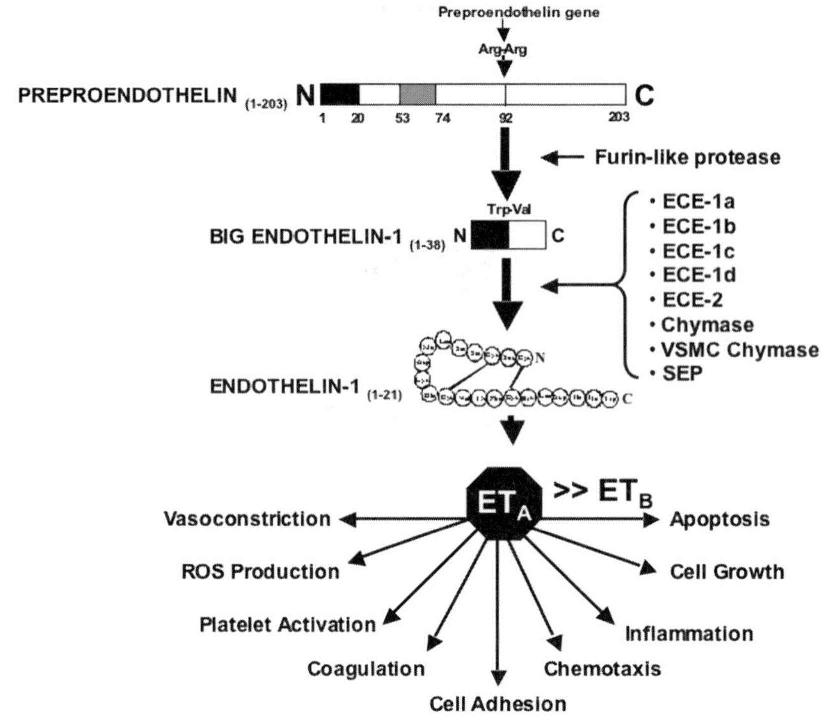

Figure 1 Biosynthesis and multiple functions of endothelin-$1_{(1-21)}$ in the cardiovascular system. Preproendothelin-1 messenger RNA is translated into preproendothelin-1, a 203 amino acid peptide, which is further cleaved by furin convertases to the inactive precursor, big-endothelin-$1_{(1-38)}$. Several enzymes, including endothelin-converting enzymes (ECE), cleave big-endothelin-1 to the active 21-amino acid peptide, endothelin-1. ECE, endothelin-converting enzyme; SEP, secreted soluble endopeptidase; VSMC, vascular smooth muscle cell; ROS, reactive oxygen species; ET_A, endothelin ET_A receptor; ET_B, endothelin ET_B receptor. Reproduced from reference [5] with permission.

and release mature endothelin-1 upon stimulation. In vitro work suggests that endothelin also has important functions for capillary formation under normal and pathological conditions. Furthermore, endothelin-1 leads to activation of cytokines in circulating blood cells and also in resident cells such as endothelial cells or vascular smooth muscle. This activation includes induction of interleukin-6, nuclear factor kappa B, and C-reactive protein, effects that in turn can be enhanced by other cytokines under certain conditions. Thus, inflammatory changes and subsequent increases in vascular permeability are key factors determining microvascular injury mediated by endogenous endothelin-1.

Endothelin and the Microvasculature in Disease

Carcinogenesis and Angiogenesis

Autocrine endothelin receptor signaling controls growth of different types of tumor cells, including ovarian cancer melanoma, Kaposi sarcoma, colon carcinoma, prostate cancer, and bone metastases of tumors [8]. In most cancer cells, growth responses involve activation of ET_A receptors. In addition, angiogenesis provides an important means to increase blood supply of the tumor growth and metastasis. Proliferation and wound healing of endothelial cells is mediated by ET_B receptors. Therefore, endothelin-1 via ET_B receptors may act as an angiogenic peptide, providing additional avenues for cancer cells to increase blood supply and disseminate. By controlling expression of vascular endothelial cell growth factor (VEGF) in an autocrine fashion, endothelin-1 also acts as an important inducer of angiogenesis, thereby stimulating tumor cell proliferation. In consequence, endothelin-1 can be considered an essential factor for tumor growth, neovascularization, and metastasis. ET_A receptor blockade has been shown to inhibit tumor growth directly and to exert antiangiogenic effects by inhibiting VEGF production and by acting on vascular smooth muscle cell and on microvascular channels lined by tumor cells (Figure 2). Moreover, because ET_A receptors are expressed by microvascular endothelial cells, inhibiton of these receptors is likely to play a role for capillary formation and growth as well. It is therefore not surprising that targeting of endothelin receptors currently appears to be a promising new anticancer strategy.

Inflammation and Vascular Permeability

Hypoxia represents a strong stimulus inducing both transcription and protein synthesis of endothelin-1. Moreover,

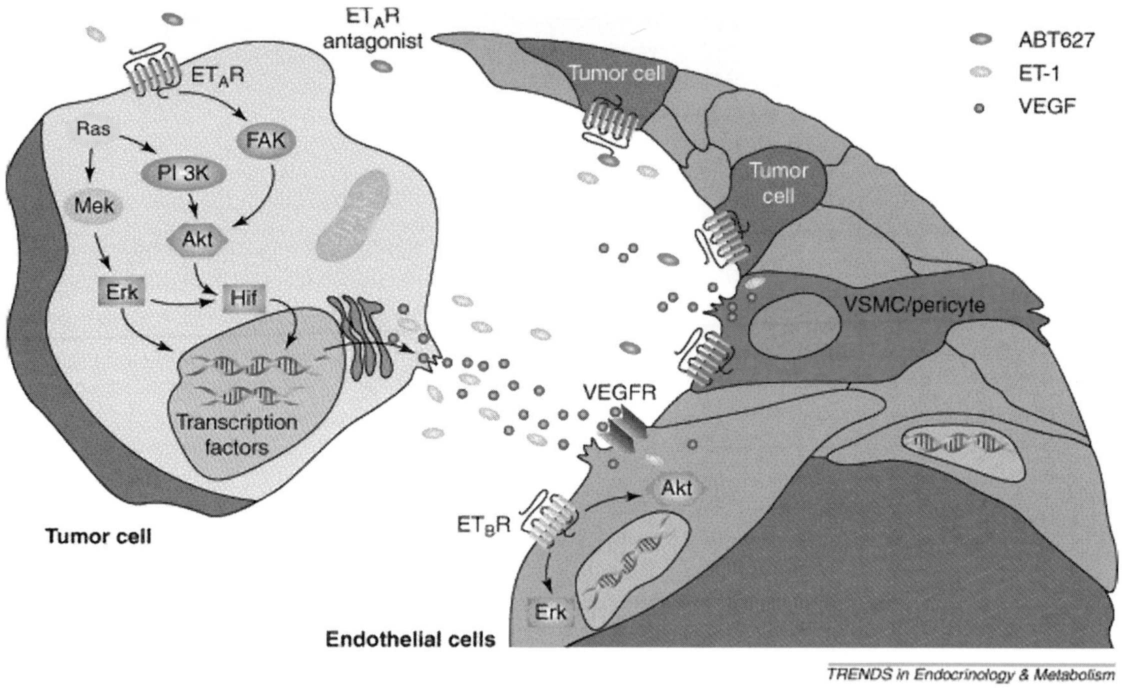

Figure 2 Proposed Overview of the Possible Role of Endothelin Signaling in Tumor Angiogenesis. This process involves a series of linked signaling pathways that typically begins with upregulation of the synthesis of angiogenic factors, such as ET-1 and VEGF, that are released by tumor cells. ET_A receptor (*ET_AR*) activation by ET-1 could promote angiogenesis by increasing VEGF production through HIF-1-dependent mechanisms. ET-1 binds to specific ET_B receptors (*ET_BR*) on endothelial cells and to ET_A receptors on pericytes/vascular smooth muscle cells or on tumor cells lining vessels (vasculogenic mimicry). ET_A receptor blockade (*ABT627*) could exert antiangiogenic effects by inhibiting VEGF production and by acting on VSMC and on microvascular channels lined by tumor cells. Another cascade of signaling events is then initiated, which leads to the formation of new tumor vessels. *Erk*, extracellular signal-related kinase; ET-1, endothelin-1; Mek, mitogen-activated protein kinase kinase; *Akt*, protein kinase B; Hif, hypoxia-inducible factor-1; VEGF, vascular endothelial cell growth factor; VEGFR, vascular endothelial cell growth factor receptor. Reproduced from reference [8] with permission. (see color insert)

oxidative stress associated with reperfusion further increases endothelin synthesis. There is now evidence suggesting that endothelin-1 directly contributes to vascular inflammation and microvascular permeability [9, 10]. Although the latter function is usually attributed to mediators such as histamine or VEGF, endothelin may not only contribute to tissue edema but also induces rolling and adherence of leukocytes to the capillary wall [11] (Figure 3). This results in inflammation, blood hyperviscosity, and microcirculatory stasis, leading to a substantial decrease in microcircular O_2 extraction and cell dysfunction. These changes can be prevented to a large extent by endothelin receptor blockers targeting the ET_A receptor, a treatment that has been shown to substantially reduce microcirculatory dysfunction in the myocardium [10, 12], brain [11], and intestine. Activation of the microvascular endothelin system and its direct and indirect effects are therefore responsible for cell injury, particularly in conditions associated with ischemia.

Atherosclerosis and Hypercholesterolemia

Cardiovascular risk factors such as increases in blood pressure, increased plasma levels of cholesterol or glucose, or lack of female sex hormones—all of which accelerate the development of atherosclerosis—result in activation of the endothelin system in vitro and in vivo [5, 6]. Recently, several studies have demonstrated that changes in microvasculature may importantly contribute to atherogenesis. Experimental evidence from John Cooke's laboratory suggests that enhancing angiogenesis may accelerate rather than limit the atherosclerotic disease process. Elegant studies by Lerman and coworkers [10, 13] have investigated some of the regulatory processes by which endothelin-dependent microvascular changes may contribute to early atherogenesis [13]. Even in the absence of structural injury, hypercholesterolemia, one of the cardinal risk factors for atherosclerosis, results in vascular inflammation, increased microvascular permeability [10], and adventitial neovascularization (Figure 5B). These alterations are associated with induction of VEGF expression (Figure 5D). Interestingly, increased microvascular permeability, adventitial neovascularization, as well as VEGF expression can be fully prevented by blocking ET_A receptors [10, 13]. Similarly, prevention of myocardial inflammation, infarction, and fibrosis in atherosclerotic mice by endothelin ET_A receptor blockade has been demonstrated [12], and beneficial effects on structural changes in mesenteric resistance vasculature of

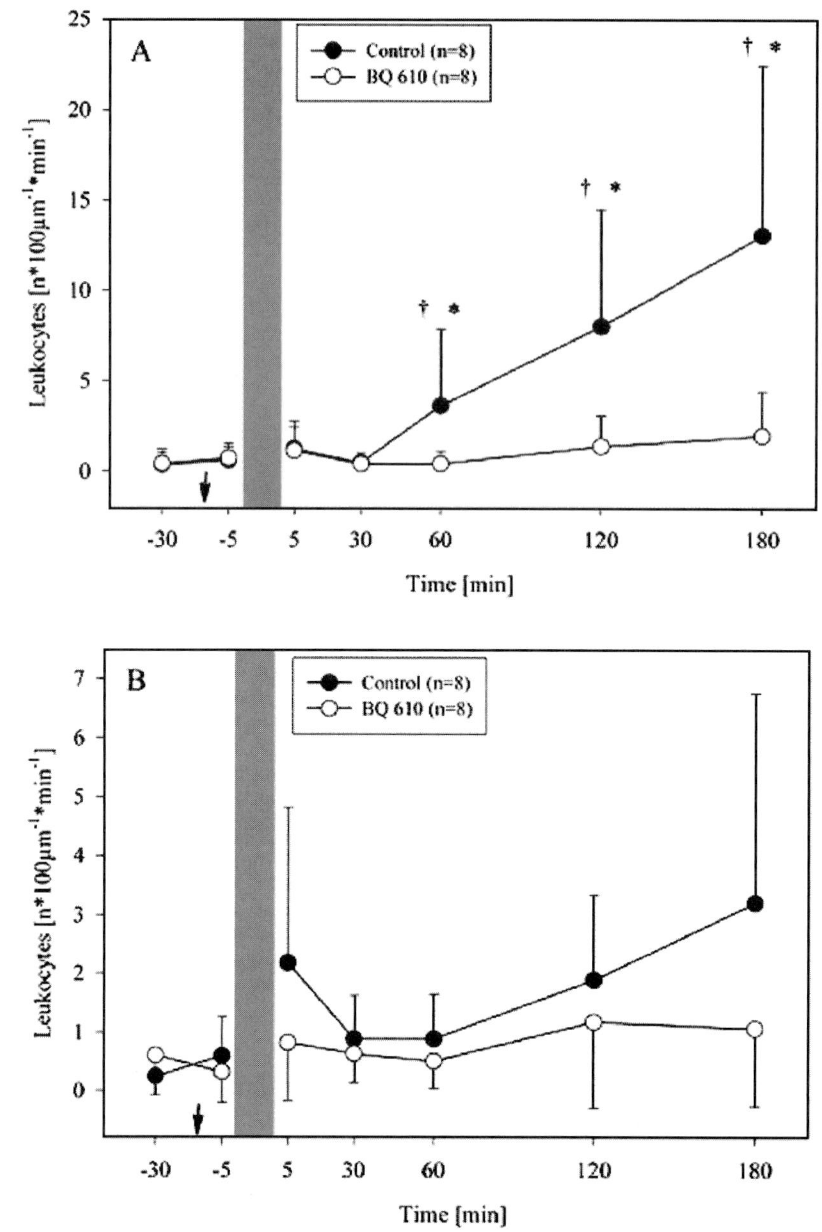

Figure 3 Role of Endothelin for Leukocyte–Endothelium Interactions in Gerbil Pial Venules. Frequency of rolling **(A)** and firm adherent leukocytes at the vessel wall of pial venules **(B)** in the absence (○) or presence (●) of the ET_A antagonist BQ-610 before and after global ischemia in gerbils. Arrow indicates time point of drug administration, bar indicates time of ischemia. *$p < 0.05$ versus control. From reference [11] with permission.

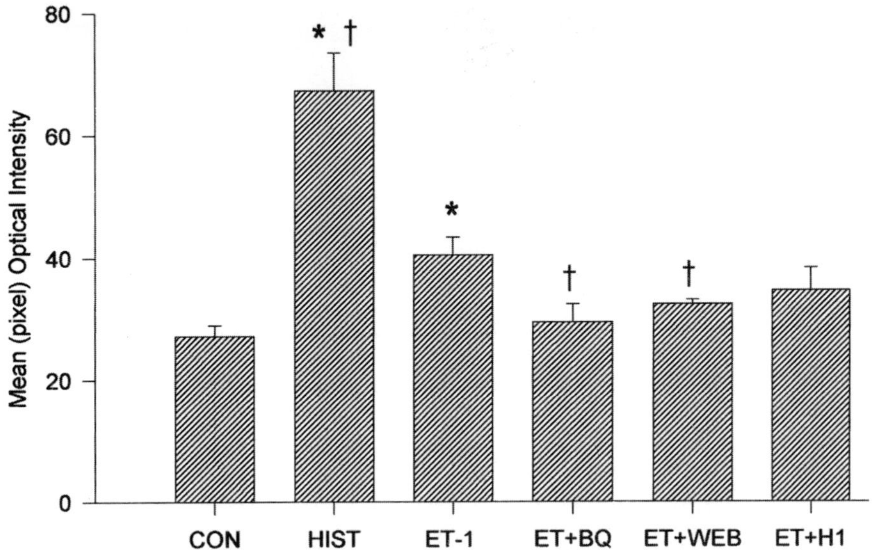

Figure 4 Vascular Permeability Measured by FITC-albumin in Guinea Pig Intestinal Submucosal Microcirculation. Effects of histamine (HIST) or endothelin-1 (ET-1) in absence or presence of treatment with antagonists of the ET_A receptor (BQ-123, BQ), platelet-activating factor (WEB-2086, WEB), or H_1 histamine receptors (diphenhydramine, H1). * $p < 0.05$ versus CON; † $p < 0.05$ versus ET-1. From reference [9] with permission.

Table I Effects of atherosclerosis and endothelin ET_A receptor blockade with darusentan for 30 weeks on vascular structure in mesenteric resistance arteries in apolipoprotein E-deficient mice ($apoE^0$) and wild-type controls (C57BL6/J) measured after treatement in isolated vessels in a arteriograph system in vitro under perfused and pressurized conditions.

Group	C57	C57 + Darusentan	$apoE^0$	$ApoE^0$ + Darusentan
Lumen diameter (μm)	222 ± 4	235 ± 7	217 ± 5	220 ± 8
Intima-media thickness (μm)	18.9 ± 0.2	18.3 ± 0.4	20.9 ± 0.7*	17.0 ± 0.4†
Intima-media CSA (×10³ μm²)	14.3 ± 0.2	13.9 ± 0.5	15.6 ± 0.6*	12.9 ± 0.5†

$ApoE^0$, apolipoprotein E deficient; C57, wild-type control; darusentan, ET_A receptor antagonist (Knoll LU135252); CSA, cross-sectional area; *$P < 0.05$ versus C57; †$P < 0.05$ versus $apoE^0$. Modified from reference [15] with permission.

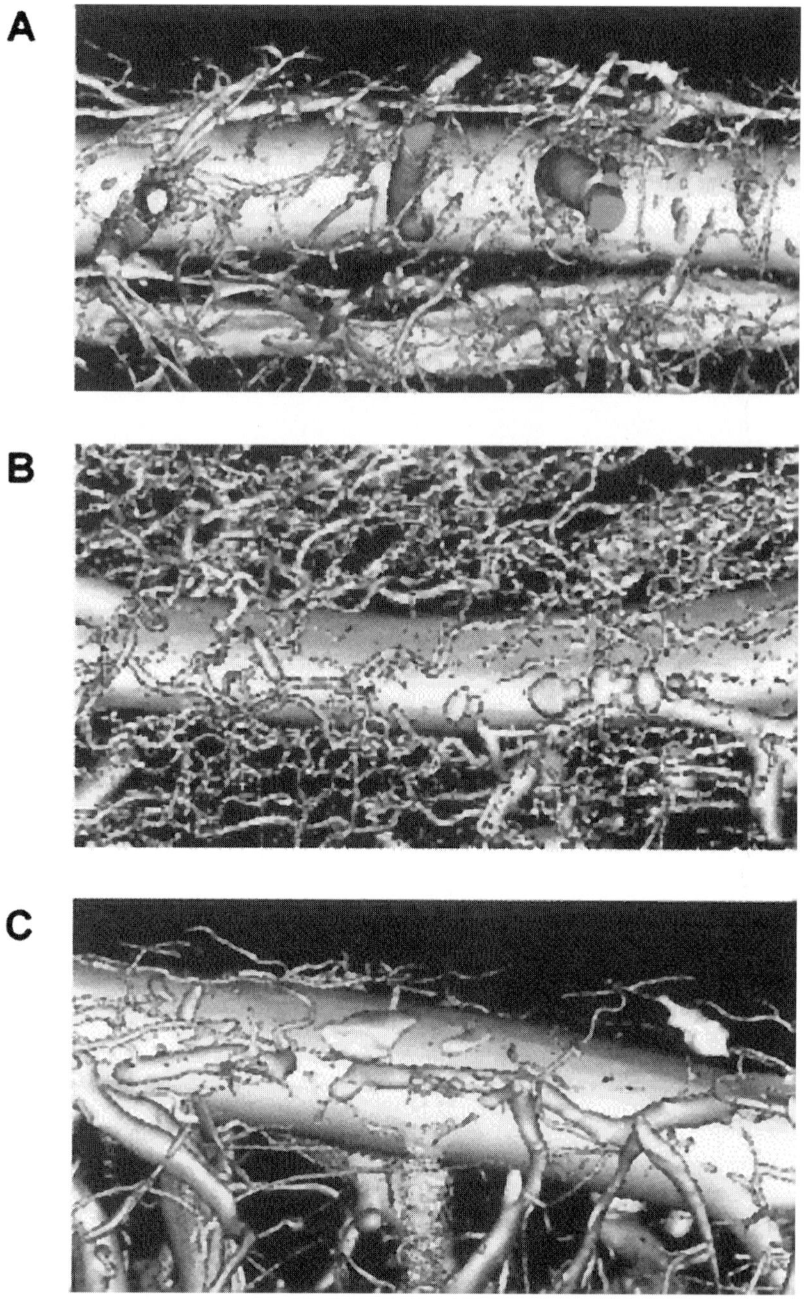

Figure 5A–C Effects of normal diet **(A)**, high-cholesterol diet alone **(B)**, or in combination with endothelin ET_A receptor antagonist treatment **(C)** on formation and spatial distribution of coronary vasa vasorum in pigs. Concomitant endothelin blockade essentially prevented adventitial neovascularization. From reference [13] with permission.

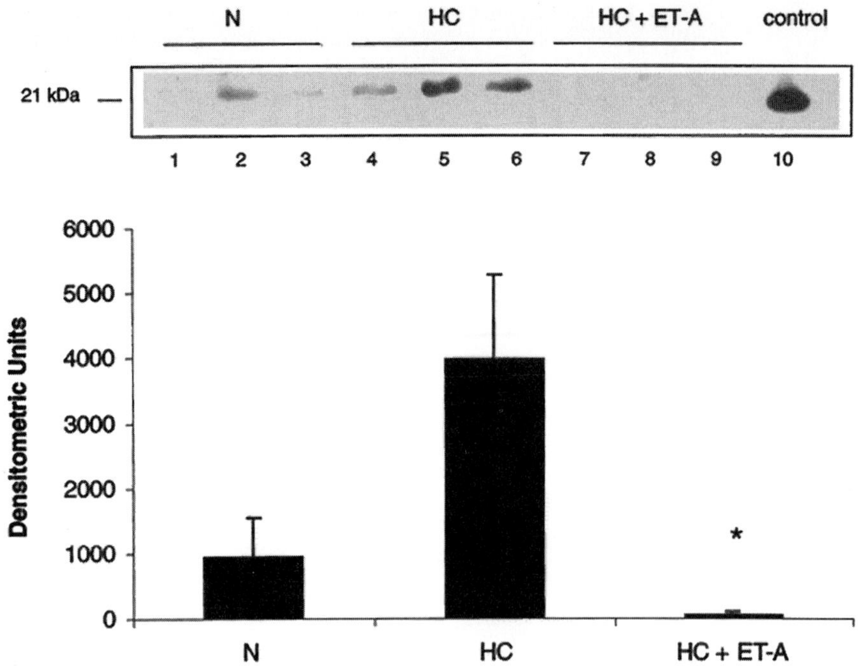

Figure 5D Coronary artery wall expression of VEGF protein in pigs after 12weeks of high-cholesterol diet. Note that endothelin ET_A receptor blockade (ET-A) fully prevented vascular upregulation of VEGF induced by hypercholesteremia. N, normal diet; HC, high-cholesterol diet; HC + ET-A, high-cholesterol diet and endothelin ET_A receptor antagonist ABT-627. From reference [13] with permission.

mice with atherosclerosis have been reported (Table I). It is not known whether the microvascular changes described mechanistically translate into the inhibition of advanced human-like atherosclerotic lesions observed after ET_A receptor blockade [14], but it is likely that anti-inflammatory and antiangiogenic effects of endothelin blockade play an essential role.

References

1. Furchgott, R. F., and Vanhoutte, P. M. (1989). Endothelium-derived relaxing and contracting factors. *FASEB J.* **3**, 2007–2018.
2. Highsmith, R. F. (1992). From endotensin to endothelin: The discovery and characterization of an endothelial cell-derived constricting factor. In *Endothelin* (G. M. Rubanyi, ed.), pp. 1–192. New York: Oxford University Press.
3. Yanagisawa, M., Kurihara, H., Kimura, S., Tomobe, Y., Kobayashi, M., Mitsui, Y., Yazaki, Y., Goto, K., and Masaki, T. (1988). A novel potent vasoconstrictor peptide produced by vascular endothelial cells. *Nature* **332**, 411–415.
4. Inoue, A., Yanagisawa, M., Kimura, S., Kasuya, Y., Miyauchi, T., Goto, K., and Masaki, T. (1989). The human endothelin family: Three structurally and pharmacologically distinct isopeptides predicted by three separate genes. *Proc. Natl. Acad. Sci. USA* **86**, 2863–2867.
5. Barton, M., Traupe, T., and Haudenschild, C. C. (2003). Endothelin, hypercholesterolemia, and atherosclerosis. *Coron. Artery Dis.* **14**, 477–490.
6. Lüscher, T., and Barton, M. (2000). Endothelins and endothelin receptor antagonists: Therapeutic considerations for a novel class of cardiovascular drugs. *Circulation* **102**, 2434–2440.
7. Kuruvilla, L., and Kartha, C. C. (2003). Molecular mechanisms of endothelial regulation of cardiac function. *Mol. Cell. Biochem.* **253**, 113–123.
8. Bagnato, A., and Spinella, F. (2002). Emerging role of endothelin-1 in tumor angiogenesis. *Trends Endocrinol. Metab.* **14**, 44–50.
A thorough review on the paracrine and autocrine effects of endothelin in cancer development and angiogenesis.
9. King-VanVlack, C. E., Mewburn, J. D., Chapler, C. K., and MacDonald, P. H. (2003). Hemodynamic and proinflammatory actions of endothelin-1 in guinea pig small intestine submucosal microcirculation. *Am. J. Physiol. Gastrointest. Liver Physiol.* **284**, G940–G948.
10. Bonetti, P. O., Best, P. J., Rodriguez-Porcel, M., Holmes, D. R., Lerman, L. O., and Lerman, A. (2003). Endothelin type A receptor antagonism restores myocardial perfusion response to adenosine in experimental hypercholesterolemia. *Atherosclerosis* **168**, 367–373.
11. Lehmberg, J., Putz, C., Fürst, M., Beck, J., Baethmann, A., and Uhl, E. (2003). Impact of the endothelin-A receptor antagonist BQ 610 on microcirculation in global cerebral ischemia and reperfusion. *Brain Res.* **961**, 277–286.
12. Caligiuri, G., Levy, B., Pernow, J., Thoren, P., and Hansson, G. K. (1999). Myocardial infarction mediated by endothelin receptor signaling in hypercholesterolemic mice. *Proc. Natl. Acad. Sci. USA* **96**, 6920–6924.
Important work demonstrating a causal role of endothelin for endothelin receptors in myocardial inflammation and infarction in mice with advanced atherosclerosis.
13. Herrmann, J., Best, P. J., Ritman, E. L., Holmes, D. R., Lerman, L. O., and Lerman, A. (2002). Chronic endothelin receptor antagonism prevents vasa vasorum neovascularization in experimental hypercholesterolemia. *J. Am. Coll. Cardiol.* **39**, 1555–1561.
An interesting article demonstrating a role of endothelin for adventitial neovascularization during hypercholesterolemia in pigs.
14. Barton, M., Haudenschild, C. C., d'Uscio, L. V., Shaw, S., Munter, K., and Luscher, T. F. (1998). Endothelin ETA receptor blockade restores NO-mediated endothelial function and inhibits atherosclerosis in apolipoprotein E-deficient mice. *Proc. Natl. Acad. Sci. USA* **95**, 14367–14372.

15. d'Uscio, L. V., Barton, M., Shaw, S., and Lüscher, T. F. (2002). Chronic ET(A) receptor blockade prevents endothelial dysfunction of small arteries in apolipoprotein E-deficient mice. *Cardiovasc Res.* **53**, 487–495.

Capsule Biography

Dr. Barton is Associate Professor of Cardiology at the University of Zürich School of Medicine, Switzerland. He is currently Attending Physician at the Department of Medicine and Director of Research of the Medical Policlinic at the University Hospital Zurich. The research in his laboratory focuses on atherogenesis, obesity, as well as vascular and renal biology. His research is supported by grants from the Swiss National Foundation (SNF), the Deutsche Forschungsgemeinschaft, and the University of Zürich.

CHAPTER 2

Control of Cell Motility by the Cytoskeleton and Extracellular Matrix

Amy Brock and Donald E. Ingber
Harvard Medical School and Children's Hospital, Boston, Massachusetts

Introduction

Most past work in the area of angiogenesis and vascular development has focused on the importance of cell growth control. However, directional motility—the process by which endothelial cells migrate in a spatially oriented manner—is equally critical for tissue expansion and vascular pattern formation. For example, angiogenic factors can stimulate 2 millimeters of capillary outgrowth in normally avascular tissues (e.g., cornea) under conditions in which cell proliferation is inhibited; this occurs exclusively through endothelial cell elongation and directed movement. Repair of the endothelium after endarterectomy also involves migration of large vessel endothelium, and failure to reconstitute the continuous cell monolayer can lead to thrombus formation and vessel occlusion. Thus, it is critical that we unravel the mechanism by which local cues from the cell microenvironment drive cell migration in order to understand angiogenesis and other vascular disease processes. But how do cells interpret, integrate, and respond to information from their local environment and decide in which direction to move? This is a complex problem in light of the fact that their environment contains multiple soluble, insoluble, and mechanical stimuli that may provide conflicting signals to individual cells.

In this chapter, we review the current understanding of how cells sense these local signals and choose a direction in which to execute purposeful locomotion, with particular focus on the role of the cytoskeleton and mechanical interactions between cells and their extracellular matrix (ECM). Insights into this mechanism have been made possible by recent development of new microtechnologies and tools that allow analysis of how microscale changes in physical parameters, such as ECM structure, topography, and mechanical compliance, impact directional cell motility and associated cytoskeletal signaling mechanisms.

Directional Motility

Directional cell migration on ECM is critical during all phases of embryonic development. For example, neural crest cells emigrate along a path defined by basement membrane fibrils during development of the nervous system. The ureteric bud epithelium migrates from the nephric duct toward the metanephric mesenchyme to induce the formation of the adult kidney. Formation of the vascular system follows a similar paradigm and is characterized by tandem cell migration during angiogenesis and migration of cell sheets during the expansion and repair of large vessels.

Angiogenesis relies on the tight control of endothelial cell migration through the balance of angiogenesis inducers and inhibitors. This dynamic, multistep process involves retraction of pericytes from the abluminal surface of the capillary, release of proteases that degrade the ECM surrounding the preexisting vessels, and endothelial cell migration in the direction of an angiogenic stimulus that is mediated by ongoing deposition (and degradation) of new ECM components. This is followed by cell proliferation, which facilitates further extension of the growing capillary sprouts. Eventually, these structures locally slow down their rate of ECM turnover, accumulate a basement membrane, and reorganize into quiescent, hollow, capillary tubes. Recruitment of pericytes and smooth muscle cells further stabilizes these newly formed blood vessels.

In the adult, cell migration during angiogenesis is critical for normal function of the female reproductive system as well as wound healing and the immune response. Directed cell movement is also a feature of the tissue remodeling that occurs during postnatal developmental processes, such as branching morphogenesis of the mammary epithelial ductal system and elongation of the ureters in the renal system. In the subventricular zone, an area of the brain containing neural stem cells, neurogenesis occurs throughout adulthood, with the majority of these new neurons migrating anteriorly into the olfactory bulb, traveling a distance of several millimeters in a highly directed manner to reach their destination. Thus, understanding of the mechanism by which directional cell movement is controlled has important implications for many developmental systems.

Cell motility also is a central component of various pathological conditions and as such may represent a common target for drug development. Tumor expansion requires the proliferation and directional migration of endothelial cells as blood vessels are recruited to supply a solid tumor with oxygen. If this angiogenic process is inhibited, the tumor must rely on preexisting blood vessels, and its growth will be checked at a maximum size of approximately 1 to 2 millimeters in diameter because of diffusion limitations. Metastatic tumor cells also acquire the ability to migrate out of the primary tumor into the vasculature and subsequently invade a secondary location. In the vascular system, misdirected endothelial cell migration is a key factor in vascular anomalies, intimal hyperplasia secondary to endarterectomy, and chronic inflammatory diseases such as arthritis and atherosclerosis.

The Migratory Process

Migration is initiated as a cellular response to chemotactic gradients of soluble mitogens and haptotactic gradients of insoluble adhesive (ECM) molecules in the local microenvironment. Cell movement is mediated by the extension of thin, flat, membrane protrusions from the leading edge of the cell (Figure 1). These outward extending membrane processes are filled with a loose network of actin filaments and are known as *lamellipodia*; they are also sometimes referred to *membrane ruffles* when they are oriented perpendicular to the plane of the adhesive substrate. In some cells, formation of these membrane processes is accompanied or preceded by extension of long, thin protrusions filled with rigidified (cross-linked) bundles of actin filaments, known as *filopodia*, that act like moving cantilevers or fingers that explore the surrounding substrate. In fact, cells may extend and retract multiple filopodia and lamellipodia as they explore their microenvironment, suggesting that this process is a key regulatory event in directional migration and pathfinding.

First identified as the primary organelles of cell motility by Abercrombie in 1971, lamellipodia are sites of active actin polymerization, and their formation requires the local recruitment of a whole array of molecular components to mediate cytoskeletal reorganization. These include actin filament nucleating factors such as Arp 2/3 and SCAR; filament depolymerizing proteins such as cofilin; barbed-end capping proteins; and actin monomer binding factors such as profilin. Filopodia also form as a result of actin polymerization; however, these filaments are cross-linked into long, stiff bundles that can extend to up to tens of micrometers in length.

Other key structural elements in migrating cells are the adhesive contacts where the cells tightly attach to the underlying ECM. In migrating cells, the formation of these local

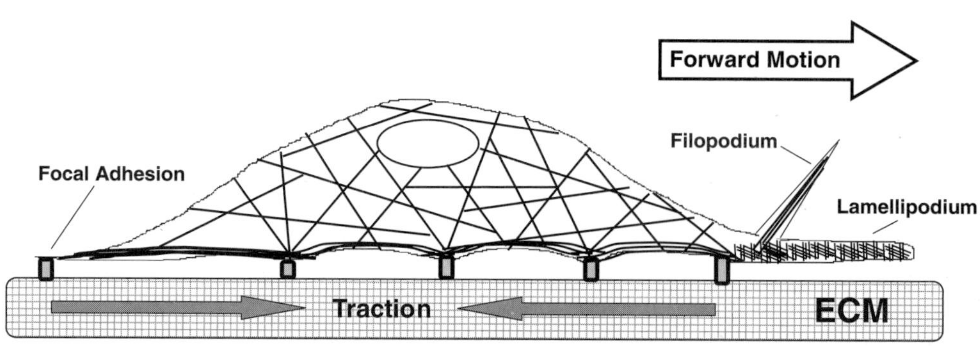

Figure 1 Diagram of a Migratory Cell. As the attachment sites between the intracellular cytoskeleton and the extracellular matrix (ECM), focal adhesions (*gray ovals*) transmit cell tractional forces to the substrate. In migrating cells, the formation of these local adhesive structures behind the leading edge stabilizes the extension of actin-based protrusions, such as lamellipodia and filopodia. The thin, sheet-like lamellipodium is characterized by a tight network of actin filaments and may be accompanied or preceded by the extension of filopodia containing rigid, cross-linked actin bundles. Forward motion requires direction extension of these processes from the leading edge, as well as exertion of cell tractional forces on the ECM substrate that help both propel the cell forward and release the cell's trailing adhesions.

adhesive contacts, known as *focal adhesions*, at the leading edge of the cell helps stabilize the actin-based membrane protrusions. ECM regions beneath these spot weld-like focal adhesions also act as traction sites that resist tensional forces that are generated via actomyosin interactions in the cytoskeleton. This buildup of isometric forces restructures and stiffens the actin cytoskeleton behind the leading edge of the cell, thereby providing a stable platform from which the actin filaments that form filopodia and lamellipodia can assemble.

Importantly, cells also must be able to exert tractional forces and pull against their fixed ECM adhesions in order to propel the cell body forward and produce purposeful locomotion. During this process, the cell must generate enough motive force to overcome the resistance of the stiffness of its own cytoskeleton and of the ECM adhesions at its trailing edge (Figure 1). The extension of lamellipodia and the formation of new adhesions at the leading edge must also be spatially and temporally coordinated with detachment of adhesions in the trailing edge and release of the tail to result in net forward movement of the entire cell.

Control of Motility by ECM

It is well known that ECM plays a central role in microenvironmental control of cell motility based on its ability to chemically mediate cell adhesion. In addition, capillary endothelial cells must increase ECM degradation to initiate cell outgrowth and also must maintain ongoing ECM synthesis and deposition to sustain progressive cell migration during both angiogenesis and healing of large vessel endothelial monolayers. Importantly, recent work has revealed that the physical properties of the ECM, including its surface topography and mechanical compliance, also can significantly impact cell movement.

Compared with cells on rigid ECM substrates, cells on flexible substrates coated with the same ECM protein display increased rates of both lamellipodial activity (protrusion and retraction) and locomotion. Cells also prefer to move from regions of low to high mechanical rigidity. In addition, they move faster along edges of adhesives surfaces than in their central regions, and along microengineered grooves or thin lines than on flat ECM substrates. Furthermore, when individual cells are constrained to polygonal ECM-coated adhesive islands with angular edges (e.g., squares, triangles, pentagons, hexagons, trapezoids), they preferentially extend lamellipodia and filopodia from their corners. Thus, cells apparently can sense micrometer-scale changes in topography and subtle alterations in ECM mechanics, and they respond by changing both their cytoskeleton and migratory behavior. This knowledge could have important implications for engineering of artificial matrices for tissue engineering applications.

Studies of endothelial cell clusters cultured in a three-dimensional collagen gel have demonstrated directly that force application can promote capillary outgrowth. These experiments also showed that matrix-transduced tensional forces in stretched collagen gels are sufficient to control *directional* outgrowth as the growing capillaries extend along the tension field lines that stretch between neighboring cells. Part of this response is based on the ability of the outgrowing endothelial cells to sense and respond to the direction of ECM fibrils that become aligned by the cell-generated forces.

The mechanism by which changes in ECM mechanics influence cell motility is less clear, largely because it has been difficult to study this process until recent years. Qualitative analysis of the traction forces exerted by migrating cells was first carried out by observing the pattern of wrinkles produced by fibroblasts cultured on deformable substrates (e.g., fibrin clots, collagen gels, silicon rubber substrates). Since that time, increasingly sophisticated methodologies have improved the quantification and spatiotemporal resolution of cell-based forces. For example, polyacrylamide gels with different degrees of cross-linking now allow substrate stiffness to be varied as an independent parameter in cell culture studies. The added use of fluorescent microbeads embedded in the gels as fiducial markers permits the direct quantitation of traction forces that are exerted by adherent cells. Combination of this method of "traction force microscopy" with micropatterning techniques also has allowed control of the position where cells exert these tractional forces, as well as analysis of the impact of varying this position on cell behavior.

These techniques have generated new and sometimes surprising insights. For example, it has recently been demonstrated that the small nascent focal adhesion complexes at the leading edge of a cell can exert stronger traction forces than larger, more mature focal adhesion plaques. In individual cells that are physically constrained to single square ECM adhesive islands, the localized distribution of focal adhesions corresponds precisely to the corner regions where the cell exerts greatest tractional forces on the substrate, deposits ECM (fibronectin) fibrils, and extends new migratory processes (lamellipodia and filopodia). Moreover, lamellipodia extension can be inhibited by dissipating cytoskeletal tension generation.

Biochemical Signaling Mechanisms

Over the past decade, great advances have been made in our understanding of the chemical factors that drive cell motility, as well as the biochemical signaling cascades that mediate cellular responses, as described in several excellent recent reviews (see Bibliography). In particular, the Rho family of small GTPases, consisting of Rho, Rac, and Cdc42, have been shown to regulate signal transduction pathways that link extracellular signals through membrane receptors to changes in cytoskeletal structure and organization that drive cell locomotion. For instance, in 1992, Hall and coworkers demonstrated that constitutively active (GTPase-deficient) mutants of Rac induce lamellipodia assembly in fibroblasts, whereas active Rho and Cdc42 mutants promote formation of stress fibers and filopodia, respectively. Similar effects have been observed in various endothelial cells as well.

In fact, Rac has been found to be essential for the migration of all cells examined thus far. Studies in macrophage, for example, have demonstrated that downregulation of Rac activity stalls cell migration while downregulation of Cdc42 renders cells unresponsive to the direction of a chemotactic gradient, causing them instead to move via a random walk. Studies utilizing fluorescence resonance energy transfer (FRET) microscopy have revealed that while Rac protein is distributed homogeneously throughout migratory fibroblasts, the active GTP-bound form is localized to the leading edge of the cell. Rac and Cdc42 may therefore act locally to govern where actin polymerization is activated at the periphery of the cell.

Analysis of Rho GTPase signaling has revealed a vast array of downstream effector proteins (at least 60 at present) that regulate many different cell processes. Rac is activated by guanine nucleotide exchange factors (GEFs) such as Trio, Vav, and Sos-1 and repressed by GTPase-activating proteins (GAPs) such as chimerin. Several factors downstream of Rac act on cytoskeletal structure and other aspects of cell migration. Perhaps the best characterized of these effectors is p21-activated kinase (Pak), a serine/threonine kinase that is activated upon binding to the GTP-bound form of Rac or Cdc42. Pak provides a direct link from Rac to cell motility through phosphorylation and activation of LIM kinase, which phosphorylates and represses the actin depolymerizing factor, cofilin. Pak may also regulate cell traction forces by promoting myosin light-chain phosphorylation and increasing cytoskeletal contractility, although these results have been controversial and may depend on the cell type examined.

The Rho GTPases also impact directional migration through the regulation of focal adhesion dynamics. In migrating cells, Rac is required for the formation of new focal adhesion complexes at the leading edge of the cell, whereas Rho is essential for the maturation of these nascent adhesive contacts into fully developed focal adhesions. Rac may regulate adhesion turnover indirectly by antagonizing Rho through its downstream effector Pak1, which localizes to focal adhesions upon activation. The mechanism by which Pak1 regulates adhesion dynamics is not yet clear but may involve the action of LIM kinase or the phosphorylation of myosin light chain.

Mechanical Signaling Mechanisms

But how do changes in ECM mechanics or in the level of tractional forces cells exert on their focal adhesions influence cell motility? Focal adhesions are critical to the migratory process because they are sites of attachment between the ECM, transmembrane adhesion receptors (known as integrins), and the actin cytoskeleton, which stabilize lamellipodia and transmit propulsive forces. In addition, the cytoskeletal backbone of the focal adhesion acts as an orienting scaffold for a vast array of signaling molecules. They also are sites where the mechanical and biochemical signals that regulate cell migration are integrated inside the cell.

The temporal and spatial organization of focal adhesion assembly is only partially understood; nevertheless, it is clear that physical forces significantly impact this process. The application of force to newly formed integrin-cytoskeleton linkages promotes additional focal adhesion assembly, and hence mechanically reinforces the connection, by inducing force-dependent signaling events inside the cell. For example, cytoskeletal tension exerted on the ECM via new adhesion sites results in the maturation of small nascent adhesive contacts ("focal adhesion complexes") into larger, more highly organized anchoring structures (classic "focal adhesions"). The earliest events in the assembly process include activation of receptor and nonreceptor tyrosine phosphatases, and the subsequent sequential recruitment of talin and paxillin, followed by slower recruitment of vinculin and FAK. The presence of tensin and zyxin may indicate a late stage of adhesion assembly or a transition to a different type of adhesion site that assembles only after the leading edge has stabilized.

To date, more than 50 different signaling molecules have been reported to be associated with focal adhesion sites, and further investigation into the functional consequences of this molecular heterogeneity is ongoing. In addition, the dynamic aspect of adhesion assembly and disassembly will need to be examined in the future. The lifetime of an adhesion is on the order of tens of minutes, whereas the exchange rates of its individual structural components appear to be on the order of seconds to minutes. The local mechanical compliance of the ECM also may spatially regulate the strength, size, number, and molecular composition of the cell-ECM adhesions, and thus it may be a major source of spatial heterogeneity in adhesion complexes. For example, treatment of cells with HL-7 to inhibit actomyosin contractility results in the rapid loss of phosphotyorosine from focal adhesions, followed by disassembly of the complex on a slower time scale. These findings suggest that tension applied through the actomyosin system may trigger local tyrosine phosphorylation events that are required for the subsequent assembly of adhesion complexes.

Micromanipulation studies utilizing magnetic or optical micromanipulation techniques have unequivocally demonstrated that changes in the balance of forces transmitted across cell surface integrins play a key role in organization of the focal adhesions that mediate cell motility. Application of physical stresses to integrins results in the recruitment of focal adhesion proteins, actin filaments, signaling molecules, and mRNA to the site of force application. Focal adhesions also respond rapidly to mechanical perturbations and may actively regulate their dynamics to control their strength, size, and spatial distribution. For example, application of fluid shear stresses to the apical membranes of confluent endothelium results in almost immediate focal adhesion remodeling at the cell base, with preferential addition of new components in the direction of the applied stress.

It is now clear that directed motility within developing tissues, such as vascular networks, is controlled by local changes in cellular force balance. Mechanical perturbation of tissues or their underlying ECM results in the transmission of forces across integrin receptors, restructuring of the CSK, and activation of various signal transduction cascades in a force-dependent manner. Recent work in Drosophila has confirmed that the migratory motion of cell layers similarly generates mechanical forces that shape the digestive tract. During development of this tissue, movements of the posterior and anterior mesoderm cells compress neighboring stromodeal precursor cells. This mechanical stress results in the translocation of Armadillo transcription factor to the nucleus, where it upregulates expression of *Twist*, a gene required for invagination of the stromodeum. In cells expressing a mutant form of *Twist*, these events can be rescued by the application of an external mechanical force (using a micropipette). Thus, mechanical force per se clearly can impact cell migration in a developmentally relevant way.

Future Directions

A great deal is known about the signaling molecules that invoke the motile response, but the mechanism by which this response becomes spatially localized within the cell to determine the site where motile structures, such as lamellipodia and filopodia, are generated is less well characterized. Future work in this area will rely on advances in imaging technology and new probes to permit the real-time visualization of activated states of signaling molecules with improved spatial resolution. As directional migration also requires the integration of microenvironmental information, including ECM composition, topography, and mechanics, systems that maintain or reproduce these parameters in vitro will greatly facilitate future study of migration within the physical and spatial context in which it normally proceeds.

Glossary

Cytoskeleton: the network of protein filaments in the cytoplasm of eukaryotic cells that gives the cell shape, guides intracellular transport, and coordinates its movement and growth. Its most abundant components are actin microfilaments, microtubules, and intermediate filaments.

Focal adhesion: the localized, spot weld-like attachment site between the cytoskeleton and extracellular matrix that anchors the cell to its adhesive substrate. Attachment is mediated by tightly clustered, transmembrane integrin receptors and a variety of structural and signaling protein components.

Lamellipodia: flat, sheet-like, membrane protrusions supported by an intracellular meshwork of actin filaments that are extended outward at the leading edge of a migrating cell.

Further Reading

Etienne-Manneville, S., and Hall, A. (2002). Rho GTPases in cell biology. *Nature* **420**, 629–635. *A comprehensive overview of the role of the small GTPases in a variety of cellular processes including movement, proliferation, morphology and contraction, phagocytosis, and secretion.*

Geiger, B., Bershadsky, A., Pankov, R., and Yamada, K. M. (2001). Transmembrane crosstalk between the extracellular matrix–cytoskeleton crosstalk. *Nat. Rev. Mol. Cell. Biol.* **2**, 793–805. *An in-depth treatment of the molecular and mechanical signaling events that occur at sites of cell-matrix adhesion, with a special emphasis on the importance of local physical forces.*

Ingber, D. E. (2002). Mechanical signaling and the cellular response to extracellular matrix in angiogenesis and cardiovascular physiology. *Circ. Res.* **91**, 877–887.

Kaverina, I., Krylyshkina, O., and Small, J. V. (2002). Regulation of substrate adhesion dynamics during cell motility. *Intl. J. Biochem. Cell. Biol.* **34**(7), 746–761.

Sheetz, M. P., Felsenfeld, D. P., and Galbraith, C. G. (1998). Cell migration: Regulation of force on extracellular-matrix-integrin complexes. *Trends Cell. Biol.* **8**, 51–54.

Stossel, T. P., Hartwig, J. H., Janmey, P. A., and Kwiatkowski, D. J. (1999). Cell crawling two decades after Abercrombie. *Biochem. Soc. Symp.* **65**, 267–280.

Webb, D. J., Parsons, J. T., and Horwitz, A. F. (2002). Adhesion assembly, disassembly and turnover in migrating cells—over and over and over again. *Nat. Cell Biol.* **4**, E97–E100. *The authors review the coordination of focal adhesion assembly, disassembly and turnover in migrating cells, as well as recent advances in imaging technology that have improved the characterization of these complexes.*

Capsule Biographies

Amy Brock, a graduate student in the Biomedical and Biosciences (BBS) program at Harvard University, is completing her Ph.D. dissertation on cell motility in the laboratory of Dr. Ingber.

Dr. Ingber has made significant contributions to the fields of angiogenesis, matrix biology, and mechanoregulation, and cellular engineering. He is the Judah Folkman Professor of Vascular Biology in the department of Pathology at Harvard Medical School, and Senior Investigator in the Vascular Biology Program at Children's Hospital in Boston, Massachusetts.
His work is supported by grants from the NIH. NASA, DARPA, ARO and DOD.

CHAPTER 3

Ultrastructural Morphometry of Capillary Basement Membrane Thickness

Edward C. Carlson
School of Medicine and Health Sciences, University of North Dakota

Introduction

Capillary Basement Membrane Morphometry Defined

Morphometry is a precise method of measurement that employs conventional morphological methods to estimate the physical dimensions of an unbiased population of biological structures. The structures may be quite large and visible to the naked eye or they may be small and require substantial magnification. In order to measure capillary basement membranes (CBMs), transmission electron microscopic (TEM) methods are applied to tissue samples. The samples must be selected without bias and carefully prepared, accurately measured, corrected for plane of section, and statistically analyzed to provide a faithful numerical description of the measured CBM population.

Historical Background

Basement membranes (BMs) were identified first in the mid-19th century. They were described as homogeneous and sometimes fibrillar extracellular sheets that supported the digestive and respiratory systems, and as continuous matrix barriers between mucous membranes and their underlying capillaries. These remarkably accurate accounts formed the bases of textbook descriptions of BMs until the 1950s, when the advent of TEM provided substantially increased image resolution of biological samples.

Subsequent TEM studies coupled with the development of new methods to isolate BMs led to biochemical and immunochemical analyses, which showed that BMs were comprised largely of collagen, noncollagenous glycoproteins, and proteoglycans assembled in a network that resulted in the electron-dense layer (*basal lamina*) imaged by TEM [1]. Studies of isolated BMs substantially increased our understanding of the production, assembly, and degradation of their molecular components. Importantly, it was shown that although most BMs comprised similar molecular subunits, they were not identical, and a nonunitary concept of BM composition was generally adopted.

BMs also varied widely in their location and morphological presentation, including width (thickness). Some (e.g., the lens capsule and Descemet's membrane in the eye) were extremely thick and could be removed from their underlying tissues with a dissecting microscope. Others, including those of most blood capillaries, were very thin and could not be imaged clearly by light microscopic (LM) techniques. It is now recognized that regardless of location and width, TEM techniques are required to provide adequate images for accurate BM measurements, and this is especially true for the microvasculature where CBMs are particularly narrow.

Although BMs exhibit slow turnover rates, they are nevertheless dynamic structures, and their thickness is altered with age, physiological state, or pathological environment. For example, many microvascular BMs are thickened in diabetes mellitus, leading to renal, retinal, and neurological disorders. Interestingly, BMs of nonvascular tissues (e.g., renal tubules) also show thickness increases in diabetes. Accordingly, it is believed that although BMs from various tissue types are differentially thickened during normal aging or in disease states, the mechanisms leading to such thickening may be similar.

Because CBM thickness is widely regarded as a biomarker for aging and a hallmark for chronic complications of diabetes, numerous studies have centered on this parameter. The purpose of this chapter is to provide the reader with background information regarding the morphological heterogeneity of CBMs, and specifically with some methods by which their thickness may be accurately measured.

Ultrastructural Definitions

Basement Membranes

BMs were first defined by LM as extracellular layers subjacent to epithelia. Because the practical limit of LM resolution is approximately 0.2 μm, these BMs necessarily included considerable quantities of stained ground substance and reticular (small collagen) fibers in addition to the narrow (~100 nm) electron-dense basal lamina of TEM. Nevertheless, these early descriptions were accurate, and subsequent TEM studies confirmed their histological locations and demonstrated that basal laminae consistently colocalized with BMs described by LM.

Subsequently, the term BM was taken over by the electron microscopists, who widened the definition considerably to include a ubiquitous network of basal laminae always closely associated with cell surfaces. Moreover, BMs and basal laminae became synonymous terms for extracellular membranes that formed physical boundaries for a large and uncompartmented connective tissue space. BMs separated the space from nonconnective tissue cell types including epithelia, muscle, and nerve, while connective tissue cells (except adipocytes) were free in the tissue space uncovered by BMs. TEM also provided substantial substructural BM detail, including three distinct layers not recognizable by LM. These included a central, dense layer *(lamina densa)* flanked by two layers that were more electron-lucent *(laminae rarae).*

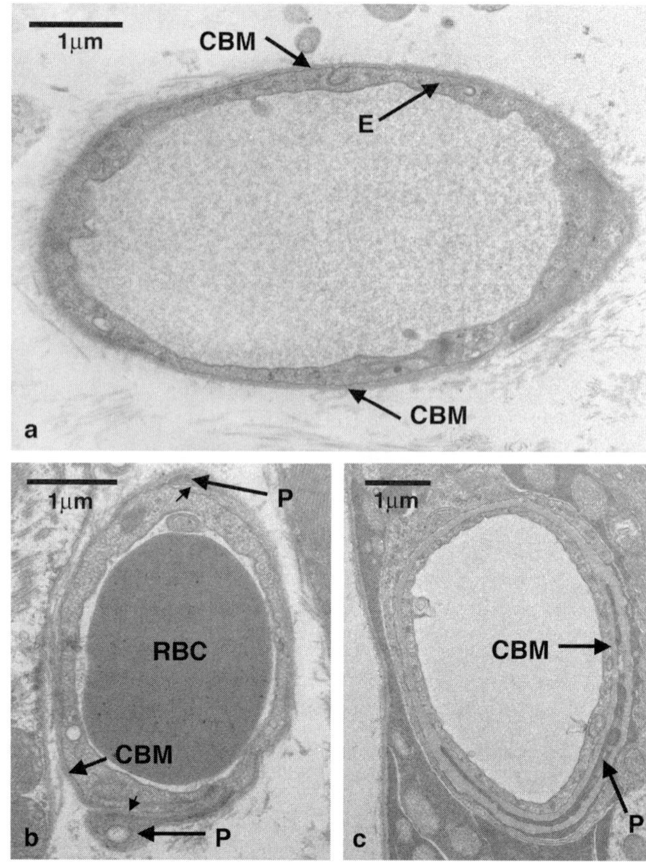

Figure 1 Transmission electron micrographs of sections through capillaries. **(A)** Simple capillary from rat skeletal muscle. A continuous capillary basement membrane (CBM) subtends the endothelium (E) and separates it from the connective tissue space. **(B)** Bovine skeletal muscle capillary surrounded by basement membrane (CBM) and decorated by pericytes (P). A shared basement membrane (*arrows*) separates pericytes from the endothelium. RBC, red blood cell, (c) Rat retinal capillary near the vitreo-retinal border. A pericyte (P) shares the basement membrane (CBM) with the endothelium for nearly half of the capillary circumference.

Capillary Basement Membranes

TEM analyses of many tissues show that all capillaries are surrounded by CBMs that closely subtend their endothelial cell layers. Simple capillaries are surrounded by BMs (basal laminae) composed of a central lamina densa separated from the endothelium by a lamina rara (Figure 1a). Capillaries decorated by perivascular cells (pericytes) are more complex. As in simple capillaries, basal laminae surround the endothelium, but they often split to enclose one or more pericytes as well. In these, CBMs separate endothelial cells from the connective tissue space and subendothelial BMs shared by perictyes, and endothelial cells often fill the narrow space between them (Figures 1b, 1c and 2).

Even more structural sophistication is demonstrated by renal glomerular capillaries, pulmonary alveolar capillaries, and capillaries of the central nervous system (including the retina). In these, both CBM surfaces are demarcated by cell membranes. In the kidney, glomerular BMs are flanked by capillary endothelial cells and podocytes; in the lung, pulmonary alveolar BMs intervene between capillary endothelium and the alveolar epithelium, and in the CNS (retina), CBMs are sandwiched between endothelial and glial (Müller) cells in a highly complex microvasculature that often includes single or multiple layers of perictyes (Figures 1c and 2).

With these structural complexities, an important question relates to the definition of CBM. In this regard, most investigators consider the CBM as the BM subjacent to the capillary endothelium, regardless of whether it is shared by other cell types.

CBMs have membranous or sheet-like shapes and are relatively narrow in one dimension (width) compared to the other two (area). Although both dimensions are subject to changes that may have physiological or pathological significance, width (thickness) is by far the simplest to estimate. Accordingly, investigators have applied various

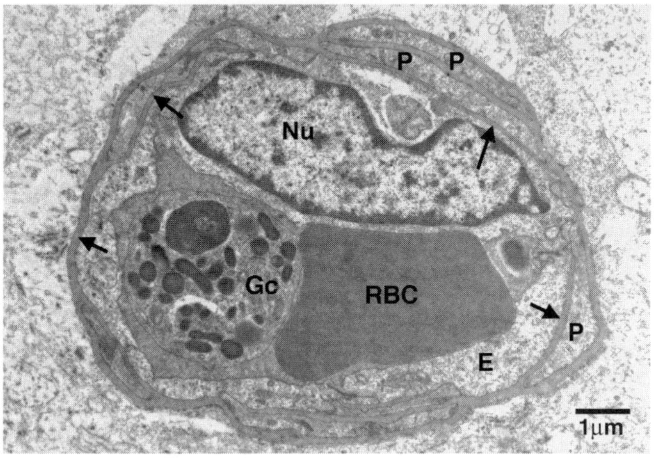

Figure 2 Transmission electron micrograph of section through complex bovine retinal capillary surrounded by numerous pericytes (P). Capillary basement membrane (*arrows*) surrounds the endothelium and often is shared by pericytes. Within the lumen are a red blood cell (RBC) and a granulocyte (Gc). Nu, endothelial cell nucleus.

morphometric techniques in an effort to determine BM thickness in normal and diseased tissues.

Capillary Basement Membrane Thickness in Health and Disease

TEM studies of CBMs in humans and other vertebrates show that regardless of animal age or health, they are continuous along the length of the microvessel and closely adhere to their parent cell types as described previously. CBM thickness, however, is widely variable within and between tissue types both in health and disease.

It is now clear that in many tissues, CBM thickness is a normal function of age and is considered a biomarker for aging. However, thickness increases with age are not seen in all tissues and all animal models, and few rigorous studies have been carried out to address this issue. Nevertheless, most investigators believe modest increases in CBM thickness are a normal age-related process. In certain disease states, however, CBM thickening is excessive and is considered a diagnostic hallmark. This is especially true in chronic stages of diabetes mellitus where most CBMs are thickened, and glomerular and retinal CBMs show remarkable thickness increases in humans and animal models of the disease [2]. In fact, it is fair to say that most CBM thickness studies have been carried out in diabetics in an effort to identify possible patterns that could point to potential molecular regulators of BM thickening. Of these investigations, a large percentage has focused on capillaries in the retina and renal glomerulus, where their diabetic sequelae often lead to retinal degeneration and blindness and/or renal decompensation and death. Moreover, because diabetes often has been cited as a disease of premature aging, recent studies have sought metabolic mechanisms that might possibly explain both normal, benign, age-related CBM thickness increases as well as those of diabetic BM disease.

Several mechanisms are implicated in the pathogenesis of microangiopathy leading to diabetic CBM thickening. These have been reviewed recently [3] and include hyperglycemia-induced increases in Type IV collagen synthesis, decreased expression of matrix metalloproteinases (MMP-2 and 3), and increased tissue inhibitors of metalloproteinase (TIMP). Vascular endothelial growth factor (VEGF) also may be involved because treatment with anti-VEGF antibodies reduces glomerular BM thickening. Also, oxygen radicals/oxidative stress and advanced glycation end products (AGEs) may play a role because aminoguanidine (which inhibits AGE formation, but also has antioxidant properties) attenuates diabetic nephropathy. In addition, the polyol enzymatic pathway is stimulated by hyperglycemia and has been implicated in the chronic sequelae of diabetes. In this regard, hyperpermeability is an early feature of diabetic microangiopathy and is reduced by aldose reductase inhibitors. Because hyperglycemia apparently increases the generation of reactive oxygen species (ROS), activates aldose reductase, and induces AGE-formation, it has been postulated that ROS production may represent a common underlying element in several pathways leading to diabetic microvascular damage, including increased CBM thickness.

Clearly, CBM thickness is an important parameter in a wide variety of biological and medical investigations. Moreover, because thickness changes necessarily alter CBM functional capacities, it is important to accurately measure their physical dimensions in order to identify patterns of change that might offer insights into molecular mechanisms regulating their form and function.

Capillary Basement Membrane Morphometric Methods

The goal of any morphometric study is to produce a numerical description that is as faithful as possible to the object described. This requires close attention to several techniques that potentially could bias the outcome.

Data Sampling

Whenever structural data are sought, a very important question regards sampling procedures. This begins at the level of selecting patients or animals for study, and rules for independent sampling must be strictly obeyed. If bias is introduced at any of the sampling levels (e.g., populations of patients or animals, blocks of tissue, visual fields, sites of measurement), then the study will be flawed and of no value. For example, with some CBM measuring techniques, it is essential that tissue sections be cut perpendicular to the long axis of the capillary. Unfortunately, this has not been closely monitored in many studies, which may explain conflicting data derived from different laboratories.

Tissue Collection and Preparation

Because CBM morphometry requires resolution at the level of ultrastructure, careful TEM techniques must be employed to consistently prepare the tissue for observation [2]. In this regard, whenever possible it is most appropriate to fix the tissue by intravascular perfusion. This should be done as quickly as possible following anesthesia and should include freshly prepared buffered fixatives. Tissues should be harvested and minced quickly, post-fixed in buffered OsO_4, and further dehydrated before embedding. A blinded (unbiased) method should be employed for choosing cured tissue blocks for sectioning. Also, the technical person responsible for generating sections, making electron micrographs, and measuring CBMs should be unaware of the source of the tissue.

Regardless of the method utilized for CBM measurement, it is advantageous to make ultrathin sections that are normal to the long axis of capillaries. In this regard, it is possible to embed some tissues (e.g., peripheral nerve, skeletal muscle, retina, ocular choriocapillaris) in epoxy blocks so that most of its microvessels are oriented perpendicular to the cutting surface. For others, however (e.g., kidney, lung, pancreas), this is not possible, and they must be sectioned randomly. Nevertheless, an effort at perpendicularity should be made by cutting thick (1 μm) sections first to determine tissue orientation and position by LM before proceeding with ultramicrotomy.

Microscopy and Micrography

As an initial preparatory step, the TEM should be stabilized (constant beam current) and calibrated using a standard cross-grating grid (e.g., 2,160 lines/inch) photographed under the same conditions and magnification used for the tissue sections.

Before CBM micrography is carried out, care must be taken to randomize visual fields selected for TEM. Consistency is critical in this regard, and whatever methods are chosen to eliminate selection bias should be applied rigorously to all sections. For example, for selecting skeletal muscle CBMs, it might be decided to utilize only those grid squares that are completely filled with tissue and only those in which the capillaries are cut primarily in cross section (i.e., where they are circular or only slightly elliptical).

Either TEM photomicrographs or digital images may be utilized for ultrastructural morphometry, but regardless of the mode chosen, all parameters (e.g., TEM and enlarger magnifications) must remain constant throughout the study.

Measuring Capillary Basement Membranes

Irrespective of the method used to physically measure CBMs, a most important point regards randomization of measurement sites. A clear plastic grid is helpful in this process (Figure 3). It can be superimposed on the photographic paper during the enlarging procedure [4], and decisions regarding sites of measurement can be determined by noting where the endothelium intersects with the sampling line grid (Figure 2). This eliminates site selection bias because measurements are made only at preselected points, and it also ensures remarkably consistent numbers of measurements per linear CBM distance.

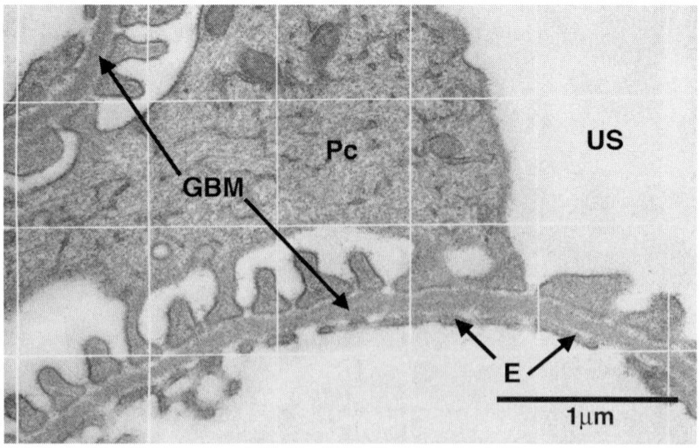

Figure 3 Transmission electron micrograph of renal glomerulus superimposed with a sampling grid. Perpendicular measurements are made at points where the grid intersects with the endothelium. Pc, podocyte; E, endothelium; GBM, glomerular basement membrane; US, urinary space.

Removing Right-Sided Skew and Calculating True CBM Thickness

By definition, CBMs are tube-like, and any series of parallel cuts through them generally will not cross precisely perpendicular to the long axis of the vessel. In fact, because many CBMs are highly folded (e.g., glomerular BM), many cuts will be partially *en face*. As a result, distributions of CBM measurements usually exhibit a right-sided skew. This phenomenon can be eliminated by measuring CBMs only where they are cut in cross section (i.e., only where adjacent endothelial cells show crisp plasmalemmae). However, this process is painstakingly slow and unnecessary because of a major advance in membrane morphometry called the orthogonal intercept method for determining membrane thickness. It was developed in 1979 by Jensen and associates and employs a measuring ruler [5] with a logarithmic scale indicating nine classes of increasing length. The ruler can be a simple transparent plastic device or it may be numerically defined using a digitizing tablet interfaced with a computer designed to measure length [2]. In either case, numbers of measurements in each class are recorded, and their harmonic mean is calculated and then multiplied by $8/3\pi$ [6]. This final calculation is designed to remove the expected right-sided measurement distribution skew [7] and to yield a true CBM thickness.

Statistical Analysis

Statistical studies of true CBM thickness derived from orthogonal intercepts are somewhat problematic because the data from many CBM measurements tend to collapse into a single number and the sample size becomes very small. Accordingly, data from different CBM types can be presented simply as ranges and medians. Alternately, true CBM thickness values can be derived from a specific number of animals, tissue samples, or micrographs, with each set representing one n. In a recent study [2], the latter method was chosen, with one set of 20 electron micrographs (approximately 100 measurements) representing one n (true CBM thickness data point). Standard deviations of these can be calculated, and distributions of control and experimental samples statistically compared using an unpaired t-test. In those instances where data fail to meet parametric testing assumptions of equal variance and normal distribution, nonparametric tests may be employed.

Interpreting the Data

The goal of all scientific data collection should be to generate credible information that can be interpreted in light of previous investigations. In the case of CBM morphometry, however, this is difficult because, as pointed out by Osterby [8], most early CBM thickness studies did not employ unbiased sampling methods—an obvious drawback to their credibility. Nevertheless, attempts must be made to compare and contrast new data with that from previous studies.

Future CBM thickness studies must be centered on unbiased samples and carried out by rigorously controlled methods. However, correlation with previous work will continue to be extremely difficult for reasons cited previously. Moreover, recent data strongly suggest that large differences exist in CBM thickness between and within individuals and between and within tissue types. This real CBM thickness variation demands that conclusions be drawn from investigations with large sample sizes prepared and analyzed under rigorous conditions. Fortunately, excellent progress has been forthcoming in the efficiency of TEM techniques and in development of appropriate morphometry software, which should be advantageous in future experimental design.

Conclusions

CBM thickness is altered with age, physiological conditions, and in several disease states—most notably diabetes mellitus. This important parameter can best be derived from well-controlled, state-of-the-art TEM morphometry studies carried out on capillaries selected by unbiased sampling methods. Given the known wide variation in CBM thickness within and between tissue types, large sample sizes will be critical for accurate estimation of CBM thickness. Recent technical advances make such studies possible, and it is hoped that new investigations will identify patterns of CBM thickness change from which can be inferred possible molecular mechanisms of metabolic dysfunction, and ultimately that these can be controlled or eliminated through therapeutic intervention.

Glossary

Basement membrane: Continuous electron-dense sheets of extracellular materials composed of collagenous and noncollagenous glycoproteins and proteoglycans, and located wherever nonconnective tissue cells border the connective tissue space. Many basement membranes thicken with age as well as in several disease states, notably diabetes mellitus.

Capillary: A small microvascular channel, generally less than 10 μm in diameter, that connects arterioles and venules; they comprise endothelial lining cells occasionally overlaid by perivascular cells (pericytes) and are surrounded by basement membrane.

Morphometry: A precise method of measurement employing conventional morphological methods to estimate the physical dimensions of an unbiased population of biological structures. The structures may be quite large and visible to the naked eye or they may require substantial magnification for measurement.

Ultrastructure: Structural elements of cells and tissues that require electron microscopic observation; it generally includes structures smaller than 0.2 μm.

References

1. Kalluri, R. (2003). Basement membranes: Structure, assembly and role in tumour angiogenesis. *Nature Reviews* **3**, 422–433. *Reviews molecular components of basement membranes with emphasis on vascular basement membranes.*
2. Carlson, E., Audette, J., Veitenheimer, N., Risan, J., Laturnus, L., and Epstein, P. (2003). Ultrastructural morphometry of capillary basement membrane thickness in normal and transgenic diabetic mice. *Anat. Rec.* **271A**, 332–341. *Includes statistical data on basement membrane thickness changes in capillaries from ten different tissues from normal and transgenic diabetic mice.*
3. Tsilibary, E. (2003). Microvascular basement membranes in diabetes mellitus. *J. Pathol.* **200**, 537–546. *Reviews the major molecular mechanisms by which capillary basement membrane thickness may be altered.*
4. Dische, F. (1992). Measurement of glomerular basement membrane thickness and its application to the diagnosis of thin-membrane nephropathy. *Arch. Pathol. Lab. Med.* **116**, 43–49. *Contains an excellent discussion of the practical use of the orthogonal intercept method for capillary basement membrane morphometry.*
5. Hirose, K., Osterby, R., Nozawa, M., and Gundersen, H. (1982). Development of glomerular lesions in experimental long-term diabetes in the rat. *Kidney Intl.* **21**, 689–695.
6. Jensen, E., Gunderson, H., and Osterby, R. (1979). Determination of membrane thickness distribution from orthogonal intercepts. *J. Microsci.* **115**, 19–33. *Describes a major breakthrough method for determining true thickness of sections of a folded membrane.*
7. Gundersen, H., and Osterby, R. (1972). Statistical analysis of transformations leading to normal distribution of measurements of the peripheral glomerular basement membrane. *J. Microsci.* **97**, 293–399.
8. Osterby, R. (1990). Basement membrane morphology in diabetes mellitus. In *Diabetes Mellitus* (R. Rifkin and D. Porte, Jr., eds.), pp. 220–233. Stamford, CT: Appleton & Lange. *Review of basement membrane morphology and its alterations in diabetes mellitus.*

Further Reading

Carlson, E., Audette, J., Klevay, L., Nguyen, H., and Epstein, P. (1997). Ultrastructural and functional analyses of nephropathy in calmodulin-induced diabetic transgenic mice. *Anat. Rec.* **247**, 9–19.

Siperstein, M., Unger, R., and Madison, L. (1968). Studies of muscle capillary basement membranes in normal subjects, diabetic, and prediabetic patients. *J. Clin. Invest.* **47**, 1973–1999. *The first ultrastructural morphometry investigation of relatively large numbers of capillary basement membranes in human tissue.*

Capsule Biography

Edward Carlson is professor and chairman of the Department of Anatomy and Cell Biology at the University of North Dakota School of Medicine and Health Sciences, in Grand Forks, North Dakota. He received his Ph.D. from the University of North Dakota and previously served as a faculty member in Departments of Human Anatomy at the University of Arizona, and the University of California at Davis, Schools of Medicine. The author's principal research interests include the morphology and cell biology of basement membranes in health and disease.

CHAPTER 4

IL-7 Receptor and Microvascular Endothelial Cells

[1]Danuta Dus, [1]Agnieszka Krawczenko, and [2]Claudine Kieda

[1]Ludwik Hirszfeld Institute of Immunology and Experimental Therapy, Polish Academy of Sciences, Wroclaw, Poland

[2]Centre de Biophysique Moléculaire (CNRS), Orléans, France

Interleukin 7 (IL-7) is a stromal cell–derived cytokine that is critical for B- and T-lymphocytes development. It was originally isolated as a murine pro-B cells growth factor and called lymphopoietin-1. Interleukin 7 functions primarily as a growth- and antiapoptotic factor in the early stages of B- and T-cell development. When bound by proteoglycans, it could exert a local biological activity. It was also shown that IL-7 is a potent endothelial growth factor and has an impact on the process of lymphangiogenesis, enhancing the expression of lymphatic markers on human endothelial cells.

Receptor for IL-7

Structure

Interleukin-7 can act on various cells through its receptor IL-7R. IL-7R is a heterodimer that consists of an α chain (CD127) that specifically binds IL-7 and the common γ chain γ_c (CD132), which is also a component of several other cytokine receptors: IL-2, IL-4, IL-9, IL-15, and IL-21. Interleukin-7 receptor is a member of the hematopoietin/cytokine receptor superfamily; both IL-7Rα and γ_c have a pair of conserved extracellular cysteine residues and an extracellular Trp-Ser-X-Trp-Ser motif and lack intrinsic tyrosine kinase activity (Figure 1). The α chain of IL-7R can also bind thymic stromal-derived lymphopoietin (TSLP), a cytokine that is important in early stages of B- and T-cell development. IL-7Rα is a type 1 membrane glycoprotein. It has a 220-amino acid extracellular domain, a 25-amino acid transmembrane fragment, and a 195-amino acid cytoplasmic tail. The mature protein has an observed molecular weight of about 75 kDa, although the predicted molecular weight is 49.5 kDa; the difference is probably caused by post-translational modification of one or more of the six potential N-linked glycosylation sites. The α chain exists in membrane-bound and soluble forms. Soluble forms of the IL-7R are produced by alternative splicing of the IL-7R gene, which is located on human chromosome 5p13 and mouse chromosome 15. The murine IL-7 receptor shows 64 percent sequence identity with the human protein. For high-affinity binding of IL-7, necessary for cell survival or proliferation signal transduction, both α and γ_c chains are required (Figure 1). On T cells, high- and low-affinity receptors are expressed according to the state of activation of the cell.

Signal Transduction

Interleukin 7 can signal through several nonreceptor tyrosine kinases associated with the cytoplasmic tail of the receptor. This includes the JAK/STAT pathway, Src family kinases ($p59^{fyn}$ in mice and $p59^{fyn}$ and $p53/p56^{lyn}$ in humans), and PI-3 kinase. In T lymphocytes, the MAPK family (p38) is also involved. IL-7 activates JAK1 and JAK3 kinases, which are also activated by the cytokines that share the γ_c receptor subunit. JAK kinases can further activate signal transducers and activators of transcription (STAT1, STAT3, and STAT5), which in turn translocate to the nucleus and activate DNA transcription. It has been demonstrated that stimulation of the pro-T cells by IL-7 results in clathrin phosphorylation. Because clathrin is involved in internalization of many receptors, its phosphorylation by IL-7 may affect the internalization of the IL-7 receptor. The disorder in JAK-mediated signal transduction is a cause for the immunodeficiency syndrome severe combined immunodeficiency (SCID). The mutation in the JAK3 kinase gene as well as in the α chain gene of the IL-7 receptor results in autosomal, recessive SCID, whereas mutation in the γ_c chain gene causes X-linked SCID (SCID-X).

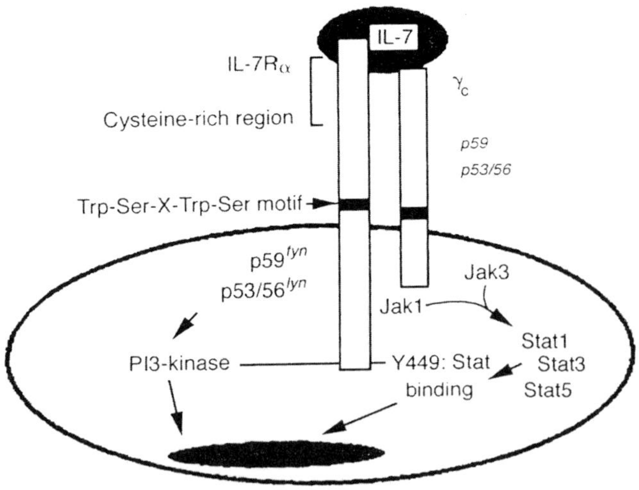

Figure 1 The Schematic Structure of IL-7R. The receptor for IL-7 consists of a specific α chain and a common γc chain. The IL-7Rα is also a part of thymic stromal lymphopoietin (TSLP), whereas the γc chain is present in other cytokines such as IL-2, IL-4, IL-9, IL-15, and IL-21. The cellular signaling upon binding of IL-7 to IL-7R can be mediated by the JAK/STAT pathway, PI-3 kinase, or Src kinases family. Reprinted from "Biological and clinical implications of interleukin-7 and lymphopoiesis" by P. M. Appasamy from *Cytokines, Cellular & Molecular Therapy*, www.tandf.no/ccmt, 1999, **5**, 25–39, by permission of Taylor & Francis AS.

Other details describing the structure of IL-7R and IL-7 signaling have been reviewed elsewhere [1–3].

IL-7R Distribution

The presence of IL-7R was reported mainly in hematopoietic human and murine cell systems and selected myeloid cells. The IL-7Rα chain is present on bone marrow progenitor cells, lymphoid T and B precursors, and on most mature T cells. Recently, its presence on epithelial cells and neurons, as well as on human microvascular endothelial cells and some tumors, was reported.

After IL-7 binding, IL-7R transduces several biological signals, playing an important role in the function of the immune system. The cytokine, *via* its receptor, supports the maturation and augment proliferation of pre-B and pre-T cells. Murine B lymphocyte development is critically dependent on IL-7 availability. The interleukin-7 receptor is essential for Peyer's patches (PP) development. Lymphoid organ-specific mesenchymal tissue that possesses IL-7R is involved in the initial phase of PP formation. Accumulating evidence indicates that both PP and lymph nodes (LN) originate from a common precursor, characterized by the presence of IL-7R. The precursor differentiation into LN or PP, and their further functional properties, are controlled by the local environment.

The Presence of IL-7R on Endothelial Cells

The endothelium is a type of highly specialized, structurally and functionally heterogenous mesodermal-derived epithelium. Microvascular endothelium controls the body compartmentation and homing of lymphocytes into lymphoid as well as nonlymphoid sites, in a tissue-specific manner. Selectins (E- and P-) and other endothelium-specific adhesive molecules called addressins (ligands for selectins) initiate the process of leukocytes' adhesion to the vascular wall. Under the chemoattractive influence of endothelium-presented chemokines, activated leukocytes adhere firmly and finally transmigrate through the vascular wall to the surrounding tissues. Endothelial cells play fundamental roles not only in normal processes such as wound healing but also in pathological ones such as atherosclerosis, inflammation, and cancer progression and metastasis.

In 1993 it was shown that murine endothelial cells from peripheral lymph nodes could be specifically activated by IL-7. In cells grown in the presence of IL-7, expression of lectin adhesion molecule and a selective induction of addressin (MECA 79 antigen) expression were noticed. These observations strongly suggested the presence of IL-7R on murine endothelial cells. Further study revealed that human endothelial cells possess the receptor for IL-7. The α chain of the receptor (IL-7R) was detected in five endothelial cell lines, derived from peripheral or mesenteric lymph nodes, lung, and appendix.

It has been postulated that there is a close association of the hematopoietic and endothelial developmental lineages because they both arise from a common precursor, the hemangioblast. Hence, the fact of the presence of an IL-7 receptor on hematopoietic as well as on endothelial cells may support this hypothesis.

Activity of IL-7 on endothelial cells is achieved through activation of its specific receptor, resulting in induction of adhesion molecules and/or cytokines. It could be hypothesized that IL-7, activating endothelial cells, may act indirectly, inducing the production of another growth factor such as VEGF or bFGF; however, no evidence supports this hypothesis. It has also been shown that IL-7 can induce signal transduction events in cells that do not express IL-7R because of the ability to involve other surface receptors such as FLT3 and c-kit.

IL-7R in Cancer

Because IL-7 is a known lymphocyte growth and differentiation factor, it also participates in the development of several hematological malignancies. However, the IL-7 receptor mRNA has also been detected in many nonhematological tumor in vitro cell lines. IL-7 was found to be a potent growth factor for breast cancer cells. Breast cancer cell lines, cancer tissue biopsies, and normal breast tissue express IL-7R mRNA. Moreover, the expression of IL-7R and its signaling molecules is higher in tumor tissue of patients with a poor clinical outcome. The aberrant expression of IL-7, IL-7R, and their signaling complex may influence the proliferation of breast cancer cells and disease progression, supporting lymphangiogenesis of the solid tumors.

References

1. Appasamy, P. M. (1999). Biological and clinical implications of interleukin-7 and lymphopoiesis. *Cytokines. Cell. Mol. Ther.* **5**, 25–39. *An extensive review about IL-7 and IL-7R, their biological functions, and potential clinical applications of IL-7.*
2. Hofmeister, R., Khaled, A. R., Benbernou, N., Rajnavolgyi, E., Muegge, K., and Durum, S. K. (1999). Interleukin-7: Physiological roles and mechanisms of action. *Cytokine Growth Factor Rev* **10**, 41–60. *This publication is a very good review describing two main aspects of the biological activity of IL-7 toward lymphoid cells: the trophic effect and an effect supporting V(D)J recombination.*
3. Fry, T. J., and Mackall, C. L. (2002). Interleukin-7: From bench to clinic. *Blood* **99**, 3892–3904. *An excellent review focused on biological activity of IL-7 from medical and clinical point of view.*

Further Reading

Al-Rawi, M. A. A., Rmali, K., Watkins, G., Mansel, R. E., and Jiang, W. G. (2004). Aberrant expression of interleukin-7 (IL-7) and its signalling complex in human breast cancer. *Eur. J. Cancer* **40**, 494–502.

Bizouarne, N., Denis, V., Legrand, A., Monsigny, M., and Kieda, C. (1993). A SV-40 immortalized murine endothelial cell line from peripheral lymph node high endothelium expresses a new alpha-L-fucose binding protein. *Biol. Cell.* **79**, 209–218.

Denis, V., Dupuis, P., Bizouarne, N., de O. Sampaio, S., Hong, L., Lebret, M., Monsigny, M., Nakache, M., and Kieda, C. (1996). Selective induction of peripheral and mucosal endothelial cell addressins with peripheral lymph nodes and Peyer's patch cell-conditioned media. *J. Leukocyte Biol.* **60**, 744–752.

Dus, D., Krawczenko, A., Zalecki, P., Paprocka, M., Wiedlocha, A., Goupille, C., and Kieda, C. (2003). IL-7 receptor is present on human microvascular endothelial cells. *Immunol Lett.* **86**, 163–168. *This paper describes for the first time the presence of IL-7R on human microvascular endothelial cells.*

Goodwin, R. G., Friend, D., Ziegler, S. F., Jerzy, R., Falk, B. A., Gimpel, S., Cosman, D., Dower, S. K., March, C. J., and Namen, A. E. (1990). Cloning of the human and murine interleukin-7 receptors: Demonstration of a soluble form and homology to a new receptor superfamily. *Cell* **60**, 941–951.

Kieda, C., Paprocka, M., Krawczenko, A., Zalecki, P., Dupuis, P., Monsigny, M., Radzikowski, C., and Dus, D. (2002). New human microvascular endothelial cell lines with specific adhesion molecules phenotypes. *Endothelium* **9**, 247–261.

Yoshida, H., Naito, A., Inoue, J-I., Satoh, M., Santee-Cooper, S. M., Ware, C. F., Togawa, A., Nishikawa, S., and Nishikawa, S-I. (2002). Different cytokines induce surface lymphotoxin-$\alpha\beta$ on IL-7 receptor-α cells that differentially engender lymph nodes and Peyer's patches. *Immunity* **17**, 823–833.

Capsule Biographies

Danuta Dus, Ph.D. hab., is a head of Laboratory of Cellular Interactions at the Institute of Immunology and Experimental Therapy at the Polish Academy of Sciences in Wroclaw, Poland; Agnieszka Krawczenko, Ph.D., is an assistant professor at the Laboratory of Cellular Interactions. The scientific activity of the laboratory concerns predictive markers of metastasis. The aims of the study are molecular mechanisms engaged in the process of mutual interactions of tumor cells with endothelial cells at the site of tumor cell extravasation, which determine the organ specificity of metastatic secondary growth. The work is supported by the Ministry of Scientific Research and Information Technology, The State Committee for Scientific Research (KBN), and by The Foundation for Polish Science (FNP).

Claudine Kieda, Ph.D., is heading the Cell Recognition: Endogenous Lectins team in the Centre for Molecular Biophysics in Orléans, France. The laboratory is focusing on molecular mechanisms of cellular interaction between endothelium and circulating cells leading to invasion. She first discovered endogenous lectins in lymphocytes; her actual special interests are glycoconjugates and their receptors as key recognition signals in inflammation and metastasis. She was awarded the CNRS bronze medal, and her work is supported by ARC, CNRS, and private foundations such as Jérôme Lejeune.

CHAPTER 5

Matrix Metalloproteinases, Tissue Inhibitors of Metalloproteinases, and the Microvasculature

Cecilia A. Fernández and Marsha A. Moses

Vascular Biology Program, and Department of Surgery, Children's Hospital, Boston, and Department of Surgery, Harvard Medical School, Boston, Massachusetts

An accumulating body of evidence suggests that the remodeling of the extracellular matrix (ECM) during angiogenesis is accomplished largely through the activity of the matrix metalloproteinases (MMPs). MMPs form a large family of metal-dependent proteolytic enzymes, which have been classified on the basis of their substrate (i.e., the ECM components that they degrade) specificity. There are four major classes of MMPs: collagenases, gelatinases, stromelysins, and membrane-type MMPs. The activity of MMPs is the rate-limiting step in ECM turnover. MMP activity is regulated at three different levels: transcription, activation of MMPs from their inactive (latent) proenzyme form, and by the activity of their endogenous inhibitors, the tissue inhibitors of matrix metalloproteinase (TIMPs) [1].

Matrix Metalloproteinases

MMPs have been implicated in promoting angiogenesis as well as tumor growth and metastasis. Capillary endothelial cells (EC), under the influence of angiogenic mitogens, secrete MMPs, and the proteolytic activity of these enzymes facilitates the sprouting of EC out from the parent venule as well as the migration of these cells at the leading edge of the new capillary sprout through the ECM during angiogenesis (Figure 1). For example, MMP-2 activity has been shown to be concentrated at the invadopodia of a variety of cell lines. In addition, MMP activity is required for cancer cell migration and invasion through extracellular spaces as well as for the cell's ability to invade a nearby blood vessel, extravasate at a distant site, and then to finally invade the distant tissue in order to seed a new metastatic lesion. MMPs have also been shown to cleave membrane-bound growth factors, such as HB-EGF and TGFβ, as well as to release angiogenic mitogens (bFGF) deposited in the matrix.

Recent studies have demonstrated that MMPs are also important at the earliest stages of the angiogenic program called the *angiogenic switch*. In a model of tumor progression that recapitulates the earliest events associated with the onset of neovascularization, our laboratory has shown that MMP-2 is a critical component of the angiogenic switch. In this study, both MMP-2 protein and activity were significantly upregulated during the transition from a preangiogenic to an angiogenic tumor [2]. Since that report, other groups have shown that MMP-9 may be a regulator of the angiogenic switch in a pancreatic tumor model [3]. Taken together, these data support the conclusion that MMP activity is one of the earliest and most sustained activities in angiogenesis. Because MMP activity is required for angiogenesis and tumor growth, it is not surprising that MMPs can be found in the urine of cancer patients and that they are independent predictors of disease status [4]. MMPs were reproducibly detected in the urine of cancer patients and correlated with disease status but were not found in the urine of healthy age-matched, sex-matched controls or in the urine of patients who showed no evidence of disease at the time of sampling and testing.

These studies highlight the importance of MMPs as regulators of angiogenesis and subsequent tumor progression. Based on these data, MMP activity represents a potentially powerful target for cancer therapy. In fact, it has been shown that a shift in the proteolytic balance in favor of the MMP inhibitors would effectively block angiogenesis, and several synthetic inhibitors of MMPs have been tested in clinical

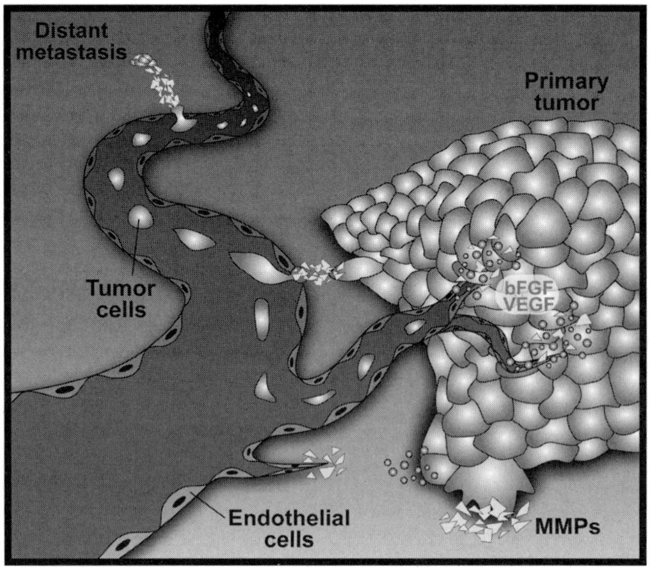

Figure 1 Role of MMPs in angiogenesis, tumor growth and metastasis.

trials designed to determine their efficacy in the treatment of solid tumors and metastasis. Unfortunately, the indiscriminate use of synthetic MMP inhibitors has proven to be less successful than originally anticipated and has also resulted in side effects that preclude their use at therapeutically effective doses. Current clinical trials are testing certain of these inhibitors in combination therapies. However, it is possible that their broad specificity and, in some cases, limited bioavailability may ultimately mean that this generation of synthetic MMP inhibitors will not be as promising as once hoped. In light of these disappointing outcomes, there is renewed interest in the endogenous MMP inhibitors, which, although larger in size, may offer advantages over synthetics, with respect to both a reduction of cytotoxicity and their lack of immunogenicity as a function of their being a part of a naturally occurring protein.

Although MMPs are routinely regarded as being proangiogenic molecules, MMPs have also recently been shown to process antiangiogenic fragments from their inactive parent molecule, suggesting that they can also function as negative regulators of angiogenesis. For example, we have shown that MMP-2 can proteolytically cleave plasminogen to yield the endogenous inhibitor, angiostatin. Other groups have reported that, depending on tumor type, additional MMP family members also share this activity. Similarly, MMPs have been shown to participate in the processing of endostatin from collagen XVIII.

Tissue Inhibitors of Metalloproteinases

There are four known endogenous inhibitors of MMPs that have been cloned to date: TIMP-1, -2, -3, and -4. These four members of the TIMP family have been shown to inhibit active MMPs equally well, although some limited degree of specificity exists. For example, TIMP-2 binds preferentially to MMP-2, whereas TIMP-1 binds to MMP-9 with higher affinity than to other MMPs. The amino acid sequence of TIMPs is highly conserved (~45%) and includes 12 cysteine residues that form 6 disulfide-bonded loop structures. The junction of the first three loops has been identified as the MMP inhibitory site by both biochemical and structural studies. In fact, the first three loops expressed alone as a truncated protein have been shown to be sufficient to inhibit MMP activity. These three loops have been popularly referred to as the N-terminal domain. The C-terminal domain, comprising the remaining three disulfide-bonded loops, is more variable among the TIMP family members in terms of its primary structure and is believed to confer some of the specificity observed for certain TIMPs [5]. For example, interactions between the C-terminal domain of TIMP-2 and the PEX domain of MMP-2 are thought to stabilize the enzyme inhibitor complex, as well as to promote pro-MMP-2 activation in a trimeric complex with MT1-MMP.

It is becoming increasingly clear that TIMPs are multifunctional proteins that affect a variety of cellular activities, such as cell growth and apoptosis. Most prominent among these functions is the ability of certain TIMPs to regulate vascular and tumor cell growth. For many years, this potential has been predicated on their ability to directly inhibit the MMP activity by binding to the enzymes' active site. Since the first report that TIMPs could inhibit angiogenesis in vitro and in vivo [6], it has largely been assumed that all TIMPs inhibit angiogenesis. However, recent studies now illustrate important differences in the ability of these family members to be bona fide inhibitors of angiogenesis in vivo. Although TIMP-1, TIMP-2, and TIMP-3 are all inhibitors of capillary EC migration, only TIMP-2 has been shown to inhibit the proliferation of normal capillary endothelial cells [7]. TIMP-1 has actually been found to be a modest stimulator of capillary EC proliferation as well as angiogenesis in vivo, whereas TIMP-3 has no significant effect on normal capillary EC proliferation. TIMP-4 has not yet been tested in these systems.

These results led us to hypothesize that the antiproliferative effects of TIMP-2 are, in fact, a unique feature of this TIMP and represent a second antiangiogenic activity in this molecule. Recently, it has been reported that this second antiangiogenic site which is responsible for directly inhibiting capillary EC proliferation, is housed in the C-terminal domain of TIMP-2 and, more specifically, in Loop 6. (Figure 2) Interestingly, while the antiproliferative domain of TIMP-2 was a potent inhibitor of corneal neovascularization when angiogenesis was stimulated by the exogenous addition of an angiogenic mitogen, the MMP-inhibitory domain resulted in only modest inhibition that was not statistically different from a mutant form of the protein that lacked MMP inhibitory activity [8]. In fact, synthetic MMP inhibitors, such as BB-94, do not inhibit angiogenesis in this system, nor do monospecific, MMP-2 immuno-

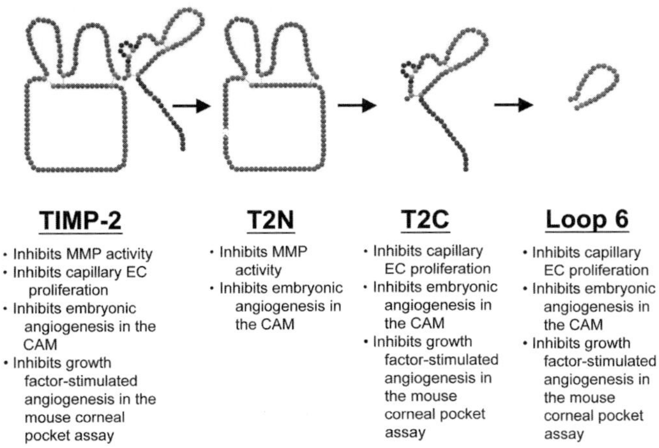

Figure 2 Uncoupling of the anti-angiogenic domains of TIMP-2. TIMP-2 possesses two structurally independent anti-angiogenic domains, T2N and T2C. T2N inhibits MMP activity but not capillary endothelial cell proliferation, while T2C inhibits capillary endothelial cell proliferation but not MMP activity. T2N inhibits embryonic angiogenesis *in vivo* as demonstrated in the chick chorioallantoic membrane (CAM) assay, while T2C inhibits *both* embryonic and mitogen-stimulated angiogenesis, as demonstrated in the CAM and in the mouse corneal pocket assays. The anti-proliferative activity of T2C has been further isolated to Loop 6 of TIMP-2. This 24 amino acid domain is, in and of itself, a potent inhibitor of angiogenesis *in vivo* as shown in the two distinct *in vivo* models.

neutralizing antibodies. These results suggest that direct MMP inhibition alone may not be sufficient to inhibit the robust angiogenesis associated with pathological neovascularization[8].

It is possible th.at the antiproliferative effects of TIMP-2 are the result of its interaction with the PEX domain of MMP-2. As mentioned previously, TIMP-2 has been shown to participate in the cell surface activation of pro-MMP-2, and the C-terminal domain alone has been shown to inhibit this process, presumably by hijacking pro-MMP-2 from the activating complex. This interaction has now been shown to be mediated by the C-terminal tail of TIMP-2, in that a mutant form of the TIMP-2 carboxy-terminal domain containing the C-terminal tail of TIMP-4 did not result in the inhibition of pro-MMP-2 activation [9]. Given these results, it is unlikely that the antiangiogenic effects of Loop 6, which is not involved in these interactions, is the result of indirect MMP inhibition.

Much remains to be learned about the multifunctional and sometimes contradictory activities of TIMPs. For example, TIMP overexpression in cancer cell lines has been shown to result in decreased tumor growth and tumor cell colonization in some systems, while stimulating tumor growth in other systems. Similarly, whereas certain TIMPs inhibit apoptosis, others have been shown to promote it. These effects often vary depending on the specific TIMP and cell type tested and the method of gene delivery utilized. These studies continue to teach us that the physiological *context* in which these inhibitors and their cognate enzymes function plays a key role in determining their ultimate effect in vivo both with respect to regulating the vasculature as well as other important physiological systems.

Glossary

Angiogenesis: The process of new capillary formation from a pre-existing vessel.

Matrix metalloproteinases (MMPs): A multigene family of metal-dependent enzymes, whose activity is considered to be a rate-limiting step of extracellular matrix degradation. The growing family of MMPs is currently composed of 28 members.

Tissue inhibitors of metalloproteinases (TIMPs): A family of endogenous inhibitors of metalloproteinases. Four family members have been cloned and expressed to date.

Acknowledgments

The authors are grateful for the support of NIH-2PO1CA455, NIH P50DK065298 and The Fortin Charitable Foundation. The authors also acknowledge Kristin Gullage for figure preparation.

References

1. Visse, R., and Nagase, H. (2003). Matrix metalloproteinases and tissue inhibitors of metalloproteinases: Structure, function, and biochemistry. *Circ. Res.* **92**, 827–839. *A recent and comprehensive review of MMPs and TIMPs.*
2. Fang, J., Shing, Y., Wiederschain, D., Yan, L., Butterfield, C., Jackson, G., Harper, J., Tamvakopoulos, G., and Moses, M. A. (2000). Matrix metalloproteinase-2 is required for the switch to the angiogenic phenotype in a tumor model. *Proc. Natl. Acad. Sci. USA* **97**, 3884–3889.
3. Bergers, G., Brekken, R., McMahon, G., Vu, T. H., Itoh, T., Tamaki, K., Tanzawa, K., Thorpe, P., Itohara, S., Werb, Z., and Hanahan, D. (2000). Matrix metalloproteinase-9 triggers the angiogenic switch during carcinogenesis. *Nat. Cell Biol.* **2**, 737–744.
4. Moses, M. A., Wiederschain, D., Loughlin, K. R., Zurakowski, D., Lamb, C. C., and Freeman, M. R. (1998). Increased incidence of matrix metalloproteinases in urine of cancer patients. *Cancer Res.* **58**, 1395–1399.
5. Brew, K., Dinakarpandian, D., and Nagase, H. (2000). Tissue inhibitors of metalloproteinases: evolution, structure and function. *Biochim. Biophys. Acta* **1477**, 267–283.
6. Moses, M. A., Sudhalter, J., and Langer, R. (1990). Identification of an inhibitor of neovascularization from cartilage. *Science* **248**, 1408–1410.
7. Murphy, A. N., Unsworth, E. J., and Stetler-Stevenson, W. G. (1993). Tissue inhibitor of metalloproteinases-2 inhibits bFGF-induced human microvascular endothelial cell proliferation. *J. Cell Physiol.* **157**, 351–358.
8. Fernandez, C. A., Butterfield, C., Jackson, G., and Moses, M. A. (2003). Structural and functional uncoupling of the enzymatic and angiogenic inhibitory activities of tissue inhibitor of metalloproteinase-2 (TIMP-2): loop 6 is a novel angiogenesis inhibitor. *J. Biol. Chem.* **278**, 40989–40995. *A recent structure/function analysis that identifies and characterizes the antiangiogenic domains of TIMP-2.*
9. Kai, H. S., Butler, G. S., Morrison, C. J., King, A. E., Pelman, G. R., and Overall, C. M. (2002). Utilization of a novel recombinant myoglobin fusion protein expression system to characterize the TIMP-4 and TIMP-2 C-terminal domain and tails by mutagenesis: The importance of acidic residues in binding the MMP-2 Hemopexin C domain. *J. Biol. Chem.* **8**, 8.

Capsule Biography

The authors are research scientists in the Vascular Biology Program at Children's Hospital and Harvard Medical School. Dr. Moses is an Associate Professor in the Department of Surgery at Harvard Medical School. Work from this group demonstrated for the first time that the inhibition of MMP activity resulted in the inhibition of angiogenesis and subsequent tumor growth, and that MMP activity was critical for the switch to the angiogenic phenotype. They have published extensively on the role of MMPs and TIMPs in the regulation of vascular growth and angiogenesis.

CHAPTER 6

Matrix Metalloproteinases and Their Inhibitors

Iwona Fijalkowska and Rubin M. Tuder

Divisions of Cardiopulmonary Pathology and Pulmonary and Critical Medicine, Departments of Pathology and Medicine, Johns Hopkins University School of Medicine, Baltimore, Maryland

Introduction

As important as the inner organelles of the cell, the extracellular matrix (ECM) plays fundamental roles in cell function. The molecular framework of the extracellular matrix serves several functions in cell-to-cell communication and integration of environmental cues to a cell's differentiation and fate. We have learned that the extracellular matrix is as plastic and dynamic as the cells that produce them. Critical in this molecular and structural plasticity is the ability to cleave the underlying molecular matrix units, which is accomplished by a vast array of enzymes with a wide spectrum of substrate specificities. Metalloproteases (MMPs) or matrixins refer to a wide superfamily of enzymes that requires the presence of a highly conserved zinc-binding motif for catalysis [1, 2]. MMPs are functionally, structurally, and at the DNA sequence homology related to a collagenase found in the tail of a tadpole undergoing metamorphosis, which was the first member to be described within this rapidly expanding superfamily of enzymes. By cleaving the molecular extracellular framework surrounding cells, MMPs contribute to cell renewal and tissue formation, both in the embryo and in the adult system. Their action continues in critical biological processes, such as tissue remodeling, would healing, and angiogenesis. Furthermore, there is growing evidence that, despite their enzymatic activity, MMPs also act as cellular regulatory factors in areas of cellular signaling, growth, differentiation, and death. These broad functions modulate or control interactions of a cell with its immediate surroundings and with its external environment.

In this chapter, the authors aim at summarizing the newest knowledge on the role of MMPs in physiological and pathophysiological processes, with an emphasis in vascular and pulmonary vascular biology.

Molecular Structure

Currently a total of 23 enzymes out of 29 MMPs have been identified in humans. Five MMPs are shorter isoforms of full-length enzymes, and one is annotated as MMP-like1 protein. Additionally, several plant and nonvertebrate MMPs have been found and characterized.

All MMP family members share a similar multidomain structure. The general structural pattern consists of a prodomain, a catalytic domain, a hinge region, and a hemopexin domain. The prodomain comprises a signal peptide [of about 20 amino acids (aa)] and a propeptide region, of about 80 aa. The signal peptide is cleaved off the molecule during its transport through the endoplasmic reticulum for cell release. The propeptide contains two specific regions, which as based on their biochemical properties, mechanistically account for the activation of the zymogen (i.e., latent form) to the active MMP. These two regions consist of a protease-sensitive sequence located in the second out of three helices and a unique, cysteine-switch motif PRCGXPD, within the peptide tail.

The catalytic domain (of about 170 aa) consists of five β-sheet and three α-helix structures linked by a connecting loop. Three calcium ions and one structural zinc atom strengthen and stabilize the structure. The active site with its characteristic metal-binding motif HEXGHXXGXXH is located near the C-terminus, a highly conserved (56% to 64%) portion of the protein. The second zinc atom, also known as the active zinc, is coordinated by three histidines present within this motif. The substrate binding region is located adjacent to the active site. Unlike the active site, the binding pocket reveals a high level of sequence variability among the different MMP classes, and thus accounts for the specificity of substrate recognition by the MMPs.

The hemopexin-like domain consists of four hemopexin-like repeats (containing about 210 aa) and occupies the C-terminal part of the molecule. Each repeat has a four-bladed β-propeller fold, with a single stabilizing disulfide bond between blades I and IV. Because these domains are also found in many other proteins, for instance in integrins, it is thought that they can also be involved in protein–protein interactions. The hemopexin domain connects to the catalytic domain via a proline-rich polypeptide hinge region of 10 to 70 aa. The hemopexin-like domains are determinants of substrate specificity of each enzyme The function of the hinge region is not fully understood.

Although all MMP family members have a similar overall structural motif, their molecular weight ranges from 28,000 to 92,000 da and suggests the existence of different levels and patterns of glycosylation, as well as some structural diversity. Whereas some enzymes lack certain domains, others have additional inserts that modulate but not drastically alter their functions. For instance, MMP-7 and MMP-26 lack the hemopexin domain, whereas MMP-2 and MMP-9 have an insert of three fibronectin type II domains in the vicinity of the catalytic site. NMR studies have indicated that the inserts consist of two antiparallel β-sheets, connected with a short α-helix. and are stabilized by two disulfide bonds. Domains 2 and 3 are thought to be structurally flexible, which can facilitate simultaneous interactions with multiple sites in the ECM.

Several members of the MMP family exhibit additional structural features that allow classifying them into a separate subgroup of membrane-type MMPs (MT-MMPs). Within their polypeptide linker that connects the prodomain with the catalytic domain, there is a sequence R-X-R/K-R, which seems to be essential for specific post-translational processing of these proteinases. Also, their C-terminal region has a 70 to 100 aa hydrophobic polypeptide tail that inserts in the cell membrane, terminating with a short cytoplasmic tail. These enzymes are anchored in a cell membrane. So far, six membrane-type proteases have been characterized [2].

Each MMP exhibits a unique set of structural and functional characteristics. For instance, MMP-23 lacks the cysteine switch motif and the hemopexin domain. Instead, it has a unique cysteine and proline-rich region, the IL-1 type II receptor-like domain, and a furin-susceptible region. MMP-23 is activated in the Golgi-like MT-MMP, but it does not have a transmembrane domain. It is released in active form, but it may associate to the cell membrane via the N-terminal part of the propeptide.

MMPs can be grouped according to different common properties, varying from primary sequence homologies, substrate specificity, or site of action. One such approach is based on grouping by DNA homology using the resources of the human genome project database, or as based on the search for proteins with homologous catalytic domain of MMPs. Relevant amino acid sequences were detected after analyses of alignments that revealed 29 different proteins stemming out from the same ancestor. Based on this approach, MMPs could be ordered into five subfamilies: nonfurin-regulated MMPs, gelatinases, transmembrane MMPs, GPI-anchored, and others. A similar clustering, with minor differences, was obtained when catalytic domain or hemopexin sequences were compared [3].

Based on the differences in substrate specificity, MMPs can be classified into six subgroups: collagenases, gelatinases, stromyelisins, matrilysins, membrane-type MMPs, and others [2]. Based on the site of action, there are two groups of enzymes: secreted into the extracellular matrix and transmembrane proteins. Table I summarizes groups of MMPs.

Activation of Metalloproteinases

The final products of MMP gene translation are inactive forms of enzymes such as zymogens or proMMPs. Functionally, the active form of MMPs requires exposure of the active catalytic center, which occurs via a series of conformational changes.

The active site containing a zinc-binding motif and surrounding amino acids consists of a cleft along the catalytic domain. In the latent form of the enzyme, the polypeptide tail of the prodomain region covers the catalytically competent cleft. The inhibitory cysteine with its sulfhydryl group acts as a fourth ligand and thus blocks accessibility to the active zinc atom. Cleavage of the protease-sensitive region in the prodomain causes translocation of the polypeptide tail away from the cleft, which upon removal of the cysteine from the active region, exposes the active zinc in coordination with a water molecule. The removal of the inhibitory prodomain exposes the substrate binding region located close to the active site. When the substrate binds and comes close to the active site, the carbonyl group of the substrate peptide bond replaces the water molecule and coordinates the active zinc. The glutamic acid located within the MMP active domain and adjacent to active zinc acts as a proton acceptor from water and provokes a nucleophilic attack on the peptide bond of substrate [5].

MMP activating factors include proteolytic enzymes that digest a protease-sensitive region in the prodomain. Glutathione and other thiol-modifying agents in general, chaotropic agents, SDS, reactive oxygen, NO, heat, or low pH can also perturb the molecular interaction between the inhibitory cysteine and the active zinc, causing activation of MMPs. MMP activation is a multistep process, involving initially the partial removal of the peptide from the prodomain by any of the above listed agents, culminating with the full removal of activation peptide of intermediate molecules.

Membrane-type MMPs contain a specific sequence between the propeptide and the catalytic domain that is susceptible for furin or furin-like proprotein convertases

Table I Groups of Human MMPs.

Enzyme	GENE	Domains	Substrate
Collagenases			
MMP-1, Interstitial collagenase, collagenase 1	11q22–q23	Pre-Pro(SH)-Cat(Zn)-Hinge-Hemopexin(S-S)	Aggrecan, collagens I, II, III, VII, VIII, X, XI, enactin, fibronectin, gelatin, IGFBPs, laminin, link protein, myelin basic, tenascin, vitronectin, α1-AC, α2-M, α1-PI, casein, C1q, fibrin, fibrinogen, IL1β, ProTNFα
MMP-8, Neutrophil collagenase, collagenase 2	11q21–q22	Pre-Pro(SH)-Cat(Zn)-Hinge-Hemopexin(S-S)	Aggrecan, collagens I, II, III, α2-M, α1-PI, C1q, fibrinogen, substance P
MMP-13 Collagenase 3	11q22.3	Pre-Pr(SH)-Cat(Zn)-Hinge-Hemopexin(S-S)	Aggrecan, collagens I, II, III, VI, IX, X, XIV, fibrillin, fibronectin, gelatin 1, osteonectin, α2-M, casein, factor XII, fibrinogen, proMMP2, ProTNFα
Gelatinases			
MMP-2, Gelatinase A	16q13	Pre-Pro(SH)-Cat-3xFnII(Zn)-Hinge-Hemopexin(S-S)	ADP-ribose polymerase (in vitro) [ref 64 S1], collagen I, III, IV, V, VII, IX, X, XI, decorin, elastin, entactin/nidogen, fibrillin, fibronectin, fibulins, gelatin1, IGFBPs, laminin, α1-AC, α1-PI, C1q, fibrin, fibrinogen, IL1β, ProTNFβ, plasminogen substance P
MMP-9, Gelatinase B	20q11.2–q13.1	Pre-Pro(SH)-Cat-3xFnII(Zn)-Hinge-Hemopexin(S-S)	Aggrecan, collagens IV, V, XI, XIV, decorin, elastin, fibrillin, gelatin1, laminin, link protein, myelin basic, osteonectin, vitronectin, α2-M, α1-PI, casein, C1q, fibrin, fibrinogen, IL1β, ProTNFα, Pro TNFβ, plasminogen, substance P
Stromelysins			
MMP-3, Stromelysin 1	11q23	Pre-Pro(SH)-Cat(Zn)-Hinge-Hemopexin(S-S)	Aggrecan, collagens III, IV, V, VII, IX, X, XI, decorin, elastin, entactin/nidogen, fibrillin, elastin, fibronectin, gelatin1, IGFBPs, laminin, link protein, myelin basic, osteonectin, tenascin, vitronectin, α1-AC, α1-PI, C1q, E-cadherin, fibrin, fibrinogen, IL1β, ProTNFα, plasminogen,substance P, T-kininogen
MMP-10, Stromelysin 2	11q22.3–q23	Pre-Pro(SH)–Cat(Zn)-Hinge-Hemopexin(S-S)	Aggrecan, collagens III, IV, V, elastin, fibronectin, gelatin1, link protein, casein, fibrinogen
MMP-11, Stromelysin 3	22q11.2	Pre-Pro(SH)FurinSite-Cat(Zn)-Hinge-Hemopexin(S-S)	IGFBPs, α2-M, α1-PI
Matrilysins			
MMP-7, Matrilysin 1, pump-1	11q21–q22	Pre-Pro(SH)-Cat(Zn)	Aggrecan, collagens I, IV, decorin, elastin, entactin/nidogen, fibronectin, fibulins, gelatin1, laminin, link protein, myelin basic, osteonectin, tenascin, vitronectin, α1-PI, casein, E-cadherin, fibrinogen, Pro-α-defensin, Fas-ligand, Pro(TNF)-α
MMP-26, Matrilysin 2, endometase	11p15	Pre-Pro(SH)-Cat(Zn)	Collagen IV, fibronectin, gelatin1, α1-PI, fibrinogen
Membrane-Type MMPs			
MMP-14, MT1-MMP	14q11–q12	Pre-Pro(SH)FurinSite-Cat(Zn)-Hinge-Homopexin(S-S)-TM-Cytoplasmic tail	Aggrecan, collagens I, II, III, elastin, entactin/nidogen, fibronectin, gelatin 1, laminin, vitronectin, α2-M, α1-PI, Factor XII, fibrin, fibrinogen, ProMMP2, ProTNFα
MMP-15, MT2-MMP	15q13–q21	Pre-Pro(SH)FurinSite-Cat(Zn)-Hinge-Homopexin(S-S)-TM-Cytoplasmic tail	
MMP-16, MT3-MMP	8q21	Pre-Pro(SH)FurinSite-Cat(Zn)-Hinge-Homopexin(S-S)-TM-Cytoplasmic tail	Collagen III, fibronectin, ProMMP2
MMP-17, MT4-MMP	12q24.3	Pre-Pro(SH)FurinSite-Cat(Zn)-Hinge-Homopexin(S-S)-GPI anchor	
MMP-24, MT-MMP	20q11.2	Pre-Pro(SH)FurinSite-Cat(Zn)-Hinge-Homopexin(S-S)-TM-Cytoplasmic tail	
MMP-25, MT6-MMP	16p13.3	Pre-Pro(SH)FurinSite-Cat(Zn)-Hinge-Homopexin(S-S)-GPI anchor	

Table I Continued

Enzyme	GENE	Domains	Substrate
Others			
MMP-12, macrophage metalloelastase	11q22.2–q22.3	Pre-Pro(SH)–Cat(Zn)-Hinge-Hemopexin	Aggrecan, collagens I, IV, decorin, elastin, entactin/nidogen, fibronectin, gelatin 1, laminin, myelin basic, tenascin, α2-M, α1-PI, Factor XII, fibrinogen ProTNFα, ProTNFβ, plasminogen
MMP-19, RASI-1	12q14	Pre-Pro(SH)–Cat(Zn)-Hinge-Hemopexin	Collagens I, IV, fibronectin, gelatin 1, tenascin, casein
MMP-20, enamelysin	11q22.3	Pre-Pro(SH)–Cat(Zn)-Hinge-Hemopexin	amelogenin
MMP-23, cysteine array MMP (CA-MMP)	1p36.3	Pre-Pro(SH)-FurinSite-Cat(Zn)-Cys-rich-Ig-like	
MMP-27, CMMP	11q24	Pre-Pro(SH)–Cat(Zn)-Hinge-Hemopexin	
MMP-28, epilysin	17q21.1	Pre-Pro(SH)FurinSite-Cat(Zn)-Hinge-Hemopexin(S-S)	

Enzymes MMP-22 from chicken, MMP-18 and MMP-21 from Xenopus have been omitted.

Abbreviations: Pre, preprodomain, signal peptide; Pro(SH), prodomain with a zinc-ligating thiol (SH) group; FurinSite, specific sequence susceptible for furin cleavage located within Prodomain; Cat(Zn), catalytic domain with active zinc; 3xFnII, three Fibronectin-like domains inserted into a catalytic domain; Hinge, hinge region, linker polypetide; Homopexin(S-S), hemopexin domain with four propeller domains and a disulfide bridge between the first and fourth domain; TM, transmembrane region of protein; C, cytoplasmic tail, intracellular region of MMp expressed on the cell surface; GPI, glycophophatidyl inositol anchoring domain; Cys-rich, cystein and proline rich domain; Ig-like, immunoglobulin like domain. This table is based on [2, 4].

cleavage. During post-translational processing, these enzymes recognize and cleave within this sequence, thus activating MT-MMPs. Once transported through the Golgi network, activated MT-MMPs remain as surface enzymes at the cell membrane. Internalization of the enzyme can occur within 60 minutes after initial cell membrane exposure. This might represent a rapid response for relocalizing active MT1-MMP at the leading edge of the cell during migration.

Active MMPs and MT-MMPs degrade several components of ECM molecules. Their targets include other proteinases, proteinase inhibitors, clotting factors, chemotactic molecules, latent growth factors, growth factor–binding proteins, cell surface receptors, and cell–cell adhesion molecules. In addition, they may act via cell receptors, as for instance α2β1 integrin, thus affecting cell survival. Hence, the mechanisms of in vivo activation and inhibition of proMMPs are of high physiological importance. It was suggested that activation may take place on the external cell surface, where plasma plasminogen, together with its uPA/uPAR complex, is converted into plasmin, and plasmin then removes part of the propeptide of a freshly secreted MMP zymogen. Membrane-type MMP also plays a role in MMP activation by cleaving the rest of the propeptide. There is some evidence that this process is mediated by a specific MMP inhibitor (TIMP) that has high affinity to both the MT-MMP and MMP zymogen, keeping the complex docked on the cell surface. The adjacent MT-MMP, which is not occupied by a TIMP, then partially cleaves MMP zymogen, which in turn can auto-activate to the fully active form. Recently, a new activating agent was revealed that offers an additional control of MMP function called the SIBLINGS proteins. Also, it was found that blood proteins involved in coagulation/fibrinolysis cascade, such as Activated Protein C, can directly activate MMPs.

Specific Inhibitors of Metalloproteinases: TIMPs

MMPs can be inhibited by α-2 macroglobulin, a general proteinase inhibitor of very broad spectrum, by inhibitors of complexed progelatinase and TIMP, and by TIMPs themselves. Tissue inhibitors of metallo-proteinases (TIMPs), the key inhibitors of MMPs in vivo, have four members (TIMPs 1 to 4) that share 41 percent to 54 percent sequence identity. They are small proteins with molecular weight that ranges between 27-kDa and 29-kDa. Each TIMP is built with one polypeptide chain that folds into two domains: N-terminal domain (about 125 aa) and C-terminal domain (about 65 aa). The interaction between TIMP and MMP or MT-MMP occurs via the binding of its four N-terminal residues to the catalytic cleft in MMP, one of them being the inhibitory cysteine. Additionally, the C-terminal region of TIMP binds with high affinity to the C-terminal part of MMP, thus interfering, which intensifies the interaction so much that the complexes are stable even in 0.1 percent SDS. The stoichiometry of complexes is 1 : 1.

The role of TIMP in the activation and inhibition of MMP seems to be dual. Experimental data show that low or moderate levels of TIMP2 promote activation of MMP2 in breast tumor cells, whereas higher levels inhibit it. These contrasting findings may be explained by the fact that high levels of TIMP saturate MT-MMP, thus blocking its ability to activate proMMP. It was found that TIMPs additionally have some cell–growth–promoting or cell growth–inhibiting activities. Molecules of TIMP were found in nuclei of fibroblasts and breast carcinoma cells. By analogy to their interaction with MMPs, their role in other physiological or pathophysiological processes can also be dual [6].

Because overactive MMPs are associated with many pathological states, the equilibrium between active enzymes and their inhibitors plays a crucial role in maintaining the proper level of metalloproteinase activity.

MMP Genes and Regulation of Expression

Genes coding MMPs were localized on chromosomes 11, 14, 16, 20, and 22. Genes coding several secreted, soluble proteases form clusters. For instance, the long arm of chromosome 11 harbors sequentially (oriented from the most centromeric to the most telomeric regions) genes coding for MMP-8, MMP-10, MMP-1, MMP-12, MMP-7, and MMP-13. In contrast, the genes coding for membrane-type MMPs were localized on different chromosomes; MMP14 (MT1-MMP) maps to human chromosome 14, MMP15 (MT-2-MMP) to human chromosome 16, and MMP16 (MT-3-MMP) to human chromosome 8.

A paradigm of MMP gene structure and organization is the MMP-12 (human macrophage elastase) gene. It is similar to the genes of stromelysins and collagenases because it has the same location to chromosome 11q22.2–22.3 and similar, highly conserved exon size and intron–exon borders. It is 13-kilobase (kb) long, composed of 10 exons and 9 introns. It was determined that MMP-12 (macrophage elastase) and MMP-3 (stromelysin 1) genes are physically linked within the 62-kb region. The promoter region of the MMP-12 gene contains several features common to other MMP genes, including a TATA box 29 nucleotides upstream to the transcription initiation site, an AP-1 motif located about −70 kb upstream, and a PEA3 element. The PEA3 is a conserved Polyoma Enhancer A binding protein site, and the AP-1 is an Activation Protein 1 binding element. Both are regions of DNA that bind directly to these transcription factors. Other regulatory elements within different MMP genes include GC-Sp-1 binding site, SBE-SIAT binding element, c/EBP-β-CCATT/enhancer binding protein β, OSE-2 (osteoblast specific element-2), SPRE (stromelysin-1 PDGF responsive element), TRE (octamer binding protein), Sil (silencer sequence), NF-kβ (nuclear factor kβ), NF-1 (nuclear factor-1), and RARE (retinoic acid responsive element). Some of these sites overlap. For instance, the phorbol ester-responsive element is located within the AP-1 binding region.

Under normal conditions, the level of MMPs is rather low and increases with the need for ECM remodeling. Thus, the MMP genes are susceptible to external induction, which occurs with participation of the regulatory elements listed above. Regulation at the translational level may occur via modulation of mRNA or via second messenger signaling.

The agents affecting gene expression are inflammatory cytokines, growth factors, hormones, phorbol esters, oxidant stress, and mechanical injury or aging. They act through the whole variety of intracellular signaling pathways, including mitogen-activated protein kinases. These pathways may cross-react with each other, resulting in a broad spectrum of intermediating proteins (enhancers, co-regulators, or inhibitors) that in turn form complexes with transcription factors. These complexes eventually bind to AP-1 or other regulatory elements in the promoter region of the MMP gene.

A good example of the level of complexity is transforming growth factor (TGF)-β. TGF-β is a potent regulator of many genes that carry AP-1 motifs in their promoters. It generally induces expression of MMP inhibitors (TIMPs) and expression of genes coding for ECM proteins that are substrates for MMPs, but suppresses synthesis of MMPs 1, 3, and 9. For its suppressive activity, TGF-β binds to its transmembrane receptor and activates the cascade of phosphorylation of Smad family proteins. These proteins can then be translocated to the nucleus, where they bind to a TGF-β inhibitory element (TIE). This is a sequence: 5′-GAATTGGAGA-3′, located 245 bp upstream of the transcription site and upstream of the AP-1 site in the MMP-1 promoter. TIE is considered to be a constitutive repressor of the MMPs. In carcinoma cells, TGF-β has an opposite effect, in that by acting through a protein kinase C and tyrosine kinase signaling pathways, it enhances expression and synthesis of MMP-13 [7, 8].

Angiogenesis

MMPs play critical roles in different areas related to vascular biology, particularly angiogenesis, which involves intimate interaction between the vascular cells and their surrounding matrix. MMP-mediated remodeling with degradation of ECM components by proteolytic cleavage rearranges or removes mechanical barriers, thus allowing cell migration and cell growth. Because of the presence of MMPs in virtually any tissue and because of activity directed to a broad spectrum of substrates, MMPs play a crucial role in embryonic development and organogenesis. For review, see recent publications [9].

Vascularization, de novo vessel formation, begins when fibroblast growth factor isoforms produced by endodermal cells induce paraxial and lateral plate mesoderm to form angioblasts and hematopoietic cells. Cells differentiated into angioblasts can either remain at a specific site of origin or migrate to distant locations to form a regional vascular plexus. Within the plexus, endothelial cells proliferate and form capillaries, which may foster further vessel formation, known as angiogenesis. These processes are regulated on several levels, one of which is the endothelial receptors tyrosine kinase TIE-2 and TIE-1 and interactions of vascular endothelial growth factors with their receptors VEGF-R2 and VEGF-R1 located on the cell surface. It is believed that the VEGF-R2 receptor mediates early-stage endothelial cell differentiation, whereas VEGF-R1 is responsible for formation of lumen in newly created capillaries. Knock-out animals lacking either of these factors died in embryonal stage from abnormal vascularization.

Angiogenesis consists of formation of new capillaries by two processes: one by splitting of a native vascular structure, also known as *intussusception*, and the other by sprouting. Splitting requires the proliferation of endothelial cells inside the existing vessels. This process can be induced either by formation of transcapillary pillars (folding capillary walls) that eventually tear the native vessels or by direct action of proteases that degrade the wall enzymatically. In the site of injury, a new vessel can be formed.

Sprouting occurs when endothelial cells from the original cluster migrate through the basement membrane of the vessel wall, invade the surrounding tissue, anchor in a new place, and proliferate. In both instances, before any move, endothelial cells release proteases that remodel ECM, thus facilitating translocation of cells from the parent venule. Lack of synthesis or inhibition of collagen-degrading enzyme prevents new vessel formation. Indeed, mice deficient in MMP-2 (gelatinase A, degrades collagen) had reduced retinal angiogenesis.

It has been recently proposed that postnatal vasculogenesis originates from endothelial progenitor cells differentiated from bone marrow stem or precursor cells. Angioblasts were found in and isolated from the peripheral blood of adult species, and the series of transplantation experiments revealed that they maintained specific endothelial cell features and were eventually incorporated into foci of neovascularization. Matrix metalloproteinase-9 (MMP-9, gelatinase B), induced in bone marrow cells, mediated the transfer of endothelial and hematopoietic stem cells from the quiescent to proliferative niche by releasing soluble Kit-ligand, a specific signal that induced stem cells to differentiate and mobilize to circulation [10].

Although the nature of the triggering signal in the angiogenic sprouting remains unknown, it is known that angiogenic stimuli, such as fibroblast growth factor (FGF), vascular endothelial cell growth factor (VEGF), or vascular permeability factor (VPF), are the most potent inducers of MMP synthesis in endothelial cells. Additionally to growth factors and cytokines, modifying action of other molecules has been reported. These molecules are side products or intermediates of proteolytic cleavage of ECM proteins (matrikines), and cryptic segments within larger proteins revealed by conformational changes or partial hydrolysis (matricryptins). Among this group of mediators of angiogenesis, some are worth mentioning: angiostatin and kringle 5 (proteolytic fragments of plasminogen), endostatin and restin (proteolytic fragments of collagen XVIII an XV), the 16 kDa N-terminal fragment of prolactin, an N-terminally truncated platelet factor 4 (PF4), vasostatin originating from calreticulin, kininostatin corresponding to the high-molecular-weight kininogen domain 5, the cleaved form of the serpin antithrombin (AT), a noncatalytic carboxy-terminal hemopexin-like domain of MMP-2 (also known as PEX), several noncollagenous domains from type IV collagen, thrombospondins, secreted protein acidic rich in cysteine (SPARC), and anastellin corresponding to the first type III repeat in fibronectin [11].

Under physiological conditions, new vessel formation is a process triggered by signals from neighbor cells or ECM, balanced by MMP synthesis, activation and inactivation mechanisms, and modulated by tissues or blood flow or blood products, depending on or benefiting from a well-maintained circulatory system. The lack of appropriate inhibition at any level results in excessive expression or overactivity of the MMPs and causes pathological states. The most severe pathological states are tumors, cancer, arthritis, atherosclerosis, nephritis, tissue ulcers, and fibrosis. For instance, there is evidence that MMP-2 is involved in the switch to the angiogenic phenotype, one of the earliest events during the transition of a tumor from the preneoplastic to the tumorigenic phenotype.

MMPs and Pulmonary Vascular Diseases

MMPs have been extensively involved in pulmonary diseases, including interstitial lung disease, emphysema, asthma, and in the context of this review, pulmonary hypertension. Pulmonary hypertension concerns the elevation of pulmonary artery pressures above 25 mmHg mean pulmonary artery pressures. Pulmonary hypertension may occur as an idiopathic disease, known historically as primary pulmonary hypertension (or, currently, as idiopathic pulmonary hypertension), or associated with underlying clinical conditions such as congenital heart malformations with left-to-right shunting, collagen vascular diseases, HIV infection, liver diseases, interstitial lung processes, or chronic obstructive pulmonary disease. Clinically, pulmonary hypertension can be divided as mild to moderate, when the pulmonary artery pressures rise to levels not more than 40 to 45 mmHg, and severe, if the pressures are equivalent to those in the systemic circulation. The severe forms of pulmonary hypertension ultimately kill as a result of right heart failure. The underlying morphological basis of pulmonary hypertension is an alteration in the number or size of medial, smooth muscle cells (hyperplasia or hypertrophy, respectively), and in growth of myofibroblasts into the intima. In a selected group of patients with severe disease, endothelial cells undergo growth, forming plexiform lesions. In aggregate, these cellular alterations are known as *pulmonary vascular remodeling*.

Degradation of components of ECM releases matrix-bound growth factors or creates fragments originated from degraded matrix proteins, resulting in biologically active mediators. The integrated action of these growth factors or active protein fragments may induce vascular cell growth and vascular remodeling. Increased elastolytic activity in pulmonary arteries was postulated as a potential mechanism leading to pulmonary vascular remodeling. Experimental pulmonary hypertension caused by monocrotaline, an alkaloid that causes pulmonary artery endothelial cell injury and pulmonary hypertension, was associated with enhanced elastolytic activity. Similar observations were then extended to the chronic hypoxia model of pulmonary hypertension. The mediator of this elastolytic activity appeared to be a serine protease called endogenous vascular elastase, and if blocked with an elastase inhibitor, pulmonary hypertension was prevented [12].

The interaction between the action of the vascular elastase and MMPs was further dissected in studies employing pulmonary arteries of pulmonary hypertensive lungs caused by monocrotaline, grown in organ cultures. These explants demonstrated increased gelatinase activity (mostly MMP2), which could be blocked by treatment with the broad-spectrum MMP inhibitor GM6001. GM6001 also inhibited tenascin (a growth factor for smooth muscle cells) expression, increased vascular cell apoptosis, decreased cell proliferation, and finally, caused a reduction of medial thickness [12]. This line of work agrees with other hypotheses. It is proposed that, during chronic hypoxia, pulmonary arteries undergo cycles of extracellular collagen synthesis and degradation, ultimately contributing to pulmonary vascular remodeling. Pulmonary vascular oxidative stress with formation of peroxynitrite can induce MMP activation, such as of MMP-1, -2, -8, and -9. There is evidence of oxidative stress in experimental hypoxic pulmonary hypertension and in pulmonary arteries of patients with idiopathic pulmonary hypertension [13]. Also, there is evidence that chronic hypoxia activates MMP-2 and MMP-13, which might degrade elastin and fibrillar collagen, respectively. The corollary of this paradigm is the protection against pulmonary hypertension and pulmonary vascular remodeling afforded by the MMP inhibitor Batismat. In summary, these data ascribe a pathogenetic role for serine proteases and MMPs in the process of pulmonary vascular remodeling, which is most pertinent to the monocrotaline model of pulmonary hypertension.

However, other investigations revealed a diametrically opposite function for MMPs in pulmonary vascular remodeling associated with pulmonary hypertension. Armed with the rationale that regression of pulmonary vascular remodeling requires proteolytic processing of the ECM in pulmonary arteries, Vieillard-Baron et al. demonstrated that TIMP-1 inhibition of MMPs worsened hypoxia-induced pulmonary hypertension. It is difficult to pinpoint the cause of this discrepancy with regard to the pathobiological importance of matrix proteases in pulmonary vascular remodeling, but it is conceivable that TIMP-1 may have an unforeseen effect on pulmonary vascular remodeling that is independent of its MMP-inhibitory effect.

Conclusion

MMPs and matrix proteases are critical effectors of cell responses. This function is ever more pertinent to dynamic states in which cells have to interact with their surrounding matrix to perform physiologic functions, or in disease states. Blood vessels undergo dynamic changes of their component cells to adapt to stresses or maladapt and cause disease. MMPs are intimately involved in these processes, but as apparent with the data presented herein, manipulation of MMP activity to modify the course of diseases may lead to adverse and unforeseen consequences.

Glossary

Metalloproteases: Protein molecules (enzymes) that require the presence of metal ion bound to their sequence to maintain a proper structure that allows their functional activity. The main function of metalloproteases is to activate or to inactivate (inhibit) other proteins (substrates), usually by cutting off one or more of their fragments.

Multidomain structure: Domains are regions in a protein molecule that are determined by amino-acid sequence and that form defined, functionally distinct units. Thus, multidomain structure is specific for and characterizes a particular protein. If several proteins have similar multidomain structure, it means these proteins may also have similar functions.

References

1. Stocker, W., Grams, F., Baumann, U., Reinemer, P., Gomis-Ruth, F. X., McKay, D. B., and Bode, W. (1995). The metzincins—topological and sequential relations between the astacins, adamalysins, serralysins, and matrixins (collagenases) define a superfamily of zinc-peptidases. *Protein Sci.* **4**, 823–840.
2. Visse, R., and Nagase, H. (2003). Matrix metalloproteinases and tissue inhibitors of metalloproteinases: Structure, function, and biochemistry. *Circ. Res.* **92**, 827–839. *An updated review of biochemical structure of metaloproteinases and their specific inhibitors.*
3. Andreini, C., Banci, L., Bertini, I., Luchinat, C., and Rosato, A. (2004). Bioinformatic comparison of structures and homology-models of matrix metalloproteinases. *J. Proteome. Res.* **3**, 21–31.
4. Sternlicht, M. D., and Werb, Z. (2001). How matrix metalloproteinases regulate cell behavior. *Ann. Rev. Cell Develop. Biol.* **17**, 463–516. *An overview of the role of metaloproteinases in physiological processes that involve tissue remodeling.*
5. Morgunova, E., Tuuttila, A., Bergmann, U., Isupov, M., Lindqvist, Y., Schneider, G., and Tryggvason, K. (1999). Structure of human pro-matrix metalloproteinase-2: Activation mechanism revealed. *Science* **284**, 1667–1670.
6. Lambert, E., Dasse, E., Haye, B., and Petitfrere, E. (2004). TIMPs as multifacial proteins. *Crit. Rev. Oncol./Hematol.* **49**, 187–198.
7. Hall, M. C., Young, D. A., Waters, J. G., Rowan, A. D., Chantry, A., Edwards, D. R., and Clark, I. M. (2003). The comparative role of activator protein 1 and Smad factors in the regulation of Timp-1 and MMP-1 gene expression by transforming growth factor-beta 1. *J. Bio. Chem.* **278**, 10304–10313. *An excellent example of modern, basic research on how tissue growth factor beta regulates expression of metaloproteinase and specific metaloproteinase inhibitor genes to control cell behavior.*
8. Yuan, W., and Varga, J. (2001). Transforming growth factor-beta repression of matrix metalloproteinase-1 in dermal fibroblasts involves Smad3. *J. Bio. Chem.* **276**, 38502–38510.
9. Stamenkovic, I. (2003). Extracellular matrix remodelling: The role of matrix metalloproteinases. *J. Pathol.* **200**, 448–464.
10. Heissig, B., Hattori, K., Dias, S., Friedrich, M., Ferris, B., Hackett, N. R., Crystal, R. G., Besmer, P., Lyden, D., Moore, M. A. S., Werb, Z., and Rafii, S. (2002). Recruitment of stem and progenitor cells from the bone marrow niche requires MMP-9 mediated release of Kit-ligand. *Cell* **109**, 625–637.

11. Bellon, G., Martiny, L., and Robinet, A. (2004). Matrix metalloproteinases and matrikines in angiogenesis. *Crit. Rev. Oncol./Hematol.* **49**, 203–220.
12. Rabinovitch, M. (1999). EVE and beyond, retro and prospective insights. *Am. J. Physiol.* **277**, L5–L12.
13. Novotna, H. J. (2002). Possible role of matrix metalloproteinases in reconstruction of peripheral pulmonary arteries induced by hypoxia. *Physiol. Res.* **51**, 323–334.

Capsule Biographies

Dr. Rubin M. Tuder, Professor of Pathology and Medicine, in the Department of Pathology, Johns Hopkins University School of Medicine. His interests are focused on pulmonary pathology and molecular mechanisms responsible for pulmonary diseases. His grants are supported by the NIH/NHLBI, AHA, and other foundations.

Dr. Iwona Fijalkowska is a Research Associate in Cardiopulmonary Pathology in the Department of Pathology, Johns Hopkins University School of Medicine. Her scientific interests cover protein chemistry and protein–protein interactions in angiogenesis and hemostasis.

CHAPTER 7

VEGF-A and Its Isoforms

Eric B. Finkelstein and Patricia A. D'Amore
Schepens Eye Research Institute, Harvard Medical School, Boston, Massachusetts

Vascular endothelial growth factor-A (VEGF) is one of the most important regulators of physiological and pathological angiogenesis, as demonstrated by genetic studies in which deletion of a single VEGF allele resulted in embryonic lethality caused by defective blood vessel formation. VEGF is an endothelial cell (EC) mitogen that serves as a mediator of angiogenesis and vessel permeability. As such, VEGF is essential for tumor growth and angiogenesis and is involved in ocular neovascularization characteristic of diabetic retinopathy and retinopathy of prematurity. VEGF belongs to a family of growth factors that includes VEGF-A, VEGF-B, VEGF-C, VEGF-D, VEGF-E, and placental growth factor (PlGF). This chapter focuses on isoforms of VEGF-A resulting from alternative splicing. Therefore, throughout the text, VEGF-A will be referred to as VEGF.

VEGF is active as a homodimer by binding two receptor tyrosine kinases: VEGF receptor-1 (VEGFR1; Flt-1) and VEGF receptor-2 (VEGFR2; Fkl-1; KDR). VEGFR2 is primarily associated with the mitogenic and angiogenic activities of VEGF, whereas VEGFR1 is thought to be a reservoir for VEGF in solution, thereby regulating its availability. VEGF activity is regulated on many levels, including transcriptionally by cytokines and oxygen tension, and by alternative splicing of the VEGF gene to produce multiple isoforms that differ in their bioavailability and bioactivity. Increasing evidence suggests that VEGF isoform levels may be a major mechanism by which VEGF activity can be regulated physiologically and pathologically.

VEGF Isoform Structure

The VEGF gene is comprised of eight exons separated by seven introns (Figure 1). Exons one through five and eight are common to all isoforms. The first five exons encode for domains involved in α-helical dimerization, a signal peptide cleaved following secretion (exons 1 and 2), and sites for VEGFR1 (exon 3) and VEGFR2 (exon 4) binding (Robinson and Stringer, 2001). VEGF is alternatively spliced to produce several isoforms that vary in amino acid number in all species studied to date (Table I). Variability among the isoforms results from the inclusion of exons six and/or seven (Figure 1).

The most abundant human isoform in most tissues is VEGF165, and as a result, the most is known about this isoform. There have been several studies of VEGF145 and VEGF121, but few have focused on VEGF189 or VEGF206 because of their low abundance and reported difficulties producing them in pure form. Many of the studies described in the following sections have utilized recombinant human VEGF121 or VEGF165, whereas studies of VEGF189 or VEGF206 have utilized in vitro–produced extracellular matrix (ECM), to which these isoforms are bound.

Several additional VEGF isoforms have been described, including murine VEGF115 (Robinson and Stringer, 2001), expressed in immortalized fibroblasts, with 37 novel carboxyl-terminal amino acids relative to VEGF120. Human VEGF183 uses an alternative splice donor site 18 nucleotides upstream of the common splice site between exons 6a and 6b, which results in six less amino acids than VEGF189 [1]. Other isoforms have been reported, including VEGF162 with angiogenic activity, VEGF165b that may negatively regulate VEGF165, and VEGF148 with unknown biological activity. The significance of these isoforms awaits further investigation.

VEGF and the ECM

Interactions with HSPG and the ECM

VEGF isoforms have variable affinities for heparan sulfate proteoglycans (HSPGs) on the cell surface and/or within the ECM (Robinson and Stringer, 2001). VEGF121 has lowest affinity, VEGF145 and VEGF165 have similar

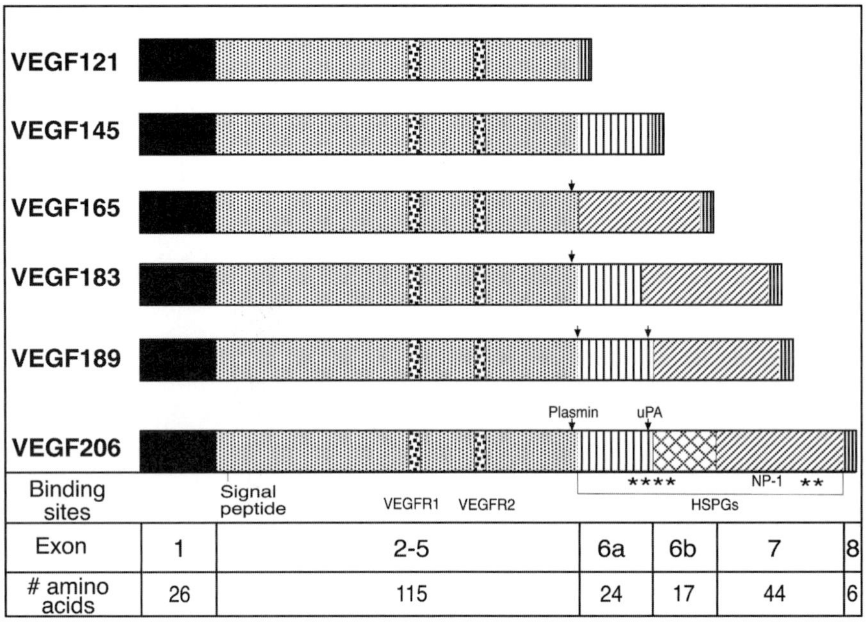

Figure 1 Isoforms of Human VEGF-A Generated by Alternative Splicing. Alternative splicing of human VEGF-A has been reported to generate several isoforms. These isoforms differ by the inclusion or exclusion of domains encoded by exons six and seven. The location of a signal peptide and VEGF receptor (VEGFR1 and VEGFR2) binding sites common to all isoforms are indicated on VEGF206. HSPG binding sites common to isoforms with exons six and seven are indicated (*), as is the neuropilin-1 binding site (NP-1). Finally, arrows indicate sites of potential proteolytic processing by either plasmin or uPA. The exons and number of amino acids in each exon is indicated at the bottom of the figure.

Table I VEGF Isoforms from Alternative Splicing in Species Studied to Date.

Species	VEGF-A Isoforms (number of amino acids)
Human	121, 145, 165, 183, 189, 206
Mouse	115, 120, 164, 188
Quail	122, 146, 166, 190
Rat	120, 164, 188
Bovine	120, 164

affinities, and VEGF189 and VEGF206 have the highest affinities for HSPG (and heparin in vitro) (Table II). The high affinity of VEGF189 and VEGF206 is a result of the presence of exons 6 and 7, which mediate binding to HSPGs (Figure 1). Exon 6a in VEGF145 mediates ECM binding independent of HSPGs because of a cell surface retention sequence. The affinity of a VEGF isoform for HSPGs is directly correlated with its diffusibility (Figure 2). As a result, the smallest isoform (VEGF121) is secreted and freely diffusible. Approximately 50 percent of intermediate-sized isoforms (VEGF145 and VEGF165) diffuse from producing cells (based on their presence in conditioned media), whereas the remainder are cell and/or ECM associated. Finally, the largest isoforms (VEGF189 and VEGF206) are secreted from producing cells but completely sequestered within the ECM with low amounts bound to the cell surface

Table II Biochemical Properties of Human VEGF Isoforms.

Isoform	Exons	MW (kDa)	Acidic or basic	Heparin binding
121	1–5, 8	34	Weakly acidic	no
145	1–5, 6a, 8	41		yes
165	1–5, 7, 8	46	basic	yes
183	1–5, 6a*, 7, 8	46		yes
189	1–5, 6a, 7, 8	52	basic	yes
206	1–5, 6a, 6b, 7, 8	~60	basic	yes

Molecular weight (MW) based on the sum of monomer band sizes.
* indicates that exon 6a is truncated.

(Figure 2). Similar to VEGF189 and VEGF206, most of VEGF183 remains bound to the cell surface [1], suggesting it has a high affinity for HSPGs. Although most of the isoforms can bind HSPG and/or the cell surface, they can be released as soluble, biologically active molecules.

Release of VEGF from the ECM

Addition of heparin to cells stably expressing VEGF165 resulted in a two- to fourfold increase in the level of VEGF in conditioned media. Therefore, VEGF165 can be released from the cell surface and/or ECM by heparin. A similar

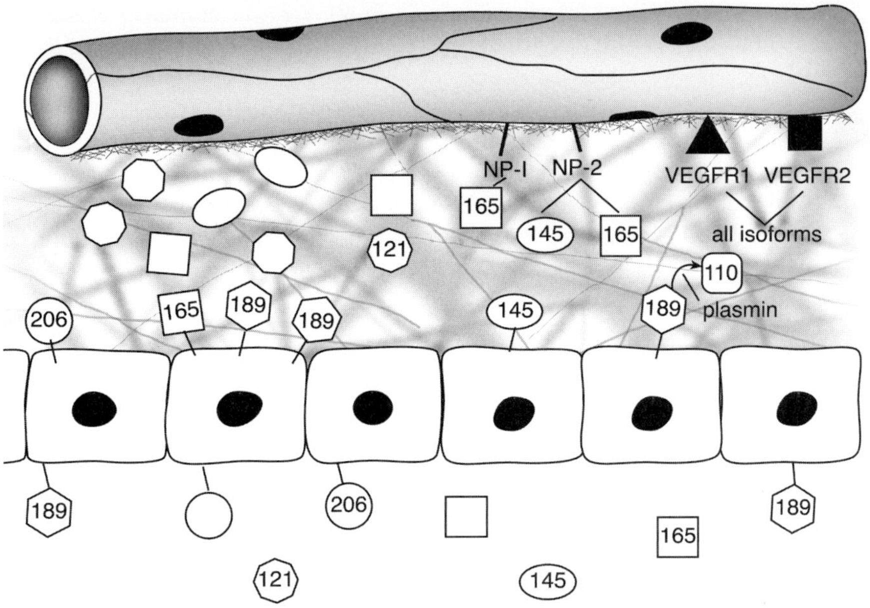

Figure 2 Diffusibility or Cell-Association of VEGF Isoforms. VEGF is depicted as being produced by an epithelial cell layer and diffusing towards endothelial cells, where it binds to the neuropilins (NP-1 or NP-2) or to VEGF receptors (VEGFR1 or VEGFR2). VEGF isoforms are diffusible to different degrees, with VEGF121 completely diffusible, while VEGF189 (or VEGF183) and VEGF206 remain bound to the cell surface and/or within the extracellular matrix. VEGF189 (and 206) can be cleaved by plasmin to generate a diffusible form, VEGF110.

effect was observed for heparinase. All VEGF isoforms within the ECM or on the cell surface can be rapidly released by proteolytic cleavage with plasmin, a serine protease (Robinson and Stringer, 2001). Plasmin cleavage produces a 110 amino acid (aa) fragment (Figure 2) that can bind VEGFR2 but not heparin. This 110aa fragment is active as an EC mitogen and vascular permeability factor similar to VEGF121, with activity reduced fiftyfold relative to VEGF165. Plasmin cleavage results in a second smaller, heparin-binding fragment without mitogenic activity because the VEGF receptor-binding domains are absent. Thus, the carboxyl-terminal heparin-binding domains of VEGF are essential for full mitogenic activity, as their deletion by proteolytic cleavage or alternative splicing significantly reduces activity. VEGF189 and perhaps VEGF183 and VEGF206 are also processed by the serine protease urokinase type plasminogen activator (uPA) near the carboxyl terminus of the domain encoded by exon six. This results in a truncated VEGF (uPA-VEGF189) with activity similar to native VEGF165. These processing steps are required for mitogenic activity and binding toVEGFR2, but not for binding to VEGFR1.

VEGF–Receptor Interactions

In addition to modulating isoform bioavailability, heparin can alter bioactivity through the VEGF receptors. Low concentrations of heparin can enhance binding of VEGF165 to VEGFR2, while high concentrations of heparin can inhibit this interaction. Binding of VEGF121 to VEGFR2 is unaffected by heparin. Heparin can inhibit interactions of VEGF121 and VEGF165 with VEGFR1, suggesting that HSPGs are necessary for VEGFR1 activity. In addition to heparin, exon seven of VEGF binds a class of receptors, known as neuropilins (Figure 1; neuropilin-1: NP-1; neuropilin-2: NP-2). NP-1 binds VEGF165 but not VEGF121, and is thought to serve as a VEGF coreceptor to potentiate signaling through VEGFR2. As such, NP-1 has a lower affinity for VEGF than VEGFR2. In support of the coreceptor concept, a peptide corresponding to exon seven can inhibit VEGF165 activity by approximately 60 percent (to a level similar to VEGF121) likely by blocking VEGF–NP-1 interactions [2]. NP-2 binds VEGF165 and VEGF145 and can mediate VEGF145 signaling.

Isoform Expression

In human tissues, VEGF121, VEGF165, and VEGF183 are expressed in virtually all cell types examined, with the notable exception of EC. VEGF145 is restricted to cell lines derived from female reproductive tract carcinomas and the placenta, and VEGF206, originally isolated from a fetal liver cDNA library, has been found in the placenta.

Investigation of VEGF isoform expression during murine development suggested that the isoforms are developmentally regulated. Quantitative analysis of isoform expression in developing and adult murine tissues revealed that while all VEGF isoforms are expressed in embryonic mouse organs, the relative levels of each isoform varied from organ to organ over developmental time [3]. In the developing lung, heart, and liver, VEGF188 levels increased from embryonic day 13 to day 18 to account for more than 50 percent of the total VEGF mRNA, and remained high in the adult. In contrast, there were only slight increases in VEGF120 and VEGF164. Little VEGF188 was expressed in the developing or adult brain, with no change detected. In adult organs, lung had the highest relative level of VEGF188, while eye, small intestine, and ovary had the lowest VEGF188 level. VEGF164 was predominant in brain, muscle, eye, and kidney. Finally, VEGF120 was highest in uterus, skin, and ovary [3]. Differences in expression levels of VEGF isoforms in developing and adult organs are an indication that the isoforms have specific functions during developmental and adult angiogenesis.

Isoform Function

In Vitro Bioactivity

All VEGF isoforms, including the cell-associated forms, are active as EC mitogens by signaling through VEGFR2. Recombinant VEGF189 and VEGF206 are unable to stimulate endothelial cell mitogenesis, because of protein folding that masks regions necessary for receptor binding. However, ECM derived from cells expressing VEGF189 or VEGF206 induces EC to proliferate, demonstrating that VEGF trapped in the ECM is an active mitogen.

In Vivo Bioactivity: Developmental

The first definitive demonstration that the various VEGF isoforms serve distinct functions in vivo was provided by analysis of mice that were engineered to express single VEGF isoforms [4]. Earlier analysis of VEGF isoform expression in adult tissues indicated that heart and lung expressed the highest levels of VEGF188. Consistent with this observation, mice that express only the soluble VEGF120 are born at a reduced frequency (indicating embryonic lethality), and those that develop to term die rapidly after birth as a result of defective pulmonary and/or cardiac development. Analysis of vessel density in a limited number of mice that lived up to two weeks revealed that whereas vessel density in wild-type mice increased threefold within a week following birth, there was no increase in vessel density in VEGF 120/120 over the same time frame. These results demonstrate that the isoforms are in fact functionally distinct. In particular, these data illustrate that VEGF120 is not able to substitute for the missing VEGF188. Although the mechanistic explanation for the resulting developmental abnormalities has not been elucidated, it is appealing to speculate that expression of an isoform that is entirely soluble (i.e., has no capacity to bind to matrix) may be ineffective in establishing a gradient that is necessary to induce postnatal angiogenesis.

Pulmonary development was also markedly defective in VEGF120/120 mice. These mice have delayed airspace development, and quantitation of air–blood barriers revealed that the 120/120 mice not only had as few as 30 percent of the air–blood barriers as their wild-type littermates, but that the architecture of barrier structures was abnormal. Whereas in the normal air–blood interface the pulmonary epithelium is closely apposed to the capillary endothelium, the pulmonary capillaries in the VEGF120/120 mice were separated from the alveolar lumens by up to three cell layers. Interestingly, the heterozygous (120/+) mice display intermediate phenotypes.

Although these observations indicate distinct functions for the individual isoforms, they do not clarify the mechanisms whereby these different functions are mediated. Some insight into the possible means by which the isoforms act is provided by localization studies. VEGF isoform production in the lung appears to be developmentally regulated [3], with VEGF188 levels relatively low until it dramatically increases at around embryonic day 16, the late canalicular and late saccular phases of lung development, when alveolar structures are forming. In addition to the temporal correlation between VEGF synthesis and pulmonary capillary formation, there is a clear spatial correlation. In situ hybridization has shown that VEGF is produced specifically by the type 2 pneumocytes, cells intimately associated with the forming vasculature. These findings strongly indicate that the intimate association between the source of the VEGF and the target cells directs the formation of the blood–air barrier. Furthermore, the fact that a majority of the VEGF produced by the pulmonary epithelium is VEGF188, an isoform that is not freely diffusible but remains locally sequestered, adds further strength to this hypothesis of the paracrine regulation of alveolar formation.

Abnormalities in the formation of other tissues and organs have been described in the isoform-specific mice. Retinal vascularization in mice that express only VEGF120 is severely impaired; there is a dramatic reduction in the formation of the primitive vascular network, reduced remodeling, and a persistence of the hyaloid vasculature [5]. Mice expressing only VEGF188 also exhibit abnormal retinal vascularization. Although the primitive vessel plexus is normal at postnatal day 3 (P3), vessel number is reduced by half only two days later (P5), with ephrin staining indicating a loss of the arterial component. As for the VEGF120 mice, a persistent hyaloid vasculature appears to migrate into the retina by P9, perhaps compensating for the regression of the arteries. More recent investigations into the relative contribution of VEGF isoforms to pathologic retinal neovascularization have revealed that retinal vessel development is normal in VEGF120/188 (VEGF164-deficient) mice [6].

The relative contribution of the various VEGF isoforms to skeletogenesis has also been investigated. Analysis of mice expressing only VEGF120 leads to the suggestion that VEGF is involved in at least three aspects of bone formation, including vessel ingrowth into the perichondrium and primary ossification center, stimulation of vessel growth and osteoclast migration into hypertrophic cartilage, and induction of osteoblast activity in both intramembranous and endochondronal bones [7].

Thus, VEGF isoforms appear to play critical and rate-limiting roles in the formation of a wide range of tissues and organs. The continued, tissue-specific expression of VEGF isoforms in adult organs [3] suggests that these functions may persist in mature organs and tissues.

In Vivo Bioactivity: Pathologic Models

To date, there have been few comparative studies on the function of VEGF isoforms, but differences clearly exist. Overexpression of VEGF121 and VEGF165 in a brain tumor model led to rapid vessel growth and breakdown (hemorrhage) around tumors, whereas VEGF189 overexpression led to tumors with a similar extent of vascularization as the other isoforms, but no hemorrhage [8].

Tumor cell lines expressing a single mouse VEGF isoform have been used to study the role of the isoforms during tumorigenesis [9]. VEGF-null fibroblasts were immortalized by transfection with SV40 large T antigen and transformed by infection with *H-ras*. The fibroblasts were then infected with plasmids encoding individual VEGF isoforms under control of the CMV promoter and used to generate fibrosarcomas in mice. Relative to VEGF-null cells that formed small, poorly vascularized tumors, VEGF164-expressing transformed fibroblasts completely rescued tumor growth, VEGF188 failed to rescue tumor growth, and VEGF120 had a partial effect. However, although VEGF188 did not rescue tumor growth, tumors were more vascularized compared to VEGF120 tumors [9]. The authors proposed a model in which the isoforms act cooperatively during tumor vascularization with soluble forms acting at a distance to promote blood vessel recruitment, while ECM-bound forms act locally to expand the capillary bed within the tumor [9].

These results and others suggest that VEGF-A isoforms have different abilities to induce vascularization, tumor growth, and blood vessel leakiness in multiple models. However, all models to date have utilized overexpression systems with VEGF under the control of nonphysiological promoters such as the CMV promoter. Therefore, although these results strongly indicate differential functions for VEGF isoforms, the absence of endogeneous VEGF regulatory elements makes the data difficult to extrapolate. Development of models in which VEGF is under the control of its endogeneous regulatory elements would greatly enhance our understanding of the contribution of VEGF isoforms to vascularization during developmental and pathological processes.

Overexpression is not an issue in studies that have utilized mice that express single VEGF isoforms. A second set of pathologies in which VEGF isoform function has been studied is ocular neovascularization, including diabetic retinopathy and retinopathy of prematurity, two leading causes of blindness in which ischemia leads to aberrant retinal blood vessel proliferation. One study utilized mice that express single VEGF isoforms (see above). VEGF164 is derived from leukocytes, is more proinflammatory than VEGF120, and is preferentially induced during ocular neovascularization [6]. Using a mouse model of retinopathy of prematurity, it was found that administration of a VEGF164-specific neutralizing reagent (aptamer) blocks leukocyte adhesion and pathologic neovascularization, with no effect on retinal revascularization or physiological (developmental) neovascularization [6]. When all VEGF isoforms were blocked, revascularization and physiological neovascularization were also affected. In mice with only VEGF120 and VEGF188 (VEGF164-deficient), retinal development was normal, similar to the aptamer results, suggesting that VEGF164 has a primary role in pathologic but not developmental vessel growth [6]. These data suggest that VEGF164 is essential for pathological retinal neovascularization, but not physiological revascularization, which can progress under the influence of only VEGF120 and VEGF188. Thus, in this ocular angiogenesis model, an inhibitor specific for VEGF164 has been developed, which eliminates the features of pathological neovascularization (i.e. leukocyte adhesion, vascular tufts) while allowing the ischemic retina to become revascularized normally.

Conclusion

It is clear that the VEGF isoforms differ in their biochemical properties. Analysis of several model systems has demonstrated that individual isoforms can differentially affect both developmental and pathological angiogenesis. Further work is necessary to elucidate the precise roles of each isoform. Such work may eventually make it possible to specifically target individual isoforms during pathological angiogenesis.

Glossary

CMV: Cytomegalovirus promoter, a ubiquitous, strong promoter often used in overexpression studies.

Exons: The coding sequences of genes. In mammalian cells, exons within genes are interrupted by introns, which are noncoding sequences.

HSPGs: Heparin sulfate proteoglycans.

Ischemia: A condition in which tissues become oxygen deficient as a result of several pathological conditions. As a result, the tissue is relatively hypoxic, leading to upregulation of the VEGF gene.

Isoforms: Different forms of a protein, derived from a single gene that result from alternative splicing.

References

1. Jingjing, L., Srinivasan, B., and Roque, R. S. (2001). Ectodomain shedding of VEGF183, a novel isoform of vascular endothelial growth factor, promotes its mitogenic activity in vitro. *Angiogenesis* **4**, 103–112.
2. Soker, S., Gollamudi-Payne, S., Fidder, H., Charmahelli, H., and Klagsbrun, M. (1997). Inhibition of vascular endothelial growth factor (VEGF)-induced endothelial cell proliferation by a peptide corresponding to the exon 7-encoded domain of VEGF165. *J. Biol. Chem.* **272**, 31582–31588.

3. Ng, Y. S., Rohan, R., Sunday, M. E., Demello, D. E., and D'Amore, P. A. (2001). Differential expression of VEGF isoforms in mouse during development and in the adult. *Dev. Dyn.* **220**, 112–121. *In this paper, it was demonstrated that VEGF isoform levels vary during development and in the adult mouse. Data suggest that VEGF isoforms have different roles in developing and adult tissues to form and maintain vascular patterns.*
4. Carmeliet, P., Ng, Y. S., Nuyens, D., Theilmeier, G., Brusselmans, K., Cornelissen, I., Ehler, E., Kakkar, V. V., Stalmans, I., Mattot, V., Perriard, J. C., Dewerchin, M., Flameng, W., Nagy, A., Lupu, F., Moons, L., Collen, D., D'Amore, P. A., and Shima, D. T. (1999). Impaired myocardial angiogenesis and ischemic cardiomyopathy in mice lacking the vascular endothelial growth factor isoforms VEGF164 and VEGF188. *Nat. Med.* **5**, 495–502. *One of the first reports that VEGF isoforms in vivo are not functionally equivalent. Mice that express only the smallest isoform (VEGF120) were generated, and were shown to die shortly after birth, compared to mice with a wild-type complement of VEGF.*
5. Stalmans, I., Ng, Y. S., Rohan, R., Fruttiger, M., Bouche, A., Yuce, A., Fujisawa, H., Hermans, B., Shani, M., Jansen, S., Hicklin, D., Anderson, D. J., Gardiner, T., Hammes, H. P., Moons, L., Dewerchin, M., Collen, D., Carmeliet, P., and D'Amore, P. A. (2002). Arteriolar and venular patterning in retinas of mice selectively expressing VEGF isoforms. *J. Clin. Invest.* **109**, 327–336.
6. Ishida, S., Usui, T., Yamashiro, K., Kaji, Y., Amano, S., Ogura, Y., Hida, T., Oguchi, Y., Ambati, J., Miller, J. W., Gragoudas, E. S., Ng, Y. S., D'Amore, P. A., Shima, D. T., and Adamis, A. P. (2003). VEGF164-mediated inflammation is required for pathological, but not physiological, ischemia-induced retinal neovascularization. *J. Exp. Med.* **198**, 483–489.
7. Zelzer, E., Mclean, W., Ng, Y. S., Fukai, N., Reginato, A. M., Lovejoy, S., D'Amore, P. A., and Olsen, B. R. (2002). Skeletal defects in VEGF(120/120) mice reveal multiple roles for VEGF in skeletogenesis. *Development* **129**, 1893–1904.
8. Cheng, S. Y., Nagane, M., Huang, H. S., and Cavenee, W. K. (1997). Intracerebral tumor-associated hemorrhage caused by overexpression of the vascular endothelial growth factor isoforms VEGF121 and VEGF165 but not VEGF189. *Proc. Nat. Acad. Sci. USA* **94**, 12081–12087.
9. Grunstein, J., Masbad, J. J., Hickey, R., Giordano, F., and Johnson, R. S. (2000). Isoforms of vascular endothelial growth factor act in coordinate fashion to recruit and expand tumor vasculature. *Mol. Cell. Biol.* **120**, 7282–7291. *A thorough analysis of the influence of different VEGF isoforms on tumor formation. It was shown that the different isoforms vary in their abilities to vascularize tumors and promote tumor growth when compared with VEGF-null or wild-type cells.*

Robinson, C. J., and Stringer, S. E. (2001). The splice variants of vascular endothelial growth factor (VEGF) and their receptors. *J. Cell. Sci.* **114**, 853–865.

Capsule Biographies

Patricia A. D'Amore received her Ph.D. from Boston University and completed a postdoctoral fellowship at Johns Hopkins Medical School. She is currently Professor of Ophthalmology and Pathology at Harvard Medical School, a Research Associate at Children's Hospital, and a Senior Scientist at the Schepens Eye Research Institute in Boston.

Eric B. Finkelstein earned his Ph.D. in 2001 at the SUNY Upstate Medical University in Syracuse, New York, working in Dr. Thomas J. Poole's laboratory on the role of VEGF in blood vessel morphogenesis. He is currently a postdoctoral fellow in Dr. D'Amore's laboratory in Boston, studying endothelial cell differentiation and the function of VEGF isoforms.

CHAPTER 8

Ephrins and the Unveiling of Distinct Arterial and Venous Microcirculations

Jeffrey M. Gelfand and Guillermo García-Cardeña

Center for Excellence in Vascular Biology, Departments of Pathology, Brigham and Women's Hospital and Harvard Medical School, Boston, Massachusetts

Introduction

When William Harvey boldly challenged centuries of medical orthodoxy in 1628 by proclaiming that blood circulated throughout the body, he still had trouble explaining one critical concept: how arteries actually connected to veins. For with no way to magnify the vessels, all the great experimenter could conjure up was the old idea that there must be some sort of vascular pores arterial blood seeped through [1]. The Italian anatomist, Marcello Malpighi, was therefore quite surprised just 30 years later when he trained the lens of his new-found microscope onto the fine vascular architecture of a frog's lung and found not pores, as he had expected, but rather an entirely new class of veritable microvessels—which became known as "capillaries"—as the definitive connection shuttling blood between arteries and veins [2].

For more than 300 years, scientists have understood capillaries to be just as Malpighi described them that day in Bologna—as a third class of blood vessels distinct in form and function from their arterial and venous counterparts. However, recent research on a novel class of cell-signaling molecules known as ephrins, some of which reliably mark either arteries or veins on a molecular level, suggests that this traditional view of capillaries may no longer be sufficient, as capillary vessels seem to exhibit clear arterial-venous identities of their own. These identities are defined by the differential expression of members of the Ephrin-B family down to the level of the smallest microvessels, demonstrating that the microvasculature, far from just being a homogenous collection of nonarterial, nonvenous capillaries linking arterioles to venules, consists rather of two discrete arterial and venous continuums: the *arterial microcirculation* and the *venous microcirculation.* These new observations of arterial and venous microcirculatory identities likely reflect previously uncharacterized genetic, morphologic, and functional differences within the microvasculature, and should provide a novel conceptual framework for the study of vascular biology and the treatment of angiogenesis-dependent disease.

Cell Signaling via Ephrins and Eph Receptors

The ephrins, and their Eph receptors, are a novel class of proteins that enable cells to send signals to one another. Although long known to be key mediators in brain formation, ephrin signaling is also essential for the proper development of new blood vessels. Because some subsets of these important signaling molecules exhibit the additional property of localizing exclusively either to arteries or veins, ephrins have also begun to serve as the first reliable molecular marker capable of distinguishing arterial from venous identity.

In 1987 researchers working with *e*rythropoietin-*p*roducing *h*epatocellular carcinoma cells happened to discover a new kind of tyrosine kinase receptor, which they accordingly named the *Eph* receptor [3]. With 15 subtypes subsequently identified (EphA1–9 and EphB1–6), Eph receptors now constitute the largest family of tyrosine kinase receptors in vertebrates [4]. Tyrosine kinase receptors participate broadly in many pathways controlling cell growth and differentiation. Unlike traditional tyrosine kinases, however, where soluble hormones and growth factors travel as ligands from cells much farther away, ephrin ligands are bound to the cell surface and can activate Eph receptors on neighboring cells only when their mutual cell membranes draw close enough to interact. As a result, Ephrin/Eph receptor signaling is particularly well-suited for mediating physical cell-to-cell interactions such as repulsion and attraction.

Ephrin ligands are classified into two groups: Ephrin-As and Ephrin-Bs. The six known Ephrin-As are tethered to the cell membrane by a chemical moiety known as the glycosylphosphatidyl inositol (GPI) anchor. The three Ephrin-B ligands, on the other hand, extend both inside and outside the cell as transmembrane proteins with active cytoplasmic tails. In general, EphA receptors tend to bind Ephrin-As, and EphB receptors bind to Ephrin-Bs.

Even more unusually for tyrosine kinase–based systems, ephrin ligands can also function as receptors unto themselves, thereby enabling a unique mode of two-way, bidirectional signaling between neighboring cells. The bipartite structure of the Ephrin-B protein particularly lends itself to this duel function, with its extracellular portion serving as the ligand to activate "forward" signaling in Eph-expressing cells, while its cytoplasmic tail controls "reverse" signaling within its own cell [5]. The mechanism for how Ephrin-A ligands may also participate in reverse signaling is, however, unknown.

In neural development, where most Ephrin/Eph research originally concentrated, Ephrin/Eph cell-to-cell signaling has proven to be fundamental to the way axons, the signal-carrying end of nerve cells, navigate their way to specific locations in the nervous system. The best-studied model is one called the retinal-tectal projection, in which axons from visual neurons in the retina migrate to an important visual relay center in a part of the brainstem known as the tectum (the superior colliculus in humans). The axons appear to navigate by way of particular gradients, where, for instance, different EphA receptor concentrations influence anterior/posterior mapping and various EphBs affect dorsal/ventral patterning. Ephrin/Eph signaling exerts its effects by mediating repulsion and attraction between neurons within these gradients. Axonal migration in other parts of the brain also relies, in part, on Ephrin/Eph signaling, as does neural crest cell migration in the early embryo [6]. The very recent observations that EphB receptors also function in the adult brain beside excitatory NMDA glutamate receptors, and may even affect synaptic plasticity, could suggest an additional, and as-of-yet uncharacterized, role for Ephrin/Eph receptors in higher-order processes such as memory formation as well [7].

Angiogenesis Requires Ephrin Signaling

In the vasculature, ephrins are essential to the proper development of new blood vessels, particularly in the process of angiogenesis, where new vessels form from existing vascular templates. Although the mechanism remains unknown for how ephrin signaling actually influences angiogenesis, a role mediating attraction and repulsion between developing arteries and veins is strongly suspected.

Blood vessels form embryonically in two discrete stages. During vasculogenesis, endothelial cell precursors first form into a crude template known as the primary capillary plexus, which serves as the first branches of the future vascular tree. In the subsequent process of angiogenesis, this plexus then remodels into a mature vascular network by sprouting off new vessels and pruning others, with the balance between pro- and antiangiogenic factors as the guide. This angiogenic developmental program can be physiologically reactivated when needed, as in wound repair and menstruation, but it can also be pathologically co-opted in many diseases. As Judah Folkman first proposed in the early 1970s, most solid tumors, for instance, must recruit their own blood supply from the host using angiogenic factors in order to grow larger than 1 to 2 millimeters in size [8]. Many other diseases are also increasingly being recognized as angiogenic-dependent, including macular degeneration, endometriosis, and rheumatoid arthritis, and others like psoriasis also appear to have strong angiogenic components to their pathophysiology. Ephrin-signaling may play a role in most, if not all, of these physiologic and pathologic forms of angiogenesis, and may therefore prove to be an important and novel target for the treatment of angiogenic-dependent diseases.

Most notably, in a 1998 report by Wang, targeted mutagenesis of vascular Ephrin-B2 in a mouse model proved to be lethal at embryonic day 11 because of defects in angiogenesis. Because Ephrin-B2 localizes to arteries as opposed to veins, whereas its EphB4 receptor sits primarily on venous cells, the hypothesis began to emerge that Ephrin-B2/EphB4 signaling between nascent arteries and veins is required for angiogenesis to proceed properly in murine development [9]. This model received even more experimental support when Gerety and Wang subsequently revealed that EphB4 mutant mice had exactly the same defective-angiogenic phenotype as the Ephrin-B2 mutants [10]. The question remained, however, whether Ephrin-B2 on endothelial cells as well as smooth muscle mesenchymal cells were both necessary for mediating angiogenesis. To distinguish between their relative roles, Gerety and Anderson crossed Tie2-Cre mice with Ephrin-B2/lox mice to selectively knock out Ephrin-B2 in endothelial cells. What they discovered was that the endothelial cell–specific knockout led to angiogenic defects which were "indistinguishable" from the conventional combined endothelial and mesenchymal Ephrin-B2 knockout, demonstrating that the endothelial expression of Ephrin-B2 alone is necessary and sufficient for its angiogenic funciton [11]. The role of Ephrin-B2 signaling in the pathologic angiogenesis of tumors is as of yet unknown, but if it mediates signaling between developing arteries and veins in the nascent tumor vasculature, much like its postulated role in development, blocking arterial Ephrin-B2 may be a potent antiangiogenic therapeutic target.

Members of the Ephrin-A family also appear to play a direct role in mediating angiogenesis, although through very different mechanisms. In 1995 Pandey demonstrated that TNF-α, a potent proinflammatory as well as proangiogenic factor, exerts its downstream angiogenic effects by inducing Ephrin-A1 ligands to bind EphA2 receptors on endothelial cells as part of its signal transduction [12]. More recent studies suggest that tumor cells often develop the ability to express Ephrin-A1 ligands, which then bind to EphA2 receptors on endothelial cells to help induce angiogenesis. Accordingly, antibodies targeted to the EphA2 receptor have been shown to inhibit tumor angiogenesis and tumor growth in vivo in models of RIP-Tag pancreatic islet cell carcinoma and 4T1 mammary adenocarcinoma, specifically by blocking endothelial cell migration [13]. Whether any other members of the Ephrin-A family in the vasculature also contribute to angiogenesis remains to be determined.

Because metastastic potential in most tumors partially depends on angiogenic potential, Ephrins and their Eph receptors are also being studied as possible markers of tumor behavior. Several studies report associations between certain Eph and Ephrin subtypes in a variety of tumor cell lines, but more research needs to be done in this area before any estimations of their utility can be made.

Ephrins Reveal Distinct Arterial and Venous Microvascular Identities

Perhaps the most important conceptual role for Ephrins, however, stems not from what they do, but rather from on which cells they act. While studying the functional role of Ephrins in the nervous system, Wang et al. happened to notice that the Ephrin-B2 ligand localizes to distinct subsets of blood vessels in the developing mouse embryo. Further analysis revealed that this staining pattern was a result of the differential expression of Ephrin-B2 in the arterial, as opposed to venous, endothelium of the vasculature [9]. Subsequent studies by Shin and García-Cardeña et al. and Gale et al. went on to demonstrate that this Ephrin-B2 expression pattern persists into adulthood and extends not just within the endothelium, but into smooth muscle cells and pericytes as well [14, 15]. Most important, they also observed that differential arterial Ephrin-B2 expression extends into even the smallest diameter vessels, suggesting for the first time that the microvessels of capillary networks may have arterial and venous identities of their own. This differential expression was evident in vessels of multiple tissues, including pancreas, muscle, fat, kidney glomeruli, brain, liver, adrenal cortex, and adrenal medulla. Intriguingly, Ephrin-B2 also appears to differentiate arterial from venous vessels in several models of both normal and pathological angiogenesis, including marking subsets of vessels sprouting into the cornea from the limbus artery in a model of VEGF-induced angiogenesis, as well as subsets of microvessels in wounded tissue undergoing healing. Strikingly, approximately half of the vessels present in the microvasculature of different types of mouse tumor models also expressed Ephrin-B2, suggesting that the tumor vasculature may also consist of distinct arterial and venous microcirculatory components. Taken together, these studies for the first time challenged the classical view that quiescent as well as remodeling capillaries have neither arterial nor venous identities of their own, and strongly suggested that the expression of Ephrin-B2 may also be important for the formation of new vascular circuitries.

The Implications of a New Microvascular Paradigm

Since the time of Malpighi in the late 17th century, scientists have understood capillary networks to consist of a unique class of tiny "hair-like" microvessels linking arterioles to venules. Part of what distinguished capillaries operationally from all other vessel types was that they were neither arterial nor venous and could therefore function as a vascular crossroads for gas and metabolite exchange at the tissue level. In the late 1970s and early 1980s, however, a series of studies out of Czechoslovakia on enzyme histochemistry began to challenge this long-standing presumption of capillary arterial-venous neutrality. In 1979, Lojda realized that because the enzyme alkaline phosphatase histochemically stained arterioles, and dipeptidylpeptidase IV (DPPIV) highlighted venules, staining for both of these enzymes together would help illuminate the entire capillary bed [16]. With this new staining method, Lojda was able to divide the capillary bed histologically into three distinct portions—an alkaline phosphatase section, a DPPIV section, and a third "transitional" zone in which both enzymes were present—revealing for the first time molecular differences within the microvasculature. Throughout the 1980s and early 1990s, a handful of studies revealed additional evidence of capillary heterogeneity with even more molecular markers, but these markers only lasted during specific windows of development. Not until the late 1990s did Ephrin-B2 prove to be the first reliable molecular marker of arterial identity from the largest down to the smallest vessels and persisting from development into adulthood. Much like Lojda's enzymes, Ephrin-B2 even more clearly maps out specific subsets of capillaries as arterial versus venous, providing the strongest challenge yet to the long-standing idea of capillaries as its own class of nonarterial, nonvenous blood vessel.

The new picture that is now emerging of microvascular identity is one of distinct arterial and venous microcirculations. These circulations are defined by the differential expression of molecular markers such as Ephrin-B2, which delineate an arterial continuum from arteries to arterioles, all the way down to corresponding microarterial capillary segments, and a venous continuum extending from veins to venules to corresponding microvenous capillary segments. These continuums display their own unique genetic, morphologic, and, perhaps, functional identities, and raise a host of new questions for the study of vascular biology and treatment of many human diseases: Why, for instance, do activated immune cells in the blood escape only through postcapillary venules and not through arterioles? Many vasculitic processes, on the other hand, exhibit specific arterial versus venous predispositions for the site of inflammation. Both giant cell and Takayasu's arteritis, for instance, occur only in large arterial vessels, and Kawasaki's disease in medium-sized arteries, but the many small-vessel vasculitic processes, such as microscopic polyangiitis and Wegener's granulomatosis, have much less exclusive propensities. Would more directed studies of these many vasculitic processes reveal any segregration between microarterial and microvenous capillary segments and perhaps provide new clues to their mysterious pathogenic processes? In genetic disorders such as CADASIL, where mutations in Notch-3 lead to migraines, mood changes, strokes, and, ultimately, vascular dementia, why are only small arterioles affected? Are there any corresponding syndromes defined by defects in the genes of the venous microcirculation? And as the tumor microvasculature appears to consist of roughly half microarterial and half microvenous vessels, how important in the end is arterial and venous identity in forming a functioning tumor vasculature? Similarly, what would happen if either the arterial or venous development of the tumor vasculature were blocked? And do chemotherapeutic or antiangiogeneic agents specifically target the arterial or venous microcirculations? Research into Ephrins is providing the first window into answering these questions.

Much like Malpighi's microscopy and its ability to reveal an entire class of vessels that even William Harvey did not imagine, the study of Ephrins has begun to serve as the first example of a 21st-century molecularly-guided microscopy powerful enough to reveal unexpected heterogeneity within even Malpighi's own capillary microvessels. The missing link in Harvey's circuit appears to be defined neither by pores, nor even liminal capillaries, but rather by "microarteries" and "microveins" bridging together distinct arterial and venous circulations.

References

1. Harvey, W. (1958). *Exercitatio Anatomica De Motu Cordis Et Sanguinis in Animalibus*. Trans. Chauncey D. Leake. Springfield: Charles C. Thomas.
2. Malpighi, M. (1929). "Malpighi's 'De Pulmonibus.'" Trans. James Young. *Proc. Royal Soc. Med., Hist. Med.* **1**, 1–11. *One of the only English translations of the Italian microscopist's famous 1661 manuscript, in which he first revealed the existence of capillary microvascular networks. It takes the form of two letters to his long-time mentor in Pisa.*
3. Hirai, H., Maru, Y., Hagiwara, K., Nishida, J., and Takaku, F. (1987). A novel putative tyrosine kinase receptor encoded by the Eph gene. *Science* **238**, 1717–1720.
4. Drescher U. (2002). Eph family functions from an evolutionary perspective. *Curr. Opin. Genet. Dev.* **12**, 397–402.
5. Himanen, J. P., Rajashankar, K. R., Lackmann, M., Cowan, C. A., Henkemeyer, M., and Nikolov, D. B. (2001). Crystal structure of an Eph receptor-ephrin complex. *Nature* **414**, 933–938.
6. Wilkinson, D. G. (2001). Multiple roles of Eph receptors and ephrins in neural development. *Nat. Rev. Neurosci.* **2**, 155–164.
7. Murai, K. K., and Pasquale, E. B. (2002). Can Eph receptors stimulate the mind? *Neuron* **33**, 159–162.
8. Folkman, J. (1971). "Tumor Angiogenesis: Therapeutic Implications." *New England Journal of Medicine* **285**(21), 1182–1186.
9. Wang, H. U., Chen, Z. F., and Anderson, D. J. (1998). Molecular distinction and angiogenic interaction between embryonic arteries and veins revealed by Ephrin-B2 and its receptor Eph-B4. *Cell* **93**, 741–753. *The first report of the functional role of a member of the Ephrin family during developmental angiogenesis, and of the expression of Ephrin-B2 in endothelial cells of developing arteries but not veins.*
10. Gerety, S. S., Wang, H. U., Chen, Z. F., and Anderson, D. J. (1999). Symmetrical mutant phenotypes of the receptor *EphB4* and its specific transmembrane ligand *Ephrin-B2* in cardiovascular development. *Molecular Cell.* **4**, 403–414.
11. Gerety, S. S., and Anderson, D. J. (2002). Cardiovascular ephrinB2 function is essential for embryonic angiogenesis. *Development* **129**, 1397–1410.
12. Pandey, A., Shao, H., Marks, R. M., Polverini, P. J., and Dixit, V. M. (1995). Role of B61, the ligand for the Eck receptor tyrosine kinase, in TNF-alpha-induced angiogenesis. *Science* **268**, 567–569.
13. Brantley, D. M., Cheng, N., Thompson, E. J., Lin, Q., Brekken, R. A., Thorpe, P. E., Muraoka, R. S., Cerretti, D. P., Pozzi, A., Jackson, D., Lin, C., and Chen, J. (2002). Soluble Eph A receptors inhibit tumor angiogenesis and progression in vivo. *Oncogene* **21**, 7011–7026.
14. Shin, D., Garcia-Cardena, G., Hayashi, S., Gerety, S., Asahara, T., Stavrakis, G., Isner, J., Folkman, J., Gimbrone, M. A. Jr., and Anderson, D. J. (2001). Expression of EphrinB2 identifies a stable genetic difference between arterial and venous vascular smooth muscle as well as endothelial cells, and marks subsets of microvessels at sites of adult neovascularization. *Dev. Biol.* **230**, 139–150. *This in-depth study of Ephrin-B2 expression in endothelial and vascular smooth muscle cells of the adult mouse vasculature provides the first demonstration that quiescent as well as remodeling capillaries possess arterial and venous identities of their own.*
15. Gale, N. W., Baluk, P., Pan, L., Kwan, M., Holash, J., DeChiara, T. M., McDonald, D. M., and Yancopoulos, G. D. (2001). Ephrin-B2 selectively marks arterial vessels and neovascularization sites in the adult, with expression in both endothelial and smooth-muscle cells. *Dev. Biol.* **230**, 151–160.
16. Lojda, Z. (1979). Studies on dipeptidyl (amino) peptidase IV (glycylproline napththylamidase). II. Blood Vessels. *Histochemistry* **59**, 153–166.

Capsule Biographies

Dr. García-Cardeña is an Assistant Professor of Pathology at Harvard Medical School and the Director of the Laboratory for Systems Biology at the Center for Excellence in Vascular Biology, Brigham and Women's Hospital, Boston, Massachusetts.

Jeffrey M. Gelfand is a medical student at Harvard Medical School and a member of Dr. García-Cardeña's Laboratory.

CHAPTER 9

Endothelial Heterogeneity

Mary E. Gerritsen
Frazier HealthCare Ventures, Palo Alto, California

Introduction

The following paragraph written by a Malaysian student studying in America aptly captures the concept of regio-specific heterogeneity, the topic of the present chapter:

America is not homogeneous, neither is its heterogeneity immediately apparent. Rather, as so many of my American friends have been so keen to enlighten me, American people differ from state to state. The mainstream American culture of fast-food chains, shopping malls and pop music is there for the world to see. But it is the only uniting factor across their vast homeland. A person from Iowa is vastly different from a person from California. Even in Virginia alone, those who lived in the northern part of the state near Washington, DC are totally different from those who grew up in central Virginia where Sweet Briar is located. Taken as a whole, America is bursting with ethnic and cultural diversity, but in reality, it is small pockets of homogeneity that combine to form a picture of unparalleled heterogeneity.[1]

The vascular endothelium lines the blood vessels of the body, and in humans, the estimated number of individual cells constituting this lining is somewhere on the order of $1–6 \times 10^{13}$. Like Americans, the mainstream "endothelial culture" of blood, tissue, and nutrients is there for all to note. These uniting factors do not, however, reveal the vast differences of specific circulations [1, 2]. Several vascular human diseases are exquisitely restricted to specific types of vessels. Vasculitis often shows a marked predilection for specific arteries, veins, or capillaries; tumor cells may metastasize to selective vascular beds, and atherosclerosis is generally restricted to the larger arteries.

Although scientists initially thought the endothelial cell layer was essentially an inert barrier between the blood and the tissue, this extensive "organ" is now known to carry out a diverse array of specialized functions which can vary markedly from one organ to another. Today the endothelium is viewed as a dynamic, mutable, heterogeneous, distributed organ with essential secretory, synthetic, metabolic, and immunologic functions. The picture of "unparalleled heterogeneity" is revealed as follows.

Endothelial Architecture

Endothelial heterogeneity is a term that has been coined to describe the diversity and regional specificity of endothelial cells at different sites in the vasculature. The concept that endothelial cells are not identical is not new; variations in the structure of capillary endothelium led to the first subclassification of endothelial cells into "continuous," "fenestrated," and "discontinuous" [3, 4] (Figure 1). Endothelia of the continuous type are the most prevalent and are found in the walls of arterioles, capillaries, and venules of skeletal, smooth and cardiac muscle, the mesentery, skin, connective tissue, lung, brain, and eye, as well as lining the major conduit vessels (large arteries, veins). These endothelial cells are characterized by occluding (tight) junctions. Fenestrated endothelia are found in the exchange vessels of secretory and excretory organs such as endocrine and exocrine glands, in the capillaries lining the gastric mucosa, the glomerular and peritubular capillaries of the kidney, as well as in the synovium and choroid plexus. (Actually, the capillary endothelial cells of the synovium exhibit a continuous morphology on the side facing away from the synovial fluid and a fenestrated morphology on the opposite side.) Fenestrated endothelial cells are characterized by numerous small "windows" (i.e., small transcellular openings ranging from 50 to 80 nm in diameter). At the ultrastructural level, differences in the fine structure of the fenestrae are revealed. In some organs, the fenestrae appear open; in others the fenestrae are

[1] Thoughtfully penned by a Citizen of the World; http://www.jellybeans.blogspot.com/2002_12_01_jellybeans_archive.html

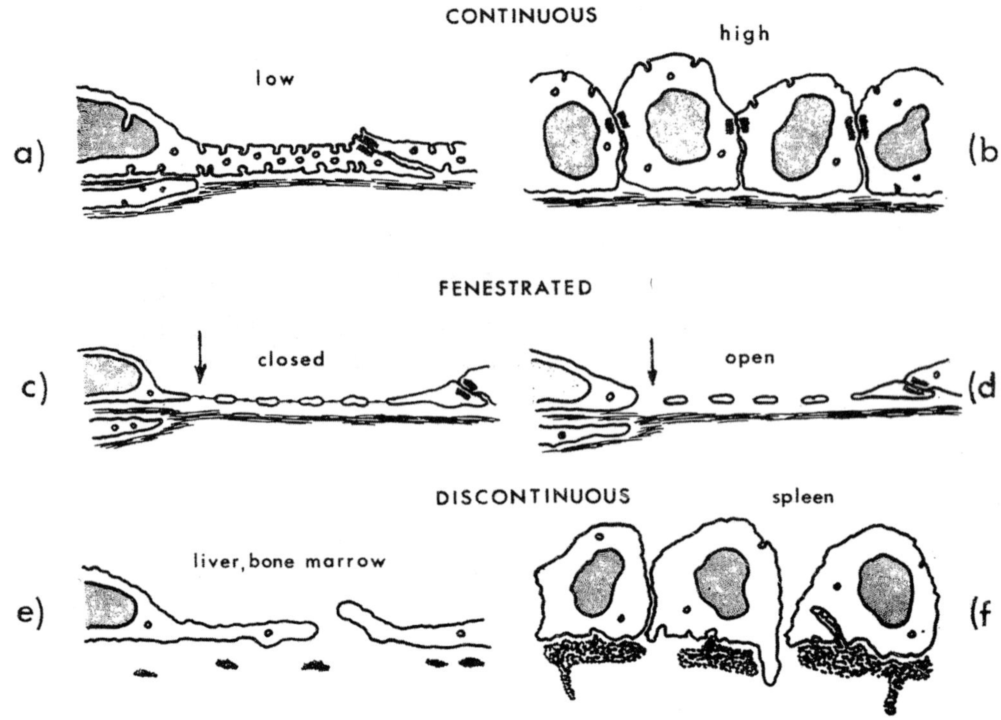

Figure 1 Classification of the endothelial cells of different capillaries according to their continuity. Three main types are distinguished (i.e., continuous, fenestrated, and discontinuous) and for each, two main varieties are represented. Little detail is shown because considerable variations exist from organ to organ; in fact, almost each organ can be said to have its own type of capillary endothelium. The scheme is based on information derived from mammalian studies. **Panels a, b:** The endothelium has no recognizable openings. The low variety **(a)** is found in striated muscle, myocardium, central nervous system, smooth muscle of digestive and reproductive systems, and subcutaneous and adipose tissue. The high variety **(b)** is typical of the postcapillary venules of the lymph nodes and thymus; a similar endothelial appearance is also seen in large arteries when they are contracted. **Panels c, d:** The endothelium has intracellular fenestrae (*arrows*) either closed **(c)** as in endocrine glands, choroids plexus, ciliary body, and intestinal villus, or open **(d)** as in the renal glomerulus. **Panels e, f:** The endothelium are discontinous, containing intercellular gaps. These vessels are often refered to as "sinusoids" and are typical of liver, bone marrow, and spleen; in each of these sites, they differ in structural detail. (Reprinted with permission from Figure 1 [4].)

"closed" by thin, membranous diaphragm-like structures. Fenestrae are not fixed structures; their frequency and location can be influenced by cytokines, ischemia reperfusion, growth factors, and other stimuli. An extreme form of fenestrated endothelia, termed *discontinuous endothelia*, are found in the sinusoids of the liver, spleen, and bone marrow. These endothelia have irregular outlines and calibers and highly fenestrated structures; in the liver, the fenestrae are often clustered to form sieve plates, and none of the fenestrae are closed with diaphragms (Figure 1).

Endothelial Glycocalyx

Endothelial diversity goes far deeper than appearance alone. In recent years, scientists have begun to appreciate the breadth of endothelial specialization, at the biochemical, metabolic, and immunological level. The endothelial surface is a highly specialized feature. This is the "face" the endothelial cell presents to the nutrients and formed elements of the blood, and not surprisingly, this surface exhibits remarkable specificity. One exquisitely localized property is the composition of the endothelial glycocalyx, in which differences have been revealed by the binding patterns of labeled lectins. For example, wheat germ agglutinin (WGA), when perfused through the vasculature of mice or rats, will bind to the luminal surface of endothelial cells in capillaries and arterioles, but not in venules (Figure 2). This selectivity is impressively abrupt; capillary endothelial cells will bind WGA strongly, whereas immediately adjacent venular cells do not. WGA interacts with N-acetylglucosamine and sialic acid glycoconjugates; thus the display of these glycoconjugates is highly regional [6].

Endothelial Lymphocyte and Leukocyte Interactions

The expression of homing receptors involved in lymphocyte homing illustrates another aspect of endothelial surface specificity, providing the so-called "zip code" for cell trafficking. For example, in the mouse, lung-specific endothelial adhesion molecule (Lu-ECAM-1) is expressed exclusively by pulmonary postcapillary endothelial cells and in some venules in the spleen; mucosal addressin cell

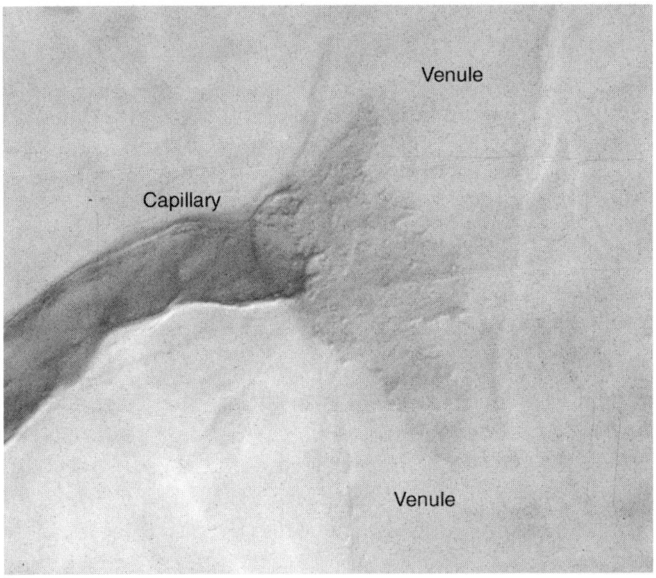

Figure 2 Staining of microvessels in rat trachea by postfixation perfusion of biotinylated wheat germ agglutinin (WGA) lectin, followed by reaction with DAB peroxidase. WGA lectin binds to the luminal surface of capillary endothelial cells, but not to that of venular endothelial cells, showing abrupt heterogeneity of cell surface glycosylation. Micrograph shows WGA binding to an endothelial cell that extends from the capillary into the postcapillary venule. (Photomicrograph kindly provided by Dr. Gavin Thurston, Regeneron Pharmaceuticals, Inc.)

Table I Different Markers of Endothelial Specialization [1, 5].

Marker	Preferential Expression
Ephrin-B2	Arterial endothelial cells
P-glycoprotein (mdr 1a)	Brain endothelial cells
Glut-1	Brain endothelial cells
CD73/transferrin receptor	Brain endothelial cells
LuECAM	Lung venular endothelial cells
GlyCAM-1	HEVs
NF-HEV	HEVs
MECA79	HEVs, chronically inflamed EECs
VAP-1	HEVs, endothelial cells in large vessels
VAP-2	HEVs, venules
CD34/Sgp90	One form, specifically glycosylated is restricted to HEVs
MadCAM-1	Peyer's patches, mesenteric lymph node HEVs, venules in intestinal lamina propria, spleen sinus
EphB4	Venous endothelial cells
Flt-4	Lymphatic endothelial cells
LYVE-1	Lymphatic endothelial cells
Podoplanin	Lymphatic endothelial cells
Prox-1	Lymphatic endothelial cells
Desmoplakin	Lymphatic endothelial cells

adhesion molecule-1 (Mad-CAM-1) is expressed primarily on high endothelial venules in Peyer's patches; GlyCAM-1 is also restricted to high endothelial venules. Postcapillary venules are the site of leukocyte adhesion and emigration, partly because of regio-specific cell adhesion molecule expression. For example, P-selectin is constitutively expressed by venular endothelial cells in most organs; however, it is absent or present only at very low concentrations in arteriolar and capillary endothelial cells. In vivo, the leukocyte adhesion molecules VCAM-1 and E-selectin are also selectively expressed by venular endothelial cells when activated by inflammatory cytokines or other mediators. Other factors contributing to this "local zip code" include the selective secretion of certain chemokines, cytokines, and differences in shear stress (see following sections for further discussion of physical forces). Table I summarizes the expression of several vascular bed–specific endothelial cell markers.

Vascular Permeability

The early topographic studies of Majno [7] were the first to recognize that the topical application of histamine or serotonin to the rat cremaster muscle selectively induced permeability in the venules. The list of inflammatory mediators that selectively induce venular responses is impressive and includes substance P, leukotriene B4, bradykinin, tumor necrosis factor, lipopolysaccharide, and ischemia-reperfusion injury. Moreover, these venular-specific responses have been observed in many organs, including cremaster muscle, diaphragm, trachea, cheek pouch, skin, bladder, stomach, pancreas, and intestine. At least some of this specificity is a result of different densities of the specific receptors for these mediators on venular endothelial cells; there may be additional contributing factors such as differences in downstream signal transduction. There are remarkable differences in the baseline permeability of different vascular beds to low-molecular-weight compounds as well as to large proteins. The exchange of these molecules across the endothelium can occur through the endothelium (by specific transporters, or by "vesicular transport") or between endothelial cells (i.e., at the endothelial junction). Obviously, the architecture of the endothelial junction plays a critical role in the roles of these different exchange pathways. There are also differences in the expression of different transporters and plasmalemmal vesicles from one vascular bed to another. For example, plasmalemmal vesicles are rarely seen in brain endothelia, yet are relatively frequent in the endothelial cells of the lung and diaphragm. The endothelium of the brain, which forms the so-called blood–brain barrier, plays a particularly important role in regulating the transport of solutes from the apical to basal surface; brain endothelial cells have specific transport systems for certain nutrients such as glucose (Glut-1) and amino acids. The endothelium of the brain also expresses the multidrug-resistance protein P-glycoprotein (*mdr1a*), an active transporter that transports low-molecular-weight molecules from the basal to apical surface, thus protecting the cells of the central nervous system from accumulating potentially neurotoxic molecules. Organs with discontinuous endothelia, such as the liver capillaries, are quite permeable, and albumin escapes at a rate several fold greater than from less permeable continuous-type capillaries of muscle and skin.

Vascular Reactivity

Endothelial cells are the source of some potent molecules that influence the contractile state of vascular smooth muscle cells, including the vasodilators nitric oxide (NO), prostaglandin (PG)PGI_2, endothelial-derived hyperpolarizing factor (EDHF), adrenomedullin, and the potent vasoconstrictor, endothelin-1. PGE_2, made predominantly in microvascular endothelial cells, may be a vasodilator or vasoconstrictor, depending on the concentration, interactions with other autocoids, and the vascular bed. The balance of these factors is an important determinant of the control of vascular tone and tissue perfusion; considerable heterogeneity of responses exists among vessels of different size from different anatomic origins and different species.

Endothelial Junctions

As discussed previously, differences in endothelial cell-to-cell junctions in different vascular beds has long been recognized. During the last two decades, much has been learned about the biochemical and molecular constituents of the endothelial junction and their regulation. The inter-endothelial adhesive structures include tight, adherens, and gap junctions, in which the surface proteins occludin, claudins, cadherins, platelet endothelial cell adhesion molecule (PECAM), CD99, and junctional adhesion molecules (JAMs) are specifically incorporated. Tight junctions, defined structurally as closely opposed neighboring plasma membranes that appear to be tightly fused, are especially well developed in the brain and retina; occludin and claudin are the cellular marker proteins of this type of junction.

Adherens junctions, defined morphologically by a cytoplasmic plaque structure of electron-dense material on the plasma membrane of adjacent cells, are ubiquitously expressed in endothelia of all vascular beds; cadherins, in particular vascular endothelial (VE)-cadherin, are the endothelial marker protein for this type of junction. Gap junctions are characterized by their morphological appearance as a patch on which the membranes of two adjacent cells are separated by a uniform narrow gap of 2 to 4 nm; connexins, in particular connexins (Cx) Cx37, Cx40, and Cx43, are the marker protein for this type of junction. This "gap" is spanned by channel-forming proteins (the connexins) that allow inorganic ions and other small water-soluble molecules to pass from the cytoplasm of one cell to another; gap junctions thus couple cells both electrically and metabolically.

There is considerable heterogeneity in the expression of some of these junction-associated molecules. For example, occludin, a structural component of the tight junction, is highly expressed in epithelial cells and in brain endothelial cells, yet barely detectable in endothelial cells of most other organs. Junctional adhesion molecule (JAM)-B (JAM-2, VE-JAM) is highly localized to the intercellular boundaries of high endothelial venules; JAM-A (JAM-1) is more highly expressed in brain endothelia, whereas JAM-C (JAM-3) is more broadly expressed in endothelia of different vascular beds. JAM-1 may be an organizer of occludin clustering, thus contributing to the establishment of tight junctions in certain endothelial cells. In the kidney, the expression of claudin 5, a component of tight junctions, is restricted to arterial endothelial cells but undetectable in veins and capillaries. Other junctional molecules such as vascular endothelial cadherin (VE-cadherin) and the connexins seem to be more ubiquitously expressed.

Metabolism

The total surface area of the vascular endothelium is impressive: 100 to 700 m^2. This surface provides an interface for nutrient transport, metabolism of bloodborne substances, and secretion of chemokines, cytokines, vasoactive peptides, enzymes, and other biologically active factors. The lung, because of its large surface area and its position as recipient of the entire cardiac output, is an important site of metabolism of both native molecules and xenobiotics. The architecture of the pulmonary capillaries, sandwiched between adjacent alveolae, provides a system in which blood flows through these vessels as if between sheets, providing for maximal exposure of the capillary blood to alveolar gases. Surprisingly, relatively little is known about the potential expression of specific metabolic enzymes in pulmonary capillary endothelia.

Endothelial Cells as a Source of Growth Factors

Endothelial cells from the bone marrow are capable of promoting long-term multilineage hemopoiesis, in particular myelopoiesis and megakariocytopoiesis. The ability of these cells to support hemopoeisis may be a result of their ability to constitutively produce certain growth factors such as colony stimulating factors 1 and 2, fibroblast growth factor, leukemia inhibitory factor, transforming growth factor β, platelet-derived growth factor, interleukin-6, and kit ligand.

Endothelial Cells and Physical Forces

In addition to biochemical activation, the endothelial lining is constantly subjected to biomechanical stimuli (i.e., shear stresses, strains, and hydrostatic pressures), generated by the pulsatile flow of the blood. The fluid mechanics of blood flow in the vasculature is complex; in arteries it involves pulsatility at high pressures, yet in veins it is quasi-steady and at low pressure. Flow patterns in the vicinity of curvatures and branch points create regions of turbulence in arteries; however, turbulence in veins is generally not seen. Flowing blood subjects the endothelium to two orthogonal components: viscous friction and distension. The frictional force per unit area acts tangential to the vascular wall and is known as wall shear stress. The distending force acts perpendicular to the wall and is essentially equal to the fluid pressure. The shear stress and the pressure stretch the wall, producing both longitudinal and circumferential strain. Endothelial cells are known to sense and respond to all of these hemodynamic challenges. Flow patterns dictated by vascular anatomy are now thought to play a crucial role in the origin of atherosclerotic lesions. These lesions have a predeliction to occur at or near vessel bifurcations, regions exposed to nonuniform, complex flow. Interestingly, in vitro and in vivo studies comparing the effects of nonuniform complex flow to time-averaged, uniform flow, such as laminar shear stress, have revealed a remarkable difference in phenotype. Disturbed or turbulent flow is associated with the generation of reactive oxygen species by endothelial cells, followed by the expression of endothelial inflammatory markers, augmented endothelial turnover, and increased endothelial cell death. In contrast, the more uniform flow is associated with minimal endothelial cell turnover, low levels of apoptosis, and the expression of antioxidants and anti-inflammatory genes. There are quite likely inherent differences in the ability of endothelial cells at different vascular sites to respond to biomechanical forces; this is a relatively understudied aspect of endothelial heterogeneity.

Endothelial Cells and the Basement Membrane

The basal lamina is a flexible, thin (40 to 120 nm thick) mat of specialized extracelluar matrix that separates the endothelium from the underlying tissue; it is composed mainly of type IV collagen, heparan sulfate proteoglycan, laminin, and nidogen/entactin (although the total number of proteins known to make up the basement membrane is at least 50). This so-called mat is far from inert or just structural; the basal lamina can influence cell metabolism, organize proteins in adjacent plasma membranes, induce proliferation, serve as a filter, act as a depot for various growth factors and other proteins, and importantly, serve as a specific highway for migrating leukocytes. Most basement membranes appear similar at the level of the electron microscope, but their composition is unique to each tissue. This molecular inequality is an important contributor to the differentiated state of the endothelial cell. The basement membrane of bone marrow sinusoids shows a distinctive structure: it is discontinous, lacks a network organization, and consists of irregular masses of amorphous material. It is also unusually abundant in chondroitin sulfate proteoglycans and poor in heparan sulfate proteoglycan. This composition may facilitate disassembly and reassembly of basement membrane material as maturing hemopoietic cells pass through the sinusoidal wall. Tumors produce basement membrane components that promote angiogenesis, and additionally, cryptic domains within basement membrane molecules (such as the collagens) can be exposed during proteolytic degradation; some of these possess antiangiogenic activity [8].

Endothelial Cells and Pericytes

Pericytes, a type of mural cell associated abluminally with endothelial cells in capillaries and in some postcapillary venules, exhibit considerable variability in their relative association with endothelial cells in different tissues (for a review, see Ref. [9]. For example, capillaries of the retina, lung, and skeletal muscle exhibit more pericytes and "coverage" of endothelial cells than capillaries found in endocrine tissues such as the adrenal gland. Pericyte coverage of newly formed blood vessels is also thought to play a role in the stabilization of these structures; the vasculature of tumors is poorly invested with pericytes.

Endothelial Cell Proliferation

Endothelial cells isolated from various microvascular beds have been traditionally more difficult to maintain in culture than their counterparts from larger vessels such as the aorta, saphenous vein, and umbilical vein. This is likely a consequence of multiple contributing factors: differences in growth factor synthesis and in growth factor receptor expression and signal transduction, harsher experimental conditions used to isolate the cells, and potential differences in the nutritional requirements (e.g., amino acids, carbohydrates, vitamins). A perusal of PubMed will quickly identify all of these as candidates, and more, in one or several publications. There are regional differences in endothelial "proliferative" states in vivo as well. For example, endothelial cells at sites of disturbed flow (e.g., branching of blood vessels, areas of curvature) exhibit a higher rate of BrDU labeling or proliferating cell nuclear antigen expression (indices of proliferation) compared to cells in adjacent regions of the vasculature exposed to laminar flow. Endothelial cells at the site of a wound, ischemia-reperfusion injury, or in reproductive organs such as the ovary or mammary gland undergo phenotypic changes consistent with the development of an angiogenic phenotype. The phenotypic characteristics of the endothelial cells in tumors are a hot topic of current research: the vision of specifically targeting the tumor vasculature offers an incredible therapeutic opportunity for the treatment of many solid tumors. A major contributor to the switch to the "angiogenic phenotype" is vascular endothelial growth factor (VEGF), a potent mitogen, morphogen, and motogen for endothelial cells. Moreover, VEGF is a potent inducer of vascular permeability. Proliferating or angiogenic endothelial cells display different antigens, altered basement membrane composition, altered junctional morphology, and numerous transendothelial channels. Recent studies now indicate that there may also be vascular bed–specific endothelial cell growth factors; for example, endocrine gland–derived VEGF (EG-VEGF) is an endothelial cell mitogen selective for endocrine gland endothelial cells.

Genetic Diversity of Endothelial Cells

Several approaches have been used to identify endothelial differences on a genomic scale, including differential display, suppressive subtractive hybridization, serial analysis of gene expression, and DNA and oligonucleotide microarrays [10–14]. Such studies have identified potential tissue-specific endothelial cell transcripts from cerebral, lymphatic, arterial versus venous, lung, nasal polyp, skin, and myometrial cells. At present these studies are too preliminary to cite the markers as fact, but they offer considerable promise for an improved molecular understanding of endothelial heterogeneity.

References

1. Garlanda, C., and Dejana, E. (1997). Heterogeneity of endothelial cells. Specific markers. *Arterioscler. Thromb. Vasc. Biol.* **17**, 1193–1202.
2. Gerritsen, M. E. (1987). Functional heterogeneity of vascular endothelial cells. *Biochem. Pharmacol.* **36**, 2701–2711.
3. Bennett, B., Luft, J., and Hampton, J. (1959). Morphological classification of vertebrate blood capillaries. *Am. J. Physiol.* **197**, 381–390. *The first attempt to distinguish endothelial cells based on their ultrastructural features.*
4. Majno, G. (1965). Ultrastructure of the vascular membrane. In W. Hamilton, ed., *Handbook of Physiology*, pp. 2293–2375. Washington, DC: American Physiological Society. *An early but thoughtful and insightful description and discussion of capillary structures and their diversity.*
5. Wang, H., Chen, Z., and Anderson, D. (1998). Molecular distinction and angiogenic interaction between embryonic arteries and veins revealed by ephrin-B2 and its receptor Eph-B4. *Cell* **93**, 741–753.
6. Thurston, G., Baluk, P., and McDonald, D. (1999). Determinants of endothelial cell phenotype in venules. *Microcirculation* **7**, 67–80. *A review of endothelial diversity in the setting of inflammation.*
7. Majno, G., Palade, G., and GL, S. (1961). Studies on inflammation. II. The site of action of histamine and serotonin along the vascular tree: A topographic study. *J. Biophys. Biochem. Cytol.* **11**, 607–625.
8. Kalluri, R. (2003). Basement membranes: structure, assembly and role in tumor angiogenesis. *Nat. Rev. Cancer* **3**, 423–433. *A detailed review of basement membranes.*
9. Shepro, D., and Morel, N. (1993). Pericyte physiology. *FASEB J.* **7**, 1031–1038.
10. Chi, J.-T., Chang, H., Haraldsen, G., Jahnsen, F., Troyanskaya, O., Chang, D., Wang, Z., Rockson, S., van de Rijn, M., Botstein, D., et al. (2003). Endothelial cell diversity revealed by global expression profiling. *Proc. Natl. Acad. Sci. USA* **100**, 10623–10628.
11. Girard, J.-P., Baekkevold, E., Yamanaka, T., Haraldsen, G., Brandtzaeg, P., and Amalric, F. (1999). Heterogeneity of endothelial cells. The specialized phenotype of human high endothelial venules characterized by suppression subtractive hybridization. *Am. J. Pathol.* **155**, 2043–2055.
12. Kallman, B.-A., Wagner, S., Hummel, V., Buttmann, M., Bayas, A., Tonn, J., and Rieckmann, P. (2002). Characteristic gene expression profile of primary human cerebral endothelial cells. *FASEB J.* **10**, 589–591.
13. Podgrabinska, S., Braun, P., Velasco, P., Kloos, B., Pepper, M., Jackson, D., and Skobe, M. (2002). Molecular characterization of lymphatic endothelial cells. *Proc. Natl. Acad. Sci. USA* **99**, 16069–16074.
14. St. Croix, B., Rago, C., Velculescu, V., Traverso, G., Romans, K., Montgomery, E., Lai, A., Riggins, G., Lengauer, C., Vogelstein, B., et al. (2000). Genes expressed in human tumor endothelium. *Science* **289**, 1197–1202.

Capsule Biography

At the time this review was written, Dr. Gerritsen was working as an independent consultant for several biotech and venture capital firms. Currently she is the Senior Director of Molecular and Cellular Pharmacology at Exelixis Inc. She has held the positions of Senior Director, Vascular Biology, at Millennium Pharmaceuticals (2001–2002); Associate Director, Cardiovascular Research, at Genentech Inc. (1997–2001); Group Leader, Inflammation, at Bayer Corporation (1990–1997) and Associate Professor, Physiology, New York Medical College, 1985–1990. Her work has focused on the biology of the endothelium.

CHAPTER 10

Portholes on a Vessel: Endothelial Fenestrae

Sofia Ioannidou, Steven J. Samuelsson, and David T. Shima
Eyetech Research Center, Eyetech Pharmaceuticals, Inc.

Precise regulation of blood–tissue interchange is critical for proper integration of organ physiology with the cardiovasculature. Accordingly, microvascular endothelial cells, the primary barrier to free blood–tissue exchange, adopt highly specialized features and behaviors to mediate and monitor the flux of macromolecules and fluids across the vascular wall. For example, the vascular endothelium of the blood–brain and the blood–retinal barriers is fortified by tight junction components that restrict access of fluid and blood components to sensitive neural tissues. The other extreme of endothelial barrier differentiation is that of lymphatic vessels, an endothelium with numerous gaps between cells enabling uptake of extravascular fluid for clearance, transport, and return to the circulation. Finally, organs mediating endocrine, absorptive, or filtrating functions have a demand for abundant but more regulated fluid and macromolecule exchange and thus are supported by a highly specialized vascular endothelium containing numerous plasma membrane pores, the fenestrae (Figure 1). Derived from the Latin word for window, the term *fenestra* is used in this context to describe an opening in the vessel wall that connects the luminal and extravascular space. In this chapter, fenestrae structure, function, and pathophysiological relevance are discussed in detail.

Fenestrae Structure

Morphology and Architecture

Ultrastructural studies have described fenestrae as transcellular circular pores with an average diameter of approximately 60 nm (although they can be as large as approximately 125 nm within the liver sinusoidal endothelium). Fenestrae are encountered in the most attenuated regions of the endothelium, where the cell profile is as little as 40 nm, and span the entire thickness of the cell without disrupting the continuity of the cell membrane (Figure 2). The substances that traverse the pore never encounter the contents of the cytoplasm and are transported in a rapid and presumably energy-efficient manner. This is in contrast to transcytosis, which involves the coupling of energy-rich endocytic and exocytic events. In most vascular beds, the fenestrae contain a diaphragm composed of approximately eight radial fibrils converging in a central knob, which further dissects the pore into 5- to 6-nm openings (Figure 3).

Fenestrae are known to occur in clusters of approximately 50 to 100, termed *sieve plates*, which are encircled by a microtubule-rich border. Within a sieve plate, fenestrae are found in a near-linear arrangement, with precise spacing between each pore, implying the presence of a complex intracellular scaffolding to support such order. Whether individual fenestrae, or the sieve plates, are stable structures that persist throughout the lifetime of a differentiated cell or dynamic structures that are rapidly turned over, is currently unknown.

Chemical and Molecular Composition

Palade and the Simionescus pioneered the study of fenestrae composition in the 1960s, 1970s, and 1980s by demonstrating that cationized ferritin (CF) preferentially deposited within a glycocalyx visible on the luminal aspect of the fenestral diaphragm. Capitalizing on this initial observation, they used the CF interaction as a probe for the molecular nature of fenestrae, by monitoring its disappearance following treatment with enzymes of defined speci-

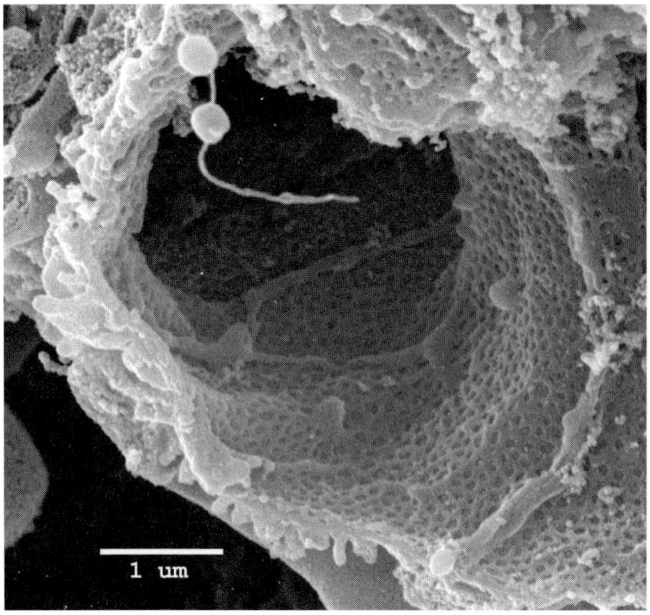

Figure 1 A scanning electron micrograph from a freeze-fractured glomerulus shows a view down the lumen of a fenestrated capillary. The endothelium appears to have a colander-like structure that permits passage of materials from the blood to the glomerular capsule. (We thank Steve Gschmeissner at Cancer Research UK for this data.) (see color insert)

ficity. Sensitivity to proteases and certain glycosidases as well as affinity for lectins suggested that acidic glycoproteins and proteoglycans could account for CF-decorated anionic sites on the diaphragms. However, the differential sensitivity to heparinase and heparitinase in the fenestral diaphragms of the intestine and choriocapillaris, respectively, together with variable results in lectin-binding studies in these tissues, highlight organ-specific differences in the glycocalyx composition: heparan sulfate proteoglycans presumably form part of the diaphragms in the intestine, while the closely related molecule heparin is thought to localize on diaphragms of the choriocapillaris. Moreover, binding of CF to fenestral diaphragms is absent altogether in the bone marrow and the fetal liver.

Recent studies by Stan and colleagues identified an endothelial cell-specific protein, Plasmalemmal Vesicle 1 Protein (PV-1) as the first known component of the fenestral diaphragm. PV-1 is a 60-kDa Type II transmembrane glycoprotein that is believed to form homodimers that constitute the primary structural component of the diaphragm. It should be noted that PV-1 and the diaphragm are not unique to fenestrae but also reside within endothelial cell caveolae and transendothelial channels.

Fenestrae Biogenesis

The inability to maintain fenestrated endothelium in tissue culture and the difficulty in quantifying and manipulating the appearance of fenestrae both in vivo and in vitro have greatly hampered the study of their biogenesis. Until recently, the induction of fenestrae in cultured endothelial cell lines has been reported to yield numbers that are three to four orders of magnitude lower in density than normally observed in vivo. Nevertheless, a small number of key studies published throughout the last three decades have highlighted several extracellular and intracellular determinants that may be involved in the differentiation program of a fenestrated endothelial cell.

Endothelial Cell Microenvironment

Fenestrated microvessels are in constant contact with the extracellular matrix, both in the form of routine basal lamina or, in some cases, elaborate, thickened matrix such as the basement membrane of the kidney glomerulus or the multilamellate Bruch's membrane, which separates pigmented retina from the choriocapillaris (Figure 4). Extracellular matrix could simply provide a structural scaffold to facilitate the extreme shape changes and cell attenuation that accompanies the fenestrated phenotype. However, a more active, instructive role for the extracellular matrix has been proposed to explain, for example, the presence of fenestrae only in the regions of the choriocapillaris that are immediately adjacent to Bruch's membrane. In vitro experiments examining the effects of a variety of extracellular matrices suggest that fenestrae formation is supported by specific matrix components, which mimic the situation observed in vivo.

Signal Transduction Pathways

The first attempts to reprogram the fenestrae differentiation program in cultured endothelial cells relied on the use of potent and relatively nonspecific initiators of intracellular signaling cascades on isolated bovine adrenal cortex endothelial cells. Phorbol myristate acetate, an activator of protein kinase C isoforms and a potent differentiation agent for some cell types, promoted a change in the shape of endothelial cells accompanied by a five-fold increase in the frequency of diaphragmed fenestrae (approximately six fenestrae per $100\,\mu m^2$). Similarly, treatment of cultures with retinoic acid led to a threefold increase in the surface density of fenestrae, while transforming growth factor β led to a sevenfold decrease in their density. The physiological relevance of these signaling pathways remains to be established.

The angiogenic growth factor, vascular endothelial growth factor (VEGF), is the strongest candidate for a signaling protein that induces fenestrae formation. VEGF is an endothelial-specific mitogen and motogen, and was originally identified as a vascular permeability factor, approximately 50,000 times more potent than histamine. Although VEGF is greatly downregulated after the completion of embryonic vasculogenesis and angiogenesis, it is continuously and highly expressed in epithelial cells adjacent to fenestrated endothelium. Moreover, fenestrae have been

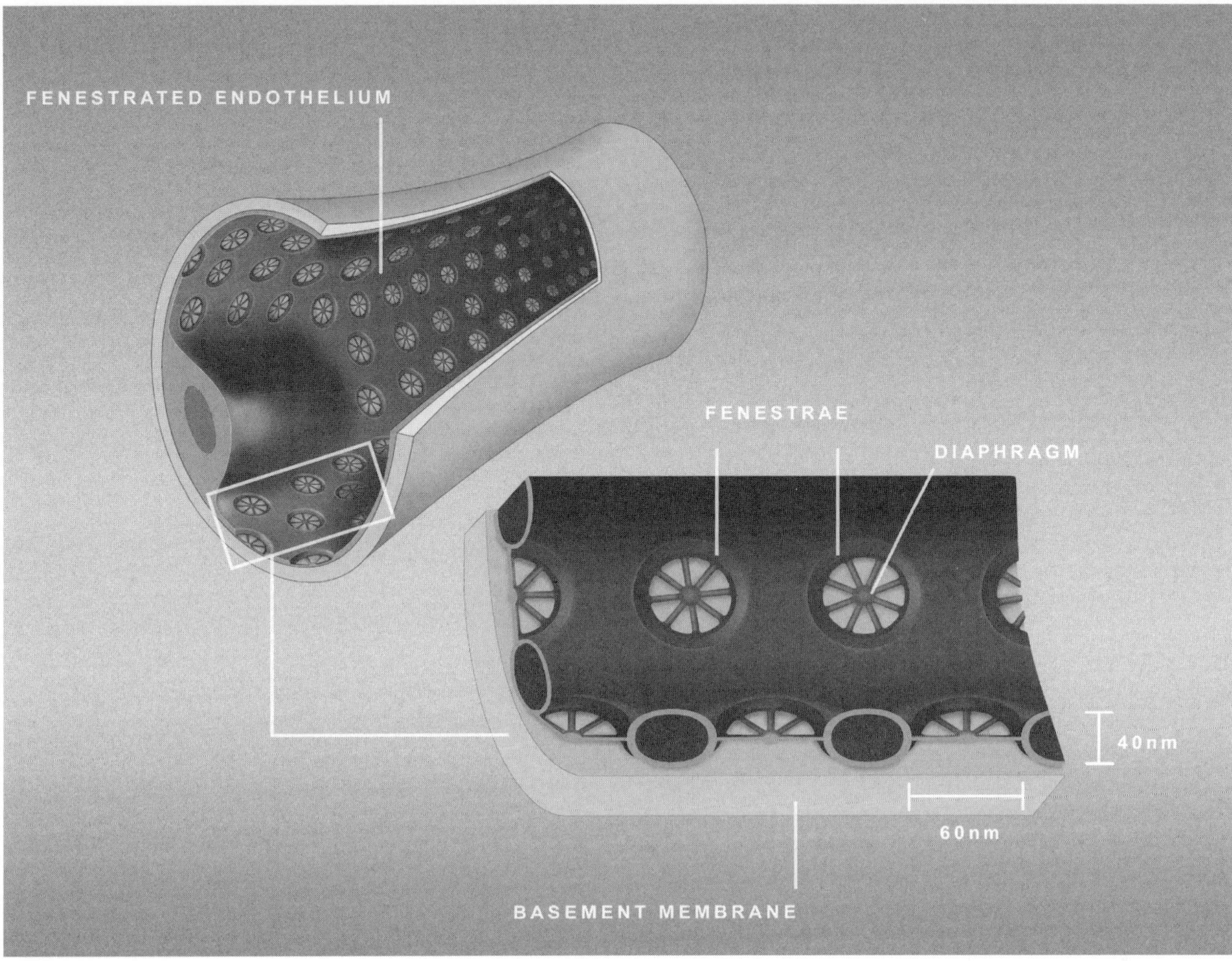

Figure 2 Schematic representation of a capillary with fenestrated endothelium and underlying basement membrane. Fenestrae are found only at the most attenuated regions of endothelial cells, where the nucleus and organelles are excluded and the distance from apical to basal plasma membrane is as little as 40 nm. They occur in groups and are arranged in a near-linear fashion with precise spacing between them. *Inset* shows the fenestral pore to be 60 nm in diameter, with an effective size of 5 to 6 nm when apertured by a diaphragm. [Adapted from Rhodin (1962), *J. Ultrastructure Res.* **6**, 171–185.] (see color insert)

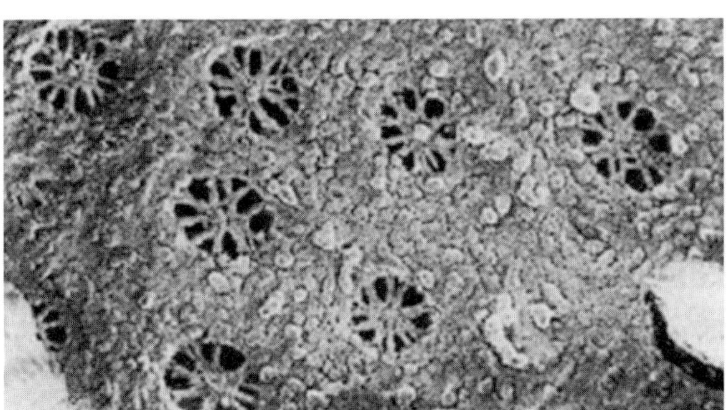

Figure 3 Fenestral diaphragms revealed by quick-freezing and deep-etching of rat kidney peritubular capillaries. The intertwining fibrils bridging the pore and converging into a central mesh are apparent in this luminal view. Magnification: 120,000×. [Courtesy of Drs. Bearer and Orci (1985), *J. Cell. Biol.* **100**, 418–428.] (see color insert)

observed in the neovasculature of tumors and in the normally continuous endothelium of the retina during diabetic microangiopathy, both pathological situations that are functionally linked to the local upregulation of VEGF. In fact, in vivo studies have demonstrated that VEGF can induce fenestrae within 10 minutes in the continuous endothelium of skeletal muscle and skin, when applied topically or injected intradermally, and in vitro studies using capillary endothelial cells further reinforce the ability of VEGF to promote fenestrae biogenesis. It is still unknown to what degree overall VEGF-induced permeability can be attributed to fenestrae formation, since VEGF also triggers the appearance of other structures implicated in permeability, such as caveolae and vesiculo-vacuolar organelles (VVOs), and regulates the gating properties of endothelial junctions.

Cytoskeleton

The importance of cytoskeletal remodeling in fenestrae biogenesis has been suggested from in vitro studies examin-

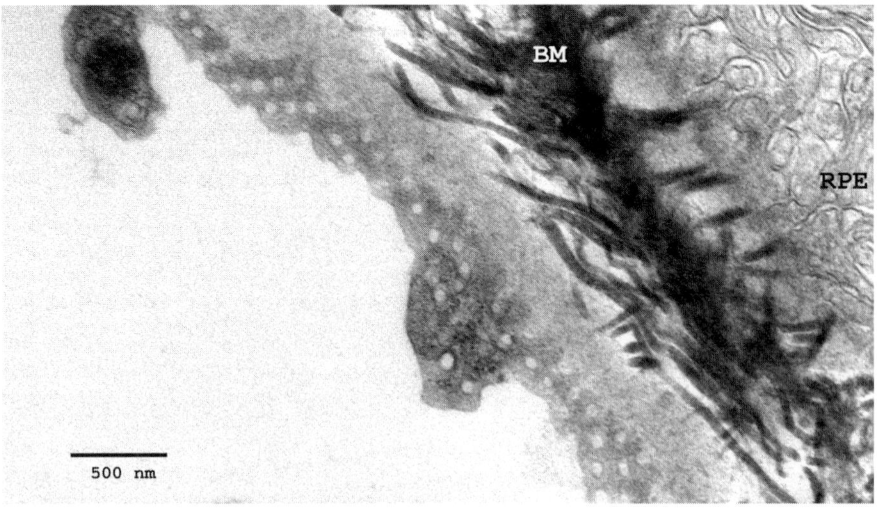

Figure 4 Fenestrated endothelium of choriocapillaris. This image reveals the close association of diaphragmed pores to Bruch's membrane (BM) and microvillar base of retinal pigmented epithelium (RPE). (We thank associate Eunice Cheung for this data.) (see color insert)

ing the effect of agents that disrupt actin microfilament assembly on highly fenestrated liver sinusoidal endothelial cells. Latrunculin A, which depolymerizes actin filaments through the sequestration of actin monomers, as well as Cytochalasin B, which leads to disassembly by capping the fast-growing end of actin filaments, both led to a two- to threefold increase in the number of fenestrae within 30 to 60 minutes. The physiological relevance of this data was recently supported by experiments showing that a dominant negative version of the small GTP-binding protein Rac could block VEGF-driven fenestrae formation during corneal angiogenesis, putatively through interfering with the reorganization of the actin cytoskeleton.

General Concepts for Fenestrae Formation

Researchers have used the limited amount of experimental data to generate several conceptual models (not mutually exclusive) to guide research in the area of fenestrae biogenesis.

Caveolae Give Rise to Fenestrae

Fenestrae and caveolae share structural features, common tissue distribution, a relationship to the VEGF signaling cascade, and a common putative role in the regulation of vascular permeability. In one of the earliest hypotheses for fenestrae formation, investigators postulated that caveolae may give rise to fenestrae in a process involving fusion of a budding caveolar vesicle with the adjacent plasmalemma. The discovery that PV-1 was common to the diaphragms that reside in both fenestrae and the caveolae of endothelial cells further supported this model. Current evidence, however, emphasizes more differences than similarities in the nature of the two organelles. General compositional differences between the two organelles were first highlighted in tracer perfusion studies showing that caveolar diaphragms (also referred to as stomatal diaphragms) lacked anionic sites among other molecular determinants that were present within the fenestral diaphragm. Furthermore, exclusion of the main structural component of caveolae, caveolin-1, from fenestrae in vivo and in vitro, and the recent finding that knockout mice completely lacking caveolae still have fenestrae, provide compelling evidence for distinct origins of caveolae and fenestrae. However, the existence of a common structural precursor that differentiates to give rise independently to fenestrae and caveolae still remains a valid point for consideration.

A Putative Role for the Diaphragm in Fenestrae Formation

The presence of a diaphragm in fenestrae is variable; it is found in the fenestrated capillaries of the intestine, the choriocapillaris, the choroid plexus, and endocrine organs, while it is absent from the more permeable microvessels of the kidney glomerulus and the liver sinusoids. Diaphragmed fenestrae, however, do initially appear within the vessels of developing fetal glomerulus, which has raised speculation on a role for the diaphragm in fenestrae biogenesis, in addition to its putative role in gating the pore. Achieving a precise circular opening and facilitating the extreme membrane curvature at the rim of fenestrae, where apical and basal plasma membranes come together, could be functions of the protein-rich diaphragm.

Apical-Basal Plasma Membrane Fusion Following Actin Displacement

Data demonstrating that actin microfilament disassembly triggers fenestrae biogenesis have spawned a model whereby removal of the actin-rich cortex beneath the plasma membrane is required to allow close apposition and fusion of apical and basal plasma membranes. Cortical actin removal has been shown to be an important prerequisite to membrane fusion during exocytosis, while removal of organelle-bound actin has been suggested to accelerate mammalian endosome and yeast vacuole membrane fusion. Whether the cytoskeleton also plays an instructive role in fenestrae formation remains to be established.

Fenestrae Function

Selectivity of the Barrier

Attempts to investigate the permeability properties of fenestrae were initiated by Clementi, Palade, Simionescu, and Pino using tracer perfusion studies. Horseradish peroxidase (~4 nm in diameter) proved to be readily permeable through diaphragmed fenestrae of the intestinal mucosal capillaries and the ocular choriocapillaris, whereas larger tracers such as ferritin (~11 nm in diameter), or dextrans and glycogens, were variably permeable. These findings fit remarkably well with data developed by Bearer and Orci, who determined that the presence of a diaphragm transected the 60-nm fenestrae into multiple channels of approximately 5 nm in diameter.

Whether permeability through fenestrae is a passive process or is facilitated by specific molecular interactions between components of the pore and traversing substances remains an open question. Although somewhat speculative, future studies in this area may benefit by examining the structure and function of the nuclear pore, a remarkably analogous structure consisting of two lipid bilayers fused to create a seamless circular opening. Transport of molecules between the cytoplasm and the nucleoplasm is chaperoned by proteins of the nuclear pore that first recognize specific export or import signals on potential cargo molecules and then act as shuttling factors in a process regulated by the GTPase, Ran. So far, selectivity within the fenestrae has only been shown with respect to molecular charge, as inferred from studies demonstrating the particularly high affinity of cationic substances for fenestral diaphragms.

Fenestrae and Organ Function

Although there is sparse direct evidence of a major physiological role for fenestrae, several functions can be safely attributed to them simply on the basis of their physical properties and their distribution within normal vasculature and during neovascular disease. Fenestrae are postulated to mediate the bidirectional exchange of water, solutes, and small macromolecules between blood and tissues. More specifically, fenestrae are believed to function (1) in the filtration of blood within the choroid plexus that gives rise to cerebrospinal fluid; (2) in the access of endocrine hormones to the bloodstream, such as the release of steroid hormones by the adrenal cortex; (3) in the supply of nutrients to, and removal of waste from, the outer retina by the underlying choriocapillaris; (4) in the ultrafiltration of blood to create primary urine within the kidney glomerulus; and (5) in the filtration of potentially hazardous substances in the blood at the liver sinusoids.

Fenestrae are also observed in normally nonfenestrated vascular beds in association with unwanted angiogenesis and other cardiovascular pathologies. The appearance of fenestrae in the neovasculature of tumors, retinal vessels in diabetic retinopathy, and capillaries of inflamed tissue, such as arthritic joints, coincides with clinical and experimental findings for vascular leakage and edema and implies that fenestrae contribute to the deregulation of vascular permeability. Fenestrae diameter has also been found to alter in conjunction with liver dysfunction, such as cirrhoses, and in kidney disease, such as pre-eclampsia.

Challenges for the Future

The lack of molecular tools, quantitative methods, and suitable model systems have slowed progress on fenestrae research and, to date, the bulk of our knowledge is still largely anecdotal and rich with assumption. A primary challenge for the future is to begin relating definitive functional properties of the fenestrae to the abundant ultrastructural information that has been collected over the past five decades. Recent advances in the fields of genomics and proteomics combined with the development of promising in vitro models for fenestrae formation are expected to open the way for (1) fascinating endothelial cell biology, (2) the potential to create light microscopic and biochemical methods to supplement ultrastructural analyses, and (3) the development of specific antagonists of fenestrae function to elucidate its contribution to cardiovascular function.

Glossary

Caveolae: Flask-shaped invaginations (50 to 100 nm in diameter) at the cell membrane implicated in endocytosis, cholesterol trafficking, and signal transduction. Also known as *plasmalemmal vesicles*.

Diaphragm: Variable feature of fenestrae and caveolae consisting of radial fibrils converging in a central knob. It is thought to consist of proteins, and its only known component to date is plasmalemmal vesicle 1 (PV-1) protein.

Fenestrae: Transcellular circular pores (60 to 70 nm in diameter) that occur in clusters in attenuated endothelia and are implicated in capillary permeability.

Vascular endothelial growth factor (VEGF): Growth factor expressed as several spliced variants and involved in vasculogenesis, angiogenesis, and vascular permeability.

Further Reading

Bearer, E. L., and Orci, L. (1985). Endothelial fenestral diaphragms: A quick-freeze, deep-etch study. *J. Cell. Biol.* **100**, 418–428.

Braet, F. et al. (1996). Microfilament-disrupting agent latrunculin A induces an increased number of fenestrae in rat liver sinusoidal endothelial cells: Comparison with cytochalasin B. *Hepatology* **24**, 627–635.

Clementi, F., and Palade, G. E. (1969). Intestinal capillaries. I. Permeability to peroxidase and ferritin. *J. Cell. Biol.* **41**, 33–58. *This article describes one of the earliest attempts to define the permeability proper-*

ties of fenestrae and to distinguish them from those of other endothelial structures. It is selected because the central dogmas postulated are still valid today.

Esser, S. et al. (1998). Vascular endothelial growth factor induces endothelial fenestrations in vitro. *J. Cell. Biol.* **140**, 947–959.

Roberts, W. G., and Palade, G. E. (1995). Increased microvascular permeability and endothelial fenestration induced by vascular endothelial growth factor. *J. Cell. Sci.* **108**, 2369–2379. *This study is selected because it provides the first direct evidence implicating VEGF signaling in fenestrae formation.*

Simionescu, N. et al. (1981). Differentiated microdomains on the luminal surface of the capillary endothelium. I. Preferential distribution of anionic sites. II. Partial characterization of their anionic sites. *J. Cell. Biol.* **90**, 605–621.

Simionescu, M. et al. (1982). Differentiated microdomains on the luminal surface of capillary endothelium: Distribution of lectin receptors. *J. Cell. Biol.* **94**, 406–413.

Sorensson, J. et al. (2002). Glomerular endothelial fenestrae *in vivo* are not formed from caveolae. *J. Am. Soc. Nephrol.* **13**, 2639–2647.

Stan, R. V. et al. (1999). PV-1 is a component of the fenestral and stomatal diaphragms in fenestrated endothelia. *Proc. Natl. Acad. Sci. USA* **96**, 13202–13207. *This article is selected because it describes the identification of the first, and to date the only, known structural component of fenestrae.*

Capsule Biography

Ms. Ioannidou, a Ph.D. student registered at University College London, United Kingdom; Dr. Samuelsson, former head of Cell and Molecular Imaging at Procter & Gamble Pharmaceuticals; and Dr. Shima, former head of the Endothelial Cell Biology Laboratory at the London Research Institute, Cancer Research UK, are currently studying various aspects of fenestrae biology at the Eyetech Research Center, Eyetech Pharmaceuticals, Inc.

CHAPTER 11

Neuropilins and the Vasculature

Roni Mamluk and Michael Klagsbrun
Departments of Surgical Research and Pathology, Vascular Biology Program, Children's Hospital and Harvard Medical School, Boston, Massachusetts

The development, patterning, and stabilization of the vasculature depends on several key steps, including the proliferation and migration of endothelial cells (EC) and the differentiation and recruitment of their supportive mural cells. These processes are regulated by the surrounding cells and tissues that secrete various growth factors, cytokines, and chemokines that activate EC receptors. It has been recently recognized that one of these EC receptors is neuropilin (NRP). Neuropilins (NRP1 and NRP2) are receptors for the vascular endothelial growth factor (VEGF) family, members of which are potent regulators of EC migration, EC proliferation, and angiogenesis. Previously, it was shown that VEGFs act via VEGF receptor tyrosine kinases, but it now appears that VEGF activity is also modulated by NRPs, which have no kinase activity. This article focuses on the role of NRPs in the vasculature and describes NRP structure, gene expression, regulation, and biological function.

Introduction

NRP1 and NRP2 are mediators of neuronal guidance and angiogenesis. NRPs bind members of the class 3 Semaphorin family, regulators of neuronal guidance, and members of the VEGF family of angiogenesis factors. Fujisawa and colleagues first identified NRP1 in 1991 as an adhesion molecule in developing nervous tissue; it was shown to be a 130- to 140-kDa highly conserved type 1 transmembrane glycoprotein. A second gene, NRP2, was identified in 1998 that had a 44 percent amino acid homology with NRP1.

A possible role of NRP1 in blood vessel development was first noted when NRP1 was overexpressed in mice. Besides ectopic sprouting and defasciculation of neurons, the mice also exhibited excess blood vessels. Subsequently, it was reported that NRP1 knockout mice exhibited various vascular abnormalities along with neuronal defects. The molecular connection of NRP1 to angiogenesis was noted when Klagsbrun and colleagues reported that VEGF, a major angiogenesis factor, was a ligand for NRP1 and NRP2.

Neuropilin Structure

Domains

NRPs are glycoproteins with a relatively large extracellular domain (860 amino acids), a transmembrane domain, and a relatively short cytoplasmic domain of about 40 amino acids. The extracellular domain consists of five subdomains, which are referred to as a1, a2, b1, b2, and c. The a1a2 and b1b2 domains are involved in ligand binding and cell adhesion. The c domain is thought to function as a site for homo- or hetero-dimerization of NRP1 and NRP2. The function of the short cytoplasmic domain, which is highly conserved, is not clear. A neuropilin interacting protein (NIP) has been identified that contains a PDZ domain. It binds to the C-terminal three amino acids of NRP1 (S-E-A-COOH).

In addition to full-length NRPs, some cell types express truncated NRP isoforms. These naturally occurring proteins contain only the a1a2 and b1b2 subdomains, are soluble, and are secreted by cells. The soluble NRP molecules are produced by premature truncation within introns and as a result are characterized by having intron-derived 3′ nucleotides and C-terminal amino acid sequences.

Neuropilin Binding Sites for Ligands

NRPs can bind multiple ligands, which can be categorized into three groupings: the class 3 semaphorins, the VEGF family, and various transmembrane proteins. Semaphorins have a sema domain that binds to the NRP a1a2 domain and an Ig-basic C-terminal domain that binds to the b1b2 domain. VEGF family members bind NRPs with a high degree of specificity. Placental growth factor-2 (PlGF-2), VEGF-B, and VEGF-E bind NRP1 but not NRP2. On the other hand, $VEGF_{145}$ and VEGF-C bind NRP2 but not NRP1. Domain binding sites have been determined for $VEGF_{165}$ and PlGF-2, both of which bind to the b1b2 domain. The binding sites of VEGF-B, VEGF-C, and VEGF-E on NRP have not been determined.

Heparin, which enhances both $VEGF_{165}$ and PlGF-2 binding to NRP1, binds the b1b2 domain. Heparin and heparan sulfate may play a critical role in EC NRP function by forming a complex of $VEGF_{165}$ or PlGF-2, NRP1b1b2, and heparin that facilitates ligand binding to NRP1. Because the b1b2 domain is a binding site for Sema3A, $VEGF_{165}$, and PlGF-2, there may be a structural basis for competitive inhibition of the different NRP1 ligands. For example, Sema3A competes $VEGF_{165}$-induced EC migration and $VEGF_{165}$ inhibits Sema3A-induced growth cone collapse, suggesting overlapping binding sites in the b1b2 domain.

The third category of NRP1 ligands are transmembrane proteins, including plexins A, L1, VEGFR-1, VEGFR-2, and heparan sulfate proteoglycans (HSPG). These bind to the NRP1 extracellular domain, but the subdomains involved have not yet been determined.

Endothelial Cell Neuropilin

Neuropilin Expression

Cultured EC express both NRP1 and NRP2. However, NRPs are expressed in vivo on specific EC types. For example, it has been demonstrated that there are differential embryonic blood vessel expression patterns for NRP1 and NRP2 in the avian vascular system. In the avian vasculature, NRP1 and NRP2 are both expressed in blood islands, which are the earliest vascular structures. However, once arteries and veins differentiate, NRP1 is expressed in arterial EC and in mesenchyme surrounding developing arteries, whereas NRP2 is expressed only in the venous EC. Similar expression patterns were reported in mice. In the developing mouse skin and retina, NRP1 is predominantly expressed in EC of arteries and arterioles. NRP2 expression was restricted to EC of veins at E10, but from E13 and on, NRP2 was downregulated in veins and highly expressed on lymphatic EC.

Neuropilin-Mediated VEGF Activity in Endothelial Cells

NRP1 appears to be a co-receptor for VEGFR-2 in cultured EC. When co-expressed in cells with VEGFR-2, NRP1 enhances the binding of $VEGF_{165}$ to VEGFR-2 and $VEGF_{165}$-mediated chemotaxis as compared to cells expressing VEGFR-2 only. Conversely, inhibition of $VEGF_{165}$ binding to NRP1 inhibits its binding to VEGFR-2 and its mitogenic activity for EC. $VEGF_{165}$ binds NRP1 via its exon 7–encoded peptide, whereas it binds VEGFR-2 via its exon 4–encoded peptide. $VEGF_{165}$ may form a bridge between the two receptors, facilitating a better presentation of $VEGF_{165}$ to VEGFR-2 (Figure 1). Support for this model was shown in immunoprecipitation studies where NRP1 and VEGFR-2 were co-immunoprecipitated only in the presence of $VEGF_{165}$. However, other reports have suggested that NRP1 can directly interact with VEGFR-2.

Semaphorin-Mediated Neuropilin Activity in Endothelial Cells

NRPs bind class 3 semaphorins, regulators of neuronal guidance. Semaphorin 3A (Sema3A), which is the best characterized semaphorin, repels axons, collapses dorsal root ganglion neuronal growth cones, and regulates migration of cortical neurons in an NRP1-dependent manner. These effects are mediated by small GTPases such as Rho A and Rac, which induce actin depolymerization. NRPs do not appear to directly activate signaling pathways in neurons. Instead, signaling is mediated by the interactions of a Sema3A/NRP1 complex with plexins, which are transmembrane signaling receptors. NRP1/plexin complex formation enhances Sema3A binding to NRP1. L1, a neuronal adhesion molecule, has also been demonstrated to be a component of the Sema3A receptor complex (Figure 1).

Sema3A binds EC via NRP1 and inhibits the motility of EC and capillary sprouting from rat aortic ring segments in an in vitro angiogenesis assay. The inhibition of EC motility by Sema3A is competed by $VEGF_{165}$. $VEGF_{165}$ and Sema3A are also antagonists in neuronal survival/apoptosis assays. Thus, a balance of semaphorins and $VEGF_{165}$ can modulate both EC and neuronal activities.

The role of semaphorins in vascular development has been analyzed in several models, and they have been shown to be regulators of vascular development. In chick limb development, overexpression or sequestering of Sema3A abrogated both vascular and neuronal patterning. In addition, it has recently been shown that in the chick embryo, EC express Sema3A, which autoregulates EC motility and vascular morphognesis via NRP1 and Plexin A1 complexes.

In zebrafish, semaphorins regulate the pathway of dorsally migrating angioblasts, which are NRP1-positive endothelial precursor cells that migrate to generate the dorsal aorta. Ubiquitous overexpression or knockdown of Sema3a1 protein interrupted dorsal migration of angioblasts and retarded development of the dorsal aorta, resulting in severely diminished blood circulation. Thus, Sema3a1 is a key regulator of early zebrafish vascular development.

In mice, however, transgenic studies seem to indicate that Sema3A does not play a role in regulating the vasculature. An initial report demonstrated that blood vessels developed normally in the limbs of Sema3A knockout mice. Consistent with this observation, transgenic mice overexpressing a

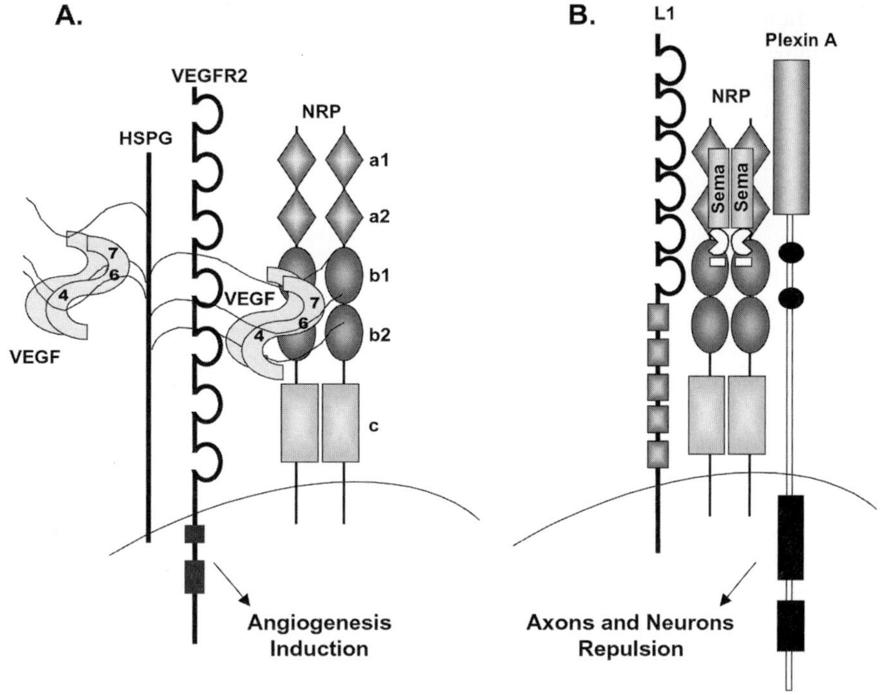

Figure 1 A Schematic of Neuropilin Signaling Complexes. (**A**) In this model, $VEGF_{165}$ bridges NRP1 (via axon 7) and the tyrosine kinase receptor VEGFR-2 (via exon 4). There are no direct NRP1/VEGFR-2 interactions. $VEGF_{165}$ and NRP1 also bind HSPG. The NRP1 binding site for $VEGF_{165}$ and HSPG is the b1b2 domain. It is suggested that the formation of a complex between $VEGF_{165}$ and the three receptors enhances $VEGF_{165}$ binding to VEGFR-2, which initiates the cellular signaling cascade, and as a consequence enhances $VEGF_{165}$ activity such as chemotaxis of endothelial cells and angiogenesis. (**B**) Sema3A binds to the a1a2 and b1b2 domains of NRP1. NRP1 binds to the transmembrane proteins plexin A and L1. The complex formation enhances the binding of Sema3A to NRP1, and subsequently plexin A induces the signaling events that lead to repulsion of axons and of neuronal cells.

mutated NRP1 that cannot bind Sema3A exhibited normal vasculature. On the other hand, in a different mouse background, Sema3A knockouts showed vascular defects in the head and had abnormal trunk blood vessels. These discrepancies as to whether Sema3A does or does not regulate the vasculature could be a result of the use of different developmental models and methodologies and will hopefully be reconciled in the future.

Neuropilin Function in Developmental Angiogenesis

Mice

There is strong evidence from transgenic mouse studies that NRPs mediate angiogenesis. Mice overexpressing NRP1 were embryonic lethal and displayed several vascular abnormalities, such as excess capillaries and blood vessels, dilation of blood vessels, hemorrhage, and malformed hearts. The chimeric embryos appeared redder than their normal counterparts, suggesting that blood vessels were leaky, which was possibly caused by enhanced vascular permeability activity of $VEGF_{165}$. It was concluded that expression of NRP1 was essential not only for neuronal development but also for development of the cardiovascular system.

The physiological role of NRPs in angiogenesis has also been determined using knockout mice. It was demonstrated that NRP1-deficient mutant mice were embryonic lethal between E12.5 to E13.5. In yolk sacs and embryos the vascular networks of large and small vessels were disorganized, the capillary networks were sparse, and normal branching did not occur. In the central nervous system (CNS), capillary invasion into the CNS was delayed and the capillary networks that were in the CNS were disorganized and had degenerated. The mutant embryos showed abnormal heart development, including lack of some of the branchial arch–related great vessels, and dorsal aorta and transposition of the aortic arch. The development of heart outflow tracts was also disturbed.

In another approach, the role of NRP1 in the vascular system was demonstrated by deleting NRP1 specifically from EC. Mutant mice were embryonic lethal by mid- to late gestation, with abnormal vasculature throughout the embryo. The larger vessels were intact, but medium and small vessels were missing. In the developing brain the vessels appeared larger and underdeveloped with very little branching, suggesting a defect in remodeling and branching of the primary vessel plexus. Taken together all of these data indicate that NRP1 is a critical receptor required for angiogenesis. On the other hand, NRP2 knockouts were viable into adulthood and did not display any abnormal vascular development.

More recently, double NRP1/NRP2 knockouts were reported. Transgenic mice, in which both NRP1 and NRP2 were targeted, died in utero at E8.5 and their yolk sacs were totally avascular. Mice that were homozygous for one gene but heterozygous for the other were also embryonic lethal and survived to E10 to E10.5. The vascular phenotypes of these mice were abnormal. The yolk sacs, while of normal size, displayed the absence of branching arteries and veins, the absence of a capillary bed, and the presence of large avascular spaces between the blood vessels. The embryos displayed blood vessels that were heterogenous in size, large avascular regions in the head and trunk, and unconnected blood vessel sprouts. The embryos were about 50 percent the length of wild-type mice and had multiple hemorrhages. These double NRP1/NRP2 knockout mice had a more severe abnormal vascular phenotype than either NRP1 or NRP2 single knockout mice. Their abnormal vascular phenotype resembled those of VEGF and VEGFR-2 knockouts. It appears that NRPs are early genes in embryonic vessel development, with overlapping functions that are required for normal blood vessel formation.

Zebrafish

The zebrafish is an excellent system for analyzing vascular development. Zebrafish intersegmental vessels correspond to mammalian capillary sprouts, whereas the axial vessels correspond to larger blood vessels, such as arteries and veins (Figure 2A, control). The zebrafish NRP1 gene (*znrp1*) was isolated, and the zNRP1 protein was shown to be a functional receptor for human $VEGF_{165}$. Whole-mount in situ hybridization showed that transcripts for *znrp1* during embryonic and early larval development were detected mainly in neuronal and vascular tissues. Knockdown of zNRP1 by using specific antisense oligos (Morpholino) in embryos resulted in severe defects in angiogenesis, including impaired circulation in the intersegmental vessels and dorsal longitudinal anastomotic vessel (DLAV) (Figure 2). However, circulation via the trunk artery and vein axial vessels, which are formed by vasculogenesis, was not affected. When zNRP1 and VEGF morpholinos were co-injected into embryos at concentrations that individually did not significantly inhibit blood vessel development, the result was a potent inhibition of blood cell circulation via both intersegmental and axial vessels. These results demonstrated that VEGF and NRP1 act synergistically to promote a functional circulatory system. These results may provide a physiological demonstration that NRP1 regulates angiogenesis through a VEGF-dependent pathway.

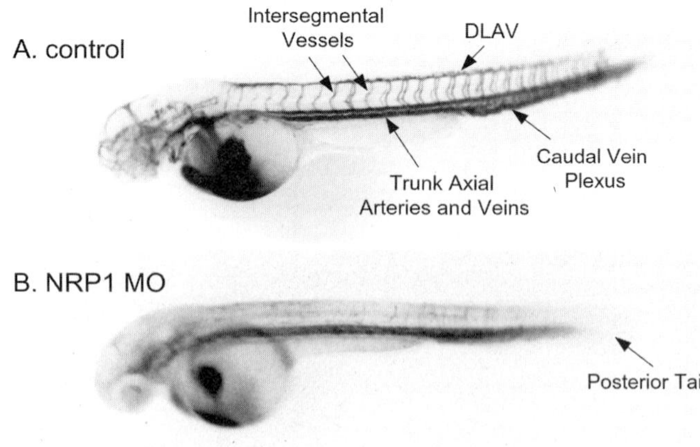

Figure 2 NRP1 Knockdown in the Zebrafish: Microangiography. Fish embryos were injected with antisense morpholino (MO) or controls at the one- to four-cell stage. To visualize blood vessels flow, FITC-dextran was injected into the cardinal vein 56 hours postfertilization. **(A)** Normal circulation of the zebrafish. Injected with the four-base mismatch morpholino control. **(B)** In zebrafish injected with anti-zNRP1 morpholino, blood flow via intersegmental vessels, DLAV, caudal vein plexus, and posterior vein is diminished. Axial vessel flow, however, is not affected.

Neuropilin in Pathological Angiogenesis

Tumor Angiogenesis

Many tumor cell types express NRP1 and NRP2 and bind $VEGF_{165}$. NRPs are the only VEGF receptors expressed by these tumor cells, so that any $VEGF_{165}$ activity must be mediated by NRPs. In several clinical studies, NRP1 and NRP2 expression was correlated with increased aggressiveness, malignancy, and/or hypervascularity. For example, NRP1 was upregulated in primary sporadic prostate tumors at different clinical stages. The correlation between NRP1 overexpression with advanced disease suggested that NRP1 overexpression might be a marker of aggressiveness.

The function of NRP1 in tumor cells has been analyzed directly by expressing NRP1 in tumor cells under the control of a tetracycline-inducible promoter. Concomitant with increased NRP1 expression in response to tetracycline, tumor cell migration and $VEGF_{165}$ binding was increased. However, induction of NRP1 did not affect tumor cell proliferation. When rats injected with rat prostate carcinoma cells were fed tetracycline, NRP1 synthesis was induced in vivo and tumor size was increased by up to sevenfold, compared to control. The tumors were characterized by markedly increased microvessel density, increased EC proliferation, dilated blood vessels, and notably less tumor cell apoptosis compared to noninduced controls. In addition, tumors overexpressing NRP1 retained higher levels of VEGF, possibly because of "trapping" by NRP1. It was concluded that NRP1 expression results in enlarged tumors associated with substantially enhanced tumor angiogenesis.

On the other hand, sNRP1 is a tumor antagonist. Tumors of rat prostate carcinoma cells overexpressing recombinant sNRP1 in vivo were characterized by extensive hemorrhage, damaged vessels, and apoptotic tumor cells. Because sNRP1 inhibits ^{125}I-VEGF$_{165}$ binding to EC and VEGF$_{165}$-induced tyrosine phosphorylation of VEGFR-2 in EC in vitro, this tumor phenotype may be caused by VEGF$_{165}$ withdrawal and lack of bioavailability. Withdrawal of VEGF$_{165}$ from tumors has previously been shown to result in vascular damage, EC apoptosis, hemorrhage, and extensive tumor necrosis.

Vascular Injury and Disease

NRPs are more highly expressed in the developing embryo as compared with the normal adult, but are induced following injury such as ischemia. Several pathologies that are characterized by ischemia-induced angiogenesis have shown NRP1 upregulation. For example, in the adult mouse, NRP1 expression in ischemic brain was significantly upregulated as early as two hours and persisted at least 28 days after focal cerebral ischemia. There was a marked increase in NRP1 expression in EC of cerebral blood vessels at the border and in the core of the ischemic lesion after seven days. These results suggest that upregulation of NRP1 may contribute to neovascular formation in the adult ischemic brain. In another mouse ischemia model system, very little NRP2 expression was observed in normal blood vessels after birth. However, NRP2 expression was induced in newly sprouting blood vessels in response to ischemia in a hind limb model in which the femoral artery was occluded. Increased NRP1 expression was also detected in pathological retinal neovascularization induced by ischemia. In this case, expression of NRP1 and VEGFR-2 was co-localized in the area of neovascularization.

Summary

Substantial evidence based on cell culture and transgenic mouse studies indicates that NRPs are novel and significant regulators of blood vessel development. During embryonic development, NRP expression occurs early, in the blood islands of the yolk sac. NRP expression is required for the normal branching and organization of large vessels and capillaries in the developing embryo. In the developing heart, NRP1 is required for normal formation of the large vessels. In the adult, NRP expression is generally reduced, but it is strongly upregulated in blood vessels in response to vascular injury. Tumor cells are among the highest expressers of NRPs, and overexpression of NRP1 enhances tumor angiogenesis and progression. The significance of direct VEGF binding to tumor cells is unknown but might involve enhancement of tumor cell migration and survival. On the other hand, whereas overexpression of full-length NRP1 in tumor cells enhances tumor angiogenesis, overexpression of sNRP1 suppresses it.

It is now well established that NRPs are crucial for vascular development and angiogenesis. In the future, goals will include determining the molecular mechanisms by which NRP regulates EC functions and elucidating the role of NRPs in normal physiology and in pathology.

Glossary

Angiogenesis: The formation of new blood vessels that sprout from preexisting blood vessels.

Axonal guidance: Also know as *axonal pathfinding* and *neuronal guidance*. The process by which neuronal axons move to their destinations in response to chemorepellants and chemoattractants, such as the semaphorins.

Vasculogenesis: The formation of blood vessels from endothelial cell precursors, for example, arteries and veins.

Further Reading

Bagnard, D., ed. (2002). *Neuropilin: from nervous system to vascular and tumor biology*. Vol. 515. New York and Texas: Kluwer Academic/Plenum Publishers and Landes Bioscience. *This book is a comprehensive review of neuropilin structure and function. Some of the leading investigators in the neuropilin field have contributed chapters focusing on their own scientific contributions.*

Ferrara, N. (1999). Molecular and biological properties of vascular endothelial growth factor. *J. Mol. Med.* **77**, 527–543.

Giger, R. J., Urquhart, E. R., Gillespie, S. K., Levengood, D. V., Ginty, D. D., and Kolodkin, A. L. (1998). Neuropilin-2 is a receptor for semaphorin IV: Insight into the structural basis of receptor function and specificity. *Neuron* **21**, 1079–1092.

Gu, C., Rodriguez, E. R., Reimert, D. V., Shu, T., Fritzsch, B., Richards, L. J., Kolodkin, A. L., and Ginty, D. D. (2003). Neuropilin-1 Conveys Semaphorin and VEGF Signaling during Neural and Cardiovascular Development. *Dev. Cell.* **5**, 45–57. *This work elegantly dissects the individual contributions of VEGF$_{165}$ and Sema3A to the development of the vascular and the neuronal systems, respectively. In this study, an Npn-1 (NRP1) construct was produced that could bind VEGF, but not Sema3A. Knock-in of Npn-1 in mice showed that VEGF$_{165}$ was needed for vascular development but Sema3A was not. Sema3A was needed, however, for normal axonal guidance.*

Kawasaki, T., Kitsukawa, T., Bekku, Y., Matsuda, Y., Sanbo, M., Yagi, T., and Fujisawa, H. (1999). A requirement for neuropilin-1 in embryonic vessel formation. *Development* **126**, 4895–4902.

Kitsukawa, T., Shimono, A., Kawakami, A., Kondoh, H., and Fujisawa, H. (1995). Overexpression of a membrane protein, neuropilin, in chimeric mice causes anomalies in the cardiovascular system, nervous system and limbs. *Development* **121**, 4309–4318.

Lee, P., Goishi, K., Davidson, A. J., Mannix, R., Zon, L., and Klagsbrun, M. (2002). Neuropilin-1 is required for vascular development and is a mediator of VEGF-dependent angiogenesis in zebrafish. *Proc. Natl. Acad. Sci. USA* **99**, 10470–10475. *Zebrafish is an excellent animal model for studying vascular development. The zebrafish is transparent so that the cardiovascular system can be readily visualized. The zebrafish can survive without a functional circulation for at least four days, thereby enabling analysis of the molecular basis of vascular defects that are lethal in mice. This paper demonstrates that NRP1 is needed for normal angiogenesis but not for vasculogenesis. Furthermore, the interdependence of VEGF and NRP1 for normal vascular function is described.*

Matsumoto, T., and Claesson-Welsh, L. (2001). VEGF receptor signal transduction. *Sci STKE* 2001, RE21. Accessed at: www.stke.org/cgi/content/full/OC_sigtrans;2001/112/re21

Miao, H. Q., Soker, S., Feiner, L., Alonso, J. L., Raper, J. A., and Klagsbrun, M. (1999). Neuropilin-1 mediates collapsin-1/semaphorin III inhibition of endothelial cell motility: Functional competition of collapsin-1 and vascular endothelial growth factor-165. *J. Cell. Biol.* **146**, 233–242.

Neufeld, G., Cohen, T., Shraga, N., Lange, T., Kessler, O., and Herzog, Y. (2002). The neuropilins: Multifunctional semaphorin and VEGF receptors that modulate axon guidance and angiogenesis. *Trends Cardiovasc. Med.* **12**, 13–19.

Soker, S., Takashima, S., Miao, H. Q., Neufeld, G., and Klagsbrun, M. (1998). Neuropilin-1 is expressed by endothelial and tumor cells as an isoform-specific receptor for vascular endothelial growth factor. *Cell* **92**, 735–745.

Takashima, S., Kitakaze, M., Asakura, M., Asanuma, H., Sanada, S., Tashiro, F., Niwa, H., Miyazaki Ji, J., Hirota, S., Kitamura, Y., Kitsukawa, T., Fujisawa, H., Klagsbrun, M., and Hori, M. (2002). Targeting of both mouse neuropilin-1 and neuropilin-2 genes severely impairs developmental yolk sac and embryonic angiogenesis. *Proc. Natl. Acad. Sci. USA* **99**, 3657–3662.

Capsule Biography

Dr. Klagsbrun is a professor in the Vascular Biology Program at Children's Hospital Boston and Harvard Medical School. He is well known for his work on vascular growth factors and their receptors. His studies on vascular molecules such as basic FGF, HB-EGF, neuropilin, and semaphorin have greatly furthered our understanding of how the vascular system is regulated and how tumor growth may be inhibited. His work is supported by grants from the National Cancer Institute.

Dr. Mamluk completed her post doctorate in the laboratory of Professor Klagsbrun. She focused her research on understanding the role of neuropilins, soluble NRPI and VEBFR-2 in vascular function and development. Currently she is conducting research for a biotechnology company aimed at developing new protein therapeutics to inhibit tumor angiogenesis.

CHAPTER 12

Receptor Tyrosine Kinase Signal Transduction in the Microvasculature

Marie B. Marron and Nicholas P. J. Brindle
Department of Cardiovascular Sciences, University of Leicester, United Kingdom

Introduction

Receptor tyrosine kinases (RTKs) have essential roles in the formation, maintenance, and function of the microvasculature. Vasculogenesis and angiogenesis, involving differentiation, migration, proliferation, and survival of endothelial cells, as well as microvascular regression, inflammation, and permeability are all controlled by RTK signaling pathways. Being able to understand how these pathways work will provide new opportunities for therapeutic manipulation of the microvasculature relevant to a wide range of diseases, from inflammatory diseases to tumor growth and tissue ischemia. In this chapter, signaling pathways utilized by four of the key families of RTKs of the microvasculature are outlined. Because of space constraints, we have focused on RTKs in the endothelium. Signaling pathways controlling the specialized functions of the microvasculature are best studied in the physiologically appropriate cell types. Although data from nonvascular cells can be extremely valuable in guiding understanding, such cells may lack key components of signaling or effector pathways, co-receptors, and signaling cross-talk important in the mechanisms by which the RTKs exert their effects in vascular cells. We have, therefore, focused on RTK signaling pathways demonstrated in endothelial cells rather than extrapolated from findings in nonvascular cells.

RTKs are activated by ligand-induced clustering. This permits transphosphorylation of regulatory tyrosine residues in the RTKs, activating kinase activity and inducing phosphorylation of additional tyrosine residues. These phosphotyrosines act as docking sites for SH2 and PTB domains in intracellular signaling molecules, leading to their recruitment to the RTKs. Such signaling molecules may then become activated by phosphorylation or protein-to-protein interactions and initiate a signaling cascade linking the RTK to effector mechanisms. In the following sections, these signaling cascades are reviewed with respect to the microvasculature. Where amino acid numbers are given, they refer to positions in the human sequences.

Signaling by VEGF Receptors

The vascular endothelial growth factors (VEGFs) have central roles in initiating blood vessel formation by both stimulating angioblast differentiation and activating angiogenesis from existing vessels. There are six members of the VEGF ligand family: VEGF-A to VEGF-E and placental growth factor (PlGF). These ligands signal via three VEGF-RTKs: VEGF-receptors-1 to 3. VEGF-A is the predominant form of the ligand and can bind all three receptors, whereas PlGF and VEGF-B recognize only VEGFR-1, and VEGF-E is specific for VEGFR-2. VEGF-C and D bind VEGFR3 and to some extent VEGFR2. All three RTKs have a similar overall structure, with a ligand binding extracellular domain containing seven immunoglobulin-like repeats and intracellular domain comprising a juxtamembrane sequence, tyrosine kinase domain with kinase insert, and carboxy-terminal tail. VEGFR-3 is mainly confined to lymphatic endothelia, where it has important roles in lymphangiogenesis. The principal VEGF-RTKs of the microvasculature are VEGFR-1 and VEGFR-2, which are the focus of this section. Because most endothelial cells express both VEGFR-1 and -2, defining the specific roles and signaling events mediated by each receptor in response to VEGF has been challenging. One way to address this issue has been by overexpression of chimeric receptors, allowing specific activation of each receptor, or the expression of each receptor individually in cells that do not normally express VEGFRs. Such approaches are useful, but care is required in extrapolating to the situation in primary endothelial cells expressing physiological levels of receptors. Some of the discrepancies reported in signaling by VEGFRs may be a result of these limitations.

VEGFR-1 Signaling

Of the two microvessel VEGF-RTKs, VEGF-A binds with highest affinity to VEGFR-1. However, this receptor appears to have relatively low kinase activity, and VEGF-A-activated increase in kinase activity or autophosphorylation of VEGFR-1 is difficult to detect in endothelial cells. One function of the receptor may therefore be to act as a negative regulator of VEGF-A activity by sequestering the ligand and preventing it from activating VEGFR-2.

In addition to any role in regulating VEGF-A availability, VEGFR-1 does have signaling activity. Transgenic mice expressing a truncated form of VEGFR-1, lacking the intracellular domain, exhibit normal developmental angiogenesis but have compromised pathological angiogenesis, indicating involvement of the intracellular VEGFR-1 signaling domain. VEGFR-1-specific ligands also activate distinct patterns of gene expression, again consistent with the receptor having discrete signaling activity. The signaling pathways utilized by VEGFR-1, however, are poorly understood. Phosphorylation has been observed in baculovirus expressed intracellular domains of the receptor and in cells overexpressing VEGFR-1 in response to VEGF-A. The principal phosphorylation sites identified in these studies were Y1169, Y1213, and Y1333. VEGF-A activation of nonoverexpressing endothelial cells increases Y1213 phosphorylation. PLCγ, Nck, and Crk bind to phosphopeptides containing the Y1333 site, and SHP2, phospholipase Cγ (PLCγ), and Grb_2 bind to Y1213 phosphopeptides. Increased tyrosine phosphorylation of PLCγ, Crk, and SHP_2 has also been observed in response to VEGF-A stimulation in cells overexpressing VEGFR-1. These data suggest that the receptor could modulate Ca^{++}/DAG (via PLCγ) and Ras/Raf pathways (via SHP_2/Grb_2), although more work is necessary to establish interaction of the receptor with these intermediates in endothelial cells expressing physiological levels of receptor in their normal cellular background.

VEGF-A and the VEGFR-1-specific ligand PlGF appear to have some distinct effects on VEGFR-1 in endothelial cells expressing normal levels of the receptor. The two ligands induce different patterns of gene expression and activate phosphorylation of different tyrosine residues in the receptor. In contrast to the effects of VEGF-A on Y1213, PlGF activates phosphorylation of Y1309. The signaling and functional consequences of Y1309 phosphorylation have yet to be defined. Such differential effects of ligands on the receptor provide a means for distinct activities of the ligands on endothelial cells and raise the important question of the mechanism for ligand-specific phosphorylations in VEGFR-1.

There is cross-talk between VEGFR-1 and VEGFR-2. Specific activation of VEGFR-1 by PlGF enhances VEGF-A-activation of VEGFR-2 by a mechanism involving transphosphorylation of VEGFR-2 by VEGFR-1. The two receptors exist as homomeric complexes and preformed heteromeric complexes in endothelial cells.

In summary, VEGFR-1 has apparently low kinase activity and can act to sequester VEGF-A and regulate signaling by VEGFR-2. Specific ligands for VEGFR-1 can modulate signaling via VEGFR-1 by displacing sequestered VEGF-A and enhancing VEGFR-2 activity by transphosphorylation. Such activities are likely to be involved in the enhancement of VEGF-induced angiogenesis by PlGF. In addition, VEGFR-1 does have distinct signaling capability responsible for regulating specific gene expression profiles, and probably other effects in endothelial cells.

VEGFR-2 Signaling

The limited ability to detect VEGFR-1 activation has led to the suggestion that the principal receptor mediating the effects of VEGF-A in endothelial cells is VEGFR-2. The receptor has been implicated in VEGF-induced endothelial migration, proliferation, and differentiation, as well as the pro-inflammatory and permeability effects of the ligand. Reported sites of tyrosine phosphorylation in VEGFR-2 include Y951, Y996, Y1054, Y1059, Y1175, and Y1214. Of these, Y1175 and Y1214 appear to be primary sites of phosphorylation, and Y1175 has been confirmed in endothelial cells expressing physiological levels of receptor.

Specific activation of VEGFR-2 in nonoverexpressing endothelial cells provides an antiapoptotic signal requiring phosphatidylinositol-3-kinase (PI-3K) and Akt. The activity of Akt is stimulated by VEGF in endothelial cells in response to VEGFR-2-specific ligands, and PI-3K and Akt are both activated via VEGFR-2 in cells overexpressing the receptor. The precise way in which PI-3K is activated by VEGFR-2 is not clear, but in some studies the receptor was found to associate directly with the p85 subunit of PI-3K (pY801, pY1175), whereas in others p85 was bound to focal adhesion kinase (FAK) recruited to pY1214 of VEGFR-2, or VEGFR-associated protein (VRAP), which is recruited to pY951. Intriguingly, the antiapoptotic activity of VEGF appears to involve the interendothelial adhesion molecule VE-cadherin. In cells from mice lacking the intracellular domain of VE-cadherin, coupling between VEGFR-2 and Akt is impaired and interaction between VEGFR-2 and p85 is decreased. Further work will be required to define the mechanism by which VE-cadherin modulates VEGFR-2 association with p85.

The signaling pathway by which VEGF induces endothelial migration via VEGFR-2 is unclear. The pathway is likely to involve FAK and PLCγ, both of which interact with VEGFR-2 and have been implicated in VEGF-stimulated endothelial migration in studies using mutant receptors and inhibitors. VEGF activation of endothelial proliferation occurs via a signaling pathway involving VEGFR-2 activation of PLCγ following its recruitment to pY1175. The resulting generation of diacylglycerol (DAG) activates protein kinase C (PKC), which in turn stimulates Raf and the Erk pathway.

VEGF markedly increases vascular permeability, and use of receptor-specific ligands indicates this is mediated via VEGFR-2. Again, however, the signaling mechanisms involved are not yet understood. There is good evidence that PLCγ-mediated DAG generation and Ca^{++} mobilization contribute to increased nitric oxide (NO) generation, leading to elevation of cGMP and increased vascular permeability. In addition, Akt, activated via VEGFR-2 as described previously, has also been implicated. One action of the Akt is to phosphorylate endothelial nitric oxide synthase (eNOS) on serine1177 and increase NO production. The PLCγ pathway appears to predominate in the first few minutes of VEGF-induced vessel permeability, after which the Akt pathway becomes more important. The downstream events mediating the effects of cGMP on permeability have yet to be clearly defined. The lack of effects of VEGF on vascular permeability in transgenic mice lacking p60Src and p62Yes indicate that these members of the Src family of intracellular tyrosine kinases also participate in the permeability signaling pathway, although exactly how is not yet known. Acute increased permeability of microvessels in response to VEGF can occur through formation of fenestrae, which is stimulated by VEGF, and increased transcellular flux. Clear definition of the mechanisms by which fenestrae form and transcellular flux increases will allow delineation of the missing links between the VEGFR-2-NO signaling pathway and increased permeability. Figure 1 summarizes in schematic form some of the best elucidated VEGFR-2 signaling pathways.

Signaling by FGF Receptors

There are more than 20 FGFs, and the best studied of these are FGF-1 and FGF-2. The FGFs signal via four FGF RTKs, FGFR-1 to FGFR-4, and variants of these receptors arise by alternative splicing. These receptors have a similar overall structure, with extracellular domains containing two or three immunoglobulin repeats and an alternatively spliced sequence rich in acidic residues and serines, the acidic box. The intracellular domains comprise a juxtamem-

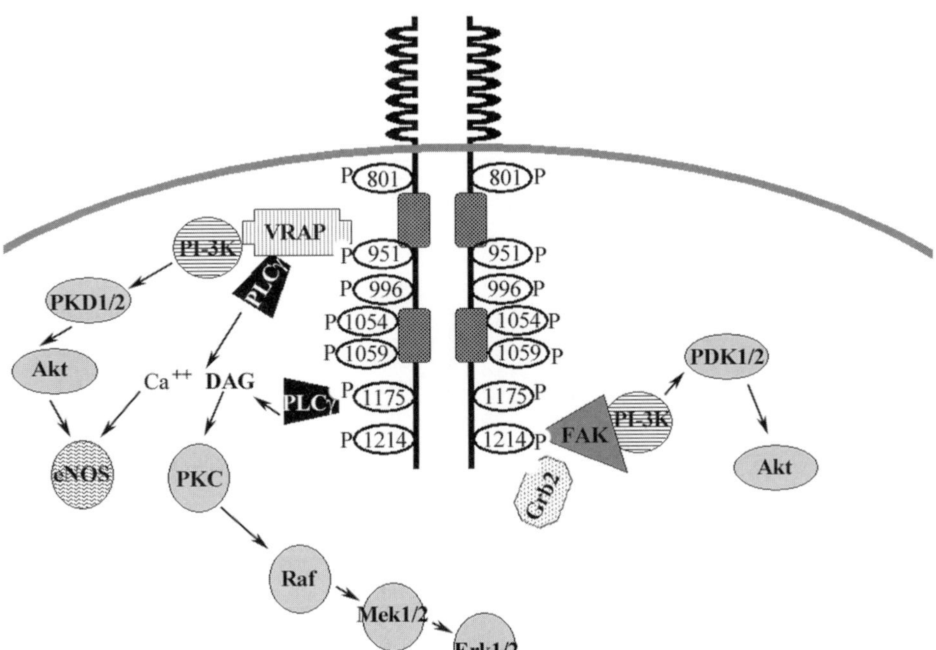

Figure 1 Overview of VEGFR-2 Signaling. Schematic representation of VEGFR-2 illustrating main sites of tyrosine phosphorylation. Shaded boxes in the intracellular portion of the receptor denote the kinase domain. Signaling intermediates, PLCγ, Grb2, FAK, and VRAP, known to interact with VEGFR-2, are shown adjacent to the phosphotyrosine residues to which they bind. Arrows indicate downstream pathways reported in endothelial cells to be initiated from signaling molecules recruited to VEGFR-2. VEGFR-2-activated PLCγ acts on phosphatidylinositol 4,5-bisphosphate to generate DAG and the Ca^{++} mobilizing second messenger inositol 1,4,5-trisphosphate (IP_3). Increased intracellular Ca^{++} may also occur via an IP_3-independent route.

brane region, tyrosine kinase domain with kinase insert, and carboxy-terminal tail. In contrast to the VEGFR family, receptors for FGF are expressed by a wide range of cell types. FGFR-1 has been shown clearly to have a role in microvessel formation and maintenance. FGFs stimulate endothelial proliferation, migration, and organization into capillary tubes.

Activation of FGFR-1 results in phosphorylation of the juxtamembrane Y463, Y583/585 in the kinase insert, Y653/654, Y730, and in the carboxy-tail Y766. Several signaling pathways are activated by FGFR-1, including the PI-3K pathway and stimulation of Src family kinases. The best-elucidated pathways are those leading to PKC activation and ERK1/2 stimulation, linking FGFR-1 to endothelial proliferation. Phosphorylation of Y766 is important for both of these pathways. This phosphotyrosine provides a site for recruitment and activation of PLCγ, leading to generation of inositol 1,4,5-trisphosphate (IP_3) and DAG with subsequent PKC stimulation.

FGFR-1 is linked to the Ras/ERK1/2 pathway and proliferation via a multisubstrate adaptor FRS2. In some studies, this adaptor has been found constitutively associated with the juxtamembrane region of FGFR-1 via the FRS2 PTB domain, although in a phosphotyrosine-independent manner. Activation of FGFR-1 results in markedly increased phosphorylation of FRS2, creating recruitment sites for Grb2 and Shp2. Grb2 is associated with the guanine nucleotide exchange factor SOS that activates Ras, thereby initiating the Raf/ERK1/2 cascade. In endothelial cells, stimulation of FRS2 phosphorylation by FGFR-1 requires the adaptor protein Shb that binds to phosphorylated Y766 in FGFR-1. This adaptor does not interact directly with FRS2 but associates with Shp2, providing an indirect link. Further work will determine the mechanism by which Shb mediates FGFR-1 phosphorylation of FRS2. These signaling events are illustrated in Figure 2. Additional, undefined pathways exist in endothelial cells for FGFR-1 activation of the ERK1/2 pathway.

In several cell types, the adaptor protein Gab1 is recruited to FGFR-1-activated FRS2 via its interaction with the SH3 domain of Grb2. FGFR-activated phosphorylation of Gab1 then creates a binding site for the SH2 domain of the p85 regulatory subunit of PI-3K, resulting in activation of PI-3K and Akt. Although not yet directly demonstrated, it is likely that a similar pathway is responsible for the FGF-activation of PI-3K/Akt in endothelial cells.

The Ephs and Ephrins

The Eph family of RTKs comprise 15 members. They are activated by ligands, the ephrins (of which there are 9), which are cell surface bound. This system functions in short-range cell-to-cell communication. The ephrins are

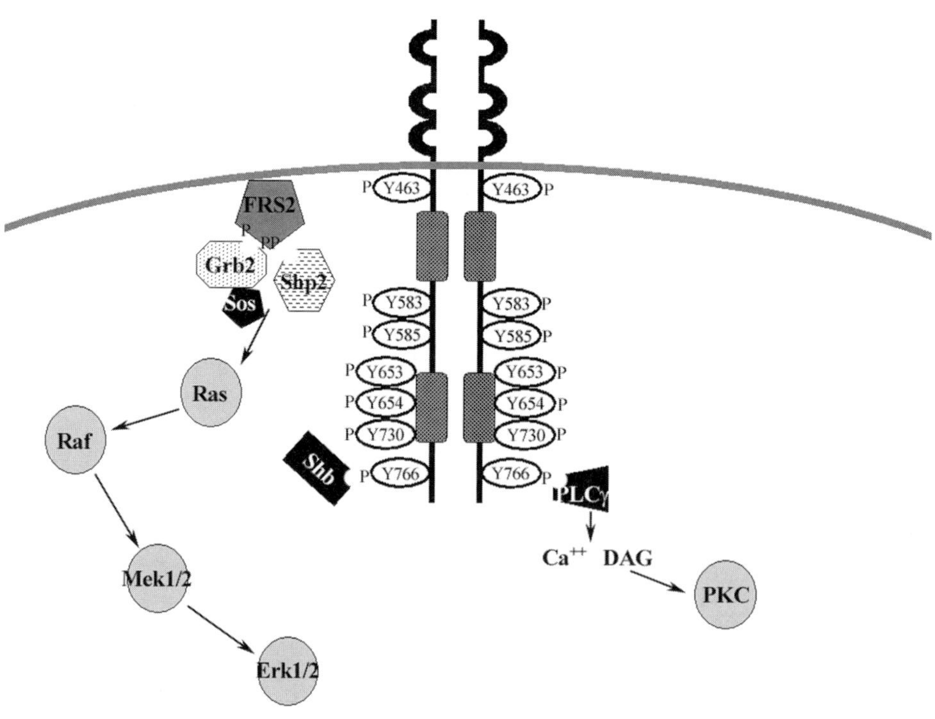

Figure 2 FGFR-1 Signaling in the Endothelium. This schematic figure shows the main tyrosine phosphorylation sites in FGFR-1. Shaded boxes in the intracellular portion of the receptor denote the kinase domain. Signaling intermediates are shown adjacent to phosphotyrosine residues to which they bind. The adaptor protein FRS2 may be constitutively associated with the juxtamembrane region of FGFR-1. Maximal phosphorylation of FRS2 requires binding of Shb to pY766. Shp2 has been found to be constitutively associated with Shb and may act to link Shb with FRS2 phosphorylation. The pathways shown are those for which clear evidence exists in endothelial cells.

divided into two subtypes: (1) those bound to the cell surface by glycosylphosphatidyl inositol (GPI) linkages, class A ephrins, and (2) those with transmembrane domains, class B ephrins. Two subtypes of Eph have also been defined, generally EphAs bind class A ephrins and EphBs class B ephrins. EphB2, EphB3, and EphB4 as well as the ligands ephrinB1 and ephrinB2 have been implicated in microvessel formation. Transgenic mice deficient in EphB4 or ephrinB2 have defects in vasculogenesis and compromised angiogenic remodeling. Deficiency of EphB2 and EphB3 also results in defective vessel remodeling, although because of some functional compensation, the single knockouts do not have a defective vascular phenotype. These phenotypes are consistent with the known involvement of the Eph/ephrin system in regulating cellular repulsion, adhesion, and migration. The broadly reciprocal expression of EphB4 in venous endothelium and its ligand ephrinB2 in arterial endothelium suggest involvement in suppression of mixing between the two cell types, again in accord with roles in regulation of adhesive and repulsive interactions.

The Eph receptors all possess a glycosylated extracellular region consisting of an N-terminally located ligand-binding site, a cysteine-rich domain, and a dimerization motif contained within two fibronectin III–like repeats. The intracellular region follows the single transmembrane spanning domain and consists of a juxtamembrane region and a single tyrosine kinase domain, each of which contain tyrosines that when phosphorylated act as docking sites for downstream signaling molecules. The C-terminus has a PDZ binding motif to which PDZ motif containing proteins bind. These may act as scaffolds for the assembly of multiprotein signaling complexes at the membrane. There is a sterile alpha motif (SAM) domain just before the C-terminus, which may regulate dimerization of the receptors.

The membrane-bound ephrin ligands present in a clustered state to the Eph receptors. The extent of clustering may determine the level of activation of the Eph receptor. Importantly, the ephrins also have signaling capacity, and the class B ligands become tyrosine phosphorylated on binding their receptors. Thus, both EphB receptors and ephrin-B ligands are involved in bidirectional signaling.

Eph/Ephrin Signal Transduction

A large number of adaptor proteins, namely, SLAP, Grb2, Grb10 Crk, and Nck, together with cytoplasmic signaling proteins such as RasGAP, Src, Abl, LMW-PTP, PLCγ, and PI-3K, have been shown to interact with Eph receptors and their ligands in studies on patterning in the nervous system. These studies have established Eph/ephrin signaling in controlling neural cell morphology and architecture, attachment, and motility. In contrast, very little is known about Eph/ephrin signaling pathways in the endothelium and microvasculature. However, the adaptor protein Crk has been strongly implicated in ephrinB1-induced membrane ruffling and focal complex assembly in endothelial cells. EphrinB1 induces phosphorylation of EphB1 in human endothelial cells and activates both Rac1, resulting in membrane spreading, and Rap1, which mediates stabilization of focal complexes. These events depend on the adaptor function of Crk, which has been shown to associate with Ephs following their activation in other cell types. Ephrin-B1 transduces signals to modulate integrin-mediated cell attachment and migration in endothelial cells and promotes angiogenesis in vivo, although the signaling pathway has yet to be clearly defined. EphB2 and EphB4 have been shown to recruit p120RasGap on activation and suppress VEGFR- and Tie2-mediated endothelial migration, but again further work will be required to establish the precise signaling cascades involved.

Signaling by the Tie Receptor Tyrosine Kinases

The Tie family comprises two members: Tie1 and Tie2. Both of these RTKs are essential for blood vessel formation and maintenance. There have been no ligands yet identified for Tie1, and relatively little is known about the cellular functions and signaling pathways utilized by this receptor. Tie1 does inhibit endothelial apoptosis and promote vessel survival. Several ligands, the angiopoietins, have been identified for Tie2. Of these, angiopoietin-1 (Ang1) and angiopoietin-2 (Ang2) have been best characterized. Ang1 is an activator of Tie2, and Ang2 can activate Tie2 or antagonize the effects of Ang1 depending on the cellular context. Tie2 inhibits endothelial death, promotes endothelial migration, suppresses microvessel regression, inhibits vascular inflammation, and promotes microvessel integrity.

Structurally, Tie1 and Tie2 share several key features. The extracellular domain consists of two immunoglobulin-like domains separated by three epidermal growth factor motifs and followed by three fibronectin III–like repeats. Intracellularly, the receptors each contain a tyrosine kinase domain interrupted by a kinase insert region.

Tie1 Signaling

Studies examining Tie1 signaling have utilized chimeric receptors in which the extracellular domain of Tie1 is replaced by that of a receptor with known ligand. Using this approach, Tie1 exhibits a very limited ability to undergo phosphorylation and appears to have a low kinase activity in comparison with Tie2. Nevertheless, the receptor does interact with the adaptor Shp2 and the p85 subunit of PI-3K. The latter interaction initiates an antiapoptotic signaling cascade via Akt. Tie1 also interacts with Tie2, and a pool of preformed Tie1:Tie2 hetero-oligomers exist in endothelial cells.

Tie undergoes regulated ectodomain cleavage in which the extracellular domain of the receptor is proteolytically released. This generates an intracellular fragment containing the tyrosine kinase domain. This fragment may have signaling functions. Ectodomain cleavage of Tie1 is stimulated by VEGF.

Tie2 Signaling

Several phosphorylation sites have been identified in Tie2, including Y992 in the activation loop of the kinase and Y1102, Y1108, and Y1113 in the carboxy-terminal tail of the receptor. Y1108 has been confirmed as a phosphorylation site in Ang1-activated endothelial cells. Ligand-activated phosphorylation of Y1108 results in recruitment of the multiple adaptor protein Dok-R and its tyrosine phosphorylation. This recruitment entails localization of Dok-R to the membrane via interaction of Dok-R PH domain with membrane phosphoinositide lipid products generated by PI-3K, and binding of the adaptor to pY1108 via its PTB domain. Maximal activation of Dok-R tyrosine phosphorylation by Tie2 therefore depends on Y1108 phosphorylation and activation of PI-3K. Phosphorylated Dok-R recruits the adaptor Nck and p21-activating kinase (Pak) that constitutively associates with it. Pak activity is stimulated in this complex of Tie2/Dok-R/Nck/Pak. This cascade mediates Tie2-stimulated endothelial motility.

The antiapoptotic activity of activated Tie2 is mediated via the PI-3K/Akt pathway. Phosphorylation of Y1102 on Tie2 creates a binding site for the p85 subunit of PI-3K. This leads to activation of PI-3K and Akt and suppression of apoptosis. Tie2 activation also stimulates tyrosine phosphorylation of FAK, and this depends on PI-3K activity. Other signaling intermediates have been reported to interact with phosphorylated Tie2, including Grb2, ShP2 Grb7, and Grb14. Further work will be required to delineate the precise involvement of these molecules in Tie2 activity. Tie2 also recruits the NFκB regulatory protein ABIN-2. This recruitment is dependent on Tie2 phosphorylation and is activated by Ang1 in endothelial cells. Interaction between Tie2 and ABIN-2 appears important in the anti-inflammatory effects of Tie2, although the signaling pathway involved has yet to be clearly defined. Tie2 signaling pathways are summarized in Figure 3.

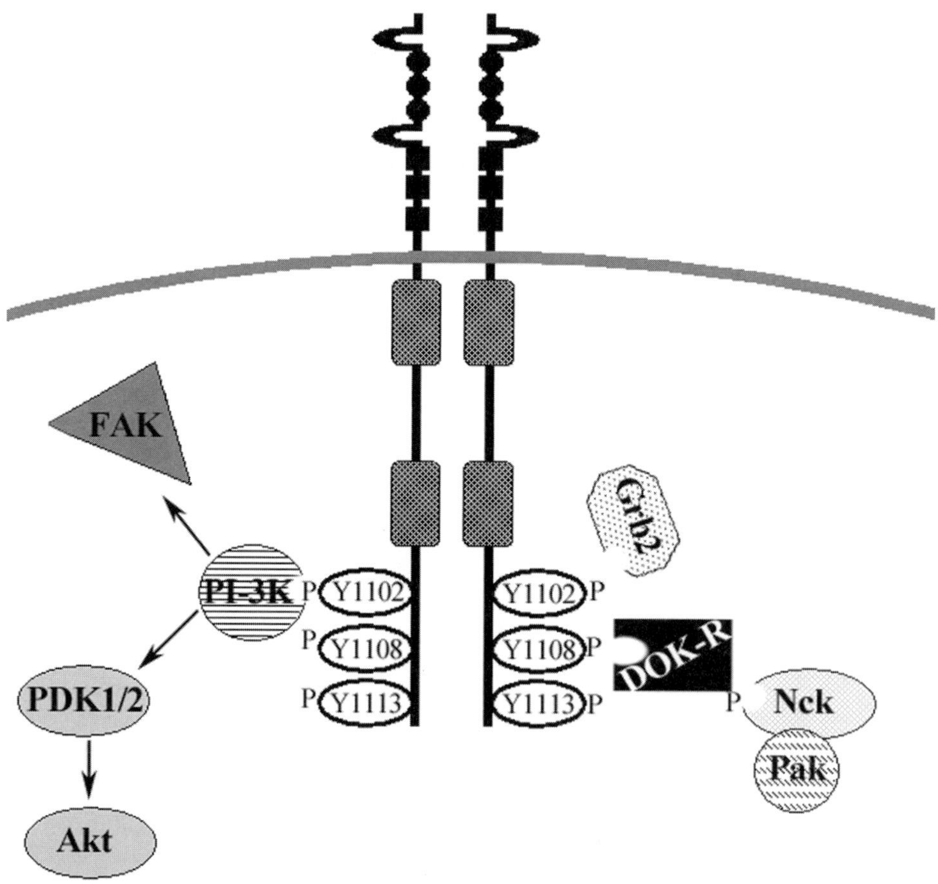

Figure 3 Tie2 Signaling Pathways. In this schematic figure, the three main phosphorylation sites in the carboxy-terminal of Tie2 are shown together with signaling intermediates reported to interact with the receptor. Shaded boxes in the intracellular portion of the receptor denote the kinase domain. Optimal DOK-R recruitment requires interaction with pY1108 in Tie2 via the DOK-R PTB domain and membrane localization via DOK-R PH domain binding to phosphoinositide lipid products generated by PI-3k action. Tie2-mediated phosphorylation of DOK-R provides a recruitment site for Nck binding. Pak is constitutively associated with Nck. Other intermediates have been found to bind Tie2 in a phosphotyrosine-dependent manner; however, their roles in Tie2 action and downstream Tie2-initiated signaling cascades have yet to be defined in endothelial cells.

Conclusion

Despite the importance of RTK signal transduction cascades in microvessel development and function, it is clear that there are large gaps in our understanding of these cascades in microvascular cells for each of the RTKs reviewed. A more comprehensive knowledge of events at the receptor, downstream pathways, and linkage with effector mechanisms is needed for each of the microvascular RTKs. Important questions also remain on the specificity of signaling cascades in the microvasculature, how different RTKs can activate the same signaling pathway but have different effects on the cell. For example, VEGF and Ang1 both activate the Akt pathway, but VEGF stimulates pro-inflammatory gene expression and Ang1 suppresses this via a mechanism that requires Akt activation. Differences in spatial, temporal, and quantitative aspects of signal cascades as well as the profile of signaling events elicited by each receptor is likely to underlie some of the specificity in signaling. The cell is required to integrate many signals into the physiological setting. An important challenge is to understand how the different RTK signaling pathways coordinate, both with each other and with non-RTK pathways. Such coordination is likely to involve interactions between cascades within the cell as well as regulation at the level of ligand and receptor. For example, VEGF impacts Tie signaling by inducing proteolytic cleavage of Tie1, and hetero-oligomeric complexes of VEGFR1:VEGFR2 and Tie1:Tie2 provide mechanisms for cross-talk between receptors of the same family. The specialized functions and architecture of the microvasculature provide an opportunity to gain fundamental insights into mechanisms of signal transduction and how they regulate tissue morphogenesis and maintenance. Such insights may provide new strategies for treatment of ischemic disease, cancer, and inflammatory conditions.

Glossary

Adaptor protein: (also called adapter protein): A protein that can act as a binding intermediate. Adaptor proteins bind to one protein, for example an activated receptor, and provide binding sites for the recruitment of other proteins.

Autophosphorylation: The phosphorylation of a protein by itself. Receptor tyrosine kinases are said to undergo autophopshorylation following ligand activation. In fact, ligand binding enhances RTK dimerization or oligomerization, allowing individual RTKs within the complex to be phosphorylated by other RTKs also in the complex (transphosphorylation).

PTB domain: Phosphotyrosine binding domains are between 100 and 170 amino acids and bind phosphotyrosine, usually flanked by an asparagines-proline-X amino peptide. PTB domains in some proteins appear to have broader binding specificity, including nonphosphorylated sequences.

SH2 domain: SH2 domains comprise approximately 100 amino acids and bind phosphotyrosine-containing peptides. The specificity of SH2 binding is determined by the amino acid residues flanking the phosphorylated tyrosine.

SH3 domain: A protein module binding proline-rich peptides.

Further Reading

Autiero, M., Waltenberger, J., Communi, D., Kranz, A., Moons, L., Lambrechts, D., Kroll, J., Plaisance, S., De Mol, M., Bono, F., et al. (2003). Role of PlGF in the intra- and intermolecular cross talk between the VEGF receptors Flt1 and Flk1. *Nat. Med.* **9,** 936–943.

Cheng, N., Brantley, D. M., and Chen, J. (2002). The ephrins and Eph receptors in angiogenesis. *Cytokine Growth Factor Rev.* **13,** 75–85.

Claesson-Welsh, L. (2003). Signal transduction by vascular endothelial growth factor receptors. *Biochem. Soc. Trans.* **31,** 20–24. *An excellent, concise review of the key features of VEGF-receptor signaling.*

Cross, M. J., Lu, L., Magnusson, P., Nyqvist, D., Holmqvist, K., Welsh, M., and Claesson-Welsh, L. (2002). The Shb adaptor protein binds to tyrosine 766 in the FGFR-1 and regulates the Ras/MEK/MAPK pathway via FRS2 phosphorylation in endothelial cells. *Mol. Biol. Cell* **13,** 2881–2893.

Jones, N., Iljin, K., Dumont, D. J., and Alitalo, K. (2001). Tie receptors: New modulators of angiogenic and lymphangiogenic responses. *Nat. Rev. Mol. Cell. Biol.* **2,** 257–267. *A very good summary of the Tie receptor biology and signaling, concentrating on roles in angiogenesis.*

Master, Z., Jones, N., Tran, J., Jones, J., Kerbel, R. S., and Dumont, D. J. (2001). Dok-R plays a pivotal role in angiopoietin-1-dependent cell migration through recruitment and activation of Pak. *EMBO J.* **20,** 5919–5928.

Nagashima, K., Endo, A., Ogita, H., Kawana, A., Yamagishi, A., Kitabatake, A., Matsuda, M., and Mochizuki, N. (2002). Adaptor protein Crk is required for Ephrin-B1-induced membrane ruffling and focal complex assembly of human aortic endothelial cells. *Mol. Biol. Cell.* **13,** 4231–4242. *A research article illustrating some of the approaches used for delineating intracellular signaling pathways activated by receptor tyrosine kinases.*

Zachary, I., and Gliki, G. (2001). Signaling transduction mechanisms mediating biological actions of the vascular endothelial growth factor family. *Cardiovasc Res.* **49,** 568–581. *This article is valuable in linking the signaling events elicited by VEGF to downstream functional effects of the growth factor.*

Capsule Biography

Marie Marron and Nick Brindle work on receptor tyrosine kinase signaling with particular focus on RTKs in vascular formation, regression, and maintenance. They are supported mainly by the British Heart Foundation and Wellcome Trust.

Chapter 13

Eicosanoids and the Pulmonary Microvasculature

Ivan M. Robbins and Brian W. Christman

Department of Medicine, Division of Allergy, Pulmonary and Critical Care Medicine, Vanderbilt University School of Medicine, Tennessee

Introduction

All cell membranes, including those of the pulmonary endothelium, are composed in large part of phospholipids structurally derived from glycerol. In addition to serving a barrier function, these compounds are metabolized by a family of phospholipases, which liberates precursors of biologically active compounds that act in both an endocrine and paracrine fashion to alter physiologic responses. These phospholipids are enriched in the sn-2 position with arachidonic acid, a 20-carbon alkane with four double bonds. Metabolism of arachidonic acid proceeds by oxygen insertion catalyzed by several different enzyme families, including cyclooxygenase 1 (COX-1) and cyclooxygenase 2 (COX-2), yielding an unstable intermediate prostaglandin (PG) H2. Further metabolism of this compound depends on the complement of available enzymes in a given cell type (e.g., prostacyclin synthetase in endothelial cells and thromboxane synthase in platelets), but can result in formation of a variety of prostaglandins, including PGD_2, PGE_2, $PGF_{2\alpha}$, prostacyclin (PGI_2), and TxA_2. In the vascular bed of the lung, the predominant prostaglandins made by vascular endothelium are prostacyclin and PGE_2. Smooth muscle cells in lung vessels are net producers of both thromboxane A_2 and prostacyclin.

It is also possible for arachidonic acid to be metabolized by lipoxygenases (yielding leukotrienes and lipoxins) and specific cytochrome P450 enzymes [yielding epoxyeicosatrienoic acids (EETs), and stereospecific hydroxyeicosatetraenoic acids, (HETEs)]. Alternately, during situations of enhanced oxidant stress, reactive oxygen species can directly interact with arachidonic acid esterified to membrane phospholipids to form isoprostanes (IsoPs), compounds analogous to prostaglandins but with different stereochemistry (Figure 1). All of these molecules derived from arachidonic acid are known as *eicosanoids*.

In vitro studies document that specific enzymes critical for synthesis of unique prostaglandins and leukotrienes are restricted in their cell distribution. As such, the appearance of end products in biological samples allows inference about which cells have been activated in vivo. In the case of intact tissues and organs perfused with blood containing cells with different metabolic capacities, additional opportunities for mediator synthesis are created. For example, early in the pulmonary inflammatory process, upregulation of adhesion molecules facilitates neutrophil binding to pulmonary vascular endothelial cells. Neither cell in isolation is capable of metabolizing arachidonic acid into cysteinyl leukotrienes (LTC_4, LTD_4, LTE_4). However, an unstable epoxide intermediate, LTA_4, can be formed by the actions of 5-lipoxygenase in the neutrophil and transferred to the endothelial cell because of intimate cellular apposition. The endothelial cell lacks 5-lipoxygenase but does contain abundant LTC_4 synthase and is able to complete formation of a cysteinyl leukotriene [1]. Such transcellular biosynthesis with transfer of arachidonic can also occur between platelets and endothelium to facilitate formation of prostacyclin. This is part of the biochemical basis for the beneficial effects of antiplatelet agents, such as aspirin, in prevention of myocardial infarction and stroke. Such processes emphasize the dynamic and complex nature of pulmonary vascular eicosanoids biosynthesis and, with waxing and waning of cell apposition, allows some plasticity in the eicosanoid synthetic capability of the lung.

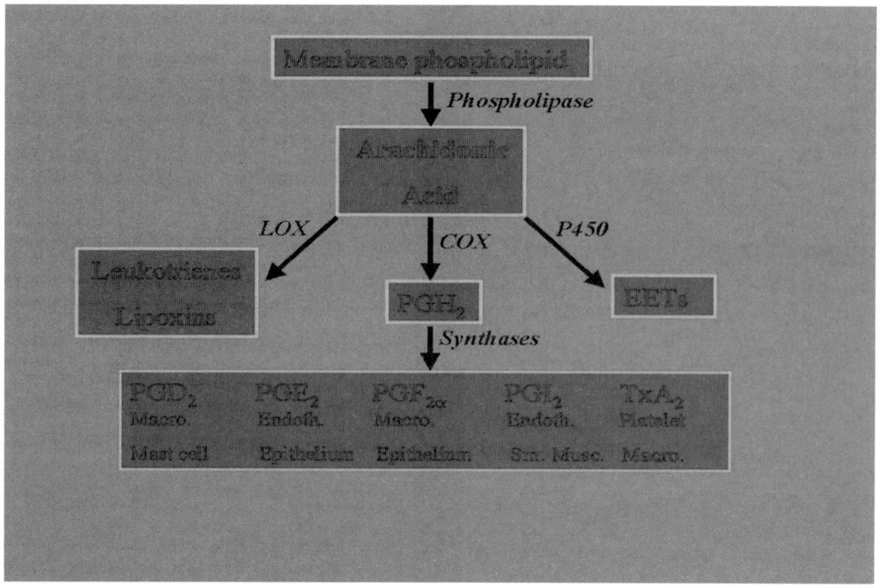

Figure 1 Enzymatic metabolism of membrane phospholipid, via arachidonic acid, to form eicosanoids. Specific enzymes are shown in italics (LOX = lipoxygenase; COX = cyclo-oxygenase; P450 = cytochrome P450; "Synthases" includes PGD synthase, PGE synthase, PGF2 isomerase, prostacyclin synthase, and thromboxane synthase. (see color insert)

The pulmonary microvasculature serves not only as an important source of prostaglandins but also as a major site of metabolism of these molecules. The standard bisenoic prostaglandins (PGE_2, PGD_2, $PGF_{2\alpha}$) are rapidly metabolized (60% to 90%) on first passage through the lung. This proceeds via oxidation of the 15-hydoxyl function followed by reduction of the 13,14-double bond, resulting in the respective 15-keto-13,14-dihydro metabolites [2]. Presumably such rapid metabolism by the lung limits systemic effects of prostaglandins generated in the splanchnic circulation. However, during periods of oxidant stress, such as in prolonged hyperoxia, or when lung endothelial function has been compromised (e.g., sepsis with ARDS), transpulmonary clearance of PGE_2 has been shown to be markedly reduced. Such loss of the normal, homeostatic clearance function of the pulmonary vascular bed may result in significant alteration in function of systemic organs during times of prostaglandin overproduction. Interestingly, metabolism of both prostacyclin and thromboxane A_2 tends to occur via both spontaneous hydrolysis, which curtails the circulating half-life of both autocoids to a few minutes, and hepatic metabolism by beta oxidation pathways in the liver.

The biological effects of prostaglandins depends on binding to specific cell surface receptors. There are subtypes of many of the prostaglandin receptors, which may signal by entirely different mechanisms. For example, of the four different E-prostaglandin (EP) receptors, one signals by altering calcium flux, two increase cellular cyclic adenosine monophosphate (AMP) levels, and one decreases c-AMP [3]. Relative regional differences in density of such receptors on target tissues and effector cells modulates the effects of eicosanoids on vascular tone and microvascular fluid and solute exchange. In addition, the structural similarity that exists between the different eicosanoids can result in "receptor promiscuity"—high levels of a structurally similar eicosanoid ligand can activate a related receptor. In vivo and in vitro studies have documented such cross-talk between PGD_2 and the thromboxane (TP) receptor and between the prostacyclin and the EP receptor.

As described previously, a wide array of eicosanoid metabolites, often with opposing actions and synthesized by multiple cell types, are produced in, and act on, the pulmonary microvasculature. The predominant endothelial cell product is prostacyclin, a key endogenous vasodilator that results in rapid elevation of intracellular c-AMP after binding to its receptor. This in turn activates protein kinase A and decreases intracellular calcium concentration, leading to relaxation of smooth muscle. Both vasoconstrictors (angiotensin II) and vasodilators (nitric oxide) have been shown to increase the synthesis of prostacyclin. In addition, increased shear force is a potent stimulus for prostacyclin synthesis by endothelium. Of note, in precapillary PAH, the ordinary pulsatile capillary flow in the lung is lost. This may contribute to the decrease in synthesis observed in many forms of pulmonary arterial hypertension (PAH).

Prostacyclin has additional properties that are important. These include inhibition of platelet aggregation and effects on muscle growth and cardiac contractility [4]. PGI_2 is a well documented inhibitor of platelet aggregation, which is mediated, similar to its vasodilatory actions, through an increase in intracellular c-AMP. Importantly, prostacyclin does not inhibit initial platelet adhesion at the site of vessel wall injury and thus allows hemostasis without vessel occlusion. When used in pharmacological doses for the treatment of PAH (see following discussion), although bleeding time increases, clinical evidence of excessive bleeding in patients does not occur. PGI_2 also modulates leukocyte adhesion to abnormal vascular surfaces, preventing local leukostasis and limiting inflammation. The intracellular increase in c-AMP resulting from engagement of PGI_2 to its receptor has also been shown to inhibit DNA synthesis in vascular smooth muscle cells. Direct inotropic effects of prostacyclin have been demonstrated in animal studies evaluating isolated ventricular muscle preparations, and clinically important improvement in cardiac output following administration of epoprostenol, the synthetic salt of prostacyclin, has been reported in patients with pulmonary hypertension receiving chronic therapy.

Thromboxane A_2 (TxA_2) is the principle eicosanoid produced by platelets, but significant production in alveolar macrophages and monocytes occurs as well, especially when they are activated [5]. TxA_2 has properties opposite to those of PGI_2 in that it is a potent vasoconstrictor and promotes platelet aggregation and smooth muscle growth, the latter possibly through a protein kinase C-linked pathway. Numerous triggers of platelet activation, including acute lung injury, endotoxemia, shear force, and procoagulant alterations of the endothelial surface, lead to increased TxA_2 synthesis.

It is well known that PGD_2 causes bronchoconstriction. Its effects on the pulmonary vasculature are less well documented, but in sheep PGD_2 causes vasoconstriction, likely through binding to the TxA_2 receptor. Conversely, and similar to prostacyclin, PGD_2 also inhibits DNA synthesis in vascular smooth muscle cells, but this is independent of an increase in c-AMP. In humans, whether PGD_2 has a beneficial or deleterious effect on the development, maintenance, or regression of pulmonary arteriopathy is unknown. PGD_2 also may tend to oppose platelet aggregation.

Leukotrienes, eicosanoid products resulting from the action of lipoxygenases on arachidonic acid, are potent mediators of inflammation with well-documented involvement in the pathogenesis of asthma. Recently, increased levels of 5-lipoxygenase and 5-lipoxygenase activating protein have been found in small and medium-sized pulmonary arteries by immunohistochemical staining of lung tissue from patients with pulmonary hypertension [6]. These results, in combination with the finding of increased leukotriene E4 in lavage fluid from newborn infants with persistent pulmonary hypertension, suggests a possible role for leukotrienes in pulmonary vasculopathies.

Isoprostanes, the result of nonenzymatic, free-radical catalyzed peroxidation of arachidonic acid, are excellent markers of in vivo oxidant stress. Evidence of lipid peroxidation has been found in numerous vascular disorders, including pulmonary vascular disease. Isoprostanes may also cause vasoconstriction and modulate platelet function, possibly by binding to receptors for TxA_2. Furthermore, they can stimulate proliferation of endothelial cells and synthesis of endothelin-1, an exceptionally potent endogenous vasoconstrictor and smooth muscle mitogen. Previously believed to be exclusively formed by cyclo-oxygenases, a recent study indicates that as much as 30 percent of PGD_2 is synthesized via the isoprostane pathway. This suggests that oxidant stress can increase prostaglandin production by two independent mechanisms: (1) facilitation of enzymatic catalysis by COX, and (2) direct peroxidation of arachidonate-containing phospholipids with phospholipase (or PAF acetylhydrolase) induced release of prostanoids.

A large body of information has examined the role of eicosanoids in endotemia and acute lung injury [7]. Increased levels of the major metabolite of thromboxane, a powerful vasoconstrictor and platelet aggregating cyclo-oxygenase product produced exuberantly by platelets, have been demonstrated in experimental models of endotoxemia. However, neither inhibition of thromboxane A_2 synthesis nor blockade of the TP receptor abrogates the decrease in vascular barrier function (10#26,69,74). Furthermore, platelet depletion does not significantly modify the pulmonary vascular permeability response to endotoxin in chronically instrumented, nonanesthetized sheep. Both humans and large animals develop pulmonary hypertension following endotoxemia. Brigham and coworkers administered *E. coli* endotoxin to awake, chronically instrumented sheep and noted a two-phased response consisting of an early phase of acute pulmonary hypertension, characterized by increased flow of lung lymph that was relatively low in protein. This was followed by a longer phase of increased vascular permeability (i.e., high lymph/plasma protein ratio), during which mean pulmonary pressure remained only slightly elevated. Both analytical and pharmacologic studies using cyclo-oxygenase inhibitors suggest that thromboxane A_2 mediates the early pulmonary vasopressor response but does not play a role in the later alteration in microvascular permeability.

Pulmonary Arterial Hypertension

A major disorder involving the pulmonary microvasculature and eicosanoids is pulmonary arterial hypertension (PAH). PAH primarily affects the small pulmonary arteries with a diameter of less than 100 microns. Pathological changes include medial hypertrophy, intimal proliferation, in situ thrombosis, and plexiform lesions, the latter likely a disordered attempt at neovascularization. PAH may occur without an associated disorder, primary pulmonary hypertension (PPH), or in association with connective tissue disease, congenital heart disease, portal hypertension, HIV infection, or use of appetite suppressants. Disruption of the endothelial lining of the microvasculature in PAH is associated with changes in the local milieu of protein and lipid mediators. Whether these changes are causative or secondary to the pathological changes is uncertain. It is noteworthy that similar pathological changes have been found in the microvasculature of patients undergoing thromboendarterectomy for treatment of large-vessel thromboembolic pulmonary hypertension, suggesting that small-vessel vasculopathy can result from multiple processes.

More specifically, alterations in the production of eicosanoid mediators have been demonstrated in animal models of pulmonary hypertension and in PAH. Although several eicosanoid products can affect the pulmonary microvasculature, PGI_2 and TxA_2 appear to play the largest role in vascular homeostasis. Both compounds are rapidly metabolized within seconds to minutes, and therefore, circulating levels of both prostacyclin and TxA_2 are extremely low (< 4 pg/mL). This implies that these eicosanoids are primarily synthesized locally in the pulmonary vasculature, and perturbations in the local environment are associated with changes in mediator production.

In an animal model of pulmonary hypertension using calves exposed to hypoxia at high altitude, investigators found decreased synthesis of PGI_2 compared to control animals. More recently, studies using in situ hybridization and Western blots have demonstrated a decrease in prostacyclin synthase in both large and small pulmonary arteries from patients with PAH. Additional studies in mouse models support a pivotal role for prostacyclin in the development of pulmonary hypertension. In a prostacyclin receptor knockout model, mice exposed to hypoxia developed more severe pulmonary hypertension and vascular remodeling compared to control animals. Conversely, when exposed to hypobaric hypoxia, transgenic mice with overexpression of prostacyclin synthase produced more prostacyclin and histologically exhibited nearly minimal medial hypertrophy in precapillary arteriole vessels compared to significant changes in control animals.

In patients with pulmonary vascular disease, with PPH or related to connective tissue disease, we found increased urinary metabolites of TxA_2 when corresponding metabolites of prostacyclin were decreased, so the overall ratio of TxA_2/PGI_2 was significantly increased compared to control subjects [8]. These findings have subsequently been confirmed in patients with PAH related to congenital heart disease. Although TxA_2 is usually synthesized by platelets (and to a lesser extent by other circulating cells), vascular synthesis of TxA_2 has been demonstrated, at least in systemic arteries. This abnormal production of TxA_2 may be even more marked in PAH where an abnormal vascular lining promotes platelet binding and activation. Combined, these studies, in humans and in animal models, suggest that the pathogenesis and/or maintenance of pulmonary hypertensive vascular disease may involve alterations in the production of PGI_2 and TxA_2.

Clinical Modulation of Eicosanoids in Pulmonary Vascular Disease

Attempts to treat PAH with pharmacological doses of epoprostenol (the synthetic salt of prostacyclin), 100 to 1,000 times that normally made endogenously, began in the early 1980s based on epoprostenol's vasodilatory properties. Because of the short half-life in vivo, epoprostenol must be continuously infused via a central venous catheter; acute clinical deterioration has been reported with acute interruption of the drug. Nevertheless, use of epoprostenol has been approved for the long-term treatment of PAH based on improvement in pulmonary hemodynamics, quality of life, and survival documented in a randomized, clinical trial [9]. Currently, epoprostenol is used to treat all forms of intrinsic PAH as well as other disorders that can affect the pulmonary microvasculature, such as distal chronic thromboembolic disease that is not amenable to operative treatment or sarcoidosis with diffuse vascular involvement.

Intravenous epoprostenol has dramatically improved the outcome of patients with PAH, most impressively in patients with PPH. Results in patients with the scleroderma spectrum of disease are not as impressive, perhaps because they have a systemic disease and are generally older than patients with PPH. Compared to historical controls or to survival predicted by the NIH PPH Registry survival equation, survival with epoprostenol is significantly improved [10]. However, not all patients respond to therapy, and one third of patients with PPH still die within three years of starting treatment. This may be explained by the fact that although most patients improve clinically, hemodynamics remain markedly abnormal. After one year of therapy, mean pulmonary artery pressure only decreases 10 to 15 mmHg, and pulmonary vascular resistance remains five to six times normal. Rarely do patients demonstrate normalization of pulmonary hemodynamics. This indicates that while most patients improve clinically, the fundamental pathological remodeling process in the precapillary pulmonary vessels is not significantly affected.

Since the approval of epoprostenol, other prostacyclin analogues have been evaluated in the treatment of PAH. Beraprost, an oral analogue of prostacyclin, was originally used in uncontrolled studies in Japan that purported to show efficacy in PPH. More recently, two randomized, placebo-controlled studies have been performed. One was a three-month study that demonstrated improvement in six-minute walk distance. The other was a 12-month study that initially showed similar clinical improvement at 3 and 6 months. However, by 9 months of treatment, no clinical benefit was apparent, indicating that long-term benefit can not be maintained with the oral preparation. Iloprost is another longer-acting prostacyclin analogue, given intravenously and by inhalation, the latter of which requires 6 to 10 treatments per day. A recent multicenter, placebo-controlled three-month trial involving patients with PAH and inoperable chronic thromboembolic disease demonstrated improvement in a combined endpoint with inhaled iloprost. Although the treatment was effective in some patients, the frequency of the treatments makes this therapy difficult to use. Finally, treprostinil, a subcutaneously delivered analogue, has been studied in the largest randomized, placebo-controlled study in PAH. Modest improvement was found in six-minute walk distance with treprostinil. Local pain at the infusion site, related to binding of the prostacyclin analogue to pain receptors, limited the dose escalation in most patients in the study and has been an impediment to widespread use of treprostinil; however, those able to increase to greater doses experienced greater clinical improvement.

The mechanism of action of prostacyclin and its analogues remains uncertain. As noted previously, most patients continue to exhibit significant pulmonary hypertension despite treatment, indicating that vasodilation is unlikely to play a major role. Much of the benefit from epoprostenol may result from inotropic effects leading to improved right ventricular contractility. Increases in c-AMP, while important during acute use of epoprostenol, are unlikely to be sustained during long-term use. In cell culture studies of both proximal and distal pulmonary artery, smooth muscle growth was inhibited by a variety of prostacyclin analogues, which correlated with increases in c-AMP. However, levels peaked by 30 minutes and declined by 4 hours. We have measured c-AMP levels in plasma in patients receiving therapy with epoprostenol and found no increase compared to baseline values.

Provision of pharmacologic doses of prostacyclin has effects on other mediator systems. Investigators have shown that chronic infusion of epoprostenol decreases circulating levels of endothelin-1, which may account for some of the beneficial action of prostaglandins. Others have shown improvement in platelet aggregation and markers of endothelial injury, such as circulating levels of von Willebrand factor with long-term epoprostenol treatment. Interestingly, in preliminary studies, despite epoprostenol's potent effects on platelet aggregation, the drug does not appear to decrease excretion of thromboxane metabolites in patients with PPH. This suggests that much of the increase in thromboxane generation in patients with PPH may derive from cells other than platelets. More recently, epoprostenol has been reported to decrease markers of oxidative damage in lung tissue samples from patients with PPH.

Use of thromboxane receptor antagonists or synthase inhibitors have been unsuccessful, perhaps because TxA_2 is not a pivotal mediator in the maintenance of the microvascular remodeling in PAH. Inhibition of thromboxane synthase with an oral agent was initially studied in the late 1980s and showed very modest improvement in resting hemodynamics. More recently, use of a combined thromboxane synthase inhibitor/receptor antagonist, terbogrel, was evaluated in a randomized, placebo-controlled study. Unfortunately, the study was stopped prematurely because of an unexpected adverse event, severe leg pain, that developed in approximately 25 percent of patients. In addition, there was no clinical or hemodynamic improvement in patients who completed the study, although TxA_2 metabolite levels in serum (reflecting platelet capacity) and in urine (reflecting total body thromboxane production) decreased significantly with drug treatment. Furthermore, urinary prostacyclin metabolites increased, supporting the concept of endoperoxide shunt from platelet to endothelium at the vascular interface. Inhibition of PGD_2 or modulation of leukotriene synthesis (or receptor blockade) or treatment with antioxidants (e.g., vitamins C and E) has not been studied in PAH.

A combination of empiric clinical trials of modulation of prostaglandin balance, translational research in patients with pulmonary vascular disease, and ongoing basic investigations of endothelial and smooth muscle biology have yielded clinically useful results over the last decade. Epoprostenol, and perhaps stable analogues of prostacyclin, have a beneficial effect in patients with PAH, particularly those with PPH. However, despite improvement in both quality of life and survival with pharmacological doses of prostacyclin, it is clear that the fundamental abnormalities of growth and disordered angiogenesis are not substantially altered by this therapy. Future studies spanning the spectrum from basic to clinic research are needed to determine more precisely the role of eicosanoids in pulmonary vascular disease.

Glossary

Cyclo-oxygenase: An enzyme protein complex present in most tissues that catalyses two steps in prostaglandin biosynthesis and produces prostaglandins and thromboxanes from arachidonic acid.

Eicosanoid: A class of oxygenated, endogenous, unsaturated 20-carbon fatty acids derived from arachidonic acid after it is cleaved from membrane phospholipids. They include prostaglandins, leukotrienes, thromboxanes, and hydroxyeicosatetraenoic acid (HETE) compounds. They exert their actions by binding to specific cellular receptors.

Epoprostenol: A prostaglandin that is biosynthesized enzymatically from prostaglandin endoperoxides generated by the action of cyclo-oxygenase in human vascular tissue. It is a vasodilator and a potent inhibitor of platelet aggregation. The sodium salt has also been used as a pharmaceutical (Flolan), to treat primary pulmonary hypertension.

Pulmonary arterial hypertension: A disorder primarily affecting the small precapillary pulmonary arteries and occurring either as an idiopathic illness [primary pulmonary hypertension (PPH)], or in association with other disorders including congenital heart disease, connective tissue diseases (especially scleroderma), liver disease, human immunodeficiency virus infections, and use of appetite suppressants. Characteristic pathologicial changes include smooth muscle medial hypertrophy, intimal proliferation, in situ thrombosis, and the development of plexiform lesions.

References

1. Claesson, H. E., and Haeggstrom, J. (1987). Metabolism of leukotriene A4 by human endothelial cells: Evidence for leukotriene C4 and D4 formation by leukocyte-endothelial cell interaction. In Samuelsson, B., Paoletti, R., and Ramwell, P. W. (eds.), *Advances in Prostaglandin, Thromboxane, and Leukotriene Research*. New York: Raven Press; 17: 115–119. *One of the earliest articles documenting transcellular biosynthesis of eicosanoids during cellular apposition. This paper documents donation of leukotriene A4, formed in neutrophils, to endothelial cells with subsequent metabolism to cysteinyl leukotrienes.*
2. Nomura, T., Lu, R., Pucci, M. L., and Schuster, V. L. (2004). The two-step model of prostaglandin signal termination: In vitro reconstitution with the prostaglandin transporter and prostaglandin 15 dehydrogenase. *Mol. Pharmacol.* **65**, 973–978.
3. Imig, J. D., Breyer, M. D., and Breyer, R. M. (2002). Contribution of prostaglandin EP(2) receptors to renal microvascular reactivity in mice. *Am. J. Renal Physiol.* **283**, F415–F422.
4. Smyth, E. M., and Fitzgerald, G. A. (2002). Human prostacyclin receptor. *Vit. Horm.* **65**, 149–165.
5. Cheng, Y., Austin, S. C., Rocca, B., Koller, B. H., Coffman, T. M., Grosser, T., Lawson, J. A., and Fitzgerald, G. A. (2002). Role of prostacyclin in the cardiovascular response to thromboxane A2. *Sci-*

ence **296**, 539–541. *Important study used transgenic murine models of acute lung injury to examine the relative importance of activation of prostacyclin (IP) and thromboxane (TP) receptors during inflammation. Animals lacking the IP receptor had more injury-induced platelet activation and vascular proliferation. Both were decreased in animals genetically deficient in the TP receptor or pretreated with a TP receptor antagonist.*

6. Wright, L., Tuder, R. M., Wang, J., Cool, C. D., Lepley, R. A., and Voelkel, N. F. (1998). 5-lipoxygenase and 5-lipoxygenase activating protein (FLAP) immunoreactivity in lungs from patients with primary pulmonary hypertension. *Am. J. Respir. Crit. Care Med.* **157**, 219–229.
7. Robbins, I. M., Newman, J. H., and Brigham, K. L. (1998). Sepsis induced pulmonary edema: Cellular and physiological mechanisms. In Matthay, M. A., and Ingbar, D. H. (eds.), *Pulmonary Edema*, 116: 203–245 in *Lung Biology in Health and Disease* series. New York: Marcel Dekker.
8. Christman, B. W., McPherson, C. D., Newman, J. H., King, G. A., Bernard, G. R., Broves, B. M., and Loyd, J. E. (1992). An imbalance between the excretion of thromboxane and prostacyclin metabolites in pulmonary hypertension. *N. Engl. J. Med.* **327**, 70–75.
9. Barst, R. J., Rubin, L. J., Long, W. A., McGoon, M. D., Rich, S., Badesch, D. B., Groves, B. M., et al. for the Primary Pulmonary Hypertension Study Group. (1996). A comparison of continuous intravenous epoprostenol (prostacyclin) with conventional therapy for primary pulmonary hypertension. *N. Engl. J. Med.* **334**, 296–301. *This was the pivotal randomized clinical trial that established the efficacy of intravenous epoprostenol in PPH. Patient received either conventional therapy (oxygen, diuretics, inotropes, etc.) or conventional therapy plus continuous infusion of epoprostenol via an indwelling catheter. Both exercise tolerance and short-term mortality were significantly improved in the group receiving prostaglandin therapy.*
10. Galie, N., Manes, A., and Branzi, A. (2003). Prostanoids for pulmonary arterial hypertension. *Am. J. Respir. Crit. Care Med.* **2**, 123–137.

Capsule Biographies

Dr. Robbins has directed the Adult Pulmonary Hypertension Center at Vanderbilt University since 2001. His research has focused on the mechanism of action of epoprostenol in the treatment of pulmonary hypertension, and he has been the lead investigator at Vanderbilt in numerous clinical trials evaluating new medications for the treatment of pulmonary hypertension. His work is supported by grants from the NIH.

Dr. Christman trained in internal medicine and pulmonary/critical care medicine at Vanderbilt University, where he is currently Vice-Chair of Medicine at Vanderbilt University and Chief of the Medical Service for the VA Tennessee Valley Health System. His laboratory uses both basic and clinic research approaches to decipher contribution of the urea/nitric oxide cycle and eicosanoid pathways in modulating both liver/lung injury after endotoxemia and pulmonary hypertension. His work is supported by an NIH Program Project Grant from the NIH National Heart Lung and Blood Institute and by a grant from the NIH National Cancer Institute.

SECTION B

Vascular Development: Vasculogenesis and Angiogenesis

CHAPTER 14

Role of platelet-derived growth factors in the microvasculature

Christer Betsholtz and Holger Gerhardt
Department of Medical Biochemistry, University of Göteborg, Medicinaregatan 9A, Box 440, SE 405 30 Göteborg, Sweden

Introduction

Platelet-derived growth factor (PDGF) is a family of dimeric ligands that bind to and activate two cell surface receptor tyrosine kinase subunits, PDGFRα and PDGFRβ. Four different ligand chains (PDGF-A, B, C and D) assemble into five different dimeric combinations (AA, AB, BB, CC and DD). Binding of bivalent ligands leads to receptor subunit dimerization. The apposition of the receptor kinase domains leads to reciprocal receptor subunit phosphorylation, further activating the tyrosine kinases and creating docking sites for down-stream signal transducers.

One of the first biological functions demonstrated for PDGF was its mitogenic activity on cultured vascular smooth muscle cells (vSMC) [1]. The PDGF preparations used for these early studier were derived from platelets and consisted mainly of PDGF-AB heterodimers. Subsequently it was demonstrated that vSMC carry both PDGFRβ and PDGFRα. In vitro studies also demonstrated that cultured vascular endothelial cells produce PDGF A and B. Thus, early in vitro findings provided indications that PDGFs and their receptors play a role in the vasculature. However, the focus was placed on large vessels and the potential function of PDGF as a "wound healing" hormone for the vascular wall and surrounding connective tissue. For example, PDGF was suggested to be of central importance in the "response to injury" model of atherosclerosis championed by Russell Ross and co-workers.

Subsequent in vitro studies argued for PDGF signaling to have a role also in the microvasculature. In co-culture studies, endothelial cells promoted pericyte migration—a function that was attributed to PDGF. Moreover, pericyte contact with the endothelial cells inhibited proliferation of the latter, and TGFβ was implicated in this function [2, 3].

Roles of PDGFs in the Microvasculature Revealed by Genetic Studies

Lessons from PDGF-B and PDGFRβ Knockouts

PDGF-B and PDGFRβ null mice both die at late gestation from microcirculatory deficiency involving generalized microvascular hemorrhage and edema [4, 5]. Closer analysis revealed that these mice fail to recruit pericytes to their microvasculature in many but not all organs [6, 7]. The acute cause of death of the PDGF-B and PDGFRβ null mice is likely circulatory failure. Live born mutants die from respiratory distress within minutes to hours after birth.

CNS Pericytes

The central nervous system (CNS) microvasculature has a remarkably high density of pericytes in comparison with most other capillary beds. The CNS is also one of the organs most strongly affected by genetic PDGF-B or PDGFRβ deficiency. CNS vessels develop by angiogenic sprouting from the perineural vascular plexus. In this process, the invading endothelial cells express PDGF-B and the pericytes PDGFRβ. In the absence of either ligand or receptor, pericyte co-recruitment largely fails, resulting in <5% of the normal number of brain pericytes. In spite of this cellular deficiency, the endothelial sprouting process proceeds relatively normally, establishing an intracerebral circulation that allows prenatal CNS growth and differentiation. At early stages (E10–14), only mild abnormalities in vessel patterns are observed in associated with a 2-fold increase in endothelial cell number [8]. However, at late gestation, the microvasculature displays more advanced morphological aberrations. Focal dilations and microaneurysms rapidly expand and rupture, leading to multiple sites of hemorrhage and edema. At the cellular level, this correlates with endothelial hyperplasia and abnormal differentiation. Notably, the endothelial luminal membrane surface is oversized and multiple cytoplasmic folds protrude into the vascular lumen. There are also abnormalities in the endothelial adherence junctions and multiple signs of increased vascular permeability. These observations suggest that PDGF-B signaling through PDGFRβ is critical for pericyte recruitment to the CNS microvasculature. Moreover, they show that pericytes, in turn, deliver important signals within the microvessel wall that regulate endothelial proliferation and differentiation. At the level of integrated functions, pericytes thereby support microvascular integrity and homeostasis. At a gross anatomical level, pericytes appear to exert control of the capillary diameter.

Mesangial Cell Recruitment

Mesangial cells are pericyte-like cells of the kidney glomerulus. PDGF-B and PDGFRβ null embryos both fail to recruit pericytes into the developing glomerulus [4, 5, 9]. As a result, the single capillary loop that invades the nascent glomerular epithelial cleft fails to develop into a complex tuft of vessels. Instead, it dilates into an aneurysm-like structure filling out the glomerular space. During glomerulus development, PDGF-B is expressed in the invading endothelium, whereas the mesangial cells express PDGFRβ. A paracrine PDGF-B signal from the endothelium thus promotes mesangial cell co-recruitment into the glomerulus, and the mesangial cells in turn deliver signals that promote glomerular capillary branching and splitting (intussusception). The role of the mesangial cells thus differ somewhat from the brain pericytes, since the overall vessel branching pattern in the brain does not seem to be influenced by the pericytes. However, there are also similarities in that the pericyte/mesangial cell absence in both organs leads to loss of capillary diameter control.

Placenta Pericytes

Placenta pericytes, like glomerular mesangial cells, appear to function in intussusceptive vessel splitting, leading to the formation of complex high-density vascular tufts. Possibly, such specialized pericytes are of particular importance where high vascular densities are needed to fulfill systemic functions (filtration, excretion, absorption etc). Both the kidney and the placenta exert such functions, and the vessels densities are therefore tailored to serve demands of the whole organism, rather than local needs. PDGF-B and PDGFRβ null mice show about 50% reduction in the number of placenta pericytes, and an associated impairment of the formation of fetal vessel tufts in the labyrinthine layer of the placenta (where fetal and maternal vessels meet) [10].

Other Locations

Other organs that show a marked (>50%) pericyte deficiency in PDGF-B or PDGFRβ null embryos include skin, heart and lung. Interestingly, the major population of liver pericytes, the perisinusoidal cells (Ito cells), is unaffected by PDGF-B or PDGFRβ deficiency, and therefore has to develop through other mechanisms. Ito cells also do not express appreciable levels of PDGFRβ (mRNA), at least prenatally, and the sinuisoidal endothelium does not detectably express PDGF-B. Also the axial arteries constitute a site at which the number of vSMC/pericytes appears to be unaffected by PDGF-B or PDGFRβ deficiency. The origin of these vSMC is believed to be the immature mesenchyme surrounding these vessels. Upon an *inductive* cue, presumably presented by the endothelial cells, the mesenchymal cells condensate around the vessel and turn on a SMC differentiation program. This *circumferential* recruitment of vSMC/pericytes is thus independent of PDGF-B/Rβ, whereas the subsequent *longitudinal* recruitment to new vessels formed by angiogenic sprouting requires paracrine endothelial-to-pericyte signaling using PDGF-B and PDGFRβ. Although dispensable for the vSMC *induction* around the axial arteries, PDGFRβ is required for the postnatal maintenance and/or expansion of this vSMC population, as shown by chimeric analysis [11].

Lessons from Conditional PDGF-B Knockouts

Using Cre-lox techniques, vascular endothelium- and neuronal-specific PDGF-B knockouts have been generated [12, 13]. The latter did not reveal observable defects in the brain microvasculature in spite of neurons being a major source of PDGF-B in the CNS. However, the endothelium-specific knockout resulted in pericyte deficiency, confirming the endothelium as the critical source of PDGF-B in pericyte recruitment [12, 14].

Evidence for the Local Importance of Pericytes

The endothelial PDGF-B gene deletion was not complete. Instead, it resulted in a chimeric situation where wild-type vessel segments alternated with PDGF-B deficient segments. In contrast to the PDGF-B null mutants, the endothelium-specific knockouts survived into adulthood, allowing analysis of postnatal angiogenesis. The chimeric nature of the mutant made it possible to spatially correlate pericyte deficiency with microvascular abnormality. This was important, as the role of systemic influences (e.g. heart and placenta defects) on microvessels formation could not be ruled out in the studies of the null mutants. It could therefore be concluded that pericytes control the function of neighboring endothelial cells. Possibly this control involves paracrine growth factors, cell adhesion and junction molecules, and pericyte-derived extracellular matrix molecules.

Retinal Angiogenesis

The survival of the endothelium-specific knockout made it possible to analyze retinal angiogenesis, an early postnatal process. As in other parts of the CNS, pericyte recruitment in the retina was shown to be dependent on endothelial PDGF-B. Studies using neutralizing PDGFRβ antibodies confirmed this conclusion [15].

PERICYTES AND DIABETIC RETINOPATHY

The chimeric nature of the endothelial PDGF-B deletion was inter-individually variable, allowing for analyses of the retinal vascular abnormalities at different states of pericyte density ranging from almost normal (≈70% of normal) to near complete deficiency (<10% of normal). Animals where the average pericyte density in the CNS was lower than 50% invariably developed proliferative lesions in the retinas that were reminiscent of diabetic retinopathy. This result was striking since diabetic retinopathy in both humans and animals has been associated with pericyte loss within the microvessel wall. It is not known, however, whether the pericyte loss constitutes a causal event in the pathogenesis of the retinopathy, or whether it is just correlative with the process. The results from the endothelium-specific PDGF-B knockouts show that independently of a diabetic challenge, pericyte deficiency may cause progressive vascular changes that mimic diabetic retinopathy. The data may also provide an explanation as to why diabetic rodents only develop mild signs of retinopathy and never proliferative disease; pericyte loss in these models is usually substantially lower that 50%.

Lessons from PDGF-B Retention Motif Knockouts

PDGF-B carries a basic stretch of amino acids in its C-terminus referred to as the "retention motif". A similar motif is found in PDGF-A and in different members of the VEGF family, where it is included or excluded by alternative splicing. In PDGF-B the retention motif may remain or be deleted from the precursor depending on proteolytic processing. The basic motifs of the PDGF-B/VEGF family have affinity for heparin and heparan sulphate proteoglycans (HSPGs) and probably serve the same function as analogous motifs in other growth factor families, namely to help generating gradients or depots of factor in the extracellular space. This may ensure graded signaling, or sequestering of the factor in the ECM for release or redistribution by proteolytic cleavage in situations of inflammation.

PDGF-B RETENTION IS REQUIRED FOR PROPER INTEGRATION OF PERICYTES IN THE VESSEL WALL

Deletion of the retention motif in PDGF-B resulted in viable mice (PDGF-B ret/ret mice), however, with severe retinopathy [16]. Other signs of microvascular dysfunction were evident as well, including brain microaneurysms and breakage of the blood brain barrier, as well as glomerular defects and proteinuria. All these problems appear to relate to abnormal pericyte and mesangial cell recruitment. Although there was a significant reduction in the number of pericytes in these mice, the degree of reduction did not seem to explain the severity of the phenotype; endothelium-specific knockouts with a similar decrease in pericyte numbers had much milder retinal and glomerular pathology. The severity of the PDGF-B ret/ret mutants was instead proposed to result from abnormal pericyte integration within the microvessel wall. Notably, pericytes are normally very tightly associated with the abluminal endothelial cell surface. In PDGF-B ret/ret mutants, pericytes were instead partially or fully detached from the endothelial cells. Likely this abrogates the normal communication between the pericytes and the endothelium, leading to a state that resembles the absence of pericytes. In the eye, the dissociated pericytes also form sheets of fibroblast-like cells at the retinal surface, which may contribute to contraction and detachment of the retina.

PDGF-B AND THE RECRUITMENT OF TUMOR PERICYTES

There are several possible explanations for the detachment of pericytes from the microvessels in PDGF-B ret/ret mice. One possibility is that a graded presentation of PDGF-B is required from the endothelium in order to guide pericyte migration along the abluminal endothelial surface. If the pericyte fails to recognize that the source of PDGF-B is endothelial, they may migrate more randomly, leading eventually to their detachment from the endothelial surface. Another possibility is that a directional presentation of PDGF (i.e. from the endothelium) is needed to polarize the pericyte, forcing unequal distribution of cell and matrix adhesions molecules, or deposition of matrix itself. Such polarization may promote adherence of the pericyte to the endothelium and/or ensure that the pericyte is properly embedded within the microvascular basement membrane. These possibilities, which are not mutually exclusive, are attractive also considering the uneven expression of PDGF-B in the vascular endothelium. The recent finding that the expression is concentrated to specialized endothelial cells situated at the sprout tip (endothelial tip-cells) [17] suggests that the location of the PDGF-B source is critical.

Studies of pericyte recruitment to tumor vessels have provided additional strong support for the idea that directional presentation of PDGF-B is important in the process of pericyte recruitment. In tumors transplanted on PDGF-B ret/ret mice, few pericytes were recruited, and they were partially or fully detached from the endothelium [18]. However, when PDGF *levels* were compensated by secretion from the tumor cells rather than the endothelium, the number of pericytes in the tumor vessels increased significantly, but the integration of the pericytes in the vascular wall failed. Since the tumors produced the wildtype form of PDGF-B (including the retention motif) it could also be excluded that the integration defect depended on a putative change in the signaling properties of the mutant PDGF-B. Instead, the most plausible explanation is that the PDGF-B protein has to be presented directionally from the endothelium, and that the retention motif helps associating the secreted PDGF-B protein to the endothelial surface or within the periendothelial matrix to facilitate such presentation.

Lessons from PDGFRβ Signaling Mutants

Using homologous recombination in ES-cells, Soriano and collaborators have generated an extensive allelic series of endogenous signaling mutations in PDGFRβ. Mice carrying these mutations show a different range of cardiovascular defects. Taken together, these analyses confirm the importance of PDGFRβ signaling in vSMC and pericyte recruitment and in the development of a functional cardiovascular and renal excretory system. However, the analyses also came up with several surprises concerning the relative importance of the different signaling pathways downstream of PDGFRβ.

PDGFRβ Signaling and Cell Function in Vitro

PDGFRβ has been the subject of intense biochemical analysis. It is extensively autophosphorylated upon ligand binding, leading to the formation of up to thirteen phosphotyrosine residues that engage in coupling to at least ten different SH2-domain containing proteins. These molecules include Src family kinases, PI3 kinase (PI3K), Shc, RasGAP, STATs, Grb2, Grb7, SHP-2, PLCγ, and Nck. Their engagement leads to the initiation of several different cellular functions in cultured mesenchymal cells, including proliferation, migration, and extracellular matrix synthesis. In vitro, a certain degree of specificity has been noticed between the signaling pathway engaged and the cellular response. For example, PI3K signaling has been strongly linked to cytoskeletal rearrangements and cell migration. Since analysis of the PDGF-B and PDGFRβ null mutants suggests a role for these proteins in pericyte migration and proliferation, it was hoped that the distinct signaling mutations in the endogenous PDGFRβ gene would separate between the putative multiple roles of PDGFRβ signaling in pericyte/vSMC recruitment. However, as discussed in more detail below, it appears that the various signaling pathways engaged downstream of PDGFRβ exert additive rather than distinctive effects in microvascular development.

PDGFRβ Unable to Signal Through PI3K and PLCγ

Surprisingly, mice homozygous for a mutation that encodes a PDGFRβ protein lacking the two PI3K-binding phosphotyrosine residues displayed no overt phenotype [19]. When challenged, however, they display a reduction in the capacity to resolve an experimentally induced tissue edema. Thus, PDGFRβ signaling may be implicated in the control of tissue interstitial fluid pressure (IFP). Increased IFP is a general feature of solid tumors and PDGF-B or PDGFRβ inhibiting compounds seem to be able to reduce tumor IFP, which in turn facilitates the delivery of substances, including cytostatic drugs, to the tumor [20]. The mechanism by which PDGFRβ regulates IFP is not clear, however it is plausible to assume that it involves pericyte or fibroblasts mediated contraction of the ECM. Hence, PDGFRβ-mediated PI3K signaling may have its major role in cell contraction in the adult tissues, rather than in cell migration in embryonic development. Recent analysis using embryonic CNS pericyte density as readout also demonstrates that PDGFRβ-mediated PI3K signaling takes part in determining the rate of expansion of the pericyte population [21, 22]. Hence PI3K signaling may also be involved in the control of pericyte progenitor proliferation in vivo.

Mice carrying mutant PDGFRβ unable to signal through both PI3K and PLCγ are likewise without an overtly abnormal phenotype [23]. However, when challenged with an experimental injury to the glomerulus, they display exaggerated pathology. Although it is unclear how the change in pathogenesis relates to defective PDGFRβ signaling, it likely involves recruitment or function of mesangial cells, since these are the only cells in the developing glomerulus known to express PDGFRβ. Since the overall growth rate of the pericyte population is decreased in the combined PI3K/PLCγ mutants, one might speculate that the proliferation of the mesangial cells is likewise negatively affected, and that this may slow down repair of the injured glomerulus and, in turn, enhance a pathological response.

PDGFRβ With Multiple Pathways Defects

Additional mutations removing 5 and 7 phosphotyrosine residues, respectively, in PDGFRβ led to overt vascular defects in several organs, in particular when combined with a PDGFRβ null allele to reduce the level of PDGFRβ expression [21, 22]. These phenotypic abnormalities coincided with a substantial (>50%) reduction in pericyte numbers. The PDGFRβ allelic series generated by Soriano and co-workers hence provides a series of mutant animals with a progressive decrease in the pericyte density. This decrease is even further aggravated in animals where the mutant alleles occur in combination with the PDGFRβ null allele.

A mutant has also been created in which the intracellular domain of PDGFRβ was replaced with that of PDGFRα [24]. While homozygous carriers of this mutation lack overt phenotype, mice carrying one hybrid receptor and one null allele, showed spontaneous pathology similar to that observed in the mutant lacking 7 phosphotyrosines. This demonstrates that PDGFRα signaling can only partially compensate for PDGFRb signaling in vSMC/pericytes, in agreement with biochemical analysis.

Pericyte Density and Its Relationship to Organ Pathology—Concluding Remarks

Analogous Phenotypic Consequences of PDGF-B and PDGFRβ Mutagenesis

Several interesting conclusions may be drawn from comparing the various mouse lines that have been generated by mutagenesis at the PDGF-B or PDGFRβ loci. Not surprisingly, perhaps, the range of phenotypes that is seen in the two mutant series is similar. Both the ligand and receptor series display a gradual reduction in pericyte density, and a phenotypic outcome ranging from normal (no phenotype in the unchallenged state) to overt pathology in the eye, heart and kidney, with the eye displaying the most profound pathology. Interestingly, the relationship between overall degree of pericyte reduction and the resulting pathology was also similar. Both series produced mutants with pericyte reduction ranging from <50 to >90%, and overt phenotypic changes in the eye appear in mutants with more than 50% reduction in the pericyte density. Thus, substantial pericyte reduction appears to be tolerated in organ development, with most organs (in which pericytes depend on PDGF-B/PDGFRβ signaling) tolerating 50% reduction or more. The exception is the eye, which has the highest known normal pericyte density, suggestive of a critical and perhaps unique function for these cells. Intriguingly, development of many organs, including the brain, can be completed relatively normally, even when there is up to 90% reduction in pericyte density. However, these mice have many signs of brain microvessel dysfunction and pathological responses in the astroglia. When mice with 50% reduction or less are exposed to a pathological challenge, such as glomerulonephritis or diabetes, there is an exaggerated pathological microvascular response, however.

Not only pericyte numbers, but also their association with the endothelium seems to matter, as illustrated by the PDGF-B retention deficient mice. Probably, a graded, or directional, presentation of PDGF-B by the endothelial cells is required to ensure that the pericytes become correctly associated with the endothelial cells and their basement membrane. Thus the extracellular compartmentalization of PDGF-B seems to be as important for pericyte recruitment as the formation of gradients or depots of VEGF-A in the guidance of endothelial sprouts [17, 25].

Do the Other PDGF/Receptors Play a Role in the Microvasculature?

No other PDGF/PDGFR member is as strongly implicated in microvascular assembly or function as PDGF-B and PDGFRβ. PDGFRα knockouts display defects in the cardiac outflow tract [26, 27] that may relate to a primary defect in the neural crest. Likewise, other neural crest-derived populations of vSMC and pericytes may be affected in the PDGFRα knockouts, particularly in the rostral part of the embryo. Again the critical signaling events involving PDGFRα likely take place before the crest-derived vSMC/pericyte populations are specified. An additional location where PDGFRα signaling might determine the fate of a progenitor of the vSMC/pericyte population is the kidney. Here, PDGFRα null mutants become progressively depleted of interstitial mesenchymal cells. However, these mutants also partially lack mesangial cells in the kidney glomeruli in spite of the fact that PDGFRβ is the dominant receptor in these cells. Thus, it is possible that the mesangial progenitors are in part (or fully) recruited from the PDGFRα positive interstitial mesenchyme [28].

While PDGF-A and -C signaling through PDGFRα may have indirect effects on microvascular development, the function of PDGF-D remains elusive. The strong similarities between PDGF-B and PDGFRβ null mutants may suggest that PDGF-D has no or little effect during development. However, it is possible that PDGF-D serves mainly postnatal functions, and that some of these functions may involve the microvasculature. Detailed expression analysis and genetic ablation of PDGF-D in mice will help answering these questions.

Glossary

PDGF: Platelet-derived growth factor. A family of five dimeric ligands (AA, BB, CC, DD and AB) composed of four constituent chains (A, B, C, D) encoded by different genes. PDGF was originally found in platelets and assumed to exert a wound healing function, but is currently know to be expressed by a multitude of cell types and exert numerous important roles, particularly during organogenesis.

PDGFR: Platelet-derived growth factor receptor. A family of two transmembrane receptors (α and β). They are structurally related receptor tyrosine kinases with 5 extracellular Ig domains and a split intracellular kinase domain. The PDGFRs dimerize upon ligand binding, autophosphorylated, and thereby engage several classical signaling pathways, including ••.

Pericyte: A mural cells of the vessel wall that shares basal lamina with the endothelium. Pericytes are vascular smooth muscle lineage cells that express several of the markers of smooth muscle.

References

1. Ross, R., Glomset, J. A., Kariya, B., and Harker, L. (1974). A platelet-dependent serum factor that stimulates the proliferation of arterial smooth muscle cells in vitro. *Proc. Natl. Acad. Sci. USA* **71**, 1207–1210.
2. Hirschi, K. K., Rohovsky, S. A., and D'Amore, P. A. (1998). PDGF, TGF-beta, and heterotypic cell-cell interactions mediate endothelial cell-induced recruitment of 10T1/2 cells and their differentiation to a smooth muscle fate. *J. Cell. Biol.* **141**, 805–814.
3. Hirschi, K. K., Rohovsky, S. A., Beck, L. H., Smith, S. R., and D'Amore, P. A. (1999). Endothelial cells modulate the proliferation of mural cell precursors via platelet-derived growth factor-BB and heterotypic cell contact. *Circ. Res.* **84**, 298–305.
4. Levéen, P., Pekny, M., Gebre-Medhin, S., Swolin, B., Larsson, E., and Betsholtz, C. (1994). Mice deficient for PDGF B show renal, cardiovascular, and hematological abnormalities. *Genes Dev.* **8**, 1875–1887.
5. Soriano, P. (1994). Abnormal kidney development and hematological disorders in PDGF beta-receptor mutant mice. *Genes Dev.* **8**, 1888–1896.
6. Lindahl, P., Johansson, B. R., Levéen, P., and Betsholtz, C. (1997). Pericyte loss and microaneurysm formation in PDGF-B-deficient mice. *Science* **277**, 242–245.
7. Hellström, M., Kalén, M., Lindahl, P., Abramsson, A., and Betsholtz, C. (1999). Role of PDGF-B and PDGFR-beta in recruitment of vascular smooth muscle cells and pericytes during embryonic blood vessel formation in the mouse. *Development* **126**, 3047–3055.
8. Hellström, M., Gerhardt, H., Kalén, M., Li, X., Eriksson, U., Wolburg, H., and Betsholtz, C. (2001). Lack of pericytes leads to endothelial hyperplasia and abnormal vascular morphogenesis. *J. Cell. Biol.* **153**, 543–553.
9. Lindahl, P., Hellstrom, M., Kalen, M., Karlsson, L., Pekny, M., Pekna, M., Soriano, P., and Betsholtz, C. (1998). Paracrine PDGF-B/PDGF-Rbeta signaling controls mesangial cell development in kidney glomeruli. *Development* **125**, 3313–3322.
10. Ohlsson, R., Falck, P., Hellstrom, M., Lindahl, P., Bostrom, H., Franklin, G., Ahrlund-Richter, L., Pollard, J., Soriano, P., and Betsholtz, C. (1999). PDGFB regulates the development of the labyrinthine layer of the mouse fetal placenta. *Dev. Biol.* **212**, 124–136.

11. Crosby, J. R., Seifert, R. A., Soriano, P., and Bowen-Pope, D. F. (1998). Chimaeric analysis reveals role of Pdgf receptors in all muscle lineages. *Nat. Genet.* **18**, 385–388.
12. Enge, M., Bjarnegård, M., Gerhardt, H., Gustafsson, E., Kalén, M., Asker, N., Hammes, H.-P., Shani, M., Fässler, R., and Betsholtz, C. (2002). Endothelium-specific platelet-derived growth factor-B ablation mimics diabetic retinopathy. *EMBO J.* **21**, 4307–4316.
13. Enge, M., Wilhelmsson, U., Abramsson, A., Stakeberg, J., Kühn, R., Betsholtz, C., and Pekny, M. (2003). Neuron-specific ablation of PDGF-B is compatible with normal central nervous system development and astroglial response to injury. *Neurochem. Res.* **28**, 271–279.
14. Bjarnegård, M., Enge, M., Norlin, J., Gustafsdottir, S. M., Fredriksson, S., Abramsson, A., Takemoto, M., Gustafsson, E., Fässler, R., and Betsholtz, C. (2004). Endothelium-specific ablation of PDGF-B leads to pericyte loss and glomerular, cardiac and placental abnormalities. *Development* **131**, 1847–1857.
15. Uemura, A., Ogawa, M., Hirashima, M., Fujiwara, T., Koyama, S., Takagi, H., Honda, Y., Wiegand, S. J., Yancopoulos, G. D., and Nishikawa, S. (2002). Recombinant angiopoietin-1 restores higher-order architecture of growing blood vessels in mice in the absence of mural cells. *J. Clin. Invest.* **110**, 1619–1628.
16. Lindblom, P., Gerhardt, H., Liebner, S., Abramsson, A., Enge, M., Hellström, M., Bäckström, G., Fredriksson, S., Landegren, U., Nyström, H., Bergström, G., Dejana, E., Östman, A., Lindahl, P., and Betsholtz, C. (2003). Endothelial PDGF-B retention is required for proper investment of pericytes in the microvessel wall. *Genes Dev.* **17**, 1835–1840.
17. Gerhardt, H., Golding, M., Fruttiger, M., Ruhrberg, C., Lundkvist, A., Abramsson, A., Jeltsch, M., Mitchell, C., Alitalo, K., Shima, D., and Betsholtz, C. (2003). VEGF guides angiogenic sprouting utilizing endothelial tip-cell filopodia. *J. Cell. Biol.* **161**, 1163–1177.
18. Abramsson, A., Lindblom, P., and Betsholtz, C. (2003). Endothelial and non-endothelial sources of PDGF-B regulate pericyte recruitment and influence vascular pattern formation in tumors. *J. Clin. Invest.* **112**, 1142–1151.
19. Heuchel, R., Berg, A., Tallquist, M., Ahlen, K., Reed, R. K., Rubin, K., Claesson-Welsh, L., Heldin, C.-H., and Soriano, P. (1999). Platelet-derived growth factor receptor regulates interstitial fluid homeostasis through phsophatidylinositol-3′ kinase signaling. *Proc. Natl. Acad. Sci. U S A* **96**, 11410–11415.
20. Pietras, K., Rubin, K., Sjoblom, T., Buchdunger, E., Sjoquist, M., Heldin, C.-H., and Östman, A. (2002). Inhibition of PDGF receptor signaling in tumor stroma enhances antitumor effect of chemotherapy. *Cancer Res.* **62**, 5476–5484.
21. Hoch, R. V., and Soriano, P. (2003). Roles of PDGF signaling in animal development. *Development* **130**, 4769–4784.
22. Tallquist, M. D., French, W. J., and Soriano, P. (2003). Additive effects of PDGF receptor beta signaling pathways in vascular smooth muscle cell development. *PLoS Biol.* **1**, E52.
23. Tallquist, M. D., Klinghoffer, R. A., Heuchel, R., Meuting-Nelsen, P. F., Corrin, P. D., Heldin, C.-H., Johnson, R. J., and Soriano, P. (2000). Retention of PDGFR-beta function in mice in the absence of phosphatidylinositol 3′-kinase and phospholipase C signaling pathways. *Genes Dev.* **14**, 3179–3190.
24. Klinghoffer, R. A., Mueting-Nelsen, P. F., Faerman, A., Shani, M., and Soriano, P. (2001). The two PDGF receptors maintain conserved signaling in vivo despite divergent embryological functions. *Mol. Cell* **7**, 343–354.
25. Ruhrberg, C., Gerhardt, H., Golding, M., Watson, R., Ioannidou, S., Fujisawa, H., Betsholtz, C., and Shima, D. (2002). Spatially restricted patterning cues provided by heparin-binding VEGF-A control blood vessel branching morphogenesis. *Genes Dev.* **16**, 2684–2698.
26. Soriano, P. (1997). The PDGF alpha receptor is required for neural crest cell survival and for normal patterning of the somites. *Development* **124**, 2691–2700
27. Morrison-Graham, K., Schatteman, G. C., Bork, T., Bowen-Pope, D. F., and Weston, J. A. (1992). A PDGF receptor mutation in the mouse (patch) perturbs the development of a non-neuronal subset of neural crest-derived cells. *Development* **115**, 133–142.
28. Li, X., Ponten, A., Aase, K., Karlsson, L., Abramsson, A., Uutela, M., Backstrom, G., Hellstrom, M., Bostrom, H., Li, H., Soriano, P., Betsholtz, C., Heldin, C. H., Alitalo, K., Ostman, A., and Eriksson, U. (2000). PDGF-C is a new protease-activated ligand for the PDGF alpha-receptor [see comments]. *Nat. Cell Biol.* **2**, 302–309.

Further Reading

Hoch, R. V., and Soriano, P. (2003). Roles of PDGF signaling in animal development. *Development* **130**, 4769–4784.

Betsholtz, C. (2004). Insight into the physiological functions of PDGF through genetic studies in mice. *Cytok Growth Fact Rev* **15**, 215–228.

Capsule Biography

Dr Christer Betsholtz is professor of Medical Biochemistry at Göteborg University since 1993. His laboratory has explored the physiological roles of the PDGF family members for more than 20 years, which has recently led into the areas of angiogenesis and, in particular, the mechanism of recruitment and function of pericytes. His work is supported by several research councils and foundations in Sweden and Europe.

Dr Holger Gerhardt is a former post doctoral fellow in Betsholtz' group and newly appointed junior group leader at Cancer Research UK, London. His previous work on avian neural development, together with a more recent focus on angiogenesis, has led him into new exciting areas of angiogenic sprout guidance in the central nervous system. He is supported by the Cancer Research UK.

CHAPTER 15

Basic Concepts of Intussusceptive Angiogenesis

Valentin Djonov and Peter H. Burri
Institute of Anatomy, University of Berne, Switzerland

Sprouting and Intussusceptive Angiogenesis

Angiogenesis is a fundamental process of microvascular growth and expansion implicated in different physiological and pathological processes. It can occur by two main mechanisms: sprouting and intussusception. Although both mechanisms lead to creation of new vascular segments, they involve different morphogenic mechanisms and probably are regulated by different factors.

Sprouting angiogenesis is characterized by local vasodilation, followed by increased vascular permeability. After proteolytic degradation of the basement membrane, the endothelial cells migrate into the extracellular matrix and proliferate. After internal reorganization, the sprouts form a vascular lumen and get connected to other capillary segments. Sprouting is a relatively sluggish process. At least three to five days are needed before the newly formed capillary loop is perfused and integrated into the vascular system.

In intussusceptive angiogenesis, the capillary system expands and grows by the insertion of new capillary meshes into the existing network.

The Basic Concept of Intussusception

Nonsprouting angiogenesis by intussusception was first described in the capillary bed of neonatal rat lungs by Caduff and coworkers [1]. Analyzing scanning electron micrographs of lung vascular corrosion casts, the authors detected a multitude of tiny holes with diameters up to 2 μm. Serial sectioning of lung tissue followed by transmission electron microscopy revealed these structures to correspond to transluminal tissue pillars [1, 2]. Four consecutive steps for pillar formation have been postulated:

- *Phase I:* Simultaneous protrusions of opposite capillary walls into the vessel lumen lead to the creation of an interendothelial contact zone across the capillary lumen.
- *Phase II:* Reorganization of the interendothelial cell junctions followed by central perforation of the sandwich-like structure composed of two basal membranes and two endothelial leaflets.
- *Phase III:* Formation of the interstitial pillar core by invasion of cell processes of pericytes and myofibroblasts; appearance of collagen fibrils. During these first three phases, the transluminal pillars are still very slender, mostly with diameters less than 2.5 μm.
- *Phase IV:* The pillars increase in girth by further deposition of material and cells, but without changes in their basic structure.

The concept of intussusception is schematically represented in Figure 1.

In the meantime, it could be demonstrated that transcapillary tissue pillars could be produced by several alternative mechanisms to the one described above. For details, the reader is referred to a recent review [3].

To describe the new angiogenic process, the authors coined the term *intussusception* to convey the meaning that creation of new capillary segments occurs by insertion of transcapillary pillars (i.e., by growth "within itself"). They postulated that the microvasculature of the developing lung was predominantly growing by intussusception, a phenomenon that had not been described hitherto and termed it intussusceptive microvascular growth (IMG) [1, 2].

The complex spatial structure of transluminal pillars, their small size, and the need of three-dimensional analysis

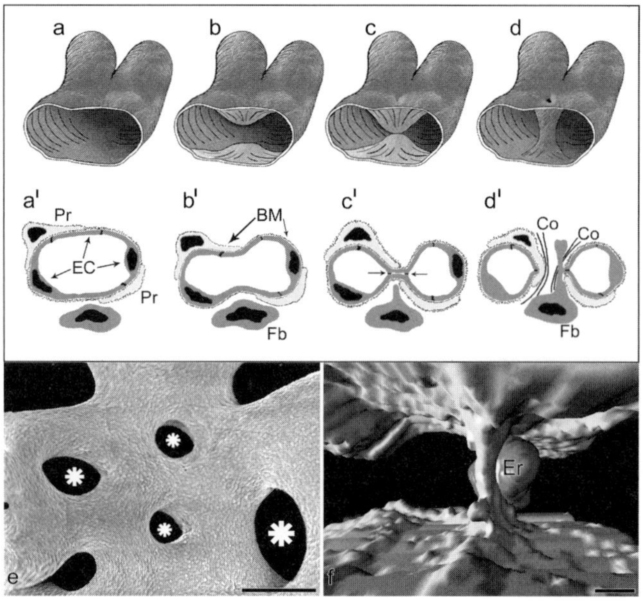

Figure 1 Three-dimensional **(a–d)** and two-dimensional **(a′–d′)** drawings illustrating intussusceptive pillar formation. The process begins with the protrusion of opposite capillary walls into the vessel lumen **(a, b, a′, b′)**. After establishing an interendothelial contact **(c, c′)**, the endothelial bilayer and the basement membranes (BM) are perforated centrally. The newly formed pillar increases in girth after being invaded by pericytes (Pr) and fibroblasts (Fb), which produce collagen fibrils (Co) **(d, d′)**. Corresponding three-dimensional visualization of transluminal pillars in capillary plexus of chicken CAM. In corrosion casts of blood vessels, newly formed transcapillary pillars appear as small holes (*asterisks*) **(e)**; three-dimensional reconstruction of a pillar based on transmission electron microscopic analysis of serial sections **(f)**. An erythrocyte (Er) is seen behind the pillar. Bars: e and f = 5 μm. (a–d′). Reproduced from Kurz et al. 2003 [8] and (e, f) from Djonov et al. 2000 [4] with the authors' and publisher's permission. (see color insert)

for their visualization may explain why intussusception was overlooked in the past. The only reliable methods to recognize transluminal pillars safely are vascular corrosion casting visualized by scanning electron microscopy, serial sectioning for light or transmission electron microscopy followed by three-dimensional reconstruction (Figure 1) or, alternatively, confocal laser microscopy [4–7]. Other methods, such as nuclear magnetic resonance, microcomputer tomography, angiography, and ultrasonics even adapted for small objects do not have the resolution (below 1 μm) needed for visualizing pillars.

Hallmarks of Intussusception

Intussusception is a widespread phenomenon that occurs in the vascular systems of all species and organs investigated so far. Some essential characteristics, which distinguish it from sprouting, are discussed in the following sections.

Intussusception Can Occur Only in a Perfused Vascular Bed

All commonly used in vitro and in vivo experimental tools, such as three-dimensional collagen gels, corneal implants, tumor implantation, wound healing, and embryonic grafting, will originally induce capillary sprouting during early tissue neovascularization. Intussusception would not be expected to occur initially in these models because (1) the facts that the tissue has first to be "colonized" by endothelial cells and (2) the absence of blood flow. Our own investigations indicated that hemodynamic factors are essential players in regulatory mechanisms of intussusception [7]. This could be an additional reason why intussusception has not been recognized until recently.

Blood Vessel Segments Are Generated More Rapidly by Intussusception than by Sprouting

As mentioned previously, sprouting is a relatively sluggish process, requiring a few days for initiation, implementation, and finally integration of the newly formed segments into the vascular system. Using improved in vivo monitoring combined with histological and ultrastructural analyses of serial tissue sections, pillar formation has been demonstrated to require a period of four to five hours for completion. Under conditions of artificially accelerated blood flow, this period was reduced to one hour only [4, 5, 7]. This swiftness could explain how the capillary network of the eye choroid or of the chorio-allantoic membrane (CAM) of the chicken could increase 25 and 100 times, respectively, in disk surface just within a few days. The augmentation of the vascular exchange surface area and vascular volume was much higher even. Similarly, in the lung vasculature during the early postnatal period, the capillary volume and surface area increase thirty-fivefold and twentyfold, respectively [4, 5, 7–9].

Intussusceptive Microvascular Growth Is Achieved with a Low Rate of Endothelial Proliferation

Our own morphometric investigations on the lung and CAM, as well as data from the literature, indicate that the intussusceptive process is associated with low endothelial proliferation. This means that capillary beds growing by intussuception achieve their rapid expansion by redistributing the preexisting total endothelial cell volume by thinning and spreading of the cytoplasmic leaflets. This was confirmed by our comparative studies of different organs before and after the onset of intussusception, as well as by reports in the literature [3–5, 8, 9].

Capillaries Formed by Intussusception Are Less Leaky

Intussusception is characterized by physiological levels of vascular transpermeability. This is an essential condition for uncompromised tissue and organ function. This important benefit could be the reason why after the formation of the primary capillary plexus by vasculogenesis or sprouting, additional vascular growth and remodeling during both embryonic development and postnatal life occurs by intussusception. The low transpermeability could furthermore be an essential factor for the resistance to chemotherapy treatment in tumors growing predominantly by intussusception [6].

Intussusception Is Energetically and Metabolically More Cost Effective

In contrast to sprouting, massive cell proliferation, extensive basement membrane degradation, and invasion of the surrounding tissue is not required in intussusception. Intussusception represents a phenomenally simple angiogenic mechanism possessing a high morphogenic potential and plasticity, and which is simultaneously energetically and metabolically more economical. This would explain why intussusceptive microvascular growth is triggered immediately after the formation of the primitive capillary plexus by vasculogenesis or sprouting.

Three Facets of Intussusceptive Angiogenesis

The term *intussusceptive angiogenesis* circumscribes a host of processes that are involved in generation, growth, and final remodeling of vascular entities with diverse morphological and functional outcomes. Although chronologically sequential, the processes overlap both in space and time, affecting different components of the vascular bed.

Intussusceptive Microvascular Growth: Expansion of Capillary Plexuses

Intussusceptive microvascular growth (IMG) represents a widespread phenomenon, by which capillary plexuses expand by simultaneous scattered pillar formation and successive pillar enlargement. By these means, new microvascular segments arise without dramatic changes in the dimensions and functional characteristics of their components (Figure 2). IMG favors the rapid expansion of the primitive capillary plexuses, thus providing a large surface area for the exchange of oxygen, carbon dioxide, and nutrients. As mentioned previously, IMG was first observed in the growing postnatal lung [1] and then in the microvasculature of many other tissues and organs, both in normal and pathological microvascular growth, such as retina, kidney, CAM, liver, intestines, stomach, excretory glands, spleen, skeletal muscle, heart, brain, in a model of tissue repair, during the cyclic vascular changes in the endometrium, in cerebral vascularization after stroke, and during tumor angiogenesis. Intussusception is a widespread, practically ubiquitous phenomenon that occurs in the vascular systems

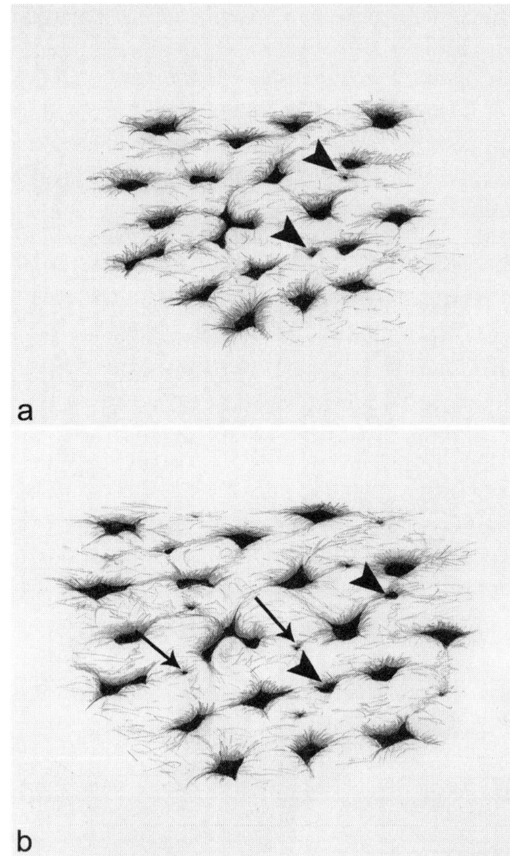

Figure 2 Drawing representing the process of intussusceptive microvascular growth (IMG). The capillary plexus expands from its initial stage in **(a)** to status **(b)** by insertion of new pillars (*arrows*) and by enlargement of already existing ones (arrowheads).

of all species thus far investigated, including rat, mice, rabbit, chick, fish, and human. Actually, in many older reports dealing with different angiogenic aspects, it is possible to recognize the well-documented signs of intussusception; they were, however, not recognized as a mechanism of angiogenesis. It is now evident that IMG represents a general and ubiquitous mechanism of capillary growth, by which the capillary beds of organs, which arise initially by sprouting and/or vasculogenesis, can undergo rapid expansion without any compromise in vascular physiology or function thanks to low vascular permeability and a low rate of endothelial cell proliferation [3, 6, 8, 9].

Intussusceptive Arborization: Formation of a Feeding Vascular Tree

Intussusceptive arborization (IAR) represents a mechanism whereby preferentially perfused segments of a capillary plexus can be transformed into terminal arterioles or collecting venules, patterning in this way a hierarchic vascular tree. IAR is initiated by the formation in the capillary bed of double rows of lined-up "vertical" tissue pillars, which demarcate future feeding vessels. These pillars undergo reshaping into narrow tissue septa that progressively fuse to delineate a new vascular entity. The remaining connecting bridges are "severed" by the formation of "horizontal" folds, the feeding vessels being thereby definitively separated from the capillary plexus. As a result of this process, a complex arterial and venous vascular tree arises to form a second layer of draining and feeding vessels (Figure 3).

IAR provides the vasculature with an important angio-adaptive mechanism. Any capillary plexus expansion is associated with an increase in the mean capillary path length between arteries and veins, with deleterious consequences for gas and nutrient exchange. The formation of new terminal supplying and draining branches is necessary, and this is achieved via IAR. The IAR process could be observed in all investigated organs and systems, but for easy demonstration it is mostly shown in nearly two-dimensional capillary beds, such as the chloroid of the eye and the chicken CAM. In these models, the de novo segregated arterial and venous vascular trees form an easily detectable second layer atop the capillary plexus [3–5, 8–9].

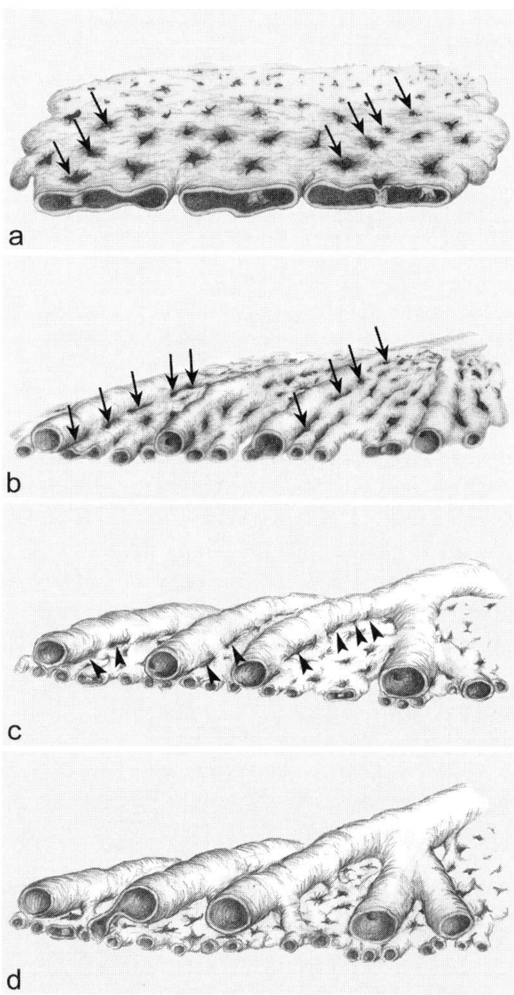

Figure 3 Drawing representing the process of intussusceptive arborization (IAR). Within a capillary plexus, a series of "vertical" pillars arise (*arrows in **a***), demarcating the pathway of future feeding vessels. These pillars undergo reshaping and fusion to form narrow tissue septa (*arrows in **b***). "Horizontal" pillars and folds are then formed (*arrowheads in **c***), which separate the feeding vessels from the capillary plexus (**d**). Adapted from Djonov et al. 2000 [5], with the authors' and publisher's permission.

Intussusceptive Branching Remodeling: Optimization of Branching Geometry and Vascular Pruning

Intussusceptive branching remodeling (IBR) refers to the process by which the branching geometry of supplying vessels is modified, thereby optimizing the pre- and post-capillary flow properties. IBR can also lead to the removal of putative supernumerary branches (vascular pruning), thereby optimizing the efficiency of the blood supply and the hierarchy of the vascular tree. Implementation of IBR is accomplished via transluminal pillars and folds, which are inserted close to the bifurcation sites of arteries and veins of up to 120 μm in diameter. These structures appear de novo and are capable of rapidly changing the vascular geometry and the hemodynamic properties at the affected branching points.

The pillars located close to bifurcation points enlarge (alternately elongate into flat longitudinal folds) until they merge with connective tissue in the branching angle (Figure 4). Thus, IBR narrows the branching angle by relocating the branching point more proximally. This may represent an important adaptive response to the continually increasing blood flow and blood pressure during embryogenesis and growth. Second, IBR optimizes the hemodynamic conditions at bifurcation sites by remodeling the diameter of one or both branches (mainly by "pillar augmentation"). Consequently, IBR yields a branching pattern that approximates the ideal hemodynamic condition predicted by Murray's Law of minimal power consumption and constant shear stress. Third, IBR is implemented in the process of vascular pruning by the successive asymmetric formation of pillars. The subtotal lumen obstruction of one of the daughter branches followed by reduction in blood flow probably contributes to the regression, retraction, and ultimate atrophy of the affected branch (Figure 4) [7–9].

From the mechanisms of IAR and IBR, detected more recently, it appears that the intussusceptive angiogenic principle represents an important morphogenetic machinery affecting all components of vascular beds: as the expansion of capillary beds (IMG), the segregation of small arteries and veins (IAR), and finally the optimization of the branching geometry and vascular pruning by IBR.

Sprouting and Intussusception: Two Complementary Angiogenic Mechanisms

Structural investigations have revealed that the vasculature of different organs of chicks and rats, such as CAM, lung, heart, endometrium, eye, kidney, and yolk sac, undergoes two main phases of development. The *early, sprouting phase* is characterized by the appearance of multiple capillary sprouts, which are responsible together with

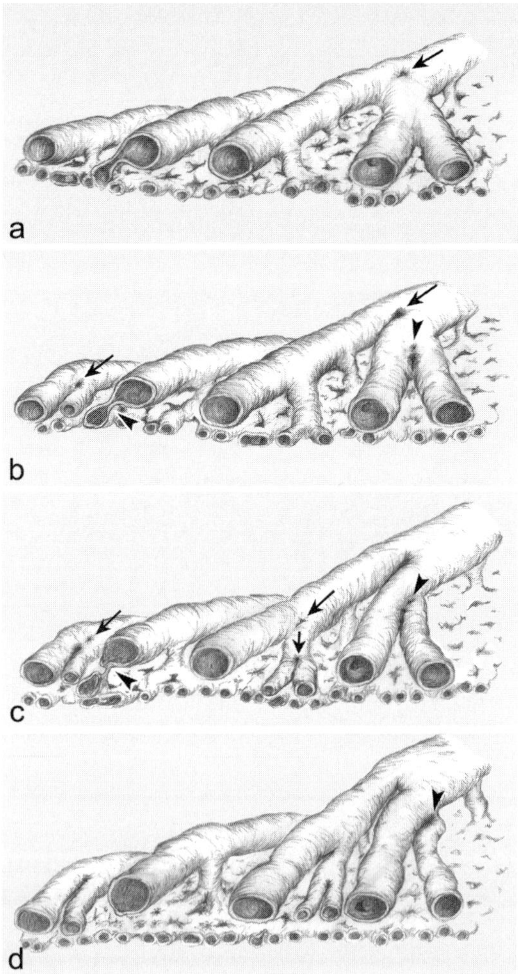

Figure 4 Drawing representing intussusceptive branching remodeling (IBR). The insertion of transluminal pillars at branching points is followed by their merging with the connective tissue in the branching angle (*arrows*). This process allows to adapt the depth of the branching angle and the diameters of the two daughter vessels. Moreover, intussusceptive branching remodeling leads to vascular pruning by repetitive eccentric formation, thickening, and fusion of pillars (*arrowheads*). Adapted from Djonov et al. 2000 [5], with the authors' and publisher's permission.

vasculogenesis for tissue neovascularization, forming the primary capillary plexuses. During the *second, intussusceptive phase*, capillary sprouting is superseded by pillar formation. Further vascular growth and remodeling occurs by intussusception as a result of advantages listed in the section on hallmarks of intussusception.

Summary

Intussusceptive angiogenesis is a general and ubiquitous formative process of the vasculature. The formation of tissue pillars across the vascular lumen has three distinct consequences depending on the pillar location. It results in rapid expansion of the capillary networks (= IMG), it implements vascular tree formation (= IAR), and/or it leads to dynamic adaptations and remodeling of the vascular bed (= IBR). It plays a crucial role in embryonic blood vessel formation, growth, tissue repair, and tumor angiogenesis.

Taking into consideration that numerous pro- and antiangiogenic agents act in complex vascular beds formed mainly by intussusception, it is evident that better understanding of the intussusceptive angiogenesis, and especially its regulatory mechanisms, will open the way for development of novel therapeutic strategies.

References

1. Caduff, J. H., Fischer, L. C., and Burri, P. H. (1986). Scanning electron microscope study of the developing microvasculature in the postnatal rat lung. *Anat. Rec.* **216**, 154–164. *In this study investigating casts of the lung microvasculature by scanning electron microscopy, the authors put forward the idea that a capillary network can grow by the insertion of slender transcapillary tissue pillars. They coined and explained the term of* intussusception.
2. Burri, P. H., and Tarek, M. R. (1990). A novel mechanism of capillary growth in the rat pulmonary microcirculation. *Anat. Rec.* **228**, 35–45. *This paper demonstrates by electron microscopy of serial ultrathin sections that transcapillary tissue pillars with diameters around 1 μm postulated for the intussusception process really exist. A mode of formation is proposed on the basis of the pillars' ultrastructure.*
3. Burri, P. H., and Djonov, V. (2002). Intussusceptive angiogenesis: the alternative to capillary sprouting. *Mol. Aspects Med.* **23**, 1–27.
4. Djonov, V., Schmid, M., Tschanz, S. A., and Burri, P. H. (2000). Intussusceptive angiogenesis: Its role in embryonic vascular network formation. *Circ. Res.* **86**, 286–292. *First demonstration of the involvement of intussusceptive pillar formation in the morphogenesis of peripheral vascular trees.*
5. Djonov, V., Galli, A. B., and Burri, P. H. (2000). Intussusceptive arborization contributes to vascular tree formation in the chick chorioallantoic membrane. *Anat. Embryol.* **202**, 347–357.
6. Djonov, V., Andres, A. C., and Ziemiecki, A. (2001). Vascular remodelling during the normal and malignant life cycle of the mammary gland. *Microsc. Res. Tech.* **52**, 182–189.
7. Djonov, V. G., Kurz, H., and Burri, P. H. (2002). Optimality in the developing vascular system: branching remodeling by means of intussusception as an efficient adaptation mechanism. *Dev. Dyn.* **224**, 391–402. *First demonstration of the involvement of intussusceptive pillar formation in the remodeling and pruning process of small arteries as adaptive optimization of structures to hemodynamics.*
8. Kurz, H., Burri, P. H., and Djonov, V. G. (2003). Angiogenesis and vascular remodeling by intussusception: From form to function. *News Physiol. Sci.* **18**, 65–70.
9. Djonov, V., Baum, O., and Burri, P. H. (2003). Vascular remodelling by intussusception. *Cell Tissue Res.* **314**, 107–117.

Capsule Biographies

Peter H. Burri is Professor Emeritus of Anatomy. He was the head of the Section for Developmental Biology at the Institute of Anatomy of the University of Berne, Switzerland, and became the Chairman of the Institute from 1993 to 2003. His primary research interest was lung development, growth, and regeneration. His research on angiogenetic mechanisms developed directly from the lung studies. Dr. Burri's work has constantly been supported by the Swiss National Science Foundation.

Valentin Djonov is Assistant Professor at the Institute of Anatomy of the University of Berne, Switzerland. After completing research in tumor biology in the lab of Professor Friis, he joined the group of Dr. Burri in 1995 and started work on intussusceptive angiogenesis. He is now leading the research group in this field. Dr. Djonov's work is supported by the Swiss National Science Foundation and the Bernese Cancer League.

CHAPTER 16

Vascular Permeability Factor/Vascular Endothelial Growth Factor (VPF/VEGF, VEGF-A)

Harold F. Dvorak

Beth Israel Deaconess Medical Center and Harvard Medical School, Boston, Massachusetts

VEGF-A is the founding member of the VPF/VEGF family of proteins that also includes VEGFs B, C, and D as well as placenta growth factor (PlGF) and a related viral protein, VEGF-E. VEGF-A, the subject of this chapter, has critical roles in vasculogenesis and pathological and physiological angiogenesis, acting through receptors (VEGFR-1, VEGFR-2, and neuropilin) that are expressed on vascular endothelium as well as on certain other cell types. The product of a single gene, VEGF-A is alternatively spliced to form several proteins of different lengths, properties, and functions. Originally discovered as a potent vascular permeabilizing factor (VPF), VEGF-A is also an endothelial cell motogen and mitogen, profoundly alters the pattern of endothelial cell gene expression, and protects endothelial cells from apoptosis and senescence. Recently, VEGF-A has been found to have additional critical roles in hematopoiesis, in expansion and differentiation of bone marrow endothelial cell precursors, and in development and maintenance of the nervous system. The functions of other VPF/VEGF family members have been less well characterized. However, VEGFs C and D are essential for development of the lymphatic system, VEGF-B has a role in the development of coronary arteries, and PlGF has important roles in pathological angiogenesis.

VEGF-A Structure

VEGF-A is a highly conserved, disulfide-bonded dimeric glycoprotein of Mr ~45 kD. It shares low but significant sequence homology with platelet-derived growth factor (PDGF) and, like PDGF, has cysteines that form integral inter- and intra-chain bonds. Crystal structure reveals that the two chains that comprise VEGF-A are arranged in antiparallel fashion with binding sites at either end. Upon reduction, VEGF-A separates into its individual chains and loses all biological activity.

The human VEGF-A gene is located on the short arm of chromosome 6 and is differentially spliced to yield predominant isoforms that encode polypeptides of 206, 189, 165, and 121 amino acids in human cells (corresponding murine proteins are one amino acid shorter) (Figure 1). Other splice variants (183, 165b, 145) have also been described. The several major VEGF-A isoforms have distinct physical properties. VEGF-A$^{120/1}$ is acidic, freely soluble, and does not bind heparin. By contrast, the 164/5 and 188/9 isoforms have increasing basic charge and bind heparin with increasing affinity; in fact, VEGF-A^{165} was originally purified on the basis of its affinity for heparin. Heparin, heparan sulfate, and heparinase all displace the larger VEGF-A isoforms from proteoglycan binding sites in tissues; proteases such as plasmin have a similar net effect, cleaving the C-terminal portion of bound VEGF-A to generate a biologically active peptide. In several situations in vivo, liberation of bound VEGF-A from cells or cell matrix is the necessary trigger that initiates angiogenic activity. The different VEGF-A isoforms have largely identical biological activities in vitro, but there is increasing evidence for distinctive functions in vivo; for example, mice

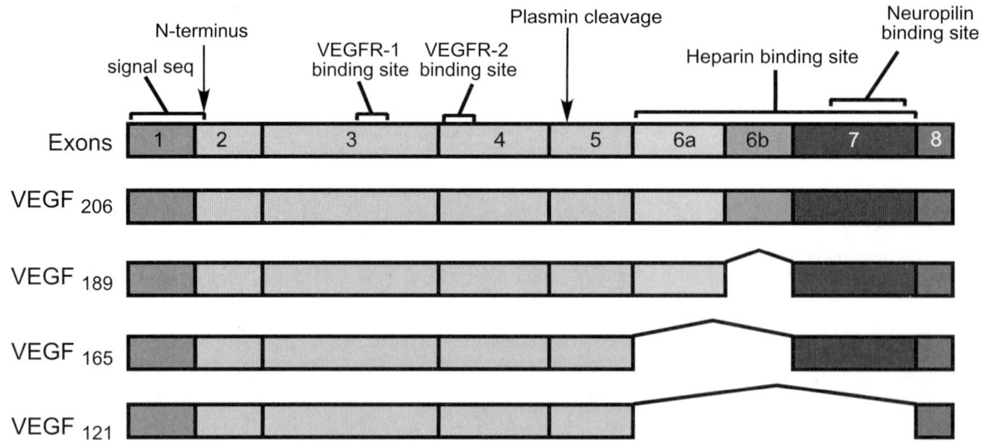

Figure 1 VEGF-A Gene Structure. VEGF-A has eight exons, the first of which encodes a hydrophobic leader sequence, typical of secreted proteins. The 206 amino acid VEGF-A isomer is encoded by all eight exons. VEGF-A^{189} lacks a portion of exon 6, whereas VEGF-A^{165}, frequently the predominant isoform, lacks all of exon 6. VEGF-A^{121} lacks both exons 6 and 7. VEGF receptor and heparin binding sites are as indicated. [Republished with permission from Cross et al. (2003), *Trends in Biochemical Sciences* **28**, 488.] (see color insert)

expressing only the 120 or 188 isoforms develop severe vascular anomalies.

VEGF-A Receptors and Signaling Pathways

VEGF-A mediates its effects primarily by interacting with two high-affinity transmembrane tyrosine kinase receptors: VEGFR-1 (Flt-1) and VEGFR-2 (KDR, Flk-1) (Figure 2). Cultured endothelial cells have approximately 3,000 copies of VEGFR-1, which binds VEGF-A^{165} with a K_d of approximately 10 pM; they have more numerous copies of VEGFR-2, which binds VEGF-A^{165} with somewhat lower affinity (K_d of 75 to 125 pM). A truncated soluble form of VEGFR-1 (sFlt) that results from alternative splicing is found in serum and retains VEGF-A binding activity. sFlt has recently been implicated in pre-eclampsia, a serious complication of pregnancy. A third, nonkinase receptor, neuropilin (NRP-1) has been found; NRP-1 potentiates VEGF-$A^{164/5}$'s binding to VEGFR-2. Neuropilin had been known as a receptor for the semaphorin/collapsin family of neuronal guidance mediators, and it is also expressed widely on nonendothelial cells. VEGF-$A^{164/5}$, B, E, and PlGF bind to NRP-1, but VEGF-$A^{120/1}$ does not (Figure 2); this inability may explain some of the abnormalities in mice engineered to express only the VEGF-A^{120} isoform. Two other members of the VPF/VEGF family (VEGF-B and both isoforms of PlGF) bind to VEGFR-1, whereas VEGFs C, D, and E bind to VEGFR-2.

Mice null for any of the three VEGF-A receptors are embryonic lethals. VEGFR-1 null mice die at day 8.5 to 9.0 as a result of vessel obstruction by an overgrowth of endothelial cells. VEGFR-2 knockout mice die at a similar stage of development because of failure of endothelial and

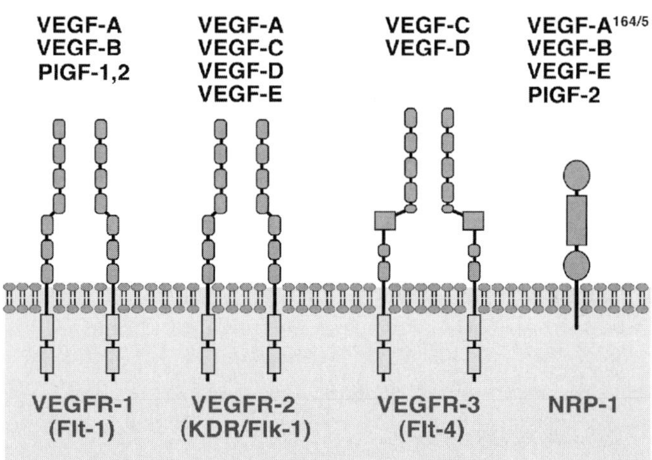

Figure 2 Binding Specificities of the VPF/VEGF Ligand Family Members and Their Receptors. VEGFR-1 and VEGFR-2 are expressed on vascular endothelial cells as well as on some other cell types; VEGR-3 is expressed on lymphatic endothelium in the adult but also on some tumor vessels. NRP-1 is expressed on vascular endothelium, neurons, and some tumor cells (see text for further details). [Republished with permission from H. Dvorak (2002), *Journal of Clinical Oncology* **20**, 4368.] (see color insert)

hematopoietic precursor cells. NRP-1 null mice also die from a failure of vascular development.

VEGFR-1 and VEGFR-2 are widely expressed on normal vascular endothelium, and expression is upregulated in tumor blood vessel endothelium and in that of other examples of pathological angiogenesis induced by VEGF-A. The mechanisms responsible for receptor overexpression are not yet fully understood, but hypoxia, which stimulates VEGF-A expression (see following discussion), may also upregulate VEGF receptor expression.

Recent reports indicate that VEGF-A receptors may also be expressed on tumor cells, raising the possibility of an autocrine loop that might stimulate tumor cell growth and migration. VEGFR-1 is additionally expressed on a population of hematopoietic stem cells and myeloid progenitors as well as on monocytes/macrophages, spermatogenic and Leydig cells, uterine smooth muscle cells, and osteoclasts. VEGFR-1 has an important role in mediating hematopoietic cell development and recruitment, particularly that of monocytes, and mediates VEGF-A-induced monocyte (but not endothelial cell) motility. As noted previously, VEGFR-1 null embryos die as a result of an overgrowth of endothelial cells, implying, along with data from cultured endothelial cells, that this receptor exerts a negative regulatory effect on the VEGF-A activities governed by VEGFR-2 signaling. VEGFR-2 is expressed on bone marrow endothelial cell progenitors, megakaryocytes, uterine smooth muscle cells, and lymphatic endothelium.

Although binding to VEGF-A with high affinity, VEGFR-1 induces only minimal stimulation of kinase activity in vascular endothelium; as a result, downstream signaling pathways have had to be worked out in endothelial cells that were engineered to overexpress this receptor. Much more is known about the signaling pathways initiated through VEGFR-2 (Figure 3). Upon binding to VEGFR-2, VEGF-A initiates a cascade of events that begins with receptor dimcrization and autophosphorylation followed by phosphorylation of numerous downstream proteins. Very recently, G proteins have been implicated in VEGF-A signaling. Most of the biological activities mediated by VEGF-A on endothelial cells (e.g., proliferation, migration, vascular permeability, antiapoptosis) are mediated through VEGFR-2 signaling. VEGFR-2 also has an important role in the development of endothelial cell precursors present in the bone marrow and circulating in blood.

Biological Activities

VEGF-A is essential for the development and organization of the vascular system (vasculogenesis) and for both physiological and pathological angiogenesis. Mice lacking even one copy of the VEGF-A gene are embryonic lethals. VEGF-A mediates its complex functions by exerting specific biological activities.

Increased Microvascular Permeability

VEGF-A was originally discovered in the late 1970s because of its ability to increase the permeability of microvessels (primarily postcapillary venules) to circulating plasma and plasma proteins. It is responsible for the nearly

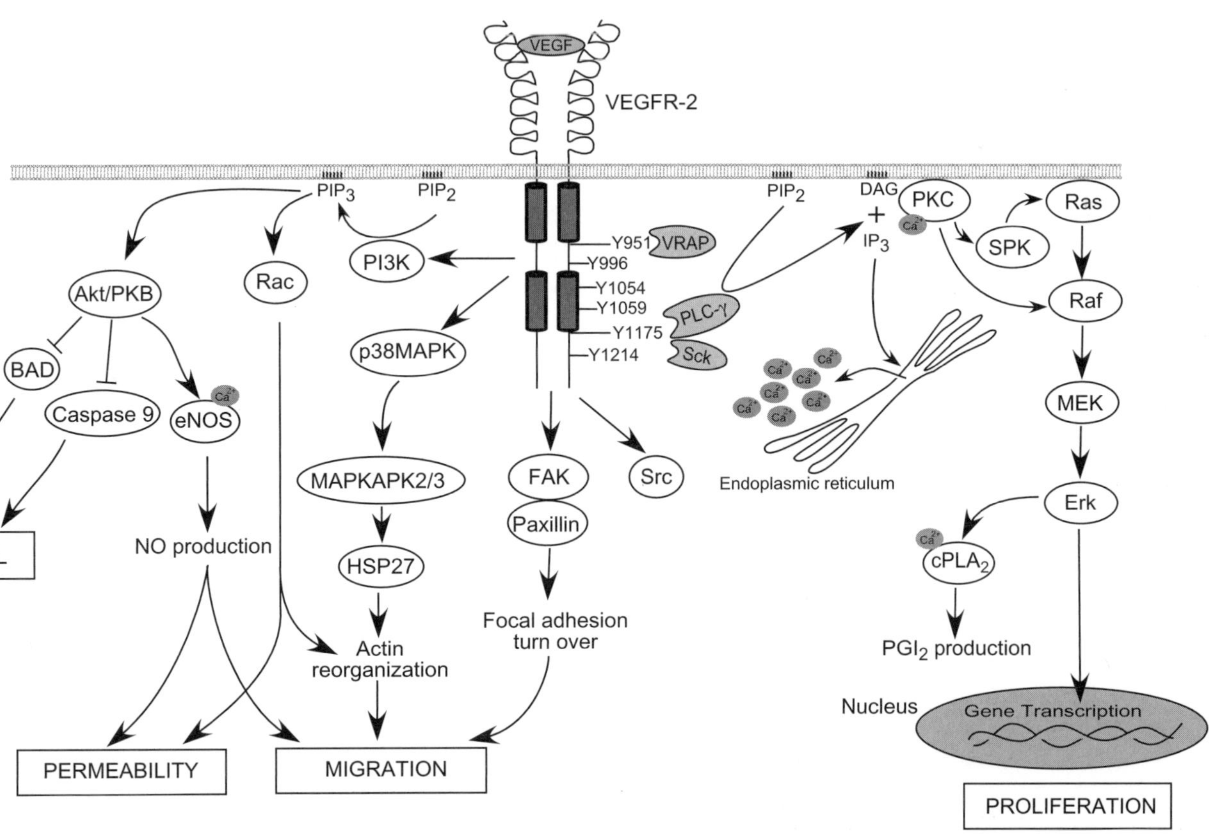

Figure 3 Schematic Diagram of VEGFR-2 Signaling. Ligand binding to the extracellular domain induces dimerization, autophosphorylation of specific intracellular tyrosine residues, and a series of downstream phosphorylations and other events that lead to endothelial cell survival, migration, proliferation, and increased permeability. [For further details, see Cross et al. (2003), *Trends in Biochemical Sciences* **28**, 488, from which this diagram is republished with permission.] (see color insert)

universal hyperpermeability of tumor blood vessels. Permeability becomes evident within a minute following injection of VEGF-A protein into skin and continues for approximately 20 minutes. VEGF-A is among the most potent vascular permeabilizing agents known, acting at concentrations below 1 nM and with potency some 50,000 times that of histamine on a molar basis.

VEGF-A induces venular permeability by its action on endothelial cells, the primary barrier to the extravasation of plasma proteins, but there has been debate about the pathway that circulating macromolecules follow in traversing endothelium. The earlier view had been that vasoactive agents cause endothelial cells to pull apart, creating an interendothelial cell gap through which macromolecules could extravasate. Although supported by data using cultured endothelial cells, evidence for an intercellular extravasation pathway in vivo is not convincing. More recent studies have shown that macromolecules cross tumor and normal venular endothelium primarily by means of a transendothelial cell pathway that involves vesiculo-vacuolar organelles (VVOs) (see Chapter 46 by A. Dvorak and D. Feng for a detailed discussion of VVOs). VEGF-A also induces endothelial fenestrations that provide an additional transcellular pathway for solute extravasation.

The increased microvascular permeability induced by VEGF-A leads to tissue edema, a characteristic of tumors, healing wounds, and other pathologies in which VEGF-A is overexpressed. Extravascular fluid accumulation is particularly prominent in tumors growing in body cavities such as the peritoneum (ascites tumors). Plasma protein leakage has several consequences. One of these is activation of the clotting system via the tissue factor pathway, leading to deposition of a fibrin gel that retards clearance of edema fluid and results in locally increased interstitial tissue pressure, a characteristic feature of many solid tumors. Deposited fibrin also provides a provisional stroma for endothelial cell and fibroblast migration that supports the angiogenesis and fibrogenesis necessary for generating mature stroma (desmoplasia in the case of tumors or scar formation in wound healing).

Other Biological Activities on Vascular Endothelium

VEGF-A has multiple effects on vascular endothelium that become apparent over the time frame of hours to days. These include striking changes in cell morphology and cytoskeleton, accompanied by stimulation of endothelial cell migration and division. At the molecular level, VEGF-A reprograms endothelial cell gene expression, causing the increased production of several different proteins, including the procoagulant tissue factor, fibrinolytic proteins (urokinase, tPA, PAI-1, urokinase receptor), matrix metalloproteases, the GLUT-1 glucose transporter, nitric oxide synthase, numerous mitogens, and antiapoptotic factors (e.g., bcl-2, A1, survivin, XIAP). VEGF-A also serves as an endothelial cell survival factor, protecting endothelial cells against apoptosis (L. Benjamin) and senescence. In addition, the Jain laboratory has shown that VEGF-A upregulates the expression of endothelial adhesion molecules on vascular endothelium, including E-selectin, ICAM-1, and VCAM; this favors inflammatory cell attachment and extravasation. Recently, the Ferrara laboratory has identified another activity of VEGF-A, that of paracrine release of hepatotrophic molecules from sinusoidal liver endothelium, a function mediated through VEGFR-1.

Effects of VEGF-A on Cells Other Than Vascular Endothelium

There is increasing interest in activities that VEGF-A exerts on nonendothelial cells that express VEGF receptors as VEGF-A stimulates monocyte chemotaxis and proliferation of uterine smooth muscle. VEGF-A also has reported effects on lymphocytes, granulocyte-macrophage progenitor cells, osteoblasts, Schwann cells, mesangial cells, and retinal pigment epithelial cells. In development, VEGF-A drives angioblasts and primitive vessels toward arterial differentiation and attracts filopodia of both endothelial cells and neurons, causing these cells to move in the direction of a VEGF-A gradient. VEGF-A receptors have now been reported on some tumor cells, raising the possibility that VEGF-A exerts autocrine effects that enhance tumor cell motility and survival.

Expression of VEGF-A in Normal Physiology and in Pathology

VEGF-A is constitutively expressed in the cells of many normal tissues and at higher levels in several types of epithelium (renal glomerular podocytes, adrenal cortex, breast, and lung), in cardiac myocytes, and in activated macrophages. It is also expressed at high levels in the physiological angiogenesis of ovarian follicular development, in corpus luteum formation, and in endochondral bone formation.

VEGF-A is overexpressed by the vast majority of solid human and animal carcinomas, where it is thought to be the prime mover of tumor angiogenesis. Most VEGF-A in solid tumors is expressed by the malignant cells, but tumor stroma can also synthesize VEGF-A. More recently, VEGF-A has been found to be expressed in lymphomas and hematological malignancies. VEGF-A is also overexpressed in at least some premalignant lesions (e.g., precursor lesions of breast, cervix, and colon cancers); furthermore, expression levels increased in parallel with malignant progression. An association between VEGF-A expression and benign tumors is less well established, in part because the latter have been less carefully studied. However, pituitary adenomas and benign hemangiomas rarely overexpress VEGF-A, whereas uterine leiomyomas and malignant vascular tumors do so.

VEGF-A is overexpressed in many other examples of pathological angiogenesis, including wound healing (e.g., in healing skin wounds, myocardial infarcts, and strokes), chronic inflammation (e.g., delayed hypersensitivity, rheumatoid arthritis, psoriasis), and various retinopathies. In all of these examples, VEGF-A is thought to be largely responsible for the accompanying angiogenesis.

Regulation of VEGF-A Expression

Several mechanisms regulate VEGF-A expression, including oxygen concentration, low pH, oncogenes and tumor suppressor genes, cytokines, hormones, and a variety of other mediators.

Oxygen Concentration and Low pH

Hypoxia potently upregulates VEGF-A expression, both by stabilizing its message and by increasing message transcription. Transcriptional regulation is mediated through hypoxia-inducible factor 1 (HIF-1), a heterodimeric protein transcription factor. One HIF-1 component, HIF-1α, is rapidly degraded under normoxic conditions by the ubiquitin pathway; however, when stabilized by hypoxia, HIF-1α dimerizes with HIF-1β, and the complex binds to and activates a hypoxia-responsive element in the VEGF-A promoter. Hypoxic regulation of VEGF-A expression is likely important in healing wounds and has been demonstrated in some tumors. However, many tumors express VEGF-A constitutively at high levels even under normoxic conditions, and therefore, regulation is achieved by other means, such as those discussed as follows.

Low pH, another hallmark of tumors and ischemic tissues, also upregulates VEGF and does so independently of HIF-1.

Oncogenes and Tumor Suppressor Genes

Several oncogenes (*src*, *ras*) and tumor suppressor genes (*p53*, *p73*, von Hippel Lindau [*vHL*]) modulate VEGF-A expression and in this way can modulate tumor growth. Cells transfected with mutant *src* or *ras* oncogenes express increased amounts of VEGF-A mRNA and protein. The *vHL* tumor suppressor gene is part of the protein complex that targets specific proteins, including HIF-1α, for ubiquitinylation and proteolysis. Therefore, when *vHL* is absent or inactivated, HIF-1α is stabilized even under normoxic conditions with resulting upregulation of VEGF-A and perhaps other members of the VPF/VEGF family. The *p53* and *p73* genes suppress VEGF-A transcription.

Cytokines and Other Mediators

Numerous growth factors, cytokines, and lipid mediators upregulate VEGF-A expression in different cells, including EGF, TGF-α, FGF-2, TGF-β, PDGF, keratinocyte growth factor, TNF, interleukins 1 and 6, insulin-like growth factor 1, HGF, and prostaglandins E1 and E2. These findings are likely to be important in autocrine regulation of VEGF-A expression in vivo in that many tumors that express VEGF-A also express other cytokines and their receptors (e.g., TGF-α, FGF-2, EGF).

Hormones

VEGF-A is expressed by many cells that make steroid hormones (adrenal cortex, corpus luteum, Leydig cells) and by cells that are under hormonal regulation (e.g., the cycling uterus and ovary). Circulating VEGF-A levels and presumably VEGF-A expression correlate with estrogen receptor positivity in breast cancer. VEGF-A is also expressed by peptide hormone-producing cells such as thyroid follicular cells, and its production in culture is upregulated by agents such as insulin, dibutyryl c-AMP, and the IgG of Graves' disease. Thyroid-stimulating hormone upregulates VEGF-A mRNA expression in human thyroid follicles and promotes VEGF-A secretion in several thyroid cancer cell lines.

Other Agents

Several other chemicals, proteins, and processes can augment VEGF-A expression or activity by direct or indirect means, including thrombin, platelet aggregation, shear stress, acidosis, lysophosphatidic acid, and adenosine. On the other hand, dexamethasone downregulates or prevents cytokine- (but not hypoxia-) induced upregulation of VEGF-A expression.

The New Blood Vessels Induced by VEGF-A

VEGF-A, and particularly its 164/5 isoform, is thought to be the most important mediator of pathological angiogenesis, including that induced by tumors, wounds, and inflammation. However, all of these complex entities involve the activities of multiple cytokines. Recently, it has been possible to tease apart those aspects of the angiogenic response that are attributable solely to VEGF-A by expressing VEGF-A^{164} in an adenoviral vector (Ad-VEGF-A^{164}). Upon introduction into mouse tissues, infected host cells express VEGF-A^{164} protein within a few hours and continue to secrete it at fairly constant levels for approximately two weeks. VEGF-A^{164} induces a highly reproducible, time-ordered sequence of events that is qualitatively similar in all tissues studied. Local microvessels become hyperpermeable to plasma proteins, resulting in tissue edema and deposition of extravascular fibrin. By 18 hours, enlarged, thin-walled, pericyte-poor, hyperpermeable, and strongly VEGF receptor-positive "mother" vessels develop from preexisting venules (Figure 4). Such mother vessels are characteristic of many tumors, healing wounds, and so forth and form according to a three-step process: (1) proteolytic degradation of the noncompliant (nonelastic) vascular basement membrane; (2) detachment of pericytes from basement membrane; and (3) transfer of VVO membranes to the

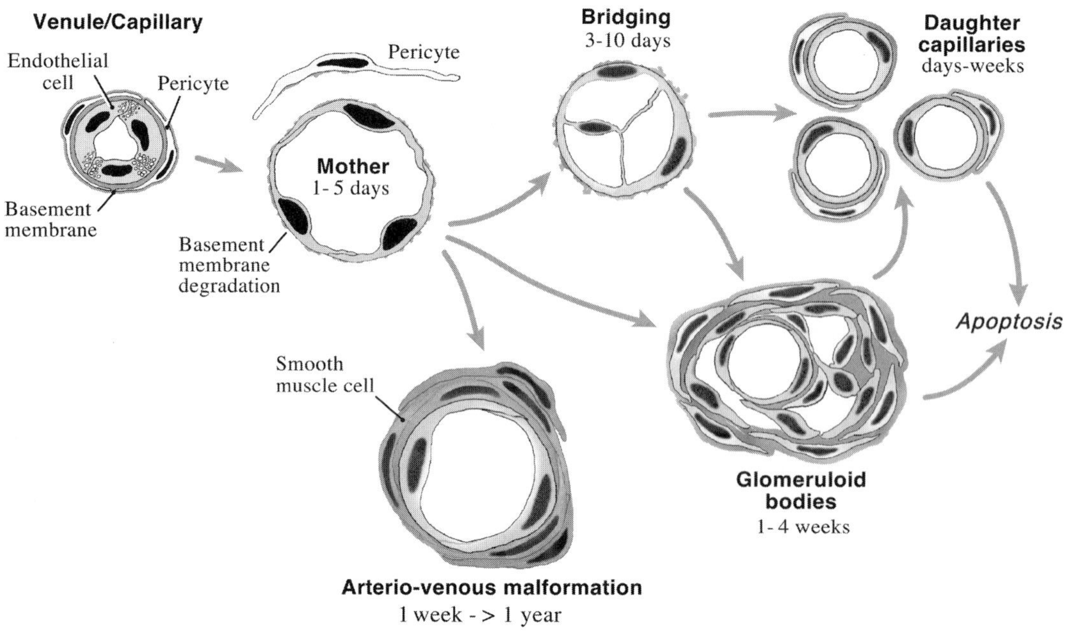

Figure 4 Schematic diagram summarizing the progression of the angiogenic response that follows introduction of a VEGF-A^{164}-expressing adenoviral vector in vivo. Mother vessels develop and may evolve into glomeruloid bodies, vascular malformations, and daughter capillaries. Finally, as VEGF-A^{164} expression wanes, glomeruloid bodies undergo apoptosis, whereas malformations achieve VEGF-A independence and persist indefinitely. [Republished in modified form from Pettersson et al. (2000), *Laboratory Investigation* **80**, 99]. (see color insert)

plasma membrane, allowing endothelial cells to thin and spread over the enlarged surface area that was made possible by basement membrane degradation.

Mother vessels are transitional structures that generally persist as such for only a few days (Figure 4). They may divide into smaller channels by projecting transluminal endothelial cell "bridges" into and across mother vessel lumens; these divide blood flow into smaller channels that separate from each other over time, forming individual smaller-caliber "daughter" capillaries. Mother vessels also evolve to form glomeruloid bodies, poorly organized conglomerates of endothelial cells and pericytes that are also found in tumors such as glioblastoma multiforme. Other mother vessels acquire a coating of smooth muscle cells and/or fibrosis and take on the appearance of arteriovenous vascular malformations. Taken together, these findings show that VEGF-A is sufficient to generate the mother vessels, glomeruloid bodies, and vascular malformations found in various human pathologies.

In addition, VEGF-A^{164} is also able to induce a modest degree of arteriogenesis and to generate abnormal "giant" lymphatics that are characterized by very large size and poor function. Lymphatics of this description occur in patients with Crohn's disease and in lymphangiomas (lymphatic malformations).

Clinical Significance of VEGF-A

Patients with large tumor burdens and widespread metastatic disease have increased levels of circulating VEGF-A, often multiples of those found in normal individuals. Although not useful as a screening tool, increasing serum VEGF-A levels may signify increased tumor growth, recurrence, or metastatic spread in individual patients. There is debate about whether plasma or serum levels are more meaningful because platelets sequester VEGF-A and because plasma $\alpha 2$ macroglobulin binds it and makes it unavailable to at least some antibodies. Also, both megakaryocytes and leukocytes synthesize VEGF-A. Therefore, serum levels reflect not only VEGF-A of tumor origin but also that released from platelets and leukocytes, making it difficult to establish a range of normal values. VEGF-A levels are also elevated in malignant effusions and in the urine of patients with bladder cancer.

Treatments that reduce VEGF-A expression or inhibit its action have been advanced as an antiangiogenesis approach to cancer therapy. Antibodies that neutralize VEGF-A strikingly inhibit tumor growth in mice and have recently shown success in prolonging the life of colon cancer patients when used in combination with chemotherapy. Conversely, VEGF-A has been used in attempts to induce the growth of new blood vessels in ischemic tissues (e.g., myocardial infarctions, peripheral vascular disease), but there are many reasons for caution. Very recently, reduced VEGF-A has been implicated in amyotrophic lateral sclerosis and in certain neurovascular disorders.

Glossary

Angiogenesis: The development of new blood vessels from preexisting blood vessels.

Vasculogenesis: The development of blood vessels from angioblasts as in development.

Vesiculo-vacuolar organelles (VVOs): Interconnected chains of uncoated cytoplasmic vesicles and vacuoles that span the entire thickness of venule endothelial cell cytoplasm from lumen to albumen. VVOs serve as a pathway for macromolecular extravasation and as an intracellular store of plasma membrane that can be transferred to the cell surface to allow rapid plasma membrane expansion and vessel enlargement.

Further Reading

Carmeliet, P. (2003). Blood vessels and nerves: Common signals, pathways and diseases. *Nat. Rev. Genet.* **4**, 710–720. *VEGF-A biological activities.*

Cross, M. J., Dixelius, J., Matsumoto, T., and Claesson-Welsh, L. (2003). VEGF-receptor signal transduction. *Trends Biochem. Sci.* **28**, 488–494. *VEGF-A receptors and signaling.*

Dvorak, H. F. (2002). Vascular permeability factor/vascular endothelial growth factor: A critical cytokine in tumor angiogenesis and a potential target for diagnosis and therapy. *J. Clin. Oncol.* **20**, 4368–4380. *VEGF-A structure, receptors and signaling, biological activities, expression, regulation of expression, clinical significance.*

Dvorak, H. F. (2003). Rous-Whipple Award Lecture. How tumors make bad blood vessels and stroma. *Am. J. Pathol.* **162**, 1747–1757. *New blood vessels induced by VEGF-A.*

Dvorak, H. F., Nagy, J. A., Feng, D., Brown, L. F., and Dvorak, A. M. (1999). Vascular permeability factor/vascular endothelial growth factor and the significance of microvascular hyperpermeability in angiogenesis. *Curr. Top. Microbiol. Immunol.* **237**, 97–132. *VEGF-A biological activities, expression.*

Feng, D., Nagy, J. A., Pyne, K., Hammel, I., Dvorak, H. F., and Dvorak, A. M. (1999). Pathways of macromolecular extravasation across microvascular endothelium in response to VPF/VEGF and other vasoactive mediators. *Microcirculation* **6**, 23–44. *VEGF-A biological activities (vascular permeability mechanisms).*

Ferrara, N., Gerber, H. P., and LeCouter, J. (2003). The biology of VEGF and its receptors. *Nat. Med.* **9**, 669–676. *VEGF-A structure, biological activities, expression clinical significance.*

Melder, R. J., Koenig, G. C., Witwer, B. P., Safabakhsh, N., Munn, L. L., and Jain, R. K. (1996). During angiogenesis, vascular endothelial growth factor and basic fibroblast growth factor regulate natural killer cell adhesion to tumor endothelium. *Nat. Med.* **2**, 992–997. *VEGF-A biological activities.*

Mukhopadhyay, D., and Zeng, H. (2002). Involvement of G proteins in vascular permeability factor/vascular endothelial growth factor signaling. *Cold Spring Harbor Symp. Quant. Biol.* **67**, 275–283. *VEGF-A receptors and signaling.*

Nagy, J. A., Dvorak, A. M., and Dvorak, H. F. (2003). VEGF-A(164/165) and PlGF: Roles in angiogenesis and arteriogenesis. *Trends Cardiovasc. Med.* **13**, 169–175. *New blood vessels induced by VEGF-A.*

Nagy, J. A., Vasile, E., Feng, D., Sundberg, C., Brown, L. F., Manseau, E. J., Dvorak, A. M., and Dvorak, H. F. (2002). VEGF-A induces angiogenesis, arteriogenesis, lymphangiogenesis, and vascular malformations. *Cold Spring Harbor Symp. Quant. Biol.* **67**, 227–237. *New blood vessels induced by VEGF-A.*

Rafii, S., Lyden, D., Benezra, R., Hattori, K., and Heissig, B. (2002). Vascular and haematopoietic stem cells: Novel targets for anti-angiogenesis therapy? *Nat. Rev. Cancer* **2**, 826–835. *Biological activities of VEGF-A.*

Rak, J., Filmus, J., Finkenzeller, G., Grugel, S., Marme, D., and Kerbel, R. S. (1995). Oncogenes as inducers of tumor angiogenesis. *Cancer Metastasis Rev.* **14**, 263–277. *Regulation of VEGF-A expression by oncogenes and tumor suppressor genes.*

Stalmans, I., Ng, Y. S., Rohan, R., Fruttiger, M., Bouche, A., Yuce, A., Fujisawa, H. Hermans, B., Shani, M., Jansen, S., Hicklin, D., Anderson, D. J., Gardiner, T., Hammes, H. P., Moons, L., Dewerchin, M., Collen, D., Carmeliet, P., and D'Amore, P. A. (2002). Arteriolar and venular patterning in retinas of mice selectively expressing VEGF isoforms. *J. Clin. Invest.* **109**, 327–336. *VEGF-A structure.*

Capsule Biography

Dr. Dvorak has served as chair of the Department of Pathology at Beth Israel Deaconess Medical Center and as Mallinckrodt Professor of Pathology at the Harvard Medical School since 1979. The recipient of the 2002 Rous-Whipple Award from the American Society for Investigative Pathology, his laboratory focuses primarily on the angiogenic response in health and disease. His work is supported by grants from the NIH and the National Foundation for Cancer Research.

CHAPTER 17

Endothelial Progenitor Cells, Vasculogenesis, and Their Contribution to New Blood Vessel Formation in the Adult

John Glod and Howard A. Fine

Neuro-Oncology Branch, National Cancer Institute, National Institute of Neurological Disorders and Stroke, National Institutes of Health, Bethesda, Maryland

Introduction

It is well established that the de novo formation of blood vessels in the embryo is accomplished by the process of vasculogenesis. Endothelial progenitor cells (EPCs) or angioblasts are formed from more primitive mesodermal precursors and migrate into areas of vessel formation, where they are incorporated into the vascular wall and differentiate into mature endothelial cells (ECs). Vasculogenesis is contrasted by angiogenesis, a process classically thought to be the exclusive means of new vessel formation in the adult. Angiogenesis consists of the replication of mature endothelial cells that are resident in established blood vessels. These cells provide a supply of endothelium for the budding and branching of new vessels from the old. However, evidence has accumulated over the past decade that vasculogenesis is not limited to the prenatal period but is also an important component of new vessel growth in the adult. This chapter summarizes the current body of work on circulating endothelial progenitor cells as building blocks of adult neovasculature.

Despite the recent burst of interest in the concept of vasculogenesis in the adult animal, evidence that an endothelial precursor is present in the adult circulation has a long history. In 1932, Hueper and Russell reported on "Capillary Tube Formation in Tissue Cultures of Leukocytes." This was followed by several reports in the 1950s of hematopoietic cells adopting characteristics of endothelial cells in vitro, and numerous publications have described the participation of blood-derived cells in the re-endothelialization of vascular grafts. Several studies suggested that the new endothelial layer was not a result of migration of cells from preexisting vessels. These early reports suggested the presence of circulating cells that could form endothelium. Several possible hypotheses could explain this finding.

Mature Endothelial Cells in the Circulation

One possibility is that mature endothelial cells are present in the circulation and can participate in neovascularization. In 1970, Gaynor and colleagues identified endothelial cells in the circulation of rabbits after exposure to endotoxin. A body of evidence has since accumulated, demonstrating the presence of mature endothelial cells in the circulation after various types of vascular injury. However, it is unlikely that these cells contribute significantly to vasculogenesis. Two pieces of data that argue against mature ECs participating in vasculogenesis are their extremely low prevalence in the blood (1 to 3 per milliliter of blood) and their limited proliferative capacity based on in vitro assays.

The prevailing evidence indicates that these circulating mature ECs are not the cells responsible for vasculogenesis in the adult. Nevertheless, the presence of increased mature ECs in the circulation may be useful as a marker for vascular injury.

Endothelial Progenitor Cells from the Bone Marrow

Another hypothesis regarding the potential reservoir for circulating endothelial progenitors is that they arise from the bone marrow. The potential for hematopoietic cells to contribute to endothelial formation is supported by the close relationship between endothelial cells and hematopoietic cells in development. Endothelial progenitors (angioblasts) and hematopoietic progenitors both arise from mesodermal precursors in the blood islands during early embryonic development. The angioblasts develop at the perimeter of blood islands in the yolk sac of the embryo, whereas hematopoietic precursors are formed in the center of the blood island. A common precursor for both hematopoietic precursors and angioblasts has also been described by Risau et al. in 1995 using in vivo studies in zebrafish and in vitro differentiation experiments. Embryonic angioblasts and hematopoietic stem cells also share several markers, including Flk-1, Tie-2, and CD34. These hint at the close ties between ECs and hematopoietic cells, suggesting that circulating endothelial progenitors may be generated from bone marrow cells. This line of reasoning ultimately led to the isolation and characterization of circulating EPCs.

Isolation and Characterization of Circulating Endothelial Progenitors from Adult Peripheral Blood

Circulating endothelial progenitors have been defined as highly proliferative cells that are derived from the bone marrow with the capacity to adopt endothelial characteristics in vitro and incorporate into neovasculature and differentiate into mature endothelial cells in vivo. Asahara, Isner, and colleagues published the first description of circulating endothelial progenitor cells in an adult in 1997. In this initial description, human peripheral blood mononuclear cells enriched for expression of the markers Flk-1 (VEGFR2) and CD34 were found to have the capacity to develop traits of mature endothelial cells when cultured. The conditions used to produce this phenotype attempted to recapitulate an angiogenic environment by including bovine brain extract in the culture medium. The cells expressed endothelial characteristics such as the markers CD31, Tie-2, Flk-1, von Willebrand Factor (vWF), and endothelial nitric oxide synthetase (ecNOS). The cells also took up DiI-labeled acetylated low-density lipoprotein (DiI acLDL) and lost the hematopoietic cell surface marker CD45. This initial work was a prelude to several studies in the area that have used a variety of culture conditions to induce the expression of endothelial traits in bone marrow–derived cells. The starting populations of cells that have been used in these experiments can be roughly divided into two categories: hematopoietic stem cells and monocytes. (See Table I for a summary of these experiments.)

Several groups have reported that cells with one or more of a constellation of markers CD133, CD34, and the vascular endothelial growth factor receptor flk1 (VEGFR2) identifies a subpopulation of circulating cells with endothelial potential. These cells have been isolated both from the bone marrow and peripheral blood and been found to adopt endothelial characteristics in culture. Upon culture with angiogenic cytokines such as vascular endothelial growth factor (VEGF), basic fibroblast growth factor (bFGF), and others, the precursor endothelial cells lose the expression of hematopoietic stem cell markers CD133 and CD34 and begin expressing the endothelial markers VE-Cadherin, von Willebrand factor, and *Ulex europaeus* agglutinin-1 (UEA-1) binding. Additionally, these cells are able to show some functional endothelial characteristics in vitro, including uptake of acLDL and tube formation in matrigel.

Another hematopoietic cell population that has been shown to adopt endothelial characteristics in vitro and participate in neovasculature formation in vivo are peripheral blood monocytes. The ability of a monocyte fraction of cells to assume endothelial characteristics in vitro was first demonstrated in 2001 by Haaraz and colleagues and has been subsequently repeated in other studies. As with the CD34+/CD133+/FLK1+ cells, monocytes require culture under angiogenic conditions to express endothelial traits. These cells were found to express the EC markers von Willebrand factor, VE-Cadherin (CD144), endoglin (CD105), thrombospondin receptor (CD36), VEGFR1, and VEGFR2. However, unlike the CD34+/CD133+/FLK1+ derived cells, the monocyte-derived cells continue to express myeloid markers such as CD45 and CD14, even as they begin showing endothelial characteristics.

Contribution of Circulating Cells to Neovasculature

In their initial report of endothelial progenitor cells Asahara and colleagues showed that following systemic injection, endothelial progenitors incorporate into sites of vessel growth and repair in rodent hind limb ischemia models. Other investigators have since repeated this finding. Several experimental designs have been used to document this process, although each design has possible pitfalls. The supply of endothelial progenitors has varied between experiments. Both undifferentiated early progenitor cells and cells that have been differentiated in vitro to a more endothelial-like phenotype have been delivered systemically and directly into areas of vessel growth. Additionally, bone

Table I. Characteristics of Endothelial Progenitor Preparations. Hematopoietic cells of different types and from different sources have been shown to express endothelial characteristics when cultured with pro-angiogenic cytokines. Starting cell populations and sources are shown in addition to culture additives; vascular endothelial growth factor (VEGF), basic fibroblast growth factor (bFGF), insulin-like growth factor I (IGF-I), and epidermal growth factor (EGF). The expression of a set of endothelial and myeloid markers is indicated for each study. Endothelial-like cells derived from hematopoietic progenitors and monocytes are grouped separately.

Starting cell population	Culture additives	EPC characteristics	Reference
Cells with Progenitor Markers			
Total human PB mononuclear cells	VEGF, bFGF, IGF-1, EGF, hydrocortisone, heparin, FBS	P1H12+, vWF+, CD36+, VE-Cad+, WP bodies, CD14–	Lin et al. (2000). *J. Clin. Invest.* **105**, 71–77
CD34+ human periperal blood	Bovine brain extract FBS	EcNOS+, Tie2+, CD34+, CD31+, CD45–	Asahara et al. (1997). *Science* **5302**, 964–967
CD34+ human bone marrow, fetal liver, cord blood, peripheral blood	VEGF, bFGF, IGF-1	AcLDL, vWF	Shi et al. (1998). *Blood* **92**, 362–367
CD133+ GCSF mobilized human peripheral blood	FBS, VEGF, SCGF, hydrocortisone	CD34+, CD31+, VE-Cad+, Flk-1+, Tie2+, UEA-1, vWF+, CD1a–, CD14–	Gehling, U. M. et al. (2000). *Blood* **95**, 3106–3112
CD133+ human bone marrow followed by selection with UEA-1 binding	FBS, bFGF, heparin	VE-Cad, vWF, P1H12, UEA-1, CD105, KDR, WP bodies, CD14–, CD45–, CD34–	Quirici et al. (2001). *Br. J. Haematol.* **115**, 186–194
CD34+/CD133+/VEGFR-2+ human peripheral blood, fetal liver, and cord blood	FBS, bFGF, heparin	AcLDL, VE-Cad, CD13, CD31	Peichev et al. (2000). *Blood* **95**, 952–958
Monocytes			
Unselected human peripheral blood mononuclear cells	VEGF, bFGF, IGF-1, EGF, FBS	AcLDL, UEA-1, CD45+, CD14+, CD11b+, CD11c+, CD31, low VE-Cad, low CD34	Rehman et al. (2003). *Circulation* **107**, 1164–1169
CD14+ human peripheral blood monocytes	FBS, VEGF, bFGF, IGF-1	VWF, VE-Cad, ecNOS, CD68	Schmeisser et al. (2001). *Cardiovasc. Res.* **49**, 671–680
Human CD34+mobilized peripheral blood	FBS, insulin, VEGF, bFGF	VE-Cad, ecNOS, vWF	Nakul-Aquaronne et al. (2003). *Cardiovasc Res* **57**, 816–823
Human CD34- peripheral blood monocytes	FBS, bovine brain extract	EcNOS, vWF, VE-Cad, Tie-2, MUC18, CD105, CD1a, CD45, acLDL	Harraz et al. (2001). *Stem Cells* **19**, 304–312
CD14+ human peripheral blood	VEGF, bFGF, EGF, IGF-1, hydrocortisone, heparin, FBS	Ac-LDL, vWF, VE-Cad, CD105, CD36, Flt-1, Flk-1, CD1a–, CD83–, CD68, HLA-DR	Fernandez Pujol et al. (2000). *Differentiation* **65**, 287–300

marrow transplant studies have been performed to document the incorporation of bone marrow–derived cells into a growing endothelium. The methods of marking the endothelial progenitors have also differed and include dye labeling, green fluorescent protein (GFP) expression, use of transgenic donor animals expressing GFP or β-galactosidase, and identification of the Y chromosome in sex-mismatched donor/recipient pairs. Some criticisms of these studies have been the possibility of dye uptake by other native cells and the inherent difficulty in demonstrating that a cell is functionally incorporated into the vasculature as opposed to spatially located in the vasculature. Despite these criticisms, the ability of several groups to derive essentially the same conclusions using varied experimental methodologies make the data compelling.

Although the potential for circulating cells to incorporate into new vasculature appears to be established, there continues to be debate on the relative importance of angiogenesis versus vasculogenesis in the adult. An effort has been made to quantitate the relative amount of endothelium derived from circulating cells versus mature local endothelial cells using rodent bone marrow transplant models. Several groups have used this strategy and found wide ranges of donor versus recipient contribution to new vascular endothelium. In different experimental situations, the donor contribution ranges from 0 to 95 percent (see Table II). The comparison of these experiments is complicated by the use of different models of adult vascular formation ranging from tumors to limb ischemia and by other factors such as differences in the extent of bone marrow engraftment and accuracy with which transplanted cells are differentiated from host cells. Several studies in human transplant patients have also been published. Host contributions to endothelium in donor hearts ranging from 0 to –25 percent have been

Table II. Contribution of Bone Marrow–Derived Cells to Neovasculature in Mouse Transplant Models. The contribution of bone marrow–derived cells in various models of neovascular growth in the adult mouse has been determined using murine transplant models. The wide range of bone marrow–derived cells illustrates possible differences in the contributions of angiogenesis versus vasculogenesis with different experimental conditions.

Model of vascular growth	Percentage of bone marrow–derived cells	Reference
Regenerative lung growth	rare	Voswinckel et al. (2003). *Circ. Res.* **93**, 372–379
Cerebral ischemia	Cells present at infarct border	Zhang et al. (2002). *Circ. Res.* **90**, 284–288
Brain injury	0	Vallieres et al. (2003). *J. Neurosci.* **23**, 5197–5207
Ischemic cardiac injury	3.3%	Jackson et al. (2001). *J. Clin. Invest.* **107**, 195–402
Sarcoma xenograft	10%	Bolontrade et al. (2002). *Clin. Cancer Res.* **8**, 3622–3627
B6RV2 lymphoma xenograft	95%	Lyden et al. (2001). *Nat. Med.* **7**, 1194–1201
Glioma xenograft	10–25%	Ferrari et al. (2003). *Gene Therapy* **10**, 647–656

demonstrated, and recipient-derived endothelial cells have been reported in transplanted kidneys and livers. Transplanted bone marrow has also been found to contribute to endothelium in recipients. There are wide variations in the reported relative contribution of circulating cells to new vascular endothelium even among different patients in the same study. This variation in transplant patients as well as rodent bone marrow transplant models suggests that many factors influence the interplay between angiogenesis and vasculogenesis and also illustrates the complexity and variability of these experiments.

Factors That Influence the Numbers of Circulating Endothelial Progenitors

The wide range of estimates on the relative contribution of bone marrow–derived cells to neovasculature suggests that this contribution may vary depending on the conditions in a particular experiment or the biological/clinical situation. One prevailing theme in the literature is the absence or low level of bone marrow–derived cells incorporated into quiescent endothelium, while higher percentages of marrow–derived cells are reported in proliferating vasculature. Other than this broad generalization, factors that may influence the relative contributions of angiogenesis and vasculogenesis to neovessel formation in different circumstances remains unclear. One variable that may play a role is the number of circulating endothelial progenitors.

In a broad sense, the number of circulating endothelial progenitors is increased by vascular injury. More specifically, the levels of several cytokines have been found to influence the numbers of circulating EPCs. Not surprisingly, VEGF was one of the first substances identified that increases EPC number. Increased plasma levels of VEGF in adult mice and humans has been shown to increase EPCs. Other chemokines such as platelet-derived growth factor (PDGF), fibroblast growth factor (FGF), and interleukin-8 (IL-8) have also been shown to promote the migration of endothelial cells to sites of injury. G-CSF, GM-CSF, and erythropoietin, cytokines already in use clinically to promote mobilization of hematopoietic cells, can also increase EPCs. EPC mobilization can also be increased by HMG-CoA reductase inhibitors (statins). Further dissection of the mobilization of circulating EPCs by Heissig and colleagues shows that mobilization of both hematopoietic and endothelial precursors is dependent on MMP-9-mediated release of soluble kit ligand (sKitL) in the bone marrow.

Are Circulating Cells Critical to the Formation of New Vasculature?

Although numerous studies have demonstrated that circulating endothelial progenitors may be involved to a greater or lesser extent in vessel formation in the adult, few have unequivocally demonstrated their necessity in this process. Mice with reduced gene dosages of the Id transcription factors are unable to support neoangiogenesis in tumors. Rafii and colleagues demonstrated that bone marrow transplantation with wild-type marrow could reconstitute tumor angiogenesis in $Id1^{+/-}Id3^{-/-}$ mice. Not only were these studies the first to imply that marrow–derived cells are involved in new vessel formation in the adult, but they also demonstrated that these cells are critical for this process of new vessel formation. In this work, approximately 90 percent of the tumor vessels were found to contain cells derived from the donor bone marrow.

Clinical Studies Using Bone Marrow–Derived Cells to Augment Neovascularization

Although our knowledge is still in its infancy, the biology of circulating EPCs has begun to show some therapeutic promise. In early animal experiments, the injection of human bone marrow–derived EPCs improved revascularization in rodent hind limb ischemia models. Local injection of EPC or CD34+ cells into the ischemic myocardium of rats has also been shown to improve the recovery of cardiac function after the injury. The use of EPCs in revascularization has progressed to clinical trials. The delivery of autologous bone marrow cells was shown to improve revascularization in patients with ischemic limbs in the Therapeutic Angiogenesis using Cell Transplantation Study. Several clinical trials have also shown benefit to the local delivery of EPCs in the treatment of myocardial infarction.

Using Bone Marrow–Derived Cells in Antiangiogenic Strategies

Using EPCs to impede vascular growth is also being investigated. Ferrari and colleagues used bone marrow–derived cells to deliver a conditional cytotoxic gene to tumor vasculature. This approach resulted in a significant decrease in tumor growth. Other laboratories have used similar methods to impede tumor growth. The depletion of circulating endothelial progenitors is also being studied as an approach for the treatment of cancer. Preclinical experiments demonstrated the necessity of endothelial progenitors in tumor growth, and low-dose chemotherapy has been shown to decrease the number of circulating endothelial precursors. This treatment strategy is currently being evaluated in clinical trials.

Conclusion

Significant advances in our understanding of the role of bone marrow–derived cells in the formation of new vasculature have occurred in the last decade. Despite these advances, however, large gaps in our understanding of this biology remain. The population(s) of cells involved continue to be unresolved, and the methods of isolation of circulating endothelial progenitor cells remain to be standardized. There continues to be controversy regarding the importance of bone marrow–derived cells in neovasculature and the relative contribution of vasculogenesis versus angiogenesis in the adult. Other important questions such as a detailed description of the mechanisms of homing to areas of vascular growth and the role and ultimate fate of the endothelial progenitor cells that have incorporated into neovasculature in vivo remains to be elucidated. It is becoming clear that understanding the biology of circulating endothelial progenitor cells is critical to understanding the mechanisms of vessel formation, a process important in many pathological and physiological processes. Continued advances in this area are likely to pay significant clinical dividends in the future.

Glossary

Angiogenesis: The growth of new blood vessels from preexisting vasculature.

Endothelial progenitor cell: A precursor cell with the potential to develop into a mature, functional endothelial cell.

Vasculogenesis: The formation of new blood vessels from the aggregation of endothelial progenitors.

Further Reading

Asahara, T., and Isner, J. M. (2002). State-of-the-art reviews on vascular stem cells and angiogenesis. *J. Hematotherapy Stem Cell Res.* **11**, 171–178.

Asahara, T., Murohara, T., Sullivan, A., Silver, M., van der Zee, R., Li, T., Witzenbichler, B., Schatteman, G., and Isner, J. M. (1997). Isolation of putative progenitor endothelial cells for angiogenesis. *Science* **275**, 964–967. *The initial description of circulating endothelial progenitor cells in the adult.*

Lyden, D., Hattori, K., Dias, S., Costa, C., Blaikie, P., Butros, L., Chadburn, A., Heissig, B., Marks, W., Witte, L., Wu, Y., Hicklin, D., Zhu, Z., Hackett, N. R., Crystal, R. G., Moore, M. A. S., Hajjar, K. A., Manova, K., Benezra, R., and Rafii, S. (2001). Impaired recruitment of bone-marrow-derived endopthelial and hematopoietic precursor cells blocks tumor angiogenesis and growth. *Nat. Med.* **7**, 1194–1201. *An important study demonstrating the necessity of bone marrow–derived cells for the formation of tumor vasculature.*

Moldovan, N. I. (ed.) (2003). *Novel Angiogenic Mechanisms: Role of Circulating Progenitor Endothelial Cells.* New York: Klumer Academic/Plenum Publishers. *A complete monograph detailing the role EPCs derived from monocytes in vascular growth.*

Rabbany, S. Y., Heissig, B., Hattori, K., and Rafii, S. (2003). Molecular pathways regulating mobilization of marrow-derived stem cells for tissue revascularization. *Trends Mol. Med.* **9**, 109–117. *Excellent review of the known factors that effect EPC mobilization.*

Rafii, S., Lyden, D., Benezra, R., Hattori, K., and Heissig, B. (2002). Vascular and hematopoietic stem cells: Novel targets for anti-angiogenesis therapy. *Nat. Rev. Cancer* **2**, 826–835.

Szmitko, P. E., Fedak, P. W. M., Weisel, R. D., Stewart, D. J., Kutryk, M. J. B., and Verma, S. (2003). Endothelial progenitor cells: New hope for a broken heart. *Circulation.* **107**, 3093–3100. *Describes current clinical trials using EPCs.*

Capsule Biographies

Dr. Fine is Chief of the Neuro-Oncology Branch of the National Cancer Institute and The National Institutes of Neurological Disorder and Stroke at the NIH. His laboratory focuses on the development of novel therapeutic strategies, including antiangiogenic approaches, for primary brain tumors.

Dr. Glod is Assistant Professor of Pediatrics in the Division of Pediatric Hematology/Oncology at The Cancer Institute of New Jersey and The University of Medicine and Dentistry of New Jersey. His research focuses on the role of bone marrow–derived cells in the growth and repair of vasculature in the brain.

CHAPTER 18

Pericytes and Microvascular Morphogenesis

Ira M. Herman

Tufts University, School of Medicine, Boston, Massachusetts

Overview

During the past few years, the cellular mechanisms regulating microvascular morphogenesis have come under close scrutiny. In consideration of the important role that cell-to-cell and cell-to-matrix interactions play in regulating these developmental and disease-related processes, researchers have taken advantage of the technology and approaches founded in genetics and cell biology in their efforts aimed at revealing the molecular mechanisms controlling developmental and pathologic angiogenesis. Although significant inroads have been made with regard to the molecular and cellular details that govern endothelial seeding of early organ primordia or the regulatory role that pericytes play in microvascular physiology, there is currently little knowledge regarding the molecular and cellular events regulating vascular morphogenesis, including the signals regulating the recruitment and local differentiation of pericytes. Presumably, the early inductive events that stimulate vascular endothelial growth are also linked to the subsequent rounds of vascular remodeling seen in association with vascular proliferative disorders (e.g., tumor-induced angiogenesis, diabetic retinopathy, age-related macular degeneration). Collectively, these vascular morphogenetic events are characterized by the (1) growth/differentiation of vascular endothelial cells, (2) subsequent immigration and/or colonization of microvessels with pericytes, and (3) incipient dedifferentiation, loss, or death of pericytes that initiates or sustains the vascular pathophysiologic state. Based on our work and experiments carried out in our colleagues' labs, it seems quite likely that pericyte growth and differentiation are critically dependent on autocrine signals, cues that emanate from the microvascular endothelium and information arising from the associating extracellular matrix components present in the capillary basement membrane. In turn, these homotypic and heterotypic cellular interactions (i.e., endothelial–endothelial, endothelial cell–pericyte, and pericyte–astrocyte) and the interactions that occur between these cells and the surrounding extracellular matrix dictate where, when, and how vascular cell growth and differentiation ensue. Moreover, acute or chronic perturbations in these molecular and cellular signaling pathways will play pivotal roles in developmental or pathologic angiogenesis.

Microvascular Morphogenesis

The vasculature forms via two processes: vasculogenesis and angiogenesis. In vasculogenesis, endothelial cell (EC) precursors called *angioblasts* associate to form early vessel tubes. Tissues that are vascularized by this process are generally of endodermal origin and include lung and pancreas as well as the heart tube and dorsal aorta. In the process of angiogenesis, small blood vessels form by budding and sprouting from larger, extant vessels. Tissues of ectodermal and mesodermal derivation, such as the kidney, brain, and retina, are thought to be vascularized primarily via angiogenesis. In addition, angiogenesis appears to be the predominant means of neovascularization during events such as wound healing and during pathologies such as proliferative diabetic retinopathy. Whether vessels form by angiogenesis or vasculogenesis, the primitive vessels are subsequently remodeled. Remodeling is a poorly understood event that involves growth of new vessels and the regression of others, as well as changes in lumen diameter and vessel wall thickness, to suit the local tissue needs.

Although there are fundamental differences between angiogenesis and vasculogenesis, these two processes share several common regulatory features.

Until recently, information regarding the regulation of angiogenesis and vasculogenesis came from classic embryological studies. During the past decade, cell and molecular analyses have divulged several vessel-specific receptors and their corresponding high-affinity ligands. These include the isoforms of VEGF and its receptors, flk1, flt1, and neuropilin-1; platelet-derived growth factors A and B (PDGF) and their receptors α and β; angiopoietins (ang) 1 and 2 and their receptor tie-2; and transforming growth factor β (TGF-β) and the TGF-β receptors 1 and 2 [1–3]. This database, when combined with the ability to genetically manipulate vascular development and differentiation using mouse genetic approaches, has begun to provide important insights into the molecular mechanisms that regulate vessel growth and remodeling. Although our knowledge of the molecules involved in the assembly of mature, quiescent blood vessels is far from complete, observations to date do permit some speculation regarding how vascular cells take advantage of these developmental and differentiative signaling pathways to assemble a physiologically functional vasculature [1, 3].

Receptor Kinases Regulate Microvascular Growth and Differentiation

One specific area of interest in vascular signal transduction has been the recent identification of receptor tyrosine kinases (RTKs), which coordinate signaling cascades integral to angiogenic phenomena. At least five transmembrane receptor kinases have been shown to possess an expression pattern that is predominantly endothelial, and these can be conceptually subdivided into two groups. One group, or subfamily, encodes the VEGF receptors, flt-1, flk-1, and flt-4, whereas the second group encodes the TEK and Tie receptors. Both groups share structural homology with extracellular Ig and cytoplasmic split-kinase motifs. Because expression patterns vary with developmental time and space, there has been considerable speculation as to whether the differential expression of each RTK is important in dictating when or how the vascular endothelial lineage is specified. Based on the very early and predominantly overlapping expression patterns of flk-1 compared with the subsequent onset of TEK and tie-1 expression during embryonic development, it has been suggested that these signaling pathways are necessary for establishing the embryonic vasculature. Analysis of dominant-negative mutations in the murine TEK gene reveal its essential role in fostering endothelial proliferation in vivo, but not in the derivation of the endothelial lineage from angioblastic precursors. When considering the recurrent rounds of neovascularization seen during normal development or accompanying adult-onset, ischemia-induced ocular angiogenesis, it is also quite likely that the regulated expression of the fibroblast, platelet-derived growth factor, as well as the vascular endothelial growth factor (VEGF) families are also of pivotal importance for endothelial as well as pericyte proliferation [3, 4].

Receptor Tyrosine Kinases: Endothelial–Pericyte Interactions?

Biochemical analyses and gene cloning studies have clearly demonstrated the importance of the fibroblast growth factor family in fostering endothelial growth and angiogenesis. Although it has been established that bFGF (FGF-2) is a component of select subendothelial basement membranes in vivo (as well as a component of the endothelial extracellular matrix in vitro), its mode of release as well as its accumulation and function in the matrix remain somewhat equivocal. Because matrix (heparin)-bound FGF envelops populations of quiescent endothelial cells, others and we have suggested that it acts as a wound healing or survival agent, being released following tissue injury. However, if FGF-2 acts only in an autocrine or paracrine role to stimulate endothelial proliferation, then it is difficult or impossible to explain the exceedingly low rate of endothelial turnover observed in microvascular beds that possess FGF-2-enriched basement membranes, let alone its presence in the extracellular compartment despite any convincing evidence of local injury or cell death. These important observations have led us to postulate other critical roles for FGF signaling in vascular cells [5]. Specifically, is it possible that the recruitment and/or local differentiation of retinal pericytes are regulated by endothelial-produced and matrix-associated growth regulators (i.e., FGF-2).

As has been well studied, the pericyte resides within the basement membrane in association with the microvascular endothelial cell. Although the pericytes' contribution to the basement membrane, per se, has not been critically established, the strategic location of this mural cell population within the basement membrane places it in a position capable of participating in unique heterocellular signaling, events that may be essential for their sustained growth and/or survival [6, 7]. Work in the field has demonstrated a role for PDGF in potentiating vascular smooth muscle and pericyte proliferation. Vascular endothelial cell (EC) production of PDGF is seemingly regulated by injury and repair, because quiescent EC populations fail to produce measurable PDGF levels in vitro. In vivo, PDGF and PDGF RTK mRNA have been localized within pericyte-deficient capillaries, but it is unknown whether such an autocrine signaling loop is responsible for stimulating endothelial cell cycle progression. Recently, work from our laboratory has revealed that FGF-2 acts as a principal regulator of pericyte growth during the development and/or remodeling of the microvasculature [5]. These observations demonstrate a critical role for FGF signaling not only in microvascular development but in the pathological morphogenesis occurring during disease as well.

Modulation of pericyte contractile phenotype is by serine/threonine kinase signal transduction, TGFβ-1, and the extracellular matrix. TGFβ-1, which is a member of the transforming growth factor gene family and represents the prototype of this multimember family, coordinates a variety of activities, including the regulation of growth, differentiation, and extracellular matrix production. Its widespread expression during embryonic development points to the important role that TGFβ-1 plays in orchestrating epithelial–mesenchymal interactions, angiogenesis, chrondro-, and osteogenesis. Only recently, through the targeted disruption of TGFβ-1, have these essential role(s) been mapped. TGFβ-1 signals through two transmembrane serine/threonine kinases, STKI and STKII, which function together and share about 40 percent homology with each other [2]. Each STK possesses an extracellular cysteine-rich domain responsible for fostering ligand binding, a single transmembrane domain, and an intracellular serine/threonine kinase domain responsible for intracellular signaling. It has been demonstrated that binding of TGFβ-1 to STKII causes receptor phosphorylation, allowing association and phosphorylation of STKI, therein forming a functional heteromeric complex capable of downstream signaling. In vivo phosphorylation occurs within a stretch of five clustered threonines and serines (TTSGSGSG) within the GS domain, a conserved stretch of 30 amino acids that are unique to STKI. Mutations that either (1) inactivate the STKI kinase domain, (2) eliminate the STKI phosphorylation sites, or (3) alter the STKII cysteine-rich region involved in recognition of TGFβ-1 all prevent STKI phosphorylation and TGFβ-1 signal transduction. Elucidation of the downstream targets for STKI and STKII may unveil important new insights in TGFβ-1 signaling during embryonic development, cellular differentiation, and the response to injury. Our recent discovery that TGFβ-1 signaling may orchestrate the onset of pericyte differentiation as well as control pericyte extracellular matrix production not only indicates the putative importance of STK signaling during retinal microvascular development, but, in consideration of the known alterations that occur in retinal capillary permeability, basement membrane composition, and vascular cell growth/differentiation during diabetes, it seems evident that aberrations in TGFβ-1 signal transduction may also be critically involved in regulating pericyte de-differentiation during disease onset [5, 8].

Despite recent in vitro demonstrations that pericytes can dramatically inhibit endothelial growth in a TGF-β-dependent pathway, little evidence indicates that the reverse is also true (i.e., that pericyte growth or contractile phenotype can be modulated by endothelial-produced, matrix-associated growth regulators). Early work helped establish the important role that the endothelial-derived extracellular matrix plays in modulating retinal pericyte growth state and contractile phenotype [5, 9]. Most interesting and perhaps critical to our more current understanding of matrix dynamics and pericyte signal transduction is the demonstration that collagen IV–enriched, endothelial-synthesized extracellular matrix fosters pericyte proliferation. In these growth-potentiated pericyte cultures, smooth muscle contractile proteins cannot be detected [5, 9]. Similarly, when pericytes are stimulated to proliferate under the control of exogenously added FGF-2 and heparin (both of which are components of the endothelial-derived matrix), α-smooth muscle actin expression is also repressed (as cells proliferate).

Not only do these observations suggest that the growth state of the cells and their resultant contractile phenotype are inversely related, but the data also suggest the phenotypic similarity that exists between vascular smooth muscle and pericytes, both with respect to the molecular events controlling contractile phenotype as well as cell cycle progression. It is tempting to speculate that control of either pericyte growth or differentiation is the sum total of the positive and negative signals that originate in the vascular basement membrane, with each signaling pathway having as targets the regulators of cell cycle progression. In fact, reports in the literature state that stimulation of growth by FGF-2 is mediated by phosphorylation of the retinoblastoma susceptibility gene product, RB, which is thought to restrict cell cycle progression through late G1. It has also been demonstrated that TGFβ-1 may inhibit cell proliferation by preventing RB phosphorylation. We presume that both of these important signaling pathways are in place in the microvasculature (Figure 1). Furthermore, in consideration of the role that TGFβ-1 signal transduction plays in fostering retinal pericyte differentiation (presumably via induction of smooth muscle gene transcription), it seems possible that other downstream targets are also affected. For example, TGFβ-1 signaling in pericytes and/or endothelial cells could account for the excessive accumulation of type IV collagen in diabetic basement membranes. In turn, reduplication of the basement membrane may cause the loss of contact of endothelial cells from pericytes. Concomitantly, pericyte phenotypic modulation would ensue, resulting in de-differentiation, loss of smooth muscle contractile proteins, and a failure to restrain endothelial cells in their growth-arrested state. These de-repressed endothelial cells would then begin proliferating and retinal angiogenesis would ensue. In this model (Figure 1), we would postulate that pericyte phenotypic modulation (de-differentiation), together with local alterations in the extracellular matrix (rather than pericyte death), serve as early signals that elicit the recurrent rounds of endothelial proliferation observed during diabetes-induced neovascularization. This hypothesis is supported by the observations that, in humans, only a modest decrement in pericyte number is seen in proliferative diabetic retinopathy (~20% to 30%). Our collective data would strongly suggest that the pericytes are present in the proliferative retinopathic vasculature, but these cells are neither fully contractile nor in intimate contact with their endothelial neighbors (Figure 1).

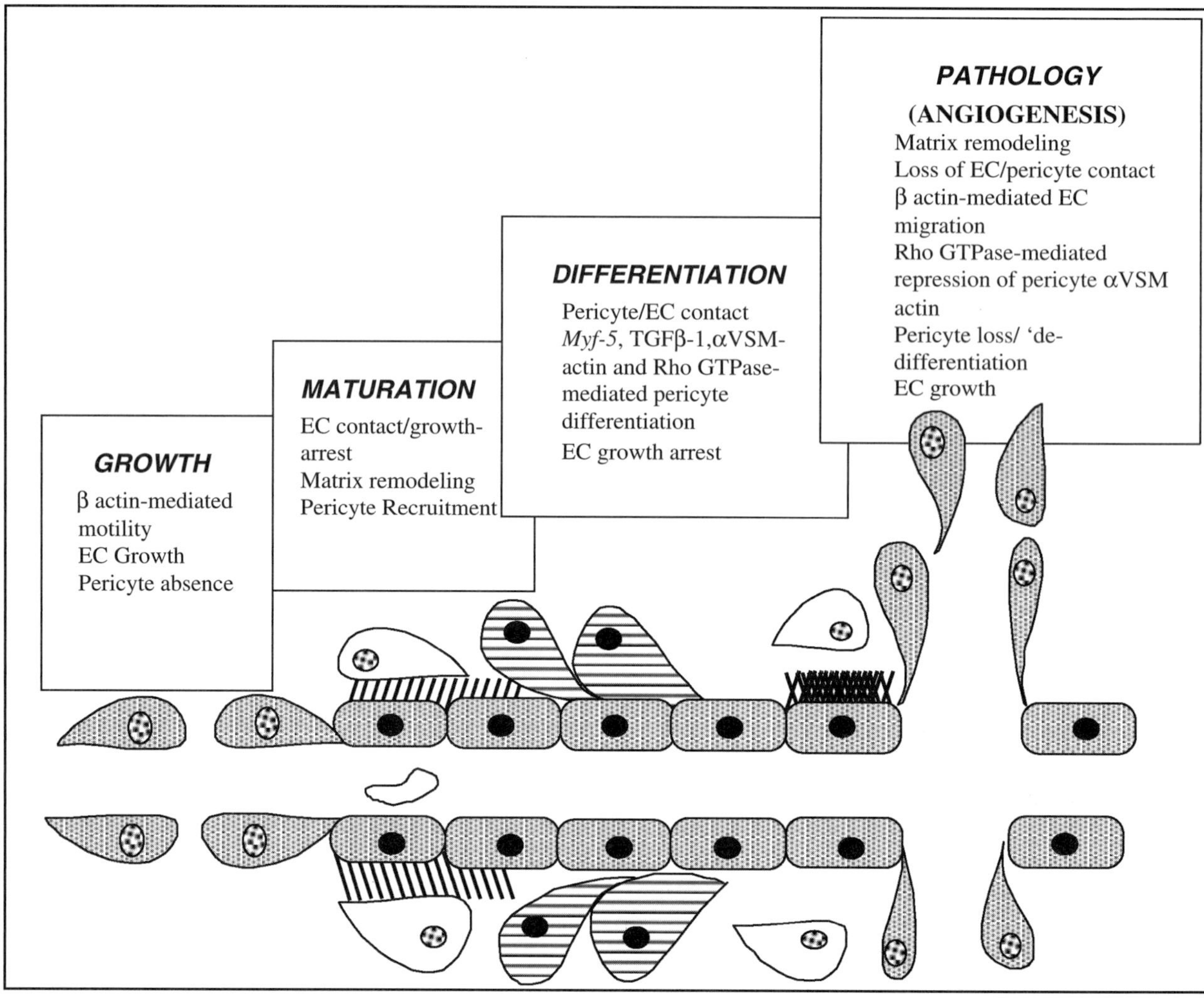

Figure 1 Model depicting microvascular morphogenesis during developmental and pathologic angiogenesis.

Pathologic Angiogenesis: Recapitulation of Microvascular Development

Development of the retinal vasculature is tightly regulated and highly dependent on cell-to-cell and cell-to-matrix interactions. For example, during retinal development, astrocytes transiently express VEGF as they migrate across the ganglion cell layer, preceding the formation of the superficial layer of retinal microvessels. Müller cells also express VEGF and direct a deeper layer of retinal microvascular morphogenesis. Blinding diseases, such as diabetic retinopathy, age-related macular degeneration, or neovascular glaucoma are directly related to aberrant angiogenic responses, which deploy common features of developmental angiogenesis, including ischemia and increased vascular permeability. For example, in diabetic retinopathy, pathological changes in the retinal vasculature yield proliferation of endothelial cells, and the loss or de-differentiation of pericytes is believed to be causally linked to the pathologic angiogenic cascade.

The correlation between the absence of pericytes and retinal neovascularization in diabetic retinopathy led to the hypothesis that pericytes exert a suppressive influence in capillary growth. Pericytes and endothelial cells share and coproduce a common basement membrane through which direct cell-to-cell contact can be achieved. Consistent with this concept was the ultrastructural observation that pericyte association with the developing capillary marked the cessation of vessel growth. Pericyte growth and recruitment is positively regulated by PDGF-β, and recently it was reported that endothelium-restricted ablation of PDGF-β generates viable mice with extensive inter- and intra-individual variation in the density of pericytes throughout the CNS, including retina. This strong inverse correlation between pericyte density and the formation of a range of retinal microvascular abnormalities suggests that pericyte depletion may be sufficient to cause retinopathy, at least, in mice. Evidence suggests that at least some of the retinal pericytes are lost during nonproliferative diabetic retinopathy as a result of apoptosis, presumably secondary to hyperglycemic episodes. However, based on earlier work carried out in the lab, we also suspect that pericyte loss is caused by their de-differentiation and concomitant dissociation from the retinal endothelium. We further speculate that pericyte apoptosis might ensue as a result of either mechanism; alterations in the retinal microvascular basement membrane are also likely to be contributory, all events leading to a similarly dysfunctional retinal microvasculature. Although these mechanisms by which pericytes suppress endothelial cell proliferation and/or stabilize capillaries have not been completely elucidated, coculture of pericytes with endothelial cells indicate that growth inhibition is achieved in a contact- and TGF-β-dependent process. It is our working hypothesis that signaling through the retinal microvascular isoactin network orchestrates the retinal microvascular remodeling required for developmental and pathologic retinal angiogenesis. Ongoing experiments will not only directly test this hypothesis, but results will also reveal the molecular switches controlling developmental and pathologic angiogenesis.

The Isoactin Network Drives Retinal Microvascular Morphogenesis

Coordination of cell shape and motility during retinal microvascular morphogenesis occurs principally through molecular cascades that signal isoactin expression, isoprotein sorting, and cytoskeletal reorganization. Both the assembly state as well as the molecular composition of the isoactin network expressed in vascular cells is tightly regulated. Furthermore, actin cytoskeletal control mechanisms arise through specific protein–protein interactions that sequester isoactin monomers, foster filament cross-linking, or alter polymerization kinetics [10]. Clearly, selective subcellular or cell-specific functions are afforded through the specific expression and sorting of multiple actin isoforms encoded by distinct, structurally related genes [11].

There are multiple lines of genetic, cellular, and biochemical data demonstrating functional diversity of the actin isoforms. Functional distinction is evidenced by altered expression profiles, biophysical properties, cellular localization, and overexpression studies. For example, in retinal pericytes or vascular smooth muscle cells that are actively dividing, nonmuscle isoactins are prevalent; however, when the cells are contractile, α-VSM actin (among other smooth muscle–specific proteins) prevails [9]. Expression of αVSM actin is strongly regulated during wound contraction, as seen in myofibroblasts, during mesangial cell damage in hypertension, and in de-differentiated, intimal vascular smooth muscle cells present in atherosclerotic lesions. This regulated pattern of vascular cell isoactin expression mediates the developmental and disease-related motile- and contractile-mediated events.

Rho GTPases Signal Isoactin-Dependent Cytoskeletal Remodeling

More recently, we have recognized the pivotal role that the Rho GTPases and their downstream signaling effectors play in regulating retinal microvascular shape, motility, and contractile potential [12, 13]. As has been well documented, activation of Rho GTPase in other cells has been shown to cause the bundling of actin filaments into stress fibers, along with the clustering of focal adhesion complex proteins. Like all GTPases, Rho GTPases act as molecular switches, which cycle between an inactive GDP-bound and an active GTP-bound state, and the ratio between these two forms depends on the activity of regulatory factors. GTPase-activating proteins (GAPs) promote the inactive state of the GTPase by increasing the GTPase's intrinsic rate of nucleotide hydrolysis, while guanine nucleotide dissociation inhibitors (GDIs) interfere with both the exchange of GDP for GTP and the hydrolysis of bound GTP. Guanine nucleotide exchange factors (GEFs) promote the active GTP-bound state and tether the GTPases to specific subcellular locations in order to generate an active signal. Biological implications of these findings are wide ranging, and Rho GTPases, together with family members, rac and Cdc42, play key regulatory roles in cell movement, axonal guidance, as well as multicellular morphogenetic processes involving changes in cell polarity and angiogenesis [3].

Two protein families appear to be required for contractility through Rho-induced assembly of stress fibers and focal adhesions: the Rho-associated kinases (ROCK) and the Dia members of the formin family. These include ROCK1 and ROCK2 as well as mDia1 and mDia2. Activation of ROCK appears to be necessary, but is not sufficient, for stress fiber formation. Inhibition of ROCK using the inhibitor Y-27632 prevents stress fiber formation, whereas a constitutively activated mutant of ROCK1 promotes the formation of stellate actomyosin filaments, which do not resemble stress fibers. Through its downstream phospho-substrates, ROCK can signal isoactin reorganization by multiple mechanisms (e.g., ROCK has been shown to phosphorylate and thereby inactivate MLC phosphatase, thus resulting in an increase in phosphorylated MLC). In addition, ROCK1 can also play a pivotal role in generating contractile force by stabilizing filamentous actin cross-links through a LIM-kinase-dependent phosphorylation of the actin depolymerizing protein, ADF-cofilin, which is inactivated by phosphorylation. In turn, this may also influence cellular ATP levels, which may also be important in regulating cellular contractility, as has recently been demonstrated for retinal pericytes [14]. For the microvasculature, we believe that these molecular signaling cascades are essential elements controlling developmental and pathologic angiogenesis and offer a newfound awareness for the molecular and cellular mechanisms regulating microvascular morphogenesis.

Glossary

Capillary: A microvascular blood vessel interposed between the arterial and venous microcirculation, which has a lumenal diameter between 5 to 8 micrometers (i.e., wide enough to permit the passage of one red blood cell to pass, one at a time).

Cytoskeleton: The structural organelle of living cells, which comprises structural and contractile protein components. The cytoskeleton is usually comprised of three distinct filament systems: the thin filaments (actin filaments), the intermediate filaments (desmin or vimentin filaments), and the microtubules (made up of the protein, tubulin).

Morphogenesis: A term referring to the structural development of an organism or a part thereof (e.g., in the formation of the blood vascular system).

Pericyte: The mural (wall) cell of the capillary and postcapillary venule, which has been shown to possess contractile and structural properties similar to the vascular smooth muscle cell and which has been demonstrated to influence vascular endothelial cell growth.

References

1. Nguyen, L. L., and D'Amore, P. A. (2001). Cellular interactions in vascular growth and differentiation. *Int. Rev. Cytol.* **204**, 1–48.
2. Attisano, L., and Wrana, J. L. (2002). Signal transduction by the TGF-beta superfamily. *Science* **296**, 1646–1647.
3. Papetti, M., and Herman, I. M. (2002). Mechanisms of normal and tumor-derived angiogenesis. *Am. J. Physiol. Cell Physiol.* **282**, C947–C970.
4. Darland, D. C., Massingham, L. J. et al. (2003). Pericyte production of cell-associated VEGF is differentiation-dependent and is associated with endothelial survival. *Dev. Biol.* **264**, 275–288. *Interesting report revealing the role that pericytes play in regulating capillary morphogenesis.*
5. Papetti, M., Shujath, J., Riley, K., and Herman, I. M. (2003). FGF-2 antagonizes the TGF-beta1-mediated induction of pericyte alpha-smooth muscle actin expression: A role for myf-5 and Smad-mediated signaling pathways. *Invest. Ophthalmol. Vis. Sci.* **44**, 4994–5005. *Thorough accounting of the up- and downstream signaling pathways responsible for regulating pericyte growth and contractile phenotype.*
6. Abramsson, A., Lindblom, P. et al. (2003). Endothelial and non-endothelial sources of PDGF-B regulate pericyte recruitment and influence vascular pattern formation in tumors. *J. Clin. Invest.* **112**, 1142–1151.
7. Cho, H., Kozasa, T. et al. (2003). Pericyte-specific expression of Rgs5: Implications for PDGF and EDG receptor signaling during vascular maturation. *FASEB J.* **17**, 440–442.
8. Sieczkiewicz, G. J., and Herman, I. M. (2003). TGFβ-1 signaling controls retinal pericyte contractile protein expression. *Microvasc. Res.* **66**, 190–196.
9. Newcomb, P. M., and Herman, I. M. (1993). Pericyte growth and contractile phenotype: Modulation by endothelial-synthesized matrix and comparison with aortic smooth muscle. *J. Cell. Physiol.* **155**, 385–393.
10. Carlier, M. F. (1998). Control of actin dynamics. *Curr. Opin. Cell. Biol.* **10**, 45–51.
11. Herman, I. M. (1993). Actin isoforms. *Curr. Opin. Cell. Biol.* **5**, 48–55.
12. Etienne-Manneville, S., and Hall, A. (2002). Rho GTPases in cell biology. *Nature* **420**, 629–635.
13. Kolyada, A. Y., Riley, K. N., and Herman, I. M. (2003). Rho GTPase signaling modulates cell shape and contractile phenotype in an isoactin-specific manner. *Am. J. Physiol. Cell Physiol.* **285**, C1116–C1121. *Demonstration that the Rho GTPase signaling pathway is active in pericytes and that modulation of cell shape and contractile potential is regulating in an isoactin-specific manner.*
14. Kawamura, H., Sugiyama, T. et al. (2003). ATP: A vasoactive signal in the pericyte-containing microvasculature of the rat retina. *J. Physiol.* **551**, 787–799. *Recent findings revealing the role that ATP plays in controlling pericyte contractility.*

Capsule Biography

Dr. Herman is currently Professor of Cellular and Molecular Physiology at Tufts University School of Medicine, where he also holds appointments as Professor of Cell Biology and Anatomy as well as Professor of Ophthalmology. For more than two decades, work in the Herman laboratory has been instrumental in revealing the molecular and cellular mechanisms regulating vascular remodeling during development and disease. For this work, Dr. Herman has been the recipient of numerous awards, including grants from the NIH, American Heart Association, and pharma.

CHAPTER 19

PDGF-B and Pericyte Recruitment

Michael J. Jarzynka and Shi-Yuan Cheng
University of Pittsburgh Cancer Institute, Pennsylvania

Vasculogenesis and Angiogenesis

Blood vessels are formed by two processes: vasculogenesis and angiogenesis. Vasculogenesis involves the de novo formation of blood vessels that occurs mostly during embryogenesis. Angiogenesis is a process comprising the development of new blood vessels from pre-existing vessels. Angiogenesis takes place during embryogenesis and is often associated with physiological conditions including wound healing and the reproductive process as well as pathological conditions such as arthritis, diabetic retinopathy, periodontal disease, psoriasis, and cancer.

Blood vessel formation is closely linked with hematopoiesis. The hemangioblast is a common precursor to endothelial and hematopoietic cells. During vasculogenesis, hemangioblasts aggregate to create blood islands. These blood islands are composed of hematopoietic stem cells and angioblasts that further differentiate into hematopoietic cells and endothelial cells, respectively. Endothelial cells form the inner lining of blood vessels and generate tube-like structures that connect to form an primitive network of capillaries. Next, vascular remodeling such as pruning and sprouting occur, resulting in a mature vessel network.

The vasculature is composed of capillaries, arterioles, and venules. The microvasculature is made up of endothelial cells and mural cells. Mural cells are vascular support cells in which pericytes constitute a major portion (Figure 1). Larger blood vessels such as arteries and veins differ slightly from the microvasculature in that they contain endothelial cells, smooth muscle cells (differentiated pericytes), fibroblasts, collagen, and elastic fibers. Pericytes and smooth muscle cells form a layer surrounding the developing vascular tube, facilitating the formation of a mature vessel.

Various growth factors modulate a variety of pathways during vasculogenesis and angiogenesis. For example, fibroblast growth factor (FGF) is responsible for the generation of hemangioblasts during vasculogenesis. Vascular endothelial growth factor (VEGF) initiates the differentiation of angioblasts into endothelial cells. Additionally, VEGF stimulates endothelial cell proliferation, migration, and sprouting, leading to endothelial tube development and eventually new vessel formation. These newly formed vessels are immature and leaky, requiring further processing. Growth factors such as angiopoietin-1 (Ang1), transforming growth factor-β (TGF-β), and platelet-derived growth factor (PDGF) are responsible for the maturation and stabilization of the newly formed vessels. PDGF facilitates the recruitment of pericytes to encase the endothelial cell layer, reinforcing the vessel. Lack of pericytes leads to leaky blood vessels that are structurally defective.

Platelet-Derived Growth Factor

Platelet-derived growth factor (PDGF) family plays important roles in angiogenesis. PDGF is an approximately 30 kilodalton (kDa) protein consisting of four related polypeptide chains (-A, -B, -C, and -D) that exist in dimeric form as a result of disulfide bonds between cysteine residues contained within the peptides. PDGF can form homodimers (-AA, -BB, -CC, and -DD) as well as a heterodimer (-AB). The c-sis/PDGF-2 gene on chromosome 22q that encodes PDGF-B in humans is homologous to the v-sis of the Simian sarcoma virus, which is significant because it is an oncogene.

The PDGF family of proteins binds to and signals through membrane-bound receptor tyrosine kinases (RTKs), of which there are two isoforms: PDGF receptor (PDGFR)-α (170 kDa) and -β (190 kDa). Similar to their ligands, the PDGFR also form homo- or heterodimers: PDGFR-αα, -αβ, and -ββ. PDGFR-α and -β arise from two different gene products and have various binding affinities for their PDGF ligands. For example, PDGF-AA binds only the -αα

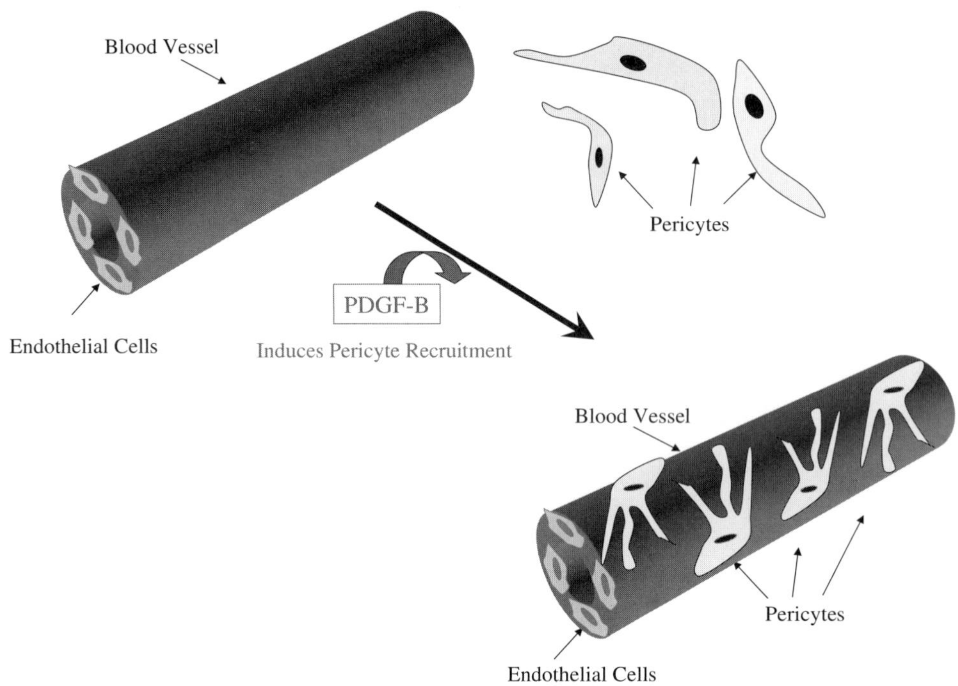

Figure 1 Schematic Representation of Pericyte Recruitment to a Newly Formed Vessel. PDGF-B released from endothelial and other types of cells acts as a chemoattractant by stimulating PDGFR-β on vascular support cells called pericytes. This stimulus causes the recruitment of pericytes that encase and structurally stabilize the growing vessel. (see color insert)

receptor homodimer, whereas PDGF-BB interacts with -αα and -αβ receptors but has the highest binding affinity for the -ββ receptor. The PDGF-AB heterodimer preferentially binds to -αα and -αβ receptors. PDGF-CC and PDGF-DD interact with the -αα and -ββ receptors, respectively. All of the receptor subtypes are comprised of five immunoglobulin-like extracellular domains containing a ligand-binding site, a single hydrophobic transmembrane domain, and an intracellular domain containing residues with tyrosine kinase activity (Figure 2).

Binding of ligand induces receptor homo- or heterodimerization depending on cellular expression of the receptor subtype and ligand present. In a process known as *autophosphorylation*, the kinase portion of each receptor monomer phosphorylates tyrosine residues within the intracellular domain of its dimeric partner. Afterward, additional phosphorylation events occur at other locations within the intracellular domain of the receptor. These phosphorylated tyrosine residues serve as docking sites for signaling molecules that contain SH2 domains such as PI-3 (phosphatidylinositol-3) kinase, Src (cellular counterpart of the Rous sarcoma virus gene, v-src), Grb2/Sos1 (growth factor receptor-bound protein 2/son of sevenless 1), Stat (signal transducer and activator of transcription), GAP (GTPase activating proteins), and PLC-γ (phospholipase C-γ), leading to modulation of cellular processes and gene expression. Cellular functions including mitogenesis, chemotaxis, and apoptosis are regulated based on the broad range of signaling cascades to which the PDGF receptors couple.

PDGF-B, Pericyte Recruitment, and Vessel Maturation

PDGF-B is the most characterized member in the PDGF family. Although first discovered as a secretory product of platelets during coagulation, PDGF-B is also expressed in many other cell types, such as endothelial cells, macrophages, smooth muscle cells, fibroblasts, glial cells, neurons, tumor cells, and possibly others. Expression of PDGF-B is constitutive in immune and nerve cells, whereas inducible expression of PDGF-B most likely occurs in other cell types. The actions of PDGF-B can occur via paracrine and autocrine pathways. PDGF-B is involved in many physiological functions, such as wound healing, inflammation, gestation, and differentiation.

PDGF-B triggers mural cell recruitment to sprouting vessels and proliferation of certain cell types during angiogenesis. Pericyte recruitment is believed to occur through the secretion of PDGF-B by endothelial cells, stimulating the PDGFR-β on pericytes, leading to migration of these cells to nascent capillaries of the developing microvasculature. In vitro, PDGF-B induces smooth muscle cell migration and proliferation. In addition, PDGF-BB has been shown to increase the expression of PDGFR-β in endothelial cells, eliciting endothelial cell proliferation, resulting in angiogenesis amplification in vitro.

In 1994, two studies revealed the roles of PDGF-B and PDGFR-β in vasculogenesis and angiogenesis during murine development. Soriano demonstrated that PDGFR-β gene knockout impaired kidney development in mice

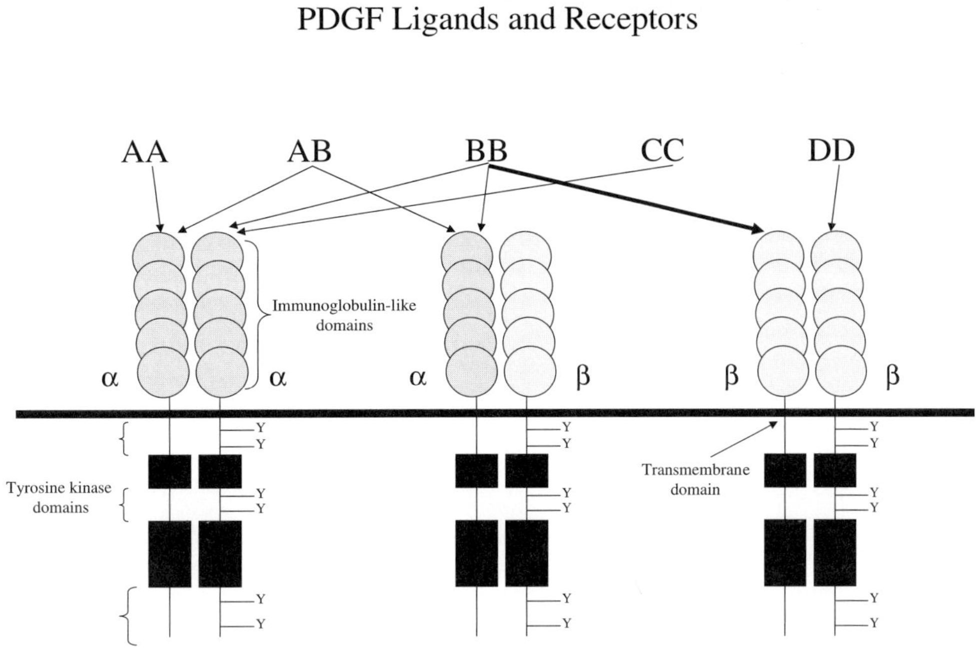

Figure 2 Schematic Representation of the Binding of PDGF Ligands to Their Respective Receptors. PDGF-AA binds only the -αα receptor homodimer. PDGF-BB binds to the -αα and -αβ receptors but has the highest binding affinity for the -ββ receptor. PDGF-AB preferentially binds to -αα and -αβ receptors. PDGF-CC and PDGF-DD interact with the -αα and -ββ receptors, respectively. The binding of PDGF to the extracellular immunoglobulin-like domains of its cognate receptors induces receptor homo- or heterodimerization, stimulating tyrosine phosphorylation in the intracellular kinase domains. The phosphorylated tyrosines trigger downstream signaling events to occur cellular events such as mitogenesis, chemotaxis, and apoptosis. (-Y, tyrosine residues) (see color insert)

because of glomerular development irregularities secondary to an absence of mesangial cells, cells that are structurally and functionally related to pericytes. In addition, hematologic abnormalities such as anemia, thrombocytopenia, erythroblastosis, and arteriolar dilation were also present. The other study performed by Betsholtz and group utilized PDGF-B-deficient mice that displayed a similar phenotype and died perinatally as a result of hemorrhage. In 1997, the effects of PDGF-B deficiency were further characterized, indicating that lack of pericytes during vessel growth produces microvascular aneurysms and related these findings to possible pathological complications seen in diabetes mellitus. These studies demonstrated the importance of PDGF-B signal-ing through the PDGF-β receptor in promoting the recruitment of pericytes to developing vessels thereby increasing their stability.

The Role of PDGF-B in Tumor Angiogenesis and Progression

Normal PDGF-B function is critical in physiological events such as embryonic development and wound healing. However, aberrant PDGF-B or PDGFR-β function has been implicated in several disease states depicted by excessive cell growth, including atherosclerosis, fibrosis, glomerulonephritis, liver cirrhosis, rheumatoid arthritis, and tumor progression.

Various types of human cancers have been shown to have increases in PDGF-B and/or PDGFR-β expression, affecting tumor growth and angiogenesis. Overexpression of PDGF and PDGF receptors has been observed in brain tumors, gastric carcinoma, lung cancer, melanomas, neuroendocrine tumors, ovarian cancer, pancreatic cancer, and prostate cancer. For example, in glioma pathogenesis, the degree of expression of PDGF-B correlates with increasing glioma tumor grade. In addition, increased expression levels of PDGF-B and PDGFR-β have been observed in tumor-associated endothelial cells in grade IV glioblastomas. Interestingly, the PDGFR-β was not expressed in the glioma cells. Additional studies comparing PDGFR-β expression in normal brain tissue and glioma tissue have demonstrated increased expression only in the tumor-associated endothelial cells. However, no detection of PDGFR-β was seen in the normal brain tissues and glioma cells. From these data, it can be presumed that PDGF-B acts on the PDGFR-β on the tumor-associated endothelial cells, stimulating glioma angiogenesis and progression.

Several in vitro studies have attempted to decipher the mechanisms behind PDGF actions on endothelial cells. PDGF-BB has been shown to have an increased potency to induce chemotaxis in rat brain capillary endothelial cells as compared to PDGF-AA. Further studies have demonstrated that reorganization of the cytoskeletal protein actin is responsible for this chemotactic effect. Additional mechanisms are involved in PDGF-B effects on endothelial cells, specifically the effect on VEGF. PDGF-B augments VEGF expression and secretion by stimulating the PI-3 kinase signal transduction cascade on PDGFR-β-expressing endothelial cells. In addition, a variety of human glioma cell lines stimulated with PDGF-B have been shown to increase VEGF secretion. Thus, PDGF-B enhances glioma angiogenesis and growth by stimulating VEGF expression in the tumor endothelia, increasing endothelial cell migration and mitogenesis, and by promoting recruitment of pericytes into growing vessels, thus facilitating vessel assembly.

The overexpression of PDGF-B in tumor cells and PDGFR-β in endothelial cells appears to play important roles in glioma development and progression, corroborating both paracrine and autocrine pathways of PDGF-B on endothelial cell proliferation and migration. Additionally, PDGF-B may act as an angiogenic switch that upregulates VEGF and enhances vessel growth, thus promoting glioma angiogenesis. Continuous high levels of expression of PDGF-B and PDGFR-β in gliomas further upregulate VEGF in the tumor endothelium in addition to increasing the recruitment of pericytes. These events would allow rapid neovascularization in malignant gliomas to occur.

In summary, the PDGF family of proteins is critical in physiological and pathological events because of their direct effects on the state of the vasculature. As demonstrated by genetic studies, lack of PDGF-B and PDGFR-β expression disrupts normal vascular development during embryogenesis, leading to premature death. In vitro analysis reinforced the importance of normal expression levels of these proteins in cell culture. In addition, in vivo tumor xenograft models proved that overexpression of PDGF-B increased glioma tumorigenicity and angiogenesis in mice. Overall, the effects of PDGF in homeostasis and pathology are becoming clearer. Researchers will continue to develop methods to induce or abrogate PDGF signaling and apply these methods to novel therapeutic treatments of disease.

Glossary

Angiogenesis: The process of new blood vessel formation from preexisting vessels.

Chemotaxis: The chemical-induced movement of cells toward or away from other cells.

Glioma: Malignant neoplasms of the glial cells (support cells) in the central nervous system.

Pericytes: Cells associated with the walls of capillaries and other small blood vessels.

Vasculogenesis: The de novo formation of new blood vessels.

Further Reading

Carmeliet, P. (2003). Angiogenesis in health and disease. *Nat. Med.* **9**, 653–660. *An outstanding review outlining the role of angiogenesis in physiological and pathological states. This review provides detailed insight into the molecular processes governing angiogenesis.*

Dunn, I. F., Heese, O., and Black, P. M. (2000). Growth factors in glioma angiogenesis: FGFs, PDGF, EGF, and TGFs. *J. Neurooncol.* **50**, 121–137.

Guo, P., Hu, B., Gu, W., Xu, L., Wang, D., Huang, H. J., Cavenee, W. K., and Cheng, S. Y. (2003). Platelet-derived growth factor-B enhances glioma angiogenesis by stimulating vascular endothelial growth factor expression in tumor endothelia and by promoting pericyte recruitment. *Am. J. Pathol.* **162**, 1083–1093. *A detailed examination of the role of PDGF-B in glioma angiogenesis and progression. This paper demonstrated the promotion of neovessel formation and pericyte recruitment induced by PDGF-B in vitro and in vivo using a mouse glioma tumor model.*

Heldin, C-H., Ostman, A., and Ronnstrand, L. (1998). Signal transduction via platelet-derived growth factor receptors. *Biochimica et Biophysica Acta* **1378**, F79–F113.

Hirschi, K. K., and D'Amore, P. A. (1997). Control of angiogenesis by the pericyte: Molecular mechanisms and significance. *Experientia Supplementa* **79**, 419–428.

Hirschi, K. K., Rohovsky, S. A., and D'Amore, P. A. (1998). PDGF, TGF-beta, and heterotypic cell-cell interactions mediate endothelial cell-induced recruitment of 10T1/2 cells and their differentiation to a smooth muscle fate. *J. Cell Biol.* **141**, 805–814.

Jain, R. K. (2003). Molecular regulation of vessel maturation. *Nat. Med.* **9**, 685–693.

Leveen, P., Pekny, M., Gebre-Medhin, S., Swolin, B., Larsson, E., and Betsholtz, C. (1994). Mice deficient for PDGF B show renal, cardiovascular, and hematological abnormalities. *Genes Devel.* **8**, 1875–1887.

Lindahl, P., Johansson, B. R., Leveen, P., and Betsholtz, C. (1997). Pericyte loss and microaneurysm formation in PDGF-B-deficient mice. *Science* **277**, 242–245. *A mechanistic study examining the pathological effects of PDGF-B deficiency during murine embryogenesis. The lack of pericyte recruitment to newly sprouting capillaries resulted in the formation of capillary microaneurysms.*

Soriano, P. (1994). Abnormal kidney development and hematological disorders in PDGF beta-receptor mutant mice. *Genes Devel.* **8**, 1888–1896. *An investigation probing the effects of PDGFR-β deficiency in mice during embryonic development. This study demonstrated that PDGFR-β deficient mice died perinatally due to renal and hematologic disorders proving a critical role that PDGFR-β signaling plays in vascular development.*

Capsule Biographies

Michael J. Jarzynka, Ph.D., is a postdoctoral fellow in the laboratory of Dr. Shi-Yuan Cheng at the University of Pittsburgh Cancer Institute. He is studying the mechanisms underlying tumor angiogenesis and invasion.

Shi-Yuan Cheng, Ph.D., is an Associate Professor of Pathology at the University of Pittsburgh Cancer Institute. His laboratory examines the molecular mechanisms by which angiogenic factors promote tumor angiogenesis/invasion and develops novel therapeutic approaches to target molecular pathways involved in brain, breast, and lung tumorigenesis and angiogenesis.

CHAPTER 20

VEGF Regulation of Morphogenesis

[1]Thomas J. Poole and [2]Eric B. Finkelstein

[1]SUNY Upstate Medical University, Syracuse, New York

[2]Schepens Eye Research Institute and Harvard Medical School, Boston, Massachusetts

Vascular endothelial growth factor (VEGF) expression and its regulation are critical for normal vascular development. Gene knockout studies in mice resulted in embryonic lethality from a failure of endothelial cell specification and blood vessel formation. Such studies have revealed an essential role for VEGF in vascular development. Animals heterozygous or homozygous for the VEGF-null mutation are embryonic lethal, demonstrating a critical role for dose regulation of VEGF during vascular morphogenesis. Additional studies in which mice were engineered to overexpress VEGF [1] or downregulate VEGF specifically in yolk sac endoderm [2] also lead to disruption of vascular patterns, a further demonstration that precise control of VEGF levels is critical for normal vascular development. Studies in the quail embryo, zebrafish, and frog have led to similar conclusions. Such studies have utilized delivery of soluble growth factor (quail, frog), delivery of soluble inhibitors (quail), and genetic knockdown with antisense or morpholinos (zebrafish, frog). All studies have demonstrated that increased or decreased VEGF levels result in abnormal vascular patterns.

Origins of the Embryonic Vascular Pattern

Vascular development is studied in several model vertebrate embryos, with different species having distinct advantages for understanding various aspects of vascular morphogenesis. Avian embryos of the chicken and Japanese quail have the advantage of accessibility in the egg or whole embryo culture, quail/chick chimeras, and a monoclonal antibody (QH-1) that labels quail angioblasts and endothelial cells. Mouse embryos from a variety of transgenic and knockout lines have been examined for vascular patterning defects, leading to the discovery of unexpected roles for several genes. Frog embryos have the advantage of ease of injection of DNA constructs for overexpression and classic tissue transplantation studies. Zebrafish embryos are nearly transparent, allowing for easy observation of the developing vasculature, and large-scale genetic screens have identified mutants with vascular pattern formation defects.

The formation of embryonic blood vessels occurs through a series of events (Figure 1). First, angioblasts (endothelial cell precursors) are induced from the embryonic mesoderm (Figure 1A), which at this stage is an epithelial cell sheet. The new angioblast initially undergoes an epithelial to mesenchymal transformation (EMT; Figure 1B), and the new mesenchymal cell migrates (Figure 1C) and coheres to other angioblasts, forming an initial solid cord of cells (Figure 1D) at sites of vessel formation (vasculogenesis). This solid cord further differentiates through unclear mechanisms to form a lumen in which a blood vessel tube is lined with endothelial cells (Figure 1E). In the case of the dorsal aorta, the first major vessel formed in the developing embryo, angioblasts originate from mesoderm directly adjacent to where they assemble in a process called *vasculogenesis type I*. In contrast, angioblasts that form the endocardium and cardinal veins migrate as individual cells or clusters that assemble at a distance from sites of angioblast origin in a process called *vasculogenesis type II*. Angiogenesis is the sprouting of new vessels from preexisting vessels. The first example of angiogenesis in the embryo

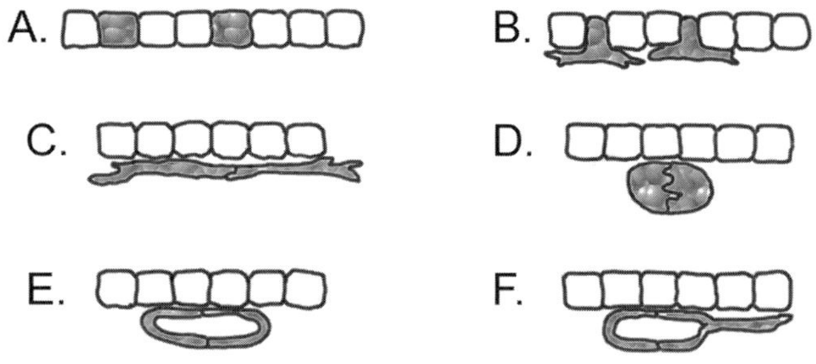

Figure 1 Initial Steps in Dorsal Aorta Morphogenesis. Embryonic blood vessels form from angioblasts that originate from the embryonic mesoderm by epithelial to mesenchymal transformation (EMT). **(A)** Two cells in a simple cuboidal epithelium acquire gene expression changes (*shaded*) that distinguish them as angioblasts. **(B)** These two angioblasts undergo EMT. **(C)** The angioblasts migrate and cohere to one another. **(D)** Two angioblasts have formed a portion of a solid cord. In the case of the dorsal aorta, a thin cord runs in the cranial-to-caudal direction. **(E)** A lumen is formed between two endothelial cells that continue to adhere at their junctions. **(F)** A sprout has formed off the vessel with the formed lumen.

is morphogenesis of the intersomitic arteries that sprout from the dorsal aorta (Figure 1F) after a lumen has formed.

Angioblast Induction

Although the role of VEGF in angioblast induction remains unclear, fibroblast growth factor 2 (FGF-2) has been implicated in angioblast induction in the quail embryo [3]. The expression of VEGF receptor 2 (VEGFR2, flk-1) resembles the expression of the quail antibody QH-1, which labels angioblasts as they undergo EMT. VEGFR2 is the first VEGF receptor expressed, suggesting that VEGF is not involved in angioblast induction. The idea that VEGF is not involved with angioblast induction fits with the phenotypes of VEGF and VEGFR2 knockout mice. In receptor knockout mice, endothelial cell precursors formed, but there were no mature endothelial cells, while in the VEGF ligand knockout mice, endothelial cells formed but failed to differentiate into a vascular pattern. Similarly, VEGFR2 mutations in zebrafish suggest a role for VEGF in sprouting and remodeling but not in initial specification of angioblasts [4].

Vasculogenesis

Morphogenetic Events

The dorsal aortae are formed by vasculogenesis. The role of VEGF in the assembly of the dorsal aorta was examined by perturbation studies in which VEGF was added or inactivated. VEGF has been inactivated by injection of soluble VEGF receptor-1/Fc hybrid protein in quail embryos [5] and by mouse or zebrafish knockout or mutation. In the quail embryo, VEGF has been added through direct injection of recombinant protein [5] or by surgical implantation of heparin chromatography beads preincubated with recombinant human VEGF165 [6], which serve as a slow-release source of VEGF (Figure 2). VEGF was delivered to somites of quail embryos with six or seven somite pairs, because somites give rise to few angioblasts. This was followed by seven hours of culture, until the embryos had 9 to 12 somite pairs. Embryos were fixed and stained as whole mounts with the QH-1 monoclonal antibody (Figure 2A). At early stages of vasculogenesis, ectopic VEGF delivery resulted in increased vascular densities such that the dorsal aorta (DA) and capillary plexus (CP) just lateral to the embryo appeared fused together, with obliteration of the avascular area (AVA) that normally separates the DA from the CP (Figures 2B and C).

The embryo and its vasculature develop in a rostral-caudal (head-to-tail) fashion, such that the most highly developed vascular pattern is located rostrally. Ectopic VEGF can affect morphogenetic events of vasculogenesis along the entire rostral-caudal axis of the embryo, resulting in hypervascularization [6]. However, qualitative differences in the vascular changes were observed at different levels. Ectopic VEGF at rostral levels resulted in an increased number of vessel branches between the DA and CP, similar to intersomitic arteries branching from the tubular dorsal aortae at this anatomic level. Moving caudal (closer to the tail), VEGF delivery at the level of a middle somite or the most recently formed somite resulted in a more generalized expansion of the vasculature. There was complete loss of the AVA, such that the DA and CP fused together (Figures 2B and C). The effect of ectopic VEGF can be seen by comparing of the side where VEGF was delivered to the opposite side, or to a control embryo (Figure 2A). These data suggest that VEGF influences multiple events of vascular morphogenesis as they occur simultaneously along the rostral-caudal axis. These include cohesion of angioblasts to form a solid cord caudally, further differentiation of the cord as a lumen forms at the middle somite level, and finally sprouting from the tubular DA at rostral levels, likely through angiogenesis (see Figure 1).

Mechanisms

There are several cellular mechanisms by which VEGF can influence endothelial cells as they form networks in vitro or in vivo. These include induction of endothelial cell proliferation or chemotaxis toward a VEGF source, where it can influence the size and pattern of blood vessels. Although it is known that increased or decreased VEGF levels in the developing embryo can profoundly affect vascular morphogenesis, to date the precise mechanism of VEGF activity remains unclear. VEGF does not induce the formation of new angioblasts from uncommitted mesoderm. Rather, it is believed to act on existing populations of angioblasts by guiding them to form a vascular pattern with vessels specifically located and of defined size. VEGF is not believed to influence the developing vasculature in our experiments by inducing endothelial cell or angioblast proliferation. The effects of VEGF were observed in as little as four hours,

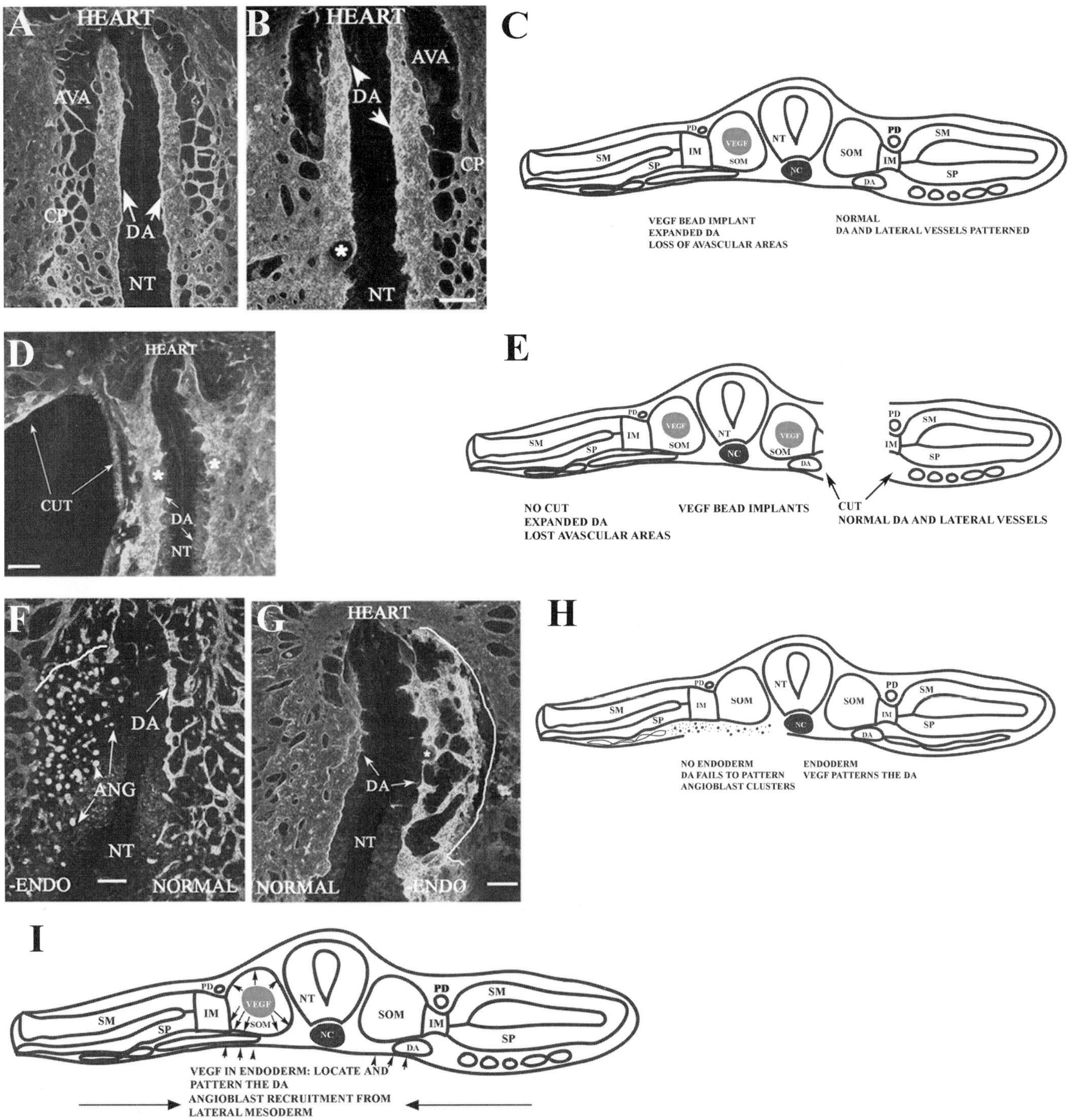

Figure 2 The Effect of VEGF and Endoderm on Angioblast Recruitment from Lateral Mesoderm to Pattern the Dorsal Aorta. Quail embryos are shown following experimental methods to study vascular morphogenesis. For each panel, the head of the embryo is up, the scale bar represents 100 µm, and asterisks (*) indicate VEGF delivery locations. **(A)** A normal quail embryo stained with QH-1. **(B)** An embryo with VEGF delivered to somite 7. **(C)** A schematic representation of the effect of ectopic VEGF on the vascular pattern, as seen in cross section. **(D)** An embryo with VEGF delivery to both sides, followed by a cut (made as indicated approximately 100 µm lateral to the midline) to separate the lateral mesoderm (a source of angioblasts) from the embryo proper, resulting in ablation of the VEGF delivery effect. **(E)** The effect of lateral mesoderm removal on the ability of ectopic VEGF to alter vascular patterns. VEGF is a chemoattractant for angioblasts from the lateral mesoderm to form the dorsal aortae. **(F)** An embryo without a VEGF bead, in which the endoderm was removed from one side. On the side without endoderm (-ENDO), the DA failed to form; rather, angioblasts (ANG) remain as single cells or clusters, compared to the unoperated (NORMAL) side of the embryo. **(G)** An embryo in which endoderm was removed from one side (-ENDO), followed by VEGF delivery on the same side. In this case, VEGF was able to partially restore the normal vascular pattern. **(H)** The effect of endoderm removal on vascular patterning. VEGF expressed within the endoderm is necessary for the proper localization and patterning of the dorsal aorta. **(I)** The role of VEGF in vascular morphogenesis. Ectopic VEGF delivery causes hypervascularization *(left)* by recruiting angioblasts from lateral mesoderm. Endogeneous VEGF within the endoderm recruits angioblasts from lateral mesoderm *(arrows)* and acts to localize and pattern the DA. In all diagrams: AVA, avascular area; CP, capillary plexus; DA, dorsal aorta; IM, intermediate mesoderm; NC, notochord; NT, neural tube; PD, pronephric duct; SM, somatic mesoderm; SP, splanchnic mesoderm; SOM, somite; VEGF, delivery site.

while the proliferation rate of quail embryo cells at the stages examined is approximately ten hours [6]. Rather, VEGF is believed to induce endothelial cells and angioblasts to undergo directed migration (chemotaxis) towards a VEGF source. Evidence for this mechanism in quail is from experiments with VEGF delivered to both sides of the embryo, followed by removal of lateral mesoderm (a tissue rich in angioblasts). After removal of lateral mesoderm, the effect of ectopic VEGF on vascular pattern formation was reduced (Figures 2D and E), suggesting that cells recruited by the VEGF to form the dorsal aorta were removed.

VEGF is known to regulate the behavior of endothelial cells during vasculogenesis, such that adding VEGF increases protrusive activity, while soluble VEGFR1 (a VEGF inhibitor) reduces protrusive activity. Protrusive activity is the extension of filopodia from endothelial cells as they extend and migrate to form a vascular pattern. This activity is necessary for endothelial cell shape changes that take place during vascular morphogenesis [5].

The Role of Endoderm

In the developing quail embryo, high levels of VEGF are expressed within the planar endoderm that underlies the DA. VEGF within the endoderm may provide a chemotactic cue to recruit angioblasts and endothelial cells from the lateral mesoderm to the midline, thereby positioning the DA. Directed cell migration of angioblasts toward an area of VEGF expression was demonstrated during development of the dorsal aorta in the frog, *Xenopus laevis* [7]. In *Xenopus*, the hypochord (an endoderm-derived midline structure that lies below the notochord) expresses predominantly the lowest molecular weight isoform of VEGF (VEGF122), a diffusible form of the growth factor. Diffusion of VEGF122 stimulates angioblast migration toward the midline of the embryo. Expression of ectopic VEGF in lateral mesoderm can divert angioblasts away from the hypochord. Vital dye labeling studies confirmed lateral mesoderm as the source of angioblasts destined to form the DA [7]. VEGF is thought to play essentially the same role in dorsal aorta formation in zebrafish [4].

In the quail embryo, removal of endoderm resulted in failure of the dorsal aorta to form (Figures 2F and H). However, unlike the situation in *Xenopus*, angioblasts were recruited to the midline, where they remained as single cells or clusters in the absence of endoderm, without undergoing cohesion to form the DA. On the contralateral side of the embryo, where the endoderm was left intact, a DA formed (Figure 2F). VEGF delivery to the side where endoderm was removed partially restored the vascular pattern and hypervascularized the control side as previously described (Figure 2G). Therefore, VEGF within the endoderm of the quail embryo locates and patterns the DA (Figure 2H). This may differ from the results in *Xenopus* because of the developmental stage when endoderm was removed in the quail (earlier removal may prevent angioblast migration), or the planar quail endoderm may function differently from the *Xenopus* hypochord. Taken together, data from lateral mesoderm removal and endoderm removal experiments in quail suggest that VEGF within the endoderm recruits angioblasts to the midline, where they undergo cellular changes under the continued influence of VEGF to form the DA (Figure 2I).

Embryonic Angiogenesis

VEGF can affect the process of embryonic angiogenesis in addition to its roles in vasculogenesis described previously. The effect of VEGF on angiogenesis has been examined in the mouse eye, where the retinal vasculature expands by sprouting as the retina grows [8]. The vascular sprout contains two types of endothelial cells: tip cells and stalk cells, which can be distinguished by differences in gene expression and morphology [9]. Filopodia of tip cells direct cell migration by sensing a gradient of VEGF, while stalk cells respond to VEGF by proliferation. VEGF induces filopodia to extend from tip cells, which are guided toward VEGF sources. VEGF antagonists inhibit these activities. This is similar to the protrusive activity of endothelial cells in the developing quail embryo, as disruption of VEGF gradients inhibits protrusive activity and migration of sprouting tip cells.

Other Functions of VEGF in Morphogenesis

In addition to its role in blood vessel morphogenesis, VEGF has been implicated as being mitogenic for other cell types, which include lymphocytes, retina pigment epithelial cells, and Schwann cells. VEGF can act as a neuron survival factor. Nerves and blood vessels are both branched structures, forming intricate networks that are often associated with each other in the same anatomic space. However, it is unknown if they form their patterns in an independent or coordinated fashion. Arteries, but not veins, specifically align with peripheral nerves. VEGF is expressed by peripheral nerves and can induce expression of arterial markers (including ephrinB2) when added to primary embryonic endothelial cells in vitro [10]. A similar effect was seen when neural cells were cocultured with endothelial cells, suggesting that neurons and Schwann cells induce ephrinB2 in primary embryonic endothelial cells by secretion of VEGF. In erbB3-/- mutant mice, Schwann cells are absent and blood vessels fail to arterialize. This is correlated with reduced nerve-associated VEGF levels, suggesting that VEGF from nerves is an arterializing factor in vivo [10]. VEGF has also been found to promote arterial fate in zebrafish [4]. Genetic studies demonstrated that VEGF acts downstream from the hedgehog signaling pathway, but upstream of Notch to specify arterial endothelial cell fate.

Use of VEGF antisense morpholinos resulted in loss of arterial markers from trunk vessels and ectopic venous marker expression in arterial vessels. This was similar to the phenotypes observed in animals with defective Notch signaling. Activated Notch induces arterial specification in the absence of VEGF, a further indication that VEGF is active upstream of the Notch pathway.

Summary

VEGF is essential for vascular morphogenesis, with precise control of its level required for normal vascular pattern formation. Results to date have focused on the role of VEGF in many organisms, using a variety of techniques. VEGF is known to be essential for forming a vascular pattern, perhaps by recruiting angioblasts and specifying the size and location of blood vessels within the embryo. Size and location may be predetermined by a VEGF gradient, before angioblast migration. However, the mechanism of VEGF activity and the signaling pathways active during vascular morphogenesis are less clear. Further studies of VEGF action in the embryo will lead to a better understanding of VEGF activity during pathological neovascularization.

Glossary

Angioblasts: The precursor cells of endothelial cells that arise from embryonic mesoderm.

Angiogenesis: The formation of new blood vessels by sprouting from existing vessels.

EMT: The process of epithelial to mesenchymal transformation.

Morphogenesis: The change in form that gives rise to the organization of cells into tissues in embryos.

Vasculogenesis: The process of vessel morphogenesis by de novo assembly of angioblasts.

References

1. Miquerol, L., Langille, B. L., and Nagy, A. (2000). Embryonic development is disrupted by modest increases in vascular endothelial growth factor gene expression. *Development* **127**, 3941–3946.
2. Damert, A., Miquerol, L., Gertsenstein, G., Risau, W., and Nagy, A. (2002). Insufficient VEGFA activity in yolk sac endoderm compromises hematopoietic and endothelial differentiation. *Development* **129**, 1881–1892.
3. Poole, T. J., Finkelstein, E. B., and Cox, C. M. (2001). The role of FGF and VEGF in angioblast induction and migration during vascular development. *Dev. Dyn.* **220**, 1–17.
4. Weinstein, B. M. (2002). Plumbing the mysteries of vascular development using the zebrafish. *Sem. Cell. Dev. Biol.* **13**, 515–522.
5. Drake, C. J., LaRue, A., Ferrara, N., and Little, C. D. (2000). VEGF regulates cell behavior during vasculogenesis. *Devel. Biol.* **224**, 178–188.
6. Finkelstein, E. B., and Poole, T. J. (2003). Vascular endothelial growth factor: A regulator of vascular morphogenesis in the Japanese quail embryo. *Anat. Rec.* **272A**, 403–414. *A thorough analysis of the influence of ectopic VEGF on vascular morphogenesis, in which it was shown that VEGF can affect multiple processes of vasculogenesis.*
7. Cleaver, O., and Krieg, P. A. (1998). VEGF mediates angioblast migration during development of the dorsal aorta in Xenopus. *Development* **125**, 3905–3914. *This paper used a variety of techniques to demonstrate the essential role of VEGF and the novel role of the Xenopus hypochord in determining dorsal aorta localization.*
8. Ruhrberg, C. (2003). Growing and shaping the vascular tree: Multiple roles for VEGF. *BioEssays* **25**, 1052–1060.
9. Gerhardt, H., Golding, M., Fruttiger, M., Ruhrberg, C., Lundkvist, A., Abramsson, A., Jeltsch, M., Mitchell, C., Alitalo, K., Shima, D., and Betsholtz, C. (2003). VEGF guides angiogenic sprouting utilizing endothelial tip cell filopodia. *J. Cell Biol.* **161**, 1163–1177. *This is the first demonstration that as a vascular network forms, there are two types of endothelial cells, which can sense a VEGF gradient, or proliferate, thereby determining the location and size of vessel networks.*
10. Mukouyama, Y. S., Shin, D., Britsch, S., Taniguchi, M., and Anderson, D. J. (2002). Sensory nerves determine the pattern of arterial differentiation and blood vessel branching in the skin. *Cell* **109**, 693–705.

Capsule Biographies

Thomas J. Poole has been studying vascular development since 1983. He has published numerous articles on the use of quail embryos and quail/chick chimeras in deciphering the origins of the embryonic vascular pattern. The current interests of his laboratory are the roles of FGF and VEGF in angioblast induction and vessel morphogenesis, and the use of arsenic in perturbation of embryonic vasculogenesis and angiogenesis.

Eric B. Finkelstein earned his Ph.D. in 2001 in Dr. Poole's laboratory, studying the role of VEGF in blood vessel morphogenesis. He is currently a postdoctoral fellow in Dr. Patricia D'Amore's laboratory in Boston, studying endothelial cell differentiation and the function of VEGF isoforms.

CHAPTER 21

The Two-Phase Model for Angiogenesis Regulation by the Extracellular Matrix

James D. San Antonio and Renato V. Iozzo
Thomas Jefferson University, Philadelphia, Pennsylvania

Introduction

Angiogenesis, the development of blood vessels from the preexisting vasculature, is recognized to play a central role in organismal development and pathologies such as tumor growth and metastasis. Two of the primary classes of molecular players in angiogenic regulation include growth factors and the extracellular matrix (ECM), the latter being the special focus of this review. The ECM scaffold that EC produce during capillary morphogenesis and in the differentiated microvasculature comprises 10 or more macromolecules, and many of these have been reported to exhibit pro- or antiangiogenic activities in their native or processed forms. Thus, at first glance, the literature presents a bewildering collection of information that appears difficult to integrate into a coherent model for the ECM regulation of angiogenesis. However, results from numerous studies suggest that a simple way to categorize the ECM control of angiogenesis is in the form of a model, wherein the intact basement membrane that promotes quiescence and differentiation comprises the first phase, and that certain conditions stimulate the endothelium to degrade the basement membrane, and encounter the second phase, the stromal or provisional matrices, including fibrillar collagens and fibrin, that promote capillary tube morphogenesis (Figures 1 and 2). The focus of this chapter is on discussing this two-phase model of angiogenesis and on deciphering the role the ECM plays in controlling and modulating this morphogenic event. Finally, we will discuss how this information may be applied to ECM bioengineering, which may be used to influence angiogenesis in vivo.

Phase I: Maintenance of Quiescence by the Basement Membrane

Insofar as angiogenesis is the process of growth of a preexisting vasculature, it is appropriate to begin our discussion with the ECM status of the quiescent, differentiated blood vessel. Basally, the capillary secretes and assembles a basement membrane scaffold, which exists in close contact with the EC that make up the capillary network (Figure 2, upper panel). The basement membrane contains several classes of glycoproteins and proteoglycans. The basement membrane conglomerate is considered to play multiple roles in the vascular system, from partitioning EC from the surrounding stroma and providing structural support to the vasculature, to creating a charge barrier between the capillary and adjacent cells and tissues. However, in addition to playing these roles, it is also clear that many of the basement membrane components may also act individually or in concert to exert significant activities on angiogenesis.

Basement Membrane Structure

The basement membrane is a highly organized, net-like molecular conglomerate. At least two modes of basement membrane construction may take place: self-assembly driven by the molecules themselves and cell-assisted processes. The two major components comprising the network are type IV collagen and laminin. Type IV collagen is proposed to self-assemble through multimerization via its N-terminal S domains and C-terminal noncollagenous domains, and lateral associations between its central col-

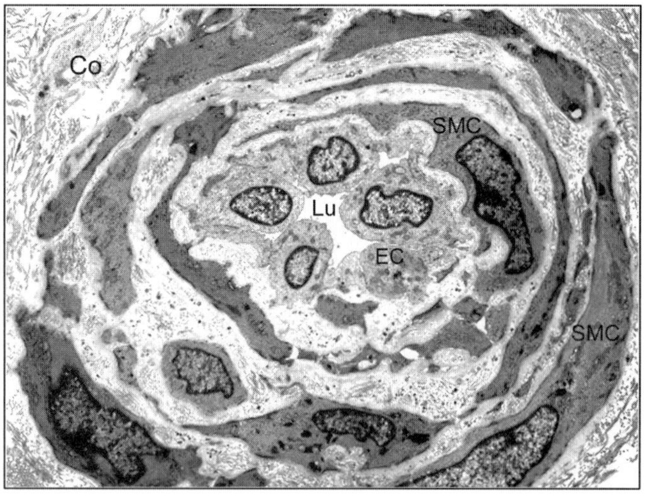

Figure 1 The Microvasculature and the ECM are Intimately Associated. This electron micrograph of an arteriole illustrates that EC that line the vessel lumen (Lu) are surrounded closely by circumferentially arranged smooth muscle cells (SMC) and interstitial ECM. A basement membrane, deemed to be phase I of angiogenic regulation in this review, lies between the EC and the underlying vessel components, but it is too thin to be resolved at this magnification. Interstitial collagen fibrils (Co), comprising phase II of angiogenic regulation, can be seen in abundance underneath the EC layer and throughout the media.

lagenous regions. Similarly, the laminin scaffold is thought to form through oligomerization of laminin monomers ($M_r \cong 900$ Kd), likely driven by low-affinity interactions between the N-terminal regions of the α, β, and γ arms of the monomers, and may be further modulated by the local protein and calcium concentration. The other major components of basement membranes include the glycoprotein nidogen ($M_r \cong 80$ Kd) and at least three proteoglycans, including perlecan ($M_r \cong 750$ Kd), agrin, and type XVIII collagen. In addition, type XV collagen harbors chondroitin sulfate side chains. Nidogen is proposed to serve as the molecular glue that holds the basement membrane together. Thus, the C-terminal domain of nidogen binds with high affinity to the short arm of the laminin γ chain, and through weaker affinity interactions, via its other domains, with type IV collagen and perlecan. The heparan sulfate (HS) chains of perlecan and, presumably, those of collagen XVIII and agrin, may also stabilize basement membrane structure because heparin-binding is exhibited by laminin and type IV collagen preparations. Depending on the tissue, other nonstructural components of basement membranes may include type XVIII or XV collagens, the agrin proteoglycan, or nonstructural components associated with the basement membranes such as basic fibroblast growth factor or matrix metalloproteinases (MMPs), proposed to bind to and be sequestered by the HS chain components of perlecan. All of the components of basement membranes are synthesized and secreted by EC, which may help regulate or direct its assembly. For example, disruption of the function of the cell surface β1 integrin receptor interferes with the ordered deposition of type IV collagen during basement membrane formation.

The Basement Membrane and EC Quiescence

In the quiescent capillary, EC are tightly associated with a basement membrane. It is proposed that the basement membrane may promote EC quiescence in two ways: (1) the intact matrix scaffold may exert strong differentiation-promoting activities on EC, and (2) individual basement membrane components or their proteolytic products may exert strong antiproliferative and antiangiogenic activities (Figure 2, upper and lower left panels).

Role of the Intact Basement Membrane

The most compelling evidence supporting the role of the intact basement membrane in promoting EC quiescence and differentiation is circumstantial in nature (i.e., the fact that quiescent capillaries assemble basement membranes and exist in tight association with them). Moreover, a hallmark of other differentiated cell types that, like the microvasculature, differentiate to form polarized, morphologically distinct tissues, is the production of and association with basement membranes. Not surprisingly, EC and many other cell types differentiate in vitro when cultured within or on top of basement membrane mixtures such as Matrigel. Further evidence in support of the role of the basement membrane in angiogenesis is supplied by studies that identified genes upregulated by at least two-fold in various angiogenesis systems [1]. Coincident with capillary tube differentiation in response to either type I collagen or fibrin, mRNA for various basement membrane components including laminin chains α4, β1, and β2, type IV collagen, the type XVIII collagen αI chain, and nidogens 1 and 2, are increased along with other molecules directly or indirectly related to basement membrane structure and function. The latter include collagen synthesis enzymes, MMPs 1, 2, and 9, disintegrin, and ADAMs (A disintegrin-like and metalloproteinase domain) –9, 10, and 17. More work is needed to understand how the complex basement membrane scaffold acts to promote EC quiescence. In this regard, it may be useful to consider how the basement membrane regulates differentiation in mammary epithelial cells that respond to the three-dimensional matrix environment in very defined ways, as elucidated by Mina J. Bissel and co-workers. Thus, when these epithelial cells interact with laminin, they change their cell morphology, undergo growth arrest, and in the presence of prolactin, express the differentiation marker, the milk protein β-casein. A multiplicity of cell surface receptors, including the α6β4 integrin, β1 integrins, and an E3 laminin receptor, were found to mediate these cellular responses. Whether the intact basement membrane promotes EC quiescence and the expression of EC-specific differentiation in a similar manner remains to be determined.

Role of Individual Basement Membrane Components

Most of the research probing the function of basement membranes in angiogenesis has taken the reductionistic approach of assaying the activities of individual basement

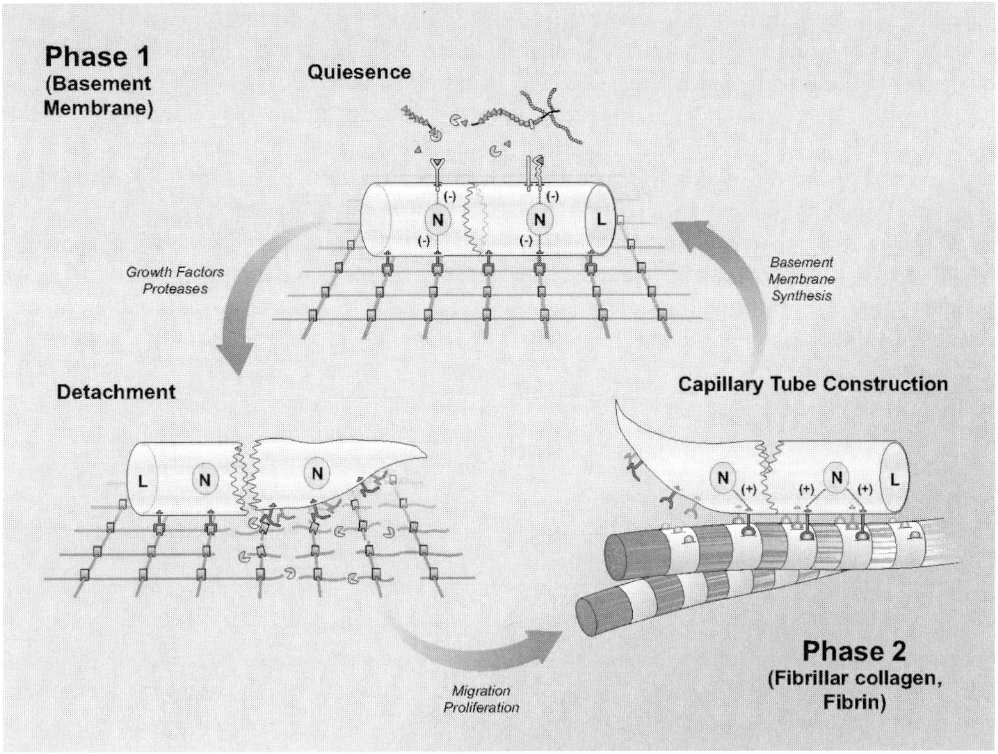

Figure 2 The Two-Phase Matrix Model of Angiogenesis Regulation. *(Phase I, Top of Diagram)* It is proposed that the basement membrane phase, shown here on an orange background, promotes EC quiescence and maintains capillary integrity in two ways: (1) via stable interactions between cell surface receptors and various basement membrane components on the basal EC side, and (2) by the release of antiangiogenic fragments of basement membrane components through the continuous low-level proteolysis and basement membrane turnover that may occur during quiescence. For the sake of clarity, this latter activity is shown only on the EC apical side, although it is presumed to occur virtually anywhere the EC has contact with the basement membrane. Antiangiogenic fragments may inhibit EC growth and migration by binding to unique receptors (cell on left) or by binding to and interfering with integrin receptor function (cell on right). *(Phase I, Bottom Left of Diagram)* Upon growth factor stimulation, or during tissue injury, massive proteolysis, likely mediated by MMPs, causes the degradation and collapse of the basement membrane, and integrins or other cell surface receptors may mediate the migration of EC into the second matrix phase (intussusception). It is proposed that the fibrillar collagen or fibrin phase, shown on a blue background, catalyzes capillary tube construction in two ways: (1) by presenting a linear, rigid scaffold that helps align EC into presumptive tubes, and (2) by presenting regularly spaced ligand-binding sites that induce clustering of EC receptors such as the $\alpha2\beta1$ integrin in the case of type I collagen, or VE-cadherin in the case of fibrin, that induce capillary lumen formation and perhaps the expression of genes required for the maintenance of EC differentiation. (see color insert)

membrane components, or their various domains on EC behavior. It is difficult to reconcile the relevance of these findings to angiogenesis, because it is not clear whether the various constituents of basement membranes are accessible to the EC during quiescence or throughout the various stages of capillary development. Nonetheless, these investigations will be summarized, and it will be assumed here, for the sake of simplicity, that the following findings are most relevant to two aspects of EC biology: (1) that basement membrane components may be exposed during the constant, low-level turnover of this ECM compartment, which may occur during EC quiescence (Figure 2, upper panel), or (2) alternatively, during the massive proteolysis and collapse of the matrix scaffold at the initiation of angiogenesis or blood vessel remodeling (Figure 2, lower left panel). Early approaches examined how EC responded to culture on a reconstituted basement membrane gel. It was found that tubes formed well within one day and that, although blocking laminin function in various ways inhibits tube morphogenesis, neither laminin nor type IV collagen gels alone support tube formation. These findings imply that a multiplicity of interactions among various cell surface receptors and basement membrane components are required for tube formation in response to the basement membrane, but again, it is questionable whether in vivo EC ever encounter these matrix components as individual entities rather than as complex supramolecular assemblies.

Laminin

Several groups have mapped the regions of laminin that influence various EC behaviors. For example, an RGD peptide in the α chain of laminin promotes EC adhesion. The

YIGSR sequence in the β1 chain, when substrate-bound, induces cell-to-cell interactions and promotes tube formation, and soluble peptide inhibits that process; the IKVAV sequence from the α1 chain also promotes angiogenesis. Moreover, it has been shown that MMP-2 cleavage of laminin-5 exposes a cryptic site in the γ2 subunit that promotes cell migration. Others examined the activity of a laminin-nidogen mixture and showed that it can promote EC sprouting in vitro at very low concentrations but is inhibitory at high concentrations, consistent with a quiescence-promoting function of the intact basement membrane.

Type IV Collagen

Type IV collagen contains multiple sites for EC surface binding, including to α1β1, α2β1, and α3β1 integrins. Noncollagenous (NC)-1 domains from the α1(IV), α3(IV), and αβ6(IV) chains of collagen IV support integrin-dependent EC adhesion and migration. NC1 domains from α1(IV), α4(IV), α3(IV), and α6(IV) are potent inhibitors of angiogenesis in vivo. Angiogenesis inhibitors derived from these fragments include arresten, generated from α1(IV); canstatin, generated from α2(IV); and tumstatin, a product of α3(IV) degradation. A site exists in collagen IV that is normally cryptic but may be unmasked through exposure to MMP-2. Blocking the function of this site in vivo disrupts angiogenesis and tumor growth.

Nidogen

The α3β1 integrin has been reported to bind directly to this basement membrane component, and its fragments regulate cellular adhesion. Because nidogen may be critical in stabilizing basement membrane structure, it can be speculated that once it undergoes proteolysis, the basement membrane scaffold may fall apart or may be easily breached by EC, and that nidogen's fragments may regulate EC migration into the tissue stroma.

Perlecan

This predominant proteoglycan of the basement membrane can interact with cell surfaces via its core protein, as well as modulating cell proliferation through its sequestration or release of various growth factors. EC are proposed to bind to the protein core via the α1, β1, and β3 integrin receptors. Moreover, the activity of basic fibroblast growth factor (FGF2) on cell growth and vascular morphogenesis is potentiated through its associations with perlecan and other proteoglycans. Thus, perlecan, but not syndecans or glypicans, extracted from fibroblast cultures facilitates high-affinity FGF2 binding to Chinese hamster ovary cells deficient in endogenous HS and engineered to express the FGF receptor-1, and soluble FGF receptors. Perlecan encapsulated in alginate beads is capable of binding FGF2 and promoting extensive angiogenesis in vivo. It was proposed that perlecan functions as a low-affinity co-receptor by delivering FGF2 to its high-affinity cell surface receptors. Consistent with this model, it was shown using an antisense approach that blocking of perlecan expression by colon carcinoma cells attenuated their growth, which correlated with a reduction in their responsiveness to FGF-7. Furthermore, antisense-mediated inhibition of perlecan expression in tumor xenografts of colon carcinoma cells or allografts of mouse melanoma cells showed a reduced capacity for tumorigenesis and the promotion of neovascularization in nude mice. Perlecan-mediated potentiation of growth factor action may occur via sites on either or both of the glycosaminoglycan chains and the core protein. At the cell surface, cell-associated proteoglycans such as perlecan are proposed to contribute HS chains in the formation of a ternary complex with two FGF1 molecules, and one growth factor receptor chain, to activate the growth factor receptor. Perlecan's HS chains can also serve as reservoirs or binding sites for other angiogenic factors such as MMP-7, which, when released, may initiate or help sustain EC migration during angiogenesis.

Investigations into the functional role played by the perlecan protein core in vascular development led to the discovery that the C-terminus potently inhibits EC migration and adhesion to type I collagen and capillary morphogenesis in vitro and in vivo. The C-terminal portion of human perlecan was named endorepellin to designate its antiendothelial activity. This domain blocks EC-matrix adhesion without interacting directly with the matrix components themselves. Endorepellin also binds with high affinity to EC cell surfaces, as well as to endostatin, and counteracts its antiangiogenic activity. These latter observations imply that perlecan-type XVIII collagen interactions could play a structural role in promoting basement membrane stability in some tissues, but that during basement membrane dissolution, the presence of both endorepellin and endostatin would likely not interfere with the subsequent EC migration, growth, and capillary morphogenesis, because these factors may interact and thus neutralize each other. Iozzo and coworkers have shown that endorepellin may be liberated from the perlecan core by BMP-1/Tolloid-like MMPs, and various lines of evidence are consistent with the liberation of endorepellin by cell cultures *in vitro*, and of its presence in the circulation of humans in various pathological states.

Type XVIII and XV Collagens

These collagens are members of the mutiplexin collagen family, because they carry multiple triple-helical domains interrupted by noncollagenous regions. Type XVIII collagen is a hybrid collagen-proteoglycan molecule and is a constituent of basement membranes of blood vessels, kidney, skeletal muscle, and retina, and some other tissues. This collagen carries up to four HS chains, each attached to one of four nontriple-helical domains. The widely distributed basement membrane molecule type XV collagen is highly similar to type XVIII in primary structure and domain arrangement. Type XV collagen is also a collagen-PG hybrid molecule, but carries chondroitin sulfate, not HS chains.

An 18 Kd protein fragment of type XVIII collagen was isolated from hemangioma cell conditioned media and shown to potently inhibit EC growth in vitro, and tumor growth in vivo, and was thus named endostatin by its discoverers, Judah Folkman and coworkers. It is proposed to be liberated from the parent molecule through the action of proteases such as cathepsin L and MMPs. Endostatin was shown to potently inhibit FGF2-induced EC growth in vitro, and in the chick chorioallantoic membrane (CAM), and to nearly abolish the growth and associated angiogenesis of a variety of tumors implanted in mice. Examination of residual carcinomas in endostatin-treated mice revealed it to act not by inhibiting tumor cell growth but by increasing apoptosis sevenfold, via its inhibition of tumor-associated angiogenesis. Discontinuation of endostatin treatment resulted in the regrowth of the implanted tumors. Thus, endostatin was proposed to hold great clinical promise as a tumor dormancy factor. A later report, however, failed to observe an effect of endostatin on the FGF2-induced growth of bovine or human EC in vitro. However, these authors found endostatin to potently inhibit the VEGF-induced migration of human EC and, at very low doses, to inhibit the growth of human renal cell carcinomas implanted subcutaneously in nude mice. A third group reported an inhibition of FGF2 but not VEGF-induced angiogenesis by endostatin. Finally, drastically different results were reported when the roles of the zinc- and heparin-binding domains of endostatin to its antiangiogenic action were examined. Notably, type XV collagen also contains a C-terminal noncollagenous domain with about 60 percent sequence homology to endostatin, that does not interact with heparin or zinc, but nonetheless inhibits FGF2- or VEGF-induced angiogenesis in the CAM. In experiments where endostatin from type XVIII collagen and ES-XV were compared for their abilities to inhibit FGF2- or VEGF-induced angiogenesis, strikingly different activities were observed. Thus, the type XV collagen-derived fragments, ES-XV, and NC1 XV similarly inhibited VEGF-induced angiogenesis, but angiogenic stimulation by FGF2 was potently inhibited by NC1-XV but not by ES-XV. In contrast, whereas type XVIII endostatin and NC1 domains failed to inhibit VEGF-induced angiogenesis, endostatin, but not the type XVIII collagen NC1 domain, were active against FGF2-induced angiogenesis.

MODE OF ANGIOGENESIS INHIBITION BY ECM FRAGMENTS

Angiogenesis inhibitors are likely to act via distinct pathways primarily because of their diversity in structure and receptors used. For example, endostatin is proposed to interact with the $\alpha5\beta1$ integrin that triggers a signaling cascade leading to inactivation of RhoA GTPase; this subsequently may cause disruption of the actin cytoskeleton and disassembly of focal adhesions. Thus, endostatin blocks the mobility of EC, thereby preventing the early stages of angiogenesis that require cells to migrate. This process requires an intact cytoskeletal organization. Cytoskeleton dynamics are of fundamental importance for the actions of another antiangiogenic factor derived from the C-terminus of perlecan, endorepellin. Endorepellin is proposed to interact with another integrin, the $\alpha2\beta1$ that is the major receptor for collagen I, and this may lead to an intracellular increase in c-AMP, activation of protein kinase A, and subsequent dissolution of actin stress fibers and focal adhesions. Notably, collagen I exerts the opposite effect, further indicating that the $\alpha2\beta1$ integrin is directly involved in endorepellin-related signaling. Interestingly, tumstatin, a C-terminal fragment of collagen type IV, is believed to act differently by binding yet another integrin, $\alpha v\beta3$, thereby triggering a signaling cascade that leads to inhibition of EC proliferation. Therefore, a key question is whether common effector pathways are similarly modulated by the various antiangiogenic inhibitors. Endpoints for these pathways would presumably contain the most essential molecular targets for inhibition of angiogenesis. There is, thus, the need to investigate not only intracellular signaling pathways, but also the initial ECM/growth factor interactions with the surface receptors of the EC. In addition, we need to further understand the factors that alter or modulate the cell surface receptors that comprise the first lines of sensing of the angiogenesis environment by the EC.

The Transition Between the Matrix Phases: Growth Factors and MMPs Provide the One-Two Punch

It is proposed that insults such as proteolysis and wounding, or growth factor action, may provide the stimulus for basement membrane degradation or disruption; these events may induce quiescent EC to enter the second ECM phase and initiate the angiogenesis program (Figure 2, upper and lower left panels). In the case of proteases, it is widely held that MMPs, which comprise a family of at least 23 distinct endoproteinases, may be the likely players; their expression can be induced by growth factor or cytokine activity or through tissue injury. Significantly, more than half of the known MMPs have substrate specificities for one or more basement membrane components, and several of the phenotypes of mice null for specific MMPs include defective angiogenesis. Depending on the physiologic state of the vasculature, MMPs can be released by the EC or by perivascular stromal or tumor cells. Moreover, the level of MMP activity would likely determine the outcome on angiogenesis (i.e., low activities may cause limited release of antiangiogenic basement membrane fragments), but if disruption of the ECM scaffold is not significant enough to allow EC to breach the basement membrane, EC quiescence may be sustained. In contrast, high proteolytic activity would cause significant dissolution of the basement membrane, as well as promoting the release of factors, such as the heparin-binding FGFs or VEGFs sequestered by the various basement membrane components, which would in turn stimulate the proliferative phase of angiogenesis. Notably, these and many others of the proangiogenic growth factors exhibit matrix-binding properties, and in some cases, require the function of cell surface or ECM proteoglycans as co-receptors for growth factor–receptor binding and activation. Although the role of proteases and growth factor action is likely crucial in promoting the progression of EC between the two ECM phases of angiogenesis, our review will not further examine these aspects because they are widely discussed elsewhere in the current literature and in this book.

Phase II: Induction of Capillary Morphogenesis by the Interstitial Matrices, Fibrillar Collagens or Fibrin

Wounding or proteolysis may be envisioned to perforate the basement membrane, a relatively thin scaffold approximately several hundred nanometers thick. Subsequently, EC would have to breach the basement membrane and migrate only small distances or, depending on the circumstances, even not at all, to encounter fibrillar collagens or fibrin (Figure 2, lower left and right panels). Thus, in many tissues the EC layer of the microvasculature or small vessels exist within several μm of light to sometimes heavy arrays of extracellular fibrillar collagens (e.g., Figure 1). During the wounding process, or in the vicinity of some tumors, fibrin may be heavily deposited throughout the tissue, and there may be cases where fibrin pools literally form adjacent to the resident ECs. Fibrillar collagens and fibrin each exert powerful proangiogenic activities on EC and are thus considered to comprise the second phase of angiogenesis regulation, and function to catalyze capillary tube construction (Figure 2, lower right panel). Because angiogenesis can occur during development in tissues lacking fibrillar collagen or fibrin (e.g., in the brain), we speculate that angiogenesis is not absolutely dependent on the presence of such polymers. Instead, we propose that fibrillar collagens or fibrin may greatly accelerate or catalyze capillary tube formation during instances when rapid revascularization is required to support tissue survival and regeneration, as in wound healing.

Type I Collagen

Type I collagen is the most abundant protein in the human body and in other vertebrates. It is synthesized by vascular cells and most other cell types and is a common and, sometimes, predominant extracellular component of many interstitial tissues. Type I collagen is synthesized as a soluble triple-helical procollagen precursor of ≅ 300 Kd, is proteolytically processed upon secretion, and assembles in a staggered fashion into the type I collagen fibril, the form most commonly found in tissues. A role for type I collagen in angiogenesis has been known for many years; initial observations showed that inhibition of collagen metabolism and cross-linking disrupted angiogenesis in vivo, and subsequently, type I collagen was shown to be an ideal angiogenic scaffold in vitro. Many investigators have developed in vitro collagen-induced angiogenesis assays. The most common type involves culturing EC within a collagen gel sandwich into which cells migrate, proliferate, undergo cell-to-cell interactions, and then form capillary-like tubes, usually within two to three days. A modification of this system was devised by our group, where an apical collagen gel is placed on a confluent EC monolayer in the presence of a serum-free medium, and tubes form as rapidly as 12 hours. Because morphogenesis occurs in the absence of significant cell migration and proliferation, we consider this to be an endpoint or tube formation assay. Using this system, we have shown that angiogenesis in response to a type I collagen gel is surprisingly specific and relies largely on interactions between the EC $\alpha 2\beta 1$ integrin receptor and only one of the three potential glycine-phenylalanine-hydroxyproline-glycine-glutamic acid-arginine (GFPGER) integrin-binding sites on the collagen fibril. This stands although the collagen fibril contains numerous sites for potential interactions with cell surfaces and their associated molecules, including those for the binding of $\alpha 1\beta 1$ or $\alpha 2\beta 1$ integrins, discoidin domain receptors, various proteoglycans, and fibronectin. Further evidence from our work suggests that $\alpha 2\beta 1$-collagen ligation correlates with activation of the intracellular signaling molecule p38MAPK and downregulation of FAK, associated with sites of focal adhesion-matrix interactions. Ongoing work attempts to determine if these signaling pathways may transduce the signal from the level of cell–collagen interactions to the intracellular signaling events that induce tube formation.

Fibrin

Fibrin is an extracellular polymer generated during thrombogenesis from the precursor fibrinogen [2]. Fibrinogen has an M_r of ≅ 340 Kd and comprises two identical subunits, each of which consists of three distinct polypeptides called the Aα, Bβ, and γ chains; the two halves of the molecule are associated by several disulfide bonds at the N-terminal regions. During clot formation, thrombin cleaves the A and B fibrinopeptides, and the resultant product of ≅45 nM in length assembles in a half-staggered fashion into fibrin polymers or fibrils. As with type I collagen, fibrin has been shown by Jose Martinez and coworkers to be a potent inducer of angiogenesis in vitro. EC monolayers exposed apically to fibrin gels form capillary-like tubes within 24 hours. It was found that an interaction between the EC VE-cadherin and the beta 15–42 sequence of fibrin was necessary for fibrin-induced tube morphogenesis.

Common Features of Type I Collagen and Fibrin Polymers

When considering whether type I collagen and fibrin act in common ways to promote angiogenesis, it may be useful to compare the molecular and supramolecular features of these polymers. On a gross level, both polymers form classic gels, which when placed apically on EC monolayers induce capillary morphogenesis. At the level of the primary protein sequence, the two molecules share no significant homology; however, both are high-molecular-weight proteins that assemble into cable-like polymers exhibiting regular periodicities. The average type I collagen fibril diameter is 82.5 nm, the lateral repeat distance is 67 nm, and the spacing between monomers is approximately 1.0 to 1.5 nm. For fibrin, the average fibril diameter is 85 nm, the lateral repeat distance is 22.5 nm, and spacing between monomers is about 5 nm. It may be significant that the average fibril diameters are quite similar and that the other dimensions are well within an order of magnitude of each other.

We envision that these molecules may play at least two roles in promoting angiogenesis. First, both polymers present a rigid, linear substrate for cell interactions that may catalyze the side-by-side positioning of EC into tube-like arrays, a function reminiscent of that suggested by the matrical track theory of ECM-induced angiogenesis, put forth by Vernon and colleagues [3]. This theory proposes that EC may exert tractional forces on a planar, relatively homogeneous fibrillar matrix scaffold, which results in a gathering of alignment of fibrillar matrix cables between cells. These cables in turn function as matrical tracks, which guide ECs to extend processes and join together to ultimately form a polygonal cellular or capillary network. In the case of the induction of angiogenesis by a collagen gel, it is possible that the EC must first act on the three-dimensional random array of collagen fibrils within the gel to gather or align multiple fibrils together, which in turn may promote linear cell-to-cell associations. In addition, we propose that a second distinct role these polymers may play is to promote the proper spatial interactions between specific EC surface receptors and regularly repeating ligand-binding sites on the polymers, resulting in cell surface receptor clustering and the ensuing activation of signaling pathways necessary for capillary morphogenesis. It can be speculated that this event may trigger EC responses such as capillary lumen formation and the expression of genes necessary to initiate and maintain capillary differentiation. In the case of type I collagen, the $\alpha2\beta1$ integrin is likely involved, and for fibrin, VE-cadherin may be the relevant receptor. Such receptor clustering may be dictated by as-yet-undetermined molecular repeat distances, which may be either common to these two polymers or unique to each type of receptor–ligand interaction. In this regard, we will briefly consider only the potential mechanism of $\alpha2\beta1$ integrin clustering by the type I collagen fibril.

Assuming anywhere from one to three of the proposed sites for $\alpha2\beta1$-binding are available on the fibril surface, it may be calculated to contain from approximately 10 to 40 $\alpha2\beta1$ integrin receptor-binding sites across its width, and along its length, to contain such sites at 67-nm intervals. Because the region of the integrin receptor that engages type I collagen is $\cong$ 7.0 nm in diameter, across the fibril of $\cong$ 80 nm in width there is only enough space for a maximum of 10 receptors to occupy their ligand-binding sites; in this scenario the receptors clearly have the potential to be tightly arranged or clustered at the same cell surface location. On the other hand, receptors bound to sites available every 67 nm along the length of the fibril may be too sparsely distributed to promote their clustering and activation.

Another level of complexity that must be investigated is the matter of whether EC must interact with multiple collagen fibrils to initiate angiogenesis, and if so, how fibril arrangement may influence their EC interactions and capillary morphogenesis. Such information will contribute to the rational design of angiogenic polymers for various applications in tissue engineering, as discussed following. Last, it is worth addressing the fact that some other molecules, such as laminin and type IV collagen mixtures, have been reported to support tube formation in vitro, which at first glance seems inconsistent with the model described previously. However, in vivo, EC may never be exposed to high concentrations of only one or two basement membrane components, and moreover, in the intact basement membrane, many of the biologically active domains of the matrix scaffold may be inaccessible to EC. Finally, some ECM molecules, although incapable of supporting angiogenesis when present as components of the native macromolecular assembly, may be capable of engaging the appropriate cell surface receptors and thus promoting angiogenesis, when isolated and presented to EC at high concentrations.

Tissue Engineering of the Microvasculature: The ECM Holds the Key

One of the greatest current challenges in biomedical science is to refine and apply tissue-engineering technologies to correct a variety of human diseases. Among the most significant limitations to generating tissue equivalents are the difficulties inherent in endowing such implants with suitable vasculatures. Although much progress has been made in the arena of large-diameter vascular conduits, technologies to develop the microvasculature are in their infancy. However, the necessity of the microvasculature is evidenced by the observation that most cells must be within $\cong$ 100 µm or so from a capillary, and in some cases (e.g., those tissues with high metabolic rates such as skeletal or cardiac muscle myocytes), may require even greater proximity (e.g., $\cong$ 50 µm). Difficulties in engineering of the microvasculature include its complex arrangement, diminutive size, and fragility. For example, the smallest module of the microvasculature is the capillary bed, which typically consists of a thoroughfare channel that passes between an arteriole and a venule, but which also feeds a fine honeycomb-like arrangement of capillary tubes. Depending on the metabolic needs of the tissue, precapillary sphincters, arranged at the junctures between the arterioles and the capillary network, may control the flow into the capillaries. The diameters of these vessels range from $\cong$ 50 µm for arterioles (Figure 1) down to 5 or less µm for true capillaries, which do not contain associated smooth muscle cells or pericytes.

Examination of the microvasculature of even relatively modest-sized tissues (e.g., the wing of the little brown bat *Myotis lucifigus*) reveals a remarkably complex structure, containing more than 2,500 distinct blood vessels ranging anywhere from ≅4 to 75 μm in diameter. To fashion such a vasculature within an engineered tissue in a directed way seems daunting, and it would be of great advantage to tissue engineers if the assumption that the development of the vascular bed is dictated by the parenchymal tissues that surround it proves to be correct. In other words, if a tissue equivalent is seeded with the proper parenchymal and supportive cells (i.e., vascular cells and others such as fibroblasts), the capability can be created to sort and undergo morphogenesis into the proper tissue arrangement. For instance, mixtures of EC and fibroblasts cultured on cross-linked mats of chitosan-chondroitin sulfate-collagen form dermis-like tissues, complete with what appeared to be microvessels, which developed in the vicinity of fibroblast- and ECM-rich regions. However, it is currently unclear if such tissues are endowed with native-type microvasculatures. Moreover, if such a tissue equivalent were transplanted into a host animal, it is uncertain whether the vasculature of the implant would be capable of anastomosing with that of the host. As with this latter example, to date, attempts to engineer the microvasculature have largely involved use of the ECM-binding growth factors and/or ECM scaffolds as key tools.

A predominant approach toward engineering the microvasculature involves attempting to stimulate vascular regeneration by promoting the growth or regeneration of the endogenous vasculature in ischemic tissues. Such methods have mainly involved delivering angiogenic growth factors, such as FGF-2 or VEGF, to tissues either by injection or sustained polymer-based release; as mentioned earlier, these factors work in concert with extracellular or cell surface proteoglycans to exert their activities. The current literature demonstrates that such approaches hold much promise in promoting microvascular regeneration in ischemic tissues, and many reviews discuss these findings, so this topic will not be further discussed.

ECM Templates for Microvascular Engineering

Another approach attempts to provide an engineered tissue with a preformed structural template of the desired vasculature. One series of studies performed by Donald E. Ingber and coworkers has defined how the two-dimensional geometry of the ECM substrate influences EC growth and differentiation. Thus, EC were cultured on self-assembled monolayers generated by stamping microscopic patterns of alkanethiols on gold-coated substrata, resulting in micropatterns of alkanethiols that were further modified by the addition of various ECM molecules such as fibronectin. This approach allows for the generation of ECM substrata in the forms of islands, stripes, or geometric patterns of various dimensions, on which EC can be cultured and examined for their adhesive, growth, and differentiation responses. A main finding was that substrate geometry and area dictates whether an EC would adhere at all, proliferate, or undergo apoptosis. Thus, on single ECM islands of 10 μm or less, cells could not adhere; those ranging in sizes between 10 to 25 μm supported attachment but also promoted apoptosis, and on ECM islands of 50 μm or more, cells proliferated. Of particular relevance to the engineering of angiogenic scaffolds, it was further shown that EC cultured on 10-μm-wide fibronectin stripes formed vascular-like cords complete with contiguous lumens, but on 30-μm-thick stripes, the cells proliferated and did not form lumens.

Some investigators have proposed the development of three-dimensional templates for the generation of the microvasculature. One such template is described as a micro-bioreactor or angiochip. The device has as its basic design a solid-state silicon capsule or wafer containing thousands of microscopic chambers measuring 1 μm or less, which can be loaded with nanoliter volumes of bioactive agents. After filling, chambers are sealed with a thin gold layer to retain their contents until a later time. After implantation in a host, the contents of individual chambers on such chips could be selectively released by electro-dissolving the gold layers limiting the chambers. For some applications, the device may be surface-tooled and treated with an appropriate ECM scaffold to render them amenable to EC seeding before implantation. Potential uses for non-EC-seeded variety of implants would be for the sustained delivery of angiogenic or angiostatic drugs, targeted to expand or reduce the endogenous microvasculature. Angiochips preseeded with EC may be useful in promoting the outgrowth of the exogenous EC to form a new microvascular network in the host tissue, which also might anastamose with the microvascular network of the host.

Another potential approach to creating microvascular conduits was discovered using a mouse transgenic model with a targeted overexpression of chemoattractant protein-1 in the myocardium. The cardiac muscle was found to contain thin channels presumably formed by the enzymatic boring of macrophages, via their release of metalloelastase, through the tissue interstitium. To apply this phenomenon toward the rational engineering of the microvasculature will require directing the formation of channels to generate a desired microvascular configuration. Some possible approaches may include imprinting a matrix scaffold with a chemotactic gradient and seeding one end of a scaffold with macrophages. The cells may respond by boring channels through the ECM in a directed fashion, in response to the gradient. A variant of this approach would be to drive the cell-boring activities actively, using an external stimulus such as a magnetic field to spatially direct the migration of the macrophages; perhaps iron-conjugated beads may be affixed to the cells to confer their response to the magnetic field. Subsequently, the macrophage-generated channels would have to be populated with the appropriate complement of vascular cells before its use as a tissue implant.

Rational Engineering and Use of Angiogenic Polymers

From our discussion of the basic biology of ECM-induced angiogenesis, and of the tissue engineering of the microvasculature, it is obvious that a crossroads of both bodies of knowledge is the matter of how one may define and rationally create an angiogenic polymer. The simplest approach to this end may be to design a polymer that has all of the appropriate, proangiogenic characteristics of type I collagen or fibrin, but that lacks any of their undesirable characteristics (Table I). For example, to replicate some of the common features of type I collagen and fibrin, the polymer should assume a rigid, linear shape approximately 100 nm or greater in diameter and sufficiently long (i.e., ≥ 100 μm) to simultaneously interact with or bridge two or more EC. The ideal polymer might carry a basic charge, because this will confer affinity for the strongly anionic character of the EC surface. Moreover, the polymer should contain ligand-binding sites for an appropriate EC surface receptor, such as the $\alpha 2\beta 1$ integrin (i.e., GFPGER sequences), which should be distributed appropriately to promote cell–polymer interactions. From this discussion, such integrin-binding sites might be spaced no closer than 10 nm and no farther apart than 50 nm on the polymer. If the polymer is to be presented apically to cells, it should be capable of forming a classic gel or, if it must be affixed as stripes on a substrata on which EC will be seeded, one would limit their width to 10 μm or less to promote proper lumen formation. To create such a polymer, either recombinant type I collagen or fibrin could be used, in which unsuitable protein epitopes are deleted (e.g., Table I). For example, in the case of type I collagen, one might remove binding sites for ligands that have the potential to interfere directly, or through steric hindrance, with integrin receptor binding, such as fibronectin and the various collagen-binding proteoglycans such as decorin. Another approach would be to design polymers having no significant homologies to collagen or fibrin, but in which multiple integrin-binding sites are included at appropriate spacing intervals. Such super polymers may someday provide an ideal angiogenic template for a variety of tissue engineering and human therapeutic applications.

Table I Angiogenic Super-Polymers.

Physical characteristics
- Form a rigid, cable-like polymer
- Assume appropriate secondary and tertiary conformations
- Assume dimensions of ≅ 100 nm wide and ≥ 100 μm long
- Gel at physiological temperature and pH
- Carry a net positive charge

Biologically Active Determinants

Include sites:
- For binding of appropriate *integrins, distributed 10–50 nm apart, or for binding of other cell surface receptors.
- For binding of angiogenic growth factors (e.g., bFGF/VEGF)
- Necessary for polymer fibrillogenesis/assembly

Exclude binding sites for:
- Ligands that sterically disrupt EC–polymer interactions
- Proteinases, pathogens, or toxins

Exclude epitopes that are:
- Immunogenic
- Pro- or anticoagulant
- Antiangiogenic

Modes of presentation to EC
- As apical or pericellular gel
- Affixed to EC substrata in stripes/islands ≤ 10 μm wide
- Injected into tissues with vascular insufficiency/ischemia
- Incorporated into tissue equivalent scaffolds for implantation in vivo
- Genetically engineered to be expressed by perivascular cells or vascular stem cells

*Integrin-binding sites in type I collagen or collagen mimetic polymers must be triple helical.

Features and uses of angiogenic super-polymers are proposed based on common characteristics of the proangiogenic type I collagen and fibrin polymers, and on various aspects of EC biology outlined in this review. Super-polymers may include genetically engineered forms of type I collagen or fibrin, or various synthetic polymers.

Glossary

Angiogenesis: Establishment of new blood vessels by branching, budding, or remodeling of the existing vasculature. Occurs during embryogenesis and in the adult (e.g., during wound healing).

Basement membrane: Extracellular matrix scaffold secreted by EC and other cell types, comprised of laminin, type IV collagen, perlecan, and nidogen, as well as other minor components that vary depending on the tissue source. The basement membrane lies in close contact with the basal side of the EC and provides structural support and functions as a barrier separating them from the underlying perivascular cells and tissue stroma.

EC: Mesenchymal cells that comprise the lining of the cavities of the heart, blood, and lymphatic vessels.

Fibrin: A polymer that forms the fibrous stroma of blood clots. It is generated during the clotting cascade by thrombin cleavage of fibrinogen, which liberates fibrin monomers that assemble in an overlapping fashion into the cable-like fibrin polymer.

Type I collagen: A ubiquitous ECM component comprised of triple-helical monomers that assemble in a staggered fashion into cable-like collagen fibrils in the extracellular space, which confer tensile strength on many tissues of the body.

References

1. Gerritson, M. E. et al. (2003). *Microcirculation*, **10**, 63–81.
2. Weisel, J. W. (1986). *Biophys. J.*, **50**, 1079–1093.
3. Vernon, R. B. et al. (1995). *In Vitro Cell Dev. Biol.* **31**, 120–131.

Further Reading

D'amore, P. A., and Ng, Y. S. (2002). Tales of the cryptic: Unveiling more angiogenesis inhibitors. *Trends in Molec. Med.* **8**, 313–315.

DiLullo, G., Sweeney, S. M., Korrko, J., Ala-Kokko, L., and San Antonio, J. D. (2002). Mapping the ligand-binding sites and disease associated mutations on the most abundant protein in the human- type I collagen. *J. Biol. Chem.* **277**, 4223–4231.

Iozzo, R., and San Antonio, J. D. (2001). Heparan sulfate proteoglycans: Heavy hitters in the angiogenesis arena. *J. Clin. Invest.* **108**, 349–355.

Moldovan, N., and Ferrari, M. (2002). Prospects for microtechnology and nanotechnology in bioengineering of replacement microvessels, *Arch. Pathol. Lab. Med.* 126, 320–324. *This review outlines many of the aspects of microvascular tissue engineering discussed here and provides their primary literature citations.*

Mongiat, M., Sweeney, S. M., San Antonio, J. D., Fu, J., and Iozzo, R. (2003). Endorepellin, a novel inhibitor of angiogenesis derived from the c terminus of perlecan. *J. Biol. Chem.* **276**, 4238–4249.

Somerville, R. P. T., Oblander, S. A., and Apte, S. S. (2003). Matrix metalloproteinases: Old dogs with new tricks. *Genome Biol.* Accessed at: http://genomebiology.com/2003/4/6/216.

Sudhakar, A., Sugimoto, H., Yang, C., Lively, J., Zeisberg, M., and Kalluri, R. (2003). Human tumstatin and human endostatin exhibit distinct antiangiogenic activities mediated by $\alpha v\beta 3$ and $\alpha 5\beta 1$ integrins. *Proc. Natl. Acad. Sci. USA* **100**, 4766–4771.

Sweeney, S. M., DiLullo, G. D., Slayter, S., Martinez, J., Iozzo, R. V., Lauer-Fields, C., Fields, G. B., and San Antonio, J. D. (2003). Type I collagen-induced angiogenesis requires $\alpha 2\beta 1$ ligation of a single GFP*GER integrin-binding collagen sequence, possibly in concert with p38 MAPK activation and focal adhesion disassembly. *J. Biol. Chem.* **278**, 30516–30524.

Weideman, M. P., Tuma, R. F., and Mayrovitz, H. N. (1981). *An Introduction to Microcirculation*, Vol. 2. New York: Academic Press. *This book describes the anatomy and function of the microvasculature and provided facts cited here regarding the structure of capillary beds, average dimensions of capillaries, and the anatomy of the bat wing vasculature.*

Xu, J., and Brooks, P. C. (2002). *Extracellular Matrix in the Regulation of Angiogenesis in Assembly of the Vasculature and Its Regulation* (R. J. Tomanek, ed.), pp. 67–95. Boston: Birkhauser Press. *This article presents a comprehensive review of ECM structure and its many roles in angiogenesis, and provides primary citations for many of the facts discussed here. Although the authors categorize the ECM into the basement membrane and interstitial compartments, they do not present them as having two distinct functions in angiogenesis regulation as we have done here.*

Capsule Biographies

Dr. James D. San Antonio is an Associate Professor of Medicine at Thomas Jefferson University in Philadelphia. His research focuses on elucidating mechanisms of type I collagen–induced angiogenesis; defining the role of proteoglycan–collagen interactions in connective tissue disorders; and invention of heparin and proteoglycan-binding peptides to use as protamine substitutes for the neutralization of anticoagulant heparin in humans, or as carriers to target drugs to proteoglycan-rich tissues such as the vasculature. His work is supported by grants from the National Institutes of Health, the Pennsylvania American Heart Association, and the Department of Defense.

Dr. Renato V. Iozzo is a Professor of Pathology and Cell Biology, and Director of the Extracellular Matrix Program at the Kimmel Cancer Center, at Thomas Jefferson University, in Philadelphia, Pennsylvania. He is the winner of several awards and was the past Chair of the Gordon Research Conference on Proteoglycans. He is the editor of two books and has published more than 200 research articles focused on extracellular matrix, tumor growth, and angiogenesis. His current work is supported by funds from the National Institutes of Health and the Department of Defense.

CHAPTER 22

Pericytes and Glial Cells and Vascular Development

[1]Stanley A. Vinores and [2]John D. Ash

[1]*The Wilmer Eye Institute, Johns Hopkins University School of Medicine, Baltimore, Maryland*
[2]*University of Oklahoma Health Sciences Center and the Dean A. McGee Eye Institute, Oklahoma City, Oklahoma*

Introduction

The retinal vasculature consists of a single layer of vascular endothelial cells, arranged to form a lumen, which are ensheathed by pericytes. In the superficial retina, these vessel structures are then ensheathed by astrocytic processes. Each cell layer is separated by a negatively charged basement membrane (Figure 1). In the deeper retina, Müller cells, rather than astrocytes, participate in the formation of the perivascular sheath. Through mechanisms that are not entirely clear, the growth, maturation, and permeability of retinal vessels are dependent on the growth and maturation of pericytes and astrocytes. A significant amount of work has been directed at determining the roles of these supporting cell types in retinal development and homeostasis, and what follows is a brief summary of the current progress. To focus this discussion, we have only included what is known for humans, mice, and rats.

Pericytes

Role of Pericytes in Maintaining Vascular Integrity

Pericytes are the vascular supporting cell type most intimately associated with the retinal vascular endothelium. They are elongated, polymorphic cells of mesenchymal origin that are sandwiched between the vascular endothelial cells and the perivascular astrocytes and are separated from each by a basement membrane, except where pericytes and endothelial cells make direct contact (Figure 1). Pericytes surround the vascular endothelium and help regulate endothelial cell growth, vessel stability, and vascular permeability. Pericyte loss, which occurs in diabetic retinopathy, is accompanied by neovascularization and vascular leakage. Pericytes contain extensive networks of intermediate filaments, which, depending on the species and tissue, consist of vimentin with or without desmin. Pericytes also contain contractile proteins, such as smooth muscle–specific isoforms of actin (α-SMA) and myosin, cyclic GMP-protein kinase, and tropomyosin, which may help in regulating blood flow [1, 2]. α-SMA, which is also found in vascular smooth muscle cells, is the most commonly used marker for the recognition and immunolocalization of pericytes [3, 4], but an RNA probe directed against PDGFβ-receptors can also be used to visualize pericytes [5].

Pericyte Development in the Retinal Vasculature

During vascularization of the rat retina, pericytes are recruited after the formation of the primary endothelial cell plexus. Pericytes migrate into the eye from arterioles branching from the hyaloid artery in the optic nerve. As the plexus matures, pericytes migrate from the arterioles toward the venules. In rats and mice, the recruitment of pericytes lags behind the formation of the endothelial plexus by several days to a week [3, 5], although some pericytes have been found in small vessels of the mouse as early as postnatal day 1 (P1)[6]. They are specifically associated with areas of the vascular network that contain collagen IV [5]. By P10, pericytes cover the primary vessel branches of the superficial retinal vascular bed in rats and by P18, they have

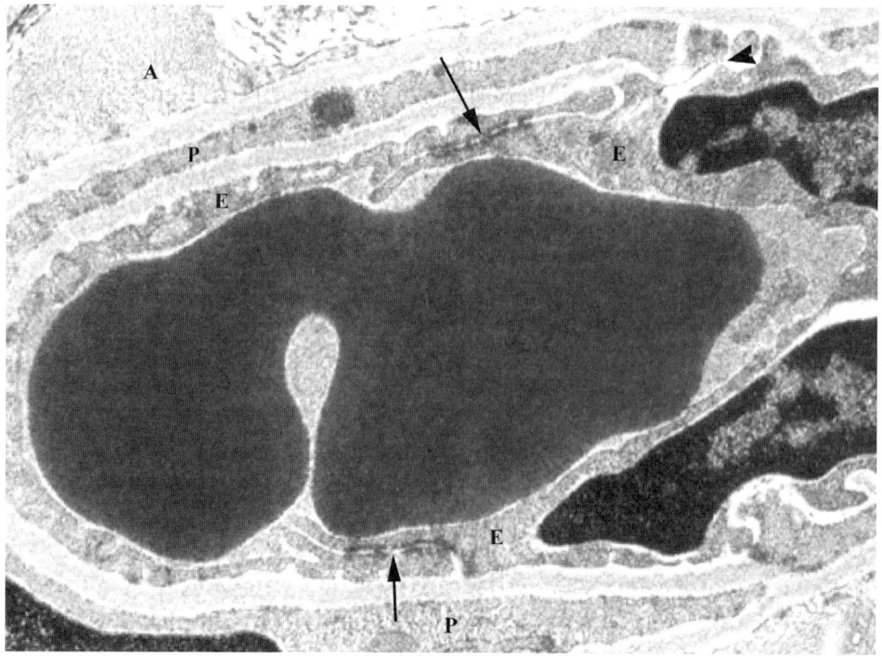

Figure 1 Ultrastructure of a normal inner retinal vessel. The tubular configuration of the vascular endothelial cells (E) forms a lumen, which contains a red blood cell. The endothelial cells are connected by tight junctions (zonula occludens) (*arrows*), which help establish the blood–retinal barrier. Pericytes (P) envelop the vascular endothelium and astrocytes (A) cover the pericytes. A basal lamina separates the astrocytes from the pericytes and the pericytes from the endothelial cells, except where there is direct cell-to-cell contact. The *arrowhead* shows a point of direct contact between a pericyte and a vascular endothelial cell. (see color insert)

reached the secondary and tertiary branches. Even though the deep capillary bed is completely formed by P10, pericytes do not cover these vessels until more than three weeks after birth, when no further vascular remodeling occurs [3]. It appears that pericytes are more involved in the maturation and remodeling of retinal vessels than in their formation. After the vasculature has formed, many of the vessels regress in more hyperoxic areas. Covering of the vascular endothelium by pericytes arrests endothelial cell proliferation and appears to stabilize the enveloped capillaries, protecting them from regressing. In mice, pericytes are always found on capillaries adjacent to the regressing segment, but never on the regressing arm [6].

The Role of PDGF-B in Pericyte Recruitment and Development

Platelet-derived growth factor-B (PDGF-B) is necessary for pericyte development, because PDGF-B-deficient mice lack pericytes [7]. Not only is PDGF-B essential for pericyte survival, but it is also critical for their migration along capillary endothelial sprouts and for normal pericyte–endothelial interactions. Pericytes are the predominant cell type in the developing retina that expresses PDGF-β receptors, which specifically bind PDGF-B. In studies where PDGF-B has been elevated during the time of vascular remodeling, either by endogenous administration [3] or because of overexpression in transgenic mice [4], the increased PDGF-B leads to a disruption of normal pericyte–endothelial cell interactions and disrupted retinal vascular development. Following intravitreal injection of PDGF-B to mice during the time of vessel remodeling, pericytes, which normally are tightly wrapped around vessels, are disorganized and often connected to more than one vessel. This leads to excessive hemorrhage followed by capillary obliteration and the widening of capillary-free zones and the formation of irregularly spaced vascular loops and capillary tangles. The retinal vasculature of these PDGF-B-treated mice is also more vulnerable to oxygen-induced damage. When PDGF-B is administered during the remodeling phase of the deep capillary plexus, but after the remodeling of the superficial plexus is complete (P18), only the deep vessels are affected [3]. Overexpression of PDGF-B in transgenic mice leads to hyperproliferation of pericytes, astrocytes, and endothelial cells. It also results in the disassociation of pericytes and vascular smooth muscle cells from vascular endothelial cells and an impairment of the formation of the deep capillary bed. These effects result in the formation of an underdeveloped vasculature in the retina. The αSMA-positive pericytes in the retinas of these mice are generally found in clusters and are usually round or oblong with a high nucleus-to-cytoplasm ratio and abundant mitochondria and rough endoplasmic reticulum [4].

Astrocytes

The Role of Astrocytes in the Development of the Blood–Retinal Barrier

Recent studies have shown that retinal astrocytes play a significant role in retinal vascular development by angiogenesis and in the induction and maintenance of the blood–retinal barrier (BRB). Glial fibrillary acidic protein (GFAP)-positive astrocytes are found in all vertebrate species that have a retinal vasculature, and the temporal-spatial development of astrocytes precedes the onset of angiogenesis in the retina [2, 5, 8]. In adults, astrocytes ensheathe the superficial retinal blood vessels (Figure 2), and their association with the vasculature is spatially and temporally correlated with the establishment of the BRB [8]. The functional role of astrocytes in establishing a blood–tissue barrier has been more clearly defined in the central nervous system and in culture (Reviewed in [9]). The end feet of astrocytes make an extensive network of cell processes that lie in close apposition to the basal surface of the microvascular endothelium. The intimate association between astrocytes and endothelial cells suggests that astrocytes may provide an inductive signal to promote the establishment of the blood–brain barrier (BBB) [9], which is analogous to the BRB. Although the exact mechanism for this induction is unknown, it can be mimicked in vitro and has been shown to require close proximity and correct cell polarity between astrocytes and vascular endothelial cells. As with the BBB, the BRB is likely to be induced by glial cells, with astrocytes providing this function in the superficial blood vessels and Müller cells in the deep capillary network. In addition to their role in establishing the BRB, retinal astrocytes also appear to play an essential role in driving the angiogenic phase of retinal vascular development through the expression of VEGF. Differentiated astrocytes are located in the retina in advance of the formation of patent vessels. In the absence of adequate blood circulation, astrocytes experience relative hypoxic conditions and have been shown to upregulate VEGF, which drives retinal vascular development by angiogenesis [2]. The association of astrocytes to retinal vessels also appears to contain their growth and provide vascular stability, and a breech of this association may result in unrestrained neovascularization and hemorrhage [10]. In order to understand the role of astrocytes in retinal vascular development, it is necessary to understand the migration, growth, and development of astrocytes in the retina.

Development of Retinal Astrocytes

Although there are significant differences between humans and mice, astrocytes in both species migrate into the retina through the optic nerve as immature astrocyte progenitors [8]. The progenitors have been defined as cells expressing Pax2 and vimentin, but not GFAP or S100 [8]. The progenitors then migrate through the nerve fiber layer toward the anterior margins of the retina. This creates a spatial and temporal pattern in which the progenitors, which are the least differentiated, are found at the leading edge of migration and the older astrocytes, which have begun to differentiate and mature, are found in the central retina. At least three stages of maturation have been described. The first stage has been defined as immature perinatal astrocytes that express vimentin, Pax2, S100, and GFAP. This is followed by the appearance of mature perinatal astrocytes that lose vimentin expression, but retain Pax2, S100, and GFAP expression. After the final stage of development, adult astrocytes are found throughout the retina and have robust expression of GFAP and S100, but have lost expression of Pax2 [8]. In mice, the interaction with the vascular endothelium seems to increase the expression of GFAP and S100 in retinal astrocytes, suggesting that this interaction may be important for the final stage of astrocyte maturation [5]. This

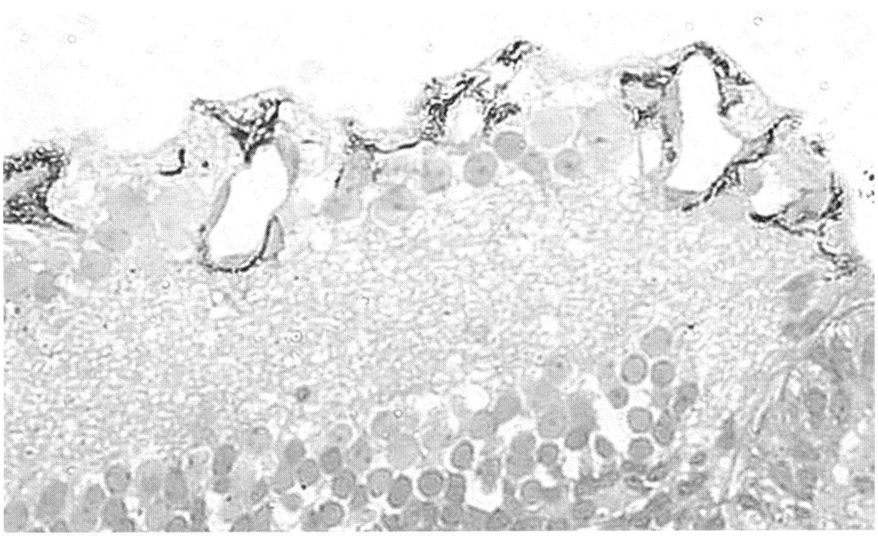

Figure 2 Immunocytochemical staining for glial fibrillary acidic protein in a P15 mouse retina decorates the astrocytic processes that encircle the vessels in the inner retina. (see color insert)

is most apparent in flat-mounted retinas from P5 mice that have been immunostained for S100 and griffonia simplicifolia isolectin B4 (GSA; Figure 3). At this stage of mouse development, S100-positive astrocytes have spread throughout the nerve fiber layer of the retina, while the developing GSA-positive vasculature has only spread throughout the central region of the retina. S100 expression is much higher in the vascular-covered central retina than in the avascular periphery.

The Role of PDGF-A in Astrocyte Migration and Proliferation

Although the inductive signals for astrocyte development have not been fully identified, the current data suggest that platelet-derived growth factor A (PDGF-A) plays an important role in their migration and proliferation. Astrocyte progenitors enter the retina well after the differentiation of retinal ganglion cells. This is perhaps a necessary sequence as astrocytes migrate into the retina through the optic nerve along the ganglion cell axons. Once in the retina, astrocyte progenitor migration may be independent of the presence of ganglion cells [11]. However, other studies have shown a direct role for ganglion cells in astrocyte development. Ganglion cells secrete PDGF-A, and astrocytes can respond to PDGF-A through the expression of PDGF receptor α (PDGFRα) [5, 12]. The simplest model suggests that astrocytes and/or their progenitors migrate into and proliferate within the retina following the chemoattractant and mitogenic signal, PDGF-A. In support of this model, studies have shown that overexpression and ectopic expression of PDGF-A from the lens results in an astrocytic hyperplasia known as a hamartoma near the optic nerve head of the retina [12]. PDGF-A has also been overexpressed in transgenic mouse retina using the neuron-specific enolase promoter and the opsin promoter [5, 13]. In both models, there was excessive accumulation of astrocytes in the retina. In each of the overexpression models, the ectopic expression of PDGF-A led to mislocalized astrocytes, suggesting that the site of PDGF-A expression is important for migration cues in normal development. PDGF-B, which can also bind and activate the PDGFRα on astrocytes, can also induce their mislocalization and hyper-proliferation [4]. Further studies have shown that disruption of PDGF-A signaling by neutralizing antibodies to PDGFRα disrupts the normal development of retinal astrocytes [5]. These studies demonstrate that PDGF-A is a potent chemoattractant and mitogen for astrocytes and that its expression is required for normal astrocyte migration in the retina. It is interesting to note that retinal vascular development is dramatically impaired in all of the mouse models overexpressing PDGF-A. These findings clearly demonstrate that growth and development of retinal astrocytes is important for the development of a functional retinal vasculature.

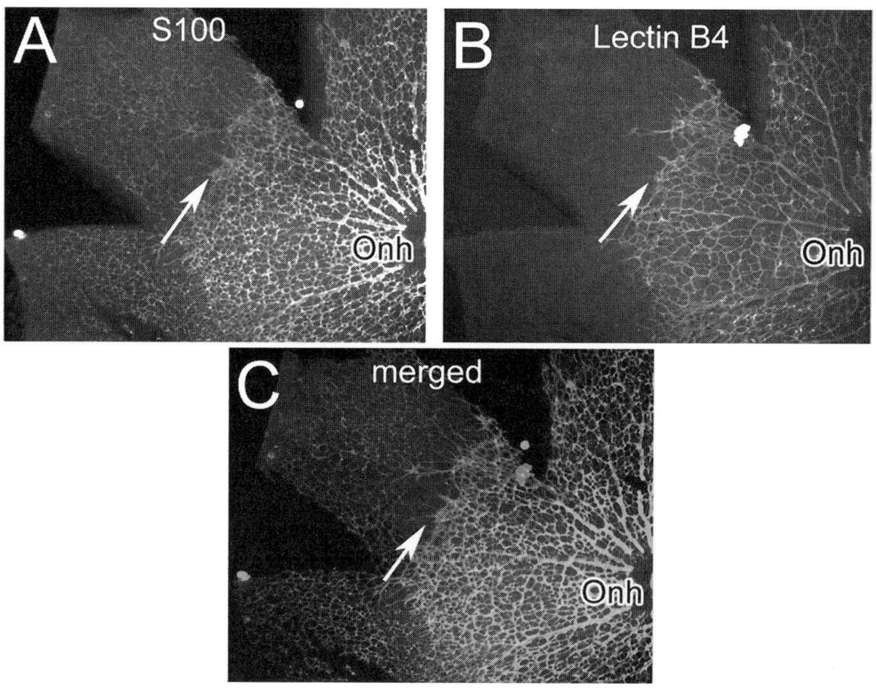

Figure 3 Immunofluorescent staining using antibodies to S100 (**A**), which is expressed only in astrocytes in the P5 mouse retina, and fluorescence staining of the developing vasculature using Griffonia simplicifolia isolectin B4 (**B**). S100 positive astrocytes are located throughout the retina; however, the staining intensity is higher in the central retina (**A**). At this age, the developing vasculature covers only the central region of the retina (**B**). The *arrow* in all panels marks the leading edge of the developing vasculature. In the merged image (**C**), it is clear that astrocytes in contact with the vasculature have higher expression of S100 than astrocytes in the avascular periphery. (see color insert)

Glossary

Astrocyte: A neuroglial cell of ectodermal origin with processes that ensheathe vessels in the inner retina and brain, helping to establish the blood–retinal or blood–brain barrier; they represent a supporting cell type for neurons as well as the vasculature.

Pericyte: A vascular supporting cell of mesenchymal origin with contractile properties that surrounds the vascular endothelium and helps regulate vascular integrity.

Platelet-derived growth factor A: A chemoattractant and mitogen for astrocytes that is important in retinal vascular development.

Platelet-derived growth factor B: A proangiogenic factor that is necessary for pericyte survival and migration and for normal retinal vascular development.

Vascular endothelial cells: The cells that line the lumen of blood vessels.

References

1. Diaz-Flores, L., Gutierrez, R., Varela, H., Rancel, N., and Valladares, F. (1991). Microvascular pericytes: A review of their morphological and functional characteristics. *Histol. Histopathol.* **6**, 269–286. *This is an excellent review describing the morphological and functional properties of pericytes.*
2. Provis, J. M. (2001). Development of the primate retinal vasculature. *Progress Retinal Eye Res.* **20**, 799–821. *This review describes the roles of the various cell types in the development of the retinal vasculature in primates and compares the findings to those of other species.*
3. Benjamin, L. E., Hemo, I., and Keshet, E. (1998). A plasticity window for blood vessel remodelling is defined by pericyte coverage of the preformed endothelial network and is regulated by PDGF-B and VEGF. *Development* **125**, 1591–1598. *This study describes the role of the pericyte in the development of the retinal vasculature and how PDGF-B and VEGF affect pericyte–endothelial interactions.*
4. Vinores, S. A., Seo, M. S., Derevjanik, N. L., and Campochiaro, P. A. (2003). Photoreceptor-specific overexpression of platelet-derived growth factor induces proliferation of endothelial cells, pericytes, and glial cells and aberrant vascular development: An ultrastructural and immunocytochemical study. *Dev. Brain Res.* **140**, 169–183. *This study shows the effect of overexpression of PDGF-B in the retinas of transgenic mice on endothelial cells, pericytes, and retinal glia and how this affects retinal vascular development.*
5. Fruttiger, M. (2002). Development of the mouse retinal vasculature: Angiogenesis versus vasculogenesis. *Invest. Ophthalmol. Vis. Sci.* **43**, 522–527. *This is the most recent study describing the relationship between prior astrocyte and pericyte development and vascular development in the mouse retina.*
6. Connolly, S. E., Hores, T. A., Smith, L. E. H., and D'Amore, P. A. (1988). Characterization of vascular development in the mouse retina. *Microvasc. Res.* **36**, 275–290.
7. Lindahl, P., Johansson, B. R., Leveen, P., and Bertsholtz, C. (1997). Pericyte loss and microaneurysm formation in PDGF-B-deficient mice. *Science* **277**, 242–245.
8. Chu, Y., Hughes, S., and Chan-Ling, T. (2001). Differentiation and migration of astrocyte precursor cells and astrocytes in human fetal retina: Relevance to optic nerve coloboma. *FASEB J.* **15**, 2013–2015. *This study carefully describes four unique stages of astrocyte development in the human retina.*
9. Abbott, N. J. (2002). Astrocyte-endothelial interactions and blood-brain barrier permeability. *J. Anat.* **200**, 629–638. *This review describes our current understanding of the role of astrocytes in the establishment and maintenance of the blood–brain barrier.*
10. Zhang, Y., and Stone, J. (1997). Role of astrocytes in the control of developing retinal vessels. *Invest. Ophthalmol. Vis. Sci.* **38**, 1653–1666.
11. Chan-Ling, T. (1994). Glial, neuronal and vascular interactions in the mammalian retina. *Prog. Retin. Eye Res.* **13**, 357–389. *This is an excellent review of the interactions between retinal glial cells, neurons, and vascular endothelium in normal development.*
12. Reneker, L. W., and Overbeek, P. A. (1996). Lens-specific expression of PDGF-A in transgenic mice results in retinal astrocytic hamartomas. *Invest. Ophthalmol. Vis. Sci.* **37**, 2455–2466.
13. Yamada, H., Yamada, E., Ando, A., Seo, M. S., Esumi, N., Okamoto, N., Vinores, M., LaRochelle, W., Zack, D. J., and Campochiaro, P. A. (2000). Platelet-derived growth factor-A-induced retinal gliosis protects against ischemic retinopathy. *Am. J. Pathol.* **156**, 477–487.

Capsule Biographies

Dr. Vinores has been a member of the faculty of the Wilmer Eye Institute at the Johns Hopkins University School of Medicine in Baltimore, Maryland, since 1991. His research primarily focuses on the blood–retinal barrier, angiogenesis, and immunocytochemistry. He is supported by funds from the National Institutes of Health (grants EY10017 and EY05951) and Research to Prevent Blindness (Lew R. Wasserman Merit Award).

Dr. Ash has been an affiliate of the Dean A. McGee Eye Institute and on the faculty at the University of Oklahoma Health Sciences Center since 1999. His research is primarily focused on the development of the vasculature in the eye and the role of astrocyte–endothelial cell interactions during retinal angiogenesis. He is supported by funds from the National institutes of Health (grants EY14206-01, EY012190, and RR017703-01), Research to Prevent Blindness, and the Oklahoma Center for Science and Technology (HR02-140RS).

SECTION C

In Vivo Models

CHAPTER 23

Hemangioma: A Model System to Study Growth and Regression of Blood Vessels

Joyce Bischoff

Vascular Biology Program, Children's Hospital Boston and Department of Surgery, Children's Hospital Boston and Harvard Medical School, Boston, Massachusetts

Hemangioma is a tumor of endothelium that occurs in 5 to 10 percent of caucasian infants. Often referred to as infantile hemangioma, it follows a unique biological course: it proliferates rapidly in the first year of life, regresses slowly during early childhood, and never recurs. These features of hemangioma distinguish it from other vascular tumors, which do not regress and can occur in adults. Like many vascular tumors, the endothelial cells that comprise hemangioma express a multitude of endothelial markers and angiogenic growth factors. Endothelial cells from hemangiomas are clonal and exhibit abnormal properties when compared to normal human microvascular endothelial cells. These findings strongly suggest that hemangiomas arise when a somatic mutation occurs, within a single endothelial cell, in a gene that controls endothelial cell proliferation and possibly responses to angiogenic stimuli. The molecular mechanisms underlying the rapid growth and the spontaneous involution of hemangioma are yet to be discovered, but when elucidated, are likely to yield unique insights into the regulation of blood vessel formation and growth.

Clinical Features of Hemangioma

Growth and Regression

Hemangiomas typically appear soon after birth, grow rapidly during the first year of life (*proliferating phase*), regress slowly from age 1 to 7 (*involuting phase*), and thereafter become a fibrofatty residuum (*involuted phase*). Hence, the life cycle of hemangioma encompasses both proliferation and regression of blood vessels and, as such, presents a unique model system to study the regulation of angiogenesis. The most common form of hemangioma appears within two weeks after birth, yet some appear fully grown at birth. These "congenital hemangiomas" fall into two subgroups: one group regresses rapidly, within months, whereas the second group fails to involute, suggesting that the life cycle of hemangioma can be compressed or extended. Understanding these variations on the more common form of infantile hemangioma will be fundamental to deciphering the mechanisms that control involution and may eventually pave the way for development of fast-acting therapies that either speed up the involution process or prevent the aberrant growth from occurring in the first place.

Location of Hemangiomas

Hemangiomas are most often cutaneous, found on the head and neck (60%), trunk (25%), and the extremities (15%). See examples of proliferating hemangiomas in Figure 1. About 20 percent of hemangiomas occur as multiple tumors. The relationship of these multifocal lesions to each other and how they differ from the more common single lesion is not understood. Visceral hemangiomas also occur, but these are rare and are likely to go undetected because most hemangiomas are small and pose no threat or complications. However, a subset of hemangiomas can grow

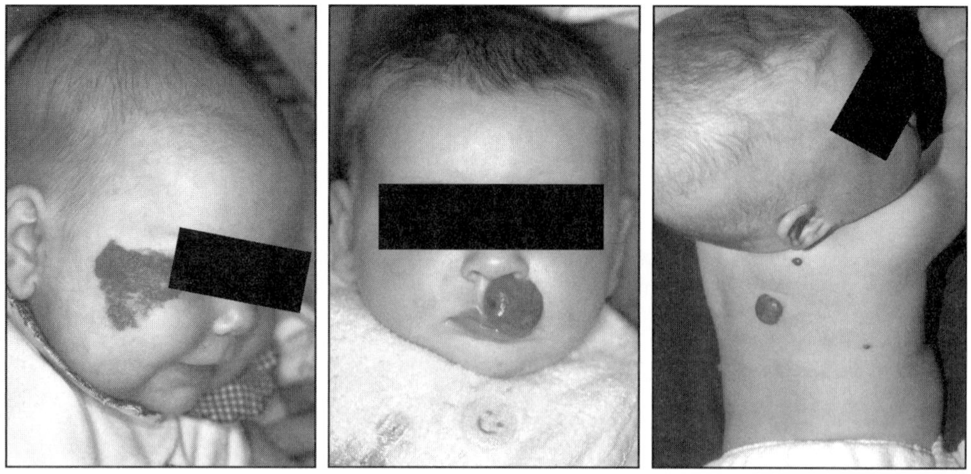

Figure 1 Three Examples of Hemangioma in the Proliferating Phase. The third panel is an example of multiple hemangiomas. (see color insert)

dramatically during infancy. For these children, the normally benign tumor can be devastating, causing tissue destruction, altering or impairing tissue function, and can become life-threatening. Even a relatively small hemangioma can cause significant problems for the infant depending on the location. Children with endangering or life-threatening hemangiomas are treated with corticosteroids, surgical resection, or with interferon-α. However, some hemangiomas do not respond to these drug therapies or respond so slowly that it is not clear if the drug accelerated the normal involution process. Furthermore, neurologic toxicity has been associated with interferon-α administration to infants with hemangioma.

Hypothyroidism in Hemangioma

Huang and colleagues discovered that severe hypothyroidism can be associated with large hemangiomas [1]. They showed that the likely underlying cause was ectopic expression of type 3 iodothyronine deiodinase (D3) in proliferating hemangiomas. D3 is widely expressed in fetal tissues and some tumors, but after birth, its expression is restricted to placenta and brain. D3 inactivates thyroid hormone and, in the case described by Huang and colleagues, can lead to consumption of therapeutic doses of hormone in a child with a large hemangioma. The impact of hypothyroidism during infancy is severe and irreversible; neurological damage in particular can lead to loss of three to five IQ points per month in children under one year of age with untreated hypothyroidism. Hence, the aberrant proliferation of endothelium that characterizes hemangioma can have deleterious systemic effects for a child in addition to the local tissue destruction. The discovery of hypothyroidism in hemangioma by Huang and colleagues, and subsequent case reports by others, affirms the critical importance of understanding the cellular and molecular alterations in hemangioma and the need for better fast-acting therapies.

Genetics

Most hemangiomas occur sporadically, with only rare examples of a hereditary predisposition. A large study of identical and fraternal twins, conducted by Cheung and colleagues [2], found no evidence for Mendelian inheritance, yet the authors speculated that a positive family history could reflect susceptibility to other "factors"—such as environmental or genetic factors—which in turn might lower the threshold for development of hemangiomas. Examples of factors that predispose one to hemangioma are being female (hemangiomas are three times more common in females compared to males), being fair-skinned, and being born prematurely. However, despite the preponderance of sporadic hemangiomas, Marchuk and colleagues reported evidence for inherited hemangiomas in a small number of kindreds in which hemangioma appears to be autosomal dominant [3]. An alternative scenario that could account for familial hemangiomas would be an autosomal recessive mutation, followed by acquisition of a somatic mutation in an endothelial cell. This would explain how an inherited mutation that would presumably affect all endothelial cells would only be expressed in a localized fashion.

Other Vascular Tumors

Hemangioma is often used as a generic term and applied to a wide array of vascular neoplasms and even vascular malformations. Examples of vascular tumors that have been called hemangioma are angiosarcoma, Kaposiform hemangioendothelioma, and pyogenic granulomas. Additionally, congenital tumors, such as tufted angioma and hemangiopericytoma, can be confused with hemangioma. Hemangioma is also a clearly distinct entity from vascular malformations. Several reviews and chapters have been written to clarify and classify human vascular anomalies, including infantile hemangioma, and to establish a logical nomenclature based on clinical features, biological properties, and when possible, molecular markers. In 2000, North and colleagues discovered that infantile hemangiomas could be distinguished from other vascular tumors and vascular malformations with high confidence by immunostaining tissue sections with an anti-glucose transporter-1 (GLUT1) antibody [4]. In a retrospective study, they reported intense endothelial immunoreactivity with anti-GLUT1 antibody in 139 of 143 hemangiomas obtained from infants and young children. In contrast, endothelial GLUT1 was not detected in vascular malformation specimens, nor pyrogenic granulomas or granulation tissue specimens. Furthermore, GLUT1 was not detected in hemangioendotheliomas, a low-grade malignant endothelial tumor, and only in a few cases of angiosarcoma, a highly aggressive endothelial tumor. In a subsequent study by North and colleagues, GLUT-1 was also not detected in congenital hemangiomas, suggesting a distinct eitiology from the more common form of hemangioma. Hence, the anti-GLUT1 antibody has proven to be a valuable tool for characterizing hemangiomas among a wide array of vascular tumors.

Life Cycle of Hemangioma

Proliferating Phase: Endothelial Progenitor Cells

Proliferating hemangiomas are highly cellular, with little connective tissue and barely discernible vascular lumens; proliferating endothelial cells are easily detected by histological methods. Because the tumor comprises endothelium, it is not surprising that many studies have found high levels of endothelial-expressed proteins. In Figure 2, tissue sections from proliferating and involuting hemangioma sections were double-labeled, with antibodies directed against two well-known endothelial markers. Vascular endothelial growth factor-receptor-2 (VEGF-R2), also known as kinase insert domain-containing receptor (KDR), is the major mediator of the angiogenic effects of VEGF on endothelial cells. In addition to being expressed on mature endothelial cells, KDR is also expressed by endothelial progenitor cells (EPC). CD31, also known as platelet-endothelial cell adhesion molecule-1 (PECAM-1), is a cell-cell adhesion molecule expressed on virtually all continuous endothelium. As seen in Figure 2, KDR is abundantly expressed in proliferating hemangioma; the green fluorescent signal is evident on nascent vascular structures and on cells lining vascular channels. CD31 is also expressed in proliferating phase hemangioma; the red fluorescent signal strongly highlights vascular channels. When the two signals are viewed simultaneously, it is clear that KDR and CD31 are co-localized, providing strong evidence for endothelial cells within the proliferating hemangioma. The same immunostaining procedure was performed on involuting hemangioma specimens. KDR-positive/CD31-positive endothelial cells are evident, but the cellular density is decreased and vascular channels are more prominent. This change in endothelial morphology suggests that nascent vascular structures present in the proliferating phase differentiate into more well-defined vascular channels as the involuting phase proceeds. The speculation that proliferating hemangiomas contain primitive angioblast-like cells has been put forth by many investigators. In support of this concept, we recently reported the detection of endothelial progenitor cells (EPC) from 11 out of 12 proliferating phase hemangiomas [5]. No EPC were detected in involuting hemangiomas or in venous malformation or lymphatic malformation tissue, suggesting that the presence of EPCs in hemangioma is unique and that EPC might contribute to the pathogenesis of hemangioma. To our knowledge, our study was the first to

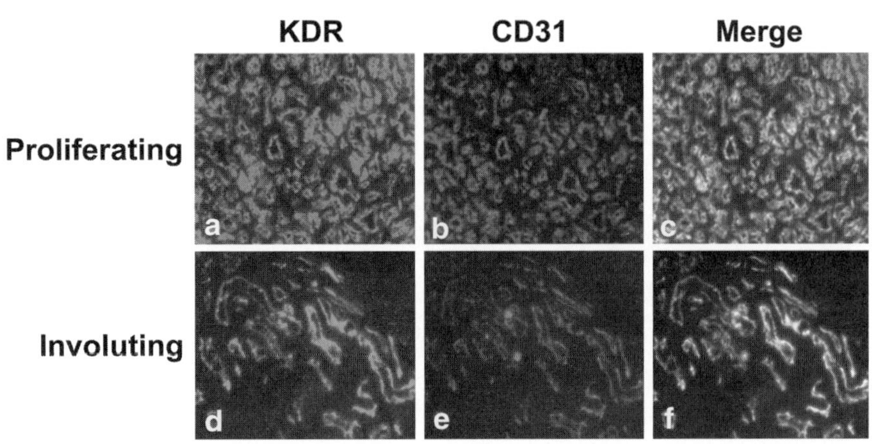

Figure 2 Expression of KDR/VEGF-R2 and CD31 in Hemangioma. Indirect immunofluorescence detection of KDR (panels a, c, d, f) and CD31 (panels b, c, e, f) in frozen sections from proliferating phase (a–c) and involuting phase (d–f) hemangiomas. In panels **a** and **d**, the green fluorescence (FITC) corresponding to KDR is seen. In panels **b** and **e**, the red fluorescence (Texas-Red) corresponding to CD31 is seen. In panels **c** and **f**, both fluorescence signals were viewed simultaneously; cells expressing both proteins are seen as yellow or yellow/orange. (Yu and Bischoff, unpublished data.) (see color insert)

show direct evidence of human EPC in a human vascular tumor.

Proliferating Hemangioma: Clonal Endothelial Cells

Two experimental approaches demonstrate that hemangioma is caused by an intrinsic defect in the endothelial cells. First, Boye and colleagues isolated hemangioma-derived endothelial cells from 10 different hemangioma tissue specimens and showed that the endothelial cells were clonal by examining X-chromosome inactivation patterns [6]. Nonendothelial fibroblast-like cells that were co-isolated from the tissue and separated from the endothelial cells were not clonal, demonstrating that the clonal expansion was restricted to the endothelial cells. In addition, the hemangioma-derived endothelial cells exhibited abnormal properties in in vitro assays that measure angiogenic activity. These results substantiate the hypothesis that hemangioma is caused by a somatic mutation in a single endothelial cell (i.e., an intrinsic defect), in a gene that controls endothelial growth and differentiation. Two other studies also support this hypothesis. Evidence for loss of heterozygosity on chromosome 5q in three out of six hemangiomas suggested a somatic mutation in this region may be the cause of sporadic hemangiomas. More recently, Walter and colleagues showed that 12 of 14 informative hemangioma lesions had nonrandom X-chromosome inactivation patterns—evidence for clonal endothelial cells—and thus, a somatic mutational event [7]. Importantly, their clonality assays were performed on DNA isolated from tissue sections, which avoids changes in cell composition that might occur during cell culture. In summary, these studies provide strong support for the hypothesis that a somatic mutation in a single endothelial cell leads to clonal expansion of dysregulated endothelial cells, resulting in hemangioma. Formal proof of the hypothesis awaits identification of the somatic mutation(s) and functional demonstration that such mutation(s) cause hemangioma.

Involuting Phase: Onset of Apoptosis

The involuting phase of hemangioma coincides with a transformation to a more lobular architecture with organized and recognizable vascular channels, which can be appreciated in Figure 2. As involution proceeds, vascular channels become larger and more prominent, and few interstitial cells are evident. Increases in perivascular cells expressing α-smooth muscle actin are consistent with the appearance of mature blood vessels. Mast cells appear to be increased in the involuting phase, and changes in extracellular matrix occur. Interferon-β in the epidermis overlying involuting and involuted hemangiomas has been proposed as a mechanism for attenuating the abnormal proliferation in hemangioma [8]. The final, involuted phase is characterized by a few remaining thin-walled vessels that resemble normal capillaries, surrounded by fibrofatty tissue.

The cellular and biochemical mechanisms that drive the involuting phase have been difficult to study because endothelial cells from involuting hemangioma do not grow in vitro [9]. Based on studies showing that increased apoptosis can offset cellular proliferation in murine tumor models, we examined cellular proliferation and apoptosis in a series of 16 hemangioma specimens from infants to children up to nine years of age, as well as normal infant skin for comparison [10]. Consistent with previous studies, cellular proliferation was high in hemangioma specimens obtained from children up to 24 months of age. Apoptosis was increased fivefold in hemangioma specimens from children one to four years of age compared to proliferating hemangioma specimens from children under one year of age, late-involuting/involuted hemangioma specimens from children over four years of age, and normal skin from infants (see Figure 3). Thus, the increase in apoptosis coincided precisely with the onset of the involuting phase. Double-labeling tissue sections for apoptosis and the endothelial marker von Willebrand Factor showed that 27 to 39 percent of the apoptotic cells were endothelial. In summary, this study suggests that regression of hemangioma involves increased apoptosis even as cellular proliferation continues. One might speculate that an agent that triggers apoptosis in proliferating endothelial cells would be an effective treatment for hemangioma.

Summary

The pathogenesis of hemangioma—its life cycle of growth and involution—is still a mystery. The studies described here support the hypothesis that sporadic heman-

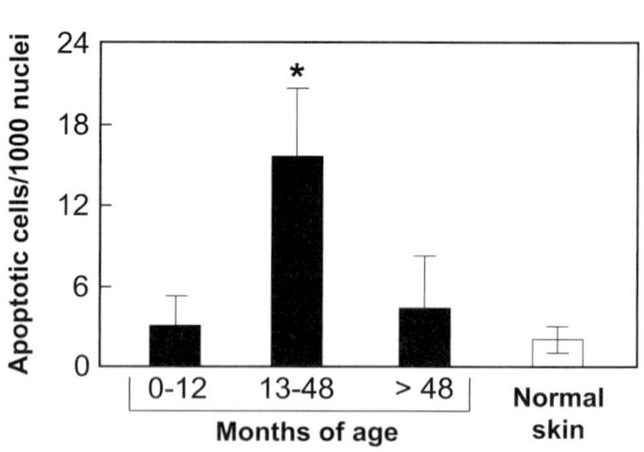

Figure 3 Increased Apoptosis Coincides with the Onset of the Involuting Phase. Apoptotic cells were quantitated using the TUNEL assay on 16 hemangioma specimens from children ranging from two months to nine years of age. Normal skin from infants and young children (n = 8) was analyzed for comparison. On average, apoptotic nuclei in eight fields, viewed at 630X magnification, were counted. Hemangiomas from children 13 to 48 months of age had a higher mean count of apoptotic nuclei compared to hemangiomas from younger children (0 to 12 months) ($p < 0.001$) and compared to hemangiomas from older children (> 48 months of age) ($p = 0.002$). There was no significant difference between younger (0 to 12 months) versus older (> 48 months) patients ($p = 0.68$). This figure is a graphical summary of data from Razon et al. (1998). *Microcirculation* **5**, 189–195.

giomas are caused by a somatic mutation in a single endothelial cell, possibly an immature endothelial cell or endothelial progenitor, that leads to clonal expansion of abnormal endothelial cells. It is likely that several somatic mutations, in one or more genes, will be discovered in individual hemangiomas, and that the assortment of mutations will reflect the differences in severity and life span of the individual tumors. Much progress has been made toward distinguishing hemangioma from the many other types of human vascular anomalies using cellular and biochemical markers. The goals now are (1) to identify the genetic defects that cause hemangiomas, so alterations in cellular signaling and function can be studied at the molecular level; (2) to develop animal models that recapitulate hemangiogenesis, so that growth and regression can be studied in vivo; and (3) to use the animal models to evaluate anti-angiogenic therapies that will either prevent growth or accelerate involution, so that better, fast-acting therapies can be provided to children with endangering hemangiomas.

Glossary

Apoptosis: A mechanism by which cell death can occur. It is characterized by morphologic changes in the nucleus and cytoplasm and chromatin cleavage at regularly spaced sites resulting in DNA fragmentation. This mode of cell death is contributes to the regulation of the size of animal tissues and in mediating pathologic processes associated with tumor growth.

Clonal expansion: A single cell undergoes multiple cell divisions to give rise to a population of descendent cells.

Endothelial progenitor cells: immature endothelial cells that co-express stem cell markers and endothelial lineage specific markers.

Immunohistochemistry: An indirect means of determining the expression and localization of a protein or antigenic determinant within a tissue. Histological tissue sections are incubated with an antibody directed against the protein of interest, followed by detection of bound antibody using a secondary antibody conjugated to a reporter.

Somatic mutations: Alterations in DNA sequence that cause defective gene products in any cell in the body except germ cells.

References

1. Huang, S. A., Tu, H. M., Harney, J. W., Venihaki, M., Butte, A. J., Kozakewich, H. P., Fishman, S. J., and Larsen, P.R. (2000). Severe hypothyroidism caused by type 3 iodothyronine deiodinase in infantile hemangiomas. *N. Engl. J. Med.* **343**, 185–189.
2. Cheung, D. S., Warman, M. L. & Mulliken, J. B. (1997). Hemangioma in twins. *Ann. Plast. Surg.* **38**, 269–274.
3. Walter, J. W., Blei, F., Anderson, J. L., Orlow, S. J., Speer, M. C., and Marchuk, D. A. (1999). Genetic mapping of a novel familial form of infantile hemangioma. *Am. J. Med. Genet.* **82**, 77–83.
4. North, P. E., Waner, M., Mizeracki, A., and Mihm, M. C., Jr. (2000). GLUT1: a newly discovered immunohistochemical marker for juvenile hemangiomas. *Hum. Pathol.* **31**, 11–22.
5. Yu, Y., Flint, A. F., Mulliken, J. B., Wu, J. K., and Bischoff, J. (2004). Endothelial progenitor cells in infantile hemangioma. *Blood* **103**, 1373–1375.
6. Boye, E., Yu, Y., Paranya, G., Mulliken, J. B., Olsen, B. R., and Bischoff, J. (2001). Clonality and altered behavior of endothelial cells from hemangiomas. *J. Clin. Invest.* **107**, 745–752.
7. Walter, J. W., North, P. E., Waner, M., Mizeracki, A., Blei, F., Walker, J. W., Reinisch, J. F., and Marchuk, D. A. (2002). Somatic mutation of vascular endothelial growth factor receptors in juvenile hemangioma. *Genes Chromosomes Cancer* **33**, 295–303.
8. Bielenberg, D. R., Bucana, C. D., Sanchez, R., Mulliken, J. B., Folkman, J., and Fidler, I. J. (1999). Progressive growth of infantile cutaneous hemangiomas is directly correlated with hyperplasia and angiogenesis of adjacent epidermis and inversely correlated with expression of the endogenous angiogenesis inhibitor, IFN-beta. *Int. J. Oncol.* **14**, 401–408.
9. Yu, Y., and Bischoff, J., unpublished observations.
10. Razon, M. J., Kraling, B. M., Mulliken, J. B., and Bischoff, J. (1998). Increased apoptosis coincides with onset of involution in infantile hemangioma. *Microcirculation* **5**, 189–195.

Further Reading

Cohen, M. M., Jr. (2002). Vasculogenesis, angiogenesis, hemangiomas, and vascular malformations. *Am. J. Med. Genet.* **108**, 265–274. *This excellent review article integrates clinical information on hemangiomas and vascular anomalies with the cellular and molecular regulation of vascular development.*

Ezekowitz, R. A., Mulliken, J. B., and Folkman, J. (1992). Interferon alfa-2a therapy for life-threatening hemangiomas of infancy. *N. Engl. J. Med.* **326**, 1456–1463.

Ferrara, N., Gerber, H. P., and LeCouter, J. (2003). The biology of VEGF and its receptors. *Nat. Med.* **9**, 669–676.

Holmdahl, K. (1955). Cutaneous hemangiomas in premature infants. *Acta Pediatrics* **44**, 370–379.

Mulliken, J. B., Fishman, S. J., and Burrows, P. E. (2000). Vascular anomalies. *Curr. Prob. Surg.* **37**, 517–584. *This comprehensive overview provides a clinical and pathophysiological overview of hemangiomas and other related vascular anomalies by internationally recognized leaders in the clinical diagnosis and treatment of vascular anomalies.*

North, P. E., Waner, M., James, C. A., Mizeracki, A., Frieden, I. J., and Mihm, M. C., Jr. (2001). Congenital nonprogressive hemangioma: A distinct clinicopathologic entity unlike infantile hemangioma. *Arch. Dermatol.* **137**, 1607–1620.

Rafii, S., and Lyden, D. (2003). Therapeutic stem and progenitor cell transplantation for organ vascularization and regeneration. *Nat. Med.* **9**, 702–712. *This article provides a comprehensive review of endothelial progenitor cells (EPCs). EPCs play important roles in normal vascularization and repair, but they can also contribute to tumor angiogenesis. The role of EPCs in disease, as well as potential applications for revascularization, is an exciting new area in vascular biology.*

Smoller, B. R., and Apfelberg, D. B. (1993). Infantile (juvenile) capillary hemangioma: A tumor of heterogeneous cellular elements. *J. Cutan. Pathol.* **20**, 330–336. *Although not the first to propose that hemangioma contains primitive angioblastic-like cells, these authors were the first to show expression of an endothelial marker (CD34) on interstitial cells not associated with vascular channels in proliferating phase hemangiomas.*

Takahashi, K., Mulliken, J. B., Kozakewich, H. P., Rogers, R. A., Folkman, J., and Ezekowitz, R. A. (1994). Cellular markers that distinguish the phases of hemangioma during infancy and childhood. *J. Clin. Invest.* **93**, 2357–2364. *This study was one of the first to perform a comprehensive analysis of the expression patterns of angiogenic factors in proliferating, involuting, and involuted hemangiomas.*

Capsule Biography

Dr. Bischoff is an Associate Professor in the Vascular Biology Program/Department of Surgery at Children's Hospital Boston and Harvard Medical School. Her laboratory has been investigating cellular and molecular mechanisms that cause hemangioma since 1997. In addition, her laboratory studies growth and differentiation of endothelial progenitor cells and endothelial cells that line cardiac valves.

CHAPTER 24

Window Models

Dai Fukumura
Massachusetts General Hospital and Harvard Medical School, Boston, Massachusetts

Introduction

The tissue preparations used for *in vivo* microscopy can be grouped into three classes: (1) *in situ* preparations (e.g., mouse ear skin, bat wing, human nail bed, mouse tail lymphatics), (2) acute (exteriorized) tissue preparations (e.g., hamster cheek pouch; rat cremaster muscle; mouse, rat, cat, or rabbit mesentery; internal organs such as the heart, lung, stomach, intestine, pancreas, liver, kidney, ovary, breast, spleen, lymph node, or Peyer's patch), and (3) chronic transparent windows (chambers). Each preparation has its strengths and weaknesses. *In situ* preparation does not require any invasive preparation, but accessible tissue is very limited. Acute exteriorization can be applied to virtually any tissue with variable degrees of difficulty, but the duration and frequency of the observation is limited, and the preparation procedures may affect the microcirculatory parameters.

Window models allow sufficient recovery time after the implantation and permit chronic observation thereafter. Various window models are available today, and more are being developed. This chapter discusses chronic window preparations and their usage for microcirculation research.

Chronic Window Preparations

Chronic windows developed to date include the rabbit ear chamber, dorsal skin chamber, cranial window, hamster cheek pouch, spinal window, rabbit bone chamber, mouse abdominal window, mammary window, and lung window (Table I). Each of these chronic windows has its advantages and disadvantages. For example, the rabbit ear chamber is perhaps the most optically clear. However, rabbits are expensive to purchase and maintain, and it takes four to six weeks to mature the granulation tissue before the study begins. Mice, hamsters, and rats are less expensive and require smaller quantities of reagents because of their smaller body weight. From the surgical point of view, rats and hamsters are easier to work with than mice, but the latter have many advantages. The availability of comprehensive genetic information on mice, immunodeficient and genetically engineered mice, and murine reagents such as surface markers or antibodies has made mice the most commonly used laboratory animals for modern biological research. The dorsal skin chamber in mice is the most widely used chamber preparation because the surgery is less involved than some of the other preparations and because of its longer history. The cranial window can be kept for up to a year, compared to 30 to 40 days for the dorsal skin chamber, and (along with the cheek pouch) is an immunoprivileged site. The main disadvantage of the cranial window is that, in most cases, the visualization of microvessels requires both epi-illumination and the injection of a contrast agent such as a fluorescent marker.

The Rabbit Ear Chamber

In 1924, Sandison developed the first transparent window (chamber) for implantation in the ear of a rabbit. Since then, various modifications of this model have been used for microcirculation studies. The model requires granulation tissue formation within the chamber, and during this period it is very useful for the study of the wound healing process. The chambers can be used either for normal tissue studies or for tumor implantation after the maturation of granulation tissue.

Typically, a rabbit ear chamber is surgically implanted in the 15-cm-long ears of New Zealand white rabbits (~3 kg body weight). Four holes are punched in the shaved ear: three outer perforations (3.5 mm in diameter) for the chamber positioning and a central puncture (5.4 mm diameter) for the transparent window housing. A molded plate is placed on the inside of the ear and aligned with the existing holes,

Table I Examples of Window Models for Microcirculation Studies.

Model	Species	Application
Ear chamber	Rabbits	Wound healing, tumor growth
Dorsal skin chamber	Rabbits, rats, hamsters, mice	Skin microcirculation, ischemia reperfusion, local cytokine application, wound healing, tumor growth, angiogenesis assay, tissue growth
Cranial window	Pigs, cats, rats, mice	Cerebral and pial microcirculation, local cytokine application, wound healing, tumor growth, angiogenesis assay, tissue growth
Cheek pouch window	Hamsters	Cheek pouch microcirculation, wound healing, tumor growth
Spinal window	Rats, mice	Spinal microcirculation
Bone chamber	Rabbits	Bone microcirculation, ischemia reperfusion, wound healing, stem cell homing
Abdominal window	Mice	Abdominal wall and pancreas microcirculation, tumor growth
Mammary window	Rats, mice	Tumor growth
Lung window	Rabbits, mice	Lung and pleural microcirculation, tumor growth

while a thin (~200 μm) cover of mica glass is positioned on the outside of the cartilage. A thin granulation tissue bed (with a thickness of ~40 μm and a diameter of ~5 mm) grows in the chamber, sandwiched between the molded plate and the mica glass. The granulation tissue develops at an average of 8 days after implantation, and reaches maturity at approximately 40 days after implantation.

For tumor implantation, the cover glass, which forms the top plate of the transparent chamber, is carefully removed. A tumor (such as VX2 carcinoma) is excised from the flank of a tumor-bearing host, minced, placed in 0.9 percent NaCl solution, and spread uniformly over the cover glass. The cover glass is replaced flush against the intact normal tissue. Angiogenic response is observed 3 to 4 days post-implant, and the tumor-bearing chamber is ready for intravital microscopy approximately 10 days after implantation.

For intravital microscopy, the animal is placed in a dorsal recumbent position in a cradle that restricts head movement while still maintaining proper circulation to the chamber. The ear containing the chamber is extended horizontally to the specimen plane of an intravital microscope. The chamber is secured to the microscope stage with an aluminum adapter.

The Dorsal Skin Chamber

In the 1940s, Algire adapted the Sandison chamber to the dorsal skin in mice and carried out pioneering studies of angiogenesis during wound healing and tumor growth. Algire's original design and its modifications for other species require granulation tissue growth similar to that in the rabbit ear chamber model. Although this type of dorsal skin chamber has better optical accessibility, the effects associated with the wound healing process may hamper the observation. Later development of new dorsal skin chamber design minimized such effects by using intact skin, as described below. This type of dorsal skin chamber has been used for immunocompetent as well as immunodeficient mice, rats, and hamsters. With the availability of genetically engineered animals, the mouse dorsal skin chamber (Figure 1) has the broadest range of applications.

For the implantation of a transparent chamber into mouse dorsal skin, the entire back of the animal is shaved and depilated and two symmetrical titanium frames (weight 3.2 g) are implanted so as to sandwich the extended double layers of skin. One layer of the skin is removed in a circular area approximately 15 mm in diameter, and the remaining layer (consisting of epidermis, subcutaneous tissue, and striated muscle) is covered with a circular cover glass (11 mm in diameter) incorporated into one of the frames. Following implantation of the dorsal skin chamber, animals are allowed to recover from microsurgery for 48 hours before *in vivo* microscopy studies or subsequent procedures are conducted.

To implant tissue, such as a tumor, the animals are positioned in a transparent polycarbonate tube (inner diameter: 25 mm). The cover glass is carefully removed and a small piece of tissue (approximately 1 mm in diameter) or 2 μl of dense cell suspension (~2×10^5 cells) is implanted at the center of the dorsal skin chamber. A new cover glass is then placed on the chamber.

For microscopy on a dorsal skin chamber–bearing mouse, the mouse is positioned in the polycarbonate tube. A heating tape is wrapped around the plastic tube to maintain the animal's core temperature at a constant ~37°C during the observations. To obtain microcirculatory parameters, chambers are observed under an intravital microscope, which is typically equipped with a high-sensitivity camera for fluorescence imaging.

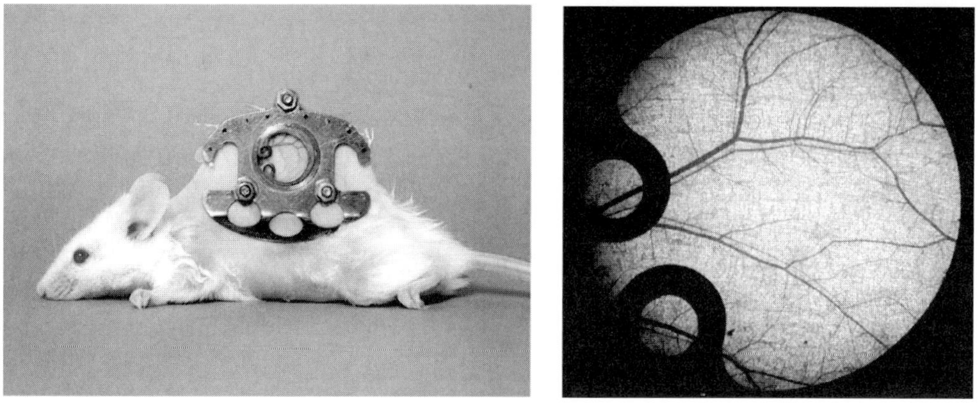

Figure 1 Mouse Dorsal Skin Chamber Model. An SCID mouse bearing a dorsal skin chamber (left) and a macroscopic image of the mouse dorsal skin chamber (right). The frame is made of titanium (3.2 g). The window is 11 mm across. (see color insert)

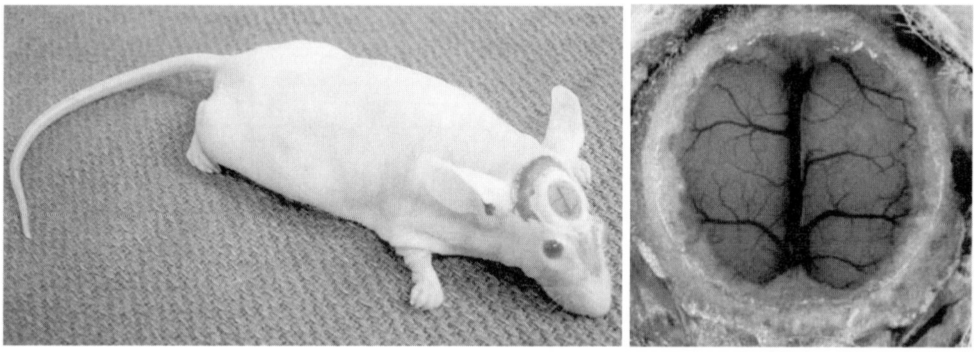

Figure 2 Mouse Cranial Window Model. Nude mouse bearing a cranial window (left) and a macroscopic image of the mouse cranial window (right). The window is 7 mm across. (see color insert)

The Cranial Window

The closed transparent cranial window model has been extensively used for cerebral microcirculation studies. After the recovery of central nervous system fluid and intracephalous pressure to normal levels, this model provides direct optical access to cerebral blood vessels under physiological condition. The cranial window is the most stable window model to date, allowing observation for as long as the entire natural life span of the animal; it is therefore very useful for studies requiring relatively long duration. Cranial windows are used for angiogenesis, vessel remodeling, and maturation studies. Furthermore, this model provides a natural microenvironment for primary and metastatic brain tumors. Various species, including mice, rats, cats, and pigs, are used for the cranial window model. We will describe the preparation of the mouse cranial window (Figure 2), but similar procedures are used for other species.

For cranial window implantation, a stereotactic apparatus holds the head of the animal. The skin is cut in a circular manner on top of the skull, and the periosteum underneath is scraped off to the temporal crests. A 6-mm circle is drawn over the skull, and a groove is made on the drawn circle using a high-speed air-turbine drill with a 0.5-mm-diameter burr tip. Drilling of the groove continues until the bone flap becomes loose. Using a malis dissector, the bone flap is separated from the dura mater underneath. A nick is made close to the sagittal sinus. Iris microscissors are passed through the nick. The dura and arachnoid membranes are cut completely from the surface of both hemispheres, avoiding any damage to the sagittal sinus. The window is sealed with a 7-mm cover glass, which is glued to the bone with histocompatible cyanoacrylate glue.

For implantation of tumors or tissue constructs, the cover glass is removed one week after the window implantation. A small piece of tissue is inoculated at the center of the window. Alternatively, 3 to 5 μl of a thick single cell suspension ($1–5 \times 10^5$ cells/μl) are implanted with a 28-gauge microsyringe (using a needle tip angle of 55 degrees and a depth of 1.75 mm).

For intravital microscopy, an animal is anesthetized and put on a polycarbonate plate, with the head fixed by means of a metal ring upper frontal tooth holder and a bilateral plastic ear holder. A heating pad is placed under the animal to maintain its temperature at ~37°C during *in vivo* microscopy.

Other Window Models

A spinal window has been made in rats (C5) and mice (C1 to C7) with a laminectomy. The windows were sealed with dental acrylic resin and a 7-mm cover glass (rats) or gas-impermeable transparent membrane (mice). The dorsal surface of the cervical cord was observed for leukocyte, T-cell, and platelet interaction with spinal vessels.

The rabbit bone chamber is a hollow titanium-alloy screw with quartz lenses embedded in its core. Vessels, followed by bone tissue, grow in a 100-µm-thick chamber between two flat quartz elements implanted in tibial bone. Intravital microscopy in the bone chamber has been used to study bone microcirculation, ischemia reperfusion injury in bone, and stem cell homing to bone marrow.

An abdominal window model for the mouse pancreas has been reported. A portion of the pancreas is gently exteriorized through a small laparotomy and kept within a space surrounded by the outer side of the abdominal wall, a titanium ring, and a circular glass cover slip (11 mm in diameter) held by the titanium ring. This model permits repeated observation of microcirculation in the pancreas and of the growth and treatment of pancreatic tumors.

The mammary window in rats and mice allows chronic observation of tumors growing in the breast tissue. This model requires removal of skin, nipple, and epithelium before injection of tumors and placement of an acrylic disk. Alternately, microcirculation in intact breast tissue and/or tumors grown in the breast can be observed acutely by placing a surgically prepared tissue flap in a specially designed microscope stage or chronically by using a mammary chamber similar to the mouse dorsal skin chamber.

Lung windows in New Zealand White rabbits have been used to study pulmonary microcirculation via intravital microscopy. The smaller intact pleural window involves the removal of the skin, intercostal muscles, and endothoracic fascia to expose the costal pleura, through which the lung surface is observed. The larger type of window requires the removal of the thoracic wall, including ribs, over a 3-cm distance and the implantation of a transparent window with a hollow cylinder to apply negative pressure. The lungs are inflated by positive air pressure of 6 mmHg, and the lung surface is sucked onto the window. Hemodynamics and short-term leukocyte/platelet-endothelial interactions were studied using these rabbit lung window models.

On the other hand, the chronic thoracic window model in mice allows temporal observation of the microcirculation of tumors implanted in the lung. For the implantation, a 6-mm ventral portion of the chest wall is removed, covered with a circular glass cover slip (8 mm in diameter), and sealed with biocompatible cyanoacrylate adhesive and cement mixture. Withdrawal of the remaining air in the chest cavity allows the lung to inflate and attach to the cover slip. One week later, a tumor fragment is implanted on the pulmonary pleural surface of the window. For intravital microscopy, the animal is placed in the lateral position and the chest wall is attached to adjustable arms on the stage by cranial and caudal sutures.

Utilization of Window Models

Window models generally allow long-term continuous and/or repeated observation and are thus ideal for time-course studies of physiological and pathological events. Window models are best suited for intravital microscopy and are used to study angiogenesis, microcirculation, tumor and engineered tissue growth, disease process, and treatment response.

Intravital Microscopy

Intravital microscopy is a powerful optical imaging technique that allows noninvasive monitoring of molecular and cellular processes in intact living tissue with 1 to 10 µm resolution. Sophisticated animal models, such as window models, are a prerequisite for this technique. With the use of appropriate tracers, optics, and analysis algorithms, a wide array of morphological and functional vascular parameters—such as blood and lymph vessel architectures, blood and lymph flow, vasomotion, vascular permeability, and leukocyte/platelet-endothelial interactions—can be determined by intravital microscopy. In addition, endogenous and exogenous probes, as well as genetically manipulated cells and animals, can provide other important parameters such as gene expression (promoter activity via GFP imaging), interstitial/intracellular pH, interstitial and microvascular pO_2, interstitial diffusion, convection and bindings, and extracellular matrix structure (second harmonic generation of collagen fiber).

Treatments and Manipulations

By observing the same vessels and/or regions repeatedly, window models provide accurate dynamic information on microcirculatory responses. In addition to monitoring the therapeutic effects of systemic treatments, window models are often used to assess the local effects of treatments or manipulations. For example, the occlusion of feeding vessels to the windows or the physical compression of tissues within the windows induces ischemia, followed by reperfusion after release; thus, windows are useful as models of local ischemia reperfusion. For local treatments, test compounds can be applied locally, either by topical application after removal of the cover glass or by injection through a port embedded in the window models. Alternately, a drug-containing matrix can be implanted within the window for continuous release of local treatments. Furthermore, the application of specific optical powers or wavelengths can activate certain drugs or carriers within the light-exposed area.

Cancer Research

A solid tumor is an organ-like structure that consists of cancer cells and nonneoplastic host stromal cells embedded in an extracellular matrix and nourished by a vascular network. The application of intravital microscopy techniques to tumors grown in window models (Figure 3) has provided unprecedented insights into the inner workings of this organ. It has been unequivocally demonstrated that the structure and function of tumor vessels is heterogeneous. The chaotic nature of tumor vessels forms a physiological barrier to the delivery of therapeutic molecules to tumors and hinders the efficacy of various tumor therapies. In vivo microscopy of tumor window models has revealed that certain antiangiogenic treatments can normalize the abnormal tumor vessels so that they become more efficient. Normalization of tumor vessels by antiangiogenesis treatment may overcome physiological resistance to drug delivery and accelerate treatment efficacy.

Angiogenesis Assay

To study angiogenesis, vessel remodeling, and maturation, an extracellular matrix construct containing either angiogenic factor(s) or vascular cells (such as endothelial cells, pericytes, vascular smooth muscle cells, and their precursors) are implanted in window models (Figure 4). Angiogenic matrix implants in window models allow for noninvasive, real-time measurements of angiogenic vessel structure and function, from early development to subsequent remodeling and maturation. Other types of animal models also permit chronic monitoring of vessel formation (e.g., the corneal micropocket assay in rabbits, rats and mice; the anterior chamber/iris assay in frogs, rabbits, guinea pigs, swines, goats and sheep; and the chick chorioallantoic membrane model), but these models do not typically use intravital microscopy and are thus unable to provide detailed functional information.

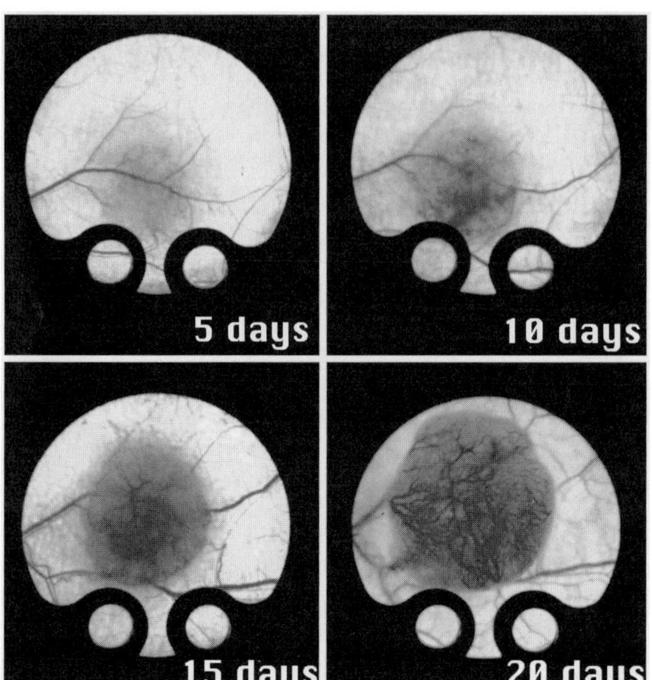

Figure 3 Angiogenesis and Tumor Growth in a Window Model. LS174T human colon cancer was implanted in an SCID mouse dorsal skin chamber. At day 5 after tumor cell implantation, enlargement of host vessels was observed, and by day 10 occasional hemorrhage and sprout formation occurred. At day 15, tumor growth and further angiogenesis became apparent. By day 20, the tumor was fully vascularized. (Adapted from Reference [3].) (see color insert)

Tissue Engineering

Tissues and organs, or their precursors (stem cells), can be implanted into window models for the study of organogenesis and/or the growth, microcirculation, and biology of the respective tissues. For example, preadipocyte implants exhibit *de novo* fat tissue formation and reveal reciprocal regulation of angiogenesis and adipose differentiation. Bone implants permit the monitoring of bone regeneration and angiogenesis as well as bone preferential metastasis such as prostate cancer. A major challenge in tissue engineering and regenerative medicine is the creation of stable vasculature. Window models provide an ideal platform for the determination of vessel development and function in an engineered tissue.

Future Perspectives

Intravital microscopy in window models has been providing useful insights into angiogenesis and microcirculation. However, two key challenges remain. First, the most widely used microscopy techniques are surface-weighted. Multi-photon microscopy can provide images at depths of up to ~700 μm, depending on the tissue and tracer used. Other optical methods, such as optical coherence tomography, can image further deeper regions. Systems and applications of these techniques have been rapidly increased recently. With more research in this area, we may someday be able to obtain dynamic images of whole organs or tissues with high spatial resolution.

The second limitation of window models is that they are currently limited to organs accessible from outside of the body. To study internal organs, surgical manipulation is required, which allows only short-term observations. The adaptation of microscopy techniques to use a fiberscope system will enable internal organs to be studied with minimum invasion. Alternately, the future advancement of optics and transponder devices may allow for both the implantation and chronic monitoring of internal organs.

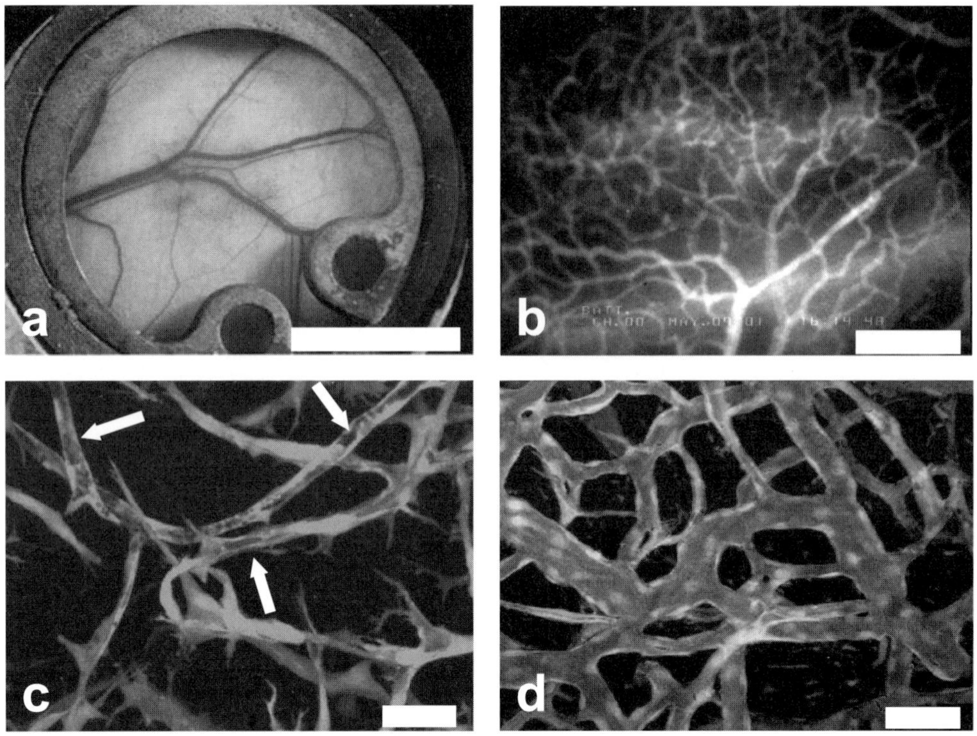

Figure 4 Vessel Formation in a Window Model. **(a, b)** Angiogenesis and vessel remodeling during adipogenesis in the mouse dorsal skinfold chamber after 3T3-F442A cell implantation. **(a)** Macroscopic image 9 days after implantation. **(b)** Intravital microscopy image of fluorescence contrast enhanced blood vessels at 28 days after implantation. **(c, d)** HUVECs and 10T1/2 cells were seeded in a 3-D extracellular matrix and implanted in a mouse cranial window. 3-D images of engineered vessels (EGFP expressing HUVEC, green; functional blood vessels contrast-enhanced with rho-dextran, red) were taken using intravital multi-photon laser-scanning microscopy. **(c)** Four days after implantation of HUVEC + 10T1/2 construct. Large vacuoles in the tubes resemble the lumens of capillaries (arrows) but they are not perfused (no red). **(d)** Four months after implantation of HUVEC + 10T1/2 construct. Engineered vessels are stable and functional. Bars indicate 5 mm (*a*), 100 μm (*b*), 50 μm (*c, d*), respectively. (Adapted from Fukumura, D., Ushiyama, A., Duda, D. G., Xu, L., Tam, J., Chatterjee, V. K. K., Garkavtsev, I., and Jain, R. K. (2003). Paracrine regulation of angiogenesis and adipocyte differentiation during *in vivo* adipogenesis. *Circ. Res.* **93**, e88–e97; and Koike, N., Fukumura, D., Gralla, O., Au, P., Schechner, J. S., and Jain, R. K. (2004). Creation of long-lasting blood vessels. *Nature* **428**, 138–139.) (see color insert)

With these improvements, window models will continue to offer new opportunities for discovery in microcirculation research.

Summary

Transparent window (chamber) models are widely used in microcirculation research. Because window models allow direct optical access without further surgical manipulation, they are especially useful for long-term monitoring of microcirculation by intravital microscopy. These techniques allow us to dissect physiological and pathophysiological processes during blood and lymph angiogenesis, wound healing, tissue engineering, tumor growth, and various therapies. The rabbit ear chamber, the rodent dorsal skin chamber, and the cranial window are the most commonly used window models. In addition, windows for various organs have been or are being designed. The continuing development of window models, along with progress in imaging techniques, analysis algorithms, molecular probes, and genetically engineered animals, increases the variety, accuracy, and efficacy of our data collection and further expands our knowledge of microcirculation.

Glossary

Cranial window: A transparent window in cranium of an animal.

Dorsal skin chamber: A transparent window model in dorsal skin of an animal.

Intravital microscopy: Microscopy technique for *in situ* observation of a living tissue.

Rabbit ear chamber: A transparent window model in a rabbit ear.

Transparent window (chamber) model: An animal model typically consists of a frame (e.g., titanium, mold, endogenous bone), a transparent window (e.g., glass, mica, quartz, acryl), and a tissue of interest.

Bibliography

Ishikawa, M., Sekizuka, E., Sato, S., Yamaguchi, N., Shimizu, K., Kobayashi, K., Bertalanffy, H., and Kawase, T. (1999). In vivo rat closed spinal window for spinal microcirculation: Observation of pial vessels, leukocyte adhesion, and red blood cell velocity. *Neurosurgery* **44**, 156–161. *Detailed description of the spinal window model.*

Jain, R. K., Brown, E. B., Munn, L. L., and Fukumura, D. (2004). Intravital microscopy of normal and diseased tissues in the mouse. *In Live Cell Imaging: A Laboratory Manual.* Cold Spring Harbor, NY: Cold Spring Harbor Laboratory Press, *in press. Comprehensive review article with detailed procedures including window models and intravital microscopy.*

Jain, R. K., Munn, L. L., and Fukumura, D. (2002). Dissecting tumour pathophysiology by intravital microscopy. *Nat. Rev. Cancer* **2**, 266–276. *Comprehensive review article on the use of window models and intravital microscopy for tumor study.*

Jain, R. K., Schlenger, K., Hockel, M., and Yuan, F. (1997). Quantitative angiogenesis assays: Progress and problems. *Nat. Med.* **3**, 1203–1208. *Comprehensive review article on the method of angiogenesis assay.*

Kuhnle, G. E., Leipfinger, F. H., and Goetz, A. E. (1993). Measurement of microhemodynamics in the ventilated rabbit lung by intravital microscopy. *J. Applied Physiol.* **74**, 1462–1471. *Detailed description of the rabbit lung window model.*

Leunig, M., Yuan, F., Menger, M. D., Boucher, Y., Goetz, A. E., Messmer, K., and Jain, R. K. (1992). Angiogenesis, microvascular architecture, microhemodynamics, and interstitial fluid pressure during early growth of human adenocarcinoma LS174T in SCID mice. *Cancer Res.* **52**, 6553–6560. *Detailed description of the dorsal skin chamber model.*

Shan, S., Sorg, B., and Dewhirst, M. W. (2003). A novel rodent mammary window of orthotopic breast cancer for intravital microscopy. *Microvasc. Res.* **65**, 109–117. *Detailed description of the mammary window model.*

Winet, H. (1989). A horizontal intravital microscope-plus-bone chamber system for observing bone microcirculation. *Microvasc. Res.* **37**, 105–114. *Detailed description of the rabbit bone chamber model.*

Yuan, F., Salehi, H. A., Boucher, Y., Vasthare, U. S., Tuma, R. F., and Jain, R. K. (1994). Vascular permeability and microcirculation of gliomas and mammary carcinomas transplanted in rat and mouse cranial window. *Cancer Res.* **54**, 4564–4568. *Detailed description of the cranial window model.*

Zawicki, D. F., Jain, R. K., Schmid-Schoenbein, G. W., and Chien, S. (1981). Dynamics of neovasculaization in normal tissue. *Microvasc. Res.* **21**, 27–47. *Detailed description of the rabbit ear chamber model.*

Capsule Biography

Dai Fukumura, M.D., Ph.D. (Associate Professor, Edwin L. Steele Laboratory, Department of Radiation Oncology, Massachusetts General Hospital and Harvard Medical School), is a research scientist in the field of cancer and vascular biology. He has documented expertise in various animal window models and intravital microscopy. His current research interests include the role of nitric oxide in angiogenesis and microcirculation, the role of host-tumor interactions in tumor angiogenesis, growth, and metastasis, and the mechanism of vessel maturation.

CHAPTER 25

The Cremaster Muscle for Microcirculatory Studies

Robert Hester
University of Mississippi Medical Center, Jackson, Mississippi

Studies of the microcirculation have typically used thin muscle preparations that are easily dissected. The cremaster muscle has proven to be an easily prepared tissue for the study of microvascular reactivity and mechanisms for the control of striated muscle blood flow. The following will provide information about the preparation of this muscle and the associated cardiovascular parameters under both normal and pathological conditions.

Animal Models

Rats and hamsters have been used for most microcirculatory studies involving the cremaster. The first mention of the cremaster muscle for the study of blood vessels *in vivo* was presented by Grant [1]. Subsequent studies provided detailed descriptions of the dissection techniques including "hints" [2]. This manuscript also presented information about preparing the rabbit cremaster, although it was mentioned that the size of the scrotum in 2 kg rabbits was similar to 250 g rats. Therefore, although the technique is applicable to the rabbit, it is more suitable in the rat. The report by Baez [3] provided the first study using an opened cremaster, which has become the classic paper for this microcirculatory preparation. With the development of transgenic mice, the mouse cremaster has become a popular model for studying the mechanisms of inflammatory responses or control of blood flow in an animal with a modified genotype. In mice the cremaster appears to be the only microcirculatory model for skeletal muscle because of the small size of the animal.

Anatomy

Anatomically the cremaster is a muscle covering the spermatic cord and testes. The cremaster originates from the internal oblique muscle as the spermatic cord passes from the abdomen to the scrotum through the inguinal canal. The physiological role of the cremaster is to retract the testes and keep the testes warm and protect it from injury. The cremaster muscle in rodents is a thin muscle, 200 to 400 μm in thickness, comprising two muscle layers.

In adults rats or hamsters, the cremaster muscles comprise 60 to 80 percent of the total area as Type IIb fibers, which are fast-twitch glycolytic, while 10 to 16 percent are Type I slow-twitch fibers and 8 to 20 percent are Type IIa fast-twitch oxidative-glycolytic fibers [4]. This percentage has been shown to occur in both hamsters and rats, irrespective of age. The average size of the muscle fibers in young rats and hamsters averages approximately 500 μm^2 and increases to 12 to 1500 μm^2 in adult animals.

Nervous Supply

The major nerve supply to the cremaster is by both the femoral and genital branches (L1 and L2) of the genitofemoral nerve. The ilio-inguinal, the ilio-hypogastric, and the lateral cutaneous nerves also supply the cremaster, and sectioning of all four nerves is required to fully denervate the muscle. The genitofemoral nerve can be accessed by an abdominal incision and stimulated directly (0.1 ms 1–5V) to elicit contraction of the cremaster muscle. Direct field stimulation using electrodes placed across the cremaster can also be used to elicit muscle contraction.

Blood Supply

The blood flow to the cremaster is supplied by the cremasteric artery, a branch of the inferior epigastric artery. The cremaster muscle is normally supplied by a paired arteriole (~100 μm) and venule (~150 μm). These vessels branch into smaller vessels, which also give off side branches. The paired arrangement, arteriole and venule, may continue to the level of 30 to 40 μm arterioles. There is no apparent relationship between the arterioles and the orientation of the muscle fibers. Arterioles continue to branch until terminal arterioles give rise to capillaries. Groups of capillaries (modules) appear to be the basic unit of vascular architecture; that is, the flow to a group of capillaries is controlled by the diameter of the upstream arteriole, and an increase in capillary flow occurs throughout the module, not within a single capillary. Normal vascular diameters are presented below.

Dissection and Preparation

The ease of preparing the cremaster muscle is a strength of this preparation. There are only minor differences across species in the preparation of this muscle; therefore, the following directions for dissection can be easily adapted for different species.

The animal should be anesthetized with an appropriate agent, for example, sodium pentobarbital intraperitoneally (40 to 60 mg/kg, for rats and hamsters), and then supplemented with intravenous infusion, either through a jugular or femoral vein. A tracheostomy should be performed to allow removal of bronchial secretion but not to support mechanical ventilation. As with any anesthetized animal, monitoring of blood gases should be considered. Placement of a femoral or carotid arterial catheter is needed only if blood pressure is required. Core body temperature should be continually monitored with an esophageal temperature probe.

The preparation and study of the cremaster muscle requires a continuous flow of a physiological salt solution over the muscle. This is accomplished by having a small glass dripper placed over the muscle, allowing for a small continuous flow of superfusate solution during the entire dissection and experiment. A warm physiological salt solution (PSS), pH 7.35 at 34°C, containing (in mM) 131.9 NaCl, 4.7 KCl, 2.0 $CaCl_2$, 1.2 $MgSO_4$, and 20 $NaHCO_3$ is appropriate for superfusion. Because the solution is bicarbonate buffered, the superfusion solution must be equilibrated with 5 percent CO_2. The amount of oxygen in the superfusion solution has varied between different laboratories, but the oxygen delivery to the cremaster tissue should be from the blood supply. Therefore, the superfusion solution should be equilibrated with a low percentage of oxygen, 0 percent or 5 percent. Thus, the superfusion solution should be equilibrated with 5 percent CO_2, 0 percent to 5 percent O_2, with a balance of N_2.

The animal is placed on its back on a clear acrylic (Lucite® or Plexiglass®) board. The animal's feet should be secured to the board with tape and then either the left or right groin is shaved. To minimize urine contamination of the cremaster, the bladder should be emptied by gentle pressure, with a tie being placed around the penis. A round acrylic pedestal or bath with a rim of silastic or silicone elastomer (Sylgard®) allows for the cremaster to be pinned to the silastic using insect pins.

Figure 1 illustrates the basic surgical procedure. First a 5-0 suture is placed on the distal portion of the scrotum, and the suture is pulled tight to secure the cremaster for surgery. The tip of the scrotum is elevated, and the skin is cut in a distal to proximal direction. At this time the PSS should be flowing over the preparation. The connective tissue between the testicle and the skin is then cut to free the testicle. The testicle can either be removed (after ligating the blood supply) or the freed testicle is gently pushed through the inguinal canal into the abdominal cavity. The solution flowing over the cremaster will hydrate the connective tissue, allowing for a better visualization of the connective tissue for removal. The skin is then pulled proximal and either removed, secured by ties, or pinned to the silastic around the acrylic pedestal. The distal tip of the cremaster is then

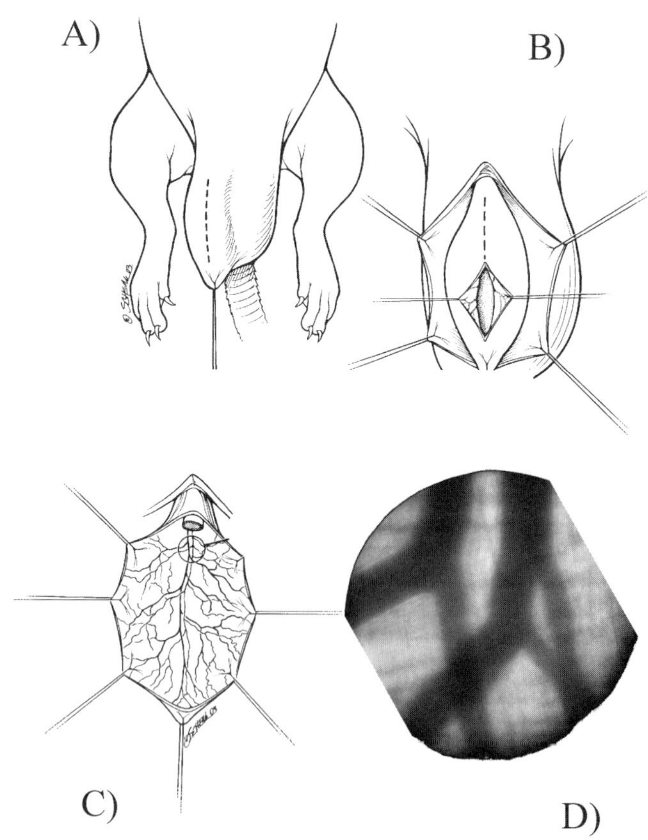

Figure 1 This figure shows the basic surgical techniques for preparation of the cremaster muscle. **(A)** A suture is placed in through the distal skin, and a distal to proximal cut is made. **(B)** The skin is retracted, the cremaster is exposed, and a distal to proximal cut is made through the cremaster to expose the testicle. **(C)** Following removal of the testicle, the cremaster is stretched over a plexiglass pedestal. Care should be taken not to overly stretch the cremaster muscle. **(D)** A microscope image of arterioles and venules.

pinned to the silastic and the cremaster is cut in a distal to proximal direction. This will free the cremaster except for a small set of blood vessels between the cremaster and the testicle. These can be occluded with forceps and gently pulled in the middle with minimal bleeding. The connective tissue is then cleared from the cremaster using fine-tipped scissors. The cremaster is then gently stretched and pinned out, starting from the distal direction. The entire animal can then be moved to the microscope for observation at higher magnification. The cremaster should be superfused with the PSS through the experimental period.

An intact cremaster preparation can be prepared by an abdominal removal of the testicle and replacement with a clear Lucite insert. Although this technique has not been used by a large number of investigators, the major advantage is that disruption of the vasculature is minimized.

Cardiovascular Parameters

Tissue Blood Flow

The preparation of the cremaster muscle requires minimal surgery. Even so, the surgery can result in a disruption of the normal blood flow. Several studies have measured total cremaster blood flow for rats and hamsters using radioactive microspheres in both the surgically dissected tissue and the contralateral undisturbed cremaster muscle (Table I). Blood flow in the dissected cremaster muscle was similar to flow in the intact cremaster and similar to flow in the biceps and gastrocnemius muscles [5]. Other studies reported higher values in the dissected cremaster muscle compared to the undisturbed contralateral muscle. From these studies, it appears that surgery results in an elevated blood flow in the cremaster, although the increased blood flow may be limited to the outside areas exposed to surgical trauma (Table I). At the current time, no measurements of blood flow in the mouse cremaster muscle have been made.

Table I Cremaster Muscle Tissue Blood Flow.

Animal	Dissected tissue blood flow ml/min/100 g	Intact tissue blood flow ml/min/100 g	Reference
Rat	10 ± 2	9 ± 2	Morff and Granger 1980 [5]
	43 ± 3	24 ± 7	Proctor and Busija 1985 [6]
Hamster	7.7 ± 2.3	3.3 ± 0.5	Klitzman and Duling 1979 [7]

Microvascular Parameters

The range of diameters of arterioles and venules are presented in the Table II. The size of each vessel will vary with the age and size of the animal. The most common method for identifying vessels is based on the feed vessel being classified as first order, and each branching vessel with a decreasing size results in an increase in order number. There are varying numbers of arterioles before capillaries, but there are at least four branching orders of arterioles before the capillaries. The terminal arterioles are located before the capillaries and because of their small size, regulate flow through the capillaries. The arterial vessels supplying the cremaster provide significant resistance such that the pressure in the first-order arterioles averages 40 percent of the systemic pressure. Arteriolar diameter, red cell velocity, and calculated blood flow for representative vessels are presented in Table III.

Models of Hypertension

There are numerous hypertensive rodent models, and the cremaster muscle has been used to determine the effects of

Table II Normal Microvessel Diameters and Pressures.

Animal	Systemic blood pressure	Vessel	Resting diameter μm	Max diameter	Microvascular pressure	Reference
Rat	91 ± 4	1st order	119 ± 7		46 ± 3	Meininger et al. 1984 [8]
		2nd order	86 ± 5		43 ± 2	
		3rd order	31 ± 3		37 ± 3	
		3rd order vein	53 ± 7		11 ± 1	
		2nd order vein	141 ± 15		10 ± 1	
		1st order vein	218 ± 7		8 ± 1	
Mouse	70–80 mmHg	1st order	52 ± 3	67 ± 3	+	Hungerford et al. 2000 [9]
		2nd order	18 ± 2	40 ± 2		
		3rd order	8 ± 1	23 ± 2		
Hamster	+	1st order	+	+	+	
		2nd order	35 ± 6	51 ± 3*		Hester and Duling 1988 [10]
		3rd order	22 ± 2	37 ± 3*		

* arteriolar diameter response to tetanic stimulation.
\+ not measured.

Table III Normal Microvessel Red Cell Velocity and Blood Flow.

Animal	Vessel	Tissue blood flow ml/min/100 gm	Vessel diameter μm	Red cell velocity mm/sec	Blood flow nl/sec	Reference
Rat	Arteriole	43	40 ± 4	9.0 ± 1.5	8.7 ± 2.0	Proctor and Busija 1985 [6]
Hamster	Arteriole		44 ± 6	6.9 ± 2	11 ± 7	Hester and Duling 1988 [10]
		8	30 ± 3	5.1 ± 0.8	5 ± 2	Klitzman and Duling 1979 [7]
Rat	Capillary		5.1 ± 0.1	100 μm/sec		Klitzman and Duling 1979 [7]

Table IV Cardiovascular and Microvessel Parameters in Hypertension Models.

Animal	Condition	Systemic blood pressure	Vessel	Diameter μm	Microvascular pressure mmHg	Reference
Rat	1 kidney 1 clip hypertension	164 ± 11	1st order	77 ± 5	60 ± 5	Meininger et al. 1984 [8]
			2nd order	53 ± 4	50 ± 6	
			3rd order	26 ± 3	39 ± 3	
Rat	2 kidney 1 clip hypertension	130 ± 8	1st order	92 ± 10	53 ± 6	Meininger et al. 1984 [8]
			2nd order	58 ± 6	41 ± 5	
			3rd order	24 ± 3	34 ± 4	
Rat	DOCA-salt hypertension	134 ± 17	1st order	67 ± 7	62 ± 12	Meininger et al. 1984 [8]
			2nd order	48 ± 6	55 ± 5	
			3rd order	18 ± 3	47 ± 3	
Rat	SHR hypertension	118 ± 12	1st order	88 ± 4		Falcone et al. 1993 [11]
			3rd order	26 ± 2		

hypertension on the microcirculation. Table IV provides data for baseline arterial pressures, diameters, and microvascular pressures in four models of hypertension. Venular data is not presented because the values are similar to normal values (Table I).

Advantages

As a result of the ease of preparation, the cremaster muscle preparation has been used extensively to study the microcirculation. The total time for preparation of the muscle is approximately one hour. With suitable surgical training, it is easy to have preparations that have substantial vascular tone. The muscle will contract vigorously in response to field stimulation, resulting in an increase in oxygen consumption, blood flow, and arteriolar diameter.

Disadvantages

The cremaster is not a true postural muscle, and the arrangement of the muscle fibers is different from most postural muscles. The muscle fibers run oblique to each other, and even though there are discreet areas supplied by capillaries, this delineation is not readily observable. The outside areas of the cremaster should be avoided for study because these vessels are dilated as a result of the surgical procedure.

Conclusion

The cremaster muscle has proven to be an excellent tissue for the study of the microcirculation. Its ease of dissection, along with good visibility, provide a tissue that all laboratories can use for the study of the microcirculation. The availability of different hypertension models, along with different species, allows investigators a wide range of preparations to study microcirculatory function. The use of the mouse cremaster in the study of microcirculatory function will undoubtedly lead to the development of new transgenic models.

Glossary

Arteriole: A small arteriole, one which, at its distal end, leads to a capillary.

Cremaster: One of the fascia-like muscles suspending and enveloping the testicles and the spermatic cord.

Red cell velocity: The average speed of the blood in blood vessels. The centerline velocity is measured by various optical techniques, and average blood velocity is calculated as $\text{Velocity}_{\text{average}} = \text{Velocity}_{\text{centerline}} / 1.6$.

Venule: A tiny vein which, at it most proximal end, collects blood from capillaries.

References

1. Grant, R. T. (1964). Direct observations of skeletal muscle blood vessels (rat cremaster). *J. Physiol.* **172**, 123–137.
2. Majno, G., Gilmore, V., and Leventhal, M. (1967). A technique for the microscopic study of blood vessels in living striated muscle (cremaster). *Circ. Res.* **21**, 823–832.
3. Baez, S. (1973) An open cremaster muscle preparation for the study of blood vessels by in vivo microscopy. *Microvasc. Res.* **5**, 384–394. *This was the first paper describing the open cremaster preparation. This manuscript is the classic paper that all others reference for the methodology.*
4. Sarelius, I. H., Maxwell, L. C., Gray, S. D., and Duling, B. R. (1983). Capillarity and fiber types in the cremaster muscle of rat and hamster. *Am. J. Physiol.* **245**, H368–H376. *The cremaster muscle is used as a model of skeletal muscle. This manuscript provides information about the muscle types in different animal models for the investigator to compare to other animal models.*
5. Morff, R. J., and Granger, H. J. (1980). Measurement of blood flow with radioactive microspheres in the intact and surgically exposed rat cremaster muscle. *Microvasc. Res.* **19**, 366–373.
6. Proctor, K. G., and Busija, D. W. (1985). Relationships among arteriolar, regional, and whole organ blood flow in cremaster muscle. *Am. J. Physiol.* **249**, H34–H41.
7. Klitzman, B., and Duling, B. R. (1979). Microvascular hematocrit and red cell flow in resting and contracting striated muscle. *Am. J. Physiol.* **237**, H481–H490. *This manuscript provides information about capillary hematocrit along with muscle blood flow. It is also important because it presents the hypothesis about the role of an unstirred layer in capillaries as a source of the low capillary hematocrit.*
8. Meininger, G. A., Harris, P. D., and Joshua, I. G. (1984). Distributions of microvascular pressure in skeletal muscle of one-kidney, one clip, two-kidney, one clip, and deoxycorticosterone- salt hypertensive rats. *Hypertension* **6**, 27–34.
9. Hungerford, J. E., Sessa, W. C., and Segal, S. S. (2000). Vasomotor control in arterioles of the mouse cremaster muscle. *FASEB J.* **14**, 197–207.
10. Hester, R. L., and Duling, B. R. (1988). Red cell velocity during functional hyperemia: Implications for rheology and oxygen transport. *Am. J. Physiol.* **255**, H236–H244.
11. Falcone, J. C., Granger, H. J., and Meininger, G. A. (1993). Enhanced myogenic activation in skeletal muscle arterioles from spontaneously hypertensive rats. *Am. J. Physiol.* **265**, H1847–H1855.

These references were used because they provide normal cardiovascular values for microvessel diameters, red cell velocities, and pressures. The three references by Morph and Granger, Proctor and Busija, and Klitzman and Duling provide normal values for whole tissue blood flows. These references will allow readers to decide the usefulness of this preparation for their experiments.

Capsule Biography

Dr. Robert Hester is a Professor in the Department of Physiology at the University of Mississippi Medical Center, Jackson, Mississippi. Dr. Hester received his Ph.D. at the University of Mississippi Medical Center and completed a postdoctoral fellowship at the University of Virginia. He has been using the cremaster muscle for the study of functional hyperemia for approximately 20 years.

CHAPTER 26

Zebrafish: A Model for Studying Microvascular Development and Function

Kameha Kidd and Brant Weinstein

Laboratory of Molecular Genetics, National Institute of Child Health and Human Development, National Institutes of Health, Bethesda, Maryland

The zebrafish is a relatively recent addition to the field of microvascular research, but it possesses a number of distinct advantages for studies of blood vessel formation in vivo including experimental and genetic accessibility and an optically clear embryo. Because of its useful attributes, the fish has already begun to contribute significantly to our understanding of key factors involved in vessel growth, differentiation, and maturation, as well as the mechanisms that guide the patterning and function of the vasculature. In this chapter we provide a historical background and brief overview of the use of the fish for studies of blood vessel formation.

Emergence of the Zebrafish as a Vertebrate Genetic Model

The zebrafish has become a popular model for research in developmental and vascular biology during the past decade, but it has actually been evolving as a model organism for more than 30 years (see Figure 1 for a timeline of important events). The late George Streisinger was probably responsible more than anyone else for developing the zebrafish into the useful model organism that it is today.

From Phage to Fish

In the late 1960s George Streisinger began working with zebrafish at the University of Oregon with the goal of developing a vertebrate genetic model system comparable to the fly. Streisinger had worked with bacteriophage at Caltech during the critical foundation period of the molecular era, when the basic DNA–RNA–protein story was being established. His studies at Caltech involved a mutational approach to evaluate the genetic code, including its structure and the process of translation in the phage. These studies helped show that in vitro proposed codons did translate to in vivo specification of amino acid sequences based on specific base sequences. The success of genetic studies such as these in phage led Streisinger and others to wonder whether more complex genetic and biological processes could also be deciphered through mutational analysis. Taking this idea forward, Streisinger and some of his Caltech colleagues decided to attempt to study the nervous system using genetic analysis. Sydney Brenner turned to *Caenorhabditis elegans* for this purpose, while Seymour Benzer turned to the fruit fly to address the questions of nervous system function and behavior. Streisinger, however, was determined to find a vertebrate genetic model and set upon the zebrafish as a vertebrate organism in which it would be feasible to use mutational analysis to dissect biological (neurological in particular) development. It was joked that he wanted to find a phage with a backbone.

Why Zebrafish?

The zebrafish possesses some of the genetic and experimental accessibility of the fly and nematode worm, but is much more closely related to mammals. Some of the general

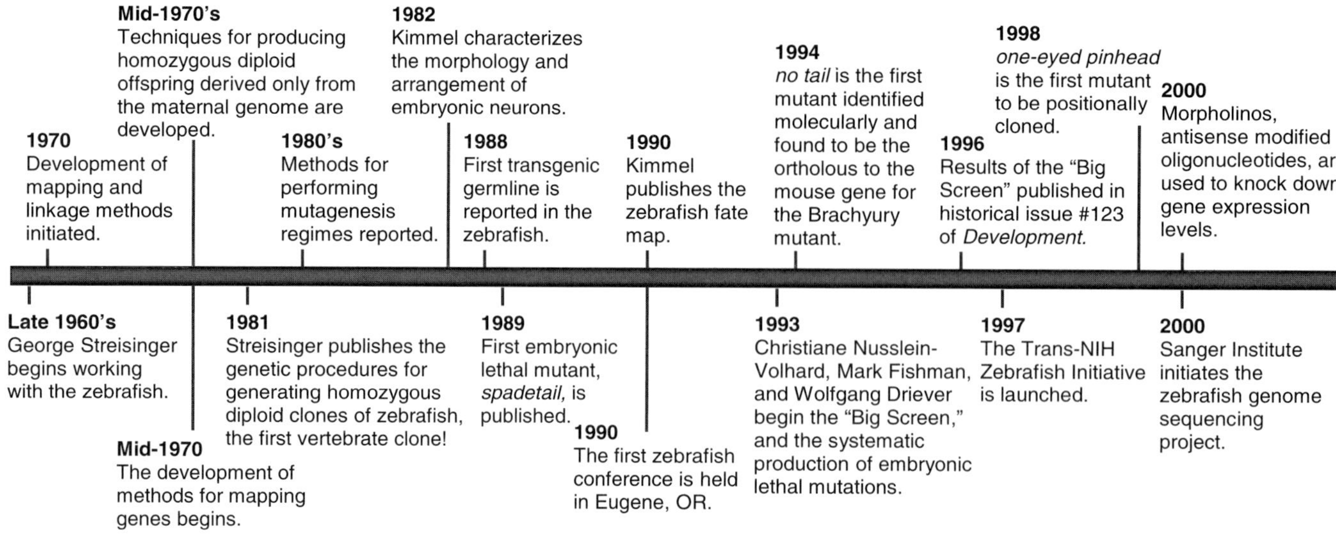

Figure 1 Major events that contributed to the success of the zebrafish. The zebrafish has been around for a number of years, but was not considered an established model for biological study until the past decade. This timeline highlights some of the major events that have contributed to the establishment of the zebrafish as a model organism. (see color insert)

features of the zebrafish that initially attracted Streisinger were the large numbers of adult fish that could be maintained in a relatively small space (adult zebrafish are only an inch long) and their year-round breeding in the laboratory. Zebrafish mature in 3 months' time, are capable of breeding every 1 to 2 weeks, and can produce hundreds of eggs per clutch. The external fertilization of the eggs permits in vitro manipulation of ploidy and fertilization for genetic analysis. Streisinger also realized the tremendous advantage afforded by the optically clear embryo of the zebrafish, facilitating screening for developmental phenotypes of interest even in organs and tissues deep within the animal.

Streisinger's Success

Streisinger's efforts to establish the zebrafish as a genetic model were not without risk. In the 1970s little was known about the conservation of regulatory pathways, and the question of translation from the fish to other vertebrates was a serious concern. Fortunately, the environment at the University of Oregon was such that Streisinger had liberty to pursue his interests without the risk of losing the support and resources provided by the "community" at Oregon, and he slowly began to make progress. He developed methods to produce haploid and diploid gynogenotes, permitting recessive mutations to be uncovered in a single generation without elaborate and time-consuming inbreeding schemes. He established a breeding program to remove lethal background mutations from strains, and developed methods to efficiently induce and recover new germ-line mutations. In 1981, Streisinger published a paper in *Nature* [1] describing the production of homozygous diploid clones of zebrafish. Meanwhile, work in other model organisms was leading to growing acceptance for forward-genetic analysis as a means of dissecting developmental processes. Arguably the most notable of these was the report published by Christiane Nüsslein-Volhard and Eric Wieschaus describing large-scale genetic analysis of *Drosophila* development [2]. As Streisinger's work with the zebrafish progressed, other faculty members at the University of Oregon also became interested in the zebrafish. In 1990 Charles Kimmel and colleagues published a cellular fate map for the zebrafish, a project that took nearly 10 years to complete [3]. This report laid the groundwork and provided tools for further studies of cell lineage and cell fate in the zebrafish. Monte Westerfield helped open the door further to other groups interested in working with zebrafish when he generated a protocol book covering the housing, handling, and laboratory use of zebrafish [4]. In 1988, Westerfield and colleagues also published the first report of stable germ-line transmission of foreign DNA in the zebrafish [5], laying the foundations for the subsequent very fruitful use of transgenic technologies in the fish.

Two additional events helped to solidify the fish as a vertebrate genetic model, by demonstrating its "relevance" to mammals and other "higher" vertebrates. The first was the molecular cloning of the zebrafish *no tail* mutant, which was shown to result from a mutation in the zebrafish homolog of the mouse *T* (*Brachyury*) gene and to cause phenotypically similar defects [6]. This demonstrated the fundamental conservation of developmental processes between these divergent vertebrate species. The second was publication of the results of the "Big Screens" carried out by two groups in Tübingen, Germany, and in Boston. Some 4,000 mutant phenotypes were identified in these screens, whose results were published in a single, historic volume of *Development* [7]. Mutants were uncovered affecting a wide variety of vertebrate developmental processes, including many modeling human congenital diseases, and analysis of these mutants has led to many insights into the genetic pathways regulating early development. These screens demonstrated the feasibility of large-scale forward-genetic analysis of vertebrate development. Today the process of setting up and running a zebrafish screen is almost routine, and numerous genetic screens are being carried out targeting a wide variety of developmental processes.

The Postscreen Era

Interest in the use of the zebrafish greatly increased after publication of the results of the first large-scale screens in 1996, and development of additional tools and financial investment accelerated. Efforts began to develop dense genetic and physical maps of the genome, work that has led to a much more rapid turnaround time between identification of an interesting mutant phenotype and identification of the gene responsible for the phenotype. In 1997 the NIH established the Trans-NIH Zebrafish Coordinating Committee (TZCC) to help provide funding for zebrafish as a model organism for the study of vertebrate development, physiology, and disease. The efforts of this committee have been instrumental in facilitating increased and better-targeted channeling of NIH resources to the zebrafish community. A whole-genome sequencing project was initiated by the Sanger Institute, a project that has reached approximately fourfold coverage of the genome to date. Anther important advance was the development of targeted gene "knockdown" technology using antisense, morpholino-modified oligonucleotides (morpholinos, MOs) to inhibit translation or splicing of target genes of interest [8]. Suspected mutant sequences can now be confirmed with MO injections, conservation of gene function between fish and mice can be easily tested, and single and multiple gene knockdowns open the door to epistasis experiments and serve as a valuable complementary method to mutants for studying vertebrate development and disease.

The Zebrafish as a Model Organism for Studying Vessel Formation

The zebrafish is a useful model for studies of development in general, but it offers particular advantages for studies of blood vessel development, including the ability to isolate vascular-specific mutants by forward genetics and use optical imaging methods to visualize blood vessels within the living animal with very high resolution. Recently the zebrafish has begun to yield novel insights into mechanisms of blood vessel formation in vivo.

Cardiovascular Mutants in the Zebrafish

Among the mutants uncovered in the Tübingen and Boston screens were a large number with circulatory system defects. Mutations that significantly impair cardiovascular development are readily identifiable in zebrafish because of the transparency and small size of the developing zebrafish embryo, which permits easy visualization of the heart and blood vessels and allows the animal to receive enough oxygen by passive diffusion to survive for 4 to 5 days in the absence of a functional circulation. The cardiovascular mutants from these and subsequent mutant screens included a number of mutants specifically defective in the formation of blood vessels. Table I provides a list of some of the vascular-specific mutants that have been published to date, while Figure 2 illustrates the vascular defects associated with a few of these mutants. Although vascular mutants represent a relatively small subset of the mutants reported thus far, several large-scale genetic screens specifically targeting blood vessel mutants have recently been undertaken [9; also our unpublished results], and these new screens are now yielding many new vascular-specific loci, providing an important resource for future studies of vascular development in the fish.

Table I

Mutant	Phenotype	Gene	Reference
cloche	Lack endothelial and circulating blood cells	unknown	*Development* (1997) **124**, 381–389.
schwentine	Loss of angioblasts (endothelial cell precursors) and failure to undergo angiogenesis	Zebrafish ortholog of *flk 1*	*Curr. Biol.* (2002) **12**(16), 1405–1412.
gridlock	Lack trunk and tail circulation due to improper assembly of the dorsal and lateral aortae	Vascular bHLH factor related to mammalian HRT2	*Science* (2000) **287**, 1820–1824.
violet beauregarde	Massive enlargement of central cranial vessels and improper arterial–venous connections	*Alk1 (acvrl1)*	*Development* (2002) **129**, 3009–3019.
plcg(y10)	Deficient in VEGF-mediated angiogenesis and arterial differentiation	Phospholipase C-γ1	*Genes Dev.* (2003) **17**(11), 1346–1351.
kurzschluss	Branchial arteries fail to form properly and arterial–venous shunts lead to loss of circulation in the trunk	unknown	*Development* (1996) **123**, 293–302.
Out-of-bounds	Display premature sprouting and mispatterned growth of the trunk intersegmental vessels	PlexnD1	*Development* (2002) **129**, 973–982.

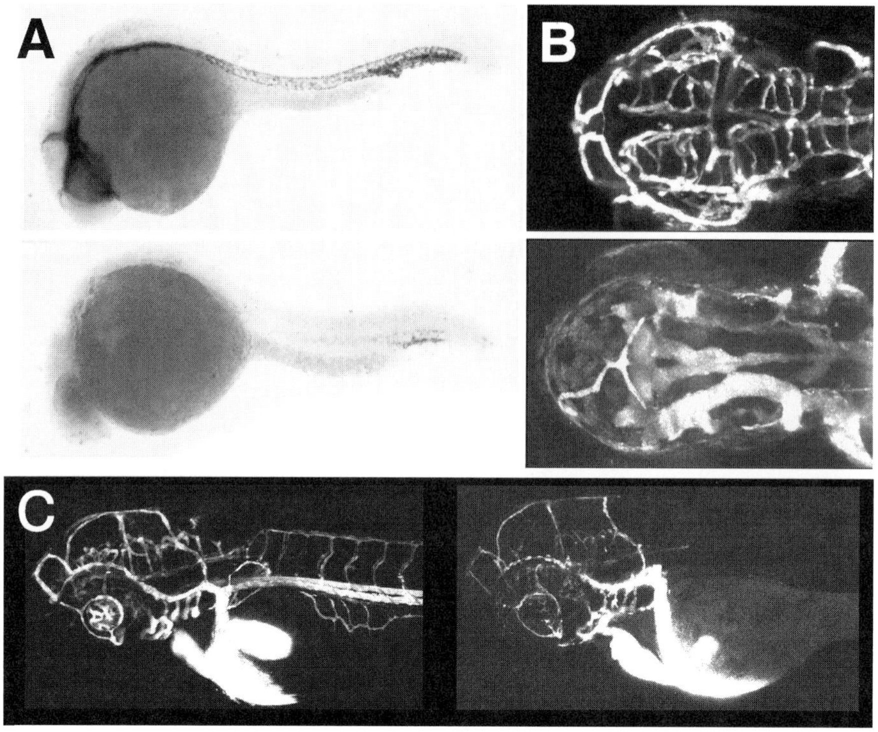

Figure 2 Vascular-specific mutants in the zebrafish. A number of different mutants have been isolated in zebrafish with defects in blood vessel formation. *Cloche* mutants lack virtually all blood endothelium (A). *In situ* hybridization of wild type (top) and mutant (bottom) embryos with vascular endothelial-specific probes such as *Fli-1* reveals a lack of vascular staining except in a small patch in the posterior trunk (24 hpf embryos, lateral views). *Violet beauregarde* mutants have defects in cranial vessel patterning (B). Mutants (bottom) display highly enlarged primary cranial vessels and reduced perfusion of secondary cranial vessels compared to wild type (top) embryos, as shown in these angiographic images (2.5 dpf embryos, dorsal views of head). *Gridlock* mutants lack caudal circulation due to a defect in proper formation of the dorsal aorta C). Mutants (right) have a normal pattern of cranial vessels but lack perfusion of more caudal vessels in comparison to wild type (left) animals (2.5 dpf embryos, lateral views of head and anterior trunk). (see color insert)

Transgenesis and Vascular Imaging in the Zebrafish

The application of transgenic technology to zebrafish cardiovascular research has provided a number of powerful tools for in vivo vascular imaging. Transgenic zebrafish lines expressing green fluorescent protein [GFP or enhanced (E)GFP] within vascular endothelial cells have been particularly useful for studying the formation of the vasculature in vivo. A germ-line transgenic expressing EGFP in the vasculature was generated using the zebrafish *fli1* promoter (Figure 3; *fli1* is a transcription factor expressed in the presumptive hemangioblast lineage, and later restricted to vascular endothelium, cranial neural crest derivatives, and a small subset of myeloid derivatives). This line expresses EGFP at high levels in vascular endothelium, permitting very high resolution long-term time-lapse analysis of developing blood vessels in vivo. Multiphoton confocal time lapse imaging of Fli-EGFP transgenic zebrafish has enabled detailed analysis of both normal vascular development and defective vessel formation due to genetic or experimental perturbations. Lawson and Weinstein [10] showed in vivo that growing blood vessels are extremely active, extending and retracting filopodial processes up to tens of microns in all directions. In a separate study Isogai et al. [11] used time-lapse imaging of Fli-EGFP animals to study how the earliest network of angiogenic vessels (consisting of intersegmental and parachordal vessels) assembles in the trunk. This study revealed a novel "two-step" mechanism for vascular network formation, with a primary vascular network forming first from artery-derived sprouts followed by emergence of a set of vein-derived secondary sprouts that interact dynamically with the primary network to generate the final functional "wiring" of the trunk network. The results of this study showed that blood flow does not play an important role in the initial assembly and gross anatomical patterning of this network, but is likely a critical determinant of the interconnections between primary and secondary vessels and their eventual arterial–venous identity. The Fli-EGFP transgenic line also permits imaging of blood vessels in adult zebrafish, as revealed in another recent study, of vascular network reassembly in the regenerating zebrafish caudal fin [12].

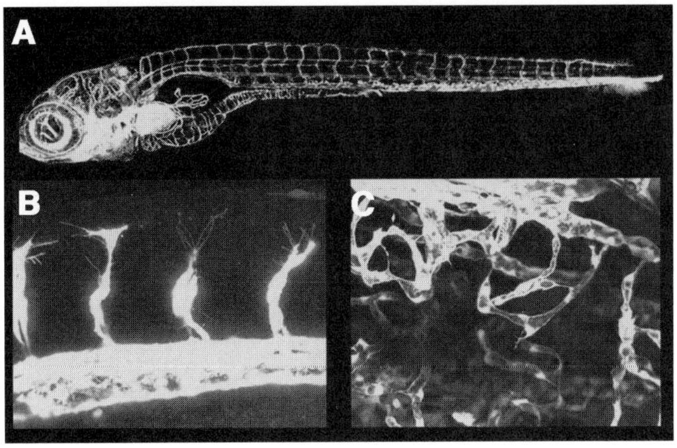

Figure 3 Blood vessels imaged in Fli-EGFP transgenic zebrafish. Vascular endothelial cells and their angioblast precursors are brightly green fluorescent in Fli-EGFP transgenic animals (**A**). Image shown is a confocal microangiogram of a 7 day-old transgenic zebrafish. Fine cellular details can be discerned, including thin filopodial processes extending from growing intersegmental vessels in the trunk (**B**) and central arteries assembling in the zebrafish hindbrain (**C**). Images shown are confocal microangiograms from approximately 1.5 day-old transgenic zebrafish. (see color insert)

Studying Arterial–Venous Fate Determination Using the Zebrafish

One of the recent important contributions to the study of microvasculature using the zebrafish has been the identification of molecular pathways responsible for arterial–venous (A-V) differentiation of endothelial cells. In the past, the arterial–venous fate of endothelial cells was believed to follow from physiological parameters such as differences in blood flow and pressure, but recent work has shown that early endothelial arterial–venous differentiation is in fact genetically programmed. Initial evidence that arterial and venous endothelial cells possess distinct molecular identities came from work with ephrin and Eph genes in mice. Wang et al. [13] described the expression of ephrin B2 (*Efnb2*), a member of the ephrin family of membrane ligands. Prior to the onset of flow, *Efnb2* is expressed specifically in arterial endothelial cells and is absent in venous endothelial cells, whereas the ephrin B2 receptor, Ephb4, is preferentially expressed in veins. Targeted gene deletion of each member of this ligand–receptor pair resulted in similar cardiovascular abnormalities, demonstrating their necessity, and likely direct interaction, for normal vascular development. Although these previous studies demonstrated the existence of functionally important molecular differences between arterial venous endothelial cells, they did not reveal how this fate choice was initially made. Zebrafish studies have been critical in uncovering and dissecting the functional roles of the upstream factors specifying arterial and venous endothelial cell fates, identifying a signaling cascade consisting of sequential hedgehog, vascular endothelial growth factor, and notch signaling. The discussion that follows reviews these findings.

A variety of studies in mammals and other vertebrates had revealed the specific expression of Notch signaling genes (Notch, Delta, Jagged, and so or) in arterial but not in venous endothelial cells. Murine knockout studies showed that these molecules play an important functional role in the vasculature, and their arterial-specific expression suggested they might be playing a specific role in artery formation. A number of recent studies in the zebrafish [14–16] have demonstrated that Notch signaling promotes arterial differentiation at the expense of venous differentiation during vascular development. Notch signaling was repressed in zebrafish embryos either genetically, using the neurogenic *mindbomb* (*mib*) mutant, or experimentally, by injecting mRNA encoding a dominant-negative DNA binding mutant of *Xenopus suppressor of hairless* protein. In either case, repression of Notch signaling resulted in loss of *ephrinB2a* expression from arteries and ectopic expansion of normally venous-restricted markers such as *ephb4* and *flt-4* into the arterial domain. Conversely, activation of Notch signaling [either by heat-shock promoter-driven ubiquitous expression of the Notch1a intracellular domain (Notch1a-ICD) or by Fli1-promoter driven vascular-specific expression of Notch5-ICD] suppressed expression of vein-restricted markers and promoted ectopic expression of *ephrinB2a* and other arterial markers in venous vessels. The vascular specific expression of Notch demonstrated the vascular endothelial cell-autonomy of Notch-ICD effects, confirming that Notch is in fact acting at the level of the vascular endothelial cell itself and not via indirect signals from some other, adjacent Notch-responsive cells or tissues.

Lawson and colleagues further dissected the A-V differentiation signaling hierarchy by demonstrating that sonic hedgehog (shh) and vascular endothelial growth factor (vegf) act upstream of Notch [15]. As in embryos lacking Notch signaling, embryos lacking shh or vegf fail to express ephrin-B2a within their blood vessels. Overexpression of shh promotes ectopic arterial vessel formation in the trunk, whereas overexpression of vegf via injection of vegf mRNA suppresses expression of vein-restricted markers and results in expression of ephrinB2a and other arterial markers in venous vessels. By combining activation or inhibition of each of the different signaling pathways in a series of "molecular epistasis" experiments, it was found that shh activity induces expression of vegf in the somites, which then activates notch signaling in the adjacent endothelial cells of the developing dorsal aorta, promoting arterial differentiation. Genetic screening methods were also used to identify genes functioning downstream from vegf in the zebrafish [17]. A zebrafish mutant was uncovered that was deficient in both angiogenesis and arterial differentiation as a result of a defect in phospholipase C gamma-1 (plcg1). Phospholipase C genes are known effectors of signaling via receptor tyrosine kinases such as the vegf receptor Flk1, and the vascular expression of plcg1 and vascular-specific phenotype of the mutant in this gene suggested that it might be functioning downstream of vegf signaling. Indeed, further experiments showed that plcg1 mutants were insensitive to both angiogenic and arterial differentiation responses to vegf overexpression. In support of the zebrafish findings regarding roles for hedgehog and vegf signaling in the vasculature, recent studies in mice have also implicated shh and vegf signaling in regulating blood vessel growth and arterial differentiation [18–21], demonstrating the conservation of genetic programs between vertebrate species.

Conclusions

The lessons learned in the zebrafish are likely to be readily transferred to other vertebrate organisms, including mammals, since the complex circulatory system of the zebrafish is in most respects quite similar to that of other vertebrates. A comparison between the blood vessels of zebrafish and other vertebrates presents a striking degree of anatomical and functional conservation of vascular pattern and anatomy, suggesting that vascular development is directed by genetically programmed, evolutionarily conserved control mechanisms. Indeed, it is clear from the analysis of a variety of zebrafish orthologs of vascular-specific genes first described in other species that most of these genes have very similar spatial and temporal expression patterns in the fish. Some of the genetic control mechanisms responsible for regulating the expression of these genes and for vascular differentiation and patterning have now begun to be studied in the zebrafish. The specific advances in our understanding of the A-V differentiation pathway will be important in designing methods for directing organ-specific vessel growth for tissue regeneration, and for targeting specific vessel components for inhibition of growth. It is anticipated that our knowledge and understanding in this area will only continue to grow with the zebrafish as an integral component, resulting in the identification of novel factors and the more subtle characterization of the various vessel types and their growth.

Glossary

Clone: An exact copy of biological material such as DNA, a whole cell, or a complete organism.

Diploid: A complete set of chromosomes, present in somatic (non-sex-determining) cells, consisting of two copies of each chromosome. Human beings have 46 chromosomes in their diploid cells.

Epistasis: The ability of a gene to mask the phenotypic effects of another gene. A downstream gene is epistatic to any upstream genes that produce the same effect.

Fate map: A map of an embryo showing areas that are destined to develop into specific adult tissues and organs.

Genetic linkage map: A linear map of the relative positions of genes along a chromosome. Gene distances are determined by the frequency at which two gene loci are separated during chromosomal recombination.

Germ-line: The propagation of genetic information from one generation to the next via the germ cell. Germ-line mutation is one that has occurred in a germ cell and will be passed to the next generation. Same logic pertains to the term *germ-line transmission*.

Gynogenotes: Individuals that derive their chromosome number solely from the maternal contribution, rather than half from each parent.

Haploid: A single set of chromosomes, present in the egg and sperm cells of animals. Humans have 23 chromosomes in their reproductive cells.

Homolog: One member of a chromosome pair, or a gene similar in structure and evolutionary origin to a gene in another species.

Morpholino oligonucleotides: Short, antisense sequences of DNA (oligos) with a morpholino backbone that provides stability, and an increased life span in vivo.

Ploidy: Refers to the number of single sets of chromosomes in a given cell or organism.

Recessive mutations: Mutations affecting those genes require two identical copies to be expressed.

References

1. Streisinger, G., Walker, C., Dower, N., Knauber, D., and Singer, F. (1981). Production of clones of homozygous diploid zebra fish (Brachydanio rerio). *Nature* **291**, 293–296.
2. Nusslein-Volhard, C., and Wieschaus, E. (1980). Mutations affecting segment number and polarity in Drosophila. *Nature* **287**, 795–801.
3. Kimmel, C. B., Warga, R. M., and Schilling, T. F. (1990). Origin and organization of the zebrafish fate map. *Development* **108**, 581–594.
4. Westerfield, M. (2001). *The Zebrafish Book*, 4th ed. University of Oregon Press, Eugene, OR.
5. Stuart, G. W., McMurray, J. V., and Westerfield, M. (1988). Replication, integration and stable germ-line transmission of foreign sequences injected into early zebrafish embryos. *Development* **103**, 403–412.
6. Schulte-Merker, S., van Eeden, F. J., Halpern, M. E., Kimmel, C. B., and Nusslein-Volhard, C. (1994). No tail (ntl) is the zebrafish homologue of the mouse T (Brachyury) gene. *Development* **120**, 1009–1015.
7. *Development* (1996). Vol.123 [special issue].
8. Nasevicius, A., and Ekker, S. C. (2000). Effective targeted gene 'knockdown' in zebrafish. *Nat. Genet.* **26**, 216–220.
9. Habeck, H., Odenthal, J., Walderich, B., Maischein, H., and Schulte-Merker, S. (2002). Analysis of a zebrafish VEGF receptor mutant reveals specific disruption of angiogenesis. *Curr. Biol.* **12**, 1405–1412.
10. Lawson, N. D., and Weinstein, B. M. (2002). In vivo imaging of embryonic vascular development using transgenic zebrafish. *Dev. Biol.* **248**, 307–318.
11. Isogai, S., Lawson, N. D., Torrealday, S., Horiguchi, M., and Weinstein, B. M. (2003). Angiogenic network formation in the developing vertebrate trunk. *Development* **130**, 5281–5290.
12. Huang, C. C., Lawson, N. D., Weinstein, B. M., and Johnson, S. L. (2003). reg6 is required for branching morphogenesis during blood vessel regeneration in zebrafish caudal fins. *Dev. Biol.* **264**, 263–274.
13. Wang, H. U., Chen, Z. F., and Anderson, D. J. (1998). Molecular distinction and angiogenic interaction between embryonic arteries and veins revealed by ephrin-B2 and its receptor Eph-B4. *Cell* **93**, 741–753.
14. Lawson, N. D., Scheer, N., Pham, V. N., Kim, C. H., Chitnis, A. B., Campos-Ortega, J. A., and Weinstein, B. M. (2001). Notch signaling is required for arterial-venous differentiation during embryonic vascular development. *Development* **128**, 3675–3683.
15. Lawson, N. D., Vogel, A. M., and Weinstein, B. M. (2002). Sonic hedgehog and vascular endothelial growth factor act upstream of the Notch pathway during arterial endothelial differentiation. *Dev. Cell* **3**, 127–136.
16. Zhong, T. P., Rosenberg, M., Mohideen, M. A., Weinstein, B., and Fishman, M. C. (2000). Gridlock, an HLH gene required for assembly of the aorta in zebrafish. *Science* **287**, 1820–1824.
17. Lawson, N. D., Mugford, J. W., Diamond, B. A., and Weinstein, B. M. (2003). Phospholipase C gamma-1 is required downstream of vascular endothelial growth factor during arterial development. *Genes Dev.* **17**, 1346–1351.
18. Pola, R., Ling, L. E., Silver, M., Corbley, M. J., Kearney, M., Blake Pepinsky, R., Shapiro, R., Taylor, F. R., Baker, D. P., Asahara, T., and Isner, J. M. (2001). The morphogen Sonic hedgehog is an indirect angiogenic agent upregulating two families of angiogenic growth factors. *Nat. Med.* **7**, 706–711.
19. Mukouyama, Y. S., Shin, D., Britsch, S., Taniguchi, M., and Anderson, D. J. (2002). Sensory nerves determine the pattern of arterial differentiation and blood vessel branching in the skin. *Cell* **109**, 693–705.

20. Visconti, R. P., Richardson, C. D., and Sato, T. N. (2002). Orchestration of angiogenesis and arteriovenous contribution by angiopoietins and vascular endothelial growth factor (VEGF). *Proc. Natl. Acad. Sci. USA* **99**, 8219-24.
21. Springer, M. L., Ozawa, C. R., Banfi, A., Kraft, P. E., Ip, T. K., Brazelton, T. R., and Blau, H. M. (2003). Localized arteriole formation directly adjacent to the site of VEGF-Induced angiogenesis in muscle. *Mol. Ther.* **7**, 441–449.

Further Reading

Grunwald, D. J., and Eisen, J. S. (2002). Headwaters of the zebrafish—emergence of a new model vertebrate. *Nat. Rev. Genet.* **3**, 717–724.

Isogai, S., Horiguchi, M., and Weinstein, B. M. (2001). The vascular anatomy of the developing zebrafish: an atlas of embryonic and early larval development. *Dev. Biol.* **230**, 278–301.

Isogai, S., Lawson, N. D., Torrealday, S., Horiguchi, M., and Weinstein, B. M. (2003). Angiogenic network formation in the developing vertebrate trunk. *Development* **130**, 5281–5290. *This reference exemplifies one of the key attributes of the zebrafish, the ability to image the developing embryo, and how this attribute translates to an increased understanding of microvascular development and formation.*

Kimmel, C. B., Warga, R. M., and Schilling, T. F. (1990). Origin and organization of the zebrafish fate map. *Development* **108**, 581–594.

Lawson, N. D., and Weinstein, B. M. (2002a). Arteries and veins: making a difference with zebrafish. *Nat. Rev. Genet.* **3**, 674–682. *This paper reviews of much of the literature detailing our knowledge of blood vessel differentiation and how the zebrafish as a model has contributed to this knowledge.*

Lawson, N. D., and Weinstein, B. M. (2002b). In vivo imaging of embryonic vascular development using transgenic zebrafish. *Dev. Biol.* **248**, 307–318.

Nasevicius, A., and Ekker, S. C. (2000). Effective targeted gene "knockdown" in zebrafish. *Nat. Genet.* **26**, 216–220.

Rasooly, R. S., Henken, D., Freeman, N., Tompkins, L., Badman, D., Briggs, J., and Hewitt, A. T. (2003). Genetic and genomic tools for zebrafish research: the NIH zebrafish initiative. *Dev. Dyn.* **228**, 490–496.

Schulte-Merker, S., van Eeden, F. J., Halpern, M. E., Kimmel, C. B., and Nusslein-Volhard, C. (1994). No tail (ntl) is the zebrafish homologue of the mouse T (Brachyury) gene. *Development* **120**, 1009–1015.

Stainier, D. Y., Fouquet, B., Chen, J. N., Warren, K. S., Weinstein, B. M., Meiler, S. E., Mohideen, M. A., Neuhauss, S. C., Solnica-Krezel, L., Schier, A. F., Zwartkruis, F., Stemple, D. L., Malicki, J., Driever, W., and Fishman, M. C. (1996). Mutations affecting the formation and function of the cardiovascular system in the zebrafish embryo. *Development* **123**, 285–292. *This paper describes many of the original mutations that were identified as affecting the vasculature of the zebrafish. Many of these mutations provided the foundation for current cardiovascular research in the zebrafish.*

Capsule Biography

Brant M. Weinstein obtained his Ph.D. in 1992 from the Massachusetts Institute of Technology. He has headed the Unit on Vertebrate Organogenesis within the Laboratory of Molecular Genetics at the National Institute of Child Health and Human Development (NICHD) since 1997. His laboratory, supported by the intramural program of the NICHD, studies vascular development using the zebrafish.

Kameha Kidd received her doctorate at the University of Arizona in 2002 under the tutelage of Drs. Stuart K. Williams and James B. Hoying. She is currently a postdoctoral fellow, under the direction of Dr. Brant Weinstein, at the National Institute of Child Health and Human Development at the NIH. Her research interest is in microvascular development and organogenesis for application in therapeutic angiogenesis and tissue engineering.

CHAPTER 27

The Significance of the Nail Fold in Clinical and Experimental Studies

Hildegard Rand Maricq
Medical University of South Carolina, Charleston, South Carolina

Definition and Description of the Nail Fold

Anatomical Description

Nail fold (NF) is the skin fold surrounding the nail; it is divided into proximal and lateral portions. Since only the proximal NF has been the site of microvascular studies, the adjective is usually dropped. A sagittal section of this fold, which covers the root of the nail, is shown in Figure 1 [1].

NF has a dorsal epidermis, similar to the usual description of skin anatomy, having epidermal ridges and dermal papillae alternating along the dermo-epidermal junction. The ventral epidermis of the NF is flat, loosely adhering to the nail. Between the two epidermal layers we find the dermal connective tissue with blood vessels, nerves, and other structures, such as sweat glands. In the proximal part of the NF the dermal papillae, containing capillaries, are perpendicular to the skin surface; distally they gradually become slanted and become parallel to the ventral epidermis at the edge of the NF (Figure 1; see also Figure 4b).

The thickness of the NF, the slope of the decreasing distance between the two layers of the epidermis from the proximal NF to its edge, and the length of the NF vary considerably between individuals and also from finger to finger.

View by in Vivo Microscopy

Superficial blood vessels in the skin can be visualized by depositing oil on its surface. In an average healthy adult, there is a relatively uniform distribution of capillaries in the NF, as seen by in vivo capillary microscopy (Figure 2, bottom). Only the tips of the capillary loops are visible in the proximal NF, where they are perpendicular to the skin surface. Gradual sloping of dermal papillae in more distal NF allows visualization of capillaries partially from the side and at the edge of the NF they can be seen the best, now horizontal and showing their arterial, apical, and venular portions of the loop.

In about 5 percent of the normal young adult population, a highly visible and extensive subpapillary venous plexus can also be observed below the superficial papillary capillaries. This visible venular plexus is often associated with longer capillary loops, especially at the edge of the NF. The overall length of the NF tends to be longer in these individuals. Small sections of this plexus can sometimes also be seen in other subjects near the edge of the NF.

Place of the Nail Fold in Human Microvascular Studies

Why the Nail Fold?

Most human studies of microvasculature have used the technique of in vivo microscopy, taking advantage of the fact that superficial microvessels of the skin can be visualized by rendering the skin surface smooth by application of oil. This noninvasive method, discovered in the beginning of the 20th century, appealed to many investigators. O. Müller in 1922 published his extensive observations from a great number of areas of human skin, in both health and disease, illustrated by careful drawings that demonstrate the considerable variability of skin microvasculature from site to site.

Over time, however, NF became the favorite site because it was easy to place a finger under the microscope, and serial studies were facilitated, because the exact location of initial observations could be easily identified.

The pattern of NF capillaries varies with age. In a normal newborn baby only a plexus of microvessels can be seen;

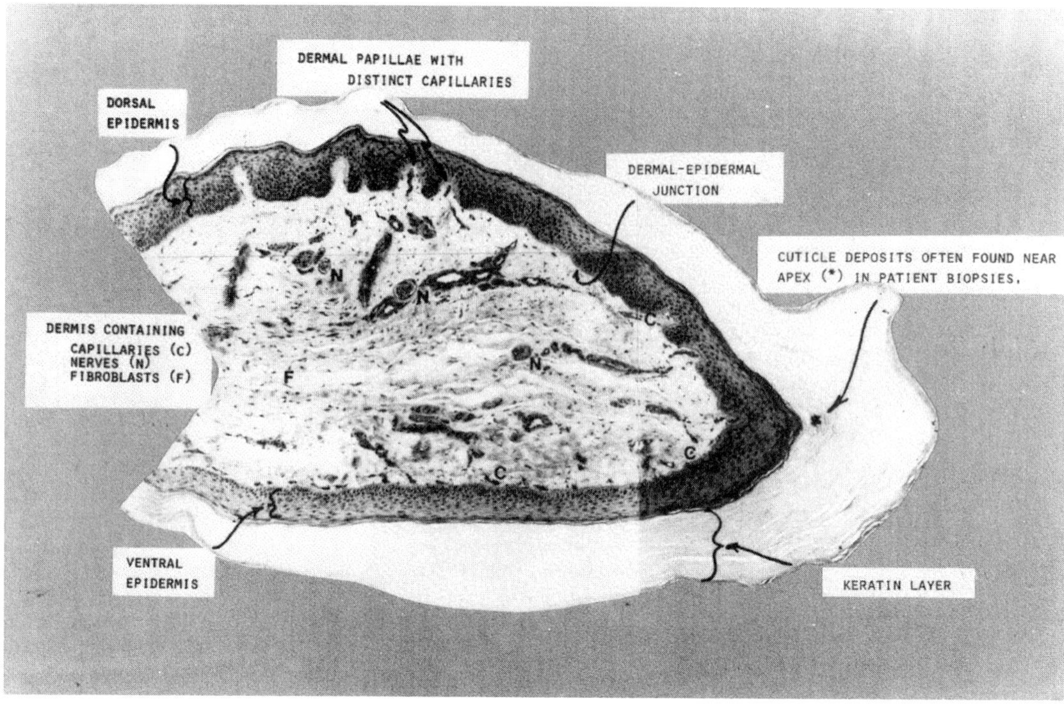

Figure 1 Sagittal section of a finger nail fold in a normal healthy subject.

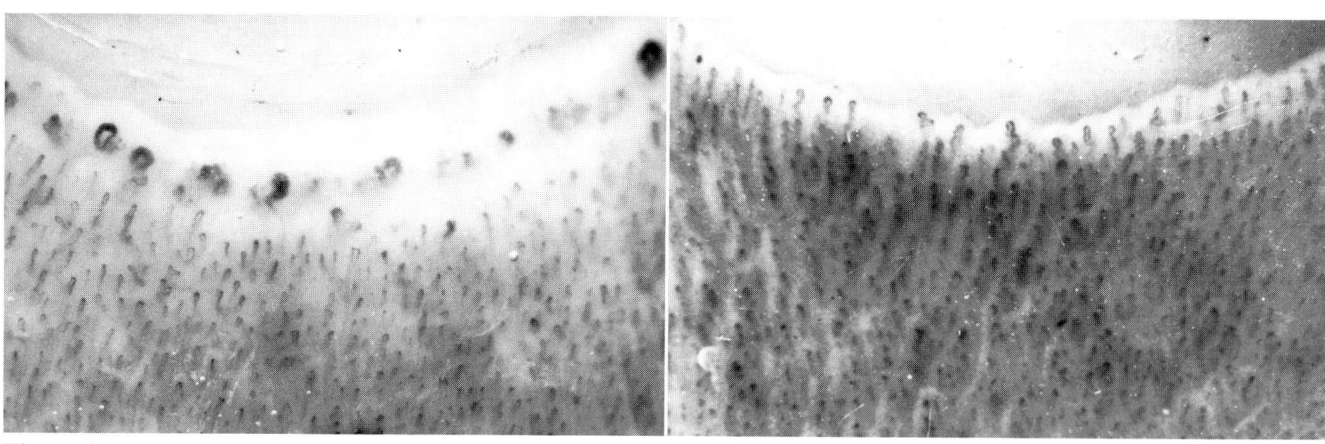

Figure 2 The wide-field view of the nail fold in a scleroderma patient and a normal control. (*Left*) Characteristic scleroderma-type capillary loops of various sizes and moderate to extensive loss of capillaries. (*Right*) Capillary distribution in the NF of a normal subject.

there are no papillary capillaries. Gradually over the first months of life a papillary layer of microvessels appears. Their density is less than in an adult and the subpapillary plexus can still frequently be seen. It appears as a polygonal network, different from the subpapillary venual plexus seen in later life. The latter appears to be a genetic trait, transmitted as Mendelian dominant, but may also be related to disease or to drugs used for treatment. In aged subjects, a decreased density of NF capillaries can occur, often associated with disease rather than aging per se.

The endrow capillaries (those in the terminal row) tend to have larger diameters than more proximally located capillaries. Because of their horizontal position, they can be seen to a greater length. Thus, it is possible to perform measurements of capillary diameters, observe red blood cell (RBC) flow, and also use minimally invasive procedures, such as capillary pressure recordings in this location.

If not otherwise indicated, NF refers to the NF of the finger, where the majority of the human microvascular studies have been performed. Not much can be found in the literature on the toe NF, despite the fact that a large number of publications are devoted to the lower leg and foot. The closest observations reported are usually from the dorsal skin of the distal phalanx of the toes, close to, but not necessarily involving the NF and the endrow loops. Reportedly the mean width of the toe NF loops is significantly larger than in the finger.

Limiting Factors

The drawback of capillary observations in the skin, including the NF, relates to the problem of illumination: Only reflected light can be used here. Compared to microscopy performed with transmitted light, conventional capillaroscopy allows capillaries to be seen only by the presence of RBC columns within them.

Only a few investigators have been able to obtain other details. Ehring and Schumann [2] succeeded in visualizing the capillary wall of NF capillaries by using a specialized optical and illumination system of Vonwiller. Mahler et al. [3] used fluorescein-tagged human albumin that helped to visualize plasma space and thus obtain a better estimate of the total intraluminal dimensions. Conventional measurements of NF capillary dimensions have been based on RBC columns.

In addition, reflections from the oil deposited on the skin interfere with photography, as does the very limited depth of field especially at low magnification. It should also be remembered that NF is a very specialized area of the skin that may not necessarily closely reflect microvascular structure and function elsewhere in the body. Despite all these shortcomings, NF capillary microscopy has yielded considerable information on human microvasculature in health and disease.

Methods

In Vivo Microscopy

Recording of Capillaries and Their Blood Flow

Most studies of the human NF have been performed by in vivo microscopy, also called capillaroscopy, because usually the focus has been on capillaries only. A variety of microscopes, illumination techniques, and photographic systems have been used over the years to obtain well-focused, sharp still pictures.

To record capillary blood flow, by measuring RBC velocity, additional instrumentation was developed: first cinematographic, then video systems. Computer technology now allows automated analysis of these data. Normal RBC velocity ranges from 0.1 to 2.8 mm/sec, with an average of 0.6 mm/sec.

Unfortunately, the measurements are often taken from selected, nonrandom, capillaries: those that happen to be well focused. Therefore, the number of capillaries per NF and the number of nail folds studied vary.

Capillary Pressure

Instrumentation for capillary pressure measurement needs to be used in combination with capillaroscopy to properly introduce the micropipette into the capillary loop. Unlike the completely noninvasive approach just discussed, minimal trauma is necessary, because the hard keratin layer is difficult to penetrate. The normal capillary pressure ranges from 10 to 22 mmHg.

Capillary Permeability

Sodium fluorescein has been used most often to estimate capillary permeability. Although this agent also diffuses out in normal subjects, pathological changes can be estimated by the speed and pattern of this diffusion compared to healthy subjects.

Laser Doppler Blood Flow Measurements

This technique has often been performed on the dorsum of the distal phalanx, or on the proximal part of the NF concurrently with capillaroscopy. It is believed to measure skin blood flow in deeper layers and not the capillary blood flow.

More detailed information on methodology can be found in the book by Bollinger and Fagrell [4] and in a review article by Shore [5].

Biopsy

NF biopsy has been rarely performed, but, as seen in Figure 1, it can show NF structure better than the schematic illustrations often used. At higher magnification, finer details of capillary wall, dermoepidermal junction, and so on can be studied and compared to in vivo observations obtained in the same NF just before the biopsy [6].

Data Obtained

Most of the data on the NF microvasculature has been obtained from the endrow capillaries, in both morphological and function studies. Only a relatively few studies focus on the total area of NF.

Morphology

Endrow Capillaries

NF endrow capillaries show variations in shape, size, and density even in normal healthy subjects; these variations increase in pathological states. Variation is seen not only between individuals, but also between fingers. Also, within an NF not all capillaries are of the same size and shape. However, over time, in a normal subject, the overall picture of NF capillaries in a given finger has been reported to remain as constant as a fingerprint.

(a) *Shape.* A normal capillary is usually described as a hairpin-like loop. The deviant capillary loops have been described as tortuous, meandering, branched, bushy, coiled, and so on. Often the same term is used for different shapes and different terms for the same shape by different investigators. (A better definition with adequate illustration should be used to make comparisons of results possible.) These data have been analyzed semiquantitatively by counting the number of deviant capillary loops per NF.

It should be noted here that vigorous "pushing back" of the cuticle during manicure can considerably distort the "normal" U-shaped capillaries at the edge of the NF.

(b) *Size.* Endrow loops have been classified subjectively as normal, slightly or moderately enlarged, markedly or extremely enlarged, and giant. Measurement data include diameters of arterial and venular limbs of the capillary loop, the caliber of the apical portion, the length of the capillary loop, and the total width of the loop. The ranges for normal measurements reported are arterial, 6 to 19 μm, and venular, 7 to 20 μm.

The problem here is the sampling of capillary loops: Usually they are not all studied. For any measurements, the capillaries have to be sharply focused, but because of technical difficulties this is not always the case. The other selection criteria may be the location within the edge of NF or a subjective preclassification into normal, enlarged, or giant loops.

(c) *Density.* The density is expressed as number of loops/mm. The normal range is reported as 12 to 17/mm.

(d) *Capillary hemorrhages.* These are frequently encountered in certain diseases. In normal persons, they are present only in response to obvious injury or microtrauma.

2. Total Area of NF

(a) *Density.* Capillary distribution proximal to the endrow is relatively uniform and can be measured as number of capillaries per mm^2. The normal density has been reported as 40 to 50/mm^2. A uniform decrease in capillary density in certain conditions can be measured the same way.

(b) *Localized loss of capillaries.* In some pathological conditions the loss of capillaries is not uniform: There may be an extensive loss, including the endrow and extending to a considerable distance proximally, or there may be discrete areas near the endrow. The loss may be complete, resulting in completely avascular areas, or may be relatively avascular, i.e., showing a few capillary loops but less than 8/mm^2. These areas can be measured quantitatively and expressed in mm^2.

A semiquantitative scale has also been used: slight, moderate, and extensive. A blind comparison of this scale with quantitative measurement has shown the following: slight 0.4 to 2 mm^2, moderate 2 to 4 mm^2, and extensive more than 4 mm^2 per NF.

Some authors have used the term "drop out" to note the local loss of capillaries. They refer to endrow loops only and require two "dropped-out" loops to define capillary loss or avascularity.

(c) *Visibility of subpapillary venous plexus (SPVP).* The extent of this feature (SPVP) can be measured by a semiquantitative scale, i.e., the plexus visualization score (PVS). PVS ranges from 0 to 4 per finger and 0 to 40 per subject. Normal range is 0 to 10.

(d) *Pattern.* The view of the total NF (magnification 12 to 14×) allows us to appreciate the overall pattern of change from the normal one. The loss of capillaries is especially striking and sometimes associated with "bushy formations" (Figure 3 [7]) near the junction of avascular and normal NF capillary bed.

This total view helps to locate and identify special features to be studied with higher magnification and help to relocate them on serial studies. Observation of the total NF also makes it easier to notice the abnormal appearance of the cuticle: Normally completely transparent, it may show yellow-orange discoloration and varied patterns of outgrowing capillary hemorrhages in the cuticle.

(e) *Detection of microinjury.* The total NF observation helps to determine whether certain microvascular features are due to local trauma and inflammation rather than related to disease.

Experimental Studies

An Overview

Data presented in the previous section were obtained from NF by microvascular techniques in baseline conditions, i.e., without intervention by experimental stimuli or change in environmental conditions. Here we summarize data of microvascular responses to selected agents likely to affect the microvasculature of the NF.

Most human studies have been clinically oriented and often used for diagnosis, differential diagnosis, or correlated with laboratory data. Investigators aimed to understand pathophysiology by assuming that the same microvascular abnormalities seen in NF may be present in internal organs. They have correlated NF data with those of internal organs that were involved in a given disease, and attempted to study the prognostic significance of microvascular changes.

Selected Stimuli

COLD EXPOSURE

(a) *Whole-body exposure.* In normal subjects exposed to cold in an experimental chamber, NF capillary blood flow may slow down but will not come to a complete standstill, even after more than 30 minutes. In certain disease states, severe slowing of capillary flow occurs in NF loops within a matter of minutes and rapidly comes to a standstill in all the endrow loops and beyond. Temperatures used range from 10 to 20°C between investigators. Other factors are also important, such as time of the day, clothing, and meals.

(b) *Local cold exposure*. Cooling of the finger can be accomplished by blowing cold air on the NF, using Peltier element and other techniques, while controlling the finger skin temperature carefully to keep it in desired range and to avoid injury. A short flow stop (measured in seconds) can also occur in normal subjects.
(c) *Immersion of contralateral hand in cold water*. This has also been used but is less effective.

REACTIVE HYPEREMIA

An arterial occlusion of finger blood flow in normal subjects produces cessation of blood flow followed by reactive hyperemia overshooting the basal flow rate. This response curve can be modified in certain disease states. This test has been used with a variety of microvascular techniques, such as RBC velocity measurement and laser Doppler flux.

RESPONSE TO DRUGS

(a) Responses to vasomotor agonists injected IA or IV and to their antagonists.
(b) Responses to drugs used in treatment of certain diseases in which microcirculation is deemed important.

The reader can find more details on experimental studies from works of Bollinger and Fagrell and of Shore [4, 5].

Clinical Studies

An Overview

NF capillary microscopic studies have been performed in a large number of disease groups by a variety of microvascular techniques. The authors wished to determine the reason for the development and/or persistence of microvascular changes in disorders in which NF abnormalities were present. Serial studies to document progression or regression of pathological changes in the NF, as related to the evolution of the disease or to the effect of therapy, have also been performed. Data obtained from the NF studies in various diseases are presented here as a short summary.

Selected Disorders

Since NF capillaroscopic studies in human disease range from every subspecialty of medical and surgical fields to mental disorders, genetics, and epidemiology, only some selected disorders are listed here: disorders in which some concordance exist between the findings of different investigators and in which high specificity has been found for diagnosis, or microvascular findings have helped to subdivide a clinical entity or help with differential diagnosis.

1. DIABETES MELLITUS

Capillary density is within normal range in both type I and II diabetes, but capillary pressure has been reported to be increased in type I, especially in those with poor glycemic control. Abnormalities in capillary perfusion have also been reported in both types of diabetes, especially in response to arterial occlusion producing reactive hyperemia. Transcapillary and interstitial diffusion of IV injected sodium fluorescein has been shown to be significantly faster in long-term diabetes than in normal controls.

2. HYPERTENSION

Capillary density is reportedly lower (rarefaction) and capillary pressure higher in patients with hypertension; capillary flow velocity may be reduced.

3. RAYNAUD PHENOMENON

(a) *Primary RP*. Capillary morphology is within normal range in this disorder and so is the capillary density. Capillary flow, as measured by RBC velocity, is lower than normal in baseline conditions and responds strongly to cold stimuli (local or whole body) leading to a flow stop of variable duration (measured in seconds or minutes, "complete" standstill or stop being defined differently by different investigators). During an RP attack, NF capillaries usually remain filled by RBC. Transcapillary diffusion of sodium fluorescein in NF capillaries is within normal limits.
(b) *Secondary RP*. There are a large number of disorders reportedly associated with RP, i.e., these disorders are considered primary, leading to a secondary RP. They are too numerous to review here and may or may not include microvascular studies. The most important group of secondary RP is found in connective tissue disease (CTD), especially in scleroderma (SD).

4. SCLERODERMA SPECTRUM

Scleroderma spectrum (SDS) disorders: Scleroderma (SD) and related disorders, such as CREST (calcinosis, RP, esophageal dysfunction, sclerodactyly, and telangiectasis), MCTD (mixed connective tissue disease), and UCTD (undifferentiated connective tissue disease), form a spectrum with many similarities in clinical and laboratory parameters including microvascular pathology. Dermatomyositis (DM, a closely related disease) shows similar but more acute, microvascular pathology. An overlap between SD and DM sometimes occurs.

Only in this disease group is the nail fold microvascular pathology highly specific and helpful for diagnosis. Enlarged NF capillaries have been described in other disorders, but what is characteristic here is the presence of capillaries in the NF that are enlarged in all three portions of the loop (arterial and venous limbs and the apex), resulting in a considerable increase in the total width of the loop (Figure 2, left). Other capillaries with less specific changes in size and shape may also be present.

These characteristic loops are located mainly along the edge of the NF but in an irregular manner: sometimes between quasi-normal-looking capillaries or in groups of variable size. The dimensions of these capillary loops vary considerably from slightly enlarged to giant loops; the size

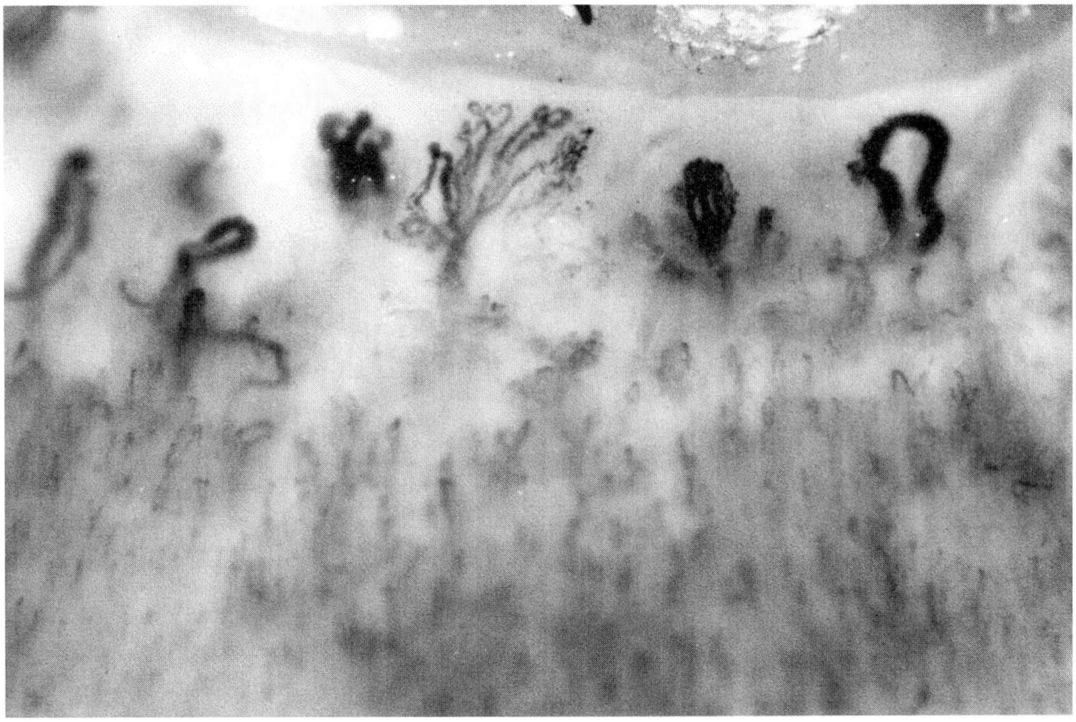

Figure 3 A "bushy formation" in a patient with dermatomyositis in the active phase of the disease.

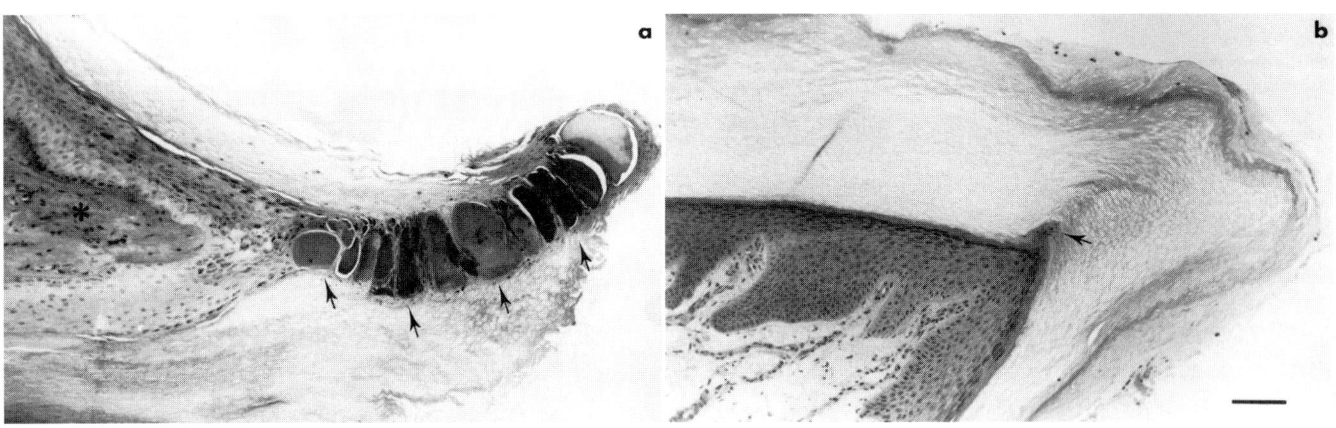

Figure 4 Cuticle deposits in a scleroderma patient compared to a normal control. (**a**) A scleroderma patient showing severe cuticle deposits (grade 6, arrows). (**b**) A normal control subject showing a grade 1 cuticle deposit (arrow), the most extreme encountered in the normal population. Bar = 100 μm.

and number of such loops vary within an NF or between NFs of the same patient. The number of capillaries per millimeter is greatly decreased because of their size, but also because they are often associated with small avascular areas along the NF edge, sometimes called dropout. However, capillary loss often extends into the proximal part of the NF and can be measured in mm^2. This characteristic pattern can be recognized even without measurements. (Pattern recognition is well known to be quite reliable.)

Other specific abnormalities, not always present but more often seen during an active phase of the disease, are "bushy formations" (Figure 3) adjacent to avascular areas (seen more often in DM than in SD).

They can be recognized even at low magnification. At higher magnification, they show many thin capillary loops springing from an arteriole or the deep end of the arterial limb. On serial photographs, they seem to be associated with filling in of an avascular area by capillaries [8]. Therefore, they are believed to be indicators of capillary regeneration. These "bushy formations" should not be confused with ramified or branched capillaries described by other authors in other disorders and even in normal subjects.

A yellow-orange discoloration of the cuticle is also associated with SD and DM. It corresponds to deposits of macromolecules (immunoglobulins, fibrinogens) in the cuticle demonstrated by biopsy (Figure 4 [6]) and reflects the permeability change of both the capillary wall and the dermoepidermal junction. In these cases, the capillary hemorrhages, often seen in SD and DM patients, do not remain self-limited when gradually growing out with the cuticle as in normal subjects, but the heme products diffuse widely throughout the cuticle.

Studies with video microscopy after injecting sodium fluorescein show an abnormally rapid diffusion of this dye, but not with a homogenous pattern as seen in long-term diabetes, but diffusing out through only certain parts of the capillary wall.

The capillary blood flow can be slow even at an average room temperature in the classical SD and has a striking response to local or whole-body cooling, leading to a complete standstill in the whole NF area at 16°C. This can actually be better appreciated by continued watching and scanning the field of NF than by automated systems. A complete stop of capillary flow can be easily recognized by standstill of all RBCs in an NF loop by observing it for at least 1 minute. When this is repeated in other loops and the whole endrow, as well as more proximal NF, scanned for any RBC movement for a total time of about 3 minutes, a conclusion of total shutoff the NF capillaries can be reached. This has not been observed in other vasospastic conditions, such as RP; its observed flow stop is of much shorter duration, intermittent, and not present in the whole NF.

CREST, a more benign variant of SD, shows similar NF capillary abnormalities but in a slightly different pattern: Giant loops at the edge of the NF are more prominent and numerous while avascular areas are quite limited or nonexistent. These patients also show characteristic capillary telangiectases elsewhere on the fingers, especially on the finger pads. These telangiectases can be microscopic, but in most cases many are large enough to be detected with the naked eye. "Bushy formations" are absent. Their capillary blood flow response to cold is similar to that of SD.

MCTD shows SD-pattern NF capillary abnormalities and more "bushy formations" than in classical SD.

UCTD also has SD-type NF capillary patterns, but usually fewer fingers are involved and show fewer abnormalities. There are some patients, however, with active disease who show marked loss of capillaries and rapid progression to definite diagnosis of SD. Both MCTD and UCTD can show capillary flow dysfunction when exposed to cold.

DM shows more extreme avascular areas and more numerous "bushy formations" than any of the conditions described earlier; the cuticular deposits and capillary hemorrhages with diffuse extension into the cuticle are also frequent. These lesions can vary from finger to finger; patients may show even completely normal NF capillaries in some fingers, and abnormalities can return to normal in remission [8].

5. OTHER CONNECTIVE TISSUE DISEASES

(a) *Rheumatoid arthritis*. An extensive visible subpapillary plexus may be present in a relatively large proportion of these patients.

(b) *Systemic lupus erythematosus*. Most patients with this disorder have NF capillaries within the normal range. In the acute phase, they may show a peculiar pattern of disappearance of all superficial (papillary) capillaries in most of the NF area, with only subpapillary venous plexus visible. The capillaries reappear in remission. SD–SLE overlaps also occur.

(c) *Discoid lupus erythematosus*. In this form of lupus, the proximal NF may be completely devoid of capillaries, showing only a prominent venular plexus; at the edge of the NF there are large looplike formations, which are probably part of the edge of the venular plexus rather than capillaries. These microvascular lesions are not present on all NFs and correspond most likely to discoid lesions occurring in the NF area.

6. HEREDITARY HEMORRHAGIC TELANGIECTASIA

This disease is mentioned here because numerous capillary telangiectases are present on skin, lips, fingers, and NF, which may sometimes need differential diagnosis with CREST.

7. SCHIZOPHRENIA

This disorder was studied by NF capillaroscopy soon after Müller's work. Many of these patients show an extensive subpapillary plexus on most of their fingers having a PVS of more than 10 (often 20 to 40 on a semiquantitative scale). This subgroup of schizophrenia also has a significantly lower finger blood flow and appear to correspond clinically to so-called process schizophrenics (those with negative symptoms). It is interesting to note that, although there is a higher than normal frequency of high PVS in both schizophrenia and RA, these disorders appear to be mutually exclusive.

Conclusions

The in vivo microvascular studies have been most fruitful clinically in SDS, not only for diagnosis and subgrouping of patients, but also in early differential diagnosis in RP subjects between primary RP and those at risk for SD. Over 90 percent of SD patients have RP, but RP usually precedes the diagnosis of SD by years. SD pattern NF capillary changes are present early and allow detection of patients at risk for SD. It has even been possible to pick out SDS patients from random samples from the general population "blindly" with the technique of in vivo microscopy and have independent confirmation of diagnosis later by clinical and laboratory examinations. This technique also allows distinction between the slowly developing CREST, acute SDS cases, and early SD with severe prognosis.

In schizophrenia research, NF capillaroscopy has been helpful to distinguish between two types of schizophrenia (those with negative versus positive symptoms) with possibly different etiopathology and prognosis.

Glossary

Capillaroscopy: In vivo microscopic examination of capillaries in the skin

Microvessels: Arterioles, capillaries, and venules

Nail fold: Skin fold surrounding the finger or toenail

References

1. Maricq, H. R. (1988). Nailfold capillary abnormalities in patients with connective tissue diseases. In: *Microcirculation in Circulatory Disorders* (Manabe, Zweifach, and Messmer, eds.), pp. 389–394, Springer-Verlag, Tokyo. *This reference was selected for the list because, to my knowledge, the nail fold has not been illustrated by a biopsy section before.*
2. Ehring and Schumann (1964). *Bibliotheca Anat.* **7**, 310–313.
3. Mahler, F., and Kneubühl, F. (1981) Application of fluorescein-tagged albumin in human nailfold capillaries. *Bibliotheca Anat.* **20**, 684–687.
4. Bollinger, A., and Fagrell, A. (1990). *Clinical Capillaroscopy*. Hogrefe and Huber.
5. Shore, A. C. (2000) Capillaroscopy and the measurement of capillary pressure. In: Research Methods in Human Cardiovascular Pharmacology (S. Maxwell and D. Webb, eds.), *Br. J. Pharmacol.* **50**, 501–513.
6. Thompson, R. P. et al. (1984). Nailfold biopsy in scleroderma and related disorders. *Arthritis Rheum.* **27**, 97–103. *To my knowledge, quantitative comparison of NF biopsy results (of the total width of the NF) and in vivo observations, including data on capillary density and cuticular deposits, in addition to histological sections and electron microscopy of capillaries (obtained in the same nail fold before biopsy in normal subjects, primary RP and SDS patients), has not been reported before.*
7. Maricq, H. R. (1982). The microcirculation in scleroderma and allied diseases. *Adv. Microcirc.* **10**, 17–52. *I found it necessary to add this article showing a specific "bushy formation" associated with capillary regeneration, not to be confused with bushy or branched capillaries also seen in other conditions and normal subjects.*
8. Maricq, H. R. (1992). Capillaroscopy in Raynaud's syndrome. *Vasc. Med. Rev.* **3**, 3–20. *This article includes serial studies of NF microvasculature in SD and DM, not often published.*

Capsule Biography

Dr. Hildegard R. Maricq has been involved with interdisciplinary research centering around her interest in microcirculation since 1961. She has combined it with other disciplines, such as genetics, epidemiology, psychophysiology, and peripheral vascular physiology. She has studied microcirculation and peripheral vascular circulation in a variety of medical disorders. Dr. Maricq has headed the Microvascular Laboratory at the Medical University of South Carolina since 1975. Now a Professor Emeritus, she is still active in international epidemiological studies that include microvascular components. She was nominated Master of the American College of Rheumatology in 2001, mainly for her work in microcirculation.

CHAPTER 28

Capillaries in the Chicken Chorioallantoic Membrane

[1]Domenico Ribatti and [2]Marco Presta

[1]Department of Human Anatomy and Histology, University of Bari

[2]Unit of General Pathology and Immunology, Department of Biomedical Sciences and Biotechnology, University of Brescia, Italy

The chick embryo chorioallantoic membrane (CAM) is an extraembryonic membrane. It serves as a gas exchange surface through the eggshell and its function is supported by a dense capillary network. Because of its extensive vascularization and easy accessibility, the CAM has been broadly used to study the morphofunctional aspects of the angiogenesis process in vivo and to investigate the efficacy and mechanisms of action of pro-angiogenic and anti-angiogenic natural and synthetic molecules. Also, because of the lack of a fully developed immunocompetence system in the chick embryo, the CAM represents a host tissue for tumor engrafting suitable to study various aspects of the angiogenic and metastatic potential that characterizes human malignancies.

Developmental Anatomy of the Blood Vessels of the CAM

The allantois of the chick embryo appears at about 3.5 days of incubation as an evagination from the ventral wall of the endodermal hind gut. During the fourth day, it pushes out of the body of the embryo into the extraembryonic coelom. Its proximal portion lies parallel and just caudal to the yolk sac. When the distal portion grows clear of the embryo it becomes enlarged. The narrow proximal portion is known as the allantoic stalk, the enlarged distal portion as the allantoic vesicle. Fluid accumulation distends the allantois so that its terminal portion resembles a balloon in complete embryos (Figure 1).

The allantoic vesicle enlarges very rapidly from days 4 to 10: an extensive morphometric investigation has shown rapid extension of the CAM surface from 6 cm^2 at day 6 to 65 cm^2 at day 14. In this process the mesodermal layer of the allantois fuses with the adjacent mesodermal layer of the chorion to form the CAM. An extremely rich vascular network connected to embryonic circulation by the allantoic arteries and veins develops between the two layers. Immature blood vessels scattered in the mesoderm and lacking a complete basal lamina and smooth muscle cells grow very rapidly until day 8, giving rise to a capillary plexus. The plexus associates with the overlaying chorionic epithelium and mediates gas exchange with the outer environment.

Capillary proliferation continues rapidly until day 10 (mitotic index equal to 23 percent). Then, the endothelial cell mitotic index declines rapidly to 2 percent and the vascular system attains its final arrangement on day 18, just before hatching. Besides sprouting angiogenesis that characterizes the early phases of CAM development, late CAM vascularization is supported by intussusceptive microvascular growth in which the capillary network increases its complexity and vascular surface by insertion of transcapillary pillars. At day 14, the capillary plexus is located at the surface of the ectoderm adjacent to the shell membrane. The blood circulation and the position of the allantois immediately adjacent to the porous shell confer a respiratory function to the highly vascularized CAM. In addition to the respiratory interchange of oxygen and carbon dioxide, the allantois also serves as reservoir for the waste products excreted by the embryo—mostly urea at first, and chiefly uric acid later.

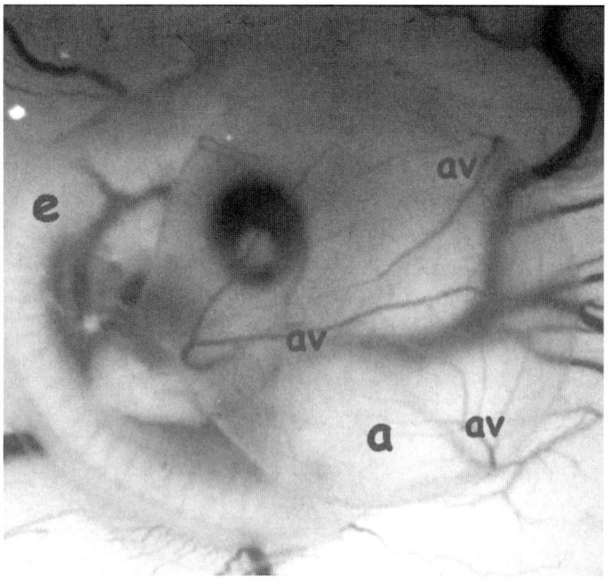

Figure 1 Allantoic sac (a) of a 5-day embryo (e) showing in ovo distribution pattern of allantoic vessels (av). Original magnification: ×25.

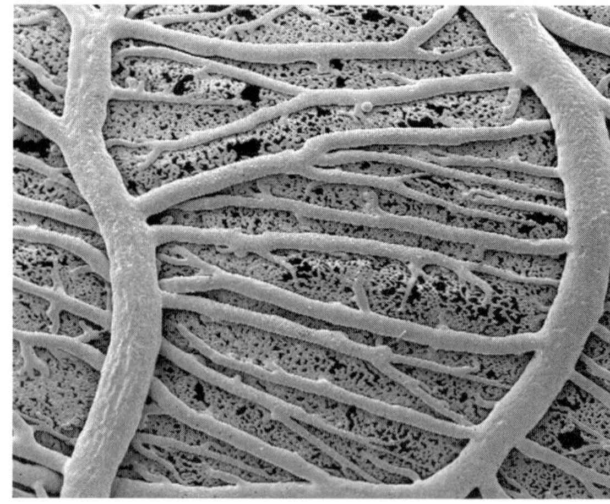

Figure 2 Image of the CAM arteriovenous and capillary systems after casting of the vasculature (provided by Dr. V. Djonov, University of Berne, Switzerland). Original magnification: ×100.

Structure of the Blood Vessels of the CAM

On day 4, all CAM vessels have the appearance of undifferentiated capillaries. Their walls consist of a single layer of endothelial cells lacking a basal lamina. By day 8, the CAM displays small (10 to 15 µm in diameter), thin-walled capillaries beneath the chorionic epithelium and larger vessels (10 to 115 µm in diameter) in the mesodermal layer. Large vessels have a layer of mesenchymal cells surrounding the endothelium and are completely wrapped by a basal lamina. Starting from day 12, capillaries contain endothelial cells and a few mesenchymal cells (presumptive pericytes) closely applied to the endothelial abluminal side. Desmin-positive cells are evenly distributed all over the CAM and are located in close association with the capillary plexus. The mesodermal vessels are now distinct arterioles and venules (Figure 2). In addition to the endothelium, the walls of arterioles (10 to 85 µm in diameter) contain one or two layers of mesenchymal cells and increased amounts of connective tissue. Venules (10 to 115 µm in diameter) are surrounded by an incomplete investment of mesenchymal cells. Mesenchymal cells are presumed to be developing smooth muscle cells, and the walls of arterioles also develop a distinct adventitia containing fibroblast-like cells.

At days 4 to 8 the endothelial cells form punctuate junctional appositions and a few plasmalemmal vesicles are observed. Between days 9 and 13 the arteriolar endothelium displays more extensive junctional complexes with multiple membrane contact points. In contrast to the arterioles, endothelial junctional appositions of the CAM venules remains punctuate. The venules possess multiple sites of interendothelial contact with areas of junctional dilations, while the arterioles display complex interdigitating cell junctions.

CAM arterioles and venules are accompanied by a pair of interconnected lymphatics. Veins are also associated with lymphatics and larger veins are surrounded by a lymphatic plexus. Lymph is drained by trunks of the umbilical stalk into the coccygeal lymphatics and the lymph hearts of the embryo. CAM lymphatic capillaries have no basal lamina and an extremely thin endothelial lining that specifically expresses vascular endothelial growth factor receptor-3 (VEGFR-3, Quek-2, flt-4) whereas VEGFR-2 (Quek-1, kdr, flk-1) is expressed by both vascular and lymphatic endothelial cells.

Endogenous CAM Modulators of Blood Vessel Formation

The development of the vascular system of the CAM is a complex, highly regulated process that depends on genetic and epigenetic factors expressed by endothelial and non-endothelial cells.

Endothelial cell interaction with extracellular matrix (ECM) plays an important role in angiogenesis. Accordingly, ECM modifies its properties during CAM development in terms of expression of several components (e.g. fibronectin, laminin, type IV collagen, Secreted Protein Acidic Rich Cysteine (SPARC)), distribution of specific glycosaminoglycans, and production of lytic enzymes.

Fibronectin accumulates in the ECM beneath the chorion at early stages of development when the subepithelial capillary plexus is not yet formed, possibly promoting the migration of endothelial cells merging by sprouting from the mesodermal blood vessels. In contrast, type IV collagen appears in the late stages of CAM vascular development concomitantly with the terminal differentiation of endothelial cells and maturation of basement membrane. Laminin immunoreactivity is present instead during all stages of CAM vessel formation, in keeping with its role in the early formation and later differentiation of the subendothelial basement membrane. Finally, ialuronic acid plays a crucial role in the formation, alignment, or migration of the capillary plexus, while heparan sulfate, chondroitin sulfate, and dermatan sulfate are important in the differentiation and development of arterial and venous vessels.

In spite of the evidence that several growth factors, such as fibroblast growth factor-2 (FGF-2), VEGF, placenta growth factor (PlGF), platelet-derived growth factor (PDGF), transforming growth factor beta-1 (TGF β-1), angiogenin, and erythropoietin (Epo), stimulate blood vessel formation when applied to the CAM (see later discussion), the role of their endogenous counterparts in the development of the CAM vascular system is poorly understood. The expression of VEGFRs suggest that, in analogy with other developmental models, endogenous VEGF may play a role in CAM vascularization. Accordingly, neutralizing anti-VEGF antibodies prevent CAM vessel development. Also, direct experimental evidences implicates endogenous FGF-2 in this process.

FGF-2 protein levels change in the CAM during development and peak between days 10 and 14. At early stages of development, FGF-2 is expressed by chorionic epithelial cells and may trigger a paracrine loop of stimulation by inducing an angiogenic response in undifferentiated vessels of the CAM mesoderm. At later stages, FGF-2 mRNA expression predominates in endothelial cells forming the capillary plexus, suggesting an autocrine function in late vessel growth and maintenance. Accordingly, neutralizing anti-FGF-2 antibodies fully prevent neovascularization when applied to the CAM at day 8 of incubation. They also decrease fibroblast density within the mesoderm, but do not affect epithelial cells of the chorion and allantois. Thus, FGF-2 plays a rate-limiting role in the maturation of stroma and blood vessels during CAM development.

The proteolytic plasmin/plasminogen activator system is also involved in CAM vascularization, possibly affecting the invasive behavior of sprouting endothelium and by modulating the activity of endogenous modulators. For instance, endogenous urokinase plasminogen activator (uPA) mediates the formation of angiogenic SPARC-derived peptides, whereas exogenous uPA can induce neovascularization by increasing the mobilization of endogenous FGF-2 from ECM-associated reservoirs.

CAM as a Model to Study the Modulation of Angiogenesis

A. Tumor Angiogenesis and Metastasis

The CAM is a suitable site for transplanting tissues because the chick embryo immunocompetence system is not fully developed and the conditions for rejection have not been established. Transplants survive and develop by peripheral anastomoses between graft and original CAM vasculature or by new angiogenic vessels grown from the CAM and invading the graft. The formation of peripheral anastomoses between host and preexisting donor vessels is the main, and the most common, mechanism involved in the revascularization of embryonic grafts. At variance, the growth of CAM-derived vessels into the graft is stimulated by tumor implants.

The CAM has long been a favored system for the study of tumor angiogenesis and metastasis. Tumor grafts remain avascular for 72 hours, after which they are penetrated by new blood vessels and a phase of rapid growth begins. The rate of growth during this vascular phase is greater for implants on days 5 and 6 to decrease at later days of implantation. The CAM may also be used to assess the ability of antiangiogenic molecules to inhibit growth and neovascularization of tumor xenografts.

Studies using the tumor cells/CAM model have also focused on the invasion of the chorionic epithelium and the blood vessels by tumor cells. The cells invade the epithelium and the mesenchymal connective tissue below, intravasate into the dense bed of blood vessels, and may invade the chick embryo.

Finally, delivery of tumor cells onto the CAM allows the fine study of the effects of tumor-derived angiogenic growth factors on blood vessel structure and functionality. For instance, tumor cell lines overexpressing FGF-2 or VEGF induce a quantitatively similar vasoproliferative response when grafted onto the CAM. However, in keeping with the different ability of FGF-2 and VEGF to modulate endothelial cell morphology and vascular permeability, an increased endothelial fenestration characterizes the blood vessels of the CAM stimulated by VEGF transfectants.

B. Molecules with Angiogenic and Anti-angiogenic Activity

The CAM is used to study molecules with angiogenic and anti-angiogenic activity following their delivery in ovo (Table I). Many protocols have been envisaged to deliver macromolecules and low-molecular-weight compounds onto the CAM by using silostatic rings, methylcellulose disks, silicon rings, filters, and plastic rings. Also, collagen and gelatin sponges treated with stimulators or inhibitors of blood vessel formation have been implanted on growing CAM. The gelatin sponge is also suitable for the delivery of cell suspensions onto the CAM surface and the evaluation of their angiogenic potential. This latter experimental condition allows the slow, continuous delivery of growth factors released by few implanted cells (20,000 cells per sponge or fewer). As compared with the application on the CAM of large amounts of a pure recombinant angiogenic cytokine in a single bolus, implants of cells overexpressing angiogenic cytokines enables the continuous delivery of growth factors, following a more "physiological" mode of interaction with the CAM vasculature.

Besides in ovo experimentation, a number of shell-less culture techniques have been devised, involving cultures of avian embryos with associated yolk and albumin outside

Table I Pro-Angiogenic and Anti-Angiogenic Molecules in the CAM Assay.

Pro-angiogenic molecules	Anti-angiogenic molecules
Fibroblast growth factor-2 (FGF-2)	Anti-FGF-2 antibodies
Vascular endothelial growth factor (VEGF)	Anti-VEGF antibodies
Placenta growth factor-1 (PlGF-1)	Anti-PlGF-1 antibodies
Angiogenin	Anti-angiogenin antibodies
Tumor necrosis factor alpha (TNF-α)	Interleukin-2 (IL-2)
Transforming growth factor beta (TGF-β)	Angiostatin
Platelet derived growth factor (PDGF)	Endostatin
Erythropoietin (Epo)	Heparan sulfate
Osteogenic protein-1 (OP-1)	Modified heparins
Leptin	Heparanase inhibitors
Adrenomedullin	Protamine sulfate
CC chemokine I-309	Platelet factor-4 (PF-4)
Osteopontin/Eta-1	Pentosan polysulfate
	Non- or low-sulfated saccharides
	Arylsulfatase inhibitors
	Sulfated polysaccharide peptidoglycan
	α-, β-, γ-cyclodextrin
	Suramin
	Spironolactone
	Tyrosine kinase receptor inhibitors
	Adhesion molecule antagonists
	Matrix metalloproteinase inhibitors

of the eggshell. Shell-less cultures facilitate experimental access and continuous observation of the growing embryo.

Typically, an angiogenic response occurs 72 to 96 hours after stimulation in the form of an increased density of vessels that converge radially toward the implant like spokes in a wheel. Conversely, when an angiostatic compound is tested, the vessels become less dense around the implant and eventually disappear (Figure 3). Alternatively, the molecules can be directly inoculated into the cavity of the allantoic vesicle so that their activity reaches the whole vascular area in a uniform manner.

Several semiquantitative and quantitative methods are used to evaluate the extent of vasoproliferative response or angiostatic activity at macroscopic and microscopic levels. Quantification of the CAM vasculature has been performed with the use of morphometric point-count methods, radiolabeled proline incorporation to measure collagenous protein synthesis, and fractal analysis of digital images.

Many techniques can be applied within the constraints of paraffin and plastic embedding, including histochemistry and immunohistochemistry. Electron microscopy can also be used in combination with light microscopy. Moreover, unfixed CAM can be utilized for biochemical studies, such as the determination of DNA, protein, and collagen content, and for reverse transcriptase–polymerase chain reaction (RT-PCR) analysis of gene expression. Recently, studies of intracellular signaling pathways mediating the angiogenic response to growth factors and cytokines have been successfully performed. This has allowed the demonstration of the role of extracellular signal-regulated kinases (ERKs) and $\alpha_v\beta_3$ integrin engagement in FGF-2-mediated angiogenesis and of JAK2/STAT-3 pathway during neovascularization induced by GM-CSF.

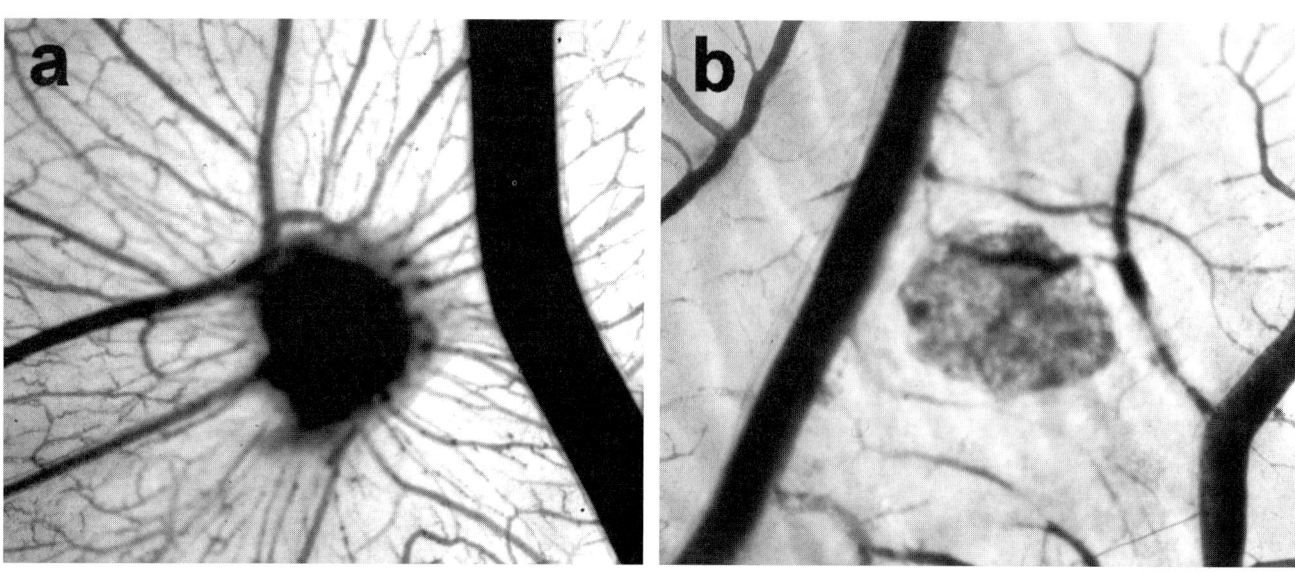

Figure 3 Modulation of CAM vascularization: numerous vessels develop radially in a "spoked-wheel" pattern toward the stimulus triggered by an angiogenic cytokine (**a**); in contrast, very few vessels are recognizable around the implant of an angiostatic molecule (**b**). Original magnification: ×50.

Concluding Remarks

One of the most important technical problems facing the study of angiogenesis and anti-angiogenesis is the difficulty of obtaining meaningful assessments of efficacy. In vivo angiogenesis assays, such as those performed in the CAM, have allowed important progress in elucidating the mechanisms of action of several angiogenic factors and inhibitors. Nevertheless, some limitations characterize this experimental model. Nonspecific inflammatory reactions may develop as a result of grafting, inducing a secondary vasoproliferative response hardly distinguishable from a direct angiogenic activity of the test material; rearrangement of existing vessels following membrane contraction may mimic neovascularization; and vasodilatation/vasoconstriction may result in apparent pro/anti-angiogenic effects, respectively (all these drawbacks may be overcome by detailed histological analysis). Finally, species-specific differences and the lack of avian-specific reagents (as well as limited genomic information) may represent serious disadvantages.

However, recent progress has contributed to improving the usefulness of the CAM as a model. For instance, retroviral, lentiviral, and adenoviral vectors have been used in the past few years to infect the CAM (as well as the whole chick embryo), leading to the expression of the viral transgene. This allows the long-lasting presence of the gene product that is expressed directly by CAM cells, which makes feasible the study of the effects of intracellular or membrane-bound proteins as well as of dominant-negative gene products. Such an approach will shed new light on the mechanisms of blood vessel formation and inhibition in physiologic and pathologic conditions.

Glossary

Angiogenesis: The formation of new blood microvessels from preexisting ones.

Chorioallantoic membrane: Extraembryonic membrane originated by the fusion of the mesodermal layer of the allantois with the adjacent mesodermal layer of the chorion. It mediates gas exchange between the avian embryo and the outer environment.

FGF-2: A pleiotropic growth factor endowed with a potent angiogenic activity in vitro and in vivo. It interacts with signaling tyrosine kinase receptors and heparan-sulfate proteoglycans on target cells.

VEGF: Endothelial cell growth factor crucial for embryonic vasculogenesis and angiogenesis. It acts also as a potent vasopermeability factor. It is involved in physiologic and pathologic angiogenesis.

Further Reading

Ausprunk, D. H., Knighton, D. R., and Folkman, J. (1974). Differentiation of the vascular endothelium in the chick chorioallantois: A structural and autoradiographic study. *Dev. Biol.* **38**, 237–247. *This fundamental study describes the morphological characteristics of the CAM vasculature.*

Djonov, V. G., Galli, A. B., and Burri, P. H. (2000). Intussusceptive arborization contributes to vascular tree formation in the chick chorioallantoic membrane. *Anat. Embryol.* **202**, 347–353.

Elicieri, B. P., Klemke, R., Stromblad, S., and Cheresh, D. A. (1998). Integrin $\alpha_v\beta_3$ requirement for sustained mitogen-activated protein kinase activity during angiogenesis. *J. Cell Biol.* **140**, 1255–1263. *Integrin/growth factor receptor-mediated signaling is investigated during angiogenesis in the CAM.*

Hagedorn, M., Javerzat, S., Gilges, D., Meyre, A., de Lafarge, B., Eichmann, A., and Bikfalvi. A. (2005). Accessing key steps of human tumor progression in vivo by using an avian model. *Proc. Natl. Acad. Sci. U. S. A.* **102**, 1643–1648.

Knighton, D., Ausprunk, D., Tapper, D., and Folkman, J. (1977). Avascular and vascular phases of tumour growth in the chick embryo. *Br. J. Cancer* **35**, 347–356.

Papoutsi, M., Tomarev, S. I., Eichmann, A., Prols, F., Christ, B., and Wilting, J. (2001). Endogenous origin of the lymphatics in the avian chorioallantoic membrane. *Dev. Dyn.* **222**, 238–251.

Ribatti, D., Gualandris, A., Bastaki, M., Vacca, A., Iurlaro, M., Roncali, L., and Presta, M. (1997). New model for the study of angiogenesis and antiangiogenesis in the chick embryo chorioallantoic membrane: The gelatin sponge/chorioallantoic membrane assay. *J. Vasc. Res.* **34**, 455–463.

Ribatti, D., Urbinati, C., Nico, B., Rusnati, M., Roncali, L., and Presta, M. (1995). Endogenous basic fibroblast growth factor is implicated in the vascularization of the chick embryo chorioallantoic membrane. *Dev. Biol.* **170**, 39–49. *This paper describes for the first time the role of an endogenous growth factor in CAM vascularization during development.*

Ribatti, D., Vacca, A., Roncali, L., and Dammacco, F. (2000). The chick embryo chorioallantoic membrane as a model for *in vivo* research on anti-angiogenesis. *Curr. Pharmacol. Biotechnol.* **1**, 73–82.

Scher, C., Haudenschild, C., and Klagsbrun, M. (1976). The chick chorioallantoic membrane as a model system for the study of tissue invasion by viral transformed cells. *Cell* **8**, 373–382.

Schughart, K., and Accart, N. (2003). Use of adenovirus vectors for functional gene analysis in the chicken chorioallantoic membrane. *Biotechniques* **34**, 178–183.

Capsule Biography

Dr. Ribatti is Full Professor in Human Anatomy at the School of Medicine of the University of Bari. Researches in his laboratory primarily focus on the study of angiogenesis during embryonic development and tumor progression.

Dr. Presta is Full Professor in General Pathology and has headed the Angiogenesis Laboratory at the School of Medicine of the University of Brescia since 1990. Researches in his laboratory primarily focus on the mechanisms of action of angiogenic growth factors and the development of angiogenesis inhibitors.

His work has been supported in part by grants from the European Union, AIRC, and MIUR.

CHAPTER 29

The Cremaster Muscle as a Microvascular Research Model

Maria Siemionow and Andrea Moreira-Gonzalez
Department of Plastic Surgery, The Cleveland Clinic Foundation, Cleveland, Ohio

Background

Despite major advances in the field of reconstructive microsurgery, complications and failures after free tissue transfer and replantation surgery remain as a significant clinical problem. The nature of these failures is often multifactorial and the pathophysiology is still not completely understood. Different research approaches and animal models were developed to improve our understanding of changes occurring at the tissue microvascular level [1].

Previously, most studies applied techniques that allowed quantifying microcirculatory perfusion by implementation of the indirect methods of measurement, such as the use of staining techniques. These methods, however, did not reveal accurate information about the dynamics of the microvascular network, which is in a constant state of change. Research discoveries improved significantly after application of modern microcirculatory techniques and models that allowed for direct in vivo observation of hemodynamics of vascular pathophysiology. These models included hamster cheek pouch, rabbit ear chamber, the cat tenuissimus muscle, bat wing, and various mesenteric preparations [1].

The rat cremaster muscle preparation has a relatively recent history of application to microcirculatory studies [2–4]. One of its advantages is that the cremaster muscle flap is structurally and functionally similar to other skeletal muscles. Thus, all changes observed in the microcirculation of this small flap could be applied to larger muscle flaps used in clinical microsurgery. Majno and Palade were the first to introduce the cremaster muscle model in their histological and electron microscopy studies on inflammation [2]. This model was accepted for microcirculatory observation by in vivo microscopy after modifications made by Grant [3]. Baez was the first to open the cremaster muscle and to detach the testicle from the flap [4]. He was then able to spread out the flap and transilluminate it under a microscope. Miller and Wiegman introduced a tissue bath to provide better control of the microcirculatory environment [5]. This standard cremaster preparation was further developed in our laboratory in such a way that the cremaster muscle was completely isolated on its neurovascular pedicle, as an island flap [6]. In this model the cremaster muscle circulation is solely dependent on one feeding artery and one outflowing vein, excluding all other sources of vascular supply. This procedure simulates the conditions encountered during surgery of free tissue transfers, where the main vessels of the flap undergo microsurgical anastomosis and changes occurring downstream from the site of vascular repair can be monitored. These microcirculatory observations can add to our understanding of variables that can directly affect flap failure and survival.

The cremaster muscle flap isolated on the neurovascular pedicle expands significantly our ability to study microcirculation in relation to (1) reactivity of the microvessels [1, 6], (2) ischemia-reperfusion injury [7], (3) assessment of microcirculatory responses to muscle denervation [8], (4) chronic observation of the microcirculation using a chamber model [7], (5) observation of the microvascular events following tumor implantation, and (6) monitoring of microvascular responses to isograft and allograft transplantation [9, 10].

Surgical Technique

Male rats weighing between 80 and 150 g are used, because at this weight range the cremaster fascia is thin enough to allow for optimal visualization of the microcirculation of the transilluminated muscle. Also, the iliac artery is large enough to allow for controlled dissection, transection, and repair under surgical microscope magnification.

Anesthesia is induced with intraperitoneal administration of pentobarbital (60 to 70 mg/kg) with additional supplements as needed. Core body temperature is maintained between 35° and 37°C. The animal's genital area and thighs are shaved; the rat is prepped and placed in a supine position.

Isolation of the Cremaster Muscle Flap

During flap dissection the tissues should be kept moist using saline solution. An operating microscope (Zeiss OPM6-SD, Zeiss, Germany) is used for pedicle dissection and isolation. A longitudinal incision is made from the tip of the scrotum to its base. From this point the incision is carried laterally toward the anterior iliac spine. The subcutaneous tissue over the groin area and the scrotum is then carefully dissected. The cremaster is visualized as a thin tissue layer covering the testis. This dissection is continued carefully on the ventral as well as dorsal surface of the muscle until the muscle is completely detached from the scrotum. A traction suture is placed through the distal tip of the cremaster muscle to facilitate atraumatic dissection (Figure 1A). The pedicle of the flap is visualized on the dorsal aspect of the muscle. A longitudinal incision is then made on the ventral surface of the cremaster muscle flap origin including the full thickness of the muscles of the abdominal wall and is directed parallel to the inguinal ligament (IL). The vascular pedicle of the testicle and the deferens duct are exposed and ligated with 5-0 silk sutures. The testicle is pulled from the scrotum and is carefully separated from the muscle. Next the testicle is removed, and the empty cremaster muscle tube flap is now ready for dissection of the neurovascular pedicle.

Isolation of the Neurovascular Pedicle

The dissection is performed in the groin area, with exposure of the femoral artery and vein and the IL. The medial insertion of the IL is then divided from the pubic bone attachment, and the ligament is mobilized from medial to lateral, exposing the genito-femoral nerve, and the external iliac artery and vein (EIA and EIV). All nonrelevant branches of the EIA are either coagulated or ligated with 9-0 nylon sutures, leaving the intact pudic-epigastric trunk, which now provides the only blood supply to the cremaster muscle flap. The frontal wall of the cremaster muscle tube flap is then opened from proximal to distal. The remaining attachments connecting the flap to the groin region are transected and the cremaster muscle is totally isolated on its vascular pedicle, as an island muscle flap.

Preparation of the Cremaster Flap for Microcirculatory Monitoring

The animal is placed on a Plexiglas observation platform. The prepared cremaster muscle is extended and is held in place with 5-0 silk sutures (Figure 1B). The muscle is irrigated with saline solution and covered with a plastic film to avoid dehydration (Saran Wrap presoaked with distilled water overnight). Figure 1C shows the schematic map of the microvessels that can be monitored in this preparation.

Direct In Vivo Microcirculatory Observations

The cremaster muscle is then transilluminated from below and the microcirculation is observed and monitored using an intravital microscope (Nikon Optiphot-2, Japan) equipped with a color video camera (Sony CCD-IRIS, Japan), a 19-inch monitor (Sony Trinitron, Japan), and a videotape recorder (Panasonic AG-1730, Japan). Using this standardized setup, the following microcirculatory parameters can be directly observed, measured, and recorded on the videotape over time for repeated observations (Figure 2):

1. *Vessel diameter*. The diameter and the endothelial wall thickness of the cremaster arterioles and venules are measured by using a video image measurement system (VIA-150, Boeckeler, Tucson, AZ).
2. *Red blood cell (RBC) velocity*. A custom-made optical Doppler velocimeter (Texas A&M, College Station, TX) is adapted to the system. It allows the measurement of RBC velocity in millimeters per second in the main arterioles and venules of the cremaster muscle flap.
3. *Capillary density*. The capillary perfusion or functional capillary density is defined as the number of capillaries in which RBCs are constantly flowing. Three regions of the cremaster muscle flap with a good capillary perfusion and clear visualization are chosen, in the proximal, medial, and distal area of the flap. Each visual field—as seen in the monitor—represents an area of tissue of $0.18\,mm^2$. Nine fields in each of three flap regions are monitored and counted for a total of 27 fields, and the average number from all fields represents the overall state of capillary perfusion of the cremaster muscle microcirculation.
4. *Leukocyte–endothelial interactions*. The postcapillary venules are chosen for the observation of leukocyte behavior. These vessels have a diameter of 20 to 40 μm and are identified in the proximal, medial, and distal regions of the flap. Rolling, "sticking" (remaining stationary for more than 20 seconds), transmigrating neutrophils and lymphocytes are observed and counted during a 2-minute period in each flap region using a hand counter (Figure 2).
5. *Endothelial-edema index*. The external and internal diameters of the postcapillary venules are measured using the VIA-150. The ratio between the two numbers is defined as endothelial edema index and is used in some studies to indicate vascular occlusion caused by leukocyte aggregation within the venular lumen, causing vessel wall damage, and microvascular injury leading to increased tissue permeability.

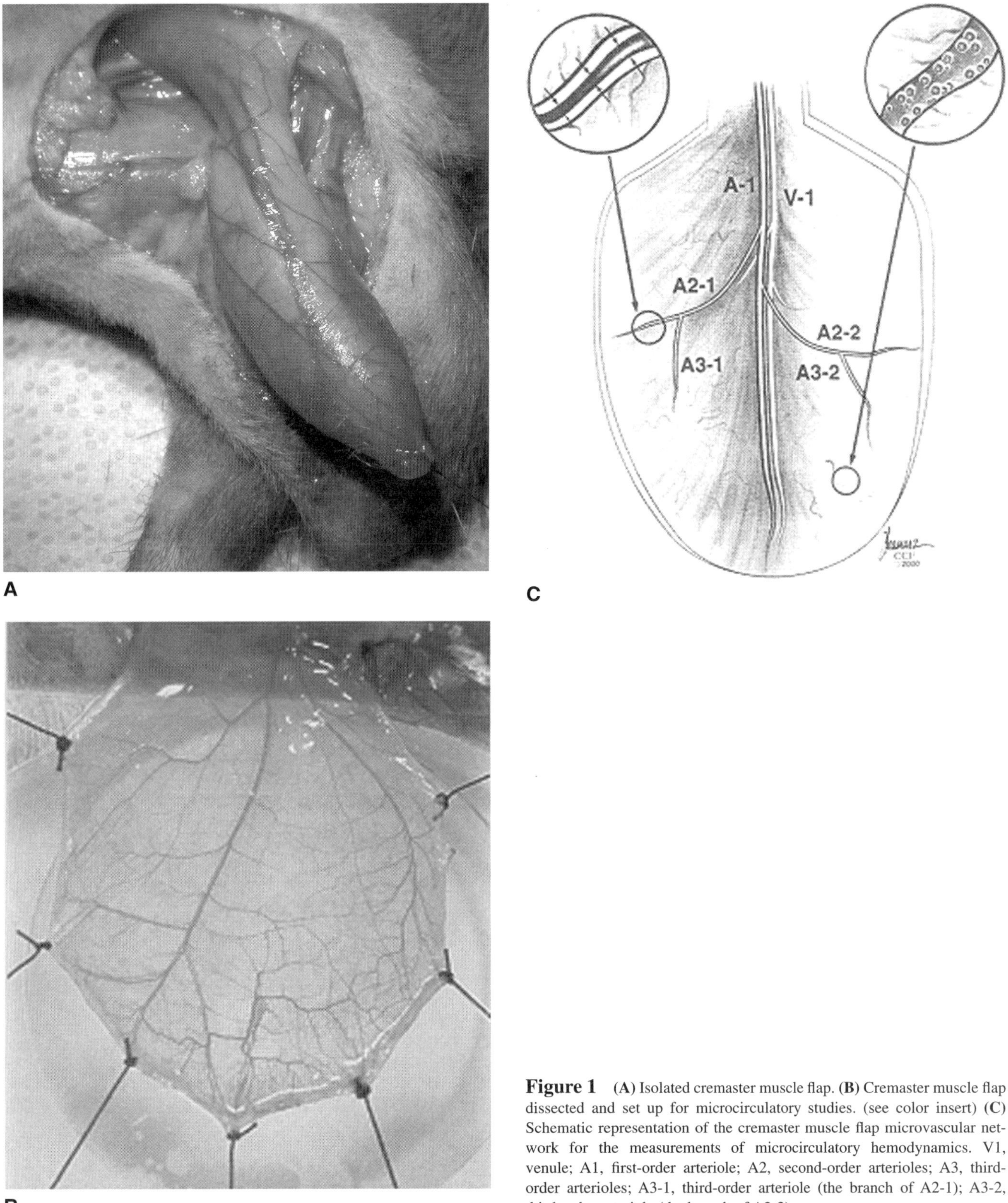

Figure 1 **(A)** Isolated cremaster muscle flap. **(B)** Cremaster muscle flap dissected and set up for microcirculatory studies. (see color insert) **(C)** Schematic representation of the cremaster muscle flap microvascular network for the measurements of microcirculatory hemodynamics. V1, venule; A1, first-order arteriole; A2, second-order arterioles; A3, third-order arterioles; A3-1, third-order arteriole (the branch of A2-1); A3-2, third-order arteriole (the branch of A2-2).

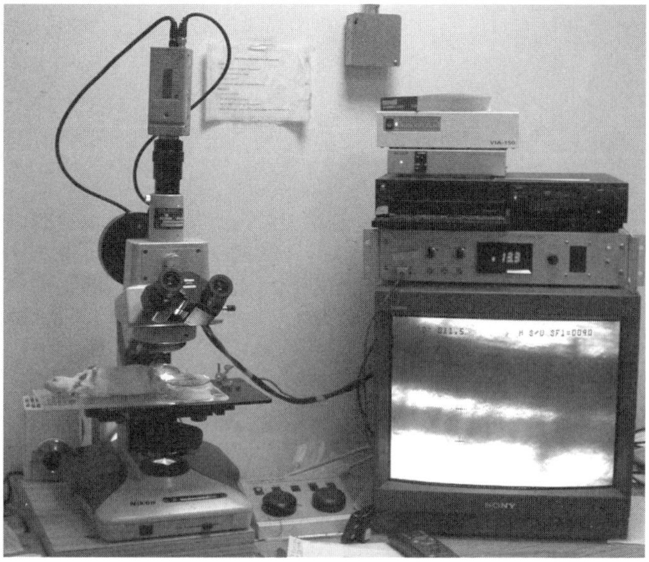

Figure 2 Intravital microscopy setup for microcirculatory monitoring. (see color insert)

6. *Thrombus formation.* The transilluminated cremaster muscle can be used to directly visualize and quantify thrombus formation. The pedicle artery is viewed with an inclinable head stereoscope (Zeiss OpMi6, Zeiss, Germany) and at the same time the downstream cremaster is observed with a compound microscope. By incorporating the two systems it is possible to image the upstream and the downstream microcirculation side by side on the same video monitor, simultaneously. This allows the correlation of thrombus formation upstream with the appearance of the dislodged platelet emboli in the downstream microcirculation.

Evaluation of the Microvascular Permeability

1. *Fluorescein microscopy setup.* A Zeiss 20T fluorescein microscope (Carl Zeiss, Germany) is used. An illumination system with a 460 to 490 nm blue light band excitation filter from a mercury arc lamp (Mercury Power Supply, Model 1200, OpticQuip Inc, NY) is used to stimulate the fluorescein isothiocyanate. The closed circuit video recording system consists of an MTI silicon-intensifying target camera (MTT SIT-68), a Panasonic AG-1290 video recorder, and a 19-inch monitor. The gain, sensitivity, and black level of the SIT camera allow the detection of very low light levels.
2. *Microvascular permeability assessment technique and analysis of macromolecule leakage.* Macromolecular leakage reflects vascular permeability. Albumin conjugated with fluorescein isothiocyanate (FITC) is injected intravenously to assess the microcirculatory permeability in the cremaster muscle flap. Postcapillary venules are selected for measurements, which are taken at zero time and after 30 minutes. The albumin appears as bright white on the video monitor. Venules without leaks look like white columns against a black background. Leaks appear as white streaks outside of the vessels. These images are digitized using a Kontron Elektronik KS 300 V2.00 image analysis system (Kontron Elektronik GmbH, Eching, Germany). Three segments in the lumen of the venule and of the interstitium are selected. An index of vascular leakage (permeability index, PI) is computed from the images. A relatively high positive PI indicates increased leakage from the postcapillary venule as result of endothelial injury. If the Kontron system is not available, the live images recorded can be captured by the computer using the Studio DC Plus software (version 3.1, Pinnacle System, Mountain View, CA) and a capture card. The images are then analyzed with the Image Pro-Plus software (version 2.1, Media Cybernetics, Silver Spring, MD).

Application of the Cremaster Muscle Island Flap for Ischemia-Reperfusion and Related Studies

There has been an increased awareness and interest in reperfusion injury or the no-reflow phenomenon. Unfortunately, the pathogenesis of this phenomenon is still not completely understood. Many hypotheses have been postulated, such as vasospasm, oxygen free radicals, platelet-plugging microthrombus formation, and changes in the vascular permeability leading to the interstitial edema. The direct observation of these highly dynamic mechanisms is important since it may explain many of the clinical events happening during postoperative follow-up in many microvascular and vascular surgery procedures.

All important parameters involved in ischemia/reperfusion injury can be evaluated using the cremaster muscle flap. They include measurements of vasospasm, vessel diameters, thrombus formation, presence of microembolism, functional capillary density, leukocytes–endothelial interaction, and microvascular leakage. Furthermore, the cremaster chamber model allows for chronic observation of the microcirculatory events, which is critical for monitoring consequences of ischemia/reperfusion.

Technique. The cremaster muscle flap dissection described earlier can be modified for application to the ischemia/reperfusion injury studies.

In order to match the events of reperfusion injury associated with the vasospastic phenomenon and decrease in capillary perfusion, the cremaster flap has been applied as a microsurgery ischemia model. Technically, the cremaster vascular pedicle is transected and re-anastomosed, and the flap is further assessed by microcirculatory monitoring An important observation from this ischemia model was the description of concept of the cremaster flap risk zones by simultaneous assessment of the site of vascular anastomosis and peripheral microcirculation. The zones were divided into the upstream risk zone (anastomotic site) and a downstream risk zone (the microcirculation) [1]. The events monitored within these zones were directly related: A process occurring within the upstream zone such as thrombus formation was influencing events at the downstream zone, causing decrease in capillary perfusion. Other observations from this model have shown that the ischemia time that is the period needed to perform the microvascular anastomosis induced a profound vasospasm peripherally, specifically in the third-order arterioles of the cremaster flap. Some authors suggested that this mechanism is most probably related to microcirculatory response to anesthetic drugs, poor microsurgical technique of vessel repair, and tissue hypoxia and acidosis.

The cremaster flap can be also used as a clamp-ischemia model. In this model homodynamic changes occurring at the microcirculatory level following clamp application to the flap feeding artery can be compared to the ischemia-induced events during free tissue transfer procedures. The technique involves clamping the cremaster vascular pedicle for a different periods ranging from 2 to 6 hours with subsequent monitoring of the microcirculatory conditions. This was specifically of interest in studies related to ischemic preconditioning and its effects on the microcirculation. This model is also applicable for simulation of reduced flow states, in which instead of clamp application a silk suture is placed around the pedicle, reducing blood flow to the flap.

The response of leukocyte–endothelial interactions to ischemia/reperfusion injury has been extensively studied. Upon reperfusion, a cascade of events takes place, leading to activation, adherence, and transmigration of the leukocytes, ultimately resulting in decreased blood flow, increased capillary permeability, and subsequent tissue damage and microcirculatory failure. Substances such as anti-ICAM and anti-PECAM antibodies, growth factors, interferon gamma, dehydroepiandrosterone, and cyclosporin and many more have been tested in our laboratory in different experimental scenarios of tissue ischemia, reperfusion, denervation, and preconditioning. The ability to reduce the effects of the ischemia/reperfusion injury after administration of different pharmacological agents has opened new horizons in the field of reconstructive microsurgery leading to better understanding and improvement of the free tissue transfer flap survival.

Application of the Cremaster Muscle for Chronic Observation of the Microcirculation

Previously, the cremaster muscle preparation was applied only for acute microcirculatory observation lasting no longer than 6 hours [1]. However, it is well known that some of the events altering the flow hemodynamics in the flap feeding artery and peripheral microcirculation of transferred tissues can occur later most often within 72 hours and no longer than 7 days after flap transfer. Therefore, the cremaster muscle flap model was modified to allow for observation of the microcirculation for prolonged periods of time, still maintaining the accuracy of the measurements [7]. Two models for chronic observation were developed.

The Cremaster Muscle Chamber for Continuous Observation of the Microcirculation

The main advantage of the cremaster chamber application is the ability of repeated measurements of hemodynamic changes within the same muscle flap for a period of up to 3 days. This period is accepted as critical time for flap survival or failure after microvascular repair [7]. After 3 days this model is less applicable since neovascularization from the surrounding tissues takes place and can obscure visibility under intravital microscopy. Also, with prolonged chamber application there is a possibility of the cremaster pedicle stretching or damage.

Technique. The cremaster muscle island flap is dissected and spread on the bottom of the Plexiglas chamber. The muscle is bathed in saline solution, then covered with a round cover glass and with the top frame of the chamber. This keeps the flap moist and free from atmospheric oxygen throughout the experiment. The base of the chamber is fixed with a suture to the body of the animal along the inguinal canal (Figure 3). Observations start at 30 minutes following chamber application to eliminate the immediate effects of surgical trauma on the flap microcirculatory response. Hemodynamic measurements can be taken at 1-hour intervals for the first 24 hours and as needed thereafter up to 3 days after chamber application. Extension of time between the measurement intervals permits the rats to recover from the anesthesia.

In our experience, the chamber is well tolerated by the animals, and the cremaster muscle preparation remains viable during the entire observational period. There is no significant pressure exerted by the use of the chamber, as evidenced by the flow measurements. Furthermore, the

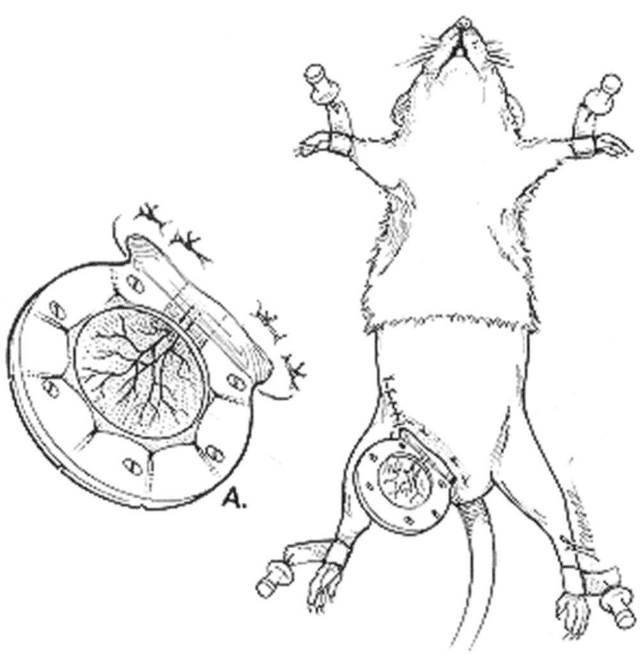

Figure 3 Schematic drawing of the cremaster muscle flap chamber model.

chamber model allowed to perform microcirculatory measurements comparable with standard cremaster muscle island and tube flap model recordings [7].

The Cremaster Muscle Tube Flap Model for Chronic Observation of the Microcirculation

If even longer periods of observation are necessary for the understanding of the microcirculatory mechanisms, the cremaster muscle tube flap is used. With this model the microcirculatory events can be observed for periods up to 14 days.

Technique. The cremaster is dissected as described before. However, after removal of the testis the empty cremaster muscle tube raised on the neurovascular pedicle is preserved (Figure 4). Next a subcutaneous tunnel is created in the anteromedial aspect of the hind limb. The flap is buried in this tunnel, secured distally with a pull-out suture, and kept in the tunnel until the day of microcirculatory measurements. At this time the flap is opened, spread over the top of the Plexigas platform, and set up for the measurements. This tube flap model is well tolerated by the animal, and the muscle viability is not affected by the staged surgical procedure required for monitoring.

This model has been successfully applied in research requiring longer periods of observation [7], such as studies on the long-term effects of flap denervation, evaluation of hemodynamic effects of pharmacological agents, and studies monitoring acute phases of allograft rejection.

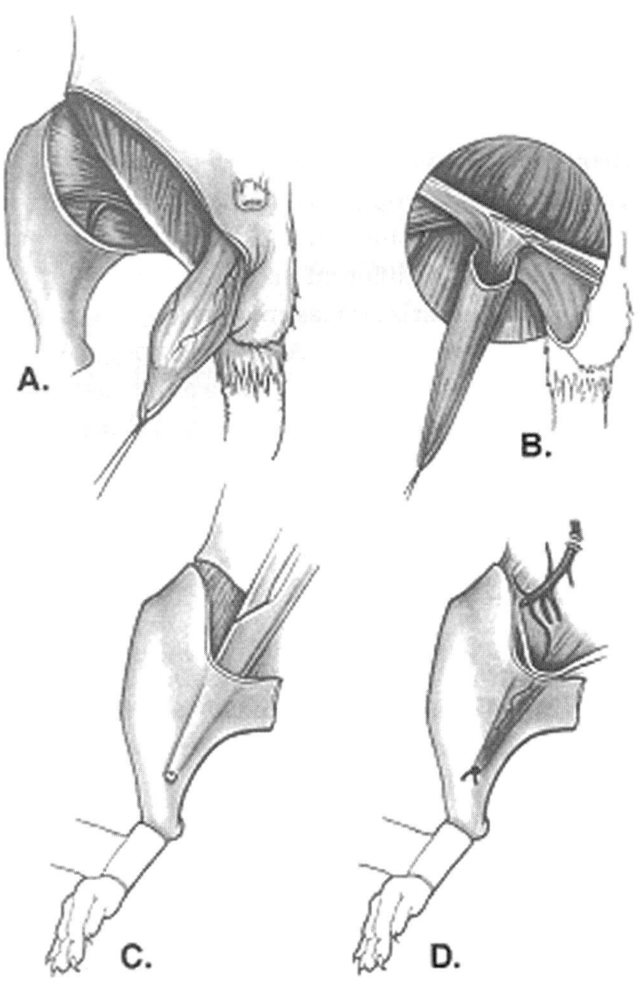

Figure 4 Steps in the hind limb–cremaster transplantation procedure: **(A)** Extraction of the testis from the cremaster muscle, **(B)** dissection of the cremaster tube flap on the neurovascular pedicle, **(C)** preparation of the subcutaneous tunnel for the preservation of the muscle flap for chronic observations, **(D)** insertion of the muscle flap into the tunnel, transection of the vascular pedicle at the iliac level, and limb amputation at the mid-femoral level.

Microcirculatory Responses to Cremaster Muscle Denervation

Since the cremaster muscle flap is an isolated axial pattern flap in which all collateral vessels are transected, it is the ideal model to study the effects of muscle denervation including influence of neurogenic factors on the microcirculatory response. Applying this model it is also possible to test the microvascular response to the topical application of the vasoactive substances. In such scenarios the cremaster muscle flap can be applied as an acute or as a chronic model.

1. *Technique—acute model*. After cremaster muscle flap dissection on the neurovascular pedicle, the denervation procedure is performed by resecting a 1-cm segment of the genitofemoral nerve in the groin area leading to flap sympathectomy. Next, the adventitia of both the iliac and femoral artery and vein are excised circumferentially just above and below the pudic-epigastric pedicle resulting in somatic denervation of the flap. The flap is then prepared for microcirculatory evaluation.
2. *Technique—chronic models*. We have developed chronic models for long-term microcirculatory recordings. The cremaster muscle tube flap could be modified by the two-stage denervation procedure. In the first stage an incision is made in the groin, and the iliac and femoral vessels and genitofemoral nerve are exposed. Next, the standard denervation procedure of 1 cm of nerve segment resection and excision of the femoral artery and vein adventitia is performed. The flap denervation can be carried out from 24 hours up to 14 days. At this point the second-stage procedure of standard surgical dissection of the cremaster muscle flap on the tissue bath is performed and microcirculatory measurements revealing the effects of flap denervation are recorded. As a result of these observations a triphasic sequence was observed, starting with (1) an acute hyperadrenergic phase, followed by (2) a non-adrenergic phase (vasodilatation) and (3) a sensitized phase where increased capillary perfusion and hyper-responsiveness to vasoactive substances was observed. Based on these observations it was evident that microcirculation can be significantly improved by muscle sympathectomy and somatic denervation [8].

A Model of Cremaster Muscle Isograft and Allograft Transplant

After transplantation, two patterns of tissue damage can occur during allograft rejection. One is caused by cell-mediated or antibody-mediated damage. The second is due to the tissue ischemia and surgical trauma causing microvascular destruction [9]. Since these patterns are dependent on the interactions between the allograft endothelium as well as allograft and recipient immune systems, a quantitative assessment of the microvascular events occurring during allograft rejection is crucial. The cremaster muscle flap was introduced to transplantation studies to serve as a "window" for direct microcirculatory measurements of the hemodynamic events occurring due to ischemia/reperfusion injury and the acute phase of allograft rejection following muscle transplantation. The cremaster tube flap can be transplanted alone as a free flap, or it can be incorporated into the donor's limb as a part of a composite tissue graft. This allows a monitoring sequence and timing of cremaster muscle allograft transplant rejection at the microcirculatory level. Further, the advantages of this cremaster flap preparation include the possibility of comparing differences between hemodynamic and cellular responses among the transplant donor and transplant graft recipient and to investigate the changes occurring within the intra- and extravascular compartments.

1. *Technique of the cremaster muscle free flap transplantation model.* The cremaster muscle tube flap is designed based on the iliac and femoral vessels, which are transected above the cremaster vascular pedicle take-off. The dissected cremaster muscle free flap is next transplanted to the recipient animal, and the pedicle is anastomosed to the femoral vessels of the recipient. Next a tunnel is created in the medial border of the hind limb, into which the flap is inserted for long-term observations. The transplanted tube flap allograft is then removed from the limb, opened, and spread on the tissue bath for the microcirculatory observation of allograft acceptance/rejection at the designated times.
2. *Technique of the combined hind limb–cremaster muscle transplantation.* This model allows study of microcirculatory response to composite tissue allografts across the MHC barrier. The rat hind limb is dissected together with the cremaster muscle, which is incorporated as a tube flap into the medial border of the limb. Transplantation from donor to recipient involves the four-step procedure. First, the hind limb of the recipient animal is amputated [10], followed by preporation of the hind limb–cremaster graft of the donor animal and preparation of the cremaster muscle tube flap implantation for chronic observations, as outlined in Figure 4. Next, the iliac vessels are exposed, ligated, and transected. The hind limb is then amputated. The hind limb–cremaster composite tissue graft is transplanted into the designated recipient (iso- or allograft), and the cremaster muscle tube graft is dissected and prepared for microcirculatory measurements.

The cremaster flap has proven to be a reliable tool for monitoring microcirculatory hemodynamics during composite tissue graft rejection. At the microcirculatory level the rejection process after allotransplantation begins as early as 24 hours posttransplant. It is characterized by a decrease of capillary perfusion and marked increase in the activation of adherent and transmigrating leukocytes, proceeding clinical the signs of allograft rejection [9, 10]. These observations were significant and added to our understanding of composite tissue allograft rejection as well as allowing us to distinguish between events occurring as a result of the transplantation trauma and rejection phenomenon.

Cremaster Muscle Model for Tumor Implantation

Since the cremaster muscle flap is a well-vascularized tissue it is applicable to solid tumor implantation. Following implantation, tumor growth and angiogenesis, along with molecular transport within the tumor and the delivery of genetic vectors and anticancer therapeutic agents, can be studied. Also, this model allows direct control of the inflow and outflow of the substances into the cremaster microcirculation, which is important during evaluation of tumor biology and drug testing.

Technique. Through an incision on the anterior surface of the exposed cremaster muscle, the testicle pedicle and spermatic cord are isolated and dissected. The testis is removed from the cremaster pouch and retracted into the abdominal cavity. The tumor cell suspension is then injected directly into the cremaster muscle. The incision is then closed, and the cremaster tube flap is implanted into the medial border of the limb. It is withdrawn from the limb only at the time of microcirculatory measurements. Intraarterial injection of substances of interest is done directly into the pudic-epigastric pedicle after clamping of the iliac and femoral vessels.

To summarize, the cremaster muscle flap model has been proven to be a novel model for cancer research with broad applicability and with specific value for the study of tumor biology and angiogenesis.

Glossary

Blood rheology: The study of the flow and deformation of red blood cells.

Cremaster muscle: Thin (200 microns) skeletal muscle covering testicles. Innervated by the genitofemoral nerve and vascular supply provided by pudic-epigastric vessels. Because of its thin muscle layers, the cremaster may be transilluminated and used for intravital microscopic observations.

Hemodynamics: The study of physical aspects of blood circulation, including cardiac function and peripheral vascular physiology characteristics.

Intravital microscopy observation: Observation performed under magnification of the light microscope upon a living tissue, in this case the cremaster muscle flap observed under 1,800× magnification.

Leukocyte–endothelial interaction: The relation between leukocytes present within the vessels and the vascular endothelium that can be altered in physiological and pathological states.

Microcirculation: Circulation of blood in the most distal part of the circulatory system, within the smallest vessels of the body such as arterioles, venules, and capillaries.

References

1. O'Shaughnessy, M., Jin-Mei, G., Wyllie, F., Acland, R., Anderson, G. L., Banis, J. C., and Barker, J. H. (1994). Microcirculatory consequences of microvascular surgery. *Microsurgery* **15**, 405–412.
2. Majno, G., and Palade, G. E. (1961). Studies on inflammation. 1. The effect of histamine and serotonin on vascular permeability: An electron microscopy study. *J. Biophys. Biochem. Cytol.* **11**, 571–605.
3. Grant, R. T. (1964). Direct observations of skeletal muscle blood vessels (rat cremaster). *J. Physiol.* **172**, 123–132.
4. Baez, S. (1973). An open cremaster muscle preparation for the study of blood vessels by in-vivo microscopy. *Microvasc. Res.* **5**, 384–394.
5. Miller, F. N., and Wiegman, D. L. (1977). Anesthesia-induced alteration of small vessel responses to norepinephrine. *Eur. J. Pharmacol.* **44**, 331–337.
6. Anderson, G., Acland, R. D., Siemionow, M., and McCabe, S. J. (1988). Vascular isolation of the rat cremaster. *Microvasc. Res.* **36**, 56. *This paper introduces a new model of the cremaster muscle flap isolated on the neurovascular pedicle for microsurgery studies. This model allows monitoring of effects of ischemia–reperfusion injury by application of vascular clamps to the feeding artery and vein of the muscle flap.*
7. Siemionow, M., and Nanhekhan, L. V. (1999). Introduction of cremaster muscle chamber technique for long-term intravital microscopy. *Ann. Plast. Surg.* **43**, 161–166. *This article describes the ability to perform chronic measurements of microcirculatory hemodynamics by application of the cremaster muscle chamber model.*
8. Siemionow, M., Andreasen, T., Chick, L., and Lister, G. (1994). Effect of muscle flap denervation on flow hemodynamics: A new model for chronic in vivo studies. *Microsurgery* **15**, 891–894.
9. Siemionow, M., Kimori, K., and Lister, G. (1998). Microcirculatory hemodynamics during the acute phase of free vascularized muscle allograft rejection. *Ann. Plast. Surg.* **41**, 275–282. *This is a first report on transplantation of the cremaster muscle allograft across a major histocompatibility barrier. The acute phase of allograft rejection is recorded by recording of leukocyte–endothelial interactions and hemodynamic changes of vessel diameters and capillary perfusion.*
10. Siemionow, M., Ozer, K., Porvasnik, S., Zins, J. (2000). Microcirculatory window for detection of allograft rejection. *Ann. Plast. Surg.* **44**, 637–643.

Capsule Biography

Maria Siemionow, M.D., Ph.D., is board certified in Rehabilitation Medicine, Orthopedic Surgery, and Traumatology and Microsurgery. Dr. Siemionow did her orthopedic surgery and hand surgery training in Poznan, Poland, and Helsinki, Finland, respectively. Later she became a Hand and Microsurgery Fellow at the Christine M. Kleinert Institute for Hand and Microsurgery (Louisville, KY).

Currently she is the Director of Plastic Surgery Research, Head Microsurgery Training, Department of Plastic Surgery at The Cleveland Clinic Foundation. She is a full staff member of the Plastic Surgery Department and holds joint appointments with the Department of Orthopedic Surgery and the Transplant Center and with the Department of Immunology, at the Cleveland Clinic Foundation. She has adjunct appointments at Ohio State University and Case Western Reserve University.

Dr. Siemionow has received several awards for her contributions in the fields of hand surgery, microsurgery and transplantation. She is the author of 122 peer-reviewed articles and 319 abstracts.

Andrea Moreira-Gonzalez, M.D., is currently a Research Fellow at the Department of Plastic Surgery at the Cleveland Clinic Foundation. Dr. Gonzalez is a fully trained plastic surgeon that did her general and plastic surgery training in Sao Paulo, Brazil. After she did a Craniofacial and Reconstructive Fellowship at the Institute for Craniofacial and Recoonstructive Surgery in Southfield, Michigan. Dr. Gonzalez received several awards for her research in Reconstructive Plastic Surgery and is the author of 15 peer-reviewed articles and 47 abstracts.

Chapter 30

The Hamster Cheek Pouch as a Research Model of Inflammation

Erik Svensjö

Instituto Biofísico Carlos Chagas Filho, Universidade Federal do Rio de Janeiro, Rio de Janeiro, Brazil

Acute inflammation is characterized by increased microvascular permeability to plasma proteins and leukocyte recruitment into inflammatory sites at postcapillary venules but not in arterioles or capillaries. The hamster cheek pouch as prepared for intravital microscopy has become a versatile model for studies of the very first events (within seconds, minutes, and hours) of inflammation as indicated by instant changes in blood flow (arteriolar constriction or dilation), number of blood-perfused capillaries, increased leukocyte rolling and adhesion, and macromolecular leakage in postcapillary venules following the application of a great variety of inflammatory stimuli. Its value for studying the effects of proinflammatory mediators on plasma exudation or macromolecular leakage as well as leukocyte adhesion in postcapillary venules has been established by a large number of studies.

Morphology

Hamster cheek pouches are bilateral invaginations of the oral mucosa and can be everted from the anesthetized hamster with their blood flow intact. They are therefore well suited for intravital microscopy. The cheek pouch tissue is transparent and the anatomical elements of the microcirculation, arterioles, capillaries and venules can be easily identified (Figure 1). Cheek pouches, as prepared for intravital microscopy, have been used for studies of inflammation, tumor growth, vascular smooth muscle function, blood flow regulation, and leukocyte behavior—rolling and adhesion—at the microcirculatory level. The hamster cheek pouch preparation (HCP) was described for the first time as a "natural window" for microscopic observation and motion picture recording of blood flow, vessel wall caliber, and intravascular behavior of blood cells by Fulton, Jackson, and Lutz in 1946. The early observation that the cheek pouch of an anesthetized hamster can be brought out and applied on a stage for intravital microscopy with minimal interference with their blood flow resulted in its application for physiological studies of arterioles, capillaries, and venules. The cheek pouch can also be studied in situ by introducing a light rod through the mouth into the pouch and applying an acrylic chamber on top of the light rod after a skin incision. It appears that these two methods of preparing the HCP for intravital microscopy give equivalent results. The way to prepare the HCP for physiological studies of the microcirculation was developed and refined by Duling in 1973 using the everted cheek pouch. Ever since dextrans labeled with fluorescein isothiocyanate—FITC dextrans—became available as intravital markers of vascular permeability in 1970, the HCP has been used in numerous studies of inflammation as induced by a large number of different inflammatory mediators or procedures, such as bradykinin, histamine, adenosine diphosphate (ADP), serotonin, prostaglandins, or leukotrienes; by temporary ischemia of the pouch (IR); or by application of live parasites (*Trypanosoma cruzi, Leishmania donovani*) (Table I).

Detailed descriptions of the morphology of the cheek pouch have been given by Priddy and Brodie in 1948 and by Handler and Shepro in 1968. Briefly, the cheek pouches of the golden hamster are invaginations of the oral mucosa, extending beneath the subcutaneous tissue down to the shoulder region, and they can be characterized histologically as skinlike. The length of the distended cheek pouch is approximately 2.5 to 5.5 cm, and the width is about 1 cm. All parts except the distal portion of the pouch have longitudi-

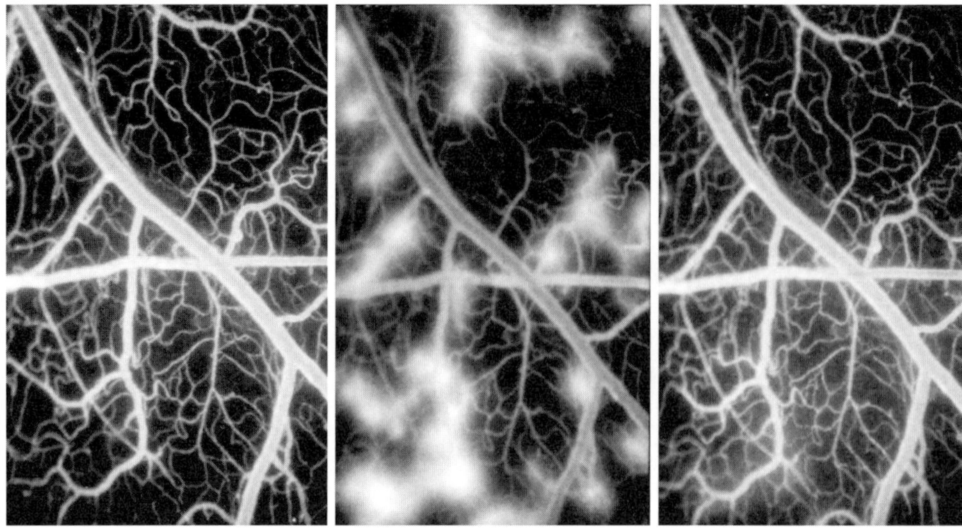

Figure 1 Fluorescent micrograph of the hamster cheek pouch after IV injection of FITC-dextran. Same area before (**A**), 5 minutes after (**B**), and 30 minutes after (**C**) topical application of bradykinin (4×10^{-7} M). More than 25 leakage sites at postcapillary venules are shown by extravasation of FITC-dextran at postcapillary venules. (see color insert)

Table I Mediators or Procedures That Have Been Shown to Increase FITC-Dextran Leakage in Postcapillary Venules.

Mediator or procedure	Effective concentration
Histamine, serotonin	10^{-6} M
Bradykinin, substance P	10^{-7} M
ADP, adenosine, inosine	10^{-5} M
Prostaglandins E_1, E_2 and E_{2a}	$<10^{-8}$ M
Leukotrienes C_4, D_4, E_4, B_4	$<10^{-9}$ M
Complement C3a, C5a	$<10^{-9}$ M
Cytokines – IL-1β	1 μg/ml
Platelet activating factor (PAF) Fibrin-derived peptides	$<10^{-9}$ M
VEGF (vascular endothelial growth factor)	$<10^{-9}$ M
Polyarginine, polylysine, major basic protein (MaBP)	$<5 \cdot 10^{-9}$ M
oxLDL, apolipoprotein B derived peptides	0, 1 mg/ml
Ischemia/Reperfusion (I/R)	30 min ischemia
ROS (reactive oxygen species)	
Oxidant injury (tertiary-butyl-hydroperoxide, TBOOH)	10^{-4} M
Immune aggregates (ovalbumin [OA] immunized hamsters)	<1 μg/ml of OA
Phorbol ester (PDBu)	10^{-6} M
Smokeless tobacco extract	
Endotoxin	7 μg/ml
Trypsin, cruzipain, gingipain	$>10^{-8}$ M
Live parasites—*Trypanosoma cruzi, Leishmania donovani*	10^7/ml

nal muscle fibers. The pouch is provided with a long retractor muscle and a sphincter-like arrangement around its aperture. There are no specific regional lymph nodes, and the nonmuscular part of cheek pouch lacks lymphatic drainage and has been characterized as an immunologically privileged site. Three layers are distinguishable in the pouch membrane. The epithelium is devoid of hair follicles and glands. The pouch is attached to the subcutaneous tissue by loose areolar connective tissue. The cheek pouch vasculature is mainly supplied by branches of the external carotid artery, via three saccular arteries. Blood is also supplied via the arteries of the retractor muscle. Terminal arterioles supply capillaries, which drain into postcapillary venules, which in turn empty into collecting venules. Arteriole-to-arteriole and venule-to-venule anastomoses are often seen, but there are no arteriole-to-venule anastomoses. The arterioles and, to a lesser extent, the venules are surrounded by numerous mast cells that appear to promote the oriented migration of leukocytes from the venules into the extravascular space.

Cheek Pouch Preparation

For microscopic observation, the single layer pouch preparation is performed essentially as described by Duling (1973). Alternatively the pouch may be studied in situ after insertion of a light rod and an acrylic skin chamber. There are no discrepancies reported for results achieved with either of these two procedures. However, the everted cheek pouch offers better optical conditions and a larger area (about 1 cm^2) for observations of the microvasculature. A hamster is anesthetized by intraperitoneal injection of pentobarbital and is then placed on a heating pad controlled by a rectal thermistor. A tracheal tube is inserted to facilitate breathing and a femoral vein is catheterized for administration of supplemental anesthetic or other intravenous drugs. The experimental setup of the hamster cheek pouch preparation is illustrated in Figure 2. The animal is placed on an acrylic stage in the middle of which there is a well with a circular silicon rubber ring surrounding a transillumination window. The cheek pouch is everted with the aid of a moist cotton stick, the distal nonmuscular part of the pouch is identified and pinned to the silicon ring, and the pouch is dissected under a stereomicroscope at 10× magnification. A central area is selected with no major arterioles or venules and an incision is made in the top layer, which then is moved aside and pinned to the silicon ring. Now the loose areolar connective tissue is exposed and can be removed to have a single-layer preparation with the best optical conditions. During the preparation, and throughout the experiment, the cheek pouch is constantly superfused with a bicarbonate-buffered saline solution that is continuously bubbled with 5 percent CO_2 in N_2 at 37°C, and thus provides a physiological level of oxygen, temperature, and acidity (pH). For studies of vascular permeability changes, an intravascular marker such as fluorescein-labeled albumin can be used. Alternatively, an intravascular marker with a molecular weight similar to that of albumin namely fluorescein-labeled dextran (FITC-dextran), is used in a dose that has minor if any effect on platelet function and blood clot formation. FITC-dextran 70,000 or 150,000 Da is dissolved in normal saline and injected i. v. in a dose of 50 to 100 $mg\,kg^{-1}$ body weight. Properly mounted and dissected, the cheek pouch is a noninflamed preparation as judged from the presence of vascular tone and absence of sites with FITC-dextran leakage (as seen in Figure 1), and absence of excessive leukocyte rolling and adhesion in postcapillary venules; the preparation can be utilized for intravital microscopy for several hours.

Changes in Vascular Permeability

In a classical electron microscopy study by Majno and Palade (1961), it was shown that postcapillary venules of 20 to 30 μm diameter leaked carbon particles through gaps formed between endothelial cells as a result of histamine stimulation and endothelial cell contraction. Such gaps have been confirmed to exist in the HCP after bradykinin stimulation. Studies combining intravital microscopy with subsequent electron microscopy of the same cheek pouch showed that interendothelial gaps had formed at sites where extravasation of FITC-dextran already had been observed in vivo, as in Figure 1B. The advantage of FITC-dextran as a marker is that it can first be observed in living tissue and then can be traced as dark precipitates after the fixation of the tissue, which demonstrates the pathway through interendothelial gaps taken by macromolecules in inflammation. The gaps are formed as a result of mediator-induced (e.g., bradykinin) contraction of endothelial cells in the postcapillary venules. Supportive evidence for endothelial cell contraction as a mechanism for gap formation can be found in the pharmacological studies of smooth muscle relaxant drugs that maintain the intracellular level of cAMP (adenyl cyclase activators, β_2-adrenoreceptor stimulants) or block its break-

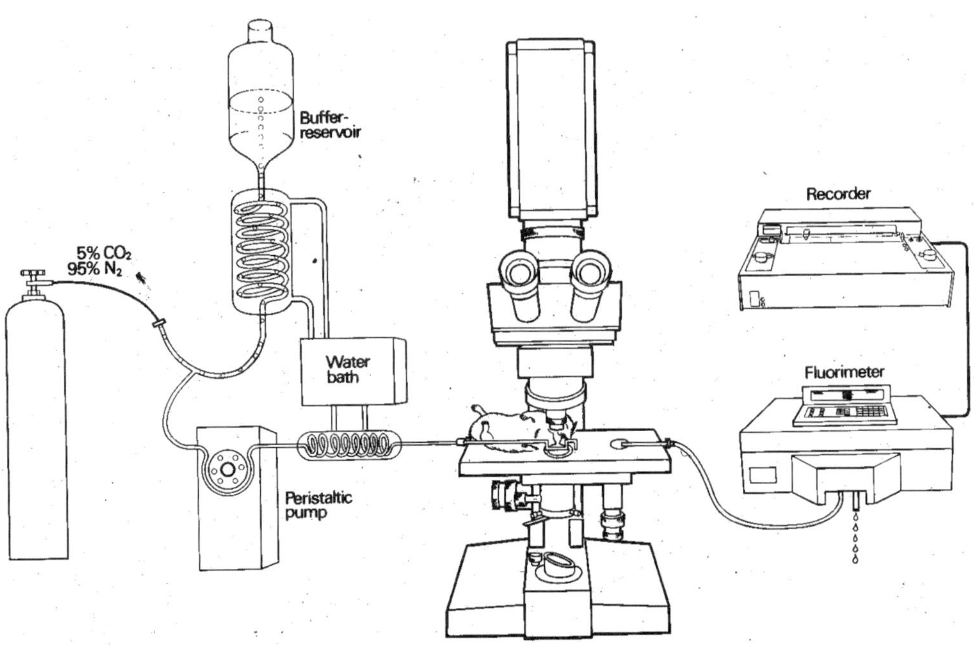

Figure 2 Schematic illustration of the experimental setup for the hamster cheek pouch model. The bicarbonate buffer superfusing the cheek pouch is heated and equilibrated with 5 percent CO_2 and 95 percent N_2. A peristaltic pump keeps buffer flow to microscope stage and fluorimeter constant.

down (phosphodiesterase inhibitors), which counteracts the mediator-induced contraction, the subsequent gap formation, and thus, the macromolecular leakage (Table II).

There has been some controversy in the past on the nature of endothelial gaps, intracellular or intercellular, but the controversy was resolved in a study by McDonald et al. (1999) on the formation of interendothelial gaps and their function as a route for plasma leakage in inflamed rat trachea. Using five different methods to examine normal and inflamed tissue samples, the researchers concluded that most of the openings in leaky venules were intercellular gaps, not transcellular holes, and that the formation and closure of gaps are likely to be energy dependent, while the process of plasma leakage is not, provided there is adequate driving force for extravasation. The cellular mechanisms of gap opening and closure still remain to be elucidated.

The mean diameter of postcapillary venules in the HCP most prone to leak on mediator stimulation varies between 8 and 15 μm, and the interendothelial gaps formed after stimulation with bradykinin vary between 0.08 and 1.4 μm in width.

For studies of macromolecular leakage at postcapillary venules, it is sufficient to use a magnification of 40× to 50× and for measurements of arteriolar diameter and for studies of leukocyte rolling and adhesion a water-immersible lens is recommended with a magnification of 200× to 400×. Illumination with a 50- to 100-W mercury lamp and filtering of the light with proper excitation and emission filters clearly shows the intravascular FITC-dextran and its extravasation after stimulation (see Figure 1). Epi-illumination gives better contrast between fluorescent and nonfluorescent tissue. Inflammatory mediators and drugs that may affect macromolecular leakage can be added to the superfusion buffer before it flows over the cheek pouch. The microvascular permeability increase or macromolecular leakage following the addition of bradykinin to the superfusion buffer is immediate; can be seen as early as at 30 seconds after its application; and reaches a maximum between 2 and 5 minutes after its application. The macromolecular leakage is measured by counting the number of leaking postcapillary venules (leaks) per cm^2 with 2-minute intervals until the maximal value is reached. The maximal number of leaks per cm^2 correlates with the fluorescent light intensity as measured with a photomultiplier on top of the microscope and also with the amount of FITC-dextran eliminated by the superfusing buffer and measured by in fluorimetry. The HCP is cleared

Table II Inhibitors of Mediator-Induced Macromolecular Leakage in the HCP.

Inhibitor	Concentration, M or dose	Mediator or procedure
β_2-adrenoceptor agonist (terbutaline, salbutamol)	10^{-6} M	Histamine, bradykinin, LTB_4, adenosine, phorbol ester, oxidant injury, I/R
Phophodiesterase inhibitors (Rolipram, milrinone)	10^{-5} M	Bradykinin
Theophylline (adenosin receptor agonist)	10^{-5} M	Histamine
Glucocorticoids (budesonide, methylprednisolone, dexamethasone)	10^{-7} M	Histamine, bradykinin, LTB_4, LTC_4, PAF, immune aggregates, oxidant injury, I/R, phorbol ester, endotoxin
Calcium antagonist (verapamil)	10^{-5} M	Histamine, bradykinin
Bradykinin-2-receptor antagonists (HOE 140, NPC 17647)	10^{-5} M	Bradykinin
Triglycylvasopressin, 1-desamino-8D-argininvasopressin	10^{-8} M	Histamine, bradykinin
H_1-receptor antagonist (mepyramine, dimethpyrindene)	$2 \cdot 10^{-6}$ M	Histamine
$5HT_2$-receptor antagonist (ketanserin)	$5 \cdot 10^{-7}$ M	Serotonin, histamine
Staurosporin (protein kinase C inhibitor)	10^{-9} M	Phorbol ester (PDBu)
NOS-inhibitors (L-NA, L-NAME, L-NMMA)	10^{-5} M	Histamine, bradykinin, I/R, PAF, endotoxin
Prostacyclin, iloprost	10^{-10} M	I/R
CuZn-SOD, EC-SOD (superoxid dismutases)	25 mg/kg, i.v.	I/R, oxidant injury
Tocopherol	10^{-5} M 1 mg/kg/day	I/R, oxidant injury (TBOOH)
Ascorbic acid	10^{-5} M	I/R, oxidant injury (TBOOH)
Flavonoids:hydroxylrutoside, diosmin, ruscus extract (Cyclo 3 Fort)	5–80 mg/kg/day	I/R, oxidant injury (TBOOH)
Glibenclamide, gliclazide (sulfonylureas, diabetes drugs)	10^{-8} M	I/R
WEB 2170 (PAF-antagonist)	10^{-5} M	PAF, oxLDL
Ropivacain (local anestetic)	10^{-5} M	LTB_4
Melatonin		I/R
Ketoprofen (COX-inhibitor)	$> 10^{-5}$ M	IL-1β, bradykinin
Nedocromil		Histamine, LTB_4
Lipoxin		LTB_4
Dextran sulfate	1, 75 mg/kg	LTB_4, polylysine

from the extravasated FITC-dextran at 30 minutes after a permeability increase and appears as before the increase (see Figure 1C). A linear dose–response relationship has been shown to exist between the number of leaks and the logarithmic dose of bradykinin, histamine, and several leukotrienes and also between the maximal number of leaks and the amount of FITC-dextran eliminated by the superfusion buffer over the course of 30 minutes.

Another characteristic of the response to certain inflammatory mediators such as bradykinin, histamine, and leukotriene B_4 is that on repeated stimulation with the same submaximal dose at 30-minute intervals, there is apparently no downregulation of receptor function and thus only slight reduction in the macromolecular leakage response over a time period when measured as number of leaks. Therefore, the HCP can be used with the first application of the mediator as a control followed by as many as seven subsequent applications together with different inhibitors. Certain mediators and procedures can make the HCP resistant to a secondary stimulation of the same kind. Thus a secondary application of ischemia/reperfusion (I/R), platelet activating factor (PAF), or parasites to the cheek pouch will result in a response which is 30% or less of the first. There is no complete understanding of the mechanisms behind this state of preconditioning to secondary stimulation, but a detailed study on the subject has been published by Korthuis et al. (2003).

Altogether, these data suggest that the HCP can be very useful in physiological and pharmacological studies on the regulation (inhibition or potentiation) of macromolecular leakage via interendothelial gaps induced by a mediators acting on the endothelial cells in postcapillary venules.

Inhibition of Mediator-Induced Macromolecular Leakage

Pharmacological inhibition of mediator (bradykinin) induced macromolecular leakage in the HCP was first shown with the β_2-adrenoreceptor terbutaline. Since then a variety of drugs have been studied for their potential to inhibit plasma leakage in the HCP as listed in Table II. The listed drugs represent different physiological and pharmacological principles as exemplified by selective receptor blockers (mepyramin, HOE 140, theophylline, ketanserin), calcium antagonists (verapamil), glucocorticoids (budesonide, methylprednisolone, dexamethasone), nitric oxide synthetase (NOS) inhibitors (L-NA, L-NAME, L-NMMA), antioxidants (ascorbic acid, tocopherol, flavonoids), potassium channel blockers (glibenclamide, gliclazid, for treatment of diabetes) and drugs interfering with leukocyte adhesion or activation (antioxidants, prostacyclin, NOS-inhibitors, superoxide dismutase, dextran sulfate).

β_2-Adrenoreceptor stimulants (terbutaline, salbutamol) and phosphodiesterase inhibitors (rolipram, milrinone) will cause arteriolar dilation on application to the HCP. Nevertheless they all inhibit macromolecular leakage in postcapillary venules, thus emphasizing that in spite of increased blood flow and perfusion pressure in the arterioles that should favor plasma leakage from the venules, the effect of the drug localized to receptors on the endothelial cells of the postcapillary venules is the dominant factor in the regulation of plasma leakage in inflammation. Neutrophil-induced changes in vascular permeability are the result of a complex interaction between adhesive proteins expressed on the surface of leukocytes and endothelial cells, selectins, and integrins, which finally results in the firm adhesion of leukocytes in postcapillary venules. It is possible to interfere with the leukocyte–endothelial cell interaction at different steps. Monoclonal antibodies against adhesion glycoproteins have been used to elaborate mechanisms of leukocyte rolling and adhesion. Dextran sulfate, and possibly also heparin, acts by neutralizing charged peptides released from the LTB_4-activated neutrophil, thereby inhibiting macromolecular leakage by 85 percent. Ropivacain, a local anesthetic with anti-inflammatory properties, may reduce the rolling and firm adhesion of LTB_4-stimulated neutrophils and also the macromolecular leakage, see Table II. These effects could be explained by a shedding of selectins on the surface of leukocytes. Studies on neutrophil or leukocyte induced macromolecular leakage illustrate that the final result of inhibitor action is localized to the venular endothelial cells in the postcapillary venules. Antioxidants such as ascorbic acid, tocopherol, flavonoids, and SODs may not exert any constriction or dilation, but they may neutralize products released from adhering and activated neutrophils. Asthma and diabetes are today recognized as inflammatory diseases, and asthma drugs, which have been studied in the HCP (β_2-adrenoreceptor stimulants, theophylline, glucocorticoids, chromoglycate), all have a significant inhibitory effect on histamine-induced macromolecular leakage. The antidiabetic drugs, glibenclamide and gliclazide, may counteract I/R-induced inflammatory effects in the HCP.

Conclusions

Increased macromolecular leakage through widened gaps between contracted endothelial cells of postcapillary venules is an important event in inflammation. It is induced by a direct action on the venular endothelium of mediators released from mast cells or from activated leukocytes. Cell-mediated (polymorphonuclear leukocyte) macromolecular leakage, like that caused by leukotriene B_4 is reversible and can be reproduced at 30-minute intervals but that induced by I/R-, PAF-, or parasite-application may cause a state of preconditioning. The HCP is a practicable model for studies on the very first events of inflammation as indicated by the increased rolling/adhesion of leukocytes and macromolecular leakage in postcapillary venules for several reasons: (1) the effects of inflammatory mediators and anti-inflammatory drugs in the HCP have been confirmed in other animal models; (2) measurement of rolling/adhesion of leukocytes and macromolecular leakage can be made simultaneously; and (3) monitored superfusion of the HCP facilitates mediator administration and drug treatment with great accuracy.

Bibliography

Granger, D. N., and Kubes, P. (1994). The microcirculation and inflammation: Modulation of leukocyte-endothelial cell adhesion. *J. Leukocyte Biol.* **55**, 662–675. *A comprehensive description and guide for the understanding of leukocyte–endothelial interactions that is still useful although it was published in 1994.*

Korthuis, R. J., Dayton, C., and Yamaguchi, T. (2003). Early and late preconditioning prevents ischemia/reperfusion injury: Signaling pathways mediating the adaptive metamorphosis to a protective phenotype in preconditioned tissues. In *Molecular Basis for Microcirculatory Disorders* (G. W. Schmid-Schönbein and D. N. Granger, eds.), pp. 343–363. France, Paris: Springer-Verlag. *This review is very useful for the understanding of the phenomenon of preconditioning, an important protective mechanism in inflammation.*

Ley, K., and Sperandio, M. (2003). Molecular mechanisms of leukocyte adhesion. In *Molecular Basis for Microcirculatory Disorders* (G. W. Schmid-Schönbein and D. N. Granger, eds.), pp. 1–71. France, Paris: Springer-Verlag. *A very complete review covering all aspects of leukocyte rolling, firm adhesion, and transmigration, and also details of mechanisms of adhesive proteins with references to more than 400 publications.*

McDonald, D. M., Thurston, G., and Baluk, P. (1999). Endothelial gaps as sites for plasma leakage in inflammation. *Microcirculation* **6**, 7–22.

Persson, C. G. A., and Svensjö, E. (1985). Vascular responses and their suppression: Drugs interfering with venular permeability. In *Handbook of Inflammation* (I. L. Bonta, M. A. Bray, and M. J. Parnham, eds.), Vol. 5, *The Pharmacology of Inflammation*, pp. 61–82. Amsterdam, Netherlands: Elsevier. *Describes the basis for the concept of the endothelial cells in the postcapillary venules as the site for physiological and pharmacological regulation of vascular permeability increase in inflammation.*

Capsule Biography

Dr. Svensjö has worked as senior pharmacologist at two major pharmaceutical companies in Sweden and is now at Laboratorio Imunologia Molecular, Instituto Biofísica Carlos Chagas Filho, Universidade Federal do Rio de Janeiro. The laboratory focuses on mechanisms of parasite infection (*Trypanosoma cruzi*, leishmaniasis). The lab is supported by grants from FAPERJ, CNPq, and WHO.

SECTION D

Permeability, Tone and Hemodynamics

CHAPTER 31

Regulation of Microvascular Permeability

D. O. Bates

Microvascular Research Laboratories, Department of Physiology, Preclinical Veterinary School, University of Bristol, Bristol, United Kingdom

Introduction

The rate at which fluid and solutes cross the vascular wall is determined by the driving forces for exchange (pressure or diffusion gradients), the surface area available for exchange, and the permeability of the vascular wall to fluid or solute. This permeability is regulated by the cells forming the capillary walls—the endothelium—and is dynamic in both time and space.

Microvascular Permeability to Water: Hydraulic Conductivity (L_p)

Although cell membranes are freely permeable to the diffusion of water, they have a high resistance to fluid flow down a hydrostatic pressure gradient. L_p is given by the fluid flow per unit pressure per unit area.

What Is the Site of the Resistance to Fluid Flow?

The main pathway of fluid flow across the walls of continuous capillaries is the intercellular cleft, and in fenestrated endothelium through fenestrae. The conductance is determined the viscosity of the fluid (η), the path length (Δx), and the fourth power of the radius of the pathway (r^4). The cleft of most endothelial cells in the capillary and postcapillary beds is approximately 20 to 30 nm wide throughout its length (see Figure 1) with at least one tight junction. Breaks in the tight junction strand, approximately every 2 to 3 µm and usually 50 to 200 nm long and overlapping junctional strands further into or out of the cleft, give rise to a tortuous pathway for fluid flow through the cleft. The path length is therefore greater than the cleft length, and the radius is smaller than the cleft width, but greater than the space between the tight junction molecules.

How Can L_p Be Measured?

Cell Culture Models

L_p can be measured of endothelial cells in culture. A monolayer grown on polycarbonate filters is pressurized and the flow rate through the monolayer measured. The area of the monolayer and the pressure difference are known, so L_p can be calculated. Two main problems with this technique are the lack of formation of a perfect monolayer and the compression of endothelial cells and the underlying basement membrane onto the filter (sealing). The L_p of monolayers is up to two orders of magnitude greater than in vivo.

Measurement of L_p in Single Vessels in Vivo

L_p can be measured in single arterioles, capillaries, and postcapillary venules using the Landis–Michel technique. A single microvessel in a thin flat tissue such as the mesentery is cannulated and perfused with a solution of known oncotic pressure and containing flow markers (erythrocytes). The vessel is then occluded by a glass rod, and the rate of filtration per unit area of vessel wall per unit pressure (L_p) is determined from the rate of movement of the flow markers, the radius of the vessel, and the length of the vessel.

Measurement of Filtration Capacity in Whole Organs

L_p can be estimated in whole tissues by measuring the gain of weight or volume induced by an increase in fluid filtration after an increase in capillary pressure. The venous

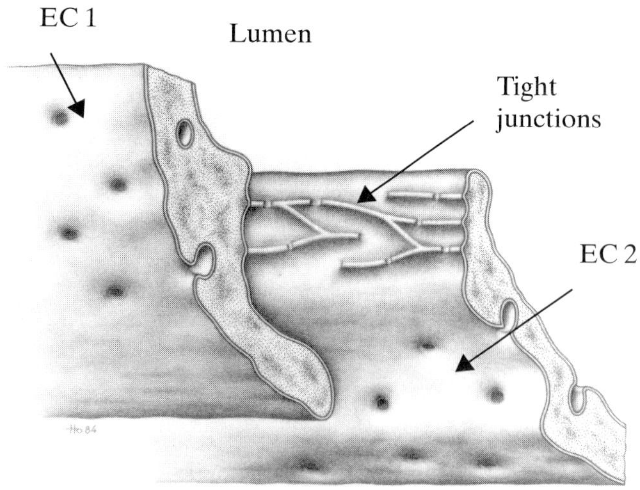

Figure 1 Three-dimensional reconstruction of an endothelial cell cleft from ultrathin serial sections. Endothelial cell 1 (EC1) has been cut away to reveal the tight-junction structure of endothelial cell 2 (EC2) to be visible in the cleft. Tight junction strands run along the cleft and have small breaks in them that allow fluid and solutes through.

pressure of the tissue can be increased, either by cannulating the draining vein or by inflating a cuff. The increase in weight or volume of the tissue per unit pressure is the capillary filtration coefficient (CFC, or K_f), which is the product of the mean L_p and the surface area available for exchange.

Microvascular Permeability to Solutes

Lipophilic versus Hydrophilic Solutes and the Importance of Flow Limitation

The solute permeability (P_s) of the capillary wall is dependent on the oil:water partition coefficient, the diffusion coefficient of the solute, Δx, the radius of the solute relative to the pore, and the pore area. Fat-soluble molecules are therefore very permeable across the capillary wall (P_s for oxygen is $1\text{–}10 \times 10^{-3}\,\text{cm}\,\text{s}^{-1}$). The concentration gradient across the capillary wall can only be maintained by relatively high blood flows. Highly permeable solutes are therefore flow limited rather than diffusion limited. Small hydrophilic molecules are also flow limited, even though their transport is predominantly through extracellular pathways. Larger molecules such as plasma proteins have a P_s that is normally so low that the concentration gradient along the vessel is maintained independently of the rate of flow for most perfused vessels, and is therefore diffusion limited. Intermediate-sized solutes such as glucose move from flow limitation to diffusion limitation as the P_s of the vessel changes.

The Sites of Solute Permeability

In continuous capillaries the location of the barrier to solute movement is in the endothelial cell cleft. There are at least three possible barriers to solute flux—the glycocalyx, the tight junctional strands, and the adherens junctions of the cleft. There is still some support for the concept of transcellular movement of protein and other large solutes through the vesicular system of endothelial cells. Fused vesicle clusters attached to the lumen have been seen to cross the entire width of the cell and fuse with vesicle clusters attached to the abluminal side of the cell, thereby providing a potential route for solute movement.

Measurement of Permeability to Water-Soluble Solutes

Problems Common to Methods of Solute Permeability Measurement

P_s can be determined by measuring the solute flux across the vascular wall and calculating P_s based on known driving forces for solute flux, in a manner analogous to that for L_p. The P_s of a membrane to a particular solute is set by Fick's law—the rate of diffusive flux per unit concentration difference per unit surface area. However, the total solute flux across the vascular wall depends upon diffusive and convective fluxes—that is, the rate of solute movement carried along in the flow of fluid that is also crossing the vessel wall. Furthermore, the concentration gradient across the vessel wall is also set by the balance of the diffusive and convective solute fluxes. The interstitial solute concentration depends upon solute flux relative to water flux. This ratio of diffusive to convective solute flux must be used to calculate the true driving force for diffusive solute movement. It is only possible to use the measurement of solute flux to accurately calculate P_s when all these parameters are known. P_s has been measured in cell culture models, in whole organs, and in single vessels. Most methods for measuring P_s use labeled solutes. These include radiolabeled (e.g., ^{125}I-albumin, ^{57}Cr-EDTA), or fluorescently labeled molecules (e.g., FITC-albumin), or dyes that bind to specific molecules (e.g., Evans blue, which binds albumin). All these methods suffer from two potential problems—dissociation of the marker from the test molecule and alteration of the test molecule by the marker.

1. *Tracer dissociation.* This can lead to a marked overestimation of P_s. For instance, Evans blue has a molecular weight of 960, albumin 66,000. Therefore for every molecule of EB that dissociates, P_s is 69-fold overestimated—if 1 percent of the EB dissociates, then 69 percent of the P_s will be that of EB, not that of albumin. Moreover, free EB is flow limited rather than diffusion limited. Therefore changes in EB transport occur far more easily by changing blood flow. This is also true for FITC and radiolabeled molecules that undergo radiolysis—^{125}I-albumin usually has 0.01 to 2.5 percent dissociated ^{125}I. This can be avoided by the use of tracers that do not readily dissociate, such as TRITC, or by nonnative tracers that are themselves detectable, such as green fluorescent protein or fluorescent dextrans.

2. *Alteration of the test molecule.* The process of labeling molecules may also alter their characteristics. Albumin is denatured when labeled with FITC, resulting in a change of charge and pH, and it has a different P_s than when it is labeled with other fluorescent molecules. Although this may have a small effect on P_s under normal conditions, it may be highly significant in conditions where albumin is endogenously modified (e.g., diabetes).

CELL-CULTURE MODELS

There are numerous problems measuring P_s of endothelial monolayers in cell culture. These include those outlined earlier concerning measurement of L_p. Because P_s of the monolayer is so much higher than in vivo, basement membrane can significantly contribute in vitro. This can be controlled by measuring P_s after removal of endothelial cells. Since cells are grown on a synthetic insert, the path length between the junction and the nearest pore in the membrane can also change. The average endothelial cell surface area and shape can be determined to eliminate this possibility. It is also necessary to ensure that there is not a significant decrease in the solute concentration difference between the endothelial cell and the pore, or the endothelial cell and the lower chamber. Calculation of true P_s (in cms^{-1}) should be carried out rather than solute flux (mgs^{-1}).

MEASUREMENT OF SOLUTE PERMEABILITY IN SINGLE VESSELS

P_s can accurately and effectively be measured in individual microvessels by determining solute flux under known conditions of diffusive and convective driving forces. A vessel is cannulated with a pipette divided by a septum. The two halves are perfused at separate pressures and one side filled with a fluorescent tracer. The perfusate is switched rapidly to the fluorescently labeled one, thereby giving a known concentration difference. The total fluorescence of and around the vessel is measured. With small molecular weight solutes (convective flux being negligible) the rate of increase of fluorescence can be used to calculate solute flux per unit concentration gradient per unit area—P_s. For larger solutes (or for small solutes in very tight vessels such as the blood–brain barrier), solute flux can be measured at a variety of pressures such that convective flux can be measured.

MEASUREMENT OF SOLUTE PERMEABILITY IN WHOLE TISSUES

Permeability–surface area product (*PS*) is usually measured in whole tissues, since it is difficult to measure the surface area. *PS* can be measured using radiolabeled, colorimetric, or fluorescent tracers—bearing in mind the measurement problems discussed earlier. Assuming that the tracers are effectively labeled, do not dissociate, and do not denature the compound, *PS* can be accurately assessed using a double-labeled tracer technique. This involves perfusing animals for increasing amounts of time with a tissue tracer (e.g., ^{125}I-albumin), followed by injection of a second reference tracer (e.g., ^{131}I-albumin) for 1 minute to allow circulation. The animal is then killed, a blood sample is taken, and the tissue is excised and weighed. The lungs are then dried completely and water content determined. The reference tracer (V_r) distribution volume is calculated from the plasma concentration and the reference tracer tissue counts. Tissue distribution volume (V_t) is calculated from the tissue tracer counts and the blood. The albumin distribution volume (V_a) can be calculated from V_r and V_t. V_a will increase over time as solute moves from vascular to tissue so the time experiments are used to calculate initial solute flux. V_a is measured under normal conditions and during increased filtration rate. The relationship between filtration rate and solute flux can be used to calculate the *PS*. A common mistake in this method is to divide rather than subtract the first tracer from the second. This results in the calculation of a meaningless ratio dependent on blood flow rather than *PS*.

Reflection Coefficient (σ)

The reflection coefficient of a membrane to a solute is the osmotic pressure exerted by that solute across the membrane as a fraction of the osmotic pressure across an ideal membrane (it is a function of the mean pore area relative to the mean cross-sectional area of the solute). σ for albumin, for instance, in normal capillaries is close to 1: It exerts almost its full osmotic pressure. An increase in the pore diameter (which would also increase solute P_s) would drastically reduce σ to albumin.

MEASUREMENT OF σ IN SINGLE MICROVESSELS

Reflection coefficient to macromolecules can be measured using the Landis–Michel technique [1]. A single vessel is cannulated and perfused as described for L_p. σ can be calculated as the root of the ratio of the pressure required to balance filtration to that of the colloid osmotic pressure determined across an ideal membrane by a colloid osmometer. σ has been determined in a number of capillary beds using these techniques.

MEASUREMENT OF σ IN WHOLE ORGANS

Under conditions of high filtration rates the interstitial protein concentration, when expressed as a fraction of the plasma protein concentration, approximates to the permitted fraction, or $1 - \sigma$. Thus if filtration rate is raised and interstitial fluid can be sampled, σ can be estimated. One way of doing this is to cannulate a lymphatic draining the tissue and measure the protein concentration under conditions of high filtration rate, after equilibration.

Normal—or Baseline—Vascular Permeability

Hydraulic Conductivity—L_p

L_p of microvascular beds is heterogeneous under normal conditions, and the mechanisms that regulate baseline permeability are not clear. The number of tight junction strands is greater in vascular beds with lower L_p and P_s (e.g., blood–brain barrier). Some signaling pathways have been shown to be able to increase junctional strand number in endothelial cells in culture and to decrease L_p in vivo. Intracellular cAMP can stimulate the formation of tight junction strands by PKA-mediated phosphorylation of junctional associated proteins. The degree of filling of Ca^{2+} stores in the endothelial cells may regulate baseline L_p, although the link between Ca^{2+} stores and cAMP mediated pathways has not yet been determined. Nitric oxide has been implicated, since inhibition of NO synthase results in a transient increase of baseline L_p. The presence of fenestrations will result in a much greater baseline L_p, as is seen in renal, salivary, and synovial capillary beds, to name a few.

Solute Permeability—P_s

Little is known of the intracellular signaling cascades that regulate normal P_s, but for small solutes this is likely to be the same as that governing L_p. For larger molecular weight solutes, despite many decades of research, it is still not clear what are the contributions to basal P_s of the different signal transduction pathways.

Oncotic Reflection Coefficient—σ

Regulation of reflection coefficient is dependent on the functional pore radius. Diaphragms in fenestrae decrease the pore radius such that fenestrated endothelium has a high σ despite a high L_p. Little is known concerning the signaling pathways that regulate fenestrations, although the isolation of the protein that forms diaphragms across fenestrae of caveolae—PV-1—should mean that regulation of fenestrations will become more clearly understood.

Agonist-Mediated Increased Vascular Permeability

Mediators of Increased Permeability

G-Protein-Coupled Receptor Agonists

Vascular permeability to both water and solutes can be increased by inflammatory mediators released by neurons (e.g., substance P), mast cells (e.g., histamine), and platelets (e.g., ATP). These agonists act through stimulation of seven transmembrane domain G-protein coupled receptors, resulting in phospholipase C-β activation, inositol 1,4,5-trisphosphate production, release of Ca^{2+} from intracellular stores, and subsequent Ca^{2+}-mediated Ca^{2+} entry across the cytoplasm. This Ca^{2+} increase is common to many agonists that increase P_s. It leads to a series of intracellular signaling events, the relative roles of which are still being established. It appears that activation of nitric oxide synthesis, production of NO, and subsequent activation of guanylyl cyclase is critical in many agonist-mediated increases in both L_p and P_s.

Tyrosine Kinase Receptor Agonists

Other factors that increase permeability include growth factors such as VEGF that act through receptor tyrosine kinases and result in activation of phospholipase C-γ, and hence diacylglycerol production and Ca^{2+} entry but in this case, independently of Ca^{2+} stores.

Physical Forces

P_s to small and intermediate-sized solutes, such as K^+ and glucose-sized molecules, can be increased by shear stress. P_s to albumin and L_p are not affected. These increases in permeability are NO dependent and can be abolished by increasing cAMP concentrations. Recent evidence has implicated the activation of NO through a VEGF-R2 mediated signaling pathway possibly involving an integrin-mediated activation of Src.

The Site of Increased Permeability

There are a number of ultrastructural alterations in endothelial cells in vessels in which permeability is increased. These include intercellular gaps (between separate endothelial cells), transcellular gaps (through individual endothelial cells), induction of fenestrations, collections of vesicle-like structures known as vesiculo-vacuolar organelles (VVOs), vacuole formation, and reduction in cleft length and tight junction overlap. Each of these ultrastructural events has been linked to increased permeability stimulated by one or more agonists, but the molecular mechanisms regulating these are not yet clear.

The Cleft

The molecules that contribute to the formation of the cleft—the occludins in the tight junction, and the cadherins in the adherens junctions—can be regulated by anchoring proteins on the cytoplasmic side of the membrane. These proteins, such as ZO-1 and the catenins, are phosphorylated in response to many permeability enhancing agonists. Furthermore, prevention of their phosphorylation can prevent permeability increases in some model systems—particularly in vitro. Although there is some evidence that this may also be true in vivo, the significance of phosphorylation of these proteins is still not clear.

Gaps

Endothelial gaps were for many years thought to be intercellular, indicating a breakdown in the cleft. However, in the vast majority of these studies only single sections have been examined using transmission electron microscopy or scanning electron microscopy. Serial section reconstruction is necessary to prove that a gap is between rather than through the cell. When serial sections have been carried out, both intercellular and transcellular gaps have been shown, and the relative frequency of these appears to alter with the agonist. Some agonists, such as VEGF, result primarily in intracellular gaps, whereas others, such as substance P, appear to create intercellular gaps.

Fenestrations, Vacuoles, and Vesiculovacuolar Organelles

The induction of fenestrations by release of paracrine factors has been shown in response to VEGF, but does not appear to be a widespread mechanisms for inducing increased permeability. The induction of clusters of vesicles (or caveolae), or the formation of large vacuoles that span the cell occurs in response to VEGF and in tumor vessels, presumably formed from a Ca^{2+}-mediated fusion of vesicles to form continuous pathways through the cell. Ferritin tracer experiments show that these caveolae are open to perfusing substances. VVOs, vacuoles, fenestrations, and transcellular gaps may all be part of a single cascade, resulting from Ca^{2+}-mediated activation of vesicle fusion.

Glossary

Hydraulic conductivity (L_p): The convective permeability of the vessel wall to fluid flow. Can be defined as the filtration rate per unit pressure difference per unit area. Dependent on relative viscosity of fluid as it flows through the cleft, the functional pore radius (to the power 4), and the path length.

Oncotic reflection coefficient (σ): The ratio of the osmotic pressure exerted by a molecule across the vascular wall to that exerted by the solute across an ideal semipermeable membrane. Is a function of the solute area relative to the pore area, or the true solute concentration in the pore relative to the free concentration.

Solute flux (J_s): The rate of solute movement across the vessel wall. This is driven by both convection (dependent on filtration rate, reflection coefficient, and plasma solute concentration), and diffusion (dependent on the concentration gradient, the surface area, and the solute permeability).

Solute permeability (P_s): The diffusive permeability to a solute. Can be defined as the diffusive solute flux per unit concentration gradient per unit area for small solutes. Depends on the restricted diffusion coefficient, the equilibrium partition coefficient, the mean pore area, and the path length.

Reference

1. Michel, C. C., and Phillips, M. E. (1987). Steady-state fluid filtration at different capillary pressures in perfused frog mesenteric capillaries. *J. Physiol.* **388**, 421–435.

Bibliography

Adamson, R. H., and Michel, C. C. (1993). Pathways through the intercellular clefts of frog mesenteric capillaries. *J. Physiol. (Lond.)* **466**, 303–327. *The first, and still most clear, description of the tight junctional strands in the endothelial cell cleft.*

Bates, D. O., Hillman, N. J., Williams, B., Neal, C. R., and Pocock, T. M. (2002). Regulation of microvascular permeability by vascular endothelial growth factors. *J. Anat.* **200**, 581–597. *A comprehensive review of how vascular growth factors increase permeability.*

Feng, D., Nagy, J. A., Pyne, K., Hammel, I., Dvorak, H. F., and Dvorak, A.M. (1999). Pathways of macromolecular extravasation across microvascular endothelium in response to VPF/VEGF and other vasoactive mediators. *Microcirculation* **6**, 23–44. *A review of pathways through the cleft and the article that crystallizes the unifying hypothesis of transcellular transport during endothelial activation.*

Levick, J. R., and Smaje, L. H. (1987). An analysis of the permeability of a fenestra. *Microvasc. Res.* **33**, 233–256.

Michel, C. C., and Curry, F. E. (1999). Microvascular permeability. *Physiol. Rev.* **79**, 703–761. *This is the essential description of vascular permeability states. It describes, in close detail and a concise and easy-to read-style, what permeability is and how it is regulated.*

Renkin, E. M., and Tucker, V. L. (1998). Measurement of microvascular transport parameters of macromolecules in tissues and organs of intact animals. *Microcirculation* **5**, 139–152. *A review of how to correctly measure vascular permeability.*

Capsule Biography

Dr. David Bates is a British Heart Foundation Lecturer, University of Bristol Senior Research Fellow, and Director of the Microvascular Research Laboratories in the Department of Physiology at the University of Bristol, United Kingdom. He received his Ph.D. in Physiology from the University of London and postdoctoral training in the Department of Genetics, University of Glasgow, and Human Physiology, University of California at Davis. He was a lecturer in Medicine at the University of Leicester, before moving to Bristol in 1999. His research focuses on the vascular endothelial growth factors and their effects on permeability, angiogenesis, and lymphangiogenesis.

CHAPTER 32

Regulation of Vascular Endothelial Cell Signal Transduction and Phenotype by Mechanical Factors

Konstantin G. Birukov

Department of Medicine, Johns Hopkins University, Baltimore, Maryland.

Vascular endothelial cells play significant roles in regulating vascular functions in health and disease. Mechanical forces such as shear stress and cyclic stretch sequentially activate mechanosensors, intracellular signaling pathways, specific transcription factors, cytoskeletal remodeling, and the expression of genes and proteins and, as result, profoundly modulate endothelial cell functions. This article describes the nature of mechanical forces acting on vascular endothelium in physiological conditions and pathological settings and discusses signaling pathways of mechanotransduction, as well as the effects of physiological and pathological mechanostimulation on gene expression patterns in vascular endothelium.

Mechanical Forces Acting on Vascular Endothelium

Mechanical forces in the form of shear stress imposed by blood flow and mechanical strain resulting from heart propulsions and hydrostatic pressure affect all vasculature in the organism and play an important role in physiological and pathological vascular responses. However, differences in hemodynamics and additional mechanical forces resulting from respiratory cycles form unique mechanical environment experienced by pulmonary vascular endothelial cells (ECs). In systemic circulation, vascular endothelium experience higher hydrostatic pressure and shear rates (15–40 dyn/cm^2), and estimated amplitude of vessel distension caused by heart propulsions ranges within 5 to 10 percent. Pulmonary circulation is characterized by lower hydrostatic pressure and shear stress. Whereas a large body of literature exists on the effects of shear stress in endothelial cells, very little is known about effects of mechanical stretch on pulmonary endothelium. Although the degree of lung cell stretching is not known precisely in critically ill patients submitted to mechanical ventilation, many in vitro studies have used cyclic uniaxial strain in the range of 5 to 30 percent elongation to mimic physiological and pathological reactions of cell stretching, which correlated with vascular stretching caused by mechanical ventilation at high tidal volumes (20–25%). Thus, cyclic strain (or stretch) is more prominent factor in the cells of the alveolar capillaries in the lung, whereas shear forces may have differential effects on various potions of the vascular bed with gradient decline from the arterial endothelium in systemic circulation to the capillaries from systemic and pulmonary vascular beds.

Shear Stress

In large arteries, the mean wall shear stress is typically in the range of 20 to 40 dyn/cm^2 in the regions of uniform geometry and away from branch vessels. Estimated shear rates in the pulmonary artery are 10 to 20 dyn/cm^2. The actual values of shear stress for different segments of human pulmonary vascular tree have not been yet determined; however, calculations using data obtained from animal models suggest that the range of shear stress in lung arterial tree is 0.1 to 4 dyn/cm^2.

The pathological role of disturbed flow in atherogenesis is manifested by the focal distribution of atherosclerotic lesions in the bifurcations and curved regions of the arterial tree where blood flow is disturbed by flow separation, and the rates of shear stress are low and unsteady in these lesion-prone areas. Direct correlation was found between decreased shear rates and turbulent flow with increased local permeability of ECs. Experimental manipulations with flow rates and patterns in situ and in cell culture models suggest increased endothelial permeability and lipid deposition in the areas of reduced or disturbed laminar flow and completely preserved barrier properties of endothelial monolayer in the areas with normal or elevated flow. Although both pulmonary and systemic circulation are constantly exposed to varying levels of shear, lung endothelial cells experience variable flow patterns in vivo under physiological and pathological conditions that are dictated by unique features of the pulmonary circulation. For example, the distribution of blood flow throughout the human pulmonary vasculature is nonuniform and decreases from the base to the apex of the lung. In pathological situations, such as severe hypovolemia or mechanical ventilation at excessive airway pressure, apical pulmonary arterial pressures may fall below alveolar pressures, resulting in capillary collapse and cessation of blood flow. Cessation of pulmonary flow through defined segments is observed after thromboembolism, or in response to hypoxic vasoconstriction, a condition unique to the pulmonary circulation, which may result in capillary collapse. Thus, the differences in systemic and pulmonary flow patterns may dictate different mechanisms of flow-induced regulation of endothelial functions in these systems.

Mechanical Strain

Mechanical strain experienced by endothelial cells from systemic and pulmonary circulation is a superposition of pulsatile and tonic components. Tensile stress is imposed on the vascular wall by hydrostatic pressure counteracted by tonic contraction of vascular smooth muscle cells and elastic components. In addition, cyclic stretch is imposed by heart propulsions. Pulsatile distension of the arterial wall in systemic circulation normally does not exceed 10 percent to 12 percent, whereas various vasomotor reactions may change diameter of smaller caliber "resistance" arteries may reach 60 percent of initial diameter or more and last minutes or hours. Chronically increased blood pressure and vascular transmural stress activates vascular cell proliferation and collagen and fibronectin synthesis, which results in thickening of the vascular wall as a feature of hypertension-induced vascular remodeling. Direct measurements of interstitial/vascular distension in the mechanically ventilated lung are not currently available because of the complexity of local distension patterns in the lung parenchyma further complicated by uneven regional lung distension observed during inflammation and lung injury. However, studies by S. Margulies' group [1] suggest that if lung volume increases from 40 to 100 percent of total lung capacity, alveolar epithelial cell basal surface area increases by 34 to 35 percent. Pathological lung distension observed in ventilator-induced lung injury (VILI) induces alveolar and vascular barrier dysfunction, increased inflammatory cytokine production, macrophage activation, and acute inflammation that may culminate in pulmonary edema or acute respiratory distress syndrome. Several groups including ours have established cell culture models related to in vivo VILI conditions and reproduced in cell culture models cellular responses such as cytokine production and exacerbation of agonist-induced endothelial barrier dysfunction by high-amplitude cyclic stretch observed in the injured lung (Birukov et al., 2003; Dos Santos and Slutsky, 2000; Pugin et al., 1998; Vlahakis et al., 1999) [2–5]. These models are now intensively utilized in studies of pathophysiological mechanotransduction and gene expression, which will be described later.

Mechanosensors and Signal Transduction Pathways

Cell membranes, cell attachment sites, and cytoskeletal network directly experience hemodynamic forces, and most likely serve as primary mechanosensors, as was shown in pioneering work by the groups of P. Davis and J. Shyy (Davies, 1995; Shyy and Chien, 2002) [6, 7]. Cells adhere to neighboring cells and to the extracellular matrix via transmembrane receptors of cadherin (cell-to-cell) and integrin (cell-to-substrate) families. In the cytoplasmic domain, these receptors are coupled to protein complexes, which link receptors with cytoskeleton and also mediate mechanical signal transduction via activation of signaling molecules such as tyrosine (focal adhesion kinase, p60Src) kinases, serine (Erk-1,2, JNK, and p38 MAP kinases) protein kinases, inositol lipid kinases (phospholipase C), and some growth factor receptors (VEGF and PDGF receptors). Activation of mechanoreceptors triggers multiple signal cascades with ion channels (Na^+ channel, K^+ channel, chloride-selective channel) and heterotrimeric G-proteins ($G\alpha q$) being activated within seconds of mechanical stimulation, and protein kinases (protein kinase C, MAP kinases, nonreceptor protein tyrosine kinases) activated within minutes of stimulation. Protein kinase–mediated phosphorylation of specific cytoskeletal and cell contact proteins, other enzymes, and transcription factors induces cytoskeletal remodeling and stimulates gene expression in vascular cells. Figure 1 summarizes major signal pathways and cellular responses induced by shear stress and cyclic stretch.

Although most putative mechanosensors and mechanotransduction pathways are stimulated by both shear stress and stretch, the nature of mechanical forces and amplitude may still differentially affect certain signaling systems. For example, shear stress exposure selectively activates small GTPase Rac, which results in peripheral translocation of actin polymerization proteins and specific cortical actin remodeling, whereas cyclic stretch stimulates small GTPase Rho without affecting Rac and induces cytoskeletal

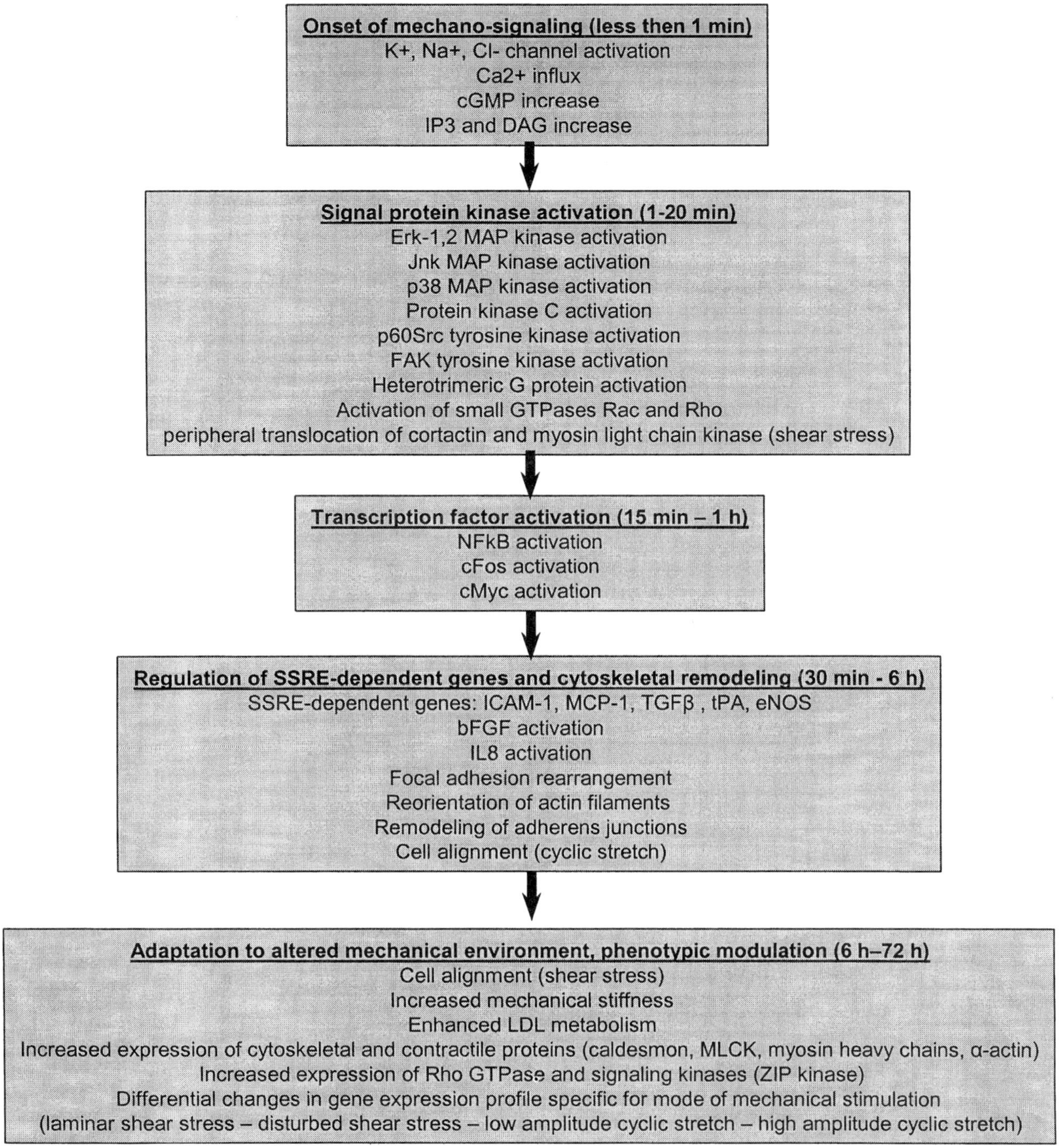

Figure 1 Major signal pathways and cellular responses induced by shear stress and cyclic stretch. Activation of mechanoreceptors triggers multiple signal cascades with ion channels (Na^+ channel, K^+ channel, chloride-selective channel) and heterotrimeric G-proteins ($G\alpha q$) being activated within seconds of mechanical stimulation, and protein kinases (protein kinase C, MAP kinases, nonreceptor protein tyrosine kinases) activated within minutes of stimulation. Protein kinase–mediated phosphorylation of specific cytoskeletal and cell contact proteins, other enzymes, and transcription factors induces cytoskeletal remodeling and stimulates gene expression in vascular cells.

remodeling distinct from shear stress. Furthermore, most protein kinases and ion channels involved in mechanochemical signaling exhibit amplitude-dependent activation that may alternatively regulate gene expression described hereafter.

Effect of Mechanical Stimulation on Endothelial Phenotype and Gene Expression Profile

Shear Stress–Regulated Gene Expression

Research on the regulation of endothelial gene expression by flow began more than 10 years ago, when expression of specific proteins and transcription of individual genes was compared between static cultures and cells exposed to laminar or disturbed flow. These studies revealed major groups of shear stress–responsive genes, which include transcription factors, cytokines, growth factors, adhesion molecules, enzymes, and cytoskeletal proteins. Analysis of shear responsive promoters identified to date showed that ability to respond to applied shear stresses can be directly linked to the presence of specific, *cis*-acting sequence elements, termed shear stress responsive elements (SSRE), which were discovered by M. Gimrone's group. Two negative and two positive SSREs within promoter regions of shear stress–regulated genes have been identified so far. In addition, two distinct groups of transcription factors mediate flow-induced induction and suppression of specific gene expression. More recently, endothelial gene regulation by shear stress was documented by several groups using DNA microarray technology [8–11]. These studies compared gene expression profiles in endothelial cells exposed to high-shear steady laminar flow, which represents "atheroprotective" flow conditions, and low-shear, nonsteady, nonunidirectional flow (disturbed flow) that simulates conditions in the atherosclerosis-prone areas of the arterial circulation. As result, thousands of endothelial genes were screened, and more than 100 shear stress–regulated genes including genes with not-yet-known function have been detected. The principal conclusions of these studies were as follows: (1) Acute shear stress transiently activates the expression of genes related to the activation of endothelium; (2) chronic laminar shear stress promotes expression of a subset of endothelial genes that can exert potent antithrombotic, antiadhesive, antiproliferative, and antioxidant effects in endothelial cells; (3) disturbed flow is not simply the absence of laminar flow, but in fact represents a distinct biomechanical stimulus that has profound impact upon the gene expression profile of endothelial cell culture; and (4) under chronic (24 hours) conditions of laminar flow, more genes are suppressed than induced, whereas disturbed flow dramatically induces proatherogenic, proinflammatory genes such as E-selectin, VCAM-1, ICAM1, PECAM-1, thrombospondin, and MCP-1. These findings support the notion about a specific shear stress–mediated EC phenotype that may play an important role in the development of vascular pathologies and responsiveness of the EC monolayer to external chemical and mechanical stimuli.

Cyclic Stretch–Regulated Gene Expression

Similar to flow, mechanical strain is also a potent regulator of gene expression. Phenotypic responses of vascular cells exposed to cyclic stretch in vitro include increased

Table I Selected Genes Differentially Regulated by Low- and High-Amplitude Cyclic Stretch.

	Fold change at 5%	Fold change at 18%
Signal transduction		
Inducible T-cell kinase	—	17.1
Nuclear receptor subfamily 1, group D, member 2	—	3.2
Angiopoietin 2	1.8	2.1
HMG-CoA-synthase	NC	4.0
Proteinase-activated receptor 2	1.4	2.1
Proteinase-activated receptor 1	NC	−1.2
Rho B GTPase	1.87	2.1
Rho C GTPase	NC	1.4
Cell adhesion		
Gap junction protein, alpha 5	—	2.8
Intercellular adhesion molecule 1 (CD54)	NC	2.0
Beta 3 Integrin	1.5	2.1
Beta-catenin	NC	2.1
Cadherin-13	8.5	1.7
Cell–cell signaling		
Placental growth factor,	1.8	2.0
VEGF-related		
Ephrin A1	1.2	2.1
Ephrin B2	1.7	2.0
Cytoskeleton		
Smooth muscle myosin heavy chain	NC	1.4
Filamin	NC	1.7
Inflammation/remodeling		
Human cyclooxygenase-2	NC	3.0
TGF-β superfamily protein	1.6	2.3
Proteinase-activated receptor 2	1.4	2.1
ZIP-kinase	1.3	1.7

Expression profiling was performed using the Affymetrix GeneChip system. Samples obtained from cells exposed to 5 percent or 18 percent cyclic stretch (48 hours) were hybridized to the Affymetrix HGU95Av2 Array (~12,000 full-length genes). Affymetrix Microarray Suite software was used to determine relative gene expression. GeneSpring and MAPPFinder software was used for microarray data analysis. Results represent fold increase in cDNA signal in cyclic stretch-preconditioned cells over static control. NC, no change; —, signal absent.

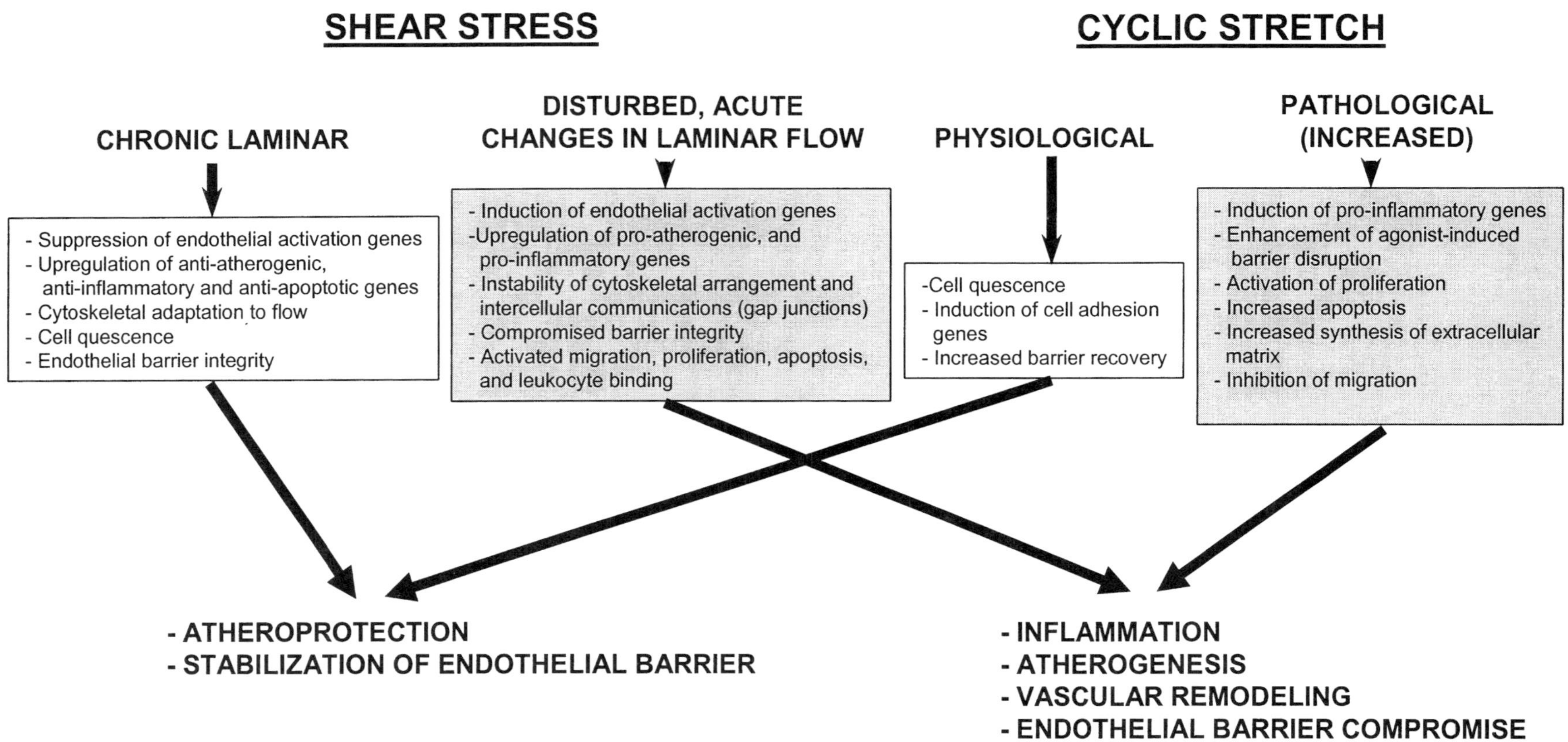

Figure 2 The differential responses of the endothelium to various patterns of shear stress and cyclic stretch. Physiologically relevant laminar shear stress and cyclic stretch promote endothelial cell quiescent state, induce expression of antiatherogenic and anti-inflammatory genes, and promote endothelial barrier integrity and recovery after edemagenic stimuli such as thrombin. In contrast, disturbed flow and pathologically increased amplitudes of cyclic stretch induce endothelial activation, extracellular matrix production, and expression of proinflammatory and apoptotic genes, which may lead to vascular remodeling and propagate endothelial barrier dysfunction, atherogenesis, and inflammatory processes in the vasculature.

expression of contractile and cytoskeletal proteins (myosin light chain kinase, smooth muscle myosin heavy chains, desmin, h-caldesmon) and increased expression of thrombin receptor PAR1 in vascular smooth muscle cells. A number of bioactive proteins regulated by cyclic stretch have been also identified in endothelial cells and macrophages and include IL-8, TGF-beta, VEGF, and monocyte chemotactic protein-1. Analysis of vascular gene expression regulated by mechanical strain reveals differential responses to physiological and pathophysiological (increased) levels of mechanical strain. For example, release of FGF-2, a growth factor involved in cellular repair after injury, is induced in vascular smooth muscle cells stretched at 14 percent and 33 percent elongation, but not at 5 percent elongation. Significant increase in IL-8 production is observed in endothelial cells exposed to cyclic stretch at 15 percent elongation, whereas stretch at 6 percent elongation did not affect IL-8 levels. In pulmonary circulation, pathological overdistension of the lung may induce inflammatory process–triggered mechanical activation of macrophages and epithelial and endothelial cells, which may cause alveolar and endothelial barrier dysfunction and vascular leak and culminate in pulmonary edema. Recently, microarray DNA technologies have been applied to assess time and amplitude dependence of cyclic stretch effects on gene expression profile in human pulmonary endothelial cells [2]. The results showed that cyclic stretch at physiologically relevant and pathological amplitudes (5% and 18% elongation, respectively) induced distinct expression patterns of genes involved in signal transduction, cytoskeletal remodeling, cell adhesion, inflammatory responses, and regulation of endothelial barrier function (Table I).

Concluding Remarks

Shear stress and tensile forces are now well recognized as factors that regulate endothelial signaling, cytoskeletal remodeling, gene expression, and physiological responses. A rapidly growing body of evidence indicates that endothelial cell discriminate between laminar and spatial gradients of shear stress, steady and pulsatile laminar flow, and low- and high-amplitude cyclic stretch. Moreover, the pattern of mechanical stimulation determines whether endothelial cells will develop pro- or anti-inflammatory cell response and also may differentially regulate endothelial barrier properties (Figure 2). Experimental and analytical tools are being developed to assess the stress distribution throughout cell structures that might be involved in mechanotransduction. Further studies will address the role of specific patterns of mechanical forces experienced by endothelium in physiological and pathological conditions (acute injury, inflammation, hypertension, ventilator-induced lung injury) and will identify key cellular targets for drug design and gene therapy.

Glossary

Mechanical strain or stretch: Change in length in relation to initial length.

Shear stress: Force per unit surface area in the direction of flow exerted at the fluid–surface interface.

Stress: Force per unit area.

References

1. Tschumperlin, D. J., and Margulies, S. S. (1999). Alveolar epithelial surface area–volume relationship in isolated rat lungs. *J. Appl. Physiol.* **86**, 2026–2033.
2. Birukov, K. G., Jacobson, J. R., Flores, A. A., Ye, S. Q., Birukova, A. A., Verin, A. D., and Garcia, J. G. (2003). Magnitude-dependent regulation of pulmonary endothelial cell barrier function by cyclic stretch. *Am. J. Physiol. Lung Cell. Mol. Physiol.* **285**(4), L785–L797.
3. Dos Santos, C. C., and Slutsky, A. S. (2000). Invited review: Mechanisms of ventilator-induced lung injury: A perspective. *J. Appl. Physiol.* **89**, 1645–1655.
4. Pugin, J., Dunn, I., Jolliet, P., Tassaux, D., Magnenat, J. L., Nicod, L. P., and Chevrolet, J. C. (1998). Activation of human macrophages by mechanical ventilation in vitro. *Am. J. Physiol.* **275**, L1040–L1050.
5. Vlahakis, N. E., Schroeder, M. A., Limper, A. H., and Hubmayr, R. D. (1999). Stretch induces cytokine release by alveolar epithelial cells in vitro. *Am. J. Physiol.* **277**, L167–L173.
6. Davies, P. F. (1995). Flow-mediated endothelial mechanotransduction. *Physiol. Rev.* **75**, 519–560. *This is a comprehensive review of endothelial biology under shear stress, which summarizes data about putative mechanosensors and signaling pathways and describes sequence of cellular events in response to flow from activation of flow-sensitive ion channels and second messenger signaling systems to cytoskeletal remodeling and regulation of gene expression.*
7. Shyy, J. Y., and Chien, S. (2002). Role of integrins in endothelial mechanosensing of shear stress. *Circ. Res.* **91**, 769–775. *This review shows a role for integrin-dependent signaling via downstream activation of protein tyrosine kinases in mechanotransduction.*
8. Brooks, A. R., Lelkes, P. I., and Rubanyi, G. M. (2002). Gene expression profiling of human aortic endothelial cells exposed to disturbed flow and steady laminar flow. *Physiol. Genomics* **9**, 27–41.
9. Chen, B. P., Li, Y. S., Zhao, Y., Chen, K. D., Li, S., Lao, J., Yuan, S., Shyy, J. Y., and Chien, S. (2001). DNA microarray analysis of gene expression in endothelial cells in response to 24-h shear stress. *Physiol. Genomics* **7**, 55–63. *This study demonstrates transactivation of growth factor receptor tyrosine kinases by mechanical forces. This mechanism converges signaling pathways triggered by mechanical and chemical factors.*
10. Garcia-Cardena, G., Comander, J., Anderson, K. R., Blackman, B. R., and Gimbrone, M. A., Jr. (2001). Biomechanical activation of vascular endothelium as a determinant of its functional phenotype. *Proc. Natl. Acad. Sci. USA* **98**, 4478–4485. *This is the first gene profiling report on the effects of laminar and disturbed shear stress in EC cultures.*
11. McCormick, S. M., Eskin, S. G., McIntire, L. V., Teng, C. L., Lu, C. M., Russell, C. G., and Chittur, K. K. (2001). DNA microarray reveals changes in gene expression of shear stressed human umbilical vein endothelial cells. *Proc. Natl. Acad. Sci. USA* **98**, 8955–8960.

Capsule Biography

Dr. Konstantin Birukov has been studying the role of mechanical factors in vascular endothelial and smooth muscle cell biology since 1989. His group primarily focuses on mechanochemical regulation of signaling, phenotypic expression and endothelial permeability. His work is supported by grants from the NHLBI and NIH (HL075349, HL076259).

CHAPTER 33

Autacoid Production by Hemodynamic Forces

Ingrid Fleming
Johann Holfgang Goethe-Universität, Frankfurt, Germany

Autacoid Production by Hemodynamic Forces

In all higher organisms the cardiovascular system represents an elaborated transport network, which is essential for the maintenance of vital functions by supplying oxygen and nutrients to tissue and removing by-products of metabolism. In order to adapt to the varying demands of the tissues, the circulatory system has evolved central and local control mechanisms that act in concert to maintain an adequate blood flow. At a given blood pressure, the blood flow to each organ is determined by the peripheral vascular resistance of this organ, which is adjusted by a variety of local mechanisms affecting the tone of the smooth muscle cells in the so-called resistance vessels, that is, small terminal arteries and large and small arterioles. Endothelial cells, situated at the interface between blood and the vessel wall, play a crucial role in controlling vascular tone and homeostasis, particularly in determining the expression of pro- and anti-inflammatory genes. Many of these effects are mediated by the generation and release of endothelium-derived autacoids, which are generally short-lived and locally acting vasoactive compounds. In vivo, endothelial cells are constantly subjected to mechanical stimulation by the viscous drag generated at the luminal surface of endothelial cells by the streaming blood (wall shear stress) and the cyclic strain of the vascular wall, which results from the pulsatile changes in blood pressure. Such stimuli not only determine acute autacoid production but can also affect the expression of autacoid-producing enzymes.

Endothelium-Derived Autacoids

The potential role of endothelial cells in flow-dependent dilation gained considerable interest after the pioneering observation by Furchgott and Zawadzki [1] that the endothelium can actively induce changes in vascular tone by the release of a labile relaxing factor. In fact, an obligatory role of endothelial cells in sensing changes in blood flow signals and transducing them into vasodilator responses was demonstrated in large conduit arteries in situ as well as in arteries in vitro. Furthermore, clinical studies revealed that the flow-dependent dilation, which occurs in different vascular beds in humans, is reduced or even abolished by hypercholesterolemia and arteriosclerosis. Although there are a number of additional endothelium-derived vasodilator and vasoconstrictor autacoids [endothelin-1, prostacyclin (PGI_2), prostaglandin H_2, the superoxide anion (O_2^-), and the endothelium-derived hyperpolarizing factor: EDHF], none of these autacoids play as central a role in the regulation of vascular tone and homeostasis as the free radical nitric oxide (NO).

Nitric Oxide

In endothelial cells NO is generated from the amino acid L-arginine by the endothelial NO synthase (eNOS), which is a constitutively expressed membrane-bound enzyme. Functional eNOS is a homodimer whose activity is determined by its interaction with a group of eNOS-associated proteins that include calmodulin, caveolin-1, and heat shock protein 90 (hsp90), as well as by intracellular levels of Ca^{2+} ($[Ca^{2+}]_i$), and the phosphorylation of several regulatory serine and threonine residues.

Distinct signaling pathways modulate the activity of eNOS in response to agonists and hemodynamic stimulation. Whereas the former is largely a Ca^{2+}-dependent process, the activation of eNOS by fluid shear stress does not require a maintained increase in $[Ca^{2+}]_i$ but is dependent on the activation of the phosphatidylinositol 3-kinase and the subsequent activation of the serine kinase Akt, which phosphorylates eNOS on Ser^{1177} (Figure 1). This step increases enzyme activity two- to four-fold and thus increases NO production. Provided the stimulus is maintained, Akt remains activated for several hours, as does the production of NO. The NO-dependent relaxation of vascular smooth muscle cells involves the activation of the soluble guanylyl cyclase and the subsequent increase in cyclic GMP levels, which, via the activation of G kinase, elicits a decrease in vascular smooth muscle cell $[Ca^{2+}]_i$.

NO is, however much more than a vasodilator and is generally thought to play a central role in maintaining the endothelium in a noninflammatory phenotype by influencing the activity of several transcription factors and thus attenuating the expression of proinflammatory genes such as the adhesion molecules E-selectin and vascular cell adhesion molecule-1 (VCAM-1). With the progression of vascular disease, NO-dependent vasodilatation becomes impaired, partly as a consequence of the interaction of NO with O_2^- generated by enzymes such as NADPH oxidase, to generate the strong oxidant peroxynitrite ($ONOO^-$) and partly as a consequence of eNOS uncoupling. The latter phenomenon relates to a switch from the production of NO to the generation of O_2^- by eNOS and has been linked to decreased cellular levels of the essential cofactor tetrahydrobiopterin. The consequence of these changes is a decrease in the bioavailability of NO and a reduction in the vasodilator response to agonists such as acetylcholine as well as the activation of redox-sensitive transcription factors (e.g., nuclear factor kappa B [NF-κB]), the activation of which is normally suppressed by NO.

Prostacyclin

Cyclo-oxygenase (COX) is the key enzyme in the synthesis of prostanoids, and two COX isoforms (COX1 and COX-2) have been identified in endothelial cells. Although these enzymes catalyze the same reaction, they have distinct pharmacological profiles and biological roles. COX-1 is constitutively expressed in endothelial cells, whereas COX-2 is expressed in other cell types in response to proinflammatory stimuli. The availability of the COX substrate, arachidonic acid, which is released from membrane phospholipids by phospholipase A_2 (PLA_2), is the rate-limiting step in prostaglandin synthesis, and PGI_2 is the major product of the subsequent COX pathway in all arterial and venous endothelial cells studied so far. As a consequence of the Ca^{2+} sensitivity of PLA_2, the synthesis of PGI_2 by endothelial cells is a Ca^{2+}-dependent process. This codependency on Ca^{2+} means that in many circumstances NO, PGI_2, and EDHF can be generated simultaneously.

The apparent redundancy in the major vasoactive autacoid generating pathways means that the loss of one vasodilator can be compensated for by one of the others. For

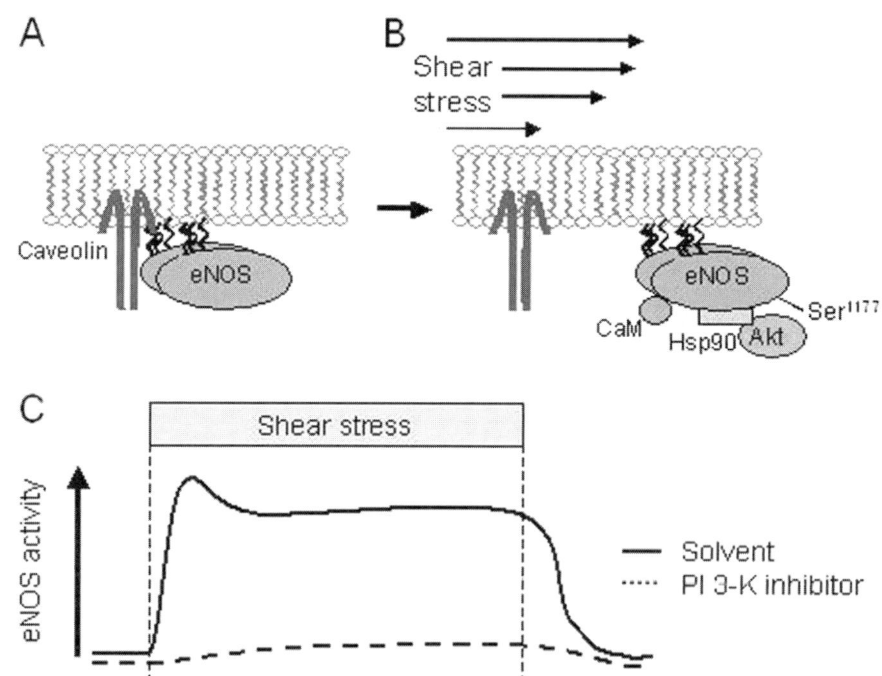

Figure 1 Endothelial cell activation by fluid shear stress increases eNOS phosphorylation and activity. In unstimulated endothelial cells the eNOS dimer is associated with caveolin, which inhibits its activation, which means that NO output is low **(A)**. In response to the application of fluid shear stress, the phosphatidylinositol 3-kinase (PI 3-K) is activated and in turn activates Akt, which associates with the eNOS scaffolding protein hsp90 and phosphorylates eNOS on Ser^{1177} **(B)**. This step elicits the dissociation from caveolin and increases NO production. Shear stress–induced increases in endothelial NO production can be abrogated by preventing the activation of PI 3-K **(C)**. (see color insert)

example, in wild-type mice flow-dependent vasodilatation in gracilis muscle arterioles appears to be mediated by NO. However, in male eNOS$^{-/-}$ mice, flow-induced vasodilatation of these arterioles can be attributed to the generation of PGI_2.

The Endothelium-Derived Hyperpolarizing Factor

In various blood vessels, endothelium-dependent relaxations can be accompanied by the endothelium-dependent hyperpolarization of smooth muscle cells. These endothelium-dependent relaxations and hyperpolarizations can be partially or totally resistant to inhibitors of COX and NO synthases and could be observed without an increase in intracellular levels of cyclic nucleotides in the smooth muscle cells. Therefore the existence of an additional pathway that involves smooth muscle hyperpolarization was suggested and attributed to an endothelium-derived hyperpolarizing factor. The contribution of EDHF-mediated responses as a mechanism for endothelium-dependent relaxation increases as the vessel size decreases, apart from the coronary and renal vascular beds in which EDHF also plays a major role in conduit arteries. Different mechanisms appear to underlie EDHF-mediated responses in different arteries; however, experimental evidence favors three explanations for the EDHF phenomenon (Figure 2):

1. K^+ ions are released from stimulated endothelial cells through Ca^{2+}-activated K^+ (K_{Ca}) channels and subsequently elicit relaxation by activating K^+ channels and/or the Na^+/K^+-ATPase on vascular smooth muscle cells, thus inducing hyperpolarization and decreasing the open probability of voltage-operated Ca^{2+} channels, through which the Ca^{2+} required for contraction enters the cell.
2. An increase in endothelial $[Ca^{2+}]_i$ triggers the activation of a phospholipase A_2 that liberates arachidonic acid from membrane phospholipids. The arachidonic acid is then metabolized by cytochrome P450 epoxygenase(s) to epoxyeicosatrienoic acids (EETs), which either contribute to the hyperpolarization of endothelial cells or are released from endothelial cells to activate K^+_{Ca} channels on underlying vascular smooth muscle cells.
3. Direct transfer of an electrical signal (i.e., endothelial cell hyperpolarization) to vascular smooth muscle cells via myo-endothelial gap junctions. This mechanism appears to be particularly important in the microcirculation.

The majority of the work performed to date that has led to the characterization of EDHF-mediated responses has involved the use of receptor-dependent agonists such as acetylcholine and bradykinin. However, the limited studies that have addressed the response to hemodynamic stimuli suggests that pulsatile stretch is a particularly important stimulus for the generation of EDHF. This phenomenon may be related to the fact that pulsatile stretch elicits much larger and more prolonged increases in $[Ca^{2+}]_i$ in endothelial cells than fluid shear stress does. Indeed, K_{Ca} channel-dependent hyperpolarization is preferentially stimulated by pulsatile flow as opposed to steady flow, and pulsatile stretch/cyclic strain has been reported to increase the generation of EETs by native porcine coronary endothelial cells as well as to increase the expression of the EET-generating cytochrome P450 epoxygenase.

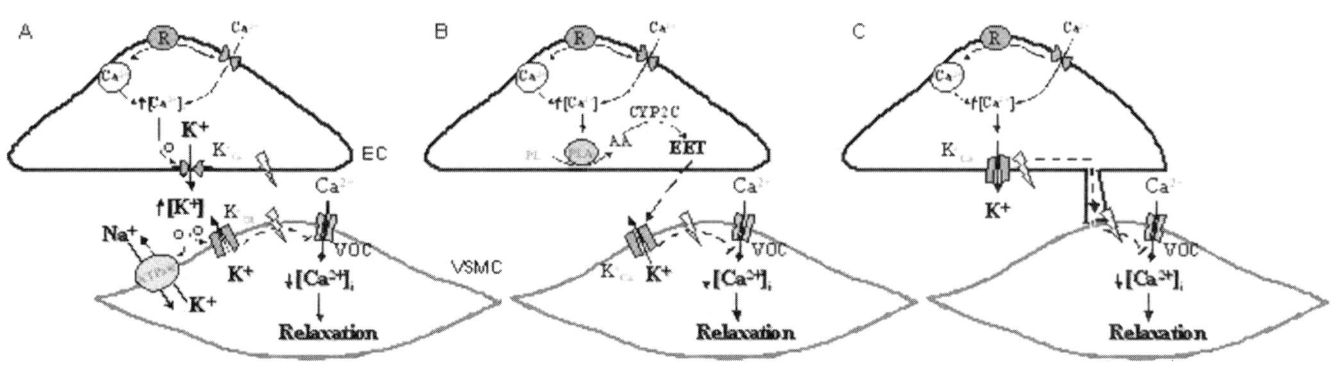

Figure 2 Three models contributing to endothelium-derived hyperpolarizing factor-mediated relaxation. **(A)** The K^+ ion hypothesis. In response to the receptor (R)-dependent activation of endothelial cells (EC), intracellular concentration of Ca^{2+} ($[Ca^{2+}]_i$) increases and activate Ca^{2+}-dependent K^+ (K^+_{Ca}) channels, which results in the efflux of K^+ ions into the subintimal space and endothelial cell hyperpolarization (lightning symbol). The increase in extracellular K^+ activates inwardly rectifying K^+ channels (K^+IR) and/or the Na-K-ATPase which elicit vascular smooth muscle cell (VSMC) hyperpolarization and decrease the open probability of voltage-dependent Ca^{2+} (VOC) channels. As a consequence $[Ca^{2+}]_i$ in the smooth muscle cells decreases and relaxation occurs. **(B)** The cytochrome P450 hypothesis. An increase in $[Ca^{2+}]_i$ activates a phospholipase A_2, which liberates arachidonic acid (AA) from membrane lipids. Arachidonic acid is then metabolized to epoxyeicosatrienoic acids (EETs) by a cytochrome P450 2C (CYP 2C) epoxygenase. EETs then increase the open probability of K^+_{Ca} in the underlying smooth muscle cells to elicit hyperpolarization and relaxation. **(C)** The gap junction hypothesis. Endothelial cell activation as described elicits endothelial cell hyperpolarization by activating K^+_{Ca} channels. This hyperpolarization is, however, transferred to smooth muscle cells via myo-endothelial gap junctions rather than by the release of an endothelium-derived factor. The final steps are the same, that is, inhibition of Ca^{2+} entry into smooth muscle cells and relaxation. (see color insert)

Hemodynamic Forces and the Endothelium

Shear Stress

Blood flow through a vascular segment generates a viscous drag at the luminal surface of endothelial cells (Figure 3), which can be expressed as shear stress (τ) and calculated according to Poiseuille's law:

$$\tau = 4\eta Q/r4\pi$$

(where η = viscosity, Q = blood flow, r = internal vessel radius).

This relation highlights the fact that relatively small decreases in vessel diameter at constant flow can markedly increase shear stress at the endothelial surface. It should be noted, however, that shear stress is calculated under the assumption that the flow profile is parabolic, a situation that is never realized in vivo because of the pulsatile nature of blood flow and the fact that blood vessels are distensible. Branching and bifurcations along the vascular tree together with a gradual decrease in diameter also disfavor the establishment of a stable flow profile.

Pulsatile Stretch/Cyclic Strain

Although it is sometimes assumed that pressure per se is a proper physiological stimulus for the vascular wall, it must be emphasized that the application of pressure in the physiological range (below 1 atm) to a tissue that is essentially incompressible because of its high water content has no direct effect on cell function. Rather, the pressure exerted on a compliant tissue elicits deformation, that is, either distension (stretch) or compression. As a result of pulsatile pressure changes, cells within the vascular wall are subjected to three-dimensional cyclic strain in the radial, longitudinal, and circumferential directions. There are also cell–cell-generated forces that are influenced by hemodynamic stimuli but that cannot be expressed by a simple physical relationship. One example is the isometric contraction in which the development of contractile force within the smooth muscle cell layer counteracts the distending transmural pressure. Under such conditions there is a relative displacement of opposing cell layers within the vascular wall (e.g., smooth muscle cells versus elastic lamina and endothelial cells), despite the fact that no net movement occurs. Although the displacement induced may be subtle, the close physical arrangement of endothelial focal adhesion contacts and the smooth muscle would tend to suggest that the forces developed at the abluminal surface of endothelial cells may be greater than those generated by shear stress on the luminal surface.

In principle, the relative contribution of pulsatile stretch and wall shear stress to the adjustment of local vascular tone is likely to depend on a number of factors including vessel

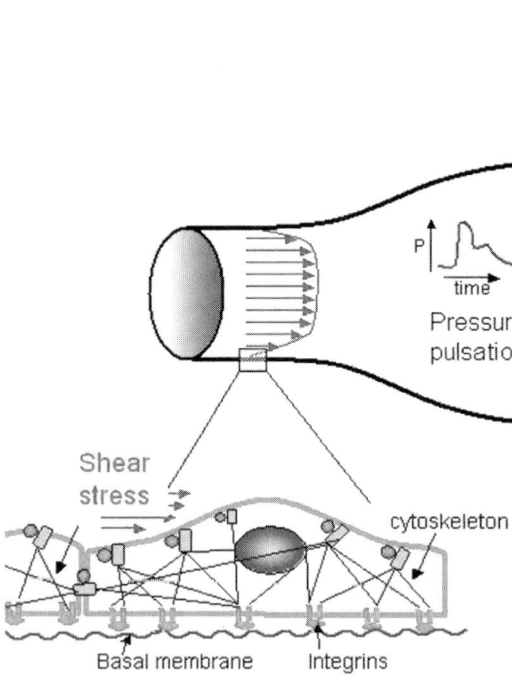

Figure 3 Scheme illustrating the hemodynamic forces affecting endothelial cells. Note that, because of the pulsatile nature of flow, the flow profile is flattened compared to the parabolic flow profile present under steady-state conditions, resulting in a greater shear stress at the endothelial cell surface. The simultaneous pressure (P) pulse distends the vessel diameter (D) and stretches endothelial cells in the range of 1–5% strain. These stimuli result in a redistribution of force across the endothelial cell cytoskeleton and the activation of signaling molecules (including integrins) at the abluminal and luminal cell surfaces as well at sites of cell-cell contact. Adapted from Busse, R., and Fleming, I. (2003). Regulation of endothelium-derived vasoactive autacoid production by hemodynamic forces. *Trends Pharmacol. Sci.* **24**, 24–29, reprinted with permission from Elsevier Science Ltd. (see color insert)

architecture, moment-to-moment changes in smooth muscle activity, and the ability of endothelial and smooth muscle cells to sense hemodynamic stimuli and to produce vasoactive factors. Although it is tempting to break down hemodynamically generated forces into singular physical components such as circumferential stretch/strain and wall shear stress, it must be emphasized that such a procedure is purely theoretical, as changes in pulsatile stretch and shear stress are inextricably linked.

Mechanochemical Coupling

The term mechanochemical coupling describes how endothelial cells translate physical stimuli into intracellular signals and ultimately into changes in autacoid production. Initially, one or more mechanoreceptors were suggested to sense changes in fluid shear stress at the endothelial cell surface; however, the endothelial cell cytoskeleton itself can be regarded as a mechanoreceptor. The cytoskeletal frame of an endothelial cell is composed of actin filaments, intermediate filaments, and microtubules that transverse the cells and end in characteristic adhesion complexes at the luminal cell surface (caveolae), at the abluminal surface (focal adhesion points), or at cell–cell boundaries. Signaling molecules are clustered around and inherent in these sites so that it is conceivable that fluid shear stress acting on the luminal surface of the endothelial cell is transmitted through the entire cell by the cell cytoskeleton to activate signal transduction cascades in specific signaling hot spots.

Other possible mechanisms by which endothelial cells may sense shear stress are through mechanosensitive ion channels and the glycocalyx, a layer of glycoproteins extending into the extracellular space, which may be displaced by shear stress to elicit an intracellular response. Alternatively, flow induces changes in the concentrations of endothelial agonists (e.g., ATP, ADP, and bradykinin) in the unstirred boundary layer at the cell surface, and thus may affect intracellular signaling pathways.

Integration of the Vascular Response (Ascending Dilatation)

The resistance network that controls blood flow to skeletal muscle comprises terminal arterioles, as well as larger arterioles and the small (feeder) arteries from which they derive. In small (terminal) arterioles the products of energy metabolism are effective dilators, but the accumulation of metabolites alone has relatively little effect on vascular resistance. The reason for this is that in order to achieve optimal vascular conductance, both small arterioles and the larger arterioles feeding them must dilate in concert. More than 70 years ago the German physician A. Schretzenmayer provided the first experimental evidence for flow-induced dilation. In the hind legs of anesthetized cats he showed that whenever blood flow to the leg was increased, there was a concomitant increase in the diameter of the feeding femoral artery. It was concluded that this flow-dependent dilator response, which improves conductivity of feeding vessels, was due to a tissue-derived signal transmitted along the vascular tree. Since distal transection did not impede the dilation of conduit arteries in response to flow, it became evident that this dilator response was a locally generated phenomenon of the vascular wall. Further studies led to the concept of an "ascending or conducted dilation" in conduit arteries under conditions of high tissue oxygen demand. This coordinated longitudinal transmission of vasomotor responses is essential for the optimization of vascular conductivity and organ perfusion.

Experimentally, ascending dilatation can be studied in vivo and in vitro by assessing the response to a vasoactive substance at the point of application (local response) as well as at a remote site. This conducted vasodilatation can travel bidirectionally and the amplitude of the conducted vasodilatation is generally smaller than that of the local response. Although there is a gradual decline in the conducted vasodilatation along some arterioles, there is no obvious decay of the conducted response along feed arteries.

Conducted responses have been intensively investigated, but the exact mechanism remains to be clarified. Most researchers agree that NO does not play a major role in this response. Indeed, conducted vasodilatation in response to a number of stimuli is not affected by NOS inhibitors and the response is apparently intact in eNOS-deficient mice. Changes in membrane potential appear to be central to the phenomenon of ascending dilatation, and responses are generally attributed to the propagation of a hyperpolarization along the vascular wall that is linked either to the actions of an EDHF or to the direct transmission of an electrical signal between vascular cells. Over the past few years evidence has accumulated to suggest that homocellular as well as heterocellular gap junctional communication are involved in the phenomenon of conducted dilatation.

At this point, it is necessary to note that not all vessels within one vascular bed respond in the same way in response to an accumulation of vascular metabolites or to mechanical stimuli (transmural pressure and shear stress). Metabolic control exerts a dominant influence on the smallest arterioles ($< 20\,\mu m$), but is a generally less important stimulus in more upstream vessels. The pressure-induced myogenic response dominates in middle-sized arterioles, whereas its influence wanes upstream in relatively large arterioles (100–$200\,\mu m$) in which fluid shear stress predominantly governs tone.

Glossary

Ascending dilation: In a contracting skeletal muscle, low PO_2 and vasoactive metabolites elicit a local response and initiate a conducted vasodilatation that "ascends" the vascular tree to induce the simultaneous vasodilatation of the feed arteries as well as branch arteries and thus increase blood flow. This coordinated longitudinal transmission of vasomotor responses is essential to achieve optimal organ perfusion.

Autacoid: From the Greek *autos* [self] and *akos* [remedy]. Endothelium-derived autacoids such as NO, PGI_2, and O_2^- are generally short-lived and locally acting.

Endothelium-derived hyperpolarizing factor: Describes endothelium-dependent relaxation that is not mediated by either nitric oxide or prostacyclin but is temporally correlated with endothelial cell hyperpolarization and followed shortly thereafter by smooth muscle cell hyperpolarization and relaxation.

Myogenic tone: An intrinsic mode of control of activity in which the stretch of the vascular smooth muscle cell membrane results in the activation of stretch-sensitive channels. The result is depolarization and contraction.

Reference

1. Furchgott, R. F., and Zawadzki, J. V. (1980). The obligatory role of endothelial cells in the relaxation of arterial smooth muscle by acetylcholine. *Nature* **288**, 373–376.

Further Reading

Busse, R., Edwards, G., Feletou, M., Fleming, I., Vanhoutte, P. M., and Weston, A. H. (2002). EDHF: Bringing the concepts together. *Trends Pharmacol. Sci.* **23**, 374–380. *This article is the first serious attempt by three different groups to take all of the current concepts relating to EDHF-mediated relaxation and combine them in one (more or less) unifying concept. This paper is an excellent overview and a great help to those confused by the apparent inconsistencies in the literature.*

Davidge, S. T. (2001). Prostaglandin H synthase and vascular function. *Circ. Res.* **89**, 650–660.

Fleming, I., and Busse, R. (2000). Activation of NOS by Ca^{2+}-dependent and Ca^{2+}–independent mechanisms. In *Nitric Oxide* (L. J. Ignarro, ed.), San Diego: Academic Press, pp. 621–632.

Fleming, I., and Busse, R. (2003). Molecular mechanisms involved in the regulation of the endothelial nitric oxide synthase. *Am. J. Physiol. Regul. Integr. Comp. Physiol.* **284**, R1–R12.

Govers, R., and Rabelink, T. J. (2001). Cellular regulation of endothelial nitric oxide synthase. *Am. J. Physiol. Renal. Physiol.* **280**, F193–F206.

Ingber, D. E. (2003). Tensegrity I. Cell structure and hierarchical systems biology. *J. Cell Sci.* **116**, 1157–1173.

Ingber, D. E. (2003). Tensegrity II. How structural networks influence cellular information processing networks. *J. Cell Sci.* **116**, 1397–1408.

A pair of state-of the-art articles that describe in detail how the architecture of endothelial cells can influence signaling in response to physical stimuli.

Segal, S. S. (2000). Integration of blood flow control to skeletal muscle: key role of feed arteries. *Acta Physiol. Scand.* **168**, 511–518.

An excellent overview of the significance of ascending dilation and the integrated control of vascular networks.

Capsule Biography

Ingrid Fleming (although from Northern Ireland) is a Professor of Physiology at the Johann Wolfgang Goethe University in Frankfurt. Her research interests center on signal transduction in vascular cells, in particular on the regulation of nitric oxide synthase activity and the phenomenon of nitric oxide/prostacyclin or endothelium-derived hyperpolarizing factor (EDHF)-mediated vascular relaxation.

CHAPTER 34

Ion Channels and Arteriolar Tone

William F. Jackson
Michigan State University, East Lansing, Michigan

Introduction

Arteriolar tone refers to the degree of contraction of smooth muscle cells that wrap around these microvessels. Smooth muscle contractile activity (i.e., arteriolar tone) determines the diameter of the arteriolar lumen, which, in turn, sets the hydraulic resistance of these microvessels to blood flow. Thus, arteriolar tone importantly contributes to regulation of systemic vascular resistance and blood pressure; the distribution of blood flow to and within the body's tissues and organs; and regulation of solute and water exchange in downstream capillaries and venules. Arteriolar smooth muscle cells integrate a myriad of excitatory and inhibitory stimuli from blood pressure, circulating hormones, neurotransmitters, local metabolites, and endothelium-derived factors to determine arteriolar tone, and hence arteriolar diameter and resistance. Ion channels, proteins that form ion-selective, water-filled pores in the plasma membrane of cells, in both smooth muscle and endothelial cells, play a central role in determining and regulating arteriolar tone. Ion channels provide the major source of intracellular Ca^{2+} that activates contractile proteins in smooth muscle and stimulates the synthesis and release of vasoactive substances from endothelial cells. In addition, ion channels significantly contribute to the determination and regulation of membrane potential. Membrane potential, in turn, modulates Ca^{2+} influx through ion channels, the release of Ca^{2+} from intracellular stores, and cell Ca^{2+} sensitivity. Ion channel–mediated changes in membrane potential also are transmitted along microvessels to modulate arteriolar tone at distant sites. Thus, ion channels participate in all aspects of regulation of arteriolar tone.

Ion Channels Expressed in Arterioles

Vascular smooth muscle and endothelial cells in the wall of arterioles express a large number of ion channels (Table I). These include one or more varieties of voltage-gated Ca^{2+} channels, at least four different classes of K^+ channels, several types of nonselective cation channels and at least two classes of Cl^- channels. All of these channels arise from multigene families that also display splice variation. Table I shows the isoforms of channel proteins that are thought to underlie the ion channels present in arteriolar vascular smooth muscle and endothelium. In addition, blockers or inhibitors of the channels are shown. Please note that with the exception of a few substances, such as the scorpion toxin, iberiotoxin, that selectively blocks large-conductance Ca^{2+}-activated K^+ channels, most of the substances listed show only limited selectivity. For example, all of the inhibitors of Ca^{2+}-activated and swelling-activated Cl^- channels also inhibit voltage-gated Ca^{2+} or nonselective cation channels, such that their use in functional studies is very limited.

Voltage-Gated Ca^{2+} Channels

Voltage-gated Ca^{2+} channels in smooth muscle cells play a major role in the regulation of arteriolar tone. Importantly, they transduce changes in vascular smooth muscle membrane potential (produced by other ion channels and transporters) into changes in Ca^{2+} influx and intracellular Ca^{2+}. Signals that depolarize smooth muscle open voltage-gated Ca^{2+} channels, allowing Ca^{2+} to diffuse into cells down their electrochemical gradients, raising intracellular Ca^{2+}, leading to smooth muscle contraction and increased arteriolar tone. Conversely, hyperpolarization of smooth muscle closes these channels, reducing intracellular Ca^{2+} and leading to decreases in arteriolar tone.

The dominant voltage-gated Ca^{2+} channels expressed in vascular smooth muscle cells are nifedipine-sensitive, L-type Ca^{2+} channels that begin to activate at −50 mV under physiological conditions. Smooth muscle cells in some arterioles also express other types of voltage-gated Ca^{2+} channels. These include T-type channels that are insensitive to

Table I Ion Channels in Arteriolar Smooth Muscle and Endothelial Cells.

Ion channel[a]	VSM[b]	ENDO[c]	Additional subunits	Conductance (pS)	Primary stimuli	Inhibitor(s)[d]
L-type Ca^{2+}	CaV 1.2b	— or ?[e]	β, $\alpha_2\delta$	25	Depolarization (> −50 mV)	Nifedipine, diltiazem, verapamil, $Cd^{2+} > Ni^{2+}$
T-type Ca^{2+}	CaV 3.1 or 3.2?	— or CaV 3.1	$\alpha_2\delta$, γ?	8	Depolarization (> −70 mV)	Kurtoxin, mibefradil, pimozide, flunarizine $Ni^{2+} > Cd^{2+}$
R-type Ca^{2+}	CaV 2.X?[e]	— or ?	β, $\alpha_2\delta$?	13	Depolarization (> −50 mV)	Nimodipine, amiloride, mibefradil, $Cd^{2+} > Ni^{2+}$
SOC	TRPC 1	TRPC 1?	TRPC 4 or 5?	16	Release of Ca^{2+} from internal stores, PKC	2-APB, Ni^{2+}, Cd^{2+}, Gd^{3+}, La^{3+}
SAC	TRPC 6 or TRPM4	?	?	25–30	Membrane stretch, PKC	*G. spatulata* venom, SKF 96365, amiloride, Gd^{3+}, La^{3+}
ROC	TRPC 6?	?	?	25–30	DAG, PKC?	SKF 96365, amiloride, Gd^3, La^{3+}
BK_{Ca}	Slo1	— or ?	Slo β	250	Depolarization, $\uparrow[Ca^{2+}]_{in}$, NO, CO, EETs	Iberiotoxin, charybdotoxin, penitrem A, paxillin, 1 mM TEA
sK_{Ca}	— or ?	SK3	Calmodulin	4–20	$\uparrow[Ca^{2+}]_{in}$	Apamin, d-tubocurarine, TBA
IK_{Ca}	—	IK1	Calmodulin	20–80	$\uparrow[Ca^{2+}]_{in}$	TRAM-39, TRAM-34, charybdotoxin, clotrimazole, TBA
K_V	KV 1.5, 1.6?	KV 1.5, 1.6?	K_V β	5–8	Depolarization	4-aminopyridine, correolide, Agitoxin-2
K_{ATP}	KIR 6.1	KIR 6.1 or 6.2?	SUR2b	50–100?	$\downarrow$ATP, $\uparrow$ADP, PKA, PKG	Glibenclamide, tolbutamide, TPA, Ba^{2+}
K_{IR}	KIR 2.1	KIR 2.1	—	20	Hyperpolarization, $\uparrow[K^+]_{out}$	Ba^{2+}, quinidine, phencyclidine
Cl_{SW}	— or CLC 3?	?	?	1?	$\uparrow$Cell volume, membrane stretch	Tamoxifen, DIDS, niflumic acid, IAA-94, 9-AC, NPPB
Cl_{Ca}	ClCA 4?	?	?	1–3	$\uparrow Ca^{2+}$	Niflumic acid, flufenamic acid, 9-AC DIDS, NPPB

[a] See text for definitions of channel abbreviations.
[b] Vascular smooth muscle.
[c] Endothelium.
[d] Inhibitor abbreviations: 2-APB, 2-aminoethoxydiphenyl borate; TEA, tetraethylammonium; TBA, tetrabutylammonium; TPA, tetrapentylammonium; DIDS, 4,4′-diisothiocyanatostilbene-2,2′-disulfonic acid; IAA-94, indanyloxyacetic acid; 9-AC, 9-anthracenecarboxylic acid; NPPB, 5-nitro-2-(3-phenylpropylamino)benzoic acid.
[e] —, Not present; ?, present, but specific isoform unclear, or mechanism unclear.

nifedipine, activate at very negative membrane potentials (−70 mV), and inactivate rapidly, and nifedipine-insensitive, high voltage-activated Ca^{2+} channels (R-type Ca^{2+}) with activation properties similar to those of L-type channels.

The presence of voltage-gated Ca^{2+} channels in endothelial cells remains controversial. Studies of endothelial cells from conduit arteries have uniformly failed to demonstrate their presence. However, a few studies of microvascular endothelial cells have demonstrated the presence of both L-type and T-type voltage-gated Ca^{2+} channel currents. The generality of these findings has not been established and the physiological function of endothelial cell voltage-gated Ca^{2+} channels in the regulation of arteriolar tone is unknown.

K^+ Channels

Arteriolar smooth muscle cells express at least four different classes of K^+ channels in their membranes. These include large conductance, Ca^{2+}-activated K^+ (BK_{Ca}) channels, several types of voltage-gated K^+ (K_V) channels, ATP-sensitive K^+ (K_{ATP}) channels, and inward rectifier K^+ (K_{IR}) channels. All appear to contribute to the regulation of arteriolar tone in the microcirculation. As in other cells in the body, K^+ channels play a major role in determining membrane potential. Opening of K^+ channels causes K^+ ions to diffuse out of cells, down their electrochemical gradients. This loss of positive charge causes membrane hyperpolarization, closure of voltage-gated Ca^{2+} channels, and decreased arteriolar tone. Conversely, closure of open K^+ channels leads to membrane depolarization and opening of voltage-gated Ca^{2+} channels, leading to increased arteriolar tone.

Microvascular endothelial cells also express multiple types of K^+ channels including two types of Ca^{2+}-activated K^+ channels, K_V channels, K_{ATP} channels, and K_{IR} channels. These K^+ channels also significantly contribute to the regulation of membrane potential in endothelial cells that, itself, can act as a signal to regulate arteriolar tone (see later discussion). In addition, membrane potential determines the electrochemical gradient for Ca^{2+} influx through nonselective cation channels, that contributes to the production of endothelium-derived vasodilators such as NO, prostacyclin, and epoxides of arachidonic which also participate in the regulation of arteriolar tone (see Figure 4). Potassium channels also may allow changes in membrane potential to be conducted, in a nondecremental fashion, along the length of arterioles, further contributing to the regulation of arteriolar tone.

BK_{Ca} CHANNELS

Large-conductance, Ca^{2+}-activated K^+ channels represent one of the most abundant K^+ channels expressed in smooth muscle membranes. These channels open with membrane depolarization and increases in intracellular Ca^{2+} and are modulated by a number of cell signaling pathways. They play a major role in the negative feedback regulation of smooth muscle membrane potential during vasoconstriction (see Figure 2) and participate in the mechanism of action of several vasodilators (see Figure 3). Like many ion channels, BK_{Ca} channels exist in signaling complexes with voltage-gated Ca^{2+} channels, protein kinases (protein kinase A, cGMP-dependent protein kinase, protein kinase C), phosphatases, and other signaling proteins (Figure 1). These macromolecular signaling complexes also appear to be located adjacent to smooth endoplasmic reticulum Ca^{2+}-activated, Ca^{2+}-release channels (ryanodine receptors, RYR) such that focal release of Ca^{2+} from RYR triggered by Ca^{2+} influx through voltage-gated Ca^{2+} channels (i.e., Ca^{2+} sparks) can activate BK_{Ca} channels in smooth muscle. Microvascular endothelial cells likely do not express BK_{Ca} channels in vivo.

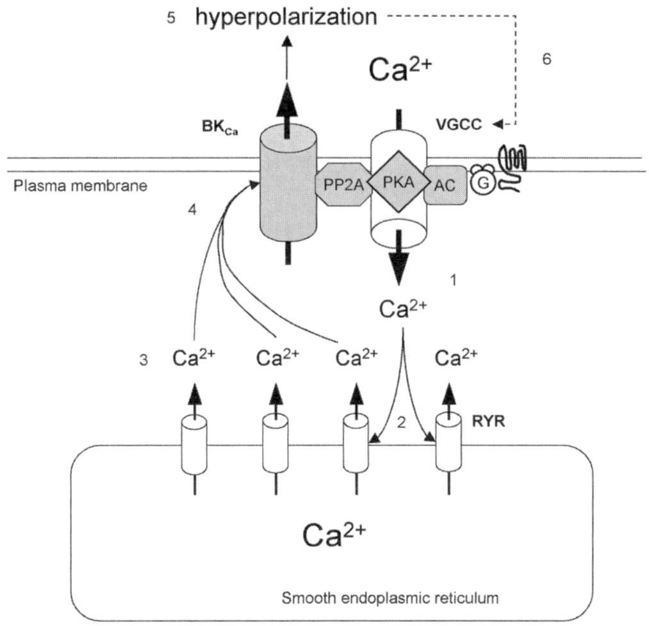

Figure 1 Ion channels form signaling complexes. Simplified schematic diagram of a signaling complex in a smooth muscle cell consisting of voltage-gated Ca^{2+} channels (VGCC), G-protein-coupled receptors, adenyl cyclase (AC), protein kinase A (PKA), protein phosphatase 2A (PP2A), and large-conductance Ca^{2+}-activated K^+ channels (BK_{Ca}). The VGCC and BK_{Ca} channels are functionally coupled to ryanodine receptors (RYR) in the membranes of smooth endoplasmic reticulum (SER) adjacent to the plasma membrane. **1**, Ca^{2+} influx through VGCC increases local Ca^{2+} concentration in the sub-plasmalemmal space. **2**, This local increase in Ca^{2+} triggers Ca^{2+}-induced Ca^{2+} release from the SER through RYR. **3**, This Ca^{2+} release results in a focal increase in subplasmalemmal Ca^{2+} (a Ca^{2+} spark). **4**, The Ca^{2+} spark, along with Ca^{2+} entry through VGCC activates BK_{Ca} channels. **5**, The resultant K^+ efflux hyperpolarizes the membrane. **6**, The hyperpolarization closes VGCC providing negative feedback control of the system.

sK_{Ca} AND IK_{Ca} CHANNELS

Native endothelial cells in arterioles express two types of Ca^{2+}-activated K^+ channels: small conductance, sK_{Ca} channels, and intermediate conductance, IK_{Ca} channels that are distinct from BK_{Ca} channels. These channels are responsible for agonist-induced hyperpolarization of endothelial cells and thus play an important role in regulation of arteriolar tone. In addition to having much smaller single channel conductances, both channels are insensitive to the highly selective BK_{Ca} channel blocker, iberiotoxin, and are not voltage-gated such that they are easily distinguished from BK_{Ca} channels. Both sK_{Ca} and IK_{Ca} channels require calmodulin to display Ca^{2+}-dependent modulation, unlike BK_{Ca} channels that bind Ca^{2+} directly.

K_V CHANNELS

Arteriolar smooth muscle cells express a number of different K_V channels and there are species and regional differences in the specific gene products that are expressed. As their name implies, K_V channels are activated by membrane depolarization. These channels participate in the regulation of resting membrane potential, the negative feedback regulation of membrane potential during vasoconstriction (Figure 2), and may participate in the mechanism of action of both vasoconstrictors (Figure 2) and vasodilators (Figure 3). As noted in Table I, most K_V channels expressed in arteriolar smooth muscle can be blocked by aminopyridines such as 4-aminopyridine. In addition, KV 1 family members can be selectively inhibited by the triterpene correolide. Agitoxin-2 also blocks KV 1 family channels with the notable exception of KV 1.5 and can be used to distinguish between KV 1.5 and other KV 1.X–mediated responses. Microvascular endothelial cells also may express K_V channels that function in the negative feedback regulation of membrane potential to limit depolarization.

K_{ATP} CHANNELS

Both arteriolar smooth muscle and endothelial cells express K_{ATP} channels. These channels open when intracellular ATP levels decrease; hence their name. However, they also may be activated by elevated nucleotide diphosphates, reduced intracellular pH, and phosphorylation by protein kinase A (PKA) and cGMP-dependent protein kinase (PKG). In both smooth muscle and endothelial cells, they may act as sensors of the metabolic status of cells, opening during ischemic or hypoxic conditions in response to

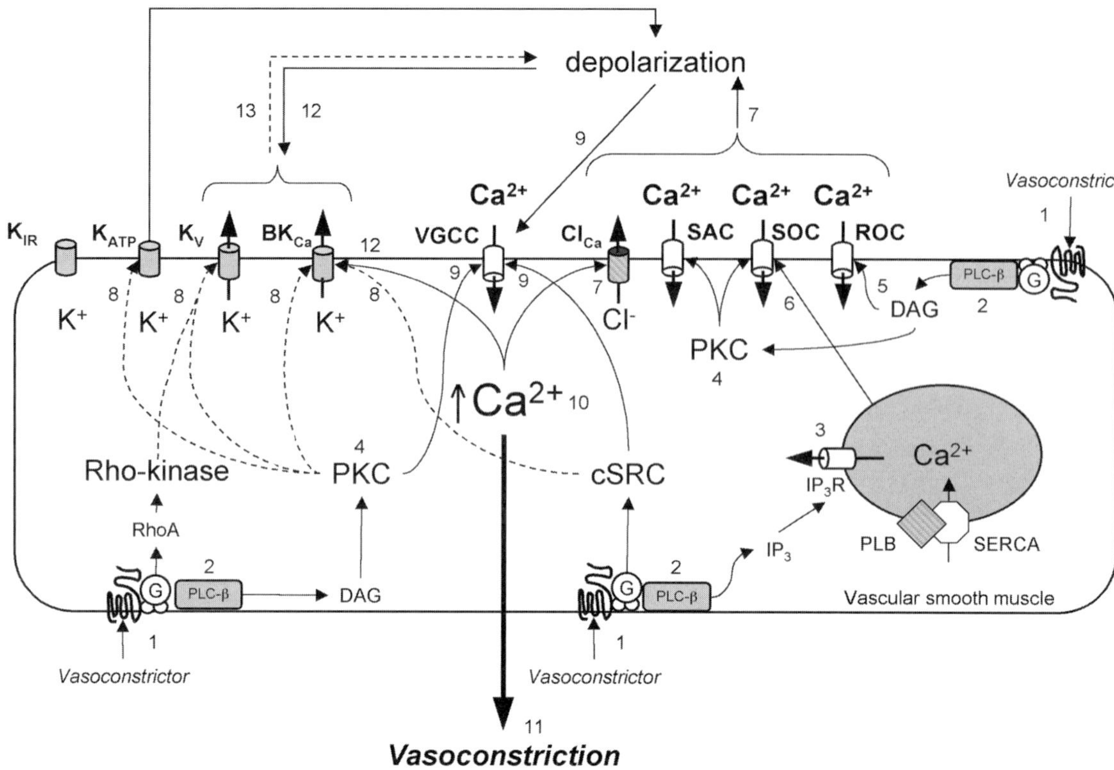

Figure 2 Ion channels and vasoconstriction. Simplified schematic diagram showing the role played by ion channels in contraction of arteriolar smooth muscle cells. Solid lines and arrows indicate stimulatory effects, whereas dashed lines represent inhibitory effects. **1**, Vasoconstrictors such as norepinephrine, angiotensin II, vasopressin or endothelin, bind to G-protein-coupled receptors to activate effector proteins such as phospholipase C-β (PLC-β). **2**, Activated PLC-β then acts on membrane phospholipids to form inositol 1,4,5-triphosphate (IP_3) and diacylglycerol (DAG). **3**, IP_3 binds to IP_3 receptors (IP_3R) on the smooth endoplasmic reticulum, releasing stored Ca^{2+}, and raising intracellular Ca^{2+}. **4**, The DAG formed and the presence of increased $[Ca^{2+}]_{in}$ activate protein kinase C, which then phosphorylates a number of ion channels including store-operated channels (SOC), stretch-activated channels (SAC), voltage-gated Ca^{2+} channels (VGCC), ATP-sensitive K^+ channels (K_{ATP}), voltage-gated K^+ channels (K_V), and large-conductance Ca^{2+}-activated K^+ channels (BK_{Ca}). **5**, In addition to activating PKC, DAG also can activate receptor-operated channels (ROC). **6**, The release of Ca^{2+} from intracellular stores, through, or in addition to, activation by PKC, opens SOC channels. **7**, The influx of Ca^{2+} and Na^+ through SAC, SOC, and ROC, along with Cl^- efflux through Ca^{2+}-activated Cl^- channels (Cl_{Ca}), depolarizes the smooth muscle cell membrane. **8**, This depolarization is supported by PKC-dependent closure of K_{ATP} channels, K_V channels, and BK_{Ca} channels, Rho-kinase-dependent inhibition of K_V channels, and downregulation of BK_{Ca} channels by the tyrosine kinase cSRC. **9**, Membrane depolarization along with stimulatory effects of PKC and cSRC on VGCC open these channels and provide a major source of steady-state Ca^{2+} influx. **10 and 11**, The released Ca^{2+}, along with Ca^{2+} influx through SAC, SOC, ROC, and VGCC, raise intracellular Ca^{2+} leading to smooth muscle contraction and vasoconstriction. **12**, The increase in $[Ca^{2+}]_{in}$ along with membrane depolarization activate BK_{Ca} channels, and the depolarization also activates K_V channels. **13**, The resulting efflux of K^+ provides a negative feedback signal that limits membrane depolarization and appears to prevent overactivation of the muscle and vasospasm. Other abbreviations: PLB, phospholamban; SERCA, smooth endoplasmic reticulum Ca^{2+} ATPase.

elevated ADP, reduced ATP, and reduced intracellular pH. Endogenous vasodilators such as adenosine, prostacyclin, epinephrine acting through β-adrenergic receptors, and calcitonin-gene-related peptide (CGRP) open these channels (Figure 3), whereas vasoconstrictors tend to close K_{ATP} channels (Figure 2). In skeletal muscle and coronary arterioles they play an important role in the regulation of smooth muscle resting membrane potential and tone, participate in functional hyperemia, and mediate, in part, hypoxia-induced vasodilation in these tissues.

K_{IR} CHANNELS

Arteriolar smooth muscle and endothelial cells also express K_{IR} channels. These channels open with small, physiological increases in extracellular K^+ concentration (Figure 3) and may participate in functional hyperemia in skeletal muscle, cardiac muscle, and the brain. As implied by their name, K_{IR} channels show strong inward rectification due to intracellular block of the outward current flow through the ion-conductive pore by Mg^{2+} and polyamines at membrane potentials positive to the K^+ equilibrium potential. More importantly, these channels may display an "N"-shaped current–voltage relationship such that between the potassium equilibrium potential and potentials 10 to 20 mV more positive, K_{IR} channels conduct small outward K^+ currents. It is these small outward currents that allow K_{IR} channels to contribute to the regulation of arteriolar tone. Endothelial K_{IR} channels also are stimulated by increased

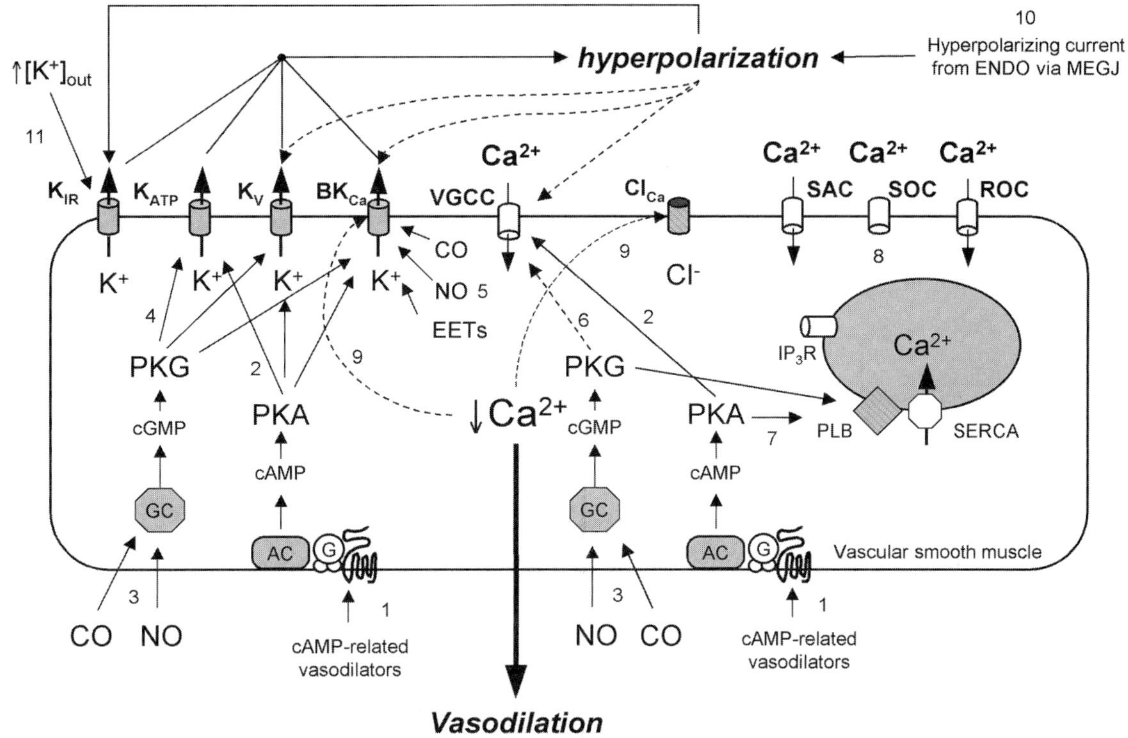

Figure 3 Ion channels and vasodilation. Simplified schematic diagram of the role played by ion channels in the mechanism of action of vasodilators on arteriolar smooth muscle cells. Solid lines and arrows indicate stimulatory effects, whereas dashed lines represent inhibitory effects. **1**, Vasodilators such as adenosine, prostacyclin (PGI_2), epinephrine, and calcitonin-gene-related peptide (CGRP) all bind to G-protein-coupled receptors and activate adenyl cyclase (AC) to form cAMP, which then activates protein kinase A (PKA). **2**, PKA phosphorylates and activates K_{ATP} channels, K_V channels, BK_{Ca} channels, and VGCC. The activation of the K^+ channels tends to hyperpolarize smooth muscle, closing VGCC, reducing Ca^{2+} influx, lowering $[Ca^{2+}]_{in}$, and leading to smooth muscle relaxation and vasodilation. **3**, Vasodilators such as NO, CO, and atrial natriuretic peptide (ANP) stimulate guanylyl cyclase (GC) to form cGMP, which then activates cGMP-dependent protein kinase (PKG). **4**, PKG targets K_{ATP}, K_V, and BK_{Ca} channels, leading to hyperpolarization and, ultimately, vasodilation. **5**, In some instances NO, CO, and epoxides of arachidonic acid may directly activate BK_{Ca} channels. **6**, Phosphorylation of VGCC by PKG is inhibitory and contributes to cGMP-PKG induced vasodilation. **7**, Both PKA and PKG also phosphorylate other proteins such as phospholamban (PLB) and phospholipase C-β (not shown in figure). Phosphorylated PLB disinhibits the smooth endoplasmic reticulum Ca^{2+} ATPase (SERCA) reducing $[Ca^{2+}]_{in}$ and refilling SER Ca^{2+} stores. **8**, This latter effect closes store-operated channels (SOC) as indicated. Phosphorylation of phospholipase C-β reduces production of DAG and IP_3, leading to reduced activity in both stretch-activated channels (SAC) and receptor-operated channels (ROC) contributing to the reduction in $[Ca^{2+}]_{in}$. **9**, The reduced $[Ca^{2+}]_{in}$ also closes Ca^{2+}-activated Cl^- channels, removing another depolarizing current, and tends to close BK_{Ca} channels, limiting their effect. **10**, In some arterioles, smooth muscle cells are electrically coupled to endothelial cells such that hyperpolarization of endothelial cells (ENDO) can also hyperpolarize smooth muscle leading to vasodilation. **11**, Inward-rectifier K^+ channels (K_{IR}) may be recruited during membrane hyperpolarization. These channels are also activated by increases in extracellular K^+ ($[K^+]_{out}$).

shear stress (Figure 4). In both smooth muscle and endothelial cells, K_{IR} channels are probably modulated by protein kinases, although this has not been studied in the microcirculation.

Nonselective Cation Channels

Vascular smooth muscle cells and endothelial cells express a number of nonselective cation channels that likely all arise from the transient receptor potential (TRP) family of ion channels originally described in the photoreceptors of the fruit fly, *Drosophila*. All of the TRPC isoforms expressed in smooth muscle and endothelial cells (see Table I) form nonselective cation channels that conduct Na^+ and Ca^{2+} into cells, and none appear to be voltage-gated. This large family of proteins forms channels that open when intracellular Ca^{2+} stores are depleted (store-operated channels, SOC), open with membrane stretch (stretch-activated channels, SAC), or open upon receptor activation (receptor-operated channels, ROC). They contribute to the regulation of arteriolar tone by providing a regulated pathway for entry of Ca^{2+} into both smooth muscle and endothelial cells. In smooth muscle, Ca^{2+} entry through SOC and ROC contributes to vasoconstrictor-induced increases in arteriolar tone, and Ca^{2+} influx through SAC importantly contributes to pressure-induced changes in arteriolar tone (Figure 2). In endothelial cells, SOC are the source of maintained increases in intracellular Ca^{2+} that contribute to the steady-state production of endothelium-derived autacoids (Figure 4).

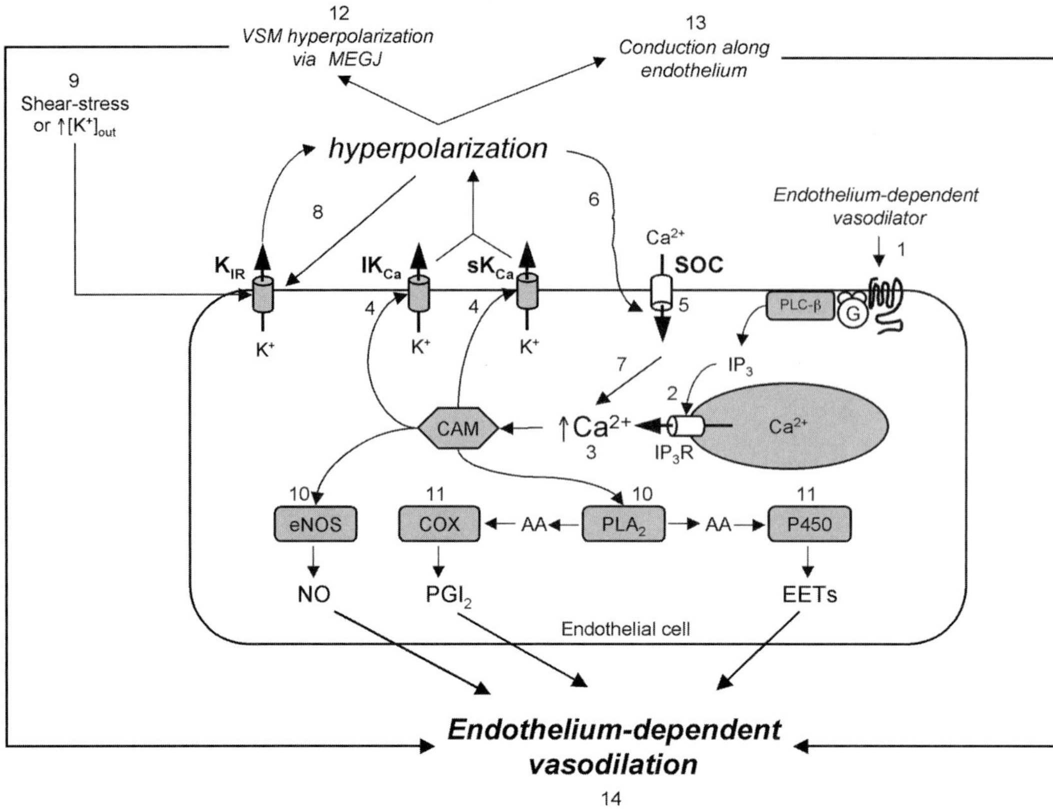

Figure 4 Ion channels and endothelial hyperpolarization. Simplified schematic diagram of the role played by ion channels in the mechanism of action of endothelium-dependent vasodilators, shear-stress and elevated extracellular K^+ (($[K^+]_{out}$). **1**, Endothelium-dependent vasodilators such as acetylcholine, bradykinin, ATP, and histamine bind to G-protein-coupled receptors to activate phospholipase C-β (PLC-β) leading to the formation of inositol 1,4,5-triphosphate (IP_3) and diacylglycerol (DAG, not shown). **2**, IP_3 binds to IP_3-receptors (IP_3R) on the smooth endoplasmic reticulum, releasing stored Ca^{2+}. **3**, This causes a rapid rise in intracellular Ca^{2+}. **4**, The increased $[Ca^{2+}]_{in}$ then binds to calmodulin (CAM) associated with a number of proteins including intermediate (IK_{Ca}) and small (sK_{Ca}) conductance Ca^{2+}-activated K^+ channels, leading to membrane hyperpolarization. **5**, The release of Ca^2 from intracellular stores activates store-operated channels (SOC). **6**, The membrane hyperpolarization caused by opening of IK_{Ca} and sK_{Ca} channels increases the electrochemical gradient for diffusion of Ca^{2+} into the cells through SOC, and amplifies this current. **7**, This amplified Ca^{2+} influx yields a sustained elevation in $[Ca^{2+}]_{in}$. **8**, The hyperpolarization induced by IK_{Ca} and sK_{Ca} channels can activate inward rectifier K^+ channels, augmenting the hyperpolarization. **9**, This effect may be enhanced by increases in shear-stress or elevated $[K^+]_{out}$, both of which activate K_{IR} channels. **10**, The elevated Ca^{2+}, through Ca^{2+}-CAM then activates proteins such as NO-synthase (e-NOS) and phospholipase A_2 (PLA_2). Activation of e-NOS leads to production of NO, while activation of PLA_2 liberates arachidonic acid (AA) from membrane phospholipid stores. **11**, Released arachidonic acid is then rapidly metabolized by cyclooxygenase (COX) to prostacyclin (PGI_2) and other vasodilator prostaglandins, and to epoxides (EETs) by cytochrome P-450 (P450). **12**, In some arterioles, smooth muscle cells (VSM) are electrically coupled to endothelial cells through myoendothelial gap junctions (MEGJ), such that endothelial hyperpolarization, per se, in the absence of another mediator, can lead to vasodilation. **13**, In addition, endothelial cells in arterioles are electrically coupled so that hyperpolarization can be conducted for long distances along the length of arteriolar endothelium. **14**, All of these signals (hyperpolarization, NO, PGI_2, EETs) are integrated by overlying smooth muscle cells to yield vasodilation.

Cl^- Channels

Vascular smooth muscle and endothelial cells also express chloride channels. These include Ca^{2+}-activated Cl^- (Cl_{Ca}) channels and swelling or volume-activated Cl^- (Cl_{SW} channels). The molecular identity of the Cl^- channels expressed in vascular smooth muscle and endothelial cells remains unclear as indicated in Table I. In both smooth muscle and endothelial cells, the intracellular concentration of Cl^- is such that opening of channels conducting these anions leads to diffusion of Cl^- out of the cells, and membrane depolarization. Calcium-activated Cl^- channels are activated by increases in intracellular Ca^{2+}. In smooth muscle, they contribute to vasoconstrictor-induced membrane depolarization (Figure 2). Their function in arteriolar endothelial cells has not been established. In endothelial cells and in smooth muscle cells, Cl_{SW} channels play an important role in the regulation of cell osmolarity and volume and likely participate in the regulation of resting membrane potential. Smooth-muscle Cl_{SW} channels may play a role in pressure (stretch)-induced membrane depolarization and increases in tone.

Membrane Potential and Arteriolar Tone at Rest

Arterioles in the microcirculation at rest display substantial tone largely due to the impact of blood pressure on the smooth muscle. Under these conditions, smooth muscle and endothelial cells have a membrane potential of about −30 mV. This relatively depolarized membrane potential results from the outward flow of K^+ through K^+ channels (a hyperpolarizing current) that is balanced by efflux of Cl^- ions and inward flow of Na^+ and Ca^{2+} ions (all depolarizing currents). There appear to be substantial regional and species differences in the specific ion channels that contribute to resting membrane potential. However, general patterns appear.

In arteriolar smooth muscle cells at rest, membrane potential is determined by K^+ efflux through K_V channels, cation (Na^+ and Ca^{2+}) influx through SOC, and probably Cl^- efflux through Cl_{SW} channels. In skeletal muscle and coronary vascular beds, K_{ATP} channels also contribute to resting membrane potential. In vivo studies suggest that voltage-gated Ca^{2+} channels and BK_{Ca} channels may not be active at rest. The mechanisms responsible for the apparent lack of activity of voltage-gated Ca^{2+} channels at rest in vivo have not been established. However, it is likely that the low resting activity of arteriolar BK_{Ca} channels in vivo reflects the low activity of voltage-gated Ca^{2+} channels, with which the BK_{Ca} channels are functionally coupled (Figure 1), along with a high Ca^{2+} threshold of microvascular BK_{Ca} channels. Inward rectifier K^+ channels do not contribute to resting potential because of the relatively depolarized membrane potential (−30 mV) found in arteriolar smooth muscle cells. Calcium-activated Cl^- channels appear silent at rest due to their high Ca^{2+} threshold.

The ion channels that contribute to resting membrane potential in arteriolar endothelial cells have not been established. However, similar to the overlying smooth muscle cells, K_V channels, Cl_{sw} channels and nonselective cation channels probably contribute to the relatively depolarized membrane potential in endothelial cells in resting arterioles (−30 mV). In addition, blood flow over the endothelium may activate endothelial K_{IR} channels through a shear-stress dependent mechanism, such that these K^+ channels may also contribute to resting membrane potential of arteriolar endothelial cells. Also, because smooth muscle and endothelial cells may be electrically coupled by myoendothelial gap junctions in some arterioles, both smooth muscle and endothelial ion channels may contribute to the membrane potential of the other cell type.

Ion Channels and Vasoconstriction

In general, endogenous vasoconstrictors such as norepinephrine, angiotensin II, vasopressin, and the endothelins tend to activate smooth muscle cation influx through voltage-gated Ca^{2+} and nonselective cation channels, activate Cl^- efflux through Cl_{Ca} channels, and inhibit K^+ efflux through K^+ channels, resulting in membrane depolarization, net Ca^{2+} influx, and an increase in intracellular Ca^{2+} leading to smooth muscle contraction (Figure 2). In addition to the mechanisms outlined in Figure 2, the vasoconstrictor-induced increase in intracellular Ca^{2+} bolsters depolarization by activation of the Ca^{2+}-dependent phosphatase 2B (PP2B, calcineurin), which dephosphorylates K_V and K_{ATP} channels to reduce their activity (not shown in Figure 2). Although the activity of K_V and BK_{Ca} channels is somewhat blunted by the inhibitory mechanisms outlined in Figure 2, the membrane depolarization and, in the case of BK_{Ca} channels, the elevated intracellular Ca^{2+} lead to net activation of both K_V and BK_{Ca} channels. This K^+ channel activity importantly limits the depolarization and provides an effective negative feedback signal so that overstimulation and vasospasm do not occur. Thus, ion channels play a key role in vasoconstriction.

Ion Channels and Vasodilation

Vasodilators tend to have effects on smooth muscle ion channels that oppose those of vasoconstrictors (Figure 3). In general, they reduce Ca^{2+} influx by a number of mechanisms (see Figure 3), leading to a reduction in intracellular Ca^{2+}, smooth muscle relaxation, and vasodilation. The only exceptions to this pattern are vasodilators that act through the cAMP-PKA signaling cascade. Voltage-gated Ca^{2+} channels are phosphorylated by PKA, leading to their activation. Although counter to the general Ca^{2+}-lowering trend induced by most dilators, this increased activity in voltage-gated Ca^{2+} channels may provide local Ca^{2+} to maintain the activity of BK_{Ca} channels (see Figure 1), despite a general lowering of intracellular Ca^{2+} by other means.

In skeletal muscle, cardiac muscle, and the brain, increases in tissue activity lead to release of K^+ ions from the active cells, resulting in elevation of the extracellular concentration of K^+ from 5 mM to between 8 and 20 mM. Such increases in extracellular K^+ can activate smooth muscle K_{IR} channels leading to membrane hyperpolarization and vasodilation (Figure 3). In arterioles where smooth muscle cells are electrically coupled to endothelial cells by myoendothelial gap junctions, endothelial K_{IR} channels may contribute to K^+-induced hyperpolarization and vasodilation through a similar mechanism (Figure 4).

Vasodilators such as acetylcholine, bradykinin, ATP, and histamine lead to relaxation of vascular smooth muscle and decreases in arteriolar tone in intact arterioles by stimulating the release of vasodilator substances from endothelial cells. Ion channels play an important part in this process, as outlined in Figure 4. Calcium-dependent activation of sK_{Ca} and IK_{Ca} channels hyperpolarizes endothelial cells and augments Ca^{2+} influx through SOC, providing a maintained increase in intracellular Ca^{2+} to support endothelial autacoid production. In addition, the endothelial hyperpolarization, per se, may be transmitted through myoendothelial gap junctions to hyperpolarize and relax the overlying smooth muscle (Figure 4). Shear stress and increases in extracellular K^+ may also hyperpolarize endothelial cells by activation of K_{IR} channels. These channels may also be recruited by membrane hyperpolarization caused by activation of other K^+ channels serving to amplify the initial hyperpolarization. Hyperpolarization-induced activation of K_{IR} channels also may allow conduction of hyperpolarization from endothelial cell to endothelial cell, that are coupled by gap junctions, providing a means to transmit hyperpolarization for long distances along arterioles. Similar to their effects on smooth muscle, vasodilators such as adenosine also may activate K_{ATP} channels on endothelial cells, providing another pathway for endothelial cell hyperpolarization and regulation of arteriolar tone (pathway not shown in Figure 4). Thus, ion channels significantly contribute to the mechanism of action of vasodilators both in smooth muscle and in endothelial cells.

Ion Channels and Disease States

Diseases such as hypertension and diabetes appear to affect ion channel expression or function and contribute to altered arteriolar tone associated with these pathologies. In hypertension there appears to be an upregulation of voltage-gated Ca^{2+} channels, nonselective cation channels, and BK_{Ca} channels, but decreased function of K_V and K_{ATP} channels that likely contribute to altered arteriolar tone and reactivity observed in this disease. Similarly, in diabetes there is a depression of BK_{Ca}, K_V, and K_{ATP} channel function that may contribute to diabetic vascular dysfunction. Conversely, sepsis is associated with a large increase in the activity K_{ATP} channels that may underlie the systemic vasodilation and reduced arteriolar tone associated with this condition. Thus, ion channels participate in both physiological and pathophysiological regulation of arteriolar tone.

Glossary

Depolarization: Refers to cell membrane potential becoming more positive.

Electrochemical gradient: The driving force that determines the direction of net diffusion of ions through ion channels. Its magnitude is determined by the difference between the cell membrane potential (which determines the electrical force on the ion) and the concentration difference for the ion that exists across the membrane (which determines the chemical "force" on the ion). Its magnitude can be calculated as the difference between the membrane potential and the equilibrium potential for a given ion.

Equilibrium potential: Theoretical membrane potential that would be required to exactly oppose diffusion of an ion down a given concentration gradient. It is also referred to as the Nernst potential.

Hyperpolarization: Refers to cell membrane potential becoming more negative.

Myoendothelial gap junctions: Gap junctions between endothelial cells and smooth muscle cells that allow ions and small molecules to diffuse between the two cell types. They allow electrical activity in one cell type to be transmitted to the other cell type.

Bibliography

Archer, S. L., and Rusch, N. J., eds. (2001). *Potassium Channels in Cardiovascular Biology*. New York: Kluwer Academic/Plenum. *This is an outstanding compilation of in-depth reviews of K^+ channels in the cardiovascular system. Very comprehensive treatment from the molecular and structural biology of K^+ channels to their function and pharmacology. Recommended reading for anyone interested in K^+ channels and the cardiovascular system.*

Beech, D. J., Xu, S. Z., McHugh, D., and Flemming, R. (2003). TRPC1 store-operated cationic channel subunit. *Cell Calcium* **33**(5–6), 433–440.

Busse, R., Edwards, G., Feletou, M., Fleming, I., Vanhoutte, P. M., and Weston, A. H. (2002). EDHF: Bringing the concepts together. *Trends Pharmacol. Sci.* **23**(8), 374–380. *An outstanding brief synopsis of endothelium-dependent hyperpolarization of vascular smooth muscle, and the role played by sK_{Ca} and IK_{Ca} channels in this process.*

Hille, B. (2001). *Ionic Channels of Excitable Membranes*. Sunderland, MA: Sinauer. *This is an excellent, comprehensive overview of all types of ion channels, membrane physiology, and biophysics. It provides a very good starting point to learn about ion channels and their function.*

Jackson, W. F. (2000). Ion channels and vascular tone. *Hypertension* **35**(1, Pt. 2), 173–178.

Jaggar, J. H., Porter, V. A., Lederer, W. J., and Nelson, M. T. (2000). Calcium sparks in smooth muscle. *Am. J. Physiol. Cell Physiol.* **278**(2), C235–256.

Jentsch, T. J., Stein, V., Weinreich, F., and Zdebik, A. A. (2002). Molecular structure and physiological function of chloride channels. *Physiol Rev.* **82**(2), 503–568.

Keef, K. D., Hume, J. R., and Zhong, J. (2001). Regulation of cardiac and smooth muscle Ca(2+) channels (Ca(V)1.2a,b) by protein kinases. *Am. J. Physiol. Cell Physiol.* **281**(6), C1743–1756.

Nilius, B., and Droogmans, G. (2001). Ion channels and their functional role in vascular endothelium. *Physiol Rev.* **81**(4), 1415–1459. *An excellent review of ion channels and their physiological function in endothelial cells. Provides a good overview of signaling in endothelial cells and the role played by ion channels in this process.*

Welsh, D. G., Morielli, A. D., Nelson, M. T., and Brayden, J. E. (2002). Transient receptor potential channels regulate myogenic tone of resistance arteries. *Circ. Res.* **90**, 248–250. *This is the landmark study demonstrating that TRPC 6 likely underlies SAC in smooth muscle.*

Capsule Biography

Dr. Jackson is a professor in the Department of Biological Sciences at Western Michigan University and in 1998 won the Distinguished Faculty Scholar Award, Western's highest honor for research accomplishments. His laboratory focuses on ion channel function and expression in arterioles, their modulation by oxygen, and their function in regulation of microvascular blood flow.

Dr. Jackson's research is supported by Public Health Service grant HL 32469 from the National Heart, Lung and Blood Institute.

CHAPTER 35

Roles of Nitric Oxide in the Microcirculation

Michelle I. Lin and William C. Sessa
Yale University School of Medicine, Boyer Center for Molecular Medicine, New Haven, Connecticut

Introduction

Nitric oxide (NO), a free radical gas produced by the NO synthase (NOS) family of proteins, is a ubiquitous second messenger in diverse physiological responses in the cardiovascular system including vasodilation, anticoagulation, vascular remodeling and angiogenesis. In vitro and in vivo studies using NOS inhibitors or more recently, using mice deficient in each of the three mammalian isoforms, namely (1) endothelial NOS (eNOS), (2) cytokine-inducible NOS (iNOS), or (3) neuronal NOS (nNOS), have shown definitive roles of NO and its sources in the cardiovascular system. The purpose of this review is to highlight the evidence supporting various functional roles of NO in the microcirculation.

NO and Vasodilation

NO is clearly the major endothelium-dependent vasodilator produced by large blood vessels (Figure 1). In the aorta or carotid arteries of mice deficient in eNOS (eNOS –/– mice), endothelium-dependent vasodilation is eliminated, thus supporting eNOS derived NO as the source of the relaxing factor that led to the Nobel Prize–winning experiments by Furchgott, Murad, and Ignarro awarded in 1998. However, the vasodilatory roles for NO in arterioles and venules appear less unanimous, and the importance of NO as a vasodilatory agent (deciphered by the sensitivity of the response to blockage of NOS) depends on the stimulus (flow, acetylcholine, bradykinin, and so on), the vascular bed studied (coronary, cremaster, gracilis, mesentery), the species (human, rat, mouse, hamster), and more recently, the number of endothelial–smooth muscle interconnections (i.e., myo-endothelial gaps). In eNOS (–/–) mice, the local vasodilatory actions of acetylcholine or flow induced changes in blood flow are not diminished, because of compensation by upregulation of nNOS, endothelium-derived hyperpolarizing factor (EDHF), and vasodilatory prostaglandins [1, 2]. However, the local vasodilatory action of histamine was absent from second order arterioles in eNOS (–/–) mice [3]. These are only a few of the many examples highlighting some of the differential response to NO as aforementioned. Another theory imparting NO as an important factor in microvascular blood flow control is the concept that NO bound to hemoglobin in red blood cells can serve as a stable NO adduct for delivery at sites of resistance. Clearly NO binds to the heme moiety of hemoglobin while a reactive thiol, cysteine β 93, and hemoglobin can undergo allosteric changes in structure initiated by a drop in arteriolar pO_2, and release NO from these sites [4]. Although this principle can be demonstrated in model systems, the physiological role of NO via its release from hemoglobin in regulating blood flow in the microcirculation remains controversial [5]. An alternative, emerging theory of how the endothelium may control microvascular blood flow is through another endothelium-dependent vasodilator, EDHF. In systems where NOS is blocked or absent, EDHF may subserve a role as a key regulator of microvascular blood flow control.

NO and the Homeostasis of the Blood Cell Interaction with the Vessel Wall

Preserving a Nonthrombogenic Surface of the Endothelium

In addition to the role of NO as a vasodilator, NO in the microcirculation can prevent platelets from adhering to the endothelium and can assist with the disaggregation of activated platelets to the endothelium or underlying basement membrane (Figure 1). More recently, this concept was unequivocally demonstrated in mice deficient for eNOS [6, 7]. Bleeding times were decreased in eNOS (–/–) mice and platelet adhesion to venules was increased in response to bacterial lipopolysaccharide (LPS) when a NOS inhibitor was given. Similarly, mice deficient in eNOS, but not iNOS or nNOS, also exhibited increased platelet adhesion. In both instances, eNOS in the endothelium as well as in the platelet per se contributed to the antiplatelet actions of eNOS-derived NO.

Maintaining an Anti-inflammatory State of the Vessel Wall

Inflammation can be described as the physiological response of the microcirculation to injury or infections. Designed to close off or destroy injured tissue, inflammatory response is hallmarked by increases in leukocyte–endothelium interactions. Recently, NO has emerged as a crucial endogenous anti-inflammatory mediator in a number of pathophysiological states including hypercholesterolemia and ischemia-reperfusion injury. More specifically, NO can act by downregulating cytokines, resulting in the downregulation of endothelial cell adhesion molecules (ECAMs). The mechanism of leukocyte recruitment is mediated by these ECAMs such as the selectin family (P- and E-selectin), which are important modulators of leukocyte–endothelium interaction via leukocyte rolling along the endothelium and adhesion to the endothelium. Once the cells begin to roll, they can then firmly attach to the endothelium via integrin interaction with endothelial intercellular adhesion molecules (ICAMs) to promote leukocyte adhesion.

Pharmacological studies have shown that NOS inhibitors could increase leukocyte adhesion, an effect that can be reversed by large amounts of exogenous L-arginine. Similar studies using eNOS (–/–) mice confirmed that upon activation of an inflammatory response, eNOS derived NO is critical for reducing leukocytes adhesion and the extent of tissue injury. Recent data suggest that NO is derived from eNOS and nNOS, but interestingly, iNOS may be an endogenous inhibitor of P-selectin. eNOS and nNOS (–/–)

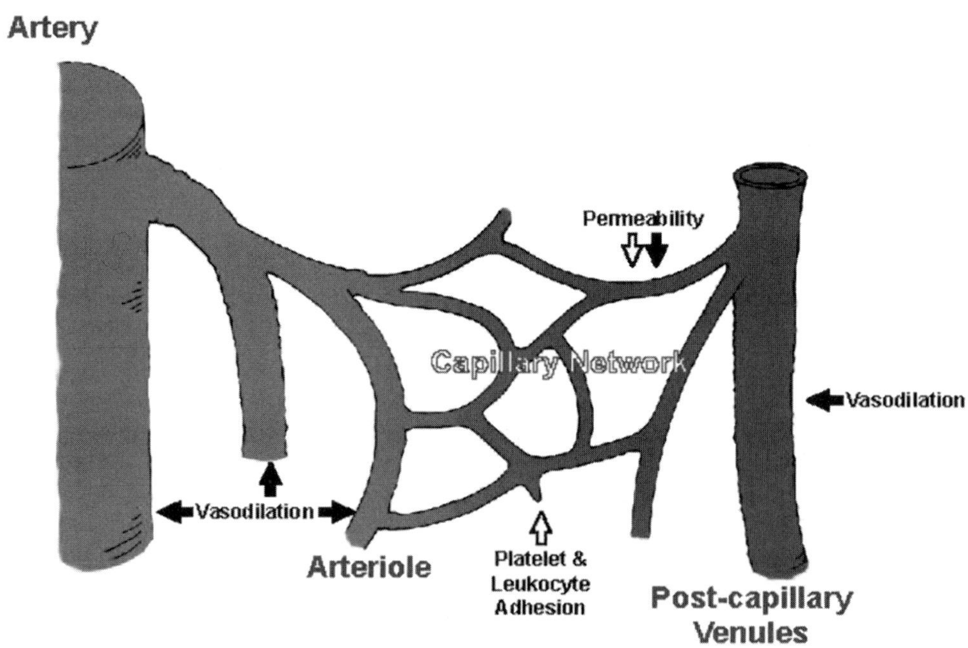

Figure 1 The role of NO in the microcirculation. NO, a gaseous free radical, can exert multiple effects in the microcirculatory bed and is a major player in regulating the cardiovascular system. Studies have shown the vasodilatory effect of NO in large blood vessels as well as in the arterioles and venules. This mainly occurs when NO synthesized by the endothelium diffuses to the surrounding vascular smooth muscle, activating guanylyl cyclase (GC), thereby increasing cytosolic cGMP to produce smooth muscle relaxation. NO can also regulate homeostasis of the blood vessel wall by preventing platelet adhering and disaggregation from the endothelium. Furthermore, NO can negatively regulate leukocyte-endothelium interaction by downregulating cytokines and ECAMs such as P-selectin that are important for leukocyte rolling and adhesion to endothelial cells. Thus, NO serves as an anti-inflammatory mediator in pathological conditions such as hypercholesterolemia or as a cardioprotectant in myocardial ischemia-reperfusion injuries. In addition, NO is a crucial regulator of microvascular permeability. Controversy exists to whether NO promotes or inhibits basal versus stimulated changes in vascular permeability. (see color insert)

mice exhibited increased expression of P-selectin in their mesenteric venules as well as increased leukocyte rolling and adhesion that can be blocked by both the P-selectin neutralizing antibody and a high-affinity P-selectin ligand [8]. This supports previous findings that NOS inhibition by NOS inhibitors caused a significant increase in P-selectin expression in the postcapillary venules. Using another inflammatory model, namely the myocardial ischemia-reperfusion injury, NO was also shown to provide a protective role in the injury cascade leading to inflammation. In particular, other studies by the Lefer group have shown that eNOS-deficient mice exhibited exacerbated reperfusion injury compared to wild-type mice, an effect associated with upregulated P-selectin expression and enhanced infiltration of neutrophils. Recently, transgenic mice overexpressing eNOS in the endothelium have been generated. In accordance with the findings from eNOS (–/–) mice, overexpression of eNOS in these transgenic mice exhibited an ameliorated response to myocardial infarction with a 33 percent reduction in infarct size without changes in systemic hemodynamics or differences in baseline ventricular morphology and function. Taken together, these findings are in agreement with previous observations of the cardioprotective role by NO in the myocardial microcirculation using exogenous NO donors and the exacerbated injury when animals were treated with NOS inhibitor.

Role of NO in Regulating Vascular Permeability

One of the initial events in an acute inflammatory response is increased extravasation of fluid and protein from postcapillary venules at sites of tissue injury. Under physiological conditions, microvascular fluid exchange is primarily determined by blood flow, Starling forces, and the perfused surface area. The perm-selectivity barrier to solute flux is determined by multiple factors including the identity of the endothelium (tissue bed), composition of tight versus adherens junctions, and extracellular matrix.

Using cannulated, venular microvessels in the mesentery, a variety of inflammatory agents such as ATP, ionomycin, bradykinin, histamine, and VEGF stimulated increases in $[Ca^{2+}]_i$ slightly preceding the initial peak in vascular leakage. The delay of the onset of changes in permeability suggests that in addition to the initial increase of $[Ca^{2+}]_i$, calcium-dependent signaling cascades (i.e., eNOS) may contribute to vascular permeability. Furthermore, NOS inhibitors can reduce ATP-induced increases in single vessel permeability, while 8-Br-cGMP, a membrane-permeable cGMP analog, can potentiate ATP driven increases in permeability. These agents, however, did not affect the ATP induced rise in $[Ca^{2+}]_i$, which further suggests that the increase in $[Ca^{2+}]_i$ was necessary for an agonist-induced increase in permeability but not entirely sufficient.

Over the years, studies on the action of NO in regulating vascular permeability have yielded conflicting observations (Figure 1). Early studies using NO donors and NOS inhibitors demonstrated the negative regulatory effect of NO in microvascular permeability. For example, administration of a NOS inhibitor, N^{ω}-nitro-L-arginine-methyl ester (L-NAME), increased transvascular fluid and protein flux, whereas subsequent NO donors such as sodium nitroprusside (SNP) reversed this effect in the intestinal circulation. The increase in permeability was not due to hemodynamic alteration, a phenomenon in which L-NAME has been suggested to play a role. Many other studies using NOS inhibitors or NO donors also demonstrated that NO negatively regulates permeability in response to a host of inflammatory agents including bradykinin, carrageenan, substance P, and mustard oil. In addition, increases in intracellular levels of cGMP by NO donors or activators of guanylate cyclase (sGC) have also been associated with a decrease in endothelial permeability.

On the other hand, the NO-sGC-cGMP pathway can positively regulate microvascular permeability. Upon receptor activation and subsequent stimulation of phospholipase C (PLC), increase in cytosolic calcium can lead to activation of eNOS and NO production. NO activates guanylate cyclase (GC) and increases cGMP production, resulting in protein kinase G (PKG) activation and phosphorylation of proteins that regulate the contractile apparatus or cytoskeleton. In cannulated porcine coronary venules, VEGF and histamine can induce increases in permeability, as measured by the rate of FITC-albumin transfer into the extravascular space in a cGMP-PKG dependent manner. Administration of the NO donor SNP increased permeability similar to that induced by histamine, and this increase can be blocked completely by an inhibitor of guanylate cyclase.

Recent studies have examined the effect of eNOS in modulating vascular permeability using eNOS (–/–) mice or using a synthetic peptide that inhibits eNOS activity. A pivotal study by Jain's laboratory [9] demonstrated that eNOS (–/–) mice, but not iNOS (–/–) mice showed impairment in VEGF-induced increases in permeability, further confirming the importance of eNOS as the predominant NOS isoforms in the regulation of VEGF induced vascular permeability. To further examine the role of eNOS in regulating vascular permeability, our laboratory has generated a synthetic fusion protein containing the eNOS inhibitory domain of caveolin-1 and a cell permeable peptide that facilitates the uptake of cargo proteins/oligonucleotides [10]. This peptide, which we termed cavtratin, can inhibit acetylcholine-induced, endothelium-dependent relaxation, thereby reducing the amount of NO generated block eNOS and not iNOS activity. Furthermore, it can specifically. Administration of this peptide in vivo can block both carrageenan- and mustard oil–induced inflammation and edema formation. Furthermore, this peptide, which did not have an antiangiogenic or cytostatic effect, can selectively block tumor microvascular permeability, leading to decreased tumor progression. Recently, cavtratin has been shown to block PAF-mediated increases in hydraulic conductivity in venules, thus validating the antileak actions of the peptide [11]. Combined with our observations that eNOS (–/–) mice exhibited decreased tumor permeability and progression, these studies may provide novel insights into development of new antipermeability drugs to target tumor growth.

Conclusions

The discovery of NO has led to the identification of multiple physiological roles for this gaseous second messenger in the microcirculation. We surmise that the local autocrine actions of NO that govern inflammation and vascular permeability in the microcirculation may be as important as its vasodilatory role in conduit vessels. Furthermore, the diverse function of NO may extend beyond the vascular bed. Recent findings using NOS inhibitors and eNOS (–/–) mice revealed an important role of endothelial derived nitric oxide in mediating lymphatic fluid flow in the microlymphatic network (new reference #12). Therefore, future mechanistic studies and the development of novel reagents to faithfully manipulate the NO-sGC-PKG pathway will help dissect the physiological importance of NO in the microcirculation and microlymphatic beds.

Glossary

Endothelium: The innermost layer of cells lining all blood vessels.

Nitric oxide: A free radical gas produced by the enzyme nitric oxide synthase.

Permeability: The leakage of fluid or proteins from the circulation after tissue injury.

Vasodilation: Dilation of a blood vessel to increase blood flow.

References

1. Meng, W., Ma, J., Ayata, C., Hara, H., Huang, P. L., Fishman, M. C., and Moskowitz, M. A. (1996). ACh dilates pial arterioles in endothelial and neuronal NOS knockout mice by NO-dependent mechanisms. *Am. J. Physiol.* **271**, H1145–1150.
2. Sun, D., Huang, A., Smith, C. J., Stackpole, C. J., Connetta, J. A., Shesely, E. G., Koller, A., and Kaley, G. (1999). Enhanced release of prostaglandins contributes to flow-induced arteriolar dilation in eNOS knockout mice. *Circ. Res.* **85**, 288–293.
3. Payne, G. W., Madri, J. A., Sessa, W. C., and Segal, S. S. (2003). Abolition of arteriolar dilation but not constriction to histamine in cremaster muscle of eNOS–/–mice. *Am. J. Physiol. Heart Circ. Physiol.* **285**, H493–498.
4. McMahon, T. J., Moon, R. E., Luschinger, B. P., Carraway, M. S., Stone, A. E., Stolp, B. W., Gow, A. J., Pawloski, J. R., Watke, P., Singel, D. J., Piantadosi, C. A., and Stamler, J. S. (2002). Nitric oxide in the human respiratory cycle. *Nat. Med.* **8**, 711–717.
5. Nase, G. P., Tuttle, J., and Bohlen, H. G. (2003). Reduced perivascular PO2 increases nitric oxide release from endothelial cells. *Am. J. Physiol. Heart Circ. Physiol.* **285**, H507–515.
6. Freedman, J. E., Sauter, R., Battinelli, E. M., Ault, K., Knowles, C., Huang, P. L., and Loscalzo, J. (1999). Deficient platelet-derived nitric oxide and enhanced hemostasis in mice lacking the NOSIII gene. *Circ. Res.* **84**, 1416–1421.
7. Cerwinka, W. H., Cooper, D., Krieglstein, C. F., Feelisch, M., and Granger, D. N. (2002). Nitric oxide modulates endotoxin-induced platelet-endothelial cell adhesion in intestinal venules. *Am. J. Physiol. Heart Circ. Physiol.* **282**, H1111–1117.
8. Lefer, D. J., Jones, S. P., Girod, W. G., Baines, A., Grisham, M. B., Cockrell, A. S., Huang, P. L., and Scalia, R. (1999). Leukocyte-endothelial cell interactions in nitric oxide synthase-deficient mice. *Am. J. Physiol.* **276**, H1943–1950.
9. Fukumura, D., Gohongi, T., Kadambi, A., Izumi, Y., Ang, J., Yun, C. O., Buerk, D. G., Huang, P. L., and Jain, R. K. (2001). Predominant role of endothelial nitric oxide synthase in vascular endothelial growth factor-induced angiogenesis and vascular permeability. *Proc. Natl. Acad. Sci. USA* **98**, 2604–2609.
10. Bucci, M., Gratton, J. P., Rudic, R. D., Acevedo, L., Roviezzo, F., Cirino, G., and Sessa, W. C. (2000). In vivo delivery of the caveolin-1 scaffolding domain inhibits nitric oxide synthesis and reduces inflammation. *Nat. Med.* **6**, 1362–1367.
11. Zhu, L., Schwegler-Berry, D., Castranova, V., and He, P. (2004). Internalization of caveolin-1 scaffolding domain facilitated by Antennapedia homeodomain attenuates PAF-induced increase in microvessel permeability. *Am. J. Physiol. Heart Circ. Physiol.* **286**, H195–201.
12. Hagendoorn, J., Padera, T. P., Kashiwagi, S., Isaka, N., Noda, F., Lin, M. I., Huang, P. L., Sessa, W. C., Fukumura, D., and Jain, R. K. (2004). Endothelial nitric oxide synthase regulates microlymphatic flow via collecting lymphatics. *Circ. Res.* **95**, 204–209.

Bibliography

Dvorak, H. F. (2002). Vascular permeability factor/vascular endothelial growth factor: A critical cytokine in tumor angiogenesis and a potential target for diagnosis and therapy. *J. Clin. Oncol.* **20**, 4368–4380.

Cirino, G., Fiorucci, S., and Sessa, W. C. (2003). Endothelial nitric oxide synthase: The Cinderella of inflammation? *Trends Pharmacol. Sci.* **24**, 91–95. *Review examining the concept that the endothelial isoform of nitric oxide synthase can promote inflammation.*

Hobbs, A. J., Higgs, A., and Moncada, S. (1999). Inhibition of nitric oxide synthase as a potential therapeutic target. *Ann. Rev. Pharmacol. Toxicol.* **39**, 191–220.

Liu, L., and Kubes, P. (2003). Molecular mechanisms of leukocyte recruitment: Organ-specific mechanisms of action. *Thromb. Haemos.* **89**, 213–220.

Michel, C. C., and Curry, F. R. (1999). Microvascular permeability. *Physiol. Rev.* **79**, 703–761. *Thorough review describing the experimental evidence for the mechanisms of vascular permeability.*

Papapetropoulos, A., Rudic, R. D., and Sessa, W. C. (1999). Molecular control of nitric oxide synthases in the cardiovascular system. *Cardiovasc. Res.* **43**, 509–520. *Analysis of the roles of nitric oxide synthases in the cardiovascular system based on phenotypes observed in genetically deficient mice.*

Walford, G., and Loscalzo, J. (2003). Nitric oxide in vascular biology. *J. Thromb. Haemost.* **1**, 2112–2118.

Capsule Bibliography

Dr. Sessa is a Professor in the Department of Pharmacology at Yale University and the Director of the Vascular Cell Signaling and Therapeutics Program. His laboratory research focuses on identifying cell biological and molecular pathway that influence the function of blood vessels during vessel remodeling, angiogenesis, and inflammation. His work is funded from the NIH.

Ms. Michelle I. Lin is a Ph.D. student examining how nitric oxide regulates vascular permeability.

CHAPTER 36

Rheology of Blood Flow in the Microcirculation

Herbert H. Lipowsky

Department of Bioengineering, The Pennsylvania State University, University Park, Pennsylvania

Introduction

It is widely acknowledged that the resistance to flow within individual microvessels (R) results from the effective viscosity (η) of blood and geometric factors, where the latter is often referred to as vascular hindrance (z). Under idealized conditions of fluid flow with a constant viscosity (i.e., a Newtonian fluid) and Poiseuille flow (parabolic velocity profile), z is proportional to vessel length (L) and inversely proportional to D^4, yielding the relationship that $R \sim \eta z$. It has been shown that over the broad range of diameters encountered as blood traverses the microvascular network, the principal determinant of resistance in the normal flow state is the relationship that $R/L \sim 1/D^4$ and that substantial departures from this behavior in the low-flow state and pathological conditions arise from the rheological properties of blood [1].

Rheology (from Greek *rheo*, flow) is the study of the flow and deformation of materials. The general aim of rheological studies is to characterize the intrinsic mechanical properties of a fluid or solid in terms of the resistance it offers to deformation under a given load, or shear at a prescribed rate. The viscous properties of blood in large-bore tubes and viscometric instruments have provided a foundation for understanding the rheology of blood in microvessels. With the assumption that blood is a homogenous fluid with an intrinsic viscosity, these devices have revealed that blood viscosity falls as shear rates ($\dot{\gamma}$) rise (shear thinning) from on the order of 0.1 to 1,000 seconds^{-1}, in contrast to the behavior of a Newtonian fluid with constant viscosity. The intrinsic viscous properties of bulk suspensions are typified by a parametric set of curves of viscosity (η) versus shear rate ($\dot{\gamma}$), as illustrated in Figure 1 for cat blood using a cone-plate viscometer. The shaded area represents the general regime of $\dot{\gamma}$ and hematocrit for the microcirculation in the normal flow state. At a given shear rate, blood viscosity rises exponentially with increasing red blood cell (RBC) concentration (hematocrit), to an extent dependent upon prevailing $\dot{\gamma}$. Viscosity of the suspending medium (plasma) has been shown to be invariant, with $\dot{\gamma}$ (Newtonian) and is dependent mainly upon protein content and temperature.

Within the circulation, in large diameter vessels representative of the macrocirculation (i.e., > 100 μm), blood may be treated as a homogeneous continuum with intrinsic properties characterized by an "apparent viscosity." The term apparent viscosity is used since viscosity of a homogenous fluid (e.g., water, molasses) is a material property that may be dependent upon shear rate, is invariant with the size of the vessel through which it flows and is dependent mainly on temperature. In vitro viscometric studies have revealed that apparent viscosity (η) rises as shear rate ($\dot{\gamma}$) is reduced. The decrease in η with increasing $\dot{\gamma}$ depends strongly on levels of hematocrit. A comparison of this "shear thinning" of blood in the presence and absence of aggregating agents suggests that about 75 percent of the decrease is a result of the disruption of red cell aggregates, and 25 percent is due to red cell deformation in response to increased shear stresses. In vivo, the particulate nature of blood affects this relationship. The dominance of noncontinuum effects in the smallest microvessels result in an effective blood viscosity that is strongly dependent upon microvessel diameter.

Microvascular Hematocrit

It is evident that microvascular hematocrit may be the dominant determinant of apparent viscosity in microvessels. Because of the particulate nature of blood, three hematocrits

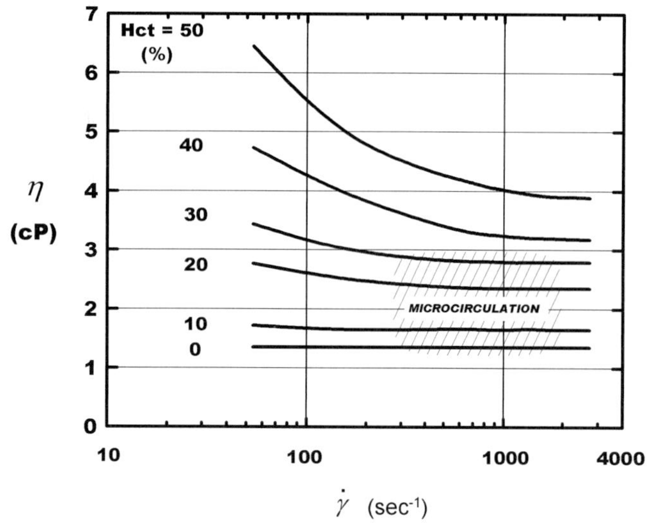

Figure 1 Apparent viscosity (η) versus shear rate ($\dot{\gamma}$) at various levels of hematocrit was obtained in a large scale cone-plate viscometer. The cross-hatching delineates the typical range in microvessel hematocrit, η, and $\dot{\gamma}$ found in the microcirculation in the normal flow state. Redrawn from a study of cat blood by Lipowsky et al. (1980). *Microvasc. Res.* **19**, 297–319.

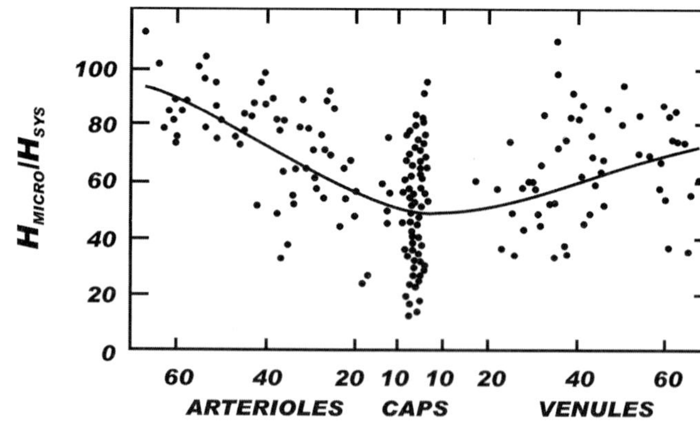

Figure 2 Representative distribution of microvessel hematocrit (H_{MICRO}), normalized with respect to systemic values (H_{SYS}), measured in the cremaster muscle of the rat for microvessels ranging from arterioles to venules of the indicated luminal diameter. H_{MICRO} falls below H_{SYS} because of the Fåhraeus effect and plasma skimming at arteriolar branches. From the data of House and Lipowsky (1987). *Am. J. Physiol.* **252**, H211–H222.

must be considered when describing blood flow in small-bore tubes or microvessels: (1) the feed hematocrit (H_{FEED}), (2) the discharge hematocrit (H_{DISCH}), and (3) the tube or microvessel hematocrit (H_{TUBE} or H_{MICRO}). The feed hematocrit represents the packed cell fraction contained in the suspension that supplies the small blood vessels and is equivalent to the red cell fraction present in large vessels of the macrocirculation, typically larger than 100 μm in diameter. The discharge hematocrit is the volume fraction of red cells found in a hypothetical collection container that receives flow from the exit of a microvessel or small tube. In the absence of any cell screening events at the tube entrance, $H_{FEED} = H_{DISCH}$. The tube or microvessel hematocrit is defined as the packed cell fraction resident within the lumen of the tube at any instant of time and is determined by suddenly stopping the flow and assaying the packed cell fraction of RBCs within the tube. Given that RBCs flow through a tube with an average velocity of V_{RBC}, it has been shown that $V_{RBC}H_{TUBE} = V_{MEAN}H_{DISCH}$, where V_{MEAN} is the mean velocity of blood (cells plus plasma) calculated by dividing the volumetric flow rate (Q) by the cross-sectional area (A) of the tube, that is, $V_{MEAN} = Q/A$. These relationships lead to a hypothetical minimum value for H_{MICRO} if it is assumed that Poiseuille's law governs the flow of blood. That is, for Poiseuille flow with a parabolic velocity profile within the tube, V_{MEAN} is $^1/_2$ the maximum velocity that occurs along the tube center line, V_{CL}. Hence, if all RBCs were infinitesimally small and traveled along the vessel centerline, the lowest value of H_{MICRO} attainable would be $^1/_2$ that of H_{DISCH}. In contrast, the theoretical maximum value of H_{MICRO} that could be attained within a tube is governed by packing considerations for rigid particles of volume equivalent volume to that of red cells. For a rigid spherical particle, the maximum volume fraction would be about 53 percent, whereas for a rigid RBC it would be about 64 percent.

Within the microvasculature, H_{MICRO} has been shown to span a much broader range of values under normal and pathological conditions. Direct measurements of H_{MICRO} reveal that as microvessel diameter falls throughout successive divisions, H_{MICRO} falls to on the order of 25 percent of systemic values (H_{SYS}) as blood approaches microvessels comparable in diameter to the characteristic size of an RBC. The average capillary hematocrit has been found to vary among tissues as a result of different topographical patterns of blood vessels and luminal diameters within a given microvascular division, as illustrated in Figure 2 for the cremaster muscle (rat). Two processes give rise to this fall: the Fåhraeus effect and plasma skimming.

Since the pioneering studies of Robin Fåhraeus (1929; see Ref. [2]) it has been recognized that hematocrit in tubes comparable in size to that of the red cell may be reduced dramatically below the feed hematocrit that supplies a tissue. This behavior has been attributed to rapid changes in the mean velocity of red cells relative to that of plasma, as summarized by Cokelet [3]. As tube diameter is diminished toward the size of a red cell, the velocity of red cells increases relative to the mean velocity of plasma. Hence, to maintain conservation of red cell flux, the tube hematocrit falls and fewer red cells travel through the tube at a faster than normal speed (relative to plasma) to maintain the same red cell volumetric flow. With further reductions in tube diameter to the point where red cells must undergo large deformation to gain entry to the tube, the Fåhraeus effect reverses, as evidenced by increasing H_{TUBE} with diminishing diameter as the motion of RBCs becomes hindered and their sequestration in the tube leads to increases in H_{MICRO} above H_{SYS}.

Direct measurements of H_{MICRO} by cell counting or spectrophotometry within single microvessels have revealed a heterogeneous distribution of H_{MICRO} within any given division of the network [4] indicative of hematocrit reductions due to other effects. The principal mechanism leading to further reductions in H_{MICRO} is that of skimming off of plasma by daughter branches at bifurcation points within the network, as first observed by Poiseuille [5]. It has been suggested that H_{MICRO} is relatively uniform across the lumen (radial direction) of an arteriole or venule, compared with the heterogeneity associated with successive branchings. However, the presence of a thin annulus of plasma surrounding the core of RBCs within a microvessel facilitates an uneven distribution of RBCs at arteriolar branchings [3]. The proportions of RBCs from the core, and the cell-free plasma layer that is captured by an arteriolar branch at a bifurcation, is dependent upon the relative magnitudes of total volumetric flow from parent to daughter branch at a bifurcation. At the final ramifications of the arteriolar network, red cell entry into capillaries is dependent upon the presence of a sufficient pressure gradient that can sustain red cell deformations at the capillary entrance and hence the capillary branch with the fastest stream (greatest pressure gradient) captures the majority of red cells. In concert with the Fåhraeus effect, plasma skimming contributes to the markedly lower than systemic hematocrits in the capillary network.

Other mechanisms of hematocrit reduction have been implicated as a source of the low capillary hematocrit. It has been suggested that irregularities in the capillary lumen, due to departures from a circular cross-section, introduce substantial errors in computing H_{CAP} that may result in systematically low values. Uncertainties in observing the luminal surface of the capillary endothelium may also introduce significant errors in estimating H_{CAP}. The presence of a thick (~ 0.5 µm) surface layer of carbohydrates and proteins on the surface of the endothelium (the glycocalyx) may also affect estimation of H_{CAP}. Removal of the endothelial surface layer by enzymatic degradation of the glycocalyx (see review by Pries et al. [6]) has resulted in a substantial rise in capillary hematocrit.

The observed low capillary hematocrits have raised critical questions on the role of hematocrit in oxygen transport to tissue. It is generally accepted that the potential for oxygen transport from blood to tissue is related to the concentration of RBCs within the microvasculature. However, the low levels of microvessel hematocrit are inconsistent with the fact that for a given level of oxyhemoglobin, oxygen content is proportional to the product of mean corpuscular hemoglobin concentration and H_{MICRO}. Attempts to account for all of the factors contributing to reduced hematocrit (Fåhraeus effect, plasma skimming, uncertainties in capillary diameter, and so on) have not resolved the disparity between H_{SYS} and H_{CAP}. This situation may have persisted because hematocrit measurements do not consider the flow-weighted flux of red cells within individual microvessels. In contrast, the flow-weighted estimates of red cell fraction obtained using techniques of indicator dilution (with fluorescently labeled RBCs and plasma) have revealed average tissue hematocrits approaching systemic values, which are substantially greater. Nonetheless, the importance of H_{MICRO} rests in that it remains a major determinant of the apparent viscosity of blood in microvessels, and hence of resistance to blood flow.

In addition to the dynamic processes that affect RBC apportionment at branch points, RBC mechanical properties may dramatically affect capillary hematocrit. Pathologically stiff red cells (e.g., in sickle cell disease, the thalassemias, or hereditary spherocytosis) may affect the distribution of cells. Because of cell–cell interactions, less deformable RBCs have a tendency to become trapped within networks with nontube-like vessels, such as the lung, spleen, liver, or bone marrow. The permeability characteristics of the microvessel wall may also affect hematocrit. In the classical study of disruption of the permeability of the small blood vessels leading to complete vascular stasis, as first described by August Krogh [1], the hyperpermeable vessel wall may allow compaction of RBCs to hematocrits approaching 100 percent.

Apparent Viscosity

The bulk viscosity of blood obtained by in vitro viscometry (Figure 1) does not fully address the viscous properties that determine microvessel perfusion for a given pressure gradient. In microvessesls larger than 20 µm in diameter, it has been demonstrated that at low physiological hematocrits, and in the normal flow state, the apparent viscosity (η) varies linearly with H_{MICRO} (see Zweifach and Lipowsky [1]). In vitro studies of blood behavior in small-bore tubes have demonstrated that reductions of H_{MICRO} attendant to diameter reductions result in a decrease in effective viscosity as well, which has been referred to as the Fåhraeus-Linqvist effect [2]. In vivo, increased viscosity occurs with reductions in $\dot{\gamma}$, suggesting that the heterogenous perfusion of the microvasculature in the low flow state may arise in part due to substantial elevations in η. However, in vitro studies with small tubes do not reflect these trends and suggest that additional mechanisms may contribute to shear-dependent resistance increases in small blood vessels [7].

Red Cell Aggregation

The effect of red cell aggregation (RCA) on the resistance of flow in vivo has also been fraught with controversy. Infusing aggregating agents such as high-molecular-weight (500 kDa) dextran into animals results in some studies in systemic effects indicative of increased resistance to flow caused by RCA. However, other studies demonstrate a reduction in regional resistance in isolated tissues, suggestive of reductions in systemic hematocrit attendant to RBC sequestration in other regions. In vitro studies of RCA in vertically positioned small-bore glass tubes have shown that the apparent viscosity of blood decreases with increasing degrees of RCA. However, in vivo measurements in the low-flow state suggest a dramatic rise in apparent viscosity with reductions in $\dot{\gamma}$. The sequestration of red cell aggregates in regional networks has been shown to dramatically decrease systemic hematocrit, the extent of which is strongly dependent on the strength the aggregating agents.

Two mechanisms have been implicated in aggregate formation: osmotic depletion, which results in hydrostatic pressures that tend to force apposing cells together in the presence of macromolecules, and molecular cross-bridging, whereby solutes cause the binding of one cell to another. Modeling and experimental studies reveal that with weaker aggregating forces, red cells assume the configuration of rouleaux (Figure 3A), which may be disrupted by shearing forces at branch points in the arteriolar network. With stronger aggregating forces, clumps of cells are formed that appear to be more easily trapped at microvascular branch points (Figure 3B).

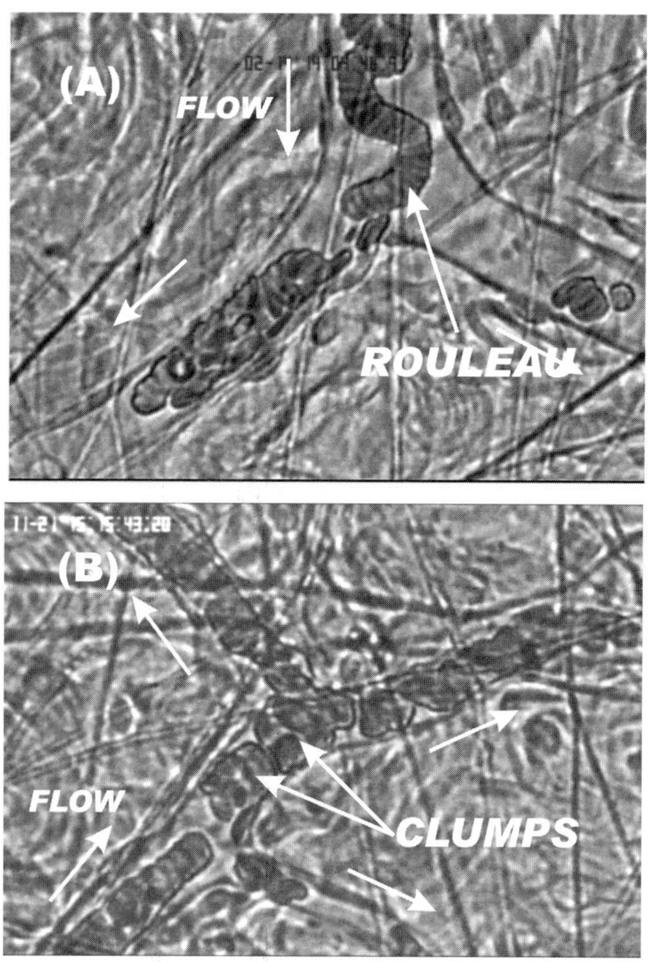

Figure 3 Red cell aggregates at the entrance of precapillary vessels in the low-flow state. Two levels of aggregation are shown, corresponding to **(A)** rouleaux formation in the presence of elevated fibrinogen (0.7 g%) and **(B)** clumping in response to 3 g% high-molecular-weight dextran (500 kDa). The apparently stronger clumps are formed as apposing RBC membranes are linked by dextran cross-bridging with the result that clumps are readily trapped within the entrance to small microvessels. From Pearson and Lipowsky (2004). *Microcirculation* **11**(3), 295–306.

Blood Cell Deformability

Blood cell deformability is a major source of the resistance to flow. Increases in bulk viscosity of RBCs in suspension clearly result from reductions in the ability of RBCs to deform under shear ($\dot{\gamma}$). However, these abnormalities in η are small compared with the dramatic rise in resistance to flow as individual cells enter the true capillaries. In most species, blood cellular elements (erythrocytes and leukocytes) are similar in size or slightly larger than the true capillaries, and they must be deformed by prevailing pressure gradients to gain entry into a capillary. The pressure gradient required to propel RBCs and WBCs into a capillary of a given diameter is strongly dependent on the ratio of cell diameter (D_{CELL}) to capillary diameter (D_{CAP}), that is, D_{CELL}/D_{CAP}. As D_{CELL}/D_{CAP} is reduced, small increases in cell stiffness may inhibit its entry into a capillary. As a result, RBCs and WBCs tend to traverse the microvascular network through pathways of least resistance, with the larger and less deformable WBCs bypassing the smallest diameter capillaries. As shown in Figure 4, plasma, red cells, and leukocytes traverse the capillary network through different pathways that are dictated by their ability to enter a vessel of a given diameter.

Resistance to Blood Flow in Microvessels

The behavior of small blood vessels in response to vasoregulatory stimuli and pathological disturbances affect the rheological state of blood within the microvasculature. The principal determinants of the apparent viscosity (hematocrit and shear rate) are directly influenced by levels of flow and alterations in vascular diameter. These physiological responses are inseparable from the viscous properties of blood. Several shear-dependent mechanisms of vasoregulation (e.g., those mediated by prostacycline and nitric oxide) are strongly dependent on levels of shear stress (τ) which is related to η by Newton's law of friction, $\tau = \eta\dot{\gamma}$. Thus, changes in vascular diameter may affect the apportionment of flow (Q) throughout successive microvascular divisions and the fraction of the overall arteriovenous pressure drop that appears along individual microvessels. These pressure gradients, in turn, dictate flow and wall shear rates within a single vessel. If blood behaves as a Newtonian fluid, and Poiseuille's law governs the relationship between pressure gradient and flow, then wall shear rate would be given by $\dot{\gamma} = 32Q/\pi D^3$. This relationship serves to couple the intrinsic properties of blood to microvessel flow (Q) and diameter (D) via the relationship between η and $\dot{\gamma}$.

The relationship among hemodynamic resistance, flow and viscosity has suggested that an optimal hematocrit may exist that maximizes the delivery of oxygen to tissue. As H_{MICRO} is increased, elevations in η will result in increased resistance and hence reductions in flow. Although RBC content (and hence oxygen carrying capacity) would be increased, the convective flux of oxygen ($\sim Q \times H_{MICRO}$)

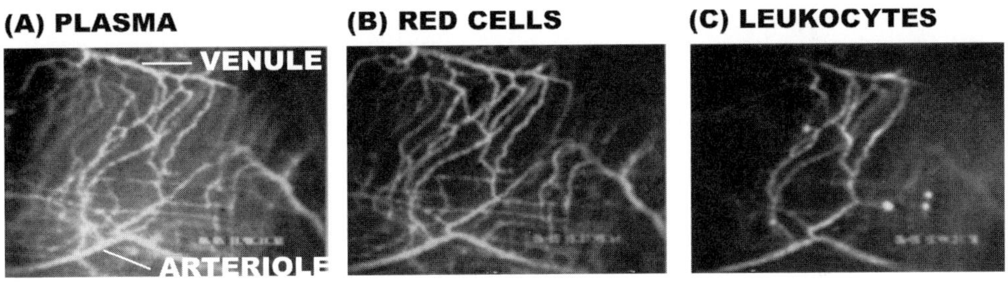

Figure 4 Blood cell deformability is an important determinant of the ability of RBCs and WBCs to pass through the capillary. In many networks the true capillaries are often smaller than the characteristic size of a blood cell. Shown are the pathways of perfusion taken by fluorescently labeled plasma (**A**), RBCs (**B**), and WBCs (**C**). Compared to plasma, the lesser intensity of RBCs in many vessels suggests that many RBCs traverse the network from arteriole to venule through larger diameter pathways. This behavior is much more striking for the larger and stiffer WBC that appears to travel through more centrally situated thoroughfare channels. From Eppihimer and Lipowsky (1994). *Am. J. Physiol.* **267**, H1122–H1134.

may be reduced. Conversely, reductions in H_{MICRO} may decrease resistance (η) and thus increase Q, while diminishing the red cell content (H_{MICRO}). It has been demonstrated that there exists an optimum level of hematocrit that serves to maximize red cell flux through the microvascular network. With changes in H_{MICRO} above or below its optimum value, red cell flux may fall dramatically. Measurements of red cell flux in various organs reveal a paraboloid maximum dependent on H_{MICRO}, with a relationship that is affected by the state of vasodilation or constriction with the arteriolar network.

The long-standing hypothesis that decreased viscosity will enhance oxygen transport by improving flow has recently been challenged by developers of blood substitutes [8]. It has been shown that elevations in blood substitute viscosity serve to elicit a vasodilatory response to increased shear stresses and thus enhance microvascular blood flow. The inverse relationship between resistance to blood flow and the fourth power of vascular diameter (Poiseuille's law; see Sutera and Skalak [5]) serves as a dominant determinant of microvascular perfusion [1].

Interactions of the RBCs and WBCs with the microvessel wall may also affect the resistance to flow. In some pathological disorders, such as sickle-cell disease, RBCs have a tendency to adhere to the microvessel wall, and if levels of flow are low enough, their adhesion may impair perfusion of the network by obstruction of the microvessel lumen. The receptor-mediated adhesion of WBCs to the venular walls of postcapillary venules may also have a dramatic effect on resistance to flow. During the inflammatory process, WBC adhesion may proceed to an extent where it can obliterate the venular lumen. It has been demonstrated that as few as 12 WBCs adhering to the wall of a postcapillary venule per 100 μm of venular length may double the resistance to flow within that vessel. Indirect studies of hindlimb perfusion have suggested that the presence of WBCs within the capillaries may affect the distribution of red cells and thus adversely increase resistance. However, removal of WBCs from the circulation has shown only a 5 percent decrease in resistance at the capillary level. In the low-flow state, the entrapment of WBCs at the entrance to capillaries may also adversely affect resistance to flow and diminish the throughput of flow from arteriole to venule. The extent to which these effects impair microvascular perfusion has been shown to be dependent upon the rate of accumulation of WBCs, the stiffness of the WBC, and the influence of inflammatory mediators that affect WBC deformability.

Although a wealth of in vitro and in vivo rheological studies have served to characterize the intrinsic mechanical properties of blood cell suspensions and blood cellular elements, new mechanisms have come to light that may affect the resistance to flow in microvessels. It has been suggested that low levels of red cell aggregation (as may occur in fibrinogenemia) may enhance the contact of WBCs with postcapillary endothelium and thus promote elevation of leukocyte–endothelium adhesion [9], leading to increased resistance to flow. Studies of the function of the endothelial glycocalyx have suggested that its removal may result in diminished resistance to blood flow [6]. Recent studies have demonstrated that components of the endothelial glycocalyx may be shed during inflammation and ischemia, resulting in diminished resistance due to increases in the vascular lumen, or increased resistance due to elevations in WBC adhesion to venular endothelium [10]. It is clearly evident that in order to understand the rheological determinants of the resistance to flow within the microcirculation proper, greater attention needs to be given to blood interactions with the endothelial cell surface at the molecular level.

References

1. Zweifach, B. W., and Lipowsky, H. H. (1984). Pressure and flow relationships in blood and lymph microcirculation. In *American Physiological Society Handbook on Microcirculation* (E. M. Renkin and C. C. Michel, eds.), Vol. I, pp. 251–307. Am Physiol Soc, Bethesda, MD. *Overview of the distribution of intravascular pressures and flows in the microcirculation proper with an emphasis on in vivo viscous properties of blood and the determinants of resistance within successive microvascular divisions.*
2. Chien, S., Usami, S., and Skalak, R. (1984). Blood flow in small tubes. In *American Physiological Society Handbook on Microcirculation*

(E. M. Renkin and C. C. Michel, eds.), Vol. I, pp. 217–249. Am Physiol Soc, Bethesda, MD. *Thorough overview of the mechanics of blood cell behavior in small bore tubes. Establishes a necessary framework to understand in vivo determinants of the resistance to flow.*
3. Cokelet, G. R. (1999). Poiseuille Award Lecture. Viscometric, *in vitro* and *in vivo* blood viscosity relationships: How are they related? *Biorheology* **36**, 343–358. *Comprehensive summary of the rheology of blood flow as assessed in viscometric instruments and its relevance to in vivo blood viscosity.*
4. Lipowsky, H. H. (1986). In situ measurement of microvascular hematocrit. In *Microvascular Technology* (C. H. Baker and W. L. Nastuk, eds.), New York: Academic Press, pp. 161–178. *Overview of techniques for measurement of microvascular hematocrit in small blood vessels and the trends routinely observed in successive divisions of the microvasculature.*
5. Sutera, S., and Skalak, R. (1993). The history of Poiseuille's law. *Ann. Rev. Fluid Mechanics* **25**, 1–20. *Historical account of the seminal studies of this pioneering scientist, who discovered the laws governing fluid flow in small-bore tubes in his quest to understand the mechanics of blood flow in microvessels.*
6. Pries, A. R., Secomb, T. W., and Gaehtgens, P. (2000). The endothelial surface layer. *Pflügers Arch. Eur. J. Physiol.* **440**, 653–666. *Overview of molecular composition and structure of the endothelial cell glycocalyx in the microvasculature.*
7. Reinke, W., Gaehtgens, P., and Johnson, P. C. (1987). Blood viscosity in small tubes: effect of shear rate, aggregation and sedimentation. *Am. J. Physiol.* **253**: H540–H547. *Illustrates the difficulties in performing in vitro simulations of blood flow in tubes comparable in size to microvessels.*
8. Tsai, A. G., and Intaglietta, M. (2001). High viscosity plasma expanders: Volume restitution fluids for lowering the transfusion trigger. *Biorheology* **38**: 229–237. *Seminal paper hypothesizing the vasodilatory response to increased viscosity (shear stress) of blood substitutes that results in greater blood flow. See the chapter by M. Intaglietta in this volume for additional details.*
9. Pearson, M. J., and Lipowsky, H. H. (2004). Effect of fibrinogen on leukocyte margination and adhesion in post-capillary venules. *Microcirculation* **11**(3), 295–306.
10. Mulivor, A. W., and Lipowsky, H. H. (2004). Inflammation and ischemia induced shedding of the venular glycocalyx. *Am. J. Physiol. Heart Circ. Physiol.*, **286**: H1672–H1680. *Characterization of the endothelial cell glycocalyx in postcapillary venules suggests that its composition reflects a balance between biosynthesis of glycans and their shear-dependent removal.*

Further Reading

Lipowsky, H. H. (1995). Shear stress in the circulation. In *Flow Dependent Regulation of Vascular Function* (J. Bevan and G. Kaley, eds.), New York: Oxford University Press, pp. 28–45. *Review of basic mechanics and hydrodynamic factors that contribute to flow induced shear stresses on the microvessel wall. Summary data depict the systematic variation of flow, hematocrit, viscosity, and shear stress throughout successive microvascular divisions.*

Author's Biographical Sketch

Dr. Lipowsky is currently Professor and Head of the Department of Bioengineering at Penn State University. He has been studying the in vivo rheology of blood flow in the microcirculation for the past 30 years and has pioneered several techniques for making rheological measurements in the living animal using techniques of intravital microscopy.

CHAPTER 37

Endothelial Ca^{2+} and Endothelium-Derived Hyperpolarizing Factor (EDHF)-Mediated Vasodilatation

Sean P. Marrelli

Assistant Professor, Department of Anesthesiology, Baylor College of Medicine, Houston, Texas

The Endothelium as a Modulator of Vascular Tone

A single thin layer of endothelial cells comprises the innermost layer of all blood vessels from capillaries to large arteries and veins. One of the more recently discovered functions of the endothelial cell layer is that of a modulator of vascular tone. Endothelial cells are capable of responding to numerous physiological stimuli (such as circulating humoral factors, sheer stress, and so on) by producing vasoactive substances that are then released to the adjacent smooth muscle cells. Two of the most well known endothelium-derived vasoactive substances, prostacyclin (PGI_2) and nitric oxide (NO), are capable of producing vasodilatation upon reaching the smooth muscle. PGI_2 is one of several potential products produced by the cellular metabolism of arachidonic acid. NO is a gaseous factor produced from the conversion of L-arginine to L-citrulline by nitric oxide synthase. NO (or an NO-containing compound) was identified as endothelium-derived relaxing factor (EDRF). The physiological importance of these two vasodilatory factors was recognized by the awarding of the Nobel Prize in Physiology/Medicine in 1982 and 1998 for work leading to the discovery and physiological role for PGI_2 and NO, respectively.

An Additional Mechanism of Vasodilatation, EDHF

In the mid-1980s, it became apparent that endothelium-dependent vasorelaxation could not always be completely accounted for by the actions of PGI_2 and NO. It had been known for some time that isolated arteries could be relaxed (or dilated) by factors such as acetylcholine (ACh), bradykinin (BK), and adenosine diphosphate (ADP). However, whereas the relaxation could be abolished by the removal or damaging of the endothelial layer, the relaxation was not always completely blocked by the combined inhibition of PGI_2 and NO synthesis. It was further noted that the PGI_2/NO-independent relaxation was associated with smooth muscle hyperpolarization. These findings prompted some investigators to conclude that an additional relaxing factor must exist; they termed it endothelium-derived hyperpolarizing factor (EDHF) [1].

Mechanism of EDHF-Mediated Vasodilatation

Although the proposed mechanism of EDHF-mediated vasodilatation is continually evolving, the following represents a general understanding at this time (see Figure 1). (1)

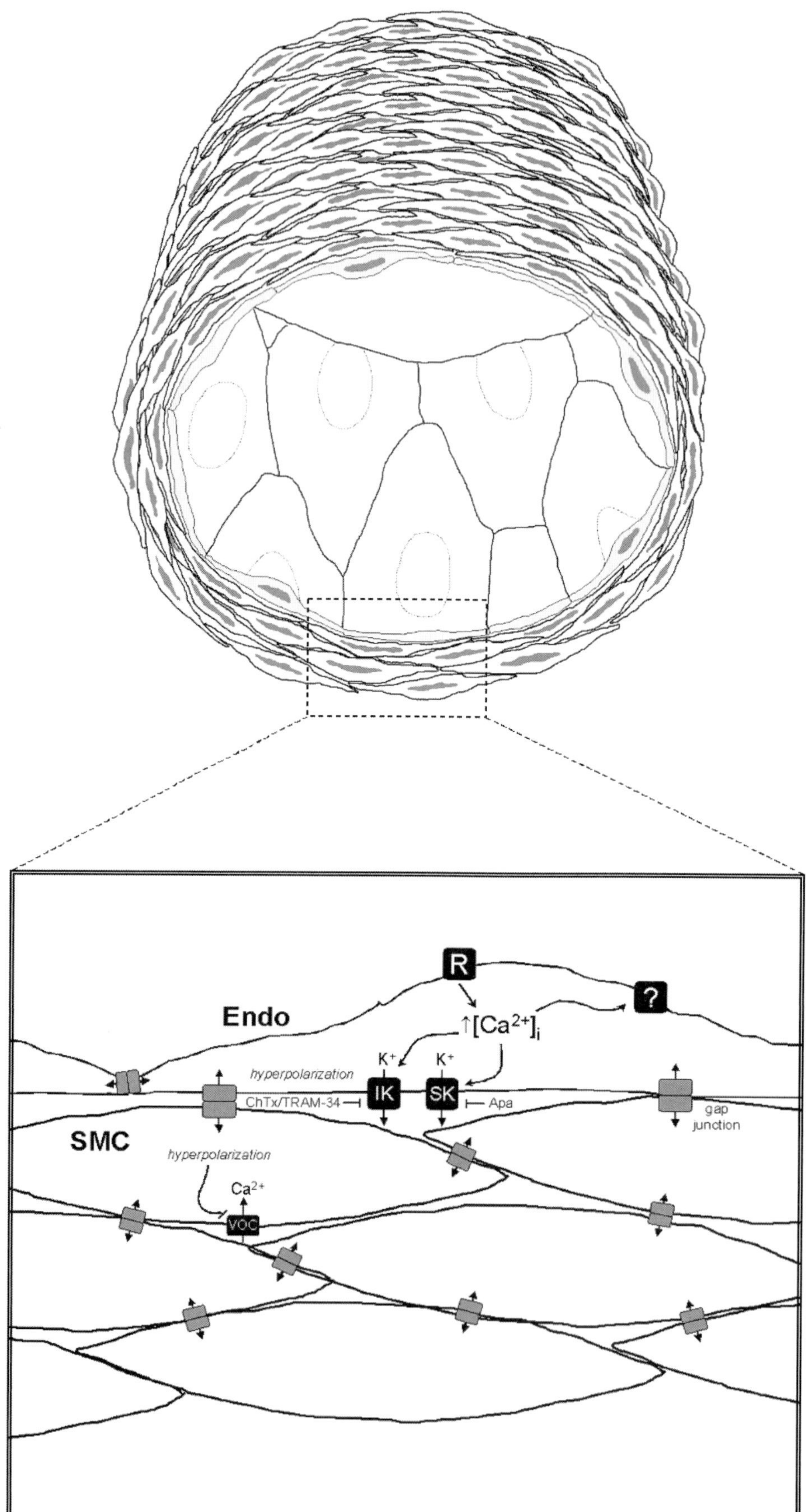

Figure 1 Cartoon depiction of the role of endothelial cytosolic Ca^{2+} ($[Ca^{2+}]_i$) in EDHF-mediated vasodilatation. Abbreviations: Endo, endothelial cell; SMC, smooth muscle cell; R, receptor; IK, intermediate-conductance K_{Ca} channel; SK, small-conductance K_{Ca} channel; VOC, voltage-operated Ca^{2+} channel; ChTx, charybdotoxin; Apa, apamin. Possible Ca^{2+}-sensitive mechanisms that could modulate the EDHF-dependent mechanism are represented by "?."

An agonist binds to its receptor on the endothelium. This binding initiates a cascade of events that leads to an elevation of endothelial cytosolic free calcium ($[Ca^{2+}]_i$). Calcium ionophores, such as A23187, have also been demonstrated to initiate EDHF-mediated dilations by producing an increase in endothelial $[Ca^{2+}]_i$. (2) Elevated endothelial $[Ca^{2+}]_i$ stimulates endothelial calcium-sensitive K channels (K_{Ca}) and promotes hyperpolarization of the endothelium. Elevated endothelial $[Ca^{2+}]_i$ also results in activation of other pathways that may be important modulators of the EDHF-dependent mechanism. (3) The endothelial hyperpolarization then promotes smooth muscle hyperpolarization by a mechanism that is still quite controversial. One of the prevailing ideas is that hyperpolarization is conducted directly via myoendothelial gap junctions, which form electrical couplings between adjacent smooth muscle and endothelial cells. There is pharmacological, electrophysiological, and histological evidence in support of such a mechanism. Another idea is that K^+ ions extruded from basolateral K_{Ca} channels stimulate smooth muscle inwardly rectifying K channels (K_{ir}) and/or Na^+/K^+-ATPase. A small elevation in extracellular K^+ has been shown to stimulate smooth muscle K_{ir} channels and Na^+/K^+-ATPase, and could thus promote smooth muscle hyperpolarization in this way. Although the evidence in support of the latter mechanism is much less prevalent, it could be partly due to the technical difficulties in measuring K^+ in the small intercellular space between endothelial and smooth muscle cells. (4) Smooth muscle hyperpolarization promotes the closure of voltage-operated Ca^{2+} channels (VOC). The smooth muscle VOCs are inactivated at hyperpolarized membrane potentials and thus significantly reduce smooth muscle Ca^{2+} influx. The reduced Ca^{2+} influx combined with Ca^{2+} extrusion and resequestration results in a reduction in smooth muscle $[Ca^{2+}]_i$ and subsequent relaxation or vasodilatation.

Although it was initially believed that EDHF-mediated dilations were the result of a transferable *factor* (similar to NO and PGI_2-mediated responses), recent studies suggest otherwise. Instead, it appears much more likely that EDHF-mediated dilations reflect a *process* by which endothelial hyperpolarization is translated into smooth muscle hyperpolarization (see previous discussion). For this reason, some investigators have begun referring to the mechanism as endothelium-dependent hyperpolarization (EDH). It should be noted that more than one EDHF-dependent mechanism may exist. For instance, the mechanism in coronary and renal arteries may critically involve the cytochrome P450-dependent production of epoxyeicosatrienoic acids (EETs). For the purpose of this article, however, only the former (and seemingly more prevalent) mechanism will be reviewed.

Role of K_{Ca} Channels in EDHF-Mediated Dilation

There are three classes of K_{Ca} channels, large-conductance K_{Ca} (BK_{Ca}), intermediate-conductance K_{Ca} (IK_{Ca}), and small-conductance K_{Ca} (SK_{Ca}), named based on their respective unitary channel conductance [2]. The conductances for BK_{Ca}, IK_{Ca}, and SK_{Ca} channels are 200–250 pS, 20–80 pS, and 2–20 pS, respectively. All of the channels are activated by Ca^{2+}, though the IK_{Ca} and SK_{Ca} channels are activated at much lower Ca^{2+} concentrations than the BK_{Ca} channels (250–500 nM versus 1–10 μM). The channels also differ in that the BK_{Ca} channels are voltage sensitive (activated upon depolarization) whereas the IK_{Ca} and SK_{Ca} channels are voltage insensitive. There are three known isoforms of SK_{Ca} channels, SK1, SK2, and SK3, whereas the IK_{Ca} channel (IK1) does not have any additional known isoforms.

One of the hallmarks of EDHF-dependent responses is the sensitivity to certain K_{Ca} channel inhibitors. For most peripheral arteries, complete inhibition of the response requires combined inhibition of IK_{Ca} (charybdotoxin) and SK_{Ca} channels (apamin) [3, 4]. In the cerebral circulation, complete inhibition appears only to require inhibition of IK_{Ca} channels. Inhibition of the BK_{Ca} channels alone (iberiotoxin) or in combination with SK_{Ca} channels does not affect the EDHF-dependent response.

Because charybdotoxin is a nonspecific blocker (inhibits BK_{Ca}, IK_{Ca}, $K_v1.2$, $K_v1.3$), earlier studies inferred the involvement of IK_{Ca} channels based on sensitivity to charybdotoxin and lack of sensitivity to iberiotoxin (BK_{Ca} selective) or other K_v channel blockers. The recent development and application of apparently IK_{Ca}-specific inhibitors (such as TRAM-34) has provided direct confirmation of the critical involvement of IK_{Ca} channels in EDHF-mediated responses.

Role of Endothelial $[Ca^{2+}]_i$ in EDHF-Mediated Vasodilatation

The critical role for endothelial $[Ca^{2+}]_i$ in EDHF-mediated responses initially gained wide support based on the following evidence: (1) agonists known to produce EDHF-mediated responses produced elevations of endothelial $[Ca^{2+}]_i$ in cultured endothelial cells, (2) Ca^{2+} ionophores (such as A23187) produced endothelium-dependent smooth muscle hyperpolarization, and (3) removal of Ca^{2+} from the bathing medium reduced or eliminated endothelium-dependent smooth muscle hyperpolarization. The evidence supporting the critical involvement of endothelial Ca^{2+} is discussed further hereafter.

Chen and Suzuki were among the first to demonstrate the Ca^{2+}-dependency of EDHF-mediated responses [5]. By measuring rabbit carotid artery smooth muscle membrane potential by sharp electrode, they were able to demonstrate that ACh-mediated smooth muscle hyperpolarization consisted of two components (see Figure 2). When ACh was applied to endothelium-intact arteries in PSS with a standard Ca^{2+} concentration ($[Ca^{2+}]_o$) of 2.5 mM, ACh produced a significant smooth muscle hyperpolarization that was sustained for several minutes (top trace). However, in the pres-

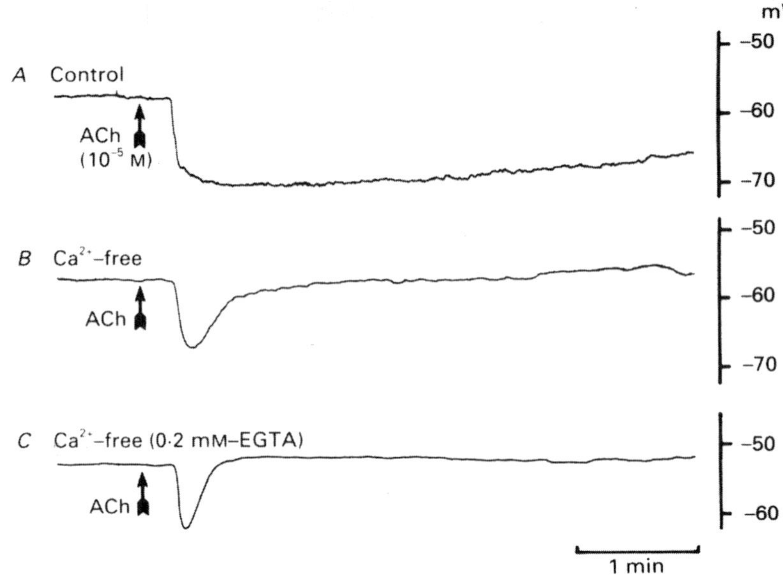

Figure 2 Acetylcholine (ACh)-induced hyperpolarization generated in different $[Ca^{2+}]_o$ solutions. $[Ca^{2+}]_o$ = 2.5 mM (**A**), 0 mM (**B**), and 0 mM with 0.2 mM EGTA (**C**). ACh (10^{-5} M) was applied at the arrow in each trace. Reproduced with permission from Chen and Suzuki (1990). *J. Physiol.* **421**, 521–534.

ence of nominally Ca^{2+}-free (no Ca^{2+} added to the PSS) or Ca^{2+}-free/EGTA solution (Ca^{2+} chelator added to remove any residual Ca^{2+}), ACh produced only a transient hyperpolarization (middle and bottom traces). Application of ACh to arteries from which the endothelium was destroyed or remove produced no smooth muscle hyperpolarization. Additionally, application of the Ca^{2+} ionophore, A23187, promoted smooth muscle hyperpolarization that was similarly dependent upon an intact endothelium. Thus, it was concluded that endothelium-dependent hyperpolarization of the smooth muscle consisted of two components, a transient hyperpolarization that utilized intracellular store Ca^{2+} and a sustained hyperpolarization that was dependent on Ca^{2+} influx from the extracellular medium.

Early on in the study of EDHF-mediated responses, it was noted that the involvement of an EDHF-dependent mechanism varied greatly in the vessels studied. For instance, whereas some vessels appeared to utilize EDRF/NO and EDHF-mediated mechanisms, others appeared to utilize only the EDRF/NO-mediated mechanism. It was therefore questioned why EDHF-mediated responses occurred in some vessels but not others. One hypothesis that was put forth by Nagao et al. speculated that the reason was due in part to differences in endothelial $[Ca^{2+}]_i$ attained in a particular artery [6]. The principal evidence for this conclusion was from an experiment in which A23187 was administered to rat mesenteric arteries in the absence or presence of an inhibitor of nitric oxide synthase (NOS). The authors found that A23187 was able to promote vasorelaxation in both conditions; however, it required a higher concentration of A23187 to do so in the presence of the NOS inhibitor. Thus, assuming that the increased concentration of A23187 resulted primarily in an increase in endothelial $[Ca^{2+}]_i$, it appeared that the EDHF-mediated response required a higher endothelial $[Ca^{2+}]_i$ than the EDRF/NO-mediated response. Thus, one possible explanation for the heterogeneity of EDHF-mediated responses was that the ability of the agonists to increase endothelial $[Ca^{2+}]_i$ varied in the different preparations.

A subsequent study by Marrelli investigated a possible Ca^{2+} threshold for EDHF-mediated responses more directly [7]. In that study, Marrelli selectively measured endothelial $[Ca^{2+}]_i$ in response to agonists that have been shown to produce or not produce EDHF-mediated responses in rat cerebral arteries. It was demonstrated that EDRF/NO-mediated dilations were initiated at a threshold of 220 nM Ca^{2+} (resting = 145 nM), whereas EDHF-mediated dilations were not initiated until 340 nM Ca^{2+} (see Figure 3). The threshold for EDHF-mediated dilation was similar for both a receptor-dependent agonist (uridine triphosphate; UTP) or a receptor-independent mechanism to increase endothelial $[Ca^{2+}]_i$ (Br-A23187). Interestingly, the study also demonstrated that a "non-EDHF-dependent" agonist could be made to elicit an EDHF-mediated dilation if the Ca^{2+} response was augmented. For instance, stimulation of endothelial $P2Y_1$ receptors with 2-methylthioadenosine triphosphate (2MeS-ATP) produced maximal dilations through an exclusively EDRF/NO-dependent mechanism. Dilations were completely abolished in the presence of a NOS inhibitor. However, if the baseline endothelial $[Ca^{2+}]_i$ was elevated or primed with a low concentration of UTP or Br-A23187 (though not enough to elicit EDHF-mediated dilation by itself), subsequent addition of 2MeS-ATP resulted in a greater increase in endothelial $[Ca^{2+}]_i$ and an EDHF-mediated dilation. These experiments indicated that EDHF-mediated dilations are not inherently linked to certain

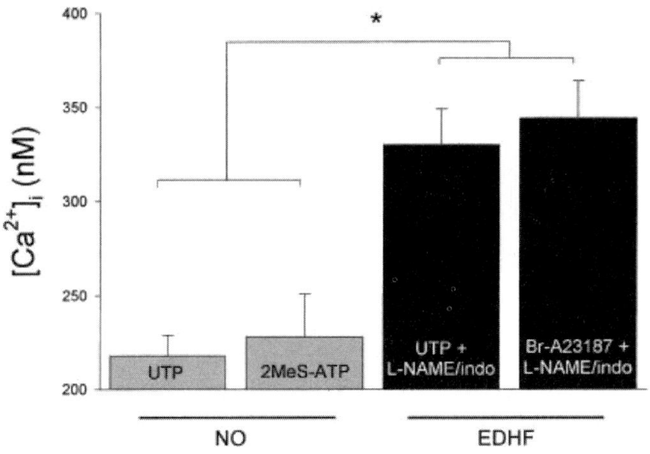

Figure 3 Comparison of the endothelial $[Ca^{2+}]_i$ threshold for nitric oxide (NO)- and EDHF-mediated vasodilatation. Data are presented as the mean SE $[Ca^{2+}]_i$ required to elicit a 15% dilation. In the concentrations used and in the absence of L-NAME (nitric oxide synthase inhibitor), UTP and 2-MeS-ATP produced vasodilatation through the production of NO (gray bars). In the presence of L-NAME plus indomethacin (cyclooxygenase inhibitor), UTP and Br-A23187 produced a dilation through the production of EDHF (black bars). There is a significantly higher threshold for EDHF-mediated vasodilatation compared with that of NO. $*P < 0.05$ between the indicated groups by one-way ANOVA. Reproduced with permission from Marrelli (2001). *Am. J. Physiol. Heart Circ. Physiol.* **281**, H1759–H1766.

receptor systems that either are or are not capable of eliciting EDHF-mediated responses. Rather, the ability of a receptor system to produce an EDHF-mediated response depends more on its potential to increase endothelial $[Ca^{2+}]_i$.

How Does Elevated Endothelial $[Ca^{2+}]_i$ Produce Vasodilatation?

The interpretation of the role of endothelial Ca^{2+} in EDHF-mediated responses has recently undergone significant revision. At one time, it was widely held that elevated endothelial $[Ca^{2+}]_i$ served primarily to activate a putative "EDHF synthase" (or to trigger release of a stored EDHF). More recent studies, however, indicate that the essential role of elevated endothelial $[Ca^{2+}]_i$ may instead be to promote endothelial hyperpolarization via stimulation of endothelial K_{Ca} channels, in particular the IK_{Ca} and SK_{Ca} channels. This idea is further supported by the fact that the endothelial $[Ca^{2+}]_i$ required to initiate EDHF-mediated dilations (340 nM Ca^{2+}) falls right within the range of $[Ca^{2+}]$ reported to activate IK_{Ca} and SK_{Ca} channels (~250–500 nM Ca^{2+}) [2, 7].

The type and location of the K_{Ca} channels involved in EDHF-mediated responses is still being worked out. However, recent data points to involvement of IK_{Ca} channels alone (cerebral arteries) or in combination with SK_{Ca} channels (peripheral arteries) located on the endothelium. Endothelial cells have been demonstrated to possess both IK_{Ca} and SK_{Ca} channels by pharmacological, electrophysiological, and molecular techniques. Native endothelial cells do not appear to express BK_{Ca} channels. Native smooth muscle cells express BK_{Ca} channels as well as certain SK_{Ca} channels, but they may not express IK_{Ca} channels under normal conditions. From these data, it would appear likely that the K_{Ca} channels involved in the EDHF-mediated response are located on the endothelial cells.

Admittedly, activation of K_{Ca} channels is not the only outcome of elevated endothelial $[Ca^{2+}]_i$. One Ca^{2+}-dependent enzyme that has been demonstrated to be involved in the EDHF-mediated response of some arteries is cytosolic phospholipase A2 ($cPLA_2$). The $cPLA_2$ enzyme is a Ca^{2+}-dependent lipase that catalyzes the hydrolysis of the *sn*-2 linkage of diacyl glycerophosphates, releasing fatty acids such as arachidonic acid. Significant activation of $cPLA_2$ has been demonstrated at Ca^{2+} concentrations of 230 to 450 nM, a concentration range that brackets the endothelial $[Ca^{2+}]_i$ concentration for initiating EDHF-mediated responses. One possible explanation of the $cPLA_2$ dependence of EDHF-mediated dilations in some arteries is simply that arachidonic acid or one or more of its numerous metabolites is the EDHF. In coronary arteries in particular, epoxygenase metabolites of arachidonic acid may act as hyperpolarizing factors. However, an alternative explanation is that arachidonic acid or its metabolites *modulate* the EDHF-dependent mechanism—but are not the EDHF themselves. For instance, arachidonic acid has been demonstrated to activate endothelial transient receptor potential cation channels (TRPV4), which are capable of carrying inward Ca^{2+} currents. Thus, one conceivable mechanism could involve $cPLA_2$-dependent release of arachidonic acid, activation of TRPV4 by arachidonic acid, increased Ca^{2+} influx, and subsequent sustained activation of endothelial IK_{Ca} and SK_{Ca} channels. In this hypothetical scenario, the sensitivity of the mechanism to $cPLA_2$ inhibition could vary depending on several tissue-specific variables such as TRPV4 channel prevalence, relative contribution of other Ca^{2+} regulation mechanisms, K_{Ca} channel Ca^{2+} sensitivity, and so on. If one therefore considers that $cPLA_2$ can participate in a modulating role, the degree to which it is able to modulate the EDHF-dependent mechanism could vary considerably between arteries and might thus explain the heterogeneity of effects seen with PLA_2 inhibitors. Although the preceding scenario is still hypothetical, it is intended to emphasize the point that a number of EDHF candidate factors should also be considered as potential modulators of endothelial K_{Ca} channels—and therefore EDHF-mediated vasodilatation.

A Relationship between Endothelial Membrane Potential (V_m) and Endothelial $[Ca^{2+}]_i$?

A number of studies have been performed to evaluate the relationship between endothelial membrane potential (V_m) and endothelial $[Ca^{2+}]_i$. The bulk of these studies have been performed with cultured endothelial cells while only a handful of studies have evaluated the relationship in endothelium from intact arteries. In cultured endothelial cells, Ca^{2+} influx has been shown to be notably modulated by endothelial V_m. As the endothelial cell becomes more hyperpolarized, the driving force for Ca^{2+} entry is increased and greater Ca^{2+} influx results following agonist stimulation [8]. If this relationship were to hold true for endothelium within intact arteries, then one possible effect of stimulating endothelial K_{Ca} channels (and thus hyperpolarizing the endothelium) would be to promote greater Ca^{2+} influx in response to EDHF-dependent agonists. In this scenario, the effect of K_{Ca} channel blockers on the EDHF-mediated response could be to prevent endothelial $[Ca^{2+}]_i$ from reaching a critical threshold. However, from the few studies that have addressed this issue in intact arteries, the relationship between endothelial V_m and endothelial $[Ca^{2+}]_i$ is uncertain. When high K^+ has been used to nonselectively inhibit all K^+ channels, agonist-mediated increases in endothelial $[Ca^{2+}]_i$ were shown to be either reduced [9, 10] or unaltered [11]. However, when charybdotoxin plus apamin was administered to block the K_{Ca} channels, agonist-mediated increases in endothelial $[Ca^{2+}]_i$ were shown to be undiminished, despite complete inhibition of the EDHF-mediated relaxation/dilation [9, 11] (see Figure 4). These findings support the conclusion that the essential role of endothelial K_{Ca} channels is not to promote or maintain adequate Ca^{2+} entry for EDHF-mediated dilation. Instead, the essential role of the endothelial K_{Ca} channels appears to be to produce endothelial hyperpolarization, entirely downstream of endothelial Ca^{2+}, and thus promote smooth muscle hyperpolarization more directly. This conclusion is supported by a study by Marrelli et al. in which an IK_{Ca} channel agonist (1-EBIO) was administered selectively to the endothelium in intact pressurized cerebral arteries [11]. 1-EBIO produced significant hyperpolarization of the endothelium and vasodilatation despite no change in endothelial $[Ca^{2+}]_i$. Thus, endothelium-dependent hyperpolarization of the smooth muscle can be produced without any elevation of endothelial $[Ca^{2+}]_i$ if endothelial K_{Ca} channels are activated by some other means. Taken as a whole, it appears that the primary role of endothelial Ca^{2+} in the EDHF-dependent mechanism may be that of the principal physiological activator of the IK_{Ca} and SK_{Ca} channels.

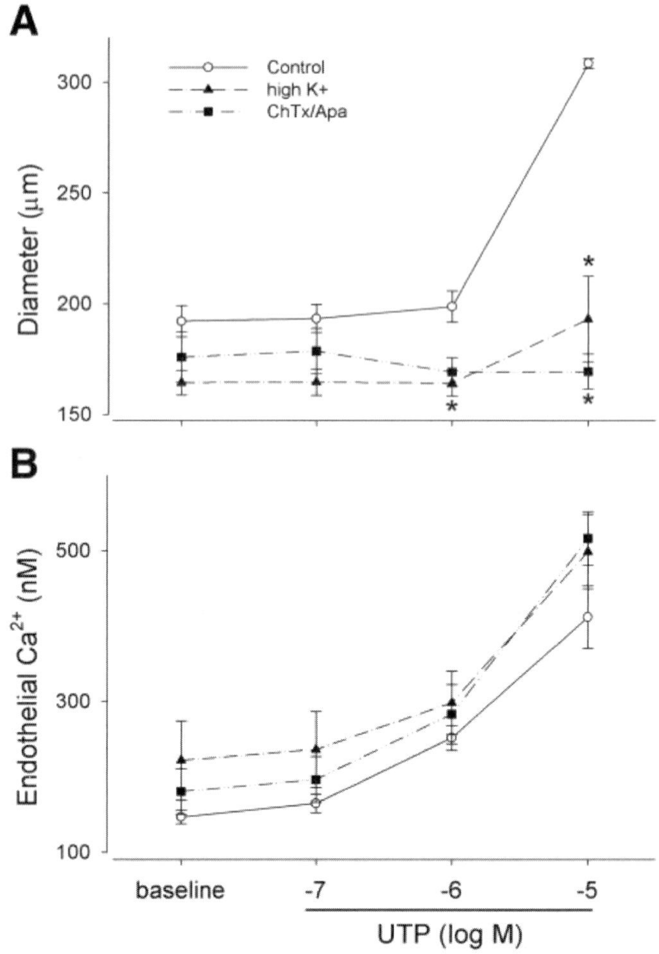

Figure 4 Role of endothelial K^+ channels in endothelial $[Ca^{2+}]_i$ response. Summary of simultaneous measurements of diameter **(A)** and endothelial $[Ca^{2+}]_i$ **(B)** in response to UTP (10^{-7} to 10^{-5} M). All experiments were performed in the presence of L-NAME and indomethacin. Responses to UTP were performed in the presence of L-NAME/indomethacin alone (control), with luminal 60 mM K^+ (high K^+, nonselective K^+ channel inhibitor), and in the presence of charybdotoxin/apamin (ChTx/Apa; combination blocks all K_{Ca} channels). Diameter responses for both high K^+ and ChTx/Apa were significantly different compared with control (two-way repeated-measures ANOVA). *Individual differences ($P < 0.05$, Tukey's test). Reproduced with permission from Marrelli et al. (2003). *Am. J. Physiol. Heart Circ. Physiol.* **285**, H1590–H1599.

Role of Endothelial $[Ca^{2+}]_i$ in EDHF-Mediated Responses in Pathology

An increasing number of studies demonstrate potentiated EDHF-mediated responses following a variety of pathological conditions, including ischemia/reperfusion (I/R), hypercholesterolemia, and congestive heart failure. The mechanism by which this potentiation occurs remains largely unknown. However, in cerebral I/R, the potentiated EDHF-mediated dilations are accompanied by augmented endothelial $[Ca^{2+}]_i$ responses [12]. In this case, for a given agonist concentration, there was a greater increase in endothelial $[Ca^{2+}]_i$ in the I/R endothelium compared with control. It was speculated that the greater Ca^{2+} responses following I/R might have resulted from some dysfunction in endothelial Ca^{2+} handling. It remains to be determined if endothelial Ca^{2+} is similarly altered in other pathological situations in which EDHF-mediated responses are potentiated.

Summary of Endothelial $[Ca^{2+}]_i$ in EDHF-Mediated Vasodilatation

EDHF-mediated vasodilatation appears to require the activation of endothelial IK_{Ca} and SK_{Ca} channels in order to promote smooth muscle hyperpolarization and subsequent vasodilatation. The role of endothelial $[Ca^{2+}]_i$ in this mechanism appears to be as the principal physiological activator of those channels. Therefore, factors that significantly affect the endothelial $[Ca^{2+}]_i$ response should be able to variably modulate IK_{Ca} and SK_{Ca} channel activation, and thus EDHF-mediated vasorelaxation.

References

1. Chen, G., Suzuki, H., and Weston, A. H. (1988). Acetylcholine releases endothelium-derived hyperpolarizing factor and EDRF from rat blood vessels. *Br. J. Pharmacol.* **95**, 1165–1174. *This article first described the nonnitric oxide, nonprostacyclin mediator of vasorelaxation as EDHF.*
2. Nilius, B., and Droogmans, G. (2001). Ion channels and their functional role in vascular endothelium. *Physiol. Rev.* **81**, 1415–1459.
3. Garland, C. J., and Plane, F. (1996). Relative importance of endothelium-derived hyperpolarizing factor for the relaxation of vascular smooth muscle in different arterial beds. In *Endothelium-Derived Hyperpolarizing Factor* (P. M. Vanhoutte, ed.), Vol. 1, pp. 173–179. Harwood Academic Publishers.
4. Zygmunt, P. M., and Hogestatt, E. D. (1996). Role of potassium channels in endothelium-dependent relaxation resistant to nitroarginine in the rat hepatic artery. *Br. J. Pharmacol.* **117**, 1600–1606.
5. Chen, G. F., and Suzuki, H. (1990). Calcium dependency of the endothelium-dependent hyperpolarization in smooth muscle cells of the rabbit carotid artery. *J. Physiol (Lond)* **421**, 521–534. *This article first described the role of endothelial Ca^{2+} in EDHF-mediated smooth muscle hyperpolarization.*
6. Nagao, T., Illiano, S., and Vanhoutte, P. M. (1992). Heterogeneous distribution of endothelium-dependent relaxations resistant to NG-nitro-L-arginine in rats. *Am. J. Physiol.* **263**, H1090–H1094. *This article is among the first to propose a higher Ca^{2+} threshold for EDHF-mediated responses compared with NO-mediated responses.*
7. Marrelli, S. P. (2001). Mechanisms of endothelial P2Y(1)- and P2Y(2)-mediated vasodilatation involve differential $[Ca^{2+}]_i$ responses. *Am. J. Physiol. Heart Circ. Physiol.* **281**, H1759–H1766. *This article provided direct evidence for an endothelial Ca^{2+} threshold for initiating EDHF-mediated vasodilatation.*
8. Schilling, W. P. (1989). Effect of membrane potential on cytosolic calcium of bovine aortic endothelial cells. *Am. J. Physiol.* **257**, H778–H784.
9. Ghisdal, P., and Morel, N. (2001). Cellular target of voltage and calcium-dependent K(+) channel blockers involved in EDHF-mediated responses in rat superior mesenteric artery. *Br. J. Pharmacol.* **134**, 1021–1028.
10. Knot, H. J., Lounsbury, K. M., Brayden, J. E., and Nelson, M. T. (1999). Gender differences in coronary artery diameter reflect changes in both endothelial Ca^{2+} and ecNOS activity. *Am. J. Physiol.* **276**, H961–H969.
11. Marrelli, S. P., Eckmann, M. S., and Hunte, M. S. (2003). Role of endothelial intermediate conductance K_{Ca} channels in cerebral EDHF-mediated dilations. *Am. J. Physiol Heart Circ. Physiol.* **285**, H1590–H1599. *This article demonstrates the importance of endothelial Ca^{2+} as the physiological stimulus for endothelial hyperpolarization via activation of endothelial K_{Ca} channels.*
12. Marrelli, S. P. (2002). Altered endothelial Ca^{2+} regulation following ischemia/reperfusion produces potentiated endothelium-derived hyperpolarizing factor-mediated dilations. *Stroke* **33**, 2285–2291.

Capsule Biography

Dr. Marrelli is an assistant professor in the Department of Anesthesiology at Baylor College of Medicine where he studies endothelial regulation of vascular tone. In particular, Dr. Marrelli's research focuses on the blood vessels of the cerebral circulation and the effects of ischemia/reperfusion (or stroke) on subsequent vascular function.

CHAPTER 38

Role of the Endothelial Cell Cytoskeleton in Microvascular Function

Laura Linz McGillem and Joe G. N. Garcia

Center for Translational Respiratory Medicine, Division of Pulmonary and Critical Care Medicine, Johns Hopkins University School of Medicine, Baltimore, Maryland

Introduction

No single cellular organizational element has commanded so much attention in the endothelial cell recently as the group of proteins known as the cytoskeleton. As depicted in Figure 1, the endothelial cytoskeleton is involved in multiple critical endothelial biologic processes including angiogenesis, mechanotransduction, apoptosis, and critical elements of the inflammatory response including leukocyte diapedisis and barrier regulation. The cytoskeleton is composed of three primary elements—actin microfilaments, intermediate filaments, and microtubules. Each will be discussed in the context of microvascular endothelial activation and dysfunction.

Microfilament Cytoskeleton Overview

Endothelial cells (ECs) contain an abundance of the molecular machinery necessary to generate tension via an actomyosin motor—actin and myosin, the key components of the microfilamentous cytoskeleton, represent a major portion of the total endothelial cellular protein content (15–20%). G-actin is a globular monomer that assembles reversibly to form polymerized actin fibers (or F-actin) conferring strength on structural elements regulating cell shape, particularly when accompanied by phosphorylated myosin. F-actin filaments within peripherally distributed cortical bands are essential for maintenance of endothelial integrity and basal barrier function. The actin microfilament cytoskeleton is a dynamic structure that undergoes rearrangement under the control of various actin binding, capping, nucleating, and severing proteins and is focally linked to multiple membrane adhesive proteins such as cadherin molecules, glycocalyx components, functional intercellular proteins of the zona occludens and zona adherens, and focal adhesion complex proteins. More than 80 actin-binding proteins have been identified and are critical participants in cytoskeletal rearrangement and tensile force generation, and serve to provide a high level of fine tuning of cell shape, adhesion, and orchestrated cell migration as well as regulation of endothelial junctional stability [1]. Edemagenic agents, such as thrombin, initiate cytoskeletal rearrangement characterized by the loss of peripheral actin filaments with a concomitant increase in organized F-actin cables that span the cell as "stress fibers" (Figure 2). One actin binding protein, cofilin for example, exerts actin-depolymerizing activity critical to cortical actin rearrangement that is inhibited by Rho GTPase pathway activation during stress fiber formation [1]. In addition, reduction in either expression or activity of the abundant actin-severing protein gelsolin significantly decreases stress fiber–dependent contraction in cultured cells. Another actin-binding protein involved in cellular contraction is the 27-kDa heat shock protein, hsp27, whose actin binding properties are altered by phosphorylation through a p38 MAP kinase-driven pathway. Reduction of hsp27-induced inhibition of actin polymerization alone can produce stress fiber formation.

Manjno and Palade first observed ultrastructurally that lung EC exhibit a rounded morphology producing paracellular gaps during inflammatory edema [2], a finding similarly observed after thrombin or histamine challenge. This dramatic cell shape change was an early implication of direct involvement of endothelial structural components, particularly the dynamic actin-containing microfilament EC cytoskeleton. Focally distributed changes in tension/relaxation can be accomplished by regulation of the level of myosin light chain (MLC) phosphorylation and actin stress fiber formation. Formation of cytoplasmic stress fibers, critical to cellular contraction and increased intracellular tension, occurs via the coordinate activation of the small GTPase Rho and Ca^{2+}/CaM-dependent myosin light chain kinase, which together increase the level of phosphorylated myosin light chains in a spatially distinct manner. The resultant increases in actin stress fiber formation and actomyosin cellular contraction disrupt the barrier-regulatory balance with tethering forces and destabilize cytoskeletal-junctional linkages culminating in increased vascular permeability [1]. Direct inhibition of either MLCK or Rho kinase as well as Ca^{2+}/calmodulin antagonism attenuates thrombin-induced MLC phosphorylation, gap formation, and barrier dysfunction as well as the increased vascular permeability observed in many but not all models of lung edema.

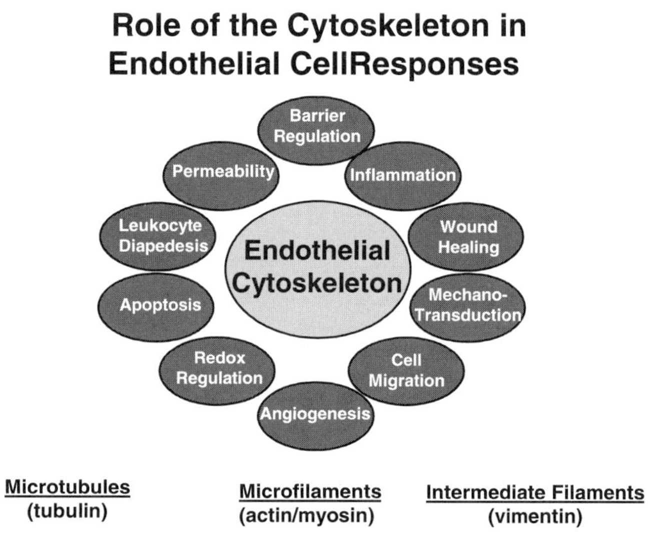

Figure 1 Role of the Cytoskeleton in Endothelial Cell Responses. (see color insert)

Endothelial Cell MLCK Isoform

EC MLCK is an ATP- and Ca^{2+}/calmodulin-dependent enzyme essential for generation of centripetal cellular tension via its ability to enhance actomyosin motor activity [1]. EC MLCK appears to be enriched in microvascular endothelium compared to macrovascular cells, likely a reflection of the diversity of function within that tissue. Gene expression studies have noted that expression of the EC MLCK gene distinguishes micro- from macrovascular endothelium. Myosin light chain kinase (MLCK)-dependent stress fiber formation results in EC contraction and intercellular gap formation with decreased transmonolayer endothelial resistance (TER) and increased fluid and solute translocation [1]. Nonskeletal-muscle MLCK exists as high-molecular-weight (~210kDa) and low-molecular-weight (130–160kDa) isoforms that are derived from a single gene. Human ECs express only the high-molecular-weight form (EC MLCK) cloned by our group, with multiple splice variants also detected by RT-PCR. The two most predominantly expressed variants (EC MLCK 1 and 2) differ only by a single exon deletion (encoding AA# 436–505), which results in EC MLCK 2 containing 69 fewer amino acids than EC MLCK 1. This deleted region contains two sites for $p60^{src}$-catalyzed tyrosine phosphorylation (Y^{464},

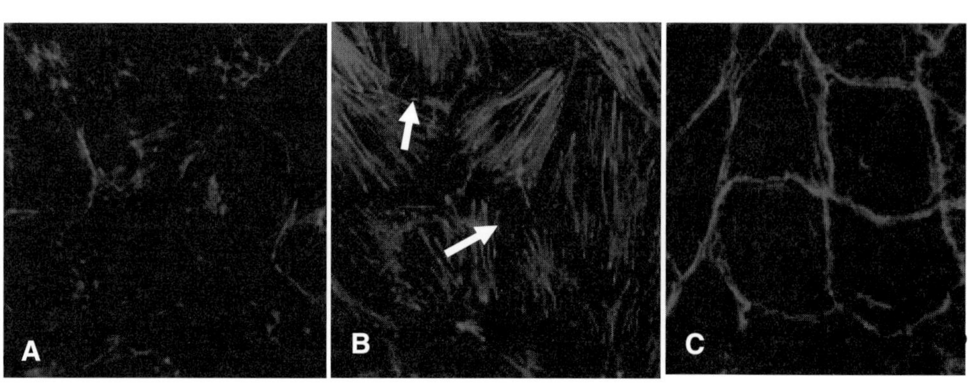

Figure 2 Actin microfilaments distribution is altered by agents that affect vascular endothelial cell permeability. Human lung endothelial cells contain randomly distributed filamentous actin (red) in a stable monolayer (A). Edemagenic agents, such as thrombin, cause cytoskeletal rearrangement resulting in thick actin bundles "stress fibers" which span the cell. Increase intracellular tension and contraction results in formation of intercellular gaps (B, arrows). Platelet derived-phopholipids such as sphingosine 1-phophate are barrier protective and result in an increase in cortical actin at the cell periphery (C). (see color insert)

Y^{471}) that serve to produce differential regulation of EC MLCK splice-variant activity.

The EC MLCK isoform, like the endothelial cytoskeleton in general, is highly multifunctional and serves as an important effector in numerous endothelial processes. In multiple in vitro and in vivo models of EC permeability, increased MLCK activity produces increased MLC phosphorylation within newly formed stress fibers, intracellular tension, cell rounding, paracellular gap formation, and subsequent EC barrier disruption. Inhibition of EC MLCK attenuates or prevents vascular leak produced by ischemia reperfusion, neutrophils, TGFB, thrombin, and mechanical stress as well as neutrophil influx in response to LTB4 or FMLP. More recently, we have described increased MLC phosphorylation in a cortical distribution during EC barrier enhancement [3], suggesting possible spatially-defined MLCK activation that can differentially regulate permeability. EC MLCK has been implicated in TNF-induced endothelial cell apoptosis, mechanotransduction, and calcium signaling.

Rho Family GTPases and the Cytoskeleton

The Rho family of small GTPases are intimately involved in cytoskeletal rearrangement and the distribution and assembly intercellular adherens complexes and focal adhesions. We and others have shown that over expression of constitutively active Rac enhances peripheral actin polymerization in the cortical ring [3]. Rac activation induces lamellipodia formation, membrane ruffling, the formation of cortical actin filaments, and the spreading of ECs, whereas inhibition of Rac GTPase leads to increased monolayer permeability and enhances the thrombin-mediated barrier dysfunction response through a variety of signaling proteins. We recently reported the critical importance of Rac GTPase-dependent cortical actin rearrangement in the augmentation of pulmonary endothelial cell (EC) barrier function by sphingosine 1-phosphate (S1P), HGF, shear stress, simvastatins, and oxidized phospholipids. Ligation of cell surface S1P receptors triggers a complex signaling cascade mediated by intracellular G-proteins that activate Rac GTPase and modulate molecular trafficking to and enzymatic activity at the cell periphery, resulting in peripheral cytoskeletal enhancement and the formation of functional adherens junction complexes.

Furthermore, Rac GTPase is essential for translocation of cortactin, an F-actin binding protein that stimulates actin polymerization and stabilizes the filamentous actin network, an event necessary for the peak barrier enhancing response to S1P in vitro [3]. Conversely, dominant negative Rac prevents the translocation of cortactin and subsequent actin polymerization in the cell periphery, and attenuates barrier enhancement by simvastatin, sphingosine 1-phosphate, and shear stress. Src kinase-mediated phosphorylation of cortactin alters the binding of cortactin, via its SH-3 domain, to MLCK, a molecule potentially important for the organization of filamentous actin in the cortical ring. Finally, Rac inactivates the actin severing protein, cofilin, in the cell periphery through a signaling cascade involving PAK-1 and LIM kinase. Clearly, multiple signaling proteins contribute to modulate dynamic cytoskeletal arrangements, which play a key role in the maintenance or disruption of endothelial barrier integrity.

Microtubule Cytoskeleton Overview

Microtubules (MTs) are cylindrical polymeric 25-nm tubes composed of parallel bundles of 13 linear protofilaments made up of α/β tubulin heterodimers. Uniform orientation of the protofilaments within the tubule conveys polarity to the entire microtubule. Microtubules continually undergo cyclic polymerization and depolymerization, known as dynamic instability. Post-translational modifications of tubulin molecules, such as acetylation and detyrosination, as well as capping of the microtubules' plus ends, are thought to stabilize and mature microtubules. Microtubules have critically important roles in mitosis, cell migration, and intracellular transport of a large number of proteins and organelles via microtubule-associated proteins (MAPs).

Multiple aspects of endothelial cell motility, including migration, morphological changes, and proliferation as well as control of endothelial cell intracellular tension and contractility, involve cross-linking between the MT and actin microfilament networks. Microtubule distribution and dynamic instability regulate the activity of small GTPase-mediated signal transduction cascades that control microfilament network dynamics due in part to nonmotor MAPs known as MT-associated guanine-nucleotide exchange factors (GEFs), a family of proteins that regulate activity of Rho GTPases. Furthermore, MT depolymerization results in activation of Rho, stress fiber formation and apoptosis. Inversely, multiple signaling molecules associate with MT and may regulate MT dynamics including Rac1, proteins upstream of Rac1, and mDia1 [4]. It is clear that microtubule distribution and state of assembly/disassembly has a significant impact on the state of endothelial cell intracellular tension and contractility.

Microtubules and Vascular Barrier Function

Control of vascular endothelial cell barrier function results from a delicate balance between contractile and tethering forces that is significantly regulated by cross-linking of MT and actin cytoskeletal networks. Disruption of the MT network appears to set off a cascade of downstream effects on Rho-dependent mechanisms, including release of MT-bound GEFs and subsequent activation of the small GTPase Rho and Rho kinase, which ultimately results in a significant increase in MLC phosphorylation and subsequent contractile force.

Multiple factors regulate the actomyosin and MT-dependent effect on EC permeability. These include the barrier-protective effect of the cAMP-dependent protein kinase A (PKA) which likely stabilizes the MT network and retards Rho kinase activation by RhoGEFs [5].

Inflammatory cytokines such as tumor necrosis factor (TNF)-α that are secreted by macrophages and endothelial cells increase EC permeability. Presence of TNF-α also induces destabilization of microtubules. TNF-α induces MLC phosphorylation accompanied by microfilament rearrangement; however, subsequent EC permeability appears to be independent of MLCK and Rho kinase. Microtubule disassembly most likely affects TNF-α induced actin network changes and EC permeability.

Microtubules and Response to Shear Stress

Vascular endothelial cells are exposed to constant physiological shear stress from blood flow. Microtubules contribute to the stability of microvascular cells via intracellular distribution changes concomitant with EC migration and motility. Within the endothelial cell in an intact quiescent monolayer, microtubules are distributed in an extensive array radiating from the centrosome, or MT organizing center (MTOC), which is randomly oriented around the nucleus. Stress on the monolayer, such as wounding or fluid shear, is accompanied by alterations in the MT network distribution such as temporary displacement of the MTOC and realignment of the microtubules [6]. Shear stress also causes EC to undergo morphological changes such as lamellipodia extension and directional migration that appear to require active MT dynamics. Chemical stabilization of MT depolymerization causes a reduction in EC migration, as well as attenuated lamellipodial protrusion in response to flow.

Microtubules and Tube Formation

Angiogenesis is a multistep process including the directed migration of endothelial cells (EC) in response to chemotactic signals. Each of the key steps in angiogenesis (sprouting, branching, lumen formation, and barrier stabilization) involves a major role for the cytoskeleton in regulation of the signaling events, especially via the Rho GTPase pathway. MT are involved in cell motility dynamics during tube and lumen formation as well as maintenance of the 3-D nascent tube postangiogenesis (Figure 3). MT destabilizing drugs that limit tumor growth via inhibition of mitosis in both tumor cells and vascular EC are an important part of antitumor therapy. Interference with MT dynamics by stabilization or depolymerization results in capillary tube collapse due to loss of EC structural integrity. This phenomenon involves the Rho GTPase pathway since inhibitors of Rho activity results in blockage of capillary tube collapse [7].

Intermediate Filament Cytoskeleton Overview

Intermediate filaments (IFs) represent the third major element involved in EC cytoskeletal structure. Despite greater diversity than the highly conserved components of either actin microfilaments or microtubules, IF proteins share a common dimer structure containing two parallel α-helices which combine to form apolar fibrils that associate with an array of IF-binding proteins while connecting to the nuclear envelope, peripheral cell junctions, and other cytoskeletal components. IF proteins are expressed in a highly cell-specific manner. Vimentin is the primary IF protein found in EC and other cells of mesenchymal origin. These data suggest that potential roles for IF in EC cytoskeletal structure, and more specifically barrier function, are likely to be subtle and subject to compensation by biologic redundancy.

IFs were defined on the basis of their 7- to 12-nm filament structure, which distinguished them from microfilament and microtubules. Helical polymers of IF proteins are made up of conserved central rod domains flanked by variable C- and N-terminal domains. In vivo studies of vimentin assembly resulted in characterization of three distinct, progressive structural assembly intermediates (particles, squiggles, and long filaments) that exist in a state of equilibrium between subunits and longer polymers. IF distribution in spread, interphase cells is predominantly arranged in long filaments extending from the nucleus to the cell surface, with particles and squiggles apparent at the extreme periphery of the cell. Evidence shows that the assembly of IFs is a complex process, which most likely is highly regulated by signaling cascades associated with cell motility.

Vimentin IFs are very dynamic structures undergoing constant assembly/disassembly, as well as anterograde and retrograde movements. Microtubule-based movement of IF is thought to be critical for assembly and maintenance of the vimentin IF network [8]. Most likely, the dynamic properties of vimentin IF network in the vascular endothelial cell is important in regulation of cell shape and resistance to hemodynamic stress that accompanies blood flow.

Mechanical Strain, Shear Stress, and Flow

Vascular endothelial cells withstand constant mechanical shear force of blood flow and can adapt in response to acute and chronic changes in hemodynamic forces. These physiological changes in the vascular wall are regulated by the IF cytoskeleton and IF-associated proteins, which not only serve as internal scaffolding for the endothelial cell, but are also linked to the plasma membrane, contacts to adjacent cells and to the basement membrane. Vimentin protein expression is higher in macrovascular endothelial cells lining vessels subjected to the highest hemodynamic strain,

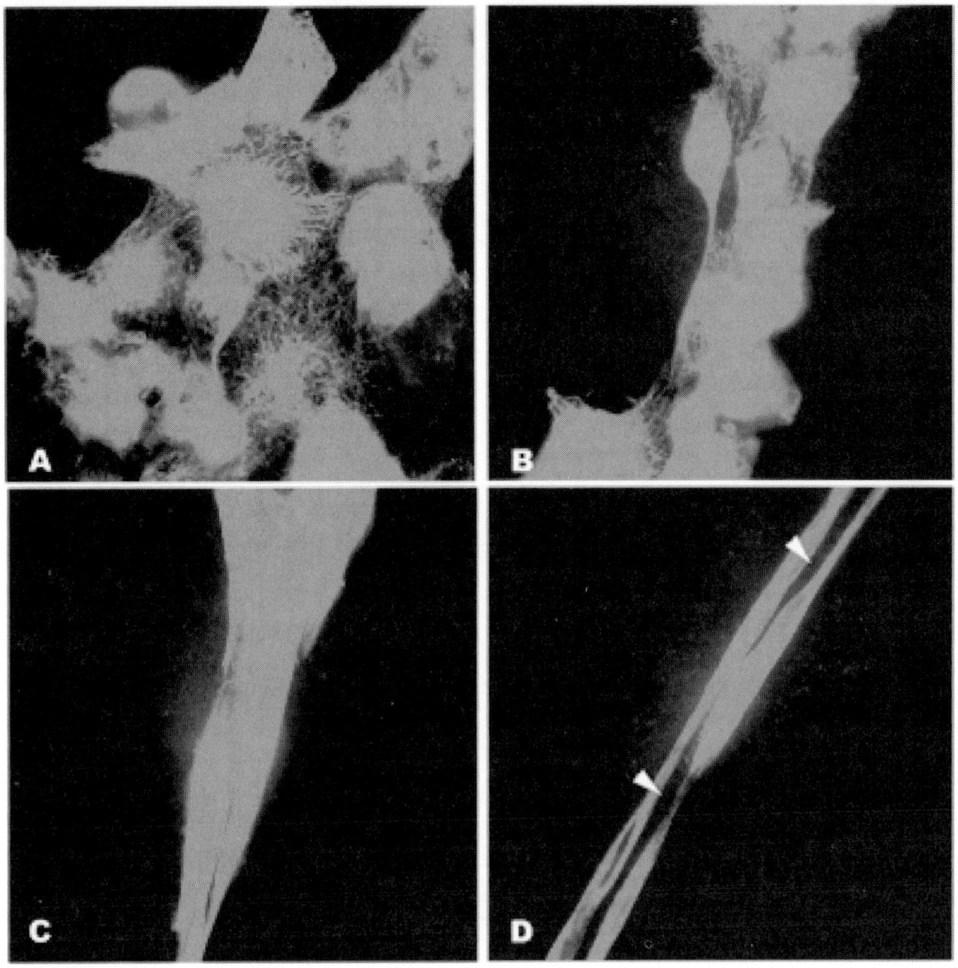

Figure 3 Microtubules are involved in endothelial cell dynamics during capillary-like tube and lumen formation. Human lung microvascular (HLMV) endothelial cells were plated on extracellular matrix (Matrigel, BD Biosciences) and labeled with antibodies specific for tubulin (green). In the early stages of capillary-like tube formation (1–4 hr) HLMV EC are grouped together and display a spread morphology with microtubules arrayed in a radial pattern (A, arrowheads) and begin to migrate into a linear arrangement (B). In the middle stage of tube formation (5–9 hrs), HLMV EC have elongated into capillary-like cords of cells with more elongated microtubule arrays (C). In late capillary-like tube formation (>12 hrs), a lumen is visible (D, arrowheads) between the HLMV EC with compressed linear microtubule distribution. (see color insert)

such as the aorta, compared to EC lining vessels under less shear stress. The physical properties of vimentin IF suggest that they are responsible for resistance to shear strain. Although rigid microfilaments are first to break or fluidize under high strain, vimentin IFs are a more resilient protein polymer with bimodal elastic properties such that vimentin IFs have a higher elasticity under conditions of low strain but become harder and more resistant under higher strains. Presence of a flexible vimentin intermediate filament network is important for maintaining not only the structural integrity of individual cells but also the integrity of the endothelial cell monolayer lining the microvessel. Specifically, vimentin IFs are critical for maintaining the structural integrity of endothelial cells under shear stress, and may also be a conduit for signaling cascades triggered by mechanical force.

Vimentin phosphorylation occurs rapidly in thrombin- or phorbol-stimulated endothelium; however, the role of vimentin in EC structure and resultant barrier function remains unclear. Although vimentin IF disassembly results in dramatic alteration of actin and microtubule filaments in cultured cells, fibroblasts derived from vimentin knockout mice displayed normal actin and microtubule architecture. While the vimentin knockout animals themselves developed normally without gross blood vessel abnormalities, absence of vimentin in walls of mesenteric arteries reduced vessel dilation in response to flow [8]. Downstream responses to flow may be the result of intracellular mechanosignaling events triggered by deformation of the IF cytoskeleton. Adaptation of the EC vimentin IF network occurs in response to changes in flow. Vimentin IF networks display rapid directional displacement within minutes of initial exposure to unidirectional laminar flow. Over a period of hours, cytoskeletal filaments align themselves in the direction of flow. Primary IF network displacement due to the onset of laminar flow imposes a significantly larger change in the vimentin IF above the nucleus compared with displacement that is occurring in the cytosol closer to the substrate. These observed spatial changes may be a means of distribution of local shear force transmission throughout the cell and therefore convey cell signaling messages via a mechanosignaling pathway.

Intermediate Filaments and Microvascular Endothelial Cell–Cell Junctions

Microvascular permeability and barrier function are dependent on maintenance of tight cell–cell intercellular junctions. Adjacent cells assemble junctions composed of transmembrane proteins linked to cytoplasmic multiprotein plaques that anchor microfilament or IF cytoskeletal networks to the plasma membrane. Microvascular EC junctional adhesive plaques maintain cell morphology, tissue integrity, and a semipermeable barrier and participate in cell signaling pathways. Endothelial cells contain microfilament-associated adherens junctions and tight junctions. Vascular endothelial cadherin (VE-cad) is an important component of the Ca^{2+}-dependent EC adherens junction and, along with other junctional proteins such as desmoplakin and β- and γ-catenin (plakoglobin), link the actin and vimentin intermediate filament networks of adjacent cells.

Inflammatory factors affect vascular permeability by disturbing cell–cell junctions and the vimentin IF network. Histamine decreases EC barrier function, most likely by rapidly increasing phosphorylation of multiple junctional proteins including VE-cadherin and vimentin and inducing a rapid reorganization of the vimentin IF. Biochemical studies reveal that the histamine treatment decreases association between the vimentin cytoskeleton and VE-cadherin, but not actin and VE-cadherin [9]. The IF network may then participate structurally and also appears to be involved in assembly and function of microvascular endothelial cell–substrate and cell–cell attachment that effects endothelial cell function.

Intermediate Filaments and Microvascular Endothelial Cell–Matrix Junctions

Endothelial cell–extracellular matrix adhesion is mediated by focal contacts and vimentin-associated matrix adhesions (VMAs) at the basal cell surface where the cytoskeleton is linked to heterodimeric clusters of α/β integrin proteins. Heterodimeric integrin pairing appears to convey functional diversity. Focal contacts are not only mechanical anchors but also participate in cell–matrix signaling. Along with microfilaments, vimentin IF can also insert into focal contacts composed of α6β4 laminin binding integrins in microvascular endothelial cells. Bone marrow microvascular cells are attached to the basement membrane via αvβ3 integrin-containing vimentin-associated matrix adhesions. Vimentin IFs are likely to have a structural/functional role in EC focal contact organization. Vimentin IFs appear to stabilize and regulate cell–matrix adhesions since VMAs became larger, greater in number, and showed increased association with vimentin IFs in ECs subjected to hemodynamic shear stress or flow. Functionality of IF attachment to focal contacts has been further demonstrated by the reduction of the size of VMAs in ECs with reduced vimentin IFs by means of vimentin-silencing RNA. These vimentin knockdown cells display drastically reduced adhesion under physiological flow rates. Vimentin IFs in endothelial cells both stabilize and regulate the size of FCs, which allows regulation of cell–matrix adhesion and resistance to flow [10].

Summary

The roles of actin microfilaments, microtubules, and intermediate filaments in EC function are becoming increasingly defined and appreciated as increasingly complex. The actin cytoskeleton is intimately involved in numerous endothelial cell biologic processes of critical importance to the function of the organ and the entire organism as well. Historically viewed as separate and distinct cytoskeletal systems, microtubules and actin filaments are now known to interact functionally during dynamic cellular processes. Microtubule disruption induces rapid assembly of actin filaments and focal adhesions, isometric cellular contraction that correlates with the level of MLC phosphorylation, increased permeability across endothelial cell monolayers, and increased transendothelial leukocyte migration, while microtubule stabilization attenuates these effects. The mechanisms involved in these effects are poorly understood but are likely to be mediated through interaction with actin filaments, suggesting significant microfilament–microtubule crosstalk and an intriguing role for the microtubule cytoskeleton in EC barrier regulation. The exact roles of intermediate filaments such as vimentin await further exploration.

Glossary

Cadherin: A cell type specific calcium-dependent transmembrane protein present in adherens junctions.

Cortactin: An actin-bundling protein that is phosphorylated by tyrosine kinase $pp60^{src}$.

Endothelial permeability: Dynamic process that involves alteration and contraction of cytoskeletal structure as well as loosened adhesive junctions to allow intercellular passage of fluid.

MLCK: ATP- and Ca^{2+}/calmodulin-dependent kinase that phosphorylates myosin light chain.

Rac GTPase: Members of large family of signaling proteins, which hydrolyze GTP to GDP and have roles in regulation of motility.

Rho GTPase: Members of large family of signaling proteins which hydrolyze GTP to GDP and are involved in signaling cascades linking extracellular stimuli to dynamic actin cytoskeletal rearrangement including stress fiber formation.

Tubulin: Globular protein that exists as a heterodimer in polar linear polymers that compose microtubules.

Vimentin: Type III intermediate filament protein expressed in cells of mesenchymal origin.

References

1. Dudek, S. M., and Garcia, J. G. (2001). Cytoskeletal regulation of pulmonary vascular permeability. *J. Appl. Physiol.* **91**(4), 1487–1500. *Maintenance of the pulmonary vascular barrier is critically important to lung function. Regulation of vascular permeability is dependent on the integrity of the endothelial cell layer, which is a complex balance between intracellular contractility and cell–matrix tethering forces mediated by the dynamics of the actomyosin-based endothelial cell cytoskeletal network. Actin network contractility is mediated in part by Rho kinase/myosin light chain kinase (MLCK) dependent and independent pathways. The actin network is inserted into cell–cell and cell–matrix adhesive protein linkages that are highly regulated to produce a dynamic endothelial cell network.*
2. Manjno, G., and Palade, G. (1961). Studies on inflammation. 1. Effect of histamine and serotonin on vascular permeability: An electron microscopic study. *J. Biophys. Biochem. Cytol.* **11**, 571–605.
3. Garcia, J. G. N., Liu, F., Wang, P., Verin, A. D., Dechert, M. A., Gerthoffer, W. T., and English, D. K. (2001). Sphingosine 1-phosphate promotes endothelial cell barrier integrity by Edg-dependent cytoskeletal rearrangement. *J. Clin. Invest.* **108**(5), 689–701.
4. Rodriguez, O. C., Schaefer, A. W., Mandato, C. A., Forscher, P., Bement, W. M., Waterman-Storer, C. M. (2003). Conserved microtubule–actin interactions in cell movement and morphogenesis. *Nat. Cell Biol.* **5**(7), 599–609.
5. Birukova, A. A., Liu, F., Garcia, J. G. N., Verin, A. D. (2004). Protein kinase A attenuates endothelial cell barrier dysfunction induced by microtubule disassembly. *Am. J. Physiol. Lung Molec. Physiol.* **287**, L86–L93.
6. Galbraith, C. G., Skalak, R., and Chien, S. (1998). Shear stress induces spatial reorganization of the endothelial cell cytoskeleton. *Cell. Motil. Cytoskeleton* **40**(4), 317–330.
7. Bayless, K. J., and Davis, G. E. Microtubule depolymerization rapidly collapses capillary tube networks in vitro and angiogenic vessels in vivo through the small GTPase Rho. *J. Biol. Chem.* Apr 20;**276**(16), 13483–13489. Epub 2001 Jan 25.
8. Helfand, B. T., Chang, L., and Goldman, R. D. (2004). Intermediate filaments are dynamic and motile elements of cellular architecture. *J. Cell Sci.* 2004. **117**(Pt 2), 133–141. *Intermediate filaments (IFs) are dynamic motile cytoskeletal elements that assemble in three distinct structural forms, but are most frequently observed as an extensive network of IFs extending from the perinuclear region to the endothelial cell surface. Vimentin IFs in each of the structural forms are motile and travel via microtubule-based motors along microtubule tracks (MT) tracks. IF motility speed appears to be based on both the type of IF and the state of assembly. Cross-linking between the MT and IF network appears to be critical to cell function, since disruption of IF motility results in significant pathological conditions in multiple cell types including neurons.*
9. Dejana, E. (2004). Endothelial cell–cell junctions: Happy together. *Nat. Rev. Mol. Cell. Biol.* **5**(4), 261–270. *Endothelial cells contain cell–cell attachment sites known as adherens and tight junctions that not only have structural roles, but also participate in endothelial cell specific functions such as control of vascular permeability, angiogenesis, and leukocyte diapedisis. Multiple transmembrane proteins such as VE-cadherins, claudins, and their associated intracellular proteins compose junctional complexes that trigger intracellular signaling cascades, which mediate a balance between endothelial cells composing a stable vascular barrier and activated endothelial cells undergoing angiogenesis or leukocyte diapedisis.*
10. Tsuruta, D., and Jones, J. C. (2003). The vimentin cytoskeleton regulates focal contact size and adhesion of endothelial cells subjected to shear stress. *J. Cell Sci* **116**(Pt 24), 4977–4984.

Capsule Biography

Dr. Joe G. N. Garcia is the Dr. David Marine Professor of Medicine, Director of the Division of Pulmonary and Critical Care Medicine, and Director of the Center for Translational Respiratory Medicine at The Johns Hopkins University School of Medicine in Baltimore, Maryland. One of his primary research interests include the regulation of the pulmonary microvascular endothelial barrier by the endothelial cytoskeleton.

Dr. Laura Linz McGillem is a postdoctoral fellow in the department of Pulmonary and Critical Care Medicine. Her thesis work at Louisiana State University Health Sciences Center included isolation and characterization of a novel nonmuscle myosin from bovine retina. Her interests include endothelial cell physiology, cell–cell adhesions, and the cytoskeleton.

CHAPTER 39

Microvascular Permeability

C. C. Michel
Biomedical Sciences Division, Faculty of Medicine, Imperial College, London

Introduction

Microvascular permeability is a collective phrase to describe the barrier properties of microvascular walls in terms of the ease with which fluid and solute molecules are transported between blood and tissues. The higher the microvascular permeability to a particular solute, the lower the barrier of the microvascular walls to that solute.

Permeability is measured quantitatively in terms of permeability coefficients whose values are determined by the properties of pathways through the endothelium. In many experimental studies, changes in permeability are inferred from changes in solute transport from blood to tissues. When this is done the results should be examined critically, because several factors in addition to permeability may influence blood tissue transport. Only changes in the permeability coefficients yield information that may be interpreted unambiguously in terms of changes in permeability.

Permeability Coefficients

For practical purposes, there are three permeability coefficients. Two of these describe the permeability to a particular solute (the diffusional permeability and the solute reflection coefficients) and the third describes fluid permeability (the hydraulic permeability).

The Solute Diffusional Permeability Coefficient

The solute diffusional permeability, P_d, is defined under conditions where there are no net movements of fluid through microvascular walls. It is the transport of the solute through unit area of microvascular wall divided by the concentration difference of that solute between the fluid inside and that immediately outside the microvessel. P_d has units of velocity (cm sec^{-1}). Writing the definition of P_d as an equation:

$$P_d = \frac{J_{Sd}/S}{(C_C - C_i)} \qquad (1)$$

where J_{Sd} is the net transport of solute from the microvascular blood into the tissues in the absence of net fluid flux through the vessel wall, S is the area of microvascular wall through which transport occurs, C_c is the mean solute concentration inside, and C_i is the mean solute concentration immediately outside the microvessel. The condition that net fluid movements between the blood and the tissue should be zero is especially important when P_d is being estimated for macromolecules. It is of less importance when permeability is high, as it is for small molecules (see later discussion). The number of perfused capillaries in a tissue determines S, and this varies independently of permeability. C_i is often unknown and estimates of P_d are often made under conditions where C_i can be assumed to be approximately zero. C_c is a function of the arterial and venous concentrations and also of microvascular blood flow and of the permeability itself (see later discussion).

P_d is dependent on the lipid solubility, molecular size, and charge of the solute. Lipid-soluble molecules have high permeabilities, as they can pass through cell membranes and so can exchange through the entire area of the microvascular walls. Oxygen, nitrogen, and carbon dioxide have high enough fat solubilities to fall into this category. The P_d values of lipid-soluble molecules may be greater that 0.1 cm sec^{-1}. Water-soluble molecules are confined to specialized pathways through the endothelium and their values for P_d are nearly always less than 10^{-3} cm sec^{-1}. In most tissues (e.g., skin, connective tissues, muscle, nerves) the specialized pathways are located between the endothelial cells and pass through occasional breaks in the junctional strands. In tissues associated with secretory and absorptive epithelia (e.g., intestinal mucosae, kidney, exocrine and endocrine glands) the permeability pathways are fenestrations, disk-like openings (400 to 700 Å in diameter) through thinned areas of endothelium covered with a fine diaphragm with an overlying layer of glycocalyx (cell coat). In intact microvessels, values for P_d of water-soluble molecules fall more rapidly than their values for diffusion coefficient as molecular size increases (see Figure 1A).

Solute Reflection Coefficient

Solutes are also transported through microvascular walls by fluid filtration and reabsorption. This *convective* component of solute transport is proportional to net fluid flow and to the solute reflection coefficient, σ. If net fluid filtration from plasma to tissues is J_V and the mean plasma concentration of the solute inside the microvessel is C_c, the convective component of transport, J_{Sc}, is

$$J_{Sc} = J_V C_c (1 - \sigma) \tag{2}$$

The reflection coefficient is therefore the fraction of solute that is reflected or rejected at microvascular walls during ultrafiltration and $(1 - \sigma)$ is the fraction of the solute molecules that is transported. The reflection coefficient also determines the effective osmotic pressure that the solution of a particular solute can exert across a particular

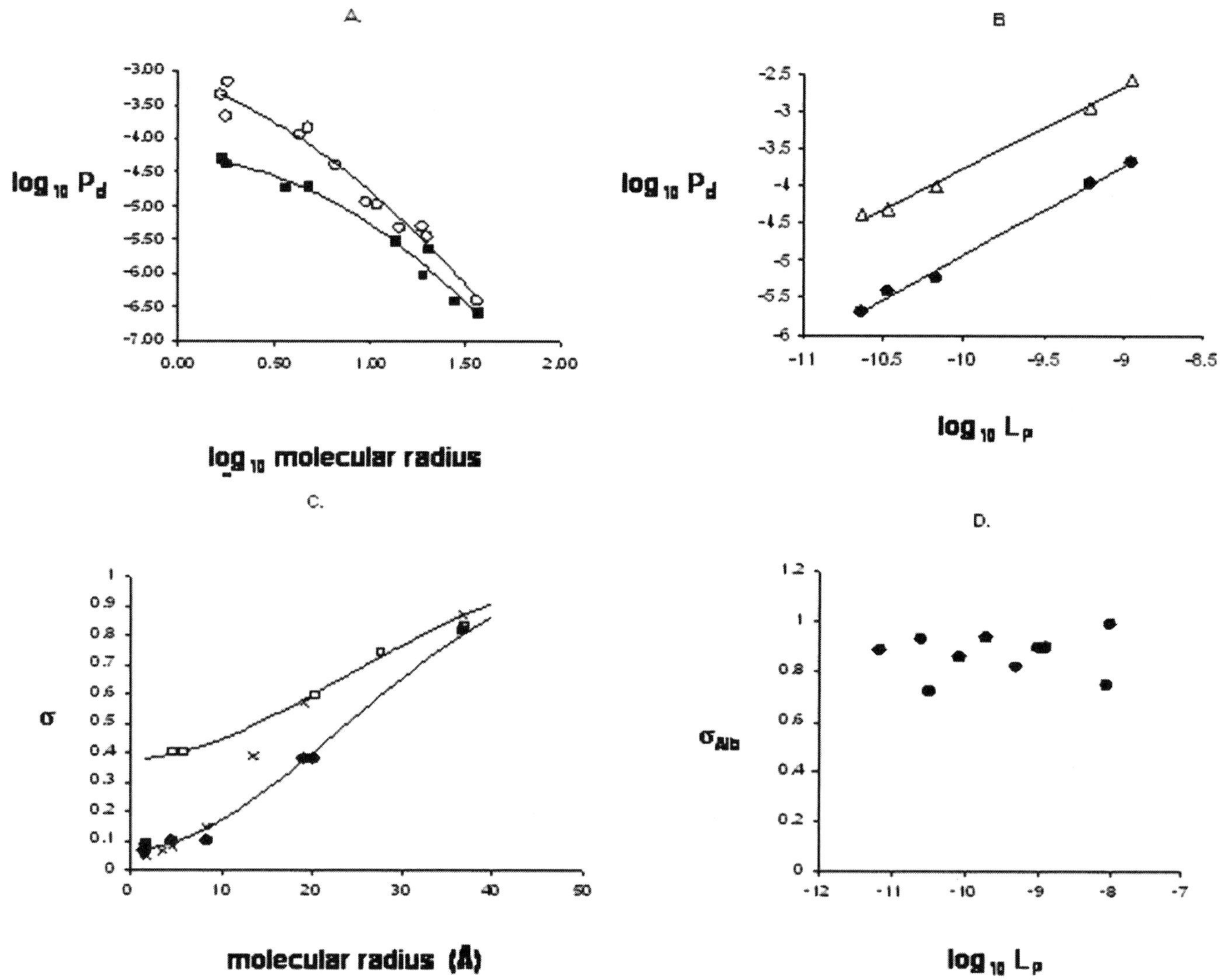

Figure 1 Microvascular permeability to fluid and hydrophilic molecules. **(A)** Diffusional permeability (P_d) to solutes of varying molecular radii in mesenteric capillaries (open circles) and muscle capillaries (closed squares). Data are plotted as $\log_{10}$ values to cover the range of P_d. Note how P_d for the the two capillary types differ by > 10-fold for the smallest molecules but converge as log radius approaches 1.6 (~40 Å). **(B)** Log P_d to NaCl (open triangles) and inulin (closed circles) plotted against log L_P. The slopes of the relations do not differ significantly from direct proportionality. [Modified from Michel and Curry (1999). *Physiol. Rev.* **79**, 703–761.] **(C)** Reflection coefficient (σ) for hydrophilic molecules for capillaries of cat skeletal muscle (open squares), rat limb capillaries (crosses), and frog mesentery (solid circles) plotted against molecular radius. The high values of σ to the smallest molecules differ in cat muscle capillaries suggest a large aquaporin pathway in these vessels [data of Wolf and Watson (1989). *Am. J. Physiol.* **256**, H282–H290]. **(D)** Reflection coefficient to serum albumin (σ_{Alb}) against $\log_{10} L_P$ for capillaries of different tissues. The lowest values of L_P are for skeletal muscle and the highest are for glomerular capillaries [based on Michel, C. C. (1988). *J. Physiol.* **404**, 1–29].

membrane. Thus the osmotic pressure of a 5-mM solution of glucose may be 127 cm H_2O in an osmometer fitted with a membrane impermeable to glucose molecules in agreement with van't Hoff's law. A 5-mM difference in glucose concentration across the walls of a mesenteric capillary, however, exerts an effective osmotic pressure less than one tenth of this. This is because 90 percent of the glucose molecules pass relatively freely through capillary walls and only 10 percent are reflected, that is, σ for glucose is just less than 0.1. This leads to an alternative definition of σ in terms of the total (or van't Hoff) osmotic pressure and the effective osmotic pressure that a solution of a particular solute exerts across a particular membrane, that is,

$$\sigma = \frac{\text{Effective osmotic pressure}}{\text{Theoretical osmotic pressure}} \tag{3}$$

Strictly speaking, σ defined in Eq. (3) is the osmotic reflection of the solute (σ_d) and that defined in Eq. (2) is the solvent drag reflection coefficient (σ_f). The values of σ_d and σ_f are identical for solutions where the relations between concentration and osmotic pressure are linear. Small differences are present for σ_f and σ_d of macromolecules where osmotic pressure rises as a second order power function of concentration. For all solutes at all membranes, σ has a maximum value of 1.0 when the membrane is completely impermeable to solute but permeable to solvent. For water-soluble molecules of biological solutions, σ has a minimum value of zero. Because σ describes a fraction or a ratio, it has no units.

Hydraulic Permeability

The hydraulic permeability, L_P, is also referred to as the hydraulic conductivity, hydraulic conductance, water permeability, and filtration coefficient. It is defined as the net fluid flow through unit area of microvascular wall per unit difference in pressure between the lumen of the vessel and its abluminal surface. Thus if net fluid filtration or reabsorption, J_V, occurs between blood and tissue through an area of capillary wall, S, then:

$$L_P = \frac{J_V/S}{(\Delta P - \sigma \cdot \Delta\Pi)} \tag{4}$$

where ΔP and $\sigma\Delta\Pi$ are the differences in hydrostatic pressure and effective osmotic pressure, respectively, between the inside and outside of the microvessel. The hydraulic permeability has units of velocity per unit pressure difference, such as cm sec^{-1} cm H_2O^{-1}. In microvessels with fenestrated endothelia, water passes through the fenestrations with water-soluble molecules. Similarly, in vessels with nonfenestrated endothelia, water shares the pathways between the cells with hydrophilic solutes, but here there is an additional (water only) pathway through cell membranes via aquaporin channels. In most nonfenestrated vessels the water-only pathways contribute 10 percent or less to L_P, though there is evidence in some skeletal muscle capillaries their contribution may be as high as 40 percent.

Permeability Coefficients of Microvessels in Different Tissues and Their Interpretation

Permeability coefficients have been measured in single microvessels mainly in mesentery but also in other tissues including skeletal muscle, brain, lung, and kidney. Estimates of microvascular permeability have also been made for capillary beds in different organs and tissues. Because many millions of capillaries may contribute to net transport here, values are obtained for the products P_dS and L_PS. The value of S, the area of microvascular wall, may be estimated from the histology of the tissue assuming that the number of perfused vessels per unit volume is same in the tissue prepared for histology and in that where the permeability measurements were made. (For methods of permeability measurement, see article in this volume on "Regulation of Vascular Permeability" by D. O. Bates.)

There is reasonably good agreement between estimates of P_d and L_P obtained from whole tissues and organs and for single vessels in the same tissue. There is, however, great variation between the mean values of L_P and P_d to the smaller hydrophilic solutes in different microvascular beds. In vessels with fenestrated endothelium, L_P and P_d for low-molecular-weight water-soluble solutes vary directly with the number of fenestrations per unit area of endothelium. Values for the L_p of fenestrated vessels are different in different tissues. Thus in the fenestrated ascending vasa recta in the renal medulla, L_p may exceed 10^{-5} cm sec^{-1} cm H_2O^{-1}, whereas in fenestrated vessels of the intestinal mucosa, L_p is in the range of 10^{-6} cm sec^{-1} cm H_2O^{-1}, reflecting the lower density of the endothelial fenestrations on the intestinal capillaries. The L_P of vessels with continuous nonfenestrated endothelium varies between 10^{-6} and 10^{-9} cm sec^{-1} cm H_2O^{-1}, depending on the tissue of origin, and similar variations are seen in P_d. Here, the ultrastructural basis of variable permeability is less obvious. Although electron micrographs of muscle capillaries and mesenteric capillaries look similar, both types of vessel appearing as tubes of flattened endothelial cells lacking fenestrations and joined by junctions of similar ultrastructure, L_P values for mesenteric capillaries are 20 to 50 times greater than L_P for muscle microvessels. Similar differences are seen for values of P_d to NaCl in mesenteric and muscle microvessels (see Figure 1A). The differences in permeability reflect different frequencies of breaks in the strands of tight junctions that join adjacent cells. At present, detailed ultrastructure of the intercellular junctions is known only for heart muscle and mesenteric capillaries. The sieving properties of these two types of vessel, however, are very similar and in both vessel types, P_d falls rapidly as solute molecular radius approaches 40 Å.

Although microvascular L_P and P_d for small water-soluble molecules vary greatly from one tissue to another, L_P bears a constant relation to P_d for a given solute (e.g., inulin). Figure 1B shows that variations in P_d to NaCl and inulin in different tissues are directly proportional to variations in L_P. This is strong evidence that the permeability pathways for small hydrophilic solutes are principally those used for fluid exchange.

By contrast, σ to macromolecules is very similar in all normal microvessels. The value for serum albumin is between 0.85 and 0.99 in nearly all microvascular beds (see Figure 1D). Such values are essential for normal fluid balance (see later discussion). This is indicative of a common structure in all microvascular walls that is responsible for molecular sieving. At present, the favored candidate for this structure is the luminal glycocalyx of the endothelial cells, and this hypothesis has been boosted by several recent reports.

A fall in σ and a rise in L_P is characteristic of the increase in permeability seen in acute inflammation.

While σ for macromolecules approaches unity as molecular radius approaches 40 to 50 Å, finite permeabilities have been measured for all naturally occurring macromolecules. These lie in the range of 10^{-9} to 10^{-8} cm sec^{-1}. Whereas smaller hydrophilic molecules pass through channels in microvascular walls that have a limiting size of 40 to 50 Å (molecular radius), macromolecular permeability involves different pathways. There is evidence for two mechanisms: transcytosis by endothelial caveolae that act as shuttles across the cell; or convection through a minute number of large pores (radius ~250 Å) consisting of transcellular channels formed by fusion of caveolae or by occasional open intercellular junctions.

Permeability and Exchange of Fluid and Solute between Blood and Tissues under Physiological Conditions

Fluid Exchange

Fluid transport is determined by the differences in hydrostatic and the effective osmotic pressures across microvascular walls. This is Starling's principle and when written as an equation it is:

$$J_V = L_P S\left(\Delta P - \sum_n \sigma_n \cdot \Delta\Pi_n\right) \quad (5)$$

where J_V is net fluid transport (filtration) from blood to tissue, S is surface area of microvascular wall through which filtration is occurring, ΔP is the mean hydrostatic pressure difference across the walls of the exchange vessels, and $\Sigma_n \sigma_n \Delta\Pi_n$ is the summation of the effective osmotic pressures set up across the exchange vessel walls by all the solutes present in the plasma. Note that when fluid is being reabsorbed into the blood, J_V has a negative sign, indicating that the tissues are losing fluid.

Starling, in 1896, first appreciated that the effective osmotic pressure difference was determined by macromolecules. He pointed out that while these solutes make a small contribution to the total osmotic pressure of the plasma, they represent the only solutes that are normally present at different concentrations in the plasma and in the interstitial fluids. For all other solutes, $\Delta\Pi$ is zero across most microvascular walls. Furthermore, we now know that unlike the other solutes of the plasma, the macromolecules normally exert more than 85 percent of their full osmotic pressure ($\sigma > 0.85$).

Microvascular walls are slightly but measurably permeable to all the macromolecules that contribute to the effective osmotic pressure term. This means that the difference in macromolecular concentration between the plasma and the tissue fluid (and hence the effective osmotic pressure term) is itself a function of the permeability and the fluid filtration rate. In most tissues, the concentration difference depends on fluid entering the tissue spaces faster than macromolecules, diluting the pericapillary macromolecules to concentrations well below those of the plasma. Drainage of tissue fluid by the lymph at the same rate as it is being formed by microvascular ultrafiltration keeps the interstitial fluid volume constant. Net fluid uptake from the tissues into the plasma (reabsorption) accelerates the equilibration of macromolecules across microvascular walls, diminishing their effective osmotic pressure. Thus fluid reabsorption can occur for only limited periods in most tissues before the effective osmotic pressure difference across the microvessel walls is reduced to less than the hydrostatic pressure difference. Reabsorption then ceases and a low level of filtration develops. One important consequence of this is that the popular textbook diagram illustrating Starling's principle of microvascular fluid balance, where fluid is filtered from the arterial end of a capillary and is reabsorbed at the venous end, bears little relation to reality.

Continuous reabsorption of fluid does occur in tissues such as the intestinal mucosa and the kidney. Here the interstitial concentration of macromolecules is kept low by the continuous secretion of protein-free fluid into the interstitium by the adjacent epithelial cells. For a full discussion of microvascular fluid exchange, the reader should consult the references by Levick (1991) or Michel (1997), listed at the end of this chapter.

Solute Exchange

CONVECTION AND DIFFUSION

The transport of solute through microvascular walls is the sum of transport of the solute by diffusion and by convection. This addition is slightly complicated for when solute and fluid are being carried in the same channels through microvascular walls; convection of solute in the same direction as diffusion reduces the concentration gradient at the entry to the channels so that the mean gradient is no longer the difference in concentration across the wall divided by wall thickness. Convection in the opposite direction to diffusion increases the gradient. This effect of convection on the diffusion gradient is expressed in terms of the ratio of the velocity of solute transport by convection to its velocity by diffusion through the membrane. This ratio is the Péclet number, *Pe,* and here it is defined as:

$$Pe = \frac{J_V(1-\sigma)}{P_d S} \tag{6}$$

Using Pe to modify the expression for the diffusional component, the total solute transport by diffusion and convection from micro-vessel to tissues becomes (see Curry 1984):

$$J_S = P_d S(C_C - C_i)\frac{Pe}{e^{Pe}-1} + J_V(1-\sigma)C_c \tag{7}$$

where $(C_C - C_i)$ is the mean concentration difference across the walls of the exchange vessels. Taylor, Renkin, and their colleagues have used expressions of the general form of Eq. (7) to analyze macromolecular transport between the plasma and lymph.

Mean Concentration in the Exchange Vessels

In many microvascular beds, small highly diffusible molecules have very low Péclet numbers and convective solute transport is negligible. Under these conditions, J_S may be described by the simple diffusion expression

$$J_S = P_d S(C_C - C_i) \tag{8}$$

C_C, the mean concentration in the microvessels, lies between the arterial concentration, C_a, and the venous concentration, C_V, by an amount that is determined by the blood flow, F, and the permeability–surface area product, P_dS. Writing this as an equation we find

$$C_c = C_i + (F/P_d S)(C_a - C_v) \tag{9}$$

When C_i is constant, Renkin showed that C_c falls exponentially between C_a and C_v. For the special case of $C_i = 0$, C_v is related to C_a by the expression

$$C_v = C_a e^{-P_d S/F} \tag{10}$$

Various forms of Eqs. (9) and (10) have been used to develop methods for measuring permeability to small molecules. Because microvascular permeability to macromolecules is usually low, their concentrations fall negligibly between entry and exit of exchange vessels and for these solutes, C_C approximates to C_a. This approximation may not be valid when permeability is increased.

Clearance and Extraction

Renkin also pioneered the use of Eq. (10) for examining how F and P_dS influence blood–tissue exchange. At the same time he clarified the terms solute clearance and solute extraction showing how they would vary with blood flow and solute permeability.

The clearance of a solute from the blood into the tissues supplied by a microcirculation is the net rate of solute transport divided by its arterial concentration, that is, J_S/C_a. Clearance from the tissues to blood is J_S/C_i. Renkin argued that blood–tissue clearance could be thought of being the product of blood flow and extraction where the extraction was the arteriovenous concentration difference divided by either C_a (for clearance blood to tissue) or C_i for clearance between tissues and blood. Using Eq. (10), he argued that

$$Clearance = F(1 - e^{-P_d S/F}) \tag{11}$$

where the extraction is $(1 - e^{-P_dS/F}) = (C_a - C_v)/C_a$. From this it follows that when P_dS is large compared with F, clearance is determined largely by blood flow, but when F exceeds P_dS, clearance approximates more and more closely to P_dS as F increases. These predictions are consistent with experiments where the blood and tissue clearances of diffusible solutes have been investigated over a wide range of blood flows.

The permeability of microvascular walls to macromolecules under normal conditions is very low compared with F. The unidirectional clearance of macromolecules from blood to tissue does not, however, yield a value of P_dS. From Eq. (7), unidirectional clearance (i.e., clearance when C_i is close to zero) is

$$Clearance = \frac{J_S}{C_a} = \frac{J_V(1-\sigma_f)}{(1-e^{-Pe})} \tag{12}$$

Only when J_V is small relative to P_dS does clearance approximate to P_dS.

Clearance is sometimes used as a measure of microvascular permeability, but Eqs. (11) and (12) show that this is valid for small molecules only when blood flow is large relative to P_dS and for large molecules only when filtration rate is small relative to P_dS. Furthermore, it is important to remember that P_dS is not permeability but the product of permeability and surface area. This means that when looking for changes in permeability, P_d and S have to be separated before a change in permeability can be inferred from a change in P_dS.

Glossary

Clearance of a solute from blood to tissues: The net transport of solute per unit time (flux) divided by the arterial (inflowing) concentration. Clearance of a solute from tissues to blood is the net flux divided by the tissue concentration. Clearance is also the product of extraction and blood flow.

Extraction, E, of a solute from blood into tissues: The difference between the arterial and venous blood (or plasma) concentrations divided by the arterial concentration. Extraction of a solute from tissues to blood (when the arterial concentration is zero) is the venous concentration of that solute divided by the tissue concentration.

Hydraulic permeability, L_p: Net fluid flow per unit area of microvascular wall divided by the difference in hydrostatic pressure across the wall. It is determined by both the frequency and the dimensions of the channels that transmit fluid through the wall.

Solute diffusional permeability, P_d: Net transport per unit time of a specific solute through unit area of microvascular wall divided by the concentration difference of that solute across the wall. For hydrophilic solutes, P_d is determined by both the number of channels per unit area of wall and the molecular radius of the solute relative to the dimensions of the channels.

Solute reflection coefficient, σ: Has two definitions. It is either the fraction of molecules of a specific solute that are reflected at (i.e., not carried through) the microvascular wall during ultrafiltration of fluid through the wall, or it is the ratio of the effective osmotic pressure that a given concentration difference of the solute exerts across the microvascular wall to the osmotic pressure that the same concentration difference would exert across a membrane that was impermeable to the solute but permeable to water. For hydrophilic molecules, it is determined by the size of the channels to solute and to water and is independent of the frequency of channels.

Bibliography

Bates, D., Lodwick, D., and Williams, B. (1999). Vascular endothelial growth factor and microvascular permeability. *Microcirculation* **6**, 83–96. *This has an excellent introductory section on the assessment of changes in permeability.*

Curry, F. E. (1984). Mechanics and thermodynamics of transcapillary exchange. In *Handbook of Physiology* (E. M. Renkin and C. C. Michel, eds.), Section 2, Vol. 4, pp. 309–374. Bethesda, MD: American Physiological Society. *This is still the best comprehensive review of the biophysics of microvascular permeability.*

Levick, J. R. (1991). Capillary filtration–absorption balance reconsidered in the light of dynamic extravascular factors. *Exp. Physiol.* **76**, 825–857. *An excellent compilation of the experimental measurements of hydrostatic and effective osmotic pressure differences across microvascular walls with important conclusions.*

Michel, C. C. (1997). Starling: The formulation of his hypothesis of microvascular fluid exchange and its significance after 100 years. *Exp. Physiol.* **82**, 1–30. *This is partly historical but also addresses recent work and recent questions about microvascular fluid exchange arising after Levick's review.*

Michel, C. C., and Curry, F. E. (1999). Microvascular permeability. *Physiol. Rev.* **79**, 703–761. *This is the most comprehensive modern review of the ultrastructural basis of microvascular permeability and of recent work on the regulation of permeability.*

Renkin, E. M. (1984). The control of the microcirculation and blood tissue exchange. In *Handbook of Physiology* (E. M. Renkin and C. C. Michel, eds.), Section 2, Vol. 4, pp. 627–687. Bethesda, MD: American Physiological Society. *Although this review does not include some new hypotheses for the regulation of microvascular blood flow, it is the best treatment of the principles of blood tissue exchange.*

Rippe, B., Rosengren, B. I., Carlsson, O., and Venturoli, D. (2002). Transendothelial transport: The vesicle controversy. *J. Vasc. Res.* **39**, 375–390. *This is an excellent review interpreting recent evidence concerning the mechanisms of transport of macromolecules through microvascular walls.*

Taylor, A. E., and Granger, D. N. (1984). Exchange of macromolecules across the microcirculation. In *Handbook of Physiology* (E. M. Renkin and C. C. Michel, eds.), Section 2, Vol. 4, pp. 467–520. Bethesda, MD: American Physiological Society. *Many applications of convection diffusion equations in the interpretation of macromolecular transport.*

Capsule Biography

C. Charles Michel is Emeritus Professor of Physiology and Senior Research Investigator at Imperial College, London. His main research interest for over 30 years has been microvascular permeability and microvascular exchange and his contributions have been recognized by various awards including the Malpighi Medal of the European Society for Microcirculation (!984), the Annual Prize Lecture of The Physiological Society (1987), and his election to Honorary Membership of the American Physiological Society (1993), The Physiological Society (2001), and the British Society for Microcirculation (2001).

CHAPTER 40

TNF-α-Induced Pulmonary Endothelial Permeability: The Role of the Microtubule Cytoskeleton

Irina Petrache, Anna A. Birukova, and Alexander D. Verin
The Johns Hopkins University

The endothelial cell cytoskeleton was long believed to merely constitute a scaffolding cellular structure. Recent research highlights an active role of cytoskeletal structures in the cellular metabolism, as well as in its response to external stresses. In this chapter we show how the endothelial cell microtubule cytoskeleton participates in endothelial barrier dysfunction induced by a proinflammatory cytokine, tumor necrosis factor-α.

Endothelial Barrier Function

The main role of the pulmonary vascular tree is in respiration: the uptake of oxygen and the elimination of carbon dioxide. At the alveolar level, blood vessels are lined only by endothelial cells that come in close proximity with epithelial cells, on the luminal side of the airways. The endothelial cells lining the vessels are in contact with each other, making the vascular wall a tight barrier. Any breach in the endothelial cells barrier results in leakage of fluid from the lumen of the vessels into the interstitial tissue and/or alveolar lumen, severely impairing gas exchange. Disruption of the vascular barrier is a prominent feature of acute lung injury syndrome and results in pulmonary edema formation and subsequent respiratory dysfunction or failure. The endothelial barrier integrity is the result of a balance between the tethering and contractile forces acting on endothelial cells, which are critically dependent upon cytoskeletal components, including the actin-based microfilaments, intermediate filaments, and microtubules [1]. Endothelial cell activation by inflammatory stimuli is often accompanied by enhanced cellular contraction and formation of intercellular gaps, major events leading to increased vascular permeability and pulmonary interstitial edema.

Tumor Necrosis Factor-α

Tumor necrosis factor-α (TNF-α) is a major proinflammatory cytokine secreted in increased amounts by macrophages and endothelial cells in situations linked pulmonary edema formation, such as pneumonia, sepsis, systemic inflammation and injury. TNF-α has been implicated in endothelial cell activation, endothelial cell death by apoptosis, and increased vascular permeability. The mechanisms by which TNF-α triggers endothelial barrier disruption include contraction of the intracellular actin microfilaments and formation of inter-cellular gaps that parallel in time the development of trans-endothelial permeability [2]. The rearrangement of the actin-based cytoskeleton is the result of actin–myosin movement, which, as in muscle cells, follows the phosphorylation of the myosin light chains. TNF-α induces significant actin microfilaments contraction via myosin light chain phosphorylation, a reaction catalyzed by both myosin light chain kinase (MLCK) and Rho kinase. However, inhibition of either enzyme does not prevent TNF-α-induced endothelial barrier dysfunction [2], suggesting that MLCK-independent microfilament changes and/or other cytoskeletal structures, such as intermediate filaments, microtubules, and adherens junctions, may be involved.

Microtubules

Microtubules are highly dynamic cytoskeletal polymers that reside in all eukaryotes and are essential for a wide variety of processes, including intracellular signal transduction, vesicular and organelle transport, development and maintenance of cell shape and cell locomotion, and regulation of cell division. The individual microtubule is formed by a linear association of protofilaments, which are composed of tubulin heterodimers α and β assembled in "head to tail" structure. Thus, each microtubule itself has a polarity, and the two ends have structural and kinetic differences. The "plus end" is characterized by a much more extensive duration of growing and shortening than the opposite "minus end"; thus the processes of polymerization and depolymerization of microtubules occur preferentially at the plus end. Microtubules form a lattice network of rigid hollow rods and undergo continual assembly and disassembly by the reversible addition and loss of tubulin dimers at the ends of microtubules. The linkage between the microtubule network and the contractile cytoskeleton has not been fully explored. However, an active crosstalk between microtubules and actin is required to control microtubule dynamics. This interplay has a key role in determining the cell polarity and regulation of migration. The microtubule–actin interaction is essential in regulation of endothelial integrity and wound repair [3–5]. The microtubule cytoskeleton is emerging as an important modulator of TNF-α-induced pulmonary endothelial permeability (Figure 1). A role for microtubule dynamics in permeability has been also demonstrated for other agents known to cause endothelial barrier damage such as nocodazole and hydrogen peroxide. The mechanisms by which TNF-α induces changes in the microtubule cytoskeleton are only beginning to be elucidated. TNF-α treatment induces microtubule destabilization and a decrease in the acetylated tubulin [6]. Microtubule stabilization by taxol or epothilone B inhibits actin rearrangement, vascular endothelial (VE)-cadherin redistribution, and

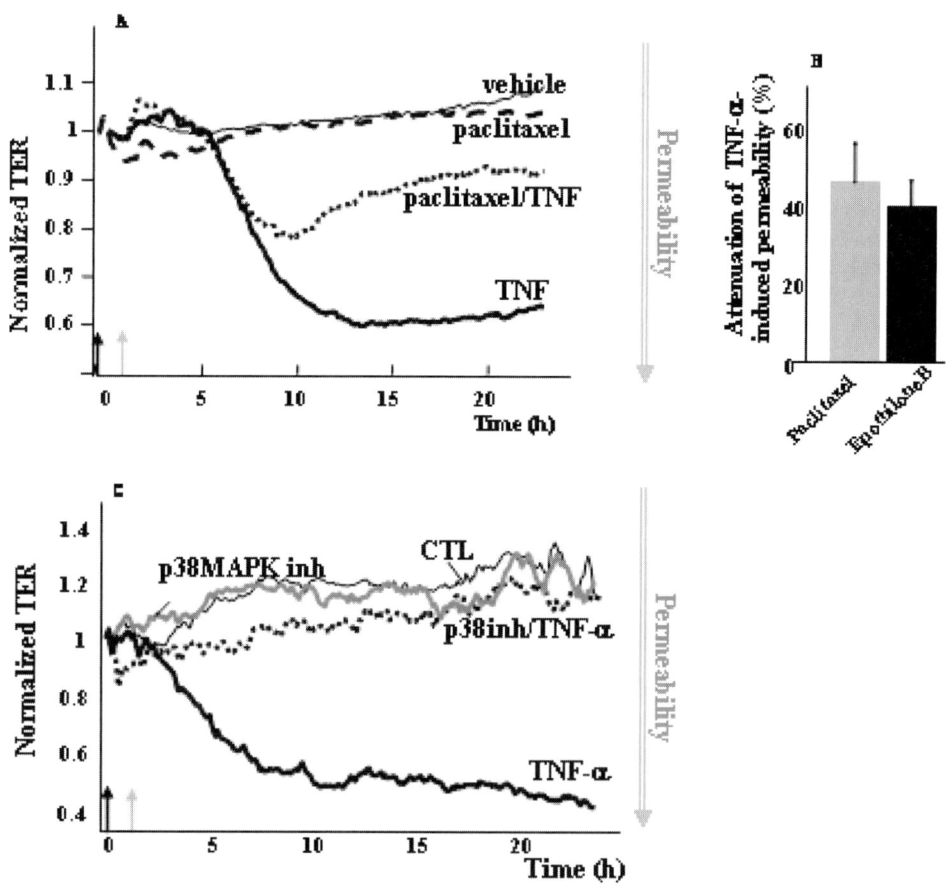

Figure 1 The effects of microtubule stabilization and p38 MAPK inhibition on TNF-α-induced endothelial permeability. **(A, C)** Representative tracings of normalized electrical resistance measured across endothelial cell monolayers (TER). Endothelial cells were exposed to vehicle (control) and TNF-α [20 ng/ml], where the gray arrow on the abscissa indicates the time of challenge. TNF-α induced TER reduction was evident at 4 hours with a maximal effect at 10 hours of exposure. **(A)** Microtubule stabilization by paclitaxel (black arrow) significantly inhibits TNF-α-induced decreases in resistance. **(C)** P38 MAPK inhibition by SB203580 inhibits TNF-α induced permeability. **(B)** Bar graph demonstrating the effect of the microtubule-stabilizing agents paclitaxel end epothilone B on the maximal TNF-α-induced decrease in TER. No significant difference was noted between the two agents ($p = 0.44$). (see color insert)

intercellular gap formation in response to TNF-α [6], suggesting first important crosstalk among the microtubule cytoskeleton, the actin microfilaments, and the zonula adherens proteins and second, that microtubules are opposing cellular contraction. This effect has been confirmed by the finding that microtubule disassembly leads to instant Rho kinase activation and cell contraction, potentially triggered by the release of Rho activating factors bound to microtubules ([7] and unpublished data).

The Role of the Mitogen Activated Protein Kinase (MAPK) Pathway

(a) The mechanisms by which TNF-α triggers microtubule network disassembly, followed by actin cytoskeleton rearrangement and intercellular junction changes, include the activation of the MAPK pathways. The MAPK pathways are important intracellular signaling pathways that are activated by TNF-α-induced ligation of the TNF receptors, TNFR1 (p55) and to a lesser extent TNFR2 (p75). Each component of the MAPK pathways is differentially recruited by specific stimuli resulting in kinase-specific signaling and regulation of cell growth, differentiation, and apoptosis. There are three main MAPK signaling pathways, the p42/p44 MAPK, the (Jun activated) JNK MAPK, and the p38 MAPK.

(b) The p38 MAPK is a stress-activated MAPK family that consists of four isoforms: p38-α, p38-β, p38-γ, and p38-δ. P38-α and β, isoforms inhibitable by cytokine-suppressive anti-inflammatory drugs, appear to be critical mediators of TNF-α-induced microtubule disassembly, actin microfilament rearrangement, and endothelial permeability. How p38 MAPK regulates microtubule dynamics in not well understood, but may involve direct phosphorylation of microtubule-associated proteins. Potential candidates are *tau*, a substrate for p38-γ, which when phosphorylated has decreased ability to promote microtubule assembly [8], and stathmin, a substrate for p38-δ, which can regulate the dynamics of microtubules [9]. In addition, p38 MAPK inactivates kinesin, an ATPase that mediates plus end–directed transport of organelles along microtubules.

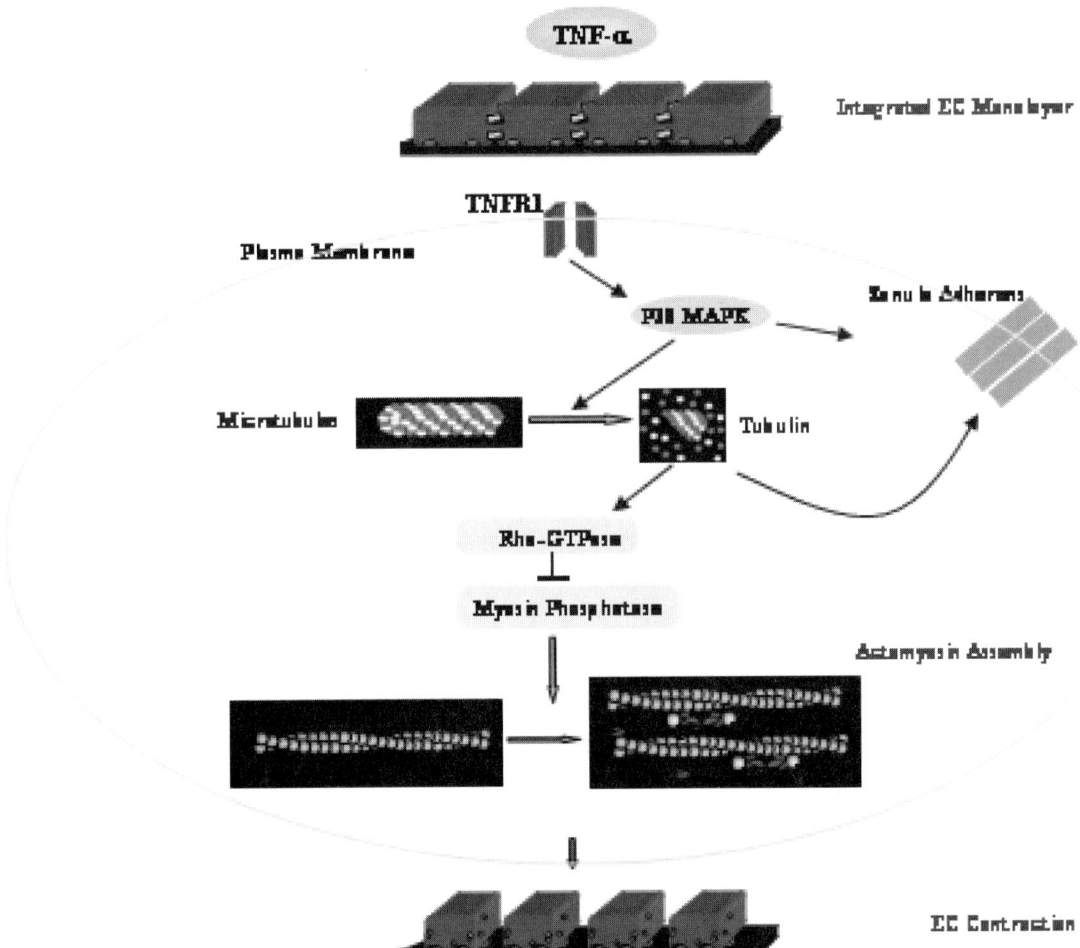

Figure 2 Schematic representation of putative pathways of TNF-α-triggered endothelial cell permeability. TNF-α acts via p38 MAPK to trigger microtubule disassembly and subsequent actin microfilament rearrangement and cell–cell contact disruption. Additionally, the p38 MAPK engages microtubule-independent cytoskeletal mechanisms of barrier dysfunction. (see color insert)

(c) The effects of p38 MAPK activation may encompass multiple cellular pathways, in addition to the microtubule cytoskeleton (Figure 2). The actin cytoskeleton function is regulated by p38 MAPK via direct activation of MAPK-activated protein kinase-2 (MAPKAP kinase-2), which in turn phosphorylates and activates an actin-binding protein, heat shock protein-27 (HSP-27). Nonphosphorylated HSP-27 normally exists in high-molecular-weight multimers that serve as chaperones. Serine phosphorylation of HSP-27 results in the dissociation of HSP-27 into monomers and dimers with redistribution of HSP-27 to the actin cytoskeleton and subsequent actin reorganization into stress fibers.

This brief review of endothelial cytoskeletal pathways engaged by TNF-α and their crosstalk with signaling pathways that regulate vital endothelial cell responses highlights the complex regulation of pulmonary vascular permeability. Elucidation of the complex orchestration of the endothelial barrier function will lead to potential therapies for the pulmonary edema seen in acute lung injury syndromes.

Glossary

Microtubules: Components of the intracellular cytoskeleton. They are organized intro a network that spans the cytoplasm and connects with other cytoskeletal structures. The dynamic function of the microtubule regulates the position and function of other intracellular proteins and organelles.

Mitogen activated protein kinases (MAPK): Important intracellular serine–threonine kinases. These enzymes are organized into signaling pathways that are activated by specific stimuli and lead to the downstream activation of specific cellular responses, such as gene transcription and cellular apoptosis.

TNF-α: A cytokine secreted by cells in increased concentrations in situations of stress, injury, and inflammation. It acts on cells by ligating specific receptors, which then trigger intracellular proinflammatory responses.

References

1. Dudek, S. M., and Garcia, J. G. (2001). Cytoskeletal regulation of pulmonary vascular permeability. *J. Appl. Physiol.* **91**, 1487–1500.
2. Petrache, I., Verin, A. D., Crow, M. T., Birukova, A., Liu, F., and Garcia, J. G. (2001). Differential effect of MLC kinase in TNF-alpha-induced endothelial cell apoptosis and barrier dysfunction. *Am. J. Physiol. Lung Cell. Mol. Physiol.* **280**, L1168–L1178.
3. Gundersen, G. G., and Cook, T. A. (1999). Microtubules and signal transduction. *Curr. Opin. Cell Biol.* **11**, 81–94.
4. Small, J. V., Geiger, B., Kaverina, I., and Bershadsky, A. (2002). How do microtubules guide migrating cells? *Nat. Rev. Mol. Cell Biol.* **3**, 957–964. *This review describes the key role of MT in cell migration, cell asymmetry, polarization maintenance, and involvement in contractility and traction. The authors discuss the crosstalk among the actin cytoskeleton, adhesion complexes, and MT and its essential role in the regulation of cell motility.*
5. Nogales, E. (2000). Structural insights into microtubule function. *Annu. Rev. Biochem.* **69**, 277–302. *This review discusses the crystal structure of tubulin and MT, the tubulin isoforms, folding and dimer formation, and MT dynamics. It provides an updated characterization of MT-associated proteins and motors, stabilizers and destabilizers of MT, including new antimitotic tubulin ligands, and provides comprehensive description of MT self-regulation in the cell.*
6. Petrache, I. B. A., Ramirez, S. I., Garcia, J. G. N., and Verin, A. D. (2003). The role of the microtubules in TNF-α-induced endothelial cell permeability. *Am. J. Respir. Cell. Mol. Biol.* **28**, 574–581.
7. Verin, A. D., Birukova, A., Wang, P., Liu, F., Becker, P., Birukov, K., and Garcia, J. G. (2001). Microtubule disassembly increases endothelial cell barrier dysfunction: Role of MLC phosphorylation. *Am. J. Physiol. Lung Cell. Mol. Physiol.* **281**, L565–574.
8. Goedert, M., Hasegawa, M., Jakes, R., Lawler, S., Cuenda, A., and Cohen, P. (1997). Phosphorylation of microtubule-associated protein tau by stress-activated protein kinases. *FEBS Lett.* **409**, 57–62.
9. Parker, C. G., Hunt, J., Diener, K., McGinley, M., Soriano, B., Keesler, G. A., Bray, J., Yao, Z., Wang, X. S., Kohno, T., and Lichenstein, H. S. (1998). Identification of stathmin as a novel substrate for p38 delta. *Biochem. Biophys. Res. Commun.* **249**, 791–796.

Capsule biography

Dr. I. Petrache has been an assistant professor in the Pulmonary and Critical Care Division at the Johns Hopkins University School of Medicine since 2000. During her fellowship training at Johns Hopkins, she focused on studies of endothelial cell permeability induced by TNF-α. Her work is supported by grants from the NHLBI and NIH.

Dr. A. Birukova was recently appointed a Research Associate in the same Division. Her research interests include molecular mechanisms of microtubule-mediated endothelial cell barrier regulation. Her work is supported by Scientist Development grant from American Heart Association.

Dr. A. Verin is currently an associate professor in the Pulmonary and Critical Care Division at the Johns Hopkins University. His primary research interest includes involvement of cytoskeletal components (particularly microtubules) in the regulation of endothelial permeability. His work is supported by grants from the NHLBI and NIH.

CHAPTER 41

Vasoactive Signals and Pericyte Function in the Retina

Donald G. Puro
University of Michigan

Although constituting the largest portion of the circulatory system and the sites of greatest resistance, capillaries traditionally have been thought to play only a passive role in the regulation of blood flow. However, evidence is accumulating that blood flow can be actively controlled at the capillary level in at least in some vascular beds. Experimental support for the concept that capillaries actively regulate local perfusion is based chiefly on studies of the retinal vasculature.

Role of Pericytes in Blood Flow Regulation

The microcirculation of the retina is anatomically and physiologically specialized to distribute oxygen and nutrients to areas of metabolic need within a tissue that must be translucent.

Specialized Features of the Retinal Vasculature

One adaptation that minimizes interference with light passing through the retina is the low density of capillaries. In the retina, the intercapillary distance is at least 50 μm, which is less dense than in other tissues. Although the relatively paucity of retinal capillaries is a successful compromise between visual function and nutritive needs, a consequence is that there little functional reserve. As a result, a tight link between local perfusion and neuronal metabolism is essential in the retina.

Having blood flow controlled exclusively by local conditions facilitates efficiency in the distribution of oxygen and nutrients within the retina. Local control is ensured by the absence of autonomic innervation, which limits extrinsic oversight by the central nervous system. Also, the presence of an endothelial barrier restricts the effects of circulating vasoactive molecules. In addition, the lack of precapillary smooth muscle sphincters, which control local perfusion in most other tissues, suggests that retinal blood may be regulated within capillaries, rather than exclusively at precapillary sites.

Pericytes as Regulators of Capillary Perfusion

Candidates for regulating retinal blood flow at the capillary level are the pericytes. As in nearly all vascular beds, these abluminally located cells envelop the capillaries and postcapillary venules of the retina. Suggestive of the particular importance of pericytes in the retinal microcirculation, the density of these cells is greatest in the retina. Their expression of molecules such as α-smooth muscle actin supports the hypothesis that pericytes are contractile elements of the microvasculature. By contracting or relaxing, pericytes may adjust lumen size and thereby control capillary perfusion. However, even though vasoconstriction is observed in isolated retinal microvessels at sites adjacent to contracting pericytes, it remains to be definitively demonstrated that these cells regulate capillary blood flow in vivo.

Regulation of Pericyte Contractility

Assay Systems

Knowledge of the regulation of pericyte contractility is based on several experimental assay systems.

CULTURED RETINAL PERICYTES

Many studies testing the effects of vasoactive molecules use cells grown on a silicone surface. The laboratories of

D'Amore and Shepro pioneered in developing this assay system for retinal pericytes. Contraction of these cells is detected by the induction of wrinkles in the underlying silicone. An advantage of using a culture system is that only pericytes are present. Thus, indirect effects mediated by other types of cells are excluded. However, one disadvantage is that pericytes are likely to change during long-term culture. Certainly the exquisite morphology of in vivo pericytes is lost, as the cells quickly become rather nondescript flat cells in vitro; molecular and physiological changes may also occur. Furthermore, the tens of minutes that are often required for the induction of wrinkles in the silicone preclude assessment of pericyte responses within shorter time periods that may be more physiologically relevant.

Isolated Retinal Microvessels

More recently, Puro and his associates have used differential interference optics and time-lapse photography to detect pericyte contraction and relaxation in microvessels freshly isolated from the rat retina. Pericytes in the isolated microvessels retain their morphology and maintain a close association with the underlying endothelial tube. Because retinal pericytes are coupled via gap junctions to dozens of neighboring vascular cells, the ability to study pericytes that are an integral component of a multicellular functional unit can reveal a more complete picture of how the retinal microvasculature responds to vasoactive signals. Another advantage is the ability to detect pericyte contraction or relaxation, as well as microvascular constriction or dilation, within seconds after the onset of exposure to vasoactive substances. However, a caution is that findings based on studies of isolated microvascular complexes must ultimately be confirmed in vivo.

Retinal Capillaries in Situ

Funk and his colleagues have shown that it is feasible to monitor the effects of various vasoactive chemicals on the diameters of capillaries in isolated rat retinas. This in situ assay system should be useful, although a direct assessment of pericyte contraction or relaxation in situ has yet to be reported. A deficiency in both the in situ system and the use of isolated microvessels is that the vascular lumens are not perfused during an experiment. As a result, the role of shear stress and other intraluminal forces in determining the microvascular response to vasoactive molecules cannot be assessed. A challenging goal is to visualize in vivo the changes in pericyte shape, lumen diameter, and blood flow induced by vasoactive signals.

Contractile Responses to Vasoactive Molecules

Putative vasoactive molecules, which are known to affect the contractile tone of retinal pericytes, originate from a number of sources (Figure 1). Sites include the underlying endothelial cells that form the capillary tube, the glia that ensheathe the pericyte/endothelial complex, and the neurons that release vasoactive neurotransmitters. In addition, a

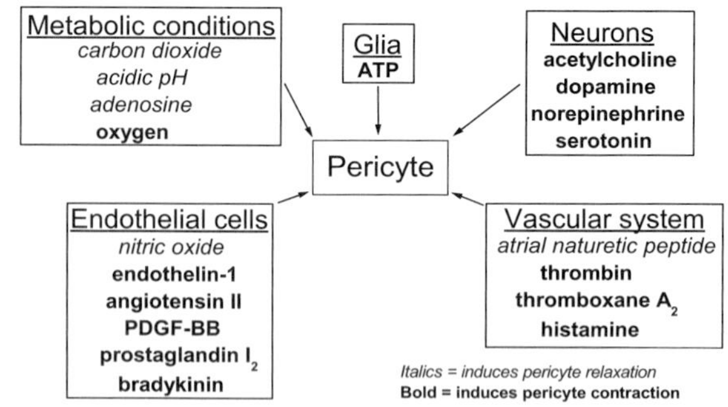

Figure 1 Schematic diagram of the likely sources of the vasoactive signals that are known to regulate the contractility of retinal pericytes. PDGF-BB, platelet-derived growth factor-BB.

breakdown of the blood–retinal barrier exposes pericytes to molecules from the vascular system.

Effects of Vasoactive Signals on Pericyte Physiology

The study of the mechanisms by which vasoactive molecules regulate pericyte contractility is in its infancy. Because ion channels are likely to be important in mediating the responses of pericytes to vasoactive signals, several laboratories have used electrophysiological methods to begin the task of determining the effect of these molecules on the ionic currents of pericytes. Also, because calcium regulates the contractile apparatus of pericytes, the effects of vasoactive molecules on the intracellular concentration of this divalent cation are of importance. However, at present, only a minority of the molecules known to alter pericyte contractility have been evaluated for effects on ion channel function and intracellular calcium levels.

Regulation of Ion Channels

Five molecules that induce retinal pericytes to contract have been analyzed for their effects on ion channel activity. These putative vasoactive signals include (1) acetylcholine, which activates muscarinic receptors to elicit pericyte contraction; (2) ATP acting via $P2X_7$ and $P2Y_4$ receptors; (3) endothelin-1, whose contractile effect is via ET_A receptors; (4) platelet-derived growth factor-BB (PDGF-BB), which is the ligand for PGDF-β receptors; and (5) angiotensin II, for which the subclass of receptor mediating the contractile response of retinal pericytes remains to be determined.

For each of these putative vasoconstrictors, a depolarization of pericytes is caused by the opening of calcium-activated chloride channels (Cl_{Ca}) and also, except for acetylcholine, the activation of calcium-permeable nonspecific cation (NSC) channels. In addition, an inhibition of ATP-sensitive potassium (K_{ATP}) channels, as has been demonstrated for endothelin-1, provides an additional mechanism by which vasoactive molecules can cause depolarization. With a decrease in the membrane potential, there is an opening of the L-type voltage-gated calcium channels (VGCCs), which are expressed by retinal pericytes (Figure 2).

Only two vasodilators have been tested for their effects on the ion channel activity of retinal pericytes. Adenosine, acting at A_1 and A_{2a} receptors, activates K_{ATP} channels. In contrast, nitric oxide (NO) does not affect K_{ATP} channels, but causes pericyte relaxation by a mechanism involving the cGMP-mediated inhibition of VGCCs.

Regulation of Intracellular Calcium

There are two general mechanisms by which putative vasoconstrictors cause the intracellular concentration of calcium to increase in retinal pericytes (Figure 2). One is an influx of this divalent cation via calcium-permeable NSC channels and VGCCs. Another is the release of intracellular stores of calcium. These pathways for increasing cytoplasmic calcium are interrelated. For example, the release of calcium stores is associated with the activation of the Cl_{Ca} channels in retinal microvessels. In addition, it is likely that an influx of calcium triggers the release of stored calcium.

Both influx and release appear to contribute to the increase in pericyte calcium observed during exposure to the five tested vasoconstrictors, that is, acetylcholine, ATP, angiotensin II, endothelin-1, and platelet-derived growth factor BB. At present, the effects on pericyte calcium levels of the vasodilators adenosine and NO are not known. Future experimental study is needed to clarify the relative functional roles in pericytes of the NSC channels and the VGCCs. Also, the relative importance of calcium influx versus calcium release in regulating pericyte contractility is not well understood.

Regulation of Microvascular Gap Junctions

In the retinal microvasculature, pericytes are coupled by gap junctions to dozens of their neighboring vascular cells. Recently, Puro and colleagues discovered that molecules such as endothelin-1, angiotensin II, and extracellular ATP potently and reversibly close these gap-junction pathways. Thus, vasoactive signals not only affect individual microvascular cells, but also regulate the effective size of the multicellular functional units that play a role in the control capillary blood flow. The regulation of intercellular communication within the pericyte-containing microvasculature may be an important, previously unappreciated mechanism by which local perfusion is dynamically affected by vasoactive molecules. Finally, the effects of putative vasoactive signals on the calcium sensitivity of the pericyte contractile apparatus remain to be determined.

Pericyte Pathobiology

Diabetes

The microvessels of the retina are particularly vulnerable to damage induced by diabetes. An early histological sign of diabetic retinopathy is the loss of pericytes. Currently, the mechanisms causing this sight-threatening complication remain uncertain.

PURINOCEPTORS

Recently, Sugiyama and his colleagues proposed that vasoactive molecules, such as extracellular ATP, can become vasotoxic in the diabetic retina. They found that activation of $P2X_7$ purinoceptors not only plays a role in transducing the contractile response of pericytes to ATP, but also can induce the formation of lethal transmembrane pores. Although high agonist doses are required to open $P2X_7$ pores in nondiabetic retinal microvessels, normally nonlethal concentrations trigger apoptosis in diabetic capillaries. Thus, a diabetes-induced increase in the vulnerability of retinal microvessels to the lethal effect of $P2X_7$ receptor activation may be a previously unrecognized mechanism by which diabetic retinopathy progresses.

ENDOTHELIN-1

Because epidemiological studies indicate that hyperglycemia is a key factor associated with the development of

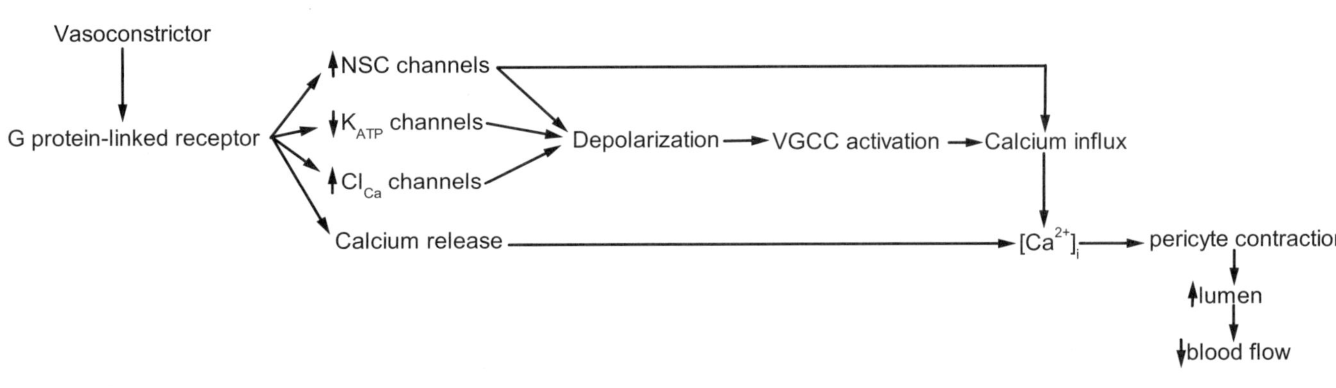

Figure 2 Schematic diagram of putative pathways by which vasoactive signals induce pericyte contraction and thereby decrease blood flow. Not shown are possible interactions between calcium release and calcium influx, that is, calcium induced calcium release and store-operated calcium channels. NSC, calcium-permeable nonspecific cation channels; K_{ATP}, ATP-sensitive potassium channels; Cl_{Ca}, calcium-activated chloride channels; VGCC, voltage-gated calcium channels; $[Ca^{2+}]_i$, intracellular calcium concentration.

diabetic retinopathy, the effect of glucose on pericyte responses to vasoactive molecules is of interest. Using cultured bovine pericytes, investigators have found that the contractile response of these cells to endothelin-1 is attenuated when the culture medium contains a high concentration of glucose. Experiments by McGinty and her coworkers indicate that a glycation-induced decrease in the function of L-type VGCC accounts for a reduction in the endothelin-induced influx of calcium and consequently the diminished contractile response of pericytes to this vasoactive molecule. An alteration in the vasoconstrictive effect of endothelin-1 may contribute to the dysfunction of autoregulatory mechanisms observed in the retinal circulation well before the onset of pericyte death and clinical signs of diabetic retinopathy.

MICROVASCULAR GAP JUNCTIONS

Soon after the onset of experimental diabetes, the intercellular communication system linking pericytes with their neighboring microvascular cells is disrupted by a mechanism involving the upregulation of protein kinase C. This closure of gap junction pathways disrupts the multicellular organization of the retinal microvasculature and may compromise mechanisms to match local blood flow to the needs of neurons. Not only could the loss of cell-to-cell communication adversely affect neuronal function, but it may metabolically isolate pericytes and contribute to their demise early in the course of diabetic retinopathy.

BLOOD–RETINAL BARRIER BREAKDOWN

In many retinal disorders, including diabetic retinopathy, the blood–retinal barrier is compromised. At sites where this barrier is defective, serum-derived molecules leak from the blood vessels into the retina. Because pericytes are located on the abluminal surface of the vascular endothelium, they are among the first cells to be exposed to molecules leaking from the circulatory system. As a result, the responses of pericytes to blood-derived molecules may determine how the retina functions when the vascular endothelial barrier is leaky.

In one of the first studies to establish that pericytes are contractile, Kelley and coinvestigators demonstrated that serum causes these cells to contract. More recently, Sakagami and coworkers showed that exposure of isolated pericyte-containing microvessels to serum activates calcium-permeable NSC and Cl_{Ca} channels. Associated with the opening of these channels, pericytes depolarize, contract, and cause the adjacent lumens to constrict.

Serum-induced contraction of pericytes may be a successful adaptive response to a breakdown of the blood–retinal barrier. For example, contraction of leaky microvessels would shunt blood away from areas with a defective vascular endothelium. On the other hand, extensive shunting of blood may cause ischemic damage, contribute to the demise of pericytes, and facilitate the progression of diabetic retinopathy.

Ischemia

Under physiological conditions, PDGF-BB activates NSC and Cl_{Ca} channels. With the opening of these channels retinal pericytes depolarize and contract (Figure 2). However, during a prolonged inhibition of ATP synthesis, some pericytes hyperpolarize and relax in response to PDGF-BB. This PDGF-induced increase in membrane potential is due to the activation of K_{ATP} channels, which close when the intracellular ATP concentration falls to an extremely low level. The capability of a vasoactive signal to elicit vasoconstriction when energy supplies are ample and vasodilation under ischemic conditions provides an efficient mechanism to link the function of the microcirculation to the local metabolic needs. Future studies may reveal other dual-action signals that play a role in the regulation of the retinal microcirculation.

Glossary

ATP-sensitive potassium (K_{ATP}) channels: Potassium-selective ion channels that are inhibited by intracellular ATP ($[ATP]_i$). By causing a decrease in $[ATP]_i$, ischemia activates these channels, which cause pericytes to hyperpolarize. Although activated as $[ATP]_i$ falls from millimolar concentrations, K_{ATP} channels require ~10 μM ATP to remain open. As a result, as $[ATP]_i$ declines below 10 μM, these channels again close.

Nonspecific cation (NSC) channels: Ion channels that are permeable to sodium and potassium and, in some cases, calcium. Opening these channels causes pericytes to depolarize. Membrane hyperpolarization increases the influx of calcium through calcium-permeable NSC channels.

$P2X_7$ purinoceptors: A member of the family of receptor-operated channels activated by extracellular ATP. Unlike nearly all other types of ion channels, the sustained activation of $P2X_7$ purinoceptors is associated with the formation of transmembrane pores, which can cause cell death.

Pericytes: Cells located on the outer walls of capillaries and postcapillary venules. Thought to play a role in regulating blood flow, maintaining vessel structure, and inhibiting endothelial proliferation. The loss of these cells is an early histological sign of diabetic retinopathy.

Voltage-gated calcium channels: Ion channels that are selectively permeable to calcium and activated by membrane depolarization.

Further Reading

Kawamura, H., Sugiyama, T., Wu, D. M., Kobayashi, M., Yamanishi, S., Katsumura, K., and Puro, D. G. (2003). ATP: A vasoactive signal in the pericyte-containing microvasculature of the rat retina. *J. Physiol.* **551**, 787–799. *The online version of this paper, which is at DOI:10.1113/jphysiol.2003.047977, contains a link to a time-lapse movie showing induced contractions of a pericyte-containing microvessel.*

Kelley, C., D'Amore, P., Hechtman, H. B., and Shepro, D. (1987). Microvascular pericyte contractility in vitro: Comparison with other cells of the vascular wall. *J. Cell Biol.* **104**, 483–490. *An early paper using a silicone substrate to detect cellular contraction in cultures of retinal pericytes.*

McGinty, A., Scholfield, C. N., Liu, W. H., Anderson, P., Hoey, D. E., and Trimble, E. R. (1999). Effect of glucose on endothelin-1-induced calcium transients in cultured bovine retinal pericytes. *J. Biol. Chem.* **274**, 25250–25253.

Oku, H., Kodama, T., Sakagami, K., and Puro, D. G. (2001). Diabetes-induced disruption of gap junction pathways within the retinal microvasculature. *Invest. Ophthalmol. Vis. Sci.* **42**, 1915–1920.

Rhinehart, K., Zhang, Z., and Pallone, T. L. (2002). Ca^{2+} signaling and membrane potential in descending vasa recta pericytes and endothelia. *Am. J. Physiol. Renal Physiol.* **283**, F852–F860. *Although pericytes of the retina have received the most attention, there is a substantial literature on pericytes from other vascular beds. In this paper, the authors studied pericytes of the kidney.*

Sugiyama, T., Kobayashi, M., Kawamura, H., Li, Q., and Puro, D. G. (2004). Enhancement of $P2X_7$-induced pore formation and apoptosis: an early effect of diabetes on the retina microvasculature. *Invest. Ophthalmol. Vis. Sci.* **45**, 1026–1032.

Capsule Biography

Dr. Puro is professor of Ophthalmology and Visual Science and of Molecular and Integrative Physiology at the University of Michigan. He was named in 2001 as the recipient of the Harrington RBP Senior Scientist Award. His research focuses on the physiology and pathobiology of retinal capillaries; his laboratory is supported by grants from the NIH, the American Diabetes Association, and Research to Prevent Blindness, Inc.

CHAPTER 42

Regulation of Microvascular Hydraulic Conductivity

Rolando E. Rumbaut and Virginia H. Huxley
[1]*Baylor College of Medicine, Houston, Texas*
[2]*University of Missouri–Columbia, Columbia, Missouri*

Introduction

Transport of water and hydrophilic solutes across microvascular walls (capillaries and postcapillary venules) is an essential mechanism for delivery of nutrients to tissues. The ease of water movement across microvessel walls is described by the coefficient hydraulic conductivity (L_p). L_p reflects net volume flux ($cm^3 sec^{-1}$) per unit surface area (cm^2) per unit pressure (cmH_2O). L_p was assumed previously to be a fixed property of microvessels; however, it is now evident that L_p may be regulated actively under both physiologic and pathologic conditions. In this chapter, we review current aspects of regulation of L_p, common methods for assessing this coefficient in vivo, and implications of alterations in L_p. We will limit our scope to data derived from in vivo models; techniques for in vitro assessment of L_p of cultured endothelial cell monolayers are exemplified by the work of Tarbell and colleagues [1].

Forces Regulating Transvascular Water Movement

Movement of water across microvascular walls is dependent upon pressures acting upon both sides of the microvessel. Starling first described this principle in 1896, by proposing that plasma was retained in the circulation by the osmotic pressure exerted by plasma proteins (oncotic, or colloid osmotic pressure), which opposes the outward hydrostatic pressure. In addition, the hydrostatic and oncotic pressures of the extravascular compartment (interstitium) affect net water movement, which is governed by the difference between hydrostatic and oncotic pressure gradients (intravascular minus extravascular). This balance, referred to as "Starling equilibrium," is illustrated in Figure 1. The modern form of the Starling equation for fluid filtration is also shown in Figure 1. It includes hydraulic conductivity (L_p), defined in the introduction, and a second coefficient, the osmotic reflection coefficient (σ, a dimensionless coefficient ranging from 0 to 1), an index of the selectivity of the microvascular barrier to an osmotically active solute. In the majority of the circulation, small solutes such as NaCl have σ approaching 0, whereas in vivo measures of σ to albumin under physiologic conditions typically range between 0.8 (heart) and 0.99 (skeletal muscle). Based on this equation, positive values of J_v denote filtration, whereas negative values denote water reabsorption into the microvessels.

Pathways for Volume Flux across Microvascular Walls

In microvessels with continuous endothelium, the majority of transvascular J_v is presumed to occur via the interendothelial cell junctions. Numerous mathematical models have been proposed to describe the geometry of these pathways, including cylindrical pores, rectangular slits, and a fiber matrix model [2]. Each of these models alone is unable to fully account for in vivo measures of volume and hydrophilic solute flux across microvessels. Of these models, the fiber matrix model appears to describe many of the transport coefficients measurable in vivo. The fiber matrix model is consistent with experiments involving disruption of the endothelial surface layer (glycocalyx). For example, perfusion of microvessels with cationized ferritin demonstrates a layer up to 100 nm thick on the endothelial cell surface. Exposure of microvessels to an enzyme, pronase, results in reduction of the thickness of this surface layer and a significant increase in microvessel L_p. Similarly,

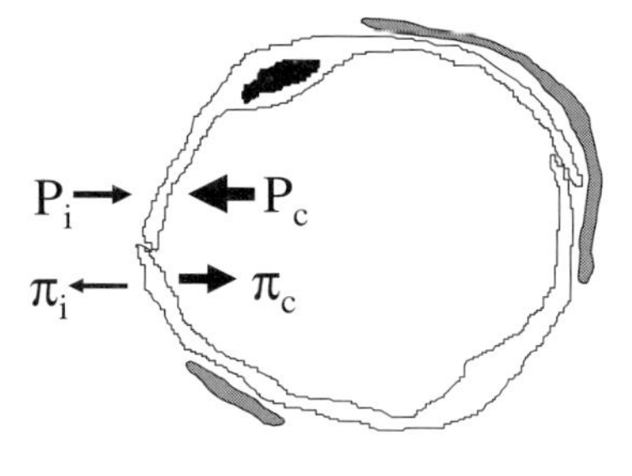

$$J_v = L_p\, S\, [\,(P_c - P_i) - \sigma\,(\pi_c - \pi_i)\,]$$

J_v: volume flux
L_p: hydraulic conductivity
S: surface area
P_c: capillary hydrostatic pressure
P_i: interstitial hydrostatic pressure
π_c: capillary oncotic pressure
π_i: interstitial oncotic pressure
σ: reflection coefficient

Figure 1 Pressures responsible for fluid movement across a capillary, and modern form of the Starling equation for fluid filtration. See text for details.

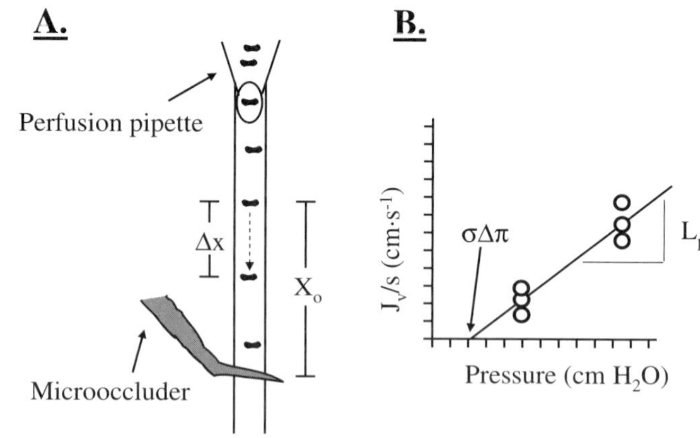

$$J_v/S = (\Delta x/\Delta t) \times (r/2X_o) \qquad J_v/S = L_p\,(\Delta P - \sigma\Delta\pi)$$

Figure 2 Landis–Michel technique for measurement of hydraulic conductivity (L_p) in individually perfused microvessels. **(A)** Relative placement of microinstruments and calculation of fluid filtration rate per unit surface area (J_v/S) based on vessel radius and erythrocyte location and velocity. **(B)** Calculation of L_p based on measures of J_v/S at different hydrostatic pressures.

microvessel perfusion with a solution devoid of proteins reduces the thickness of the surface layer and induces a significant increase in L_p (known as the "protein effect"). These studies support the notion that the glycocalyx (which may be modeled as a fiber matrix) contributes significantly to the microvascular barrier to volume flux. Based on recent electron microscopic observations suggesting the presence of a thick endothelial surface layer (200 to 500 nm), elaborate models of a fiber matrix overlying interendothelial cell clefts with junctional strand gaps have been proposed [3]. Although these models describe many in vivo measures of transvascular volume and solute flux, their validity remains to be confirmed.

Assessment of Transvascular Water Movement in Vivo

Single Perfused Microvessels

Eugene Landis first reported measures of L_p of individual microvessels in the early 20th century [4]. By occluding capillaries of the frog mesentery, he described that on occasions erythrocytes moved toward the micro-occluder and on other occasions they moved away from the occlusion site. Cells moving toward the occluder were interpreted as representing fluid filtration, whereas those moving away denoted reabsorption. He calculated the rate of filtration or reabsorption per unit surface area (J_v/S) based on the erythrocyte velocity. By correlating J_v/S to capillary pressure (P_C) measured with a micropipette, he determined a linear relation between P_C and J_v/S. The slope of that relation is the L_p, and the x pressure axis intercept in Landis' experiments approximated the oncotic pressure of frog plasma. Those experiments provided evidence in vivo of Starling's hypothesis for fluid filtration. The Landis technique to assess L_p, modified by Michel and colleagues [5], is illustrated in Figure 2A. A microvessel is perfused with a micropipette connected to a water manometer, to measure microvascular hydrostatic pressure, containing marker erythrocytes in a physiological salt solution of predetermined composition. Downstream from the site of cannulation, a micro-occluder is used to stop flow intermittently in the vessel. Assuming cylindrical geometry, and negligible interstitial hydrostatic pressure (P_i), L_p is the slope of the measures of fluid filtration at the set of hydrostatic pressures (see Figure 2B). This technique allows for paired measures of L_p in individual microvessels under control conditions and following exposure to a variety of experimental conditions in the luminal, abluminal, or both surfaces of the microvessel. Further, given that the oncotic pressure is known, the intercept on the pressure axis can be used to determine σ for the specific vessel and colloid (Figure 2B). Much of our current understanding of physiologic and pathologic regulation of hydraulic permeability is based on experiments using the Landis–Michel technique to measure L_p.

Whole Organs

Experimental evidence of fluid filtration consistent with Starling's hypothesis was described in whole organ preparations in the mid-20th century [6] when hind limbs from cats and dogs were isolated, perfused, and the weight of the hind limb recorded continuously. Using this technique, fluid filtration was determined from the rate of gain of limb weight, and fluid reabsorption from the rate of loss of limb weight. The arterial and venous hydrostatic pressures were adjusted experimentally, and values that resulted in neither filtration nor reabsorption were established, resulting in an "isogravimetric state." The hydrostatic pressures required for the isogravimetric state varied according to the oncotic pressure of the perfusate solution, as predicted by the Starling equation. The relationship between filtration rate and "capillary" pressure (calculated from the arterial and venous hydrostatic pressures) was linear, with the slope equal to the filtration coefficient (K_f). K_f is the product of L_p and surface area, the latter coefficient being unknown in whole-organ experiments but measurable in single-perfused microvessels (assuming cylindrical geometry).

Regulation of Hydraulic Conductivity

Control Conditions

Figure 3 shows the distribution of control values of L_p determined in individual microvessels of the frog ($n = 1111$) and rat mesentery ($n = 89$) in our laboratories. Median L_p for frog capillaries is 3.3 and for rat venules is 1.8 ($\times 10^{-7}$ cm sec^{-1} cmH$_2$O). Several features are noteworthy. First, the shapes of the distribution are similar despite being derived from different vessel types (capillaries and postcapillary venules) from two phylogenetically distinct species. Second, the broad spread of the distribution under physiologic conditions precludes describing J_v by a single value. Third, the distributions in each case do not fit a normal distribution; they are normalized by a fourth root transformation [7]. Of interest, a similar frequency distribution was noted by Pappenheimer's group of K_f values in the feline hind limb [8]. The authors speculated that if the pathways responsible for fluid filtration were "pores" with a circular cross-section, a normal distribution of pore radii would result in a skewed distribution of filtration coefficient, varying according to the fourth power of the pore radius. However, ultrastructural evidence of cylindrical pores as a pathway for volume flux is lacking.

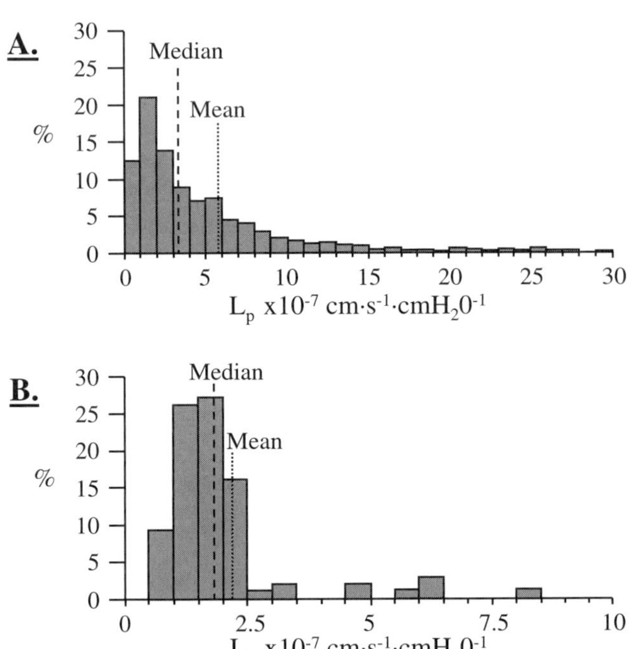

Figure 3 Frequency distribution of L_p of single perfused frog mesenteric capillaries (**A**, $n = 1111$) and rat mesenteric venules (**B**, $n = 89$). Note nonnormal distribution of values with similar shape in both cases.

Control L_p values in frog mesenteric capillaries demonstrate a gradient of values across the capillary network. Capillaries near arterioles (arteriolar capillaries) typically have the lowest values, followed by midstream capillaries (true capillaries) and then capillaries near postcapillary venules (venular capillaries). This phenomenon is illustrated by our median L_p values (in units of 10^{-7} cm sec^{-1} cmH$_2$O) in the frog mesentery of arteriolar capillaries ($L_p = 1.8$, $n = 192$), true capillaries ($L_p = 2.9$, $n = 580$), and venular capillaries ($L_p = 5.1$, $n = 339$). Since hydrostatic pressure across the capillary network has a decreasing gradient, the gradient of L_p values may attenuate variations in net volume flux across the capillary network. Despite this, the relative L_p and hydrostatic pressure gradients across the network tend to favor filtration across true and venular capillaries.

Active Changes in Hydraulic Conductivity

In Landis' initial experiments, he reported that frog mesenteric capillary L_p increased in response to a variety of insults including ischemia, hypoxia, and increased acidity. More recent work has shown acute increases in microvascular L_p with less injurious stimuli. For example, agonists that increase intracellular levels of cyclic guanosine monophosphate (cGMP) such as atrial natriuretic peptide (ANP), sodium nitroprusside, and 8-bromo-cGMP induce reversible, dose-dependent increases in capillary L_p. The endothelial cell structures responsible for these reversible changes in L_p remain to be determined, though in the case of ANP, the change in permeability is independent of the "protein effect" [9]. These data suggest that the microvascular barrier for volume flux may be regulated independently by different mechanisms. Other agents shown to increase L_p include histamine, bradykinin, serotonin, vascular endothelial growth factor (VEGF), adenosine triphosphate and calcium ionophores (see Ref. [2]). Many of these agents enhance intracellular endothelial calcium; temporal correlations between changes in endothelial calcium and microvascular L_p have been described [10]. In addition to increases in L_p, several agents have been reported to decrease microvascular L_p, including nitric oxide synthase inhibitors [11] as well as agents that increase endothelial cell cyclic adenosine monophosphate or activate adenylyl cyclase (rolipram, isoproterenol, forskolin; [12]). These studies demonstrate that microvascular volume flux may be subject to regulation by a variety of physiologic as well as pathologic stimuli. Under physiologic conditions, changes in L_p may influence the rate of delivery of water and nutrients to tissues. Pathologic increases in permeability, or hyperpermeability, would favor enhanced fluid filtration and may lead to tissue swelling, or edema. Of interest, on occasion a single inflammatory mediator may result in different changes in permeability in different portions of the microvasculature. For example, the inflammatory agonist bradykinin induced changes in L_p with distinct temporal and spatial variations in frog capillaries [13]. Thus, in addition to the heterogeneous control values of L_p across the network described earlier, the changes in L_p in response to the same agonist may vary across the microvascular network.

Summary

Delivery of water and hydrophilic nutrients to tissues may be regulated at the level of the microvasculature, primarily in capillaries and postcapillary venules. Hydraulic conductivity (L_p) is a coefficient that reflects the ease of passage of water across microvascular walls. Data derived from single-perfused microvessels demonstrate that control values of L_p have a broad distribution, with evidence of spatial differences within microvascular networks. Further, L_p is not a fixed coefficient of individual microvessels, but instead is subject to active regulation by both physiologic and pathologic stimuli. The cellular molecular basis for active changes in the endothelial pathways that regulate microvascular L_p is an area of ongoing research.

Glossary

Microvascular hydraulic conductivity (L_p): The ease of passage of water across microvascular walls, reflecting net volume flux ($cm^3 sec^{-1}$) per unit surface area (cm^2) per unit pressure (cmH_2O).

Protein effect: An effect of plasma proteins that reduces L_p, presumably at the level of the endothelial surface layer. Removal of proteins from microvessel consistently induces significant elevations in L_p.

Reflection coefficient (σ): An index of the selectivity of the microvascular barrier to an osmotically active solute; it is a dimensionless coefficient, theoretically ranging from 0 to 1.

References

1. Sill, H. W., Chang, Y. S., Artman, J. R., Frangos, J. A., Hollis, T. M., and Tarbell, J. M. (1995). Shear stress increases hydraulic conductivity of cultured endothelial monolayers. *Am. J. Physiol.* **268**, H535–H543. *A description of a cultured endothelial cell monolayer model to measure endothelial hydraulic conductivity in vitro. The control values of hydraulic conductivity of the endothelial cell monolayers in this model are comparable to those of single-perfused microvessels depicted in Figure. 3.*
2. Michel, C. C., and Curry, F. E. (1999). Microvascular permeability. *Physiol. Rev.* **79**, 703–761. *A comprehensive review of microvascular permeability under physiologic conditions as well as under conditions of hyperpermeability. Includes extensive modeling of the pathways responsible for volume and solute flux across microvessels and a discussion of signal transduction of changes in permeability.*
3. Weinbaum, S., Zhang, X., Han, Y., Vink, H., and Cowin, S. C. (2003). Mechanotransduction and flow across the endothelial glycocalyx. *Proc. Natl. Acad. Sci. USA* **100**, 7988–7995. *An intriguing model of the role of the endothelial cell glycocalyx as a primary barrier to microvascular volume flux, based on current electron microscopic observations of a complex structure on the endothelial surface.*
4. Landis (1927). *Am. J. Physiol.* **82**, 217–238.
5. Michel, C. C., Mason, J. C., Curry, F. E., and Tooke, J. E. (1974). A development of the Landis technique for measuring the filtration coefficient of individual capillaries in the frog mesentery. *Q. J. Exp. Physiol.* **59**, 283–309. *Original description of the modified Landis technique used currently for assessment of hydraulic conductivity of single perfused microvessels in vivo. Includes extensive discussion of the assumptions and limitations of the technique.*
6. Pappenheimer and Soto-Rivera (1948). *Am. J. Physiol.* **152**, 471–491.
7. Hargrave et al. (1995). *Am. J. Physiol.* **268**, R468–R474.
8. Pappenheimer et al. (1951). *Am. J. Physiol.* **167**, 13–46.
9. Huxley and Meyer (1990). *Am. J. Physiol.* **259**, H1351–H1356.
10. He et al. (1996). *Am. J. Physiol.* **271**, H2377–H2387.
11. Rumbaut et al. (1995). *Am. J. Physiol.* **268**, H1856–H1861.
12. Adamson et al. (1998). *Am. J. Physiol.* **274**, H1885–H1894.
13. Williams and Huxley (1993). *Am. J. Physiol.* **264**, H1575–H1581.

Further Reading

Levick, J. R. (1995). Circulation of fluid between plasma, interstitium and lymph. In *An Introduction to Cardiovascular Physiology*, 2nd ed., pp. 158–187. Butterworth-Heinemann, Oxford. *A practical overview of transvascular fluid filtration with an emphasis on fluid across the interstitium and the mechanisms responsible for formation of edema. Includes descriptions of capillary pressure and capillary filtration measurements in humans, as well as a discussion on lymph flow relative to transvascular water movement.*

Capsule Biography

Dr. Rumbaut earned an M.D. degree from the Instituto Tecnológico y de Estudios Superiores de Monterrey and a Ph.D. in physiology from the University of Missouri–Columbia. He has been an Assistant Professor of Medicine and Pediatrics at Baylor College of Medicine in Houston, Texas, since 2000. His laboratory focuses on microvascular permeability and microvascular thrombosis; his research is funded by the National Institutes of Health.

Dr. Huxley earned a Ph.D. in biophysics from the University of Virginia and was a postdoctoral fellow at the University of California at Davis. She has been a Professor of Medical Pharmacology and Physiology at the University of Missouri-Columbia since 1994. Her laboratory focuses on regulation of microvascular permeability and gender influences on microvascular physiology; her research is funded by the National Institutes of Health.

CHAPTER 43

The Role of Gap Junctions in the Regulation of Arteriolar Tone

N. M. Rummery and C. E. Hill

John Curtin School of Medical Research, Australian National University, Canberra, Australia

Gap junctions are low-resistance channels which connect adjacent cells in almost all mammalian tissues, including blood vessels, allowing the direct transfer of electrical current and molecules of < 1 kDa in molecular weight. Because of their small size, many important metabolites and signaling molecules elicited by receptor activation, such as calcium, inositol trisphosphate, or cyclic AMP, would be expected to pass through gap junctions from one cell to another. Since arteriolar tone relies on the balance of vasoconstrictor and vasodilator pathways, which involve such signaling molecules or the opening of ion channels in the cell membrane and subsequent changes in membrane voltage, it is expected that gap junctions will play a pivotal role in coordinating this process.

Structure and Regulation of Gap Junctions

Structure of Gap Junctional Channels

Gap junctions comprise a family of membrane proteins known as connexins (Cxs). Six individual Cx proteins oligomerize in the plasma membrane of one cell to form a connexon or hemichannel. A complete gap junctional channel linking adjacent cells results from the docking of two connexons, one from each of the two opposing cells (Figure 1). To date, 20 different types of mammalian Cx proteins, commonly named after their molecular weight in kilodaltons (kDa), have been identified. All of these Cxs have the same basic molecular structure, consisting of four α-helical membrane-spanning domains, two extracellular loops, and three cytoplasmic regions including one intracellular loop. Gap junctional channels may be either homo- or heteromeric in nature, depending on whether a single or multiple types of Cxs are present in each connexon (Figure 1). Gap junctions can also be homo- or heterotypic depending on the Cx content of the two connexons (Figure 1).

Gap junctional channels demonstrate unique electrical and biochemical properties depending on the constituent Cx isotypes. In turn, the distinct functional properties of each Cx protein arise from specific amino acid sequences found in the intracellular loop and carboxy terminus. The formation of heterotypic and heteromeric channels, which express a mixture of more than one Cx, has been shown to confer channel properties that are distinctly different from those of homotypic channels formed from each of the individual Cxs. For example, the ability to permit the passage of particular second messengers has been shown to vary with the Cx composition of gap junctions. Thus, it is important to determine the specific Cx composition of the gap junctional channels formed in any particular artery or arteriole, since this is likely to have a significant impact on vascular function.

Regulation of Gap Junctions

The function of gap junctions can be regulated by factors that act in either the short or long term. Short-term regulation is predominantly concerned with rapid and reversible changes in channel conductance resulting from alterations in the opening probability of the channels or alterations in channel turnover. On the other hand, long-term regulation of gap junctions involves alterations in gene expression at the mRNA and protein levels. In blood vessels, changes in Cx expression have been observed at branch points where blood flow is disturbed and shear stress turbulent. Such forces may also be expected to be encountered within the branching of resistance networks.

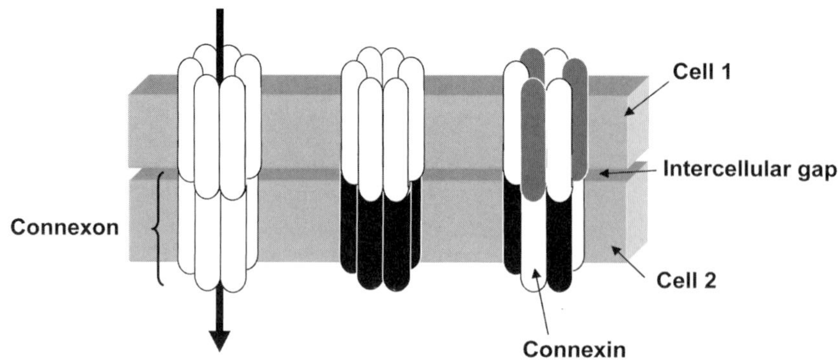

Figure 1 Structure of gap junctional channels. Gap junctions comprise two connexons, each of six Cx proteins, which may be of the same or different types, in adjacent cells.

Conductance of gap junctional channels can be modulated by a number of different factors, such as phosphorylation, changes in intracellular Ca^{2+} concentration and pH. The result of the phosphorylation of Cx proteins on conductance appears to depend on the component Cx isotypes and presumably relates to the number and location of specific phosphorylation sites, both of which vary among the different Cx proteins. Since many important vasoactive stimuli activate intracellular pathways resulting in increases in intracellular calcium or in the activation of protein kinases, it is likely that these stimuli will also modulate gap-junctional communication. Thus, cellular communication within the vascular wall may be dynamically controlled by increased or decreased channel open probability and gating, as well as Cx synthesis, assembly, trafficking, and degradation.

Expression of Gap Junctions in Vascular Tissue

The vascular wall is composed of a single inner layer of endothelial cells surrounded by layers of smooth muscle cells called the media (Figure 2). Both the endothelial and smooth muscle cells are spindle shaped but they are arranged with their longitudinal axes at right angles, the endothelial cells along the vessel axis and the smooth muscle cells transversely around the vessel (Figure 2). Large conduit arteries, such as the thoracic aorta, have many layers of smooth muscle cells that are interspersed with layers of elastic tissue to provide rigidity for the pressures experienced in these vessels. The so-called muscular arteries lack these elastic laminae, but still show considerable variation in the number of smooth muscle cell layers within the media. In the microcirculation, the number of smooth muscle cell layers is reduced to one or two.

At the ultrastructural level, gap junctions have been identified by their pentalaminar structure in which the proximity

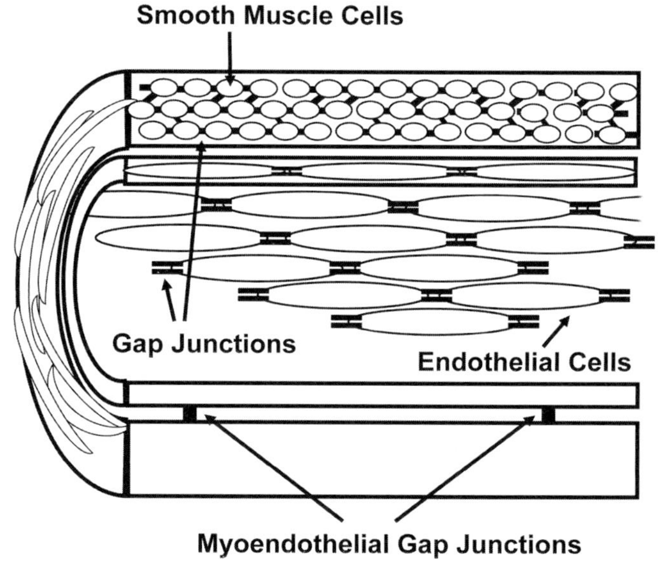

Figure 2 Anatomical structure of the arterial wall. Endothelial and smooth muscle cells are arranged at right angles to each other with adjacent cells being coupled by gap junctions.

of the outer membrane leaflets of the two opposing cells is seen as a single dark line (Figure 3A). In some cases, a 2-nm gap can be seen at this point. Such gap junctions have been commonly identified between adjacent endothelial cells and, more occasionally, between adjacent smooth muscle cells as well as between the two cell layers.

Within vascular tissue, four Cxs have been identified using subtype selective antibodies against Cxs 37, 40, 43, and 45. Cxs 37, 40, and 43 have been detected in the vascular endothelium of most vessels, whereas the situation in the smooth muscle appears to be more variable (Figure 3B). At least some of this variability may be due to heterogeneity in Cx expression among different animal species, as well as to variation within and between different vascular beds. Given

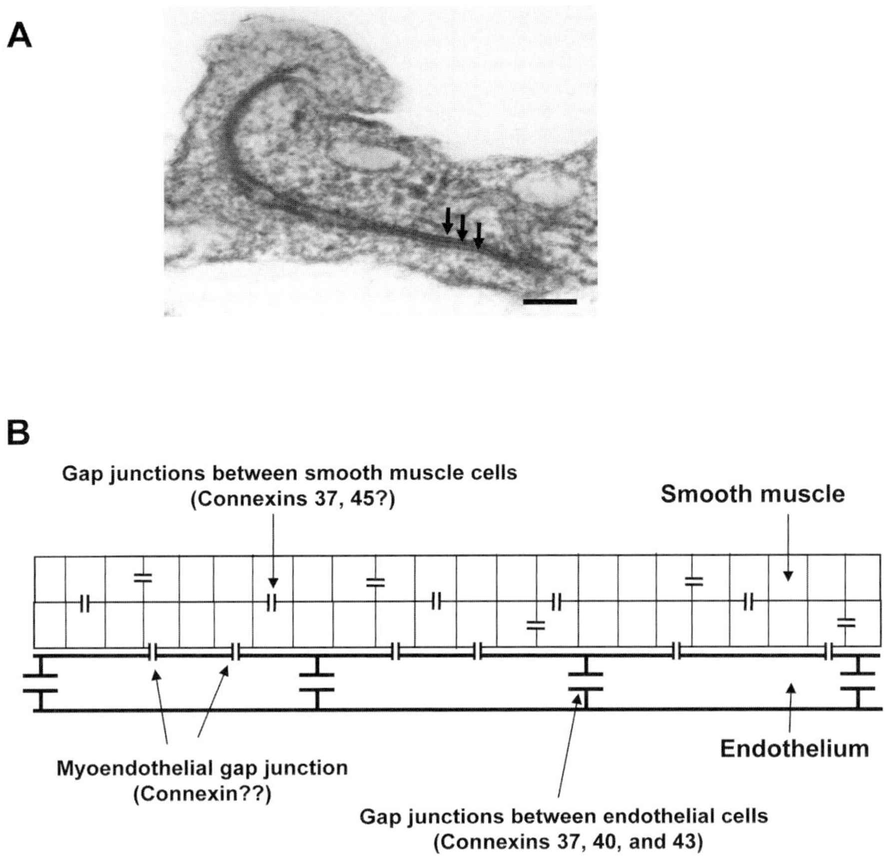

Figure 3 Gap junctions in the arterial wall. **(A)** Pentalaminar structure (arrows) of a gap junction connecting two endothelial cells of the rat basilar artery (kindly supplied by Dr. Shaun Sandow). Calibration bar 50 nm. **(B)** Cxs vary within the cell layers, as do the size of the gap junctional plaques.

that different Cx isotypes may form channels with unique electrical and biochemical properties, the heterogeneity in Cx expression among different vessels may contribute to heterogeneity in vascular responses.

Gap Junctions in the Endothelium

In the vascular endothelium, Cxs 37, 40, and 43 have been commonly found in both large and small arteries. At the ultrastructural level, large areas of intact pentalaminar gap junctions, up to and sometimes larger than 1 μm in length, are readily found in vessels of all sizes, confirming the view that endothelial cells are well coupled throughout the vasculature (Figure 3A). At the light microscope level, using immunohistochemistry and subtype selective antibodies, gap junctions appear as puncta that delineate the cell borders of endothelial cells (Figure 4A,B).

In the endothelium of many large vessels, Cxs 37, 40, and 43 have been described to be colocalized in the same gap junctional plaque, suggesting an ability to form heteromeric gap junctional channels. Although some variation in the relative expression of the three Cxs has been observed among different arteries, Cx40 appears to be the most highly expressed, whereas the expression of Cx43 may be the most variable in large vessels. Indeed, Cx43 appears to be specifically associated with areas where flow is turbulent, for example, at arterial branch points. Expression of Cxs is also subject to variation between species and in disease, development, and aging.

In the small vessels of the microvasculature, as in the larger muscular arteries, expression of all three Cxs has been reported in the endothelium (Figure 3B). Once again, variability exists in the relative expression within and between vascular beds and in the same vessel in different species, although Cx40 still appears to be the most highly expressed of the endothelial Cxs, whereas Cx37 expression may be more variable in these smaller vessels. Thus, Cx37 was consistently found along with Cxs40 and 43 in the endothelium of mesenteric resistance arteries, but was absent from the arterioles of the rat mesentery.

Gap Junctions in the Media

For many years Cx43 was considered to be the Cx linking smooth muscle cells. This view largely arose through investigations involving large elastic arteries such as the aorta, where Cx43 expression is indeed robust and abundant, and through the use of smooth muscle cell lines derived from the aorta. Comparative studies of Cx43 expression in a range of vessels including large and small

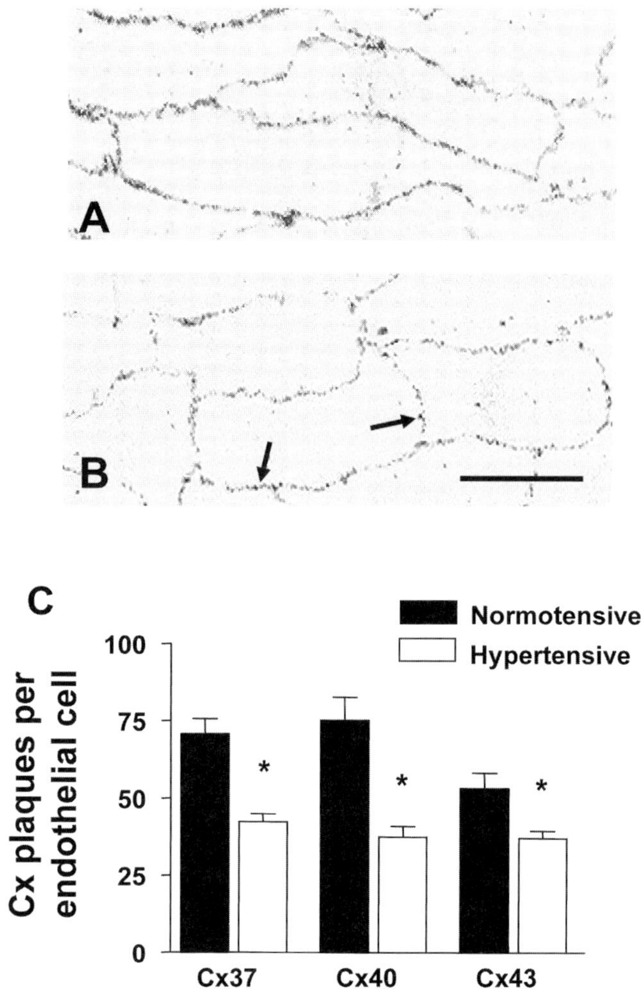

Figure 4 Changes in Cx expression in endothelial cells in hypertension. Punctate Cx40 expression (arrows) in caudal artery of normotensive **(A)** and spontaneously hypertensive rats **(B)**. Calibration bar 20 μm. **(C)** Expression of Cxs37, 40, and 43 per endothelial cell is reduced in hypertension ($*P < 0.05$). [Rummery et al. (2002). *J. Hypertens.* **20**, 247–253; permission granted from Lippincott, Williams & Wilkins.]

muscular arteries, however, demonstrated that expression of Cx43 was absent from the muscle layers of the latter vessels and that, even in the large conduit vessels, the expression of Cx43 decreased with distance from the heart.

The development of Cx45 knockout mice provided the first anatomical evidence that this Cx may play a role in the media of murine arteries. Subsequently Cx45 was detected in the media of the thoracic aorta and caudal arteries of rats and in smaller cerebral vessels. Interestingly, expression of Cx45 was reported to be inversely correlated with that of Cx43 in the aorta, and the suggestion was made that these two Cxs may be associated with different states of cell activity and differentiation. Although reports of Cxs37 and 40 expression in the media of large arteries also exist in the literature, these have not always been confirmed and may have arisen through the use of incompletely characterized Cx antibodies or may constitute further examples of variation in expression across animal species.

In muscular arteries, both large and small, there are few reports of the expression of Cx43. In general, Cx40 is also rarely described in the media of these vessels, although it is convincingly expressed in the renin-secreting smooth muscle cells of the juxtaglomerular apparatus. More recently, a role for Cxs37 and 45 has been identified in the media of a number of muscular arteries of rodents, including small microcirculatory cerebral vessels.

The intensity and extent of the Cx staining within the media of muscular arteries and arterioles suggest that the incidence of gap-junctional channels may be less than that observed in the endothelium or in the media of conduit vessels. At the ultrastructural level, pentalaminar gap junctions between smooth muscle cells are rarely observed and, where described, are much smaller (~100 nm) than those in the endothelium. Dyes, such as Lucifer yellow, that readily pass between endothelial cells do not pass between smooth muscle cells in arterioles of the microcirculation, again suggesting major differences between the coupling within the two cell layers. Interestingly, in larger muscular arteries such as the caudal artery, Lucifer yellow passes between smooth muscle cells in a radial but not a longitudinal direction, and nerve-evoked responses elicited in adventitial smooth muscle cells can be recorded in luminal smooth muscle cells. Taken together, these data suggest that coupling within the media of arteries and arterioles may be significantly less than that within the endothelium and may often be more effective in the radial compared with the longitudinal direction.

Myoendothelial Gap Junctions

Unlike homocellular coupling within the blood vessel wall, less is known regarding the nature of heterocellular coupling via myoendothelial gap junctions (MEGJs) between the endothelial and smooth muscle cell layers, and the Cxs comprising these junctions remain unidentified. Several studies have confirmed electrophysiologically that MEGJs exist in the microcirculation through the transfer of current pulses, dyes, and calcium from one cellular layer to the other. However, few structural reports of MEGJs existed in the literature until serial section electron microscopy was employed to elucidate their incidence. These latter studies have revealed that MEGJs are very small (~100 nm) and that in general the incidence of MEGJs increases as vessel size decreases. Thus, MEGJs might be expected to play an important role in the function of the microcirculation and in the control of vascular resistance.

Involvement of Gap Junctions in Responses Affecting Arteriolar Tone

The maintenance of arteriolar tone is achieved through a balance between vasoconstriction and vasodilation. Alterations to this balance have been implicated in disease states such as hypertension and atherosclerosis.

Vasoconstrictor Responses

THE MYOGENIC RESPONSE

The myogenic response refers to the nerve- and endothelium-independent vasoconstriction in response to increased transmural pressure and vasodilation in response to reduced pressure. This autoregulatory response is considered to be a major determinant of vascular tone and to reduce downstream damage to microcirculatory elements from significant and sudden changes in systemic pressure. The myogenic response is more prominent in smaller vessels, but is also differentially effective among vascular beds. It is now well established that the myogenic response is initiated by depolarization of vascular smooth muscle cells and an increase in intracellular calcium concentration via the opening of voltage-dependent calcium channels. The involvement of gap junctions in this response is suggested from the attenuation of myogenic tone by putative inhibitors of gap junctions. Gap-junctional coupling among the smooth muscle cells would be expected to facilitate the spread of the depolarizing current to coordinate constriction, particularly if the pressure stimulus was insufficient to activate all the intramural muscle cells—for example, those more distant from the lumen. Thus, it may be predicted that gap junctions would be less important in the myogenic response of arterioles.

VASOMOTION

Rhythmical contractions, or vasomotion have been demonstrated in many different vascular beds, including microcirculatory vessels of the systemic and cerebral circulations both *in vivo* and *in vitro*. Vasomotion has been suggested to be important for the maintenance of vascular resistance and blood flow. The common underlying mechanism involves release of intracellular calcium, which initiates a depolarization of the cell membrane and subsequent oscillation in membrane voltage due to sequential opening and closing of calcium and potassium channels. Gap-junctional coupling has been suggested to be critical to coordinate these oscillations between adjacent smooth muscle cells, and putative gap-junction inhibitors have been shown to prevent vasomotion in a number of vascular beds.

Vasodilatory Responses

The endothelium plays an important role in the control of vascular tone through the release of several mediators that act on the underlying smooth muscle to cause a relaxation. These factors, which are released in response to hemodynamic forces such as shear stress, include nitric oxide (NO), prostaglandins, and endothelium-derived hyperpolarizing factor (EDHF). NO and prostaglandins are both released from the endothelial cells and diffuse to the smooth muscle cells, where they activate intracellular pathways. Although it is possible that in large vessels gap junctions within the media might be involved in the spread of vasodilatory responses, this is unlikely to be so critical in the microcirculation where the media is made up of only one or two cell layers.

As its name suggests, EDHF produces a hyperpolarization of the endothelium, followed by hyperpolarization and relaxation of the smooth muscle due to a reduced calcium influx through voltage-dependent calcium channels. The precise identity of EDHF is controversial, with chemical factors, which can diffuse from the endothelium to the smooth muscle, as well as simple electrical coupling of the two cell layers via MEGJs being proposed as the likely candidates.

In general, the importance of NO and EDHF are inversely related. Thus, NO is more important in large arteries with multiple layers of smooth muscle, whereas EDHF has been found to be more important in smaller vessels with fewer smooth muscle cell layers. EDHF would therefore be seen to play a significant role in the microcirculation. Correlation of the incidence of MEGJs and the importance of EDHF as a vasodilatory factor has led to the suggestion that EDHF activity could be more simply explained by heterocellular coupling via MEGJs. Data from a number of laboratories, using putative gap-junctional uncouplers to inhibit EDHF activity and serial section electron microscopy to quantify MEGJs, have provided convincing support for this hypothesis in a variety of vessels, including the mesenteric microcirculation. Other studies have indicated that the myogenic response is antagonized by EDHF, since they share a common target of calcium-dependent potassium channels, whose activity will reduce membrane potential and decrease the activity of the voltage-dependent calcium channels.

Conducted Vascular Responses

An important function of the microcirculation is to regulate tissue perfusion by conducting vasodilatory responses from capillaries and small arterioles to upstream feed vessels to coordinate tissue blood flow to metabolic demands. Conducted vasoconstriction has also been reported in microvessels, and in the kidney this may be important for the operation of the tubuloglomerular feedback mechanism.

Conduction of vasomotor responses has been shown to rely on the electrotonic spread of current through gap-junctional channels within the vascular wall. Selective disruption of the continuity of one or the other of the two cellular layers has confirmed that the endothelium is the principal site for conduction of vasodilation, although in some vessels vasodilatory responses can be conducted in the smooth muscle if the endothelial pathway is impaired. On the other hand, vasoconstrictor responses are conducted in the smooth muscle and not in the endothelium in some vessels, whereas in others these responses are not conducted along the vessel at all. These data are in line with the immunohistochemical results suggesting that coupling between smooth muscle cells along the length of arteries may be less effective than coupling within the endothelium.

Acetylcholine evokes a conducted vasodilation that appears to be mediated by EDHF, whereas NO, prostaglandins, and EDHF are all thought to modulate vasodilation at the local site of drug application. MEGJs are therefore considered to play an important role at the distal site in order for the hyperpolarization to be transferred to the smooth muscle and effect a relaxation through the inactivation of voltage-dependent calcium channels. The reduced conduction of endothelium-dependent vasodilatory responses in the skeletal microcirculation of mice lacking Cx40 further highlights the importance of endothelial gap junctions in these responses.

Conclusions

The preceding discussion has indicated that gap junctions play an important role in microvascular functions that can affect arteriolar tone and, in turn, systemic pressure and effective tissue perfusion. Since these functions are often compromised during vascular disease, such as hypertension, diabetes, and atherosclerosis, it may be expected that changes in expression of particular Cxs will accompany these states. Indeed, mice lacking Cx40 are hypertensive, while changes in expression of Cxs have been reported in the endothelium of large arteries in hypertensive animals (Figure 4). Increases in Cx43 have also been described early in the formation of atherosclerotic lesions, suggesting that upregulation of Cx43 may be required for the initial smooth muscle cell proliferation, while inflammatory mediators appear to modulate Cx expression and compromise conduction of microvascular responses.

Taken together, these data point to gap junctions as potential therapeutic targets for the treatment of cardiovascular diseases. More precise data regarding the control of gap-junctional function in intact tissues, the distribution of particular Cx isoforms in different parts of the microcirculation, and changes during disease will shed further light on the feasibility of this proposal.

Glossary

Conduction: Propagation of physiological responses to distal sites within the vasculature.

Cxs: Protein components of gap-junctional channels.

Endothelium-derived hyperpolarizing factor (EDHF): Putative factor producing endothelial hyperpolarization and relaxation of smooth muscle.

Gap junction: Intercellular channel connecting the cytoplasm of adjacent cells, allowing the transfer of electrical current and small ions.

Myoendothelial gap junction: Gap junction connecting endothelial and smooth muscle cells.

Myogenic response: Nerve- and endothelium-independent vasoconstriction in response to increased transmural pressure and vasodilation in response to reduced blood pressure.

Vasomotion: Rhythmical contractions of a blood vessel.

Further Reading

Christ, G. J., Spray, D. C., el-Sabban, M., Moore, L. K., and Brink, P. R. (1996). Gap junctions in vascular tissues. Evaluating the role of intercellular communication in the modulation of vasomotor tone. *Circ. Res.* **79**, 631–646. *Comprehensive review identifying the importance of gap junctions in the physiology of vascular responses.*

Davis, M. J., and Hill, M. A. (1999). Signaling mechanisms underlying the vascular myogenic response. *Physiol. Rev.* **79**, 387–423.

de Wit, C., Roos, F., Bolz, S.-S., Kirchhoff, S., Krüger, O., Willecke, K., and Pohl, U. (2000). Impaired conduction of vasodilation along arterioles in connexin 40-deficient mice. *Circ. Res.* **86**, 649–655. *First study to demonstrate a role for a specific connexin in the conduction of vasodilatory responses in the vascular wall of microcirculatory vessels.*

Hill, C. E., Phillips, J. K., and Sandow, S. L. (2001). Heterogeneous control of blood flow amongst different vascular beds. *Med. Res. Rev.* **21**, 1–60.

Hong, T., and Hill, C. E. (1998). Restricted expression of the gap junctional protein connexin 43 in the arterial system of the rat. *J. Anat.* **193**, 583–593. *Demonstration that Cx43 is absent from the media of muscular arteries.*

Sandow, S. L., and Hill, C. E. (2000). The incidence of myoendothelial gap junctions in the proximal and distal mesenteric arteries of the rat is suggestive of a role in EDHF-mediated responses. *Circ. Res.* **86**, 341–346. *First study to correlate the incidence of MEGJs with the role of EDHF in vasodilatory responses.*

Segal, S. S., Emerson, G. G., and Bartlett, I. S. (2000). Endothelium and smooth muscle pathways for conduction along resistance microvessels. In: *EDHF 2000* (P. M. Vanhoutte, ed.). London: Taylor & Francis.

Severs, N. J. (1999). Cardiovascular disease. In: *Novartis Foundation Symposium*, Vol. 219, *Gap Junction–Mediated Intercellular Signaling in Health and Disease* (N. B. Giluda, ed.), pp. 188–211. New York: Wiley. *First to question the nonspecificity of many connexin antibodies.*

Willecke, K., Eiberger, J., Degen, J., Eckardt, D., Romualdi, A., Guldenagel, M., Deutsch, U., and Sohl, G. (2002). Structural and functional diversity of connexin genes in the mouse and human genome. *Biol. Chem.* **383**, 725–737.

Capsule Biography

Dr Nicole Rummery completed her Ph.D. on the expression of vascular Cxs in arteries from normotensive and hypertensive rats under the supervision of Professor Caryl Hill, who heads the Blood Vessel laboratory at the John Curtin School of Medical Research. Research in the laboratory is focused on the involvement of gap junctions in vascular function, particularly the role played in vasomotion and in the mechanism of action of EDHF. Work is supported by the National Health & Medical Research Council and the National Heart Foundation of Australia.

CHAPTER 44

Vascular Control Mechanisms in Skeletal Muscle

Ingrid H. Sarelius
University of Rochester, Rochester, New York

In skeletal muscle, blood flow is very tightly coupled to metabolism. That is, as the metabolic rate of the muscle tissue changes (for example, during exercise) blood flow through the tissue changes proportionately as a result of changes in tone (i.e., the relative dilation) of the arterioles. In resting muscle, arterioles are partially constricted; generally this constriction is measured at 40 to 60 percent in arterioles of resting muscle. Thus there is considerable capacity in most skeletal muscle beds to change the resistance to flow through the tissue, or through some regions with respect to others. Studies such as those of Folkow and Halicka in the late 1960s established that the metabolically related increase in blood flow was greater in oxidative versus glycolytic muscles; this response differential can occur between different regions in a single muscle where the metabolism is different. These kinds of studies, especially when undertaken in isolated muscles, established that there must be direct coupling between arteriolar responses and metabolic rate, and a great deal of effort, over many decades, has identified roles for a variety of metabolically related products in vascular control. These include, for example, changed PO_2, purines (especially adenosine), K^+, H^+, lactate, and many others. Duling's group, in the 1970s, showed, using direct electrically induced contraction of individual skeletal muscle fibers, that arterioles can indeed be directly and locally coupled to a metabolic stimulus: Interestingly, it is only recently that new advances have been made in identifying mechanisms underlying this response (see Section IV). The goal of this chapter is to summarize what is known about the mechanisms underlying principally the local control of skeletal muscle arteriolar tone. Where appropriate, contributions to these local responses by integrating signals (such as those from neural inputs, or from flow) will be discussed, and important gaps in our knowledge will be identified.

Not all arterioles respond similarly to the same stimuli, and an important aspect of understanding the control of these vessels is to identify how different inputs differentially affect different parts of the microcirculation, and how this is integrated into a coordinated vascular response subserving changes in muscle blood flow. Topics that will be covered in this chapter therefore include the organization of skeletal muscle microvasculature with particular reference to the responses of different parts of the arteriolar tree; mechanisms of communication among the cells of the arteriolar wall; local metabolic response mechanisms; and control of arteriolar function via capillaries and venules.

I. Microvascular Network Organization in Skeletal Muscle

As in many other organ systems, skeletal muscle microcirculation consists of, in series, a divergently branching arteriolar network, a system of branching capillaries (the principal exchange bed for respiratory gases and nutrients), and a converging venular system. Until the early 1980s, most people viewed skeletal muscle microcirculation as a simple branching hierarchy, with each branch level being in series with those proximal and distal to it, and with all elements at any specified branch level being functionally equivalent and arranged in parallel with each other. During the 1980s and subsequently, rekindled interest in understanding the organization of skeletal muscle microcirculation, particularly arterioles and capillaries, has led to a better understanding of the organization of this system.

First, it became clear that like many other organs, skeletal muscle has prominent arcading arteriolar systems. These arteriolar arcades presumably serve to equalize the driving pressure for flow through subsequent branches across wide regions of the tissue; each arcade gives rise to several smaller arterioles that each branch two or three times and then give rise to capillaries. In general, the arcading arterioles are larger (greater than about 80 to 100 μm in diameter) than the terminally branching arteriolar systems. These larger vessels are also the arterioles that are most likely to be "paired" with collecting venules: The functional consequences of this anatomical arrangement for arteriolar control will be discussed later.

Second, the terminally branching arteriolar trees (diameter usually less than about 65 μm, down to approximately 10 μm) are not paired anatomically with venules and are the controllers of capillary perfusion (see Section VI). Capillaries arise from the distal ramifications of these arteriolar trees. They are arranged in groups (variously called networks, units, or modules), and capillary recruitment occurs when these groups of capillaries are perfused as a result of dilation of the small arterioles from which they arise.

II. The Cells of the Arteriolar Wall

All but the very largest arterioles consist of a single layer of smooth muscle cells (SMCs) wrapped circumferentially around the vessel, which is lined by a single layer of endothelial cells (ECs) oriented axially along the vessel. In addition to these major cell types, the adventitial (abluminal) aspect of the arteriole supports perivascular nerves, pericytes, fibroblasts, and a substantial extracellular matrix. All of these elements have the capacity to contribute to arteriolar control; the role of neural inputs, both central and perivascular, is best documented, but it is also now established that signals generated by the extracellular matrix can influence arteriolar tone through integrin coupling to Ca^{2+} channels, although how this might contribute to control of skeletal muscle arterioles is not known. Similarly, the contribution of pericytes (or other cell types located in close apposition to the arteriolar wall) to responses of arterioles in skeletal muscle remains largely undefined. The arteriolar wall is exposed to inputs from both the vessel lumen and the tissue space. It is established that ECs respond to mechanical inputs, of which the most prominent is flow, but other luminal stimuli have been identified. For example, red blood cells, in addition to their role in oxygen delivery, have been postulated to act as modifiers and/or sources of ATP and NO. On the abluminal side of the arteriole, skeletal muscle myocytes are a primary source of vasoactive agents, but additionally, the arteriole can respond to external mechanical stimuli, such as compression of microvessels induced by muscle contraction, and will be of course be influenced by neural inputs. Overall, these arterioles are exquisitely responsive to a very wide range of inputs; redundancy of control mechanisms clearly confers survival advantages on this important system but requires that the microcirculatory network also have the capacity to integrate all this information to produce appropriately coordinated responses. Thus, an important emerging issue is that of identifying the mechanisms by which many inputs are integrated locally, and perhaps of even more significance, how they are integrated over the length of individual arteriolar segments, and across the microvascular network.

III. Differential Responses of Large versus Small Arterioles

Skeletal muscle arterioles, like those of other organs, display myogenic tone, that is, in response to the transmural (distending) pressure exerted on the wall by the blood, the SMCs increase their degree of constriction. How this is manifest at any level of the microcirculation depends on the transmural wall stress, which is a function of both the pressure and the vessel radius. Thus the myogenic effect should be less in smaller vessels; despite this, smaller arterioles typically have higher tone those located more proximally. This reflects longitudinal differences in the ability of arterioles to respond to other regulatory mechanisms (local, neural). Differential sensitivities of larger versus smaller arterioles to a variety of stimuli is an important way in which integration of responses is achieved in the arteriolar microcirculation. In skeletal muscle microcirculation, perivascular nerves and varicosities are found in association with arterioles of all sizes, but arteriolar sensitivity to adrenergic inputs is not uniform across the network. Both α_1 and α_2 adrenergic receptors have been identified on large arterioles, while α_2 receptors appear to predominate on smaller arterioles. Furthermore, most studies suggest that larger vessels (sometimes classified as first- and second-order vessels, in contrast to the smaller third- through fifth-order vessels) are more susceptible to neural inputs than are smaller arterioles; third- through fifth-order arterioles are in turn more responsive to metabolic stimuli than are larger arterioles. In like manner, oxygen sensitivity appears to predominate in smaller arterioles, while flow-dependent responses are more prominent in first- and second-order vessels. Notwithstanding these gradations in response characteristics as one moves distally through the network toward smaller arterioles, most arterioles do exhibit some degree of responsiveness to all these inputs. An important unanswered issue is to identify which of these differences are an acute reflection of the local conditions, and which reflect phenotypic differences in response capability.

IV. Metabolic Regulation of Arteriolar Tone

A. Oxygen

Oxygen has consistently been implicated in regulation of arteriolar tone, and although involvement of changes in P_{O_2} in regulation of skeletal muscle function is intuitively compelling, no obvious "oxygen sensor" has been identified despite decades of searching. Thus it has seemed unlikely that oxygen has a direct effect on the arteriolar wall. However, recent work has revisited this issue, and although the signaling mechanisms are still not clarified, it is clear that at least in some arterioles, including those in skeletal muscle, oxygen can directly affect the cells of the arteriolar wall via a cytochrome P450–dependent pathway, and also possibly by mitigation of adrenergic vasoconstrictor tone at a step in the SMC signaling cascade that is distal to an increase in Ca^{2+}. There is compelling evidence that K_{ATP} channels are implicated in local metabolic responses, but this does not appear to be via a direct effect of oxygen on the function of these channels in SMCs.

Most investigators have concluded that oxygen modulates local arteriolar responses indirectly—that is, as muscle metabolism changes, local PO_2 will also change, as will the balance of metabolites released from the contracting muscle. It is thus easy to conceptualize a control system in which some or all such products feed back on the arteriole to produce changes in force development by the SMCs. Using this logic, a variety of possible mediators have been identified. The challenge is to demonstrate directly what role any such agents play on the cellular mechanisms regulating tone in the blood vessel wall.

B. Adenosine

By reason of its release from contracting muscle fibers during decreased Po_2, adenosine (ADO) has long been implicated in metabolic control. There is evidence from direct observations of small arterioles that ADO is indeed involved in the local metabolic response. However, it accounts for only a fraction of the dilation in response to near-maximal muscle contraction and appears to have a greater effect in smaller arterioles. This might explain why studies of whole organs or in intact animals have not invariably implicated this purine in metabolic dilation. It is widely assumed that ADO acts on smooth muscle, and it is established that SMCs in skeletal muscle arterioles display both P_1 and P_2 purinergic receptors; however, purinergic receptors are also found on ECs, and recent work suggests that local metabolic responses in small arterioles may be coupled through P_1 receptors on ECs. Purines are known to increase EC Ca^{2+}, and this increase in EC Ca^{2+} has also been implicated in mediation of metabolic vasodilation in small arterioles. In some circulations (e.g., coronary) ADO acts via K_{ATP} channels, but in skeletal muscle arterioles this does not appear to be the case.

C. Ions, Particularly K^+

Many ions have vasoactive properties and have been identified as possible mediators of metabolic vasodilation. These include potassium, hydrogen, and overall osmolarity, as well as charged moieties such as lactate, all of which have been detected in venous blood during changes in muscle metabolism. Of these, raised extracellular $[K^+]$ has received the most attention, and it is established that small increases in $[K^+]$ of the order of 5 to 10 mM can lead to vasodilation in a variety of arterioles, including those from skeletal muscle. The mechanism(s) by which this dilation is brought about are still unclear: Evidence supporting a role for K_{IR} channels and for Na^+/K^+ ATPase activation has been published, but more definitive work in this area is needed.

D. Other Molecules Contributing to Functional Hyperemia

There is considerable evidence implicating nitric oxide (NO) as a signaling intermediate—local arteriolar dilation to muscle contraction can be significantly attenuated in the presence of blockers such as L-NAME. Several different actions of NO have been identified. It appears to support vascular dilation directly, possibly via a K_{ATP} channel-dependent signaling pathway. Studies in NOS-deficient mouse models suggest that at least part of the NO is derived from nNOS located in skeletal muscle fibers; contributions from eNOS have also been identified, but there is a lot still to be learned about the mechanisms underlying this response. In many systems, NO release from ECs has been implicated in flow-dependent responses, and it is possible that this stimulus also contributes to NO release during blood flow changes associated with changes in muscle metabolism. NO also attenuates sympathetic vasoconstricion in an action known as "sympatholysis": Again, this mechanism is not fully understood, but this aspect of the contribution of NO to vasodilation also appears likely to involve K_{ATP} channels. Whether they are located on SMCs, ECs, or the skeletal muscle fibers remains to be determined.

It has also been shown that the arteriolar dilation to muscle contraction is partially blocked by indomethacin, implicating prostaglandins in the regulation of this function. As both NO and prostaglandins have been implicated in flow-dependent responses in skeletal muscle, a challenge will be to sort out which aspects of this pathway are primary versus secondary as a consequence of local changes in the prevailing flow that are themselves a consequence of local changes in arteriolar tone. Recent work has shown, not surprisingly, that multiple local vasoactive pathways can be activated simultaneously upon contraction of skeletal muscle, and at least some of these are independent of any local hemodynamic changes.

V. Integration of Metabolic Response Pathways

A. Myoendothelial Communication

In addition to considerations about which control mechanisms for arteriolar responses reside on SMCs versus ECs, it has become apparent that these two cell types can be closely coupled, thus enabling the arteriolar wall to act as a syncytium. In small arterioles there are gap junctions between adjacent ECs and between adjacent SMCs; in addition the two cell types are connected by myoendothelial gap junctions. This arrangement allows signals to spread axially along the vessel, and also to be communicated directly between ECs and SMCs. Local paracrine signaling from ECs to SMCs is, of course, well established (NO, EDHF), but it is now clear that intracellular transfer of signaling intermediates can also occur from SMCs to ECs. It has recently been shown that the increased SMC Ca^{2+} that occurs when arterioles constrict can be transferred to ECs, raising Ca^{2+} in these cells and contributing to NO release, which in turn ameliorates the original vasoconstriction. How this response pattern contributes to in situ arteriolar responses is not yet clear, but it indicates that vessels in their native environment can respond in an even more closely integrated way than previously thought.

B. Axially Propagated Responses

In the past two decades, a great deal of work, much of it in skeletal muscle arterioles, has established that in these vessels, responses can be propagated axially over considerable distances (millimeters) from their point of local origin. Most of this work has focused on understanding the mechanisms underlying dilations and constrictions conducted axially as a result of local stimulation by acetylcholine and phenylephrine, respectively. These agents induce hyper- and hypopolarizations, respectively, that are transmitted axially along the vessel in a manner that appears to be dependent largely (although not necessarily entirely) on gap-junction communication. Although the principal pathway for this transmission appears to be ECs, signals can also be transmitted via a SMC pathway, and an important task will be to identify what control mechanisms determine whether transmission is via ECs or SMCs or both cell types. Not all local vasomotor events result in a propagated signal; for example, it has been shown that contraction of skeletal muscle fibers causes a local dilation that involves ADO, NO, and a separate, K_{ATP} channel–dependent pathway, but that the NO-dependent component of this response is not propagated axially, whereas the ADO and K_{ATP} channel–dependent components are. Similarly, locally applied purines induce a mixed response, constriction followed by dilation, but only the dilator component is propagated axially. Not all of these responses appear to depend entirely on gap-junctionally mediated transmission, but what other signaling pathways are involved is not known at this time. Initial work suggests that the remote response to purines is dependent on increased EC Ca^{2+} at the remote site, but how this relates to the signaling cascade is entirely unknown. Despite the fact that there is much to be learned about the mechanisms underlying these communicated responses, their potential contribution to integration of skeletal muscle responses is clear. If only local dilations were produced in muscles in response to contraction of individual motor units, then at submaximal motor unit recruitment one might expect arterioles to exhibit a series of randomly located local dilations that would not support changes in flow through the vessel: Conduction of these dilations along the arteriolar wall allows the vessel to respond as a coordinated entity, thus facilitating decreased resistance to flow across the vessel (and network region) as a whole.

C. Flow-Dependent Dilations

Skeletal muscle arterioles are sensitive to flow, and, interestingly, larger rather than smaller vessels are more responsive to this stimulus (compared to the conducted responses discussed earlier, which appear to be most prominent in smaller arterioles). It has also been shown that ascending flow-dependent dilations can be propagated proximally to encompass the inflow vessels to the tissue, thus enabling flow to be increased to the entire tissue. Thus the picture that is emerging is one in which small arterioles, by reason of their electrical and/or local paracrine coupling, are able to exhibit responses that are conducted beyond their region of origin to enable the vessel to respond as a coordinated whole, while larger arterioles, being flow sensitive, exhibit an ascending dilatory behavior that subserves increased flow into the tissue under conditions of increased metabolism.

VI. Control of Capillary Recruitment

Capillaries, having no smooth muscle, are not generally thought to actively participate in blood flow control. However, in a series of studies in the 1980s and 1990s, Tyml and others showed that capillaries in skeletal muscle are capable of responding to vasoactive stimuli such as norepinephrine by producing vasoconstriction elsewhere in the network and hence decreasing capillary flow. The implication was that these signals were axially propagated along the vessel wall. In the 1990s, Sarelius and coworkers showed that capillaries could sense contraction of underlying skeletal muscle fibers and send a dilator signal to upstream arterioles, thus increasing flow through the "active" capillary network. They showed that this signal was specific to capillaries associated with actively contracting muscle fibers. This mechanisms explains how capillary recruitment can be matched closely with metabolic requirements, despite the fact that the controllers of these vessels are located upstream of the capillaries that they control, and often remote from the contracting fibers.

VII. Venular–Arteriolar Communication

As noted in Section I, larger arterioles are located anatomically in a "pair" with larger collecting venules. This raises the possibility that control mechanisms may be coupled from these outflow venules to their neighboring arterioles—an attractive idea that conceptualizes how changes in venous outflow due to changes in metabolism could be coupled back locally within the tissue to change inflow capacity. The idea that venous Po_2 may be sensed in this way by inflow arterioles has been raised repeatedly for many decades. In recent years, veno-arteriolar coupling of responses to metabolites such as adenosine has been clearly demonstrated, indicating that this means of communication within the muscle has the capacity to modify inflow to the tissue in a way that, at least conceptually, should serve to match blood flow to metabolism more closely.

VIII. Summary

Blood flow control in skeletal muscle achieves very close coupling between flow and the metabolism of the tissue. Integration of control is achieved by several means. There are differences in response capacity to many stimuli between large and small arterioles. Communication of signals among the cells of the arteriolar wall, both axially and between SMCs and ECs, enables the vessel and network to respond as a coordinated whole. Ascending flow-dependent dilation has also been identified as a mechanism supporting rapid modulation of the inflow to the tissue. Cellular signaling mechanisms underlying these responses are still being worked out: In addition to roles for such metabolically related substances as oxygen and adenosine, contributions from NO, and one or more EDHFs, have been identified. A role for K_{ATP} channels is strongly indicated, but where in the signaling cascade these channels contribute to the response is not clear.

Glossary

Arcading arteriolar system: Arterioles that form a continuous loop from which smaller arterioles branch, in contrast to a bifurcating arteriolar system, in which an arteriole diverges into two smaller branches.

Gap-junction communication: Transfer of signals between the cytosols of two adjacent cells via a channel in their membranes.

NOS: Nitric oxide synthase, the enzyme responsible for production of nitric oxide. (eNOS is an isoform found in endothelial cells; nNOS, an isoform found in skeletal muscle.)

Transmural wall stress: The (normalized) tension in the vessel wall that resists the pressure exerted by the blood in the vessel lumen.

Further Reading

Bolz, S. S., Fissithaler, B., Pieperhoff, S., De Wit, C., Fleming, I., Busse, R., and Pohl, U. (2000). Antisense oligonucleotides against cytochrome P450 2C8 attenuate EDHF mediated Ca^{2+} changes and dilation in isolated resistance arteries. *FASEB J.* **14**, 255–260.

Dora, K. A. (2001). Intercellular Ca^{2+} signalling: The artery wall. *Semin. Cell. Dev. Biol.* **12**, 27–35. *A good current source of how Ca^{2+} signaling is involved in communicated responses in the walls of larger arterioles and small arteries.*

Gorczynski, R. J., and Duling, B. R. (1978). Role of oxygen in arteriolar functional vasodilation in hamster striated muscle. *Am. J. Physiol.* **235**, H505–H515. *This is included because it is a very elegant study showing direct coupling between skeletal muscle contraction and arteriolar responses.*

Lash, J. M. (1996). Regulation of skeletal muscle blood flow during contractions. *Proc. Soc. Exp. Biol. Med.* **211**, 218–235. *A comprehensive review.*

Laughlin, M. H., and Korzick, D. H. (2001). Vascular smooth muscle: Integrator of vasoactive signals during exercise hyperemia. *Med. Sci. Sports. Exerc.* **33**, 81–91. *Another comprehensive review with a different emphasis than the Lash review.*

Murrant, C. L., and Sarelius, I. H. (2000). Coupling of muscle metabolism and muscle blood flow in capillary units during contraction. *Acta Physiol. Scand.* **168**, 531–541.

Segal, S. S., Damon, D. N., and Duling, B. R. (1989). Propagation of vasomotor responses coordinates arteriolar resistances. *Am. J. Physiol.* **256**, H832–H837.

Segal, S. S., and Jacobs, T. L. (2001). Role for endothelial cell conduction in ascending vasodilatation and exercise hyperaemia in hamster skeletal muscle. *J. Physiol. (London)* **536**, 937–946.

Capsule Biography

Dr. Sarelius is a Professor of Pharmacology and Physiology and of Biomedical Engineering at the University of Rochester. Her laboratory focuses on cell communication in the arteriolar wall, and on signaling between blood cells (primarily leukocytes) and the microvessel wall. Her work is supported by grants from the National Institutes of Health.

CHAPTER 45

Protein Kinases and Microvascular Permeability

Mack H. Wu and Sarah Y. Yuan
University of California School of Medicine, Sacramento, CA

Blood–tissue exchange occurs in the downstream segments of microvasculature, namely capillary and postcapillary venules. The wall of these exchange microvessels consists of a sheet of cells, endothelial cells, which connect to each other with closely opposed intercellular junctions. This endothelial sheet is tethered to a network of extracellular matrices in the basement membrane, forming a barrier structure that controls the transvascular movement of blood components. The permeability of microvascular endothelium is maintained by an equilibrium between the contractile force generated at the endothelial cytoskeleton and the adhesive forces produced at endothelial cell–cell junctions and cell–matrix focal contacts (Figure 1). A dynamic interaction among these structural elements controls the opening and closing of the paracellular pathways for fluid and cells to move across the vessel wall and thus serves as a fundamental mechanism in the physiological and pathological regulation of blood–tissue exchange.

Aberrations of microvascular barrier function lead to an abnormal extravasation of blood components, resulting in tissue edema and dysfunction. This cellular process is involved in inflammation, ischemia-reperfusion injury, trauma, sepsis, diabetes, and atherosclerosis. Under such disease conditions, many inflammatory mediators are released, and most of them possess a potent permeability-increasing effect. These mediators, including histamine, bradykinin, platelet activating factor, growth factors, glycation products, cytokines, reactive oxygen species, and activated leukocytes, can interact with the microvascular endothelium and activate an array of intracellular second messengers, which in turn catalyze a series of biochemical reactions at the endothelial cytoskeleton and cell–cell and cell–matrix adhesive structures (Figure 2). Ultimately, endothelial cells contract or undergo conformational changes, resulting in endothelial hyperpermeability and microvascular barrier dysfunction.

Protein kinases, classified as serine/threonine kinases or tyrosine kinases based on their substrate specificity, constitute an important category of intracellular second messengers that mediate the structural and functional changes occurring in the endothelium under physiological conditions as well as during inflammatory stimulation. Several groups of protein kinases have been characterized and their involvement in the cellular response to inflammation or injury has been documented. The following discussion focuses on the protein kinases that are frequently implicated in the modulation of microvascular permeability.

Myosin Light-Chain Kinase (MLCK)

Actin and myosin are the major contractile components in the cytoskeleton. The cross-bridge movement between actin and myosin provides a mechanical basis for not only the maintenance of centripetal tension but also the development of contractile force in cells during physical or chemical stimulation. In vascular endothelial cells, the interaction between actin and myosin is mainly governed by the phosphorylation status of the regulatory myosin light chain (MLC). Two major mechanisms control the activity of MLC. On one hand, myosin light-chain kinase directly phosphorylates MLC at Thr-18 and/or Ser-19, resulting in MLC activation and actin–myosin binding. On the other hand, myosin-associated protein phosphatase dephosphorylates MLC and thus counteracts the MLCK effect, leading to cell relaxation. Certain types of agonists and cells can

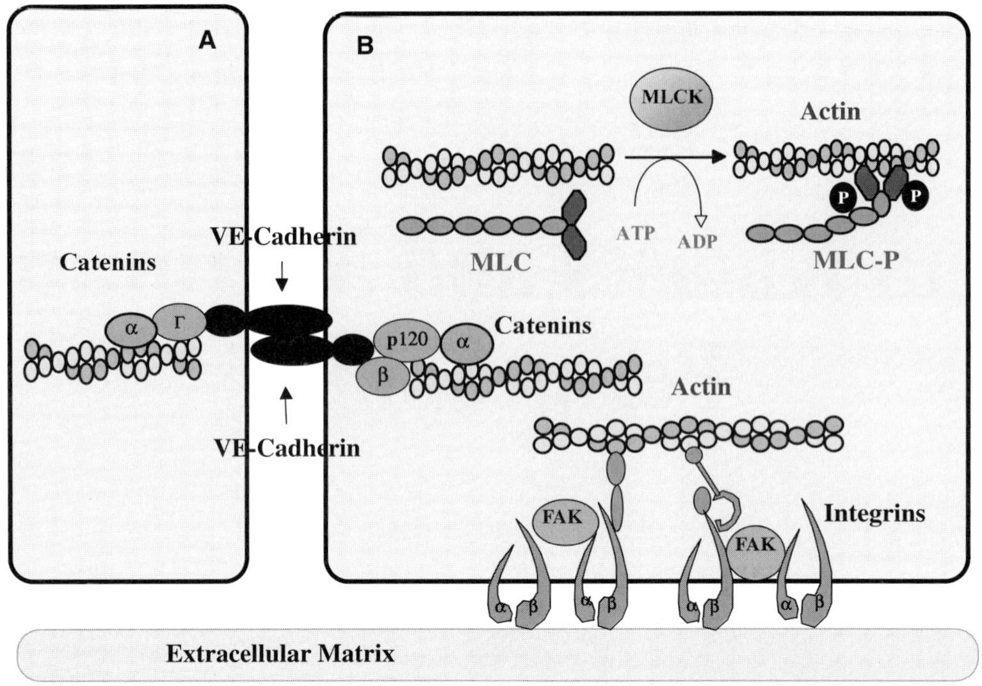

Figure 1 A schematic diagram of microvascular endothelial barrier structure. The barrier is formed by a layer of endothelial cells that connect to each other (e.g., cell A to cell B) through the junctional adhesive molecule VE-cadherin, which binds to another VE-cadherin in the junction and connects to actin cytoskeleton via a family of catenins (αβγ, and p120). This endothelial lining is tethered to the extracellular matrix through the binding of transendothelial receptor integrins (composed of α and β subunits) and a family of cytoskeleton-linking proteins including focal adhesion kinase (FAK). The integrity of this barrier structure is maintained by VE-cadherin-mediated cell–cell adhesions and focal adhesion-supported cell–matrix attachment, whereas myosin light chain kinase (MLCK)-catalyzed myosin light chain (MLC) phosphorylation promotes cross-bridge movement between actin and myosin leading to cell contraction. A dynamic interaction among these structural elements controls the opening and closing of the paracellular pathways for fluid, proteins, and cells to move across the endothelium.

activate the contractile process by increasing MLCK activity through calcium/calmodulin signaling or by directly phosphorylating MLCK. Other agents are able to stimulate actomyosin contraction by inhibiting MLC phosphatase activity. In particular, a family of Rho-like small GTPases and their downstream effector, Rho kinase, have been implicated in the regulation of MLC phosphorylation through inhibition of MLC phosphatase.

The functional importance of the contractile mechanism in controlling microvascular permeability has been subject to extensive investigation. Recent studies provide strong evidence for a link between MLC phosphorylation and stress fiber formation or tension development in the endothelial cells. A causal role of MLCK in the endothelial hyperpermeability response to inflammatory mediators has been established through experiments in cultured endothelial monolayers as well as in intact isolated venules, where administration of pharmacological agents or transference of synthetic peptides that specifically block MLCK function can prevent the increase in permeability caused by thrombin, histamine, cytokines, oxygen radicals, and activated neutrophils. Furthermore, inhibition of MLCK or knockout of endothelial MLCK genes has proven to effectively attenuate microvascular leakage under clinically relevant conditions such as severe burns and sepsis. These findings confirm that endothelial MLCK plays a critical role in mediating microvascular barrier dysfunction during inflammation and injury.

Protein Kinase C (PKC)

Protein kinase C represents a family of at least 10 serine/threonine kinases (α, β1, β2, γ, δ, ε, ν, θ, μ, ξ, and λ). These isozymes show distinct expression patterns and are responsible for diverse cellular responses. Most of them can be activated by phorbol esters and diacylglycerol (DAG), as experimentally indicated by membrane translocation and increased phosphorylating activities. PKC activity is negatively regulated by serine/threonine protein phosphatases.

Functionally, despite its perhaps minor effect on the basal barrier property of the endothelium, PKC has been frequently referred to as a key mediator of microvascular hyperpermeability under stimulated conditions. Direct activation of PKC with DAG or phorbol esters increases the flux of fluid and macromolecules across the microvascular endothelium, whereas PKC inhibitors reduce the increases

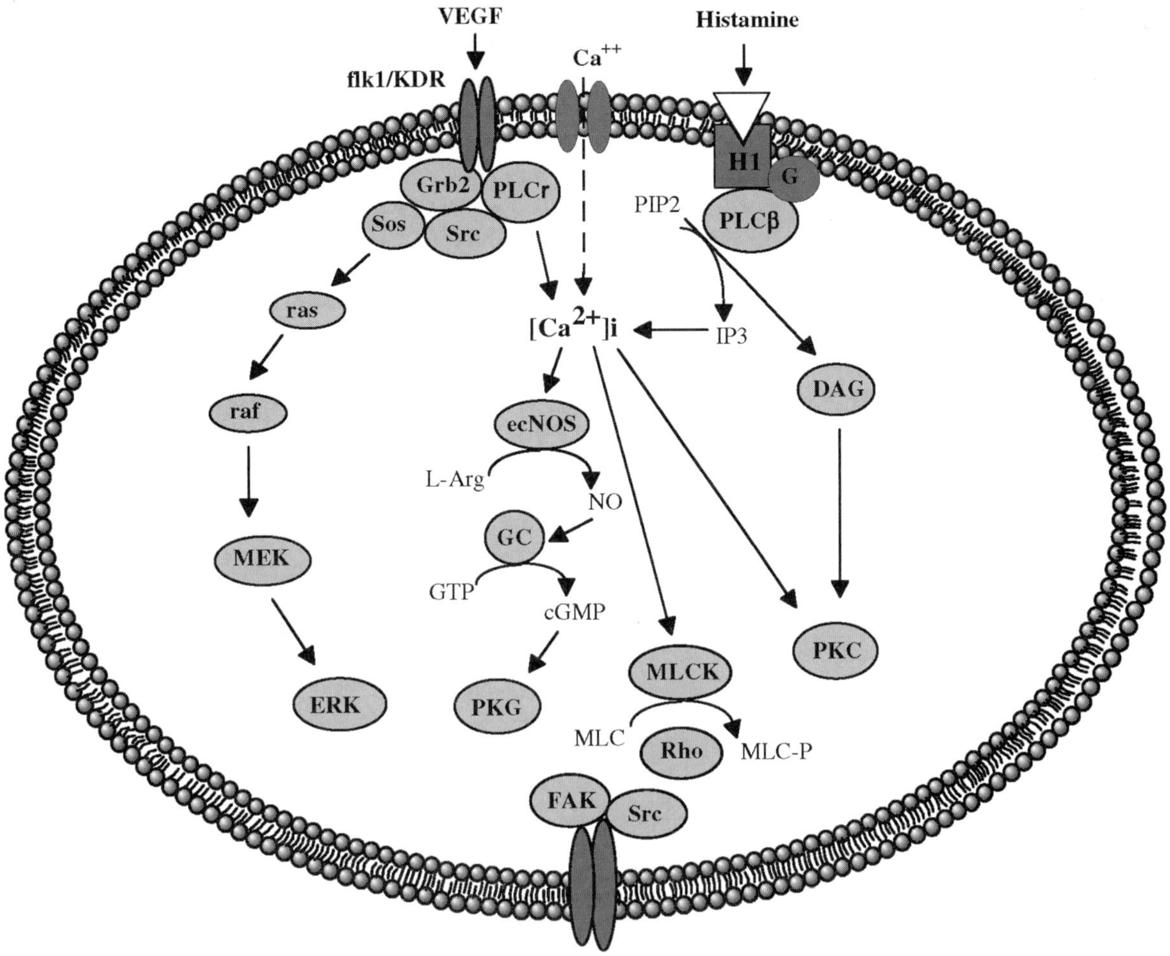

Figure 2 A model for the signal transduction of agonist-induced microvascular hyperpermeability. Histamine and vascular endothelial growth factor (VEGF) are used to represent permeability-increasing agonists. Histamine binding to its receptor (H) activates phospholipase C beta (PLCβ) via a G protein (G), catalyzing the production from inositol bisphosphate (PIP_2) to inositol trisphosphate (IP_3) and diacylglycerol (DAG). DAG directly activates protein kinase C. In parallel, IP_3 stimulates an internal release of Ca^{2+} followed by an influx of extracellular Ca^{2+}, leading to upregulation of nitric oxide synthase (ecNOS) and nitric oxide (NO) production from its precursor Larginine (L-Arg). Nitric oxide stimulates guanylate cyclase (GC) to produce cGMP, a potent activator of cGMP-dependent protein kinase (PKG). The elevated intracellular calcium also stimulates PKC and myosin light-chain kinase (MLCK); the latter phosphorylates myosin light chain (MLC-P). The hyperpermeability effect of VEGF is triggered by its tyrosine kinase receptor flk1/KDR and mediated through two major signaling pathways. On one hand, activated phospholipase C gamma (PLCγ) elevates the intracellular calcium and subsequently activates the PKG, PKC, and MLCK cascades. On the other hand, KDR receptor occupancy results in the recruitment of several adaptor proteins, such as Grb2, Sos, and Src, which further produces the ras-raf-MEK-ERK cascade. In addition, the nonreceptor tyrosine kinases FAK and Src as well as the Rho-family of GTPases are also involved in the endothelial response to inflammatory stimulation. All of these protein kinases—PKG, PKC, MLCK, MAPK, and tyrosine kinases—may directly or indirectly target the endothelial barrier structure to cause an opening of the paracellular pathway for transendothelial flux of fluid and macromolecules. (see color insert)

in endothelial permeability caused by thrombin, bradykinin, platelet activating factor, hydrogen peroxide, VEGF, or neutrophils. Importantly, clinical and experimental evidence is accumulating that PKC activation and subsequent microvascular barrier dysfunction may underlie the initiation and progress of circulatory disorders associated with diabetes, atherosclerosis, and ischemia-reperfusion injury.

Recent advances in molecular technologies enable direct targeting of particular PKC isoforms by selective inhibitors, synthetic peptides, or antisense oligonucleotides. Experiments employing these approaches reveal the relative importance of PKCα, PKCβ, PKCδ, and PKCμ (also known as PKD) in the regulation of microvascular permeability. Some studies demonstrate that endothelial hyperpermeability resulting from hyperglycemia, ischemia, and angiogenesis or inflammatory stimulation is a PKCα-dependent event. In contrast, others suggest that PKCβ overexpression in endothelial cells potentiates agonist-induced transendothelial flux of macromolecules, whereas pharmacological or antisense inhibitors of PKCβ diminish the increase in endothelial permeability caused by phorbol esters or oxygen radicals. Oral administration of LY333531 or LY290181, selective pharmacological antagonists of PKCβ, prevents microvascular leakage in the retina and kidney of diabetic animals. More recently, evidence is emerging that PKCδ and PKD may be required for the microvascular permeability response to phorbol ester and diacylglycerol, typical exogenous and endogenous activators of PKC, respectively.

How PKC alters endothelial barrier function remains an interesting question. Both the cytoskeleton and cell–cell junctions are potential targets of this potent kinase. More specifically, it has been reported that PKC activates the endothelial contractile apparatus by inducing MLC phosphorylation and actin polymerization. Recent experiments suggest that the endothelial contractile response can be triggered by a PKC-dependent activation of the Rho pathway. Furthermore, disassembly of endothelial adherens junctions has been linked to PKC activation. In addition to its direct effect on the cytoskeletal and junctional structures, PKC may indirectly alter the endothelial barrier function through crosstalk with other intracellular signaling molecules. In this regard, the nitric oxide (NO) pathway has been increasingly recognized as a potential downstream target of PKC. An in vivo experiment in the hamster cheek pouch demonstrates that phorbol ester–induced macromolecular transport in the microcirculation is reduced when NO synthesis is blocked. This is further supported by evidence of enhanced nitric oxide synthase (NOS) phosphorylation and NO production in microvessels or endothelial cells exposed to PKC activators. A comparative analysis of the permeability response to PKC activators in isolated venules before and after NOS inhibition indicates that the barrier loosening effect of PKC is mediated, at least in part, through the endothelial production of NO.

Cytosolic calcium is an important cofactor required for the regulation of PKC function. Within this context, PKCα and PKCβ are the predominant isozymes that are dependent on calcium for activation. Many permeability-increasing agonists are able to stimulate the release of calcium from intracellular stores followed by influx from extracellular space. This process, in addition to directly activating PKC, can upregulate the activity of other enzymes and second messengers in endothelial cells, such as MLCK and NO, which more likely serve as intracellular signals secondary to the PKC pathway in the transduction of endothelial permeability response. Currently, studies are ongoing to define the precise role of cytosolic calcium in the complex interactions among multiple protein kinases in microvascular endothelial cells upon inflammatory stimulation.

cAMP- and cGMP-Dependent Protein Kinases (PKA and PKG)

In general, an increase in the intracellular level of cAMP promotes endothelial barrier integrity, and the effect is most likely mediated through PKA. Studies have been carried out in cultured cells, perfused microvessels, and intact tissues in which application of PKA activators, such as β-adrenergic agonists or cAMP analogs, reduces endothelial permeability and prevents microvascular leakage during inflammation and ischemia-reperfusion injury. Recently, an endogenous PKA competitor, PKI, has been identified in vascular endothelial cells and is being used for selective inhibition of PKA. Infection of human dermal microvascular endothelial cells with adenovirus containing PKI gene results in overexpression of PKI and abrogates the cAMP-mediated protection against increased endothelial permeability.

The barrier protection effect of PKA may be related to its ability to stabilize endothelial cytoskeletal and adhesive structures. There are reports that PKA causes dephosphorylation of MLC, dissociation of F-actin from myosin, stabilization of cytoskeletal filaments, and strengthening of cell–matrix adhesions. Moreover, PKA is well known for inhibiting leukocyte adhesion and platelet aggregation. This effect may play an indirect role in preventing the endothelial barrier from being damaged by activated leukocytes or platelets and their metabolites. Finally, it is possible that the PKA pathway functions through interactions with other signaling molecules such as PKG.

In contrast to the general consensus on the barrier-tightening effect of PKA, there is considerable controversy as to whether the cGMP-PKG cascade acts as a barrier protector or a permeability-increasing factor. Most in vitro experiments using cultured endothelial cells derived from large vessels or nonexchange microvessels show that an elevation of intracellular cGMP by NO donors or guanylate cyclase activators is associated with a decrease in endothelial permeability. On the other hand, in vivo observation and in vitro studies using microvascular endothelial cells report a PKG-dependent increase of microvascular permeability in response to a variety of agonists, including NO donors, histamine, bradykinin, tumor necrosis factor, platelet activating factor, and VEGF. These apparently contradictory findings may result from variations in cell types and experimental conditions. With regard to experimental models, cultured cells may not fully mimic the in vivo state of microcirculation. For example, it has been reported that PKG expression is dramatically reduced in cultured endothelial cells during serial passage. Whether the cells are exposed to a physiological range of shear stress and a normal calcium environment also affects the expression and activity of the kinase. Recent experiments in intact perfused microvessels show that selective inhibition of NOS, guanylate cyclase, or PKG suppresses venular hyperpermeability caused by shear stress, histamine-type agonists, and VEGF. It has been postulated that certain types of physical and chemical stimuli increase microvascular permeability by activating the NO-cGMP-PKG signaling cascade. Such a mechanism may account for the direct effect of the stimuli on microvascular endothelium where leukocytes and platelets are absent.

It seems that PKA and PKG form a pair of yin-and-yang-like functional antagonists in controlling microvascular permeability. Protein kinase A may play a dominant role in the maintenance of basal barrier property and counteract the effect of PKG in response to inflammation. The interaction between the two protein kinases can occur at several levels: competing for a common target at the cytoskeleton or junctions, counteracting to modulate leukocyte and platelet function, or feedback regulating through phosphodiesterases (PDE). Three forms of PDE, namely, cGMP-stimulated PDE II and IV and cGMP-inhibited PDE III, have been shown to affect PKA-associated alteration in endothelial permeability.

Mitogen-Activated Protein Kinases (MAPKs)

Mitogen-activated protein kinases are involved in multiple cascades of serine/threonine and tyrosine phosphorylating reactions that mediate diverse cellular responses to growth factors, physical stress, and cytokines. The most studied MAPKs in mammalian cells are extracellular signal-related kinases (ERK 1/2), c-jun N-terminal kinases (JNK), and p38 proteins. Although ERK 1/2 is mainly known for its role in cell growth and p38 MAPK plays a major role in cellular responses to stress and injury, both have been implicated in the regulation of vascular endothelial permeability. A typical case is seen in endothelial cells or intact microvasculature subjected to VEGF stimulation: Inhibition of ERK 1/2 with the ERK kinase inhibitor PD98059 or inhibition of p38 with SB203580 blocks VEGF-induced increases in permeability. In addition to the growth factor, many inflammatory agonists, including histamine, thrombin, hydrogen peroxide, and intracellular calcium elevating agents, are able to phosphorylate ERK 1/2. In parallel, the same types of substances have been found to induce endothelial cytoskeletal contraction and junctional barrier failure through a p38-related mechanism. Clearly, activation of either the ERK 1/2 or the p38 pathway can cause changes in endothelial barrier function; however, their relative contribution may vary depending on the form of stimulation, type of vessels, and duration of observations.

Nonreceptor Protein Tyrosine Kinases (PTKs)

Phosphorylation of proteins on serine/threonine residues is often initiated by or coupled with tyrosine phosphorylation. In fact, tyrosine phosphorylation has so far been considered the primary or even the exclusive indicator of signal transduction in multicellular organisms. Enhanced PTK activity has been implicated in not only cancer and proliferative diseases, but also in inflammatory disorders. Inhibition of PTKs effectively blocks the increase in endothelial permeability induced by a wide spectrum of inflammatory mediators, whereas upregulation of tyrosine phosphorylation with tyrosine phosphatase inhibitors exerts opposite effects.

Protein tyrosine kinases can be categorized into receptor tyrosine kinases, such as the VEGF receptor KDR, and nonreceptor tyrosine kinases, such as Src kinases and focal adhesion kinase (FAK). The Src family, containing c-Src, Lyn, Fyn, Lck, Hck, Fgr, Blk, and Yes, exerts potent phosphorylating and transforming effects through interaction with the SH2 and SH3 domains of effector proteins. The activity of Src is upregulated by phosphorylation at Tyr-416 (located in the catalytic domain) and negatively regulated by phosphorylation at Tyr-527 (near the carboxyl terminus). It has been documented that the Src signaling is required in multiple cellular processes, such as cell contraction and migration, angiogenesis, and vascular leakage. Blockage of Src activity or Src deficiency reduces cerebral edema associated with stroke in animal models. Recent experiments show that Src mediates vascular endothelial permeability responses to tumor necrosis factor and reactive oxygen metabolites. In addition, specific blockage of Src activity by pharmacological agents or Src-inhibiting peptides abolishes the increase in albumin permeability caused by C5a-activated neutrophils in isolated venules as well as in cultured venular endothelial monolayers. Further biochemical analysis confirms that activated neutrophils stimulate Src phosphorylation at Tyr-416 and decrease phosphorylation at Tyr-527, two events that are known to upregulate Src activity.

Several pathways may be activated through Src signaling. For example, MAPK has been identified as a downstream effector of Src. More importantly, Src may directly alter endothelial barrier structure by phosphorylating contractile proteins and adhesion proteins. Supporting this is biochemical evidence of coimmunoprecipitation and immunocolocalization between Src and the cytoskeletal and junctional components. Myosin light chain kinase, β-catenin, and FAK have been identified as potential Src substrates. Therefore, it is highly possible that Src activation serves as a common signal in coordinating cytoskeletal contraction, junctional disorganization, and focal adhesion redistribution in response to multiple protein kinase cascades that are triggered by inflammatory mediators.

Another nonreceptor tyrosine kinase known to modulate endothelial barrier structure and function is FAK. Focal adhesions are referred to as transmembrane structures that anchor the cells to extracellular matrices in the basement membrane. Major components of the focal structure include a family of transmembrane receptors, namely integrins, and their associated intracellular proteins, such as talin, α-actinin, and paxillin, which link the cytoskeleton to integrins. Not only does this structure support the physical attachment of endothelial lining to the basement membrane, it also serves as a transducer in mediating cellular response to physical stress or chemical stimuli. Because of the lack of catalytic activity in integrin receptors, such signaling reactions are transduced via a network of integrin-associated proteins located in focal adhesions. Of these proteins, FAK is a major protein kinase capable of catalyzing various downstream signaling cascades leading to a dynamic regulation of focal adhesion morphology and distribution. The activity of FAK is mainly regulated through tyrosine phosphorylation, preferentially by the Src-family tyrosine kinases. FAK-mediated focal adhesion assembly and reorganization play a critical role in cell contraction and migration during stimulation by integrins or nonintegrin factors including shear stress, growth factors, and permeability-increasing agents such as thrombin, platelet activating factors, phorbol esters, and even leukocyte adhesion molecules. Recent evidence is accumulating in support of the functional importance of FAK to the microvascular barrier effect. For example, there are reports that protein tyrosine kinase blockades abolish FAK phosphorylation-coupled endothelial barrier dysfunction. Downregulation of FAK expression using antisense approaches modulates agonist-induced increases in endothelial monolayer permeability. More importantly, inhibition of FAK activity through direct delivery of FAK-related nonkinase (FRNK), an endogenously expressed FAK competitive inhibitor, into intact isolated venules significantly attenuates the hyperpermeability response to VEGF or activated leukocytes.

Several pathways may be considered for potentially underlying the effect of FAK on endothelial barrier function. On one hand, FAK-signaled focal adhesion disassembly and redistribution may reduce cell–matrix adhesion forces, leading to rounding of the endothelial cell and widening of interendothelial channels. This is consistent with an emerging view that FAK activation leads to disassembly, rather than assembly, of focal adhesion complexes. On the other hand, activated FAK may directly participate in the development of contractile force in the endothelial cells resulting in intercellular gap formation. In support of this notion, FAK phosphorylation has been linked to actin polymerization and stress fiber formation in endothelial cells treated with VEGF. Alternatively, FAK could indirectly affect endothelial barrier function by coordinating secondary intracellular signaling molecules, including the Src family of tyrosine kinases and the Rho family of GTPases. In this regard, FAK may function as a scaffolding protein that recruits and activates Src, Rho, and other focal adhesion-associated proteins, which in turn cause changes in the cytoskeleton or intercellular junctions, leading to endothelial hyperpermeability.

Summary

Endothelial barrier integrity is maintained by adhesive interactions occurring at the cell–cell and cell–matrix contacts via junctional proteins and focal adhesion complexes that are anchored to the cytoskeleton. Under physiological conditions, cyclic AMP and PKA counteract with the NO-PKG pathway to protect the basal barrier function. Upon stimulation by physical stress, growth factors, or inflammatory agents, endothelial cells undergo a series of intracellular signaling reactions involving activation of PKC, PKG, MAPKs, and PTKs. These protein kinases further trigger biochemical and conformational changes in the barrier structure and ultimately lead to an opening of the paracellular pathway. In particular, MLCK activation and subsequent MLC phosphorylation in endothelial cells directly result in cell contraction and shape changes. The PTK-mediated phosphorylation of junctional proteins causes disorganization of adherens junctions or dissociation of VE-cadherin–catenin complex from its cytoskeletal anchor, leading to loose or opened intercellular junctions. Additionally, FAK phosphorylation-coupled focal adhesion assembly and redistribution provide an anchorage support for the conformational changes occurring in the cells and at the cell junctions. The Src-family tyrosine kinases may serve as a common signal that coordinates these molecular events to facilitate the paracellular transport of macromolecules. The critical roles of protein kinases in the endothelial hyperpermeability response implicate the therapeutic significance of protein kinase inhibitors in the prevention and treatment of diseases and injuries that are associated with microvascular barrier dysfunction.

Further Readings

Baldwin, A. L., and Thurston, G. (2001). Mechanics of endothelial cell architecture and vascular permeability. *Crit. Rev. Biomed. Eng.* **29**, 247–278.

Dudek, S. M., and Garcia, J. G. (2001). Cytoskeletal regulation of pulmonary vascular permeability. *J. Appl. Physiol.* **91**, 1487–1500. *Dr. Joe Garcia and his colleagues are leading investigators in the molecular control of pulmonary vascular permeability. This review paper presents a refined model that they have proposed for decades describing the structural basis of endothelial barrier function.*

Duran, W. N., Takenaka, H., and Hobson 2nd, R. W. (1998). Microvascular pathophysiology of skeletal muscle ischemia-reperfusion. *Semin. Vasc. Surg.* **11**(3), 203–214. *Dr. Joe Garcia and his colleagues have been studying the molecular mechanisms that regulate endothelial barrier funciton. This review paper presents a refined model that they have proposed for decades describing the structural basis of endothelial barrier.*

Huxley, V. H., and Rumbaut, R. E. (2000). The microvasculature as a dynamic regulator of volume and solute exchange. *Clin. Exp. Pharmacol. Physiol.* **27**, 847–854.

Lum, H., and Malik, A. B. (1996). Mechanisms of increased endothelial permeability. *Can. J. Physiol. Pharmacol.* **74**, 787–800. *Dr. Asrar Malik's work has been focusing on the signal transduction in pulmonary microvascular endothelial hyperpermeability. This paper provides a thorough review on the signaling molecules that are involved in the modulation of endothelial permeability.*

Michel, C. C., and Curry, F. E. (1999). Microvascular permeability. *Physiol. Rev.* **79**, 703–761. *This is a classic paper presenting important information on the physiological basis of microvascular exchange. The review covers both the physical and biochemical aspects of permeability regulation.*

Shepro, D. (1988). Endothelial cells, inflammatory edema, and the microvascular barrier: Comments by a "free radical." *Microvasc. Res.* **35**, 246–264.

van Nieuw Amerongen, G. P., and van Hinsbergh, V. W. M. (guest eds.) (2002). Special Issue: Endothelial hyperpermeability in vascular leakage. *Vasc. Pharmacol.* **39**(4–5), 171–272. *The entire issue discusses the molecular mechanisms of vascular endothelial hyperpermeability. It includes eight review articles by leading investigators in the field, addressing the signaling effects of calcium, Rho GTPases, tyrosine kinases, vascular endothelial growth factors, and interendothelial adhesion molecules as well as some pharmacological perspectives regarding the treatment of endothelial hyperpermeability.*

Capsule Biography

Dr. Mack H. Wu is Associate Professor at University of California, Davis, School of Medicine, Department of Surgery. Previously he worked at the Texas A&M University Health Science Center, Department of Medical Physiology and Cardiovascular Research Institute led by Dr. Harris J. Granger. Dr Wu has been investigating the signaling mechanisms underlying the microvascular permeability response to growth factors and inflammatory stimuli. His research work is supported by the National Institute of Health.

Dr. Sarah Y. Yuan is Peal Stewart Professor and Director of Research at University of California at Davis School of Medicine, Department of Surgery. Previously, she worked as Professor and held the Raleigh White Chair of Surgery at the Texas A&M University Health Science Center. Dr. Yuan has been leading research programs focusing on the cellular and molecular control of microvascular barrier function in inflammation, trauma, sepsis, and diabetes. Her work is supported by the National Institute of Health, American Heart Association, and Scott & White Foundation.

SECTION E

Transport, Junctions, Adhesion Molecules

CHAPTER 46

Vesiculo-vacuolar Organelles Are Permeability Structures in the Endothelium of Normal and Tumor Microvessels

Ann M. Dvorak and Dian Feng

Beth Israel Deaconess Medical Center, Boston, Massachusetts, and Harvard Medical School

Introduction

The microcirculation functions to provide rapid exchange of nutrients and waste products between blood and tissues. Large molecules, such as plasma proteins, exit most capillaries at low rates and by two routes: (1) vesicular transport via the shuttling of 50- to 70-nm cytoplasmic vesicles present in endothelial cells (or by interconnected vesicles that form transendothelial cell channels), and (2) interendothelial cell gaps.

We identified a new endothelial cell organelle, which we termed the vesiculo-vacuolar organelle (VVO) (Figure 1) [1]. This organelle provides the major route of extravasation of macromolecules at sites of augmented vascular permeability induced by vascular permeability factor/vascular endothelial growth factor (VPF/VEGF), a tumor-derived cytokine, in venules associated with experimental tumors [1, 2]. We also reported extensive VVOs in venular endothelium in an animal model of allergic inflammatory eye disease, characterized by histamine secretion from augmented mast cell expression. Further studies used multiple ultrastructural methods to determine (1) the substructure and contents of VVOs, (2) the formation of VVOs, (3) the mechanism of upregulated function of VVOs, and (4) whether tumor cytokines and mediators of inflammation with permeabilizing properties could recapitulate in normal vessels the upregulated VVO function of tumor vessels [3].

VVOs Are the Major Site of Hyperpermeability of Tumor Blood Vessels

Tumor microvessels are typically fourfold to tenfold more permeable to circulating macromolecules than comparable normal vessels. We injected several different macromolecular tracers intravenously into mice or guinea pigs bearing solid or ascites tumors and followed the extravasation of these tracers from tumor and normal vessels over time by light and electron microscopy [1, 2]. These studies led to the identification of VVOs in the endothelium of tumor microvessels. The vesicles and vacuoles of individual VVOs extend across endothelial cells, interconnecting with each other and with the luminal, abluminal, and often the lateral plasma membranes by stomata that may be open or that are guarded by thin diaphragms. These stomata and their closing diaphragms likely provide the structural basis for regulation of tracer passage across the microvascular endothelium.

VVOs were found to provide a transcytotic pathway by which soluble macromolecular tracers extravasated from leaky tumor blood vessels. Within seconds of intravenous

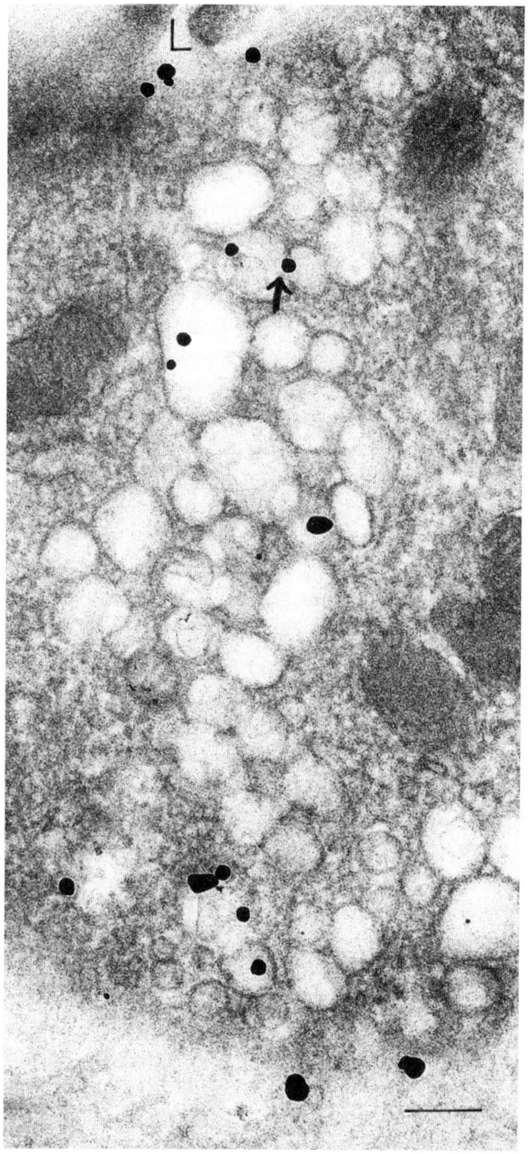

Figure 1 Platelet endothelial cell adhesion molecule (PECAM, CD31) localization to a VVO and plasma membranes (luminal and abluminal) of a mouse skin venule. In this thickened portion of venular endothelium the characteristic VVO stretches from the luminal to the abluminal surfaces of this endothelial cell. Localization is with an immunonanogold method. Particles are associated with stomata and their diaphragms (arrow), which typify the ultrastructural anatomy of VVOs. L, Lumen. Bar = 170 nm.

injection into tumor-bearing mice, macromolecular tracers were found in VVO vesicles that opened to the vascular lumen. Initially, tracers were found in VVO vesicles that opened to the vascular lumen, but VVO vesicles and vacuoles were rapidly labeled at all levels of endothelial cell cytoplasm. Vesicles that opened to the endothelial cell ablumen spilled both tracers into the underlying basal lamina. Ferritin ($d \sim 11$ nm) was not found to exit tumor vessels through interendothelial cell junctions at any time interval up to 1 hour. The small protein tracer HRP ($d \sim 5$ nm) did exit tumor vessels through normally apposed interendothelial junctions, but only after a delay of at least 5 minutes, at a time when extensive HRP extravasation had already taken place through VVOs.

Subsequent study revealed that morphologically similar VVOs were present, and with equal frequency, in the venular endothelium supplying skin and many other tissues in normal animals as well as remote from tumor sites in tumor-bearing animals. However, normal venules extravasated only very small amounts of macromolecules and their VVOs were minimally labeled with ferritin and HRP.

Vessels supplying some tumors were fenestrated and prominent collections of ferritin were also found in and beneath individual fenestrae, suggesting that these structures had provided a second pathway for tracer extravasation [1]. In contrast, some vessels supplying other tumors were not fenestrated, and tracer extravasated exclusively by the transcellular VVO route.

VVO Structure

Clusters of VVOs are deployed at intervals in the cytoplasm of endothelial cells. They are often concentrated near lateral borders of endothelial cells, that is, parajunctionally, and provide a direct link between the vascular lumen and ablumen. Approximately 12 percent of VVOs extend from luminal to the abluminal plasma membranes when viewed in single, random 80-nm electron microscopic sections, and in serial sections, this value approaches 100 percent [3]. The individual vesicles and vacuoles comprising VVOs are bounded by membranes and interconnect with each other and with the endothelial cell plasma membranes by means of stomata that are closed by thin diaphragms. The stomatal diaphragms measure less in diameter than the diameter of individual VVO vesicles and vacuoles. The presence of tracers within adjoined vesicles–vacuoles, as well as within the stomata connecting them, provides prima facie evidence that such vesicles and vacuoles are in open communication with each other. However, in some cases, the stomata joining individual vesicles and vacuoles were closed by a diaphragm, and passage of macromolecular tracers was restricted such that one vesicle or vacuole contained tracer whereas the other did not. It would seem, therefore, that the diaphragms that separate individual stomata serve to restrict the passage of cargo and may be opened and closed individually.

VVO structure cannot be fully appreciated by electron microscopy of standard 70- to 100-nm sections. In serial 12- to 14-nm ultrathin sections for electron microscopy and computer-assisted three-dimensional reconstructions, a network of interconnecting vesicles and vacuoles was revealed and established that VVOs provide a continuous, often serpentine pathway across venular endothelium, extending to both the lumen and ablumen at multiple sites [3]. Subsequent studies have revealed that the VVOs of adjacent endothelial cells also open to the interendothelial cleft, and in some instances, VVOs from adjacent endothelial cells connect across the intercellular cleft between adjacent endothelial cells, raising the possibility that plasma may extravasate by a VVO pathway that extends across overlapping endothelial cells.

The large size and complexity of VVOs suggest that they are sessile structures and that their component vesicles and vacuoles do not shuttle back and forth across endothelial cell cytoplasm. In evaluation of ultrathin (12- to 14-nm-thick) serial sections, only 9 of 1,395 (0.65%) uncoated vesicles were found to be single entities, free and unattached in the cytoplasm, that is, > 99 percent were attached to other vesicles and vacuoles as parts of VVOs [3].

VVO Function

The venules of normal tissues permit only minimal extravasation of circulating macromolecules; consistent with this finding, the VVOs of normal tissues, although structurally similar to those of tumor vessels, differ functionally from tumor endothelial cell VVOs in that they permit only minimal entry and passage of macromolecular tracers. To account for the functional differences between the VVOs of tumor vessels and those of normal venules, we postulated that VVO function was regulated by vasoactive mediators that in some way opened the stomata that connected individual VVO vesicles and vacuoles with each other and with the venular lumen and ablumen. We also postulated that one such mediator, VPF/VEGF, was likely responsible for opening these stomata in hyperpermeable tumor vessels, thereby accounting for the relatively free passage of macromolecular tracers through tumor vessel VVOs. Consistent with this hypothesis, the tumors we studied synthesize and secrete large amounts of VPF/VEGF. Moreover, tumor cell-secreted VPF/VEGF was found to localize on the surface of tumor microvascular endothelial cells as well as in association with the VVO vesicles and vacuoles in their cytoplasm [3].

In addition, another such mediator, histamine, could act similarly in opening stomata in hyperpermeable vessels in allergic inflammation. In fact, using a new enzyme-affinity-gold postembedding method to detect histamine, we localized histamine, which is secreted from mast cells in the allergic eye disease mouse model, to VVOs in involved vessels.

If VVO function is regulated by VPF/VEGF or other vasoactive mediators, then these mediators would be expected to increase the microvascular permeability of normal venules by opening VVO stomata to the passage of macromolecules. In fact, this proved to be the case. Injection of small amounts of VPF/VEGF, histamine, or serotonin into the normal flank or scrotal skin of guinea pigs, rats, and mice greatly increased the permeability of local venules [3]. Electron microscopy demonstrated that the circulating tracer (anionic ferritin) exited such venules primarily by way of VVOs, just as in tumor microvessels. Rapidly after intradermal injection of these mediators, and continuing for some minutes thereafter, increasing amounts of circulating ferritin entered VVO vesicles contiguous with the venular lumen and proceeded across the endothelium through a succession of interconnecting VVO vesicles and vacuoles to reach the vascular ablumen and underlying basal lamina, that is, tracers followed the same pathway across normal dermal venule endothelial cells as in leaky tumor microvessels. Most stomata connecting adjacent VVO vesicles and vacuoles to one another and to the luminal and abluminal plasma membranes were functionally open in that the passage of ferritin was not restricted. However, diaphragms closed some stomata to the passage of ferritin since ferritin molecules accumulated in immediately proximal vesicles or vacuoles. Thus, stomatal diaphragms were able to serve as barriers that limited the further transcellular passage of macromolecular tracers. Endothelial cell junctions remained intact, and ferritin was never observed in them. Together, these data establish VVOs as the major pathway by which soluble plasma proteins exit nontumor venules in response to several mediators that increase venular hyperpermeability.

VPF/VEGF induces its biological effects by binding to two tyrosine kinase receptors, VEGFR-1 (fms-like tyrosine kinase receptor or Flt) and VEGFR-2 (fetal liver kinase 1 or Flk-1 in rodents; kinase insert domain-containing receptor or KDR in man), which are selectively expressed in vascular endothelium and are both strikingly upregulated in tumors, wounds, and inflammation in which VPF/VEGF is overexpressed. We used ultrastructural preembedding immunoperoxidase and immunonanogold methods to localize VEGFR-2 (flk-1/KDR) in vascular endothelium in model systems in which VPF/VEGF is highly expressed: (1) glomerular and peritubular capillaries of normal mouse kidney; (2) microvessels supplying a well-characterized mouse mammary carcinoma; and (3) new vessels induced by an adenoviral vector, engineered to overexpress VPF/VEGF (adeno-vpf/vegf) [4]. Microvascular endothelial cells were positive for VEGFR-2 in all three models and, in the latter two, could be localized to the luminal and abluminal surfaces and to the membranes of cytoplasmic VVOs. The stomatal diaphragms of some VVOs and caveolae were VEGFR-2-positive, best seen with the peroxidase reporter when the entire vesicle membrane was not stained. Localization of VPF/VEGFR-2 to VVO membranes, to the luminal and abluminal plasma membranes of vascular endothelium, but not to the lateral plasma membranes at interendothelial cell junctions is consistent with the mechanisms that we have proposed for the increased microvascular permeability that is induced by VPF/VEGF and other vasoactive mediators [4].

Stomatal diaphragms are likely the structures that regulate VVO permeability. Support for this concept is based on the strategic localization of VEGFR-2 on this diaphragm which closes stomata in VVOs and of VPF/VEGF bound to VVOs in endothelia present in animal tumor models [4].

VVO Formation

Given the similarities between caveolae and VVO vesicles and vacuoles, it is possible that VVOs might form from the linking together of individual caveolae. It is also possible that the larger vesicles and vacuoles of VVOs might form from the fusion of two or more caveola-sized vesicles. We did ultrastructural preembedding immunoperoxidase to localize caveolin in VVOs in vitro and in vivo [5] to investigate the first possibility. These studies showed, both in vivo and in vitro and as in caveolae, that endothelial cell VVOs were caveolin-positive, lending support for formation of caveolin-positive VVO clusters from the joining of individual caveolin-positive caveolae. We used an ultrastructural morphometric approach [3] to investigate the second possibility. For example, a fusion mechanism has been proposed in the generation of several types of cytoplasmic secretory granules in which small progranules of unit size fuse with each other in varying combinations to form larger mature granules whose volumes represent multiples of the volume of the unit progranule [6]. By analogy, the volume of the various VVO vesicles and vacuoles would not be expected to fall on a continuum but instead would represent multiples of the volume of the smaller unit vesicle. Measurements of the volumes of vesicles and vacuoles comprising VVOs revealed a heterogeneous distribution, where the volumes of individual vesicles and vacuoles were not continuous but exhibited a periodic modal distribution [3]. The most frequently occurring vesicle–vacuole had a unit volume of ~0.00015 μm^3; this value corresponds to a spheroid of diameter ~60 nm, that is, the size of typical capillary caveolae. Vacuoles corresponding to the fusion of as many as 10 unit vesicles were detected. The data, therefore, are consistent with the hypothesis that larger VVO vesicles and vacuoles arise from the fusion of different numbers of caveola-sized unit vesicles.

We also have investigated the substructural localization of other proteins of importance to vesicular trafficking and fusion, generally. One such protein, vesicle-associated membrane protein (VAMP), is localized to the cytoplasmic side of caveolar and VVO membranes [3]. The presence of this vesicle-SNARE protein in these locations lends support to the proposed construction of VVOs from fused caveolae.

VVO Fate

Once formed, VVOs are sessile, multichambered structures occupying vast expanses of venular cytoplasm. They restrict the passage of macromolecules from the blood vascular spaces at their narrow points in stomata that are closed by thin diaphragms. These diaphragms are opened, making VVOs porous—a process that is rapidly induced by exposure to well-known permeabilizing molecules, such as VPF/VEGF, histamine, and serotonin.

We envision at least three possible fates for VVOs in venular endothelium: (1) opened, leaky VVOs could close, thereby returning vessels to their prior nonleaky state; (2) opened, enlarged stomata in permeabilized VVOs could open further, allowing retention of opened vesicles and vacuoles in their fused state to form large transcellular holes through which circulating endogenous and exogenous particulates could freely enter the extravascular space [3]; or (3) VVOs could serve as an extensive intracellular store of membranes that, in certain circumstances, could rapidly and greatly expand the endothelial plasma membrane. This expansion could be instrumental in the rapid formation of large vessels ("mother" vessels) in VPF/VEGF angiogenesis models [4] and in the enhanced endothelial surface process and migration sac formation that accompanies endothelial thinning and loss of VVOs at points of neutrophil transmigration in an animal model of acute inflammation [3].

Presently, there are no data to confirm the first possibility, but long-term studies of VVO architecture following transient permeability events are in progress. Considerable data have accrued in support of the second proposed fate of VVOs [3]. In studies of VPF/VEGF and other vasoactive mediators in combination with soluble macromolecular tracers, we encountered relatively few openings across venular endothelium. In such experiments, ferritin and HRP extravasated across venular endothelium primarily by way of VVOs. However, when we used colloidal carbon ($d \sim$ 50 nm) as a tracer, a particulate that is, for the most part, too large to enter or pass through the narrow necks of stomata in VVOs, we observed a substantial (threefold to thirtyfold) increase in endothelial cell openings [3], suggesting that, in response to vasoactive mediators, holes may develop from a rearrangement of VVO vesicles and vacuoles to form larger membrane-lined vacuolar structures and, eventually, channels of sufficient size to allow the passage of particulate tracers as large as erythrocytes.

The third possible fate of VVOs as a membrane store for rapidly expanding endothelial plasma membrane has gained some support from studies of transendothelial cell neutrophil migration in acute inflammation [3]. Here, we noted that VVOs became less numerous in thinned endothelium through which neutrophils were traveling, and that a markedly expanded endothelial cell luminal membrane covered many extended endothelial cell cytoplasmic processes and provided additional membrane to form migration sacs that enclosed migrating neutrophils. By analogy, the markedly and rapidly enlarged "mother" vessels in tumor and VPF/VEGF-induced angiogenesis models lend support to the third possibility, since VVOs were not a conspicuous component of the cytoplasm of the enlarged mother vessels and the formation of mother vessels occurred rapidly in an angiogenesis model. VVOs may provide membrane for the rapid formation of these vessels in angiogenesis.

Thus, the fate of VVOs includes at least three possibilities—to recover in place by reforming diaphragms between their components or to persist by remaining in place without reclosure of stomata (e.g., in acute and chronic inflammation); to expand further to form large transcellular holes (to facilitate transcellular trafficking in inflammation); or to provide rapid expansion of endothelial plasma membranes by externalizing their membrane in acute inflammatory and angiogenesis models.

Summary

A newly defined endothelial cell permeability structure, termed the vesiculo-vacuolar organelle (VVO), has been identified in the endothelia of normal venules, in the microvasculature that accompanies tumors, and in venules associated with allergic inflammation. This organelle provides the major route of macromolecular extravasation at sites of increased vascular permeability induced by vascular permeability factor/vascular endothelial growth factor (VPF/VEGF), serotonin, and histamine in animal models. Continuity of these large sessile structures between the vascular lumen and extracellular space has been demonstrated in kinetic studies with ultrastructural electron-dense tracers and by ultrathin serial sections. Ultrastructural enzyme-affinity cytochemical and immunocytochemical studies have identified histamine and VPF/VEGF bound to VVOs in vivo in animal models where these mediators of permeability are released from mast cells and tumor cells, respectively. The high-affinity receptor for VPF/VEGF, VEGFR-2, and platelet endothelial cell adhesion molecule (PECAM-1, CD31) were localized to VVOs and their substructural components by pre-embedding ultrastructural immunonanogold and immunoperoxidase techniques. Similar methods were used to localize caveolin and vesicle-associated membrane protein (VAMP) to VVOs and caveolae, indicating a possible commonality of formation and function of VVOs to caveolae.

Glossary

Caveola: A small, flask-shaped substructural organelle attached to the plasma membranes of endothelial cells. It is distinguished by suborganellar anatomic structures including stomata, knobs, diaphragms, and necks and functions in transendothelial cell transport of macromolecules, either by motion of vesicles through endothelial cells or by fusion of vesicles to form short, transendothelial cell channels.

Enzyme-affinity-gold electron microscopy: A specialized ultrastructural technique developed to identify subcellular sites of specific enzyme substrates by binding of gold–enzyme complexes to sample substrates. An example is the localization of the substrate histamine, by the specific enzyme diamine-oxidase-gold probe.

Vesiculo-vacuolar organelle: A newly identified endothelial cell structure that connects all plasma membrane domains of endothelial cells and provides an anatomical route of transendothelial cell passage of macromolecules. They are particularly prominent in thick venular endothelium, and their functions are upregulated by potent permeability mediators. This organelle is composed of fused small, caveola-sized vesicles and larger vacuoles that display suborganellar anatomic structures including stomata, knobs, diaphragms, and necks.

Acknowledgments

Supported by NIH grants AI-33372 and AI-44066. Portions of this text are reprinted with permission from Dvorak, A. M., and Feng, D. (2001). The vesiculo-vacuolar organelle (VVO): A new endothelial cell permeability organelle. *J. Histochem. Cytochem.* **49**, 419–431.

Further Reading

1. Kohn, S., Nagy, J. A., Dvorak, H. F., and Dvorak, A. M. (1992). Pathways of macromolecular tracer transport across venules and small veins. Structural basis for the hyperpermeability of tumor blood vessels. *Lab. Invest.* **67**, 596–607. *This report is the first description of a system of interconnecting vesicles and vacuoles termed vesiculo-vacuolar organelles (VVOs) that are prominent in both tumor-associated and control vessel endothelial cells and provide the primary pathway for macromolecular extravasation. The large increase in permeability characteristic of tumor vessels is likely attributable to upregulation of VVO function.*
2. Dvorak, A. M., Kohn, S., Morgan, E. S., Fox, P., Nagy, J. A., and Dvorak, H. F. (1996). The vesiculo-vacuolar organelle (VVO): A distinct endothelial cell structure that provides a transcellular pathway for macromolecular extravasation. *J. Leukoc. Biol.* **59**, 100–115. *This is a report of the high-magnification ultrastructural analysis of electron-dense tracer passage across endothelial cell VVOs. In it the passage of macromolecules was revealed to be regulated at stromatal diaphragms, thereby demonstrating a mechanism for controlling the passage of macromolecules across endothelial cells.*
3. Feng, D., Nagy, J. A., Dvorak, H. F., and Dvorak, A. M. (2002). Ultrastructural studies define soluble macromolecular, particulate and cellular transendothelial cell pathways in venules, lymphatic vessels and tumor-associated microvessels in man and animals. *Microsc. Res. Tech.* **57**, 289–326. *This review extensively documents what is currently known about the structure and function of VVOs. Studies reviewed here include ultrastructural serial sections and computer-assisted, three-dimensional reconstructions of soluble macromolecular, particulate, and cellular transendothelial cell pathways.*
4. Feng, D., Nagy, J. A., Brekken, R. A., Pettersson, A., Manseau, E. J., Pyne, K., Mulligan, R., Thorpe, P. E., Dvorak, H. F., and Dvorak, A.M. (2000). Ultrastructural localization of the vascular permeability factor/vascular endothelial growth factor (VPF/VEGF) receptor-2 (FLK-1, KDR) in normal mouse kidney and in the hyperpermeable vessels induced by VPF/VEGF-expressing tumors and adenoviral vectors. *J. Histochem. Cytochem.* **48**, 545–555.
5. Vasile, E., Qu-Hong, Dvorak, H. F., and Dvorak, A. M. (1999). Caveolae and vesiculo-vacuolar organelles in bovine capillary endothelial cells cultured with VPF/VEGF on floating matrigel-collagen gels. *J. Histochem. Cytochem.* **47**, 159–167.
6. Hammel, I., Lagunoff, D., Bauza, M., and Chi, E. (1983). Periodic, multimodal distribution of granule volumes in mast cells. *Cell Tissue Res.* **228**, 51–59.

Capsule Biography

Dr. Dvorak, a pathologist, heads the electron microscopy unit in the Department of Pathology at Beth Israel Deaconess Medical Center, a major teaching hospital at Harvard Medical School in Boston, MA. Her laboratory primarily focuses on the cell biology of endothelial cells, mast cells, basophils, and eosinophils in health and disease. Her work is supported by grants from the NIH.

CHAPTER 47

The Multistep Cascade of Leukocyte Extravasation

[1,2]Britta Engelhardt and [1]Dietmar Vestweber

[1]Max Planck Institute for Molecular Biomedicine

[2]Theodor Kocher Institute, Bern, Switzerland

Introduction

Leukocyte extravasation from the blood into tissues plays a key role in innate and adaptive immunity and in inflammation. In order to understand the task of extravasation of a leukocyte traveling in the blood stream, Paul Kubes [1] recently suggested that we picture ourselves plunged into the water of a "roaring river," in order to imagine the high shear forces white cells are exposed to in the mainstream of the blood. In either scenario stopping and exiting the flow becomes an incredibly daunting task. Whereas we might struggle by various means using our hands and feet to exit the river, in the case of the white cell, a well-coordinated and extremely efficient sequence of leukocyte endothelial interactions has evolved allowing the leukocyte to marginate from the main bloodstream, to slow down and to "hold on" (adhere) to the endothelial vascular wall, and eventually to extravasate. Here, we will summarize the sequence of molecular steps known to govern leukocyte extravasation across the vascular wall to date.

The Multistep Paradigm

In order to ensure immunosurveillance of the body the recruitment of leukocytes from the blood into the different tissues needs to be a well controlled process directing the extravasation of the appropriate cell at the proper location at the right time. The multistep paradigm for leukocyte extravasation [2] has proven to be a valid framework for understanding the sequence of dynamic interactions between the circulating leukocyte and the endothelial cell during this process. The multistep paradigm postulates that an initial transient contact of the circulating leukocyte with the vascular endothelium, generally mediated by adhesion molecules of the selectin family and their respective carbohydrate ligands, slows down the leukocyte in the bloodstream. After an initial capture the leukocyte rolls along the vascular wall with greatly reduced velocity. The rolling leukocyte can now "sense" chemotactic factors from the family of chemokines presented on the endothelial surface. Chemokines bind to their respective serpentine receptors on the leukocyte surface. These receptors deliver a G protein–mediated pertussis toxin–sensitive signal into the cell resulting in the functional activation of adhesion molecules of the integrin family present on the leukocyte surface. Activation of integrins leads to an increase in their affinity and/or avidity. Only "activated" integrins are able to mediate the firm adhesion of leukocytes to the vascular endothelium by binding to their endothelial ligands from the immunoglobulin (Ig) superfamily of adhesion receptors. This ultimately leads to leukocyte diapedesis, through endothelial cell contacts involving molecules of the endothelial adherens and tight junctions [3]. Some reports even suggest that leukocytes extravasate by transcytosis through endothelial cells. Successful recruitment of circulating leukocytes into the tissue depends thus on the productive leukocyte/endothelial interaction during each of these sequential steps.

Capture and Rolling: Selectins and Selectin Ligands

Extravasation of leukocytes takes place in postcapillary venules. The process known as capture or tethering hereby represents the very initial contact of a white cell with the endothelium. The transiently captured leukocyte might begin to roll along the endothelial surface. Rolling velocities are between 5 and 40 µm/sec and thus below that of freely flowing cells within the same postcapillary venules where mean blood flow velocities range from 1,000 to 4,000 µm/sec (Figure 1).

Capture and rolling are usually mediated by the selectin family of adhesion molecules, which has only three members. Selectins recognize via their N-terminal lectin domain fucosylated and sialylated glycoprotein ligands. Leukocyte (L)-selectin is constitutively expressed on all granulocytes and monocytes and most lymphocytes. Platelet (P)-selectin is stored in α-granules of platelets and in Weibel-Palade bodies of endothelial cells and translocated to the cell surface within minutes upon their activation during inflammation. Endothelial (E)-selectin is not constitutively expressed but induced on endothelial cells by inflammatory stimuli.

As post-translational modifications rather than the protein itself are recognized by the selectins, there are many candidate ligands for selectins. P-selectin glycoprotein ligand 1 (PSGL-1) has been most extensively characterized at the cellular, molecular, and functional level. PSGL-1 is constitutively expressed on all lymphocytes, monocytes, eosinophils, and neutrophils. It is the major ligand for P-selectin but can also bind to E-selectin. L-selectin glycoprotein ligands have been identified in high endothelial venules

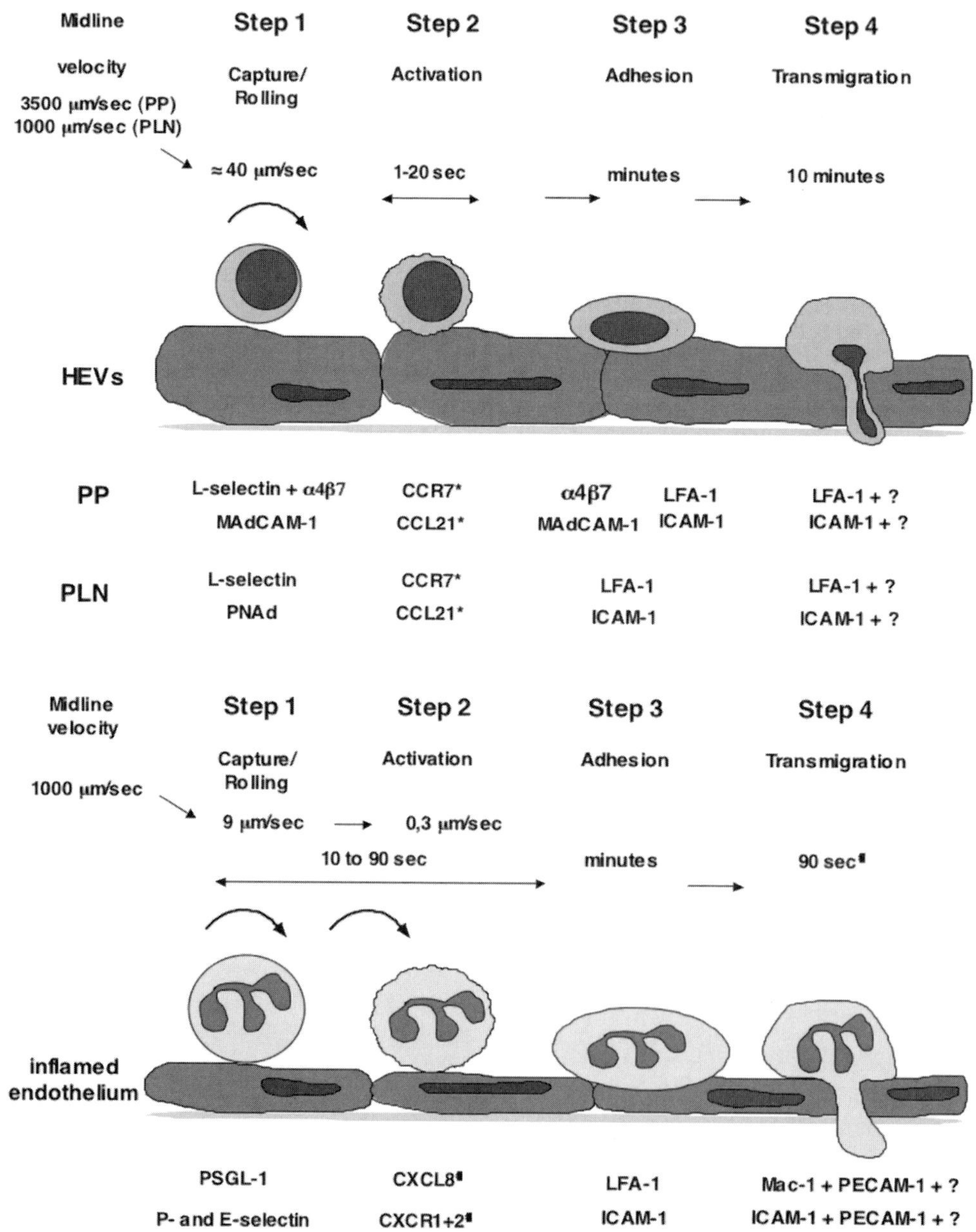

Figure 1 The multistep model of leukocyte endothelial interaction. **Top**: Lymphocyte/HEVs interactions in Peyer's patches (PP) or peripheral lymph nodes (PLN). **Bottom**: Neutrophil interaction with inflamed endothelium. *, Only T cells, unknown for B cells; #, only demonstrated in vitro. The velocities of free-flowing (noninteracting) cells and rolling leukocytes as well as other time estimates are derived from intravital microscopic observations published from the laboratories of Ulrich von Andrian, Eugene Butcher, and Klaus Ley. (see color insert)

(HEVs) of secondary lymphoid organs and are collectively known as the peripheral node addressins (PNAd) and recognized by the monoclonal antibody MECA-79. L-selectin mediates the rolling of naive lymphocytes on PNAd in HEVs of peripheral lymph nodes. Mice lacking L-selectin have very small peripheral lymph nodes due to the failure of their lymphocytes to roll on HEVs at this site. Evidence for the importance of selectins for effective neutrophil recruitment into inflamed tissues is derived from mice lacking either selectins or selectin ligands and from patients suffering from leukocyte adhesion deficiency II (LADII). These patients fail to incorporate fucose into selectin ligands and thus lack functional selectin ligands. Interestingly, the defect does not affect a biosynthetic enzyme. The defective gene in LAD II is a GDP fucose transporter in the Golgi membrane that provides the fucosyltransferases with their substrate [4]. As a consequence of this defect, leukocytes lacking selectin ligands cannot be captured to endothelium and patients suffer from bacterial infections.

Besides the adhesion receptor family of selectins and their ligands, on lymphocytes, α4-integrins have been shown to support capture and rolling. α4-integrin mediated rolling is rather slow and was measured to be around 10 μm/sec. α4β1 (VLA-4) mediated capture of T lymphoblast on its vascular ligand VCAM-1 is required for successful recruitment of activated T cells into the central nervous system [5], whereas α4β7 (LPAM-1) mediated capture and rolling via MAdCAM-1 on HEVs in mucosal associated lymphatic tissue is required for the recruitment of α4β7-integrin^{positive} gut homing lymphocytes to these sites. One key to the participation of α4-integrins in capture and rolling of lymphocytes is their localization on the microvillous processes on the surface of circulating cells, which they share with L-selectin.

Activation: Chemokines and Chemokine Receptors

In order to stop a rolling leukocyte, further signals are required. This is illustrated by in vivo observations that both neutrophils and lymphocytes can use L-selectin to roll in HEVs of peripheral lymph nodes; however, only lymphocytes have the ability to stop [6]. This points to the necessity of selective signals that specifically trigger adhesion of lymphocytes but not of neutrophils in HEVs.

Chemokines have been recognized to trigger adhesion of leukocytes to the endothelium under shear. Chemokines are a family of about 50 low-molecular-weight chemotactic cytokines (8 to 14 kDa) that can bind to and signal through seven transmembrane spanning G-protein coupled receptors expressed on leukocytes but also on other cell types. Chemokines share sequence homologies and are divided into four groups (CC, CXC, CX_3C, and C) based on the orientation of conserved cysteines in their amino termini [7]. The chemokine receptors are grouped accordingly into four receptor families (CCR, CXCR, CX_3CR, and CR). Chemokine receptors preferentially signal via $G_{\alpha i}$-proteins, which can be blocked by pertussis toxin. The expression of chemokine receptors varies greatly among different leukocyte subsets and can change upon lymphocyte activation. Furthermore, chemokines are differentially expressed in different tissues. Many chemokines have been shown to bind to the surface of endothelial cells. This is due to the fact that most chemokines have highly charged amino acid residues that can mediate binding to heparan sulfate and other glycosaminoglycans (GAGs).

In order to examine the ability of chemokines to induce adhesion in the context of the multistep recruitment cascade, many investigators have employed in vitro flow chamber assays that mimic blood flow. Using these assay systems, chemokines have been demonstrated to trigger arrest of rolling lymphocytes on purified ligands or on endothelial cells under shear and have therefore been called "arrest chemokines" [8]. The chemokine CCL21 (SLC) was shown to trigger the arrest of T lymphocytes rolling on purified L-selectin ligands on ICAM-1. CCL21 is constitutively expressed on HEVs in peripheral lymph nodes. As only T lymphocytes but not neutrophils possess CCR7, the receptor for CCL21, it becomes plausible that only lymphocytes but not neutrophils can be arrested in HEVs in vivo. The importance of the CCL21/CCR7 chemokine/chemokine receptor interaction for successful T-cell homing to peripheral lymph nodes is exemplified by the lack of T cells within peripheral lymph nodes of CCR7-deficient mice or in mice that, because of a spontaneous mutation, lack expression of CCL21. Other chemokines have been characterized triggering the arrest of monocytes on inflamed endothelium [8]. Thus the chemokine/ chemokine receptor system has a great impact in the specificity of leukocyte extravasation at the level of cell arrest.

Firm Adhesion: Integrins and Ig Superfamily Members

Leukocyte integrins form a family of noncovalently linked αβ-heterodimeric transmembrane proteins and are normally expressed on circulating leukocytes in a relatively inactive state. Upon chemokine receptor engagement, integrins become functionally activated within seconds, triggering leukocyte arrest on the endothelium. With the exception of the fact that a pertussis toxin–sensitive G-protein signal is delivered into the cell by the chemokine receptors, the downstream signals leading to integrin activation, commonly referred to as "inside-out-signaling," are as yet unknown. "Functional activation" of integrins on the leukocyte surface is achieved by an affinity increase through conformational changes, an avidity increase by integrin clustering, or a combination of both effects [9].

Each class of leukocytes displays a particular pattern of integrins that can change in a signal- and time-dependent fashion. For example resting human T lymphocytes express β1-, β2-, and β7-integrins on their surface, whereas neutrophils express mostly β2-integrins, and to a lesser degree also β1- and β3-integrins. All leukocytes, however, express at least one member of the β2-integrin family, which is restricted to the leukocyte lineage and therefore referred to as the leukocyte integrins. The β2-integrins are a family of four receptors, namely LFA-1 (αLβ2; CD11a/CD18), Mac-1 (αMβ2; CD11b/CD18), p150,95 (αxβ2; CD11c/CD18), and αDβ2 (CD11d/CD18). Especially LFA-1 and Mac-1 are involved in leukocyte extravasation by mediating adhesion to intercellular adhesion molecule-1 (ICAM-1) and ICAM-2 on endothelial cells, whereas αDβ2-integrin was shown to bind to vascular cell adhesion molecule 1 (VCAM-1).

Involvement of LFA-1 in lymphocyte homing to peripheral lymph nodes was first demonstrated by Alf Hamann (in 1988) by antibody inhibition studies and was more recently confirmed by intravital microscopy, where it could be demonstrated that LFA-1 mediates the activation-dependent adhesion of lymphocytes on HEVs at this site [6].

The importance of β2-integrins in leukocyte recruitment is exemplified by an inherited autosomal recessive disease in man, called leukocyte adhesion deficiency-I (LAD-I). LAD-I is the result of mutations in the β2 integrin subunit inhibiting cell surface expression and/or function of all β2-integrins. Patients suffer from recurring soft-tissue infections, impaired wound healing, and gingivitis. The CD18 "knockout" mouse has similar problems to human LAD-I patients. The disorder is primarily considered a failure of neutrophil function, extravasation of which relies most heavily on β2-integrins, because they lack significant levels of other integrins.

Monocytes and lymphocytes can use other integrins, such as the α4-integrins, when the β2-integrins are missing. There are two α4-integrins, VLA-4 (α4β1; CD49d/CD29) and α4β7-integrin. α4β7 is the lymphocyte receptor for mucosal addressin cell adhesion molecule -1 (MAdCAM-1). MAdCAM-1 is expressed on postcapillary venules in the lamina propria of the gut, and on HEVs of the intestine associated lymphoid tissues, especially the Peyer's patches and the mesenteric lymph nodes, thus directing trafficking of gut homing lymphocytes expressing its lymphocyte receptor α4β7 το τηεσε σιτεσ. MAdCAM-1 is also found to be upregulated above constitutive levels in the intestinal lamina propria of patients suffering from ulcerative colitis or Crohn's disease and contributes to the maintenance of these chronic inflammatory diseases. Constitutive expression of VCAM-1 on endothelial cells is low or absent with the exception of the central nervous system, where constitutive VCAM-1 is involved in mediating the firm adhesion of α4-integrin positive T lymphoblasts in postcapillary venules [5].

Transendothelial Migration

Firm adhesion of the leukocyte to the endothelial surface is a prerequisite for diapedesis. The β2-integrins LFA-1 and Mac-1 and their endothelial ligand ICAM-1 have been implied to be involved in transendothelial migration (TEM) of T lymphocytes. In fact, endothelial cells lacking both ICAM-1 and ICAM-2 no longer support TEM of T cells in vitro. The apparent dual role of ICAM-1 in mediating firm adhesion and TEM can be assigned to different parts of the molecule. Whereas the extracellular domain of endothelial ICAM-1 suffices to mediate T-cell adhesion, the cytoplasmic domain is required to mediate TEM of T cells, probably by inducing Rho-signaling within the endothelial cells. Activation of Rho leads to cytoskeletal rearrangements within the endothelium, which are necessary to allow the passage of leukocytes across the endothelial cell wall. The actin cytoskeleton within the endothelium is anchored at cell–matrix interaction sites but also in cell-to-cell contacts forming adherens and tight junctions. Migrating leukocytes have been found to induce a delocalization of the vascular endothelial (VE)-cadherin from the endothelial adherens junctions by a yet unidentified mechanism, which potentially disrupts the endothelial adherens junction in a defined region and allows the leukocyte to exit via the opened gap. On the other hand, direct manipulation of adherens junctions causing molecular disorganization and an increase in vascular permeability shows no influence on TEM of leukocytes in vitro, demonstrating that TEM of leukocytes is not directly correlated to changes in endothelial permeability. Besides VE-cadherin other molecules were found to be concentrated at the lateral borders of endothelial cells and have been implicated in the process of transendothelial migration of leukocytes. These include the Ig-superfamily members platelet endothelial cell adhesion molecule 1 (PECAM-1) and the junctional adhesion molecules (JAMs: JAM-A, JAM-B, and JAM-C) and probably a unique molecule CD99. All of these molecules are potentially capable of homophilic interactions, which are thought to be involved in the establishment of the endothelial cell-to-cell contacts. PECAM-1, CD99, JAM-A, and JAM-C have also been demonstrated to be expressed on leukocytes. Therefore leukocyte extravasation through the endothelial cell junctions is often pictured as a zipper-like model, where the leukocyte on its passage transiently replaces the homophilic molecular interactions usually occurring in between the endothelial cells. This concept is supported by the findings that antibodies blocking homophilic interactions of PECAM-1 arrest leukocytes at the apical surface of endothelial cells. On the other hand, PECAM-1 deficient mice do not reveal any major defect in leukocyte extravasation, indicating that a requirement of PECAM-1 is not obligatory for TEM. Additionally, antibody inhibition studies have implied the JAMs to be involved in leukocyte extravasation. Interestingly, all members of the JAM family are able to bind to either β2– or β1–integrins on activated leukocytes, further supporting their possible involvement in leukocyte endothelial interactions.

Observations arguing for an alternative transcellular route for extravasating leukocytes through endothelial cells may suggest that their might be more than one molecular mechanism or route for leukocytes to move through the endothelial cell barrier. A number of very well performed studies documented leukocyte extravasation by sparing endothelial cell-to-cell contacts, suggesting that leukocytes traverse the endothelial cell proper [10]. In fact, Gowans and colleagues, who discovered the postcapillary venules within the lymphoid organs as the exit sites of circulating lymphocytes in the early 1960s [11] were even convinced that the specialized endothelial cells in HEVs engulf lymphocytes by mechanisms resembling phagocytosis, thus allowing their passage from the bloodstream to the lymphoid tissue. Thus, although certain stimuli may trigger a transcytotic pathway of leukocyte TEM, based on the current experimental evidence the general belief is that leukocytes traverse the endothelium through small gaps at intercellular junctions. The orchestration of molecular signals exchanged between the leukocyte and the endothelium during this process remains to be determined.

Outlook

Since the multistep cascade of leukocyte recruitment was put forward [2], we have come a long way in our understanding of the traffic signals involved in leukocyte recruitment from the blood into tissue. However, although many of the players involved have been characterized, we still lack detailed mechanistic insights into their orchestrated function. How can chemokines trigger integrin activation and leukocyte arrest in seconds? Where and how do leukocytes penetrate the vascular wall without increasing vascular permeability? How are shear forces involved in leukocyte adhesion and TEM? Filling in the gaps in our understanding of the multistep cascade of leukocyte/endothelial interaction in the near future promises exciting new insights into this well-regulated process, which directs the trafficking of the correct cells at the right time to the right place in order to fight infections and disease.

References

1. Kubes, P. (2002). Introduction: The complexities of leukocyte recruitment. *Semin. Immunol.* **14**, 65–72.
2. Butcher, E. C. (1991). Leukocyte–endothelial cell recognition: Three (or more) steps to specificity and diversity. *Cell* **67**, 1033–1036. *In this minireview Eugene Butcher puts forward the idea of the multistep cascade of leukocyte–endothelial interaction, based on his observations and ideas.*
3. Vestweber, D. (2002). Regulation of endothelial cell contacts during leukocyte extravasation. *Curr. Opin. Cell Biol.* **14**, 587–593.
4. Lühn, K., Wild, M. K., Eckhardt, M., Gerardy-Schahn, R., and Vestweber, D. (2001). The gene defective in leukocyte adhesion deficiency II encodes a putative GDP-fucose transporter. *Nat. Genet.* **28**, 69–72.
5. Vajkoczy, P., Laschinger, M., and Engelhardt, B. (2001). Alpha4-integrin-VCAM-1 binding mediates G protein-independent capture of encephalitogenic T cell blasts to CNS white matter microvessels. *J. Clin. Invest.* **108**, 557–565.
6. Warnock, R. A., Askari, S., Butcher, E. C., and von Andrian, U. H. (1998). Molecular mechanisms of lymphocyte homing to peripheral lymph nodes. *J. Exp. Med.* **187**, 205–216.
7. Zlotnik, A., and Yoshie, O. (2000). Chemokines: A new classification system and their role in immunity. *Immunity* **12**, 121–127. *This review focuses on several areas of chemokine biology of particular interest to immunologists and gives a comprehensive list of known human chemokines. Most importantly, it introduces the new nomenclature for chemokines, which is based on the chemokine receptor nomenclature in use.*
8. Ley, K. (2003). Arrest chemokines. *Microcirculation* **10**, 289–295.
9. Kim, M., Carman, C. V., and Springer, T.A. (2003). Bidirectional transmembrane signaling by cytoplasmic domain separation in integrins. *Science* **301**, 1720–1725.
10. Feng, D., Nagy, J. A., Pyne, K., Dvorak, H. F., and Dvorak, A. M. (1998). Neutrophils emigrate from venules by a transendothelial cell pathway in response to FMLP. *J. Exp. Med.* **187**, 903–915.
11. Gowans, J. L., and Knight, E. J. (1964). The route of recirculation of lymphocytes in the rat. *Proc. R. Soc. Lond. B Biol.* **159**, 257. *This paper describes the discovery of the specialized high endothelial venules (HEVs) in lymph nodes as the sites where circulating lymphocytes leave the bloodstream and enter the lymphatic tissue.*

Capsule Biography

After heading a research group at the Max-Planck Institute in Münster, Germany, Dr. Engelhardt became director of the Theodor Kocher Institute at the University of Bern, Switzerland, in November 2003. Her research focuses on leukocyte trafficking into the central nervous system.

Dr. Vestweber became head of the Institute of Cell Biology at the University of Münster in 1994 and in 2001 became the founding director of the new Max-Planck-Institute for Molecular Biomedicine in Münster. His work focuses on the extravasation of leukocytes into inflamed and lymphoid tissue.

CHAPTER 48

Vascular Endothelial Cadherin and Neutrophil Transmigration

Bastien Hermant, Marianne Weidenhaupt, Stéphanie Heyraud, and Danielle Gulino-Debrac

Laboratoire d'Ingénierie des Macromolécules, Institut de Biologie Structurale Jean-Pierre Ebel, Grenoble, F-38027 and UMR-CNRS-CEA-UJF 5075, France

This review focuses on the major proteins involved in the various stages of the transmigration of leukocytes across the endothelium and on the molecular mechanisms underlying this process.

Introduction

He Endothelium

The endothelium is a monolayer of cells covering the inside of the entire vascular tree. It is also a semipermeable barrier that, by modulating its intercellular junctions, allows the transmigration of various blood constituents (Figure 1). Morphologically, three types of organelles constitute endothelial cell junctions. At the apical side of the cells, tight junctions seal the cells. Gap junctions, which are sometimes intercalated with tight junctions, allow the exchange of ions and small molecules between adjacent cells. Adherens junctions, located at a more basal position, mediate the physical contacts between cells and are anchored in the actin cytoskeleton.

The endothelium is also the point of entry of circulating leukocytes into inflamed tissue. The mechanism of leukocyte entrance into tissues is a complex, multistep event involving leukocyte adhesion to the endothelium and subsequent migration across the blood vessel wall [1]. Whereas the early steps leading to leukocyte adhesion are well understood, it is still unclear how leukocytes actually crawl through the endothelial wall. A recent study suggested that leukocytes could pass across the body of endothelial cells using a transcellular route, but it is generally accepted that most leukocyte transendothelial migration occurs via paracellular rather than transcellular routes.

Neutrophils

Leukocytes are circulating blood cells involved in defending the body against infective organisms and foreign substances. During normal immunosurveillance, leukocytes circulate in search of foreign antigens. Injury or infections cause inflammatory responses that lead to the recruitment of granulocytes, monocytes, and T cells to the sites of inflammation. Responding to the signals induced by infections or tissue damages, sentinel cells stationed in the tissues, mast cells and macrophages, release histamine, eicosanoids, and chemokines, which cause vasodilatation of blood vessels and consequently a decrease of the blood flow. This facilitates the capture of leukocytes on the surface of the endothelium. In parallel, inflammatory cytokines such as TNFα stimulate the endothelium to express adhesion receptors on its apical surface with which leukocytes can interact.

Following transmigration into the tissue, leukocytes undergo directed migration through a concentration gradient of chemotactic cytokines to the site of inflammation. According to their type and their specialized function, leukocytes release different inflammatory cytokines that, in turn, activate other white blood cells and enhance leukocyte recruitment to the inflammatory site.

Neutrophils form the primary defense against bacterial infections. Similarly to the other cells of the immune

*Current address: European Molecalar Biology Laboratory (EMBL), Grenoble, France

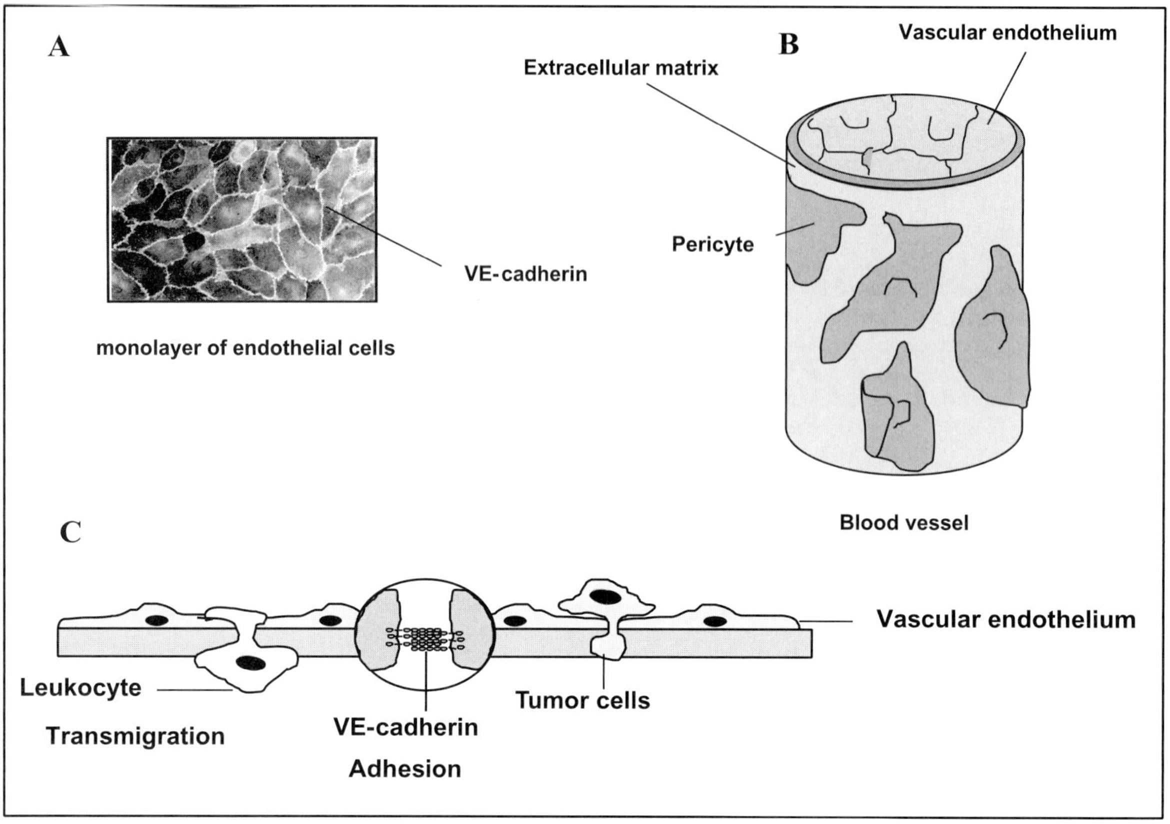

Figure 1 VE-cadherin is involved in the maintenance and restoration of the endothelium integrity. The endothelium is made up of a monolayer of endothelial cells (**A**) that separates blood components from underlying tissues (**B**). VE-cadherin is expressed at the lateral junctions between endothelial cells (**C**). Different types of leukocytes and tumor cells transmigrate at the intercellular junctions where VE-cadherin is expressed (**C**).

system, they are produced in the bone marrow and circulate in the bloodstream. However, they can move out of blood vessels into infected tissue where they attack the foreign substances mainly through phagocytosis which allows them to "eat" other cells and foreign substances. Both the transmigration of neutrophils across the endothelium and the intracellular degradation of antigens are at least in part mediated by proteases, of which there are four classes: serine proteases, metalloproteases (MMPs), thiol proteases, and aspartate proteases. The first two classes play a critical role in the degradation of extracellular components, whereas the two others are involved in the intracellular degradation of antigens. Serine proteases, characterized by the presence of a catalytic triad made of histidine, aspartic acid, and serine, are packed into the azurophil granules of neutrophils. They include elastase, cathepsin G, proteinase 3, and Cap37, the last being enzymatically inactive. Upon stimulation of neutrophils by proinflammatory mediators, the content of the azurophil granules is translocated toward the external membrane of neutrophils. These surface-bound proteases are catalytically active and remarkably resistant to naturally occurring protease inhibitors [2].

Markers Expressed at Endothelial Cell–Cell Junctions

In addition to proteins that are encountered at interepithelial cell contacts, such as JAMs and CD99, the endothelial junctions express specific marker proteins such as PECAM-1 and VE-cadherin (vascular endothelial cadherin, also known as cadherin 5). All these adhesive receptors, being localized at the endothelial cell–cell junctions, are on the path of leukocyte emigration.

Platelet Endothelial Cell Adhesion Molecule-1 (PECAM-1)

PECAM-1 is a transmembrane glycoprotein of the immunoglobulin gene superfamily expressed at the surface of monocytes, neutrophils, and naive T lymphocytes and also concentrated at endothelial cell–cell junctions. It possesses an extracellular region consisting of six Ig-like domains and a cytoplasmic tail exhibiting a variable size due to differential splicing [3] (Figure 2).

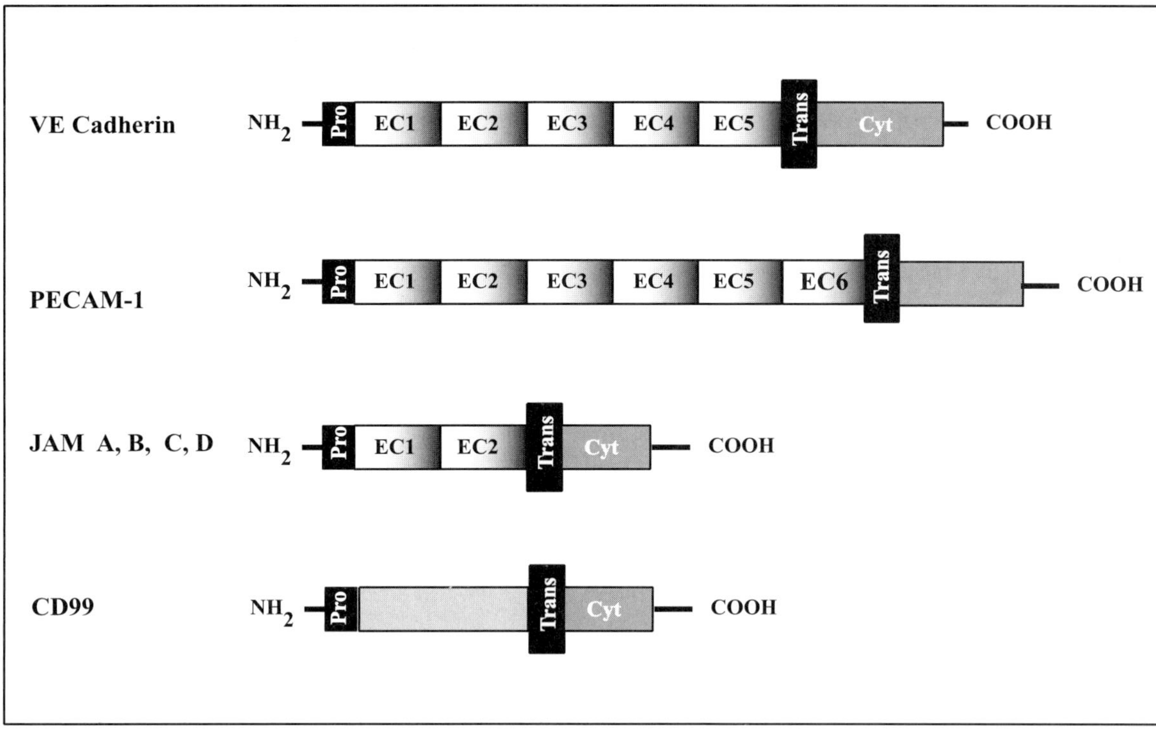

Figure 2 Schematic structures of the adhesion receptors expressed at endothelial cell–cell junctions. VE-cadherin belongs to the cadherin superfamily and PECAM-1 and JAM A, B, C, D to the immunoglobulin family, whereas CD99 is a unique molecule. ECi: Ig domains. Pro and Trans correspond to the propeptide and the transmembrane domain of each molecule. Cyt designates their cytoplasmic parts.

The extracellular domain of PECAM-1 binds very weakly to PECAM-1-expressing cells because of its low affinity for its cellular target. Once expressed at the surface of liposomes, however, PECAM-1 is able to promote the association of these vesicles. It can be deduced that PECAM-1 is able to interact homophilically when it is highly concentrated at the surface of the membrane. To define the immunoglobulin domains responsible for mediating homophilic binding, several chimeric proteins with one, two, three, four, or five Ig domains deleted were elaborated. Using an adhesion assay, these studies establish that the two N-terminal Ig domains EC1 and EC2 of PECAM-1 (Figure 2) are both necessary and sufficient to mediate homophilic interactions, but maximal stable binding requires all six extracellular Ig domains of PECAM-1. These homophilic interactions occur at the junctions between endothelial cells and may be involved in the diapedesis of leukocytes by creating heterotypic links between leukocytes and endothelial PECAM-1 molecules. However, PECAM-1-deficient mice suffer from only limited effects on the transendothelial migration of leukocytes. This probably results from an existing redundancy between adhesion molecules involved in leukocyte transmigration.

Similarly to other adhesive receptors of the immunoglobulin superfamily, PECAM-1 is also able to establish heterophilic interactions with integrins, particularly with $\alpha V\beta 3$ expressed at endothelial cell–cell junctions (Figure 4).

Cadherins

Endothelial cells express both N-cadherin and VE-cadherin at their surface. Despite similar expression levels, only VE-cadherin is expressed at cell–cell junctions, whereas N-cadherin is spread over the cell membrane. In fact, whereas VE-cadherin mediates homotypic interactions resulting in endothelial cell–cell attachment, N-cadherin seems to essentially participate in the anchorage of other N-cadherin-expressing cells, such as pericytes, to the endothelium. These N-cadherin-mediated interactions may allow the formation of the cellular multilayer structure required for the elaboration of shear stress–resistant vessels. This review will focus mainly on VE-cadherin as this is the major cadherin molecule of endothelial cell junctions.

VE-cadherin belongs to the superfamily of cadherins which are Ca^{2+}-dependent cell–cell adhesion receptors. Cadherin molecules consist of five extracellular modules designated as EC1 to EC5, a single transmembrane domain, and a cytoplasmic tail (Figure 2). The major property of cadherins is to mediate, via their extracellular parts, homophilic interactions. This means that cadherin X selectively interacts with cadherin X. Based on these homophilic interactions, cells expressing identical cadherins interact together by means of homotypic interactions. In contrast, cells expressing various cadherins segregate into like groups when mixed. To generate strong cell–cell adhesion,

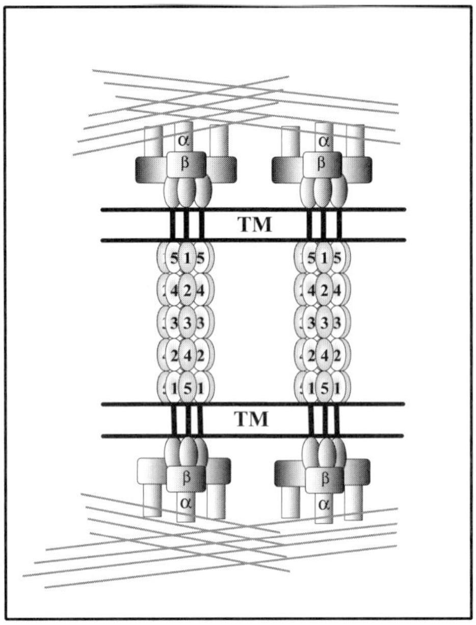

Figure 3 VE-cadherin and its intracellular partners elaborate the adherens junctions between endothelial cells. VE-cadherin molecules elaborate homophilic interactions that are able to link neighboring endothelial cells. Their extracellular domains self-associate to form a hexameric structure at the endothelium surface while their cytoplasmic tails bind proteins such as β catenin. VE-cadherin is anchored to the actin cytoskeleton (—) by the fixation of α catenin on VE-cadherin-bound β catenin.

homophilic and homotypic interactions between the cadherin ectodomains must be strengthened by intracellular interactions involving the cytoplasmic tails of cadherins. This domain exhibits two highly conserved regions: the first one, located at the C terminus of the cytoplasmic domain, binds β-catenin. The second one, close to the membrane, interacts with another catenin protein named p120. The anchorage of cadherins to the actin cytoskeleton is indispensable for the development of strong cell–cell adhesion. This connection is mediated by α-catenin, which interacts on one hand with cadherin–bound β-catenin and on the other with the actin cytoskeleton (Figure 3).

We demonstrated that VE-cadherin participates in the maintenance and restoration of the endothelium integrity. Moreover, we also proved that VE-cadherin elaborated hexameric probably antiparallel structures absolutely required for the elaboration of stable endothelial adherens junctions [4] (Figure 3).

The existence of an unknown cadherin at the surface of T cells was discovered in 1996. Similarly to classical cadherins, this new cadherin is able to bind α, β, and γ catenins and may interact with VE-cadherin, thus facilitating the homing of T cells.

CD99

CD99 is a 32-kDa glycoprotein, discovered in 1988 and expressed at the surface of all human T cells and many other hematopoietic cell types such as platelets, neutrophils, and red blood cells. It is also found at the lateral borders of confluent monolayers of human endothelial cells similarly to PECAM-1 and VE-cadherin. CD99 shares no structural homology with any known family of proteins.

Recent studies established that CD99 is involved in the control of the transport of newly synthesized MHC class I molecules to the plasma membrane in B lymphocytes, suggesting that it may be associated with the post-Golgi trafficking machinery.

In T cells, CD99 exists under two different isoforms, 32 kDa and 28 kDa, respectively, generated by alternative splicing. The 32-kDa form corresponds to the full-length protein, whereas the 28-kDa form exhibits a deletion within its cytoplasmic domain. These two CD99 forms can elaborate homophilic associations, leading to the formation of homodimers and heterodimers. Recently, it was established that the heterodimeric form, created via the formation of a covalent disulfide bridge between the cysteines 155, is able to induce T-cell apoptosis. It can be deduced that the elaboration of these dimers modulates the fate of T cells via the transduction of a proapoptotic signal.

Analysis of the cytoplasmic tail of CD99 reveals a lysine triplet (K151K152K153) that may function as a putative binding site for ezrin/radixin/moesin. These proteins are known to link surface proteins, such as E selectin and ICAM-2, to the actin cytoskeleton.

Junctional Adhesion Molecules (JAMs)

JAMs form a new Ig-like superfamily of adhesion receptors whose first member, called JAM A, was only discovered in 1998. Since that time, three other members have been described as JAM B, C, and D [5]. (JAM D is also named as JAM 4 in the literature.) They contain two extracellular immunoglobulin-like domains, a transmembrane domain and a relatively short cytoplasmic tail (Figure 2). Although all JAMs localize at endothelial cell–cell junctions and more precisely at tight junctions, they display different cellular distributions. Thus, the expression of JAM B is restricted to some endothelial subpopulations including high endothelial venules or lymphatic cells, whereas JAM A is expressed in platelets and in endothelial and epithelial cells. Following their transfection into CHO or MDCK cells, JAM A and B molecules are specifically enriched at cell–cell contact sites at the level of tight junctions, suggesting that they are capable of promoting cell–cell adhesion. In vitro studies indicate that the extracellular part of JAM-A self-associates as a dimer and this homodimerization seems to be required for its adhesive function and the organization of the junctional endothelial structure.

JAMs are also able to mediate heterophilic interactions. For instance, although JAM C is unable to interact with JAM A, it binds firmly to JAM B. Moreover, JAMs can interact with some integrins generally expressed at the surface of leukocytes. Thus, whereas JAM-A is a ligand of the β2 integrin LFA-1, JAM-B is the counterreceptor for the leukocyte β1 integrin α4β1, and JAM C interacts with the

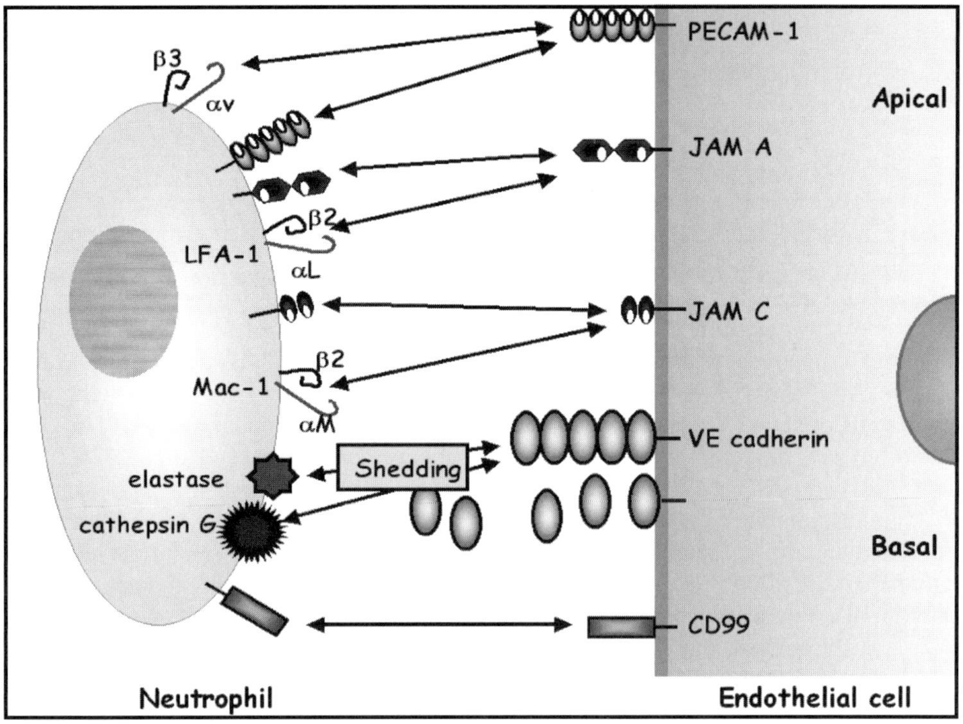

Figure 4 Adhesive interactions between endothelial cells and transmigrating neutrophils. The transmigration process involves several adhesive receptor pairs. By means of homophilic and heterophilic interactions, neutrophils progress through the endothelial junctions. The first transmigration step implicates homophilic and heterophilic PECAM-1 interactions. These homophilic interactions are elaborated between PECAM-1 molecules expressed at the surface of both endothelial cells and neutrophils, whereas the heterophilic ones involve the leukocyte integrin αVβ3. These PECAM-1 interactions are relayed by homophilic CD99-mediated interactions. Several interactions mediated by different JAMs receptors also participate in the transmigration, but their time course is still not known. When migrating neutrophils reach the adherens junctions where VE-cadherin is expressed, elastase and cathepsin G bound to the neutrophil cell surface can cleave its extracellular domain, thus opening their way through the adherens junctions.

β2 integrin Mac-1. These interactions significantly expend the repertoire of contacts between endothelial cells and leukocytes.

Mechanism of Leukocyte Diapedesis

The mechanism of leukocyte entrance into tissues requires, first, leukocyte adhesion to the endothelium and, subsequently, migration across the blood-vessel wall.

Leukocyte Adhesion

The leukocyte adhesion cascade is a sequence of adhesion and activation events that precede the extravasation of leukocytes. This multistep event is mediated by the engagement of adhesion receptors on leukocytes and their counter-receptors on endothelial cells. First, leukocytes escape from the bloodstream by capture or tethering to the endothelium. Indeed, during inflammation, endothelial cells are activated by signals emanating from inflammatory mediators such as TNFα, nitric oxide, and complement factors. Within minutes of stimulation, P- and E-selectins are expressed at the surface of endothelial cells and interact with their primary ligand PSGL-1 (P-selectin glycoprotein ligand-1), constitutively found on all leukocytes. This allows the capture of leukocytes on the surface of endothelial cells and permits neutrophil celerity to be slowed. Following a rolling phase, neutrophils are arrested on the endothelial cell surface. This adhesion step is mostly mediated by interactions involving β2-integrins on the leukocyte surface and immunoglobulins such as ICAM-1 and VCAM-1 on the endothelium. Subsequently, neutrophils migrate on the endothelial surface by amoeboid movement toward the borders of endothelial cells where transmigration occurs.

For more details on the coordinated interplay between the adhesion and signaling proteins involved in the process of firm adhesion of leukocytes on the endothelium, the reader is referred to an excellent recent review [6].

Leukocyte Transmigration

The multiple steps implicated in the transmigration process require the formation and destruction of homophilic and heterotypic interactions between receptors expressed at the surface of leukocytes and endothelial cells. To investigate the role of each adhesion receptor described earlier, attempts were made to block transendothelial migration of leukocytes using specific antibodies.

Role of PECAM-1 and CD99

Pretreating monocytes or neutrophils with antibodies specific for PECAM-1 inhibited their emigration across endothelial cell monolayers in in vitro assays. Reciprocally, blockage of endothelial cell PECAM-1 results in an effective inhibition of leukocyte transmigration. It was also found that PECAM-1 is required for in vivo leukocyte recruitment, because antibodies against PECAM-1 were able to block the accumulation of neutrophils in the peritoneal cavity in a rat model system. This means that homophilic interactions of PECAM-1 on leukocytes with PECAM-1 on endothelial junctions are involved in transmigration. In fact, PECAM-1 is constitutively recycled along the endothelial cell borders. During the transendothelial migration process, recycling is targeted to the part of the junction across which monocytes are squeezing. Indeed, anti-PECAM-1 antibodies block the recruitment of PECAM-1 to the zones of leukocyte migration and thereby the leukocyte transmigration process.

The role of CD99 in leukocyte transendothelial migration was more recently established by a similar method using anti-CD99 antibodies. Thus, whereas anti-CD99 antibodies have no effect on monocyte adhesion, they are able to block their transmigration significantly. Inhibition of either leukocyte or endothelial cell CD99 blocked leukocyte transmigration with equal efficiency, suggesting that homophilic interactions between CD99 expressed on both monocytes and endothelial cells are required for transmigration [7]. This effect is more drastic than the inhibition induced by anti-PECAM-1 antibodies. Simultaneous use of anti-PECAM-1 and anti-CD99 antibodies completely blocks the ability of monocytes to transmigrate. This additive effect suggests that PECAM-1 and CD99 possess distinct roles in the transmigration process. In fact, leukocytes arrested by anti-PECAM-1 antibodies adhere to the apical surface of the endothelium, whereas those arrested by anti-CD99 antibodies are trapped inside the endothelial cell–cell junctions. This indicates that PECAM-1 and CD99 are differentially located along the interendothelial junction and are sequentially involved in the transmigration process. PECAM-1-mediated homotypic interactions are probably relayed by CD99-mediated ones during leukocyte migration. To date, it is difficult to ascertain whether such a sequential process can be extended to the other leukocyte subfamilies, the surface expression of CD99 of the various types of leukocytes being only partially known.

Role of JAMs

Concerning the role of JAMs on the transendothelial migration of leukocytes, several antibodies directed against JAM A were described to have the capability to inhibit the transmigration of monocytes, T cells, and neutrophils in vitro and in vivo. Thus, in a model of meningitis in mice, it was demonstrated that the addition of anti-JAM A antibodies strongly reduced monocyte and neutrophil infiltration into the cerebrospinal fluid and brain parenchyma. Moreover, anti-JAM B or a soluble fragment of JAM B were also able to block chemokine-induced lymphocyte transmigration.

In fact, JAM A plays a critical role in the diapedesis of leukocytes. After stimulation of endothelial cells with both TNFα and IFNγ, JAM-A is redistributed from the junctions to the endothelial surface. Subsequently, it becomes available to interact with the activated form of the β2-integrin LFA-1 (β2αL), thus contributing to the adhesion of leukocytes on the activated endothelium. At this stage, homophilic interactions involving JAM A expressed both on leukocytes and on endothelial cells are not crucial to firm adhesion of leukocytes. This is particularly true for monocytes and neutrophils that express very low amounts of JAM A at their surfaces. The interaction of JAM A with LFA-1 contributes first to the chemokine-triggered adhesion and second to the transmigration of memory T cells, neutrophils, and lymphocytes.

The binding of endothelial JAM A to LFA-1 on leukocytes may disrupt junctional homophilic JAM A interactions, thereby unlocking interendothelial junctions. Complementary data are needed to understand the interplay between the homophilic and heterophilic JAM A–mediated interactions occurring during leukocyte transmigration.

Moreover, according to the subpopulation of leukocytes, the transmigration step involving the JAM family implicates different couples of adhesion receptors. For example, JAM B expressed at the surface of high endothelial venules interacts with JAM C of T cells and NK and dendritic cells. This molecular pair seems to play a critical role in the trafficking of T, NK, and dendritic cells into and out of high endothelial venules.

Role of VE-Cadherin

In contrast to antibodies specific for PECAM-1, JAMs, or CD99, antibodies directed against the extracellular domain of VE-cadherin increase the recruitment of leukocytes to the infection sites. In fact, the anti-VE-cadherin antibodies have the ability to disrupt the VE-cadherin homophilic interactions. This results in the formation of gaps between endothelial cells through which leukocytes can easily squeeze. In physiological conditions, alterations of VE-cadherin-mediated adhesion also occur at endothelial junctions during leukocyte transmigration. Indeed, immunofluorescence experiments reveal that VE-cadherin and α-, β-, and γ- catenins disappear from cell-to-cell contacts following adhesion of neutrophils to endothelial monolayers. This effect appears only where neutrophils firmly adhere while VE-cadherin-based complexes remain intact in areas devoid of adherent neutrophils. By contrast, PECAM-1 distribution remains unaffected.

The mechanism leading to the formation of gaps at cell–cell junctions during leukocyte transmigration is still controversial. To cast some light on this mechanism, movements of VE-cadherin occurring during leukocyte transmigration were observed using real-time microscopy. Thus, transmigration of fluorescently labeled leukocytes was followed in real time by two-color fluorescence microscopy using HUVECs expressing endogenous VE-cadherin and a VE-cadherin protein fused with its C-terminal part to Green Fluorescent Protein (VE Cad-GFP) [8]. This study showed that transmigration occurs both through preexisting gaps and through de novo gap formation. Gap widening accommodates to the size of the transmigrating leukocyte, allowing a narrow contact between the endothelial cell and the leukocyte. The widening of the gaps seems to be accompanied by a clustering of the VE Cad-GFP molecules on the edge of the forming clefts. After transmigration, the displaced molecules diffuse back to reconstitute the junctions. This leads to a rapid resealing of the junctions within 5 minutes after leukocyte transmigration. These results were confirmed by a real-time study following the differential movements of VE-cadherin in living endothelial cells during transmigration of neutrophils with anti-VE-cadherin antibodies that do not interfere with transendothelial migration. The lateral movement of VE-cadherin was hypothesized to be a consequence of its decoupling from the cytoskeleton, probably initiated by an intracellular signal due to leukocyte adhesion to the endothelium. In both of the studies, only the movement of the cytoplasmic tail of VE-cadherin was followed in real time. It is not known whether the extracellular domain of VE-cadherin remains intact during the lateral movement.

Real-time fluorescent imaging shows that neutrophils only move laterally underneath the vascular endothelium instead of deeply into the vascular tissue. Thus, the question as to how neutrophils go across the endothelium before penetrating into the surrounding tissues remains unanswered. The curtain effect that moves VE-cadherin is necessary but not sufficient to completely support leukocyte transmigration.

The event initiating the lateral movement of VE-cadherin may correspond to the cleavage of this adhesive receptor by proteases expressed at the surface of transmigrating neutrophils. This hypothesis is based on the fact that elastase, at the surface of transmigrating neutrophils, mainly localizes to the migration front [9]. Thus, after the adhesion to the endothelial cell surface, elastase is expressed on the apical face of neutrophils, avoiding contact between proteases and the endothelial cell surface. When neutrophils reach endothelial cell–cell junctions, elastase moves from the apical face to the basal face where it comes in close proximity with junctional components. An in vivo study suggested that elastase may contribute to lung neutrophil accumulation and microvascular injury during intestinal ischemia-reperfusion (IR). Moreover, following IR, elastase and VE-cadherin fragments were found in the bronchoalveolar fluid.

Despite results obtained in vitro and in vivo, the participation of proteases in the transmigration of leukocytes is still controversial. Indeed, a study published in 2002 suggested that elastase and MMP9 are not essential for neutrophil transmigration. This assumption is based on the use of neutrophils depleted in both elastase and MMP9 that show no defect in their transmigration capacity. These results contradict our own study as we assessed that VE-cadherin is cleaved following adhesion of neutrophils to endothelial cell monolayers [10]. Using specific inhibitors of neutrophil proteases, we were able to identify elastase and cathepsin G bound to the neutrophil surface as the major proteases involved in the cleavage of VE-cadherin. Furthermore, inhibition of both membrane-bound proteases inhibited neutrophil transmigration in in vitro assays. In vivo, the surface-bound elastase and cathepsin G may cleave VE-cadherin on very restricted areas close to the sites of neutrophil adhesion. On the endothelium, the redundancy between these two proteases can explain why neutrophils still transmigrate across endothelial monolayers in the presence of specific inhibitors of elastase and also why neutrophils from elastase-deficient mice show no defect in transendothelial migration. Indeed, cathepsin G can replace elastase, thus allowing the neutrophil transmigration to occur despite elastase inhibition or deficiency.

The involvement of a paracrine mechanism in neutrophil-mediated alteration of the endothelial barrier function was suggested by Gautam and coworkers. Upon neutrophil adhesion to the endothelium, leukocyte β2-integrin signaling triggered the release of the neutrophil CAP37, a serine protease without catalytic activity that induced a cytoskeletal rearrangement and the formation of gaps in the endothelium in vitro. Alteration of the endothelium may result from the interaction between CAP37 on the surface of neutrophils and proteoglycans present at the endothelial cell surface. This interaction could stimulate endothelial cell contraction by a yet unidentified mechanism.

Conclusions

During the past 5 years, it has been established that new receptors such as JAMs and their integrin counterreceptors play major roles in leukocyte transmigration. Also, the fate of VE-cadherin during leukocyte transmigration is now better understood. In particular, it has been shown that membrane-bound neutrophil elastase and cathepsin G facilitate neutrophil transmigration by cleaving the extracellular part of VE-cadherin.

These adhesive receptors are definitely more than just a molecular glue, because it is suspected that they mediate intracellular signaling. Thus, some data from our laboratory suggest the existence of VE-cadherin-mediated signaling pathways triggered by the destabilization of cadherin junctions. This pathway leads eventually to the restoration of endothelium integrity by a neosynthesis of VE-cadherin molecules. The molecules involved in this pathway remain to be determined. Nevertheless, several elements suggest that β-catenin and possibly the Wnt signaling pathway might play a role in the restoration of endothelium integrity.

A better understanding of the sequential events occurring during transendothelial migration of leukocytes may lead to the discovery of new classes of target molecules for the design of novel drugs that are able to interfere with the transmigration process. Such drugs could have major applications in several important human illnesses such as polyarthritis or cancer.

Glossary

Endothelium: Monolayer of endothelial cells that line the inside of vascular tree.

Neutrophils: A subtype of white cells.

Transmigration: Cellular process allowing white cells to migrate across the endothelial barrier.

VE cadherin: Ca++-dependent adhesive receptor exclusively expressed in endothelium.

References

1. Worthylake, R. A., and Burridge, K. (2001). Leukocyte transendothelial migration orchestrating the underlying molecular machinery. *Curr. Opin. Cell Biol.* **13**, 569–577.
2. Owen, C. A., Campbell, M. A., Sannes, P. L., Boukedes, S. S., and Campbell, E. J. (1995). Cell surface-bound elastase and cathepsin G on human neutrophils: A novel, non-oxidative mechanism by which neutrophils focus and preserve catalytic activity of serine proteinases. *J. Cell. Biol.* **131**, 775–789. *This paper describes the translocation of serine proteases such as elastase and cathepsin G from azurophil granules to the cell surface after stimulation of neutrophils by chemokines. These cell surface–bound proteases are catalytically active and resistant to proteinase inhibitors.*
3. Newman, P. J. (1997). The biology of PECAM-1. *J. Clin. Invest.* **99**, 3–8.
4. Bibert, S., Jaquinod, M., Concord, E., Ebel, C., Hewat, E., Vanbelle, C., Legrand, P., Weidenhaupt, M., Vernet, T., and Gulino-Debrac, D. (2002). Synergy between extracellular modules of vascular endothelial cadherin promotes homotypic hexameric interactions. *J. Biol. Chem.* **277**, 12790–12801.
5. Muller, W. A. (2003). Leukocyte–endothelial-cell interactions in leukocyte transmigration and the inflammatory response. *Trends Immunol.* **24**, 327–334.
6. Weber, C. (2003). Novel mechanistic concepts for the control of leukocyte transmigration: Specialization of integrins, chemokines, and junctional molecules. *J. Mol. Med.* **81**, 4–19.
7. Schenkel, A. R., Mamdouh, Z., Chen, X., Liebman, R. M., and Muller, W. A. (2002). CD99 plays a major role in the migration of monocytes through endothelial junctions. *Nat. Immunol.* 143–150.
8. Shaw, S. K., Bamba, P. S., Perkins, B. N., and Luscinskas, F. W. (2001). Real-time imaging of vascular endothelial-cadherin during leukocyte transmigration across endothelium. *J. Immunol.* **167**, 2323–2330. *Using a VE-cadherin/green fluorescent protein fusion construct (VEcadGFP), Shaw et al. visualized alterations in endothelial junctional structures in real time during transmigration of human neutrophils and monocytes in an in vitro flow model. They show that migrating leukocytes appeared to push aside VEcadGFP in the plane of the junction, and this displaced material subsequently diffused back to refill the junction.*
9. Cepinskas, G., Sandig, M., and Kvietys, P. R. (1999). PAF-induced elastase-dependent neutrophil transendothelial migration is associated with the mobilization of elastase to the neutrophil surface and localization to the migrating front. *J. Cell Sci.* **112**, 1937–1945.
10. Hermant, B., Bibert, S., Concord, E., Dublet, B., Weidenhaupt, M., Vernet, T., and Gulino-Debrac D. (2003). Identification of proteases involved in the proteolysis of VE cadherin during neutrophil transmigration. *J Biol Chem* **278**, 14002–14012. *This paper established that neutrophil elastase and cathepsin G, bound at the surface of transmigrating neutrophils, are able to cleave VE-cadherin at endothelial cell–cell junctions, thus facilitating neutrophil diapedesis.*

Capsule Biography

Dr. Gulino-Debrac is a senior scientist from the Centre National de la Recherche Scientifique (CNRS). Since 1996 she has been in charge of a group working on VE-cadherin in Institut de Biologie Structurale Jean-Pierre Ebel (Grenoble, France). Her work is supported by grants from the CNRS, the Commissariat à l'Energie Atomique (CEA), and the Ligue contre le Cancer.

CHAPTER 49

Vascular Gap Junctions

[1]Michael Koval and [2]Jahar Bhattacharya

[1]*Department of Physiology and Institute for Environmental Medicine, University of Pennsylvania School of Medicine, Philadelphia, Pennsylvania*

[2]*Lung Biology Laboratory, College of Physicians and Surgeons, St. Luke's-Roosevelt Hospital Center, Columbia University, New York, New York*

Introduction

Gap junctions are intercellular channels that allow diffusion of small solutes and ions between adjoining cells. This intercellular exchange enables intercellular communication of specific signals and metabolic products that promote coordination of multicellular function. In blood vessels, gap junctions have been implicated in such diverse functions as regulation of vascular tone, transendothelial cell migration, and the pathogenesis of atherosclerosis. Here we focus on some essential principles of gap junctional function in the vascular bed.

Connexins

Gap junction channels are composed of proteins in the connexin family. There are almost two dozen mammalian connexins. "Cx" followed by a number, for example, Cx43, denotes each connexin. The number predicts molecular mass, although this prediction may not always be accurate. A gap junction channel consists of two hexameric hemichannels, one in each cell. For some connexins, e.g., Cx32, hemichannels are formed in the endoplasmic reticulum (ER), while for others, such as Cx43 and Cx46, the hemichannels are assembled in the Golgi apparatus [1]. Regulated trafficking of connexin hemichannels to the plasma membrane enables modulation of channel number at the intercellular interface.

The major vascular connexins are Cx37, Cx40, and Cx43, which are expressed differing extents depending on tissue type and location [2]. These connexins, when expressed by the same cell, have been shown to form heteromeric (mixed) gap junction channels including Cx40 : Cx43, Cx37 : Cx43, and Cx37 : Cx40 channels.

The ability of connexins to intermix defines permeability and gating characteristics of the gap junction channel. Connexins also have multiple conductance states. The interpretation of cell–cell coupling by gap junctions requires consideration of both the extent of compatibility between connexin subtypes and the physical–chemical properties of the probe being used. A good approach is to obtain electrophysiological quantifications of net ionic flux across gap junctions simultaneously with those of fluorescent dye transfer. This dual approach normalizes channel permeability against channel surface area. Using this approach, Burt et al. found that communication through heteromeric Cx40 : Cx43 channels increased with an increase in the amount of Cx43 relative to Cx40 [3]. Other examples of heteromeric gap junction channels with unique permeability characteristics include the following connexin pairings: Cx37 : Cx43, Cx40 : Cx43, and Cx43 : Cx45. Given that different classes of vascular cells express different ratios of these connexins, this suggests that intercellular communication is regulated through differential connexin expression.

Gap Junctional Regulation

Increasing evidence indicates that proteins that associate with connexins may also regulate channel function. For example, the association of zonulae occludens-1 with both connexins and tight junction proteins suggests that crosstalk may exist between gap and tight junctions. Gap junctions are also modulated by the cystic fibrosis transmembrane conductance regulator, which may also be a part of the junctional complex, and by wnt-1, suggesting a link with β-catenin and cadherins in gap junctional regulation [4]. However, the functional implications of these interactions are not well understood.

Connexin-specific gating mechanisms modify gap junctional communication. Acidic intracellular pH gates channels formed with Cx43 and Cx46, although not all connexins—for example, Cx37—are pH sensitive. Differential phosphorylation of connexins is another potential mechanism for gap junctional gating. Kinases implicated in this regulation include protein kinase C, which inhibits Cx43 assembly and thereby decreases Cx43-mediated gap junctional communication. A similar inhibitory effect on Cx43 gap junctions is attributed to MAP kinases and src kinases. These considerations indicate that connexin expression alone does not control the level of gap junctional communication; instead, crosstalk between connexins and other signaling pathways also needs to be considered.

Vascular Connexin Expression

Connexin expression is sensitive to cell state and varies with cell and vessel type. Connexin protein expression in situ is typically defined by immunohistochemical localization or by immunoblot. However, cross-reactivity between antibodies, particularly cross-reactivity of anti-Cx40 antibodies to Cx43, is a potential concern when considering analysis of connexin expression in the vasculature.

Gap junctions are more prominent in large than small vessels. Interendothelial and inter–smooth muscle junctions are homologous, whereas endothelial–smooth muscle junctions are heterologous. Cx37 and Cx40 are uniformly expressed by vascular endothelial cells in situ. However, endothelial expression of Cx37 and Cx43 is greater in large than in small vessels. Endothelial Cx43 expression might be species-specific, since the expression is pronounced in rat and bovine aortic endothelium, but less prominent in mouse aortic endothelium. Also, Cx43 expression is higher in areas with turbulent flow, such as at vessel branch points.

Cx43 is the most prominent connexin expressed by vascular smooth muscle. Low levels of Cx37 expressed by resting vascular smooth muscle may be upregulated during new vessel growth. Other connexins expressed by vascular smooth muscle include Cx45, which is limited to specific vascular beds such as cerebral vessels and ascending aorta of rat, and the recently identified Cx31.9.

Heteromer formation is likely to be important for formation of functional myo-endothelial gap junctions since endothelial Cx37 : Cx40 hemichannels are linked to vascular smooth muscle hemichannels containing mainly Cx43 to form well-coupled heterotypic gap junctions. In the absence of Cx37, homomeric Cx40 hemichannels expressed by endothelium and Cx43 hemichannels expressed by vascular smooth muscle would form poorly conducting heterotypic gap junction channels.

Knockout models indicate that the coordinated expression of connexin subtypes is complex in the vascular bed. In one study, endothelial cell–cell coupling was maintained in Cx40-deficient mice by compensatory upregulation of endothelial Cx37, but not Cx43 [5]. However, Cx37 deficiency had little effect on Cx40 or Cx43 expression or on cell coupling. Although Cx37 or Cx40 deficiency alone has little effect on vessel morphology, mice doubly deficient in Cx37 and Cx40 die soon after birth and show abnormally dilated blood vessels, particularly in skin and testis [6]. Vessels in Cx37 –/– Cx40 –/– mice also have permeability defects and localized hemorrhages, suggesting a role for endothelial gap junctions in regulating vessel barrier function. Cx43 deficiency does not affect global vascular development, although coronary artery development is partially defective [7]. Although Cx45 expression in the vascular bed is limited, it plays a critical, yet undetermined, role in vessel development during embryogenesis since Cx45-deficient mice have defective vascular branching and do not develop smooth muscle in large vessels.

Connexins and Vascular Tone

Gap junctions contribute to regulation of vascular tone. Thus, pharmacologic agents and connexin mimetic peptides, which inhibit gap junctional communication, also inhibit vasodilation. In fact, the well-documented phenomenon of conducted vasodilatation in conduit arteries is explained by longitudinal NO conduction along the vascular wall through gap junctions. Further, connexin-mimetic peptides inhibit the transmission of endothelial hyperpolarization to underlying smooth muscle cells, consistent with a role for heterologous endothelial–vascular smooth muscle communicating junctions in so-called endothelium-derived hyperpolarizing factor (EDHF)-induced vasodilation [8]. EDHF-type vessel relaxation mediated by gap-junctional communication is more prominent for small than large vessels. This is consistent with the finding that functional gap-junctional communication is present in small vessels as indicated by the presence of Ca^{2+} waves that propagate between endothelial cells [9].

Consistent with the prominent expression of Cx40 by endothelial cells, endothelial electrical conduction, vasodilation, and smooth muscle activation are defective in Cx40-deficient mice. These mice also have increased blood pressure, indicating a predominant role for Cx40, rather than Cx43, in establishing gap junctional connections between vascular smooth muscle and endothelium [10].

The neonatal lethal phenotype of Cx43-deficient mice has hampered the understanding of Cx43 in vascular function. To address this, Liao et al. [11] used *cre* expression driven by the Tie 2 promoter to preferentially excise the Cx43 gene from endothelial cells. These mice develop normally and their endothelial cells express two other connexins, Cx37 and Cx40, but they have abnormally low blood pressure and a lowered heart rate. Lowered blood pressure was due to increased plasma NO production, which was most likely due to increased eNOS activity. However, since endothelial NO production is linked to gap-junctional coupling between vascular smooth muscle and endothelial cells, this indicates that Cx43 depletion actually enhanced coupling between these cells, perhaps through enhanced formation of homomeric Cx40 gap junction channels.

Vascular Connexins in Pathologic Conditions

Vascular endothelium has the capacity to form gap junctions with circulating cells such as neutrophils expressing Cx43 and macrophages expressing Cx37. Such interactions may be significant in the development of inflammation, particularly since inflammation increases Cx43 expression in endothelial cells and leukocytes [12]. In contrast, myoendothelial cell–cell communication is downregulated in inflammation. The gap-junctional role in inflammation is also indicated in that gap junction inhibitors promote vascular leak and transvascular cell migration [13], suggesting that gap junctions help maintain vascular barrier function. Tumor cell extravasation might also be regulated by gap-junctional communication, since melanoma cells expressing high levels of Cx26 are able to communicate with vascular endothelium and are more metastatic than cells that lack connexin expression. Interestingly, since Cx26 is believed to be incompatible with Cx37, Cx40, and Cx43, the melanoma cell–endothelial interaction may be attributable to a novel mode of gap-junctional communication, or to an unknown endothelial connexin that pairs with Cx26.

Gap-junctional communication between foam cells and the vascular wall is also likely to promote atherosclerotic plaque formation [14]. Consistent with a role for Cx43 in plaque formation, Cx43 +/– deficient mice that are deficient for LDL receptor expression show significantly less endothelial Cx43 expression and fewer atherosclerotic plaques than mice expressing normal levels of Cx43. Cx43 upregulation in response to disturbed flow during plaque formation or in response to vessel injury may exacerbate this effect. Whether other connexins are involved in plaque formation remains to be determined; however, there are some hints that Cx37 may also play such a role. One possibility is that while some gap-junctional communication is important for transvascular cell migration, excessive cell–cell coupling can be inhibitory, which, in turn, could stabilize a foam cell precursor at sites of plaque formation.

Conclusion

In summary, vascular connexins are central to the structural and functional organization of the vascular bed. Functionally, they are overwhelmingly important for normal vascular development and in the regulation of vascular tone. The formation of gap-junctional channels is a complex process that is subject to plasticity and depends upon cell type and connexin subtype specificity. However, several questions remain. For instance, more needs to be understood with respect to the role of vascular connexins in pathological states. This role may be extensive and highly facilitatory to disease progression, especially in the vascular involvement in inflammation, in atherosclerotic plaque formation, and in abnormalities of blood coagulation. The emerging role of gap junctions in the consideration of vascular pathobiology promises to be fertile ground for future research.

Glossary

Connexin: Channel-forming transmembrane protein subunit. A complete gap junction channel consists of six connexins in one cell pairing with six connexins in an adjacent cell that allow direct transfer of small cytoplasmic molecules from one cell to an adjacent cell.

Gap junction: Originally defined by electron microscopy as a class of cell–cell contact sites with a uniform ~16-nm "gap" between the cells as opposed to tight junctions that show no gap. Gap junctions consist of an array of connexin-based channels that mediate cell–cell communication.

Heteromeric: A gap junction channel composed of two or more different types of connexins completely intermixed.

Heterotypic: Head-to-head interaction between one type of connexin in one cell and a different type of connexin in an adjacent cell.

Homomeric: A gap junction channel composed of a single type of connexin.

References

1. Das Sarma J., Wang, F., and Koval, M. 2002. Targeted gap junction protein constructs reveal connexin-specific differences in oligomerization. *J. Biol. Chem.* **277**, 20911–20918.
2. Severs, N. J., Rothery, S., Dupont, E., Coppen, S. R., Yeh, H. I., Ko, Y. S., Matsushita, T., Kaba, R., and Halliday, D. 2001. Immunocytochemical analysis of connexin expression in the healthy and diseased cardiovascular system. *Microsc. Res. Tech.* **52**, 301–322.
3. Burt, J. M., Fletcher, A. M., Steele, T. D., Wu, Y., Cottrell, G. T., and Kurjiaka, D. T. 2001. Alteration of Cx43:Cx40 expression ratio in A7r5 cells. *Am. J. Physiol. Cell Physiol.* **280**, C500–C508.
4. Xu, X., Li, W. E., Huang, G. Y., Meyer, R., Chen, T., Luo, Y., Thomas, M. P., Radice, G. L., and Lom C. W. 2001. Modulation of mouse neural crest cell motility by N-cadherin and connexin 43 gap junctions. *J. Cell Biol.* **154**, 217–230.
5. Kruger, O., Beny, J. L., Chabaud, F., Traub, O., Theis, M., Brix, K., Kirchhoff, S., and Willecke, K. 2002. Altered dye diffusion and upregulation of connexin37 in mouse aortic endothelium deficient in connexin40. *J. Vasc. Res.* **39**, 160–172.
6. Simon, A. M., and McWhorter, A. R. 2002. Vascular abnormalities in mice lacking the endothelial gap junction proteins connexin37 and connexin40. *Dev. Biol.* **251**, 206–220.
7. Li, W. E., Waldo, K., Linask, K. L., Chen, T., Wessels, A., Parmacek, M. S., Kirby, M. L., and Lo, C. W. 2002. An essential role for connexin43 gap junctions in mouse coronary artery development. *Development* **129**, 2031–2042.
8. Sandow, S. L., and Hill, C. E. 2000. Incidence of myoendothelial gap junctions in the proximal and distal mesenteric arteries of the rat is suggestive of a role in endothelium-derived hyperpolarizing factor-mediated responses. *Circ. Res.* **86**, 341–346.
9. Ying, X., Minamiya, Y., Fu, C., and Bhattacharya, J. 1996. Ca2+ waves in lung capillary endothelium. *Circ. Res.* **79**, 898–908.
10. de Wit, C., Roos, F., Bolz, S. S., Kirchhoff, S., Kruger, O., Willecke, K., and Pohl, U. 2000. Impaired conduction of vasodilation along arterioles in connexin40-deficient mice. *Circ. Res.* **86**, 649–655.
11. Liao, Y., Day, K. H., Damon, D. N., and Duling, B. R. 2001. Endothelial cell-specific knockout of connexin 43 causes hypotension and bradycardia in mice. *Proc. Natl. Acad. Sci. USA.* **98**, 9989–9994.
12. Navab, M., Liao, F., Hough, G. P., Ross, L. A., Van Lenten, B. J., Rajavashisth, T. B., Lusis, A. J., Laks, H., Drinkwater, D. C., and

Fogelman, A. M. 1991. Interaction of monocytes with cocultures of human aortic wall cells involves interleukins 1 and 6 with marked increases in connexin43 message. *J. Clin. Invest.* **87**, 1763–1772.

13. Zahler, S., Hoffmann, A., Gloe, T., and Pohl, U. 2003. Gap-junctional coupling between neutrophils and endothelial cells: a novel modulator of transendothelial migration. *J. Leukoc. Biol.* **73**, 118–126.
14. Kwak, B. R., Veillard, N., Pelli, G., Mulhaupt, F., James, R. W., Chanson, M., and Mach, F. 2003. Reduced connexin43 expression inhibits atherosclerotic lesion formation in low-density lipoprotein receptor-deficient mice. *Circulation* **107**, 1033–1039.

Further Reading

Brink, P. R. (1998). Gap junctions in vascular smooth muscle. *Acta Physiol Scand.* **164**, 349–356.

Griffith, T. M., Chaytor, A. T., Taylor, H. J., Giddings, B. D., and Edwards, D. H. (2002). cAMP facilitates EDHF-type relaxations in conduit arteries by enhancing electrotonic conduction via gap junctions. *Proc. Natl. Acad. Sci. USA* **99**, 6392–6397. *Study using connexin mimetic inhibitory peptides to show that calcium transmission through gap junctions was required for EDHF-type vascular relaxation.*

Hill, C. E., Rummery, N., Hickey, H., and Sandow, S. L. (2002). Heterogeneity in the distribution of vascular gap junctions and connexins: Implications for function. *Clin. Exp. Pharmacol. Physiol.* **29**, 620–625.

Kruger, O., Plum, A., Kim, J.S., Winterhager, E., Maxeiner, S., Hallas, G., Kirchhoff, S., Traub, O., Lamers, W. H., and Willecke, K. (2000). Defective vascular development in connexin 45-deficient mice. *Development* **127**, 4179–4193.

Liao, Y., Day, K. H., Damon, D. N., and Duling, B. R. (2001). Endothelial cell-specific knockout of connexin 43 causes hypotension and bradycardia in mice. *Proc. Natl. Acad. Sci. USA* **98**, 9989–9994. *Mice where connexin43 was specifically depleted from endothelial cells using the cre/lox recombinase system were used to demonstrate a role for endothelial connexin43 in regulating blood pressure.*

Saez, J. C., Berthoud, V. M., Branes, M. C., Martinez, A. D., and Beyer, E. C. (2003). Plasma membrane channels formed by connexins: Their regulation and functions. *Physiol Rev.* **83**, 1359–1400.

Capsule Biography

Dr. Koval's laboratory studies the molecular mechanisms of membrane protein assembly and roles for intercellular communication in regulating pulmonary function. His work on gap junctions started more than a decade ago and is supported by grants from the NIH and American Heart Association.

Dr. Bhattacharya's laboratory investigates intercellular connectivity and coordination in proinflammatory signaling in microvessels. Using lung inflammation as a model, the Bhattacharya group has developed optical imaging methods to quantify signaling mechanisms in cells in situ.

CHAPTER 50

Adhesion Molecules and the Recruitment of Leukocytes in Postcapillary Venules

Klaus Ley

Cardiovascular Research Center and Departments of Biomedical Engineering, Molecular Physiology and Biological Physics, University of Virginia, Charlottesville, Virginia

Postcapillary Venules

Postcapillary venules start where two capillaries meet and discharge their blood into a microvessel of peculiar morphology that rapidly increases its diameter as more capillaries enter. At the beginning, postcapillary venules have a diameter of less than 10 μm, and further downstream, postcapillary venules can be as large as 80 μm. Wall shear rate, the rate of change in velocity near the wall, and wall shear stress, the force exerted per unit area of venule parallel to the vessel axis, are lower in postcapillary venules than in precapillary arterioles, and this contributes to, but does not entirely account for, preferential leukocyte adhesion in these vessels [1]. Recently, the presence of a significant endothelial glycocalyx surface layer has been discovered [2], which is up to 0.5 μm thick and effectively reduces shear stress on the endothelial surface to zero [3]. The shear forces experienced by adherent or rolling leukocytes can be calculated from the cell geometry and an interfacial shear stress present at the interface between glycocalyx and lumen. This interfacial shear stress is, on average, five times higher than the Newtonian wall shear stress estimated for a homogenous fluid and no glycocalyx [3a].

Leukocytes

In human blood, most leukocytes are neutrophils, with a smaller fraction of lymphocytes, monocytes, eosinophils, and basophils. In mouse blood, lymphocytes are more abundant than neutrophils. The lymphocyte compartment contains B cells, various types of T cells, natural killer cells, and others. Blood lymphocytes comprise hundreds of phenotypes distinguishable by flow cytometry.

Under most conditions, leukocyte interactions with postcapillary venules are dominated by neutrophils. Typically, more than 90 percent of all rolling and adherent cells are neutrophils, with the balance monocytes and eosinophils. Lymphocytes rarely roll or adhere in postcapillary venules of nonlymphatic tissues, but roll vigorously in high endothelial venules of lymph nodes and other secondary lymphatic organs. Therefore, reports of leukocyte rolling and adhesion observed by intravital microscopy in nonlymphatic organs almost always refer to neutrophils, although this is not always specifically mentioned. Recently, mice have become available in which certain leukocyte subsets express green fluorescent protein (GFP), so that specific intravital microscopic studies become possible for neutrophils, monocytes, and certain T cells.

Adhesion Molecules

Leukocyte adhesion molecules are defined as transmembrane receptor–ligand pairs that transmit mechanical force. Adhesion molecules are required for leukocyte capture or initial tethering, rolling, firm adhesion, and transendothelial migration (Figure 1). All known adhesion molecules relevant for leukocyte adhesion are type I single-spanning transmembrane molecules (N terminus outside the cell), some are homo- or heterodimers. Although it is formally possible that type II or multispanning or GPI-linked molecules could be involved in leukocyte adhesion, there is no evidence that these molecules actually mediate leukocyte rolling or adhesion under in vivo conditions. Notably, a chemokine–chemokine receptor interaction, a variant of fibroblast growth factor receptor, a GPI-anchored receptor, and a secreted molecule have all been proposed to serve as leukocyte adhesion molecules, but none of these proposed interactions have been substantiated in vivo.

Selectins

The selectins are a class of three C-type (calcium-binding) lectins that are expressed on leukocytes (L-selectin), inflamed endothelial cells (E-selectin), and activated platelets and endothelial cells (P-selectin) [4]. They share a common genomic organization and protein structure with an N-terminal lectin domain followed by an epidermal growth factor (EGF) domain, a number (two to nine) of consensus repeats, a transmembrane domain, and a cytoplasmic domain that is homologous among selectins from different species, but unique to each type of selectin [4]. Selectins bind glycosylated selectin ligands (see later discussion) with high to intermediate affinities (nanomolar range) and high on-rates and off-rates. This unique combination of properties enables selectins to mediate capture between cells in the flowing blood and the endothelium, and rolling through formation and breakage of reversible bonds. All three selectins are important in inflammation. In addition, L-selectin mediates rolling of (naive) lymphocytes in high endothelial

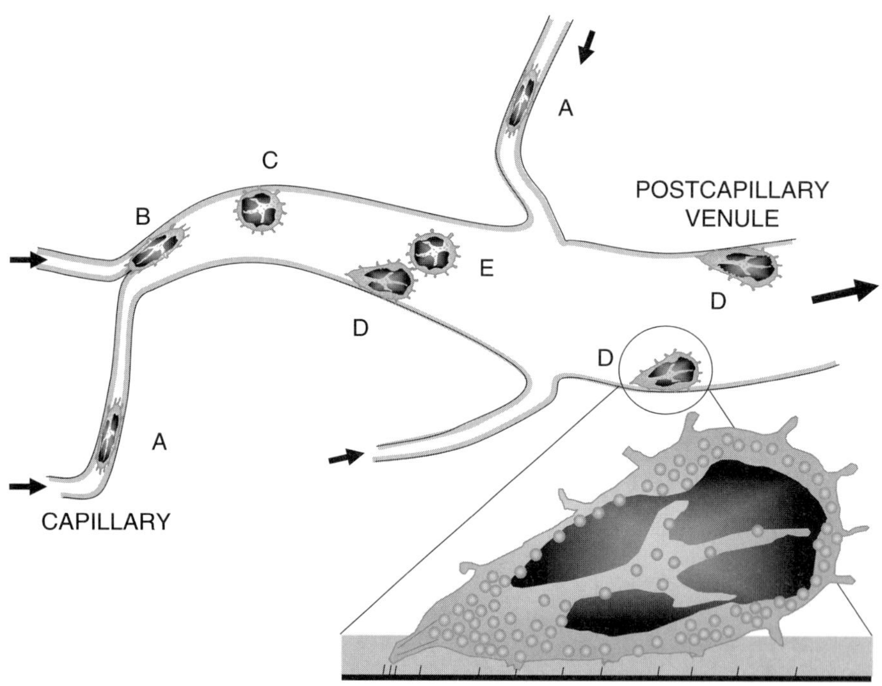

Figure 1 Neutrophils interacting with the endothelium lined with a 500-nm glycocalyx (blue) in a postcapillary venule. In capillaries, neutrophils are deformed into a near-cylindrical shape (A) and deform the endothelial glycocalyx. Almost all leukocyte rolling is initiated at the beginning of postcapillary venules, where leukocytes are in close physical contact with the endothelial surface (B). Noninteracting deformed leukocytes eventually recover their spherical shape (C), but rolling leukocytes, especially neutrophils, acquire a characteristic teardrop shape that reflects the effect of adhesive forces balanced by shear forces on the cytoskeleton (D). Rolling or adherent cells can nucleate L-selectin-PSGL-1-dependent secondary capture or tethering events (E). A close-up of a rolling neutrophil (bottom right) shows the endothelial glycocalyx (grey 500 nm) with endothelial adhesion molecules (E-selectin, 30 nm, and P-selectin, 40 nm). The black hairlike lines represent the length of a P-selectin-PSGL-1 pair (100 nm) completely buried in the glycocalyx. Rolling leukocytes probably continuously deform the glycocalyx as they roll while selectin (and integrin) bonds are formed at the leading edge (right) and broken at the trailing edge (left). Arrows indicate direction of blood flow. (see color insert)

venules of secondary lymphatic organs, P-selectin mediates platelet interactions with leukocytes and endothelial cells, and E-selectin mediates homing of certain lymphocytes to the skin. L-selectin is constitutively expressed on all neutrophils, monocytes, eosinophils, and most lymphocytes. P-selectin is stored in Weibel-Palade bodies of endothelial cells and platelet alpha granules and released upon stimulation by secretagogues. E-selectin is transcriptionally regulated through nuclear factor kappaB (NFκB)-dependent pathways. It is suspected that all three selectins and other adhesion molecules also serve signaling functions. Leukocyte activation upon ligation of L-selectin has been demonstrated, as has endothelial cell remodeling secondary to E-selectin engagement. There is little evidence for signaling through P-selectin, but engagement of endothelial or platelet P-selectin appears to result in activation of the bound cells via selectin ligands.

Selectin Ligands

Many selectin ligands have been proposed, but only P-selectin glycoprotein ligand-1 (PSGL-1) has been fully characterized and confirmed to be relevant in vitro and in vivo. PSGL-1 is a homodimer with functional groups near the N terminus, including three (in mouse, two) sulfated tyrosine residues and a critical threonine decorated with an O-linked carbohydrate chain terminating in the tetrasaccharide sialyl Lewis^x (sLex). sLex biosynthesis requires fucosyl transferase VII and a sialyl transferase (ST), possibly ST3GalIV, and is most likely directly involved in selectin binding. The critical O-linked side chain is located on a core2 structure so that core2 *N*-acetylglucosaminyl transferase is required for biosynthesis of fully functional PSGL-1. PSGL-1 functions much better as a dimer than as a monomer, possibly because P-selectin is also present as a dimer on the cell surface. Cocrystals between PSGL-1 and P-selectin show that both the tyrosine sulfate and the carbohydrate interact with the lectin domain simultaneously.

Although PSGL-1 was initially discovered as a P-selectin ligand, it was later shown to also bind L-selectin and E-selectin. PSGL-1 knockout mice have a significant defect in P-selectin mediated rolling, a complete absence of transient leukocyte–leukocyte interactions known as secondary tethering, and a moderate defect in E-selectin-dependent rolling. This suggests that PSGL-1 may be the only relevant L-selectin ligand on leukocytes, although other L-selectin ligands have been postulated. PSGL-1 is clearly the major P-selectin ligand, accounting for 80 to 90 percent of P-selectin mediated rolling. PSGL-1 appears to mediate leukocyte capture on E-selectin, but is not the E-selectin ligand responsible for the characteristic slow rolling.

Other proposed selectin ligands include mouse GlyCAM-1, which mediates L-selectin-dependent leukocyte rolling when coated on a solid substrate, but is normally secreted from high endothelial venules in secondary lymphatic organs, and in milk. No ortholog of GlyCAM-1 has been found in humans so far. The GlyCAM-1 knockout mouse showed no evidence of defective lymphocyte homing. E-selectin glycoprotein ligand-1 (ESL-1) was shown to bind E-selectin with high affinity, but most ESL-1 is present in the Golgi system and not accessible from the cell surface. CD24 is a GPI-linked glycoprotein that can function as a P-selectin ligand and can mediate rolling of cancer cells on P-selectin, but a function in leukocyte adhesion or rolling was not demonstrated. Other molecules on neutrophils including the integrin Mac-1 express sLex, but the function of these molecules as selectin ligands in leukocyte recruitment remains unclear. A number of other putative L-selectin ligands have been characterized in high endothelial venules, but it is unclear which of these molecules are physiologically relevant.

Integrins

Integrins are heterodimeric transmembrane molecules expressed by almost all cells except erythrocytes. Most if not all integrins exist in two conformations, active (ligand-binding) and inactive. In some integrins, dramatic changes in molecular shape have been demonstrated upon activation [5], but some of these changes may be secondary to isolation conditions. Although these and other results are suggestive of a conformational change, the exact nature of integrin activation remains an area of ongoing investigation. In addition to conformational activation, integrins cluster into groups upon cell activation, leading to a change in avidity of interaction with clustered ligands that may be present on interacting cells.

Integrins are classified by their (smaller) β chain. For leukocyte adhesion, β_1, β_2, β_3, and β_7 integrins are relevant. β_1 integrins are expressed on the surface of most cells including endothelial and smooth muscle cells. On memory lymphocytes, several β_1 integrins are expressed, with an increased expression after cell activation. $\alpha_4\beta_1$ integrin (VLA-4) is also expressed on monocytes and appears to be important in monocyte rolling, adhesion, and recruitment. $\alpha_4\beta_7$ is involved in lymphocyte homing to gut-associated tissues β_2 integrins are almost leukocyte-specific, the only exception being glial cells. All leukocytes express LFA-1 ($\alpha_L\beta_2$). Monocytes, neutrophils, and some effector lymphocytes also express Mac-1 ($\alpha_M\beta_2$). Dendritic cells express $\alpha_x\beta_2$, and some lymphocytes express $\alpha_d\beta_2$. Whereas LFA-1 serves important functions in leukocyte adhesion (see later discussion) and in forming the immunological synapse, Mac-1 is mostly engaged in phagocytic functions of macrophages and other phagocytes. The functions of $\alpha_x\beta_2$ and $\alpha_d\beta_2$ in leukocyte adhesion are unknown.

Immunoglobulins

Most cellular integrin ligands are immunoglobulins. Intercellular adhesion molecule-1 (ICAM-1) is a five-domain immunoglobulin-like molecule expressed on endothelial cells, most lymphocytes, and most other cells after exposure to proinflammatory cytokines such as interleukin-1 (IL-1) or tumor necrosis factor-α (TNFα). ICAM-1 binds LFA-1 and Mac-1 through its first and third immunoglobulin domain, respectively. ICAM-2 contains two immunoglobulin domains and is expressed on resting endothelial cells and platelets. It also binds LFA-1 and Mac-1. Vascular cell adhesion molecule-1 (VCAM-1) is the main endothelial ligand for VLA-4 and is also expressed on inflamed smooth and cardiac muscle cells.

Another class of immunoglobulins is engaged in homophilic interactions. Platelet-endothelial adhesion molecule-1 (PECAM-1) is a six-domain immunoglobulin expressed by endothelial cells, where it localizes to interendothelial borders, and on platelets and most leukocytes. Antibodies to PECAM-1 inhibit monocyte and leukocyte transendothelial migration, but PECAM-1 knockout mice show a clear defect in transendothelial migration only on some backgrounds. The junctional adhesion molecule (JAM) family encompasses three members, JAM1, JAM2, and JAM3, that also localize to interendothelial borders and regulate endothelial monolayer permeability. In addition, JAM1 interacts with LFA-1, JAM2 (also called JAM-3) with Mac-1, and JAM3 (also called JAM-2) with VLA-4. The JAMs are likely involved in regulating transendothelial migration.

Physiology of Leukocyte–Endothelial Interactions

In postcapillary venules, leukocytes, endothelial cells, shear forces, and adhesion molecules interact to allow leukocyte recruitment to sites of inflammation and lymphocyte recruitment to secondary lymphatic organs. The physiological processes discussed next have been observed in many model tissues and probably apply to skeletal muscle, cardiac muscle, skin, and lymph nodes, but not to the liver, spleen, or lung. In these latter organs the physiology of leukocyte recruitment is not well understood.

Capture or Tethering

Leukocyte adhesion is initiated by leukocyte–endothelial, leukocyte–platelet, and leukocyte–leukocyte contact. Direct capture of leukocytes by endothelial cells is rarely observed, probably because the endothelial cells are covered by a thick glycocalyx layer (500 nm) that dwarfs the adhesion molecules, the longest of which (P-selectin) extends a mere 40 nm from the endothelial plasma membrane. In postcapillary venules, most leukocyte–endothelial interactions are initiated at the point where leukocytes enter the venules from capillaries. Since capillary diameters are smaller than leukocyte diameters in most organs, leukocytes must deform to pass, and in the process squeeze down the endothelial surface layer, which brings them in close contact with the endothelial adhesion molecules. Once engaged, these adhesion molecules support stable rolling, during which the endothelial surface layer is probably continuously flattened. The layer appears to recover after the leukocyte has passed. Most primary leukocyte capture or tethering is mediated by P-selectin on the endothelial cell and PSGL-1 on the leukocyte.

Capture in other parts of postcapillary venules is rare. The few interactions initiated along the venule and not at its beginning involve leukocytes attaching to already adherent or rolling leukocytes in a process called secondary capture or tethering. This process requires L-selectin on the flowing leukocytes and PSGL-1 (and possibly other molecules) on the adherent leukocytes. Sometimes, adherent or rolling leukocytes may leave behind fragments in the form of tethers or cytoplasts, which can nucleate more leukocyte rolling through an L-selectin-dependent process similar to secondary tethering. Endothelial E-selectin does not support much tethering, but its limited tethering function appears to require PSGL-1. Integrins and immunoglobulins have no known role in capture, but LFA-1 stabilizes adhesion events initiated by selectins in plate-and-cone model systems.

Rolling

Under mild inflammatory conditions, P-selectin is rapidly expressed on the surface of endothelial cells and supports stable neutrophil rolling at a characteristic velocity of 20 to 50 μm/sec. During rolling, molecular bonds form and break continuously, and the cell is rolled forward by the torque resulting from the blood flow until the next available bond is pulled taut. In ideal rolling, cells remain in continuous contact with the postcapillary endothelium without ever breaking free, but occasional hops and skips with reattachment are also observed. The velocity of neutrophil, but not lymphocyte, rolling is regulated by proteolytic cleavage of L-selectin from the leukocyte surface.

Lymphocyte rolling in high endothelial venules of the inguinal lymph node, Peyer's patches, and probably other lymphatic organs is achieved by transient L-selectin binding to sulfated and glycosylated endothelial ligands. Although early studies identified GlyCAM-1 and CD34 as potential L-selectin ligands, later studies in gene-targeted mice did not confirm this. Potential L-selectin ligands relevant for rolling include podocalyxin and an incompletely defined molecule called sgp200. In mesenteric lymph nodes, interaction of $\alpha_4\beta_7$ integrin on lymphocytes with MAdCAM-1 on endothelial cells also contributes to rolling.

When inflammatory conditions persist, for example, in the presence of TNFα, E-selectin is also expressed on endothelial cells. E-selectin supports rolling at a markedly reduced velocity of about 5 μm/sec by binding through an unknown leukocyte ligand distinct from PSGL-1. This ligand requires sialylation by sialyl transferase 3-Gal-IV for full functionality. E-selectin engagement may also serve to activate rolling neutrophils.

During severe inflammation induced by TNFα or other cytokines, the leukocyte integrins participate in the rolling process, presumably by engaging in transient bonds with ligands on endothelial cells [6]. Specifically, Mac-1 has been shown to engage ICAM-1 during rolling, LFA-1 engages an unknown ligand other than ICAM-1 or ICAM-2 [6a]. There is a strong synergy between E-selectin and β_2 integrins in leukocyte rolling [6] so that in the absence of both almost no rolling occurs, and leukocyte recruitment is severely compromised.

Arrest

Rolling leukocytes must arrest to become firmly adherent. Interestingly, this arrest mechanism is fundamentally different for neutrophils and lymphocytes [7]. Lymphocyte arrest was studied in postcapillary venules of Peyer's patches and lymph nodes. It requires activation by the chemokine CCL21 (SLC) through the G-protein-coupled receptor CCR7. In flow chamber systems, CXCL12 (SDF-1) interacting with its receptor CXCR4 has also been shown to mediate arrest of lymphocytes and monocytes. In large blood vessels such as the carotid artery, monocyte arrest is mediated by CXCL1 (Gro-α) binding to its receptor CXCR2 and by CCL5 (RANTES) binding to CCR1,3,5 or a combination thereof.

For neutrophils, the arrest situation is much less clear. Although the textbook paradigm proposes activation through chemokines, and there are plenty of chemokines immobilized on inflamed endothelial cells and at least two relevant chemokine receptors, CXCR1 and 2 (only CXCR2 in mice) expressed on resting neutrophils. Recently, CXCR2 was identified as a ventrophil arrest chemokine receptor in postcapillary venules [8]. Neutrophils roll for a long time (1 to 3 minutes) and distance (several hundred micrometers), apparently scanning the endothelial surface for activating signals, before they arrest. This is in contrast to lymphocyte, which arrest within less than a second of activation [9]. The difference may be found in other activating signals that neutrophils, but not lymphocytes, may receive while rolling. Most prominently, such signals have been shown to emerge from engagement by E-selectin, P-selectin, and L-selectin.

Transendothelial Migration

Transendothelial migration proceeds through interendothelial clefts and through the body of endothelial cells (transcellular). The molecular requirements for each pathway are likely different. β_2 integrins such as LFA-1 and Mac-1 have long been known to be required for transendothelial migration. Recent evidence suggests that they bind junctional adhesion molecules. Certainly, transendothelial migration requires an active participation by endothelial cells, because it can be blocked by chelating intracellular free calcium or by blocking other signaling pathways. Transendothelial migration appears to involve retraction of vascular endothelial (VE)-cadherin. Another candidate molecule possibly involved in transendothelial migration is PECAM-1. Antibody blockade blocks transmigration, possibly by endothelial signaling, and PECAM-1-deficient mice have a significant defect in transendothelial migration.

Migration in the Tissue

Transmigrated neutrophils (and possibly other leukocytes, although this has not been investigated) acquire a different phenotype after transmigration. They express $\alpha_4\beta_1$, $\alpha_2\beta_1$, $\alpha_6\beta_1$, and other β_1 integrins on their surfaces that enable them to engage extravascular matrix components. For migration, $\alpha_2\beta_1$ appears to be most important.

Unresolved Questions

Postcapillary venules are formidable machines for leukocyte recruitment. More than 90 percent of all neutrophils in capillaries begin rolling in postcapillary venules, and more than 90 percent of rolling leukocytes (neutrophils) arrest successfully in cytokine-activated venules. Yet, the overall extraction efficiency during inflammation is less than 50 percent, suggesting that some arresting neutrophils may detach again, or that transmigration may be inefficient. This process is not understood at the quantitative level. The role of the endothelial surface glycocalyx is just beginning to be explored. Rolling, arrest, and transmigration of many lymphocyte subsets including activated Th1 and Th2 CD4 helper cells, CD8 cytotoxic cells, natural killer cells, $\gamma\delta$ T cells, and NKT cells have not been studied in vivo. Very little is known about eosinophil and basophil recruitment and the resolution of inflammation, a process that likely also involves cell interactions in postcapillary venules. Finally, the mechanisms mentioned previously are valid for some tissues, but not the lungs, liver, spleen, kidney, or brain. Suitable intravital microscopy systems and other models need to be developed to address these issues.

Reference

1. Ley, K., and Gaehtgens, P. (1991). Endothelial, not hemodynamic differences are responsible for preferential leukocyte rolling in venules. *Circ. Res.* **69**, 1034–1041.
2. Vink, H., and Duling, B. R. (1996). Identification of distinct luminal domains for macromolecules, erythrocytes, and leukocytes within mammalian capillaries. *Circ. Res.* **79**, 581–589. *Discovery of the endoluminal endothelial surface layer.*
3. Smith, M. L., Long, D. S., Damiano, E. R., and Ley, K. (2003). Near-wall micro-PIV reveals a hydrodynamically relevant endothelial surface layer in venules in vivo. *Biophys. J.* **85**, 637–645.

3a. Long, D. S., Smith, M. L., Pries, A. R., Ley, K., and Damiano, E. R. (2004). Microviscometry reveals reduced blood viscosity and altered shear rate and shear stress profiles in microvessels after hemodilution. *Proc. Natl. Acad. Sci. USA* **101**, 10060–10065.

4. Kansas, G. S. (1996). Selectins and their ligands: Current concepts and controversies. *Blood* **88**, 3259–3287. *Comprehensive review on selectins and selectin ligands.*
5. Shimaoka, M., Xiao, T., Liu, J. H., Yang, Y., Dong, Y., Jun, C. D., McCormack, A., Zhang, R., Joachimiak, A., Takagi, J., Wang, J. H., and Springer, T. A. (2003). Structures of the alphaL I domain and its complex with ICAM-1 reveal a shape-shifting pathway for integrin regula-

tion. *Cell* **112**, 99–111. *Discovery of major conformational shape change upon integrin activation.*

6. Jung, U., Norman, K. E., Ramos, C. L., Scharffetter-Kochanek, K., Beaudet, A. L., and Ley. K. Transit time of leukocytes rolling through venules controls cytokine-induced inflammatory cell recruitment in vivo. *J. Clin. Invest.* **102**, 1526–1533.

6a. Dunne, J. L., Collins, R. G., Beaudet, A. L., Ballantyne, C. M., and Ley, K. (2003) Mac–1, but not LFA–1, uses intercellular adhesion molecule-1 to mediate slow leukocyte rolling in TNF-alpha-induced inflammation. *J. Immunol.* **171**, 6105–6111.

7. Alon, R., Grabovsky, V., and Feigelson, S. (2003). Chemokine induction of integrin avidity on rolling and arrested Leukocytes: Local signaling events or global stepwise activation? *Microcirculation* **10**, 297–311. *Recent review showing two different mechanisms of arrest for T cells and neutrophils.*

8. Smith, M. L., Olson, T. S., and Ley, K. (2004) CXCR2- and E-selectin-induced neutrophil arrest during inflammation in vivo. *J. Exp. Med.* **200**, 935–939.

9. Campbell, J. J., Hedrick, J., Zlotnik, A., Siani, M. A., Thompson, D. A., and Butcher, E. C. (1998). Chemokines and the arrest of lymphocytes rolling under flow conditions. *Science* **279**, 381–384. *Discovery of arrest chemokines.*

Capsule Biography

Dr. Ley has headed the Cardiovascular Research Center at the University of Virginia since 2001 and was the winner of the 1986 Abbott Microcirculation Award and the 2001 Curt A. Wiederhielm Award of the Microcirculatory Society. His laboratory primarily focuses on inflammatory processes in the microcirculation, in atherosclerosis and inflammatory bowel disease. His work is supported by grants from the NIH.

CHAPTER 51

Cell Adhesion Molecules: Structure, Function, Organization, and Role in Leukocyte Trafficking through the Lung

[1]Lina H.K. Lim and [2]Bruce S. Bochner

[1]Department of Physiology, National University of Singapore, Singapore

[2]Department of Medicine, Division of Allergy and Clinical Immunology, The Johns Hopkins University School of Medicine, Baltimore, Maryland

Introduction

The endothelial cells (ECs) lining the walls of blood vessels in the microcirculation have a number of important functions. These include contributions to tissue structure and growth, and mediation of leukocyte recruitment from the circulation into the tissue, either for homeostasis or during inflammation. ECs can express many structures on their luminal surfaces including adhesion molecules (AMs) that are key players in leukocyte infiltration from the circulation to the tissue. This chapter will focus on the description of AMs and their ligands that mediate leukocyte-EC circulating interactions, relating to cell recruitment. Further, it will address tissue-specific expression of AMs and their related functions, particularly in the lung and pulmonary circulation [1].

General Dynamic Events Leading to Leukocyte Migration

Leukocyte recruitment from the blood into tissue sites of inflammation involves initial contact with the vascular endothelium under shear flow. Leukocytes in the circulation generally passage harmlessly through in the center stream of the vessel, where the velocity is greatest. Diffusive forces caused by the microcirculatory network (increasing diameters from the capillary to the venule) allow smaller, faster erythrocytes to pass the more slowly moving leukocytes, driving them toward the vascular wall. This encourages leukocyte–endothelial interactions, resulting in rolling, where leukocytes move more slowly than flowing erythrocytes. Under these conditions, leukocyte rolling velocities on postcapillary venules are reduced to approximately 10 μm/sec. Leukocytes can then overcome shear forces and attach to the endothelium through a sequence of adhesive events, mediated by specific AMs present on both the leukocyte and the endothelium. Subsequently, transmigration across the EC can occur, followed by further migration through the basement membrane and into the surrounding interstitium (Figure 1).

Leukocytes generally do not interact with the endothelium on the arteriolar wall, but adhere and migrate at the site of the postcapillary venule. The diffusive forces and shear rates created through the microvascular network can partly explain this typical location of leukocyte extravasation. However, arteriolar and venular ECs exposed to the same shear rates in vitro demonstrated dramatically different leukocyte–EC interactions and adhesion properties, indicating that shear forces alone are not enough to sustain leukocyte trafficking. Venular ECs are specialized in supporting

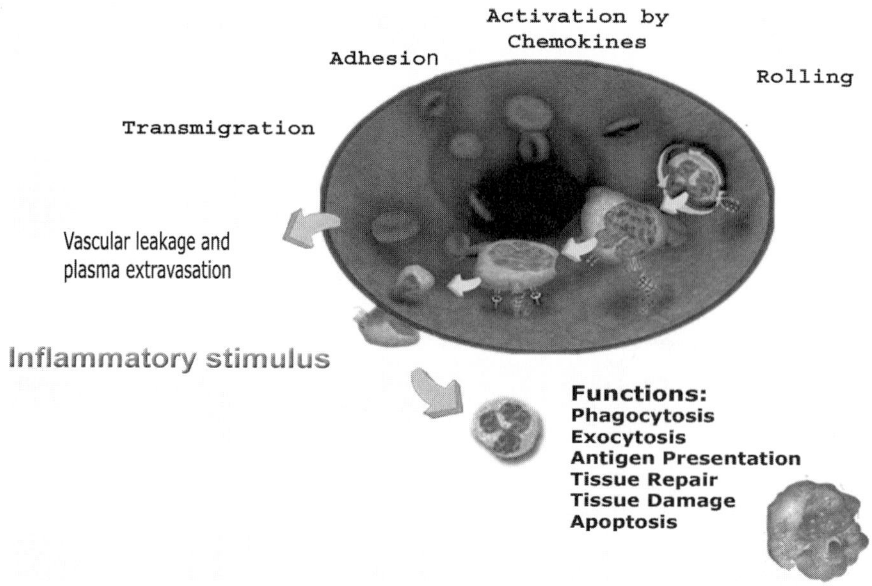

Figure 1 General mechanisms involved in leukocyte trafficking in the microcirculation. Shown is a schematic representation of the multistep paradigm of leukocyte recruitment. Leukocytes undergo rolling and activation by EC-displayed chemokines, followed by adhesion onto and transmigration across the endothelium to perform their extravascular functions. Illustration by A. K. Khaw. (see color insert)

leukocyte adhesion, distinctly different from the arteriolar ECs, because of particular AMs (e.g., E-selectin) located exclusively on the venular endothelium. In addition, lymphocytes exit the circulation and enter lymph nodes through specialized high endothelial venules that express their own unique, specific AMs for lymphocyte homing.

Techniques Used to Observe Leukocyte Trafficking

Many indirect and endpoint experimental systems have been used to study inflammation and leukocyte infiltration, such as lavages and histology. However, these do not serve to determine the sequences and events leading to leukocyte extravasation. Consequently, a state-of-the-art in vivo technique, known as intravital microscopy, was established to this end, and provides direct visual observation of living circulation in vivo (e.g., in mice, rats, guinea pigs, hamsters, bats and cats). Here, all the events of leukocyte trafficking, that is, rolling, adhesion, and transmigration through the vascular endothelium, can be visualized in real time. Furthermore, the development of in vitro rolling assays, exerting physiologically relevant shear forces on leukocytes over a layer of ECs, allows intricate dissection of the molecules and mediators involved in endothelial activation and leukocyte recruitment. Further, the relative importance of all the AMs involved in leukocyte recruitment has been demonstrated using blocking reagents (e.g., antibodies and small molecule antagonists), gene targeting, or knockout animals. However, because of compensatory mechanisms, overlapping functions and regulatory processes, data from knockout (double or triple) animals need to be more carefully examined.

Endothelial AMs Involved in Leukocyte Trafficking

The AMs found on vascular ECs, with their corresponding ligands on the leukocyte, are usually distinguished by their contribution to rolling or firm adhesion (see Figure 2) [2]. The selectins (E-selectin, P-selectin) on the EC, and L-selectin on the leukocyte, are involved in initial tethering and rolling, while members of the immunoglobulin (Ig) superfamily of AMs, such as intercellular adhesion molecule-1 (ICAM-1) and cadherins, are mainly involved in adhesion and migration, respectively. However, it is possible that under certain circumstances, ICAMs and integrins can mediate rolling. The time course and magnitude of EC AM expression coordinates the recruitment of leukocytes to sites of inflammation. Thus, the rapid recruitment of rolling leukocytes results from L-selectin activation on leukocytes and rapid mobilization of preformed P-selectin to the EC surface. Leukocyte adhesion can also be fast because of the activation of β_2-integrins on rolling leukocytes, which can interact with constitutively expressed ICAM-1, leading to firm adhesion and subsequent migration (see Figure 1). However, tissue-specific adhesion characteristics are apparent, such as in the liver sinusoidal microcirculation and the lung pulmonary circulation, where selectins are not required for leukocyte recruitment.

Rolling

Rolling is a transitory process that occurs in both noninflamed and inflamed vessels. As a result, many leukocytes "patrolling" the microvasculature will come into brief

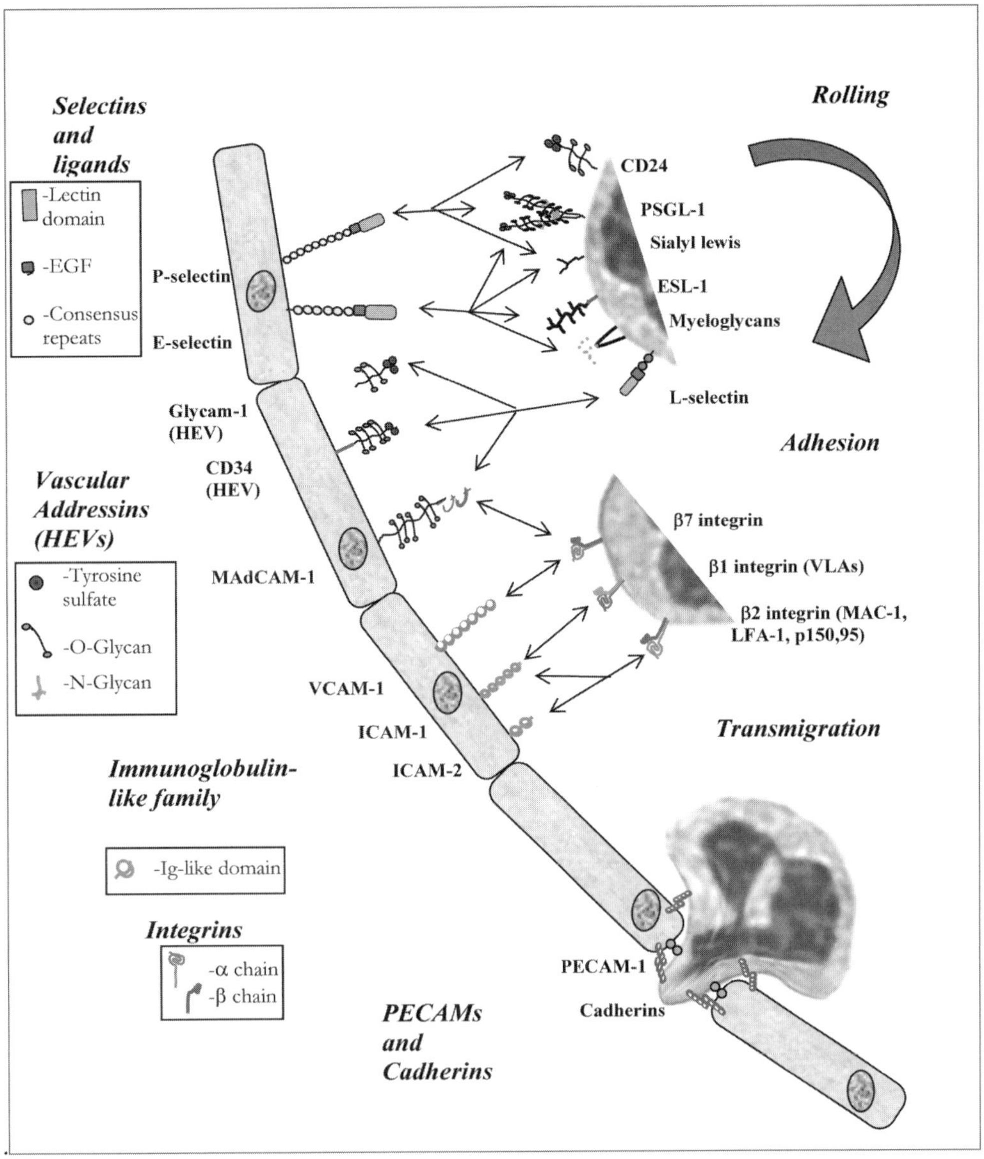

Figure 2 Endothelial cell adhesion molecules and counterligands on leukocytes involved in cell trafficking. Activated EC express selectins that can bind to their carbohydrate receptors to allow leukocyte rolling under physiologic shear forces. Addressins expressed on high endothelial venules (HEVs) in lymph node vessels are expressed for lymphocyte homing; their peripheral vascular counterparts are less well defined. If activation occurs, leukocytes adhere to the endothelium via Ig-like superfamily proteins and integrins. Transendothelial migration is regulated by PECAM-1, cadherins, and other structures. (see color insert)

contact with the endothelium. Sometimes, rolling leukocytes interact with other attached leukocytes in a process known as tethering. Approximately 80 percent of these cells will gently roll and detach from the endothelium and rejoin the circulation, while the remainder will roll more slowly and may adhere and flatten onto the endothelium. The first response to inflammatory stimuli that can be observed using intravital microscopy is an increase in the number of rolling cells, with a concurrent reduction in leukocyte velocity. Leukocytes will integrate chemokine and other signals on the EC surface while rolling along the endothelium until they reach a critical level of activation, resulting in firm adhesion.

SELECTINS AND THEIR LIGANDS

The selectins are C-type lectins characterized by the presence of an N-terminal calcium-dependent carbohydrate recognition domain [3]. They are single-chain membrane glycoproteins that recognize counter-ligands usually containing sialylated and fucosylated carbohydrate residues, such as sialyl lewis x (Figure 2). There are three selectin AMs important in mediating initial tethering and rolling. Two selectins (E-selectin and P-selectin) are found on ECs, and one, L-selectin, is found on leukocytes.

L-selectin (Leukocyte-selectin) is involved mainly in the tethering of leukocytes to the endothelial surface, and proteolytic loss of expression of L-selectin, especially on granulocytes, occurs in conjunction with the activation of other AMs such as integrins. This soluble L-selectin can be detected in the serum and can competitively bind to endothelial ligands, blocking the interactions with leukocyte-bound L-selectin. The interaction of L-selectin with the EC is short-lived and usually with one of several ligands, such as CD34, mucosal addressins (e.g., MAdCAM-1), or vascular addressins.

P-selectin (platelet-selectin), originally named granule-external membrane protein, is a 140-kDa glycoprotein constitutively expressed in α-granules of platelets and the Weibel-Palade bodies of ECs. P-selectin is thought to be the predominant AM responsible for the rolling phase of the leukocyte adhesion cascade. Stimulation of ECs with inflammatory mediators such as histamine and thrombin results in translocation and expression of P-selectin on the cell surface within minutes. Reactive oxygen species (ROS), sulfidopeptide leukotrienes, and complement products can also induce fast mobilization of P-selectin onto the EC surface. P-selectin surface expression is transient and is recycled by reinternalization. P-selectin expression can also be transcriptionally regulated by cytokines such as interleukin-4 (IL-4) and IL-13. Thus, it is possible that P-selectin is important in Th2-mediated inflammatory diseases. In contrast, tumor necrosis factor (TNF), IL-1, and lipopolysaccharide (LPS), which induce other AMs on ECs, do not induce P-selectin gene transcription or expression. Like all the selectins, P-selectin binds to sialyl lewis x molecules but more specifically binds to P-selectin glycoprotein ligand-1 (*PSGL-1*), a homodimeric structure and member of the sialomucin family of AMs found on all circulating leukocytes. P-selectin can also bind to CD24 and other ligands. The interactions between P-selectin and PSGL-1 cause leukocyte–EC contacts and rolling along the venular endothelium. The P-selectin–EC associations are longer than the L-selectin–EC interactions and result in slower leukocyte rolling velocities. The signal transduction pathways that follow P-selectin binding to its ligand involve the activation of mitogen-activated kinases, Src kinases, and increases in intracellular calcium. All these pathways can lead to other downstream events such as integrin activation, shape change, and further ROS production.

E-selectin (endothelial-selectin) expression is protein synthesis dependent, occurring on the cell surface approximately 4 hours after activation with TNF, IL-1, or LPS. Eosinophils and neutrophils bind equally well to P-selectin, whereas eosinophils form fewer primary tethers to E-selectin. E-selectin has similar, possibly redundant functions to P-selectin, but more importantly, may be important in converting rollers to adherent cells, as E-selectin mediates even slower rolling than P-selectin. The off-rate of E-selectin is similar to that of P-selectin; thus, it is likely that E-selectin or its ligand is more highly expressed than P-selectin and PSGL-1. In addition, because its peak expression is around 4 hours, E-selectin could functionally support rolling when P-selectin expression is reduced or when P-selectin is internalized. E-selectin is capable of binding to PSGL-1 and other sialic acid–containing glycoproteins, but the primary E-selectin ligands in humans are glycolipid structures termed myeloglycans. In mice, the identified ligand of E-selectin is a glycoprotein known as E-selectin ligand-1 (ESL–1).

Studies in knockout animals have revealed the importance of L-selectin and P-selectin in the initial induction of rolling, while all selectins can support stable rolling. However, in E-selectin and P-selectin double-deficient mice, L-selectin can sufficiently produce rolling that results in activation and adhesion, suggesting that overlap and redundancy occur in selectin function. Interestingly, E-selectin and β_2 integrin AMs (together with ICAMs) cooperate to control the time that a leukocyte takes to roll and firmly adhere. Further, these double-deficient mice exhibit signs of early lethality. This suggests that E-selectin operates "downstream" from P-selectin, more toward the firm adhesion step of the cascade.

In summary, selectins are either constitutively expressed, or inducible with inflammatory mediator or cytokine exposure. These AMs all play roles in early stages of extravasation involving tethering and rolling, with varying degrees of attachment. This allows temporary interaction of patrolling leukocytes in the circulation with vascular ECs, allowing them to react to stimuli (e.g., chemokines) presented on the EC surface. This, in turn, causes leukocyte activation and subsequent extravasation.

ADDRESSINS

In secondary lymphoid tissues (such as lymph nodes and Peyer's patches), the entry site for naive T and B lymphocytes is known as the High Endothelial Venule (HEV), because of its cuboidal shape. The ECs of HEV express specific AMs that allow lymphocyte homing into the lymphoid tissue, while other cells, such as granulocytes, may enter the HEV but do not stop. Murine HEVs selectively bind to an antibody, MECA-79, that recognizes a carbohydrate epitope not found on ECs of postcapillary venules or large vessels in the spleen or thymus. These carbohydrate epitopes bind to L-selectin, are partly responsible for the tissue-specific trafficking of lymphocytes, and are called vascular addressins [4, 5]. The peripheral node addressins (PNAds) include GlyCAM-1 (glycosylation-dependent cell AM), CD34, podocalyxin, and Sgp200, whereas mucosal or mesenteric addressin is thought to be MAdCAM-1 (Mucosal Addressin cell AM-1). The adhesion molecules are important in the homing of lymphocytes and dendritic cells to lymphoid tissues for antigen presentation.

Firm Adhesion

Firm adhesion of leukocytes on the vascular endothelium occurs after the activation of leukocyte AMs by chemokines and other stimuli presented on the EC surface. All firmly adherent leukocytes are exclusively recruited from the rolling pool. The AMs involved in firm adhesion are mainly the members of the Ig superfamily of AMs and their integrin ligands.

IG SUPERFAMILY OF AMS

This is a wide family of AMs, consisting of Ig-like domains characterized by intrachain disulfide bonds. Together with their ligand counterparts, they play major roles in leukocyte adhesion and the regulation of firm adhesion. Members include ICAM (intercellular AM)-1, -2, and -3 and VCAM-1 (vascular cell AM-1).

ICAM-1 is found constitutively at low levels on the EC surface, but is upregulated upon exposure to inflammatory cytokines (e.g., TNF, IFN-γ, and IL-1) with peak expression after about 24 hours. Of note, treatment of ECs with IFN-γ selectively increases the expression of ICAM-1, with no effect on other AMs. ICAM-1 contains five Ig-like extracellular domains, where the first Ig-like domain (on the amino-terminal end) is the recognition site for fibrinogen, an extracellular matrix protein, and the integrin lymphocyte associated function antigen (LFA-1 or CD11a/CD18). The third Ig-like domain recognizes another integrin, macrophage antigen-1 (Mac-1 or CD11b/CD18). Mac-1, found on granulocytes and monocytes, binds ICAM-1 with high affinity. As described, LFA-1 can also bind to ICAM-1, but with lower affinity than Mac-1. ICAM-1 has another interesting function, in that it is the exploited receptor of most strains of rhinovirus and the malarial parasite. The ligation of ICAM-1 can cause structural changes in junctional EC cytoskeleton proteins without causing EC injury or retraction. These structural proteins have been shown to dissociate within minutes of leukocyte–EC adhesion. This could be linked to subsequent events in transmigration.

ICAM-2 is a truncated form of ICAM-1, with only two Ig-like domains. LFA-1 binds to ICAM-2 with higher affinity than to ICAM-1. ICAM-2 is highly and constitutively expressed on ECs, without further expression after stimulation. It is found preferentially at endothelial borders, and thus could be important in transmigration. *ICAM-3* is found on resting leukocytes, but can be induced in ECs. It contains five Ig-like extracellular domains, similar to ICAM-1, and binds to LFA-1 and $\alpha d\beta_2$, another β_2 integrin found on most leukocytes.

VCAM-1 is found at very low levels, if at all, on nonactivated ECs, but is upregulated substantially after long-term exposure to cytokines. It is expressed on the surface of IL-1 and TNF activated EC, and similar to the ICAMs, binds to integrins. However, it binds to a different class of integrins with β_1 and β_7 subunits. The most common ligand for the binding of VCAM-1 is very late antigen-4 (VLA-4, CD49d/CD29), found on eosinophils, basophils, mast cells, lymphocytes, and monocytes, but not neutrophils. VLA-4-VCAM-1 binding has been shown to mediate leukocyte rolling and adhesion. In particular, VCAM-1 ligation is essential for eosinophil and monocyte migration, and exposure of ECs to the Th2 dependent cytokines IL-4 and IL-13 (and TNF and IL-1 synergistically) selectively increases VCAM-1 expression, indicating the potential importance of VCAM-1 in Th2-mediated diseases such as allergy and asthma.

In summary, the Ig superfamily of adhesion molecules and integrins plays an extremely important role in the adhesion step of the recruitment cascade. ICAM-1 is important in adhesion of essentially all leukocytes, while VCAM-1 is the Ig-like adhesion molecule involved in adhesion of all leukocytes except neutrophils.

Transmigration

The final step of cell recruitment involves transmigration between adjacent ECs at tricellular junctions, a pause in the wall, and subsequent penetration through the basement membrane. During the adhesion phase, a disruption of EC-cadherin complexes occurs, inducing a loss of lateral junction localization, leading to migration between ECs [6]. *PECAM-1* (platelet endothelial cell adhesion molecule-1, CD31) is another member of the Ig super family and is vitally important in the transmigration step of neutrophil recruitment. It has six Ig-like domains as well as immunoreceptor tyrosine-based inhibitory motifs. PECAM-1 is in especially high density at the gap junction area of ECs, but is evenly distributed on platelets and leukocytes. Because of its concentration at the intercellular junctions of the vascular endothelium, it is thought to be a crucial AM to guide adherent cells to the migration site. PECAM-1 mediates endothelial transmigration through binding to itself (homophilic) or to other ligands (glycosaminoglycans such as CD44). PECAM-1 is also involved in the migration through the basement membrane and subendothelial matrix. PECAM-1 is critical for neutrophil (but not eosinophil) transendothelial migration, and the first Ig-like domain is important for this step. Cells treated with anti-PECAM-1 antibodies, or PECAM-IgG constructs comprising domain 1, arrest on ECs but do not transmigrate. Furthermore, the 6th Ig-like domain is thought to be important in the migration across the basement membrane. In this case, as there is no PECAM-1 expressed in the basement membrane, PECAM-1 is thought to bind to an as yet unidentified structure in the basal lamina. It may be possible that PECAM-1 is more involved in migration across the basement membrane, as leukocytes from PECAM-1–deficient mice do migrate through ECs but are delayed in their migration across the basement membrane. Altogether the process of leukocyte recruitment from the circulation into the tissue involves many AMs on the leukocyte as well as the EC. Thus, the process requires a step-by-step activation and deactivation (or deadhesion) of each and every constituent for successful leukocyte extravasation.

Tissue-Specific Expression of AMs in the Lung Microvasculature

The aforementioned sequence of events and mechanisms leading to the transmigration of leukocytes out of the circulation has been shown to occur in vitro and in vivo in tissues such as the mesentery. However, several recent studies have shown differences in leukocyte trafficking characteristics in the pulmonary circulation when compared to the systemic circulation [1]. The location of leukocyte extravasation and the requirements for AM expression have also proven to be different, possibly because of the unique structural features of the lung microcirculation [7]. This suggests that tissue-specific leukocyte trafficking and differential expression of AMs exist in the lung, and thus, the mechanisms of leukocyte trafficking must not be thought of as a generalized set of occurrences. We have recently developed a rat model of intravital microscopy in the airway circulation or tracheal microcirculation [8] and demonstrated that leukocyte recruitment in the circulation of the trachea is similar to the systemic circulation. Other intravital microscopy techniques to study microvessel behavior and leukocyte trafficking in the pulmonary circulation have been created. One utilized an implanted window in the thoracic wall of rabbits and dogs. In addition, transplantation of neonatal lung into skin folds of nude mice has been used to examine leukocyte recruitment in revascularized lung tissues [9].

Margination and Migration in the Pulmonary Circulation

In the systemic circulation of the airways and other tissues, migration occurs in postcapillary venules, following the typical sequence of leukocyte trafficking described previously. However, several distinct characteristics are seen in the pulmonary circulation. For example, leukocytes marginate and transmigrate in alveolar capillaries, rather than the normal site of migration, the postcapillary venule. Many leukocyte–EC interactions take place in the pulmonary capillary bed via a process known as sequestration. Here, the network consists of capillary segments that connect several alveolar walls and generally contain many more leukocytes compared to other tissue vascular beds. In fact, the concentration of leukocytes (neutrophils or polymorphonuclear [PMN] cells in this case) in pulmonary capillary blood is 35 to 100 times more than that of systemic vessels. This could be an active process, or simply due to small capillary diameters (2 to 5 μm). Leukocytes are approximately 8 μm in diameter, so they would need to deform to squeeze through such small vessels, taking more time to travel through this microcirculation. Although selectin-mediated rolling occurs in pulmonary postcapillary venules, most migrated neutrophils (97%) are found in the capillary network, where rolling does not occur, as the leukocyte arrests without rolling because of the size of the capillaries. Similarly, leukocytes transmigrate out of the capillary network in the lung, because of their slow transit and interaction with the capillary EC, possibly independent of PECAM-1.

SELECTINS

Due to the small diameters of pulmonary capillaries, arrest and migration in the pulmonary circulation can occur without rolling. P-selectin is expressed on pulmonary venular and arteriolar EC, but not capillary EC in rabbits, whereas in the rat, P-selectin is completely nonexpressed. Thus, it can be proposed that in this case, selectins are not required for leukocyte trafficking in the pulmonary circulation. Further, from experiments using E-selectin and P-selectin-deficient mice, pulmonary migration to *Streptococcus pneumoniae,* a Gram-positive bacterium (see next section), was increased, indicating that selectin expression actually suppresses leukocyte migration out of the pulmonary circulation. This contrasts with other, allergen-driven cell recruitment models, where selectin deficiency or blockade reduces T cell and eosinophil recruitment. The depletion of L-selectin from neutrophils does not affect the rate and response of sequestration in the lung induced by complement fragments. This suggests that selectins may not be required for migration to Gram-positive bacteria. However, selectin-mediated processes are involved in migration to Gram-negative bacteria or endotoxin (*Escherichia coli,* LPS), where selectin deficiency protects against endotoxin-induced septicemia, lung injury, and death.

During acute lung injury, blockage of all selectins can result in approximately 40 percent inhibition of leukocyte infiltration and vascular permeability. High tidal volume ventilation activates lung microvessels in a P-selectin-dependent process. Furthermore, E-selectin is upregulated in lungs of fatal acute respiratory distress patients infected with Gram-negative bacteria. Thus the role of selectins in acute respiratory distress and acute lung injury appears somewhat variable depending on the inflammatory condition.

IG SUPERFAMILY AND INTEGRIN LIGANDS

ICAM-1 is expressed at constitutively high levels on pulmonary capillary and venular ECs, and at low levels on ECs in the systemic circulation [10]. This could be one reason why activated leukocytes (with increased expression of activated β2 integrins) are sequestered in the lungs. However, pulmonary migration of granulocytes can occur independently of β2 integrins depending on the inflammatory stimuli. Using neutralizing antibodies to CD18, leukocyte migration from activation by Gram-negative bacteria (*Pseudomonas aeruginosa* and *E. coli*), IL-1, LPS, and phorbol esters can induce CD18-dependent lung migration pathways, involving ICAM-1, while Gram-positive bacteria (*S. pneumoniae*) and complement products induce CD18-independent pulmonary recruitment. Accordingly, ICAM-1 expression (as high as it is on pulmonary ECs) is further upregulated by Gram-negative bacteria and LPS, but not by Gram-positive bacteria in these animal models. Interestingly, expression of CD18 and L-selectin in neutrophils exposed to *E. coli* was altered only after migration and not while in the circulation. In contrast, in response to *S. pneumoniae*, CD18 and L-selectin levels were increased and reduced, respectively, before migration, indicating their involvement. In these cases, the expression and requirement for CD18 before and during migration is inversely correlated and the limiting factor could be the increased expression of ICAM-1. In patient samples of sepsis-induced lung injury, ICAM-1 as well as VCAM-1 expression is highly upregulated. Hydrochloric acid instillation into the trachea, an animal model of gastric acid inspiration, causes lung injury and endothelial activation. It does not require CD18 for leukocyte accumulation, but subsequent lung injury and injury to other organs is CD18 dependent. It is possible that VCAM-1 or other AMs can be involved in Gram-positive bacterial infections. The mechanism underlying CD18-independent leukocyte recruitment in the pulmonary circulation is still unknown.

Migration in the Systemic Bronchial Circulation

Leukocyte trafficking and endothelial migration in the systemic circulation of the lungs is not well studied. Only 1 percent of the cardiac output is directed to the airway circulation to supply the central airways and bronchi. These vessels form a plexus network around the airways, supplied by the bronchial artery, sending branches into the muscle and submucosa. It remains unknown whether leukocyte infiltration close to the bronchi is occurring from the bronchial circulation or the pulmonary circulation. It seems unlikely that tissue-infiltrating leukocytes such as eosinophils found in and around the bronchi in airway conditions such as asthma exit through the pulmonary circulation. Thus, it is important to know the recruitment properties of this circulation. As mentioned previously, we have developed a model of intravital microscopy in the upper airways to address this issue [8]. Although this model uses the tracheal circulation, the function and regulation of tracheal muscle and tissue mirror the smaller airways. Leukocyte–endothelial rolling and adhesion induced by Gram-negative bacterial LPS and the leukocyte-activating bacterial peptide fMLP (formylmethionyl-leucyl-phenylalanine) occurred in postcapillary venules, with parameters similar to that of the systemic circulation. Furthermore, using the same model, high-pressure mechanical ventilation induced EC activation and leukocyte recruitment in tracheal microvessels, which was sensitive to selectin blockage. In a lung transplant model [9], where intravital microscopy was done in revascularized vessels of the lung, TNF-induced leukocyte rolling and adhesion occurred in the venules and arterioles, and this, too, was dependent on selectins. It was interesting that no arrest or sequestration was seen in the capillaries. However, rolling leukocytes in the postcapillary venules were elongated, perhaps because of shape change from traversing through the capillary network. It is possible that because these are newly revascularized vessels, it may be a model of angiogenesis rather than pulmonary circulation per se. In any case, this indicates that selectins are likely important in the recruitment of leukocytes in the airways. Similarly, other studies have observed that lymphocyte homing to the bronchial endothelium is P-selectin-PSGL-1 mediated, and VCAM-1, E-selectin, and CD18 independent. Furthermore, strong expression of P-selectin and ICAM-1 was found on bronchial ECs, with low or negligible expression of E-selectin.

Summary

This review has described the general mechanisms involved in leukocyte recruitment, focusing on the important roles of AMs. Recruitment characteristics in the lung microcirculation have been outlined. The inflammatory stimuli and disease condition may play a role in the site and regulation of leukocyte recruitment and EC activation in the lungs. Various conditions that affect the central conducting airways may induce recruitment in the systemic airway circulation, while the pulmonary network may be more important in other more acute lung injuries from invading organisms, septicemia, or chemicals. The mechanisms of leukocyte recruitment in this circulation are not well elucidated, but it appears that the bronchial microvasculature is similar to other systemic microcirculation. This suggests CD18-dependent recruitment, with selectin-mediated rolling and transmigration across ECs in bronchial/tracheal vessels. This also indicates that the events leading to leukocyte infiltration from the bronchial systemic circulation may differ from those in the more well-studied pulmonary circulation. It is here where cells are sequestered without rolling, and CD18-independent migration occurs. It can be suggested that in the lung, two different endothelial phenotypes may exist, depending on the structure and function of the circulation they support, i.e., gaseous exchange or blood supply, for the pulmonary and airway circulation, respectively. Future studies should elucidate the differences between these two circulations in the same functioning organ, enhancing the understanding of the pathophysiology of lung disorders, and possibly validating novel targets to treat diseases that primarily affect the lungs.

Glossary

Adhesion molecule: A cell surface glycoprotein or glycolipid that functions in the interaction between cells or between cells and the extracellular matrix. Adhesion molecules are found on virtually all cell types.

Alveoli: The air sacs in the lungs where oxygen and carbon dioxide are exchanged between blood and inhaled air.

Bacterial endotoxin (LPS): A component of Gram-negative bacterial cell wall released from bacteria. Once released, it stimulates many inflammatory reactions.

Capillaries: The smallest of vessels, with the thinnest walls, where gas exchange takes place between the blood and the surrounding tissues.

Endothelial cells: These exist as a monolayer of cells lining the insides of blood vessels. They act together (endothelium) to form a selective barrier that regulates movement of cells, proteins, and fluids from the circulation to the tissue. They are also important in blood vessel growth, tissue repair, and coagulation.

Gram-negative bacteria: Bacteria characterized by two outer membranes that fail to stain with Gram's stain. Organisms include *E. coli*, *Salmonella*, and cholera.

Gram-positive bacteria: Bacteria characterized by two outer membranes that do stain with Gram's stain. Organisms include *Staphylococcus*, *Streptococcus*, and *Pneumococcus*.

Intravital microscopy: A technique used in live animals to directly observe the microcirculation of tissues in real time using video microscopy.

Leukocytes: Also known as white blood cells, these migratory cells function in immune surveillance and host defense.

Nude mice: Mice lacking thymus development and T-lymphocytes.

Polymorphonuclear cells (PMNs): Also called granulocytes, these are leukocytes with a characteristic multilobed nucleus and cytoplasmic granules. They can be classified as neutrophils, eosinophils, and basophils. Neutrophils are the most abundant circulating leukocyte important in responses to bacterial infections, whereas eosinophils and basophils are more rare in the circulation and important in allergic and parasitic diseases.

Pulmonary circulation: This is the circulation carrying blood from the heart to the lungs and back for oxygen and carbon dioxide exchange.

Systemic circulation: This is the circulation carrying blood from the heart to the whole body for oxygen consumption and back to the heart and lungs for reoxygenation.

Transgenic (or knockout) mice: These are mice with specific mutations (or deletions) and altered genetics used for study of specific gene function.

References

1. Wagner, J. G., and Roth, R. A. (2000). Neutrophil migration mechanisms, with an emphasis on the pulmonary vasculature. *Pharmacol. Rev.* **52**, 349–374.
2. Bochner, B. S. (2003). Cellular adhesion in inflammation. In: *Middleton's Allergy Principles and Practice*, 6th ed. (N. F. Adkinson, Jr., J. W. Yunginger, W. W. Busse, B. S. Bochner, S. T. Holgate, and F. E. R. Simons, eds.). St. Louis, MO: Mosby, pp. 117–134. *This is a complete review on the function and regulation of adhesion molecules found on the EC and the leukocyte, with a focus on allergic inflammatory responses.*
3. Patel, K. D., Cuvelier, S. L., and Wiehler, S. (2002). Selectins: Critical mediators of leukocyte recruitment. *Semin. Immunol.* **14**, 73–81.
4. von Andrian, U. H., and M'Rini, C. (1998). In situ analysis of lymphocyte migration to lymph nodes. *Cell. Adhes. Commun.* **6**, 85–96.
5. Girard, J. P., and Springer, T. A. (1995). High endothelial venules (HEVs): Specialized endothelium for lymphocyte migration. *Immunol. Today* **16**, 449–457.
6. Muller, W. A., and Randolph, G. J. (1999). Migration of leukocytes across endothelium and beyond: Molecules involved in the transmigration and fate of monocytes. *J. Leukoc. Biol.* **66**, 698–704.
7. Burns, A. R., Smith, C. W., and Walker, D. C. (2003). Unique structural features that influence neutrophil emigration into the lung. *Physiol. Rev.* **83**, 309–336. *This is a comprehensive review on the recruitment of neutrophils in the lungs, through the pulmonary circulation and migration in the extracellular matrix, focusing on the structural features of the lungs that differ from other tissues.*
8. Lim, L. H., Bochner, B. S., and Wagner, E. M. (2002). Leukocyte recruitment in the airways: An intravital microscopic study of rat tracheal microcirculation. *Am. J. Physiol. Lung Cell Mol. Physiol.* **282**, L959–L967.
9. Sikora, L., Johansson, A. C., Rao, S. P., Hughes, G. K., Broide, D. H., and Sriramarao, P. (2003). A murine model to study leukocyte rolling and intravascular trafficking in lung microvessels. *Am. J. Pathol.* **162**, 2019–2028.
10. Hogg, J. C., and Doerschuk, C. M. (1995). Leukocyte traffic in the lung. *Annu. Rev. Physiol.* **57**, 97–114. *This is a good clear review on previous experiments done to elucidate CD18-dependent and independent mechanisms. It also discusses leukocyte traffic in the bronchial circulation.*

Capsule Biography

Dr. Bochner is Professor of Medicine at the Johns Hopkins University School of Medicine and became the Director of the Division of Allergy and Clinical Immunology at the Johns Hopkins Asthma and Allergy Center in 2003. Dr. Bochner's laboratory focuses on mechanisms of human allergic inflammation, with a particular interest in eosinophil, basophil, and mast cell adhesion, migration, survival, and apoptosis. His work is supported by grants from the NIAID branch of the NIH.

Dr. Lim is an assistant professor in the National University of Singapore. Her laboratory focuses on cellular recruitment during inflammation and cancer. Her research is currently supported by grants from the University Research Committee and the National Medical Research Council of Singapore.

CHAPTER 52

The Endothelium from a T-Cell Point of View

Federica Marelli-Berg
Imperial College, London

Lymphocyte recirculation and the localized recruitment of antigen-specific T cells to sites of inflammation are critical to the surveillance and effector functions of the immune system. A tightly regulated series of molecular interactions with the microvascular endothelium (extravasation) allows T lymphocytes to continuously recirculate from the bloodstream into lymphoid and nonlymphoid tissue scanning for their cognate antigen. During these frequent encounters the endothelium itself has the opportunity to display antigen to transiting T lymphocytes with the potential to modify their functions.

Regulation of T-Cell Trafficking by the Endothelium

The Multistep Paradigm of Lymphocyte Extravasation

Lymphocyte extravasation is initiated by transient interactions with the blood vessels through adhesion receptors including selectins and their ligands, members of the integrin superfamily, and the vascular hyaluronan-binding CD44. By virtue of these interactions T cells tether the vascular wall where they are exposed to chemokines (chemotactic cytokines), bound to the endothelial cell (EC) surface by glycosaminoglycans (GAGS). Chemokines trigger pertussis toxin–sensitive G-protein coupled receptors on the T cells. This "activation" event leads to conformational changes in the T-cell integrin adhesion molecules (including members of the β2 and β1 family of integrins, such as LFA-1 and VLA-4) that increase their affinity for their ligands. Chemokines also induce changes in the T cell shape (from round and villous to polarized), which create an ideal surface of interaction between T cells and ECs. As a consequence T cells flatten over and firmly adhere to the endothelium. Additional molecular interactions involving adhesion molecules (including integrins and junction adhesion molecules) and chemokines guide the migration of the adherent lymphocyte through the endothelial junctions (transendothelial migration or diapedesis). Some EC molecules such as CD31 appear to optimize the efficiency of this last step of the migratory process. A summary of the T cell–EC interactions leading to T-cell extravasation is provided in Figure 1. Together with intrinsic properties of the T cells (such as the high expression of certain adhesion molecules, as observed in memory T cells), signals mediated by these relatively well-characterized ligand–receptor interactions are likely to determine whether a T cell that is engaging in adhesive interactions with EC will eventually migrate into the underlying tissue.

T Lymphocyte Phenotype and Function is Modified By Transendothelial Migration

Once T cells have crossed the endothelial barrier, they remain localized for a variable length of time between the endothelial layer and the basal membrane and eventually enter the tissue. Tissue infiltration is a slow and complex process, which involves T-cell interactions with the basal membrane and the extracellular matrix and migration induced by tissue-derived chemokines. Thus, once a T cell has crossed the endothelial barrier and is committed to migration, it undergoes a genetic reprogramming leading to functional changes including induction of matrix metalloproteinase (MMP) expression, and responsiveness to tissue-derived chemoattractants that enable them to invade the

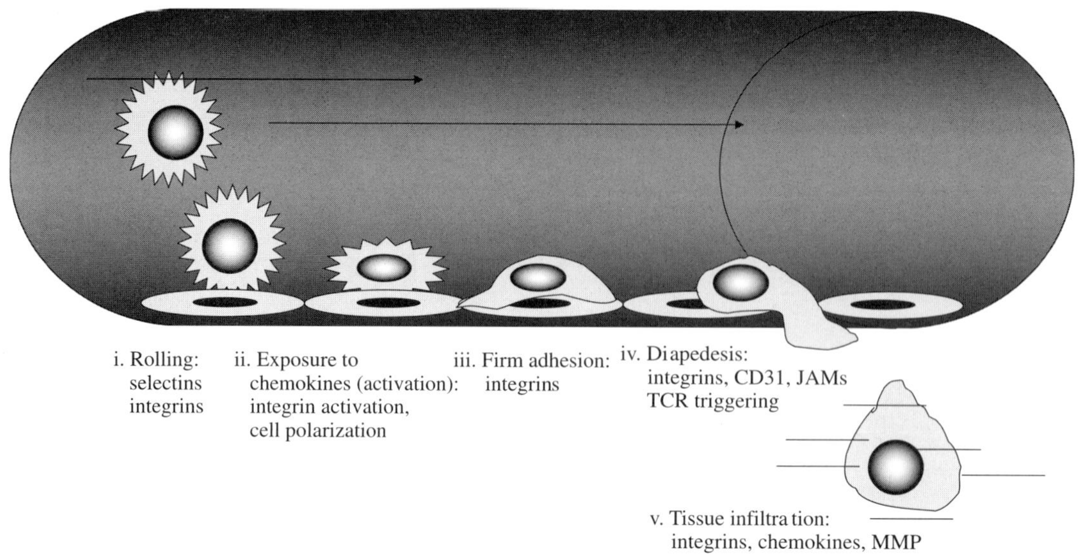

Figure 1 The multistep paradigm of lymphocyte extravasation. T cells are first engaged in selectin-mediated transient interactions with the endothelium. In this way, they become exposed to endothelium-bound chemokines. This induces increased affinity of T-cell integrin for their endothelial ligands and causes T-cell polarization (i.e., change of shape and redistribution of surface molecules). T cells thus firmly adhere to the endothelium and eventually migrate through the endothelial cell junctions. Once in the tissue, T lymphocytes respond to stimuli that lead them to the site of inflammation. (see color insert)

tissue and reach the inflammatory site. It is likely that the endothelium itself delivers the necessary signals to transiting T cells, which favor T-cell motility and further infiltration of the underlying tissue. In this context, upregulation of certain molecules (including at the transcriptional level), such as CD86 and CD69, has been observed in T cells following noncognate interactions with the EC. These changes are likely to help further activation of the T cells once they have reached their antigenic sites within the tissue. In line with these studies, we have observed that T-cell transendothelial migration is accompanied by activation of the transcription factor AP-1, without induction of NF-κB. This genetic reprogramming correlates with the upregulation of certain adhesion (LFA-1, VLA-4), activation (CD69), and costimulatory (CD86) receptors and with increased T-cell motility and antigen responsiveness to tissue invasiveness. These effects were dependent on $\alpha_L\beta_2$ integrin/CD54 interactions during adhesion to the endothelium.

Thus, T-cell extravasation is accompanied by phenotypic and functional changes induced by the interactions with the EC, which favor tissue infiltration by T cells and their further activation once they reach the antigenic site. The nature of the endothelium-derived signals that induces "proinvasive" genetic reprogramming in T lymphocytes and the molecular changes required for tissue invasion by T cells remain poorly characterized.

Recirculation of Memory and Naive T Cells and Tissue-Specific Homing

The ability of T lymphocytes to reach antigenic tissue is regulated at different levels. Naive T cells recirculate through secondary lymphoid organs (spleen, lymph nodes, and Peyer's patches) through the expression of L-selectin (CD62L) and the chemokine receptor CCR7. Expression of their ligand is restricted to the microvasculature of secondary lymphoid organs. Following antigen encounter and activation in the lymphoid tissues, changes are induced in the array of surface-expressed adhesion receptors, which allow effector memory T-cell access into nonlymphoid tissue. These changes include loss of CD62L and CCR7 expression and upregulation of other adhesion (LFA-1) and chemokine (CXCR4, CCR5) receptors. It has recently been described that a subset of memory T cells (central memory T cells) retain expression of the chemokine receptor CCR7 and the ability to traffic through secondary lymphoid organs. Finally, a third subset of memory T lymphocytes differentiates following priming in the lymph nodes, namely the follicular homing T cells, which specialize in delivering help to B cells. These cells are retained in the lymphoid tissue characterized by the expression of the chemokine receptor CXCR5, which mediates their recruitment in the B cell areas of the lymph node.

Memory and effector T cells generally display selective tropism for specific peripheral tissue such as the skin or the gut (tissue-specific homing). This is mediated by the use of different combinations of adhesion and chemokine receptors (molecular "area code") at distinct anatomical sites that facilitate the T lymphocyte return to that site. For example, high-level expression of $\alpha4\beta7$ integrin, whose ligand mucosal addressin cell adhesion molecule (MAdCAM)-1 is expressed on postcapillary venules in the intestinal lamina propria, targets one population of memory T cells to this site. In addition, a subset of these intestinal $\alpha4\beta7$hi cells also expresses CCR9, the specific receptor for the chemokine CCL25, or thymus-expressed chemokine (TECK), which is produced by small intestinal epithelium. In contrast, memory and effector cell recruitment to inflamed skin requires expression of lymphocyte surface ligands for vascular selectins, such as the E-selectin ligand cutaneous lymphocyte antigen (CLA) in humans and P-selectin ligand in mice. In addition, skin-homing T cells express CCR4, the specific receptor for the chemokine CCL17, or thymus and activation-regulated chemokine (TARC), expressed by keratinocytes. T cells acquire these receptors during priming in the lymphoid tissue. The genetic reprogramming leading to the acquisition of a molecular area code is determined by the local lymphoid organ microenvironment.

Immunogenicity of the Microvascular Endothelium

The properties of endothelial cells as antigen-presenting cells are likely to be important during local inflammatory responses and following transplantation of vascularized tissues, where EC are the first alloantigen-bearing target encountered by transiting host T cells. Endothelial cells constitutively express major histocompatibility complex (MHC) class I molecules, which enables them to present antigen to $CD8^+$ (cytotoxic) T lymphocytes. In addition, vascular EC often acquire the potential to present antigen to $CD4^+$ (helper) T cells, since MHC class II molecule expression on small vessels is a common finding in a variety of inflammatory conditions and human capillary ECs have been shown to express MHC class II molecules in heart, kidney, and liver. Triggering of the T-cell antigen receptor (TCR) by the endothelium is likely to occur during the sustained adhesive interactions leading to T-cell extravasation. The functional consequences of antigen presentation by the endothelium, particularly its ability to initiate T-cell responses, are still controversial. A key point in these issues is that ECs (at least in the human system) do not express the B7 family of costimulatory molecules that, when triggered in conjunction with the antigen receptor, allow autocrine production of the T cell growth factor inteleukin-2 and cell division. Other costimulatory molecules that have been found to be expressed by endothelial cells, such as inducible T-cell costimulator-ligand (ICOS-L), are unable to induce IL-2 gene transcription, although they can support the production of other cytokines, such as IL-4. Finally, negative regulators of T-cell activation, such as programmed death-ligand 1 (PD-L1) and PD-L2, have been identified on ECs.

$CD4^+$ (Helper) T Cells

The conclusions from some studies on the functional effect of cognate recognition of the endothelium by $CD4^+$ T cells are to date controversial. Some groups have reported that EC can support mitogen-induced T-cell proliferation, antigen-specific T-cell proliferation, and primary alloresponses. These authors hypothesized that during antigen presentation by ECs, costimulation can be provided by surface molecules other than B7, such as LFA-3. Other groups, including ourselves, have failed to detect alloproliferation of human and mouse $CD4^+$ T lymphocytes induced by either resting or cytokine-treated endothelium. Possible explanations for these discrepancies include a contamination with bone-marrow-derived antigen presenting cells (such as dendritic cells) of either the EC or the T-cell preparations.

An interesting observation that arose from these studies is that, unlike what has been observed with other costimulation-deficient parenchymal cells, memory T cells were not rendered hyporesponsive by cognate recognition of B7-negative endothelium. It is possible that the availability of accessory molecules displayed on the EC surface and coengaged during cognate recognition might modulate the signal delivered through TCR engagement. Thus antigen presentation in the absence of costimulation, but in the presence of strong cell–cell adhesion, mediated by the large array of accessory molecules displayed on the EC surface, might fall below the threshold required for T-cell activation but above that leading to T-cell silencing.

$CD8^+$ (Effector) T Cells

The few studies that have investigated the effects of cognate recognition of EC by $CD8^+$ T cells have focused on the development of proliferative responses, and, as for $CD4^+$ T cells, contradictory results have been reported. Some studies have concluded that EC can induce human and murine $CD8^+$ T lymphocyte differentiation, although with a low efficiency compared with conventional antigen-presenting cells (APC), and with impaired ability to secrete γ-IFN. In contrast, other investigations suggested that human ECs are unable to stimulate a functional response by freshly purified $CD8^+$ T cells.

In Vivo Studies

Similarly to all the in vitro studies described, the few studies analyzing antigen presentation by EC in vivo have led to conflicting conclusions. A caveat that must be borne in mind is that the microvascular endothelium has different phenotypic characteristics in different species. For example, murine EC express the costimulatory molecule CD80 but not CD40, whereas human ECs are totally B7-negative but express functional CD40. A recent study has suggested that EC allorecognition by adoptively transferred $CD8^+$ T cells in the absence of $CD4^+$-mediated T cell help and professional hematopoietic antigen-presenting cells leads to heart allograft rejection in MHC class I recipients. However it has clearly been shown that secondary lymphoid organs are required for transplant rejection, implying that antigen presentation by graft-resident dendritic cells migrated in the host lymphoid tissue is required for alloreactive T cells to be activated. The reason for these discrepancies remains unresolved.

A possible compromise of these opposite views is the recent definition of EC as "semiprofessional" APC, due to their ability to enhance T-cell responsiveness and cytokine production without eliciting full T-cell activation. Given that ECs might display tissue-specific as well as foreign peptides, full competence as APC would be potentially dangerous in terms of self-tolerance, and induction of nonresponsiveness following cognate recognition of ECs would be likewise undesirable for virus-specific T cells entering an infected tissue. Thus the neutral effects on subsequent T-cell behavior, described earlier, could be advantageous to overall immunoregulation. Of course the primary immunological function of EC is to regulate lymphocyte trafficking. Recent developments in vascular immunology suggest that antigen presentation by EC might induce a different activation response, inducing T-cell transendothelial migration.

TCR-Dependent T-Cell Trafficking

The possibility that Ag location might in some way mediate specific T-cell localization and tissue retention has been suggested by some studies looking at the migration of autoantigen-specific T cells in experimental models of inflammation. The participation of TCR-derived signals in the regulation of T-cell migration was suggested long ago by the observation that TCR triggering can induce integrin activation and immobilize migrating T cells. A role for TCR-triggering by the endothelium (during extravasation) in the regulation of T cell trafficking has also emerged from in vitro studies showing that antigen display by the endothelium leads to the recruitment of specific human and murine $CD4^+$ and $CD8^+$ T cells into the tissue. A recent study has provided the first in vivo evidence that insulin-specific $CD8^+$ T cells require antigen presentation by pancreatic microvascular endothelium to gain access to pancreatic islets [1]. These data have been confirmed by our observation that TCR-triggering by the endothelium induces T cell diapedesis without affecting adhesion [2]. The mechanism whereby cognate recognition promotes transmigration has yet to be fully defined, and it is likely to involve TCR-mediated activation of signaling pathways that control adhesion molecule activation and cytoskeletal rearrangements. In this context, recent studies have established a role for adapter proteins such as Vav, ADAP (FYB), and SKAP-55 in the activation of integrins following TCR triggering. Downstream events in these pathways are likely to involve Ras/Rho family GTPases, including Ras, Rho, Rac, and Cdc42.

Although it is clear that the majority of T-cell recruitment into tissues is regulated by noncognate mechanisms, antigen display by the endothelium can enhance the specificity and efficiency of lymphocyte recruitment into and retention at antigenic sites of inflammation, thus minimizing collateral

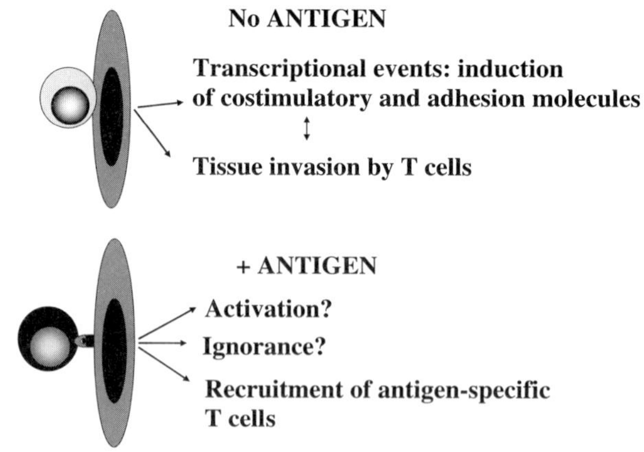

Figure 2 Functional consequences of T cell–endothelial cell interactions. Sustained interactions with the endothelium during extravasation induce a genetic reprogramming in T lymphocytes leading to upregulation of adhesion and costimulatory molecules. As a consequence, T cells become more tissue-invasive and hyperreactive to antigenic stimuli. The effects of antigen presentation by the endothelium to transiting T cells during extravasation are less clear; however, this event appears to favor the extravasation of antigen-specific T lymphocytes. (see color insert)

damage caused by excessive antigen-nonspecific inflammatory cells. A summary of the functional effect of T cell–EC interactions discussed here is provided in Figure 2.

References

1. Savinov et al. (2003). *J. Exp. Med.* **197**, 643.
2. Marelli-Berg et al. (2004). *Blood* **103**, 3111.

Bibliography (Further Reading)

Fabbri, M., Bianchi, E., Fumagalli, L., and Pardi, R. (1999). Regulation of lymphocyte traffic by adhesion molecules. *Inflamm. Res.* **48**, 239–246.

Kunkel, E. J., and Butcher, E. C. (2002). Chemokines and the tissue-specific migration of lymphocytes. Immunity **16**, 1–4. This review clearly highlights the role of chemokines/adhesion molecules in controlling homeostatic lymphocyte trafficking, including the localization of cutaneous and intestinal memory T cells.

Luscinskas, F. W., Ma, S., Nusrat, A., Parkos, C. A., and Shaw, S. K. (2002). The role of endothelial cell lateral junctions during leukocyte trafficking. *Immunol. Rev.* **186**, 57–67. *This article analyzes the endothelial-dependent mechanisms that coordinate transendothelial migration in and highlights the role of certain lateral junctional proteins and protein complexes.*

Marelli-Berg, F. M., and Jarmin, S. J. (2004). Antigen presentation by the endothelium: A green light for antigen-specific T cell trafficking? *Immunol. Lett.* **93**, 109–113. *This is a critical overview of the often conflicting reports on the functional effects of antigen presentation by endothelial cells to T lymphocytes.*

Capsule Biography

Federica Marelli-Berg is a PI in the T Cell Motility Laboratory, Department of Immunology, Imperial College London. The focus of her research revolves on the regulation of T lymphocyte motility, particularly by the interactions with the endothelium.

CHAPTER 53

Integrins and Formation of the Microvasculature

[1]Joseph H. McCarty and [2]Richard O. Hynes

[1]*Department of Cancer Biology, University of Texas MD Anderson Cancer Center, Houston, Texas*

[2]*Howard Hughes Medical Institute and Center for Cancer Research, Massachusetts Institute of Technology, Cambridge, Massachusetts*

Introduction

The formation of the microvasculature is a complex process requiring proper cell–cell and cell–extracellular matrix (ECM) communication events. A vascular basement membrane rich in many ECM proteins surrounds all blood vessels and regulates multiple vascular cell functions. Endothelial cells and pericytes, and indeed most eukaryotic cells, recognize ECM ligands primarily via members of the integrin family of proteins [1]. Integrins are heterodimeric cell surface receptors consisting of noncovalently associated α and β subunits. Most integrins recognize multiple ECM ligands, and many ECM proteins can bind to more than one integrin receptor. In vertebrates there are 26 integrin genes, 18 encoding α subunits and 8 encoding β subunits, yielding 24 distinct integrin heterodimers (Figure 1). At least five integrins (α1β1, α2β1, α5β1, αvβ3, and αvβ5) are expressed at some point on endothelial cells or pericytes [2]. Here we summarize pertinent data linking integrins and their ECM ligands to various microvascular events.

Microvascular Integrins and Their Extracellular Matrix Ligands

α5β1 Integrin

Various genetic data support the notion that the integrin α5β1 and its ECM ligand, fibronectin (FN), are absolutely essential for normal blood vessel function. Mouse embryos harboring an ablated α5 gene die by embryonic day (E) 10.5, displaying severe embryonic and extraembryonic vascular defects, as well as posterior trunk and somitic abnormalities. Additionally, α5-null embryonic stem (ES) cells injected into syngeneic mice form teratocarcinomas with reduced size and a poorly developed vasculature. Also, embryoid bodies derived from α5-null ES cells display delayed vascular differentiation and organization. In concordance with the α5 genetic data, deletion of the FN gene leads to embryonic lethality associated with severe vascular defects, and similar angiogenic abnormalities are observed in FN-null embryoid bodies [2].

Other experimental models also support a necessary role for α5β1 and FN in angiogenesis. For example, protein expression of α5β1 and FN is upregulated on angiogenic tumor blood vessels. Likewise, in the chick chorioallantoic membrane (CAM) assay, angiogenic growth factors upregulate α5β1 expression. In both systems anti-α5β1 or anti-FN function-blocking antibodies or peptide antagonists perturb neovascularization [3].

α1β1 and α2β1 Integrins

The collagen-binding integrins α1β1 and α2β1 are both upregulated on angiogenic blood vessels following vascular endothelial growth factor (VEGF) stimulation (reviewed in Ref. [2]). Function-blocking anti-α1 and -α2 antibodies inhibit angiogenesis in vivo using Matrigel plugs or transplanted tumor models. Interestingly, α1-null mice are viable and fertile, yet are less susceptible to tumor growth. This resistance correlates with increased levels of circulating matrix metalloproteinase (MMP) activity involving MMP-2, MMP-7, and MMP-9. Elevated proteolysis in α1-null mice leads to the conversion of plasminogen to angiostatin, a potent inhibitor of endothelial cell neovascularization. Mice harboring a deletion of the α2 gene are viable and fertile and do not display any gross vascular abnormalities. Obviously, analysis of mice null for both α1β1 and α2β1

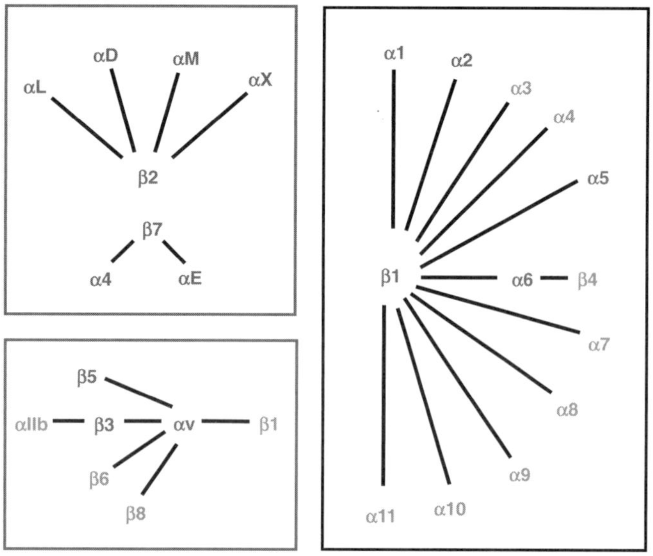

Figure 1 The integrin family of ECM receptors. The 26 integrin α and β subunits, separated into three groups, are shown. Lines connecting subunits indicate the 24 known heterodimeric pairs. The leukocyte-specific integrins are shown in blue. Integrins with proposed roles in various microvascular functions are highlighted in red. (see color insert)

expression should be useful in understanding the exact functions of these integrins during physiological and pathological angiogenesis.

β1 Integrins

The β1 integrin subunit can pair with at least 12 α subunits (Figure 1). Various β1-containing integrins are expressed in virtually all vertebrate cells and are involved in most aspects of development. Not surprisingly, β1 genetic ablation leads to very early (E7.5) embryonic lethality, prior to the onset of vascular development. Analyses of β1-null ES cells, embryoid bodies, and teratocarcinomas reveal an essential role for β1 integrins in blood vessel formation. Whereas β1 integrins are dispensable for endothelial cell differentiation, they are required for VEGF-induced endothelial cell proliferation and blood vessel elaboration.

As mentioned previously, integrins α1β1 and α2β1 are also important for microvascular formation. It is likely that some of the defects observed in the absence of β1 integrin can be explained by loss of α1β1 and α2β1 function. However, ES cells or embryoid bodies null for β1 integrin, α5 integrin, or fibronectin display many strikingly similar defects. Thus, many of the β1-null defects can be explained by loss of integrin α5β1. Again, these data emphasize that fibronectin and its primary receptor, α5β1 integrin, play essential roles in the regulation of vascular development [4].

αv Integrins

The αv integrin subunit can pair with β1, β3, β5, β6, and β8, yielding five distinct integrin heterodimers (Figure 1). An extensive body of published data exists concerning αvβ3 and αvβ5 during various types of physiological and pathological angiogenesis. Various angiogenic growth factors or cytokines upregulate αvβ3 or αvβ5 protein expression on cultured endothelial cells or on vessels using the CAM assay. Likewise, many angiogenic blood vessels display increased levels of αvβ3 or αvβ5. Function-blocking antibodies or peptide antagonists inhibit this integrin-mediated neovascularization in these various systems, ultimately leading to endothelial cell death. Along these lines, αvβ3 has been intensely pursued as a potential anti-angiogenic target, and αvβ3 antagonists are currently being tested in human clinical trials.

Surprisingly, ablation of the β3 and/or β5 genes does not lead to angiogenic abnormalities. Mice null for both αvβ3 and/or αvβ5 are viable and fertile, and developmental vasculogenesis and angiogenesis proceed normally in these mutants. Physiological angiogenesis occurs normally in β3-null or mice null for both β3 and β5. No experimental differences are observed using hypoxia-induced retinal angiogenesis or growth factor–mediated corneal neovascularization models. Interestingly, tumor growth and neovascularization are actually enhanced in β3-/- or β3/β5-null mice [5]. Thus, it is difficult to infer the significance of αvβ3 and αvβ5 in neovascularization based on the antibody and peptide inhibition results versus the genetic ablation data.

One could argue that, in mice lacking β3 and/or β5 gene expression, other αv-associated integrins are compensating for normal αvβ3 and/or αvβ5 functions. As mentioned earlier, αv can indeed pair with five different β subunits. However, mice null for the αv gene, and thus lacking the expression of all five αv integrins, do not display general vascular defects. Approximately 70 percent of αv-nulls survive to E10.0, having developed a normal vasculature, but die by E10.5 because of placental defects. The remaining αv-null embryos that survive the mid-gestation crisis subsequently develop to term with a largely normal vasculature. However, they do develop cerebral hemorrhage and a cleft palate, and they die shortly after birth. The hemorrhage is not due to endothelial or pericyte defects, but rather involves defective associations between cerebral vessels and central nervous system glia [6]. Interestingly, the cerebral hemorrhage in the αv-nulls is specifically due to the loss of integrin αvβ8; mice null for the β8 gene display phenotypes that are essentially identical to those seen in the αv-nulls [2].

Interpreting the Conflicting Data

It is difficult to reconcile the conflicting experimental data regarding the necessity of αvβ3 function during pathological, and especially tumor, neovascularization. It is important to note that the antagonism of an existing integrin function is certainly quite different from ablation of the integrin gene product. Thus, one should exercise caution when interpreting data using such different experimental approaches. However, there are cases where genetic ablation of integrin expression and acute inhibition of integrin function lead to quite similar results. For example, various data using β3 antagonists implicate αvβ3 as important for bone reabsorption. This function is also revealed after ablation of the β3 gene. Antibody or peptide inhibition of the platelet integrin αIIbβ3 leads to defective platelet aggregation. Mutations in the human β3 gene also lead to abnormal platelet aggregation and the bleeding disorder Glanzmann thrombasthenia. β3-null mice develop abnormalities that are very similar to this human disease. Similarly, genetic ablation of the α5 gene, as well as acute inhibition of α5 function, both clearly show that this integrin is important for neovascularization.

Accumulating data support the concept of cross-talk pathways existing among different integrins. Inhibition of one integrin signaling cascade can in turn affect other integrin-mediated signaling events. Thus, antagonists that perturb αvβ3 function may be affecting other integrins involved in neovascularization, possibly including α5β1 and/or α1β1. Another likely possibility, discussed in more detail later, is that in some cases a given integrin can be pro- or antiangiogenic. For example, growing evidence suggests that αvβ3 is a negative regulator of pathological angiogenesis. Ablation of the β3 gene would remove this negative regulation, leading to enhanced neovascularization. This is indeed the result during pathological angiogenesis in the β3-null mouse model. In any case, it is likely that the acute inhibitors of angiogenesis are affecting microvascular events in a far more complicated manner than originally expected, and more extensive reviews discussing these possibilities have recently been published [5, 7].

Cross Talk between Vascular Integrins and Receptor Tyrosine Kinases

Integrin–ECM adhesion regulates many intracellular signaling events, affecting cellular responses such as migration, differentiation, and proliferation. Initiation of integrin-ECM adhesion occurs via two primary mechanisms, generally termed *outside-in* and *inside*-out signaling. Inside-out signaling involves the modulation of integrin–ECM affinity via signaling events originating within the cell. Outside-in signaling results from direct integrin-ECM binding, leading to the activation of downstream signaling cascades. Whereas integrin activation can occur via a variety of stimuli, growth factors and their cognate receptor tyrosine kinases (RTKs) are established synergistic signaling partners [8]. Importantly, cross talk between integrin and RTK pathways is necessary for microvascular development and maintenance [9].

Generally, signaling cross talk between vascular integrins and RTKs positively regulates each other's function. For example, the integrin αvβ3 can physically interact with Flk-1, the VEGF-A receptor tyrosine kinase. αvβ3-Flk1 association augments VEGF-A-induced Flk-1 phosphorylation and enhances endothelial cell migration on vitronectin, an αvβ3 ligand. A similar physical link between αvβ3 and the PDGFR-β has been reported. αvβ3 interaction with both Flk-1 and PDGFR-β occurs via the αvβ3 extracellular domain and is independent of growth factor stimulation.

Growth factor regulation of integrin signaling also occurs at the transcriptional level. VEGF treatment of endothelial cells stimulates the transcription of multiple integrin subunits including αv, β3, β5, α1, and α2. In other cases, growth factor stimulation of integrin expression can be quite specific. For example, basic fibroblast growth factor (bFGF)-induced angiogenesis leads to the specific upregulation of endothelial αvβ5 integrin.

The Role of the ECM in Regulating Vascular Integrin Functions

The *angiogenic switch* model [10] posits that a delicate balance exists between pro- and antiangiogenic factors. Changes in the levels of such factors regulate whether a blood vessel remains in a quiescent state, or becomes activated and angiogenic. Many ECM proteins within the vascular basement membrane, acting via integrins, are major regulators of the angiogenic switch. An emerging theme is that the exact function of a given ECM protein, as well as the integrin receptor(s) it binds, can be greatly affected by proteolytic events [11]. Several examples of how such endogenous "cryptic" fragments affect microvascular function are described next.

Type IV collagen, a major component of the vascular basement membrane, is normally a ligand for the integrins α1β1 and α2β1. MMP9 cleavage of collagen IV yields the potent angiogenesis inhibitor *tumstatin*. In vitro tumstatin induces apoptosis of endothelial cells by inhibiting the protein translation machinery. Likewise in vivo, tumstatin is a potent inhibitor of tumor neovascularization. Tumstatin's antiangiogenic effects are mediated by binding to the αvβ3 integrin.

Proteolysis of the C-terminal NC1 domain of collagens XV or XVIII generates the antiangiogenic factor endostatin. Murine *endostatin* inhibits VEGF- or bFGF-induced endothelial cell proliferation and migration in vitro, and also represses pathological angiogenesis in some transplantable tumor models. Endostatin is reportedly a ligand for both α5β1 and αvβ3. Intact collagen XV and XVIII do not bind to these integrins. Importantly, while the exact molecular mechanisms by which endostatin exerts its antiangiogenic effects remain unknown, it is clear that they are different from those regulated by tumstatin and do not require αvβ3.

Proteolysis of many other ECM proteins also generates antiangiogenic fragments, although the mechanistic details of how they exert these effects remain unclear. For example, when proteolytically cleaved, thrombospondin-1 "cryptic fragments" bind to the scavenger receptor CD36 and serve as potent repressors of angiogenesis [12]. Similarly, degra-

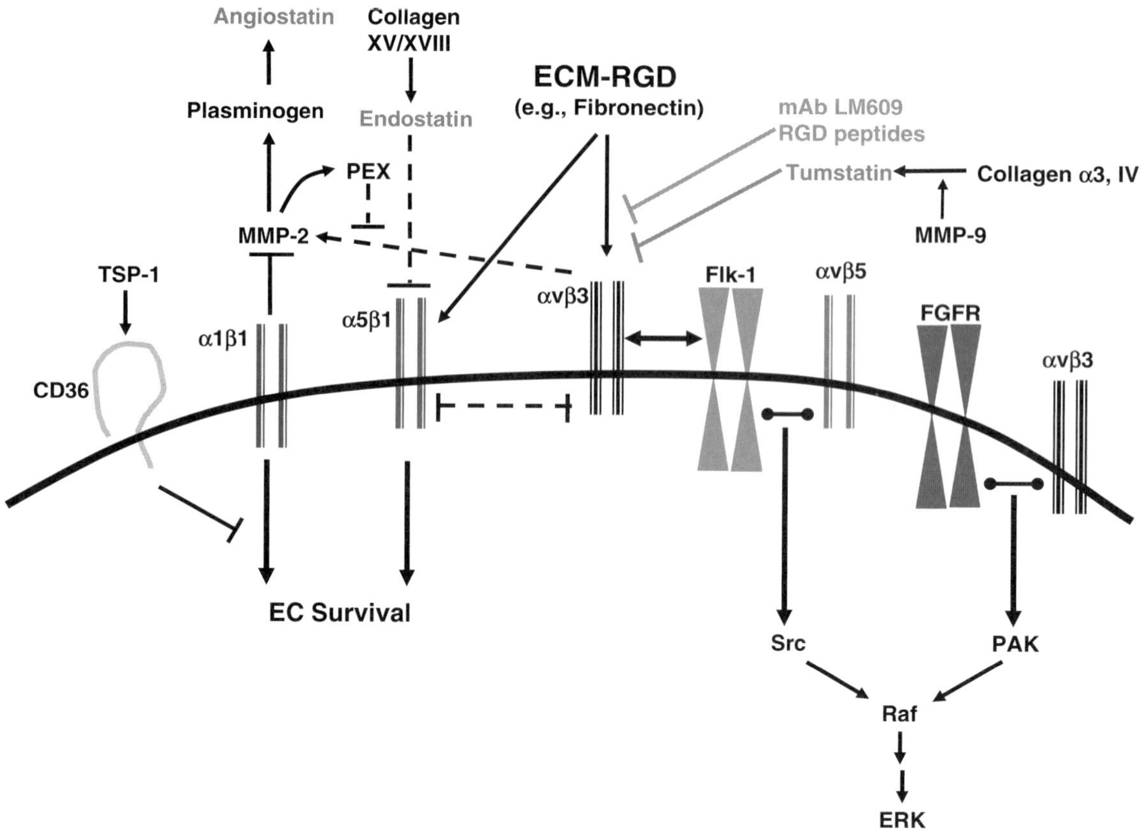

Figure 2 Integrin signaling cascades during angiogenesis. Various integrins, extracellular matrix proteins, and receptor tyrosine kinases with roles in endothelial cell function are shown. Pink lettering indicates endogenous inhibitors of integrin-mediated angiogenesis. Orange lettering indicates blocking antibodies or peptide antagonists that inhibit integrin function. Lines with double-headed arrows indicate direct physical connections between proteins. Double-headed ball-ended lines denote signaling cross-talk between receptors. (see color insert)

dation of the proangiogenic factor fibronectin generates the angiogenesis inhibitor anastelin.

Conclusion

Integrins and their various ECM ligands play essential roles in virtually every aspect of physiological and pathological microvasculature formation. This is controlled largely through integrin-mediated signal transduction and involves extensive cross talk with other signaling cascades, including those involving growth factor receptor tyrosine kinases (Figure 2). Additionally, ECM proteolysis regulates integrin functions, in many cases switching integrin regulation of neovascularization from a positive to a negative state. These various data highlight microvascular integrins and their ECM ligands as attractive candidates for therapeutic intervention to selectively enhance or repress various forms of physiological or pathological neovascularization.

Glossary

Cell Adhesion: The process by which cells adhere to each other and to components of the extracellular environment. Precise regulation of cell adhesion is necessary for normal developmental and physiologic maintenance. Abnormal regulation of cell adhesion events can lead to various disease states, including cancer.

Extracellular matrix: A complex mixture of secreted macromolecules that comprise the noncellular material of all tissues. Besides providing structural support, the extracellular matrix is biologically active and influences a host of dynamic cellular responses.

Integrin: Heterodimeric cell surface molecules composed of α and β subunits. Integrins serve as the cell's primary adhesion receptors for extracellular matrix ligands and play important roles in regulating cellular functions, ranging from differentiation to proliferation and migration.

References

1. Hynes, R. O. (2002). Integrins: Bidirectional, allosteric signaling machines. *Cell* **110**(6), 673–87. *A review of the molecular models for integrin adhesion, activation, and signaling.*
2. Hynes, R. O., Lively, J. C., McCarty, J. H., Taverna, D., Francis, S. E., Hodivala-Dilke, K., and Xiao, Q. (2002). The diverse roles of integrins and their ligands in angiogenesis. *Cold Spring Harbor Symp Quantitat. Biol.* **67**, 143–153. *A detailed summary of the molecular genetic data supporting roles for different integrins in angiogenesis.*
3. Kim, S., Bell, K., Mousa, S. A., and Varner, J. A. (2000). Regulation of angiogenesis in vivo by ligation of integrin alpha5beta1 with the central cell-binding domain of fibronectin. *Am. J. Pathol.* **156**(4), 1345–1362.
4. Bouvard, D., Brakebusch, C., Gustafsson, E., Aszodi, A., Bengtsson, T., Berna, A., and Fassler, R. (2001). Functional consequences of integrin gene mutations. *Circ. Res.* **89**(3), 211–223.

5. Hynes, R. O. (2002). A reevaluation of integrins as regulators of angiogenesis. *Nat. Med.* **8**(9), 918–921.
6. McCarty, J. H., Monahan-Earley, R. A., Brown, L. F., Keller, M., Gerhardt, H., Rubin, K., Shani, M., Dvorak, H. F., Wolburg, H., Bader, B. L., Dvorak, A. M., and Hynes, R. O. (2002). Defective associations between blood vessels and brain parenchyma lead to cerebral hemorrhage in mice lacking alphav integrins. *Mol. Cell. Biol.* **22**(21), 7667–7677.
7. Stupack, D. G., and Cheresh, D. A. (2002). Get a ligand, get a life: Integrins, signaling and cell survival. *J. Cell Sci.* **115**(Part 19), 3729–3738.
8. Giancotti, F. G., and Ruoslahti, E. (1999). Integrin signaling. *Science* **285**(5430), 1028–1032. *This article provides an excellent summary of the complexity of integrin signaling and how these events regulate cell proliferation, differentiation, and migration.*
9. Hood, J. D., Frausto, R., Kiosses, W. B., Schwartz, M. A., and Cheresh, D. A. (2003). Differential αv integrin-mediated Ras-ERK signaling during two pathways of angiogenesis. *J. Cell Biol.* **162**(5), 933–943.
10. Hanahan, D., and Folkman, J. (1996). Patterns and emerging mechanisms of the angiogenic switch during tumorigenesis. *Cell* **86**(3), 353–364.
11. Kalluri, R. (2003). Basement membranes: Structure, assembly and role in tumour angiogenesis. *Nat. Rev. Cancer* **3**(6), 422–433. *This article describes in detail the importance of various basement membrane components and their cryptic fragments in regulating different aspects of developmental and pathological angiogenesis.*
12. Lawler, J. (2000). The functions of thrombospondin-1 and -2. *Curr. Opin. Cell Biol.* **12**(5), 634–640.

Capsule Biography

Dr. Joseph McCarty was a postdoctoral researcher in the Center for Cancer Research at MIT. He is currently an Assistant Professor in the Department of Cancer Biology at the University of Texas MD Anderson Cancer Center. His research focuses on the development and maturation of the brain vasculature.

Dr. Richard Hynes was a Daniel K. Ludwig Professor of Biology at MIT and an Investigator of the Howard Hughes Medical Institute. His research interests concern the role of cell adhesion in physiology and pathology.

CHAPTER 54

Platelet Phospholipids Tighten the Vascular Endothelial Barrier

Fred L. Minnear

Department of Physiology and Pharmacology, Robert C. Byrd Health Sciences Center, West Virginia University, Morgantown, West Virginia

Platelets are key to the physiological processes of thrombosis and hemostasis and are repositories for biologically active mediators that influence many cellular functions such as aggregation, adhesion, growth, and mitogenesis as well as matrix formation and fibrinolysis. Platelets (or thrombocytes) also support the vascular endothelium, maintaining the semipermeable barrier to the passage of water and protein. Recently, two platelet phospholipids, lysophosphatidic acid (LPA) and sphingosine 1-phosphate (S1P), have been demonstrated to have a myriad of biological activities including cell proliferation, migration, survival, mitogenesis, and chemotaxis as well as matrix assembly, tumor cell invasion, and angiogenesis. LPA and S1P also enhance the barrier function of the vascular endothelium. This review will focus on the interaction of platelets and the vascular endothelium in promoting endothelial barrier function and will identify possible signaling pathways and cellular targets activated by LPA and S1P.

Platelets Support the Endothelial Barrier

Thrombocytopenia in patients is often characterized by petechial hemorrhages due to capillary leakage. Thrombocytopenic animals demonstrate gross purpura and alterations in the microvasculature as evidenced by gaps between endothelial cells and leakage of radioactive tracers into the surrounding tissues. The vascular endothelia are thinner and have open pores and membranous diaphragms. Evidence of a dysfunctional endothelial barrier and tissue edema due to thrombocytopenia have been observed in many experimental models such as the frog hindlimb, the thyroid lobe of dogs, the ear of postirradiated rabbits, the thoracic duct lymph of dogs, the coronary microvasculature of rats, and the lungs of sheep. Sheep chronically depleted of platelets with an anti-platelet antibody develop petechial hemorrhages, an increase in the transendothelial clearance of protein in the lungs, and pulmonary edema at a left atrial pressure of 20 mmHg.

Transfusion with platelet-rich plasma reverses these abnormalities. When comparing an isolated, thyroid lobe of a dog perfused with platelet-poor plasma and its counterpart lobe perfused with platelet-rich plasma, the lobe with platelet-poor plasma had gross purpura, endothelial gaps, and increased leakage of protein into the tissue. Transfused platelets labeled with radiosulfate rapidly interact with the vascular endothelium of thrombocytopenic animals and incorporate the label within the endothelium. Interestingly, a common finding is that the number of platelets required to reverse hemorrhage or capillary leakage is much less than that required to correct bleeding times. Thus, platelets in some way nurture the microcirculation.

Three theories have been proposed to explain this endothelial barrier-supporting function of platelets. The first and obvious explanation is that platelets physically block pores or gaps in the vessel wall. The second theory is that platelet components promote endothelial cell growth. The third, and most studied, theory is that platelets continually release a novel humoral factor that decreases endothelial permeability.

The measurement of diffusive permeability across confluent, endothelial cell monolayers developed in the 1980s provided a novel methodology to directly test these three theories. Initial studies addressed the first theory that platelets simply act as plugs between leaky endothelial cells. Washed human platelets, but not paraformaldehyde-fixed platelets, both of which attached sparingly to endothelial cells, reduce the flux of ^{125}I-labeled albumin across endothelial cell monolayers. Subsequently, a number of studies demonstrated that the permeability-decreasing activity of platelets resides in a soluble factor as platelet-conditioned medium (PCM) consistently decreases albumin permeability across endothelial cell monolayers derived from bovine aorta and bovine and human pulmonary arteries and microvessels. Thus, platelets decrease endothelial permeability via the release of a soluble factor and not by plugging of endothelial pores or gaps.

Accepting the notion that this activity of platelets resides in a soluble factor, investigators focused on a number of familiar platelet components such as serotonin, norepinephrine, and cyclooxygenase metabolites. Proven antagonists or inhibitors of these metabolites did not prevent the decrease in albumin permeability induced by platelets, platelet lysate, or PCM. At the time, norepinephrine (predominately a β_1-adrenergic agonist) was an intriguing possibility because it was known that β-agonists decrease endothelial permeability. This activity of β-agonists, however, favors the β_2-subtype, the predominant receptor subtype on vascular endothelial cells.

When my laboratory entered this area of research, adenosine was proposed to be the active platelet factor. We [1] ruled out adenosine as the likely candidate by the following experimental findings. PCM contains micromolar concentrations of AMP, ADP, and ATP, but adenosine is below detectable levels as measured by high-performance liquid chromatography. Although adenosine directly decreases endothelial permeability, inactivation of adenosine with adenosine deaminase, which metabolizes adenosine to inactive inosine, does not block the activity of PCM. Furthermore, the adenosine-receptor antagonist BW-A1433U83 does not block the decrease in endothelial permeability induced by platelets or PCM. A fraction of PCM smaller than 3 kDa that contains micromolar concentrations of AMP and ADP has no activity, whereas a fraction larger than 3 kDa that contains much reduced levels of AMP and ADP significantly reduces permeability. Thus, adenosine and adenine nucleotides are not the primary factors responsible for the platelet activity; instead the activity resides in a fraction of PCM larger than 3 kDa.

As early as 1956, a protein fraction extracted from platelets was reported to reduce capillary permeability in the rat hind limb. Haselton and Alexander [2] and Patil and others [1] demonstrated 35 years later that the active platelet factor is heat-stable, trypsin-sensitive, and >3 kDa. Some of these findings point to a protein as the active factor. An obvious difficulty with this notion is how platelets could replenish the store of a protein continuously released into the plasma, considering platelets have a limited ability to synthesize proteins. This criticism prompted us to consider viable alternatives. There were also conflicting pieces of data in the literature. The fact that the active factor is sensitive to trypsin and, in addition, is precipitated in saturated ammonium sulfate implicates a large charged protein, yet the activity is heat stable.

Still assuming that the permeability-decreasing activity of PCM is associated with a protein fraction, Patil et al. [1] performed anion- and cation-exchange chromatography to increase the specific activity of a protein fraction. These experiments provided evidence that the activity resides with a negatively charged protein such as albumin because the active fraction binds to and requires a high salt concentration to elute the activity from an anionic exchange column. Applying PCM to a calibrated Sephacryl S-200 gel filtration column yields three major protein fractions with activity present only in the fraction with an elution volume of albumin. Activity also resides with the albumin immunoprecipitate following removal of albumin from the PCM with an anti-albumin antibody. It was at this time that Alexander and coworkers [3] provided convincing evidence that the active factor is a phospholipid. Their findings pointed to lysophosphatidic acid (1-acyl-2-hydroxyl-3 phosphoglyceride, LPA).

That LPA might be the active factor was somewhat of a surprise because LPA had been reported to increase the permeability of endothelial cell monolayers derived from brain capillaries. However, the findings of Alexander and coworkers [3] are compelling. Activity is present in the methanol extract of PCM and is eliminated by enzymatic treatment with phospholipase B, which cleaves at the *sn*-1,2 positions, and with alkaline phosphatase, which cleaves at the phosphomonoester bond. Phospholipase A_2, which cleaves at the *sn*-2 position where a hydroxyl group but no lipid is present in a phospholipid, has no effect. Using a similar protocol, Minnear et al. [4] extracted lipids with methanol from albumin fractions obtained from PCM either by immunoprecipitation with an anti-albumin antibody or by passing the PCM through a Blue-Sepharose column. Successive extractions with methanol yield activity in the methanol extract and loss of activity in the extracted albumin fraction.

Lysophosphatidic Acid (LPA)

LPA was originally reported to be a growth factor released from activated platelets and other injured cells. LPA and thrombin, another mitogen and well-known permeability-increasing agent in vitro, initiate similar cellular responses and signaling pathways such as mobilization of intracellular calcium, stimulation of phospholipase C and protein kinase C, formation of actin stress fibers, and contraction of fibroblasts. These known actions of LPA would link this phospholipid to an increase in endothelial permeability, as has been observed in endothelial cells from brain microvessels and human umbilical veins. In contrast, LPA decreases endothelial permeability in bovine endothelial cells from the lung and aorta. Thus, the effects of LPA on vascular permeability may be organ, tissue, or species specific.

Sphingosine 1-Phosphate (S1P)

S1P came to prominence a few years after LPA and was found to mimic many of the biologic activities of LPA. A seminal study from the laboratory of Garcia [5] demonstrated that S1P decreases endothelial permeability across cell monolayers derived from bovine and human pulmonary arteries and human umbilical vein. In a number of cell types, S1P appears biologically to be much more potent than LPA. For example, S1P is more active than LPA in human umbilical vein endothelial cells with regard to the intracellular mobilization of calcium, cytoskeletal reorganization, and cell migration. In our hands, S1P increases the barrier function of endothelial cell monolayers derived from pulmonary arteries to an approximately twentyfold greater extent than does LPA.

S1P and LPA are the leading candidates for the soluble factor released from platelets that maintains the barrier function of the vascular endothelium. The plasma concentrations of S1P and LPA are 0.2 μM and negligible, respectively, and the serum concentrations are 0.5 to 1.0 and 10 μM, respectively. Both S1P and LPA are present in platelets and to a lesser extent in other cells such as leukocytes, fibroblasts, erythrocytes, and adipocytes and are released upon activation during blood coagulation, inflammation, tissue injury, and neoplasia. Platelets have the potential to synthesize and store S1P because platelets have an abundance of sphingosine kinase, which converts sphingosine to S1P, and relatively no lipases to degrade S1P. S1P and LPA bind to albumin on which they retain biological activity and are protected against hydrolysis by serum phospholipases. Cellular responses of S1P and LPA have been grouped into two categories: (1) growth-related activities such as proliferation, differentiation, and suppression of apoptosis; and (2) cytoskeletal functions such as shape change, aggregation, adhesion, chemotaxis, contraction, and secretion. Micromolar concentrations are required for growth-related functions and nanomolar concentrations stimulate cytoskeletal responses.

Although S1P has been proposed to be the major factor in PCM that imparts the barrier-enhancing property, no one has detected S1P in active conditioned medium. We attempted to identify the lipids in two batches of PCM and in the albumin immunoprecipitate from one of the batches of PCM by using electrospray mass spectrometry (Table I); all three samples were equally potent in enhancing endothelial electrical resistance. A number of lipids were identified, and there were 10 unknown lipids. Lipid profiles were almost identical between PCM and the methanol extract of the albumin immunoprecipitate, indicating the importance of albumin as a lipid-carrier. Interestingly, LPA was detected as a small peak of palmitoyl (16:0)-LPA only in one batch of conditioned medium. Palmitoyl-LPA, stearoyl-LPA, and arachidonoyl-LPA are the major molecular species of LPA produced from thrombin-stimulated platelets. Surprisingly, S1P was not present in any of the three samples tested.

Table I Electrospray Mass Spectrometry of Bligh & Dyer and Modified Folch Extracts of Platelet-Conditioned Medium.

Bligh & Dyer extraction	Folch extraction
Decanoic (10:0)	Octanoic (8:0)
Myristic (14:0)	Unknown
Pentadecanoic (15:0)	Unknown
Palmitic (16:0)	Myristic (14:0)
Linoleic (18:2)	Palmitoleic (16:1)
Oleic acid (18:1)	Palmitic (16:0)
Stearic (18:0)	Linolenic (18:3)
C18 sphingosine	Linoleic (18:2)
C18 sphingonine	Oleic (18:1)
Arachidonic (20:4)	Stearic (18:0)
Homogamma linolenic (20:3)	C18 dihydrosphingosine
Eicosadienoic (20:2)	C18 sphingosine
D-Ribo-phytosphingosine	C18 sphingonine
Unknown	Arachidonic (20:4)
Unknown	Homogamma linolenic (20:3)
C20 ceramide, 16:0–18:1 DG	Unknown
16:0–18:0 DG	Lignoceric acid (24:0)
Unknown	16:0 LPA
Unknown	Unknown
Unknown	Unknown C20 ceramide, 16:0–18:1 DG 18:1–20:4 PA

Regulation of Vascular Endothelial Permeability and Proposed Cellular Targets

A working hypothesis for the regulation of endothelial permeability, which was popular in the 1990s, states that signaling pathways that balance competing adhesive and contractile forces control the paracellular flux of macromolecules. Endothelial cell–cell and cell–matrix contacts tether endothelial cells to each other and to the extracellular matrix, respectively, and act against centripetal tension generated by actin–myosin motors. Thus, agents that enhance barrier function would do so by decreasing isometric tension; and the opposite would occur for barrier-decreasing agents. This hypothesis is supported by studies that have demonstrated that cAMP-enhancing agents increase barrier activity in association with decreases in myosin light-chain phosphorylation. The additional measurement of cellular isometric tension, however, provides a different conclusion in that cAMP-enhancing agents promote barrier function independently of cellular tension. In our hands, S1P and LPA increase endothelial barrier activity in association with increases in myosin light-chain phosphorylation and cellular isometric tension. S1P increases isometric cellar tension to a slightly lesser extent than can be induced by thrombin, which decreases endothelial barrier function. Thus, changes in cellular tension are not solely responsible for changes in barrier function as both S1P and cAMP increase barrier activity yet have opposite effects on myosin light-chain phosphorylation and cellular isometric tension.

cAMP-enhancing agents and S1P affect cellular targets that initiate cell spreading and alter the adherens junction proteins, cell-specific cadherins and catenins that mechanically hold cells together. An increase in intracellular cAMP causes the remodeling of actin filaments to the cell periphery and a concomitant loss of actin stress fibers. Spreading of adjacent cells resulting in closer apposition or overlap would increase the pathway for diffusive permeability. The simple act of cell spreading would counteract the cell retraction/contraction induced by barrier-loosening agents such as thrombin. S1P profoundly affects the actin cytoskeleton, inducing both stress fibers and an increase in peripheral actin within minutes. Whether this remodeling of actin influences cell spreading remains to be determined. Alternatively, changes in cortical actin may facilitate the stabilization of the adherens junction, and cAMP-enhancing agents and S1P may function by stabilizing junctions. An increase in intracellular cAMP has been shown to maintain within 10 minutes the transvascular flux of ^{3}H inulin, transepithelial resistance, and the peripheral localization of E-cadherin and ZO-1 in the presence of low calcium in epithelial cells and to increase by 2 hours the peripheral localization of E-cadherin in brain microvascular endothelial cells. S1P has been shown to increase within 1 hour the localization of the adherens junction proteins, VE-cadherin and the catenins, α, β, and γ, at the cell periphery and to increase the amount of VE-cadherin in the actin-associated cell fraction in human umbilical vein endothelial cells. These observations by Lee from the research laboratory of Hla [6] have led to the conclusion that S1P induces the formation of adherens junctions. Therefore, S1P and cAMP-enhancing agents may function to tighten the endothelial barrier at the level of the actin cytoskeleton and/or the adherens junction.

Cell-Signaling Pathways

Although much effort has been generated to identify the active platelet factor, there is a paucity of information concerning the cellular mechanism by which the active factor functions to decrease endothelial permeability. Two signaling pathways have been studied in depth with regard to endothelial permeability. It is well known that the cAMP/protein kinase A pathway prevents and reverses an increase in vascular permeability induced by a variety of mediators, diseases, and syndromes. The effect of the cGMP/protein kinase G pathway on endothelial permeability is controversial and may be cell, tissue, and/or organ specific. Considering the overwhelming list of publications involving the above two signaling pathways, Gainor and coworkers [7] asked whether one of these two pathways is responsible for the permeability-decreasing activity of PCM. Two approaches were taken, using a desensitization protocol with three different cAMP/protein kinase A-enhancing agents and using inhibitors of protein kinases A and G. The cAMP-enhancing agents isoproterenol, forskolin, and 8-bromo-cAMP were incubated individually with endothelial cell monolayers as two separate challenges that were 45 minutes apart. Cells respond to the first challenge of each of these agents with an increase in electrical resistance. Following the second challenge with each agent, electrical resistance does not increase, indicative of desensitization. In contrast, desensitization is not observed when PCM is administered as the second challenge. Pharmacological inhibition of protein kinase A with KT-5720 blocks the increased endothelial electrical resistance induced by 8-bromo-cAMP but has no effect on the activity of PCM, LPA, or S1P. Similar findings are obtained upon inhibition of protein kinase G with KT-5823. The conclusion is that PCM, LPA, and S1P tighten the endothelial barrier via a cellular mechanism independent of protein kinases A and G.

Since cAMP/protein kinase A and G are not involved in the platelet activity, the next logical approach was to turn to the proposed signaling pathways elicited by LPA. LPA as well as S1P activate endothelial differentiation gene (Edg now LPA or S1P) receptors with nanomolar affinities. LPA and S1P bind to a number of Edg receptors, most notably Edg 2 (LPA 1) and Edg 4 (LPA 2) and Edg 1 (S1P 1), Edg 3 (S1P 3), and Edg 5 (S1P 2), respectively. In neurites and fibroblasts, LPA initiates signaling via activation of the G proteins, G_i, G_q, and $G_{12/13}$. The mitogenic response induced by LPA in NIH 3T3 fibroblasts is pertussis toxin sensitive, indicating involvement of the G_i protein, and also includes Ras and mitogen-activated protein kinase. LPA also profoundly influences the actin cytoskeleton via activation of $G_{12/13}$, Rho, and Rho kinase. Rho signaling has been shown to increase actin stress fibers. LPA as well as thrombin reorganizes the cortical actin and causes cell rounding of N1E-115 mouse neuroblastoma cells via a signaling pathway independent of calcium mobilization, activation of protein kinase C, or altered levels of intracellular cAMP. Inactivation of Rho by C3 exoenzyme or *Clostridium difficile* toxin B and subsequent disruption of actin causes an increase in endothelial and epithelial permeability. These observations indicate that novel pathways independent of cAMP/protein kinase A exist that influence mitogenesis, cell shape, cell motility, and endothelial permeability, and that the G_i and $G_{12/13}$ signaling pathways are prime signaling candidates.

Based upon the foregoing literature, initial studies with PCM, LPA, and S1P focused on the G_i signaling pathway because the pathway is well known, it can be manipulated with selective pharmacological inhibitors, and the G_q signaling pathway has been implicated in increasing endothelial permeability. Endothelial cell monolayers were treated first with genistein to generally inhibit tyrosine kinases because there are a number of proteins in the G_i signaling pathway that require tyrosine phosphorylation. Genistein prevents the increase in endothelial electrical resistance induced by PCM and S1P. Next, the Edg-1 receptor was depleted with antisense oligonucleotide methodology and the G_i protein was inhibited with pertussis toxin. Edg 1 couples primarily with G_i and Edg 3 and Edg 5 with G_q and $G_{12/13}$. Endothelial cells appear to express Edg 1 more abundantly than Edg 3, and Edg 5 is undetectable in human umbilical veins. Depletion of Edg-1 and Edg-3 receptors inhibits the activity of S1P, and pertussis toxin inhibits the activities of both LPA and S1P. Interestingly, pertussis toxin delayed by 10 to 15 minutes the characteristic, rapid rise in endothelial electrical resistance induced by PCM. These different findings with pertussis toxin would suggest that there are other active components in PCM, other than LPA and S1P. Also, pertussis toxin does not prevent the increase in endothelial electrical resistance induced by isoproterenol, providing additional evidence that S1P- and cAMP-enhancing agents function via different cellular targets.

Phosphatidylinositol 3-kinase (PI-3 kinase) is a downstream effector of G_i in many cells and has been linked to other cellular responses induced by LPA and S1P. Inhibition of PI-3 kinase with two mechanistically different inhibitors, wortmannin and LY-294002, also causes a delay of 10 to 15 minutes in the initial, rapid rise in endothelial electrical resistance induced by PCM, LPA, and S1P. Because inhibitors of PI-3 kinase only delay the increase in electrical resistance, other G proteins and signaling pathways could be involved. One possible explanation is that the active platelet factor initiates a fast signaling pathway dependent on PI-3 kinase and a slower-acting pathway independent of this kinase. Another very plausible explanation is that the rapid signaling pathway elicited by S1P requires the interaction of the lipid by-products of PI-3 kinase with pleckstrin homology domains of the signaling proteins. In other words, activation of PI-3 kinase positions the essential signaling proteins in a complex that facilitates a rapid response. Inactivation of PI-3 kinase somewhat uncouples the signaling proteins and temporally delays the signaling process. Taken together these results indicate that PCM, LPA, and S1P rapidly increase endothelial electrical resistance via a novel signaling pathway involving tyrosine kinases, the G_i protein, and PI-3 kinase.

Since inhibitors of G_i and PI-3 kinase significantly modify the permeability-decreasing activity of PCM, LPA, and S1P, the next logical step was to determine if the downstream extracellular signal-regulated kinase (ERK) was involved. PCM and LPA rapidly and transiently increase ERK phosphorylation, indicative of an increase in ERK activity. Inhibition of MEK, the upstream kinase of ERK, with U-0126 blocks the increase in ERK phosphorylation but has no effect on the activity of PCM. Therefore, platelets do not function via ERK to increase endothelial electrical resistance.

Rac and Rho

The function and localization of both Rac1 and Rho in the cell have recently attracted the attention of those investigators studying the barrier function of the endothelium and epithelium. Kaibuchi and coworkers [8] have put forth an intriguing hypothesis that activation of Rac1 and/or Cdc42 strengthens the adherens junction of epithelial cells. Their findings support the notion that active Rac1 and Cdc42 can sequester IQGAP1, a proposed negative regulator of the junction, away from the junction allowing for strong cell–cell adhesion. The opposite occurs when Rac1 activity is decreased in Madin-Darby canine kidney II cells treated with a phorbol ester or hepatocyte growth factor, the latter inducing cell scattering. Now IQGAP1 is associated with β-catenin and not Rac1 at the cell periphery, and there is a loss of α-catenin from cell–cell borders. Expression of a constitutively active Rac1 blocks this disappearance of α-catenin. The binding site on β-catenin for IQGAP1 (amino-terminal 1–183 amino acids) contains the α-catenin binding domain (120–151 amino acids). α-Catenin is the key intermediary that links VE-cadherin and β-catenin to the actin cytoskeleton to form a strong cell–cell adhesion. The presence of IQGAP1 disrupts this linear linkage and results in weak adhesion between cells. Thus, IQGAP1 may contribute to the dynamic nature of the adherens junction.

S1P induces the translocation of Rac1 and its guanine nucleotide exchange factor Tiam1 to the cell periphery in human umbilical vein endothelial cells but does not alter the spacial movement of Rho. However, inhibition of Rac1 or Rho reduces the immunolocalization of VE-cadherin and β-catenin at sites of cell–cell adhesion. Experiments with antisense oligonucleotides, have suggested that Rac1 and Rho function via Edg 1 and Edg 3, respectively. Recruitment of actin to epithelial junctions is dependent on Rac1 activation. The involvement of Rac1 and Rho in the integrity of the adherens junction of endothelial cells, however, is controversial. Inhibition of Rho kinase with Y-27632 has been reported to attenuate the increase in endothelial electrical resistance induced by S1P.

S1P increases the activity of Rac1, and this activation is sensitive to pertussis toxin. Rac1 activity is determined with an affinity precipitation assay using glutathione-*S*-transferase fused to the Rac1/Cdc42 binding domain of p^{21}activated kinase (PAK). Treatment of endothelial cell monolayers with S1P results in an increase in active Rac1 by 10 minutes. Thus, S1P-induced increases in Rac1 activity and endothelial electrical resistance occur within the same time frame and both responses are sensitive to pertussis toxin, indicating the involvement of the G_i protein. As stated earlier, S1P remodels actin within minutes into stress fibers and an increased peripheral actin. Rac1 induces the formation of cortical actin at the leading edge of lamellipodia, and Rho forms stress fibers. Formation of actin stress fibers by S1P is abolished with C3 exoenzyme, an inhibitor of Rho, and with Y-27632, the Rho kinase inhibitor. A dominant-negative Rac1 inhibits both the changes in stress fibers and cortical actin.

Wojciak-Stothard and coworkers in the laboratory of Ridley [9] reported that inhibition of Rho with C3 exoenzyme or adenoviral expression of a dominant-negative Rho and inhibition of Rho kinase with Y-27632 prevent the increase in endothelial permeability and the decrease in transendothelial electrical resistance induced by thrombin and histamine. These inhibitors also prevent the disassembly of the adherens junctions induced by these inflammatory agents. Surprisingly, adenoviral expression of either a dominant-active or a dominant-negative Rac1 increases the permeability of unstimulated monolayers of human umbilical vein endothelial cells. Faced with the same data induced by either activation or inhibition of Rac1, the authors concluded that the activity of Rac1 must be tightly regulated to maintain the integrity of the adherens junction between endothelial cells. According to the hypothesis of Kaibuchi, both Rac1 and Cdc42 can enhance the barrier function of epithelial cells. Therefore, Rac1 and Cdc42 may be redundant proteins with regard to the regulation of the endothelial barrier.

Beyond Rac1, one study by Garcia and coworkers [5] has looked at the downstream kinase PAK, and cofilin, an actin-severing protein inactivated by LIM kinase. A Rac1-bound PAK phosphorylates LIM kinase causing the subsequent inactivation of cofilin. Both PAK and cofilin translocate to the cell periphery after treatment with S1P. Expression of a dominant-negative PAK-1 or wild-type cofilin reduces the increase in cortical actin. The latter also blunts the increase in endothelial electrical resistance induced by S1P. These authors concluded that a thickened, cortical actin plays a prominent role in the enhanced endothelial barrier activity of S1P. Therefore, inactivation of cofilin via Rac1 to PAK signaling is key to this activity of S1P.

Conclusions and Future Studies

S1P induces cell migration, but as emphasized in this review, S1P also promotes the barrier function of the vascular endothelium. S1P activates both Rac1 and Rho, and both small GTPases have been implicated in the physiological processes of cell migration and endothelial barrier function (see Figure 1). At least for Rac1, this apparent discrepancy may depend on whether Rac1 is localized at cell–cell junctions and involved in barrier function or at the leading edge of a wounded cell and promoting cell migration. To complicate our understanding, these two GTPases have also been associated with processes that loosen the endothelial barrier. Therefore, the current take-home message is that the integrity of the adherens junction, at least in endothelial

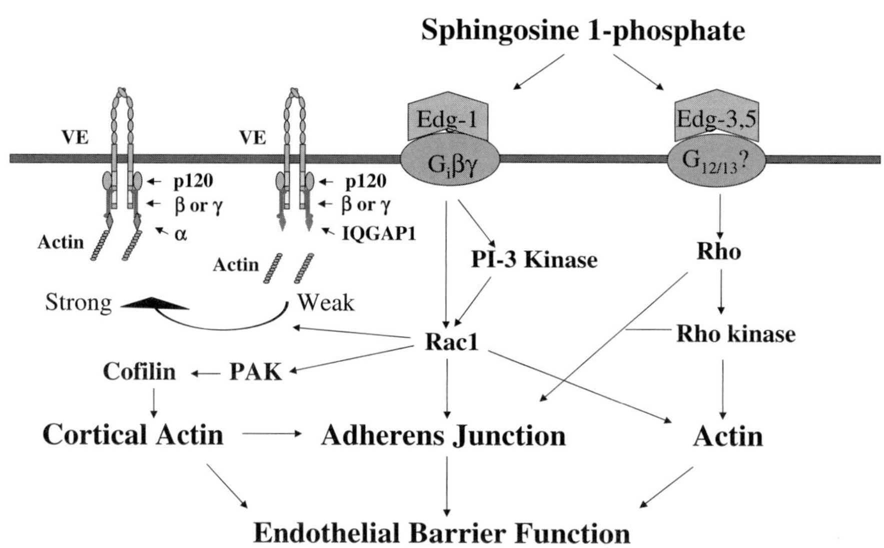

Figure 1 Proposed cellular targets and signaling pathways by which sphingosine 1-phosphate (S1P) tightens the endothelial barrier. S1P binds to endothelial differentiation gene receptors (Edg). Edg 1 couples primarily with G_i and Edg 3 and Edg 5 with G_q and $G_{12/13}$. S1P activates Rac1, and many of the physiological outcomes of S1P and Rac1 are linked to G_i. Down-regulation of Edg 1 or Edg 3 or inhibition of G_i prevents the S1P-induced enhancement of the endothelial barrier. In epithelial cells, Rac1 sequesters IQGAP1 and removes this proposed negative regulator from adherens junction proteins. The resultant linkage to the actin cytoskeleton facilitates the integrity of the adherens junction, although this sequence of events has not been described in endothelial cells. S1P and active Rac1 also remodel the actin cytoskeleton as evidenced by a thickened, cortical actin and stress fibers. Active Rac1 appears to affect cortical actin via activation of PAK and the subsequent inactivation of cofilin. The formation of actin stress fibers involves Rho and Rho kinase, and inhibition of Rho kinase also prevents the barrier-enhancing property of S1P. The linkage of Rho with $G_{12/13}$ and Edg 3 and Edg 5 requires demonstration. To complicate matters, both Rac1 and Rho have also been associated with loosening of the endothelial barrier. Thus, subtle changes in the activation of Rac1 and Rho may profoundly influence endothelial barrier function. VE, vascular endothelial cadherin; p120, p120-catenin; α, α-catenin; β, β-catenin; γ, γ-catenin or plakoglobin; p^{21}-activated kinase. (see color insert)

cells, may be influenced by subtle changes in the Rho family of GTPases, Rho, Rac1, and Cdc42. The cellular targets that regulate endothelial barrier function, whether adherens junction proteins or the actin cytoskeleton located at adhesions between cells and cell to substratum, need to be elucidated and require further study.

Glossary

Barrier function: The semipermeable nature of the endothelium to the passage of water and protein. Includes the movement of water and protein by diffusion and by convection (or bulk flow of water).

Permeability: The process whereby water or proteins diffuses from a higher concentration in the blood through the junctions between endothelial cells to a lower concentration in the tissue.

Rac and Rho: Members of the Rho family of small GTPases that function as intracellular signaling molecules.

Thrombocytopenia: Platelet count in the blood of 50,000/μL or less. Normal platelet count is 150,000 to 300,000 cells per microliter of blood.

Acknowledgments

This work was supported by a National Institutes of Health grant, HL-68079, and by an American Heart Association grant, AHA-97-127A.

Further Reading

1. Patil, S., Kaplan, J. E., and Minnear, F. L. (1997). Protein, not adenosine or adenine nucleotides, mediates platelet decrease in endothelial permeability. *Am. J. Physiol. Heart Circ. Physiol.* **273**, H2304–H2311.
2. Haselton, F. R., and Alexander, J. S. (1992). Platelets and a platelet-released factor enhance endothelial barrier. *Am. J. Physiol. Lung Cell Mol. Physiol.* **263**, L670–L678.
3. Alexander, J. S., Patton, W. F., Christman, B. W., Cuiper, L. L., and Haselton, F. R. (1998). Platelet-derived lysophosphatidic acid decreases endothelial permeability in vitro. *Am. J. Physiol. Heart Circ. Physiol.* **274**, H115–H122. *The first report that a phospholipid(s), possibly lysophosphatidic acid (LPA), is the active platelet factor.*
4. Minnear, F. L., Patil, S., Bell, D., Gainor, J. P., Morton, C. A. (2001). Platelet lipid(s) bound to albumin increases endothelial electrical resistance: Mimicked by LPA. *Am. J. Physiol. Lung Cell Mol. Physiol.* **281**, L1337–L1344.
5. Garcia, J. G. N., Liu, F., Verin, A. D., Birukova, A., Dechert, M. A., Gerthoffer, W. T., Bamburg, J. R., and English, D. (2001). Sphingosine 1-phosphate promotes endothelial cell barrier integrity by Edg-dependent cytoskeletal rearrangement. *J. Clin. Invest.* **108**, 689–701. *S1P increases endothelial permeability across cell monolayers derived from bovine and human pulmonary arteries and human umbilical vein. Depletion of Edg-1 and Edg-3 receptors and pertussis toxin inhibit the activity of S1P. Both p^{21}-activated kinase (PAK) and cofilin translocate to the cell periphery after treatment with S1P. Expression of a dominant-negative PAK-1 or wild-type cofilin reduces the increase in cortical actin, and the latter also blunts the increase in endothelial electrical resistance induced by S1P.*
6. Lee, M.-J., Thangada, S., Claffey, K. P., Ancellin, N., Liu, C. H., Kluk, M., Volpi, M., Sha'afi, R. I., and Hla, T. (1999). Vascular endothelial cell adherens junction assembly and morphogenesis induced by sphingosine-1-phosphate. *Cell* **99**, 301–312.
7. Gainor, J. P., Morton, C. A., Roberts, J. T., Vincent, P. A., and Minnear, F. L. (2001). Platelet-conditioned medium increases endothelial electrical resistance independently of cAMP/PKA and cGMP/PKG. *Am. J. Heart Circ. Physiol.* **281**, H1992–H2001. *Platelet-conditioned medium (PCM) and LPA rapidly increase endothelial electrical resistance via a novel, signaling pathway involving tyrosine kinases, the G_i protein, and PI-3 kinase. This pathway is independent of protein kinases A and G.*
8. Fukata, M., Nakagawa, M., Itoh, N., Kawajiri, A., Yamaga, M., Kuroda, S., and Kaibuchi, K. (2001). Involvement of IQGAP1, an effector of Rac1 and Cdc42 GTPases, in cell–cell dissociation during cell scattering. *Mol. Cell Biol.* **21**, 2165–2183.
9. Wojciak-Stothard, B., Potempa, S., Eichholtz, T., and Ridley, A. J. (2001). Rho and Rac but not Cdc42 regulate endothelial permeability. *J. Cell Sci.* **114**, 1343–1355.

Capsule Biography

Dr. Minnear is Professor of Physiology and Pharmacology, Assistant Dean of Graduate Studies, and Director of the M.D./Ph.D. Scholars Program at West Virginia University School of Medicine. His research focuses on regulation of the vascular endothelial barrier to the passage of water and protein with emphasis on the adherens junction and the actin cytoskeleton. His work is supported by grants from the National Heart, Lung and Blood Institute and the American Heart Association.

CHAPTER 55

Role of Gap Junctions in Capillary-Arteriolar Communication

Karel Tyml
University of Western Ontario, London, Canada

Introduction

Gap junctions, found in nearly all mammalian cells, are intercellular channels that permit a direct passage of ions and small molecules (up to 1 kDa molecular mass) between the cytosols of adjacent cells. Gap junctional communication between the cells of vascular wall has been proposed to coordinate microvascular function, including participation in the local control of blood flow.

Conceptually, the local control serves to match the local tissue blood flow to the tissue's metabolic demand. Vasoactive metabolites generated within the tissue during periods of demand are envisioned to reach upstream arterioles to cause adjustment of arteriolar diameter and blood flow to meet this demand. However, for tissues where the distance between the site of demand and upstream arterioles is large (e.g., distances along skeletal muscle fibers can be of the order of 1 mm), it is not clear how metabolic signals reach these arterioles. Since simple diffusion of metabolites across these distances may be too slow in relation to the speed of observed vasomotor responses, a new mechanism of matching the arteriolar response to the local demand has been proposed [1]. This mechanism is hypothesized to involve metabolically induced electrical signals that spread along the capillary endothelium to the arteriole via gap junctions (i.e., capillary–arteriolar communication). The present chapter discusses the available evidence regarding this mechanism and examines how the pathological condition of systemic inflammation could affect this communication.

Gap Junctions in the Microvasculature

Gap junctions (GJ) are formed by a family of integral membrane proteins termed connexins (Cx). Vascular endothelial cells express three connexin family members Cx37, Cx40, and Cx43. In each gap junction, six connexins oligomerize to form a connexon hemichannel at one cell membrane; this connexon docks with a connexon from another cell's membrane to establish an intercellular channel. At the arteriolar level, the presence of gap junctions between endothelial cells, between smooth muscle cells, and between endothelial and smooth muscle cells has been established structurally (i.e., by electron microscopy, immunohistochemistry), and functionally (i.e., by dye coupling and electrical coupling). In mammalian capillaries, gap junctions have been detected by electron microscopy and immunohistochemistry (Cx37 and Cx40 appear to dominate here; [2]).

Direct intercellular communication via gap junctions can be regulated by numerous mechanisms (e.g., gating, connexin protein expression, rates of connexin insertion and removal from the cell membrane) and can be affected by a number of intracellular signaling pathways. At the "macroscopic" level, communication has been shown to be modulated by changes in cytosolic pH and calcium, and by changes in trans-GJ and transmembrane voltages. Thus, capillary–arteriolar communication and, consequently, the local blood flow control, could be modulated acutely (e.g., via altered GJ gating) and chronically (via altered connexin expression) during disease processes whose targets include gap junctions.

Evidence for Capillary–Arteriolar Communication

In Vivo Work

The majority of experimental evidence for communication along the capillary is indirect. Using amphibian and mammalian skeletal muscle preparations, our laboratory has shown that micropipette application of a minute amount of various vasoactive agents (norepinephrine, phenylephrine, acetylcholine, bradykinin, adenosine analog NECA, KCl) on capillaries 300 to 500 μm away from the feeding arteriole causes constriction/dilation of this arteriole and ensuing reduction/increase in blood flow in capillaries fed by this arteriole (Figure 1). Simple diffusion failed to explain these responses since interventions that do not alter the diffusional process [e.g., treatment of the capillary midpoint with GJ uncoupler, injury at this midpoint (Figure 1), or systemic application of GJ uncoupler] inhibited these responses. These experiments suggested that the capillary itself could sense vasoactive agents and that it could provide a communication feedback path between the site of locally produced/applied agents and the arteriole. Further, stimulation of two capillaries fed by the same arteriole yielded additive/subtractive responses at the arteriolar level, suggesting that the capillary bed could function as a dispersed sensor within the tissue that is able to integrate metabolic signals within this tissue into an "overall" response at the arteriolar level [5].

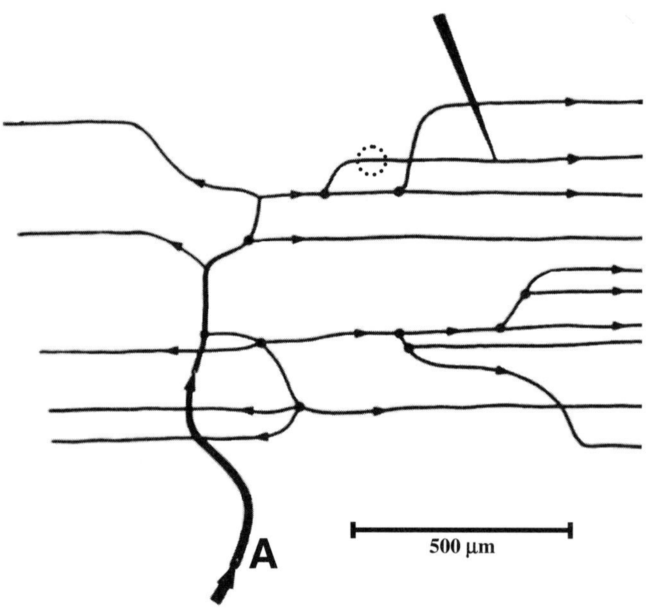

Figure 1 Schematic diagram of experimental arrangement for distal stimulation of capillaries. In this diagram, capillaries are oriented horizontally, run in parallel to each other, and are fed by arteriole "A." Arrows indicate direction of blood flow while dots represent physical junction between microvessels. During the experiment, a minute amount of vasoactive agent is placed on capillary by means of a glass micropipette, 300 to 500 μm away from the arteriole. Responses are measured in terms of changes in arteriolar diameter or changes in red blood cell flow in the stimulated capillary (or any capillary supplied by the terminal end of the arteriole). Local damage of the stimulated capillary at midpoint (i.e., indicated by dotted circle), or pretreatment of the midpoint by gap junction uncoupler, inhibits the arteriolar/blood flow response to the distal capillary stimulus [3, 4]. (Adapted from Ref. [3].)

Although the capillary sensing/communication phenomenon characterized by the above observations gave rise to a new concept of local blood flow control, the physiological significance of this phenomenon is not clear. To this end, two laboratories provided data relevant to this issue. First, Berg and coworkers [6] reported that, in hamster cremaster muscle, locally induced contraction of muscle bundle unit supplied by capillary bed via a single terminal arteriole caused conducted dilation along this arteriole. A local application of a GJ uncoupler on this arteriole prevented this dilation. Since mechanical tugging of the muscle unit did not result in conducted dilation, it was speculated that the capillary bed was involved in the transmission of contraction-induced vasoactive metabolite(s) to the arteriole. Thus, capillary–arteriolar communication could be involved in matching of the dilation-mediated increase in blood flow to the contraction-induced metabolic demand. Second, Collins and coworkers [7] reported that intravenular application of ATP (1 μM) via micropipette caused an upstream arteriolar dilation. ATP at this concentration has been reported to be released from red blood cells in response to low pH and hypoxia in venous blood. In order to explain the upstream dilation, these workers proposed that ATP stimulated endothelial P_{2Y} and P_{2U} receptors in venules to initiate conducted vascular response across the capillary bed to the arteriole. Thus, capillary endothelium could participate in communication of oxygen sensing signals between the venular and arteriolar ends of the microvasculature.

Unlike in capillaries, conduction of local agonist-induced responses is well characterized for the arteriole. Using recording electrodes inserted into the arteriolar wall at the local site of agonist application and at an upstream site, agonist-induced hyper/depolarizations are seen to spread electrotonically along the arteriolar endothelium. Since the capillary endothelium forms a continuous layer with the arteriolar endothelium, the mechanism of capillary sensing/communication has been assumed to be of a similar nature, that is, involving an electrotonic spread of locally induced hyper/depolarizations. Because of technical difficulties, however, direct electrical recordings from the capillary endothelium have not yet been carried out to confirm this mechanism. To date, an indirect approach has been used instead. Using a voltage-sensitive dye loaded into the capillary endothelium, hyperpolarization and depolarization of endothelial cells were observed after local application of agonists [8]. Employing this method, local depolarization initiated at the upstream arteriole site conducted to a downstream capillary site, demonstrating that electrotonic communication can occur along the capillary endothelium [9].

Another group has used a calcium-sensitive dye in venular capillaries of perfused lung. Spontaneous calcium waves were seen to spread along the capillary wall [10]. That this spread was inhibited by a GJ uncoupler indicated that capillary endothelial cells could communicate via gap junctions in vivo.

In Vitro Work

There is considerable in vitro evidence that microvascular endothelial cells (i.e., the major component of the capillary wall) can communicate electrically via gap junctions. When a single endothelial cell grown in a culture dish is connected to a patch-clamp electrode, its measured input electrical conductance is much smaller than that obtained for a confluent cell monolayer (i.e., indicating substantial intercellular coupling within the monolayer). Alternatively, when a cell in a monolayer is injected with brief pulses of electrical current, rapid electrotonic spread of these pulses through the monolayer is seen (i.e., pulse amplitude decreases with distance along the monolayer). Cultured microvascular endothelial cells have been shown to hyper/depolarize in response to application of vasoactive agents. (Note: with increased cell passaging, the responsiveness to many agents disappears, most likely due to the loss of expression of the appropriate receptors.) We have shown that this hyper/depolarization also communicated along the monolayer, but with a substantial spatial decay (only 20 to 30% of the local membrane potential change was seen at 300 μm distance) [11].

Recently, our laboratory has employed an alternative in vitro model to study endothelial cell-to-cell communication. Instead of monolayers, microvascular endothelial cells (origin: rat skeletal muscle) were grown in Matrigel matrix as "capillary-like" structures (i.e., 300 to 500 μm long with about four cells per 100 μm). A local, agonist-induced hyper/depolarization communicated electrotonically along these structures with a much smaller decay (i.e., 70% of the local response at 300 μm) than in monolayers. Thus, communication along these structures better mimicked the one-dimensional communication along in vivo microvessels than the two-dimensional communication seen in monolayers [11]. In both models, GJ uncouplers inhibited communication. Since, in general, all three vascular connexins have been found in cultured endothelial cells, the available functional and structural evidence indicates that endothelial cells in vitro have the fundamental ability to respond to a variety of vasoactive stimuli and to communicate this response (e.g., hyper/depolarization) rapidly along the microvascular endothelium, including the capillary endothelium.

Communication along the Microvasculature during Sepsis

Sepsis is a systemic inflammatory response to a local infectious insult. The release of lipopolysaccharide (LPS) (or comparable bacterial product) into the bloodstream initiates activation of immune cells and subsequent release of inflammatory cytokines that precipitate a number of circulatory disorders. These include decreased systemic vascular resistance, reduced responsiveness to vasoconstrictors and vasodilators at the microvascular level, impaired oxygen utilization, and maldistribution of flow within the capillary bed. Since sepsis can affect endothelial function (e.g., biosynthesis of nitric oxide, permeability), it is possible that sepsis could also affect the capillary–arteriolar communication and, consequently, the feedback pathway of the local blood flow control.

Using the micropipette application approach (Figure 1) in septic rat skeletal muscle (i.e., 24-hour cecal ligation and perforation model of sepsis), we found that sepsis attenuated the arteriolar response to vasodilators applied distally on capillaries. Since this attenuation was more pronounced than the attenuation seen after application of vasodilators directly on the arteriole, we concluded that the capillary sensing/communication was affected by sepsis [12].

Because of the technical limitations of our *in vivo* model, we used our *in vitro* models of communication to tease out the effect of sepsis on capillary sensing and/or communication. Based on the "capillary-like" structure model, LPS (an initiating factor in sepsis) was found to attenuate electrotonic communication along "capillary" length but not alter the local agonist-induced hyper/depolarization [13]. Thus, capillary communication, rather than capillary sensing, was more likely to be affected by sepsis in vivo. Attenuated communication along the "capillary" could occur via reduced GJ conductance between adjacent endothelial cells or via increased transmembrane conductance (i.e., conductance across cell membrane to extracellular space) of these cells. Based on our endothelial cell monolayer model, LPS was found to reduce GJ conductance, but not alter the transmembrane conductance. Further, this reduction was found to be reversible, protein tyrosine kinase dependent, but protein synthesis and mitogen-activated protein kinase independent [14]. Our subsequent study pinpointed Cx43 as the likely GJ protein to be responsible for the reduced GJ conductance. LPS caused phosphorylation of Cx43 tyrosine residue(s), an event that is generally associated with GJ closure and reduced coupling. Altered tyrosine kinase and phosphatase activities modulated in parallel both LPS-induced phosphorylation of Cx43 and reduction in GJ conductance. Thus these studies provided evidence that at least one agent of sepsis (i.e., LPS) could affect the capillary–arteriolar communication via reduced gap junctional coupling of endothelial cells [15].

Summary

Although GJ-mediated intercellular communication between cells of the vascular wall has, in general, been recognized to play a key role in the coordination of vascular function, the role of GJ in capillary endothelial cells has not yet been completely clarified. Available indirect in vivo evidence implicates capillary GJ participation in capillary–arteriolar communication and the feedback pathway of local blood flow control. Microvascular endothelial cells grown in vitro retain their ability to respond to vasoactive stimuli and to communicate these responses via gap junctions. Thus, concerted research efforts should be undertaken to provide the necessary direct in vivo evidence to firmly establish participation of capillary GJs in local blood flow control under normal and pathophysiological conditions.

Glossary

Capillary: The smallest blood vessel (~5 μm in diameter in mammalian tissues) lacking continuous muscular layer.

Electrotonic spread: Passive spread of electrical current (e.g., carried by the ions of the cell cytosol) whose amplitude decays with distance.

Gap junction: Intercellular channel directly connecting the cytosols of two adjacent cells.

Sepsis: Systemic inflammatory response to local infection.

Acknowledgements

This research was supported by the Canadian Institutes of Health Research and Heart and Stroke Foundation of Ontario.

References

1. Dietrich (1989). *Microvasc. Res.* **38**, 125–135.
2. Cai et al. (2001). *J. Mol. Cell. Cardiol.* **33**, 957–967.
3. Dietrich and Tyml (1992). *Microvasc. Res.* **43**, 87–99.
4. Yu et al. (2000). *Microvasc. Res.* **59**, 162–168.
5. Song, H., and Tyml, K. (1993). Evidence for sensing and integration of biological signals by the capillary network. *Am. J. Physiol.* **265**, H1235–H1242.
6. Berg, B. R., Cohen, K. D., and Sarelius, I. H. (1997). Direct coupling between blood flow and metabolism at the capillary level in striated muscle. *Am. J. Physiol.* **272**, H2693–H2700. *This paper implies physiological relevance of capillary–arteriolar communication.*
7. Collins et al. (1998). *Microvasc. Res.* **56**, 43–53.
8. McGahren, E. D., Beach, J. M., and Duling, B. R. (1998). Capillaries demonstrate changes in membrane potential in response to pharmacological stimuli. *Am. J. Physiol.* **274**, H60–H65. *Using a voltage-sensitive dye, this work shows for this first time that capillary endothelial cells can sense pharmacological stimuli in terms of hyper/depolarization of endothelial cells.*
9. Beach, J. M., McGahren, E. D., and Duling. B. R. (1998). Capillaries and arterioles are electrically coupled in hamster cheek pouch. *Am. J. Physiol.* **275**, H1489–H1496. *Based on measurements of voltage-sensitive dye loaded in capillary endothelial cells in vivo, this work shows for this first time that these cells are electrically coupled.*
10. Ying et al. (1996). *Circ. Res.* **79**, 898–908.
11. Ouellette et al. (2000). *Microvasc. Res.* **60**, 222–231.
12. Tyml et al. (1998). *J. Appl. Physiol.* **84**, 837–844.
13. Lidington, D., Ouellette, Y., and Tyml, K. (2002). Communication of agonist-induced electrical responses along "capillaries" in vitro can be modulated by lipopolysaccharide, but not nitric oxide. *J. Vasc. Res.* **39**, 405–413. *This work demonstrates electrical communication in capillary-like structures grown in vitro.*
14. Lidington, D., Ouellette, Y., and Tyml, K. (2000). Endotoxin increases intercellular resistance in microvascular endothelial cells by a tyrosine kinase pathway. *J. Cell. Physiol.* **185**, 117–125.
15. Lidington et al. (2002). *J. Cell. Physiol.* **193**, 373–379.

Bibliography

Gustafsson, F., and Holstein-Rathlou, N. H. (1999). Conducted vasomotor responses in arterioles: Characteristics, mechanisms, and physiological significance. *Acta Physiol. Scand.* **167**, 11–21.

Lidington, D., Ouellette, Y., Li, F., and Tyml, K. (2003). Conducted vasoconstriction is reduced in a mouse model of sepsis. *J. Vasc. Res.* **40**, 149–158.

Mitchell, D., Yu, J., and Tyml, K. (1997). Comparable effects of arteriolar and capillary stimuli on blood flow in rat skeletal muscle. *Microvasc. Res.* **53**, 22–32. *This paper shows that arteriolar responses to distant capillary pharmacological stimuli cannot be explained by direct diffusion of agents from capillary to arteriole.*

Saez, J. C., Berthoud, V. M., Moreno, A. P., and Spray, D. C. (1993). Gap junction. Multiplicity of controls in differentiated and undifferentiated cells and possible functional implications. In: *Advances in Second Messenger and Phosphoprotein Research* (S. Shenolikar and A. C. Nairn, eds.), Vol. 27, New York: Raven Press, pp.163–199.

Capsule Bibliography

K. Tyml is a professor at the Department of Medical Biophysics, University of Western Ontario. His research interests include local microvascular blood flow control, arteriolar reactivity, and the role of capillaries and vascular cell-to-cell communication in this control. Recently, he has examined the pathophysiological effects of skeletal muscle disuse, and sepsis, and the protective effect of vitamin C, on this local control.

Part II

Organ Microvascular Adaptations

Section A

Cardiology

CHAPTER 56

Cardiac Microvasculature

William C. Aird

Department of Medicine, Beth Israel Deaconess Medical Center and Harvard Medical School, Boston, Massachusetts

Introduction

The goal of this chapter is to review the structural and functional properties of the cardiac microcirculation. The primary focus will be the endothelial lining of the microvessels. The endothelium may be viewed as a subsystem of the blood vessel wall. The blood vessel, in turn, may be considered a subsystem of the cardiovasculature. In serving as a frame of reference, the endothelium readily lends itself to a consideration of other cellular and noncellular components of the cardiac microvasculature.

Structural Topology

The coronary circulation, like all vascular beds, has a treelike or fractal structure. Two major arteries arise from the aorta, the left main coronary artery (which divides into the left anterior descending and the circumflex arteries) and the right coronary artery. These large vessels lie on the epicardial surface of the heart and function as distribution or conduit vessels, providing rapid mass transport of blood to the myocardium. Once the large arteries penetrate the myocardium, they branch into resistant vessels or arterioles. The small arteries and arterioles (75 to 200 μm in diameter) are the primary sites of coronary vascular resistance, and thus the principal regulators of coronary blood flow. These arterioles give rise to a densely packed capillary network, which runs parallel to sheets of cardiomyocytes. Capillaries serve as exchange vessels, transferring oxygen and nutrients to the underling tissue. The mean intercapillary distance is 17 μm at rest, thus providing a mean maximal diffusion distance of 8.5 μm. There are approximately 2,500 cardiomyocytes and 2,500 capillaries per mm^2. In rodents, the capillary–cardiomyocyte ratio ranges between 0.91 and 1.12. At rest only a third of the capillaries are perfused at any one time. During exercise, coronary vasodilation results in increased flow and capillary recruitment. Blood drains from the capillaries into postcapillary venules, cardiac veins, and ultimately the coronary sinus and right atrium. The microcirculation consists of vessels smaller than 300 μm in diameter, and includes small arteries, arterioles, capillaries, and postcapillary venules. The endocardium, which represents the endothelial lining of the atria and ventricles and heart valves (endocardium), is a developmentally and phenotypically distinct vascular bed and is not discussed in this chapter.

Regulation of Coronary Flow

The heart provides bulk flow of blood to the various tissues of the body, contracting approximately $2^1/_2$ billion times during the lifetime of a human. Compared with other organs, the heart has high basal oxygen consumption (8 to 10 mL O_2/min/100 g) and a large A-VO_2 difference (10 to 13 mL/100 mL). The resistance arterioles (and to a far lesser extent the large coronary arteries) are responsible for mediating vasomotor tone and blood flow. Coronary flow is closely matched to metabolic demands and myocardial oxygen consumption. The precise mechanisms that couple oxygen demand with blood flow remain unknown. Several factors appear to be important, including metabolic changes (termed *metabolic dilation*), flow characteristics, and fluctuations in intramural pressure. Each of these factors is differentially regulated in space and time. For example, metabolic-mediated changes in arteriole diameter and blood flow depend, at least in part, on the initial diameter of the vessel and location within the myocardial wall. The arterial system in the human heart is interconnected by collateral vessels that can enlarge and rescue ischemic areas of myocardium.

The Endothelium as an Input–Output Device

Much of endothelial cell biology can be understood—at least conceptually—by considering each and every endothelial cell in the body as an adaptive input–output device (Figure 1). The input arises from the extracellular milieu and may include biochemical or biomechanical signals. The output is manifested as the cellular phenotype and includes a number of structural and functional properties. Some of these properties are expressed at the level of individual cells or cell culture (e.g., protein, mRNA, proliferation, apoptosis, migration, and permeability), whereas other properties are expressed at higher levels of organization—for example the blood vessel (e.g., leukocyte trafficking, vasomotor tone, fibrin deposition) or organism (e.g., redistribution of blood flow, vascular bed-specific phenotypes). Each endothelial cell may have unique intrinsic properties, the so-called set point. Differences in set point are brought about by environmentally induced epigenetic changes in the "hard wiring" of the cell (see later discussion). Differences in signal input and set point conspire to generate phenotypic differences in space and time—a phenomenon known as endothelial cell heterogeneity or vascular diversity.

If one accepts the analogy of each endothelial cell representing its own input–output device, it is not a stretch to consider the endothelium as a circuit board—one that is hard wired (to some extent) to meet the demands of the tissue, and one that is highly vulnerable to short-circuiting as a mechanism of vasculopathic disease. There are important caveats to this model. First, the input–output device is not a black box, but rather a multiorganelle cell with highly complex nonlinear signaling networks. Second, the system displays properties of emergence. In other words, while the individual capillary endothelial cell follows simple local rules and at any point in time and space has a definable (albeit nonlinear) input–output relationship, the properties of the single cell do not predict for behavior of the whole capillary, heart or organism. An important challenge in vascular biology—whatever the vascular bed of interest—is to learn how to integrate the results of studies at the reductionist and holistic ends of the biomedical spectrum to understand the role for the endothelium in health and disease.

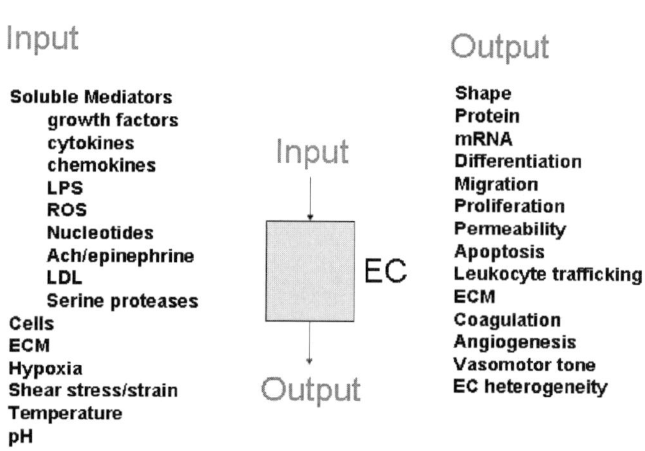

Figure 1 The endothelial cell as an input–output device. Input arises from the extracellular environment and includes a combination of biomechanical and biochemical signals (*left*). The output, or cellular phenotype (*right*), may be expressed at the level of single endothelial cells or at higher levels of organization (e.g., blood vessel, organ, or whole organism). The device itself—depicted here as a "black box"—contains a highly complex array of signal transduction pathways that couple input with output. (see color insert)

Applying the Input–Output Analogy to Cardiac Microvasculature in Health

The notion that the endothelium represents a series of input–output devices is helpful when considering the vasculature or endothelium during development and in the postnatal period. A consideration of cardiac development is beyond the scope of this chapter. The following section will focus on the adult cardiac endothelium.

Conduit and Resistance Vessels

The endothelium of the coronary and resistance vessels has presumably evolved to optimize the transport blood (oxygen and nutrients) to the myocardium. Compared with capillaries or veins, the arterial/arteriolar endothelium is exposed to higher oxygen concentration and shear stress. Like other segments of the cardiac vasculature, the arterial endothelium also experiences forces generated by contraction of the heart. On the abluminal side the endothelium receives signals from extracellular matrix and vascular smooth muscle cells. Based on studies of the heart and other organs, the arterial endothelium differs from its capillary and venous counterparts in morphology, gene and protein expression profiles, and cellular function. As one example, endothelial cells lining the resistance vessels express the highest amounts of endothelial nitric oxide synthase (eNOS), a pattern that is consistent with the established role of nitric oxide (NO) in regulating coronary vascular tone.

The tight coupling of myocardial oxygen consumption and coronary flow may be attributed, in part, to cross talk between cardiomyocytes and resistance vessels. Indeed, it has been proposed that coronary blood flow involves a dynamic, metabolically regulated balance in the production of cardiomyocyte-derived vasodilators (e.g., adenosine, NO, prostacyclin, and bradykinin) and vasoconstricting molecules (e.g., angiotensin II). Whether these and/or other cardiomyocyte-derived factors directly engage vascular smooth muscle cells or communicate indirectly via the endothelium is not known.

Capillaries

Whereas the endothelial lining of arteries and arterioles has evolved to optimize conduit function, the capillary endothelium is adapted to meet the unique and high metabolic needs of the myocardium. If one considers the arteries and arterioles as analogous to train tracks, the capillaries represent the train station—the site where circulating molecules and cells leave and enter the blood. As an input–output device, the endothelium senses numerous mechanical forces, including pressure generated during the cardiac cycle. With each heart contraction, the pressure may result in reduced diameter and flow, particularly in the intramural and subendocardial vessels. The capillary endothelial cell is exposed to multiple biochemical stimuli arising from the blood and abluminal surface (Figure 2). The cardiomyocyte secretes many factors that potentially act in paracrine pathways to maintain or modulate endothelial cell phenotype, including vascular endothelial growth factor (VEGF), basic fibroblast growth factor (bFGF), and platelet-derived growth factor (PDGF). VEGF may play a role not only in new blood vessel formation, but also in maintaining survival and adaptive function of the endothelium. On the output side, capillary endothelial cells express a unique pattern of genes and proteins. Moreover, these cells secrete myriad factors that interact with underlying cardiomyocytes, thus contributing to a bidirectional communication between the two cell types. For example, endothelial cells in the myocardium produce NO, endothelin, angiotensin, and prostacyclin, each of which may influence cardiac metabolism and growth, contractile function and conduction. At a functional level, the endothelium participates in the local balance of hemostasis, leukocyte trafficking, and survival/apoptosis. Endothelial cells (together with pericytes) play a critical role in mediating new blood vessel formation. Angiogenesis depends on the sprouting of preexisting capillaries or postcapillary venules, whereas collateral growth (variably termed arteriogenesis) arises from an outward remodeling of preexisting arterioles or small arteries. Angiogenesis normally results in blood vessels of capillary diameter (5 to 8 μm), while arteriogenesis gives rise initially to 10 to 20 μm resistance arterioles and ultimately to conductance arteries. The formation or remodeling of blood vessels depends on the concerted activity of soluble factors released by cardiomyocytes (e.g., VEGF and bFGF), endothelium (e.g., NO), and monocytes (e.g., urokinase-type plasminogen activator, metalloproteinases). Interestingly, there is increasing evidence for the role of endothelial progenitor cells in postnatal blood vessel formation (vasculogenesis) in the setting of cardiac ischemia. The elucidation of the molecular control of angiogenesis, arteriogenesis, and vasculogenesis represents an important foundation for novel cardiovascular therapies.

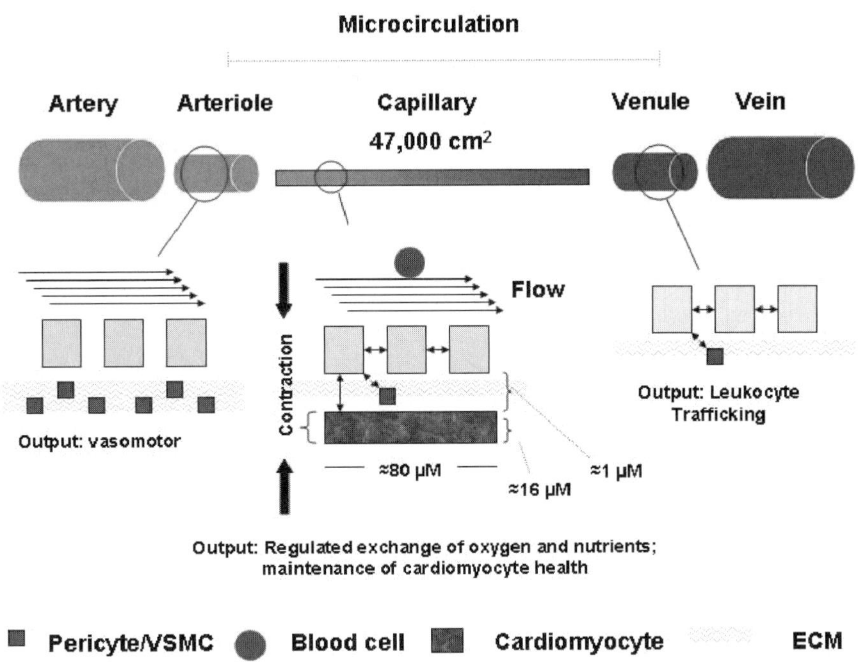

Figure 2 Schematic of cardiac endothelium in health. The endothelium of the artery and arteriole is exposed to comparatively high shear stress and communicates with neighboring smooth muscle cells in the blood vessel wall. In addition to maintaining blood fluidity and barrier function, the endothelial lining of the arteries and arterioles contributes to the control of vasomotor tone. The cross-sectional area of the capillaries is far greater than that of the arteries and veins, thus reducing shear stress at the level of the endothelium and facilitating exchange of oxygen and nutrients with underlying tissue. Multiple input signals arise from the luminal surface (shown is blood flow) and abluminal surface (shown is communication with pericytes and cardiomyocyte, extracellular matrix [ECM], and contractile force). The capillary endothelium plays a critical role in maintaining the viability and function of the myocardium. The endothelium lining the postcapillary venules is the primary site for leukocyte trafficking. (see color insert)

Endothelial cells from the capillaries of thc hcart (as well as those in the arteries, arterioles, venules, and veins) may possess a distinct "set point," defined as epigenetically programmed switches in phenotype. According to this hypothesis, certain properties of the capillary cells are irreversibly "locked in" by site-specific environmental factors—either during development or in the postnatal period. Although there is compelling evidence that endothelial cells from the epicardial arteries and the endocardium are derived from different subpopulations of precursor cells, little is known about the embryonic origin of capillary endothelial cells. The observation that capillary endothelial cells undergo phenotypic drift in tissue culture, and that this effect is partially reversed by coculture with cardiomyocytes, suggests that at least certain vascular bed–specific properties of these cells are reversibly regulated by the microenvironment. It is important to note that the retention of certain vascular bed–specific properties in vitro does not provide evidence for "genetic predetermination"—after all, each and every cell in the human body has an identical genome. Rather, such findings point to the influence of an epigenetic process at some point during the life of that cell. The distinction between endothelial cell phenotypes dependent on signal input and those that are governed by set point has important therapeutic implications. For example, diabetes is associated with phenotypic changes in microvascular endothelium. Are these changes reversible with correction of the metabolic disease? Or are certain phenotypes irreversibly and indelibly imprinted on the endothelium through modification of the genome? If the latter is correct, then the reversal of endothelial cell "dysfunction" associated with diabetes will be considerably more challenging.

As a result of differences in input signal and set point, the properties of the endothelium differ between different segments of the vasculature within the heart, and even between neighboring endothelial cells within the same vessel. For example, von Willebrand factor (vWF) is expressed at higher levels within the veins of the heart, compared with the arteries or capillaries. eNOS expression is highest in the endocardial and coronary artery endothelium. As a final example, the receptor-type protein tyrosine phosphatase μ is expressed predominantly on the arterial side of the cardiac vasculature.

Venules

Venules and veins function as conduit vessels to deliver deoxygenated blood to the right atrium. Postcapillary venules are the primary site for leukocyte–endothelial interactions and transmigration of leukocytes into the subendothelial space. The propensity for trafficking to occur in postcapillary venules may be explained by the low flow state and/or differential expression of cell adhesion molecules.

Applying the Input–Output Analogy to Cardiac Microvasculature in Disease

Endothelial cell activation is a term that was originally coined to describe the increased adhesiveness of cultured endothelial cells that had been pretreated with cytokines. The term today may be more broadly defined as the phenotypic response of the endothelium to an inflammatory stimulus, usually consisting of some combination of procoagulant and proadhesive properties and loss of barrier function. The term endothelial cell dysfunction was—and to some extent continues to be—used to describe reduced vasodilator reserve in conduit vessels and microvascular resistance vessels in the heart, as observed in atherosclerosis, transplantation, and heart failure. However, the term may be more broadly applied to states in which the endothelial phenotype—whether or not it meets the definition of activation—poses a net liability to the host. Thus, whereas endothelial cell activation describes a definable phenotype of the endothelium, endothelial cell dysfunction describes the cost of the phenotype to the organism.

Endothelial cell dysfunction may arise from any number of changes in the extracellular environment and may in turn contribute to progression in pathophysiology (Figure 3). For example, local differences in blood flow (hemodynamics) may conspire with coronary risk factors to modulate endothelial cell phenotype in the coronary arteries as a prelude to atherosclerosis. Takayasu's arteritis is a necrotizing vasculitis that may be associated with aneurismal dilation of the coronary arteries. Recent studies point to a possible role for endothelial-derived metalloproteinases in this syndrome. Many of the risk factors for conduit vessel disease have also been shown to affect microvascular endothelium, including diabetes and hypertension. Microangiopathic hemolytic anemia, as exemplified by thrombotic thrombocytopenic purpura (TTP), may be complicated by widespread occlusion of small arteries in the heart and recurrent cycles of ischemia–reperfusion. TTP is associated with increased apoptosis of endothelial cells in affected organs. The phenotype of cardiac microvascular endothelial cells has been shown to change in patients with heart failure. Patients with dilated cardiomyopathy have reduced VEGF expression and capillary density. Ischemia–reperfusion injury of the heart is associated with increased ROS generation, decreased NO bioactivity, generation of cytokines and chemokines, complement activation, increased expression of cell adhesion molecules, trafficking of neutrophils and monocytes, mast cell activation, increased permeability, and changes in the ECM. Patients with cardiac syndrome X have reduced coronary microvascular dilatory responses and increased coronary resistance. Sepsis, which is associated with systemic activation of inflammatory and coagulation cascades, may result in abnormalities in systolic and diastolic function. The mechanism does not appear to be related to reduced coronary flow, but rather to the effect of soluble mediators (e.g., input signals) on cardiac endothelium and cardiomyocytes.

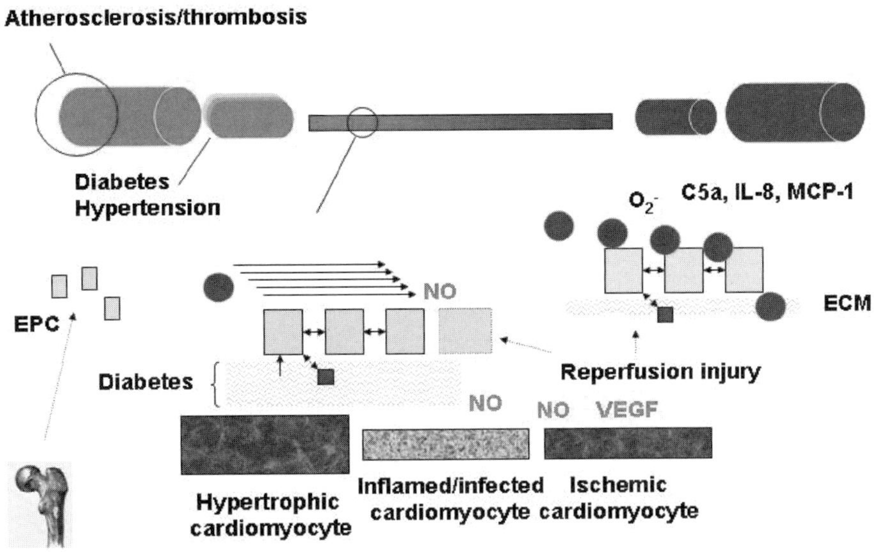

Figure 3 Schematic of cardiac endothelium in disease. The endothelium is both affected by and contributes to cardiac pathophysiology. Atherosclerosis affects the large coronary arteries. Diabetes and hypertension affect large- and small-vessel endothelium. Abnormalities in cardiomyocyte function (e.g., hypertrophic cardiomyopathy, myocarditis, and ischemia) may result in altered paracrine signaling in the endothelium and secondary endothelial cell activation and/or dysfunction. Shown is cardiomyocyte production of nitric oxide (NO) and VEGF. Ischemia/reperfusion results in the exposure of the endothelium to myriad signals, including free oxygen radicals (O_2^-), complement, and cytokines [e.g., interleukin (IL)-8 and monocyte chemoattractant protein (MCP)-1]. Cardiac ischemia may result in increased mobilization and/or uptake of circulating endothelial precursor cells (EPCs) in the cardiac vasculature. (see color insert)

As a final example of microvascular disorders, endothelial dysfunction plays a critical role in mediating cardiac allograft vasculopathy following heart transplantation.

Conclusions

Endothelium and blood vessels carry out certain common core functions: Arteries and arterioles transport blood (oxygen and nutrients); capillaries regulate the exchange of oxygen, and other blood constituents to the underlying tissue; and veins drain the organs of carbon dioxide and metabolic by-products. Different vascular beds employ different strategies to accomplish these tasks. Moreover, each vascular bed—particularly at the level of the capillary—is tightly coupled to the local tissue environment. In the case of the heart, coronary blood flow is highly coupled to the metabolic needs of the cardiomyocytes and is protected by multiple "backup" systems, including the presence of collateral vessels. The microvascular endothelium of myocardial capillaries is exposed to a spatially and temporally regulated input of heart-specific signals—including those that arise from the cardiomyocyte and the hemodynamic forces associated with the cardiac cycle. The dynamic and malleable nature of the endothelium renders these cells vulnerable to pathophysiological changes in the composition of the extracellular milieu. The altered signal input may be transduced by the endothelial cell in ways that perpetuate or accentuate the underlying pathophysiology. Important goals for the future will be to understand how the endothelium contributes to cardiac microvascular disease, and how its privileged location and communication network within the local microenvironment can be exploited for therapeutic purposes.

Glossary

Endothelial cell activation: The phenotypic response of the endothelium to an inflammatory stimulus, usually consisting of some combination of procoagulant and proadhesive properties, and/or alteration in permeability. The criteria for activation depend on the location within the vascular tree.

Endothelial cell dysfunction: An endothelial cell phenotype that represents a net liability to the host. Common examples include coronary artery disease (local dysfunction) and sepsis (systemic dysfunction).

Input–output device: A device that senses the environment and responds in a way that is usually adaptive (function) or nonadaptive (dysfunction). Each endothelial cell behaves like an input–output device, coupling changes in the extracellular milieu (input) to alterations in phenotype (output) via nonlinear networks of intracellular signaling pathways.

Bibliography

Brutsaert, D. L. (2003). Cardiac endothelial-myocardial signaling: Its role in cardiac growth, contractile performance, and rhythmicity. *Physiol. Rev.* **83**(1), 59–115. *This is a comprehensive review of cardiac endothelium and the role for the environment in regulating endothelial phenotypes.*

Carmeliet, P. (2003). Angiogenesis in health and disease. *Nat. Med.* **9**(6), 653–660. *This review provides an overview of differences in the molecular control of angiogenesis and arteriogenesis in the heart.*

Rosenberg, R. D., and Aird, W. C. (1999). Vascular-bed–specific hemostasis and hypercoagulable states. *N. Engl. J. Med.* **340**(20), 1555–1564. *This article reviews the molecular basis for a cardiac-specific signaling pathway that involves cross-talk between microvascular endothelial cells and cardiomyocytes.*

Stevens, T., Rosenberg, R., Aird, W., et al. (2001). NHLBI workshop report: Endothelial cell phenotypes in heart, lung, and blood diseases. *Am. J. Physiol. Cell. Physiol.* **281**(5), C1422–C1433. *This paper provides an update on endothelial cell heterogeneity in the heart, lung, and blood and includes a summary of the embryonic origin of the coronary endocardium.*

Capsule Biography

Dr. Aird is Chief of the Division of Molecular and Vascular Medicine at the Beth Israel Deaconess Medical Center and is the Deputy Director of the institution-wide Center for Vascular Biology Research. Dr. Aird's laboratory focuses on understanding the spatial and temporal regulation of endothelial cell phenotypes. His work is supported by grants from the National Institutes of Health.

SECTION B

Central Nervous System and Eye

CHAPTER 57

The Vasculature of Choroid

Imran A. Bhutto and Gerard A. Lutty
Wilmer Ophthalmological Institute, Johns Hopkins Hospital, Baltimore, Maryland

The choroidal microvasculature lies immediately under or posterior to the retina. Although the gross choroidal anatomy is well known, little information is available on the arrangement of arterioles and venules and their relationship to the choriocapillaris. This chapter will focus on mammalian choroids because avian choroids are uniquely different.

Gross Anatomy

The choroid is considered to be the ocular homolog of the pia arachnoid in brain. It is a long, thin, vascular, and pigmented tissue, which forms the posterior portion of the uveal tract (the iris, ciliary body, and choroid). Because of its great vascularity, the choroid has some of the properties of erectile tissue. Its capillaries form a very unusual pattern, being arranged in a single layer restricted to the inner portion of the choroid; this arrangement enables the capillary layer to provide nutrition for the outer retina (Figure 1). The choroidal stroma contains a considerable number of pigmented melanocytes, which are distributed through all areas of the choroid, although they are more sparse in the innermost layers. The pigment in the melanocytes gives the choroid its characteristic dusky brown color. The choroid has an extensive nerve supply, and many ganglion cells can be found in the stroma and suprachoroidea. The choroid is easily detached from the sensory retina and from the sclera (the collagenous coat of the eye), but it is firmly bound to the optic nerve, because its connective tissue is continuous with that of the optic nerve.

Internally, the choroid is so closely attached to the retina that the retinal pigment epithelium adheres more to Bruch's membrane than to the retinal photoreceptors. Bruch's membrane is the anatomical border between retina and choroid (Figure 1). The area around the choriocapillaris has little or no pigment (Figure 1). Externally the pigmented suprachoroidal connective tissue lamellae are closely bound to those of the lamina fusca of the sclera, the border of choroid and sclera.

The thickness of the choroid is difficult to determine because it diminishes after enucleation or fixation; it has been estimated to be approximately 0.1 mm anteriorly and 0.22 mm posteriorly, with greatest thickness under the macula. The choroidal vasculature in primates is composed of the choriocapillaris (an internal or anterior layer), medium-sized vessels (Sattler's layer), and large arteries measuring 40 to 90 μm, large veins measuring 20 to 100 μm (Haller's layer), nerves, and lymphatics (Figure 2).

The Choroidal Vasculature and Associated Structures

Arteries

There are three main arterial sources of blood to the choroid. The long posterior ciliary arteries (LPCA, temporal, and nasal) follow long, oblique intrascleral courses and, therefore, are easily affected by scleritis. The LPCAs travel in the potential suprachoroidal space and send branches from the ora serrata region posteriorly to supply the choroid as far posterior as the anatomical equator. A ciliary nerve accompanies each LPCA. The second arterial source is the short posterior ciliary arteries (SPCA, 15 to 20 in number) that supply the choroid from equator to optic nerve. The distribution of their entry is perifoveal, peripapillary, or in a compromise pattern, the papillomacular oval. The arteries surround the optic nerve (the circle of Zinn), penetrating and then branching peripherally in a wheel-shaped arrangement. The radial areas supplied by the arteries are separated by watersheds and are triangular, with the apices directed toward the fovea. A watershed zone in choroid is an area that normally fills slowly with blood. Hayreh [1], one of the pio-

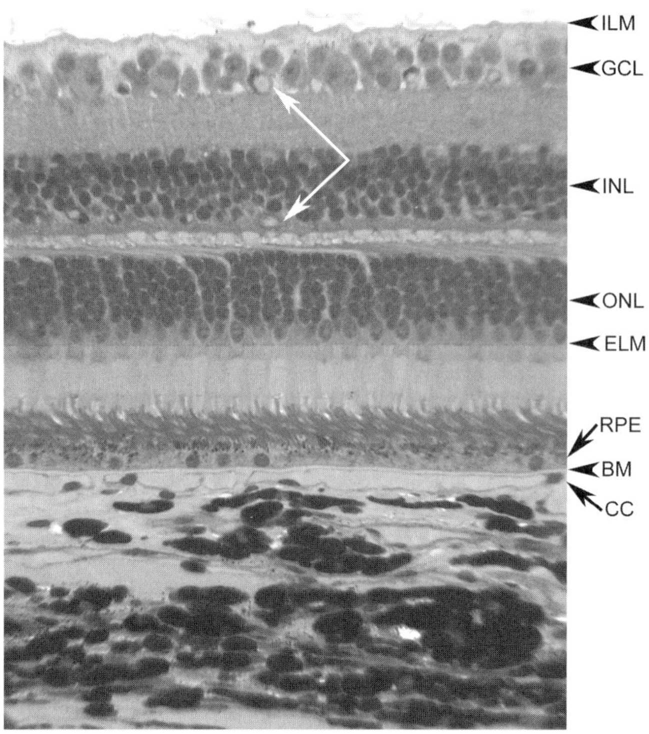

Figure 1 Section from the eye of a rhesus monkey showing retina from the internal limiting membrane (ILM) at the interface of retina and vitreous to the retinal pigment epithelial cells (RPE). Retinal blood vessels (white arrows) are in the ganglion cell layer (GCL) of inner retina, and also a secondary or deep plexus is associated with the inner nuclear layer; their inner and outer segments are below the external limiting membrane (ELM). The nuclei of the photoreceptors are in the outer nuclear layer (ONL). Below Bruch's membrane (BM) is the choroid with choriocapillaris (CC), immediately posterior to Bruch's membrane and large choroidal vessels posterior in the densely pigmented outer choroid. (see color insert)

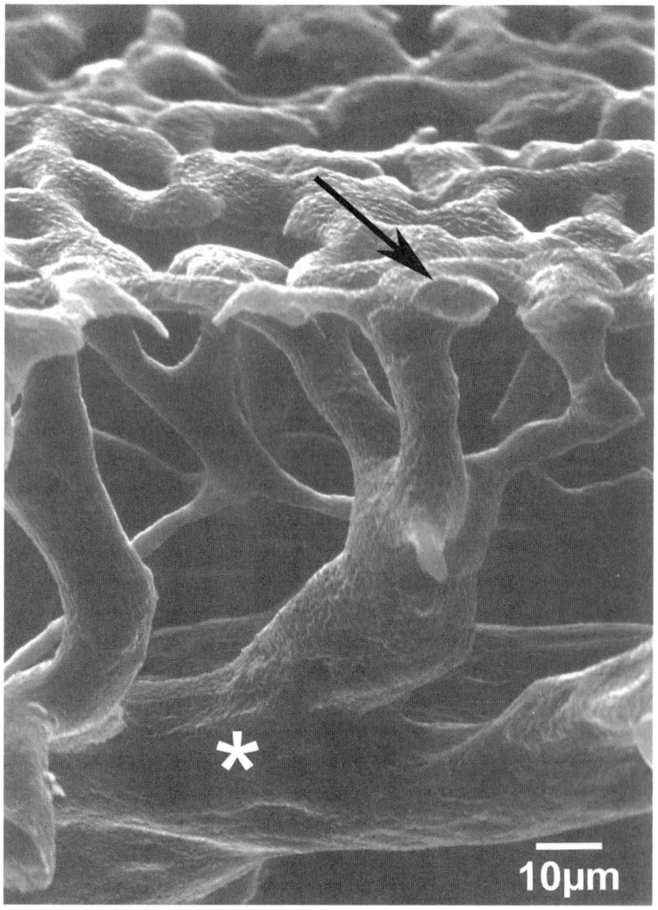

Figure 2 A vascular cast of the choroidal vasculature of a dog. The oval-shaped choriocapillaris (arrow) are at the top with draining venules below connecting to a large vein (asterisk). This section of the cast is the area of the tapedum lucidum, so the venules are more elongated to traverse the tapedum than they would be in a primate.

neers in the study of the choroidal vasculature, has done extensive in vivo experimental studies on choroidal circulation and its watershed zones in human. He has shown that the choroidal vascular bed is a strictly segmental and end-arterial system and has watershed zones situated between the various PCAs, the short PCAs, the choroidal arteries, the arterioles, and the vortex veins. The nature of the choroidal vasculature and the existence of watershed zones in the choroid are of great clinical importance and play a significant role in the production of various ischemic lesions in the choroid. The final arterial source of blood is the anterior ciliary arteries, which send recurrent branches posteriorly to supply the choroid at 3 o'clock and 9 o'clock soon after they pierce the anterior sclera. The arteries of all three vessel systems rapidly extend internally via arterioles to supply blood to the choriocapillaris. Mast cells are intimately associated with most choroidal arteries.

Choroidal Veins (Vortex Veins)

The main venous drainage of the choroid occurs through four to six vortex veins that drain into superior and inferior ophthalmic veins. Postcapillary venules are closely arranged in the macular region. In this area, the venous portion predominates over the arterial one. The meshwork of the venous plexus becomes less dense with increasing distance from the macula. In the extramacular region, the vessels are straighter, losing the tortuousity that is characteristic of the macular region. Vessels of larger lumen form the subcapillaris plexus and eventually flow into the vortex veins. Venous drainage is segmentally organized into quadrants, with watersheds oriented horizontally through the disc and fovea and vertically through the papillomacular region. The macula is centered over both arterial and venous watersheds, which may either predispose it to relative ischemia or prevent ischemia through multiple submacular blood supplies.

Choriocapillaris

The choriocapillaris, located solely in the internal portion of the choroid, appears as a nonhomogenous network of large (20 to 50 μm) capillaries (Figure 2). This monolayer vascular network, flattened in the anterior–posterior aspect, changes from a dense, honeycomb-like, nonlobular structure in the peripapillary area to a lobule-like pattern in submacular areas and most of the posterior pole and equatorial areas (Figure 3). In the peripheral area, choriocapillaries form more elongated, palmlike or fanlike vascular networks, and finally form arcades that terminate at the ora serrata [2]. The network of choriocapillaris is supplied by feeding arterioles derived from the short posterior ciliary arteries and drained by the collecting venules (Figure 2). These arterioles and venules form the medium-sized vessels of the choroid occupying the choroidal stroma (Sattler's layer). The majority of these vessels in the peripapillary and submacular areas form a 90-degree angle with the posterior aspect of the choriocapillaris (Figure 2). The choriocapillaris lobules measure 0.6 to 1.0 mm (Figure 3).

There is controversy over the idea that "lobules" exist and subdivide the choroid into many functional islands. One hypotheses suggests that the choriocapillaris is a single, continuous capillary vascular layer; although lobules were anatomically separated, they were functionally interconnected. The other hypothesis proposed by Hayreh [1] has advocated the presence of noncommunicating lobules. On the basis of fluorescein angiography findings he described a mosaic of lobules, each one containing an arteriole in the middle and a venule at its periphery.

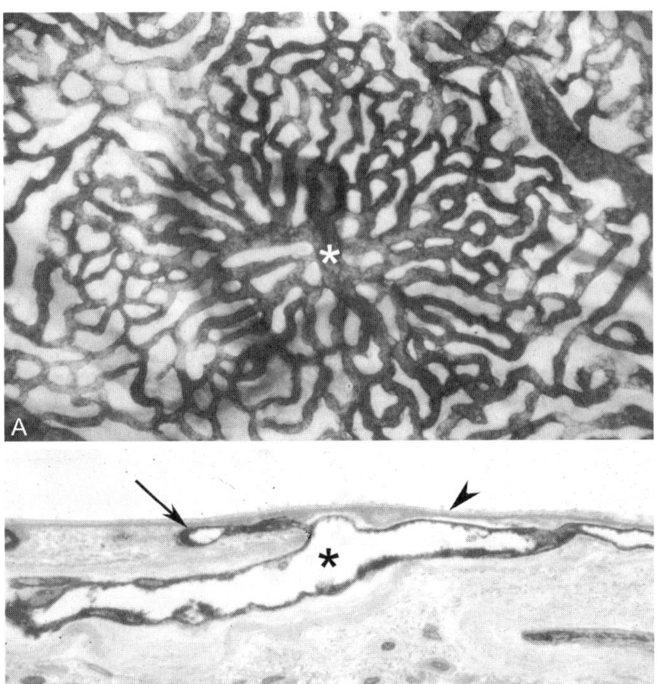

Figure 3 **(A)** Alkaline phosphatase incubated normal human choroid after bleaching. In this area of choroid near the macula, the lobular organization of the choriocapillaris is apparent. There is a draining venule (asterisk) in the center of this lobule. **(B)** Section through the center of the lobule shown in (A) shows blue APase activity in choriocapillaris lumens (arrow) and in the draining venuole (asterisk). The internal limiting membrane (arrowhead) is PAS positive. RPE cells have been removed from Bruch's membrane in processing. (Blue APase reaction product, PAS, and hematoxylin stain.) (see color insert)

There is disagreement as well about the location of arterioles and venules in the lobule. Shimizu and Ujiie [3] confirmed the central location of the artery and the peripheral location of the venule. On the contrary, McLeod and Lutty [2], who performed histologic studies of the choriocapillaris stained with alkaline phosphatase reaction product, described a lobular organization of the choroid with arterioles and venules located peripherally and centrally, respectively, in most lobules.

Bruch's Membrane (Lamina Vitrea)

Bruch's membrane is a thin noncellular lamina separating the choriocapillaris from the retinal pigment epithelium (RPE). Ultrastructurally, it is composed of the basement membrane of the retinal pigment epithelium (inner basal lamina; 0.3 μm thick); inner collagenous zone (1.5 μm thick); elastic fibers; outer collagenous zone (0.7 μm thick); and basement membrane of the choriocapillaris (0.14 μm thick). Bruch's membrane is eosinophilic and PAS-positive. Mallory and other connective tissue stains show its collagen, and elastic tissue stains show that the membrane contains a well-developed layer of elastic tissue. The inner collagenous layer is thick and may degenerate and split with age. The elastic layer distribution is much more diffuse posteriorly. The basal lamina of the choriocapillaris is incomplete in that it is limited to the actual capillaries and absent at the intercapillary septa.

Retinal Pigment Epithelial (RPE) Cells

RPE cells form the outer blood retinal barrier (BRB). RPE cells measure 16 μm in height and 10 to 60 μm in diameter; they feature apical zonulae occludens (the outer BRB). The RPE nucleus is basal, and its pigment is apical and therefore distant from the choriocapillaris (Figure 1). Inner choroidal vessels (choriocapillaris and medium-sized vessels) are sandwiched between apical RPE pigment of neuroepithelial origin and outer choroidal pigment of neural crest origin.

Visualization of the Choroidal Vasculature

In vivo visualization of the choroid in humans is difficult because of pigmentation of the RPE and choroidal melanocytes. Angiography can be accomplished with sodium fluorescein in lightly pigmented or albino eyes, but not in darkly pigmented fundi as observed in humans of African descent. Angiography with indocyanine green dye, which absorbs and fluoresces in the infrared, has provided the best tool to view the choroidal vasculature in any pigmented fundus. ICG angiography clearly shows blood flow in large choroidal blood vessels, but only high-speed ICG angiography with pulsed laser light source provides enough resolution to study the choriocapillaris [4].

McLeod and Lutty developed an alkaline phosphatase technique to study histologically the human, cat, and dog choroidal vasculature postmortem [2]. By incubating the excised choroid for alkaline phosphatase histochemical activity and then bleaching the choroid, the choroidal vasculature can be studied at high resolution. The choroid can then be flat-embedded in glycol methacrylate and viewed in dual perspective: choroidal vessel pattern in the flat perspective; vessels of interest sectioned and viewed in cross section as well (Figure 3). Only viable choroidal vessels have alkaline phosphatase activity, so sites of vascular dropout can be studied in detail.

The corrosion cast technique is also an excellent method for studying the three dimensional properties of the choroidal vasculature postmortem [3]. A polymer is injected into the vasculature and allowed to polymerize, and then the tissue is digested away from the polymer. After coating the cast with gold palladium, the vasculature can be viewed in three dimensions with scanning electron microscopy (Figure 2).

Physiology

The choroid lacks autoregulation, that is, it lacks metabolic regulation. We have found that when neonatal dogs are exposed to 100 percent oxygen for 4 days the choroidal vasculature does not constrict. In cat the Po_2 in choroid is normally around 70 mmHg, whereas in hyperoxia it increases to 250 mmHg [5]. As a consequence, systemic hypoxia and elevated intraocular pressure lead to decreases in choroidal Po_2. The choroidal vasculature supplies oxygen to outer retina while retinal vasculature supplies oxygen to inner retina. From the measurements and modeling of Linsenmeier's lab [5], the photoreceptors live at the edge. The retina consumes more oxygen per gram of tissue than many other tissues, including brain, barely receiving enough oxygen for maintenance of its normal metabolic function. All of the oxygen for photoreceptor metabolism is supplied by the choroidal vasculature in light conditions and 90 percent of the oxygen consumed by photoreceptors in the dark [5]. As a consequence, choroidal nonperfusion is extremely detrimental to photoreceptor function and viability and retinal detachment from choroid also results in photoreceptor death.

Blood flow rate in cat choroid is said to be at least 10 times higher than retinal blood flow as measured by radioactive microspheres [6]. High blood flow was thought necessary to provide enough oxygen to the retina. However, our recent study using fluorescently labeled erythrocytes demonstrates that, at least in rat, the RBC velocity in choriocapillaris is actually four times slower than in retinal capillaries. The difference in these two values may be due to species differences, but it is more likely due to techniques where the former values are based on blood flow in all choroidal vessels whereas, the RBC velocities where measured directly in capillaries.

Cell Biology

It has been found that rat choriocapillaris had 50 percent fewer pericytes than retinal capillaries. Also, unlike retinal capillaries, pericyte loss does not occur in rat experimental diabetes. Choriocapillaris is unique in that only the retinal side of the capillary has fenestrae, making transport of nutrients to RPE and photoreceptors and removing waste from photoreceptor disk shedding and RPE digestion of these disks more efficient. The same sidedness is observed in the location of VEGF receptors. It has been reported that both VEGF receptor-1 (FLT-1) and VEGF receptor 2 (FLK-1 or KDR) are found on endothelial cells on the retinal side of the choriocapillaris. Perhaps this is related to the basal production of VEGF by RPE. RPE was actually one of the first cells shown to produce VEGF and upregulate production during hypoxia. Perhaps the release of VEGF on the retinal side encourages maintenance of fenestrae on the retinal side of the choriocapillaris.

Pathology

Age-Related Macular Degeneration

The choroidal vasculature is at the heart of age-related macular degeneration. There are many forms of AMD, but the two major types are exudative or wet and nonexudative or dry AMD. In nonexudative AMD eyes, there are macular drusen and sharply defined focal areas of RPE atrophy, which are associated with varying degrees of loss of the choriocapillaris. Drusen is a form of deposit on Bruch's membrane, which is believed to be incompletely digested material from the RPE that cannot traverse Bruch's membrane for removal by the choriocapillaris. One hypothesis states that the deposit of the debris on Bruch's membrane is a result of choriocapillaris insufficiency. Alternatively, debris may accumulate on Bruch's membrane and choriocapillaris atrophy results since it is not needed any more. Drusen appear as refractile structures when viewing the ocular fundus, and the ophthalmologist refers to them as hard drusen when they are sharply demarcated and highly refractile. We have found that hard drusen are present almost always over intercapillary septa and not over choriocapillaris lumens, suggesting that this material accumulates on Bruch's membrane where transport is least likely.

There is considerable controversy over the role of the choroidal vasculature in AMD. Our recent study on geographic atrophy, a form of dry AMD, demonstrated that the RPE cells degenerated first and then the choriocapillaris atrophied. Interestingly, even in areas with complete RPE atrophy, some choriocapillaris segments remained viable (alkaline phosphatase positive) but severely constricted [7].

Exudative AMD is characterized by fluid and hemorrhage beneath the RPE or sensory retina due to choroidal neovascularization (CNV). In exudative AMD, we have observed areas with choriocapillaris atrophy in advance of the growing choroidal neovascularization. Every example of CNV in our study had surviving RPE associated with it. It is tempting to speculate that choriocapillaris loss results in ischemic RPE, and the RPE cells in turn produce angiogenic factors such as vascular endothelial cell growth factor, which stimulated growth of CNV. CNV grows through Bruch's membrane and spreads under the retinal pigment epithelium in most cases (Figure 4).

Whether choriocapillaris is responsible in part for AMD and which forms it initiates remains to be determined. However, there can be no question that there is a dynamic symbiosis between choriocapillaris and RPE and this synergy is disrupted in AMD.

Systemic Diseases

HYPERTENSION

Stenosis, focal narrowing, and occlusion of choroidal arteries accompanied by hypertrophy of smooth muscle cells, endothelial cell proliferation, and reduplication of basement membrane occurs frequently in benign hypertension. These findings led to the assumption that arterial occlusion may precede rarefaction of the choriocapillaris. Long-standing hypertension leads to severe arteriosclerosis, with marked thickening and hyalinization of choroidal arterioles, irregular thickened basement membrane, and reduction in caliber of the lumen. In addition, necrosis and fibrinoid deposition occurs in the choroidal vessel wall. Based on morphologic study of vascular casts on spontaneously hypertensive rats, the early changes in the choroidal vasculature were a decrease in number of draining venules, especially in the peripapillary choroid. As the systolic blood pressure increased progressively, the choroidal vessels were often sparse and tortuous. Some of the venules showed localized aneurismal dilatation and irregular calibers. In long-term hypertension, severe tortuosity, caliber irregularity, and generalized narrowing of the choroidal arteries; considerably decreased number of the draining venules converging from all directions into the vortex veins; narrow ampullae of the vortex veins; and elongated and engorged choriocapillaris were the most extensive morphological changes in the choroid [8].

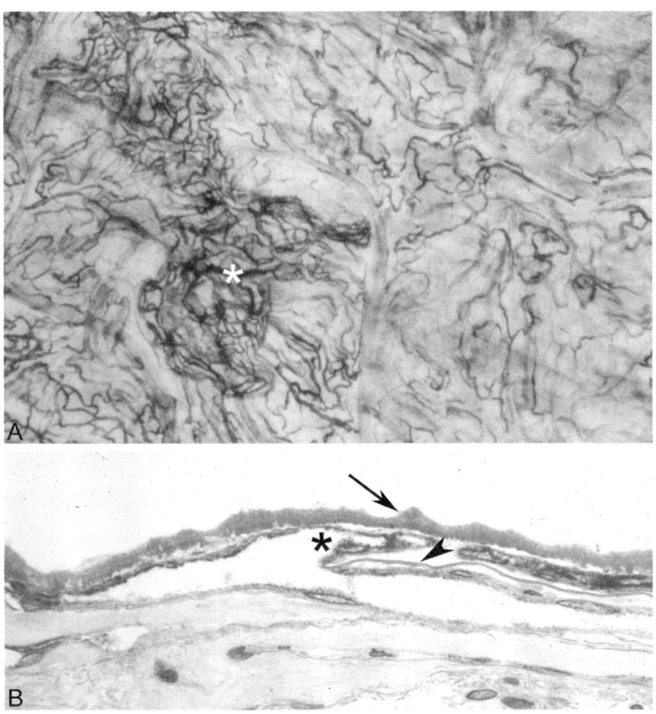

Figure 4 **(A)** Alkaline phosphatase incubated choroid from a diabetic subject. Choroidal neovascularization (CNV) is apparent (asterisk) adjacent to an area with severe choriocapillaris dropout (lack of viable, APase-positive capillary lumens). The yellow areas are drusen. **(B)** Section through the CNV formation in (A) shows that the new blood vessels (asterisk) are between a basal laminar deposit (arrow), which is characteristically positive for PAS, and Bruch's membrane (arrowhead). The origin of the new vessels appears to be choriocapillaris. (Blue APase reaction product, PAS, and hematoxylin stain.) (see color insert)

DIABETIC CHOROIDOPATHY

In both hypertensive and diabetic choroids, we have observed intrachoroidal neovascularization lying deep in choroids near the interface with sclera. These structures, which we termed intrachoroidal microangiopathy or ICMA, were the only sites where we found microaneurysms in human choroids. They were present in 20 percent of the diabetic choroids we have analyzed.

Using the alkaline phosphatase flat-embedding technique, we have observed a fivefold greater loss in choroidal capillaries in diabetic subjects compared to older nondiabetic subjects. This capillary dropout is a hallmark of diabetic retinopathy but had never been quantified in diabetic choroid before [9]. The cause of the choriocapillaris loss may be related to neutrophil-initiated occlusions in choriocapillaris. We documented a twofold greater number of PMNs in diabetic choriocapillaris, compared to normal subjects, and there was a significant increase in PMNs in areas of diabetic choroid with capillary loss as compared to normal-appearing areas of choriocapillaris. The retention of PMNs may be due to increased expression of both ICAM-1 and P-selectin in diabetic choroid. ICAM-1 is constitutively made in normal choriocapillaris, but levels were increased in diabetic choriocapillaris and immunoreactivity was also observed in other choroidal vessels. P-selectin, responsible for leukocyte rolling, was also observed in diabetic choriocapillaris and platelet aggregates in choroid, whereas in normal subjects it was confined to postcapillary venules [10].

Summary

The choroidal vasculature lies between sclera and retina and supplies the majority of nutrients to the outer retina. It is necessary for normal visual function because it removes waste from photoreceptor disk shedding and supplies 90 to 100 percent of the oxygen photoreceptors require. Dysfunction of the choroidal vasculature occurs in systemic diseases such as diabetes and hypertension and in AMD.

Glossary

Age-related macular degeneration (AMD): Degeneration of choriocapillaris, photoreceptors, and retinal pigment epithelium in macula.

Choriocapillaris: Capillary system of choroid.

Fenestrations: Apertures in the walls of choroidal capillaries that exclude molecules by size.

Acknowledgments

The authors acknowledge and are very grateful to D. Scott McLeod, Wilmer Ophthalmological Institute, who created the figures, critically evaluated this chapter, and provided many of the insights on the choroidal circulation that are mentioned herein. They also thank Stephen Wajer, Johns Hopkins Applied Physics Lab, Laurel, MD, who provided the image of the vascular cast.

References

1. Hayreh, S. S. (1990). In vivo choroidal circulation and its watershed zones. *Eye* **4**, 273–289. *This is a landmark paper describing watershed zones and their importance in choroid.*
2. McLeod, D. S., and Lutty, G. A. (1994). High resolution histologic analysis of the human choroidal vasculature. *Invest. Ophthalmol. Vis. Sci.* **35**, 3799–3811.
3. Shimizu, K., and Ujiie, K. (1978). *Structure of Ocular Vessels.* New York: Igaku-Shoin. *This textbook has the most extraordinary vascular casts of the eye.*
4. Flower, R. W., Frychowski, A. W., McLeod, D. S. (1995). Variability in choriocapillaris blood flow distribution. *Invest. Ophthalmol. Vis. Sci.* **36**, 1247–1258.
5. Wangsa-Wirawan, N. D., and Linsenmeier, R. A. (2003). Retinal oxygen. Fundamental and clinical aspects. *Arch. Ophthalmol.* **121**, 547–557. *This is an excellent review about oxygen in retina and choroid and how to measure it. Linsenmeier is one of the pioneers in making direct measurements of oxygen in tissue.*
6. Alm, A., and Bill, A. (1972). The oxygen supply to the retina, II: Effects of high intraocular pressure and of increased arterial carbon dioxide tension on uveal and retinal blood flow in cats. *Acta Physiol. Scand.* **84**, 306–319.
7. McLeod, D. S., Taomoto, M., Otsuji, I., Green, W. R., Sunness, J. S., and Lutty G. A. (2002). Quantifying changes in RPE and choriocapillaris in eyes with age-related macular degeneration. *Invest. Ophthalmol. Vis. Sci.* **43**, 1986–1993.
8. Bhutto, I. A., and Amemiya, T. (2002). Choroidal vascular changes in spontaneously hypertensive rats—transmission electron microscopy and scanning electron microscopy with casts. *Ophthalm. Res.* **34**, 54–62.
9. Cao, J., McLeod, S., Merges, C. A., Lutty, G. A. (1998). Choriocapillaris degeneration and related pathologic changes in human diabetic eyes. *Arch. Ophthalmol.* **116**, 589–597. *This was the first study to quantify the loss of choriocapillaris in diabetic choroid.*
10. McLeod, D. S., Lefer, D. J., Merges, C., Lutty, G. A. (1995). Enhanced expression of intracellular adhesion molecule-1 and P-selectin in the diabetic human retina and choroid. *Am. J. Pathol.* **147**, 642–653.

Capsule Biography

Dr. Imran Bhutto was awarded an M.D. degree from Sind University, Pakistan and a Ph.D. in Ophthalmology from the University of Nagasaki, Japan, and was a Research Associate in Department of Ophthalmology, University of Nagasaki, Japan, from 1996 to 2001. He was the winner of the Medical Research Award of University of Nagasaki School of Medicine in 2000. His research work primarily focused on ocular blood vessels and systemic diseases, and he has recently become a Research Fellow with Dr. Gerard Lutty at the Wilmer Ophthalmological Institute, Johns Hopkins Hospital, Baltimore, Maryland.

Dr. Gerard A. Lutty was awarded a masters of science degree from Catholic University of America, Washington, DC, and a Ph.D. from the Johns Hopkins School of Medicine Biochemistry, Cellular and Molecular Biology. The Wilmer Ophthalmological Institute is the Department of Ophthalmology but it seemed redundant to say that. He is the recipient of an American Heart Association Established Investigator and a Research to Prevent Blindness Lew Wasserman Merit Award. His research focuses on development of the retinal and choroidal vasculatures and how they change in disease states like diabetes, sickle cell disease, retinopathy of prematurity and age-related macular degeneration. He is an Professor at the Wilmer Ophthalmological Institute, Johns Hopkins Hospital, Baltimore, Maryland.

CHAPTER 58

The Role of the Cytoskeleton in Barrier Function of the Microvasculature

Steven M. Dudek and James H. Finigan
Johns Hopkins University School of Medicine, Baltimore, Maryland

Introduction

The endothelium functions as a semipermeable barrier between circulating vascular contents and the interstitium of the surrounding tissues, and maintenance of this barrier is essential for optimal organ function. For example, the lung is particularly sensitive to perturbations of barrier function because of the enormous surface area of the pulmonary vasculature. Disruption of the pulmonary endothelial lining leads to alveolar flooding during inflammatory conditions such as adult respiratory distress syndrome (ARDS) and sepsis and contributes significantly to the morbidity of these clinical syndromes. As a result, a comprehensive understanding of the regulatory mechanisms involved in endothelial permeability has important clinical implications. Current evidence supports an integral role for the endothelial cytoskeleton in dynamic modulation of vascular barrier function, with recent work highlighting the contribution of the microvasculature to this process. In this chapter, we will focus on the pulmonary vasculature as a model system to outline the cytoskeletal mechanisms that regulate endothelial permeability. Our discussion will begin with a general overview of regulatory mechanisms inherent to endothelial cells (ECs) throughout the vascular bed, and then we will proceed to highlight the permeability properties unique to the microvessel circulation.

Overview of the Cytoskeleton

The cellular cytoskeleton consists of three primary structural elements—actin microfilaments, intermediate filaments, and microtubules (for detailed discussion, see Chapter 38, "Role of the Endothelial Cell Cytoskeleton in Microvascular Function"). The actin cytoskeleton is a dynamic structure that undergoes rearrangement under the control of various actin binding, capping, nucleating, and severing proteins that are intimately involved in regulating the cell shape, motility, and contractile status of cells. Early studies revealed the primary role of actin filaments in EC permeability as demonstrated by the observations that cytochalasin D, which disrupts the actin cytoskeleton, increases EC permeability, while phallacidin, an actin stabilizer, prevents barrier disruption by various agonists. Providing a structural framework to the cell, the actin microfilament system is focally linked to multiple membrane adhesive proteins that connect it to cell–cell (adherens junctions, tight junctions) and cell–matrix (focal adhesion) junctions that anchor the endothelium. In addition, actin filaments, through interaction with myosin, are critical to EC tensile force generation necessary for cell shape changes and barrier regulation (as discussed in detail later).

The functional roles of microtubule and intermediate filament cytoskeletal components in EC barrier regulation are less well defined. Microtubules, polymers of α and β-tubulin forming a lattice network of rigid hollow rods spanning the cell that undergo frequent assembly and disassembly, interact functionally with actin filaments during dynamic cellular processes. Microtubule disruption with agents such as nocodazole or vinblastine induces rapid rearrangement of actin filaments and focal adhesions, cellular contraction, and increased permeability across EC monolayers, while microtubule stabilization with paclitaxel attenuates these effects. These results suggest that actin microfilament microtubule crosstalk plays an important, but still poorly defined, role in EC barrier regulation.

Although the third major element involved in EC cytoskeletal structure, intermediate filaments (IFs), exhibit greater diversity than the highly conserved components of actin microfilaments or microtubules, IF proteins share a common dimer structure containing two parallel α-helices that associate with IF-binding proteins that in turn connect to the nuclear envelope, peripheral cell junctions, and other cytoskeletal components. Vimentin, the primary IF protein found in EC and other cells of mesenchymal origin, does not appear to have an essential function in EC barrier regulation. Fibroblasts derived from vimentin knockout mice have normal actin and microtubule architecture, while the animals themselves develop normally without any obvious phenotypic abnormalities, suggesting that functional roles for IF in EC barrier function are subject to compensation by biologic redundancy.

Regulation of Permeability by the Endothelial Cell Cytoskeleton

Because the actin microfilament system is well studied and known to be essential for EC barrier regulation, we will explore its functional role in greater detail in this section. The current paradigm of EC barrier regulation suggests that a balance exists between barrier-disrupting cellular contractile forces and barrier-protective cell–cell and cell–matrix tethering forces. As outlined in Figure 1, both competing forces in this model are intimately linked to the actin-based EC cytoskeleton by a variety of actin-binding proteins that are critical both to tensile force generation and to linkage of the actin cytoskeleton to adhesive membrane components. Although less structured than in skeletal or smooth muscle cells, ECs contain similar molecular machinery for generation of tension via an actomyosin motor (actin and myosin represent more than 15% of total EC protein). Focally distributed changes in tension within EC are accomplished by regulation of the level of myosin light chain (MLC) phosphorylation, which promotes actomyosin interaction, and subsequent actin stress fiber formation. There is a good association between the development of transcellular actin stress fibers, increased MLC phosphorylation, enhanced tension development, paracellular gap formation, and increased EC permeability. A key regulator of this EC contractile apparatus in this process is the Ca^{2+}/calmodulin (CaM)-dependent enzyme myosin light chain kinase (MLCK).

Balancing these EC contractile forces are cell–cell and cell–matrix contacts that provide tethering forces essential for mechanical stability and barrier maintenance (Figure 1). Adherens junctions are the primary cell–cell contacts along the EC monolayer and are composed of cadherins bound together in a homotypic and Ca^{2+}-dependent fashion to link adjacent ECs. Cadherins interact through their cytoplasmic tails with the catenin family of intracellular proteins, which provide anchorage directly to the actin cytoskeleton. The primary cadherin in human EC adherens junctions, vascular endothelial (VE)-cadherin, is critical to EC barrier integrity since infusion of VE-cadherin blocking antibody increases lung vascular permeability in cultured ECs and mice. Focal adhesions provide the primary tethering sites of ECs to the underlying extracellular matrix through transmembrane integrin receptors connected to the actin cytoskeleton via multiprotein focal adhesion plaques. These linkages are also essential for EC barrier integrity as blocking antibodies to β_1 integrin alter EC attachment and permeability. Unliganded integrins are not associated with the cytoskeleton, but binding to the ECM induces the attachment of integrins to intracellular actin fibers and stimulates tyrosine phosphorylation of multiple proteins and Ca^{2+} influx. Moreover, intracellular signaling pathways that regulate cytoskeletal rearrangement can also modulate cell–matrix contacts. For example, inhibition of the small GTPase Rho dissociates stress fibers from focal adhesions, decreases phosphotyrosine content of focal adhesion proteins, and enhances EC barrier function. Thus, the tethering forces of both focal adhesion and adherens junction complexes are essential components of EC barrier integrity.

Barrier Disruption

Activation of the contractile apparatus is a critical step in many models of agonist-induced EC barrier dysfunction. For example, one well-characterized model utilizes the central coagulation regulatory protein, thrombin, to induce significant pulmonary vascular leakage. The observation that microthrombi occur in the pulmonary microvasculature of patients with acute lung injury suggests this model has biologic relevance to EC barrier regulation in human disease. Thrombin induces a profound increase in EC permeability through rapid actin cytoskeletal rearrangement and force generation dependent on actomyosin interaction catalyzed by the phosphorylation of regulatory MLC by MLCK. Direct activation of MLCK is sufficient to produce EC contraction and barrier disruption, whereas MLCK inhibition abolishes barrier dysfunction in rat lung models of ischemia–reperfusion injury and ventilator-induced lung permeability. Tyrosine phosphorylation status appears to play an important role in regulation of EC permeability since tyrosine phosphorylation of EC MLCK evokes significant increases in MLCK activity, EC contraction, and subsequent EC barrier dysfunction.

The small GTPase Rho also plays an important role in regulation of the EC contractile apparatus in several models of agonist-induced EC barrier dysfunction. Linking extracellular stimuli to dynamic actin cytoskeletal rearrangement, Rho activation induces increased MLC phosphorylation, actomyosin interaction, stress fiber formation, and subsequent EC barrier dysfunction. The relative contributions of the EC MLCK and Rho pathways in regulating EC permeability are not completely understood since inhibition of either MLCK or Rho activation attenuates

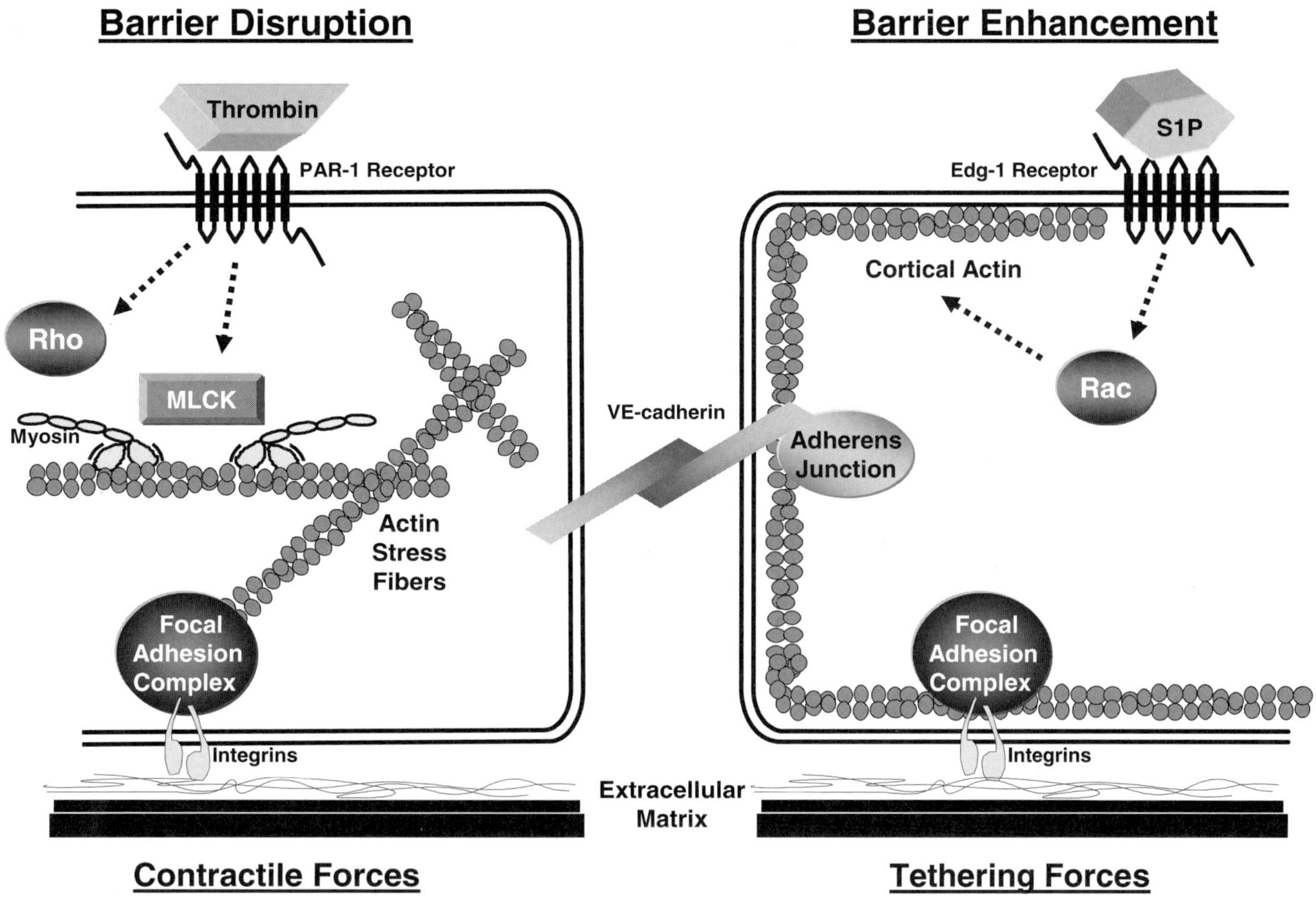

Figure 1 Actomyosin contractile elements compete with cellular tethering forces to regulate EC permeability. The left-hand cell illustrates cytoskeletal changes associated with agonist-induced EC barrier disruption. Thrombin binding to its PAR-1 receptor leads to Rho and MLCK activation along actin stress fibers, actomyosin interaction, contraction, and increased permeability. Conversely, the right-hand cell shows the barrier-enhancing cytoskeletal changes produced by sphingosine 1-phosphate (S1P), which strengthens the cortical actin ring through Edg receptor binding and subsequent downstream Rac activation, cortical actin formation, and adherens junction and focal adhesion rearrangement to decrease permeability. Vascular barrier integrity is dependent upon a balance of these two competing forces.

thrombin-induced EC barrier dysfunction. Recent studies suggest that Rho and MLCK may differentially regulate MLC phosphorylation according to spatial localization within ECs.

Barrier Enhancement

Cytoskeletal rearrangements are also essential for production of EC barrier enhancement by various agonists, such as the platelet-derived phospholipid sphingosine 1-phosphate (S1P), which dramatically reduces pulmonary vascular permeability. S1P binds to G-protein coupled Edg receptors on EC to initiate a series of cytoskeletal protein rearrangements that decreases EC permeability (Figure 1). Activation of the small GTPase Rac results in formation of a prominent cortical actin ring that accompanies S1P-induced EC barrier enhancement. In addition to inducing this cortical actin cytoskeletal rearrangement, S1P alters cell–cell and cell–matrix contacts to reduce EC permeability. S1P dramatically increases localization and interaction of VE-cadherin and catenin proteins at EC cell–cell junctions (adherens junctions). Moreover, while the barrier disrupting agent thrombin rearranges cell–matrix contacts so that focal adhesion proteins assemble at the ends of massive actin stress fibers to anchor cell contraction that pulls ECs apart, S1P induces differential focal adhesion protein rearrangement that associates with cortical actin ring formation that enhances EC barrier function.

Ongoing studies of these contrasting models of EC barrier disruption and enhancement continue to provide insights into potential targets for therapeutic modulation of vascular permeability. This recent work has described additional agonists with impressive EC barrier-enhancing properties. For example, hepatocyte growth factor (HGF) significantly increases pulmonary EC barrier function through enhanced cortical cytoskeleton linkage to cell–cell tethering junctions. Although HGF-induced barrier enhancement requires actin rearrangement and rapidly activates Rac in a manner similar to S1P, the mechanistic pathways are not identical since HGF requires PI-3′-kinase activity whereas S1P does not. Thus, multiple signaling pathways exist for promotion of EC barrier function. The 3-hydroxy-3-methylglutaryl coenzyme A (HMG CoA) reductase inhibitor and cholesterol lowering agent, simvastatin, induces actin cytoskeletal cortical rearrangement and gene expression alterations over time that protect pulmonary EC barrier integrity from thrombin disruption without affecting basal permeability. In addition, the importance of mechanical forces in regulation of EC permeability is being increasingly recognized. The application of physiologic levels of shear stress to cultured pulmonary EC results in enhanced cortical actin and other cytoskeletal rearrangements that promote barrier function. Conversely, stretch forces similar to those produced during mechanical ventilation make pulmonary EC more sensitive to the barrier disrupting effects of thrombin.

Differential Regulation of Permeability in the Microvasculature

Having described the general mechanisms through which the cytoskeleton regulates endothelial permeability, we will now outline specific characteristics intrinsic to microvascular barrier function. Although the endothelium was initially regarded as relatively homogeneous throughout the vasculature, important phenotypic and functional differences are now appreciated at various sites along the vascular tree. These differences account for the tremendous physiologic variability observed in vascular function along various EC segments in different organ systems. Studies of endothelial heterogeneity typically divide ECs into two primary groups, macrovascular cells from large conduit vessels and microvascular cells from small arterioles, venules, and capillaries.

Isolation and culture of the microvasculature pose a particular challenge, in large part because it is not possible to cannulate the microvasculature. Isolation of microvascular ECs requires physical dissociation of ECs via homogenizing, mincing, or cutting the tissue into 3- to 5-mm^3 pieces before incubation with proteolytic enzymes to release the microvascular ECs. However, these techniques often yield other cell types in addition to microvascular ECs (most frequently fibroblasts); therefore, purification is a key step in the isolation of microvascular ECs. Visual identification of cells based on morphologic characteristics followed by manual separation techniques was an early method of purification but is labor intensive and time consuming. Microvascular ECs can also be removed with specially coated glass beads that specifically adhere to ECs when added to the culture and then are removed using a magnet. Other techniques include fluorescence activated cell sorting (FACS) and gradient centrifugation to separate cells differentially by weight. Regardless of the technique employed, the possibility that other cell types are present in microvascular EC cultures needs to be considered when evaluating experimental results.

Multiple factors contribute to the differential barrier properties of microvascular and macrovascular EC (Figure 2), including cell-specific attributes such as surface protein expression, cell morphology, and permeability. Morphologic differences include a fourfold larger surface area for microvascular ECs relative to pulmonary artery ECs, while electron microscopy reveals tighter intercellular connections, fewer visible gaps, and increased focal adhesion sites in microvascular cells compared to macrovascular ECs. In addition, microvessel ECs contain a more abundant array of intercellular junctional complexes. All of these microvascular EC characteristics combine to decrease basal permeability in these cells relative to large conduit ECs. Specifically, cultured microvascular ECs have tenfold higher barrier integrity than macrovascular ECs as measured by electrical resistance across monolayers. Even within each of these broad categories of macro- and microvascular ECs, significant variability is present. EC morphologic appearance, cell surface glycoproteins, cell–cell interactions, and protein and mRNA expression are all highly variable throughout the vasculature. The functional importance of this variability remains a highly active area of research.

Studies comparing the responses of various EC phenotypes to specific agonists have provided insight into differential barrier function along the vasculature. In general, microvascular ECs are more resistant to agonist-induced lung permeability. Cultured bovine and sheep pulmonary artery ECs have been shown to be more sensitive than microvascular ECs to barrier disruption by agents such as endotoxin or TNFα. However, this relative resistance of microvascular ECs is not universal to every model of agonist-induced permeability since thrombin produces more dramatic barrier disruption in microvascular ECs compared to pulmonary artery ECs. The mechanisms underlying this differential EC barrier regulation remain poorly understood, but it appears that Ca^{2+}-dependent signaling events play a role. For example, macrovascular ECs are more sensitive to barrier disruption induced by store-operated Ca^{2+} entry, whereas microvascular ECs have increased levels of intracellular cAMP, a determinant of focal adhesion complex formation that contributes to barrier integrity. Differential adherens junction protein expression may also account for some of the variability in EC barrier function along the vascular tree. Microvascular ECs express significantly more VE-cadherin protein than macrovascular cells, an observation that appears to have functional importance for cell–cell integrity since the infusion of anti-VE-cadherin antibodies causes increased permeability primarily in alveolar capillaries. Finally, variability in the extracellular environment likely plays an important role in differential EC barrier function. Mechanical forces such as shear stress and cyclic stretch significantly affect EC cytoskeletal organization in ways that alter permeability (as discussed earlier) and are differentially distributed along the vasculature. For example, understanding the effects of shear stress on pulmonary EC barrier integrity must take into account regional differences in blood flow that affect the amount of shear applied

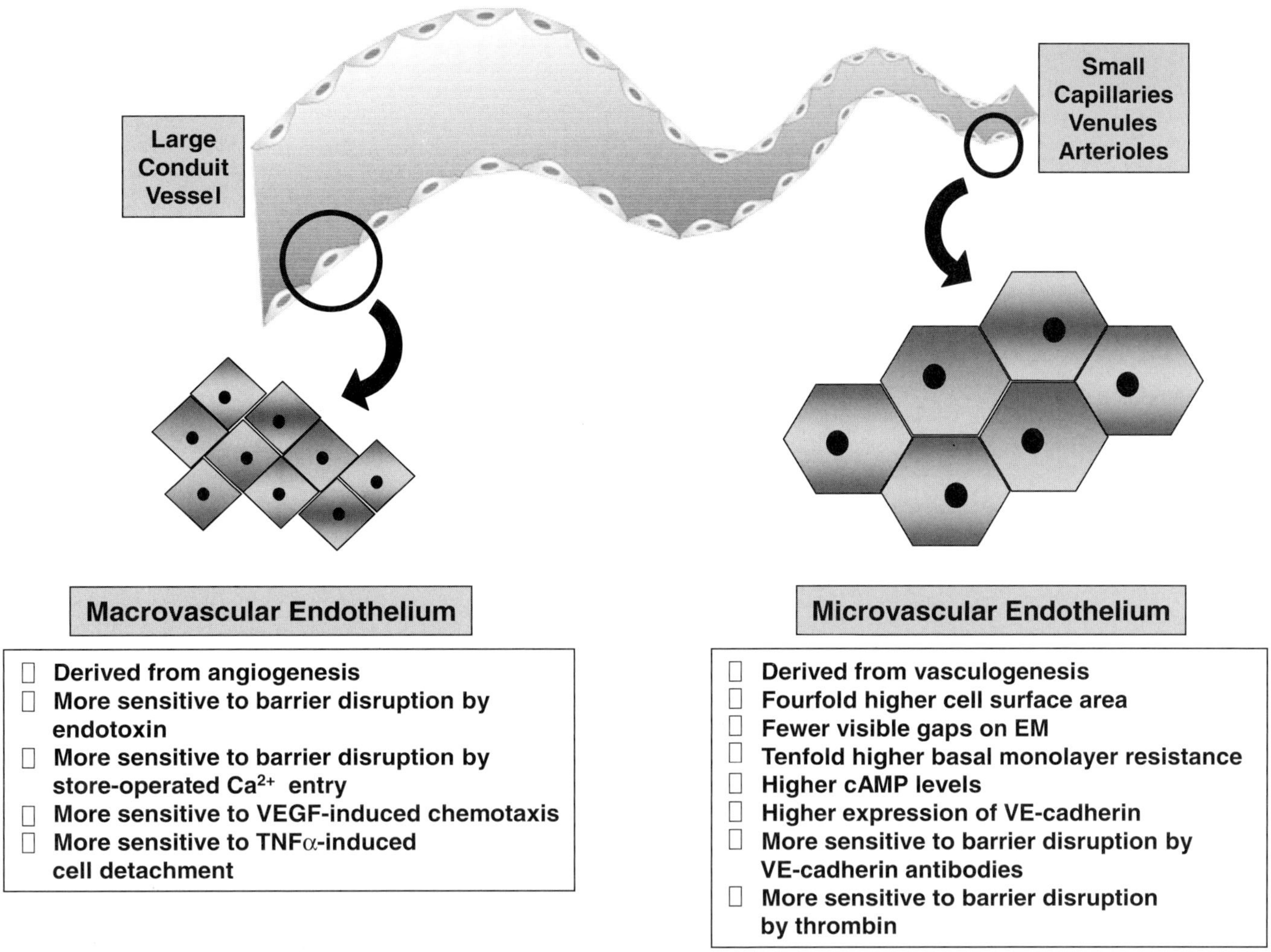

Figure 2 Differential barrier properties of endothelial phenotypes. The diagram shows the cross section of a blood vessel extending from the proximal large arteries to the small microvessels. Listed beneath each broad category of endothelium are specific characteristics related to differences in barrier function between these vascular areas.

to macrovascular ECs relative to microvessel ECs. The contribution of these mechanical forces to differential EC permeability is still incompletely understood.

Glossary

Cytoskeleton: Framework of structural proteins necessary for cell shape and movement. The three primary components are actin microfilaments, microtubules, and intermediate filaments.

Myosin light chain kinase (MLCK): A Ca^{2+}/calmodulin, ATP-dependent enzyme that phosphorylates myosin light chains on serine-19 and threonine-18, which results in increased actomyosin interaction. MLCK activity is a necessary step in many models of vascular permeability.

Sphingosine 1-phosphate (S1P): A biologically active phospholipid generated by hydrolysis of membrane lipids, S1P is produced in significant quantities by circulating platelets and has potent EC barrier-enhancing and chemotactic properties.

Thrombin: A central regulatory molecule in the coagulation cascade, thrombin produces potent EC barrier disruption through stimulation of actomyosin interaction, stress fiber formation, contraction, and intercellular gap formation.

Bibliography/Further Reading

Aird, W. C. (2003). Endothelial cell heterogeneity. *Crit. Care Med.* **31**, S221–S230. *This article describes structural and functional differences in EC subpopulations and applies this knowledge to an understanding of vasculopathic processes, especially acute lung injury.*

Amano, M., Chihara, K., Kimura, K., Fukata, Y., Nakamura, N., Matsuura, Y., and Kaibuchi, K. (1997). Formation of actin stress fibers and focal adhesions enhanced by rho-kinase. *Science* **275**, 1308–1311.

Corada, M., Mariotti, M., Thurston, G., Smith, K., Kunkel, R., Brockhaus, M., Lampugnani, M. G., Martin-Padura, I., Stoppacciaro, A., Ruco, L., McDonald, D. M., Ward, P. A., and Dejana, E. (1999). Vascular endothelial-cadherin is an important determinant of microvascular integrity in vivo. *Proc. Natl. Acad. Sci. USA* **96**, 9815–9820.

Dudek, S. M., and Garcia, J. G. N. (2001). Cytoskeletal regulation of pulmonary vascular permeability. *J. Appl. Physiol.* **91**, 1487–1500. *A recent review by this chapter's first author, this article explores EC cytoskeletal components and mechanisms of barrier regulation in greater detail than is possible here.*

Garcia, J. G., Siflinger-Birnboim, A., Bizios, R., Del Vecchio, P. J., Fenton, J. W., and Malik, A. B. (1986). Thrombin-induced increase in albumin permeability across the endothelium. *J. Cell Physiol.* **128**, 96–104.

Garcia, J. G. N., Liu, F., Verin, A. D., Birukova, A., Dechert, M. A., Gerthoffer, W. T., Bamburg, J. R., and English, D. (2001) Sphingosine

1-phosphate promotes endothelial cell barrier integrity by Edg-dependent cytoskeletal rearrangement. *J. Clin. Invest.* **108**, 689–701. *This novel work represents the initial report describing the potent barrier-enhancing effects of S1P and the cytoskeletal changes involved in producing this response.*

Goeckeler, Z. M., and Wysolmerski, R. B. (1995). Myosin light chain kinase-regulated endothelial cell contraction: The relationship between isometric tension, actin polymerization, and myosin phosphorylation. *J. Cell Biol.* **130**, 613–627.

Kelly, J. J., Moore, T. M., Babal, P., Diwan, A. H., Stevens, T., and Thompson, W. J. (1998). Pulmonary microvascular and macrovascular endothelial cells: Differential regulation of Ca^{2+} and permeability. *Am. J. Physiol.* **274**, L810–L819.

Michel, C. C., and Curry, F. E. (1999). Microvascular permeability. *Physiol. Rev.* **79**, 703–761. *This extensive review discusses classical structural questions about microvascular permeability using recent experimental work on intact microvascular beds, single perfused microvessels, and EC cultures.*

Wojciak-Stothard, B., and Ridley, A. J. (2002). Rho GTPases and the regulation of endothelial permeability. *Vascul. Pharmacol.* **39**, 187–199. *This article summarizes the antagonistic effects of Rho and Rac small GTPases on EC barrier function.*

Capsule Biography

Dr. Dudek is a faculty member of the Division of Pulmonary and Critical Care Medicine of the Johns Hopkins School of Medicine. His research focuses on the cytoskeletal regulation of pulmonary vascular permeability in acute lung injury syndromes and is supported by grants from the NIH.

Dr. Finigan is a research fellow performing similar studies in the Division of Pulmonary and Critical Care Medicine of the Johns Hopkins School of Medicine.

CHAPTER 59

Microcirculation of the Brain

Frank M. Faraci and Donald D. Heistad
Departments of Internal Medicine and Pharmacology, Cardiovascular Center, University of Iowa Carver College of Medicine and VA Medical Center, Iowa City, Iowa

The microcirculation of the brain is unique with respect to both its complexity and its heterogeneity. Beginning more than a century ago, studies by Ehrlich and others led to the development of the concept of the blood–brain barrier (BBB). Later work established that cerebral blood vessels have an endothelial structure that is unique and a critically important component of the cerebral circulation. Thus, this present chapter will focus on basic characteristics of the cerebral microcirculation with an emphasis on endothelium. An overview of the role of endothelium under normal conditions will be presented, as well as examples of mechanisms that contribute to microvascular dysfunction in disease states.

Endothelium and the Blood–Brain Barrier

In most regions of brain, tight junctions are present between adjacent endothelial cells. These cells, which constitute the blood–brain barrier (BBB), do not express/possess fenestrations, contain few pinocytotic vesicles, and exhibit very high transendothelial electrical resistance. These unique features are due in very large part to induction of BBB properties by astrocytes. Once induced during later embryological development, the BBB plays many important functions including protecting the brain from fluctuations in changes in concentrations of ions, neurotransmitters, and other substances that could affect neuronal and/or glial function. The genomic and proteomic expression patterns that account for these unique characteristics are only beginning to be defined. Recent studies have begun to explain at the molecular level how permeability of the BBB is regulated. For example, studies in gene-targeted mice have shown that deficiency in claudin-5 (a major cell adhesion protein expressed within endothelium and at tight junctions) increases permeability of the BBB to small molecules.

In addition to the presence of tight junctions between cells, the BBB exhibits other unique permeability characteristics. While the BBB is very selective in relation to substances are allowed to pass from blood to brain, it also expresses a host of transporters for key substances such as glucose and some amino acids.

Many studies have examined the pathophysiology of changes in permeability of the BBB. This work has identified veins and venules as major sites of disruption of the barrier, and found that many conditions (including ischemia with reperfusion, acute hypertension, proinflammatory stimuli, subarachnoid hemorrhage, and meningitis) disrupt the BBB. Although several mechanisms may by involved, a common mediator of this disruption in disease states may be reactive oxygen or reactive nitrogen species (see later discussion).

In contrast to most regions of the brain, some sites within the central nervous system do not have tight junctions between endothelial cells. These sites include the choroid plexus, which is the major site of formation of cerebrospinal fluid (CSF), and the circumventricular organs. Resting levels of blood flow and microvascular permeability within these regions are very high compared to regions expressing the BBB. Regulation of blood flow is also unique in these structures. For example, humoral stimuli such as circulating angiotensin II and vasopressin have marked effects on blood flow to choroid plexus and production of cerebrospinal fluid but little effect on blood flow to other regions of the brain.

The cerebral microcirculation is also unique in that resistance of large arteries appears to be greater in the cerebral circulation than in other vascular beds. Large arteries contribute importantly to total cerebral vascular resistance and are major determinants of local microvascular pressure (arteriolar pressure), which is lower than in other vascular beds.

Endothelium and Vascular Tone

In addition to its fundamental role in controlling vascular permeability, endothelium is also a major regulator of vascular tone. The primary mechanism by which endothelium regulates tone of underlying vascular muscle under normal conditions is by release of endothelium-derived relaxing factor(s) (EDRFs). These factors include nitric oxide (NO), endothelium-derived hyperpolarizing factor(s)(EDHFs), and prostacyclin (Figure 1).

Under normal conditions, most studies indicate that nitric oxide is the predominant EDRF in the cerebral circulation. In both pial (vessels on the surface of the brain) and parenchymal arterioles, many lines of evidence illustrate that nitric oxide from endothelium influences basal tone and mediates the majority of the response to acetylcholine (the classic endothelium-dependent agonist), other receptor-mediated agonists, and increased shear stress (Figure 1). Once it diffuses to vascular muscle, nitric oxide produces relaxation predominantly via activation of soluble guanylate cyclase and increased production of cGMP (Figure 1). This mechanism of endothelium-dependent relaxation may differ from that seen in select peripheral microvascular beds where some studies have suggested that EDHF is the major EDRF in microvessels.

The cerebral microcirculation is unusual in that responses to bradykinin, another endothelium-dependent agonist, are mediated by reactive oxygen species (ROS) in normal cerebral arterioles. It is not entirely clear which reactive oxygen species specifically mediates this response, but hydrogen peroxide may be responsible for bradykinin-mediated vasodilatation (Figure 1).

In addition to modulation of microvascular function, nitric oxide and reactive oxygen species may affect microvascular structure. For example, preliminary findings indicate that genetic deficiency in superoxide dismutase-1 (SOD-1), the CuZn isoform of SOD, produces increases in superoxide and hypertrophy of cerebral arterioles. Increases in cross-sectional area of the vessel wall (hypertrophy) may have functional consequences because hypertrophy of vascular muscle can impair maximal vasodilator capacity.

Although nitric oxide is the primary mediator of microvascular responses to many endothelium-dependent stimuli, EDHF may mediate a major portion of the response of cerebral microvessels to other endothelium-dependent stimuli including selected purines (Figure 1). To date, there has been no convincing evidence that EDHF or reactive oxygen species influence resting tone of cerebral arterioles under normal conditions.

In summary, endothelium-derived nitric oxide influences resting microvascular tone and is a major EDRF in brain. For selected stimuli, however, reactive oxygen species and EDHF can also be important EDRFs in the cerebral circulation.

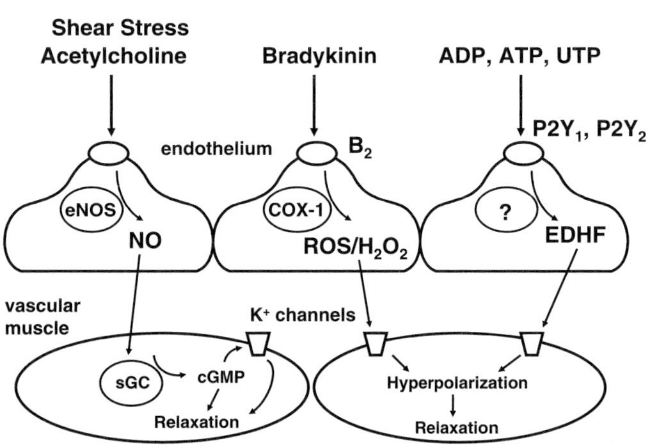

Figure 1 Schematic representation of major mechanisms of endothelium-dependent relaxation of vascular muscle in cerebral microvessels. Nitric oxide (NO) is produced by the endothelial isoform of NO-synthase (eNOS) in response to activation of the M_5 subtype of muscarinic receptor and other endothelium-dependent stimuli. NO diffuses to vascular muscle, where it activates soluble guanylate cyclase (sGC), causing increased production of cyclic GMP (cGMP), activation of cGMP-dependent protein kinase I, and relaxation. In contrast, bradykinin (via activation of the B_2 subtype of receptor) activates cyclooxygenase-1 (COX-1), resulting in production of reactive oxygen species (ROS) including hydrogen peroxide (H_2O_2). H_2O_2 activates potassium channels (K^+ channels) in vascular muscle, producing membrane hyperpolarization and relaxation. In response to some purines (ATP, UTP, ADP), endothelium can produce EDHF through an unknown pathway. By definition, EDHF produces relaxation of vascular muscle by activation of potassium channels and membrane hyperpolarization. In some vessels, cGMP can also activate potassium channels.

Potassium Channels

In blood vessels, there is a close relationship between membrane potential and intracellular calcium, and between membrane potential and vascular diameter. Thus even small changes in membrane potential of vascular muscle produce significant changes in vascular tone. Opening or closure of potassium channels has important effects on membrane potential and thus vascular tone.

Studies utilizing pharmacological approaches, and more recently genetically altered mice, suggest that potassium channels play a major role in mediating microvascular dilatation in response to diverse stimuli. These stimuli include elevations in concentrations of extracellular potassium ion, receptor-mediated agonists (including calcitonin gene-related peptide, prostacyclin, and adenosine), second messengers (cAMP and cGMP), EDRFs, reactive oxygen species (hydrogen peroxide and superoxide anion), chemical stimuli (hypoxia, hypercapnia, acidosis), and calcium sparks.

Endothelium-Derived Contracting Factor

Endothelium has the potential to produce and release a diverse group of contracting factors, named endothelium-derived contracting factors (EDCFs). These factors include endothelin, products of cyclooxygenase activity, and reactive oxygen species. In normal vessels, production of EDCFs appears to be absent or minimal. However, EDCFs may be important mediators of vascular dysfunction in disease states. For example, a cyclooxygenase-derived EDCF produces contraction by activation of PGH_2-thromboxane A_2 receptors in cerebral arterioles of spontaneously hypertensive rat and in a rat model of type I diabetes. In these arterioles, indomethacin or inhibition of prostaglandin H_2/thromboxane A_2 receptors restores endothelial function to normal, suggesting that endothelial dysfunction is mediated by a cyclooxygenase-derived EDCF.

Reactive Oxygen and Nitrogen Species

Nitric oxide and superoxide anion react very efficiently and inactivate each other as they form peroxynitrite. Thus, the bioactivity of nitric oxide depends, in part, on its interaction with superoxide. Increased levels of superoxide impair relaxation of cerebral arteries in response to endothelium-dependent stimuli. Kontos and Wei were the first to provide evidence that superoxide inactivates nitric oxide (EDRF) in vivo.

In addition to reducing bioavailability of nitric oxide, the reaction of nitric oxide with superoxide forms peroxynitrite, a potent oxidant that has the potential to produce endothelial dysfunction through a variety of mechanisms (Figure 2). These mechanisms include nitrosylation and reduction in function of prostacyclin synthase and the mitochondrial isoform of superoxide dismutase, oxidation of tetrahydrobiopterin (a critical cofactor for normal function of NO-synthase), and activation of poly (ADP-ribose) synthetase (PARS).

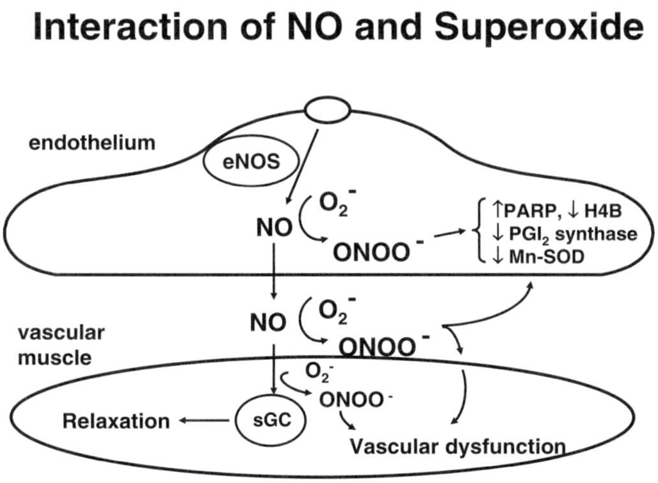

Figure 2 Schematic representation of the interaction of nitric oxide (NO) and superoxide (O_2^-). NO is produced by eNOS and can diffuse to vascular muscle and activate soluble guanylate cyclase (sGC) resulting in vasorelaxation. NO can also rapidly react with superoxide to form peroxynitrite ($ONOO^-$). This interaction can occur in different cell type and subcellular compartments—wherever superoxide is produced. Superoxide can be formed within blood vessels by several sources including NAD(P)H oxidase, cyclooxygenase, xanthine oxidase, and mitochondria. Once formed, $ONOO^-$ can produce vascular dysfunction through a variety of potential mechanisms (see text).

Superoxide may mediate impairment of endothelial function in the cerebral microcirculation in response to angiotensin II, β-amyloid peptide, or amyloid precursor protein (APP), alcohol, nicotine, and homocysteine, as well as in models of acute hypertension, traumatic brain injury, and ischemia. Endothelium, smooth muscle, and adventitia all have the potential to produce superoxide. There are multiple sources of superoxide, including NAD(P)H oxidase, xanthine oxidase, cyclooxygenase, mitochondria, and NO-synthases, as well as autooxidation of substances such as hemoglobin, tetrahydrobiopterin, and β-amyloid. Thus, inactivation of nitric oxide by superoxide contributes to impaired endothelial and nitric oxide–mediated dilatation of cerebral blood vessels under several pathophysiological conditions.

Angiotensin II

It has been known for many years that the renin–angiotensin system contributes importantly to some forms of hypertension. Recent work suggests that activation of this system may contribute to vascular disease under other pathophysiological conditions, including atherosclerosis and diabetes. Angiotensin II has many effects on vascular cells. In aorta, for example, angiotensin II produces oxidative stress, activates components of the inflammatory cascade, and produces endothelial dysfunction.

In the cerebral microcirculation, angiotensin II produces superoxide-mediated endothelial dysfunction and impairs local increases in blood flow in response to somatosensory activation. In stroke-prone spontaneously hypertensive rats (SHRSP) that have endothelial dysfunction, local inhibition of angiotensin-converting enzyme restores endothelial function toward normal, suggesting that endogenous angiotensin II may be a mediator of vascular dysfunction. Using transgenic mice that overexpress both human renin and human angiotensinogen, we have provided evidence that the renin–angiotensin system is an important determinant of microvascular "remodeling" during chronic hypertension.

Acknowledgments

Original studies by the authors that are summarized in this review were supported by National Institutes of Health grants HL-38901, NS-24621, HL-62984, HL-14388, and HL-16066, as well as by funds from the VA Medical Center and from the Carver Trust.

Further Reading

Faraci, F. M., and Heistad, D. D. (1998). Regulation of the cerebral circulation: Role of endothelium. *Physiol. Rev.* **78**, 53–97.

Iadecola, C., Zhang, F., Niwa, K., Eckman, C., Turner, S. K., Fischer, E., Younkin, S., Borchel, D. R., Hsiao, K. K., and Carlson, G. A. (1999). SOD1 rescues cerebral endothelial dysfunction in mice overexpressing amyloid precursor protein. *Nat. Neurosci.* **2**, 157–161. *A fine example of* *how genetically altered mice can be used as models of human disease and to study mechanisms of vascular dysfunction.*

Kamii, H., Kato, I., Kinouchi, H., Chan, P. H., Epstein, C. J., Akabane, A., Okamoto, H., and Yoshimoto, T. (1999). Amelioration of vasospasm after subarachnoid hemorrhage in transgenic mice overexpressing CuZn-superoxide dismutase. *Stroke* **30**, 867–872.

Mayhan, W. G. (2001). Regulation of blood–brain barrier permeability. *Microcirculation* **8**, 89–104.

Nag, S. (2003). *The Blood-Brain Barrier. Methods in Molecular Medicine.* Totowa, NJ: Humana Press.

Nitta, T., Hata, M., Gotoh, S., Seo, Y., Sasaki, H., Hashimoto, N., Furuse, M., and Tsukita, S. (2003). Size-selective loosening of the blood–brain barrier in claudin-5-deficient mice. *J. Cell. Biol.* **161**, 653–660. *An excellent example of using modern molecular approaches to study mechanisms that control permeability of the BBB.*

Rosenblum, W. I. (1998). A review of vasomotor responses of arterioles on the surface of the mouse brain: The necessary prelude to studies using genetically manipulated mice. *Microcirculation* **5**, 129–138.

Capsule Biography

Dr. Faraci is Professor of Internal Medicine and Pharmacology at the University of Iowa College of Medicine. His research deals with the biology of the carotid artery and cerebral circulation with a focus on defining mechanisms that produce vascular dysfunction in disease states. He is an Established Investigator of the American Heart Association and has received the Lamport Award from the American Physiological Society. He is Assistant Editor for Stroke, and Associate Editor for Arteriosclerosis, Thrombosis and Vascular Biology, serves on numerous editorial boards, and is the lead investigator on the Cerebral Vascular Biology Program at the University of Iowa.

Dr. Heistad is Zahn Professor in the Departments of Internal Medicine and Pharmacology at the University of Iowa and staff physician at the VA Medical Center. A major focus of Dr. Heistad's laboratory is physiology and pathophysiology of cerebral blood vessels. He has received the Wiggers Award from the American Physiological Society, the Landis Award from the Microcirculatory Society, and the Research Achievement Award from the American Heart Association. He is Editor-in-Chief of *Arteriosclerosis, Thrombosis, and Vascular Biology*. His work is supported by grants from the NIH and VA.

CHAPTER 60

Model Systems for Studying Blood–Brain Barrier Function

K. L. Hallene, G. Dini, and D. Janigro
The Cleveland Clinic Foundation, Cleveland, Ohio

The blood–brain barrier (BBB) has been described in detail over the years, both physiologically and morphologically [1]. At the cellular level, the BBB is composed of glia and microvascular endothelial cells (ECs). The latter are characterized by the presence of tight junctions, lack of fenestrations, and minimal pinocytotic vesicles. ECs, together with astrocytes, form a selectively permeable barrier protecting the brain from systemic influences, while still providing pathways of transport for nourishment to neurons in the parenchyma. Today, it is generally accepted that movement of ions and molecules across the BBB is restricted due to endothelial cell tight junctions. However, under certain conditions, some toxins may actually increase BBB permeability, allowing entry of potentially noxious substances into the brain. It is therefore recognized that the blood–brain barrier plays a crucial role in the determination of neurotoxicity and its prevention by specific transport mechanisms [2].

Because of the anatomical and topographic obstacles associated with the direct investigation of BBB in vivo, its structure and physiological properties are often inferred from studies of isolated microvessels and primary cultures of brain microvascular endothelial cells. This is true for both the normal and disease states.

Experimental observations first made over a century ago showed that the central nervous system (CNS) is not stained by intravascular water-soluble dyes, providing the first demonstration of the existence of a barrier to the passage of polar compounds from the blood to the brain. Pioneering studies of the BBB were performed in vivo using intracarotid injection single-pass techniques. Further characterization of the BBB at the cellular level more recently has led to the development of in vitro experimental approaches. Isolated brain microvasculature preparations, as well as tissue culture systems using brain endothelial cells, have proven to be a promising methodology to define the characteristics of the brain capillary endothelium at the molecular and cellular level. The purpose of this review is to describe the virtues and pitfalls of cell culture-based models of the BBB.

Monodimensional Models of the Blood–Brain Barrier: Endothelial Cells

The most commonly used model to study the blood–brain barrier in vitro stems from a rather simplified view of the BBB and consists of the so-called "Transwell system," where cells are grown on porous membranes and immersed in growth media (Figure 1). A variety of cells have been successfully grown under these conditions, including, but not limited to, endothelial and epithelial cells from rodents, primates, or human origin. The most attractive features of this model are its simplicity and the ability to perform several experiments at the same time. Many Transwells can be simultaneously loaded with cells or cell lines, and experiments can therefore be rapidly performed, minimizing the cost and need for an experimenter's skills. However, the major pitfalls of this system are that cells are grown in the absence of physiological stimuli, such as perivascular glia, shear stress, or interaction with blood cells present in the cerebral circulation under normal conditions, develop an abnormal permeability across the EC layer. Another limitation of this system is that cells are grown in the presence of serum, on both the luminal and abluminal sides. Conversely, in vivo, only the luminal side is exposed to serum proteins and the abluminal part is exposed to glial influences or cere-

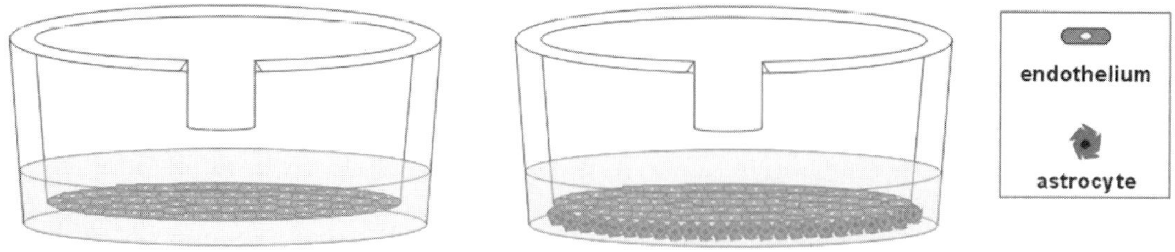

Figure 1 Schematic representation of the Transwell system. Either monodimensional endothelial cultures (*left*) or bidimensional endothelial-astrocyte coculture can be established (*right*). (see color insert)

Table I Advantages and Disadvantages of Each Model Used to Study the BBB in Vitro.

	Monodimensional (Transwell)	Bidimensional (coculture)	Tridimensional (DIV-BBB)
Pros	Ease of use Ideal for Michaelis-Menten kinetics of transport due to fixed volumes in each compartment Can do several experiments at same time Minimal skills Cost efficient	Coculture allows induction of BBB properties Molecule flux across is lower than monodimensional Cost efficient	Artificial capillaries connected by gas-permeable tubing allows source of growth medium, exchange of O_2+CO_2, exposure to flow Induction of BBB properties High TEER Low permeability to sucrose Stereoselective transport Can mimic intravenous/intra-abdominal delivery versus administration by mouth Long term studies possible Presence of drug extrusion mechanisms
Cons	Lacks physiological flow High permeability to sucrose Low TEER Exposure to serum on both sides (luminal + abluminal)	Lacks physiological flow High permeability to sucrose Low TEER (to a lesser extent)	Linear kinetic studies more difficult, because of flow Cost Skills Number of cells to be loaded/used

brospinal fluid. In conclusion, monodimensional cultures based on endothelial cells lack true physiological "barrier function." However, Galla and his colleagues have shown that with the addition of hydrocortisone to serum, the barrier function greatly improves [3]. The physiopathological significance of these findings is unclear, but the modification provided by these investigators greatly improved the usefulness of these models.

Bidimensional Models of the Blood–Brain Barrier: Co-culture of Endothelial Cells and Glia

Bidimensional models of the BBB have been created through the addition of glia to the Transwell system with various levels of success (Figure 1). In doing so, investigators were able to mimic at least one of the fundamental aspects of the BBB, that is, glial–endothelial interactions. It has been well known for some time that glial cells exert a yet unknown influence on their neighboring endothelial counterparts. The nature and significance of this interaction is poorly understood, but it is clear however, that at least in vitro these cells do interact and that glial cells induce BBB properties in endothelial cells. Interestingly, the efficacy of glia to induce BBB properties in endothelial cells is not limited to cerebral endothelial cells since investigators were able to induce BBB properties in nonbrain endothelial cells as well [4].

The main advantage of the bidimensional model, compared to monodimensional, is the attempt to closely mimic the in vivo situation. Thus, the addition of glial cells has led to a decrease in transendothelial leakage. However, this model may also experience some "nonphysiological leak" as well, thus making it more difficult to predict or estimate permeation across the BBB of pharmacological agents or other molecules. Other than those already stated, there are no obvious pitfalls, besides that a source of glial cells is essential, which slightly increases the required skills of the investigator/scientist (Table I).

Tridimensional Dynamic Models of the Blood–Brain Barrier (Flow-Based)

Both short- and long-term changes occur in cerebral arterioles in response to intraluminal flow [5]. Endothelial cells in vivo are continuously exposed to shear stress generated by blood flow. A new in vitro model of the BBB has recently been developed and is characterized by a tridimensional, pronectin-coated hollow fiber structure that enables coculturing of ECs with glia (Figure 2) [6]. In the hollow-fiber apparatus, ECs are seeded intraluminally and are exposed to

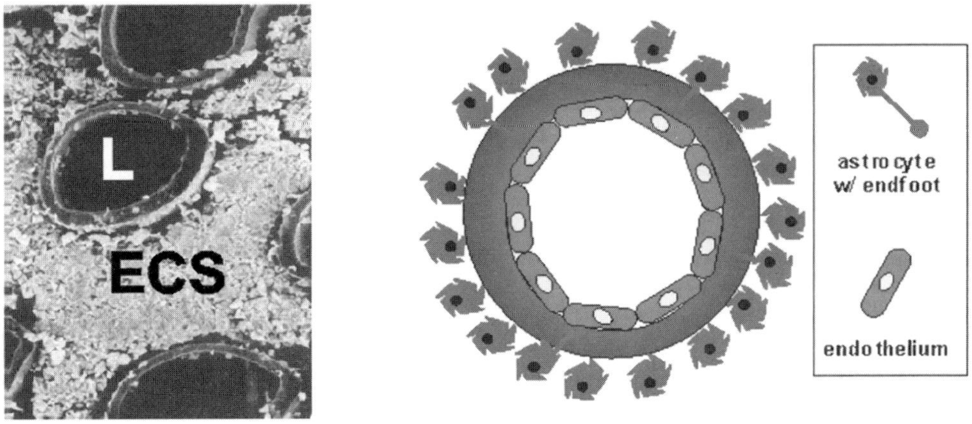

Figure 2 Tridimensional in vitro model of the BBB: Endothelial cells are seeded intraluminally in a pronectin-coated hollow fiber, while astrocytes are cultured extraluminally. *Left panel*: Hollow fiber section (L, lumen; ECS, extracellular space); *right panel*: schematic representation. (see color insert)

flow. Under these conditions, endothelial cells develop a morphology that closely resembles the endothelial phenotype in situ, demonstrating that endothelial cells that are grown under flow conditions develop a greater differentiation than conventional culture. We reported the induction of BBB properties in endothelial cells grown in hollow fibers in the presence of extraluminally seeded glia; this induction of a BBB-specific phenotype included low permeability to intraluminal potassium, negligible extravasation of proteins, and the expression of a BBB-specific glucose transporter. In addition, coculturing of EC with glia affected the overall morphology of these cells and induced the expression of BBB-specific ion channels [6, 7].

The model system used for these studies resulted from a modification of a traditional cell culture system normally used for expansion of hybridoma cells, and the general design of the hollow-fiber apparatus was derived from attempts toward the development of a "cell factory." Cell culture on hollow fibers has been extensively exploited for mass production of rare cell types, antibody production, and modeling of organlike structures such as the blood–brain barrier. Ott et al. [8] used a hollow fiber cell culture apparatus for studies of flow-mediated effects on endothelial cell growth, while Stanness and her colleagues [4] further developed this system by performing permutations of intra- and extraluminal cell growth to study the effects of glia on endothelial cells.

The dynamic in vitro BBB model (DIV-BBB) is constituted by a plastic support containing a variable number of artificial capillaries (approximately 300 μm cross diameter with a lumen of approximately 75 μm). These capillaries bear 0.5-μm transcapillary pores that allow free diffusion of solutes from the extraluminal compartment to the intraluminal space and vice versa. The capillaries are intraluminally perfused at various shear stress rates by pulsatile flow (Figure 3). It was shown that induction of BBB-like characteristics occurs following prolonged coculture of glia and bovine

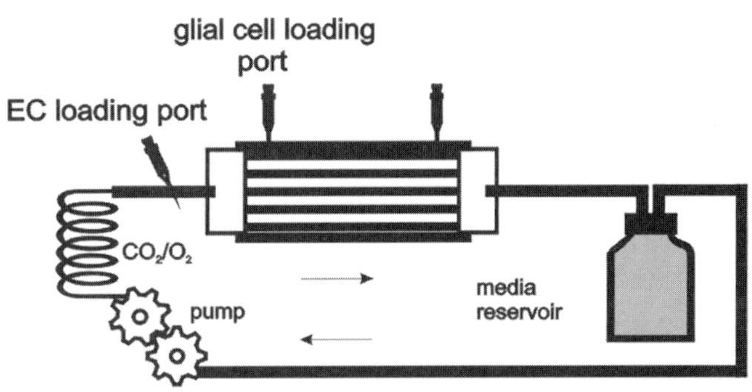

Figure 3 Schematic representation of the tridimensional dynamic in vitro model (DIV-BBB); cells are injected into the system through the loading ports and the flow is maintained by a pump. (see color insert)

aortic endothelial cells and that glia can induce the expression of BBB-specific ion channel proteins in non-BBB endothelial cells [6]. Interestingly, genetically altered astrocytes lacking intermediate filaments were not capable of these induction properties [9].

Several improvements in model design have allowed for better understanding and study of the central nervous system and neurological diseases. This model in particular has several advantages compared to mono- or bidimensional models, such as closely mimicking the conditions of pre- and postcapillary vessels in vivo, and allowing for experimental manipulations and reutilization of the device. In this model, high transendothelial electrical resistance (TEER), low permeability to sucrose, and stereoselective transport also prevail. It is also possible to perform long-term studies (months), and the presence of drug extrusion mechanisms allows for meaningful drug permeation studies. However, there are obvious limitations to this model, including, but not limited to, lack of blood cells, no intraluminal pressure changes, and a lack of neuronal influences. Some of these issues have been addressed recently by Krizanac-Bengez and co-workers [10].

Table II Advantages and Disadvantages of Cell Lines versus Primary Human Cultures.

	Cell lines	Primary human cultures
Pros	Cost effective Easy to use Readily available	Variety of etiologies available Not manipulated Don't divide uncontrollably
Cons	Some conditions/diseases may be unable to be mimicked by these cells May become hyperproliferative or "immortalized"	Obtaining human specimens Skills Number of cells isolated

Cell Lines versus Primary Cultures

Once the "ideal model of the blood–brain barrier" has been designed, one must decide what kind of cells to use for a particular experiment. There are no definitive rules, but one must keep in mind that some conditions are only present in a particular species, as, for example, in the case of humans. However, experiments with dynamic modeling have shown that one of the initiating factors in endothelial cell differentiation consists of cell cycle arrest induced by flow [11]. It is therefore imperative, whatever cells one may decide to use, that these cells themselves be capable of being distracted from the cell cycle by environmental cues.

In fact, if cell lines are to be used, one has to carefully assess their tendency to become either hyperproliferative or immortalized since both conditions will make their differentiation into a mature endothelial phenotype either hard or impossible (Table II). Obviously, cell lines offer numerous advantages, including cost-effectiveness and ease of use. However, with the advent of stem cell research, and our increasing understanding of how blood cells differentiate into endothelial cells, one can envision a future where differentiating factors will be used to induce blood–brain barrier properties in omnipotent stem cells.

Many factors must be taken into consideration if the desire exists to use primary human cultures (Table II). The availability and ease of obtaining human specimens to isolate cells can be a challenging obstacle, unless experiments are performed in proximity to a hospital or similar environment where specimens are readily available. The numbers of cells to be isolated are crucial to running a successful experiment and therefore increase the techniques and time required of the scientist. Although these pitfalls seem major, the benefits of using primary human cultures far outweigh the costs. The main advantages in using primary human cultures are that these cells are not manipulated, allowing the study of the actual disease and its progress. Also, cells can be manipulated as needed for any given experiment. Obtaining human specimens provides a variety of etiologies that otherwise may not be available in cell lines or commercial products. Under normal conditions, primary cells also tend not to divide excessively or take on an "immortal" phenotype, as often occurs in cell lines.

In Vitro Models of the Human Blood–Brain Barrier

The main goals of in vitro studies of the blood–brain barrier are to reproduce as many aspects of the physiology and biology of the BBB in vivo as possible and to understand human disease. Therefore, it is imperative to stress the need for human cell-based models since several pathological conditions cannot be reproduced in cell-line-derived or rodent-based BBB models.

The current understanding of brain pathophysiology has led to the hypothesis that numerous diseases of the CNS are associated with failure of BBB structural integrity or function, as an altered BBB permeability may be observed in several diseases—epilepsy, ischemia, trauma, etc.

Extravasation of plasma proteins with BBB dysfunction can occur through any number of different transcellular or paracellular routes, including altered tight junctions, disruption of EC membranes, or formation of transendothelial channels. These pathways, however, are not mutually exclusive. Regardless of whether BBB disruption or consequences of the disease(s) themselves are the main factors leading to dysfunction or changes in structural integrity, the understanding of the mechanisms of the BBB is limited. This is perhaps due to the lack of comprehensive BBB models.

Conclusions

This chapter has described several properties of BBB models along with the advantages and disadvantages of each apparatus or technique. In the future, new and improved models may be invented and contribute to a better understanding of the etiology of both chronic and acute neurodegenerative brain disorders and disease. In conclusion, an individual cannot clearly define the ideal blood–brain barrier model, since in vivo approaches have obvious pitfalls, and in vitro models are pliable, yet different from in vivo conditions. Given that the ideal model does not exist, caution and skill should be exercised to mimic as closely as possible the specific condition that is to be studied. For example, the use of newborn human cells is a poor model of Alzheimer dementia.

Glossary

Blood–brain barrier (BBB): A membrane characterized by tight junctions, presence of glia, lack of fenestrations, and minimal pinocytotic vesicles separating the brain tissue from circulating blood and possible harmful substances.

Cell culture: A technique for growing cells outside of the organism in nutrient medium under laboratory conditions.

In vitro: In an artificial environment outside the organism.

In vivo: In an environment inside the organism.

Stem cell: A cell that has the ability to divide (self-replicate) for indefinite periods, often throughout the life of the organism. Under appropriate conditions, stem cells can give rise (differentiate) into many different cell types that compose the organism.

References

1. Rubin, L. L., and Staddon, J. M. (1999). The cell biology of the blood–brain barrier. *Annu. Rev. Neurosci.* **22**, 11–28. *There is an ever-increasing number of review articles on various aspects of blood–brain barrier function. This article will appeal to both the expert and the non-BBB savvy. The main aspects of BBB physiology are clearly outlined and current controversies are addressed in a balanced framework.*
2. Schinkel, A. H. (1999). P-Glycoprotein, a gatekeeper in the blood–brain barrier. *Adv. Drug Deliv. Rev.* **36**, 179–194. *This important article describes in detail the molecular and physiologic mechanisms that regulate the passage of xenobiotics into the brain. While a more recent set of review articles has expanded our view on these "gatekeepers," P-glycoprotein remains the archetype molecule on which our knowledge of drug extrusion at the BBB hinges.*
3. Hoheisel, D., Nitz, T., Franke, H., Wegener, J., Hakvoort, A., Tilling, T., and Galla, H. J. (1998). Hydrocortisone reinforces the blood–brain properties in a serum free cell culture system. *Biochem. Biophys. Res. Commun.* **247**, 312–315.
4. Stanness, K. A., Westrum, L. E., Fornaciari, E., Mascagni, P., Nelson, J. A., Stenglein, S. G., Myers, T., and Janigro, D. (1997). Morphological and functional characterization of an in vitro blood–brain barrier model. *Brain Res.* **771**, 329–342. *Many of the aspects of this mini-review are introduced in this early paper describing the advantages of dynamic modeling of the BBB.*
5. Yuan, Y., Granger, H. J., Zawieja, D. C., and Chilian, W. M. (1992). Flow modulates coronary venular permeability by a nitric oxide-related mechanism. *Am. J. Physiol.* **32**, H641–H646.
6. Cucullo, L., McAllister, M., Kight, K., Krizanac-Bengez, L., Marroni, M., Mayberg, M., Stanness, K., and Janigro, D. (2002). A new dynamic in vitro model for the multidimensional study of astrocytes–endothelial cell interactions at the blood–brain barrier. *Brain Res.* **951**, 243.
7. McAllister, M. S., Krizanac-Bengez, L., Macchia, F., Naftalin, R. J., Pedley, K. C., Mayberg, M. R., Marroni, M., Leaman, S., Stanness, K. A., and Janigro, D. (2001). Mechanisms of glucose transport at the blood–brain barrier: An in vitro study. *Brain Res.* **904**, 20–30.
8. Ott, M. J., Olson, J. L., and Ballermann, B. J. (1995). Chronic *in vitro* flow promotes ultrastructural differentiation of endothelial cells. *Endothelium* **3**, 21–30.
9. Pekny, M., Stanness, K. A., Eliasson, C., Betsholtz, C., and Janigro, D. (1998). Impaired induction of blood–brain barrier properties in aortic endothelial cells by astrocytes from GFAP-deficient mice. *Glia* **22**, 390–400.
10. Krizanac-Bengez, L., Kapural, M., Parkinson, F., Cucullo, L., Hossain, M., Mayberg, M. R., and Janigro, D. (2003). Effects of transient loss of shear stress on blood–brain barrier endothelium: Role of nitric oxide and IL-6. *Brain Res.* **977**, 239–246.
11. Desai, S. Y., Marroni, M., Cucullo, L., Krizanac-Bengez, L., Mayberg, M. R., Hossain, M. T., Grant, G. G., and Janigro, D. (2002). Mechanisms of endothelial survival under shear stress. *Endothelium* **9**, 89–102.

Capsule Biography

Kerri Hallene, B.S., is a research technologist who has been involved in human and animal studies of the blood–brain barrier in vitro and in situ.

Gabriele Dini, Ph.D., is a biomedical engineer studying various aspects of shear stress and glial involvement in organ models including the blood–brain barrier.

Damir Janigro, Ph.D., is Professor of Molecular Medicine and Director of the Cerebrovascular Center at the Cleveland Clinic Lerner College of Medicine.

CHAPTER 61

Adrenomedullin and Blood–Brain Barrier Function

Béla Kis
Wake Forest University, Winston-Salem, North Carolina

Adrenomedullin (AM), a recently discovered vasodilator peptide, was originally isolated from human pheochromocytoma, and the initial reports suggested that the adrenal medulla, ventricle, kidney, and lung have the highest levels of AM expression. However, since the discovery that the AM gene expression is twentyfold to fortyfold higher in endothelial cells than even in the adrenal medulla, this peptide has been regarded as an important secretory product of vascular endothelium, together with nitric oxide, endothelin-1, and other vasoactive metabolites. Recent data from our laboratory indicate that adrenomedullin, as an endothelium-derived autocrine/paracrine hormone, plays an important role in the regulation of specific barrier properties of brain endothelial cells. In this chapter, we describe current knowledge about the possible role of AM in blood–brain barrier (BBB) functions.

AM, a Vasoactive Hormone

AM is a multifunctional peptide, which is present ubiquitously in the body. The normal plasma concentration of AM is in the range of 1 to 15 pmol/L, and AM is derived mostly from endothelial cells. Its plasma half life is about 20 minutes. It is degraded initially by metalloproteases followed by an aminopeptidase. The lungs are the main clearance site of AM in humans.

AM molecule shows some homology with calcitonin gene-related peptide (CGRP) and amylin, suggesting that AM belongs to the CGRP superfamily. AM consists of 52 amino acid residues in humans and 50 amino acid residues in rats. It has two characteristic structures, a ring structure formed by an intermolecular disulfide bridge and a C-terminal amide structure, which are essential for the receptor binding and the biological activity of AM.

The biological actions of AM are mediated via two specific adrenomedullin receptors, and at higher concentrations AM also can cross-react with the CGRP type 1 (CGRP1) receptor. Although the pharmacology of these receptors is distinct, the main component of both is the calcitonin-receptor-like receptor (CRLR). Whether CRLR functions as an AM or CGRP receptor is dependent on the presence of one of the three receptor activity modifying proteins (RAMP). When CRLR combines with RAMP1 it defines the CGRP1 receptor, whereas the combination of CRLR with either RAMP2 or 3 forms the AM receptors.

CRLR is a member of the type II G-protein coupled receptor family, and like other members of this receptor family it signals via Gs and adenylate cyclase activation. Increased intracellular cyclic adenosine 3′,5′-monophosphate (cAMP) level is the usual result of AM challenge. Intracellular calcium concentrations appear to be unaffected by AM in many different cell types expressing endogenous AM receptors, except endothelial cells where AM can also increase intracellular calcium concentration. When injected intravenously, AM acts predominantly in organs in which the AM gene is highly expressed. These observations suggest a local autocrine and/or paracrine role for AM.

AM has been shown to have a wide range of physiological effects. It is implicated in the regulation of fluid/electrolyte homeostasis, insulin secretion, and glucose metabolism. AM has an important role in pregnancy and fetal growth, inhibits apoptosis, and regulates cell proliferation.

The major role of AM is in the paracrine control of vascular function, supported by the high levels of AM secreted by endothelial cells and vascular smooth muscle cells. AM is a powerful hypotensive peptide. Its vasodilator effects are mediated in part by an elevation of cytoplasmic cAMP leading to relaxation of vascular smooth muscle cells. In addition, AM acts on endothelial cells; it stimulates nitric oxide production via calcium-dependent activation of endothelial nitric oxide synthase, which also contributes to vasodilatation. AM regulates vascular smooth muscle cell proliferation, inhibits endothelial apoptosis, promotes angiogenesis, and regulates blood coagulation and fibrinolysis. Transgenic overexpressing and knockout models further emphasize that AM is crucial to vascular morphogenesis and function.

Numerous studies defining the circulating AM plasma concentration in different clinical settings showed an elevated AM plasma concentration in a variety of pathological conditions including essential hypertension, chronic heart failure, acute myocardial infarction, pulmonary hypertension, chronic renal failure, diabetes, hyperthyroidism, hyperparathyroidism, sepsis, schizophrenia, autism, bipolar affective disorders, surgical interventions, or during normal pregnancy. Because such a broad range of conditions are associated with elevated AM levels, it seems likely that increases in AM are not causative of disease but rather are compensatory to these pathological events. The results from transgenic mice overexpressing AM gene support the idea of a protective role for AM.

Role of AM in the Microcirculation

Endothelial cells have been shown to contain an elaborate microfilament system allowing active actin- and myosin-based cell contraction. In peripheral endothelium, activation of cell contraction and disturbance of junctional organization subsequently result in the induction of interendothelial gaps followed by enhanced paracellular endothelial permeability. An increased intracellular cAMP level in endothelial cells has been known to decrease the basal permeability properties of the barrier and attenuate the increase in permeability when the endothelial cells are exposed to inflammatory agents. AM, as an autocrine/paracrine hormone, is a potent stimulator of cAMP elevation in endothelial cells. AM can stabilize the barrier function of endothelial cells by a cAMP-dependent relaxation of the microfilament system, thereby preventing endothelial cell contraction and paracellular fluid flux.

Results on transgenic mice show the importance of AM in the regulation of microcirculation. Mice overexpressing AM in their vasculature have turned out to be resistant to lipopolysaccharide-induced shock. On the other hand, AM knockout homozygous mice die at mid-gestation with extreme hydrops fetalis and cardiovascular abnormalities including severe hemorrhages and pericardial effusions. These suggest a general role for AM as an endothelium-derived autocrine hormone in the regulation of endothelial permeability.

AM Production in the Cerebral Microcirculation

Cerebral endothelial cells (CECs) are the most important anatomical and functional constituents of the cerebral microcirculation. The finding of high AM production in peripheral endothelial cells made it probable that CECs could be an important source of AM in the cerebral microcirculation. However, the first report did not support this hypothesis, because Sugo et al. [1] found very low AM production by cultured bovine CECs, which was only a few percent of the AM production of rat aortic endothelial cells. Later this observation was supported by Ladoux and Frelin [2], who described weak AM mRNA expression in clones of rat CECs. It should be mentioned, however, that CECs were passaged a large number of times (10 to 20 passages) in both studies, which is known to deteriorate the original phenotype of the primary cells.

In contrast to these experiments we have recently found unexpectedly high AM production in primary cultures of rat CECs both at the peptide and at the mRNA levels [3]. Rat CECs had about one magnitude higher AM production than was published for other primary cells. Thus, available data indicate that rat CECs have the highest rates of AM synthesis and secretion among the cells studied.

The high AM production of rat CECs was further induced by astrocyte-derived factors; significantly elevated AM production was detected in the culture medium of primary rat CECs cocultured with astrocytes or cultured in astrocyte-conditioned medium. These results suggest that the in vivo AM production may be even higher than that detected in vitro. AM production by rat CECs, however, could not be induced by cytokines, bacterial lipopolysaccharide, and thrombin, which are the most powerful inducers of AM release in peripheral endothelial cells. Our studies also revealed that AM is secreted primarily but not exclusively at the luminal (blood) side of CEC monolayers. Contrary to AM, endothelin-1, a vasoconstrictor peptide, is secreted mostly toward the abluminal (brain) side of bovine CECs.

In addition to CECs, other cellular elements of the cerebral vessels can produce AM, that is, vascular smooth muscle cells, pericytes, or even astrocytes and neurons. All of these cells, however, are on the brain side of the BBB; therefore AM production by both these cells and the choroid plexus seems more likely to contribute to the AM level of the cerebrospinal fluid. AM concentration in the cerebrospinal fluid is comparable to or lower than that in plasma, and an independent regulation of AM level in these two compartments was suggested.

In vivo an approximately 50 percent higher AM concentration was found in the jugular vein than that in the carotid artery, and similarly about a 50 percent higher AM concentration was detected in the venous plasma effluxed from the brain than that from peripheral organs [3]. Previously no significant difference was observed between AM levels in the venous blood from different organs.

These observations strongly support the in vitro findings that the cerebral circulation has an exceptionally high AM concentration due to the significantly elevated basal AM secretion by brain endothelial cells that is induced by astrocyte-derived factors.

AM Receptors in the Cerebral Microcirculation

The fact that AM behaves as a local autocrine and/or paracrine hormone raises the possibility that AM released by CECs acts primarily on AM receptors present in the cerebral endothelium itself and on neighboring cells. The expression of CRLR, RAMP-1, -2, -3, was characterized on isolated rat cerebral microvessels [4], in rat CECs and cerebral pericytes [3], and in rat astrocytes [2]. In cultured rat CECs and pericytes RAMP-2 showed the highest expression, followed by RAMP-3 and RAMP-1, and exogenous AM increased the intracellular cAMP concentration suggesting the existence of functional AM receptors on these cells [3]. Although astrocyte-derived factors increased the AM production of rat CECs, they did not change the expression of AM receptor components in rat CECs. Oliver et al. [5] reported the same expression pattern of RAMPs in human cerebral vasculature.

Role of AM in Blood–Brain Barrier Functions

Among the endothelial cells of different tissues, the endothelium of the brain is very specific. Although cerebral endothelial cells (CECs) share many common properties of the peripheral endothelium, they have a unique morphological and functional feature, the formation of the BBB. The BBB contributes to the stability of the brain parenchymal microenvironment by strictly controlling the traffic of molecules and cells between the blood and the central nervous system. The BBB has specific, epithelial cell-like morphological characteristics and physical barrier properties, such as tight intercellular junctions, the absence of intercellular clefts, low rate of pinocytosis, many mitochondria, continuous basement membrane, or high transendothelial electrical resistance. Specific, polarized, highly discriminatory membrane transport systems, such as glucose transporter-1, nucleoside transport system, or large neutral amino acid transporter, operate to supply the brain. Efflux transporters, such as P-glycoprotein or members of multidrug-resistance protein family (MRP-1, -3, -4, -5, -6), can eliminate toxic substances from endothelial compartments and brain. The BBB phenotype of CECs is induced by the astroglial environment. Although the effect of neurons on the induction of some specific CEC properties have been described, the effect of astroglia on CECs in vitro has been more extensively studied.

The involvement of cAMP in the regulation of BBB functions has long been known; for example, cAMP elevates transendothelial electrical resistance and decreases paracellular permeability, reduces the rate of fluid-phase endocytosis, and increases P-glycoprotein function. It has been proved that AM has cAMP-like effects on specific BBB functions in vitro [6]. Exogenous AM increased transendothelial electrical resistance and reduced endothelial permeability for the low-molecular-weight sodium fluorescein, which suggests a tightening of intercellular junctions. AM also decreased endothelial fluid-phase endocytosis, a type of nonspecific endothelial transport, although it did not change the transendothelial albumin permeability. AM activated the P-glycoprotein efflux pump, while it did not change the activity of MRP-1 efflux transporter in cultures of rat CECs. Treatment with either the AM receptor antagonist AM_{22-52} or the AM antisense oligonucleotide decreased the basal intracellular cAMP level in rat CECs. Michibata et al. [7] have reported that neutralization of endogenous AM by monoclonal antibodies reduced the basal cAMP production in bovine aortic endothelial cells. In primary rat CECs antisense treatment significantly reduced the AM production and decreased transendothelial electrical resistance [6]. It is remarkable that the basal intracellular cAMP concentration is the highest in rat CECs followed by GP8-immortalized rat CECs and human umbilical vein endothelial cells, which corresponds to the AM production of these cells. These observations suggest that AM, as an autocrine mediator, plays an important role in the maintenance of basal intraendothelial cAMP levels, and AM appears to be an autocrine inducer of BBB functions of CECs via the activation of adenylate cyclase enzyme. Moreover, astrocyte-derived factors increased the AM production by primary rat CECs, suggesting that AM is involved in the astrocytic regulation of the BBB phenotype. Therefore, AM can be a physiological link between astrocyte-derived factors, cAMP, and the induction and maintenance of the BBB properties by rat CECs (Figure 1).

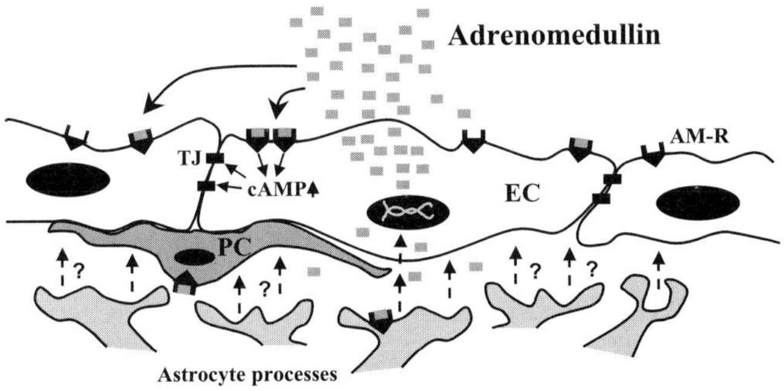

Figure 1 Adrenomedullin and the BBB. Adrenomedullin production in cerebral endothelial cells is stimulated by unknown astrocyte-derived factors. The synthesized adrenomedullin is secreted mostly at the luminal surface of the cerebral endothelial cells into the blood, but also at the abluminal surface. The released peptide acts as an autocrine/paracrine hormone and stimulates cAMP production in the cells. The elevated intracellular cAMP will improve the barrier functions of the cerebral endothelial cells. EC, cerebral endothelial cell; PC, pericytes, TJ, tight interendothelial junction; AM-R, adrenomedullin receptor.

Summary

AM seems to be an important hormone in the cerebral circulation. The concentration of AM is approximately 50 percent higher here than in other regional circulations because of the astrocyte-induced elevated AM production of CECs. AM causes vasodilation in the cerebral circulation and may be important in the maintenance of the resting cerebral blood flow and protective against ischemic brain injury. Recent data from our laboratory indicate that AM, as an endothelium-derived autocrine/paracrine hormone, plays an important role in the regulation of specific BBB properties. AM can be one of the physiological links between astrocyte-derived factors, cAMP, and the induction and maintenance of the BBB. Moreover, the role of AM in the differentiation and proliferation of peripheral endothelial cells and in angiogenesis suggests a more complex function for AM in the cerebral circulation and in the development of the BBB.

Glossary

Gs: A G protein that stimulates adenylate cyclase, leading to increased intracellular cAMP level.

P-Glycoprotein: A 170-kDa integral plasma membrane glycoprotein that functions as an adenosine triphosphate–dependent efflux pump, causing multidrug resistance.

Pheochromocytoma: Hormone-producing (mostly catecholamine) neoplasm of the adrenal medulla.

References

1. Sugo, S., Minamino, N., Kangawa, K., Miyamoto, K., Kitamura, K., Sakata, J., Eto, T., and Matsuo, H. (1994). Endothelial cells actively synthesize and secrete adrenomedullin. *Biochem. Biophys. Res. Commun.* **201**, 1160–1166.
2. Ladoux, A., and Frelin, C. (2000). Coordinated Up-regulation by hypoxia of adrenomedullin and one of its putative receptors (RDC-1) in cells of the rat blood-brain barrier. *J. Biol. Chem.* **275**, 39914–39919.
3. Kis, B., Kaiya, H., Nishi, R., Deli, M. A., Ábrahám, C. S., Yanagita, T., Isse, T., Gotoh, S., Kobayashi, H., Wada, A., Niwa, M., Kangawa, K., Greenwood, J., Yamashita, H., and Ueta Y. (2002). Cerebral endothelial cells are a major source of adrenomedullin. *J. Neuroendocrinology* **14**, 283–293.
4. Kobayashi, H., Minami, S., Yamamoto, R., Masumoto, K., Yanagita, T., Uezono, Y., Tsuchiya, K., Mohri, M., Kitamura, K., Eto, T., and Wada, A. (2000). Adrenomedullin receptors in rat cerebral microvessels. *Brain. Res. Mol. Brain. Res.* **81**, 1–6.
5. Oliver, K. R., Wainwright, A., Edvinsson, L., Pickard, J. D., and Hill, R. G. (2002). Immunohistochemical localization of calcitonin receptor-like receptor and receptor activity-modifying proteins in the human cerebral vasculature. *J. Cereb. Blood. Flow. Metab.* **22**, 620–629.
6. Kis, B., Deli, M. A., Kobayashi, H., Ábrahám, C. S., Yanagita, T., Kaiya, H., Isse, T., Nishi, R., Gotoh, S., Kangawa, K., Wada, A., Greenwood, J., Niwa, M., Yamashita, H., and Ueta Y. (2001). Adrenomedullin regulates blood-brain barrier functions in vitro. *Neuroreport* **12**, 4139–4142.
7. Michibata, H., Mukoyama, M., Tanaka, I., Suga, S., Nakagawa, M., Ishibashi, R., Goto, M., Akaji, K., Fujiwara, Y., Kiso, Y., and Nakao, K. (1998). Autocrine/paracrine role of adrenomedullin in cultured endothelial and mesangial cells. *Kidney Int.* **53**, 979–985.

Further Reading

Caron, K. M., and Smithies, O. (2002). Multiple roles of adrenomedullin revealed by animal models. *Microsc. Res. Tech.* **57**, 55–59. *A review of the adrenomedullin transgenic animal models.*

de Boer, A. G., van der Sandt, I. C. J., and Gaillard, P. J. (2003). The role of drug transporters at the blood–brain barrier. *Annu. Rev. Pharmacol. Toxicol.* **43**, 629–656.

Hinson, J. P., Kapas, S., and Smith, D. M. (2000). Adrenomedullin, a multifunctional regulatory peptide. *Endocr. Rev.* **21**, 138–167. *The best available detailed review of adrenomedullin research.*

Hippenstiel, S., Witzenrath, M., Schmeck, B., Hocke, A., Krisp, M., Krull, M., Seybold, J., Seeger, W., Rascher, W., Schutte, H., and Suttorp N. (2002). Adrenomedullin reduces endothelial hyperpermeability. *Circ. Res.* **91**, 618–625. *This paper demonstrates a role for adrenomedullin in the regulation of permeability of peripheral endothelial cells.*

Rubin, L. L., and Staddon, J. M. (1999). The cell biology of the blood–brain barrier. *Annu. Rev. Neurosci.* **22,** 11–28.

Capsule Biography

Béla Kis earned his M.D. in 1994 and his Ph.D. in 1998 at Albert Szent-Györgyi Medical University, Szeged, Hungary. He spent 2 years as postdoctoral research fellow at the University of Occupational and Environmental Health, Kitakyushu, Japan, where most of his adrenomedullin-related work was accomplished. Now he is a postdoctoral research fellow at Wake Forest University, Winston-Salem, North Carolina. His research interest is vasoactive mediators and blood–brain barrier function as well as neuronal preconditioning.

CHAPTER 62

Cerebral Microvascular Blood Oxygenation and Neuronal Activity

Ute Lindauer
Experimental Neurology, Charité, Berlin, Germany

Introduction: Definition of the Subject

A tight temporal and spatial relationship exists among neuronal activity, metabolism, and blood flow in the brain. Changes in neuronal activity induce localized changes in metabolism and regional cerebral blood flow (rCBF), a phenomenon termed *neurometabolic and neurovascular coupling*. Consequently, consumption and delivery of oxygen and glucose, the metabolic substrates that fuel the brain, change at the same time. Different temporal and spatial kinetics of oxygen consumption and blood flow lead to complex changes in regional cerebral blood oxygenation. This chapter presents a detailed description of the dynamic changes of blood flow and blood oxygenation during functional activation and relates these to our current knowledge on activation-induced changes of oxygen metabolism in the brain.

Cerebral Blood Oxygen Tension and Hemoglobin Oxygen Saturation at Rest

The brain is a highly oxidative organ. It consumes approximately 20 percent of the body's oxygen and 25 percent of the body's glucose supply. Compared to its small fraction of total body weight (only 2%), a disproportionately large percentage of oxygen and glucose is consumed in the brain. The supply of oxygen to the brain depends on the vasculature, and the brain receives approximately 15 percent of the cardiac output. By reversibly binding oxygen to hemoglobin located in erythrocytes, oxygen transport capacity within the vasculature is significantly increased compared to the amount of oxygen physically dissolved in plasma. Blood oxygen tension (Po_2), hemoglobin oxygen saturation (So_2), and blood flow of brain microvessels on the one hand, and tissue oxygen tension on the other hand are the basic parameters determining oxygen delivery to tissue. Under resting conditions, a tight but nonlinear relationship exists for brain interstitial oxygen tension and regional cerebral blood flow. The difference in oxygen partial pressure between capillary blood and the tissue determines the PO_2 gradient and, hence, the oxygen flux from microvasculature to nerve and glial cells.

In systemic arterial blood, Po_2 values of 80 to 90 mmHg result in an oxygen saturation of hemoglobin of 95 percent. Studies in barbiturate-anesthetized rats have shown that following brain passage Po_2 values are reduced to approximately 40 mmHg, leading to an oxygen saturation of hemoglobin of around 60 percent within the blood taken from the sagittal sinus, the venous outflow system of the brain. Blood oxygen tension decreases significantly from pial arteries to intracerebral arterioles, resulting in a comparably low Po_2 of approximately 60 mmHg in precapillary resistance vessels. Blood oxygen saturation within the capillary bed drops from approximately 80 percent (Po_2 60 mmHg) to 60 percent (Po_2 40 mmHg), and the major part of oxygen supply to brain tissue occurs within capillaries.

However, PO_2 on the venous side of the capillary tree further decreases in small venules. Surprisingly, oxygen saturation of approximately 55 percent in postcapillary venules is significantly lower than in the blood of larger intracerebral veins and the superior sagittal sinus [1]. This higher PO_2 in larger venules and veins compared to postcapillary venules can be explained by the well-known fact that arterial O_2 is lost to countercurrent venous exchange before reaching the tissue. In general it can be concluded that the bulk of oxygen delivery to brain tissue occurs at the level of the microcirculation, and not only capillaries but also small arterioles and venules significantly contribute to oxygen supply to brain tissue during normoxia.

Microvascular oxygenation strongly depends on erythrocyte flux within the vasculature. Whereas all capillaries in the brain are perfused by plasma, erythrocyte perfusion under resting conditions is characterized by a pronounced heterogeneity, suggesting that many capillaries are not maximally used. Rapid fluctuations and spatial heterogeneity of density and velocity of red blood cells occur within the capillary network. During global stimulation of cerebral blood flow following systemic hypercapnia, mean blood cell flux and velocity increase, capillary perfusion with red blood cells becomes more homogenous, and the number of poorly blood-cell perfused capillaries decreases. In addition there is evidence for a dilation of capillaries. "Capillary recruitment" within brain microcirculation exists only in the sense of a dynamic adjustment of red blood cell perfusion in continuously plasma-perfused capillaries, but not as opening or closing of capillaries [2].

Cerebral Microvascular Blood Flow Changes during Functional Activation

During increased neuronal activity, vasodilation occurs in the active brain areas. The concept of neurovascular coupling has already been postulated in the 19th century by Roy and Sherrington (for reference see [3]). From their observations Roy and Sherrington postulated that "unknown signals" during increased activity of neurons in the brain induce regional vasodilation and increase of blood flow. Despite extensive research these signals are still not fully understood. In addition, relevant aspects of the different temporal and spatial kinetics of cellular energy metabolism and oxygen consumption leading to the observed complex changes in regional cerebral blood oxygenation have remained enigmatic. Nevertheless, the vascular changes based on the phenomenon of neurometabolic and neurovascular coupling are widely utilized under experimental and clinical conditions by modern techniques providing noninvasive functional brain imaging, where they serve as surrogate markers for neural activity.

The regional CBF is spatially as well as temporally coupled to brain activity. Studies on neurovascular coupling are therefore mainly focused on changes within the cerebral microcirculation.

When neurons are further activated from resting state activity, within the first 1 to 2 seconds regional CBF increases. It reaches a peak response after 3 to 5 seconds, followed by a plateau somewhat lower than the peak response during ongoing neuronal activity, as shown for the cortical blood flow response during somatosensory stimulation of the whisker system in rats [4]. When stimulation stops, CBF returns to baseline within several seconds, and a transient small undershoot can irregularly be detected. A typical example of microvascular blood flow changes due to functional activation by 60 seconds (**A**) or 4 seconds (**B**) of whisker hair deflection in the anesthetized rat is provided in Figure 1.

Changes of red blood cell flow within single capillaries can be investigated by two-photon laser scanning imaging,

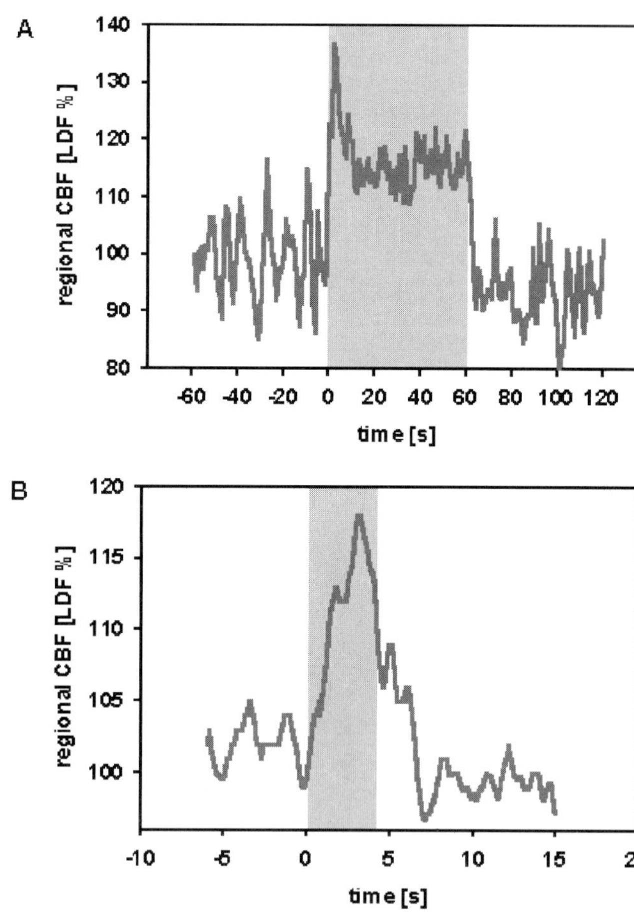

Figure 1 Typical example of microvascular blood flow changes due to functional activation by 60 seconds (**A**) or 4 seconds (**B**) of whisker hair deflection (3 Hz) in the anesthetized rat. Regional cerebral blood flow is measured by laser Doppler flowmetry (LDF); percentage changes from baseline flow at rest (100%) are presented. The stimulation period is indicated by the gray bar within each figure. The regional cerebral blood flow change is temporally tightly coupled to the somatosensory stimulation. Note that during prolonged stimulation (A) the peak blood flow increase is followed by a plateau response slightly lower than the peak response immediately after stimulation onset. No plateau response occurs when short stimulations are performed (B). (see color insert)

providing higher spatial resolution. In olfactory bulb glomeruli of anesthetized rats, it has been shown that during short periods of odor-evoked neuronal activation, a transient and reproducible increase in red blood cell flow in capillaries within the odor specific glomerulus occurs. This increase is largely caused by an increase in red blood cell velocity and only partially accompanied by an increase in linear density of red blood cells. Vascular responses occur as early as 1 to 2 seconds after the neuronal response within the glomerulus. In the olfactory bulb, the arteriolar network penetrating into glomeruli from the surface does not exhibit any precise organization that could provide an arteriolar-based regulation of the glomerulus-specific vascular responses to odors. These findings provide first functional evidence for the hypothesis of a local capillary regulation of microcirculatory blood flow within activated brain areas [5].

Cerebral Microvascular Blood Oxygenation Changes during Increased Neuronal Activity

Cerebral Microvascular Blood Oxygenation Changes Induced by the Regional Cerebral Blood Flow Response during Increased Neuronal Activity

Changes of hemoglobin oxygenation during neuronal activity can be investigated by methods measuring reflection changes followed by spectroscopic analysis of light received from the cortex illuminated with light in the visible wavelength range in experimental animals or in the near-infrared wavelength range in humans. The principle features of changes in oxy- and deoxyhemoglobin during increased brain activity is comparable in animals and humans. Counterintuitively, the stimulation-induced changes in oxy- and deoxyhemoglobin result in regional hyperoxygenation of the blood and, hence, an increase in hemoglobin oxygen saturation. Oxyhemoglobin concentration starts to increase within the first 1 to 2 seconds of stimulation onset and reaches its maximum 2 to 3 seconds later followed by a plateau during ongoing activity. When stimulation stops, oxyhemoglobin concentration returns to baseline within several seconds followed by a transient small undershoot. Deoxyhemoglobin concentration starts to decrease approximately 2 to 3 seconds after stimulation onset and peaks 2 to 3 seconds later. At the end of stimulation deoxyhemoglobin also returns to baseline followed by a transient overshoot. Beside an increase in hemoglobin oxygenation during functional activation, a significant increase in partial molecular oxygen pressure within the microcirculation can be detected. The time course of microvascular Po_2 changes exactly match that of oxyhemoglobin during increased neuronal activity [3, 6]. With neuronal activation the fractional change in regional CBF leading to blood hyperoxygenation is approximately twice as large as the fractional change in oxygen metabolism, and both are coupled through a linear relationship that does not depend on stimulus type. It can be suggested that oxygen delivery from brain microcirculation to tissue is limited at rest. Therefore a larger increase in CBF is required to support the smaller increase of oxygen metabolism, because the oxygen extraction fraction is significantly dropping with increased red blood cell velocity [7]. A typical example of microvascular blood oxygenation (**A**) and hemoglobin oxygen saturation (**B**) changes due to functional activation by whisker hair deflection in the anesthetized rat is provided in Figure 2.

Early Cerebral Microvascular Blood Oxygenation Changes during Increased Neuronal Activity

Despite intense research efforts a number of unresolved issues exist regarding functional activation-induced changes in oxygen consumption and delivery. Most investigators agree on the principal features of blood oxygenation changes more than 2 seconds after stimulation onset. However, early oxygenation changes, preceding the rCBF

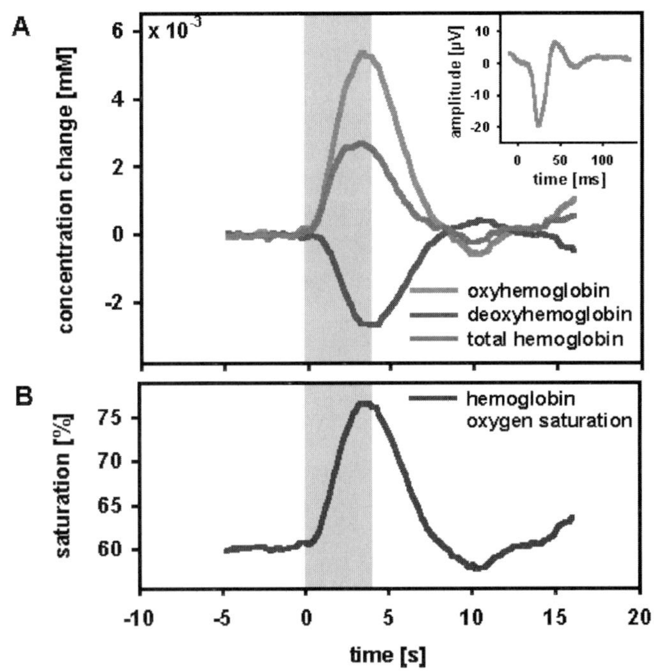

Figure 2 Typical example of microvascular blood oxygenation (**A**) and hemoglobin oxygen saturation (**B**) changes due to functional activation by 4 seconds of whisker hair deflection (3 Hz) in the anesthetized rat. Regional cerebral oxygenation changes are measured by microfiber spectroscopy (wavelength range 510 to 805 nm, wavelength-dependent differential path-length corrected analysis, estimated hemoglobin concentration 20 µM, mean saturation 60 percent as determined for microcirculatory areas). Total hemoglobin and hemoglobin oxygen saturation changes are calculated from measured oxy- and deoxyhemoglobin changes. Neuronal activity was recorded as somatosensory evoked potentials (SEP). The SEP (average of 4 seconds of 3-Hz stimulation) responsible for the hyperoxygenation response is provided in the small inset in part A of the figure. During functional activation, oxyhemoglobin increases and deoxyhemoglobin decreases, leading to microvascular hyperoxygenation and an increase in hemoglobin saturation. The blood oxygenation changes are temporally tightly coupled to the somatosensory stimulation. (see color insert)

response leading to hyperoxygenation, are controversial. The existence of an early hemoglobin deoxygenation within the first seconds after stimulation onset seems highly elusive. The phenomenon of an early deoxyhemoglobin increase has been termed *initial dip*, since it is detected as a decrease in the magnetic resonance imaging signal induced by an increase of deoxyhemoglobin, which has—in contrast to oxyhemoglobin—paramagnetic properties and can be used as an endogenous paramagnetic contrast agent. Whereas some studies have shown an early increase in deoxyhemoglobin preceding the onset of the CBF response [8, 9], other studies argue against the occurrence of an early loss of hemoglobin oxygenation and decrease in microvascular Po_2 that precede the rise in CBF [3, 10]. From further studies in cats and rats it can be concluded that the detection of an early dip mainly depends on the hemodynamic properties. Strong evidence exists for a correlation of the occurrence of an early increase of deoxyhemoglobin and the temporal onset of the regional CBF response, which very much parallels the oxyhemoglobin increase. Hypothetically, a lower baseline CBF and thus lower mean transit time of red blood cells through the cerebral microvasculature may favor a condition in which oxygen delivery lags oxygen extraction. In addition, a correlation of an early deoxyhemoglobin increase and the systemic arterial pco_2 may exist in humans as well as in animal experiments. The obvious variability of occurrence, amplitude, and rise time of the dip in experiments in cats and monkeys may also depend on the anesthetic and on the depth of anesthesia. Consequently, the early dip has a low overall contrast-to-noise ratio. Its occurrence seems to be highly susceptible to subtle changes of basal physiological condition, and it may thus not be detectable under physiological conditions. In principle, the existence of an early increase in deoxyhemoglobin at the onset of functional activation would suggest that early metabolic changes in the brain during increased neuronal activity are oxidative. An early desaturation of hemoglobin might support the hypothesis of oxygen tension–related mechanisms within the tissue triggering yet unknown signals to the vasculature to induce vasodilation. However, it is still unresolved whether increases in regional CBF and oxygen consumption are dynamically coupled following neural activation or occur in parallel and independent from each other.

Spatial Resolution of Cerebral Microvascular Blood Flow and Blood Oxygenation Changes during Neuronal Activation

During increased neuronal activity, early changes of regional CBF are more localized than later changes at time points when increased blood flow reaches the area of the draining veins. Deoxyhemoglobin concentration is larger in the venous compartment. Therefore, imaging techniques based on deoxyhemoglobin changes during the phase of hyperoxygenation are more affected by the draining vein effect than techniques using early oxyhemoglobin or regional CBF changes for brain mapping. However, as long as the mechanisms and mediators of neurovascular coupling are not known in detail, the question remains whether the area of increased blood flow exactly matches neuronally active areas of the brain or whether cerebral blood flow is controlled on a more coarse spatial scale.

Cerebral Microvascular Blood Oxygenation Changes during Increased Inhibition or Cortical Deactivation

As previously described, noninvasive functional imaging of the activated brain relies on the elevation of glucose and oxygen metabolism, blood flow, and oxygenation due to increased neuronal activity, leading to, for example, an increase of the blood oxygenation level-dependent (BOLD) MRI signal. Traditionally, functional activation is equated with increased excitatory synaptic activity, accompanied by an increase in energy metabolism and CBF and blood hyperoxygenation. On the other hand, reduced activity can be caused either by a decrease in excitatory input to the region of interest (deactivation) or by active synaptic inhibition within the region of interest. Active inhibition is based on the excitation of γ-aminobutyric acid (GABA)-releasing interneurons followed by the interaction of GABA, the main inhibitory neurotransmitter in the brain, with its target receptors. Whereas deactivation is accompanied by a decrease in energy metabolism and glucose consumption within the deactivated area, inhibition by activation of inhibitory interneurons is suppose to be associated with an increase in energy consumption. In line with these considerations, hypermetabolic areas can be defined as areas where excitatory as well as inhibitory neurons are activated. On the other hand, hypometabolic regions are regions with inactivation of both excitatory and inhibitory neurons.

Consequently, different patterns of cerebral microvascular blood oxygenation changes can be expected during active inhibition compared to deactivation. Until now no convincing data have been shown regarding the characteristics of cerebral microvascular blood flow and oxygenation changes during purely or predominantly inhibitory GABAergic neuronal activity. However, during deactivation of the visual cortex it was demonstrated that, compared to classical activation, inverse changes for the microvascular oxygenation parameters occur. A decrease in cortical neuronal activity is accompanied by a decrease in regional blood flow, a decrease in the concentration of oxyhemoglobin, and an increase in the concentration of deoxyhemoglobin. In addition, a decrease in BOLD MRI signal can be detected [6]. This finding implies that the transition from "deactivation" to "rest" to "activation" is a continuous equilibrium of cerebral microvascular blood flow and oxygenation.

Summary

Within the brain, the regional CBF is spatially as well as temporally coupled to neuronal activity. Increased neuronal activity is accompanied by an increase in CBF, whereas during deactivation CBF is reduced. After more than a century of research, we still do not fully understand the physiological meaning of the regional CBF response to changes in cerebral activity: Vascular coupling may provide a constant glucose and oxygen supply to the activated brain tissue—on the other hand, removal of tissue metabolites as lactate may be the driving force for vasodilation. In addition, it remains unclear which signals initiate the very early blood flow response, and which factors or mechanisms mediate the sustained increase in CBF during increased neuronal activity. The CBF response to functional activation or deactivation of brain tissue induces characteristic changes in hemoglobin oxygenation, leading to microvascular hyper- or hypo-oxygenation, respectively. Stimulation-induced metabolic and vascular responses are utilized to visualize human or animal brain at work and to map brain activity. Neurometabolic and neurovascular coupling thus form the physiological basis for modern functional brain imaging techniques such as functional MRI, near infrared spectroscopy, and optical imaging spectroscopy. Therefore we still need to better understand the physiology of neurometabolic and neurovascular coupling to fully exploit the potential of modern functional brain imaging.

Glossary

Capillary recruitment: Opening or closing of capillaries as a mechanism regulating microvascular blood flow according to demand.

Functional brain imaging: Investigation of stimulation-induced active brain areas by utilizing changes of metabolic or vascular parameters.

Initial dip: Early increase of the concentration of deoxyhemoglobin as evidence for early desaturation of hemoglobin and oxidative tissue metabolism at the onset of functional activation.

Neurovascular coupling: Tight temporal and spatial relationship between regional cerebral blood flow and neuronal activity.

References

1. Vovenko, E. (1999). Distribution of oxygen tension on the surface of arterioles, capillaries and venules of brain cortex and in tissue in normoxia: An experimental study on rats. *Pflugers Arch.* **437**, 617–623.
2. Villringer, A., Them, A., Lindauer, U., Einhäupl, K., and Dirnagl, U. (1994). Capillary perfusion of the rat brain cortex. An in vivo confocal microscopy study. *Circ. Res.* **75**, 55–62.
3. Lindauer, U., Royl, G., Leithner, C., Kuhl, M., Gold, L., Gethmann, J., Kohl-Bareis, M., Villringer, A., and Dirnagl, U. (2001). No evidence for early decrease in blood oxygenation in rat whisker cortex in response to functional activation. *NeuroImage* **13**, 988–1001. *In this publication it was shown that the method of analysis of spectroscopic data from optical imaging studies may have a considerable impact on the calculation of hemoglobin oxygenation changes under physiological as well as pathophysiological conditions.*
4. Lindauer, U., Villringer, A., and Dirnagl U. (1993). Characterization of CBF response to somatosensory stimulation: Model and influence of anesthetics. *Am. J. Physiol.* **264**, H1223–H1228.
5. Chaigneau, E., Oheim, M., Audinat, E., and Charpak, S. (2003). Two-photon imaging of capillary blood flow in olfactory bulb glomeruli. *Proc. Natl. Acad. Sci. USA* **100**, 13081–13086.
6. Wenzel, R., Wobst, P., Heekeren, H. R., Kwong, K. K., Brandt, S. A., Kohl, M., Obrig, H., Dirnagl, U., and Villringer A. (2000). Saccadic suppression induces focal hypooxygenation in the occipital cortex. *J. Cereb. Blood Flow Metab.* **20**, 1103–1110.
7. Buxton, R. B. (2001). The elusive initial dip. *NeuroImage* **13**, 953–958. *This article is an excellent commentary on the issue of the "initial dip," which integrates data from experimental studies with current models of the coupling of cerebral blood flow and oxygen metabolism.*
8. Malonek, D., and Grinvald, A. (1996). Interactions between electrical activity and cortical microcirculation revealed by imaging spectroscopy: Implications for functional brain mapping. *Science* **272**, 551–554. *In this publication on functional brain mapping, optical imaging spectroscopy is introduced as an experimental technique that offers high spatial, temporal, and spectral resolution. The data for the first time suggest that vascular responses within the first seconds after stimulation onset are significantly better localized than in the later phase of the vascular response.*
9. Mayhew, J., Zheng, Y., Hou, Y., Vuksanovic, B., Berwick, J., Askew, S., and Coffey, P. (1999). Spectroscopic analysis of changes in remitted illumination: The response to increased neural activity in brain. *NeuroImage* **10**, 304–326.
10. Silva, A. C., Lee, S. P., Iadecola, C., and Kim, S. G. (2000). Early temporal characteristics of cerebral blood flow and deoxyhemoglobin changes during somatosensory stimulation. *J. Cereb. Blood Flow Metab.* **20**, 201–206.

Capsule Biography

Dr. Lindauer is heading the research group on cerebrovascular regulation within the Department of Experimental Neurology at Charité Hospital in Berlin, Germany. Her research mainly focuses on the investigation of mediators and mechanisms of neurovascular coupling as the basis for functional brain imaging.

CHAPTER 63

The Retina Microvasculature

Ming Lu and Anthony P. Adamis
Eyetech Research Center, Woburn, Massachusetts
Foundation of the American Academy of Ophthalmology

Anatomy

Retinal Circulation

Microangiopathies of the retina, including diabetic retinopathy and retinal vessel occlusive diseases, are the leading cause of visual loss. The central retinal artery and its branches perfuse the inner retina extending to the inner portion of the inner nuclear layer. The outer retina is perfused by the choriocapillaris, a capillary system of the choroidal arteries that branch from the ciliary arteries. Both the central retinal artery and the ciliary artery are derived from the ophthalmic artery branch of the internal carotid artery. The larger retinal vessels lie in the innermost portion of the retina near the inner limiting membrane. Astrocytes constrain the retinal vessels to the retina. The retinal blood vessels usually do not extend deeper than the middle limiting membrane, formed by a zone of desmosome-like attachments in the region of the synaptic bodies of the photoreceptor cells. In 30 percent of eyes, a cilioretinal artery also supplies part of the inner retina, with contributions to some portions of the macular circulation in approximately 15 percent of eyes. The venous blood of the retina is drained by the central retinal vein into the cavernous sinus. The interposed capillary bed links precapillary arterioles and postcapillary venules (Figure 1), except in the peripheral retina and the perifoveal area where large anastomotic capillary communications connect the terminal arterioles and venules. The fovea thus contains a capillary-free zone of 400 to 600 μm in diameter. The extreme retinal periphery measuring approximately 1.5 mm in width is avascular.

Retinal Capillary

The retinal capillary plexus is arranged in a two-layer pattern. The superficial layer resides in the nerve fiber and ganglion cell layer and the deeper one in the inner portion of the inner nuclear layer. The peripapillary area contains an additional capillary network in the superficial portion of the nerve fiber layer, called the radial peripapillary capillaries. The inner blood–retinal barrier is retained with tight junctions between capillary endothelial cells, whereas the outer barrier is formed by the apical zonulae occludentes of the retinal pigment epithelium (RPE). The endothelial cells are oriented along the axis of the capillary, with cytoplasmic extensions encircling the lumen. A basement membrane covers the outer surface of the endothelium. There is an interrupted layer of pericytes (mural cells) in the basement membrane, with a ratio of 1 : 1 to the number of endothelial cells. Pericytes regulate microvascular growth and function. Interaction with glial cells including astrocytes is required for the endothelial cells to develop tight junctions and to maintain the inner blood–retinal barrier. The retinal capillaries do not leak fluorescein, glucose, amino acids, or small ions, but are highly permeable to water and lipid-soluble substances such as O_2 and CO_2. Metabolic substrates are transported through the endothelial cells through specific channels, pinocytosis, and other carrier-mediated transport systems. Diabetes, hypertension, and surgical trauma compromise the inner blood–retinal barrier.

Retinal Arterioles

When compared to the same size arteries of other organs in humans, retinal arteries lack an internal elastic lamina and contain gradually decreased layers of smooth muscle cells from the optic disk to the periphery. Retinal arteriolar precapillary annuli are also absent. Where arterioles and venules cross, they share a common basement membrane. Venous occlusive disorders commonly occur at an arteriovenous (AV) crossing.

Innervation

In contrast to the choroidal vessels, the retinal vasculature lacks sympathetic fibers. The retinal blood flow

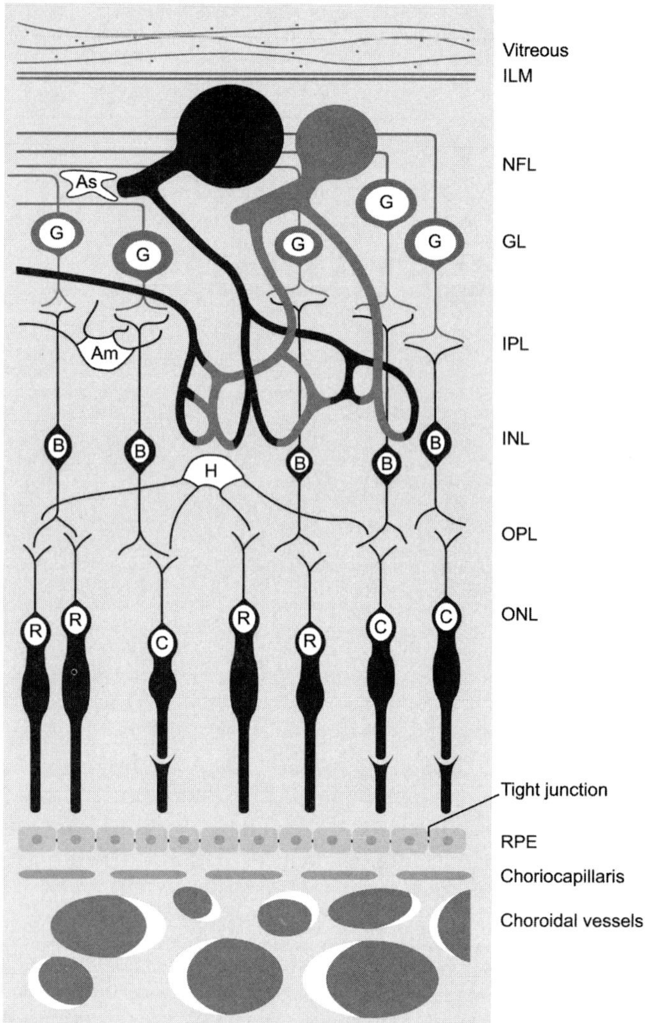

Figure 1 Schematic representation of the retinal and choroidal circulations. The photoreceptor layer is avascular, receiving oxygen from choriocapillaris by diffusion. ILM, internal limiting membrane; NFL, nerve fiber layer; GL, ganglion cell layer; IPL, inner plexiform layer; INL, inner nuclear layer; OPL, outer plexiform layer; ONL, outer nuclear layer; RPE, retinal pigment epithelium; As, astrocyte; Am, amcrine cell; H, horizontal cell; G, ganglion cell; B, bipolar cell; R, rod photoreceptor cell; C, cone photoreceptor cell. (see color insert)

is autoregulated through modifications of vascular resistance.

Physiology

Blood Flow

Eighty-five percent of the total blood supply to the eye is distributed to the choroid and only 4 percent to the retina. The remaining is distributed to the ciliary body and the iris. The human retinal blood flow has been calculated to be between 35 and 80 μL/min. The oxygen extraction for the retinal blood is about 38 percent. In contrast, the oxygen extraction from the uveal blood is very low, with the arteriovenous difference for the choroidal blood about 3 percent. Choroidal blood flow is extremely high, probably because of the large caliber of the vascular lumen. The choroidal blood flow near the fovea and around the optic nerve head is much higher than the blood flow in the periphery of the eye. Despite the low oxygen extraction rate, about 60 percent of oxygen and 75 percent of glucose are delivered by the choroidal circulation. The high volume of choroidal blood flow may contribute to the removal of heat generated during visual transduction, to the removal of fluid from the outer retina, and to meeting the metabolic needs of the retina. Proteins and other substances of high molecular weight can enter the interstitial space through the fenestration of the choroidal capillaries. Retinal binding protein, vitamin A, and many other micronutrients and ions become available to the RPE for transport to the outer layers of the retina. Both active and passive transport mechanisms facilitate the movement of selected nutrients and waste products across the outer blood–retinal barrier. During aging, lipoprotein accumulates in Bruch's membrane from rod outer segment degradation, which may obstruct the movement of water and waste products across the RPE.

Regulation of Blood Flow

The retinal blood flow is determined by the perfusion pressure and the diameter of the retinal capillaries. Local myogenic responses, metabolic factors, and endothelium-derived substances influence the vascular resistance. The retinal blood flow is autoregulated through modifications of the vascular resistance, that is, changes of the contractile state of the retinal arterioles. Because of the lack of precapillary sphincter in retinal arterioles, metabolic and myogenic stimuli are the major modifiers of the retinal arteriolar diameter. The retinal blood flow is maintained relatively constant despite moderate variations in perfusion pressure, up to 41 percent of baseline values of systemic blood pressure or during an increase of intraocular pressure up to 30 mmHg. As a result, the inner retina tissue partial pressure of O_2 (Po_2) is maintained at constant values during moderate reductions of the perfusion pressure. In contrast, there is no autoregulation in the choroidal vasculature. Choroidal blood flow changes little during sudden increments in blood pressure, as an increased sympathetic activity and consequent vasoconstriction of the choroidal vessels maintain a constant choroidal blood flow. The parasympathetic system seems to play little role in the regulation of the choroidal blood flow. Loss of sympathetic innervation may cause accumulation of fluid in the retina and retinal edema. This is important in diseases such as diabetes and hypertension, in which autonomic control is altered. The choroidal vessels contain α-adrenergic vasoconstrictor receptors but no β-adrenergic vasodilator receptors.

Mechanical stretch and increases in the transmural pressure stimulate the endothelial cells to release contracting factors. The factors released by the retinal metabolism also optimize retinal blood flow according to metabolic needs. These factors include Po_2, the tissue partial pressure of CO_2 (Pco_2), nitric oxide (NO), prostaglandins (PGs, including PGE_2, PGF_2, PGH_2, and fo forth), and lactate, released either in the blood or in the surrounding retinal arteriole glial or neuronal tissue. Hyperoxia induces vasoconstriction, whereas hypoxia induces vasodilatation of retinal arterioles. The Po_2 values measured in the inner retina up to half of its thickness remained stable during hypoxia. In contrast, the Po_2 measured near the choroid and in the outer retina decreased in a linear manner according to the variations of systemic Po_2. High oxygen delivery by the choroidal circulation is thus necessary for photoreceptors to function properly. Systemic hypoxia induces an increase in the retinal lactate release, which dilates the arteriolar wall. A rise in Pco_2 induces an increase in blood flow.

The endothelium influences the vascular tone by releasing either endothelium-derived relaxing factors, such as NO and PGE_2 and PGF_2, or contracting factors, such as endothelin-1, thromboxane A_2, and PGH_2. These molecules induce relaxation or contraction of vascular smooth muscle cells and capillary pericytes, thus affecting the diameters of arterioles and capillaries. NO released by endothelial cells acts on the vascular smooth muscle cells and accounts for the biologic properties of the endothelium-derived relaxing factor. There is a gradient and a continuous production of NO by the retinal tissue. The effect of endothelin on a vascular bed is difficult to predict, because apart from causing vasoconstriction, it can also release the vasodilators prostacyclin and NO. PGE_2 and PGF_2 are the predominant PGs produced by the retina and choroid, and they induces vasodilatation of the retinal arterioles.

Retinal Oxygen Distribution

The retina has a high rate of oxygen consumption and metabolism for visual transduction. The PO_2 is heterogeneously distributed close to the vitreoretinal interface. O_2 diffusion from the retinal arterioles affects the juxta-arteriolar preretinal and inner retinal layer's Po_2, which is higher than that of the intervascular areas due to oxygen consumption by retinal cells. In contrast, the preretinal and inner retinal (30% depth) Po_2 far beyond the retinal vessels remains constant in all retinal areas. The intraretinal Po_2 gradually decreases from both the retinal surface and the choroid toward the midretina, with the minimal value at 50 percent of retinal depth. The Po_2 near RPE is significantly higher than it is at the inner limiting membrane level, due to much higher O_2 delivery by the choroidal circulation than by the retinal circulation.

Pathophysiology

Retinal Ischemia

Occlusion of retinal arterioles leads to an acute decrease of the inner retinal Po_2 and subsequent injury to the inner retinal cells. Systemic hyperoxia but pure O_2 increases inner retina Po_2. This is due to the vasoconstrictive effect of 100 percent O_2. The ischemic retina may require other metabolites in addition to O_2 to resume normal function. In chronic retinal ischemia such as diabetic retinopathy and retinal vein occlusion, modifications of the retinal vascular bed lead to the formation of ischemic areas. The ischemic areas produce vasoproliferative factors such as vascular endothelial growth factor (VEGF), stimulating new vessel formation. The new vessel is leaky because of the scarce intercellular junctions between the endothelial cells. Clinical trials are under way to test the use of intravitreal or subtenon steroids, as well as oral and local anti-VEGF agents, in preventing retinal neovascularization due to ischemia.

Glossary

Autoregulation of retinal blood flow: The ability of the retina to maintain its blood flow relatively constant despite moderate variations of perfusion pressure.

Further Reading

Foundation of the American Academy of Ophthalmology. (2000). Fundamentals and principles of ophthalmology. In *Basic and Clinical Science Course*, Section 2, pp. 76–88. San Francisco, CA: The Foundation of the American Academy of Ophthalmology. *This chapter offers detailed anatomical description of the retinal and choroidal microvasculature.*

Harris, A., Bingaman, D. P., Ciulla, T. A., and Martin, B. J. (2001). Retinal and choroidal blood flow in health and diseases. In *Retina*, 2nd ed. (S. J. Ryan, ed.), pp. 68–88. St. Louis, MO: Harcourt-Mosby. *This chapter has good coverage of the pathophysiology of retinal and choroidal microvasculature.*

Pournaras, C. J., and Donati, G. (2000). Retinal and choroidal circulations. In *Principles and Practice of Ophthalmology*, 2nd ed. (D. M. Alberts and F. A. Jakobiec, eds.), pp. 1804–1817. Philadelphia: Saunders. *The chapter has in-depth discussion on the level of oxygen at various retinal and choroidal layers, as well as its regulation.*

Capsule Biography

Dr. Ming Lu is a Heed Ophthalmic Fellow at the University of Texas Southwestern Medical Center. Winner of the Leslie Nesmith Award from the Schepens International Society in 1999, he is interested in diabetic retinopathy and retinopathy of prematurity.

Dr Adamis is cofounder and Chief Scientific Officer of Eyetech Pharmaceuticals. His research is focused on the mechanisms of diabetic retinopathy and age-related macular degeneration.

CHAPTER 64

Nitric Oxide and Blood–Brain Barrier Function

William G. Mayhan
Department of Physiology and Biophysics, University of Nebraska Medical Center, Omaha, Nebraska

The endothelium of cerebral blood vessels forms the interface between the blood and brain tissue (blood–brain barrier), and is the main physical barrier that minimizes the entry of blood constituents into brain tissue. This restriction by the blood–brain barrier arises by the presence of tight junctions (zonulae occludens) between adjacent endothelial cells and a relative paucity of pinocytotic vesicles within endothelium of cerebral arterioles, capillaries, and venules. Physical, chemical, and metabolic stimuli can alter the permeability characteristics of cerebral endothelium. For example, acute increases in arterial blood pressure beyond the autoregulatory capacity of cerebral blood vessels, application of hyperosmolar solutions, cerebral hypoxia, cerebral ischemia–reperfusion, synthesis/release of inflammatory mediators during brain injury, and/or activation of blood-borne elements can dramatically alter the permeability characteristics of the blood–brain barrier. However, the precise cellular mechanisms that account for maintaining the basal integrity of the blood–brain barrier and stimulus-induced increases in permeability of the blood–brain barrier remain poorly defined. Recent evidence suggests that the synthesis/release of nitric oxide may play a role in the regulation of permeability of the blood–brain barrier. Thus, this review will focus on studies that have implicated an important role for nitric oxide in the integrity of the blood–brain barrier during physiologic and pathophysiologic conditions.

Concept of a Blood–Brain Barrier

Before a discussion of the role of nitric oxide in the permeability characteristics of the blood–brain barrier, it is important to understand the evolution of our understanding of the blood–brain barrier. More than a century ago Ehrlich found that intravascular injection of dye distributed relatively equally throughout the most tissues of the body, with the exception of the brain. From these studies, Ehrlich concluded that the brain had a lower affinity for the dye than peripheral tissues. However, studies by Goldmann found that dye injected into the circulation stained all tissues except that of brain and spinal cord. Further, when the dye was injected into the cerebrospinal fluid, it stained the brain but did not enter the peripheral circulation. Thus, these discoveries led to the concept that the central nervous system must be separated from the peripheral circulation by some type of barrier, that is, the blood–brain barrier. However, the identity of the ultrastructure of this barrier remained unknown for nearly 50 years.

Since astrocytes were shown to encapsulate cerebral capillaries, it was initially thought that astrocytes composed the blood–brain barrier. With the arrival of the electron microscope, it became clear that structures within cerebral blood vessels, and not astrocytes, constituted the blood–brain barrier. These structures consist of tight junctions between adjacent endothelial cells (i.e., zonulae occludens). In addition, endothelial cells of cerebral vessels are devoid of fenestrations and transendothelial channels and have a paucity of pinocytotic vesicles. Because of its morphology, it was initially thought that the blood–brain barrier was a very static and rigid structure separating the brain from the peripheral environment. However, this view of the blood–brain barrier has evolved over the years, and it is now known that the blood–brain barrier is a dynamic and multifaceted structure that contains important transport pathways and complex enzyme systems that participate in the regulation of the brain's microenvironment.

Nitric Oxide Synthases

Nitric oxide is generated from the metabolism of L-arginine by nitric oxide synthase (NOS). Functionally, NOS can be separated into constitutive and inducible isoforms. The constitutive isoforms of NOS [endothelial NOS (eNOS) and neuronal NOS (nNOS)] appear to be active under basal conditions and can be stimulated by receptor-mediated increases in intracellular calcium. These isoforms of NOS (eNOS and nNOS) produce small quantities of nitric oxide for relatively short periods of time. In addition to the constitutive isoforms of NOS, an inducible isoform of NOS (iNOS) was first identified in macrophages. Unlike that found for the constitutive isoforms of NOS, iNOS is calcium-independent, and when stimulated can produce large, unregulated quantities of nitric oxide for relatively long periods of time.

Many cell types contained within the brain may be responsible for the synthesis/release of nitric oxide under physiologic and pathophysiologic conditions. It appears that a small portion of parenchymal neurons stain for NADPH diaphorase (a marker for nNOS), and thus may be capable of releasing nitric oxide during the activation of nNOS. Since these neurons may be adjacent to cerebral blood vessels, nitric oxide produced during activation of these neurons may influence vascular tone and the transport of molecules across the blood–brain barrier (Figure 1). Endothelium contained within cerebral blood vessels have been shown to express eNOS, and in some instances iNOS. In vivo and in vitro studies of cerebral blood vessels have shown that nitric oxide released by cerebral endothelium contributes to basal tone, agonist-induced changes in diameter of cerebral blood vessels and in reactivity of cerebral blood vessels in response to physiologic stimuli. Thus, the synthesis/release of nitric oxide by activation of eNOS is important in the control of diameter of cerebral blood vessels, and could conceivably be important in the regulation of permeability of the blood–brain barrier (Figure 1). Glial cells (astrocytes, microglia, and oligodendrocytes) are the predominant cell types in the brain and have been shown to release nitric oxide by activation of iNOS. Astrocytes, for example, not only appear to express a constitutive form of NOS in response to a number of vasoactive agonists including neurotransmitters, but also are capable of the synthesis/release of nitric oxide from iNOS in response to lipopolysaccharide and a number of other important cytokines (Figure 1).

Thus, it is conceivable that the synthesis/release of nitric oxide by these cells contained within the brain could have a major impact on the basal integrity of the blood–brain barrier and on increases in permeability of the blood–brain barrier during brain trauma. Unfortunately, very few studies have examined the role of these cellular elements and the various isoforms of NOS on the regulation of permeability of the blood–brain barrier.

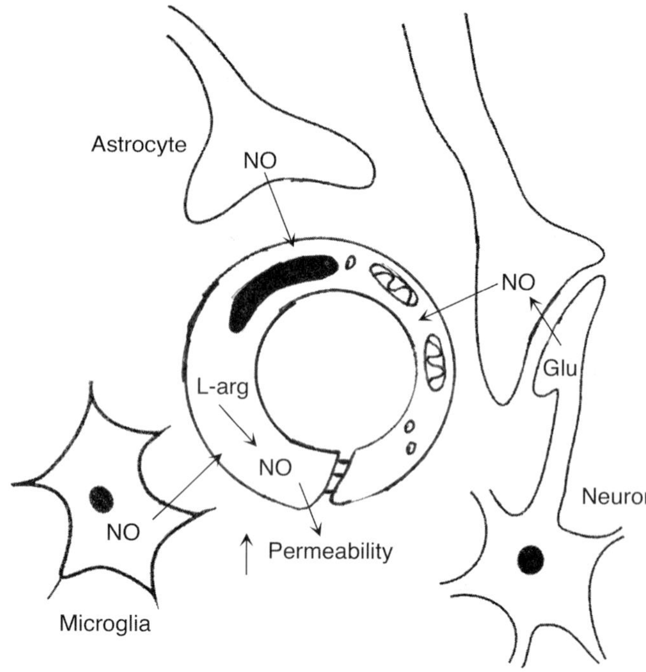

Figure 1 Schematic showing possible sources of nitric oxide (NO) in the brain. NO can be synthesized/released by cerebral endothelium, neurons, astrocytes, and microglia. Activation of glutamate receptors on neurons and stimulation of astrocytes and microglia can stimulate nNOS and/or iNOS, causing local production of NO that could diffuse to cerebral endothelium and influence the permeability of the blood–brain barrier via activation of a number of cellular pathways. Endothelial generated NO from L-arginine (L-arg) in response to inflammatory mediators also may stimulate a number of cellular pathways to influence the permeability of the blood–brain barrier.

Role of Nitric Oxide in Basal Integrity of the Blood–Brain Barrier

Although it is clear that the endothelium, by the synthesis/release of nitric oxide, plays a critical role in basal tone of peripheral and cerebral blood vessels, the role of nitric oxide in maintaining the integrity of the endothelial barrier in peripheral and cerebral vessels remains controversial. A few investigators have suggested that inhibition of NOS (using L-NMMA or L-NAME) increases basal levels of permeability in the peripheral circulation and potentiates agonist-induced increases in peripheral vascular permeability. In contrast, a majority of studies including studies from our laboratory have suggested that inhibition of nitric oxide synthesis using L-NMMA and L-NAME does not influence or decreases basal levels of peripheral microvascular permeability, and inhibits increases in venular permeability of the peripheral microcirculation in response to mechanical stimuli and/or inflammatory mediators. The precise reasons for discrepancies between studies regarding a role for nitric oxide in peripheral microvascular permeability remain elusive, but may relate to regional differences in responses of the microcirculation.

Since many cell types of the central nervous system are capable of the synthesis/release of nitric oxide from the various isoforms of NOS it is conceivable that nitric oxide may play a critical role in maintaining the integrity of the blood–brain barrier under physiologic conditions. Unfortunately, no studies have systematically examined the precise role of the various isoforms of nitric oxide on basal integrity of the blood–brain barrier, and studies that have attempted to examine the potential role of nitric oxide, in general, in basal permeability of the blood–brain barrier have reported conflicting results. Studies from our laboratory found that L-NMMA (a nonspecific inhibitor of NOS) and aminoguanidine (a specific inhibitor of iNOS) did not alter the basal transport of molecules across the blood–brain barrier. In addition, others have not reported any differences in the basal permeability characteristics of the blood–brain barrier between wild-type and nNOS knockout mice. In contrast, others have reported that inhibition of NOS with L-NAME produced an increase in permeability of the blood–brain barrier. However, treatment with L-NAME also produced a significant increase in blood pressure, and the contribution of changes in blood pressure to an increase in the permeability of the blood–brain barrier was not examined. Thus, although the basal synthesis/release of nitric oxide may play a critical role in basal tone of cerebral blood vessels, the role of nitric oxide in maintaining the integrity of the blood–brain barrier during physiologic conditions remains uncertain. Future studies, using a variety of specific inhibitors of NOS and genetically altered models, will be required in order to precisely determine the role of the various isoforms of nitric oxide in basal integrity of the blood–brain barrier.

Role of Nitric Oxide in Stimuli-Induced Changes in the Blood–Brain Barrier

As discussed for the role of nitric oxide in maintaining the basal integrity of the peripheral circulation, the role of nitric oxide in stimuli-induced changes in peripheral microvascular permeability is also controversial. Some have suggested that the generation of nitric oxide inhibits, whereas an increasing number of studies now suggest that the synthesis/release of nitric oxide accounts for increases in permeability of peripheral blood vessels in response to a variety of stimuli. Although the role of nitric oxide in permeability of the peripheral circulation is becoming increasingly clear, the precise role and cellular mechanisms by which nitric oxide may alter the permeability of the blood–brain barrier in response to various stimuli remains uncertain. The following discussion focuses on studies that have begun to implicate a role for nitric oxide in permeability of the blood–brain barrier under several conditions, that is, stimulation with inflammatory mediators, cerebral hypoxia, cerebral ischemia, and bacterial infections.

Inflammatory Mediators/Cytokines

Before a precise role for nitric oxide in stimuli-induced changes in permeability of the blood–brain barrier could be advanced, it was important to determine whether nitric oxide could directly alter the permeability of the blood–brain barrier. Several lines of evidence suggest that nitric oxide can influence the permeability of the blood–brain barrier. Initial studies found that application of a nitric oxide donor, sodium nitroprusside, could reduce electrical resistance across cultured brain endothelium. In addition, application of nitric oxide increased paracellular permeability of isolated cerebral endothelium. Further, studies from our laboratory have shown that application of donors of nitric oxide (Sin-1 and SNAP) to the cerebral microcirculation produced an increase in permeability of the blood–brain barrier. Thus, it appears that nitric oxide has the ability to produce an increase in the permeability of the blood–brain barrier.

Over the past several years, it has become apparent that many cerebrovascular diseases have an inflammatory component. Since nitric oxide has been shown to contribute to changes in peripheral vascular permeability in response to inflammatory mediators, a number of studies have begun to examine the contribution of nitric oxide to changes in permeability of the blood–brain barrier during stimulation with inflammatory mediators/cytokines. Unfortunately, few studies have examined the precise role of the various isoforms of nitric oxide in this process. For example, it has been difficult to examine the precise role of eNOS in the permeability of the blood–brain barrier since there are no specific inhibitors of eNOS. Thus, most studies that have attempted to implicate a role of eNOS in stimuli-induced changes in permeability of the blood–brain barrier have used a rationale similar to that used for studies of cerebrovascular reactivity, that is, the use of nonspecific inhibitors of NOS (L-NAME, L-NMMA, and L-NNA). Using this experimental regimen, we and others have suggested that the synthesis/release of nitric oxide, presumably via activation of eNOS, contributes to increases in permeability of the blood–brain barrier, primarily in cerebral venules and veins, in response to several important inflammatory mediators (i.e., histamine, bradykinin) that may be released during brain trauma and during acute arterial hypertension.

Cytokines are a group of polypeptides that are involved in the activation of the immune system and in the inflammatory response. Cytokines are classified into several categories including interleukins, tumor necrosis factors, chemokines, interferons, growth factors, and neurotrophins. Cytokines are synthesized/released by many cell types within the brain (microglia, astrocytes, endothelial cells, macrophages, and T-cells) and may play a crucial role in the development of inflammation in the brain during cerebrovascular diseases. For example, there is an increase in the levels of interleukins and tumor necrosis factor in the cerebrospinal fluid of human subjects following acute stroke, subarachnoid hemorrhage, cerebral ischemia, or traumatic brain injury. In addition, neurological diseases such as Alzheimer's disease, Parkinson's disease, multiple sclerosis, and AIDS-related dementia also appear to have an inflammatory component in their pathology that may be related to the synthesis/release of cytokines. Thus, although a number of studies have shown that cytokines can produce dramatic changes in the permeability of the blood–brain barrier, only recently has the role of nitric oxide in this process been acknowledged. However, the precise role of nitric oxide, the contribution of the various cellular sources of nitric oxide, and potential mechanisms by which nitric oxide can alter the permeability of the blood–brain barrier during pathophysiologic conditions remain unknown.

Cerebral Hypoxia

A number of metabolic changes to the brain can influence reactivity of cerebral blood vessels, regulation of cerebral blood flow, and perhaps the permeability of the blood–brain barrier. Hypoxia produces dramatic increases in the diameter of cerebral blood vessels that appears to be related, in part, to the synthesis/release of nitric oxide, presumably via the activation of nNOS. However, the effects of hypoxia on the blood–brain barrier are controversial. Some investigators have shown that hypoxia produces an increase in permeability of the blood–brain barrier that can be inhibited by exogenous generation of nitric oxide. In contrast, others have shown that hypoxia-induced changes in permeability of the blood–brain barrier can be markedly reduced by inhibition of eNOS or iNOS. Further, others have shown that hypoxia-induced alterations in permeability of the blood–brain barrier are related to changes in the expression of tight-junctional proteins between adjacent endothelial cells. Thus, future studies will be required to precisely examine the role of nitric oxide, the cellular source of nitric oxide, and the influence of nitric oxide on pathways that account for changes in permeability of the blood–brain barrier during hypoxia.

Cerebral Ischemia

Cerebral ischemia has dramatic effects on the nitric oxide biosynthetic pathway. NOS-dependent reactivity of cerebral arteries is impaired following cerebral ischemia/reperfusion via a mechanism that appears to be related to the inactivation of nitric oxide by oxygen radicals. In addition, studies have shown that nitric oxide concentration in the ischemic area increases within several minutes during cerebral ischemia. This increase in nitric oxide production appears to involve activation of eNOS and nNOS during the onset of ischemia, and iNOS during the progression of cerebral ischemia. Although there is little debate regarding the effects of cerebral ischemia on reactivity of cerebral blood vessels and the production of nitric oxide, there is considerable debate regarding the role nitric oxide plays in the mechanisms of ischemic brain injury. Early studies that examined the effects of nonspecific inhibition of nitric oxide on ischemic brain damage found contradictory results, that is, enhanced, reduced, or no change in cerebral ischemic damage. Since these early studies, the use of more specific inhibitors of NOS has yielded more consistent results. Selective inhibition of nNOS has been shown to reduce cerebral ischemic damage, including cerebral edema. Similarly, the use of specific inhibitors of iNOS has been shown to reduce postischemic infarct size following cerebral ischemia. Further, the use of NOS-deficient mice has strengthened the position that eNOS activity might be protective (due to the effects of eNOS on cerebrovascular hemodynamics), while nNOS and iNOS activity may be detrimental to the ischemic brain. Thus, an understanding of the precise roles for the various isoforms of NOS in ischemia-induced damage to the brain, including an increase in permeability of the blood–brain barrier, may lead to new therapeutic approaches for the treatment of many cerebrovascular disorders, including stroke.

Bacterial Infection

Bacterial meningitis remains a life-threatening illness with high morbidity and mortality rate in infants, children, and adults. A characteristic feature of bacterial meningitis, which appears to contribute to the adverse neurological outcome, is an inflammatory response to endotoxins that produces disruption of the blood–brain barrier. Investigators have examined the pathology of meningitis by examining the effects of lipopolysaccharide on the permeability of the blood–brain barrier. Lipopolysaccharide is a major constituent of the outer membrane of Gram-negative bacteria and plays a pivotal role in initiating inflammation of the brain. Treatment of the cerebral circulation with lipopolysaccharide produces an increase in the permeability of the blood–brain barrier. Further, treatment of the cerebral microcirculation with aminoguanidine, to inhibit iNOS, inhibited lipopolysaccharide-induced increases in permeability of the blood–brain barrier. Thus, it appears that the synthesis/release of nitric oxide plays a critical role in changes in permeability of the blood–brain barrier during bacterial infections.

Cellular Mechanisms for the Effects of Nitric Oxide on the Blood–Brain Barrier

There are several potential mechanisms by which nitric oxide could contribute to changes in permeability of the blood–brain barrier (Figure 2). It appears that changes in permeability of the blood–brain barrier during stimulation with inflammatory mediators, for example, may involve the activation of the cGMP pathway by nitric oxide. Activation of this pathway may, in turn, activate a cascade of cellular pathways, including protein kinase G, protein kinase C, and/or mitogen activated protein kinase (MAPK). In addition, it is possible that the synthesis/release of nitric oxide may stimulate the contraction of endothelial cytoskeletal proteins to produce contraction of adjacent endothelial cells,

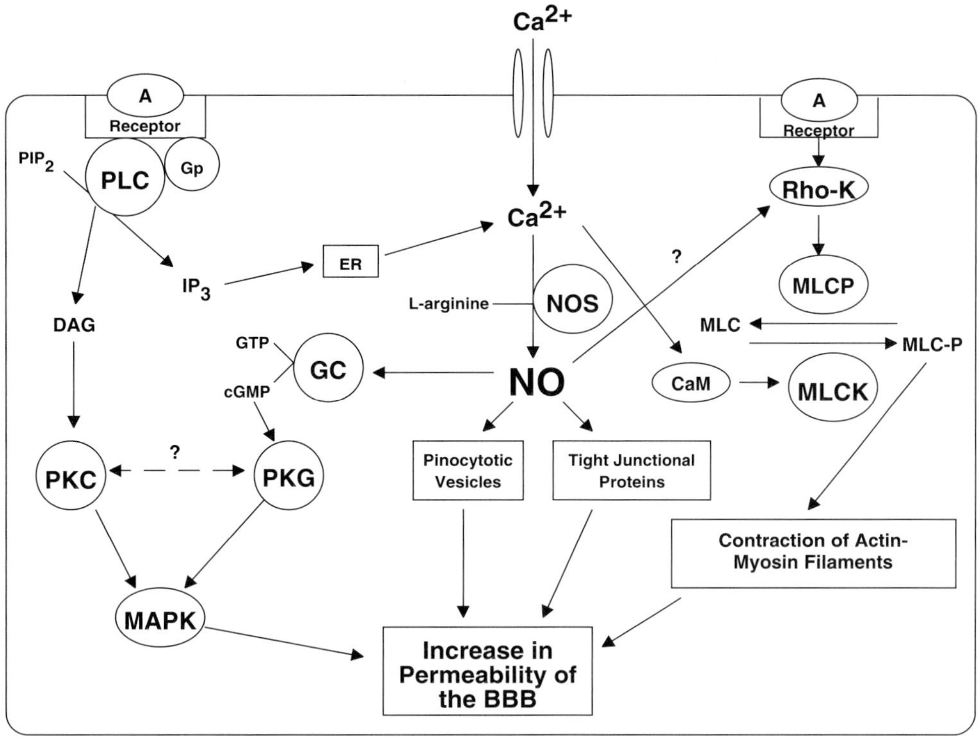

Figure 2 Schematic representation of potential pathways by which nitric oxide (NO) may influence the permeability of the blood–brain barrier. It appears that an agonist (A) binding to a receptor may activate phospholipase C (PLC) via a G protein (Gp) to catalyze the production of inositol trisphosphate (IP_3) and diacylglycerol (DAG) from inositol bisphosphate (PIP_2). IP_3 stimulates the release of calcium from the endoplasmic reticulum (ER) and also increases the influx of extracellular calcium, leading to activation of nitric oxide synthase (NOS) to produce NO from L-arginine. NO appears to stimulate guanylate cyclase (GC) to produce cyclic GMP (cGMP), a potent activator of protein kinase G (PKG). PKG may directly increase the permeability of the blood–brain barrier, may interact with protein kinase C (PKC), and/or may stimulate mitogen activated protein kinase (MAPK) to produce an increase in permeability of the blood–brain barrier. In parallel, DAG may stimulate PKC, may interact with PKG, and/or may stimulate MAPK to produce an increase in permeability of the blood–brain barrier. In addition to this pathway, NO may have direct effects on the formation of pinocytotic vesicles within endothelial cells and/or on tight junctional proteins to produce a pathway for the movement of molecules from the vasculature into brain tissue. Further, changes in permeability of the blood–brain barrier may be related to the effects of NO on endothelial cytoskeletal components. Thus, an increase in intracellular calcium via a Ca-calmodulin interaction (CaM) activates myosin light chain kinase (MLCK) to phosphorylate myosin light chain (MLC), leading to an increase in actinomyosin interaction, and subsequent endothelial cell contraction to increase permeability of the blood–brain barrier. It is also possible that an agonist and/or NO may stimulate myosin light chain phosphorylation (MLC-P) through Rho kinase (Rho-K) pathway inhibition of myosin light chain phosphatase (MLCP) activity resulting in contraction of actinomyosin filaments and an increase in permeability of the blood–brain barrier.

and thus provide a pathway for the movement of molecules and/or influence the formation of pinocytotic vesicles within cerebral endothelium (Figure 2). Further, it is possible that that nitric oxide does not directly produce an increase in vascular permeability, but plays an obligatory role in macromolecular transport, as has been shown for peripheral blood vessels. Future directions to determine the precise role of nitric oxide and mechanisms by which nitric oxide produces changes in permeability of the blood–brain barrier in response to various stimuli, including inflammatory mediators and mechanical and metabolic stimuli, will certainly advance our understanding of this process and will contribute to new therapeutic approaches for the treatment of inflammatory cerebrovascular diseases.

Summary and Conclusions

Although a significant number of studies have clearly demonstrated an important role for nitric oxide in the regulation of diameter of cerebral blood vessels, there is an overall lack of information regarding the potential role of nitric oxide in permeability of the blood–brain barrier. Future studies must continue to focus on identifying the contribution of the nitric oxide biosynthetic pathway to the permeability of the blood–brain barrier. In addition, while many cellular second-messenger systems have been implicated in contributing to the effects of nitric oxide on peripheral microvascular permeability, the role of these pathways in changes in permeability of the blood–brain barrier under physiologic and pathophysiologic conditions remains unknown. Thus, future studies directed at identifying the precise cellular role of nitric oxide in the permeability of the cerebral microcirculation will be critical. The complementary use of in vitro and in vivo models using innovative pharmacological approaches and unique genetic models will certainly advance our understanding of the role of nitric oxide in basal integrity and pathophysiologically induced changes in permeability of the blood–brain barrier. The answers provided by these unique studies will provide insights into new therapeutic approaches for the treatment of many neurological and cerebrovascular diseases that produce an increase in the permeability of the blood–brain barrier.

Glossary

Blood-brain barrier: An anatomical and physiological barrier. It is formed by complex interendothelial cell tight junctions that form a continuous sealing. The barrier serves to protect the brain from agents in the blood, but also controls the influx and efflux of numerous substances to maintain proper homeostasis in the brain.

Cerebrovascular disease: Disease processes that involve blood vessels of the brain.

Inflammation: A condition in which tissues react to an injury. Inflammation is marked by vasodilation, leukocyte infiltration, redness, heat and pain.

Nitric oxide: A free radical gas derived from a guanidino nitrogen of L-arginine. Nitric oxide is a major secretory product of many cell types and is a pluripotent biological mediator.

Permeability: A process by which substances penetrate or pass across or through a cell membrane.

Further Reading

Brian, J. E., Faraci, F. M., and Heistad, D. D. (1996). Recent insights into the regulation of the cerebral circulation. *Clin. Exp. Pharmacol. Physiol.* **23**, 449–457.

Faraci, F. M., and Heistad, D. D. (1998). Regulation of the cerebral circulation: Role of endothelium and potassium channels. *Physiol. Rev.* **78**, 53–97.

Iadecola, C. (1997). Bright and dark sides of nitric oxide in ischemia brain injury. *Trends Neurol. Sci.* **20**, 132–139.

Krizbai, I. A., and Deli, M. A. (2003). Signalling pathways regulating the tight junction permeability in the blood–brain barrier. *Cell. Mol. Biol.* **49**, 23–31.

Mayhan, W. G. (2001). Regulation of blood–brain barrier permeability. *Microcirculation* **8**, 89–104.

Pardridge, W. M. (1986). Blood–brain barrier: Interface between internal medicine and the brain. *Ann. Intern. Med.* **105**, 82–95.

Yuan, S. Y. (2000). Signal transduction pathways in enhanced microvascular permeability. *Microcirculation* **7**, 395–403.

Capsule Bibliography

Dr. William G. Mayhan has been at the University of Nebraska Medical Center since 1989 and is currently a Professor and the Vice-Chair of the Department of Physiology. Recipient of the Young Investigator Award from the Microcirculatory Society in 1983, his research primarily focuses on endothelial control of the cerebral microcirculation during physiologic and pathophysiologic conditions. His work is supported by grants from the National Institutes of Health, American Heart Association and American Diabetes Association.

CHAPTER 65

Fatty Acids Induce Tight Junctions in Brain Capillary Endothelial Cells

[1]Kazuo Yamagata, [2]Motoki Tagami, [3]Yasuo Nara, [4]Shingo Itoh, [1]Hidehiko Yokogoshi, and [5]Yukio Yamori

[1]*Laboratory of Preventive Nutrition, Department of Food Science and Technology College of Bioresource Sciences, Nihon University, Fujisawa-shi, Kanagawa-ken, Japan*

[2]*Department of Internal Medicine, Sanraku Hospital, Chiyoda-Ku, Tokyo, Japan*

[3]*Laboratory of Health Science, Department of Biological Pharmacy, School of Pharmacy, Shujitu University, Okayama-shi, Okayama, Japan*

[4]*Laboratory of Food Science, Department of Food Science and Technology College of Bioresource Sciences, Nihon University, Fujisawa-shi, Kanagawa-ken, Japan*

[5]*World Health Organization Collaborating Center for Research on Primary Prevention of Cardiovascular Diseases, Kyoto-shi, Kyoto, Japan*

Introduction

The brain is highly enriched in fatty acids such as docosahexaenoic acid (DHA) and arachidonic acid. These fatty acids are major constituents of the phospholipid bilayer of the cell membrane, and modulate the tight junction permeability and occludin protein expression in brain capillary endothelial cells (ECs). This review summarizes the main findings of the in vitro approach in blood–brain barrier (BBB) research, describes the unique tight-junction features of the brain endothelium, and provides a short overview on how these BBB characteristics can be induced by the fatty acids in cerebral ECs.

Overview: Tight Junction and Fatty Acids

Brain capillary ECs lining the cerebral microvasculature form the BBB. The BBB is a complex of tight junctions between brain ECs. This barrier between the blood and the central nervous system (CNS) is relatively impermeable to ions, many amino acids, small peptides, and proteins. The maintenance of such an efficient barrier between the blood and brain compartments is essential for normal brain function. The barrier formed by the tight junctions has the important function of protecting the brain from fluctuations in the composition of the plasma. Accordingly, disruption of the BBB can be a relatively major part of the pathology following cerebral ischemia. Physiological and morphological studies have indicated that the relationship between the transepithelial electrical resistance (TER) and intramembrane particle strand number is logarithmic. In fact, the TER across the BBB in vivo is believed to be $1{,}000\,\Omega\cdot cm^2$. In freeze-fracture electron micrographs, tight junctions appear as a set of continuous, anastomosing intramembrane particle strands in the cytoplasmic face with complementary grooves in the extracellular face. Various tight junction–associated proteins have been identified, including zonula occludens (ZO)-1, ZO-2, ZO-3, cingulin, 7H6 antigen, occluding, and claudin. It has been shown that ZO-1 binds to occludin in vitro and is colocalized with F-actin in cultured cells. In particular, occludin and claudin-5 are integral membrane proteins localized within tight junction strands that have been shown to serve as functional components of the tight junction of vascular ECs. Therefore, both occludin and claudin-5 contribute to the electrical barrier function of tight junctions and possibly to the formation of aqueous pores within tight junction strands in brain ECs. These proteins together with the actin cytoskeleton are major determinants of tight-junction structure and also play a role in the regulation of tight junctions. Tight-junction formation appears to be regulated in part by signal transduction pathways involving G proteins, release of intracellular Ca^{2+}, and activation of protein kinase C. Fatty acids are the major constituents of the phospholipid bilayer of the epithelial cell membrane and reportedly modulate the permeability of the epithelial cells. Various reports have demonstrated the effects of fatty acids on tight junction permeability and occludin protein expression in an umbilical vein endothelial cell line. It has also been demonstrated that exposure to fatty acids significantly increased TER in ECs derived from the brain. On the other hand, the absorption-enhancing effects of the sodium salts of two medium-chain fatty acids have been studied in monolayers of intestinal epithelial cell lines derived from colonic carcinoma. Medium fatty acids induced a rapid increase in epithelial permeability in the intestinal cell lines. These reports suggest that the fatty acids contribute to the regulation of tight junctions in ECs. Essential fatty acids are structural components of all tissues and are indispensable for cell membrane synthesis. Brain, retinal, and other neural tissues are particularly rich in long-chain polyunsaturated fatty acids (PUFAs). These fatty acids serve as specific precursors for eicosanoids, which regulate numerous cell and organ functions. It is generally accepted that PUFAs are essential for growth and development, and their crucial role in the development of the CNS has been the subject of many studies. The important role of fatty acids, including PUFAs, for tight junction assembly in brain capillary ECs is well established. Furthermore, tight junction formation appears to be regulated in part by the release of PUFA from astrocytes in vivo.

Metabolism and Supply of Polyunsaturated Fatty Acids to Brain Capillary Endothelial Cells

The brain is highly enriched in the long-chain PUFAs such as DHA (22:6ω-3) and arachidonic acid (20:4ω-6). In particular, DHA is the most abundant ω-3 fatty acid in the brain. These PUFAs can be synthesized from fatty acid precursors linoleic acid (18:2ω-6) and linolenic acid (LA,18:3ω-3) through a process of fatty-acid chain elongation and desaturation. By contrast to the PUFAs, their respective precursors, linoleic acids and LA, are present at only low levels in the brain. The liver is a major site where shorter-chain dietary PUFA is converted to arachidonic acid and DHA, although some retinal and cerebral cells, such as retinal pigment epithelial cells and astrocytes, have the ability to produce DHA. Moreover, low levels of DHA and several of its ω-3 PUFA precursors normally are present in the plasma, and all of these substrates can be utilized in the brain. The brain capillary endothelial cells are thought to take up PUFA precursors and target them preferentially into the brain, performing some elongation and desaturation in the process. In rats, after incubation of cultured brain endothelial cells with [3–^{14}C] docosapentaenoic acid (22:5ω-3) in the presence of serum, radioactivity is primarily recovered in eicosapentaenoic acid (EPA, 20:5ω-3); however, DHA, 24:5ω-3, and 24:6ω-3 were also labeled. Therefore, brain ECs can synthesize and secrete DHA and it 24-C precursors. In addition, one study demonstrated that mouse cerebral endothelium converted 18:3ω-3 into 22:5ω-3, but could not accomplish the last desaturation. The brain capillary ECs are highly enriched in DHA and arachidonic acid and important for the supply of nutrients to the brain, but it is thought that ECs do not play a major role in DHA synthesis. On the other hand, it has been demonstrated that rat astrocytes appear to be a major site for DHA and arachidonic acid synthesis. Indeed, astrocytes, not neurons, elongated and desaturated the 18- and 20-carbon precursors and released DHA and arachidonic acid into the culture medium. It has also been shown that the amount of DHA synthesis from ω-3 PUFA precursors in astrocytes may be regulated by the availability of DHA or other PUFA in the brain tissue or cerebral circulation. Furthermore, in vitro data indicate that DHA synthesis from ω-3 PUFA precursors takes place primarily in astrocytes and that these cells supply some of the newly formed DHA to the neurons and BBB endothelium. In order to examine the participation of astrocytes in this synthesis, cocultures of endothelial cell and astrocytes mimicking the in vivo BBB were used. Endothelial cells cultured alone weakly converted the precursor fatty acids into 20:4ω-6 and 22:6ω-3. Astrocytes play a major role in the delivery of essential PUFAs to the barrier itself and to the brain. It was discovered based on experiments using cocultures of astrocytes and ECs that astrocytes synthesize DHA from ω-3 PUFA precursors and contribute to the ability of brain ECs in the reconstituted BBB to synthesize DHA and arachidonic acid from their 18-carbon precursors. When endothelial cells were cocultured with astrocytes, their content of PUFA increased dramatically. These fatty acids were released by astrocytes after they were synthesized from the precursor fatty acids that passed through the endothelial cell monolayer into the astrocyte medium. Astrocytes play an important role in the brain by elongation of free fatty acids and desaturation action of ω-3 and ω-6 of fatty acid precursors of arachidonic acids and DHA. Astrocytes may thereafter complete the conversion of precursors to DHA, releasing it for uptake by neurons and brain ECs. Studies of cultured cells suggest that astrocytes are the main site of DHA and arachidonic acid synthesis in the brain. Astrocytes may contribute positively to the high level of fatty acid desaturation necessary for capillary endothelial cell functions such as those carried out in the BBB.

Regulation of Tight Junction Formation by Fatty Acids

Fatty acids are major constituents of the phospholipid bilayer of the epithelial cell membrane, and there have been a few reports on the effects of fatty acids on tight junction permeability; fatty acids have been shown to affect occludin protein expression in the umbilical vein endothelial cell line ECV304 and in brain capillary ECs. It has been demonstrated that the TER and the level of occludin mRNA were increased by fatty acids in capillary ECs isolated from brain, and the expression of occludin protein paralleled the change in the pattern of its mRNA expression. The first reported such finding was that the changes in tight junction permeability induced by EPA and gamma LA (GLA) in brain capillary ECs are due to the increase in TER and expression of occludin protein, respectively. Subsequent reports demonstrated the effect of a range of oleic acids, LA, GLA, alpha LA, arachidonic acids, eicosatrienoic acids, and EPA on the permeability of tight junctions and the protein level of occludin in the human umbilical vein endothelial cell line ECV304. GLA, alpha LA, and EPA significantly increased the TER and protein expression, whereas LA significantly decreased the TER of ECV304 cells. These changes in TER were related to alterations in paracellular permeability, which was significantly reduced by EPA and GLA, and significantly increased by LA in ECV304. Furthermore, it was recently shown that PUAC induces tight junctions to form in brain capillary ECs. In brain capillary ECs, the effects of GLA, EPA, and LA on tight junctions were studied in cell monolayers. In those studies the mRNA expression of occludin and the involvement of tyrosine phosphorylation, protein kinase C, and PI 3-kinase were investigated. In vitro studies have shown that fatty acids, EPA, and GLA affect TER and apical-basol permeability. The effect of LA, however, was different from that reported in the umbilical vein endothelial cell line ECV304. Namely, in the in vitro study using ECV304 cells, it was demonstrated that LA decreased the TER. This finding suggests that the difference in the effect of LA is due to a difference of response to fatty acids between umbilical vein ECs and brain capillary ECs, although the mechanism of the response remains unknown. In addition, the rise in TER induced by EPA and GLA was inhibited by tyrosine kinase inhibitors and protein kinase C inhibitor in brain capillary ECs. In contrast, the rise in TER induced by EPA and GLA was not inhibited by the PI 3-kinase inhibitor. EPA and GLA have the ability to increase the TER and the expression of occludin mRNA in brain capillary ECs. This GLA- and EPA- induced assembly of tight junctions is likely to be regulated by protein kinase C and tyrosine kinase activity. The tight junction barrier is regulated by phosphorylation. Various reports have shown that phosphorylation of occludin is a key step in tight junction assembly, which in turn affects the localization of occludin in the cell. The high-molecular-weight forms of occludin are the functional forms that participate in the formation of the tight-junction barrier. The localization of different phosphorylated forms of tight junction–associated proteins in the cell may be especially important. It has been demonstrated that in Madin-Darby canine kidney (MDCK) cells, the less phosphorylated low-molecular-weight forms of occludin are found in the basolateral membrane and cytosol, whereas the more highly phosphorylated high-molecular-weight forms are concentrated exclusively in the tight junction. Differential phosphorylation of occludin might be important for generating tight junctions with different permeability properties. Indeed tyrosine phosphorylation of occludin is known to occur during the assembly of the tight junctions in MDCK cells. Recently the effects of tyrosine kinase inhibitors, genistein, and PP2 on the response to polyunsaturated fatty acids have been reported. It was found that tyrosine kinase inhibitors inhibited the EPA- and GLA-induced increases in TER in brain capillary ECs. The data suggested that EPA and GLA stimulated the tyrosine phosphorylation of multiple proteins. Furthermore, the results indicated that GLA and EPA act through the phosphorylation of protein tyrosine residues. On the other hand, the absorption-enhancing effects of the sodium salts of two medium-chain fatty acids, capric acid and lauric acid, have been studied in monolayers of intestinal epithelial Caco-2 cells derived from colonic carcinoma. Both fatty acids induced a rapid increase in epithelial permeability to the hydrophilic marker molecule sodium fluorescein in the Caco-2 cell line. These reports showed that diverse reactions of free fatty acid depend on differences of the organization. However, the role of fatty acids in tight junction assembly in brain capillary ECs in vivo is not yet well understood.

Conclusion

In summary, in vitro studies have demonstrated that fatty acids induce tight junction formation in brain capillary ECs. On the other hand, brain endothelial cells take up preformed PUFA and transfer them into the brain. Furthermore, astrocytes can complete the conversion of precursors to DHA and arachidonic acid. Astrocytes may modulate tight junctions through provision of fatty acids.

Further Reading

Abbott, N. J. (2002). Astrocyte–endothelial interactions and blood–brain barrier permeability. *J Anat.* **200**, 629–638. *This article provides a detailed review of the role of barrier induction and modulation in normal physiology and in pathology.*

Bernoud, N., Fenart, L., Benistant, C., Pageaux, J. F., Dehouck, M. P., Moliere, P., Lagarde, M., Cecchelli, R., and Lecerf, J. (1998). Astrocytes are mainly responsible for the polyunsaturated fatty acid enrichment in blood-brain barrier endothelial cells in vitro. *J. Lipid Res.* **39**, 1816–1824. *This article demonstrated that astrocytes could play a major role in the delivery of essential polyunsaturated fatty acids to the barrier itself and to the brain.*

Calderon, V., Lazaro, A., Contreras, R. G., Shoshani, L., Flores-Maldonado, C., Gonzalez-Mariscal, L., Zampighi, G., and Cereijido,

M. (1998). Tight junctions and the experimental modifications of lipid content. *J. Membrane Biol.* **164**, 59–69.
Contreras, M. A., and Rapoport, S. I. (2002). Recent studies on interactions between ω-3 and ω-6 polyunsaturated fatty acids in brain and other tissues. *Curr. Opin. Lipidol.* **13**, 267–272. *This article summarizes recent studies on interactions between ω-3 and ω-6 polyunsaturated fatty acids in brain and other tissues.*
Innis, S. M., and Dyer, R. A. (2002). Brain astrocyte synthesis of docosahexaenoic acid from ω-3 fatty acids is limited at the elongation of docosapentaenoic acid. *J. Lipid Res.* **43**,1529–1536.
Jiang, W. G., Bryce, R. P., Horrobin, D. F., and Mansel, R. E. (1998). Regulation of tight junction permeability and occludin expression by polyunsaturated fatty acids. *Biochem. Biophys. Res. Commun.* **244**, 414–420.
Lindmark, T., Kimura, Y., and Artursson, P. (1998). Absorption enhancement through intracellular regulation of tight junction permeability by medium chain fatty acids in Caco-2 cells. *J. Pharmacol. Exp. Ther.* **284**, 362–369.
Moore, S. A. (2001). Polyunsaturated fatty acid synthesis and release by brain-derived cells in vitro. *J. Mol. Neurosci.* **16**, 195–200.
Williard, D. E., Harmon, S. D., Kaduce, T. L., and Spector, A. A. (2002). Comparison of 20-, 22-, and 24-carbon ω-3 and ω-6 polyunsaturated fatty acid utilization in differentiated rat brain astrocytes. *Prostaglandins Leukot. Essent. Fatty Acids* **67**, 99–104.
Yamagata, K., Tagami, M., Takenaga, F., Yamori, Y., Nara, Y., and Itoh, S. (2003). Polyunsaturated fatty acids induce tight junctions to form in brain capillary endothelial cells. *Neuroscience* **116**, 649–656.

Capsule Biography

Dr. Yamagata is at the Laboratory of Preventive Nutrition, Department of Food Science and Technology College of Bioresource Sciences, Nihon University. His current research includes control of tight junctions and brain and nutrients.

Dr. Tagami is at the Department of Internal Medicine, Sanraku Hospital. He is studying the pathology of stroke.

Dr. Nara is with the Laboratory of Health Science, Department of Biological Pharmacy, School of Pharmacy, Shujitu University.

Dr. Itoh is at the Laboratory of Food Science, Department of Food Science and Technology College of Bioresource Sciences, Nihon University. He is studying fatty acids analysis and brain nutritional relation.

Dr. Yamori is with the World Health Organization Collaborating Center for Research on Primary Prevention of Cardiovascular Diseases studying preventive nutrition and food.

Dr. Yokogoshi is with the Laboratory of Nutritional Biochemistry, School of Food and Nutritional Sciences, The University of Shizuoka, and his current research is Nutritional Biochemistry.

SECTION C

Gastrointestinal Tract

CHAPTER 66

Microcirculation in Peritoneal Exchange

Michael F. Flessner
University of Mississippi Medical Center, Jackson, Mississippi

Introduction

The importance of the peritoneal microcirculation is dependent on its use as an exchange organ, and this chapter will focus on those aspects of the transport barrier that affect the use of the peritoneal cavity in therapy. Therapy can be divided into two broad categories: (a) removal of waste metabolites and water by peritoneal dialysis from the body of a renal failure patient and (b) regional drug delivery, exemplified by intraperitoneal (IP) chemotherapy for metastatic ovarian or colorectal cancer. Whereas the goals of intraperitoneal chemotherapy are different from those of peritoneal dialysis, the transport mechanisms are basically the same. In intraperitoneal chemotherapy drugs are dissolved within a 2- to 3-L solution of dialysis fluid, which is instilled into the peritoneal cavity and allowed to absorb into the surrounding tissue. While in the cavity, the drugs diffuse across the peritoneum, through the tissue space, and into the blood vessels. A prime concern is maximizing the concentration in the targeted peritoneal metastases. In contrast, the goal of dialysis is to maximize transfer of water and solutes from the blood of anephric patients to the solution in the cavity. As illustrated in Figure 1, blood circulates through microvessels in the tissue below the peritoneum, and solutes and water must transport across the endothelium of the capillary, through the tissue interstitium, and across the peritoneum to the dialysis fluid, which is drained every 2 to 6 hours and replaced with fresh solution.

This functional view of the peritoneal microcirculation calls for an integrative approach that includes discussion of the lymphatic drainage of the cavity, the microvasculature, the interstitium surrounding the microvasculature, and the anatomic peritoneum. The system elements are made up of cells that are highly reactive to immunologic stimuli and which in turn affect transport across the peritoneum. In addition to the elements of the transport system, the area of contact between a solution in the peritoneal cavity and the tissue determines the route of absorption, metabolism, and rate of transfer. Each of these elements will be integrated to provide a conceptual model of the transport of solutes and water across the peritoneum.

Description of the System

Peritoneal Development and Anatomy

The peritoneal cavity has not been designed as an efficient organ of solute or water exchange, in the same fashion as the kidney. The yolk sac of the primitive embryo is the progenitor of the peritoneal cavity. Analogous to the collapse of an inflated balloon around a tube, the peritoneal cavity develops as a potential space due to intricate folding of the yolk sac into which grow associated elements of the digestive and genitourinary systems. The surface area of the peritoneum approximates the body surface area, and therefore there is potentially a large area of transfer with a multitude of well-perfused organs in the abdominal and pelvic cavities. The peritoneum covers a portion of the diaphragm and liver, a large part of the stomach and spleen, the small and large intestines, the uterus, ovaries, fallopian tubes, retroperitoneal muscles, kidneys, and a broad expanse of the anterior abdominal wall. The multiple tissues with their different blood circulations lead to the question of which parts are most important to the transport processes.

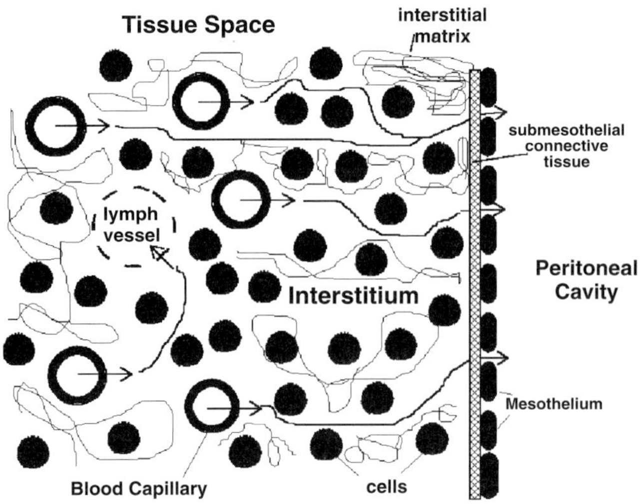

Figure 1 Distributed model concept of peritoneal transport in dialysis. Solutes in the blood transport across the wall of the capillaries distributed in the subperitoneal tissues. Solutes must follow a tortuous path through the interstitium, restricted by matrix molecules and by cells. The final barrier is the peritoneum, made up of a single layer of mesothelial cells and several layers of connective tissue. All solutes must pass across the peritoneum to the therapeutic solution in the cavity, which is replaced with fresh solution periodically to maintain the blood-to-peritoneal cavity concentration gradient that drives diffusion.

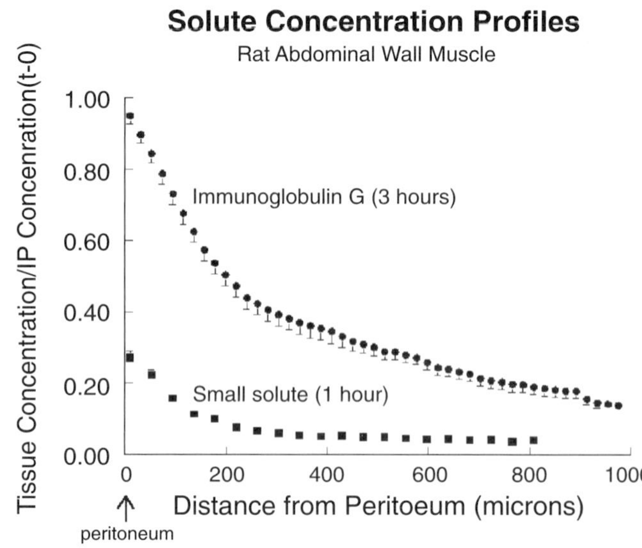

Figure 2 Solute concentration profiles measured by quantitative autoradiography in the abdominal wall of rats after 1 to 3 hours of dialysis with an isotonic solution containing either radiolabeled immunoglobulin G (MW = 150,000 Da) or a small solute (MW = 340 Da). Intraperitoneal pressure was 3 to 4 mmHg. The tissue concentrations have been normalized by the concentration in the original solution infused into the cavity. The degree of penetration of the profiles (0.5–0.6 mm for small solutes and >1 mm for macromolecules) demonstrate that transperitoneal transport involves much more than the peritoneum itself.

The Anatomic Peritoneum

Careful dissection of the human intestinal peritoneum reveals six layers with a total thickness of somewhat less that 100 μm. The most superficial layer is the mesothelium, a continuous monolayer of cells that closely adhere to each other, which is a potential resistance to transport. Separating the mesothelial layer from the underlying connective tissue is a basement membrane. Under the basement membrane are networks of collagenous and elastic fibers. These five layers, making up the most superficial 30 to 40 μm, are devoid of blood and lymphatic vessels. The final layer is the deep latticed collagenous layer that ranges from 50 to 60 μm thick and contains most of the blood vessels of the anatomic peritoneum. The total thickness of the intestinal peritoneum of the human is 80 to 95 μm. Thus after passing the mesothelium, a solute transporting from the peritoneal cavity towards the blood must pass through 25 to 35 μm of tissue before reaching any blood vessels, which are scant in the deep collagenous layer. In the intestine, the layer of tissue under the peritoneum is the smooth muscle of the gut. There are at least 1,000 to 3,000 μm of muscle between the peritoneum and the extensive blood vasculature of the human intestinal villi. As discussed in the following section, the blood supply available to the cavity is chiefly from the smooth muscle of the gut.

The rat has been used as a model of peritoneal transport for more than 30 years, and the normal rat peritoneum resembles that of the human in many respects. Differences are that there is not a collagenous layer with blood vessels and that the peritoneum has a total thickness of 25 μm. Because the rat has very few blood vessels in the normal peritoneum, transport occurs between blood vessels in the parenchymal tissue below the peritoneum and the peritoneal cavity. Recent research has shown that the rat anatomic peritoneum has very little resistance to the passage of small solutes and macromolecules. Viral vectors, such as adenovirus, are the exception to this; they are completely absorbed into the mesothelial cell layer and do not pass beyond the normal anatomic peritoneum.

The Transport System Extends Well Beyond the Anatomic Peritoneum

The microcirculation of the underlying tissues is essential to transport of solutes and fluid between the peritoneal cavity and the circulating blood. The microcirculation of the anatomic peritoneum makes up only a small part of the microcirculation that is involved in transport of solute and water across the peritoneum. The blood supply to the mesentery exemplifies the microcirculation of the intrinsic peritoneum. The membranous-like structure of the mesentery with the sparse population of blood capillaries is a convenient means of study of the microvasculature, but it may not be representative of the microcirculation involved in transperitoneal exchange. Evidence for the amount of tissue involved with transport across the peritoneum is best obtained through the study of solute penetration into this tissue from the peritoneal cavity. Figure 2 demonstrates the degree of penetration of small solutes (MW = 200 to

5,000 Da) and a larger molecule such as immunoglobulin G (IgG, MW = 150,000 Da). In both of these experiments, a rat was dialyzed with large volume in the peritoneal cavity with a radiolabeled compound of mannitol or IgG. At 60 to 180 minutes after the introduction of the solution into the peritoneal cavity the animal was rapidly euthanized, the fluid was drained, and the tissue was rapidly frozen to preserve the concentration profile. Tissue was subsequently collected and analyzed with quantitative autoradiography. The profiles demonstrate that in the case of diffusion-dominated, small molecules (MW less than 5,000 Da), the penetration is on the order of 500 to 600 μm or approximately 1 mm. This is well beyond the thickness of the peritoneum, which in the rat is approximately 25 μm. On the other hand, the profile of a large molecule such as a protein, which is dominated by hydrostatic pressure-driven convection, demonstrates a much greater degree of penetration within the abdominal wall, because of tissue binding and relative lack of lymphatics that could remove the protein from the tissue. From this figure, it can be concluded that the tissue underlying the peritoneum that is involved with transport is a minimum of 0.5 to 1 mm thick and involves a significant amount of parenchymal tissue below the anatomic peritoneum.

The Subperitoneal Blood Circulation

Despite its continuous surface, there is an anatomical division of the peritoneum into the *visceral portion*, covering the intra-abdominal and pelvic organs, and the *parietal peritoneum*, which adheres to the retroperitoneal structures, the diaphragm, and the anterior abdominal wall. The abdominal aorta is the large vessel that passes through the diaphragm and supplies all of the abdominal tissues. The vena cava is the large vein to which all abdominal vessels ultimately drain. The visceral circulation supplies all of the internal organs including the liver, spleen, stomach, and intestines. The celiac artery divides into the splenic, left gastric, and hepatic arteries supplying respectively the spleen, stomach, and liver. Two other major vessels, the superior and inferior mesenteric arteries, form the arcade of vessels supplying the small and large intestines. Branches of the inferior mesenteric artery anastomose with branches of the superior mesenteric, which in turn anastomose with branches of the celiac. The venous drainage of the subperitoneal tissues is a distinguishing feature of the visceral and parietal peritoneum. The veins that correspond to the arteries for these tissues combine to form the portal vein, through which blood is delivered first to the liver prior to reaching the vena cava. Therefore substances that transport into the blood from either the luminal side or the peritoneal side of the hollow organs will transit through the liver before reaching the central circulation and experience a "first-pass effect" of hepatic metabolism.

In contrast to the venous system draining the visceral tissues, blood from the parietal tissues drains directly into the vena cava. Separate arteries supply the kidneys and the ovaries, while their respective veins drain to the vena cava. The right and left deep circumflex iliac arteries supply the sublumbar muscles of the abdominal wall, and at the level of the sacrum approach the skin. Five lumbar arteries branch from the abdominal aorta and supply the para-axial and lumbar muscles, the skin, and the spinal cord, while the epigastric arteries supply the remaining muscles of the abdominal wall. The veins corresponding to these arteries that supply the retroperitoneal tissue and the inferior abdominal wall are close to the arteries and essentially drain into the vena cava directly without going through the liver. The phrenic arteries supply the diaphragm, and the corresponding veins drain to the ipsilateral renal veins, which lead directly to the vena cava.

From a pharmacokinetic point of view, the important information to note from this discussion is that the parietal peritoneal circulation is separate from the visceral circulation and that the drainage from the parietal circulation does not transit through the liver before joining the general circulation. Thus substances that transport into the tissue of the diaphragm, lumbar muscles, or the anterior abdominal wall are transferred directly to the vena cava and to the central circulation without passing through the liver. On the other hand, substances that transport into the tissues of the gut, liver, spleen, and pancreas will transit through the liver prior to entering the central circulation. Figure 3 is a diagram of a

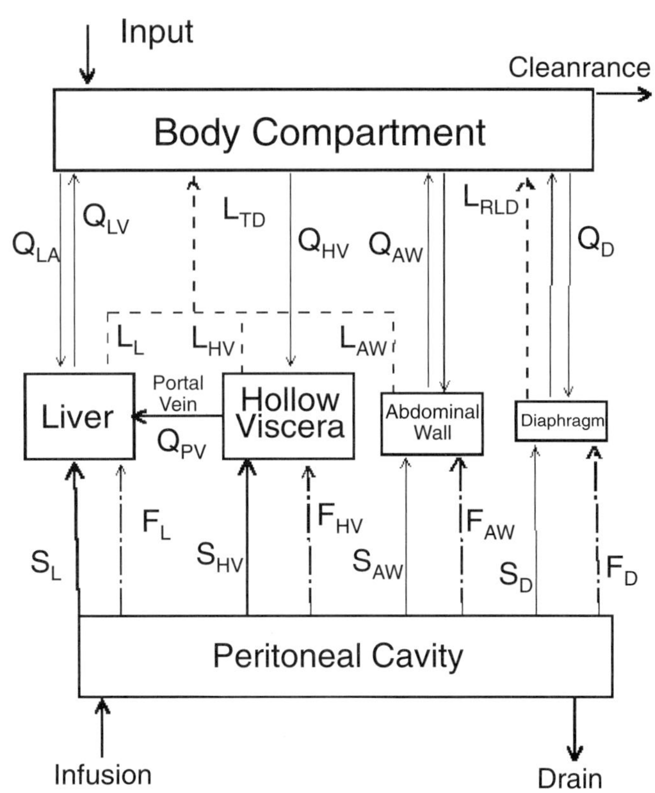

Figure 3 Multicompartment model of intraperitoneal drug delivery. The four tissue compartments result from the anatomy of the cavity and the blood and lymph drainage of the tissues. In the figure: Q, blood flow; L, lymph flow; S, solute transport; F, volume flow. Subscripts: AW, abdominal wall; D, diaphragm; HV, hollow viscera; L, liver; LA, liver artery; LV, liver vein; PV, portal vein; RLD, right lymph duct; TD, thoracic duct. (Data replotted from M. F. Flessner, *Am. J. Physiol.* 1985, *Cancer Res.* 1994.)

multicompartment model that includes this division of the circulation and which is particularly useful in consideration of drug absorption and mathematical modeling.

Peritoneal Microcirculation Involved in Transport

The classic description of the peritoneal microvasculature was provided by Chambers and Zweifach more than 50 years ago. The typical capillary network consists of arterioles, terminal arterials, precapillary sphincters, arterial venules and anastomotic thoroughfare channels, capillaries, postcapillary venules, and venules. Blood flows into the capillary system through the arterioles and exits through venules. Arterioles and thoroughfare channels control actual blood flow to the exchange vessels, which are the capillaries and the venules. Although most data have been collected in studies of the mesenteric bed because of the convenience of imaging these vessels, they may not be the best representation of the major microcirculation involved in transperitoneal transport. As discussed earlier, the transport involves vessels within approximately a millimeter of the mesothelial surface. The mesentery is a double-walled fold of the mesothelium with a single layer of sparse vessels making up the mesenteric circulation. The vessels tend to be much greater distances apart than the vessels in the abdominal wall or in the wall of the gut. Because the largest percentage of the anatomic peritoneum is made up of muscle (either smooth muscle of the hollow viscera or the skeletal muscle of the abdominal wall or of the retroperitoneum), it may be better to think of the peritoneal circulation involved with transport to be similar to that of muscle. This circulation is only bathed by solution in the cavity on one side, whereas the mesentery is bathed on both sides of a single layer of vessels. Although nonfenestrated capillaries of the liver and spleen are extremely permeable and are likely a source of protein loss during peritoneal dialysis, the amount of direct exchange is limited because their surface exposure is a small fraction (about 10%) of the potential surface area.

Transendothelial Transport: The Three-Pore Model

Because of the importance of muscle capillaries, the remainder of this discussion will be focused on this type of nonfenestrated capillaries. Rippe and colleagues have recently published a detailed description of the microcirculation and has described the endothelial barrier in terms of a three-pore model as illustrated in Figure 4. One to two percent of the total pore surface area is made up of transcellular pores or "aquaporins" of radius 0.1 to 0.5 nanometers (nm) in diameter that permit only water to cross. Most of the pore surface area (about 95%) is made up of "small pores" with a radius of 4.0 to 6.0 nm. The remaining 2 to 3 percent of the total pore area is made up of "large pores" on the order of 20 to 30 nm, across which the hydrostatic pressure dominates the transport forces. The figure also depicts the endothelial cells coated on the luminal side with a glycocalyx, which is typically 0.1 to 0.2 μm thick but can be as thick

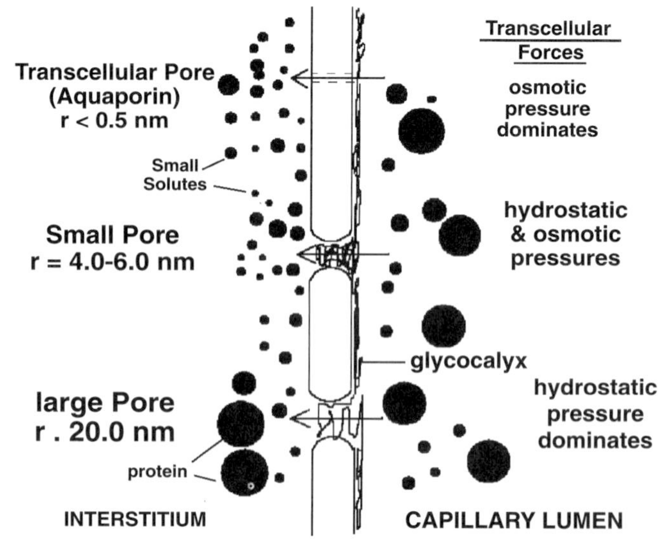

Figure 4 Pore-matrix theory of transendothelial transport. See text for details.

as 0.5 μm. The pore-matrix theory hypothesizes that this glycocalyx restricts the passage of molecules across the endothelium and may be responsible for the membrane properties of the "pores." In this theory, the pore sizes, which have been estimated from the transfer of different molecular size solutes across the capillary wall, may be in fact made up of similar-sized interendothelial gaps that are filled with different densities of glycocalyx. Mathematically, each theory can be shown to be equivalent to the other.

Water, which transports during dialysis in response to an osmotic pressure difference across the capillary wall, flows chiefly through endothelial aquaporins, unique structures in the cell membrane that allow only water to pass through the membrane. The portions of endothelium containing these very small pores are perfect semipermeable membranes, and therefore the entire osmotic pressure difference is effectively exerted. In experiments in an aquaporin-knockout mouse, approximately 50 percent of the osmotic filtration during dialysis was eliminated, demonstrating that these water-only pores are responsible for half of the osmosis during dialysis. The remaining 50 percent of the water transport occurs across the small pores and the large pores.

Small pores make up approximately 95 percent of the total pore area and are primarily responsible for transendothelial transport of substances with a molecular weight of less than 5,000. Permeability of the small pore decreases with increasing molecular size up to a radius of 3.0 nm. Larger solutes are excluded from these pathways and pass through the large pore. Since these openings are too large to restrict macromolecules, there is essentially no osmotic pressure difference across these pores. Because the hydrostatic pressure in the capillary or venule lumen is typically higher than in the interstitium, solutes and water are driven out of the bloodstream under hydrostatic pressure–driven convection. Although some have postulated that there is a vesicular transport of macromolecules across the endothelium, this has been discounted by functional studies in which macromolecular transport continued to be observed under conditions in which vesicular transport should be near zero. Additional evidence comes from direct histologic observation in serial thin sections, which demonstrated that the so-called vesicles were actually invaginations in the cell wall.

Blood Flow and Limitations of Transport

In the transfer of solutes and water, there has been disagreement on whether blood flow limits the transport. In theory, very low blood flows may limit transfer of very small solutes such as urea (MW = 60 Da). However in experiments in rats, the blood flow perfusion to individual organs was decreased to 20 to 30 percent of its original level without change in the mass transfer rates of urea across the peritoneum of the cecum, stomach, or abdominal wall. The transfer across the liver was significantly altered with a decrease in blood flow. In analogous experiments, water transport was decreased across all of these surfaces, but only approached statistical significance across the liver peritoneum. In another approach to the problem, animals were bled to decrease their blood pressure to shock levels, which resulted in minor but statistically significant changes in rates of mass transfer in small solutes. However, even under these extreme conditions the rate of transfer was sufficient to support dialysis. From these studies it can be concluded that under normal circumstances, there are no significant blood flow limitations for water-soluble molecules, but there may be limitations in removal of water during states of systemic shock.

Peritoneal Lymphatic System

The lymphatic system that drains the peritoneal cavity can be divided up into two parts. In the subdiaphragmatic system, which drains 70 to 80 percent of the lymphatic flow from the peritoneal cavity, the diaphragm acts as a pumping mechanism that pulls fluid from the pelvic regions of the cavity toward itself. As the diaphragm moves upward in expiration, the lymphatic plexus expands and a negative pressure is established in the lymphatic vessels. Penetrations in the basement membrane called lacunae open to take in fluid, solutes, and particles up to 25 μm in diameter. When the diaphragm contracts, the tension in the lymphatic wall is released and the lacunae are closed. The fluid is then propelled upward toward the right lymphatic duct or into the thoracic duct. The remaining 20 to 30 percent of lymph flow from the peritoneal cavity travels through the visceral lymphatics. These drain to the mesenteric lymphatics and ultimately to the cisterna chyli at the base of the thoracic duct. In healthy peritoneal dialysis patients, the total lymph flow has been determined to be between 7 and 21 mL/hour. The capacity of the lymphatic system can increase significantly in states of massive cirrhotic ascites to rates of greater than 30 mL/hour or decrease to near zero with carcinomatous obstruction of the lymph channels.

Subperitoneal Interstitium and Effects of Intraperitoneal Pressure

Although some transport models neglect the presence of interstitium and portray the peritoneal barrier as a capillary endothelium, the interstitial space, which surrounds the blood vessels and parenchyma cells of the subperitoneum, presents a significant barrier to transport. This was demonstrated most clearly by the absorption of inert gases from the peritoneal cavity of pigs, which demonstrated a one-hundredfold range of clearances that correlated with the log of the aqueous diffusivity. If the resistance in the barrier were equivalent to just the blood capillary wall, then the transfer of gas would be limited by blood flow as it is in the lung, and the rates of inert gas absorption would have been the same. Correlation with diffusivity of water implies that the resistance in the tissue interstitium limits the transport of these inert gases.

The space between the cells, termed interstitium, is a very ordered structure of the tissue. At one time the interstitial matrix of molecules was considered an amorphous mixture of interstitial ground substance. However, now it is known to be a highly ordered matrix made up of collagen fibers that are anchored to the surrounding cells by adhesion molecules called integrins. Collagen fibers have been demonstrated to hold the cells and extracellular matrix in active tension that can be disrupted in states of inflammation or tissue damage. Wrapped around the collagen molecules are large hyaluronan molecules, which imbibe water and bind proteoglycans such as chondroitin sulfate, keratin sulfate, and heparin sulfate. The hyaluronan molecules are negatively charged and together with the collagen present a major barrier to the passage of negatively charged proteins. These molecules also restrict proteins to approximately 20 percent of the extracellular space. These interstitial molecules in the subperitoneal tissues link the parenchymal cells, fibroblasts, and other cells and force solutes to take a very tortuous path when transiting from the blood to the peritoneal cavity or vice versa. The presence of the interstitium and cells surrounding the exchange blood vessels in the tissue has the effect of decreasing the rates of diffusive transport from the blood to the peritoneal cavity 30 to 100 times from the rate of free diffusion of the same molecule through water. The greater the distance between the capillary and the peritoneal surface, the more significant the resistance of the interstitium becomes. The portion of resistance attributable to the interstitium in the diffusion of a small solute such as sucrose is estimated to be 20 percent if the capillary is located 50 μm from the peritoneum. However, it increases to more than 80 percent of the total peritoneal resistance if the capillary is located 600 μm from the peritoneum.

During a therapeutic procedure in which a volume of 2 to 3 L is infused into the cavity, pressures are 2 to 20 mmHg, depending on the size of the patient, the volume administered, and the position of the patient (sitting position has the highest pressure; supine, lowest). In the abdominal wall, even a relatively low pressure of 4 mmHg causes the extracellular volume of the abdominal wall of the rat to expand by 100 percent, which results in a marked decrease in the resistance to diffusion or convection. Therefore, the interstitial portion of the peritoneal transport system during a large abdominal dwell of fluid can potentially be quite variable in its characteristics.

Quantitative Approach to Peritoneal Exchange

Distributed Model of Peritoneal Transport

Because of the complex relationships between the various parts of the peritoneal transport system, an integrative approach needs to be taken in order to quantify the system. The distributed model was devised as a simplification of the actual system. In this concept, the blood capillaries are distributed uniformly throughout the tissue space, surrounded by parenchymal cells and the interstitium, which is also assumed to have uniform concentrations of interstitial matrix molecules. A layer of mesothelial cells and the underlying connective tissue of the peritoneum overly the tissue space. These principles are portrayed in a simplified fashion in Figure 1. Lymphatic vessels, which are typically at the tissue planes in the parietal tissue or distributed within the multiple layers of smooth muscle of the visceral tissues, are modeled as a separate flow from each tissue compartment (see Figure 3). Within this system, diffusion of small molecules (MW less than 6,000 Da) between the blood and the peritoneal cavity is a symmetrical process, that is, small molecules from blood to the peritoneal cavity or from the peritoneal cavity to the blood follow an equivalent path. An asymmetry in transport exists in the case of larger molecules, which can transfer across capillary walls from the lumen to the interstitium via "large pores" but can only return to the blood via the lymphatics.

An engineering approach to the transfer of small solutes across the peritoneum is illustrated in the following equation:

$$\begin{aligned} \textit{rate of mass transfer} &= \frac{d(C_{pc}V_{pc})}{dt} \\ &= \sum_i MTC_i \bullet A_i (C_{plasma} - C_{pc}) \quad (1) \end{aligned}$$

where C_{pc}, C_{plasma} are concentrations in the peritoneal cavity and the plasma respectively. V_{pc} is the volume in the peritoneal cavity and in this version is assumed to remain constant. The terms MTC_i and A_i are, respectively, the mass transfer coefficient and the contact surface area of each tissue element "i" of the transport system. In this equation, the peritoneal volume is assumed to be well mixed with a uniform concentration, and the total transport resistance of each tissue element has been lumped into the term MTC_i.

The MTC_i can be related to the tissue diffusivity and the capillary permeability with the following equation:

$$MTC_i = \sqrt{D_i (pa)_i} \quad (2)$$

where D_i is the effective solute diffusivity for each tissue and $(pa)_i$ is the capillary permeability (p) times the area density (a) for the tissue. The term p incorporates the contributions from the two pores that permit passage of solute. This equation implicitly assumes that there are no blood flow limitations.

For a diffusion-limited solute (MW less than 6,000 Da), the concentration profile in the tissue can defined as

$$\frac{C_i - C_{plasma}}{C_{pc} - C_{plasma}} = \exp^{-\sqrt{\frac{(pa)_i}{D_i}}x} \quad (3)$$

where C_i is the tissue concentration and x is the distance into the tissue.

Peritoneal Surface Contact Area

In consideration of Equation (1), a major factor that determines the rate of mass transfer across the peritoneum is the peritoneal surface contact area (A_i) between the peritoneal solution containing a therapeutic drug or dialysate and the peritoneum. As illustrated in Figure 1, transfer will not occur between the vessels located in the tissue and the cavity without contact of fluid on the overlying surface. Although some authors have discounted the three-dimensional nature of the transport barrier and equate the perfused vascular area to the peritoneal surface area, animal and human studies have demonstrated that rate of solute transfer is directly dependent on the peritoneal contact area. In human studies, increasing the volume of fluid within the peritoneal cavity increases the surface contact area, which increases the creatinine mass transfer proportionately. With the use of relatively large volumes in rodents and human dialysis patients, only 30 to 40 percent of the total anatomic peritoneal surface area is in contact with the solution.

Contact surface area is likely not distributed proportionally between parietal and visceral tissues. Although the visceral peritoneum makes up approximately 60 to 70 percent of the mammalian anatomic peritoneum, evisceration in rodents has been shown to decrease the peritoneal mass transport only 10 to 30 percent. Transfer rates of small solutes across different surfaces of the visceral and parietal peritoneum, independent of surface area, have been shown to be essentially equivalent. Therefore, the actual distribution of the fluid contact area may not make a difference in determination of the overall average mass transfer rate. Pathological changes in the peritoneum, such as the development of adhesions that restrict free flow of fluid in the cavity, may alter the distribution.

Modulation of Transport Barrier: Normal Physiology

ALTERING SURFACE CONTACT AREA

A problem may develop in IP chemotherapy if the residence time of the solution in contact with the carcinomatous target is short. The goal of intraperitoneal chemotherapy is to expose 100 percent of the peritoneal surface to the therapeutic solution for a maximal duration. Typically, this is performed in humans by infusing 2 L of peritoneal dialysis fluid into the cavity and allowing it to be absorbed. Rates of fluid transfer are typically unimportant to the chemotherapist, since most of these patients have intact kidneys and can excrete the absorbed fluid. Increased surface area during intraperitoneal chemotherapy would be very advantageous to treatment of metastatic carcinoma. If the solution containing the drug is in contact with the surface of the tumor for 10 minutes out of a 24-hour period there will likely be very little drug deposited into the tumor. Use of relatively high concentrations of diacetyl sodium sulfosuccinate (DSS) in rodents resulted in 100 percent of the peritoneal surface area in contact with the fluid, but toxicity was observed. A further problem occurs in some patients with extensive carcinomatoses who develop severe adhesions, which bind together portions of the visceral and parietal peritoneum, restricting the movement of fluid to all parts of the peritoneal surface and limiting therapy. Surface-active agents may be useful in patients with end-stage carcinoma, but testing in animals should be undertaken prior to use in humans to elucidate potential toxicity.

In Equation (1), which describes the rate of mass transfer across the peritoneum in quantitative terms, the rate is directly proportional to the surface contact area. Enhancement of the peritoneal surface area by even 20 percent would increase the rate of mass transfer by 20 percent and could provide enough additional dialytic therapy to improve dialysis efficacy. Increasing the volume from 2 to 3 L in average-sized humans has been shown to improve the rate of transfer of creatinine by 25 percent, proportional to the increased surface area in contact with the fluid. In animals, the use of surface-active agents such as DSS has been shown to increase the surface contact area and to proportionally increase the rate of mass transfer. In small rodents, this was shown to also increase the rate of protein loss during the dialysis. Since in chronic dialysis, the fluid is removed from the body and discarded, this additional loss of protein could be detrimental to the patient. At this time, there is no data on the use of surface active substances in the human peritoneal cavity. Low doses of surface-active substances should first be tested in animal models and demonstrated to be safe for chronic use prior to their utilization in human beings.

Use of larger volumes to cover more of the peritoneal surface is currently the simplest method of increasing surface contact area. The major factors that determine the tolerable peritoneal volume are the size of the patient, the expansion capability of the abdominal wall, the length of time of the procedure, and the position of the patient. For ambulatory peritoneal dialysis, the volumes are typically 2 to 3.5 L. Very few people can tolerate more than this volume in their peritoneal cavity, while carrying out their activities of daily living. On the other hand, a patient undergoing radical resection of metastases in the peritoneum and who is anesthetized might tolerate very large volumes in the cavity while supine and anesthetized. During experimental IP chemotherapy, rates of transfer have also been markedly enhanced by placing two catheters into the cavity after surgical debulking of tumor and perfusing at a very high flow rate into one catheter and removing the fluid from the other. By maintaining a very high drug concentration in the peritoneal cavity and by perfusing the peritoneal surface with a rapid flow rate, the surface area and the concentration difference between the cavity and the blood in the underlying microcirculation were maximized, and these measures caused a doubling of the mass transfer rate of low-molecular-weight drugs from the cavity.

ALTERING THE MASS TRANSFER COEFFICIENT

A second method of altering transperitoneal transport is adjustment of the mass transfer coefficient (MTC). As shown in Equation (2), the MTC is equal to the square root of tissue diffusivity times the capillary permeability–area density product. By altering any of these terms, one can obtain a modest increase or decrease in the MTC in proportion to the square root of the factor. In the setting of dialysis where maximal rates of transfer are desired, many studies have demonstrated that the use of drugs such as nitroprusside does increase the clearance of urea, creatinine, inulin, and protein in a dose-dependent fashion. Although there are no data on which term within the (pa) is actually altered, it is assumed that vasodilators increase the vascular surface area per unit volume of tissue or a. Vasodilator substances diffuse into the tissue from the peritoneal cavity and set up a concentration profile similar to the small solute profile in Figure 2. Thus the most marked effect would be close to the peritoneal surface, whereas the vasodilation more than 500 μm from the surface would be much less. Unfortunately, intraperitoneal nitroprusside appears to be limited by a loss of effect after approximately five exchanges with the drug. Removal of the drug from the dialysis solution causes a decrease to the baseline of the permeability. In addition, there may be effects on systemic blood pressure that limit the use of vasodilators in the peritoneal cavity.

In contrast, vasoconstrictors cause an acute decrease in the rate of mass transfer. It is presumed that they work in the opposite fashion by decreasing the perfused vascular surface area. These might be useful in chemotherapy where local absorption into tissue should be maximized, but systemic absorption into general circulation should be minimized. Experiments in animals have shown that these drugs acutely will cause a change, but there is little data in humans. Theoretically none of the vasodilators or vasoconstrictors should change the intrinsic capillary permeability, nor should they change any characteristics of the tissue diffusivity.

The tissue diffusivity can be altered by changing the structure of the extracellular space. As described earlier, a large volume dwell tends to increase the extracellular space of the subperitoneal tissue. Even at very low intraperitoneal pressures of 4 mmHg, the extracellular space of the rat abdominal wall doubles. This will translate to a doubling of the effective diffusivity for small solutes, which are restricted to the extracellular space. This may also cause an increase in the rate of diffusion of larger solutes, such as immunoglobulins, which distribute to a much smaller portion of the tissue space because of interactions with the negatively charged matrix molecules. Alteration of the extracellular matrix by decreasing the concentration of hyaluronan or collagen will lead to less macromolecular restriction and a higher rate of transport.

Pathological States and Alterations in the Peritoneal Transport System

Metastatic Cancer in the Peritoneum

In the patient undergoing intraperitoneal chemotherapy, the peritoneum has been altered by metastasis of the primary tumor. The subdiaphragmatic lymphatics may become obstructed, and the outflow of cellular materials, protein, and fluid may be markedly decreased. In addition, metastasis to the liver may cause an increase in transhepatic ascitic fluid production, which, coupled with the loss of outflow through subdiaphragmatic lymphatics, will result in accumulation of fluid in the peritoneal cavity and marked distention of the abdominal wall with an increase in intraperitoneal volume and pressure. Additionally, metastasis to specific areas of the peritoneum will cause the mesothelium to be altered and will result in adhesions between the parietal and visceral peritoneum or between tumor and normal peritoneum. This leads to reduction of peritoneal volume, dysfunction of the gastrointestinal tract, decreased access of a therapeutic solution to the entire peritoneum, and nonuniform distribution of the peritoneal fluid in the cavity. The presence of ascitic fluid and extensive adhesions decreases the effectiveness of intraperitoneal chemotherapy. Only with radical peritoneal surgery can the cavity be restored to a state appropriate for treatment of metastatic abdominal cancer. Recent evidence has shown that peritoneal stripping of intra-abdominal carcinomatosis results in restoration of the distribution of the therapeutic solution and the anticipated rates of mass transfer across the peritoneum.

Chronic Inflammation from Sterile Dialysis Solutions

With the use of the peritoneal cavity as a dialysis system, the requirements are significantly different from those of chemotherapy, and the types of pathology that alter the tissue barrier are somewhat different. Although some patients have had prior cases of peritonitis or abdominal surgery that have produced adhesions, the vast majority of peritoneal dialysis patients do not have the same problem that patients with peritoneal carcinomatoses have. Over the past 10 years knowledge has grown concerning the chronic state of inflammation that the peritoneum experiences with long-term, continuous peritoneal dialysis. The solutions, which typically have a low pH, a high osmotic pressure due to glucose, and glucose degradation products from heat sterilization, have been shown to cause significant changes in the peritoneum, subperitoneal interstitium, and transport microvasculature. The high glucose content, which varies from 15 to 42.5 g/L, sets up a diabetic state within the cavity. Twenty-four hours a day the mesothelium is bathed with these solutions. The resident macrophages in the peritoneal cavity and the mesothelial cells have both been shown to react to these glucose solutions. Human studies have demonstrated that in as little as 8 months, the peritoneal mesothelium changes from an epithelioid cell type to a fibroblastic phenotype. This is not due to repeated infective peritonitis but merely constant exposure to sterile, hypertonic solutions. Although the low pH and hyperosmolality have been implicated in these changes, current evidence points to glucose degradation products and glucose itself. Biopsy studies of long-term peritoneal dialysis patients have shown that the submesothelial compact zone, which is typically less than 60 to 80 μm thick, increases in thickness to as much as 2,000 μm after 6 to 7 years. Simultaneously, the degree of angiogenesis and vasculopathy in the layer directly under this subcompact mesothelial layer increases in approximately 90 percent of patients after 6 years of exposure. The rate of solute transfer is often enhanced while the rates of fluid removal with osmotic ultrafiltration are markedly depressed. This results in loss of capacity to remove fluid from the patient and necessitates an alternative mode of therapy.

Peritoneal Inflammatory Response

Figure 5 illustrates the peritoneal cellular system and some of the mediators of the inflammatory cascade. Mesothelial cells overlie the connective tissue of the peritoneum, and within the tissue space, there are fibroblasts, muscle (parenchymal) cells, and endothelial cells. The glucose solutions stimulate the production of interleukin 1 (IL-1) and tumor necrosis factor alpha (TNFα) and the peroxidatic activity of macrophages that reside in the peritoneal cavities of dialysis patients. Mesothelial cells also respond directly to glucose solutions and spontaneously release hydrogen peroxide, IL-1, and TNFα. Whereas in vitro exposure of mesothelial cells to low pH dialysis solutions causes marked toxicity, these solutions are buffered in 10 to 15 minutes in vivo. Therefore the solution pH may not be a major factor in a chronic inflammation. However, bicarbonate-buffered solutions have been shown to decrease production of transforming growth factor-1 (TGFβ-1) and to improve the overall health of the mesothelium. The

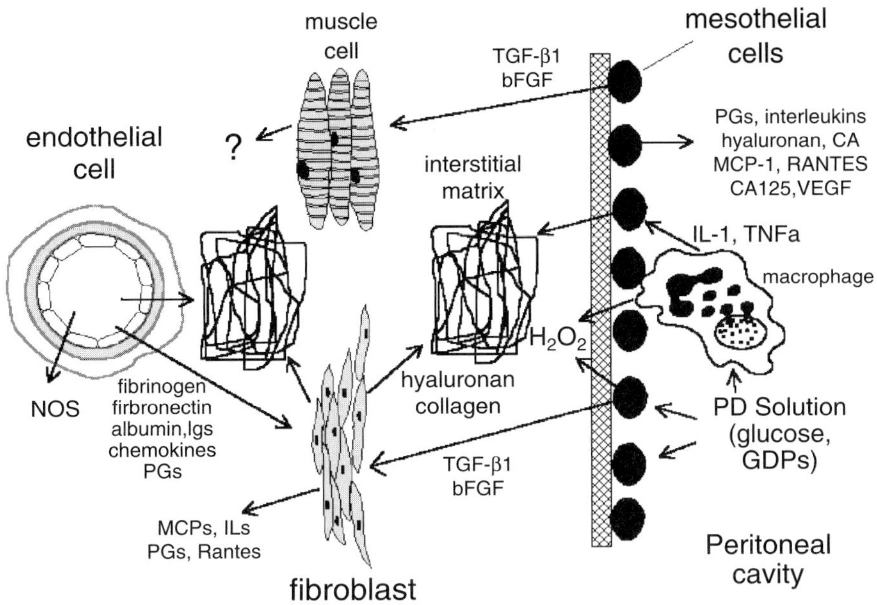

Figure 5 Cellular system in peritoneal inflammation. Dialysis solutions in the peritoneal cavity contain glucose and glucose degradation products (GDPs) that activate macrophages and mesothelial cells to secrete numerous cytokines. These in turn stimulate fibroblasts and attract leukocytes to migrate through activated endothelium. This chronic inflammatory process results in scarring of the subperitoneal tissue and vasculopathy in the parenchymal vessels. See text for definitions of cytokine acronyms. (see color insert)

mesothelial cells, once stimulated by glucose or by the IL-1 and TNFα from stimulated macrophages, produce prostaglandins, interleukins, hyaluronan, cancer antigen 125 (CA125, a marker for mesothelial cell number), monocyte chemoattractant protein (MCP-1), RANTES, vascular endothelial growth factor (VEGF), basic fibroblast growth factor (bFGF), and TGFβ-1. In vivo gene transfer of an active TGFβ-1 into rat mesothelium via an adenoviral vector resulted after 7 days in a markedly thickened peritoneum with increased vascularization and extensive collagen deposition. The infected animals demonstrated an increased transport of solutes and decreased ultrafiltration in response to hypertonic solutions.

Stimulated fibroblasts produce a series of cytokines as well as hyaluronan and collagen. It is likely that they play a major role in the fibrosis in the subcompact area of the peritoneum during long-term peritoneal dialysis. Coupled to cells in the interstitium are activated endothelial cells in the vessels through which white blood cells pass through the endothelium into the tissue space. Nitrogen oxide synthetase (NOS) has been shown to be upregulated and to result in a state of vasodilation. A constant state of hyperosmolality due to high glucose concentrations essentially sets up a diabetic state within the tissue, with the possible formation of advanced glycation end products. These are correlated with many of the fibrotic and angiogenic changes within the tissue. Whether parenchymal muscle cells or other cells such as pericytes associated with endothelium within the tissue take part in this product state of inflammation is unknown.

Intravital microscopy has recently been used to analyze the vasoactive effects of peritoneal dialysis solutions on the mesenteric vessels. Low-GDP solutions have caused transient vasodilation, whereas conventional 4.25 percent dextrose solutions result in maximal vasodilation, doubling of the arterial flow, and a 20 percent increase in perfused capillary length/area. Neutral, lactate-buffered solutions had only transient effects. Peritoneal NOS has been shown to rise fivefold in long-term dialysis patients. It also correlated with increased vascular density and endothelial area in biopsies. VEGF was colocated in the peritoneal endothelium with advanced glycation end products. Filtered dialysis solutions contain far less of the glucose degradation products that occur with heat sterilization. Treatment of patients with these more biocompatible solutions typically produces lower concentrations of hyaluronan (a marker of inflammation) in the peritoneal dialysate, higher concentrations of CA125, and higher concentrations of procollagen-1-C peptide and procollagen–3-N-terminal peptide.

In summary, mesothelial cells that are continually exposed to the high concentrations of heat-sterilized, glucose-based peritoneal dialysis undergo a cellular transition from an epithelial cell type to a fibroblastic phenotype. High levels of TGFβ-1 and VEGF appear to promote fibrosis and angiogenesis in these subperitoneal tissue spaces. Attempts to block specific modulators of this inflammatory system have resulted in decreased proliferation in vitro. In vivo investigation will likely proceed in the next several years.

Integration of the Peritoneal Transport System

The complexity of the peritoneal transport barrier requires an integrated approach toward its description. Figure 3 is a useful conceptual model when the system is modeled mathematically. Quantitative flows of blood and lymph must be assigned to all of the various pathways in this illustration. Rates of mass transfer with individual mass transfer coefficients and surface contact areas must be assigned to each tissue group. The system in Figure 3 must then be integrated into a total-body system with multiple compartments for each organ system. Quantitative values for the various parameters have been suggested by Flessner and Dedrick (see Further Reading).

The distributed model concept of Figure 1 requires tissue-specific transport coefficients that cannot be obtained from the compartmental model of Figure 3. Solutes transport from the cavity through the peritoneum, which has been shown to be an insignificant barrier, through the tissue interstitium and are absorbed either directly into the distributed blood system throughout the subperitoneal tissues or into the lymphatic vessels, if macromolecules. Characteristics of the interstitium and its variability during a large-volume dialysis or chemotherapeutic procedure must be taken into account. In addition, the vascular system is highly reactive to hypertonic solutions often used in dialysis. Therefore in the early stages of dialysis, there may be a localized vasodilation in the layers of the tissue closer to the anatomic peritoneum. However, after 15 to 20 minutes, this tends to disappear and a rather uniform rate of blood flow appears to be the rule. Rates of small-solute transport appear to be the same whether in the direction from the cavity to the blood or from the blood to the cavity. Larger solutes, however, experience an asymmetry in their transport. They leave the blood vessels via large pores and transport through the tissue interstitium to the peritoneal cavity. However, if administered into the peritoneal cavity, they transport across the peritoneum and through the interstitium, but cannot be reabsorbed directly across the endothelium; they must be taken up by lymphatics. The restrictive properties of the peritoneal interstitium must be taken into account when describing the transport. Although a small portion of the peritoneal surface area is made up of solid organs such as the liver and spleen, the vast majority of the tissue microcirculation can be represented by vessels of the muscle.

The cells illustrated in Figure 5 need to be considered as a system. Although each of these cell types can be studied in vitro, it is only through their interaction that the clinically observed pathology is created. It is also only through this interaction that an improved understanding of the complex process of chronic inflammation will come about. The integrated cellular system within the peritoneal tissue reacts to multiple stimuli. In bacterial peritonitis, the interaction of the organisms with the macrophages and the mesothelium results in a rapid response of the local inflammatory cascade. This affects the underlying tissue space in the short term but appears to be rapidly reversed if the infection is brought under control. On the other hand, chronic inflammation occurs at a very slow rate and is due to the exposure of the peritoneum to noncompatible dialysis solutions. It is not known whether any solutions other than an isotonic salt solution could be truly compatible with these cells. The macrophages and the mesothelial cells are both stimulated by dialysis solutions to produce a myriad of cytokines, prostaglandins, and matrix proteins, such as collagen and hyaluronan. Hyaluronan is secreted into the peritoneal cavity and is deposited within the tissue. This along with fibronectin and other macromolecules that escape the inflamed endothelium results in a thickening and scarring of the submesothelial peritoneum. Stimulation of tissue fibroblast results in greater deposition of matrix materials with further scarring. Parenchymal cells such as skeletal muscle may just be innocent bystanders, but their true role has not been elucidated. It is anticipated that the system in Figure 5 will become more complex and more specific over the next several years.

To truly understand the peritoneal transport system in living human beings who utilize this for solute and water excretion over many years, one must combine Figures 1 and 5. Overlying the physiologic transport resistances of the various portions of the tissue space is a highly reactive cellular system of resident macrophages, mesothelial cells, fibroblasts, and endothelial cells. All of these combine in an integrated system of inflammatory response, which has the capacity to bring about over time marked alterations in the peritoneal transport system and its functional characteristics. It is anticipated that many advances will be made in the description of these changes and how to combat them and improve intraperitoneal therapies.

Glossary

Distributed model: A conceptual and integrative model of transperitoneal transport in which the microvasculature is distributed uniformly in the tissue.

Intraperitoneal chemotherapy: Regional treatment of peritoneal metastases with intraperitoneal solutions containing high concentrations of chemotherapeutic agents.

Peritoneal dialysis: A therapeutic technique used to remove waste metabolites and excess water from patients with kidney failure.

Three-pore model: A classic conceptual model of transendothelial transport.

Further Reading

Chambers, R., and Zwiefach, B. W. (1946). Functional activity of the blood capillary bed, with special reference to visceral tissue. *Ann. N. Y. Acad. Sci.* **46**, 683–694. *This is the classic description of the topography of the mesenteric circulation.*

Flessner, M. F. (1991). Peritoneal transport physiology: Insights from basic research. *J. Am. Soc. Nephrol.* **2**, 122–135. *This integrative review of the peritoneal transport system includes discussion of the role of the peritoneum, interstitium, lymphatics, and blood circulation. It includes quantitative data obtained in a rat model.*

Flessner, M. F., and Dedrick, R. L. (2000). Intraperitoneal chemotherapy. In *Textbook of Peritoneal Dialysis* (R. Gokal, R. Khanna, R. Th. Krediet, and K. D. Nolph, eds.), pp. 809–827. Dordrecht: Kluwer Academic.

This chapter reviews the details of the multicompartment model of intraperitoneal drug delivery and presents equations and parameter values for various drugs of interest.

Rippe, B., Rosengren, B. I., and Venturoli, D. (2001). The microcirculation in peritoneal dialysis. *Microcirculation* **8**, 303–320. *This is a detailed review of the three-pore model of transcapillary transport in the peritoneum. Included are the detailed equations used to describe the pore model.*

White, R., and Granger, D. N. (2000). The peritoneal microcirculation in peritoneal dialysis. In *Textbook of Peritoneal Dialysis* (R. Gokal, R. Khanna, R. Th. Krediet, and K. D. Nolph, eds.), pp. 107–134. Dordrecht: Kluwer Academic. *This is a detailed description of the peritoneal microvasculature, including effects of vasoactive substances on blood flow and leukocyte interactions with the endothelium during states of inflammation.*

Capsule Biography

Dr. Flessner's background combines biomedical engineering with medicine. He has been the Director of Nephrology and the John Bower Professor of Nephrology and Hypertension, University of Mississippi Medical Center, since 2001. He is supported by grants from the NIH and the American Heart Association to study intraperitoneal immunotherapy and chronic inflammation in the peritoneal cavity.

CHAPTER 67

Peritoneal Microvessels and High Peritoneal Solute Transport Rate (PSTR)

[1]Miwako Numata, [2]Makio Kawakami, [3]Bengt Lindholm, and [1]Masaaki Nakayama

[1]Division of Kidney and Hypertension, Department of Internal Medicine, Jikei University School of Medicine, Tokyo, Japan

[2]Department of Pathology, Jikei University School of Medicine, Tokyo, Japan

[3]Divisions of Renal Medicine and Baxter Novum, Karolinska Institutet, Huddinge University Hospital, Stockholm, Sweden

Introduction

It is well known that peritoneal solute transport rate (PSTR) increases during peritonitis in peritoneal dialysis (PD) patients. However, there was great concern recently over the possibility that long-term PD itself increases PSTR. An increasing number of studies report that an increased PSTR may have a negative impact on both the technique survival and the patient survival in PD patients. Furthermore, it was reported that encapsulating peritoneal sclerosis, which is a serious complication for PD patients, is related to high PSTR.

Several factors such as increases in the peritoneal contact area, tissue diffusivity, capillary surface area, and capillary permeability could conceivably explain an increase in PSTR. The contact area between the dialysate and the peritoneum is the first barrier to solute transport between the underlying microvessels and the dialysis solution. The wall of microvessels is considered to be the main functional barrier to the peritoneal solute transport as demonstrated by Flessner et al. It is conceivable that increasing vascular surface area could be caused by both increase of the amount and dilatation of peritoneal perfused capillary.

Peritoneal Microvascular Architecture

Like other tissues, a series of systemic vascular flows can be recognized in the biopsy specimens of peritoneal tissue. Specifically, the series is composed of a small artery, arteriole, precapillary arteriole, capillary, postcapillary venule, venule, and small vein.

The small artery has a triple-layered structure including intima that is covered with endothelial cells, media constituting a layer composed of smooth muscle cells, and adventitia that is composed mainly of collagen and a few fibroblasts. The first two layers are distinguished by the formation of internal elastic lamina; the adventitia, however, is an aggregate of collagen fibers that does not form a membranous structure like an internal elastic lamina. The arteriole is composed of two layers—one of endothelial cells and the other of the media. The latter is made up of one or two layers of smooth muscle cells, but the growth of the elastic lamina is not well developed. The inner vascular diameter is less than 15 μm. The morphological characteristics of the capillary are low muscular volume and composition of only endothelial cells, basement membrane, and a few pericytes that form a reticular periphery; they also lack an intima and media. The inner diameter is less than 10 μm for the capillary. More peripherally located postcapillary venules are of diameter 10 to 30 μm and have a richer loose media component composed of reticular frame lamella and smooth muscle cells. The larger venule and small vein are composed of endothelial, intima that is made up of cells, and smooth muscle cell–based media. The inner diameter is less than 50 μm for venules and 100 μm for small veins.

Three-Pore Theory

Three-pore theory explains water and solute transport across the peritoneal microvessel during peritoneal dialysis. For this transport process, there are three kinds of pores: a large number of small pores (radius 40 to 50 Å) for low molecular weight, a very low number of large pores (radius 250 Å) for macromolecules, and ultrasmall pores (radius 3 to 5 Å) for water transport but rejecting the transfer of solutes. Aquaporins, especially aquaporin-1, which was identified recently, are the proteins constituting these transendothelial water channels; they are assumed to be ultrasmall pores.

Estimation of Peritoneal Solute Transport Rate

The peritoneal transport of low-molecular-weight solutes has traditionally been investigated with the peritoneal equilibration test (PET). Parameters that can be calculated from such a standardized test are the mass transfer area coefficient (MTAC) of a solute, its dialysate-to-plasma (D/P) ratio, and the clearance of that solute. The MTAC or the D/P ratio is determined mainly by the peritoneal vascular surface area, because this solute transport is the major process in PD. It is suggested that under basal circumstances only about 25 to 50 percent of the peritoneal capillaries are perfused. Based on D/P creatinine, practically used as small solute transport rate, patients have been classified into four transport categories: low, low-average, high-average, and high.

The Relationship between PSTR and Alterations of Peritoneal Microvessels in PD Patients

As described earlier, many reports have been published with the theory that peritoneal transport is related to peritoneal vascular surface area; however, there is only one report, from the present authors, that investigated this phenomenon as the clinical fact. We reported alterations in the peritoneal capillary and postcapillary venule (i.e., the main part of water and solute transport; named "microvessel" in the report) in biopsy specimens of peritoneum obtained from PD patients.

The parietal peritoneal tissue was obtained from PD patients and stained by the modified Masson-Noguchi method (Figure 1) before using an image analysis program. The relative microvessels area (RVA, calculated as total area of microvessels/total area of peritoneal field) and relative microvessels number (RVN, calculated as number of microvessels/total area of peritoneal field) were determined for each specimen. The ratio RVA/RVN was used to assess the average area of microvessels. The D/P for creatinine showed a significant positive correlation with both RVA (Figure 2) and RVA/RVN (Figure 3), but not with RVN (Figure 4). This indicates that the dilatation of each perfused microvessel has an impact on PSTR. The D/P for β2 microglobulin (middle size molecule) correlated with RVA (Figure 5) with weaker correlation than the D/P for creati-

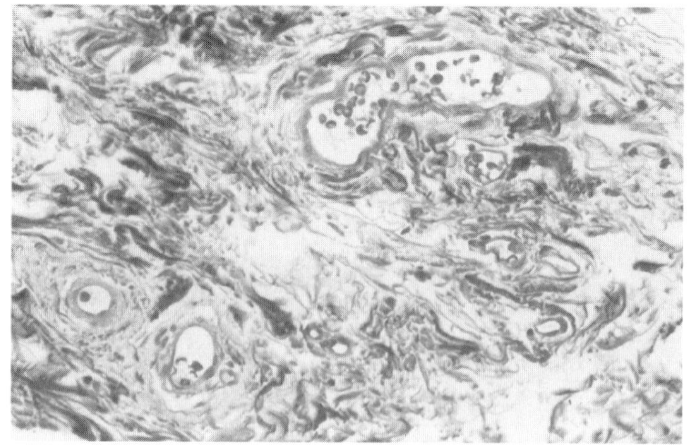

Figure 1 Human peritoneal membrane stained with Masson-Noguchi modified stain (original magnification 200×). (see color insert)

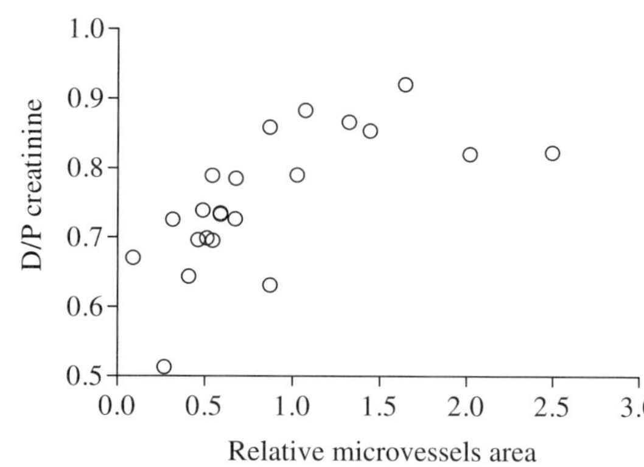

Figure 2 Correlation between the relative microvessels area and D/P creatinine (rho = 0.77, $p < 0.001$, $n = 22$).

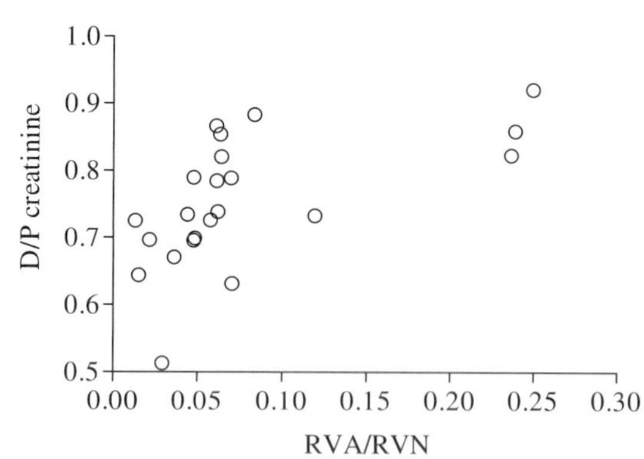

Figure 3 Correlation between D/P creatinine and microvessel cross-section (RVA/RVN) (rho = 0.66, $p = 0.003$, $n = 22$).

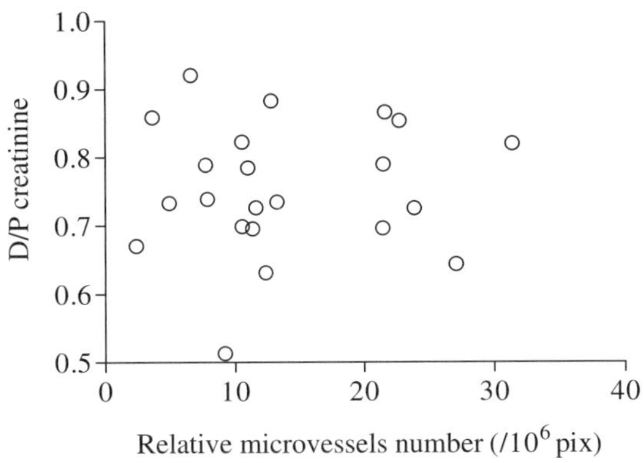

Figure 4 Relationship between D/P creatinine and relative microvessels number (rho = 0.02, p = NS, n = 22).

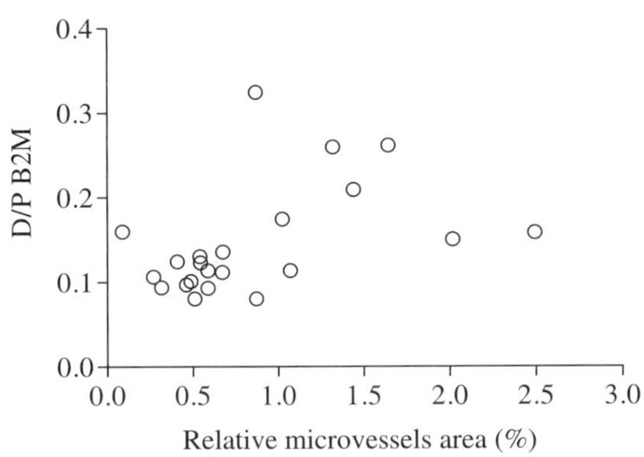

Figure 5 Correlation between the relative microvessels area and D/P β2 microglobulin (rho = 0.52, p = 0.018, n = 22).

nine, but not with RVN and RVA/RVN. No correlation was found between D/P albumin (large molecule) and RVA (Figure 6) or RVN. These observations cannot be explained only by the presence of microvessel dilatation, and therefore other pathology may be present in the PD peritoneum as well.

Increased microvessel surface area may be well reflected by the perfusion volume. In the study mentioned earlier, it was not possible to take into consideration the possible influence of the actual perfusion state of the microvessels. The various microvascular networks are open or closed at different times, and this might have an impact on the transport properties of the membrane. Recently intravital microscopy was introduced in the PD field. It may be helpful to clarify the peritoneal microvessels status during PD.

In PD animal models, the peritoneum has been reported to gain in thickness, and presumably this occurs also in patients on PD. Thus, it can be speculated that the thickening of the peritoneum may provide an increase in the total capillary length, while the density of microvessels does not decrease. However, this has not been elucidated in the clinical study.

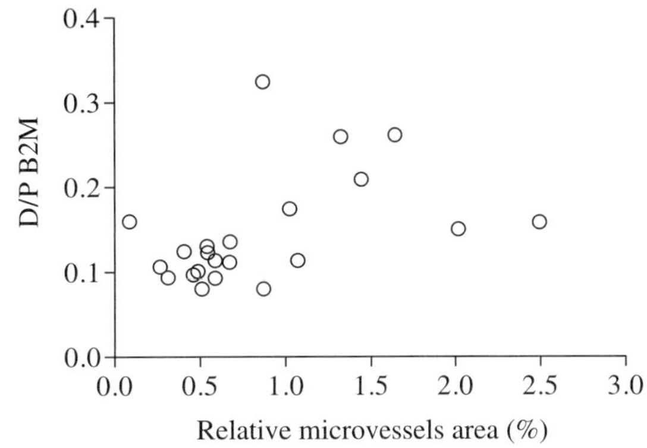

Figure 6 Relationship between the relative microvessels area and D/P albumin (rho = –0.01, p = NS, n = 22).

Candidate Molecules Related to Angiogenesis and/or Dilatation

The main pathology that alters the vasculatures of the PD peritoneum has not been clarified in patients on PD. However, accumulating studies have reported the important role of some cytokines.

Based on the experimental study, vascular endothelial growth factor (VEGF) is assumed to be the main factor enhancing angiogenesis and vascular permeability. It is expressed in the endothelium lining peritoneal capillaries in the human peritoneal membrane. VEGF is detected in the dialysate from both PD rat model and PD patients. In the rat model it was reported that the dialysate level of VEGF correlates with small-solute transport. Furthermore, VEGF concentration in the effluent is increased as a function of time on CAPD, and it is very closely related with PSTR in PD patients. It was reported that VEGF blockade normalized PSTR in streptozotosin-induced diabetic rats.

VEGF stimulates nitric oxide synthase (NOS) production, leading to the production of nitric oxide (NO). Endothelial NOS (eNOS) is required for angiogenesis and vascular permeability driven by VEGF, and VEGF upregulates eNOS production. Therefore, NO plays an important role in the regulation of vascular tone and permeability. In fact, it is reported that NOS production is increased in the peritoneum of long-term CAPD patients. There is an interesting recent report that eNOS4(a/b) gene polymorphism is associated with PSTR at the start of PD in uremic Chinese patients.

Vasculopathy

It is suggested that change of peritoneal vascular characteristics develops in PD patients. Honda et al. has reported that peritoneal vascular hyalinization is observed especially in the media of postcapillary venules to small venules in long-term PD patients with high PSTR. They observed that formation of carboxymethyl lysine (CML), an advanced glycation end product (AGE), in the peritoneum correlates with the development of microvascular sclerosis in PD patients, and this vascular change is associated with high PSTR. They also pointed out that the morphological changes of the peritoneal vasculature are similar to those associated with diabetic microangiopathy.

AGE accumulation in CAPD peritoneum, especially in the vascular area, increases with time on CAPD, and it may be directly related with PSTR and peritoneal fluid absorption rate. AGE formation could be driven by several factors such as high-glucose solutions, glucose degradation products (GDPs), which are present in the unused PD solution as well as the uremic circulation. It was reported that GDPs could induce VEGF production; on the other hand, NO inhibits AGE (pentosidine) formation by scavenging free radicals and by inhibiting carbonyl compound formation.

AGEs influence cellular properties that allow the binding of AGEs to specific receptors, of which the best characterized is the receptor for AGEs (RAGE). RAGE is a member of the immunoglobulin superfamily and is expressed by the endothelium, smooth muscle cells, and mononuclear phagocytes, cells central to both vascular homeostasis and pathogenesis of vascular lesions. Recently it was reported that RAGE is expressed in peritoneal mesothelial cells and that anti-RAGE antibody does not affect the PSTR, but does affect submesothelial fibrosis in diabetic rat. These findings suggest the possible role of mesothelium for developing high PSTR through AGE–RAGE interaction.

In summary, it is suggested that the dilatation of microvessel, the increasing number of microvessels, and the progression of vasculopathy play a role in increased PSTR. However, at present, there is little clinical information about cause(s) and mechanisms for the observed alterations in the peritoneal microvessels in PD patients. Further studies are needed in this area.

Glossary

Advanced Glycation Endproduct: The products by non enzymatic chemical reaction between glucose or glucose degradation products and proteins.

Peritoneal dialysis: Dialysis method using peritoneal membrane. Dialysate are infused in and remain in the peritoneal cavity for a time and are drained. The wall of microvessels is considered to be the main functional barrier to the solute transport between dialysate and blood.

Peritoneal solute transport rate: It is associated with an increased surface area of peritoneal microvessels. At the same time, other factors such as increases in the peritoneal contact area and tissue diffusivity could conceivably explain an increase in PSTR.

References

Combet, S., Miyata, T., Moulin, P., Pouthier, D., Goffin, E., and Devuyst, O. (2000). Vascular proliferation and enhanced expression of endothelial nitric oxide synthase in human peritoneum exposed to long-term peritoneal dialysis. *J. Am. Soc. Nephrol.* **11**, 717–728.

De Vriese, A. S., Tilton, R. G., Stephan, C. C., and Lameire, N. H. (2001). Vascular endothelial growth factor is essential for hyperglycemia-induced structural and functional alterations of the peritoneal membrane. *J. Am. Soc. Nephrol.* **12**(8), 1734–1741.

Flessner, M. F. (1991). Peritoneal transport physiology: Insights from basic research. *J. Am. Soc. Nephrol.* **2**(2), 122–135. *This article provides a detailed review of the mechanisms of peritoneal solute transport using some models.*

Honda, K., Nitta, K., Horita, S., Yumura, W., and Nihei, H. (1996). Morphological changes in the peritoneal vasculature of patients on CAPD with ultrafiltration failure. *Nephron* **72**(2), 171–176.

Honda, K., Nitta, K., Horita, S., Yumura, W., and Nihei, H. (1999). Accumulation of advanced glycation end products in the peritoneal vasculature of continuous ambulatory peritoneal dialysis patients with low ultra-filtration. *Nephrol. Dial. Transplant.* **14**(6), 1541–1549. *This article elucidates the relationship between peritoneal transport and vasculopathy.*

Nakayama, M., Kawaguchi, Y., Yamada, K., Hasegawa, T., Takazoe, K., Katoh, N, et al. (1997). Immunohistochemical detection of advanced glycosylation end-products in the peritoneum and its possible pathophysiological role in CAPD. *Kidney Int.* **51**(1), 182–186.

Nakayama, M., Yamamoto, H., Ikeda, M., Hasegawa, T., Kato, N., Takahashi, H., Otsuka, Y., Yokoyama, K., Yamamoto, R., Kawaguchi, Y., and Hosoya, T. (2002). Risk factors and preventive measures for encapsulating peritoneal sclerosis—Jikei experience 2002. *Adv. Perit. Dial.* 18: 144–148.

Numata, M., Nakayama, M., Nimura, S., Kawakami, M., Lindholm, B., and Kawaguchi, Y. (2003). Association between an increased surface area of peritoneal microvessels and a high peritoneal solute transport rate. *Perit. Dial. Int.* **23**, 116–122. *This article elucidates the relationship between peritoneal solute transport rate and vascular surface area with human subjects.*

Wong, T. Y.-H., Szeto, C.-C., Szeto, C. Y.-K., Lai, K.-B., Chow, K.-M., and Li, P. K.-T. (2003). Association of ENOS polymorphism with basal peritoneal membrane function in uremic patients. *Am. J. Kidney Dis.* **42**, 781–786.

Zweers, M. M., de Waart, D. R., Smit, W., Struijk, D. G., and Krediet, R. T. (1999). Growth factors VEGF and TGF-beta1 in peritoneal dialysis. *J. Lab. Clin. Med.* **134**, 124–132.

Capsule Biography

Miwako Numata, M.D., PhD, is a nephrologist in Jikei University Hospital, Tokyo, Japan. She worked at Baxter Novum, Karolinska Institutet, as guest scientist in 2001–2003. Her research focuses on alteration of peritoneal function and peritoneal structure in peritoneal dialysis patients.

Masaaki Nakayama, M.D., Ph.D., is a lecturer in the Department of Nephrology and Hypertension, Jikei University School of Medicine, which is one of the leading centers of peritoneal dialysis in Japan. His research focuses on the pathology of peritoneal injury in CAPD patients.

Bengt Lindholm, M.D., Ph.D., has been head of the Division of Baxter Novum, Karolinska Institutet, Stockholm, Sweden, since 1993. His research focuses on peritoneal dialysis, malnutrition, inflammation, and atherosclerosis and the role of genetics in these conditions in end-stage renal disease patients. His work is supported by Baxter Healthcare and Karolinska Institutet.

Makio Kawakami, M.D., PhD is a pathomorphologist in the Department of Pathology, Clinical Service, at Jikei University Hospital. He has long worked on the functional unit of the various organs from the viewpoint of structural alteration. Now, his interest is focusing on the peritoneal device as an analog of the nephron.

CHAPTER 68

The Gastrointestinal Microcirculation

Thorsten Vowinkel and D. Neil Granger

Department of Molecular and Cellular Physiology, Louisiana State University Health Sciences Center, Shreveport, Louisiana, and Department of General Surgery, University of Münster, Münster, Germany

More than 25 percent of cardiac output normally flows through the gastrointestinal tract. However, the rate of blood perfusion through the gastrointestinal (GI) tract is dynamic, responding to both the metabolic requirements of local tissue and the needs of vital organs. For example, the ingestion of a meal is associated with large increases in GI blood flow, while profound reductions in flow are elicited by stresses (e.g., hemorrhage) that threaten to deprive the brain and heart of their normal blood supply. Although blood vessels in the different organs comprising the GI tract share common mediators and mechanisms of blood flow regulation, some differences related to tissue function have been demonstrated. The vascular supply to the gastrointestinal mucosa is particularly well suited for the absorptive and secretory functions of this tissue in that it allows for a high rate of blood flow, has a large exchange surface area, and exhibits characteristics that allow for easy permeation of nutrients and water, yet proteins are largely retained within the plasma compartment. Disease processes affecting the GI circulation can be associated with either impaired blood perfusion (e.g., ischemia) or over-perfusion (portal hypertension).

Gastric Microcirculation

Gastric blood flow plays an important role in sustaining the normal physiologic functions of the stomach, and it helps to protect the gastric mucosa against ulcer formation. Intrinsic regulatory mechanisms have evolved to ensure that blood flow is adjusted to meet the energy-demanding processes of gastric secretion and motility. Gastric blood flow also helps to maintain a barrier against back-diffusion of luminal acid, thereby preventing mucosal damage and ulceration. Impairment of gastric blood flow renders the mucosa vulnerable to the damaging actions of gastric juice as well as ingested agents, such as ethanol, aspirin, and bacteria (e.g., *Helicobacter pylori*).

The parallel-coupled capillary networks of the gastric muscular layer and the mucosa are under separate control, responding independently to tissue metabolism, other local factors, and extrinsic neural input. Between meals, blood flow in the mucosal layer is about six times higher than that of the muscle layer, and approximately 75 percent of total gastric blood flow is distributed to the mucosa, with 25 percent directed to the muscle layer. This intramural distribution of blood flow is altered when either of the two layers becomes functionally active—that is, when the mucosa is stimulated to produce acid, mucosal blood flow (and its percent of total flow) preferentially increases.

Gastric blood flow is controlled by neural, humoral, and metabolic factors. Sympathetic activation elicits reductions in total gastric blood flow and mucosal flow through arteriolar constriction. Parasympathetic nerves exert a tonic vasodilatory influence on gastric arterioles, with vagotomy resulting in a reduction in blood flow. Gastrin and histamine, both powerful stimulants of gastric acid secretion, increase mucosal blood flow. Oxygen consumption by the stomach increases in proportion to acid production. Changes in both blood flow and oxygen extraction assist in meeting the demand for additional oxygen in the acid-secreting stomach. When gastric blood flow is reduced, however, acid secretion and blood flow may fall in parallel, owing to the fact that the rate of oxygen delivery to the parietal cells is limited by blood flow.

The gastric microcirculation contributes to gastric ulcer formation in several ways. Capillary transport of parietal cell–derived bicarbonate normally plays an important role in protecting the surface epithelium against acid-induced injury and ulceration. The mucosal capillaries originating near the gastric pits transport bicarbonate toward the mucosal surface, where it can diffuse into the interstitial compartment beneath the surface epithelial cells. The latter cells transport bicarbonate into the gastric lumen where it can buffer luminal acid. As a consequence, there is an inverse relationship between gastric mucosal injury and the rate of vascular bicarbonate delivery to the mucosal surface.

Intestinal Microcirculation

The blood circulation plays an important role in the support of intestinal functions, such as propulsion of chyme and assimilation of ingested nutrients. Intrinsic regulatory mechanisms allow the intestine to adjust the distribution of blood flow between the muscular and mucosal layers in accordance with local metabolic needs. An extensive network of collateral channels within and external to the gut wall helps to ensure adequate intestinal blood flow.

Intestinal blood flow accounts for 10 to 15 percent of the resting cardiac output (500 to 750 mL/min) in the adult human. There appears to be an oral-to-anal gradient in blood flow (mL/g tissue) along the small intestine. In the resting state, approximately 65 percent of the total intestinal blood flow is directed to the mucosa, 25 percent to the muscularis, and the remainder to the submucosa. This distribution of flow within the bowel wall is usually attributed to the greater metabolic demand of the mucosa. Stimulation of epithelial transport processes favors improved mucosal perfusion while enhanced motor activity redistributes blood flow toward the muscle layers.

Extrinsic control of intestinal blood flow is exerted by neural and humoral factors. Activation of parasympathetic nerves usually results in vasodilation (increased blood flow) mediated by acetylcholine. Sympathetic nerve stimulation elicits vasoconstriction (decreased blood flow) that is mediated by norepinephrine. This α-adrenergic vasoconstriction is short-lived because intestinal arterioles escape from the constrictor influence of norepinephrine, resulting in partial restoration of normal blood flow (autoregulatory escape). Local release of adenosine appears to mediate this autoregulatory escape. Hormones such as VIP, cholecystokinin, and secretin can induce vasodilation and increase blood flow, while angiotensin II and vasopressin are potent constrictors of intestinal arterioles. Indeed, a large proportion of basal vascular tone in the intestine can be attributed to circulating angiotensin II and vasopressin.

Intrinsic control of intestinal blood flow is mediated by both metabolic and nonmetabolic factors. Ingestion of a meal results in an increase in both intestinal oxygen consumption and blood flow. The postprandial hyperemia is directly coupled to the increase in intestinal oxygen consumption. For any given increase in oxygen consumption, a greater initial oxygen extraction results in a larger postprandial hyperemic response. If the initial oxygen extraction is low, then the postprandial hyperemia is minimal and the increased oxygen demand is met primarily by an increase in oxygen extraction. The opposite holds if the initial oxygen extraction is high.

The postprandial hyperemia is confined to that segment of intestine directly exposed to chyme; segments distal to the bolus of chyme have normal resting blood flow. Of the hydrolytic products of food digestion, luminal glucose and oleic acid are capable of eliciting an intestinal hyperemia. Intraluminal glucose presumably elicits a hyperemia due to stimulation of absorptive processes, since 2-deoxyglucose (which is not absorbed) does not elicit a hyperemia. The glucose-induced hyperemia is mediated by metabolic factors, such as low tissue po_2, and adenosine release. The same metabolic factors contribute to the oleic acid–induced functional hyperemia; however, a portion of the hyperemia can be attributed to oleic acid–induced irritation of the mucosa, which is linked to local release of vasoactive intestinal peptide. The importance of active transport of nutrients to the postprandial hyperemia is best exemplified by the differential responses of the jejunum and ileum to luminal bile or bile salts. In the jejunum, bile does not elicit a hyperemia, whereas in the ileum (where bile salts are actively transported), luminal bile produces a profound hyperemic response.

Pathophysiology of the GI Circulation

Ischemia–Reperfusion (I/R)

Ischemic damage to the GI tract occurs when splanchnic blood flow falls to a level at which delivery of oxygen and other nutrients is insufficient to maintain oxidative metabolism and hence cell integrity. Reduced blood flow to the gastrointestinal tract may occur during generalized nonocclusive ischemia (e.g., circulatory shock and congestive heart failure, especially in those treated with cardiac glycosides) and in occlusive disorders (e.g., emboli, atherosclerosis, thrombosis) that primarily involve the mesenteric circulation. Occlusion of a major intestinal artery does not result in the expected reduction in intestinal blood flow. In adult animals, occlusion of a branch of the superior mesenteric artery results in only a 30 to 50 percent reduction in intestinal blood flow, which is attributed to the extensive network of intramural and extramural collateral channels. In neonatal animals, which have less developed collateral channels, a similar arterial occlusion reduces intestinal blood flow by 70 percent. This may explain why the neonatal intestine is more vulnerable to ischemic necrosis than adult intestine.

Increases in the capillary filtration coefficient and a reduction in the osmotic reflection coefficient have been demonstrated in the GI tract after I/R. The permeability response is derived from an increase in the number of large (200 Å radius) pores, while the small-pore (50 Å radius) population is unaffected. In the intestine, the increase in the capillary filtration coefficient observed after I/R is not solely a result of increased capillary surface area.

Ischemic injury in the intestine appears to be related, either primarily or secondarily, to the effects of tissue hypoxia. Possible mechanisms of mucosal injury induced by tissue hypoxia include depletion of high-energy phosphates necessary to produce protective substances; accumulation of histamine, leading to increased microvascular permeability; production of metabolic acidosis, leading to release of lysosomal enzymes and cellular digestion; conversion of xanthine dehydrogenase to xanthine oxidase, an enzyme that can produce cytotoxic oxygen-derived free radicals during reoxygenation; and the attraction and activation of circulating granulocytes that can injure tissue by producing proteases and oxidants. There is considerable evidence that implicates both reactive oxygen species and infiltrating granulocytes in I/R injury in the GI tract. The I/R-induced microvascular responses in the GI tract are comparable to those observed during an intense inflammatory response. With the reintroduction of molecular oxygen at reperfusion endothelial cells assume both a proinflammatory and prothrombogenic phenotype that is characterized by the recruitment of adherent leukocytes and platelets in postcapillary venules. The platelets act to amplify the neutrophil activation response while neutrophils per se appear to mediate the endothelial barrier dysfunction associated with I/R. Extravasated neutrophils then contribute to the mucosal dysfunction and epithelial cell necrosis after I/R by releasing a variety of proteases and reactive oxygen metabolites.

Chronic Portal Hypertension

Whereas an acute elevation in portal venous pressure is likely to elicit a myogenic-mediated constriction of splanchnic arterioles, chronic portal hypertension tends to exert a vasodilatory influence on the splanchnic vasculature. Furthermore, chronic portal hypertension has a significant impact on other regional vascular beds and on systemic hemodynamics. Blood flow to the gastrointestinal tract, kidneys, and skeletal muscle is significantly elevated. This presumably results from an increase in circulating vasodilators (e.g., glucagon) and a decrease in vascular sensitivity to vasoconstrictors (e.g., norepinephrine). The widespread dilation of arterioles results in a reduction of peripheral vascular resistance and a corresponding reduction of arterial blood pressure. In addition, cardiac output is elevated as a consequence of the increased venous return associated with the splanchnic and peripheral vasodilation. The elevated portal pressure results in the opening of portosystemic shunts to divert portal blood from the liver and reduce portal pressure. The increase in portal pressure impairs venous drainage from the spleen into the portal vein. This results in the accumulation of blood within, and distension of, the spleen (splenomegaly).

Organ blood flow is determined by the arterial–venous pressure gradient and vascular resistance. It follows then that portal pressure is determined by portal venous inflow and portal venous resistance. When portal vascular resistance is normal, an increase in portal venous flow will produce a proportional increase in portal pressure. However, when portal vascular resistance is increased, the relationship between portal pressure and portal venous flow is shifted upward and to the left. At any given portal venous inflow, an increased portal vascular resistance will result in an increase in portal pressure. Portal pressure can be further increased when there is a concomitant increase in portal venous flow and portal vascular resistance. Indeed, the latter instance appears to reflect the vascular changes that account for the elevated portal pressure that is observed in some experimental models of chronic portal hypertension and is likely to account for the portal hypertension associated with some forms of liver disease. With a portal vascular resistance that is 40 percent higher in the portal hypertensive than in the control state, it is predicted that increased portal inflow and increased portal vascular resistance account for 40 percent and 60 percent of the increase in portal pressure, respectively.

The portal hypertensive state leads to the development of collaterals (mostly along the esophagus; esophageal varices) to shunt blood from the congested portal vein, around the liver, to the systemic circulation (portosystemic shunting). Since a large proportion of portal venous blood bypasses the liver via portosystemic shunting, the hepatic degradation of different compounds, including circulating vasodilators, such as glucagon, is reduced. The diminished catabolism of circulating vasodilators increases their concentration in the plasma, allowing these agents to relax arteriolar vascular smooth muscle and reduce splanchnic vascular resistance. Another important action of some of the vasodilators that accumulate in chronic portal hypertension (e.g., glucagon) is to reduce the sensitivity of splanchnic arterioles to vasoconstrictors such as norepinephrine, vasopressin, and angiotensin. The net result of the direct and indirect actions of the accumulated circulating vasodilators is an increased splanchnic blood flow, which serves to perpetuate the portal hypertensive state.

Glossary

Chyme: Hydrolytic products of food within the lumen of the intestine.

Portal hypertension: An abnormal elevation of hydrostatic pressure within the portal veins that drain the GI tract and empty into the liver.

Portosystemic shunting: Diversion of blood flow from the portal vein away from the liver, through either existing or newly formed channels (shunts).

Postprandial hyperemia: Increase in gastrointestinal blood flow that results from ingestion of a meal.

Reperfusion: The restoration of tissue blood flow following a period of ischemia.

Further Reading

Abdel-Salam, O. M., Czimmer, J., Debreceni, A., Szolcsanyi, J., and Mozsik, G. (2001). Gastric mucosal integrity: Gastric mucosal blood flow and microcirculation. An overview. *J. Physiol. Paris* **95**, 105–127. *This article summarizes evidence implicating the gastric microcirculation in protecting the gastric mucosa against ulcer formation.*

Carden, D. L., and Granger, D. N. (2000). Pathophysiology of ischemia–reperfusion injury. *J. Pathol.* 190, 255–266. *This article provides a detailed review of the mechanisms that have been implicated in the pathogenesis of ischemia–reperfusion in the gastrointestinal tract.*

Crissinger, K. D., and Granger, D. N. (1999). Gastrointestinal blood flow. In *Textbook of Gastroenterology* (T. Yamada, ed.), pp. 519–546. Baltimore: Lippincott, Williams & Wilkins.

Granger, D. N., Kvietys, P. R., Korthuis, R. J., and Premen, A. J. (1989). Microcirculation of the intestine. In *Handbook of Physiology. Section 6, The Gastrointestinal System*, Vol. 1: *Motility and Circulation*, Part 2, pp. 1405–1474.

Holzer, P., Livingston, E. H., and Guth, P. H. (1994). Neural, metabolic, physical, and endothelial factors in the regulation of gastric circulation. In: *Physiology of the Gastrointestinal Tract* (L. R. Johnson, ed.), pp. 1311–1330. New York: Raven Press. *An excellent summary of the factors that regulate blood flow in the stomach.*

Tepperman, B. L., and Jacobson, E. D. (1994). Circulatory factors in gastric mucosal defense and repair. In: *Physiology of the Gastrointestinal Tract* (L. R. Johnson, ed.), pp. 1331–1352. New York: Raven Press

Capsule Biography

Thorsten Vowinkel earned his M.D. degree from the University of Würzburg, Germany. He is a resident at the Department of General Surgery, University of Münster, Germany, and currently works as a postdoctoral fellow at the Department of Molecular and Cellular Physiology at Louisiana State University Health Sciences Center in Shreveport. His research interests include microvascular responses to intestinal ischemia–reperfusion and inflammation.

D. Neil Granger, Ph.D., is Boyd Professor and Head of the Department of Molecular and Cellular Physiology at Louisiana State University Health Sciences Center in Shreveport. He has served as President of the Microcirculatory Society, Editor-in-Chief of *Microcirculation*, and was elected to serve as President of the American Physiological Society.

SECTION D

Hematopoietic System

CHAPTER 69

The Bone Marrow Vasculature

Jorge Domenech and Pierre Charbord
Faculty of Medicine, University François Rabelais, Tours, France

The bone marrow (BM) is the site of hematopoiesis in the adult mammal. The BM vasculature that comprises a specific structure, the sinus where exchange between BM parenchyma and circulating blood takes place, not only has a structural role, but also is actively regulating hematopoietic stem cell (HSC) self-renewal/differentiation and hematopoietic cell trafficking. The molecular mechanisms accounting for HSC homing and engraftment and for HSC mobilization from BM to blood begin to be understood. The BM vasculature contributes to the physiopathology of a number of blood neoplastic diseases.

Hematopoiesis and Its Microenvironment

Hematopoiesis is the process of blood cell formation. In adult mammals it takes place in the bone marrow (BM). In normal humans hematopoietic BM is located in flat bones (skull, sternum, ribs, pelvis) and vertebrae. It may extend in case of increased hematopoietic demand (e.g., in anemias) or in some blood neoplastic diseases (e.g., chronic or acute leukemias) to long bones (humerus, femur, tibia). In rodents long bones are hematopoietic under normal conditions.

During development hematopoiesis occurs at different sites. In the embryo, hematopoiesis is first detected in the yolk sac and on the ventral floor of the dorsal aorta. In the fetus, hematopoiesis develops in the liver. By the second half of gestation, hematopoiesis finds its final location, that is, the BM. Spleen also constitutes a late fetal site of hematopoiesis, transient in humans, but persisting into adulthood in the mouse.

All blood cells derive from a small compartment of hematopoietic stem cells (HSCs). HSCs are characterized by two cardinal properties: They are multipotent and they self-renew. HSCs give rise to hematopoietic progenitors, in turn the progeny of precursors. The mature cells having completed their differentiative cycle must enter the circulating blood to be delivered to the different peripheral tissues. The finely tuned process of release into blood of mature cells is taking place in a vascular structure specific for the BM, the BM sinus.

In the adult mammal, HSCs are located in the BM. During development HSCs migrate from one site to the other. For example, in the fetus HSCs migrate from the liver to the BM via the circulating blood where HSC concentration is high. In the adult such HSC migration is restricted and the HSC blood concentration is small. When patients lacking BM hematopoiesis because of a blood disease (aplastic anemia) or following myelotoxic treatment (heavy radio- or chemotherapy) are given allogeneic HSCs by intravenous injection, the stem cells circulate until they settle within the marrow tissue. This process called engraftment is experimentally duplicated by injecting allogeneic or syngeneic BM HSCs to animals that have been lethally irradiated and can be rescued by the marrow graft. Instead of being recovered from the BM, HSCs can be mobilized from BM to peripheral blood using myelotoxic chemotherapeutic agents and/or hematopoietic growth factors (HGFs), in particular granulocyte colony-stimulating factor (G-CSF). This procedure is in wide use since it allows the collection of HSCs from the peripheral blood, avoiding the surgical act of marrow collection under general anesthesia.

The BM tissue is made essentially of hematopoietic cells packed in the hematopoietic parenchyma (the marrow cords or "logettes"). The nonhematopoietic cells found in the BM constitute the hematopoietic microenvironment, consisting of endosteal cells lining bone trabeculae, of scattered fat-laden adipocytes, and of vascular cells forming the various vascular structures. The description of long-term marrow cultures where HSCs are maintained for months in mice and weeks in humans has shown that "stromal" cells generated in the adherent layer are required for survival, proliferation, and differentiation of HSCs. A subset of stromal cells forms the hematopoietic niche where HSCs in physical contact with stromal cells maintain the adequate balance between self-renewal and differentiation. In vivo, the BM counterpart for stromal cells appears to be a subpopulation of microenvironmental cells with vascular smooth muscle cell characteristics. Such a cell population belongs to the BM vasculature. Cells from the BM vasculature are therefore essential for blood cell formation since they regulate both HSC self-renewal/differentiation and hematopoietic cell trafficking.

The Vascular Network

As for other vasculatures throughout the body, the BM vasculature is a network of arteries, arterioles, capillaries, and veins. However, there is a structure highly specific for the BM, which is the marrow sinus that connects capillaries to the venous system.

The Network

In mouse long bones, the arterial blood supply comes from two sources. The first source is the nutrient artery that penetrates the bony cortex through the nutrient canal. In the marrow cavity it bifurcates into ascending and descending arteries, which give rise to radial branches that travel to the inner surface of the cortex and penetrate it. In the cortex, they become capillaries forming the bone canalicular system, then reenter the marrow parenchyma where they give rise to the branching network of marrow sinuses. Blood from the sinuses is collected in a larger, central sinus from which emissary veins depart, crossing back the cortex to enter the systemic venous circulation. The second source of blood supply is derived from muscular arteries giving rise to periosteal capillaries that enter the cortex, where they anastomose with the radial capillaries issuing from the nutrient artery.

In long bones (in particular in rodents where they have been extensively studied), the concentric structure of the vascular network is conspicuous, with vessels running along the bone longitudinal axis, that is, the nutrient artery and the central sinus. However, in flat bones and in vertebrae, the marrow vascular network appears to be more complex, bony trabeculae of significant thickness being found within the central region and large sinuses being observed in the immediate juxtaosseous region.

BM arteries and arterioles are similar to structures found in other tissues, comprising an intima, a media, and an adventitia. The intima is made of endothelial cells (ECs). The media comprise one or several layer(s) of vascular smooth muscle cells (VSMCs). The adventitia is made of loose connective tissue, with often nerve structures. In mice, a particular cell type, the periarterial adventitial cell, concentrically surrounds both nerves and arterioles.

BM capillaries comprise, as for other capillaries throughout the body, an endothelial lining and, on the abluminal or adventitial side of the ECs, pericytes forming a discontinuous cover.

Sinuses are usually small (5 to 30 μm in diameter in mouse long bones, but they may be larger in human flat bones). They are made of an endothelial lining flanked on the abluminal side by cells with VSMC characteristics. These abluminal cells are called by different names—adventitial reticular cells, myoid cells, barrier cells—depending on the author who first described them. Abluminal cells are separated from ECs by a thin and interrupted basement membrane.

The Cells

ECs express, as usual, the membrane antigens CD31 and CD34 and, in the cytoplasm, the von Willebrand factor. Sinusal ECs are broad and flattened with often overlapping and interdigitated junctions. In some cells, away from the prominent nucleus, the cytoplasm thickness may attenuate to a diameter approaching that of a double plasma membrane, forming fenestrae with a diaphragm. The looseness of the junctions, the presence of fenestrae, and the interrupted underlying basement membrane are features specific for the BM sinus endothelial lining.

VSMCs contain, as usual, myofibers expressing alpha-SM actin (ASMA). Pericytes also express ASMA. Abluminal cells from BM sinuses are all believed to be contractile cells akin to VSMCs: Adventitial reticular cells contain bundles of microfilaments, myoid cells are ASMA-positive, and barrier cells are similar to wound-healing myofibroblasts. Abluminal cells are flattened cells covering, in the unperturbed mouse, approximately 65 percent of the endothelial abluminal surface. However, in situations of stress or in leukemias the coverage may substantially decrease. Abluminal cells also present long cytoplasmic extensions within the cord, being in contact with many surrounding hematopoietic cells. Some myoid cells might even be observed within the parenchyma, without obvious relation to a BM sinus. Hence, myoid or barrier cells form a network joining one sinus to the other. The network is reinforced in case of stress. In particular, in the mouse after administration of the proinflammatory cytokine interleukin-1, the network extends almost continuously from bone endosteal surfaces to medullary sinuses, coursing within the parenchyma and enveloping hematopoietic cells.

In conclusion, the BM sinus is characterized not only by a particular endothelial lining, but also by a sheath of abluminal contractile cells forming a regulated outer vascular coat. The highly specialized cell types found in these specific structures are involved in the regulation of hematopoiesis, that is, both proliferation/differentiation of HSCs and hematopoietic cell trafficking.

The Vasculature during Development

In human long bones, at 6 to 8 gestational weeks (gw), the bone rudiments are entirely cartilaginous. Numerous CD34-positive capillaries are found in the perichondral limb mesenchyme, running parallel to the rudiments. At 8 to 9 gw BM cavities appear: chondrolysis is actively proceeding and capillaries invade the rudiments. At 9 to 10.5 gw the vascular bed develops between ossifying trabeculae. In the diaphyseal region loose connective-tissue chambers are observed. They are centered by an arteriole with an intima of CD34-positive cells and a media of ASMA-positive VSMCs. They are limited from the surrounding vascular space by an endothelial lining of CD34-positive cells flanked by ASMA-positive myoid cells. At this stage, these structures, called *primary logettes*, do not contain hematopoietic cells. At 11 to 15 gw hematopoietic cells appear within the primary logettes. The chambers then extend considerably within the diaphyseal vascular space, being attached to the juxtaosseous tissue by a short pedicule. They are now centered by an artery. The number of myoid cells lying in an abluminal position to the outer endothelial lining or observed within the logette increases. Maximal extension of the primary logette is observed by 15 gw. The central artery is on some sections connected to a perforating artery, a situation similar to what is observed in long bones of adult rodents. From gw 16 onward the hematopoietic parenchyma fills almost all the space within the diaphyseal medullary area, hampering the visualization of logettes. It is only after birth that BM human hematopoiesis declines in long bones. In rodents the sequence of events before and after the onset of BM hematopoiesis (by day 16 post-coitus) has not been studied in detail.

In conclusion, vascular structure formation precedes the onset of hematopoiesis and vascular organization appears to be a prerequisite for hematopoiesis.

The BM Blood Flow

Measurements made after injection of flow tracers (in particular microspheres) whose distribution to the various organs mirrors organ blood flows indicate that in humans the BM blood flow is approximately 10 mL/min/100 g tissue; the hematopoietic BM receives therefore approximately 2 to 4 percent of the cardiac output. Similar values have been found in animals.

Various factors regulate the BM blood flow: cytokines conveyed by the bloodstream such as erythropoietin (Epo) or secreted locally such as G-CSF, endothelium-derived vasoactive substances such as nitric oxide (NO), neurotransmitters, and neuropeptides released by efferent nerve terminals or by vascular or hematopoietic cells.

BM blood flow increased threefold in rats with 50 percent blood loss and in rats supplemented with high doses of Epo. The administration of G-CSF to rats also caused a twofold increase in the BM blood flow. Following injection of either cytokine, cardiac output, mean arterial blood pressure, and ventilation rate remained unchanged, which excludes a generalized systemic effect of the cytokines on the vasculature. The increase in blood flow after Epo was found not only in BM, but also in bone. The increase in blood flow after G-CSF was restricted to the BM; blood flows in nonhematopoietic organs (muscle, small bowel, skin, kidneys), in spleen, and even in the bone itself were unchanged. Epo and G-CSF appear therefore to cause different perfusion patterns, with different effects on the blood supplies of the nutrient artery versus the muscular arteries, as well as on venous collection. Inhibition of NO formation in BM greatly decreased the increase in blood flow observed after bleeding or administration of G-CSF, indicating that the cytokines may induce the synthesis and/or release of the vasodilatory substance by ECs. NO may then affect the tone of nearby VSMCs.

Nerve fibers are observed in BM of humans or rodents, most conspicuously in association with the vasculature. In the mouse, approximately 92 percent of the efferent nerve terminals with many synaptic vesicles have been found nearby arteries and arterioles (in contact with VSMCs or periarterial adventitial cells), while only 3 percent were in contact with sinuses and 5 percent within the hematopoietic parenchyma. In humans, approximately 75 percent of the arteries and arterioles and 6 percent of the capillaries are associated to nerve fibers. In spite of so widespread a BM innervation, whether there is neuronal control of BM blood flow remains controversial; noticeably, the increase in BM blood flow induced by bleeding or administration of G-CSF has not been modified by denervation.

The Marrow Sinus

The sinus is the exclusive site of exchange between the hematopoietic and the circulating blood. It is at this sole location that mature hematopoietic cells (polymorphs, platelets, and red cells) enter the circulation to finalize their life cycle as blood cells and that hematopoietic stem cells (HSCs) enter the marrow parenchyma during engraftment.

BM Sinus ECs: Molecular Aspects

BM sinus ECs express in a specific way a number of molecules. Some cell adhesion molecules (CAMs), such as E-selectin or vascular cell adhesion molecule-1 (VCAM-1), are expressed constitutively while in ECs from nonhematopoietic tissues, their expression is usually induced by proinflammatory cytokines. The heparan sulfate (HS) proteoglycans present on the luminal face of BM ECs bind, with an affinity depending on the sulfatation patterns of the composing disaccharides, chemotactic cytokines (chemokines), such as interleukin-8, macrophage inhibitory protein-1α (MIP-1α), growth-related activity-α(GRO-α), and stromal-derived factor 1 (SDF-1), known to be the major chemotactic factor for HSCs. A current model hypothesizes the existence of a SDF-1 gradient within the hematopoietic parenchyma, with increasing concentration from the sinus wall to the depth of the parenchyma, which would explain how HSCs are maintained within the marrow cords.

Release of Mature Cells

As opposed to diapedesis, where leukocytes migrate between two endothelial cells to enter peripheral tissues (interendothelial migration), the egress of mature cells from BM occurs through pores within endothelial cells (transendothelial migration). When cell maturation is achieved, reticulocytes or polymorphs go through a fenestra with a diaphragm creating a pore whose diameter (of approximately 2 μm) is far less than the cell size. Active movements of ECs favor the cell release into the vessel lumen. In the case of platelet release, pseudopods arising from the megakaryocyte membrane flow through pores into the sinus lumen where platelet fragmentation takes place.

The release of hematopoietic cells must occur when cells are fully differentiated and not before. One mechanism is via extracellular matrix (ECM) to integrin receptor recognition. Fibronectin is such an ECM protein widely expressed in the BM cords. Precursors of red cells express fibronectin receptors (integrins VLA-5) up to the stage of reticulocytes. It is generally assumed that the lack of adhesion of reticulocytes to fibronectin is a critical factor for the release of these cells into circulating blood. Similar, but less known, interactions involving other integrins and ECM molecules might be at play for leukocytes, or even for immature cells in pathological conditions. Other mechanisms might also be operative, in particular those involving abluminal cells whose number and endothelial coverage are tightly regulated.

HSC Homing and Mobilization

Homing of circulating HSC to BM requires sequential specific signals for, first, attachment to sinus ECs; second, transendothelial migration; and third, anchoring within the hematopoietic parenchyma. This multistep process, comparable to transendothelial migration of leukocytes in response to inflammatory stimuli, involves interactions of HSCs with CAMs expressed on sinus vascular cells and with ECM adhesive molecules. These interactions are modulated by HGFs and SDF-1. Present data, provided mainly by studies in vitro, suggest the following scenario (Figure 1A). In the first step, HSCs must reduce their speed by tethering to, and rolling along, the endothelial lining. The sinusal blood flow is relatively slow, allowing weak interactions between P-selectin glycoprotein ligand-1 (PSGL-1) expressed on HSCs and E- and P-selectins expressed on BM ECs. Chemokines, such as SDF-1, or HGFs, such as Flt3-ligand or thrombopoietin, located on the endothelial membrane or bound to proteoglycans, may then activate VLA-4 and LFA-1 integrins on the rolling HSCs. VLA-4 and LFA-1 activation would convert to firm adhesion the interactions of these molecules with CAMs of the Ig superfamily (VCAM-1 and ICAM-1) on the endothelial cells, resulting in HSC arrest. Actin polymerization would then be activated in HSCs, leading to their migration through the endothelium. The transendothelial migration would be mediated by VLA-4 and VLA-5 expressed on HSCs and fibronectin present in the sinusal ECM. The platelet endothelial cell adhesion molecule-1 (PECAM-1/CD31) expressed on both ECs and HSCs could also be involved in this process through homotypic interactions. In the last step, HSCs would polarize and migrate following a SDF-1 gradient to reach the hematopoietic niche. The final anchoring might depend on interactions of HSC VLA-4 with microenvironmental cell VCAM-1 and of HSC VLA-5 with parenchyma ECM fibronectin; the c-Kit ligand produced locally might enhance the anchoring; moreover, the continuous production of SDF-1 by stromal cells would allow HSCs to be confined within the niche.

Mechanisms leading to the mobilization of HSC are less well understood. Among the multiple CAMs involved in HSC adhesion to the microenvironment, HSC VLA-4 interacting with microenvironmental cell VCAM-1 appears critical. It is now well established that HSC mobilization is associated with an increased local production of proteases within the BM, such as leukocyte serine proteases (elastase, cathepsin G) or matrix metalloproteinases (MMP-9), able to degrade the ECM. Elastase is known to disrupt VLA-4/VCAM-1 interactions (by degrading VCAM-1) and CXCR-4/SDF-1 interactions (by degrading both CXCR-4, the SDF-1 receptor, and SDF-1), while MMP-9 can also cleave SDF-1. Collectively, the increase in BM protease activity can explain the decrease, observed after G-CSF-induced HSC mobilization, in BM SDF-1 levels and in CXCR-4 expression on circulating CD34-positive hematopoietic cells. A model showing the main factors involved in HSC trafficking is proposed in Figure 1B.

Pathology

Increased BM blood flow has been described in cases of increased hematopoietic demand such as that observed in hemolytic anemia, which is in agreement with experimental data on animals after bleeding. Similarly, some authors have related the bone pain experienced by patients administered with G-CSF to the increased blood flow associated to an increase in the interstitial fluid pressure in the BM.

The myeloproliferative syndrome associates neoplastic blood cell diseases (polycythemia vera, chronic myelogenous leukemia, myelofibrosis) affecting HSCs and resulting in overproduction of blood cells, fibrosis of the hematopoietic parenchyma, and reactivation of liver and spleen hematopoiesis. In this syndrome the BM blood flow is increased. In myelofibrosis there is novel vessel formation (angiogenesis).

Angiogenesis is also a major anatomo-pathological feature of acute leukemias and multiple myeloma. Neoplastic cells (leukemic blasts or plasma cells), by producing angiogenic factors -cytokines such as fibroblast growth factor-2 (FGF-2) or vascular endothelial growth factor (VEGF) and/or metalloproteases (MMP2, MMP9), induce endothelial cell proliferation and organization into capillary tubes that provide oxygen and help tumor development.

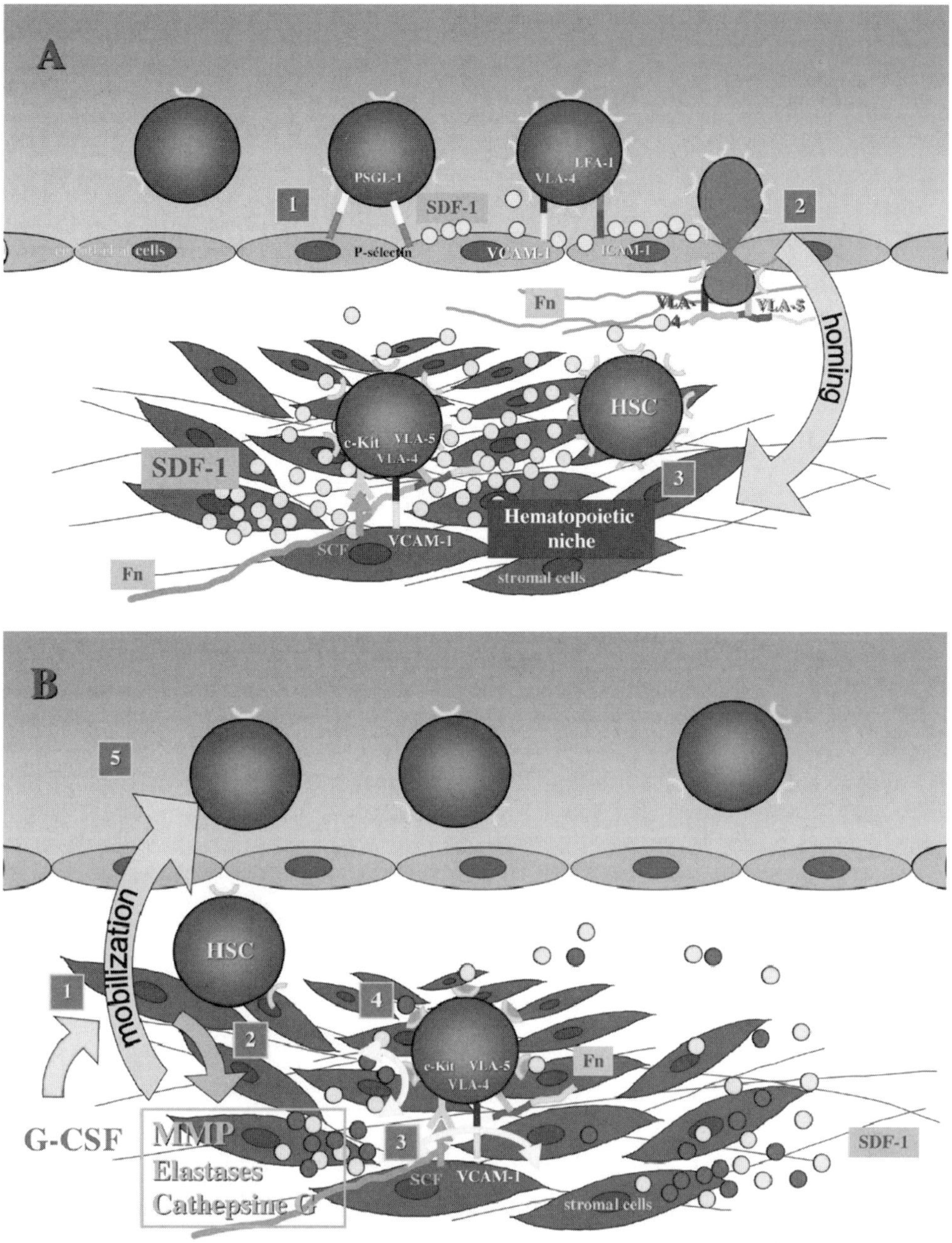

Figure 1 Molecular modeling of HSC trafficking. Data supporting the following models are provided mainly by studies in vitro using HSCs, BM ECs, and stromal cells generated from long-term marrow cultures. **(A) HSC homing.** **(1)** Weak interactions between P-selectin glycoprotein ligand-1 (PSGL-1) expressed on HSCs and E- and P-selectins constitutively expressed on BM ECs lead HSCs to tether to, and roll along, the BM endothelium. SDF-1 located on the endothelial surface and bound to heparan sulfates induces activation of HSC integrins VLA-4 and LFA-1. VLA-4 and LFA-1 activation converts to firm adhesion their interactions with VCAM-1 and ICAM-1 constitutively expressed on endothelial cells. **(2)** This activation induces the arrest of HSCs and stimulates actin polymerization that favors HSC transendothelial migration mediated by VLA-4 and VLA-5 in the presence of fibronectin (Fn). **(3)** HSCs polarize, migrate along the local gradient of SDF-1 produced by stromal cells, and reach the hematopoietic niche. This anchoring depends mainly on close interactions of HSC VLA-4 with stromal cell VCAM-1 and of HSC VLA-5 with Fn of the ECM. Moreover, the continuous production of SDF-1 by stromal cells allows the confinement of HSC within the hematopoietic niche. **(B) HSC mobilization.** **(1)** Addition of G-CSF induces **(2)** a local production of proteases (MMPs, elastases, and cathepsin G). **(3)** These molecules are able to degrade the ECM. Moreover, elastase can cleave VLA-4/VCAM-1 and **(4)** CXCR-4/SDF-1 interactions by degrading both CXCR-4 and SDF-1. The loss of attachment to stromal cells and to the ECM together with the loss of SDF-1 activity **(5)** favors the release of HSCs from the hematopoietic niche. (see color insert)

Angiogenesis is characteristic of diseased BM since in normal conditions there is little, if any, novel vessel formation. Antiangiogenic compounds are therefore promising novel therapeutics for leukemias.

Glossary

Cell adhesion molecules (CAMs): These molecules, expressed on the cell membrane, are involved in cell-to-cell and cell-to-extracellular matrix interactions. They include integrins (e.g., VLA-4 and VLA-5), selectins (e.g., P-, E-, and L-selectins), sialomucins (e.g., CD34 antigen), and immunoglobulin superfamily molecules (e.g.,VCAM-1 and ICAM-1). Interactions between selectins and sialomucins are of low affinity while integrin activation transforms interactions between integrins and immunoglobulin superfamily molecules into the high-affinity state. Integrins can be transiently activated by growth factors, such as c-Kit- and Flt3-ligands and thrombopoietin, or by chemokines, such as SDF-1 and IL-8. This induces changes in the cytoskeleton structure that affects cell motility.

Chemokines: Chemotactic cytokines with low molecular mass (8 to 17 kDa) showing approximately 20 to 50 percent sequence homology among each other at the protein level. These molecules mainly exert a chemotactic activity on leukocyte subpopulations induced during inflammatory reactions. They are divided into four groups (CXC, CX_3C, CC, and C) according to the positioning of the first two closely paired and highly conserved cysteines of the amino acid sequence. Chemokine receptors belong to the large group of G-protein-coupled seven-transmembrane-domain receptors.

Hematopoiesis: The process of blood cell formation. Blood cells derive from hematopoietic stem cells (HSCs) that are multipotent: They give rise to cells of all blood lineages, and that self-renew. In the progeny of a mother HSC there is a daughter HSC with identical capacity for proliferation and differentiation. HSCs give rise to hematopoietic progenitors with great proliferative potential but that are unable to self-renew, being already committed to a hematopoietic lineage, in turn giving rise to precursors (differentiating cells). The process is extremely dynamic since several billion hematopoietic cells need to be generated from a small compartment of HSCs each day to replace dying blood cells.

Microenvironment: The set of nonhematopoietic cells found in the BM parenchyma (cells from the vasculature, fat cells, and endosteal cells). Cells from this compartment derive from a specific population of mesenchymal stem cells (MSCs). One finds therefore in the BM parenchyma two different types of stem cells: HSCs giving rise to blood cells and MSCs giving rise to microenvironmental cells. In long-term marrow cultures hematopoiesis can be maintained for several weeks to months, provided there is generation of adherent stromal cells, therefore considered to be the population associated to HSCs and critical for their maintenance. A subset of stromal cells forms the hematopoietic niche where HSCs in physical contact with stromal cells maintain the adequate balance between self-renewal and differentiation. In vivo, the BM counterpart for stromal cells appears to be a subpopulation of microenvironmental cells with vascular smooth muscle cell characteristics.

Migration: A fundamental process in cell biology that allows cell locomotion. This property is essential during development, particularly embryogenesis, but also throughout adult life. It is a mandatory process for cell renewal such as occurs in hematopoiesis. Two classical types of migration are known: *chemotaxis*, where cell migration is directed by concentration gradients of soluble extracellular agents, and *chemokinesis*, characterized essentially by an undirected movement of cells. The main molecular factors involved in the migration process are cell adhesion molecules, chemotactic factors such as chemokines, and proteases, particularly matrix metalloproteinases (MMPs).

Further Reading

Charbord, P., Tavian, M., Humeau, L., and Péault, B. (1996). Early ontogeny of the human marrow from long bones: An immunohistochemical study of hematopoiesis and its microenvironment. *Blood* **87**, 4109–4119.

Dennis, J. E., and Charbord, P. (2002). Origin and differentiation of human and murine stroma. *Stem Cells* **20**, 205–214. *A review on the hematopoietic microenvironment.*

Iversen, P. O., Nicolaysen, G., and Benestad, H. B. (1993) The leukopoietic cytokine granulocyte colony-stimulating factor increases blood flow to rat bone marrow. *Exp. Hematol.* **21**, 231–235.

Levesque, J. P., Takamatsu, Y., Nilsson, S. K., Haylock, D. N., and Simmons, P. J. (2001). Vascular cell adhesion molecule-1 (CD106) is cleaved by neutrophil proteases in the bone marrow following hematopoietic progenitor cell mobilization by granulocyte colony-stimulating factor. *Blood* **98,** 1289–1297.

Lichtman, M. A. (1981) The ultrastructure of the hemopoietic environment of the marrow: A review. *Exp. Hematol.* **9**, 391–410.

Peled, A., Petit, I., Kollet, O., Magid, M., Ponomaryov, T., Byk, T., Nagler, A., Ben-Hur, H., Many, A., Shultz, L., Lider, O., Alon, R., Zipori, D., and Lapidot, T. (1999). Dependence of human stem cell engraftment and repopulation of NOD/SCID mice on CXCR4. *Science* **283**, 845–848. *This paper provides strong evidence for the pivotal role of CXCR-4/SDF-1 interaction for HSC homing into the bone marrow. Models of in vitro migration along a SDF-1 gradient and of engraftment in mice were used. Both migration and homing of HSCs were prevented by antibodies blocking SDF-1 or downregulation of CXCR-4, but increased by CXCR-4 upregulation.*

Peled, A., Kollet, O., Ponomaryov, T., Petit, I., Franitza, S., Grabovsky, V., Slav, M. M., Nagler, A., Lider, O., Alon, R., Zipori, D., and Lapidot, T. (2000). The chemokine SDF-1 activates the integrins LFA-1, VLA-4, and VLA-5 on immature human CD34(+) cells: Role in transendothelial/stromal migration and engraftment of NOD/SCID mice. *Blood* **95,** 3289–3296.

Perez-Atayde, A. R., Sallan, S. E., Tedrow, U., Connors, S., Allred, E., and Folkman, J. (1997). Spectrum of tumor angiogenesis in the bone marrow of children with acute lymphoblastic leukemia. *Am. J. Pathol.* **150**, 815–821. *A princeps paper showing the involvement of angiogenesis in a neoplastic blood disease.*

Petit, I., Szyper-Kravitz, M., Nagler, A., Lahav, M., Peled, A., Habler, L., Ponomaryov, T., Taichman, R. S., Arenzana-Seisdedos, F., Fujii, N., Sandbank, J., Zipori, D., and Lapidot, T. (2002). G-CSF induces stem cell mobilization by decreasing bone marrow SDF-1 and up-regulating CXCR4. *Nat. Immunol.* **3**, 687–694. *G-CSF-induced HSC mobilization leads to a dramatic decrease of marrow SDF-1 levels in mice. This phenomenon is related to local proteolytic degradation of SDF-1 by neutrophil elastase and is correlated to HSC mobilization. Antibodies neutralizing CXCR-4 and SDF-1 inhibited human and murine HSC mobilization, demonstrating the role of SDF-1/CXCR-4 signaling in cell egress.*

Weiss, L., and Geduldig, U. (1991) Barrier cells: Stromal regulation of hematopoiesis and blood cell release in normal and stressed murine bone marrow. *Blood* **78**, 975–990.

Capsule Biography

Dr. Jorge Domenech is a university scientist and physician with clinical and experimental experience in hematopoiesis. His research is focused on the mechanisms leading to the migration of hematopoietic stem cells.

Dr. Pierre Charbord is an Inserm scientist with experience in experimental hematology. His research is focused on the hematopoietic microenvironment (stromal and hematopoietic stem cells).

SECTION E

Kidney

CHAPTER 70

Renal Medullary Microcirculation

Thomas L. Pallone, Janos Pittner, and Whaseon Lee-Kwon
University of Maryland School of Medicine, Baltimore

Introduction

The purpose of this chapter is to review the current state of knowledge concerning the microcirculation of the renal medulla. This is a broad topic that touches on many areas. Perhaps the best recognized function of descending and ascending vasa recta (DVR and AVR) is to trap solute through countercurrent exchange. The classical notions of diffusive exchange have been revised to account for endothelial expression of aquaporin-1 water channels and the UTB-facilitated urea carrier. In addition to transport-related functions, DVR are contractile microvessels that regulate regional perfusion of the medulla. Medullary perfusion is of great importance, not only to the urinary concentrating mechanism but also to the maintenance of oxygen tension and renal salt handling. Evidence has accumulated that reduction of medullary perfusion can be accompanied by ischemic insult because oxygen tensions in the renal medulla are particularly low. Also of great interest is the finding that reduction of perfusion can lead to experimental forms of hypertension.

Tubular Vascular Relationships in the Renal Medulla

Blood flow to the renal medulla is provided by descending vasa recta, microvessels approximately 15 μm in diameter that are predominantly derived from efferent arterioles of juxtamedullary glomeruli. As efferent arterioles cross the corticomedullary junction to the outer stripe of the outer medulla, the muscular layer is diminished and replaced by smooth muscle–like cells called "pericytes" (Figure 1). Pericytes impart contractile function and are less prominent as DVR penetrate to the inner medulla. The outer medulla is characterized by its separation into vascular bundles and the interbundle capillary plexus. The "simple" vascular bundle (rabbit, guinea pig, dog, cat, monkey, man) is made up of DVR and AVR separated by a minimum of interstitium. Close apposition of DVR and AVR in vascular bundles provides opportunity for exchange of solutes and water. The capillary plexus of the interbundle region arises from DVR that peel off from the periphery of the vascular bundles in the inner stripe, suggesting a role for DVR to regulate regional blood flow within the medulla. Based on this arrangement it seems likely that modulation of perfusion of the interbundle capillary plexus regulates oxygen delivery to the thick ascending limb in the hypoxic medulla. Below the inner–outer medullary junction, vascular bundles disappear and vasa recta become more evenly dispersed among loops of Henle and collecting ducts. DVR terminate at various levels to form a sparse capillary plexus that coalesces to form AVR. DVR have a continuous endothelium with tight junctions. Fenestrations appear in the DVR wall as they terminate and fenestrations cover the majority of capillary plexus and AVR surface area. AVR return to the renal cortex from the inner medulla by traversing outer medullary vascular bundles. AVR that arise from the interbundle plexus reach the cortex independent of the bundles [1].

Countercurrent Exchange and Transport Properties

General Concepts

It is broadly accepted that the microcirculation of the renal medulla is a countercurrent exchanger that traps NaCl and urea deposited to the interstitium by the loops of Henle and collecting ducts. Textbook illustrations generally show parallel vessels that equilibrate by diffusion. Most simply, DVR blood flow is concentrated by diffusive influx of NaCl and urea as blood flows from the corticomedullary junction toward the papillary tip. Conversely, blood flowing toward the corticomedullary junction in AVR is diluted by diffusive efflux so that solute is trapped and recycled. Recent studies have shown that this depiction of a purely diffusive "U-tube" exchanger is oversimplified. Osmotic removal of water from DVR across water channels occurs, sieving NaCl and urea to concentrate these solutes in DVR plasma. Thus, both molecular sieving and diffusion contribute to equilibration of DVR plasma with the interstitium. As discussed later, shunting of water from DVR to AVR in vascular bundles might play a role in optimization of urinary concentrating ability by reducing blood flow rate to the deep medulla.

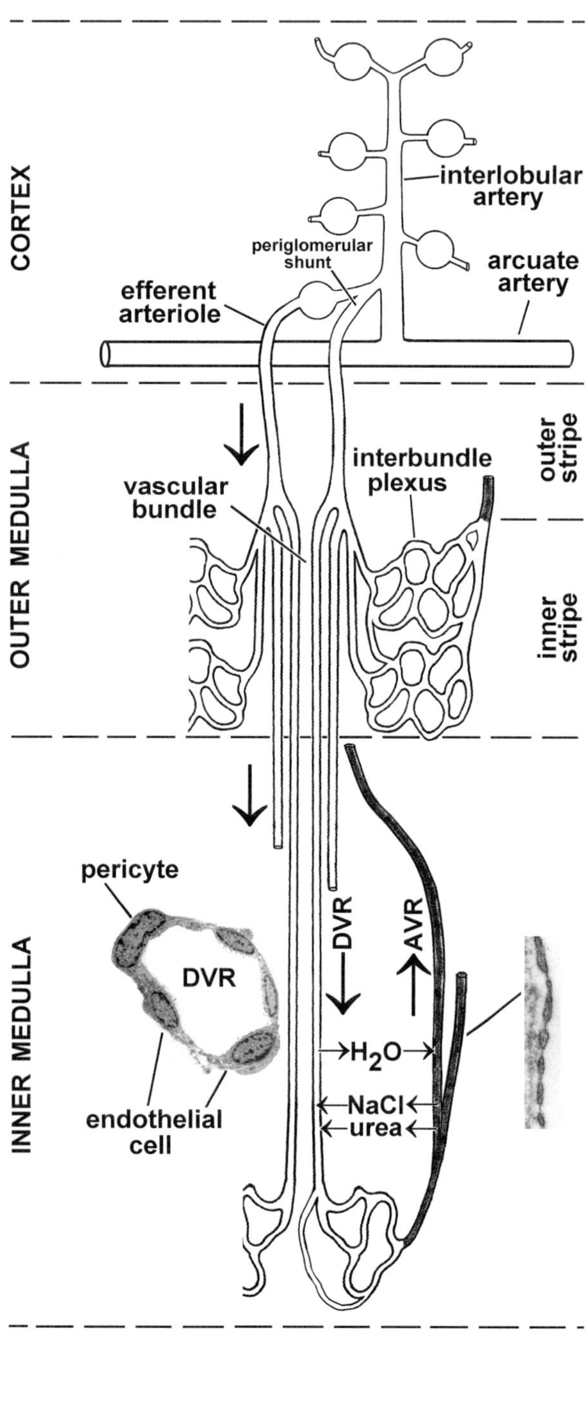

Figure 1 Anatomy of the medullary microcirculation. **Cortex:** Interlobular arteries arise from the arcuate artery and ascend toward the cortical surface. Juxtamedullary glomeruli arise at a recurrent angle from the interlobular artery. The majority of blood flow reaches the medulla through juxtamedullary efferent arterioles; however, some may also be from periglomerular shunt pathways. **Outer medulla:** In the outer stripe, juxtamedullary efferent arterioles give rise to DVR that coalesce to form vascular bundles in the inner stripe. DVR on the periphery of vascular bundles give rise to the interbundle capillary plexus that perfuses nephrons (thick ascending limb, collecting duct, long looped thin descending limbs, not shown). DVR in the center continue across the inner–outer medullary junction to perfuse the inner medulla. Thin descending limbs of short looped nephrons may also associate with the vascular bundles in a manner that is species dependent (not shown). **Inner medulla:** Vascular bundles disappear in the inner medulla and vasa recta become dispersed with nephron segments. AVR that arise from the sparse capillary plexus of inner medulla return to the cortex by passing through outer medullary vascular bundles. DVR have a continuous endothelium (inset) and are surrounded by contractile pericytes. The number of pericytes decreases with depth in the medulla. AVR are highly fenestrated vessels (inset). As blood flows toward the papillary tip, NaCl and urea diffuse into DVR and out of AVR. Transmural gradients of NaCl and urea abstract water across the DVR wall across aquaporin-1 water channels. Reproduced with permission from Ref. [2]. Renal Medullary Microcirculation in Encyclopedia of the Microcirculation, edited by David Shepro, Elsevier Inc.

Efflux of Water from DVR

DVR plasma protein concentration rises along the direction of blood flow, indicating water loss from plasma to renal medullary interstitium. Thus, DVR and nephrons deposit water to the renal medulla and AVR take up that water, accounting for overall mass balance. Water efflux from DVR raises two paradoxes. First, the purpose of depositing water from DVR plasma to the hypertonic medullary interstitium seems enigmatic. Second, Starling forces (hydraulic and oncotic pressure) do not account for the direction of DVR water transport because intraluminal oncotic pressure that favors water uptake exceeds the hydraulic pressure that favors efflux. NaCl and urea gradients generated by the lag in equilibration of DVR blood with interstitium favor water efflux and could account for osmotic water abstraction from DVR, but this requires the presence of a "small pore" pathway across which such small solutes exert effective osmotic driving force. The discovery of the aquaporins led to the molecular identification of that route (Figure 2A). Blockade of aquaporin-1 (AQP1) with mercurial agents or AQP1 knockout eliminates water efflux driven by abluminal hypertonic NaCl [3, 5]. Thus, transport of water across the DVR wall must be described by at least two parallel pathways, the properties of which are summarized in Table I. One is the highly water-selective AQP1 molecule. Another is a "large pore" route, likely paracellular, that conducts the majority of water movement driven by Starling forces. Evidence is consistent with the notion that the paracellular pathway offers little or no restriction to convective small solute flux (small solute reflection coefficient nearly 0, $\sigma_{SS} \approx 0$) while the AQP1 pathway is completely restrictive ($\sigma_{SS} \approx 1$). In addition to AQP1, another mercurial insensitive pathway conducts water efflux across the DVR wall when the driving solute is urea, glucose, or raffinose.

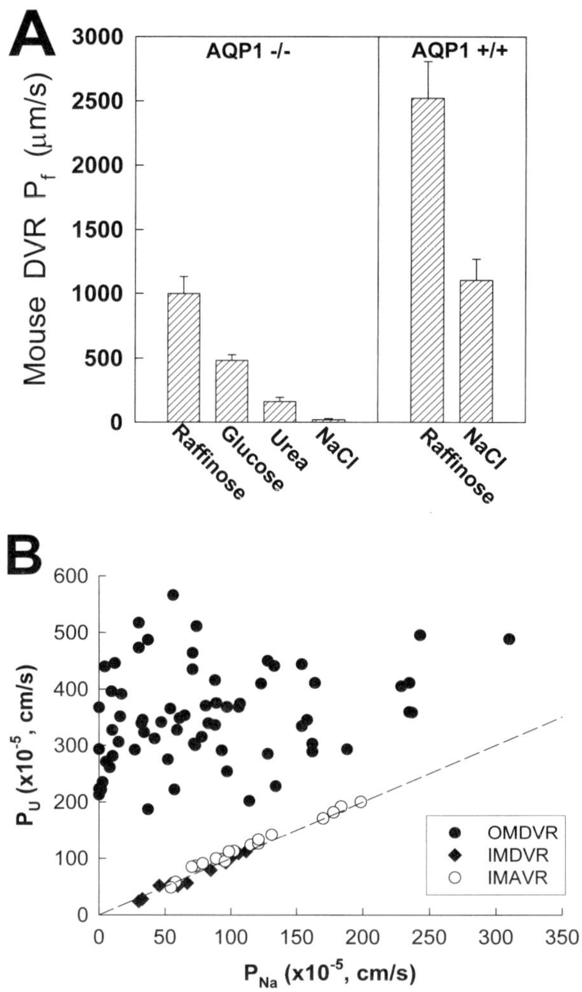

Figure 2 Osmotic water permeability of murine DVR. **(A)** All P_f measurements are summarized (ordinate, mean SE) for AQP1 –/– (*left*) and +/+ mice (*right*). The solute used to drive water movement is shown on the abscissa. Transmural NaCl gradients failed to induce water flux across AQP1-deficient DVR; however, larger solutes (glucose, urea, raffinose) were effective. Reproduced with permission from Ref. [3]. **(B)** Simultaneous measurement of ^{14}C-urea (P_U) and ^{22}Na (P_{Na}) permeability was performed by dual isotope microperfusion of outer medullary (OMDVR), inner medullary descending vasa recta (IMDVR), or inner medullary ascending vasa recta (IMAVR). OMDVR were studied by in vitro perfusion while inner medullary vessels were perfused on the surface of the exposed papilla in vivo. OMDVR P_U was always high, independent of P_{Na}. This is due to expression of the UTB urea carrier in DVR endothelia. Reproduced with permission from Ref. [4]. Renal Medullary Microcirculation in Encyclopedia of the Microcirculation, edited by David Shepro, Elsevier Inc.

The UTB urea transporter is expressed by DVR endothelia and exhibits mercurial insensitive water channel activity. The role of UTB to conduct water as well as urea flux across the DVR wall has yet to be examined. Simulation of these various transport pathways has been the underpinning of recent mathematical models of medullary microvascular exchange [7].

Mathematical simulations showed that DVR AQP1 might improve medullary concentrating ability by providing a route through which DVR water is shunted to AVR. The effect of that activity in the superficial medulla is to reduce blood flow to the deep medulla where interstitial gradients of NaCl and urea are most steep. A broadly unsettled question concerns regulation that might shift DVR equilibration between diffusive influx and water efflux/molecular sieving. High DVR solute permeability favors diffusive influx while low permeability favors water efflux across endothelial AQP1 water channels. If DVR permeability is acutely or chronically regulated, effects on "solute washout" and perfusion of the deep medulla would be expected to occur.

AVR Water Uptake

As required for overall mass balance in the medulla, AVR must remove the water deposited to the interstitium by nephrons, collecting ducts, and DVR. Transmural oncotic pressure gradients favor water uptake across the AVR wall and AVR hydraulic conductivity is very high [8, 9]. In vivo, transmural gradients in the AVR generated by the osmotic lag between blood and interstitium are directed to favor water uptake (luminal concentration greater than interstitial concentration). For AVR water uptake to be augmented by those gradients, the AVR wall must have nonzero reflection coefficients to NaCl and/or urea and transmural gradients of those solutes must be of significant magnitude. Rigorous measurements of AVR NaCl and urea reflection coefficients have been thus far impossible to obtain, but the general hypothesis that NaCl might augment transmural volume flux has been tested. In contrast to similar experiments in DVR, in vivo microperfusion of AVR with buffers made hypertonic or hypotonic to the papillary interstitium yielded no measurable water flux [10], suggesting that AVR reflection coefficients to small hydrophilic solutes is negligible ($\sigma_{SS} \approx 0$).

Vasa Recta Solute Permeability

As blood flows from the corticomedullary junction toward the papillary tip, rising interstitial concentrations of NaCl and urea are encountered. Those solutes equilibrate with the DVR lumen; however, the lag creates transmural gradients so that interstitial NaCl and urea concentrations exceed their respective concentrations in DVR plasma. Diffusive influx of NaCl and urea is favored. Additionally, the transmural gradient abstracts water across AQP1 water channels leading to molecular sieving of NaCl. Thus both diffusion and sieving across AQP1 contribute to DVR plasma equilibration. Quantification of diffusive permeabilities of the DVR wall to NaCl has been achieved by measurement of the rate of efflux of radiolabeled tracers from microperfused vessels. Those experiments have been performed both in vivo, on the surface of the exposed papilla, and in vitro, in isolated microperfused DVR. A summary of reported permeability measurements is provided in Table II. In vivo perfusion can underestimate permeabilities if the rate of diffusion of the isotopes away from the vessel in the surrounding interstitium is too low. In that case, ^{22}Na or [^{14}C]urea concentrations on the abluminal surface accumu-

Table I Hydraulic Conductivity (L_p), Osmotic Water Permeability (P_f), and Reflection Coefficients of Vasa Recta.

Parameter	OMDVR	IMDVR	IMAVR	Driving force
$L_p \times 10^{-6}$ (cm sec^{-1} mmHg^{-1})		1.4[a]		Albumin gradient
$L_p \times 10^{-6}$ (cm sec^{-1} mmHg^{-1})	1.6			Albumin gradient
$L_p \times 10^{-6}$ (cm sec^{-1} mmHg^{-1})	0.12[b]			NaCl gradient
$L_p \times 10^{-6}$ (cm sec^{-1} mmHg^{-1})			12.5	Hydraulic pressure
Parameter	**OMDVR**	**IMDVR**	**IMAVR**	**Method**
$\sigma_{albumin}$	0.89[c]			Sieving
$\sigma_{albumin}$			0.78	Sieving
$\sigma_{albumin}$			0.70	Osmotic
σ_{Na}		<0.05[d]	0.00[d]	Osmotic
σ_{Na}	~0.03[d]			Osmotic
σ_{Na}	~1.0[e]			Sieving
$\sigma_{Raffinose}$	~1.0[e]			Sieving

[a] Assumes a reflection coefficient to albumin of 1.0.
[b] Evidence shows that transmural NaCl gradients drive water flux exclusively through water channels, whereas albumin drives water flux predominantly through water channels along with a small component via other pathway(s).
[c] Not significantly different from 1.0.
[d] Measurement of σ_{Na} for the vessel wall as a whole.
[e] σ_{Na}, $\sigma_{Raffinose}$ for the aquaporin-1 water channel pathway through which NaCl gradients drive water flux. References to original data in Ref. [6].

late, violating the assumption of zero abluminal concentration. In vitro perfusion, due to the presence of a continuously flowing bath, is less likely to yield errors from such boundary layer effects, but necessitates the trauma of isolation and exposes the vessel to artificial buffers that could alter transport properties. In addition, DVR permeability is strongly dependent upon perfusion rate. Whether this rate dependence exists in vivo is uncertain but has important implications. If the true DVR NaCl permeability is very low, then abstraction of water across AQP1 might be the dominant mode of NaCl equilibration. That mode of equilibration may reduce blood flow to the deeper regions of the medulla and enhance interstitial osmolality. DVR urea permeability is the sum of transport via phloretin-sensitive transcellular, carrier-mediated route(s) and other, for example pericellular, pathways (Figure 2B). Histochemical evidence and in situ hybridization have identified the DVR urea transporter as the same as that expressed by the RBC-UTB.

AVR solute permeability has not been as thoroughly evaluated as that in DVR because AVR have not been isolated and perfused in vitro, owing to technical difficulties. Transport properties have been measured only by the difficult approach of in vivo microperfusion of vessels on the surface of the exposed papilla of rats and hamsters (Table II). The values so obtained exceed DVR permeabilities but, even so, are probably underestimated because the ^{22}Na and ^{14}C-urea tracers might have accumulated in the interstitium to significant levels.

Vasoactivity of DVR

Constriction and Dilation of DVR by Vasoactive Agents

DVR are contractile microvessels (Figure 3). A large number of mediators have been shown to constrict or dilate DVR, and a number of receptors have been identified in medullary vascular bundles by ligand binding, autoradiography, RT-PCR, or immunochemistry. Table III summarizes the findings. The entries listed in Table III include observations of pharmacological effects on vasomotion of microperfused DVR isolated from vascular bundles of the rat as well as receptor studies that have employed a variety of methods. We attribute constriction to the action of pericytes. Vasopressin (compared to angiotensin II and endothelin) is a weak DVR vasoconstrictor. Endothelins constrict primarily via the ET_A receptor and are thought to exert a self-limiting vasodilatory effect through ET_B receptor stimulation. ET_1 and ET_2 isoforms are ET_A and ET_B receptor agonists and have proven to be the most potent vasoconstrictors of DVR thus far observed [11]. Angiotensin II (AngII) also reliably constricts isolated DVR. AT_2 receptor antagonists enhance AngII constriction and AT_2 receptor expression has been verified in DVR by RT-PCR. Effects of adenosine effects are concentration dependent. At low concentration, adenosine A_1 receptor stimulation induces DVR constriction. At high concentration, A_2 effects predominate

Table II Solute Permeability of Vasa Recta.

Permeability $\times 10^{-5}$ cm/sec	OMDVR[a]	IMDVR[b]	IMAVR[b]	Species
P_{Na}		28	51	Hamster
P_{Na}	76	75	115	Rat
P_{Na}		67	116	Rat
P_{Urea}		47		Rat
P_{Urea}	360	76	121	Rat
P_{Urea}	343 → 191[c]			Rat
P_D	476[d]			Rat
$P_{raffinose}$	40			Rat
Permeability ratio	**OMDVR[a]**	**IMDVR[b]**	**IMAVR[b]**	**Species**
P_{Urea}/P_{Na}		1.09	0.98	Rat
P_{Cl}/P_{Na}	1.33			Rat
$P_{raffinose}/P_{Na}$	0.35			Rat
P_{Inulin}/P_{Na}	0.22			Rat

Abbreviations: OMDVR, outer medullary descending vasa recta; IMDVR, inner medullary descending vasa recta; IMAVR, inner medullary ascending vasa recta.

[a] Values obtained with in vitro microperfusion are highly dependent upon perfusion rate.

[b] Values obtained with in vivo microperfusion in the exposed papilla are probably underestimated due to boundary layer effects.

[c] Values are before and after inhibition with 50 mM thiourea.

[d] Diffusional water permeability measured with ^3H_2O efflux. References to original data in Ref. [6].

Table III Mediators That Constrict and Dilate DVR.

	Constriction[a]	Dilation	Receptor studies[b]
Angiotensin	+++ AT_1	+ AT_2	AT_1, AT_2
Endothelins	++++ $ET_A > ET_B$	–	ET_A, ET_B
Vasopressin	++ V_1	+ V_2	V_1
Adenosine	++ A_1	+ A_{2a}, A_{2b}	A_1, A_{2a}, A_{2b}
Prostaglandin E_2	–	+	–
Nitric oxide	–	+	–
Kinins	+ B_1	+ B_2	–
Acetylcholine	++	+	–
Norepinephrine	++	–	α_{2B}

[a] Table entries refer to observations of constriction or dilation in response to receptor specific or nonspecific agonists. The intensity of constriction is graded from (–) to (++++).

[b] Table entries show the receptor subtypes expressed in DVR. Information is derived from studies that employed various methods, including RT-PCR, radioligand binding, and immunochemistry.

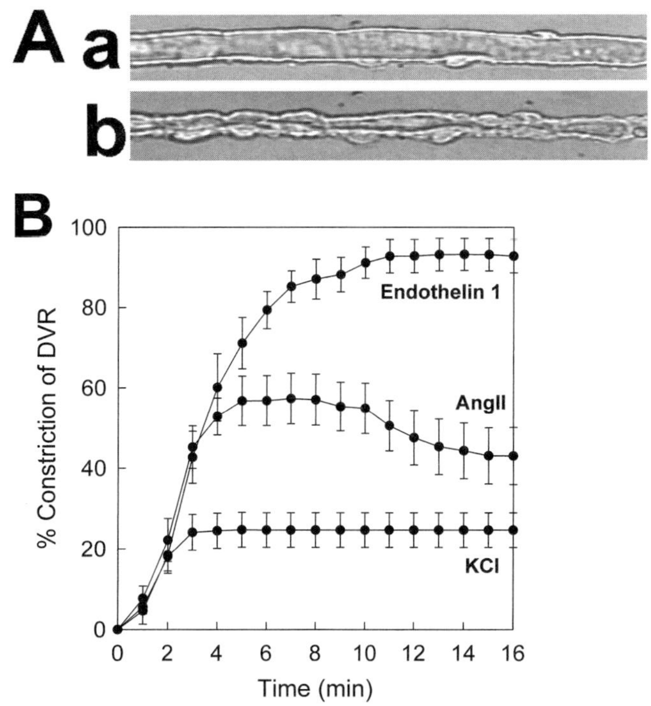

Figure 3 Vasoconstriction of outer medullary DVR. **(A)** DVR isolated and microperfused in vitro is exposed to AngII (10 nM) by abluminal application from the bath. Panels **a** and **b** show the vessel prior to and after constriction. Two cell types can be seen. Pericyte cell bodies project from the abluminal surface and endothelia line the lumen. **(B)** The graph shows quantification of DVR constriction through measurement of luminal diameter. Results are expressed as percent constriction = $100 \times (D_o - D)/D_o$, where D_o is basal diameter and D is after constriction. The mean luminal diameter of perfused DVR is ~14 µm. Constriction has been induced by abluminal exposure to endothelin 1 (0.1 nM, $n = 6$) or AngII (10 nM, $n = 15$), or by raising extracellular K^+ concentration from 5 to 100 mM by isosmotic substitution for NaCl ($n = 6$). Reproduced with permission from Ref. [2]. Renal Medullary Microcirculation in Encyclopedia of the Microcirculation, edited by David Shepro, Elsevier Inc.

and adenosine dilates preconstricted DVR. Kinins increase DVR endothelial intracellular calcium concentration and promote nitric oxide generation through the bradykinin, B_2 receptor. The cataloging of agents in Table III does not provide an integrated hypothesis of renal medullary function; however, the large number of agents to which DVR respond is clear. We assume that, in vivo, DVR vasomotor tone is governed by the integrated response to many hormonal and paracrine influences.

Pericyte Ca^{2+} Signaling and Channel Architecture

Recently, the mechanisms by which AngII induces vasoconstriction have been evaluated using fluorescent probes of intracellular calcium concentration and membrane potential and by electrophysiological recording. As expected for signaling via the AngII AT1 receptor, a classical peak and plateau $[Ca^{2+}]_i$ response is elicited in globally fura2-loaded pericytes. Both electrophysiological recording and measurements with a potential-sensitive fluorescent probe showed that AngII depolarizes the pericyte, mediated primarily through activation of a Ca^{2+}-sensitive Cl^- conductance that shifts membrane potential away from the equilibrium potential of K^+ ion toward that of Cl^-. An 11 pS Ca^{2+}-sensitive Cl^- channel has been identified in DVR pericytes. Membrane potential of AngII-treated pericytes often oscillates and voltage-clamped cells held at −70 mV exhibit classical spontaneous transient inward currents (STICs) typical of various smooth muscle preparations [12, 13].

The role of membrane depolarization to gate Ca^{2+} entry had been well established in the afferent arteriole. Until recently, however, the existence of voltage-gated calcium entry pathways in the efferent circulation and DVR pericyte was uncertain. RT-PCR, immunochemistry, and examination of vasoreactivity in isolated arterioles verified expression of T-type and L-type calcium channel α subunits in efferent arterioles of juxtamedullary (but not superficial) glomeruli and in DVR [14]. Indeed, the L-type channel blocker diltiazem vasodilates AngII constricted DVR and reduces $[Ca^{2+}]_i$ of AngII-treated pericytes. Both high external K^+ concentration and the L-type agonist BAYK8644 are weak DVR vasoconstrictors. Finally, agents that repolarize pericytes, bradykinin and the K_{ATP} channel opener pinacidil, are effective vasodilators (Figure 4) [15]. The many downstream effects of pericyte AngII and endothelin receptor activation remain unknown, however actions independent of $[Ca^{2+}]_i$ elevation must occur. Principally, depolarization in the absence of agonist induces far less intense constriction than does AngII or endothelins. Phosphorylation events that sensitize the intracellular contractile machinery to the effects of Ca^{2+} are likely to be implicated.

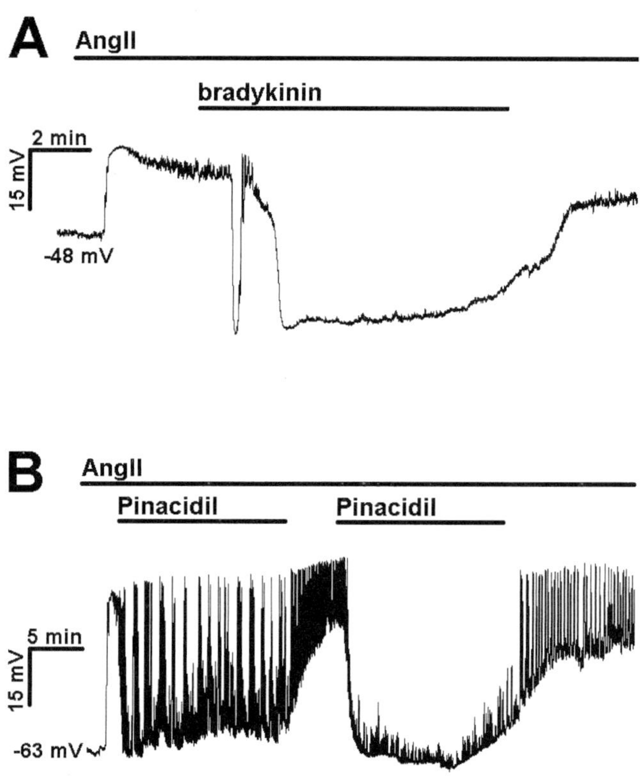

Figure 4 Repolarization of AngII depolarized pericytes by vasodilators. **(A)** Recording of membrane potential from a DVR pericyte successively exposed to AngII (10 nM) and then bradykinin (100 nM). A biphasic repolarization occurs after exposure to bradykinin. **(B)** Similar recording from a DVR pericyte exposed to AngII (10 nM) and then the K_{ATP} channel opener, pinacidil (10 μM). Both bradykinin and pinacidil repolarize pericytes and vasodilate preconstricted DVR. Reproduced with permission from Ref. [2]. Renal Medullary Microcirculation in Encyclopedia of the Microcirculation, edited by David Shepro, Elsevier Inc.

Endothelial Ca^{2+} Signaling and NO Production

The calcium-sensitive fluorophore fura2 loads avidly into DVR endothelia (to the near exclusion of pericytes) and so it has been possible to examine global intracellular Ca^{2+} transients generated by endothelium dependent vasodilators. Bradykinin (BK) generates a peak and plateau calcium response, enhances NO generation, and induces vasodilation. An unexpected finding is that the vasoconstrictor AngII suppresses basal calcium and inhibits BK, acetylcholine, thapsigargin, and cyclopiazonic acid induced calcium responses in DVR endothelia [16]. This is surprising because the effect is inhibited by high concentrations of the AT1 blocker losartan and modulated by AngII AT2 receptors [17]. AngII and AT1 receptors mediate the vast majority of effects by signaling through IP_3 generation and calcium mobilization. Second, infusion of AngII has been observed to lead to secondary enhancement of NO levels within the medulla and in isolated cortical microvessels. Given that eNOS/NOS3 is a calcium-dependent isoform of nitric oxide synthase (NOS), suppression of calcium would be expected to block rather than enhance endothelial NO generation. Possibly, adjacent nephrons that also express NOS isoforms might be responding to generate NO on the vascular bundle periphery, providing a feedback loop through which those structures regulate their own perfusion. It has been hypothesized that AngII might suppress DVR endothelial Ca^{2+} signaling as a means of turning regulation of DVR vasoactivity away from the endothelium to the medullary thick ascending limb (mTAL).

The physiological roles of NO cannot be thoroughly evaluated without considering interactions with oxygen free radicals. Oxygen radicals result from reductions of O_2 to generate superoxide ($O_2^{\cdot-}$), hydrogen peroxide (H_2O_2), hypochlorous acid, and hydroxyl radical ($^{\cdot}OH$), the "reactive oxygen species" (ROS). $O_2^{\cdot-}$ reacts with NO to form peroxynitrite ($ONOO^-$), a product that is a weak vasodilator. ROS are generated by "leak" of electrons from the mitochondrial electron transport chain and a variety of enzymatic processes. Intrinsic mechanisms limit cellular levels of ROS. Several isoforms of superoxide dismutase (SOD) convert $O_2^{\cdot-}$ to O_2 and H_2O_2. The SOD mimetic tempol has been shown to enhance medullary perfusion. NO production by, and AngII constriction of, isolated DVR are both enhanced by tempol. Given the importance of NO in the maintenance of medullary blood flow, it is inviting to speculate that the level of "oxidative stress" in the renal medulla has physiological and pathophysiological regulatory roles.

Medullary Oxygen Tension and Perfusion of the Medulla

Oxygen tension in the medulla of the kidney is low, 10 to 25 mmHg. This is predicted to be a consequence of the countercurrent arrangement of vasa recta because oxygen in DVR blood diffuses to AVR to be shunted back to the cortex. Several hormonal systems play a role in the protection of the medulla from ischemic insult. Each shares the ability to enhance medullary blood flow and inhibit salt reabsorption by nephrons. Hypothetically this should have a dual effect to enhance the supply of oxygen and simultaneously reduce the demand for O_2 consumption. One example is cyclooxygenase (COX) production of vasodilatory prostaglandins. Medullary perfusion is sensitive to COX inhibition, and renomedullary interstitial cells release prostaglandin E_2 (PGE_2) in response to AngII and bradykinin. Apparently, the cylooxygenase-2 isoform is responsible because AngII reduces medullary blood flow in COX2- but not COX1-deficient mice. In most vascular beds, ischemia favors generation of adenosine, a paracrine agent that enhances blood flow through local vasodilation. Within the renal medulla (but not the cortex) adenosine acts as a vasodilator and inhibits salt reabsorption by the medullary thick ascending limb of Henle (mTAL). It is reasonable to hypothesize that adenosine produced by the mTAL diffuses to and dilates DVR on the periphery of vascular bundles. Adenosine A_1 and A_2 receptor stimulation constricts and dilates DVR, respectively.

The Importance of Medullary Blood Flow and Nitric Oxide in Hypertension

Renal medullary NOS activity and NO production exceeds that in the cortex. NO acts in autocrine and paracrine fashion to modulate both vasoconstriction and epithelial NaCl reabsorption. NOS isoforms have specific effects [18]. NOS1 inhibition reduces NO levels in the medulla and induces salt-sensitive hypertension without altering medullary perfusion. Global inhibition of NOS1, NOS2, and NOS3 isoforms decreases medullary NO levels, medullary blood flow, and tissue oxygen tension and leads to salt-dependent hypertension [19]. NO generation may be important to abrogate tissue hypoxia that would otherwise arise from release of vasoconstrictors. AngII, norepinephrine, and vasopressin stimulate release of NO in the medulla. Subpressor infusion of N(ω)-nitro-L-arginine methyl ester (LNAME) into the renal interstitium does not affect medullary blood flow or pO_2 but enables otherwise ineffective doses of AngII, norepinephrine, or vasopressin to reduce perfusion. Data broadly support the conclusion that medullary NO production has a tonic effect to maintain perfusion, favor saliuresis, and protect from ischemic injury and hypertension.

Glossary

Hydraulic conductivity: Proportionality constant generally denoted L_p that relates the rate of vectorial water flux that occurs across a membrane in response to driving forces imposed by osmotic, oncotic, and hydraulic pressures.

Juxtamedullary: Refers to the deepest region of the renal cortex. Glomeruli that lie in the deep, juxtamedullary cortex have efferent arterioles that extend into the outer stripe of the outer medulla where they break up like a horse's tail to form descending vasa recta.

Osmotic water permeability: Proportionality constant generally denoted P_f that relates the rate of vectorial water flux that occurs across a membrane in response to driving forces imposed by osmotic, oncotic, and hydraulic pressures. Can be converted to hydraulic conductivity by the relationship $L_p = (P_f V_w)/(R T)$, where V_w is the partial molar volume of water, R is the universal gas constant, and T is absolute temperature.

Solute permeability: Proportionality constant for the ith solute generally denoted P_i that relates the rate of diffusive solute flux across a membrane that occurs when a transmural concentration difference of that solute exists across the membrane.

Vasa recta: System of parallel microvessels that traverse the renal outer and inner medulla. Descending vasa recta carry blood flow from the juxtamedullary cortex to the medulla and are sometimes referred in older literature to as arteriolar vasa recta. Ascending vasa recta carry blood flow from the inner and outer medulla back to the cortex and are sometimes referred to as venous vasa recta.

References

1. Lemley, K. V., and Kriz, W. (1987). Cycles and separations: the histotopography of the urinary concentrating process. *Kidney Int.* **31**, 538–548.
2. Pallone, T. L., Zhang, Z., and Rhinehart, K. (2003). Physiology of the renal medullary microcirculation. *Am. J. Physiol.* **284**, F253–F266.
3. Pallone, T. L., Edwards, A., Ma, T., Silldorff, E. P., and Verkman, A. S. (2000). Requirement of aquaporin-1 for NaCl-driven water transport across descending vasa recta. *J. Clin. Invest.* **105**, 215–222.
4. Pallone, T. L., Work, J., Myers, R. L., and Jamison, R. L. (1994). Transport of sodium and urea in outer medullary descending vasa recta. *J. Clin. Invest.* **93**, 212-222.
5. Pallone, T. L., Kishore, B. K., Nielsen, S., Agre, P., and Knepper, M. A. (1997). Evidence that aquaporin-1 mediates NaCl-induced water flux across descending vasa recta. *Am. J. Physiol.* **272**, F587–F596.
6. Pallone, T. L., Turner, M. R., Edwards, A., and Jamison, R. L. (2003). Countercurrent exchange in the renal medulla. *Am. J. Physiol.* **284**, R1153–R1175.
7. Edwards, A., Delong, M. J., and Pallone, T. L. (2000). Interstitial water and solute recovery by inner medullary vasa recta. *Am. J. Physiol.* **278**, F257–F269.
8. Pallone, T. L. (1991). Resistance of ascending vasa recta to transport of water. *Am. J. Physiol.* **260**, F303–F310.
9. MacPhee, P. J. and Michel, C. C. (1995). Fluid uptake from the renal medulla into the ascending vasa recta in anaesthetized rats. *J. Physiol.* **487**, 169–183.
10. Pallone, T. L. (1991). Transport of sodium chloride and water in rat ascending vasa recta. *Am. J. Physiol.* **261**, F519–F525.
11. Silldorff, E. P., Yang, S., and Pallone, T. L. (1995). Prostaglandin E2 abrogates endothelin-induced vasoconstriction in renal outer medullary descending vasa recta of the rat. *J. Clin. Invest.* **95**, 2734–2740.
12. Zhang, Z., Huang, J. M., Turner, M. R., Rhinehart, K. L., and Pallone, T. L. (2001). Role of chloride in constriction of descending vasa recta by angiotensin II. *Am. J. Physiol.* **280**, R1878–R1886.
13. Pallone, T. L. and Huang, J. M. (2002). Control of descending vasa recta pericyte membrane potential by angiotensin II. *Am. J. Physiol.* **282**, F1064–F1074.
14. Hansen, P. B., Jensen, B. L., Andreasen, D., and Skott, O. (2001). Differential expression of T- and L-type voltage-dependent calcium channels in renal resistance vessels. *Circ. Res.* **89**, 630–638.

15. Zhang, Z., Rhinehart, K., and Pallone, T. L. (2002). Membrane potential controls calcium entry into descending vasa recta pericytes. *Am. J. Physiol.* **283**, R949–R957.
16. Pallone, T. L., Silldorff, E. P., and Cheung, J. Y. (1998). Response of isolated rat descending vasa recta to bradykinin. *Am. J. Physiol.* **274**, H752–H759.
17. Rhinehart, K., Handelsman, C. A., Silldorff, E. P., and Pallone, T. L. (2003). ANG II AT2 receptor modulates AT1 receptor-mediated descending vasa recta endothelial Ca2+ signaling. *Am. J. Physiol.* **284**, H779–H789.
18. Mattson, D. L., and Wu, F. (2000). Control of arterial blood pressure and renal sodium excretion by nitric oxide synthase in the renal medulla. *Acta Physiol. Scand.* **168**, 149–154.
19. Nakanishi, K., Mattson, D. L., and Cowley, A. W., Jr. (1995). Role of renal medullary blood flow in the development of L-NAME hypertension in rats. *Am. J. Physiol.* **268**, R317–R323.

Bibliography

Bankir, L., and de Rouffignac, C. (1985). Urinary concentrating ability: Insights from comparative anatomy. *Am. J. Physiol. Regul. Integr. Comp. Physiol.* **249**, R643–R666.

Cowley, A. W., Jr., Mori, T., Mattson, D., and Zou, A. P. (2003). Role of renal NO production in the regulation of medullary blood flow. *Am. J. Physiol. Regul. Integr. Comp. Physiol.* **284**, R1355–R1369. *This review describes the important role of nitric oxide to maintain renal medullary perfusion. Important references that demonstrate the hypertensive effect of inhibiting medullary NO generation are cited.*

Jamison, R. L., and Kriz, W. (1982). *Urinary Concentrating Mechanism*. New York: Oxford University Press.

Knepper, M. A., Saidel, G. M., Hascall, V. C., and Dwyer, T. (2003). Concentration of solutes in the renal inner medulla: Interstitial hyaluronan as a mechano-osmotic transducer. *Am. J. Physiol. Renal. Physiol.* **284**, F433–F446.

Mattson, D. L. (2003). Importance of the renal medullary circulation in the control of sodium excretion and blood pressure. *Am. J. Physiol. Regul. Integr. Comp. Physiol.* **284**, R13–R27.

Michel, C. C. (1995). Renal medullary microcirculation: Architecture and exchange. *Microcirculation* **2**, 125–139.

Pallone, T. L., and Silldorff, E. P. (2001). Pericyte regulation of renal medullary blood flow. *Exp. Nephrol.* **9**, 165–170.

Capsule Biography

Thomas Pallone obtained the bachelor's degree in engineering at the Massachusetts Institute of Technology (1977) and the Doctor of Medicine degree from the Pennsylvania State University (1981). Subsequently, his graduate studies in the Health Science and Technology program at M.I.T. were mentored by William M. Deen, Ph.D. in Chemical Engineering. His interest in the renal medullary microcirculation was kindled as a thesis project (S.M., 1982). After completing medicine residency training at the University of Maryland (1985), he obtained clinical and laboratory fellowship training in the Nephrology Division at Stanford University under Rex L. Jamison, M.D. (1988). He has held staff positions at Penn State University (through 1995) and the University of Maryland at Baltimore, where he is currently Professor of Medicine and Physiology.

CHAPTER 71

Renal Cortical Microcirculation

William J. Welch
Georgetown University, Washington, D.C.

Introduction

The kidney maintains homeostasis by regulation of fluid and electrolyte balance, as well as the elimination of nitrogenous and toxic products. The kidney has adapted an intricate vascular network that contains as much as 20 percent of the cardiac output at any given time. This high throughput is critical to the function of the glomerular capillaries, which filter the fluid component of blood. Fluid and electrolytes filtered at the glomerulus are reabsorbed by the nephron and mostly returned to the circulation. The glomerular capillaries are located exclusively in the cortex and give rise to cortical (superficial) and juxtamedullary nephrons. Cortical nephrons have short loops of Henle passing only a short distance into the medulla. Juxtamedullary nephrons, with long loops of Henle, pass deep into the medulla to the papilla. Cortical glomeruli give rise to short efferent arterioles that pass into a peritubular capillary that bathes the proximal tubule of cortical nephrons and is structurally similar to nonrenal capillaries. Juxtamedullary glomeruli also give rise to efferent arterioles, some of which lead to peritubular capillaries and some that lead directly to small venules, the vasa recta. The main determinant of filtration, glomerular capillary hydrostatic pressure, is tightly regulated by the highly resistant afferent and efferent arterioles, which are subject to humoral, neural, and physical factors. In this review, I will summarize studies that make new and important observations on renal cortical microcirculation and are representative of recent research in this area.

Control of Renal Cortical Microvascular Resistance

Methods of Evaluation

Much of the recent focus on cortical microcirculation research has been on identification of the systems that regulate vascular tone, affecting segmental and overall renal vascular resistance. Access to the important resistance vessels is limited, but recent methodological gains have provided new information. Both in vivo and in vitro observations continue to be made in hydronephrotic kidneys, developed by prolonged ureteral occlusion. However, there is concern that responses in this nonfiltering kidney model are not fully representative of normal function. The recent use of intravital microscopy, with improved optics and imaging analysis, shows promising results suggesting that in vivo inspection of renal cortical microcirculation in normal kidneys is possible. Isolated segments of the interlobar, arcuate, and interlobular arteries continue to provide data, though they often are similar to nonrenal vessels. However, more studies have emerged from direct perfusion of the isolated afferent arterioles and in some cases the efferent arterioles, which provide more renal specific responses. These extremely small (15 to 30 μm) vessels have been isolated from rabbits, mice and more recently rats and microperfused in temperature-controlled baths. Though many of these observations may be limited to the experimental species, they provide a sensitive resistance vessel model with which to test several hypotheses. Renal micropuncture also continues to provide data on the control of cortical vascular tone.

Angiotensin II

Angiotensin II is the major end product of the proteolytic enzyme renin, which is primarily produced in the kidney in the afferent arteriole. Ang II is a potent vasoconstrictor in vascular beds mediated by abundant type 1 (AT_1) receptors. Type 2 receptors (AT_2), which are much less abundant, may cause vasodilation when activated, but supporting evidence for a physiological role is currently lacking. Ang II may have unique vascular control in the kidney, since it is well established that the level of Ang II–dependent constriction varies between afferent and efferent arterioles. Several new studies have focused on the relative contributions of Ang II–dependent constriction of the AA and EA. In an attempt to characterize Ang I and Ang II actions in these arterioles, Marchetti et al. [1] measured changes in intracellular Ca^{2+} (Ca_i) concentrations as a parameter of vasoactive responses in isolated rat juxtaglomerular arterioles. They studied three populations of arterioles isolated from juxtamedullary glomeruli, the afferent arterioles (AA), efferent arterioles terminating as peritubular capillaries (EA-T), and EA terminating as vasa recta (EA-M). Ca_i was increased in all arterioles by both Ang I and II and was blocked in all arterioles by losartan, an Ang II receptor blocker. Lisinopril (an angiotensin-converting enzyme inhibitor) blocked the increase in Ca_i due to Ang I in the AA and the EA-M, yet had no effect in EA-T. The ratio of EC_{50} values for Ang I to Ang II was also much higher in EA-M. The authors suggest that local ACE acts to convert Ang I to Ang II, which then activates AT receptors and initiates Ca_i mobilization and constriction. However, in the EA-M, which may have no ACE activity, Ang I either activates the Ang II receptors or is converted to some other active form. The authors fail to identify a physiological role for this arteriolar heterogeneity. The same group also assessed expression of AT_{1a}, AT_{1b}, and AT_2 receptors in these microdissected arterioles. All three receptors were expressed in AA and EA-M. Yet, only modest levels of mRNA for AT_{1a} and AT_2 were expressed in EA-T. The expression of the receptor subtypes correlates with the Ang II effects on Ca^{2+} mobilization.

The arteriolar diameter responses to systemic infusion of Ang II were also evaluated in the dog kidney using intravital microscopy [2]. This improved imaging method may provide direct evidence of differential responses to Ang II. However, renal artery injections of Ang II reduced diameters in EA and AA similarly in superficial glomeruli. Total constriction differed only modestly between the AA and EA in the juxtamedullary glomeruli. Inhibition of prostaglandins enhanced the Ang II response only in the juxtamedullary AA, suggesting that PG vasodilating products offset Ang II in the juxtamedullary AA, but not in the cortical AA. However, nitric oxide may also contribute to the Ang II response in juxtamedullary AA. This study was able to demonstrate zonal heterogeneity of Ang II control of vascular tone. The lack of sufficient resolution made it difficult to distinguish major differences between AA and EA responses to Ang II, which has been demonstrated more clearly in isolated perfused arterioles. Another weakness of this technique is the inability to inject vasoactive agents directly into the microvasculature in the video field to ascertain local responses.

Adenosine Receptors

Adenosine, acting on type 1 receptors (A1-R) in the afferent arteriole, constricts the AA, reduces the GFR, and also mediates tubuloglomerular feedback (TGF). Local adenosine levels are increased during elevated Na^+ transport in the nephron and subsequent metabolism of adenosine triphosphate (ATP). The relationship between Ang II and adenosine on the regulation of afferent arteriolar tone has been the focus of several studies. These agents act through AT_1 and A1 receptors that have been localized in the AA. Both acute and chronic suppression of the renin–angiotensin system (RAS) reduces A1-R vasoconstriction. The dependence of Ang II on A1-R is not as clear. In the hydronephrotic kidney, A1-R inhibition had no effect on Ang II constriction. However, in isolated perfused rabbit AA, A1-R inhibition reduced Ang II constriction by 50 percent. With the development of A1-R knockout mice, this question was readdressed to test the chronic effects of A1-R inhibition. Hansen et al. [3] measured renal function in anesthetized mice and in separate experiments tested the tone of the isolated perfused AA in response to Ang II. The GFR did not differ between wild type and mice lacking A1-R during the control period. An acute pressor dose of Ang II raised MAP similarly in both groups (+11 to +13 mmHg). However, Ang II reduced GFR by 40 to 50 percent in the wild-type (+/+) mice, but only by 20 percent in the A1-R (–/–) mice (Figure 1). Ang II reduced the renal blood flow (RBF) more in the knockout mice, similar to its effect on GFR. In the microperfused AA the baseline diameters were similar between the groups. However, Ang II was more effective in the A1-R (+/+) than in the A1-R (–/–) in reducing AA diameter. At a physiological dose (10^{-10} M) Ang II reduced the diameter in A1-R (+/+) by 50 percent, yet had no effect in A1-R (–/–). These differences were probably not due to differences in AT_1 receptor density, since the expression of mRNA for AT_1 was not different between AA from the two

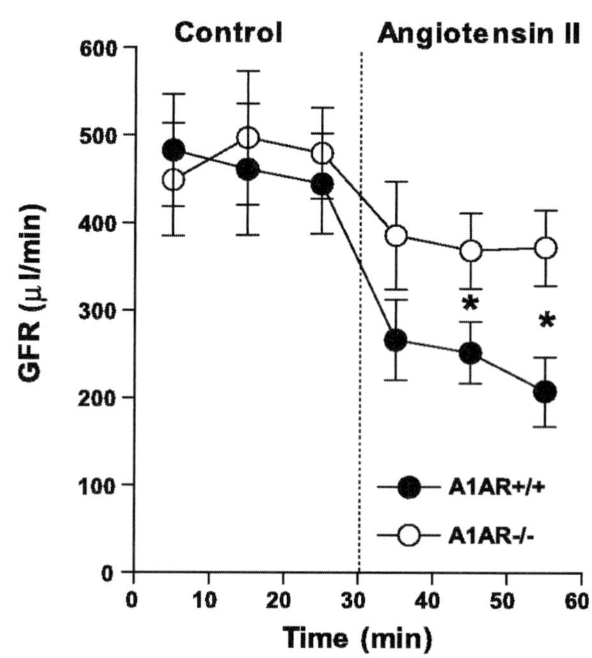

Figure 1 Glomerular filtration rate (GFR) of wild-type and A1 adenosine receptor (A1R) knockout mice in three consecutive 10-minute periods during control and during intravenous infusion of Ang II at 1.5 ng/min. Values are means of 6 experiments SE. *Significance between A1R +/+ and –/– for a given period ($p < 0.05$).

groups. This study provides clear evidence that the full effect of Ang II constriction in the AA requires a functioning A1-R, and presumably activation by increased adenosine release. Alternately, the level of interaction could be mediated by G-protein coupled signaling of these two receptors.

Purinergic Receptors

The vascular control mediated by adenosine and its A1-R may be closely related to ATP-linked purinergic receptors, particularly in the renal cortical circulation where both families of receptors are abundantly expressed. The P2X (ATP) and P2Y (UTP) families of receptors, specifically $P2X_1$, $P2X_3$, and $P2Y_2$, mediate increases in intracellular calcium concentration in vascular smooth muscle cells and perhaps vascular tone in the afferent arterioles. Activation of both types of receptors in freshly harvested preglomerular smooth muscle cells increases intracellular calcium concentration, presumably through different mechanisms. To test this hypothesis in a vascular preparation, Inscho and Cook [4] measured the diameters of rat juxtamedullary AA in response to perfusion of ATP and UTP with and without treatment of diltiazem, a calcium channel blocker. The P2 agonists α,β-methylene ATP, ATP, and UTP reduced the diameters of AA by 8 to 30 percent (Figure 2). The constrictive response to the specific $P2X_1$ agonist α,β-methylene ATP was completely blocked by diltiazem, which suggests that this receptor acts to increase calcium concentration from extracellular sources. The constrictive response to ATP at doses less than 10 μM was also blocked by diltiazem. However, the response to UTP was similar before and during diltiazem. These physiological observations confirm previous studies in cells that ATP and UTP increase intracellular calcium, which is consistent with the increased tone of the AA elicited in this study. Further, the increases to ATP appear to be mediated via L-type channels from extracellular sources and the increases to UTP from release of intracellular stores.

Arachidonic Metabolites

The major product of cyclooxygenase activity in the kidney is prostaglandin E_2 (PGE_2), which mediates vasodilation through activation of the EP_2 receptor. Four G protein–coupled EP receptors have been identified in the kidney, and the physiological effects of these have not been fully established. In an attempt to clarify these roles, Imig et al. [5] tested the effects of PGE_2 in mice with disrupted EP_2 receptors. They used a preparation in the isolated perfused mouse kidney that surgically exposes the juxtamedullary glomeruli. PGE_2 added to the perfusate dilated the preconstricted afferent arteriole in EP_2 (+/+) mice, but further constricted the AA in EP_2 (–/–) mice (Figure 3). This suggests that the renal vasodilation associated with PGE_2 is mediated specifically by the EP_2 receptor. In the absence of EP_2, PGE_2 activates other EP receptors, which causes vasoconstriction. Selective inhibition of EP_1 and EP_3 receptors prevented the PGE_2-associated constriction. In addition this constriction was also blocked with ACE inhibition, demonstrating the link between PGE_2 and Ang II. EP_2 receptors may be critically involved in maintenance of renal blood flow or glomerular filtration, specifically by mediating the vasodilation associated with PGE_2.

Endothelially Derived Hormones

The relative resistance of preglomerular arteries and the afferent arteriole is often considered when assessing total renal vascular resistance and its hormonal control. Larger arteries are more accessible and are often used for isolated in vitro studies, yet the well-established dominant resistance

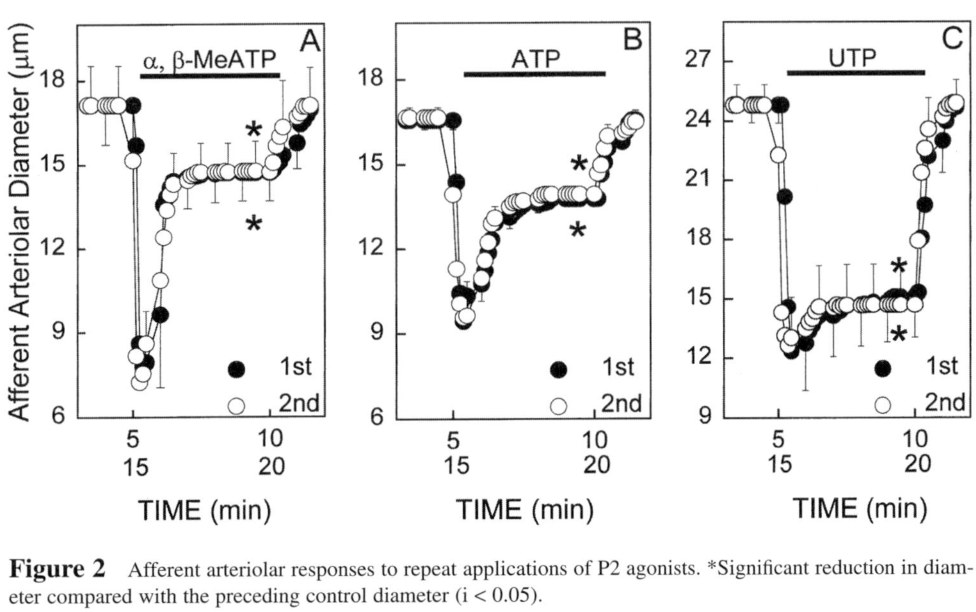

Figure 2 Afferent arteriolar responses to repeat applications of P2 agonists. *Significant reduction in diameter compared with the preceding control diameter (i < 0.05).

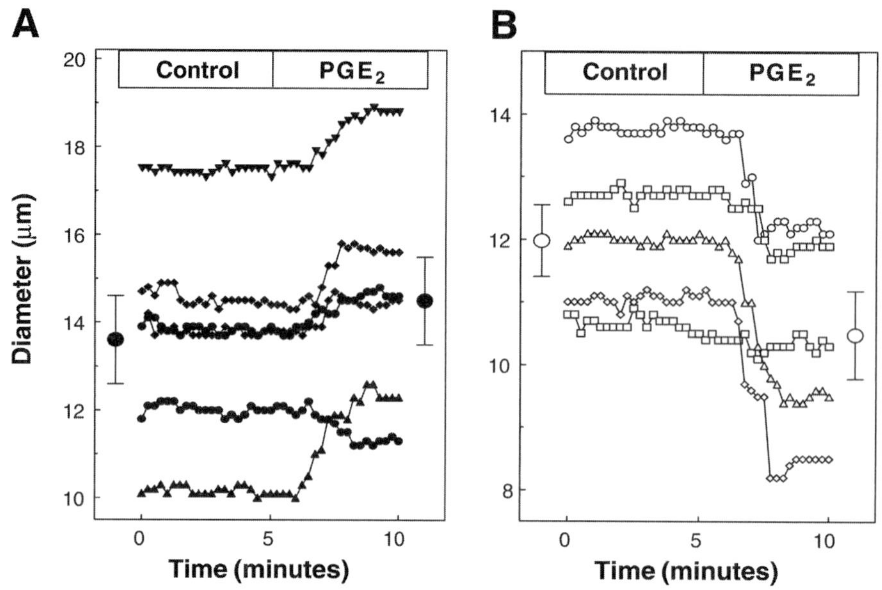

Figure 3 Afferent arteriolar diameter responses to PGE_2 in EP_2 +/+ and EP_2 −/− mice. The afferent arteriolar PGE_2 responses in EP_2 +/+ mice are shown in **A**, and PGE_2 −/− mice are shown in **B**. Diameter measurements at 15-second intervals are depicted under control conditions (first 5 minutes) and after addition of PGE_2 (second 5 minutes).

capacity of the afferent arteriole makes this tissue more applicable in control of glomerular hydrostatic pressure and RBF. However, because of the small size and fragility of this tissue, there have been only a few in vitro observations in this key resistance vessel, and these have been limited to rabbits and mice. Several hormones have been implicated in the control of larger artery resistance in studies using isolated preglomerular vessels. Wang et al. [6] recently evaluated several vasodilators in the isolated perfused rabbit AA. Nitric oxide (NO) and endothelium-derived hyperpolarizing factors (EDHF) mediate acetylcholine-induced vasodilation in small vessels, and the relative effects of these have been fully investigated in the larger preglomerular arteries. EDHF may be more important in smaller resistance vessels, yet the relative roles have not been studied extensively in the AA. The epoxide product epoxyeicosatrienoic acid (EET) may mediate vasodilation in these vessels, but another epoxide 20-hydroxyeicosatetraenoic acid (20-HETE) acts as a vasoconstrictor and may offset its effect. Wang et al. tested the roles of EDHF, NO, and prostaglandins on the Ach-induced dilation in isolated preconstricted AA [6]. COX inhibition had no effect on the Ach-induced vasodilation, whereas more than 50 percent could be blocked with inhibition of NO. Most of the remaining dilation was blocked by inhibitors of Ca^{2+} activated K^+ channels or by high extracellular K^+, which suggests EDHF. In an effort to characterize the EDHF, the tissue was treated with a competitive inhibitor of 14,15-EET, the vasodilating epoxide identified in other renal arteries. During NO inhibition 14,15-epoxyeicosa-5(*Z*)-enoic acid (14,15-EEZE) blocked much of the remaining dilation, equal to K^+ channel blockade. Both inhibitors, however, did not account for all of the Ach-induced vasodilation in this model. Though this study identified a novel dilator in the AA, the level of NO-dependent dilation was greater than in studies with other small-resistance vessels.

Rho/Rho Kinase

More novel pathways that may regulate renal vascular tone continue to be described and investigated. In vascular smooth muscle cells (VSMC), one of the final steps for contraction is the increased intracellular Ca^{2+} activation of myosin light chain (MLC) kinase, which leads to MLC phosphorylation (MLCP). However, several studies have shown that MLCP inhibition increases Ca^{2+} sensitivity and contraction in VSMC. One endogenous mechanism that checks MLCP is the activation of Rho kinase (ROK) by the G protein RhoA. Recently developed ROK inhibitors have been used to characterize its role in constriction. The early use of these compounds reduced in vitro vascular contraction, which is consistent with the theoretical control of ROK on Ca^{2+} sensitivity. However, since ROK increases stress fiber formation and actin polymerization, the vasoconstriction could occur via these mechanisms. Cavarape et al. [7] were the first to test the effects of ROK inhibitors on the renal vascular tone. In the in vivo hydronephrotic rat kidney, they showed that ROK inhibition increased glomerular blood flow and dilated interlobular arteries, and afferent arterioles more than efferent arterioles. They further tested agonists to three small renal artery constrictors that all increase intercellular Ca^{2+}: endothelin B receptors (ETB), adenosine 1 receptors (A1-R), and guanylyl cyclase (GC). ETB and A1-R are more effective on AA and GC is more effective in EA. ROK inhibitors prevented the constriction associated with each agonist in nearly all vessels. ROK inhibitors had no effect on expression of actin; therefore the authors conclude that the action of ROK inhibition is on renal VSMC Ca^{2+} sensitivity.

Ischemia

The recent development of miniaturized video imaging and improved image analysis has spurred the study of cortical microcirculation under normal and experimental conditions. With a pencil-like lens of 1 mm diameter, this video system is capable of visualizing the glomerular and pre- and postglomerular circulation, as well as interlobular arteries. This improved instrument has been validated by confirming microcirculatory responses to Ang II and TGF stimulation originally observed by in vivo renal micropuncture. Early use of this instrument was made in anesthetized dogs, but now applications are made in the rat, as well. Evaluations of microcirculatory changes due to pathophysiology and responses to vasoactive agents are now emerging.

In a study that demonstrates the capabilities of this new tool, Yamamoto et al. [8] observed peritubular capillaries, glomerular arterioles and interlobular arteries before and after 45 minutes of renal artery occlusion in the rat (Figure 4). By measuring the velocity of red blood cells (RBC), they showed that blood flow quickly returned to normal levels during the first 1 to 2 minutes after release of the occlusion. However, capillary flow dropped to levels 10 to 20 percent of normal during the first 2 to 5 minutes after release and gradually improved over the next 60 minutes (Figure 5). Similar observations were made in the glomerular blood flow, except that flow was fully restored after 60 minutes. This could not be explained by interlobular artery flow, since these vessels tended to dilate after occlusion. A second finding was that flow through the peritubular capillaries was always orthograde during the control period, but after ischemia the authors noted a mixture of orthograde, retrograde and no flow. The lack of perfusion of all peritubular capillaries confirms and strengthens the "no-flow" theory of postischemic vascular damage, directly demonstrating profound and sustained reduction of flow. The authors conclude that 45 minutes of renal ischemia can induce vascular nephropathy.

Vascular Resistance in Transplanted Kidneys

Chronic allograft failure of transplanted kidneys is often related to hemodynamics rather than immunological rejection, which usually occurs more acutely. Therefore the vascular compliance of transplanted kidneys is now being considered as another predictive parameter in long-term survival of renal transplants. Using an animal model to assess the role of changes in vascular resistance in transplanted kidneys, Gabriels et al. [9] measured the responses of excised small renal arteries to a series of vasoactive agents. Small arteries (200 to 270 μm) were obtained from kidneys donated from Fischer rats (Fl) transplanted into nephrectomized Lewis rats. Several control animals were also studied: rats with syngeneic transplants, uninephrectomized, uninephrectomized with denervated kidneys, and kidneys made ischemic, as well as native kidneys. All animals were treated with cyclosporine. After 18 weeks, the coronary, femoral, and mesenteric arteries and small arteries from the kidney were excised and placed in a myograph. The reactivity was assessed by the responses to constrictors (norepinephrine, epinephrine, Ang II, vasopressin) and dilators (Ach, sodium nitroprusside). Histology showed that only the kidneys from FL rats had glomerular lesions and vasculopathy. However, renal vessels from both allogeneic and syngeneic transplants were more sensitive to all

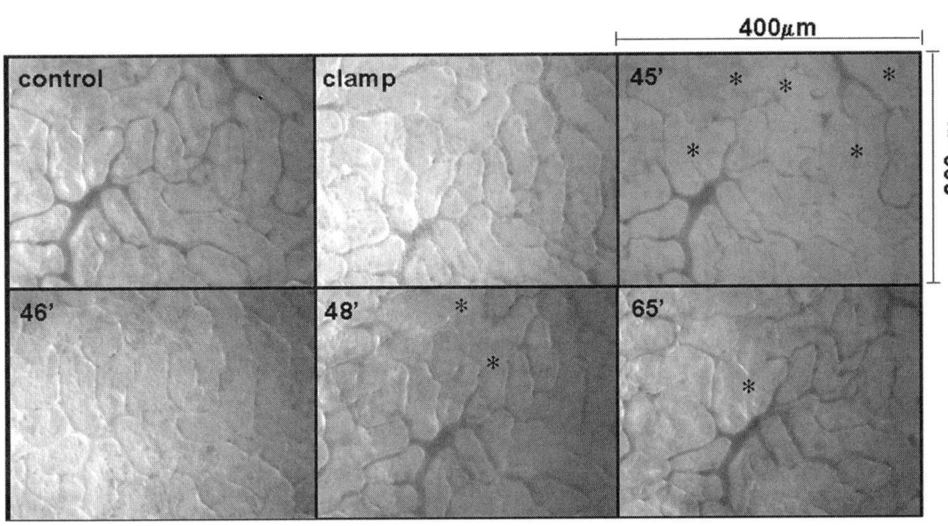

Figure 4 Sequential images obtained from a representative field of superficial peritubular capillaries before renal artery occlusion and after release of the clamp. Note the initially interrupted blood flow abated after the renal artery was clamped and 1 to 2 minutes after release of the clamp (frame 46′). Of note, not all capillaries seen at the initial image remain perfused on the subsequent postischemia images.

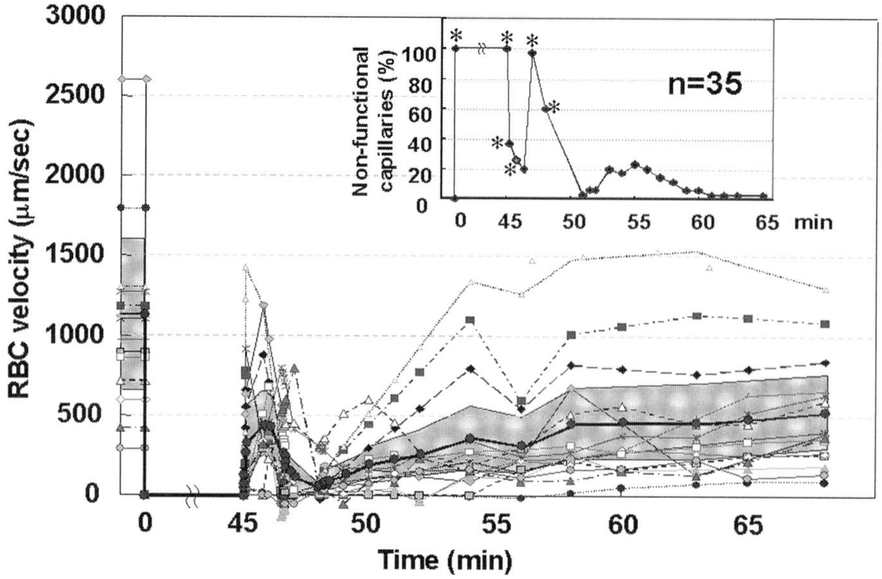

Figure 5 Time course of changes in red blood cell (RBC) velocity in postischemic peritubular capillaries. Shaded area, means SE. Individual experimental sequences are represented by different symbols. **Inset:** Proportion of nonfunctional peritubular capillaries in the postischemic kidney.

vasoconstrictors, compared to the control groups. Similarly, the vessels from transplanted kidneys were less responsive to the vasodilators than was the control tissue. There were no vessel reactivity differences in nonrenal arteries. This study demonstrates that small-artery reactivity is reduced in transplanted kidneys, regardless of their source. Further, the small arteries in all the transplanted kidneys appeared more constricted, which may contribute to reduced function and survival in transplanted kidneys.

A recent clinical study used a similar hypothesis to identify the predictive value of vascular resistance on allograft survival. In a prospective study of 601 transplanted patients, renal segmental arterial index (RI) was measured at least 3 months after surgery in order to exclude acute rejection causes [10]. RI was determined by the percent reduction of the end-diastolic flow compared with the systolic flow in renal segmental arteries, measured by Doppler ultrasonography. This value reflects primarily renal cortical vascular resistance. They divided the subjects into RI of greater than 80 (122 patients) and less than 80 (479 patients) and tested the hypothesis that lower renal resistances (an RI of less than 80) predicted long-term allograft survival. The combined end point that defined survival was a reduction in creatinine clearance of 50 percent. Patients were followed for up to 3 years. Sixty-nine percent of the high-RI patients showed a 50 percent reduction in creatinine clearance compared to only 12 percent of the low-RI patients. In addition, 47 percent of the high-RI patients and only 9 percent of the low-RI patients required dialysis. Though this study did not evaluate nonresistance issues in rejection, it confirmed that high vascular resistance predicted poor allograft performance.

Summary

The renal cortical microcirculation delivers blood to the glomerular capillaries, where blood is filtered into the nephron. Pre- and postglomerular arteriolar tone is closely regulated to maintain a sufficient filtering pressure. The systems that control resistance of these arterioles remain a rich source of investigation. The goals of these studies are to better understand the physiology of glomerular blood flow and the potential dysfunctions that may contribute to renal failure and hypertension. Advances have been made both in methods to study these small-resistance vessels, such as intravital microscopy and isolated perfused afferent arterioles, and in studies of the hormones that regulate arteriolar resistance.

Glossary

Afferent arteriole: The small artery leading into the glomerular capillary, providing the major vascular resistance in the kidney.

Efferent arteriole: The smallest artery in the cortex, which emerges from the glomerular capillary and leads to either a peritubular capillary (in the outer cortex) or vasa recta (in the juxtamedullary cortex). This vessel has a relatively high resistance capacity.

Glomerular capillary: A capillary containing an intricate network of branches and anastomoses, where the filtration of the fluid component of blood passes into the nephron. Located only in the renal cortex.

Renal vascular resistance: The change in hydrostatic pressure due to the tone of the downstream vasculature.

References

1. Marchetti, J., Helou, C. M., Chollet, C., Rajerson, R., and Alhenc-Gelas, F. (2003). ACE and non-ACE mediated effect of angiotensin I on intracellular calcium mobilization in rat glomerular arteries. *Am. J. Physiol. Heart Circ. Physiol.* **284**(6), H1933–H1941.
2. Matsuda, H., Hayashi, K., Arakawa, K., Kubota, E., Honda, M., Tokuyama, H., Suzuki, H., Yamamoto, T., Kajiya, F., and Saruta, T. (2002). Distinct modulation of superficial and juxtamedullary arterioles by prostaglandin in vivo. *Hypertens. Res.* **25**(6), 901–910.
3. Hansen, P. B., Hashimoto, S., Briggs, J., and Schnermann, J. (2003) Attenuated renovascular constrictor responses to angiotensin II in adenosine 1 receptor knockout mice. *Am. J. Physiol. Regul. Integr. Comp. Physiol.* **285**(1), R44–R49. *This study combined whole kidney function and isolated perfused isolated afferent arterioles from the adenosine type 1 receptor knockout mouse to assess the chronic absence of adenosine vasoconstriction control. The authors showed that the full vasoconstriction associated with Ang II required adenosine 1 receptors.*
4. Inscho, E. W., and Cook, A. K. (2002) P2 receptor-mediated afferent arteriolar vasoconstriction during calcium blockade. *Am. J. Physiol. Renal Physiol.* **282**(2), F245–F255.
5. Imig, J. D., Breyer, M. D., and Breyer, R. M. (2002). Contribution of prostaglandin EP_2 receptors to renal microvascular reactivity in mice. *Am. J. Physiol. Renal Physiol.* **283**, F415–F422.
6. Wang, D., Borrego-Conde, L. J., Falck, J. R., Sharma, K. K., Wilcox, C. S., and Umans, J. G. (2003). Contributions of nitric oxide, EDHF, and EETs to endothelium-dependent relaxation in renal afferent arterioles. *Kidney Int.* **63**, 2187–2193.
7. Cavarape, A., Endlich, N., Assaloni, R., Bartoli, E., Steinhousen, M., Parekh, N., and Endlich, K. (2003). Rho-kinase inhibition blunts renal vasoconstriction induced by distinct signaling pathways in vivo. *J. Am. Soc. Nephrol.* **14**(1), 261–264. *This study introduces a novel regulator to the control of renal cortical microcirculation, the Rho/Rho kinase system. Though previously studied in VSMCs and other nonrenal tissue, this is the first attempt to identify the role of these agents in the renal cortex.*
8. Yamamoto, T., Tada, T., Brodsky, S. V., Tanaka, H., Moiri, E., Kajiya, F., and Goligorsky, M. S. (2002). Intravital videomicroscopy of peritubular capillaries in renal ischemia. *Am. J. Physiol. Renal Physiol.* **282**(6), F1150–F1155. *These investigators report new observations with intravital microscopy in the rat kidneys. These results demonstrate both the advantages and limitations of this improved tool.*
9. Gabriels, G., August, C., Grisk, O., Steinmetz, M., Kosch, M., Rahn, K. H., and Schlatter, E. (2003). Impact of renal transplantation on small vessel reactivity. *Transplantation* **75**(5), 689–697.
10. Radermacher, J., Mengel, M., Ellis, S., Stuht, S., Hiss, M., Schwarz, A., Eisenberger, U., Burg, M., Luft, F. C., Gwinner, W., and Haller, H. (2003). The renal arterial resistance index and renal allograft survival. *N. Engl. J. Med.* **349**(2), 182–184.

Capsule Biography

William J. Welch has worked extensively on intrarenal hormones that regulate tubuloglomerular feedback and its effect on renal function. He has used in vivo micropuncture to access the renal cortical vasculature and cortical nephrons. The interaction between single nephron function and vascular consequences has been the primary focus of his more than 60 published studies. He has used both normal and hypertensive models to show modulatory roles for angiotensin II, thromboxane, nitric oxide, and superoxide in the control of GFR.

SECTION F

Liver

CHAPTER 72

Hepatic Microvascular Responses to Inflammation

Claudine S. Bonder and Paul Kubes
University of Calgary, Calgary, Alberta, Canada

Introduction

A hallmark feature of hepatic inflammation is the activation of resident nonparenchymal cells, the upregulation of proinflammatory mediators, and the recruitment of leukocytes from the blood to the afflicted site. The liver is a unique organ that houses an elaborate vasculature to handle an enormous volume of blood and monitors for any pathogens that may have breached the gut wall. In this article we highlight cells and substances that mediate hepatic inflammation. We also discuss the adhesion molecules that are key for leukocyte recruitment to the liver during an inflammatory response.

The Hepatic Microvascular System

The liver receives as much as 25 percent of the cardiac output, and much of this blood has already passed through the intestinal tract. In conjunction with handling such a large quantity of incoming blood, it also responds to potential infectious agents that have entered the blood. To do all this, the liver has evolved to maximize vascular spaces and developed mechanisms to fight foreign entities.

The hepatic microvascular system consists of a portal circulation including portal venules, sinusoids, and central venules and an arterial circulation (hepatic arterioles). Blood leaving the gut via the portal venules enters the sinusoids via the sphincter-like cells. Arteriole blood enters the sinusoids through branches of the hepatic arterioles. Both portal and arterial vessels can contract and therefore regulate blood flow within the sinusoids. The sinusoids take blood through the lobule from the portal tract to the central (terminal) vein. Close to the portal tract the sinusoids form an interconnecting network that becomes organized into parallel vessels. As blood leaves the sinusoids it flows through outlet sphincters. These sphincters control the rate of blood flow entering the central veins.

The portal blood represents a route through which infectious gut-derived organisms can enter the liver. Specific mechanisms have evolved to instigate and regulate inflammatory responses against invading pathogens. The close proximity and cross-talk between the different cells within the liver dictates that many of the hepatic cells are involved in an inflammatory response. Unfortunately the development of an inflammatory response, if uncontrolled, can manifest in permanent liver injury. Indeed, there is irrefutable evidence for the involvement of inflammatory cells in a large number of known liver diseases, including alcoholic hepatitis and cirrhosis, viral hepatitis, ischemia–reperfusion injury (transplantation, tumor resection, shock), sepsis- or endotoxin-induced liver injury, acute and chronic rejection, primary biliary cirrhosis, and primary sclerosing cholangitis. Following is an overview of the inflammatory cells, molecules and mechanisms used by the liver to combat and regulate liver injury and dysfunction. As this review is brief, we strongly encourage the readers to turn to our list of suggested reading, and elsewhere, for more information.

Liver Constitution

The liver is formed by parenchymal cells, that is, hepatocytes, and nonparenchymal cells. In contrast to hepatocytes that occupy almost 80 percent of the total liver volume, nonparenchymal liver cells occupy only 6 percent. Notably, it is the nonparenchymal cells of the liver that regulate an inflammatory response. For the following section we recommend reviews [1–3].

Endothelial Cells

Liver endothelial cells line the portal venules as well as the sinusoids. Whereas the endothelium of the portal venules can be likened to endothelium in other parts of the body, the sinusoidal endothelial cells are morphologically unique. The sinusoidal endothelial cells are characterized by an absence of tight junctions between cells, by a lack of basal lamina, and by the presence of open fenestrations that are arranged into characteristic sieve plates. These cells perform an important filtration function, as these small fenestrations allow free diffusion of many small substances between the blood and the hepatocyte surface. Sinusoidal endothelial cells have also been shown to endocytose many ligands including glycoproteins, components of the extracellular matrix (such as hyaluronate, collagen fragments, and fibronectin) and immune complexes. Sinusoidal endothelial cells may function as antigen-presenting cells in the context of both MHC-I and MHC-II restriction and are also active in the secretion of cytokines, eicosanoids, (i.e., prostanoids and leukotrienes), endothelin-1, nitric oxide, and some extracellular matrix components.

Kupffer Cells

Kupffer cells are intrasinusoidally located tissue macrophages (Figure 1) and have endocytic and phagocytic capabilities. Hepatic macrophages are in constant contact with gut-derived portal blood. Exposure of Kupffer cells to bloodborne pathogens (e.g., endotoxin) can lead to the intensive production of inflammatory mediators and the formation of liver injury. Liver macrophages modulate immune responses via antigen presentation and interaction with T cells. Among the many inflammatory mediators produced by Kupffer cells are reactive oxygen species, eicosanoids, nitric oxide, carbon monoxide, cytokines (TNFα), and chemokines [CXCL10 (IP-10) and CXCL9 (Mig)]. The activation of these cells identifies them as regulators of the early phase of liver inflammation and an important part in innate immune defense.

Hepatic Stellate Cells

Hepatic stellate cells are present in the perisinusoidal space. They are positioned with their well-branched cytoplasmic processes in close proximity to the endothelial cells (Figure I). In the normal liver, hepatic stellate cells control turnover of extracellular matrix and are important in regulating the contractility of sinusoids. In the event of hepatocyte damage, stellate cells transform into myofibroblast-like

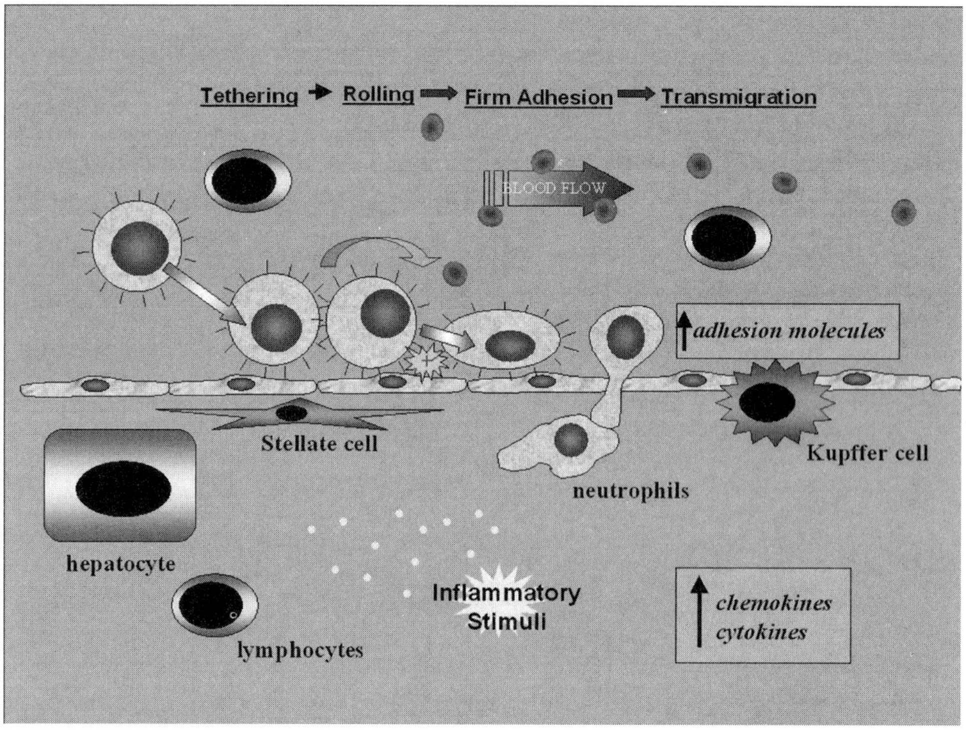

Figure 1 The leukocyte recruitment cascade. Fast-moving leukocytes in the bloodstream tether and roll on activated endothelium via interactions between adhesion molecules. Chemokines or other proinflammatory mediators released by various sources within the tissue are presented on the endothelium to rolling leukocytes, resulting in firm adhesion. Leukocyte transmigration across the endothelium and entry into inflamed tissue ensues. (see color insert)

Table I Adhesion Molecules and Chemokines Upregulated during Hepatic Inflammation.

	Inflamed portal vessels	Inflamed sinusoids
ICAM-1	++	+
VCAM-1	+	+
VAP-1	+	+
PECAM-1	+	–
P-selectin	+	–
E-selectin	+	–
CCL2 (MCP-1)	++	+
CCL3 (MIP-1α)	++	+
CCL4 (MIP-1β)	++	+
CCL5 (RANTES)	++	+
IL-8	++	++
CXCL9 (Mig)	+	++
CXCL10 (IP-10)	+	++
CXCL11 (I-TAC)	+	++

cells and mediate the fibrotic response during inflammation. Activated stellate cells, like Kupffer cells, are potent sources of several chemokines including CXCL-10 (IP-10) and CCL-2 (MCP-1).

Lymphocytes

Intrahepatic lymphocytes comprise 25 percent of nonparenchymal cells in the normal liver (Figure 1). In contrast to peripheral blood, the major lymphoid population in the normal liver are pit cells (i.e., NKT cells, 30%), then T cells (20% TCR-αβ and 10% TCR-γδ), NK cells (20%), and very few B lymphocytes (5%). Natural killer T cells are a unique population of cells that express both CD3 and NK1 on their cell surface, secrete IL-2, IL-4, and IFNγ, and recognize antigen in association with CD1. After NKT cells, the predominant intrahepatic T cells are CD8+. Notably, this predominance of CD8+ T cells in the liver differs from the CD4+ T cell majority observed in peripheral blood and spleen. CD8+ T cells, NKT cells, and NK cells all possess cytotoxic activity. Activated CD8+ T cells can be retained by the liver via their interaction with sinusoidal endothelial cells. These "trapped" activated T cells can then cause liver injury by releasing inflammatory cytokines and by apoptosing. Alternatively, circulating lymphocytes can be activated within the liver by Kupffer cells or hepatocytes. Prior to their "activation-induced" cell death, these nonspecifically trapped T cells can also mediate liver damage by secreting mediators of inflammation (e.g., IL-12 and IFNγ).

Neutrophils

Neutrophils are an essential component of innate host resistance and the adaptive responses to inflammation. Although neutrophils constitute a relatively small percentage (1 to 2%) of the nonparenchymal liver cells in a normal (uninfected) mouse, a dramatic tenfold to twentyfold increase in neutrophils occurs within 2 hours following viral infection. Previous observations suggest that neutrophils are the first cell type to arrive at sites of inflammation and as such form the first line of immune defense. In response to chemotactic factors released at inflammatory sites, neutrophils migrate from the bloodstream through the vascular endothelium and adhere to parenchymal cells (Figure 1). Once at the inflammatory site, neutrophils can act as effector cells in an attempt to resolve the inflammation. Alternatively, neutrophils may be cytotoxic and cause necrotic tissue destruction by releasing proteases or reactive oxygen species. Neutrophils can also release chemokines such that more inflammatory cells are recruited to the afflicted site. A detrimental effect of neutrophils in acute and chronic liver disease processes has been shown in hepatic ischemia–reperfusion injury, endotoxemia, sepsis, and alcoholic hepatitis.

Taken together, the unique structure and constitution of the liver, such as the open fenestrations and position of the Kupffer cells, dictate that the barriers against inflammation differ from those in other parts of the body. Indeed, it has repeatedly been shown that inflammatory mediators have easy access to hepatocytes.

Mediators of Hepatic Inflammation

When microbes and other inflammatory stimuli first enter the hepatic circulation, they activate sinusoidal endothelial cells and Kupffer cells. These cells then produce a cocktail of proinflammatory mediators such as prostanoids, nitric oxide, cytokines, chemokines, many growth factors, and reactive oxygen species. With the increased expression of these inflammatory mediators comes (1) the recruitment of leukocytes into the inflamed area, (2) the killing of the invading microbes, and in some cases (3) damage to liver parenchymal cells. For more elaborate discussions on the mediators of hepatic inflammation, we recommend the following reviews: Ref. [4–7].

Vasocontriction and Vasodilation

Alterations to blood flow within the liver have a significant impact on liver function. Several vasoactive substances can affect constriction and relaxation of hepatic stellate cells. These mediators have been shown to come from stellate cells themselves (endothelin-1, NO) as well as hepatocytes (carbon monoxide, leukotrienes), endothelial cells (endothelin, nitric oxide, prostaglandins), and Kupffer cells (prostaglandins, NO). Notably, during the early phase of sepsis both nitric oxide (NO) and carbon monoxide (CO) act as antagonists of vasoconstrictors (e.g., endothelin-1) by mediating relaxation of sinusoidal vessels such that there is an increase in vasodilation. In acute and chronic hepatic inflammation, dysregulation of the vasoactive substances can result in a circulatory collapse within the liver.

Reactive Oxygen Species

Reactive oxygen species (ROS) are important cytotoxic and signaling mediators in the pathophysiology of inflammatory liver diseases. They can be generated by resident and nonresident liver cells following stimulation. ROS are essential for host-defense functions of phagocytes. In addition, ROS have been shown to indirectly affect vasoconstriction and liver regeneration. Although ROS are able to induce cell damage, they also modulate signal transduction pathways by regulating transcription factors such as NF-κB. By regulating NF-κB, ROS can indirectly affect the pathophysiology of the liver by (1) directly inducing and/or regulating apoptotic and necrotic cell death and (2) modulating the expression of inflammatory mediators and adhesion molecules.

Eicosanoids

Prostaglandins and leukotrienes are ubiquitous mediators of liver function and disease. In the inflamed liver, Kupffer cells produce prostaglandins to modulate hepatic glucose metabolism. Upregulation of glucose correlates with the increasing demand for energetic fuel by the inflammatory cells such as leukocytes, sinusoidal endothelial cells, and Kupffer cells. Studies have also demonstrated a protective role of PGE_2 in viral hepatitis as it abrogates the induction of TNFα, attenuating the expression of antigens on hepatocytes, and may inhibit viral replication. By contrast, in cirrhosis and fulminant hepatic failure a dysregulation of prostaglandin metabolism has been noted. Leukotrienes (e.g., LTB4) have vasoconstrictive and metabolic effects in the liver. These mediators have been shown to cause liver injury in ischemia–reperfusion and endotoxemia.

Cytokines

Experimental data have demonstrated a critical role for proinflammatory cytokines (e.g., TNFα, IL-1, and IFNγ in the development of liver injury. Although TNFα is required for normal hepatocyte proliferation during liver regeneration, it is also involved in inflammation. TNFα is known to induce the transcription factor nuclear factor-kappaB (NF-κB) which regulates many proinflammatory mediators (e.g., TNFα and IL-1). Overexpression of TNFα (from Kupffer cells or infiltrating neutrophils) has been shown to correlate with hepatotoxicity in endotoxemia, alcoholic liver disease, and viral hepatitis. Another cytokine noted for its involvement in liver injury is IFNγ. In alcoholic hepatitis, overexpression of IFNγ mediates liver injury. By contrast, in viral hepatitis it is a lack of functioning IFNγ that permits the liver damage.

Chemokines

The recruitment of inflammatory cells to the liver is a critical step to the development of liver disease. Chemokines have emerged as one of the most important regulators of leukocyte trafficking and activation. In the noninflamed liver resident cells express and secrete low levels of CCL5 (RANTES), CCL2 (MCP-1), CCL3 (MIP-1α), CCL4 (MIP-1β), and IL-8. These chemokines have been detected on the vascular endothelium in portal tracts as well as the sinusoids. Interestingly, inflammation-induced chemokine expression varies between portal and sinusoidal vessels. Chemokines CXCL9 (Mig), CXCL10 (IP-10), and CXCL11 (I-TAC) are detected on sinusoids in association with liver inflammation, whereas relatively little induction of CCL5 (RANTES), CCL2 (MCP-1) or IL-8 is observed. By contrast, elevated levels of CCL5 (RANTES), CCL3 (MIP-1α), CCL2 (MCP-1), and CCL4 (MIP-1β) have been detected on the portal vessels. Although the reasons for these differences are still unclear, it may reflect that the sinusoidal endothelium is fundamentally distinct from the portal endothelium. In fact, these observations may suggest a selective process for regional leukocyte recruitment during liver injury.

Cellular Adhesion Molecules

Cellular adhesion molecules (CAMs) are cell surface glycoproteins involved in cell–cell and cell–matrix interactions. CAMs are critical for leukocyte adhesion to endothelium, transmigration, binding to target cells, and cytotoxicity. Following is a review of CAMs that have been shown to be relevant to inflammatory liver disease, and we suggest the following articles to readers: Ref. [6, 8–10].

Integrins

Resident liver cells (e.g., Kupffer cells and pit cells) as well as newly recruited monocytes, T lymphocytes and neutrophils have β_1 and β_2-integrins on their surface. LFA-1 ($\alpha_L\beta_2$) is highly expressed on all leukocytes. By contrast, Mac-1 ($\alpha_M\beta_2$) is present primarily on neutrophils, Kupffer cells, and monocytes, and to a small degree on pit cells. VLA-4 ($\alpha_4\beta_1$) is expressed mainly by T cells, pit cells, and monocytes, and to some degree on Kupffer cells and under limited conditions neutrophils. The activation of integrins has been observed in situations akin to liver injury in vivo. Indeed, increased Mac-1 expression on neutrophils was observed during reperfusion after hepatic ischemia, endotoxemia, and sepsis.

Intercellular Adhesion Molecules

Intercellular adhesion molecules (ICAM-1, -2 and -3), vascular adhesion molecule (VCAM-1), and platelet endothelial adhesion molecules-1 (PECAM-1) are also important in an inflammatory response. In the normal liver, no ICAM-1 is found on hepatocytes or stellate cells and low expression of ICAM-1 can be found on the entire hepatic endothelium and Kupffer cells. During inflammation, ICAM-1 is strongly upregulated by TNFα, IL-1β, and IFNγ on all liver cells (Table I). This induction is promptly followed by the recruitment of leukocytes into the liver, where they attach to and subsequently damage the liver cells. In normal livers there is low VCAM-1 expression on venule, arteriole, and sinusoidal endothelial cells. No constitutive or inducible VCAM-1 has been observed on hepatocytes. With exposure to TNFα or IL-1β both Kupffer cells and endothelial cells upregulate VCAM-1 (Table I). Vascular adhesion protein (VAP)-1 is constitutively expressed on human sinusoidal endothelial cells and has been shown to mediate lymphocyte recruitment and possibly transendothelial transmigration. PECAM-1 has been shown to be important for leukocyte transendothelial migration. Studies have shown PECAM-1 to be constitutively expressed on large vessel endothelial cells but not sinusoidal endothelial cells. PECAM-1 is not inducible with inflammatory mediators (Table I).

Selectins

The selectin family includes three molecules, that is, L-, E-, and P-selectin. L-selectin is constitutively expressed on neutrophils and is shed from the cell surface during activation. Notably, this shedding coincides with Mac-1 upregulation on neutrophils. There are important differences in the expression of selectin adhesion molecules between vascular and sinusoidal endothelium. P-selectin is stored in secretory granules (Weibel-Palade bodies in endothelial cells). Liver sinusoidal endothelial cells do not contain WP bodies and consistent with this lack of WP bodies, sinusoidal endothelial cells do not express P-selectin (Table I). By contrast, under inflammatory conditions, P- and E-selectin can be strongly induced on the portal vascular endothelium (Table I). These differences may reflect a lack of requirement for selectin-mediated tethering in the low-flow environment of the sinusoid. Indeed, leukocyte recruitment to the sinusoidal area has been shown, in vivo, to occur in the absence of selectins.

Conclusion

In conclusion, the liver is unique in its histoarchitecture compared to other organs. Additional differences exist within the liver itself, and these become more obvious during an inflammatory response. Herein we have touched on the concept that the disruption of the normal liver occurs through a multitude of cells and proinflammatory mediators. The challenge in the future is to define various models that best describe the mechanisms that are critical to induction and/or maintenance of liver damage. Understanding the precise mechanisms of hepatic injury or inflammation may permit the development of specific anti-inflammatory strategies. Only then can we aim at preventing effector cell recruitment while leaving general homeostatic liver functions intact.

Glossary

Adhesion molecules: Molecular determinants that regulate homeostatic recirculation and tissue-specific trafficking of cells.

Inflammation: A localized protective response elicited by injury or destruction of tissues, which serves to destroy, dilute, or sequester both the injurious agent and the injured tissue. Histologically, it involves a complex series of events, including dilation of arterioles, capillaries, and venules, with increased permeability and blood flow, exudation of fluids, including plasma proteins, and leukocyte migration into the inflammatory focus.

Liver: A solid organ located in the right upper quadrant of the abdomen that plays a major role in metabolism, digestion, detoxification, and elimination of substances from the body.

Mediators: Substances released from cells as a result of the interaction between antigen and antibody or by the action of antigen with a sensitized lymphocyte.

Recruitment: The collection or increased abundance of cells into a defined area, such as antigen-experienced lymphocytes into inflammatory effector sites via specific vascular addressins and chemokines.

Further Reading

*Papers of interest have been marked for their special relevance to this topic and as outstanding reviews.

1. Ekataksin, W., and Kaneda, K. (1999). Liver microvascular architecture: An insight into the pathophysiology of portal hypertension. *Semin. Liver Dis.* **19**, 359–382.
2. Knolle, P. A., and Gerken, G. (2000). Local control of the immune response in the liver. *Immunol. Rev.* **174**, 21–34.
3. Sheth, K., and Bankey, P. (2001). The liver as an immune organ. *Curr. Opin. Crit. Care* **7**, 99–104.
4. Bone-Larson, C. L., Simpson, K. J., Colletti, L. M., Lukacs, N. W., Chen, S. C., Lira, S., Kunkel, S. L., and Hogaboam C. M. (2000). The role of chemokines in the immunopathology of the liver. *Immunol. Rev.* 177, 8–20.* *This particular review addresses the current understanding of the manner in which chemokines contribute to both acute and chronic liver injury by examining a variety of clinical and experimental studies.*
5. Jaeschke, H. (2000). Reactive oxygen and mechanisms of inflammatory liver injury. *J. Gastroenterol. Hepatol.* **15**, 718–724.
6. Jaeschke, H. (2003). Molecular mechanisms of hepatic ischemia-reperfusion injury and preconditioning. *Am. J. Physiol Gastrointest. Liver Physiol* **284**, G15–G26.* *This review is an excellent source for understanding the mechanisms that are key in the development of hepatic inflammation after an ischemic event. It discusses the role of cytokines, reactive oxidant species, complement, adhesion molecules, and neutrophil trafficking in the inflamed liver.*
7. Ramadori, G., and Armbrust, T. (2001). Cytokines in the liver. *Eur. J. Gastroenterol. Hepatol.* **13**, 777–784.
8. Jaeschke, H. (1997). Cellular adhesion molecules: Regulation and functional significance in the pathogenesis of liver diseases. *Am. J. Physiol* **273**, G602–G611.* *Hartmut Jaeschke is world renowned for his work in studying liver inflammation. This review focuses on CAMs that have been shown to be relevant to inflammatory liver disease processes and discusses their regulation and functional significance.*

9. Lalor, P. F., Shields, P., Grant, A., and Adams, D. H. (2002). Recruitment of lymphocytes to the human liver. *Immunol. Cell Biol.* **80**, 52–64.* *Many studies have identified lymphocytes to be key mediators of liver inflammation. This review is an excellent source for understanding and comparing the molecular mechanisms that regulate the entry of lymphocytes into the normal and inflamed liver.*
10. Wong, J., Johnston, B., Lee, S. S., Bullard, D. C., Smith, C. W., Beaudet, A. L., and Kubes, P. (1997). A minimal role for selectins in the recruitment of leukocytes into the inflamed liver microvasculature. *J. Clin. Invest* **99**, 2782–2790.

Capsule Biography

Dr. Bonder is a Canadian Association of Gastroenterology postdoctoral fellow with Dr. Kubes at the University of Calgary. Her work focuses on the recruitment of leukocytes to splanchnic organs during inflammation.

Dr. Kubes is chair of the Immunology Research Group in the Department of Physiology and Biophysics. His laboratory is supported by grants from the Canadian Institutes for Health Research (CIHR) and a CIHR group grant.

CHAPTER 73

Liver Microcirculation

Robert S. McCuskey
University of Arizona, Tucson, Arizona

The purpose of this chapter is to provide an overview of what is known about the dynamic structure, function, and regulation of the hepatic microvascular system in health and some of its basic responses to disease.

Blood Supply to the Liver

It is important to recognize that the liver has a dual blood supply. Approximately 80 percent of the blood entering the liver is poorly oxygenated blood supplied by the portal vein containing venous blood flowing from the intestines, pancreas, spleen, and gallbladder. The remaining 20 percent of the blood supply is well oxygenated and delivered by the hepatic artery. Both the hepatic portal vein and hepatic artery enter the liver at the hilus, where efferent bile ducts as well as lymphatics also exit the organ.

Branches of the hepatic artery, hepatic portal vein, bile duct, and lymphatic vessels travel together in portal tracts through the liver parenchyma. After repeated branching, terminal branches of the blood vessels (portal venules and hepatic arterioles) supply blood to the sinusoids, which are the exchange vessels in the liver. Branches of hepatic arterioles also supply the peribiliary plexus of capillaries nourishing the bile ducts and then drain into sinusoids or occasionally into portal venules. After flowing through the sinusoids, blood is collected in small branches of hepatic veins termed central venules (central veins, terminal hepatic venules). These course independently of the portal tracts and drain via hepatic veins, which leave the liver on the dorsal surface and join the inferior vena cava. Occasionally, branches of hepatic arterioles bypass the sinusoidal bed to supply the walls of hepatic veins.

Lymphatic vessels originate as blind-ending capillaries in the connective tissue spaces of portal tracts. The fluid contained in these lymphatics flows toward the hepatic hilus and eventually into the cisternae chyli.

Microvascular Functional Units

The organization of each liver lobe into structural or functional units related to function and/or disease has been the subject of considerable debate during the past century. Several models, none of which are mutually exclusive, have been proposed as follows and as illustrated in Figure 1.

The classic hepatic lobule is a polygonal structure having as its central axis a central venule, with portal tracts distributed along its peripheral boundary. The peripheral boundaries of these lobules are poorly defined in most species, including humans. (In some species, e.g., the pig, there is considerably more connective tissue present in the liver and the connective tissue is distributed along the peripheral boundary of classic lobules, thus making them very distinct.) Considerable sinusoidal anastomoses occur between adjacent lobules, and thus the blood collected by each central venule is supplied by several portal venules. For these reasons, and because of intralobular regional differences in oxygenation, metabolic functions, and responses to some disease, an acinar concept was proposed to define the hepatic functional unit.

The hepatic acinus is a unit having no distinct morphologic boundaries. Its axis is a portal tract and its peripheral boundary is circumscribed by an imaginary line connecting the neighboring terminal hepatic venules (central hepatic venules of the classic lobule), which collect blood from sinusoids. Contained within the acinus are three zones, each having different levels of oxygenation and metabolic function. Although the acinar concept has been widely accepted, it fails to account for those mammalian species (e.g., pig, seal) that have connective tissue boundaries circumscribing the classic lobule. Additional inconsistencies increasingly have been identified in three-dimensional studies of metabolic heterogeneity and microvascular structure.

In yet another model of hepatic organization, the unit is defined by bile drainage. So-called portal lobules have at

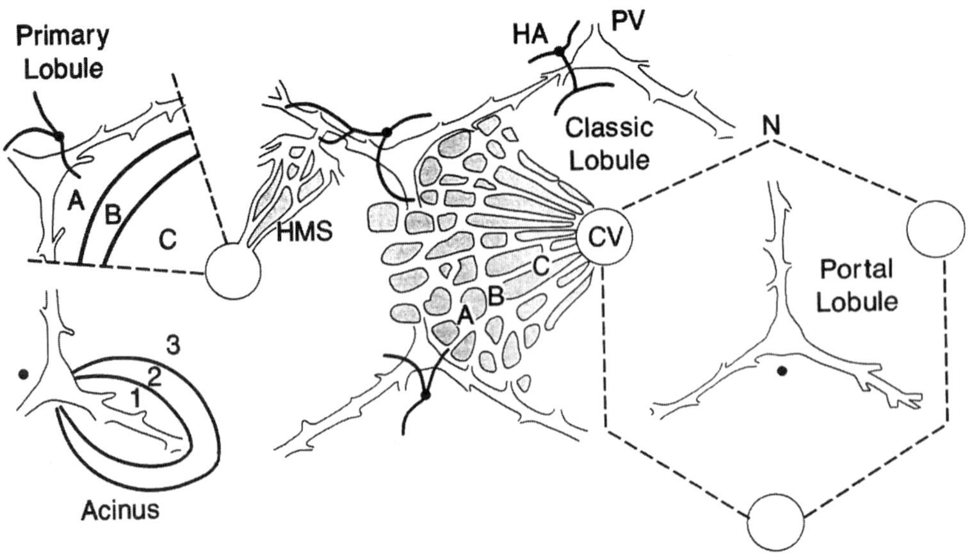

Figure 1 Contiguous hepatic lobules illustrating the interconnecting network of sinusoids derived from two portal venules (PV). Note that the sinusoids become more parallel as they course toward the central venule (CV), which forms the axis of the classic lobule. Hepatic arterioles (HA) supply blood to sinusoids near the periphery of the lobule, usually by terminating in inlet venules or terminal portal venules. As a result, three zones (1, 2, 3) of differing oxygenation and metabolism have been postulated to compose a hepatic acinus, with its axis being the portal tract (*lower left*). Several acini would compose the portal lobule (*lower right*). Each classic lobule contains several cone-shaped subunits having convex surfaces fed by portal and arterial blood at the periphery and apices at the central venule (*upper left*). A, B, and C represent hemodynamically equipotential lines in a "primary lobule." A recent modification further subdivides lobules into conical hepatic microcirculatory subunits (HMS), each being supplied by a single inlet venule. (see color insert)

their center a portal tract, with central veins present around the periphery of each lobule. This concept has received little support.

Currently, the concept of subunits of the classic lobule forming functional units is the most consistent with existing evidence. In this model, each "classic" lobule consists of several "primary lobules." Each primary lobule is cone-shaped, having its convex surface at the periphery of the classic lobule supplied by terminal branches of portal venules and hepatic arterioles, and its apex at the center of the classic lobule drained by a central (terminal hepatic) venule. These "primary lobules" were renamed *hepatic microvascular subunits* (HMS) and were demonstrated to consist of a group of sinusoids supplied by a single inlet venule and its associated termination of a branch of the hepatic arteriole from the adjacent portal space. Further confirmation of this HMS concept was obtained by studying their development in neonatal livers. Accompanying the HMS are hepatic parenchymal cells and the associated cholangioles and canaliculi. Hepatocellular metabolic gradients also have been demonstrated to conform to this proposed functional-unit concept.

Studies using three-dimensional reconstruction of sectioned livers, scanning electron microscopic examination of corrosion casts, and in vivo microscopy of several species support this concept of the functional unit being a conical microvascular subunit of the classic lobule.

Hepatic Microvascular System

The hepatic microvascular system comprises all blood and lymphatic vessels immediately involved in the delivery and removal of fluids to and from the hepatic parenchyma, namely, portal venules, hepatic arterioles, sinusoids, central venules, and lymphatics. Figure 2 is a diagram illustrating the afferent and efferent microvascular connections to the sinusoids within a single hepatic lobule.

Most blood enters the sinusoids from portal venules. These inlets are reported to be guarded by sphincters composed of sinusoidal lining cells termed the afferent or inlet sphincters. Arterial blood enters some of the sinusoids, principally through branches of the hepatic arterioles. These vessels, arterio-sinus twigs, terminate in sinusoids near their origins from portal venules. In addition, occasional direct connections (arterio-portal anastomoses, APA) have been observed with the terminal portal venules. The frequency of these APAs appears to be species-dependent. Because all of these structures are independently contractile, the sinusoids receive a varying mixture of portal venous and hepatic arterial blood. Finally, some evidence suggests that the fraction of blood delivered to the sinusoids by the hepatic artery differs between the hilus and periphery of hepatic lobes. Within the network of sinusoids, blood flow is reportedly regulated by contractile sinusoidal lining cells that control not only the velocity of flow but also its distribution within the network.

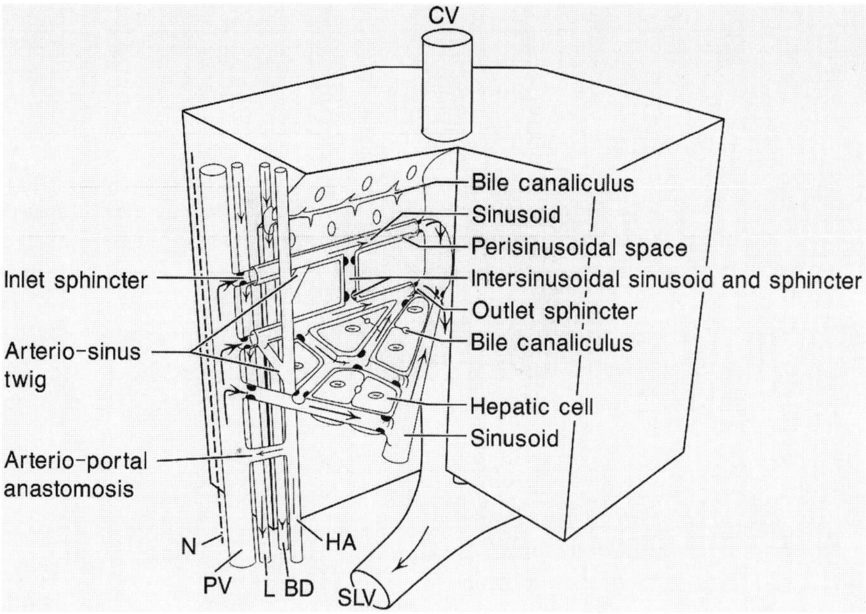

Figure 2 Hepatic microvasculature as determined by in vivo microscopic studies. PV, portal venule; HA, hepatic arteriole; L, lymphatic; BD, bile ductule; N, nerve; CV, central venule; SLV, sublobular hepatic vein. Arrows indicate direction of flow. (see color insert)

Blood leaves the sinusoids by flowing into central (terminal hepatic) venules, reportedly by passing through outlet or efferent sphincters composed of sinusoidal lining cells.

Structure and Function of Hepatic Sinusoids

The sinusoids are unique exchange vessels composed of specialized nonparenchymal cells that exhibit structural and functional heterogeneity.

Structure

The structure of the sinusoid is illustrated in Figure 3. The endothelial cells are highly fenestrated and lack a supporting basal lamina. The fenestrae are organized in clusters known as sieve plates. As a result, there is continuity between the plasma in the sinusoid lumen and the perisinusoidal space (of Disse). The sinusoidal endothelial cells contain numerous endosomes and scavenge a number of substances including breakdown products of connective tissue. They also are the source of several cytokines, eicosanoids, nitric oxide, and endothelins.

Stellate cells (fat-storing cells of Ito) lie external to the endothelium in the space of Disse. They are pericytes that frequently contain lipid droplets serving as storage sites for vitamin A. Multiple cytoplasmic projections of these cells surround and embrace the abluminal surfaces of the endothelial cells. When activated, these cells produce collagen and become contractile. As a result, they are thought to play a role in the regulation of sinusoidal blood flow.

Kupffer cells are attached to the luminal surfaces of the endothelium. These are highly phagocytic, specialized fixed macrophages of the liver and contain numerous lysosomes and phagosomes. Kupffer cells are involved in a number of host defense mechanisms and immune functions and are the source of a number of cytokines, eicosanoids, free radicals, and nitric oxide.

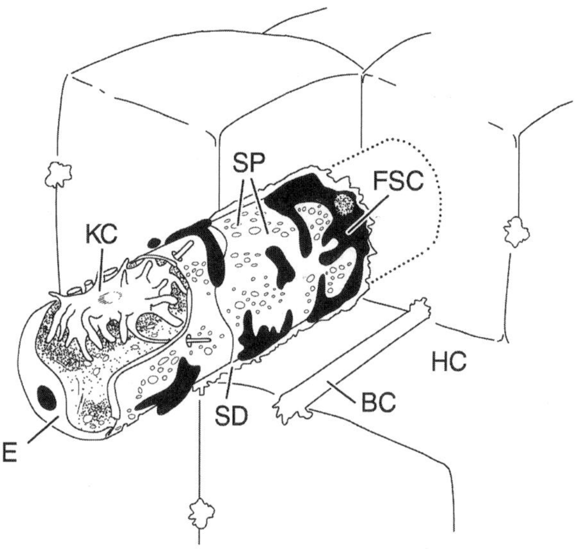

Figure 3 Sinusoid and contiguous hepatic parenchymal cells (HC). E, endothelium; KC, Kupffer cell; SD, space of Disse; SP, sieve plate composed of endothelial fenestrae; FSC, fat storing cell (stellate) cell; BC, bile canaliculus.

Heterogeneity

The organization of the sinusoid network exhibits heterogeneity. Near portal venules and hepatic arterioles, sinusoids

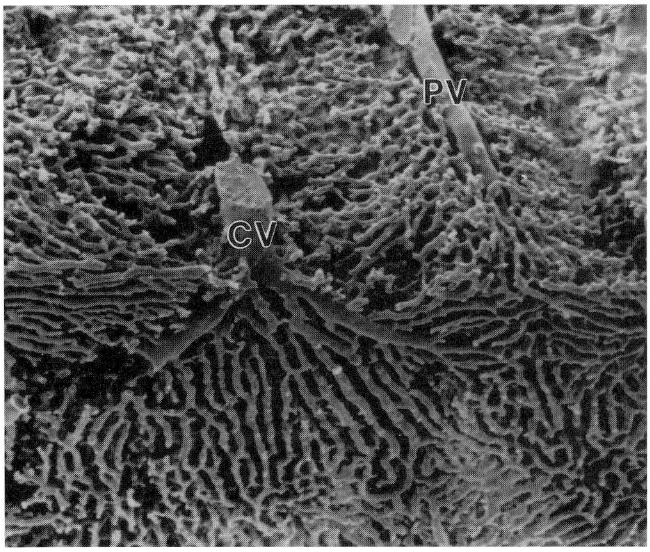

Figure 4 Vascular cast of the hepatic microvasculature illustrating the tortuous, anastomotic sinusoids adjacent to the portal venule (PV) and the more parallel and larger sinusoids near the central venule (CV). (see color insert)

are arranged in interconnecting polygonal networks; farther away from the portal venules the sinusoids become organized as parallel vessels that terminate in central venules (terminal hepatic venules). Short intersinusoidal sinusoids connect adjacent parallel sinusoids. Figure 4 is a microvascular cast illustrating these regional differences.

In the periportal area, the volume of liver occupied by sinusoids is greater than that surrounding central venules. However, because of the smaller size and anastomotic nature of the periportal sinusoids, the surface available for exchange in this area (surface/volume ratio) is greater than in centrilobular sinusoids. The size and pattern of distribution of endothelial fenestrae differs along the length of the sinusoid. At the portal end, the fenestrae are larger but comprise less of the endothelial surface area than they do in the pericentral region. The functional significance of these regional differences is unclear but relates to the functional metabolic heterogeneity that has been demonstrated for hepatocytes in different regions of the lobule. This, in turn, may depend on the recognized portal-to-central intralobular oxygen gradient and the unique microcirculation in the liver.

Morphologic Sites for Regulating the Hepatic Microcirculation

There are several potential morphological sites for regulating blood flow through the sinusoids. These include the various segments of the afferent portal venules and hepatic arterioles, the sinusoids themselves, and central and hepatic venules. These vessels contain several potentially contractile cells—smooth muscle cells in arterioles and venules, and in sinusoids, endothelial, stellate, and Kupffer cells.

Portal venules and central venules contain limited amounts of smooth muscle in their walls relative to their luminal size, but nevertheless are contractile and respond to pharmacologic agents. Hepatic arterioles are more responsive because of a complete investment of smooth muscle and relatively small lumens. The principal site of regulation of blood flow through the sinusoids, however, is thought to reside in the sinusoid itself, where the major blood pressure drop occurs in the liver.

The sinusoidal lining cells are responsive to a wide variety of pharmacodynamic substances. By contracting (or swelling), they may selectively reduce the patency of the sinusoid lumen, thereby altering the rate and distribution of blood flow. The relative roles of Kupffer versus endothelial cells in this process are not yet resolved, but both appear to be involved. The participation of perisinusoidal, stellate cells (fat-storing, Ito cells) in regulating sinusoidal diameter also has been reported. All three cell types contain filaments, tubules, and contractile proteins suggestive of contractile activity.

Because of these structures, blood flow through individual sinusoids is variable. At sites where the lumen is narrowed by the bulging, nuclear regions of sinusoidal lining cells, flow may be impeded by leukocytes that transiently plug the vessel and obstruct flow. Transient leukocyte plugging is more frequent in the periportal sinusoids, which are narrower and more tortuous than those in the centrilobular region. The more plastic erythrocytes usually flow easily through such sites unless the lumen is reduced to near zero. Some sinusoids, however, may act as thoroughfare channels and have relative constant rates of blood flow, while others have more intermittent flow. This may depend on not only the distribution of intrasinusoidal sphincter cells but also on the distribution of arterio-sinus twigs (AST) and the contribution of arterial blood flowing to individual sinusoids. For example, arterial blood flowing into an individual sinusoid through a dilated AST may increase the rate of sinusoidal blood flow. Because of the delivery of arterial blood at higher pressure, some arterial blood may even reverse the entry of portal blood into the sinusoids. As a result, the AST in concert with the initial segment of the sinusoid in which it terminates may form a "functional" arterio-portal anastomosis so that arterial blood is delivered into the portal venules. In the anesthetized, healthy animal, however, terminal branches of the hepatic arteriole containing flow are seen infrequently so that most blood delivered to the sinusoids is derived from the portal venules. Consistent with this is the in vivo microscopic observation that the velocity of flow in sinusoids and portal and central venules located near the capsule of the liver is not significantly altered by hepatic artery occlusion in healthy anesthetized rats. However, arterial inflow to the sinusoids may be more significant in regions near the hepatic hilum.

The frequency distribution of the wide variations in blood flow in the sinusoids exhibits a polymodal pattern composed of several Gaussian distributions. These wide variations in flow are due to the structural features previously described for sinusoids and also are due to intermittent arterial inflow into the sinusoids. Blood pressures in portal and central venules have been measured to be about 6 to

7 cmH_2O and 1.5 to 3.0 cmH_2O, respectively. Arterial blood enters the sinusoid at pressures ranging from 12 to 25 cmH_2O.

Pathophysiology of the Hepatic Microcirculation

Significant interactive roles for endotoxin, cytokines, chemokines, reactive free radicals, nitric oxide (NO), endothelin (ET-1), carbon monoxide (CO), sinusoidal lining cells, leukocytes, and platelets have been demonstrated in the pathophysiology of hepatic microvascular disturbances and parenchymal injury resulting from infection, toxicants, and ischemia/reperfusion following hemorrhage or liver transplantation. The responses of the hepatic microvasculature are of two basic types: (a) an inflammatory response involving paracrine activation of the sinusoidal endothelial cells (SEC) by mediators released from adjacent Kupffer cells and/or hepatic parenchymal cells following stimulation by toxicants, leading to the upregulation of adhesion molecules and the subsequent adhesion of leukocytes to the SEC, as well as swelling of the SEC, both of which restrict sinusoidal blood flow; and (b) direct injury of the SEC, resulting in loss of fenestrae, formation of gaps, penetration of the sinusoidal lining by blood cells, destruction of SEC, and obstruction of the sinusoid by SEC debris. The inflammatory response results from endotoxemia, sepsis, ischemia–reperfusion injury, and acute alcohol ingestion while direct injury is elicited by acetaminophen (APAP) or during hepatic venoocclusive disease by pyrrolizidine alkaloids. Both types of injury may occur together or sequentially, as is seen when ethanol sensitizes sinusoidal lining cells to other toxicants such as APAP or endotoxin, which exacerbates the extent of injury. It should be noted that because of the highly anastomotic nature of the hepatic sinusoid bed, plugging of single or scattered segments of sinusoids results in redirection of blood flow into adjacent unplugged vessels. However, unless the injury is highly localized, these latter vessels eventually become plugged, resulting in microcirculatory failure.

Glossary

Hepatic lobule: A polygonal structure having as its central axis a central venule, with portal tracts distributed along its peripheral boundary; major functional unit in the liver.

Hepatic microvascular system: Comprises all blood and lymphatic vessels immediately involved in the delivery and removal of fluids to and from the hepatic lobule and its parenchymal cells, namely, portal venules, hepatic arterioles, sinusoids, central venules, and lymphatics.

Hepatic sinusoids: Unique exchange vessels in the liver composed of specialized nonparenchymal cells (sinusoidal endothelial cells, Kupffer cells, and stellate cells) that exhibit structural and functional heterogeneity.

Further Reading

Bloch, E. (1955). The in vivo microscopic vascular anatomy and physiology of the liver as determined with the quartz-rod method of transillumination. *Angiology* **6**, 340–349. *Bloch was involved in the earliest hepatic microcirculatory studies, the results of which are contained in this paper.*

Clemens, M., and Zhang, J. (1999). Regulation of sinusoidal perfusion: In vivo methodology and control by endothelins. *Semin. Liver Dis.* **19**, 383–396. *A review of some of the current concepts concerning regulation of sinusoidal blood flow.*

Ekataksin, W., Zou, Z., Wake, K., Chunhabundit, P., Somana, R., Nishida, J., and McCuskey, R. (1997). The hepatic microcirculatory subunits: An over-three-century-long search for the missing link between the exocrine unit and an endocrine unit in mammalian liver lobules. In *Recent Advances in Microscopy of Cells, Tissues and Organs* (P. Motta, ed.), pp. 407–412. Rome: University of Rome "La Sapienza." *A review of the various functional unit concepts and the evidence for hepatic microvascular subunits.*

Matsumoto, T., and Kawakami, M. (1982). The unit-concept of hepatic parenchyma—a reexamination based on angio architectural studies. *Acta Pathol. Japon.* **32**, 285–314. *Matsumoto first described subdividing the lobule into primary lobules.*

McCuskey, R. (1994). The hepatic microvascular system. In *The Liver: Biology and Pathobiology* (I. Arias, J. Boyer, N. Fausto, W. Jakoby, D. Schachter, and D. Shafritz, eds), 3rd ed., pp. 1089–1106. New York: Raven Press. *Comprehensive review of the structure and function of hepatic microvascular system in health and disease.*

McCuskey, R. (2000). Morphologic mechanisms for regulating blood flow through hepatic sinusoids. *Liver* **20**, 3–7. *Review of morphologic sites for regulating sinusoidal blood flow and controversies.*

McCuskey, R., and Reilly F. (1993). Hepatic microvasculature: Dynamic structure and its regulation. *Semin. Liver Dis.* **13**, 1–12. *Review of the effects of various substances and nerve stimulation of the hepatic microcirculation.*

Rappaport, A. (1973). The microcirculatory hepatic unit. *Microvasc. Res.* **6**, 212–228. *Rappaport originated the concept of the acinus as the functional unit in the liver.*

Wisse, E., Braet, F., Luo, D., DeZanger, R., Jans, D., Crabbe, E., and Vermoesen, A. (1996). Structure and function of sinusoidal lining cells in the liver. *Toxicol. Pathol.* **24**, 100–111. *Review of the structure and function of the cells that comprise the sinusoidal lining.*

Wisse, E., DeZanger, R., Jacobs, R., Charels, K., Van der Smissen, P., and McCuskey, R. (1985). The liver sieve: Consideration concerning the structure and function of endothelial fenestrae, the sinusoid wall and the space of Disse. *Hepatology* **5**, 683–692. *Review of the structure and function of hepatic sinusoidal endothelium.*

Capsule Biography

Robert S. McCuskey is Professor and Head of Cell Biology and Anatomy, as well as Professor of Physiology and Professor of Pediatrics, at the University of Arizona College of Medicine. During 2003–2005, he also is President of the American Association of Anatomists. Dr. McCuskey has been studying the hepatic microvasculature in health and disease for more than 40 years.

CHAPTER 74

Gaseous Signal Transduction in Microcirculation

Makoto Suematsu and Nobuhito Goda
Department of Biochemistry and Integrative Medical Biology, School of Medicine, Keio University, Tokyo, Japan

Gaseous signal transduction is an event where gaseous molecules produced or utilized in the body transfer biological signals to their receptors. Molecular oxygen (O_2) is not only necessary to maintain aerobic ATP synthesis in mitochondria but also serves as a primary substrate to synthesize the signaling gases, such as, nitric oxide (NO) and carbon monoxide (CO). These gaseous monoxides are generated by oxygenases, such as, NO synthase (NOS) and heme oxygenase (HO), respectively, and both play important roles in regulation of microvascular functions. Aerobic consumption of molecular oxygen occurs concurrently with glucose oxidation and thereby upregulates synthesis of CO_2 in the Krebs cycle. On the other hand, 3-phosphoglyceric acid, a product of glycolysis, serves as a substrate for serine, which interacts with methionine to generate cysteine. Besides its roles in synthesis of glutathione, taurine, and sulfate ion, this amino acid serves as a substrate for cystathionine β-synthase and/or cystathionine γ-lyase to generate hydrogen sulfide (H_2S) (Figure 1). Among these gases, this article focused on the roles of O_2, NO, and CO in microvascular signal transduction where molecular mechanisms for their reception have been revealed with physiologic implications.

Signal Transduction by O_2

Cytochrome *c* Oxidase (CCO)

Molecular oxygen (O_2) functions primarily as a terminal acceptor of electrons on mitochondria and is consumed through this enzyme to generate H_2O. This reaction is coupled with active transport of protons from the inner to the outer side of mitochondrial inner membrane. Reflux of the electron across the membrane through ATP synthase is necessary to maintain oxidative phosphorylation. Considering that more than 95 percent of oxygen consumed in our body is used for this reaction, it is not unreasonable to hypothesize that the enzyme per se serves as an oxygen sensor. Experiments using the photochemical action spectrum by Wilson et al. [1] first suggested that CCO serves as an oxygen sensor in the carotid body. The hypoxia-induced activation of the afferent neural burst of glomus cells in the body was CO sensitive, suggesting that the sensor is a heme protein, and the sensitivity profile of the CO effect to monochromatic light followed the absorption spectrum of CCO as a function of the wavelength of the light. However, mechanisms by which the CCO changes transfer the hypoxic signal to glomus cells are quite unknown in these experiments. Hypoxia-induced alterations in the CCO reaction trigger a number of events leading to remodeling of cell functions. Active transport of protons across the membrane is automatically reduced. NADH in the matrix is increased to inhibit dehydrogenase reactions in the Krebs cycle; this event secondarily decreases ATP and increases AMP to accelerate anaerobic glycolysis to compensate ATP synthesis. Cessation of oxygen consumption and accumulation of NADH causes generation of reactive oxygen species from the midpoint of the mitochondrial electron transfer system. In addition, a reduction of ATP synthesis inversely increases adenosine, a potent vasodilator. At present, it is unknown which of these events could play a critical role in transducing the hypoxic signal toward critical intracellular components to regulate vascular functions.

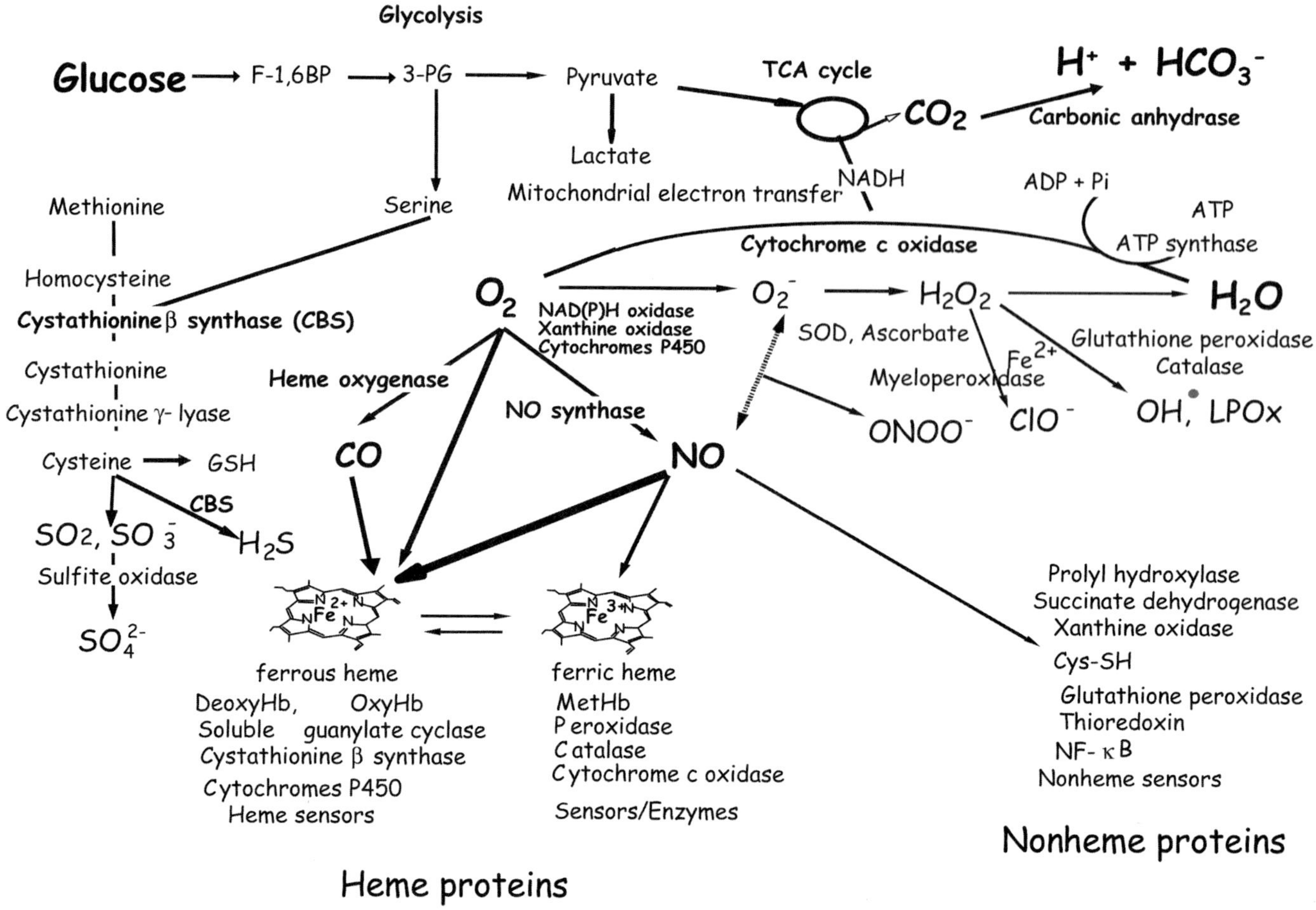

Figure 1 Landscape view of gas mediators and their receptor systems. From Ref. [17] with permission.

Oxygenases and Hemoglobin as a Class of Oxygen Sensors

Oxygenases constitute another class of oxygen sensors, since some of these enzymes accept molecular oxygen as a substrate and generate biologically active compounds as messenger molecules. Cyclooxygenases, cytochromes P450, NO synthase (NOS), and heme oxygenase (HO) are such candidate oxygenases that produce potent vasoactive mediators for hypoxic remodeling. NOS has largely been examined for its roles in regulation of neurovascular systems, but few studies showing its role in oxygen sensing have been available. Although a reduction of molecular oxygen in the cells could decrease the generation of the gas from the enzyme, it is unlikely that hypoxia causes a decrease in NO, since the half-life of NO by itself in vivo greatly depends on local oxygen tension [2]. Recent studies suggest that not only endothelial cells but also erythrocyte-dependent recycling mechanisms for NO serve as resources of the gas available in microcirculation. Under hypoxic conditions, erythrocytes alter their hemoglobin allostery toward the T-state, and thereby increase the conductance of band III on their membranes; this reaction causes an increase in outflow of NO-glutathione into circulation [3]. Such sequential reactions in and around the erythrocyte membrane also trigger the release of ATP which could secondarily stimulates NO in endothelial cells [4]. Besides endothelial and neural NOS in situ, these circulating resources of NO or NO stimulators could explain an endothelial cell-independent fraction of NO supplied into microvascular beds in vivo [5].

Prabhakar et al. [6] suggested that HO-2 serves as an oxygen sensor in the carotid body; CO constitutively generated by HO-2 in glomus cells is decreased upon hypoxia and secondarily causes depolarization of the cells, leading to neural burst of their dopaminergic fibers. Interestingly, NOS and HO can be induced upon hypoxia as a result of transcriptional upregulation of the inducible enzymes, and compensate blood supply in hypoxic regions. Such events are mediated by nuclear translocation of hypoxia-inducible factor (HIF-1) upon hypoxia as described next.

Hypoxia-Inducible Factor (HIF)-1α and HIF Prolyl-hydroxylase

In the early 1990s, a novel hypoxia-inducible transcriptional factor, HIF-1, was identified. This factor binds at the 3′ region of the hypoxia response element (HRE) of the erythropoietin gene after exposure to hypoxia [7]. HIF-1 is a heterodimeric transcription factor that consists of two

Further Reading

Semenza, G. L. (2001). HIF-1, O2, and 3PHDs: How animal cells signal hypoxia to the nucleus. *Cell* **107**, 1–3. *An excellent review of hypoxia-sensing mechanisms.*

Suematsu, M., and Ishimura, Y. (2000). The heme oxygenase-carbon monoxide system: A regulator of hepatobiliary function. *Hepatology* **31**, 3–6. *A concise summary to understand CO-mediated regulatory mechanisms for hepatic microcirculation.*

Capsule Biography

Dr. Suematsu has headed the Department of Biochemistry and Integrative Medical Biology at the Keio University School of Medicine since 2001. He was the winner of the first Lafon Microcirculation Award in 1995. His laboratory primarily focuses on gas biology, which handles the molecular basis of gaseous signal transduction and its pathophysiologic implications.

Dr. Goda is an assistant professor in the same department who has organized a project for analyses of HIF-1α conditional knock out mice.

Dr. Suematsu is a leader of the National Leading Project for Biosimulation assisted by metabolome analyses.

SECTION G

Lung

CHAPTER 75

Lung Endothelial Cell Heterogeneity and the Microvasculature

Abu-Bakr Al-Mehdi, Kane L. Schaphorst, and Troy Stevens
Departments of Pharmacology and Medicine, Center for Lung Biology, University of South Alabama, Mobile, Alabama

Introduction

Although all individual endothelial cells share some properties in common, the vascular endothelium taken as a whole represents a conglomerate cell system that exhibits diverse biological functions. All endothelial cells lie on the luminal surface of a vascular basement membrane, thereby forming a continuous interface between blood (or lymph) and the tissue interstitium. In this position the endothelial cell is strategically positioned to fulfill an essential bidirectional signaling role. A critical function of the vascular endothelium is the formation of a semipermeable barrier that limits bulk water and protein transport from the vascular space, while specifically trafficking particular protein molecules, such as hormones and immunoglobulins, across the barrier. Under homeostatic conditions, the endothelial cell barrier is antiadhesive but may, in response to certain chemical stimuli, become proadhesive to facilitate the directed recruitment of white blood cells to sites of injury or infection. An additional role for the vascular endothelium apart from its ability to regulate permeability is its role in regulating vascular tone and distribution of flow. Endothelial cells produce multiple vasoactive autocoids that regulate vascular tone moment by moment. For example, nitric oxide and prostacyclin are well-recognized vasodilators, whereas endothelin and angiotensin-converting enzyme are well-recognized vasoconstrictors. These physiologic roles, which are essential vascular functions, clearly establish that the vascular endothelium is a key regulator of homeostasis.

While endothelial cells in all vascular beds or segments share certain features in common, there is a growing recognition of their phenotypic diversity as well. For example, endothelial cells lining different organs exhibit a discernible histological structure. Endothelium that lines the blood–brain barrier, skin, muscle, and connective tissue is continuous, whereas endothelium that lines the sinusoids of the hepatic or splenic circulations, red bone marrow, adrenal and parathyroid glands, and the carotid body is not continuous. Indeed, fenestrated endothelial cells (with transcytoplasmic vacuoles) are observed in vessels of glomeruli, some endocrine glands, pancreas, and intestinal villi. Whereas fluid readily permeates renal glomerular and liver sinusoidal endothelium, brain endothelium is nearly fluid impermeable. Similarly, striking diversity may also be found within a single organ and, in some instances, within the same vascular segment such as described by Majno and Palade [1], who observed histamine-induced intercellular gaps between the endothelial cells of postcapillary venules, but not in all immediately adjacent cells. Although the notion of heterogeneity of endothelial cells and its potential role in organ specific function is appreciated, the origins and mechanisms of heterogeneity are poorly understood.

Heterogeneity may be broadly defined as secondary diversity in a grossly "homogeneous" population. Morphological and functional cellular heterogeneity among a "homogenous" type of cell population can be related to developmental origins, anatomical and histological location, local environmental influences, and, most importantly, the pattern of gene expression. For example, in any particular cultured cell type, considerable variation is seen with regard to cellular morphology and function. Some of the variations can be attributed to different stages of cell cycle, but others may be due to a differential organization, regulation, or expression of genes (i.e., due to epigenetic controls). The study of cellular heterogeneity and its origins not only is interesting from a theoretical biological viewpoint, but also is important in predicting the fate of a cell and, consequently, its contribution to the development of disease. Individual susceptibility to a particular disease, vulnerability to particular disease risk factors, or responsiveness to a particular treatment are all factors that likely represent manifestations of cellular heterogeneity.

Endothelial Cell Heterogeneity in the Pulmonary Circulation

The pulmonary circulation is unique among vascular beds in that it is the only organ system to receive 100 percent of the cardiac output through a low-resistance, low-pressure circuit. Within this unique circulation, different anatomical compartments can be distinguished between the macro- and microvascular segments and between the bronchial and pulmonary circulations. Whereas the bronchial circulation delivers oxygenated arterial blood from the left ventricle to support the bronchial and large airway tissues, the pulmonary circulation delivers mixed venous blood from the right ventricle through the pulmonary artery to the alveolar septal network for gas exchange. This oxygenated blood is then returned through pulmonary venules and veins to the left atria and ventricle for delivery to the systemic circulation. Considerable structural and functional diversity has been described among endothelial cells in these various compartments within the pulmonary circulation.

Heterogeneity among pulmonary endothelial cells of large vessels and capillaries is now well recognized both morphologically and physiologically. Morphologically, cells in larger vessels under laminar shear rates are elongated and aligned in the direction of flow, whereas those in the capillaries are not aligned in the direction of flow. Because of their exposure to different magnitude and types of shear stress, part of the heterogeneity may be attributed to adaptation to blood flow, suggesting that lung macro- and microvascular endothelial cells differ in their interpretation of and response to mechanical forces.

In addition, physiologic heterogeneity has also been demonstrated by various groups who have examined segment-specific permeability features of lung endothelium. Early work in this area illustrated most of the baseline fluid flux occurred across capillary (46%) and venous segments (38%). Segment measurements rely on subtracting arterial or venule permeability from total filtration. To ascertain these values, high airway pressure is typically used to collapse alveolar capillaries segments, stopping continuity of blood flow. Although it is unlikely that flow is completely abolished through corner vessels, these findings are broadly reflective of segment-specific responses. In fact, a series of subsequent studies have borne out these initial results, demonstrating pulmonary artery/arteriole segments account for 16 to 24 percent of basal water permeability, while capillary (36 to 50%) and venule/vein (33 to 46%) segments each account for the bulk of the circulation's total basal water flux [2]. Although this work illustrates that capillary endothelial cells account for a substantial portion of basal permeability, they do not suggest capillary endothelial cells are highly water and protein permeable. Indeed, when standardized to relative surface area, it becomes immediately apparent that the capillary endothelial cell barrier is 58 and 26-fold more restrictive to flux than either the pre- or post-capillary barrier segments, respectively.

It is of particular importance that precapillary (artery/arteriole), capillary, and postcapillary (venule/vein) vascular segments are differentially targeted by physiological and pathophysiological stimuli. In the case of increased airway pressure, which is relevant to ventilator-induced lung injury, increased permeability occurs primarily across capillary segments [2]. Similar results were observed by West and colleagues [3] in capillaries subjected to high vascular pressure that illustrated endothelial cell "stress failure" by electron microscopy. In the case of oxidant-mediated injury, which is relevant to lung transplantation, and in the case of hydrogen peroxide- and xanthine oxidase–induced vascular injury, postcapillary venule/vein segments are the preferential targets [4]. Increased permeability arising from either segment can eventually cause frank pulmonary edema if the rate of fluid accumulation exceeds the rate of reabsorption and clearance through the lymphatics. Thus, it is critical to understand the basis for such site-specific heterogeneity.

In 1997 our group demonstrated that the plant alkaloid thapsigargin activates store operated calcium entry and increases lung endothelial cell permeability. These findings indicate that calcium entry across the cell membrane is sufficient to produce intercellular gaps that form a paracellular pathway for water and protein transudation, similar to earlier work using G_q agonists such as histamine, thrombin, and substance P. However, upon close inspection of the pulmonary circulation, we resolved that not all segments were equally effected by thapsigargin [5]. Indeed, this agonist only induced intercellular gaps in arteries and veins larger than approximately 75 to 100 μm (Figure 1A). These findings suggested that rises in cytosolic calcium, through store-operated calcium entry channels, selectively targeted large vessel segments, prompting a more sophisticated evaluation of mechanisms of endothelial cell heterogeneity.

Evidence that thapsigargin increases permeability in pulmonary artery and vein segments, but not in capillary segments, suggests that neither drug concentration nor site-specific "environmental" influences underlie the heterogeneity in response. Another possible explanation is that endothelial cells within the large and small blood vessels arise from different progenitor cells and become imprinted by their unique (site-specific) environments to achieve a differentiated phenotype. As these endothelia are genetically identical, multiple epigenetic (i.e., beyond the gene sequence) mechanisms likely account for the stable modifications in cell phenotype, including methylation of promoter sequences that inhibit gene expression, and acetylation of histone proteins that control access of transcription factors to promoter sequences [8].

Lung developmental studies also provide evidence that the endothelia in particular segments are phenotypically distinct. DeMello and coworkers [6] mapped the origin of lung blood vessels using a Mercox casting technique (Figure 1B). Their findings indicated the rudimentary pulmonary circulation was first apparent in the early pseudoglandular phase of lung development. However, the circulation was incomplete, and only large blood vessels could be resolved to be intact with the right ventricle (Figure 1B, left panel). Parallel studies revealed simultaneous development of blood islands with immature capillaries, apparently not contiguous with the pulmonary artery or vein. Late in the pseudoglandular phase of lung development, an intact circulation was observed with large vessel segments contiguous with capillaries (Figure 1B, right panel). These authors suggested that large vessels arise from angiogenesis, the sequential branching of new blood vessels from existing ones, whereas capillaries (alveolar septal network) arise by vasculogenesis, the coalescence of blood lakes made up of mesenchymally derived cells. They further suggested at mid-gestation these different segments fuse to form a continuous vascular network. Although certain aspects of these findings have been debated, it appears clear that large and small blood vessel endothelium is derived from different mesenchymal cells that interact with site-specific environmental cues and are stably imprinted to become differentiated phenotypes.

If this idea is correct, then macro- and microvascular cells should share global endothelial cell behaviors but pos-

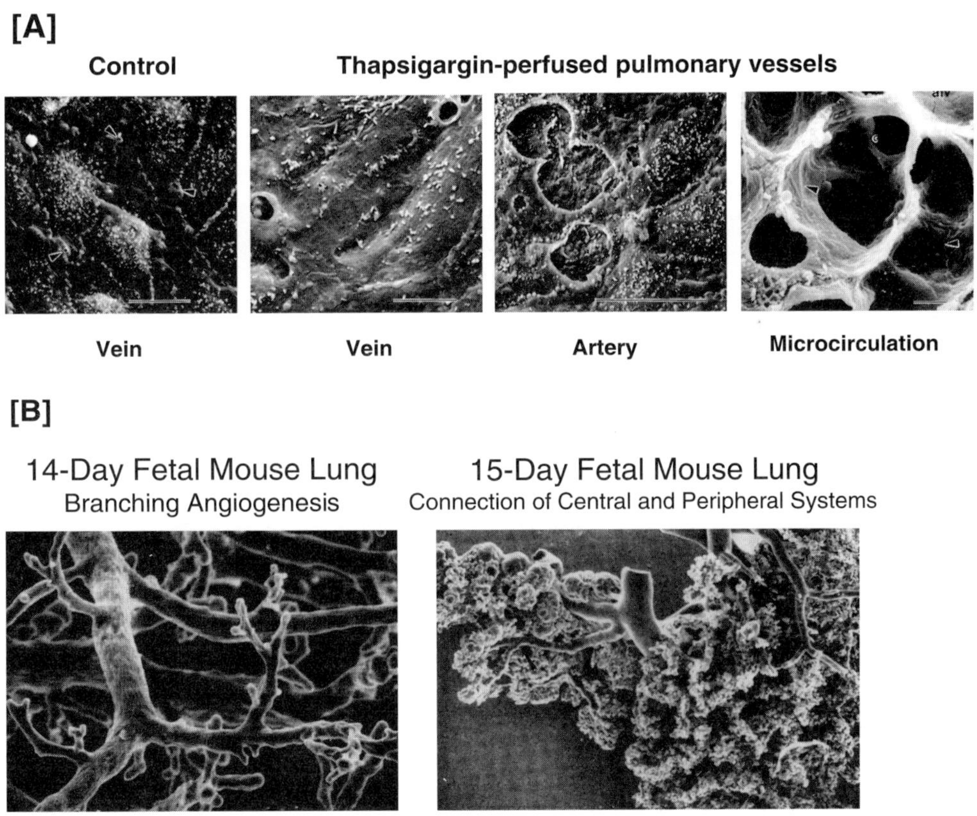

Figure 1 Functional heterogeneity along the lung vascular tree can be resolved by segment-specific responses to agonists such as thapsigargin and is related to the unique differentiation of endothelial cells during their normal embryological development. **(A)** Thapsigargin induces intercellular gaps in artery/arteriole and venule/vein segments, but not in capillary segments. For experimental details see Ref. [5]. **(B)** Scanning electron microscopy evaluation of vascular casting reveals an incomplete pulmonary circulation in the early pseudoglandular phase of lung development (*left panel*), even though blood islands can be resolved in parallel by transmission electron microscopy (not shown). Near the end of the pseudoglandular phase of lung development, capillaries can be resolved intact with the pulmonary artery (*right panel*). For experimental details see Ref. [6]. **(C)** Rat pulmonary macro- and microvascular endothelial cells can be distinctly resolved in the intact circulation by their interaction with *Griffonia simplicifolia* and *Helix pomatia* (*top panel*). (see color insert) Since lectins are agglutinins, cells isolated from each segment agglutinate with exposure to lectin even in the presence of trypsin (bottom panel). For experimental details see Ref. [7].

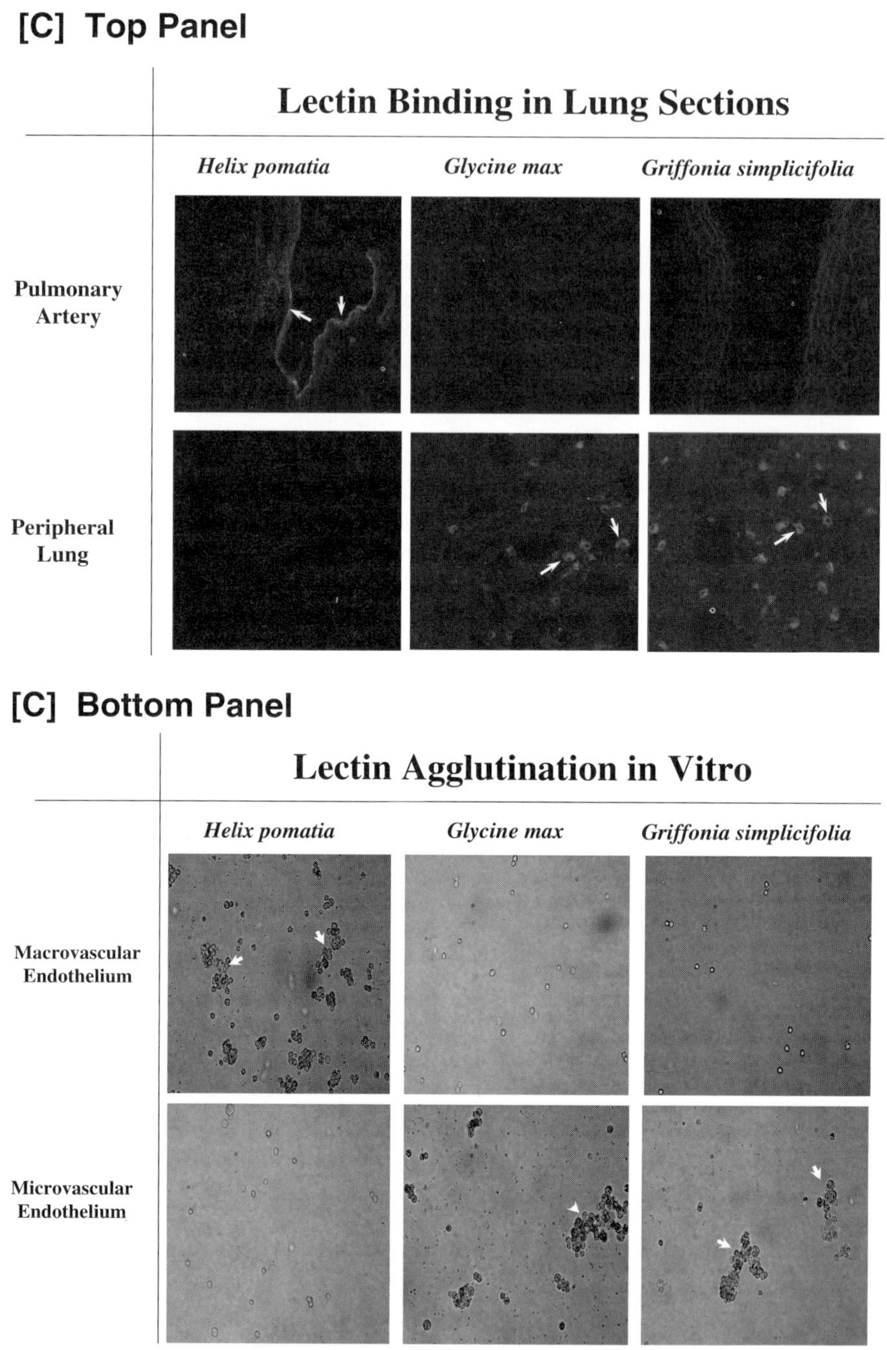

Figure 1 *Continued.*

sess specialized functions. It is not surprising then that pulmonary artery endothelial cells flow-align in vivo, but microvascular endothelial cells do not flow-align. Macro- and microvascular endothelial cells in vivo each fulfill a barrier function, but the microvascular barrier is more restrictive. To more rigorously divulge the environment-independent features of macro- and microvascular endothelial cells, we sought approaches to isolate and culture cells from these different vascular segments.

Ryan and colleagues [9] approached the problem of isolating lung capillary (microvascular) endothelial cells in the 1970s. They (and other groups) developed methods for enriching endothelial cells from capillaries using minimal vascular protease digestion and perfusion of small beads. As an alternative, peripheral lung segments were cut (pleural cut technique), minced, exposed to minimal protease digestion, and expanded from cell colonies. Both approaches were successful, and have been adapted to further improve the isolation success rate. One major advance was recognition that endothelial cells from different vascular segments uniquely interact with plant lectins. *Griffonia simplicifolia* and *Glycine max* have been shown by many different laboratories to recognize pulmonary microvascular endothelial cells and not pulmonary artery endothelial cells, whereas *Helix pomatia* (in rat) recognizes pulmonary artery endothelial cells [7]. Thus, by incubating cells isolated using the pleural cut technique with *Griffonia simplicifolia* coated beads, microvascular endothelial cells can be efficiently ascertained (Figure 1C, top panel) and isolated (Figure 1C, bottom panel).

Use of purified pulmonary artery and microvascular endothelial cells has greatly contributed to our understanding of heterogeneity along the vascular tree. As in the intact circulation, pulmonary artery endothelial cells in vitro flow-align in response to shear stress, whereas microvascular endothelial cells do not flow-align (Al-Mehdi, unpublished). Similarly, cultured pulmonary artery endothelial cells align tangential to the axis of mechanical perturbation, whereas microvascular endothelial cells do not (Parker, unpublished). These findings are consistent with conclusions from the intact circulation, suggesting that these cell types possess unique mechanotransduction signaling programs.

Because oxidant-induced lung injury occurred principally in postcapillary (venule/vein) segments, Gillespie and others [10] examined potential mechanisms that may account for such site-specific oxidant sensitivity. Using endothelial cells isolated from each of the three vascular segments, their results indicated that pulmonary vein endothelial cells were most susceptible to xanthine oxidase and hypoxanthine-induced DNA damage. This DNA damage corresponded with DNA repair rates in the respective cell types, where vein endothelium exhibited the lowest rate of DNA repair. These findings therefore support the results from the intact circulation, suggesting that the propensity for oxidants to increase permeability in postcapillary segments could result from the inability of these cells to mitigate the oxidant signal.

Since thapsigargin initiated site-specific responses in situ, our group examined its effect in vitro. Activation of store-operated calcium entry triggered reorganization of the actin cytoskeleton into stress fibers, increased myosin light chain phosphorylation, decreased adhesive forces, and induced intercellular gap formation in pulmonary artery endothelial cells. Gap formation corresponded to increased macromolecular permeability. Although thapsigargin increased myosin light chain phosphorylation in pulmonary microvascular endothelial cells, it did not induce intercellular gaps and was not sufficient to increase micromolecular permeability. Thus, these findings are also consistent with results from the intact circulation.

Mechanisms accounting for the enhanced barrier function of pulmonary microvascular endothelial cells are not well understood, but microarray analysis comparing the two cell types has begun to reveal some insight. Microvascular endothelial cells express a number of cell–cell adhesion molecules not typically observed in pulmonary artery endothelial cells, including activated leukocyte cell adhesion molecule (ALCAM/CD166; immunoglobulin family), N-cadherin (cadherin family), and ZO-2 (ZO/MAGUK family). We have confirmed our results from microarray studies using either quantitative PCR or Western blot analysis, but the functional contribution of any of these molecules to barrier enhancement has not been established. Nonetheless, these findings provide a step toward understanding the distinct molecular anatomy of the macro- and microvascular endothelial cell barrier.

A similar but more comprehensive approach has been undertaken by Brown and colleagues [11], when they performed global expression profiling experiments using mRNA from 53 different endothelial cell phenotypes. More than 2.4 million gene expression measurements were made. Using stable gene profiles in culture, they were able to discriminate macrovascular from microvascular endothelial cells and, further, artery from vein endothelial cells. Relevance of these unique expression profiles in vitro has not been confirmed in the intact circulation. Nonetheless, their work makes the important point that endothelial cells in culture are able to retain a stable memory of their origin, even under otherwise similar environmental conditions.

Whereas the wealth of in vivo, in situ, and in vitro evidence supports the notion of distinct endothelial phenotypes among precapillary, capillary, and postcapillary segments, an even more puzzling diversity exists within highly confined vascular segments. The idea of pacemaker cells was introduced in cell culture experiments to acknowledge that single cells control in synchrony the signal transduction events of near neighbors. Bhattarcharya and colleagues [12] have recently illustrated this concept in the pulmonary circulation in situ. They showed that calcium oscillations originate first in pacemaker endothelial cells that control the calcium response of their surrounding neighbors, indicating the presence of heterogeneity within the microvascular cells themselves in the pulmonary circulation. More recently, the same group has shown that endothelial cells at branch points of lung venular capillaries possess twofold more mitochondrial content by MitoTracker Green staining than their midsegment counterparts. The spatial pattern of the TNF-α-induced cytosolic Ca^{2+} transition and oxygen radical generation corresponded to the mitochondrial distribution pattern [13]. A major difference in the environment of midsegment and branch point cells is the magnitude and pattern of shear stress. In midsegment, flow is laminar and shear stress is thought to be of consistent magnitude; in the branch points flow is turbulent and shear stress is thought to be highly variable. This suggests that shear stress regulates the endothelial cell phenotype, perhaps partly by altering mitochondrial mass and function in individual cells.

Conclusion

For many years structural and functional differences in endothelial cells have been recognized among organs, and between specialized sites, such as in high endothelial venules. More recently, however, phenotypic variation in endothelial structure and function along a single blood vessel has been examined. These results illustrate a heterogeneity that arises not only because of the environment in which the cell resides, but also because of the cell's origin. Indeed, endothelial cells become imprinted in the course of development, as do all cells, based upon their interactions with surrounding tissue environments to achieve differentiation. As this differentiated phenotype is stable through mitotic cell divisions, the cell retains a memory from whence it came even through multiple passages in tissue culture. We are only now beginning to appreciate the significance of these epigenetic modifications in endothelial function and their relevance to site-specific disease susceptibility. Subsequent investigation will provide significant insight into how normal lung vascular development imprints on macro- and microvascular endothelial cells a phenotype that is uniquely suited to fulfill their site-specific functions.

Summary

Although all endothelial cells form a metabolically active, semipermeable interface between blood (or lymphatics) and tissue, considerable heterogeneity in endothelial structure and function exists even within a single vascular segment. Part of this heterogeneity is due to the environment in which a cell resides. However, endothelial cells are also imprinted during their development to impart a stable differentiated phenotype. Accordingly, endothelial cells from different sites within a blood vessel can be discriminated based upon their active gene expression profiles, even when their environments are the same. Interaction between environmental stimuli (as in blood flow patterns or transmural pressure) and the cell's origin (precapillary, capillary, postcapillary) therefore coordinately regulate site-specific endothelial cell behavior.

References

1. Majno, G., Palade, G. E., and Schoefl, G. I. (1961). Studies on inflammation. II. The site of action of histamine and serotonin along the vascular tree: A topographic study. *J. Biophys. Biochem. Cytol.* **11**, 607–626.
2. Parker, J. C., and Yoshikawa, S. (2002). Vascular segmental permeabilities at high peak inflation pressure in isolated rat lungs. *Am. J. Physiol. Lung Cell. Mol. Physiol.* **283**, L1203–L1209.
3. West, J. B., Mathieu-Costello, O. , Jones, J. H., Birks, E. K., Logemann, R. B., Pascoe, J. R., and Tyler, W. S. (1993). Stress failure of pulmonary capillaries in racehorses with exercise-induced pulmonary hemorrhage. *J. Appl. Physiol.* **75**, 1097–1109.
4. Khimenko, P. L., and Taylor, A. E. (1999). Segmental microvascular permeability in ischemia–reperfusion injury in rat lung. *Am. J. Physiol.* 276, L958-L960.
5. Chetham, P. M., Babal, P., Bridges, J. P., Moore, T. M., and Stevens, T. (1999). Segmental regulation of pulmonary vascular permeability by store-operated Ca^{2+} entry. *Am. J. Physiol.* **276**, L41–L50. *This paper demonstrates that a calcium agonist (thapsigargin) increases permeability in arteries and veins, and not in capillary segments, suggesting endothelial cells exhibit site-specific responses to environmental stimuli.*
6. deMello, D. E., Sawyer, D., Galvin, N., and Reid, L. M. (1997). Early fetal development of lung vasculature. *Am. J. Respir. Cell. Mol. Biol.* **16**, 568–581. *This paper demonstrates that lung macro- and microvascular segments arise from different cell origins, potentially providing the important developmental cues that impart stable cell memories.*
7. King, J., Hamil, T., Creighton, J., Wu, S., Bhat, P., McDonald, F., and Stevens T. (2004). Structural and functional characteristics of lung macro- and microvascular endothelial cell phenotypes. *Microvasc. Res.* (in press).
8. Herman, J. G., and Baylin, S. B. (2003). Gene silencing in cancer in association with promoter hypermethylation. *N. Engl. J. Med.* **349**, 2042–2054.
9. Habliston, D. L., Whitaker, C., Hart, M. A., Ryan, U. S., and Ryan, J. W. (1979). Isolation and culture of endothelial cells from the lungs of small animals. *Am. Rev. Respir. Dis.* **119**, 853–868.
10. Grishko, V., Solomon, M., Wilson, G. L., LeDoux, S. P., and Gillespie, M. N. (2001). Oxygen radical-induced mitochondrial DNA damage and repair in pulmonary vascular endothelial cell phenotypes. *Am. J. Physiol. Lung Cell. Mol. Physiol.* **280**, L1300–L1308.
11. Chi, J. T., Chang, H. Y., Haraldsen, G., Jahnsen, F. L., Troyanskaya, O. G., Chang, D. S., Wang, Z., Rockson, S. G., van de Rijn, M., Botstein, D., and Brown, P. O. (2003). Endothelial cell diversity revealed by global expression profiling. *Proc. Natl. Acad. Sci. USA* **100**, 10623–10628. *This article reports on the global mRNA expression patterns of 53 different endothelial cell populations. Hierarchical cluster analysis revealed that artery-, microvascular-, and vein-derived cells were reproducibly distinguishable based upon their selective expression patterns, indicating that a memory of their origin was retained in culture.*
12. Ying, X., Minamiya, Y., Fu, C., and Bhattacharya, J. (1996). Ca^{2+} waves in lung capillary endothelium. *Circ. Res.* **79**, 898–90813.
13. Parthasarathi, K., Ichimura, H., Quadri, S., Issekutz, A., and Bhattacharya, J. (2002). Mitochondrial reactive oxygen species regulate spatial profile of proinflammatory responses in lung venular capillaries. *J. Immunol.* **169**, 7078–7086.

Capsule Bibliography

Dr. Al-Mehdi has held academic appointments at the University of Pennsylvania (Institute for Environmental Medicine) and, more recently, at the University of South Alabama (Assistant Professor of Pharmacology, Center for Lung Biology). His work focuses on lung vascular biology, with specific attention given to endothelial cell responses to oxygen transitions, and the role of oxygen in lung metastasis. Work in the Al-Mehdi lab is supported by the NIH.

Dr. Schaphorst has held academic appointments at Johns Hopkins University (Pulmonary and Critical Care Medicine) and, more recently, at the University of South Alabama (Associate Professor of Medicine, Center for Lung Biology). His work has addressed signal transduction networks that interplay to control endothelial cell barrier function, with emphasis on hepatocyte growth factor activation of c-Met signaling.

Dr. Stevens has held academic appointments at the University of Colorado Health Sciences Center (CVP Research Laboratory) and, more recently, at the University of South Alabama, where he is the Director of the Center for Lung Biology. His work seeks to understand the interplay between genetic and environmental factors that control endothelial cell phenotype. A major emphasis in his laboratory is to reveal the unique signal transduction pathways that control macro- and microvascular endothelial cell function. Work in the Stevens laboratory is supported by the NIH.

CHAPTER 76

Lung Edema and Microvascular Permeability

Richard D. Minshall and Stephen M. Vogel
University of Illinois at Chicago, Chicago, Illinois

Introduction

Pulmonary edema, or fluid accumulation in the lung extravascular spaces, results in impaired gas exchange and hypoxemia. When the lung fails to maintain fluid balance, liquid initially accumulates in the interstitium and ultimately invades the alveolar airspaces. The restrictive barrier properties of the pulmonary microvascular endothelium are established by tight junctions and adherens junctions between neighboring endothelial cells. Breakdown of the endothelial barrier leads rapidly to lung edema, the hallmark of acute lung injury and ARDS (acute respiratory distress syndrome). This chapter will focus on the mechanisms that regulate the barrier properties of the lung, highlighting the role of the transcellular and paracellular permeability pathways in edema formation.

Starling Forces Underlying Pulmonary Edema Formation, Accumulation, and Resolution

Formation of Edema Fluid

Edema formation can be understood by consideration of the Starling forces across the microvessel wall. The mathematical relationship between fluid filtration rate (J_v) and transmural hydrostatic and oncotic pressures is written as

$$J_v = L_p S[(P_c - P_i) - \sigma(\Pi_c - \Pi_i)],$$

where σ is the osmotic reflection coefficient, S the vessel surface area, L_p the hydraulic conductivity, P the capillary hydrostatic pressure, Π the oncotic pressure, and the subscripts c and i, capillary and interstitial compartments. The term $(P_c - P_i)$ gives the transmural hydrostatic pressure difference and the product $\sigma(\Pi_c - \Pi_i)$ the effective transmural oncotic pressure drop; the difference between these parenthetical terms defines the "driving pressure" for net fluid filtration or reabsorption through pulmonary microvessel walls. Note that higher $(P_c - P_i)$ favors fluid filtration (increased J_v), whereas higher $\sigma(\Pi_c - \Pi_i)$ favors fluid reabsorption. Normally, an overall equilibrium is achieved between fluid filtration and fluid reabsorption at proximal and distal segments of capillaries, and little or no net filtration occurs through capillary walls. The lymphatic system in the lung maintains a negative interstitial pressure by continuously withdrawing fluid from interstitium. Edema develops when fluid filtration substantially exceeds its reabsorption and the capacity of the lymphatic system to remove fluid from the pulmonary interstitium.

UNDERLYING CAUSES OF EDEMA FLUID FORMATION: ROLE OF STARLING FORCES

An elevated net driving pressure without a marked increase in permeability underlies "pressure edema" (i.e., edema resulting from increase in the pulmonary capillary hydrostatic pressure). Hydrostatic edema, when it is not associated with frank barrier breakdown, is generally protein-poor, at least in early stages of the syndrome, because the barrier properties tending to exclude large molecules are preserved. In hydrostatic edema the ratio of plasma to alveolar fluid protein concentration is usually less than 0.6. Critical capillary pressure for formation of edema due strictly to elevated hydrostatic pressure is a P_c above 25 mmHg. Fluid accumulation in the lung is minimized by "safety factors" that are activated below this critical capillary pressure (see later discussion). The pulmonary extravascular water content increases progressively as a result of the inability of these safety factors to reduce fluid filtration rate when capillary hydrostatic pressure increases above the critical value. Most clinical manifestations of pulmonary edema can be understood in terms of changes in Starling forces through pulmonary microvessel walls. A decrease in the plasma protein concentration, such as in hypoalbuminemia, reduces the transmural oncotic pressure difference, thus favoring increased fluid filtration. In this case, the critical capillary pressure at which lungs begin to gain water decreases in direct proportion to the reduction in plasma oncotic pressure.

CAUSES OF LUNG EDEMA: INCREASED VASCULAR PERMEABILITY

Barrier rupture or breakdown underlies "permeability edema," which creates a protein-rich fluid because of the loss of normal protein-excluding properties of the alveolar–capillary barrier. Some evidence suggests that stimulation of protein transport could make a contribution to formation of protein-rich edema fluid. For example, a vigorous transcellular albumin transport process involving vesicular carriers is well established in pulmonary microvascular endothelial cells [1]. These carriers are predominantly caveolae that occupy a remarkably high percentage (15%) of the endothelial cell volume. An important area of investigation is whether pathologic conditions can substantially stimulate this active transport process leading to secretion of a protein-rich edema fluid.

Lung vascular permeability can increase as a result of a direct injury of endothelial cells, alterations in the dimensions of interendothelial junctions (i.e., paracellular pathways), or a combination of these factors [2]. The increase in lung vascular permeability is operationally defined in the Starling equation by an increased capillary filtration coefficient ($K_{f,c}$), which is equivalent to the L_pS term in the equation. An increase in the $K_{f,c}$ value corresponds to decreased barrier resistance to the movement of liquid across the capillary wall barrier. The albumin reflection coefficient (ϕ_{Alb}), which describes the albumin permeability of the vascular endothelial barrier, provides a widely used measure of the protein permeability of the barrier. In high-permeability pulmonary edema, the alveolar fluid protein concentration approximates the plasma protein concentration. The increase in lung vascular permeability shifts the relationship between left atrial pressure and pulmonary extravascular water content toward lower pressures, indicating that edema will occur at a reduced driving pressure in the face of an increased vascular permeability.

CAUSES OF PULMONARY EDEMA: ROLE OF THE LYMPHATIC SYSTEM

Lymphatics are capable of removing excess extravascular fluid because of their effectiveness as a pump. Lymphatic propulsion is determined by the intrinsic contractility of lymphatic vessels, by inspiration and expiration, and by unidirectional lymphatic valves. The extent to which lymphatic insufficiency serves as an important mechanism of fluid accumulation in the lung is not clear. Some studies have indicated that surgical removal of the lymphatics predisposes lung to edema, although the increase in water content is usually transient.

Fluid Accumulation in the Lung

Newly formed edema fluid initially distends the interstitial compartment and then disrupts the interstitial lattice; proteolysis of interstitial structural proteins may occur, leading to increased interstitial compliance [3]. Fluid that cannot be cleared by lymphatics accumulates in the connective tissue surrounding smaller vessels and bronchioles. The fluid then migrates down the interstitial fluid pressure gradient to interstitial spaces around larger vessels and airways. If lymphatics in the connective tissue sheaths are unable to remove the excess fluid, undrained fluid becomes compartmentalized and forms perivascular cuffs. Normally the interstitial hydrostatic pressure in the lung is a negative value (−9 mmHg; [4]). Because of the low interstitial compliance, excess fluid accumulation within the interstitium will rapidly increase tissue pressure to slightly positive values. The alveolar barrier breaks down at a pressure of 2 mmHg, corresponding to an increase in the interstitial fluid volume of 35 to 50 percent; tissue pressure values above this threshold will cause a precipitous alveolar edema during which individual alveoli begin to flood in an all-or-nothing manner. Initially, the distribution of alveolar flooding is patchy, but rapid severe flooding follows this. The exact route by which fluid moves into the alveoli is not known. Fluid movement may involve bulk flow through large epithelial pores or channels or may be the result of increased transport through intercellular pathways in respiratory epithelium of terminal bronchioles. There is also the possibility of epithelial injury involving detachment of epithelial cells from the underlying matrix, resulting in movement of fluid directly into the alveoli.

Removal of Accumulated Edema Fluid

Resolution of pulmonary edema involves both removal of already accumulated fluid and protein from alveoli and termination of the conditions causing edema. Mechanisms governing edema resolution following lung injury are poorly understood. A restructuring of the dilated pulmonary interstitium following the degradation of matrix proteins appears to be required. Stimulation of transport processes that drive fluid and protein from distal alveolar epithelium including the (Na, K)-ionic pump [5] and albumin transport [6] may be critical removal processes.

Role of "Safety Factors" in Lung Fluid Homeostasis

Several safety factors protect the lung against edema; these are the decrease in albumin exclusion volume, the lymphatic system, and the increase in interstitial hydrostatic pressure.

In high-pressure edema, protein-poor fluid begins to accumulate in the pulmonary interstitial space by ultrafiltration. A decrease in the exclusion volume for albumin (or an increase in its actual volume of distribution) becomes important in decreasing the interstitial protein concentration and thereby decreasing π_c. Such a decrease in π_c, according to the Starling equation, further reduces net fluid filtration and augments fluid reabsorption through pulmonary microvessel walls.

The pulmonary lymph flow is capable of increasing by a large factor in response to increased interstitial fluid volume. Lymph flow is actually dependent on the interstitial pressure, which in turn is a function of interstitial volume and compliance. Beyond a critical fluid volume, pulmonary lymph flow can no longer increase in proportion to the increase in interstitial pressure. Until this saturation value is attained, lymphatic drainage tracks the rate of edema fluid formation and thereby limits fluid accumulation.

An increase in tissue pressure also represents the lung's short-term protective mechanism to limit edema formation. The low interstitial compliance in the lung reflects an unusually low interstitial volume plus a rigid protein infrastructure. This means that pulmonary tissue pressure (P_i) undergoes a large rise for a relatively small increase in interstitial volume; such an increase in P_i favors fluid reabsorption, according to the Starling equation, and in this sense qualifies as an important safety factor.

Cellular Mechanisms of Permeability Edema: Endothelial Barrier Dysfunction

Overview

Capillary endothelial cells form the primary barrier between the plasma and interstitial fluid. Intercellular contacts between endothelial cells and cellular adhesion to the underlying subendothelial matrix are responsible for the characteristic barrier properties of endothelium, including a low permeability to plasma proteins. The manner in which pathological conditions foster barrier dysfunction is not well understood. Barrier dysfunction results in a loss of contact between microvascular endothelial cells and/or weakening of their adhesion to the basement membrane. Certain substances elaborated during inflammatory processes such as thrombin, histamine, and bradykinin disrupt barrier function by a direct action on the endothelium to increase vascular permeability by opening intercellular junctions.

Activation Endothelial Retraction and Disruption of Cell–Cell Junctions

The endothelial cell is a target for many proinflammatory and thrombogenic mediators and growth factors. These agents may disrupt interendothelial junctions, increasing endothelial permeability and allowing the passage of plasma proteins through intercellular gaps. Inflammatory mediators such as thrombin result in increased endothelial permeability by causing an intense endothelial cell retraction and shape change. The signal transduction pathways activated by thrombin that promote loss of barrier function involve a complex series of biochemical events leading to a rapid and sustained phosphorylation of myosin light chain (MLC) and a simultaneous inhibition of MLC phosphatase that prevents dephosphorylation of MLC and prolongs the response. The disruption of endothelial junctions is likely precipitated by increased interaction of endothelial contractile proteins actin and myosin. Phosphorylation of MLC by calcium-dependent myosin light chain kinase (MLCK) is required for actin–myosin interaction. Filamentous actin within endothelial cells is known to associate with the cytoplasmic tail of the major adherens-junction protein VE-cadherin. Contractile force may "unhinge" adherens junctions between endothelial cells, resulting in formation of interendothelial gaps [7]. Such gaps, which thrombin induces within minutes in pulmonary endothelial cells in vitro, provide a plausible structural basis for increased paracellular permeability.

Ca^{2+}-Dependent and Ca^{2+}-Independent Permeability Increasing Mechanisms

Several key signaling and effector proteins in this important pathway have been identified (see Figure 1). The agonist thrombin activates PAR-1 (the proteinase activated receptor). GTP-binding protein G_q signals calcium release from intracellular stores, and calcium-store depletion, in turn, signals calcium entry (see later discussion). Free calcium in the cytosol is thought to bind to calmodulin (a calcium-binding protein); the calcium–calmodulin complex activates MLCK, which induces the phosphorylation of MLC. In parallel with G_q, the $G_{12/13}$ G-protein pathway, acting through cytoplasmic Rho GTPase and its effector Rho kinase, inhibits MLC dephosphorylation. The combined effect of MLCK plus Rho kinase activity is to strongly induce and maintain MLC phosphorylation, resulting in formation of contractile units (stress fibers) that exert force upon interendothelial junctional complexes [8]. The model provided is probably an oversimplification of the biochemical process underlying endothelial cell retraction, since there is considerable cross talk and lateral regulation of the seemingly parallel pathways.

Considerable attention has been devoted to the regulation of endothelial calcium entry by thrombin. Inositol (1,4,5)-trisphosphate (IP_3) formation induced by thrombin is known to cause release of sequestered calcium and elicit calcium entry via store-operated channels (SOC). Tiruppathi and associates have identified *TRPC4* (Transient Receptor Potential Channel 4) as an essential constituent of the SOC in the mouse lung [9]. Their data support a causal relationship between increases in calcium entry and elevated pulmonary microvascular permeability. Thus, increased

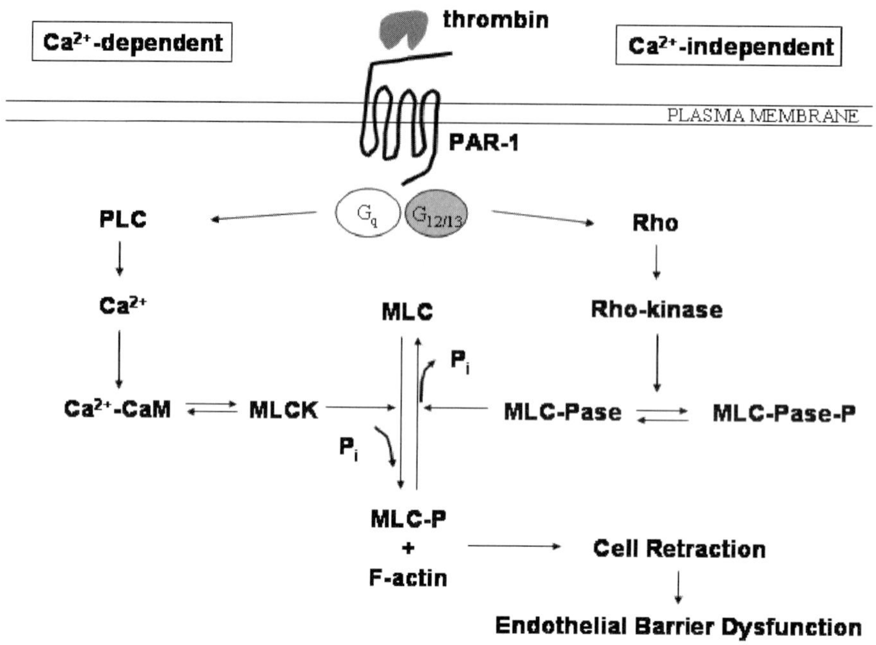

Figure 1 Signaling mechanism of thrombin-induced endothelial cell retraction and vascular barrier dysfunction. The protease thrombin activates the G protein coupled protease activated receptor 1 (PAR-1) on the endothelial cell surface. The Ca^{2+}-dependent signaling pathway is activated by G protein G_q stimulation of phospholipase C, the generation of lipid intermediates that induce the release of Ca^{2+} from an intracellular storage compartment, and the regulation of myosin light chain kinase (MLCK) by Ca^{2+}–calmodulin (Ca^{2+}-CaM). Phosphorylated MLC interacts with F-actin, forming stress fibers that contract and pull apart the cell–cell junctions. The Ca^{2+}-independent component of thrombin signaling is mediated by G_{12} and G_{13} activation of Rho and Rho kinase, which inhibits myosin light chain phosphatase (MLC-Pase), thereby preserving MLC phosphorylation and the contractile event in the face of declining Ca^{2+} levels. Thus, endothelial barrier dysfunction caused by inflammatory mediators such as thrombin is the result of Ca^{2+} induced actin–myosin based contraction of individual endothelial cells, which generates gaps between cells and increases albumin leakage into perivascular spaces.

calcium influx leading to activation of endothelial retraction may be a fundamental, underlying cause of pulmonary edema.

Summary

The pulmonary microvessels serve as primary sites for fluid and solute exchange within the lung tissue. Starling forces govern fluid filtration from microvessels into the surrounding perimicrovascular interstitial space. The pulmonary lymphatics collect fluid and protein leaking into the interstitium and return the fluid and dissolved solute to the vascular system. Pathophysiologic events and mediators that substantially perturb the Starling forces culminate in pulmonary edema.

Glossary

Adherens junction: An anchoring junction between two cells (made up of cadherin and associated proteins) that is linked to the actin cytoskeleton.

Edema: Swelling or excessive buildup of fluid in the tissues, or an increase in tissue mass due to fluid accumulation.

Fluid filtration: The passage of a liquid through a porous membrane resulting from a hydrostatic pressure difference.

Reflection coefficient: A measure of the relative permeability of a particular membrane to a particular solute calculated as the ratio of observed osmotic pressure to that calculated from van't Hoff's law.

Vascular permeability: The velocity of transport through the endothelial barrier through any or all of the following pathways: (1) between the cells, (2) through the cells, via pores (diaphragms or fused vesicles), and (3) transcellularly, via shuttling vesicles and specific receptors.

References

1. John et al. (2003). *Am. J. Physiol. Lung Cell. Mol. Physiol.* **284**, L187–L196.
2. Lum and Malik (1994). *Am. J. Physiol.* **267**, L223–L241.
3. Negrini et al. (1996). *Am. J. Physiol.* **270**, H2000–H2007.
4. Negrini and Fabbro (1999). *J. Physiol.* **520**, 761–769.
5. Rutschman et al. (1993). *J. Appl. Physiol.* **75**, 1574–1580.
6. John et al. (2001). *J. Physiol.* **533.2**, 547–559.
7. Sandoval et al. (2001). *J. Physiol.* **533.2**, 433–445.
8. Birukova et al. (2004). *Microvasc. Res.* **67**, 64–77.
9. Tiruppathi et al. (2002). *Circ. Res.* **91**, 70–76.

Further Reading

Dudek, S. M., and Garcia, J. G. N. (2001). Cytoskeletal regulation of pulmonary vascular permeability. *J. Appl. Physiol.* **91**, 1487–1500. *This article provides an outstanding overview of the endothelial cell signaling mechanisms and effector proteins that mediate thrombin-induced increase in endothelial permeability as well as mechanisms that stabilize or maintain endothelial barrier integrity.*

Flick, M. R., and Matthay, M. A. (2000). Pulmonary edema and acute lung injury. In *Textbook of Respiratory Medicine* (J. F. Murray and J. A. Nadel, eds.), 3rd ed., pp. 1575–1629. Philadelphia: W.B. Saunders.

Malik, A. B., Vogel, S. M., Minshall, R. D., and Tiruppathi, C. (2000). Pulmonary circulation and regulation of fluid balance. In *Textbook of Respiratory Medicine* pp. 119–154. 3rd edition, (J. F. Murray and J. A. Nadel, eds.), Philadelphia: W.B. Saunders.

Minshall, R. D., Tiruppathi, C., Vogel, S. M., and Malik, A. B. (2002). Vesicle formation and trafficking and its role in regulation of endothelial barrier function. *Histochem. Cell Biol.* **117**, 105–112. *This review outlines the signaling mechanisms involved in caveolae-mediated endocytosis and transcytosis of albumin in endothelial cells. This transcellular transport mechanism is an important determinant of basal albumin permeability and nutrient/hormone delivery into the tissue when cell–cell junctions are intact.*

Taylor, A. E. and Parker, J. C. (1985). Pulmonary interstitial spaces and lymphatics. In *Handbook of Physiology*. Section 3: *The Respiratory System. Circulation and Nonrespiratory Function* (A. P. Fishman and A. B Fisher, eds.), Vol. 1, pp. 167–230. Bethesda, MD: American Physiological Society.

Tuma, P. L., and Hubbard, A. L. (2003). Transcytosis: Crossing cellular barriers. *Physiol. Rev.* **83**, 871–932.

Tiruppathi, C., Minshall, R. D., Paria, B. C., Vogel, S. M., and Malik, A. B. (2003) Role of Ca^{2+} signaling in the regulation of endothelial permeability. *Vasc. Pharmacol.* **39**, 173–185. *This review summarizes the mechanisms which regulate intracellular calcium homeostasis (release mechanisms and influx pathways) in endothelial cells and their role in thrombin-induced increase in endothelial permeability.*

Capsule Biographies

Dr. Richard Minshall is an Assistant Professor of Pharmacology and Anesthesiology at The University of Illinois at Chicago. A Parker Francis Fellow in Pulmonary Research from 2001 to 2004, his NIH-funded research focuses on the physiological and pathological roles transcellular and paracellular permeability pathways of the endothelium.

Dr. Stephen Vogel is an Assistant Professor of Pharmacology at The University of Illinois at Chicago investigating the mechanisms of protein and fluid permeability regulation in the lung. Dr. Vogel is regarded as a significant contributor in the lung biology field for the development of ex vivo isolated mouse and rat lung models used to investigate the physiological importance of caveolae-mediated albumin transport processes in the epithelial and endothelial cell barriers.

CHAPTER 77

Forms of Reserve in the Pulmonary Capillary Bed

Wiltz W. Wagner, Jr. and Robert G. Presson, Jr.
Indiana University School of Medicine, Indianapolis, Indiana

Basic lung anatomy is simple. Branching airways terminate in alveoli, and the arterial and venous trees interconnect via capillaries that are embedded in alveolar walls. The complexity of the structure comes from its ~3×10^8 alveoli and their accompanying ~3×10^{10} capillaries. These very large numbers have caused anatomists and physiologists to simplify the lung to a one-alveolus, one-capillary model. With time, this model has tended to be thought of as the real lung; some senior scientists express considerable surprise when they realize that pulmonary capillary pathways cross more than one alveolus.

Although pulmonary microvascular anatomy and its effect on perfusion are extremely complex, they are not hopelessly so. We have considerable information about how the pulmonary circulation utilizes the extensive reserve in the pulmonary capillary bed when the demand for oxygen increases. In trying to understand how these reserves come into play when the lung reaches its full potential as a gas exchanger during heavy exercise, we consider three kinds of changes that occur in the capillary bed as cardiac output increases from rest to exercise.

Capillary Transit Time

As red blood cells pass through the capillary network during basal conditions, oxygen and hemoglobin combine with sufficient rapidity to completely saturate hemoglobin in about the first quarter of its transit. The results of a number of studies of the saturation of hemoglobin with oxygen, each using differing assumptions, have been reviewed by Wagner [1]. Even with the wide range of variables used by the investigators, 0.25 seconds consistently appeared as a reasonable bound for the minimum time required for red blood cell P_{O_2} to increase from 40 to 100 torr. With exercise, capillary transit times decrease as cardiac output increases, and more of the capillary length is utilized for oxygen uptake. Only if capillary transit time becomes too rapid, that is, less than 0.25 seconds, will red blood cells leave the capillaries without becoming fully saturated.

The predominant method for measuring pulmonary capillary transit time is based on the diffusing capacity of the lung for carbon monoxide to compute pulmonary capillary volume. Capillary transit time can then be calculated by dividing capillary blood volume by cardiac output measured concomitantly. For example, 75 mL (capillary volume) ÷ 75 mL/second (a 4.5 L/minute resting cardiac output) = 1.00 second (capillary transit time). Based on this kind of calculation, it is believed that, in the resting human, red blood cells spend an average of 0.75 to 1.0 seconds in the pulmonary capillaries [1], well above the calculated 0.25-second minimum transit time. During maximal exercise, however, hypoxemia has been shown to be present in horses, in exercising humans at simulated high altitude, and even in normal humans exercising maximally at sea level. Too rapid a transit time for the fastest moving red cells is implicated in these findings. Naturally, not every red cell traverses the pulmonary capillaries at the mean time. Rather, there is a range of transit times distributed about the mean with some times significantly shorter than the mean, but the shape of that important distribution has escaped measurement with nearly all methods. During the past two decades, there has been sufficient development of intensified television cameras and computer video processing systems to permit pulmonary capillary transit times to be measured directly. Video fluorescence microscopic recordings of the passage of a fluorescent dye bolus through the subpleural pulmonary microcirculation have been used to obtain dye dilution curves from an arteriole and venule that shared a common capillary network. Initially, the experiments were restricted to calculation of the mean capillary transit time. Using this technique, rabbits have a more rapid mean transit time than dogs at basal cardiac outputs, (rabbits, 0.41 to 0.60 seconds and dogs, 1.37 to 2.0 seconds). These experiments also showed a substantial vertical gradient of transit times in the dog lung, 10 to 20 seconds in the upper lung and 1.5 to 2.0 seconds in the lower lung. These video microscopic measurements have been confirmed by Hogg and colleagues using independent techniques.

Although the measurements of mean capillary transit time have been important, the accurate measurement of the entire distribution of transit times, including the minimum and maximum, has eluded investigators. Recently, however, Presson et al. have made these measurements using a fluorescent plasma marker. As blood flow was increased in this study, both the skew and kurtosis of the curves changed so that the longest times shortened, and the curves approached the ideal shape, that is, a spike at 0.25 seconds such that all red blood cells would become saturated exactly at the downstream end of the capillaries. The more homogeneous transit time distribution (a higher, sharper peak with a narrower base) along with some perfusion of previously unperfused capillaries that increased capillary blood volume, both acted to prevent even the most rapidly moving plasma from crossing in less than 0.25 seconds.

In the latest report from Presson and colleagues [2], the transit time distribution of fluorescently labeled red blood cells was reported. In intact dogs, they found that fluorescently labeled plasma boluses required about half again as long to traverse the capillaries as red blood cells (Figure 1), probably because of a combination of factors including the plasma-skimming Fahraeus effect in perfused capillaries, and the possibility of plasma alone perfusing some capillaries. This means that data obtained from the fluorescently labeled plasma dilution curves overestimate red cell transit time. Nevertheless, at the highest cardiac outputs, capillary recruitment and increasing red blood cell transit time homogeneity combined, as they did for plasma capillary transit time, to prevent even the most rapidly transiting red blood cells from going below 0.25 seconds in the dog lung at a cardiac output increased fourfold over basal.

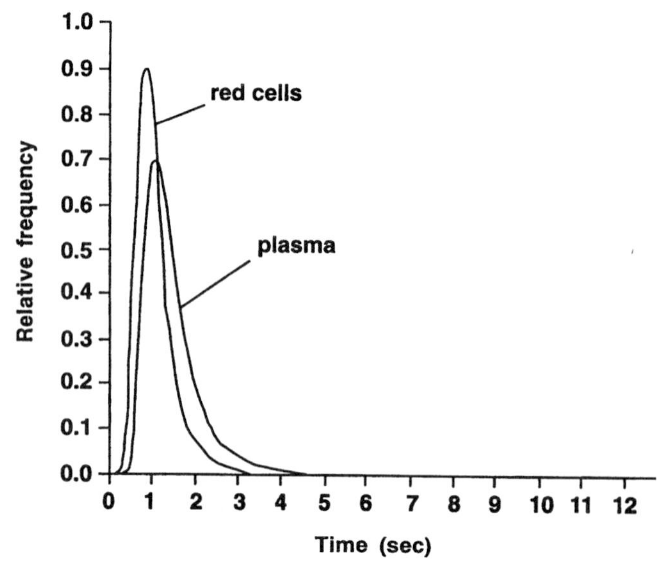

Figure 1 In intact dogs, fluorescently labeled plasma boluses require about half again as long to traverse the capillaries as red blood cells. This causes the plasma dilution curves to overestimate red blood cell capillary transit times. The mean transit times (plasma = 1.4 seconds, red blood cell = 1.0 seconds), the minimum transit times (plasma = 0.6 seconds, red blood cell = 0.4 seconds), and the maximum transit times (plasma = 3.5 seconds, red blood cell = 2.5 seconds) were all significantly longer for plasma. At higher cardiac outputs, the curves become more narrow with higher peaks; the minimum transit time is < 0.25 seconds, even at very high cardiac outputs. (Reproduced with permission from Ref. [2].)

Distention

Distention of the capillaries, which increases the number of red blood cells in the gas exchanging vessels, has been studied extensively by Fung and Sobin [3] and colleagues. Much of their work has been based on careful measurements of capillary diameters filled under various pressure conditions by a polymer that, after having flowed into all the vessels, formed a cast of the capillary bed. The appearance of the extraordinarily dense capillary network embedded in thin alveolar walls suggested that perfusion of the pulmonary capillaries functioned more like a sheet of flowing blood rather than flow through a network of tubes. Measurements of the polymer casts produced the relationship shown in Figure 2. In this preparation, once transmural capillary pressures exceeded alveolar pressure, the capillary bed opened suddenly to form a sheet about 4 μm thick. Additional pressure increases produced a linear increase in thickness. The sheet flow model permitted mathematical modeling of capillary hemodynamics, a significant advantage. The complexity of other models, composed of a seemingly infinite number of tubes nearly as wide as they were long, made it virtually impossible to manage the mathematical analysis of flow.

Other workers have also found that the capillaries distended as pressure increased. In their classic paper, Glazier et al. [4] studied isolated perfused greyhound lungs that were rapidly frozen. The lungs were frozen under three vascular pressure conditions: (1) where alveolar pressure was greater than pulmonary arterial pressure, which in turn, was greater than pulmonary venous pressure, and there was little or no flow (PA > Ppa > Ppv = zone 1); (2) where arterial pressure exceeded alveolar pressure, but alveolar pressure was greater than venous pressure (Ppa > PA > Ppv = zone 2); and (3) where arterial pressure exceeded venous pressure, and venous pressure was greater than alveolar pressure (Ppa > Ppv > PA = zone 3). In lungs frozen under zone 2 and zone 3 conditions, the capillaries distended linearly as vascular transmural pressure increased.

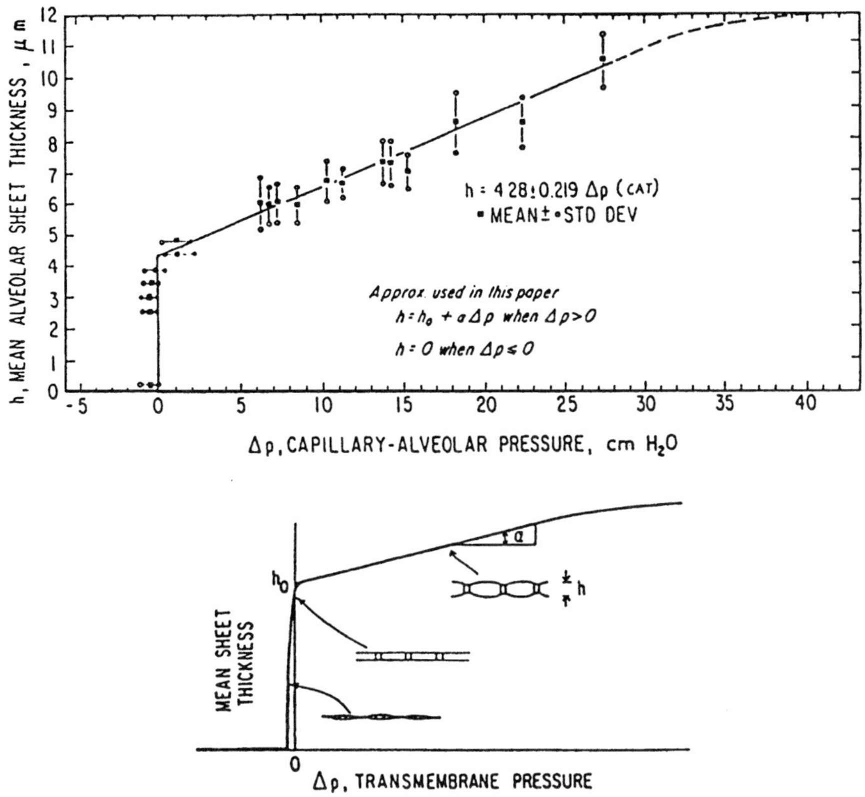

Figure 2 (*Upper panel*) The distensibility curve of the pulmonary capillaries in the cat lung. Transmural capillary pressure is shown on the *x*-axis and sheet thickness on the *y*-axis. The sheet remains empty until capillary pressure is ≈ 0, then the sheet opens suddenly to a thickness of ~4 μm and distends linearly thereafter. (*Lower panel*) Theoretical capillary shapes are shown as transmural pressure increases. (Reproduced with permission from Ref. [3].)

As a final note about distending capillaries, the situation becomes more complex when hematocrit is considered. In individual capillaries, hematocrit varies from moment to moment, as can readily be demonstrated by in vivo microscopy. As the capillaries distend sufficiently to admit red blood cells edge on, rouleau formations become possible. In that extreme, capillary hematocrit can reach very high values. One the other hand, large plasma gaps between red cells can reduce individual capillary segmental hematocrit to 0 percent momentarily. Other combinations of red cells and plasma cause the entire gamut of hematocrits to be run in a matter of seconds. These variations have been quantitated; their implications for gas exchange have recently been explored and suggest that hematocrit, red blood cell orientation, and the cross-sectional shape of the capillary may have an important effect on gas exchange.

Recruitment

At baseline, not all capillaries are perfused by red blood cells. As more capillaries become perfused by red blood cells, they are "recruited" and add to the gas exchange surface area. There are two models of capillary recruitment: segment-by-segment recruitment over a wide range of distending pressure, and the sheet flow model in which all capillaries suddenly become perfused when alveolar pressure is exceeded. This disparity between capillary recruitment models has been a difficult and important issue to resolve. The evidence supporting each idea appears solid, yet the physiological implications of each model are significantly different. Recent work by Godbey et al. [5] may have resolved the issue, because they found a way in which capillaries could be predictably recruited either segment by segment or as a sheet. The pattern of recruitment did not depend on capillary pressure or on alveolar pressure per se; rather, the recruitment pattern depended in a surprising way on the state of distention of the alveoli. Distended alveoli stretch capillaries into oval cross sections, which seemed to cause the individual segments to have a range of opening pressures. This, in turn, led to segment-by-segment recruitment. Whether alveolar distention was caused by positive airway pressure or by the pull of gravity on a lung suspended in the thorax by negative intrapleural pressure as in life (0 mmHg airway pressure), progressive capillary recruitment occurred over a wider range of capillary pressures as the alveoli enlarged. Further, the oval cross sections of the capillaries constrained the red blood cells to flow single file with their broad sides facing alveolar gas, an orientation favorable for gas exchange. Smaller, less distended alveoli permit capillaries to become circular; the "relaxed" capillaries in smaller alveoli opened suddenly as alveolar pressure was exceeded, that is, as a sheet. The circular cross sections permitted the red blood cells to flow in any orientation with respect to alveolar gas. However, many of these orientations are less favorable for gas exchange.

These experiments showed that each model of capillary recruitment was possible; the appropriate model depended on the amount of alveolar distention (Figure 3). The sheet flow model seems more predictive for the lower lung of large animals where alveoli are small, and in small animals in which the alveoli are relatively small in diameter. Segment-by-segment recruitment seems more appropriate in the upper lung of large animals where alveoli are large. That the state of alveolar distention could so dramatically affect capillary recruitment characteristics was unexpected and surprising. One explanation is based on the anatomical observation that connective tissue fibers are interlaced throughout the capillary network. When the lobes are inflated to low levels of distention (airway pressure = 2 mmHg), the connective tissue fibers are slack. Under these conditions, all of the capillaries could be readily perfused as microvascular pressure exceeds alveolar pressure. As the alveoli are distended, the connective tissue fibers are stretched. Individual fibers become progressively taut, but not equally so. As taut fibers cross each capillary, they tend to pinch the segments. The amount of pinching would be likely to vary among segments and lead to a range of opening pressures and therefore recruitment over a range of pressures. Further increases in alveolar distention would make these variations more pronounced. In this way, capillary recruitment on a segment-by-segment basis or on a sheet flow basis can be explained (Figure 3).

Although the work of Godbey explained how average patterns of recruitment (sheet versus segment-by-segment) were a function of the degree of alveolar distension, there was significant variability in the recruitment pattern among neighboring alveoli, all of which were inflated with the same pressure. Baumgartner and colleagues [6] designed a study to investigate this variability. In isolated, pump-perfused canine lung lobes, fields of six neighboring alveoli were recorded with video microscopy as pulmonary venous pressure was raised from 0 to 40 mmHg in 5-mmHg increments. The largest group of alveoli (42%) recruited gradually. Another group (33%) recruited suddenly (sheet flow). Half of the neighborhoods had at least one alveolus that paradoxically derecruited when pressure was increased, even though neighboring alveoli continued to recruit capillaries. At pulmonary venous pressures of 40 mmHg, 86 percent of the alveolar–capillary networks were not fully recruited. They

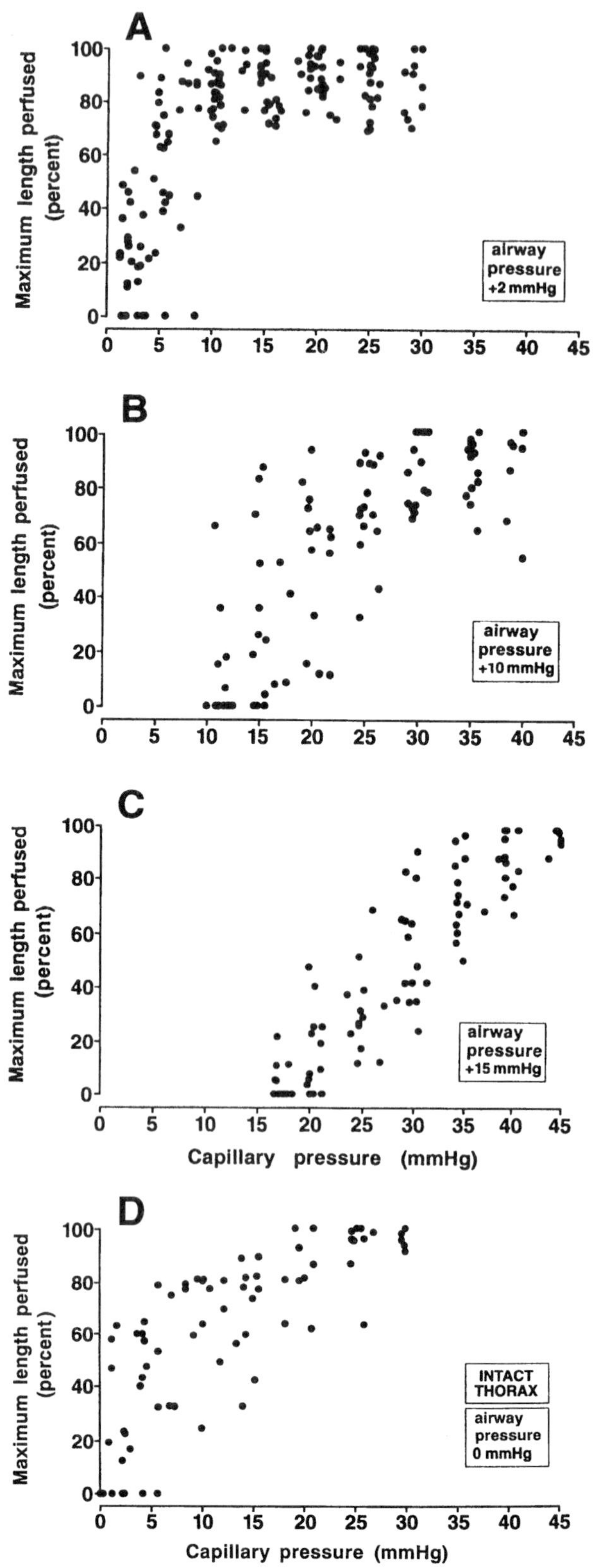

Figure 3 In excised perfused dog lobes (panels **A**, **B**, and **C**), when alveolar pressure is low (**A**), the capillary bed opens suddenly and completely as alveolar pressure is exceeded by perfusion pressure; the capillaries are perfused as a sheet. As alveolar pressure is increased (**B**) and alveoli enlarge, capillary recruitment occurs more gradually. Once the alveoli are distended (**C**), recruitment is linear. Recruitment patterns are different in panel **D**. The lungs were inflated by negative intrapleural pressure and perfused in an intact canine thorax (alveolar pressure = 0 mmHg). Observations were made on the uppermost lung where highly distended alveoli had capillary networks that were distended and recruited gradually on a segment-by-segment basis. (Reproduced with permission from Ref. [5].)

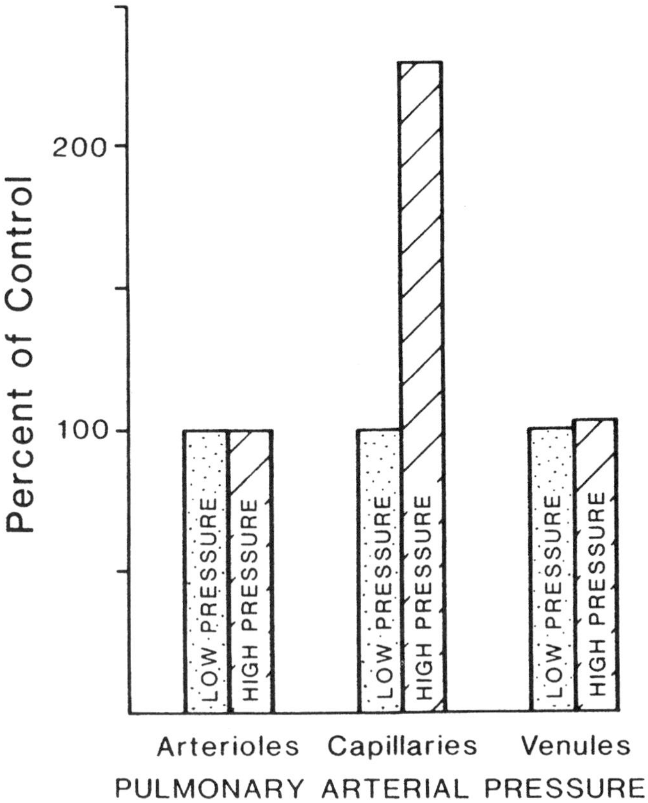

Figure 4 Average capillary recruitment for alveoli ($n = 54$) in six isolated, pump-perfused dog lung lobes. Although there were a variety of modes of recruitment among alveoli, the average recruitment was gradual. The arterio-venous pressure gradient at a venous pressure of zero was 8 mmHg, which decreased to 3 mmHg at a venous pressure of 40 mmHg, which suggests that capillary transmural pressure was close to venous pressure at the higher perfusion pressures. (Reproduced with permission from Ref. [6].)

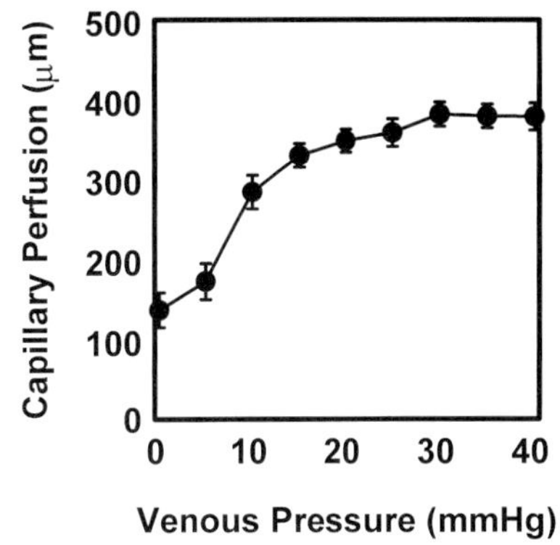

Figure 5 Experiments in intact dogs in which pulmonary arterial pressure was raised either by airway hypoxia or by fluid loading with dextran. In both cases, pulmonary arterial pressure rose from 8 to 23 mmHg. The only vessels that recruited in response to these hemodynamic loads were capillaries, which showed that recruitment is a capillary event. (Reproduced with permission from Ref. [7].)

concluded that the pattern of recruitment among neighboring alveoli was complex, was not homogeneous, and may not have reached full recruitment, even during extreme pressures. Taken as a group, however, the capillaries recruited gradually (Figure 4).

Although these investigations have increased our knowledge about capillary recruitment, it was not known whether larger vessels in the circulation could be recruited. If arterioles or arteries could be recruited, then the recruitment of a single precapillary vessel would add significantly to capillary volume. Hanson et al. [7] investigated this issue using in vivo microscopy to study capillary segments and branches of arterioles and venules in the dog lung. With each vessel serving as its own control, observations were made during low and high vascular pressures. The only vessels to recruit were capillaries, indicating that, in the dog, recruitment was exclusively a capillary event (Figure 5). A similar conclusion was reached in the study of Warrell et al., who studied frozen dog lungs.

Much of the data discussed thus far have involved static or very nearly static conditions: injected vessels, frozen lungs, or vessels perfused at very low flow rates all sampled at one point in time. As valuable as those data are, they lack the dynamic quality of perfusion that exists in the living lung. The complexity of the pulmonary microcirculation can only be measured by performing studies in vivo. Data from in vivo microscopy produced the first direct evidence for a recruitable reserve of pulmonary capillaries. Those classic observations were made by Wearn et al. [8], who studied the pulmonary microcirculation on the surface of the cat lung. Not only did they demonstrate that the capillary bed was only partially perfused by red blood cells at any given time, but they also showed that the perfusion pattern varied with time within a single alveolar wall. Wearn et al. speculated that the changes in the capillary perfusion pattern could be explained by changes in pulmonary blood flow or pressure, but in 1934, they lacked the techniques necessary to study these hemodynamic variables.

These important findings remained unconfirmed by direct observation for 60 years until Okada et al. [9] repeated Wearn's experiment using modern cardiovascular monitoring combined with in vivo microscopy of an intact canine preparation. During a 45-minute study period, 1-minute observations were made every 5 minute to determine which capillaries were perfused. On average, Okada et al. found about half of the capillaries to be perfused over the course of a total of nine observations, confirming that not all of the capillaries were perfused all of the time. The level of recruitment fluctuated, but the variations in perfusion did not correlate with the small fluctuations in pulmonary arterial pressure or cardiac output that occurred during the study. In most animals, the correlation was poor (r^2 less than 0.06), which was a surprise, because previous work by this group [10] had shown that pulmonary arterial pressure and capillary recruitment were tightly correlated (Figure 6), at least when the pressure changes were large. These findings imply that capillary perfusion alterations were being influenced by factors more subtle than small variations in pulmonary arterial pressure or cardiac output.

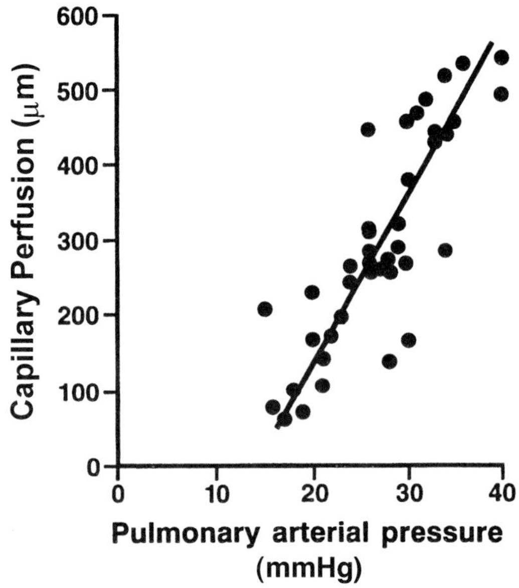

Figure 6 In intact dogs, capillary recruitment was closely linked to pulmonary arterial pressure. Airway hypoxia was used to raise pulmonary arterial pressure. (Reproduced with permission from Ref. [9].)

Two characteristics emerged when the perfusion of individual capillary segments was studied in detail. First, of the perfused segments, more than half were perfused by red blood cells during at least eight of the nine observations. This finding showed that some capillary segments had hemodynamic characteristics that were sufficiently stable to cause them to be continuously or nearly continuously perfused. These capillaries generally were interconnected to form pathways across entire alveolar walls. Second, the remaining half of the segments were not perfused stably. Some were perfused only once or twice during the nine 1-minute observation periods. The frequency of perfusion of the unstably perfused segments was evenly dispersed among observations and resulted in considerable switching of flow among the segments within each alveolar network.

The complexity of the recruitment process is even more apparent when it is considered that capillary recruitment generally refers to perfusion of new segments by a combination of red cells and plasma. However, capillaries can be perfused (recruited) by plasma alone. Recent work by Conhaim and Rodenkirch [10a] has shown that capillaries in the rat lung even under zone 1 conditions have an estimated functional diameter of 1.7 μm, sufficiently large to admit plasma, but an impediment to red blood cell perfusion. Consistent with these findings, König et al. [10b] injected colloidal gold nanospheres (8 nm diameter) into the pulmonary circulation of rabbits. After 2 minutes, the circulation was stopped, and the lungs were fixed and examined both by electron and light microscopy. They found that the entire pulmonary capillary bed was perfused by gold particles after 2 minutes, and in later experiments, complete dispersion occurred in 10 seconds. These studies suggest that all of the capillaries were perfused by some of the plasma all of the time. This work suggests the following possibilities. The entirety of the capillary endothelium could act continuously on the plasma to perform its many metabolic functions. Red blood cell recruitment, however, would occur only when capillary pressure increased sufficiently to distend capillary segments enough for the passage of red cells.

At the outset, we pointed out that the lung is conceptualized in an overly simplistic way: a single-airway, single-alveolus, single-capillary model. Much evidence demonstrates that the pulmonary microcirculation behaves in a complex manner with perfusion rapidly switching between capillary segments and red blood cell velocity and hematocrit continually varying. As cardiac output increases with exercise, however, a number of changes occur that cause the microcirculation to behave more homogeneously. First, as capillary pressure increases, capillary recruitment approaches 100 percent. Presumably, capillary resistance becomes more uniform as segments that were closed open and distend toward the plateau of their compliance curve. These changes lead to a diminution of red blood cell velocities, and the heterogeneity of capillary transit times lessens. The complexity of perfusion of the pulmonary microcirculation during increasing exercise, therefore, decreases and the simplistic one-alveolus, one-capillary model becomes progressively more realistic.

Glossary

Capillary recruitment: perfusion of previously unperfused capillary
Capillary transit time: time to vessel from arteriods to venules
Pulmonary capillaries: gas exchange vessel in alveolar wall

References

1. Wagner, P. D. (1977). Diffusion and chemical reaction in pulmonary gas exchange. *Physiol. Rev.* **57**, 257–312. *Although somewhat dated, this superb review will provide excellent grounding in the pulmonary physiology of gas exchange.*
2. Presson, R. G., Jr., Hanger, C. C., Godbey, P. S., Graham, J. A., Sidner, R. A., and Wagner, W. W., Jr. (1995). Distribution of pulmonary capillary red blood cell transit times. *J. Appl. Physiol.* **79**, 382–388.
3. Fung, Y. C., and Sobin, S. S. (1972). Pulmonary alveolar blood flow. *Circ. Res.* **30**, 470–490. *These investigators did classic work showing that all of the pulmonary capillaries opened suddenly and completely once capillary pressure exceeded alveolar pressure, a phenomenon known as sheet flow. This work started a long-running controversy between the sheet flow school and the recruitment school, which was finally settled by the paper in Ref. [5].*
4. Glazier, J. B., Hughes, J. M. B., Maloney, J. E., and West, J. B. (1969). Measurements of capillary dimensions and blood volume in rapidly frozen lungs. *J. Appl. Physiol.* **26**, 65–76. *This group led the way in showing that blood flow tended to perfused the base of the lung rather that the apex (gravitationally dependent versus independent). In this paper they show that more capillaries are perfused with blood (recruited) in the lower lung.*
5. Godbey, P. S., Graham, J. A., Presson, R. P., Jr., Wagner, W. W., Jr., and Lloyd, T. C., Jr. (1995). Effect of capillary pressure and lung distention on capillary recruitment. *J. Appl. Physiol.* **79**, 1142–1147.
6. Baumgartner, W. A., Jr., Jaryszak, E. M., Peterson, A. J., Presson, R. G., Jr., and Wagner, W. W., Jr. (2003). Heterogeneous capillary recruitment among adjoining alveoli. *J. Appl. Physiol.* **95**, 469–476.

7. Hanson, W. L., Emhardt, J. D., Bartek, J. P., Latham, L. P., Checkley, L. L., Capen, R. L., and Wagner, W. W., Jr. (1989). The site of recruitment in the pulmonary microcirculation. *J. Appl. Physiol.* **66**, 2079–2083.
8. Wearn, J. T., Ernstene, A. C., Bromer, A. W., Barr, J. S., German, W. J., and Zschiesche, L. J. (1934). The normal behavior of the pulmonary blood vessels with observations on the intermittence of the flow of blood in the arterioles and capillaries. *Am. J. Physiol.* **109**, 236–256.
9. Okada, O., Presson, G. R., Jr., Godbey, P. S., Capen, R. L., and Wagner, W. W., Jr. (1994). Temporal capillary perfusion patterns in single alveolar walls of intact dogs. *J. Appl. Physiol.* **76**, 380–386.
10. Wagner, W. W., Jr., Latham, L. P., and Capen, R. L. (1979). Capillary recruitment during airway hypoxia: Role of pulmonary arterial pressure. *J. Appl. Physiol.* **47**, 383–387.
10a. Conhaim, R. L., and Rodenkirch, L. A. (1997). Estimated functional diameter of alveolar septal microvessels at the zone I–II border. *Microcirc.* **4**: 51–59.
10b. König, M. F., Lucocy, J. M., and Weibel, E. R. (1993). Demonstration of pulmonary vascular perfusion by electron and light microscopy. *J. Appl. Physiol.* **75**: 1870–1876.
11. Wagner, W. W., Jr., Latham, L. P., Hanson, W. L., Hofmeister, S. E., and Capen, R. L. (1986). Vertical gradient of pulmonary capillary transit times. *J. Appl. Physiol.* **61**, 1270–1274. *This group was the first to demonstrate that not only is there less capillary recruitment in the upper lung, but also the transit times are much longer than in the lower lung. This work was controversial until verified by independent techniques by other investigators.*

Capsule Biographies

Dr. Wagner has been the Director of Research for the Anesthesia Department at Indiana University since 1985. His laboratory has focused on the pulmonary microcirculation using in vivo microscopy with the aim of understanding how blood flow is controlled at rest and during exercise. His work is supported by the NIH.

Dr. Presson has been a close collaborator with Dr. Wagner for the past 17 years. He has focused on mathematical analysis and modeling of pulmonary microcirculatory flow data. In addition he works with animal models of emphysema. He is a full-time practicing anesthesiologist. His work is supported by NIH.

Fragility of Pulmonary Capillaries

John B. West
University of California San Diego, La Jolla, California

Introduction

Only in the past 15 years has it been appreciated that pulmonary capillaries are remarkably mechanically fragile, and that ultrastructural changes can occur in their walls under normal though unusual physiological conditions. We now know that many pulmonary and cardiovascular disorders are associated with structural failure of the walls of the capillaries in response to high mechanical stresses ("stress failure"). In retrospect it is extraordinary that it took so long to realize the vulnerability of pulmonary capillaries because recognition of the extreme thinness of their walls goes back to the first electron micrographs in 1952. A simple calculation on the back of an envelope using the Laplace relationship will show that any substantial increase in transmural pressure of the capillaries must result in extremely high hoop or circumferential stresses. Perhaps people were misled by early measurements suggesting that pulmonary capillary pressures remained low under all physiological conditions, although we now know that this is a serious misconception. This brief account summarizes information on the structure and strength of pulmonary capillaries, the stresses that can develop in their walls under both physiological and pathological conditions, the phenomenon of stress failure, and recent work on how the structure of the capillary wall is regulated.

Pulmonary versus Systemic Capillaries

Those of us who work on the pulmonary circulation tend to bristle at the term "lesser circulation." After all the lung is the only organ of the body that receives the whole of the cardiac output. It is true that the pulmonary circulation has much lower pressures than the systemic circulation, but of course its functions are equally important.

A cardinal feature of a pulmonary capillary as shown in Figure 1 is that parts of its wall are extraordinarily thin. For example in the human lung the wall is of the order of 0.2 μm in thickness over much of its area of 50 to 100 m^2 [2]. This extreme thinness is essential to allow the enormous amounts of gas exchange of oxygen and carbon dioxide to occur by passive diffusion. The pulmonary capillary is essentially naked in the sense that it is directly exposed to the alveolar gas.

By contrast, systemic capillaries are embedded in tissue and receive substantial mechanical support from this [3]. Baez and colleagues measured the diameters of various small blood vessels in the exteriorized meso-appendix of rats and showed that the smallest (8 to 10 μm), that is, the capillaries, behaved essentially as rigid tubes in that their diameter remained constant over a range of pressures from 20 to 200 mmHg. Fung calculated that the surrounding tissue of capillaries in mesentery contributed over 99 percent of their rigidity with less than 1 percent coming from the endothelium and basement membrane. Pressures of the magnitude of 100 mmHg will certainly cause damage to the walls of pulmonary capillaries. It is critically important to appreciate the very different mechanical environment of pulmonary capillaries compared with systemic capillaries.

Structure of Pulmonary Capillaries

Pulmonary capillaries typically have a thick and thin side as shown in Figure 1. On the thin side, the blood–gas barrier consists only of a thin epithelium, this being the protoplasmic extension of a type I alveolar epithelial cell, an interstitium or extracellular matrix, and a thin capillary endothelial layer. There is also a very thin layer of surfactant lining the alveolar space, but this is not shown in Figure 1 because it was lost during preparation of the section.

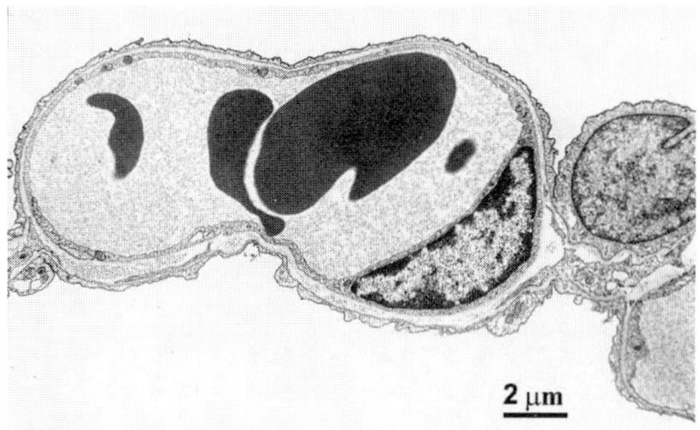

Figure 1 Electron micrograph of a pulmonary capillary. The thin side of the blood–gas barrier is in the upper part of the micrograph while the thick side is below. The micrograph emphasizes the extreme thinness of some portions of the barrier and the fact that it is completely unsupported by surrounding tissue. (From Ref. [1].)

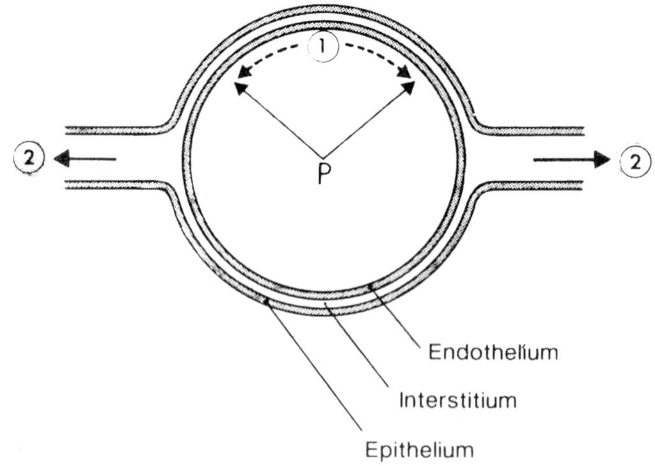

Figure 2 Diagram showing the two mechanisms that can increase the mechanical stress in the capillary wall. (1) shows the hoop or circumferential stress as a result of an increase in the capillary transmural pressure. (2) indicates the longitudinal tension in the alveolar wall as the lung is inflated to a high volume. Part of this tension is transmitted to the capillary wall. (From Ref. [5].)

By contrast, the thick side of the pulmonary capillary contains additional structures in the interstitium, notably fibers of type I collagen that contribute to the structural framework of the lung, and also interstitial cells including fibroblasts and other pericytes. This anatomy means that the thin side of the blood–gas barrier is mainly responsible for diffusional gas exchange, whereas the thick side is concerned with fluid exchange and also mechanical stability. Discussion about the fragility of pulmonary capillaries mainly concerns the thin side where the stresses will be highest, other things being equal.

Strength of Pulmonary Capillaries

There is now good evidence that the strength of the thin side of the blood–gas barrier comes not from the cellular layers themselves, but from the interstitium or extracellular matrix, and in particular the type IV collagen in the basement membranes. Here it is important to appreciate that in the thin side of the blood–gas barrier, the basement membranes of the alveolar epithelium and capillary endothelium fuse in the center of the extracellular matrix to give a central electron-dense band rich in type IV collagen perhaps only 50 nm thick. It is believed that this extremely thin sheet of type IV collagen is mainly responsible for the mechanical strength of the thin side of the blood–gas barrier.

The evidence for this can be summarized as follows. We have found that in animal studies where the capillary pressure is raised to high levels, disruptions can be seen in the epithelial and endothelial layers but often the basement membrane remains intact [4, 5]. Additional evidence comes from studies on isolated rabbit renal tubules that show that the distensibility of the tubules is determined by their basement membrane [6]. In addition, studies of the distensibility of frog mesentery capillaries show that this is attributable to the mechanical properties of the basement membrane. Further studies show that the thickness of the basement membrane of systemic capillaries increases down the body as the pressure within them rises. Finally, it is well known that renal glomerular capillaries have a thick basement membrane consistent with their high transmural pressure.

What is known about the structure and ultimate tensile strength of type IV collagen? It has a triple helix structure similar to that of other matrix collagens, but is unusual in that two molecules join at the C-terminal end to give a doublet about 800 nm long, and four molecules join at the N terminal to give a matrix configuration rather like chicken wire. There are few studies of the ultimate tensile strength of basement membrane, but measurements on cat lens capsule together with studies on isolated rabbit renal tubules suggest that the ultimate tensile strength is approximately $1 \times 10^6\,\mathrm{Nm}^{-2}$, a very large value. Thus type IV collagen is ideally suited to confer the high tensile strengths needed in the wall of the summary capillary.

Mechanical Stresses in the Walls of Pulmonary Capillaries

Figure 2 shows two mechanisms by which mechanical stresses develop in the capillary wall. The first is the hoop or circumferential stress that results from the difference of pressure between the inside and outside of the capillary (transmural pressure) acting across the curved capillary wall according to the Laplace relationship. We can think of the capillary as part of a thin-walled cylindrical tube, and in such a structure the wall stress S is given by Pr/t, where P is the transmural pressure, r is the radius of curvature, and t is the thickness of the load-bearing structure.

It is informative to make an approximate calculation of this hoop or circumferential stress. First, how high can the transmural pressure in human capillaries rise during exercise? Pulmonary arterial wedge pressures (a measure of pulmonary venous pressure) as high as 21 mmHg have been recorded. Consistent with this, mean pulmonary artery pressures have been shown to be as high as 37.2 mmHg during exercise. Micropuncture studies of the pressures in small pulmonary blood vessels in anesthetized cats have shown that mean capillary pressure is about halfway between arterial and venous pressures, but that much of the fall in pressure occurs in the capillary bed. Therefore a conservative estimate for capillary pressure at midlung during maximal exercise is about 29 mmHg, although the pressures seen by capillary segments in the upstream regions of the bed will be higher. If we add the hydrostatic gradient for capillaries at the bottom of the upright human lung, this gives a capillary pressure there of about 36 mmHg [5].

There are no good data on the radius of human pulmonary capillaries at high capillary pressures, but in rabbit and dogs the value is about 3.5 μm [7]. The most elusive number is the thickness of the load-bearing structure, but if we assume that this is the thin band of type IV collagen in the center of the extracellular matrix of the thin side of the blood–gas barrier, the value is of the order of 50 nm. Inserting these numbers into the Laplace relationship gives a tensile stress in the layer of type IV collagen of about $3 \times 10^5\,\mathrm{Nm}^{-2}$, which is approaching the ultimate tensile strength of type IV collagen as discussed earlier. The upshot of all this is that the normal lung does not appear to have a great deal of reserve in terms of the strength of the capillary wall, and this is consistent with evidence for changes in the structure of the wall that we have seen in elite athletes at high levels of exercise as discussed later.

The second mechanism for increasing mechanical stresses in the capillary wall is increased tension in the alveolar wall as a result of inflating the lung to high volumes. We can think of the alveolar wall as a string of capillaries with part of the longitudinal tension of the wall being transmitted to the capillary walls. As we shall see later, there is strong evidence that when the lung is inflated to high volumes, for example as a result of high levels of positive end-expiratory pressure, the capillary wall is damaged.

Stress Failure in the Wall of the Pulmonary Capillary

The ultrastructural changes that occur in the walls of pulmonary capillaries when the wall stresses are increased have been extensively studied. Measurements have been made in anesthetized rabbits where the pulmonary artery and left atrium were cannulated so that the capillary transmural pressure could be raised and accurately measured. The lungs were prepared for electron microscopy using intravascular fixation with buffered glutaraldehyde at the required pressure. An example of the ultrastructural changes seen in such experiments is shown in Figure 3. Note that the alveolar epithelial layer is disrupted near the top of the micrograph and, in addition, there is disruption of the capillary endothelial layer near the bottom. Interestingly, a blood platelet appears to be adhering to the exposed basement membrane, which is not surprising because the membrane is electrically charged and highly reactive.

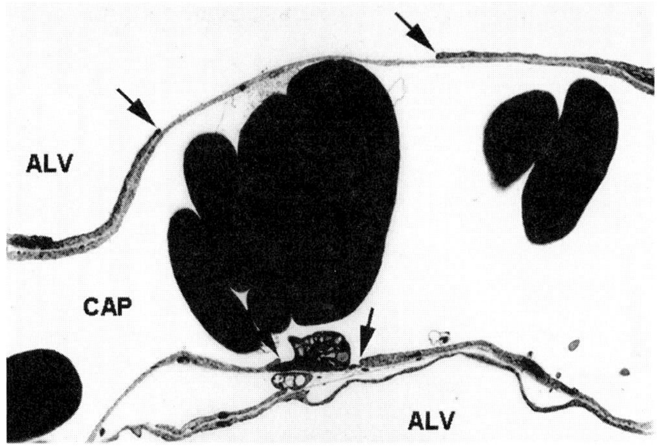

Figure 3 Electron micrograph showing some effects of raising the capillary transmural pressure in rabbit lung. Note that there is disruption of the alveolar epithelial layer (top arrows) and also a disruption of the capillary endothelial layer (bottom arrows). A platelet is adhering to the exposed basement membrane below. ALV, alveolus; CAP, capillary. (From Ref. [5].)

The pressures required to cause these ultrastructural changes have been investigated and it has been shown that the first indications of stress failure in rabbit lung occur at capillary transmural pressures of about 24 mmHg, but the number of breaks is much increased at a pressure of 39 mmHg [4]. It is perhaps remarkable that pressures as low as 24 mmHg result in some changes in view of the fact that in the human lung capillary transmural pressures as high as 36 mmHg apparently occur during heavy exercise. Of course, we cannot assume that the strength of the capillary walls is the same in rabbit and human lung, and in fact we have shown that different pressures are required for stress failure in rabbit, dog, and horse lung [7].

An important feature of these ultrastructural changes is that many are rapidly reversible when the capillary transmural pressure is reduced. This has been shown in our rabbit preparation by first raising the pressure to a high level, and then reducing it to a low level for several minutes with intravascular fixation at the low pressure. The result was that about 70 percent of both the endothelial and epithelial breaks closed within a few minutes [8]. The breaks that closed were mainly those that were initially small and associated with an intact basement membrane.

The micromechanics of stress failure are not fully understood, but one possibility is that the high tensile stress in the type IV collagen layer causes this to stretch with distortion of the matrix arrangement. It is known that type IV collagen molecules have sites that allow bending to occur. Thus it is possible that the basement membrane elongates under stress, causing the overlying cell to disrupt. When the tensile stress is removed the matrix reestablishes its normal configuration and the disruptions in the cells close. However, there is no direct evidence for this hypothesis.

Physiological Conditions Associated with Stress Failure of Pulmonary Capillaries

The most remarkable physiological situation where stress failure of pulmonary capillaries is seen is in galloping thoroughbred racehorses. There is now evidence that all thoroughbreds in training bleed into their lungs. This condition, called exercise-induced pulmonary hemorrhage, has been known since Elizabethan times, and all sorts of explanations have been offered such as moldy hay. However, recently it has become apparent that exercising thoroughbreds have extremely high pulmonary vascular pressures. For example, studies on animals galloping on a treadmill show that a left atrial pressure measured directly with a catheter can be as high as 70 mmHg, and this is associated with a mean pulmonary artery pressure as high as 120 mmHg. Other vascular pressures are equally astonishing, with a mean systemic arterial pressure as high as 240 mmHg and a mean right atrial pressure up to 40 mmHg.

The reason for these high pressures is that these animals have been selectively bred for hundreds of years for very high aerobic performances. As examples, their maximal oxygen consumptions reach 180 $mLmin^{-1}kg^{-1}$ and cardiac outputs are as high as 750 $mLmin^{-1}kg^{-1}$. In order to develop these extraordinary cardiac outputs the filling pressures of the left ventricle must be very high, and this in turn translates to high pulmonary venous and capillary pressures. We have looked at the lungs of thoroughbreds immediately after they have been galloped on a treadmill and shown unequivocal evidence of breaks in the pulmonary capillaries [9].

These extraordinary findings in racehorses raise the question of whether elite human athletes ever develop ultrastructural changes in their pulmonary capillaries during maximal exercise. There is now strong evidence for this. Six elite cyclists sprinted uphill over several minutes at maximal effort sufficient to give a mean heart rate of 177 beats $minute^{-1}$, and within an hour they voluntarily underwent bronchoalveolar lavage (BAL). The results were compared with those of normal sedentary subjects who did not exercise before BAL, and it was found that the athletes had significantly higher concentrations of red blood cells, total protein, albumin, and leukotriene B_4 (LTB_4) in their BAL fluid compared with the control subjects [10]. Thus brief but very intense exercise in elite athletes apparently causes changes in the integrity of the capillary wall.

Would the same changes occur if the athletes worked for a longer period at slightly lower levels of exercise? This was tested by making additional measurements on a similar group of six elite cyclists who exercised at 77 percent of their maximal work level for 1 hour and then underwent BAL. Again the controls were eight normal nonathletes who did not exercise before BAL. In contrast to the results of the previous studies, these athletes showed no differences compared with the controls. Thus we can conclude that only extremely high levels of exercise cause changes in the capillary walls, and indeed this is what might be expected on evolutionary lines. In general organisms evolve to cope with all but the most extreme stresses to which they are subjected.

Table I Pathological Conditions Causing Stress Failure of the Blood–Gas Barrier.

1. High capillary pressure resulting in high-permeability edema, e.g., high-altitude pulmonary edema, neurogenic pulmonary edema
2. High capillary pressure causing edema and hemorrhage, e.g., mitral stenosis, left ventricular failure
3. High state of lung inflation, e.g., positive end-expiratory pressure in the intensive care unit
4. Abnormal extracellular matrix, e.g., Goodpasture's syndrome

Pathological Conditions Causing Stress Failure

The fact that normal subjects develop stress failure of their pulmonary capillaries at extremely high but nevertheless physiological pressures immediately suggests that any pathological condition that raises the transmural pressure of the capillaries to unphysiologically high levels will result in stress failure. This is indeed the case, and Table I summarizes the pathological conditions under which stress failure of the capillaries has been described. There is insufficient space here to discuss these conditions. However, attention should be drawn to the third group, namely high states of lung inflation. It was pointed out earlier that increased longitudinal tension in the alveolar wall associated with high lung volumes would be expected to increase the wall stress of the capillaries. We have tested this in animal preparations where lung volume was increased to high levels while keeping the capillary transmural pressure constant. Such experiments show a great increase in the number of disruptions of both the capillary endothelial and alveolar epithelial layers at the high lung volumes. Consistent with this, trials of low versus traditional high tidal volumes during mechanical ventilation in intensive care units have shown reduced mortality with the low tidal volumes [11], although there is still some controversy in this area.

Regulation of the Structure of Pulmonary Capillaries

It is clear from the above that the blood–gas barrier has a bioengineering dilemma. On the one hand it needs to be extremely thin for adequate gas exchange, but on the other it must be immensely strong to withstand the high mechanical stresses that develop when the capillary pressure rises on exercise, or the lung is expanded to high volumes. We have seen that the human blood–gas barrier maintains its integrity under all but the most exceptional physiological conditions, that is, maximal exercise in elite athletes. However, pathological conditions associated with an unphysiologically high capillary pressure inevitably cause stress failure.

A central question in lung biology is how the structure of the blood–gas barrier is regulated to optimize these conflicting requirements. The most likely hypothesis is that the capillary wall senses wall stress in some way, and then the wall structure, presumably particularly the amount of type IV collagen in the extracellular matrix, is regulated accordingly. Much of our research over the past few years has been devoted to testing this hypothesis.

Remodeling of the walls of larger pulmonary blood vessels in response to increased stress is well known, and there is an extensive literature on the subject; see Stenmark and Mecham [12] for a review. As an example, Tozzi and colleagues stretched explanted rings of rat main pulmonary artery and showed increases in collagen synthesis, elastin synthesis, mRNA for pro-α1(I) collagen, and mRNA for proto-oncogene V-sis within 4 hours. A feature of these changes is that they were endothelium dependent because they did not occur when the endothelium was removed.

However, it is remarkable that in contrast to the large literature on vascular remodeling in larger pulmonary blood vessels, remodeling of pulmonary capillaries has been almost completely neglected. There is certainly evidence that it occurs because marked thickening of the basement membranes of the capillary endothelial and alveolar epithelial cells is seen in the pulmonary capillaries of patients with mitral stenosis where the capillary pressure is raised over months or years. The same appearances have been described in patients with pulmonary veno-occlusive disease who also have an increased pulmonary capillary pressure.

To investigate mechanisms of pulmonary capillary remodeling, experiments have been designed in which the capillary wall stress is raised either by increasing the transmural pressure of the capillaries, or by inflating the lung to high volumes. One of the challenges in these experiments is to measure only the changes associated with increased wall stress of pulmonary capillaries, and not those associated with the larger pulmonary blood vessels. This is difficult to achieve because both experimental maneuvers have effects on all the pulmonary blood vessels. We have attempted to overcome this by analyzing tissue from only the outer few millimeters of lung, which is predominantly occupied by capillaries. There is not space to describe the results in detail here, and indeed no clear consensus has emerged as yet. Suffice it to say that experiments have shown increases in mRNA for procollagens α1(I), α2(II), and α2(IV), fibronectin, laminin, basic fibroblast growth factor, transforming growth factor β1, and platelet-derived growth factor-B. The cell that senses the increased stress has not yet been identified, nor the cells in which the increased gene expression occurs. Much more work needs to be done on this central problem in lung biology.

Glossary

Fragility: In this context, the tendency of pulmonary capillaries to develop breakages in part or all of their walls when exposed to increased stresses.

Laplace relationship: Expression relating the pressure across a curved surface such as the wall of a tube to the radius of curvature of the tube, the wall thickness, and the stress in its wall.

Remodeling: In this context, histological changes in the walls of blood vessels in response to some physiological stimulus, often increased pressure.

Stress: The force per unit area, in this case in the wall of the capillary, which tends to cause breakages.

Stress failure: Ultrastructural changes in the walls of capillaries as a result of their exposure to high mechanical stresses.

References

1. Weibel, E. R. (1973). Morphological basis of alveolar–capillary gas exchange. *Physiol. Rev.* **53**, 419–495.
2. Gehr et al. (1978). The normal human lung: ultrastructure and morphometric estimation of diffusion capacity. *Respir. Physiol.* **32**, 121–140.
3. Fung, Y. C., et al. (1966). Elastic environment of the capillary bed. *Circ. Res.* **19**, 441–461.
4. Tsukimoto, K., et al. (1991). Ultrastructural appearances of pulmonary capillaries at high transmural pressures. *J. Appl. Physiol.* **71**, 573–582.
5. West, J. B., et al. (1991). Stress failure in pulmonary capillaries. *J. Appl. Physiol.* **70**, 1731–1742.
6. Welling and Grantham (1972). Physical properties of isolated perfused renal tubules and tubular basement membranes. *J. Clin. Invest.* **51**, 1063–1075.
7. Birks, E. K., et al. (1994). Comparative aspects of the strength of pulmonary capillaries in rabbit, dog and horse. *Respir. Physiol.* **97**, 235–246.
8. Elliott, A. R., et al. (1992). Short-term reversibility of ultrastructural changes in pulmonary capillaries caused by stress failure. *J. Appl. Physiol.* **73**, 1150–1158.
9. West, J. B., et al. (1993). Stress failure of pulmonary capillaries in racehorses with exercise-induced pulmonary hemorrhage. *J. Appl. Physiol.* **75**, 1097–1109. *This article presents the evidence that all thoroughbred racehorses in training break their pulmonary capillaries and discusses the cause.*
10. Hopkins, S. R., et al. (1997). Intense exercise impairs the integrity of the pulmonary blood-gas barrier in elite athletes. *Am. J. Respir. Crit. Care Med.* **155**, 1090–1094. *This is a study showing that high levels of exercise in elite human athletes apparently alter the structure of the pulmonary blood–gas barrier.*
11. Brower et al. (2000). Ventilation with lower tidal volumes as compared with traditional tidal volumes for acute lung injury and the acute respiratory distress syndrome. *N. Engl. J. Med.* **342**, 1301–1308.
12. Stenmark, K. R., and Mecham, R. P. (1997). Cellular and molecular mechanisms of pulmonary vascular remodeling. *Ann. Rev. Physiol.* **59**, 89–144.

Further Reading

West, J. B. (2003). Thoughts on the pulmonary blood-gas barrier. *Am. J. Physiol. Lung Cell. Mol. Physiol.* **285**, L501–L513. *This is an extensive review of the topic with many references.*

Capsule Biography

Dr. West has been a Professor of Medicine and Physiology in the School of Medicine, University of California, San Diego, since 1969. He is a Member of the Institute of Medicine of the National Aademy of Sciences, a Fellow of the American Academy of Arts and Sciences and has received many other awards. His research covers many aspects of respiratory physiology, high-altitude physiology, space physiology, and the history of physiology. Most of his work is support by grants from the NIH and NASA.

SECTION H

Lymphatics

CHAPTER 79

Lymphatic Endothelial Cells: Heterogeneity from Blood Endothelium and Unique Properties in Health and Disease

J. Steven Alexander
LSU Health Sciences Center, Shreveport, Louisiana

Lymphatic versus Blood Endothelial Biology

The field of endothelial cell biology has existed for more than four decades, during which time enormous advances have been made in our understanding of endothelial cell structure and function. Once regarded as only the "passive" lining of blood vessels and constituent of capillaries, the endothelium is now a well recognized participant in angiogenesis, inflammation, vasomotion, and the metabolism of bloodborne substances. Endothelial biology has made such vast progress due in part to the identification of specific blood endothelial cell (BEC) markers and growth factors/receptors that govern blood vessel development which have led to the development of the many models and therapies that are based on, and which exploited these findings. Although BEC and lymphatic endothelial cells (LECs) share many structural and functional characteristics, several recently described markers, growth factors, and receptors are now known that distinguish LEC structurally, functionally, and pathophysiologically from BEC, and in the past 5 years, our understanding of LEC biology has gone through a dramatic revival because of the discovery of these LEC markers and their unique functions.

Unique Structural/Functional Markers: Lymphatic versus BEC

LECs share some markers expressed by most BECs, including factor VIII, anti-thrombin 3 and MHC I, receptors for low-density lipoprotein (LDL-R), and angiotensin-converting enzyme (ACE). However, some studies suggest that LEC may only modestly express some other classical BEC markers. For example, Weibel-Palade bodies and their constituents (von Willebrand factor/factor VIII-related antigen, alpha-1 fucosyltransferase VI, P-selectin) are expressed in LEC [1], but may exhibit quantitative differences based on anatomical location and species. Similarly, levels of expression of 5′-nucleotidase and 5′-nucleotidase/alkaline phosphatase have also been suggested as markers that can qualitatively distinguish LECs and BECs [1]. Extracellular matrix (e.g., heparan sulfate proteoglycan, collagen type IV) proteins are clearly less densely expressed on LECs compared to BECs. However, on the basis of these structural markers alone, BECs and LECs cannot be clearly distinguished. Another structural marker, Pathologische Anatomie Leiden-Endothelium (PAL-E), a caveolae-related marker associated with leaky vessels, does appear to be a highly specific BEC marker and is absent on LECs. Unlike these markers, which were preliminarily used to distinguish BECs from LECs, the more important and highly specific/selective

LEC markers [1] now include podoplanin, Prox-1, LYVE-1, and VEGF-R3 (Flt-4), which are described below.

Podoplanin

Podoplanin is expressed by LECs and has been used as an LEC specific marker. It was originally described as a 43 kDa mucoprotein on kidney podocytes (hence the name *podoplanin*), but is not entirely specific for LEC; podoplanin is also densely expressed in lung alveolar epithelium, osteoblasts, and several other cell types. The main function of podoplanin in LEC is not known, but it may help to regulate LEC vessel structure, or govern solute exchange in lymphatics. Podoplanin has a large extracellular domain with 6 *O*-glycosylation sites and a shorter cytoplasmic tail with two sites for phosphorylation that may be important in regulating its function. In disease, the altered expression of podoplanin has been reported during the lymphatic proliferation seen in IBD, cancer, and Kaposi's sarcoma (KS).

Prox-1

The transcription factor Prox-1 is an important regulator of lymphatic system development, but also regulates development of several tissues including the lens, retina, and liver [2]. Prox-1 is probably the most important determinant for endothelial progression/commitment to the lymphatic endothelial phenotype. The activation of Prox-1 appears to help commit lymphatic progenitors both by activating LEC transcripts and by suppressing BEC transcripts. Some of LEC-specific transcripts that are activated by Prox-1 activity include integrins $\alpha 9$ and $\alpha 1$, desmoplakins I/II, adducin-γ, plakoglobin, matrix Glα protein, TIMP-3, macrophage mannose receptor, alpha-actinin-2 associated LIM protein, IL-7, SDF-1b, and the cell cycle regulators, Cdkn-1b and Cdkn-1c.

Interestingly, Petrova et al. [3] report that Prox-1 transfection into BEC led to the induction of numerous lymphatic endothelial specific genes including p57kip2 and VEGF-R3. Prox-1 expression in BEC also suppressed 40% of BEC-specific genes, including STAT-6, laminin, VEGF-C, neuropilin-1, ICAM-1, MCP-1, IL-6, and P-selectin. It is interesting to note that VEGF-C, which is a potent LEC growth factor (see later discussion), is a normal transcript for BEC, which may be suppressed when cells commit to the LEC lineage to avoid a persistent autocrine growth stimulation. The suppression of some of these genes may also reflect Prox-1 dependent inhibition of STAT-6 (which is required to sustain BEC transcripts such as MCP, IL-6, and P-selectin). When Prox-1 is expressed in nonendothelial cells it upregulates message for cyclins E1 and E2 and activates the cyclin E promoter, but does not induce LEC genes. Therefore Prox-1 is necessary for the LEC phenotype, but does require a prior commitment to the endothelial lineage, a prerequisite for the final "programming" in the lymphatic endothelial lineage [3].

LYVE-1

LYVE-1 (lymphatic vascular endothelial hyaluronan receptor) is a homolog of CD44 that shares approximately 41% homology with CD44 [4]. LYVE-1 participates in endocytic processing of hyaluronic acid (HA) in LECs, like CD44 in BECs, but it may also perform other functions. In lymphatic vessels, LYVE-1 on the cell surface appears to colocalize with HA, which lines the lymphatic lumen where it has been suggested to help modulate leukocyte/immune cell trafficking. The exact role of the hyaluronan matrix for lymphatic function is still not well understood, but may be important to signaling, lymphocyte migration, or differentiation [5]. Compared to CD44, LYVE-1 may exhibit greater specificity for HA than CD44 [4]. LYVE-1 is expressed on the cell surface as a 60-kDa protein; 20 kDa of its mass appears to be produced by dense glycosylation. LYVE-1 expression appears to be generally restricted to lymphatic endothelium within the spleen, lymph nodes, heart, lung, and fetal liver. In adult tissues, LYVE-1 is low in the liver, muscle, bone marrow, and appendix and is generally absent in BECs, hematopoietic cells, and lymphocytes. However, while LYVE-1 is expressed mainly by LECs, some LYVE-1 expression is retained by BECs in the spleen and liver, in syncytiotrophoblasts and in macrophages [2, 5, 6].

VEGF-R3 (Flt-4)

The growth factors and growth factor receptor involved in LEC proliferation, maturation, and survival include VEGFs-C and D and the VEGF receptor, Flt-4 (VEGF-R3). Flt-4 was perhaps the first specific lymphatic endothelial marker described and plays an important role in the development of the embryological capillary system; in the adult, Flt-4 action regulates mainly the growth and maintenance of lymphatic vessels. Flt-4 is a 210-kDa receptor for VEGFs-C and D that is expressed at high levels on LECs, but usually not on BECs. Some exceptions when Flt-4 is induced on BECs include wound healing and expression in some tumor blood vessels and in fenestrated endothelium (e.g., bone marrow, spleen and hepatic sinusoids, kidney glomeruli, and endocrine gland endothelium). Flt-4 is also expressed in some nonendothelial cells (e.g., dendritic cells). Flt-4 is a member of the class III receptor tyrosine kinases like Flt-1 and Flk-1/KDR that have seven Ig-like domains and 12 potential glycosylation sites. At least two Flt-4 isoforms are known that are derived from a single transcript.

Animals made genetically deficient in Flt-4 are ultimately embryonic lethals because of defective fluid drainage (chylous ascites), and some abnormal vascular assembly. In human disease, primary (hereditary) lymphedema can be produced by single amino acid substitutions in Flt-4 (VEGFR-3) that inactivate the receptor kinase activity and prevent the maintenance/development of a normal lymphatic system.

VEGFs C and D

VEGFs C and D are closely related growth factors produced by several tissues, such as lung, skeletal muscle, colon, small intestine, tumor cells, and BECs (umbilical vein ECs). VEGF-C is a potent ligand for Flt-4, but also binds Flk-1/KDR (VEGF-R2) but not VEGF-R1, and induces lymphangiogenesis and LEC migration. Interestingly, although thought to be an LEC-specific growth factor, viral induction of VEGF-C is one of the most powerful stimuli for angiogenesis and lymphangiogenesis [7]. Although both VEGF-C and D can bind Flk-1/KDR and Flt-4, experiments with Flt-4 specific forms of these VEGFs show that Flt-4 binding alone promotes lymphatic vessel growth. VEGF-D is also a ligand for Flt-4 (and VEGF-R2), but not VEGF-R1 (Flt-1). In mice VEGF-D appears to be a selective ligand for Flt-4 [8]. Stimulation of lymphatic endothelial cells with VEGF-C is a strong stimulus for the increased expression of Ang-2 (via VEGFR-2) and in lymphatic development, angiopoetin-2 (Ang-2) appears to control lymphatic development, since Ang-2 knockout mice show errors in the maturation of lymphatic vessels. Interestingly, Ang-1 appears to correct the lymphatic, but not the angiogenesis defects, suggesting that Ang-2 acts as a Tie-2 agonist in the former setting, but as an antagonist in the latter setting. Interestingly, LEC have been shown to express Ang-2, and it has been suggested that the lack of pericytes/support cells in lymphatic microvessels could be related to this LEC Ang-2 expression.

Junctions and Solute Barrier Function in Blood and Lymphatic Vessels

BECs express well-developed adherens and tight junctions that contain both cadherin-5 (vascular endothelial (VE) cadherin) and N- and P-cadherins. VE-cadherin appears to be expressed by both BECs and LECs. While BEC maintain conventional adherens and tight junctions at endothelial borders that regulate solute exchange, LEC junctions form a specialized type of adherens junction termed the *complexus adherentes*, that contains high levels of plakophilin-2, a desmosomal protein found in nonclassical adherens junctions. Besides plakophilin-2, cadherin-13 and zona occludens 2 (ZO-2) also appear to be elements of the LEC complexus adherentes junction. Desmoplakins have been reported in LECs, but have also been reported in venous BECs and may be an unreliable marker for discriminating BEC from LEC [6]. Despite the proposition that LECs lack the same level of junctional integrity as BECs, podoplanin found within the overlapping cell borders of LEC may help create the LEC capillary "valves," which support the opening and closing of these vessels without the need to remodel tight or adherens junctions [1]. Junctional adhesion molecule-2 (JAM-2) is a junctional component within the CTX [cortical thymocyte Xenopus (CTX) molecular family] that is shared by both high endothelial venule EC and LEC. JAM-2 supports lymphocyte extravasation from the blood vascular space [9], but its function in lymphatics is not clear. Most likely JAM-2 expressed on HEV and lymphatics plays a role in lymphocyte targeting in that JAM-2 on these endothelium binds to JAM-3 expressed on T, NK, and dendritic cells. Unlike JAM-2, JAM-1 appears to be associated with tight junctions in conjunction with occludin and claudins.

Endothelial-Cell Adhesion Molecules (ECAMs) in BECs versus LECs

Both LECs and BECs express several types of adhesion molecules including those in the Ig-superfamily and selectins family.

BECs basally express CD54/ICAM-1 and CD102/ICAM-2 as well as CD31/PECAM. BECs also increase their expression of several adhesion molecules, such as VCAM-1, ICAM-1, and E- and P-selectin, after stimulation with several proinflammatory cytokines (TNF-α, IL-1β) or lipopolysaccharide in a tissue- and anatomy-specific manner. Unlike BECs, LEC may not densely express ICAM-1 and -2. LEC expression of ICAM-1 may also be anatomy specific; lymphatic and submandibular lymph node capillaries only weakly express ICAM-1, and ICAM-1 is among the BEC transcripts that are suppressed by the LEC fate determinant Prox-1. In cancer, ICAM-1 is however elevated in peritumoral lymphatics and involved lymph nodes.

PECAM-1 is constitutively expressed by most LECs, BECs, and HEV, but generally, PECAM-1 expression in LECs also appears lower than in other endothelial types.

VCAM-1 is not expressed basally, by either BEC or LEC but VCAM-1 (and ICAM-1, ICAM-3) are expressed by LEC during active inflammation (especially gut inflammation), in response to inflammatory cytokines. E-selectin is also mobilized in BEC during inflammation, and some tissues (e.g., tongue) even show constitutive LEC expression of E-selectin.

CD34 is a 90- to 120-kDa cell surface sialomucin expressed by BECs, but not by LECs. CD34 is expressed on early lymphoid hematopoietic stem and progenitors, small-vessel endothelial cells, embryonic fibroblasts, some fetal cells, and in adult nervous tissue. The function of CD34 is probably to control interactions of BEC with lymphocytes via L-selectin in high endothelial venules.

α9-Integrin also appears to participate in the development of lymphatic vessels, since α9-integrin knockout mice die from respiratory failure/chylothorax resulting from a failure in normal lymphatic development. α9-Integrin is densely expressed on several cell types including smooth muscle, epithelia, and neutrophils. α9-Integrin is also a ligand for VCAM-1, which may play a role in lymphocyte trafficking in forms of filariasis, an infestation of the lymphatic system. Although α9-integrin appears to play an important regulatory role in the maturation of the lymphatic system, it is not clear if it is expressed by LECs, or if it is involved in signaling that supports lymphatic maturation.

Chemokines Released by LECs and BECs

LECs and BECs express several chemokines that modulate leukocyte adhesion, migration, and inflammation. LECs apparently release several CXC, CC, and C chemokines. LECs express mRNAs for KC, IP10, Mig-1, BCL, MIP-2, SLC, RANTES, MCP-1, C10, and Lptn, but not MIP-1α,β or γ. The release of C-10 is especially unusual for LEC since C-10 expression had long been thought to be limited to cells of the hematopoietic lineage. LECs, but not BECs, constitutively secrete the CC chemokine receptor CCR-7 ligand secondary lymph node–associated chemokine (SLC)/CCL21 at their basolateral surface, and secretion is suppressed by TNF-α and IL-1β.

Lymphatic Development

Embryologically, lymphatics are derived from endothelial progenitors that also give rise to venous endothelium under the influence of several growth factors, and their receptors that are differentially expressed during development. Lymphatic develop as off-"buds" derived from the venous system and was originally described in the classical studies by Florence Sabin. The first step in progenitor commitment to lymphatic phenotype is induction of the lymphatic specific transcription factor, Prox-1, followed by the release of secondary lymphoid chemokines (SLCs). LECs expressing these markers may further increase VEGF-R3 expression [10], which increases their sensitivity to VEGF-C/D and helps maintain their commitment to the LEC lineage.

This second step in lymphatic development appears to require signaling through the receptor tyrosine kinase Syk and its substrate, SLP-76 [11]. Signals from the Syk/SLP-76 system appear to be crucial for maintaining the separation of the blood and lymphatic systems, since deletion of either gene leads to anastomosis of blood and lymphatic vessels. It is interesting that while Syk/SLP-76 are certainly determinants of lymphatic development, it is not clear that they are proteins that are expressed by endothelium. It has therefore been suggested that Syk/SLP-76 could (1) activate lymphoid progenitor cells (which lose Syk/SLP-76 during maturation), (2) modulate the release of IL-2, or (3) be related to platelet degranulation, which is affected by Syk/SLP-76 [12].

In the adult, venous endothelial cells in lymphatic buds express VEGF-R3, which in endothelium is generally specific for lymphatic endothelial cells, indicating that some venous endothelial cells retain an intermediate phenotype, at least within lymphatic buds. In the process of lymphatic budding, the lymphatic specific transcription factor, Prox-1, and Syk/SLP-76 must be activated for these cells to mature into lymphatics, which express VEGF-C/VEGF-R3 signaling, LYVE-1, and podoplanin.

Lymphatics in Disease: Lymphedema and Cancer

Cancer

The role of lymphatics in cancer is complex. Although functional lymphatics in tumors have long been disputed, immunochemical analysis of lymphatic EC specific markers (e.g., VEGF-R3) can show lymphatics at the periphery of solid tumors. The functional significance of these vessels may be less effective for the nutrition/drainage of tumors but rather important as conduits for the intravasation/metastasis of tumors in cancer. There is a very strong clinical correlation between expression of lymphatic growth factors/receptors, metastasis, and a poor prognosis in several types of malignancies.

Lymphatic vessels are known to play an essential role in the expansion and dissemination of several forms of cancer. Whereas malignancies of mesenchymal origin frequently spread via the blood vascular system, carcinomas more often migrate and pass through the lymphatic system. Many models now describe a near absence of metastasis in animals devoid of or deficient in lymphatic vessels and establish that many forms of cancer, most notably breast cancer, are lymphatic dependent. Many tumors (e.g., gastric carcinoma and pancreatic and invasive lung adenocarcinomas) secrete lymphatic specific growth factors such as VEGF-C and -D and represent significant risk factors for tumor progression. These studies showed a high risk associated with VEGF-C expression in tumor status. Besides its role in solid tumor progression, Flt-4 may also play roles in leukemia as well, since several leukemia cell lines express Flt-4. Furthermore, findings that many lymphatic markers such as VEGF-R3 and podoplanin are detectable on KS lesions have prompted consideration of KS as a lymphoproliferative disorder, and it may be possible that VEGF-R3 antagonists could be useful treatments for several types of cancer and KS.

Lymphedema

Lymphedema is a debilitating and disfiguring swelling of tissues caused by insufficient lymphatic drainage. Lymphedema is divided into primary (or hereditary lymphedema), and several forms of secondary lymphedema. The genetic basis of primary lymphedema is now thought to reflect alterations in several genes that control the development of the lymphatic system including VEGFs C and D, Prox-1, FOXC2 (forkhead transcription factor), and VEGF-R3 [5]. Some forms of primary lymphedema reveal mutations in the VEGF-R3 receptor (Flt-4), particularly within the 5q35 locus. These mutations diminish Flt-4 kinase activity and also appear stabilize the receptor against proteolysis.

Secondary lymphedema generally represents environmental, parasitic, or surgical blockade or damage to the lymphatics. In the setting of cancer, secondary lymphedema often results from the surgical removal of lymph nodes and vessels leading to poor tissue drainage. Adjuvant radiation treatments will typically worsen this condition by further limiting the regrowth of lymphatic vessels and decreasing lymphatic drainage. Filariasis is an infestation of lymphatics by parasites and a significant cause of lymphedema, especially in tropical regions. Filariasis blocks lymphatics and injures the lymphatics, leading to profound forms of lymphedema. We have already described the role of lymphoproliferation in cancer and lymphedema; however, lymphoproliferation may participate in several other diseases as well. For example, there may be an important role of lymphatic vessels in many chronic inflammatory syndromes including inflammatory bowel disease, arthritis, and lupus (SLE).

Therapeutic Manipulation of Lymphatics?

Gene therapeutic adenoviral constructs that encode VEGFs A, B, C, and D and pharmacological antagonists for VEGF-Rs have recently been used to enhance/regress both blood and lymphatics through their effects on BEC and LEC growth and survival, and both may be therapeutic under different circumstances. Although VEGF-C/D might promote the spread of cancer cells to distant nodes, the induction of VEGF-C or -D may be beneficial in lymphedema by creating or restoring lymphatic drainage and relieving tissue injury caused by edema. As the several key growth factors, receptors, and signaling systems that induce, maintain, and reduce lymphatics become increasingly better understood, they hold out enormous promise for (1) creating/enhancing the lymphatics in primary lymphedema and (2) the reestablishment of lymphatics after surgery, cancer, or radiation therapy.

Summary

The molecular mechanisms that create, separate, and regulate the lymphatic and blood vascular systems have important implications for our understanding of the development of the vascular system, events in some forms of vascular disease, and possible new avenues for therapy for several diseases. Specific and selective markers (LYVE-1, VEGF-R3, Prox-1) now exist that can distinguish LEC from BEC differentiation at several levels, particularly the genetic and molecular levels and permit the separation of these different populations for examination of their properties. Several of these new structural and functional LEC markers permit selective targeting of LECs, and examination of their unique roles in development and their contributions to several important pathologies. These studies on LEC markers have in only a few years already provided remarkably important new evidence for the roles of lymphatics in several diseases including cancer, lymphedema, wound healing, and inflammation. These results promise to provide additional information on the roles of lymphatics in many other diseases of the immune system, development, and iatrogenic complications. Most importantly, these markers not only provide prognostic signs of specific features of disease, but also provide new mechanisms for disease processes, and consequently important new targets for prophylactic and therapeutic intervention.

Glossary

Chemokines: Soluble polypeptide factors secreted by immune system cells that stimulate some activity of other cells. Chemokines are often chemoattractants and act as tertiary messengers between cells.

Lymphedema: Swelling in the extremities caused by interstitial fluid buildup in tissues due to congenital malformation of lymphatics (primary) or surgical/traumatic (secondary) loss of lymphatic drainage (e.g., when nodes are removed or blocked).

VEGFs: A family of structurally related proteins that stimulate growth of blood and lymphatic endothelial cells and several other cell types. VEGF actions are stimulated through binding to VEGF receptors VEGFRs 1, 2, and 3. Placental growth factor (PlGF) and VEGF-B bind primarily to VEGFR-1. PlGF modulates angiogenesis and may participate in inflammation. VEGF-C and VEGF-D bind primarily to VEGFR-3 and stimulate lymphangiogenesis rather than angiogenesis.

References

1. Sleeman, J. P., Krishnan, J., Kirkin, V., and Baumann, P. (2002). Markers for the lymphatic endothelium: in search of the holy grail? *Microsc. Res. Tech.* **55**(2), 61–69.
2. Oliver, G., and Detmar, M. (2002). The rediscovery of the lymphatic system: Old and new insights into the development and biological function of the lymphatic vasculature. *Genes Dev.* **16**(7), 773–783.
3. Petrova, T. V., Makinen, T., Makela, T. P., Saarela, J., Virtanen, I., Ferrell, R. E., Finegold, D. N., Kerjaschki, D., Yla-Herttuala, S., and Alitalo, K. (2002). Lymphatic endothelial reprogramming of vascular endothelial cells by the Prox-1 homeobox transcription factor. *EMBO J.* **21**(17), 4593–4599.
4. Banerji, S., Ni, J., Wang, S. X., Clasper, S., Su, J., Tammi, R., Jones, M., and Jackson, D. G. (1999). LYVE-1, a new homologue of the CD44 glycoprotein, is a lymph-specific receptor for hyaluronan. *J. Cell Biol.* **144**, 789–801.
5. Baldwin, M. E., Stacker, S. A., and Achen, M. G. (2002). Molecular control of lymphangiogenesis. *Bioessays* **24**(11), 1030–1040.
6. Sleeman, J. P., Krishnan, J., Kirkin, V., and Baumann, P. (2001). Markers for the lymphatic endothelium: In search of the holy grail? *Microsc. Res. Tech.* **55**(2), 61–69.
7. Rissanen, T. T., Markkanen, J. E., Gruchala, M., Heikura, T., Puranen, A., Kettunen, M. I., Kholova, I., Kauppinen, R. A., Achen, M. G., Stacker, S. A., Alitalo, K., and Yla-Herttuala, S. (2003). VEGF-D is the strongest angiogenic and lymphangiogenic effector among VEGFs delivered into skeletal muscle via adenoviruses. *Circ Res.* **92**(10), 1098–1106.
8. Baldwin, M. E., Catimel, B., Nice, E. C., Roufail, S., Hall, N. E., Stenvers, K. L., Karkkainen, M. J., Alitalo, K., Stacker, S. A., and Achen, M. G. (2001). The specificity of receptor binding by vascular endothelial growth factor-d is different in mouse and man. *J. Biol. Chem.* **276**(22), 19166–19171.
9. Johnson-Leger, C. A., Aurrand-Lions, M., Beltraminelli, N., Fasel, N., Imhof, B. A. (2002). Junctional adhesion molecule-2 (JAM-2) pro-

motes lymphocyte transendothelial migration. *Blood* **100**(7), 2479–2486.

10. Kriehuber, E., Breiteneder-Geleff, S., Groeger, M., Soleiman, A., Schoppmann, S. F., Stingl, G., Kerjaschki, D., and Maurer, D. (2001). Isolation and characterization of dermal lymphatic and blood endothelial cells reveal stable and functionally specialized cell lineages. *J. Exp Med.* **194**(6), 797–808.
11. Abtahian, F., Guerriero, A., Sebzda, E., Lu, M. M., Zhou, R., Mocsai, A., Myers, E. E., Huang, B., Jackson, D. G., Ferrari, V. A., Tybulewicz, V., Lowell, C. A., Lepore, J. J., Koretzky, G. A., and Kahn, M. L. (2003). Regulation of blood and lymphatic vascular separation by signaling proteins SLP-76 and Syk. *Science* **299**(5604), 247–251.
12. Jain R. K., and Padera, T. P. (2003). Development. Lymphatics make the break. *Science* **299**(5604), 209–210.

Further Reading

Aurrand-Lions, M., Johnson-Leger, C., Wong, C., Du Pasquier, L., and Imhof, B. A. (2001). Heterogeneity of endothelial junctions is reflected by differential expression and specific subcellular localization of the three JAM family members. *Blood* **98**(13), 3699–3707.

Breiteneder-Geleff, S., Matsui, K., Soleiman, A., Meraner, P., Poczewski, H., Kalt, R., Schaffner, G., and Kerjaschki, D. (1997). Podoplanin, novel 43-kd membrane protein of glomerular epithelial cells, is down-regulated in puromycin nephrosis. *Am. J. Pathol.* **151**(4), 1141–1152.

Gumkowski, F., Kaminska, G., Kaminski, M., Morrissey, L. W., and Auerbach, R. (1987). Heterogeneity of mouse vascular endothelium. In vitro studies of lymphatic, large blood vessel and microvascular endothelial cells. *Blood Vessels* **24**(1–2), 11–23.

He, Y., Karpanen, T., and Alitalo, K. (2004). Role of lymphangiogenic factors in tumor metastasis. *Biochim. Biophys. Acta* **1654**(1), 3–12. *This review addresses the current understanding of how tumor cells penetrate the lymphatics to reach lymph nodes from primary tumor sites and summarizes recent findings in tumor/lymphatic dissemination.*

Podgrabinska, S., Braun, P., Velasco, P., Kloos, B., Pepper, M. S., and Skobe, M. Molecular characterization of lymphatic endothelial cells. *Proc. Natl. Acad. Sci. USA* **99**(25), 16069–16074.

Sabin, F. R. (2002). Preliminary note on the differentiation of angioblasts and the method by which they produce blood-vessels, blood-plasma, and red blood-cells as seen in the living chick. *J. Hematother. Stem Cell Res.* **11**(1), 5–7. *This paper is a reprinting of some of Dr. Rena Sabin's classical findings on the development of the vascular system from The Anatomical Record* ***13****, 199–204 (1917).*

Scavelli, C., Vacca, A., Di Pietro, G., Dammacco, F., and Ribatti, D. (2004). Crosstalk between angiogenesis and lymphangiogenesis in tumor progression. *Leukemia* **18**(6), 1054–1058. *This brief review considers chemical crosstalk between elements of the blood and lymphatic vessels in the context of cancer progression.*

Veikkola, T., Lohela, M., Ikenberg, K., Makinen, T., Korff, T., Saaristo, A., Petrova, T., Jeltsch, M., Augustin, H. G., and Alitalo, K. (2003). Intrinsic versus microenvironmental regulation of lymphatic endothelial cell phenotype and function. *FASEB J.* **17**(14), 2006–2013. *This manuscript describes diverse differences in genes activated by growth factor receptors in blood and lymphatic endothelium and how their responses are affected by other cells in the vasculature.*

Capsule Biography

Dr. Alexander has been a member of the department of Molecular and Cellular Physiology in Shreveport, LA. since 1993 and the codirector of the cell biology core, and the director of research for the department of Gastroenterology. His laboratory primarily focuses on the roles of endothelial cells in chronic inflammation, cancer, and cardiovascular disease. The NIH has supported his work since 1993.

CHAPTER 80

Lymphocyte Interaction with Microvascular Endothelium

Sirpa Jalkanen and Marko Salmi
University of Turku and National Public Health Institute, Turku, Finland

Physiological Lymphocyte Recirculation

A subset of vascular endothelial cells in peripheral lymph nodes and in organized lymphatic tissues in the gut (Peyer's patches and appendix) have differentiated to optimally support continuous lymphocyte trafficking into these tissues. They do not mediate extravasation of granulocytes in normal conditions. These vessels are postcapillary venules of the vascular tree and their endothelial cells are morphologically plump. To discriminate them from the flat-walled venules present in all organs, they are called high endothelial venules (HEV). The expression patterns of homing-associated molecules on HEV in peripheral lymph nodes and mucosa-associated lymphatic tissues differ, resulting in functional differences in the entry of lymphocyte populations into these two tissues. Lymphocytes also efficiently enter the spleen, but the mechanisms and molecules and even the contribution of vascular endothelium is not clear in lymphocyte homing into this organ. Vessels in nonlymphoid tissues support lymphocyte entrance only at a very low level and therefore, only occasional lymphocytes in search of microbes can be found in these tissues.

HEV are the most important site for lymphocyte entrance to organized lymphoid tissues. However, about 15 percent of lymphocytes enter the lymph nodes via afferent lymphatics. Lymphocytes leave the nodes via efferent lymphatic vessels. The vascular network guiding lymphocyte recirculation is schematically presented in Figure 1. Endothelium in lymphatics differs morphologically from endothelium in blood vessels, since lymphatic endothelium is flat and discontinuous. Although both of these two types of endothelia interact with lymphocytes, they have largely different molecular phenotypes.

Inflammation-Induced Changes

Inflammatory mediators induce molecular and morphological changes in endothelium, optimizing its capacity to efficiently mediate entry of different leukocyte populations to sites of inflammation. Several homing-associated molecules not normally present on flat endothelium appear on the surface of these vessels upon inflammation. In chronic inflammations the flat endothelial vessels morphologically transform to HEV-like vessels. Different leukocyte populations sequentially accumulate at the site of inflammation, lymphocytes being the last cell type to enter. Their main wave of entry is 2 to 3 days after the initiation of inflammation. In extreme chronic cases a nonlymphatic tissue can in fact start to resemble organized lymphatic tissues.

Molecular Events in Lymphocyte–Endothelial Cell Interaction

Lymphocyte interaction with endothelium is a dynamic process that takes place in a multistep fashion involving binding between several molecular pairs on lymphocyte and endothelial cell surface. The adhesive steps can be divided to the following phases: (1) tethering and rolling, (2) activation, (3) firm adhesion, and (4) transmigration. These steps are schematically depicted in Figure 2. All these steps have to be appropriately executed before a lymphocyte can enter the tissue. Molecular mechanisms mediating these steps have been traditionally divided into unique categories. However, depending on the hemodynamic conditions and on the vascular bed where the interaction takes place, one receptor–ligand pair can be involved in other phases as well.

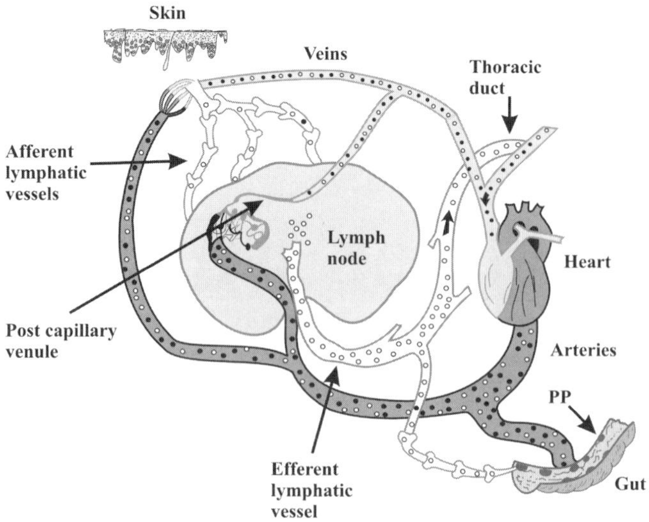

Figure 1 Vascular network guiding lymphocyte recirculation. Transport of foreign material (pieces of microbes) into lymphoid organs takes place, for example, through the epithelium of the gut and via afferent lymphatics draining the skin. Bloodborne lymphocytes enter the organized lymphatic tissues (lymph nodes and Peyer's patches; PP) from the blood via the arterial tree, flow through the capillaries, and extravasate in the postcapillary high endothelial venules. Thereafter, the lymphocytes migrate through the tissue parenchyma in search of their cognate antigens. If the antigen is found, the lymphocytes get activated within the germinal centers (B cells) or outside the centers (T cells). Activated cells and cells that did not find their antigen leave the node by entering the lymphatic vessels and return via the efferent lymphatics back to the blood circulation. Activated cells then preferentially leave the blood in peripheral tissues, where the microbial insult took place, whereas the nonactivated lymphocytes continue their journey between blood and lymphoid organs. (see color insert)

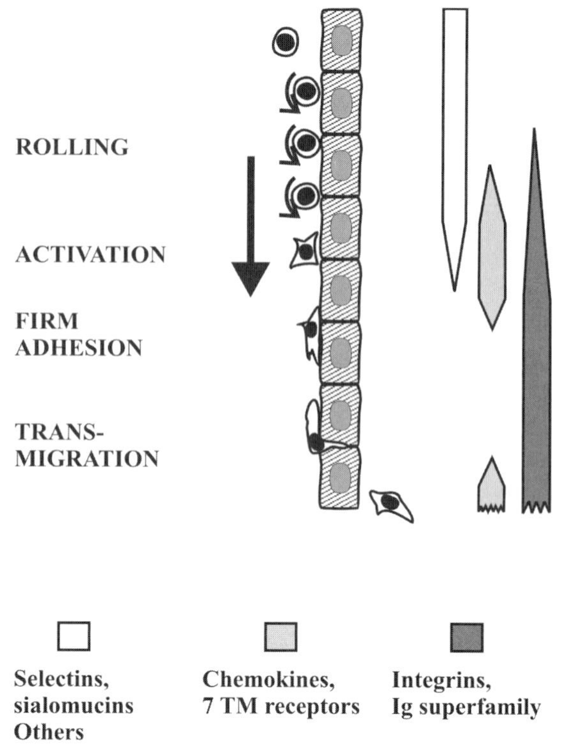

Figure 2 The multistep cascade of lymphocyte extravasation. The bloodborne lymphocyte first makes transient and weak contacts with endothelial cells (rolling) using selectins and sialomucins. If the lymphocyte has the right chemokine receptor to interact with a chemokine presented by the endothelium, this molecular interaction leads to activation of integrins. Endothelial immunoglobulin superfamily counterparts bind to activated integrins, which results in firm adhesion of the lymphocyte to the vessel wall. Thereafter, the lymphocyte transmigrates into the tissues between or through the endothelial cells.

The key molecular interplayers at each step are described in the following and summarized in Figure 3.

Tethering and Rolling

Freely flowing lymphocytes in the blood randomly make collisions with the vessel wall. They form transient and weak contacts with endothelium, which results in slowing of their velocity. This initial interaction is predominantly mediated by selectins and their sialomucin ligands. A typical feature in this interaction is that it involves binding of the lectin domain of a selectin to carbohydrates on sialomucin molecules. There are three members in the selectin family. They are L-selectin (CD62L) expressed on leukocytes, E-selectin (CD62E) present on endothelium, and P-selectin (CD62P) present on both platelets and endothelium.

The counter-receptors of L-selectin are sulfated, fucosylated, and sialylated carbohydrates (sialyl Lewis X-like structures) present on several protein backbones [for example, on glycosylation-dependent cell adhesion molecule-1 (GlyCAM-1), CD34, podocalyxin, and even on a subset of mucosal addressin cell adhesion molecule-1 (MAdCAM-1) in Peyer's patches]. These molecules are collectively called peripheral lymph node addressins (PNAd). HEV in peripheral lymph nodes strongly express PNAd and thus guide L-selectin expressing lymphocytes efficiently into the nodes.

E-selectin is one of the inducible molecules found at sites of inflammation. Its expression becomes prominent within few hours after initiation of inflammation. E-selectin mediates lymphocyte entry, especially into inflamed skin. Cutaneous lymphocyte antigen CLA (a variant of P-selectin glycoprotein ligand, PSGL-1) serves as the lymphocyte ligand for E-selectin in the skin. In addition, E-selectin is also able to bind E-selectin ligand-1 (ESL-1) on leukocyte surface.

P-selectin is translocated within minutes from specific granules (Weibel-Palade bodies) to the surface of endothelium. Although it mainly binds granulocytes at an early phase of inflammation, it is also present in chronic inflammations due to new protein synthesis and binds to PSGL-1 on lymphocytes.

Endothelial cells at mucosa-associated lymphatic tissues have a unique phenotype, because they, unlike other normal endothelium in the body, express MAdCAM-1. MAdCAM-1 binds $\alpha 4\beta 7$ integrin on lymphocytes and selectively mediates traffic of the $\alpha 4\beta 7$ positive lymphocytes to mucosal lymphoid tissues. MAdCAM-1 also contributes to the firm adhesion step.

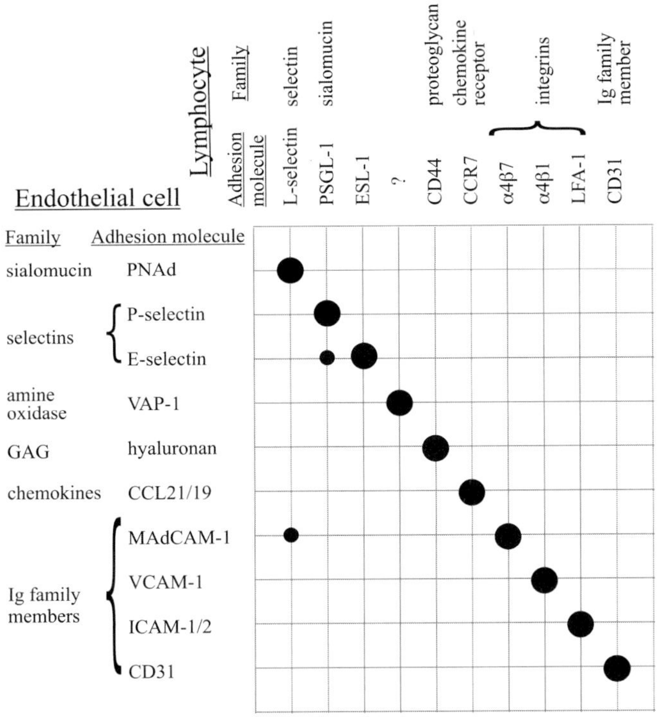

Figure 3 Molecules involved in lymphocyte extravasation. The most relevant endothelial glycoproteins mediating lymphocyte–endothelial cell interactions and their best characterized lymphocyte receptors are shown as receptor–ligand pairs. Their molecular families are also indicated, as well as the relative contribution of each receptor–ligand pair (the size of the black dot) in lymphocyte–endothelial cell interaction. (In addition to the molecules presented in this figure, there are individual reports describing involvement of many other molecules in the extravasation process). Ig, immunoglobulin; GAG, glucosaminoglycan.

Other molecules, such as vascular adhesion protein-1 (VAP-1) and hyaluronan presented on endothelium, are also involved in the lymphocyte rolling step, especially in inflammatory conditions. VAP-1 is an endothelial adhesin that also possess enzymatic activity catalyzing oxidation of amines to hydrogen peroxidase, aldehyde, and ammonium. These end products regulate the functional status of endothelial cells. Hyaluronan, on the other hand, serves as an endothelial cell counterstructure for lymphocyte CD44.

Activation

Chemokines present on endothelial cells and their receptors on lymphocytes are thought to be the key players in the activation step. Their interaction leads to activation of integrins. Typically chemokines are small soluble molecules that need to be presented by endothelial proteoglycans to be available for rolling lymphocytes. Chemokines can be classified structurally into four groups: CXC, CC, CX3C, and C (C is cysteine and X can be any amino acid). The most important chemokine presented by HEV-like vessels is CCL21. Also CCL19 is involved in activating lymphocytes binding to HEV. Both bind CCR7 on lymphocytes. Endothelial cells in different vascular beds have certain degree of tissue specificity regarding chemokine expression. One example is the skin, where endothelial CCL17 attracts CCR4 bearing lymphocytes.

Firm Adhesion

Members of the immunoglobulin superfamily, intercellular adhesion molecules (ICAM-1, CD54, and ICAM-2, CD102) and vascular adhesion molecule (VCAM-1, CD106) on vascular endothelium serve as ligands for activated integrins on lymphocytes. ICAM-1 and ICAM-2 bind to lymphocyte activation associated antigen (LFA-1, CD11a/CD18) and VCAM-1 adheres to α4β1 (CD49d/CD29) on lymphocytes. Expression of ICAM-1 and VCAM-1 is upregulated at sites of inflammation, whereas synthesis of ICAM-2 is not influenced by inflammatory mediators. ICAM-1, ICAM-2, and VCAM-1 do not confer tissue specificity in the homing process.

Transmigration

In successful cases firm adhesion leads to the transmigration phase. The active contribution of endothelial cells at this step can be seen as rapid changes in their cytoskeletal organization, which are controlled, for example, by small GTPases. It is still under debate whether lymphocytes transmigrate through the endothelial cell or whether they penetrate through intercellular junctions.

Molecules preferentially localized at the junction, such as, CD31 and junctional adhesion molecules A, B and C (JAM-A, B and C), are involved in lymphocyte transmigration. Also ICAM-1/LFA-1 and VCAM-1/α4β1 pairs as well as VAP-1 can contribute to this process. It is obvious that a transmigrating lymphocyte needs to use proteolytic mechanisms to get through the vessel wall. Those as well as rapid repair mechanisms required to close the transmigration path remain to be elucidated.

Importance of the Endothelial Homing-Associated Molecules

Modern technology has provided us with the ability to produce mice in which one or more endothelial adhesion molecule(s) have been deleted. Phenotypes of the knockout mice lacking ICAM-1, ICAM-2, P-selectin, E-selectin, CD31, VCAM-1, and carbohydrate decorations of PNAd have been reported. All these mice (excluding VCAM-1 deficient mice) show variable defects either in normal lymphocyte homing or leukocyte trafficking to sites of inflammation or in both of these functions. Mice lacking VCAM-1, in contrast, are embryonic lethal because of the essential role of VCAM-1 in formation of umbilical cord and placenta. In general, the mice lacking only one endothelial homing-associated molecule do not show very severe phenotype. However, when more molecules are simultaneously deleted, the mice are no longer able to respond efficiently to inflammatory stimuli. This illustrates the marked redundancy in the function of the endothelial homing-associated molecules.

Endothelial Homing-Associated Molecules as Drug Targets

As endothelial cells in different vascular beds are at a key position to control lymphocyte accumulation in the inflamed tissues, homing-associated molecules on endothelial cells are potential targets when new anti-inflammatory drugs are developed. All homing-associated endothelial molecules mentioned in this chapter have been targeted with promising results in in vivo animal studies utilizing a variety of inflammatory models. Unfortunately, in clinical trials many of them have not fulfilled the expectations set based on the animal studies. However, many of them are still being tested in clinical trials. The excellent results obtained by targeting α4-integrin (present mainly on lymphocytes) in multiple sclerosis suggest that by targeting endothelial molecules equally good efficacy may be obtained in harmful inflammations.

Summary

Endothelial cells at different vascular beds are important in regulating lymphocyte entrance into the organs. Therefore, they are at a key position determining what types of immune reactions take place in different tissues of the body. Adhesion molecules present on the endothelium serve as counter-receptors for homing receptors expressed on circulating lymphocytes. The molecular signature of vascular endothelial cells in different tissues vary, and only those lymphocytes having the right repertoire of homing receptors for the given endothelial counter-molecules are able to enter a particular tissue. Inflammation induces and upregulates several homing-associated adhesion molecules on endothelium, allowing extensive traffic of massive number of lymphocytes into the tissues. Accumulation of lymphocytes can cause devastating symptoms in harmful inflammations as for example in diabetes, rheumatoid arthritis, and inflammatory bowel diseases. Therefore, homing-associated molecules at vascular endothelium are potential targets for drug development aiming at blocking abnormal lymphocyte trafficking.

Glossary

Chemokines: Small attractant molecules.

Extravasation: The process in which a cell migrates from the vasculature into the tissue site.

High endothelial venules: Specialized postcapillary vessels mediating lymphocyte entrance into the lymphoid organs.

Homing: Migration of recirculating lymphocytes from the bloodstream to particular lymphoid sites.

Lymphocyte recirculation: Continuous trafficking of a subset of white blood cells between the blood and lymphoid organs.

Further Reading

Butcher, E. C. (1991). Leukocyte–endothelial cell recognition: Three (or more) steps to specificity and diversity. *Cell* **67**, 1033–1036. *This classical review introduces the concept of the multistep adhesion cascade in leukocyte–endothelial cell interaction.*

Gowans, J. L., and Knight, E. J. (1964). The route of re-circulation of lymphocytes in rat. *Proceedings of the Royal Society of London.* **159**, 257–282. *This is the pioneering work that demonstrated the recirculation routes of lymphocytes and is considered as the beginning of the modern lymphocyte homing field.*

Hogg, N., Henderson, R., Leitinger, B., McDowall, A., Porter, J., and Stanley, P. (2002). Mechanisms contributing to the activity of integrins on leukocytes. *Immunological Reviews* **186**, 164–171.

Johnston, B., and Butcher, E. C. (2002). Chemokines in rapid leukocyte adhesion triggering and migration. *Seminars in. Immunology* **14**, 83–92.

Lowe, J. B. (2003). Glycosylation in the control of selectin counter-receptor structure and function. *Immunological Reviews* **186**, 19–36.

Miller, D. H., Khan, O. A., Sheremata, W. A., Blumhardt, L. D., Rice, G. P., Libonati, M. A., Willmer-Hulme, A. J., Dalton, C. M., Miszkiel, K. A., and O'Connor, P. W. (2003). International Natalizumab Multiple Sclerosis Trial Group. A controlled trial of natalizumab for relapsing multiple sclerosis. *New England Journal of Medicine* **348**, 15–23. *The first clinical breakthrough using antibody therapy against an adhesion molecule.*

Muller, W. A. (2003). Leukocyte–endothelial-cell interactions in leukocyte transmigration and the inflammatory response. *Trends in Immunology.* **24**, 327–334, 2003.

Salmi, M., and Jalkanen, S. (1997). How do lymphocytes know where to go: Current concepts and enigmas of lymphocyte homing. *Advances in. Immunology* **64**, 139–218. *A comprehensive review of the history and progress in the field of lymphocyte—endothelial cell interactions.*

http://www.ncbi.nlm.nih.gov/PROW/guide/45277084.htm *A data bank and systematic nomenclature for molecules (CD classification).*

http://cytokine.medic.kumamoto-u.ac.jp/ *A data bank for chemokines.*

Capsule Biography

Sirpa Jalkanen, M.D., Ph.D., is Professor of the Finnish Academy and Professor of Immunology at University of Turku.

Marko Salmi, M.D., Ph.D., is Laboratory Director (Cell Trafficking) at National Public Health Institute in Turku. They have long experience in the field of leukocyte trafficking.

They lead the National Center of Excellence called Cell Trafficking that studies different aspects of cell migration in physiological and pathological conditions such as in harmful inflammations and cancer spread.

CHAPTER 81

The Lymphatic System

Michael S. Pepper
NetCare Molecular Medicine Institute, Unitas Hospital, Lyttelton, Pretoria, South Africa

The lymphatic vascular system regulates tissue fluid balance and pressure, facilitates interstitial protein transport, and serves immunological functions by regulating the movement of cells between tissue interstitium and lymph nodes. Lymphatics are also responsible for absorption of fat from the gut. Given its central role in maintaining normal tissue homeostasis, it is indeed surprising that the lymphatic vascular system has until recently been very poorly characterized from a molecular point of view. The recent surge of interest in this field can be attributed principally to two factors: the discovery of growth factors that induce the formation of new lymphatic capillaries (lymphangiogenesis), and the identification of molecular markers that allow one to differentiate between blood vascular and lymphatic endothelium. This in turn has led to the establishment of techniques for the isolation of relatively pure populations of blood vascular and lymphatic endothelial cell (BEC and LEC) populations.

Growth and Structure of the Lymphatic Tree

Work on the lymphatic system began in the 17th century, and by the beginning of the 19th century, the anatomy of most of the lymphatic system had been described. Much of the work carried out late in the 19th century and in the first half of the last century was aimed at determining the embryonic origin of lymphatic endothelium. Two theories were proposed. The first suggested that lymphatic endothelium derives by sprouting from blood vascular (venous) endothelium, the so-called centrifugal theory. The second, the so-called centripetal theory, suggested that lymphatic endothelium differentiates in situ from primitive mesenchyme, and secondarily acquires connections with the blood vascular system at a limited number of sites. More recent studies have provided support for both hypotheses. Thus, the homeobox transcription factor Prox-1, which is a master switch for inducing the lymphatic phenotype, is expressed at sites of budding of new lymphatics from veins during development. There is also evidence for the existence of lymphangioblasts that give rise to the de novo differentiation of lymphatic endothelial cells in tissue interstitium during development.

In postnatal life, new lymphatic capillaries grow by lymphangiogenesis during acute and chronic inflammation. This occurs by sprouting from preexisting lymphatic capillaries in much in the same way as new blood capillaries arise by sprouting from preexisting blood capillaries or postcapillary venules during angiogenesis. The appearance of new lymphatic capillaries is secondary to that of blood capillaries, although linear growth occurs at a comparable speed. Lymphatic capillaries are less labile than blood capillaries: They send out fewer sprouts, anastomose less frequently, and show much less tendency to retract or undergo changes in size or form. Although temporary lymphedema occurs following lymphatic disruption, this usually resolves due to spontaneous regeneration or reconnection of lymphatics. In resolving inflammation, newly formed lymphatic capillaries tend to involute more rapidly than do blood capillaries.

The lymphatic vascular system originates in the interstitial space of virtually all tissues of the body as a series of blind-ending thin-walled capillaries (Figure 1). Lymphatic capillaries lack a continuous basement membrane and pericytes, but contain anchoring filaments that aid in maintaining luminal patency in the face of increased interstitial pressure (Table 1). Tissues that are purported to be devoid of lymphatic vessels include, among others, the central nervous system, placenta, bone, and islets of Langerhans. Careful immunohistochemical studies with the newly described lymphatic markers (see later discussion) will be required to carefully map the distribution of lymphatics throughout the body. Lymphatic capillaries drain into larger collecting lymphatics that contain valves. Collecting lymphatics also contain smooth muscle. Lymph nodes are interposed along the

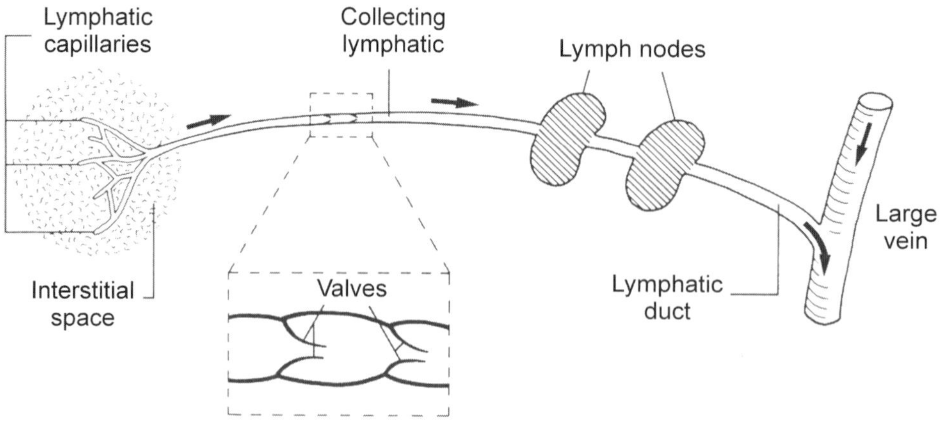

Figure 1 Schematic representation of the lymphatic vascular tree.

path of collecting lymphatics. Segments of collecting lymphatics between valves are called lymphangions. Collecting lymphatics progressively converge and ultimately end up as two large trunks, the thoracic duct and the right lymphatic duct, which anastomose with the large veins at the base of the neck (Figure 1). Lymphatic ducts are reinforced by smooth muscle cells, and the adventitia of large lymphatic ducts contains vasa vasorum and a rich neural network. Unlike the blood vascular system, in which flow is ensured by the rhythmic contractile activity of the heart, flow of lymph within the lymphatic tree is aided initially by external mechanical forces (contraction of surrounding skeletal muscle) and later by the contraction of smooth muscle cells in the walls of larger lymphatics. Unidirectional flow of lymph and cells is ensured by the presence of valves. The blood and lymphatic vascular systems very rarely anastomose.

Lymphatic Endothelial Cells

Lymphatic Endothelial Markers

A number of markers have been identified that distinguish blood from lymphatic vascular endothelium (Table 1). These include podoplanin, a glomerular podocyte membrane mucoprotein; Prox-1, a homeobox gene product involved in regulating early lymphatic development; and LYVE-1, a lymphatic endothelial receptor for the extracellular matrix/lymphatic fluid glycosaminoglycan, hyaluronan. None of these markers are expressed entirely specifically by LECs. LECs also express vascular endothelial growth factor-3 (VEGFR-3); with the exception of fenestrated blood vascular endothelium, VEGFR-3 is expressed exclusively by lymphatics in normal adult tissues. However, it is widely expressed in embryonic blood vascular endothelium and is reexpressed in tumor blood vessels. In wound healing VEGFR-3 is confined to lymphatic endothelium (quiescent and regenerating), and is not expressed in blood vessels.

5′-Nucleotidase has also been used to distinguish lymphatic from blood vascular endothelium, and several apparently lymphatic-specific monoclonal antibodies have been reported, although none appear to have found widespread use. Finally, since lymphatic capillaries lack a continuous basement membrane, lack of immunoreactivity for extracellular matrix components (collagen IV, laminin, collagen XIII) has also been used to distinguish them from capillaries of the blood vascular system. However, this is unlikely to be reliable in tumors, since angiogenic blood vessels also appear to be partially or completely devoid of a basement membrane.

Lymphatic Endothelial Growth Factors

Two growth factors have been identified that induce growth of new lymphatic capillaries, and both belong to the vascular endothelial growth factor (VEGF) family. These are VEGFs -C and -D. Members of the VEGF family are highly conserved secreted glycoproteins that regulate vasculogenesis, hematopoiesis, angiogenesis, lymphangiogenesis, and vascular permeability and are implicated in many physiological and pathological processes. The VEGF family comprises VEGFs-A, -B, -C, -D and placental growth factor (PlGF). Of the three VEGF tyrosine kinase receptors identified thus far (VEGFR -1, -2, and -3), VEGFR-1 binds VEGFs-A and -B as well as PlGF, VEGFR-2 binds VEGFs-A, -C, and -D, and VEGFR-3 binds VEGF-C and -D. VEGFRs differ with respect to mechanisms of regulation and patterns of expression. For example, VEGFRs-1 and -2 are expressed almost exclusively by vascular endothelial cells and hematopoietic precursors, whereas VEGFR-3 is widely expressed in the early embryonic vasculature but becomes restricted to lymphatic endothelium at later stages of development and in postnatal life. VEGF-C also binds neuropilin-2, a receptor for class III semaphorins, which regulate chemorepulsive guidance of developing axons. Neuropilin-2 appears to be required for normal lymphatic development.

VEGFs-C and -D display a high degree of similarity to VEGF-A in their so-called VEGF homology domain, including conservation of the eight cysteine residues

Table I Characteristics of Lymphatic and Blood Vascular Endothelium.

	Lymphatic capillaries	Blood capillaries
Morphological features		
Lumen diameter	10–50 μm	10 μm
Pericytes	Absent	Present
Basement membrane	Absent or discontinuous	Continuous
Anchoring filaments	Present	Absent
Intercellular junctions	Overlapping, narrow	Well developed
Weibel-Palade bodies	Present	Present
Caveolae	Present	Present
Molecular markers		
CD31/PECAM-1	+	+
CD34	+ (low)	+ (high)
Von Willebrand factor	+	+
VE-cadherin	+	+
VEGFR-1	+	+
VEGFR-2	+	+
VEGFR-3	+ (high)	+ (low)
Tie-2	–/+	+
PAL-E	–	+
LYVE-1	+	–
Podoplanin	+	–
Prox-1	+	–
Mannose receptor	+	–
CCBP2/D6	+	–

characteristic of the VEGF family. However, VEGFs-C and -D also contain N- and C-terminal extensions that are removed by cell-associated proteolytic processing following secretion. Processing increases the affinity of the ligands for VEGFR-3 and also allows them to bind to VEGFR-2. In addition to their lymphangiogenic effect, under certain conditions VEGFs-C and -D also stimulate angiogenesis. The respective roles of VEGFR-2 and -3 in mediating the lymphangiogenic and angiogenic effects of VEGF-C and -D are incompletely understood.

Other molecules that have been implicated in the formation of the lymphatic vascular system include angiopoietin-2, the adaptor protein SLP76 and the tyrosine kinase syk, both of which are expressed primarily in hematopoietic cells, the transcription factors Prox-1 and net, and chemokines (CCL21/SLC) and their receptors (CCBP2/D6). α9 integrin has also been implicated.

Isolation and Culture of Lymphatic Endothelial Cells

Although many attempts have been made in the past to isolate and culture lymphatic endothelial cells from a variety of species, all of these studies have described isolation of the cells from large lymphatic vessels and have employed crude mechanical methods of cell separation. Identification of cell surface markers that allow easy distinction between lymphatic and blood vascular endothelium has recently led to the development of superior techniques for the isolation of pure endothelial cell populations. LECs have been isolated by positive selection using antibodies to podoplanin, VEGFR-3, or LYVE-1, and by a negative selection with antibodies to CD34. Molecular characterization of these cells has revealed many important differences in their molecular profiles. Of major importance is the observation that LECs express Prox-1, since Prox-1 appears to be a master switch for the lymphatic phenotype. In the adult organism, lymphatic endothelium has been shown to expresses VEGFR-2 and -3. Cultured LECs express VEGFR-1, -2, and -3, while BECs express VEGFR-1 and -2 and low levels of VEGFR-3. Many of the characteristic in vivo patterns of gene expression are altered with time in culture. For example, LYVE-1 is progressively lost from cultured LECs.

A Renewed Perspective on the Physiopathology of the Lymphatic System

Regulation of Interstitial Pressure

Lymph is essentially protein-rich interstitial fluid, which in turn is an ultrafiltrate of plasma. Unique to lymphatic capillaries are overlapping intercellular junctions that are formed by the extensive superimposition of adjacent LECs. LECs also have anchoring filaments, which link the basolateral plasma membrane to the adjacent extracellular matrix. As interstitial pressure rises, collagen fibers and other matrix components are forced apart. This in turn pulls on anchoring filaments, the result of which is an opening up of interendothelial junctions. Microarray analysis has revealed significant differences between BECs and LECs with respect to the expression of molecules that are involved in intra- and transcellular transport. Particularly highly represented in LECs were genes encoding proteins that control specificity of vesicle targeting and fusion. It is likely that LECs will display selectivity (quantitative and qualitative) with regard to regulating the formation of lymph.

Lymphedema is the term used to describe pathological conditions in which there is excessive, regional interstitial accumulation of protein-rich fluid. It can be either primary or secondary (i.e., acquired). Genetic studies have identified mutations in at least three genes that are associated with primary lymphedema: VEGFR-3, SOX18, and FOXC2. Inactivating mutations in the catalytic domain of VEGFR-3, a receptor tyrosine kinase for the lymphangiogenic factors VEGF-C and -D, are associated with Milroy's disease, an autosomal dominant form of early-onset lymphedema. SOX18 is a transcription factor that has been implicated in dominant and recessive forms of hypotrichosis–lymphedema–telangiectasia. Late-onset primary lymphedema has been attributed to truncating mutations in the forkhead transcription factor FOXC2. Secondary lymphedema occurs as a result of disruption or obstruction of preexisting lymphatics following surgery, radiotherapy, trauma, and neoplastic or inflammatory conditions.

Regulation of Immune Function

The lymphatic system plays an important role in inflammatory and immunological responses. With regard to dendritic cells, following antigen capture, these cells migrate into lymph nodes via afferent vessels. Within the lymph nodes they present major histocompatibility (MHC)-bound antigens to lymphocytes. It is very likely that chemokines produced by LECs are required for stimulating migration of mature dendritic cells (and possibly lymphocytes) from the tissue interstitium into lymphatic capillaries, although the nature of these chemokines remains to be fully elucidated. However, previous studies have shown that CCL21 is expressed by lymphatic endothelium in the small intestine and liver. Cultured LECs produce CCL21. CCL21 is a potent chemoattractant for dendritic cells in vitro and enhances the emigration of dendritic cells from skin to regional lymph nodes. This effect is likely to be mediated by CCR7 on mature or cytokine-activated dendritic cells.

Tumor Cell Dissemination

VEGF-C and -D, which bind to and activate VEGFR-2 and -3, have been shown to induce lymphangiogenesis in vivo. Of particular importance is the discovery of a tight correlation between levels of VEGF-C in primary human tumors, and the presence of regional lymph node metastases. Experimental studies have demonstrated a direct role for VEGF-C and -D in the induction of lymphangiogenesis and the formation of regional lymph node metastases. This has been demonstrated in mouse transgenic and tumor xenotransplantation models. Both lymphangiogenesis and the formation of lymph node metastases could be efficiently inhibited by antagonists of VEGF-C and -D. Similarly, overexpression of a soluble form of VEGFR-3 in the skin of transgenic mice selectively inhibited spontaneous lymphangiogenesis, which in turn resulted in lymphedema. However, one of the most important questions concerning lymphatic metastases still remains to be answered: Do tumor cells employ existing lymphatics for their dissemination, or does dissemination require the de novo formation of lymphatic capillaries (lymphangiogenesis)? With regard to human tumors, there is at present very little evidence for peri- or intratumoral lymphangiogenesis. This then begs the question: Why are high levels of VEGF-C in primary tumors associated with regional lymph node metastases?

Tumor cell dissemination via the lymphatic system requires the intravasation of tumor cells into lymphatic capillaries. Very little is known about how this process is regulated, but it is conceivable that LEC-derived chemokines may play a role. Apart from the fragmentary data on cultured LECs (see earlier discussion), very little is known about which chemokines are produced by LECs in vivo. However, evidence is accumulating that tumor cells express chemokine receptors (CXCR4, CCR7, and CCR10) that mediate their dissemination to lymph nodes. Signaling via these receptors mediates intracellular calcium flux, actin polymerization, changes in integrin expression, formation of pseudopodia, and tumor cell chemotaxis. Neutralization of the CXCL12/CXCR4 interaction in vivo significantly reduces metastasis of breast cancer cells to regional lymph nodes and lung.

Summary and Perspectives

The lymphatic network subserves almost every organ in the body, and along its pathway are several hundred lymph nodes, the majority of which are in the head and neck region. Because of their strategic localization, LECs are intimately involved in regulating the movement of lymph and cells of the immune system between tissue interstitium and the lymphatic system. The recent ability to isolate and culture LECs has provided tools with which to study these functions.

From the patient's perspective, there are settings in which both stimulation and inhibition of lymphangiogenesis could be beneficial. In the case of the former, the feasibility of using VEGF-C gene therapy for the treatment of primary lymphedema has been clearly demonstrated in pre-clinical studies. In the case of the latter, in preclinical studies, inhibition of tumor lymphangiogenesis has been shown to reduce the formation of metastases in regional lymph nodes. However, it will be important to clearly define an appropriate therapeutic window in cancer patients, since it is likely that tumor cell dissemination will already have occurred in many patients at the time of initial presentation.

Glossary

Endothelium: The layer of flattened epithelial-like cells that lines the inner luminal surface of blood and lymphatic vessels (also known as intima) as well as the heart (also known as endocardium). The endothelium is in direct contact with blood in blood vessels and the heart, and with lymph in lymphatic vessels.

Lymph: A transparent fluid often containing lymphocytes that is found within lymphatic vessels; originates from tissue interstitium throughout the whole body; is returned to the systemic circulation via the lymphatic system; may be a milky white in the small intestine following fat absorption.

Lymphangiogenesis: The growth of new lymphatic capillaries by a process of sprouting from preexisting veins or lymphatic capillaries; may also include the differentiation of lymphatic endothelium from lymphangioblasts.

Lymphedema: Swelling of a part of the body (usually a limb) due to the accumulation of and inability to drain interstitial fluid as a consequence of absent, hypoplastic, or obstructed lymphatic vessels.

Lymphogenous metastasis: Secondary tumors that develop at sites distant from the primary tumor as a consequence of tumor cell dissemination via the lymphatic system; usually involve lymph nodes but may include other sites as well.

Bibliography

Alitalo, K., and Carmeliet, P. (2002). Molecular mechanisms of lymphangiogenesis in health and disease. *Cancer Cell* **1**, 219–227.

Baldwin, M. E., Stacker, S. A., and Achen M. G. (2002). Molecular control of lymphangiogenesis. *BioEssays* **24**, 1030–1040. *This is a comprehensive and up-to-date review on the molecular mechanisms of lymphangiogenesis.*

McClure, C. F. W. (1921). The endothelial problem. *Anat. Rec.* **22**, 219–237. *This paper discusses much of the early work done on the development of the lymphatic system, and contrasts the so-called "centrifugal" and "centripetal" hypotheses of lymphatic endothelial origin.*

Oliver, G., and Detmar, M. (2002). The rediscovery of the lymphatic system: Old and new insights into the development and biological function of the lymphatic vasculature. *Genes Dev.* **16**, 773–783.

Pepper, M. S., and Skobe, M. (2003). Lymphatic endothelium: Morphological, molecular and functional properties. *J. Cell Biol.* **163**, 209–213.

Pepper, M. S., Tille, J. C., Nisato, R., and Skobe, M. (2003). Lymphangiogenesis and tumor metastasis. *Cell Tissue Res.* **314**, 167–177.

Rockson, S. G. (2001). Lymphedema. *Am. J. Med.* **110**, 288–295.

Swartz, M. A. (2001). The physiology of the lymphatic system. *Adv. Drug Deliv. Rev.* **50**, 3–20. *This review deals in depth with the structure and function of the lymphatic vascular system.*

Wilting, J., Neeff, H., and Christ, B. (1999). Embryonic lymphangiogenesis. *Cell Tissue Res.* **297**, 1–11.

Witte, M. H., Way, D. L., Witte, C. L., and Bernas, M. (1997). Lymphangiogenesis: Mechanisms, significance and clinical implications. *EXS* **79**, 65–112.

Capsule Biography

Michael Pepper graduated M.B.Ch.B from the University of Cape Town Medical School in 1982, and obtained his Ph.D. and M.D. degrees at the University of Geneva Medical School in 1990 and 1992, respectively. His research interests over the past 20 years have included angiogenesis and more recently lymphangiogenesis. Dr. Pepper's principal objective has been to foster the transfer of information between basic and clinical sciences.

CHAPTER 82

Foundations of Microlymphatic Function

Geert W. Schmid-Schönbein

Department of Bioengineering, Whitaker Institute of Biomedical Engineering, University of California, San Diego, La Jolla, California

Introduction

Lymphatics are an integral part of the tissue transport system and the site of adipogenesis and immunological control. All organs, with the exception of the brain, tendons, and ligaments, have a lymphatic system. They carry fluids containing a wide variety of organic and inorganic molecules, colloids, and cells. In tumors, lymphatics have special significance as one of the main pathways for metastatic cells. Lymphatic fluid moves unidirectionally from the tissue through afferent channels toward strategically located lymph nodes. After passage through the lymph node, lymphatic fluid is carried through confluent efferent channels into the two central lymphatic ducts and returned through the subclavian junctions back into the venous circulation. In the following review, we will describe the main microanatomical, cellular, and molecular features of microlymphatics common among different tissues and organs, and we will outline mechanisms that serve the transport of lymph fluid and lymph angiogenesis.

Structure of Microlymphatics

Compared with the high density of the microvasculature, the microlymphatic network is sparse. Individual microlymphatics originate in the tissue and form bifurcating trees and in some organs, meshworks. Their detailed network morphology depends on the tissue and organ involved.

The lymphatics can be divided into two general classes, the *initial* lymphatics and the *contractile* lymphatics (Figure 1). Preceding the initial lymphatics there are also prelymphatic channels. We will describe the three microstructures in the sequence through which tissue fluid is transported.

Prelymphatic Channels

In mesentery, extracellular tissue fibers are connected to the terminal endings of the initial lymphatics. The prelymphatic channels are formed by tissue fibers oriented in the direction towards the lymphatic endothelial cells. In the brain, the delineation of tracers injected into the cerebrospinal space identifies prelymphatic channels. Such tracers tend to follow the perivascular space of arterioles and eventually drain into the initial lymphatics.

The Initial Lymphatics

These consist of a continuous but highly attenuated endothelial lining, and they have no smooth muscle media. In organs such as the skin and intestine, the initial lymphatics form a highly regular hexagonal branching pattern. Initial lymphatics are frequently associated with structures that have smooth muscle (arterioles, respiratory bronchioles, and so on). They are located in the adventitia of arterioles in close proximity to the smooth muscle cells, nerve fibers, adipocytes, and what is a continuous line of mast cells (Figure 2). In contrast, other parts of the tissue parenchyma may be relatively depleted of initial lymphatics. For example, the capillary space in skeletal muscle has no initial lymphatics. Instead, initial lymphatics form a meshwork that is tightly paired with the arcade arterioles and larger venules in this organ [1].

In contrast to the cylindrical geometry of most blood vessels, the lumen cross section of the initial lymphatics has an irregular shape in many tissues; the initial lymphatics almost never have circular cross sections. The lumen may, in fact, be completely collapsed. Yet in normal tissue, the initial lymphatics are readily identified based on morphology and specific location in the tissue. In tumor tissues, initial lymphatics may be more difficult to detect on histological

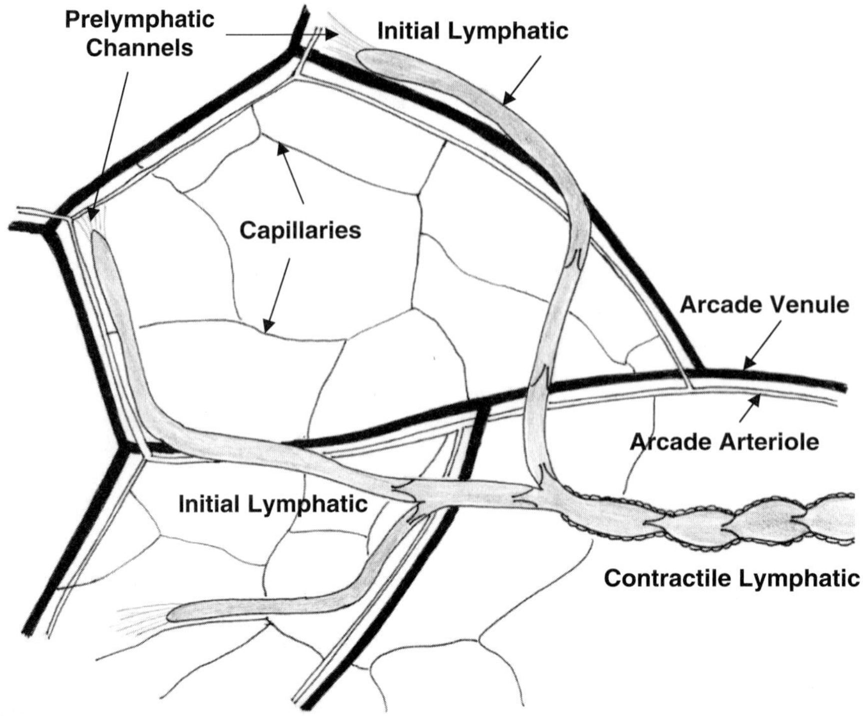

Figure 1 Schematic of initial lymphatics without smooth muscle and contractile lymphatics with smooth muscle in the mesentery microcirculation. The vessel dimensions are not drawn to scale. The initial lymphatics tend to be associated with paired arcade arterioles and venules.

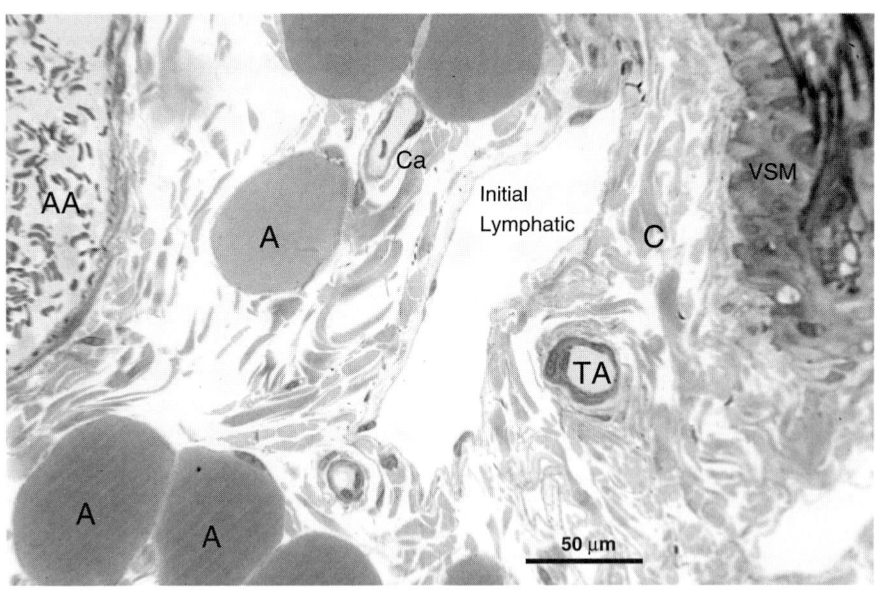

Figure 2 Light microscopic cross section of initial lymphatic in rat skeletal muscle. The vessel has a single endothelial lining, is positioned in vicinity of adipose cells (A), a terminal arteriole (TA), embedded in collagen fibers(C), and positioned between two arcade arterioles (AA) with vascular smooth muscle media (VSM). (see color insert)

sections and therefore require immunological labels (such as an antibody against LYVE-1, a lymphatic specific receptor for hyaluronan [2], or the lymphatic marker podoplanin) for identification as lymphatic endothelium.

The endothelium in initial lymphatics has a discontinuous basement membrane with openings in the vicinity of the interendothelial junctions. In contrast to capillary endothelium, there are no pericytes on lymphatic endothelium. Short interstitial fibers (anchoring filaments) connect the abluminal side of the lymphatic endothelium to the adjacent connective tissue. In the organs investigated to date, no adrenergic innervation of the initial lymphatics has been encountered. In the interlobular connective tissue of the rat liver, neuropeptide Y- and substance P–containing nerve fibers around initial lymphatics have been observed.

Initial lymphatics have the ability to collect and transport interstitial fluid, proteins, lipid particles, colloids, and cells, including metastatic cells. In tissues that are resting, peristaltic motion sufficient to expand or compress the lumen cross section as a fluid pump mechanism has yet to be observed in initial lymphatics. To date, no human lymphatic system has been described that is without initial lymphatics. Compared with contractile lymphatics, the initial lymphatics are considerably more numerous and make up the major portion of the lymphatic network.

The Contractile Lymphatics

In contrast to the initial lymphatics, the contractile lymphatics have a smooth muscle media and exhibit regular peristaltic motion with periodic compression and expansion of the lymphatic lumen. Contractile lymphatics frequently have circular cross sections so that a circumferential mechanical stress can be generated by the smooth muscle. The contractile lymphatics undergo spontaneous peristaltic contractions, a phenomenon not seen in the initial lymphatics. They are rarely associated with vascular smooth muscle and frequently form vessels that pass through the tissue in the form of isolated channels unpaired with the arterioles or large veins. Contractile lymphatics are positioned downstream of (i.e., proximal to) the initial lymphatics and serve to drain fluid from the initial lymphatics into the lymph nodes and the central thoracic ducts. Contractile lymphatics have pacemaker cells, and smooth muscle contraction travels in the form of peristaltic waves along the contractile lymphatic ducts.

The lymphatic smooth muscle cells exhibit many of the characteristics of arteriolar smooth muscle, with smooth muscle myosin heavy chain isoforms as well as cardiac alpha-actin, vascular alpha-actin, enteric gamma-actin, and skeletal alpha-actin.

The contractile lymphatics exhibit a myogenic response accompanied by generation of membrane action potentials and calcium influx into the endothelial cytoplasm. They also show nitric oxide–dependent vasodilation. The flow-dependent dilation can be mimicked by nitric oxide. The smooth muscle in contractile lymphatics has among many other receptors, stimulatory α-adrenergic and β-inhibitory adrenoreceptors [3], histamine receptors, and ATP-sensitive K^+ channels. The contractile lymphatics are also responsive to a variety of inflammatory mediators, such as oxygen free radicals and substance P. Contractile lymphatics may have a noradrenergic innervation [4] with unmyelinated nerve fibers that reach into the subendothelial connective environment in close association with endothelial cells. These nerve fibers react with calcitonin gene–related peptide or substance P.

In mammalian organs, like the intestine or skeletal muscle, all lymphatics inside the tissue parenchyma are initial lymphatics. Contractile lymphatics originate only at strategic transition points where the lymphatics exit the tissue parenchyma [5].

Lymphatic Endothelium

Lymphatic endothelial cells are continuous but highly attenuated and have a discontinuous basement membrane. Their cytoplasm contains an assortment of organelles, caveolae, and mitochondria. The cytoplasm contains typical cytoskeletal proteins, including actin, intermediate filaments, and microtubules with the ability to form actin-bundle (stress) fibers. The cytoskeletal proteins facilitate active changes in cell shape, such as pseudopod projection, phagocytosis, and formation of adhesion site or cell migration, as in lymph angiogenesis. There is no conclusive evidence that such cytoplasmic motion by the endothelial cells can be translated into a contractile mechanism for the lumen of a lymphatic channel comparable to the smooth muscle action in contractile lymphatics. The major part of the lymphatic endothelial cells is firmly attached to the surrounding collagen matrix.

Lymphatic endothelial cells in culture express remarkably high levels of genes implicated in protein metabolism, sorting, and trafficking. Current analysis indicates that genes with high representation are those encoding proteins that control specificity of vesicle targeting and fusion, such as members of the SNARE family, rab GTPases, AAA ATPases, and sec-related proteins, reflecting the existence of an extensive caveolar system [6].

Lymphatic Endothelial Junctions

Junctions between lymphatic endothelial cells have a paucity of tight junction desmosomes (macula adherents) and interendothelial membrane adhesion molecules, such as VE-cadherins, desmoplakin, or plakoglobin. The cell extensions at the junctions form overlapping cellular flaps, which have the ability to separate from each other because of the lack of interendothelial adhesion molecules. These cell extensions are not attached to the underlying tissue matrix; they form the part of an endothelial cell that can form a small flap (see the following section).

The Primary Lymphatic Valves

Investigation of lymphatic endothelial sections fixed during lymph pumping shows open interendothelial junctions not found in resting initial lymphatics. These results suggest that initial lymphatics have a (primary) valve system at the level of the endothelial junctions. Cross sections through the initial lymphatics show open lymphatic endothelial flaps in muscle that during tissue fixation has been periodically compressed to preserve lymph flow into the lymphatic channel. No open junctions can be observed on cross sections of initial lymphatics fixed in a resting state without oscillatory muscle compression. Colloidal tracer particles can readily enter into the lumen of the initial lymphatics. But inside the lumen of the initial lymphatics, no significant escape of tracer particles out of the lymph lumen occurs. Escape of tracer particles from the lymphatic lumen is small, even after lowering the fluid pressure in the interstitium adjacent to the initial lymphatics.

The primary lymphatic valves can open and close rapidly, possibly by a mechanical flap action of overlapping cell extensions at the endothelial junctions [7]. The initial lymphatic endothelium serves as a unidirectional transport barrier. This particular endothelium is leaky when fluid enters from the interstitial space, but it is tight for fluid transport in the reverse direction.

The Secondary Intralymphatic Valves

Lymphatics have a system of intraluminal (secondary) valves that prevent reflow along the length of the lymphatic channels. The valves have bileaflet morphology with the individual valves made up of a thin collagen sheet sandwiched between two layers of endothelial cells. The valve leaflets are relatively long (typically about twice the length of the lymph cross-sectional dimension), forming a funnel inside the lymphatic channels. This feature facilitates the opening and closing of the secondary valves by fluid pressure drops across the leaflets generated by viscous pressure gradients inside the funnel. Secondary valves also operate in lymphatics with highly irregular lumen cross sections.

The spacing of secondary valves along the initial lymphatics is organ dependent. For example, in the intestinal wall there are relatively few secondary valves, whereas in skeletal muscle they frequently arise at convergent bifurcations of two lymphatic channels into a single daughter vessel, preventing reflow into the two parent channels. Contractile lymphatics in general have secondary valves, arranged in a ladder configuration, designated as lymphangions. Each lymphangion consists of a contractile compartment with an inlet and an outlet valve.

The Interplay between Primary and Secondary Lymphatic Valves

In the presence of two valve systems, the unidirectional transport in lymphatics is easy to understand. Consider two periods during a pump cycle, an expansion period and a compression period of an initial lymphatic (Figure 3). During lymphatic expansion, an initial lymphatic channel becomes filled with interstitial fluid. During this period the primary lymphatic valves are open and the secondary valves inside the lumen are closed to prevent reflow of fluid along the lymphatic duct. During lymphatic compression, the primary valves close and prevent escape of fluid back into the interstitium while the secondary valves in the lymphatic lumen are open. Thus fluid is transported along the lumen of the initial lymphatics in a proximal direction toward the contractile lymphatics and nodes. During each compression and expansion of initial lymphatics, the cycle repeats.

Compartments that rely on volume expansion and compression as the primary mode of fluid movement require two valves for unidirectional transport. For example, a lymphangion along the contractile lymphatics has two valves, one at the entry and one at the exit from the lymphangion. Puncturing one of the valves compromises the fluid transport in and out of the lymphangion. A chain of lymphangions, as seen along the contractile lymphatics, serves as entry and exit valves from one such compartment to the next.

Leakage of the Primary Valves and Tissue Edema

The interendothelial junctions can be readily separated if tension is applied to the lymphatic wall, for example by overinflation of the lymphatic lumen [8]. There are local regions along the interendothelial junctions where neighboring endothelial cells can be separated when tension is applied to them. When these junctions are stretched open, one can identify endothelial attachment to the underlying

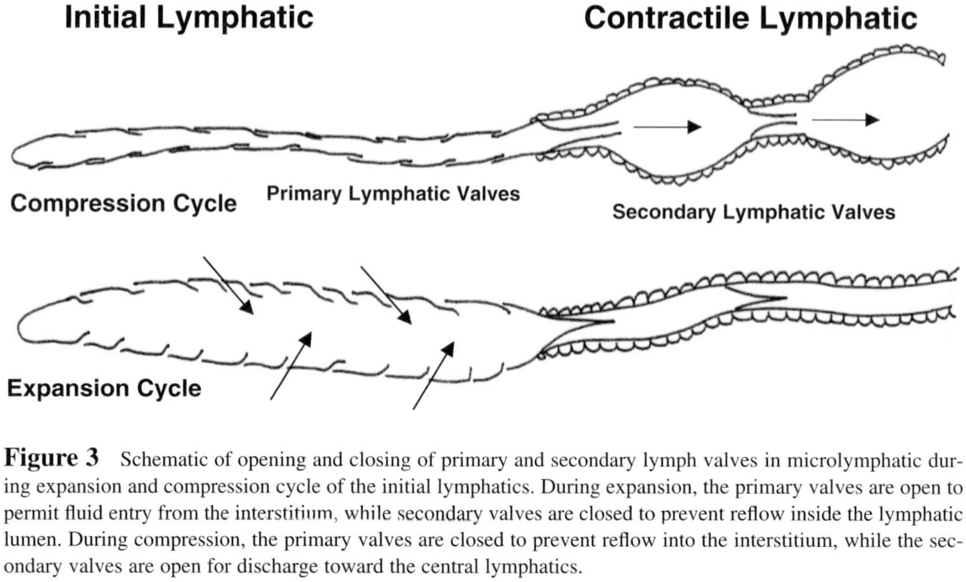

Figure 3 Schematic of opening and closing of primary and secondary lymph valves in microlymphatic during expansion and compression cycle of the initial lymphatics. During expansion, the primary valves are open to permit fluid entry from the interstitium, while secondary valves are closed to prevent reflow inside the lymphatic lumen. During compression, the primary valves are closed to prevent reflow into the interstitium, while the secondary valves are open for discharge toward the central lymphatics.

basement membrane and to anchoring filaments, and can detect large openings between endothelial cells that permit free fluid entry in and out of the initial lymphatics. The primary lymphatic valves may cease functioning under edematous conditions.

Mechanisms for Periodic Compression of Initial Lymphatics

How can initial lymphatics expand and compress their lumens? Since these lymphatics have no smooth muscle media, their lumen compression and expansion depends on periodic motion of tissue structures positioned in proximity to the initial lymphatics. A variety of organ-dependent arrangements exist. For example, in skeletal muscle the initial lymphatics are located in the adventitia of the arterioles and in close apposition to each other. Expansion of the arterioles during vasodilation leads to compression of the adjacent lymphatic, whereas active contraction of the arterioles leads to expansion of the initial lymphatic vessel [1]. Thus, both vasomotion and pulse pressure serve to promote lymph fluid [9]. In addition, skeletal muscle contractions may also control the expansion and compression of the initial lymphatics. Thus, lymphatic pumping is a carefully controlled process closely linked to the physiological activity of the organ.

The pressure pulse leads to only low amplitudes of arteriolar expansion and compression, so that the lymph flow generated in this fashion is relatively small. Lymph flow rates are greatly enhanced by either passive muscle stretching or active contractions. Cessation of pressure pulsations and vasomotion in a resting skeletal muscle leads to undetectable, near-zero lymph flow rates.

Many physical activities that generate cyclic compression and expansion of initial lymphatics also increase lymph transport. In addition to arterial pulse and vasomotion, rhythmic tissue deformations such as walking and running, heart muscle contractions, respiration, intestinal peristalsis, or skin massage enhance lymph flow. The faster the rate at which the initial lymphatics are pumped, the higher the lymph flow. The actual lymph flow rates achieved during such activities depend on the frequency of the cyclic tissue motions.

Lymph flow can also be enhanced by the elevation of capillary fluid pressure (e.g., elevation of venous pressure) and capillary filtration (e.g., elevation of endothelial permeability) as well as the elevation of interstitial fluid pressure, especially if the fluid pressure reaches values above those inside the initial lymphatics.

Lymph Fluid Collection and Transport

In addition to plasma proteins, colloidal materials (up to micrometer size) readily enter into the initial lymphatics across the primary valves. Cells the sizes of lymphocytes or of metastatic cells are carried along the lymphatics in the skin during periodic tissue compression, but not without a periodic compression of the initial lymphatics [10]. This evidence suggests that in addition to the junctions between endothelial cells in initial lymphatics, which can open up to let micrometer-size particles pass, the lymphatic endothelium may be able to open interendothelial and transendothelial pores of a size sufficient for entire cells to migrate across. Initial lymphatics may carry lymphocytes and, under conditions of ischemia and shock, erythrocytes, platelets, and in cancer metastatic cells as well.

Lymphatic Fluid Pressures

In resting tissues after general anesthesia, the fluid pressures in initial lymphatics are close to zero and tend to be similar to the fluid pressure measured in the interstitial space. There are currently no direct measurements of fluid pressure in pumping initial lymphatics. In contractile lymphatics, the fluid pressures rises during each contraction. The pressure may also rise from one lymphangion to the next if there is an outflow resistance in proximal segments of the lymphatics. Plasma oncotic pressures are lower in lymphatic fluid compared with blood plasma.

Lymphatic Angiogenesis

During embryonic development, lymphatic growth is initiated after establishment of vascular endothelial cells, a process that may require a transcription factor. Lymphatic channels develop in the arteriolar adventitia after the arterioles have become innervated and exhibit active smooth muscle contraction. The lymphatic growth is accompanied by the development of adipose tissue. Lymphatics and lymph nodes in many tissues are surrounded by adipose tissue and control the growth of adipocytes [11].

The homeobox gene transcription factor Prox-1 appears to be required for differentiation of endothelial cells to a lymphatic phenotype [12]. Two peptide lymphangiogenic factors that are members of the VEGF family have been identified to date: VEGF-C and VEGF-D. They have the ability to induce lymphangiogenesis and vasculogenesis. Binding of these growth factors to the vascular endothelial growth factor receptor 3 regulates lymphangiogenesis, in contrast to signaling via the vascular endothelial growth factor receptor-2, which regulates angiogenesis [13]. Overexpression of soluble vascular endothelial growth factor receptor 3 in transgenic mice inhibits lymph channel formation [14]. Other molecules that may play a role in the development of the lymphatic system include angiopoietin-2 and neuropilin-2.

Lymphangiogenesis in wound healing of the skin shows that interstitial fluid channels, detected by fluorescent fluid tracers, form before lymphatic endothelial cell organization. Lymph endothelial cell migration, vascular endothelial growth factor-C expression, and lymphatic capillary network organization are initiated primarily in the direction of lymph flow. These data suggest that lymph channel growth is determined not only by tissue growth factors but also by stresses in the tissue surrounding the lymphatics [15]. Lymphangiogenesis is currently an active research field because of its significance for understanding of lymph edema and metastatic cell dissemination.

Lymphatics in Tumors

In both animal and human tumors, lymphatic structures that stain for LYVE-1/Prox1 within the cancer tissue have been detected. But these structures do not function properly as lymphatics and thus contribute the interstitial hypertension in tumors. Metastatic cells that are carried along lymph channels and may accumulate in the nodes enter the lymphatics in the margin of tumors [16].

Glossary

Contractile lymphatics: Lymphatic channels with their own smooth muscle media draining fluid from the initial lymphatics.

Initial Lymphatics: Lymphatic channels embedded in the tissue parenchyma without smooth muscle media collecting fluid from the interstitial fluid space.

Lymphatic valves: Primary valves at the junction between endothelial cells in initial lymphatics, secondary valves in the lumen lymphatic channels.

Lymphatic Endothelial Growth Factor: Receptor agonist (the first one discovered is vascular endothelial growth factor – C, VEGFC) with affinity to the vascular endothelial growth factor receptor (e.g. VEGFR 3) on lymphatic endothelium that stimulates lymphatic endothelial migration with lymphatic channel growth and enlargement.

Lymphatic Smooth Muscle: Smooth muscle layer in the media of contractile lymphatics with innervation, vasomotor activity, and myogenic activity.

Vascular Endothelial growth factor (VEGF): Family of growth factors binding to endothelial VEGF receptors on vascular endothelium (e.g. VEGFR 2), enhancing vascular endothelial permeability and stimulating endothelial cell migration.

Acknowledgment

Supported by NSF Grant IBN 9876379 and in part by NIH Grant HL 43026. I thank Drs. Thomas C. Skalak, Michelle Mazzoni, Fumitaka Ikomi, Mr. J. Trzewik, and Ernesto Mendoza for numerous discussions about lymphatics. Special thanks to Frank A. DeLano for assistance with the experiments.

References

1. Skalak, T. C., Schmid-Schönbein, G. W., and Zweifach, B. W. (1984). New morphological evidence for a mechanism of lymph formation in skeletal muscle. *Microvasc. Res.* **28**, 95–112. *A demonstration of the tight association between arteriolar motion and lymphatic pumping in the skeletal muscle microcirculation.*
2. Banerji, S., Ni, J., Wang, S. X., Clasper, S., Su, J., Tammi, R., Jones, M., and Jackson, D. G. (1999). LYVE-1, a new homologue of the CD44 glycoprotein, is a lymph-specific receptor for hyaluronan. *J. Cell Biol.* **144**, 789–801. *Description of a molecular marker relatively specific for lymphatics.*
3. Ohhashi, T., and Azuma, T. (1984). Sympathetic effects on spontaneous activity in bovine mesenteric lymphatics. *Am. J. Physiol.* **247**, H610–H615.
4. McHale, N. G., Roddie, I. C., and Thornbury, K. D. (1980). Nervous modulation of spontaneous contractions in bovine mesenteric lymphatics. *J. Physiol.* **309**, 461–472.
5. Unthank, J. L., and Bohlen, H. G. (1988). Lymphatic pathways and role of valves in lymph propulsion from small intestine. *Am. J. Physiol.* **254** (Gastrointest., Liver Physiol 17), G389–G398. *A comprehensive three-dimensional reconstruction of the microlymphatic network in the intestine of several species.*
6. Podgrabinska, S., Braun, P., Velasco, P., Kloos, B., Pepper, M. S., and Skobe, M. (2002). Molecular characterization of lymphatic endothelial cells. *Proc. Natl. Acad. Sci. USA* **99**, 16069–16074. *Gene expression profile of lymphatic endothelial cells in culture.*
7. Mendoza, E., and Schmid-Schönbein, G. W. (2003). A model for mechanics of primary lymphatic valves. *J. Biomech. Eng.* **125**, 407–414.
8. Castenholz, A. (1984). Morphological characteristics of initial lymphatics in the tongue as shown by scanning electron microscopy. *Scan. Electron Microsc.* **1984**, 1343–1352. *Scanning electron microscopic images of lymphatic endothelial junctions before and after partial separation by mechanical tension.*
9. Intaglietta, M., and Gross, J. F. (1982). Vasomotion, tissue fluid flow and the formation of lymph. *Int. J. Microcirc. Clin. Exp.* **1**, 55–65. *The report demonstrates that arteriolar vasomotion serves to pump lymphatics.*
10. Ikomi, F., and Schmid-Schönbein, G. W. (1996). Lymph pump mechanics in the rabbit hind leg. *Am. J. Physiol.* **271**, H173–H183.
11. Macqueen, H. A., Waights, V., and Pond, C. M. (1999). Vascularisation in adipose depots surrounding immune-stimulated lymph nodes. *J. Anat.* **194**(Pt 1), 33–38.
12. Wigle, J. T., and Oliver, G. (1999). Prox1 function is required for the development of the murine lymphatic system. *Cell* **98**, 769–778.
13. Lymboussaki, A., Partanen, T. A., Olofsson, B., Thomas-Crusells, J., Fletcher, C. D., de Waal, R. M., Kaipainen, A., and Alitalo, K. (1998). Expression of the vascular endothelial growth factor C receptor VEGFR-3 in lymphatic endothelium of the skin and in vascular tumors. *Am. J. Pathol.* **153**, 395–403. *The study shows vascular endothelial growth factor C receptor-3 labeling of lymphatic endothelial cells.*
14. Makinen, T., Jussila, L., Veikkola, T., Karpanen, T., Kettunen, M. I., Pulkkanen, K. J., Kauppinen, R., Jackson, D. G., Kubo, H., Nishikawa, S., Yla-Herttuala, S., and Alitalo, K. (2001). Inhibition of lymphangiogenesis with resulting lymphedema in transgenic mice expressing soluble VEGF receptor-3. *Nat. Med.* **7**, 199–205.
15. Boardman, K. C., and Swartz, M. A. (2003). Interstitial flow as a guide for lymphangiogenesis. *Circ. Res.* **92**, 801–808. *An illustration of the initial lymphatics in the mouse skin and their growth in the direction of the lymph flow.*
16. Padera, T.P., Katambi, A., di Tomaso, E., Mauta Carreira, C., Brwon, E. B., Boucher, Y., Choi, N. C. Mathisen, D., Wain, J., Mark, E. J., Munn, L. L., and Jain, R. K. (2002). Lymphatic metastasis in the absence of functional intratumor lymphatics. *Science* **296**, 1883–1886. *The report illustrates that although tumors may have lymphatics, these lymphatics have no normal function. Metastatic cells enter lymphatics that surround the tumor.*

Capsule Biography

Dr. Schmid-Schönbein has headed the Microcirculation Laboratory at the University of California, San Diego, since 1979. He was President of the Microcirculatory Society in 2003. His laboratory primarily focuses on cell mechanics and transport in living tissues; his work is supported by grants from the NIH and NSF.

CHAPTER 83

Morphology and Physiology of Lymphatic Endothelial Cells

H. Zöltzer
University of Kassel

Introduction

The lymphatic system of man and mammals is in close functional interaction with the blood circulation and is connected with it by means of large lymph trunci in the venous system of the heart region. So the lymphatic vessels are part of a comprehensive circulatory system serving the substance exchange within the body and between the body and environment. Whereas the blood vessel system is closed, with continuous circulation of the blood and with capillaries by which small molecules can be quickly exchanged, the lymphatic vessel system in its initial region serves for the uptake of fluid, larger molecules, and cells from the interstitium, which cannot be reabsorbed by the blood vessels. Depending on their function, different morphological parts of the lymphatic vessel system exist: initial lymphatics, precollectors, collectors, and lymph stems (Table I). All these vessels are lined with a functionally adapted endothelium. Whereas the endothelium of the collectors and trunci is similar to the endothelium of the venous system, the endothelium of initial lymphatics and precollectors displays a special differentiation.

History

"White blood" was first described in ancient times (Hippocrates, ca. 460–377 BC; Aristotle, 384–322 BC). Herophilus (about 300 BC) and Erasistratos (about 330–250 BC) also knew something about gut lymphatics, marked with natural fatty substances, without understanding their function. Pecquet (1651) was the first to discover the lymphatics and the ductus thoracicus in human. The first comprehensive overviews of the lymphatic vessels originated from T. Bartholinus (1653, 1657), O. Rudbeck (1653, 1654), and Hunter (1752). Bartholinus established the term "lymph" and described it as a clear watery liquid. In the 17th and 18th centuries, lymph was understood as a nutritional fluid and a source of harm or illness ("Fons aegritudinum," Mascagni, 1787). Each swelling of lymph nodes was diagnosed as a stagnation of the "bad nutritive fluids." Von Recklinghausen (1860) and Hoyer (1865) demonstrated the endothelial nature of the inner surface of lymphatics by means of silver nitrate solutions. Ludwig (1858) and Starling (1908) described the lymph as a filtrate from the blood via the capillaries and demonstrated the connection between hydrostatic and osmotic pressure in the exchange of substances between blood and lymph. The school of C. K. Drinker ([1–3] and so on) and the group around I. Rusznyák ([4] and so on) created the basis of modern research in the lymphatic system.

The Endothelium of the Initial Lymphatic

Initial lymphatics have a large diameter (20 to 70 μm) and an irregular luminal shape. They start in the interstitial spaces as plexus or blind-opening tubes, demonstrable with only few exceptions in all parts of the body. Valvelike structures and trabecules can divide the vascular lumen in a labyrinthine pattern. The occurrence of initial lymphatics is, as a rule, at the utmost distance from blood capillaries. In organs with an epithelium, initial lymphatics are located under the blood capillary network. Only precollectors, collectors, and trunci run parallel to blood vessels.

Table I The Human Lymph Vessel System.

Structural Elements	
Interstitial space	
Initial lymphatics	
Precollectors	
Collectors	
Lymph nodes	
Lymph stems	
Lympho-venous anastomoses	
Fluid and cells of the lymph	
Functions	
Retransportation of proteins, other macromolecules, and water to the blood vessel system	
Recirculation of lymphocytes	
Elimination of macromolecular substances and antigenous material from the body fluids	
Metabolism and turnover of extracellular matrix constituents	
Prevention of edema = maintenance of the inner balance of body	
Transport of long-chain fatty acids, lipid-soluble vitamins, and other food particles reabsorbed by the gut via the gut lymphatics	

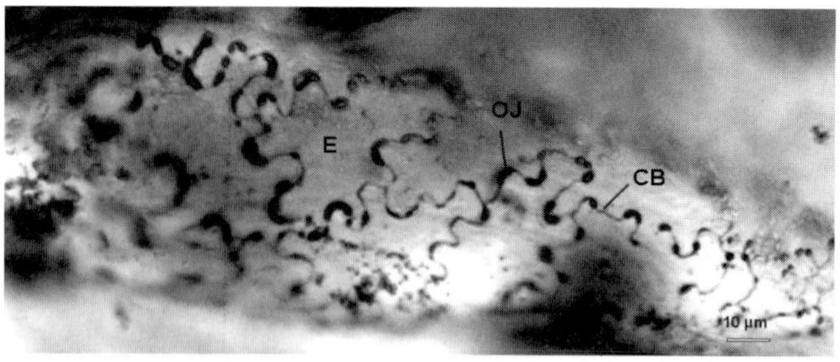

Figure 1 Typical cell (E) circumference similar to an oak leaf in light microscope. The cell border line (CB) is contrasted with silver. Open-junction formations (OJ) are visible as dark half-moons. (see color insert)

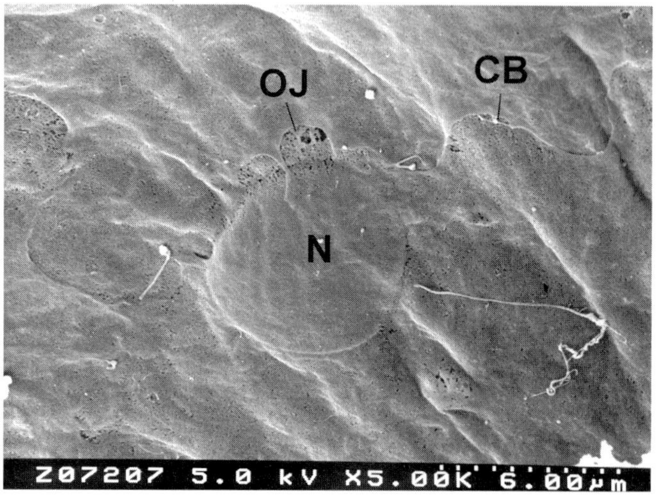

Figure 2 Initial lymphatic endothelium in SEM. N = nucleus, CB = cell border, OJ = open-junction formation. Triton-X-100 method.

Shape of the Initial Lymphatic Endothelium Cell

The inner surface of initial lymphatics is lined by a single layer of endothelial cells. With the exception of the perinuclear zone the cytoplasm is very thin, from 0.1 to 1.2 µm. Those thin areas are zones with very small distances for the diffusion or the transport of molecules. In the nuclear zone the cell is 0.9 up to 1.4 µm and the cell diameters are about 60 × 30 µm. Cell density is 500 per μm^2 in an area of 1 mm^2. The cell circumference is similar to an oak leaf and it has contact structures to four to eight neighboring cells (Figure 1). The diameter of the inlets is about 4 to 8 µm. Neighboring cells overlap each other and are connected with zonulae adherentes. The peripheral cytoplasm is electron-dense and partially interrupted by pores. Sometimes endothelial trabecular structures running in the lumen connect with another endothelial cell or produce structures like incomplete valves. In the lumen of initial lymphatics one can find fine granular material (coagulated lymph), in transmission electron microscope (TEM) a gray flocky material.

Organelles of the Initial Lymphatic Endothelium Cell

The pattern of organelles corresponds mostly to the expected occurrence. The endothelial nucleus forms a prominent structure, bulging the cell in direction of the lumen (Figure 2). All over the cell many vesicles and invaginations exist, but in spite of the very thin endothelium, no fenestrations are formed. The vesicles occur mostly under the luminal and abluminal surface and have an average diameter of 30 to 100 nm. Most of the vesicles are uncoated; only rarely are coated vesicles demonstrable. In larger vesicles of deeper zones of the cytoplasm a positive acid phosphatase activity exists. Often merging zones of the vesicles with the plasmalemma are visible (Figure 3), both up to the luminal or abluminal direction, with openings from 50 to 150 nm. These vesicles are transport systems into the cell, out of the cell, and through the cell (endocytosis, transcytosis, and exocytosis).

Great numbers of fine filaments are typical for the initial endothelium cell. The diameters are about 6 nm; sometimes they are dispersed, and sometimes aggregated to bundles, then running mostly in the longitudinal axis of the cell. The length of the filaments is from one to several micrometers. Following treatment with Triton X-100, elements of the cytoskeleton are visible in SEM. Above the nucleus, its texture is relatively compact with a mesh about 75 nm wide. In contrast, in the peripheral zone it is wider with meshes of about 100 to 750 nm. Often bundles of filaments are running along the cell border (Figure 4). This dense peripheral band (DPB) can be demonstrated in TEM and SEM, clearly visible after application of the Triton X-100 method. In endothelial cells of blood venules Wong and Gottlieb [5] determined actin, myosin, tropomyosin, α-actinin, and vinculin as components of the DPB. Thus, it is able to contract, and it is likely that the DPB controls cell shape.

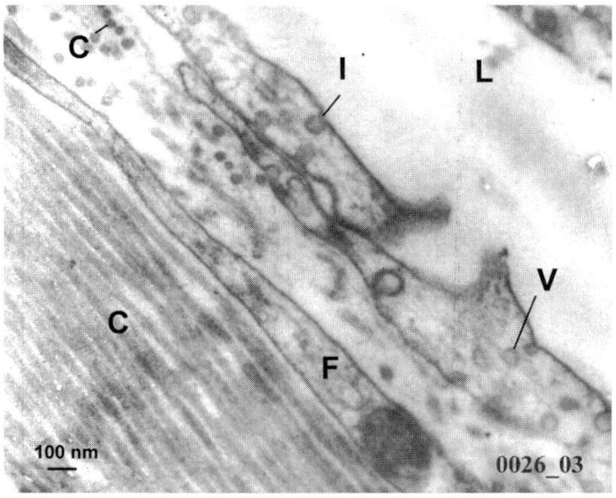

Figure 3 Vesicles (V) and invaginations (I) of initial lymphatic endothelium (E) in TEM. F, fibrocyte; C, collagenous fibers.

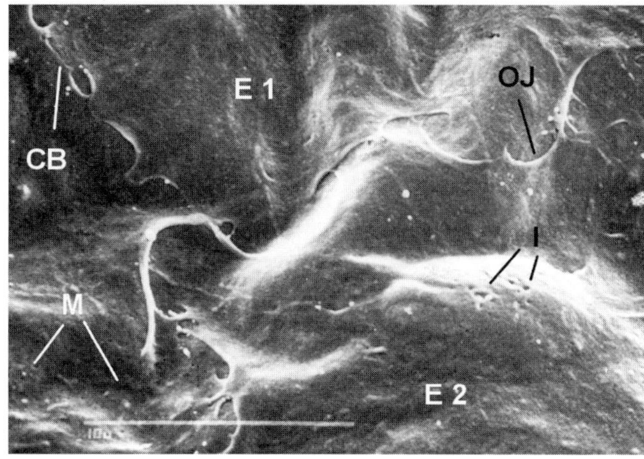

Figure 5 Luminal surface of initial lymphatic endothelium (E) in SEM. I, invagination; OJ, open-junction formations; M, microvilli; CB, cell border.

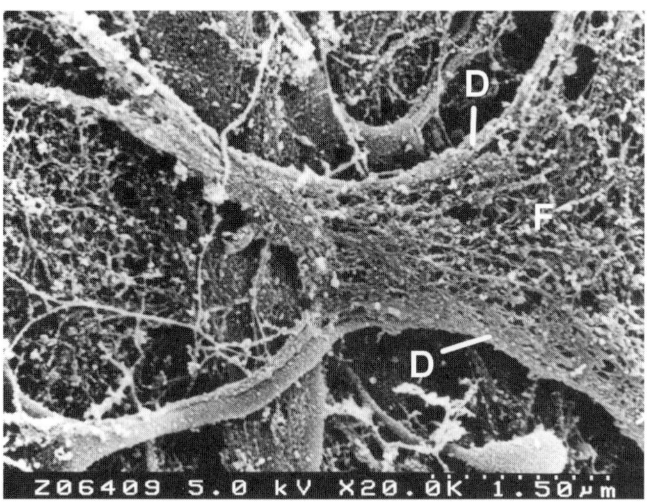

Figure 4 Cytoskeleton of initial lymphatic endothelium in SEM following treatment with Triton X-100. D, dense peripheral band; F, filaments of the cytoskeleton.

Luminal Surface of the Initial Lymphatic Endothelium Cell

On the surface of the initial endothelium cells partly some microvilli, microridges, and microfolds are visible. Numerous invaginations are signs for active transport phenomena. Fenestrations of the cells could not be demonstrated (Figure 5). Otherwise the surface is relatively smooth, covered with a so-called glycocalyx like every cell. It consists of glycoprotein, glycolipids, free polysaccharides, and glycosaminoglycans as part of the cell membrane. This glycocalyx has been the subject of research in the past 10 years as researchers have tried to find specific markers for the lymphatic endothelium to be able to demonstrate the distribution of the lymphatic vessel system in healthy and pathologically altered organs.

Table II Markers for Enzyme- and Immunohistochemical Characteristics of the Initial Lymphatic Endothelium.

5′-Nucleotidase (+ +)
Adenylate cyclase and guanylate cyclase activity (+ –)
Anti-desmoplakin-1 2.17 (+ + +)
Beta-chemokine receptor D6 (+ –)
CD31 (+ –)
CD34 (+ –)
D2-40 (+ –)
Factor VIII associated antigen (+ –)
ICAM-1, -3 (Intercellular adhesion molecule) (+ –)
Lectin binding reaction for UEA I, PNA, DBA, GS I, MPA, and RCA I (+ –)
LyMAb, a G(1) immunoglobulin (+ –)
LYVE-1 (lymphatic vessel endothelial hyaluronan receptor) (+ + +)
Platelet–endothelial cell adhesion molecule-1 (PECAM-1) (+ –)
Podoplanin (+ + +)
Prox-1 (+ –)
VCAM-1 (vascular cell adhesion molecule-1) (+ –)
VEGF-C (+ +)
VEGFR-3 (vascular endothelial growth factor receptor-3) (+ +)
VEGFR-2 (+ –)

+ and –: Valuation for usefulness and distinctness.

Enzyme- and Immunohistochemical Characteristics of the Initial Lymphatic Endothelium Cell

Different enzyme- and immunohistochemical methods have been used to describe the numerous more or less specific characteristics, but only few are typical and selective for the initial lymphatic endothelium to differentiate it from the blood vessel system (Table II). As the lymphatic sinus lacks a continuous basement membrane, immunohistochemistry for extracellular matrix components has also been applied to distinguish them from capillaries of the blood vascular system. Since about 1995, antibodies have been established for a selective immunohistochemical characterization of lymphatics. At present three antibodies seem to be best suitable for detecting lymphatic vessels in human: anti-podoplanin, a polyclonal mouse IgG antibody against a 43-kDa glycoprotein of the lymphatic endothelial cell membrane; anti-desmoplakin-1 2.17, a monoclonal mouse-IgG that binds a 220-kDa protein of the complexes adherentes of the lymphatic endothelia and other epithelial structures; and LYVE-1, a lymph specific receptor for hyaluronan, homolog to the CD 44 glycoprotein.

Cell Culture

Most studies describe the isolation of cells from mesenteric collecting or thoracic ducts, that is, they originate from large vessels. To date, isolation of endothelial cells from initial sinus has failed. Studies of lymph angiogenesis from a cultured large-vessel lymphatic endothelium show spontaneous reorganization into a branching and anastomosing network of capillary-like tubes in planar cultures of bovine and ovine lymphatic endothelium cells. The lumina of such tubes contained extracellular matrix and cell debris. Spontaneous sprouting has been observed from aggregates of cultured lymphatic endothelial cells in three-dimensional collagen gels as well.

Junctions of the Initial Lymphatic Endothelium Cell

The cell branches of the oak leaf–like initial lymphatic endothelium cell clearly overlap each other. Partly they form closed-junction formations, partly open-junction formations, acting as inlet valves for the uptake of tissue fluids and for reflux prevention of the lymphatic fluid respectively.

Closed Junctions

In closed-junction formations, the endothelial cells of the initial lymphatics were connected with zonulae adherentes, electron-dense structures in TEM. The silver deposits of the silver method correlate with these junction structures. Zonulae adherentes, on the one hand, serve as anchoring structures for the dense peripheral band (DPB) of the cytoskeleton and, on the other hand, they are the structural basis for the contact between neighboring cells. The intercellular cleft is about 15 nm wide. In the cell membrane there exist transmembrane molecules (E-cadherin and N-cadherin). The neighboring cadherins are connected to each other in the intercellular space by means of calcium ions (Ca^{2+}). Within the cell, transmembrane molecules are bound to protein linkers on the actin filaments of the DPB.

The closed-junction formations possess a great morphological variability (Figure 6). The elementary form is a vertical contact of two cells with two zonulae adherentes, one in the luminal and the other in the abluminal position. Another most common cell–cell contact structure is a simple overlap of two neighboring cells. In this course two or three zonulae adherentes can be demonstrated, mostly at a distance of 0.3 to 6.5 μm, binding the cells between the apical side of one cell and the basal side of the other. Often a cell border digitates into the neighboring cell border like a finger in an imprint and so the closed junction is single fold. Mostly more than two zonulae adherentes build up. Less

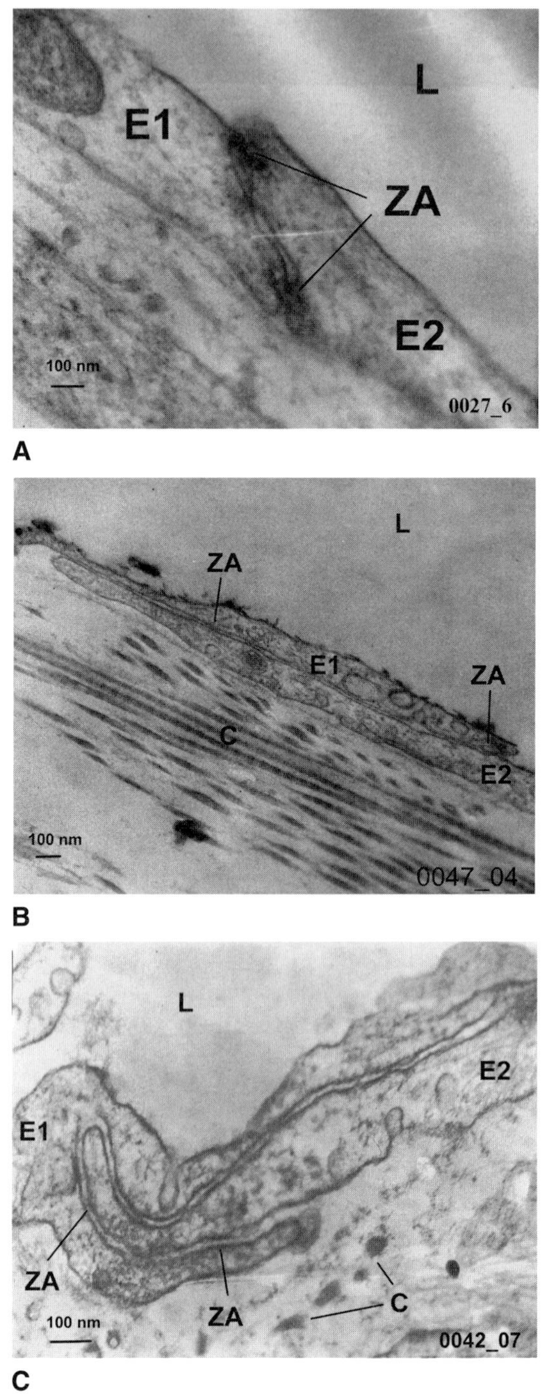

Figure 6 Closed-junction formations. E, endothelial cell; ZA, zonula adherens. TEM. (**A**) Elementary form. (**B**) Simple overlap. (**C**) Single fold.

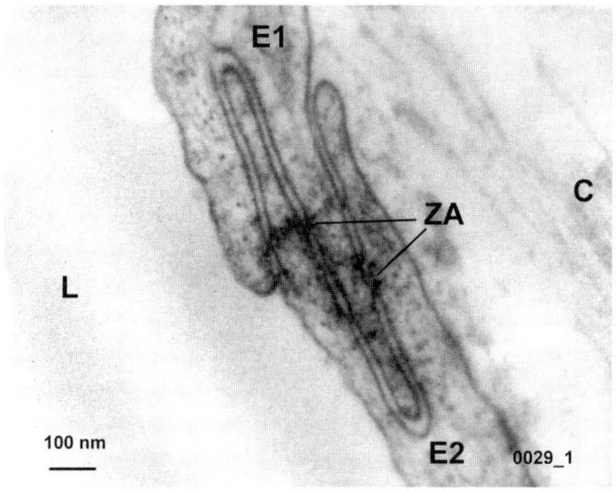

D

Figure 6—*Cont'd.* (**D**) Multifold.

common are closed-junction formations with a multifold system of cell border interdigitations.

OPEN JUNCTIONS

Beneath the zonulae adherentes the cells of the initial lymphatics typically possess inlets where neighboring cells only overlap one another in a distance of about 0.2 up to 1.4 μm, thus forming inlet valves (Figure 7). The diameter of an open-junction formation ranges from 3 to 6 μm. Each cell possesses up to 15 inlet valves in common with its neighbor. So the number of open-junction formations is about 3,750/mm^2. On the abluminal surface anchoring filaments connect the endothelium cell with the filament system of the surrounding tissue via the sparse basement membrane.

Morphological Specialties

In the region of the cell borders spaces or *caveolae* occur without or with an opening toward either the luminal or the abluminal surface. Within these caveolae one can find a clear fluid, membrane-surrounded vesicles, condensed material, and cell detritus or whole cells such as eosinophils. So the caveolae can be developed by migration of cells from the interstitium into the lymphatics, and/or they are spaces for the removal and destruction of older lymph cells due to the initial lymphatics as a part of the reticular endothelial system.

A further specialty of the initial lymphatics is the occurrence of open-interface formations. These are specific structures arising from initial lymphatics, branching and tapering into the surrounding connective tissue, presumably acting as pressure relief valves.

Sometimes sproutlike formations are visible, consisting of spindle-shaped endothelial cells in the beginning region of initial lymphatics. They are 3 μm in width and 12 to 14 μm in length. In this region a basement membrane is not demonstrable. This arrangement may be able to build up new open interfaces or new initial lymphatics within a short time under different conditions.

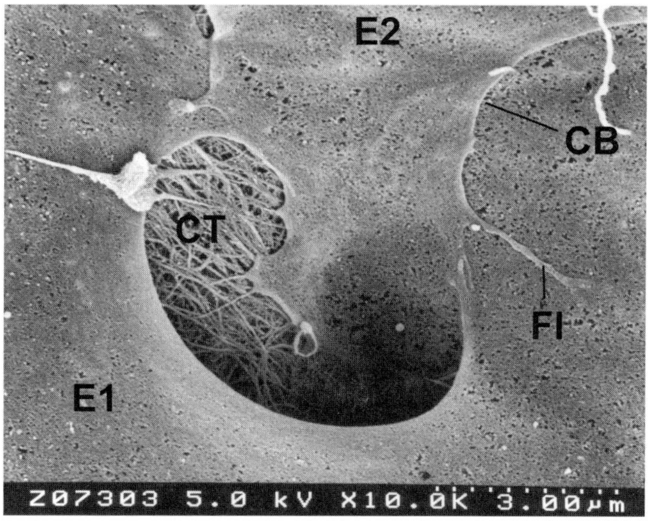

A

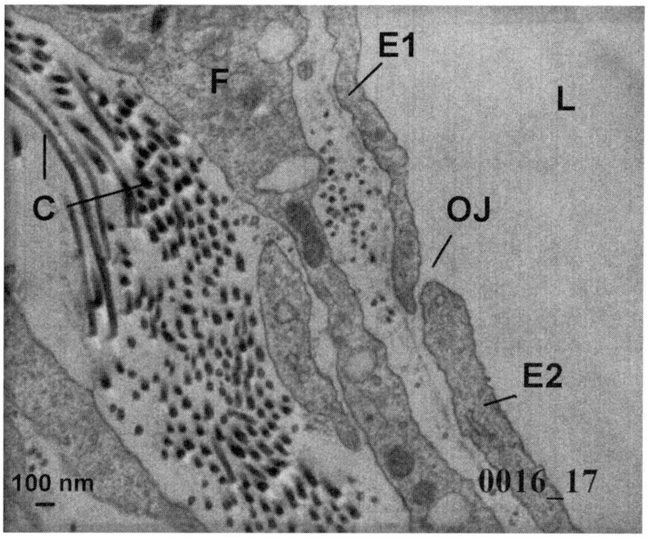

B

Figure 7 Open-junction formation. E, endothelial cell; ZA, zonula adherens; CB, Cell border; FI, filopodium; CT, connective tissue; L, lumen; OJ, open-junction formation; F, fibrocyte. TEM. (**A**) Stretched open-junction formation, due to perfusion pressure. SEM. (**B**) Normal situation. TEM.

Basement Membrane

In contrast to blood capillaries, the basement membrane of the initial lymphatics is sparse and reduced to a few reticular fibers. Interruptions in the basement membrane and their absence over long distances can be demonstrated. In SEM, on the abluminal surface of initial lymphatics a fine network of woven fibers is visible, allowing particles to pass this membrane. Periendothelial cells cannot be demonstrated (Figure 8).

Interrelation to the Connective Tissue

Interwoven in the reticular network of the basement membrane or in direct connection with the lymphatic

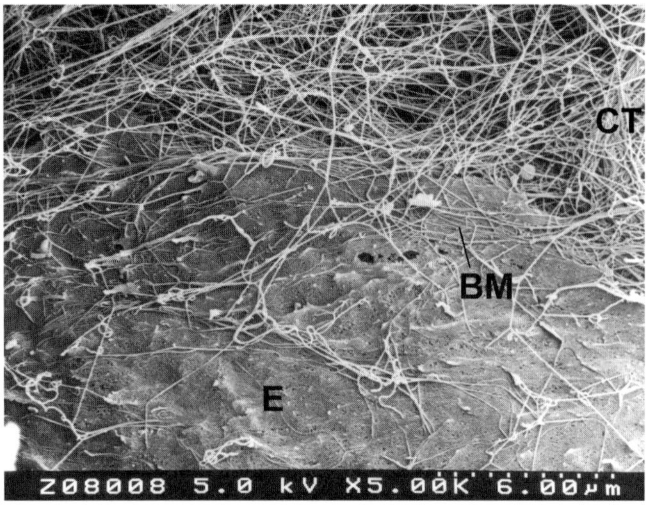

Figure 8 Basement membrane (BM) and connective tissue (CT) at the backside of an initial lymphatic endothelium cell. SEM.

endothelium, there are anchoring filaments with a diameter of 50 to 60 nm, radiating over a distance of 10 µm in the connective tissue around the initial lymphatic. Such anchoring filaments obviously consist of elastic and collagenous components and have a diameter of about 10 nm. The fixation of the anchoring filaments at the outer surface of the initial lymphatics is performed with the aid of adhesion molecules, such as kinase, vinculin, talin, and β-actin. Thus, mechanical signals of the interstitium could be transformed into biochemical signals within the endothelial cell. This elastic fiber system could probably generate intraendothelial stimuli for adaptation, for example in modifying the number of open-junction formations.

In SEM the filaments of the connective tissue around the initial lymphatic build up an irregular network. They can be aggregated to bundles along the vessel wall and extend in some distance between collagenous bundles and fibrocytes of the connective tissue.

Elastic fibers in the neighborhood of the initial lymphatics form longitudinally orientated networks mainly in a distance of 5 µm, whereas blood vessels do not have such elastic networks.

Lymph Formation

Exchange of Fluids between the Blood- and Lymph-Vessel System

In the interstitial space most of the extracellular matrix is in a gelatinous phase. Within this gel-interstitium finer and larger channels and spaces branch, filled with a watery phase of the extracellular matrix. Von Recklinghausen (1862) and Kihara [6] described these structures as "Saftkanälchen" (little juice channels), and Hauck [7] as "tissue channels" or "low-resistance pathways," underscoring their function of facilitating the transport of molecules and particles through the interstitium. This prelymphatic system normally contains about 1 percent free fluid; 99 percent is bound by the gelatinous phase of the extracellular matrix. So in TEM and SEM regular spaces are visible, lined only by a net of fibers or incompletely lined with widespread endothelium-like fibrocytes. Within these tissue channels the so-called "lymphatic load" is transported toward the initial lymphatics with their open-junction and partly open-interface formations.

Function of the Open-Junction Formations

The open-junction formations build up a free transport path for substances from the tissue spaces directly into the initial lymphatic, guaranteeing the quick removal of the lymphatic load under normal conditions. Of the surface of an undilated initial lymphatic, 2.3 percent is part of the inlet zones. Because of the close embedding in the fiber network of the connective tissue, in the case of an increasing lymphatic load the initial lymphatics are stretched in all directions by the transferred forces over the anchoring and collagenous fibers interwoven in the basement membrane. The resulting dilatation leads to a widening of the open-junction formations of the initial lymphatics. Under such physiological conditions, a better uptake of the lymphatic load results, because these structures prevent edema under normal conditions. When the pressure within the initial lymphatic is higher than in the surrounding tissue, the open-junction formations act under normal conditions as a one-way system and the valves will close. The elastic fiber system around the lymphatics stores the energy during dilatation and restores this energy in the relaxation phase. So the lymph will be propelled in proximal direction and the cycle starts again (Figure 9).

Vesicular Transport Phenomena

Lymph formation is a result not only of the number and function of open-junction formations but also of active transport phenomena. On both the luminal and abluminal surfaces of the initial lymphatic endothelium fine pores are visible, interpreted as vesicular uptake or delivery of substances. So a protein enrichment of the lymph with selectively, actively transported substances results. This leads to a higher protein concentration within the lymph, supporting the uptake of lymphatic load along the oncotic gradient. In all these processes ER, the Golgi-apparatus, and lysosomes are involved. The coated pits in TEM are responsible for the receptor-controlled endocytosis for the uptake of, for example, cholesterin, hormones, toxins, virus, carrier proteins, and antibodies.

Conclusions to the Phenomena of Lymph Formation

Several mechanism are responsible for lymph formation

1. Fluid movement along a hydrostatic pressure gradient (theory of Starling, 1894 and 1896)
2. Fluid movement along a colloid osmotic gradient (theory of Casley-Smith, 1977 and 1982)
3. Open-junction system as passive and active structures
4. Open interface formations

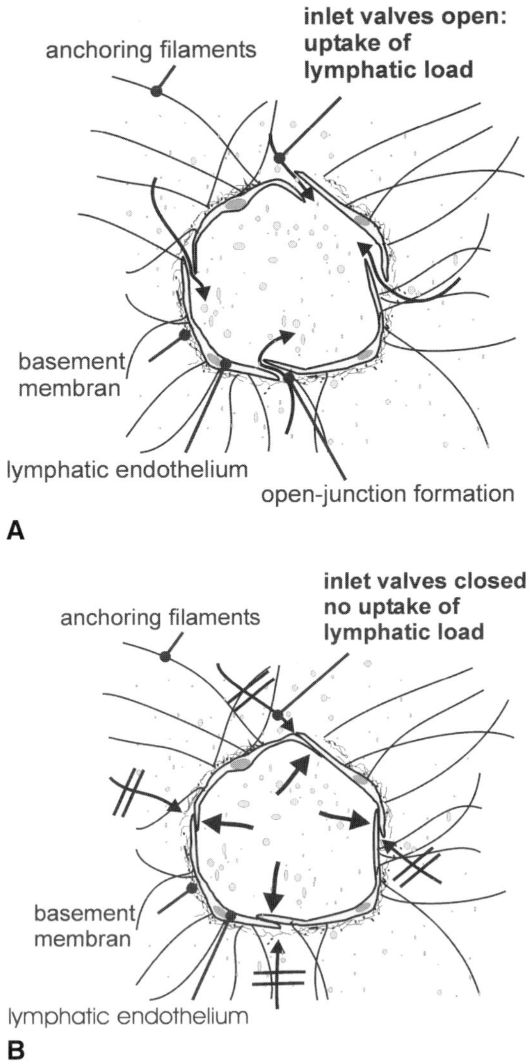

Figure 9 Function of open-junction formations. **(A)** The pressure in the initial lymphatic is lower than in the connective tissue. Inlet valves are open: uptake of lymphatic load. **(B)** Increased pressure in the initial lymphatic. Inlet valves are closed: transport of lymph toward the precollector. (see color insert)

5. Vesicular transport phenomena of the initial lymphatic cell
6. Phagocytosis and deposition of particular material

These functional properties of the lymphatic endothelium cell and the sparse basement membrane, the lack of pericytes, and all the other properties described make it clear that the initial lymphatics are not capillaries in the common sense. The term "capillary" should be applied only for the finest vessels of the blood vessel system, which possess totally different properties than the initial lymphatics or much better the "initial lymph sinus." So far, the endothelium of initial lymphatics is characterized as a flexible structure with the ability to respond to higher tissue fluid demands and the need of transporting more lymph, if necessary.

The Endothelium of the Precollectors

Precollectors are structures of different length between initial lymphatics and lymph collectors. Their irregular diameter can be from 20 to 100 μm, and often they are in close contact with arterial vessels.

Two shapes of endothelial cells are present. In cells of the distal part an oak-leaf pattern dominates, whereas proximal cells are more stretched in the flow direction of the lymph and are preferentially rhombic, similar to the venous endothelium. Open-junction formations can regularly be seen in the more distal part, acting as inlet valves. A specialty of the precollector endothelium is the development of trabecular and valvelike structures, giving the lumen a labyrinthine structure, looking like an eel basket. The building of these structures is managed by spindle-shaped modified endothelial cells based in the endothelium and running through the lumen of the precollector. Distal valves are more incomplete structures like ridges in the luminon; proximal to the collector they become sufficient and more are visible. Here, the distance between these valves is about 130 μm. This valve system ensures a more proximal flow direction of the lymph. Additionally to the incomplete valves, the trabecular system generates turbulences of the lymph stream so that all substances within the lymph (molecules, particles, cells) contact the endothelial surface, which possesses immunological and resorptive properties. In this way, there exists a controlling system for the prenodal lymph, proofing this fluid, and with the capacity of unspecific as well as specific immunological reactions (Figure 10).

The nuclei of the precollector endothelium are oval and oriented longitudinally. Their diameter is about 1.2 to 1.9 μm. Numerous vesicles, particularly near the cell borders, are typical. The endothelium is rich in pinocytotic vesicles, and the luminal surface of the endothelium often possesses some microvilli, indicating active transport processes. The cytoskeleton of the precollector endothelium, in contrast to the initial lymphatic endothelium, is directed more along the length of the cells. Partially interendothelial caveolae are visible, including especially multilamellar bodies, perhaps relics of destroyed cells or material. A thin discontinuous basement membrane is regularly visible, as are anchoring filaments. Sometimes smooth muscle cells adjacent to the basement membrane can be seen. Their occurrence is not related to the site of valves. Because of the lack of a continuous muscular wall like the collectors, they use different forces for the propulsion of lymph fluid. Often precollectors are arranged between two arterial vessels or an arterial vessel is arranged via a mesangion in the lumen of the precollectors. So, because of the pulsation of the bloodstream, the rhythmical compression displaces the lymph fluid, which can flow only in the proximal direction because of the valvelike structures or complete valves. The typical fiber system of the connective tissue surrounds the precollector. Thus, the precollector endothelium is a particular structure, neither initial nor collector endothelium, involved in the fluid absorption and lymph propulsion processes.

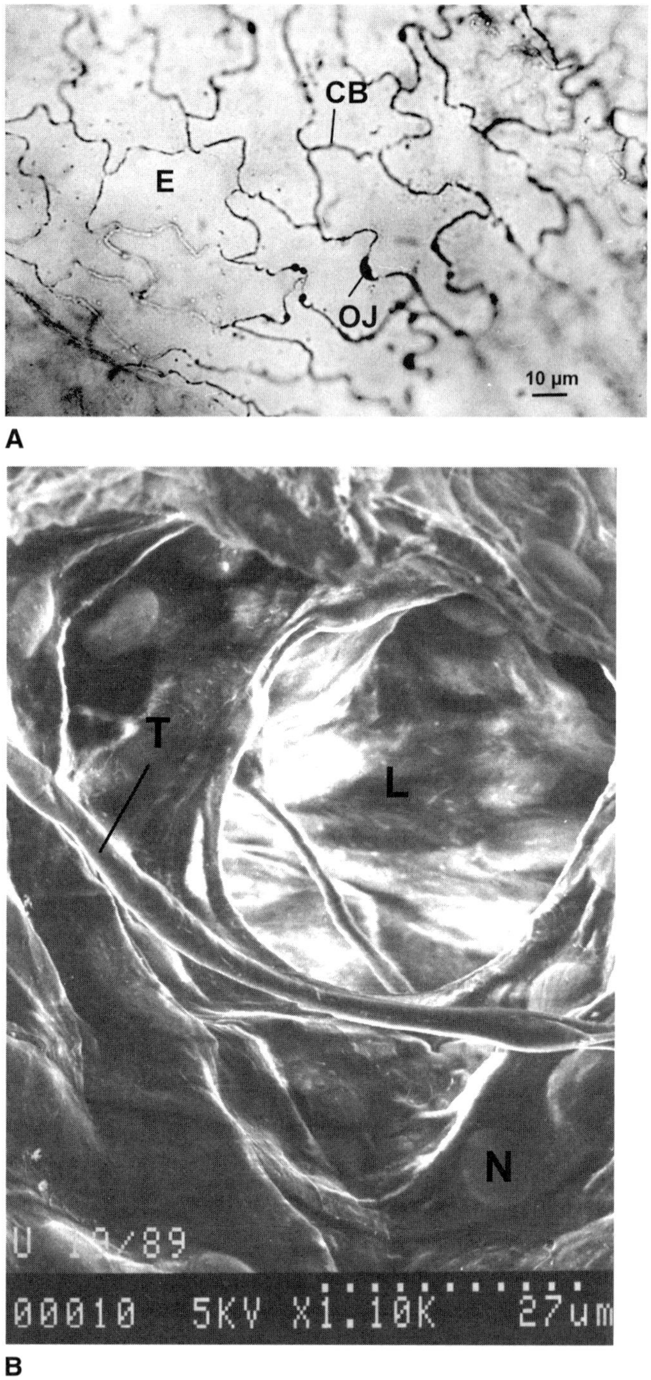

Figure 10 Precollector. E, endothelial cell, CB, cell border, N, nucleus, T, trabecule, L, lumen. **(A)** The cell border line (CB) is contrasted with silver. LM. **(B)** Inside view of the lumen of a precollector. SEM.

The Endothelium of the Collectors and Lymphatic Stem Vessels

These parts of the lymphatic vessel system possess regular valves, an intima, a media, and an adventitia. Their diameter is about 40 to 100 µm, the thickness of the wall about 2 to 3 µm. The rhombic shape of the endothelial cells is similar to that in the venous system. All collectors possess a well-developed continuous basement membrane, often in close contact with collagenous and elastic fibers. Collectors have to carry the lymphatic fluid into the venous vessel system. The greatest lymph stem is the ductus thoracicus, which transports about 2.5 L lymph per day. The lymph is propelled by intrinsic forces of these vessels, which have their morphological correlation in the lymphangion structure. Each of those segments consist of parts between two valves (from 0.2 to several millimeters) and a strong muscular cuff. A slight dilatation of the vessel wall of a lymphangion produces a contraction of the vessel wall. So the lymph is propelled in both directions, backward and forward. The act of closure of the distal and the opening of the proximal valves carries the lymph into the nearest proximal lymphangion. This autonomically repeated rhythmic action can modify by supplying a network of nervous fibers. The frequency of this action is 6 to 10 per minute, but values of 15 to 20 per minute are also possible. The resulting pressure is between 1 and 25 mmHg, normally 3 to 5 mmHg. This process was first described by Mislin and Rathnow [8]. Pulsation of neighboring arteries and muscle and respiratory action are additional amplifying factors. So insufficient lymph flow can be supported by massage for treatment of edema.

The Endothelium in the Lymph Nodes

Lymph nodes are interconnected in collectors and are controlling stations for the lymph fluid. Their number is estimated as 600 to 700; morphology and size are variable. Lymph nodes are encapsulated structures and incompletely subdivided by trabecules, between which a cellular network exists containing white blood cells. These can react to antigens via humoral and cellular immune responses, and small particles in the lymph are eliminated as well. Lymph nodes consist of a cortical zone (B-cell zone with primary and secondary follicles), a paracortical zone (T-cell zone), and a medullary region. Vasa afferentia lead the primary lymph to the marginal sinus; from there it runs through the trabecular sinus via intermediary sinus into the medullary sinus, draining to vasa efferentia. An exact and complete description of the endothelium of the lymphatic parts of the lymph nodes does not yet exist. The marginal sinus possesses a morphology similar to that of the precollectors. The intermediary lymph sinus endothelium has gaps so that the lymph fluid can be in contact with the reticular cells of the lymph node. Additional lymph nodes can concentrate the lymph proteins three- to fourfold. In the lumen of the lymphatics of the lymph node, preferentially macrophages (particularly in the medullary sinus), a variable number of erythrocytes, neutrophils, lymphocytes, and plasma cells occur.

Glossary

Anchoring filaments: A characteristic feature of initial lymphatic vessels. They connect the abluminal membrane of endothelial cells to the surrounding elastic fibers. The main molecular component is fibrillin.

Basement membrane: Extracellular matrix characteristically found under epithelial cells. There are two distinct layers: the basal lamina, immediately adjacent to the cells, is a product of the epithelial cells themselves and contains collagen type IV and the reticular lamina is produced by fibroblasts of the underlying connective tissue and contains fibrillar collagen.

Caveolae: Small invaginations of the plasma membrane in many cell types, especially in endothelial cells. These flask-shaped structures are rich in proteins and lipids and are used for several functions in signal transduction (Anderson, 1998) They are also believed to play a role in endocytosis, oncogenesis, and the uptake of pathogenic bacteria.

Connective tissue: Any type of biological tissue with an extensive extracellular matrix. There are several basic types. Loose connective tissue holds organs and epithelia in place, and has a variety of proteinaceous fibers, including collagen and elastin.

Dense peripheral band (DPB): Dense 150–300 nm broad band, following the circumference of the cell, built up by cytoskeletal filaments (mostly actin-filaments).

Elastic fibers: Fibers which are capable of returning to their original length after being stretched. The protein molecules which compose these fibers are synthesized in fibroblasts and smooth muscle cells. They are not found in bundles but occur as solitary fibers.

Elementary form: Very simple formation of cell-cell-connection, without any peculiarities. Vertical contact zone between two cells.

Filaments: Cytoskeletal elements like intermediate filaments, thin actin filaments and microtubules. Frequently the three components work together to enhance both structural integrity, cell shape, and cell and organelle motility.

Multifold system: Cell-cell-connection with interdigitating structure.

Nucleus: Largest cell organelle, found in the majority of eukaryotic cells, which contains most of the cell's genetic material. Nuclei have two primary functions: to control chemical reactions within the cytoplasm and to store information needed for cellular division.

Open-interface formations: Pressure relief valve with direct connection between tissue channels and initial lymphatics.

Open-junction formations: Cell borders of neighboring cells overlap one another, thus forming inlet valves.

Shapes of endothelial cells: Initial lymphatics possesses the circumference similar to an oak leaf.

Simple overlap: Cell-cell-connection. The apical part of one cell is connected with the basal part of another cell

Single fold: Cell-cell-connection. The cell border of one cell digitates into the neighboring cell border like a finger in an inprint.

Sproutlike formations: Bud-like cell formation, extended from an initial lymphatic, consisting of circularly arranged spindle-shaped endothelial cells.

Tissue channels: 1% of extracellular matrix, containing free fluid otherwise bound on the gelatinous phase.

Trabecular and valvelike structures: Trabecular structure: spindle-shaped modified endothelial cells, based in the endothelium and running through the lumen to connect with an endothelial cell in an opposite position. Valvelike structure: spindle-shaped modified endothelial cells, forming a ridge, acting as uncompleted valve.

Vesicles and invaginations: Vesicles are relatively small and enclosed cell compartments, separated from the cytosol by at least one lipid bilayer. Vesicles store, transport, or digest cellular products and wastes. Invaginations are the folding in of the plasmalemma so as to form a pocket in the surface (surface vesicle).

References

1. Drinker and Field (1933).
2. Drinker and Yoffey (1941).
3. Yoffey and Courtice (1970).
4. Rusznyák, Földi, and Szabó (1969).
5. Wong and Gottlieb (1986).
6. Kihara (1956).
7. Hauck (1973).
8. Mislin and Rathnow (1961).

Further Reading

Azzali, G. (1992). Morphological features of absorbing peripheral lymphatic vessels studied by TEM, SEM, and three-dimensional models. In *Scanning Electron Microscopy of Vascular Casts: Methods and Applications*, pp. 83–98. Dordrecht: Kluwer Academic.

Cahill, R. N. P., Wayne, W. G., and Hay, J. B. (2001). Lymphatic system. In *Encyclopedia of Life Sciences*. Nature publishing group: www.els.net. *A good overview, easily obtained on the Internet.*

Castenholz, A. (1998). Functional microanatomy of initial lymphatics with special consideration of the extracellular matrix. *Lymphology* **31**(3), 101–118. *This study is based on scanning electron microscopy and confocal laser scanning microscopy and gives insights into the architecture of the fibrous network of the extracellular matrix and its functional features. The author proposes a concept that considers the histomechanics of the initial lymphatics with the surrounding connective fiber tissue describing the structural basis for the permeability of the lymphatic vascular wall.*

Földi, M., and Casley-Smith, J. R. (eds.) (1983). *Lymphangiology.* Stuttgart: Schattauer. *The aim of this book is to build bridges between basic scientists, radiologists, and clinicians. It is a classical handbook with more than 800 pages. Themes of the first ten chapters are phylogeny; structure and functioning of the blood vessels, interstitial tissues, and lymphatics; the lymphangion; enzymes in lymph; insufficiency of lymph flow; general pathology of the lymph vascular system; injury and the lymphatic system; the lymphatic system in burns; and special pathology of the lymph vascular system with anatomical, physiological, and pathological aspects. Twelve chapters deal with anatomy, physiology, pathophysiology, and therapy of special organs. Further chapters are concerned with lymphedema; surgical therapy for lymphedema; pharmacology of the lymphatics; nuclear medicine techniques; thoracic duct cannulation; lymphatic filariasis; malformation of the lymphatic system; and pediatric lymphangiology. It seems to be the most actual book on these issues about the lymphatic system in English. In German the 5th edition of* Lehrbuch der Lymphologie, *by Földi, M., and Kubik, S., was published in 2002 (München: Urban & Fischer ISBN: 3-437-45321-1).*

Gerli, R., Solito, R., Weber, E., and Aglianó M. (2000). Specific adhesion molecules bind anchoring filaments and endothelial cells in human skin initial lymphatics. *Lymphology* **33**(4), 148–157. *Anchoring filaments connect the abluminal membrane of endothelial cells to the surrounding elastic fibers and possibly enable transmission of chemical and/or mechanical stimuli from the extracellular matrix to the endothelial cells. So they could contribute to the initial formation of lymph.*

Kato, S. (2000). Organ specificity of the structural organization and fine distribution of lymphatic capillary networks. Histochemical study. *Histol. Histopathol.* **15**, 185–197. *The histochemical method of analyzing the topography of lymphatics. The lymphatics and blood vessels are characterized by an enzyme-histochemical method and the results are discussed in relation to their ability to demonstrate the organ specificity of vascular networks under normal and pathological conditions.*

Pepper, M. S. (2001). Lymphangiogenesis and tumor metastasis. *Clin. Cancer Res.* **7**, 462–468.

Podgrabinska, S., Braun, P., Velasco, P., Kloos, B., Pepper, M. S., Jackson, D. G., and Skobe, M. (2002). Molecular characterization of lymphatic endothelial cells. *Proc. Natl. Acad. Sci. USA* **99**, 16069–16074. *Immunoselection of primary lymphatic and blood microvascular endothelial cells from human skin with the lymphatic marker LYVE-1. Description and classification of the differently expressed genes into functional groups. The characteristic gene expression profile of human lymphatic microvascular endothelial cells indicates a more active role of lymphatic endothelium in uptake and transport of molecules than previously anticipated.*

Schmidt-Schönbein, G. W. (1990). Microlymphatics and lymph flow. *Physiol. Rev.* **70**, 987–1028. *Review of the morphological and physiological basis of lymph production and lymph transport. Description of the mechanism that causes expansion and compression of the initial lymphatics.*

Zöltzer, H. (2003). Initial lymphatics—morphology and function of the endothelial cells. *Lymphology* **36**, 7–25. *Lymph formation usually is considered as a passive process. The cells of the initial lymphatics, however, hold a key position in absorbing fluid from the interstitial space. This study suggests that the forming and closing of the open-junction formations is an active component of the lymphatic endothelium.*

Capsule Biography

Dr. Zöltzer has been engaged in lymphological research since 1985. He is head of the working group "Theoretical Lymphology" of the German Association of Lymphology (DGL).

Color Insert

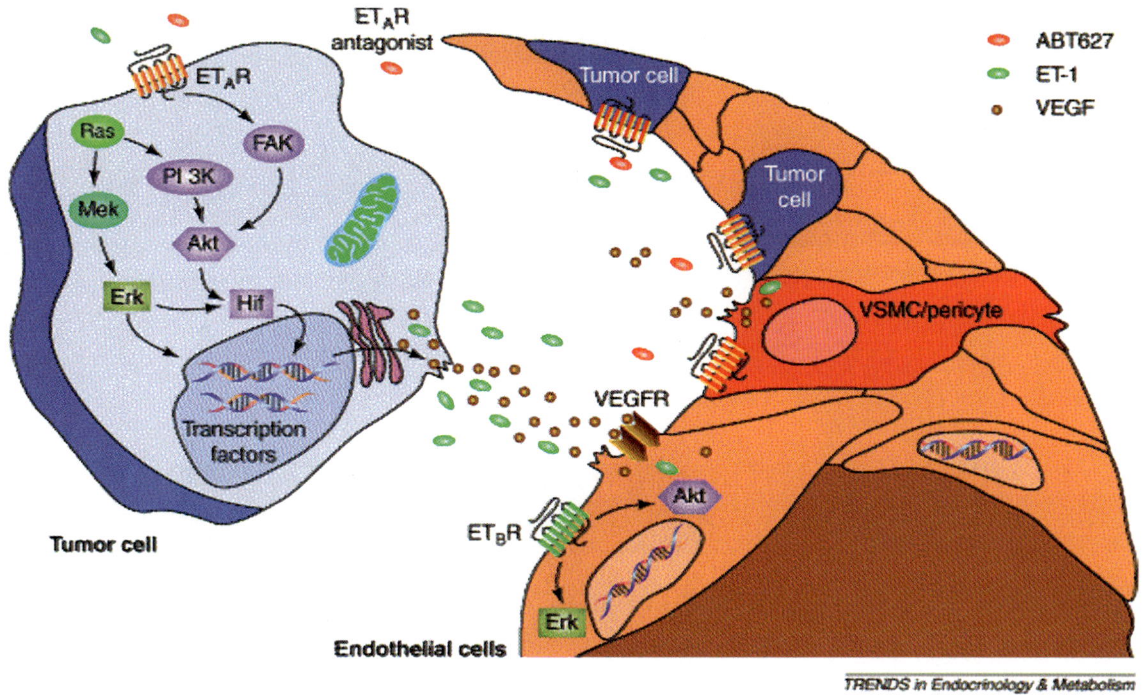

Chapter 7 Figure 2 Proposed Overview of the Possible Role of Endothelin Signaling in Tumor Angiogenesis. This process involves a series of linked signaling pathways that typically begins with upregulation of the synthesis of angiogenic factors, such as ET-1 and VEGF, that are released by tumor cells. ET_A receptor (ET_AR) activation by ET-1 could promote angiogenesis by increasing VEGF production through HIF-1-dependent mechanisms. ET-1 binds to specific ET_B receptors (ET_BR) on endothelial cells and to ET_A receptors on pericytes/vascular smooth muscle cells or on tumor cells lining vessels (vasculogenic mimicry). ET_A receptor blockade (*ABT627*) could exert antiangiogenic effects by inhibiting VEGF production and by acting on VSMC and on microvascular channels lined by tumor cells. Another cascade of signaling events is then initiated, which leads to the formation of new tumor vessels. *Erk*, extracellular signal-related kinase; ET-1, endothelin-1; Mek, mitogen-activated protein kinase kinase; *Akt*, protein kinase B; Hif, hypoxia-inducible factor-1; VEGF, vascular endothelial cell growth factor; VEGFR, vascular endothelial cell growth factor receptor. Reproduced from reference [8] with permission.

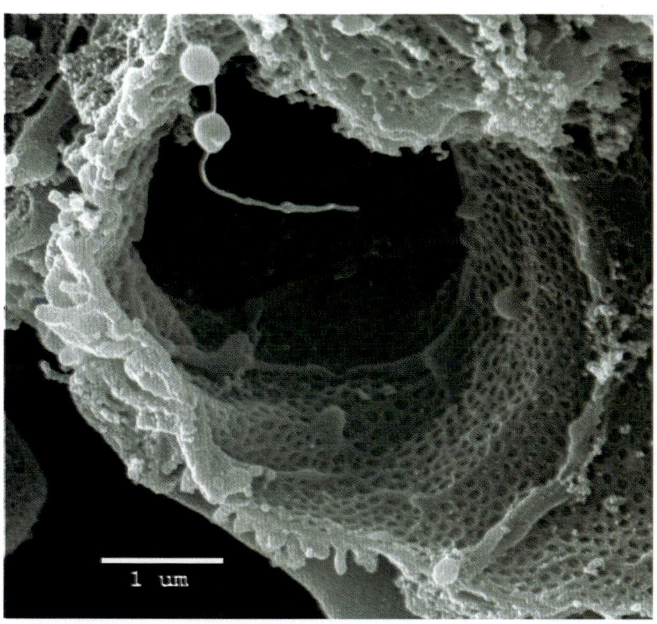

Chapter 10 Figure 1 A scanning electron micrograph from a freeze-fractured glomerulus shows a view down the lumen of a fenestrated capillary. The endothelium appears to have a colander-like structure that permits passage of materials from the blood to the glomerular capsule. (We thank Steve Gschmeissner at Cancer Research UK for this data.)

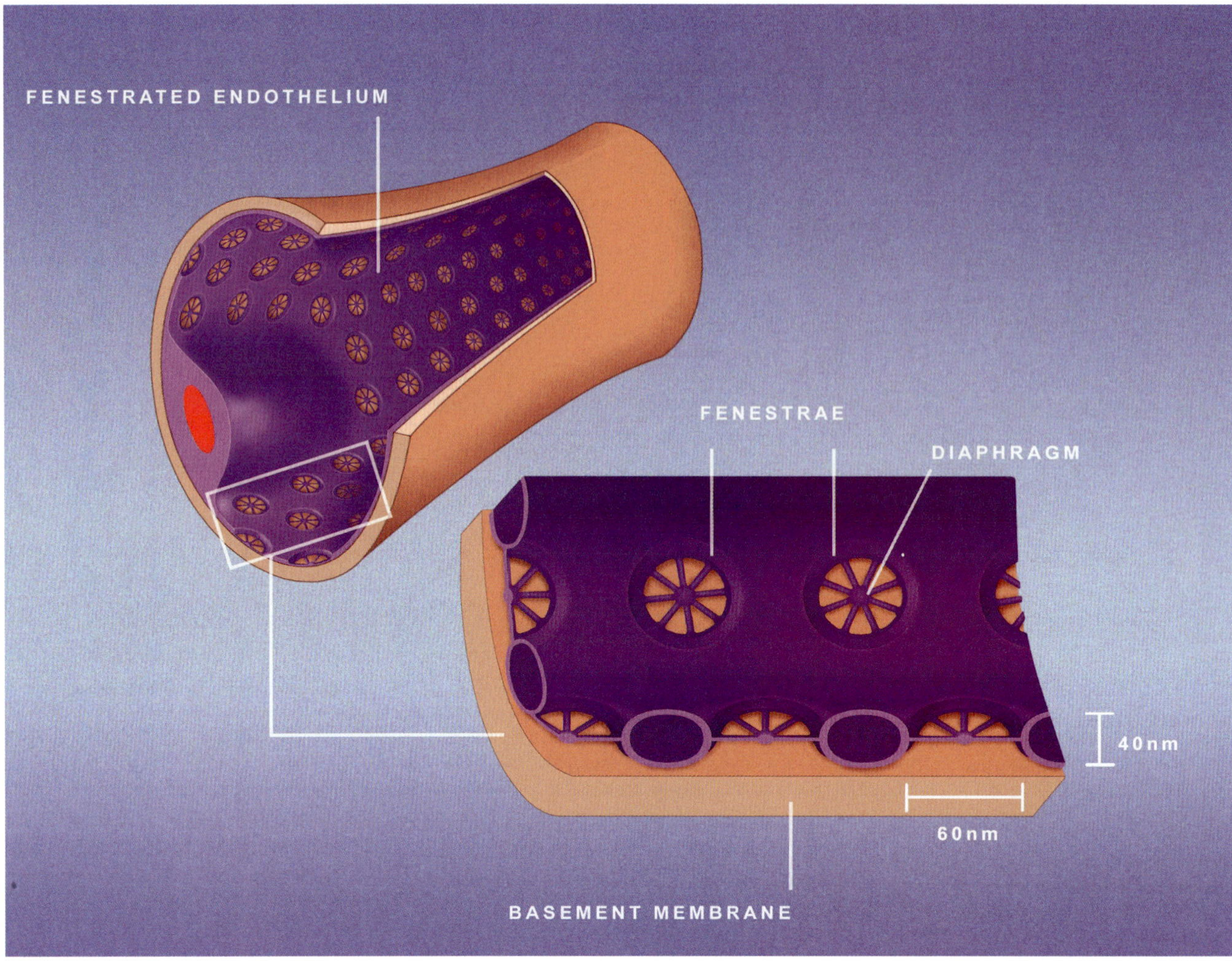

Chapter 10 Figure 2 Schematic representation of a capillary with fenestrated endothelium and underlying basement membrane. Fenestrae are found only at the most attenuated regions of endothelial cells, where the nucleus and organelles are excluded and the distance from apical to basal plasma membrane is as little as 40 nm. They occur in groups and are arranged in a near-linear fashion with precise spacing between them. *Inset* shows the fenestral pore to be 60 nm in diameter, with an effective size of 5 to 6 nm when apertured by a diaphragm. [Adapted from Rhodin (1962), *J. Ultrastructure Res.* **6**, 171–185.]

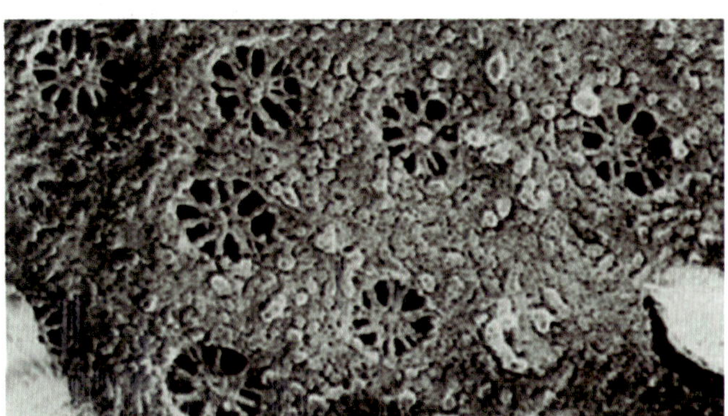

Chapter 10 Figure 3 Fenestral diaphragms revealed by quick-freezing and deep-etching of rat kidney peritubular capillaries. The intertwining fibrils bridging the pore and converging into a central mesh are apparent in this luminal view. Magnification: 120,000×. [Courtesy of Drs. Bearer and Orci (1985), *J. Cell. Biol.* **100**, 418–428.]

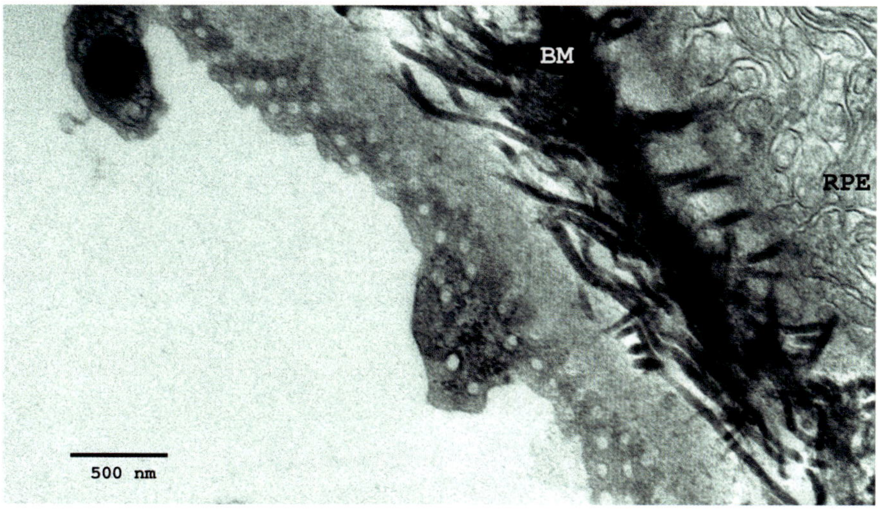

Chapter 10 Figure 4 Fenestrated endothelium of choriocapillaris. This image reveals the close association of diaphragmed pores to Bruch's membrane (BM) and microvillar base of retinal pigmented epithelium (RPE). (We thank associate Eunice Cheung for this data.)

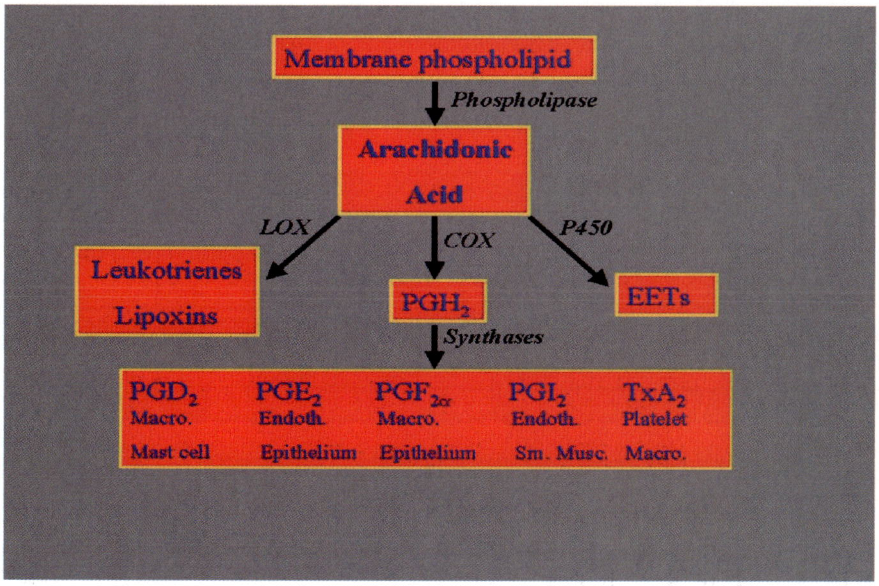

Chapter 13 Figure 1 Enzymatic metabolism of membrane phospholipid, via arachidonic acid, to form eicosanoids. Specific enzymes are shown in italics (LOX = lipoxygenase; COX = cyclo-oxygenase; P450 = cytochrome P450; "Synthases" includes PGD synthase, PGE synthase, PGF2 isomerase, prostacyclin synthase, and thromboxane synthase.

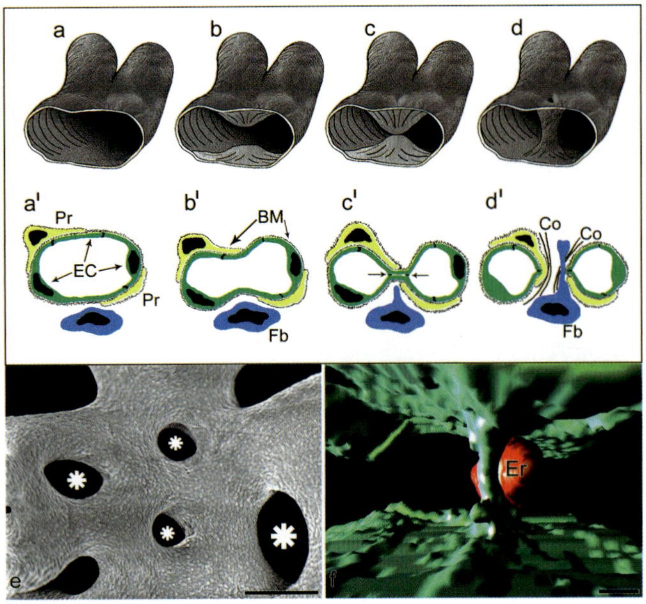

Chapter 15 Figure 1 Three-dimensional **(a–d)** and two-dimensional **(a′–d′)** drawings illustrating intussusceptive pillar formation. The process begins with the protrusion of opposite capillary walls into the vessel lumen **(a, b, a′, b′)**. After establishing an interendothelial contact **(c, c′)**, the endothelial bilayer and the basement membranes (BM) are perforated centrally. The newly formed pillar increases in girth after being invaded by pericytes (Pr) and fibroblasts (Fb), which produce collagen fibrils (Co) **(d, d′)**. Corresponding three-dimensional visualization of transluminal pillars in capillary plexus of chicken CAM. In corrosion casts of blood vessels, newly formed transcapillary pillars appear as small holes (*asterisks*) **(e)**; three-dimensional reconstruction of a pillar based on transmission electron microscopic analysis of serial sections **(f)**. An erythrocyte (Er) is seen behind the pillar. Bars: e and f = 5 µm. (a–d′). Reproduced from Kurz et al. 2003 [8] and (e, f) from Djonov et al. 2000 [4] with the authors' and publisher's permission.

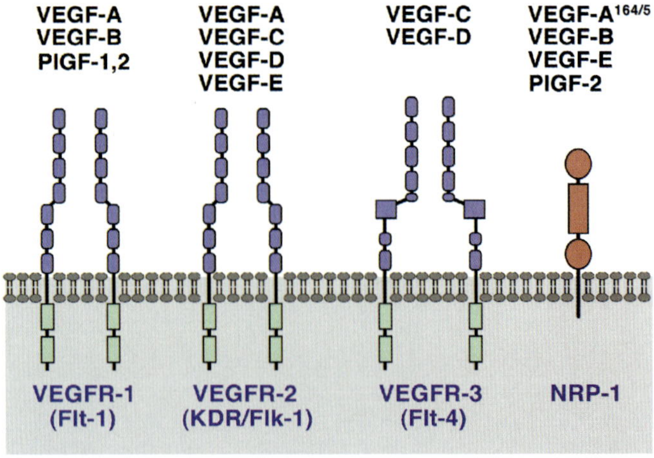

Chapter 16 Figure 2 Binding Specificities of the VPF/VEGF Ligand and Family Members and Their Receptors. VEGFR-1 and VEGFR-2 are expressed on vascular endothelial cells as well as on some other cell types; VEGR-3 is expressed on lymphatic endothelium in the adult but also on some tumor vessels. NRP-1 is expressed on vascular endothelium, neurons, and some tumor cells (see text for further details). [Republished with permission from H. Dvorak (2002), *Journal of Clinical Oncology* **20**, 4368.]

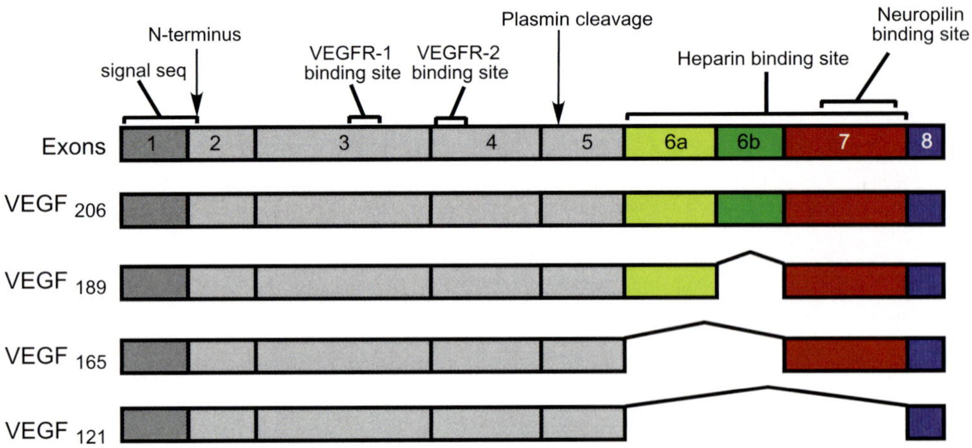

Chapter 16 Figure 1 VEGF-A Gene Structure. VEGF-A has eight exons, the first of which encodes a hydrophobic leader sequence, typical of secreted proteins. The 206 amino acid VEGF-A isomer is encoded by all eight exons. VEGF-A^{189} lacks a portion of exon 6, whereas VEGF-A^{165}, frequently the predominant isoform, lacks all of exon 6. VEGF-A^{121} lacks both exons 6 and 7. VEGF receptor and heparin binding sites are as indicated. [Republished with permission from Cross et al. (2003), *Trends in Biochemical Sciences* **28**, 488.]

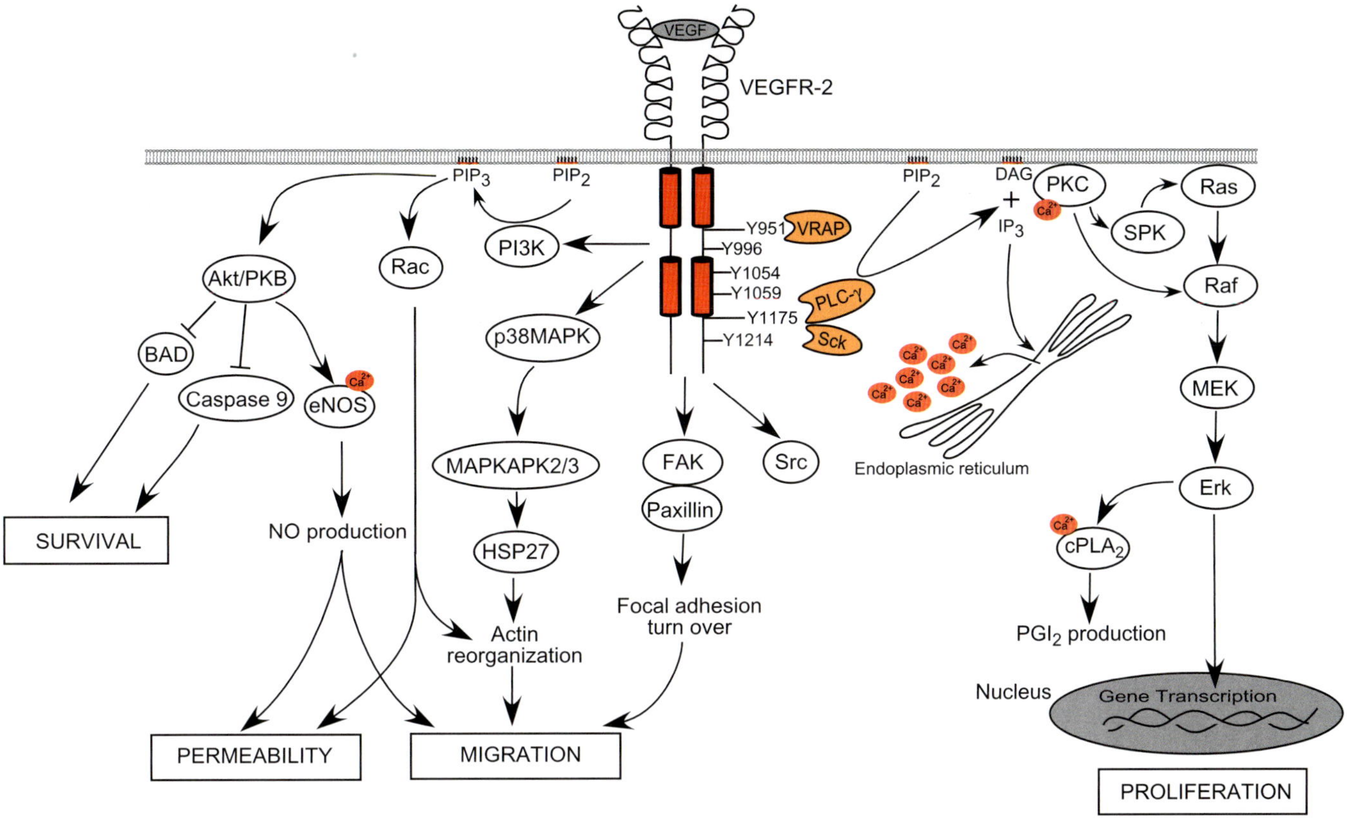

Chapter 16 Figure 3 Schematic Diagram of VEGFR-2 Signaling. Ligand binding to the extracellular domain induces dimerization, autophosphorylation of specific intracellular tyrosine residues, and a series of downstream phosphorylations and other events that lead to endothelial cell survival, migration, proliferation, and increased permeability. [For further details, see Cross et al. (2003), *Trends in Biochemical Sciences* **28**, 488, from which this diagram is republished with permission.]

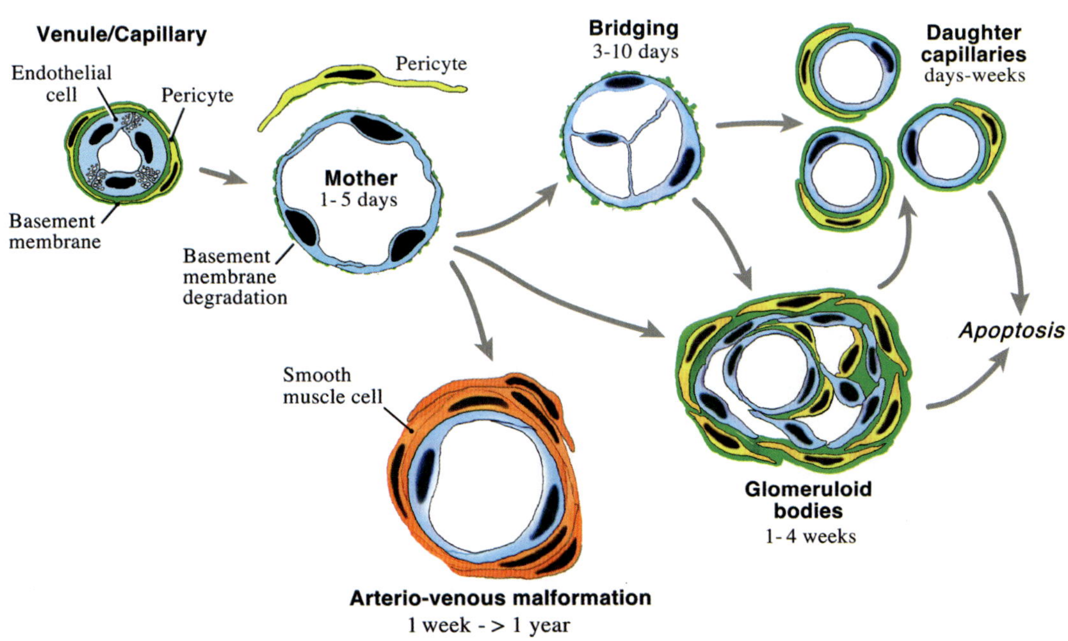

Chapter 16 Figure 4 Schematic diagram summarizing the progression of the angiogenic response that follows introduction of a VEGF-A^{164}-expressing adenoviral vector in vivo. Mother vessels develop and may evolve into glomeruloid bodies, vascular malformations, and daughter capillaries. Finally, as VEGF-A^{164} expression wanes, glomeruloid bodies undergo apoptosis, whereas malformations achieve VEGF-A independence and persist indefinitely. [Republished in modified form from Pettersson et al. (2000), *Laboratory Investigation* **80**, 99].

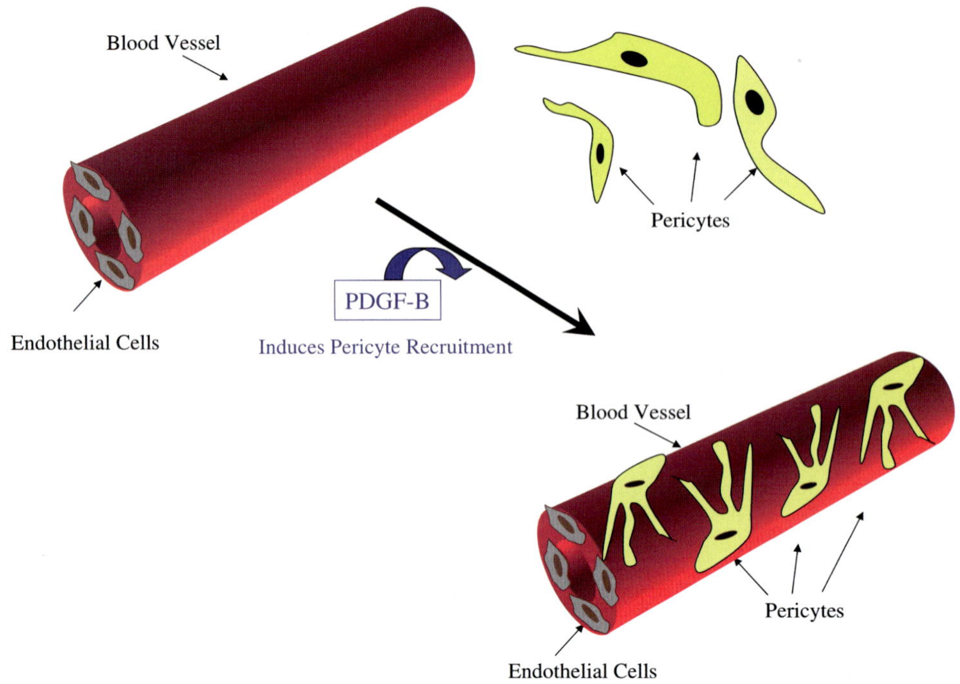

Chapter 19 Figure 1 Schematic Representation of Pericyte Recruitment to a Newly Formed Vessel. PDGF-B released from endothelial and other types of cells acts as a chemoattractant by stimulating PDGFR-β on vascular support cells called pericytes. This stimulus causes the recruitment of pericytes that encase and structurally stabilize the growing vessel.

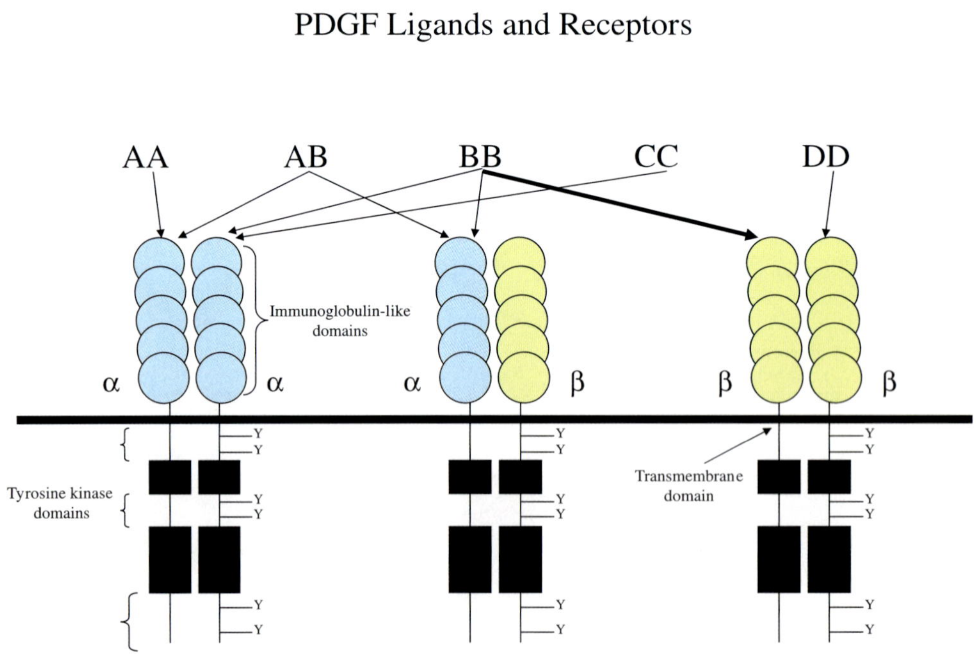

Chapter 19 Figure 2 Schematic Representation of the Binding of PDGF Ligands to Their Respective Receptors. PDGF-AA binds only the -αα receptor homodimer. PDGF-BB binds to the -αα and -αβ receptors but has the highest binding affinity for the -ββ receptor. PDGF-AB preferentially binds to -αα and -αβ receptors. PDGF-CC and PDGF-DD interact with the -αα and -ββ receptors, respectively. The binding of PDGF to the extracellular immunoglobulin-like domains of its cognate receptors induces receptor homo- or heterodimerization, stimulating tyrosine phosphorylation in the intracellular kinase domains. The phosphorylated tyrosines trigger downstream signaling events to occur cellular events such as mitogenesis, chemotaxis, and apoptosis. (-Y, tyrosine residues)

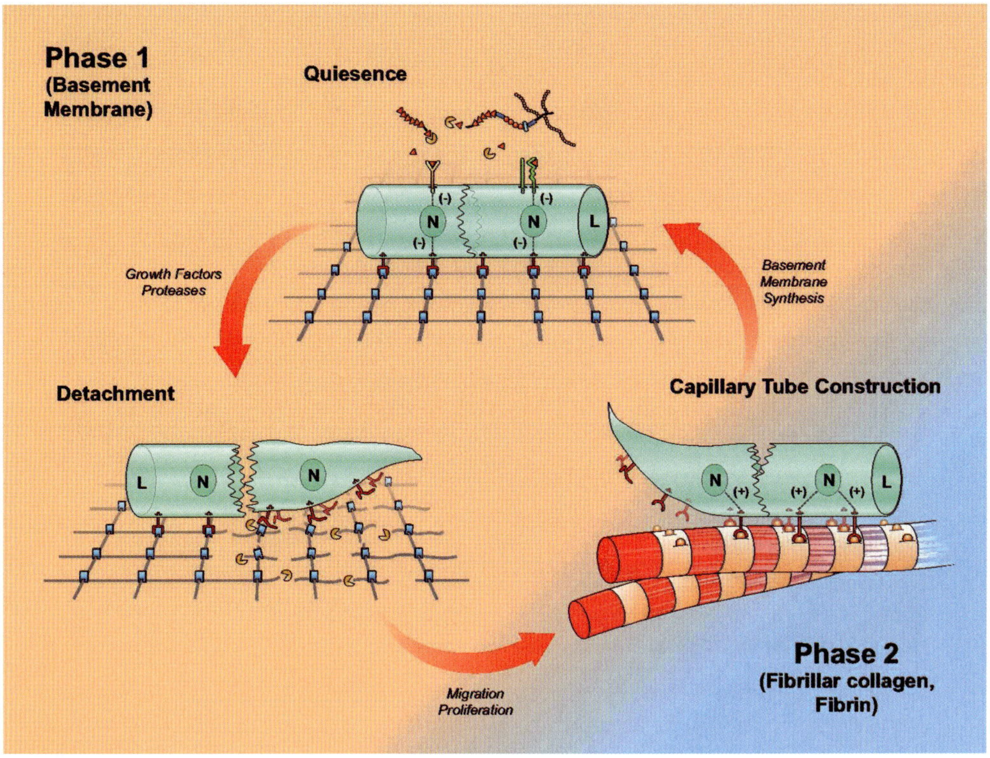

Chapter 21 Figure 2 The Two-Phase Matrix Model of Angiogenesis Regulation. *(Phase I, Top of Diagram)* It is proposed that the basement membrane phase, shown here on an orange background, promotes EC quiescence and maintains capillary integrity in two ways: (1) via stable interactions between cell surface receptors and various basement membrane components on the basal EC side, and (2) by the release of antiangiogenic fragments of basement membrane components through the continuous low-level proteolysis and basement membrane turnover that may occur during quiescence. For the sake of clarity, this latter activity is shown only on the EC apical side, although it is presumed to occur virtually anywhere the EC has contact with the basement membrane. Antiangiogenic fragments may inhibit EC growth and migration by binding to unique receptors (cell on left) or by binding to and interfering with integrin receptor function (cell on right). *(Phase I, Bottom Left of Diagram)* Upon growth factor stimulation, or during tissue injury, massive proteolysis, likely mediated by MMPs, causes the degradation and collapse of the basement membrane, and integrins or other cell surface receptors may mediate the migration of EC into the second matrix phase (intussusception). It is proposed that the fibrillar collagen or fibrin phase, shown on a blue background, catalyzes capillary tube construction in two ways: (1) by presenting a linear, rigid scaffold that helps align EC into presumptive tubes, and (2) by presenting regularly spaced ligand-binding sites that induce clustering of EC receptors such as the $\alpha 2\beta 1$ integrin in the case of type I collagen, or VE-cadherin in the case of fibrin, that induce capillary lumen formation and perhaps the expression of genes required for the maintenance of EC differentiation.

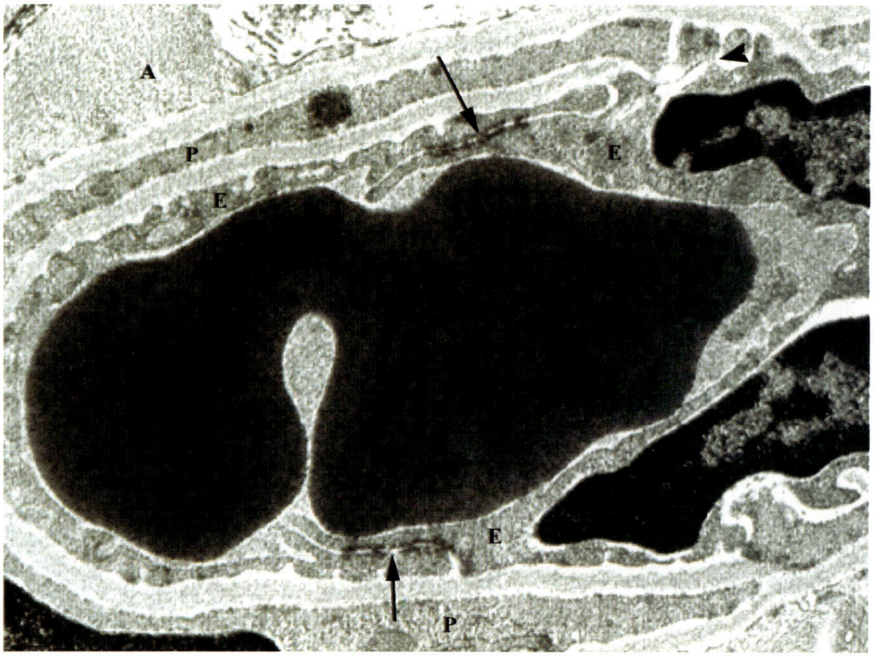

Chapter 22 Figure 1 Ultrastructure of a Normal Inner Retinal Vessel. The tubular configuration of the vascular endothelial cells (E) forms a lumen, which contains a red blood cell. The endothelial cells are connected by tight junctions (zonula occludens) (*arrows*), which help establish the blood–retinal barrier. Pericytes (P) envelop the vascular endothelium and astrocytes (A) cover the pericytes. A basal lamina separates the astrocytes from the pericytes and the pericytes from the endothelial cells, except where there is direct cell-to-cell contact. The *arrowhead* shows a point of direct contact between a pericyte and a vascular endothelial cell.

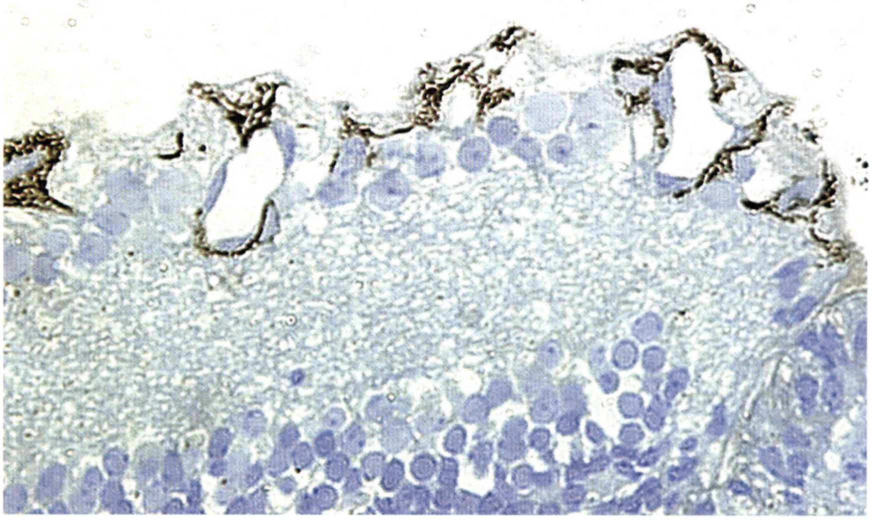

Chapter 22 Figure 2 Immunocytochemical staining for glial fibrillary acidic protein in a P15 mouse retina decorates the astrocytic processes that encircle the vessels in the inner retina.

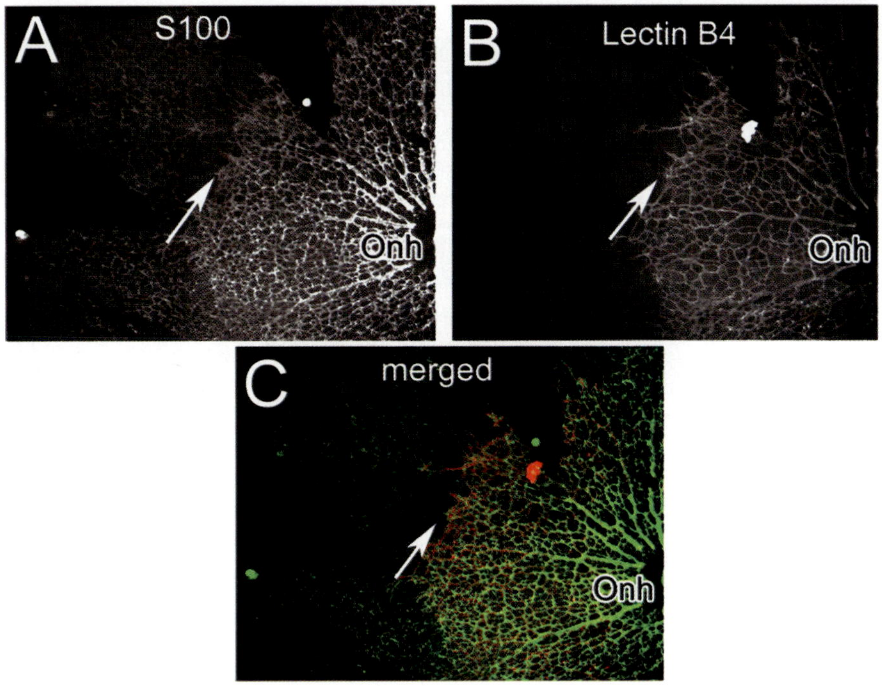

Chapter 22 Figure 3 Immunofluorescent staining using antibodies to S100 (**A**), which is expressed only in astrocytes in the P5 mouse retina, and fluorescence staining of the developing vasculature using Griffonia simplicifolia isolectin B4 (**B**). S100 positive astrocytes are located throughout the retina; however, the staining intensity is higher in the central retina (**A**). At this age, the developing vasculature covers only the central region of the retina (**B**). The *arrow* in all panels marks the leading edge of the developing vasculature. In the merged image (**C**), it is clear that astrocytes in contact with the vasculature have higher expression of S100 than astrocytes in the avascular periphery.

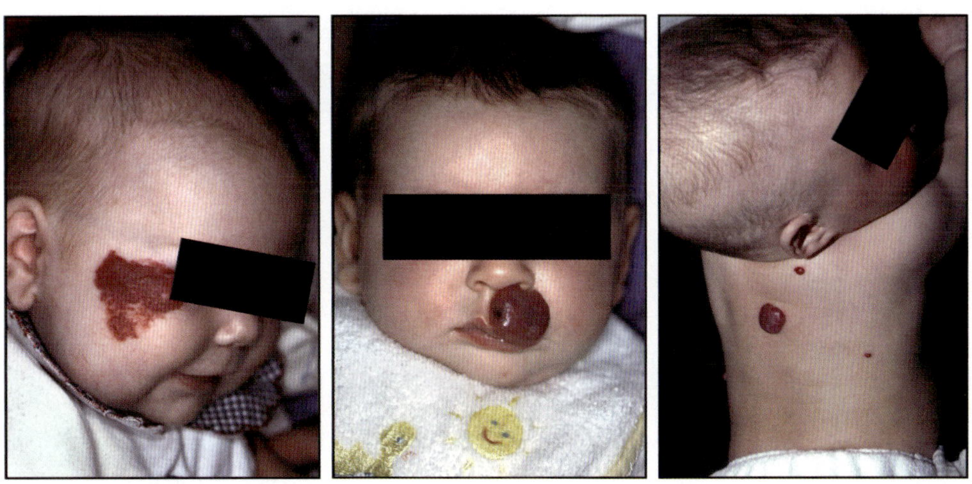

Chapter 23 Figure 1 Three Examples of Hemangioma in the Proliferating Phase. The third panel is an example of multiple hemangiomas.

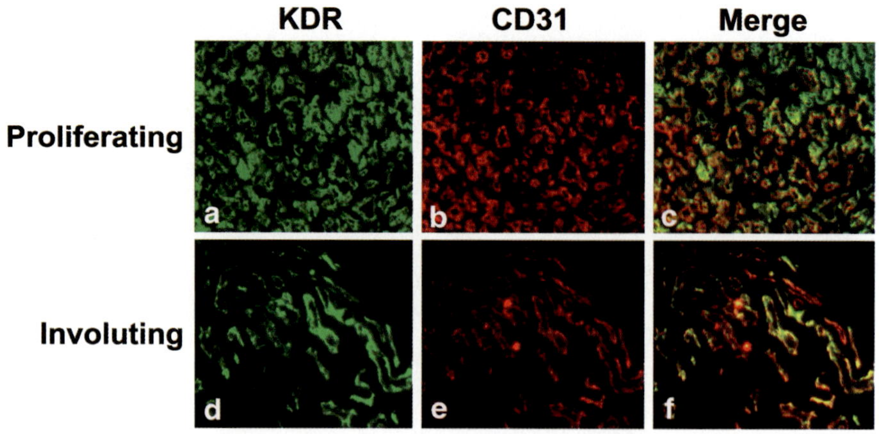

Chapter 23 Figure 2 Expression of KDR/VEGF-R2 and CD31 in Hemangioma. Indirect immunofluorescence detection of KDR (panels a, c, d, f) and CD31 (panels b, c, e, f) in frozen sections from proliferating phase (a–c) and involuting phase (d–f) hemangiomas. In panels **a** and **d**, the green fluorescence (FITC) corresponding to KDR is seen. In panels **b** and **e**, the red fluorescence (Texas-Red) corresponding to CD31 is seen. In panels **c** and **f**, both fluorescence signals were viewed simultaneously; cells expressing both proteins are seen as yellow or yellow/orange. (Yu and Bischoff, unpublished data.)

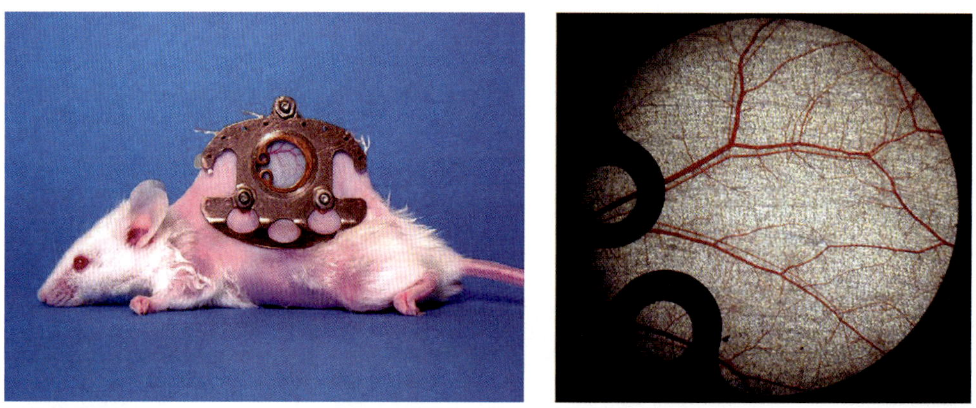

Chapter 24 Figure 1 Mouse Dorsal Skin Chamber Model. An SCID mouse bearing a dorsal skin chamber (left) and a macroscopic image of the mouse dorsal skin chamber (right). The frame is made of titanium (3.2 g). The window is 11 mm across.

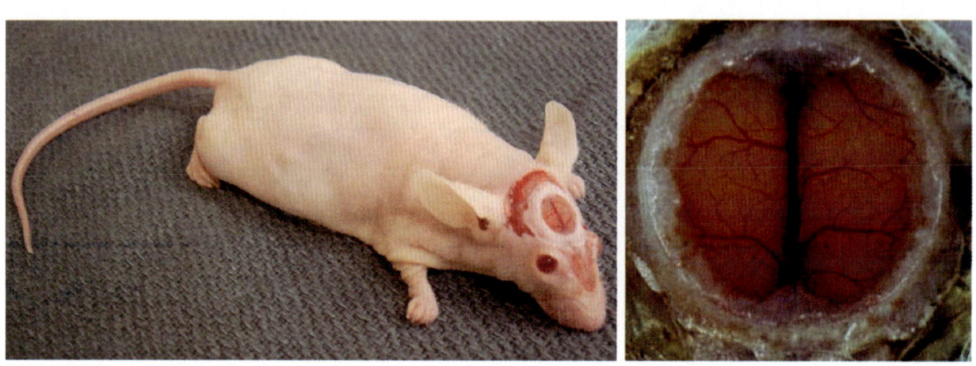

Chapter 24 Figure 2 Mouse Cranial Window Model. Nude mouse bearing a cranial window (left) and a macroscopic image of the mouse cranial window (right). The window is 7 mm across.

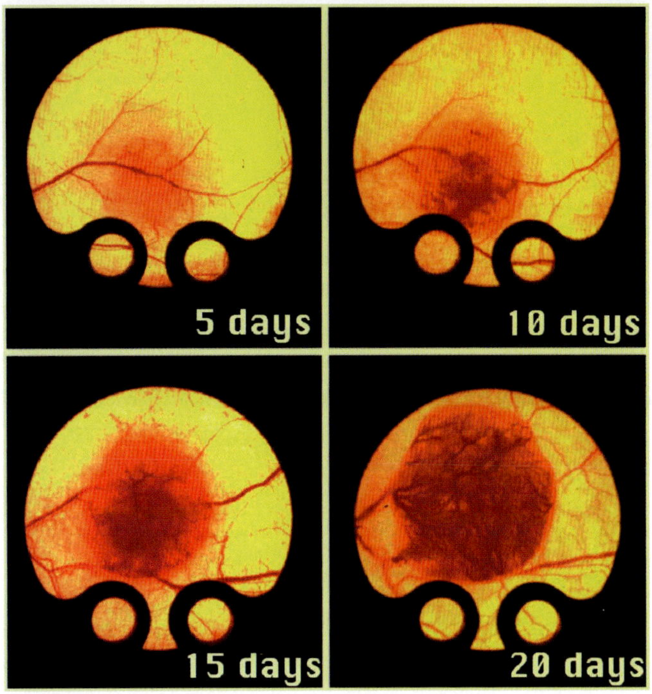

Chapter 24 Figure 3 Angiogenesis and Tumor Growth in a Window Model. LS174T human colon cancer was implanted in an SCID mouse dorsal skin chamber. At day 5 after tumor cell implantation, enlargement of host vessels was observed, and by day 10 occasional hemorrhage and sprout formation occurred. At day 15, tumor growth and further angiogenesis became apparent. By day 20, the tumor was fully vascularized. (Adapted from Reference [3].)

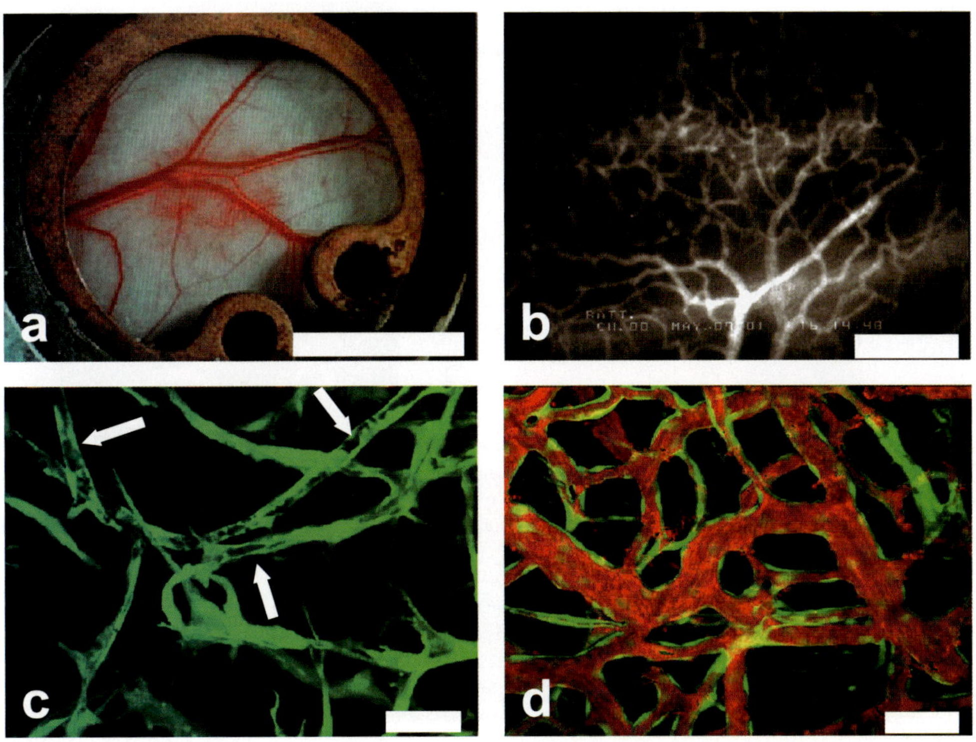

Chapter 24 Figure 4 Vessel Formation in a Window Model. **(a, b)** Angiogenesis and vessel remodeling during adipogenesis in the mouse dorsal skinfold chamber after 3T3-F442A cell implantation. **(a)** Macroscopic image 9 days after implantation. **(b)** Intravital microscopy image of fluorescence contrast enhanced blood vessels at 28 days after implantation. **(c, d)** HUVECs and 10T1/2 cells were seeded in a 3-D extracellular matrix and implanted in a mouse cranial window. 3-D images of engineered vessels (EGFP expressing HUVEC, green; functional blood vessels contrast-enhanced with rho-dextran, red) were taken using intravital multi-photon laser-scanning microscopy. **(c)** Four days after implantation of HUVEC + 10T1/2 construct. Large vacuoles in the tubes resemble the lumens of capillaries (arrows) but they are not perfused (no red). **(d)** Four months after implantation of HUVEC + 10T1/2 construct. Engineered vessels are stable and functional. Bars indicate 5 mm (*a*), 100 μm (*b*), 50 μm (*c*, *d*), respectively. (Adapted from Fukumura, D., Ushiyama, A., Duda, D. G., Xu, L., Tam, J., Chatterjee, V. K. K., Garkavtsev, I., and Jain, R. K. (2003). Paracrine regulation of angiogenesis and adipocyte differentiation during *in vivo* adipogenesis. *Circ. Res.* **93**, e88–e97; and Koike, N., Fukumura, D., Gralla, O., Au, P., Schechner, J. S., and Jain, R. K. (2004). Creation of long-lasting blood vessels. *Nature* **428**, 138–139.)

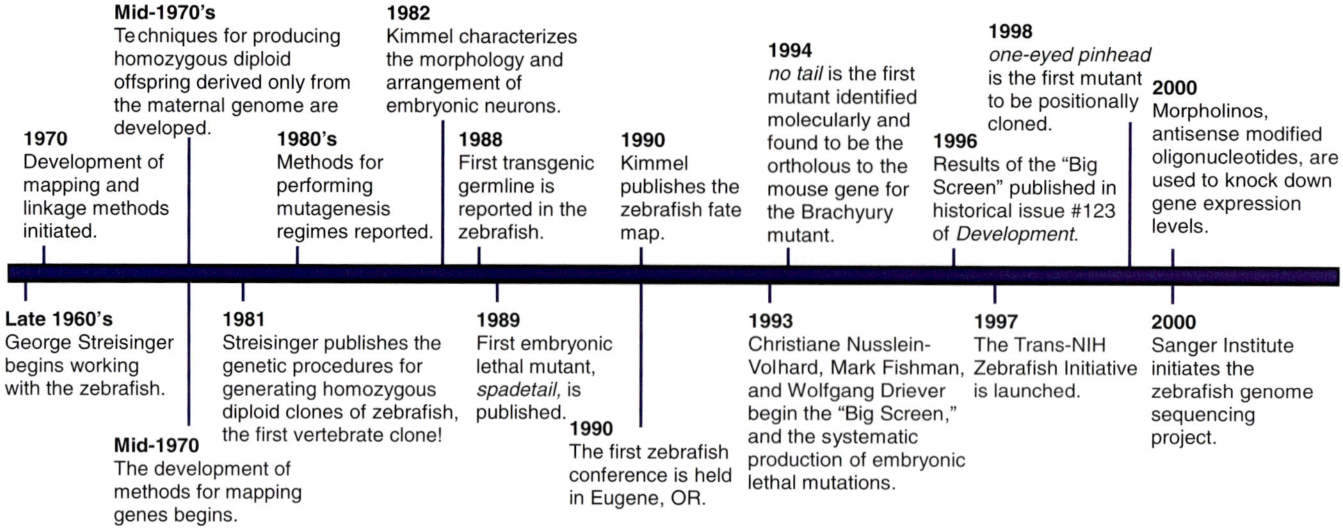

Chapter 26 Figure 1 Major events that contributed to the success of the zebrafish. The zebrafish has been around for a number of years, but was not considered an established model for biological study until the past decade. This timeline highlights some of the major events that have contributed to the establishment of the zebrafish as a model organism.

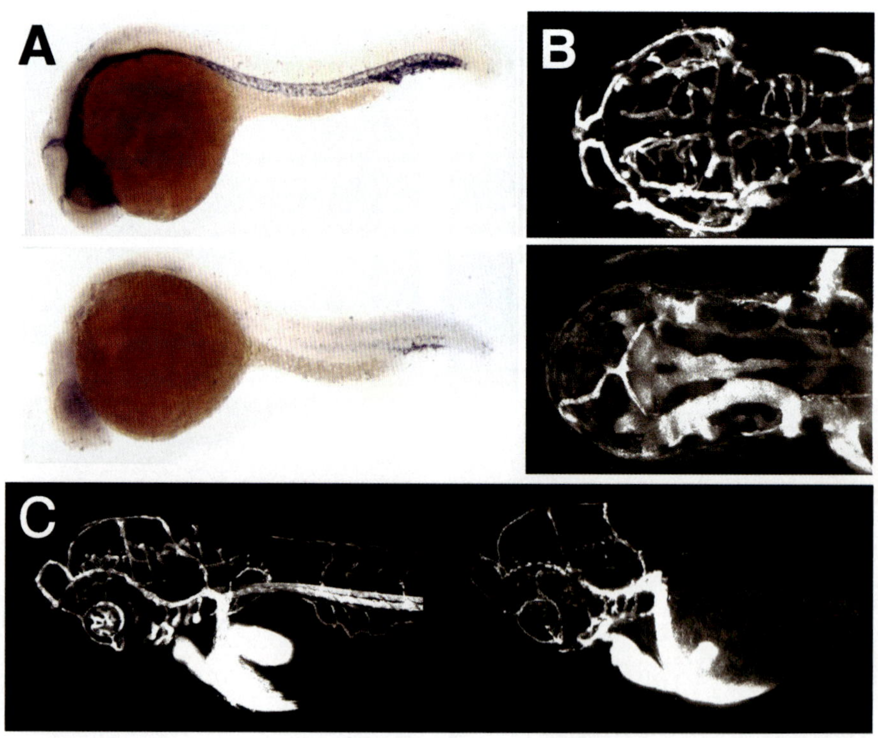

Chapter 26 Figure 2 Vascular-specific mutants in the zebrafish. A number of different mutants have been isolated in zebrafish with defects in blood vessel formation. *Cloche* mutants lack virtually all blood endothelium (A). *In situ* hybridization of wild type (top) and mutant (bottom) embryos with vascular endothelial-specific probes such as *Fli-1* reveals a lack of vascular staining except in a small patch in the posterior trunk (24 hpf embryos, lateral views). *Violet beauregarde* mutants have defects in cranial vessel patterning (B). Mutants (bottom) display highly enlarged primary cranial vessels and reduced perfusion of secondary cranial vessels compared to wild type (top) embryos, as shown in these angiographic images (2.5 dpf embryos, dorsal views of head). *Gridlock* mutants lack caudal circulation due to a defect in proper formation of the dorsal aorta C). Mutants (right) have a normal pattern of cranial vessels but lack perfusion of more caudal vessels in comparison to wild type (left) animals (2.5 dpf embryos, lateral views of head and anterior trunk).

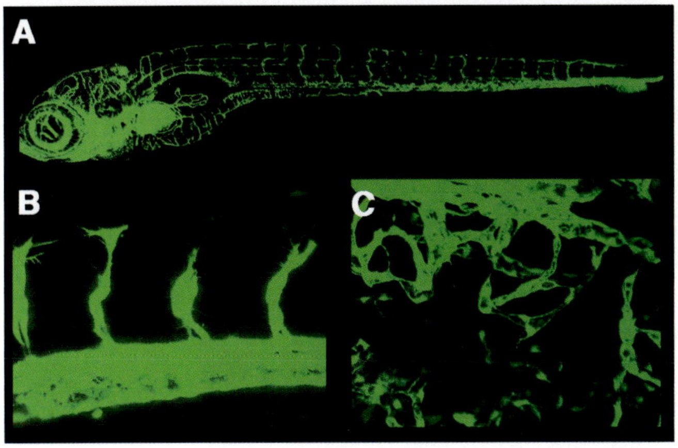

Chapter 26 Figure 3 Blood vessels imaged in Fli-EGFP transgenic zebrafish. Vascular endothelial cells and their angioblast precursors are brightly green fluorescent in Fli-EGFP transgenic animals (**A**). Image shown is a confocal microangiogram of a 7 day-old transgenic zebrafish. Fine cellular details can be discerned, including thin filopodial processes extending from growing intersegmental vessels in the trunk (**B**) and central arteries assembling in the zebrafish hindbrain (**C**). Images shown are confocal microangiograms from approximately 1.5 day-old transgenic zebrafish.

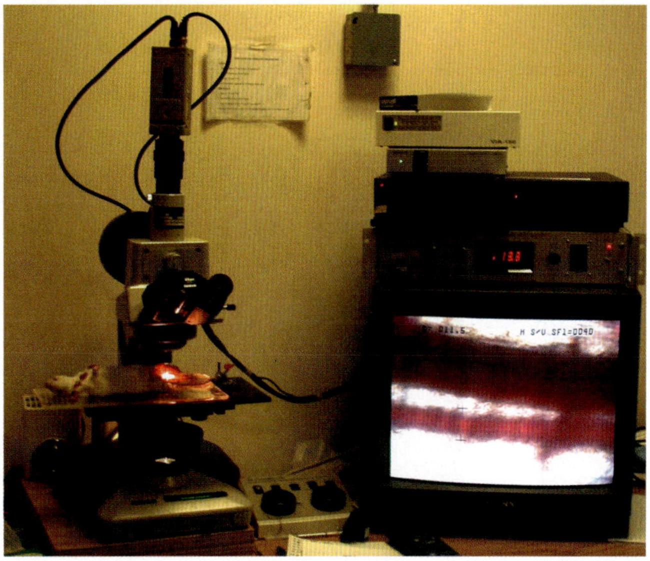

Chapter 29 Figure 2 Intravital microscopy setup for microcirculatory monitoring.

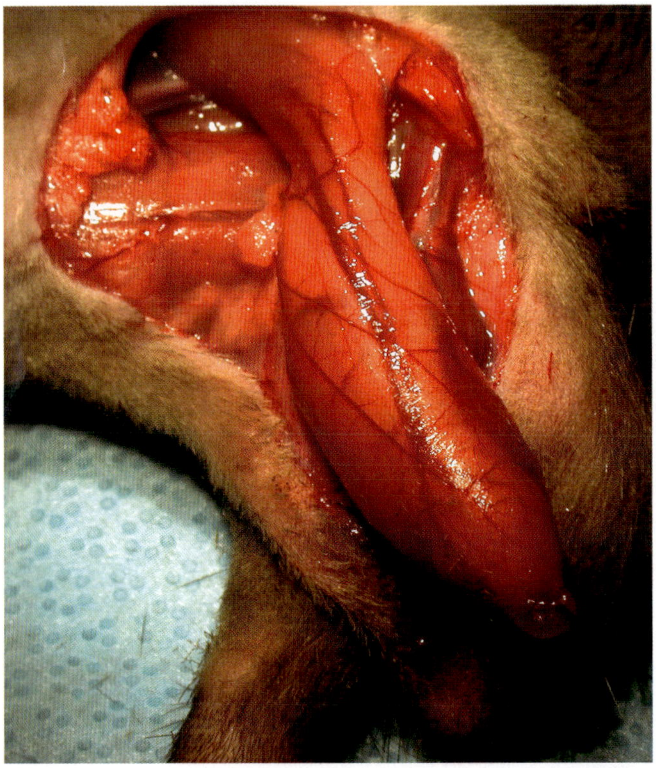

A

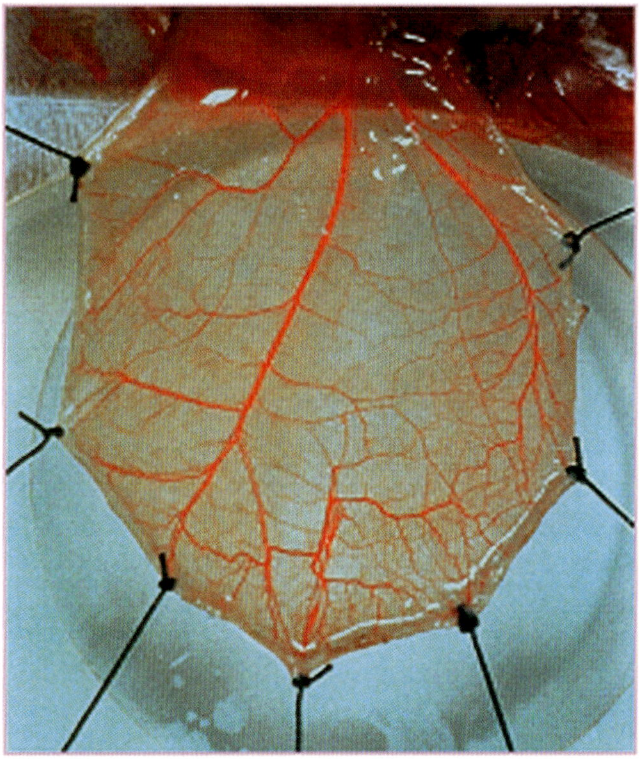

B

Chapter 29 Figure 1 (**A**) Isolated cremaster muscle flap. (**B**) Cremaster muscle flap dissected and set up for microcirculatory studies. (**C**) Schematic representation of the cremaster muscle flap microvascular network for the measurements of microcirculatory hemodynamics. V1, venule; A1, first-order arteriole; A2, second-order arterioles; A3, third-order arterioles; A3-1, third-order arteriole (the branch of A2-1); A3-2, third-order arteriole (the branch of A2-2).

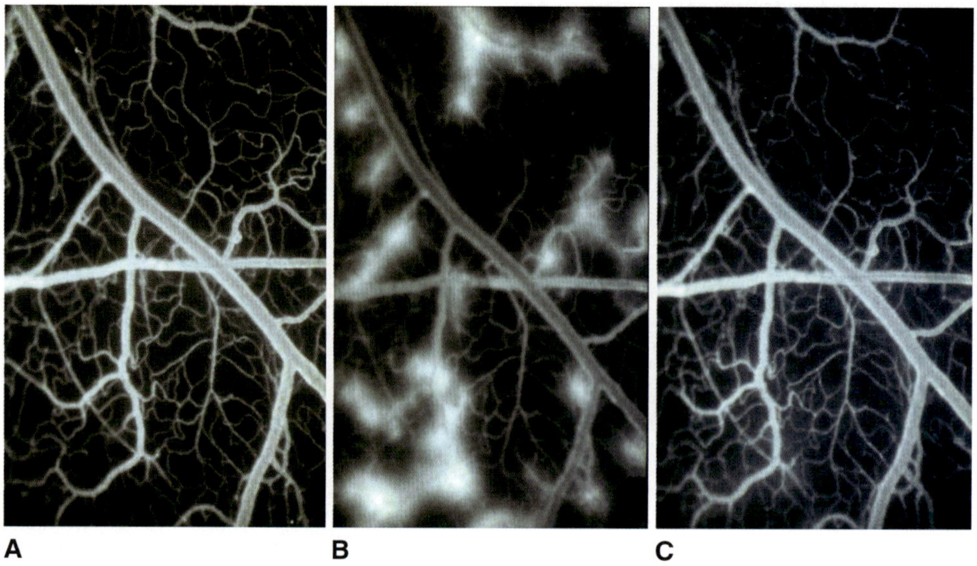

Chapter 30 Figure 1 Fluorescent micrograph of the hamster cheek pouch after IV injection of FITC-dextran. Same area before **(A)**, 5 minutes after **(B)**, and 30 minutes after **(C)** topical application of bradykinin (4×10^{-7} M). More than 25 leakage sites at postcapillary venules are shown by extravasation of FITC-dextran at postcapillary venules.

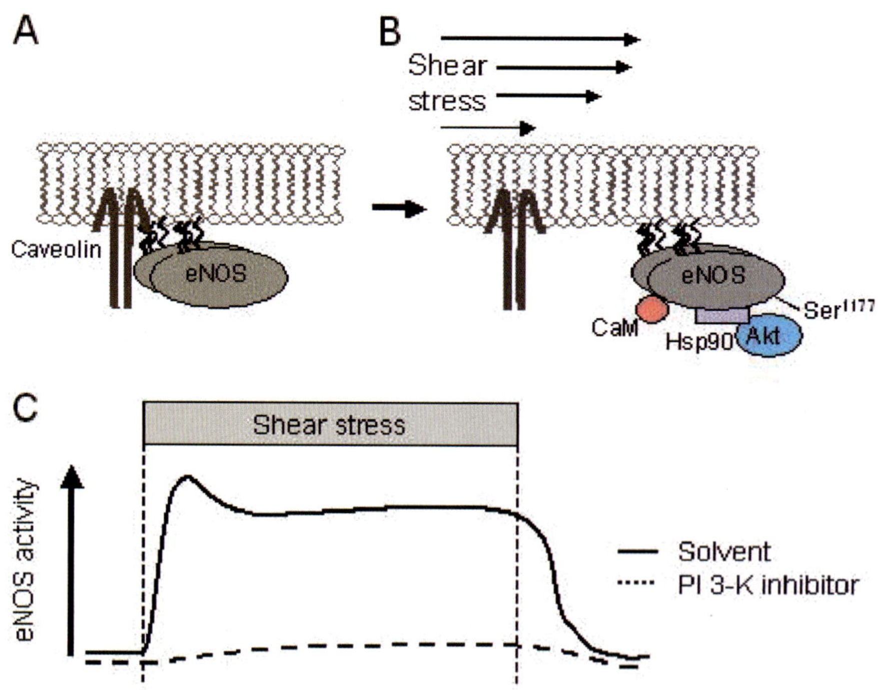

Chapter 33 Figure 1 Endothelial cell activation by fluid shear stress increases eNOS phosphorylation and activity. In unstimulated endothelial cells the eNOS dimer is associated with caveolin, which inhibits its activation, NO output is low **(A)**. In response to the application of fluid shear stress, the phosphatidylinositol 3-kinase (PI 3-K) is activated and in turn activates Akt, which associates with the eNOS scaffolding protein hsp90 and phosphorylates eNOS on Ser^{1177} **(B)**. This step elicits the dissociation from caveolin and increases NO production. Shear stress–induced increases in endothelial NO production can be abrogated by preventing the activation of PI 3-K (C).

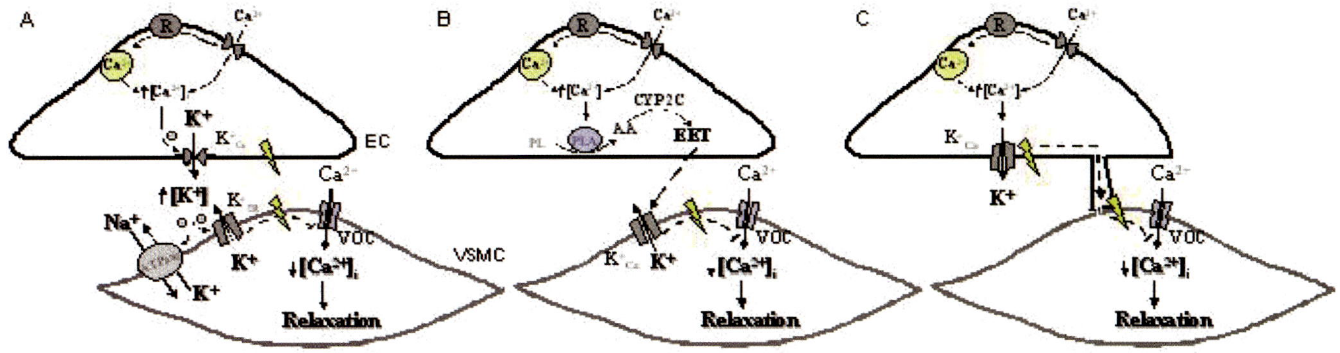

Chapter 33 Figure 2 Three models contributing to endothelium-derived hyperpolarizing factor-mediated relaxation. **(A)** The K^+ ion hypothesis. In response to the receptor (R)-dependent activation of endothelial cells (EC), intracellular concentrations of Ca^{2+} ($[Ca^{2+}]_i$) increase and activate Ca^{2+}-dependent K^+ (K^+_{Ca}) channels, which results in the efflux of K^+ ions into the subintimal space and endothelial cell hyperpolarization (lightning symbol). The increase in extracellular K^+ activates inwardly rectifying K^+ channels (K^+IR) and/or the Na-K-ATPase which elicit vascular smooth muscle cell (VSMC) hyperpolarization and decrease the open probability of voltage-dependent Ca^{2+} (VOC) channels. As a consequence $[Ca^{2+}]_i$ in the smooth muscle cells decreases and relaxation occurs. **(B)** The cytochrome P450 hypothesis. An increase in $[Ca^{2+}]_i$ activates a phospholipase A_2, which liberates arachidonic acid (AA) from membrane lipids. Arachidonic acid is then metabolized to epoxyeicosatrienoic acids (EETs) by a cytochrome P450 2C (CYP 2C) epoxygenase. EETs then increase the open probability of K^+_{Ca} in the underlying smooth muscle cells to elicit hyperpolarization and relaxation. **(C)** The gap junction hypothesis. Endothelial cell activation as described elicits endothelial cell hyperpolarization by activating K^+_{Ca} channels. This hyperpolarization is, however, transferred to smooth muscle cells via myo-endothelial gap junctions rather than by the release of an endothelium-derived factor. The final steps are the same, that is, inhibition of Ca^{2+} entry into smooth muscle cells and relaxation.

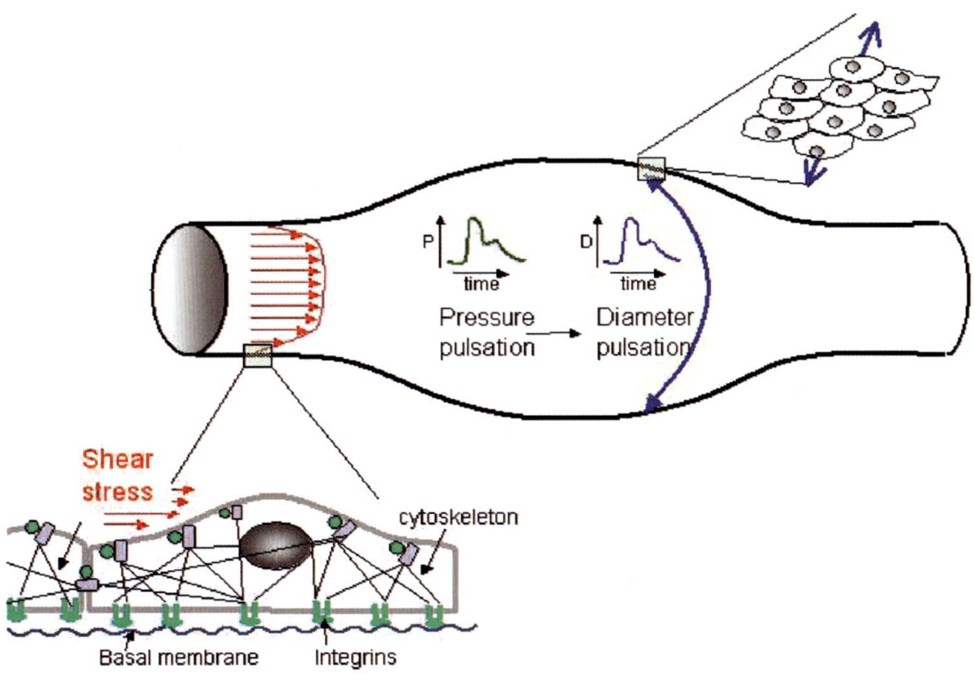

Chapter 33 Figure 3 Scheme illustrating the hemodynamic forces affecting endothelial cells. Note that, because of the pulsatile nature of flow, the flow profile is flattened compared to the parabolic flow profile present under steady-state conditions, resulting in a greater shear stress at the endothelial cell surface. The simultaneous pressure (P) pulse distends the vessel diameter (D) and stretches endothelial cells in the range of 1–5% strain. These stimuli result in a redistribution of force across the endothelial cell cytoskeleton and the activation of signaling molecules (including integrins) at the abluminal and luminal cell surfaces as well at sites of cell-cell contact. Adapted from Busse, R., and Fleming, I. (2003). Regulation of endothelium-derived vasoactive autacoid production by hemodynamic forces. *Trends Pharmacol. Sci.* **24**, 24–29, reprinted with permission from Elsevier Science Ltd.

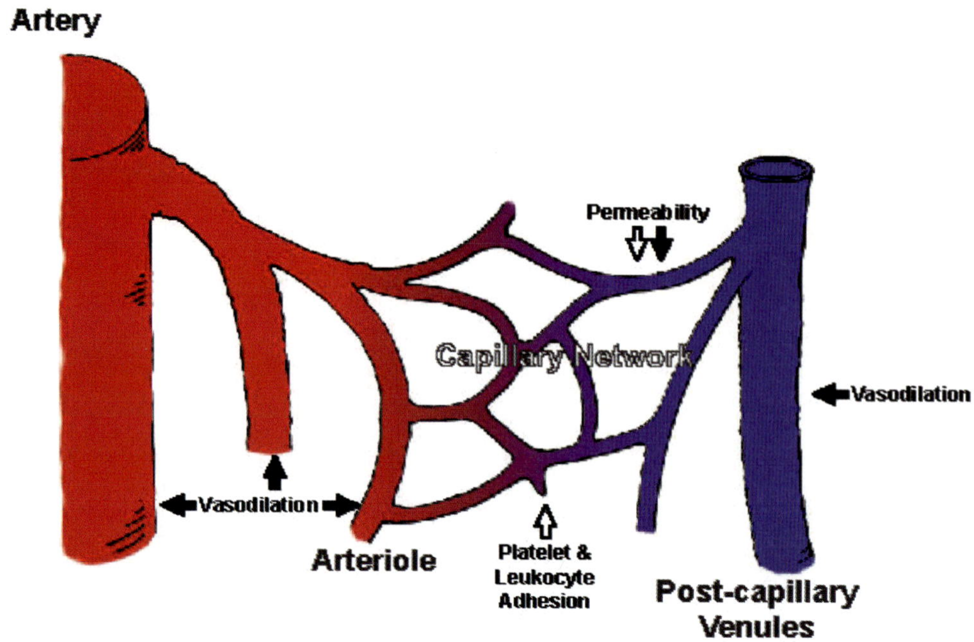

Chapter 35 Figure 1 The role of NO in the microcirculation. NO, a gaseous free radical, can exert multiple effects in the microcirculatory bed and is a major player in regulating the cardiovascular system. Studies have shown the vasodilatory effect of NO in large blood vessels as well as in the arterioles and venules. This mainly occurs when NO synthesized by the endothelium diffuses to the surrounding vascular smooth muscle, activating guanylyl cyclase (GC), thereby increasing cytosolic cGMP to produce smooth muscle relaxation. NO can also regulate homeostasis of the blood vessel wall by preventing platelet adhering and disaggregation from the endothelium. Furthermore, NO can negatively regulate leukocyte-endothelium interaction by downregulating cytokines and ECAMs such as P-selectin that are important for leukocyte rolling and adhesion to endothelial cells. Thus, NO serves as an anti-inflammatory mediator in pathological conditions such as hypercholesterolemia or as a cardioprotectant in myocardial ischemia-reperfusion injuries. In addition, NO is a crucial regulator of microvascular permeability. Controversy exists to whether NO promotes or inhibits basal versus stimulated changes in vascular permeability.

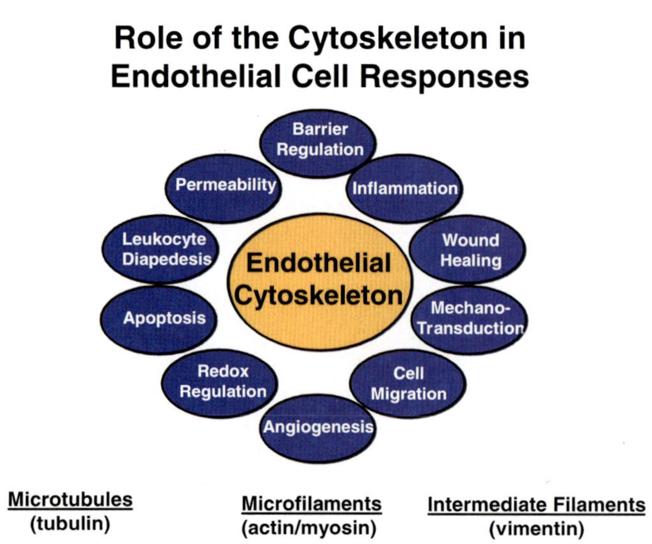

Chapter 38 Figure 1 Role of the Cytoskeleton in Endothelial Cell Responses.

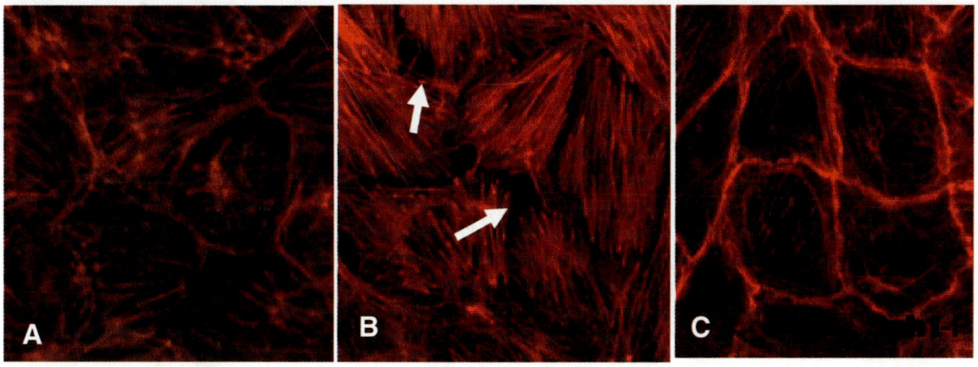

Chapter 38 Figure 2 Actin microfilaments distribution is altered by agents that affect vascular endothelial cell permeability. Human lung endothelial cells contain randomly distributed filamentous actin (red) in a stable monolayer (A). Edemagenic agents, such as thrombin, cause cytoskeletal rearrangement resulting in thick actin bundles "stress fibers" which span the cell. Increase intracellular tension and contraction results in formation of intercellular gaps (B, arrows). Platelet derived-phopholipids such as sphingosine 1-phophate are barrier protective and result in an increase in cortical actin at the cell periphery (C).

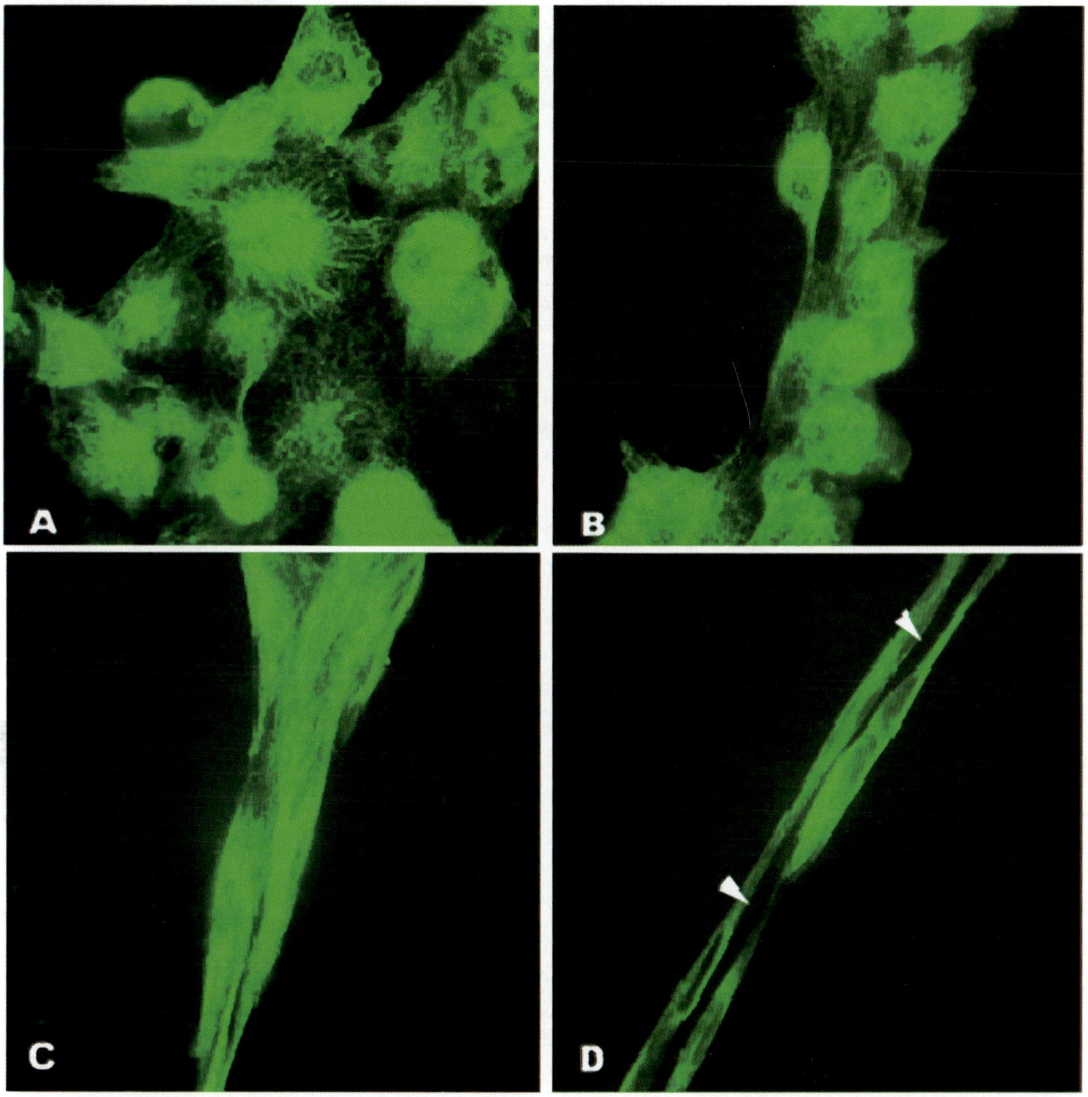

Chapter 38 Figure 3 Microtubules are involved in endothelial cell dynamics during capillary-like tube and lumen formation. Human lung microvascular (HLMV) endothelial cells were plated on extracellular matrix (Matrigel, BD Biosciences) and labeled with antibodies specific for tubulin (green). In the early stages of capillary-like tube formation (1–4 hr) HLMV EC are grouped together and display a spread morphology with microtubules arrayed in a radial pattern (A, arrowheads) and begin to migrate into a linear arrangement (B). In the middle stage of tube formation (5–9 hrs), HLMV EC have elongated into capillary-like cords of cells with more elongated microtubule arrays (C). In late capillary-like tube formation (>12 hrs), a lumen is visible (D, arrowheads) between the HLMV EC with compressed linear microtubule distribution.

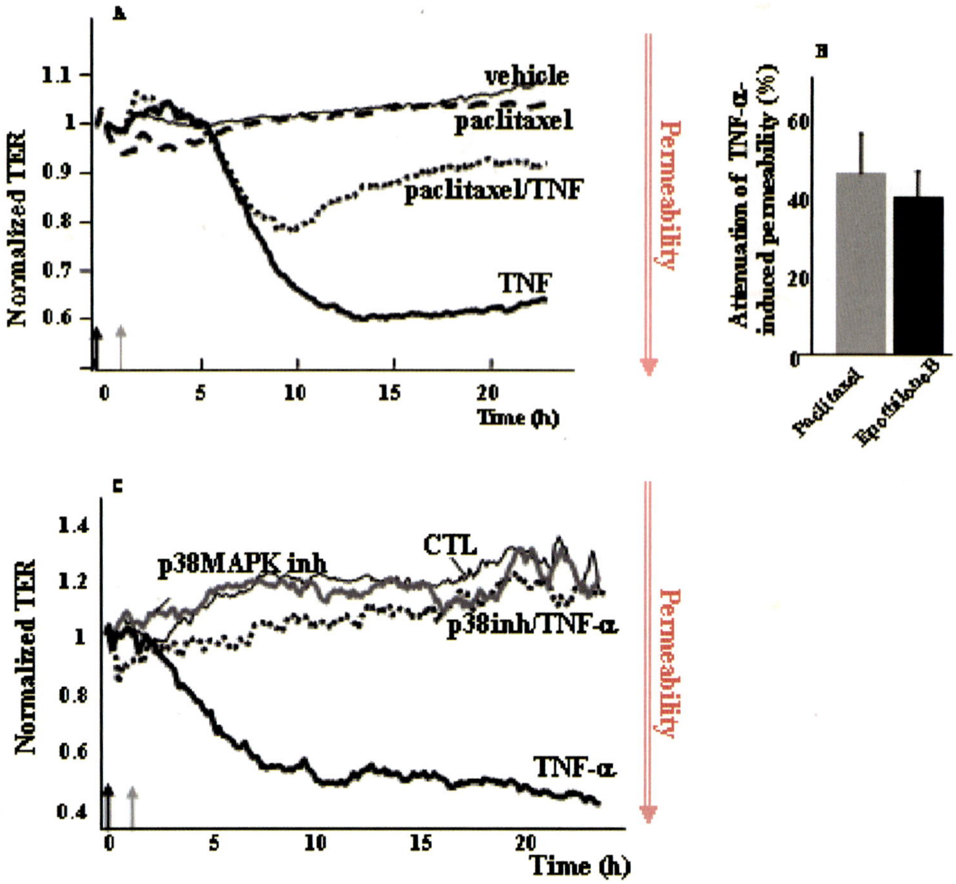

Chapter 40 Figure 1 The effects of microtubule stabilization and p38 MAPK inhibition on TNF-α-induced endothelial permeability. (**A**, **C**) Representative tracings of normalized electrical resistance measured across endothelial cell monolayers (TER). Endothelial cells were exposed to vehicle (control) and TNF-α [20 ng/ml], where the gray arrow on the abscissa indicates the time of challenge. TNF-α induced TER reduction was evident at 4 hours with a maximal effect at 10 hours of exposure. **(A)** Microtubule stabilization by paclitaxel (black arrow) significantly inhibits TNF-α-induced decreases in resistance. **(C)** P38 MAPK inhibition by SB203580 inhibits TNF-α induced permeability. **(B)** Bar graph demonstrating the effect of the microtubule-stabilizing agents paclitaxel end epothilone B on the maximal TNF-α-induced decrease in TER. No significant difference was noted between the two agents ($p = 0.44$).

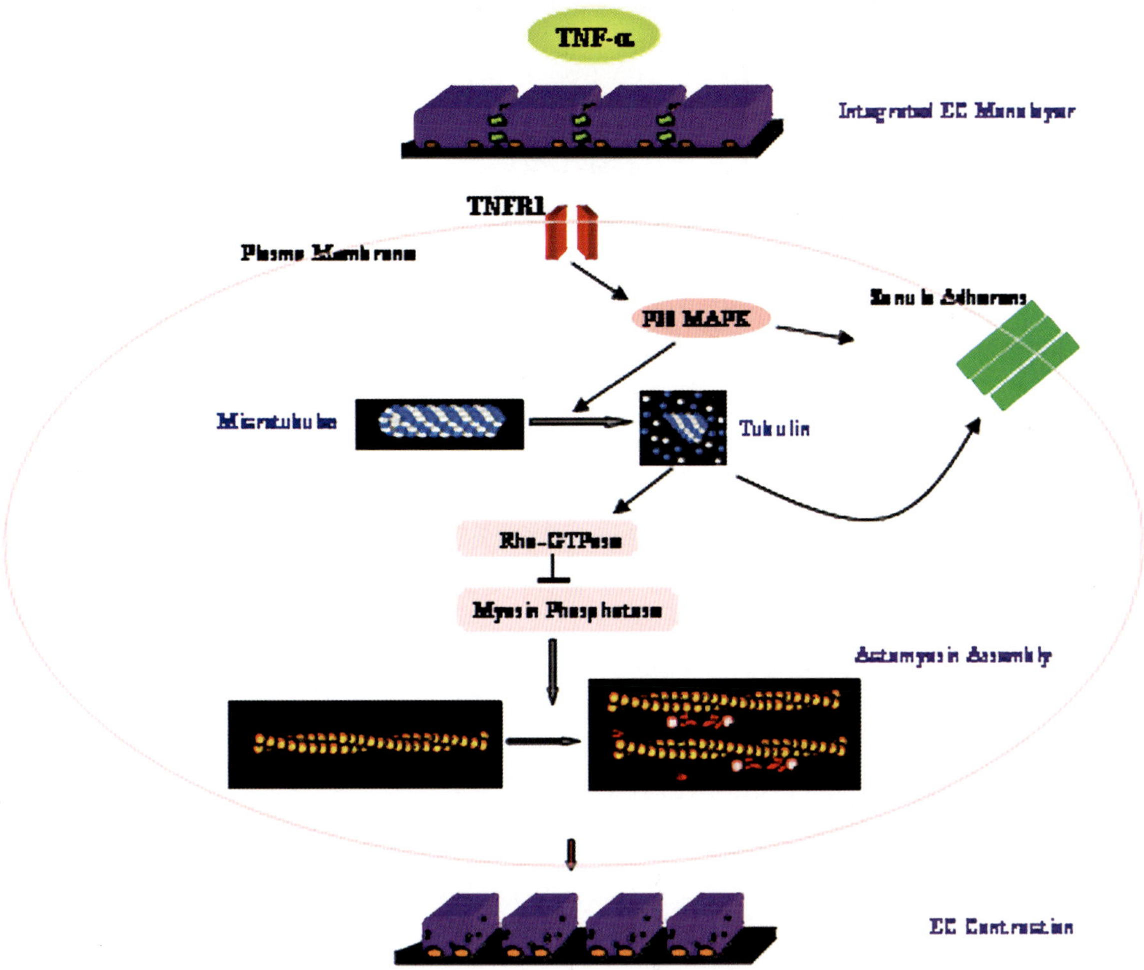

Chapter 40 Figure 2 Schematic representation of putative pathways of TNF-α-triggered endothelial cell permeability. TNF-α acts via p38 MAPK to trigger microtubule disassembly and subsequent actin microfilament rearrangement and cell–cell contact disruption. Additionally, the p38 MAPK engages microtubule-independent cytoskeletal mechanisms of barrier dysfunction.

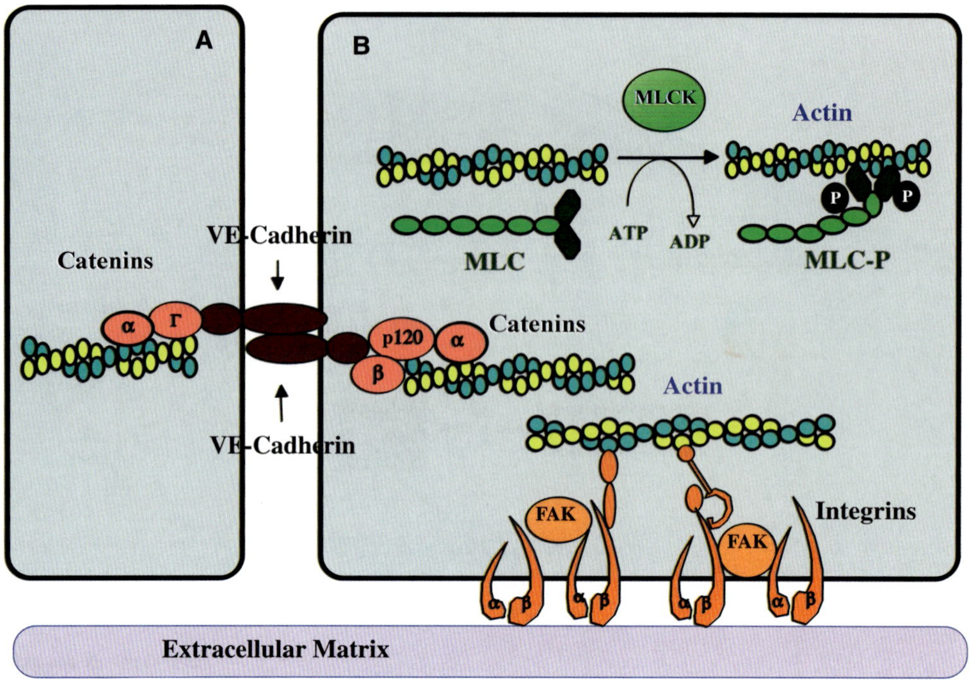

Chapter 45 Figure 2 A model for the signal transduction of agonist-induced microvascular hyperpermeability. Histamine and vascular endothelial growth factor (VEGF) are used to represent permeability-increasing agonists. Histamine binding to its receptor (H) activates phospholipase C beta (PLCβ) via a G protein (G), catalyzing the production from inositol bisphosphate (PIP_2) to inositol trisphosphate (IP_3) and diacylglycerol (DAG). DAG directly activates protein kinase C. In parallel, IP_3 stimulates an internal release of Ca^{2+} followed by an influx of extracellular Ca^{2+}, leading to upregulation of nitric oxide synthase (ecNOS) and nitric oxide (NO) production from its precursor L-arginine (L-Arg). Nitric oxide stimulates guanylate cyclase (GC) to produce cGMP, a potent activator of cGMP-dependent protein kinase (PKG). The elevated intracellular calcium also stimulates PKC and myosin light-chain kinase (MLCK); the latter phosphorylates myosin light chain (MLC-P). The hyperpermeability effect of VEGF is triggered by its tyrosine kinase receptor flk1/KDR and mediated through two major signaling pathways. On one hand, activated phospholipase C gamma (PLCγ) elevates the intracellular calcium and subsequently activates the PKG, PKC, and MLCK cascades. On the other hand, KDR receptor occupancy results in the recruitment of several adaptor proteins, such as Grb2, Sos, and Src, which further produces the ras-raf-MEK-ERK cascade. In addition, the nonreceptor tyrosine kinases FAK and Src as well as the Rho-family of GTPases are also involved in the endothelial response to inflammatory stimulation. All of these protein kinases—PKG, PKC, MLCK, MAPK, and tyrosine kinases—may directly or indirectly target the endothelial barrier structure to cause an opening of the paracellular pathway for transendothelial flux of fluid and macromolecules.

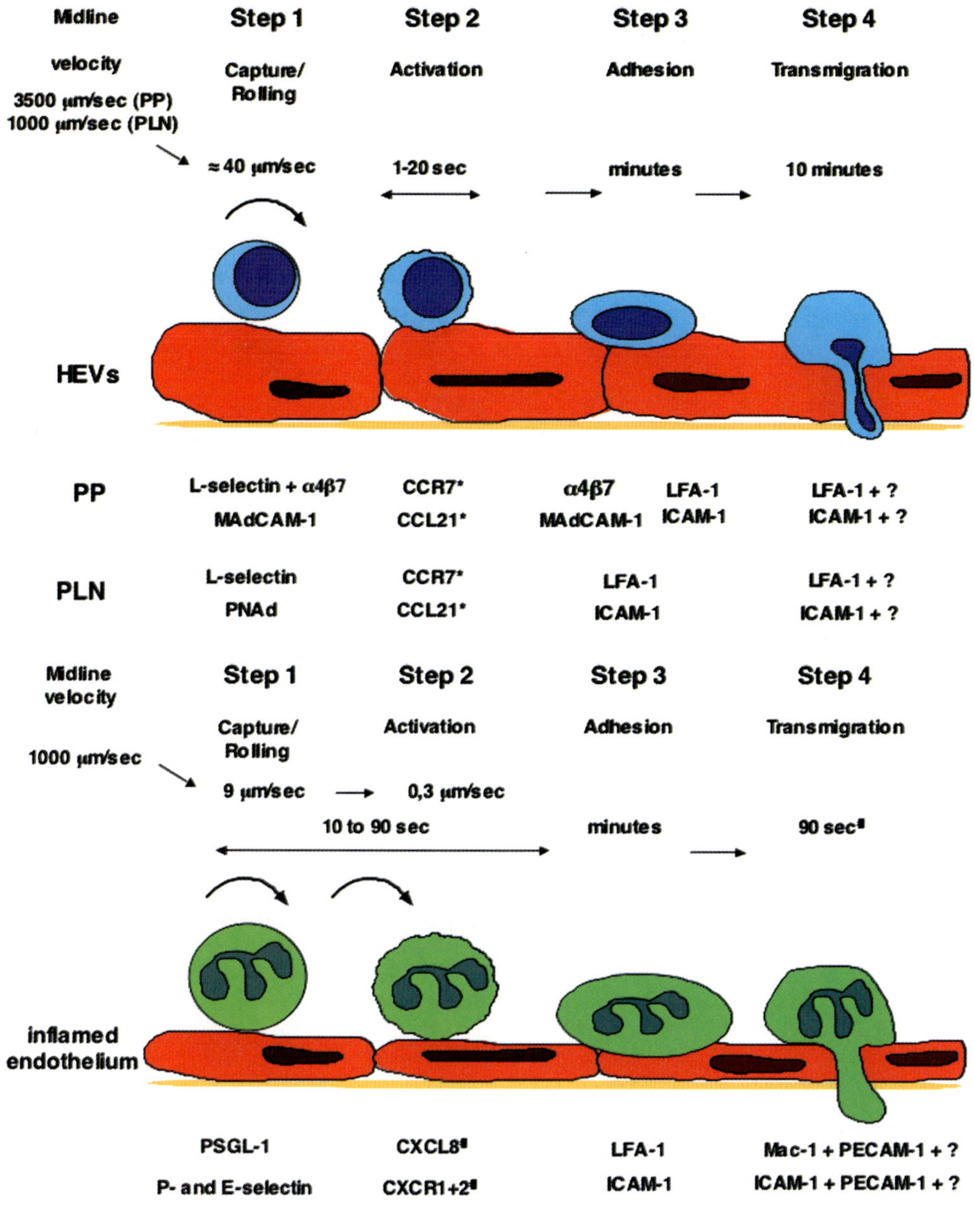

Chapter 47 Figure 1 The multistep model of leukocyte endothelial interaction. **Top**: Lymphocyte/HEVs interactions in Peyer's patches (PP) or peripheral lymph nodes (PLN). **Bottom**: Neutrophil interaction with inflamed endothelium. *, Only T cells, unknown for B cells; #, only demonstrated in vitro. The velocities of free-flowing (noninteracting) cells and rolling leukocytes as well as other time estimates are derived from intravital microscopic observations published from the laboratories of Ulrich von Andrian, Eugene Butcher, and Klaus Ley.

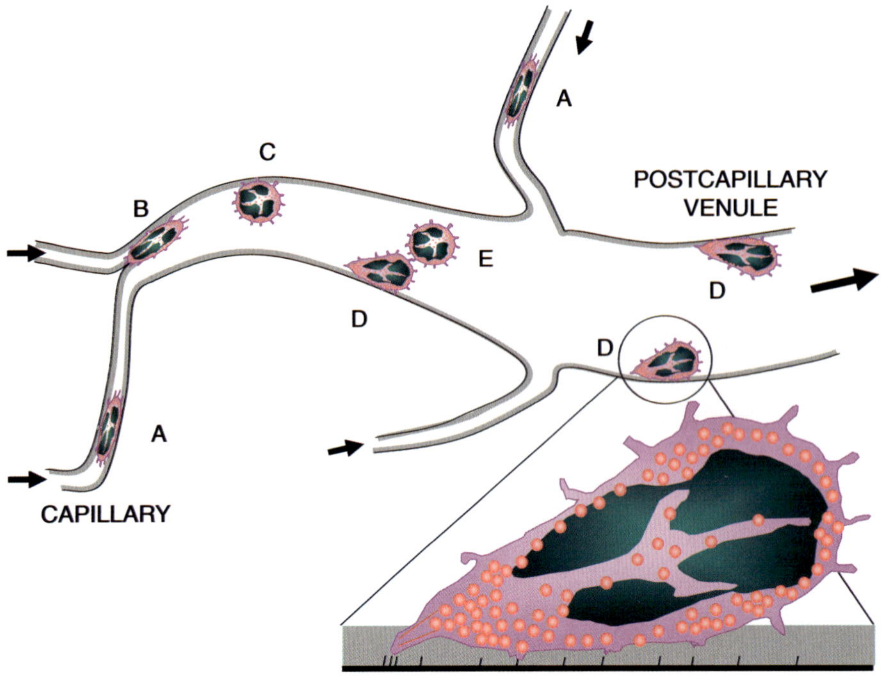

Chapter 50 Figure 1 Neutrophils interacting with the endothelium lined with a 500-nm glycocalyx (blue) in a postcapillary venule. In capillaries, neutrophils are deformed into a near-cylindrical shape (A) and deform the endothelial glycocalyx. Almost all leukocyte rolling is initiated at the beginning of postcapillary venules, where leukocytes are in close physical contact with the endothelial surface (B). Noninteracting deformed leukocytes eventually recover their spherical shape (C), but rolling leukocytes, especially neutrophils, acquire a characteristic teardrop shape that reflects the effect of adhesive forces balanced by shear forces on the cytoskeleton (D). Rolling or adherent cells can nucleate L-selectin-PSGL-1-dependent secondary capture or tethering events (E). A close-up of a rolling neutrophil (bottom right) shows the endothelial glycocalyx (blue, 500 nm) with endothelial adhesion molecules (E-selectin, 30 nm, and P-selectin, 40 nm). The black hairlike lines represent the length of a P-selectin-PSGL-1 pair (100 nm) completely buried in the glycocalyx. Rolling leukocytes probably continuously deform the glycocalyx as they roll while selectin (and integrin) bonds are formed at the leading edge (right) and broken at the trailing edge (left). Arrows indicate direction of blood flow.

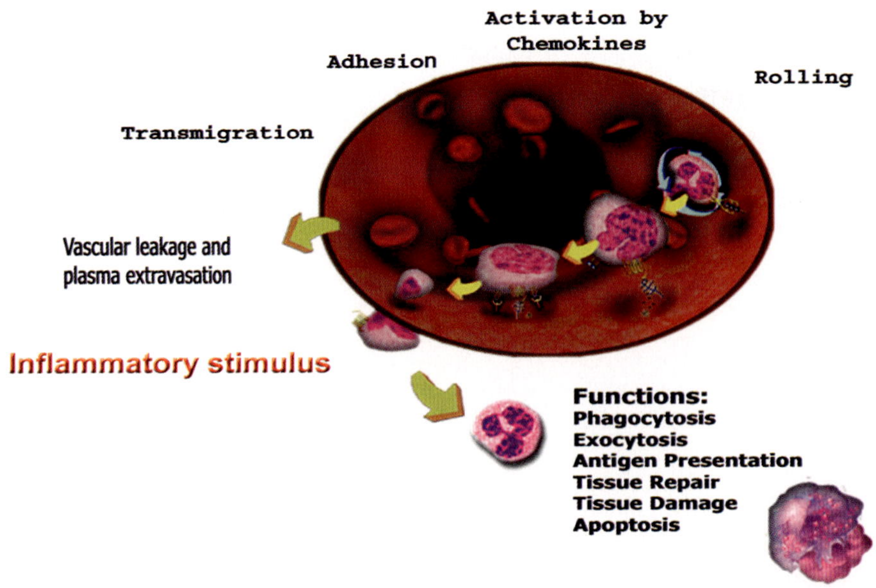

Chapter 51 Figure 1 General mechanisms involved in leukocyte trafficking in the microcirculation. Shown is a schematic representation of the multistep paradigm of leukocyte recruitment. Leukocytes undergo rolling and activation by EC-displayed chemokines, followed by adhesion onto and transmigration across the endothelium to perform their extravascular functions. Illustration by A. K. Khaw.

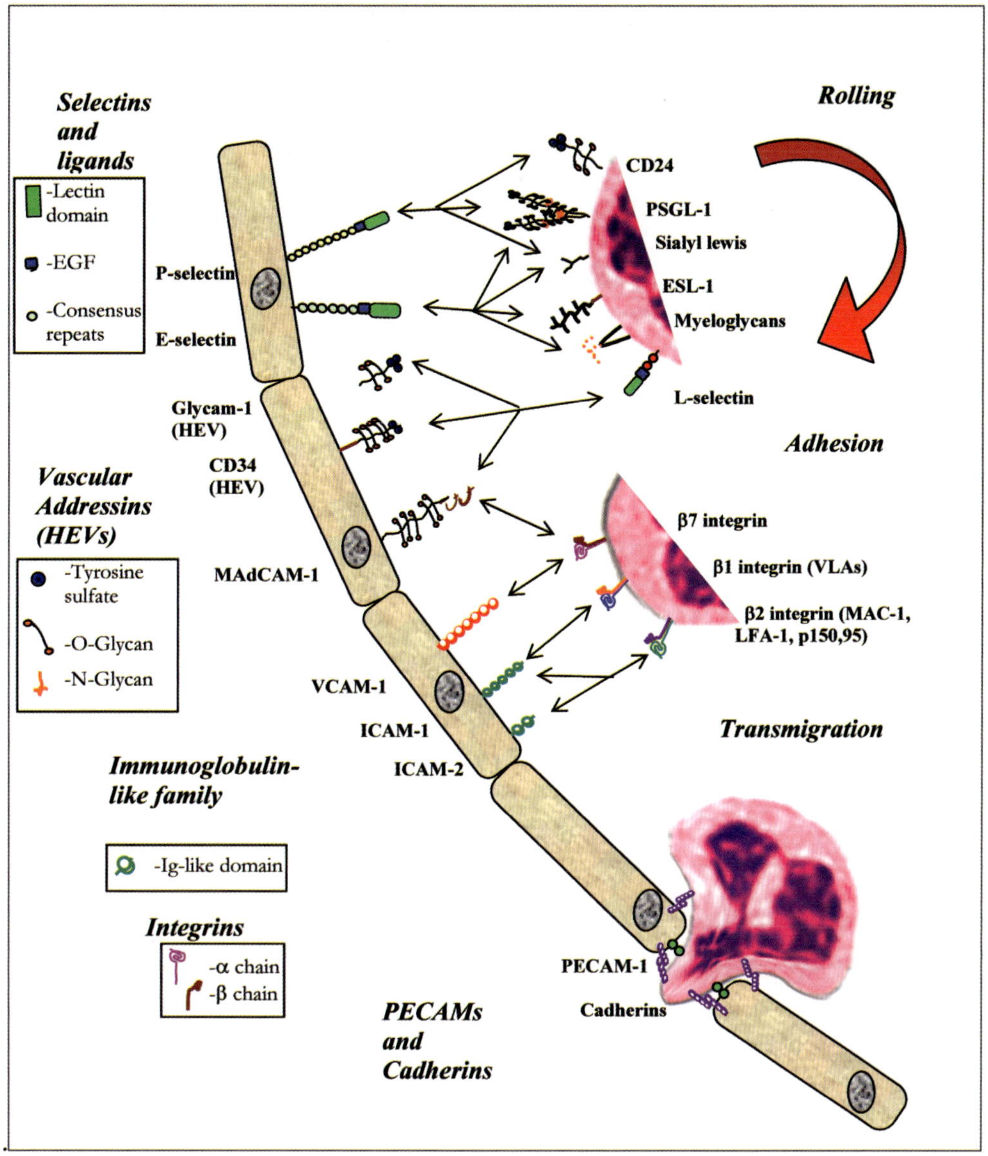

Chapter 51 Figure 2 Endothelial cell adhesion molecules and counterligands on leukocytes involved in cell trafficking. Activated EC express selectins that can bind to their carbohydrate receptors to allow leukocyte rolling under physiologic shear forces. Addressins expressed on high endothelial venules (HEVs) in lymph node vessels are expressed for lymphocyte homing; their peripheral vascular counterparts are less well defined. If activation occurs, leukocytes adhere to the endothelium via Ig-like superfamily proteins and integrins. Transendothelial migration is regulated by PECAM-1, cadherins, and other structures.

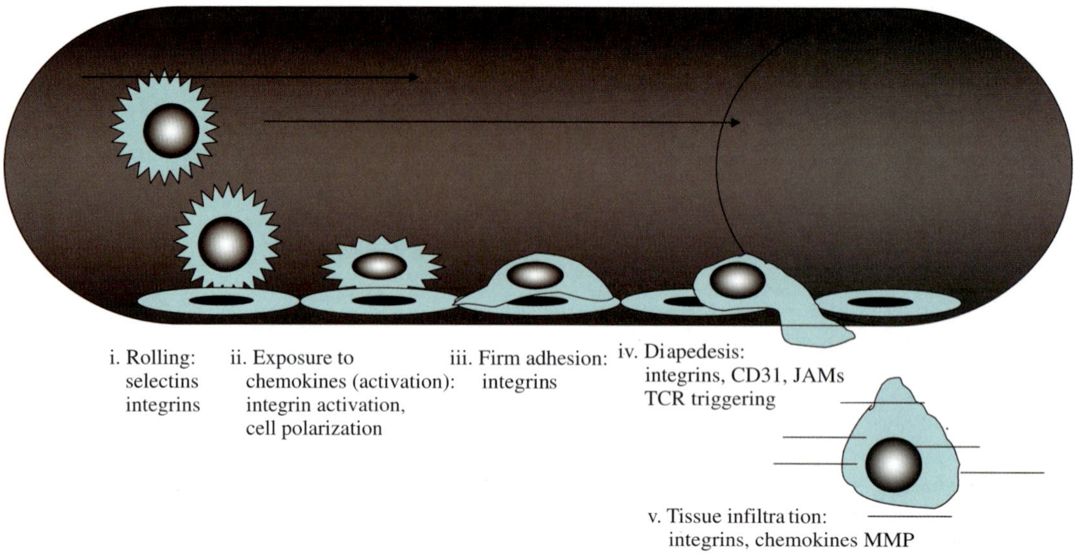

Chapter 52 Figure 1 The multistep paradigm of lymphocyte extravasation. T cells are first engaged in selectin-mediated transient interactions with the endothelium. In this way, they become exposed to endothelium-bound chemokines. This induces increased affinity of T-cell integrin for their endothelial ligands and causes T-cell polarization (i.e., change of shape and redistribution of surface molecules). T cells thus firmly adhere to the endothelium and eventually migrate through the endothelial cell junctions. Once in the tissue, T lymphocytes respond to stimuli that lead them to the site of inflammation.

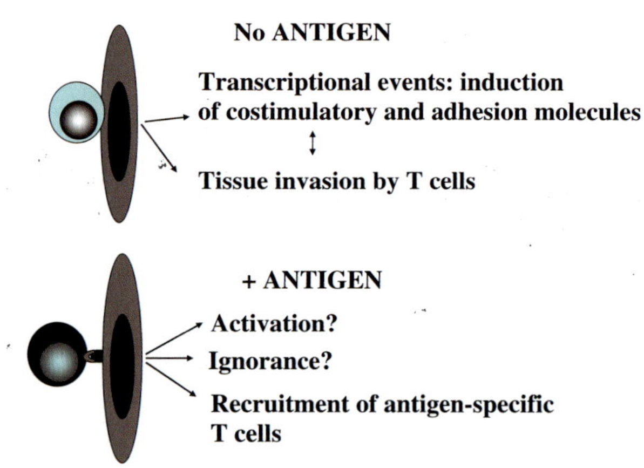

Chapter 52 Figure 2 Functional consequences of T cell–endothelial cell interactions. Sustained interactions with the endothelium during extravasation induce a genetic reprogramming in T lymphocytes leading to upregulation of adhesion and costimulatory molecules. As a consequence, T cells become more tissue-invasive and hyperreactive to antigenic stimuli. The effects of antigen presentation by the endothelium to transiting T cells during extravasation are less clear; however, this event appears to favor the extravasation of antigen-specific T lymphocytes.

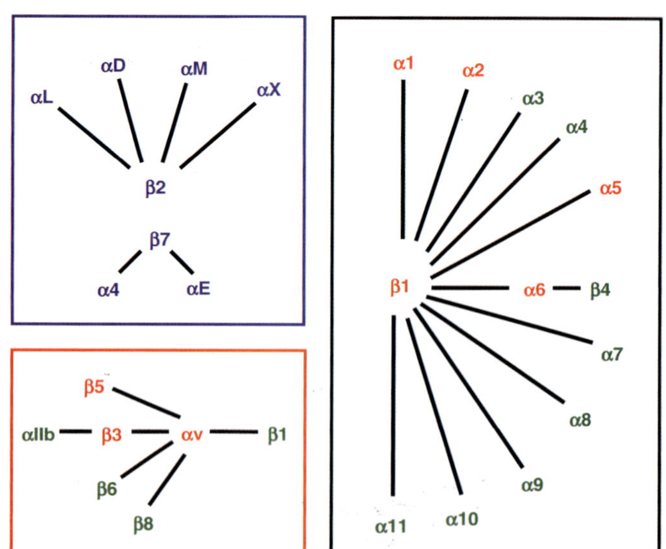

Chapter 53 Figure 1 The integrin family of ECM receptors. The 26 integrin α and β subunits, separated into three groups, are shown. Lines connecting subunits indicate the 24 known heterodimeric pairs. The leukocyte-specific integrins are shown in blue. Integrins with proposed roles in various microvascular functions are highlighted in red.

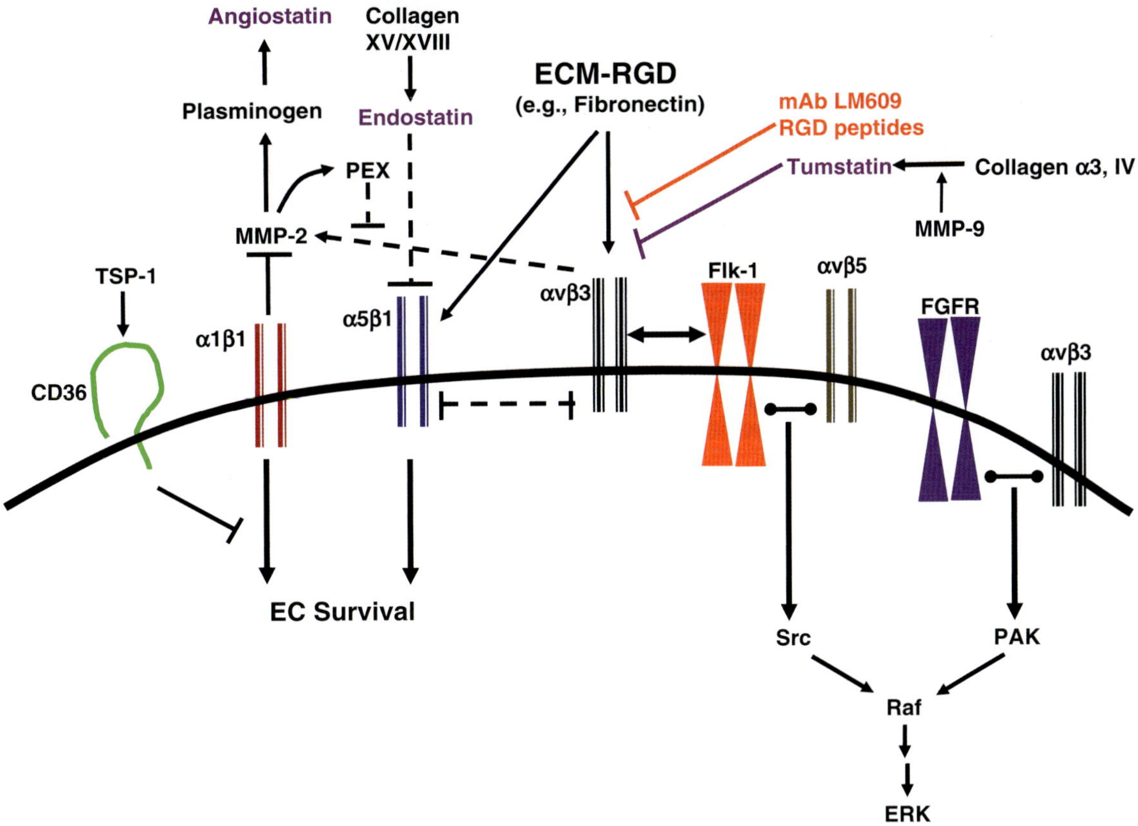

Chapter 53 Figure 2 Integrin signaling cascades during angiogenesis. Various integrins, extracellular matrix proteins, and receptor tyrosine kinases with roles in endothelial cell function are shown. Pink lettering indicates endogenous inhibitors of integrin-mediated angiogenesis. Orange lettering indicates blocking antibodies or peptide antagonists that inhibit integrin function. Lines with double-headed arrows indicate direct physical connections between proteins. Double-headed ball-ended lines denote signaling cross-talk between receptors.

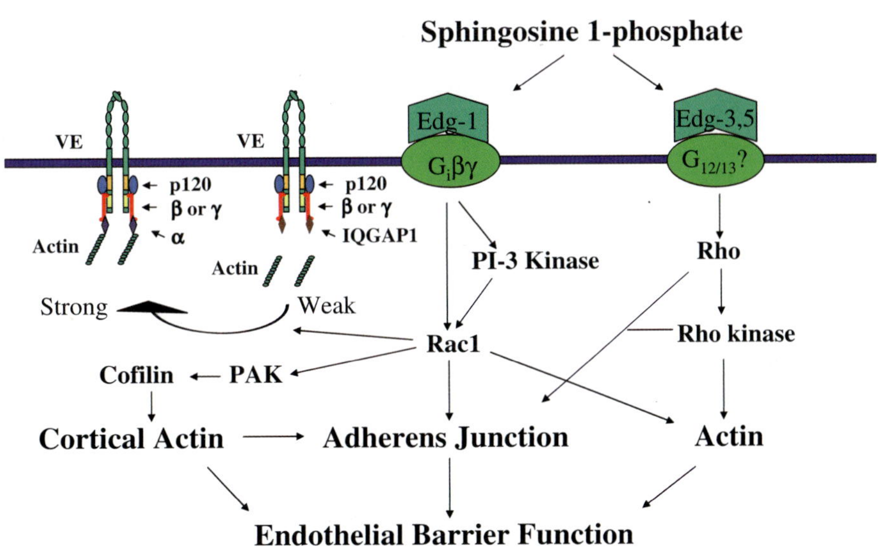

Chapter 54 Figure 1 Proposed cellular targets and signaling pathways by which sphingosine 1-phosphate (S1P) tightens the endothelial barrier. S1P binds to endothelial differentiation gene receptors (Edg). Edg 1 couples primarily with G_i and Edg 3 and Edg 5 with G_q and $G_{12/13}$. S1P activates Rac1, and many of the physiological outcomes of S1P and Rac1 are linked to G_i. Down-regulation of Edg 1 or Edg 3 or inhibition of G_i prevents the S1P-induced enhancement of the endothelial barrier. In epithelial cells, Rac1 sequesters IQGAP1 and removes this proposed negative regulator from adherens junction proteins. The resultant linkage to the actin cytoskeleton facilitates the integrity of the adherens junction, although this sequence of events has not been described in endothelial cells. S1P and active Rac1 also remodel the actin cytoskeleton as evidenced by a thickened, cortical actin and stress fibers. Active Rac1 appears to affect cortical actin via activation of PAK and the subsequent inactivation of cofilin. The formation of actin stress fibers involves Rho and Rho kinase, and inhibition of Rho kinase also prevents the barrier-enhancing property of S1P. The linkage of Rho with $G_{12/13}$ and Edg 3 and Edg 5 requires demonstration. To complicate matters, both Rac1 and Rho have also been associated with loosening of the endothelial barrier. Thus, subtle changes in the activation of Rac1 and Rho may profoundly influence endothelial barrier function. VE, vascular endothelial cadherin; p120, p120-catenin; α, α-catenin; β, β-catenin; γ, γ-catenin or plakoglobin.

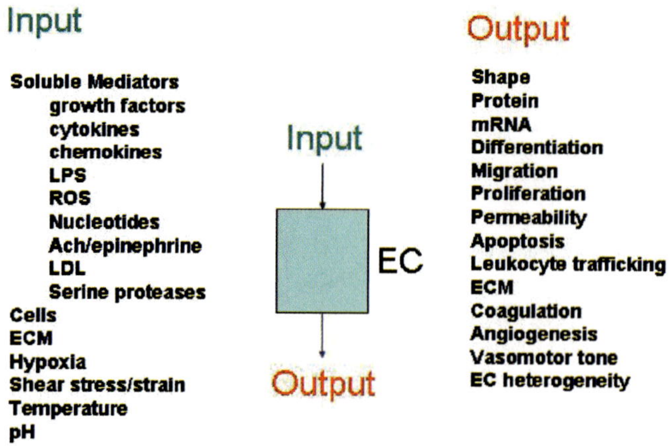

Chapter 56 Figure 1 The endothelial cell as an input–output device. Input arises from the extracellular environment and includes a combination of biomechanical and biochemical signals (*left*). The output, or cellular phenotype (*right*), may be expressed at the level of single endothelial cells or at higher levels of organization (e.g., blood vessel, organ, or whole organism). The device itself—depicted here as a "black box"—contains a highly complex array of signal transduction pathways that couple input with output.

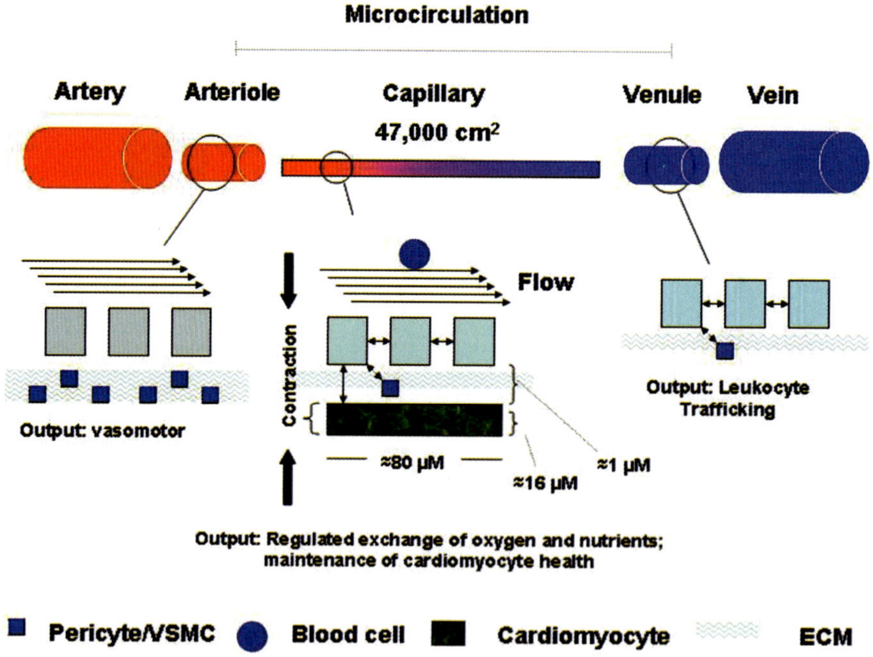

Chapter 56 Figure 2 Schematic of cardiac endothelium in health. The endothelium of the artery and arteriole is exposed to comparatively high shear stress and communicates with neighboring smooth muscle cells in the blood vessel wall. In addition to maintaining blood fluidity and barrier function, the endothelial lining of the arteries and arterioles contributes to the control of vasomotor tone. The cross-sectional area of the capillaries is far greater than that of the arteries and veins, thus reducing shear stress at the level of the endothelium and facilitating exchange of oxygen and nutrients with underlying tissue. Multiple input signals arise from the luminal surface (shown is blood flow) and abluminal surface (shown is communication with pericytes and cardiomyocyte, extracellular matrix [ECM], and contractile force). The capillary endothelium plays a critical role in maintaining the viability and function of the myocardium. The endothelium lining the postcapillary venules is the primary site for leukocyte trafficking.

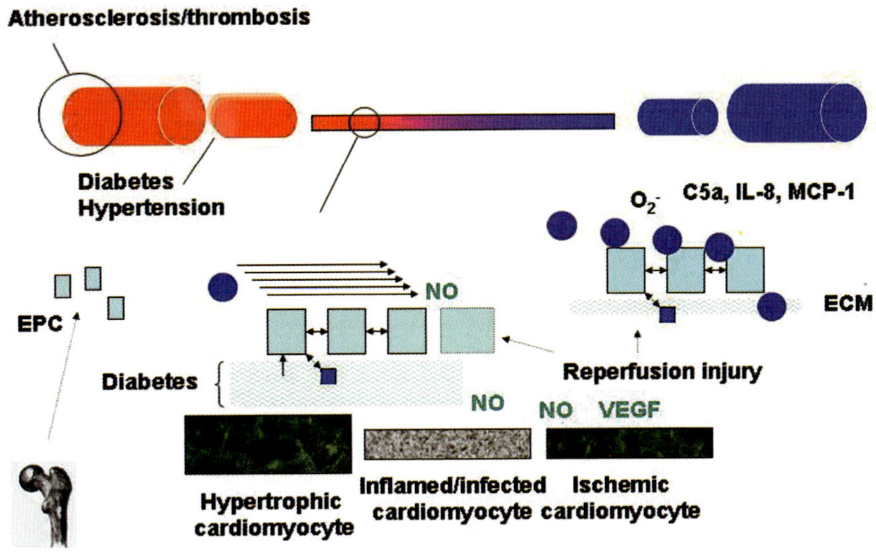

Chapter 56 Figure 3 Schematic of cardiac endothelium in disease. The endothelium is both affected by and contributes to cardiac pathophysiology. Atherosclerosis affects the large coronary arteries. Diabetes and hypertension affect large- and small-vessel endothelium. Abnormalities in cardiomyocyte function (e.g., hypertrophic cardiomyopathy, myocarditis, and ischemia) may result in altered paracrine signaling in the endothelium and secondary endothelial cell activation and/or dysfunction. Shown is cardiomyocyte production of nitric oxide (NO) and VEGF. Ischemia/reperfusion results in the exposure of the endothelium to myriad signals, including free oxygen radicals (O_2^-), complement, and cytokines [e.g., interleukin (IL)-8 and monocyte chemoattractant protein (MCP)-1]. Cardiac ischemia may result in increased mobilization and/or uptake of circulating endothelial precursor cells (EPCs) in the cardiac vasculature.

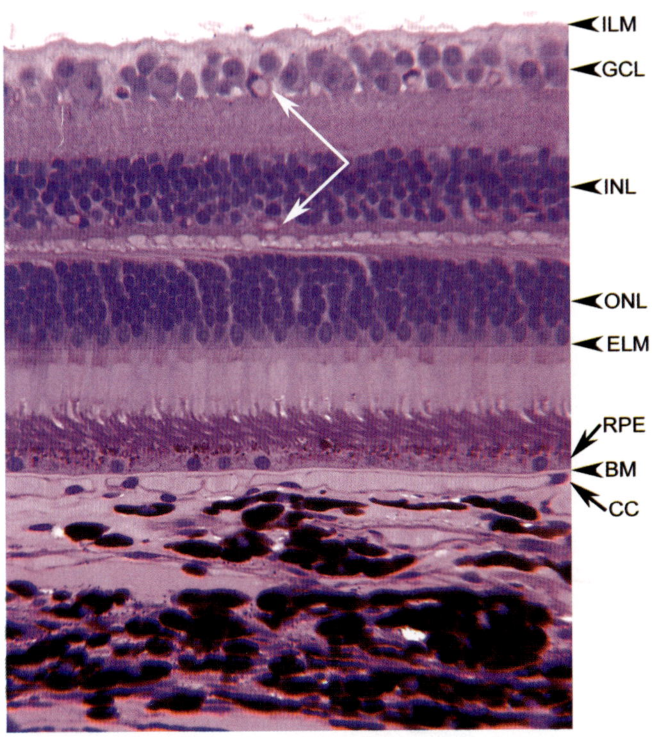

Chapter 57 Figure 1 Section from the eye of a rhesus monkey showing retina from the internal limiting membrane (ILM) at the interface of retina and vitreous to the retinal pigment epithelial cells (RPE). Retinal blood vessels (white arrows) are in the ganglion cell layer (GCL) of inner retina, and also a secondary or deep plexus is associated with the inner nuclear layer; their inner and outer segments are below the external limiting membrane (ELM). The nuclei of the photoreceptors are in the outer nuclear layer. Below Bruch's membrane (BM) is the choroid with choriocapillaris immediately posterior to Bruch's membrane and large choroidal vessels posterior in the densely pigmented outer choroid.

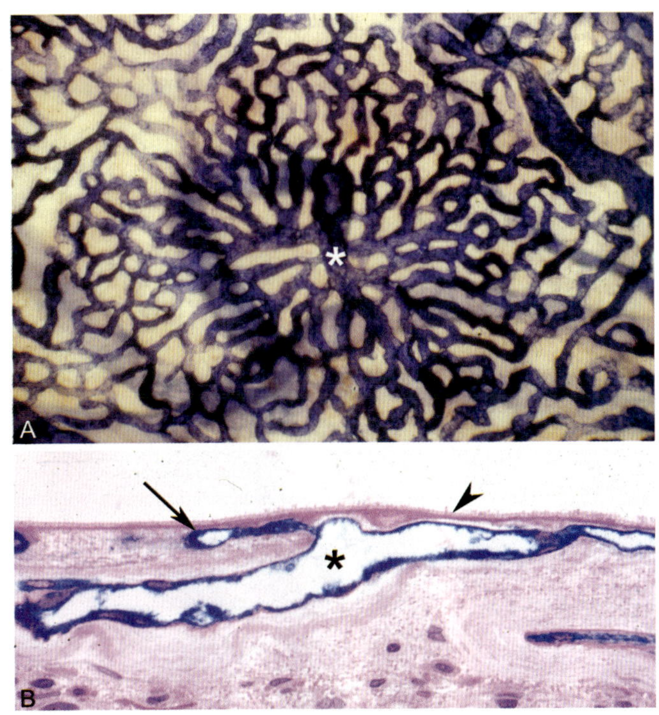

Chapter 57 Figure 3 **(A)** Alkaline phosphatase incubated normal human choroid after bleaching. In this area of choroid near the macula, the lobular organization of the choriocapillaris is apparent. There is a draining venule (asterisk) in the center of this lobule. **(B)** Section through the center of the lobule shown in (A) shows blue APase activity in choriocapillaris lumens (arrow) and in the draining venuole (asterisk). The internal limiting membrane (arrowhead) is PAS positive. RPE cells have been removed from Bruch's membrane in processing. (Blue APase reaction product, PAS, and hematoxylin stain.)

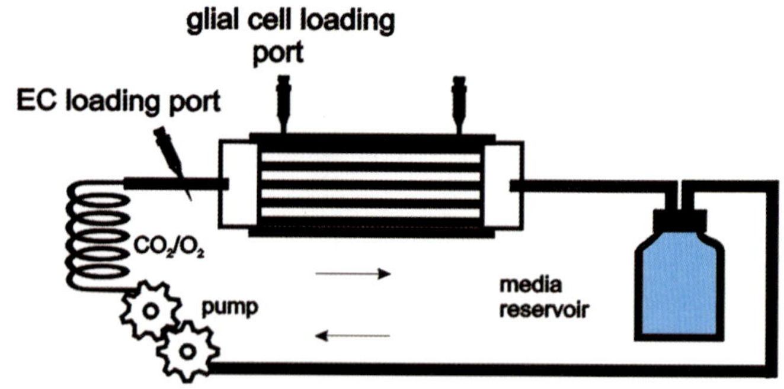

Chapter 60 Figure 3 Schematic representation of the tridimensional dynamic in vitro model (DIV-BBB); cells are injected into the system through the loading ports and the flow is maintained by a pump.

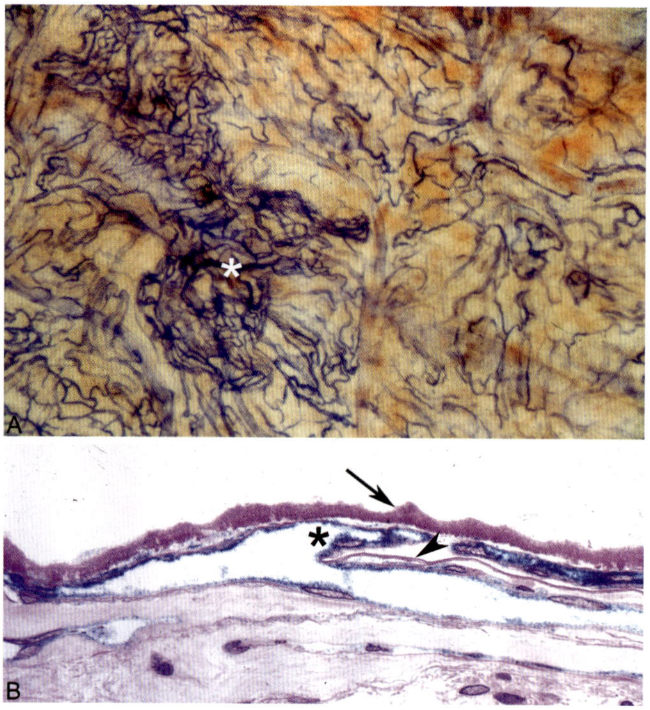

Chapter 57 Figure 4 **(A)** Alkaline phosphatase incubated choroid from a diabetic subject. Choroidal neovascularization (CNV) is apparent (asterisk) adjacent to an area with severe choriocapillaris dropout (lack of viable, APase-positive capillary lumens). The yellow areas are drusen. **(B)** Section through the CNV formation in (A) shows that the new blood vessels (asterisk) are between a basal laminar deposit (arrow), which is characteristically positive for PAS, and Bruch's membrane (arrowhead). The origin of the new vessels appears to be choriocapillaris. (Blue APase reaction product, PAS, and hematoxylin stain.)

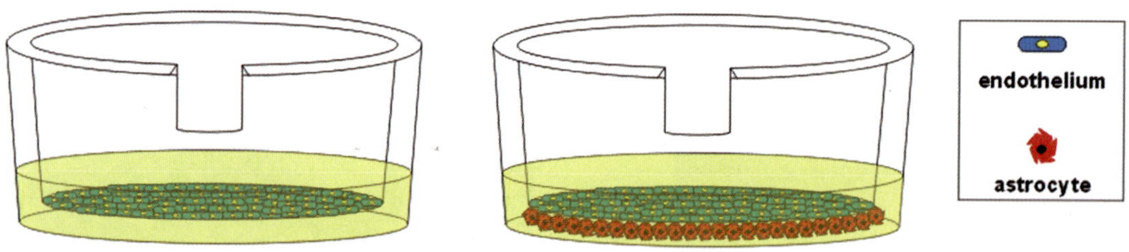

Chapter 60 Figure 1 Schematic representation of the Transwell system. Either monodimensional endothelial cultures (*left*) or bidimensional endothelial-astrocyte coculture can be established (*right*).

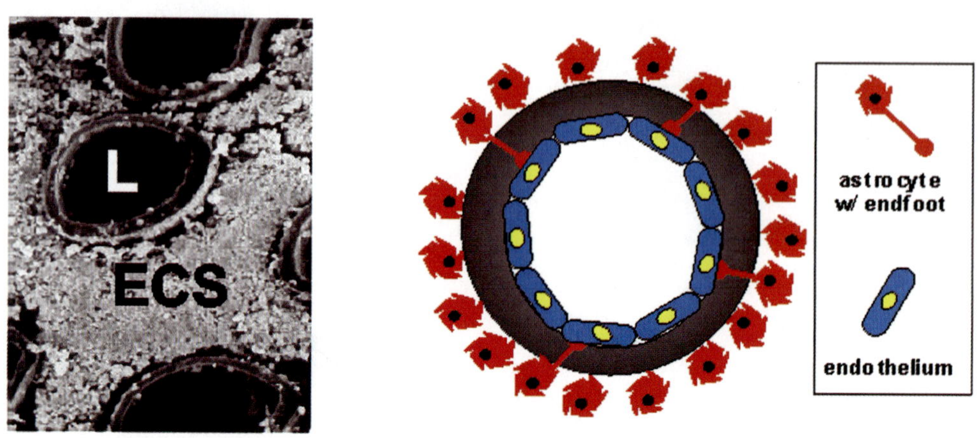

Chapter 60 Figure 2 Tridimensional in vitro model of the BBB: Endothelial cells are seeded intraluminally in a pronectin-coated hollow fiber, while astrocytes are cultured extraluminally. *Left panel*: Hollow fiber section (L, lumen; ECS, extracellular space); *right panel*: schematic representation.

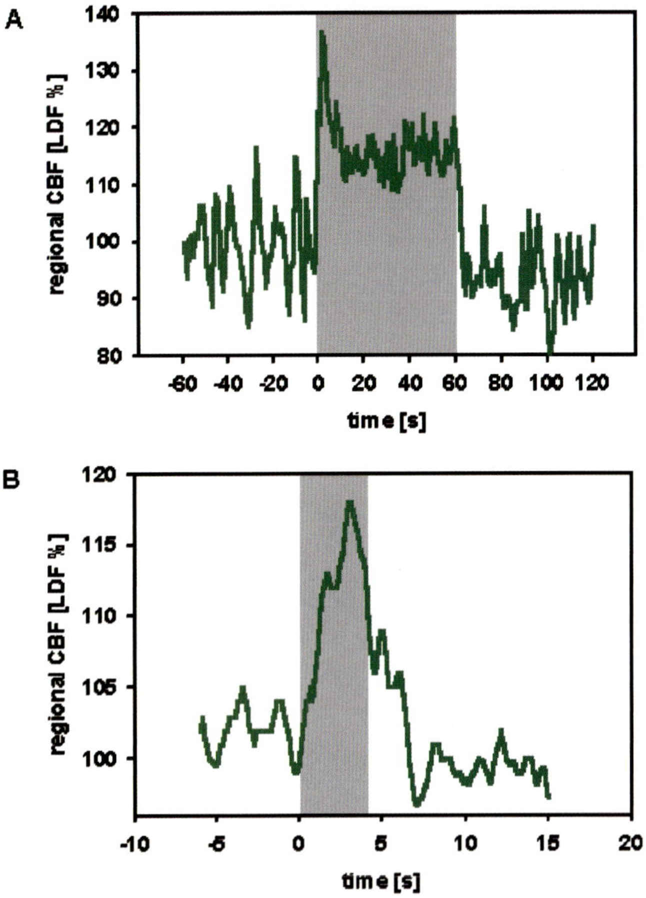

Chapter 62 Figure 1 Typical example of microvascular blood flow changes due to functional activation by 60 seconds **(A)** or 4 seconds **(B)** of whisker hair deflection (3 Hz) in the anesthetized rat. Regional cerebral blood flow is measured by laser Doppler flowmetry (LDF); percentage changes from baseline flow at rest (100%) are presented. The stimulation period is indicated by the gray bar within each figure. The regional cerebral blood flow change is temporally tightly coupled to the somatosensory stimulation. Note that during prolonged stimulation (A) the peak blood flow increase is followed by a plateau response slightly lower than the peak response immediately after stimulation onset. No plateau response occurs when short stimulations are performed (B).

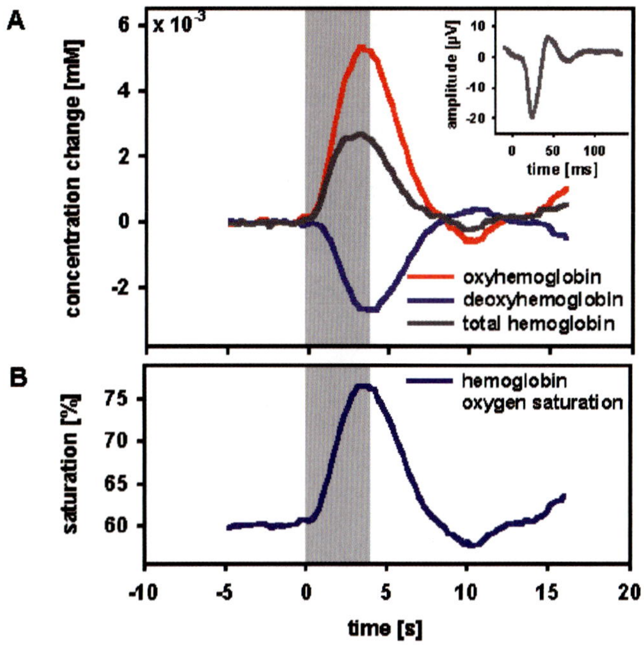

Chapter 62 Figure 2 Typical example of microvascular blood oxygenation **(A)** and hemoglobin oxygen saturation **(B)** changes due to functional activation by 4 seconds of whisker hair deflection (3 Hz) in the anesthetized rat. Regional cerebral oxygenation changes are measured by microfiber spectroscopy (wavelength range 510 to 805 nm, wavelength-dependent differential path-length corrected analysis, estimated hemoglobin concentration 20 µM, mean saturation 60 percent as determined for microcirculatory areas). Total hemoglobin and hemoglobin oxygen saturation changes are calculated from measured oxy- and deoxyhemoglobin changes. Neuronal activity was recorded as somatosensory evoked potentials (SEP). The SEP (average of 4 seconds of 3-Hz stimulation) responsible for the hyperoxygenation response is provided in the small inset in part A of the figure. During functional activation, oxyhemoglobin increases and deoxyhemoglobin decreases, leading to microvascular hyperoxygenation and an increase in hemoglobin saturation. The blood oxygenation changes are temporally tightly coupled to the somatosensory stimulation.

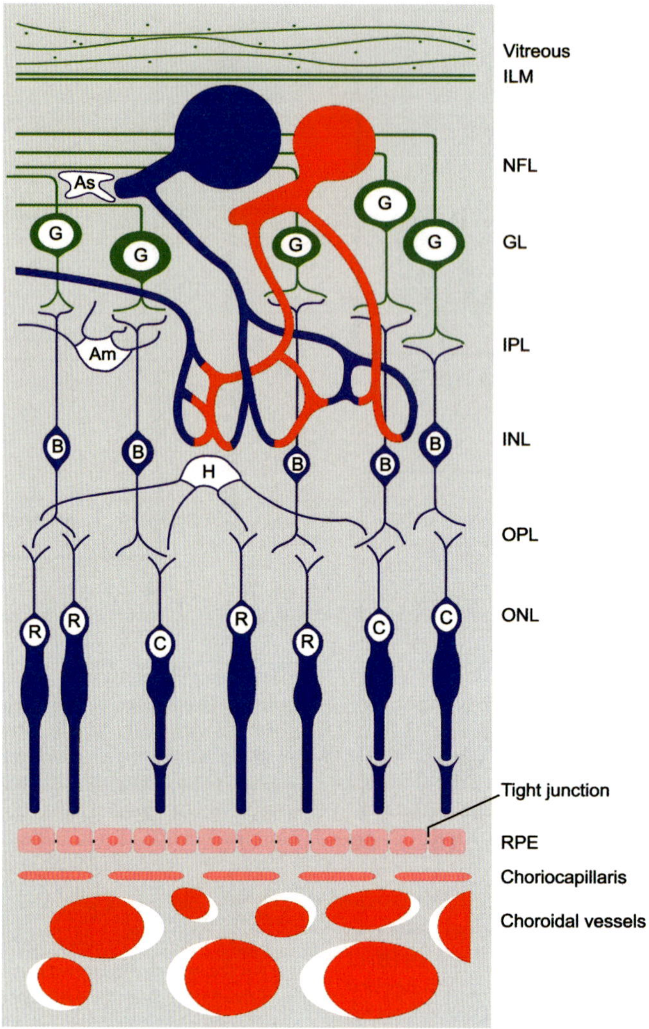

Chapter 63 Figure 1 Schematic representation of the retinal and choroidal circulations. The photoreceptor layer is avascular, receiving oxygen from choriocapillaris by diffusion. ILM, internal limiting membrane; NFL, nerve fiber layer; GL, ganglion cell layer; IPL, inner plexiform layer; INL, inner nuclear layer; OPL, outer plexiform layer; ONL, outer nuclear layer; RPE, retinal pigment epithelium; As, astrocyte; Am, amcrine cell; H, horizontal cell; G, ganglion cell; B, bipolar cell; R, rod photoreceptor cell; C, cone photoreceptor cell. From: Figure 121–2A, p. 1806 of Pournaras, C. J., and Donati, G. (2000). Retinal and choroidal circulations. In *Principles and Practice of Ophthalmology*, 2nd ed. (D. M. Albert and F. A. Jakobiec, eds.). Philadelphia: WB Saunders.

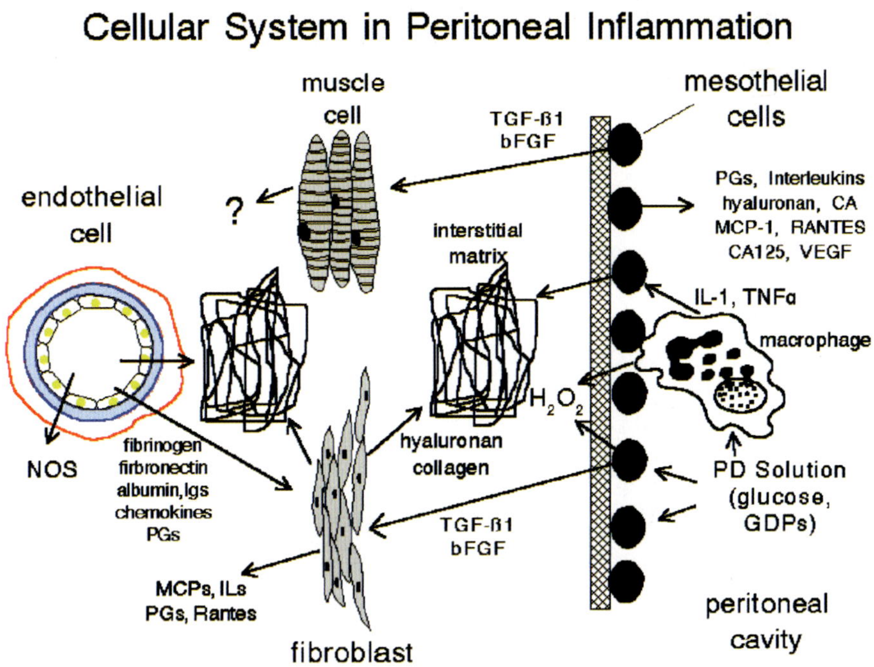

Chapter 66 Figure 5 Cellular system in peritoneal inflammation. Dialysis solutions in the peritoneal cavity contain glucose and glucose degradation products (GDPs) that activate macrophages and mesothelial cells to secrete numerous cytokines. These in turn stimulate fibroblasts and attract leukocytes to migrate through activated endothelium. This chronic inflammatory process results in scarring of the subperitoneal tissue and vasculopathy in the parenchymal vessels. See text for definitions of cytokine acronyms.

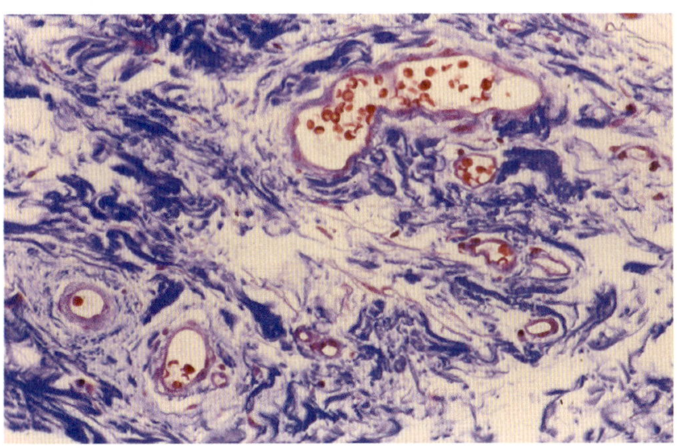

Chapter 67 Figure 1 Human peritoneal membrane stained with Masson-Noguchi modified stain (original magnification 200×).

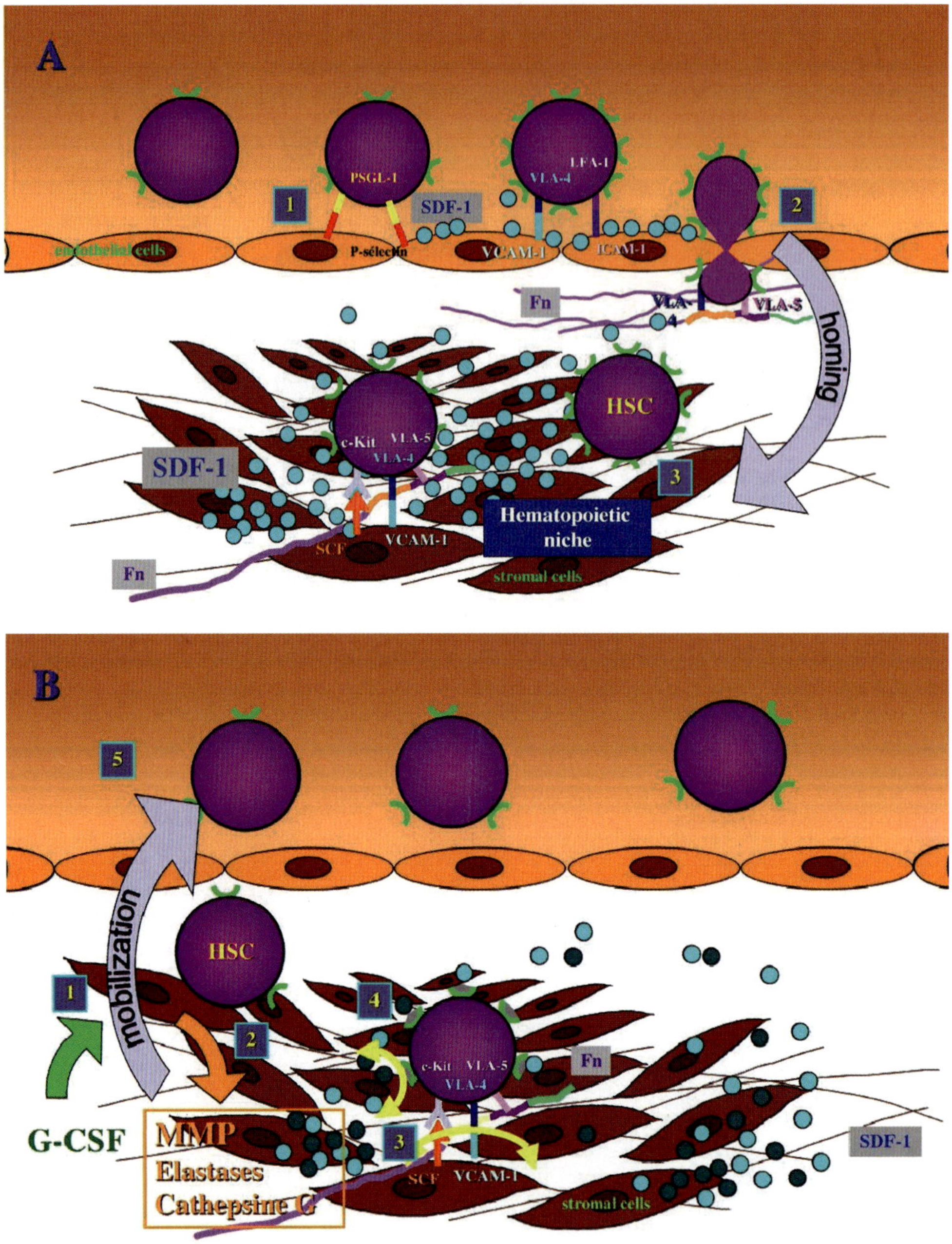

Chapter 69 Figure 1 Molecular modeling of HSC trafficking. Data supporting the following models are provided mainly by studies in vitro using HSCs, BM ECs, and stromal cells generated from long-term marrow cultures. **(A) HSC homing. (1)** Weak interactions between P-selectin glycoprotein ligand-1 (PSGL-1) expressed on HSCs and E- and P-selectins constitutively expressed on BM ECs lead HSCs to tether to, and roll along, the BM endothelium. SDF-1 located on the endothelial surface and bound to heparan sulfates induces activation of HSC integrins VLA-4 and LFA-1. VLA-4 and LFA-1 activation converts to firm adhesion their interactions with VCAM-1 and ICAM-1 constitutively expressed on endothelial cells. **(2)** This activation induces the arrest of HSCs and stimulates actin polymerization that favors HSC transendothelial migration mediated by VLA-4 and VLA-5 in the presence of fibronectin (Fn). **(3)** HSCs polarize, migrate along the local gradient of SDF-1 produced by stromal cells, and reach the hematopoietic niche. This anchoring depends mainly on close interactions of HSC VLA-4 with stromal cell VCAM-1 and of HSC VLA-5 with Fn of the ECM. Moreover, the continuous production of SDF-1 by stromal cells allows the confinement of HSC within the hematopoietic niche. **(B) HSC mobilization. (1)** Addition of G-CSF induces **(2)** a local production of proteases (MMPs, elastases, and cathepsin G). **(3)** These molecules are able to degrade the ECM. Moreover, elastase can cleave VLA-4/VCAM-1 and **(4)** CXCR-4/SDF-1 interactions by degrading both CXCR-4 and SDF-1. The loss of attachment to stromal cells and to the ECM together with the loss of SDF-1 activity **(5)** favors the release of HSCs from the hematopoietic niche.

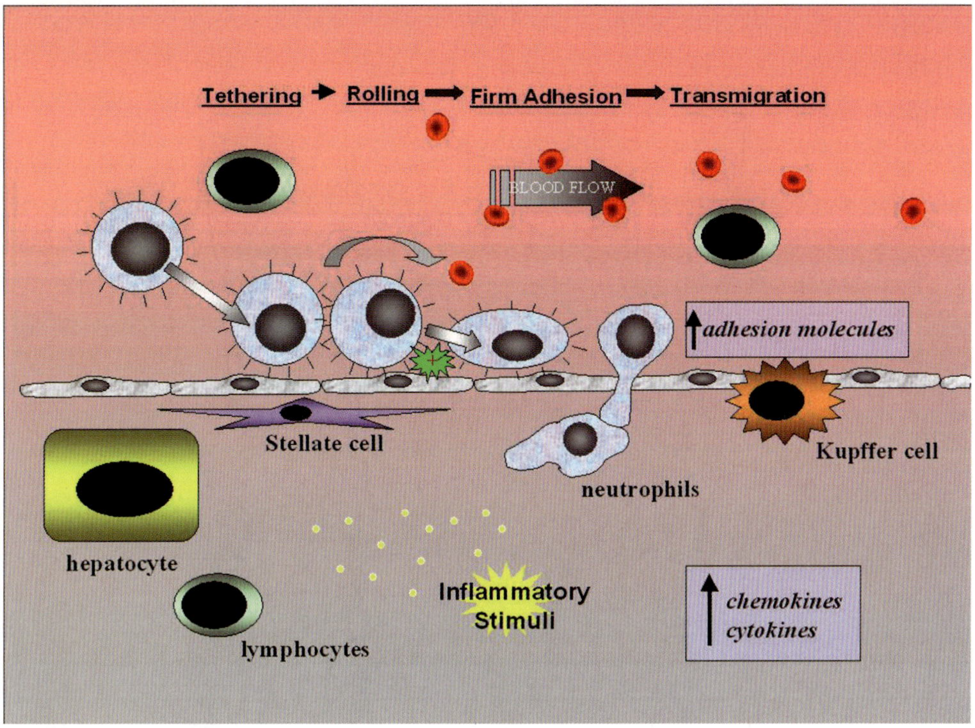

Chapter 72 Figure 1 The leukocyte recruitment cascade. Fast-moving leukocytes in the bloodstream tether and roll on activated endothelium via interactions between adhesion molecules. Chemokines or other proinflammatory mediators released by various sources within the tissue are presented on the endothelium to rolling leukocytes, resulting in firm adhesion. Leukocyte transmigration across the endothelium and entry into inflamed tissue ensues.

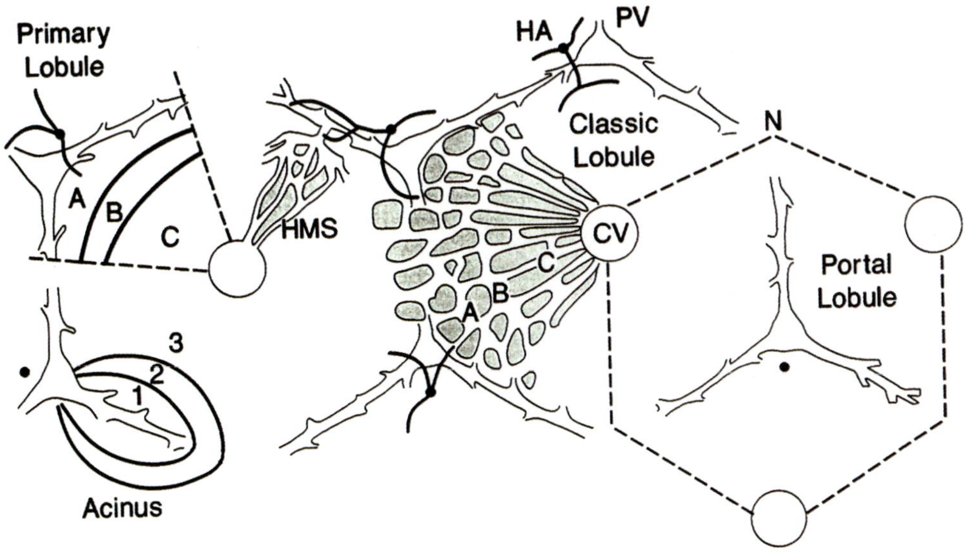

Chapter 73 Figure 1 Contiguous hepatic lobules illustrating the interconnecting network of sinusoids derived from two portal venules (PV). Note that the sinusoids become more parallel as they course toward the central venule (CV), which forms the axis of the classic lobule. Hepatic arterioles (HA) supply blood to sinusoids near the periphery of the lobule, usually by terminating in inlet venules or terminal portal venules. As a result, three zones (1, 2, 3) of differing oxygenation and metabolism have been postulated to compose a hepatic acinus, with its axis being the portal tract (*lower left*). Several acini would compose the portal lobule (*lower right*). Each classic lobule contains several cone-shaped subunits having convex surfaces fed by portal and arterial blood at the periphery and apices at the central venule (*upper left*). A, B, and C represent hemodynamically equipotential lines in a "primary lobule." A recent modification further subdivides lobules into conical hepatic microcirculatory subunits (HMS), each being supplied by a single inlet venule.

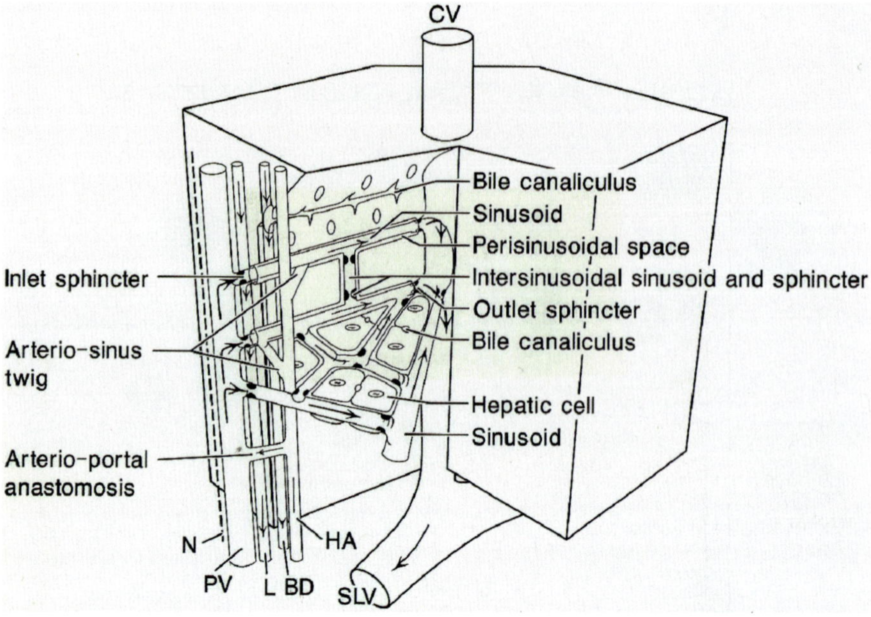

Chapter 73 Figure 2 Hepatic microvasculature as determined by in vivo microscopic studies. PV, portal venule; HA, hepatic arteriole; L, lymphatic; BD, bile ductule; N, nerve; CV, central venule; SLV, sublobular hepatic vein. Arrows indicate direction of flow.

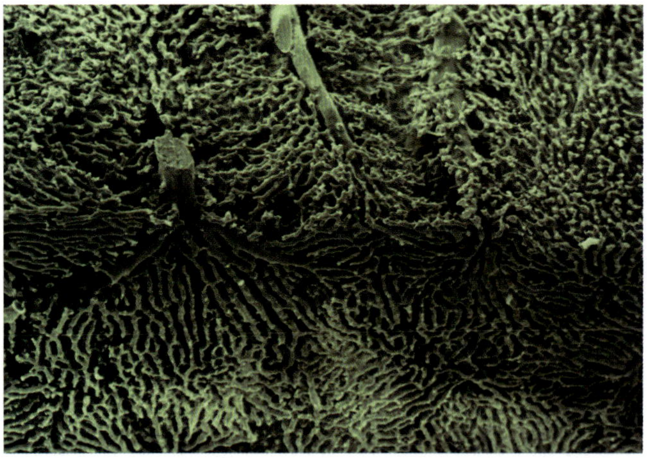

Chapter 73 Figure 4 Vascular cast of the hepatic microvasculature illustrating the tortuous, anastomotic sinusoids adjacent to the portal venule (PV) and the more parallel and larger sinusoids near the central venule (CV).

[C] Top Panel

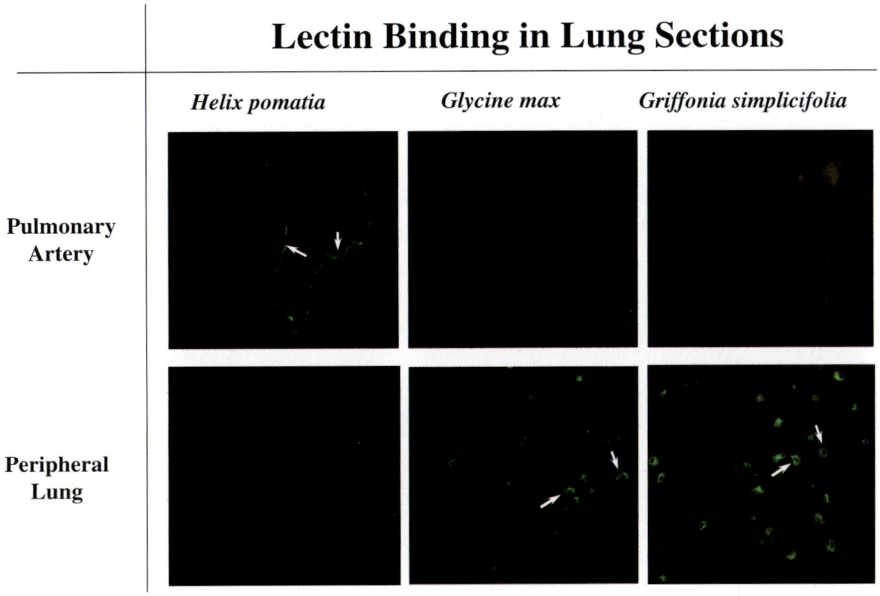

Chapter 75 Figure 1 (C) Rat pulmonary macro- and microvascular endothelial cells can be distinctly resolved in the intact circulation by their interaction with *Griffonia simplicifolia* and *Helix pomatia* (*top panel*).

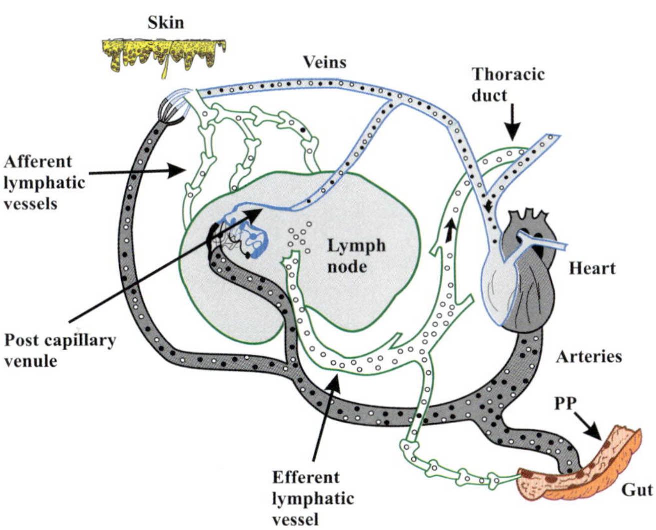

Chapter 80 Figure 1 Vascular network guiding lymphocyte recirculation. Transport of foreign material (pieces of microbes) into lymphoid organs takes place, for example, through the epithelium of the gut and via afferent lymphatics draining the skin. Bloodborne lymphocytes enter the organized lymphatic tissues (lymph nodes and Peyer's patches; PP) from the blood via the arterial tree, flow through the capillaries, and extravasate in the postcapillary high endothelial venules. Thereafter, the lymphocytes migrate through the tissue parenchyma in search of their cognate antigens. If the antigen is found, the lymphocytes get activated within the germinal centers (B cells) or outside the centers (T cells). Activated cells and cells that did not find their antigen leave the node by entering the lymphatic vessels and return via the efferent lymphatics back to the blood circulation. Activated cells then preferentially leave the blood in peripheral tissues, where the microbial insult took place, whereas the nonactivated lymphocytes continue their journey between blood and lymphoid organs.

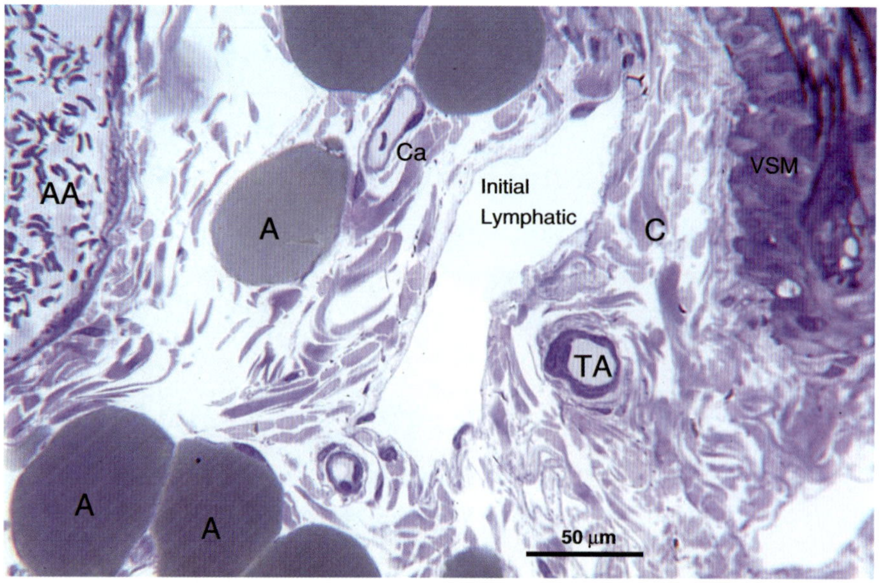

Chapter 82 Figure 2 Light microscopic cross section of initial lymphatic in rat skeletal muscle. The vessel has a single endothelial lining, is positioned in vicinity of adipose cells (A), a terminal arteriole (TA), embedded in collagen fibers(C), and positioned between two arcade arterioles (AA) with vascular smooth muscle media (VSM).

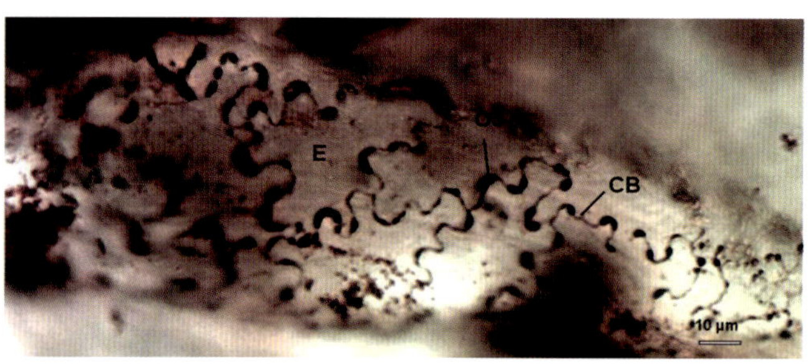

Chapter 83 Figure 1 Typical cell (E) circumference similar to an oak leaf in light microscope. The cell border line (CB) is contrasted with silver. Open-junction formations (OJ) are visible as dark half-moons.

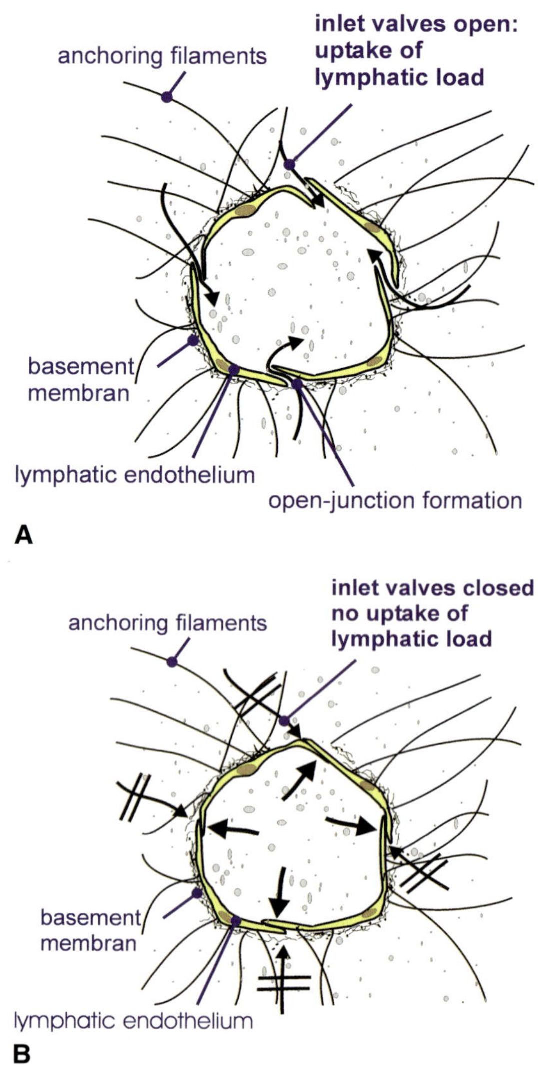

Chapter 83 Figure 9 Function of open-junction formations. **(A)** The pressure in the initial lymphatic is lower than in the connective tissue. Inlet valves are open: uptake of lymphatic load. **(B)** Increased pressure in the initial lymphatic. Inlet valves are closed: transport of lymph toward the precollector.

Microvascular Research

VOLUME 2

Microvascular Research

VOLUME 2

Editor-in-Chief

David Shepro, Ph.D.
Boston University
Boston, MA, USA

ELSEVIER
ACADEMIC
PRESS

Amsterdam Boston Heidelberg London New York Oxford
Paris San Diego San Francisco Singapore Sydney Tokyo

Elsevier Academic Press
30 Corporate Drive, Suite 400, Burlington, MA 01803, USA
525 B Street, Suite 1900, San Diego, California 92101-4495, USA
84 Theobald's Road, London WC1X 8RR, UK

This book is printed on acid-free paper. ♾

Library of Congress Cataloging-in-Publication Data
Application Submitted.

British Library Cataloguing in Publication Data
A catalogue record for this book is available from the British Library

ISBN 13: 978-0-12-639510-5 (set)
ISBN 10: 0-12-639510-1 (set)
ISBN 10: 0-12-639511-X (Volume 1)
ISBN 13: 978-0-12-639511-2 (Volume 1)
ISBN 10: 0-12-639512-8 (Volume 2)
ISBN 13: 978-0-12-639512-9 (Volume 2)

For all information on all Elsevier Academic Press publications
visit our Web site at www.books.elsevier.com

Printed in the United States of America
05 06 07 08 09 10 9 8 7 6 5 4 3 2 1

Cover image based on an Aboriginal rock painting showing what may be the first depiction of the circulatory system. The original artwork dates from circa 6000 B.C.

Foreword I

Dr. David Shepro has compiled an outstanding book on Microvascular Research that summarizes the state-of-the-art advances in this exciting field and points to the future horizons in this interdisciplinary area of great important in health and disease.

The book starts with a Section on Basic Science that covers the molecular and cellular bases of endothelial cell structure and function; roles of various molecules and cells in regulating vascular development; *in vitro* models ranging from zebrafish to mammalian and human systems; physiological regulation of permeability, vascular tone and hemodynamics, and the process of transendothelial transport in relation to endothelial junctions and adhesion molecules. This Basic Science Section provides an excellent foundation and underpinning for the subsequent Sections.

The second Section presents the current knowledge on Microvascular Adaptation in Organs, including the heart, central nervous system and eye, gastrointestinal tract, hematopoietic system, kidney, liver, lung, lymphatics, pancreas, skeletal muscle, skin, and reproductive systems. The comprehensive presentations in this Section point out the uniqueness of each organ/tissue and at the same time provide insights into the similarity in microvascular adaptation among different organs and tissues. This organ-level treatment of microvascular physiology leverages on the molecular and cellular approaches in the Basic Science Section and set the stage for the Sections on Pathology and Therapy that follow.

The third Section on Pathology covers the microvascular basis of pathological changes seen in various disease states. Some of the clinical conditions are well-documented as stemming from microvascular abnormalities, e.g., thrombosis and hemostasis, diabetes, inflammation, wound healing, etc. With the expansion of knowledge on microvascular research, there is now increasing evidence that the microvasculature also plays a significant role in Alzheimer's disease, arthritis, transplant rejection, tumor growth and metastasis, etc. This book provides an in-depth analysis of the role of microvasculature in a variety of pathological conditions.

The fourth Section on Therapy provides new insights into the great potential of applying the fruits of microvascular research to clinical conditions such as fluid treatment of hemorrhagic shock, the betterment of delayed preconditioning, therapeutic use of statins, therapeutic angiogenesis, and the delivery of molecular and genetic therapeutic agents to vascular endothelium. This Section points to the translational potential of microvascular research to enhance human health and combat disease.

The fifth Section on New Research Modes and Procedures presents a stimulating and inspiring view of the vista of microvascular research. It covers important recent advances such as DNA microarray, tissue engineering, endothelial biomarkers, knockout/transgenic models, mathematical models, and proteomics, as well as stem cells and their therapeutic promises. This Section provides a great ending of the book with the exciting future of microvascular research.

In summary, this book provides a comprehensive and logical progression from basic science through organ adaptation to pathology and therapy, ending with new research approaches. It is the most definitive book ever written on Microvascular Research and it is a must for all researchers and students in this field.

Dr. Shepro's superb background and experience made him uniquely qualified to edit such a state-of-the-art treatise on Microvascular Research. He has made important contributions to microvascular research, publishing over 200 peer-reviewed articles and four books. He is a Founding Editor and the Editor-in-Chief of *MICROVASCULAR RESEARCH: An International Journal*. In this capacity, Dr. Shepro has a marvelous knowledge of the leading scientists working at the frontiers of microvascular research, and hence he was able to gather such a stellar group of authors contributing to this wonderful book. As a member of the microvascular research community, I wish to congratulate Dr. Shepro and the contributors to this book for their outstanding accomplishments and to thank them for disseminating their precious knowledge to everyone working on this subject. This book undoubtedly will have a major impact on the further advancement of this important field.

Shu Chien, M.D., Ph.D.
University Professor of Bioengineering and Medicine
Chair, Department of Bioengineering
University of California, San Diego
La Jolla, California, U.S.A.
March 2005

Foreword II

The present compilation of articles on the topic of Microvascular Research is both the most comprehensive and most current of the works ever written on this subject. Its breadth is dazzling and the assembled cast of contributors is a Who's Who of the field's best minds. There is virtually no area of scientific discovery relevant to the structure, function and pathology of the microvasculature that is not addressed in this text. Together Drs Shepro and D'Amore have logged more that 75 years of research in vascular biology and their knowledge of the field and of the top scientists in this field is unparalleled.

The volume starts with the basic biology of the cells that comprise the microvascular bed and moves from there to vascular development and permeability, *in vivo* models, peri-vascular transport and on to the role of the microvasculature in specific organ beds as well as in human disease. What is noticeable is the extremely large list of human diseases that are caused or influenced by alterations in the microvasculature. One begins to think that the list of diseases that are *not* influenced by microvascular alterations must be considerable shorter. Of course, a large amount of the research on microvessels over the past 35 years has been conducted on the topic of tumor angiogenesis and this topic is well covered in this volume.

It is remarkable to see how much progress has been made in microvascular research over the past few years and how much information is now known. In some cases, the text reveals new progress on topics that have been studied for decades. Among these are the relationship of the endothelium with the extracellular matrix, endothelial metabolism, endothelial heterogeneity, the distinction between developmental vessel formation and pathologic vessel formation, the traditional experimental models such as the chick chorioallantoic membrane and hamster cheek pouch, the maintenance of vascular tone and the regulation of microvascular hemodynamics, a long-term interest of Dr. Shepro.

At the same time, there are a variety of new areas of microvascular research that have burst onto the scene in just the last few years and these, too, are wonderfully represented in this volume. Among these are the role of ephrins as signaling mediators in vascular cells, the ability of zebrafish to function as model system for vascular biology, the effect of mechanical signals on microvascular structure and function, and the extraordinary new work on VEGF and VEGF antagonists that is leading to important new treatments for both cancer and for vascular diseases of the eye.

If there is a message that emanates from this book as a whole, it is the concept that blood vessels have a crucial yet differing role in the function of virtually every tissue and in the pathogenesis of a host of diseases. No longer is it sufficient to consider blood vessels as passive tubes that merely carry nutrients to tissues. Rather the microvessels are a vital part of every tissue that interact mechanically, biochemically and metabolically with the parenchymal cells of the tissues they support. The communication is elaborate and goes in both directions, from the vessels to tissue and vice versa. If you study the vasculature, you must look at it from the context of the tissue in which it occurs. If you study any specific tissue, you must understand its relationship with the microvessels that inhabit it. Reading this volume allows a new appreciation of these interactions.

It should be noted that Dr. D'Amore began her career as a graduate student in Dr. Shepro's laboratory and is now one of the leading investigators in the field of vascular biology. Perhaps they didn't realize at the time that they would one day put together a volume as definitive and thoughtful as this one. One only hopes that there is a graduate student in Dr. D'Amore's laboratory who can be counted on to update this volume in ten or twenty years.

Bruce Zetter, Ph.D.
Charles Nowiszewski Professor of Cancer Biology
Children's Hospital
Harvard Medical School
Boston, Massachusetts

Introduction

Rationale and Reader's Guide

I can safely state that no one truly comprehends the physiology and pathophysiology of the microvasculature. In a large measure, this less than desirable state is the result of an explosion of new scientific data that creates paradoxically [or maybe poetically] a concomitant explosion of ignorance. But challenges enhance curiosity. Regardless how far scientists' understanding may be from the natural laws that regulate microvessel functions, we share an incumbency to accumulate and disseminate, by whatever means, current new concepts based upon the latest tested investigations. Hence, the *Biology and Pathology Microvasculature Research* in printed and electronic formats. As Lord Florey once stated: "I do not think there is any sharp line between physiology and experimental pathology," hence the subtitle.

In 1967, at a Boston conference the *Microcirculation as Related to Shock*, (D. Shepro and G.P. Fulton, Eds. Acad. Press, 1968) Robert Ebert, then dean of Harvard Medical School, summarized the proceedings in his keynote address that: *"in shock all physiological parameters are disturbed but if there is a common denominator it would be the microcirculation."*

Endothelium is metabolically one of the most active tissues in the body . . . probably second only to nervous tissue. Microvessels bridge the subcellular to the cellular to the organ to the system. **The breaching of the microvascular barrier is part and parcel of every disease. An inflammatory response, initially localized by microvessels, is frequently a persistent parameter of every disease.** Hence, whatever the therapeutic intervention, continuous attenuation of this defense mechanism, when awry, is required. As a closing but not a final argument, one important current concept is that there are specific *microvascular diseases* that to a large extent are independent of events in other parts of the cardiovascular system; e.g. angina with patent coronary arteries, cardiac slow flow states, long flight induced edema. Some scientists have speculated that the first stage of all forms of dementia is a breakdown of the blood brain barrier.

In the past, writers and publishers at some point had to agree on an endpoint, regardless of the manuscript's blemishes, and submit what was on hand for publication. As the "*ink dried*," the accumulated responses from critics and readers and new information would form the basis to kick-off a new edition, more frequently than not, years after the initial publication. Informatics technology, in all research endeavors, adds a new dimension to publishing, namely instant editing when the need arises, whether it be daily, weekly, quarterly, biannually. Because the editors and publishers are mindful of the fact that the currency of each scientific publication devaluates rapidly, the electronic version of *Microvascular Research* will be updated frequently. If "break throughs" should occur, an epilogue can be added immediately or reverse, obsolete "facts" and speculations that do not dovetail with new documented data can be immediately deleted.

The readers will quickly note that each presentation is limited to about 10–15 printed pages, unusual especially for review articles, and with a limited annotated (author's) bibliography. This editorial decision is based upon the realization that investigators are swamped with data and with the new tools to share information, speed has become an essential parameter. Global updating is so easily and readily available that, *less can be an advantage*. Given the space constraints, our contributors are challenged to provide the readers with their best objective and subjective take on the state of their sub-discipline. The redeeming fall-out of condensation is that lengthy introductions, redundant citations, speculations beyond the call, are eliminated. The creative sifting and contracting prior to submitting an article serves as a *needle's eye*.

A Perspective on Historical Recordings

In keeping with the editorial consensus that brevity can be a virtue and in light of the speed of scientific discoveries [see Rationale], a classical account that chronologically recapitulates microcirculatory milestones is not included in the volume. Some readers will regard this decision as an intellectual miscalculation; that without a record of discoveries there is blinding ignorance. Yet, it is an accepted fact of academia that recorders of events cannot prevent their imaginations from slipping into their "factual accounts." The noted historian Pieter Geyl states "history is an argument without end". For example: W*ho was the first to posit the existence of capillaries and their connecting role: Erisistratus? Galen? Malpighi? Harvey? Leeuwenhoek? Boerhaave? Henle?* The winner of the paternity title will go to the histiographer of your choice. Fortunately, a wealth of microcirculatory data are readily available in paper and electronic format, which lessens the need for such a chapter in the Encyclopedia. However, to accent the editors' recognition and appreciation of history, an illustration may provide the wherewithal to unravel one scientific puzzle in the history of microvascular research.

Why did it take over a century for scientists to appreciate the heterogeneity and high metabolic activity of the ubiquitous endothelial cell [not to mention the pericyte]? Elie Metchnikoff (1883), at the Pasteur Institute, in describing his new theory on inflammation that laid the foundation for the modern concept of this defense mechanism, stated:

"Movement [contractility] of endothelial cells plays an important role in the formation of stomata during inflammation as Klebs thought and as I imagined and stated in my first paper on inflammation."

Another Nobel laureate, August Krogh, described and illustrated endothelial cell motility in "The Anatomy and Physiology of Capillaries" [1922], the publication of his Sterling Lectures given at Yale University. Although both works are highly quoted, the scientific concepts that they embrace were virtually ignored in endothelial research until a scant 30 years ago. In 1966, the laureate Lord Florey, in his address entitled, "The Endothelial Cell", honoring the noted pathologist Sir Roy Cameron, spoke exclusively on the morphology of these cells. Presciently, he concluded that his was an interim report and "*I should expect to see in the next ten years a rich harvest of new [functional] knowledge. . . .*"

Again the question—why the achievements of three laureates, whose collective contributions significantly codified microvascular biology, were largely ignored? One explanation certainly can be argued on the importance of technology over imagination; that in the life sciences technology has always played a dominant role (review the history of the microscope) in contrast to Newton's and Einstein's science, where imagination was dominant. In other words, the technology for accurately profiling mural cells was unavailable. But in this scientific teaser I believe the answer lies in the incandescence of another stellar scientist, Earnest H. Starling, whose hypothesis on fluid transport still remains a major physiological principal. But for years this monumental work may have put blinders on investigators regarding the true nature of the microvascular wall. The Starling quotation in part supports this opinion:

"We have no sufficient evidence to conclude that endothelial cells of capillary walls take any part in the formation of lymph" [1896]

The elegance of Starling's hypothesis and research on filtration was so convincing that for decades microvascular physiologists devoted their efforts, for the most part, to fine tuning his formula. I hasten to add that this opinion does not diminish Starling's achievements and legacy an iota. Newton's unique contributions were eventually and correctly challenged and even the theories of Einstein, the genius of the 20th century, are now under scrutiny.

I would add one very personal speculation (a chronicler's prerogative) that might explain why Starling ignored other scientific data that could have affected his unilateral view of the microvessel wall as a passive barrier. Notwithstanding their correspondence on science and other matters, Starling did not appear to appreciate Krogh's experimental data. My perception is that he viewed Krogh as a junior level scientist and I "*imagine*" he was a tad jealous of his Danish colleague's talents.

In summary, I respectfully suggest to those who are interested in the historical analysis of microcirculation to select the format that best matches their interest. The chronological approach is the most common and of course should include examples from ancient oriental cultures as well as those from western civilization. Another choice would be to select the superstars of different eras. For example, Andreas Vesalius's "**De Humani Corporis Fabrica**" could be the starting point of modern science since this publication marked the deathknell for skewing scientific data to comply with authority (Aristotle; Galen). Vesalius's work was based on careful observations. Would there have been a Harvey [17 century] without a Versalius [16 century]? Advances in technology, such as the history of the microscope, can also be an artful approach to chronicle advances in microcirculatory knowledge. Whatever your choice, the endpoint will be similar.

References

Florey, H.W. 1966. The endothelial cell. *Brit. Med. J.* 27: 487–490

Krogh, A. Anatomy and Physiology of Capillaries. New Haven: Yale University Press. 1922. pp. 47–69

Metchnikoff, E. Lectures on the Comparative Pathology of Inflammation. New York: Diver Publications, Inc., 1968. pp. 137–156.

Starling. E.H. 1896. On the absorption of fluids from connective tissue spaces. *J. Physiol. Lond.* 19: 312–326

Acknowledgements

I thank all of the contributors, the editors who developed and shaped the contents of these volumes, our Elsevier Editor, Noelle Gracy, who inaugurated the project, and provided often needed encouragement, notwithstanding a transfer to Holland and a *maternity leave*. And to Karen Dempsey, at Elsevier, San Diego, who managed to organize our input, in bits and pieces, into a professional, valuable publication—always positive in her responses and never displays an "ounce of static." To my assistant, Christina Pitcher-Cozzone, for her "can do" handling of correspondence and almost daily revisions of contributors, titles and contents. Lastly, my gratitude to my wife, Marilyn, for her encouragement whenever I whined about the problems of a head editor and who did not complain about lengthy disappearances to my computer during the final stages of production.

Editorial Board Members

Editor-in-Chief
David Shepro, Ph.D.

Dr. Shepro is Professor of Surgery and Professor of Biology at Boston University. His research interests include cellular signaling and gene regulation and was the first to culture successfully large numbers of aortic endothelial cells that provided the wherewithal to research their metabolic activity and responses to inflammatory agonists.

David Shepro is a Fellow of the American Association for the Advancement of Science and a Fellow of the American Physiological Society. He has received the national and international awards from the American Microcirculatory Society, the Landis Award and the Zweifach Gold Medal Award, respectively, and the Taylor Smith Gold Medal Award from the New England Ophthalmological Society, the latter for his research on the biology/pathology of the retinal microcirculation. He was Burroughs Welcome Visiting Professor in Basic Medical Sciences and Distinguished Visiting Professor at the University of Cagliari School of Medicine, Sardinia.

Dr. Shepro is the Chair of the Literature Selection Technical Review Committee for the National Library of Medicine and Chairman of the Scientific User Committee for the Marine Biological Laboratory—Woods Hole Oceanographic Institute Library. He is the founding editor of the journal *Microvascular Research*.

Associate Editor
Patricia A. D'Amore, Ph.D.

Dr. Patricia A. D'Amore is a Senior Scientist and the Associate Director of Research at The Schepens Eye Research Institute. She is Professor of Ophthalmology and Pathology at Harvard Medical School, and Chair of the Angiogenesis, Invasion, and Metastasis Program in Development at the Dana Farber Harvard Cancer Center. She is the author of 95 publications which have appeared in journals such as *Nature Medicine, Journal of Clinical Investigation, Development*, and *Journal of Cell Biology*. She is Co-Editor-in-Chief of the journal *Microvascular Research*.

Editorial Board
Dame Carol M. Black, Ph.D.

Professor Dame Black is currently President of the Royal College of Physicians (UK) and Professor of Rheumatology at the Centre of Rheumatology, the Royal Free and University College Medical School. Her department at the Royal Free Hospital is a major centre for scleroderma in Europe and one of the largest in the world, with an extensive programme of clinical and basic research.

Joe G. N. Garcia, M.D.

Dr. Garcia is the Lowell T. Coggeshall Professor and Chair, Department of Medicine at the University of Chicago. His research program studies endothelial cytoskeletal regulation of the pulmonary microvascular endothelial barrier with a focus on the signaling events regulating barrier function of the endothelial cell monolayers. This work has resulted in over 200 peer-reviewed publications and numerous book chapters. He is Co-Editor-in-Chief of *Microvascular Research*, and Associate Editor of the *American Journal of Respiratory Cell & Molecular Biology*, and the *Journal of Stem Cell Research*.

D. Neil Granger, Ph.D.

Dr D. Neil Granger is Boyd Professor and Head of the Department of Molecular & Cellular Physiology at LSU Health Sciences Center. He has authored over 500 publications dealing with the role of the microcirculation in health and disease, and has served as Editor-in-Chief of Microcirculation, President of the Microcirculatory Society, and President of the American Physiological Society. His areas of research interest include ischemia-reperfusion injury, inflammation, and oxidative stress.

Christian Haudenschild, M.D.

Christian Haudenschild directs the Experimental Pathology Department at the American Red Cross Jerome H. Holland Laboratory for the Biomedical Sciences. His laboratory's main area of research is angiogenesis.

Herbert B. Hechtman, M.D.

Dr. Hechtman is Professor of Surgery at Harvard Medical School and Brigham and Womens Hospital. His laboratory's areas of research include the consequences of ischemia and inflammatory reactions in the lungs.

Rakesh K. Jain, Ph.D.

Rakesh K. Jain is Andrew Werk Cook Professor of Tumor Biology, Department of Radiation Oncology, Harvard Medical School and Director of the Edwin L. Steele Laboratory of Tumor Biology at Massachusetts General Hospital. Dr. Jain is regarded as a leader in the areas of vascular and interstitial biology of tumors and is known for integrating tumor biology with bioengineering. He is a member of the Institute of Medicine and the National Academy of Engineering of the U. S. National Academies.

Joseph A. Madri, Ph.D., M.D.

Joseph A. Madri received his BS and MS in Biology from St. John's University, NY and his Ph.D. in Chemistry and an M.D. from Indiana University. He is a Professor of Pathology at Yale University with appointments in the Medical School and Graduate School of Arts & Sciences. His areas of interest include vascular and cardiac development, angiogenesis, immune cell-endothelial cell interactions, and extracellular matrix biology.

Contributors

David Abraham
Centre for Rheumatology, Royal Free Hospital, London NW3 2QG, UNITED KINGDOM

Anthony P. Adamis
Research Center, Eyetech Pharmaceuticals, Woburn, MA 01801

Alexei V. Agapitov
East Amherst, NY 14051

William Cameron Aird
Department of Medicine, Division of Molecular/Vascular Biology, Beth Israel Deaconess Medical Center, Harvard Medical School, Boston, MA 02215

J. Steven Alexander
Department of Molecular and Cellular Physiology, Louisiana State University Health Sciences Center, Shreveport, LA 71130-3932

Abu-Bakr Al-Mehdi
Department of Pharmacology, College of Medicine, University of South Alabama, Mobile, AL 36688

Arne Andersson
Department of Medical Cell Biology, University of Uppsala, Uppsala, SE-75123, Sweden

Nick Anim-Nyame
Department of Maternal and Fetal Medicine, Faculty of Medicine, Imperial College of Science, Technology and Medicine at Chelsea and Westminster Hospital, London SW10 9NH, UNITED KINGDOM

Peter B. Anning
Department of Medical Biochemistry and Immunology, University of Wales College of Medicine, Cardiff, Wales CF14 4XN, UNITED KINGDOM

John D. Ash
Departments of Ophthalmology and Cell Biology, University of Oklahoma Health Sciences Center, Oklahoma City, OK 73104

James W. Baish
Department of Mechanical Engineering, Bucknell University, Lewisburg, PA 17837

Ann L. Baldwin
Department of Physiology, University of Arizona College of Medicine, Tucson, AZ 85724-5051

William A. Banks
Geriatrics Research, Educational, and Clinical Center, Veterans Affairs Medical Center, St Louis, MO 63106

Margery A. Barrand
Department of Pharmacology, University of Cambridge, Cambridge CB2 1PD, UNITED KINGDOM

Matthias Barton
Medical Policlinic, Departement of Internal Medicine, University Hospital Zurich, Zürich, CH-8091, SWITZERLAND

David O. Bates
Microvascular Research Laboratories, Department of Physiology, Preclinical Veterinary School, University of Bristol, Clifton, Bristol BS2 8EJ, UNITED KINGDOM

John F. Beltrame
Cardiology Unit, The Queen Elizabeth Hospital, Woodville South, SA-5011, SOUTH AUSTRALIA

Monica M. Bertagnolli
Division of Surgical Oncology, Department of Surgery, Brigham & Women's Hospital, Boston, MA 02115

Christer Betsholtz
Department of Medical Biochemistry, Göteborg University, Goteburg, SE-40330, SWEDEN

Jahar Bhattacharya
Lung Biology Laboratory, College of Physicians and Surgeons, St. Luke's/Roosevelt Hospital, Columbia University, New York, NY 10019

Imran Bhutto
Ophthalmology Department, Johns Hopkins Hospital, Baltimore, MD 21287-9115

David G. Binion
Division of Gastroenterology and Hepatology, Medical College of Wisconsin, Milwaukee, WI 53226

Konstantin Birukov
Department of Medicine, Section of Pulmonary Critical Medicine, University of Chicago, Chicago IL 60611

Anna Birukova
Department of Medicine, Section of Pulmonary Critical Medicine, University of Chicago, Chicago IL 60611

Joyce E. Bischoff
Vascular Biology Program, Department of Surgery, Children's Hospital, Harvard Medical School, Boston, MA 02115

Carol M. Black
Centre for Rheumatology, Royal Free Hospital, London NW3 2QG, UNITED KINGDOM

Bruce S. Bochner
Department of Medicine—Allergy and Clinical Immunology, Johns Hopkins Unversity School of Medicine, Baltimore, MD 21224

Claudine S. Bonder
Departments of Physiology and Biophysics, University of Calgary Health Sciences Centre, Calgary, AB T2N 4N1, CANADA

Nick Brindle
Cardiovascular Research Institute and Department of Surgery, University of Leicester, Leicester LE2 7LX, UNITED KINGDOM

Amy Brock
Vascular Biology Program, Children's Hospital, Harvard Medical School, Boston, MA 02115

Peter H. Burri
Institute of Anatomy, Developmental Biology, University of Berne, Bern 9, CH 3000, SWITZERLAND

Pedro Cabrales
Microhemodynamics Laboratory, Department of Bioengineering, University of California—San Diego, La Jolla, CA 92093-0412

Edward C. Carlson
Department of Anatomy & Cell Biology, University of North Dakota, Grand Forks, ND 58203

Per-Ola Carlsson
Department of Medical Cell Biology, University of Uppsala, Uppsala, SE-75123, Sweden

Gediminas Cepinskas
Vascular Biology/Inflammation Program, Lawson Health Research Institute, London, Ontario N6A 465, CANADA

Subrata Chakrabarti
Department of Pathology, Pathologist, London Health Sc. Ctr, The University of Western Ontario, London, Ontario N6A 5A5, CANADA

Rodney K. Chan
Division of Surgery, Brigham and Women's Hospital, Boston, MA 02115

Pierre Charbord
Laboratoire d'Hématopoïèse, Bâtiment Bretonneau, Faculté de Médecine, Université François-Rabelais, Tours Cedex, F-37032, FRANCE

Shi-Yuan Cheng
Research Pavilion at the Hillman Cancer Center, University of Pittsburgh Cancer Institute, Pittsburgh, PA 15213-1863

Francesco Chiarelli
Department of Medicine, Section of Pediatrics, University of Chieti, Chieti Scalo, I-66100, ITALY

Brian W. Christman
Department of Medicine, Division of Allergy, Pulmonary and Critical Care Medicine, Vanderbilt University, Nashville, TN 37232-2650

Martin K. Church
Allergy and Inflammation Sciences, University of Southampton School of Medicine, Southampton SO16 6YD, UNITED KINGDOM

Geraldine Clough
Infection Inflammation and Repair, University of Southampton School of Medicine, Southampton SO16 6YD, UNITED KINGDOM

Cecile Colpaert
Department of Pathology, University Hospital Antwerp, Edegem-Antwerp, B-2650, BELGIUM

Hiran C. Cooray
Department of Pharmacology, University of Cambridge, Cambridge CB2 1PD, UNITED KINGDOM

Patricia A. D'Amore
Schepens Eye Research Institute, Boston, MA 02114

Sergei M. Danilov
Department of Anesthesiology, University of Illinois in Chicago, Chicago, IL 60612

Michael R. Dashwood
Department of Clinical Biochemistry, Royal Free and University College Medical School, London NW3 2QG, UNITED KINGDOM

Marc Dellian
Department of Otorhinolaryngology, Head and Neck Surgery, Klinikum Grosshadern, University of Munich, München, D-81377, GERMANY

Robert H. Demling
Brigham and Women's Hospital, Harvard Medical School, Boston, MA 02115

Michael Detmar
Cutaneous Biology Research Center, Massachusetts General Hospital, Charlestown, MA 02129

Mario Di Gioacchino
Department of Medicine and Science of Aging, "G. D'Annunzio" University School of Medicine, Chieti, I-66013, ITALY

Gabriele Dini
Department of Neurosurgery, The Cleveland Clinic Foundation, Cleveland, OH 44195

Valentin Djonov
Institute of Anatomy, University of Berne, Bern, CH 3012, SWITZERLAND

Jorge Domenech
Laboratoire d'Hématologie, Hôpital Bretonneau, CHU de Tours, Tours Cedex, F-37044, FRANCE

Steven M. Dudek
Department of Medicine, Section of Pulmonary Critical Medicine, University of Chicago, Chicago IL 60611

Michael Dunn
Department of Neuroscience, Institute of Psychiatry, King's College London, London SE5 8AF, UNITED KINGDOM

Danuta Dus
Ludwik Hirszfeld Institute of Immunology and Experimental Therapy, Polish Academy of Sciences, Centre of Excellence "IMMUNE", Wroclaw, PL-53-114

Ann M. Dvorak
Department of Pathology, Beth Israel Deasoness Medical Center, Boston, MA 02215

Harold F. Dvorak
Department of Pathology, Beth Israel Deaconess Medical Center, Harvard Medical School, Boston, MA 02215-5491

Britta Engelhardt
Max-Planck Institute for Vascular Biology, ZMBE-Institute for Cell Biology, Muenster, D-48149, GERMANY

Denis English
Experimental Cell Research Laboratory and Allied Health Sciences, Methodist Research Institute, Indiana University School of Medicine, Indianapolis, IN 46202

Frank M. Faraci
Department of Internal Medicine, Division of Cardiovascular Diseases, University of Iowa College of Medicine, Iowa City, IA 52242-1081

Susan A. Farr
Department of Internal Medicine, Division of Geriatrics, Saint Louis University School of Medicine, St Louis, MO 63104

Dian Feng
Department of Pathology, Beth Israel Deaconess Medical Center, Boston, MA 02215

Shuju Feng
Department of Medicine, Section of Thrombosis Research, Baylor College of Medicine, Michael E. DeBakey VA Medical Center, Houston, TX 77030

Cecilia A. Fernández
Laboratory of Surgical Research, Vascular Biology Program, Children's Hospital, Harvard Medical School, Boston, MA 02115

Iwon Fijolkowska
Department of Cardiopulmonary Pathology, The Johns Hopkins University, Baltimore, MD 21205

Howard A. Fine
Neuro-oncology Branch, National Cancer Institute, National Institutes of Health, Bethesda, MD 20814

James H. Finigan
Department of Medicine, Pulmonary, Johns Hopkins University School of Medicine, Baltimore, MD 21224

Eric B. Finkelstein
Schepens Eye Research Institute, Boston, MA 02114

Susan J. Fisher
Department of Stomatology, University of California, San Francisco, San Francisco, CA 94143-0512

Ingrid Fleming
Institut für Kardiovaskuläre Physiologie, Klinikum der J.W. Goethe-Universität, Frankfurt am Main, D-60596, GERMANY

Michael M. Flessner
Department of Medicine; Division of Nephrology, University of Mississippi School of Medicine, Jackson, MS 39216

Alpha A. Fowler
Division of Pulmonary and Critical Care Medicine, Department of Internal Medicine, Virginia Commonwealth University, Richmond, VA 23298

Dai Fukumura
Department of Radiation Oncology, Massachusetts General Hospital, Harvard Medical School, Boston, MA 02114

John Gamble
Department of Maternal and Fetal Medicine, Faculty of Medicine, Imperial College of Science, Technology and Medicine at Chelsea and Westminster Hospital, London SW10 9NH, UNITED KINGDOM

Joe G.N. Garcia
Department of Medicine, Section of Pulmonary Critical Medicine, University of Chicago, Chicago IL 60611

Guillermo Garcia-Cardena
Center for Excellence in Vascular Biology, Brigham and Women's Hospital, Boston, MA 02115

Giampietro Gasparini
Medical Oncology, San Filippo Neri Hospital, Rome, I-00135, ITALY

Ursula Gehling
Abteilung für Hämatologie und Onkologie, University Hospital Hamburg-Eppendorf, Hamburg, D-20246, GERMANY

Jeffrey M. Gelfand
Pathology Dept., Vascular Research, Brigham and Women's Hospital, Boston, MA 02115

Holger Gerhardt
Department of Medical Biochemistry, Göteborg University, Goteburg, SE-40330, SWEDEN

Berhane Ghebrehiwet
Department of Medicine, State University of New York at Stony Brook, Stony Brook, NY 11794-8161

Derek W. Gilroy
Department of Experimental Pathology, William Harvey Research Institute, Queen Mary University of London, London EC1M 6BQ, UNITED KINGDOM

Reda Girgis
Division of Pulmonary and Critical Care Medicine, The Johns Hopkins Medical Institutions, The Johns Hopkins Hospital, Baltimore, MD 21205

John Glod
Cancer Institute of New Jersey, New Brunswick, NJ 08903

Nobuhito Goda
Department of Biochemistry and Integrative Medical Biology, School of Medicine, Keio University, Shinjuku-ku, Tokyo, JAPAN

Andrew P. Golden
Department of Biomedical Engineering, Boston University, Boston, MA 02215

Norberto C. Gonzalez
Department of Molecular and Integrative Physiology, University of Kansas Medical Center, Kansas City, KS 66160-7401

Andrea Moreira Gonzalez
Department of Plastic and Reconstructive Surgery, The Cleveland Clinic Foundation, Cleveland, OH 44195

D. Neil Granger
Department of Molecular and Cellular Physiology, Louisiana State University Health Sciences Center, Shreveport, LA 71130-3932

Anna T. Grazul-Bilska
Department of Animal and Range Science, North Dakota State University, Fargo, ND 58105-5727

Rita Greco
Department of Medicine, Section of Pediatrics, University of Chieti, Chieti Scalo, I-66013, ITALY

Dmitry Grigoryev
Division of Pulmonary and Critical Care Medicine, Johns Hopkins University School of Medicine, Baltimore, MD 21224

Matthew B. Grisham
Department of Molecular & Cellular Physiology, Louisiana State University Health Sciences Center, Shreveport, LA 71130-3932

Danièle Gulino-DeBrac
Laboratoire d'Ingénierie des Macromolécules, Jean-Pierre Ebel Institute of Structural Biology (CEA/CNRS, UJF), Grenoble Cedex 1, F-38027, FRANCE

J. Silvio Gutkind
Oral and Pharyngeal Cancer Branch, National Institute of Dental and Craniofacial Research, National Institutes of Health, Bethesda, MD 20892-4340

Mark Haas
Department of Pathology, Johns Hopkins University School of Medicine, Baltimore, MD 21287

Kerri L. Hallene
Department of Neurosurgery, The Cleveland Clinic Foundation, Cleveland, OH 44195

Hans-Peter Hammes
Abt. für Nephrologie/Endokrinologie, 5. Medizinische Klinik, Mannheim der Universität Heidelberg, Mannheim, D-68167, GERMANY

Kevin A. Harvey
Experimental Cell Research Laboratory and Allied Health Sciences, Methodist Research Institute, Indiana University School of Medicine, Indianapolis, IN 46202

Paul M. Hassoun
Department of Medicine—Pulmonary, Johns Hopkins University School of Medicine, Baltimore, MD 21224

Ossama A. Hatoum
Division of Gastroenterology and Hepatology, Medical College of Wisconsin, Milwaukee, WI 53226

William G. Haynes
Department of Internal Medicine, University of Iowa, Iowa City, IA 52242

Herbert B. Hechtman
Division of Surgery, Brigham and Women's Hospital, Boston, MA 02115

Donald D. Heistad
Department of Internal Medicine, Division of Cardiovascular Diseases, University of Iowa College of Medicine, Iowa City, IA 52242-1081

Ira M. Herman
Program in Cellular and Molecular Physiology, Tufts University School of Medicine, Boston, MA 02111

Baskin Hermant
Laboratoire d'Ingénierie des Macromolécules, Jean-Pierre Ebel Institute of Structural Biology (CEA/CNRS, UJF), Grenoble Cedex 1, F-38027, FRANCE

Dan D. Hershko
Department of Surgery A, Rambam Medical Center, Haifa, IL-31096, ISRAEL

Robert L. Hester
Department of Physiology and Biophysics, University of Mississippi Medical Center, Jackson, MS 39216-4505

Stéphanie Heyraud
Laboratoire d'Ingénierie des Macromolécules, Jean-Pierre Ebel Institute of Structural Biology (CEA/CNRS, UJF), Grenoble Cedex 1, F-38027, FRANCE

Caryl E. Hill
Division of Neuroscience, John Curtin School of Medical Research, Australian National University, Canberra, ACT 0200, AUSTRALIA

May Ho
Faculty of Medicine, Department of Microbiology and Infectious Diseases, University of Calgary, Calgary, Alberta T2N 4N1, CANADA

Virginia Huxley
Department of Medical Pharmacology and Physiology, University of Missouri—Columbia, Columbia, MO 65211

Richard O. Hynes
Center for Cancer Research, Massachusetts Institute of Technology, Cambridge, MA 02139

Shahrul Izham Ibrahim
Division of Surgery, Brigham and Women's Hospital, Boston, MA 02115

Donald Ingber
Vascular Biology Program, Children's Hospital, Harvard Medical School, Boston, MA 02115

Marcos Intaglietta
Microhemodynamics Laboratory, Department of Bioengineering, University of California—San Diego, La Jolla, CA 92093-0412

Sofia Ioannidou
Eyetech Research Center, Eyetech Pharmaceuticals, Inc., Woburn, MA 01801

Renato V. Iozzo
Department of Pathology, Anatomy, and Cell Biology, Kimmel Cancer Center, Thomas Jefferson University, Philadelphia, PA 19107

Shingo Itou
Laboratory of Food Science, Department of Food Science and Technology College of Bioresource Sciences, The University of Shizuoka, Shizuoka-shi, JAPAN

Yuichi Izumi
Department of Periodontology, Kagoshima University Graduate School of Medical and Dental Sciences, Kagoshima 890-8544, JAPAN

William F. Jackson
Department of Biological Sciences, Western Michigan University, Kalamazoo, MI 49008-5410

Jeffrey R. Jacobson
Department of Medicine, Section of Pulmonary Critical Medicine, University of Chicago, Chicago IL 60611

Sirpa Jalkanen
MediCity Research Laboratory, University of Turku, Turku, FIN-20520, FINLAND

Damir Janigro
Department of Neurological Surgery, Cerebrovascular Research Center, The Cleveland Clinic Foundation, Cleveland, OH 44195

Leif Jansson
Department of Medical Cell Biology, University of Uppsala, Uppsala, SE-75123, Sweden

Michael J. Jarzynka
Research Pavilion at the Hillman Cancer Center, University of Pittsburgh Cancer Institute, Pittsburgh, PA 15213-1863

Jeffrey L. Jasperse
Department of Sports Medicine, Natural Science Division, Pepperdine University, Malibu, CA 90263

Jamie Y. Jeremy
Division of Cardiac, Anaesthetic and Radiological Sciences, University of Bristol, Bristol BS2 8HW, UNITED KINGDOM

Randall S. Johnson
Division of Biological Sciences, Molecular Biology, University of California, San Diego, La Jolla, CA 92093-0366

Michael Jünger
Klinik und Poliklinik fur Hautkrankheiten, der Ernst-Moritz-Arndt-Universitat Greifswald, Griefswald, D-17487, GERMANY

Örjan Källskog
Department of Medical Cell Biology, University of Uppsala, Uppsala, SE-75123, Sweden

Makio Kawakam
Department of Pathology, Jikei University School of Medicine, Minato-ku, Tokyo 105-8471, JAPAN

Usamah S. Kayyali
Pulmonary and Critical Care Division, Tupper Research Institute, Tufts University School of Medicine, New England Medical Center, Boston, MA 02111

Zia Ali Khan
Department of Pathology, Pathologist, London Health Sc. Ctr, The University of Western Ontario, London, Ontario N6A 5A5, CANADA

Kameha R. Kidd
Laboratory of Molecular Genetics, National Institute of Child Health and Human Development, National Institutes of Health, Bethesda, MD 20892

Claudine Kieda
Centre de Biophysique Moléculaire-Glycobiologie, Centre National de la Recherche Scientifique, Orleans, F-45071

Bela Kis
Department of Physiology/Pharmacology, Wake Forest University School of Medicine, Winston-Salem, NC 27157

Michael Klagsbrun
Departments of Surgical Research and Pathology, Vascular Biology Program, Children's Hospital, Harvard Medical School, Boston, MA 02115-5737

Moritz A. Konerding
Anatomisches Institut * Makroskopischer Bereich, Johannes Gutenberg-Universität Mainz, Mainz, D-55099, GERMANY

Robert D. Koos
Department of Physiology, Unviersity of Maryland School of Medicine, Baltimore, MD 21201-1559

Ronald J. Korthuis
Department of Medical Pharmacology and Physiology, University of Missouri-Columbia School of Medicine, Columbia, MO 65212

Michael H. Koval
Department of Physiology and Institute for Environmental Medicine, University of Pennsylvania School of Medicine, Philadelphia, PA 19104

Michael M. Krausz
Department of Surgery A, Rambam Medical Center, Haifa, IL-31096, ISRAEL

Michael M. Krausz
Department of Surgery A, Rambam Medical Center, Haifa, IL-31096, ISRAEL

Agnieszka Krawczenko
Ludwik Hirszfeld Institute of Immunology and Experimental Therapy, Polish Academy of Sciences, Centre of Excellence "IMMUNE", Wroclaw, PL-53-114

Michael H. Kroll
Department of Medicine, Section of Thrombosis Research, Baylor College of Medicine, Michael E. DeBakey VA Medical Center, Houston, TX 77030

Paul Kubes
Departments of Physiology and Biophysics, University of Calgary Health Sciences Centre, Calgary, AB T2N 4N1, CANADA

Rainer Kunstfeld
Cutaneous Biology Research Center, Massachusetts General Hospital, Charlestown, MA 02129

Peter R. Kvietys
Vascular Biology/Inflammation Program, Lawson Health Research Institute, London, Ontario N6A 465, CANADA

Evanthia Lalla
Division of Periodontics, Columbia University School of Dental and Oral Surgery, New York, NY 10032

Ira B. Lamster
Division of Periodontics, Columbia University School of Dental and Oral Surgery, New York, NY 10032

Robert Langley
Department of Cancer Biology, The University of Texas M. D. Anderson Cancer Center, Houston, TX 77030

Stephen Laroux
Department of Molecular & Cellular Physiology, Louisiana State University Health Sciences Center, Shreveport, LA 71130-3932

Maurice H. Laughlin
Department of Biomedical Sciences, College of Veterinary Medicine, University of Missouri-Columbia, Columbia, MO 65211

Andrew Leask
Centre for Rheumatology, Royal Free Hospital, London NW3 2PF, UNITED KINGDOM

Whaseon Lee-Kwon
Division of Medicine and Nephrology, University of Maryland Hospital—North, Baltimore, M.D. 21201-1595

Hans-Anton Lehr
Institute of Pathology, University of Mainz, Medical Center, Mainz, D-55101, GERMANY

Martin Leuwer
University Department of Anaesthesia, University of Liverpool, Liverpool L69 3GA, UNITED KINGDOM

Klaus F. Ley
Cardiovascular Research Center, University of Virginia Health System, Charlottesville, VA 22908-1394

Lina H.K. Lim
Department of Physiology, National University of Singapore, Singapore 117597, Republic of Singapore

Michelle I. Lin
Department of Pharmacology, Yale University School of Medicine, New Haven, CT 06510-0812

Ute Lindauer
Experimental Neurology, Charité, Universitätsmedizin Berlin, Berlin, D-10098

Bengt Lindholm
Baxter Novum and Renal Medicine, Department of Clinical Science, Carolinska Institutet, Stockholm, S-141 86, SWEDEN

Herbert H. Lipowsky
Department of Bioengineering, Pennsylvania State University, University Park, PA 16802

Perry Hsien-Tsung Liu
Department of Plastic Surgery, Brigham and Women's Hospital, Harvard Medical School, Boston, MA 02115

Julian H. Lombard
Department of Physiology, Medical College of Wisconsin, Milwaukee, WI 53226-1408

Raffaele Longo
Medical Oncology, San Filippo Neri Hospital, Rome, I-00135, ITALY

Ming Lu
Department of Ophthamology, Massachusetts Eye & Ear Infirmary, Boston, MA 02114

Gerard Lutty
Ophthalmology Department, Johns Hopkins Hospital, Baltimore, MD 21287-9115

Shwu-Fan Ma
Department of Medicine, Section of Pulmonary Critical Medicine, University of Chicago, Chicago IL 60611

Miho Machigashira
Department of Periodontology, Kagoshima University Graduate School of Medical and Dental Sciences, Kagoshima 890-8544, JAPAN

William MacNee
ELEGI Colt Laboratories, University of Edinburgh Medical School, Edinburgh EH8 9YG, SCOTLAND

Roni Mamluk
Departments of Surgical Research and Pathology, Vascular Biology Program, Children's Hospital, Harvard Medical School, Boston, MA 02115-5737

Philipp C. Manegold
Department of Otorhinolaryngology, Head and Neck Surgery, Klinikum Grosshadern, University of Munich, München, D-81377, GERMANY

Federica M. Marelli-Berg
Department of Immunology, Division of Medicine, Faculty of Medicine, Imperial College at Hammersmith Campus, London W6 8RF, UNITED KINGDOM

Hildegard R. Maricq
Department of Medicine, Medical University of South Carolina, Charleston, SC 29412

Sean P. Marrelli
Department of Anesthesiology, Baylor College of Medicine, Houston, TX 77030

Marie B. Marron
Department of Cardiovascular Sciences, University of Leicester, Leicester LE2 7LX, UNITED KINGDOM

Gernot Marx
University Department of Anaesthesia, University of Liverpool, Liverpool L69 3GA, UNITED KINGDOM

William G. Mayhan
Department of Physiology and Biophysics, University of Nebraska Medical Center, Omaha, NE 68198-4575

Joseph H. McCarty
Center for Cancer Research, Massachusetts Institute of Technology, Cambridge, MA 02139

Robert S. McCuskey
Department of Cell Biology and Anatomy, University of Arizona College of Medicine, Tucson, AZ 85724-5044

Laura Linz McGillem
Department of Pulmonary and Critical Care Medicine, The Johns Hopkins University, Baltimore, MD 21224

Emma McGregor
Proteome Sciences plc, Institute of Psychiatry, King's College London, London SE5 8AF, UNITED KINGDOM

Michael D. Menger
Institute for Clinical and Experimental Surgery, Universität des Saarlandes, Homburg, D-66421, GERMANY

Christopher Charles Michel
Faculty of Medicine, Division of Biomedical Sciences, Imperial College School of Medicine, London SW7 2AZ, UNITED KINGDOM

Ross Milner
Department of Surgery, Emory University School of Medicine, Atlanta, GA 30322

Fred L. Minnear
Robert C. Byrd Health Sciences Center, West Virginia University, Morgantown, NY 26506-9104

Richard D. Minshall
Department of Anesthesiology and Pharmacology, The University of Illinois College of Medicine, Chicago, IL 60612

David R. Mole
Wellcome Trust Center for Human Genetics, University of Oxford, Oxford OX3 7BN, UNITED KINGDOM

Silvia Montaner
Oral and Pharyngeal Cancer Branch, National Institute of Dental and Craniofacial Research, National Institutes of Health, Bethesda, MD 20892-4340

Francis Daniels Moore
Division of General and GI Surgery, Department of Surgery, Brigham & Women's Hospital, Harvard Medical School, Boston, MA

Guido Morgese
Dipartimento di Pediatria, Ostetricia e Medicina della Riproduzione, University of Siena, Siena, I-53100, ITALY

John E. Morley
Department of Internal Medicine, Division of Geriatrics, Saint Louis University School of Medicine, St Louis, MO 63104

Marsha A. Moses
Surgical Research, Room 1009, Children's Hospital, Harvard Medical School, Boston, MA 02115

Lance L. Munn
Department of Radiation Oncology, Massachusetts General Hospital, Harvard University, Boston, MA 02114

Kazunori Murakami
Department of Anesthesiology, University of Texas Medical Branch, Galveston, TX 77555-0591

Vladimir R. Muzykantov
Department of Pharmacology, IFEM, University of Pennsylvania Medical Center, Philadelphia, PA 19104-6068

Keigo Nakatani
Department of Pediatrics, Kobayashi city Hospital, Kobayashi city, Miyazaki 886-8503, JAPAN

Masaaki Nakayama
Division of Kidney and Hypertension, Department of Internal Medicine, Jikei University School of Medicine, Minato-ku, Tokyo 105-8471, JAPAN

Yasuo Nara
Laboratory of Health Science, Department of Biological Pharmacy, School of Pharmacy, Shujitu University, Okayama, JAPAN

Gerard B. Nash
Department of Physiology, The Medical School, The University of Birmingham, Birmingham B15 2TT, UNITED KINGDOM

Craig M. Neville
Department of Surgery, Massachusettes Generl Hospital, Boston, MA 02114

Naoko Nonaka
Department of Internal Medicine, Division of Geriatrics, Saint Louis University School of Medicine, St Louis, MO 63104

Miwako Numata
Division of Kidney and Hypertension, Department of Internal Medicine, Jikei University School of Medicine, Minato-ku, Tokyo 105-8471, JAPAN

Valerie B. O'Donnell
Department of Medical Biochemistry and Immunology, University of Wales College of Medicine, Cardiff, Wales CF14 4XN, UNITED KINGDOM

Kenji Okajima
Department of Diagnostic Medicine, Graduate School of Medical Science, Kumamoto University, Kumamoto 860-0811, JAPAN

Dennis P. Orgill
Department of Plastic Surgery, Brigham and Women's Hospital, Harvard Medical School, Boston, MA 02115

Ewa Paleolog
Kennedy Institute of Rheumatology, Faculty of Medicine, Imperial College, Hammersmith, London W6 8LH, UNITED KINGDOM

Thomas L. Pallone
Division of Nephrology, University of Maryland at Baltimore, Baltimore, M.D. 21201-1595

Carlo Palombo
Department of Internal Medicine, University of Pisa, Pisa, I-56126, ITALY

Panos N. Papapanou
Division of Periodontics, Columbia University School of Dental and Oral Surgery, New York, NY 10032

Mohan Parigi
Division of Pulmonary and Critical Care Medicine, Johns Hopkins University School of Medicine, Baltimore, MD 21224

Jeremy Pearson
Centre for Cardiovascular Biology & Medicine, School of Biomedical Sciences, King's College London, London SE1 1UL, UNITED KINGDOM

Ellinor I. B. Peerschke
Department of Pathology, Weill Medical College of Cornell University, New York, NY 10021

Michael Sean Pepper
Department of Morphology, University Medical Center, l'Université de Genève, Geneva 4, F-1211, SWITZERLAND

Mauro Perretti
Department of Biochemical Pharmacology, William Harvey Research Institute, Queen Mary University of London, London EC1M 6BQ, UNITED KINGDOM

Irina Petrache
Division of Pulmonary and Critical Care Medicine, Johns Hopkins School of Medicine, Baltimore, MD 21224

Emanuel Petricoin
Food & Drug Administration Clinical Proteomics Program, National Cancer Institute, Bethesda, MD 20892

Janos Pittner
Division of Medicine and Nephrology, University of Maryland Medical School, Baltimore, M.D. 21201-1595

Thomas J. Poole
Department of Cell and Developmental Biology, State University of New York, Upstate Medical University, Syracuse, NY 13210

Aleksander S. Popel
Department of Biomedical Engineering, Whitaker Biomedical Engineering Institute, School of Medicine, Johns Hopkins University, Baltimore, MD 21205

Marcus J. Post
Department of Physiology and of Biomedical Technology, Maastricht University, Maastricht, D-6200 MD, GERMANY

Robert G. Presson
Department of Anesthesiology, Indiana University School of Medicine, Indianapolis, IN 46202-5200

Marco Presta
Department of Biomedical Sciences and Biotechnology, The University of Brescia School of Medicine, Brescia, I-25123, ITALY

Donald G. Puro
Department of Ophthalmology and Visual Sciences, University of Michigan, Ann Arbor, MI 48105

Hamid Rabb
Kidney Transplant Program, Johns Hopkins University School of Medicine, Baltimore, MD 21205

Vineeth Rajkumar
Centre for Rheumatology, Royal Free Hospital, London NW3 2QG, UNITED KINGDOM

Peter J. Ratcliffe
Wellcome Trust Center for Human Genetics, University of Oxford, Oxford OX3 7BN, UNITED KINGDOM

Dale A. Redmer
Department of Animal and Range Science, North Dakota State University, Fargo, ND 58105-5727

Lawrence P. Reynolds
Center for Nutrition and Pregnancy, Department of Animal and Range Science, North Dakota State University, Fargo, ND 58105-5727

Domenico Ribatti
Department of Human Anatomy and Histology, University of Bari, Bari, ITALY

Sarah Riley
Centre for Cardiovascular Biology & Medicine, School of Biomedical Sciences, King's College London, London SE1 1UL, UNITED KINGDOM

Stephen M. Robbins
Faculty of Medicine, Department of Microbiology and Infectious Diseases, University of Calgary, Calgary, Alberta T2N 4N1, CANADA

Ivan M. Robbins
Department of Medicine, Division of Allergy, Pulmonary and Critical Care Medicine, Vanderbilt University, Nashville, TN 37232-2650

L. Christie Rockwell
Department of Anthropology, Temple University, Philadelphia, PA 19122-6089

Marco Rossi
Department of Internal Medicine, University of Pisa, Pisa, I-56126, ITALY

Kristofer Rubin
Department of Medical Biochemistry and Microbiology, Uppsala University, Uppsala, SE-75123

Tao Rui
Vascular Biology/Inflammation Program, Lawson Health Research Institute, London, Ontario N6A 465, CANADA

Rolando E. Rumbaut
Department of Medicine, Baylor College of Medicine, Houston, TX 77030

Nicole May Rummery
Division of Neuroscience, John Curtin School of Medical Research, Australian National University, Canberra, ACT 0200, AUSTRALIA

Hiromi Sakai
Department of Polymer Chemistry, Waseda University, Shinjuku-ku, Tokyo 169-8050, JAPAN

Marko Salmi
Director, Cell Trafficking, National Public Health Institute, Turku, FIN-20520, FINLAND

Alexei V. Salnikov
Department of Medical Biochemistry and Microbiology, Uppsala University, Uppsala, SE-75123

Steven J. Samuelsson
Eyetech Research Center, Eyetech Pharmaceuticals, Inc., Woburn, MA 01801

James D. San Antonio
Department of Medicine, Thomas Jefferson University, Philadelphia, PA 19107-5099

David C. Sane
Department of Cardiology, Wake Forest University School of Medicine, Winston-Salem, NC 27157-1045

Ingrid H. Sarelius
Department of Pharmacology and Physiology, University of Rochester School of Medicine and Dentistry, Rochester, NY 14642

Vladimir Savransky
Kidney Transplant Program, Johns Hopkins University School of Medicine, Baltimore, MD 21205

Kane L. Schaphorst
Center for Lung Biology, College of Medicine, University of South Alabama, Mobile, AL 36688

Anja Schlez
Universitats-Hautklinik, Eberhard-Karls Universitat, Tubingen, D-72076, GERMANY

Geert W. Schmid-Schoenbein
Department of Biomedical Engineering, University of California San Diego, La Jolla, CA 92093-0412

Geert W. Schmid-Schoenbein
Deparmtent of Biengineering, The Whitaker Institute for Biomedical Engineering, University of California, San Diego, La Jolla, CA 92093-0412

Ann Marie Schmidt
Department of Surgery, College of Physicians & Surgeons, Columbia University, New York, NY 10032

Tiffany N. Seagroves
Department of Biology, University of California, San Diego, La Jolla, CA 92093

William C. Sessa
Department of Pharmacology, Yale University School of Medicine, New Haven, CT 06510-0812

William L. Shao
Division of Pulmonary and Critical Care Medicine, Gene Expression Profiling Core, Johns Hopkins University School of Medicine, Baltimore, MD 21224

David T. Shima
Eyetech Research Center, Eyetech Pharmaceuticals, Inc., Woburn, MA 01801

Maria Siemionow
Department of Plastic and Reconstructive Surgery, The Cleveland Clinic Foundation, Cleveland, OH 44195

Michael Simons
Section of Cardiology, Dartmouth-Hitchcock Medical Center, Lebanon, NH 03756

Robert B. Smith
Department of Surgery, Emory University School of Medicine, Atlanta, GA 30322

Akrit Sodhi
Oral and Pharyngeal Cancer Branch, National Institute of Dental and Craniofacial Research, National Institutes of Health, Bethesda, MD 20892

Philip J. Steer
Department of Maternal and Fetal Medicine, Faculty of Medicine, Imperial College of Science, Technology and Medicine at Chelsea and Westminster Hospital, London SW10 9NH, UNITED KINGDOM

Troy Stevens
Department of Pharmacology, College of Medicine, University of South Alabama, Mobile, AL 36688

Karen Stokes
Department of Molecular and Cellular Physiology, Louisiana State University Health Sciences Center, Shreveport, LA 71130-3932

John H. Stone
Department of Medicine, The Johns Hopkins Vasculitis Center, Baltimore, MD 21224

Makoto Suematsu
Department of Biochemistry and Integrative Medical Biology, School of Medicine, Keio University, Shinjuku-ku, Tokyo Tokyo, JAPAN

Christian Sundberg
Department of Pediatrics, Uppsala University, Uppsala, SE-751 85

Kanyarat Suthin
Department of Conservative Dentistry, Faculty of Dentistry, Prince of Songkla University, Hadyai, Songkhla 90112, THAILAND

Erik Svensjö
Laboratorio de Imunologia Molecular, Instituto Biofisica Carlos Chagas Filho, Universidade Federal do Rio de Janeiro, Rio de Janeiro, RJ 21944-900, BRAZIL

Motoki Tagami
Department of Internal Medicine, Sanraku Hospital, Tokyo, JAPAN

Seiichiro Takeshita
Faculty of Education, Dept. of School Health Science, Ibaraki University, Mito, Ibaraki 310-8512, JAPAN

Min D. Tang
Department of Biomedical Engineering, Boston University, Boston, MA 02215

Miranda ten Kate
Laboratory of Experimental Surgery and Oncology, Erasmus Medical Center, Rotterdam 3015 GD, THE NETHERLANDS

Joe Tien
Department of Biomedical Engineering, Boston University, Boston, MA 02215

Daniel L. Traber
Department of Anesthesiology, University of Texas Medical Branch, Galveston, TX 77555

Amy G. Tsai
Microhemodynamics Laboratory, Department of Bioengineering, University of California—San Diego, La Jolla, CA 92093-0412

Rubin Tuder
Division of Pulmonary and Critical Care Medicine, Department of Pathology, The Johns Hopkins University School of Medicine, Baltimore, MD 21205

Karel Tyml
Department of Medical Biophysics, University of Western Ontario, London, Ontario N6A 5C1, CANADA

Joseph P. Vacanti
Department of Surgery, Massachusettes Generl Hospital, Boston, MA 02114

C.H.J. Van Eijck
Laboratory of Experimental Surgery and Oncology, Erasmus Medical Center, Rotterdam 3015 GD, THE NETHERLANDS

Alexander D. Verin
Division of Pulmonary and Critical Care Medicine, Johns Hopkins School of Medicine, Baltimore, MD 21224

Peter B. Vermeulen
Lab Pathology, AZ Sint-Augustinus, Wilrijk—Antwerp, B-2610, BELGIUM

Nicola Verna
Department of Medicine and Science of Aging, Section of Allergy, Clinical Immunology, and Occupational Medicine, "G. D'Annunzio" University School of Medicine, Pescara, I-65132, ITALY

Alberto Verrotti
Department of Medicine, Section of Pediatrics, University of Chieti, Chieti 66013, ITALY

Dietmar Vestweber
Institute of Cell Biology, ZMBE, University of Muenster, Muenster, D-48149, GERMANY

Stanley A. Vinores
Wilmer Ophthalmological Institute, Johns Hopkins Hospital, Johns Hopkins University School of Medicine, Baltimore, MD 21287

Stephen M. Vogel
Department of Anesthesiology and Pharmacology, The University of Illinois College of Medicine, Chicago, IL 60612

Brigitte Vollmar
Faculty of Medicine, Department of Experimental Surgery, University of Rostock, Rostock, D-18055, GERMANY

Thorsten Vowinkel
Department of Molecular and Cellular Physiology, Louisiana State University Health Sciences Center, Shreveport, LA 71130-3932

Wiltz W. Wagner
Department of Anesthesia, Indiana University School of Medicine, Indianapolis, IN 46202

Jean-Luc Wautier
Institut National de la Transfusion Sanguine, University Paris VII, Paris cedex 15 75739, FRANCE

Marie-Paule Wautier
Institut National de la Transfusion Sanguine, University Paris VII, Paris cedex 15 75739, FRANCE

Marianne Weidenhaupt
European Molecular Biology Laboratory (EMBL), Grenoble, F-38027, FRANCE

Brant M. Weinstein
Laboratory of Molecular Genetics, National Institute of Child Health and Human Development, National Institutes of Health, Bethesda, MD 20892

William J. Welch
Department of Medicine, Division of Nephrology and Hypertension, Georgetown University Medical Center, Washington, DC 20057

John B. West
Department of Medicine, Physiology, University of California, San Diego, La Jolla, CA 92093

Michael S. Wolin
Department of Physiology, New York Medical College, Valhalla, NY 10595

John G. Wood
Department of Molecular and Integrative Physiology, University of Kansas Medical Center, Kansas City, KS 66160

Mack H. Wu
Department of Surgery, Cardiovascular Research Institute, Texas A & M University Health Sciences Center, Temple, TX 76504

Kazuo Yamagata
College of Bioresource Science, Department of Food Science & Technology, Nihon University, Fujisawa, Kanagawa 252-8510, JAPAN

Yukio Yamori
Collaborating Center for Research on Primary Prevention of Cardiovascular Diseases, World Health Organization, Kyoto, JAPAN

Shui Qing Ye
Division of Pulmonary and Critical Care Medicine, Gene Expression Profiling Core, Johns Hopkins University School of Medicine, Baltimore, MD 21224

Bryan G. Yipp
Faculty of Medicine, Department of Microbiology and Infectious Diseases, University of Calgary, Calgary, Alberta T2N 4N1, CANADA

Hidehiko Yokogoshi
College of Bioresource Science, Department of Food Science & Technology, Nihon University, Fujisawa, Kanagawa 252-8510, JAPAN

Sarah Y. Yuan
Department of Surgery, Cardiovascular Research Institute, Texas A & M University Health Sciences Center, Temple, TX 76504

Hellmuth Zöltzer
Institut für Biologie, Abteilung Zellbiologie, Arbeitsgruppe Lymphologie, Universität Kassel, Kassel, D-34132, GERMANY

Contents

VOLUME 2

Section G: Hemorrhagic Fever Viruses

Section H: Hypercholesterolemia

Section I: Hypertension

Section J: Inflammation

Section M: Leak Syndrome

Section N: Lung Injury

Section O: Organ Transplantation

Section P: Periodontal Disease

Section Q: Pollutants

Section B: Engineered Tissue

Section C: Endothelial Biomarker

Section D: Knockout/transgenic Models

Section E: Mathematical Models

Section F: Proteomics

Section E: Transport, Junctions, Adhesion Molecules

SECTION I

Pancreas

CHAPTER 84

Pancreatic Microcirculation in Health and Disease

Leif Jansson, Arne Andersson, Örjan Källskog, and Per-Ola Carlsson
Department of Medical Cell Biology, Uppsala University, Uppsala, Sweden

Introduction

The pancreas constitutes an organ that consists of both exocrine and endocrine tissues. The latter, that is, the 1 to 2 million islets of Langerhans, are scattered throughout the exocrine parenchyma and make up only 1 to 2 percent of the volume in the adult pancreas. Studies during the recent two decades have shed more light on the complex organization of the pancreatic microvasculature and have underlined the autonomy of the microvascular networks in the endocrine and exocrine parts of the gland, both anatomically and functionally. Thus, from a microvascular point of view the pancreas consists of two different organs, with differently regulated blood perfusion.

Normal Pancreas

Arterial and Venous Supply

As is common in the gastrointestinal tract, numerous anastomoses exist between the intrapancreatic blood vessels. In general the head region of the gland receives its supply from the superior mesenteric artery, whereas the celiac trunk supplies the body and tail of the gland. This is of interest, since this region develops separately from the rest of the pancreas in the embryo. The venous drainage of the pancreas flows into the portal vein from the superior mesenteric and splenic veins. The blood contains high concentrations of the pancreatic hormones, especially insulin and glucagons, which elicit many of their functions directly on the hepatocytes. Interlobular arteries and veins run in parallel to one another and adjacent to the exocrine ducts. The intralobular arteries run centrally in the lobuli and send tributaries to the exocrine acini, islets and ductules, respectively. However, the latter organization is species dependent [1].

Microvasculature

The capillaries derived from the arterioles just mentioned differ markedly between islets and exocrine tissue in that they are wider (approximately 8 to 10 μm versus 4 to 6 μm) in the former. Furthermore, they constitute a much denser network in the islets. The capillary endothelium in the pancreas is fenestrated, but those in the islets possess almost 10 times as many fenestrae as capillaries in the exocrine and ductular compartments [2]. It seems as if vascular endothelial growth factor (VEGF), locally produced within the islets, is responsible for the formation of the fenestrations. Functionally, pancreatic capillaries are very permeable [3], and particularly so those within the islets. Thus, the vasculature is unlikely to pose any restrictions to the diffusion of the islet hormones.

The microvasculature of the islets has been described as glomerular-like, since its initial description by Paul Langerhans in 1869. A number of species from various vertebrates have been studied, and species differences exist. A general finding is, however, that all mammals, that is, including humans, have a direct and separate arteriolar flow to the islets. The islet vasculature is otherwise dependent on the size of the islets. Thus, smaller islets are located in the periphery of the lobuli, whereas larger islets (more than 250 μm in diameter) are mainly found at major branches of blood vessels or in association with major ducts. Small islets receive their blood supply from one arteriole and drain through numerous efferent capillaries into a basket-like network around the islets, which subsequently drains into intralobular venules. According to some investigators [4], these efferent capillaries communicate with those around exocrine acini and/or ducts, thereby forming a so called insulo-acinar portal system. Large islets, on the other hand, possess one to three arterioles, and the efferent capillaries drain into postcapillary venules at the edge of the islets, which then empty into intralobular veins. In addition to this, numerous small capillaries and/or venules connect the islets with capillaries in the acini and the ducts. Previously it has been debated to what extent these arrangements are species dependent, and whether all blood in the pancreas passes through the insulo-acinar network. At present this seems unlikely [4].

A summary of microvascular parameters in the endocrine and exocrine pancreas is given in Figure 1.

Lymphatic Capillaries

The exocrine pancreas possesses a network of lymphatic capillaries within the lobules, which drain into larger vessels within interlobular septa, associated with blood vessels and nerves [5]. Their major function is, as expected, to drain interstitial fluid, and they seem to play a minor role in the transportation of exocrine and endocrine secretions. Actually, it is generally agreed upon that the islets of Langerhans do not contain lymphatic capillaries, even though a network in the vicinity of the islet periphery can be seen, especially in rodents.

Blood Perfusion

The blood perfusion of the whole pancreas is in the order of 40 to 100 mL/min × 100 g tissue, and it is usually increased in association with increased demands for exocrine secretion. Both metabolic and myogenic mechanisms contribute to the blood flow regulation, and the oxygen uptake remains constant in the flow range of 40 to 100 mL/min × 100 g [6]. Administration of secretin and cholecystokinin, for example, increases the blood perfusion,

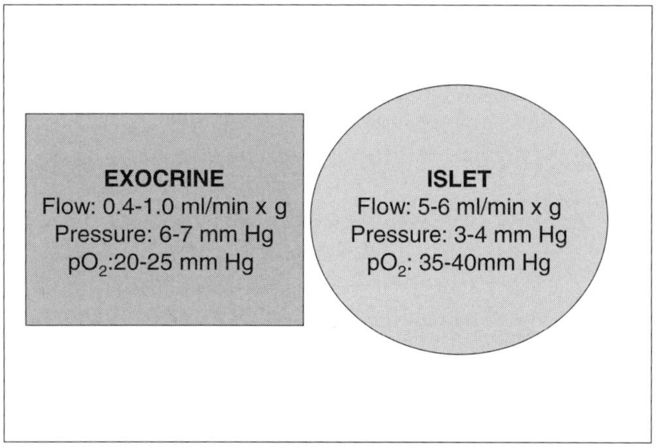

Figure 1 Summary of microvascular parameters in the endocrine and exocrine pancreas of rodents. (see color insert)

by mechanisms similar to those seen in other salivary glands. That is, nervous signals mediated through parasympathetic nerves elicit their effects by using acetylcholine to increase the exocrine secretions and vasoactive intestinal peptide to induce a simultaneous blood flow increase. In contrast, stimulation of adrenergic nerves, or administration of catecholamines, induce an initial vasoconstriction followed by a longer period of vasodilatation (e.g., Ref. [7]). In general, however, the blood flow to the whole gland normally responds to the functional state of the exocrine parenchyma. In addition to nervously mediated signals, this is ascertained by locally produced factors that may either increase (e.g., adenosine and nitric oxide) or decrease (endothelin, angiotensin II) blood flow (see Table I).

The pancreatic islets, on the other hand, have a very complex blood flow regulation, which is totally autonomous from that to the remaining parts of the pancreas [4, 8]. Despite constituting only 1 to 2 percent of the whole pancreas the islets receive from 5–10 percent (mice and rats) to 15 percent (rabbits) of the total pancreatic blood perfusion, representing a single islet blood flow of 10 to 15 nL/minutes. The blood flow is controlled through a series of complex interactions between locally produced factors, nerves and gastrointestinal hormones [8]. A summary of some of the factors affecting the blood perfusion of the total gland as well as the islets is given in Table I. In general there seems to be an association between islet glucose metabolism and blood perfusion, and thereby also a correlation between blood flow and insulin secretion. Thus, in general hyperglycemia increases islet blood flow, whereas hypoglycemia decreases islet blood perfusion.

Vasculature and Blood Flow during Development

The development of the vasculature in the embryonic pancreas has been studied only to a minor extent. Of interest is the recent observation that signals from the endothelium seems to be necessary for induction of endocrine cell development [9]. The vasculature of the islets develop late during gestation and is initially seen as a complex of blood vessels in the vicinity of the developing islets, which then sends tributaries into the clusters of endocrine cells [10]. Furthermore, it seems as if the autonomous blood flow regulation of the islets is initially absent, and matures with time.

Adaptations to Functional Demands

Stimulation of the functions and growth of the exocrine pancreas, such as after chronic hypercholecystokininemia or partial pancreatectomy, leads to increases in the blood flow to the whole gland. Chronic pancreatitis is discussed later.

Also, the pancreatic islets increase their blood perfusion when their function is stimulated, such as initially during hyperglycemia, after partial pancreatectomy, and during pregnancy. Thus, the response of the pancreatic

Table I Effects of Different Substances on Total Pancreatic and Islet Blood Flow in Anesthetized Rats.

Substance	Islet blood flow	Total pancreatic blood flow
Nitric oxide	++	+
Endothelin	–	–
Angiotensin II	–	–
Natriuretic peptides	0 or (+)	0 or (+)
Super oxide	–	(–)
Adenosine	+	+
Indomethacin	0	0
Insulin (hypoglycaemia)	–	0
C-peptide	0	0
Glucagon	+	+
Somatostatin	–	0 or (–)
Islet amyloid polypeptide	0	–
Pancreatic polypeptide	0 or –	0 or –
Gastric inhibitory polypeptide	+**	0
Glucagon-like peptide 1	+**	0
Vasoactive intestinal polypeptide	0	++
Neuropeptide Y	0	0
Leptin	–***	0
Calcitonin gene-related peptide	–	–
Corticotropin-releasing factor	+*	+*
Cholecystokinin	–	+
Secretin	0	+
Pituitary adenylate cyclase activating polypeptide	0 or +	+
Substance P	–	–
Adrenaline	–	–
Metformin	+	0
Sulfonylureas	–	0
Receptors for neurotransmittors		
β2-adrenoceptors	–	0 or (+)
α2-adrenoceptors	+	0
M3-receptors	+	0

– denotes a decerase, + an increase, and 0 no change in blood flow. * only in the tail of the pancreas. ** only in hyperglycemic animals. *** in obese mice.

compartments to changes in functional demands is similar to those of other organs.

After transplantation of the whole pancreatic gland an increase in blood perfusion is consistently seen. This is likely to reflect the functional denervation of the pancreas, and especially so the lack of sympathetic nerves. Implantation of isolated islets, on the other hand, is usually associated with a decreased blood perfusion. The islets are usually transplanted into the liver and depend on stimulation of revascularization and reinnervation for their survival. There is substantial evidence for a vascular dysfunction and lack of sufficient revascularization of transplanted islets. The reader is referred to a recent review for details [11].

Acute Pancreatitis

A complex sequence of events of multifactorial origin is inherent in the development of acute pancreatitis. Microcirculatory derangements are, however, often seen and the pancreas is highly susceptible to hypoperfusion and ischemic injury. At least graft pancreatitis, seen as a consequence of ischemia/reperfusion injuries to the implanted organ, is caused more or less exclusively by vascular factors. In the pathogenesis of acute pancreatitis of other etiologies, microcirculatory changes are likely to be aggravating factors, even though a causative role remains to be determined. Increased production of endothelin and/or decreased formation of nitric oxide have been suggested to be factors of importance for the decreased blood flow. The reader is referred to some recent reviews found in the reference list.

The islets seem to be much less susceptible to hypoperfusion, also during acute pancreatitis, and we have recently observed that islet blood flow is unchanged in an experimental model of edematous acute pancreatitis.

Chronic Pancreatitis

This disease is usually associated with atrophy of the exocrine parts of the pancreas as well as histological changes in the vasculature, and the blood flow to the whole pancreas is decreased [12]. At present the exact role of pancreatic ischemia as an initiator or perpetuator of the disease is unclear, as is the extent to which the blood flow regulation during these conditions is altered.

Islet blood flow has so far been investigated only acutely after duct ligation, which leads to atrophy of the exocrine pancreas and growth and neogenesis of islets. However, up to 4 weeks after ligation islet blood flow remains normal, despite the fact that the total islet volume increases.

Pancreatic Cancer

Like most tumors pancreatic cancers have a defect microcirculation, with regard to both morphology and function, and depend for their growth on angiogenesis. The presence of VEGF in pancreatic tumors has been demonstrated to correlate with disease progression. Indeed, in experimental studies angiogenic inhibitors have been shown to be able to suppress growth of pancreatic cancers [13].

Diabetes Mellitus

Diabetes mellitus is a syndrome characterized by hyperglycemia caused by an absolute or relative lack of insulin.

This heterogeneous disease is usually divided into two major groups: the autoimmune type 1 diabetes with a destruction of the insulin-producing β-cells, and type 2 diabetes characterized by insulin resistance and a relative insulin insufficiency.

In long-standing type 1 diabetes, the lack of local anabolic effects of insulin within the pancreas leads to a slight exocrine dysfunction, but this has rarely been studied. Since the remaining islets are few and mainly consist of glucagon- and somatostatin-producing cells, they are difficult to find, and no studies on their microcirculation have been published. In the spontaneously diabetic NOD mouse we have observed that as the islet volume decreases due to the autoimmune assault, the remaining islets initially become hyperperfused to try to cope with the increased functional demands. However, as the disease progresses blood perfusion decreases.

In type 2 diabetes there are no major changes in total pancreatic blood flow. However, islet blood flow is markedly affected. Thus, in the prediabetic and early phases of the disease an increased islet blood flow is consistently seen in all models studied so far. This increased blood flow is, at least in GK rats, associated with an islet capillary hypertension. To what extent the hyperperfusion of blood alleviates or aggravates the impaired islet endocrine function is at present unknown. It is tempting to speculate that increased shear stress on islet endothelial cells changes their expression and production of different vasoactive mediators, such as nitric oxide and endothelins, which may affect also the endocrine functions.

Conclusions

The pancreas contains both exocrine and endocrine tissues, which possess vascular systems that are autonomous of one another. Furthermore, it seems as if changes in the blood perfusion are of importance for the pathogenesis of several pancreatic disorders.

Glossary

Diabetes mellitus: A group of diseases characterized by high blood glucose concentration and lack of insulin.

Endocrine pancreas: Consists of 1 to 2 million groups of cells, each consisting of approximately 3,000 cells, scattered throughout the pancreas. These cells produce, for example, the blood glucose lowering hormone insulin.

Exocrine pancreas: Produces enzymes released into the intestines, which help to degrade food.

Pancreatitis: An inflammation of the pancreas. The acute forms can be severe and life-threatening, and the chronic forms can be associated with intractable pain. The background of these conditions is still insufficiently known.

References

1. Lifson and Lassa (1981). *Microvasc. Res.* **22**, 171–176.
2. Henderson and Moss (1985). *Q. J. Exp. Physiol.* **70**, 347–356.
3. Kvietys et al. (1983). *Am. J. Physiol.* **245**, G519–G524.
4. Brunicardi, F. C., Stagner, J., Bonner-Weir, S., Wayland, H., Kleinman, R., Livingston, E., Guth, P., Menger, M., McCuskey, R., Intaglietta, M., Charles, A., Ashley, S., Cheung, A., Ipp, E., Gilman, S., Howard, T., and Passaro, E., Jr. (1996). Microcirculation of the pancreatic islets. *Diabetes* **45**, 385–392. *Excellent and complete review on the microcirculation on pancreatic islets.*
5. O'Morchoe, C. C. (1997). Lymphatic system of the pancreas. *Microsc. Res. Tech.* **37**, 456–477. *Summarizes the scanty knowledge that exists on pancreatic lymphatics.*
6. Kvietys, P. R., McLendon, J. M., Bulkley, G. B., Perry, M. A., and Granger, D. N. (1982). Pancreatic circulation: Intrinsic regulation. *Am. J. Physiol.* **242**, G596–G602. *Seminal paper on exocrine pancreatic blood flow regulation.*
7. Barlow et al. (1974). *J. Physiol.* **236**, 421–433.
8. Jansson, L. (1994). The regulation of pancreatic islet blood flow. *Diab. Metab. Rev.* **10**, 407–16. *Summarizes the mechanisms responsible for islet blood flow regulation.*
9. Lammert et al. (2001). *Science* **294**, 564–567.
10. Chilvers and Thomas (1986). *J. Anat.* **136**, 339–347.
11. Jansson, L., and Carlsson P.-O. (2002). Graft vascular function after transplantation of pancreatic islets. *Diabetologia* **45**, 749–763. *Presents a theory on vascular dysfunction in grafted islets, which may help to explain why so many transplanted islets are needed to cure diabetic patients.*
12. Uhlmann, D., Ludwig, S., Geissler, F., Tannapfel, A., Hauss, J., and Witzigmann, H. (2001). Importance of microcirculatory disturbances in the pathogenesis of pancreatitis. *Zentralbl. Chir.* **126**, 873–878. *A summary on vascular derangements in both acute and chronic pancreatitis.*
13. Prox et al. (2003). *World J. Surg.* **27**, 405–411.

Further Reading

Bonner-Weir, S. (1986). The microvasculature of the pancreas, with special emphasis on that of the islets of Langerhans. In *The Pancreas: Biology, Pathobiology, and Disease* (V. L. W. Go, E. P. DiMagno, J. D. Gardner, E. Lebenthal, H. A. Reber, and G. A. Scheele, eds.), 2nd ed., pp. 759–768. New York: Raven Press. *Provides an excellent overview on the morphology of the microvasculature, especially in the islets. All relevant older references can be found in this article.*

Menger, M. D., Pluszyk, T., and Vollmar, B. (2001). Microcirculatory derangements in acute pancreatitis. *J. Hepatobil. Pancreat. Surg.* **8**, 187–194. *Excellent and succinct review on the theories of circulatory causes for acute pancreatitis.*

Wayland, H. (1997). Microcirculation in pancreatic function. *Microsc. Res. Tech.* **37**, 418–433. *A summary on microcirculation in both the endocrine and exocrine pancreas.*

Capsule Biography

All authors are from the Department of Medical Cell Biology, Uppsala University, Uppsala, Sweden.

Leif Jansson is Professor of Medical Cell Biology and Circulation Physiology. His main research interests have been the regulation of pancreatic islet blood flow in endogenous and transplanted pancreatic islets.

Arne Andersson is Professor of Diabetes Research. His research interests include the regulation of pancreatic islet growth and transplantation of insulin-producing cells.

The research interests of Örjan Källskog, Professor of Medical Physiology and Medical Physics, include microcirculation in kidneys and pancreas.

Per-Ola Carlsson is Associate Professor of Medical Cell Biology. His research interests include the microcirculation in endogenous and transplanted pancreatic islets.

SECTION J

Skeletal Muscle

CHAPTER 85

Exercise and Skeletal Muscle Circulation

[1]Jeffrey L. Jasperse and [2]M. Harold Laughlin

[1]*Department of Sports Medicine, Pepperdine University, Malibu, California;*

[2]*Departments of Medical Pharmacology and Physiology and Veterinary Biomedical Sciences, and Dalton Cardiovascular Research Center, University of Missouri, Columbia, Missouri*

Introduction

Blood flow to skeletal muscle is important for oxygen and nutrient delivery and for removal of metabolic waste products. With the large increases in muscle metabolism that occur during exercise, muscle blood flow must increase dramatically in order for activity to be maintained for more than a few seconds. Skeletal muscle blood flow is relatively low at rest (3 to 5 mL/min/100 g), but can increase up to 100-fold (400–500 mL/min/100 g in some muscles) during maximal, large muscle, rhythmic exercise (such as running or cycling) [1]. The rate of blood flow to skeletal muscle is determined by the perfusion pressure (roughly equivalent to arterial blood pressure) and the vascular resistance to flow within the muscle. Diastolic blood pressure changes little during this type of exercise while systolic blood pressure increases in an intensity-dependent manner, meaning that mean arterial blood pressure (and therefore mean perfusion pressure) increases only modestly during exercise. Thus the major mechanism for increasing blood flow to skeletal muscle during exercise is a reduction in vascular resistance within the exercising skeletal muscle. Elegant control mechanisms are required to mediate this large increase in blood flow and to direct it to the capillaries in regions within the muscle where it is needed without causing dramatic reductions in systemic blood pressure or excessive transcapillary fluid loss. These mechanisms include sympathetic neural control and a host of local humoral, metabolic, and mechanical factors. The purpose of this chapter is to review the role of these control mechanisms in the microcirculation in mediating exercise-induced increases in skeletal muscle blood flow.

Heterogeneity of Skeletal Muscle Blood Flow during Exercise

One of the most remarkable aspects of muscle blood flow is its close matching with muscle metabolism. Blood flow to active skeletal muscle increases at the onset of exercise and remains elevated throughout the duration of an exercise bout. For example, blood flow through the femoral artery of humans is approximately 0.3 L/minute at rest and increases in an intensity-dependent manner during leg extension exercise up to 6 to 10 L/minute [2]. One result of the close matching of blood flow to metabolism is that blood flow during exercise is not homogeneous in its distribution across different muscles or muscle regions. This regional heterogeneity of muscle blood flow appears to result from two interrelated factors: first, muscle fiber type and second, exercise intensity and recruitment order. Under resting conditions where postural maintenance is the only muscular activity, slow-twitch oxidative (SO) fibers are primarily recruited and blood flow is directed primarily to regions of muscle containing these fibers. During low-intensity exercise, SO and fast-twitch oxidative–glycolytic (FOG) fibers are recruited and blood flow is directed to regions containing these two fiber types. As exercise intensity increases up to maximal, fast-twitch glycolytic (FG) fibers are increasingly recruited and blood flow to these fibers also increases. However, even during maximal exercise, blood flow to FG fibers is never as high as flow to more oxidative fibers. In fact, the magnitude of the increase in blood flow to a particular region of muscle during exercise is directly correlated to its FOG fiber content [3]. Thus the regional distribution of blood flow within and among skeletal muscles is determined by exercise intensity and by the oxidative capacities of the fibers in various regions of muscle.

Skeletal muscle blood flow is also not constant across time, and this temporal heterogeneity takes two forms. First, blood flow increases markedly with the first contraction and then gradually levels off to a steady rate during the first 15 to 60 seconds of exercise at a given intensity. One implication of this is that the initial hyperemic response at the onset of muscle contraction may result from different vascular control mechanisms than the maintained steady-state hyperemia of continued exercise. Second, because of the mechanical effect of muscle contraction "squeezing" and occluding blood vessels located within the muscle, blood flow to contracting muscle follows a cyclical pattern in which almost all blood flow enters the muscle during the relaxation phase and very little blood flow enters the muscle during the actual contractions. Venous outflow of blood from contracting muscle follows the exact opposite pattern. Muscle contraction compresses veins, forcing blood to move, and venous valves ensure that the direction of movement is out of muscle and toward the heart. Thus most blood flow into muscle occurs during the relaxation phase of muscular contraction while most venous outflow from muscle occurs during the contractile phase.

Mechanisms Determining Blood Flow Distribution

Vascular Structure

Two major determinants control the distribution of blood flow: vascular structure and vascular control, or mechanisms that control vascular resistance by modulating smooth muscle contraction and, therefore, diameter of arterioles. Both capillary and arteriole densities are greater in muscle regions composed primarily of oxidative fibers than in those composed primarily of glycolytic fibers. This structural difference between the vascularization of different muscle fiber types allows oxidative regions to have a lower vascular resistance and a higher capacity for blood flow. Thus structural differences in the vasculature between oxidative and glycolytic regions of muscle at least partially explain the greater maximal blood flows measured in oxidative regions.

Vascular Control

There are a number of vascular control mechanisms that contribute to exercise hyperemia in skeletal muscle. These include extrinsic (from outside the tissue) or central mechanisms such as neural and humoral control and intrinsic (or local) mechanisms such as metabolic or myogenic control mechanisms. In general, extrinsic control mechanisms are more concerned with systemic blood pressure regulation, whereas intrinsic control mechanisms are more concerned with control of oxygen delivery to local areas. There is much evidence that differences in vascular control exist in regions of different fiber type and that this may help determine the heterogeneous distribution of blood flow during exercise.

Extrinsic Vascular Control Mechanisms

SYMPATHETIC NERVES

Under resting conditions blood flow to skeletal muscle is limited by sympathetically mediated constriction of arteries and arterioles in the muscle. The involvement of a neural mechanism in mediating exercise hyperemia is attractive because it would help to explain the rapid increase in blood flow at the onset of exercise. Possible mechanisms for exercise hyperemia include either sympathetic withdrawal or a sympathetically mediated vasodilation. However, neither of these mechanisms appears to play a role in exercise hyperemia. While sympathetic cholinergic nerves have been found in some vascular beds and in some species, it does not appear that sympathetic cholinergic vasodilation contributes to increases in skeletal muscle blood flow during exercise. Additionally, sympathetic nerve activity to skeletal muscle is not reduced during exercise. On the contrary, sympathetic nerve activity appears to be increased during exercise [1, 4], and this vasoconstrictor influence is thought to limit the vasodilation of exercise in order to help maintain systemic blood pressure. Thus withdrawal of sympathetic nerve activity does not mediate increases in blood flow during exercise.

HUMORAL CONTROL

A number of hormones or other bloodborne agents can cause changes in vascular diameters and therefore could help to mediate changes in muscle blood flow during exercise. The concentrations of most of these agents do not change rapidly enough to explain the large increases in muscle blood flow that occur with the onset of exercise, and most also do not appear to have much effect on blood flow during sustained exercise. Vasopressin and angiotensin II are two vasoconstrictor hormones that appear to play a role in the distribution of blood flow during exercise, but their primary effect is in constricting resistance vessels in noncontracting tissues of the gut region. One substance that may have an effect is acetylcholine, the neurotransmitter of muscle contraction and a well-known vasodilator, as there is some evidence that acetylcholine spillover from the myoneural junction may increase muscle blood flow [5]. This is an attractive hypothesis for two reasons. First, as acetylcholine release from motor nerves would presumably increase with the initiation of exercise, this would provide a mechanism for the immediate increase in blood flow that occurs at the commencement of exercise. Second, as acetylcholine release from motor nerves would presumably increase as exercise intensity increases, this would provide a mechanism for relating increases in blood flow to the intensity of exercise. However, other studies using blockers of acetylcholine receptors have failed to show any effect on blood flow to exercising skeletal muscle [6]; therefore, the role of acetylcholine in helping mediate exercise hyperemia is presently in doubt.

Intrinsic Vascular Control Mechanisms

METABOLIC VASODILATION

The most attractive hypotheses regarding the mechanisms mediating exercise hyperemia involve metabolic vasodilation. According to this idea, the rate of muscle metabolism and the rate of muscle blood flow are coupled, presumably by one or more metabolites produced by the exercising muscle. These metabolites diffuse to the neighboring arterioles and when, present in sufficient quantities, cause vasodilation. The resultant increase in blood flow will help to "wash out" or carry away the metabolites, which reduces the vasodilator stimulus. Thus the level of vascular tone is the result of a constant tension between metabolite production and metabolite "washout."

The identity of the substance or substances causing metabolic vasodilation is not currently known, though a host of substances have been proposed and studied. To contribute to metabolic vasodilation, the vasodilator substance must meet at least two criteria: It must increase in concentration at the vascular wall as exercise intensity increases, and it must have the ability to relax vascular smooth muscle, either directly or indirectly [7]. The list of potential metabolites involved includes adenosine and/or adenine nucleotides, potassium, hydrogen, carbon dioxide, reduced oxygen levels, tissue osmolality, inorganic phosphate, prostaglandins, and histamine.

Adenosine is a known vasodilator that directly causes smooth muscle relaxation and also acts indirectly via its action on the endothelium [1, 3, 7]. It is an attractive candidate for metabolic vasodilation because its production should be directly related to the rate of muscle metabolism as adenine nucleotides are broken down for their energy-bearing phosphate bonds. Indeed, adenosine appears to play a major role in dilation of the coronary circulation. But studies examining the role of adenosine in skeletal muscle hyperemia have not been definitive, with the majority of studies showing no role for adenosine [1, 3, 8]. Studies measuring adenosine concentrations in contracting muscle have generally found levels lower than that which would cause vasodilation [8], suggesting that adenosine plays at most a minor role in mediating increased blood flow, but adenosine is rapidly degraded and accurate sampling of adenosine at the smooth muscle cell is difficult [7, 8]. It appears that the inconclusive results may result at least in part from a regional heterogeneity in the action of adenosine. Studies examining adenosine's role at moderate exercise intensities in cats and miniature swine have found no effect of adenosine in glycolytic muscle but that adenosine causes vasodilation in primarily oxidative regions of muscle [1, 3, 8]. At high exercise intensities, adenosine appears to cause vasodilation in all skeletal muscle types [1, 3, 8]. Furthermore, vascular reactivity to adenosine is greater in smaller than larger arterioles, and this may also complicate data interpretation [7]. Thus the effects of adenosine in mediating exercise hyperemia appear to depend both on fiber type composition of muscle and on exercise intensity.

Potassium is a vasodilator of the microcirculation in low concentrations and it is released from both motor neurons and myocytes during membrane repolarization [7, 8]. Venous potassium levels rise during exercise, and venous potassium concentrations have been reported to correlate well with the reduction in vascular resistance in contracting muscle. However, the dilatory effect of potassium is transient, and the increase in potassium during exercise appears to be transient as well [7, 8]. This evidence suggests that potassium contributes little if at all to prolonged increases in blood flow during steady-state exercise, but that it may play a role in the initiation of the increase in blood flow at the onset of exercise.

Because one of the major functions of increasing blood flow is to increase O_2 delivery and CO_2 clearance, the notion that concentrations of these gases may directly regulate vascular tone is appealing. It is well known that in skeletal muscle reduced arterial O_2 concentration causes vasodilation while increased O_2 concentration causes vasoconstriction. Arachidonic acid metabolites, via the action of various cytochrome P-450 enzymes, may mediate at least part of the response of arteriolar smooth muscle to O_2. But the role of O_2 in mediating exercise hyperemia is unclear for at least two reasons. First, while tissue and venous Po_2 values are reduced during muscle contraction, periarteriolar Po_2 values decrease only temporarily at the onset of muscle contraction and return to resting levels after a few minutes [8]. Second, it has been difficult experimentally to measure periarteriolar Po_2 values in small arterioles where oxygen may have its largest effect [7]. Thus a direct role for O_2 in regulating vascular diameters during exercise is uncertain. That does not necessarily rule out O_2 as a mediator of exercise hyperemia. The correlation between reductions in tissue and venous Po_2 and reductions in vascular resistance suggests that O_2 may have an indirect effect on vascular diameters through production of a secondary signal that is dependent on oxygen levels [8].

Although CO_2 production is increased and venous CO_2 concentrations increase markedly during exercise, CO_2 has generally been considered to be a fairly weak vasodilator. It is possible for two reasons that this idea should be reconsidered. First, the potency of CO_2 as a vasodilator may be enhanced in conjunction with reduced O_2 levels [9]. This suggests that there is a synergistic effect of O_2 and CO_2 and that examining CO_2 alone does not provide a full picture of its effects. Second, Duling and Dora have pointed out that the CO_2 concentration around arterial smooth muscle is unknown [7]. While experiments have generally altered arterial CO_2, the effects of gas diffusion, of shunting into veins, and of CO_2 buffering are unknown, and the CO_2 concentration in the microenvironment of the vascular smooth muscle cells of small arterioles is not known in these experiments, even when arterial and venous CO_2 concentrations can be measured. Thus CO_2 may play a larger role than previously recognized.

A number of other potential vasodilatory metabolites have been studied, and clear evidence is lacking that any of them play a major role in metabolic vasodilation. Hydrogen ions appear to be produced by contracting muscle and venous and muscular pH is decreased during exercise [8]. However, it is not clear that periarteriolar pH changes significantly during exercise, and experimentally reducing periarteriolar pH fails to cause significant increases in blood flow. Inorganic phosphate is also produced during muscle contraction, but if it causes vasodilation its role is relatively minor [7]. Indeed, the available evidence does not point to any single metabolic product of exercise as the primary mediator of exercise hyperemia. However, this does not rule out metabolic vasodilation as a major contributor to exercise hyperemia. Rather, metabolic vasodilation probably results from the combined effect of many substances working in concert. Part of this effect is likely due in a small way to the effect of these metabolites on plasma osmolality. The osmolality of venous blood is increased during exercise, presumably because of the increased production of adenosine, H^+, K^+, and a number of other molecules during exercise, and increases in osmolality have been shown to elicit a small degree of vasodilation. But the majority of the metabolic vasodilation occurring during exercise is probably due to the combined effect of numerous vasodilatory metabolic products.

It should be noted that the previous discussion regarding metabolic vasodilation refers only to the sustained steady-state period of exercise. The immediate increase in blood flow at the onset of exercise is unlikely to be affected by metabolic factors for at least three reasons. First, the accumulation of metabolites occurs at a slower rate than the initial increase in blood flow. Second, measurements of arteriolar diameters during muscle contraction indicate that vasodilation occurs at a slower rate than the increase in blood flow, with blood flow increasing within the first seconds of exercise and dilation taking 5 to 20 seconds following onset [1]. Third, there is a poor correlation between work rate (and the associated rate of metabolism) and dilation at the onset of exercise, as demonstrated by the fact that changes in work rate do not necessarily alter the magnitude of the initial increase in blood flow [1]. Thus metabolic vasodilation is thought to be a major contributor to the blood flow response to sustained exercise, but it is unlikely to contribute to the initial hyperemia at the onset of exercise.

Functional Sympatholysis

Sympathetic nerve activity (SNA) increases at the onset of exercise, and the magnitude of the increase during sustained exercise is directly related to the exercise intensity [4]. The increased SNA mediates a variety of cardiovascular responses necessary to continue exercise for more than a few seconds, including an increased heart rate and stroke volume, and a redistribution of blood flow from noncontracting tissues to active skeletal muscle. The fact that there is significant vasodilation in active muscle despite the increased SNA directed to the muscle vasculature has led to the concept of functional sympatholysis. According to this idea, metabolites produced by the contracting muscle interfere with sympathetic vasoconstriction and allow the metabolic vasodilation to prevail. The mechanism of this inhibition of sympathetic vasoconstriction by metabolic products is probably twofold. First, some metabolites likely inhibit the release of norepinephrine from sympathetic nerve endings at the vessel wall (prejunctional inhibition) [1, 4]. Second, there is good evidence that metabolic products decrease alpha-adrenergic receptor sensitivity to norepinephrine. Both α-1 and α-2 receptors are located in larger arterioles, but only α-2 receptors are present in smaller arterioles, and the α-2 receptors are more affected by metabolic inhibition than the α-1 receptors [4]. The net result of functional sympatholysis is that sympathetically mediated vasoconstriction occurs in noncontracting muscle regions while it is overridden in active areas by metabolic vasodilation. This provides an efficient means of directing blood flow specifically to regions of muscle that are active during exercise. Additionally, because the muscle regions active at a given intensity of exercise are heavily influenced by the fiber type composition of the muscle, functional sympatholysis contributes to the fiber type dependence of blood flow distribution.

Despite the production of vasodilatory metabolites and the vasodilation that occurs in active regions of skeletal muscle, the overriding of SNA in these regions is not complete [1, 3, 5, 8]. Rather, there is a constant tension between metabolic vasodilation and sympathetically mediated vasoconstriction, and the level of vascular tone at any time is a result of this tension. The resulting vasomotor tone is important both for the maintenance of fluid balance in active muscle and for the maintenance of systemic blood pressure [1]. Total vasodilation in active regions would outstrip cardiac output and baroreceptor reflexes would be unable to maintain systemic blood pressure.

Myogenic Control

Because the mechanism for the initial hyperemia at the onset of exercise must be rapidly acting and because neural factors do not play a role, it is likely that mechanical factors associated with muscle contraction are involved in the early response. One mechanical factor is the myogenic response, which refers to the intrinsic property of vascular smooth muscle that causes it to contract when it is stretched. The result is that increased arteriolar distending pressures cause arterioles to constrict and decreased distending pressures cause dilation. Arterioles located within skeletal muscle are compressed during muscular contraction, which would be expected to reduce the transmural pressure in the vessel leading to decreased vascular distension, vasodilation, and increased blood flow. Studies utilizing increases in extravascular pressures to decrease vascular transmural pressure have yielded mixed results [1, 8], and the role of myogenic control in exercise hyperemia is uncertain.

CHANGING MUSCLE LENGTH

A second category of mechanical factors has to do with the distortion of arterioles that occurs during muscle contraction. This distortion as the muscle shortens can elicit both passive and active vasomotor responses in the vasculature [5, 8]. Passive changes during muscle shortening occur as arterioles are twisted, kinked, and compressed, and as interbranch vessel lengths and bifurcation angles at branch points are altered by the changing geometry of the arteriolar network. The net effect of these passive changes on vascular resistance and blood flow is uncertain. Active changes in vasomotor tone also occur, as shortening of muscle causes vasodilation and lengthening causes vasoconstriction. These active changes are apparently caused by sympathetic nerves that are sensitive to changing muscle length and respond to increasing length by releasing norepinephrine, a well-documented vasoconstrictor [5]. Both passive and active muscle length–dependent changes in vascular resistance may occur rapidly and are independent of muscle metabolism, and could thus be involved in hyperemic responses both at the onset of muscular contraction and during sustained exercise.

MUSCLE PUMP

As noted earlier, the duty cycle of contracting skeletal muscle causes a cyclical compression of blood vessels within the muscle that leads to alternating periods of arterial inflow and venous outflow during exercise. The cyclical compression of blood vessels also increases the overall rate of blood flow through contracting muscle, a phenomenon known as the "muscle pump" [1, 3]. The compression of veins during muscle contraction expels blood from veins toward the heart, thus emptying veins of their blood volume. The small veins within skeletal muscle are tethered by connective tissue to the surrounding muscle and are pulled open as the muscle relaxes again. This creates a negative pressure within the veins and effectively increases the arterial–venous pressure gradient for blood flow through the muscle. This cycle is repeated with each contraction–relaxation cycle of muscle and acts as a pump that helps drive blood flow through contracting muscle. It has been estimated that the muscle pump can account for 30 to 60 percent of the increase in blood flow during moderate intensity muscle contraction [3, 8]. The cyclical compression also appears to have important effects on fluid clearance from the interstitial spaces via pumping action on the lymphatic vessels.

There are a number of important factors to consider regarding the muscle pump. First, the pump is dependent on the contraction–relaxation cycle. Therefore, it is active only during rhythmic muscle contractions, not during isometric contractions. Second, increases in blood flow due to the muscle pump do not result from vasodilation, that is, a change in vascular conductance. Rather the increased blood flow results from decreased pressures in the small veins and increased kinetic energy imparted to the blood [1, 3]. Thus this has sometimes been called an increase in "apparent vascular conductance." Third, the efficacy of the pump mechanism is dependent on the order of activation of different muscles, a factor that separates voluntary locomotory exercise from in situ and in vitro experimental models of muscle contraction. Fourth, the effectiveness of the muscle pump appears to be greater in deeper regions of muscle than in more superficial muscle regions. Because more oxidative muscle fibers tend to be located deeper and more glycolytic fibers are located superficially, the muscle pump appears to be more effective in regions composed primarily of oxidative muscle fibers than in regions composed primarily of more glycolytic fibers [1].

Endothelium-Dependent Control

The huge importance of the endothelium in the control of blood flow has come to light in the past two decades. The endothelium plays an important role in the control of vascular tone and in the adaptation of the vasculature to chronic stimuli [3]. The endothelium produces and releases a number of vasoactive substances that exert their effects by altering vascular smooth muscle tone. Nitric oxide has been the most studied of these substances and is very important in the regulation of basal vascular tone under resting circumstances in skeletal muscle. Whether NO or other endothelium-dependent dilators contribute to the functional hyperemia of exercise is less clear. Inhibition of nitric oxide synthesis has been shown in some preparations to reduce the hyperemic response to exercise, but other studies have not shown any effect. Additionally, in some cases when NO synthesis inhibition during exercise reduces the hyperemic response, the magnitude of the reduction in blood flow is similar to the reduction seen under resting conditions. These data suggest that it is basal synthesis of NO that is being blocked in these cases rather than exercise-specific production of NO [3]. Interpretation of these data is complicated by the redundancy of control mechanisms in the arterial vasculature, that is, production of other vasodilator substances may compensate for inhibition of NO synthesis. A second complication is that although NO may not increase the magnitude of the hyperemic response to whole exercising limbs, it appears to be involved in the muscle fiber-type specific distribution of blood flow within the limb as blocking NO synthesis decreases blood flow to a greater degree in highly oxidative regions of muscle than in more glycolytic regions [3]. A third complication in interpreting data in this area is that NO may exert a greater effect on small arteries and large arterioles than on smaller arterioles [8]. Thus the current data regarding the role of the endothelium in contributing to exercise hyperemia are not definitive.

Control by Red Blood Cells

Traditional thinking about control of blood flow has usually located the origin of that control in the vascular wall or in the surrounding parenchymal tissue, but evidence collected in the past decade indicates that red blood cells (RBCs) play a role in determining their own destination [2, 10]. The oxygen saturation of hemoglobin is dependent on the surrounding Po_2, and the level of oxygen–hemoglobin saturation appears to act as a built-in sensor of oxygen levels in the surrounding environment. Low concentrations of oxygenated hemoglobin stimulate the production and release of dilator substances by RBCs. There appear to be at least two dilator substances utilized in this process. First, low levels of oxygenated hemoglobin stimulate the production of ATP by RBCs, and this ATP may act as a vasodilator at the vascular wall, either directly or through some secondary process [2]. Second, under hypoxic conditions RBCs also appear to produce and release NO [10]. NO stimulates vascular smooth muscle guanylate cyclase production of cGMP and consequent vascular smooth muscle relaxation. Thus hypoxic conditions in contracting skeletal muscle are sensed by RBCs, which produce vasodilators that increase vascular conductance specifically in the hypoxic region. Finally, another function of hemoglobin is that it normally acts as a nitric oxide scavenger [2]. Hypoxia is associated with inadequate blood flow and hemoglobin levels, and when hemoglobin concentrations are low the scavenging action of hemoglobin is also reduced. The levels of NO should, therefore, remain higher under these conditions, and the associated vasodilator stimulus should remain elevated. Thus both vasodilator production by RBCs and reduced NO scavenging by RBCs in hypoxic regions help to direct blood flow specifically to those areas that are most in need.

Feeding Arteries

Although metabolically mediated vasodilation of small arterioles within the exercising muscle is an important step in increasing blood flow, the magnitude of the increase in blood flow requires that larger arterioles and feeding arteries located outside the muscle also dilate. It has been estimated that 40 to 50 percent of vascular resistance in skeletal muscle is located in the feed arteries outside of the muscle proper [8]. Dilation of arterioles without concurrent dilation of upstream feed arteries would not allow the magnitude of increase in blood flow that occurs during exercise [5], but a coordinated dilation at all levels from small arterioles to feed arteries enables large increases in blood flow to occur. The mechanism for feed artery dilation cannot be metabolic because these feed arteries are located outside the muscle and are not directly exposed to the metabolic environment within the muscle. Three mechanisms have been proposed to mediate feed artery dilation: flow-induced dilation, retrograde propagation, and arterial–venous coupling.

Flow-induced dilation occurs when the flow velocity in a blood vessel increases, thereby increasing the shear stress of the blood against the endothelial cell wall. Theoretically, dilation of arterioles within the muscle during exercise would cause reduced resistance to flow within the muscle and increase the rate of flow through upstream vessels. However, most studies of flow-induced dilation have been done in cultured cell or in vitro experimental preparations, and it is not clear whether flow-induced dilation contributes to the hyperemia associated with exercise in vivo. First, it is not clear that wall shear stress increases during exercise, as a dilation that matches the increase in blood flow could maintain shear stress at a constant level. Second, flow-induced dilation of rat feed arteries in vitro occurs at very low levels of shear stress, levels that are lower than the calculated shear stress present in feed arteries under normal nonexercising conditions. If flow-induced dilation in vivo also occurs at shear stress levels this low, then flow-induced dilation probably helps to determine vascular tone at rest, but is most likely not involved in exercise-induced dilation of feed arteries.

Retrograde propagation of vascular signals involves the movement of electrical signals through gap junctions between cells [5, 7]. Thus a local depolarization of vascular smooth muscle or endothelial cells within the contracting muscle caused by muscle metabolites would be propagated along the arterioles out to the feed arteries. This mechanism, too, clearly occurs in cultured cell or in vitro experimental preparations and in response to electrically induced contractions in situ, although it is unclear whether the electrical signal is propagated via smooth muscle cells, endothelial cells, or both. The propagation is very rapid, but propagated vasodilation appears to travel only limited distances, and technical problems limit the study of propagated signals in vivo and under exercising conditions. Thus the contribution of a propagated vasodilation to feed artery dilation during voluntary exercise in conscious animals has not been definitively determined.

Arterial–venous coupling is a third mechanism that has been proposed to cause dilation of feeding arterioles during exercise [11]. Anatomically, skeletal muscle feeding arteries lie in close proximity to one or more veins draining the same muscle. Thus a metabolite or a group of metabolites produced in the muscle during exercise may exit the muscle via venous blood flow and activate the venular endothelium. The venular endothelium then produces a signaling molecule, most likely prostacyclin [11], which diffuses to the paired artery and causes it to dilate. Alternatively, red blood cells (RBCs) in the veins may activate the venular endothelium. As noted above, RBCs are known to release ATP and NO under hypoxic conditions, conditions that are likely to exist in the veins during exercise, and ATP or NO from the RBCs may be the signal that activates venular endothelium during exercise.

Thus the mechanisms causing feed artery dilation have not been clearly determined. It is probable that these three

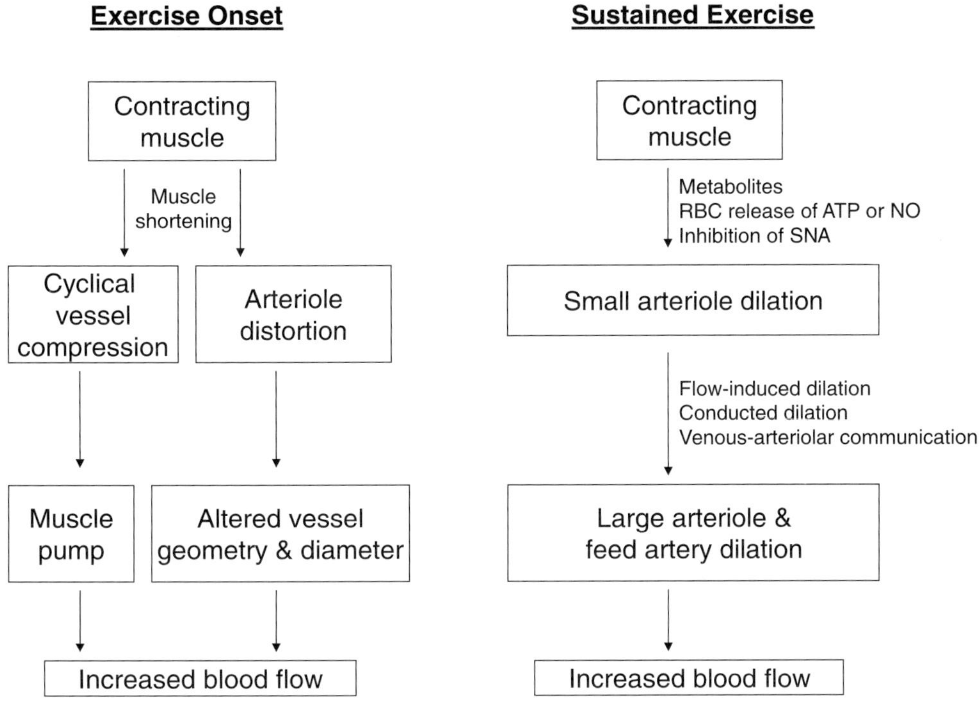

Figure 1 Schematic of potential mechanisms involved in skeletal muscle hyperemia at the onset of exercise and during sustained exercise. Exercise onset: As muscle shortens during contraction, vessels within the muscle are compressed and distorted. These physical forces may increase blood flow by increasing the kinetic energy of the blood (muscle pump), by causing passive changes in vascular geometry, or by causing active changes in vascular tone. Sustained exercise: Small arterioles are dilated by the products of muscle metabolism and by ATP or NO released from red blood cells (RBCs). The effects of sympathetic nerve activity (SNA) are reduced by metabolites from contracting muscle. Vasodilation spreads to larger arterioles and feed arteries by flow-induced dilation, upstream signal conduction, or venous–arteriolar communication. Vasodilation of all branches of the arterial vasculature combines to cause increased blood flow and oxygen delivery to contracting muscle.

factors, possibly in concert with other as yet undetermined factors, act in combination to elicit dilation. It may also be that the exact mechanism varies between different muscles or in regions of different muscle fiber type.

Integration/Conclusion

It is clear from the preceding discussion that a host of different microcirculatory control mechanisms contribute to exercise hyperemia in skeletal muscle. Although much is known about these mechanisms, the complete blood flow response to exercise is brought about by a coordinated and complex interaction of these factors, about which there is still much to learn. Figure 1 shows potential mechanisms involved at the onset of exercise and during sustained exercise. The hyperemic response to the onset of exercise is likely due to some combination of mechanical factors associated with muscle contraction, such as the muscle pump and/or vessel distortion. Whether those factors cause immediate vasodilation and an increase in vascular conductance or produce increased blood flow without vasodilation is not known. The hyperemic response to sustained exercise likely begins with metabolically induced vasodilation of small arterioles located within the contracting muscle, and vasodilators released by red blood cells in response to a hypoxic environment contribute as well. The dilation spreads upstream via one or more of the following mechanisms: flow-induced dilation, retrograde propagation, and arterial–venous coupling. The overall responses are complicated by the spatial and temporal heterogeneity described earlier.

Figure 2 illustrates another type of spatial heterogeneity—a variation in the relative importance of each vascular control mechanism throughout the vascular tree. That is, although each vascular control mechanism described appears to be present throughout the vascular tree, the relative effectiveness of each varies depending on location in the tree. For example, metabolic dilation and the myogenic mechanism both have a greater impact in smaller than in larger arterioles. Also, the efficiency of sympatholysis is enhanced by the fact that metabolic dilation is greatest in the same sized arterioles as sympathetic constriction. Finally, flow-induced dilation, a mechanism that may be important in the upstream spread of vasodilation, is more important in the larger than in the smaller arterioles. Thus the variations

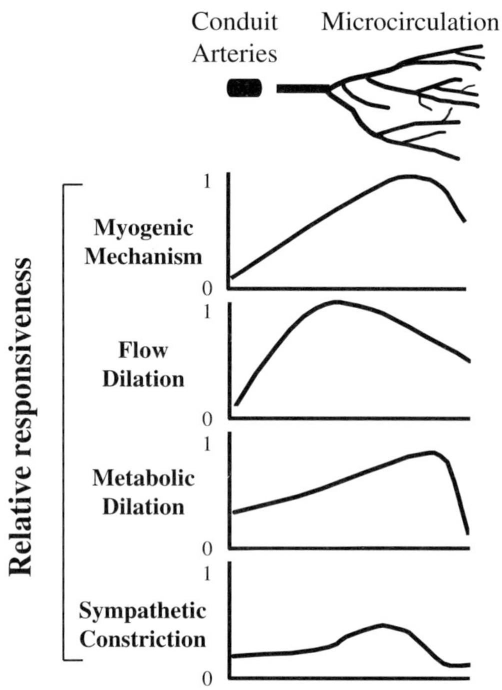

Figure 2 This figure illustrates the relative responsiveness of each section of the arterial tree (at the top) for the myogenic mechanism, flow-induced dilation (representing endothelium-dependent control), metabolic dilation, and sympathetic constriction. Redrawn based on a figure of the coronary circulation presented by Kuo et al. [12].

in the importance of each vascular control mechanism throughout the vascular tree contribute to the efficiency of the hyperemic response to exercise.

Glossary

Adenosine: A purine nucleotide consisting of adenine and ribose. It is a degradation product of ATP and thus is produced in increased quantities during increases in cellular metabolism.

Metabolic vasodilation: Refers to the relaxation of vascular smooth muscle caused by many products of metabolism. When present in increased quantities, these products cause vasodilation and help increase blood flow to active regions of the body.

Muscle pump: The cyclical compression of blood vessels that occurs within contracting skeletal muscle. This compression helps increase venous return of blood to the heart and is thought to contribute to increase blood flow through contracting muscle.

Myogenic mechanism: Refers to the property of smooth muscle that causes it to contract when it is stretched. Contributes to autoregulation of blood flow in a vascular bed because increased pressure elicits vascular constriction and increased resistance to blood flow.

Sympatholysis: Refers to the decreased effects of sympathetic nerve activity on blood vessel constriction in contracting skeletal muscle. It is thought to result from metabolic by-products interfering with some aspect of sympathetic neurotransmission or postsynaptic signaling.

References

1. Laughlin, M. H., Korthuis, R. J., Duncker, D. J., and Bache, R. J. (1996). Control of blood flow to cardiac and skeletal muscle during exercise. In *Handbook of Physiology Section 12: Exercise: Regulation and Integration of Multiple Systems* (L. B. Rowell and J. T. Shepherd, eds.), pp. 705–769. New York: Oxford University Press.
2. Saltin, B., Radegran, G., Koskolou, M. D., and Roach, R. C. (1998). Skeletal muscle blood flow in humans and its regulation during exercise. *Acta Physiol. Scand.* **162**, 421–436.
3. Delp, M. D., and Laughlin, M. H. (1998). Regulation of skeletal muscle perfusion during exercise. *Acta Physiol. Scand.* **162**, 411–419.
4. Buckwalter, J. B., and Clifford, P. S. (2001). The paradox of sympathetic vasoconstriction in exercising skeletal muscle. *Exer. Sport Sci. Rev.* **29**(4), 159–163.
5. Segal, S. S. (2000). Dynamics of microvascular control in skeletal muscle. In *Exercise and Circulation in Health and Disease* (B. Saltin, R. Boushel, N. Secher, and J. Mitchell, eds.), pp. 141–153. Champaign, IL: Human Kinetics.
6. Buckwalter, J. B., Ruble, S. B., Mueller, P. J., and Clifford, P. S. (1998). Skeletal muscle vasodilation at the onset of exercise. *J. Appl. Physiol.* **85**, 1649–1654.
7. Duling, B. R., and Dora, K. (1997). Control of striated muscle blood flow. In *The Lung: Scientific Foundations* (R. G. Crystal, J. B. West, et al., eds.), 2nd ed., pp. 1935–1943. Philadelphia: Lippencott-Raven.
8. Lash, J. M. (1996). Regulation of skeletal muscle blood flow during contractions. *Proc. Soc. Exper. Biol. Med.* **211**, 218–235.
9. Mohrman, D., and Regal, R. (1988). Relation of blood flow to Vo_2, Po_2, and CO_2 in dog gastrocnemius muscle. *Am. J. Physiol.* **255**, H1004–H1010.
10. Pawloski, J. R., and Stamler, J. S. (2002). Nitric oxide in RBCs. *Transfusion* **42**, 1603–1609.
11. Hester, R. L., and Choi, J. (2002). Blood flow during exercise: Role for the venular endothelium? *Exer. Sport Sci. Rev.* **30**(4), 147–151.
12. Kuo, L., Davis, M. J., and Chilian, W. M. (1992). Endothelial modulation of arteriolar tone. *News Physiol. Sci.* **7**, 5–9.

Further Reading

Duling, B. R., and Berne, R. M. (1970). Propagated vasodilation in the microcirculation of the hamster cheek pouch. *Circ. Res.* **26**, 163–170. *Classic paper is the first demonstration of retrograde propagation of vasodilatory responses in the microcirculation.*

Furchgott, R. F., and Zawadski, J. V. (1980). The obligatory role of endothelial cells in the relaxation of arterial smooth muscle by acetylcholine. *Nature* **288**(5789), 373–376. *Classic demonstration of the role of the endothelium in blood flow control by secretion of an endothelium-derived relaxing factor. Furchgott was one of three winners of the 1998 Nobel Prize in Medicine for this work.*

Kingwell, B. A. (2000). Nitric oxide–mediated metabolic regulation during exercise: Effects of training in health and cardiovascular disease. *FASEB J.* **14**(12), 1685–1896. *Good review of the role of nitric oxide in skeletal muscle exercise hyperemia.*

Lash, J. M. (1996). Regulation of skeletal muscle blood flow during contractions. *Proc. Soc. Exp. Biol. Med.* **211**, 218–235. *Very good and complete review of the mechanisms controlling skeletal muscle blood flow during exercise.*

Laughlin, M. H. (1987). Skeletal muscle blood flow capacity: Role of muscle pump in exercise hyperemia. *Am. J. Physiol.* **253**, H993–H1004. *Complete (and most widely cited) discussion of various factors involved in the muscle pump and its effect on skeletal muscle blood flow during exercise.*

Laughlin, M. H., and Armstrong, R. B. (1982). Muscular blood flow distribution patterns as a function of running speed in rats. *Am. J. Physiol.* **243**, H296–H306. *This is the first paper showing the muscle fiber type dependency of blood flow during exercise.*

Laughlin, M. H., Korthuis, R. J., Duncker, D. J., and Bache, R. J. (1996). Control of blood flow to cardiac and skeletal muscle during exercise. In *Handbook of Physiology Section 12: Exercise: Regulation and Integration of Multiple Systems* (L. B. Rowell and J. T. Shepherd, eds.), pp. 705–769. New York: Oxford University Press. *This is the most complete and most widely cited recent review of the various mechanisms involved in the regulation of skeletal muscle and coronary blood flow during exercise. This article also examines the effects of exercise training on these mechanisms.*

Prior, B. M., Lloyd, P. G., Yang, H. T., and Terjung, R. L. (2003). Exercise-induced vascular remodeling. *Exer. Sport Sci Rev.* **31**(1), 26–33. *Good review of structural changes in the vasculature induced by exercise training.*

Segal, S. S., and Duling, B. R. (1986). Flow control among microvessels coordinated by intercellular conduction. *Science* **234**(4778), 868–870. *Important paper demonstrating the role of propagated vasodilation in the control of blood flow in the microcirculation.*

Welsh, D. G., and Segal, S. S. (1996). Muscle length directs sympathetic nerve activity and vasomotor tone in resistance vessels of hamster retractor. *Circ. Res.* **79**(3), 551–559. *Seminal paper showing mechanical effects of muscle shortening on vascular resistance.*

Capsule Biography

Dr. Jasperse is a Seaver Fellow in the Department of Sports Medicine at Pepperdine University. His research focuses on control of blood flow to skeletal muscle and how it is altered by changes in chronic activity.

Dr. Laughlin is chair of the Department of Biomedical Sciences at the University of Missouri College of Veterinary Medicine. A Fellow in the American College of Sports Medicine (ACSM) and the Circulation Council of the American Heart Association, his research examines the effects of exercise training on the coronary circulation and skeletal muscle vascular beds. His work is funded by the National Institutes of Health and the American Heart Association, and he was winner of the 2003 ACSM Joseph Wolffe Award and the 1998 ACSM Citation Award for his career contributions to exercise science.

SECTION K

Skin

CHAPTER 86

Vascular Responses in Human Skin

Geraldine F. Clough and Martin K. Church
School of Medicine, University of Southampton

Our knowledge of microvascular function, both normal and abnormal, is derived largely from in vitro studies of isolated tissues or cells or from intravital studies of whole microvascular beds in animal models. Where investigation of the human microcirculation has been performed, it has generally been restricted to the observation of the more accessible microcirculatory beds, such as that of the skin. This short review will focus on those data obtained in humans, in vivo.

As modulation of endothelial function is an early and important event in the pathogenesis of cardiovascular disease and is common to most organs of the body, the dermal microvasculature bed is frequently used as a functional model in which to study endothelial function in humans in vivo. Also, because of its ease of access, the skin and its circulation has been used as an experimental test bed for the development of investigative tools, many of which have been transferred to a clinical setting. Some of the investigative techniques developed in the skin are now well established and widely used in both a research and a clinical setting for the investigation and diagnosis of altered cardiovascular function in vascular beds other than the superficial vessels of the skin.

Structure and Function of the Cutaneous Microvasculature

The skin is the largest organ in the body. It weighs on average 4 kg and covers an area of approximately 2 m^2. The two major functions of the skin are as a barrier protecting the body from the external environment and as a thermoregulatory organ. As a consequence, in addition to providing nutritional support and maintaining tissue homeostasis, the skin microvasculature also plays key roles in immunosurveillance, hemostasis, and tissue repair and remodeling, as well as in thermoregulation. Consequently, the small blood vessels supplying the skin have become highly organized into two horizontal plexuses (Figure 1), the upper of which lies 1 to 1.5 mm below the epidermis. This superficial vascular plexus forms the dermal papillary loops that are the major site of exchange of gases and nutrients within the skin. The three-dimensional structure of this superficial plexus has been extensively studied and painstakingly reconstructed by Braverman and his colleagues [1]. Indeed, because these vessels are the primary site of anatomical variation in a number of skin diseases, it is one of the best characterized vascular beds both in health and disease. The lower plexus is located on the dermal–hypodermal junction. These vessels supply the upper plexus and also branch to supply dermal appendages, such as hair follicles and sweat glands.

In acral or nonhairy skin, the two plexuses are additionally connected by arteriovenous anastomoses—important in the thermoregulatory shunting of blood between the upper and lower plexuses. These anastomoses, which are approximately 30 to 35 μm in diameter, occur mainly in the periphery at exposed sites such as the fingers and toes, lips, nose and ears. As much as 60 percent of the cutaneous blood can be shunted through these anastomoses away from the capillary loops and to the venous plexus in order to reduce convective heat loss.

As a result of the need to accommodate the many different and sometimes conflicting requirements of the tissue, blood flow in the skin is extremely variable. In a thermoneutral environment, flow generally ranges between 10 and 20 $mL\,min^{-1}\ 100\,g^{-1}$. It may, however, fall as low as $<1\,mL\,min^{-1}\ 100\,g^{-1}$ or reach maximal flows of 150 to 200 $mL\,min^{-1}\ 100\,g^{-1}$ at times of cold or severe heat stress, respectively. Not unexpectedly, control of this labile vascular bed is complex and involves neural, humoral, and local influences. Together this affords the investigator a rich opportunity to investigate the functional regulation of microvascular perfusion and integrity.

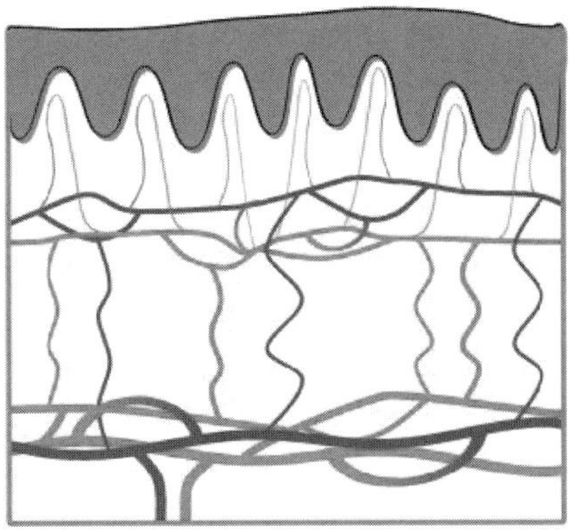

Figure 1 Diagram of the organization of the cutaneous microvascular bed. (see color insert)

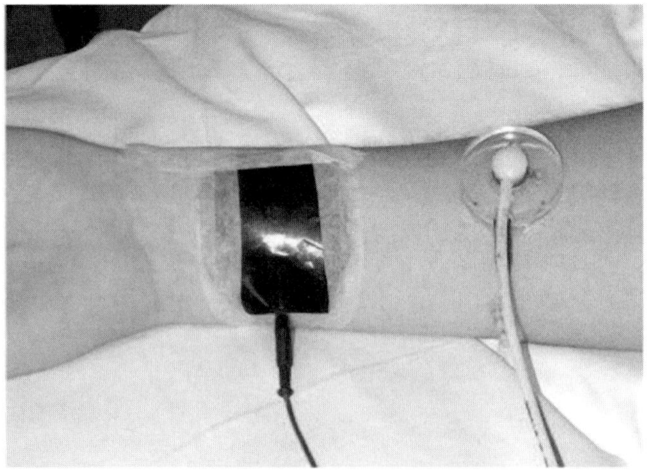

Figure 2 Ions of soluble salt are introduced into the tissue by iontophoretic current. The amount of drug delivered to the skin is proportional to the magnitude and duration of the applied current. (see color insert)

Techniques to Assess Cutaneous Microvascular Function in Vivo

Techniques to study cutaneous microvascular function will be discussed under three headings: techniques to introduce agents into the skin, techniques to assess blood flow and changes occurring therein, and techniques to recover biologically active chemicals from the dermis in order to investigate the biochemical and cellular mechanisms underlying changes in blood flow. Although many of the techniques originally developed in laboratory settings are either too invasive or technically unsuited for use in humans in vivo, over recent years a number of less invasive methods have been developed. Some of the most frequently used of these are the noninvasive imaging of blood flux and the delivery of exogenous vasoactive mediators to the skin and/or the recovery of endogenous mediators from the skin [2]. The remainder of this chapter will focus on these techniques and the type of information that has accrued using them.

Introduction of Agents into the Skin

Historically, pharmacologically active agents used to study vascular function have been introduced either into the systemic circulation, or locally via intradermal injection or topical application. Although both methods of local administration have their place, they both suffer from limitations. Intradermal injections have two potentially serious drawbacks. First, injection of even small volumes will cause major changes in local tissue pressure that may stimulate reflexes that confound or complicate the investigation under question. Second, agents diffuse a maximum of 2 to 3 mm from their site of injection and, hence, have no widespread actions. Topical application of agents to the skin also has limitations. The stratum corneum is permeable only to relatively small lipophilic molecules. Consequently, to get hydrophilic molecules into the skin requires the use of complex pharmaceutical preparations, or removal of the outer epidermis by tape stripping is necessary.

In recent years cutaneous iontophoresis has been developed as a method to deliver vasoactive agonists and antagonists to the skin, particularly in a clinical setting (Figure 2). The technique is noninvasive and avoids the confounding effects of local perturbation of the skin by intradermal injection or of systemic drug administration. Also, agents may be delivered to relatively large areas of the skin. Coupled with laser Doppler flowmetry, iontophoresis has been used to study the role of the endothelium in the modulation of blood flow in the skin. Examples of such studies are responses to acetylcholine (endothelium-dependent) and sodium nitroprusside (endothelium-independent) in both healthy volunteers and in patients with a variety of disorders including diabetes by Dr. Shore's group, hypertension, systemic sclerosis, and Raynaud's phenomenon [3]. Furthermore, introduction of agents into the skin by iontophoresis has allowed studies of the physiology of dermal microvasculature in thermoregulation under heat stress and prolonged exercise [4], the response of the dermis to xenobiotics [5], and the modulation of neurogenic inflammation in the skin [6].

Imaging of Blood Flux

Intravital imaging using transmitted light is not generally possible in solid organs and accurate measurement of blood flow has proved somewhat problematic [7]. Some of the methods currently used include measurement of transcutaneous oxygen, radionucleotide techniques, thermography, ultrasound, laser Doppler fluximetry, and infra red imaging and, most recently, the optical imaging technique of orthogonal polarization spectral analysis [8]. Of all the methods, those based on laser Doppler imaging, which use a measure of red blood cell movement to evaluate vascular flux, have proved the most suitable and best validated for use in a clinical setting.

High-resolution scanning laser Doppler imaging uses a low-power (~2 mW), 633-nm red He–Ne laser to scan the skin in a raster pattern and build up a two-dimensional image of red blood cell flux up to a depth of approximately 0.6 to 1 mm. It is able to provide a real-time output of temporal and spatial changes in skin blood flux, particularly during dermal provocation. As the scanning head may be mounted 30 cm or more above the skin surface, it also allows space for manipulation within the scanned area. Thus, scanning laser Doppler imaging has been used by many groups to study the dermal microvasculature in a wide range of disease states including hypertension, diabetes, and peripheral vascular disease, in inflamed and damaged skin following burn injury, and in nonhealing wounds.

To illustrate the usefulness of scanning laser Doppler imaging in research, we will take the specific example of its application in investigating the mechanisms weal and flare responses to intradermal injections of histamine, bradykinin, and allergen. In these studies, changes in dermal blood flux were assessed by accumulating about 16,000 data points for analysis over a scanned area of skin of 5 cm × 5 cm using scanning laser Doppler imaging (SLDI, Moor LDI, Moor Instruments Ltd, Axminster, Devon, UK) (Figure 3). From these images, the areas of the weal and flare responses were calculated using the manufacturer's software. Our experience has shown that the changes in weal and flare areas may be measured to an accuracy of 0.05 cm^2 and changes in perfusion to 5% [9].

Recovery of Vasoactive Mediators in the Skin

Until recently, the only way in which mechanistic studies could be performed in human skin in vivo was to use pharmacological agonists or antagonists to simulate or block, respectively, the response under investigation. However, such techniques yield only circumstantial evidence, which may not always be correct. What is needed is direct evidence, which may only obtained by sampling the interstitial space surrounding the microvasculature. A variety of techniques have been employed for this purpose, including needle or wick aspirate, tissue exudates such as blister fluid or wound fluid, and most recently microdialysis and ultrafiltration. Biological molecules recovered using these techniques range from metabolites such as glucose and lactate, vasoactive mediators such as histamine, nitric oxide, cyclic nucleotides, eicosanoids, and plasma proteins, and most recently cytokines, growth factors, and neuropeptides [10]. It is thus possible to identify and, if the characteristics of the recovery system are known and the assays sensitive enough, quantify changes in tissue levels of bioactive molecules.

Microdialysis, one of the most successful and widely used sampling techniques used in the skin, was originally developed to recover neurotransmitters from the brains of experimental animals. Subsequently, it has been modified to sample endogenous chemicals within the extravascular space in a number of organs in vivo and as a diagnostic tool to monitor tissue levels metabolites [11] (Figure 4). Microdialysis has the advantage over other dermal sampling techniques in that it causes minimal tissue trauma, is well tolerated, and can be used to follow the temporal variations in the generation and release of a substance at a discreet location within the tissue space. It has also been used to follow the pharmacokinetics of a wide range of xenobiotics within the dermis [12].

In an extension of its original application, the dialysis technique has also been used to investigate the balance and distribution of blood flow within tissues such as muscle and skin that are served by two different vascular beds (nutritive and nonnutritive). By measuring the extraction of highly diffusible markers from the dialysis perfusate, it is possible not only to determine changes in total blood flow but also to detect changes in the distribution of functional blood flow with a high level of sensitivity [13]. This opens the way to studies of the functional control of local tissue perfusion in humans in vivo under both physiological and pathophysiological conditions.

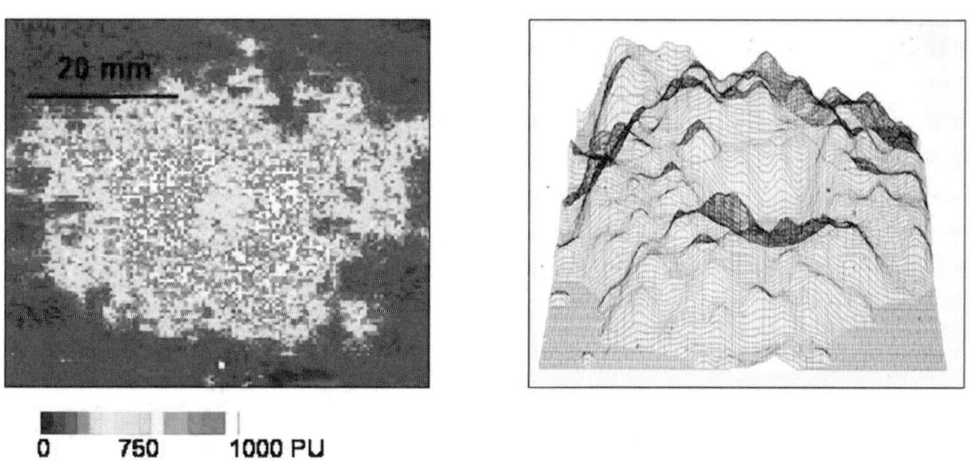

Figure 3 High-resolution scanning laser Doppler imaging uses a low-power (~2 mW), 633-nm red He–Ne laser to scan the skin in a raster pattern and build up a two-dimensional image of red blood cell flux up to a depth of approximately 0.6 to 1 mm. (see color insert)

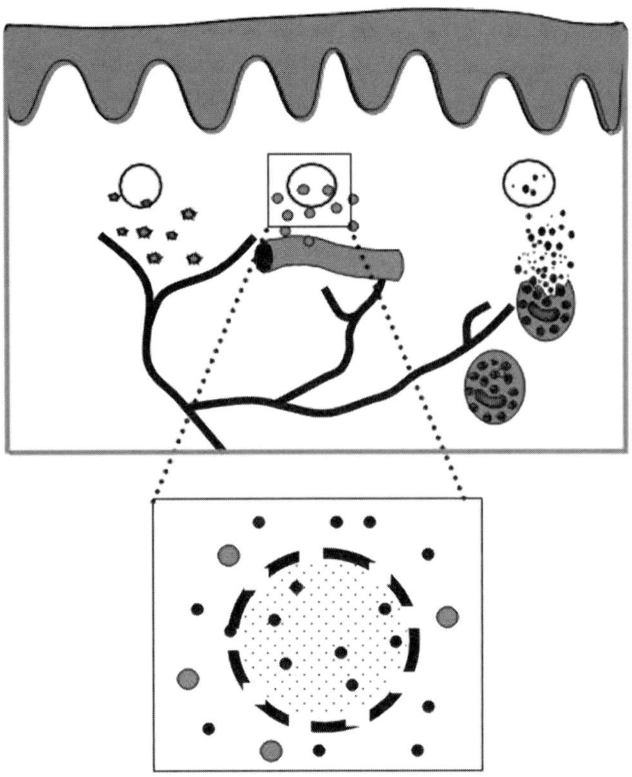

Figure 4 Principles of microdialysis. The recovery of a substance by microdialysis is governed by the same principles as that of exchange of small hydrophilic molecules across the microvascular wall, that is, passive diffusion down a concentration gradient as described by the Fick principle. It is influenced by a number of factors including the characteristics of the dialysis membrane and the recovered solute (its charge, size, solubility) and the concentration of the solute outside the membrane (which in turn may be influenced by tissue metabolism, hydration, and local blood flow). (see color insert)

Vascular Responses in Dermal Inflammation

Inflammation can be induced by a wide variety of stimuli. These include direct mechanical damage, exposure to bacterial toxins or exogenous chemical irritants, tissue ischemia, the release of endogenous inflammatory mediators, and the direct stimulation of nerves. As in other organs the inflammatory response in skin is designed to inactivate or dilute harmful stimuli and to bring about repair. In a balanced situation the response is transient and is rapidly downregulated. When the balance is lost the inflammatory response may continue unchecked, leading to chronic disease or even death. Thus it is important to understand the mechanisms of inflammation in health and disease.

The necessity of obtaining direct evidence rather than indirect circumstantial evidence when investigating the mechanism of the response of the vasculature to inflammatory mediators is exemplified by the brief examples given below. For instance, in the weal and flare response induced by intradermal injection of allergen, the combination of scanning laser Doppler imaging and microdialysis has shown the weal to be caused by the focal release of histamine and prostaglandin D_2 immediately surrounding the point of injection, no mediators being detectable more than 3 mm away. In the region of the neurogenic flare, where no histamine release is found, other studies have shown there to be a direct effect of vasodilator neuropeptides [14] on the vasculature in part mediated through the activation of NOS and the generation of NO.

Compare this with the weal and flare to bradykinin. Initial pharmacological studies showed that H_1-antihistamines blocked the effects of bradykinin. As H_1-antihistamines are not antagonists at the B_2-receptor for bradykinin expressed by endothelial cells, this indirect evidence led to the conclusion that the weal and flare response was secondary to the release of histamine by the kinin. However, when microdialysis was employed in an attempt to gain direct evidence of histamine release, none was found. The implication of this is that there appears to be a complex interrelationship between the effects of vasodilator agents, probably at the level of their receptor G-proteins [15].

Finally, as to the role of NO in the cutaneous vascular response, there is now considerable evidence from the direct measurement of tissue levels of NO that the presence of NO is required for the full expression of the dilator response, suggesting that as well as having a direct effect, NO may also act synergistically with other (as yet unknown) vasodilator substances to maintain vasodilatation [16], possibly at the level of the nerve to enhance the release of the unknown neurotransmitter which mediates the response.

Concluding Remarks

The endothelial cells lining the vessels of the cutaneous vasculature share many features in common with those elsewhere in the body. They play a pivotal role in vascular homeostasis both as the functional barrier to the movement of water, solutes, and cells between the blood and the tissues and as a rich source of a wide range of vasoactive autocoids with both autocrine and paracrine actions, many of which have been implicated in the modulation of the vascular response. The cutaneous vasculature affords a unique opportunity to study microvascular function in humans in vivo and by direct interrogation of the endothelium and surrounding tissue space to elucidate its role in both health and disease.

Glossary

Cutaneous inflammatory response: The response of the skin to an injurious stimulus, classically the weal and flare response, resulting from changes in microvascular perfusion and integrity.

Laser Doppler imaging: Intravital imaging technique that measures red blood cell movement, used to evaluate vascular flux.

Microdialysis: A sampling technique used to recover water-soluble solutes from the interstitial space.

Skin microcirculation: The small blood vessels supplying the skin, highly organized into two horizontal plexuses.

References

1. Braverman, I. M. (2000). The cutaneous microcirculation. *J. Invest. Dermatol. Symp. Proc.* **5**, 3–9.
2. Clough, G. F., and Church, M. K. (2002). Vascular responses in the skin: An accessible model of inflammation. *News Physiol. Sci.*. **17**, 170–174.
3. Khan, F., and Belch, J. J. (1999). Skin blood flow in patients with systemic sclerosis and Raynaud's phenomenon: effects of oral L-arginine supplementation. *J. Rheumatol.* **26**, 2389–2394.
4. Kellogg, D. L., Jr., Morris, S. R., Rodriguez, S. B., Liu, Y., Grossmann, M., Stagni, G., and Shepherd, A. M. (1998). Thermoregulatory reflexes and cutaneous active vasodilation during heat stress in hypertensive humans. *J. Appl. Physiol.* **85**, 175–180.
5. Boutsiouki, P., Thompson, J. P., and Clough, G. F. (2001). Effects of local blood flow on the percutaneous absorption of the organophosphorus compound malathion: a microdialysis study in man. *Arch. Toxicol.* **75**, 321–328.
6. Berghoff, M., Kathpal, M., Kilo, S., Hilz, M. J., and Freeman, R. (2002). Vascular and neural mechanisms of ACh–mediated vasodilation in the forearm cutaneous microcirculation. *J. Appl. Physiol.* **92**, 780–788.
7. Swain, I. D., and Grant, L. J. (1989). Methods of measuring skin blood flow. *Phys.Med.Biol.* **34**, 151–175.
8. Groner, W., Winkelman, J. W., Harris, A. G., Ince, C., Bouma, G. J., Messmer, K., and Nadeau, R. G. (1999). Orthogonal polarization spectral imaging: a new method for study of the microcirculation. *Nat. Med.* **5**, 1209–1212.
9. Clough, G. F., Bennett, A. R., and Church, M. K. (1998). Effects of H_1 antagonists on the cutaneous vascular response to histamine and bradykinin: a study using scanning laser Doppler imaging. *B. J. Dermatol.* **138**, 806–814.
10. Schmelz, M., and Petersen, L. J. (2001). Neurogenic inflammation in human and rodent skin. *News. Physiol. Sci.* **16**, 33–37.
11. Muller, M. (2002). Science, medicine, and the future: Microdialysis. *BMJ* **324**, 588–591.
12. Groth, L. Cutaneous microdialysis—methodology and validation. (1996). *Acta. Derm. Venereol. (Stockh.) Suppl.* **76**, 1–61.
13. Newman, J. M., Di Maria, C. A., Rattigan, S., and Clark, M. G. (2001). Nutritive blood flow affects microdialysis O/I ratio for [(14)C]ethanol and (3)H(2)O in perfused rat hindlimb. *Am. J. Physiol. Heart Circ. Physiol.* **281**, H2731–H2737.
14. Schmelz, M., Luz, O., Averbeck, B., and Bickel, A. (1997). Plasma extravasation and neuropeptide release in human skin as measured by intradermal microdialysis. *Neurosci. Let.* **230**, 117–120.
15. Leurs, R., Church, M. K., and Taglialatela, M. (2002). H1-antihistamines: inverse agonism, anti-inflammatory actions and cardiac effects. *Clin. Exp. Allergy* **32**, 489–498.
16. Wilkins, B. W., Holowatz, L. A., Wong, B. J., and Minson, C. T. (2003). Nitric oxide is not permissive for cutaneous active vasodilatation in humans. *J. Physiol.* **548**, 963–969.

Capsule Biography

Geraldine Clough is a vascular physiologist whose work over many years has focused on functional responses of the microvasculature in human skin in inflammation and its assessment in vivo. Her work is currently supported by grants from the British Heart Foundation and the Wellcome Trust, as well as from industry.

Martin Church is a pharmacologist with a special interest in the mechanisms of allergic responses and the implications of these for treatment, areas on which he has published widely.

SECTION L

Reproductive Systems

Chapter 87

Vascular Transformation of Placental Cytotrophoblasts

Susan J. Fisher
University of California-San Francisco, San Francisco, California

Survival of the eutherian embryo/fetus depends on formation of a transient but vital organ, the placenta. Placentation is the first test of the differentiative and organogenic capacity of the conceptus. During this process, the placenta's specialized cells, termed trophoblasts, acquire many unusual properties. For example, one population of mononuclear cells, termed cytotrophoblasts (CTBs), deeply invades the uterine wall. Invasion accomplishes two critical goals: attaching the embryo to the uterus, and diverting maternal blood flow to the placenta, thus enabling effective gas, nutrient, and waste exchange. The molecular underpinnings of this complex process require the precise coordination of intricately timed differentiation programs executed by both placental and uterine cells (for reviews, see Refs. [1, 2]). This chapter focuses on the profound changes in trophoblast phenotype that occur during CTB invasion of the uterus. Data gathered from the studies we review point to the central importance of molecules that control the intimate relationship that develops between fetal CTBs and the maternal vasculature. These findings have important implications for understanding mechanisms of placentation and point to possible therapeutic strategies for remodeling other vascular beds.

Morphological Aspects of Human Placentation

The complex anatomy of the placenta has made studies of the entire organ difficult. Commonly the placenta is depicted as a simple, pancake-shaped sponge connecting the embryo/fetus to the uterus. This portion of the placenta, which is expelled from the uterus during delivery, is easy to obtain and, consequently, frequently studied. But the most interesting part of the placenta is rarely seen. This portion, which lies buried within the uterine wall, separates from the rest of the placenta during pregnancy termination or delivery. As a result, the only way to obtain this tissue is by uterine biopsy of the site where the placenta attached. Thus, special procedures that are similar to the methods used to obtain any other surgical specimen are required to collect this portion of the placenta.

A full understanding of the placenta can be obtained only by studying both its parts. Figure 1 diagrams these two parts and joins them into a single unit to show how they function in conjunction with modified uterine structures during human pregnancy. The placenta is made up of individual units termed chorionic villi. Each villus has a connective tissue core that contains fetal blood vessels and numerous macrophages, termed Hofbauer cells. The macrophages often lie adjacent to a thick basement membrane, which underlies a layer of CTB progenitors that give rise to all the trophoblast lineages.

The differentiation pathway that CTB progenitors take depends on their location. In *floating villi*, the CTBs fuse to form a multinucleate syncytium, the syncytiotrophoblast, that covers the villus surface. These villi are attached at only one end to the treelike fetal portion of the placenta. The rest of the villus floats in a stream of maternal blood, which optimizes exchange, across the syncytium, of substances between the mother and fetus. In *anchoring villi*, CTB progenitors detach from the basement membrane and form a column of nonpolarized mononuclear cells that invade the uterus. As a result, these villi are attached at one end to the fetal portion of the placenta and at the other end to the

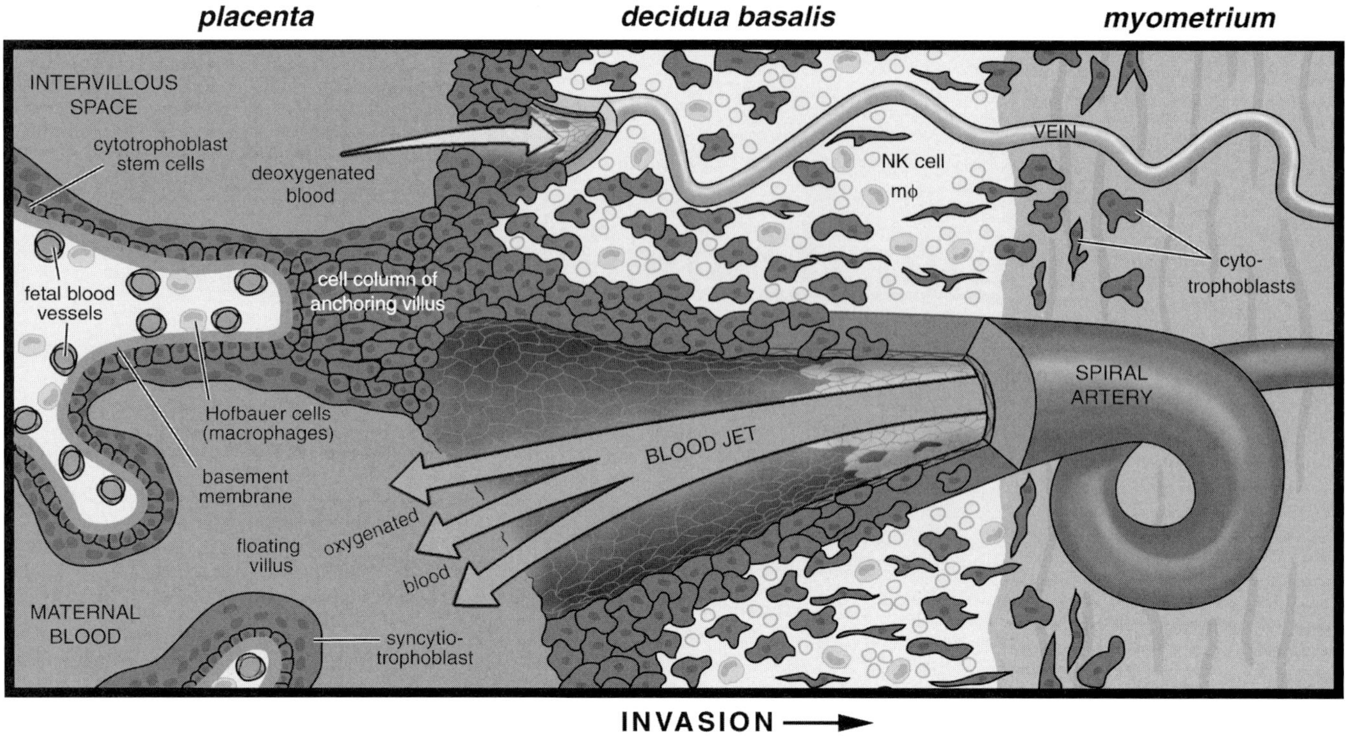

Figure 1 Diagram of a longitudinal section of the maternal–fetal interface. Fetal cytotrophoblasts from the placenta form cell columns that attach to the uterine wall. In turn, the columns give rise to a specialized subpopulation of cytotrophoblasts that invade the uterine wall (decidua and myometrium), thereby anchoring the fetus to the mother. Cytotrophoblasts also invade uterine blood vessels. In this location they replace the resident maternal endothelial and vascular smooth muscle cells. As a result they form vascular channels that carry blood to and from the placenta. (Reproduced with permission from *Biochemistry.*)

uterus. This arrangement anchors the villus to the uterine wall. Invasive CTBs rapidly traverse most of the uterine parenchyma (interstitial invasion). They also breach the uterine veins and arteries they encounter (endovascular invasion). Their interactions with veins are confined to the portions of the vessels that lie near the inner surface of the uterus, but CTBs migrate in a retrograde direction along much of the intrauterine course of the arterioles. Eventually, these fetal cells completely replace the maternal endothelial lining and partially replace the muscular wall of these vessels. This unusual process diverts uterine blood flow to the floating villi.

Molecular Regulators of CTB Pseudovasculogenesis

Our laboratory studies molecular aspects of the CTB differentiation pathway that leads to uterine invasion. Given the cells' unusual ability to line the uterine blood vessels and channel maternal blood, we hypothesized that they might replicate portions of standard vasculogenesis and/or angiogenesis programs. As a first test of this theory we examined the cells' expression of adhesion molecules during interstitial and endovascular invasion. We found that the onset of CTB differentiation/invasion is characterized by reduced staining for receptors characteristic of polarized CTB epithelial-type progenitors—integrin $\alpha6\beta4$ and E-cadherin—and the onset of expression of adhesion receptors characteristic of endothelium—VE-cadherin, IgG family members VCAM-1 and PECAM-1, and integrins $\alpha V\beta3$ and $\alpha1\beta1$. Accordingly, we termed this phenomenon pseudovasculogenesis (reviewed in Ref. [3]). Thus, as CTBs from anchoring villi invade and remodel the wall of the uterus, these epithelial cells of ectodermal origin acquire an adhesion receptor repertoire characteristic of endothelial cells. We theorize that this switch permits the heterotypic adhesive interactions that allow fetal and maternal cells to cohabit the uterine vasculature during normal pregnancy.

These findings prompted us to examine the CTB repertoire of ligands and receptors that control blood and/or lymphatic development [4]. These studies used a combination of in situ and in vitro analyses to characterize the cells' expression of vascular endothelial growth factor (VEGF) family members. CTB differentiation/invasion during the first and second trimesters of pregnancy was associated with down-regulation of VEGF receptor (VEGFR)-2. Invasive CTBs in early gestation expressed VEGF-A, VEGF-C, placental growth factor (PlGF), VEGFR-1, and VEGFR-3 and, at term, VEGF-A, PlGF, and VEGFR-1. In vitro the cells incorporated VEGF-A into the surrounding extracellular matrix; PlGF was secreted. Experiments directed toward understanding the functions of these ligand–receptor pairs showed that CTBs respond to the VEGF ligands they produce. Blocking ligand binding significantly decreased their expression of integrin $\alpha 1$, an adhesion molecule highly expressed by endovascular CTBs, and increased apoptosis. Together, the results of this study showed that VEGF family members are autocrine regulators of CTB pseudovasculogenesis and survival.

In addition, we examined the cells' expression of angiopoietin (Ang) ligands and their Tie receptors [5]. At the mRNA and protein levels, CTBs predominantly expressed Ang2. The absence of receptor expression suggested that Ang2, along with CTB-derived VEGF family ligands, could have paracrine effects on the maternal vasculature. This theory was tested by culturing uterine microvascular endothelial cells in CTB-conditioned medium, which supported their growth. Removal of VEGF-C, PlGF, and/or Ang2 from the medium caused a marked reduction in cell number as a result of massive apoptosis. We also assayed the angiogenic potential of CTB-derived factors in the chick chorioallantoic membrane assay (Figure 2). The cell-conditioned medium stimulated angiogenesis to a level comparable to that of basic fibroblast growth factor (FGF). Removal of VEGF-C, PlGF, and/or Ang2 from the medium reduced this activity by 70 to 80 percent. These data suggest that invasive human CTBs use an unusual repertoire of factors to influence the angiogenic state of maternal blood vessels and that this cross talk plays an important part in the endovascular component of uterine invasion.

Physiological Regulators of CTB Pseudovasculogenesis

We discovered that oxygen tension, which has profound effects on blood vessels, controls the switch between cytotrophoblast proliferation and differentiation/invasion [6, 7]. In these experiments, control anchoring villus explants (6 to 8 weeks of gestation) were maintained in either a 20 percent or an 8 percent O_2 atmosphere, mimicking standard culture conditions and the environment within the uterine interstitium, respectively. Other villi were cultured in 2 percent O_2, mimicking the hypoxic conditions in the fetal compartment near the uterine lumen at this time. In 20 percent and 8 percent O_2, CTBs exit the cell cycle, upregulate integrin $\alpha 1\beta 1$, and become highly invasive. In hypoxia, CTBs continue to proliferate, but they fail to express $\alpha 1\beta 1$ integrin and do not invade. These observations suggest the following model. Before CTBs access the maternal blood supply (e.g., at 10 weeks of gestation), the hypoxic environment near the uterine lumen in which early placental development occurs favors CTB proliferation, a phenomenon observed in situ. As interstitial invasion proceeds, invasive CTBs encounter a positive oxygen gradient that favors differentiation/invasion. The effects we observed are due, in part, to the actions of the von Hippel–Lindau tumor suppressor protein (pVHL), which regulates the stability of hypoxia-inducible factor-1α (HIF1α and HIF2α and thus is pivotal in cellular responses to changes in oxygen tension. The results of in vitro experiments to study VHL function in CTBs, together with the placental defects previously observed in $VHL^{-/-}$ mice, suggest that pVHL is a component of the mechanism that transduces local differences in oxygen tension at the maternal-fetal interface to changes in the biological behavior of CTBs [8].

Pre-eclampsia Is Associated with Defects in CTB Pseudovasculogenesis

Pre-eclampsia (PE) is a serious complication that affects 7 percent of first-time pregnancies in the United States (reviewed in Ref. [9]). The mother shows signs of widespread alterations in endothelial function, such as high blood pressure, proteinuria, and edema. In some cases the fetus stops growing, resulting in intrauterine growth restriction. Compounding the dangers of this condition is the fact that the maternal and fetal signs can appear suddenly at any time from mid-second trimester until term—hence the name pre-eclampsia, derived from the Greek *eklampsis*, meaning "sudden flash or development."

The PE syndrome reveals the significance of CTB pseudovasculogenesis. Floating chorionic villi in PE placentas are relatively unaffected. However, anchoring villi and the invasive CTBs they spawn show distinct anomalies. The extent of interstitial invasion by CTBs is variable, but frequently shallow. Endovascular invasion is consistently rudimentary, making it extremely difficult to find any maternal vessels that contain CTBs [10]. These anatomical defects suggested to us that in PE, CTB differentiation along the invasive pathway is abnormal. Biopsies of the uterine wall of women with this syndrome showed that invasive CTBs retain expression of adhesion receptors characteristic of the progenitor population and fail to turn on receptors that promote invasion and/or assumption of an endothelial phenotype [11].

In addition to adhesion defects, immunolocalization on tissue sections of placentas obtained from the most severely affected patients showed that CTB VEGF-A (Figure 3) and VEGFR-1 staining decreased; staining for PlGF was unaffected. Cytotrophoblast secretion of the soluble form of VEGFR-1 (sFlt-1) in vitro also increased [4], an interesting observation in light of the recent discovery that excess sFlt-1 contributes to a pre-eclampsia-like syndrome in mice [12].

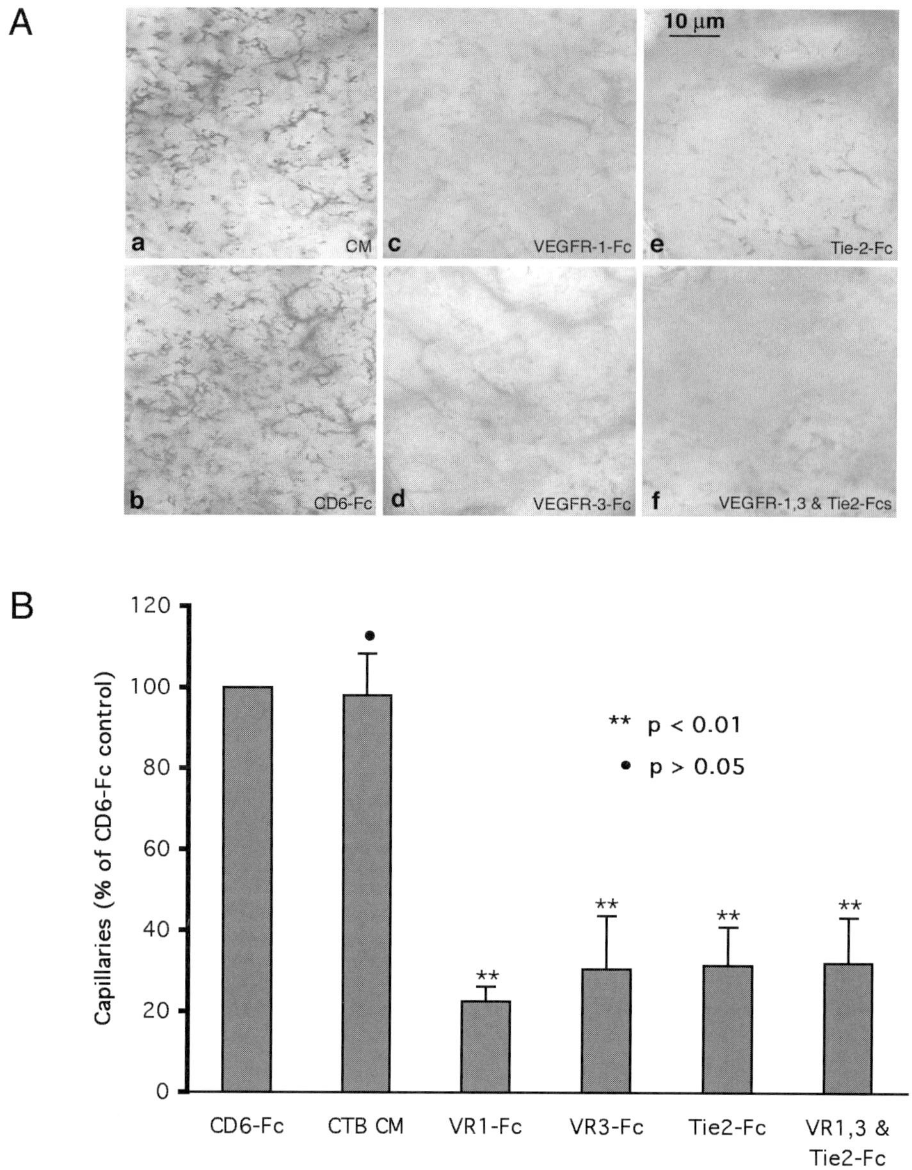

Figure 2 PlGF, VEGF-C, and Ang2 in cytotrophoblast-conditioned medium stimulate angiogenesis in vivo. The angiogenic effects of factors in cytotrophoblast-conditioned medium were assessed by using the chick chorioallantoic membrane assay. (**A**) The membranes were photographed after exposure to the samples for 48 hours. Cytotrophoblast-conditioned medium (a) and medium treated with a control CD6-Fc protein (b) induced a robust angiogenic response that was comparable to the response induced by basic FGF (data not shown). Removal of PlGF (c), VEGF-C (d), or Ang2 (e) diminished this effect by 70 to 80 percent. (f) Depletion of all three factors produced no further decrease in the angiogenic capacity of cytotrophoblast-conditioned medium. (**B**) The data were quantified by counting the number of new capillary sprouts. Data are expressed as means and standard deviations of seven experiments. The significance of the data from four separate experiments was determined by analysis of variance (ANOVA). (Reproduced with permission from *Developmental Biology*.) (see color insert)

Conclusions

In summary, recent advances have uncovered the unexpected finding that specialized fetal cells of the human placenta—CTBs—undergo a novel pseudovasculogenesis differentiation program that enables them to masquerade as the endothelial lining of maternal uterine vessels. Discovery of the regulatory mechanisms that govern this unusual transformation is offering fascinating insights into how the placenta forms. Will the results of these studies be applicable to other normal and pathogenic processes that require similar sorts of plasticity? Probably. It is now clear that some tumor cells also mimic the properties of vascular cells (reviewed in Ref. [13]), an important consideration for the design of antiangiogenesis cancer therapies. Whether or not elements of the strategy used by CTBs to remodel the uter-

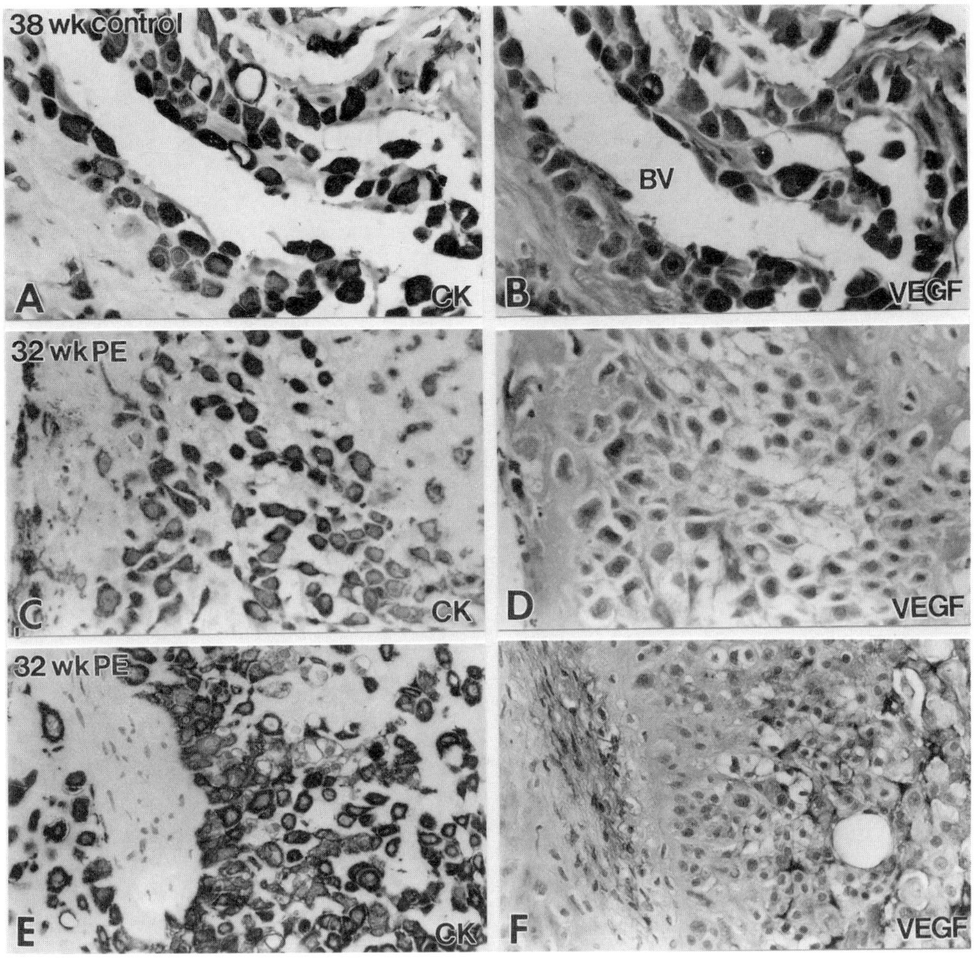

Figure 3 In pre-eclampsia, staining of invasive cytotrophoblasts with anti-VEGF-A decreased. In the third trimester of normal pregnancy (38 wk), cytotrophoblasts (CTB) in the interstitium (arrow) and in the walls of blood vessels (BV) (arrowheads) stained for VEGF-A (B). In pre-eclampsia (PE), staining of invasive cytotrophoblasts for VEGF-A was strikingly downregulated (D, F). CK, cytokeratin staining to identify trophoblasts; VS, villus stroma. (Reproduced with permission from the *American Journal of Pathology*.)

ine vasculature could be exploited for therapeutic angiogenesis of ischemic conditions is an interesting question that has yet to be addressed.

Glossary

Cytotrophoblasts: The specialized epithelial cells of the placenta.

Placenta: A transient organ that supplies food and oxygen to the fetus.

Pseudovasculogenesis: The ability to partially imitate cells that line blood vessels.

Vascular mimicry: The ability to imitate cells that line blood vessels.

Acknowledgment

Supported by HL 64597 and HD 30367.

Further Reading

1. Norwitz, E. R., Schust, D. J., and Fisher, S. J. (2001). Implantation and the survival of early pregnancy. *N. Engl. J. Med.* **345**, 1400–1408.
2. Paria, B. C., Reese, J., Das, S. K., and Dey, S. K. (2002). Deciphering the cross-talk of implantation: Advances and challenges. *Science* **296**, 2185–2188.
3. Damsky, C. H., and Fisher, S. J. (1998). Trophoblast pseudo-vasculogenesis: Faking it with endothelial adhesion receptors. *Curr. Opin. Cell Biol.* **10**, 660–666. *This article is a review of the process whereby cytotrophoblasts switch their adhesion molecule phenotype to resemble that of endothelial cells.*
4. Zhou, Y., McMaster, M., Woo, K., Janatpour, M., Perry, J., Karpanen, T., Alitalo, K., Damsky, C., and Fisher, S. J. (2002). Vascular endothelial growth factor ligands and receptors that regulate human cytotrophoblast survival are dysregulated in severe preeclampsia and hemolysis, elevated liver enzymes, and low platelets syndrome. *Am. J. Pathol.* **160**, 1405–1423.
5. Zhou, Y., Bellingard-Dubouchaud, V., Feng, K.-T., McMaster, M., and Fisher, S. J. (YEAR). Human cytotrophoblasts promote endothelial survival and vascular remodeling through secretion of Ang-2, PlGF, and VEGF-C. *Dev. Biol.* (in press).
6. Genbacev, O., Joslin, R., Damsky, C. H., Polliotti, B. M., and Fisher, S. J. (1996). Hypoxia alters early gestation human cytotrophoblast differentiation/invasion in vitro and models the placental defects that occur in preeclampsia. *J. Clin. Invest.* **97**, 540–550.
7. Genbacev, O., Zhou, Y., Ludlow, J. W., and Fisher, S. J. (1997). Regulation of human placental development by oxygen tension. *Science*

277, 1669–1672. *This article describes a role for physiological hypoxia in regulating human placental development.*
8. Genbacev, O., Krtolica, A., Kaelin, W., and Fisher, S. J. (2001). Human cytotrophoblast expression of the von Hippel–Lindau protein is down-regulated during uterine invasion in situ and upregulated by hypoxia in vitro. *Dev. Biol.* **233**, 526–536.
9. Roberts, J. M., and Lain, K. Y. (2002). Recent insights into the pathogenesis of pre-eclampsia. *Placenta* **23**, 359–372.
10. Brosens, I. A., Robertson, W. B., and Dixon, H. G. (1972). The role of the spiral arteries in the pathogenesis of preeclampsia. *Obstet. Gynecol. Annu.* **1**, 177–191.
11. Zhou, Y., Damsky, C. H., and Fisher, S. J. (1997). Preeclampsia is associated with failure of human cytotrophoblasts to mimic a vascular adhesion phenotype. One cause of defective endovascular invasion in this syndrome? *J. Clin. Invest.* **99**, 2152–2164. *This article shows that pre-eclampsia is associated with defects in the cytotrophoblast pseudovasculogenesis program.*
12. Maynard, S. E., Min, J. Y., Merchan, J., Lim, K. H., Li, J., Mondal, S., Libermann, T. A., Morgan, J. P., Sellke, F. W., Stillman, I. E., Epstein, F. H., Sukhatme, V. P., and Karumanchi, S. A. (2003). Excess placental soluble fms-like tyrosine kinase 1 (sFlt1) may contribute to endothelial dysfunction, hypertension, and proteinuria in preeclampsia. *J. Clin. Invest.* **111**, 649–658.
13. Hendrix, M. J., Seftor, E. A., Hess, A. R., and Seftor, R. E. (2003). Vasculogenic mimicry and tumour-cell plasticity: Lessons from melanoma. *Nat. Rev. Cancer* **3**, 411–421.

Capsule Biography

Susan Fisher's laboratory at the University of California–San Francisco has studied human placental development for two decades. In addition to the work described here, the group also examines mechanisms of implantation and maternal tolerance of the fetal hemi-allograft.

CHAPTER 88

The Microvasculature of the Ovary

Lawrence P. Reynolds, Anna T. Grazul-Bilska, and Dale A. Redmer
North Dakota State University, Fargo, North Dakota

The microvasculature of the ovary is relatively unique in several respects, which we will explore in this chapter. However, beyond its uniqueness, which makes studying it interesting and informative in and of itself, the ovarian microvasculature is important for practical reasons as well. Among these reasons is the fact that the ovary not only is critical to normal reproductive function but also is the site of several major pathologies, including ovarian hyperstimulation syndrome, polycystic ovary syndrome, and ovarian carcinoma. In addition, ovarian follicles and corpora lutea are some of the most rapidly growing and most vascular tissues known, and therefore they can serve as outstanding models for understanding the regulation of normal microvascular growth, development, and function. This is extremely important because as Hudlicka pointed out, "There are numerous accounts of capillary growth in developing organisms or under pathological conditions, but very little is known about capillary growth in normal adult tissues" [1]. Our discussion will be confined to the ovary of mammals; however, many similarities exist between ovarian structures across all vertebrates, although there are some striking differences as well (for an in-depth discussion of this topic, see *The Vertebrate Ovary*, which is listed under Further Reading).

General Aspects of Ovarian Function

Follicles

The ovary contains all of the female gametes, or oocytes. Mitosis of gametes is completed in females either during fetal life or shortly after birth; that is, there are no more gametogenic "stem" cells (oogonia), and the final pool of primordial oocytes is therefore present. These primordial oocytes are arrested at the diplotene, sometimes also called the dictyate, stage of the first meiotic prophase. The oocytes will remain in this arrested state until they are recruited to begin growing, and for this discussion the most important period of follicular growth occurs after puberty during the estrous/menstrual cycle.

Although not all of the primordial oocytes survive, those which do survive become surrounded by a layer of squamous epithelial cells, which are derived from the ovarian stroma and/or surface epithelium, and are termed granulosa cells. Together the granulosa cells and oocyte comprise a primordial follicle, which is enclosed by a basement membrane separating the follicular compartment from the rest of the ovary.

After the onset of puberty, a group or cohort of primordial follicles begins to grow during the follicular, or preovulatory, phase of each ovarian cycle (Figure 1). Most of the follicles that enter the growing population regress in a process known as atresia; thus, only a relatively few follicles ever reach the mature, preovulatory stage. In nonpregnant, sexually mature mammals, the ovarian cycle is usually divided into the follicular phase, which is when the follicles are growing and culminates in ovulation, and the luteal phase, which will be discussed later. The ovarian cycle is termed the menstrual cycle in primates and the estrous cycle in nonprimates, but is similar between the two groups in most respects other than the presence or absence of menses.

Soon after a primordial follicle enters the population of growing follicles, it acquires a layer of cells outside of the basement membrane. This follicular layer, termed the theca, is derived from the ovarian stroma and is composed of fibroblasts, epithelial cells, and microvessels. The theca subsequently differentiates into the theca interna, which contains most of the follicular capillaries, and the theca externa, which contains mostly arterioles and venules.

The capillaries of the theca interna form an extensive network, termed the capillary wreath, which lies close to the follicular basement membrane and completely invests the follicle (Figure 2). As ovulation approaches, the capillaries

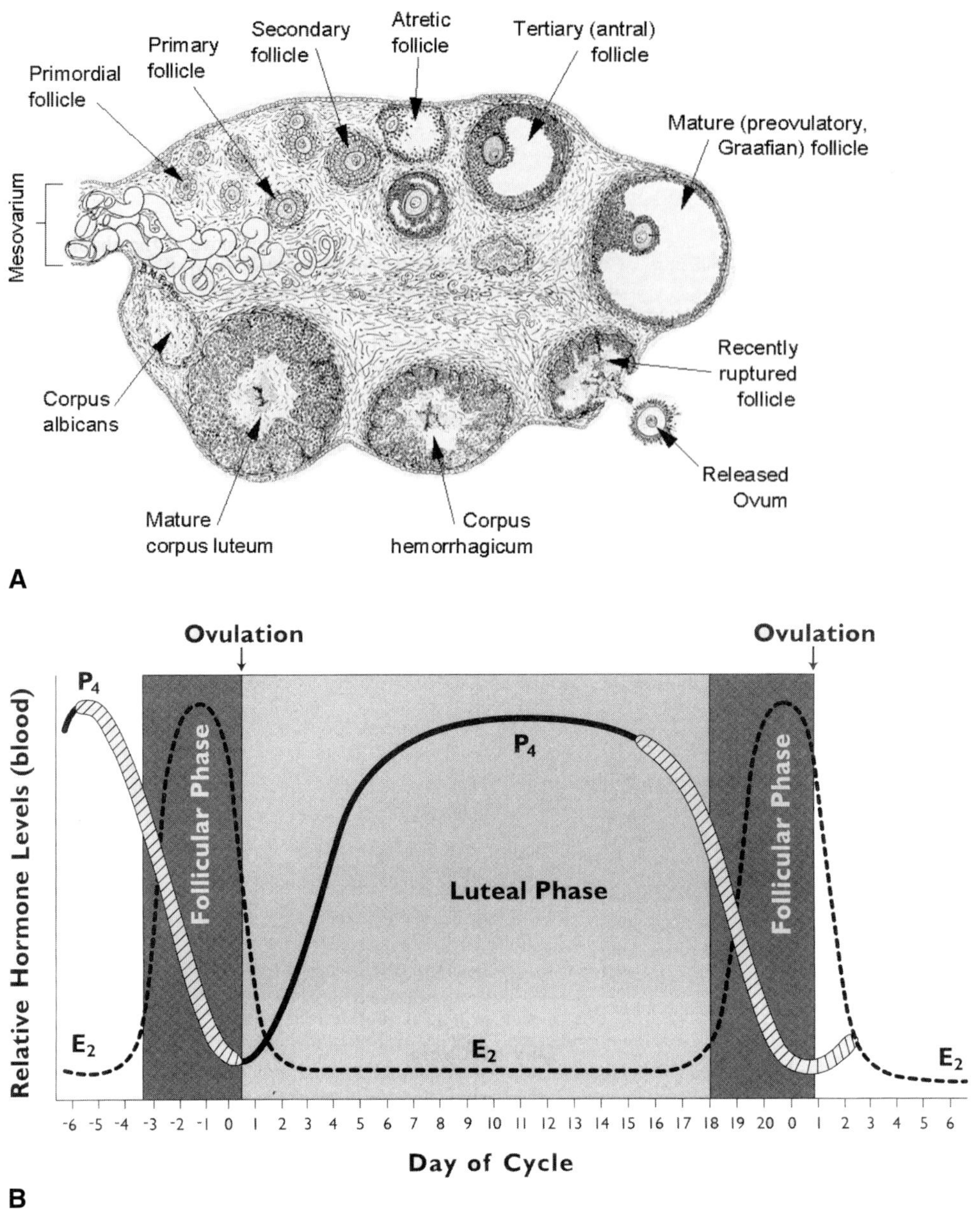

Figure 1 (**A**) Schematic depiction of the ovarian cycle, which progresses from the primordial follicle pool, to the cohort of growing follicles (primary, secondary, and tertiary or antral follicles), to the mature preovulatory follicle (also called the Graafian follicle after Regnier de Graaf, who first described them), to ovulation, to the growing corpus luteum (also called the corpus hemorrhagicum), to the mature corpus luteum, to the regressed corpus luteum (also known as the corpus albicans). The mature corpus luteum consists of several parenchymal lobules, each separated by a connective tissue tract (for further explanation, see text). An atretic follicle is shown at the top of the schematic, and the mesovarium (ovarian hilus) is shown on the left. (**B**) Schematic depiction of the stages of the estrous/menstrual cycle. During the follicular phase, the antral follicles grow and secrete estrogens (E2); during the luteal phase (after ovulation), the corpus luteum grows and secretes progesterone (P4). Note that ovulation occurs at the border between the follicular and luteal phases. Although this schematic depicts the estrous cycle of the cow, it can be applied to any estrous or menstrual cycle by adjusting the number of days and the relative length of the follicular and luteal phases; for example, in humans the menstrual cycle is approximately 28 days long and comprises a 14-day follicular phase followed by a 14-day luteal phase (menses is during the first part of the follicular phase), with ovulation again occurring at the border between the follicular and luteal phases. (**A**) Adapted with permission from Patten (*Foundations of Embryology*, 2nd edition, p. 44. New York: McGraw-Hill). (**B**) Taken with permission from Senger (*Pathways to Pregnancy and Parturition*, 1st revised ed., p. 119. Pullman, WA: Current Conceptions).

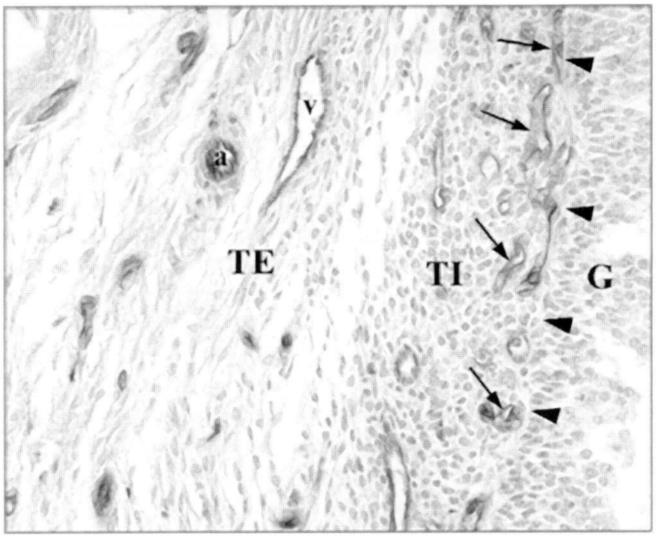

Figure 2 Histochemical staining of a preovulatory sheep follicle for BS-1 lectin, which stains microvessels. Note the dense "capillary wreath" (arrows) that is located in the theca interna (TI) adjacent to the basement membrane (delineated approximately by the arrowheads) surrounding the granulosa layer (G). Also note that the granulosa layer is avascular. An arteriole (a) and venule (v) are indicated within the theca externa (TE). The follicular fluid-filled antrum is to the right of the granulosa layer (a portion of the antrum is visible in the lower right corner of the micrograph), and the oocyte, which would be located antrally and surrounded by granulosa cells, is not visible. The section has been counterstained with nuclear fast red, and the micrograph was taken with a 20× objective. (see color insert)

forming this wreath, and especially those closest to the basement membrane, become enlarged and form extensive sinusoids (see the chapter by Ellinwood et al. in *The Vertebrate Ovary*, which is listed under Further Reading). It has been suggested that this dramatic change in morphology of the capillary wreath surrounding the follicle reflects increased permeability and preparation for invasion into the granulosa layer when the follicle ruptures at ovulation [2].

As the follicle grows the granulosa and thecal compartments become multilayered, and the follicle acquires an inner cavity, termed the antrum. The antrum is filled with follicular fluid, which is mostly a transudate of plasma but also contains secretory products from the follicular cells. The thecal microvasculature meets all of the respiratory, nutrient, hormonal, and transport demands associated with follicular development, since the granulosa layer remains avascular and is separated from the thecal compartment by the follicular basement membrane until after the follicle ruptures at ovulation.

Just before ovulation, as the growing follicles reach preovulatory size, they begin to produce large amounts of estrogens, which are ovarian steroids. In fact, the preovulatory ovarian follicles are the primary source of circulating estrogens in the female, and the theca and granulosa cells cooperate in its production. Estrogens, in turn, have an important stimulatory role in many of the primary and secondary sexual characteristics of females, including sexual behavior, and mammary and reproductive tract development, blood flow, and vascular development. For example, if one removes the ovaries, effectively removing ovarian steroids from the system, and then treats the ovariectomized animal with estrogens, uterine blood flow increases by as much as tenfold within 30 to 60 minutes, and uterine size and vascular bed volume increase by two- to fivefold within 24 hours after treatment.

The preovulatory surge of gonadotropins from the anterior pituitary, or adenohypophysis, triggers ovulation as well as the resumption of meiosis by the oocyte, a process termed meiotic maturation. Once released from the follicle, the mature but unfertilized egg is "picked up" by the oviductal fimbriae surrounding the ovary and transported to the ampulla of the oviduct where it can be fertilized.

Corpora Lutea

After ovulation, the ruptured follicular wall vascularizes and grows rapidly, and is now referred to as a corpus luteum, which reaches its mature size over a period of about 10 to 12 days. Both granulosa and thecal cells of the ruptured follicle contribute to formation of the corpus luteum and, in fact, all of the luteal microvasculature is probably derived exclusively from the thecal microvascular pericytes and endothelial cells, both of which invade the granulosa layer within a few hours after ovulation.

Early in its life span, the corpus luteum begins to secrete the ovarian steroid progesterone, and is the main source of this progestational hormone. If pregnancy does not ensue, the corpus luteum regresses at the end of the luteal phase, which lasts from a few days to a couple of weeks depending on the species. Once luteal regression occurs, progesterone secretion declines, and a new follicular phase is initiated, again culminating in ovulation (Figure 1).

In some species, such as the rat, the corpus luteum does not secrete substantial amounts of progesterone unless mating occurs. However, in all species, the corpus luteum is "rescued" by the presence of a viable conceptus in the uterus. Thus, if pregnancy is successfully established, the corpus luteum does not regress but rather is transformed into the "corpus luteum of pregnancy," and it continues to secrete progesterone, which is critical for successful establishment and maintenance of pregnancy, for the majority of the gestational period. Either somewhat before or immediately after delivery of the offspring, the corpus luteum of pregnancy regresses spontaneously.

The growth rate of the early corpus luteum is phenomenally rapid; for example, the doubling time for tissue mass and cell number is about 72 hours. Thus, during its growth phase the rate of luteal growth exceeds that of all but the most aggressive and dangerous tumors. However, unlike tumor growth, growth of the corpus luteum, including proliferation of the steroidogenic epithelial cells and the microvascular cells (vascular smooth muscle cells, endothelial cells, and pericytes), is closely coordinated and highly regulated, such that the mature corpus luteum stops growing once it reaches a species-specific size.

Thus, the corpus luteum is an outstanding model for investigating the regulation of microvascular growth in a normal, rapidly growing, highly metabolically active tissue. Understanding this process should also give us a much better understanding of what has "gone wrong" in various pathological conditions that are associated with abnormal tissue and microvascular growth, including tumor growth, various retinopathies, chronic varicose ulcers, and nonhealing fractures and wounds. In fact, we have just recently begun to apply our knowledge of microvascular growth in ovarian follicles and corpora lutea to the study of wounds in diabetics, which display delayed wound healing [3].

In addition to being species-specific, the average size of the mature corpus luteum, or the total size of the several corpora lutea present in the ovary in species in which multiple ovulations are common, is closely related to the average mature body size of the species in question. For example, the total weight of mature corpora lutea in the shrew (body weight ≈15 g) is about 1 mg, whereas that of the blue whale (body weight ≈100,000–150,000 kg) is about 1 kg. This difference in luteal size probably reflects differences in the number of granulosa and thecal cells making up the ovulatory follicle, and is thought to reflect the importance of having sufficient luteal mass to produce adequate amounts of progesterone to maintain pregnancy.

The Ovarian Microvasculature

Follicles

In the late 19th century and first half of the 20th century, several investigators noted the tremendous vascularity of ovarian follicles and corpora lutea and recognized the importance of the ovarian vasculature in supporting growth and development of these tissues. In his classic treatise, Clark [4] evaluated vascular development of the human ovary from the fetal stage through senescence and provided us with the first clear picture of the vascular architecture of the various ovarian tissues.

After evaluating vascular anatomy throughout follicular development, Clark concluded, "the vital impulse to growth in the theca interna depends not upon a maintenance of its primitive blood supply, but upon a decided increase of that supply." More recent studies have confirmed that dominant follicles (that is, growing follicles that go on to ovulate rather than regressing) have not only a more vascular theca but also an increased uptake of serum gonadotropins compared with other antral follicles, implying that they also have increased blood flow compared with other follicles. Thus, increased vascularity may be a primary determinant of follicular dominance.

These early investigators also observed the heterogeneity in follicular vascularization and suggested its importance in determining whether follicles remain healthy or become atretic. Subsequent work has supported this contention that maintenance of the follicular vasculature is important for maintaining follicular health. For example, Moor and coworkers (summarized in Ref. [5]) observed that early atretic ovine follicles will regenerate when placed in vitro, and suggested that decreased vascularity may limit access of atretic follicles to nutrients, substrates, and tropic hormones in vivo. More recently, Greenwald [6] found that reduced DNA synthesis of follicular endothelial cells was associated with reduced follicular vascularity and was one of the earliest signs of atresia. Similarly, we have observed that cessation of thecal endothelial cell proliferation, associated with a decrease in thecal vascularity, is an early event associated with atresia in bovine, ovine, and porcine follicles.

Corpora Lutea

Vascular development of the ovarian follicle becomes even more impressive after ovulation, in association with development of the corpus luteum. The capillary network of the mature corpus luteum is so extensive, in fact, that the majority of epithelial (parenchymal or steroidogenic) cells are adjacent to one or more capillaries (Figure 3). This is not surprising, since in several mammalian species most (50 to 85%) of the cells that are proliferating during growth of the developing corpus luteum are endothelial cells or capillary pericytes. In addition, morphometric studies in several species have shown that endothelial cells and pericytes constitute about 50 percent of the cells of the mature corpus luteum. In agreement with its high degree of vascularity, the mature corpus luteum also receives most of the ovarian blood supply, and not only has one of the greatest

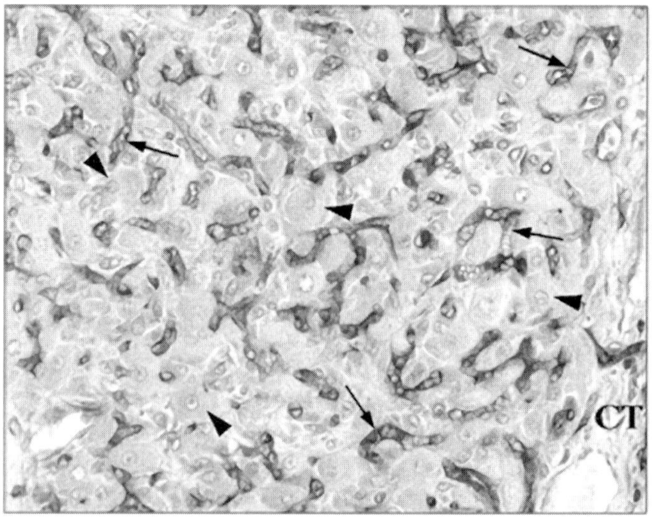

Figure 3 Histochemical staining of a mature sheep corpus luteum for BS-1 lectin, which stains microvessels. Note that nearly every steroidogenic epithelial cell (a few are indicated by arrowheads) in the parenchymal area is in contact with a capillary (a few are indicated by the arrows; some are filled with red blood cells). Also note a few larger lymphatic capillaries located in the parenchymal area (upper right and lower left of the micrograph). A portion of a connective tissue tract (CT), which is where the vast majority of the luteal arterioles and venules are located, is visible in the lower right corner. The section has been counterstained with nuclear fast red, and the micrograph was taken with a 20× objective. (see color insert)

weight-specific rates of blood flow of any known tissue but also has an extremely high rate of oxygen consumption. Ovarian blood flow and microvascular perfusion are highly correlated with progesterone secretion, which emphasizes their importance for normal luteal function, just as microvascular perfusion may be a critical factor in determining which ovarian follicles become preovulatory and acquire the ability to produce and secrete estrogens.

Conversely, inadequate luteal function has been associated with decreased luteal vascularization. Several investigators have suggested that reduced ovarian blood flow also plays a critical and active role in normal luteal regression. It is interesting to note that during luteal regression, although a portion of the capillary bed regresses, some of it remains intact and apparently functional throughout regression. This is thought to be important because the corpus luteum does not regress by necrosis but rather regresses in an orderly fashion such that the luteal tissue is resorbed within just a few days or weeks.

Unique Aspects of the Ovarian Microvasculature

The ovarian follicle and corpus luteum are relatively unique not only because they are so vascular and have such high rates of blood flow but also because of the composition of their microvasculature. For example, in the several parenchymal lobules of the mature corpus luteum, which constitute the bulk of luteal volume, the vast majority of the microvessels are blood or lymphatic capillaries, with arterioles and venules only rarely observed (Figure 3). This observation, along with physiological studies, has led to the suggestion that the luteal microvasculature comprises a "sluice flow," or relatively unregulated, system, and that blood flow is therefore regulated acutely by vasoconstriction or vasodilation of ovarian vessels external to the corpus luteum itself. This appears to be at least partially correct, but acute regulation of regional blood flow probably also occurs within the corpus luteum via intraluteal arterioles and venules that are present in the connective tissue tracts between the parenchymal lobules (Figure 3). Additionally, Niswender and colleagues (see the chapter by Ellinwood et al. in *The Vertebrate Ovary*, which is listed under Further Reading) have demonstrated substantial regional heterogeneity of luteal blood flow distribution.

Consistent with the relative lack of vasoactivity, the microvessels of the luteal parenchymal lobules do not appear to be innervated. In the follicle, nerve fibers are not present in the avascular follicular granulosa layer, whereas the arterioles and venules of both the theca interna and theca externa are densely innervated and respond to adrenergic agents. Similarly, luteal blood flow is highly responsive to adrenergic agents. In fact, ovarian blood flow is regulated primarily by periarterial sympathetic vasoconstrictor innervation within the follicular and luteal arterioles (see Ford, 1982, which is listed under Further Reading).

In addition to being composed mostly of capillaries, the microvasculature of the theca interna and corpus luteum contains a large proportion of capillary pericytes. For example, Redmer and coworkers [7] have recently found that the endothelial cell : pericyte ratio in the corpus luteum is approximately 1 : 1. In addition, although the number of lymphatic capillaries has not been quantified, the corpus luteum and ovarian follicles have an extensive lymphatic system and a relatively high rate of lymph flow per unit of tissue mass (up to 1% of blood flow), which probably reflects their high rates of tissue perfusion and blood flow.

The microvasculature of the ovarian follicle and corpus luteum also is relatively unique in that it is highly permeable. This high permeability is not surprising because of the highly fenestrated nature of the capillaries, especially in preovulatory follicles and mature corpora lutea. For example, both ferritin (molecular weight 462 kDa) and albumin (molecular weight approximately 66 kDa) readily pass through the capillaries into the interstitial fluid. A relatively high level of plasma protein in the ovarian lymphatic drainage reflects this high permeability. In fact, the ovarian lymphatic drainage transports up to one-thousandfold more protein per day, on a tissue weight specific basis, than the lymphatic drainage of the hindlimbs.

Angiogenesis in the Ovary

Ovarian Angiogenic Factors

As mentioned in the preceding sections, microvascular growth, or angiogenesis, is a critical aspect of normal follicular and luteal function. Angiogenesis in these ovarian tissues appears to be regulated ultimately by a suite of angiogenic factors common to many tissues. This suite includes the vascular endothelial growth factor (VEGF) and the fibroblast growth factor (FGF) families. Both ovarian follicles and corpora lutea produce VEGFs and FGFs and also contain receptors for these angiogenic factors. In addition, a relatively large proportion of the angiogenic activity secreted by follicles and corpora lutea can be immunoneutralized with antibodies against VEGF and FGF. Recent work by a number of investigators has confirmed that blocking VEGF or its receptors inhibits follicular and luteal growth and function (for example, see Fraser and Wulff, 2001, which is listed under Further Reading).

We also have suggested that the FGFs are involved not only in luteal angiogenesis, which occurs primarily early in the estrous cycle, but also in other aspects of luteal function. For example, we and others have shown that FGF regulates luteal progesterone production. In addition, the FGFs have been shown to inhibit cell death in several cell types. As mentioned previously, many of the larger luteal microvessels are maintained during luteal regression, and we observed an increase in FGF receptors in these vessels, which could explain how they selectively avoid cell death while the rest of the luteal tissue is resorbed [8]. We have therefore suggested that FGF may affect not only luteal cell proliferation but also luteal cell function and vascular maintenance.

As mentioned earlier, thecal pericytes invade the granulosa layer within hours after ovulation. These thecal microvascular pericytes are the major source of VEGF in the developing corpus luteum. We also have found that granulosa cells from preovulatory follicles produce a factor that promotes migration of pericytes. Thus, we have suggested that at ovulation the granulosa cells signal the pericytes to invade, and these thecal-derived pericytes may subsequently direct vascularization of the developing corpus luteum via production of VEGF.

We also have recently found that VEGF mRNA expression in cultured ovine luteal cells is increased by only about 30 percent with luteinizing hormone (LH) treatment but by 300 percent under low oxygen (O_2) conditions. These studies were based partly on the observation that gonadotropin treatment induces VEGF mRNA expression in preovulatory rat follicles and in cultured bovine luteal cells. If VEGFs were major luteal angiogenic factors, their regulation by LH would make sense because LH is an important luteotropic factor and is critical for normal luteal development and function. However, our work is consistent with the observation that O_2 is a potent stimulator of VEGF expression across a number of cell and tissue types and bolsters the concept that metabolic demand is the primary factor regulating vascular development in all tissues, including those of adults.

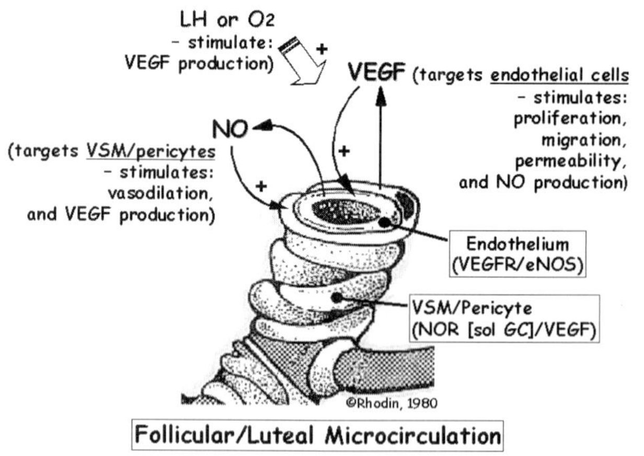

Figure 4 Working model for the regulation of ovarian follicular and luteal vascular function by VEGF and NO in sheep. This model is based on numerous in vitro and in vivo studies from our laboratories and those of others in which the major angiogenic growth factors have been not only quantified but also localized to specific tissues and cell types. In the model, vascular endothelial growth factor (VEGF) is secreted by follicular/luteal vascular smooth muscle (VSM) cells as well as capillary pericytes. The secreted VEGF acts on the endothelial cells of the follicular/luteal arterioles and capillaries, via VEGF receptor (VEGFR), to stimulate not only angiogenesis but also endothelial nitric oxide synthase (eNOS) and thus nitric oxide (NO) secretion. NO acts on the follicular/luteal VSM and pericytes, via NO receptor (NOR; soluble guanylate cyclase [sol GC]), to stimulate vasodilation and VEGF secretion. This system thereby establishes a positive feedback loop to maximize follicular/luteal angiogenesis and blood flow. Low oxygen tension seems to be the major stimulator of VEGF production, but luteinizing hormone (LH) also has a modest stimulatory effect. The schematic of the arteriole was adapted with permission from Rhodin (Architecture of the vessel wall. In *Handbook of Physiology* [D. F. Bohr, A. P. Somlyo, and H. V. Sparks, Jr., eds.], Section 2, Vol. II, p. 2. Bethesda, MD: American Physiological Society).

Angiogenic Factors in Ovarian Pathology

As mentioned earlier, a variety of pathologies of the female reproductive organs are associated with disturbances of the angiogenic process, including ovarian hyperstimulation syndrome, ovarian carcinoma, and polycystic ovary syndrome. These pathologies also are associated with altered expression of VEGFs and/or FGFs. These pathologies of the female reproductive organs represent major socioeconomic problems. For example, ovarian carcinoma often shows a poor prognosis and low survival rate and therefore is recognized as one of the most dangerous cancers in female patients. In fact, ovarian, uterine, and cervical cancers represent approximately 13 percent of new cases of cancer and 10 percent of cancer deaths in the United States, making these the fourth-leading cause of deaths due to cancer in women. In the near future, angiogenic or antiangiogenic compounds may prove to be effective therapeutic agents for treating these pathologies. In addition, monitoring of angiogenesis or angiogenic factor expression may provide not only a diagnostic tool but also a means of assessing the efficacy of these therapies.

A Model for Follicular/Luteal Angiogenesis

Recent work also has shown that nitric oxide (NO), which is primarily an endothelial product and an important local vasodilator, can stimulate VEGF production and angiogenesis. Similarly VEGF, which as mentioned previously is present in luteal perivascular cells, can stimulate endothelial nitric oxide synthase (eNOS) expression and thus NO production.

We and others recently have found that eNOS is expressed in endothelial cells of arterioles and capillaries of preovulatory follicles and developing corpora lutea. We also have shown that the follicular and luteal endothelial cells expressing eNOS often are in close physical association with perivascular cells (capillary pericytes and arteriolar smooth muscle) that express VEGF. These observations led us to propose a new model for follicular and luteal vascular function (Figure 4), involving the existence of a paracrine loop whereby endothelial cells release NO, which stimulates perivascular VEGF production, which in turn stimulates endothelial expression of eNOS. This paracrine loop would thereby serve as a feed-forward system to maximize vasodilation and angiogenesis during follicular and luteal growth and development.

Glossary

Corpus luteum: The progesterone-secreting body formed on the ovary from the wall of the ruptured follicle after ovulation; during the nonpregnant (estrous or menstrual) cycle, the corpus luteum secretes progesterone for several days but then regresses at the end of the cycle; during gestation, the corpus luteum remains viable and secretes progesterone throughout most of pregnancy.

Follicle: The ovarian structure composed of an oocyte (or egg), an avascular granulosa cell layer surrounding the oocyte and enclosed by a basement membrane, and a vascular thecal layer; both the granulosa and thecal layers are involved in biosynthesis of estrogens. At ovulation, the wall of the follicle ruptures and the oocyte is released and subsequently "picked up" by the oviduct where it is fertilized; the ruptured wall of the follicle transforms into the corpus luteum (q.v.) over the next few days.

Granulosa (stratum granulosum or granulosa layer): The avascular inner epithelial layer of the ovarian follicle, which surrounds the oocyte, is itself surrounded by a basement membrane, and is involved in biosynthesis of estrogens.

Theca (thecal layer): The outer layer of the growing ovarian follicle, which contains both epithelial and fibroblastic cells and is divided into a theca interna and a theca externa. The theca interna contains a capillary wreath adjacent to the basement membrane surrounding the granulosa layer and also contains a few arterioles and venules; the theca externa contains mostly arterioles and venules and is continuous with the ovarian stroma. The theca interna also is involved in biosynthesis of estrogens.

References

1. Hudlicka, O. (1984). In *Handbook of Physiology* (E. M. Renkin and C. C. Michel, eds.), Section 2, Vol. IV, p. 165. Bethesda, MD: American Physiological Society.
2. Redmer et al. (2001). *Biol. Reprod.* **65**, 879–889.
3. Grazul-Bilska et al. (2003). *Drugs Today* **Vol**, pp.
4. Clark (1900). *Johns Hopkins Hosp. Rep.* **9**, 593–676.
5. Moor and Seamark (1986). *J. Dairy Sci.* **69**, 927–943.
6. Greenwald (1989). *Biol. Reprod.* **41**, 175–181.
7. Redmer et al. (2001). *Biol. Reprod.* **65**, 879–889.
8. Doraiswamy et al. (1998). *Growth Factors* **16**, 125–135.

Further Reading

Augustin, H. G., Iruela-Arispe, M. L., Rogers, P. A. W., and Smith, S. K. (eds.) (2001). *Vascular Morphogenesis in the Female Reproductive System.* Cardiovascular Molecular Morphogenesis Series (R. R. Markwald, series ed.). Boston: Birkhauser. *See especially the several articles related to the ovary on pp. 107–205.*

Ford, S. P. (1982). Control of uterine and ovarian blood flow throughout the estrous cycle and pregnancy of ewes, sows and cows. *J Anim. Sci.* **55**(Suppl. 2), 32–42. *This reference provides a review of the regulation of ovarian blood flow.*

Fraser, H. M., and Wulff, C. (2001). Angiogenesis in the primate ovary. *Reprod. Fertil. Dev.* **13**, 557–566.

Jones, R. E. (ed.) (1978). *The Vertebrate Ovary.* New York: Plenum Press. *See especially the chapters by Ellinwood, Nett, and Niswender (Ovarian vasculature: Structure and function, pp. 583–614), and Jones (Evolution of the vertebrate ovary: An overview, pp. 827–840).*

Knobil, E., Neill, J. D., Greenwald, G. S., Markert, C. L., and Pfaff, D. W. (eds.) (1994). *The Physiology of Reproduction*, 2nd ed. New York: Raven Press. *See especially the chapters by Byskov and Hoyer (Embryology of mammalian gonads and ducts), Greenwald and Roy (Follicular development and its control), and Niswender and Nett (Corpus luteum and its control in infraprimate species).*

Redmer, D. A., and Reynolds, L. P. (1996). Angiogenesis in the ovary. *Rev. Reprod.* **1**, 182–192. *This reference provides a review of vascular growth and development in the ovary.*

Reynolds, L. P., Grazul-Bilska, A. T., and Redmer, D. A. (2000). Angiogenesis in the corpus luteum. *Endocrine* **12**, 1–9.

Reynolds, L. P., Grazul-Bilska, A. T., and Redmer, D. A. (2002). Angiogenesis in the female reproductive system: Pathological implications. *Int. J. Exp. Path.* **83**, 151–164. *This reference provides a review of pathologies related to microvascular dysfunction in the ovaries, uterus, and placenta.*

Reynolds, L. P., and Redmer, D. A. (1998). Expression of the angiogenic factors, basic fibroblast growth factor (bFGF) and vascular endothelial growth factor (VEGF), in the ovary. *J. Anim. Sci.* **76**, 1671–1681.

Reynolds, L. P., and Redmer, D. A. (1999). Growth and development of the corpus luteum. In *Reproduction in Domestic Ruminants IV, Proceedings of the Fifth International Symposium on Reproduction in Domestic Ruminants* (W. W. Thatcher, E. K. Inskeep, G. D. Niswender, and C. Doberska, eds.). *J. Reprod. Fertil. Suppl.* **54**, 181–191. *This reference provides a review of luteal growth and development.*

Capsule Biography

Dr. Reynolds was trained as a reproductive physiologist and works primarily in the areas of ovarian, uterine, and placental vascular function. His current research interests are the cellular and molecular regulation of follicular and luteal angiogenesis as well as uterine and placental angiogenesis; he also is involved in a collaborative effort to mathematically model development of the ovarian and utero-placental vascular beds. Dr. Reynolds is Director of the Cell Biology Center, and Codirector of the Center for Nutrition and Pregnancy, both at North Dakota State University.

Drs. Grazul-Bilska and Redmer, also at North Dakota State University, have been long-time collaborators in these research efforts.

CHAPTER 89

The Microvasculature of the Endometrium

[1]Robert D. Koos and [2]L. Christie Rockwell

[1]Department of Physiology, University of Maryland School of Medicine, Baltimore, Maryland;
[2]Department of Anthropology, Temple University, Philadelphia, Pennsylvania

Introduction

The endometrium, the inner lining of the uterus, is perhaps the single most dynamic tissue in the human body, rapidly growing, functionally differentiating, and then degenerating during each menstrual cycle (Figure 1). Growth and regression of the arteries, veins, and capillaries of the endometrial vasculature are an intrinsic part of this cycle. This occurs throughout the decades-long reproductive life span in response to a precise pattern of circulating steroid hormones produced by the ovaries. This cycling is punctuated normally only by pregnancy, during which the endometrial vasculature undergoes further extensive modification for delivery of maternal blood to the placenta. In humans and other primates, the endometrium is composed of two main layers, the functionalis and the basalis (Figure 2). The functionalis, the innermost portion surrounding the lumen, is composed of a richly vascularized stroma covered on its luminal surface by a simple epithelial cell layer. The functionalis undergoes dramatic changes in thickness (fourfold or more) during the menstrual cycle, first rapidly regenerating itself (during the first half or proliferative phase of the cycle), then functionally differentiating in preparation for implantation of a developing embryo (called the secretory phase of the cycle, in recognition of the full development and function of the epithelial cell–lined endometrial "glands"). If pregnancy does not occur, the functionalis is shed during menstruation. The basalis, the thinner, basal layer that remains after menstruation, is the foundation and tissue reservoir from which the functionalis, including its vasculature, is regenerated during the subsequent cycle.

To support the regrowth of the endometrium, and pregnancy if it occurs, a rich new vasculature regenerates within it during each proliferative phase. This contrasts to the situation in almost all other tissues of the body where blood vessels are highly stable once formed. Thus, the uterus is one of the few sites in the body in which angiogenesis, the development of new blood vessels, is a regular occurrence. In addition to the standard role of supplying oxygen and nutrients and removing wastes for surrounding cells, the endometrial microvessels also play a special role, via increases in their permeability, in the rapid regrowth of the endometrium, secretion by endometrial glands, and the process of implantation. This last role, and the factors that mediate it, will be emphasized in this review.

Anatomy of the Endometrial Microvasculature

Most studies of the endometrial vasculature have focused on the large "spiral" arteries rather than the microvasculature (Figure 2). Reference to the latter is usually limited to the statement that the spiral arteries terminate in a "capillary plexus" in the subepithelial zone. As in most tissues, however, the most fundamental function of the endometrial circulation, i.e., the delivery of nutrients and the removal of wastes, occurs at the level of the microvasculature—in the capillaries. The fact that this capillary plexus is so richly supplied by feed arteries, which penetrate the full width of the functionalis with minimal branching, is in itself an indication of its functional importance. Structurally, the subepithelial capillaries have been described as highly variable

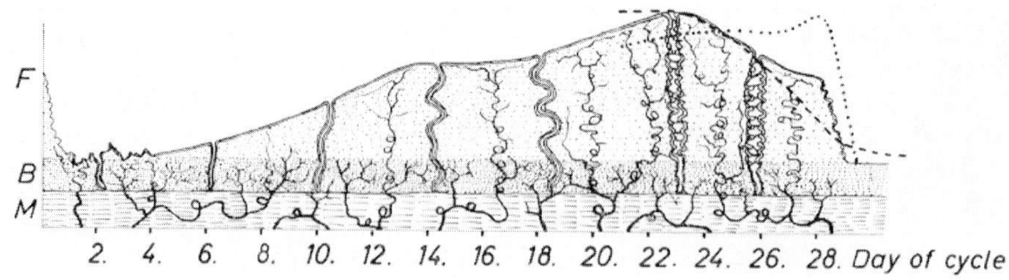

Figure 1 Diagram of endometrial development during the human menstrual cycle. F, functionalis layer; B, basalis layer; M, myometrium (modified from Ref. [1]).

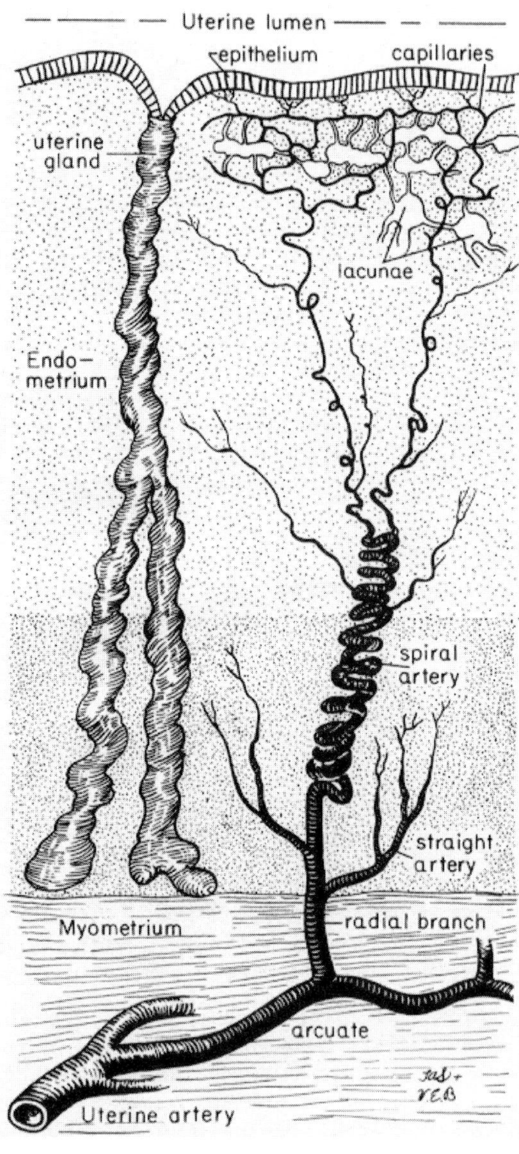

Figure 2 Diagram showing the histologic organization and major features of the endometrium and its vasculature. Although this figure shows the subepithelial capillary plexus, it does not show that a plexus also surrounds each gland (from Ref. [2]).

in form, diameter, and length [1], with numerous anastomoses interconnecting the various channels and large-diameter lacunae (Figure 3). This description suggests that flow through the plexus may be relatively slow, which would maximize exchange of bloodborne materials with the subepithelial stroma, the epithelium, and the lumen (analogous to the subcutaneous capillary plexus of the skin that facilitates the exchange of heat with the external environment). The capillaries of the plexus empty via narrow branches to larger collecting venules, which in turn empty into progressively larger veins. These eventually enter a venous plexus between the basalis and the myometrium (the outer muscular layer of the uterus).

During the cycle, epithelial cell–lined invaginations, the endometrial glands, develop and extend from the lumen deep into the functionalis layer, thereby greatly increasing the epithelial surface area of the endometrium. In humans, these glands reach their maximum development during the secretory phase of the cycle under the influence of progesterone. Glands also are present in the endometrium of nonmenstruating species, and their development is also stimulated by progesterone. Although seldom discussed, a distinct capillary plexus also envelops each gland [3]. Such a "basket" of capillaries can clearly be seen in Figure 4, surrounding a gland in the monkey endometrium [4]. Capillary branches penetrate every fold of the gland, bringing them into close proximity to the epithelial cells. This suggests that the capillaries are critical for glandular function, which is to produce and secrete or transport nutrients and other factors into the uterine lumen. Indeed, these secretions play essential roles in embryo survival, development, and pregnancy recognition even before the conceptus invades the endometrium to gain more direct access to maternal blood (see later discussion).

Mode of Endometrial Angiogenesis

Although it is clear that complete revascularization of the endometrium takes place each cycle, the exact process involved is far from clear. Histological studies to date have fallen short of clearly capturing the nature of the growth or its timing relative to the changes in the other components of the endometrium during the menstrual cycle. The regeneration of the human endometrium is a unique process, and there is little with which to compare it; menstruation occurs only in humans, some Old World primates, and a couple of other rare species, so research animal models of endometrial regeneration are quite limited (recently, Brasted et al. [5] reported on a mouse model of menstruation, based on

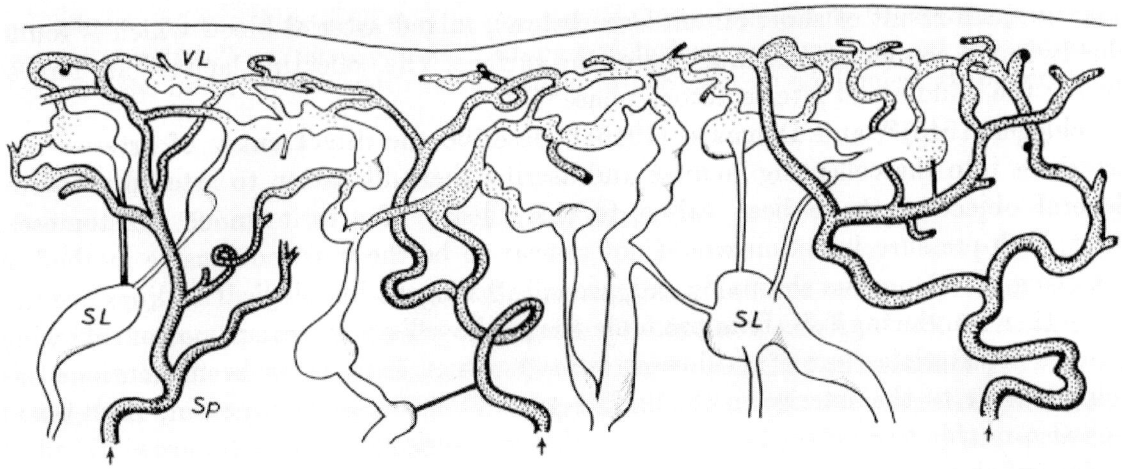

Figure 3 A diagram of the subepithelial capillary complex, showing the connecting lacunae (VL, collecting lacunae (SL), and the terminal branches of the spiral arteries (Sp) that supply the complex (modified from Ref. [1]).

Figure 4 A thick frozen section of the endometrium from a rhesus monkey at the mid-proliferative stage of the cycle immunostained with an antibody to the type 2 VEGF receptor (KDR), which is expressed almost exclusively on endothelial cells. Note the richness of small vessels in the endometrium (upper portion of image) versus the myometrium (lower quarter of the image), especially around the endometrial glands. A paraglandular capillary plexus is clearly discernible around the gland in the lower right quadrant of the endometrium (from Ref. [4]. Copyright 2000, The Endocrine Society).

progesterone treatment and rapid withdrawal, that may in part address this gap). The most extensive studies of endometrial angiogenesis have been carried out by Rogers and co-workers [6], who have concluded, somewhat surprisingly, that it does not involve the sort of progressive sprouting from existing microvessels that characterizes the angiogenesis associated with tumor growth, wound healing, and vascularization of the corpus luteum after ovulation. This is based mainly on their failure to observe sprouts or markers for sprouting endothelial cells in the endometrium, and their observation that proliferation of endothelial cells occurs primarily within the walls of existing vessels. Instead they find an elevation in average vessel segment length in the mid- to late-proliferative endometrium and an increase in the number of vessel junctions in the early to mid-secretory phase. For these reasons they suggest that the linear growth of endometrial vessels is due mainly to vessel elongation brought about by the proliferation of endothelial cells in response to mechanical (stretch) or chemical signals from the surrounding, growing endometrial tissue. At the level of the subepithelial microvasculature, they propose that new channels may arise by a process called intussusception, which, simply put, involves one vessel splitting lengthwise to give rise to two vessels. Although these processes probably are involved in the elongation of endometrial vessels, especially in menstruating species in which the endometrium becomes progressively thicker, and the remodeling of capillary beds within the endometrium, further work is needed to determine whether or not angiogenesis via endothelial cell sprouting also occurs. The subepithelial capillary plexus, after all, is completely lost with menstruation. Furthermore, injection or corrosion cast preparations of the endometrial vasculature of experimental animals, which do undergo substantial remodeling during the reproductive cycle [3], often show the presence of blind-ended capillary branches during the portion of the cycle when estrogen is rising. These are generally interpreted to be new capillary sprouts. If sprouting occurs in the endometrium of animals in which the endometrium is not lost, it seems likely that it would also occur when an entirely new vasculature is regenerated. Better methods to visualize endometrial vessels, down to the level of capillaries, in intact, high-quality endometrial samples collected at all stages of the cycle from experimental primates will be required to resolve this issue.

Factors Regulating Endometrial Angiogenesis

The development of the endometrium during the first half of the menstrual cycle—the proliferative phase—is induced by estrogen from the maturing follicle, and its full functional differentiation occurs during the second half of the cycle under the influence of progesterone, and some estrogen, from the corpus luteum. It is likely that the development of the uterine microvasculature occurs in response to the same two hormones. Although endothelial cells have been reported to express low levels of estrogen and progesterone receptors, it is unlikely that they mediate the steroid-induced angiogenesis. Rather, the growth and differentiation is induced by paracrine angiogenic factors, such as vascular endothelial growth factor (VEGF), synthesized by the epithelial cells or stromal fibroblasts in response to estrogen and/or progesterone (and possibly by hypoxia as a result of increased tissue metabolism leading to oxygen demand exceeding supply). Treatment of immature rodents with estrogen induces a very rapid upregulation of VEGF expression in the uterus [7]. Progesterone has also been shown to influence VEGF expression in several studies, and may especially induce its stromal production. As shown in Figure 4, the endometrial vessels are richly endowed with VEGF receptors.

Perhaps the best study to date of VEGF expression in the endometrium in relation to the vascular remodeling is that of Nayak and Brenner [8], in rhesus monkeys, a menstruating species. They observed a marked increase in VEGF expression in the proliferating epithelial cells during restoration of the luminal epithelium. This was matched by an increase in VEGF receptor expression in the underlying small vessels. This implicates VEGF in the repair of the endometrial microvasculature during postmenstrual healing. Hypoxia and the numerous cytokines and growth factors linked to wound healing could drive VEGF expression during this phase. Following reconstruction of the luminal epithelium, during the midproliferative stage of the menstrual cycle, there was a major peak in blood vessel growth (i.e., endothelial cell proliferation). This coincides with the rise in estrogen production by the preovulatory follicle and a peak in stromal VEGF expression. Endothelial cell labeling and VEGF expression were lower during the progesterone-dominated, secretory phase, but could still contribute to the continued lengthening of the larger feed arteries, which causes them to coil (hence, the origin of the term "spiral arteries"), as well as the final elaboration of the glandular and luminal subepithelial capillary complexes. Finally, there was an increase in VEGF expression in the glandular epithelial cells during the late secretory phase. This would be predicted to increase the permeability of the paraglandular capillaries, thereby enhancing glandular secretory capabilities.

Role of the Endometrial Vasculature in Menstruation

Menstruation is triggered by a fall in progesterone production by the ovary at the end of an infertile cycle. Constriction of blood vessels, specifically the spiral arteries, was long believed to cause the ischemic degeneration and shedding of the functionalis layer. A review of the literature, however, offers little direct support for this theory. In recent years, more emphasis has been placed on the role of proteases in the degradation of the endometrium. This theory proposes that the loss of blood vessel integrity and decreased blood flow is secondary to a large increase in protease production, triggered by the fall in progesterone, resulting in extensive proteolytic degradation of the extracellular matrix and basement membranes throughout the endometrium, including those in the walls of blood vessels [9]. In part, this theory is favored over the vasoconstriction one because of a lack of evidence that ischemia, or hypoxia, occurs prior to the major breakdown in the endometrium. However, it was recently reported that there is a large increase in VEGF expression in the superficial endometrium during the early stages of menstruation in monkeys [8]. Since this occurs in tissue that will be shed, and did not seem to be due to steroid stimulation, the most likely explanation for it is tissue ischemia and hypoxia. If so, to what degree that ischemia is due to vasoconstriction or to the general loss of structural integrity of vessels remains to be determined. The increase in VEGF at this point in time could be functionally significant. VEGF induces protease production by endothelial cells, which might further contribute to a breakdown in vessel integrity. Furthermore, the increase in VEGF was matched by an increase in stromal cell expression of VEGF receptors. Thus it is possible that VEGF may trigger a similar increase in protease production by stromal cells.

Cyclic Changes in Endometrial Microvascular Permeability

In addition to being dynamic in terms of growth and regression, the uterine microvasculature may also be one of the most variable in terms of changes in permeability in the body. These changes, which play a critical role in endometrial function, are also governed by VEGF [7, 10], which is the most potent inducer of vascular permeability known.

The first visible effect of estrogens on the rodent uterus is the rapid induction of massive stromal edema [11]. This is readily observable within a few hours as a large increase in uterine size and wet weight. These changes closely resemble those that occur in response to endogenous estrogen during proestrus in the normal adult rat. Although not as outwardly apparent as in rodents, cyclic endometrial edema also takes place in humans during both the midproliferative and midsecretory phases of the menstrual cycle [12], periods that correspond to increases in estrogen production by first the developing follicle and then the corpus luteum. Furthermore, the administration of exogenous estrogens induces stromal edema in the human endometrium, just as it does in the rat. There have been few studies of uterine edema in other species, but it is likely that it is a universal response to estrogen in mammals.

The estrogen-induced increase in permeability is preceded by a large increase in VEGF expression [7]. Its morphological basis is the formation of gaps between and fenestrae within capillary endothelial cells. VEGF can induce both gaps and fenestrations and is the only factor known to induce the latter effect [13]. Furthermore, administration of an antibody to VEGF prevents estrogen-induced uterine edema [10], conclusive evidence that VEGF is the key factor involved. Estrogen also induces a marked increase in blood flow to the uterus, probably via the induction of increased synthesis of nitric oxide, a potent vasodilator. The combination of an increase in vessel porosity (a route for fluid flow out of the blood vessel) and an increase in blood flow (which creates the pressure that drives the filtration) leads to the translocation of large amounts of fluid from the intravascular to the extravascular space, directly bathing cells of the endometrial stroma in serum.

It is now clear that this increase in permeability is far more than just an interesting biological phenomenon. Increased microvascular permeability plays a fundamental role in both normal and pathological tissue remodeling (reviewed in Ref. [14]). The infusion of the extravascular compartment of a tissue with plasma proteins, such as plasminogen and fibrinogen, leads to the breakdown of the existent extracellular matrix and formation of a provisional one upon which stromal cells, including endothelial cells, can migrate, proliferate, differentiate, and reorganize. Extravascular fibrin deposition occurs in the human endometrium during the period of maximum edema, as does a breakdown of its collagenous extracellular matrix. Increased permeability also facilitates the delivery to cells of other essential bloodborne elements, such as oxygen, essential nutrients, serum growth factors, leukocytes and platelets—with their own rich stores of growth modulators—and estrogen-carrying proteins. In the uterus, such enhanced passage of estrogen to target cells may accelerate and sustain the full range of estrogen responses.

Thus, increased permeability is probably an essential requirement for the rapid growth and differentiation of the endometrium. Consistent with this, inhibition of uterine edema with anti-inflammatory steroids significantly retards subsequent uterine epithelial cell proliferation. Furthermore, the induction of transient edema in one horn of rat uteri by ligation of the associated uterine veins (thereby increasing resistance, capillary hydrostatic pressure, and filtration) resulted in the short term in edema, which was followed within days by marked endometrial and myometrial hyperplasia [15]. The authors concluded that this cell proliferation was triggered by the experimentally-induced edema. In addition to creating the optimal environment for growth, the uterine distension caused by edema may trigger stretch-activated pathways that further stimulate the proliferation of cells.

In rodents, the interstitial fluid subsequently enters and distends the uterine lumen during estrus, which is when mating occurs. This may create the optimal environment for sperm transport and capacitation.

Structural Changes in the Endometrial Microvasculature Associated with Implantation

The initial structural change in the endometrial microvasculature upon the penetration of the implanting embryo into the stroma is actually the establishment of a totally avascular zone, called the primary decidual zone, immediately around it. Whether this involves a displacement of stromal vessels away from the embryo by decidualizing stromal fibroblasts or the regression of vessels (or a combination of the two processes) is unclear. Studies have demonstrated that this dense zone of decidual cells is an effective barrier to the diffusion of proteins larger than about 45,000 MW to the embryo. Thus, it probably serves a protective function, preventing harmful immunoglobulins and immune cells from reaching the embryo, which is immunologically foreign, until it establishes its own protective layers, while still permitting the passage of oxygen and essential nutrients from capillaries immediately outside the zone. Outside this zone, continuous growth and remodeling of the capillary bed and associated arteries and veins continues to support and accommodate the invading and growing embryo [16].

An increase in endometrial microvascular permeability and edema are critical for implantation. In rodents, this increase is triggered by the brief spike in estrogen on the fourth day after mating. The increase in permeability makes it possible to readily visualize implantation sites in rodents with the naked eye following intravenous injection of blue protein dyes [11]. Marked edema also occurs peripheral to the implanting blastocyst in primates [17]. Increased permeability immediately precedes decidualization, in which it is thought to play a causal role. Edema is probably critical for implantation for several reasons in addition to the general role it plays in the promotion of cell growth, tissue remodeling, and angiogenesis, as described earlier. The swelling of the uterine walls clasps the blastocyst in the uterine lumen, bringing it in intimate association with the epithelium. This may trigger the breakdown of the latter, placing the blastocyst into direct contact with the endometrial stroma, where it further enhances local microvascular permeability and the decidualization process. In species with multiple young, such as rats and mice, edema at the implantation sites, and not in adjacent portions of the uterus, has also been proposed to cause differential uterine elongation and growth, thereby leading to even spacing between embryos, which is likely critical as they grow to full-term fetuses.

The implantation-related increase in permeability is again VEGF dependent. Several studies have shown that VEGF and its receptors are expressed immediately around the implanting blastocyst, and we have demonstrated that administration of an antibody to VEGF to mice on the fourth day after mating completely blocks implantation [10].

The Endometrial Vasculature and Pregnancy

In humans, Old World monkeys, rodents, and rabbits, the growing placenta deeply invades the uterine wall. This invasion involves an extensive remodeling of the endometrial vasculature to create an interface with the placenta, which is itself primarily a large vascular bed (roughly the size of the liver in humans). By the 10th day of human pregnancy, the blastocyst is completely embedded in the endometrial stroma and cytotrophoblast cells, specialized stem cells that cover the surface of the conceptus, invade the endometrium, eventually penetrating it entirely, and even entering the outer third of the myometrium. During the initial stages of development, the growth of subepithelial capillaries, stimulated by factors from the embryo, are sufficient to support its growth. Further development, however, requires even greater access to maternal blood. This is accomplished when cytotrophoblasts invade the walls of the endometrial arterioles and veins, replacing endothelial and smooth muscle cells, to create new large-bore, low-resistance hybrid vessels. The arterial channels deliver maternal blood directly onto the surface of the placenta, which consists of finger-like-villi composed of a stromal core surrounded by the cytotrophoblast cells. Within the stromal core are fetal capillaries. In essence, then, the endometrial microvasculature is largely replaced by, and its function taken over by, the microvasculature of the fetal placenta. This condition persists until parturition when the placenta is shed, the endometrium is restored to the nonpregnant state, and cycling resumes as a result of the resumption of cyclic ovarian function and steroid hormone production.

Studies have shown that oxygen levels play an important role in the invasion of the cytotrophoblast cells [18]. When levels are low, due to increasing oxygen demand by the embryo, the cytotrophoblasts are highly proliferative and invasive. When they encounter and invade the blood vessel wall, however, they are exposed to the higher oxygen levels in maternal blood. This inhibits further proliferation and invasion, and rather induces the merger of the cytotrophoblasts into the multinucleate syncytiotrophoblast layer, which synthesizes large amounts of progesterone important for maintenance of pregnancy. This transformation involves the expression of endothelial cell–type adhesion molecules and other proteins typical of endothelial cells. Consistent with a vascular cell phenotype, their differentiation and survival appears to be dependent on members of the VEGF family [19].

The Endometrial Vasculature and Uterine Pathologies

The angiogenesis and changes in permeability that characterize the endometrial microvasculature may contribute to a number of serious diseases of the uterus.

Infertility: As discussed previously, changes in vascular permeability play a critical role in the preparation of the uterus for implantation and in the implantation process itself. Inadequate permeability, due to inadequacy of hormonal stimulation, the inability of the endometrium to react with appropriate VEGF expression, or failure of vessels to respond to VEGF, could give rise to a nonreceptive endometrium and failure of implantation. Infertility could also result from placental insufficiency. Hypoxia, one of the strongest inducers of VEGF expression, is believed to play an essential role in the growth of the placenta through maintenance of trophoblast cell proliferation [18]. Furthermore, knockout of HIF-1, a key regulator of VEGF expression (below), blocks placentation.

Dysmenorrhea/uterine bleeding: VEGF mediates normal estrogen-induced endometrial development and vascularization [7, 8, 10]. Imbalances in VEGF expression or activity, therefore, could contribute to abnormal uterine bleeding. This is supported by the results of a recent clinical trial of the protein hormone relaxin for the treatment of progressive systemic sclerosis [20]. Relaxin, like estrogen, is an ovarian hormone that stimulates VEGF expression in the uterus (although its role in uterine physiology is less well understood at this time). The most frequent adverse event reported during the trial was heavier-than-usual or irregular menstrual bleeding. The authors of the study attributed this effect to relaxin-induced overexpression of VEGF in the uterus. VEGF is also implicated in the formation of uterine fibroids (below), which are also a frequent cause of abnormal uterine bleeding.

Endometriosis: Endometriosis is the abnormal growth of endometrial tissue at ectopic sites in the peritoneal cavity. It is a major cause of infertility in women of reproductive age, as well as of severe pelvic pain. Establishment of the lesions presumably requires their ability to revascularize at ectopic sites. A wide range of studies have implicated VEGF in this. The increase in VEGF expression in the superficial endometrium just prior to shedding during menstruation may make fragments that enter the peritoneal cavity prone to implantation there [10]. Estrogen, relaxin, and hypoxia could all be involved in the regulation of VEGF expression by endometriotic tissue.

Leiomyoma: Uterine leiomyomas (fibroids) are benign tumors of the uterus that occur in one third of women over the age of 30. Although they originate in the smooth muscle of the myometrium rather than the endometrium, they impinge upon the endometrium and cause abnormal uterine bleeding, infertility, and pregnancy loss. They are one of the most common indications for hysterectomy. The regulation of their growth, however, is poorly understood, but estrogen is undoubtedly involved. Angiogenesis is also hypothesized to play a role in their formation, which implicates VEGF in their etiology, and the expression of VEGF has been reported to be elevated in endometrium and myometrium of leiomyoma-bearing uteri.

Uterine cancer: Expression of VEGF is significantly increased in most tumors, including uterine cancers. Circulating levels of VEGF even correlate with uterine tumor stage and burden. This strongly suggests that VEGF is an important angiogenic factor in endometrial carcinoma. Inhibition of VEGF action is being actively pursued as a means of treatment of cancer and other pathologies characterized by angiogenesis, with some agents now in phase III trials for treatment of human tumors. Such drugs could also prove useful for the treatment of noncancerous endometrial disorders such as dysmenorrhea and endometriosis.

Summary and Conclusion

It can fairly be argued that the microvasculature of the endometrium is the essential functional, as opposed to merely supportive, unit of this organ. While it certainly supports the epithelial cells and stromal fibroblasts, it also plays a special and essential role by generating extraordinary amounts of interstitial fluid at specific times—fluid that is absolutely essential for the rapid tissue repair, growth, and differentiation that prepare the endometrium to receive a fertilized egg. Capillaries also are an integral functional component of the endometrial glands, secretions of which nourish and protect the early embryo. Finally, they supply oxygen and nutrients and remove wastes for the invading, growing conceptus in early pregnancy. When they are no longer adequate for this task, the placental cells themselves take over by invading the vessels and forming low-resistance channels that open directly onto the surface of the placenta, bathing it in maternal blood. Rapid progress is being made in understanding the factors that control these complex developmental events, but much remains to be learned. Further research could reveal much about the basic mechanisms that regulate new blood vessel growth, its cessation, and vessel regression in general, knowledge that could be important for understanding and developing new treatments for several uterine pathologies as well as the angiogenesis that plays an essential role in tumor growth and numerous other diseases.

Glossary

Endometrium: The lining of the uterus that consists of the functionalis layer (lumenal surface epithelium, stroma, and glands) and the basalis layer, which is adjacent to the myometrium (the external muscular layer of the uterus) and from which a new functionalis is regenerated during each menstrual cycle.

Estrogen: The steroid hormone produced cyclically by the ovary that stimulates growth of the uterine endometrium and initiates implantation (in part, via induction of VEGF expression).

Glands: Epithelial cell–lined invaginations that extend from the uterine lumen deep into the functionalis and secrete nutrients and other substances essential for the survival, implantation, and early development of the embryo.

Spiral arteries: The major feed arteries that extend from the basalis to near the lumenal surface epithelium of the functionalis layer and give rise to the capillary networks that surround the glands and underlie the lumen; they grow, becoming progressively more convoluted, and regress again during each menstrual cycle.

VEGF: Vascular endothelial growth factor, also known as vascular permeability factor (VPF), is a protein that binds to specific endothelial cell receptors, and through them induces greatly increased microvascular permeability and new blood vessel growth (angiogenesis).

References

1. Schmidt-Matthiesen, H. (1963). Vascularization. In *The Normal Human Endometrium* (H. Schmidt-Matthiesen, ed.), pp. 208–226. New York: McGraw-Hill.
2. Blandau, R. J. (1977). The female reproductive system. In: *Histology, 4th edition*, (L. Weiss, and R. O. Greep, ed.), p 911. McGraw Hill, New York.
3. Williams, M. F. (1948). The vascular architecture of the rat uterus as influenced by estrogen and progesterone. *Am. J. Anat.* **83**, 247–308.
4. Nayak, N. R., Critchley, H. O., Slayden, O. D., Menrad, A., Chwalisz, K., Baird, D. T., and Brenner, R. M. (2000). Progesterone withdrawal up-regulates vascular endothelial growth factor receptor type 2 in the superficial zone stroma of the human and macaque endometrium: Potential relevance to menstruation. *J. Clin. Endocrinol. Metab.* **85**(9), 3442–3452.
5. Brasted, M., White, C. A., Kennedy, T. G., and Salamonsen, L. A. (2003). Mimicking the events of menstruation in the murine uterus. *Biol. Reprod.* **69**(4), 1273–1280.
6. Gargett, C. E., and Rogers, P. A. (2001). Human endometrial angiogenesis. *Reproduction* **121**(2), 181–186. *This review provides further explanation of the possible modes of endometrial angiogenesis.*
7. Cullinan-Bove, K., and Koos, R. D. (1993). Vascular endothelial growth factor/vascular permeability factor expression in the rat uterus: Rapid stimulation by estrogen correlates with estrogen-induced increases in uterine capillary permeability and growth. *Endocrinology* **133**, 829–837. *This paper was the first to demonstrate that estrogen rapidly induces the expression of VEGF in the uterus.*
8. Nayak, N. R., and Brenner, R. M. (2002). Vascular proliferation and vascular endothelial growth factor expression in the rhesus macaque endometrium. *J. Clin. Endocrinol. Metab.* **87**(4),1845–1855. *This paper is the first to clearly show the spatial and temporal patterns of VEGF expression in the primate uterus throughout the menstrual cycle, thereby providing insight into its roles in the remodeling of the endometrial vasculature.*
9. Salamonsen, L. A. (2003). Tissue injury and repair in the female human reproductive tract. *Reproduction* **125**(3), 301–311.
10. Rockwell, L. C., Pillai, S., Olsen, C. E., and Koos, R. D. (2002). Inhibition of vascular endothelial growth factor/vascular permeability factor action blocks estrogen-induced uterine edema and pregnancy in the rodent. *Biol. Reprod.* **67**, 1919–1926. *This paper shows that VEGF plays an essential role in both estrogen-induced uterine edema and implantation.*
11. Psychoyos, A. (1973). Endocrine control of egg implantation. In *Handbook of Physiology* (R. O. Greep, E. G. Astwood, and S. R. Geiger, eds.), Section 7, Vol. II, Part 2, pp. 187–215. Washington, DC: American Physiological Society.
12. Noyes, R. W., Hertig, A. T., and Rock, J. (1950). Dating the endometrial biopsy. *Fertil. Steril.* **1**, 3–25.
13. Roberts, W. G., and Palade, G. E. (1995). Increased microvascular permeability and endothelial fenestration induced by vascular endothelial growth factor. *J. Cell Sci.* **108**(Part 6), 2369–2379.
14. Dvorak, H. F., Brown, L. F., Detmar, M., and Dvoark, A. M. (1995). Vascular permeability factor/vascular endothelial growth factor,

microvascular hyperpermeability, and angiogenesis. *Am. J. Pathol.* **146**, 1029–1039.
15. Sorger, T., and Soderwall, A. (1981). The aging uterus and the role of edema in endometrial function. *Biol. Reprod.* **24**(5), 1135–1144.
16. Goodger, A. M., and Rogers, P. A. (1993). Uterine endothelial cell proliferation before and after embryo implantation in rats. *J. Reprod. Fertil.* **99**(2), 451–457.
17. Enders, A. C., Welsh, A. O., and Schlafke, S. (1985). Implantation in the Rhesus monkey: Endometrial responses. *Am. J. Anat.* **173**, 147–169.
18. Genbacev, O., Zhou, Y., Ludlow, J. W., and Fisher, S. J. (1997). Regulation of human placental development by oxygen tension. *Science* **277**(5332), 1669–1672.
19. Zhou, Y., Genbacev, O., and Fisher, S. J. (2003). The human placenta remodels the uterus by using a combination of molecules that govern vasculogenesis or leukocyte extravasation. Ann. N. Y. Acad. Sci. **995**, 73–83. *This paper reviews the process by which fetal cells called cytotrophoblasts invade endometrial blood vessels and assume the function of vascular endothelial cells.*
20. Unemori, E. N., Erikson, M. E., Rocco, S. E., Sutherland, K. M., Parsell, D. A., Mak, J., and Grove, B. H. (1999). Relaxin stimulates expression of vascular endothelial growth factor in normal human endometrial cells in vitro and is associated with menometrorrhagia in women. *Hum. Reprod.* **14**(3), 800–806.

Further Reading

Dvorak, H. F., Nagy, J. A., Feng, D., Brown, L. F., and Dvorak, A. M. (1999). Vascular permeability factor/vascular endothelial growth factor and the significance of microvascular hyperpermeability in angiogenesis. *Curr. Top. Microbiol. Immunol.* **237**, 97–132.

Hastings, J. M., Licence, D. R., Burton, G. J., Charnock-Jones, D. S., and Smith, S. K. (2003). Soluble vascular endothelial growth factor receptor 1 inhibits edema and epithelial proliferation induced by 17beta-estradiol in the mouse uterus. *Endocrinology* **144**(1), 326–334. *This recent paper directly links estrogen-induced, VEGF-mediated edema to the subsequent growth of endometrial epithelial cells.*

Capsule Biographies

Dr. Koos is a Professor and Graduate Program Director in the Department of Physiology at the University of Maryland School of Medicine. His research, funded by the NICHD and NCI, focuses on the regulation of VEGF expression in the uterus and ovary by estrogen and hypoxia. He has served on numerous NIH review panels, the editorial boards of the journals *Endocrinology* and *Biology of Reproduction*, and the FASEB Board of Directors and Public Affairs Executive Committee.

Dr. Rockwell is an Assistant Professor and Director of the Biological Anthropology Research Laboratory at Temple University. Her research includes projects on ovarian function and implantation and is supported by grants from the NIH and Temple University.

CHAPTER 90

The Vasculature of the Normal Mammary Gland: HIF-1-Independent Expansion and Regression

Tiffany N. Seagroves and Randall S. Johnson
University of California San Diego, La Jolla, California

Introduction

In adults angiogenesis is primarily confined to the female reproductive system, including the uterus, the ovary, and the breast. More recently, it has become clear that hypoxia is a potent inducer of angiogenesis, and thus can represent a physiological stimulus within tumors for neovascularization, an essential feature of tumor growth as well as metastasis. The normal rodent mammary gland is an excellent model system for studying the process of angiogenesis and vascular remodeling because it exhibits controlled expansion and regression of blood vessels over the course of pregnancy, lactation, and involution.

As the epithelial cell population of the mammary gland expands during pregnancy to prepare for lactation, there is also a concomitant expansion of the mammary-associated vasculature. During the first half of pregnancy, the vasculature expands by sprouting angiogenesis, effectively doubling the number of blood vessels. During the second half of pregnancy, the vasculature expands through nonproliferative or intussusceptive angiogenesis, dilating at parturition to facilitate nutrient exchange. Upon pup weaning (involution), the vasculature regresses along with the mammary epithelium through an undefined process that involves capillary collapse. The mechanisms responsible for the angiogenic switch during breast tumorigenesis remain undefined.

Overview of Mammary Gland Development

In contrast to other organs, the majority of mammary gland development occurs postnatally, facilitating investigation of how the mammary vasculature develops in conjunction with the epithelium. The mammary gland is primarily composed of epithelial cells, which function to synthesize milk at lactation, myoepithelial cells, which contract to express milk at lactation, and the stroma (or adipose). The relative proportions of these cell types change over the course of the development. The nulliparous (virgin) gland is comprised of approximately 9 percent epithelial cells, whereas the lactating gland contains about 90 percent epithelial cells (Figure 1). To date, the function of the epithelial cell population has been the primary focus of mammary gland studies in transgenic and gene-deleted mice. Recently, however, several laboratories have begun to analyze the contribution of other cell types within the gland using genetic models; included in these studies have been the myoepithelium and immune cells. Thus far, the effects of endothelial cell-specific gene deletion have not been analyzed in the context of normal or tumor mammary epithelial cell biology. Furthermore, it remains unclear if unique molecular markers are expressed in either normal or tumor-associated mammary gland vasculature.

Postnatal development of the epithelium consists of four tightly regulated stages: ductal outgrowth into the stromal

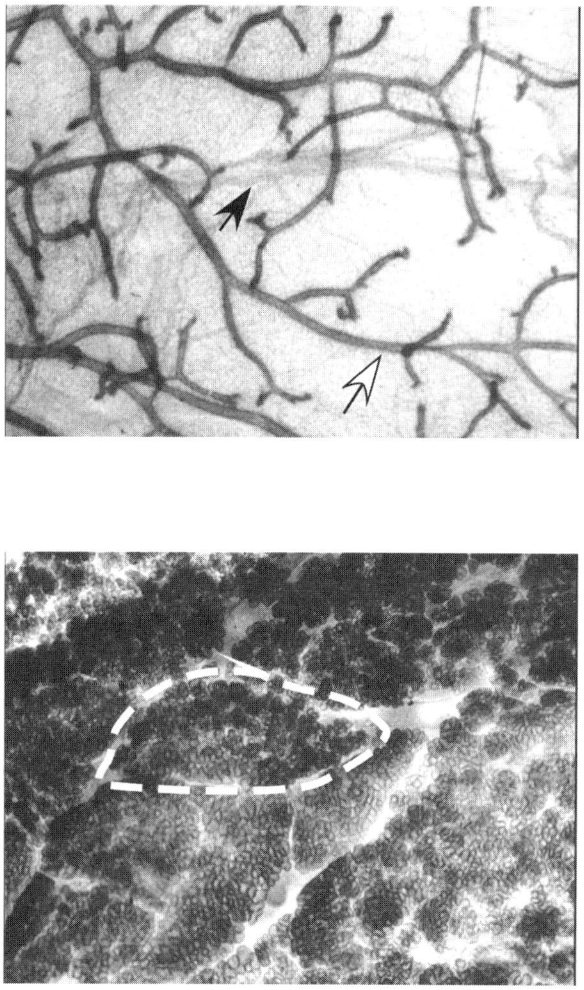

Figure 1 The gross histology of the mammary gland; hematoxylin-stained whole mount preparations from nulliparous **(a)** or lactating **(b)** glands observed at the same magnification (40×). In the mature nulliparous mammary gland, the mammary epithelium is organized into a ductal tree, which is often described as "naked" or void of alveolar development. The open arrow indicates a duct, whereas the black arrow points to the paler stained structure indicative of a large vessel. At lactation, the majority of the mammary fat pad is filled with epithelial cells, which are organized into lobules (white dotted circle) that contain individual alveoli, the polarized, secretory units of the gland that produce milk. Following involution, the mammary gland returns to a "nulliparous-like" state containing primarily ducts.

fat pad from 3 to 9 weeks of age, lobuloalveolar proliferation and differentiation during pregnancy, synthesis and secretion of milk at lactation, and involution of the secretory epithelium following weaning. Each stage depends on a critical balance among proliferation, differentiation, and apoptosis. Ductal morphogenesis progresses via a balance between proliferation and apoptosis within multilayered club-shaped structures known as terminal end buds (TEBs). As the ducts approach the edges of the fat pad, the TEBs disappear, signaling the end of ductal morphogenesis (Figure 1a). The virgin gland remains relatively quiescent until the onset of pregnancy, or the administration of exogenous hormones such as estrogen (E) and progesterone (P).

Pregnancy induces proliferation of the secretory units of the mammary gland, the alveoli, which originate from ductal progenitor cells and proliferate to fill the entire stromal fat pad at lactation (Figure 1b). During lactation, the secretory epithelium produces and secretes milk. At involution, which peaks at day four following pup removal, extensive tissue remodeling and apoptosis of the secretory epithelium occurs until the gland contains the simple ductal network observed in the virgin animal; relatively little epithelium is present by day 10 of involution. A concise review of the important genes that regulate mammary gland development at each of these stages can be found in Ref. [1].

Tools to Assay Angiogenesis and Microvessel Density in the Mammary Gland

Historically, to look at the gross structure of the mammary vasculature, vascular corrosion casts were prepared from mammary tissues isolated over the course of development, a process that destroyed the epithelium and the stroma. These corrosion casts were then subjected to analysis by scanning electron microscopy (SEM). A more recent technique preserves the relationship of the epithelium and mammary-associated vasculature. In this procedure tomato lectin, which binds to the interior of blood vessels, is injected intravenously into live mice followed by perfusion with fixative. Tissues of interest may then be harvested, followed by preparation of thick frozen sections, and subsequent staining for any other marker of interest, such as phalloidin (which stains actin networks) or E-cadherin. Following confocal microscopy a highly refined three-dimensional reconstruction of the vascularization of the mammary epithelium may be visualized, as demonstrated for a late pregnant gland in Figure 2a.

A classic morphological technique to quantify the fine microvessels surrounding alveoli was to perfuse rodents with India ink, to section the mammary gland, and to count the dark spots corresponding to the ink. More recently, staining with anti-CD31 antibodies (Figures 2b–d) and performing Chalkley counts has been accepted as a reproducible, quantitative measure of microvessel density (MVD). Finally, to look at the fine ultrastructure of individual endothelial cells, such as pinocytotic vesicles, microvilli, and fenestrations, transmission electron microscopy (TEM) has been performed on thin sections of intact fixed glands.

Angiogenesis during Normal Mammary Gland Development Is Not HIF-1 Dependent

The architecture and patterning of the vasculature shift dramatically from nulliparous to pregnant to lactating to involuting mice. Relatively few vessels are present in the nulliparous gland, and these vessels either run parallel with

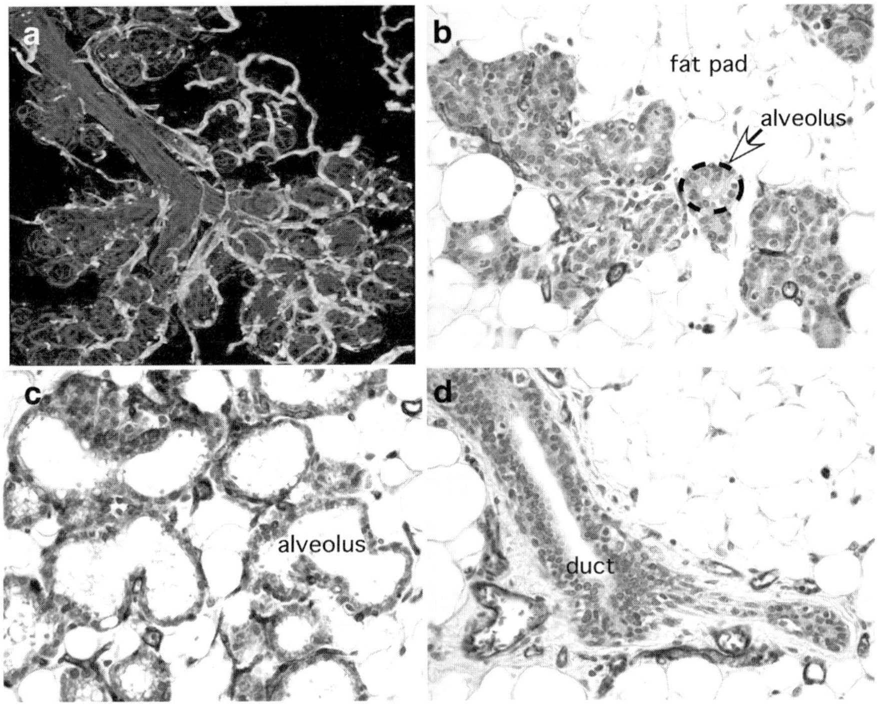

Figure 2 Tools to visualize the mammary-associated vasculature. **(a)** A three-dimensional composite image of a lectin and phalloidin-stained mammary gland harvested in the last third of pregnancy. The pregnant host was injected intravenously with lectin-FITC (green), the inguinal mammary gland was embedded in OCT and sectioned at 50 µm, and then poststained with phalloidin–Alexa Red 455 (image courtesy of Bryan Welm and Jeffrey Rosen, Baylor College of Medicine). Note the network of vessels enveloping the alveoli and the vessels that run in parallel with and encircle the ducts. To observe the fine microvessel structure, anti-CD31 immunohistochemistry (blue color) was performed on zinc-fixed paraffin sections counterstained with nuclear fast red **(b–d)** isolated from mice at day 15 of pregnancy **(b)**, day 2 of lactation **(c)**, or from nulliparous mice **(d)** (staining courtesy of Debbie Liao, UCSD). (see color insert)

or encircle the ductal tree (Figure 2d). The vasculature rapidly expands during pregnancy. Yasugi et al. demonstrated by India-ink perfusion that the vascular density of the rat mammary gland increased twofold from early (day 5) to mid-pregnancy (day 10) [2]. Matsumoto and Djonov have demonstrated that in the first third of pregnancy, vessels expand by capillary sprouting, in which endothelial cells migrate, proliferate, and form tubes [3, 4]. These sprouts ramify and anastomose with each other and will organize into distinct basket-like uniformly sized networks surrounding the alveoli by the second third of pregnancy (Figures 2a and b). Toward the end of pregnancy, capillary sprouting decreases and the vessels begin to expand instead by transluminal pillar formation that is independent of endothelial cell proliferation, a process known as intussusceptive microvascular growth (IMG), or intussusception [4]. The highly organized, homogenous structure of the mammary-associated vasculature observed during pregnancy is altered during lactation, at which time the vessels begin to dilate and to become more tortuous, in a manner similar to that observed in tumor vessels. At an ultrastructural level, Matsumoto observed by TEM that the length of microvillus processes and the number of pinocytotic vessels, indicative of active secretion, increased over the course of pregnancy compared to the nulliparous gland. Finally, there is controlled regression/collapse of the capillary network during involution. Peak capillary regression occurs at day 6 of involution, slightly delayed compared to peak epithelial cell death at day four of involution [4]. The critical steps in mammary gland angiogenesis are summarized in Figure 3.

Together these observations suggest that expansion and regression of the vasculature is under the same hormonal control as alveolar proliferation. This hypothesis is supported by the observations of Soemarwoto and Bern in 1958, and of Matsumoto in 1992, that administration of E+P to ovariectomized mice promoted development of capillaries as well as the epithelium and that prolactin augmented this effect [5, 6]. Furthermore, epithelium-free or "cleared" fat pads fail to induce angiogenesis, indicating that there is a necessary paracrine relationship between the epithelium and the vasculature. Therefore, signals from the epithelium control both the expansion and regression of the mammary-associated vasculature.

But, the question remained does hypoxia, a classical inducer of angiogenesis, also play a role in normal mammary gland vasculature development? Several laboratories

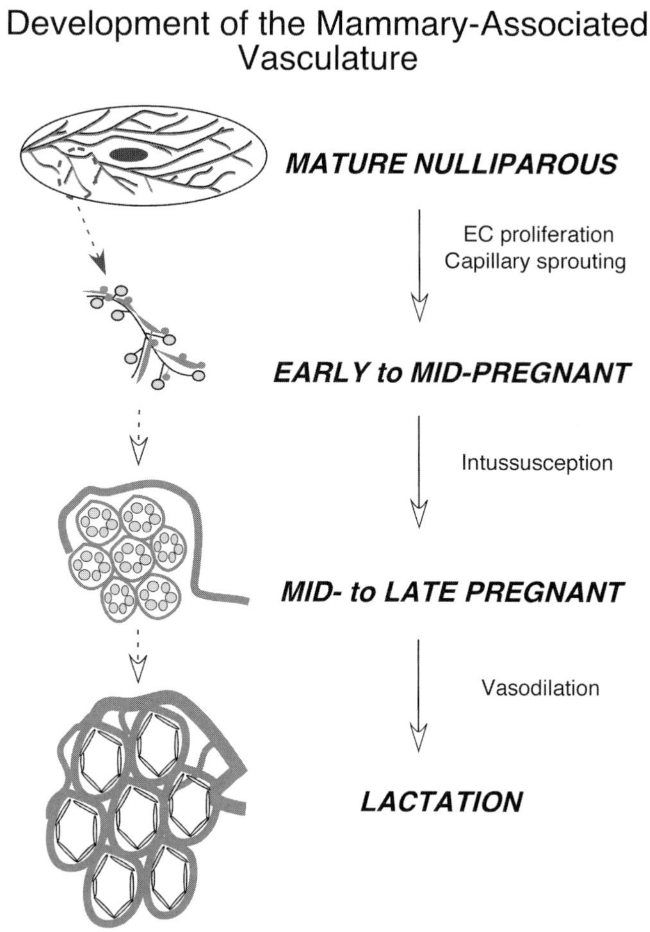

Figure 3 Cartoon representation of the various stages of mammary-associated vasculature development in conjunction with epithelial cell development. In nulliparous mice, the ducts (black lines) do not contain any alveolar structures (indicated in yellow), and the vasculature (indicated in red) runs in parallel with the ductal tree. In the first half of pregnancy, the vasculature expands by rapid endothelial cell (EC) proliferation and capillary sprouting, doubling from day 5 to day 10 of pregnancy in the rat mammary gland. In the latter half of pregnancy, when the alveoli have organized into lobuloalveolar structures, the vasculature expands by nonproliferative intussusceptive angiogenesis. At lactation, the epithelial cells become flattened since the alveoli are expanded and engorged with milk and the vessels dilate and become more tortuous. Finally, the vasculature regresses along with the epithelium during involution. The entire process will repeat over the course of each successive pregnancy. (see color insert)

have demonstrated that the transcription factor hypoxia-inducible factor-1 (HIF-1) is a master regulator of oxygen homeostasis. HIF-1 is a heterodimeric transcription factor complex induced under decreased oxygen tensions that is composed of two subunits: the aryl hydrocarbon receptor nuclear translocator (ARNT or HIF-1β) and HIF-1α, the oxygen-responsive subunit. One of HIF-1's direct transcriptional targets is vascular endothelial growth factor (VEGF). Although HIF-1α is a key component in the hypoxically induced regulation of VEGF, it should be noted that in both the mammary epithelium and stroma there is an increase in VEGF mRNA in response to estrogen and progesterone treatment. VEGF may be transcribed into three isoforms from a single promoter. This process produces three isoforms in the mouse: $VEGF_{120}$, $VEGF_{164}$, and $VEGF_{188}$. These isoforms differ in their increasing affinities for the extracellular matrix and decreasing solubility, respectively. Recent studies have shown that mammary epithelial cells express the $VEGF_{164}$ and $VEGF_{120}$ isoforms, with expression increasing twofold during pregnancy and lactation, whereas the stroma constitutively expresses $VEGF_{188}$ [7].

Given these relationships, we recently investigated the effect of mammary epithelial cell–specific conditional deletion of HIF-1α in pregnant mice using Cre/lox technology. Although deletion of HIF-1α impaired mammary epithelial cell differentiation and milk metabolism, to our surprise, there was no difference in vasculature density, patterning, vessel diameter, MVD, or VEGF mRNA expression between wild type and HIF-1$\alpha^{-/-}$ glands [8]. These results suggest that either pregnancy hormones are sufficient to induce angiogenesis, or that hypoxia is not a key driver of the angiogenesis and remodeling that occurs during normal mammary gland development. It is possible that there is a switch to a hypoxia-dependent mechanism during breast tumorigenesis.

Angiogenesis in Breast Cancer

Breast cancer is the most frequently diagnosed noncutaneous cancer of women in the United States, projected to affect 1 in 8 women over their lifetimes. All solid tumors must recruit a blood supply from neighboring vessels through angiogenesis in order to grow beyond a small size, known as the angiogenic switch. However, in contrast to normal development, which produces regular, controlled expansion of functional vessels, solid tumors develop tortuous, leaky vessels with disrupted endothelial linings, creating a hypoxic microenvironment. It has been demonstrated that the normal breast has relatively low angiogenic activity. For example, normal breast tissue implanted into nude mouse hosts induces a mild angiogenic response, whereas breast tumor tissue will induce a potent angiogenic response [9]. Therefore, understanding the factors controlling the angiogenic switch in breast cancer is clinically relevant.

Several studies have reported that increased microvessel density is an independent prognostic factor in several tumor types, including the breast. Given HIF-1's relationship to angiogenesis, it is not surprising that HIF-1α protein was found in 1999 to be upregulated in a variety of human tumors and their metastases [10]. A subsequent study by Bos et al. followed in 2001 [11], which investigated the correlation between the levels of HIF-1α overexpression and other prognostic factors of breast tumors, including proliferation rates, VEGF expression, MVD, expression of estrogen receptor (ER), and p53 expression. As in the previous study, it was found that HIF-1α overexpression was detected in the majority of ductal carcinoma in situ (DCIS) lesions, as well as in all poorly differentiated invasive carcinoma samples.

In contrast, HIF-1α expression was below detection levels in normal breast tissue or ductal hyperplasias. Moreover, in DCIS lesions, HIF-1α overexpression was statistically significantly associated with increased tumor MVD, high levels of proliferation, strong expression of VEGF, and ER positivity, but not p53 expression.

Several rodent mammary tumor models have been well characterized during the stages of mammary tumorigenesis. The most commonly used transgenic mouse models utilize the mouse mammary tumor virus (MMTV) long terminal repeat (LTR) to drive transgene expression in the mammary epithelium. Angiogenesis in transgenic mammary tumors has been investigated by corrosion casting using the MMTV-neuT model, which produces constitutive expression of activated rat ErbB2–2/Neu, producing mammary tumors with almost 100 percent penetrance [4]. Evidence of sprouting angiogenesis and intussusception was observed in these tumors, with angiogenesis occurring preferentially at the tumor border and intussusception occurring in the tumor center [4]. Because intussusception is not dependent on endothelial cell proliferation, it is likely that the centers of large breast tumors would be resistant to standard cytotoxic treatments.

Therapeutic Implications

Several lines of evidence indicate that hypoxia induces growth arrest of a variety of normal as well as transformed cell types both in vivo and in vitro. In addition, the percentage of proliferating cells in a tumor decreases as the distance from blood vessels increases. These observations are relevant to cancer therapy since sensitivity to radiation damage depends on the presence of oxygen and since most chemotherapeutic drugs are targeted against proliferating cells. Therefore, hypoxic areas of tumors in which HIF-1α is expressed should be the most resistant to cancer therapy. In support of this hypothesis, recent studies of the response of HIF-1α null transformed mouse embryonic fibroblasts to cytotoxic agents found that the HIF-1α null cells are more susceptible to treatment with carboplatin, etoposide, and ionizing radiation than HIF-1α wild-type cells both in vitro and in vivo [12].

Given the widespread nature of hypoxia in tumors, there has been impetus to search for tumor-specific and hypoxia-responsive cytotoxins to augment or replace current therapies. One possibility currently in testing is the creation of a vector in which a prodrug enzyme is placed under control of a hypoxic response element (HRE), the HIF-1 binding site. Upon subsequent delivery of the vector to the tumor site, the hypoxic environment would result in preferential expression of the construct under control of HIF-1. With respect to breast cancer treatment, other gene therapy–based approaches are being developed in which killing agents are induced under control of both an HRE and an estrogen response element (ERE), potentially increasing the specificity of the response.

Summary and Future Directions

In order to better develop targeted therapies so that they are specific for breast tumors, it is critical to know how the vasculature develops in the normal breast and how it becomes altered to induce hypoxia during breast tumorigenesis. Further studies that seek to determine the localization and timing of expression of angiogenic factors and their regulators/receptors in the epithelium, endothelium, myoepithelium, and stroma should begin to clarify the important mediators of mammary gland angiogenesis. With the widespread availability of well-described, tissue-specific conditional gene deletion and tumor mouse models, the genetic tools are now available to begin comparing the effects of deletion of each putative angiogenic factor in the mammary epithelium or the endothelium. The ultimate goal is to be able to distinguish how these two cell types interact with each other in normal development as well as during mammary tumorigenesis.

Glossary

Cre/loxP conditional gene deletion: A gene deletion strategy in which a portion of a gene, flanked by bacteriophage P1 recombination target sites (loxP sites), will be deleted upon introduction of the Cre recombinase protein. In our case, the HIF-1α conditional mouse created in our laboratory was bred to a transgenic mouse expressing Cre under control of the MMTV promoter (MMTV-Cre, Jackson Labs), which targets Cre expression preferentially to the mammary epithelium.

Ductal carcinoma in situ (DCIS): The most common early lesion observed in breast cancer in which the epithelial cells have proliferated beyond the normal single layer of epithelium, but not have not yet penetrated the integrity of the basement membrane. The analogous structure in the mouse is referred to as the hyperplastic alveolar nodule or HAN.

Ovariectomy: A surgical procedure to remove both ovaries from female mice in order to deplete circulating levels of steroid hormones responsible for initiating alveolar development.

References

1. Hennighausen et al. (2001). *Dev. Cell* **1**, 467–475.
2. Yasugi et al. (1989). *Arch. Histol. Cytol.* **52**, 115–122.
3. Matsumoto et al. (1992). *J. Vet. Med. Sci.* **54**, 937–943.
4. Djonov et al. (2001). *Microsc. Res. Techn.* **52**, 182–189.
5. Soemarwoto et al. (1958). *Am. J. Anat.* **103**, 403–435.
6. Matsumoto et al. (1992). *J. Vet. Med. Sci.* **54**, 1117–1124.
7. Hovey et al. (2001). *Mol. Endocrinol.* **15**, 819–831.
8. Seagroves et al. (2003). *Development* **130**, 1713–1724.
9. Lichtenbeld et al. (1998). *Int. J. Cancer* **77**, 455–459.
10. Zhong et al. (1999). *Cancer Res.* **59**, 5830–5835.
11. Bos et al. (2001).
12. Unruh et al. (2003). *Oncogene* **22**, 3213–3220.

Further Reading

Brown, J. M. (2000). Exploiting the hypoxic cancer cell: Mechanisms and therapeutic strategies. *Mol. Medicine Today* **6**, 157–162.

Calvo, A., Yokoyama, Y., Smith, L. E., Ali, I., Shih, S.-U., Feldman, A. L., Libutti, S. K., Sundaram, R., and Green, J. E. (2002). Inhibition of the mammary carcinoma angiogenic switch in C3(1)/SV40 transgenic mice by a mutated form of human endostatin. *Int. J. Cancer* **101**, 224–234.

Goonewardene, T. I., Sowter, H. M., and Harris, A. L. (2002). Hypoxia-induced pathways in breast cancer. *Microsc. Res. Techn.* **59**, 41–48. *Provides an up-to-date, general overview of the signaling mechanisms by which hypoxia may affect breast cancer.*

Helczynska, K., Kronblad, A., Jogi, A., Nilsson, E., Beckman, S., Landberg, G., and Pahlman, S. (2003). Hypoxia promotes a dedifferentiated phenotype in ductal breast carcinoma *in situ*. *Cancer Res.* **63**, 1441–1444.

Kato, T., Kameoka, S., Kimura, T., Nishikawa, T., and Kobayashi, M. (2003). The combination of angiogenesis and blood vessel invasion as a prognostic indicator in primary breast cancer. *Br. J. Cancer* **88**, 1900–1908.

Kurz, H., Burri, P. H., and Djonov, V. G. (2003). Angiogenesis and vascular remodeling by intussusception: From form to function. *News Physiol. Sci* **18**, 65–70. *An excellent description of the basic steps and structural mechanisms involved in intussusceptive angiogenesis.*

Malathy, P., Shekhar, V., Werdell, J., and Tait, L. (2000). Interaction with endothelial cells is a prerequisite for branching ductal-alveolar morphogenesis and hyperplasia of preneoplastic human breast epithelial cells: Regulation by estrogen. *Cancer Res.* **60**, 439–449. *Using a three-dimensional cell culture model, the authors begin to address how the interactions between the epithelium and endothelium affect formation of vascular networks and epithelial outgrowth.*

Sacco, M. G., Soldati, S., Cato, E. M., Cattaneo, L., Pratesi, G., Scanziani, E., and Vezzoni, P. (2002). Combined effects on tumor growth and metastasis by anti-estrogenic and antiangiogenic therapies in MMTV-neu mice. *Gene Ther.* **9**, 1338–1341.

Capsule Biography

Tiffany N. Seagroves is currently investigating the contribution of the hypoxia response during normal mammary gland development and mammary tumorigenesis in the laboratory of Randall S. Johnson. This work was supported by a fellowship to T. N. S. from the Department of Defense Breast Cancer Research Program (DAMD17-01-1-0186) and NIH grand CA82515 to R. S. J.

PART III

Pathology

SECTION A

Alzheimer's

CHAPTER 91

Alzheimer's Disease and the Microcirculation

[1]William A. Banks, [1,2]Naoko Nonaka, [1]Susan A. Farr, and [1]John E. Morley

[1]*Geriatric Research, Education, and Clinical Center, Veteran Affairs Medical Center—St. Louis and Department of Internal Medicine, Division of Geriatrics, Saint Louis University School of Medicine, St. Louis, Missouri.*

[2]*Department of Oral Anatomy, Showa University School of Dentistry, Tokyo, Japan, and Department of 1st Anatomy, Showa University School of Medicine, Tokyo, Japan*

The dynamic functioning of the central nervous system (CNS) is increasingly viewed to include cell types in addition to neurons, such as astrocytes and microglia. The endothelial cells that make up the microvasculature must be included in this new view. The microvasculature of the CNS is specially modified to play roles that it seldom undertakes in other tissue beds. Specifically, it forms the blood–brain barrier (BBB), which acts to prevent the unrestricted entry of circulating substances into the brain. But the BBB also is endowed with numerous saturable transport systems. All of the vitamins, minerals, glucose, amino acids, free fatty acids, and other nutrients needed by the CNS are delivered to it by the BBB. The BBB also plays a homeostatic role for the CNS, exporting toxins, regulating electrolyte levels, and controlling pH. Finally, the BBB has a role in communication between the CNS and peripheral tissues. Both blood-to-brain influx and brain-to-blood efflux systems participate in this communication. By regulating the exchange between the CNS and blood of information molecules such as cytokines and peptides, the BBB acts as a key regulatory point in an endocrine-like communication between the CNS and the peripheral tissues. Failure of the microcirculation, especially of some aspect of BBB function, has figured largely in several theories of the etiology of Alzheimer's disease (AD). It should be noted that the theories are not mutually exclusive. Indeed, if some driving force, such as amyloid beta protein (ABP) is invoked, it is easy to imagine how the various theories can be interlinked.

Perturbances in Microvascular Circulation and Endothelial Cell Function

The microvasculature is clearly perturbed in AD. With normal aging, there is an increased thinning of the endothelial wall, a decreased number of endothelial cells forming the capillary bed, a decreased number of pericytes, and an increased susceptibility to events that can disrupt the BBB. In AD, many of these findings are accentuated. In particular, the microvasculature is extremely tortuous. It has been suggested that the tortuosity is so severe that perturbed hemodynamics occur, with a shift from normal laminar flow to one of chaotic flow [1]. Such chaotic flow could perturb the exchange of solutes across the BBB, leading to deficiencies in brain of critical substances, which, in turn, would lead to cognitive impairments.

Brain endothelial cells from AD patients show a number of abnormalities. For example, they have decreased protein kinase C activity, decreased capacity for transporting glucose, an increased density of β-1 adrenergic receptors, and increased nitric oxide synthase activity. They secrete more proinflammatory cytokines in culture, and the perturbance in protein kinase C activity is associated with the secretion of an unidentified neurotoxic protein.

ABP, the leading candidate as an etiologic agent in AD [2], has a wide variety of effects that could alter the functions of brain endothelial cells. ABP acts as an ionophore, inducing channels capable of transporting calcium and potassium. ABP can generate free radicals and affect nitric oxide synthase activity. ABP decreases glucose uptake by brain endothelial cells, inhibits their proliferation, and induces apoptosis and secretion of proinflammatory cytokines, including endothelin-1. ABP-induced reductions in cerebral blood flow are likely mediated by release of endothelin-1.

Altered Permeability of the Blood–Brain Barrier

Disruption of the Blood–Brain Barrier

One of the most investigated areas of AD and the BBB is whether the BBB is disrupted to blood-borne proteins. One of the main findings supporting a disruption of the BBB is that the cerebrospinal fluid (CSF)/serum ratio for albumin is increased in AD. Albumin is virtually excluded from the CNS by the BBB, with only small amounts entering by way of the extracellular pathways. An increase in the CSF/serum ratio for albumin is a hallmark for BBB disruption. However, not all studies have found an elevation in the CSF/serum ratio for albumin. Furthermore, other causes for an increase in the CSF/serum ratio for albumin, such as a decrease in the rate at which the CSF is replaced, have not been ruled out. Overall, the majority of studies have concluded that the BBB remains intact in AD. This is largely supported to the degree it has been investigated in animal models of AD.

Alterations in the Saturable Transport Properties of the Blood–Brain Barrier

STUDIES IN HUMANS

Other CSF/serum ratios are also abnormal in AD. For example, ratios for both vitamin B_{12} and insulin are decreased in AD. Both of these substances are transported by saturable systems across the BBB. Therefore, disturbances in their levels suggests that selective impairments in the transport properties of the BBB can be perturbed in AD. This is clearly supported by work in animal models of AD.

STUDIES IN ANIMAL MODELS OF ALZHEIMER'S DISEASE

Animal models have been key in supporting a causal link between ABP and AD. Studies using animal models of AD also support the idea that transporters are altered in this disease. Whereas transgenics have been the primary model in investigating ABP as a cause of AD, a natural mutation has also been used in examining questions related to BBB. The SAMP8 is a natural mutation that develops age-related learning and memory deficits. At 4 months of age, SAMP8 mice have normal cognition, but develop severe deficits between 8 and 12 months of age. These deficits are reversed by antibodies or antisense molecules directed at ABP or its precursor. Several transporters for key peptides or regulatory proteins have been examined in the SAMP8 mouse. The transport of insulin and that of interleukin-1 show no changes with aging in the SAMP8 mouse. Tumor necrosis factor is transported more rapidly into the occipital cortex, midbrain, and striatum of aged SAMP8 mice. Pituitary adenylate cyclase activating polypeptide (PACAP) is transported into brain by peptide transport system-6, or PTS-6. Enough PACAP is transported into brain by PTS-6 to prevent apoptosis of CA-1 hippocampal neurons even when given intravenously 24 hours after four-vessel stroke. PTS-6 is present throughout the brain excepting the pons medulla of CD-1 mice. With aging, SAMP8 mice lose PTS-6 activity in the thalamus, midbrain, and olfactory bulbs.

In contrast to the selective changes in some saturable transport systems, even very old SAMP8 mice do not show disruption of the BBB. The BBB remains largely intact to both albumin and sucrose.

Transport of Amyloid Beta Protein

Blood-to-Brain Transport

Traditionally, peptides arising in the periphery have been thought to only affect peripheral receptors. But with the appreciation that biologically relevant amounts of peptides are able to cross the BBB, it has become clear that peptides arising on one side of the BBB can interact with receptors on the other side. The possibility that ABP can cross the BBB was considered early on. An immediate question that arises is which ABP should one study. ABP is cleaved irregularly from its precursor so that a series of peptides of between 39 and 43 amino acids is formed. The 1-40 and 1-42 forms are the most abundant and of greatest interest. These two forms of ABP differ substantially in their biological and chemical properties. ABP1-40 is the more abundant form in human plasma and cerebrospinal fluid. However, ABP1-42 is considered the more toxic form. It more readily forms fibrils than does ABP1-40, which is likely related to its neurotoxicity. Fibrillation makes it more difficult to work with, and so most studies have used ABP1-40. However, a single amino acid substitution can substantially alter the properties of ABP, so it cannot be assumed that work with one form of ABP can be directly applied to another form.

Another potential problem in examining the ability of a peptide to cross the BBB is the number of related interactions. For example, if a peptide is degraded rapidly in blood or by the BBB, sequestered by the vasculature, binds avidly with circulating substances, or has multiple molecular forms because of fibrillation, it may appear to have different interactions with the BBB when studied by different techniques. The various ABPs undergo a great number of these interactions (Figure 1). By comparing and contrasting results from different methods, a great deal of information had been gathered about the interactions of the ABPs with the BBB.

Radioactively labeled ABP1-40 given by intravenous injection can be recovered from both brain and cerebrospinal fluid [3], and radioactively labeled ABP1-42 has been recovered from brain after brain perfusion [4]. This is in contrast to ABP1-28, originally the only form of ABP commercially available for study, which is so rapidly degraded in blood or at the BBB as to likely preclude its

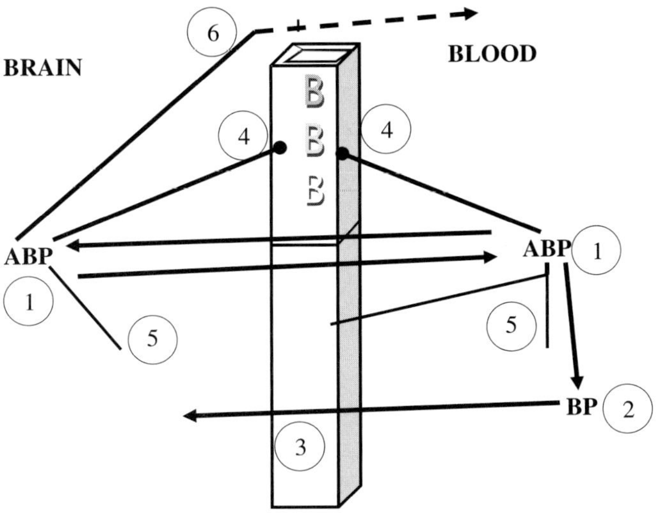

Figure 1 Postulated and proven pathways: interactions between amyloid beta protein (ABP) and the blood–brain barrier (BBB). 1, Peripheral and CNS sources of ABP are in dynamic equilibrium because of bidirectional transport of ABP across the BBB. 2, ABP binds to circulating substances in blood; 3, these binding proteins may themselves either be transported across the BBB, thus forming an alternate route for the transport of ABP across the BBB, or may be less permeable to the BBB than unbound ABP, thus retarding ABP transport across the BBB. 4, ABP can be sequestered by or accumulate on cells forming the BBB. 5, Some forms of ABP are especially rapidly degraded by enzymes in blood, at the BBB, and in the CNS. 6, ABP in the CSF may enter the circulation by way of bulk flow, the process of reabsorption of CSF back into the blood.

entry into brain. Uptake of radioactively labeled human ABP1-40 is not saturable in the mouse when given by intravenous injection but is in the guinea pig and mouse when given by brain perfusion. This discrepancy could be caused by much of the unlabeled ABP being sequestered by bloodborne binding proteins. In vitro studies with monolayers of brain endothelial cells and in vivo studies have shown that the receptor for advanced glycation end products, or RAGE, plays a major role in the transport of ABP into brain. There is also evidence that the scavenger receptor type A can bind and transport ABP.

Bloodborne ABP1-40 taken up by the BBB is largely sequestered by the capillaries. Only about 20 percent completely crosses the BBB to reach the brain parenchymal and brain interstitial fluid space, with about 80 percent being retained by the capillaries. ABP1-28 is similarly sequestered, but not the synthetic reverse peptide ABP40-1.

ABP1-40 binds to Apo J, which itself is transported by a saturable system across the BBB by glycoprotein 330/megalin. However, the circulating levels of Apo J far exceed the K_m of its saturable transporter so that the transporter is completely saturated under physiological conditions. Therefore, the net effect of Apo J binding is to retard the entry of ABP.

Brain-to-Blood Transport

Normal Physiology

Several studies with ABP1-40 have shown that it is transported rapidly out of the brain by a saturable efflux system. The efflux system has been suggested to be LDL receptor-related protein-1, and others have suggested it to be P-glycoprotein. Studies in monkeys and mice have suggested that efflux is impaired with aging and in AD. This has led to the vasculogenic hypothesis stating that ABP accumulates in brain because of an impairment in efflux transport [5].

Animal Models of AD

DeMattos et al. used transgenic mice overexpressing the precursor to ABP to dramatically demonstrate brain-to-blood efflux of ABP [6]. Rapid clearance of ABP from blood prevents serum levels from being very useful as an indicator of brain levels of ABP. DeMattos et al. peripherally administered an ABP-binding antibody to mice. After injection, serum ABP1-40 levels increased dramatically and correlated with levels of ABP and of amyloid load in the hippocampus. It is thought that because of clearance from blood was slowed, ABP was able to accumulate in blood. This accumulation would be related to the amount of ABP entering the blood from the brain, which, in turn, would be dependent on the amount of ABP in brain. This modeling of ABP efflux and clearance from blood suggests that administration of antibody to ABP could be used to determine brain levels of ABP in a diagnostic setting. If blood-to-brain influx of ABP is a significant contributor to brain levels of ABP, this model would also support the use of antibodies as a therapeutic intervention.

The preceding studies with ABP1-40 raise the question of whether ABP1-42 is also transported across the BBB and whether its transport is impaired in animal models of AD. We recently addressed both these questions [3] in the CD-1 and in the SAMP8 mouse strains. Human ABP1-40, the form almost exclusively studied to date, and mouse ABP1-42 were each transported out of the brain by a saturable process in young CD-1 mice, young SAMP8 mice, and aged SAMP8 mice. The young SAMP8 mouse had impaired transport of the ABP1-42 and ABP1-40 and the aged SAMP8 mouse had impaired transport of ABP1-42 (Figure 2). Other forms of ABP (mouse ABP1-40 and human ABP1-42) showed statistically significant differences only with aging in the SAMP8, with total loss of efflux. These results support the hypothesis that impaired efflux is associated with AD. Since decreased efflux occurred prior to the development of cognitive impairments in the SAMP8, decreased efflux could play its postulated role in leading to ABP accumulation. Finally, the results show that the BBB does not interact in the same way with all forms of ABP.

Summary

The vasculature of the central nervous system (CNS) is specially modified to form the blood–brain barrier (BBB). The BBB not only prevents the unrestricted leakage of serum proteins into the CNS, it also transports into the brain vitamins, minerals, and nutrients while transporting out of the brain toxins and other substances. The BBB also aids communication between the CNS and peripheral tissues

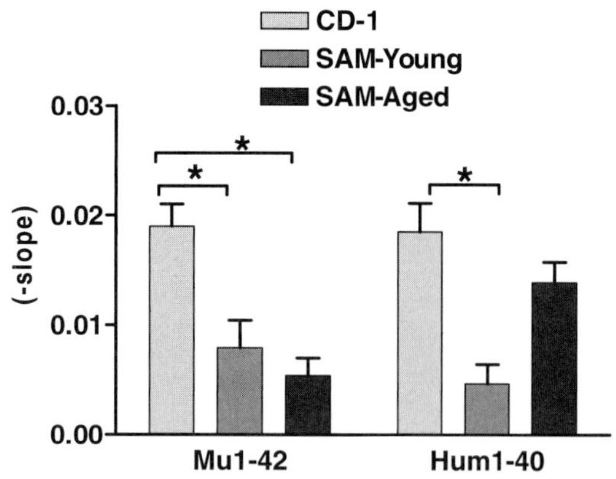

Figure 2 Decreased brain-to-blood efflux in SAMP8 mice of ABP. The SAMP8 mouse, a natural mutation used as an animal model of AD, has an impaired ability of the BBB to transport out the various forms ABP, especially the more toxic ABP1-42. Values for (–) slopes are shown, which are inversely related to half-time clearance from brain; a larger value means more rapid clearance.

through its transport of peptides and regulatory proteins. The majority of papers find the barrier function of the BBB to remain intact with Alzheimer's disease (AD). However, other aspects are altered and have been postulated to play a causal role in AD. Hemodynamic alterations in blood flow caused by vascular tortuosity could interfere with the transport of substances across the BBB. The individual brain endothelial cells that make up the BBB have alterations in their biochemical functions. Amyloid beta protein (ABP), considered widely to play a causal role in AD, induces many changes, including cell death, in brain endothelial cells. Animal models of Alzheimer's disease have shown alterations in the transporter properties of ABP. Perhaps most relevant of these are a loss of the ability to transport out the most toxic form: ABP1-42. This last finding supports the vasculogenic theory, which states that loss of the ability of the BBB to rid the brain of ABP leads to its accumulation and, ultimately, to AD.

Conclusions

This review has considered how the microvasculature is altered in AD. It has particularly addressed theories in which those alterations play a pathologic or even causal role in AD. Observations in humans has been greatly aided by work with both transgenic and natural mutation animal models of AD. Overall, work shows alterations in the morphology, cellular biology, and transport characteristics of the BBB. ABP is toxic to brain endothelial cells and could induce many of these changes. The loss of the ability of the BBB to transport ABP out of the CNS could be an immediate cause of ABP accumulation in the brain. Thus, the brain microvasculature likely plays a central role in the pathogenesis of AD.

Glossary

Alzheimer's disease: A degenerative disease of the brain characterized by the insidious onset of dementia.

Amyloid beta protein: A peptide of 39 to 43 amino acids thought to play a central role in the cause of Alzheimer's disease.

Blood–brain barrier: The modified vasculature of the CNS that prevents the unrestricted entry of blood-borne proteins into the CNS and also transports into and out of the CNS vitamins, minerals, glucose, amino acids, free fatty acids, peptides, and regulatory proteins.

CD-1: The official strain name for the standard albino outbred white lab mouse used since about 1926 and sold by Charles River.

Central nervous system: Consisting of the brain, spinal cord, and cranial nerves.

SAMP8: A mouse strain with a natural mutation that leads to an age-related decline in learning and memory. Used as an animal model of Alzheimer's disease.

References

1. de la Torre, J. C., and Mussivand, T. (1993). Can disturbed brain microcirculation cause Alzheimer's disease? *Neurol. Res.* **15**, 146–153.
2. Rosenberg, R. N. (2000). The molecular and genetic basis of AD: The end of the beginning. The 2000 Wartenberg lecture. *Neurology* **54**, 2045–2054. *In-depth review of etiologies for Alzheimer's disease with emphasis on amyloid beta protein.*
3. Banks, W. A., Robinson, S. M., Verma, S., and Morley, J. E. (2004) Efflux of human and mouse amyloid β proteins 1-40 and 1-42 from brain: Impairment in a mouse model of Alzheimer's disease. *Neuroscience*, in press. *An in-depth study of the brain-to-blood transport kinetics of the various forms of amyloid beta proteins in an animal model of Alzheimer's disease. Neurosci.* **121**, 487–492.
4. Deane, R., Yan, S. D., Submamaryan, R. K., LaRue, B., Jovanovic, S., Hogg, E., Welch, D., Manness, L., Lin, C., Yu, J., Zhu, H., Ghiso, J., Frangione, B., Stern, A., Schmidt, A. M., Armstrong, D. L., Arnold, B., Liliensiek, B., Nawroth, P., Hofman, F., Kindy, M., Stern, D., and Zlokovic, B. (2003). RAGE mediates amyloid-β peptide transport across the blood–brain barrier and accumulation in brain. *Nat. Med.* **7**, 907–913. *A classic study of the role of blood-to-brain transport in affecting brain levels of amyloid beta protein with emphasis on mediation of the receptor for advanced glycation end products.*
5. Zlokovic, B. V., Yamada, S., Holtzman, D., Ghiso, J., and Frangione, B. (2000). Clearance of amyloid β-peptide from brain: Transport or metabolism? *Nat. Med.* **6**, 718–719.
6. DeMattos, R. B., Bales, K. R., Cummins, D. J., Paul, S. M., and Holtzman, D. M. (2002). Brain to plasma amyloid-β efflux: A measure of brain amyloid burden in a mouse model of Alzheimer's disease. *Science* **295**, 2264.

Further Reading

Banks, W. A., Kastin, A. J., Maness, L. M., Banks, M. F., and Shayo, M. (1997). In *Interactions of β-Amyloids with the Blood–Brain Barrier* (J. C. de la Torre and V. Hachinski, eds.), Vol. 826, pp. 190–199. New York: New York Academy of Sciences.

Capsule Biography

Dr. William A. Banks is a staff physician and a principal investigator at the Veterans Affairs Medical Center—Saint Louis. He is also a professor in the Division of Geriatrics in the Department of Internal Medicine and in the Department of Physiological and Pharmacological Sciences. He is editor-in-chief of Current Pharmaceutical Design and has more than 250 publications in the area of blood–brain barrier.

SECTION B

Acute Respiratory Distress Syndrome

CHAPTER 92

Acute Respiratory Distress Syndrome (ARDS)

Robert H. Demling
Professor of Surgery, Harvard Medical School, Brigham and Women's Hospital, Boston, Massachusetts

Definition

The acute respiratory distress syndrome, ARDS, remains a common, often lethal form of acute lung injury prevalent in both surgical and medical patients. The term *ARDS* was recently converted from adult to acute respiratory distress because children are also affected. The definition describes a severe acute lung injury characterized by progressive hypoxemia, increased pulmonary shunt fraction, decreased lung compliance, and diffuse bilateral lung infiltrates in the presence of a normal pulmonary artery wedge pressure. An increase in lung water is also seen due to increased lung capillary permeability. The degree of shunt, however, is not directly correlated with the degree of increased water content as would be the case with cardiogenic edema. The lung insult is very complex with a significant parenchymal, interstitial, and airway component to the disease process.

The common practice of referring to ARDS as one specific clinical entity probably results in more confusion than clarification. There are probably several distinct ARDS states, each with a different initiating cause but with a common pulmonary response.

In addition, a number of qualifying statements must be added when making the diagnosis. Since ARDS is an acute syndrome, chronic states such as idiopathic pulmonary fibrosis or lymphatic spread of lung cancer are not considered to be ARDS despite similar physiologic changes.

Incidence and Outcome

The incidence today appears to be comparable to that reported several decades ago, being 65 cases per 100,000 person years. The current incidence is actually three times greater than initially predicted. This fact is of interest because avoiding risk factors such as overresuscitation and high airway pressures during ventilation is now a standard of care, which should have decreased incidence. One explanation would be that there is an increase in risk factors for ARDS now present in the critically ill population. These include an increasingly older patient population and an increase in the severity of illness. Also of interest is the fact that the mortality rate of 40 to 60 percent, described with the initial characterization of the disease, has remained at this level until very recently. Several recent studies do report a decline in mortality to about 35 percent. Possible explanations for this recent decrease in mortality include the adoption of better supportive care of critically ill patients, especially those with early signs of acute lung injury. In addition, prevention and early treatment of sepsis should attenuate the disease process.

Patients surviving ARDS typically have some degree of lung dysfunction for a number of months, often attributed to the time required to resolve residual alveolar consolidation and fibrosis. The vast majority of survivors demonstrate improvement in the 10 to 14 days after the onset of illness. However, in severe cases significant chronic lung dysfunction can develop.

Mortality 30 years ago was due mainly to progressive respiratory failure. Today however, due in part to better lung support techniques, most patients who die with ARDS have developed multisystem organ failure and a resulting refractory hemodynamic instability that is the eventual cause of death. There is a smaller group of patients who die from progressive respiratory failure as a result of an obliterative alveolar fibrosis or from recurrent pneumonia.

There are two types of clinical disorders that can lead to acute lung injury and ARDS. These include either a direct

Table I Common Causes of ARDS.

Indirect lung injury	
Sepsis	
Severe nonpulmonary trauma	
	Soft tissue damage
	Shock
	Multiple transfusions
Acute pancreatitis	
Fat emboli	
Direct Lung Injury	
Pneumonia, usually bilateral	
Lung contusions	
Aspiration	
Inhalation injury	

Table II Inflammatory Mediators in ARDS.

Proinflammatory cytokines
Products of neutrophils and macrophages
Oxygen radicals and lipid peroxides
Neutrophil proteases
Endotoxin
Activated coagulation cascade
Products of cyclo-oxygenase and lipoxygenase cascades

lung injury or more commonly an indirect lung injury caused by a systemic insult (Table I). The most common systemic insult is sepsis (Table I). Both types of insults lead to a generalized lung inflammation and a similar diffuse microvascular and alveolar injury.

Other systemic risk factors include a preexisting lung dysfunction or other organ dysfunction, especially involving the liver.

Pathogenesis

Acute Lung Injury

It is well established that the initial lung damage is the result of inflammatory mediators attacking the alveolar–capillary membrane. First endothelial cell damage occurs, leading to increased lung microvascular permeability and the leak of protein-rich fluid into the interstitium. A number of clinical studies have verified the increased protein content of edema and bronchoalveolar fluid, verifying animal studies that clearly indicate an increase in permeability. As the ARDS progresses, the alveolar epithelial cells are damaged. First the flat type I cells are damaged, leading to alveolar edema, by disrupting the normal epithelial fluid transport. This process is followed by damage to the more injury-resistant and complex type II cuboidal cells whose functions include ion transport, surfactant production, and differentiation of type I cells. Surfactant production is decreased, leading to an alveolar stability and microatelectasis. The normal local immune defenses are also severely impaired, and nosocomial pneumonia is a common complication of the later stages of ARDS.

Computed tomographic scanning has indicated that alveolar edema and atelectasis are seen mainly in dependent areas of the lung. However, selective bronchoalveolar lavage indicates that all areas of the lung are inflamed and contain protein-rich fluid.

Causes of Injury

A number of inflammatory cells and inflammatory mediators have been described which appear to cause the lung damage in ARDS. The most common causes are described in Table II.

Neutrophil-Dependent Lung Injury

Neutrophil sequestration, first along the microvascular membrane, then in the alveoli, is characteristically seen in ARDS. Adhesion molecules on the neutrophil, known as neutrophil integrins, are known to be increased in ARDS, facilitating the binding of neutrophils to endothelium. The neutrophils are then activated by cytokines or other inflammatory agents, releasing proteases and toxic oxygen radicals that damage endothelial and epithelial cells.

Modulation of these adhesion molecules using antiadherence agents has been reported in animals to decrease the initial neutrophil sequestration and the degree of acute lung injury. However, long-term studies in humans have not been performed, because of the risk of impairing lung and systemic immune defenses by impeding the neutrophils' ability to control infection.

Some new evidence would suggest that the presence of neutrophils may be the result rather than the cause of the lung injury. In addition, ARDS has been reported in neutropenic patients. This latter finding may help explain why anti-inflammatory strategies aimed at suppressing neutrophil activity have largely been unsuccessful.

Proinflammatory Cytokines

Proinflammatory cytokines, especially tumor necrosis factors (TNF) and interleukin-8 have been shown to be produced and released by the increased numbers of lung inflammatory cells, thereby amplifying the lung inflammatory response. This self-perpetuating inflammatory response then becomes autodestructive. However, selective inhibition of specific proinflammatory cytokines has not been shown to be beneficial in preventing or treating ARDS in humans. A number of new inhibitors of the proinflammatory cytokines are now available for study in ARDS. These include soluble tumor necrosis factor receptor and antibodies against tumor necrosis factor.

OXIDANTS AND ANTIOXIDANTS

Oxidant damage to the lung, measured by increased levels of lipid peroxides in lung tissue and bronchoalveolar lavage, has been well described. Also, endogenous antioxidant levels have been reported to be decreased with acute lung injury, altering the oxidant–antioxidant balance in favor of excess oxidant activity. Infusion of *N*-acetylcysteine, a precursor of the antioxidant glutathione, in patients with ARDS did transiently improve a number of hemodynamic parameters as well as lung compliance. However, there was no significant improvement in the severity of the ARDS in humans. To date, antioxidant therapy has not been shown to modify the course of ARDS.

COAGULATION ACTIVATION

Activation of the coagulation cascade, with evidence of a consumptive or disseminated intravascular coagulation, is frequently evident in patients with ARDS. Microaggregates of fibrin platelets are well described in lung microvessels, suggesting a causal relationship with the lung dysfunction. In addition, specific fibrin degradation products have been reported to be toxic to the lung microcirculation. Also, thrombosis of pulmonary arterioles has been reported, all reflecting the role of coagulation and clotting in the pathophysiology. However, anticoagulant therapy has not been shown to be effective in attenuating ARDS.

Pathophysiology and Course

The course of ARDS can be divided into a number of phases as the disease progresses through early acute lung injury to end-stage ARDS or to resolution. It is important to point out that the early phases can resolve and do not necessarily progress to severe ARDS (Table III). The potential for resolution is often dependent on the severity and longevity of the initiating insult.

Early pathologic findings are described in Table III and shown in Figure 1. An acute inflammatory response is invariably present, with increased numbers of neutrophils marginated along the endothelium as well as migrating into the interstitium and alveoli. Fibrin-platelet aggregates are also frequently found in the microvessels. At this stage, the ARDS process is reversible but only if the initiating factor, such as a septic focus, is controlled.

If the process continues to progress over the next several days, established ARDS develops, characterized by evidence of acute respiratory failure necessitating mechanical ventilation and an increased fractional oxygen concentration to treat hypoxia.

A hypermetabolic, catabolic state is also characteristically seen with the onset of ARDS. The lung inflammation, as well as any systemic inflammatory or septic foci, produces a maladaptive endocrine environment known as the

Table III Pathophysiologic Changes in Acute Respiratory Distress Syndrome.

Radiographic change	Clinical finding	Physiologic change	Pathologic change
Phase 1 (Early Changes)			
Normal radiograph	Dyspnea, tachypnea, normal chest examination	Mild pulmonary hypertension, normoxemic or mild hypoxemia, hypocarbia	Neutrophil sequestration, no clear tissue damage
Phase 2 (Onset of Parenchymal Changes)			
Patchy alveolar infiltrates beginning in dependent lung No perivascular cuffs (unless a component of high-pressure edema is present) Normal heart size	Dyspnea, tachypnea, cyanosis, tachycardia, coarse rales	Pulmonary hypertension, normal wedge pressure, increased lung permeability, increased lung water, increasing shunt, progressive decrease in compliance, moderate-to-severe hypoxemia	Neutrophil infiltration, vascular congestion, fibrin strands, platelet clumps, alveolar septal edema, intra-alveolar protein, type I epithelial damage
Phase 3 (Acute Respiratory Failure with Progression, over 2–10 days)			
Diffuse alveolar infiltrates Air bronchograms Decreased lung volume Normal heart	Tachypnea, tachycardia, hyperdynamic state, sepsis syndrome, signs of consolidation, diffuse rhonchi	Progression of symptoms, increasing shunt fraction, further decrease in compliance, increased minute ventilation, impaired oxygen extraction of hemoglobin	Alveolar consolidation from alveolar exudates and protein-rich fluid, type II cell damage, beginning fibroblast proliferation, thromboembolic occlusion
Phase 4 (Pulmonary Fibrosis–Pneumonia with Progression, > 10 days)			
Persistent diffuse infiltrates Superimposed new pneumonic infiltrates Recurrent pneumothorax Normal heart size Enlargement with cor pulmonale	Symptoms as above, recurrent sepsis, evidence of multiple system organ failure	Recurrent pneumonia, progressive lung restriction, impaired tissue oxygenation, impaired oxygen extraction, multiple system organ failure	Type II cell hyperplasia, interstitial thickening, infiltration of macrophages, fibroblasts, loculated pneumonia and/or interstitial fibrosis, medial thickening and remodeling of arterioles

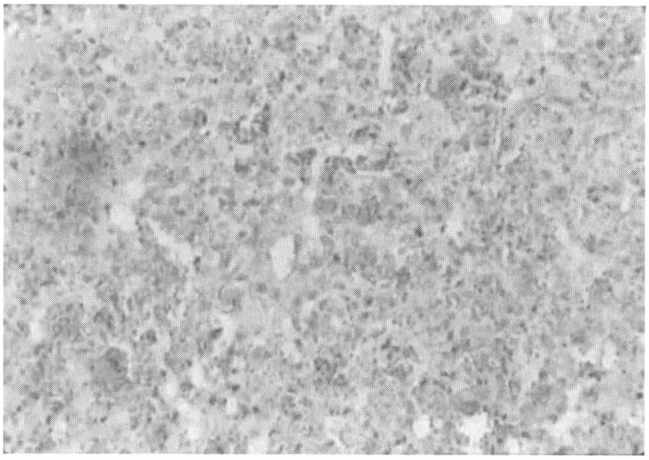

Figure 1 ARDS: Acute phase. Intense lung congestion and inflammation is evident in the interstitial and intra-alveolar space. (see color insert)

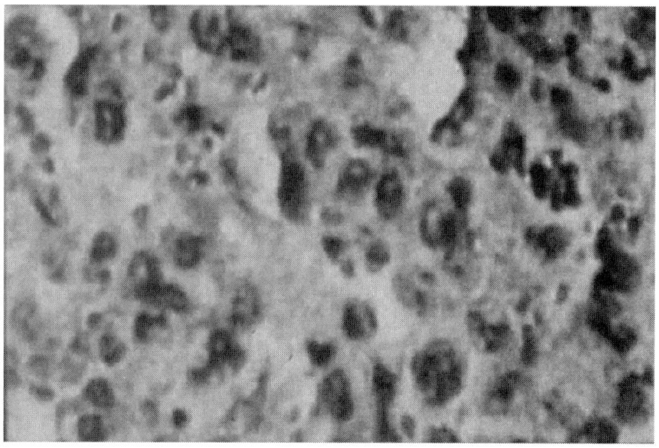

Figure 2 ARDS: Late phase. Macrophage infiltration and increased collagen (pink) is evident. Macrophages release inflammatory cytokines that increase the inflammatory response. These cytokines can produce both lung and systemic organ injury. Increased fibroblasts are also present as precursors to later lung fibrosis. (see color insert)

"stress response." The increase in catabolic cytokines and hormones leads to a progressive decrease in body protein. The increase in circulatory catechols produces an increase in metabolic rate comparable to that seen with major trauma or sepsis.

Pathologically, the lungs are much more cell dense with an increasing interstitial population of mononuclear cells and fibroblasts (Figure 2). An alveolar inflammatory exudates persists, and type II cell damage with compensatory proliferation is evident. Thrombotic or thromboembolic occlusion of precapillary arteries and loss of capillary surface area is a common finding.

The cause of death during this late phase is usually from a progressive multisystem organ failure. The mechanism of this systemic process remains unclear but it is likely due to the release of cytokines and inflammatory mediators from the lung into the systemic circulation. The lung changes from the organ being injured to being the focus of systemic organ injury. Approximately 50 percent of patients with established ARDS develop multiple system organ failure. Although the mortality rate at this stage is greater than 50 percent, complete resolution can occur over the subsequent weeks if further lung insults can be avoided.

If resolution or death does not occur, a fourth phase develops. This phase is characterized by progressive respiratory failure with pulmonary fibrosis and/or recurrent pneumonia. If progressive pulmonary fibrosis develops, there will be a significant increase in barotrauma-related complications. Increased collagen production in the interstitium and the alveoli actually begins in the earlier phases. Increased deposition of type III collagen is usually present by day 4 or 5. Also, destruction of the more elastic collagens, which are usually found in the basement membrane of the alveolar capillary junction, is well documented. The rate of collagen deposition is very rapid in ARDS as compared with other forms of pulmonary fibrosis.

The mortality rate of this phase is greater than 80 percent, again with death usually due to multiple system organ failure and systemic hemodynamic instability. The majority of patients with multiple system organ failure develop ARDS as the initial organ failure.

Treatment

Improved strategies in supportive care have developed in response to a better understanding of the pathogenesis of the lung injury in ARDS (Table IV). It is now well accepted that support modalities, especially ventilator support and fluid management, can further increase damage if not performed properly. Although current pharmacologic strategies have in large part been unsuccessful, new approaches are being developed to help correct the underlying acute lung damage (Table IV).

Mechanical Ventilation

A large number of modes of mechanical ventilation have been used over the past several decades (Table IV). Until recently, a high tidal volume (12 to 15 mL/kg) has been considered as standard for supportive treatment of ARDS, with a resultant increase in peak and mean airway pressures. More recently, lower tidal volumes (6 to 8 mL/kg) have been used to decrease airway pressure. This approach has been found to significantly reduce ARDS mortality, with a decrease in mortality from 40 percent to 31 percent in one series. A higher respiratory rate is usually required along with some permissive hypercapnia.

Fluid Management and the Level of Oxygen Delivery

Because of the report several decades ago, that oxygen consumption V_{O_2} was increased and was oxygen delivery dependent (DO_2) in ARDS, an approach of increasing DO_2 until V_{O_2} peaked was attempted in order to avoid a hidden oxygen debt. This approach was an attempt to deliver supernormal levels of oxygen in order to prevent multisystem organ failure. Both blood volume and systemic perfusion

Table IV ARDS Treatment Modalities.

Ventilator support		Pharmacotherapy	
Type	Outcome	Type	Outcome
High-level PEEP	No benefit	Glucocorticoids (early)	No benefit
High-frequency jet vent	No benefit	Glucocorticoids (during fibrosis)	Improved (small study)
Pressure control inverse I : E ratio	Inconclusive	Glucocorticoids (rescue doses)	Inconclusive
Low tidal volume	Improvement	Surfactant	No benefit to date
		Inhaled nitric oxide	No benefit
Prone position ventilation	Inconclusive	Antioxidants	No benefit
		Cyclo-oxygenase inhibition	No benefit

have been increased by using inotropes. Recent data indicate that this approach is not advantageous and may actually be harmful. Currently the approach of fluid restriction, maintaining a low to normal filling pressure, is being tested. The degree of lung edema appears to be decreased. No increase in MSOF has been reported with this approach.

Anti-Inflammatory Support

Since the lung injury is in large part inflammation induced, attempts have been made at pharmacologic anti-inflammatory therapy. The use of glucocorticoids has been the most popular. Past studies have demonstrated no benefit with high-dose glucocorticoid treatment, the exception being in progressive lung fibrosis where some benefit has been shown. Short courses of corticosteroids may be of benefit as "rescue therapy" to control excessive inflammation and allow recovery. Current trials are pursuing this approach. Other anti-inflammatory agents that have been tested have also not been found to be effective (Table IV). These include the use of anti-endotoxin antibody, a number of proinflammatory cytokine inhibitors as well as anti-neutrophil adherence agents.

Surfactant Therapy

Abnormalities in the production and composition of alveolar surfactant are well recognized in ARDS and are likely due to damage to the type II alveolar cells as well as deactivation of surfactant by oxidants. This decrease likely contributes to the microatelectasis and decreased lung compliance.

Providing exogenous surfactant by aerosol has been shown to be beneficial in treating neonatal respiratory distress syndrome. However, studies to date have not shown any benefit in the adult population with ARDS. This failure may be due to the aerosol delivery system being used as less than 5 percent of the surfactant actually reaches the surfactant-depleted alveoli using this approach. Newer preparations using more efficient delivery systems are being tested.

Vasodilator Therapy

Since increased pulmonary artery pressures are characteristically present, especially with severe ARDS, the use of a variety of pulmonary vasodilators has been attempted. Unfortunately, recent trials with inhaled nitrous oxide, an approach that does not produce systemic vasodilation, have not demonstrated a decrease in mortality. Other vasodilator agents such as prostacyclin and prostaglandin-E have also not demonstrated long-term benefit.

Nutritional Support

A major advance in the management of ARDS is the recognition of the need for nutritional support and early implementation by the enteral route. Providing the increased calorie, protein, and micronutrient needs is essential to avoid an energy deficit and the excess erosion of lean body mass. This process will lead to the impairment of immune and lung function defenses, especially the strength of the respiratory muscles needed for coughing and for weaning from mechanical support.

Summary

The acute respiratory distress syndrome remains a common source of respiratory failure and mortality. Its incidence does not appear to be decreasing and its high mortality rate has only recently been reported to be improving. Its etiology appears to be due to the combination of a variety of inflammatory mediators activated either directly by a source of lung inflammation or indirectly through a systemic focus. Activation of the various inflammatory cascades occurs, with the lung being the target organ.

The current advances in treatment appear to be focused on avoiding further lung damage caused by therapy. Low-volume mechanical ventilation and adoption of an approach to avoiding an increase in blood volume appear to be beneficial. Early aggressive nutritional support is also of benefit. The formation of an NIH-supported Multicenter Acute

Respiratory Distress Syndrome Research Team has provided the impetus for the evaluation of current treatment modalities and the development of new therapeutic strategies. Attempts at pharmacologically modifying the degree of acute destructive inflammation have as yet not been successful. However, it can be stated that prevention of further septic episodes or lung infection will lead to a more rapid resolution of an acute lung injury or established ARDS.

Glossary

Cytokines: Hormone-like low-molecular-weight proteins, secreted by many cell types, which regulate many aspects of the inflammatory immune and healing responses.

Oxidant: An unstable metabolite of oxygen that leads to oxidation of local compounds.

Syndrome: An aggregate of signs and symptoms associated with any morbid process and constituting the description of the disease.

Bibliography

Franchimont, D., Kino, T., Galon, J., Meduri, G., and Chrousas, G. (2003). Glucocorticoids and inflammation revisited: The state of the art. *NIH Clin. Staff Conf. Neuroimmune Modul.* **10**, 247–260.

Groeneveld, A. (2002). Vascular pharmacology of acute lung injury and acute respiratory distress syndrome. *Vascul. Pharmacol.* **39**, 247–256.

Lewis, J., and Jobe, A. (1993). Surfactant and the adult respiratory distress syndrome. *Am. Rev. Respir. Dis.* **147**, 218–233.

Matthay, M., Zimmerman, G., Esmon, C., and Harabin, A. (2003). Future research directions in acute lung injury; summary of a National Heart, Lung, Blood Institute working group. *Am. J. Respir. Crit. Care Med.* **167**, 1027–1035. *This article describes the various projects of the NIH study group on ARDS. Current projects are discussed as well as a sneak preview of future directions in management.*

Parodo, I., Kajikawa, O., and De Perrot, M. (2003). Injurious mechanical ventilation and end organ epithelial cell apoptosis and organ dysfunction in an experimental model of acute respiratory distress syndrome. *JAMA* **289**, 2104–2112.

Puybasset, L., Grosman, P., Muller, J., Cluzel, P., Conat, P., and Rouby, J. (2000). Regional distribution of gas and tissue in acute respiratory distress syndrome. Consequences of the effects of positive end expiratory pressure. CT scan ARDS study group. *Intensive Care Med.* **26**, 1215–1227.

Shanna, S., and Kimar, A. (2003). Septic shock, multiple organ failure and acute respiratory distress syndrome. *Curr. Opin. Pulm. Med.* **9**, 199–209.

Sprint, M., Nemery, B., and Decramer, M. (2003). Survivors of the acute respiratory distress syndrome. *N. Engl. J. Med.* **348**, 2149–2150.

Steinbrook, R. (2003). How best to ventilate? Trial design and patient safety in studies of the acute respiratory distress syndrome. *N. Engl. J. Med.* **348**, 13293–13401. *This review describes the variety of techniques used for mechanical ventilation in ARDS. Each approach is then analyzed as to its pros and cons.*

Ware, L., and Matthay, M. (2000). The acute respiratory distress syndrome. *N. Engl. J. Med.* **342**, 1334–1342. *This review presents an up-to-date discussion of the current status of ARDS. Incidence, pathophysiology, and treatment modalities are described.*

Capsule Biography

Dr. Demling is a Professor of Surgery at Harvard Medical School. The focus of his clinical and research activity has been critical care with emphasis on respiratory distress and the "stress response" to injury. He is the director of the Trauma, Burn, Critical Care Division of the Brigham and Women's Hospital.

SECTION C

Arthritis

CHAPTER 93

The Role of the Microvasculature in the Pathophysiology of Rheumatoid Arthritis

Ewa Paleolog
Kennedy Institute of Rheumatology, Faculty of Medicine, Imperial College, London, United Kingdom

Introduction

In the past, the vascular endothelial lining of the cardiovascular system, which covers the network of arteries, veins, and lymphatics of the human body, was considered to be merely an inactive barrier between the circulation and underlying tissues. However, it is now clear that despite its relatively small total mass, endothelium actively participates in vascular physiology. Endothelial cells produce mediators regulating blood flow and influence coagulation and fibrinolysis, usually presenting a nonthrombogenic surface to flowing blood. Furthermore, endothelium fulfills a gatekeeper role in cell recruitment, through expression of leukocyte activators and chemoattractants, as well as adhesion molecules. Finally, endothelial cells can act as antigen-presenting cells and play a role in the growth of new blood vessels (angiogenesis), which are vital for efficient supply of oxygen and nutrients to tissue, and for removal of waste products. These *effector* actions of endothelium fulfill a homeostatic function and act as a rapid response facility in situations of injury or infection. As the lining of blood vessels, endothelium also has the potential to act as the primary *target* for the action of mediators such as cytokines. It is therefore unsurprising that the endothelium contributes to the initiation and maintenance of many pathologies.

The focus of this review is to discuss the role of the vasculature in rheumatoid arthritis (RA), particularly in the light of observations using therapies such as anti-tumor necrosis factor α (TNFα) biologicals, and on the potential for development of vascular-targeted therapies for treatment of RA.

The Pathogenesis of Rheumatoid Arthritis

RA is a common human disease with a prevalence of about 1 percent in most parts of the world [1]. The clinical presentation can vary in terms of severity and the age of onset, although the peak occurs between the fifth and sixth decades of life. Patients display painful, stiff, and swollen joints, generally presenting with a symmetrical polyarthritis, predominantly involving the small joints of the hands and wrists, as well as the metatarsophalangeal joints, ankles, knees, and cervical spine. Periarticular structures, such as bursae and tendon sheaths, are commonly inflamed. Nonarticular features may also be seen in RA, including inflammatory nodules, vasculitis, and pericarditis, together with involvement of the lungs and nervous system. The mortality of patients with severe RA is equivalent to that of individuals with triple vessel coronary artery disease. After 2 years, joint erosions are seen in most patients, and the majority of patients become clinically disabled within 20 years. In the UK, around 387,000 adults have RA, equivalent to 0.81 percent of the adult population, and there are approximately 12,000 new cases each year. Surveys indicate that more than 9 million working days were lost because of RA in 1999–2000 in Great Britain, representing £833 million lost production. Indeed, in a recent study, RA patients of working age were found to be 32 times more likely to stop work on health grounds than matched controls. RA thus imposes a significant social and economic burden, due to loss of earnings and medical expenses, apart from adversely affecting quality of life.

In spite of many years of intensive investigation, the cause of RA remains unknown, although current thinking favors the concept of a multifactorial disease, in which contributory genetic factors combine with environmental and possibly infectious influences to initiate disease. The association between RA and genetic elements encoded within the HLA-DR region supports the importance of T cells in disease pathogenesis. More than 80 percent of Caucasian RA patients express HLA-DR1 or HLA-DR4 subtypes. The primary site of inflammation is the synovial lining of the closed spaces of articular joints, which increases greatly in mass and becomes infiltrated by blood-derived cells of lympho-hematopoietic origin, including T cells, B cells, and macrophages. The lymphocytes infiltrating the synovium are predominantly CD4+ T cells, with high expression of memory CD45R0 antigens and activation markers such as HLA-DR and CD69. A feature of the synovium in RA is an altered density of sublining capillaries and postcapillary venules. Subsequently, the synovium becomes locally invasive at the synovial interface with cartilage and bone. Progressive destruction of cartilage and bone eventually combines to produce deformities and functional deterioration and profound disability in the long term.

Over the past two decades, our understanding of the pathogenesis of RA has increased considerably, based on a range of studies using human tissue and animal models of disease. Of relevance to this review, the importance of the microvasculature in RA has become increasingly apparent, and this is discussed in subsequent sections.

Why Is the Microvasculature Important in RA?

As the lining of the vasculature, endothelium has the capacity to integrate many different signals and responses, and as a consequence is involved in the pathogenesis of many diseases. The involvement of endothelium in the pathogenesis of RA can be inferred from observations that RA is associated with vascular and hematological abnormalities. For example, the excess mortality in RA is predominantly due to coronary artery atherosclerosis [2]. A recent study showed that the adjusted relative risks of myocardial infarction and stroke in women with RA were 2.0 and 1.48, respectively, when compared to women without RA. Many studies have reported an association between RA and traditional cardiovascular risk factors such as cholesterol and low-density lipoprotein levels. The acute phase response marker C-reactive protein (CRP) is a novel risk factor for atherosclerosis and predicts future risk of coronary artery disease in initially healthy individuals. Since CRP levels are markedly elevated in RA, as part of the ongoing systemic inflammatory processes, such an augmented inflammatory burden may account for the increased cardiovascular risk. However, in addition to these somewhat circumstantial observations, the vasculature is also implicated in RA by the fact that many of the events known to occur in RA—such as leukocyte extravasation—involve the participation of endothelial cells.

Effector Role of Vascular Endothelium in RA

It was recognized more than 30 years ago that endothelial cells in RA synovium acquire the characteristic appearance of lymphatic endothelium, which controls lymphocyte emigration. Considerable evidence has now converged, documenting the responsiveness of endothelium to cytokines expressed in RA, the presence on endothelial cells of receptors for these cytokines, and expression by endothelium of adhesion molecules and chemoattractants. Endothelium can also contribute to RA pathogenesis through the formation of new blood vessels.

Recruitment and Activation of Leukocytes in RA

Adhesion of leukocytes to vascular endothelium in vivo must overcome the normal vascular mobility of circulating cells and result in a localized arrest of leukocytes at relevant sites [3]. Over the past few years there have been a number of studies using immunohistochemical analyses of RA tissue, documenting alterations in the pattern of expression of adhesion molecules. It has been shown, for example, that antibody to E-selectin stained endothelium in RA synovium. In addition, RA synovial endothelial cells were also found to express increased levels of vascular cell adhesion molecule (VCAM)-1 and intercellular adhesion molecule (ICAM)-1. As discussed previously, the typical RA synovial infiltrate is rich in memory CD45RO+ T cells. Synovial membrane and synovial fluid T cells display an enhanced capacity to interact with purified E-selectin and VCAM-1, relative to peripheral blood lymphocytes from either the same patients or from healthy donors, due to increased levels of VLA-4α, the counterligand for VCAM-1. In addition, synovial fluid lymphocytes show higher expression of other integrins such as CD29 (β1), VLA-1α, VLA-5α, and VLA-6α. Accumulation of memory T cells in RA synovium thus appears to result from elevated expression of adhesion receptors on synovial microvascular endothelium, leading to the selective emigration of memory T lymphocytes, which may bear enhanced levels of ligands for these adhesion molecules as a result of a previous activation step.

Moreover, endothelial cells are a source of a range of proinflammatory cytokines, including interleukin (IL)-1, IL-6, and granulocyte macrophage colony-stimulating factor (GM-CSF). Many of the features of the rheumatoid synovial environment suggest possible roles for chemoattractant cytokines, in that the large number of infiltrating leukocytes, especially the selective accumulation of memory T cells, could in part be a response to the elaboration of chemokines. Endothelial cells secrete and present on cell surface proteoglycans chemokines of both C-C and C-X-C subsets, in particular IL-8, monocyte chemoattractant protein (MCP)-1, RANTES, and Groα. The ability of endothelium to capture chemokines may be of particular significance in RA, in that mediators such as RANTES or MIP-1α, secreted by other participating cells and anchored on the endothelial cells surface, would ensure a relatively high concentration of chemoattractants close to the blood vessel wall, and hence temporally and spatially restricted activation of circulating cells.

ANGIOGENESIS IN RA

A consequence of the synovial hyperplasia, which occurs in RA, is an increase in the distance between the proliferating cells and the nearest blood vessels, leading to local hypoxia and hypoperfusion. It has been reported, for example, that synovial oxygen tension is low in aspirated synovial fluid samples taken from human RA knee joints. In general, the in vivo response to hypoxia is to form new blood vessels, to restore perfusion and oxygenation to the compromised area. Endothelial cells lining blood vessels within RA synovium express cell cycle–associated antigens, and indices of endothelial turnover are increased in synovia from patients with RA compared with noninflamed controls. A morphometric study also suggested that capillaries are distributed more deeply in RA synovium. Many of the cytokines and growth factors expressed in RA have the potential to stimulate angiogenesis [4]. For example, serum levels of vascular endothelial growth factor (VEGF) are markedly elevated in RA, relative to either patients with OA or normal controls, and correlate with levels of CRP. Expression of VEGF by RA lining layer cells has been reported, and microvascular endothelial cells in the vicinity of VEGF-positive cells express VEGF receptors. Since VEGF is upregulated by hypoxia, it is likely that this factor is central to the regulation of angiogenesis in RA. Other proangiogenic molecules expressed in RA include fibroblast growth factor (FGF)-1 and FGF-2, which stimulate proliferation of a variety of cell types, including endothelial cells. Members of the FGF family (FGF-1 and FGF-2) have been detected in human RA synovium. However, in RA the role of the FGFs, and other mitogens expressed in RA, such as platelet-derived growth factor, hepatocyte growth factor, and transforming growth factor-β, is unclear.

In summary, the invasive synovium in RA is highly vascularized, and numerous growth factors are expressed, which might promote new blood vessel formation.

Target Role of Vascular Endothelium in RA

As the interface between the blood and tissues, endothelium can respond to a range of cytokines and growth factors. In RA, many studies have focused on the identity of mediator(s) involved in disease pathogenesis, but a key milestone was the realization that intercellular messengers, now termed *cytokines*, may play a central role. Both TNFα and IL-1 have the capacity to activate cells in the synovium, as well as regulating cartilage turnover, stimulating bone resorption and inhibiting proteoglycan synthesis. In the context of RA pathogenesis, TNFα is produced in large quantities by enzymatically dissociated synovial cells from RA patients, and immunohistochemical analyses have demonstrated the presence in RA synovium of TNFα and its receptors. Addition of anti-TNFα antibodies to RA synovial cell cultures resulted in downregulated expression of IL-1, and reduced expression of other cytokines and angiogenic factors [5]. These studies were instrumental in the formation of a key hypothesis by Marc Feldmann and Ravinder N. Maini and their researchers—namely, that TNFα plays a vital role in the pathogenesis of RA.

In terms of vascular endothelium, TNFα is a fundamental inducer of endothelial cell responses. For TNFα to transmit a signal and exert an effect on endothelial cells, it must initially bind to specific cell surface receptors. The TNF receptor family has been studied extensively, in our own and other laboratories, and two receptors for TNFα, CD120a or TNF-R1 and 75-kDa CD120b or TNF-R2, have been cloned. Both receptors have been detected on the surface of cultured endothelium. We and others have observed that selective stimulation of endothelial cells through TNF-R1 induced responses comparable to those observed for TNFα, actions not mimicked by TNF-R2 agonists. It is likely that the presence of the higher affinity TNF-R2 on the cell surface serves to "pass" ligand to the lower affinity TNF-R1, leading to cell activation. Studies from our laboratory using sections of RA synovium have also localized TNFα and its receptors to endothelium.

TNFα has the potential capacity to regulate many of the events occurring in the RA microvasculature—leukocyte extravasation, chemotaxis, and angiogenesis [6]. E-selectin, which is not expressed on unstimulated endothelial cells, is induced in response to TNFα and IL-1β, with maximal expression observed after approximately 4 to 6 hours. Induction of ICAM-1 and VCAM-1 by TNFα and IL-1β occurs over a slower time-course, peaking at 16 to 24 hours. TNFα also induces production by endothelial cells of IL-6, GM-CSF, IL-8, monocyte chemoattractant protein-1 (MCP-1), RANTES, and Groα. The effects of TNFα on the angiogenic process are both stimulatory and inhibitory, depending on the experimental system. TNFα inhibits basal and FGF-2-stimulated endothelial cell proliferation and migration in vitro, but stimulates neovascularization in the rabbit cornea. TNFα treatment of endothelial cells has been reported to enhance urokinase-type plasminogen activator activity, which could contribute to the proangiogenic effects of TNFα. Brief exposure to TNFα induced release from endothelial cells of VEGF and FGF-2, whereas prolonged exposure of microvascular cells to TNFα inhibited capillary-like structure formation in vitro, suggesting that the net effect of TNFα on angiogenesis may reflect a balance of pro- and anti-angiogenic responses.

Another key endothelial cell activator is VEGF. Several members of the VEGF family have been described, including VEGF-A, -B, -C, and -D, which bind differentially to tyrosine kinase receptors VEGF-R1 or Flt (fms-like tyrosine kinase)-1, VEGF-R2 (kinase insert domain containing receptor or KDR/Flk-1) and VEGF-R3 (Flt-4). Signaling through VEGF-R1 appears to be important in controlling the number of endothelial cells during vessel formation, as well as in the organization of these cells during angiogenesis. Activation of VEGF-R2 enables the differentiation of endothelial cells and blood vessel formation [7]. As well as being essential for vasculogenesis and angiogenesis, VEGF is also capable of inducing procoagulant tissue factor and expression of cytokines and adhesion molecules. For example, VEGF was shown to induce expression of MCP-1 and IL-8, as well as ICAM-1, VCAM-1, and E-selectin. Essential to its proangiogenic function, VEGF can inhibit apoptosis, or programmed cell death. It has been shown, for example, that VEGF withdrawal results in obliteration of immature blood vessels. The antiapoptotic effects of VEGF are mediated via the induction of members of the Bcl family and inhibitors of apoptosis (IAP), that in turn inhibit terminal effector caspases-3 and -9. VEGF induces Bcl-2, as well as the IAPs survivin and XIAP. Thus the continued expression of VEGF in RA, as well as promoting angiogenesis, is likely to sustain the vasculature, thus further augmenting the inflammatory process.

Finally, in the context of an ongoing inflammatory response, it is likely that under conditions of flow, cytokines such as TNFα may be less important as activators of vascular endothelium than cell-bound stimuli. Although activation of endothelial cells via soluble factors has been extensively studied, cell–cell interactions can also modulate the responses of the target cells, and could thus potentially further perpetuate inflammation. T cell adhesion to the endothelial lining of blood vessels is an early event in inflammation, and cell contact–mediated signaling to endothelial cells by stimulated T-cells could be an important endogenous mechanism of endothelial activation [8]. We recently described the ability of stimulated, but not resting, human T lymphocytes to modulate endothelial cell activation by direct cell–cell contact. T cells, either activated through the T cell receptor (TCR) with immobilized anti-CD3 monoclonal antibody or incubated in the presence of a combination of TNFα, IL-6, and IL-2, were able to induce endothelial cell production of chemokines and cytokines (MCP-1, IL-8, and IL-6), although the relative amounts of different cytokines were dependent on the method of T cell stimulation. For example, TCR-activated T cells induced relatively greater amounts of endothelial cell MCP-1 than did cytokine-activated T cells. In contrast, production of IL-6 was upregulated to a greater extent in the presence of cytokine-activated T cells. Furthermore, addition of anti-CD154 (anti-CD40 ligand) and anti-TNFα antibodies reduced release of IL-6, IL-8, and MCP-1 from endothelial cells cocultured with T cells. Such direct cell–cell contact between stimulated T lymphocytes and endothelium represents a novel pathogenic mechanism in inflammatory diseases such as RA. For example, overexpression of MCP-1 induced by TCR-activated lymphocyte-endothelial cell contact interactions might preferentially regulate the influx of monocytes into the RA lesions. Cytokine-activated lymphocyte–endothelial cells contact interactions might also contribute to the systemic inflammatory response in RA patients, for example, elevated expression of IL-6.

Insights from Clinical Trials of Anti-TNFα Biologicals

The concept of TNFα as a therapeutic target in RA, put forward by Feldmann and Maini, was tested in a series of clinical trials in the UK and in other countries. The first Phase I/II study was an open-label trial of a single intravenous infusion of infliximab (Remicade), a chimeric mouse Fv, human IgG1 antibody that binds both soluble and membrane-bound TNFα with high affinity, in long-standing active RA patients who had failed all prior therapy. The results were striking, showing reductions in pain and morning stiffness, swollen and tender joint counts, and CRP levels. Currently, in addition to infliximab, two other anti-TNFα biologicals are approved for use in RA (Table I). The demonstration of the clinical efficacy of TNFα blockade, as well as providing a therapeutic benchmark, also allowed the teams of investigators to analyze the mechanism of action of this approach, and to use this knowledge to further elucidate the processes underlying RA [9].

Since trafficking into the synovium of bloodborne cells is a feature of RA, and since TNFα is one of the most potent regulators of leukocyte trafficking, it seemed reasonable to hypothesize that infliximab treatment might regulate synovial infiltration. This question has been addressed over the years with increasingly sophisticated studies, which started with the measurement of soluble adhesion molecules, which could be quantified in serially acquired serum samples. Treatment with anti-TNFα antibody significantly decreased serum E-selectin and ICAM-1 levels, and these reductions correlated with a rapid and sustained increase in the blood lymphocyte counts. Perhaps surprisingly, given that VCAM-1 is involved in mononuclear cell adhesion, no effect on serum VCAM-1 levels was detected. Moreover, a reduction was observed in synovial granulocyte counts, as well as CD3+ and CD68+ cell infiltration, and synovial expression of ICAM-1, VCAM-1, and E-selectin was also decreased. The beneficial effects of infliximab in RA are thus due in part to diminished synovial cellularity (Figure 1).

Table I TNFα Blocking Agents for the Treatment of RA.

Name	Composition	Approval	Other uses
Infliximab (Remicade)	Chimeric mouse Fv/human Fc anti-TNFα monoclonal antibody	Europe and USA	Severe Crohn's disease
Etanercept (Enbrel)	Human TNF receptor p75 dimer fused to human IgG Fc	Europe and USA	Psoriatic arthritis Juvenile RA
Adalimumab (Humira)	Human anti-TNFα monoclonal antibody	USA and Europe	

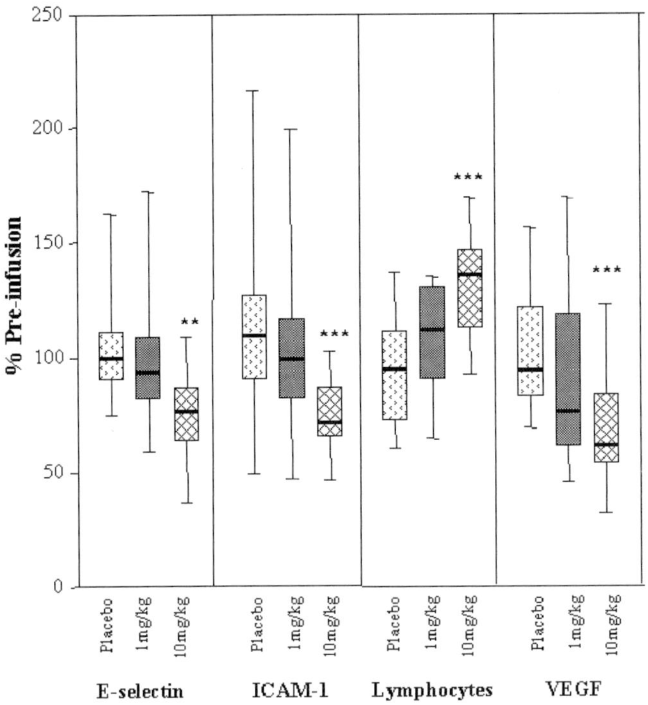

Figure 1 Anti-TNFα antibody: differential effects on vascular endothelium. Patients received a single intravenous infusion of either placebo or anti-TNFα antibody infliximab, at doses of either 1 mg or 10 mg per kilogram body weight. Whole blood and serum samples were obtained before and 4 weeks after treatment. Peripheral blood lymphocyte counts were assessed, and serum levels of E-selectin, ICAM-1, and VEGF were measured by ELISA. Results were expressed relative to preinfusion levels for each patient. Graph shows median values, interquartile ranges, and minimum and maximum values for each parameter. Data were analyzed versus placebo-treated patients by one-way ANOVA with Bonferroni correction for multiple comparisons: $^{**}p < 0.01$, $^{***}p < 0.001$.

We also measured serum VEGF concentrations following anti-TNFα treatment, to determine whether improvement in RA is associated with reduced angiogenesis. We observed that treatment of RA patients with infliximab resulted in a reduction in serum VEGF levels (Figure 1). There is now direct evidence of reduced synovial angiogenesis, as assessed by reduced expression of CD31 and von Willebrand factor. Recently, use of high-resolution ultrasound and power color Doppler has shown reduced synovial thickening and joint vascularity after infliximab treatment.

In summary, insights from studies in anti-TNFα-treated RA patients suggest that an important mechanism of action is reduced deactivation of the vascular endothelium (Table II).

Vascular Endothelium: A Possible Therapeutic Target in RA?

Increasing use of anti-TNFα biologicals has led to reports of cases of tuberculosis. Additionally, adverse effects such as non-Hodgkin-type lymphomas have been described. There is clearly scope for improvement, and cytokines other than TNFα have been proposed to be possible therapeutics goals. A recombinant, nonglycosylated form of IL-1 receptor antagonist was recently approved in the United States for the treatment of RA in combination with methotrexate, following successful clinical trials. Combined IL-1 and TNFα blockade, although effective in animal models of arthritis, might be proinfective in RA. IL-6 blockade, using anti-human IL-6 receptor antibody, resulted in improvement in signs and symptoms of RA, in parallel with reduced serum VEGF levels. Thus, like anti-TNFα antibody, IL-6 receptor antibody may exert effects on vascular endothelial activation [10]. The reduction in synovial infiltration suggest that targeting vascular endothelium might also be beneficial in RA. Results from trials of murine anti-ICAM-1 antibody enlimomab in RA were encouraging, but repeated treatments were less effective, and adverse effects such as fever, leucopenia, and skin reactions were observed.

Table II Summary of the Effects of Anti-TNFα Antibody on the Microvasculature.

Parameter	Observed effect of infliximab
Leukocyte adhesion	Reduced serum adhesion molecules Reduced synovial adhesion molecules Reduced synovial CD3+ and CD68+ cells Reduced granulocyte trafficking Increased circulating lymphocytes
Chemokine expression	Reduced synovial chemokines Reduced serum chemokines
Angiogenesis	Reduced serum VEGF Reduced synovial vascularity Reduced synovial thickening Reduced joint vascularity
Hematological markers	Reduction in elevated fibrinogen Reduction in elevated platelet counts Restoration of reduced hemoglobin

The central role of angiogenesis in RA suggests that suppression of blood vessel formation should retard arthritis progression. There is certainly considerable literature describing the ability of broadly acting angiogenesis inhibitors such as paclitaxel (Taxol) to modulate disease in animal models. Specific blockade of angiogenesis has also proved to be effective in animal models of disease, such as collagen-induced arthritis in mice. For example, we and others have reported that blockade of VEGF in collagen-induced arthritis significantly reduced disease severity and joint destruction, paralleled by significantly less joint inflammation, reduced bone and cartilage destruction, and reduced synovial vascularity. We have also shown that another angiogenesis inhibitor related to angiostatin, termed kringles 1-5 (K1-5), was also extremely effective in the murine model of collagen-induced arthritis. Angiogenesis inhibition, using molecules such as K1-5 or soluble Flt-1, is thus effective in animal models of RA.

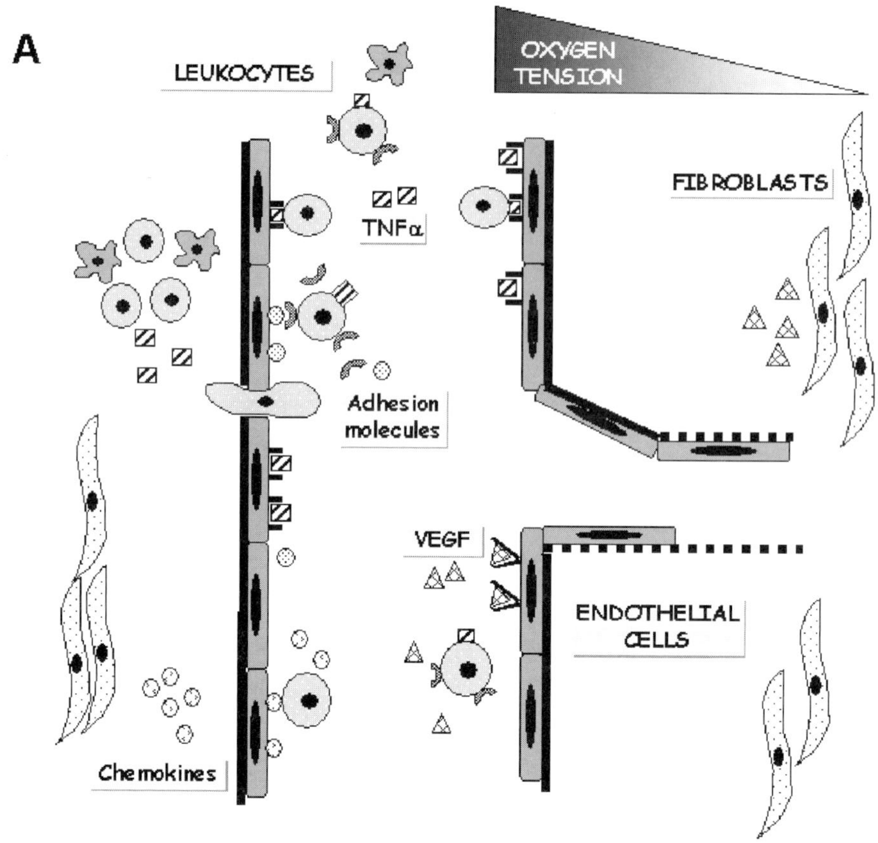

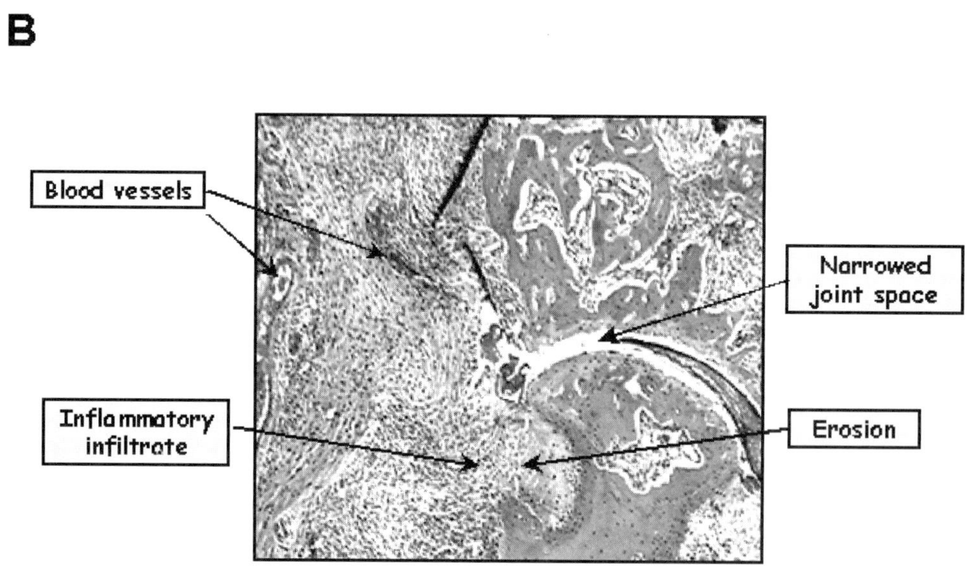

Figure 2 Role of the microvasculature in RA. Schematic representation of the target-effector role of vascular endothelium in RA. (**A**) Leukocytes secrete and express on the cell-surface cytokines including TNFα, which bind endothelial cell adhesion molecules such as E-selectin, through cognate ligand : counterligand interactions. TNFα also has the capacity to upregulate expression of adhesion molecules and secretion of chemokines. Concomitantly, VEGF, secreted either by circulating cells in response to cytokines or CD40 ligand and/or by fibroblasts under conditions of hypoxia, stimulates angiogenesis. (**B**) This process culminates in synovitis and joint erosion, as exemplified in this hematoxylin and eosin-stained section of a tarsal joint from a DBA-1 mouse, in which arthritis was induced by immunization with bovine type II collagen in complete Freund's adjuvant. Blood vessels, leukocyte infiltration, and joint space narrowing and erosions are indicated.

Therapeutic agents and strategies are being devised to either interrupt or inhibit one or more of the pathogenic steps involved in angiogenesis, and blockade of neovascularization has been effective in many tumor models. Clearly angiogenesis can be targeted at several different stages, including inhibition of production of stimuli such as VEGF, binding of proangiogenic factors (using antibodies or soluble receptors), interruption of downstream signaling, blockade of matrix degradation, or even using antiangiogenic stimuli such as angiostatin or K1-5. Many of these approaches have been used with varying degrees of success for human cancers. The first anti-angiogenic agent, bevacizumab (Avastin), is an anti-VEGF antibody and was approved in December 2003 by the US Food and Drug Administration (FDA) as first-line treatment of patients with metastatic carcinoma of the colon or rectum in combination with intra-venous 5-fluorouracil-based chemotherapy.

Concluding Remarks

In terms of the number of individuals affected or the rate of mortality, RA cannot realistically be considered in the same terms as cancer or cardiovascular disease. However, the increased mortality due to coronary artery atherosclerosis, combined with the fact that many patients are disabled at a relatively young age, make it imperative to understand the pathophysiology of RA. Interestingly, there are a number of parallels between RA and tumors, and atherosclerosis, not least the important contribution of the microvasculature to the initiation and maintenance of RA, through the regulation of leukocyte trafficking and formation of new blood vessels (Figure 2).

Some questions regarding the function of endothelium in RA still remain unanswered, including the relative roles of the different adhesion molecules and chemokines in RA. However, it is not unreasonable to suggest that targeting the vasculature in RA, in combination with other therapies such as anti-TNFα, may lead to a more persistent reduction in pannus volume and hence modify disease progression. However, confirmation of this hypothesis requires appropriate clinical trials. Importantly, targeting the inflammatory and vascular components of RA, by combining TNFα inhibition with angiogenesis blockade, could increase benefit to patients with RA, without augmenting the infection risk.

Glossary

Angiogenesis: The process of postnatal new blood vessel formation, generally considered to be regulated by a balance between proangiogenic and antiangiogenic factors, with angiogenesis occurring as the balance shifts to an angiogenic phenotype (the "angiogenic switch").

RA: Rheumatoid arthritis, a human autoimmune systemic disease, characterized by a chronic inflammatory reaction in the synovial lining joints and associated with degeneration of cartilage and erosion of bone.

TNFα: Tumor necrosis factor α, a pivotal molecule regulating cellular activation and interactions in rheumatoid arthritis, and a target for new anti-TNFα biological therapies.

VEGF: Vascular endothelial growth factor (also known as vascular permeability factor), a potent inducer of angiogenesis and endothelial cell survival, and a target for antiangiogenic therapies, especially in the field of oncology.

Acknowledgments

The Kennedy Institute of Rheumatology is a division of the Faculty of Medicine, Imperial College, and receives a Core Grant from arc (Registered Charity No. 207711).

References

1. Harris, E. D., Jr. (1986). Pathogenesis of rheumatoid arthritis. *Am. J. Med.* **80**, 4–10.
2. Goodson, N. (2002). Coronary artery disease and rheumatoid arthritis. *Curr. Opin. Rheumatol.* **14**, 115–120.
3. Haskard, D. O., and Landis, R. C. (2002). Interactions between leukocytes and endothelial cells in gout: Lessons from a self-limiting inflammatory response. *Arthritis Res* **4**(Suppl. 3), S91–S97.
4. Paleolog, E. M. (2002). Angiogenesis in rheumatoid arthritis. *Arthritis Res.* **4**(Suppl. 3), S81–S90.
5. Feldmann, M. (2002). Development of anti-TNF therapy for rheumatoid arthritis. *Nat. Rev. Immunol.* **2**, 364–371. *A historical article describing in detail the development of anti-TNFα biologicals.*
6. Pober, J. S. (2002). Endothelial activation: Intracellular signaling pathways. *Arthritis Res.* **4**(Suppl. 3), S109–S116. *An excellent review of the effect of TNFα on endothelium.*
7. Ferrara, N. (2002). Role of vascular endothelial growth factor in physiologic and pathologic angiogenesis: Therapeutic implications. *Semin. Oncol.* **29**, 10–14. *Recent review of the role of VEGF in angiogenesis.*
8. Monaco, C., Andreakos, E., Kiriakidis, S., Feldmann, M., and Paleolog, E. (2003). T cell–mediated signalling in immune, inflammatory and angiogenic processes: The cascade of events leading to inflammatory diseases. *Curr. Drug Targets Inflamm Allergy*, in press.
9. Maini, R. N., and Feldmann, M. (2002). How does infliximab work in rheumatoid arthritis? *Arthritis Res.* **4**(Suppl. 2), S22–S28. *This review summarizes the current understanding of the mechanisms of action of infliximab in RA.*
10. Taylor, P. C. (2003). Anti-cytokines and cytokines in the treatment of rheumatoid arthritis. *Curr. Pharm. Des.* **9**, 1095–1106. *This is an up-to-date review of the rationale for targeting TNFα and IL-1 in RA, and discusses the results of clinical studies designed to validate this approach.*

Capsule Biography

Dr. Ewa Paleolog is a Senior Lecturer at the Kennedy Institute of Rheumatology, Faculty of Medicine, Imperial College, London. She has more than 15 years of research experience in vascular biology, especially regarding the cytokine biology of vessels in the pathogenesis of rheumatoid arthritis.

SECTION D

Blood, Thrombosis and Homeostasis

CHAPTER 94

Review Hemoglobin-Induced Microvascular Dysfunction

Ann L. Baldwin

Department of Physiology, College of Medicine, University of Arizona, Tucson, Arizona

Introduction

Cell-free hemoglobin-based oxygen carriers have been proposed as blood substitutes for transfusions because of their plasma expansion and oxygen transport capabilities. Apart from their use after accidents or major surgery, such substitutes could be employed to alleviate anemia in patients with hematocrits too high to qualify for blood transfusions. Hemoglobin-based blood substitutes have the added advantages that they can be easily purified, stored for relatively long periods of time, and used in patients of all blood types.

Normally hemoglobin in the bloodstream is enclosed in red blood cells, and this tends to stabilize the molecule in its tetrameric form. In addition, the red cells contain a metabolite, 2,3-diphosphoglycerate (2,3-DPG), that both cross-links the hemoglobin, further promoting its stability, and reduces its oxygen affinity so that it is able to release bound oxygen to the tissue. Red blood cells also contain antioxidant enzymes, such as catalase and superoxide dismutase, which catalyze the breakdown of the reactive oxygen species (ROS), hydrogen peroxide (H_2O_2), and superoxide ($O_2^{\bullet-}$). When hemoglobin is released from red cells it rapidly splits up into dimers because these stabilizing factors are no longer present (see Figure 1, stroma-free hemoglobin, SFH). Hemoglobin contained in red cells has a low oxygen affinity, the partial pressure of oxygen needed for 50 percent oxygen saturation, P_{50}, being about 27 mmHg. Cell-free hemoglobin, however, has a high oxygen affinity ($P_{50} \sim$ 8 mmHg), meaning that it will only give up its oxygen to very anoxic tissue.

If free hemoglobin is going to be effective as an oxygen carrier and deliverer, it needs to be stabilized and its oxygen affinity must be reduced. One way to achieve these goals is to cross-link the two α subunits of the hemoglobin molecule with bis(3,5-dibromosalicyl) fumarate. Two such products are diaspirin cross-linked hemoglobin (DCLHb) and its noncommercial analog (DBBF-Hb). Alternatively, hemoglobin can be covalently cross-linked and then conjugated with macromolecules, such as polyethylene glycol (PEG), or it can be polymerized, resulting in inter- and intramolecular cross-linked polymers of various molecular sizes. Some hemoglobins that have been modified in these ways, such as DBBF-Hb, PEG-Hb, PolyHbBv (Oxyglobin), and O-R-PolyHb (Hemolink), have been used in clinical trials. In fact, PolyHbBv (Oxyglobin) is currently FDA-approved for veterinary use in the United States, and its human counterpart has recently been approved for clinical use in humans in South Africa. Studies are also underway to encapsulate modified and nonmodified hemoglobins in liposomes. Further details concerning the properties and physiological effects of these hemoglobins can be found in a review by Alayash [1]. However, transfusion of hemoglobin (Hb)-based blood substitutes, designed for their plasma expansion and oxygen transport capabilities, has resulted in some major problems, such as organ dysfunction, during clinical trials. These events might be linked to deleterious responses occurring in the microvasculature in the presence of the blood substitutes.

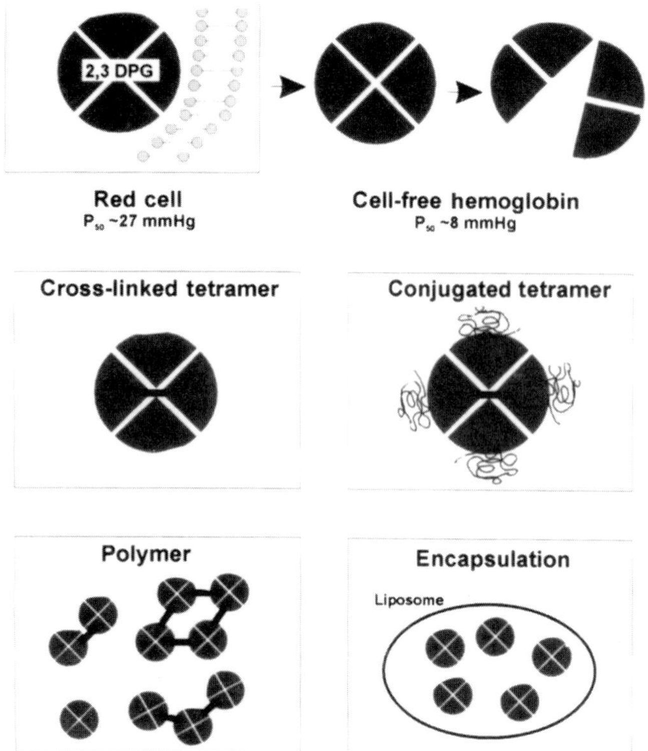

Figure 1 Development of hemoglobin solutions as blood substitutes. SFH is either cross-linked, conjugated with macromolecules, or polymerized in order to stabilize the hemoglobin tetramer and enhance its function. SFH are, in some cases, encapsulated within liposomes. (From Ref. [1].)

Problems Encountered with Hemoglobin-based Blood Substitutes during Trials

Although hemoglobin solutions have been well tolerated in human volunteers and in dialysis, septic shock, and cardiac surgery patients, a number of largely unresolved problems were found during preclinical trials and development of some of these hemoglobin-based substitutes. These include cardiovascular/hemodynamic effects, gastrointestinal changes, immune cell activation, coagulation changes, oxidative stress, and decreased host resistance to overwhelming infection. Preclinical studies reported the detection of myocardial lesions in a number of animal models infused with the Baxter Health Care Inc. product, diaspirin cross-linked Hb (DCLHb). These lesions were characterized by a mild to moderate focal myocardial degeneration and/or necrosis in a highly vascularized portion of the myocardium. Baxter has recently terminated its clinical development of this product as a result of increased fatalities in the test group [2]. In this study 46 percent of 52 patients infused with DCLHb died compared to 17 percent of 46 patients infused with saline solution. A dose–response study performed on dogs by Biopure Corporation has shown that Oxyglobin increases arterial oxygen content in the face of normovolemic anemia, but produces transient clinical signs (skin discoloration, discolored stools, nausea, vomiting). In addition, histopathology of Oxyglobin administration includes activation of tissue macrophages in multiple organs. Hemolink has recently been withdrawn from Phase III clinical trials in cardiac bypass grafting because it produced adverse cardiac events. In addition, numerous animal studies have demonstrated that the administration of extracellular hemoglobin derivatives may lead to a variety of undesirable side effects.

Problems with Hemoglobin-based Blood Substitutes in the Microcirculation

Arteriolar Constriction

Role of Hemoglobin as a Nitric Oxide Scavenger

Many experimental studies have reported an increase in blood pressure after administration of hemoglobin solutions, and pilot clinical trials have confirmed this observation. One theory to explain this phenomenon is that arterioles constrict due to removal of the endogenous vasodilator, nitric oxide (NO), by the hemoglobin. This theory is supported by studies showing that administration of the NO precursor, L-arginine, at the same time as DCLHb reduces the vasopressor effects. It is thought that when Hb is enclosed in red blood cells its ability to scavenge NO from the endothelium is diminished because of the existence of a cell-free plasma layer in the microvessels. The vasopressor effects of Hb-based blood substitutes are generally considered to be a disadvantage because vasoconstriction will reduce the blood supply to the peripheral circulation. However, in some cases the accompanying increase in blood pressure could be advantageous. For example, hemoglobin solutions may improve recovery from cardiac arrest by simultaneously increasing coronary perfusion and oxygenation. In addition, in cases of sepsis when an increase in inducible NO synthase activation causes an overproduction of NO, leading to hypotension, a scavenger of NO would be beneficial. Apart from scavenging the vasodilator, NO, modified hemoglobins may also cause the endothelium to release the vasoconstrictor, endothelin, or activate endothelin receptors.

Role of Hemoglobin as an Oxygen Deliverer

Recently, a third explanation has been proffered to explain hemoglobin-induced vasoconstriction: That is, the Hb provides excess oxygen to tissue, and in response, the arterioles narrow to reduce blood flow and hence tissue oxygen delivery. This response is known as "autoregulation." For this reason, a new blood substitute, Hemospan (Sangart, San Diego, CA), has been designed to reduce oversupply of oxygen. Hemospan is prepared by conjugating polyethylene glycol to hemoglobin tetramer from outdated human blood [3]. This product, in contrast to most of the other blood substitutes, has a high oxygen affinity (low po_2) and a high viscosity (similar to blood rather than to plasma). The high oxygen affinity ensures that the modified Hb will release oxygen only to tissue that is anoxic, and thus autoregulatory vasoconstriction, produced by excess oxygen delivery, will be eliminated. It is thought that because of the higher viscosity of this product, the shear stresses in the microvessels will be maintained, thus ensuring that the endothelium continues to release NO at a rate similar to that obtained with blood. However, this is unlikely to occur because shear stress is proportional not only to viscosity, but also to the rate of strain in the vessel (dv/dr, where v = flow velocity and r = radius). A higher viscosity will reduce the volumetric flow rate, and hence flow velocity and rate of strain. Thus the shear stress will not increase with increased viscosity unless the flow rate is maintained. At present Hemospan is in the preclinical phase of testing, but Phase I trials are planned.

Toxicity

It is important to remember that blood substitutes should not just be considered as oxygen carriers. One other important property of whole blood is that it has a high antioxidant capacity and makes a significant contribution to the total antioxidant defenses of the body. Infusion of resuscitation fluids may therefore influence the antioxidant capacity of plasma, dependent on their composition. It is vital that the antioxidant capacity of a transfusion fluid be adequate because reperfusion of the circulation after loss of blood triggers production of reactive oxygen species (ROS). The effects of dilution of blood on its antioxidant properties were demonstrated elegantly by Moison et al. [4]. They found that addition of isotonic saline, or pasteurized plasma protein solution (lacking uric acid and vitamin C with reduced sulfhydryl level), to cord blood from babies decreased the peroxyl radical trapping capacity. In contrast, fresh, frozen plasma did not lower this capacity. It was hypothesized that the use of resuscitation fluids with low antioxidant capacity may temporarily decrease the ability of the baby to inactivate reactive oxygen species.

However, not only do most modified hemoglobins have a low antioxidant capacity; there is evidence that modified hemoglobins, injected in vivo, produce highly reactive oxygen species themselves, leading to tissue damage [5]. The ROS are formed because Hb is susceptible to oxidation and auto-oxidation. Some Hb-based blood substitutes have been shown to oxidize more readily than Hb contained in red blood cells in response to chemical modifications aimed at lowering oxygen affinity. Although cell-free Hb may present a low risk to people with normal redox status, patients who are sick and have a poor antioxidant supply may be at risk. Oxidative damage is particularly dangerous in the microcirculation because gaps form between the endothelial cells resulting in excess leakage of plasma components into the interstitium. Such leakage disturbs the fluid balance between blood and tissue and alters the kinetics of delivery of intravascularly injected drugs and endogenous enzymes and hormones to various tissues.

POSSIBLE MECHANISMS RESPONSIBLE FOR HEMOGLOBIN-INDUCED TOXICITY

Three possible mechanisms can account for the propensity of modified Hbs to form oxidative by-products. First, the ferrous and ferric forms of Hb can react with hydrogen peroxide (H_2O_2), produced by neutrophils or macrophages, to form the highly reactive ferryl intermediate Hb Fe^{4+}, and this may lead to heme degradation and cytotoxicity. A second possible redox mechanism involves the release of free iron from Hb after oxidative damage. The free iron then catalyses the reaction between H_2O_2 and superoxide ($O^{-\bullet}_2$), both produced by phagocytes, to form the hydroxyl radical ($OH^\bullet$) by the Fenton reaction. Finally, NO can react with oxygen free radicals, or with Hb, to produce more ROS.

Ferryl hemoglobin. It has been postulated that the excess H_2O_2 reacts with ferrous Hb ($HbFe^{2+}$) and ferric, or "met" Hb ($HbFe^{3+}$), in vivo, to initiate further oxidation cycles resulting in the formation of highly reactive ferryl-Hb ($HbFe^{4+}$). Specifically, when $HbFe^{2+}$ reacts with H_2O_2, it donates two electrons to the H_2O_2, to form $HbFe^{4+}$ and OH^-. In $HbFe^{4+}$ the iron center is at a higher oxidation state. Despite its transient nature, Fe^{4+} heme can peroxidize lipids, degrade carbohydrates, and modify proteins. The further interaction of $HbFe^{4+}$ with H_2O_2 results in the formation of rhombic heme, which is considered to be one of the best measures of the toxicity of a blood substitute. The rhombic heme in which the geometry of the iron is distorted due, in some instances, to the chemical modification of the protein, then initiates a cascade of oxidative side reactions resulting in the formation of free iron.

Fenton reaction. In the hemoglobin molecule the hemes are normally bound in pockets from which water is largely excluded. Conformational changes that open up the heme pockets and allow greater access of water and small anions favor the conversion of the heme iron to the ferric or "met" state. The reaction (auto-oxidation), which also produces superoxide anions, occurs spontaneously in modified hemoglobins. A key study showed that 72 hours after infusion of glutaraldehyde polymerized bovine Hb into an animal, almost 40 percent of circulating hemoglobin is in the met form [6]. This is important because it has been estimated that met Hb concentrations greater than 10 percent significantly decreases the ability of Hb to deliver oxygen to tissues. The met Hb then reacts with H_2O_2 to release free iron and free heme, which decomposes H_2O_2 to form hydroxyl radical ($OH^\bullet$) and hydroxide anion (OH^-) by the Fenton reaction:

$$Fe^{2+} + H_2O_2 \rightarrow Fe^{3+} + OH^{\cdot} + OH^-$$

The Fe^{3+} that is formed can react with superoxide radicals to produce more Fe^{2+}:

$$Fe^{3+} + O_2^{-} \rightarrow Fe^{2+} + O^2$$

This Fe^{2+} then undergoes the Fenton reaction to produce more $OH^\bullet$. Thus, the combination of H_2O_2 and Fe^{2+} represents a highly toxic potential that contributes to peroxidation in hemoglobin-containing systems. One $OH^\bullet$ can result in the conversion of many hundred fatty acid side chains into lipid hydroperoxides. Iron is also capable of catalyzing the production of alkoxy and peroxy radicals from lipid peroxides, and the production of these reactive species could contribute to tissue injury.

In normal blood, the reservoir of heme iron is compartmentalized within the erythrocyte, which limits its ability to act as a catalyst. Erythrocytes are rich in antioxidant enzymes, such as superoxide dismutase (SOD) and catalase, which inactivate the $O^{-\bullet}_2$ and H_2O_2, respectively. These enzymes are in close proximity to Hb within the erythrocytes, and although Hb can release catalytic iron when exposed to oxidant stress, injury is limited because the enzymes react with the resultant ROS before they reach the cell membrane.

Reactions involving nitric oxide. Nitric oxide, which is constitutively produced by endothelial cells and is produced by macrophages when they are activated, can have either a protective or a deleterious effect on tissue in the presence of Hb. The deleterious effect can arise in two ways. One way is from its interaction with $O^{-\bullet}_2$ to form peroxynitrite anion, $ONOO^-$:

$$O_2^{\bullet-} + NO^{\bullet} \rightarrow ONOO^-$$

The amount of $O_2^{\bullet-}$ present determines the amount of $ONOO^-$ that is formed. The concentration of $O_2^{\bullet-}$ in the circulation is kept extremely low by a high concentration of superoxide dismutase (SOD). A balance between $NO^\bullet$ and $O_2^{\bullet-}$ is therefore maintained under physiological conditions, and the reaction between the two to form $ONOO^-$ is limited. When the circulation is perfused with a modified Hb, the Hb reacts very quickly with the $NO^\bullet$ and the resulting reduction in $NO^\bullet$ drives the formation of $ONOO^-$. On protonation, $ONOO^-$ decomposes to the highly reactive $OH^\bullet$. Nitric oxide can also react with oxyHb (Hb(Fe-O_2)) to produce met Hb, which is unstable and easily releases iron that catalyzes production of $OH^\bullet$. The protective effect of $NO^\bullet$ occurs when it scavenges the damaging oxidant peroxynitrite:

$$ONOO^- + NO^{\bullet} \rightarrow NO_2 + NO_2^-$$

It is obvious from these reactions that the role played by $NO^\bullet$, whether it acts as a pro-oxidant or as an antioxidant, depends on the relative amounts of $NO^\bullet$ and (ROS) that are present. The fact that Hb has a high affinity for $NO^\bullet$ has the potential for upsetting an existing balance between $NO^\bullet$ and ROS.

EFFECTS OF HEMOGLOBIN-BASED BLOOD SUBSTITUTES ON MICROVASCULAR PERMEABILITY

It is well known that excess ROS oxidize lipids of the cell membranes. Lipid peroxidation damage of membrane components is thought to play an important role in increasing microvascular permeability. When hemoglobin-based blood substitutes are injected into the circulation, excess ROS, such as H_2O_2, $O_2^{\bullet-}$, $OH^\bullet$, and Fe^{4+}, will form if the hemoglobin increases its oxidation state. Since H_2O_2 can easily diffuse across cell membranes, and $O_2^{\bullet-}$ can traverse membranes via the chloride anion channel, it is likely that these (ROS) will leave the microvasculature and gain access to other cells in the tissue. This action can have deleterious effects. For example, if the ROS reach mast cells in the interstitium they will cause the mast cells to degranulate and release a selection of inflammatory mediators, including histamine, which will increase microvascular permeability. In addition, degranulating mast cells release eosinophil and neutrophil activating factors, triggering these leukocytes to release more ROS. Surprisingly, only a few studies have been performed to determine whether injection of a hemoglobin-based blood substitute into the circulation significantly increases the accumulation of ROS in surrounding tissue. In one study [5], a bolus injection of DBBF-Hb into the circulation of rats was found to rapidly produce excess ROS, as detected by fluorescence of dihydrorhodamine-123, in the intestinal mucosa.

Although it is known that formation of the unstable form of hemoglobin, met Hb, proceeds rapidly after injection of modified hemoglobins into the circulation, and that

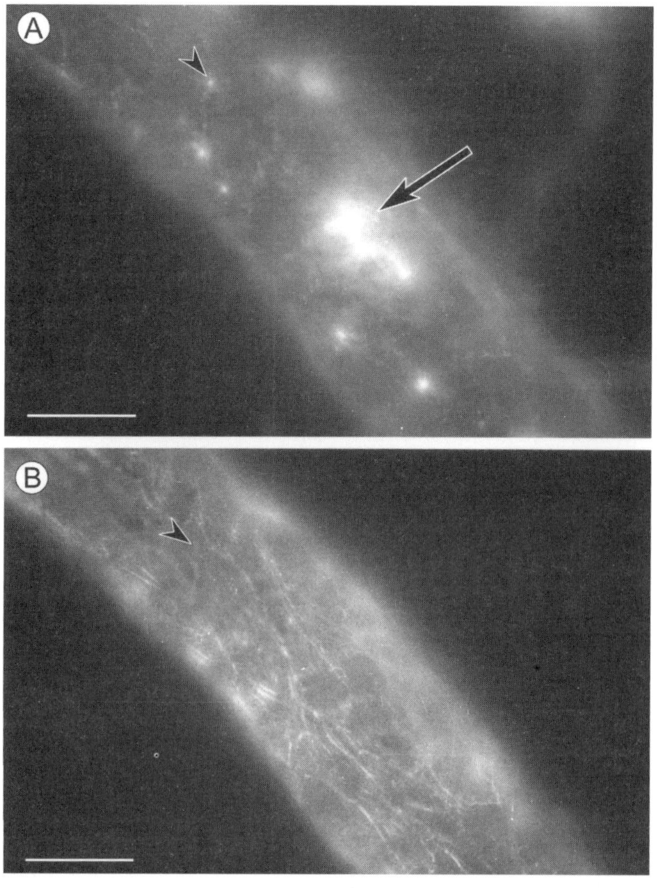

Figure 2 Image pair of FITC-albumin **(A)** and rhodamine phalloidin-stained **(B)** 10-minute DBBF-Hb preparation showing venules. Note small leaks (arrowhead in A) and extensive leak (arrow in A). The small leaks **(A)** are coincident with the endothelial junctions as identified by rhodamine-stained peripheral actin rims (arrowhead in B). Precise positions of the small leaks coincide with discontinuities in PARs. Scale bars = 25 μm. (From Ref. [7]).

ROS increase microvascular permeability, very few studies have been performed to determine whether injection of hemoglobin-based blood substitutes increase microvascular permeability. One study [7] showed that bolus injection of DBBF-Hb, in rats, increased venular leakage to fluorescently labeled bovine serum albumin (FITC-BSA) and produced mast cell degranulation in the rat mesentery. Such changes are characteristic of an inflammatory response. Focal leakage of FITC-BSA from mesenteric venules following a 10-minute perfusion with DBBF-Hb is illustrated by a figure from this study (Figure 2). The upper panel demonstrates focal leaks of FITC-BSA. Small leaks (arrowhead) and a large leak (arrow) are visible. The lower panel shows the same segment of venule, but stained for F-actin. The arrowhead indicates that small leaks are sometimes coincident with focal breaks in the peripheral endothelial actin rim. This type of focal macromolecular leakage is similar to that observed in the presence of histamine. The discrete nature of the leakage means that it is not easily detected in studies in which only the average fluorescence in the tissue is measured. It is disadvantageous for a potential blood substitute to cause microvascular leakage, because the substitute itself will rapidly leave the circulation, and in addition, alterations in transvascular exchange of plasma proteins will disturb the fluid balance between blood and tissue. Increased microvascular leakage also changes the kinetics of delivery of intravascularly injected drugs, and of endogenous enzymes and hormones, to various tissues. When transfusions are needed, for example after hemorrhagic shock, it is important that regulation of microvascular exchange not be compromised.

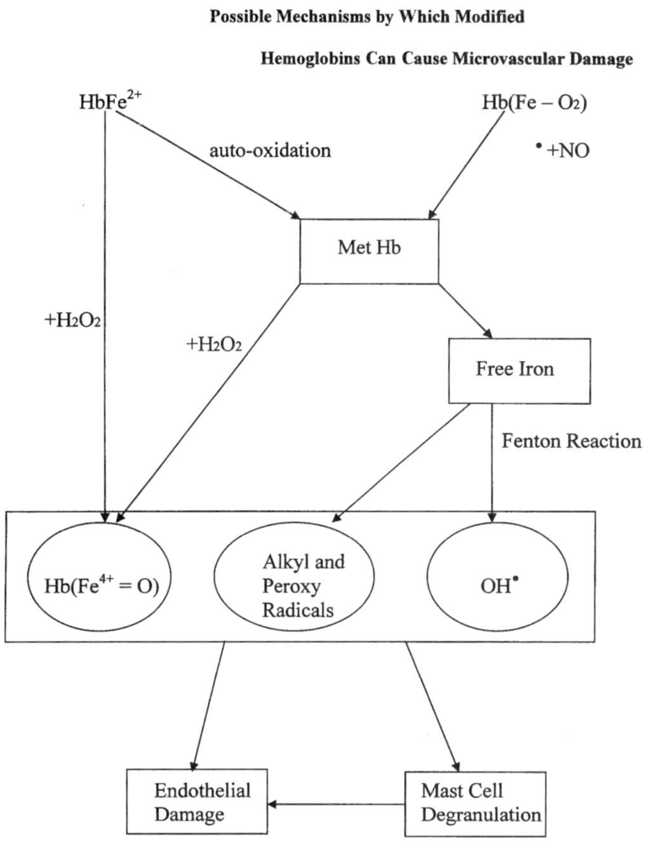

Figure 3 Diagram to show possible mechanisms by which modified hemoglobins can cause microvascular damage. Hb(Fe-O_2): oxyhemoglobin; $NO^•$: nitric oxide; $HbFe^{2+}$: ferrous hemoglobin; met Hb: ferric hemoglobin; H_2O_2: hydrogen peroxide; Hb(Fe^{4+}=O): oxyferryl hemoglobin; $OH^•$: hydroxyl radical.

METHODS TO REDUCE TISSUE DAMAGE BY BLOOD SUBSTITUTES IN VIVO

The diagram in Figure 3 depicts the various mechanisms by which modified hemoglobins can cause tissue damage. From this diagram, it is clear that there are at least three different types of agents that could be administered to reduce microvascular damage: (a) antioxidants, (b) iron chelators, and (c) mast cell stabilizers. Very little progress has been made using iron chelators and mast cell stabilizers, and so only antioxidants will be considered.

Use of superoxide dismutase and catalase. Superoxide dismutase and catalase are endogenous antioxidants for $O_2^{•-}$ and H_2O_2, respectively. Several methods currently incorporate SOD and catalase into cross-linked Hb (i.e., Ref. [8]). This technique ensures that the free-radical scavengers are in close contact with the source of the ROS, the hemoglobin, and also effectively addresses the problem that SOD has a short half-life in blood (10 to 40 minutes). An alternative way of increasing the half-lives of SOD and catalase is to bind them to PEG. Polyethylene glycol–SOD has a half-life of several days, and PEG linkage increases the half-life of catalase from 2 to 50 hours.

Use of nitroxides. Another antioxidant that has been investigated for use with blood substitutes is nitroxide. Nitroxides are able to scavenge $O_2^{•-}$, and thus can act as potent antioxidants. However, free nitroxide is cleared very rapidly from the circulation and very little is left even 5 minutes after intravenous infusion. For this reason a polynitroxylated hemoglobin-based oxygen carrier has been developed in which nitroxide molecules are covalently bound to Hb so that the circulatory half-life of the nitroxide molecules is greatly increased [9]. At this time there is no published record of the ability of hemoglobins, conjugated with antioxidants, to minimize oxidative tissue damage.

Use of selenium. Recently, the protective effect of selenium on Hb-mediated lipid peroxidation has been investigated for use with blood substitutes. The rationale for using selenium is that it is a very powerful antioxidant. Selenium is thought to act as an antioxidant in the body because it is a component of the enzyme, glutathione peroxidase, which catalyzes removal of H_2O_2. Hemoglobin's oxidative reactions are very complex, and so total protection cannot be achieved using hydroxyl radical scavengers exclusively. Sodium selenite (Na_2SeO_3), when administered orally or intravenously, has been shown to significantly reduce the microvascular leakage associated with bolus injection of DBBF-Hb in rats. In addition, Na_2SeO_3 reduces the oxidation rate of DBBF-Hb while in the presence of oxidants in vitro. Thus it appears that Na_2SeO_3 moderates hemoglobin-induced damage at least partly through its interactions with the hemoglobin molecule itself, and that there is no need for glutathione peroxidase to be involved in the process. However, the disadvantage of this treatment is that selenium compounds might only be useful adjuncts to Hb-based blood substitutes that can be reduced by interaction with selenium.

Future Possibilities

The possible mechanisms by which modified hemoglobins cause microvascular damage are summarized in Figure 3. It is obvious that in order to prevent oxidative tissue damage by hemoglobin-based blood substitutes during transfusion, the formation of, and the effects of, different nitrogen- and oxygen-derived radicals must be prevented. To prevent formation of such radicals it is necessary to reduce the tendency of the hemoglobin to oxidize, either by chemical modification during manufacture, or by adding an appropriate reducing agent at the time of infusion. Ideally, a therapeutic agent would scavenge all deleterious radicals, whether or not their production is catalyzed by iron, and would be able to cross biological membranes. One possible candidate that has been suggested is Tempol (4-hydroxy-2,2,6,6-tetramethylpiperidine-N oxyl), a stable nitroxide that attenuates the effects of peroxynitrite and oxygen-derived free radicals such as superoxide anions and hydroxyl radicals [10]. Unlike recombinant superoxide dismutase, which is not able to cross biological membranes, Tempol permeates biological membranes and accumulates in the cell cytosol. Future studies using molecules such as Tempol, in conjunction with a range of Hb-based blood substitutes with various oxygen affinities and NO scavenging capacities, may lead to a product that both is nontoxic and delivers oxygen at a rate suitable for the conditions.

Glossary

Auto-oxidation: The removal of electrons from a molecule. In the case of Hb, Fe^{2+}, or ferrous Hb, is auto-oxidized to form Fe^{3+}, or ferric (met) Hb.

Reactive oxygen species: A cluster of atoms, including oxygen, which contains an unpaired electron in its outermost shell. This is an extremely unstable configuration that reacts quickly with other molecules to achieve the stable configuration of four pairs of electrons in the outer shell.

Redox: Changes in the redox status of a molecule involve the addition (reduction) or removal (oxidation) of electrons to or from the molecule.

Shear stress: The intensity of force per unit area acting tangentially to the area.

Vasopressor: A substance that causes an increase in blood pressure by constricting the arteries.

Bibliography

1. Alayash, A. I. (1999). Hemoglobin-based blood substitutes: Oxygen carriers, pressor agents, or oxidants? *Nat. Biotech.* **17**, 545–549. *This review provides a summary of the different types of hemoglobin-based blood substitutes for which clinical trials are underway, and provides some insights into the complex redox chemistry of hemoglobin.*
2. Sloan, E. R., Koenigsberg, M., Gens, D., Cipolle, M., Runge, J., Mullory, M., and Rodman, G. (1999). Diaspirin cross-linked hemoglobin (DCLHb) in the treatment of severe traumatic hemorrhage shock: A randomized controlled efficacy trial. *JAMA* **282**,1857–1864.
3. McCarthy, M. R., Vandergriff, K. D., and Winslow, R. M. (2001). The role of facilitated diffusion in oxygen transport by cell-free hemoglobins: Implications for the design of hemoglobin-based oxygen carriers. *Biophys. Chem.* **92**, 247–117.
4. Moison, R. M. W., van Hoof, E. J. H. A., Clahsen, P. C., van Zoeren Grobben, D., and Berger, H. M. (1996). Influence of plasma preparations and donor red blood cells on the antioxidant capacity of blood from newborn babies: An *in vitro* study. *Acta Paediatr.* **85**, 220–224.
5. Baldwin, A. L., and Wiley, E. B. (2002). Selenium reduces hemoglobin-induced damage to intestinal mucosa. *Art. Cells Blood Subs. Immob. Biotech.* **30**(1), 1–22.
6. Lee, R., Neya, K., Svizzero, T. A., and Vlahakes, G. J. (1995). Limitations of the efficacy of hemoglobin-based oxygen carrying solutions. *J. Appl. Physiol.* **79**, 236–242.
7. Baldwin, A. L. (1999). Modified hemoglobins produce venular interendothelial gaps and albumin leakage in the rat mesentery. *Am. J. Physiol.* **277**, H650–H659. *This paper is one of a series in which fluorescence microscopy is used to demonstrate the focal microvascular leakage caused by intravascular injection of DBBF-Hb.*
8. D'Agnillo, F., and Chang, T. M. (1993). Cross-linked hemoglobin-superoxide dismutase-catalase scavenges oxygen-derived free-radicals and prevents methemoglobin formation and iron release. *Biomat. Artif. Cells, Immobil. Biotech.* **21**(5), 609–621.
9. Saetzler, R. K., Arfors, K. E., Tuma, R. F., Vasthare, U., Ma, L., Hsia, C. J. C., and Lehr, H.-A. (1999). Polynitroxylated hemoglobin-based oxygen carrier: Inhibition of free radical-induced microcirculatory dysfunction. *Free Radical Biol. Med.* **27**(1/2), 1–6.
10. Thiemermann, C., McDonald, M. C., and Cuzzocrea, S. (2001). The stable nitroxide, Tempol, attenuates the effects of peroxynitrite and oxygen-derived free radicals. *Crit. Care Med.* **29**(1), 223–224.

Biography

Dr. Baldwin is a Professor of Physiology at the University of Arizona and is currently Treasurer of the Microcirculatory Society. Her work on reactive oxygen species and hemoglobin-induced microvascular damage has been supported by NIH for the past 8 years.

CHAPTER 95

Determinants of Platelet Glycoprotein Ib-IX-V Mediated Microvascular Hemostasis: Are They Different from Those Regulating Thrombosis?

Michael H. Kroll and Shuju Feng

Department of Medicine, Section of Thrombosis Research, Baylor College of Medicine, VA Medical Center, Department of Biomedical Engineering, Rice University, Houston, Texas

Introduction

Physiological hemostasis and pathological thrombosis take place within a complex in vivo milieu. The complex variables that modulate prohemostatic and prothrombotic responses are probably best organized by examining them within Virchow's triad. Virchow's triad reminds us that hemostasis and thrombosis are regulated by the simultaneous interactions among blood (cells and soluble constituents), blood vessel (endothelium, subendothelium, and smooth muscle), and blood flow (related to diameter, branching, turbulence, and obstruction) (Figure 1). In considering the variables that affect—and perhaps differentiate—glycoprotein (Gp) Ib-IX-V-dependent hemostasis and thrombosis, one should begin by examining features within Virchow's triad that are unique to the microcirculation, the site of hemostasis.

The most unique feature of the microcirculation is its blood vessels. The afferent arterioles are defined as 100 μm descending to <10 μm in diameter, with little subendothelium separating the luminal endothelium from the vascular smooth muscle cell layer. The capillary bed is comprised of an extensive collateral network of tiny (about 6 μm) thin-walled vessels that function in gas, solute, or cell exchange and transport. The capillaries are in some cases discontinuous (e.g., "sinusoidal" capillaries as are found in the bone marrow, liver, and spleen) or fenestrated (e.g., "transport" capillaries as are found in endocrine organs and the renal glomeruli). In the brain, the capillaries are tightly joined to one-another to create a barrier for selective transport and exchange. Capillaries then converge into widening branches of the efferent venules, which expand in size from tens to hundreds of microns in diameter, and possess two important features: fenestrae that permit egress of cells and proteins adluminally and valves that prevent backflux into the capillary network. These features of the microcirculatory vascular network vary depending on their location, and it should be emphasized that unique anatomical and functional characteristics can be found in unique capillary beds, such as the skin, skeletal muscle, brain, liver, kidney, lungs, and heart. Such unique characteristics undoubtedly influence an organ's vulnerability to specific hemostatic and thrombotic challenges, although the mechanisms by which these influences develop are in most cases unknown.

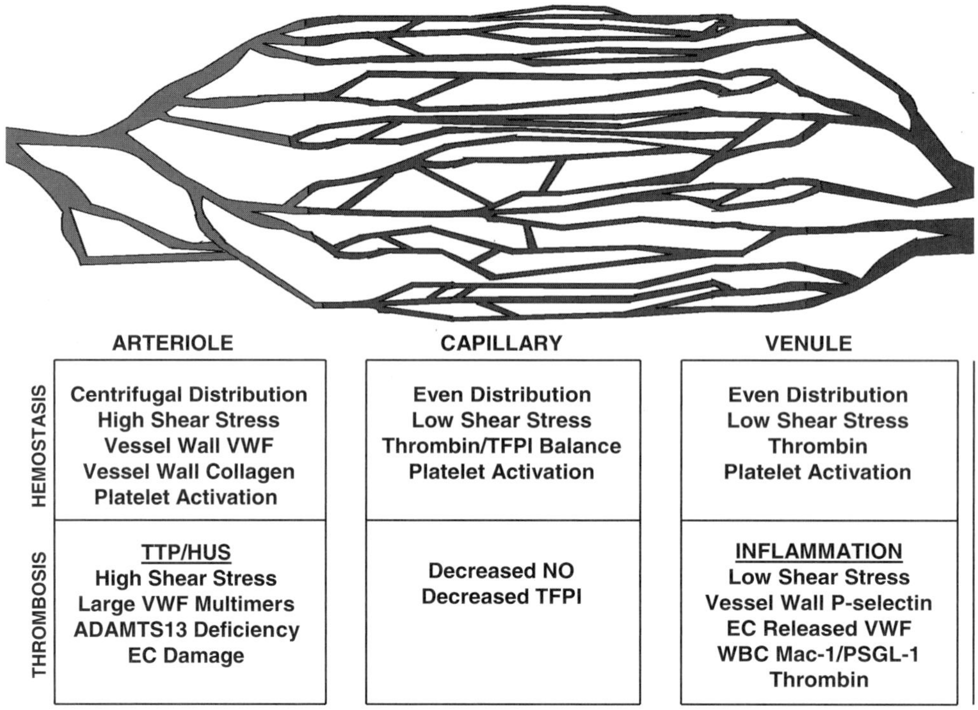

Figure 1 The elements within Virchow's triad that direct hemostasis or thrombosis in the microcirculation. The microcirculation—which is generally defined by a vascular lumen diameter less than 100 µm—encompasses three vessel beds of strikingly different morphology and rheology: arterioles, capillaries, and venules. Each possesses a unique repertoire of physiological and pathological attributes and responses. Distribution refers to where the majority of blood platelets are found within a cross-sectional vessel lumen. VWF, von Willebrand factor; TTP, thrombotic thrombocytopenic purpura; HUS, haemolytic–uremic syndrome; ADAMTS13, a disintegrin and metalloproteinase with thrombospondin type-1 motif 13; EC, endothelial cell; TFPI, tissue factor pathway inhibitor; WBC, white blood cell; PSGL-1, P-selectin glycoprotein ligand-1. (see color insert)

Rheological principles governing microvascular hemostasis and thrombosis emanate from its unique anatomic features. High shear stress (up to 60 dyn/cm^2) "feed" arterioles bifurcate into branch arterioles at intervals of about 200 µm, with declining flow velocities and shear stress as branches narrow (for example a first branch shear stress of 20 dyn/cm^2 and a fourth branch shear stress of 9 dyn/cm^2). At branch points there is turbulence, increased resistance and even backflux as cyclical blood flow overcomes afferent arteriolar autoregulatory vasoconstriction triggered with every cardiac diastolic relaxation. As blood enters the extensively branched capillary bed, shear stress decreases further. There are very few published data that provide one with precise measurements of shear stress or other rheological parameters in capillaries, but estimates based on viscosity calculations and flow and pressure measurements leading into and out of capillaries indicate that physiological capillary blood flow is low flow velocity and generates low shear stresses. The branching postcapillary venules then converge into efferent vessels of increasing diameter and capacitance and gradually increasing—but still low—shear stress (less than 5 dyn/cm^2). Venular backflux is limited by tiny two-cusped valves even when low flow and high capacitance result in venular segmental flow reversal. Flow turbulence decreases and flow velocity increases as smaller venular branches come together into larger venules and, eventually, small veins (more than 100 µm in diameter).

The unique anatomy and rheology of the microvasculature direct the activity of the platelets, coagulation proteins, red cells, and leukocytes in hemostasis and thrombosis, and the behavior of blood within the microcirculation directs vascular and rheological responses. Although the dynamics of these interactions exceed the boundaries of simplification, their complexity is best appreciated from a perspective that brings *time* into the biological framework: Every element of Virchow's triad—whether pro- or antihemostatic or pro- or antithrombotic—works within a response that occurs over time. This is important for the blood vessel (e.g., vascular smooth muscle vasoreactivity will constrict or dilate an afferent arteriole) and for blood flow (e.g., a nondeformable 7-µm red cell will obstruct a 4-µm capillary), but it is absolutely essential for the physiology and pathophysiology of the blood. In considering how the microvascular environment interacts bidirectionally and temporally with the blood to promote hemostasis and thrombosis, two fundamental observations require emphasizing. The first is that high shear stress (i.e., arteriolar) hemostasis *and* thrombosis are predominantly platelet mediated, with the coagulation proteins and other blood cells playing an important but secondary role (e.g., as vaso-occlusion causes flow and shear stress to diminish, red cells, leukocytes, and fibrin accumulate in the nascent thrombus). The second is that low shear stress thrombosis (i.e., capillary and venular) is predominantly fibrin mediated, with red cells and leukocytes playing essential primary roles, and platelets generally coming into the picture secondarily in response to coagulation factor activation (e.g., the generation of thrombin) or leukocyte recruitment (e.g., P-selectin glycoprotein ligand-1 on leukocytes bound to inflamed or damaged postcapillary venules will recruit activated platelets expressing P-selectin). There is, however, one example of a platelet-initiated low shear stress (venular) response. Platelets attach and roll along inflamed venule endothelium and the transitory contacts that mediate this response are both GpIb-IX-V mediated: platelet GpIbα binding to venular P-selectin and platelet GpIbα binding to venular EC surface VWF. It is noteworthy that these interactions do not lead to significant thrombus accrual, suggesting that elevated shearing stresses are required for platelet-dependent thrombosis to occur in the microvasculature.

Molecular Effectors of Platelet-Mediated Hemostasis and Thrombosis

The GpIb-IX-V Receptor Complex

The platelet Gp Ib-IX-V complex—universally credited as an essential component of hemostasis and as a notorious mediator of arterial atherothrombosis—is also a critically important factor in the pathogenesis of human diseases affecting the microvasculature. It plays a primary role in afferent arteriolar and capillary thrombosis in thrombotic thrombocytopenic purpura (TTP) and the haemolytic–uremic syndrome (HUS). It may also play a primary role in platelet adherence to inflamed venular endothelium and a significant secondary role in effecting venular thrombosis after leukocytes have been recruited to the injured microvasculature.

GpIb-IX-V directs specific platelet responses by organizing disparate factors operating within Virchow's triad into a single series of functional responses: adhesion, secretion, and aggregation. Its functional versatility—as a switch for turning platelet-dependent hemostasis and thrombosis "on" within different vascular and rheological microenvironments—relates to its primary structural attributes. GpIb is made up of GpIbα and GpIbβ disulfide bonded at a single perimembranous extracellular site. There are about 25,000 GpIb heterodimers per platelet noncovalently associated with GpIX and GpV, and the stoichiometry of the GpIbα/β-IX-V complex is believed to be 2:2:2:1. Each of the members of the GpIb-IX-V complex is a related member of the leucine-rich repeat protein family and each has a single transmembranous domain. The "business ends" of the complex are found mainly on GpIbα: the large extracellular N-terminal ligand binding domain and the 96 amino acid cytoplasmic C-terminal cytoskeletal-binding and signaling domain. In addition, experimental data suggest that the 34 amino acid cytoplasmic domain of GpIbβ modulates platelet activation and the extracellular domain of GpV, which is a substrate for thrombin cleavage, modulates platelet responses to thrombin.

GpIb-IX-V Ligands

Pathological shear stress-dependent platelet adherence in arteries and arterioles is triggered by platelet GpIbα binding to plasma or vessel wall *von Willebrand factor* (VWF). VWF is synthesized by vascular endothelium and by megakaryocytes. It is constitutively released adluminally (into the subendothelium) and abluminally (into the blood) by the endothelium, and it is also stored and secreted following cellular activation (stored in endothelial cell Weibel-Palade bodies and in platelets' α-granules). Subendothelial VWF is most abundant in the macrovasculature large veins and arteries (most large veins > pulmonary artery > cerebral arteries > aorta > coronary arteries > renal arteries > hepatic arteries > pulmonary vein). It is generally less abundant in microvasculature subendothelium, and its distribution in the microvasculature is noteworthy: venules > arterioles > capillaries, with very little or no VWF observed in any embryo capillary bed (in mice) and with relatively little VWF observed in adult myocardium (in pigs).

It is a multimeric protein built up of tens to hundreds of disulfide-bonded multivalent protomeric units. Larger multimers appear to have greater hemostatic and prothrombotic properties. Each protomeric subunit is composed of two disulfide-linked mature VWF polypeptides, each one of which is divided structurally and functionally into several domains: the A, B, C, and D domains. The A domains are of particular importance because the A1 domain forms the primary GpIbα recognition site and binds to type VI collagen, the A2 domain contains the recognition sequence for degradation by the VWF multimer-cleaving protease ADAMTS (a disintegrin and metalloproteinase with thrombospondin type 1 motif) 13, and the A3 domain binds to fibrillar collagens type I and III found in arterial subendothelium, thus allowing soluble VWF to tack down onto the subendothelium of ruptured atherosclerotic plaques. Two other domains of VWF are important for hemostasis and thrombosis: The C2 domain contains a RGD integrin recognition domain essential for VWF binding to platelet αIIbβ3 and an N-terminal D domain binds Factor VIII.

VWF binding to GpIbα is required for microvascular hemostasis, which appears to be triggered in the injured arteriole—a high shear stress microenvironment—by soluble and subendothelial VWF alone or bound to collagen (type VI may be the predominant microvascular collagen) attaching to platelets via GpIbα, with the subsequent activation of αIIbβ3 to a ligand-receptive conformation. The high shear stress in the arteriole limits fibrin deposition and leukocyte recruitment, and neither soluble coagulation factors nor leukocyte number or function contribute in any clinically important manner to microvascular hemostasis in the epithelium (such as skin, mucous membranes, and the urinary and GI tracts). As shear stress falls in the occlusive hemostatic plug to venular levels, fibrinogen binding to activated αIIbβ3 is important for interplatelet cohesion. The importance of fibrinogen is emphasized by clinical observations that fibrinogen deficiency causes a severe hemostatic defect. Regional regulation of the VWF-triggered hemostatic plug induced by a bleeding time wound is remarkably fine-tuned, as the platelet-rich thrombi accrue only at the mouths of the transected arterioles, and little or no platelet accumulation occurs within the adjacent arteriole lumen.

VWF-dependent arteriolar platelet thrombosis is the hallmark of TTP and HUS. In TTP the primary pathogenetic event may be an acquired deficiency of ADAMTS13, resulting in persistent "ultralarge" VWF multimers—on endothelial cell (EC) surfaces and in the blood—effecting GpIb-IX-V-dependent platelet adhesion, secretion (predominantly dense granules which contain costimulatory adenosine diphosphate), αIIbβ3 activation, and aggregation. In diarrhea-associated HUS the primary pathophysiology may involve an enterotoxin-mediated overstimulation of "ultralarge" VWF multimers from arteriolar endothelial Weibel-Palade bodies. This bolus of large VWF multimers somehow remains attached to the EC surface and thereby recruits passing platelets through the shear-dependent binding of platelet GpIbα to the EC-associated VWF. A similar effect may occur in low-shear-stress postcapillary venules when rapid EC Weibel-Palade body release is stimulated by calcium ionophore or histamine, suggesting that there may be clinical conditions associated with venular inflammation in which part of the pathophysiological process is caused by VWF/GpIb-IX-V mediated platelet recruitment within a low shear stress microenvironment.

There are several other molecules relevant to microvascular hemostasis and thrombosis that bind directly to GpIb-IX-V. These include *P-selectin*, which is expressed on inflamed venules and mediates platelet rolling through a direct interaction with GpIbα. Platelet GpIbα binding to the leukocyte integrin receptor *Mac-1*—expressed on activated neutrophils and macrophages—is involved in leukocyte recruitment to platelet thrombi and in platelet recruitment to sites of microvascular leukocyte accumulation. Platelet GpIbα binds directly to *thrombin* and *Factor XI*, and such binding appears to accelerate coagulation at the surface of activated platelets under low-shear-stress (capillary and venular) conditions. GpIbα also binds to *high-molecular-weight kininogen* and *Factor XII*, and these interactions block thrombin binding to GpIb-IX-V and thrombin-induced platelet activation. Thrombin also binds to GpV, and this interaction may be involved in attenuating thrombin-induced platelet activation.

Co-receptors

Although the GpIb-IX-V complex is required for the initiation of microvascular hemostasis and is a primary factor contributing to microvascular thrombosis in TTP and HUS, several additional platelet receptors are also involved in both microvascular physiology and pathology. In every case, the co-receptor functions to support platelet-mediated microvascular responses *after* GpIb-IX-V engages ligands (the co-receptor supports *postadhesion* responses), and in some cases the co-receptor requires activation or expression downstream of VWF/GpIb-IX-V interactions (αIIbβ3 and α1β2 are "activated," while P-selectin is expressed following α-granule secretion). Observations of human and/or mouse deficiencies of *αIIbβ3* (the human defect is called *Glanzmann's thrombasthenia*) reveal that it causes a severe bleeding disorder but protects against macrovascular arterial thrombosis. Similar observations have been made with the *protease-activated receptors* that mediate thrombin-induced platelet activation, the *thromboxane* A_2 receptor, and the $P2Y_{12}$ *receptor* that binds adenosine diphosphate (ADP), is the target of thienopyridine antiplatelet drugs, and appears to be indispensable to platelet aggregation triggered by VWF binding to GpIb-IX-V. Similar but lesser effects on hemostasis and macrovascular thrombosis are seen in mice lacking the $P2Y_1$ *receptor* or *P-selectin*. In contrast, human deficiencies of the collagen receptors—*GpVI* or integrin *α2β1*—result in a mild bleeding diathesis, while GpVI- or α2β1-deficient mice have only a small or no hemostatic defect, respectively. Either GpVI or α2β1 deficiency or pharmacological perturbation results in delayed and decreased ex vivo thrombus formation on type 1 collagen under both arteriolar and low-shear-stress conditions, indicating that GpVI and α2β1 (probably to a lesser extent in comparison to GpVI) are important but secondary mediators of microvascular thrombosis. The prototypical example of a platelet co-receptor that appears to effect hemostasis and thrombosis paradoxically is platelet endothelial cell adhesion molecule (PECAM)-1, which is expressed by the microvascular endothelium and platelets. Mice deficient in PECAM-1 have a hemostatic defect not because of a loss of platelet expression, but because they lack homotypic interactions between PECAM-1 on adjacent ECs needed to regulate the endothelial component of Virchow's hemostatic response. In fact, PECAM-1-deficient platelets are actually hyperresponsive to both VWF and collagen because PECAM-1 is a negative regulator of GpIb-IX-V- and GpVI-dependent signaling.

ADAMTS13

ADAMTS13 is the VWF multimer-cleaving protease. It is synthesized in the liver, circulates in blood, and may attach to the vascular endothelium. Although the regional distribution of ADAMTS13 activity is not yet understood, based on the end-organ pathology of TTP it appears to play an important role in processing "ultralarge" VWF multimers—released constitutively and secreted following EC stimulation—in the cerebral, mesenteric, myocardial, splenic, renal, pancreatic, and adrenal arterioles. It also appears that arteriole-level shear stress is required for ADAMTS13-mediated cleavage of "ultralarge" VWF multimers: higher levels of shear stress open up or untangle the multimers and thereby expose the ADAMTS13 cleavage site in the VWF monomer A2 domain. The hypothetical mechanism by which ADAMTS13 deficiency leads to TTP presents a tidy pathophysiological model. Under physiological conditions, ADAMTS13 is synthesized, homes to the arteriole EC, and, under arteriole levels of shear stress, breaks down prothrombotic "ultralarge" VWF multimers. This maintains blood flow within the microvasculature. When ADAMTS13 is deficient, arteriolar-level shear stress triggers platelet GpIbα binding to the unprocessed multimers, thus causing the arterioles to become occluded with both EC-attached platelets and platelet clumps that exceed the diameter of the narrowing arteriolar branches. This leads to ischemia and infarction of the involved organs.

Other Regulatory Vascular Factors

An intact vascular endothelium actively maintains blood fluidity. Recall that a bleeding time wound shows platelet thrombus formation only at the site of arteriolar transection. This is because the adjacent arteriolar, capillary, and venular ECs constitutively secrete or express on their surface molecules that prevent GpIb-IX-V-dependent platelet adhesion, secretion, and aggregation. These include *PGI_2* and *nitric oxide* (which inhibit GpIbα-mediated adhesion and signaling), the ectoADPase *CD39* (which breaks soluble ADP down), *thrombomodulin* and *heparin sulfates* (which bind and inactivate thrombin), and urokinase-type plasminogen activator (*uPA*) and tissue plasminogen activator (*tPA*) (which generate plasmin capable of degrading VWF and fibrinogen). The capillary endothelium is particularly richly endowed with *tissue factor pathway inhibitor* (TFPI), which is both secreted into the blood and retained on capillary endothelium, where it binds to and inactivates the tissue factor/Factor VIIa/Factor Xa complex, thereby eliminating thrombin generation and the assembly of a platelet GpIbα-based site of fibrin generation and Factor XI activation. This suggests that the capillary may be an important gate preventing the initiation of coagulation despite continuous low-level exposure to prothrombotic stimuli.

Only uPA or tPA deficiency is associated with spontaneous microvascular thromboses, but a deficiency of any one of these antiplatelet or anticoagulant molecules leads to exaggerated responses to thrombotic stimuli. The exception to this is CD39 deficiency, which causes a severe bleeding diathesis due to compromised hemostasis because elevated blood levels of adenine nucleotides lead to $P2Y_1$-mediated platelet desensitization. This is a useful reminder that time-dependent desensitization can affect many components of Virchow's triad.

Other Regulatory Blood Factors

Platelet GpIb-IX-V-dependent hemostasis is mainly a physiological arteriolar response to injury. In addition to the blood factors that interact directly with GpIbα (see the earlier section on GpIb-IX-V ligands), there are blood elements that modify GpIb-IX-V/vascular interactions indirectly. Perhaps most important is the red cell. Under arteriolar flow conditions the red cells flow centrally and push the platelet stream peripherally toward the arteriolar wall. Because blood flow in a tubular arteriole can be considered to be parabolic and comprised of an infinite number of infinitesimal laminae, the centripetal movement of platelets exposes them to flow laminae of lowest velocity and highest shear stress, thus slowing them and making wall collisions more efficient (or sticky). As arterioles narrow and branch into capillaries their luminal diameter narrows, red cells are excluded from the central stream (and eventually the entire stream), flow velocity falls, and platelets become evenly dispersed throughout the bloodstream. If the average pulmonary capillary diameter is calculated to be 5.8 μm, and the size of a platelet is 3 μm and a red cell is 7 μm, it becomes intuitively apparent that platelets flow *relatively* better (e.g., faster) than red cells through capillaries. This facilitates red cell–mediated oxygen uptake (in the pulmonary alveoli) and delivery (everywhere else). It also keeps platelets from slowing and sticking: Platelet/capillary interactions are minimized simply because fewer platelets are at the wall and more platelets are in the central stream with highest flow velocity. So platelet thrombosis in capillary beds adjacent to bleeding time wounds does not occur, at least in part, because of rheological factors—which are a direct consequence of capillary diameter and red cell size, shape, and deformability—that keep platelets flowing rapidly through the thin central stream of capillary blood (calculated velocity of platelet flow through a pulmonary capillary is about 500 μm per second).

A similar rheological environment is found in the venules: Until red cells that have squeezed their way through the capillary bed gradually queue back into concentric central stream laminae, platelets remain randomly dispersed throughout the venular lumen even as the flowing blood accelerates up to velocities three to four times greater than those found in the capillaries. This means that platelet and venule wall collisions are relatively rare and that efficient collisions (collisions resulting in attachment) between platelet GpIbα and inflamed venular P-selectin or surface VWF are even rarer, suggesting that other blood factors enter the milieu and participate in pathological venular thrombosis. Perhaps the most important factor is the state of platelet activation: Platelets that enter the capillaries and venules already activated are most likely to be deposited in these vascular beds. The mechanisms by which platelet activation affects their adherence to venular endothelium are unclear, although it appears that P-selectin on activated platelets is the key and that it binds to venular and arteriolar endothelium by recognizing EC PSGL-1 and related molecules. Activated platelets also bind lymphocytes, neutrophils, and monocytes, and they release growth factors and cytokines. These responses—alone or in combination, but always operating within Virchow's triad—contribute substantially to pathological venular thrombosis. Mechanisms by which platelets become activated within the venular circuit are not clear except in one case: In bacterially induced sepsis, splenic venular endothelium is activated to express tissue factor (TF), which generates thrombin through the "extrinsic" coagulation pathway (the TF/Factor VIIa complex activates Factor X, and Factor Xa—by a calcium and phospholipid-dependent reaction—cleaves prothrombin to thrombin).

Clinical Models That Separate GpIb-IX-V Mediated Thrombosis from Hemostasis

Bernard-Soulier Syndrome (BSS)

BSS is rare human genetic disorder in which platelet GpIb-IX-V is absent or dysfunctional. Such patients suffer a bleeding disorder manifested by mucosal and cutaneous bleeding following hemostatic insults that are clinically trivial when platelet GpIb-IX-V is working properly (referred to as "easy bruising or bleeding"). Their bleeding times are always elevated. The impact of BSS on the development and natural history of atherothrombosis of the mid-sized arteries—or other vasculopathies—is not of any obvious significance, but BSS platelets show poor VWF-dependent adherence and aggregation in vitro under high-shear-stress conditions. The only acquired human disorder that resembles BSS is iatrogenic: Blood-banked platelets gradually suffer a BSS-like lesion when their ligand-binding extracellular domain of GpIbα is slowly proteolyzed during storage.

Von Willebrand Disease (VWD)

VWD includes a variety of disorders defined by a deficiency or dysfunction of VWF. There are three major types of VWD. Type 1 VWD is due to a mild deficiency and type 3 VWD is due to severe or absolute deficiency. In both of these diseases VWF structure and multimerization are normal. Their clinical impact is directly proportional to the magnitude of the VWF deficiency. There are also VWD types and subtypes in which the patient suffers mild provoked bleeding because the larger normal plasma VWF multimers (not the "ultralarge" forms that participate in TTP and HUS) are deficient. Most type 2A VWD is due to the production of mutant VWF multimers that are unusually sensitive to normal proteolytic processing. Type 2A VWD is less commonly caused by mutations that interfere with the normal synthesis of plasma VWF multimers. Type 2B VWD is due to VWF mutations that render it unusually avid for platelet GpIb-IX-V; in type 2B VWD, VWF binds to GpIbα even in the absence of a modulator (e.g., ristocetin) or elevated shear stress. When this happens in vivo, larger VWF multimers become depleted from blood plasma (they bind to circulating platelets) and some patients suffer a resulting "consumptive" thrombocytopenia due to platelets with surface-bound VWF being cleared from the intravascular compartment by some unknown mechanism(s). A similar loss of large VWF multimers from blood onto platelet surface GpIb-IX-V also occurs in *pseudo-VWD*, which is caused by mutations in the extracellular ligand-recognition domain of GpIbα.

In considering the pathophysiology of VWD, two clinical observations direct two general conclusions about the function of VWF in physiology and pathology. The first conclusion is that larger plasma VWF multimers are required for normal hemostasis: Persons with type 2 or pseudo-VWD have a bleeding disorder. The second conclusion is based on the observation that direct multimeric VWF binding to GpIb-IX-V—as occurs in type 2B and pseudo-VWD—has no effect on platelet-dependent hemostasis and thrombosis in vivo independent of the fact that it causes an intravascular depletion of larger VWF multimers. This suggests that platelet-surface VWF is not an important determinant of hemostasis or thrombosis unless it becomes surface-bound by a series of specific interactions developing within the triad of Virchow.

Moderately severe underproduction thrombocytopenia (e.g., as occurs in leukemic patients who have received intensive induction chemotherapy and is defined by blood platelet concentrations of ~20,000 to 50,000/μL) rarely causes spontaneous dermal microvascular hemorrhage (petechiae). Severe underproduction thrombocytopenia (fewer than 10,000 platelets/μL blood) inevitably causes petechiae. BSS (which is accompanied by thrombocytopenia and giant platelets) and most VWD (in which platelet number and structure are normal) rarely cause spontaneous mucocutaneous bleeding. Severe VWD causes petechiae, albeit intermittently. These clinical observations reveal at least two things about how platelets serve as "guardians of the microvasculature." First, platelets' guardian function is in large part performed by their GpIb-IX-V complex and its capacity to recognize and bind VWF and thereupon signal the activation of αIIbβ3. Second, platelet guardian function is maintained reasonably well even when platelet number or the number of GpIb-IX-V complexes or VWF multimers is very small, suggesting that evolution has "overproduced" hemostatic cells and proteins under the selective pressure of preventing hemorrhage. This may have biased our evolutionary biology towards thrombosis. The observation that only severe VWD, but not mild (types 1 and 2) VWD, appears to protect against atherothrombosis is consistent with the theory that human evolution has favored hemostasis over thrombosis. If this theory is correct, the hunt to discover unique mechanisms of GpIb-IX-V mediated thrombosis versus hemostasis should begin by reexamining the old axiom that "thrombosis is simply hemostasis occurring in the wrong place." If we accept this as true, we should then begin to try to figure out what it is about the "wrong place's" vascular system and rheological features that forces "hemostasis" to occur.

TTP and HUS

TTP and HUS are thrombotic microangiopathies (small-vessel thromboses associated with microangiopathic hemolytic anemia) due to platelet thrombi developing in the microvasculature. As more clinical and in vitro data about these diseases are generated, more debate is generated about the pathophysiological relationship between TTP and HUS. Notwithstanding this fact, or the fact that HUS and TTP are distinct clinical disorders, or the fact that TTP can be treated with plasma exchange and HUS cannot, or the fact that in HUS—but not TTP—fibrin can be found deposited in the microvasculature of the renal glomerulus and peritubular regions, or the fact that no etiological factor has yet been identified in TTP, it is nonetheless clear that both diseases share a common trigger: "unusually large" VWF multimers wreak havoc because they exist where they should not. In sporadic TTP it is an acquired deficiency of ADAMTS13, usually because of an autoantibody. In sporadic HUS it is because an enterotoxin (a *shiga-* or *verotoxin*) poisons the microvascular ECs, which respond by releasing "unusually large" VWF multimers. In familial TTP and HUS, the same pathways are involved via different mechanisms: in familial TTP there are inherited mutations in ADAMTS13 that affect its secretion, stability, or activity; and in familial HUS there is a deficiency of plasma factor H, which normally prevents or dampens complement-mediated cellular toxicity, including renal glomerular microvascular toxicity. Microangiopathies are also associated with radiation, solid organ and bone marrow transplantation, and mitomycin, cyclosporine, or tacrolimus therapy. These rarely fall neatly into one category or another (i.e., TTP or HUS) and, except for bone marrow transplant recipients (who almost always have normal levels of ADAMTS13), it is difficult to predict or establish the relative importance of ADAMTS13 insufficiency or EC damage (and perhaps even VWF-GpIb-IX-V) in their pathogenesis.

Miscellaneous Conditions Where the Link between GpIb-IX-V and Pathology Is Uncertain

When one examines the molecular determinants of GpIb-IX-V-mediated hemostasis and thrombosis and then cross-references them with common diseases that affect the systemic or organ-specific microvasculature, certain hypotheses come to light. For example, patients who suffer from bacteremia and sepsis overproduce EC NO, which has pleiotropic effects. Is it possible that one of the effects, that of NO-mediated inhibition of platelet adhesion and activation resulting at least in part from GpIb-IX-V binding to VWF, contributes importantly to bleeding associated with sepsis? If we examine the question from the perspective of responses by many different factors occurring dynamically within Virchow's triad, and that each group of factors operates in unique microvascular compartments over a 24- to 48-hour time period, the answer must be "I don't know, but maybe sometimes."

As another example, consider that solid organ (e.g., kidney, liver, or heart) transplant rejection often involves a microvasculopathy with histopathological evidence of platelet deposition, including venular deposition. Is it possible that venular platelet deposition is mediated by GpIbα binding to VWF or P-selectin expressed on the venular EC inflamed by the storm of cytokines that accompanies graft rejection? How about the relationship between platelet-mediated hemostasis and thrombosis and the pathogenesis of brain plaques in Alzheimer's disease? Evidence supports the idea that platelet-secreted amyloid precursor protein (APP) is involved in the formation of brain plaques, but are platelets of primary importance or secondary importance, or is platelet APP an irrelevant epiphenomenon? Answers are possible only when we simultaneously examine changes in the vascular and rheological components of Virchow's triad within the cerebral microvasculature. Finally, what is the role of platelet GpIb-IX-V and VWF in the growth of cancers? Pretty solid evidence indicates that platelet αIIbβ3 contributes to new capillary sprouts in several cultured malignant cell lines. How does it work, and does it work at all in vivo? Does it require activation and, if so, is it activated by VWF binding to GpIb-IX-V? We can only begin to answer these questions after we have assiduously dissected molecular mechanisms of GpIb-IX-V-mediated hemostasis and thrombosis.

Summary and Conclusions

Atherosclerosis is a modern human disease caused by genetic, environmental, and degenerative (i.e., senescence) factors. It is therefore likely that a unique set of factors operate within Virchow's triad to promote thrombosis when a coronary plaque ruptures or when an internal carotid artery continues to narrow beyond a 60 percent stenosis. If we can identify these factors and rank their importance in the pathogenesis of heart attack or stroke, we might be provided with the opportunity to develop drugs that target these factors. An ideal drug will inhibit thrombosis without inhibiting hemostasis. It will be "lesion-specific."

Differentiating molecular determinants of platelet GpIb-IX-V-mediated hemostasis from those regulating thrombosis is difficult and perhaps futile. It is difficult because there are few experimental systems that provide an examination of both hemostasis and thrombosis while also controlling for the important prothrombotic and antithrombotic variables simultaneously operating within Virchow's triad. This difficulty will likely be overcome because such experimental systems—both flow chambers and whole animal preparations—are being continually revised and optimized. It is perhaps futile because the evolutionary principles of conservation of function and functional redundancy must also apply to platelet GpIb-IX-V. But such principles may relate mainly to physiology and not pathology, and it is reasonable to consider that they may not apply to GpIb-IX-V when it serves a pathological response.

Further experimentation will test the hypothesis that there are specific molecular determinants directing platelet GpIb-IX-V-mediated thrombosis but not hemostasis (i.e., "lesion-specific" determinants). If the hypothesis is validated, new drugs for heart attacks, strokes, and peripheral arterial diseases will be developed that have a therapeutic index superior to those currently in use. If the hypothesis is refuted, data generated during the process will inevitably contribute to our understanding of microvascular pathophysiology. Such a "win–win" situation reflects both the importance and the wonderfully mazelike complex heterogeneity of human microvascular circulation.

Further Reading

Andre, P., Delaney, S. M., LaRocca, T., Vincent, D., DeGuzman, F., Jurek, M., Koller, B., Phillips, D. R., and Conley, P. B. (2003). $P2Y_{12}$ regulates platelet adhesion/activation, thrombus growth and thrombus stability in injured arteries. *J. Clin. Invest.* **112**, 398–406. *$P2Y_{12}$ knockout mice had very elevated bleeding times. Provoked arterial thrombus formation was delayed and decreased. Platelet tethering to and translocation along the injured artery were unaffected. ADP-induced platelet activation resulting from a GpIb-IX-V initiated feedback loop that begins with dense granule release and terminates on the $P2Y_{12}$ receptor is an essential component of both hemostasis and thrombosis.*

Eichorn, M. E., Ney, L., Massberg, S., and Goetz, A. E. (2002). Platelet kinetics in the pulmonary microcirculation in vivo assessed by intravital microscopy. *J. Vasc. Res.* **39**, 330–339. *This paper establishes experimental measurements of an entire microvascular circuit. The subject of their paper—the rabbit pulmonary microvasculature having a luminal diameter less than about 40 μm—is particularly useful and illuminating when considering platelet determinants of hemostasis and thrombosis. This is because the investigators specifically assayed platelet movement through the entire microvasculature and determined that red cell streams and vessel size determine platelet GpIb-IX-V-dependent hemostasis (an arteriolar response) while platelet activation (which does not affect GpIb-IX-V function) determines the magnitude of platelet deposition and thrombosis in the capillaries and venules.*

Feng, S., Reséndiz, J. C., Lu, X., and Kroll, M. H. (2003). Filamin A binding to the cytoplasmic tail of glycoprotein Ibα regulates von Willebrand factor-induced platelet activation. *Blood* **102**, 2122–2129. *Shear-induced binding of VWF to GpIb-IX-V may activate platelets and stimulate thrombosis by transducing the frictional shearing force generated by pathological blood flow through a cytoskeletal tether that links the cytoplasmic domains of GpIbα and the β3 subunit of αIIbβ3. Such a tether is shown to operate at a shear stress that exceeds any within the microvasculature, suggesting that it could be a mechanism of platelet activation that develops only under conditions of pathological arterial blood flow.*

Jackson, S. P., and Schoenwaelder, S. M. (2003). Antiplatelet therapy: In search of the "magic bullet." *Nat. Rev. Drug Discov.* **2**, 775–789. *A review of clinically important platelet physiology, pathophysiology, and pharmacology. It emphasizes the feasibility and importance of hunting for platelet proteins that specifically direct thrombotic responses without affecting hemostasis. If one such protein were discovered and targeted pharmacologically, it would be an ideal drug for acute arterial occlusion.*

Liu, L., and Kubes, P. (2003). Molecular mechanisms of leukocyte recruitment: Organ-specific mechanism of action. *Thromb. Haemost.* **89**, 213–220. *This review provides an important and unique perspective from which to examine the microvasculature. It describes our state-of-knowledge of organ-specific microvasculature function focusing on the liver (where capillaries are arranged in discontinuous sinusoids) and the brain (where capillaries are fused tightly together to form the "blood–brain barrier"). Although platelets are not emphasized, information is presented in support of the hypothesis that GpIb-IX-V is a "bridge for the endothelial–leukocyte–platelet interaction."*

Moake, J. L. (2002). Thrombotic microangiopathies. *N. Engl. J. Med.* **347**, 589–600. *This clear and beautifully illustrated review article describes the biology of VWF and ADAMTS13. It also breaks down the nosology of the "TMs" into a pathophysiological framework, with TTP a unique disease resulting from ADAMTS13 deficiency and HUS a separate disease resulting from an insult (e.g., an enterotoxin) to specific microvascular compartments (e.g., the renal glomerulus).*

Nieswandt, B., and Watson, S. P. (2003). Platelet–collagen interaction: Is GpVI the central receptor? *Blood* **102**, 449–461. *This thorough review examines the question of platelet determinants of hemostasis versus thrombosis from a collagen receptor point of view. Based on an extensive body of work (much of it done in the authors' laboratories) they conclude that collagen binding to GpVI is more important than collagen binding to integrin α2β1, and that GpVI is an important co-receptor mediating hemostasis and thrombosis.*

Sadler, J. E. Contact—how platelets touch von Willebrand factor. *Science* **297**, 1128–1129. *This is a small review based on an accompanying paper by Huizinga et al. presenting the crystal structure of the VWF/GpIbα interaction. In summary, the model of platelet adhesion emphasizes that first contact involves the N-terminal hairpin of GpIbα's extracellular domain binding to VWF's A1 domain. This binding appears to be shear dependent, with elevated shear stress displacing an inhibitory N-terminal extension of the VWF A1 domain. Following this initial contact, GpIbα's extracellular domain undergoes a β-switch that directs a second contact between several of its leucine-rich repeats and another site on VWF's A1 domain. These primary and secondary contacts are probably transitory, as shear stress drives platelet rolling (or tumbling), implying that bonds between VWF and GpIbα are being rapidly formed and broken. In order for stable adherence to develop, shear-dependent platelet capture and rolling must be followed by collagen binding to GpVI and α2β1, and by VWF binding to activated αIIbβ3.*

Wagner, D. D., and Burger, P. C. (2003). Platelets in inflammation and thrombosis. *Arterioscler. Thromb. Vasc. Biol.*, **23**, 2131–2137. *This timely review presents experimental evidence of many different mechanisms by which platelets participate in the inflammatory response. It reminds us that vascular pathology should generally be considered within the context of Virchow's triad for the pathogenesis of thrombosis, even when thrombosis is an irrelevant or minor secondary response. Platelets, and platelet GpIb-IX-V, appear to be important components in the pathophysiology of many inflammatory processes in the low-shear microvasculature and may be critically important in the initiation of atherosclerosis in the high shear macrovasculature.*

Ware, J., Russell, S., and Ruggeri, Z. M. (2000). Generation and rescue of a murine model of platelet dysfunction: The Bernard-Soulier syndrome. *Proc. Natl. Acad. Sci. USA* **97**, 2803–2808. *It is possible that all future knowledge about distinct mechanisms of GpIb-IX-V-mediated hemostasis versus thrombosis will be based on this or related mouse models. This paper shows that the deficiency of GpIbα in mice recapitulates a human disease. In doing this, it raises the possibility that genetically engineered alterations of GpIbα function in mice—both decreased or increased function—will someday pinpoint molecular interactions that are specifically prothrombotic or prohemostatic.*

Capsule Biographies

Dr. Kroll is a member of the Specialized Center for Research in Thrombosis and an Associate Professor of Medicine and Molecular Physiology and Biophysics at Baylor College of Medicine. He is also an Associate Professor of Bioengineering at Rice University, where he is a member of the Cox Laboratory for Biomedical Engineering. His research, which is funded by the National Institutes of Health and the Department of Veterans' Affairs, focuses on mechanisms by which rheological factors affect platelet adhesion–activation coupling.

Dr. Feng is an Assisaent Professor of Medicine in the Section of Hematology-Oncology. His research to identify unique prothronbotic pletelet signaling pathways is funded by the Amercian Heart Assocation.

Chapter 96

Comtheplement

Francis D. Moore, Jr.
Brigham and Women's Hospital, Harvard Medical School, Boston, Massachusetts

Introduction

Vascular organisms have evolved elaborate systems to rapidly identify areas of injury and to begin the reparative process. Such systems signal through the blood to recruit selective elements to assist in this process, making the vasculature a key element in any reparative system. Collectively, these systems produce the *inflammatory response*, whereby areas of injury develop increased blood flow and leaky capillaries, augmenting an influx of cellular elements into the interstitial space to remove and repair damaged elements. Within the inflammatory response is the system of *innate immunity*. Areas of injury are also areas where barriers to microbes are breached. The system of innate immunity, a system to which *complement* belongs, recognizes and begins to check infection, at the same time attracting more host defenses to the cause. The distinction between inflammation and innate immunity is, in the end, artificial as the two processes involve a near-identical sequence of simultaneous steps and systems. Complement, as an example, plays major roles in the response to both infection and injury.

Despite the plethora of component proteins and protein fragments, there are only two functional parts to the complement system. One part includes the serum proteins and produces a cascade-like sequential activation of proteins in response to specific stimuli, "*activation*" (Figure 1). The other portion consists of cell surface proteins on immune competent cells and allows them to derive information from specific products of complement activation, "effector" (Figure 1). The link between the two parts is the pivotal step of the complement system: the production of a covalent bond between the C3b fragment of C3 and the target of complement activation, "opsonization" (Figure 1).

Complement Activation

Activation refers to the steps by which C3 is cleaved. For each activation sequence, there are corresponding specific activators, "recognition" (Figure 1). The specificity of activation is critical, as the misdirection of complement activation to vital structures would be devastating. Activation sequences are self-amplifying: complement activation potentially begets more complement activation, in sharp contrast to metabolic pathways where an excess of product generally exerts feedback inhibition on the initial enzymes of a particular pathway. Thus, understanding complement activation also requires an understanding of the mechanisms that keep it in check.

Enzymatic cleavage of C3 destabilizes an internal thioester linkage, making the thiol group briefly reactive. C3 cleavage in aqueous solution produces futile reaction with water and little biologic activity. C3 cleavage adjacent to target surfaces produces covalent binding of C3b to the surface via amide or hydroxyl groups. This binding of C3b is responsible for all of its biologic activities. Both C4 and C5 have similar internal thioester linkages.

Of the three main ways in which C3 can be activated, the *classical* (or first described) pathway is the easiest to understand by analogy to the progressive proteolytic steps that lead from vascular injury to the formation of a blood clot (Table 1). In this type of cascade, the upstream element enzymatically converts its successor protein from an inactive proenzyme into an element of the enzyme that can cleave the next proenzyme in the sequence. This produces split fragments (many with biological activities) with each cleavage and multiple potential points of control.

The triggering step for the classical pathway is the engagement of antigen by immunoglobulins IgG or IgM, as

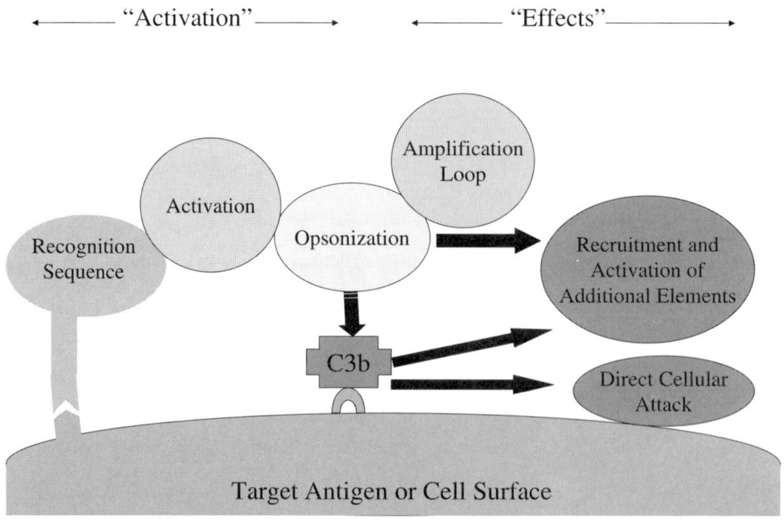

Figure 1 Schematic diagram of complement functional units. (see color insert)

would be typically the case on a surface with many antigens exposed and to which the host had been immunized previously (IgD, IgA, IgE, and IgG4 are not known to interact with complement). An example of such a surface would be that of a common microbe. "Natural" IgM antibodies cause this pathway to be activated also by certain antigens to which the host has not previously been exposed or by host antigens exposed after injury. Although the affinity for antigen by an individual IgM binding site is low compared to that of IgG, engagement of all 10 IgM binding sites at once produces a more potent activation of complement than does the bivalent IgG. As a result, there is specificity inherent in the activation of this pathway that arises from the exquisite specificity of antibodies for their antigens. Upon interacting with antigen, a change in conformation of the antibody attracts C1 and thereby creates activated C1, attached to the antigen by the antibody bridge. Activated C1 then both attracts and cleaves C4, exposing the reactive thiolester site in its C4b fragment. Although the thiol group will most frequently react futilely with water, complement activation proceeds after the covalent attachment of C4b to amino or hydroxyl groups on the activating surface adjacent to the Ig-C1 complex. C4a, a spasmogenic peptide or "anaphylatoxin," is simultaneously liberated during C4 cleavage and can be detected clinically. The surface-bound C4b attracts C2, positioning C2 for cleavage by the nearby Ig–C1 complex, leading to production the bimolecular enzyme C4b,C2a. The other portion of C2, C2b, is liberated into solution and is a by-product with kinin activity. C4b,C2a then cleaves C3, the most important of the complement proteins. C3 also has an internal thiolester site, and when activated, as C3b, binds covalently to the activating surface adjacent to the C4b,C2a complex. In that location, C3b attracts C5, which is then cleaved by the C4b,C2a and in turn bonded transiently nearby to the surface as C5b. Both C3 and C5 cleavage liberate anaphylatoxin and chemotactic moieties, C3a and C5a.

Activation of the *alternative pathway* also results in C3 and C5 cleavage, but by a much different mechanism (Table 1). This pathway does not require antibody and thus does not have the limitation of needing a prior exposure for a high-affinity antibody. In contrast, this pathway can respond innately. A small amount of reactive C3 is produced continuously in serum by spontaneous hydrolysis of the innately unstable internal thioester linkage. This does not liberate C3a, but instead leads to an opened form of C3 with weak reactivity as an autoenzyme. Each hydrolyzed C3, in combination with the other alternative pathway proteins (factor B, factor D, and properdin), briefly acquires C3 cleaving activity. This results in the continuous generation of small quantities of reactive C3b. Although such C3b moieties react with water ineffectively, a few bind to nearby surfaces. Once bound to a surface, these C3bs become the scaffold for assembly of alternative pathway activation enzyme complexes. These multimolecular assemblies are C3b,Bb, which cleaves C3, and C3b, C3b,Bb, which cleaves C5. Whether or not activation proceeds entirely depends on surface-specific equilibrium conditions for binding of factor B or factor H to the C3b initially deposited on the surface. Binding of B creates activation by producing the C3-cleaving enzyme, C3b,Bb, in concert with factor D (a serine esterase) and properdin (a ring-shaped protein that stabilizes these inherently labile enzymes), and amplified deposition of C3b ensues. In contrast, binding of H prevents activation because of H's property as a cofactor in the degradation of C3b by factor I (also called C3b, C4b inactivator protein). The binding of H to a surface-bound C3b produces a form of C3b that cannot participate in further alternative pathway activation. B and H are present in equimolar quantities in serum and extracellular fluid and have similar association constants for C3b. Whether B or H binds to a C3b is known to be a function of charge or glycosylation in the microenvironment surrounding the surface-bound C3b. Surfaces lacking sialic acid residues or heparan polysaccharides favor binding of B, and thus are strong alternative pathway activators. Bacterial surfaces are lacking in sialic acid residues, whereas mammalian cell walls are rich. Thus, this activation mechanism forms the basis by which the alternative pathway distinguishes invader from host. Furthermore, since C3b can be generated adjacent to activator surfaces by the classical pathway but not close enough to participate in classical pathway activation, the classical pathway-deposited C3b can be a source of amplification of classical pathway activation by the alternative pathway.

A third pathway of complement activation has been recognized recently, the *lectin pathway* (Table 1). This pathway is a mimic of the classical pathway, and it is suspected that the classical pathway arose as an adaptation of this lectin pathway to the presence of antibodies. This pathway reacts to the presence of a surface rich in mannose or *N*-acetylglucosamine residues, residues that characterize microbes as opposed to mammalian cells. These residues are bound by mannose-binding lectin protein (MBL). MBL shares a common structure with the C1q subunit of C1, having a collagenous central stalk with six globular heads attached by flexible links. Once bound, the MBL undergoes a conformational change allowing two enzymes to interact and then cleave C4 and C2 in exact analogy to the classical pathway. These proteases, "MBL-associated serine proteases," MASP1 and MASP2 are closely related to the C1s and C1r proteolytic subunits of C1.

There are descriptions of additional paths to complement activation of unknown significance: (1) C-reactive protein binding to bacterial polysaccharides, (2) activation of the kallikrein–bradykinin system, (3) direct activation by certain microbes and viruses, and (4) interaction with high concentrations of endotoxin.

Membrane Attack Complex

Once C5b has been produced and bound briefly to an activator surface by either pathway, a cascade of enzymatic steps resulting in membrane insertion of the membrane attack complex C5b-C9 (MAC) occurs (Table 1). The MAC assembly has the shape of a grommet. One side is hydrophobic, and it opens a physical hole through the target membrane. This allows for unregulated inflow of water and solute into the targets, causing potential hypotonic lysis of the target. Although this is clearly demonstrated to cause lysis of erythrocytes with in vitro complement assays and in some hemolytic anemias, the degree to which cytolysis contributes to cell death with nucleated cells is unclear. Cytolysis may be important in the host defense against certain encapsulated organisms, as deficiencies of the terminal complement components are associated with an increased incidence of infections with neisserial bacteria species. The noncytolytic, membrane-perturbing effects of MAC insertion may be more clinically relevant. For instance, insertion of a MAC into an endothelial cell causes endothelial activation, possibly leading to irreversible cell injury not by lysis but by attraction of activated neutrophils.

Given the mechanisms and stimuli for complement activation, it is not surprising that the clinical sites of complement activation should coincide with sites of inflammation, such as sites of acute trauma and microbial invasion. Known possible activators at such sites would include, for the alternative pathway, microbes, denatured DNA and proteins released from cell death, and injured tissue with impaired protection against autologous complement fixation (see later discussion)—for the classical pathway, primarily microbes with antibody affixed and local antigen–antibody complexes, and for the lectin pathway, microbes, and, perhaps, perturbed cell membranes. In each case, C3a and C5a are released into the blood to attract leukocytes and capillary endothelium is rendered permeable.

Clinical Assessment of Complement Activation

As products of a triggered activation system, one would expect that the anaphylatoxin fragments would not be detectable in healthy patients, with levels appearing only during complement activation. In truth, C4a, C3a, and C5a are not detectable in vivo, but their degradation products are. Their C-terminal arginines rapidly are cleaved off by a serum carboxypeptidase, generating the less active catabolites, designated desArg. The desArg peptides are stable and are the targets of detection in clinical immunoassays. In the case of C4a desArg and C3a desArg, those clinical conditions that affect clearance are not described, and the normal concentration in plasma is surprisingly high at 400 ng/mL and 100 ng/mL, respectively. In contrast, C5a is so avidly ligated to leukocyte cell surface receptors (C5aR) that levels of C5a desArg are not detectable in systems that contain leukocytes, such as any naturally occurring activation event, even with massive complement activation.

Whether complement activation has occurred and to what degree cannot be ascertained solely through analysis of serum concentrations of C3 and other intact complement proteins. C3 is a relatively abundant serum protein (1 to 2 mg/mL), which is synthesized by hepatocytes for plasma levels, by macrophages in the periphery at sites of inflammation, and by intestinal enterocytes. Concentrations of C3 at inflammatory sites cannot be measured clinically. Furthermore, serum C3 concentrations are a balance of synthesis, degradation, and serum dilution, none of which can be measured clinically, in addition to consumption by activation. Those factors that determine the rate of synthesis and degradation of C3 are not characterized well enough to predict C3 levels in a given medical condition. C3 also is an acute-phase reactant, causing concentrations to rise, despite consumption by complement activation. On the other hand, in a state of poor nutrition, concentrations might fall without any complement activation to consume C3. As a plasma protein, C3 concentrations also are dependent on degrees of hemodilution in the ill patients in whom C3 might be

Table I Complement Proteins.

Activation proteins, serum	
Classical pathway	Activates complement with antigens
IgM, IgG	Binds to antigen
C1 : C1r, C1s, C1q subunits	Binds to Ig, cleaves C2
C2	Cleaves C4
C4	With C2, cleaves C3
C3	Binds to target, cleaves C5
Alternative pathway	Activates complement with membrane charge
	Amplifies C3 deposition by other pathways
D	Cleaves B
B	Binds to C3, cleaves C3/C5
C3	With B, cleaves C3/C5
P	Stabilizes C3b, Bb enzyme
Lectin pathway	Activates complement with sugars
MBL	Binds mannose
MASP1, MASP2	With MBL, cleave C2
Terminal proteins, serum	
C3	Binds to target, cleaves C5, signals leukocytes
C5	Cleaves C6, signals leukocytes
C6–9	Form MAC, attack cell membranes
Control proteins	
C1-INH	Serum, inhibits C1r, C1s
C4bp	Serum, inhibits C4
H	Serum, serum, inhibits B
C3-INH	Serum, inhibits C3, cleaves C3
CR1	Membrane, degrades C3 cleaving enzymes
DAF, MCP	Membrane, decay C3 cleaving enzymes
CD59	Membrane, inhibits MAC formation
Signaling proteins	
C3a, C4a, C5a	Serum, anaphylatoxins, chemoattractants
CR1	Membrane receptor for C3b, phagocytosis
CR2	Membrane receptor, antibody formation
CR3	Membrane receptor, leukocyte adhesion
C3aR, C5aR	Membrane receptor for anaphylatoxin

measured but in whom hemodilution is rarely considered. Finally, complement activation reactions that consume more than 5 to 10 percent of available serum C3 would be truly massive. Assessments of the functional capability of the complement lytic system using the CH50 assay have been employed as a repetitive measure with which to monitor the activity of a chronic systemic inflammatory disorder, such as systemic lupus erythematosus. However, the CH50 is a highly contrived assay designed to measure differences in red cell hemolysis at limiting dilutions of complement proteins (typically several hundredfold). Whether undiluted serum can ever acquire a complement lytic functional deficiency is doubtful. Thus, the mainstay of assessments for complement activation is the measurement of complement activation-derived cleavage products such as C3a desArg. However, assuming that complement activation has occurred simultaneously with measured elevations of C3a desArg could be misleading, as the factors that influence the rate of C3a desArg degradation are not characterized. Isolated elevations of C3a desArg concentrations alone are thought to indicate alternative pathway activation. Elevations of both C3a desArg and C4a desArg indicate that classical pathway activation is occurring but is unable to distinguish whether there is simultaneous alternative pathway activation. In most acute injury settings, complement activation has been alternative-pathway mediated. In most chronic diseases, the classical pathway has been paramount. Additional complement cleavage products, such as C4d, Bb, and the C5b-9 neoantigen, have been reported to be clinically valuable, as well.

Intrinsic Regulation of Complement Activation

The rapid spontaneous decay of the multimolecular enzymes of each pathway that cleave C3 and C5 serves as the primary feature that prevents a complement activation event from amplifying uncontrollably (Table 1). Furthermore, several highly specific complement regulatory proteins exist that each inhibit portions of the complement activation cascades. In addition to limiting amplified complement activation, there seem to be three more functions served by these "control" proteins. One is to prevent continuous complement activation in plasma due to imperfect specificity. One is to protect cells against autologous complement attack arising from imperfect specificity. The final function is to allow leukocytes to function in a complement activation site without becoming damaged and for them simultaneously to derive information from the surrounding milieu.

C1 esterase inhibitor (C1-INH) inactivates the catalytic subunits of C1 and is the only known protein to have this function. Both subunits of C1 are serine proteinases, and C1-INH is a member of a larger class of related proteins whose function is to inhibit serine proteinases. C1-INH also can inhibit kallikrein and Hageman factor and has activity against factor XIIa. The inborn deficiency of this control protein serves as an example of uncontrolled complement activation in humans. This gives rise to *hereditary angioneurotic edema*, a disease characterized by sudden systemic complement activation via the classical pathway.

Factor I (C3b, C4b-inhibitor) produces complement inhibition by enzymatically degrading C3b and C4b, those forms of C3 and C4 that participate in generation of further complement activation, to inactive forms. Factor I requires a cofactor in this degradation reaction (factor H, membrane cofactor protein, C4-binding protein, or CR1) to bind to the C3b and C4b. Genetic deficiency of factor I produces clinical susceptibility to bacterial infection, as the natural rate of spontaneous C3b generation is much accelerated in the absence of factor I, leading to secondary critical depletion of C3 and factor B. Thus, this control protein, in analogy to C1-INH, also ensures that the proper specificity for complement activation is maintained, in this case for the alternative pathway, and that futile fluid-phase turnover is avoided.

Additional circulating proteins that down regulate complement are *Factor H* and *C4 binding protein*. Both serve to displace ligands from their targets of inhibition, Bb from

Table II Functions of Complement.

Via C3a, C4a, C5a	
	Attract leukocytes to site of complement activation
	Activate leukocytes
	Alterations in vascular smooth muscle
	Endothelial activation
	Mast cell activation
Via C5b-9	
	Cellular disruption
	Microbial killing
Via C3	
	Covalent attachment of C3b "targeting label" to activator surface
	Molecular adjuvant for antibody formation and phagocytosis
	Clearance of immune complexes
Via membrane complement receptors	
	Protection against autologous complement attack
	Mediation of molecular adjuvant roles
	Increased cell–cell adhesion

C3b in the case of H and C2a from C4b in the case of C4-binding protein. C3b and C4b then acquire susceptibility to factor I and are degraded.

A unique feature of complement activation is that the activation-dependent cleavage of C4, C3, and C5 generates their respective "b" fragments with a highly reactive thiolester group. What molecules or surfaces fix these fragments is primarily determined by spatial availability. While the covalent bonding (other than to water) generally occurs on the activating surface, as a membrane-bound Ig–C1 complex lays down C4 and C3 around itself to generate C5 cleavage by the classical pathway, and as membrane-bound C3b serves as the scaffold for alternative pathway-dependent C3 and C5 cleavage. However, activated but unreacted C3b, C4b, and C5b fragments are available in solution very briefly, and thus can also bond to a nearby surface that was *not* the initial activator. This can give rise to assembly of C3 cleaving enzymes and MACs on the membranes of cells that are not directly involved in the complement activation, are not themselves complement activators, and are therefore to be damaged as innocent bystanders. As mentioned earlier, the spontaneous decay of the C3b,Bb and C4b,C2a complexes combined with the preferential interaction of C3b with factor H (and then factor I) on nonactivating surfaces should prevent bystander injury to homologous tissue. However, this must be imperfect as a second class of complement control proteins exists to protect host cells from nearby complement activation reactions. These are, therefore, membrane proteins.

Decay accelerating factor (DAF, CD55) is expressed on the surface of essentially all cell types and accelerates the proteolytic activity of factor I on membrane-bound C4b and C3b. It has homology to the circulating proteins of similar function, C4bp and factor H, and prevents a membrane-bound C3b from becoming a nidus of assembly of a C3-cleaving enzyme. The disease, *paroxysmal nocturnal hemoglobinuria*, is complement mediated and is associated with absent erythrocyte membrane.

Membrane cofactor protein (CD46) has cofactor activity for factor-I-mediated degradation of C3b and is expressed by fibroblasts, epithelium, and endothelium. Transfection experiments into CHO cells as well as experiments with antibody directed against native MCP suggest that MCP preferentially protects against alternative pathway attack, with DAF filling the same role for the classical pathway.

The leukocyte and erythrocyte cell surface *C3b receptor (CR1 or complement receptor type 1 or CD35)* overlaps with these two proteins in its ability to capture and degrade C3b and C4b. CR1 can also block C1 interactions with antibody. CR1 has major additional important functions, as will be discussed later.

Cell surface proteins also exist that interfere with the assembly of a MAC (homologous restriction factor, protectin, CD59, C8-binding protein). CD59 is expressed by endothelium, erythrocytes, leukocytes, and epithelium. CD59 has been found to be shed from cardiac myocytes in areas of ischemia, suggesting a mechanism of complement damage in ischemic tissue.

Interactions between Leukocytes and Complement

Leukocytes have specific high-affinity cell surface receptors for the activator-bound and soluble fragments of C3, C4, and C5 that are produced by complement activation (Table 1). The complement ligand–leukocyte receptor interactions cause influx of cells to the site of activation (inflammation) with an increased cellular functionality. Such a mechanism leads to "opsonization" of complement activators with the result that activated leukocytes perform their functions with respect to the opsonized material. Receptors for C3a, C4a, C5a, C1q, factor H, C3b and its degradation products, and C4b have been described. Engagement of a specific complement fragment with its corresponding receptor has a specific effect, depending on which fragment and on which type of cell is expressing the receptor. The receptors of acute inflammation, C5aR, CR1, CR3, and CR2, are discussed next.

C5aR is a 40 to 50 M_r membrane protein of granulocytes and macrophages and has nanomolar affinity for C5a. When exposed to C5a dispersing from a site of complement activation, granulocytes migrate toward the C5a source (chemotaxis) and simultaneously become more activated, exposing, among other functions, more opsonic receptors, such as the C3b-C4b receptor (CR1) and the iC3b receptors (CR3), on their cell surfaces. When the C5aR become saturated with C5a, the cell loses its capacity to respond to a subsequent C5a exposure, as a result of internalization of engaged receptors without replacement. Thus, granulocytes exposed to a C5a solution (rather than a gradient) lose their responsiveness to a C5a gradient, a phenomenon termed *desensitization.* Clinically, neutrophils from a patient with an injury large enough to produce concentrations of C5a that saturate local neutrophil C5aR and that then penetrate the vascular compartment may lose their capacity to migrate to the point of injury. Because C5a causes both neutrophil and endothelial activation, such an injury may produce pulmonary leukocyte extravasation and acute respiratory distress syndrome (ARDS) (see later discussion). This interaction of C5a with granulocytes is the most apparent effect of complement activation in vivo and has the presumably beneficial effect of concentrating immuno- and phagocytosis-competent cells at the site of injury and complement activation. C5a has other proinflammatory immunoregulatory properties with respect to lymphocytes.

CR1 is one of the opsonic receptors for granulocytes, as well as being a potent inhibitor of complement activation. The natural ligands for CR1 are C3b and C4b that have covalently interacted through the thiolester site, for which it has nanomolar affinity. CR1 is also expressed by human erythrocytes, eosinophils, macrophages, and some lymphocytes. Although granulocytes express a limited number of CR1 molecules on their cell surfaces, a large intracellular pool is pre-synthesized and available. This pool rapidly translocates to the cell surface after exposure of cells to activators, such as C5a, FMLP, endotoxin, TNF, GM-CSF, and PDGF, causing peak numbers of about 75,000 CR1 per cell surface. CR1 in this functional state adheres C3b-coated activators to the leukocyte surfaces. As each activator surface will have many C3bs on it arrayed around the initial C3bs deposited, and, therefore, multiple sites of CR1 engagement, CR1 molecules become effectively immobilized on the granulocyte cell surface after ligating C3b. Increased diglyceride production, indicative of increased cellular function, results from this CR1 cross-linking. If antibody is also present on the activator, enhanced phagocytosis (compared to antibody alone) takes place. Thus, granulocyte CR1 participates in phagocytosis and, by its activity as a cofactor for the degradation of C3b on its own surface, protects granulocytes from being damaged by the intensity of the complement activation.

CR3 (CD18, CD11b) shares many of the properties of CR1; however, its complement ligand is iC3b, the degraded form of cell-bound C3b that follows interaction with factor I and cofactor proteins (factor H, MCP, DAF, CR1). CR3 is expressed by the same cells that express CR1 and has a constitutive level of cell-surface expression with the capacity to be quickly translocated to the cell surface by the same activators that affect CR1. CR3 is independent of CR1, as it is a member of the β2-integrin family with a two-chain structure that bears little relationship to CR1. CR3 functions to promote the attachment of granulocytes to endothelium at the site of inflammation with extravasation of granulocytes and engulfment of iC3b-opsonized complement activator. The lung injury caused by systemic complement activation and pulmonary leukosequestration appears to be CR3 dependent. Such injury can be experimentally prevented by white cell depletion, complement depletion, complement inhibition, anti-CD18 antibodies, anti-CD11b and 11a antibodies, and anti-ICAM-1 antibodies.

Since a similar spectrum of inhibitors prevent the inflammatory response to a local injury, it would appear that complement activation triggers a simple integrated response. Complement responds innately to local injury or microbial invasion by activation, simultaneously generating C5a and opsonizing the foreign material with C3b. Elaborated C5a both activates local endothelium and causes passing granulocytes to adhere and then extravasate at the site. This reaction requires iC3b on the endothelium and the C5a-induced expression of CR3 on granulocytes and ICAM-1 on endothelium. Granulocytes migrate through the interstitium to the site of C5a generation and engulf the complement activator, using the C3 fragments bound to the activator for targeting through CR1 and CR3.

CR2 (CD22) forms the link between the phagocytic response to acute injury and the lymphocyte response to produce antibody and future immunity. This cell surface protein of B-lymphocytes ligates to the activator-bound degradation fragments of C3b, iC3b, and C3dg. Experimental immunogens do not generate an IgG response either after depletion of C3 or in the presence of soluble CR2 used to compete with native lymphocyte CR2. Absence of CR2 may interfere with the development of the B-cell repertoire of antibody altogether. Just as the absence of CR2 interferes with B-cell function, the absence of CR1 interferes with the response of T-cells to antigen.

Complement Deficiency States

Spontaneous complement deficiencies give us clues to the biological functions of complement in humans (Table 2). Deficiency of C3 produces a predisposition to bacterial infection that can be lethal. Such patients also display a deficient repertoire of antibodies, possibly due to a lack of CR2-mediated lymphocyte events in the absence of C3b-mediated opsonization of new antigens. C3 deficiency occurs in several different ways: a genetic lack of C3 production, a secondary loss due to accelerated fluid-phase cleavage from an inborn deficiency of factor 1 or factor H, or acquisition of an autoantibody that causes C3 activation. Excess bacterial infections are seen in patients with granulocyte deficiencies of CR3 and other adhesion molecules or with deficiencies of properdin. Immune complex diseases result from deficiencies of erythrocyte CR1 expression or of classical pathway proteins, as (in primates) the classical pathway produces immune complex clearance. Patients deficient in C8 or C9 display increased susceptibility to infection by *Neisseria* bacterial species.

Complement Inhibitors

sCR1 refers to the naturally occurring soluble form of the C3b-C4b cell surface receptor, CR1 (see earlier discussion). Because CR1 can both bind and act as a cofactor in the degradation of these two cell bound foci of further complement activation, sCR1 is a potent complement inhibitor in solution. sCR1 has now been bioengineered into a drug. It is in Phase 2 trials as an agent that can lessen the amount of heart muscle killed during a heart attack.

CAB2 refers to a recombinant protein that combines the active sites of DAF and MCP (see earlier discussion). As such, it would be expected to have a similar activity to sCR1 and be a potent inhibitor of complement activation at the C3 step. Initial designs for clinical trials are underway, the drug having shown promising activity in rodents. Both sCR1 and CAB2 suffer a potential flaw in that the development of an antibody to the drug(s) in a patient could lead to hemolytic anemia, arising from interference with the parent, critical proteins.

Monoclonal anti-C5 is the furthest in clinical development. In theory, interference at this step would prevent neutrophil chemotaxis to a site of injury, and so prevent any deleterious effects to be caused by neutrophil activation. Furthermore, as will be noted later, C5a is thought to cause much of the lung dysfunction associated with major injury, and an anti-C5 could prevent that. Finally, this agent would prevent endothelial activation from injury by preventing the formation of MACs. In large Phase 2 trials, it may improve the outcomes after cardiac surgery.

Clinical Injury and Complement Activation

Apparently damaging complement activation has been demonstrated in a number of clinical circumstances.

Hemodialysis

The observation that simultaneous neutropenia, neutrophil aggregation, pulmonary leukosequestration, and hypoxemia occurred during hemodialysis for kidney failure led to the hypothesis that dialysis membrane-induced intravascular complement activation causes pulmonary dysfunction. The hypothesis was expanded to be the general explanation for the etiology of ventilatory failure secondary to complement-producing injuries elsewhere. By this theory, since widely adopted, any source of intravascular complement activation is capable of causing lung malfunction, through a mechanism involving activated leukocytes. Careful analysis of dialyzer membrane materials and reengineering to reduce complement activation has made these hypoxic events during dialysis much less common.

Cardiopulmonary Bypass (CPB)

Given the similarity between dialysis membranes and the pump oxygenators used in open-heart surgery, the high frequency of pulmonary dysfunction after open-heart surgery provoked an examination of complement activation in this clinical setting. Not surprisingly, complement activation was found in CPB and coincided with intravascular neutrophil activation. Patients with postoperative respiratory failure had C3a desArg concentrations immediately after surgery that were twice the mean for the patients without respiratory complication. Reengineering of CPB circuits in analogy to the recent dialysis experience has failed to produce nonactivating equipment, even in the case of heparin bonded circuits, likely reflective of the highly complex blood–plastic interfaces of modern bypass.

Thermal Injury

Burn injury was thought to cause complement activation as early as 1963. The complement activation accompanying thermal injury may unnecessarily injure the host. One mechanism is the unnecessary amplification of the inflammatory response at the burn wound. This deleterious effect is suggested by the observation that early mortality in mice receiving a burn of reproducible size was reduced from 65 percent to 10 percent by the prior crippling of complement function produced by depleting animals of C3 with cobra venom factor treatments. Complement activation also harms the host in general by causing defective neutrophil chemotaxis. It has been shown that burn patients' neutrophils have been exposed to systemic levels of C5a and have subsequently lost their capacity to migrate up a gradient of increasing C5a concentration to the source of complement activation, a possible explanation as to why burned patients have a propensity for nonburn-site infections. In addition, burned patients, in the absence of direct heat or volatile chemical injury to the tracheobronchial tree, frequently exhibit transient hypoxemia. As the burn wound produces intravascular complement activation and systemic neutrophil activation, pulmonary leukosequestration could follow, as was the case with hemodialysis.

Sepsis and Septic Shock

Septic shock causes alternative pathway complement component alterations. Initial studies documented normal levels of complement components in bacteremic patients, but reduced levels of alternative pathway proteins and terminal complement components in patients with septic shock. In many studies now, complement activation with increased concentrations of C3a desArg was found in those patients in septic shock and those patients whose C3a desArg concentrations failed to return to normal went on to multiple system organ failure. *It has been difficult to discern whether complement activation is a beneficial response to a major inflammatory event or whether it promotes sepsis and shock.* The antibacterial properties of complement suggest that complement activation would be advantageous during septic states. However, recent evidence in rodents favors the other interpretation.

Ischemia and Reperfusion Injury

The injury to muscle caused by arterial occlusion followed by reperfusion has similarities to the inflammatory reactions discussed earlier, as the degree of injury is both neutrophil and complement dependent. Myocardial infarction has been extensively studied in this regard, and the amplification of the ischemic injury occurring upon reperfusion seems to be an example of a local complement activation response that is harmful. Areas of myocardial necrosis in humans demonstrate deposition of complement proteins. Furthermore, when administered to models of coronary artery occlusion and reperfusion, sCR1 given just before reperfusion produces a substantial reduction in the size of the resulting myocardial infarct. sCR1 also has been used to attenuate the reperfusion injury to ischemic smooth muscle and mucosa of the gut in rats subjected to mesenteric arterial occlusion and reperfusion. A small dose of sCR1 given prior to reperfusion or a larger dose at reperfusion resulted in a gross change of bowel appearance after 3 hours, from purple-black in untreated animals to normal in treated animals, and the 5-day mortality was reduced from 80 percent to 45 percent. The complement and neutrophil dependency of skeletal muscle I/R, such as might occur in trauma or vascular occlusion, has been shown as well with improved indices of injury in these models is produced by sCR1. Other complement inhibitors are also active in attenuating these injuries: anti-C5 antibodies attenuate the mucosal injury in gut reperfusion, as does a C5aR antagonist. Anti-CR3 antibodies attenuate hind limb reperfusion injury. In the heart reperfusion injury models, a deficiency of C6 is protective, as is a C1s inhibitor and anti-C5a antibodies.

ARDS

Systemic or excessive complement activation may cause the development of ARDS, often the initial component of the multisystem organ failure that can follow severe injury or infection. In a study of patients at risk for developing ARDS, patients who went on to develop the syndrome manifested increased concentrations of C3a desArg, decreased chemotaxis to C5a in vitro, and evidence of systemic neutrophil degranulation. This was also noted in additional patients with multisystem organ failure of varying etiologies. A pulmonary injury is produced by skeletal muscle or gut I/R and can be lessened using anti-CD18 antibodies or sCR1. In a model of lung injury, which would not produce systemic complement activation, infusion of low doses of endotoxin and platelet-activating factor, sCR1, also protected against pulmonary edema without changing the degree of neutrophil leukosequestration.

Other Conditions

A large number of inflammatory diseases have been shown to have complement activation as at least one major component of the ongoing tissue damage. These include vasculitis, nephritis, and arthritis. In the vasculitides, complement fragments are found in the endothelium of involved blood vessels, and experimental models are highly complement dependent. In some of the nephritis categories, both immune complexes and complement fragments are found in the damaged kidneys. In systemic lupus erythematosus, the overall activity of the disease can be followed with the CH50 assay.

Summary

Complement is a highly regulated and intricate proinflammatory system that has been shown to be activated in situations of injury, sepsis, and multiple organ failure. Although these observations have long suggested a role for complement in the response to injury, it is the recent advent of the specific complement inhibitor, sCR1, that has begun to reveal the degree to which complement is involved. Furthermore, as one of the earliest inflammatory systems to respond, complement activation inevitably leads to cellular activation with elaboration of cytokines and oxidants, with worsening of injury and further complement activation.

Bibliography

Alper, C. A., Abramson, N., Johnston, R. B., Jr., Jandl, J. H., and Rosen, F. S. (1970). Increased susceptibility to infection associated with abnormalities of complement-mediated functions and of the third component of complement (C3). *N. Engl. J. Med.* **282**, 349–354. *This is the initial report linking complement activation to the defense against microbial invasion.*

Craddock, P. R., Fehr, J., Brigham, K. L., Kronenberg, R. S., and Jacob, H. S. (1977). Complement and leukocyte-mediated pulmonary dysfunction in hemodialysis. *N. Engl. J. Med.* **296**, 769–774. *In this paper, the first link between systemic complement activation and secondary lung injury is made.*

Ellison, R. T., Kohler, P. F., Curd, J. G., Judson, F. N., and Reller, L. B. (1983). Prevalence of congenital or acquired complement deficiency in patients with sporadic meningococcal disease. *N. Engl. J. Med.* **308**, 913–916. *This reports a high incidence of terminal complement pathway proteins as the basis for a common devastating infection.*

Fearon, D. T. (1978). Regulation by membrane sialic acid of β-1h-dependent decay-dissociation of amplification C3 convertase of the alternative complement pathway. *Proc. Natl. Acad. Sci. USA* **75**, 1971–1975. *This work forms the basis of understanding alternative pathway activation.*

Fearon, D. T. (1980). Identification of the membrane glycoprotein that is the C3b receptor of the human erythrocyte, polymorphonuclear leukocyte, B lymphocyte, and monocyte. *J. Exp. Med.* **152**, 20–30. *This is the initial work describing Complement Receptor Type 1, CR1.*

Gerard, N. P., and Gerard, C. (1991). The chemotactic receptor for human C5a anaphylatoxin. *Nature* **349**, 614–617. *This is the initial description of C5aR.*

Hill, J., Lindsay, T. F., Ortiz, F., Yeh, C. G., Hechtman, H. B., and Moore, F. D., Jr. (1992). Soluble complement receptor type 1 ameliorates the local and remote organ injury after intestinal ischemia–reperfusion in the rat. *J. Immunol.* **149**, 1723–1728. *This report extends the sCR1 findings to include prevention of secondary lung injury.*

Moore, F. D., Jr., Davis, C., Rodrick, M., Mannick, J. A., and Fearon, D. T. (1986). Neutrophil activation in thermal injury as assessed by increased expression of complement receptors. *N. Engl. J. Med.* **314**, 948–953. *This report links neutrophil activation to complement activation in this important injury type.*

Stevens, J. H., O'Hanley, P., Shapiro, J. M., Mihm, F. G., Satoh, P. S., Collins, J. A., and Raffin, T. A. (1986). Effects of anti-C5a antibodies on the adult respiratory distress syndrome in septic primates. *J. Clin. Invest.* **77**, 1812–1816. *This a report of the first experimental attempts to interfere with complement-activation induced pulmonary injury.*

Till, G. O., Johnson, K. J., Kunkel, R. G., and Ward, P. A. (1982). Intravascular activation of complement and acute lung injury. Dependency on neutrophils and toxic oxygen metabolites. *J. Clin. Invest.* **69**, 1126–1135. *This work further develops the relationship between systemic complement activation and secondary lung injury.*

Weisman, H. F., Bartow, T., Leppo, M. K., Marsh, H. C., Jr., Carson, G. R., Concino, M. F., Boyle, M. P., Roux, K. H., Weisfeldt, M. L., and Fearon, D. T. (1990). Soluble human complement receptor type 1: In vivo inhibitor of complement suppressing post-ischemic myocardial inflammation and necrosis. *Science* **249**, 146–151. *This is the first report of the salutary effects of a sCR1 complement inhibitor.*

Capsule Biography

Dr. Moore trained with Drs. D. T. Fearon and K. F. Austen at the time of the discovery of the mechanism of alternative pathway activation and, subsequently, CR1 and sCR1. His personal work has primarily focused on the detailed mechanisms of complement activation by injured tissue. His current clinical responsibility is as Chief of General and Gastrointestinal Surgery at Brigham and Women's Hospital.

CHAPTER 97

Regulation of Microcirculation by Antithrombin

Molecular Mechanism(s) and Possible Therapeutic Applications

Kenji Okajima

Department of Biodefense, Nagoya City University, Graduate School of Medical Sciences, Nagoya, Japan

Antithrombin (AT) is an important natural anticoagulant that inhibits serine proteases generated during activation of the coagulation cascade. The physiological significance of AT is clearly illustrated by the development of recurrent thrombosis in patients with congenital AT deficiency. In addition, AT improves microcirculatory disturbances by attenuating inflammatory responses in experimental animals given endotoxin or in those subjected to organ ischemia–reperfusion. Since both inflammation and microthrombus formation are important events that induce microcirculatory disturbances, AT might play critical roles in maintenance of proper microcirculation.

In this chapter, the author describes the molecular mechanism(s) by which AT regulates microcirculation and further mentions the possible therapeutic applications of AT to improve microcirculatory conditions in various disease states.

Anticoagulant Activity of AT

AT, a glycoprotein, has a molecular weight of 58,200 Da with 432 amino acids and is mainly synthesized in the liver. AT is a physiological serine protease inhibitor that inhibits activated coagulation factors such as thrombin and factor Xa. The reactive site loop of AT includes a P1-P'1 (Arg393-Ser394) bond (Figure 1). When thrombin cleaves this bond that resembles the substrate of thrombin, the protease is covalently linked to the P1 residue. Inhibition of these serine proteases by AT is accelerated approximately one-thousandfold by binding of heparin to arginine residues located at the heparin binding site of AT (Figure 1), with resultant conformational change leading to exposure of the P1-P'1 reactive center. Amino acid residues other than arginine shown in Figure 1 have also been found to be critical for interaction with heparin. AT is activated on the endothelial surface where thrombin generation is increased through binding to heparan sulfate molecules of ryudocan or syndecan. The physiological importance of AT is well illustrated by the presence of patients with congenital AT deficiency who developed recurrent thrombosis in their youth; 70 percent of the patients developed thrombosis before 35 years of age. Congenital AT deficiency usually presents as a heterozygous state associated with venous thrombosis, and the homozygous state is extremely rare, probably because of its presentation as lethal neonatal thrombosis. Similar observations have been reported for the pathologic sequelae of AT knockout mice [1]. The homozygous state of AT deficiency was only reported in patients who have variant AT molecules with heparin binding defects, and patients with such variant molecules have arterial thrombosis as well as venous thrombosis [2]. Possible mechanism(s) by which arterial

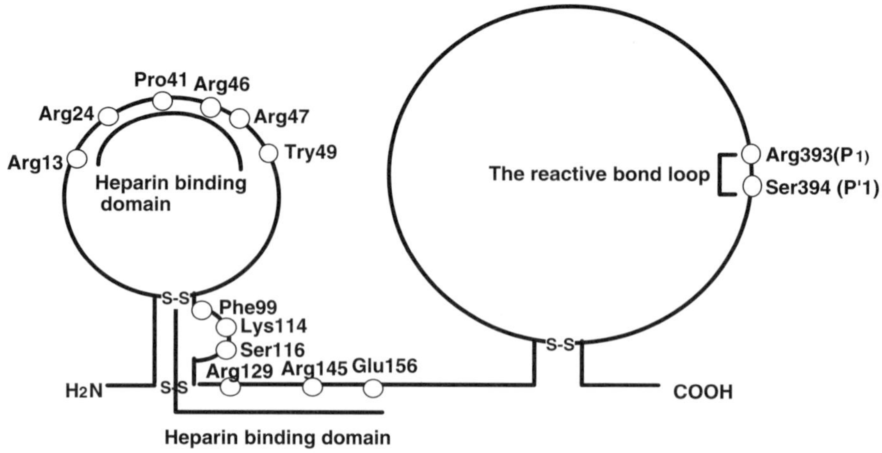

Figure 1 Localization of heparin binding domains and the reactive site in the primary structure of antithrombin.

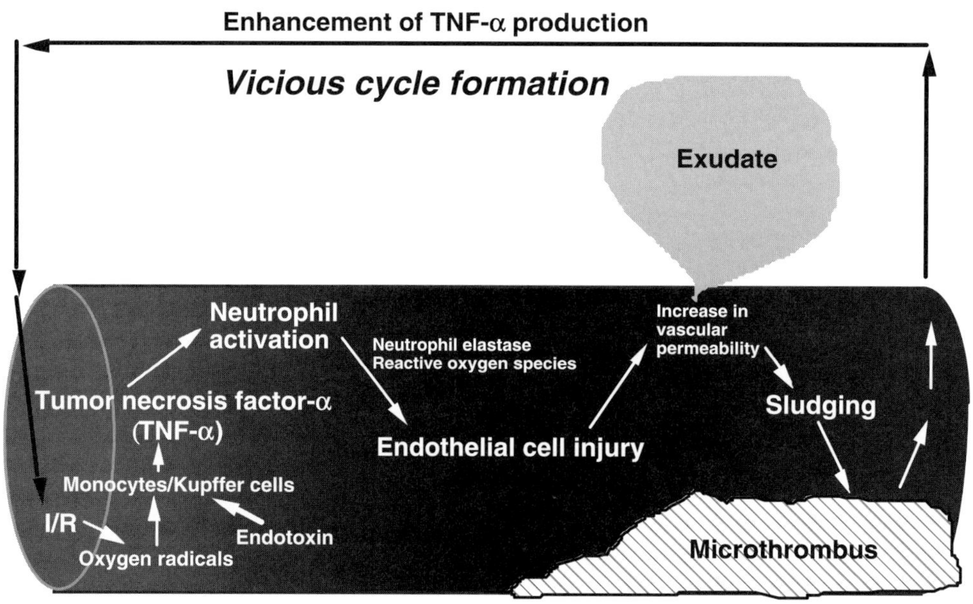

Figure 2 Pathologic mechanism(s) leading to the microcirculatory disturbance associated with sepsis or circulatory shock. I/R: ischemia-reperfusion.

thrombosis occurs in these patients will be discussed later in this chapter. These observations strongly suggested that AT plays a critical role in regulation of the coagulation system by inhibiting thrombin and other serine proteases, and the interaction of AT with the endothelial cell surface heparin-like substances is quite important for rapid and effective inhibition of such coagulation factors.

Anti-Inflammatory Activity of AT

Activation of the coagulation system leading to disseminated intravascular coagulation is frequently seen as a part of the inflammatory responses in pathologic conditions such as sepsis and circulatory shock. In these inflammatory responses, proinflammatory cytokines such as tumor necrosis factor-α (TNF-α) play an important role [3]. Although TNF-α is implicated in the activation of the coagulation system in sepsis, it also activates neutrophils, thereby promoting the release of a wide variety of inflammatory mediators such as neutrophil elastase and free oxygen radicals that are capable of damaging endothelial cells. The resultant endothelial cell damage increases microvascular permeability, leading to microcirculatory disturbance due to the hemoconcentration. Such microcirculatory disturbance precedes microthrombus formation [4]. Microthrombus formation further increases TNF-α production, thereby exacerbating the inflammatory responses to form a vicious cycle of progression of microcirculatory disturbances [5] (Figure 2).

We previously reported that AT reduced pulmonary endothelial injury by inhibiting neutrophil accumulation in the lung of rats given endotoxin. AT also reduced endotoxin-induced hypotension in rats by inhibiting TNF-α production [6]. These therapeutic effects of AT could not be explained by the anticoagulant activity, but by its promotion of endothelial production of prostacyclin (PGI_2), which potently inhibits leukocyte activation. Interaction of AT with glycosaminoglycans might be critical for promotion of endothelial production of PGI_2. Ischemia–reperfusion is an important pathologic mechanism leading to the development of coagulation abnormalities and organ failure seen in patients with circulatory shock. TNF-α is also critically involved in this pathologic process. AT increased the hepatic tissue blood flow by inhibiting neutrophil activation in rats subjected to hepatic ischemia–reperfusion. AT also increased the renal tissue blood flow by inhibiting neutrophil activation through inhibition of TNF-α production, thereby reducing renal injury [7]. These effects of AT were also independent of its anti-coagulant activity, but dependent on its capacity to promote endothelial production of PGI_2. Although AT itself has been shown to inhibit TNF-α production by monocytes stimulated with endotoxin in vitro, inhibition by AT of TNF-α production in vivo was not observed when animals were pretreated with indomethacin, which inhibits prostaglandin biosynthesis. Iloprost, a stable derivative of PGI_2, produced effects similar to those of AT in these animal models of sepsis and in those subjected to tissue ischemia–reperfusion. These observations strongly suggested that anti-inflammatory activity of AT might be mediated by PGI_2 released from endothelial cells.

Molecular Mechanism(s) of the AT-Induced Increase in Endothelial Production of PGI_2

To elucidate the precise molecular mechanism(s) by which AT exerts anti-inflammatory activity, we examined the effect of AT on the endothelial production of PGI_2 using cultured endothelial cells. However, AT did not directly increase endothelial production of PGI_2 in cultured endothelial cells. Thus, the mechanism(s) by which AT promotes endothelial release of PGI_2 in vivo might involve unknown factors other than endothelial cells.

Capsaicin-sensitive sensory neurons are nociceptive neurons that are activated by a wide variety of noxious physical and chemical stimuli. Since ablation of sensory fibers can result in a marked increase in the severity of inflammation, the sensory neurons have been shown to play a role in the maintenance of tissue integrity by regulating local inflammatory responses. On activation, the sensory neurons release calcitonin gene-related peptide (CGRP) that can increase the endothelial production of PGI_2 in vitro. Since various noxious stimuli that activate the sensory neurons to release CGRP are capable of inducing tissue damage, the CGRP-induced increase in endothelial production of PGI_2 might contribute to attenuation of local inflammatory responses, thereby reducing the tissue damage. Consistent with this hypothesis, we previously reported that capsaicin-sensitive sensory neurons were activated during hepatic ischemia–reperfusion or water-immersion restraint stress in rats, leading to an increase in the endothelial production of PGI2 via activation of both endothelial nitric oxide synthase (NOS) and cyclooxygenase (COX)-1 [8] (Figure 3).

We previously reported that AT reduced ischemia–reperfusion-induced liver injury in rats by increasing endothelial production of PGI_2. However, the mechanism(s) underlying this phenomenon remains to be fully elucidated. AT significantly enhanced the ischemia–reperfusion-induced increase in hepatic tissue levels of CGRP in rats. The increase in hepatic tissue levels of 6-keto-PGF_{1a}, a stable metabolite of PGI_2, increase in hepatic-tissue blood flow, and attenuation of both hepatic local inflammatory responses and liver injury in rats administered AT were completely reversed by administration of capsazepine, an inhibitor of the sensory neuron activation and CGRP(8-37), a CGRP antagonist. AT significantly increased CGRP release from cultured dorsal root ganglion neurons isolated from rats in the presence of capsaicin. These observations strongly suggested that AT might extravasate at the site of endothelial cell injury, thereby enhancing the activation of capsaicin-sensitive sensory neurons leading to increase in hepatic tissue levels of PGI_2 [9] (Figure 3).

Roles of Anti-Inflammation and Anticoagulation by AT in Regulation of Microcirculation

TNF-α has an important role in the development of the microcirculatory disturbance associated with sepsis or circulatory shock, primarily by activating neutrophils, thereby inducing endothelial cell injury leading to microcirculatory disturbance due to local hemoconcentration, and secondarily by inducing microthrombus formation, which precipitates the preexisting microcirculatory disturbance as described earlier (Figure 2). Thus, inhibition of both neutrophil activation and microthrombus formation might be important for maintenance of proper microcirculation. AT increases the endothelial production of PGI_2 via enhancement of the capsaicin-sensitive sensory neurons. PGI_2 is a vasodilator and has a potent inhibitory effect on platelet activation. Since platelet activation and subsequent vasospasms are critical factors in the development of arterial thrombosis, AT might prevent arterial thrombosis by increasing the endothelial production of PGI_2 through interaction with heparin-like substances on endothelial cells, contributing to maintenance of proper microcirculation in the arteries (Figure 3). This hypothesis is consistent with the observations that arterial thrombosis, as well as venous thrombosis, are frequently found in patients in a homozygous state with variant AT molecules that do not have heparin affinity. In addition, PGI_2 has inhibitory effects on monocytic TNF-α production and neutrophil activation. Activated neutrophils are critically involved in the development of microcirculatory disturbance in the postcapillary venules due to the

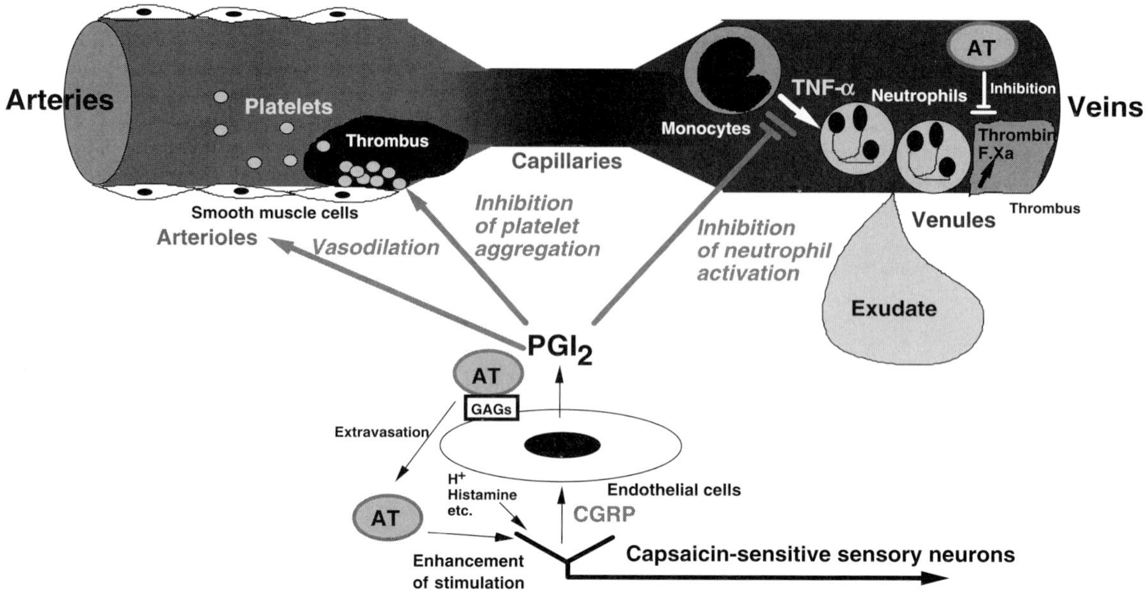

Figure 3 Regulation of the microcirculation by antithrombin. AT, antithrombin; TNF-α, tumor necrosis factor-α; CGRP, calcitonin-gene related peptide; F. Xa, activated form of coagulation factor X; PGI_2, prostacyclin.

increase in microvascular permeability, leading to reduced blood flow resulting from local hemoconcentration. In addition to stasis, activated coagulation factors such as factor Xa are essential for development of venous thrombosis (Figure 3). Thus, AT might prevent thrombus formation by promoting endothelial production of PGI_2 in postcapillary venules where activated neutrophil-induced endothelial damage is frequently induced and also in veins by its inherent anticoagulant activity. Thus, both anti-inflammatory and anticoagulant activities of AT might contribute to the maintenance of proper microcirculation.

Possible Therapeutic Applications of AT for Microcirculatory Disturbances

As described earlier, intravenous administration of AT reduced pulmonary vascular injury and hypotension because of its anti-inflammatory activity in rats given endotoxin, suggesting that AT supplementation might be useful for treatment of acute respiratory distress syndrome and shock associated with sepsis. Administration of AT was also effective in the treatment of ischemia–reperfusion-induced liver and kidney injuries and stress-induced gastric mucosal injury in rats based on its capacity to promote endothelial production of PGI_2, suggesting that AT might be useful for treatment of various organ failures associated with circulatory shock.

Ischemia–reperfusion-induced spinal cord injury is an important pathologic mechanism for the development of paraplegia after operations on the descending thoracic and thoracoabdominal aorta due to interruption of the intercostal and lumbar arteries feeding the spinal cord. Intravenous administration of AT significantly improved motor disturbances by inhibiting reduction of the number of motor neurons in rats subjected to transient spinal cord ischemia [10]. Since both local inflammatory responses and spinal cord microinfarction were significantly reduced through increases in spinal cord tissue levels of PGI_2 in animals treated with AT, both anti-inflammatory and anticoagulant activities of AT might be critical for the therapeutic effect of AT. These observations strongly suggested that AT might be a useful neuroprotective agent for prevention of spinal cord injury after surgery to repair aortic aneurysms. These possibilities should be examined in the clinical setting in the near future.

Conclusions and Perspectives

Although AT is an important natural anticoagulant, it might attenuate inflammatory responses by inhibiting the production of TNF-α. TNF-α is capable of inducing microcirculatory disturbance leading to various organ failures by activating neutrophils and by inducing microthrombus formation. Since microthrombus formation further increases TNF-α production to form a vicious cycle in the progression of microcirculatory disturbance, inhibition of both neutrophil activation and microthrombus formation by AT through promotion of endothelial production of PGI_2 and inherent anticoagulant activity might be important in improving microcirculatory disturbance. Because of these important properties, AT should be an useful therapeutic agent for microcirculatory disturbance in various disease states.

Glossary

Activated neutrophils: Neutrophils activated by various agonists such as pro-inflammatory cytokines and release neutrophil proteases and oxygen radicals that are capable of damaging endothelial cells.

Antithrombin: One of natural anti-coagulants inhibiting coagulation factors with serine protease properties.

Calcitonin-gene related peptide: A neuropeptide synthesized in dorsal root ganglion cells and released from sensory neurons on activation.

Capsaicin-sensitive sensory neurons: Nociceptive neurons that are activated by a wide variety of noxious physical stimuli and have roles in regulation of local inflammatory responses.

Endothelial cell injury: Damage of endothelial cells induced by of noxious substances such as neutrophil elastase and oxygen free radicals and leads to impairment of the endothelial function to maintain proper microcirculation.

Prostacyclin: One of prostaglandins synthesized in endothelial cells and has potent anti-platelet, vasodilatory and anti-inflammatory activities.

Tumor necrosis factor-α: One of pro-inflammatory cytokines produced by circulating monocytes and macrophages in response to endotoxin and a representative causal substance for coagulation abnormalities and organ failures observed in sepsis.

References

1. Ishiguro, K., Kojima, T., Kadomatsu, K., Nakayama, Y., Takagi, A., Suzuki, M., Takeda, N., Ito, M., Yamamoto, K., Matsushita, T., Kusugami, K., Muramatsu, T., and Saito, H. (2000). Complete antithrombion deficiency in mice results in embryonic lethality. *J. Clin. Invest.* **106**, 873–878.
2. Hirsh, J., Piovella, F., and Pini, M. (1989). Congenital antithrombin III deficiency. Incidence and clinical features. *Am. J. Med.* **87**, 34S–38S.
3. Okajima, K. (2001). Regulation of inflammatory responses by natural anticoagulants. *Immunol. Rev.* **184**, 258–274.
4. Okajima, K., Harada, N., Kushimoto, S., and Uchiba, M. (2002). Role of microthrombus formation in the development of ischemis/reperfusion-induced liver injury in rats. *Thromb. Haemost.* **88**, 473–480.
5. Okajima, K., Harada, N., and Uchiba, M. (2003). Microthrombus formation enhances tumor necrosis factor-α production in the development of ischemia/reperfusion-induced liver injury in rats. *J. Thromb. Haemost.* **1**, 1316–1317.
6. Isobe, H., Okajima, K., Uchiba, M., Harada, N., and Okabe, H. (2002). Antithrombin prevents endotoxin-induced hypotension by inhibiting the induction of nitric oxide synthase in rats. *Blood* **99**, 1638–1645.
7. Mizutani, A., Okajima, K., Uchiba, M., Isobe, H., Harada, N., Mizutani, S. R. N., and Noguchi, T. (2003). Antithrombin reduces ischemia/reperfusion-induced renal injury in rats by inhibiting leukocyte activation through promotion of prostacyclin production. *Blood* **101**, 3029–3036.
8. Harada, N., Okajima, K., Uchiba, M., and Katsuragi, T. (2003). Contribution of capsaicin-sensitive sensory neurons to stress induced increases in gastric tissue levels of prostaglandins in rats. *Am. J. Physiol. Gastrointest. Liver. Physiol.* **285**, G1214–G1224.
9. Harada, N., Okajima, K., Yuksel, M., and Isobe, H. (2005). Contribution of capsaicin-sensitive sensory neurons to antithrombin-induced reduction of ischemia/reperfusion-induced liver injury in rats. *Thromb. Haemost.* **93**, 48–56.
10. Hirose, K., Okajima, K., Taoka, Y., Uchiba, M., Nakano, K., Utoh, J., and Kitamura, N. (2004). Antithrombin reduces the ischemia/reperfusion-induced spinal cord injury in rats by attenuating inflammatory responses. *Thromb. Haemost.* **91**, 162–170.

Further Reading

Esmon, C. T. (2003). Inflammation and thrombosis. *J. Thromb. Haemost.* **1**, 1343–1348.

Lefer, A. M., and Lefer, D. J. (1993). Pharmacology of the endothelium in ischemia–reperfusion and circulatory shock. *Ann. Rev. Pharmacol. Toxicol.* **33**, 71–90. *This excellent review describes the fundamental and important mechanisms of both protection and the development of damage of endothelial cells during ischemia–reperfusion.*

Opal, S. M., and Esmon, C. T. (2003). Bench-to-bedside review: Functional relationship between coagulation and the innate immune responses and their respective roles in the pathogenesis of sepsis. *Crit. Care* **7**, 23–38. *This review article demonstrates the possible coevolution of coagulation and innate immunity from a common ancestral substrate early in eukaryotic development, thereby explaining the molecular mechanisms of pathologic sequelae of sepsis.*

Taylor, F. B., Jr., Chang, A. C. K., Peer, G. T., Mather, T., Blick, C., Catlett, R., Lockhart, M. S., and Esmon, C. T. (1991). DEGR-factor Xa blocks disseminated intravascular coagulation initiated by *Escherichia coli* without preventing shock or organ damage. *Blood* **78**, 364–368. *This paper appears to be the first demonstrating that inflammatory responses as well as coagulation abnormalities were important factors in the pathologic process for the development of microcirculatory disturbances seen in sepsis.*

Vallet, B., and Wiel, E. (2001). Endothelial cell dysfunction and coagulation. *Crit. Care Med.* **29**(7 Suppl), S36–S41. *This article shows an important role of coagulation abnormalities induced by tissue factor generation in monocytes in the development of endothelial dysfunction, thus conferring the possible role of microthrombus formation in the formation of a vicious cycle in the progression of microcirculatory disturbance seen in sepsis.*

Capsule Biography

Dr. Okajima is Professor of the Department of Biodefense and Chief Investigator of the vascular biology research group. He contributed much to the elucidation of molecular mechanism(s) in the important linkage between the coagulation and inflammation. His work is supported by grants from the Japanese Ministry of Education, Cultures, Sports, Science and Technology and by the departmental funds of Kumamoto University in which he worked before.

SECTION E

Coronary Slow Flow

CHAPTER 98

The Coronary Slow Flow Phenomenon

John F. Beltrame
The Queen Elizabeth Hospital, University of Adelaide, Adelaide, Australia

The coronary microvasculature plays a fundamental role in the provision and regulation of myocardial perfusion. Accordingly, coronary microvascular dysfunction may frequently contribute to impaired myocardial perfusion and thus play an important role in the genesis of myocardial ischemia. Thus research into coronary microvascular disorders will not only elaborate new conditions and their mechanisms but also provide insights into the mechanisms by which the microvasculature controls myocardial perfusion.

Syndrome X is the most extensively investigated coronary microvascular disorder and must be delineated from the "metabolic syndrome X." The term was first coined by Kemp in 1973, in an accompanying editorial to a study by Arbogast and Bourassa. This landmark study demonstrated both ST segment depression and myocardial lactate production during rapid atrial pacing in 11 patients with angiographic obstructive epicardial coronary artery disease (group C) and also in 10 patients with angina but no epicardial coronary disease (group X). Today, syndrome X is broadly defined as chronic angina occurring in the absence of fixed or dynamic flow-limiting epicardial coronary obstructions on angiography. The diagnosis is most frequently made in patients with exertional angina, ST segment depression on exercise stress testing, and angiographically normal epicardial coronary arteries.

Definition

The year prior to the description of syndrome X, Tambe and associates described a curious angiographic finding of slow contrast flow in six patients with angina and angiographically smooth epicardial coronary arteries. They speculated that this angiographic observation was attributable to an increased coronary microvascular resistance. Today, the coronary slow flow phenomenon (CSFP) is characterized as a delayed passage of contrast media during adequate selective coronary arteriography in the absence of obstructive epicardial coronary artery disease. It is referred to as a "phenomenon" because it is defined on the basis of an angiographic observation rather than a clinical syndrome.

What constitutes slow contrast flow has not been uniformly defined. Most published studies have utilized the subjective impression of an experienced angiographer. However, an objective definition is required if clinicians and researchers are to ensure they are referring to the same condition. Two approaches to this problem have evolved. These both utilize conventional indices developed by the Thrombolytics in Myocardial Infarction (TIMI) investigators.

The TIMI flow grade is a qualitative index of angiographic coronary flow where TIMI-3 flow is considered as normal flow. In TIMI-2 flow, distal vessel opacification is delayed and often arbitrarily defined as requiring three or more cardiac cycles to opacify the distal vasculature. Thus a simple quantifiable definition of the CSFP is the presence of TIMI-2 flow in the absence of obstructive epicardial coronary disease. The advantage of this definition is its simplicity, which allows prompt identification of the phenomenon during angiography. Furthermore, angiographic flow is normalized for heart rate, thus reducing artifacts. Its disadvantage is that it incorporates angiographic flows that are six or more standard deviations from the mean and thus may be too restrictive.

The TIMI frame count (TFC) is a quantitative angiographic flow index that assesses the number of angiographic frames required to opacify a coronary vessel. Published average TFC for the left anterior descending (LAD), circumflex (Cx), and right coronary arteries (RCA) are 36.2 ±

2.6, 22.2 ± 4.1, and 20.4 ± 3.0 frames, respectively [1]. Some investigators have defined the CSFP on the basis of TFC just above these threshold values, whereas others have required counts at least two standard deviations above these values, and yet others have utilized their own control values. Although this approach is more quantitative than those described earlier, its limitations include (a) frequent necessity for offline analysis; (b) TFC variations with heart rate, nitrate therapy, and dye injection relative to cardiac phase; (c) literature references that require adjustment for angiographic acquisition rate; and (d) clinical characteristics of the control data that are often ill-defined.

Irrespective whether the CSFP is defined by subjective impression, TIMI flow grade, or TFC, the criterion for what constitutes slow contrast flow remains an arbitrary definition. Despite this, there is clearly a group of patients with curious delayed vessel opacification in the absence of obstructive epicardial coronary disease and thus attributable to an increased microvascular resistance. The purpose of this chapter is to summarize the angiographic, pathophysiologic, and clinical characteristics associated with this angiographic phenomenon as well as describing potential therapies for the associated angina.

Primary and Secondary Coronary Slow Flow Phenomenon

The initial description of the CSFP was reported in patients undergoing angiography for the evaluation of chest pain. This same angiographic appearance has also been observed in patients immediately following technically successful coronary angioplasty. As the delayed opacification occurs in the absence of significant large vessel coronary disease in both scenarios, it is plausible that both have a similar underlying pathophysiology and therefore warrant concurrent consideration. Accordingly, CSFP occurring de novo (usually in patients undergoing angiography for evaluation of chest pain) with angiographically normal or near-normal coronary arteries should be considered as *primary CSFP*. When the phenomenon occurs following successful dilatation of a coronary stenosis (by angioplasty, Atherectomy, or stent deployment) resulting in a widely patent residual lesion and no evidence of vessel dissection, then it is referred to as *secondary CSFP*. This chapter will exclusively focus on primary CSFP, as discussion of the secondary form (often referred to as the "no reflow phenomenon") warrants a chapter of its own.

Angiographic Features

The prevalence of the CSFP among patients undergoing diagnostic angiography is 1 percent when the phenomenon is defined as TIMI-2 flow in the absence of obstructive epicardial coronary artery disease. The associated epicardial artery morphology and angiographic flow have been characterized in 65 patients with the CSFP [2]. Smooth contoured vessels were found in 74 percent of patients, coronary ectasia in 14 percent, and minor epicardial coronary artery disease (less than a 50% lesion) in 12 percent of CSFP patients. Thus the CSFP is seldom due to a capacitance effect from large ectatic epicardial coronary arteries.

Examination of the opacification rate (heart beats to fill a vessel) for the major epicardial vessels reveals that the LAD takes the longest to opacify in patients with the CSFP, followed by the right coronary and circumflex arteries (3.6 ± 1.2, 3.0 ± 1.1, and 2.4 ± 1.0 beats, respectively). This would be expected on the basis of the respective vessel lengths; however, if the angiographic flows are assessed by the TFC with normalization for their respective lengths, the LAD still exhibits a more delayed angiographic flow compared with the other vessels in the CSFP. Alternatively, if the frequency of a vessel exhibiting TIMI-2 flow is examined, the LAD is involved in 86 percent of patients, the RCA in 63 percent, and the Cx in 37 percent. This reflects the heterogeneous distribution of flow in the CSFP, although two-thirds of patients have TIMI-2 flow in multiple vessels.

Angiographic technical artifacts such as poor catheter engagement may potentially mimic the CSFP; however, these are readily recognized and excluded by experienced angiographers. A reduced perfusion pressure or myocardial oxygen demand may also be associated with reduced coronary angiographic flow. However, hemodynamic measurements recorded during angiography show no difference in central arterial pressure, epicardial artery diameter, and rate pressure product between patients with the CSFP and controls. Hence, in most patients, this angiographic phenomenon cannot be attributed to secondary changes in coronary blood flow.

Although the severity of the CSFP varies with time, serial studies in affected patients demonstrate a persistence in the delayed opacification. In a study of 12 patients with the CSFP who underwent repeat elective angiography at median interval of 7 months from the index study (range: 1 week to 10 years), the vessel opacification rates were similar (Figure 1). However, when the repeat study was compared with controls who do not have CSFP, vessel opacification remained considerably delayed (Figure 1).

Clinical Features

Although the angiographic features of the CSFP are of interest, it is the associated clinical features that are of key importance and warrant the condition being considered as a new coronary microvascular disorder. The clinical presentation prompting coronary angiography, the subsequent clinical progress, and markers of myocardial ischemia in patients with the CSFP are discussed next.

Initial Presentation

Syndrome X is clinically characterized as exertional angina that precipitates referral to a physician. An exercise

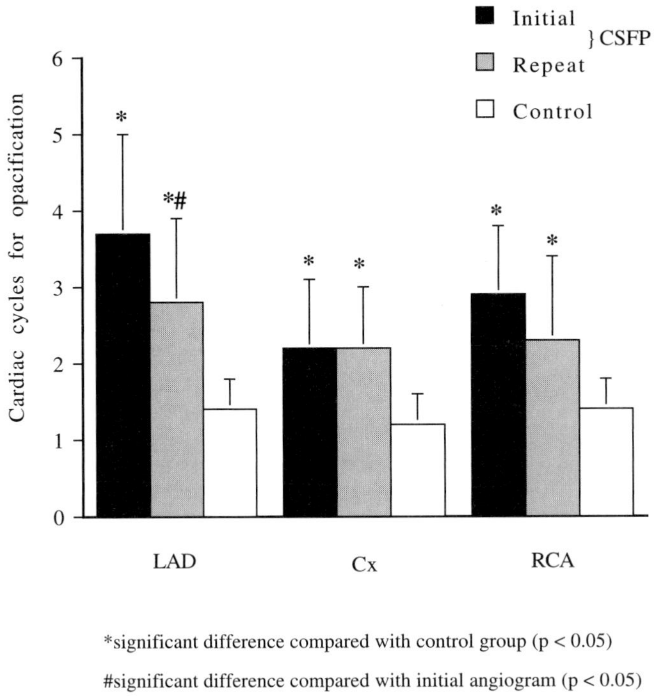

Figure 1 Serial angiographic studies in patients with CSFP. Number of cardiac cycles to opacify the LAD, Cx, and RCA in 12 patients with the CSFP who underwent a repeat angiogram (median interval = 7 months) and 47 control patients who did not exhibit this phenomenon. Although there was a trend toward improvement in the repeat study (significant for the LAD only), both studies showed delayed opacification as compared with controls.

Table I Clinical Characteristics of the CSFP.[a]

Characteristic	Control (n = 47)		Coronary slow flow (n = 47)	
Coronary risk factor				
Age	55 ± 11 years		50 ± 10 years[b]	
Males	21	45%	32	68%[b]
Current smoker	4	9%	15	32%[b]
Hypertension	22	47%	18	38%
Diabetes	3	6%	5	11%
Positive family history	19	41%	19	41%
Hypercholesterolemia	8	20%	15	37%
Chest pain features				
Recent onset (<1 month)	15	32%	29	62%[b]
Predominantly rest pain	28	59%	40	85%[b]
Pain prompting urgent admission	10	21%	35	74%[b]
CCU admission	8	17%	31	66%[b]
Urgent angiography	10	21%	33	70%[b]
Acute myocardial infarction	0	0%	3	6%
ECG findings				
ST/T wave changes on resting ECG	8	17%	17	36%[b]
Positive stress test[c]	11	39%	6	19%

[a] Clinical characteristics of patients with chest pain, smooth epicardial vessels on angiography, and absence (control) or presence (cases) of CSFP. (From Beltrame et al., 2002, *Cardiology* **97**, 197–202, with permission.)

[b] Significant difference between controls and coronary slow flow patients (chi-square or unpaired t-test, $p < 0.05$).

[c] Data available for each 47 control and coronary slow flow patients except for hypercholesterolemia (39 versus 41, respectively) and stress test (28 versus 32, respectively).

stress test is then usually performed and angiography undertaken that reveals no obstructive epicardial coronary artery disease. Patients with CSFP have a different presentation scenario. These patients usually present with an acute coronary syndrome to the emergency department and are admitted to the coronary care unit for further observation. When the case is reviewed by a cardiologist, angiography is undertaken because of the apparent unstable nature of the clinical syndrome. Hence, these patients are clinically (and angiographically) different from patients with syndrome X.

The data for the preceding clinical description is derived from observational and case-control studies. In an observational study of 65 patients with CSFP, review of their clinical records revealed that all except one patient underwent coronary angiography for the investigation of chest pain. The exceptional patient was diabetic and underwent angiography following a presentation with ventricular tachycardia requiring electrical cardioversion. Further evaluation of the remaining patients revealed that 75 percent underwent angiography following an acute coronary syndrome presentation, with most fulfilling Braunwald Class IIIB criteria.

The findings of this study prompted a more detailed case-control study to further characterize the clinical features associated with this angiographic phenomenon. Forty-seven consecutive patients with the CSFP who had smooth epicardial vessels on angiography were compared with 47 patients who also had smooth epicardial vessels on angiography but who did not have TIMI-2 flow in any vessels. The comparative findings are summarized in Table I. Compared with controls, the patients with CSFP were more often younger, male, and smokers and presented with recent onset, rest pain prompting admission to the coronary care unit and thereafter angiography. Of note, 6 percent of the CSFP patients had elevated cardiac markers consistent with myocardial infarction. However, only 19 percent had a subsequent positive stress test and, hence, the majority would not fulfill standard criteria for syndrome X.

These findings have been supported by case reports and small observational studies, in particular in the TIMI-IIIA study where angiography was undertaken in 391 patients presenting with unstable angina/non-ST elevation myocardial infarction. Of these patients, 14 percent had no significant epicardial coronary artery disease with one-third exhibiting the CSFP [3].

Clinical Progress

Follow-up of 64 patients with CSFP during a median period of 21 months (range 1 to 126 months) found that 84 percent experienced recurrent chest pain with 33 percent re-presenting to the emergency department with severe rest pain and 19 percent readmitted to the coronary care unit for intravenous nitrate therapy. Although many studies of patients with chest pain and normal angiography demonstrate recurrent chest pain, those with the CSFP appear to be particularly susceptible. Voelker and colleagues followed 88 patients with angina and normal angiography for a mean period of 9 years (range: 6 to 11 years) and examined factors that were predictive for ongoing chest pain. The best predictor of continuing chest pain at follow-up was the demonstration of CSFP on the initial angiogram.

Although CSFP is associated with recurrent chest pain and frequent hospital readmissions, the risk of a subsequent cardiac event is low. None of the 64 patients just described endured a subsequent documented myocardial infarct, although one patient experienced a fatal cardiac arrest at home following an episode of chest pain. This latter finding is of concern since ventricular tachycardia has been documented in a number of patients with CSFP. Furthermore a recent investigation has demonstrated a higher corrected QT dispersion among patients with CSFP, suggesting a predisposition to ventricular arrhythmias [4]. Further investigation is required into the relationship between CSFP and ventricular arrhythmias.

Myocardial Ischemia in the Coronary Slow Flow Phenomenon

The presence of myocardial ischemia in patients with syndrome X is controversial with studies demonstrating disparate results. The same appears to be so for CSFP, with some investigators finding a poor association with ischemic markers while others claim that the delayed contrast opacification may be considered "a marker for ischemia."

Clinical Markers of Ischemia

Table II summarizes the findings from exercise stress testing and myocardial perfusion scintigraphy in patients with the CSFP. In both investigations there is considerable heterogeneity in the findings. This may be attributable to (a) different definitions for the CSFP, (b) different exclusion criteria, and (c) variable definitions as to what constitutes a positive result. In relation to exercise stress testing, significant ST depression has been reported between 0–71 percent amongst the various studies. Five of the 7 studies (Table II) report positive stress test findings in less than 20 percent of patients suggesting that there is seldom evidence of ischemia on standard exercise stress testing in the CSFP. Myocardial perfusion scintigraphic studies are more often positive for ischemia with approximately a third demonstrating a reversible perfusion defect (Table II).

Biochemical Markers of Ischemia

Myocardial lactate production is the gold standard marker of ischemia, and two studies have evaluated its presence in the CSFP. Yaymaci and others undertook rapid atrial pacing in 34 patients with the CSFP and demonstrated ischemic lactate production in only six patients (18%). In an independent study by Beltrame and colleagues, 12 patients with documented CSFP failed to demonstrate net lactate production during rapid atrial pacing, cold pressor stimulation, or acetylcholine provocation despite many patients experiencing chest pain. Thus, consistent with the clinical markers, most patients with CSFP do not have objective evidence of ischemia on metabolic criteria. Whether this is due to an absence of myocardial ischemia or inability to detect it is open to speculation.

Pathophysiological Mechanisms

Tambe and colleagues' initial description of CSFP speculated that the phenomenon was due to an increased microvascular resistance. Structural biopsy studies and functional coronary hemodynamic studies have been under-

Table II Studies of Clinical Ischemic Markers in CSFP.

Study	n	Ischemic marker	Ischemia present
Exercise stress tests			
Ciavolella (1994)	17	↓ST (flat) ≥ 1 mm ↓ST (upsloping) ≥ 1.5 mm	9 (53%)
Cesar (1996)	17	↓ST (downsloping)	17 (12%)
Mangieri et al. (1996)	10	"Pathological ↓ST"	0 (0%)
Kurtoglu et al. (2001)	25	Not defined	0 (0%)
Goel et al. (2001)	28	↓ST ≥ 1 mm	20 (71%)
Beltrame et al. (2002)	44	↓ST ≥ 1 mm	7 (16%)
Demirkol et al. (2002)	60	↓ST (downsloping) ≥ 0.5 mm ↓ST (flat) ≥ 1 mm ↓ST (upsloping) ≥ 1.5 mm	60 (13%)
Myocardial perfusion scintigraphic studies			
Ciavolella (1994)	21	Fixed or Reversible defect	18 (86%)
Cesar (1996)	17	Reversible defect	7 (41%)
Beltrame et al. (2002)	39	Reversible defect	10 (26%)
Demirkol et al. (2002)	60	Reversible defect	17 (28%)

[a] Summary of published findings from standard exercise stress test and myocardial perfusion scintigraphic studies in patients with the CSFP.

taken to examine the role of microvascular dysfunction in CSFP.

Biopsy Studies

Mosseri and coworkers performed right ventricular biopsies in six patients with CSFP. In contrast to syndrome X, they demonstrated significant structural microvascular changes including hyperplastic fibromuscular thickening of the small arteries with endothelial cell swelling and degeneration. Whether these small vessel changes were pathogenic in the CSFP or secondary to ventricular hypertrophy was controversial as many patients also had hypertension.

In a follow-up study, Mangieri and others performed left ventricular biopsies in 10 patients with the CSFP, none of whom had concurrent hypertension, diabetes, or myocardial hypertrophy on echocardiography. As in the previous study, the small vessels had markedly thickened walls with luminal narrowing. Hence, independent studies have demonstrated structural microvascular abnormalities that could account for an increased coronary vascular resistance and thus slow distal vessel filling.

Coronary Hemodynamic Studies

Coronary hemodynamic studies have confirmed the presence of an increased coronary vascular resistance in CSFP. In 12 CSFP patients, resting coronary sinus oxygen saturation was measured and found to be abnormally low as compared with controls (23 ± 4% versus 31 ± 4%, respectively; $p < 0.001$, Figure 2), reflecting an increased resting coronary resistance.

The structural small vessel disease demonstrated in the foregoing biopsy studies could account for the observed increased coronary resistance. However, pharmacological studies would suggest that there is a significant functional component. Mangieri and colleagues demonstrated that acute administration of a small vessel vasodilator (dipyridamole) alleviated the delayed vessel opacification whereas a large vessel vasodilator (nitrates) had little impact. Kurtoglu and others extended these observations demonstrating resolution of the angiographic phenomenon with 1 month of oral dipyridamole therapy. A significant dynamic component to the increased resistance is also consistent with the clinical presentation of rest pain, since predominantly exertional pain would be anticipated with a fixed increased resistance.

Possible Biologic Mechanisms

Although evidence for structural and functional components of the increased coronary vascular resistance have been identified, the responsible biological mechanisms are unknown. Endothelin-1 is a potent small vessel vasoconstrictor that also has vascular remodeling properties and therefore is a potential candidate autacoid for this disorder. Consistent with this hypothesis, intracoronary administration of endothelin-1 in rabbits and dogs has been shown to mimic CSFP. Likewise, intracoronary neuropeptide Y has been shown to produce CSFP in patients with chest pain and normal angiography, suggesting another potential mechanism. These potential mechanisms are interesting but further studies are required to elucidate the exact mechanisms responsible for CSFP.

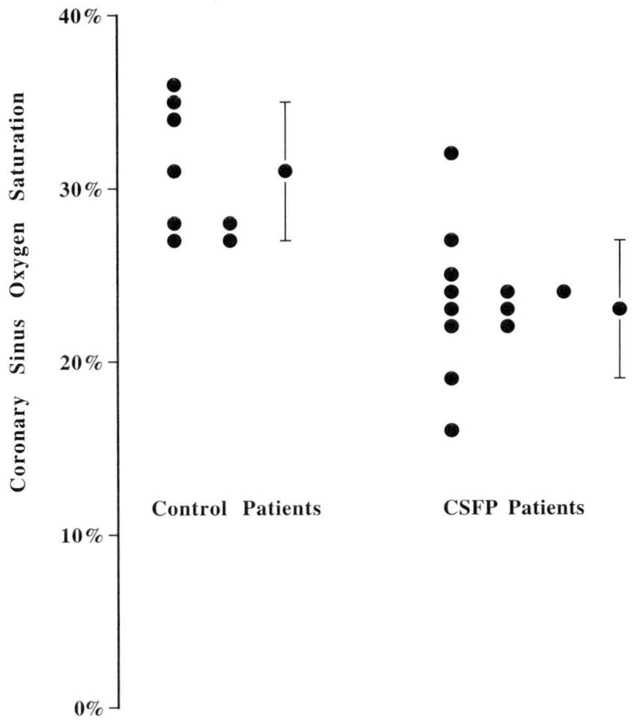

Figure 2 Resting coronary sinus oxygen saturations in CSFP. Resting coronary sinus oxygen saturations (including mean ± SD) for eight control patients with normal angiography but without CSFP and 12 patients with CSFP. (From Beltrame et al., *American Heart Journal*, **146**, 84–90, with permission.)

Therapeutic Considerations

Although the CSFP generally has a good prognosis in relation to subsequent cardiac events, it is associated with considerable morbidity with many patients experiencing recurrent chest pain and therefore requiring antianginal therapy. To date, there is only one published study examining potential therapy for this disorder. In an open-label, observational study, Kurtoglu and coworkers administered dipyridamole 75 mg three times daily to 25 CSFP patients, with angiography performed at baseline and after 1 month of therapy. Although there are limitations to this study, this group demonstrated resolution of the chest pain and angiographic phenomenon with dipyridamole.

From our own clinical experience we have found verapamil and oral nitrates to be of limited benefit. However, we have found mibefradil, a unique calcium channel blocker, to be particularly effective. We surveyed the patient's per-

Table III Summary of Angiographic, Pathophysiological, and Clinical Characteristics of CSFP.

Feature	Characteristic
Angiographic considerations	
Definition	No obstructive epicardial coronary disease with delayed distal vessel opacification (delayed opacification may be based upon TIMI flow grade or TIMI frame count)
Prevalence	1% of diagnostic angiograms
Angiographic findings	LAD most frequently affected vessel
	Often observed in multiple vessels
	Phenomenon persistent at repeat angiography
Pathophysiological considerations	
Coronary hemodynamics	Increased resting coronary vascular resistance due to
	(a) structural obstructive small vessel disease
	(b) functional small vessel constriction
Biologic mechanisms	Endothelin?
	Neuropeptide Y?
Clinical considerations	
Presentation	Acute coronary syndrome (75% of patients)
Clinical progress	Low mortality (concerns over ventricular arrhythmias)
	Low risk of subsequent myocardial infarction
	High morbidity (84% experience recurrent angina)
Exercise stress test	Ischemic ECG changes in approximately 10–20% of patients
Myocardial scintigraphy	Reversible perfusion defect in a third of patients
Potential therapies	Dipyridamole
	Mibefradil

ceived response to mibefradil in 22 patients (56 ± 14 years, 16 males) with CSFP who previously responded poorly to long-acting nitrates and conventional calcium channel blocker therapy. In contrast to the conventional antianginal therapy, all patients reported at least a moderate improvement in their angina frequency with mibefradil, including 73 percent who reported a major improvement. Randomized, double-blind controlled studies are required to corroborate this observational finding.

Synopsis

In summary, CSFP is an angiographic phenomenon which was first described more than 30 years ago but until recently has been largely considered an angiographic curiosity and thus frequently neglected. Table III summarizes our current knowledge of this disorder.

Although initially diagnosed on the basis of an angiographic observation, pathophysiological investigations have demonstrated that it is a coronary microvascular disorder with an increased resting coronary resistance due to both structural and functional abnormalities. Potential biologic mechanisms include the autacoids, endothelin, and neuropeptide Y.

The disorder differs clinically from syndrome X because patients often present initially as an acute coronary syndrome and seldom have positive stress tests. There is significant associated subsequent morbidity with most patients experiencing recurrent chest pain and thus require prophylactic antianginal therapy. Dipyridamole and mibefradil may be effective therapies but require further evaluation.

Glossary

Coronary slow flow phenomenon (CSFP): An angiographic observation characterized by the delayed opacification of the distal vasculature in the absence of obstructive epicardial coronary disease.

TIMI flow grade: A qualitative index of angiographic coronary flow (graded from 0 to 3) developed by the Thrombolytics in Myocardial Infarction (TIMI) investigators. TIMI-0 flow = occluded vessel; TIMI-1 flow = contrast penetrates the obstruction but does not opacify the distal vessel; TIMI-2 flow = delayed distal vessel opacification; and TIMI-3 flow = normal filling of the distal vasculature (i.e., within three cardiac cycles).

TIMI frame count (TFC): A quantitative angiographic flow index developed by the Thrombolytics in Myocardial Infarction (TIMI) investigators. Flow is assessed by the number of angiographic frames required to opacify a coronary vessel to predefined end points.

References

1. Gibson, C. M., Cannon, C. P., Daley, W. L., Dodge, J. T. Jr, Alexander, B. Jr, Marble. S. J., McCabe, C. H., Raymond, L., Fortin, T., Poole, W. K., and Braunwald, E. (1996). TIMI frame count: a quantitative method of assessing coronary artery flow. *Circulation* **93**(5), 879–888.
2. Beltrame, J. F., Limaye, S. B., and Horowitz, J. D. (2002). The coronary slow flow phenomenon—a new coronary microvascular disorder. *Cardiology* **97**, 197–202. *This seminal paper characterizes the clinical features of patients with the coronary slow flow phenomenon, thereby associating a clinical syndrome with the angiographic phenomenon and thus identifying a new disorder.*
3. Diver, D. J., Bier, J. D., Ferreira, P. E., Sharaf, B. L., McCabe, C., Thompson, B., Chaitman, B., Williams, D. O., Braunwald, E. (1994). Clinical and arteriographic characterization of patients with unstable angina without critical coronary arterial narrowing (from the TIMI-IIIA trial). *Am. J. Cardiol.* **74**(6), 531–537.
4. Atak, R., Turhan, H., Sezgin, A. T., Yetkin, O., Senen, K., Ileri, M., Sahin, O., Karabal, O., Yetkin, E., Kutuk, E., and Demirkan, D. (2003). Effects of Slow Coronary Artery Flow on QT Interval Duration and Dispersion. *Ann. Noninvasive Electrocardiol.* **8**(2), 107–111.

Further Reading

Beltrame, J. F., Limaye, S. B., Wuttke, R., and Horowitz, J. D. (2003). Coronary hemodynamic and metabolic studies of the coronary slow flow phenomenon. *Am. Heart J.* **146**, 84–90. *This paper identifies important pathophysiological features of the coronary slow flow phenomenon.*

Beltrame, J. F., Turner, S. P., and Horowitz, J. D. (2001). Persistence of the coronary slow flow phenomenon. *Am. J. Cardiol.* **88**, 938.

Ciavolella, M., Avella, A., Bellagamba, S., Mangieri, E., Nigri, A., and Reale, A. (1994). Angina and normal epicardial coronary arteries: radionuclide features and pathophysiological implications at long-term follow-up. *Coron. Artery Dis.* **5**(6), 493–499.

Demirkol, M. O., Yaymaci, B., and Mutlu, B. (2002). Dipyridamole myocardial perfusion single photon emission computed tomography in patients with slow coronary flow. *Coronary Artery Dis.* **13**, 223–229.

Goel, P. K., Gupta, S. K., Agarwal, A., and Kapoor, A. (2001). Slow coronary flow: A distinct angiographic subgroup in syndrome X. *Angiology* **52**, 507–514.

Kurtoglu, N., Akcay, A., and Dindar, I. (2001). Usefulness of oral dipyridamole therapy for angiographic slow coronary artery flow. *Am. J. Cardiol.* **87**, 777–779.

Mangieri, E., Macchiarelli, G., Ciavolella, M., Barilla, F., Avella, A., Martinotti, A., Dell'Italia, L. J., Scibilia, G., Motta, P., and Campa, P. (1996). Slow coronary flow: Clinical and histopathological features in patients with otherwise normal epicardial coronary arteries. *Cathet. Cardiovasc. Diagn.* **137**, 375–381. *This important study describes the structural abnormalities associated with the coronary slow flow phenomenon and provides evidence for functional abnormalities of the coronary microvasculature.*

Maseri, A. (1995). Syndrome X and microvascular angina. In *Ischemic Heart Disease. A Rational Basis for Clinical Practice and Clinical Research*, pp. 507–532. New York: Churchill Livingstone.

Mosseri, M., Yarom, R., Gotsman, M. S., and Hasin, Y. (1986). Histologic evidence for small-vessel coronary artery disease in patients with angina and patent large coronary arteries. *Circulation* **74**, 964–972.

Yaymaci, B., Dagdelen, S., Bozbuga, N., Demirkol, O., Say, B., Guzelmeric, F., and Dindar, I. (2001). The response of the myocardial metabolism to atrial pacing in patients with coronary slow flow. *Int. J. Cardiol.* **78**,151–156.

Capsule Biography

Dr. Beltrame is a Senior Lecturer in Medicine at the University of Adelaide, The Queen Elizabeth Hospital Campus. He has a strong interest in coronary vasomotor reactivity disorders and been researching the coronary slow flow phenomenon for the past 10 years.

SECTION F

Diabetes

CHAPTER 99

Endothelins and Microvascular Endothelial Responses in Diabetes

[1,2]Subrata Chakrabarti and [1]Zia Ali Khan

[1]*Department of Pathology, University of Western Ontario, London, Canada*

[2]*Department of Microbiology and Immunology, University of Western Ontario, London, Canada*

Diabetes is the leading cause of blindness, renal failure, and limb amputation in the North American population [1, 2]. In spite of improvements in therapeutic modalities, this disorder accounts for significant morbidity and mortality in diabetic patients. Long-standing diabetes leads to structural and functional alterations in both micro- and macrovasculature. The most devastating complications in terms of morbidity are, however, of microvascular origin. The determinant of these complications is sustained hyperglycemia, which leads to biochemical and structural anomalies in the eye, kidney, heart, and peripheral nerves. Microvascular endothelial damage may be a key factor in the pathogenesis of chronic diabetic complications. Endothelins, by virtue of multifunctional capability and widespread tissue distribution, may affect function and structure of microvasculature in several target organs of diabetic complications. Early events in small vessel disease include functional deficits such as blood flow alteration and increased vascular permeability [2]. These early events are essentially reversible with adequate blood glucose control. With progression, however, structural remodeling of microvessels takes place that entails thickening of capillary basement membrane (BM), loss of capillary pericytes, and microaneurysm formation. Later stages may also lead to neovascularization in some organs such as the retina. This review will outline the role of endothelins in microvascular complications of diabetes with emphasis on putative mechanisms of vascular endothelial cell damage.

Endothelins

Endothelins (ETs) are by far the most potent vasoactive peptides identified to date. These peptides exert vasoregulatory action by interacting with cell surface receptors on vascular endothelial and smooth muscle cells [1, 3]. Three structurally similar isoforms of ETs have been identified, ET-1, ET-2, and ET-3. These 21-amino-acid peptides are produced by a number of tissues with vascular endothelium being the major source. ET-1 appears to be the predominant isoform that is constitutively expressed in the vascular endothelium. ETs are regulated primarily at the transcriptional level and a number of stimulators have been identified that upregulate ET expression. These ET inducers include growth factors, cytokines, and various physiochemical factors [1]. In addition to transcriptional regulation, ET production may be regulated via destabilization of mRNA species.

ET gene products undergo two steps of enzymatic cleavage to generate biologically active ET peptides. These peptides perform vasoregulatory action by interacting with specific cell surface receptors, ET_A, ET_B, and ET_C. Among the ET receptors, only ET_A and ET_B receptor types are expressed in mammals. These receptors are coupled to phospholipase C via G proteins [1, 3]. ET_A receptors are localized primarily on vascular smooth muscle cells and are involved in sustained slow-onset vasoconstriction. Activation of ET_A receptors results in calcium influx via phospholipase C-mediated diacylglycerol (DAG) and inositol trisphosphate (IP_3) production. Elevated intracellular calcium and DAG-mediated protein kinase C (PKC) activation lead to myosin light chain kinase (MLCK) phosphorylation and smooth muscle cell contraction. ET_B receptors are involved in generation of nitric oxide (NO) by endothelial cells and thus regulate vasodilation. Endothelial-derived NO activates guanylate cyclase in smooth muscle cells and causes vasodilation by decreasing intracellular calcium levels. The net vascular effect would, therefore, depend on

ET concentration, relative density of ET receptor types, and the vascular tissue.

Mechanism of ET Alteration in Diabetes

Alteration of ETs has been demonstrated in both type I and type II diabetes. Although a number of studies can be cited that provide contradictory reports of plasma ET levels in diabetic patients, it should be noted that these peptides act in both an autocrine and paracrine fashion. Therefore, plasma levels may not provide an adequate assessment of their biological activity [1]. In both animal and human diabetes, use of ET antagonists may be more revealing in terms of the consequences of ET alteration. We and others have demonstrated that in endothelial cells and in several target organs of diabetic complications, ETs are upregulated and mediate structural and functional alterations [1].

The mechanisms by which sustained hyperglycemia leads to upregulation of ETs include activation of PKC, augmented polyol pathway and pseudohypoxia, oxidative stress, elaboration of growth factors, and alteration of vasoactive factors such as NO. Mechanisms and consequences of ET alteration in diabetes are diagrammed in Figure 1. We will briefly describe the possible mechanism by which these biochemical anomalies may lead to alteration of the ET system in diabetes.

PKC Activation

PKC activation has been demonstrated in both diabetic micro- and macrovasculopathy [2]. High glucose levels can

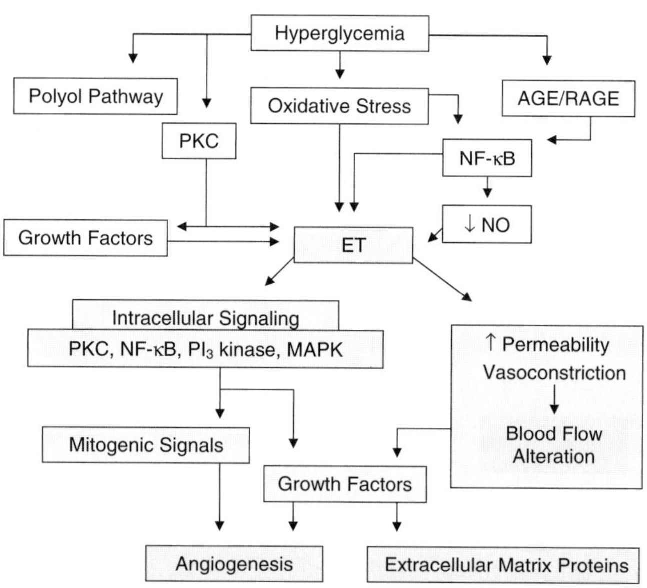

Figure 1 Putative mechanisms and consequences of ET alteration in diabetes. A schematic outlining various hyperglycemia-induced pathways leading to upregulation of endothelin levels is shown in the upper panel. Some of the major effects of increased ET levels are also presented. (see color insert)

induce de novo synthesis of DAG and activation of PKC. PKC has been implicated in mediating several important vascular functions such as regulation of blood flow, vascular permeability, expansion of extracellular matrix, and in the elaboration of various growth factors and cytokines. Studies have demonstrated an interactive relationship between PKC and ETs. We have previously demonstrated inhibition of high glucose-induced ET upregulation by both general (chelerythrine) PKC inhibitor and specific (LY379196) PKCβ inhibitor [4]. PKC activation may also regulate several other growth factors such as vascular endothelial growth factor (VEGF), platelet-derived growth factor, epidermal growth factor, insulin-like growth factor, and fibroblast growth factor. Elaboration of these growth factors may also mediate PKC-induced ET alteration in diabetes.

Polyol Pathway

Polyol pathway has been implicated in several chronic diabetic complications [1]. High intracellular glucose levels overwhelm the glycolytic pathway and lead to enzymatic conversion of glucose to sorbitol via aldose reductase (AR). Sorbitol is subsequently metabolized to fructose by sorbitol dehydrogenase. AR activity requires oxidation of NADPH whereas sorbitol dehydrogenase requires reduction of NAD^+. Therefore, increased flux through the polyol pathway leads to alteration of $NADH:NAD^+$ and $NADPH:NADP^+$. Such imbalance in redox state may cause endothelial dysfunction secondary to hyperglycemia [1]. Interestingly, the augmented polyol pathway may reduce NO synthesis, which also requires NADPH. Impairment of the NO system may lead to ET alteration, as NO has been shown to regulate ETs.

Nonenzymatic Glycation

Glucose and other reducing sugars such as glucose 6-phosphate, trioses, and fructose may react nonenzymatically with amino groups of proteins. Advanced glycation end products (AGEs) may also be produced from glycating dicarbonyl compounds such as 3-dioxyglucosone, methylglyoxal, and glyoxals. AGEs were first thought to mark senescent proteins for degradation. However, in recent years, numerous AGE receptors (RAGEs) have been identified. Binding of AGEs to AGE receptors may mediate intracellular signaling and cause upregulation of growth factors such as ET-1 and VEGF [1]. In addition, AGE formation has been shown to reduce NO, which would further lead to ET alteration.

Oxidative Stress and NO

Increased oxidative stress due to glucose autoxidation, AGE/RAGE interaction, and NO generation has been implicated in the pathogenesis of diabetic complications. In cultured endothelial cells as well as several target organs of diabetic complications, NO synthase mRNA has been shown to be upregulated. However, diabetic patients exhibit impaired endothelium-dependent relaxation. Several theories have been proposed to reconcile these contradictory results. It is interesting to note that concurrent with increased NO synthase expression is increased production of free radicals. Activation of various lipoxygenase enzymes, secondary to hyperglycemia, may promote scavenging of NO by superoxide anions. This interaction yields highly reactive peroxynitrite and hydroxyl radicals. Sequestration of NO by superoxide anions could also contribute to reduced NO bioactivity and availability leading to upregulation of ETs. Increased oxidative stress has also been demonstrated to mediate PKC activation, AGE formation, augmented polyol pathway and sorbitol accumulation, and NF-κB activation in endothelial cells. Such anomalies may further alter the ET system in diabetes. Recently, poly (ADP-ribose) polymerase (PARP), an enzyme well known for polymerizing ADP-ribose in DNA backbone synthesis, has been implicated in ET upregulation. Use of various PARP inhibitors was shown to prevent diabetes-induced alteration of ETs in kidney tissues [5]. Whether diabetes-induced PARP activation leads to ET upregulation in other target tissues, such as the retina, remains to be determined.

Hyperinsulinemia

Insulin represents another factor leading to ET alteration which could be of significance to microvascular complications in type II diabetes. Insulin has been shown to upregulate ET peptide and receptor expression in vascular endothelial and smooth muscle cells [1]. Administration of insulin also increases plasma ET levels in both humans and animals. In addition, hyperinsulinemia has been linked to accelerated macroangiopathy in diabetic patients. The role of insulin, however, in ET alteration and microangiopathy still remains to be determined.

Other Factors

Several other factors may be of importance in augmented ET-1 expression in diabetes. A large number of studies indicate the role of angiotensin II and transforming growth factor-β (TGF-β) in the development of diabetic micro- and macroangiopathy. Angiotensin II is mitogenic for smooth muscle cells and can lead to increased extracellular matrix (ECM) protein synthesis. Recent reports indicate that angiotensin II possibly mediates mitogenic and fibrogenic effects via ET system. Angiotensin II has also been shown to increase synthesis and secretion of ET-1 from vascular endothelial cells. In addition, an interactive relationship between TGF-β and ET has been established. These findings suggest multiple signaling pathways leading to alteration of ET-1 in chronic diabetes.

ETs and Microcirculatory Flow Alterations

Hemodynamic alterations in diabetes are believed to arise as a result of hyperglycemia-induced metabolic abnormalities and elaboration of vasoactive factors including ETs. There is great heterogeneity in findings from microcirculation studies in humans. Study of nailfold microcirculation has revealed elevated as well as reduced blood velocity in diabetic patients when compared to healthy subjects. Reduced blood velocity has also been observed in gastric mucosal blood flow studies. Much of the inconsistency in such studies can be attributed to duration of diabetes, interstudy variability, and limitations of techniques used for measurement of blood flow.

In parallel to in vivo blood flow studies, ex vivo measurement of blood vessel responsiveness to ETs has produced conflicting results. Depending on the relative distribution of ET_A and ET_B receptors in the vascular bed, responsiveness has been shown to be attenuated as well as exaggerated. Limited animal model studies suggest that diabetes leads to alteration of ET responsiveness. Based on existing evidence, however, diabetes is believed to cause vasoconstriction and reduced blood flow early in target organs of chronic complications. These hemodynamic alterations are mediated by increased vasoconstrictors such as ETs and reduced vasodilators including NO. Administration of ET-1 in humans has demonstrated reduced coronary and renal blood flow and increased vasoconstriction. Furthermore, we have demonstrated increased diabetes-induced vasoconstriction in the retina which was normalized with ET receptor antagonist [6].

ETs in Microvascular Endothelial Dysfunction

Endothelial cell dysfunction is increasingly being realized as the unifying mechanism of development and progression of chronic diabetic complications. Endothelial cells are critical for a complex array of functions such as providing a barrier between blood and tissues, maintaining growth and phenotypic characteristics of smooth muscle cells, balancing pro- and anti-inflammatory changes, and fluidity of blood. Alteration of endothelial function, therefore, may affect one or more of these properties. In diabetic microangiopathy, endothelial dysfunction is exhibited as increased permeation, vasoconstriction, and increased synthesis of ECM proteins. Hyalinosis of arterioles and capillaries in diabetes suggests accelerated loss of microvascular endothelial cells and increased ECM deposition. Endothelial degeneration, together with pericyte loss, may bring about a proliferative response and successive elaboration of BM proteins. ETs are implicated in several parameters of microvascular endothelial dysfunction. Administration of ET antagonists has been shown to prevent increased permeability, vasoconstriction, and BM protein expression.

Permeability

We have previously demonstrated that ETs regulate vascular endothelial permeability. Such increased permeability was normalized by treatment with ET receptor antagonist and PKC blocker [4]. The mechanisms by which ETs regulate endothelial permeability are not fully understood. Increased permeability may be arbitrated through endothelial cell contraction. Administration of calcium has been shown to cause phosphorylation of MLCK and cell contracture in endothelial cells [7]. Augmented ET expression by high glucose levels could increase endothelial permeability through interaction with ET_B receptor that is G protein coupled and increases calcium via augmented IP_3. In addition to cell contracture, ETs could also increase permeability via an MLCK-independent retraction mechanism. In such a process, ET-mediated PKC activation is of great significance. PKC has been shown to phosphorylate actin-linking proteins, talin and vanculin, producing intercellular gaps and increased permeability [8].

Mitogenic Responses

ETs are potent mitogens for vascular endothelial cells. The mitogenic property of ETs was first demonstrated in the early 1990s by DNA synthesis assays. Administration of ET-1 was shown to induce DNA synthesis in brain capillary endothelial cells. It has been demonstrated that selective ET_B receptor antagonist can prevent endothelial cell proliferation and migration. In addition to endothelial cells, ETs exhibit mitogenic property toward smooth muscle cells. One interesting difference between the signaling pathways for endothelial and smooth muscle cell proliferation is the involvement of ET receptor type. It has been demonstrated that mitogenic signals are mediated through respective predominant receptor type, that is, for endothelial cells, ET_B, and for smooth muscle cells, ET_A.

In addition to several in vitro studies, ex vivo and in vivo studies also demonstrate mitogenic effects of ETs. Several biochemical pathways may mediate such proliferative signals. ET-induced tyrosine phosphorylation of proteins, such as Src, focal adhesion kinase, and janus kinase, may be involved in these mitogenic responses. In addition, PKC-dependent activation of mitogen activated protein kinase (MAPK) family members may be important in the transduction of mitogenic signals.

ECM Protein Upregulation

A structural hallmark of diabetic microangiopathy is increased capillary BM thickening. Increased expression and decreased degradation of ECM proteins is believed to be critical in BM thickening. The major fibrogenic proteins involved in upregulation of ECM proteins are ETs, TGF-β, and angiotensin II. We have previously demonstrated that high glucose concentration in endothelial cells and hyperglycemia in diabetes leads to upregulation of ECM proteins, fibronectin (FN) and collagen, via ET alteration. Furthermore, recent studies suggest that TGF-β and angiotensin II may also cause increased expression of ECM proteins through ETs. Studies from our laboratory demonstrate that ETs activate NF-κB and AP-1 in target organs and in cultured microvascular endothelial cells leading to FN upregulation [9]. It should be noted, however, that parallel activation of PKC and MAPK family by ETs may also be involved in increased FN expression.

In addition of direct upregulation of ECM proteins such as FN, ETs may also regulate composition of ECM. Recently, we have demonstrated that ETs regulate preferential expression of oncofetal FN, a splice variant of FN [10]. Oncofetal FN is exclusively expressed in proliferating tissues such as embryos and tumors and has recently been proposed to be a marker of tumoral angiogenesis. We have also shown that ET-mediated oncofetal FN is involved in microvascular endothelial cell proliferation. The mechanism by which ET-mediated oncofetal FN regulates cellular proliferation is still obscure. However, recent studies from our laboratory suggest a potential role of oncofetal FN in VEGF expression.

ETs in Organ-Specific Microvascular Alterations in Diabetes

Diabetic Retinopathy

Diabetic retinopathy (DR) predominantly affects the vascular components of the retina. Early in the disease course, diabetes causes functional alterations such as reduced retinal blood flow [2]. With sustained hyperglycemia structural changes such as capillary BM thickening, loss of pericytes, and breakdown of intracellular endothelial cell junctions occur.

Retinal tissue is a rich source of ET-1 and ET-3 [1]. We and others have shown increased mRNA and protein expression of both ET-1 and ET-3 in retinas of diabetic animals in association with retinal vasoconstriction [1]. Blockade of ET receptor mediated signaling and ECE1, enzyme involved in ET peptide processing, prevents retinal vasoconstriction and associated structural changes.

With respect to structural changes, we have demonstrated that ET receptor blockade with dual ET_A and ET_B antagonist, bosentan, prevents diabetes-induced upregulation of FN and collagen alpha-1 (IV) mRNA, and increased capillary BM thickening in animals [1]. ETs could also arbitrate later stages of DR as selective ET_B receptor antagonists can prevent endothelial cell proliferation and migration, two fundamental steps in the process of angiogenesis. In a few recent studies, vitreous ET-1 levels were found to be significantly elevated in patients with proliferative DR as compared to nondiabetic subjects [10].

Diabetic Nephropathy

Diabetic nephropathy (DN) remains the most common cause of renal failure. Sustained hyperglycemia leads to glomerular hyperfiltration and microalbuminuria. With progression, patients develop overt macroalbuminuria and reduced glomerular filtration rate. Pathological features of DN include mesangial matrix expansion, thickening of glomerular capillary BM and tubulointerstitial fibrosis.

ETs may regulate renal blood flow and glomerular filtration. Recent studies from our laboratory have demonstrated increased expression of ET-1, ET-3, ET_A, and ET_B in the diabetic rat [1]. Increased ET-1 mRNA and increased renal ET-1 clearance in association with proteinuria has been demonstrated in human diabetes. Furthermore, treatment of diabetic animals with ET receptor antagonist has been shown to prevent microalbuminuria.

Studies in rat mesangial cells have implicated ETs in regulating ECM protein production. Diabetes-induced increased expression of ECM proteins and other fibrogenic growth factors has been shown to be completely blocked by treatment with an ET_A receptor antagonist and dual ET_A and ET_B receptor antagonist.

Diabetic Cardiomyopathy

Diabetic cardiomyopathy is a prominent cardiac complication that involves structural and functional changes in both cardiomyocytes and capillary endothelial cells. Pathological features of diabetic cardiomyopathy include myocyte hypertrophy and/or necrosis, interstitial and perivascular fibrosis, and capillary BM thickening.

ETs have been shown to be produced by both cardiomyocytes and endothelial cells. We have previously demonstrated upregulation of ET-1 along with ET_A and ET_B receptor expression in heart tissues of diabetic rats. Such alterations were associated with focal apoptosis of cardiomyocytes, scarring of the myocardium, and increased expression of ECM proteins. Inhibition of ET receptor signaling completely prevented these structural abnormalities. Furthermore, a duration-dependant alteration of chronotropic and inotropic responses to ET-1 has been demonstrated in isolated atria of diabetic rats. Recently, we have demonstrated that ET-1 may interact with sodium–hydrogen exchanger-1 (NHE-1) in mediating diabetes-induced structural and functional changes. NHE-1 may act as the downstream mediator in the development of ET-mediated functional and structural changes in diabetic myocardium.

Diabetic Neuropathy

Diabetic neuropathy is one of the most prevalent complications of chronic diabetes. The pathogenesis of diabetic neuropathy involves chronic hyperglycemic insult to both neurovasculature and neuronal parenchyma. Studies in STZ-induced diabetic rats have established a role of ETs in impairment of endoneurial blood flow. In addition, reduced NO production in the vasculature of the peripheral nerve has been demonstrated, which may further augment ET expression. ET receptor antagonism has been shown to prevent impairment of endoneurial blood flow in diabetic animals.

Neuronal parenchymal damage is believed to be due to impaired nerve conduction velocity. Impaired nerve conduction velocity has been associated PKC activity and could possibly be mediated via ETs. We have demonstrated increased immunoreactivity of ET-1 and ET-3 in peripheral nerves in diabetes. Furthermore, inhibition of ET receptor-mediated signaling has been shown to prevent early nerve conduction velocity deficits in STZ-induced diabetic rats.

Concluding Remarks

Experimental and clinical studies over the past few years indicate that ETs are of significance in several human diseases. Their predominant expression in vascular tissues and their multifunctional nature do indeed suggest that alteration of ETs may be involved in diseases affecting both the micro- and macrovasculature. In both animal and human diabetes, ETs have been shown to be upregulated. Hyperglycemia-induced biochemical anomalies such as PKC activation, nonenzymatic glycation, oxidative stress, augmented polyol pathway, and elaboration of various growth factors and cytokines may contribute to alteration of ETs. ETs, in turn, may regulate other vasoactive factors and growth factors leading to changes in both hemodynamic and structural parameters. A schematic outline of ET alteration and its consequences has been depicted in Figure 1. In support of a central role of these multifunctional peptides in diabetes-induced pathogenetic changes, it has been shown that ET-receptor antagonists prevent structural and functional abnormalities in all target organs of chronic diabetic complications in animal models. Based on the available data, ET antagonism may have a potential role in the treatment of these complications.

Glossary

Basement membrane: A ubiquitous supportive tissue that underlies an epithelium or endothelium. This tissue contains macromolecules such as collagen, fibronectin, laminin, and sulfated proteoglycans.

Extracellular matrix: A meshwork-like substance found within the extracellular space. It provides a supporting structure for cells and regulates cellular events.

Neovascularization: The development of new blood vessels. Neovascularization is an important event in tissues where circulation has been impaired by trauma or disease.

Vascular permeability: The property of the vasculature to be pervaded by water and large molecular weight proteins.

Vasoactive factors: Factors that exert an effect on the caliber of blood vessels. These factors include endothelins, angiotensin II, nitric oxide, and prostacyclins.

References

1. Khan, Z. A., and Chakrabarti, S. (2003). Endothelins in chronic diabetic complications. *Can. J. Physiol. Pharmacol.* **81**, 622–634. *This review summarizes the current literature supporting a role of endothe-*

lins in diabetic complications. Special emphasis is on the possible mechanisms of endothelin alteration in diabetes and the effects of such alteration in target organs of chronic diabetic complications.

2. Sheetz, M. J., and King, G. L. (2002). Molecular understanding of hyperglycemia's adverse effects for diabetic complications. *JAMA* **288**, 2579–2588. *This review summarizes the molecular mechanisms underlying chronic diabetic complications. Several theories are presented including oxidative stress, protein kinase C activation, and augmented polyol pathway.*
3. Levin, E. R. (1995). Endothelins. *N. Engl. J. Med.* **333**, 356–363.
4. Chen, S., Apostolova, M. D., Cherian, M. G., and Chakrabarti, S. (2000). Interaction of endothelin-1 with vasoactive factors in mediating glucose-induced increased permeability in endothelial cells. *Lab. Invest.* **80**, 1311–1321.
5. Minchenko, A. G., Stevens, M. J., White, L., Abatan, O. I., Komjati, K., Pacher, P., Szabo, C., and Obrosova, I. G. (2003). Diabetes-induced overexpression of endothelin-1 and endothelin receptors in the rat renal cortex is mediated via poly(ADP-ribose) polymerase activation. *FASEB J.* **17**, 1514–1516.
6. Deng, D., Evans, T., Mukherjee, K., Downey, D., and Chakrabarti, S. (1999). Diabetes-induced vascular dysfunction in the retina: Role of endothelins. *Diabetologia* **42**, 1228–1234.
7. Wysolmerski, R. B., and Lagunoff, D. (1991). Regulation of permeabilized endothelial cell retraction by myosin phosphorylation. *Am. J. Physiol.* **261**, C32–C40.
8. Werth, D. K., Niedel, J. E., and Pastan, I. (1983). Vinculin, a cytoskeletal substrate of protein kinase C. *J. Biol. Chem.* **258**, 11423–11426.
9. Chen, S., Mukherjee, S., Chakraborty, C., and Chakrabarti, S. (2003). High glucose-induced, endothelin-dependent fibronectin synthesis is mediated via NF-kappa B and AP-1. *Am. J. Physiol. Cell Physiol.* **284**, C263–C272.
10. Khan, Z. A., Cukiernik, M., Gonder, J. R., and Chakrabarti, S. (2004). Oncofetal fibronectin in diabetic retinopathy. *Invest. Ophthalmol. Vis. Sci.* **45**, 287–295. *A recent study that provides the first evidence of altered ECM protein composition in diabetic retinopathy. The study focuses on an aberrant fibronectin molecule, oncofetal fibronectin, which is regulated by endothelins and which may mediate angiogenesis. The authors conclude that endothelin-induced oncofetal fibronectin could be involved in endothelial cell proliferation.*

Capsule Biography

Dr. Chakrabarti is a professor in the Department of Pathology at the University of Western Ontario, and a pathologist at the London Health Sciences Centre, Canada. His laboratory primarily focuses on structural and functional alterations in diabetic microangiopathy. His work is supported by grants from the Canadian Diabetes Association, Canadian Institutes of Health Research, Heart and Stroke Foundation of Ontario, and Lawson Health Research Institute.

Mr. Zia A. Khan is a graduate student in the Department of Pathology at the University of Western Ontario, Canada.

CHAPTER 100

Pericytes and Diabetic Retinopathy

Hans-Peter Hammes
University Hospital Mannheim, University of Heidelberg, Germany

General Remarks

This review aims to select and summarize available information relating to the adult retina, and to mechanisms that are likely involved in the pathogenesis of diabetic retinopathy. Insofar as developmental aspects are important for a deeper understanding of the pathogenesis of diabetic retinopathy, and the role of pericytes herein, they will be presented. Otherwise, I refer the interested reader to relevant chapters of this encyclopedia, and to excellent recent reviews of the general role of pericytes in vascular biology (see Bibliography).

Introduction

Incipient diabetic retinopathy is characterized by increased vascular permeability and progressive vascular occlusion. The substrate of vascular occlusion is the formation of nonperfused, acellular capillaries within the retina. The primary morphologic change in the diabetic retina is the loss of pericytes that precedes the formation of acellular capillaries. Pericytes play an important role in the development and maturation of the vascular system. Under nonhyperglycemic conditions, pericytes have a survival- and growth-inhibiting effect on endothelial cells. Under hyperglycemic conditions, the protective function of pericytes on endothelial cells is traded for the increased repair capacity of the capillary endothelium. It was previously thought that pericyte loss is the result of toxic product accumulation and induction of destructive cellular signals generated within the pericyte. In contrast, new experimental data indicate that pericyte dropout may result from regulations that induce pericyte elimination as an active process. The prevention of acellular capillaries without the rescue of pericyte loss in experimental diabetic retinopathy suggests that the endothelium is the primary therapeutic target, and that restoration of pericyte coverage in retinal capillaries may not be necessary for the prevention of diabetic retinopathy.

Pericytes in Vascular Biology: A Brief Overview

Retinal capillaries consist of three elementary structures: endothelial cells, basement membrane tubes, and intramural pericytes that are located within the basement membrane (Figure 1). Pericytes both in humans and in rodents are present in the retina at an almost 1 : 1 ratio with endothelial cells. Their high numbers in the retina, even higher than the numbers in the brain, has been associated with the tightness of the blood–retinal barrier.

The recruitment of pericytes to the vessel wall has been most extensively studied in rodents. Pericytes arrive in the retinal vessel closely following the sprouting tip of the growing capillary. With improved techniques circumventing the problem that certain markers (such as SMA) are not expressed in pericyte subpopulation invading the CNS, it emerges that the existing window of plasticity of the developing retinal capillary network is not explained by pericytes lagging behind the sprouting of endothelial cells. The resistance of retinal capillaries to regressive signals such as hyperoxia marks the end of the vascular plasticity. From in vivo studies in newborn mice, it is likely a variety of growth factors such as vascular endothelial growth factor (VEGF), the angiopoietins, and others associated with pericyte recruitment such as platelet-derived growth factor (PDGF)-B and TGF-β are involved in the maturation of retinal capillaries. However, the precise mechanisms are not yet defined, and the role of pericytes is still undetermined except for the fact that their attachment to the vessel wall does not coincide with the resistance of the capillaries to regression. With regard to the pathogenesis of diabetic retinopathy, which develops almost exclusively in the mature retina, pericytes are crucially involved in promoting

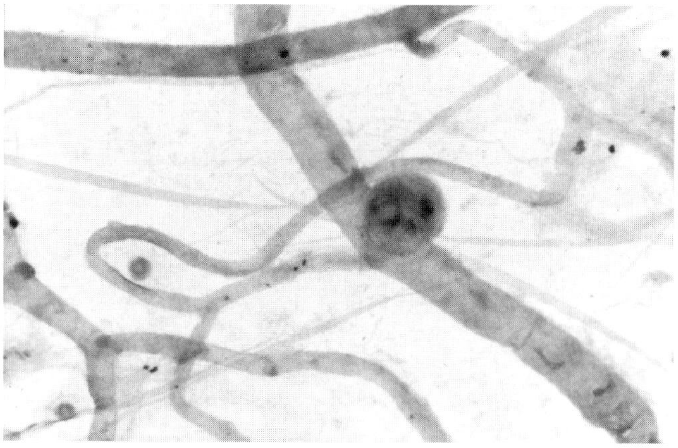

Figure 1 Retinal digest preparation of a human retina, obtained postmortem, of a patient with type 1 diabetes. A microaneurysm is positioned next to acellular capillaries. Original magnification 600×. Periodic acid–Schiff (PAS) staining.

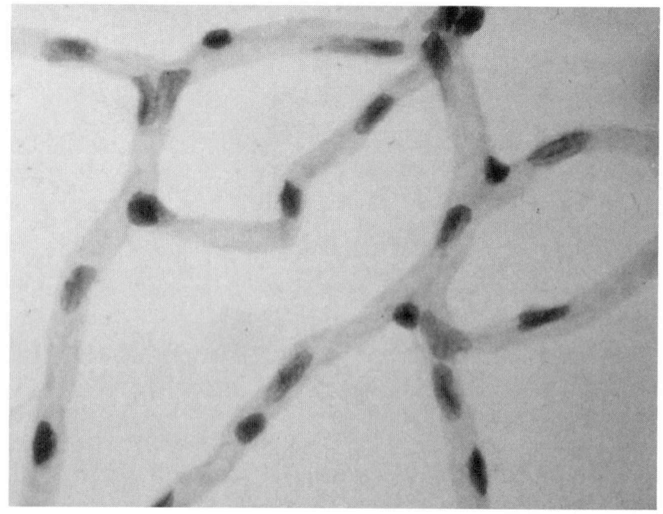

Figure 2 Retinal digest preparation (original magnification 400×; stained with PAS) showing retinal capillaries of a 12-month-old rat. Note the distribution of endothelial cells and pericytes within the basement membrane tube.

the survival of retinal capillaries. Active disruption of the cellular cross-talk between ECs and pericytes leads to aberrant remodeling. While pericyte recruitment to capillaries in the developing retina is initially PDGF-Rβ dependent, capillaries from older mice become resistant to PDGF-B depletion. From these data, it emerges that pericytes in the developing retina control sprouting and vessel remodeling, whereas in the adult/matured retina, they serve as survival-supporting cells for endothelial cells. This is also reflected by experiments showing that the adult retina is resistant to hyperoxia-induced vascular regression, even when pericyte numbers are reduced, suggesting that mechanisms independent of the physical presence of pericytes determines plasticity.

The Natural Course of Human Diabetic Retinopathy

The two most characteristic features of incipient diabetic retinopathy are increased vascular permeability and progressive vascular occlusion. The method of retinal digest preparations, developed by Kuwabara and Cogan, allowed inspection and quantitation of changes in the affected retina liberated from neuroglial tissues due to differential susceptibility against trypsin digestion. As a result of their work and that of others presented later, pericyte loss was identified as the earliest change in the diabetic retina. With time, microvascular endothelial cell loss, in part due to programmed cell death, and progressive capillary occlusion occur. Acellular capillaries are the most significant lesions in the diabetic retina, as they (1) represent the phenotype of hyperglycemia-induced vascular cell damage, which is a general feature of diabetic complications, and (2) are the likely harbingers of all subsequent changes. Microaneurysms, which are the first clinically detectable lesions in the eye of a diabetic patient, are found predominantly around areas of occluded capillaries (Figure 2). As suggested by Ashton, microaneurysms are considered a first, abortive attempt at neovascularization due to focal retinal ischemia.

With more extended nonperfused areas due to progressive capillary occlusions, larger retinal areas react forming intravascular endothelial proliferations in hypercellular vessels (not necessarily capillaries!). The subsequent formation of venous beadings indicates progressive retinal ischemia, representing the extension of the response to ischemic injury to the veins. With long-standing diabetes, and progressive vascular regression, the retina starts to form new blood vessels that enter the vitreous body. The number of newly formed vessels frequently correlates with the extent of nonperfusion in the retina. Neovascularizations tend to spread within the preretinal vitreous and are prone to rhexis and preretinal bleeding. They are often accompanied by matrix-producing cells such as fibroblasts and by inflammatory cells. Both tractive forces due to formation of membranes, which tend to shrink, and the fragility of the newly formed vessels explain the high risk of vitreous hemorrhages in patients with proliferative diabetic retinopathy.

Microaneurysms are often locations of increased vascular permeability, and leaky microaneurysms may lead to macular edema and hard exudates. Focal hard exudates and dot-blot hemorrhages are direct indicators of a breakdown of the blood–retinal barrier. Increased vascular permeability infrequently occurs when other typical retinal lesions are still absent, when glucose control is inadequate, and disappear upon normalization of hyperglycemia.

Pericyte dropout in the retina has not been clearly related to diabetes duration in humans, nor is it pathognomonic for the diabetic retina, since pericyte dropout can occur in various other ocular and systemic diseases such as venous occlusion, polycythemia, and hypertension. Permeability changes in the diabetic retina can be focal or diffuse, suggesting that predilections exist unrelated to pericyte dropout.

The time course of the initial hyperglycemic capillary damage has been most carefully studied in diabetic animal models. Here, the first appreciable sign of structural vascular damage is the loss of pericytes. Pericyte loss has been demonstrated in a variety species (mice, hamsters, rats, dogs, cats, monkeys), and with various modes of diabetes (chemical, genetic). In streptozotocin-diabetic rats, pericyte loss starts between 4 and 8 weeks of diabetes. Subsequently, there is a steady increase in the number of endothelial cells per capillary area unit, indicating random distribution of endothelial cell proliferation. This change becomes significant after 4 to 5 months of hyperglycemia. In parallel with the increase of endothelial cell numbers, acellular occluded capillaries occur, starting with a unilateral focal obstruction in the vicinity of capillaries that still contain cells. This phenotype favors the idea that capillary occlusion in the diabetic retina is the result of both endothelial cell damage (and loss) and extraluminal factors. Although in vivo experiments suggest that intraluminal obstruction may occur through cellular components, acellular capillaries in retinal digest preparations of diabetic rats are typically devoid of cellular debris or blood cells. Acellular capillaries are also present in nondiabetic animals, but are more numerous in diabetic animals, and become more frequent after 3 to 4 months of diabetes. Microaneurysms resembling very early lesions in human retinae (i.e., a unilateral outpouching of the capillary) are occasionally found in diabetic rodents, appearing after 5 to 6 months of hyperglycemia.

Although microaneurysms are occasionally observed in diabetic rodents, and much more frequently in diabetic dogs, there is no consistent evidence of sprouting vessels or proliferative changes in diabetic rats or mice despite a level of pericyte dropout that can reach 50 percent.

Abnormal vascular permeability measured by dye techniques is observed as early as 1 week after induction of chemical diabetes in rats. A relation to pericyte dropout has not been determined, but is unlikely to happen, given the discrepant time courses of both changes.

Lessons from Genetically Modified Diabetic Animals

Mice with a genetic ablation of PDGF-B exhibit several vascular phenotypes that are highly reminiscent of certain characteristics of diabetic retinopathy, including microvascular leakage and hemorrhage, pericyte deficiency, and microaneurysms in brain capillaries. A similar phenotype is observed when the PDGF-receptor β is absent, suggesting that this system is crucially involved in the recruitment of pericytes to developing vessels. Moreover, experiments in postnatal mice blocking pericyte recruitment by antibody inhibition of PDGF-receptor β revealed impaired remodeling, distortion and larger diameters of capillaries, and leakiness of retinal vessels. Of note, no signs of apoptosis or acellular occluded vessels were present. When administered to older animals, the inhibition of the PDGF-B/receptor β system had no effect on pericyte recruitment, suggesting that the system is only important for recruiting but not for maintaining pericytes in the capillaries. This is to be expected given the redundancy of factors involved in pericyte recruitment and attachment, such as angiopoietin-1, TGF-β, and tissue factor. In the absence of pericytes, changes in capillary diameters, with both increased and decreased luminal diameters, are reported. It is an open and largely disputed question whether this is attributable to the propensity of pericytes to contract. However, independent from these open questions, it is concluded that pericytes have a specific and profound role in sprouting angiogenesis in the developing retina, and that pericytes inhibit endothelial proliferation, as endothelial cell hyperplasia is seen in mice that lack pericytes.

Note that diabetic retinopathy initiates in the adult, mature retina, which is mostly resistant to manipulations such as hyperoxia or growth factor withdrawal, suggesting that important functions of pericytes and system that affect pericyte recruitment may vary.

Some indication of pericyte function in the adult vessel comes from experiments using retinal digest preparations from mice in which PDGF-B is reduced by half (heterozygous PDGF-B mice). These mice had a 28 percent reduction in pericyte numbers, and a moderate but significant increase in acellular capillaries, suggesting that pericytes function as survival factors for endothelial cells. In diabetic animals, the degree of pericyte loss was increased as expected, but the number of acellular capillaries increased exponentially. This means that with more pericyte loss, PDGF-B is not only a pericyte recruiting factor, but also an indirect survival factor for the stressed endothelium in diabetes.

In endothelial specific PDGF-B knockout mice, a heterogeneous degree of pericyte loss was observed, when assessed in several areas of the brain. In the retina, the numbers of capillary occlusions in the intermediate and deep capillary layers was inversely correlated with the numbers of pericytes on these vessels. The more pronounced the pericyte deficit, the more capillary occlusions occurred. However, mice with the least pericyte coverage did not develop any capillary layers other than the superficial, and developed proliferative signs in their remaining superficial layer. This indicates that pericytes are important cells for developmental angiogenesis. However, their inhibitory effect on endothelial proliferation needs revision, as pericytes do accompany proliferating vessels in this (pathological) and in other (physiological) settings.

Apart from the PDGF-B/PDGF-βR system, other ligand–receptor systems may be involved in the developmental recruitment of pericytes. Angiopoietin-1, which signals via the endothelial specific tyrosine kinase receptor Tie-2, determines capillary sprouting, endothelial cell survival, and vascular remodeling and has been implicated in the stabilization of vessels by recruiting pericytes. Angiopoietin-2 can act as a natural antagonist of Ang-1. In a rat model of experimental diabetic retinopathy, Ang-2 is upregulated manyfold prior to the onset of pericyte dropout. The upregulation of Ang-2 persists over time, suggesting a nontransient effect of hyperglycemia on Ang-2 transcription. Confirmatory data come from a mouse model in which a reporter construct is expressed under the control of the Ang-2 promoter (Ang-2 LacZ knockin mice). Intravitreal application of Ang-2 mimics the pericyte-depleting effect of hyperglycemia without affecting retinal capillary diameters or endothelial cell survival. When Ang-2 LacZ knockin mice were maintained hyperglycemic for 6 months, they did not develop pericyte loss, as did the wild-type controls, suggesting that Ang-2 plays a significant role in the early pericyte loss by hyperglycemia. Preservation of pericytes in this diabetic model was accompanied by a partial reduction of acellular capillaries compared with nontransgenic diabetic mice. Pericytes partially protect endothelial cells in chronic hyperglycemia, but other mechanisms, related to the newly formulated hypothesis of the unifying concept of vascular damage in diabetes by Brownlee (see later discussion), have been proposed.

Confirmation of the role of Ang-2 in pericyte regulation of retinal capillaries also comes from experiments in which Ang-2 was overexpressed in the photoreceptor using a photoreceptor-specific promoter. The resulting increase in Ang-2 in the deep layers of the retina induce a selective moderate pericyte loss of the deep capillary network, while leaving the superficial network unaffected. Hyperglycemia in these mice induces a premature damage of the deep capillary layer, as indicated by an untimely formation of acellular capillaries.

Pericytes and acellular capillaries are unevenly distributed within the retina. Whereas pericyte ghosts, which are empty basement membrane pockets suggesting pericyte loss, are similar in all four quadrants of a diabetic dog's retina, acellular capillaries are most frequent in the superior temporal area. A number of questions need to be addressed before these findings are accepted as contradiction against pericytes being survival factors for endothelial cells in the diabetic retina, such as the local production of growth factors, the main subpopulation of pericytes leaving ghosts behind, and the possible contribution of local blood flow in different areas of the retina.

Linking Biochemistry to Cell Biology of Diabetic Retinopathy: The Unifying Hypothesis

The causal link between diabetes and microvascular complications is chronic hyperglycemia. Large prospective clinical studies in both type 1 and type 2 diabetics have demonstrated a strong relationship between glycemia and diabetic microvascular complications. The link between hyperglycemia and endothelial cell damage in diabetes has been intensively studied during the past 20 years. Biochemically, four independent pathways were investigated and led to the development of pharmacological inhibitors (Figure 3): increased polyol pathway flux, increased advanced glycation end product formation, activation of protein kinase C, and increased hexosamine pathway flux. Hyperglycemia-driven mitochondrial overproduction of reactive oxygen species has recently been identified as the underlying biochemical abnormality that is not only a common denominator of the four seemingly unrelated pathways, but also a new crystallization point of concepts for the development of antiangiopathy principles in treatment of diabetes. The model is based on the findings that cells which are exposed to a high ambient glucose, such as endothelial cells, have an increased substrate flux through glycolysis and the TCA cycle. The net result is an increased production of superoxide by the mitochondrial electron transport chain. This induces a substantial reduction in the activity of the glyceraldehyde-phosphate dehydrogenase with an upstream accumulation of glycolytic intermediates such as glyceraldehyde 3-phosphate and fructose 6-phosphate. These metabolites are important initiators of the previously mentioned biochemical pathways, such as the hexosamine pathway (fructose 6-phosphate) or the protein kinase C pathway (glyceraldehyde 3-phosphate). The mechanism responsible for the inhibition of GAPDH involves ROS-mediated activation of poly-ADP-ribose polymerase.

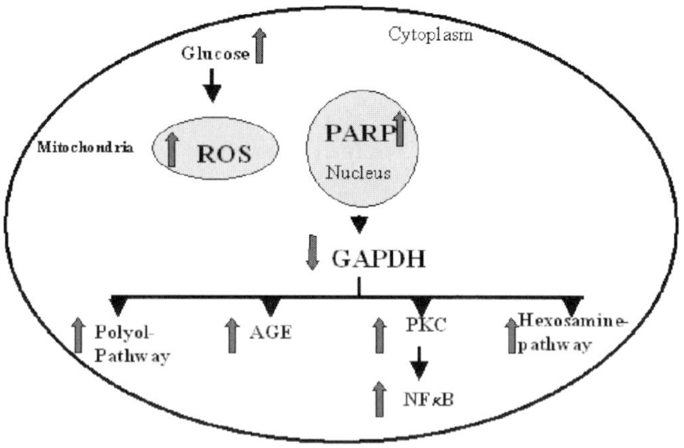

Figure 3 Biochemistry of diabetic vascular damage: Mitochondrial overproduction of reactive oxygen species (mt-ROS) induces upregulation of poly(ADP-ribose)polymerase (PARP), which inhibits a central enzyme of glycolysis (GAPDH) with subsequent effects on major pathways of hyperglycemia-induced damage. The initiator of mt-ROS is increased transport of glucose into endothelial cells followed by increased flux through glycolysis and the tricarbon cycle. (From Brownlee, Claude Bernard Lecture 2003, Paris.)

Another approach to alleviate diabetic vascular complications is to shift of the increased flux of toxic intermediates into a nontoxic pathway. Activating transketolase by the cofactor vitamin B_1 or lipid-soluble prodrugs such as benfotiamin has been shown to reduce all major biochemical pathways activated by hyperglycemia, and reduces acellular capillary formation in diabetic rats. Notably, retinal capillaries did not exhibit any change in pericytes by transketolase activation, but the well-documented increase in endothelial cells in areas not affected by acellular capillaries was reduced toward normal. These findings are consistent with the hypothesis that the endothelium modulates the balance between pericyte-recruiting and pericyte-eliminating factors in chronic hyperglycemia, and that the endothelium is the primary therapeutic target in diabetic microangiopathy.

Glossary

Acellular capillaries: Smallest vessels in the eye in which the cells (endothelial cells and pericytes) have been either lost or destroyed by chronic hyperglycemia, and the subsequent biochemical abnormalities.

Diabetic retinopathy: Most prevalent microvascular complication of common types of diabetes mellitus, affecting vascular and neuroglial structures of the retina.

Microaneurysms: Lesion typical, but not pathognomonic, for incipient diabetic retinopathy. Fusiform or saccular in appearance, microaneurysms tend to cluster around areas of acellular capillaries, suggesting that they are an early aberrant angiogenic process.

Retinal digest preparation: A method to isolate the retinal vasculature by means of enzymes that digest away neuroglial elements; subsequent staining with periodic acid–Schiff base and hematoxylin–eosin provides a selective view on the retinal vasculature.

Bibliography

Allt, G., and Lawrenson, J. G. (2001). Pericytes: Cell biology and pathology. *Cells Tissues Organs* **169**, 1–11. *A recent comprehensive overview on the diverse functions of pericytes throughout the vascular system, and their implications.*

Antonelli-Orlidge, A., Saunders, K. B., Smith, S. R., and D'Amore, P. A. (1989). An activated form of TGF-beta is produced by co-cultures of endothelial cells and pericytes. *Proc. Natl. Acad. Sci. USA* **86**, 4544–4548. *A landmark paper that established the inhibitory role of pericytes in endothelial cell proliferation.*

Benjamin, L. E., Hemo, I., and Keshet, E. (1998). A plasticity window for blood vessel remodelling is defined by pericyte coverage of the preformed endothelial network and is regulated by PDGF-B and VEGF. *Development* **125**, 1591–1598.

Brownlee, M. (2001). Biochemistry and molecular cell biology of diabetic complications. *Nature* **414**, 813–820. *The most recent integrative review on the effect of high glucose on vascular cells, presenting the new "unifying concept."*

Carmeliet, P. (2003). Angiogenesis in health and disease. *Nat. Med.* **9**, 653–660.

Drake, C. J., Hungerford, J. E., and Little, C. D. (1998). Morphogenesis of the first blood vessels. *Ann. N. Y. Acad. Sci.* **857**, 155–179.

Du, X., Matsumara, T., Edelstein, D., Rosetti, L., Zsengeller, Z., Szabo, C., and Brownlee, M. (2003). Inhibition of GAPDH activity by poly(ADP-ribose) polymerase activates three major pathways of hyperglycemic damage in endothelial cells. *J. Clin. Invest.* **112**, 1049–1057.

Gerhardt, H., and Betsholtz, C. (2003). Endothelial-pericyte interactions in angiogenesis. *Cell Tissue Res.* **17**, 1835–1840. *A review that provides new insights into the role of pericytes as modulators of angiogenesis.*

Hammes, H.-P., Du, X., Edelstein, D., Taguchi, T., Matsumara, T., Ju, Q., Lin, J., Bierhaus, A., Nawroth, P., Hannack, D., Neumeier, M., Bergfeld, R., Giardino, I., and Brownlee, M. (2003). Benfotiamine blocks three major pathways of hyperglycemic damage and prevents experimental diabetic retinopathy. *Nat. Med.* **9**, 294–299.

Hellström, M., Gerhardt, H., Kalén, M., Li, X., Eriksson, U., Wolburg, H., and Betsholtz, C. (2001). Lack of pericytes leads to endothelial hyperplasia and abnormal vascular morphogenesis. *J. Cell Biol.* **153**, 543–553.

Lindahl, P., Johansson, B. R., Levéen, P., and Betsholtz, C. (1997). Pericyte loss and microaneurysm formation in PDGF-B-deficient mice. *Science* **277**, 242–245.

Little, H. L., Jack, R. L., Patz, A., and Forsham, P. H. (1983) *Diabetic Retinopathy*. Stuttgart: Thieme.

Nehls, V., and Dreckhahn, D. (1993). The versatility of microvascular pericytes: From mesenchyme to smooth muscle? *Histochemistry* **99**, 1–12.

Sims, D. E. (2000). Diversity within pericytes. *Clin. Exp. Pharmacol. Physiol.* **27**, 842–846.

Capsule Biography

Dr. Hammes headed the laboratory of experimental retinopathy at the 3rd Medical Dept., University of Giessen, Germany, from 1989 to 2000. In that time, the laboratory focused on the understanding of the biochemical and cellular mechanisms that lead to the development of early diabetic retinopathy. One question important to Dr. Hammes has always been which mechanisms contribute to the early loss of pericytes in the course of the disease. In 2000, he moved to the Medical School Mannheim, 5th Medical Department, University of Heidelberg. His research is funded by the Deutsche Forschungsgemeinschaft, the German Diabetes Foundation, the European Foundation for the Study of Diabetes, and the Juvenile Diabetes Research Foundation.

CHAPTER 101

Von Willebrand Factor Diabetes and Endothelial Dysfunction

[1]Alberto Verrotti, [1]Rita Greco, [2]Guido Morgese, and [1]Francesco Chiarelli

[1]*Department of Medicine, Section of Pediatrics, University of Chieti, Italy*
[2]*Department of Pediatrics, University of Siena, Italy*

Introduction

Type 1 diabetes mellitus (T1DM) is associated with the development of micro- and macrovascular disease, which represents the major cause of morbidity and mortality among these patients.

Diabetic angiopathy in children and adolescents is represented mainly by microangiopathy, characterized by structural changes in the eye and renal glomeruli, whereas macroangiopathy consists of an accelerated form of atherosclerosis that increases the risk of cardiovascular disease.

The pathogenesis of diabetic angiopathy is not clear and the progression, in most cases is unpredictable. The vascular endothelium is a primary target of unbalanced glycemic metabolism in both T1DM and type 2 diabetes mellitus (T2DM). Several studies have shown that endothelial dysfunction is associated with the development of diabetic micro- and macrovascular complications and, in particular, may explain the link between microalbuminuria and the risk of atherosclerotic cardiovascular disease.

Physiopathology of Vascular Endothelium

The endothelial cells are flattened by intraluminal hydrostatic pressure, and show a wide cytoplasm with a central nucleus and, frequently, a nucleole. The cytoplasm contains the Weibel-Palade bodies, which are the depot of von Willebrand factor (vWF).

The endothelium produces a variety of mediators, such as nitric oxide (NO), prostanoids, tissue-type plasminogen activator (t-PA) and plasminogen activator inhibitor-1 (PAI-1), and vWF.

The release of endothelium-derived relaxing factors (EDRFs) must be considered. These factors include, in addition to NO and prostacyclin, the endothelium-derived hyperpolarizing factor (EDHF), not yet fully identified, which may play a role, especially in smaller vessels. The importance of these relaxing factors is supported by the fact that endothelium-dependent vasodilatation is usually used as an easily reproducible and accessible parameter to test endothelial function in various conditions, such as T1DM and T2DM.

Endothelial dysfunction can be defined as the loss of normal vascular endothelium properties (e.g., alteration of protein synthesis, increased vascular tone and permeability, acquisition of prothrombotic and antifibrinolytic properties).

Definition of the Problem

In human T1DM, impaired endothelium-dependent vasodilatation has been reported in the absence of clinical signs of vascular complications. However, in well-controlled diabetics without albuminuria no alteration of endothelium-dependent vasodilatation has been reported. This is probably true also for T2DM. Whereas elevation of both vWF and its propeptide could indicate acute perturbation, elevated vWF levels alone may reflect chronic endothelial cell activation.

Impaired endothelium-dependent vasodilatation may be due to reduction of EDRF production, increase in EDRF inhibition, alteration in diffusion of EDRFs from the endothelium to the underlying smooth muscle cells, reduction of smooth muscle cell responsiveness to EDRFs, and increase in the production of endothelium-derived constricting factors (EDCFs), most likely prostanoids [1]. Impaired endothelium-dependent vasodilatation in T1DM correlates with duration of disease, glycemic control, and small-vessel disease.

Patients with high plasma levels of t-PA, PAI, vWF, and endothelin have an adverse cardiovascular prognosis. In particular, increased plasma vWF levels are detectable in patients with microalbuminuria, but not in those with early, uncomplicated, and well-controlled T1DM, or in those with early diabetic retinopathy. Nevertheless, the initial phase of vascular endothelium damage, defined by the term *endothelial perturbation*, which refers to a reversible dysfunction of the endothelial cells, is revealed by a simultaneous increase in circulating levels of vWF and t-PA that exceed 2 SD above control values. The combined determination of vWF and t-PA may be a specific and sensitive index of endothelial cell alteration, since vWF may also originate from platelets, and t-PA is also secreted by endothelial cells. vWF is considered an available marker of endothelial dysfunction. A prospective study showed that increased plasma vWF levels started about 3 years before the onset of microalbuminuria in adult patients with T1DM.

Pathogenesis of Endothelial Dysfunction

Poor glycemic control is the major factor responsible for the developing diabetic complications. However, despite meticulous metabolic control, 10 to 15 percent of patients develop renal complications after 20 years of diabetes duration. The mechanism of the progression of nephropathy is not completely understood. There is some evidence for the potential role of genetic factors in the development of diabetic kidney disease.

There is also evidence that all patients with diabetes will develop minimal retinopathy by 20 years after the onset of the disease. Vascular endothelium is a primary target of the altered glycemic metabolism in T1DM. High glucose concentrations may alter, either directly or indirectly, the antiatherogenic and antithrombotic properties of the vascular endothelium, through the formation of advanced glycosylation end products (AGEs) and reactive oxygen species (ROS), thus setting the stage for the ensuing development of vascular complications. Increased ROS may represent a single mechanism of induction, linking elevated glucose concentrations and three major biochemical pathways involved in the diabetic damage. Several parameters, such as vWF or prorenin, have been found to be associated with the development of microangiopathy independently from glycemic control in limited studies in children.

Hyperglycemia

In patients with diabetes, glycosylated hemoglobin is generally considered the best parameter to detect the individual risk for the development of long-term vascular complications. The multicenter, randomized, prospective Diabetes Control and Complications Trial (DCCT) confirmed previous studies demonstrating glycemic control to be the most important predictor for the development and severity of complications in T1DM. In patients with T1DM more severe endothelial dysfunction is present when they have poor glycemic control compared with patients with better glycosylated hemoglobin. There is evidence that vasodilatation mediated by endothelium-derived NO is altered in patients with both T1DM and T2DM DM. Moreover, impaired NO-mediated vasodilatation during acute hyperglycemia has been documented in nondiabetic humans. Other studies, however, were not able to demonstrate a correlation between HbA1c values and degree of endothelium-dependent vasodilatation.

Polyol Pathway and Protein Kinase C Activation

Hyperglycemia activates the polyol pathway, resulting in the formation of sorbitol by aldose reductase. As aldose reductase utilizes NADPH for the reduction of glucose to sorbitol, cellular stores of NADPH may be depleted (55). NADPH is required for the functioning of several enzymes, such as NO, synthase for NO generation, and cytochromes P450, and for the activity of glutathione reductase that replenishes glutathione, one of the most important endogenous antioxidant systems. Increased polyol pathway activity is associated with the occurrence of long-term complications in patients with diabetes. This is demonstrated by the efficacy of aldose reductase inhibitors in the restoration of impaired endothelium-dependent vasodilatation, as well as in the prevention of diabetic neuropathy, albuminuria, and cataracts in animal models but so far not in humans.

Advanced Glycosylation End Products (AGEs)

Glucose binds to amino groups on proteins or to lipids, leading to the generation of weak bonds or Schiff bases by Maillard's reaction. These early nonenzymatic glycosylation products are reversible and are known as Amadori products. Through several oxidative and nonoxidative reactions, including glycoxidation and auto-oxidative glycosylation, they are converted to irreversibly modified cross-linked condensation products of glucose and lysine or arginine residues, so-called advanced glycosylation end products (AGEs). AGEs formation normally occurs at a low rate during the normal aging process, whereas it is accelerated by hyperglycemia; thus plasma levels of AGEs are increased in patients with T1DM and may precede the occurrence of microangiopathy. AGEs carry out their harmful effects by accumulating in tissue and generating ROS. AGEs bind to their respective cellular receptors (RAGEs), activate endothelial cells, monocytes, and mesangial cells, and increase oxidative stress. AGEs have been found in many proteins, such as collagen, albumin, and plasma lipoproteins, and in lipids. Moreover, they can also increase lipoprotein oxidizability and atherogenicity; they have been demonstrated in atherosclerotic lesions. AGEs can impair the properties of the basement membrane and stimulate the interaction between mononuclear cells and modified tissue; they can cause endothelial dysfunction by inhibiting type IV collagen formation. AGEs can quench NO and lead to the loss of its vasodilatating and antiproliferative properties. It has been shown that serum AGEs concentrations are increased in adolescents and young adults with T1DM and with diabetic retinopathy and nephropathy, and that the severity of diabetic angiopathy correlate with serum levels of AGEs.

Oxidative Stress

Oxidative stress represents one of the most important factors involved in the pathogenesis of endothelial dysfunction during diabetes and is characterized by an elevation of ROS as a consequence of increased free radicals production and/or reduced antioxidant systems. Diabetes mellitus has been associated with increased formation of free radicals and reduced concentrations of superoxide dismutase, catalase, glutathione, and ascorbic acid. The increased oxidative stress observed in diabetes may be due to different mechanisms, such as glucose autoxidation, PKC activation, AGE generation, increased availability of substrates through the polyol pathway, and enhanced eicosanoid metabolism. Therefore, free radicals may alter endothelium-dependent vasodilatation by inhibiting NO or behaving as an endothelium-derived constricting factor. Several studies have shown that administration of antioxidant enzymes also in combination may improve or restore the impaired endothelium-dependent response in diabetes and during high glucose exposure. However, other studies demonstrated that superoxide dismutase had less effect or no effect in improving endothelium-dependent vasodilatation compared to hydroxyl radical scavengers; therefore, hydroxyl radicals are thought to be more important in eliciting endothelial dysfunction.

Coagulation Cascade

Endothelial dysfunction is also characterized by an ongoing prothrombotic state and an increased risk of thrombotic events, and its persistence may represent a negative prognostic index. Patients with micro- and macroalbuminuria present an accentuated activation of the coagulative cascade [2]. It has been found that the simultaneous increase in circulating levels of vWF and tPA (that exceed 2 SD above control values) is associated with increased levels of prothrombin fragments 1 and 2 (F1 + 2) in children with T1DM. This finding seems to indicate that at the onset of the disease, young patients with diabetes may suffer from a subclinical coagulative, prothrombotic disorder. Increased F1 + 2 level is a marker of an unbalanced tissue factor/tissue factor inhibitor pathway, as well as an index of enhanced thrombin generation and of a hypercoagulable state. Diabetes is characterized by several coagulative disorders, such as an increase of factor VII activation and thrombin and fibrin generation, and a reduction of antithrombin III activity.

Structure and Function of von Willebrand Factor

The von Willebrand factor (vWF) is a glycoprotein involved in primary hemostasis and secreted mainly by endothelial cells and megakaryocytes. vWF is synthesized as a large 360-kDa precursor named pro-vWF. After many changes in the endoplasmic reticulum and the Golgi apparatus, this precursor undergoes cleavage into mature vWF and a large 97-kDa propeptide [3, 4]. Stimulation of exocytosis causes the release of equimolar amounts of vWF and its propeptide (vWF propeptide), which has a circulating half-time of 2 to 3 hours, compared with more than 12 hours for vWF itself. Propeptide concentrations are less influenced by factors such as blood groups, adhesive properties, and catabolism, whereas vWF concentrations can be influenced by ABO blood groups [3], exercise, and hypoxic reperfusion injury.

von Willebrand Factor and Diabetes Mellitus

Endothelial dysfunction represents an early feature of vascular disease, and several mechanisms may be involved in its pathogenesis. It has also been reported in young normotensive and normoglycemic first-degree relatives of patients with T2DM in association with insulin resistance, independent of age, sex, body mass index, serum insulin, and lipids. Abnormal markers of endothelial cell activation and impaired endothelium-dependent vasodilatation have been observed in children, adolescents, and young adults with T1DM without clinical evidence of vascular disease in the work of Elhadd, Kennedy, and others.

The increase in plasma concentration of vWF and vWF propeptide are found in patients with overt nephropathy [5] and in patients with microalbuminuria, whereas only the propeptide was increased in normoalbuminuric patients [3]. Assessment of propeptide concentrations is a valuable complement in studies of endothelial activation and also in chronic vasculopathies. Besides, it has been found that the increase of vWF and vWF propeptide in plasma precedes microalbuminuria. This increase can thus be useful to identify children with T1DM at risk of developing incipient nephropathy later in life, although not all authors are in agreement with this opinion.

Finally, the parallel determination of vWF and vWF propeptide does not add any useful information for the early prediction of the development of microalbuminuria.

It has been reported that the increase in plasma vWF precedes, and may therefore help to predict, the development of microalbuminuria in T1DM.

The type of endothelial dysfunction reflected by increased vWF and fibrinogen levels is closely related to microalbuminuria not only in T1DM [6], but also in T2DM and essential hypertension. It is not clear, however, whether the prognostic value of vWF and vWF propeptide is related to their specific functions, that is, enhancement of platelet adhesion and factor VIII availability, or whether they are simply markers of endothelial injury and dysfunction.

High plasma levels of vWF and vWF propeptide are found also in patients with T2DM, and probably in T2DM patients an increased risk of new cardiovascular events is present only in patients with vWF concentrations above the median, but not in those with lower values. From the literature it is quite difficult to draw a clear conclusion about changes of vWF in T2DM patients because the data are discordant; therefore, larger and more complete studies are needed in order to really understand the value of vWF evaluation is T2DM subjects.

Glossary

Diabetic angiopathy: Represented mainly by microangiopathy, characterized by structural changes in the eye and renal glomeruli, and by macroangiopathy that consists of an accelerated form of atherosclerosis that increases the risk of cardiovascular disease.

Endothelial dysfunction: Long-term complication of diabetes mellitus characterized by structural changes in small vessels.

T1DM: Type 1 diabetes mellitus; generally, the onset is in pediatric age. This is an autoimmune disease and the patients are insulin dependent.

T2DM: Type 2 diabetes mellitus; generally, the onset is in old age and obesity is one of the most frequent causes.

von Willebrand factor: A glycoprotein involved in primary hemostasis and secreted mainly by endothelial cells and megakaryocytes.

References

1. De Vriese, A. N. S., Verbeuren, T. J., Van de Voorde, J., Lameire, N. H., and Vanhoutte, P. M. (2000). Endothelial dysfunction in diabetes. *Br. J. Pharmacol.* **130**, 963–974. *This is a very clear and well done paper on the topic.*
2. Ceriello, A. (1993). Coagulation activation in diabetes mellitus: The role of hyperglycemia and therapeutic prospects. *Diabetologia* **45**, 471–477. *One of the first and most complete papers on the relationship between coagulation and diabetes mellitus.*
3. Vischer, U. M., Emeis, J. J., Bilo, H. J., Stehouwer, C. D., Thomsen, C., Rasmus, O., Hermansen, K., Wollheim, C. B., and Ingerslev, J. (1998). von Willebrand factor (vWF) as a plasma marker of endothelial activation in diabetes: Improved reliability with parallel determination of the vWF propeptide (vWf: AgII). *Thromb. Haemost.* **80**, 1002–1007.
4. Visher, U. M., and de Moerloose, P. (1999). von Willebrand factor: From cell biology to the clinical management of von Willebrand's disease. *Crit. Rev. Oncol. Hematol.* **30**, 93–109.
5. Zareba, W., Pancio, G., Moss, A. J., Kalaria, V. G., Marder, V. J., Weiss, H. J., Watelet, L. F., and Sparks, C. E. (2001). Increased level of von Willebrand factor is significantly and independently associated with diabetes in postinfarction patients THROMBO Investigators. *Thromb. Haemost.* **86**, 791–799.
6. Stehouwer, C. D. A., Stroes, E. S. G., Hackeng, W. H. L., Mulder, P. G. H., and den Ottolander, G. J. H. (1991). von Willebrand factor and development of diabetic nephropathy in insulin-dependent diabetes mellitus. *Diabetes* **40**, 1746.

Further Reading

Elhadd, T. A., Kennedy, G., Hill, A., McLaren, M., Newton, R. W., Greene, S. A., and Belch, J. J. (1999). Abnormal markers of endothelial cell activation and oxidative stress in children, adolescents and young adults with type 1 diabetes with no clinical vascular disease. *Diabetes Metab. Res. Rev.* **15**, 405–411.

Greaves, M., Malia, R. G., Goodfellow, K., Mattock, M., Stevens, L. K., Stephenson, J. M., and Fuller, J. H. (1997). Eurodiab IDDM complications study group. Fibrinogen and von Willebrand factor in IDDM: Relationships to lipid vascular risk factor, blood pressure, glycaemic control and urinary albumin excretion rate: The Eurodiab IDDM complications study. *Diabetologia* **40**, 698–705. *One of the largest studies about fibrinogen and von Willebrand factor in diabetes mellitus.*

Huvers, F. C., De Leeuw, P. W., Houben, A. J. H. M., De Haan, C. H. A., Hamulyak, K., Schouten, H., Wolffenbuttel, B. H. R., and Schaper, N. C. (1999). Endothelium-dependent vasodilatation, plasma markers of endothelial function, and adrenergic vasoconstrictor responses in type I diabetes under near-normoglycemic conditions. *Diabetes* **47**, 1300–1307.

Verrotti, A., Greco, R., Basciani, F., Morgese, G., and Chiarelli, F. (2003). von Willebrand factor and its propeptide in children with diabetes. Relation between endothelial dysfunction and microalbuminuria. *Pediatr. Res.* **53**, 1–5.

Capsule Biography

Prof. Alberto Verrotti is associate professor at the University of Chieti, Italy, and his scientific activity is devoted to vascular mechanisms of microvascular complications of diabetes mellitus. He works in connection with many international research institutes. His research has been published in high-level journal and has been frequently supported by national and international grants.

Rita Greco graduated in 1995. She has specialized in pediatrics since 1999 at the University of Chieti, Italy. She is an active researcher in the field of endocrinology with special interest in diabetes mellitus and microcirculation. She has published many papers in international journals and is working at the Pediatry of Pescara Hospital.

Prof. Guido Morgese is full professor of pediatrics at the University of Siena, Italy. He dedicated himself for many years to research on childhood endocrinology and, in particular, on type 1 diabetes mellitus. In the past 20 years, he has published many papers in international journals, reporting his scientific data about the main aspects of pathophysiology of long-term complications of type 1 diabetes mellitus, with particular reference to microvascular complications. He is director of the Pediatric Clinic of the University Hospital in Siena, Italy.

Prof. Francesco Chiarelli is full professor of pediatrics at the University of Chieti, Italy. He spent many years researching childhood endocrinology and type 1 diabetes mellitus. Recently, he has studied in animal models and in humans the possible relationships between oxidative stress, coagulation abnormalities, and microvascular complications; his original data have been published in many international journals. He is director of the Pediatric Clinic of the University Hospital in Chieti, Italy, and he is the Secretary of the European Society of Pediatric Endocrinology.

SECTION G

Hemorrhagic Fever Viruses

CHAPTER 102

The Vascular Endothelium as a Target for the Hemorrhagic Fever Viruses

Silvia Montaner[1,2] and J. Silvio Gutkind[3]

[1]*Department of Diagnostic Sciences and Pathology, University of Maryland, 666 W. Baltimore Street, Baltimore, MD 21201, USA*
[2]*Greenebaum Cancer Center, University of Maryland, 22 S. Greene Street, Baltimore, MD 21201, USA*
[3]*Oral and Pharyngeal Cancer Branch, National Institute of Dental and Craniofacial Research, National Institute of Health, 9000 Rockville Pike, Bethesda, MD 20892, USA*

Hemorrhagic Fever Viruses

The hemorrhagic fever viruses (HFVs) are small RNA viruses that belong to one of four families: *Arenaviridae*, *Bunyaviridae*, *Flaviviridae*, and *Filoviridae*. HFVs are transmitted to humans by contact with infected animal hosts or arthropod vectors. However, the natural history and reservoir for some HFVs remain unknown. Infection of humans can cause an acute febrile illness known as viral hemorrhagic fever (VHF) characterized by an unremarkable symptom complex that includes fever, fatigue, dizziness, muscle aches, loss of strength, and exhaustion. Patients can go on to develop systemic hemorrhages, which may be accompanied by shock, coma, seizures, and sometimes death.

Arenaviridae

The family *Arenaviridae* includes 23 viral species, of which five are included in the Category A Pathogen List established by the CDC. These five viruses are Junin [Argentine hemorrhagic fever (HF)], Machupo (Bolivian HF), Guanarito (Venezuelan HF), Sabia (Brazilian HF), and Lassa virus (Lassa fever). Rodents are the natural hosts of the arenaviruses, although person-to-person transmission can occur by direct contact with infectious blood and bodily fluids. Their clinical manifestations are almost identical and may include hemorrhagic and neurological disorders. Infection with these viruses is associated with a lethality rate of almost 20 percent.

Bunyaviridae

Human viruses in all four genera of the *Bunyaviridae* family can cause hemorrhagic fevers. Crimean–Congo hemorrhagic fever (CCHF) is a severe tick-borne nairovirus infection found in Africa, Asia, and Eastern Europe. Rift Valley fever (RVF) virus (*Phlebovirus* genus) is a mosquito-transmitted disease of cattle, sheep, and other ruminants in sub-Saharan Africa. Infection of humans can cause a range of symptoms including febrile illness, retinal disease, and hemorrhagic fever. Hantaviruses, maintained as persistent infections in rodents, are associated with two major diseases: hemorrhagic fever with renal syndrome (HFRS) and hantavirus pulmonary syndrome (HPS). The identification of a new hemorrhagic fever-causing bunyavirus called Garissa has been also recently described.

Flaviviridae

Flaviviruses include the viral etiologic agents for yellow fever and dengue hemorrhagic fever. Although there are no reported cases of person-to-person transmission or nosocomial spread of flaviviruses, humans are readily infected by yellow fever virus or dengue viruses from the bite of an infected mosquito. Historically, yellow fever, the original viral hemorrhagic syndrome, was one of the most feared lethal diseases. Today, despite the availability of an effective vaccine, the disease still affects as many as 200,000 people annually in tropical regions of Africa and South America and is associated with a mortality of up to 50 percent worldwide. Dengue virus is one of the most important emerging human viral pathogens, posing a threat to nearly one-third of the global human population. Although the primary infection with one serotype of this virus is often subclinical, a second infection with a different serotype is an important risk factor for dengue hemorrhagic fever (DHF) or dengue shock syndrome. Omsk hemorrhagic fever (OHF) and Kyasanur Forest disease (KFD) viruses are acquired from the bite of an infected tick and are responsible for a number of infections annually in Russia and India, respectively.

Filoviridae

Included among the biological agents of greatest concern are the enigmatic filoviruses (Ebola and Marburg). These viruses cause the most tissue destruction and harbor the highest fatality rates (up to 90 percent) among the HFVs. Their effective modes of transmission, rapid clinical course, and high mortality have elevated these tiny viruses to considerable threats. Indeed, the feasible use of HFVs as aerosolized biological weapons targeting large populations necessitates the development of vaccines and effective therapies against these infectious agents [1]. Unfortunately, molecular characterization of filoviruses is limited, as current studies require the direct manipulation of these lethal human pathogens within the few available biosafety level 4 facilities. Despite these restrictions, emerging evidence supports a role for direct and indirect damage to the vascular endothelium in the mechanism of action of HFVs. This chapter will focus on the pathogenesis of a member of the *Filoviridae* family, Ebola, which can serve as a prototype for understanding other HFVs.

Infection with Filoviruses: Ebola and Marburg

The first outbreak of human VHF caused by a filovirus, Marburg virus, occurred in 1967 in Germany and Yugoslavia. In 1976, simultaneous outbreaks of two distinct Ebola virus subtypes erupted near the Ebola River Valley in Zaire and southern Sudan, resulting in hundreds of deaths. Since then, additional reports of human VHFs secondary to filoviruses have appeared sporadically, mainly in Africa, with mortality rates ranging from 50 to 90 percent. Outbreaks involving the Ebola virus subtypes, however, have remained the most deadly [2].

There are currently four recognized subtypes of Ebola virus: Zaire (the most aggressive), Sudan, Reston, and Côte d'Ivoire. The Reston subtype is pathogenic in monkeys but does not cause lethal infection in humans. Epidemiological evidence suggests that transmission of Ebola virus to humans usually occurs after direct contact with blood, secretions, or tissues of affected patients or nonhuman primates. Following infection, patients initially present with an unremarkable symptom complex that includes fever, headache, muscle pain, nausea, vomiting, abdominal pain, and diarrhea. As the infection progresses, patients exhibit marked thrombocytopenia with abnormal platelet aggregation, and coagulation abnormalities, which together cause severe bleeding often involving the gastrointestinal tract and conjunctiva. Exaggerated inflammatory responses and hematological irregularities, including lymphopenia with subsequent neutrophilia, are also often observed. Although the viral infection can run its course in 14 to 21 days, damage to the liver, combined with massive viremia, can lead to disseminated intravascular coagulopathy (DIC). Death is usually associated with shock characterized by fluid distribution abnormalities (secondary to increased vascular permeability), hypotension, variable degrees of hemorrhages, and widespread focal tissue destruction, especially affecting the liver, lymphoid organs, and kidneys.

Unfortunately, current treatment options for filovirus infection consist primarily of contact isolation and supportive care. The identification of suitable therapeutic targets for effective antiviral therapies or vaccines requires a strong understanding of the pathogenesis of these enigmatic viruses [3]. In this regard, there are many questions that remain unanswered about the filoviruses. Among them are explanations for their extreme virulence, the differences in aggressiveness among the different subtypes, and the specific mechanisms that facilitate their pathogenicity. Although many of these questions remain unanswered, emerging evidence suggests that the vascular endothelium is a principle target in Ebola pathogenesis.

The Role of the Vascular Endothelium in Ebola Pathogenesis

Although we do not yet fully understand why Ebola virus is so potent, the clinical signs and symptoms of human infection with this agent point toward a dysfunction of the vascular system. The syndrome caused by this virus—and all HFVs—are acute multiorgan diseases, associated with widespread tissue damage and diffuse vascular dysfunction. Macrophages, monocytes, endothelial cells, and hepatocytes are the main cellular targets of the Ebola virus [4]. Macrophages and monocytes are considered the primary target for infection. The virus then spreads to the endothelium and the parenchyma of different organs rapidly. Several factors may enable Ebola to be such an invasive and cytopathic virus: its effective blockage of the immune response, its unusually high rate of viral replication, its disruption of normal vascularization, and the severe damage it induces in infected cells.

In Ebola infection, the vascular endothelium can be targeted in different ways: by indirect dysregulation through cytokines, chemokines, and other mediators released upon infection of primary target cells (including monocytes and macrophages); by secreted nonstructural viral proteins that may trigger the alteration of endothelial cells either directly or through the activation of primary target cells; and by direct viral infection and replication within vascular endothelial cells [5] (Figure 1).

Indirect Alterations of the Vascular Endothelium through Paracrine Secretions

Filoviruses infect macrophages and other cells of the mononuclear phagocytic system at the site of entry and in regional lymph nodes. They then spread to fixed tissue macrophages in the liver, spleen, and other tissues throughout the body. Infected monocytes and macrophages become activated and release large quantities of vasoactive agents such as cytokines (i.e., IL-1β, TNFα, and IL-6), chemokines (i.e., IL-8, GROα), histamines, and peroxidases, which may lead to endothelial cell disorganization and increased permeability of the endothelial lining of blood vessels. In particular, antibodies that neutralize TNFα block increased endothelial permeability in vitro, indicating a pivotal role of this proinflammatory cytokine in virus-induced disease. A marked elevation of IL-2, IL-10, TNFα, IFNα, and IFNγ has been also noted in fatal Ebola cases. Platelet-derived agents triggered by damaged endothelial cells might also contribute to a shock syndrome. Experimental data further provide evidence for changes in the protein organization of the endothelial cell junctions, particularly the VE-cadherin/catenin complex, although the molecular mechanisms involved have not been fully elucidated. These factors may contribute to the disruption of fluid balance between the intravascular and extravascular spaces.

Indirect Alterations of the Vascular Endothelium through Secreted Viral Proteins

Ebola virus encodes nonstructural proteins, which are released from infected host cells. These proteins may trigger the activation of endothelial cells either directly or through the activation of their target cells. Their function as potential determinants in pathogenicity remains to be investigated.

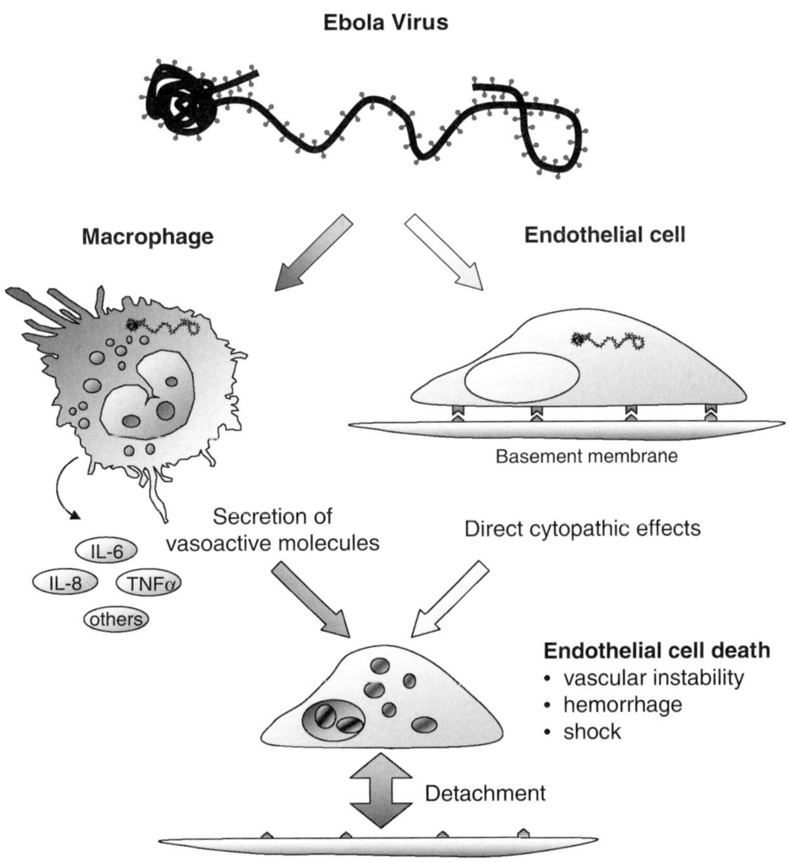

Figure 1 Effects of Ebola virus infection on the vascular endothelium. (see color insert)

Direct Infection of the Vascular Endothelium

In addition to inducing indirect damage to the vascular endothelium, Ebola virus also directly infects microvascular endothelial cells. Direct infection of endothelial cells can cause massive endothelial cell damage, compromising vascular integrity and inducing microthrombus formation. This may result in interstitial hemorrhages, fluid shifts, tissue ischemia, and cytokine dysregulation, which together may lead to development of disseminated intravascular coagulopathy (DIC), diffuse bleeding, and shock. In addition, cytopathic involvement of the parenchymal cells of multiple organs may cause tissue damage, especially involving the liver, which could further contribute to the DIC.

Experimental evidence supports a role for both direct and indirect damage to the vascular endothelium in Ebola pathogenesis. Ongoing efforts have recently focused on identifying the Ebola-encoded genes responsible for these pathogenic effects.

Ebola Pathogenic Genes

In general, filoviruses possess a noninfectious, negative-sense, single-stranded, nonsegmented RNA genome within a lipid envelope. The cellular receptors required for cell infection are not definitively clarified. The Ebola virus genome is 19 kb long and includes seven open reading frames encoding structural proteins [the glycoprotein (GP), the nucleoprotein (NP), and the matrix proteins VP24 and VP40], nonstructural proteins (VP30 and VP35), and the viral polymerase (Figure 2). Additionally, the GP open reading frame of Ebola virus, through transcriptional editing, generates two different gene products: a full-length 150- to 170-kDa protein (GP) that inserts into the viral membrane and a soluble 60- to 70-kDa protein (sGP), that is secreted from infected cells. The Ebola glycoproteins, GP and sGP, share approximately 300 amino-terminal amino acids, but contain an additional 380 and 70 carboxy-terminal amino acids, respectively. Both are thought to play important roles in viral pathogenesis [6].

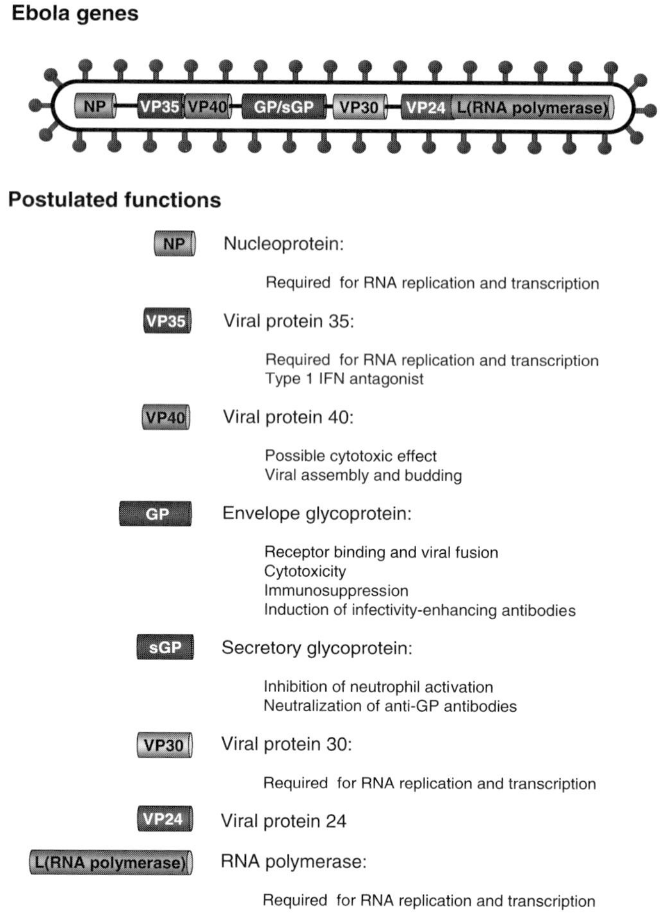

Figure 2 The Ebola genome. (see color insert)

Ebola Envelope Glycoprotein (GP)

The Ebola GP forms a trimeric complex expressed on the viral envelope and is responsible for receptor binding and viral fusion with host cells. The affinity of Ebola for endothelial cells appears to be due, in part, to GP's preferential binding to these cells [7]. In addition to mediating host cell binding, GP also appears to harbor cytopathic effects when expressed in different cell types in vitro. Specifically, GP expression in cultured human endothelial and epithelial cells causes cell rounding and detachment. These effects correlate with the downregulation of expression of important cell-surface molecules involved in matrix adhesion such as integrins [8]. Interestingly, GP from the Ebola Zaire strain—which is lethal in both humans and nonhuman primates—has the most robust activity in this cell culture assay. Ebola Zaire GP also induces severe damage and increased permeability of the endothelial cell lining in blood vessel explants derived from both human and nonhuman primates [9]. Conversely, GP encoded by the Ebola Reston strain—which is lethal only in nonhuman primates—induces similar effects only in nonhuman primate vessels. This remarkable congruence between Ebola GP cytotoxicity in vitro and Ebola strain pathogenicity in vivo suggests that the GP may be an important viral determinant of endothelial cell toxicity in Ebola infection.

Ebola Secreted Glycoprotein

Two functions have been hypothesized for the secretory glycoprotein (sGP). This dimeric, nonstructural protein has the ability to inhibit the neutralizing activity of anti-GP antiserum, suggesting that it may act as a decoy to adsorb neutralizing antibodies against GP. This might explain why GP-specific antibodies are not usually detected in patient sera during acute infection, while antibodies to the other major viral proteins are readily found. sGP may also play an important role in immune evasion. sGP has been reported to bind to human neutrophils, interfering with the physical and functional interaction between CD16b and CR3, thereby impairing early steps in neutrophil activation that ordinarily contribute to virus clearance. This may help explain why Ebola virus spreads so rapidly through the body without eliciting a proper immune response in acutely infected patients.

Other Ebola Proteins

Several of the other Ebola proteins have been shown to be essential for efficient viral infection, replication, and pathogenesis. The Ebola virus nucleoprotein, VP35, VP30, and RNA polymerase are all required for RNA replication and transcription. Recent data also indicate that VP35 can act as a type 1 IFN antagonist, suggesting that this protein may also be required for full Ebola virulence. Viral protein 40 (VP40) is another suspected cytotoxic protein, which induces particle formation when expressed in mammalian cells. This process most likely requires cellular WW-domain-containing proteins that interact with the amino-terminal proline-rich region of VP40. It was also demonstrated that VP40 interacts with an ubiquitin ligase. The biological significance of this interaction is still unclear, although it may be involved in assembly and budding of filoviruses.

Ultimately, the key to fully understanding the relative contribution of Ebola genes to VHF will rely on the development of model systems to safely and effectively examine this virus in vitro and in vivo.

Models to Study the Pathogenesis of Ebola Virus

In general, studies on Ebola virus require sophisticated yet inconvenient safety precautions, considerably limiting the investigation and understanding of Ebola pathogenesis. Fortunately, isolation of the viral cDNAs and the development of artificial expression strategies have allowed the study of Ebola virus gene products in vitro under less restrictive conditions. For example, an artificial replication system has been developed based on the vaccinia virus T7 expression model. Also, a reverse genetics system enables the generation of infectious Ebola virus from cloned cDNA. Using this strategy, cultured cells are transfected with plasmids for the expression of the Ebola proteins, a plasmid for the Ebola viral RNA controlled by T7 RNA polymerase promoter, and a plasmid for T7 RNA polymerase. This system is effective for the study of mechanisms underlying the pathogenicity of Ebola virus, as viral genes and proteins can be manipulated as desired.

Several animal models have also been used to examine Ebola virulence, the host responses to infection, and the efficacy of immunization. Although Ebola Zaire infection is traditionally examined using nonhuman primate hosts, the virus has recently been adapted to produce uniformly fatal infection in other animals including guinea pigs and mice, through serial animal-to-animal passaging. These models have provided critical insights into Ebola pathogenesis and are especially valuable for testing antiviral medications and vaccines.

More recently, a safe high-throughput avian-derived retroviral gene transfer system was engineered to specifically express candidate viral genes in mouse endothelial cells in vivo [10]. The *TIE2*-TVA transgenic mouse system expresses the avian leukosis virus (ALV) receptor, TVA, under the control of the endothelial cell-specific *TIE2* promoter. Using this system, only mammalian cells engineered to express the TVA receptor can be transduced by infection with avian-derived viruses, enabling the somatic introduction of multiple genes in vivo, in a tissue-specific manner, using a single transgenic animal. This unique animal model safely mimics the infectious process by which Ebola targets endothelial cells in vivo. To verify viral infectivity and demonstrate the potential of this model to study HFV virulence genes, an ALV-derived vector [RCASBP(A)] expressing Polyoma Middle T Antigen (PyMT)—which induces hemorrhagic diathesis when expressed in mice—was used. Endothelial cell–specific retroviral transduction with PyMT had a dramatic effect on mouse survival. Histological examination of killed animals revealed massive hemorrhaging in the liver and spleen, very similar to that seen in human patients infected with Ebola virus. These promising preliminary results suggest that this model may be uniquely suited to safely examine the contribution of Ebola genes, individually or in combination, to VHF. This system may provide fundamental insight into the molecular pathogenesis of this lethal virus and may ultimately help identify diagnostic markers and gene-product targeted therapies for VHF.

Conclusion

Infection with the Ebola Zaire virus causes a viral hemorrhagic fever syndrome with one of the highest fatality rates (up to 90 percent) among all known human viruses. This astonishing lethality can be attributed, at least in part, to the viruses ability to target the vascular endothelium through a combination of direct and indirect mechanisms. Although current efforts to elucidate the molecular pathogenesis of this virus have been inspired by its potential use as a bioweapon, examination of Ebola may further provide considerable insight into the complex relationship between inflammation and vascular dysfunction.

Glossary

Disseminated intravascular coagulopathy (DIC): DIC is a disorder of widespread systemic activation of the clotting cascade that results in diffuse fibrin deposition in small and midsize vessels and depletion of clotting factors in the blood. It occurs when the blood clotting mechanisms are activated throughout the body instead of being localized to an area of injury. This disorder is variable in its clinical effects and can result in either clotting symptoms or, more often, bleeding. Bleeding can be severe.

Glycoprotein: A protein that has carbohydrate groups attached to the polypeptide chain.

Viral hemorrhagic fevers: A group of acute febrile illnesses caused by viral infection. They are characterized by the presence of symptoms including fever, fatigue, dizziness, muscle aches, loss of strength, and exhaustion. Patients can go on to develop systemic hemorrhages, which may be accompanied by shock, coma, seizures, and sometimes death.

References

1. Bray, M. (2003). Defense against filoviruses used as biological weapons. *Antiviral Res.* **57**, 53–60.
2. Sanchez, A., Khan, A. S., Zaki, S. R., Nabel, G. J., Ksiazek, T. G., and Peters, C. J. (2001). Filoviridae: Marburg and Ebola viruses. In D. M. Knipe and P. M. Howley (eds.), *Fields Virology*, pp. 1279–1304. Philadelphia: Lippincott, Williams & Wilkins.
3. Sullivan, N. J., Geisbert, T. W., Geisbert, J. B., Xu, L., Yang, Z. Y., Roederer, M., Koup, R. A., Jahrling, P. B., and Nabel, G. J. (2003). Accelerated vaccination for Ebola virus haemorrhagic fever in non-human primates. *Nature* **424**, 681–684.
4. Colebunders, R., and Borchert, M. (2000). Ebola haemorrhagic fever—a review. *J. Infect.* **40**, 16–20.
5. Schnittler, H. J., and Feldmann, H. (2003). Viral hemorrhagic fever—a vascular disease? *Thromb. Haemost.* **89**, 967–972.
6. Takada, A., and Kawaoka, Y. (2001). The pathogenesis of Ebola hemorrhagic fever. *Trends Microbiol.* **9**, 506–511.
7. Sullivan, N., Yang, Z. Y., and Nabel, G. J. (2003). Ebola virus pathogenesis: implications for vaccines and therapies. *J. Virol.* **77**, 9733–9737.
8. Simmons, G., Wool-Lewis, R. J., Baribaud, F., Netter, R. C., and Bates, P. (2002). Ebola virus glycoproteins induce global surface protein down-modulation and loss of cell adherence. *J. Virol.* **76**, 2518–2528. *The Ebola virus envelope glycoprotein (GP) derived from the pathogenic Zaire subtype mediates cell rounding and detachment from the extracellular matrix in 293T cells. GPs from Sudan, Côte d'Ivoire, and Reston subtypes also induced cell rounding, albeit at lower levels than Zaire GP. While causing cell rounding and detachment, Ebola virus GP did not cause an increase in cell death. A global mechanism for surface molecule downregulation was observed. Integrin β1 (concurrent with the loss of cell adhesion), major histocompatibility complex class I, and epidermal growth factor receptor were among the proteins with reduced expression.*
9. Yang, Z. Y., Duckers, H. J., Sullivan, N. J., Sanchez, A., Nabel, E. G., and Nabel, G. J. (2000). Identification of the Ebola virus glycoprotein as the main viral determinant of vascular cell cytotoxicity and injury. *Nat. Med.* **6**, 886–889. *The authors show that the envelope glycoprotein (GP) of Ebola Zaire virus induces cytotoxic effects in human endothelial cells in vitro and in vivo. Gene transfer of GP into explanted human or porcine blood vessels caused massive endothelial cell loss that led to a substantial increase in vascular permeability. The Zaire GP induced endothelial cell disruption and cytotoxicity in both nonhuman primate and human blood vessels, while GP derived from the Reston strain of virus disrupted the vasculature of only nonhuman blood vessels. These findings indicate that GP is the viral determinant of Ebola pathogenicity and likely contributes to hemorrhage during infection.*
10. Montaner, S., Sodhi, A., Molinolo, A., Bugge, T. H., Sawai, E. T., He, Y., Li, Y., Ray, P. E., and Gutkind, J. S. (2003). Endothelial infection with KSHV genes in vivo reveals that vGPCR initiates Kaposi's sarcomagenesis and can promote the tumorigenic potential of viral latent genes. *Cancer Cell* **3**, 23–36. *A transgenic mouse line was engineered to express the avian leukosis virus (ALV) receptor, TVA, under the control of the vascular endothelial cell-specific TIE2 promoter. Only mammalian cells engineered to express the tva transgene can be transduced by infection with ALV, thus enabling the somatic introduction of multiple genes in vivo, in the vascular endothelium. In particular, endothelial cell–specific retroviral transduction using an ALV-derived vector expressing Polyoma Middle T Antigen (PyMT) had a dramatic effect on mouse survival, causing massive hemorrhaging, very similar to that seen in human patients infected with Ebola virus. Using this model, it is possible to examine the effects of the expression of different genes in mouse endothelial cells, by infection with the corresponding ALV-derived vectors.*

Capsule Biography

Dr. Silvia Montaner is an Assistant Professor in the Department of Diagnostic Sciences and Pathology at the University of Maryland, Baltimore and the UM Greenebaum Cancer Center. Dr. Montaner developed a vascular endothelial cell-specific retroviral gene transfer system (*TIE2-tva*) during her postdoctoral training in the laboratory of Dr. Silvio Gutkind, at the National Institute of Dental and Craniofacial Research (NIDCR), National Institutes of Health. She used the *TIE2-tva* system to identify the viral gene, encoded by HHV-8, responsible for the initiation and progression of Kaposi's sarcoma.

Dr. J. Silvio Gutkind has been the Branch Chief in the Oral and Pharyngeal Cancer Branch, National Institute of Dental and Craniofacial Research, at the National Institutes of Health since 1997. His lab has elucidated key signaling pathways by which cellular and viral receptors, coupled to heterotrimeric G proteins, can regulate gene expression. His pioneering discoveries provided novel insights into the biochemical routes by which these membrane receptors control normal and aberrant cell growth. Dr. Gutkind is currently using the *TIE2-tva* system to examine the molecular pathogenesis of the Ebola virus in vivo.

SECTION H

Hypercholesterolemia

Chapter 103

Microvascular Responses to Hypercholesterolemia

Karen Y. Stokes and D. Neil Granger
Louisiana State University Health Sciences Center, Shreveport, Louisiana

Introduction

Hypercholesterolemia is an established risk factor for the development of cardiovascular diseases. The atherosclerotic lesions that result from a sustained elevation in blood cholesterol concentration are associated with an accumulation of inflammatory cells and platelets that facilitate the deposition of lipids in the walls of lesion-prone arteries. However, long before these changes occur in large arteries, inflammatory and prothrombogenic responses are observed in arterioles and venules throughout the vascular system (Figure 1). These responses are manifested as endothelial dysfunction and the binding of leukocytes and platelets to the vessel wall. Although several mechanisms have been proposed to explain the phenotypic changes that occur in the microvasculature during hypercholesterolemia, oxidative stress and a diminished bioavailability of nitric oxide (NO) have gained much attention in recent years. This chapter describes the responses of the microcirculation to hypercholesterolemia and addresses the mechanism that underlies this systemic inflammatory condition.

Hypercholesterolemia and Arterioles

Under normal physiological conditions, basal NO production by endothelial cells maintains vascular tone and inhibits inflammation. However, during hypercholesterolemia, several events occur that negatively influence the vasodilatory role of NO in arterioles. Although the concentration of L-arginine, the substrate for NO synthase (NOS), is not reduced during hypercholesterolemia, the interaction between L-arginine and endothelial NOS may be blocked by the endogenous inhibitor asymmetric dimethylarginine (ADMA), the levels of which are increased during hypercholesterolemia. The elevated ADMA levels likely result from the diminished activity of dimethylarginine dimethylaminohydrolase (DDAH), which normally degrades ADMA, that accompanies hypercholesterolemia. Furthermore, the NOS enzyme cofactor tetrahydrobiopterin (BH_4) is also reduced during hypercholesterolemia, which would further reduce the bioavailability of NO. The net effect of these changes in both animal and human subjects during hypercholesterolemia is a diminished capacity for arteriolar endothelium to produce NO and to mediate endothelium-dependent vasorelaxation. It is important to note that these vessels are capable of responding to direct stimulation of the smooth muscle cells by NO donors, supporting the concept that there is a loss of NO bioavailability rather than a deterioration of smooth muscle cell responses to NO. However, the response of denuded coronary arterioles to vasoconstrictors is reduced in hypercholesterolemic humans, compared to their normocholesterolemic counterparts, suggesting that smooth muscle–dependent alterations do occur in this condition, albeit via an NO-independent mechanism.

Concomitant with these changes in NO production is the release of reactive oxygen species (ROS). Although several ROS-producing enzymes have been implicated in the endothelial dysfunction induced in large arteries by hypercholesterolemia, the source of excess ROS generation in arterioles remains poorly understood. We have recently demonstrated that, as in large arteries exposed to hypercholesterolemia, enhanced superoxide (O_2^-) production also contributes to the impaired relaxation responses observed in arterioles during acute hypercholesterolemia. Furthermore, using gp91^{phox}-deficient mice, we demonstrated that NAD(P)H oxidase is a major source of the O_2^- that mediates

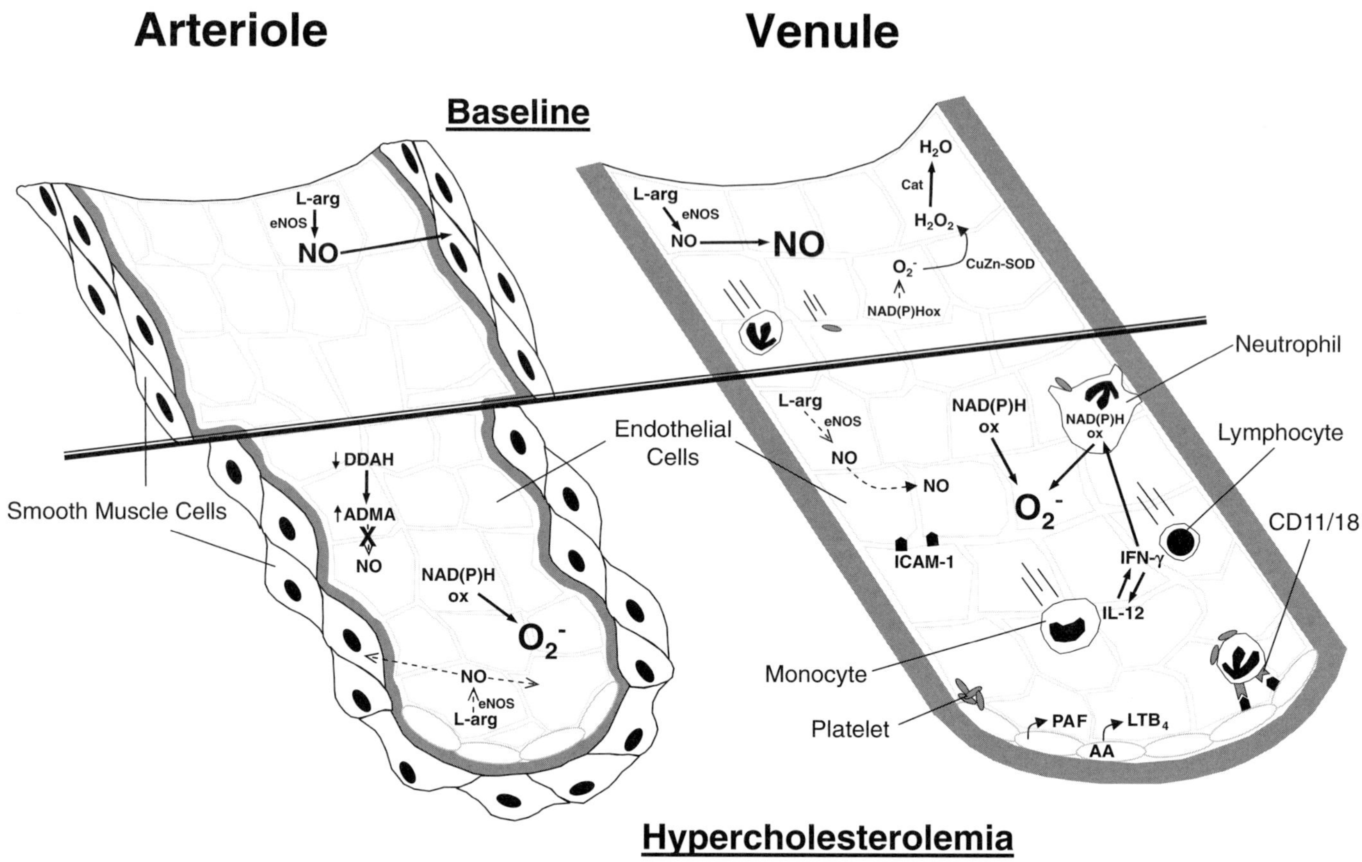

Figure 1 Inflammatory alterations in arterioles (*left*) and venules (*right*) elicited by hypercholesterolemia. Under baseline conditions (top portion of vessels), the basal release of nitric oxide (NO) maintains arteriolar smooth muscle tone and prevents cell–cell interactions in venules. Antioxidants such as catalase and CuZn-SOD minimize the levels of proinflammatory oxidants. During hypercholesterolemia (lower portions of vessels), NO bioavailability is reduced and oxidant production is enhanced. This promotes smooth muscle contraction (constriction) in arterioles. In the venular segment of the microcirculation, a proinflammatory and prothrombogenic phenotype is assumed. Adhesion molecule expression is increased resulting in the recruitment of platelets and leukocytes. The adhesion response is induced by lipid mediators and cytokines. L-arg, L-Arginine; eNOS, endothelial nitric oxide synthase; O_2^-, superoxide; cat, catalase; H_2O_2, hydrogen peroxide; SOD, superoxide dismutase; DDAH, dimethylarginine dimethylaminohydrolase; ADMA, asymmetric dimethylarginine; ICAM-1, intercellular adhesion molecule-1; IL-12, interleukin-12; IFN-γ, interferon-γ; AA, arachidonic acid; PAF, platelet-activating factor; LTB_4, leukotriene B_4; NAD(P)H ox, NAD(P)H oxidase.

the impaired endothelium-dependent vasodilation exhibited by arterioles during hypercholesterolemia.

Unlike in venules and atherosclerosis-prone large arteries, there is little evidence for the accumulation of adherent leukocytes or platelets in arterioles in animal models of diet-induced hypercholesterolemia. However, there is some evidence that oxidized low-density lipoprotein (oxLDL), which is elevated in blood of hypercholesterolemic humans, can induce the rolling and firm adhesion of leukocytes in arterioles. The leukocyte rolling response is mediated by P-selectin, while the firm adhesion of leukocytes is supported by β2-integrins.

Hypercholesterolemia and Capillaries

In response to the changes in arteriolar tone that accompany hypercholesterolemia, red blood cell velocity is reduced. This leads to erythrocyte aggregation and stasis in smaller microvessels. Humans with elevated blood cholesterol levels are said to exhibit reduced capillary perfusion, which likely reflects a diminished red blood cell velocity in capillaries. Recent work suggests that NO-dependent pathways contribute to the impaired capillary perfusion during hypercholesterolemia. There is evidence that leukocyte accumulation in downstream venules may also contribute to the impaired capillary perfusion during hypercholesterolemia, possibly through the release of inflammatory mediators.

Administration of oxLDL to otherwise normal animals promotes the degradation of the endothelial glycocalyx in capillaries. Platelets adhere to the endothelial cells of these damaged capillaries. The glycocalyx breakdown and resultant platelet adhesion can be inhibited by superoxide dismutase (SOD) and catalase. It remains unclear whether diet-induced hypercholesterolemia induces a similar injury response in capillaries. However, it has been shown that hypercholesterolemia exacerbates the capillary leak that occurs in response to acute inflammatory stimuli, such as ischemia–reperfusion (I/R). This likely reflects impaired endothelial junction integrity and occurs in a neutrophil-dependent manner, suggesting that leukocytes adherent within venules may release inflammatory mediators such as ROS that worsen the response to other stimuli.

Hypercholesterolemia and Postcapillary Venules

Although large veins appear to be relatively unaffected by acute or chronic elevations in blood cholesterol concentration, postcapillary venules in the diameter range of 20 to 40 μm exhibit profound changes in response in these conditions. Some of the alterations in signaling and inflammatory pathways induced in arterioles by hypercholesterolemia are also manifested in venules. For example, the reduced NO bioavailability and elevated ROS production are seen in both segments of the microvasculature. However, the consequences of these changes vary between the vascular segments, with NO and ROS exerting minimal influence on the diameter of venules while exerting a profound effect on arterioles. In venules, NO exerts a major influence on the expression of cellular adhesion molecules and consequently serves to minimize the adhesive interactions between circulating blood cells and venular endothelium. However, during hypercholesterolemia the expression of several adhesion molecules is increased on venular endothelium. These adhesion molecules support the leukocyte infiltration and platelet–endothelial interactions that occur in postcapillary venules when blood cholesterol concentration is elevated.

Blood Cell–Endothelial Cell Interactions

The use of blocking antibodies and mutant mice has revealed key roles for several adhesion molecules in the pathogenesis of atherosclerosis. The adhesion molecules that contribute to the binding of leukocytes and platelets in postcapillary venules during hypercholesterolemia have been less well defined. oxLDL also causes degradation of the endothelial glycocalyx in venules. It is noteworthy that this leads to loss of heparin sulfate proteoglycans, which would normally contribute to the negative charge and anti-adhesive properties of the normal endothelial cell. Hence, the increased adhesion molecule expression (and possibly the oxidatively modified surface of the endothelial cell) causes the endothelial cells to assume a phenotype that supports the adhesion of leukocytes and platelets during hypercholesterolemia.

Both intercellular adhesion molecule-1 (ICAM-1) and P-selectin are upregulated on venular endothelium when mice are placed on a cholesterol-enriched diet for 1 week. This protein expression coincides with the recruitment of leukocytes. In the early stages of leukocyte recruitment in hypercholesterolemic venules, neutrophils appear to represent the major cell population that interacts with the vessel wall. Both $CD4^+$ and $CD8^+$ T-lymphocytes participate in this response, but in an indirect manner, by producing cytokines that upregulate endothelial cell adhesion molecules.

Platelets are also recruited into postcapillary venules during acute hypercholesterolemia. It has been demonstrated using knockout mice that the platelets interact with the venular wall via P-selectin that is expressed on the surface of circulating platelets, although P-selectin on venular endothelium also contributes but to a lesser extent. The latter may participate by binding P-selectin glycoprotein ligand-1 (PSGL-1) on leukocytes, which in turn may serve as a platform for platelet binding to the venular wall. oxLDL promotes the formation of leukocyte–platelet aggregates that can interact with the venular wall. The formation of these aggregates can be inhibited using a P-selectin blocking antibody suggesting that platelet P-selectin is interacting with PSGL-1 on the leukocytes.

Nitric Oxide and Reactive Oxygen Species

Although there are very few mechanistic data available on platelet accumulation in postcapillary venules, a large body of evidence supports a role for an NO-ROS imbalance in the hypercholesterolemia-induced leukocyte–endothelial cell interactions. First, NO inhibitors fail to exacerbate leukocyte adhesion responses in postcapillary venules of hypercholesterolemic mice, unlike the greatly enhanced responses observed in normal mice. This suggests that basal NO release is impaired, possibly because of the augmented circulating levels of ADMA mentioned earlier. Exposure of normal postcapillary venules to an analog of ADMA (at levels comparable to circulating levels during hypercholesterolemia) elicits leukocyte adhesion in venules and impairs endothelial barrier function, supporting the proposal that the elevated ADMA levels during hypercholesterolemia are indeed proinflammatory. Second, many NO donors (e.g., sodium nitroprusside, spermine-NO, and L-arginine) have been successfully employed to reduce the inflammatory responses (adhesion molecule expression and leukocyte accumulation) observed both in diet-induced hypercholesterolemia and following exposure to oxLDL, supporting a role for decreased NO bioavailability in this leukocyte recruitment process.

The importance of ROS in the venular responses to hypercholesterolemia is underscored by the observation that oxidative stress, measured using an oxidant-sensitive fluorescent probe, is elevated in postcapillary venules of hypercholesterolemic mice when compared with their normocholesterolemic counterparts. This coincides with increases in leukocyte adhesion and emigration in venules. Furthermore, the leukocyte recruitment is profoundly attenuated in CuZn–SOD-overexpressing mice, suggesting that O_2^- is a major component of the ROS generated during hypercholesterolemia. Similarly, administration of CuZn-SOD has been shown to be effective in preventing oxLDL-induced venular responses. The enzymes that contribute to the increased ROS generation in hypercholesterolemic venules have not been clearly defined. However, mice that are genetically deficient in the $p47^{phox}$ subunit of NAD(P)H oxidase demonstrate a significantly lower level of leukocyte recruitment in response to hypercholesterolemia. Interestingly, when bone marrow chimeras were made to separate blood cell versus vessel wall sources of this enzyme, both sources appeared to be equally important in the generation of the inflammatory phenotype observed in postcapillary venules after 2 weeks on a cholesterol-enriched diet.

Other Inflammatory Mediators

Several mediators have been implicated in inflammatory responses of venules to oxLDL- or diet-induced hypercholesterolemia. For example, arachidonic acid metabolism appears to be important in oxLDL-induced leukocyte recruitment. Blocking leukotriene biosynthesis can prevent the leukocyte adhesion elicited by oxLDL in both arterioles and venules. Platelet-activating factor (PAF) receptor antagonists are equally effective in attenuating the inflammatory responses to oxLDL. Although the contribution of these lipid mediators to diet-induced microvascular alterations has not been assessed, a role for these factors in diet-induced atherosclerotic lesion formation is well established, supporting the possibility that they may also contribute to the early inflammatory responses elicited in venules.

The T-cell-derived cytokine interferon-γ (IFN-γ) has also been implicated in vascular responses to hypercholesterolemia. IFN-γ can promote adhesion molecule expression, and it is a potent activator of NAD(P)H oxidase. The microvasculature of IFN-γ-knockout mice exhibits reduced leukocyte adhesion and significantly lower oxidant stress in response to hypercholesterolemia, when compared with wild-type mice. This suggests that T-lymphocytes may be mediating the inflammatory responses to hypercholesterolemia via IFN-γ, and that this cytokine acts by promoting ROS generation. Another step in this inflammatory pathway may be the release of IL-12, a cytokine that is intimately linked to IFN-γ production. Like IFN-γ, IL-12 is expressed in atherosclerotic lesions and it has recently been shown to contribute to the oxidative stress and leukocyte recruitment induced in postcapillary venules by hypercholesterolemia. These observations suggest that IFN-γ and IL-12 act in concert to promote leukocyte adhesion and emigration by enhancing the production of ROS.

Exaggerated Inflammatory Responses during Hypercholesterolemia

There is a growing body of evidence that hypercholesterolemia renders microvascular endothelium more susceptible to the deleterious consequences of inflammatory stimuli such as I/R. I/R per se is known to elicit an oxidative stress and promote leukocyte adhesion in postcapillary venules. Both of these responses are exacerbated during hypercholesterolemia and can be blocked by pretreatment with either SOD or a xanthine oxidase inhibitor (allopurinol). This suggests that O_2^- generated from xanthine oxidase mediates the leukocyte accumulation elicited by hypercholesterolemic tissues exposed to I/R. Hypercholesterolemia also enhances the P-selectin upregulation that is normally elicited by I/R. In addition, the hypercholesterolemia-induced exacerbation of inflammation is seen when tissues are challenged with either lipid mediators (leukotriene B_4 and PAF) or cytokines such as TNF-α.

It has also been shown that hypercholesterolemia exacerbates the protein extravasation in venules induced by various inflammatory stimuli, and that this may be due to the enhanced leukocyte recruitment. However, administration of oxLDL in the local arterial supply of tissues exposed to I/R promotes leukocyte adhesion and emigration, without an accompanying increase in albumin extravasation. Although oxLDL is able to induce most of the microvascular alterations observed during diet-induced hypercholesterolemia, the underlying mechanisms appear to differ between these two forms of microvascular dysfunction.

Relevance of Hypercholesterolemia-Induced Microvascular Responses to Atherosclerosis

Many of the inflammatory responses and pathways that are initiated in the microvasculature by hypercholesterolemia have also been implicated in the development of atherosclerotic plaques. Whether the early inflammatory responses seen in venules influence the development of lesions in large vessels remains unclear. Since the inflammatory responses to hypercholesterolemia appear to be experienced by all tissues in the body, it appears tenable that the large endothelial surface area (> $500\,m^2$) within the microvasculature may serve as a motor that drives the systemic immune response, ultimately leading to lesion development in large arteries. It is clear, however, that this risk factor, while rendering tissue more likely to experience an ischemic episode through the development of atherosclerosis, also predisposes organs to greater microvascular dysfunction and more tissue injury following a given ischemic insult. Hence, an improved understanding of the mechanisms that underlie the inflammatory phenotype that is assumed by the microvasculature during hypercholesterolemia may reduce the morbidity and mortality associated with cardiovascular diseases.

Glossary

Adhesion molecules: Molecules expressed on the endothelial cells, leukocytes, and platelets, which bind their ligands on other cells, thereby mediating the interactions between the circulating cells and the vessel wall.

Blood cell recruitment: The adhesion of blood cells (leukocytes and platelets) to the vascular endothelium at sites of inflammation.

I/R: Ischemia/reperfusion, or the cessation and restoration of blood flow to an organ or tissue.

Oxidant stress: This usually occurs as a result of an imbalance between nitric oxide and oxidant-generating systems, resulting in an overall increase in the oxidative capacity of the tissue.

OxLDL: Low-density lipoprotein (normally responsible for carrying cholesterol to tissues) that is oxidatively modified, thereby attaining proinflammatory properties.

Acknowledgment

Supported by a grant from the National Heart, Lung and Blood Institute (HL26441).

Bibliography

Granger, D. N. (2003). Risk factors for cardiovascular disease amplify reperfusion-induced inflammation and microvascular dysfunction. In *Molecular Basis for Microcirculatory Disorders*, G. W. Schmidt-Schönbein and D. N. Granger, eds., pp. 333–342. Paris: Springer-Verlag France.

Hansson, G. K. (2001). Immune mechanisms in atherosclerosis. *Arterioscler. Thromb. Vasc. Biol.* **21**, 1876–1890.

Landmesser, U., Hornig, B., and Drexler, H. (2000). Endothelial dysfunction in hypercholesterolemia: mechanisms, pathophysiological importance, and therapeutic interventions. *Semin. Thromb. Hemost.* **26**, 529–537.

Laroia, S. T., Ganti, A. K., Laroia, A. T., and Tendulkar, K. K. (2003). Endothelium and the lipid metabolism: The current understanding. *Int. J. Cardiol.* **88**, 1–9.

Napoli, C., and Lerman, L. O. (2001). Involvement of oxidation-sensitive mechanisms in the cardiovascular effects of hypercholesterolemia. *Mayo Clin. Proc.* **76**, 619–631. *This article focuses on the pathways involved in the oxidant-mediated events that are associated with hypercholesterolemia and examines potential therapeutic strategies in this context.*

Ross, R. (1999). Atherosclerosis—an inflammatory disease. *N. Engl. J. Med.* **340**, 115–126. *This is a comprehensive review of the mechanisms involved in the development of atherosclerosis, many of which are now being addressed in the microvascular responses to hypercholesterolemia.*

Scalia, R., Appel 3rd , J. Z. and Lefer, A. M. (1998). Leukocyte–endothelium interaction during the early stages of hypercholesterolemia in the rabbit: Role of P-selectin, ICAM-1, and VCAM-1. *Arterioscler. Thromb. Vasc. Biol.* **18**, 1093–1100.

Stokes, K. Y., Clanton, E. C., Clements, K. P., and Granger, D. N. (2003). Role of interferon-gamma in hypercholesterolemia-induced leukocyte-endothelial cell adhesion. *Circulation* **107**, 2140–2145.

Stokes, K. Y., Cooper, D., Tailor, A., and Granger, D. N. (2002). Hypercholesterolemia promotes inflammation and microvascular dysfunction: Role of nitric oxide and superoxide. *Free Radic. Biol. Med.* **33**, 1026–1036. *This provides a more in-depth review of the topic discussed here, with particular reference to the role of oxidative stress in the responses to hypercholesterolemia.*

Capsule Biography

Karen Stokes earned her Ph.D. in physiology from Trinity College, Dublin. She is currently an instructor in the Department of Molecular and Cellular Physiology at Louisiana State University Health Sciences Center in Shreveport. Her research interests include the microvascular responses to ischemia–reperfusion and to hypercholesterolemia.

D. Neil Granger, Ph.D., is Boyd Professor and Head of the Department of Molecular and Cellular Physiology at Louisiana State University Health Sciences Center in Shreveport. He has served as President of the Microcirculatory Society, President of the American Physiological Societ, and Editor-in-Chief of *Microcirculation*.

SECTION I

Hypertension

CHAPTER 104

Arteriolar Responses to Arterial Hypertension

Julian H. Lombard
Medical College of Wisconsin, Milwaukee, Wisconsin

Importance of the Arterioles in Hypertension

Arterioles and the small arteries that are located immediately upstream from the arterioles are the major sites of vascular resistance in the peripheral circulation. Thus, changes in the structure and function of these vessels can play a crucial role in the development and maintenance of hypertension, since an elevation in peripheral vascular resistance is a common denominator in virtually all forms of this disease. In addition to controlling the resistance to blood flow in peripheral vascular beds, arterioles play a crucial role in determining the distribution of blood flow within the tissues. Therefore, changes in arteriolar structure, function, and microvessel density can have important implications for the supply of oxygen and nutrients to the tissues in hypertensive individuals. A variety of alterations in arteriolar structure and function can lead to an elevated vascular resistance in hypertension (Figure 1). These include increases in active resting tone; an enhanced response to vasoconstrictor stimuli; an impaired relaxation in response to vasodilator stimuli; a reduced number of arterioles and capillaries (microvascular rarefaction), and structural alterations leading to reduced lumen diameter, increases in wall/lumen ratio, and increases in vessel stiffness.

Enhanced Sensitivity of Arterioles to Vasoconstrictor Stimuli in Hypertension

One of the primary functional alterations that has been reported in arterioles during hypertension is an increase in their sensitivity to a variety of vasoconstrictor stimuli including the adrenergic neurotransmitter norepinephrine and other vasoconstrictor agonists, elevations in intravascular pressure (myogenic response), and an enhanced constriction in response to physiological stimuli such as increased oxygen availability. The increase in the sensitivity of arterioles to vasoconstrictor stimuli in hypertension may have a number of underlying causes, including intrinsic alterations in the electrophysiological responses of the vascular smooth muscle (VSM) cell membrane, changes in the nature and/or production of chemical modulators of vascular tone produced by the endothelium or by the vascular smooth muscle cells themselves, alterations in intracellular Ca^{2+} homeostasis or in other second-messenger systems regulating contractile function, and the potential effects of increased sympathetic nerve activity, humoral factors, and elevated arterial pressure per se in enhancing the sensitivity of the vessels to other vasoconstrictor stimuli.

Under normal physiological conditions, the transmembrane potential (E_m) of the vascular smooth muscle (VSM) cells is a crucial regulator of their active contractile force (and therefore the diameter of arterioles and small resistance arteries). As a result, diameters of small arteries and arterioles are tightly regulated by the dynamic interaction between Ca^{2+} and K^+ ion channels in the smooth muscle cells. Calcium influx through voltage-gated Ca^{2+} channels (activated by membrane depolarization) induces vasoconstriction, whereas the opening of K^+ channels mediates hyperpolarization and vasodilation due to the inactivation of voltage-gated Ca^{2+} channels. There is a very steep relationship between VSM transmembrane potential and contractile force generation in the E_m range between approximately −50 mV and −30 mV, so that small changes in E_m produce large changes in vascular tone. Because of the tight coupling

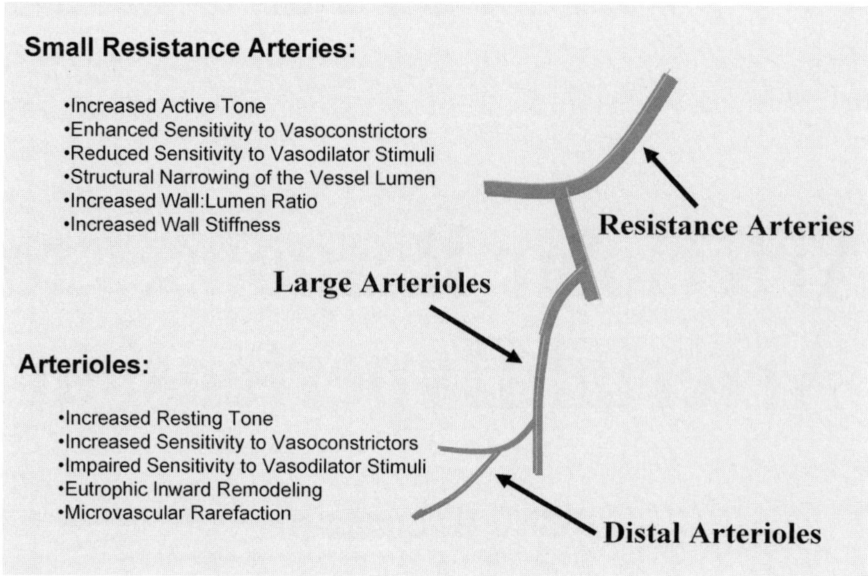

Figure 1 Summary diagram of the mechanisms by which changes in arteriolar structure and function may contribute to an elevated vascular resistance in hypertension. (Adapted with thanks from an original drawing by Dr. Francis A. Sylvester.) (see color insert)

between transmembrane potential and active tone in the smooth muscle cells, alterations in the electrophysiological responses of the VSM cell membrane could contribute to both an enhanced response of arterioles to vasoconstrictor stimuli and to an impaired relaxation in response to vasodilator stimuli in resistance vessels of hypertensive individuals. In this regard, it is important to note that resting E_m in the VSM cells of in situ arterioles and resistance arteries lies within the steep portion of the E_m–active force relationship, or very near the threshold for mechanical activation of the smooth muscle. Therefore, arteriolar tone is very sensitive to activation by vasoconstrictor stimuli such as norepinephrine and transmural pressure elevation, which would result in depolarization and VSM contraction. In hypertension, enhanced sympathetic nerve activity, elevated levels of intravascular pressure, and local or circulating humoral factors not only could activate the vessels directly via depolarization of the VSM cells, but also could bring VSM E_m closer to the threshold for mechanical activation, which would increase the sensitivity of the arterioles to other vasoconstrictor constrictor stimuli that act via membrane depolarization.

Several types of voltage-sensitive ion channels, including L-type Ca^{2+} (Ca_L) channels, voltage-gated K^+ (K_V) channels, and high-conductance voltage- and Ca^{2+}-sensitive K^+ (BK_{Ca}) channels, play a crucial role in the regulation of arteriolar tone, and many studies suggest that high blood pressure may trigger cellular signaling cascades that alter the expression of different ion channels in arterial smooth muscle, leading to further modifications of vascular tone. There is substantial evidence that calcium current through $Ca^{2+}{}_L$ channels is increased in blood vessels of hypertensive animals, which would contribute to an enhanced response to vasoconstrictor stimuli. There is also evidence that K_v channel current is reduced in vessels of hypertensive animals. This reduced K_v current would tend to depolarize the VSM cell membrane, resulting in increased resting tone and an enhanced response to vasoconstrictor stimuli. On the other hand, K_{Ca} channel current at physiological membrane potentials is significantly higher in cerebral arterial smooth muscle cells from spontaneously hypertensive rats (SHRs) compared to those of normotensive Wistar-Kyoto (WKY) controls, apparently because of an increased density of the K_{Ca} channels in the VSM cell membrane. These findings suggest that elevated levels of blood pressure lead to increased K_{Ca} channel expression in vascular smooth muscle membranes as a compensatory mechanism to offset the enhanced Ca^{2+} current and reduced K_v current in the cells. In the absence of this compensatory increase in K_{Ca} channel expression, the increased Ca^{2+} current and the reduced K_v current in the VSM cells could lead to a "vicious cycle" of positive feedback that would cause intense vasoconstriction and severe hypertension.

As noted earlier, alterations in the nature or release of chemical modulators of vascular tone, including a variety of arachidonic acid metabolites, could also contribute to an enhanced response of arterioles to vasoconstrictor stimuli. For example, there is evidence that the release of the vasoconstrictor substances endothelin, prostaglandin H_2 (PGH_2), and/or thromboxane A_2 (TxA_2) from the vascular endothelium contributes to the enhanced myogenic response to pressure elevation in arterioles of spontaneously hypertensive rats. Release of PGH_2 and/or TxA_2 also appears to contribute to an enhanced response to vasoconstrictor stimuli in SHRs and other forms of hypertension. Studies of the role of these compounds in the potentiation of the myogenic response to transmural pressure elevation in hypertensive animals suggest that these endothelium-derived constrictor substances increase the Ca^{2+} sensitivity of the contractile apparatus of arteriolar smooth muscle cells in hypertension, so that similar increases in internal Ca^{2+} concentration in the VSM cells during pressure elevation in the vessel cause an enhanced myogenic constriction of arterioles. The latter observation suggests that intracellular signaling cascades (in addition to changes in VSM E_m) can participate in the altered vascular responses to vasoconstrictor stimuli occurring in arterioles of hypertensive animals. There is also evidence that elevations in intravascular pressure can increase the formation of superoxide anion in arterioles, which could then interfere with nitric oxide (NO)-dependent vascular relaxation and potentiate myogenic responses indirectly. The enhanced myogenic response and the increased constriction of arterioles in response to elevated Po_2 that have been demonstrated in many forms of hypertension could also be due to an enhanced production of 20-hydroxyeicosatetraenoic acid (20-HETE) or to an increased sensitivity of the smooth muscle cells to the vasoconstrictor effects of 20-HETE, a metabolite of the cytochrome P450 pathway of arachidonic acid metabolism that has been implicated in mediating myogenic responses to transmural pressure elevation and arteriolar constriction in response to increased Po_2.

Impaired Relaxation of Arterioles to Vasodilator Stimuli

In addition to an enhanced response to vasoconstrictor stimuli, arterioles of hypertensive animals exhibit an impaired relaxation in response to a variety of vasodilator stimuli including hypoxia, shear stress, and endothelium-dependent vasodilators, such as acetylcholine (ACh). Impaired relaxation of arterioles to endothelium-dependent vasodilator stimuli such as ACh has also been demonstrated in human hypertensive patients. The impaired vascular relaxation in hypertensive individuals has been proposed to be due to a reduced production of endothelium-derived vasodilator compounds, such as nitric oxide (NO) or vasodilator prostaglandins, and/or to an enhanced release of vasoconstrictor factors (e.g., thromboxane or prostaglandin H_2) from the endothelium. The reduction in endothelium-dependent dilation mediated by NO also may result from increased levels of oxidative stress in the tissue, which would destroy NO and reduce its availability for mediating vascular relaxation. There is also evidence that fundamental alterations in receptor–heterotrimeric G protein coupling may contribute to impaired vasodilator responses in hypertensive animals and in normotensive animals on a high-salt diet. In the latter case, arterioles and resistance arteries of hypertensive animals and normotensive animals on high-salt diet not only exhibit impaired relaxation in response to vasoactive agonists acting through the cyclic AMP pathway of vascular relaxation, but also fail to respond to direct activation of the alpha subunit of the G_s protein with cholera toxin. Taken together, these observations suggest that hypertension and high salt diet may both be associated with fundamental alterations of signaling pathways in the vascular smooth muscle cells.

Nitric oxide–dependent relaxation and prostaglandin-mediated vasodilation are both impaired in skeletal muscle arterioles of spontaneously hypertensive rats. This appears to be due to an impaired synthesis and/or action of nitric oxide (including reduced bioavailability of NO due to increased oxidative stress) and alterations in the metabolism of arachidonic acid to favor an enhanced production of the vasoconstrictor metabolite PGH_2 and a reduced production of vasodilator prostaglandins in the arterioles. Findings such as these suggest that a simultaneous dysfunction of these two major endothelium-dependent vasodilator pathways could make a significant contribution to the elevated vascular resistance in hypertension. Agonists such as norepinephrine and acetylcholine also cause an increased release of the endothelium-dependent vasoconstrictors thromboxane A_2 and/or PGH_2 in arterioles and resistance arteries of hypertensive rats, leading to a reduced sensitivity to acetylcholine and to an enhanced vasoconstrictor response to norepinephrine.

As noted earlier, arteriolar dilation in response to the physiological stimulus of increased flow or shear stress is also impaired in arterioles of hypertensive rats. The impaired relaxation of arterioles of spontaneously hypertensive rats in response to increased flow and shear stress appears to be due to an impairment of the NO-mediated portion of flow-dependent dilation, but may also involve an enhanced release of endothelium-derived vasoconstrictor factors such as PGH_2. Current evidence suggests that augmented hemodynamic forces in the microcirculation can alter the shear stress-induced synthesis of prostaglandins and other vasoactive factors in hypertension, possibly contributing to the elevated vascular resistance in this disease.

It has been proposed that the elevated hemodynamic forces present in hypertension may initiate alterations of signaling pathways in the endothelium and smooth muscle cells of arterioles that could, in turn, enhance the release of reactive oxygen species such as superoxide. Any reduction in the availability of NO due to increased levels of superoxide released by high pressure in the arterioles (or in response to other pathophysiological alterations in hypertension) would likely cause an impaired dilation of arterioles in response to shear stress- and other NO-dependent vasodilator stimuli, leading to the maintained elevation of wall shear stress and peripheral vascular resistance that exists in hypertension. It has also been proposed that alterations in the mechanisms of functional vascular control in hypertension may eventually lead to the development of irreversible structural changes in the microcirculation. The latter hypothesis is consistent with the increasing body of evidence that elevated levels of reactive oxygen metabolites may contribute to the vascular dysfunction commonly observed in hypertension.

Structural Alterations in Arterioles during Hypertension

In addition to the changes in functional arteriolar control mechanisms in hypertension, structural alterations of arterioles and small arteries can contribute to the elevated vascular resistance in this disease. These changes include structural narrowing of the lumen, thickening of the vascular wall leading to an increased wall/lumen ratio, and altered mechanical properties of the vessel, such as increased stiffness and reduced distensibility of the vessel wall. These structural alterations of the vascular wall may be caused, at least in part, by alterations in the composition of the wall, such as changes in collagen and elastin content in the vessel wall, or by alterations in the specific types of collagen present in the vascular wall.

Most reports indicate that wall thickening is not a common response to elevated blood pressure in the smaller arterioles; however, wall thickening is prominent in the larger arteries that lie upstream from the microcirculation. There is evidence that the nature of the structural alterations occurring in the vascular wall in response to hypertension may be determined by the response of individual vessels to the increase in circumferential wall stress occurring during elevations in arterial pressure. In large arteries or in resistance arteries that do not exhibit strong myogenic responses, hypertrophy of the smooth muscle cells and the deposition of extracellular matrix thicken the walls of the vessel during the development of hypertension without reducing the size of the lumen. In contrast, small arteries and arterioles that exhibit myogenic contractile activation in response to elevated pressure show an inward remodeling that reduces lumen diameter without thickening the vessel wall. This inward remodeling is mediated through the rearrangement of the smooth muscle cells around a smaller lumen. In this case, the initial increase in circumferential wall stress that occurs in response to increased pressure in the vessel can account for inward remodeling because myogenically active small arteries and arterioles can constrict in response to an elevation of intravascular pressure, thus restoring circumferential wall stress toward control levels. In contrast, larger arteries have little or no myogenic response and respond to the increase in wall stress by initiating growth processes in the vascular wall. There is substantial evidence that the structural alterations in small arteries of hypertensive individuals reflect an adaptation to the elevated blood pressure, rather than being the primary cause of increased vascular resistance in hypertension. However, structural narrowing of the vessel lumen and thickening of the vascular wall may play a crucial role in the maintenance and exacerbation of the elevated vascular resistance in hypertension. In this respect, elevations in vascular resistance that arise from structural alterations are less responsive to therapeutic approaches than those that result from an elevated vascular smooth muscle tone, which can be treated with drugs that lead to vascular smooth muscle relaxation.

Arteriolar Rarefaction

In addition to structural narrowing and increased wall thickness, a decrease in the number of arterioles and capillaries (rarefaction) has been widely reported in many different animal models of hypertension and in hypertensive humans. Microvessel rarefaction has two components: functional rarefaction, mediated by active closure of arterioles, and anatomical or structural rarefaction, mediated by an actual reduction in the number of arterioles. Several lines of evidence suggest that functional rarefaction can eventually progress to anatomical rarefaction. Microvascular rarefaction involves both the capillaries and the smaller (third- and fourth-order) arterioles and is accompanied by structural changes in the microvessels. Mathematical models suggest that microvessel rarefaction can have substantial effects on the microcirculation, including an elevation in vascular resistance (especially in conjunction with the constriction of arterioles) and a reduction in tissue Po_2. The latter changes may be particularly significant in contributing to tissue damage under conditions of reduced perfusion. One interesting observation regarding microvessel rarefaction in salt-dependent forms of hypertension is that it has also been demonstrated to occur in normotensive animals on a high-salt diet. Arteriolar rarefaction in salt-dependent hypertension forms of hypertension and with high-salt diet in normotensive animals develops very rapidly and appears to be mediated by the angiotensin II (ANG II) suppression that occurs in response to elevated salt intake, since it can be prevented by continuous intravenous (i.v.) infusion of a low dose of ANG II to maintain normal circulating levels of ANG II without increasing blood pressure.

Oxidative Stress and Arteriolar Function in Hypertension

Enhanced production of reactive oxygen species such as superoxide anion may contribute to arteriolar dysfunction, elevated vascular resistance, and organ damage in hypertensive individuals. Studies assessing the contribution of enhanced oxidative stress to altered microvascular function have demonstrated that dihydroethidine fluorescence and tetranitroblue tetrazolium dye reduction (indicators of oxidative stress) are significantly increased in arterioles and venules of spontaneously hypertensive rats and Dahl salt-sensitive hypertensive rats on high-salt diet. This finding suggests that there is an enhanced production of oxygen free radicals in the microvasculature of hypertensive animals that could contribute to impaired relaxation in response to NO-dependent vasodilator stimuli or other vascular relaxation mechanisms, for example, prostacyclin-induced dilation. It also appears that elevated dietary salt intake alone can lead to an increase in oxidative stress in arterioles and resistance arteries of normotensive animals. As discussed next, alterations in the function of arterioles and resistance arteries due to the effects of high-salt diet alone may have important implications for the development of an elevated vascular resistance in salt-sensitive forms of hypertension.

Dietary Salt Intake and Arteriolar Function

Many individuals exhibit salt-sensitive forms of hypertension, in which elevated dietary salt intake leads to an increase in arterial blood pressure. This elevation of blood pressure is accompanied by an increase in peripheral vascular resistance. A particularly valuable genetic animal model of salt-sensitive hypertension is the Dahl salt-sensitive (Dahl S) rat, an inbred strain of rats in which elevation of dietary salt intake leads to an elevated vascular resistance and a substantial degree of hypertension. In Dahl S rats, the development of hypertension in response to elevated dietary salt intake is accompanied by a uniform increase in hemodynamic resistance throughout most of the peripheral vasculature. In the spinotrapezius muscle, this increase in resistance is largely due to an intense constriction of proximal arterioles. The mechanisms responsible for this increased arteriolar tone include increased responsiveness to oxygen and a loss of tonic nitric oxide (NO) availability caused by reduced endothelial NO production and/or accelerated degradation of NO by reactive oxygen species.

In recent years, it has become increasingly clear that elevated dietary salt intake alone can lead to profound changes in the structure and function of resistance vessels of normotensive animals, as well as vessels of salt-sensitive experimental models of hypertension such as the Dahl S rat. These changes include microvascular rarefaction and an impaired relaxation of blood vessels in response to vasodilator stimuli such as hypoxia, acetylcholine, and prostacyclin. In normotensive Dahl salt-resistant (Dahl R) rats, elevated dietary salt intake also leads to an impaired dilation during the elevated shear stress that occurs in response to increased flow in the arteriole. The impaired dilation in response to increased flow in arterioles of Dahl R rats on high-salt diet appears to be due to a salt-induced suppression of NO activity in the absence of hypertension. Emerging evidence suggests that impaired vascular relaxation in normotensive animals on a high-salt diet involves alterations in the function of both the endothelium and the vascular smooth muscle cells, and that increased levels of oxidative stress in the vasculature can contribute to the impaired vascular relaxation in animals on high-salt diet.

The impaired relaxation of blood vessels of normotensive animals on high-salt diet in response to vasodilator stimuli such as hypoxia, acetylcholine, and prostacyclin appears to be due to the suppression of angiotensin II (ANG II) levels that occurs in response to high-salt diet, because continuous i.v. infusion of a low dose of ANG II to prevent salt-induced ANG II suppression restores normal vasodilator responses without raising blood pressure in normotensive animals on a high-salt diet. The direct effect of high-salt diet itself in contributing to microvessel rarefaction and impaired vascular relaxation in normotensive rats suggests that elevated dietary salt intake may be an important initial contributor to the increased vascular resistance in salt-sensitive forms of hypertension, since it would tend to elevate vascular resistance even before the increase in arterial blood pressure. In combination with other predisposing factors for hypertension, such as impaired renal function, these changes not only could lead to the development of salt-sensitive hypertension, but also could play a major role in the maintenance and progression of the elevated vascular resistance in salt-sensitive forms of this disease.

Influence of Gender on Arteriolar Function in Hypertension

Females prior to menopause are much less susceptible to hypertension and other cardiovascular diseases than males, indicating that gender has a protective effect in these disorders and that female sex hormones can offset some of the alterations in arteriolar function that may occur with hypertension in males. For example, flow-induced arteriolar dilation is significantly reduced in male spontaneously hypertensive rats compared to females, because of the loss of the nitric oxide (NO)-mediated portion of the response. This impairment of the NO-mediated component of flow induced dilation results in a maintained elevation of wall shear stress in the male rats, suggesting that female sex hormones play an important role in maintaining NO-dependent vasodilator responses and in preserving the regulation of arteriolar shear stress by nitric oxide. Arteriolar dilation in response to increases in perfusate flow is also impaired in isolated gracilis muscle arterioles of ovariectomized female SHRs, compared with those of intact female SHRs and ovariectomized female SHRs receiving estrogen replacement. The impaired flow induced dilation in ovariectomized female SHRs appears to be due to the loss of the NO-dependent component of shear stress–induced vascular relaxation, providing additional evidence that estrogen preserves the NO-mediated portion of flow/shear stress–induced dilation in female hypertensive rats, resulting in a lower maintained wall shear stress in the female SHRs, compared to their male counterparts. The lower wall shear stress in the females may contribute to a lowering of systemic blood pressure and to the lower incidence of cardiovascular diseases in females. In contrast, the maintained elevation of shear stress in arterioles of the male rats could trigger other pathological alterations in the vascular wall, as discussed earlier.

Norepinephrine-induced constrictions are also enhanced in arterioles of ovariectomized female SHRs compared with those of intact female SHRs and ovariectomized female SHRs receiving estrogen supplementation. These differences in norepinephrine-induced constriction of arterioles are eliminated by inhibiting NO synthesis, suggesting that estrogen also preserves the modulating effect of NO on arteriolar responses to vasoconstrictor agonists in female rats.

Although female sex hormones may attenuate endothelial dysfunction in hypertensive animals by preserving endothelium-dependent vasodilation, less is known regarding the influence of ovarian hormones on the generation of contractile substances by the endothelium. However, it appears that female sex hormones attenuate the generation of vasoconstrictor prostanoids and superoxide anion ($O^{\bullet}{}_2{}^-$) by the endothelium of mesenteric microvessels from spontaneously hypertensive rats. Microvessels from ovariectomized female SHRs exhibit an increased sensitivity to norepinephrine and a reduced sensitivity to acetylcholine, compared to those from intact female SHRs. Treatment with estradiol or estradiol + progesterone restores normal reactivity to norepinephrine and acetylcholine in vessels of ovariectomized female SHRs. Inhibition of cyclooxygenase and scavenging of superoxide with superoxide dismutase (SOD) also restore normal responses to norepinephrine and acetylcholine in vessels of ovariectomized female SHRs. Norepinephrine-induced release of prostaglandin $F_{2\alpha}$ ($PGF_{2\alpha}$), a vasoconstrictor metabolite of the cyclooxygenase pathway of arachidonic acid metabolism, is also greater in endothelium-intact microvessels of ovariectomized female SHRs compared to those of intact female SHRs. This response is normalized by treatment with estrogen or estrogen + progesterone. Taken together, these findings suggest that estrogen may protect female SHRs against severe hypertension, not only by preserving NO-dependent dilation, but also by decreasing the synthesis of endothelium derived contracting factors such as PGH_2, $PGF_{2\alpha}$, and $O^{\bullet}{}_2{}^-$.

Glossary

Angiotensin II: Biologically active peptide formed from a precursor peptide (angiotensin I) by angiotensin-converting enzyme (ACE). Angiotensin II has numerous biological actions, including vasoconstriction, stimulation of aldosterone release, stimulation of sodium reabsorption by the kidney, and regulation of vessel structure, vessel funciton, and microvessel density.

Arachidonic acid: Major lipid precursor to various eicosanoids, which are fatty acid derivatives that act as signaling molecules to mediate many biological functions. Arachidonic acid is cleaved from membrane phospholipids and converted into a variety of biologically active lipid metabolites by various enzymes, such as cyclooxygenases, to form the immediate precursor (PGH_2) for various prostaglandins (e.g., prostacyclin, prostaglandin E_2, prostaglandin $F_{2\alpha}$) and thromboxane A_2; lipoxygenases to form leukotrienes; and cytochrome P450 enzymes to form vasodilator compounds such as eicosatrienoic acids (EETs) and vasoconstrictor compounds such as 20-hydroxyeicosatetraenoic acid (20-HETE).

Heterotrimeric G protein: Cell membrane spanning protein that binds guanosine triphosphate (GTP) and mediates the functional coupling of membrane receptors to downstream target enzymes or ion channels involved in cellular signal transduction.

Reactive oxygen species (ROS): Reactive chemical derivatives of oxygen, such as superoxide anion, hydrogen peroxide, hypochlorous acid, and hydroxyl radical. ROS can be formed by a variety of enzymes including xanthine oxidase, nitric oxide synthase (NOS), NAD(P)H oxidase, and cyclooxygenase. Elevated levels of reactive oxygen species in blood vessels cause increased oxidative stress and can contribute to vascular dysfunction in hypertension.

Transmembrane potential (E_m): Electrical potential difference that exists across the cell membrane. The magnitude of E_m differs among cell types, but generally ranges between −50 mV and −30 mV in vascular smooth muscle cells of in vivo microvessels and resistance arteries. A reduced magnitude of the E_m (depolarization) is associated with contraction of the smooth muscle due to increased Ca^{2+} influx into the cells via voltage activated Ca^{2+} (Ca_L) channels, while an increased magnitude of E_m (hyperpolarization) is associated with reduced Ca^{2+} influx into the cells, leading to relaxation.

Bibliography

Dantas, A. P., Scivoletto, R., Fortes, Z. B., Nigro, D., and Carvalho, M. H. (1999). Influence of female sex hormones on endothelium-derived vasoconstrictor prostanoid generation in microvessels of spontaneously hypertensive rats. *Hypertension* **34**, 914–919.

Huang, A., Sun, D., Kaley, G., and Koller, A. (1998). Superoxide release to high intra-arteriolar pressure reduces nitric oxide-mediated shear stress- and agonist-induced dilations. *Circ. Res.* **83**, 960–965. *This study demonstrated that elevated pressure in arterioles can cause increased superoxide production, which could subsequently lead to impaired dilation of arterioles in response to elevated shear stress and other NO-dependent vasodilator stimuli. These findings may be directly relevant to the impairment of vascular relaxation that occurs in hypertensive individuals not exhibiting other alterations that may impair vascular relaxation mechanisms, such as diabetes or low circulating ANG II levels.*

Izzo, J. L., Jr., and Black, H. R. (senior eds.) (2003). *Hypertension Primer*, 3rd ed. Dallas, TX: American Heart Association.

Lee, R. M. K. W. (ed.) (1989). *Blood Vessel Changes in Hypertension: Structure and Function*, Vols. I and II. Boca Raton, FL: CRC Press.

Liu, Y., Hudetz, A. G., Knaus, H.-G., and Rusch, N. J. (1998). Increased expression of Ca^{2+}-sensitive K^+ channels in the cerebral microcirculation of genetically hypertensive rats: Evidence for their protection from cerebral vasospasm. *Circ. Res.* **82**, 729–737.

Lombard, J. H., Sylvester, F. A., Phillips, S. A., and Frisbee, J. C. (2003). High salt diet impairs vascular relaxation mechanisms in rat middle cerebral arteries. *Am. J. Physiol.* **284**, H1124–H1133.

Special Topics Issue: Microcirculatory Adaptations to Hypertension. (2002). *Microcirculation* **9**(4), 221–328. *This special issue of Microcirculation features a collection of recent reviews about various aspects of microcirculatory adaptations to hypertension, including: (1) "Microvascular structure and function in salt-sensitive hypertension," by M. A. Boegehold (pp. 225–241); (2) "New expression profiles of voltage-gated ion channels in arteries exposed to high blood pressure," by R. H. Cox and N. J. Rusch (pp. 243–257); (3) "The inflammatory aspect of the microcirculation in hypertension: Oxidative stress, leukocytes/endothelial interaction, apoptosis," by M. Suematsu, H. Suzuki H, F. A. Delano, and G. W. Schmid-Schönbein (pp. 259–276); (4) "Signaling pathways of mechanotransduction in arteriolar endothelium and smooth muscle cells in hypertension," by A. Koller (pp. 277–294); (5) "Adaptation of resistance arteries to increases in pressure," by R. L. Prewitt, D. C. Rice, and A. D. Dobrian (pp. 295–304); (6) "Structural adaptation of microvascular networks and development of hypertension," by A. R. Pries and T. W. Secomb (pp. 305–314); and (7) "Adaptations of the renal microcirculation to hypertension," by J. D. Imig and E. W. Inscho (pp. 315–328).*

Stekiel, W. J., Contney, S. J., and Rusch, N. J. (1993). Altered β-receptor control of in situ membrane potential in hypertensive rats. *Hypertension* **21**, 1005–1009. *This study demonstrated that receptor-G-protein coupling is impaired in arterioles and venules of rats with reduced renal mass hypertension. VSM transmembrane potential (E_m) was measured with glass microelectrodes in first order arterioles and venules of the in situ cremaster muscle of hypertensive and normotensive rats. Arterioles of hypertensive rats failed to hyperpolarize in response either to the cAMP-dependent beta adrenergic agonist isoproterenol or to cholera toxin, a direct activator of the alpha subunit of the G_s protein coupling* *the receptor to downstream signaling events. In contrast, arterioles of normotensive controls hyperpolarized in response to both isoproterenol and cholera toxin. Subsequent studies by other laboratories demonstrated that G protein coupling is impaired in other forms of hypertension such as SHRs, and in animals on high salt diet. Winner (W. J. Stekiel) of the 1993 Harry Goldblatt Award in Cardiovascular Research, awarded by the publications committee of the American Heart Association Council for High Blood Pressure Research, to recognize the most significant new contribution to the understanding of the causes and/or consequences of hypertension.*

Ungvari, Z., and Koller, A. (2000). Endothelin and prostaglandin H_2/thromboxane A_2 enhance myogenic constriction in hypertension by increasing Ca^{2+} sensitivity of arteriolar smooth muscle. *Hypertension* **36**, 856–861.

Zweifach, B. W. (1983). The microcirculation in experimental hypertension: State of the art review. *Hypertension* **5**, I-10–I-16.

Capsule Biography

Dr. Lombard is currently Professor of Physiology at the Medical College of Wisconsin. He is a former President of the Microcirculatory Society and is a fellow of the Cardiovascular Section of the American Physiological Society, the Council for High Blood Pressure Research of the American Heart Association, and the Council on Basic Cardiovascular Sciences of the American Heart Association. His laboratory focuses on microcirculatory control under normal conditions and during pathological conditions such as hypertension. His work is currently supported by several grants from the National Institutes of Health.

CHAPTER 105

Capillary and Venular Responses to Arterial Hypertension

Geert W. Schmid-Schönbein

Department of Bioengineering
Whitaker Institute of Biomedical Engineering
University of California, San Diego, La Jolla, CA 92093-0412

Introduction

Since arterial hypertension is diagnosed by elevated blood pressure in central arteries, it is frequently regarded as a condition that is almost exclusively affected by hemodynamic resistance in the small arteries and arterioles. This point of view has lead for several decades to a focus on arteries and arterioles as mediators of the syndrome and as focus for therapeutic targets. But microvascular studies indicate that several forms of arterial hypertension may be affected by more general mechanisms, which not only involve the arteries and arterioles, but a range of pathophysiological mechanisms.

This chapter will be focused on the manifestations of hypertension in capillaries and venules, microvessels that are not exposed to blood pressure elevation. Capillaries and venules are involved in blood flow regulation and exchange functions. They are also an integral part of the inflammatory cascade, a potentially important aspect of hypertension as a vascular disease with cell and organ damage. We will summarize an array of pathophysiological phenomena in hypertension for which there is still no conclusive evidence for a pressure-mediated mechanism and which instead point towards a more general metabolic and regulatory defect.

The majority of the evidence cited here has been obtained in the spontaneously hypertensive rat (SHR) and its normotensive control, the Wistar Kyoto (WKY) rat, as well as in the salt dependent Dahl-S hypertensive rat with its normotensive control, the salt resistant Dahl-R strain. Both of these models have a strong genetic linkage. Unless indicated otherwise, these models serve as the main basis for current knowledge at the microvascular level.

Microcirculation in Hypertension

The Microvascular Network Pattern and Cell Morphology

The microvascular network topology in hypertensive and normotensive animals is overall the same but differs in quantitative terms. For example, in skeletal muscle the microvascular branching pattern formed by *feed arterioles*, by *arcade arterioles*, and by their regular side branches, the *terminal* (previously designated also as *transverse*) *arterioles*, is the same. The terminal arterioles form asymmetric dichotomous trees, which give rise to the *capillary* network. The SHR has a higher density of arcade arterioles with smaller trees forming the terminal arterioles. The capillaries form *bundles* with a modular pattern of alternating terminal arterioles and *collecting venules*, which in turn feed into the *arcade venules* and discharge into the central circulation through the *draining veins*. Compared to WKY rats, the SHR exhibits on average a lower capillary network density although individual capillaries have on average greater length (between bifurcations) and diameter. Apart from the fact that collecting venules of SHR are narrower while arcade venules are wider in lumen diameter than in the WKY rats, the two strains exhibit no differences in venular network topology.

The innervation of microvessels in skeletal muscle extends to the terminal arterioles in form of adrenergic fibers. The nerve fibers are positioned at the interface between smooth muscle media and adventitia down to the endings of the terminal arterioles. The density of adrenergic fibers in the SHR is significantly higher compared to that of the WYK rats. Capillaries or venules have no adrenergic innervation.

The walls of capillaries in hypertensives and normotensives consist of endothelial cells with pericytes. Venules have a thinner wall structure than their arteriolar counterpart, with attenuated endothelial thickness, pericytes and smooth muscle cells in the media, and fibroblast in the adventitia. Ultrastructural examination of adult capillaries and venules in hypertensives often reveals morphological damage not found in normotensives, e.g. in form of membrane bleb formation.

Microvascular Pressure

The elevated blood pressure in arteries of hypertensives is reduced in arterioles and in terminal arterioles to values which are similar to those in normotensive animals (**Figure 1**). Apart from the fact that the pressure drop on the venular side is small in both normotensives and hypertensives, there are no significant differences in blood pressure values in venules.

Microvascular Flow

Both the cardiac output and the average local flow rates in different hierarchies of microvessels in hypertensive and normotensive microvascular networks are almost indistinguishable (Zweifach et al., 1981). However, within each microvessel hierarchy the hypertensives have larger variations of flow rates among individual vessels.

Hemodynamic Resistance

Estimates of the average hemodynamic resistance derived from micro-pressure and flow measurements indicate a higher resistance in arcade and terminal arterioles of the hypertensives without such significant differences in the venular counterparts (Boegehold, 1991).

The control of the hemodynamic resistance involves smooth muscle contraction and restructuring of the arterioles. In addition, also blood rheological mechanisms serve to control the hemodynamic resistance in capillaries and venules. In spite of the relatively small number of circulating leukocytes compared to significantly faster moving erythrocytes, in capillaries with single file of blood cells the larger and stiffer leukocytes have an important influence on apparent viscosity and capillary resistance. The mechanism is due to hydrodynamic interaction of slower moving leukocytes with more flexible erythrocytes, which in capillaries displaces the erythrocytes away from their center-line position and leads to an elevated apparent viscosity. The effect is sensitive with respect to the exact erythrocyte and leukocyte counts, and does not require membrane attachment to the endothelium (Helmke et al., 1997).

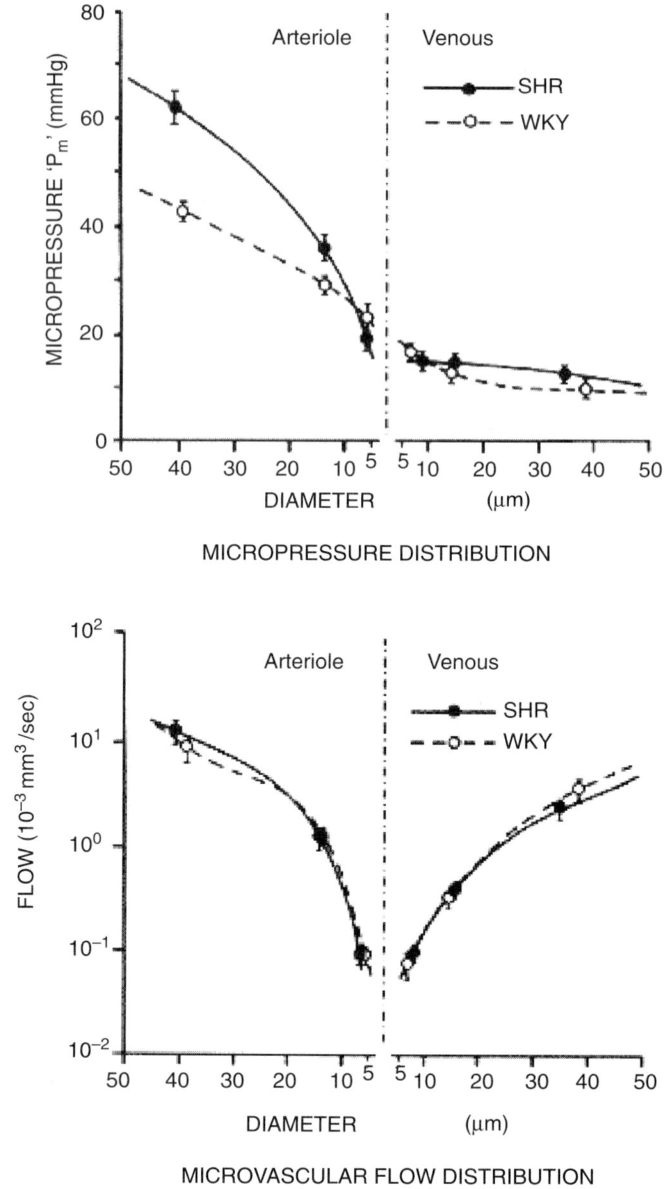

Figure 1 Micropressure (top) and flow rate (bottom) and flow grouped according to vessel diameter in mature rat spinotrapezius muscle, according to Zweifach et al., 1981, in matched WKY and SHR. Capillaries are in the diameter range of about 5 μm on the abscissa. To the left are the diameters of the arterioles and to the right the diameters of the venules. Means and standard deviations are shown. While there is an elevated blood pressure in arterioles of the SHR, their capillaries and venules have no elevation of the blood pressure. Compared with the WKY rats, the SHR exhibit no differences in average flow rates.

Reactive Oxygen Species Production in Microvessels of Hypertensives

Oxygen Free Radical Species

Evidence derived from direct measurements in the microcirculation and in blood samples, experimental results obtained by use of scavengers, and observations on isolated cells of hypertensives indicate an alteration in oxygen metabolism and overproduction of biologically active oxygen species in hypertension. Reactive oxygen species (ROS) (superoxide anion (O_2^-), hydrogen peroxide (H_2O_2), nitric oxide (NO), carbon monoxide (CO) and their derivatives) serve to regulate vascular functions but may also be toxic. In the mesentery and skeletal muscle microcirculation, reduction of nitroblue tetrazolium and fluorescent labeling of hydroethidine (superoxide dependent probes) show enhanced levels of ROS in the endothelium not only in arterioles, but also in capillaries and especially in venules. The rise of the NBT reduction in venules exceeds any enhancement on the arteriolar side in all hypertensive models investigated so far (**Figure 2**).

Superoxide anion, a primary radical product generated by one electron donation to molecular oxygen, is generated through xanthine oxidase and nicotinamide adenine dinucleotide/nicotinamide adenine dinucleotide phosphate (NADH/NADPH) oxidase. The superoxide anion has the ability to react with NO and with guanylate cyclase-dependent vasorelaxation, may activate platelets, leukocytes, and endothelial cells. After cancellation of NO-mediated relaxation and elevation of tone in the hypertensive rats, superoxide generation promotes overexpression of NO synthase mRNA. Superoxide anion inhibits soluble guanylate cyclase, the major target of NO.

Chronic overproduction of superoxide and related ROS triggers alterations in the expression of genes encoding proteins that control the tissue inflammatory responses, such as endothelial adhesion molecules (intracellular adhesion molecule-1 (ICAM-1), and P- and E-selectins) as well as superoxide dismutase, NO synthase and heme oxygenase-1. Nitric oxide depression is by itself a pro-inflammatory stimulator (Kubes et al., 1991) and a driving mechanism for superoxide mediated injury in hypertensives.

The significance of the ROS is further highlighted by the fact that the enhanced peroxide production is detectable before the blood pressure is elevated.

Enzymatic Sources of Oxygen Free Radicals

NADP/NADPH oxidase and xanthine oxidase in neutrophils, the monocyte/macrophage system and vascular endothelial cells, smooth muscle cells, as well as parenchymal cells, are involved in superoxide generation in vivo. Recent evidence derived from immuno-histochemistry indicates that both enzymes are expressed in almost all cells of the microcirculation. The population of superoxide-generating neutrophils in the circulation is greater in SHR than in WKY rats over their entire lifetime.

Xanthine Oxidase

The dehydrogenase (XD) oxidizes hypoxanthine to yield uric acid and is coupled with a reduction of NAD into NADH. Once the enzyme is converted to the oxidase form (XO), the same reaction utilizes molecular oxygen as an electron acceptor and serves as a superoxide-generating system. Endothelial cells in microvessels, but less those in larger vessels, serve as a major source of the XD/XO system. In the mesentery, where the vasculature constitutes a major source of the enzyme, both XD + XO and XO activities are elevated in SHR compared to WKY rats (Suzuki et al., 1998).

NADPH Oxidase

In addition to the role of NADH oxidase in regulation of proliferative responses in vascular smooth muscle cells, this enzyme may also be a major player in the hypertensive syndrome. Phox22, a subunit necessary for the enzyme activity, exhibits a distinct feature that is characteristic of the superoxide production from phagocytic NADPH oxidase. In vascular smooth muscle cells, this enzyme can increase superoxide generation in response to angiotensin II and regulate vascular hypertrophy.

Cytochrome P450 Monooxygenase

This set of enzymes is responsible for elevation of vascular tone and tissue inflammatory responses in hyperten-

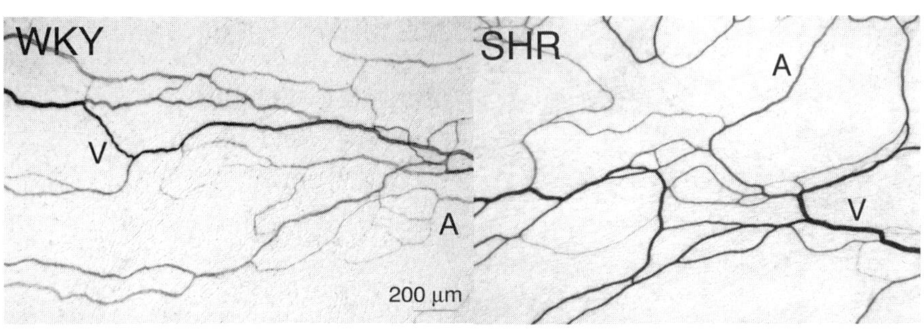

Figure 2 Microimages of the mesentery microcirculation in mature Wistar Kyoto (WKY, left) and age- and gender- matched spontaneously hypertensive rats (SHR, right). The tissue was labeled in the living state under identical conditions with nitro-blue tetrazolium (NBT), a superoxide sensitive indicator. Note the enhanced labeling in microvessels compared with the interstitial state and the strong labeling in the venules (V) compared to arterioles (A) in both animals. The SHR exhibits widespread enhancement of NBT labeling in almost all microvessels. Both rat strains have reduced labeling in capillaries. (see color insert)

sives. Cytochrome P450-derived adrenocortical hormones such as corticosterone and aldosterone play a role in the formation of hypertensive states. The metabolism of arachidonic acid by cytochrome P450 epoxygenases leads to the formation of biologically active eicosanoids for regulation of local inflammatory responses, such as epoxyeicosatrienoic acids (EETs), dihydroxyeicosatrienoic acids (DHETs), and hydroxyeicosatetraenoic acids (HETEs).

HEME OXYGENASE AND CO

Heme oxygenase may also serve as a modulator of vascular tone and smooth muscle cell hypertrophy due to the biological actions of CO, a vasorelaxing mediator (Imai et al., 2001). The potency of CO is less than that of NO. CO modestly activates cyclase when local NO levels are low.

ROS SCAVENGER DEPLETION

The enhanced oxidative stress in hypertensives may also be due to the suppression of scavenger mechanisms for oxygen radicals. In the SHR, mRNA levels and enzyme activity of superoxide dismutase and catalase are reduced in most but not all tissues. Treatment with superoxide dismutase or analogues serves in part to control blood pressure. Treatment is hampered by the limited ability to transport current scavenging agents, such as superoxide dismutase, to relevant sites in the microcirculation, including the endothelial or smooth muscle cells.

The evidence for enhanced ROS formation supports the hypothesis that the SHR may suffer from a genetic shift of the glycolytic metabolism into oxidative metabolism, in line with the close linkage of hypertension with insulin resistance in this model.

Microvascular Rarefaction

Rarefaction in the microcirculation of hypertensives has been encounntered in patients and in experimental models with hypertension (Hutchins and Darnell, 1974). *Functional* rarefaction and *structural* rarefaction can be observed (Prewitt et al., 1982). Functional rarefaction refers to a condition in which no or few blood cells are present in a microvessel due to a state of high vascular tone, but blood cells can be readily reintroduced by application of a vasodilator. Structural rarefaction refers to the physical loss of intact microvessel.

Functional Rarefaction and Blood Cell Distribution

Capillary networks, which are supplied by arterioles under high levels of vascular tone, exhibit functional rarefaction. The tone in arterioles is due in large part to smooth muscle contraction, which in small arterioles is also associated with deformation and folding of the endothelial cell. In terminal arterioles, a high tone may reach the point of complete lumen closure. Vascular tone is defined as

$$\{\text{steady state diameter minus the dilated diameter}\}/\{\text{dilated diameter}\}.$$

The dilated diameter is measured after application of a saturating dose of vasodilators, such as papaverine or adenosine.

The average tone in arterioles of hypertensives is enhanced, leading to a reduction of blood flow from the arcade arterioles into terminal arterioles. At a divergent bifurcation from an arcade to a terminal arteriole, erythrocytes and leukocytes enter preferentially into the vessels with the higher flow rates, i.e. terminal arterioles with enhanced microvascular tone receive a lower fraction of blood flow and therefore also a lower fraction of the blood cells. Therefore, if a terminal arteriole is constricted to the point at which no erythrocytes or leukocytes and only plasma and sporadic platelets enter, its downstream capillaries are filled mostly with plasma and exhibit functional rarefaction. Dilation of the terminal arteriole raises the blood flow and restores the flow of blood cells back into the capillary network. Thus functional rarefaction represents a redistribution of the microhematocit in the smallest microvessels and is reversible.

Structural Rarefaction and Endothelial Apoptosis

In contrast, structural rarefaction is not readily reversible by application of a vasodilator. Recent evidence in several forms of hypertension suggests that loss of capillaries is due to endothelial cell apoptosis (Vogt and Schmid-Schönbein, 2001).

Endothelial apoptosis may be detected in most segments of the circulation in hypertensives. In larger arterioles or venules with multiple endothelial cells along the wall perimeter, apoptosis of individual endothelial cells leads to a temporary shift in local endothelial permeability. In contrast, apoptosis of endothelial cells in true capillaries leads to actual loss of the microvessels since their wall is made up of single endothelial cells. The mesentery microcirculation of the SHR and WKY rats is subject to a non-uniform pattern of cell death, and is enhanced in selected microvascular segments by a glucocorticoid driven mechanism. Apoptosis is present in arterioles but also in capillaries and venules without elevated blood pressure. Enhanced apoptotic activity has been reported in every hypertensive model investigated to date, including the SHR, glucocorticoid-mediated hypertensives, and one kidney/one clip Goldblatt hypertensives. Apoptotic activity is observed in the kidney, heart, smooth and skeletal muscle, mesentery, and in the thymus and can be detected before blood pressure is elevated.

Leukocyte-Endothelial Adhesion in Venules

P-selectin Suppression

P-selectin mediates the rolling interaction between neutrophils and endothelial cells. But in SHR the adhesion of leukocytes to microvascular endothelium induced by inflammatory mediators under physiologic blood shear rates is reduced (Suematsu et al., 1995). The downregulation of leukocyte adhesion appears to involve both leukocyte- and endothelial cell-dependent mechanisms. The SHR has reduced levels of P-selectin on the endothelial membrane of post-capillary venules. There is also a reduction of the sialyl Lewis X-like carbohydrate structure on the leukocytes.

Attenuation of leukocyte rolling has two important consequences.

- The SHR has a chronically elevated count of circulating leukocytes with enhanced levels of free radical production and cytoplasmic degranulation. The elevation in the number of circulating neutrophils and monocytes may result from demargination of these cells in postcapillary venules by suppression of the selectin-dependent membrane interaction. A similar leukocytosis is encountered in P-selectin gene knockout mice with diminished ability for leukocyte rolling on venular endothelium.
- The SHR exhibits a diminished sensitivity to inflammatory stimuli relative to WKY rats and consequently enjoys a surprising protection against inflammatory mediators. The suppression of a P-selectin mediated adhesion pathway may compromise normal leukocyte response under physiological fluid shear conditions and early steps in tissue and lesion repair.

I-CAM Upregulation

Endothelial ICAM-1 expression under both constitutive and induced conditions is upregulated in SHR in splanchnic organs but not necessarily in heart or skeletal muscle. Circulating leukocytes adhere to and spread on endothelium with ICAM-1 overexpression under conditions of reduced fluid shear stress and thereby cause a selectin-independent margination of leukocytes.

In contrast to the resistance to inflammation, the SHR is more vulnerable than the WKY rat to hemorrhagic hypotension or acute ischemia and reperfusion (Cerwinka and Granger, 2001). Enhanced numbers of activated leukocytes trapped in the microcirculation during ischemia are associated with increased organ injury and reduced survival. Once exposed to hemorrhagic shock, activated neutrophils in the circulation are trapped in microvessels of the SHR and expose the tissue to greater oxidative stress than in the WKY rat. The SHRs display a greater extent of microvascular protein leakage upon ischemia-reperfusion than WKY.

Lymphocyte Apoptosis

The SHR has an atrophied thymus and suppressed indices of immune function with extensive lymphocyte apoptosis. Adrenalectomy in the SHR reduces apoptotic death rates of lymphocytes in the thymus. Supplementation with a glucocorticoid enhances the apoptosis in the thymus and several other organs. The process may involve DNA binding of the glucocorticoid receptor. Glucocorticoids disrupt mitochondrial transmembrane potentials, deplete nonoxidized glutathione levels, increase the production of reactive oxygen species, elevate cytosolic free Ca^{2+} levels and produce nuclear and cytoplasmic shape changes. There is evidence for oxygen free radical involvement in early apoptosis of dexamethasone-treated splenocytes and SHR smooth muscle cells.

The Glucocorticoid Pathway in the Spontaneously Hypertensive Rat

We have seen that several diverse microvascular defects in the SHR model of hypertension depend on glucocorticoids. These defects include capillary apoptosis and microvascular rarefaction, apoptosis of SHR thymocytes and lymphocytes, impaired leukocyte-endothelial interaction in post-capillary venules with central leukopenia, enhanced levels of xanthine oxidase and reactive oxygen species, and last not least increased blood pressure with elevated arterial tone. Adrenalectomy serves to normalize the blood pressure in the SHR and attenuates most of the microvascular abnormalities encountered in the SHR while supplementation with glucocorticoids restores the hypertensive state. The response in the adrenalectomized WKY rats at equal levels of glucocorticoids is significantly lower than in the SHR. The SHR suffers from a greatly increased response to adrenal glucocorticoids as well as mineralocorticoids. There is currently no conclusive evidence that the adult SHR has increased levels of glucocorticoids, although it has in the mesentery microcirculation significantly elevated density of glucocorticoid and mineralocorticoid receptors. Glucocorticoids modulate phosphorylation of the insulin receptor, and may be involved in the insulin resistance of the SHR, forming one of the links between hypertension and diabetes.

Conclusion

Hypertensives have a number of microvascular defects, which are independent of the arterioles, and affect the capillary and venular segment of the microcirculation, vessels that are not exposed to elevated blood pressure. These defects expose hypertensives to an enhanced risk for organ injury. We still have little evidence to suggest that the arterial blood pressure elevation per se induces vascular lesions similar to those encountered in hypertension. It appears that a co-factor exists that serves to enhance cell activation in the circulation. In the SHR the glucocorticoid pathway may be involved in production of reactive oxygen in endothelium, apoptosis and capillary rarefaction, inhibition of leukocyte adhesion to postcapillary venules, and suppression of the glucose receptor. These are indications of a microvascular inflammatory response, which compared to that in normotensives is blunted by the suppression of leukocyte rolling and adhesion to the postcapillary venules. The enhanced organ injury in hypertension may be associated with microvascular apoptosis.

Acknowledgements

Supported by NIH Grant HL-10881. I thank Drs. Makoto Suematsu, Hidekazu Suzuki, Fred Lacy, Allan Swei, Camille Vogt, and Dale Parks for numerous discussions and inspirations. Special thanks to Frank A. DeLano for the assistance with Figure 2.

References

Boegehold, M. A. (1991). Effect of salt-induced hypertension on microvascular pressures in skeletal muscle of Dahl rats. *American Journal of Physiology* **260**, H1819–H1825.

Cerwinka, W. H., and Granger, D. N. (2001). Influence of hypercholesterolemia, and hypertension on ischemia-reperfusion induced P-selectin expression. *Atherosclerosis* **154**, 337–344.

Helmke, B. P., Bremner, S. N., Zweifach, B. W., Skalak, R., and Schmid-Schönbein, G. W. (1997). Mechanisms for increased blood flow resistance due to leukocytes. *Am. J. Physiol.* **273**, H2884–H2890.

Hutchins, P. M., and Darnell, A. E. (1974). Observations of a decreased number of small arterioles in spontaneously hypertensive rats. *Circ. Res.* **34–35**, 161–165.

The report contains the first quantitative assessment of microvascular rarefaction in arterial hypertension after repeated qualitative descriptions of the phenomenon in hypertensive patients.

Imai, T., Morita, T., Shindo, T., Nagai, R., Yazaki, Y., Kurihara, H., Suematsu, M., and Katayama, S. (2001). Vascular smooth muscle cell-directed overexpression of heme oxygenase-1 elevates blood pressure through attenuation of nitric oxide-induced vasodilation in mice. *Circ. Res.* **89**, 55–62.

Kubes, P., Suzuki, M., and Granger, D. N. (1991). Nitric oxide: An endogenous modulator of leukocyte adhesion. *Proc. Natl. Acad. Sci. USA* **88**, 4651–4655.

The report summarizes the first detailed documentation of the inflammatory reaction after blockade of NO synthesis in the microcirculation.

Prewitt, R. L., Chen, I. I. H., and Dowell, R. F. (1982). Development of microvascular rarefaction in the spontaneously hypertensive rat. *Am. J. Physiol.* **243**, H243-H251.

Suematsu, M., Suzuki, H., Tamatani, T., Iigou, Y., DeLano, F. A., Miyasaka, M., Forrest, M. J., Kannagi, R., Zweifach, B. W., Ishimura, Y., and Schmid-Schönbein, G. W. (1995). Impairment of selectin-mediated leukocyte adhesion to venular endothelium in spontaneously hypertensive rats. *J. Clin. Invest.* **96**, 2009–2016.

An examination of the molecular mechanisms for attenuation of leukocyte adhesion to the endothelium in postcapillary venules of the SHR.

Suzuki, H., DeLano, F. A., Parks, D. A., Jamshidi, N., Granger, D. N., Ishii, H., Suematsu, M., Zweifach, B. W., and Schmid-Schönbein, G. W. (1998). Xanthine oxidase activity associated with arterial blood pressure in spontaneously hypertensive rats. *Proc. Natl. Acad. Sci. USA* **95**, 4754–4759.

Vogt, C. J., and Schmid-Schönbein, G. W. (2001). Microvascular endothelial cell death, and rarefaction in the glucocorticoid-induced hypertensive rat. *Microcirculation* **8**, 129–139.

Zweifach, B. W., Kovalcheck, S., DeLano, F. A., and Chen, P. (1981). Micropressure-flow relationship in a skeletal muscle of spontaneously hypertensive rats. *Hypertension* **3**, 601–614.

This report summarizes a comprehensive set of micro-hemodynamic measurements in young, and old SHR skeletal muscle.

Capsule Biography

Dr. Schmid-Schönbein has headed the Microcirculation Laboratory at the University of California San Diego since 1979. President of the Microcirculatory Society in 2003, his laboratory primarily focuses on cell mechanics, cell activation, mechanisms of inflammation, and tissue injury. His work is supported by grants from the NIH and NSF.

SECTION J

Inflammation

CHAPTER 106

Free Radicals and Lipid Signaling in Microvascular Endothelial Cells

Peter B. Anning and Valerie B. O'Donnell
Department of Medical Biochemistry, Cardiff University, Heath Park, Cardiff, Wales

Lipoxygenases, Prostaglandin H Synthases and Cytochrome P450s in endothelium

Endothelial cells (ECs) express several enzymes that oxidize unsaturated lipid to signaling mediators. These include both constitutive and inducible isoforms of prostaglandin H synthases (PGHS), lipoxygenases (LOX), and cytochrome P450 (CYP), with the levels of expression and isoform type being dependent on the tissue of origin and inflammatory state of the cells. The healthy endothelium generates a number of oxidized lipid mediators including prostacyclin (PGI_2) and epoxyeicosatetraenoic acids (EET). Following an inflammatory challenge, the properties of the endothelium alter with a switch from generation of vasoprotective mediators, to formation of factors that can potentiate the inflammatory response, including cysteinyl leukotrienes and hydroxyeicosatetraenoic acids (HETEs) (Figure 1). The predominant substrate utilized by all these pathways is arachidonate, hydrolyzed from the *sn*2 position of phospholipids by phospholipase A_2, in response to agonist activation. Following release, it undergoes enzymatic oxidation and isomerization forming a complex variety of signaling mediators that either are released to signal in adjacent cells or signal intracellularly in the endothelium itself. The following sections will describe each signaling pathway focusing in particular on their expression and function in the microvascular endothelium and biological actions of their lipid products on EC themselves.

Prostaglandin H Synthases-1 and -2 in ECs

Enzymology of PGHS-1 and -2

Prostaglandins (PGs) are predominantly generated through the action of PGHS, of which there are both constitutive (PGHS-1: stomach, gut, kidney, platelets) and inducible (PGHS-2: fibroblasts, macrophages) isoforms. Synthesis involves a two-step conversion of arachidonic acid. First, the enzyme oxidizes arachidonic acid to a cyclic endoperoxide, prostaglandin-G_2 (PGG_2), by a cyclooxygenase activity; then a peroxidase reduces the peroxide to a hydroxide, yielding the endoperoxide, prostaglandin-H_2 (PGH_2).

Biochemically, PGHS-1 and -2 are very similar, with 60 percent sequence homology, identical reaction mechanisms, superimposable x-ray crystal structures, and the same subcellular localization at the endoplasmic reticulum and nuclear membrane. However, PGHS isoforms function as two independent prostaglandin synthesis systems utilizing different cellular arachidonate pools in the same cell type, and with very different patterns of expression control.

PGHS in Vascular Disease

In the vasculature, PGHS isoforms regulate vascular homeostasis through generation of PGH_2, the precursor for prostacyclin (PGI_2, endothelial) or thromboxane (TXA_2, platelets). PGHS is transiently activated in platelets or endothelial cells by agonists, such as thrombin, collagen (platelets), bradykinin, or acetylcholine (endothelium). Following this, the PGH_2 is rapidly converted into PGI_2 or TXA_2 by the CYP enzymes, prostacyclin synthase or thromboxane synthase, respectively. Platelet PGHS-1 is the primary source of plasma TXA_2 in both healthy humans and patients with vascular disease, whereas endothelial PGHS-2 is the major source of PGI_2. These eicosanoids have opposing effects, with PGI_2 being vasodilatory and an inhibitor of platelet activation via elevating cAMP, and TXA_2 causing vasoconstriction and platelet activation (Figure 2).

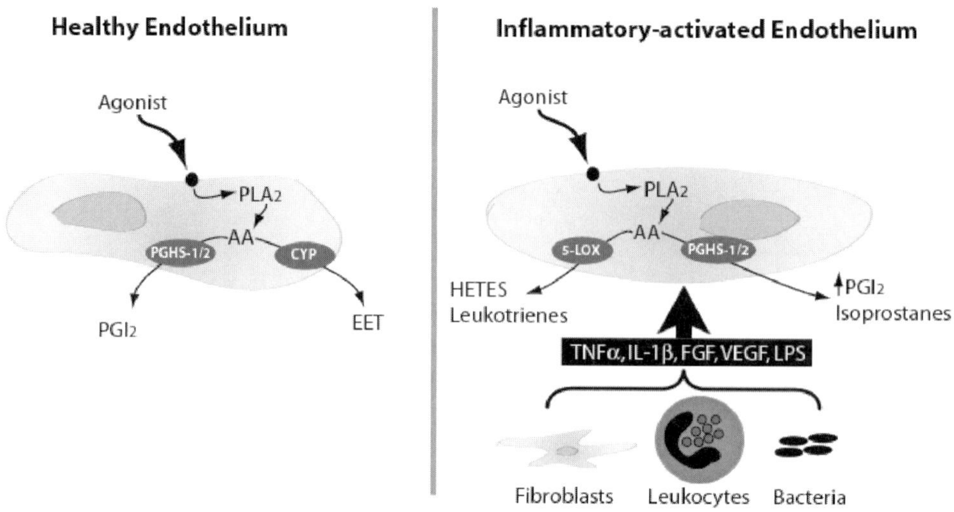

Figure 1 Generation of bioactive eicosanoids by healthy and inflammatory-activated endothelium. Following its hydrolysis from the membrane by phospholipase A_2 (PLA_2), arachidonate is oxidized to prostacyclin (PGI_2) or epoxyeicosatetraenoic acid (EET) by prostaglandin H synthase (PGHS). (see color insert)

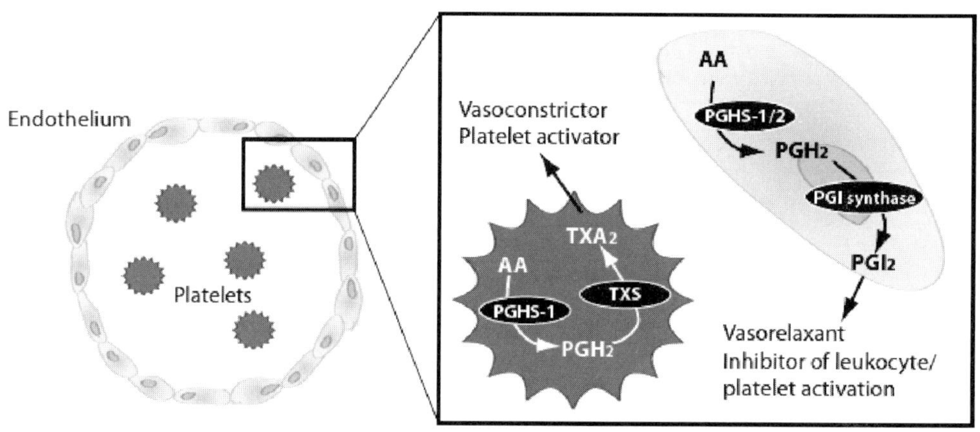

Figure 2 Localization of PGHS isoforms in vascular cells. Platelets contain PGHS-1, which forms prostaglandin H_2 (PGH_2) and is subsequently metabolized by thromboxane synthase (TXS) to thromboxane A_2 (TXA_2). Endothelial cells contain PGHS-1 and -2, both of which are responsible for providing PGH_2 for PGI_2 synthesis by PGI synthase. (see color insert)

The formation of PGHS-derived prostaglandins, including TXA_2, PGI_2, and isoprostanes, is markedly elevated in vascular disease. For example, urinary 8-epi-prostaglandin $F_{2\alpha}$ is increased 130 percent in hypercholesterolemia. Also, isoprostanes are present in human atherosclerotic lesions along with PGHS-1 and -2.

Endothelial Expression of PGHS Isoforms

It has long been considered that PGI_2 is the main prostanoid synthesized by ECs, and TXA_2 the main prostanoid from platelets. However, cultured human umbilical vein endothelial cells (HUVECs) and lung microvascular and cerebral ECs express PGHS-1 constitutively, with this enzyme being the major source of EC-derived PGH_2 precursor for low-level TXA_2 synthesis in HUVECs. Basal expression of PGHS-2 is low or absent in most ECs, but following stimulation with a number of mediators [including laminar flow, HIV-infected monocytes, platelet-derived TXA_2, hypoxia, interleukin (IL)-1β, tumor necrosis factor-α (TNFα), fibroblast growth factor, phorbol ester, lipopolysaccharide (LPS) or vascular endothelial growth factor (VEGF)], its upregulation through an immediate early gene leads to generation of PGI_2 and PGE_2 in a number of microvascular EC types (including human pulmonary, cerebral, and atherosclerotic). Interestingly, IL-1β induces PGI synthase and PGE synthase in tandem with PGHS-2, but not TX synthase. It is therefore likely that the PGHS-2-dependent generation of PGI_2 in vivo in both healthy people and patients with vascular disease requires continuous stimulation of gene expression, for example by laminar flow or proinflammatory cytokines. In contrast to HUVECs, PGHS-2 is a significant source of TXA_2 generated by human microvascular endothelial cells, which can inhibit migration and angiogenesis in vitro. The in vivo importance of this is unclear, however, since platelet PGHS-1 is the major source of TXA_2 in healthy people. PGHS-2 is also negatively regulated at the transcriptional level in ECs. For example, aspirin, sodium salicylate, or nitric oxide inhibits IL-1β-, phorbol-, or LPS-induced PGHS-2 expression in HUVECs and bovine pulmonary artery endothelial cells.

Although PGHS-1 is expressed constitutively by a number of EC types, its expression is also controlled by transcriptional regulation. For example, upregulation of PGI_2 synthesis in intrapulmonary vessels rises markedly during late fetal life, because of a developmental increase in PGHS-1 expression that occurs via estrogen stimulation of the estrogen receptor. This may also have implications for PGHS-1 expression in pre- and postmenopausal women where risk of vascular disease increases with decreased estrogen levels, and estrogen replacement is associated with decreased cardiovascular risk.

Regulation of EC Function by PGHS Products

Endothelial cell function is regulated in several ways through PGHS signaling (Figure 3). In particular, recent data have implicated the prostaglandin 15-deoxy-δ(12,14)-prostaglandin J2 (15δ-PGJ2) in mediating multiple responses through activating peroxisome proliferator–activated receptors (PPARs). These are members of the nuclear receptor superfamily of transcription factors that are important mediators of the inflammatory response. Through this pathway, 15δ-PGJ2 activation of endothelial PPARs inhibits leukocyte–endothelial interactions, IFNγ-induced expression of CXC chemokines, and TNF-induced oxidized LDL receptor (LOX-1) and induces stress proteins including heme oxygenase and plasminogen activator inhibitor type-1 (PAI-1) in a number of ECs (including brain microvascular). 15δ-PGJ2 also signals in a PPAR-independent manner in ECs, inducing apoptosis and synthesis of GSH and IL-8.

In addition to 15δ-PGJ_2, additional prostaglandins that signal in ECs include PGE_2, which induces expression of P-selectin, VEGF, and endothelial nitric oxide synthase (eNOS) through activation of ERK/JNK2 signaling pathways, and PGD_2, which can relax vessels through stimulation of eNOS activity in bovine coronary arteries (Figure 3).

In summary, PGHS isoforms expressed in ECs regulate normal vascular function and participate in the pathophysiology of vascular disease. In addition, PGHS products generated by adjacent cells are important in regulating numerous microvascular EC functions, including apoptosis, integrin expression, and eNOS activity.

Lipoxygenases in ECs

Enzymology of Lipoxygenases

Lipoxygenases (LOX) are nonheme iron-containing enzymes that catalyze arachidonate or linoleate oxidation to form a series of lipid hydroperoxides. In mammalian cells, several isoforms are known, named by their position of oxygen insertion into arachidonate. Lipoxygenases contain a single nonheme iron that alternates between Fe^{2+} and Fe^{3+} during catalysis. Resting enzyme predominantly exists as the reduced form, requiring oxidation by hydroperoxides before dioxygenation can occur.

Vascular and Endothelial Expression of LOX Isoforms

LOX enzymes are predominantly expressed by leukocytes (5- and 15-LOX in humans, rabbits, 12/15-LOX in mice, rats, pigs) and platelets (12-LOX). Under basal conditions, ECs do not appear to express significant LOX protein. However, increased protein expression of 5-LOX has been reported in pulmonary artery ECs of patients with primary pulmonary hypertension, in hypoxic rats, and in antigen-

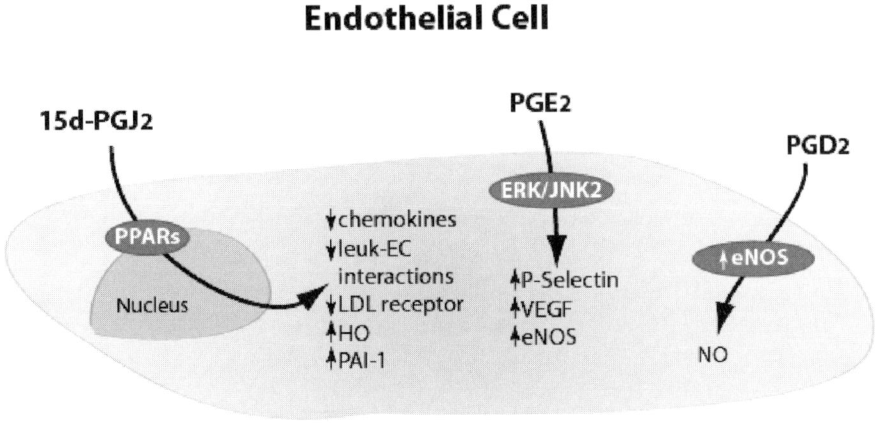

Figure 3 Effects of PGHS products on endothelial cell function. A series of PGHS metabolites have potent biological effects on endothelial cell function. PPAR, peroxisomal proliferator activating receptor; HO, heme oxygenase; VEGF, vascular endothelial growth factor; 15δ-PGJ_2, 15-deoxy-δ (12,14)-prostaglandin J_2; PAI-1, plasminogen activator inhibitor type-1. (see color insert)

challenged mice. This suggests 5-LOX may be upregulated in inflammatory-activated endothelium, although the mechanisms involved are unknown. Finally, although functional 5-LOX protein is not expressed in ECs, leukotrienes can be generated by HUVECs and pig aortic ECs following intercellular transfer of LOX products from associated granulocytes. It is unknown whether similar generation of leukotrienes can occur in microvascular ECs following granulocyte transfer of precursors.

Regulation of EC Function by LOX Products

LOX products stimulate a variety of EC functions. In particular an important role for platelet 12-LOX expressed by tumor cells in regulating ECs in cancer microvessels is emerging since 12-LOX can stimulate proliferation, migration, and tube differentiation in vitro and angiogenesis in vivo. In addition, 12-HETE upregulates expression of $\alpha v\ \beta 3$ integrin on microvascular ECs, which is required for angiogenesis of breast cancer, whereas biosynthesis of 12-HETE by B16 melanoma cells is a determinant of their metastatic potential. Finally, 12-HETE can stimulate monocyte endothelial interactions following incubation of ECs with high glucose or minimally oxidized low-density lipoprotein, suggesting a role for LOX activation of EC in inflammatory vascular disease (Figure 4).

In summary, expression of LOX in most EC is low or absent under normal conditions. Upregulation in vivo following inflammatory challenge may result in generation of low amounts of HETEs, but it is unclear whether this is of biological significance. In contrast, generation of LOX products by adjacent cells including leukocytes and tumor cells is centrally involved in regulating microvascular EC function under pathophysiological conditions.

CYP Enzymes in ECs

CYP enzymes are a ubiquitously expressed family of heme enzymes that play central roles in xenobiotic metabolism and lipid oxidation. CYP-dependent arachidonate oxidation occurs through three pathways, allylic oxidation, omega hydroxylation, and olefin epoxidation. These result in a series of oxygenated metabolites, including epoxides and fatty acid alcohols.

Nonhepatic cytochrome P450 arachidonate metabolites act as intracellular signaling molecules in vascular tissue (Figure 5). The major EC CYP isoforms are prostacyclin synthase (PGI synthase) and thromboxane synthase (TXS), which generate prostacyclin (PGI_2) or thromboxane A_2 (TXA_2), respectively, from PGHS-derived PGH_2 (described earlier). Both enzymes are controlled through transcriptional regulation, although the pathways are not well characterized. For example, TXS is inducible in pig aortic ECs by xenoreactive antibodies, whereas IL-1β elevates PGI synthase in tandem with PGHS-2 in HUVECs.

Additional EC-derived CYP products include the epoxides, 11,12-epoxyeicosatetraenoic acid (EET) and 5,6-EET, and dihydroxyeicosatrienoic acids (DHET). 11,12-EET is avidly esterified into endothelial phospholipid pools and mediates vascular relaxation, possibly accounting for a component of endothelial-derived hyperpolarizing factor (EDHF) activity. Preformed EETs in endothelial membranes can influence vascular function by altering membrane characteristics, ion transport, or lipid-dependent signaling pathways. For example, 5,6-EET mediates vasodilation by either increasing nitric oxide production through stimulating Ca^{2+} influx into ECs, including rat cerebral microvessels, or by directly activating smooth muscle K_{ca} channels. A final important vasoactive CYP product, 20-HETE is generated by CYP 4A and promotes renal vasoconstriction. However, this is generated by smooth muscle, rather than ECs.

Generation of Free Radical Species by PGHS or LOX and CYP

Lipid peroxidation enzymes generate free radical intermediates during catalysis. For example, both PGHS and LOX form enzyme-bound lipid alkyl ($L^\bullet$) and peroxyl

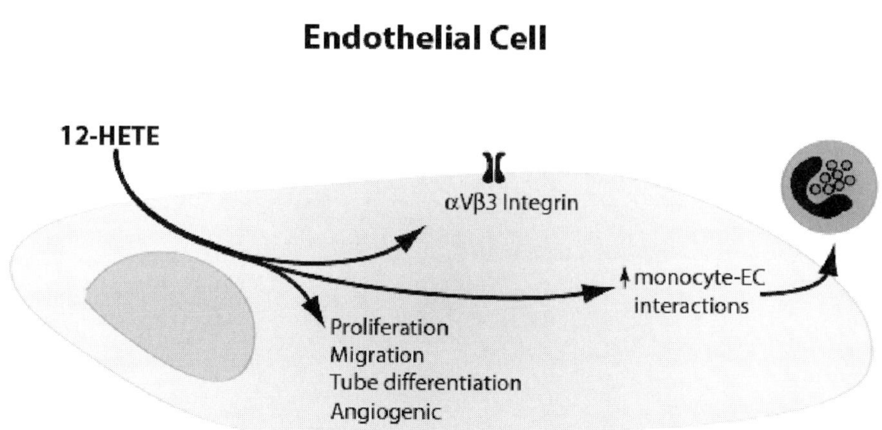

Figure 4 Signaling properties of 12-HETE in endothelial cells. The LOX product 12-HETE induces multiple biological effects in endothelium. (see color insert)

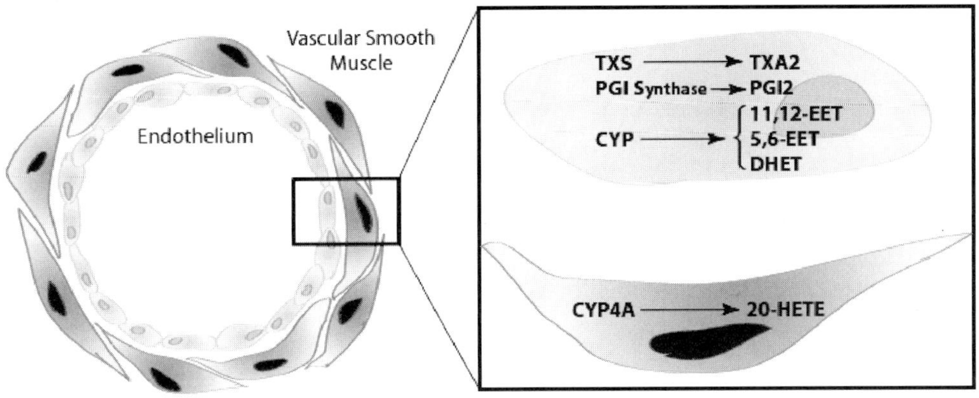

Figure 5 Localization of CYP isoforms in vascular tissue. Endothelium contains a number of CYP enzymes that generate bioactive lipid products. An additional important isoform is CYP4A, in smooth muscle that generates 20-HETE. TXS, thromboxane synthase; EET, epoxyeicosatetraenoic acid; DHET, dihydroxyeicosatetraenoic acid. (see color insert)

(LOO$^•$) radicals that are ultimately converted into hydroperoxides (LOOH) before release from the active site. At low O_2 tension, a small proportion of lipid radicals (up to 10%) escape the active site. These react with O_2 at diffusion-controlled rates to form free LOO$^•$, which can then propagate secondary nonenzymatic lipid peroxidation. During this, a proportion of racemic products is formed. This reaction may be a significant source of LOOH in late atherosclerosis where lipid peroxidation product specificity is lost.

Although they do not directly bind or activate O_2, PGHS and LOX can generate $O_2^{•-}$ through secondary side reactions involving oxidation of certain peroxidase substrates. In these reactions, substrates including NAD(P)H and GSH are oxidized to radicals [i.e., NAD(P)$^•$ and GS$^•$, respectively] that can ultimately react with O_2 either directly, or indirectly forming $O_2^{•-}$. To date, these reactions have only been observed using purified enzyme and it is unknown whether they contribute to free radical levels in intact cells or tissue. Finally, it has been suggested that CYP2C9 is a significant source of reactive oxygen species in porcine coronary arteries that may play a role in regulating vascular tone.

Regulation of PGHS, LOX, and CYP by Reactive Oxygen and Nitrogen Species

Lipid oxidation enzymes are regulated in several ways through the action of reactive oxygen and nitrogen species. In general, enzyme turnover is activated by oxidation [e.g., for LOX or PGHS by LOOH, H_2O_2, or peroxynitrite ($ONOO^-$)] and inhibited by reduction (e.g., nordihydroguairetic acid and baicalein as LOX inhibitors, or removal of LOOH or H_2O_2 by glutathione peroxidase or catalase-dependent reduction).

Nitric oxide (NO) inhibits LOX turnover through scavenging the enzyme-bound LOO$^•$, but exerts no direct effect on PGHS turnover in vitro. The lack of effect on PGHS turnover is intriguing since NO can interact with this enzyme in multiple ways including scavenging of the catalytic tyrosyl radical and acting as a reducing peroxidase substrate. In contrast to its lack of effect on purified PGHS, NO has multiple and often contradictory effects on PGHS expression and activity in intact cells. In several systems (including purified recombinant COX-2, intact platelets, endothelial cells, RAW-264.7 cells, an ex vivo model of renal inflammation, and following in vivo administration of $^•$NO donors to rats), $^•$NO highly stimulates PG production. However, other investigators have found $^•$NO either to be inhibitory towards PGHS, or to have no effect on either PGHS activity (platelets) or LPS-induced expression in RAW-264.7 cells. In some cell types however (rat microglial cells and peritoneal macrophages), $^•$NO suppresses LPS-induced COX-2 expression, resulting in apparent enzyme inhibition. Finally, $^•$NO inhibits CYP through formation of an iron–nitrosyl complex, and perhaps additional uncharacterized mechanisms. In rat renal microvessels, this attenuates EET-dependent dilation, but conversely inhibits 20-HETE-dependent vasoconstriction through inhibition of CYP4A.

Conclusions

Oxidized lipid mediators generated by PGHS, LOX, or CYP are of central importance in the normal physiology of the endothelium, with their aberrant generation playing a major role in the pathogenesis of inflammatory vascular disease. In addition, these enzymes generate a small amount of lipid radicals that may propagate nonenzymatic lipid peroxidation, a hallmark of atherosclerotic lesions. Although much is known regarding function and control of these pathways in ECs (especially PGHS and LOX), others, especially the CYP enzymes, are less studied. Studying the biological roles and signaling pathways of CYP in EC is becoming a major focus of research in vascular biology and will undoubtedly lead to a fuller understanding of their roles in both normal homeostasis and vascular pathophysiology.

Finally, although much is known regarding the biological chemistry and cell biology of these pathways, their relative importance in vessels of different origin is not clear. In particular, the role of PGHS, LOX, or CYP in control of vascular tone through regulating vascular function in large vessels, resistance vessels, and capillary beds may vary tremendously. Elucidation of tissue-specific functions and control mechanisms for lipid oxidation pathways in subtypes of EC is becoming an area of active and fruitful investigation that will yield major insights into their role in regulating vascular biology in health and disease.

Glossary

Lipoxygenases: Lipid oxidizing enzymes that play important roles in vascular function and immune regulation. There are several mammalian isoforms, with one in particular (12/15-lipoxygenase) being involved in vascular dysfunction associated with hypertension, diabetes, and atherogenesis.

Nitric oxide: Free radical signaling molecule generated by oxidation of L-arginine by nitric oxide synthases (NOS), which causes smooth muscle relaxation and inhibits platelet and leukocyte activation.

Prostaglandin H synthases: Lipid oxidizing enzymes that generate prostaglandins, signaling mediators that regulate vessel tone (e.g., prostacyclin) and platelet aggregation (e.g., thromboxane).

Acknowledgments

Research funding from the Wellcome Trust and British Heart Foundation is gratefully acknowledged.

Bibliography

Prostaglandin H Synthase

Belton, O., Byrne, D., Kearney, D., Leahy, A., and Fitzgerald, D. J. (2000). Cyclooxygenase-1 and -2-dependent prostacyclin formation in patients with atherosclerosis *Circulation* **102**, 840–845. *This study describes the relative importance of PGHS-1 and -2 in generation of prostanoids in vivo in humans with vascular disease.*

Caughey, G. E., Cleland, L. G., Penglis, P. S., Gamble, J. R., and James, M. J. (2001). Roles of cyclooxygenase (COX)-1 and COX-2 in prostanoid production by human endothelial cells: Selective up-regulation of prostacyclin synthesis by COX-2. *J. Immunol.* **167**, 2831–2838.

Smith, W. L., and Marnett, L. J. (1991). Prostaglandin endoperoxide synthase: Structure and catalysis. *Biochim. Biophys. Acta* **1083**, 1–17. *This is an excellent review on PGHS isoforms, focusing mainly on the biochemistry and enzymology of these pathways.*

Lipoxygenases

Feinmark, S. J., and Cannon, P. J. (1986). Endothelial cell leukotriene C4 synthesis results from intercellular transfer of leukotriene A4 synthesised by polymorphonuclear leukocytes. *J. Biol. Chem.* **261**, 16466–16472.

Kühn, H., Belkner, J., Zaiss, S., Fahrenklemper, T., and Wohlfeil, S. (1994). Involvement of 15-lipoxygenase in early stages of atherogenesis. *J. Exp. Med.* **179**, 1903–1911.

Liu, B., Marnett, L. J., Chaudhary, A., Ji, C., Blair, I. A., Johnson, C. R., Diglio, C. A., and Honn, K. V. (1994). Biosynthesis of 12(*S*)hydroxyeicosatetraenoic acid by B16 amelanotic melanoma cells is a determinant of their metastatic potential. *Lab. Invest.* **70**, 314.

Zhang, Y. Y., Walker, J. L., Huang, A., Keaney, J. F., Clish, C. B., Serhan, C. N., and Loscalzo, J. (2002). Expression of 5-lipoxygenase in pulmonary artery endothelial cells. *Biochem. J.* **361**, 267–276.

Cytochrome P450

Campbell, W. B., Gebremedhin, S., Pratt, P. F., and Harder, D. R. (1996). Identification of epoxyeicosatrienoic acids as endothelium-derived hyperpolarizing factors. *Circ. Res.* **78**, 415–423.

Capdevila, J. H., Falck, J. R., and Estabrook, R. W. (1992). Cytochrome P450 and the arachidonate cascade. *FASEB J.* **6**, 731–736. *This is an excellent review on the enzymology of CYP oxidation of arachidonate to bioactive mediators. Although it may be a little out-of-date, it still is a very useful starting point for the reader new to the area.*

Rosolowski, M., and Campbell, W. B. (1996). Synthesis of hydroxyeicosatetraenoic (HETEs) and epoxyeicosatrienoic acids (EETs) by cultured bovine artery endothelial cells. *Biochim. Biophys. Acta* **1299**, 267–277.

Capsule Biography

Drs. Anning and O'Donnell are based at Cardiff University, UK. Their work focuses on lipoxygenase and nitric oxide signaling in the vasculature and is funded by the British Heart Foundation and Wellcome Trust.

CHAPTER 107

Eicosanoids

Monica M. Bertagnolli
Brigham and Women's Hospital, Department of Surgery, Boston, Massachusetts

Introduction: History and Definitions

Eicosanoids (from the Greek *eicosa*, "twenty") are a large family of inter- and intracellular signaling molecules derived from arachidonic acid, a fourfold unsaturated C_{20} fatty acid sequestered in membrane phospholipids. Eicosanoid production is tightly controlled by mediators of membrane lipid mobilization, and by the cellular concentrations and activities of the enzymes involved in their metabolism. Eicosanoids are capable of mediating a great variety of cellular functions, including processes as seemingly diverse as vascular contractility, inflammatory response, protection of the gastric mucosa, and renal function. Eicosanoids are rapid responders to external stimuli, and in keeping with this role, they are not stored within the cell, but are rapidly synthesized and then quickly degraded as a result of spontaneous hydrolysis or enzyme-mediated inactivation.

The first eicosanoids identified were members of a category known as prostaglandins. Prostaglandins were discovered in the 1930s when a reproductive physiologist, von Euler, observed that a substance in human semen induced contraction of uterine smooth muscle. Because he assumed that the active agent was produced by the prostate gland, he named this substance "prostaglandin." Members of the leukotriene family were first identified in the early 1940s as a result of their effects as mediators of anaphylaxis. In 1969, Piper and Vane showed that aspirin inhibited vasoactive substances produced by rabbit aorta, and in 1971, Vane discovered that prostaglandins were the target of nonsteroidal anti-inflammatory drug (NSAID) activity. Samuelsson identified thromboxanes as distinct products in 1979. In the latter half of the 20th century, the recognition of eicosanoids as important mediators of both normal and pathologic physiological responses has added a wealth of data to this increasingly complex field.

Converting Membrane Components into Signaling Molecules

Introduction

Upon cell stimulation, arachidonic acid is released from the plasma membrane through the activation of phospholipase A_2 (PLA_2). Further metabolism of arachidonic acid, and the eventual composition of resulting eicosanoids, depends upon the availability of enzymes responsible for arachidonic acid metabolism within a specific cell. These enzymes may be classified into three major groups, including the cyclooxygenases, lipoxygenases, and P450-monooxygenases (Figure 1).

The products of arachidonic acid metabolism exert a vast range of downstream effects on cell signaling pathways. The primary mode of eicosanoid action is through specific G protein—coupled receptors. In the highly complex network of cell signaling, these mediators influence many different systems, including those governing cell proliferation and differentiation (e.g., MAP kinase and PPARs), cytoskeletal dynamics (e.g., Rho GTPases), apoptosis (e.g., Akt and PI_3K), ion transport (e.g., Ca^{2+} channels), and many others. Some of the eicosanoid downstream signaling pathways, such as those involved in inflammation, have been extensively studied. Others, such as the effects of metabolic pathways mediated by different cyclooxygenase isoforms, have only recently come under intense scrutiny.

The Cyclooxygenase Pathway

Prostaglandins and thromboxanes are bioactive substances that result from the metabolism of arachidonic acid by cyclooxygenases. These molecules are produced by most cells in the body and act as autocrine and paracrine mediators of a diverse range of cell functions, including pain

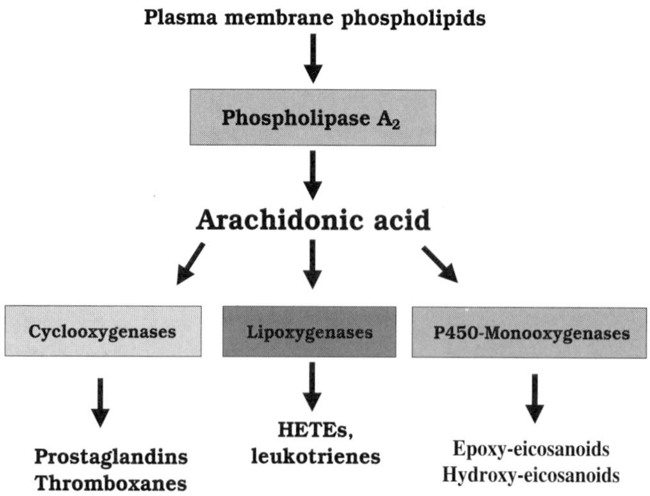

Figure 1 Arachidonic acid metabolic pathways. (see color insert)

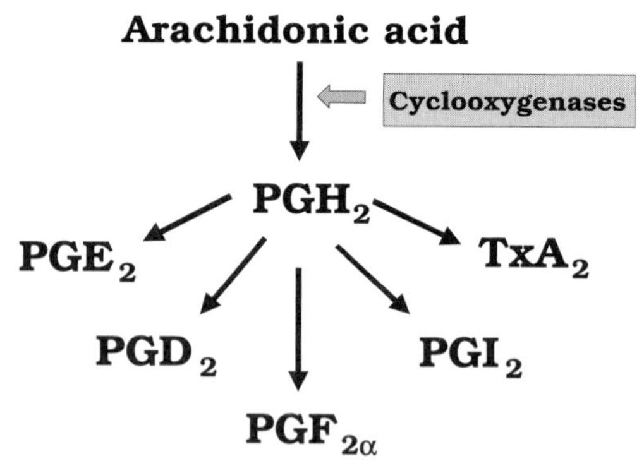

Figure 2 Tissue-specific products of cyclooxygenase activity. (see color insert)

Table I Physiologic Activities of Products of Cyclooxygenase Activity.

PGE_2	Vasodilatation, bronchodilatation, inhibition of gastric acid secretion, gastric mucosal protection, hyperalgesia, pyrexia, increased uterine contractility
PGD_2	Vasodilatation, regulation of renal blood flow, pulmonary artery constriction, bronchoconstriction
$PGF_{2\alpha}$	Pulmonary artery constriction, bronchoconstriction, increased uterine contractility
PGI_2	Vasodilatation, inhibition of platelet aggregation
TxA_2	Vasoconstriction, bronchoconstriction, promotion of platelet aggregation, increased membrane permeability, neutrophil activation

generation, vasomotor tone, vascular permeability, febrile response, and uterine contractility. Stimulation of the cell by growth factors, cytokines, or mechanical trauma leads to mobilization of arachidonic acid from the phospholipid membrane, followed by cyclooxygenase-mediated conversion of arachidonic acid to a short-lived intermediate, prostaglandin H_2 (PGH_2). This molecule is then modified by specific enzymes to produce a variety of bioactive substances (Table I) that share a similar chemical structure (Figure 2). Cell-specific expression of arachidonic metabolites exists as a result of differential expression of both downstream metabolizing enzymes and receptor isoforms. For example, epithelial cells contain prostaglandin synthetase, leading to the production of prostaglandin E_2 (PGE_2), platelets contain thromboxane synthetase and therefore produce thromboxane A_2 (TxA_2), and endothelial cells produce prostaglandin I_2 (PGI_2), also known as prostacyclin, through the activity of prostacyclin synthase. There are at least nine known prostaglandin receptor forms, conveying an additional level of tissue specificity to prostaglandin-mediated activities. Four of the receptor subtypes bind PGE_2 (EP_1-EP_4), two bind PDG_2 (DP_1, DP_2), and separate receptors bind $PGF_{2\alpha}$ (FP), PGI_2 (IP) and TxA_2 (TP). These receptors are transmembrane G protein—coupled proteins linked to a number of different signaling pathways. In complex tissues, receptors for a wide variety of prostaglandins are present on the surface of various components, such as epithelial cells, stromal fibroblasts, stromal endothelial cells, and inflammatory cells.

Until 1991, only one form of cyclooxygenase was recognized. This family of enzymes is now known to contain at least two forms, each with distinct roles in tissue regulation (Table II). Cyclooxygenase-1 (Cox-1) is constitutively expressed in the gastrointestinal mucosa, kidneys, platelets, and vascular endothelium and is responsible for maintenance of normal physiologic function of these tissues. Cyclooxygenase-2 (Cox-2) was identified in the early 1990s as a distinct enzyme associated primarily with inflammation. Cox-2 is the product of an intermediate-early response gene whose tissue expression is increased 20-fold in response to growth factors, cytokines, and tumor promoters. Cox-2 is not found in significant quantities in the absence of stimulation, which explains why it remained undetected as a distinct molecule for 20 years.

The Lipoxygenase Pathway

Lipoxygenases convert arachidonic, linoleic, and other polyunsaturated fatty acids into biologically active hydroperoxy derivatives that modulate cell signaling. In mammals, lipoxygenases are classified according to their positional specificity for fatty acid oxygenation, and are therefore designated as 5-, 8-, 12-, or 15-lipoxygenase. The 12- and 15-lipoxygenases are further differentiated according to whether they are derived from platelets (12-*S*-LOX), epidermis (12-*R*-LOX and 15-LOX-2), or reticulocytes (12-LOX-1).

The physiologic effects of lipoxygenase metabolites have not been characterized as extensively as those produced by cyclooxygenase activity. The 5-, 8-, and 12-LOX isoforms result in production of hydroxyeicosatetraenoic acids (HETEs), including 5-*S*-HETE, LTB_4, 8-*S*-HETE, and 12-HETE (Figure 3). In general, these metabolites increase cell proliferation, contribute to inflammatory changes, and

Table II Cyclooxygenases.

	Constitutive expression	Inducible expression	Effects	Inhibited by
Cox-1	Gastric epithelium, platelets	Rare or none under physiological conditions	Pain, platelet activation, protection of gastric mucosa	Most NSAIDs, including aspirin; with minimal or no inhibition by Cox-2 selective agents
Cox-2	Kidney, brain	Induced in most tissues by growth factors, inflammatory cytokines, neurotransmitters, oxidative stress	Pain, inflammation, fever, angiogenesis, tumorigenesis	Most NSAIDs

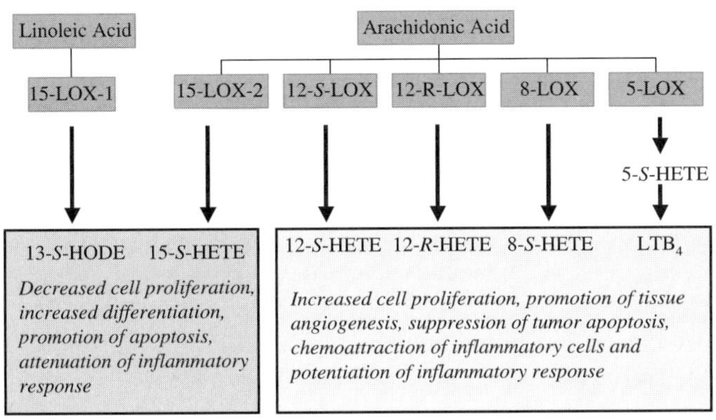

Figure 3 Products of lipoxygenase metabolism. (see color insert)

promote angiogenesis. The 15-LOX isoenzymes (15-LOX-1, 15-LOX-2) convert arachidonic acid to 15-*S*-HETE and linoleic acid to 13-*S*-hydroxy-9,11-octadecadienoic acid (13-*S*-HODE). These substances may have effects opposite to those of the 5-, 8-, and 12-LOX, as they induce epithelial cell differentiation and promote both growth inhibition and cell apoptosis.

Leukotrienes are a family of paracrine mediators derived from oxidative metabolism of arachidonic acid by 5-LOX. Leukotriene B_4 (LTB_4) is a powerful chemoattractant, responsible for the recruitment of leukocytes to sites of inflammation. 5-LOX is found primarily in inflammatory cells, such as granulocytes, monocytes, and mast cells. Leukotriene receptors include B-LT_1, a high-affinity receptor present on leukocytes, and B-LT_2, a moderate-affinity receptor that has widespread tissue distribution. A cysteinyl leukotriene receptor, CysLT_1, is found on smooth muscle cells of the bronchioles and on vascular endothelial cells.

Products of P450 Monooxygenases

Cytochrome (CYP) P450s are a large family of enzymes present in virtually all mammalian tissues. These enzymes have a variety of roles and have been most extensively studied for their ability to metabolize various exogenous substances such as xenobiotics, as well as a vast variety of drugs. A number of CYP P450s employ arachidonic acid and other fatty acids as substrates, resulting in the generation of eicosanoids. The only fatty acid—utilizing CYP to be extensively studied is the CYP4A subfamily. The CYP4A epoxygenase metabolizes arachidonic acid and linoleic acid to a set of compounds known as epoxyeicosatrienoic acids (EETs). The EETs are further metabolized by CYP4A to 19- and 20-hydroxylepoxyeicosatrienoic acids (HEETs). These mediators are vasodilators and modulators of intracellular Ca^{2+}, Na^+, and K^+ transport.

Keeping the Balance: Implications for Health and Disease

Vascular Effects

Regulation of Blood Flow

Because of their ability to modulate the balance between vasoconstriction and vasodilatation, eicosanoids provide a highly responsive mechanism for regulating organ and tissue blood flow. An excellent example of this is the effect of PGE_2 on the ductus arteriosus. Fetuses have high circulating levels of PGE_2, and in the 1970s, experiments on fetal lambs showed that the vasodilatatory effects of PGE_2 are responsible for the maintenance of ductus arteriosus patency in utero. After birth, PGE_2 levels decrease dramatically, a response associated with closure of the ductus arteriosus and establishment of postnatal patterns of pulmonary artery blood flow. Because they inhibit PGE_2 production, NSAIDs such as indomethacin are used to induce ductus closure in low-birth-weight infants who have persistent patency of the ductus arteriosus. Conversely, the synthetic agent, PGE_1, is administered to infants when maintenance of a patent ductus arteriosus is beneficial. This situation occurs in newborns with cardiopulmonary anomalies whose systemic or pulmonary blood flow depends upon shunting between the aorta and the pulmonary artery.

Angiogenesis

Angiogenesis is a process whereby new blood vessels are created in response to inducible stimuli. This feature of the microvasculature occurs in a variety of settings, both physiologic and pathologic, including chronic inflammation, embryogenesis, parturition, and tumorigenesis. Products of cyclooxygenase activity, including TxA_2, PGE_2, and PGI_2, directly stimulate endothelial cell migration and angiogenesis in vivo and may result in increased endothelial cell survival. In addition, the product of 12-LOX, 12-*S*-HETE, possesses activities contributing to angiogenesis, as it modulates both endothelial cell adhesion and motility. In in vivo studies, selective Cox-2 inhibitors effectively suppressed formation of new blood vessels in response to basic fibroblast growth factor. In in vitro model systems employing coculture of endothelial cells with epithelial tumors, cyclooxygenase inhibition reduced production of prostaglandins and proangiogenic factors and inhibited both endothelial cell migration and in vitro angiogenesis. Related data also suggested that NSAIDs have antiangiogenic properties that are independent of cyclooxygenase inhibition. As a result, NSAIDs are currently under study as both chemopreventive and cancer therapeutic agents.

Renal Function

In the kidney, eicosanoids are important regulators of blood flow and glomerular filtration rate. Consistent with this, the predominant eicosanoids produced in the kidney are PGE_2, PGI_2, and TxA_2. As a component of the body's response to stress, the synthesis of eicosanoids by renal parenchyma and endothelial cells is increased in response to vasoconstrictive stimuli such as angiotensin, vasopressin, or catecholamines. PGE_2 causes vasodilatation of the renal vasculature, and production of PGE_2 in the kidney is an important compensatory response in patients with shock, congestive heart failure, or ureteral obstruction. Administration of NSAIDs to patients with these conditions reduces prostaglandin production and is frequently associated with impairment of renal function. This failure is not associated with structural damage to the renal parenchyma and is reversible when the drugs are discontinued. In normal individuals, NSAIDs only rarely cause changes in renal blood flow or glomerular filtration rate.

Hemostasis and Coagulation

When the endothelium is injured, the resulting exposure of collagen and thrombin lead to platelet activation and adhesion to the site of injury. Following adhesion, a number of active substances are released by platelets, including TxA_2. TxA_2 contributes to hemostasis by causing local vasoconstriction, enhancing platelet aggregation, and mediating further release of TxA_2 from platelets. Although beneficial following trauma, the vasoconstrictive and platelet-aggregating effects of TxA_2 are highly detrimental in the setting of atheromatous narrowing of coronary and cerebral arteries. NSAIDs, because of their ability to inhibit Cox-1, decrease TxA_2 synthesis and reduce platelet function. As a result, NSAIDs, particularly aspirin, are beneficial preventive agents for patients at high risk of coronary artery and cerebral vascular disease. An overview of randomized trials of aspirin for the prevention of occlusive vascular disease concluded that 81 to 325 mg of aspirin daily provided protection against myocardial infarction, stroke, and death due to cardiovascular disease. This benefit was achieved at a small risk of increased hemorrhage and gastrointestinal tract ulceration due to long-term aspirin use.

Inflammation

The tissue response to inflammation is characterized by vasodilatation, increased vascular permeability, and early neutrophil accumulation. This response is largely produced by the local activity of eicosanoids that are produced by both damaged tissues and inflammatory cells. Cellular release of PGI_2 and PGE_2 causes vasodilatation, and TxA_2, leukotrienes, and histamine all increase vascular permeability. PGE_2, together with histamine and bradykinin, produces pain at the site of inflammation. Neutrophil chemoattraction and activation are caused by TxA_2 and LTB_4, as well as by complement activation. LTB_4 stimulates the synthesis and release of inflammatory cytokines, such as tumor necrosis factor and IL-1, thereby potentiating the inflammatory response.

By virtue of their ability to inhibit prostaglandin, thromboxane, and leukotriene synthesis, NSAIDs markedly attenuate the inflammatory process. As a result, naturally occurring salicylates have been used for centuries to treat pain and fever. The term "nonsteroidal anti-inflammatory drug" was coined by rheumatologists in 1949 to distinguish the activity of phenylbutazone from that of glucocorticoids, whose anti-inflammatory properties in the treatment of arthritis had recently been identified. This term came to apply to all "aspirin-like drugs" that were used clinically as antipyretics, analgesics, and anti-inflammatory agents. Recently, Cox-2 has been identified as the inducible isoenzyme responsible for inflammation. Because the beneficial effects of NSAIDs on the gastric mucosa and kidney are mediated by Cox-1, selective Cox-2 inhibitors were developed to minimize the side effects of NSAIDs while preserving their anti-inflammatory efficacy.

Protection of the Gastroduodenal Mucosa

Prostaglandins produced by constitutive activity of Cox-1 in the upper gastrointestinal tract exert important protective effects in gastroduodenal tissue. In the harsh chemical environment of the stomach and duodenum, prostaglandins are responsible for protection of the mucosa through promotion of mucus production, bicarbonate secretion, and mucosal blood flow. PGE_2 also inhibits both basal and stimulated gastric acid release. This effect may be particularly important in individuals with duodenal ulcer disease.

The use of nonselective NSAIDs (i.e., NSAIDs able to inhibit both Cox-1 and Cox-2) can produce damage to the mucosa of the stomach and duodenum and increase the complication rate of preexisting peptic ulcers. Some degree of gastrointestinal upset is present in approximately 30 percent of patients using nonselective NSAIDs on a regular basis. In addition, endoscopic surveillance of patients using NSAIDs regularly demonstrates a 20 percent prevalence of gastric ulceration, often not associated with dyspepsia. Patients with a prior history of gastroduodenal ulcers are at particular risk for serious complications, including upper gastrointestinal hemorrhage and perforation. Because of their specificity for the inducible isoenzyme, selective Cox-2 inhibitors have a significantly reduced incidence of both minor and severe gastrointestinal side effects.

Regulation of Reproductive Function

The eicosanoids, particularly prostaglandin family members, are regulators of many aspects of the reproductive process. PGE_2 stimulates LHRH secretion and may also directly stimulate ovarian follicle maturation. PGE_2 also promotes both ejaculation and implantation of the embryo in the uterine wall. PGE_2 and $PGF_{2\alpha}$ from seminal fluid promote fertility by enhancing transport of sperm into the fallopian tube. Eicosanoids also regulate gestational length and parturition. Levels of PGE_2, $PGF_{2\alpha}$, and LTB4 are elevated in the maternal circulation prior to the onset of spontaneous labor, and exogenous administration of PGE_2 or $PGF_{2\alpha}$ induces softening of the cervix and uterine contractions in both full-term and preterm labor. Although their use in pregnancy is somewhat controversial, both PGE_2 and the synthetic prostaglandin misoprostol (PGE_1) have been used successfully for induction of labor.

Role in Ischemia—Reperfusion Injury

Eicosanoids are important mediators of the harmful consequences of tissue ischemia and reperfusion. Although decreased tissue perfusion causes compensatory increases in PGI_2 levels, ischemia also stimulates thromboxane synthesis and release. Upon reestablishment of blood flow to an ischemic organ, the ratio of TxA_2 to PGI_2 is increased, producing a net vasoconstrictive effect. Together with lipoxygenase metabolites produced during ischemia, TxA_2 activates neutrophils, which become sequestered in the ischemic organ and the lung. The products of neutrophil activation include locally destructive proteases and reactive oxygen species, as well as inflammatory cytokines that contribute to increased capillary permeability and edema. As a result, tissue injury and decreased capacity for oxygenation occur both in the ischemic organ and at the diffusional surfaces of the lung. Leukotrienes produced as a result of myocardial ischemia can be particularly damaging upon restoration of coronary blood flow, as these agents may have negative inotropic and arrhythmogenic effects. There is no one pharmacologic agent able to counteract the harmful effects of ischemia—reperfusion injury. Once tissue perfusion has been reestablished, pharmacological therapy focuses upon limiting leukocyte activation and the resulting tissue damage. For example, vasodilatation can be promoted by nitrates, calcium channel blockers will limit neutrophil superoxide formation and release, and angiotensin-converting enzyme inhibitors can prevent leukocyte adhesion.

Eicosanoids and Tumorigenesis

In 1968, Williams recognized that tumors contained increased levels of prostaglandins compared to adjacent normal tissue. Since that time, data from a wide array of studies suggest that prostaglandins stimulate tumorigenesis. By-products of eicosanoid production include a number of potentially genotoxic substances, including organic free radicals, peroxides, and activated oxygen species. These substances are suspected to play a role in every stage of carcinogenesis, including activation of environmental carcinogens, direct DNA damage, stimulation of proliferation, inhibition of apoptosis, suppression of antitumor immunity, and stimulation of metastasis.

The cellular processes responsible for eicosanoid-mediated tumorigenesis are incompletely understood. There are numerous clinical associations and experimental links between inflammation and epithelial cancers. Inflammatory bowel disease, burn injuries, chronic ulcers, and long-standing cirrhosis examples of conditions that carry a cancer risk proportional to their duration in an individual. Initiation of the inflammatory response activates intracellular signaling cascades that govern cell proliferation and motility. When this condition becomes chronic, it provides a setting for selection of cells with other defects in growth control, eventually producing a clone of cells with a malignant phenotype. Recently, it was recognized that abnormal cell proliferation in a terminally differentiated epithelial cell population leads to progressive telomere shortening, resulting eventually in anaphase bridging, chromosomal instability, "telomere crisis," and the emergence of cells with unlimited proliferative potential due to reactivation of telomerase.

An interesting new observation in the field of eicosanoid biology comes from study of the peroxisome proliferated-activated receptor (PPAR) transcription factors. These receptors were initially cloned as a family of orphan receptors, but are now known to interact with a wide variety of ligands, including hypolipidemic drugs and the eicosanoids 8-*S*-HETE, LTB_4, and prostaglandins of the J series. In this capacity, certain eicosanoids resemble steroid and thyroid hormones. Cell culture data also suggests that PPARs may be a target of NSAID activity, although in vivo data confirming this have yet to be reported.

The antitumor effects of NSAIDs have been examined in both animal models and human clinical trials. Many antitumor effects have been ascribed to NSAID-mediated inhibition of cyclooxygenase activity. In particular, upregulation of Cox-2 may be a key component of epithelial tumorigenesis, and its suppression the main factor associated with the antitumor activity of NSAIDs. Tissue-selective overexpression of Cox-2 by promoter-specific targeting of murine epithelial cells induced tumorigenesis. In an animal model of FAP, intestinal tumor formation was dramatically decreased by either genetic deletion of Cox-2 or its inhibition by a Cox-2 specific NSAID. Recent studies in human tumor xenografts that constitutively expressed both Cox-1 and Cox-2 showed that selective inhibition of Cox-2 decreased intratumoral PGE_2 and reduced tumor growth. This result was also achieved by specifically inhibiting PGE_2 with a neutralizing antibody, but not by selective inhibition of Cox-1 with a new NSAID, SC-560.

In addition to enhanced expression in tumors, Cox-2 and PGE_2 are also increased in fibroblasts and endothelial cells associated with intestinal tumors. Disruption of the PGE_2 receptor, EP_2, in *Apc*-mutant mice produces tumor suppression, an effect primarily due to a positive feedback mechanism for Cox-2 expression by PGE_2 in adenoma stromal cells. Cox-2 is highly expressed in tumor-associated endothelial cells, and PGE_2 supports angiogenesis in human tumors. These observations led to the hypothesis that Cox-2 upregulation supports tumor angiogenesis, and that NSAIDs are antiangiogenic because of their ability to suppress cyclooxygenase activity. This concept is supported by data showing that selective Cox-2 inhibitors suppress angiogenesis in the FGF-rat corneal micropocket assay.

Lipoxygenase metabolites may also play a role in tumor formation. Because they promote cell proliferation and angiogenesis and suppress tumor cell apoptosis, the 5-, 8-, and 12-LOX isoforms are characterized as "tumorigenic," whereas the opposite effects of 15-LOX suggest that this enzyme may inhibit tumor formation. In support of this characterization, human epithelial tumors exhibit decreased levels of 15-LOX compared to normal tissues, and in vitro treatment of colorectal cancer cells with NSAIDs increased levels of 15-LOX, augmented tumor apoptosis, and decreased cell growth. This effect appeared to be a direct effect of NSAIDs on 15-LOX expression rather than a shift of substrate from cyclooxygenase to lipoxygenase metabolic pathways.

Activity and Specificity of Inhibitors of Arachidonic Acid Metabolism

NSAIDs

Because cyclooxygenases are the main target of NSAID therapy, the role of cyclooxygenase-derived lipid mediators has been widely studied. Inhibition of cyclooxygenase leads to a decrease in the production of all prostaglandins and thromboxanes, and this accounts for the observed effects of NSAIDs as anti-inflammatory, antipyretic, analgesic, and antithrombotic agents. It also explains their gastrointestinal and renal side effects. Enormous effort has been expended to develop NSAIDs whose specificity of action will enhance the benefits of eicosanoid inhibition yet minimize the harmful effects on gastric mucosa and renal vasculature. The discovery of Cox-2 and its role in inflammation but not gastric protection led not only to the development of specific inhibitors of Cox-2, but also to studies examining the differential effects of existing NSAIDs upon the cyclooxygenase isoforms.

Aspirin is currently the only NSAID that covalently modifies cyclooxygenase. Aspirin has greater inhibitory activity against Cox-1 than against Cox-2, and this explains its antiplatelet and cardiovascular effects, as well as its tendency to produce ulceration of gastric mucosa. Cyclooxygenase blockade by the other known NSAIDs occurs as a result of reversible binding of the drug to the cyclooxygenase molecule. The kinetics of NSAID—cyclooxygenase interactions are quite complex, with both competitive and time-dependent elements. This, together with the complexity of prostaglandin biology in vivo, makes it difficult to compare the Cox-1/Cox-2 selectivity of different NSAIDs. Depending upon dosage, cell type, and assay conditions, every NSAID exhibits some degree of inhibition of both Cox-1 and Cox-2. In general, however, most NSAIDs, such as aspirin, indomethacin, and piroxicam, are relatively nonspecific. A few, such as meloxicam, have some degree of increased specificity for Cox-2. A new class of NSAIDs, the selective Cox-2 inhibitors, include NS398, celecoxib, and rofecoxib. These agents are strong inhibitors of Cox-2 with minimal effect on Cox-1.

Leukotriene Modifiers

Leukotrienes, through their ability to modulate leukocyte—endothelial cell interactions, are thought to mediate NSAID-associated gastric mucosal damage. Leukotrienes are also potent vasoconstrictors and inducers of bronchospasm in susceptible individuals. The enzyme responsible for leukotriene synthesis, 5-LOX, is expressed only in a limited repertoire of cells, mostly leukocytes. Leukotriene receptors, however, are widely distributed among smooth muscle cells of the vasculature and respiratory tract. Leukotriene modifiers, such as zileuton and montelukast, are 5-LOX inhibitors used clinically for asthma therapy. For unknown reasons, these agents are particularly useful for exercise-induced and aspirin-intolerant asthma.

Combination Agents

A promising therapeutic approach to minimize the gastric side effects of aspirin while providing antithrombotic therapy is to concurrently suppress the activities of both cyclooxygenase and 5-LOX enzymes. Based upon the activity profiles of cyclooxygenase and 5-LOX products, these agents would be clinically useful in a wide variety of diseases, including inflammatory states, cancer prevention, and cardiovascular disorders. Several of these dual inhibitors of prostaglandin and leukotriene synthesis have been developed. A few of these, including the agent licofelone, are under evaluation in Phase III clinical trials for the treatment of osteoarthritis.

Conclusion

Studies of eicosanoid biology have provided great insight into normal physiology and the pathogenesis of disease. The rapidly responsive, tissue-localized nature of eicosanoid activities make them ideal targets for therapeutic intervention, and it is therefore easy to see why modulators of eicosanoid synthesis, such as aspirin, are among the oldest known therapeutics. Because eicosanoids play central roles in a wide range of disease states, inhibitors of eicosanoid synthesis can achieve a broad spectrum of activity. This is clearly demonstrated by the use of NSAIDs for conditions as diverse as pain relief, prevention of cardiovascular disease, and inhibition of tumor formation. In the future, the development of specific agonists and antagonists for eicosanoid receptors will yield further insight into the relevance of various pathways to disease states and provide new, more specific avenues for therapy.

Bibliography

Baker, R. R. (1990). The eicosanoids: A historical review. *Clin. Biochem.* **23**, 455–458. *A concise review of early studies of eicosanoid biology.*

Bishop-Bailey, D., and Wray, J. (2003). Peroxisome proliferator-activated receptors: A critical review on endogenous pathways for ligand generation. *Prostaglandins Other Lipid Mediat.* **71**, 1–22. *A detailed analysis of multiple signaling networks containing eicosanoids.*

Funk, C. D. (2001). Prostaglandins, and leukotrienes: Advances in eicosanoid biology. *Science* **294**, 1871–1875. *Basic current review of eicosanoid biology and potential for targeted therapies.*

Other Useful References

Kudo, I., and Murakami, M. (2003). Regulatory functions of prostaglandin E2 synthases. *Adv. Exp. Med. Biol.* **525**, 103–106.

Nie, D., and Honn, K. V. (2002). Cyclooxygenase, lipoxygenase and tumor angiogenesis. *Cell. Mol. Life Sci.* **59**, 799–807.

Thun, M. J., Henley, S. J., and Patrono, C. (2002). Nonsteroidal anti-inflammatory drugs as anticancer agents: Mechanistic, pharmacologic, and clinical issues. *J. Natl. Cancer Inst.* 94, 4–20.

Capsule Biography

Monica M. Bertagnolli is an Associate Professor of Surgery at Harvard Medical School and a member of the Division of Surgical Oncology at Brigham and Women's Hospital and the Dana Farber Cancer Institute. Her primary research interest is the impact of nonsteroidal anti-inflammatory drugs on early tumorigenesis.

CHAPTER 108

Sepsis and the Microvasculature

Alpha A. Fowler III
Division of Pulmonary Disease and Critical Care Medicine, Department of Internal Medicine, Virginia Commonwealth University

Sepsis Epidemiology

New technology and specialized medical practices have evolved over the past five decades that permit support of acute and chronic organ failure. The current era in medicine is remarkable for rapid progress in diverse fields such as cancer therapy and transplantation of bone marrow and solid organs. Consequently, patients suffering from diseases that were formerly fatal now more often than not move to "postacute" or chronic phases of illness, necessitating frequent or extended hospitalizations. A striking increase in the incidence of sepsis has accompanied medical advances. Recent studies suggest that more than 750,000 new cases of sepsis occur in the United States annually. Mortality rates attributable to sepsis range from 25 percent to 30 percent with higher mortality linked to increasing age. Preexisting or comorbid medical conditions as well as greater numbers of organs systems failed are important factors that determine outcomes. Thus, expectations are that greater than 200,000 deaths will occur annually from sepsis with annual total costs to the U.S. economy alone exceeding 16 billion dollars. Given a host of biological factors combined with aging populations and increased need for care of chronic illness, conservative projections call for a 1.5 percent increase per annum in the incidence of sepsis [1]. Worldwide incidence figures may vary, but sepsis exacts a huge toll in lost human life and productivity. Microvascular endothelial cells (ECs) are integrally involved in regulating blood flow, coagulation, leukocyte trafficking, edema formation, and angiogenesis. Insights into the pathogenesis of sepsis are gained by examining important concepts established through careful study of microcirculatory biology.

Nitric Oxide Induces Microcirculatory Dysfunction

Inducible Nitric Oxide Synthase Mediates Microvascular Dysfunction

A consensus has slowly emerged that organ failure and mortality from sepsis arise from injury and disordered circulatory homeostasis and hyperdynamic states. Hypoxemia and hypotension unresponsive to pharmacologic intervention are commonly present during sepsis. Despite the presence of enhanced oxygen delivery associated with hyperdynamic states, defects in oxygen extraction and tissue oxygen utilization produce lactic acidosis, strongly suggesting that a microcirculatory dysfunction is present. Cumulative research indicates that all anatomic compartments of the microcirculation are dysregulated. Cryer et al. demonstrated loss of vascular tone with significant dilatation of third- and fourth-order skeletal muscle arterioles (20 to 50 μM) following onset of hyperdynamic *Escherichia coli* sepsis [2]. Subsequently, workers demonstrated that resistance arterioles are hyporeactive to the vasoconstrictive effects of norepinephrine in organ-specific resistance microvasculature (e.g., liver, lung) in sepsis. Significant research indicates that the reactive nitrogen intermediate nitric oxide (NO) is a key factor producing disordered vasoregulation in sepsis. Under physiologic conditions, NO is continuously produced at low levels by endothelium and vascular smooth muscle cells through transcription of the constitutive NO synthase gene (NOSIII). However, abrupt increases of inducible NO synthase (iNOS or NOSII) expression by endothelium, vascular smooth muscle cells, and monocyte/macrophages occur following onset of sepsis, producing remarkable surges of detectable NO in the circulation.

Postarteriolar microcirculatory constituents undergo significantly reduced capillary perfusion despite the hyperdynamic conditions that exist in arterioles. Studies employing intravital microscopy in animal models of sepsis not only reveal a "stopped-flow" phenomenon in capillaries but also show reduced capillary density with increased velocity of erythrocytes perfusing capillary beds that remain patent. These phenomena result in disruption of cellular oxygen delivery, thus producing the metabolic acidosis characteristically observed in sepsis.

Lipopolysaccharide Entry into the Circulation and Cytokine Transcription Promote Microvascular Dysfunction in Sepsis

Significance of Lipopolysaccharide Binding

Lipopolysaccharide (LPS) glycolipid is a key outer cell wall component of Gram-negative bacteria released from the surface of replicating and dying bacteria into the circulation where it interacts with ECs on luminal microvascular surfaces. LPS is a fundamental factor inducing microvascular dysfunction and shock in sepsis. LPS bioactivity is detected in the circulation of septicemic patients. Studies show that levels of circulating LPS predict development of multiple organ failure, including acute respiratory distress syndrome (ARDS) [3]. A substantial body of human and animal experimentation reveals that entry of LPS into the circulation produces prompt transcription of proinflammatory cytokines such as tumor necrosis factor-α (TNFα), interleukin-1 (IL-1), and interleukin-6 (IL-6). Abrupt cytokine surges in turn promote iNOS transcription by endothelium and vascular smooth muscle cells, leading to increased nitric oxide generation and loss of microvascular responsiveness to vasoconstrictors. Further, cytokine surges produced by LPS lead to acute changes in endothelial cell physiology, resulting in disruption of essential endothelial functions such as adhesion, thrombogenesis, and permeability (see later discussion). LPS binds to both circulating monocytes and tissue-bound macrophages, mediating signaling through a complex array of cell-bound receptors. These data suggest the presence of an LPS receptor-signaling complex. The β_2 integrin CD11/CD18 and glycosylphosphatidylinositol-anchored membrane protein (CD14) bind LPS at the cell surfaces of monocytes and macrophages. This binding event facilitates interaction with toll-like receptor 4 (TLR4) and a nonmembrane-spanning molecule designated MD-2. Emerging work suggests that LPS binding to CD11/CD18 and CD14 activates TLR4, initiating signaling events that lead to transcription factor activation and cytokine secretion. Studies using TLR4 knockout mice show the appearance of explicit LPS resistance and suggest that TLR4 mediates specificity to certain bacterial lipoproteins and other components of Gram-positive bacterial cell walls [4].

Importance of Tumor Necrosis Factor-α

Current research indicates that cytokines secreted by LPS-activated mononuclear cells (e.g., TNFα, IL-1) are primary mediators of shock and microvascular dysfunction in sepsis. This is supported by multiple studies showing that (a) intravenous infusion of TNFα or IL-1 into the circulation promotes systemic responses identical to that observed following LPS infusion, (b) infusion of neutralizing antibodies to TNFα or chimeric inhibitors of the 55-kDa TNF receptor (TNF-R1) fused to an immunoglobulin heavy chain fragment attenuates LPS-induced shock in animal models, and (c) deletion of the TNF-R1 gene in mice produces increased survival following LPS infusion [5]. Both in vivo and in vitro studies suggest that TNFα is a central proinflammatory cytokine mediating activation of microvascular endothelium following onset of endotoxemia in sepsis. Binding of TNFα to its transmembrane receptor ligand, TNFR-1, induces signaling events that initiate activation of nuclear factor κB (NFκB). NFκB exists in endothelial cell cytoplasm complexed to a family of inhibitor κB proteins (IκBα-ε). Phosphorylation of IκB by the upstream kinase NFκB inducing kinase (NIK) targets IκBα for ubiquitination and proteasome degradation, thus freeing NFκB for nuclear translocation and binding to cognate sequences of key endothelial cell genes. TNFα induces coordinated activation of other important transcription factor families including activator protein-1 (AP-1), a member of the c-Jun and c-Fos family of proteins, and CREB, the cAMP-responsive element binding protein. Cooperative interactions between transcription factors such as NFκB, AP-1, and CREB lead to transcriptional activation of essential endothelial genes that induce a substantial number of pathophysiological events observed in sepsis.

New information also suggests that TNFα not only produces explosive activation of NFκB-related genes but also mediates their sustained activation. Important work by Johnson and colleagues revealed that following TNFα exposure, human endothelial cells exhibit a 50 percent reduction in IκBβ, the IκB isoform essential for retarding or downregulating subsequent NFκB translocation [6]. Johnson's work further showed that TNFα induces extended periods (i.e., 22 hours) to onset of resynthesis of IκBβ, permitting protracted NFκB translocation. Thus, TNFα not only induces immediate NFκB activation in endothelial cells but also prolongs activation of proinflammatory genes mitigating and amplifying the septic process.

Apoptosis of Endothelium in Sepsis

Endothelial Cell Apoptosis Is Widespread in Humans with Sepsis

Apoptosis is an ATP-dependent form of cell death characterized by chromatin condensation, DNA fragmentation, cell shrinkage, and "blebbing" of plasma membranes.

Significant evidence from human and experimental animal studies suggests that endothelial cell apoptosis and injury are fundamental features of the microvascular response to sepsis. Mutunga et al. showed that human subjects suffering from sepsis and septic shock exhibited increased numbers of detached endothelial cells found free floating in the systemic circulation [7]. Mutunga's study further showed that increased numbers of circulating endothelial cells correlated negatively with survival. Histological evidence of widespread microvascular EC injury is consistently observed in postmortem lung biopsies obtained from patients dying of sepsis-related acute respiratory distress syndrome. In vivo studies in mice, rats, rabbits, and baboons reveal that infusion of LPS or live Gram-negative organisms invariably leads to increased numbers of apoptotic ECs as well as pervasive EC detachment from capillary beds in lung, liver, and small bowel.

LPS and TNFα Promote Apoptosis of Microvascular Endothelium

Caspase activation is a hallmark of apoptosis. LPS and TNFα released into the circulation during sepsis activate caspases leading to key proteolytic events that produce apoptosis. Of the 14 known caspases expressed in mammalian cells, LPS and TNFα are reported to activate caspases 1, 3, 6, and 8. Controversy exists concerning the precise mechanism leading to the disseminated apoptosis observed in microvascular ECs during sepsis. In vitro, LPS induces proapoptotic signaling in ECs with resultant cell injury and death *in the absence of host-derived mediators such as TNFα.* A synthesis of prior publications shows that LPS binding directly activates signaling pathways that lead to both NFκB activation and apoptosis and that these pathways share certain signaling molecules including MyD88, IRAK-1, and TRAF-6 (Figure 1). Eloquent studies incorporating antibody and knockout strategies have produced an emerging discernment of the pathways that result in apoptosis following LPS binding to ECs in microvasculature.

Equally compelling is research that suggests LPS mediates microvascular EC apoptosis via biological mediators such as TNFα. Figure 2 shows that binding of TNFα to its 55-kDa receptor triggers formation of a multiprotein complex in which cytoplasmic proteins and the receptor interact through respective "death domain" motifs that lead to activation of a cascade of caspases. Recent studies also link induction of apoptosis by TNFα to activation of the sphingomyelin pathway. A key feature of this pathway is the generation of ceramide from the hydrolysis of sphingomyelin. Ceramide content in tissue beds such as gut and lung rises dramatically following LPS infusion. Ceramide generation is consistently associated with diffuse microvascular EC apoptosis. Once generated, ceramide signals mitochondrial membrane permeability transition, a committed step in the apoptotic process. Haimovitz-Friedman et al. employed a unique binding protein inhibitor of TNFα by constructing a

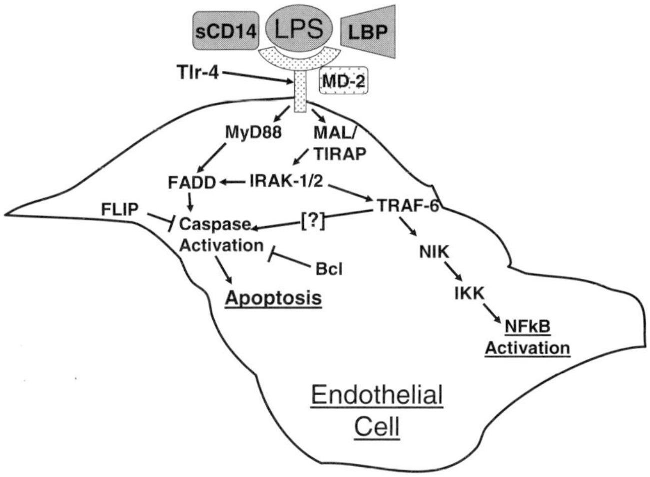

Figure 1 Schematic diagram of LPS proapoptotic signaling in endothelial cells. After LPS binding to TLR-4, EC signaling pathways leading to nuclear translocation of NFκB and apoptosis are activated. These pathways share some of the same signaling molecules, including MyD88, IRAK-1, and TRAF-6. A redundant signaling pathway involving MAL/TIRAP similarly mediates both NFκB activation and apoptosis. Downstream of TRAF-6, the signaling pathways that activate these two processes diverge. The molecules that link the upstream NFκB signaling molecules to the recruitment and activation of caspases remains unknown. LBP, LPS-binding protein; NIK, NFκB-inducing kinase; IKK, IκB kinase; sCD14, soluble CD14; FADD, Fas-associated death domain; FLIP, FADD-like interleukin converting enzyme–like inhibitor protein; TRAF-6, TNF receptor-associated factor-6; IRAK-1/2, IL-1 receptor-associated kinase-1; MyD88, myeloid differentiation factor 88. (see color insert)

chimera comprising the extracellular domain of the TNF receptor fused to an immunoglobulin heavy chain fragment [8]. The TNF-binding protein construct in Haimovitz-Friedman's study bound biologically active TNFα and protected C_{57}BL/6 against LPS-induced death. The TNF-binding protein strategy blocked LPS-induced ceramide generation and resulted in striking reductions in the intensity of EC apoptosis, leading to the schema proposed in Figure 2. These and other studies suggest that systemic TNFα is a crucial mediator that is fully capable of directly producing disseminated apoptosis and injury of microvascular ECs in systemic organ beds.

Adhesion Molecules Orchestrate Neutrophil Sequestration

Intense Cellular Sequestration Leads to Diffuse Injury in Septic Microvasculature

Pathological specimens from human and animal subjects with sepsis uniformly reveal capillary injury associated with intense sequestration of all formed blood elements. Real-time inspection of septic microvasculature in animal models via intravital microscopy has advanced understanding of the genesis of interactions that occur between blood constituents and microvascular ECs. Erythrocyte deformability

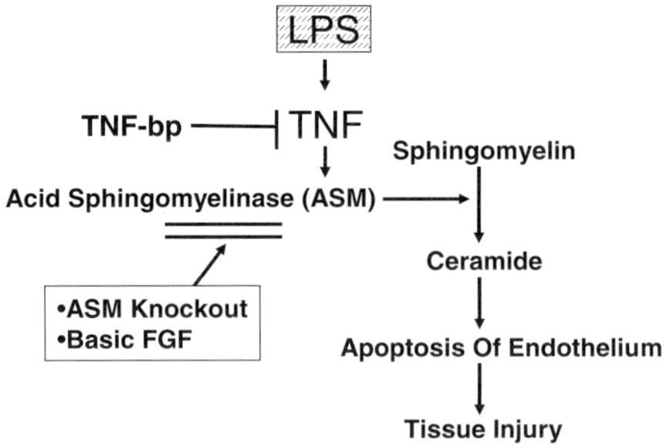

Figure 2 Entry of LPS into the circulation promotes TNFα release from monocytes and macrophages. TNFα promotes hydrolysis of sphingomyelin producing ceramide through the action of acid sphingomyelinase. Endothelial cell apoptosis results. Basic fibroblast growth factor and the use of acid sphingomyelinase knockout mice attenuate the generation of ceramide and the production of apoptosis. (see color insert)

is rapidly reduced (to less than 30 minutes) following onset of experimental sepsis (LPS infusion) giving rise to erythrocyte rouleaux formation and the appearance of spherocytes. Blood neutrophils (PMN) promptly leave axial blood flow during this period to begin rolling adhesion along venular microvascular surfaces. PMN rolling-type adhesion is followed rapidly by firm adherence that halts PMN progress through the microvasculature. Platelet activation and aggregation occurs in virtually simultaneous time frames, exacerbating obstruction and further trapping erythrocytes. Endothelial cell engorgement and the appearance of microvascular thrombi promote further obstruction to blood flow through venules and capillaries. Remarkable migration of activated, sequestered PMN occurs across microvascular walls in sepsis. Extravasation of erythrocytes and the movement of plasma proteins follow PMN migration. Ultrastructural studies reveal that these processes lead to EC vacuolization, loss of EC lining, and ultimately exposure of basal laminar structures to the circulation [9].

Endothelial and Neutrophil Adhesion Receptors Mediate Sepsis-Associated Vascular Injury

Cell adhesion molecules mediate a coordinated sequestration of PMN in the microvasculature. Endothelial glycoprotein receptors E- and P-selectins are rapidly displayed following entry of LPS and TNFα into the circulation. Rolling adhesion is mediated via selectin-to-selectin binding between endothelial-anchored selectins and the PMN affixed L-selectin receptor. A microvascular "environment for injury" is subsequently created by molecular mechanisms that tether activated PMN tightly to venular and capillary EC surfaces. Newly expressed intercellular adhesion molecule-1 (ICAM-1), a member of the immunoglobulin superfamily driven by NFκB activation, is a key participant mediating tight binding of PMN to microvascular EC surfaces. The PMN counter receptor for ICAM-1 is the β_2 integrin, CD11/CD18. ICAM-1/CD11/CD18 binding induces flattening of PMN on EC surfaces producing an interface or microenvironment where proteolytic enzymes and reactive oxygen species produced by PMN serve as effectors inducing EC injury. Studies performed in diverse animal model systems utilizing antibody and knockout strategies reveal that disruption of both rolling and tight EC/PMN adhesion attenuates microvascular injury.

Sepsis Induces Microvascular Thrombosis and Produces Poor Clinical Outcomes

Importance of Disseminated Intravascular Coagulation

Clinicians and researchers have long recognized progressive sepsis to be associated with microvascular thrombosis and the appearance of multiple organ failure [10]. LPS, other bacterial products, and proinflammatory cytokines (TNFα IL-1) generate a procoagulant environment in microvessels. Significant research across many species shows that tissue factor (TF) expressed by activated/injured microvascular ECs and monocytes potently activates extrinsic coagulation pathways. Tissue factor binds activated factor VII forming a TF/factor VIIa complex that activates factor X. Factor X together with factor Va potently generates thrombin from its precursor molecule prothrombin. Thrombin generation markedly shifts the microvascular milieu to a coagulant environment poorly capable of fibrinolysis. Dramatic increases in plasma levels of plasminogen activator inhibitor-1 (PAI-1) and tissue factor pathway inhibitor (TFPI) suppress fibrinolysis and promote fibrin deposition. Plasma levels of major physiological anticoagulants such as antithrombin III and protein C become dramatically attenuated, further promoting a prothrombotic environment. Recent studies in humans with sepsis show that plasma PAI-1 levels herald prothrombotic states in sepsis and highly correlate with lethal outcomes [11]. Procoagulant environments and widespread microvascular thrombosis lead to widespread thrombotic occlusion of microvasculature, producing tissue hypoperfusion, organ dysfunction, and ultimately multiple organ failure in septic patients. Evolving research reveals that thrombin generation augments injury significantly, amplifying generation of reactive oxygen species within microvascular endothelium. Further new information suggests that thrombin significantly alters microvascular EC barrier function, inducing enhanced permeability in capillary and venular endothelium by triggering rearrangement of the endothelial cytoskeletal elements. Signaling events that lead to formation of intracellular "stress fibers" are mediated through heterotrimeric G-proteins and key protein tyrosine kinases. These molecular events produce dramatic changes in endothelial cell shape, subsequently permitting leakage of plasma proteins and water into extravascular spaces [12].

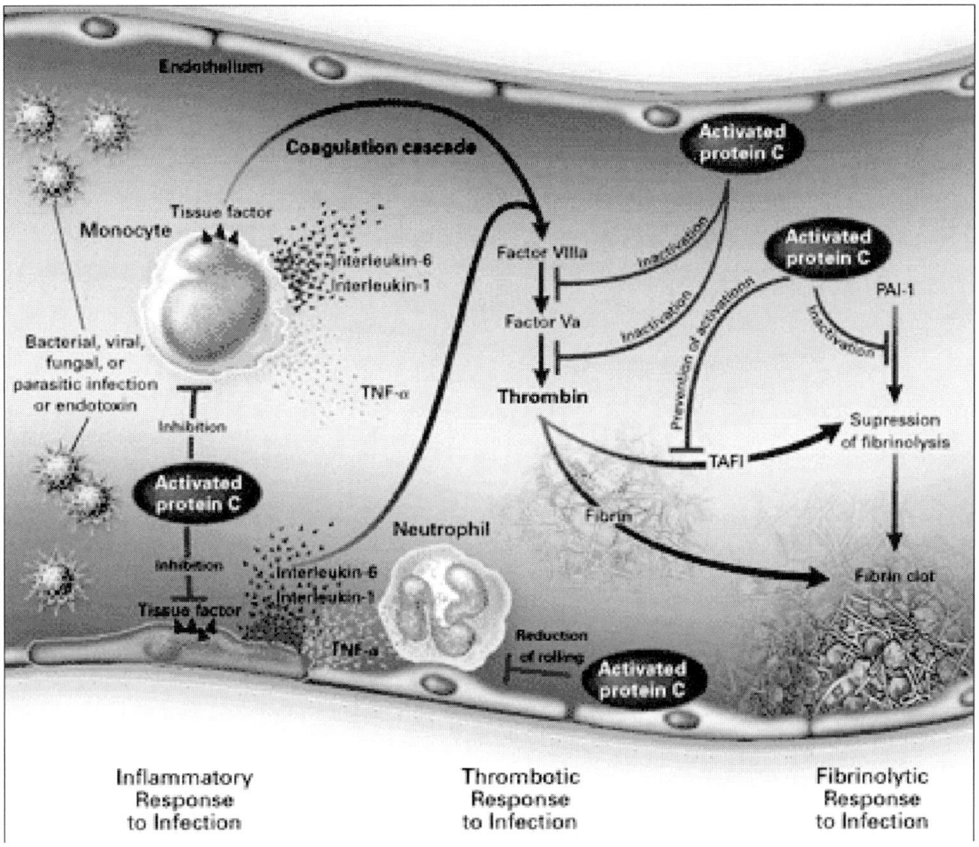

Figure 3 Proinflammatory cytokines such as TNFα and IL-1 shift the microvascular milieu to a procoagulant environment through generation of tissue factor from monocytes and the endothelium. Tissue factor generates thrombin and a fibrin deposition. Cytokines and thrombin inhibit fibrinolytic activity by promoting the release of plasminogen-activator inhibitor 1 (PAI-1). PAI-1 inhibits tissue plasminogen activator activity. Thrombin activity further suppresses fibrinolytic system by activating thrombin-activatable fibrinolysis inhibitor (TAFI). Protein C activation is secondarily dramatically reduced. Microvascular injury decreases thrombomodulin, an essential factor producing protein C activation. Widespread microvascular thrombosis results, leading to multiorgan failure and enhanced mortality. Infusion of activated protein C produces a potent antithrombotic effect by inhibiting formation of factors Va and VIIIa and the subsequent generation of thrombin. Thrombin levels in the circulation fall and anti-inflammatory effects occur by reduction of TNFα and IL-1 transcription in monocytes. Further, activated protein C infusion enhances fibrinolytic activity through its ability to inhibit PAI-1. (see color insert)

Importance of Activated Protein C in Septic Microvasculature and Potential for New Therapies

Under limited physiologic conditions activated protein C is an essential element limiting coagulation. Thrombin generation induces protein C activation by stimulating binding to both thrombomodulin and endothelial protein C receptors. Activated protein C formation is essential for swinging microvascular milieus to profibrinolytic environments. Following activation, protein C and protein S serve to minimize thrombin generation by inactivating factors Va and VIIIa. Activated protein C further promotes fibrinolysis by diminishing plasma concentrations of PAI-1 and by downregulation of monocyte proinflammatory cytokine (TNFα, IL-1) output [13]. Recent exciting results were reported from a phase III trial where activated protein C (Drotrecogin Alfa) was infused into adult patients with sepsis for 96 hours. Patients who received Drotrecogin Alfa (activated) had a significantly lower 28-day mortality rates compared to patients who received placebo (25% versus 31%, $p = 0.005$) [14].

Conclusions

Physicians have long struggled against sepsis, striving to interrupt pathological processes that produce enormous patient mortality and disability. Sepsis has claimed countless lives over the millennia. Disturbing trends that show increasing incidence bring urgency to the task of finding comprehensive therapies. Basic and clinical research performed during the past 50 years has produced significant new knowledge. Cumulative information arising from this research has provided clearer understanding of the fundamental issues and challenges that must be addressed and overcome to institute lifesaving therapies in the future.

References

1. Angus, D. C., Linde-Zwirble, W. T., Lidicker, J., Clermont, G., Carcillo, J., and Pinsky, M. R. (2001). Epidemiology of severe sepsis in the United States: Analysis of incidence, outcome, and associated costs of care. *Crit. Care Med.* **29**(7), 1303–1310. *Angus and colleagues identified 192,980 cases, yielding national estimates of approximately 751,000 cases of sepsis., of whom 51.1 percent received intensive care. Seventeen percent of patients were ventilated in an intermediate care unit or in a coronary care unit. Incidence of sepsis increased more than one-hundredfold with age. Overall mortality was 28.6 percent nationally showing a range of 10 percent in children to 38.4 percent in elderly. The average cost per case in Angus' study was $22,100 with annual total costs of $16.7 billion nationally.*
2. Cryer, H. M., Garrison, R. N., Kaebnick, H. W., Harris, P. D., and Flint, L. M. (1987). Skeletal microcirculatory responses to hyperdynamic *Escherichia coli* sepsis in unanesthetized rats. *Arch. Surg.* **122**, 86–92.
3. Brandtzaeg, P., Kierulf, P., Gaustad, P., Skulbert, A., Bruun, J. N., Halvorsen, S., and Sorensen, E. (1989). Plasma endotoxin as a predictor of multiple organ failure and death in systemic meningococcal disease. *J. Infect. Dis.* **159**, 195–204.
4. Poltorak, A., He, X., Smirnova, I., Liu, M., Huffel, C., Du,X., Birdwell, D., Alejos, E., Silva, M., Galanos, C., Freudenberg, M., Ricciardi-Castagnoli, P., Layton, B., and Beutler, B. (1998). Defective LPS signaling in C3H/HeJ and C57BL/10ScCr mice: Mutations in Tlr4 gene. *Science* **282**, 2085–2088.
5. Pfeffer, K., Matsuyama, T., Kundig, T. M., Wakeham, A., Kishihara, K., Shahinian, A., Wiegmann, K., Ohashi, P. S., Kronke, M., and Mak, T. W. (1993). Mice deficient for the 55 kd tumor necrosis factor receptor are resistant to endotoxic shock, yet succumb to *L. monocytogenes* infection. *Cell* **73**(3), 457–467.
6. Johnson, D. R., Douglas, I., Jahnke, A., Ghosh, S., and Pober, J. S. (1996). A sustained reduction in IκB-β may contribute to persistent NF-κB activation in human endothelial cells. *J. Biol. Chem.* **271**(27), 16317–16322.
7. Mutunga, M., Fulton, B., Bullock, R., Batchelor, A., Gascoigne, A., Gillespie, J. I., and Baudouin, S. V. (2001). Circulating endothelial cells in patients with septic shock. *Am. J. Respir. Crit. Care Med.* **163**(1), 195–200. *Mutunga and colleagues obtained blood samples from (1) healthy volunteers and (2) ICU "control" patients without sepsis, (3) patients with sepsis but without shock, and (4) patients with septic shock. Endothelial cells in blood were identified by immunofluorescence, using antibodies to von Willebrand factor and the vascular endothelial growth factor receptor KDR. EC counts per milliliter of blood were significantly higher ($p < 0.0001$) in patients with sepsis and septic shock when compared to healthy or ICU control patients. Thus, Mutanga showed that dramatic increases in circulating ECs can be identified during sepsis and septic shock, supporting the hypothesis that endothelial damage occurs in human sepsis.*
8. Haimovitz-Friedman, A., Cordon-Cardo, C., Bayoumy, S., Garzotto, M., McLoughlin, M., Gallily, R., Edwards, C., Schuchman, E., Fuks, Z., and Kolesnick, R. (1997). Lipopolysaccharide induces disseminated endothelial apoptosis requiring ceramide generation. *J. Exp. Med.* **186**, 1831–1841.
9. McCuskey, R. S., Urbaschek, R., and Urbaschek, B. (1996). The microcirculation during endotoxemia. *Cardiovasc. Res.* **32**, 752–763.
10. Fourrier, F., Chopin, C., Goudemand, J., Hendrycx, S., Caron, C., Rime, A., Marey, A., and Lestavel, P. (1992). Septic shock, multiple organ failure, and disseminated intravascular coagulation. Compared patterns of antithrombin III, protein C, and protein S deficiencies. *Chest* **101**(3), 816–823.
11. Mesters, R. M., Florke, N., Ostermann, H., and Kienast, J. (1996). Increase of plasminogen activator inhibitor levels predicts outcome of leukocytopenic patients with sepsis. *Thromb. Haemost.* **75**(6), 902–907. *Mesters and colleagues studied patients at high risk for development of sepsis and measured, at onset of fever, plasminogen activator inhibitor activity and PAI-1 antigen levels. They showed that levels increased significantly normal baseline and were significantly higher in patients developing septic shock compared to those with severe sepsis alone. Increase in PAI activity and antigen in patients with septic shock was accompanied by increase in tissue-type plasminogen activator antigen and total fibrin(ogen) degradation products. In patients who developed severe sepsis without shock, the variables of the fibrinolytic system remained unchanged after onset of fever. Differences observed between septic shock and severe sepsis were sustained throughout the septic episode for all variables. These findings suggest that septic shock in leukocytopenic patients is associated with significant activation of the fibrinolytic system arising from vascular endothelium responses to inflammatory injury. Finally, PAI activity measurements are sensitive markers of unfavorable outcomes in patients with sepsis.*
12. Bogatcheva, N. V., Garcia, J. G., and Verin, A. D. (2002). Molecular mechanisms of thrombin-induced endothelial cell permeability. *Biochemistry (Moscow)* **67**(1), 75–84.
13. Esmon, C. T., Ding, W., Yasuhiro, K., Gu, J. M., Ferrell, G., Regan, L. M., Stearns-Kurosawa, D. J., Kurosawa, S., Mather, T., Laszik, Z., and Esmon, N. L. (1997). The protein C pathway: New insights. *Thromb. Haemost.* **78**(1), 70–74.
14. Bernard, G. R., Vincent, J. L., Laterre, P. F., LaRosa, S. P., Dhainaut, J. F., Lopez-Rodriguez, A., Steingrub, J. S., Garber, G. E., Helterbrand, J. D., Ely, E. W., and Fisher, C. J., Jr. (2001). Recombinant human Protein C Worldwide Evaluation in Severe Sepsis (PROWESS) study group. Efficacy and safety of recombinant human activated protein C for severe sepsis. *N. Engl. J. Med.* **8;344**(10), 699–709.

Capsule Biography

Dr. Fowler is the William Taliaferro Thompson Professor of Medicine and Chairman of the Division of Pulmonary and Critical Care Medicine at Virginia Commonwealth University School of Medicine in Richmond, Virginia. Dr. Fowler's research and laboratory effort over the past two decades has focused on mechanisms of vascular inflammatory injury occurring following onset of sepsis and following reperfusion of ischemic vascular beds. A major area of focus has been the regulation of interleukin-8 expression by nitric oxide in microvascular endothelium.

CHAPTER 109

Interaction of C1q with the Endothelium: Relevance to Inflammation

[1]Berhane Ghebrehiwet and [2]Ellinor I. B. Peerschke
[1]*State University of New York, Stony Brook, New York*
[2]*Weill Medical College of Cornell University, New York, New York*

The first component of complement, C1, is a multimolecular complex comprising C1q and the Ca^{2+}-dependent tetramer $C1r_2$-$C1s_2$. The primary role of C1q within the complex is that of a recognition signal, which triggers activation of the classical pathway of complement. However, C1q is also able to bind to a variety of cell types in a manner that induces specific biological responses. Interaction of C1q with endothelial cells, for example, leads to cellular activation followed by release of biological mediators and/or expression of adhesion molecules, all of which contribute, directly or indirectly, to the inflammatory process. These specific responses are thought to be mediated by the interaction of C1q with proteins of the endothelial cell surface. Endothelial cells express four types of putative C1q binding proteins/receptors: cC1q-R/CR, or calreticulin (CR), a 60-kDa collectin receptor; gC1q-R/p33, a 33 kDa homotrimeric protein; C1q-Rp (CD93), a 129-kDa *O*-sialoglycoprotein; and CR1 (CD35), the receptor for C3b. Although the specific role of each of these molecules in the C1q-mediated biological responses is yet to be worked out, all of them may, in one fashion or another, participate in the inflammatory processes and vascular lesions that occur on the endothelium.

The Endothelium and Inflammation

The response of blood vessels to injury is made up of a series of interrelated reactions involving endothelial cells, platelets, and proteins of the coagulation, complement, kinin, and fibrinolytic systems, all of which may directly or indirectly contribute to the inflammatory process. Inflammation may in turn contribute to thrombosis and atherosclerosis through the complement system, or via inflammatory cytokines secreted by activated macrophages and lymphocytes present at inflammatory sites, atherosclerotic plaques, and sites of vascular injury. The inflammatory process, which begins as a protective response, is therefore the result of multiple interactions including those that involve the complement system and the coagulation cascade.

In addition to its role of maintaining vessel wall integrity, the endothelium plays a critical role in the pathophysiology of thrombosis, atherosclerosis, and inflammation. This is largely due to the ability of endothelial cells to respond to a wide range of signals that include microbial proteins, cytokines, growth factors, and proteins of the complement system and the kinin-forming cascade. Some of these responses are rapid and do not require new gene transcription or protein synthesis. These responses, which are the hallmark of the early nonleukocyte-dependent phase of inflammation, include expression of surface P-selectin, activation of nitric oxide (NO) metabolic pathways, and enhanced vascular permeability. A large number of agonists are capable of eliciting this type of response, including proteins of the clotting and complement systems. Vasoactive peptides derived from both of these systems can in turn recruit proinflammatory cells into the inflammatory site, thus exacerbating the disease process.

The Endothelial Cell Surface and C1q Receptors

Various receptor systems have been described on vascular cells that participate in the recognition, activation, and clearance of components involved in humoral defense. Among these molecules are receptors for the complement protein, C1q. Four C1q binding proteins or receptors have been described to date: cC1q-R/CR or collectin receptor, a 60-kDa calreticulin (CR) homolog, which binds to the collagen-like tail of C1q; gC1q-R/p33, a 33 kDa homotrimeric protein with high affinity for the globular heads of C1q; C1q-Rp (CD93), a 129 kDa *O*-sialoglycoprotein involved in phagocytosis; and CR1 (CD35), the receptor for C3b. In addition to C1q, the collectin receptor (cC1q-R/CR) is also able to bind members of the collectin family of proteins including surfactant proteins A and D (SP-A and SP-D) and mannan binding lectin (MBL). Similarly, the gC1q-R/p33 molecule can bind plasma proteins other than C1q, including high-molecular-weight kininogen (HK) and FXII, which may play a significant role in the pathogenesis of thrombosis and atherosclerosis. In addition, gC1q-R/p33 also binds a number of bacterial and viral proteins including protein A of *Staphylococcus aureus*, known to be associated with bacterial endocarditis. Whereas CD93 and CD35 are bona fide transmembrane proteins and can signal directly, cC1q-R/CR and gC1q-R/p33 each lack a transmembrane segment in their respective sequences and therefore rely on a docking/signaling partnership with other transmembrane proteins in order to relay their message across the membrane (e.g., CD91 and cC1q-R/CR; β1-integrin and gC1q-R/p33).

C1q-Mediated Endothelial Cell Responses: Role of C1q and HK

Interaction of C1q with endothelial cells induces a plethora of biological functions including adhesion and spreading; stimulation and expression of the adhesion molecules E-selectin, ICAM-1, and VCAM-1; and production of interleukin (IL)-6, IL-8, and monocyte chemoattractant protein-1 (MCP-1). Endothelial cell adhesion and spreading can be inhibited by both anti-cC1q-R/CR and gC1q-R antibodies, and production of IL-6, IL-8, and MCP-1 can be inhibited by anti cC1q-R/CR. Therefore, at sites of atherosclerotic and inflammatory and vascular lesions, where C1q is present in demonstrable quantities, C1q can trigger or contribute to the inflammatory process not only by activation of the classical pathway of complement, but also by interaction with its receptors, which induces the release of proinflammatory cytokines and recruits inflammatory cells.

In addition to its participation in C1q-mediated responses, gC1q-R, together with urokinase plasminogen activator receptor (uPAR) and cytokeratin-1, has been shown to serve as a high-affinity receptor for HK and FXII as shown by inhibition with monoclonal antibodies to gC1q-R. The binding of HK to gC1q-R is entirely dependent on the presence of 10- to 50-μM zinc. This is largely because zinc can induce exposure of hydrophobic sites in the C-terminal domain of gC1q-R, which includes those residues recognized by inhibitory monoclonal antibody. It is under these conditions that the binding of HK to the endothelial cell gC1q-R becomes greatly increased. This high-affinity interaction can then facilitate the assembly of proteins of the intrinsic coagulation/kinin-forming cascade resulting in the generation of bradykinin. Bradykinin, a potent vasoactive peptide, can in turn bind to the endothelial cell bradykinin receptor (B2) and induce a wide range of pathophysiologic responses including morphologic changes in the endothelium rendering the subendothelial matrix accessible to blood components inflammatory cells. Furthermore, bradykinin can participate in the processes of tumor metastasis and angiogenesis.

Therefore, at sites of atherosclerotic and inflammatory and vascular lesions, where gC1q-R is ubiquitously expressed and both C1q and HK are present (see Figure 1), these molecules can have an additive effect in augmenting the inflammatory process—C1q by activation of complement, and via gC1q-R or other C1q binding proteins,

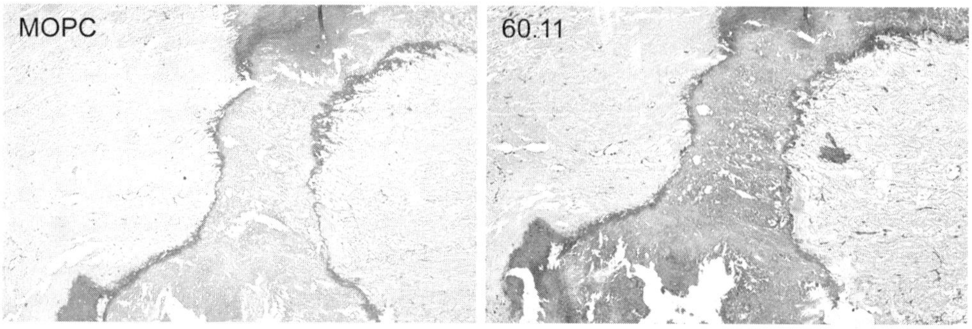

Figure 1 Immunohistochemical staining of serial sections of the necrotic core of an atherosclerotic lesion of human coronary artery stained with anti-gC1q-R mAb 60.11 (*right panel*) or with isotype- and species-matched antibody (MOPC-21). Upon closer microscopic examination, foam cells in necrotic areas in the deeper intima and the vessel walls in the adventitia showed the strongest gC1q-R staining. Moreover, the necrotic center of calcified atherosclerotic plaques stained strongly for gC1q-R. Staining of endothelial cells was also noted. (see color insert)

inducing the release of proinflammatory cytokines and the recruitment of inflammatory cells; and HK, by generation of bradykinin, which in turn induces vascular dilation and increased permeability through contraction of endothelial cells and extravascular smooth muscle.

Regulation of the Endothelium

Although the C1q binding proteins are constitutively expressed on resting nonthrombotic endothelium, and gC1q-R in particular has the potential to bind C1q, as well as HK and FXII, the mechanism by which a continuously thrombogenic state of the endothelium is averted is not completely understood. Based on the available data, however, it is postulated that efficient engagement of gC1q-R by its ligands is restricted to conditions where under chemical, physical, or infectious insult, the endothelial cell is converted to a prothrombotic and proinflammatory phenotype leading to upregulation of gC1q-R, the induction of cytokines such as IL-1 or tumor necrosis factor-α (TNFα), and/or expression of cell adhesion molecules (CAMs). Freshly recruited leukocytes, such as macrophages and T cells bound to the cell adhesion molecules, can also release cytokines, which in turn can amplify the process by upregulating the expression of C1q binding molecules in a manner that allows efficient binding of ligands. It has been shown that cytokines, such as LPS, interferon-γ (IFNγ), and TNFγ, can upregulate the expression of both cC1q-R and gC1q-R. Thus, modulation of endothelial cell function by soluble and/or immobilized proinflammatory ligands, such as C1q or HK, especially at sites of inflammatory and vascular lesions, may contribute significantly to the development of thrombosis and exacerbation of the inflammatory process.

Bibliography

Andrews, B. S., Shadforth, M., Cunningham, P., and Davis, J. S. IV (1981). Demonstration of a C1q receptor on the surface of human endothelial cells. *J. Immunol.* **127**, 1075–1080.

Feng, X., Tonnesen, M. G., Peerschke, E. I. B., and Ghebrehiwet, B. (2002). Cooperation of C1q receptors and integrins in C1q-mediated endothelial cell adhesion and spreading. *J. Immunol.* **168**, 2441–2448.

Ghebrehiwet, B., Lim, B.-L., Kumar, R., Feng, X., and Peerschke, E. I. B. (2001). gC1q-R/p33: A member of a new class of multifunctional and multicompartmental cellular proteins is involved in inflammation and infection. *Immunol. Rev.* **180**, 65–77. *This is an up-to-date and detailed summary of the structure and function of gC1q-R/p33 as well as cC1q-R/CR with 76 relevant references.*

Guo, W.-X., Ghebrehiwet, B., Weksler, B., Schweizer, K., and Peerschke, E. I. B. (1999). Upregulation of endothelial cell binding proteins/receptors for complement C1q by inflammatory cytokines. *J. Lab. Clin. Med.* **133**, 541–550.

Joseph, K., Ghebrehiwet, B., Peerschke, E. I. B., Reid, K. B. M., and Kaplan, A. P. (1996). Identification of the zinc-dependent endothelial cell binding protein for high molecular weight kininogen and factor XII: Identity with the receptor which binds to the globular "heads" of C1q (gC1q-R). *Proc. Natl. Acad. Sci. USA* **93**, 8552–8557.

Lozada, C., Levin, R. I., Huie, M., Hirschhorn, R., Naime, D., Whitlow, M., Recht, P. A., Golden, B., and Cronstein, B. (1995). Identification of C1q as a heat-labile serum cofactor required for immune complexes to stimulate endothelial expression of the adhesion molecules E-selectin and intercellular and vascular cell adhesion molecules 1. *Proc. Natl. Acad. Sci. USA* **92**, 8378–8382.

Malhotra, R., Lu, J., Thiel, S., Jensenius, J.-C., Willis, A. C., and Sim, R. B. (1992). C1q receptor (collectin receptor): Primary structure, homology, and interaction with ligands. *Immunobiology* **184**, 1437–1445.

Peerschke, E. I. B., Malhotra, R., Ghebrehiwet, B., Reid, K. B. M., Willis, A. C., and Sim, R. B. (1993). Isolation of a human endothelial cell C1q receptor (C1q-R). *J. Leukoc. Biol.* **53**, 179–184.

Peerschke, E. I. B., Smyth, S. S., Teng, E., Dalzell, M., and Ghebrehiwet, B. (1996). Human umbilical vein endothelial cells possess binding sites for the globular domains of C1q. *J. Immunol.* **157**, 4154–4158.

Silverstein, R. L. (1999). The vascular endothelium. In: *Inflammation: Basic Principles and Clinical Correlates,* J. I. Gallin and R. Snyderman, eds., p. 207. *This is an excellent reference and extensively describes the role of endothelial cells in inflammation.*

van den Berg, R. H., Faber-Krol, M. C., Sim, R. B., and Daha, M. R. (1998). The first subcomponent of complement, C1q, triggers the production of IL-8, IL-6 and monocyte chemoattractant peptide-1 by human umbilical vein endothelial cells. *J. Immunol.* **161**, 6924–6930. *This report is the first to describe the C1q-mediated release of important cytokines from endothelial cells and has a very well balanced bibliography.*

Capsule Biography

Dr. Ghebrehiwet is a Professor in the Department of Medicine at Stony Brook University, and his laboratory focuses on the biological response as a consequence of the interaction between C1q and cell surface proteins. He is credited with having discovered and characterized both cC1q-R (which is identical to calreticulin) and gC1q-R as ubiquitously distributed, multifunctional C1q binding cell surface proteins.

Dr. Ellinor Peerschke is Professor of Pathology at Weill Medical College of Cornell University in New York. Her research focuses primarily on blood platelet and vascular endothelial cell function in hemostasis, thrombosis, and inflammation.

CHAPTER 110

The Microvasculature in Inflammatory Bowel Disease

Ossama A. Hatoum and David G. Binion
Department of Medicine, Digestive Disease Center, Froedtert Memorial Lutheran Hospital, Medical College of Wisconsin, Milwaukee, Wisconsin

Introduction

The two major forms of human inflammatory bowel disease (IBD), Crohn's disease (CD), and ulcerative colitis (UC), represent classic chronic inflammatory disorders, characterized by progressive destructive inflammation in the gastrointestinal tract. Although the majority of research into IBD pathogenesis has focused on immune dysregulation, vascular involvement has also been recognized. The intestinal microcirculation and its endothelial lining contribute to normal immune homeostasis in the gut and potentially to the chronic dysregulated inflammation, which characterizes IBD. Alterations in endothelial–leukocyte interaction have been identified in the chronically inflamed intestine, including an enhanced capacity for endothelial–leukocyte interaction. The mechanisms of altered endothelial function in IBD immune dysregulation appear to involve a diminished capacity to generate nitric oxide (NO), as well as altered expression of cell adhesion molecules and vascular addressins. In addition to its role in leukocyte recruitment, the intestinal microvasculature may also contribute to chronic inflammation through its central role in the regulation of tissue perfusion. Studies have demonstrated an altered microvascular architecture characterized by stenosis, which correlates with decreased in vivo perfusion in chronically inflamed and remodeled intestine. Recent work has demonstrated an acquired microvascular dysfunction, with impaired vasodilation in IBD involved intestinal microvessels. Chronically inflamed IBD microvessels demonstrate a diminished endothelium-dependent vasodilatory capacity, a finding not present in uninvolved IBD gut arterioles. In summary, important alterations in IBD intestinal microvascular physiology and function have been identified in human IBD, which may contribute to chronic inflammatory and ischemic pathophysiology.

The Role of the Microvasculature and Endothelial–Leukocyte Interaction in Chronic Intestinal Inflammation

Endothelial cells are now understood to play an early and rate-limiting step in the inflammatory process, as endothelial activation in response to cytokines and inflammatory mediators results in expression of cell adhesion molecules and chemokines, which play a critical role in leukocyte recruitment from the circulation. Early investigation into the role of endothelial cells in IBD pathogenesis focused on histologic evaluation, characterizing the morphology of the microvasculature in chronically inflamed bowel. Using transmission electron microscopy, Dvorak et al. evaluated intestinal specimens resected from CD patients [1], demonstrating abnormalities in the endothelial cells lining the microcirculation. These included loss of monolayer integrity with tissue edema, extravasation of red blood cells, focal venular endothelial necrosis adjacent to areas of undamaged endothelial cells, and endothelial cell hypertrophy.

Research focusing on the role of microvascular endothelial cells in leukocyte recruitment to the gut has characterized increased expression of cell adhesion molecules (CAM) [2]. Immunolocalization of CAM demonstrated marked increase in E-selectin and ICAM-1 expression in IBD intestine, while VCAM-1 expression was less clearly demonstrated. Subsequent investigation by Briskin et al. has demonstrated an increase in the gut-specific homing molecule MAdCAM-1 (mucosal addressin cell adhesion molecule 1), which plays a major role in the recruitment of leukocytes expressing the alpha 4 integrin into the mucosal immune compartment [3].

Alterations in leukocyte homing patterns in IBD gut were characterized by Salmi et al., who demonstrated that naive lymphocytes are preferentially recruited to the chronically inflamed intestinal microvascular endothelium, whereas control intestinal microvessels preferentially bind memory lymphocytes [4]. These findings were confirmed by Burgio et al., who demonstrated an altered pattern of leukocyte binding in Crohn's disease, where naive monocytes and T cells were again preferentially recruited to the chronically inflamed intestine [5]. These authors also demonstrated increased expression of ICAM-1, E-selectin, and CD34 in the IBD gut microvessels.

To more fully define the contribution of microvascular endothelial cells in chronic intestinal inflammation, studies have been carried out in intestinal and disease-specific cultures of microvascular endothelial cells. Human intestinal microvascular endothelial cells (HIMECs) have been isolated and characterized, demonstrating classic endothelial markers including factor VIII associated antigen, Weibel-Palade bodies, expression of PECAM-1 (CD31), E-selectin, ICAM-1, and VCAM-1 as well as unique patterns of leukocyte adhesion and growth compared to human umbilical vein endothelial cells (HUVEC) [6, 7]. More importantly, HIMECs have also been isolated from involved and uninvolved CD and UC intestine. HIMECs isolated from both chronically inflamed CD and UC demonstrated a significantly enhanced capacity to adhere leukocytes, compared to control HIMECs, a phenomenon that was only elicited following activation with proinflammatory cytokines (interleukin-1β, tumor necrosis factor-α) and bacterial lipopolysaccharide. Leukocyte "hyperadhesion" appears to be an acquired phenomenon, as uninvolved IBD intestinal segments failed to demonstrate increased leukocyte binding [8]. The mechanisms underlying leukocyte hyperadhesion in the chronically inflamed IBD HIMECs did not appear to involve increased levels of CAM expression, compared to control cultures, which prompted investigation of possible alterations in the intracellular mechanisms that govern the downregulation of inflammatory activation in endothelial cells. NO plays a central role in the regulation of endothelial activation and the maintenance of vascular homeostasis, exerting a potent anti-inflammatory effect, downregulating the activation of vascular endothelial cells as well as their capacity to bind circulating leukocytes, normally an early and rate-limiting step in the inflammatory process. Control HIMECs displayed distinct patterns of NO generation through both constitutive endothelial (e) nitric oxide synthase (eNOS; NOS3) as well as inducible NOS (iNOS; NOS2), which was expressed following inflammatory activation [9]. In marked contrast, IBD HIMECs failed to express iNOS and increased levels of NO following inflammatory activation [10]. This loss in NO generation in IBD HIMECs was linked to enhanced leukocyte binding, as administration of NO donors restored a normal binding pattern in the activated IBD HIMECs. Further investigation demonstrated that iNOS-derived NO in the IBD HIMECs appears to function as an endogenous antioxidant, quenching superoxide anion, which is a central mediator of inflammatory activation in gut microvascular endothelial cells. The mechanisms that lead to altered iNOS expression in the chronically inflamed IBD HIMECs have not been fully defined.

The Role of Microvascular Dysfunction and Tissue Ischemia in Chronic Intestinal Inflammation

The human bowel receives a significant proportion of cardiac output, which varies with physiologic need. At rest, intestinal perfusion via the superior mesenteric artery will range from 29 to 70 mL/min/100 g intestinal tissue, whereas in the fed state, splanchnic hyperemia increases perfusion from 28 to 132 percent. Investigation of intestinal perfusion in the setting of chronic inflammation in IBD has been carried out using a variety of in vivo and in vitro techniques. Angiographic studies of the IBD intestine have demonstrated preserved anatomy in the superior and inferior mesenteric arteries, with significant abnormalities in the vasa recta, characterized by tortuous, dilated vessels together with loss of normal tapering and terminal coiling as they penetrate the bowel wall. In early stages of IBD, angiographic studies have demonstrated arteries that abruptly taper as the vessels reach the bowel wall with right-angle bifurcation, bizarre distribution, and small luminal irregularities in the peripheral branches. Furthermore, advanced IBD lesions demonstrate reduced vessel diameter, decreased vascular density, and diminished blood flow in the involved segments, which may contribute to an impaired capacity to heal and resolve the chronic inflammatory "wound."

The microvascular architecture was characterized by Wakefield et al. [11] using scanning electron micrographs of corrosion microcasts from control and CD patients, identifying occlusive fibrinoid lesions in the arterioles supplying affected intestine, which were not demonstrated in uninvolved areas of bowel. Morphologically, the chronically inflamed microvessels were tapered and stenosed compared to vessels from uninvolved and control bowel. Vascular damage was demonstrated as an early pathologic finding, which preceded the development of mucosal ulceration. Vascular damage appeared to be highest where vessels penetrated through the muscularis propria, and bursts of angiogenic vessels were seen in distal areas of the mucosal circulation. These authors concluded that multifocal gastrointestinal infarction plays a pathogenic role in the chronic inflammatory lesion in CD, with the extent of vascular damage correlating with the severity of intestinal injury. The concept of microvascular dysfunction and relative tissue ischemia was further investigated by Funayama et al., who used tissue histometry to describe remodeling in the CD intestinal microcirculation [12]. Assuming that medial atrophy is an indirect measure of decreased vascular perfusion and pressure, these investigators characterized atrophy of arterial media in the submucosal CD vessels, which indirectly suggests ischemia and increased vascular resistance in deeper submucosal arteries. Taken together, these studies suggest that the microvascular anatomy in the chronically inflamed CD intestine has undergone extensive remodeling. However, these studies did not directly define microvascular function, or potential factors that would lead to these alterations in vascular architecture.

Assessment of intestinal microvascular physiology in human IBD has been characterized using direct and indirect methods to assess microvascular blood flow. Hulten et al. used an intraoperative isotope washout technique as well as in vivo abdominal angiography to demonstrate alterations of intestinal blood flow, including perfusion patterns associated with distinct phases of disease [13]. Early fulminant IBD with severe inflammation was characterized by increased vascular perfusion, which is typical of an acute inflammatory response. In contrast, reduced regional blood flow was seen in chronically inflamed and remodeled tissues, particularly areas of CD stricture. These observations have been confirmed in subsequent studies, demonstrating diminished vascular perfusion associated with fibrosis and long-standing inflammation. Using endoscopic laser Doppler flowmetry [14], decreased mucosal blood flow was seen in the neoterminal ileum after ileocolectomy and in rectal mucosal perfusion in patients with long-standing UC [15].

The poorly healing, refractory inflammatory ulceration and damage in the IBD intestine suggests that microvascular dysfunction resulting in diminished vasodilatory capacity and tissue hypoperfusion are found in the IBD gut. The molecular physiology underlying microvascular dysfunction in IBD was assessed directly by Hatoum et al., who characterized vasodilator responses in human intestinal microvessels, by measuring in vitro vasodilatory capacity in response to acetylcholine from perfused 50- to 150-μm diameter arterioles isolated from intestinal resections [16]. Normal intestinal microvessels vasodilate in response to Ach using NO- and cyclooxygenase (COX)-dependent mechanisms, while chronically inflamed IBD arterioles (both CD and UC) demonstrated a significantly diminished vasodilatory capacity. This decreased vasodilatory capacity in the chronically inflamed IBD microvessels was directly related to a loss of NO-dependent function, and these same vessels were found to be heavily dependent on COX to maintain their vascular tone. Microvascular endothelial dysfunction in chronically inflamed IBD tissues was associated with excess levels of oxidative stress as measured by intravital dyes and confocal fluorescence microscopy, which was not present in vessels isolated from normal intestine or uninvolved areas of IBD bowel. The microvascular dysfunction identified in arterioles from chronically inflamed IBD gut was not a generalized response to inflammation, as it was not demonstrated in vessels isolated from acute inflammation (i.e., diverticulitis) and uninvolved areas of IBD. This demonstrates that medications used to treat IBD patients at the time of surgery did not contribute directly to the microvascular dysfunction, and further substantiates that intrinsic, acquired alterations in the chronically inflamed and remodeled microcirculation underlie this pathophysiology.

Summary

The vasculature plays a central role in human IBD, as studies have demonstrated an altered microvascular anatomy in the chronically inflamed bowel, which corresponds with altered immunologic function and vascular physiology. Microvascular dysfunction in IBD results in diminished perfusion capacity, which is linked to impaired wound healing and may contribute to the refractory mucosal damage that characterizes IBD. Only areas of the microvascular anatomy characterized by extensive remodeling from chronic inflammation and damage correlate with impaired vasoperfusion, as well as enhanced leukocyte interaction and sustained endothelial activation. Thus, we believe that the vasculature contributes to the initiation and perpetuation of chronic intestinal inflammation in IBD. Improved understanding of the vascular contribution in IBD pathogenesis will pave the way for the development of novel strategies targeting inflammatory and ischemic mechanisms, specifically endothelium–leukocyte interaction with antiadhesion molecule biologic agents and vasodilatory compounds to improve wound healing in areas of refractory mucosal ulceration.

Glossary

Crohn's disease: An inflammatory disease of the gastrointestinal tract whose etiopathogenesis is linked to a combination of both genetic and environmental causes. The peak incidence of onset occurs between 15 and 25 years of age, and a second peak is associated with the sixth decade of life. Common symptoms include recurrent abdominal pain, fatigue, fever, nausea, vomiting, weight loss and diarrhea. Complications include anemia, intestinal blockage from strictures, fistulas and anal fissures. Treatments target inflammation and include 5-aminosalicylate compounds, corticosteroids, immunomodulators and anti-TNFalpha agents. Surgery is frequently needed to ameliorate complications of this life-long disease.

HIMEC (human intestinal microvascular endothelial cells): acronym for human small and large intestine specific in vitro endothelial cell lines.

Inflammatory bowel disease: The group of chronic inflammatory disorders involving the gastrointestinal tract, most commonly associated with ulcerative colitis and Crohn's disease.

MAdCAM-1: Mucosal cell adhesion molecule 1 is a member of the immunoglobulin gene superfamily, and this cell surface glycoprotein is expressed on specific vascular beds in the gastrointestinal mucosa, pancreas and brain. MAdCAM-1 is increased during inflammatory bowel disease, and is the endothelial ligand for leukocytes bearing the alpha4 beta 7 integrin, mediating selective leukocyte recruitment into tissues.

Microvascular dysfunction: is the physiologic disturbance or impairment of normal endothelial vasorelaxation in response to mediators such as acetylcholine. Microvascular dysfunction has been associated with coronary artery disease, hypercholesterolemia, diabetis mellitus and inflammatory bowel disease.

Ulcerative colitis: One of the inflammatory bowel diseases, characterized by peak onset during adolescence and early adulthood, with exclusive involvement of the large intestine. Disease invariably involves the anorectum, and will extend proximally in specific patient subgroups, involving the entire colon in approximately one-third of individuals. Treatment is focused on 5-aminosalicylates, corticosteroids and immunomodulator agents. A high percentage of patients (approximately one-third) will ultimately require total abdominal colectomy due to refractory inflammation or high propensity for malignant transformation.

References

1. Dvorak, A. M., Monahan, R. A., Osage, J. E., and Dickersin, G. R. (1980). Crohn's disease: Transmission electron microscopic studies. II. Immunologic inflammatory response. Alterations of mast cells, basophils, eosinophils, and the microvasculature. *Hum. Pathol.* **11**, 606–619.
2. Podolsky, D. K., Lobb, R., King, N., Benjamin, C. D., Pepinsky, B., Sehgal, P., and deBeaumont, M. (1993). Attenuation of colitis in the cotton-top tamarin by anti-alpha 4 integrin monoclonal antibody. *J. Clin. Invest.* 92, 372–380, 1993.
3. Briskin, M., Winsor-Hines, D., Shyjan, A., Cochran, N., Bloom, S., Wilson, J., McEvoy, L. M., Butcher, E. C., Kassam, N., Mackay, C. R., Newman, W., and Ringler, D. J. (1997). Human mucosal addressin cell adhesion molecule-1 is preferentially expressed in intestinal tract and associated lymphoid tissue. *Am. J. Pathol.* **151**, 97–110.
4. Salmi, M., and Jalkanen, S. (2001). Human leukocyte subpopulations from inflamed gut bind to joint vasculature using distinct sets of adhesion molecules. *J. Immunol.* **166**, 4650–4657.
5. Burgio, V. L., Fais, S., Boirivant, M., Perrone, A., and Pallone, F. (1995). Peripheral monocyte and naive T-cell recruitment and activation in Crohn's disease. *Gastroenterology* **109**, 1029–1038.
6. Binion, D. G., West, G. A., Ina, K., Ziats, N. P., Emancipator, S. N., and Fiocchi, C. (1997). Enhanced leukocyte binding by intestinal microvascular endothelial cells in inflammatory bowel disease. *Gastroenterology* **112**, 1895–1907. *This is the first description of altered microvascular activation in chronically inflamed human intestinal microvascular endothelial cells isolated from patients with both CD and UC.*
7. Haraldsen, G., Rugtveit, J., Kvale, D., Scholz, T., Muller, W. A., Hovig, T. and Brandtzaeg, P. (1995). Isolation and longterm culture of human intestinal microvascular endothelial cells. *Gut* **37**, 225–234.
8. Binion, D. G., West, G. A., Volk, E. E., Drazba, J. A., Ziats, N. P., Petras, R. E., and Fiocchi, C. (1998). Acquired increase in leucocyte binding by intestinal microvascular endothelium in inflammatory bowel disease. *Lancet* **352**, 1742–1746.
9. Binion, D. G., Fu, S., Ramanujam, K. S., Chai, Y. C., Dweik, R. A., Drazba, J. A., Wade, J. G., Ziats, N. P., Erzurum, S. C., and Wilson, K. T. (1998). iNOS expression in human intestinal microvascular endothelial cells inhibits leukocyte adhesion. *Am. J. Physiol.* 275, G592–G603.
10. Binion, D. G., Rafiee, P., Ramanujam, K. S., Fu, S., Fisher, P. J., Rivera, M. T., Johnson, C. P., Otterson, M. F., Telford, G. L., and Wilson, K. T. (2000). Deficient iNOS in inflammatory bowel disease intestinal microvascular endothelial cells results in increased leukocyte adhesion. *Free Radic. Biol. Med.* **29**, 881–888.
11. Wakefield, A. J., Sawyerr, A. M., Dhillon, A. P., Pittilo, R. M., Rowles, P. M., Lewis, A. A., and Pounder, R. E. (1989). Pathogenesis of Crohn's disease: Multifocal gastrointestinal infarction. *Lancet* **2**, 1057–1062. *Microvascular architectural remodeling is found exclusively in chronically inflamed CD that is not seen in uninvolved areas of bowel.*
12. Funayama, Y., Sasaki, I., Naito, H., Fukushima, K., Matsuno, S., and Masuda, T. (1999). Remodeling of vascular wall in Crohn's disease. *Dig. Dis. Sci.* **44**, 2319–2323.
13. Hulten, L., Lindhagen, J., Lundgren, O., Fasth, S., and Ahren, C. (1977). Regional intestinal blood flow in ulcerative colitis and Crohn's disease. *Gastroenterology* **72**, 388–396. *The first description of relative ischemia and tissue hypoperfusion from intraoperative assessment of the chronically inflamed CD and UC intestine.*
14. Angerson, W. J., Allison, M. C., Baxter, J. N., and Russell, R. I. (1993). Neoterminal ileal blood flow after ileocolonic resection for Crohn's disease. *Gut* **34**, 1531–1534.
15. Tateishi, S., Arima, S., and Futami, K. (1997). Assessment of blood flow in the small intestine by laser Doppler flowmetry: Comparison of healthy small intestine and small intestine in Crohn's disease. *J. Gastroenterol.* 32, 457–463.
16. Hatoum, O. A., Binion, D. G., Otterson, M. F., and Gutterman, D. D. (2003). Acquired microvascular dysfunction in inflammatory bowel disease: Loss of nitric oxide-mediated vasodilation. *Gastroenterology* **125**, 58–69. *This article demonstrates a loss of nitric oxide mediated vasodilation is an acquired component of microvascular dysfunction in chronically inflamed microvessels isolated from resected CD and UC intestinal segments. Microvascular dysfunction was not identified in acute inflammation, nor was it seen in areas of uninvolved IBD intestine.*

Capsule Biography

Ossama A. Hatoum, M.D., M.Sc., is a colorectal surgeon and Research Associate in the Division of Cardiovascular Medicine at the Medical College of Wisconsin, in Milwaukee, Wisconsin. His laboratory investigation focuses on mechanisms of vasodilator response in the human intestinal microcirculation in health and IBD.

David G. Binion, M.D., is Director of the IBD Center and Associate Professor of Medicine at the Medical College of Wisconsin, in Milwaukee, Wisconsin. His laboratory has focused on the vascular biology of chronic intestinal inflammation in human inflammatory bowel disease (IBD), with a special emphasis on endothelium–leukocyte interaction. His clinical work as a gastroenterologist is focused on the development of experimental therapy for patients with Crohn's disease and ulcerative colitis.

CHAPTER 111

Pathophysiology of Edema Formation

Dan D. Hershko and Michael M. Krausz
Department of Surgery A, Rambam Medical Center, Haifa, Israel

The extracellular fluid volume is maintained within narrow limits in normal human subjects, despite day-to-day variations in the dietary intake of salt and water. Plasma volume, in turn, is determined by the total extracellular fluid volume, and the partitioning of this volume between the extravascular fluid and the intravascular compartments also remains remarkably constant. The relationship of extracellular volume and, in particular, the volume of the plasma compartment to overall vascular capacitance determine to a large extent the fundamental indices of cardiovascular performance such as mean arterial blood pressure and left ventricular filling volume.

Edema is the clinicopathological term used to describe an excessive accumulation of fluids in the extravascular (interstitial) component of the extracellular fluid volume. Depending on the etiology and mechanism leading to its formation, edema may occur locally at specific tissue and organs or it may have a more generalized distribution. Local edema may be the result of a site-specific hemodynamic derangement or it may be formed secondary to an inflammatory reaction to injury and sepsis. Generalized edema, the clinical hallmark of extracellular fluid volume expansion, usually represents the accumulation of excessive volumes of fluid in the interstitial compartment secondary to renal sodium retention, the result of activation of various compensatory mechanisms [1]. Nevertheless, recent studies have suggested that other contributory mechanisms may also have a significant role in the pathophysiology of edema formation.

In clinical practice, when edema is severe and involves all body tissues and organs, it is commonly termed *anasarca*. Accumulation of edema fluids in the peritoneal cavity is known as *ascites* and in the pleural cavity as *hydrothorax*. The composition of the edema is of importance, as it too depends on the etiology and mechanism leading to its formation. Noninflammatory edema fluid, such as accumulates in heart failure, cirrhosis of the liver, and various renal diseases, is protein-poor in nature and is termed a *transudate*. In contrast, the inflammatory edema fluid, which is termed an *exudate*, is protein- and leukocyte-rich and is associated with increased endothelial permeability. Thus, when characterizing the pathophysiology of edema formation, it is convenient to differentiate between the inflammatory and noninflammatory (hemodynamic) derangements that lead to edema formation.

Hemodynamic Aspects of Edema Formation

Approximately one-third of the total body water content is confined to the extracellular space. This compartment is composed of the plasma volume, which under normal circumstances comprises 25 percent of the extracellular space, and the remainder is interstitial fluid. Sodium content governs the total fluid volume in both intravascular and interstitial compartments while plasma proteins (mainly albumin) govern the partitioning between these two compartments. The main regulatory hemodynamic mechanism that controls the disposition of fluid between the intravascular and interstitial compartments contains two sets of opposing forces known as Starling forces [2]. The Starling principle provides a framework for analyzing fluid movement across capillaries. Thus, the direction and rate of fluid movement are determined by the balance between the hydraulic and osmotic pressures of the intravascular and interstitial compartments. The Starling equation may be expressed as follows:

$$Q_f = K_f[(P_c - P_i) - (\pi_c - \pi_i)]$$

where Q_f = rate of fluid movement across capillaries; K_f = filtration coefficient; P_c = capillary hydraulic pressure; P_i = interstitial hydraulic pressure; π_c = plasma osmotic pressure; and π_c = interstitial osmotic pressure. Hence, according to the Starling equation, the hydraulic pressure within the vascular system and the colloid osmotic pressure in the interstitial fluid promote the movement of fluid from the circulation toward the extracellular space. In contrast, the hydraulic pressure of the interstitial fluid and the intravascular colloid osmotic pressure produced by the plasma proteins tend to drive fluid back into the vascular system. Thus, any change in any one of these driving forces may alter the direction of fluid movement from one compartment to the other [3]. These forces are usually delicately balanced, resulting in a volume steady state between the intravascular and the extravascular compartments. At the arteriolar end of the microcirculation the hydraulic pressure (about 32 mmHg) is higher than the colloid oncotic pressure (25 mmHg) such that the prevailing balance of Starling forces favors the net filtration of fluid into the interstitium. Net outward movement of fluid along the length of the capillary is associated with an axial decrease in the capillary hydraulic pressure and an increase in the plasma colloid osmotic pressure. Thus, at the venular end of the capillary the intravascular hydraulic pressure (about 12 mmHg) is lower than the colloid osmotic pressure, favoring movement of fluid back to the circulation. Nevertheless, in several tissues the local transcapillary hydraulic pressure gradient continues to exceed the opposing colloid osmotic pressure gradient throughout the length of the capillary bed, such that filtration occurs along its entire length. In such capillary beds, a substantial volume of filtered fluid must therefore be returned to the circulation via lymphatics. Given this importance of lymphatic drainage in this situation, the ability of lymphatics to expand and proliferate and the ability of lymphatic flow to increase in response to increased interstitial fluid formation provide protective mechanisms for minimizing edema formation.

Other mechanisms for minimizing edema formation even under normal conditions have also been identified [4]. A rise in the venous pressure causes myogenic arteriolar constriction that will lower capillary pressure and decrease capillary surface area. Precapillary vasoconstriction tends to lower capillary hydrostatic pressure and diminish the filtering surface area in a given capillary bed. Indeed, excessive precapillary vasodilatation in the absence of appropriate microcirculatory myogenic reflex regulation appears to account for lower extremity interstitial edema associated with calcium blocker vasodilator therapy. Increased net filtration itself is associated with dissipation of capillary hydraulic pressure, dilution of interstitial fluid protein concentration, and a corresponding rise in intracapillary plasma protein concentration. The resulting change in the profile of Starling forces associated with increased filtration therefore tends to mitigate against further interstitial fluid accumulation. Furthermore, even small increases in interstitial fluid volume tend to augment tissue hydraulic pressure, again opposing further transudation of fluid into the interstitial space. Finally, as tissue pressure increases, the lymphatic network is widely opened, allowing the return of large volumes of fluid to the microcirculation. Therefore, it is evident that maintenance of adequate steady state between fluid volumes of the circulation and the interstitial space depends on the delicate balance among the microcirculatory hydraulic pressure, the level of plasma proteins, and the adequacy of the lymphatic drainage system.

The appearance of generalized edema implies one or more disturbances in the microcirculatory hemodynamics associated with expansion of the extracellular volume. There are multiple factors that may affect hemodynamic stability, such as increased venous pressure transmitted to the capillary bed, unfavorable adjustments in pre- and postcapillary resistances, reduced plasma colloid osmotic pressure, renal sodium retention, or lymphatic flow inadequate to drain the interstitial compartment and replenish the intravascular compartment. All of these factors may shift the Starling forces toward increased movement of fluids to the extracellular space and hence promote the formation of edema (Figure 1A).

Increased Hydraulic Pressure

An increase in the hydraulic pressure at the venular end of the microcirculation raises the capillary filtration pressure resulting in increased movement of transudate to the interstitial compartment. In clinical practice, this mechanism of edema formation is most commonly seen in congestive heart failure. When primarily right ventricular function is affected, the distribution of the edema is systemic, involving all body tissues. In contrast, decreased left ventricular function results in increased venous pressure that primarily affects the lungs. The factors leading to edema in congestive heart failure, however, are far more complex [5]. The decrease in cardiac output results in a decrease in the effective arterial blood volume. This in turn decreases renal blood flow, which triggers the renin–angiotensin–aldosterone axis in a compensatory attempt to increase the effective arterial blood volume through increased sodium and water retention in the kidney. In addition, reduction in the arterial renal blood flow leads to renal vasoconstriction with a marked reduction in the glomerular filtration rate. The direct consequence of the glomerular hemodynamic alterations is an increase in the fractional reabsorption of filtered sodium and water at the level of the proximal tubule. At the same time there is increased secretion of arginine vasopressin (AVP). The stimulus for increased AVP secretion in congestive heart failure appears to involve a nonosmotic drive such as attenuated compliance of the left atrium, hypotension, and activation of the renin–angiotensin system. The net effect at the kidney level is again enhanced water retention, this time at the distal nephron. The resultant increase in plasma volume from all of these compensatory mechanisms increases significantly the venous hydraulic pressure, which in turn imposes a further burden on the

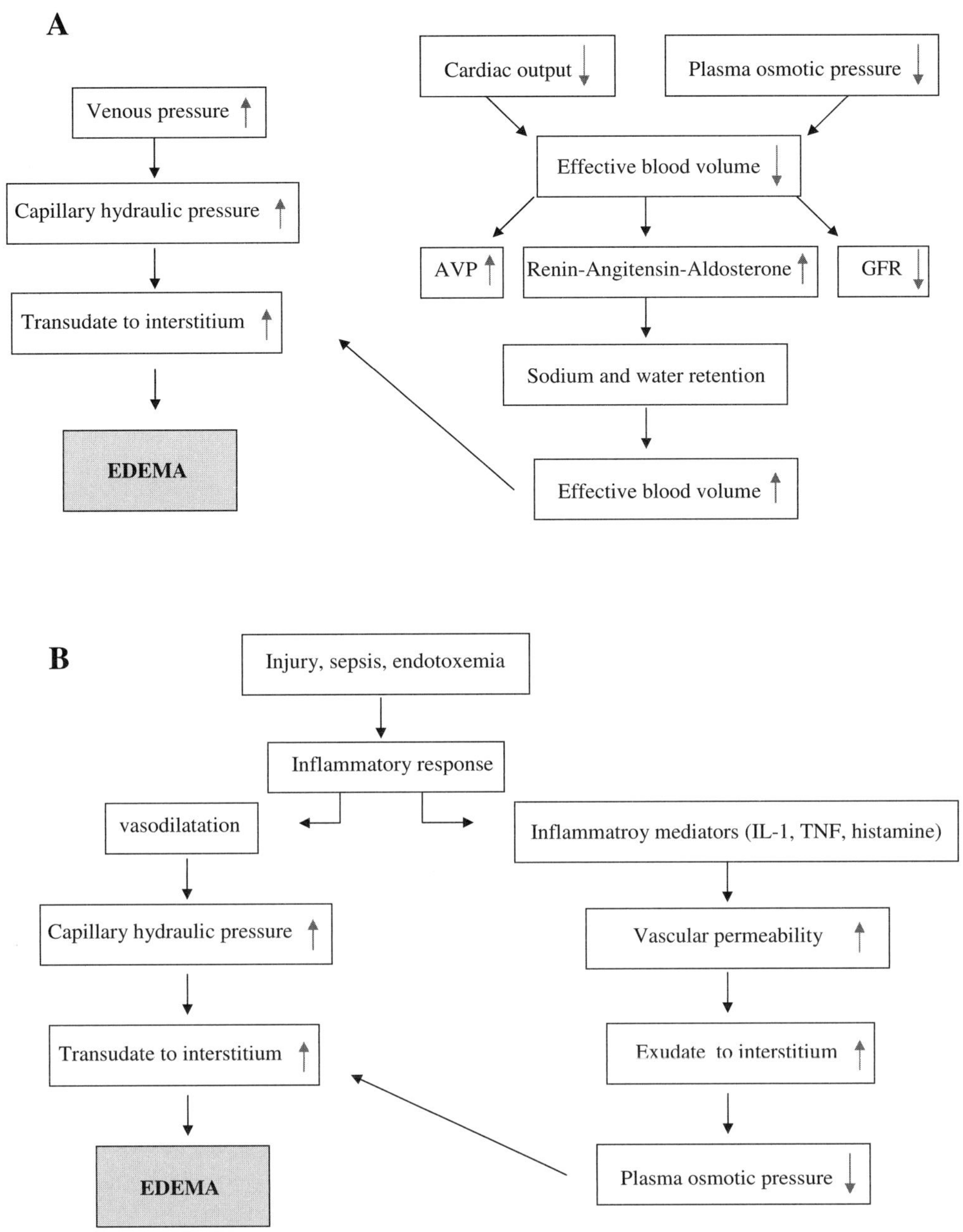

Figure 1 The hemodynamic **(A)** and inflammatory **(B)** events leading to edema formation. (see color insert)

failing heart. This leads to further edema formation inflicted by the transient increase in the central venous pressure. Thus, a vicious cycle is created whereby compensatory mechanisms are constantly activated in order to increase the effective arterial blood volume but ultimately fluid escapes to the interstitium, thereby lowering again the effective arterial blood volume and further increasing edema formation [6].

Increased hydraulic pressure may also be involved in the pathogenesis of local edema formation. For example, obstruction of venous return from a limb secondary to major vein thrombosis may lead to a marked increase in the local venous pressure that overcomes the colloid osmotic pressure gradient, resulting in local accumulation of fluid in the interstitial space of the affected limb.

Reduced Plasma Oncotic Pressure

This cause for edema is less common than increased hydraulic pressure, but when it is involved in the pathogenesis of edema formation, it tends to create a severe form of generalized edema. In most instances, the critical level of plasma protein needed to maintain adequate plasma oncotic pressure is 2.5 mg/dL. A decrease in the critical content of plasma proteins, such as may occur in several clinical settings including cirrhosis of the liver, nephrotic syndrome, malnutrition, and various protein-losing gastroenteropathies, will significantly reduce the oncotic pressure responsible for reabsorption of interstitial fluids. This favors the net movement of fluids from the intravascular compartment to the interstitial space. The reduction in the intravascular volume leads to a state of decreased effective arterial blood volume and, as in the pathogenesis described earlier for congestive heart failure, activation of multiple compensatory pathways. As such, decreased renal blood flow will decrease the glomerular filtration rate and activate the renin–angiotensin–aldosterone axis, resulting in increased reabsorption of sodium and water. However, the resultant expansion of the intravascular volume will rapidly move to the interstitium because of the low plasma osmotic pressure, resulting in the massive fluid shift. Again, this compensatory mechanism may lead to the same vicious cycle as described in the case of heart failure, whereby renal sodium and water retention are continuously activated secondary to a sustained perceived decrease in effective arterial blood volume.

Sodium Retention

The sodium body content is the main determinant of volume in the extracellular compartment. Hence, sodium retention, which is the result of an imbalance in renal tubule sodium reabsorption relative to filtered load, may result in excessive fluid accumulation in the interstitial compartment. In generalized edema, the expansion of the extracellular volume is invariably associated with renal sodium retention. Some of the causes and pathways leading to sodium retention were outlined earlier. In congestive heart failure and cirrhosis with ascites, the primary disturbance leading to sodium retention does not originate within the kidney. Instead, renal sodium retention is the response to a disturbance of the effective circulation induced by disease of the heart or liver. In the nephrotic syndrome, glomerular injury accompanied by heavy proteinuria is associated with sodium retention and leads to a profound disturbance in circulatory homeostasis. In each of these conditions, the renal effector mechanisms that normally operate to conserve sodium and protect against a sodium deficit are exaggerated and continue despite subtle or overt expansion of interstitial compartment volume.

Inadequate Lymphatic Flow

This form of edema, commonly termed *lymphedema*, is nearly always confined to a specific anatomic region. Under normal circumstances, excess of interstitial fluid is removed by the lymphatic system, eventually returning to the circulation through the thoracic duct. Lymphatic obstruction is commonly observed in the arms following axillary lymph node dissection or irradiation, and in the lower extremities following inguinal lymph node dissection or chronic infection with the filarial worm *Wuchecheria bancrofti*. This form of edema may be massive and severely impair limb function, again demonstrating the important role of the lymphatic network in minimizing edema formation.

In summary, changes in any one or more of the factors that contribute to Starling forces may result in violation of the normal body fluid partitioning and the development of edema. The magnitude of edema is therefore a reflection of the intensity of the hemodynamic derangement, and every attempt should be made to identify the cause and aim treatment strategies to counter these factors.

Inflammatory Aspects of Edema Formation

One of the common consequences of the inflammatory response is the development of edema. Although some of the pathophysiological events that lead to its formation are similar to those described for hemodynamic derangements, there are numerous events that are unique to inflammation. One of the hallmarks of the inflammatory response is the production and secretion of various proinflammatory cytokines and vasogenic mediators. Among the proinflammatory mediators and cytokines produced during inflammation, it is mainly histamine, bradykinin, platelet-activating factor, interleukin (IL)-1, and tumor necrosis factor (TNF) that are involved in the pathogenesis of inflammation-mediated edema [7]. The magnitude of their action is largely dependent on the etiology of the inflammatory response and length of exposure to offending stimuli. Furthermore, the dominance of each of these agents also depends on the offending stimuli leading to endothelial injury. For example, following minor stimuli IL-1 and TNF secretion provide cellular and tissue protection through upregulation of certain anti-inflammatory cytokines (i.e., IL-6, IL-8). In contrast, IL-1 and TNF have a cardinal role in the endothelial damage associated with sepsis and endotoxemia. Most cases of sepsis are caused by endotoxin-producing Gram-negative bacilli. Endotoxins are bacterial lipopolysaccharides (LPS) released by degradation of the cell wall. LPS binds to LPS binding proteins, and this complex binds to specific receptors (CD14) on monocytes, macrophages, and neutrophils. Exposure to high levels of LPS induces the inflammatory response of primarily mononuclear phagocytes, leading to the production and secretion of TNF, which in turn activates IL-1 synthesis. Both cytokines act on the endothelial cells to produce other cytokines and endothelial effectors. Some of these effectors such as nitric oxide, platelet-activating factors, and other eicosanoids may mediate the release of proteolytic enzymes by neutrophils that can damage the endothelium itself and surrounding tissues. Thus, the pathogenesis of edema formation during inflammation is complex and may involve the combination of multiple mechanisms.

The main events that participate in the pathogenesis of inflammatory edema are therefore, alterations in vascular hemodynamics and an increase in vascular permeability (Figure 1B).

Alterations in Vascular Hemodynamics

During inflammation there is a sequence of events in the microvasculature that leads to the formation of edema. Initially, there is transient vasoconstriction mediated by the secretion of various vasoconstricting mediators. This, in turn, is rapidly followed by vasodilatation of arterioles and expansion of microvascular beds in the involved region. Hence, the increase in the hydraulic pressure secondary to this vasodilatation results in transudation of protein-poor fluid to the extracellular compartment. At the same time, another edema-forming mechanism is activated. This mechanism, which is the hallmark of inflammatory edema, involves the induction and development of increased microvascular permeability.

Increased Vascular Permeability

Increased vascular permeability during inflammation occurs mainly at the microcirculatory level, involving the arterioles, capillaries, and venules. Unlike the forms of edema described earlier, here there is loss of protein-rich fluid (exudate) from the intravascular compartment to the interstitial space. Thus, a decrease in the intravascular oncotic pressure accompanied by an increase in the interstitial osmotic pressure result in marked movement of fluids to the interstitium and impairment of venous return.

Under normal conditions, the microvascular endothelium is composed of a thin, continuous layer of squamous epithelium with closely apposed intercellular junctions. Thus, bidirectional movement of water and small solutes is permitted between the intravascular compartment and the interstitial compartment, but larger particles, such as albumin and other plasma proteins, cannot cross freely. At the capillary level there is somewhat freer movement of proteins owing to the presence of micropinocytotic vesicles (25 nm). Despite its simple structure, the endothelium is actually metabolically active and is capable of secreting various proteins, including prostaglandins, cytokines, and collagen.

The primary factor leading to increased vascular permeability during inflammation is an injury to the endothelium. Acute inflammation may induce vascular leakiness of endothelial monolayers through a response elicited by a number of pathophysiological mechanisms. Arterioles, capillaries, and venules may be affected differently, depending on the severity of the inflammatory response and the pathways involved. These mechanisms may present distinctively or, more commonly, they tend to overlap each other. The first and most common response is observed immediately following injury and is of very short duration. This response, also known as the immediate transit response, occurs distinctively at small and medium-sized postcapillary venules. This type of response is reversible and results from the activation of histamine, bradykinin, and leukotrienes. The primary effect of these mediators is a significant contraction of the endothelial cells, leading to the formation of large intercellular gaps and escape of exudate to the interstitial space. Vascular permeability may also be increased by structural changes in the endothelial cytoskeleton. This pathway, which is mediated by the action of the proinflammatory cytokines IL-1 and TNF, is of longer duration (up to 24 hours). Here, vascular leakage results from endothelial cell retraction (not contraction) and disruption of the intercellular junctions.

Another mechanism that may lead to a massive and sustained increase in vascular permeability is severe endothelial injury that results in endothelial cell necrosis. Various insults such as severe burn injuries, sepsis, and endotoxemia are usually associated with this pathway. Cell necrosis may be the result of direct injury or may occur secondary to outpouring of noxious cellular mediators including oxygen radicals and proteolytic enzymes. These factors are mainly derived from recruited leukocytes. Because leukocytes need to adhere to become effective, this type of injury is commonly seen in venules or pulmonary capillaries. Here, vascular leakage also begins immediately, and this reaction will be sustained until the endothelium is repaired or the blood vessel is occluded by a thrombus. Another variant of this type of response is of delayed onset but again may be of long duration. It is commonly seen following thermal injury, exposure to various toxins, and irradiation. In this setting, the increase in vascular permeability occurs mainly at the capillary level. The exact mechanism leading to the delayed onset is unknown, but the possible role of some cytokines and apoptosis were recently suggested.

Idiopathic Edema

Idiopathic edema is a disorder characterized by fluid retention that cannot be attributed to the known causes or the pathogenic mechanisms of edema formation listed earlier. Although the etiology of this syndrome has not been clarified yet, it is clear that idiopathic edema overwhelmingly occurs in women. It is important to distinguish this disorder from excessive estrogen–stimulated sodium and fluid retention associated with the premenstrual state (cyclic or premenstrual edema). True idiopathic edema has been reported in postmenopausal women, and episodes of fluid retention bear no relationship to the menstrual cycle. In most series of patients, other edema-forming states such as cardiac, renal, or hepatic disease have been excluded. It is therefore likely that idiopathic edema is not a single entity, but rather the result of multiple etiologies [8].

An important consistent finding in idiopathic edema is the observation that most of the affected patients retain sodium and water in the upright posture, but diurese in the recumbent position. There is evidence to suggest that decreased plasma volume with concomitant increase in plasma renin and aldosterone activity are frequent findings. The relative importance of secondary hyperaldosteronism in sustaining the progressive fluid gain in this disorder is highlighted by the observation that treatment with angiotensin-converting enzyme inhibitors often results in fluid loss and symptomatic improvement.

The possibility that an underlying abnormality exists in capillary permeability leading to loss of intravascular volume in the upright posture has also been considered. Studies provided evidence showing a decrease in plasma volume and increased disappearance rates of albumin from the plasma. Some studies have suggested an increase in edema fluid protein concentrations, and others have reported increased catabolic rates for serum albumin. Although severe hypoalbuminemia is not a feature of the disorder, small decreases in serum albumin concentration have been noted. The mechanisms underlying the increased permeability of the capillaries are poorly understood. However, it was shown that IL-2 may have a major role in the pathogenesis of this syndrome. It was also suggested that some patients with idiopathic edema are, or have been, diuretic users. Because diuretic agents induce a degree of volume depletion, many of the mechanisms controlling sodium excretion are stimulated to compensate for the natriuresis. This "braking" phenomenon provides patients a new steady state of external sodium balance. Thus, proximal fluid reabsorption might increase and the renin–angiotensin–aldosterone system would be overstimulated.

In summary, idiopathic edema may be the end result of multiple etiologies. After exclusion of certain metabolic and other edema-forming states, it is likely that certain forms exist in which primary disturbances in capillary permeability result in decreased circulating plasma volume.

Glossary

Colloid osmotic pressure: The effective pressure between the plasma and the interstitial compartment provided by osmotically active particles (proteins) that do not pass the freely between semi-permeable membranes.

Extracellular fluid: The fluid contained in the plasma and the interstitial compartment.

Hydraulic pressure: The hydrostatic pressure formed within the intravascular compartment.

Microvascular permeability: Alterations in the vascular permeability of the microcirculation.

Starling equation: Provides the calculation of forces driving fluid movement across the capillaries.

References

1. Peters, J. P. (1948). The role of sodium in the production of edema. *N. Engl. J. Med.* **239**, 353–362.
2. Michel, C. C. (1997). Starling: The formulation of his hypothesis of microvascular fluid exchange and its significance after 100 years. *Exp. Physiol.* **82**(1), 1–30. *This review article emphasizes the important contribution of Starling to the understanding of the main mechanisms of edema formation.*
3. Kumar, V., Cortan, R. S., and Robbins, S. L. (2003). *Basic Pathology*, 7th ed. Philadelphia: W. B. Saunders.
4. Abbas, A. K., Lichtman, A. H., and Pober, J. S. (2000). *Cellular and Molecular Immunology*, 4th ed. Philadelphia: W.B. Saunders.
5. Schrier, R. W. (1988). Pathogenesis of sodium and water retention in high-output and low-output cardiac failure, nephrotic syndrome, cirrhosis, and pregnancy. *N. Engl. J. Med.* **319**(16), 1065–1072. *An excellent review on the pathogenesis of edema formation in clinical practice.*
6. Schrier, R. W., and Abraham, W. T. (1999). Hormones and hemodynamics in heart failure. *N. Engl. J. Med.* **341**(8), 577–585.
7. Kaufman, D. R., and Choi, Y. (1999). Signaling by TNF receptors: Pathways, paradigms and targets for therapeutic modulation. *Int. Rev. Immunol.* **18**(4), 405–427. *A comprehensive review on proinflammatory cytokines involved in edema formation.*
8. Streeten, D. H. (1995). Idiopathic edema. Pathogenesis, clinical features, and treatment. In: *Endocrinology and Metabolism Clinics of North America*, R. G. Dluhy, ed., Philadelphia: W. B. Saunders.

Capsule Biography

Dr. Hershko is a Senior Surgeon in the Department of Surgery A at the Rambam Medical Center in Haifa, Israel. His main research interests are the regulation of proinflammatory cytokines in sepsis and the role of cell-cycle regulatory proteins in cancer progression. He has received the Israel Cancer Research Fund (ICRF) Career Development Award.

Dr. Krausz has headed the Department of Surgery A at the Rambam Medical Center in Haifa, Israel, since 1996. His main research interests are fluid treatment of trauma casualties in civilian as well as military scenarios, and lately his research has also focused on gender differences in hemorrhagic as well as septic shock, and the role of leukocytes in pathophysiology of these phenomena. He has received grants from the Israel Academy of Sciences and the Israeli Ministry of Health.

CHAPTER 112

Endothelium in Malarial Infection: More Than an Innocent Bystander

May Ho, Bryan G. Yipp, and Stephen M. Robbins
University of Calgary, Calgary, Alberta, Canada

Introduction

The adhesion of infected erythrocytes containing mature stages of the parasite *Plasmodium falciparum* (IRBCs) to host endothelium is a key pathogenic process of the infection. The resulting sequestration of IRBCs in the microvasculature leads to tissue hypoxia, metabolic disturbances, multiorgan dysfunction, and ultimately the death of 1 to 2 million patients worldwide annually. Under physiological flow conditions, the process of cytoadherence is mediated by the parasite protein *Plasmodium falciparum* erythrocyte membrane protein 1 (PfEMP1) expressed on the surface of IRBCs, and its synergistic interactions with a number of endothelial receptor molecules, mimicking the adhesive events in the leukocyte recruitment cascade. In previous studies, we demonstrated that although IRBCs from clinical parasite isolates can tether and roll on intercellular adhesion molecule 1 (ICAM-1), vascular cell adhesion molecule 1 (VCAM-1), and P-selectin, they adhere preferentially to CD36 on microvascular endothelial cells.

In addition to serving as points of attachment, adhesion molecules are known to serve as receptor signaling molecules capable of transducing an extracellular signal leading to cellular activation. Such signaling activities have been reported for a number of adhesion molecules including the selectins, the immunoglobulin superfamily of adhesion molecules, and integrins, all of which play critical roles in leukocyte recruitment to sites of inflammation.

In this review, we will discuss recent experimental data that indicate that the adhesion of IRBCs to CD36 can also lead to endothelial intracellular signaling. Furthermore, the evidence suggests that there are both outside-in and inside-out signaling mechanisms that modulate the affinity of the IRBC–CD36 interaction through a change in the phosphorylation state of endothelial CD36. Vascular endothelium thus has a much more dynamic role in cytoadherence than previously recognized. The proposed novel intracellular signaling mechanism involving CD36, Src-family kinases, and an ecto-alkaline phosphatase could potentially be exploited to inhibit the pathophysiological chain of events seen in patients with severe falciparum malaria.

Plasmodium falciparum Malaria

Malaria, a disease caused by parasites of the genus *Plasmodium*, remains a major medical problem worldwide with an incidence of 300 million new cases and 1 to 2 million deaths each year. Of the four species of human malaria parasites, *P. falciparum* is the most lethal. The clinical manifestations of falciparum malaria are variable. They range from a mild febrile illness to severe and frequently fatal syndromes such as cerebral malaria and multiorgan failure. The mortality due to severe falciparum malaria is about 15 to 20 percent despite effective antimalarial drugs and optimal clinical care. In areas of high endemicity, such as sub-Saharan Africa, severe falciparum malaria mainly affects children less than 5 years of age, accounting for 90 percent of the annual malarial mortality worldwide. There is lower incidence in older children and adults in these areas, because of the acquisition of partial immunity. Almost all the deaths in African children have three main overlapping clinical syndromes: respiratory distress, cerebral malaria, and severe anemia. In areas of low endemicity such as Southeast Asia, severe falciparum malaria can affect all age groups. Acute renal failure, jaundice, and pulmonary edema are common in adults. Cerebral complications may occur by itself or in conjunction with other organ involvement. It generally has a poor prognosis. However, approximately 95 percent of adults and 85 percent of children who recover from cerebral malaria show no persistent neurological sequelae.

Cytoadherence

The most common and consistent pathological feature of severe falciparum malaria is the sequestration of IRBC in the capillaries and postcapillary venules of vital organs. Sequestration has been observed in all major organs such as the brain, heart, lung, kidney, and spleen. The organ distribution of sequestration varies and often reflects the clinical features of the preceding illness. For example, patients with cerebral malaria show increased cerebral sequestration compared with that in other organs. Even within the brain, the cerebral cortex and cerebellum are preferentially affected compared to the midbrain and brain stem. The sequestration of IRBCs in microvessels is thought to allow the parasites to evade splenic clearance. Unfortunately, the resulting impairment of microcirculatory blood flow may lead eventually to the demise of the host.

The molecular basis of cytoadherence in the vasculature has been studied extensively. Under static conditions, a number of adhesion molecules including CD36, ICAM-1, E-selectin, VCAM-1, PECAM-1, and thrombospondin-1 (TSP-1) have been shown to support IRBC adhesion, although the degree of adhesion to the different molecules varies by several orders of magnitude. A unique situation is the sequestration of IRBCs in the human placenta where IRBCs appear to adhere to chondroitin sulfate A (CSA) expressed on syncytiotrophoblasts that line the placental villous space. Under physiological flow conditions, we found that cytoadherence on microvascular endothelium in a parallel-plate flow chamber in vitro and in a human microvasculature in vivo is mediated by a number of adhesion molecules in a synergistic fashion (Figure 1). IRBCs can tether and roll on several host endothelial receptors such as ICAM-1, VCAM-1, and P-selectin, but not CSA. These low-affinity interactions do not by themselves lead to the arrest of the interacting cells, but enhance the subsequent adhesion of nearly all clinical isolates tested to CD36. ICAM-1 and VCAM-1 mediate their effect by increasing the percentage of rolling cells that become adherent, while P-selectin increases the absolute number of rolling and adherent IRBC, consistent with its ability to tether flowing cells from the bloodstream in leukocyte recruitment. Despite the similarities, there are important differences between leukocyte and IRBC recruitment. Whereas leukocyte recruitment occurs in a stepwise progression from selectins to members of the immunoglobulin superfamily, IRBCs can tether and roll on a number of different molecules before adhesion to CD36. In fact, CD36 by itself can mediate all three components of the adhesive interactions. If the cumulative affinity of all interactions is sufficiently high, the IRBC will adhere.

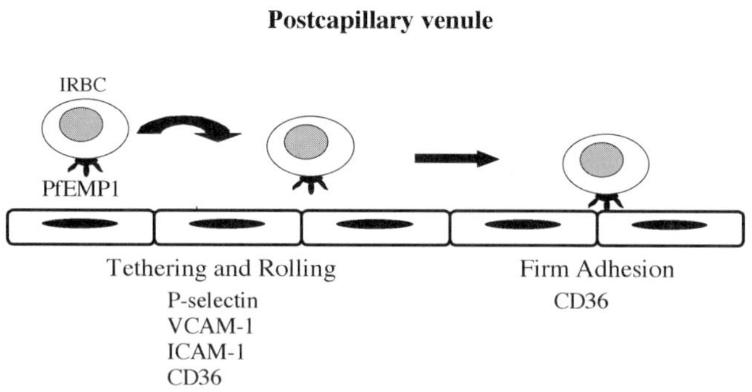

Figure 1 Schematic model of the cytoadherence cascade under flow conditions. IRBCs are observed to have three distinct types of adhesive interactions with microvascular endothelium: tethering, rolling, and firm adhesion mediated by a number of different adhesion molecules.

Otherwise, they will continue rolling. These findings underscore the pivotal role of CD36 in the pathogenesis of severe falciparum malaria.

CD36

CD36 is an 88-kDa membrane glycoprotein of 471 amino acids with two hydrophobic regions near the amino and carboxyl terminals that serve to anchor the molecule to the plasma membrane (Figure 2). It is expressed on a wide variety of cell types, such as microvascular endothelial cells, erythroblasts, monocytes, platelets, striated muscle cells, adipocytes, and mammary epithelial cells. The natural ligands of CD36 include collagen, TSP-1, both native and oxidized low-density lipoproteins (LDLs) and high-density lipoproteins (HDLs), and apoptotic neutrophils. These receptor–ligand interactions can lead to diverse cellular and biological responses including uptake of apoptotic bodies, TGF-β activation, fatty acid transport, and endothelial cell apoptosis. CD36 was first identified as a receptor for IRBCs indirectly by the inhibition of adhesion of IRBCs to C32 melanoma cells and endothelial cells by the anti-CD36 monoclonal antibody OKM5. Direct evidence was obtained subsequently by demonstrating that IRBCs adhered to immobilized purified CD36 protein from platelets and COS cells transfected with CD36. Most clinical parasite isolates studied to date bind CD36, and OKM5 can block the adhesion of all parasite isolates tested. These results imply that a CD36 binding domain is expressed in most parasite isolates and it recognizes a common region on CD36. Peptide mapping studies on CD36 reveal that residues 145 to 171 are important for IRBC adhesion, and OKM5 is thought to bind to residues 155 to 183.

PfEMP1 and Recombinant 179 Peptides

Plasmodium falciparum erythrocyte membrane protein 1 (PfEMP1) is the parasite protein directly involved in adhesive interactions with microvascular endothelium. PfEMP1

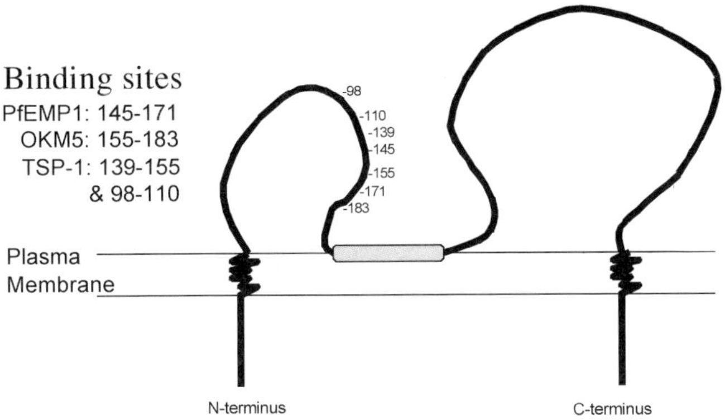

Figure 2 CD36 membrane topology and binding sites. This unique structure contains two cytoplasmic tails and two extracellular loops separated by a hydrophobic section that sits within the outer leaflet of the plasma membrane. Threonine-92 is phosphorylated in platelet CD36.

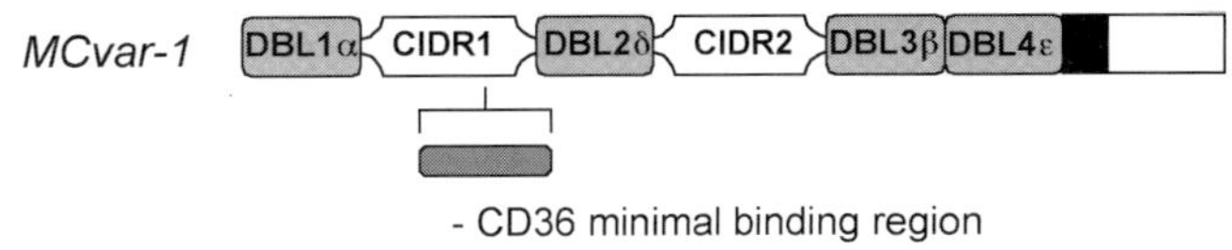

Figure 3 Structural domains of PfEMP1 from the parasite clone *MCvar1* for which the minimal CD36 binding site (179 amino acids) has been mapped to the CIDR region 1. DBL, Duffy binding-like; CIDR, cysteine-rich interdomain.

is a highly variable protein encoded by the large *var* gene family that is expressed on electron-dense protrusions on the surface of infected erythrocytes. In the early stages of the parasite cycle, many *var* genes are transcribed in a single IRBC. As the intracellular parasite matures to a trophozoite, at which stage cytoadherence occurs, only one *var* gene product is expressed, while the remaining genes are silenced by an as yet unknown mechanism. However, a single parasite clone can bind to more than one receptor molecule through its distinct binding modules: the Duffy binding-like domains (DBL) and the cysteine-rich interdomain regions (CIDR) (Figure 3). A critical region of PfEMP1 involved in binding to CD36 is localized to a 179-amino-acid sequence within the CIDR1 region. A recombinant 179 amino acid peptide in *Escherichia coli* from the parasite strain Malayan Camp (MC) *var*1 gene has been shown to bind to CD36 and inhibit and reverse the adhesion of several CD36-binding laboratory-adapted parasite strains in static and flow chamber based assays in vitro. More importantly, we recently demonstrated that a yeast recombinant peptide PpMC-179 could inhibit and reverse cytoadherence of diverse clinical isolates in human microvessels in vivo in a human–SCID mouse chimeric model.

CD36-Mediated Intracellular Signaling

CD36 is known to mediate intracellular signaling. The molecule is physically associated with Src-family kinases in membrane invaginations known as caveolae that are enriched in a variety of signal transduction molecules. Caveolae serve as a subcompartment of the plasma membrane where clustering of signaling proteins may result in more rapid cross talk and/or promote efficiency of signal transmission. The binding of one of the natural ligand TSP-1 to CD36 on dermal microvascular endothelial cells sequentially activates the Src-family kinase Fyn, caspase-3-like proteases, and the p38 MAPK pathway. As well, a CD36-mediated signaling cascade involving the Src-family kinases Lyn and Fyn is responsible for the inflammatory effects of fibrillar β-amyloid in mouse peritoneal macrophages and microglial cells. Several lines of evidence in the literature suggest that IRBC adhesion might also induce intracellular signaling. IRBC adhesion to CD36 on human monocytes induces a respiratory burst, and cross-linking of CD36 on monocytes by OKM5 activates both the ERK 1/2 and p38 MAP kinase pathways. IRBCs have also been shown to modulate the maturation of dendritic cells through an interaction with cell surface CD36.

An additional feature of platelet CD36 is that it is constitutively phosphorylated on its ectodomain on residue Thr^{92} that results in a low-affinity receptor for TSP-1. Initial CD36–TSP-1 interaction induces platelet degranulation with release of acid phosphatases capable of dephosphorylating the ectodomain of CD36, resulting in a high-affinity receptor for TSP-1. Although the phosphorylated isoform of CD36 have not been identified on endothelium, the PKC-dependent targeting sequence RGPYTYRVRFLA for Thr^{92} phosphorylation is conserved in the endothelial protein. The close proximity of CD36, dually acylated Src-family kinases, and a glycosylphosphatidylinositol (GPI)-anchored alkaline phosphatase in caveolae suggests that a change in the phosphorylation state of endothelial CD36 might also occur upon ligand binding and intracellular signal activation.

CD36 Signaling via Src-family Kinases Modulates IRBC Adhesion

We have generated pharmacological and functional data that are consistent with a role for intracellular signaling and subsequent dephosphorylation of the CD36 ectodomain in mediating firm adhesion of IRBC under flow conditions. Specifically, we have demonstrated that the binding of PpMC-179 to CD36, as well as cross-linking CD36 with OKM5, activated the ERK 1/2 MAP kinase pathway in HDMEC that was dependent on Src-family kinase activity. The PpMC-179 peptide represents the functional binding domain of the parasite cytoadherent ligand PfEMP1 to CD36. Our finding is therefore the first direct demonstration that IRBC-CD36 interaction can in fact induce an intracellular signal in the host cells.

To determine the functional consequence of the intracellular activating, we treated HDMEC with PP1, a Src-family

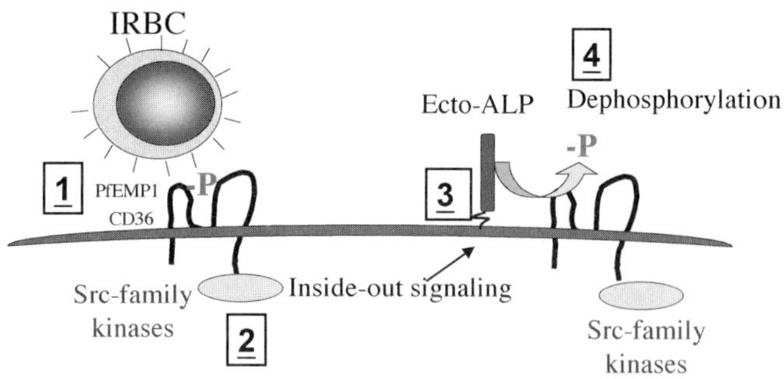

Figure 4 Proposed model of the regulation of IRBC adhesion to CD36 on microvascular endothelium under flow conditions. The initial attachment of IRBCs to CD36 (step 1) leads to a Src-family kinase-dependent intracellular signal (step 2) that is responsible for increasing subsequent IRBC adhesion to CD36 by means of an ecto-alkaline phosphatase (ecto-ALP) (step 3) that dephosphorylates and hence increases the binding affinity of CD36 for IRBCs (step 4). (From Yipp et al., 2003, with permission of The American Society of Hematology.)

kinase selective inhibitor. PP1 inhibited adhesion of IRBCs to HDMEC monolayers in a flow chamber assay and to intact human microvessels in vivo in a human–SCID mouse model. The effect of PP1 could be mimicked by levamisole, a specific alkaline phosphatase inhibitor. Furthermore, firm adhesion to PP1-treated endothelium could be restored by the addition of exogenous alkaline phosphatase. Collectively, these results strongly support a novel mechanism for the modulation of cytoadherence under flow conditions through a signaling pathway involving CD36, Src-family kinases, and an ecto-alkaline phosphatase, most likely the GPI-anchored endothelial alkaline phosphatase. Based on these findings, we propose that the initial attachment of IRBCs to CD36 under flow conditions (step 1 in Figure 4) triggers a Src-family kinase-dependent intracellular signal (step 2) that is responsible for increasing subsequent adhesion of IRBCs to CD36 by means of an ecto-alkaline phosphatase (step 3). The enzyme dephosphorylates CD36, leading to a higher binding affinity of CD36 for IRBCs (step 4). Through this mechanism, a small number of strongly adherent IRBCs in a clinical parasite isolate can activate the endothelium and promote the adhesion of the majority of IRBCs.

Downstream Signaling Events

In contrast to the increase in adhesion following the activation of Src-family kinases and possibly an ecto-alkaline phosphatase, activation of the ERK 1/2 and p38 MAP kinase pathways had no immediate effect on IRBC adhesion. However, this lack of effect of MAP kinases on adhesion does not exclude an important downstream role for MAP kinases in regulating gene expression of molecules induced as a result of PfEMP1 ligation of CD36, including cytokines and chemokines that would in turn modulate the adhesion process. The occurrence of downstream effects of MAP kinases would be consistent with the observation that the ERK 1/2 and the p38 MAPK pathways are activated in monocytes following CD36 cross-linking with OKM5, and blockade of these pathways result in inhibition of phagocytosis of IRBCs. In dendritic cells (DC), the binding of IRBC to surface-expressed CD36 leads to a switch in production of IL-12 to IL-10, and an associated inability to activate T cells. Recently, apoptosis in lung microvascular endothelial cells has been observed after the cells were cocultured with IRBCs. However, the percentage of apoptotic cells appears to be limited (~15%), which might explain the absence of widespread endothelial damage in patients with severe falciparum malaria. Collectively, these findings suggest that the interaction of IRBCs with host cells can result in profound modulation of their normal function.

Possible Role of Other Receptor Molecules

If the signaling by IRBCs through CD36 on microvascular endothelial cells is analogous to its action in other cell types, we could anticipate that an assembly of membrane receptors might be involved for endothelial cell activation as a result of IRBC adhesion. Integrins are likely candidate receptors for IRBCs on endothelium, as they promote adhesion to other cells and matrix proteins and are often associated physically and functionally with CD36. Indeed, CD36 is known to guide integrins into signaling rafts, and in so doing may regulate integrin function. The antiangiogenic effect of TSP-1 has been shown to involve its binding to endothelial CD36 as well as to $\alpha_3\beta_1$ integrin. On microglial and other myeloid cells, the activation of CD36 by β-fibrillar peptides (amyloid plaque) involves CD36, $\alpha_6\beta_1$ integrin, and the integrin-associated protein CD47. In retinal pigment epithelium, phagocytosis of outer segment (OS) fragments is mediated by CD36 and $\alpha_v\beta_5$ integrin, but the molecules appear to have different roles. Whereas $\alpha_v\beta_5$ acts as the receptor for OS, the interaction with CD36 with resulting dimerization of the molecule is necessary and sufficient to activate the internalization process. Although we have shown that intracellular signals can be generated by the direct binding of PfEMP1 to endothelial CD36, it is conceivable that additional signals could be induced through the interaction of IRBCs with integrins and/or TSP-1, which can in turn interact with CD47 or the matrix protein $\alpha_v\beta_3$ integrin. The engagement of a core of receptors will immobilize IRBCs on the cell surface, leading to focal aggregation of the receptors into a functional complex. In support of this hypothesis, an anti-α_v antibody has been shown to inhibit IRBC adhesion to HDMEC under flow conditions in vitro. As well, the uptake of apoptotic cells require both CD36 and $\alpha_v\beta_3$ on macrophages.

Other Signaling Events Triggered by *P. falciparum*

Other *P. falciparum* molecules have been reported to activate signaling in endothelial cells. The GPI anchor of two *P. falciparum* proteins MSP-1 (merozoite surface protein 1) and MSP-2 has been shown to stimulate nitric oxide production in human umbilical vein endothelial cells. The same glycolipid also upregulates the expression of ICAM-1, VCAM-1, and E-selectin on endothelial cells, leading to increased leukocyte and parasite adhesion. The latter effect of the GPI can be inhibited by the Src-family kinase antagonist herbimycin A, suggesting that the activation of nonreceptor protein tyrosine kinases in endothelial cells may constitute a critical proximal event in integrating the regulation of IRBC adhesion by diverse signals. Further understanding of the cellular consequences cytoadherence is urgently required to provide the basis for novel modes of therapy against one of the most devastating infections of humankind.

Glossary

CD36: An 88-kDa transmembrane scavenger molecule found on microvascular endothelial cells, erythroblasts, monocytes, adipose tissue, platelets, dendritic cells, and microglial cells. CD36 serves many physiological roles such as the uptake of apoptotic bodies and lipids, fatty acid transport, and apoptosis of endothelial cells to prevent neovascularization and may have significant roles in the pathogenesis of atherosclerosis, diabetes, and Alzheimer's disease.

Cytoadherence: The process whereby erythrocytes infected with mature stages of *Plasmodium falciparum* adhere to vascular endothelium. Cytoadherence is mediated by the parasite protein *Plasmodium falciparum* erythrocyte membrane protein 1 expressed on the surface of the infected red cell and a number of endothelial adhesion molecules in a cascade of interactions resembling the adhesive events in leukocyte recruitment.

Intracellular signaling: A cascade of specific molecular interactions that facilitates information transduction from the inner cytoplasmic membrane leaflet throughout the interior of the cell into the nucleus and often results in changes in DNA regulation and expression.

Src family kinases: A family of nonreceptor protein tyrosine kinases that mediate intracellular signaling. These molecules include Src, Fyn, Yes, Fgr, Lyn, Hck, Lck, Blk, and Yrk and share characteristic src-homology domains (SH) that are regulated through tyrosine phosphorylation.

Bibliography

Asch, A. S., Liu, I., Briccetti, F. M., Barnwell, J. W., Kwakye-Berko, F., Dokun, A., Goldberger, J., and Pernambuco, M. (1993). Analysis of CD36 binding domains: Ligand specificity controlled by dephosphorylation of an ectodomain. *Science* **263**, 1436–1440. *The first description of the importance of the phosphorylation state of the ectodomain of platelet CD36 as a determinant of its binding affinity for various ligands.*

Febbraio, M., Hajjar, D. P., and Silverstein, R. L. (2001). CD36: A class B scavenger receptor involved in angiogenesis, atherosclerosis, inflammation, and lipid metabolism. *J. Clin. Invest.* **108**, 785–791.

Ho, M., and White, N. J. (1999). Molecular mechanisms of cytoadherence in malaria. *Am. J. Physiol.* **276** (Cell Physiol. 45), C1231–C1242. *A comprehensive review of the roles of various adhesion molecules and parasite proteins involved in cytoadherence.*

Schofield, L., Novakovic, S., Gerold, P., Schwarz, R. T., McConville, M. J., and Tachado, S. D. (1996). Glycosylphosphatidylinositol toxin of *Plasmodium* upregulates intercellular adhesion molecules-1, vascular cell adhesion molecule-1, and E-selectin expression in vascular endothelial cells and increases leukocyte and parasite cytoadherence via tyrosine kinase-dependent signal transduction. *J. Immunol.* **156**, 1886–1896.

Shattil, S. J., and Brugge, J. S. (1991). Protein tyrosine phosphorylation and the adhesive functions of platelets. *Curr. Opin. Cell Biol.* **3**, 869–879.

Smith, J. D., Gamain, B., Baruch, D. I., and Kyes, S. (2001). Decoding the language of *var* genes and *Plasmodium falciparum* sequestration. *Trends Parasitol.* **17**, 538–545.

Yipp, B. G., Anand, S., Schollaardt, T., Patel, K. D., Looareesuwan, S., and Ho, M. (2000). Synergism of multiple adhesion molecules in mediating cytoadherence of *Plasmodium falciparum*–infected erythrocytes to microvascular endothelial cells under flow. *Blood* **96**, 2292–2299.

Yipp, B. G., Robbins, S. M., Resek, M. E., Baruch, D. I., Looareesuwan, S., and Ho, M. (2003). Src-family kinase signaling modulates the adhesion of *Plasmodium falciparum* on human microvascular endothelium under flow. *Blood* **101**, 2850–2857. *The first demonstration that IRBC adhesion to CD36 on microvascular endothelial cells leads to intracellular signaling in the host cell that could eventually result in a change in the phosphorylation state and hence binding affinity of CD36 for IRBCs. The findings also suggest the therapeutic potential of phosphatase inhibitors such as levamisole in the treatment of severe falciparum malaria.*

Capsule Biography

Dr. Ho is a Professor in the Department of Microbiology and Infectious Diseases at the University of Calgary, Canada. She is a Senior Medical Scholar of the Alberta Heritage Foundation for Medical Research, and her laboratory primarily focuses on the molecular pathogenesis of severe *P. falciparum* malaria. Her work is supported by grants from the Canadian Institutes of Health Research and the Anemia Institute of Research and Education, Canada.

Bryan Yipp is an M.D./M.Sc. student at the University of Calgary who studies the molecular pathogenesis of *Plasmodium falciparum* in Dr. May Ho's laboratory.

Dr. Robbins is an Associate Professor in the Departments of Oncology and Biochemistry and Molecular Biology at the University of Calgary. He is a Senior Scholar of the Alberta Heritage Foundation for Medical Research and currently holds a Canada Research Chair in the area of Cancer Biology. His research, which has focused on defining signal transduction pathways in various pathophysiological conditions, is supported by grants from the Canadian Institutes of Health Research, Cancer Research Society, Kids Cancer Care Foundation of Alberta, and the Alberta Cancer Board.

CHAPTER 113

Ischemia–Reperfusion Injury

Shahrul I. Ibrahim, Rodney K. Chan, and Herbert B. Hechtman
Department of Surgery, Brigham and Women's Hospital and Harvard Medical School, Boston, Massachusetts

Introduction

Over the years, scientists have been studying the phenomenon of ischemia–reperfusion injury (IR). In simple terms, it is the resultant injury to tissues that are reperfused following a period of ischemia. It can be a localized event or it can be an overwhelming global phenomenon affecting various organs in the body, so-called remote organ injury. This is rather paradoxical, as one would expect the restoration of blood supply to be beneficial, but in effect the injury sustained following reperfusion is far worse than that from the ischemic event alone.

The clinical relevance of reperfusion injury is enormous and encompasses many different disciplines of medicine. Every successful free flap, organ transplant, balloon angioplasty, or decompression fasciotomy can be threatened by ischemia–reperfusion injury. It is a common clinical problem but there has so far been no effective treatment. Early research implicated oxygen free radicals and neutrophils as central to ischemia–reperfusion injury, but this did not lead to any successful clinical trials, let alone treatment. The Harvard physiologist Walter Bradford Cannon and his colleagues stimulated much of the early research work in ischemia–reperfusion. Cannon proposed the humoral theory of disease, that all bodily functions are controlled by circulating factors [1].

Events Occurring during Ischemia–Reperfusion

Rapid changes occur during ischemia, which leads to pathway changes in signaling and surface molecule expression. There is accumulation of toxic products intracellularly, leading to apoptosis and necrosis. Ischemia can be partially or completely reversible or irreversible, depending on its severity and duration. Following reperfusion, accumulated toxic metabolites are flushed into the general circulation, which may have adverse local or remote effects on organs.

Two major concepts have been forwarded regarding the mechanisms of ischemia–reperfusion injury. Early studies implicated neutrophil stimulation and oxygen free radicals as the main culprits. Following ischemia, there is enhanced neutrophil adherence to endothelial cells secondary to increased expression of adhesion molecules. This then leads to neutrophil diapedesis, their oxidative burst, and the formation of oxygen free radicals [2].

Most recent studies, however, point toward ischemia–reperfusion as part of the inflammatory response to injury, involving the complement system. The fact that numerous studies employing novel ideas of blocking neutrophils or oxygen free radicals failed to produce any meaningful or clinically relevant findings has changed the focus of investigators toward the complement system, which was first described in myocardial infarction in the 1980s.

Overview of Inflammation

Inflammation involves a complex series of reactions including the localized accumulation and activation of leukocytes and certain plasma proteins following a toxin exposure, cell injury, or infection. Inflammation is a protective mechanism, controlling infections and promoting tissue repair; it can also be a cause of tissue damage and disease if prolonged or overly intense.

Proinflammatory cytokines and chemokines are released in response to injury or infection. These activate mast cells in the connective tissue as well as basophils, neutrophils, and platelets. These leukocytes can migrate from injured microvessels and release or stimulate the synthesis of vasodilators such as nitric oxide, histamine, bradykinins, and prostaglandins, as well as some powerful vasoconstrictor agents including leukotrienes and thromboxane A_2. Products of the complement system can also trigger mast cell and platelet release of vasoactive agents.

Cytokines are polypeptides produced in response to microbes and other antigens that mediate and regulate immune and inflammatory reactions. They are structurally diverse but share several properties: secretion is a brief, self-limiting event; the actions often are pleiotropic and redundant; they often influence the synthesis and actions of other cytokines; actions may be local or systemic; these actions are initiated by cytokine binding to specific membrane receptors; the expression of cytokine receptors and the resultant cell responsiveness are regulated by external signals; and cellular responses to cytokines are due to changes in gene expression in target cells. Examples of cytokines include the interleukins, tumor necrosis factor (TNF), and the interferons.

Neutrophils play a major role in the immune system and are recruited locally into sites of injury by chemotactic agents including lipopolysaccharide derived from bacterial cell walls, cytokines, eicosanoids produced by local tissue monocytes and endothelial cells, and complement-derived anaphylotoxins C3a and C5a. Once neutrophils accumulate at the target site, they are stimulated by chemoactivators to release oxygen free radicals and proteases whose objective is to destroy offending organisms. In severe injury, infection or ischemia–reperfusion injury, there is indiscriminate and uncontrolled neutrophil-endothelial adhesion and the release of injurious agents that damage host tissues [3].

Neutrophils and Free Radicals in Ischemia–Reperfusion

In abdominal aortic aneurysm repair, reperfusion injury occurs following the release of the aortic clamp and is mediated in part by neutrophils. Free radical and eicosanoid production occur locally in the reperfused tissue and in circulating neutrophils. This occurs faster than cytokine synthesis, which requires transcription. Neutrophils have been implicated in many types of inflammation including the local and remote effects of ischemia–reperfusion.

The recruitment of neutrophils into reperfused tissue is a self-perpetuating cycle. Oxygen radicals cause endothelial activation and the upregulation of P-selectin and other neutrophil-specific integrins [4, 5]. Activated neutrophils produce more free radicals and recruit more neutrophils. The activated complement components C3a and C5a are also potent chemotactic agents. Neutrophil presence alone, however, is often not sufficient to cause reperfusion injury. Hence, in many experimental animal models, the creation of a neutrophil-free environment does not reduce injury. However, complement inhibition even in the presence of neutrophil sequestration is sufficient to limit local injury. This calls into question the actual role of neutrophils during reperfusion injury. They may play a secondary though not an unimportant role in mediating injury to the ischemic organ.

In intestinal ischemia, use of neutrophil adhesion molecule antagonists or antibodies to the endothelial counter receptor to prevent neutrophil sequestration blunted the remote lung but not the local intestinal injury. In one study, animals made neutropenic by pretreatment with anti-neutrophil antibodies before lower torso reperfusion had significantly decreased remote pulmonary and hepatic injury.

In discussing the role of oxygen radicals in reperfusion injury, paramagnetic resonance has shown that reactive oxygen species appear in tissue beds soon after an ischemic insult. The blockade of reactive oxygen radicals has produced promising preclinical experimental results. Manson et al. demonstrated enhanced skin flap survival when superoxide dismutase (SOD) was given after arterial and venous occlusion to scavenge reactive oxygen species. However, the administration of free radical scavengers after the onset of reperfusion proved ineffective.

It thus seems that the generation of oxygen radicals is likely to be a transient early event. The relationship between the generation of oxygen free radicals and complement activation is not well described. It is postulated that oxygen radicals are generated early and derived either from parenchymal cell xanthine oxidase or neutrophil NADPH. Either one could cause endothelial activation and cell injury, which might include exposure of the so-called "ischemia antigen" and the ensuing activation of the complement system. This idea is supported by in vitro cell culture studies that showed an increase in complement activation when hydrogen peroxide was added to human umbilical vein endothelial cells in a hypoxia–reoxygenation chamber. Studies have now concentrated on the role of the complement system in ischemia–reperfusion injury, which have been boosted by the development of specific complement knockout animals that make the study of the various components of the complement system possible. Reports indicate that therapeutic inhibition of complement with sCR1, C1 esterase inhibitor or monoclonal antibodies directed against specific complement components results in decreased tissue injury, improved organ function, and increased survival in different animal models of human disease [4, 6].

Overview of the Complement System

The complement system consists of more than 30 serum and cell surface proteins. It was first discovered by Jules Bordet, the Belgian Nobel laureate, in 1899. He noted that when serum with antibacterial antibodies was added to a sample containing bacteria, the bacteria underwent lysis by the action of the antibodies. However, when the serum was heated above a certain temperature, lysis ceased to occur. He concluded that since antibodies are heat stable, there must be another substance in the serum that assists or complements the antibodies. Hence the term "complement" was coined.

There are three separate pathways of the complement system: the classical, alternative, and mannose-binding lectin pathways. They differ in the way they are activated but have a common end product, the terminal C5-9, also known as the membrane attack complex (Figure 1).

The classical pathway is activated by antigen–antibody interaction, leading to the activation of C1q, with the subsequent C2- and C4-dependent cleavage of C3 (by C3 convertase C4b2a) and cleavage of C5 by the C5 convertase (C4b2a3b).

The lectin pathway is activated when the serum mannose binding lectin (MBL), which is a homolog of C1q, recognizes microbial surface mannose and triggers activation of MBL-associated proteases (MASP1-3), resulting in the same formation of C3 and C5 as in the classical pathway.

The alternative pathway is activated by the presence of lipopolysaccharide and also spontaneously generated C3b. C3b binds to factor B and forms a complex that is cleaved by factor D to form the alternative C3 convertase, C3(H20)Bb. The complex is stabilized by properdin, which acts as an amplifying activator. This enables the cleavage product C3b to bind to it, forming the alternative C5 convertase (C3b3bBb).

All three pathways use C3 and cleave C5, which leads to the formation of the proinflammatory cleavage products C5a and C5b-9, which are believed to be mainly responsible for ischemia–reperfusion injury. There is cross talk between the three pathways in a few species including mice. The pathways augment each other. C3 is at the intersection of all three pathways and its inhibition will lead to a complete blockade of complement activation and hence the formation of the membrane attack complex. This complex results in

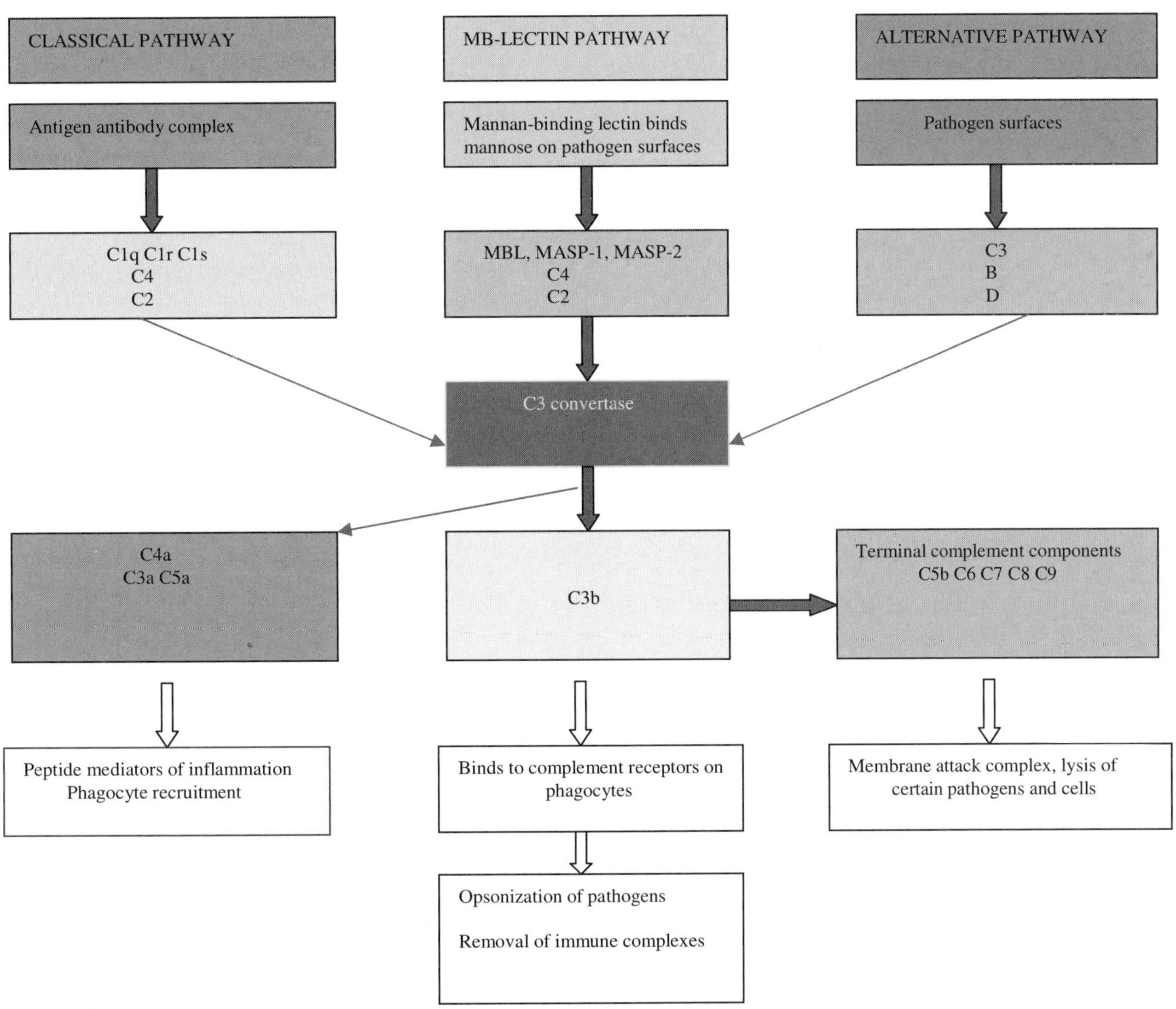

Figure 1 The three complement pathways. (see color insert)

pore formation in the cell membrane, causing abnormal ion transport, altered signaling, and possible cell lysis and eventual death.

In early studies, C3 depletion using cobra venom factor was reported to be protective during ischemia–reperfusion in kidney and heart [7]. Since then, complement inhibition in multiple animal models has been successful in limiting IR injury. In experimental myocardial ischemia, C1 esterase inhibitor administered before reperfusion prevented the deposition of C1q and significantly reduced the area of cardiac muscle necrosis. It was also shown to confer protective effects in models of lung transplantation. Animals genetically deficient in C3 have less local tissue necrosis after skeletal muscle or intestinal ischemia. The protection from local and remote injury accorded to C5 deficient animals in models of ischemia–reperfusion confirmed the central role of the membrane attack complex [8]. These preclinical results highlight the importance of complement activation.

Free radical scavengers have been used in similar studies and showed promise, but results of clinical trials were discouraging. Although the efficacy of complement antagonism in clinical trials cannot be predicted with certainty, these drugs are likely to fare better than the free radical scavengers. Free radical scavengers have never worked when given after the onset of reperfusion. In contrast, an inhibitor of complement activation, soluble *crry* (complement receptor-related gene Y), was effective against injury when given both prior to and after the onset of murine intestinal reperfusion. The use of a soluble complement receptor 1 (sCR1) in a model of skeletal muscle ischemia also reduced muscle injury when given up to an hour after the start of reperfusion. The complement system has also been shown to mediate intestinal injury after resuscitation from hemorrhagic shock, which can be viewed as global body ischemia and reperfusion. The inhibition of complement after hemorrhage reduced tissue injury and neutrophil influx and preserved nitric oxide synthase function. The exact complement pathway has yet to be ascertained, but complement inhibition might provide a novel avenue for treating patients with major hemorrhage and prevent the compounding problem of gut ischemia, as well other problems associated with the multiple organ dysfunction syndrome.

Complement Activation by Immune Complexes during Reperfusion

The idea that antibodies are involved in reperfusion injury initially came from observations after intestinal ischemia that C4-deficient animals were protected to a degree similar to those deficient in C3. Since C4 is unique to the classical pathway, which is typically activated by an immune complex, the hypothesis set forward was that immune complexes were formed during reperfusion (Figure 2). Supporting evidence has been developed using immunoglobulin-deficient RAG mice. These animals have injury reduced to levels seen in C4-deficient mice. Further, reconstitution of RAG animals with wild-type IgM will fully reproduce injury while reconstitution with wild-type IgG will not [5].

IgG, however, might play a role in augmenting IgM-mediated injury. Immunohistochemical staining for IgM and C3 on skeletal muscle demonstrates that IgM binding occurs

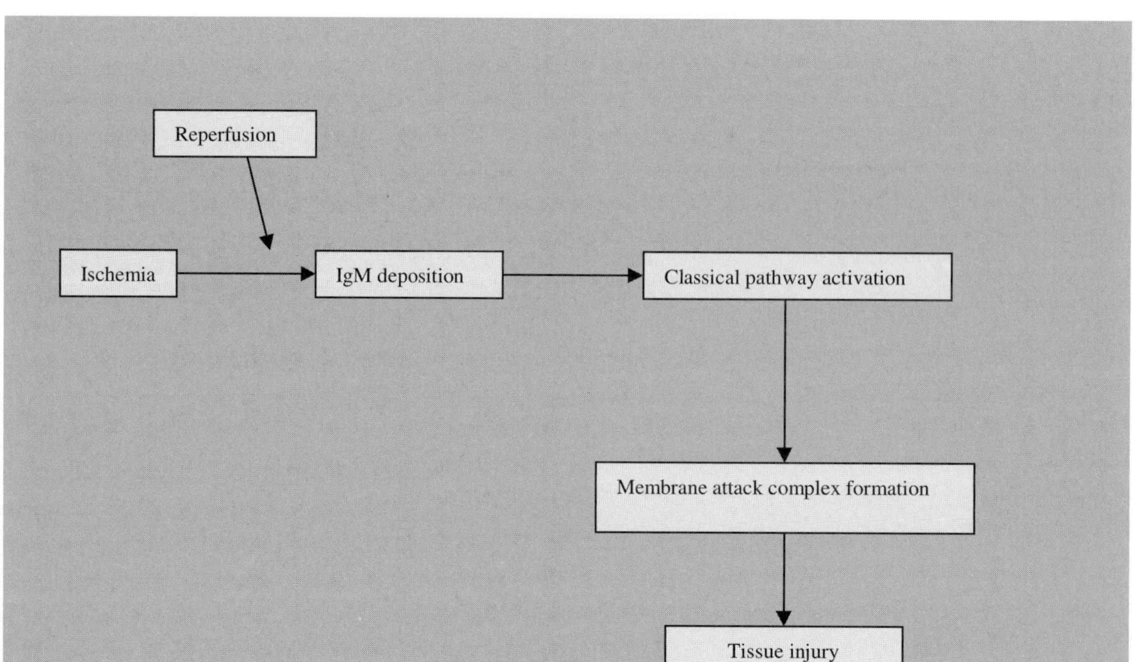

Figure 2 IR activates the classical complement pathway with formation of the membrane attack complex. (see color insert)

during ischemia while complement activation and injury occur during reperfusion [11]. This temporal sequence along with the histologic colocalization of IgM and C3 indicates that the observed IgM binding to tissue is the initiator of complement activation and subsequent reperfusion injury.

Most immunologists will agree that our understanding of IgM and its immunological interactions is at best elementary. IgM exists in circulation as a pentamer linked by covalent interactions to a J-chain. Identification of its role in mediating reperfusion injury presents a new paradigm and the opportunity to target a key mediator of this clinical phenomenon.

Binding of the Ischemia Antigen by Natural Antibodies

Reperfusion injury has been shown to be mediated by a specific category of IgM, produced by a limited subset of $CD5^+$ B-1 cells found predominantly in the pleural and peritoneal surfaces of adult humans and mice. These cells were traditionally thought of as being limited in their repertoire, and being polyreactive with low affinity for their antigens. They have been implicated in the removal of senescent red cells and in the pathogenesis of type 1 or nonobese diabetes and rheumatoid arthritis, as well as other autoimmune diseases.

Mice genetically lacking the complement receptor 2 molecule were found to have a limited production of $CD5^+$ B-1 cells, to levels 40 percent of that in wild-type mice. These CR2-deficient animals, similar to RAG mice, when subjected to either hind limb or intestinal ischemia–reperfusion, were spared injury. The injury phenotype was reproduced when these animals were supplied with a repertoire of nascent IgM or B-1 cells purified from a wild-type animal. Such experiments indicate that the causative antibody lies not just within the pool of total body IgM or even total IgM natural antibodies but within the 60 percent of natural antibodies missing from CR2–/– animals. Preliminary evidence derived from cloned peritoneal B1 cells suggests that at least one IgM natural antibody exists that is sufficient to cause injury. Although the identity of epitopes to which this "injury-causing" IgM binds is still unknown, it is believed to be a unique protein antigen.

It is widely believed that ischemia leads to the formation of an ischemia antigen on cellular surfaces that binds to circulating IgM natural antibody, forming an immune complex that activates C1 binding, complement activation, and formation of C3a and C3b. C3b then activates the rest of the complement cascade, resulting in the formation of the membrane attack complex that causes cell injury.

Drugs have been designed to treat complement and natural antibody–mediated injury by targeting the classical pathway or terminal components of complement. Two drugs in clinical trials include sCR1 and an antibody against complement component C5. Identification of the neoantigen would allow for design of more specific neutralizing antibodies, noncomplement-activating F_{AB} fragments, or short amino acid chains that might bind and block the ischemia epitope.

Current Controversies

The hypothesis that natural antibodies and the classical complement pathway are involved in all ischemia–reperfusion injuries might be oversimplified. The kidney was found to be protected in a model of kidney reperfusion injury using C3-, C5-, and C6-deficient animals, but not C4-deficient animals, implying that the classical pathway may not be the initiator of injury in all cases. In fact, another study indicated that complement might not be mediating renal ischemia–reperfusion injury at all as C3 inhibition with *crry* did not convey protection. These paradoxical results require investigation.

The classical, alternative, and lectin pathways also appear to communicate, much like the crossovers in the blood clotting system. The alternative pathway has been reported to be important in intestinal reperfusion injury using mice deficient in factor D [9]. It is possible that the classical pathway initiates injury but the alternative pathway is required for its amplification [10].

Lectin, the third pathway leading to complement activation, has been shown to attenuate myocardial reperfusion injury in rats given an MBL antibody post ischemia. These complement pathways may be species or organ specific, a phenomenon that might explain the different observations in the various animal studies.

Clinical Use of Complement Inhibitors in Reperfusion Injury

C1 esterase inhibitors have been used with positive outcome in three patients undergoing emergency coronary artery bypass after failed angioplasty. There are additional positive data from a study of 35 patients with attenuated myocardial injury: improved neurological outcome and less blood loss in the human C5 antibody treatment cohort after nonemergent cardiopulmonary bypass. Further, the use of sCR1 in patients undergoing lung transplant requiring bypass has shortened the ventilator-dependent time.

However, in the only randomized controlled clinical trial (Phase IIa) involving a complement antagonist, where sCR1 was administered to 600 patients undergoing adult cardiopulmonary bypass, there was no demonstrable mitigation of injury to the heart, brain, or other organs. It has been argued that since adult cardiopulmonary bypass has a low mortality and complication rate, averaging only 1 to 2 percent, it would be difficult for any drug to achieve a clinically significant difference. It is noteworthy that among safety trials performed with sCR1, there have been no reported complications. However, the potential side effects of increased susceptibility to infection might become evident in the future. It is important to keep in mind the beneficial actions of complement and, hence, the possible detrimental effects of its inhibition such as the susceptibility to infectious agents.

Paradoxically complement activation can be protective against reperfusion injury, especially in sublytic quantities. Infusion of 0.5 percent normal human plasma, containing small amounts of activated complement, into rabbits subjected to myocardial ischemia–reperfusion resulted in significantly smaller infarcts. This protective effect of plasma could be reversed by heat inactivation, or by pretreatment with sCR1 or anti-C5a antibody.

The Future: Mast Cells and Substance P

Mast cells have recently been shown to be activated in reperfused tissue. In the skeletal muscle of the reperfused mouse hind limb, extensive mast cell degranulation has been noted. Use of mast cell–deficient (W/W^v) mice has yielded protection against local and remote reperfusion injury. Mast cells have also been shown to mediate local injury after acid aspiration. Whether mast cells are complement-triggered effectors is uncertain. The mast cell and complement pathways likely augment one another, although evidence is still lacking.

It has been postulated that reperfusion injury could be attenuated by interrupting certain vagal nerve conduction pathway either surgically or pharmacologically because of the proximity of the mast cells to the peripheral nerve endings. Cervically vagotomized mice showed a 90 percent reduction in their remote lung injury and 40 percent reduction in local skeletal injury following hind limb ischemia–reperfusion. Rats treated with a substance P antagonist L-703606, a blocker of the NK1 receptor, have attenuated myocardial injury after ischemia. It is likely that, depending upon the timing of therapy, reperfusion injury would best be treated using a combination of anti-complement and anti-mast cell agents.

Conclusion

As we gain more understanding of the problem of ischemia–reperfusion injury, the emphasis has shifted from free radicals to complement as the main culprit. The role of the complement system will continue to be a vigorously investigated field for the foreseeable future. It is most likely that there is interaction among all three known pathways of the complement system. However, depending upon the particular organ involved and whether ischemia and reperfusion are synchronous, such as in hemorrhage, one complement pathway might dominate. The potential benefits are enormous if current animal studies examining the role of complement in diverse conditions as limb ischemia, wound healing and hemorrhagic hypotension could be translated into positive clinical trials. It would certainly change the way we look at many disease processes in the body, and would clearly change therapy.

Glossary

Complement: A group of serum and cell surface proteins that generate effectors of innate and adaptive immune responses through interactions with one another or with the rest of the immune response system. The term was first coined following its discovery by Jules Bordet, a Belgian Nobel laureate, in the late 19th century.

Ischemia antigen: The antigen thought to be formed following ischemia that binds to naturally occurring IgM, forming an immune complex that then activates C1 binding, complement activation, and formation of C3a and C3b. C3b then activates the rest of the complement cascade, resulting in the formation of the membrane attack complex that causes cell injury.

Membrane attack complex: Terminal C5b-9 common end product of the complement cascade which is inserted into cell membranes. It has both cytolytic and noncytolytic effects, with the latter thought to be more important by causing irreversible cell injury by attracting activated neutrophils.

sCR1: Naturally occurring soluble form of the C3b-C4b cell surface receptor, CR1. Now bioengineered and in Phase II clinical trials.

Acknowledgment

Supported by U.S. Public Health Service Grant P50 GM52585.

References

1. Moon, V. H. (1938). Systemic circulatory reactions to diverse agents. In: *Shock and Related Capillary Phenomena*, V. H. Moon, ed., pp. 236–251. Philadelphia: Oxford Medical Publications. *In the late 1800s, Walter Bradford Cannon, the Harvard physiologist, proposed the humoral theory of disease, arguing that all bodily functions are controlled by circulating factors. He and colleagues fueled much of the early research on ischemia–reperfusion injury.*
2. Riedeman, N. C., and Ward, P. A. (2003). Complement in ischemia reperfusion injury. *Am. J. Pathol.* **162**, 363–367. *This paper gives an overview of reperfusion injury. It also describes the various ways of blocking the complement system in experimental ischemia–reperfusion injury and reviews the paper by Stahl et al. that describes the role of the alternative complement pathway.*
3. Weiser, M. R., Gibbs, A. L., and Hechtman, H. B. (1997). Strategies to inhibit cellular adhesion molecules. In: *Adhesion Molecules in Health and Disease*, (Leendert, C. P., and Thomas, B. I., eds.) p. 55. New York: Marcel Dekker. *This paper describes the role of neutrophils and their role as a key component of the defense against infection, severe injury, and ischemia–reperfusion injury, as well as strategies employed to block them and limit the unwanted effects of their activation.*
4. Weiser, M. R., Williams, J. P., Moore, F. D., Jr., Kobzik, L., Ma, M., Hechtman, H. B., and Carroll, M. C. (1996). Reperfusion injury of ischemic skeletal muscle is mediated by natural antibody and complement. *J. Exp. Med.* **183**(5), 2343–2348.
5. Williams, J. P., Pechet, T. T., Weiser, M. R., Reid, R., Kobzik, L., Moore, F. D., Jr., Carroll, M. C., and Hechtman, H. B. (1999). Intestinal reperfusion injury is mediated by IgM and complement. *J. Appl. Physiol.* **86**(3), 938–942.
6. Weissman, H. F., Bartow, T., Leppo, M. K., Marsh, H. C. J., Carson, G. R., Concino, M. F., Boyle, M. P., Roux, K. H., Weisfeldt, M. L., and Fearon, D. T. (1990). Soluble human complement receptor type 1: In vivo inhibitor of complement suppressing post ischemic myocardial inflammation and necrosis. *Science* 249: 146–151.

7. Stein, S. H., Osgood, R. W., Barnes, J. L., Reineck, H. J., Pinckard, R. N., and McManus, L. M. (1985). The role of complement in the pathogenesis of postischemic acute renal failure. *Miner. Electrolyte Metab.* **11**, 256–261.
8. Kyriakides, C., Austen, W., Jr., Wang, Y., Favuzza, J., Kobzik, L., Moore, F. D., Jr., and Hechtman, H. B. (1999). Skeletal muscle reperfusion injury is mediated by neutrophils and the complement attack complex. *Am. J. Physiol.* **277**(6 part 1), C1263–C1268.
9. Stahl, G. L., Xu, Y., Hao, L., Miller, M., Buras, J. A., Fung, M., and Zhao, H. (2003). Role for the alternative complement pathway in ischemia reperfusion injury. *Am. J. Pathol.* **162**(2), 449–455.
10. Riedemann, N. C., and Ward, P. A. (2003). Complement in ischemia reperfusion injury. *Am. J. Pathol.* **162**(2), 363–367.
11. Chan, R. K., Ding, G., Verna, N., Ibrahim, S. I., Oakes, S., Hechtman, H. B., Moore, F. D., Jr. (2004) *IgM binding to injured tissue precedes complement activaion during skeletal muscle ischemia-reperfusion Journal of Surgical Research* **122**(1), 29–35.

Capsule Biography

Dr Shahrul Ibrahim is a graduate of The Royal College of Surgeons in Ireland and a Member of The Royal College of Surgeons of Edinburgh. He did his surgical training in Cumbria and East Yorkshire, UK, and is currently a surgical research fellow at the Brigham and Women's Hospital and Harvard Medical School, Boston, MA.

Dr Rodney Chan is a graduate of MIT and qualified from Harvard Medical School. He is on the Brigham and Women's Surgical Residency Program and currently a surgical research fellow at the Brigham and Women's Hospital and Harvard Medical School, Boston, MA.

Dr Herbert Hechtman is a graduate of Harvard Medical School and is Professor of Surgery at the Brigham and Women's Hospital and Harvard Medical School, Boston, MA. He has authored several books and published extensively on the subject of ischemia–reperfusion injury.

CHAPTER 114

Endothelial Xanthine Oxidoreductase and Vascular Disease

[1]Usamah S. Kayyali and [2]Paul M. Hassoun

[1]*Tufts-New England Medical Center, Tufts University School of Medicine, Boston, Massachusetts*

[2]*Johns Hopkins University School of Medicine, Baltimore, Maryland*

Introduction

Xanthine oxidoreductase (XOR) is the terminal enzyme in purine metabolism (xanthine dehydrogenase). Because of its abundance in milk, XOR has been the subject of extensive biochemical characterization as the leading member of the family of molybdo-flavoenzymes. Another unique property of the enzyme is its ability to generate reactive oxygen species (ROS), especially when it is converted from the original translational product, xanthine dehydrogenase, to the oxidase form. This capacity to generate ROS brought XOR to the forefront in the research fields of oxidative stress, ROS signaling, and inflammation. However, after several decades of investigation, many unanswered questions remain about the specific roles XOR might play in health and disease. This chapter will review certain aspects of XOR with special emphasis on its physiology in microvascular endothelial cells. The reader is also encouraged to consult recent comprehensive reviews related to the general biology of XOR [1–3].

Biology and Function of Xanthine Oxidoreductase

Xanthine oxidase is derived from xanthine dehydrogenase, the translational product of the XOR gene, by posttranslational modification. Biochemical characterization revealed that xanthine dehydrogenase (XDH), in its active form, exists as a homodimer of two protein molecules. Catalysis occurs by transfer of electrons along a chain of molybdo-pterin cofactor, two Fe_2S_2 sites, and an FAD site. XOR can be the target of oxidation and proteolysis, modifications that change the binding specificity of the enzyme and lead to the formation of xanthine oxidase (XO). Unlike XDH, which binds to NAD^+ as its preferred electron acceptor leading to the formation of NADH, XO can no longer bind NAD^+. However, the enzyme XO retains its activity in the conversion of hypoxanthine into xanthine and uric acid, and utilizes molecular oxygen as an electron acceptor, leading to the formation of the highly reactive oxygen species superoxide. The alteration in binding specificity for NAD^+ is irreversible in the case of proteolysis, but can be reversed by reducing agents in the case of oxidation. However, it is important to point out that the dehydrogenase form of XOR can also utilize molecular oxygen and produce ROS, albeit less efficiently than the oxidase form. Furthermore, XOR can also function as an NADH oxidase, which generates ROS without metabolizing xanthine or hypoxanthine. The catalysis in this reaction occurs at the FAD site and does not involve the molybdenum cofactor, and therefore it is not prevented by XOR inhibitors, such as allopurinol, that target the molybdenum cofactor [2].

Structural discoveries followed the biochemical characterization of XOR and largely supported the model derived from biochemical characterization. The crystal structure of XOR from bovine milk was elucidated in an elegant study which demonstrated that the conformation of the Fe_2S_2

clusters, the FAD, and the moybdo cofactor domains corresponded to those characterized biochemically [4]. Furthermore, when XOR was subjected to limited proteolysis *in vitro*, it underwent a conformational change that resulted in changing of the electrostatic charge at the opening to the FAD site, thus blocking access of NAD^+ to that site [4]. Such a conformational change supports the modifications in biochemical properties resulting from the conversion of xanthine dehydrogenase into the oxidase form.

Although its primary role is as the terminal limiting step in purine metabolism, XOR is known to metabolize other substrates as well, for example, ethanol. Of particular interest for endothelial physiology is the capacity of XOR to reduce nitrite resulting in the formation of NO. However, the physiological significance of this particular activity is not clear considering that the K_m value for this reaction is much higher than the concentration of nitrite normally found in tissues. Nevertheless, XOR has been argued to be a potentially significant source of NO, particularly in ischemic tissues.

Xanthine Oxidoreductase in Endothelial Cells

Although XOR has been implicated in the pathogenesis of different types of injury, the source of XOR causing the injury is still under debate. The complexity stems from the differences in XOR distribution between species. Another source of difficulty comes from the particular method used to study distribution, such that the level of mRNA detected by in situ hybridization might not correspond to the amount of protein detected by immunocytochemistry, which in turn might not correspond to the amount of active enzyme in lysed tissue. In rats and mice, the highest levels of XOR mRNA, protein, and activity were found in the small intestine, followed by liver and lung. In bovine tissue, XOR was immunolocalized to the epithelium and endothelium of the mammary glands as well as in the liver, heart, and lungs. It is concentrated in capillary, but not in macrovascular, endothelial cells. In humans, XOR was immunolocalized to skeletal and cardiac muscles, and its activity was reported in brain autopsy samples. However, other reports claimed very low enzyme activity in human heart autopsy samples. In another study, while XOR mRNA was detected in the lung, heart, brain, and kidney, enzyme activity was very low. Lack of significant activity in the just-mentioned organs called into question the role of XOR in ischemic injury to these organs. However, the discrepancy has been explained either by dilution or loss of XOR activity during tissue isolation, or by involvement of XOR from a remote source (e.g., released from liver) in ischemic injury to organs such as lung. In addition, it is possible that although the basal level of XOR activity is low in these organs, it might be induced under disease conditions. For example, drastic changes in XOR activity are seen in mammary tissue under specific conditions (see later discussion).

XOR is found primarily in the cytoplasm of endothelial cells, but several other reports have localized it on the cell membrane. We and others have found XOR to be localized in discrete cytoplasmic vesicles in normoxic microvascular endothelial cells. However, XOR protein expression increases in response to hypoxia, and the enzyme assumes a diffuse cytoplasmic distribution (Figure 1). In mammary epithelial cells, XOR is also found as part of the milk fat globule complex associated with proteins such as butyrophyllin. This localization, and the fact that XOR is normally undetectable in mammary epithelial cells but increases significantly in late pregnancy and during lactation, suggested a role for XOR in secretion. As previously mentioned, the fact that XOR can indeed be secreted in plasma has implicated the enzyme in damage caused to organs distant from the site of primary injury. The role of circulating XOR in disease is further discussed in a later section, "XOR and Models of Injury."

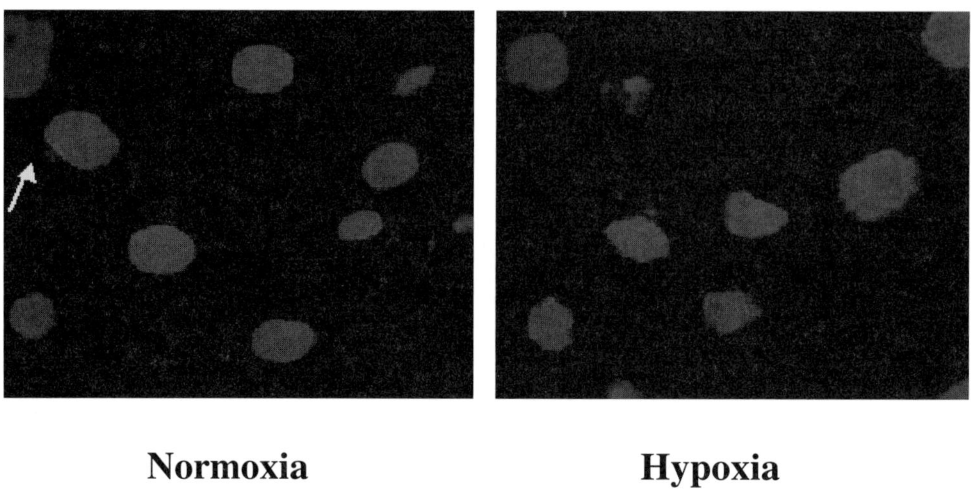

Figure 1 XOR is localized in discrete vesicles in pulmonary microvascular endothelial cells (arrow) and undergoes redistribution in response to hypoxia. (see color insert)

Regulation of Xanthine Oxidoreductase in Health and Disease

The expression of XOR varies in response to several cytokines, hormones, and stresses such as hypoxia and exposure to tobacco smoke. Studies have shown that XOR expression can be altered at the activity, protein, mRNA, and gene promoter levels (Figure 2). However, it is important to point out that there are some differences in the XOR promoter region of human versus rodent. Unlike the rat and mouse, the human XOR promoter contains TATA-like elements, which are believed to play a role in repressing the basal level of XOR in humans compared to rats. The human XOR promoter region also contains sites responsive to cytokines, such as IL-6 responsive elements as well as potential TNF-α, IFN-γ, and IL-1β responsive elements. Moreover, cytokines such as IFN-γ, IL-1β, and TNF-α have been shown to upregulate XOR in human mammary epithelial cells. IFN also upregulates XOR in the L929 fibroblast cell line. Treatment of mice with endotoxin lipopolysaccharide (LPS) activates XOR in several tissues. In a rat model of acute lung injury, LPS in combination with IL-1 upregulated XOR in the lung. This effect was also observed in cultured rat pulmonary artery microvascular endothelial cells [5].

Upregulation of XOR by LPS was further enhanced by exposing rats or endothelial cells to hypoxia. Furthermore, hypoxia alone upregulates XOR expression in bovine pulmonary artery endothelial and smooth muscle cells [5], and rat epididymal fat pad [5], and pulmonary artery microvascular endothelial cells [6]. However, hypoxia and cytokines upregulate XOR at multiple levels of expression. For example, the fact that cytokines increase the XOR mRNA and protein severalfold, albeit not enough to explain the observed increase in activity, suggested that part of the upregulation is due to post-translational modification [7]. This also applies to hypoxia, which increases mRNA and protein levels only after days of exposure [5], while the enzyme activity can be upregulated as early as four hours [6]. Post-translational modification of the protein can account for acute enzymatic activation. XOR activity is reduced by removal of the molybdo-cofactor or by desulfuration [2]. Hence, addition of the molybdo-cofactor and sulfuration of the XOR protein have been proposed as mechanisms for post-translational activation. Post-translational modification (phosphorylation) of XOR, causing acute enzyme activation in response to hypoxia, is linked to activation of p38 MAP kinase- and CK2-dependent pathways in pulmonary artery endothelial cells [6].

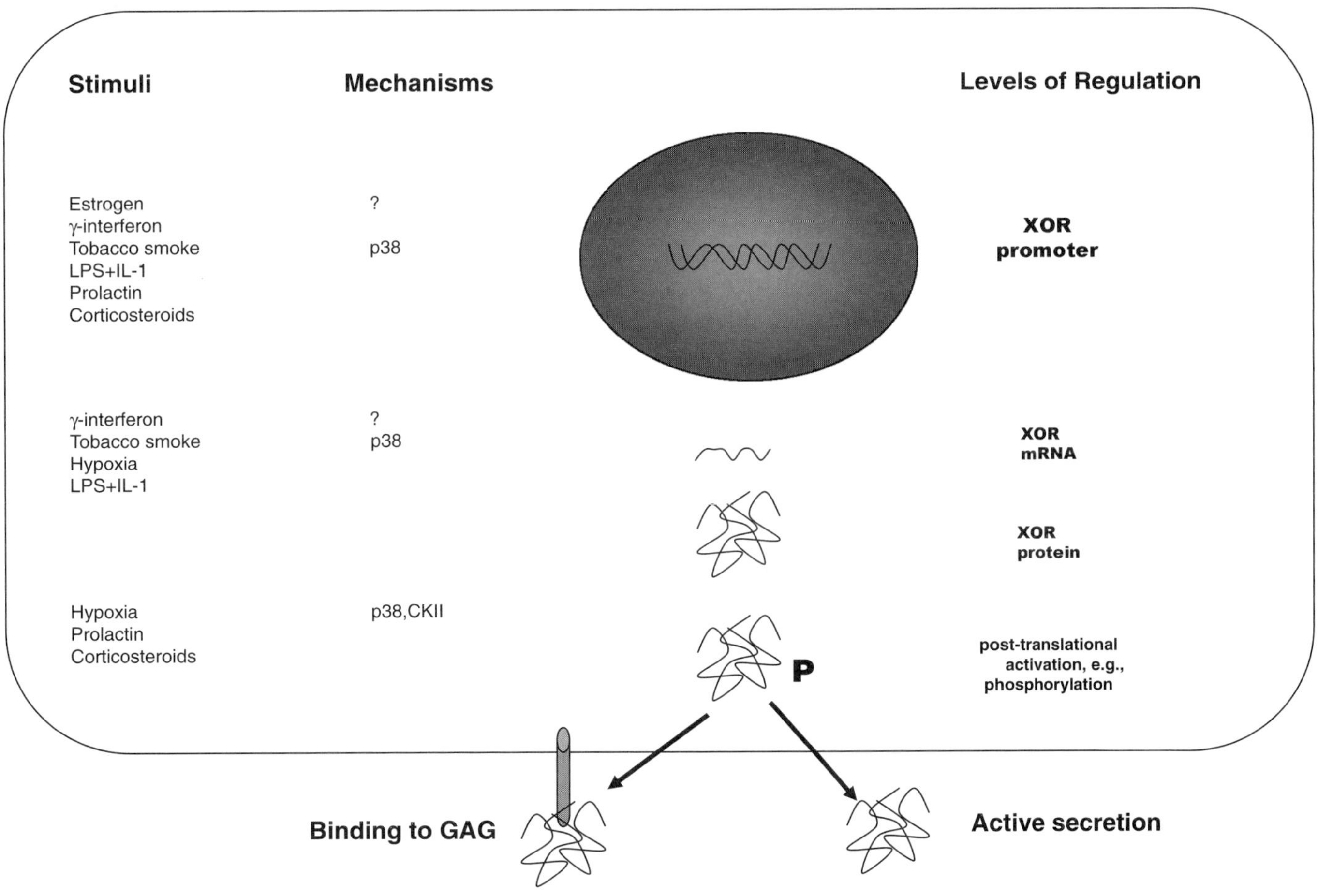

Figure 2 XOR expression is regulated at multiple levels. (see color insert)

Hormones such as prolactin and glucocorticoids are also important in regulating XOR expression. Both hormones upregulate XOR in mammary epithelial cells. XOR upregulation by prolactin and cortisone (both important for lactation) is probably relevant to the mechanisms of secretion and activation in milk, a very rich source of this enzyme. However, upregulation of XOR by glucocorticoids has also been observed in human kidney epithelial cells [8] and rat pulmonary microvascular endothelial cells (Kayyali and Hassoun, unpublished data). Since corticosteroids are important in inflammation, the latter action might be related to the putative role of XOR in the acute phase response and inflammation as discussed later. Estrogens, on the other hand, inhibit the upregulation of XOR by hypoxia in pulmonary microvascular endothelial cells. However, the finding that the nonreceptor-binding α-estradiol also inhibits XOR upregulation suggests a novel antioxidant property for estrogen-related compounds. Although the mechanisms of action of estrogens in the regulation of XOR need to be further elucidated, a hope is that nonestrogenic compounds, such as α-estradiol and possibly phytoestrogens, might represent potential therapeutic drugs having cardioprotective effects (e.g., antioxidant properties) without the deleterious effects of the active hormone.

Tobacco smoke is another form of stress that upregulates XOR in tissues such as the stomach mucosa, striated muscle blood vessels, synaptosomes, and pulmonary microvascular endothelial cells. In addition XOR has been reported to be upregulated by other toxicants such as PMA and tetrachlorodibenzodioxin (TCDD).

Signaling mechanisms that regulate the expression of XOR in different cells remain to be elucidated. One kinase that appears to be important for XOR regulation at different levels of expression is p38 MAP kinase. Inhibiting p38 and CK2 blocks the post-translational upregulation of endothelial XOR by hypoxia, a stimulus that has been shown to activate p38 [6]. In addition, inhibiting p38 blocks XOR gene promoter and transcriptional activation by tobacco smoke in pulmonary microvascular endothelial cells. Since p38 MAP kinase is a stress-activated kinase believed to be important in reactive oxygen signaling, its involvement in regulation of XOR expression merits further investigation.

Relation to Other Reactive Oxygen Producers

Although XOR has gained special attention because of its suggested involvement in ischemia–reperfusion injury, its relative importance has been challenged by other ROS producers. For example, in the mitochondria, ROS are produced as part of the respiratory oxidative phosphorylation. Mitochondria have also been reported to produce ROS under hypoxic conditions. In this so-called "reductive burst," ROS production is presumed to precede activation of other ROS-producing systems such as XOR and has been implicated in hypoxia-induced increased endothelial permeability and IL-6 production in human umbilical vein endothelial cells (HUVECs).

The NAD(P)H oxidase is another potential source of ROS in endothelial cells. This enzyme complex has been shown to produce ROS in vascular cells, including smooth muscle cells, fibroblasts, and endothelial cells, in a manner similar to the neutrophil NAD(P)H oxidase. However, unlike neutrophils, in which ROS are generated in large amounts in an "oxidative burst," vascular cell NAD(P)H oxidase generates sustained small amounts of ROS believed to be important in signaling and in the pathogenesis of vascular diseases. The NAD(P)H oxidase is also a regulated enzyme, the activity of which can be modulated by changes in one of its subunits, such as $p47^{phox}$. Activation of NAD(P)H oxidase has been implicated in a variety of vascular diseases, including hypertension.

In considering potential sources of ROS one should keep in mind the aldehyde oxidases, which bear significant homology to xanthine oxidase and are believed to be derived from copies of the XOR gene. This interesting family of oxidases has been shown to metabolize several xenobiotics. However, their natural substrates remain unknown. Since antibodies directed at XOR protein may cross-react with aldehyde oxidase protein, a note of caution should be raised about conclusions related to XOR in studies using such antibodies. Finally, an important source of ROS in the endothelium is nitric oxide synthase, which is covered in detail in other chapters.

The relative importance of each of the just-mentioned sources of ROS to different types of endothelial injury and vascular diseases needs to be elucidated. Although results obtained in a certain cell type under specific conditions might suggest a major role for one source versus others, most of these ROS-producing systems are likely to be involved in endothelial signaling or injury. Indeed, it is possible that these enzymatic systems might work in concert to mediate different phases of a specific response. For example, hypoxia might rapidly activate p38 through mitochondrial production of ROS, and p38 in turn will activate XOR, resulting in further production of ROS.

XOR and Models of Injury

Ischemia–Reperfusion Injury

Because of its ability to generate potentially noxious ROS, XOR has long been suspected to be involved in ischemia–reperfusion injury (IRI). Conversion of XOR from the dehydrogenase to the oxidase form was at the heart of a hypothesis proposed by Granger et al. and others to explain the specific involvement of XOR in IRI. In that scheme, ischemia causes an increase in intracellular calcium, stimulating a calcium-activated neutral protease [9] that irreversibly converts xanthine dehydrogenase to the oxidase form. Concomitantly, ischemia depletes intracellular ATP and leads to the accumulation of the XOR substrate hypoxanthine. With reoxygenation, xanthine oxidase converts molecular oxygen to superoxide and hydrogen peroxide, causing oxidative damage [9]. Although data from several studies indeed support a role for XOR in IRI, the hypothesis is challenged by the finding that conversion of XOR from the dehydrogenase to the oxidase form is either too slow or not significant during IRI. Furthermore, the beneficial effects on IRI of so-called specific XOR inhibitors, such as allopurinol, may be questioned since these compounds exert other effects on cellular metabolism and have anti-oxidant properties unrelated to their XOR inhibitory activity. However, as mentioned earlier, the conversion of XOR to the oxidase form is not necessary for ROS generation by the enzyme (see "Biology and Function of Xanthine Oxidoreductase").

Two lines of evidence that implicate XOR in IRI are of particular relevance to the microvasculature. The first is circulating XOR, which interacts with glycosaminoglycans on endothelial cells. Several studies have indicated that although normal plasma XOR levels are low, they increase in disease conditions such as IRI. Circulating XOR derived from XO-rich organs such as the liver has been proposed to cause damage to a variety of distant organs, particularly the lung. Such effect might explain multiorgan dysfunction in response to ischemia in a single organ such as the liver. The second line of evidence comes from studies that propose a role for XOR in mediating interactions between neutrophils and microvascular endothelial cells. Neutrophils have a significant role in tissue injury including IRI. Since interaction of neutrophils with endothelial cells involves upregulation of adhesion molecules by ROS, it is possible that, in addition to ROS generated by neutrophil NADH oxidase, XOR-derived ROS contribute to IRI by enhancing neutrophil–endothelial interactions.

XOR and Acute Lung Injury

The concept of injury related to XOR-derived ROS was originally promoted by McCord [10] and others in disease processes of the heart, kidney, lung, liver, and intestine. In animal models of acute lung injury (ALI), XOR is increased in lung parenchyma and bronchoalveolar lavage (about four-hundredfold). In humans, increased hypoxanthine and XOR levels have been demonstrated in the epithelial lining fluid from premature neonates with bronchopulmonary dysplasia, and in the serum of patients with the acute respiratory distress syndrome (ARDS), as compared to normal controls or critical-care patients with other organ diseases. Furthermore, plasma hypoxanthine levels are highest in nonsurvivors of ARDS, implicating oxidative damage (and presumably XOR) as a determinant of mortality in these mechanically ventilated patients. It is also possible that, in ARDS and ALI, XOR-derived superoxide and nitric oxide (NO) generated from inducible nitric oxide synthase (iNOS) react to form the highly toxic oxidant peroxynitrite, resulting in lung protein nitrotyrosine formation and oxidative damage. In support of oxidant-mediated toxicity by peroxynitrite is the demonstration of nitrotyrosine residues in the vascular endothelium and subendothelial tissues in patients with sepsis-induced ALI, and in the bronchoalveolar lavage of patients with ARDS. Whether the source of superoxide in this case is XOR or other lung oxidases, such as NAD(P)H oxidase, remains to be determined. Likewise, the specific source of NO may indeed be iNOS but potentially other oxidases, such as the neutrophil myeloperoxidase (through its NO oxidase activity), may also be involved.

XOR and Cardiovascular Diseases

The role of endothelial dysfunction and production of ROS in the pathogenesis of cardiovascular diseases has long been recognized. There has been renewed interest in the specific role of XOR in models of vascular diseases such as atherosclerosis and cardiomyopathy. Indeed, the endothelium has an important role in regulating local vasomotor tone. This is achieved through production of vasodilator (e.g., nitric oxide and prostacyclin) and vasoconstrictive substances (e.g., endothelin and superoxide). Therefore, tissue perfusion is quite dependent on a functional endothelium and a balance between these opposing endothelium-derived factors. An example of such imbalance is impaired flow-dependent, endothelium-mediated vasodilation in congestive heart failure (CHF), which appears to be secondary to reduced NO availability. The latter is dependent on production by endothelial nitric oxide synthase (NOS) but also on NO degradation by oxygen radicals, and in particular superoxide. Recent studies suggest that XOR is an important generator of superoxide in human vessels and may be one of the main sources of decreased NO availability, specifically through production of superoxide. Enhanced endothelium-bound XOR activity, in conjunction with decreased extracellular superoxide dismutase activity, has recently been associated with vascular oxidative stress in patients with CHF. A role for XOR in vascular oxidative stress is also suggested by reports of elevated myocardial XOR levels found in experimental heart failure and in patients with CHF, and identification of XOR in atherosclerotic plaques. Finally, an imbalance between NOS and XOR signaling pathways has been proposed in the regulation of myocardial mechanical efficiency, with upregulation of XOR relative to NOS as a leading factor in mechanoenergetic uncoupling in heart failure.

XOR and Lactation

The presence of XOR in cow and human milk has been long recognized; however, the exact function of the enzyme in the mammary gland has been unclear. The expression of XOR appears to be restricted to the mammary epithelium, and targeted disruption of XOR in mice revealed that XOR is important in the secretion of fat droplets in milk and, therefore, in maintaining lactation. In XOR +/– females, the mammary gland undergoes premature involution due to collapse of the mammary epithelium. In the secretion of the milk fat droplet, XOR translocates from a cytoplasmic position to an apical membrane location through structural interaction with two other milk fat globule proteins, butyrophilin and adipophilin. Whether this unique feature of XOR in secretory processes is also operative in the endothelium and in tissues other than the mammary gland is unclear at this time.

Xanthine Oxidoreductase: An Enigmatic Enzyme

Considerable research has been focused on XOR and its role in health and disease. Yet the extent of XOR involvement in disease processes has been debated over several decades. Questions remain concerning the low level of XOR activity in human tissues and about the exact significance of the conversion of XOR from the dehydrogenase to the oxidase form. However, there is no doubt that the enzyme can generate ROS and can be upregulated under a variety of pathological stimuli. Low basal XOR activity in normal human endothelial cells probably represents a tighter regulation in human versus other species. The debate regarding the significance of the conversion from the dehydrogenase to the oxidase form may be somewhat irrelevant since both forms of the enzyme can produce ROS. Therefore, as long as overall activity is upregulated under specific conditions, it is likely that ROS will be produced either as signaling molecules or as injurious agents depending on the situation.

The fact that uric acid, the end product of XOR catalysis in humans, is a strong antioxidant also raised speculation about the importance of XOR catalysis as an antioxidant rather than a pro-oxidant system. However, XOR continues to be a formidable generator of ROS when exogenously added to cultured cells. In vivo, XOR-derived ROS may be sufficient to directly damage cells, or to combine with NO (in the case of superoxide) in endothelial cells to produce highly reactive and damaging species such as peroxynitrite. Alternatively, XOR-derived ROS may be operative as signaling molecules mediating specific cellular responses to stress, such as interaction of phagocytes with endothelial cells.

A recent review suggested an evolutionarily conserved role for XOR in innate immunity [3]. Microvascular endothelial cells or mammary and intestinal epithelial cells are proposed to protect the organism through production of the antioxidant uric acid. Furthermore, signaling through XO-derived ROS may help recruit phagocytes or directly neutralize pathogens when ROS are produced in large amounts. This view is consistent with a role for XOR in the inflammatory or acute-phase response in which the endothelial cell might be a major protagonist. By analogy to the inflammatory process, such a protective role for XOR is not incompatible with the notion that XOR might also cause damage in disease processes. An inflammatory reaction generally offers protection to the organism but can lead to deleterious effects when unregulated, and diseases ranging from asthma to coronary artery disease involve dysregulated immune responses. Likewise, the role of XOR in disease should be viewed in terms of dysregulation leading to signaling processes that alter endothelial cell physiology and/or interaction with other cells. Rather than focusing on XOR as a simple producer of oxidants or antioxidants, future studies need to explore new avenues and search, for example, for proteins that interact with the enzyme, or identify mechanisms that regulate the enzyme or target it to specific cell compartments. The horizon for XOR research, whether or not the enzyme, like Janus, offers two faces and remains very promising.

References

1. Garattini, E., Mendel, R., Romao, M. J., Wright, R., and Terao, M. (2003). Mammalian molybdo-flavoenzymes, an expanding family of proteins: Structure, genetics, regulation, function and pathophysiology. *Biochem. J.* **372**, 15–32. *A comprehensive review of the family of molybdo-flavoenzymes to which XOR belongs. It provides insights into its function and relationship to other enzymes such as aldehyde oxidase whose function is not well understood.*
2. Harrison, R. (2002). Structure and function of xanthine oxidoreductase: Where are we now? *Free Radic. Biol. Med.* **33**, 774–797. *A comprehensive review of XOR that covers different aspects of its biochemistry and physiology. A very useful source for primary references on XOR research.*
3. Vorbach, C., Harrison, R., and Capecchi, M. R. (2003). Xanthine oxidoreductase is central to the evolution and function of the innate immune system. *Trends Immunol.* **24**, 512–517. *A discussion of the importance of XOR as a component of the immune system.*
4. Enroth, C., Eger, B. T., Okamoto, K., Nishino, T., and Pai, E. F. (2000). Crystal structures of bovine milk xanthine dehydrogenase and xanthine oxidase: Structure-based mechanism of conversion. *Proc. Natl. Acad. Sci. USA* **97**, 10723–10728. *A description of the crystal structure of xanthine oxidase that relates biochemical behavior of the different forms of XOR to structural components of the protein.*
5. Hassoun, P. M., Yu, F. S., Shedd, A. L., Zulueta, J. J., Thannickal, V. J., Lanzillo, J. J., and Fanburg, B. L. (1994). Regulation of endothelial cell xanthine dehydrogenase xanthine oxidase gene expression by oxygen tension. *Am. J. Physiol.* **266**, L163–L171.
6. Kayyali, U. S., Donaldson, C., Huang, H., Abdelnour, R., and Hassoun, P. M. (2001). Phosphorylation of xanthine dehydrogenase/oxidase in hypoxia. *J. Biol. Chem.* **276**, 14359–14365.
7. Page, S., Powell, D., Benboubetra, M., Stevens, C. R., Blake, D. R., Selase, F., Wolstenholme, A. J., and Harrison, R. (1998). Xanthine oxidoreductase in human mammary epithelial cells: Activation in response to inflammatory cytokines. *Biochim. Biophys. Acta* **1381**, 191–202.
8. Pfeffer, K. D., Huecksteadt, T. P., and Hoidal, J. R. (1994). Xanthine dehydrogenase and xanthine oxidase activity and gene expression in renal epithelial cells. Cytokine and steroid regulation. *J. Immunol.* **153**, 1789–1797.
9. Granger, D. N., Hollwarth, M. E., and Parks, D. A. (1986). Ischemia-reperfusion injury: Role of oxygen-derived free radicals. *Acta Physiol. Scand. Suppl.* **548**, 47–63.
10. McCord, J. M. (1985). Oxygen-derived free radicals in postischemic tissue injury. *N. Engl. J. Med.* **312**, 159–163.

Capsule Biography

Usamah S. Kayyali, Ph.D., M.P.H., is Assistant Professor of Medicine at Tufts-New England Medical Center/Tufts University School of Medicine. Dr. Kayyali's research interests include mechanisms of endothelial cell injury and signaling in hypoxia. His research focuses on p38 MAP kinase pathway, reactive oxygen signaling, and cytoskeletal regulation of endothelial physiology.

Paul M. Hassoun, M.D., is Associate Professor in the Department of Medicine at Johns Hopkins University School of Medicine, Division of Pulmonary and Critical Care Medicine. He is Director of the Pulmonary Hypertension Program at Johns Hopkins Hospital. Dr. Hassoun has had a long-standing interest in pulmonary vascular diseases, endothelial biology, and the role of oxidant systems in vascular diseases. His research work is focused on the regulation of certain endothelial enzymes, such as xanthine oxidoreductase and nitric oxide synthase, and their functions in disease processes.

CHAPTER 115

Antecedent Ethanol Ingestion Prevents Postischemic Microvascular Dysfunction

Ronald J. Korthuis
Department of Medical Pharmacology and Physiology, University of Missouri-Columbia

Introduction

The results of a large number of epidemiologic studies indicate that long-term, regular consumption of alcoholic beverages at low to moderate levels (one to three drinks per day for months to years) decreases the incidence of coronary artery disease and improves survival in patients suffering myocardial infarctions. Some of the first evidence in favor of this concept was derived from studies examining the relationship between cardiovascular mortality rates and dietary fat consumption in several European and North American countries. Whereas there was direct correlation noted between these variables in most Westernized societies, France was a notable exception in that this population demonstrated a much lower risk of death from cardiovascular disease than would be predicted from the levels of fat in their diets. A potential explanation for this "French paradox" became apparent when red wine consumption was factored into the analysis. Although the antioxidant properties of red wine constituents were originally thought to largely explain its cardioprotective effects, subsequent epidemiologic evidence indicated that consumption of white wine, beer, or spirits was also beneficial. The latter observations pointed to the importance of ethanol per se in the beneficial actions of alcohol intake on cardiovascular mortality.

Subsequent studies directed at the mechanistic underpinnings for the cardioprotective effects of moderate ethanol consumption focused primarily on lipoprotein and hemostatic factors. However, ethanol intake maintains a significant association with reduced cardiovascular mortality even after controlling for its effects on plasma lipids, platelet function, and fibrinolytic activity. Very recent work indicates that antecedent ethanol ingestion induces the development of an anti-inflammatory phenotype in postcapillary venules such that these vessels fail to support leukocyte adhesion and emigration in postischemic tissues. Given the critical importance of infiltrating leukocytes in the pathogenesis of atherosclerosis and ischemia–reperfusion (I/R) injury, these new observations provide novel insight regarding the mechanisms whereby ethanol ingestion reduces the likelihood and extent of I/R injury in individuals at risk for cardiovascular disease. The purpose of this review is to summarize our current understanding of the mechanisms whereby ethanol consumption induces the adaptive transformation to a protected or defensive phenotype in postcapillary venules such that these vessels fail to support I/R-induced leukocyte adhesion and emigration. Understanding the signaling pathways that are invoked by ingestion of alcoholic beverages may provide a mechanistic rationale for the development of novel treatment interventions that target both the microcirculatory and parenchymal sequelae to I/R, but do not produce the negative social and pathophysiologic consequences of ethanol ingestion.

Microvascular Consequences of Ischemia–Reperfusion

The microcirculation is particularly vulnerable to the pathologic effects of ischemia and reperfusion (I/R), with postischemic endothelial dysfunction occurring in all segments of the microvasculature and becoming evident as one of the earliest responses to reperfusion. For example, endothelium-dependent arteriolar vasoregulatory mechanisms are disrupted in postischemic tissues and contribute to the development of I/R-induced perfusion deficits in localized areas of ischemic tissues. Impaired arteriolar vasoregulation also appears to be a primary cause of variant angina and coronary vasospasm. The barrier function of endothelial cells in capillaries and postcapillary venules is also compromised as a consequence of I/R, an effect that leads to enhanced fluid and protein leakage from these vessels and the formation of interstitial edema. The accumulation of fluid in the extravascular compartment increases the path length over which oxygen must diffuse to reach parenchymal cells. This effect is exacerbated by postischemic endothelial cell swelling and edema-induced extravascular compression of the capillary lumen, processes that contribute to insufficient nutritive perfusion in I/R, a phenomenon referred to as capillary no-reflow. Postcapillary venular endothelium also participates in the production of I/R injury by coordinating the arrest and infiltration of leukocytes in ischemic regions, which then direct a focused attack upon parenchymal cells. I/R-induced platelet adhesion in postcapillary venules may facilitate this leukosequestration while their interactions with each other, leukocytes, and/or the arteriolar wall may promote thrombogenesis.

The aforementioned considerations indicate that preservation of endothelial function should be an important therapeutic goal in I/R. Of the many avenues pursued in this regard, preconditioning tissues to resist the deleterious effects of prolonged ischemia and reperfusion by antecedent exposure to ethanol represents one of the most promising interventions studied to date. For example, the effects of I/R to induce endothelial-dependent vasoregulatory dysfunction in arterioles, nutritive perfusion failure in capillaries (capillary no-reflow), adhesion molecule expression and leukocyte adherence in postcapillary venules, and endothelial barrier disruption in capillaries and postcapillary venules are completely prevented by prior ethanol ingestion. Of the beneficial actions microcirculatory actions that are induced by antecedent ethanol ingestion, our best mechanistic understanding pertains to its anti-inflammatory effects in postcapillary venules. Thus, this effect will be reviewed in the more detail.

Anti-inflammatory Effects Induced by Ethanol Ingestion

Based on the large body of evidence indicating that leukocyte infiltration into postischemic tissues plays an essential role in the production of reperfusion injury, it was suggested that ethanol may exert anti-inflammatory actions in the setting of I/R. Indeed, ethanol consumption at low to moderate levels is associated with reduced C-reactive protein levels, a plasma marker of inflammation, and decreases in the production of inflammatory cytokines such as interleukin-6. Using an intravital microscopic approach to examine postischemic leukocyte rolling, adhesion, and emigration in postcapillary venules in real time, we have demonstrated that antecedent ethanol exposure prevents adhesive interactions between circulating leukocytes and the endothelium from developing in tissues exposed to I/R. The temporal expression of this protected or preconditioned state induced by ethanol ingestion is biphasic. The first or acute phase is short-lived, first becoming apparent within 1 hour after ingestion, with peak anti-inflammatory effects occurring 2 to 3 hours after intake, and then disappearing by 4 hours after consumption. The second or delayed phase of ischemic tolerance induced by ethanol becomes evident 18 to 24 hours after ethanol consumption and is notable for its magnitude of protection, being much more powerful than the acute phase (Figure 1). In these studies, plasma ethanol levels peaked at 45 mg/dL within 30 minutes of gastric instillation by gavage and returned to control levels within 60 minutes of ingestion. This observation indicates that neither the acute phase of protection that arises within 1 to 3 hours of alcohol ingestion nor the late phase that reemerges 18 to 24 hours later are due to direct effects of ethanol. Indeed, ethanol must be absent from the blood before the cardioprotective effects of the alcohol become apparent. Interestingly, continued presence of ethanol prevents the infarct-sparing effects of other interventions that induce a preconditioned, protected state such as antecedent exposure to short bouts of ischemia or prior treatment with agents that activate mitochondrial ATP-sensitive potassium channels.

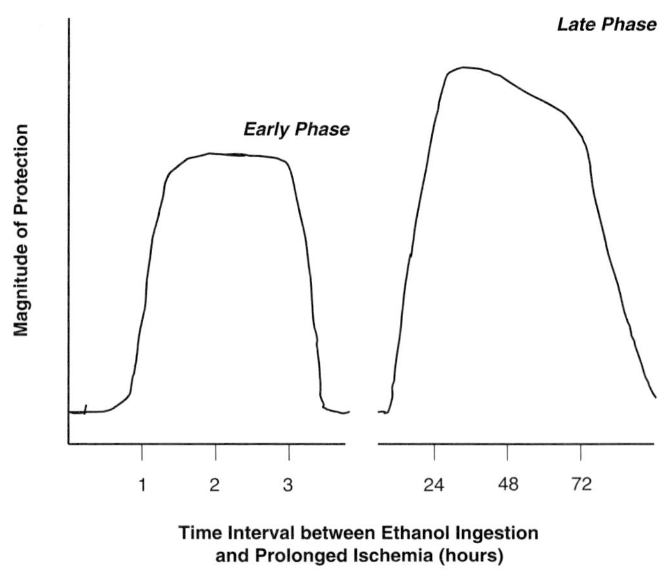

Figure 1 Antecedent ethanol ingestion produces two temporal phases of protection against the proinflammatory effects of ischemia/reperfusion. The early or acute phase develops within 1 hour after ingestion and is relatively short-lived, persisting for 1 to 3 hours and then disappearing. This is followed by the emergence of a second window of protection 12 to 24 hours later; this is more potent and persists for at least 2 days. The second wave of protection is termed late phase or delayed ethanol preconditioning.

Signaling Mechanisms That Induce the Adaptive Transformation to a Protected Phenotype in the Microcirculation after Ethanol Ingestion

The fact that the protected or preconditioned state that is induced by ethanol ingestion does not become apparent until it is removed from the tissues implies that the alcohol induces biochemical adaptations within the affected cells that render them resistant to the deleterious effects of I/R. Moreover, the differences in the time course for development and magnitudes of protection of the acute versus delayed phases of the anti-inflammatory phenotype induced by ethanol suggest that they may rely on different mechanisms.

The acute phase of protection induced by ethanol ingestion appears to arise as a result of the effect of ethanol to increase the concentration of adenosine in extracellular fluid, an effect related to the ability of the alcohol to inhibit the nucleoside transporter in cell membranes. This concept is based on the observation that the anti-inflammatory effects of superfusing the mesentery with ethanol at 45 mg/dL for 10 minutes followed by a 10-minute washout period prior to I/R are prevented by coincident treatment with adenosine deaminase or an adenosine A1/A3, but not an A2, receptor antagonist. Moreover, the acute-phase beneficial actions of ethanol in the microcirculation are mimicked by treatment with adenosine A1 receptor agonists. It appears that activation of specific protein kinase C isoforms and mitochondrial ATP-sensitive potassium (KATP) channels may play a role as downstream signaling elements in producing the protected phenotype. These results suggest that acute ethanol exposure induces ischemic tolerance by a mechanism similar to that described for early phase (or classical) ischemic preconditioning. However, one major difference in these two forms of cardioprotection appear to involve their respective mechanisms for increasing interstitial fluid adenosine concentrations. That is, decreased washout of adenosine occurs secondary to the reductions in blood flow during the bouts of preconditioning ischemia, whereas acute ethanol exposure inhibits the nucleoside transporter in cell membranes, thereby preventing adenosine reuptake.

Increased tissue adenosine levels are also required to trigger the development of the late-phase anti-inflammatory phenotype (Figure 2). However, adenosine A2 receptor occupancy is required for this process, an important mechanistic distinction as the downstream signaling elements activated by adenosine A1/A3 versus A2 receptors are very different. These findings suggest that the acute versus chronic phases of protection may involve downstream phophotidylinositol-3-kinase (PI3K)/protein kinase B (Akt)- versus adenylyl cyclase/cAMP/protein kinase A–dependent signaling mechanisms, respectively. Indeed, adenylyl cyclase or protein protein kinase A blockade, but not PI3K inhibitors, prevents the late phase of ethanol preconditioning. In addition, administration of cell-permeant cAMP analogs or adenylyl cyclase activators (e.g., isoproterenol, forskolin) mimics the postischemic anti-inflammatory effects of late-phase ethanol preconditioning.

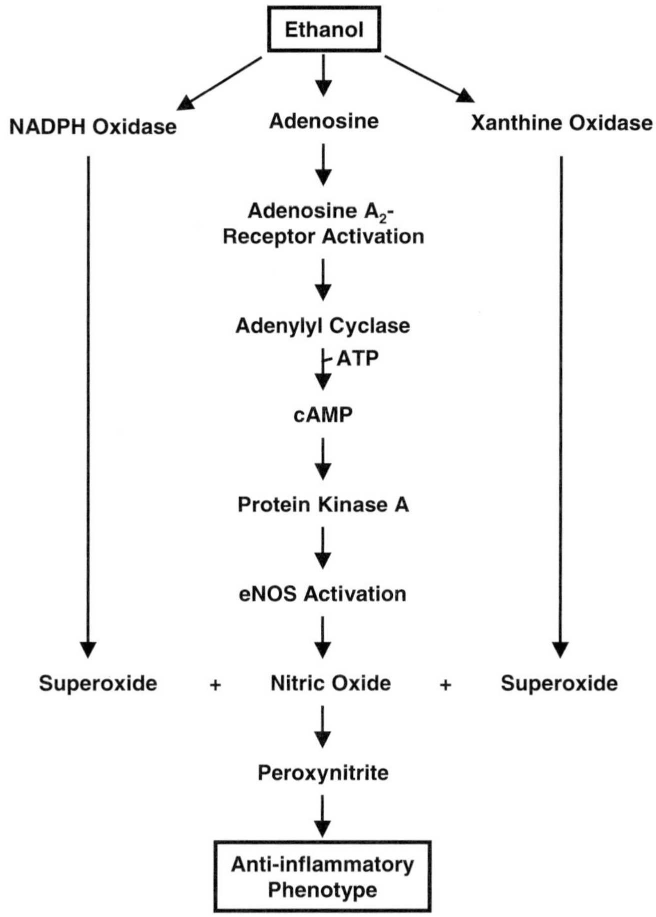

Figure 2 Factors involved in the triggering or initiation of the anti-inflammatory phenotype exhibited by postcapillary venules 12 to 24 hours after ethanol ingestion (late phase of ethanol preconditioning). See text for explanation.

Because ethanol enhances both basal and flow-stimulated nitric oxide synthase (NOS) activity and nitric oxide (NO) production in vivo and in cultured endothelial cells, it has been suggested that production of this gaseous monoxide during the period of ethanol exposure may serve as an important triggering element for the late phase of ethanol preconditioning. Support for this concept is derived from four lines of evidence. First, administration of NOS antagonists just prior to, but not 1 hour after, ethanol administration on day 1 abolishes the antiadhesive effects of late EtOH-PC on day 2. The latter observation supports the concept that NO plays an important role in instigating the development of the anti-inflammatory phenotype that becomes apparent 18 to 24 hours later. Second, plasma levels of nitrite/nitrate, a marker for NO production, are increased during the period of ethanol exposure. Third, tissues pretreated with NO donors in lieu of ethanol develop an anti-inflammatory phenotype 24 hours after administration. Finally, the anti-inflammatory phenotype induced by ethanol does not appear in mice that are genetically deficient in endothelial nitric oxide synthase (eNOS). This last finding not only provides the fourth line of evidence supporting a role for NO as a triggering element in ethanol preconditioning, but also indicates that the eNOS isoform is essential for the development of the anti-inflammatory phenotype in response to ethanol.

Although the factors responsible for increasing eNOS activity in late EtOH-PC (or any form of preconditioning for that matter) are unknown, the observations that NOS inhibition was as effective as adenosine deaminase or adenosine A_2 receptor blockade in abolishing the protective effects of late EtOH-PC suggest that adenosine and NO may serve as sequential triggering elements in the signaling cascade that induces the development of the protected phenotype rather than acting as independent initiators of this preconditioned state. Indeed, the findings that (1) NOS inhibition prevents late preconditioning induced by coadministration of an adenosine A_2 receptor agonist; (2) the development of the late-phase anti-inflammatory phenotype in response to NO donors is not prevented by adenosine A2 receptor antagonists; and (3) NO donors, but not adenosine A2 receptor agonists, induce the development of a preconditioned state in eNOS knockout animals support the concept that ethanol triggers the development of an anti-inflammatory state by a mechanism involving adenosine A2 receptor-dependent eNOS activation. This notion is supported by the observations that ligation of adenosine A2 receptors increases the activity of cAMP-dependent kinase (PKA), which in turn activates eNOS by phosphorylating Ser-1177. Moreover, adenosine stimulates l-arginine transport and NO biosynthesis by activation of A_2 receptors on human umbilical vein endothelial cells.

Another well-known effect of ethanol is to increase the generation of reactive oxygen species, including superoxide and the hydroxyethyl radical. Although they are generally considered to exert deleterious effects in biologic systems, it is becoming increasingly apparent that reactive oxygen species may participate in a number of normal physiologic phenomena by serving as second messengers in transmembrane signaling processes. Indeed, administration of a cell-permeant superoxide dismutase mimetic, MnTBAP, coincident with ethanol prevents the postischemic antiadhesive effects that become apparent 24 hours after ingestion of the alcohol. Moreover, exposing postcapillary venules to a superoxide generating system (hypoxanthine/xanthine oxidase) 24 hours prior to I/R mimicked the antiadhesive effects produced by antecedent ethanol exposure. Additional support for the concept that oxidants may participate in triggering the development of the anti-inflammatory phenotype in response to antecedent ethanol ingestion is derived from studies directed at elucidating their source of production. Inhibition of either xanthine oxidase or NADPH oxidase alone attenuated the antiadhesive effects of ethanol preconditioning by 50 percent, whereas concomitant inhibition of both oxidant-producing enzymes effectively prevented the development of the protected phenotype. The latter studies indicate that xanthine oxidase and NADPH oxidase are important enzymatic sources of the reactive oxygen species that trigger entrance into the anti-inflammatory phenotype displayed by postcapillary venules exposed to ethanol 24 hours prior to I/R.

As noted earlier, there is evidence implicating nitric oxide, formed secondary to adenosine A_2-receptor-dependent activation of endothelial nitric oxide synthase, in the beneficial actions of antecedent ethanol ingestion. This raises the possibility that NO produced during the period of ethanol preconditioning initiates the protective effects of late EtOH-PC by a mechanism that involves its interaction with xanthine oxidase- and/or NAD(P)H oxidase-derived oxidants. This is an important issue because we have obtained preliminary evidence implicating isoform-selective protein kinase C (PKC) translocation and activation as an obligatory downstream signaling element in late EtOH-PC (unpublished observations). However, NO and NO-releasing agents reversibly inactivate PKC. On the other hand, peroxynitrite, which is formed by the interaction of NO with superoxide, not only induces PKC activation, but has been implicated as a trigger for the beneficial actions of other forms of preconditioning including that induced by antecedent exposure to brief ischemia.

Although the mediators of late phase ethanol preconditioning are unclear, the time course required for its development suggests that the appearance of the protected phenotype 18 to 24 hours after ethanol ingestion requires the formation of new gene products capable of producing anti-inflammatory agents. In this regard, it is tempting to speculate that ethanol might enhance the expression of cyclooxygenase-2 (COX-2), inducible NOS (iNOS), or heme oxygenase-1 (HO-1), which in turn generate prostacyclin and other eicosanoids, NO, and carbon monoxide, respectively, all of which produce robust antiadhesive effects in postcapillary venules. In addition to carbon monoxide, HO-1 also generates the powerful antioxidants bilirubin and biliverdin, which may act to prevent the formation of oxidant-dependent chemoattractants during I/R. It is also possible that ethanol may exploit the antiadhesive properties of adenosine as a mediator of the preconditioned state secondary to enhanced production or decreased salvaging (via activation of 5′-nucleotidase or inhibition of adenosine kinase, respectively) of the nucleoside in postischemic tissues.

P-Selectin Expression as a Molecular End Target of Ethanol Preconditioning

A likely effector target of the signaling cascade induced by antecedent ethanol ingestion is P-selectin expression. This adhesive glycoprotein is upregulated in postischemic tissues, where it participates in leukocyte recruitment by supporting leukocyte rolling, a requisite step for both subsequent establishment of stationary adhesive interactions and diapedesis across postcapillary venular endothelium. Ethanol preconditioning prevents P-selectin expression in postischemic tissues by an adenosine- and NO-triggered, oxidant-dependent mechanism. These observations provide the molecular basis for the antiadhesive actions demonstrated by ethanol preconditioning in the setting of I/R.

Bibliography

Albert, M. A., Glynn, R. J., and Ridker, P. M. (2003). Alcohol consumption and plasma concentration of C-reactive protein. *Circulation* **107**, 443–447.

Das, D. K., and Ursini, F. (eds.). (2002). Alcohol and wine in health and disease. *Ann. N. Y. Acad. Sci.* **957**, 1–350. *A comprehensive selection of articles dealing with the cardioprotective actions associated with consumption of moderate levels of wine and other alcoholic beverages.*

Dayton, C., Yamaguchi, T., Warren, A., and Korthuis, R. J. (2002). Ischemic preconditioning prevents postischemic arteriolar, capillary, and postcapillary venular dysfunction: Signaling pathways mediating the adaptive metamorphosis to a protected phenotype in preconditioned endothelium. *Microcirculation* **9**, 73–89.

Dayton, C. B., Kamada, K., Yamaguchi, T., Carter, P., Gute, D. C., and Korthuis, R. J. (2004). Ethanol ingestion prevents postischemic P-selectin expression. Submitted for publication. *Microcirculation* **11**, 209–718.

Granger, D. N. (1999). Ischemia-reperfusion: Mechanisms of microvascular dysfunction and the influence of risk factors for cardiovascular disease. *Microcirculation* **6**, 167–178.

Korthuis, R. J. (ed.). (2004). Alcohol and cardioprotection. *Pathophysiology* (in press). *A Special Topics supplement to the journal with concise reviews regarding the epidemiologic evidence for the cardioprotective effects of ethanol, mechanisms underlying its infarct-sparing and anti-inflammatory actions of ethanol and the red wine constituent polyphenol resveratrol, and the effects of ethanol on platelet function, fibrinolytic activity, and lipid metabolism. Pathophysiology* **10**, 81–145.

Shigematsu, S., Ishida, S., Hara, M., Takahashi, N., Yoshimatsu, H., Sakata, T., and Korthuis, R. J. (2003). Resveratrol, a red wine constituent polyphenol, prevents superoxide-dependent inflammatory responses induced by ischemia/reperfusion, platelet-activating factor, or oxidants. *Free Radic. Biol. Med.* **34**, 810–817.

Wallerath, T., Poleo, D., Li, H., and Forstermann, U. (2003). Red wine increases the expression of human endothelial nitric oxide synthase: A mechanism that may contribute to its beneficial cardiovascular effects. *J. Am. Coll. Cardiol.* **41**, 471–478.

Yamaguchi, T., Dayton, C., Shigematsu, T., Carter, P., Yoshikawa, T., Gute, D. C., and Korthuis, R. J. (2002). Preconditioning with ethanol prevents postischemic leukocyte–endothelial cell adhesive interactions. *Am. J. Physiol.* **283**, H1019–H1030. *The first description of the effect of ethanol to promote the development of an anti-inflammatory phenotype in the microcirculation and the critical role that adenosine A2 receptor–dependent activation of endothelial nitric oxide synthase plays in this adaptive transformation.*

Yamaguchi, T., Dayton, C. B., Ross, C. R., Yoshikawa, T., Gute, D. C., and Korthuis, R. J. (2003). Late preconditioning by ethanol is initiated by an oxidant-dependent signaling pathway. *Free Radic. Biol. Med.* **34**, 365–376.

Capsule Biography

Dr. Korthuis is the Bolm Distinguished Professor and Chairman of the Department of Medical Pharmacology and Physiology at the University of Missouri-Columbia. His research is directed at elucidating the cellular mechanisms that underlie the development of protected phenotypes in the microcirculations of the small intestine and skeletal muscle in response to ischemic preconditioning and antecedent ethanol ingestion. Dr. Korthuis' work is supported by grants from the National Institutes of Health (HL-54797 and DK-43785).

CHAPTER 116

Microvascular Dysfunction in Inflammatory Bowel Disease

F. Stephen Laroux and Matthew B. Grisham

Department of Molecular and Cellular Physiology, Louisiana State University Health Sciences Center, Shreveport, Louisiana

Introduction

Active episodes of inflammatory bowel disease (IBD; Crohn's disease, ulcerative colitis) are characterized by rectal bleeding, diarrhea, exudation, fever, abdominal pain, and weight loss. Histopathological examination of the inflamed gut reveals vasodilation, venocongestion, edema, infiltration of large numbers of inflammatory cells, erosions, and frank ulcerations. There is a growing body of experimental and clinical evidence suggesting that IBD may result from a dysregulated immune response to components of the normal gut flora. Current theory suggests that the inability to regulate the normal T-cell-mediated immune responses to normal comensal bacteria gut flora results in activation of Th1-type $CD4^+$ T-cells with subsequent release of proinflammatory mediators such as interleukins-2, -12, and -1β (IL-2, IL-12 and IL-1β, respectively), interferon-γ (IFN-γ), and tumor necrosis factor-α (TNF-α), as well as platelet activating factor (PAF), nitric oxide (NO), reactive oxygen species (ROS) and certain arachidonate metabolites [1] (Figure 1). Because many of these proinflammatory mediators are vasoactive and activate the microvascular endothelium, it is not surprising to find that much of the pathophysiology observed in the chronically inflamed gut may result from alterations in the colonic and/or intestinal microcirculation [2] (Figure 2). For example, some of these inflammatory mediators (e.g., histamine, bradykinin, NO, prostaglandins) possess potent vasodilatory properties and will increase blood flow (hyperemia), thereby producing the characteristic erythema in the inflamed bowel. Increased arteriolar blood flow will also raise hydrostatic pressure in the downstream capillaries, resulting in transcapillary fluid movement across intestinal capillaries and inducing interstitial edema. Certain proinflammatory cytokines (e.g., TNF-α, IFN-γ, IL-12) expressed during T-cell and macrophage activation will activate venular endothelial cells to enhance the expression of certain vascular adhesion molecules that mediate leukocyte–endothelial interactions and subsequent emigration of leukocytes (PMNs, monocytes, and lymphocytes) into the surrounding tissue. Leukocyte emigration is often associated with vascular protein leakage with accumulation of albumin and other plasma proteins in the gut interstitium. The resultant increase in tissue oncotic pressure further promotes fluid filtration across capillaries and accelerates the development of interstitial edema. Taken together, these pathophysiological considerations suggest that the intestinal microcirculation contributes largely to the pathophysiology of chronic gut inflammation.

This brief review will summarize the current state of knowledge regarding the role of the microvasculature in chronic gut inflammation.

Organization of the Intestinal and Colonic Microcirculation: Anatomical and Functional Considerations

The microcirculation of the intestine consists of a branching network of arterioles, capillaries, and venules. Arterioles in the intestine are usually less than 500 μm in diameter wrapped by an external muscular coat consisting of two to four smooth muscle cells arranged circumferentially. The terminal (precapillary) arterioles have an internal diameter of 15 to 20 μm and are surrounded by only one layer of smooth muscle cells.

The majority of capillaries are derived from terminal arterioles and consist of a tube with an inner diameter of 4

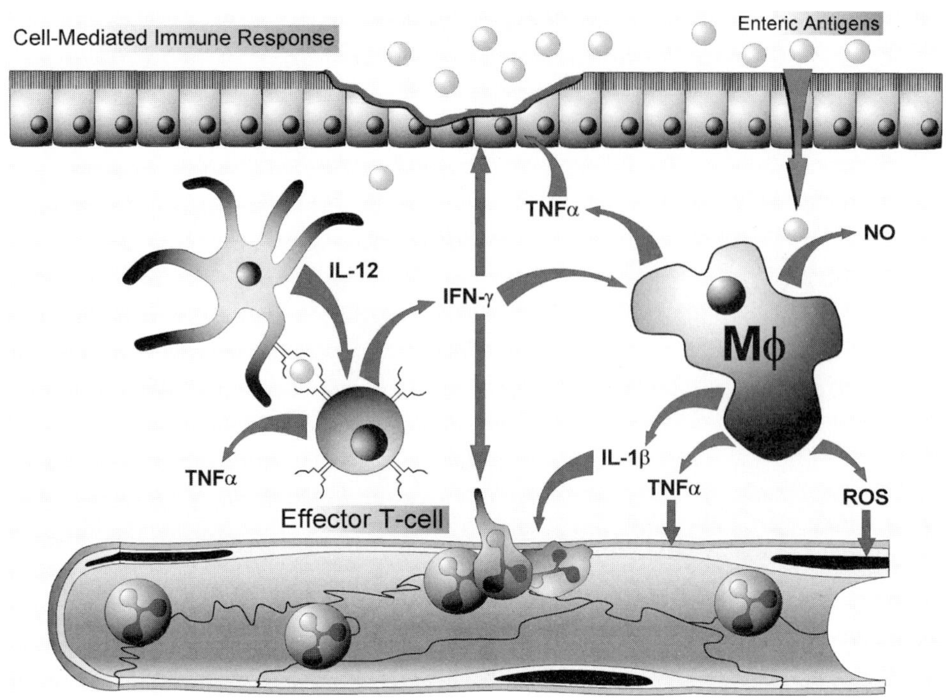

Figure 1 Intestinal immune response to enteric antigens in the absence of regulatory cells. Effector CD4$^+$ T-cells produce Th1-type cytokines in response to T-cell receptor engagement of antigens processed and presented by APCs. These cytokines (such as IFN-γ) may affect the gut epithelium directly and/or activate resident Mϕ to release large amounts of proinflammatory mediators and cytokines as well as reactive metabolites of oxygen (ROS) and NO. The net result is the recruitment of additional leukocytes and subsequent inflammation and injury. APC, antigen presenting cell; Mϕ, macrophage; NO, nitric oxide; IFN-γ, interferon-γ. (see color insert)

to 10 μm and are lined by a single layer of endothelial cells and a thin basement membrane. In the gut, only a minority (e.g., 20–30%) of the capillaries are open to perfusion under normal resting conditions. The capillary network, with its large surface area and an endothelial barrier that is highly permeable to lipid-soluble and small water-soluble molecules, appears well suited for the exchange of gases, nutrients, and water between the bloodstream and tissues.

As with all tissues, capillaries drain into postcapillary venules that are also devoid of smooth muscle cells. These vessels represent the segment of the microvasculature that is most involved in inflammatory responses and contain intercellular endothelial junctions that can open to allow plasma proteins and circulating leukocytes to escape from the blood. The different patterns of microvascular anatomy are illustrated in the small and large intestine [3] (Figure 3). For example, the arteriolar supply from the submucosal arterial plexus passes directly to the villus tip of the small intestine, where this vessel then branches into a fountain-like formation of capillaries (Figure 3B). This dense plexus of subepithelial capillaries is drained by a single venule. The capillaries of the colonic mucosa, on the other hand, are arrayed in a honeycomb-like plexus or ring pattern in which each ring of capillaries surrounds the openings to the colonic crypts (Figure 3A).

Endothelial cells and smooth muscle cells represent the major functional elements of the blood vessel wall that allow arterioles, capillaries, and venules to carry out their functions. Although the two cell types are clearly capable of functioning independently, there are processes that enable one cell type to influence the other. Pericytes and mast cells are examples of such auxiliary cells that can exert a profound influence on the function of arterioles, capillaries, and/or venules.

Mast cells are also found closely apposed to the microvasculature, particularly the postcapillary venules. These cells are exquisitely sensitive to activation by a variety of stimuli, including neuropeptides (e.g., substance P), ROS, lipid mediators (PAF, LTB4), and bacterial peptides. Upon activation, mast cells release a number of substances that can influence the function of endothelial cells and vascular smooth muscle in all segments of the microvasculature. Mast cell–derived modulators of microvascular function include histamine, adenosine, nitric oxide, cytokines (e.g., tumor necrosis factor, interleukin-1), proteases (e.g., cathepsin G), and oxidants.

Arteriolar and Capillary Alterations Associated with IBD

Blood Flow

The resting tone of arterioles appears to result from a complex interaction of metabolic, myogenic, neurohumoral, and physical (e.g., stretch or shear) signals received by the

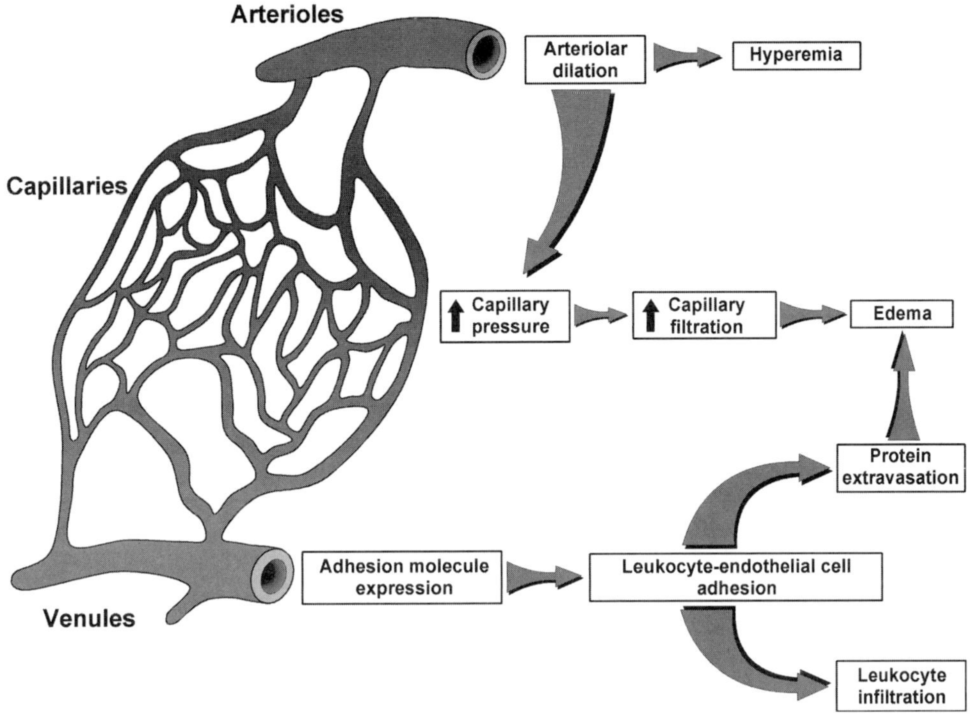

Figure 2 All segments of the microcirculation contribute to the pathophysiology of chronic gut inflammation. A variety of inflammatory mediators (e.g., histamine, bradykinin, nitric oxide, prostaglandins) produced by the affected tissue relax the vascular smooth muscle surrounding arterioles. The consequent dilation of arterioles leads to an increased blood flow (hyperemia), thereby producing erythema. Another consequence of arteriolar dilation is an increased hydrostatic pressure in the downstream capillaries. The increased capillary hydrostatic pressure alters the balance of forces that govern fluid movement across intestinal capillaries to favor net fluid filtration. The increased capillary filtration rate contributes to the interstitial edema associated with gut inflammation. Proinflammatory cytokines (TNF, IFN-γ, IL-12) released by activated mast cells, macrophages, and lymphocytes activate venular endothelial cells and increase expression of endothelial cell adhesion molecules that mediate leukocyte–endothelial cell adhesion and eventual emigration (tissue infiltration) of leukocytes. Leukocyte emigration is often associated with vascular protein leak (extravasation). Consequently, inflammation generally results in a diminished endothelial barrier function in venules that promotes the accumulation of albumin and other plasma proteins in the interstitium. The resultant increase in interstitial oncotic pressure further promotes fluid filtration across capillaries and accelerates the development of interstitial edema. (see color insert)

blood vessel wall. Historically, the responses of arterioles to physiological stimuli were believed to be initiated almost exclusively by signals sensed by VSM cells. There is now clear evidence that endothelial cells play an important role in maintaining vascular tone by releasing substances (e.g., NO, prostacyclin) that modulate the delicate balance between vasodilation and vasoconstriction. An appreciation for the contribution of endothelial cells to vascular tone comes from studies demonstrating that acetylcholine dilates arterial smooth muscle only if the endothelium is intact and viable.

In active ulcerative colitis, the submucosal arteries display a convoluted course and dilatation and congestion of the mucosal and submucosal microvessels is often striking. Studies employing microangiography, vascular casts, and mesenteric angiography have demonstrated widened arteries and a rapid venous return, suggesting that an increase in colonic blood flow occurs with active ulcerative colitis. Another frequent vascular abnormality is vasodilation of the lymph node vasculature.

In Crohn's disease, the morphologic alterations are less uniform than those described for ulcerative colitis and the vascular changes show considerable variation from one patient to another. For example, angiographic examination of the small bowel in Crohn's disease indicates that the degree of dilation and engorgement of ileal and colonic microvessels may be as conspicuous as in active ulcerative colitis, although a reduced vascularity is also commonly reported. In addition, in areas with mild alteration (deep lymphocytic infiltration but not ulcerative lesions of fissures), there is a distinct focal hypervascularity in the submucosa evidenced by numerous dilated arterial vessels that have a straight "broom-like" course [3]. Studies in patients with IBD indicate that blood flow in affected regions may increase two- to sixfold. In addition, determination of the intramural distribution of blood flow indicated that this increase is confined largely to the mucosal–submucosal layer. Based on these observations, Hulten and associates estimated that blood flow in the inflamed colon corresponds to 25 to 30 percent of the resting cardiac output and

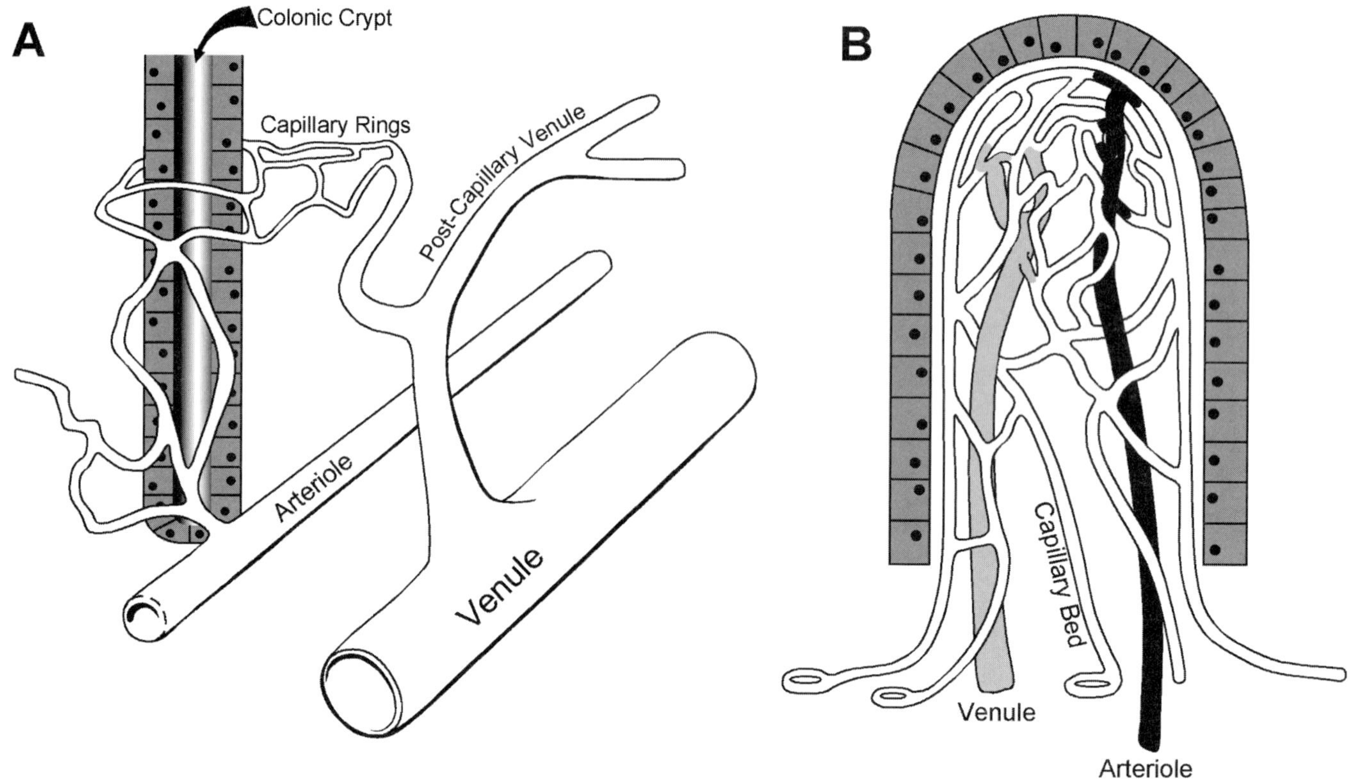

Figure 3 Microcirculatory pattern of the large and small intestinal mucosa. **(A)** Capillaries nearest the colonic mucosa are arranged in a honeycomb-like plexus of interconnecting rings in which each ring of capillaries surrounds the openings of the colonic crypts. **(B)** Fountain-like pattern of blood flow to the small intestinal villus. The pattern of arterioles and venules conforms to the shape of the villus. This capillary plexus immediately underlies the base of the epithelium. (Figure adapted from Ref. [3].)

suggested that these high flows may represent an important factor contributing to the physical deterioration often observed in these patients [4]. In contrast to the results obtained in active and exudative stages of ulcerative colitis and Crohn's disease, colonic blood flow decreases below normal in the late fibrosing stage. A similar pattern is observed for the ileum in patients with Crohn's disease affecting this segment of the bowel. These findings correlate well with the observed reduction in vascularity in this stage of disease progression. In addition to these changes, a characteristic distention or clubbing of the villi occurs and, in a later phase, epithelial denudation and destruction of the villi. Although the mechanisms underlying these alterations in colonic blood flow are unknown, sustained overproduction of NO and/or arachidonic acid metabolites may be important.

Arteriolar-Dependent Mechanisms of Interstitial Edema in IBD

Interstitial edema and mucosal exudation are cardinal histopathological signs of inflammatory bowel disease. All three segments of the intestinal microvasculature, that is, arterioles, capillaries, and venules, contribute to the interstitial edema associated with IBD. The arteriolar dilation that accounts for the intense hyperemia during inflammation may represent a major pathway for enhanced filtration of fluid across the walls of downstream capillaries. This arteriole dilation–dependent enhancement of capillary fluid filtration results from an increased capillary hydrostatic pressure. Capillary pressure (P_c) rises when arterioles dilate because a larger fraction of the prevailing arterial pressure is transmitted to the downstream capillaries. An elevated P_c alters the balance of hydrostatic and oncotic forces that govern fluid movement across capillaries. If the increment in P_c is of sufficient magnitude, the rate of fluid filtration is accelerated to an extent that produces interstitial edema, that is, the rate of fluid entry into the mucosal or submucosal interstitium exceeds the capacity of lymphatics to drain the interstitial compartment. That capillary filtration is elevated in the intestinal vasculature during IBD is supported by reports describing increased intestinal lymph flow. Although the magnitude of the increase in P_c during IBD has not been determined experimentally, published reports of blood flow changes in human subjects provide some insights into this potential P_c elevation during the inflammatory response.

Angiographic studies in patients with ulcerative colitis or Crohn's disease demonstrate widened splanchnic arteries. However, estimates of colonic blood flow (using an isotope washout technique) in patients with inflammatory bowel disorders indicate a two- to sixfold increase in colon blood flow that is largely confined to the mucosal–submucosa layers [5]. As mentioned previously, Hulten and associates have estimated that the inflamed colon, in the early "exudative" stage, may be supplied by approximately 1,500 mL/min of blood, corresponding to 25 to 30 percent of the resting cardiac output [4]. Assuming that the decrease in colonic vascular resistance occurs predominately at the arteriolar level, it can be estimated that microvascular pressure may rise by 10 to 40 mmHg. An increase in capillary pressure of this magnitude should profoundly increase the rate of capillary fluid filtration and promote a massive accumulation of interstitial fluid. Similarly, the fluid filtration that results from such a large increase in P_c should lead to disruption of the mucosal barrier and exudation of interstitial fluid into the lumen.

Angiogenesis

The intestinal vascular beds have only a small fraction (e.g., 20–30%) of the total capillary population open for blood perfusion. Physiologic stresses such as an increased metabolic demand and/or reduced blood flow in these tissues are generally associated with the recruitment of additional perfused capillaries. Consequently, the classic concept has been that alterations in functional capillary density allow for local modulation of O_2 exchange area and capillary-to-cell diffusion distances.

Capillary proliferation (e.g., angiogenesis) represents a potential mechanism whereby tissues can compensate for chronic alterations in oxygen delivery and/or metabolic demand, and to restore organ function after injury. The endothelium that lines normal capillaries is an extremely stable population of cells with very low mitotic activity; only 0.01 percent of endothelial cells in the body are dividing at any given time. Hence, capillary growth and proliferation is rarely observed in normal adult tissues except during wound healing and cyclical events in the female reproductive cycle (ovulation, menstruation). In the presence of appropriate stimuli, the process of angiogenesis (development of new blood vessels from an existing vascular network) can be initiated. Endothelial cells exposed to such stimuli first release proteases that degrade the underlying basement membrane and surrounding structural elements [6]. The cells then migrate toward the angiogenic (chemotactic) stimulus within the extravascular space, with a concomitant proliferation of the endothelial cells lining the vessel wall to replace the previously migrated cells. The migrating and proliferating endothelial cells form cordlike structures in target tissues that later canalize to functional vessels, which are further stabilized by surrounding pericytes. The initiation of angiogenesis is often associated with an increased capillary permeability that serves to enrich the adjacent interstitial compartment with plasma components. The role of angiogenesis in IBD is, at present, unclear.

Venular Alterations in IBD

It is now well recognized that postcapillary venules represent the major site of transvascular protein exchange (vascular permeability to plasma proteins) and leukocyte trafficking (leukocyte–endothelial cell adhesion) driving acute and chronic inflammation. The localization of these inflammatory functions in venules is believed to reflect the unique characteristics of endothelial cells in this segment of the microcirculation. Consequently, the literature is replete with reports that describe the responses of cultured venous endothelial cells (usually derived from human umbilical vein) to various inflammatory stimuli. These in vitro studies, coupled to data derived from experiments utilizing intravital microscopy, have improved our understanding of the potential contribution of venules to the pathogenesis of certain inflammatory diseases.

Leukocyte–Endothelial Cell Adhesion

Circulating leukocytes are recruited to sites of inflammation and tissue injury by a highly coordinated process that occurs primarily in postcapillary venules. As leukocytes exit capillaries, hemodynamic forces give rise to an outward radial movement of leukocytes toward the venular endothelium. This margination process is generally attributed to red blood cells (which normally pile up behind the larger leukocyte in capillaries) that overtake the leukocyte and tend to push it toward the venular wall. The initial adhesive interaction between the leukocyte and venular endothelium is rolling. This low-affinity (weak) interaction is subsequently strengthened such that the leukocytes attach to endothelium and remain stationary [7]. The leukocytes are then able to migrate into the interstitium through spaces between adjacent endothelial cells. These interactions are regulated by sequential activation of different families of adhesion molecules expressed on the surface of neutrophils and endothelial cells.

It is well appreciated that Th1- or macrophage-derived cytokines such as IL-1β, TNF-α, lymphotoxin-α, or IL-12 promote leukocyte adhesion and extravasation in vitro and in vivo. L-selectin of leukocytes and the P- and E-selectins of endothelial cells have been implicated as the major ECAMs involved in rolling of leukocytes along the endothelium. PMN-endothelial cell interactions predominate in acute flares of IBD, whereas lymphocyte, monocyte, and in some cases eosinophil interactions with the microvascular endothelium are more prevalent during the chronic stages of gut inflammation. The mononuclear leukocytes possess a β1 integrin called very late activation antigen-4 (VLA-4; $\alpha_4\beta_1$), which binds to the inducible VCAM-1 and MAdCAM-1 expressed on the surface of cytokine-activated endothelial cells.

Lymphocytes possess an additional ligand/counter receptor pair that is important in cell–cell adhesion, signaling, trafficking, and regulation of the immune responses in mucosal tissues, especially in the gastrointestinal tract (e.g.,

MAdCAM-1/$\alpha4\beta7$). It has recently been reported that MAdCAM-1 surface expression on venular endothelial cells in the lamina propria of the gut is enhanced in foci of inflammation in biopsies obtained from patients with active ulcerative colitis or Crohn's disease. Three recent studies using low-molecular-weight antagonists or immunoneutralizing monoclonal antibodies to either $\alpha_4\beta_7$ or MAdCAM-1 demonstrate that the $\alpha_4\beta_7$/MAdCAM-1 interaction plays an important role in the pathophysiology in three different models of colitis. Future studies may reveal that MAdCAM-1 represents a potentially important therapeutic target for the treatment of IBD.

The involvement of ECAM expression in experimental models of IBD can be readily demonstrated in the IL-$10^{-/-}$ and CD45RB^{high}/SCID transfer models. In both models it has been shown that colonic expression of E-selectin, ICAM-1, and VCAM-1 is significantly enhanced with the onset of colitis [8, 9]. Interestingly, expression of ICAM-2 was unchanged in both models. Furthermore, recent studies have demonstrated that colonic and cecal but not ileal MAdCAM-1 expression increases dramatically (approximately elevenfold) with the onset of colitis in the SCID/CD45RB^{high} and IL-$10^{-/-}$ models of IBD [8, 9].

Extravasated PMNs as well as other phagocytic leukocytes migrate through the interstitium in response to certain chemotactic stimuli including bacterial products and proinflammatory mediators released by the inflamed and/or injured tissue as well as by bacterial products that have made their way into the tissue from the lumen. However, relatively little information is available regarding the movement of PMNs through the interstitium. The formation of crypt abscesses represents the terminal step in PMN migration out of the circulation. In order for PMNs to be present within the lumen of the crypts, these inflammatory cells not only must extravasate from the circulation and move through the interstitial matrix but must interact with the basement membrane and basolateral surface of the crypt epithelium to emigrate out of the gut and into the lumen. The driving force for this directed migration out of the tissue into the lumen is provided by the bacterial gradient present in the distal bowel. Transendothelial as well as transepithelial migration of PMNs and other leukocytes may account for some of the fluid accumulation (edema) and enhanced mucosal (i.e., epithelial) permeability observed in patients with active IBD. Taken together, these studies suggest that the colonic and/or intestinal microvasculature regulates chronic gut inflammation by virtue of its ability to modulate the infiltration of different populations of leukocytes into the interstitium.

There are several factors that influence leukocyte–endothelial cell adhesion in postcapillary venules. Nitric oxide, adenosine, and prostacyclin produced by endothelial cells tend to prevent adhesion, whereas the oxygen radicals (superoxide, hydrogen peroxide) generated by activated leukocytes and endothelial cells promote leukocyte adhesion. These agents appear to exert their actions by interfering either with the production of inflammatory mediators (e.g., platelet activating factor) that induce leukocyte adhesion or with the induction of adhesion molecule expression on endothelial cells and/or leukocytes respectively. Auxiliary cells, such as mast cells (histamine), macrophages (cytokines), and platelets (leukotrienes), also produce substances that can promote leukocyte–endothelial cell adhesion.

The physical forces generated by the movement (flow) of blood in the microcirculation also play an important role in the modulation of leukocyte–endothelial cell adhesion. The prevailing shear rate exerted on the walls of postcapillary venules determines the level of leukocyte rolling and firm adherence, and it dictates the contact area between leukocytes and the endothelial cell surface. Even in the absence of an inflammatory stimulus, graded reductions in venular shear rate for brief periods (less than 2 minutes) elicit progressive recruitment of both rolling and firmly adherent leukocytes. Similarly, it has been noted that the number of adherent leukocytes recruited into venules by an inflammatory stimulus is inversely proportional to the wall shear rate, suggesting that it is easier for leukocytes to create strong adhesive bonds with endothelial cells at low shear rates and that high shear rates may prevent the creation of such bonds.

Vascular Permeability to Plasma Proteins

Vascular endothelial cells normally serve as a barrier that minimizes the movement of fluid and proteins from blood to interstitium. There are several characteristics of postcapillary venules that allow this segment of the microvasculature to regulate vascular permeability to plasma proteins. Ultrastructural analyses of the pathways for transvascular exchange have revealed that both the size and frequency of interendothelial junctions and endothelial fenestrae are higher in postcapillary venules than in either arterioles or capillaries. These pathways are normally large enough to allow for a low basal of plasma protein leakage that is driven by both diffusive and convective (coupled to fluid filtration) mechanisms. Venular endothelium also appears to possess a higher density of cell surface receptors for inflammatory mediators than their counterparts in arterioles and capillaries. Engagement of certain inflammatory mediators (e.g., histamine, platelet activating factor) with their receptors on venular endothelial cells elicit cell contraction, which results in a widening of the junctions (gaps) between adjacent endothelial cells and a consequent increase in protein extravasation. Furthermore, since postcapillary venules are the preferred site for leukocyte trafficking (because of the high density of leukocyte adhesion receptors), these endothelial cells are more frequently exposed to neutrophil products (proteases, oxidants), which can diminish barrier function, than their counterparts in arterioles and capillaries. The process of leukocyte emigration also appears to render venular endothelial cells more vulnerable to barrier dysfunction. In some models of inflammation, a strong positive correlation has been shown between venular albumin leakage and the rate of transendothelial migration of leukocytes.

Interstitial Edema: Venule-Dependent Mechanisms

Another mechanism that could explain the interstitial edema that is associated with IBD is an increased vascular permeability. This response of the vasculature to inflammation is generally localized in the postcapillary venules, which is the normal site of plasma protein extravasation in most vascular beds. It is likely that endothelial cell contraction, mediated by the variety of inflammatory mediators that are detected in diseased tissue, is responsible for the increased vascular permeability of IBD. Histamine, serotonin, bradykinin, substance P, cytokines, and other biogenic amines that are released in the inflamed bowel can elicit endothelial cell contraction and an increased permeability. The mucosal ulceration accompanying inflammatory bowel disease allows bacterial toxins to gain access to the interstitial space, where the toxins can elicit the release of cytokines and other mediators from macrophages, and subsequently increase vascular permeability. A role for bacterial endotoxins in producing an increased vascular permeability in inflammatory bowel disease is supported by reports that approximately 50 percent of patients with these diseases have elevated titers of antibody to antigens derived from *Escherichia coli* or other bacteria. Since leukocyte infiltration is another histopathological feature of IBD, these cells may also contribute to the increased vascular permeability.

The consequences of such a leakage of plasma proteins from venules can be predicted from existing principles of capillary fluid exchange. When plasma proteins accumulate in the mucosal interstitium there is a resultant increase in interstitial oncotic pressure (π_t), which alters the balance of forces governing capillary fluid exchange to favor enhanced filtration. Another consequence of the diminished endothelial barrier function in inflamed venules is a greater capacity of these vessels to transport water for a given intravascular pressure. Hence, the combined effects of increased vascular pressures, resulting from arteriolar dilation, and an increased hydraulic conductance, which reflects the endothelial barrier dysfunction, likely lead to a profound enhancement of vascular fluid filtration in the inflamed bowel.

The amount of fluid that accumulates in the intestinal interstitium is determined by the balance between the rates of capillary fluid filtration and lymph flow. In Crohn's disease, dilated lymphatic vessels are frequently present, and it has generally been considered that lymphatic obstruction plays a prominent part in the submucosal edema observed in the inflamed bowel. The results of a morphologic study suggest that structural changes in the walls of lymphatic capillaries may result in a decreased permeability of these vessels that diminishes the drainage of fluid and protein from the interstitium. The changes described include a lack of open intercellular junctions seen in normal lymphatics, development of a more extensive basement membrane, and accumulation of protein-rich fluid at the abluminal surface of the lymphatic capillaries. A causal role for lymphatic obstruction in the pathogenesis of IBD has been proposed based on the observation that chronic lymphatic obstruction (either surgically or with sclerosing agents) results in intestinal inflammation (including the infiltration of lymphocytes and monocytes), interstitial edema, and hyperemia. These changes were most marked in animals with no demonstrable reorganization of the lymphatic system (due to lymphaneogenesis) and bear some resemblance to the vascular and tissue responses seen in Crohn's disease [10].

Acknowledgment

Some of this work was supported by the NIH (DK64023) and the Yamanouchi USA Foundations.

References

1. Powrie, F. (1995). T cells in inflammatory bowel disease: Protective and pathogenic roles. *Immunity* **3**, 171–174. *A landmark review of the role of T-cells in the pathogenesis of inflammatory bowel disease.*
2. Laroux, F. S., and Grisham, M. B. (2001). Immunological basis of inflammatory bowel disease: Role of the microcirculation. *Microcirculation* **8**, 283–301. *A comprehensive review of the immunological and pathophysiological alterations of the intestinal and colonic microcirculation in experimental and clinical inflammatory bowel disease.*
3. Gannon, B. J., and Perry, M. A. (1989). Histoanatomy and ultrastructure of vasculature of alimentary tract. In: *The Gastrointestinal System* (Schultz, S. G., Rauner, B. B., Wood, J. D., and the American Physiological Society, eds.), Section 6. Bethesda, MD: American Physiological Society.
4. Hulten, L., Lindhagen, J., Lundgren, O., Fasth, S., and Ahren, C. (1977). Regional intestinal blood flow in ulcerative colitis and Crohn's disease. *Gastroenterology* **72**, 388–396.
5. Erikson, U., Fagerberg, S., Krause, U., and Olding, L. (1970). Angiographic studies in Crohn's disease and ulcerative colitis. *Am. J. Roentgenol. Radium Ther. Nucl. Med.* **110**, 385–392.
6. O'Byrne, K. J., and Dalgleish, A. G. (2001). Chronic immune activation and inflammation as the cause of malignancy. *Br. J. Cancer* **85**, 473–483.
7. Granger, D. N. (1997). Cell adhesion and migration. II. Leukocyte-endothelial cell adhesion in the digestive system. *Am. J. Physiol* **273**, G982–G986.
8. Kawachi, S., Jennings, S., Panes, J., Cockrell, A., Laroux, F. S., Gray, L., Perry, M., van der, H. H., Balish, E., Granger, D. N., Specian, R. A., and Grisham, M. B. (2000). Cytokine and endothelial cell adhesion molecule expression in interleukin-10-deficient mice. *Am. J. Physiol. Gastrointest. Liver Physiol.* **278**, G734–G743. *This was the first study to quantify the in vivo expression of different endothelial cell adhesion molecules in the colonic vasculature from a spontaneous mouse model of chronic colitis.*
9. Kawachi, S., Morise, Z., Jennings, S. R., Conner, E., Cockrell, A., Laroux, F. S., Chervenak, R. P., Wolcott, M., van der, H. H., Gray, L., Feng, L., Granger, D. N., Specian, R. A., and Grisham, M. B. (2000). Cytokine and adhesion molecule expression in SCID mice reconstituted with CD4+ T cells. *Inflamm. Bowel. Dis.* **6**, 171–180.
10. Kalima, T. V. (1971). The structure and function of intestinal lymphatics and the influence of impaired lymph flow on the ileum of rats. *Scand. J. Gastroenterol. Suppl.* **10**, 1–87.

Capsule Biography

Dr. Matthew B. Grisham is a Boyd Professor and Dr. F. Stephen Laroux received his Ph.D. from the Department of Molecular and Cellular Physiology at LSU Health Sciences Center in Shreveport, Louisiana. He is currently a postdoctoral fellow at Beth Israel Deaconess Medical Center, Harvard Medical School, in the laboratory of Dr. Cox Terhorst. His research interests include the molecular and cellular mechanisms of host defense against infections pathogens. He is a member of the American Association of Immunologists and has received young investigator and travel awards from the Society for Mucosal Immunology and the American Association of Anatomists.

Dr. Matthew B. Grisham is Professor of Molecular and Cellular Physiology and Associate Director of Research for the Arthritis Center of Excellence at LSU Health Sciences Center in Shreveport, Louisiana. Dr. Grisham currently serves as Chair of the Gastrointestinal and Liver Section of the American Physiological Society. He is also an Associate Editor of *The American Journal of Physiology: Gastrointestinal and Liver and Free Radical Biology and Medicine* and has served on the editorial boards of numerous journals, including *Inflammatory Bowel Disease* and *The Journal of Immunology*. Dr. Grisham's research is currently funded by the NIH to investigate the molecular, cellular, and immunological mechanisms responsible for chronic gut inflammation.

CHAPTER 117

Analytical Approaches to Leukocyte–Endothelium Interactions

Lance L. Munn
Harvard Medical School

Introduction

A virus infects an epithelial cell, multiplies, and soon spreads to a few hundred cells in the vicinity. At the same time, a paper cut to a finger a meter away is infiltrated by bacteria. These attacks happen every day to us or those around us, usually without complications. How do the cells of the immune system respond to these insults quickly and efficiently?

The cellular immune response is performed by a collection of cells, consisting of leukocytes (also known as white blood cells) that circulate between the blood and lymphatic systems. Some of these cells, such as neutrophils and monocytes, kill foreign or infected cells directly or when tagged with antibodies produced elsewhere in the immune system. Others, such as lymphocytes, are more specific, requiring a perfect match between cell-surface receptors and antigens on the interrogated cell before killing commences. But how do these cells, which circulate through the bloodstream, find the tissue where injury or invasion has occurred?

The answer is a sophisticated combination of fluid dynamics and adhesion molecules that bring the leukocytes to the endothelial wall in the area of insult, slow their motion, and finally arrest them. The specificity of the response comes from a signaling cascade that starts in the infected or damaged cells and propagates through "sentinel" mast cells. Cytokines produced by these cells diffuse to nearby endothelial cells. The end effect is an upregulation of the required adhesion molecules on the endothelium. These adhesion molecules collect the leukocytes.

Although the cellular components of the inflammatory response were identified 100 years ago, most of our understanding of the biophysics of this process has come from research performed in the past 20 years. This research has relied on carefully designed analytical tools and computational methods to characterize blood rheology and adhesive mechanisms responsible for leukocyte infiltration. In this article, each of the critical steps in leukocyte trafficking is discussed in light of the analytical methods that have contributed to their understanding.

Margination

The first step in leukocyte adhesion is margination—the movement of the flowing cell toward the vessel wall (Figure 1). To understand this process, we have to consider the fluid dynamics of blood. The importance of blood rheology in leukocyte adhesion illustrated the fact that in the absence of red blood cells (RBCs), leukocytes only rarely contact the vessel wall and adhere.

Red blood cells constitute approximately 40 percent of the volume of blood. There are far fewer circulating leukocytes (approximately 1,000 RBCs for every leukocyte). So blood is a dense particulate suspension of RBCs, and this imparts unique fluid dynamics. In the years 1917–1938, Robin Fåhraeus pioneered the study of blood as a flowing fluid. Among other things, he studied the decrease in viscosity as blood enters smaller vessels, known as the Fåhraeus–Lindqvist effect. This occurs due to the formation

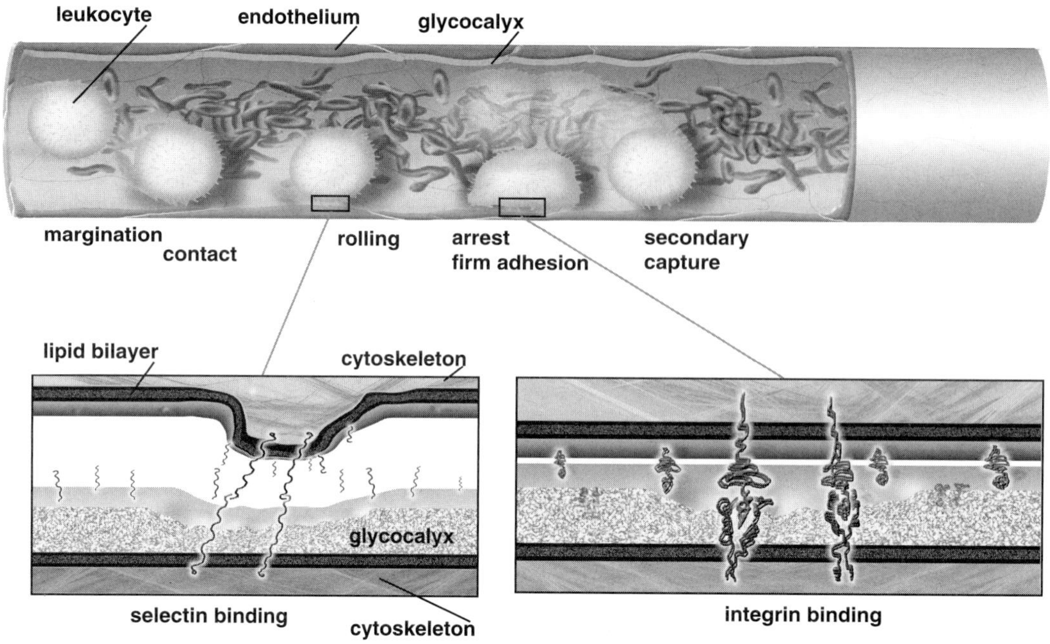

Figure 1 Leukocyte adhesion in a postcapillary venule. Leukocytes flowing in a dense suspension of RBCs first marginate toward the vessel wall, influenced by vessel geometry and the forces exerted by passing RBCs. Upon contact with the wall, selectin binding can occur, resulting in slowing of the cell velocity and commencement of rolling along the endothelium. If activation of the leukocyte and endothelium are sufficient, firm integrin binding can take over, and the cell stops rolling. Eventually, this cell will extravasate into the tissue to carry out its immune function. Secondary capture can occur downstream of the adherent leukocyte to amplify the recruitment of cells. (see color insert)

of a plasma-rich zone near the vessel wall and an enrichment of the RBCs near the center.

This is important because in flowing blood, leukocyte dispersion due to collisions with RBCs can force them toward the wall. This causes the leukocytes to travel preferentially in the plasma-rich zone in vessels larger than 30 μm, increasing the probability of contact with the endothelium.

The seminal studies in this area were performed in the Goldsmith laboratory using capillary tubes and sophisticated imaging techniques to track cell trajectories with high precision. Elegant mathematical analyses of particulate flow have also advanced the field (see, for example, Ref. [1]).

Contact

Because RBCs are more deformable and faster than the leukocytes, they tend to pass the leukocytes at branch points or expansion zones (such as in postcapillary venules). As RBCs maneuver around a leukocyte, they push it even closer to the wall, often causing a collision between the leukocyte and endothelium. Consequently, high levels of adhesion are observed in these regions.

The combined effect of the plasma-rich zone and collisions with passing erythrocytes (predominantly from the radial direction) "traps" the leukocytes near the vessel wall and at the same time moves them along in the axial direction, thus increasing the number of collisions between the leukocyte and the surface.

Because of the difficulty in quantifying rapid leukocyte–endothelium collisions, much of the work in this area has relied upon mathematical simulation. These studies conclude that capillary:postcapillary venule diameter ratio, RBC configuration, and RBC shape are critical determinants of the initiation of contact in postcapillary venules.

Similar fluid dynamics can cause "secondary leukocyte capture" in larger vessels (Figure 1). This phenomenon occurs when a leukocyte in the free stream passes another already rolling or adherent on the surface. The low-pressure zone behind the adherent cell, and leukocyte–leukocyte interactions can cause the flowing cell to contact the wall. This can lead to rapid accumulation of adherent cells in the inflamed vessel.

Rolling

When it contacts the endothelium, the leukocyte is still moving rapidly relative to the endothelium. In order to slow it down, surface adhesion molecules known as selectins engage upon contact. Selectins are glycoproteins that extend from the cell surface, have high mobility, and are able to rapidly bind to counter-receptors on the other cell. The binding is relatively weak, and the bond can quickly dissociate under applied force. But repeated selectin binding and breaking results in "rolling" of the cell along the endothelial wall and, most importantly, a slowing of its motion.

The most important selectins identified to date are E-, P-, and L- selectin (Table I). E- and P-selectin are expressed on endothelial cells, and L-selectin on leukocytes. Selectins bind to sialylated, fucosylated tetrasaccharide sialyl-Lewis^X structures. These adhesion molecules are located on microvilli, small protuberances of 300 to 700 nm on the cell surface, whereas their counterparts on the endothelial wall lie within the glycocalyx layer, which extends 50 to 500 nm above the endothelial plasma membrane. It is thought that this arrangement allows penetration of the microvilli into the negatively charged glycocalyx of the endothelium, facilitating adhesion molecule binding.

Mathematical models of cell rolling have characterized the kinetics of single adhesive bonds and the forces exerted on rolling leukocytes by the blood cells. Once rolling is achieved, the RBCs near the plasma-rich zone encourage the continued rolling of the leukocyte, adding a normal force component and torque.

Probably the most important tool in the study of cell rolling has been the parallel-plate flow chamber. This device allows direct observation of interactions between cells in a flowing fluid and a surface, which is usually coated with counter-receptors or endothelial cells. The forces experienced by cells at the wall can be calculated and varied to mimic adhesion in large and small vessels. Most importantly, this device allows observations under dynamic conditions. This is useful because many of the molecular bonds formed during leukocyte adhesion behave differently depending on the speed of the cell and the force applied. For this reason, selectin molecules could not have been characterized in static adhesion assays.

Arrest

The overall goal of inflammation is leukocyte infiltration—that is, getting the immune cells to enter the tissue at the region of insult. In order to do this, the leukocytes must adhere firmly to the endothelium, resisting the forces of the flowing blood. Firm adhesion is mediated by another set of adhesion molecules, known as integrins, with properties distinct from those of selectins (Table I). Arrest of the cell on the endothelial surface occurs through integrin binding, usually after selectin-mediated rolling.

Integrins are less nimble than selectins (they have more difficulty engaging their counter-receptors because of their relative bulk, slow diffusion, and inflexibility), but once engaged, the bond is much stronger. Integrins are classified into several subfamilies according to their common β subunits; these β subunits form heterodimers with various α subunits.

Several members of the integrin family participate in leukocyte adhesion to endothelium: VLA-4, LFA-1, Mac-1, and p150,95. LFA-1, Mac-1, and p150,95 share the β2 subunit (and are thus called β2 integrins). The β2 subunit is a very important component of the leukocyte adhesion mechanism—patients who have a mutation in this molecule have a disorder known as leukocyte adhesion deficiency (LAD), which results in recurrent bacterial infections due to the inability to effectively recruit leukocytes in response to infections.

The parallel-plate flow chamber has also contributed to our current understanding of integrins and their binding properties. By studying adhesion after genetic manipulation of adhesion molecules or blocking them with antibodies, their contributions and binding kinetics could be determined. In addition, much has been learned about endothelial adhesion molecules and their regulation using this device.

Recently, it has been shown that integrin molecules can increase their affinity in response to certain stimuli. Remarkably, this can be triggered by leukocyte binding to endothelial cells, and it can produce extremely rapid changes in receptor configuration, bringing the leukocyte to an abrupt halt. Although still not well characterized, the mechanism might involve selectin or integrin signaling. Intriguingly, binding of immobilized chemokines on the endothelium to their receptors on the leukocyte can also achieve this response. It remains to be seen what role this plays in leukocyte trafficking and inflammation.

Conclusion

Leukocyte adhesion to the vascular endothelium requires a complex and elegant sequence of events involving remark-

Table I Leukocyte Adhesion Molecules.

Adhesion molecule	Designation	Counter-receptor	Expression
Selectins			
E-selectin	CD62E	SLeX	Endothelium
P-selectin	CD62P	SLeX	Endothelium
L-selectin	CD62L	SLeX	Leukocytes
Integrins			
LFA-1 (αlβ2)	CD11a/CD18	ICAM-1, ICAM-2, ICAM-3	Neutrophils, monocytes, lymphocytes
Mac-1 (αMβ2)	CD11b/CD18	ICAM-1	Neutrophils, monocytes
P150,95 (αxβ2)	CD11b/CD18	?	Monocytes, granulocytes, NK cells, lymphocytes
VLA-4 (α4β1)	CD49d/CD29	VCAM-1	Lymphocytes

able fluid dynamics and a sophisticated array of adhesion molecules. The mechanisms involved have been elucidated using careful analytical methods and mathematical approaches, but there is still much we do not understand. For example, recent studies have shown that the glycocalyx presents a steric barrier to adhesion, and that common inflammatory agents can cause shedding of this glycocalyx, enhancing binding. This exciting finding opens up a new area of research in the field of leukocyte adhesion. Other areas that need more work include those in which the system malfunctions. This is the case in autoimmune diseases such as arthritis where leukocytes adhere to and damage normal tissue. Similarly, overzealous leukocytes are implicated in atherosclerosis in larger arteries, where adhesion is not normally expected. On the other hand, leukocyte infiltration in tumors is generally lower than needed to control tumor growth. The challenge, then, is to determine how the adhesion machinery is breaking down in these pathologies in order to devise new strategies for treatment.

Glossary

Erythrocyte: Red blood cell; any of the hemoglobin-containing cells that carry oxygen to the tissues and are responsible for the red color of vertebrate blood—called also erythrocyte, red blood corpuscle, red cell, red corpuscle.

Leukocyte: White blood cell; any of the blood cells that are colorless, lack hemoglobin, contain a nucleus, and include the lymphocytes, monocytes, neutrophils, eosinophils, and basophils.

Reference

1. Goldman, A. J., Cox, R. G., and Brenner, H. (1967). Slow viscous motion of a sphere parallel to a plane wall—II. Couette flow. *Chem. Eng. Sci.* **22**, 653–660. *This is one of the landmark papers in the mathematical analysis of the fluid dynamics of particles in shear flow.*

Further Reading

Goldsmith, H. L., Cokelet, G. R., and Gaehtgens, P. (1989). Robin Fåhraeus: Evolution of his concepts in cardiovascular physiology. *Am. J. Physiol.* **257**, H1005–H1015. *A comprehensive review of the work of Robin Fåhraeus, written by a group that also contributed greatly to the field of blood fluid dynamics.*

Hynes, R. O. (2002). Integrins: Bidirectional, allosteric signaling machines. *Cell* **110**, 673–687. *A recent review on integrin structure and function.*

Lawrence, M. B., McIntire, L. V., and Eskin, S. G. (1987). Effect of flow on polymorphonuclear leukocyte/endothelial cell adhesion. *Blood* **70**, 1284–1290. *A seminal paper describing the use of the parallel-plate flow chamber to study leukocyte adhesion under dynamic conditions.*

Lawrence, M. B., and Springer, T. A. (1991). Leukocytes roll on a selectin at physiological flow rates: Distinction from and prerequisite for adhesion through integrins. *Cell* **65**, 859–873. *The first detailed analysis of leukocyte rolling on selectins.*

Ley, K. (2001). Functions of selectins. *Results Prob. Cell Differ.* **33**, 177–200.

Schmid-Schönbein, G. W. (1987). Leukocyte kinetics in the microcirculation. *Biorheology* **24**, 139–151.

Sun, C., Migliorini, C., and Munn, L. L. (2003). Red blood cells initiate leukocyte rolling in postcapillary expansions: A lattice-Boltzmann analysis. *Biophys. J.* **85**, 208–222. *Blood flow in a "virtual vessel" elucidates the importance of RBC–leukocyte collisions in margination.*

Zhao, Y., Chien, S., and Weinbaum, S. (2001). Dynamic contact forces on leukocyte microvilli and their penetration of the endothelial glycocalyx. *Biophys. J.* **80**, 1124–1140.

Capsule Biography

Dr. Munn's laboratory in the Steele Lab for Tumor Biology, Massachusetts General Hospital, focuses on applying quantitative methods to the study of tumor physiology and blood flow. He is Director of Microscopy and Image Analysis of the Steele Lab and is supported by grants from the NCI and NHLBI.

CHAPTER 118

Microvascular Responses to Eicosanoids

Mauro Perretti and Derek W. Gilroy

The William Harvey Research Institute, Bart's and the London, Queen Mary University of London, London, United Kingdom

Eicosanoids are low-molecular-weight lipid molecules rapidly generated during normal cell hemostasis or, more often, after cell activation and in conditions of stress. The fact that their generation is rapid indicates their important function in promoting immediate alterations in cell plasticity, thus allowing rapid adaptation to new environments. In more specific terms, eicosanoids are produced following activation of cells that play a central role in modulating microvascular responses (endothelial cells, macrophages and other myeloid cells, fibroblasts, and so on), thereby having a marked impact on a series of microvascular events that characterize several experimental and clinical cardiovascular inflammatory conditions. In this review we will document the various effects eicosanoids have on the microvasculature in terms of cell trafficking, vascular tone, and edema formation.

Arachidonic Acid Metabolism

Arachidonic acid (AA) is a 20-carbon polyunsaturated lipid, with four double bonds, contained in the plasma membrane of virtually all cells. Upon cell activation, AA is released from the phospholipid pool to be metabolized by cyclooxygenase (COX), lipoxygenase (LOX), or cytochrome P450, thus giving rise to families of eicosanoids called the prostaglandins (PGs), leukotrienes (LTs), and epoxyeicosatrienoic acids (EETs), respectively (see Figure 1). PG biosynthesis arises following the metabolism of AA by COX enzymes to PGH_2, which may be further metabolized by downstream synthases to both PGs and thromboxane, collectively referred to as the prostanoids. LOXs, on the other hand, catalyze the metabolism of AA to unstable hydroperoxy intermediates that subsequently form the LTs, hydroxyeicosatetraenoic acid, and lipoxins (LXs). AA metabolism by the cytochrome P450 monooxygenase system gives rise to the less characterized EETs.

Eicosanoids have generated enormous interest mainly because of their ubiquity and pleiotropy of biological activities, which affect not only pharmacology but also physiological and pathophysiological events. For instance, it is the inhibition of prostaglandin formation that accounts, at least in part, for the pharmacological mode of action of nonsteroidal anti-inflammatory drugs, one of the most widely used families of drugs for the treatment of inflammatory mediated diseases, pain, and control of vascular tone. The use of these drugs and careful analysis of their effects have shed light on previously unappreciated pathophysiological pathways that operate during inflammation.

Prostaglandins

Metabolism

Prostaglandins are oxygenated polyunsaturated 20-carbon fatty acids containing a cyclopentane ring. Prostaglandins are designated by the letters A–J depending the nature and position of substituents on the cyclopentane ring and the presence and position of double bonds within the ring, and by a numerical subscript, which indicates the number of double bonds in the alkyl side chains [1]. For instance, metabolism of eicosatrienoic acid ($C_{20:3\ n\text{-}6}$) by COX gives rise to prostaglandins of the 1-series; those of the 2 series are from eicosatetraenoic acid ($C_{20:4\ n\text{-}6}$); and the 3 series are from eicosapentaenoic acid ($C_{20:5\ n-6}$). The synthesis of prostanoids and thromboxane occurs wholly within cells following the release of AA from membrane phospholipids by phospholipase A_2 (PLA_2; of which several isoforms exist) (Figure 1). The hydrolyzed AA is then oxidized to PGG_2, which is then rapidly reduced to PGH_2 by either COX-1 or COX-2 isoform. These two isoforms are distinctly expressed in different tissues as well as in relation to the activation (inflammation) status. The endoperoxide PGH_2 then serves as a substrate for a series of downstream synthases including PGE_2 synthase, PGD_2 synthase, prostacyclin (or PGI_2) synthase, $PGF_{2\alpha}$ synthase, and thromboxane (Tx) A_2 synthase, giving rise to PGE_2, PGD_2, PGI_2, $PGF_{2\alpha}$, and TxA_2, respectively (Figure 1). The unique profile of potently bioactive prostanoids released by a particular cell presumably depends upon the expression/enzyme activity and/or functional coupling of these downstream synthases to either COX isoform. PGG_2 and PGH_2 are relatively short-lived endoperoxide intermediates possessing a dioxygen bridge between C-9 and C-11 on the ring (Figure 1). PGI_2 has an oxygen bridge between C-6 and C-9 and is broken down to the stable metabolite, 6-keto-$PGF_{1\alpha}$. Likewise, as TxA_2 has an unstable bicyclic oxane–oxetane ring structure, it is also rapidly converted to a stable oxane derivative, TXB_2.

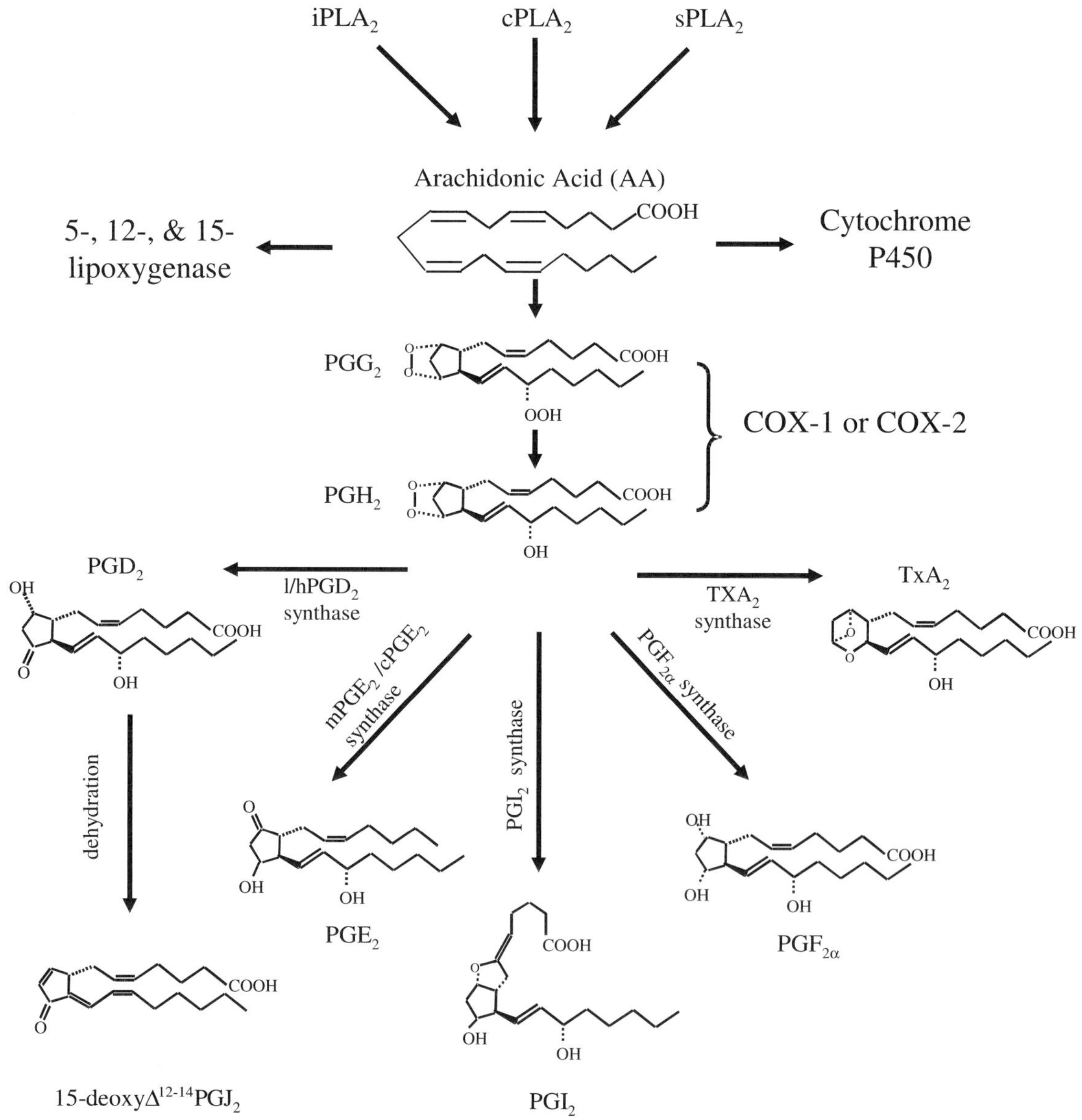

Figure 1 The arachidonic acid cascade. In resting cells arachidonic acid (AA) is esterified to membrane phospholipids. However, upon cell activation and the subsequent action of phospholipase A_2, AA can be liberated from the *sn*-2 position of the glycerol backbone. There are several types of PLA_2 including cytosolic PLA_2 (c PLA_2), calcium-independent PLA_2 (i PLA_2), and several isoforms of secretory PLA_2 (s PLA_2). AA is then oxidized to PGG_2, which is then rapidly reduced to PGH_2 by either cyclooxygenase (COX) isoform. PGH_2 acts then as a substrate for several other synthases leading to the synthesis of specific prostanoids (see main text for more details).

Biological Properties of Prostaglandins

Thrombosis

Platelet COX converts AA to TxA_2, along with comparatively smaller amounts of PGE_2 and $PGF_{2\alpha}$. TxA_2 has a half-life of about 30 seconds and breaks down to inactive TxB_2. TxA_2 constricts large blood vessels, has variable vasoconstrictor activity in the microcirculation, and is a potent stimulus for platelet aggregation. PGI_2, on the other hand, is the main AA metabolite of endothelial cell lining the larger arteries and veins, but is also formed by the endothelial cells of the microvessels. Prostacyclin (PGI_2) is a vasodilator and a potent inhibitor of platelet aggregation. It also exerts other potential protective effects on the endothelium by promoting the outflow of free cholesterol from endothelial cells, inhibiting the release of growth factors from endothelial cells, platelets, and macrophages, and reducing the release of free radicals from leukocytes [2].

A number of thrombotic diseases have been associated with an imbalance in the prostacyclin–thromboxane system. Platelets from patients with arterial thrombosis, deep vein thrombosis, or recurrent venous thrombosis produce more prostaglandin endoperoxides and TxA_2 than normal platelets. Also, platelets from rabbits made atherosclerotic by a high-fat diet or from patients who have survived myocardial infarction are abnormally sensitive to aggregating agents and upon stimulation produce higher amounts of TxA_2 than controls. In general terms, it seems that in diseases where there is a tendency for thrombosis to develop, TxA_2 production is elevated whereas PGI_2 production maybe either maintained or reduced. The opposite is found in some diseases associated with increased bleeding tendency [3, 4].

Vascular Dynamics

Prostaglandins of the E-series along with PGI_2 potently dilate arterioles in the microcirculation (Figure 2). On the other hand, prostaglandins of the A, D, and F series only produce negligible changes in vascular permeability in guinea-pig or rabbit skin even when given at high doses. Thus, in general, it is considered that prostaglandins do not contribute significantly to edema formation by a direct effect on blood vessel permeability; rather, it is believed that vasodilator prostaglandins synergize with other inflammatory mediators to cause inflammatory edema. For instance, the effects of bradykinin or histamine on plasma exudation in guinea-pig skin or edema in the rat paw are potentiated by low doses of E-series prostaglandins. Moreover, PGE_2 causes a dose-dependent increase in blood flow with little or no plasma exudation in the rabbit skin, but when mixed with bradykinin or histamine causes a marked increase in plasma exudation [5, 6].

Pain

Like their effects on the vasculature, none of the COX products cause pain directly, but some eicosanoids potentiate the pain-causing effects of histamine and bradykinin. This finding has led to the suggestion that prostaglandins can sensitize the afferent pain nerve endings, leading to the hyperalgesia characteristic of inflammation.

Cell Recruitment

Another critical component of the inflammatory response is the recruitment of inflammatory cells. However, it is generally accepted that the COX pathway is not involved in the recruitment of these inflammatory leukocytes. In fact, it is likely that PGE_2 or PGI_2 attenuates leukocyte migration by inhibiting leukocyte motility and preventing adherence to vascular endothelial cells [7].

Leukotrienes

Metabolism

In a similar manner to the generation of the prostaglandins, AA can also be metabolized entirely within the cell by 5-, 12-, or 15-LOX to the potently bioactive LTs. Although 12-LOX was the first LOX to be identified in human platelets, it was the discovery of leukocyte 5-LOX that attracted most attention as it was recognized that this enzyme was responsible for the generation of both the slow-reacting substances of anaphylaxis (LTC_4 or LTC_4, LTD_4, and LTE_4; potent mediators of allergic responses) and LTB_4, a powerful PMN chemoattractant [8]. The first step in all the LOX pathways is the formation of the hydroperoxide of the parent straight-chain C-20 polyunsaturated fatty acid. Hydroperoxides derived from AA are termed hydroperoxyeicosatetraenoic acids (HPETEs) and are converted to hydroxyeicosatetraenoic acids (HETEs). Leukotrienes are conjugated trienes produced by the action of 5-LOX. Leukotrienes of the 4-series (LTA_4, LTB_4) are derived from AA, whereas the 3-series is derived from eicosatrienoic acid and the 5-series from eicosapentaenoic acid. LTA_4 is an unstable 5,6-epoxy intermediate in the formation of LTB_4. LTC_4 is also formed from LTA_4 by the action of γ-glutamyl-*S*-transferase. LTD_4 is formed from LTC_4 by the action of γ-glutamyl transpeptidase that removes glutamic acid, and further removal of glycine by cysteinyl glycinase gives rise to LTE_4. LTF_4 can arise from LTE_4 by reincorporation of glutamic acid. Besides COX and the LOX pathways of AA metabolism, the cytochrome P450 superfamily of drug-metabolizing enzymes also generates epoxyeicosatrienoic acid (EETs) from AA. More recently, β-naphthoflavone-stimulated cytochrome P-450 in porcine coronary artery endothelial cells not only synthesized EETs, but was also shown to generate endothelium-derived hyperpolarizing factor, a critical factor in mediating vascular dilation. This discovery was particularly important, as substantial endothelium-dependent dilation to various stimuli was known to occur even after inhibition of both nitric oxide synthase and COX. Another family of eicosanoids derived from the concerted action of LOX–LOX or COX–LOX interaction are the LXs, which will be discussed in greater detail later.

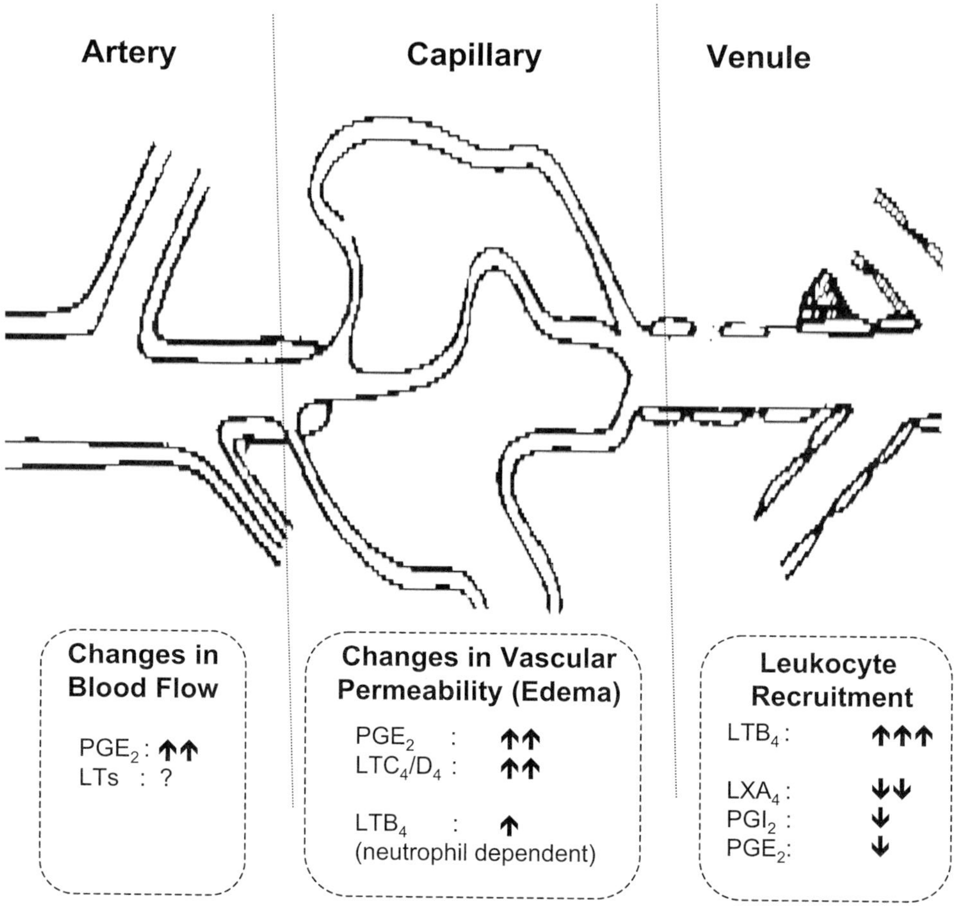

Figure 2 Eicosanoids and the microcirculation. Schematic representation of a given microcirculation showing the arterial section followed by the bed of capillaries prior to end into the venular side. Eicosanoids have been reported to produce several effects on distinct sections of the vascular tree with a great degree of species and tissue specificity (see main text for more details). In general terms, prostaglandin E_2 (PGE_2) is the major vasoactive eicosanoid able to increase blood flow, thereby facilitating the process of plasma protein extravasation, as based upon the two-mediator theory. Direct effect on the capillaries is produced by peptidoleukotrienes, including LTC_4 and LTD_4, that act on the endothelial cells to directly promote changes in vascular permeability. The effect of LTB_4 is less clear in the sense that, in line with several other chemoattractants, it can cause increases in vascular permeability indirectly, by promoting white blood cell adhesion. This latter point is particularly true at the level of the postcapillary venule, whereby LTB_4 is the most potent eicosanoid in producing leukocyte recruitment. Lipoxin A_4 (LXA_4) and some prostaglandins (including prostacyclin or PGI_2) are potent inhibitors of leukocyte recruitment, thereby producing tonic inhibition (as in the case of the antiadhesive actions of prostacyclin), or are predominantly involved during the resolution phase of inflammation (as is the case for LXA_4).

Biological Properties of LTs

Vascular Tone

For the most part, products of the LOX pathway of AA metabolism contribute little to the changes in vascular tone. Moreover, the vascular effects that have been reported differ depending on the particular metabolite [8–10]. For instance, hydroperoxides of AA cause erythema in human skin and LTC_4 and LTD_4 are vasoconstrictors in guinea pig and rat, whereas LTB_4 has no effect on blood flow in the rabbit or hamster cheek pouch model. However, it is possible that LTs influence vascular permeability, though this may be species dependent. LTC_4 and LTD_4 increase vascular permeability in human, guinea pig, and rat skin and are about 1,000 times more potent than histamine in inducing macromolecular leakage from postcapillary venules. In contrast, they have little or no activity in rabbit skin and unlike PGE_2 do not potentiate bradykinin-mediated edema formation. LTB_4, on the other hand, is a weak inducer of plasma exudation in rabbit skin but synergizes with vasodilator prostaglandins to cause enhanced plasma exudation.

Cell Recruitment

Despite its disparate and perhaps modest effects on the vasculature, LTB_4 is one of the most potent endogenous chemotactic factors known. Once added to white blood cells, LTB_4 induces chemotaxis, cell aggregation, and release of lysosomal enzymes. In vivo, LTB_4 induces leukocyte accumulation in rabbit and human skin as well as in the anterior chamber of the rabbit eye. LTB_4 superfusion over the hamster cheek pouch microvascular bed leads to leukocyte adhesion and emigration (diapedesis), whereas LTC_4 and LTD_4 are inactive. The potent effects of LTB_4 on the recruitment of circulating leukocytes to the site of inflammation contributes to the increase of vascular permeability it produces, as mentioned in the previous subsection.

INFLAMMATORY PAIN

There are conflicting opinions regarding the contribution of LTs to inflammatory pain, and this may depend on the model system investigated. For instance, LTB_4 and LTC_4 antagonize bradykinin-induced hyperalgesia in the rabbit ear. Similarly, at high doses, LTB_4, LTC_4, and LTD_4 antagonize intradental nerve excitability. However, using the rat paw withdrawal model, LTB_4 induces hyperalgesia in a manner that is dependent on circulating polymorphonuclear cells and activation of the 15-LOX pathway.

Lipoxins

Metabolism

In contrast to the synthesis of prostaglandins and LTs that follows more canonical pathways in a single cell type, the release of the LXs results from transcellular and cell–cell interaction-mediated biochemical events. In humans, the generation of LXs is the result of LOX–LOX or COX–LOX interactions [11]. There are now three defined routes for their generation. In eosinophils, monocytes, or epithelial cells, the first biosynthetic step involves the insertion of O_2 to the OH group on carbon 15 of AA by 15-LOX, yielding hydroperoxyeicosatetraenoic acids. Following their release from these cell sources and entry into either polymorphonuclear cells or monocytes, a 5,6-epoxytetraene is generated by the 5-LOX, which is then hydrolyzed within these recipient cells by either LXA_4 hydrolase or LXB_4 hydrolase to bioactive LXA_4 and LXB_4, respectively.

The second route of LX biosynthesis results from the generation of LTA_4 by the leukocyte 5-LOX, its release, and its subsequent uptake by platelets that contain a 12-LOX to generate LXA_4 and LXB_4.

The third route for LX generation results from the acetylation of COX-2 by acetylsalicylic acid in endothelial cells and other cell types. This results in the synthesis of (15*R*)-hydroxyeicosatetraenoic acid (15-*R*-HETE), which is rapidly metabolized in a transcellular manner by adherent leukocyte, vascular endothelial, or epithelial 5-LOX to 15-epiLXA_4 or 15-epiLXB_4.

Within seconds or minutes from their synthesis, LXs are converted to inactive metabolites, characterized by dehydrogenation at the C-15 position. Once formed these intermediate LXs are no longer further metabolized by PMN but are rapidly inactivated, as a result of dehydrogenation, to inactive oxo-LXs by monocytes.

Having detailed the three potential biosynthetic pathways, it is natural to ask how, in practice, are LXs formed? The three major pathways just described can operate independently or simultaneously within the vasculature. For instance, GM-CSF-primed neutrophils recruited to the inflammatory site can interact with platelets. After platelets adhere to the neutrophil surface, active leukocyte-generated LTA_4 is released and transformed by platelet 12-LOX to generate LXs. Within the vasculature, the aspirin-triggered LX pathway can also be initiated when activated endothelial cells interact with adherent neutrophils to generate 15-epi-LXs. Leukocytes interacting with epithelial cell surfaces, as in the case of respiratory, renal, or gastrointestinal inflammation, can also generate LXs through bidirectional routes, in which (15S)-HETE and (15R)-HETE are released by epithelial cells and converted to LXs by neutrophils. The other component of this bidirectional interaction can involve neutrophil-released LTA_4, which is converted by 15-LOX in epithelial cells, in particular tracheal epithelial cells, to generate LXs.

Biological Properties of LXs

VASCULAR TONE

LXs exert a range of biological effects [12]. LXA_4 and LXB_4 promote vasorelaxation and relax aorta and pulmonary arteries. LXA_4 also reverses precontraction of the pulmonary artery induced by $PGF_{2\alpha}$ and the potent vasoconstrictor mediator endothelin-1. The mechanisms of LXA_4- and LXB_4-induced vasodilation involve endothelium-dependent vasorelaxation as well as prostaglandin-dependent and -independent pathways. In endothelial cells, LXs can stimulate the synthesis of PGI_2 and nitric oxide, both of which produce vasodilation.

CELL RECRUITMENT

In terms of cell trafficking, LXs inhibit neutrophil and eosinophil chemotaxis in the nanomolar range. LXA_4 inhibits neutrophil transmigration across endothelial and epithelial cells in vitro, whereas in vivo it blocks leukocyte diapedesis from postcapillary venules; the end point would be inhibition of white blood cell entry into inflamed tissues as seen in animal models of vascular inflammation. Because of the very short half-lives of the LXs, a range of stable biologically active analogs have been designed and tested for their anti-inflammatory effects in experimental animal models. The LXA_4 analog 16-phenoxy-LXA4 and the stable analog of the aspirin-triggered epi-LXs, 15-epi-16-phenoxy-LXA_4, markedly reduced LTB_4-induced ear swelling in the mouse. In this assay, the LX analogs were as effective as dexamethasone in preventing both leukocyte infiltration and changes in vascular permeability (edema).

Interestingly, unlike neutrophils and eosinophils, LXs are potent chemoattractants from monocytes. LXA_4 and LXB_4 stimulate monocyte chemotaxis and adherence to endothelial monolayers, and these effects may be related to the recruitment of monocytes to sites of wound injury. However, such LX-recruited monocytes do not generate superoxide anions or degranulate in the presence of LXs. In fact, they exhibit a greater capacity to phagocytose effete and apoptotic leukocytes, leading to resolution of acute inflammation.

Glossary

Arachidonic acid: Polyunsaturated fatty acid, with 20 carbon atoms, liberated from the phospholipid membrane by the action of phospholipases.

Eicosanoids: This term refers to all products derived from arachidonic acid metabolism.

Nonsteroidal anti-inflammatory drugs: Also referred to as aspirin-like drugs, these therapeutic agents are effective because they block the catalytic action of COX, thereby blocking prostanoid generation.

Prostanoids: This term refers to all products derived from COX activity, and thus to prostaglandins and thromboxanes.

Bibliography

1. Samuelsson, B., Goldyne, M., Granstrom, E., Hamberg, M., Hammarstrom, S., and Malstem, C. (1978). Prostaglandins and thromboxanes. *Ann. Rev. Biochem.* **47**, 997–1029.
2. Wu, K. K. (1996). Endothelial prostaglandin and nitric oxide synthesis in atherogenesis and thrombosis. *J. Formos. Med. Assoc.* **95**, 661–666.
3. Hedqvist, P., Raud, J., Palmertz, U., Kumlin, M., and Dahlen, S. E. (1991). Eicosanoids as mediators and modulators of inflammation. *Adv. Prostaglandin Thromboxane Leukot. Res.* **21B**, 537–543.
4. Vane, J. R., and Botting, R. M. (1994). Formation by the endothelium of prostacyclin, nitric oxide and endothelin. *J. Lipid Med.* **6**, 395–404.
5. Williams, T. J., and Morley, J. (1973). Prostaglandins as potentiators of increased vascular permeability in inflammation. *Nature* **246**, 215–217. *This is the first paper describing the two-mediator theory, in which prostaglandin E_2 cooperates to promote vascular permeability and edema formation during the early stages of inflammation.*
6. Williams, T. J., and Peck, M. J. (1977). Role of prostaglandin-mediated vasodilatation in inflammation. *Nature* **270**, 530–532.
7. Williams, K. I., and Higgs, G. A. (1988). Eicosanoids and inflammation. *J. Pathol.* **156**, 101–110.
8. Samuelsson, B., Dahlen, S. E., Lindgren, J. A., Rouzer, C. A., and Serhan, C. N. (1987). Leukotrienes and LXs: Structures, biosynthesis, and biological effects. *Science* **237**, 1171–1176.
9. Ford-Hutchinson, A. W., Bray, M. A., Doig, M. V., Shipley, M. E., and Smith, M. J. (1980). Leukotriene B, a potent chemokinetic and aggregating substance released from polymorphonuclear leukocytes. *Nature* **286**, 264–265. *This is the first report on the generation of leukotriene B_4 and description of its chemotactic role for neutrophils.*
10. Dahlen, S. E., Bjork, J., Hedqvist, P., Arfors, K. E., Hammarstrom, S., Lindgren, J. A., and Samuelsson, B. (1981). Leukotrienes promote plasma leakage and leukocyte adhesion in postcapillary venules: In vivo effects with relevance to the acute inflammatory response. *Proc. Natl. Acad. Sci. USA* **78**, 3887–3891.
11. Serhan, C. N., Hamberg, M., and Samuelsson, B. (1984). Lipoxins: Novel series of biologically active compounds formed from arachidonic acid in human leukocytes. *Proc. Natl. Acad. Sci. USA* **81**, 5335–5339.
12. Serhan, C. N., and Samuelsson, B. (1988). Lipoxins: A new series of eicosanoids (biosynthesis, stereochemistry, and biological activities). *Adv. Exp. Med. Biol.* **229**, 1–14.

Capsule Biography

Both Mauro Perretti (Ph.D., Professor of Immunopharmacology) and Derek W. Gilroy (Ph.D., Senior Research Fellow) are at The William Harvey Research Institute, Queen Mary University of London. Though their research activities are independent of one another, the scientific interests of the authors can be grouped under the umbrella of anti-inflammation, that is, the study of the endogenous mediators/pathways that operate in the host to ensure the transient nature of the inflammatory response. Perretti's major research efforts have been spent in the field of annexin-1, glucocorticoids, and melanocortins, whereas Gilroy has pioneered the anti-inflammatory role of COX-2 and derived products, as well as nitric oxide, in acute inflammation. DWG: Center for Clinical Pharmacology, Rayne Institute, 5 University Street, University College London WC1 6JJ, UK.

CHAPTER 119

Oxidants and Vascular Signaling Mechanisms

Michael S. Wolin
New York Medical College, Valhalla, New York

Importance of Oxidant Signaling Mechanisms in the Microcirculation

It is rapidly becoming recognized that the formation of reactive oxygen species (ROS), reactive nitric oxide (NO)-derived species (RNS), and their interaction with cellular regulatory systems are fundamental components of signaling systems that contribute to the control of physiological and pathophysiological aspects of microvascular function in processes such as the activation of certain receptors and cellular systems which sense changes in oxygen tension, flow, and pressure. Oxidase enzymes are the primary source of generation of ROS as a result of their ability to transfer electrons to molecular oxygen, forming species including superoxide anion ($O_2^{\bullet-}$) and hydrogen peroxide (H_2O_2), as a result of the oxygen accepting one or two electrons, respectively. Although the levels of these ROS are kept extremely low by metabolizing enzymes including superoxide dismutases, peroxidases, and catalase, and antioxidants that also control the actions of RNS, each individual ROS and RNS has concentration-dependent mechanisms through which it interacts with intracellular and intercellular signaling systems that control processes within each cell type in the microcirculation (e.g., endothelium and vascular smooth muscle).

Sources of Oxidant Species

There are multiple oxidase enzymes that generate ROS, and each enzyme appears to have specific mechanisms that regulate ROS production and specific cellular localization sites. Evidence is rapidly emerging that one of the most important types of oxidases in physiological processes are NAD(P)H oxidases (containing specific nox subunits) that resemble the oxidase initially identified in phagocytic cells, which is now called nox-2. Vascular smooth muscle and endothelium appears to contain the nox-1, nox-2 and nox-4 forms of these nox oxidases. The nox oxidases appear to be composed of membrane bound nox and p22phox subunits possessing an electron transport system that transfers electrons from cytosolic NADPH and/or NADH to molecular oxygen through a flavin site and a b_{558}-type cytochrome. Electron transfer resulting in the generation of ROS by the nox oxidases appears to be controlled by stimulation of the binding of one or more cytosolic protein subunits that include or resemble rac-1, p47phox and p67phox. All cells seem to generate ROS under unstimulated conditions, and it appears that nox-containing systems are an important contributor to the basal levels of ROS production in vascular cells. Stimuli including stretch and shear and the activation of receptors linked to cell growth (e.g., angiotensin II) are thought to increase nox activity through cytosolic subunit binding triggered by mechanisms including phosphorylation by protein kinase C or other signaling processes. Some growth factors also appear to increase oxidase activity through promoting the expression of the nox and p22phox subunits. The nox enzymes that are present in vascular cells appear to have important roles in intracellular signaling events that control vascular function and cellular growth.

Endothelium contains multiple additional oxidant-generating enzyme systems, including cytochrome P450, NO synthase (NOS), cyclooxygenase (COX), and xanthine oxidase (XO). Recent evidence suggests that H_2O_2 is an important endothelium-derived relaxing factor, and receptor stimulation of its generation by cytochrome P450 and other oxidases appears to be a source of its generation. Although the major product of the NOS reaction is NO, this system has NADPH oxidase activity that is suppressed by the availability of the reduced form of its cofactor tetrahydrobiopterin, and by the availability of L-arginine for NO biosynthesis. When NOS is making both NO and $O_2^{\bullet-}$, the reaction between these molecules generates peroxynitrite ($ONOO^-$), which is a reactive species that generates other RNS (including nitrogen dioxide), which oxidizes tetrahydrobiopterin and causes thiol oxidation, nitrosation and nitration, and tyrosine nitration, whereas, when the generation of $O_2^{\bullet-}$ is the primary product of NOS, this system appears to become a major source of vasoactive endothelium-derived H_2O_2 generation. Vascular disease processes are known to promote oxidant production from NOS, COX, and XO. The generation of $O_2^{\bullet-}$ from COX seems to be associated with the availability high levels of arachidonic acid and high rates of prostaglandin generation by this enzyme. Generation of both $O_2^{\bullet-}$ and H_2O_2 by XO appears to require conversion of this enzyme to an oxidase from its dehydrogenase form by a combination of thiol oxidation and/or proteolysis, and the availability of its substrates hypoxanthine and xanthine. Prolonged hypoxia and reoxygenation generally stimulates XO activation and the accumulation of its substrates from the degradation of tissue ATP. Thus, while endothelium is a source of vasoactive levels of H_2O_2, this cell type can also become a major source of vascular oxidant production during pathophysiological processes.

Mitochondria of vascular cells also appear to have an important role in controlling oxidant generation either through producing $O_2^{\bullet-}$ or as a result of their role in controlling the redox status of cytosolic NAD(H), and perhaps NADP(H). The NADH dehydrogenase and coenzyme Q sites in the electron transport chain appear to be sites where electrons can be transferred to molecular oxygen, and processes associated with the development of apoptosis seem to markedly increase mitochondrial oxidant generation. Mitochondria appear to contribute to the sensing of physiological PO_2 levels through poorly understood redox processes that seem to be linked to the generation of ROS.

Mechanisms of Interactions of Oxidant Species with Vascular Signaling Systems

Each oxidant species has unique ways, shown in Figure 1, of interacting with the signaling systems listed in Table I as a result of its chemical and metabolic properties.

The major ROS that is generated by cellular oxidases appears to be $O_2^{\bullet-}$, which is a negatively charged free radical at physiological pH, with very selective chemical reaction properties. The levels of SOD in the cytosol (SOD-1) and mitochondria (SOD-2) appear to function in a manner that maintains intracellular $O_2^{\bullet-}$ concentrations in the picomolar

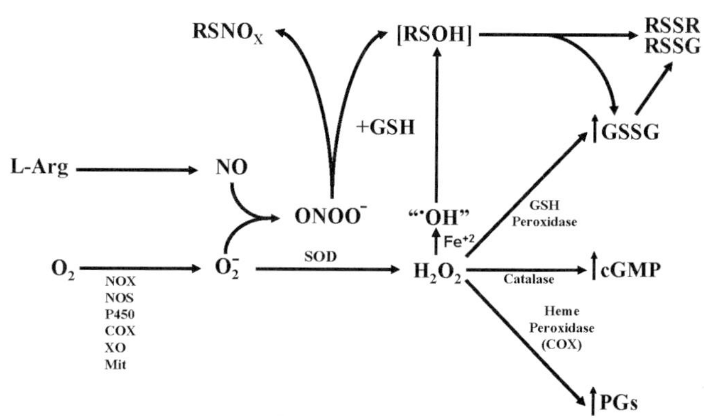

Figure 1 Scheme showing oxidases that generate ROS, and potential ways in which ROS interact with vascular signaling systems.

Table I Oxidant Species and Some of Their Interaction with Signaling Systems.

Species	Interaction	Regulatory action
$O_2^{\bullet-}$	NO	Inhibits actions of NO
	Fe	Releases iron bound to proteins as Fe^{2+}
	SOD	Converts $O_2^{\bullet-}$ to H_2O_2
H_2O_2	Cyclooxygenase	Stimulates prostaglandin production
	Glutathione peroxidase	Converts GSH to GSSG
	Catalase	Stimulates sGC and cGMP production
	Fe^{2+}	Generates reactive "·OH" species
NO	Heme binding	Stimulates sGC and cGMP production
		Reversibly inhibits mitochondrial respiration
	$O_2^{\bullet-}$	Generates peroxynitrite (ONOO)
$ONOO^-$ (RNS)	Thiols (RSH)	Oxidation and nitrosation
	Mitochondria (Fe-S)	Irreversible inhibition of respiration
	Tyrosine	Nitration, inactivation of PGI_2 synthase
Thiol	Oxidation or Nitrosation (RSSG, RSSR, RSOH, $RSNO_x$)	Opens potassium channels
		Activates tyrosine phosphorylation
		Altered Ca^{2+} reuptake mechanisms
		Inhibits calcium influx

range. These low levels of $O_2^{\bullet-}$ appear to keep it from directly interacting with most signaling systems. Because of the extremely rapid rate of reaction of NO with $O_2^{\bullet-}$, increases in $O_2^{\bullet-}$ are observed to readily attenuate physiological actions of NO [e.g., stimulation of soluble guanylate cyclase (sGC) and inhibition of tissue mitochondrial respiration] and can form $ONOO^-$ in amounts that interact with signaling systems when high levels of NO are present. Vascular smooth muscle secrete an extracellular form of SOD (SOD-3), which appears to protect NO from $O_2^{\bullet-}$ as it diffuses through the vessel wall. Elevated levels of $O_2^{\bullet-}$ are also known to damage iron–sulfur centers, associated with the inhibition of mitochondrial respiration, and release of iron. A major role for $O_2^{\bullet-}$ and SOD in signaling appears to be the generation of H_2O_2.

Hydrogen peroxide has a vast array of interactions with signaling systems, and its basal levels in cells are thought to be in the low nanomolar range. The signaling mechanisms most sensitive to the actions of H_2O_2 appear to be linked to its metabolism by heme peroxidases, catalase, and glutathione (GSH) peroxidase, and these systems participate in the generation of prostaglandins, cGMP, and oxidized glutathione (GSSG), respectively. The local concentration of GSSG and proteins present that have thiol groups (RSH) that can be S-thiolated (RS-SG) or oxidized to disulfides (RSSR) are key processes that link ROS to cellular signaling systems. Some of the cellular proteins and systems known to be regulated by thiol redox that potentially influence microvascular function are included in Table I. When ferrous iron (Fe^{2+}) is released, it promotes the formation of very reactive or hydroxyl radical ("$^{\bullet}OH$")-like species from H_2O_2, which cause cellular injury.

Evidence is emerging that RNS have important regulatory functions through interactions with signaling systems. These species readily modify proteins with reactive thiol groups (see Table I) by nitrosating (RS-NO), nitrating (RS-NO_2), or oxidizing (RS-SG, RSSR, RSOH, RSO_x) them in a manner similar to H_2O_2 (or elevated levels of other ROS). Protein tyrosine groups, iron–sulfur centers, and unsaturated fatty acids are also readily modified by RNS (and ROS). Some well-documented actions of $ONOO^-$ that potentially influence microvascular regulation include the inactivation of prostaglandin I_2 synthase by nitration of a key active-site tyrosine, and inactivation of tissue mitochondrial respiration by damaging key iron–sulfur centers. Thiols linked to zinc binding sites are also very sensitive to oxidative processes that release zinc and promote cellular signaling. While $O_2^{\bullet-}$ and peroxide have the potential for interactions with the protein thiols and other sites modified by RNS, there seems to be a need for an enhancement of the reactivity of the thiol site (e.g., by acidification) or ROS by cofactors such as iron for the observance of selective signaling-type regulation. Thus, although both peroxide metabolism and RNS cause the oxidation of GSH to GSSG, the chemical reactive properties that RNS possess enable them to have multiple additional regulatory and pathophysiological interactions, especially at activated thiol and metal-binding sites.

Tyrosine phosphorylation, potassium channel opening, and certain systems that control intracellular Ca^{2+} and phospholipid metabolism appear to be cellular regulatory systems having components that seem to be directly regulated by ROS, RNS, or changes in the redox status of cytosolic NADH, NADPH, and/or GSH. Tyrosine phosphatases have activated thiols (RSH) at their catalytic sites that appear to be readily modified or oxidized by ROS and RNS, resulting in inhibition of the tyrosine-phosphate hydrolyzing activity of these enzymes. Some tyrosine kinases appear to autoactivate when the status of their phosphorylated tyrosine groups changes. Tyrosine phosphorylation appears to control multiple processes often associated with vascular contraction, gene expression, and cell growth through changes in the activities of systems including protein kinases B and C, growth factor receptor signaling, and mitogen-activated protein (MAP) kinases. Potassium channels often open when they undergo oxidation, and the thiol groups they contain may be key sites for redox regulation. Plasma membrane Ca^{2+} channels (e.g., L-type channels) and the sarcoplasmic–endoplasmic reticulum Ca^{2+}–ATP pump or SERCA also appear to readily show inhibition of Ca^{2+} influx and alterations in Ca^{2+} reuptake, respectively, when these systems are modified by thiol oxidation. Many of these oxidant-regulated signaling systems seem to function in a coordinated manner to control phospholipid signaling, including the release of arachidonic acid, which is used for the biosynthesis of prostaglandins. The hyperpolarization-mediated relaxation induced by opening K^+ channels activated by H_2O_2 has resulted in it being considered as an endothelium-derived hyperpolarizing factor. Overall, the subcellular localization of both ROS generation and redox changes observed may have fundamental roles in controlling which oxidant-regulated signaling mechanisms are activated, and which physiological or pathophysiological responses are observed.

Physiological Importance of Vascular Oxidant Signaling

PO_2 Sensing

The sensing of changes in oxygen tension potentially originates in many circumstances from the PO_2-dependence of the substrate requirements for oxygen utilization by oxidases that generate ROS or by oxidative enzymes that use oxygen for the formation of reactive enzyme intermediates involved in the generation of vasoactive mediators including prostaglandins, NO, and cytochrome P450. There is substantial evidence that many of these systems are key processes through which changes in PO_2 regulate vascular function. It appears that ROS (e.g., H_2O_2) originating from the basal activity of nox-type oxidases and mitochondria are mediators of PO_2-linked signaling systems that may function through the redox regulation of K^+ and Ca^{2+} channels, and cGMP production by sGC.

Sensing of Physical Forces

Stretch and shear forces in the vessel wall have been demonstrated to promote activation of nox-type oxidases. The stretch component of increased blood pressure activates myogenic contractions through the generation of ROS and signaling systems potentially activated either directly or indirectly by ROS including protein kinase C and MAP kinases which appear to contribute to stretch-induced vascular contractile and growth-related adaptive remodeling responses. High levels of shear and stretch increase superoxide in endothelium, resulting in an attenuation of NO-mediated dilator responses.

Hormone and Growth Factor Receptor Function

Many receptors that possess growth-promoting actions in vascular tissue cause activation of nox-type oxidases and the stimulation of redox-linked signaling systems such as MAP kinases. Although many of the signaling events activated through growth-promoting receptors do not appear to be redox-mediated, ROS seem to have an important modulating role in controlling the growth response. Excessive generation of ROS will activate apoptosis-related signaling and processes associated with cell death as mitochondrial function deteriorates.

Roles of Oxidant Signaling Mechanisms in Microvascular Function

The fundamental roles ROS and RNS play in cellular signaling processes provide these reactive species with a wealth of roles in the control of microvascular function. A combination of the local balance between the activities of NO- and $O_2^{\bullet -}$-generating systems and ROS- and RNS-scavenging systems, localized redox perturbations caused by these species, the function of other non-ROS signaling systems, and energy metabolism-related balances originating from the delivery of O_2 and the work-related metabolic needs of the tissue are likely to all function together as an integrated interactive system that controls the expression and importance of each individual cellular signaling mechanism involving ROS and RNS in the microcirculation. Signaling systems sensitive to regulation by low levels of ROS seem to have important coordinated roles in the acute vasoactive and more chronic adaptive physiological responses to changes in PO_2, flow, and pressure. Aging and multiple vascular disease processes including hypertension, atherosclerosis, diabetes, heart failure, and ischemia activate increased ROS and RNS production in the vessel wall through multiple mechanisms, resulting in impaired endothelium-derived NO mediated relaxation and a suppression of the antithrombotic and anti-inflammatory actions of NO, increased expression of adhesion proteins, and endothelial cell permeability. Thus, at elevated levels of ROS and RNS, signaling mechanisms controlled by these species appear to actively participate in promoting the progression of microvascular disease processes.

Glossary

Oxidant signaling mechanisms: Processes involved in controlling the generation of reactive O_2 species and their interactions with specific systems that regulate cellular function.

Reactive O_2 species: Low-molecular-weight oxygen-containing biological molecules resulting from electrons being transferred to O_2, which are generally quite unstable and reactive with other specific cellular molecules or protein functional groups.

Redox signaling: Processes involved in the control of low-molecular-weight biological molecules and protein functional groups that exist in various oxidized and reduced forms having a role in regulating specific systems that control cellular function.

Bibliography

Cai, H., and Harrison, D. G. (2000). Endothelial dysfunction in cardiovascular diseases: The role of oxidant stress. *Circ. Res.* **87**, 840–844.

Dröge, W. (2002). Free radicals in the physiological control of cell function. *Physiol. Rev.* **82**, 47–95.

Griendling, K. K., Sorescu, D., Lassègue, B., and Ushio-Fukai, M. (2000). Modulation of protein kinase activity and gene expression by reactive oxygen species and their role in vascular physiology and pathophysiology. *Arterioscler. Thromb. Vasc. Biol.* **20**, 2175–2183.

Griendling, K. K., Sorescu, D., and Ushio-Fukai, M. (2000). NAD(P)H oxidase: Role in cardiovascular biology and disease. *Circ. Res.* **86**, 494–501.

Lassègue, B., and Clempus, R. E. (2003). Vascular NAD(P)H oxidases: Specific features, expression, and regulation. *Am. J. Physiol. Regulatory Integrative Comp. Physiol.* **285**, 277–297.

Thannickal, V. J., and Fanburg, B. L. (2000). Reactive oxygen species in cell signaling. *Am. J. Physiol. Lung Cell. Mol. Physiol.* **279**, 1005–1028. *This is a broad review on how reactive oxygen species can function in cellular signaling processes.*

Wolin, M. S. (2000). Interactions of oxidants with vascular signaling systems. *Arterosclet. Thromb. Vasc. Biol.* **20**, 1430–1442.

Wolin, M. S., Burke-Wolin, T. M., and Mohazzab-H., K. M. (1999). Roles for NAD(P)H oxidases and reactive oxygen species in vascular oxygen sensing mechanisms. *Respir. Physiol.* **115**, 229–238. *The focus of this review is how NAD(P)H oxidase can function in the role of a vascular oxygen sensor.*

Wolin, M. S., Gupte, S. A., and Oeckler, R. A. (2002). Superoxide in the vascular system. *J. Vasc. Res.* **39**, 191–207. *This is an in depth review on the control of vascular contractile function by superoxide anion and its derived species.*

Capsule Biography

The doctoral work of Michael S. Wolin was focused on characterizing the catalytic mechanism of the adenylate cyclase, and he received his Ph.D. in chemistry in 1982 from Yale University. He then joined the laboratory of Louis J. Ignarro in the Department of Pharmacology at Tulane University, and his postdoctoral studies involved elucidating how NO, free radicals, and heme modification regulate soluble guanylate cyclase. He then joined the faculty of the Department of Physiology at New York Medical College in 1983, where he is currently Professor of Physiology, with research interests that focus on how reactive species derived from O_2 and NO participate in signaling mechanisms regulating vascular and microvascular function, with a focus on the soluble guanylate cyclase system, O_2 sensing mechanisms, and endothelial–vascular regulation.

CHAPTER 120

Reactive Oxygen Species and the Microcirculation

John G. Wood and Norberto C. Gonzalez
University of Kansas Medical Center, Kansas City, Kansas

Oxidative stress to the vascular endothelium is now recognized to play a significant role in the pathogenesis of many vascular diseases, including ischemia/reperfusion (I/R), athlerosclerosis, diabetes, and systemic hypoxia. Reactive oxygen species (ROS) are normally produced in all cells during aerobic metabolism. Although ROS are highly reactive, multiple defenses exist within cells to prevent ROS-induced dysfunction. However, these defenses may be overwhelmed in conditions that markedly increase oxidant generation. When such a state of oxidative stress develops, a host of microvascular inflammatory responses may be initiated, resulting in impaired microvascular function and potentially contributing to the progression of cardiovascular disease. This article will examine the effects of reactive oxygen species (ROS) on the microcirculation, pathways capable of generating ROS, and endogenous defenses against oxidative stress. Finally, evidence will be presented regarding the role of ROS in the pathogenesis of microvascular inflammation resulting from I/R and systemic hypoxia.

Microvascular Responses to Oxidative Stress

Endothelial cells form a single cell layer on the inner surface of all vessels of the cardiovascular system. These cells play a critical role in physiological regulation of the microcirculation as well as in the pathogenesis of microvascular inflammation. It is now recognized that endothelial dysfunction resulting from oxidative stress is an underlying cause of many vascular diseases. In addition, oxidant-induced endothelial dysfunction can impair microvascular function in several ways (Figure 1).

Arterioles

These vessels, which consist largely of vascular smooth muscle, regulate organ blood flow through alterations in vascular resistance (via changes in vessel diameter). In arterioles, oxidative stress results in a diminished response to endothelium-dependent vasodilators. This alteration promotes arteriolar vasoconstriction and decreased organ blood flow, which may exacerbate tissue injury. Impaired arteriolar dilation is primarily due to oxidant-induced endothelial dysfunction rather than compromised vascular smooth muscle function. In conditions associated with oxidative stress, administration of antioxidants has been shown to preserve arteriolar responsiveness to endothelium-dependent dilators. In addition, oxidative stress may promote generation of vasoactive factors that promote constriction of arteriolar smooth muscle.

Capillaries

Oxidative stress has been shown to increase fluid filtration across capillaries as well as to decrease the number of perfused capillaries. Evidence indicates that the higher rate of filtration under these conditions is primarily due to increased capillary permeability rather than elevated capillary hydrostatic pressure. The reduction in capillary perfusion after oxidative stress has been attributed to several factors, including capillary plugging by activated leukocytes as well as compression of capillaries due to increased interstitial pressure resulting from edema.

Venules

Oxidative stress promotes increased leukocyte–endothelial cell adhesive interactions, leukocyte emigration

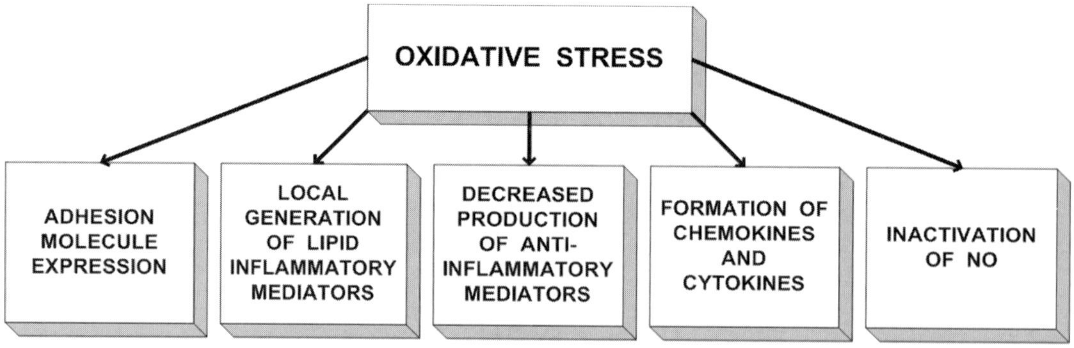

Figure 1 Oxidative stress promotes microvascular inflammation in several ways, including (1) increased adhesion molecule expression that promotes leukocyte–endothelial cell adhesive interactions; (2) local generation of lipid inflammatory mediators, such as platelet activating factor and leukotriene B_4; (3) decreased production of anti-inflammatory mediators, such as prostacyclin; (4) formation of proinflammatory chemokines and cytokines; and (5) inactivation of nitric oxide by superoxide radical.

to the perivascular space, and increased vascular permeability within postcapillary venules. These inflammatory responses result from oxidant-induced endothelial dysfunction within these vessels. Venular endothelial cells have a particularly high capacity to generate oxidants compared to endothelial cells of arterioles or capillaries. Furthermore, release of oxidants from adherent leukocytes can markedly increase the oxidative stress to venular endothelium. Oxidants have been shown to play an initiating role in the development of these inflammatory responses in a variety of conditions. For example, antioxidants attenuate leukocyte adherence and the increase in vascular permeability following I/R.

Sources of ROS

Xanthine Oxidase

Xanthine dehydrogenase is a cytosolic enzyme that catalyzes the formation of xanthine from hypoxanthine. Activation of proteases may convert this enzyme to xanthine oxidase (XO). Under normal conditions, little XO exists within cells. This enzyme uses molecular oxygen as an electron acceptor (instead of NAD^+ as in the case of xanthine dehydrogenase), which results in production of superoxide. A central role for XO in I/R-induced microvascular inflammation has been supported by the observation that XO inhibitors attenuate both venular ROS levels and leukocyte adherence after I/R.

Nitric Oxide Synthase

In some conditions, ROS may be produced by nitric oxide synthase (NOS). This enzyme catalyzes the oxidation of L-arginine to L-citrulline, which results in the production of NO. This reaction requires several cofactors, including NADH and tetrahydrobiopterin (BH_4). Limited availability of these cofactors can result in generation of $O_2^{\bullet-}$ and H_2O_2 via uncoupling of NOS. For example, BH_4 levels are insufficient for optimal NOS activity in hypercholesterolemia and in some forms of hypertension, resulting in $O_2^{\bullet-}$ production by this enzyme. Administration of BH_4 to spontaneously hypertensive rats decreased systemic blood pressure, an effect attributed to generalized arteriolar vasodilation resulting from reduced $O_2^{\bullet-}$ production by NOS.

NADPH Oxidase

This enzyme was first identified as that responsible for superoxide generation during the respiratory burst of activated neutrophils. Because this pathway is capable of producing large amounts of ROS to kill bacteria, adherence of leukocytes to the endothelium as well as emigration of leukocytes to the perivascular space can significantly enhance oxidative stress within the microcirculation. Recent evidence has demonstrated that this enzyme is localized in other cells as well, including endothelial cells, mast cells, platelets, and vascular smooth muscle cells. In these cells, NADPH oxidase generates much lower amounts of $O_2^{\bullet-}$ than does the neutrophilic enzyme, and likely plays a role in signal transduction rather than bactericidal actions.

Mitochodrial Electron Transport

Approximately 5 percent of molecular oxygen is converted to superoxide during electron transport within mitochondria under normal conditions. A growing body of evidence has shown that superoxide generation increases above this basal rate when O_2 levels are reduced. Using fluorescent dyes to detect oxidants, a reduction in PO_2 was demonstrated to rapidly increase ROS formation within endothelial cells in vitro. Evidence implicates ubisemiquinone, a free radical associated with complex III of the electron transport chain, as the major site of hypoxia-induced superoxide production. The significance of this finding is the demonstration of a pathway capable of increasing $O_2^{\bullet-}$ production when O_2 levels are reduced; superoxide formation by this route is inversely proportional to O_2 levels. In contrast, in the pathways discussed previously, the rate of ROS generation is directly proportional to the available O_2.

Endogenous Defenses against ROS

Superoxide Dismutase

This enzyme catalyzes the dismutation of superoxide to hydrogen peroxide (H_2O_2):

$$O_2^{\bullet-} + 2H^+ \leftrightarrows H_2O_2 + O_2$$

Several isozymes of SOD exist: (1) Cu/Zn-SOD, a cytosolic form; (2) Mn-SOD, which is localized within mitochondria; and (3) extracellular SOD (ecSOD). The last is released from cells and binds to sulfated polysaccharides along the cell surface. As a result, ecSOD plays a particularly important role in defense against ROS released from leukocytes adherent to the vascular endothelium. Although conflicting results have been reported regarding the ability of exogenous SOD to attenuate microvascular inflammation, these results have been attributed to rapid plasma clearance of the enzyme or a limited ability of this protein to reach intracellular sites of ROS generation. However, the fact that inhibition of endogenous SOD augments microvascular oxidative stress in various conditions clearly demonstrates the importance of this enzyme as a defense against ROS.

Catalase

As shown earlier, dismutation of superoxide results in formation of H_2O_2. This lipophilic oxidant can be disposed of by catalase, a cytosolic enzyme that degrades H_2O_2 to water according to the following reaction:

$$2H_2O_2 \leftrightarrows 2H_2O + O_2$$

Catalase was recognized to be an important endogenous antioxidant based on the following observations: Exogenous administration of catalase reduces oxidative stress as well as the degree of microvascular inflammation in various settings, whereas inhibition of endogenous catalase augments both of these responses.

Glutathione

In addition to catalase, H_2O_2 can also be degraded via glutathione peroxidase. This cytosolic enzyme utilizes reduced glutathione (GSH) as a proton donor to degrade H_2O_2 to oxidized glutathione (GSSG) and water:

$$2GSH + H_2O_2 \leftrightarrows GSSG + 2H_2O$$

Another cytosolic enzyme, GSH reductase, converts GSSG back to its reduced form. This recycling of glutathione plays an important role in microvascular defenses against oxidative stress as interventions that reduce cellular GSH augment both ROS levels and the extent of microvascular inflammation.

Nitric Oxide

Under normal conditions, the rate of NO synthesis within endothelial cells is greater than that of $O_2^{\bullet-}$. NO avidly interacts with $O_2^{\bullet-}$ at a rate approximately threefold higher than dismutation of $O_2^{\bullet-}$ by SOD. The difference between these reaction rates has led to the proposal that the most important physiological role of NO is its antioxidant action. According to this view, inactivation of $O_2^{\bullet-}$ by NO maintains low oxidant levels within cells under normal conditions. However, a marked increase in the rate of $O_2^{\bullet-}$ generation would result in NO depletion, and therefore oxidative stress.

A growing body of evidence indicates that it is the balance between the cellular levels of oxidants and NO that plays a key role in the development of microvascular inflammation. Interventions that decrease NO or increase ROS levels (i.e., nitric oxide synthase inhibitors, inhibition of endogenous antioxidants) cause arteriolar vasoconstriction, leukocyte–endothelial cell adhesive interactions, and increased vascular permeability. Conversely, administration of exogenous NO or antioxidants (which increase NO or decrease oxidant levels) enhances endothelium-dependent arteriolar vasodilation and attenuates leukocyte adhesion and increases in vascular permeability in various conditions.

Nonenzymatic Antioxidants

Ascorbic acid, vitamin E, lipoic acid, uric acid, bilirubin, and β-carotene are among a group of endogenous free radical scavengers. The antioxidant action of these compounds is due to their ability to donate an electron to an oxidant species, thereby inactivating it. Because transition metals such as iron and copper can react with $O_2^{\bullet-}$ to form the highly reactive hydroxyl radical ($OH^{\bullet}$), another group of compounds that act as metal chelators also contribute to antioxidant defenses. These compounds include ferritin, ceruloplasmin, and transferrin.

Ischemia/Reperfusion-Induced Microvascular Inflammation

I/R Promotes Oxidative Stress and Microvascular Inflammation

The microvascular alterations associated with I/R include the following: (1) impaired endothelium-dependent vasodilation in arterioles, (2) increased capillary permeability as well as a decrease in the number of perfused capillaries, and (3) enhanced leukocyte trafficking in postcapillary venules and increased venular permeability. Considerable evidence supports the involvement of ROS in the pathogenesis of these microvascular alterations. During reperfusion after prolonged ischemia, the reintroduction of oxygen to tissues increases oxidant formation in endothelial cells of all vascular segments, but particularly so in venules. Throughout the ischemic period, XO activity progressively increases within endothelial cells. When tissue O_2 levels increase during reperfusion, marked increases in oxidant generation occur via XO. The resulting oxidant stress at the onset of reperfusion causes rapid microvascular alterations through several actions, including local generation of lipid inflammatory mediators (i.e., platelet activating factor, leukotriene B_4) as well as mobilization of preformed P-selectin to the endothelial surface (which promotes leukocyte rolling). In addition, ROS also activate genes leading to upregulation of other adhesion molecules, which serves to prolong the microvascular inflammatory responses to I/R.

Superoxide/NO Balance and I/R-Induced Microvascular Dysfunction

Granger and colleagues have provided compelling evidence implicating changes in the balance between superoxide and NO levels as a critical determinant of microvascular alterations after I/R. According to this view, under normal conditions, NO production exceeds ROS generation such that NO can effectively scavenge superoxide, yet NO levels remain sufficient to maintain the endothelium in an anti-inflammatory state (Figure 2, upper panel). When superoxide generation increases markedly upon reperfusion, the balance shifts in favor of ROS: NO levels are no longer able to effectively scavenge superoxide, resulting in a progressive decrease in NO levels and corresponding increase in ROS. This shift initiates a proinflammatory state within the microcirculation, leading to the microvascular dysfunction associated with I/R (Figure 2, lower panel). In support of this view, interventions which reduce ROS levels or increase NO levels (i.e., XO inhibitors, antioxidants, NO donors) attenuate the severity of microvascular dysfunction after I/R.

Microvascular Responses to Systemic Hypoxia

ROS Are Generated during Hypoxia

Several studies have shown that hypoxia alone, in the absence of ischemia, can result in the generation of ROS. For example, graded reductions in O_2 lead to progressive increases in ROS levels within endothelial cells and vascu-

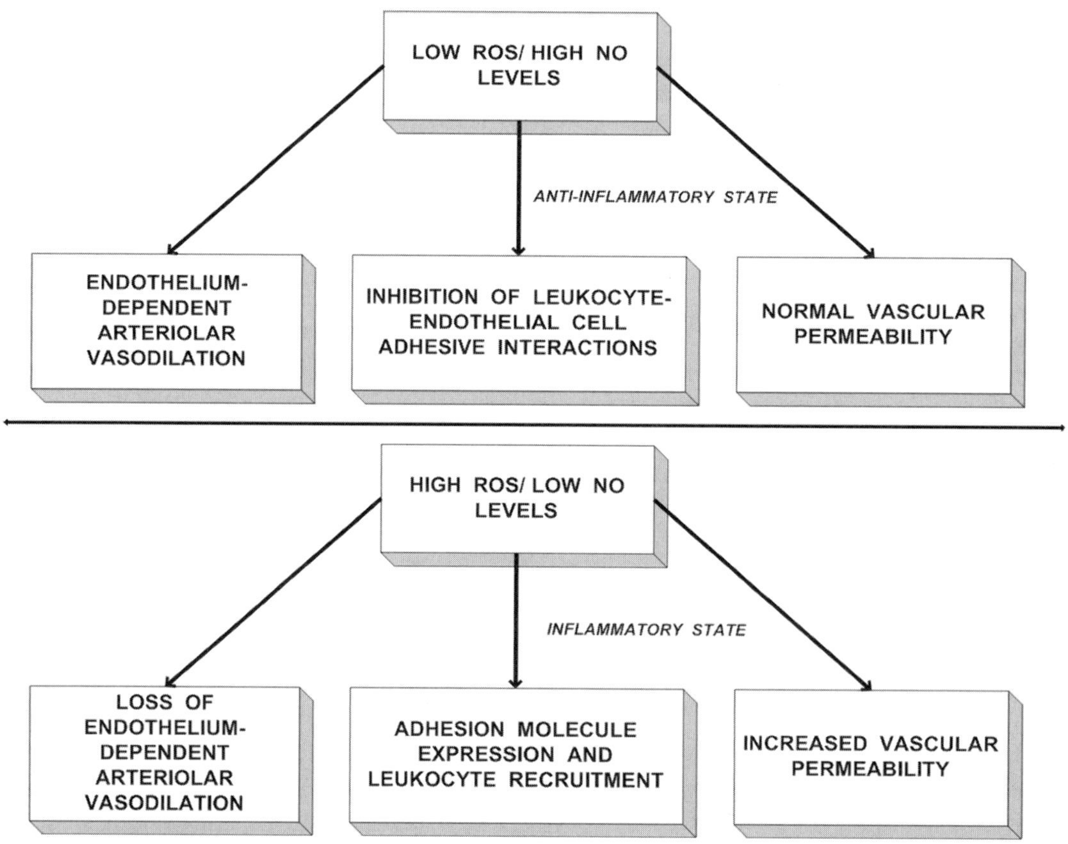

Figure 2 (*Upper panel*) Under normal conditions, low oxidant levels and high NO levels create an anti-inflammatory state within the microcirculation, as evidenced by (1) arteriolar vasodilation in response to endothelium-dependent agents, (2) inhibition of adhesive interactions of leukocytes within postcapillary venules, and (3) normal vascular permeability in capillaries and postcapillary venules. (*Lower panel*) When the balance between ROS and NO is shifted toward the former, the resulting oxidative stress promotes microvascular inflammation, as evidenced by (1) a loss of endothelium-dependent arteriolar vasodilation that reduces organ blood flow, (2) expression of adhesion molecules that promotes leukocyte recruitment into tissues, and (3) increased vascular permeability due to a loss of vascular integrity within capillaries and postcapillary venules.

lar smooth muscle cells, as well as other cell types. In addition, hypoxia reduces the levels of several antioxidants in cultured endothelial cells as well as in the liver in vivo, effects attributed to oxidative stress. In vivo studies have shown that ROS generation within mesenteric venules increases rapidly during systemic hypoxia when tissue O_2 levels are reduced, as well as decreased oxidant production upon a return to normoxia when tissue O_2 levels are restored to normal. The pattern of ROS generation in hypoxia is clearly distinct from that observed in I/R. With regard to I/R, microvascular ROS levels increase upon reperfusion when tissue O_2 levels are elevated, whereas little oxidant generation occurs during the ischemic period.

NO Levels Are Reduced during Hypoxia

Recent studies have shown that hypoxia decreases NO formation in endothelial cells in vitro and in an isolated lung preparation, suggesting a potential role for NO depletion in the microvascular inflammatory response to systemic hypoxia. NO depletion during hypoxia could result from increased inactivation of NO as a result of elevated superoxide generation, or, alternatively, from decreased O_2 availability since O_2 is a substrate for NO formation.

Antioxidants and NO Donors Attenuate Hypoxia-Induced Microvascular Inflammation

A central role for ROS in hypoxia-induced microvascular inflammation is supported by the observation that antioxidants prevent both the hypoxia-induced increases in venular ROS levels and in leukocyte adherence to postcapillary venules. Antioxidants also attenuated the increases in leukocyte emigration as well as in vascular permeability during systemic hypoxia. Interventions designed to increase tissue NO levels (NO donors or the NO precursor, L-arginine) markedly reduced leukocyte–endothelial cell adhesive interactions, leukocyte emigration, and the increase in vascular permeability during systemic hypoxia.

Summary

Oxidative stress is now recognized as playing a critical role in the initiation and progression of microvascular dysfunction associated with various cardiovascular diseases, including ischemia–reperfusion, hypercholesterolemia, atherosclerosis, diabetes, and systemic hypoxia. Considerable progress has been made in identifying the cellular sites of oxidant generation as well as the mechanisms underlying microvascular inflammatory responses associated with oxidative stress.

Further Reading

Carden, D. L., and Granger, D. N. (2000). Pathophysiology of ischaemia–reperfusion injury. *J. Pathol.* **190**, 255–266.

Cooper, D., Stokes, K. Y., Tailor, A. Z., and Granger, D. N. (2002). Oxidative stress promotes blood cell–endothelial cell interactions in the microcirculation. *Cardiovasc. Toxicol.* **2**, 165–180.

Cuzzocrea, S., Riley, D. P., Caputi, A. P., and Salvemini, D. (2001). Antioxidant therapy: A new pharmacological approach in shock, inflammation, and ischemia/reperfusion injury. *Pharmacol. Rev.* **53**, 135–159.

Gonzalez, N. C., and Wood, J. G. (2001). Leukocyte–endothelial interactions in environmental hypoxia. *Adv. Exp. Med. Biol.* **502**, 39–60.

Grisham, M. B., Granger, D. N., and Lefer, D. J. (1998). Modulation of leukocyte–endothelial interactions by reactive metabolites of oxygen and nitrogen: Relevance to ischemic heart disease. *Free Radic. Biol. Med.* **25**, 404–433. *This article provides a comprehensive discussion of the chemistry of reactive metabolites of oxygen and nitrogen relevant to physiological processes, as well as the role of these metabolites in promoting leukocyte–endothelial adhesive interactions following ischemia/reperfusion and in ischemic heart disease.*

Lum, H., and Roebuck, K. A. (2001). Oxidant stress and endothelial cell dysfunction. *Am. J. Physiol. Cell Physiol.* **280**, C719–C741.

Rathaus, M., and Bernheim, J. (2002). Oxygen species in the microvascular environment: Regulation of vascular tone and the development of hypertension. *Nephrol. Dial. Transplant.* **17**, 216–221.

Schnackenberg, C. G. (2002). Physiological and pathophysiological roles of oxygen radicals in the renal microvasculature. *Am. J. Physiol. Regul. Integr. Comp. Physiol.* **282**, R335–R342.

Stokes, K. Y., Cooper, D., Tailor, A., and Granger, D. N. (2002). Hypercholesterolemia promotes inflammation and microvascular dysfunction: Role of nitric oxide and superoxide. *Free Radic. Biol. Med.* **33**, 1026–1036. *This review critically analyzes the evidence supporting a role of reactive oxygen and nitrogen metabolites in the pathogenesis of microvascular inflammation during hypercholesterolemia, as well as the cellular and molecular mechanisms involved in this inflammatory response.*

Suematsu, M., Suzukim H., Delano, F. A., and Schmid-Schonbein, G. W. (2002). The inflammatory aspect of the microcirculation in hypertension: Oxidative stress, leukocytes/endothelial interaction, apoptosis. *Microcirculation* **9**, 259–276. *This review examines the role of ROS in microvascular inflammation in hypertension and evaluates potential molecular mechanisms involved in oxidant-dependent leukocyte–endothelial cell adhesive interactions in this setting.*

Capsule Biography

Dr. Gonzalez has studied the mechanisms of adaptation to acute and chronic hypoxia in intact animals, as produced by a reduction in oxygen levels in the inspired air. In collaboration with Dr. Wood, a major current research effort is the study of the underlying mechanisms and the physiological significance of the microvascular inflammatory response to systemic hypoxia. Their work is supported by grants from the NIH.

SECTION K

Immunology

CHAPTER 121

Autoantibodies to Endothelial Cells

Jeremy D. Pearson and Sarah E. Riley
Cardiovascular Division, King's College London, Guy's Campus, London, United Kingdom

Patients with a variety of diseases involving damage to small and large blood vessels, unlike healthy controls, have circulating antibodies that bind to endothelial cells. These autoantibodies constitute a marker of the extent of disease activity and increasingly have been implicated as contributing to the pathogenesis of several of these diseases.

Autoimmune Diseases and Autoantibodies

Autoimmune diseases are those in which clinical problems are due to dysregulation of the patient's immune system in the absence of ongoing infection, and they are characterized by the presence of autoantibodies, that is, antibodies that react with self-antigens. Autoimmune diseases arise in susceptible individuals due to variations in the ability of T and/or B lymphocytes to be activated or to undergo apoptosis—this variation presumably usually being genetically determined, and often involving inappropriate expression of cytokines—coupled with exposure to an environmental trigger, often unknown in individual cases but including microbial antigens and drugs.

Two processes are considered to be important in the generation of autoantibodies: molecular mimicry and epitope spreading. The former implies that the initial antibody formation is directed at epitopes of foreign antigens that are sufficiently similar to epitopes of self-antigens that they cross-react with them. The latter is well documented, though not fully understood, and describes how the repertoire of epitopes recognized increases with time to encompass a wider variety of self-antigens, probably due to B cell processing of the initial antigenic complexes leading to presentation of previously cryptic peptides to T cells. The identified self-antigens recognized in autoimmune diseases are commonly intracellular, often intranuclear, proteins, perhaps exposed during apoptosis. In the systemic autoimmune diseases the pattern of autoantibodies found is highly characteristic of each disease and often allows subsetting with prognostic implications. Both autoreactive T cells and autoantibodies can damage tissues, though the extent to which autoantibodies are markers of disease extent rather than pathogenic remains debatable.

Anti-endothelial Cell Autoantibodies (AECAs)

Over the past 20 years it has become apparent that AECAs are present in a significant proportion of patients with a wide variety of autoimmune diseases that include a vascular pathology. In addition, AECAs have been detected in patients with advanced atherosclerosis, and antibodies that target donor endothelium have been implicated in the process of chronic rejection of transplanted tissues and organs. One of the first suggestions that AECAs are involved in microvascular pathology was made by Shingu and Hurd in 1981, who reported that immunoglobulins (Ig) from the sera of some patients with the autoimmune disease systemic lupus erythematosus (SLE), but not Ig from healthy controls, bound selectively to cultured endothelial cells. The patients whose Ig bound to endothelium were those who had clinical evidence of inflammatory damage to blood vessels (vasculitis), indicating a possible causal relationship.

Since then, AECAs, mainly IgG or IgM, have been detected in patients with many different diseases where damage to blood vessels is occurring, ranging from primary systemic vasculitic diseases (such as Wegener's granulomatosis and microscopic polyangiitis) and connective tissue diseases (notably systemic sclerosis, also known as scleroderma) to thrombotic disorders (such as hemolytic uremic syndrome and anti-phospholipid syndrome) and multiple sclerosis. AECAs are usually detected by their ability to bind to cultured human endothelial cells, and although estimates vary it seems likely that between a third and two thirds of patients with a systemic vasculitic disease, SLE, or scleroderma have detectable AECAs.

As noted earlier, patients with systemic autoimmune diseases have autoantibodies to a wide variety of target proteins, with particular signature autoantibodies defining clinical subsets of the diseases and often being valuable in prognosis. AECAs are in general distinct from these signature autoantibodies. The endothelial cell specificity of AECAs is less rigorously defined; it is clear that AECAs do not recognize target antigens on blood cells, but it is less certain that they cannot additionally recognize antigens present in some other cell types or in the extracellular matrix. However, since they are found in the context of vascular damage, it is reasonable to assume that AECAs either contribute to causing the damage or at the least arise as a consequence of the damage and thus serve as markers of the extent of vascular pathology.

The main difficulty with research in this field has been the lack of characterization of the target antigens recognized by AECAs. In part, this stems from the fact that AECAs are clearly heterogeneous between different diseases and between individuals with a particular disease, and exhibit multiple specificities within individuals. Several investigators have tried to identify target proteins by immunoblotting with purified AECAs, and they have usually found multiple bands on gels and have not molecularly identified any band unequivocally. When IgG AECAs from a series of patients with SLE were used to compete with radiolabeled AECAs from one individual for binding to endothelial cells, some did block and others did not, indicating that binding specificities for the AECAs from different patients are only partially shared. More recent attempts to define the target antigens have used methods such as phage display libraries of endothelial proteins, with some success. For example, Frampton and colleagues showed in 2000 that AECAs from patients with active SLE included Ig that recognized an endothelial cell–selective antigen (plasminogen activator inhibitor type 1).

Potential Pathogenic Roles for AECAs

The spectrum of autoantibodies present in patients with autoimmune diseases provides diagnostic utility and must also in principle provide information about the mechanisms underlying the disease process. In several instances, it is also certain that these autoantibodies play a causal role in pathogenesis (e.g., myasthenia gravis, autoimmune thyroid disease). A pathogenic role for AECAs is less certain, and it may well be that in some diseases AECAs arise as the consequence of vascular damage, which exposes epitopes not usually visible to the immune system, and thus they are simply markers of the extent of vascular pathology. However, there is good evidence that other autoantibodies in systemic vasculitic diseases are pathogenic—notably anti-neutrophil cytoplasmic antibodies (ANCAs)—and increasing evidence that AECAs have effects that are consistent with a role in pathogenesis in the diseases where they are detected.

A pathogenic role for AECAs was first demonstrated in patients with Kawasaki disease, an acute childhood autoimmune vasculitis almost certainly triggered by infection. If not recognized and promptly treated, these children suffer from blood vessel damage that can lead to fatal aneurysm formation. In the mid-1980s Leung and colleagues discovered that sera from patients during the active phase of Kawasaki disease contained Ig that in the presence of complement were selectively cytotoxic to cultured endothelial cells that had been pretreated with proinflammatory cytokines. There is now extensive knowledge of the wide range of changes in endothelial cell functions induced by cytokines such as interleukin-1 (IL-1), tumor necrosis factor, and interferon-γ during the acute inflammatory response, notably including the upregulation of a series of cell surface molecules involved in leukocyte adhesion. Thus it is likely that the AECAs in Kawasaki disease (where the patients have raised circulating levels of these cytokines) recognize one or more newly expressed endothelial cell surface antigens and contribute directly to the vascular damage. More recent work, using a cell binding assay rather than a cell killing assay, supported this hypothesis by showing that sera from patients with Kawasaki disease contain AECAs that bind to unstimulated endothelial cells in culture, but AECA binding is significantly increased after treatment of the endothelial cells with proinflammatory cytokines.

AECAs detected in other systemic autoimmune vasculitic diseases or in scleroderma are not usually cytotoxic to cultured endothelium, whether in the presence of complement or in that of leukocytes. As in Kawasaki disease, however, AECAs from patients with Wegener's granulomatosis bind more strongly to cytokine-pretreated endothelial cells. In addition, there is increasing experimental evidence that these AECAs may be pathogenic, by themselves altering endothelial cell phenotype in a proinflammatory or prothrombotic manner similar to the effects caused by defined proinflammatory cytokines. Early indications included the demonstration that Ig from SLE patients can either directly induce or enhance the ability of tumor necrosis factor to induce the expression of the procoagulant tissue factor on endothelial cells in culture. More recently, a series of studies has demonstrated that pretreatment of cultured endothelial cells with AECAs from patients with SLE, systemic vasculitis, or scleroderma causes the subsequent upregulation of leukocyte adhesion molecules including E-selectin and VCAM on the endothelial cell surface, comparable to the effects of pretreating with IL-1 or bacterial endotoxin (see Figure 1). Indeed, Carvalho and coworkers were able to show that the majority of this effect was due to the induction by AECAs of endogenous IL-1 synthesis in the endothelial cells, which then acts in an autocrine or paracrine fashion to induce adhesion molecule expression.

As noted previously, the target antigens recognized by AECAs are not well described. However, in SLE one of the signature autoantibodies in a subset of patients recognizes

SECTION L

Kidney Transplant Rejection

Microvasculature in Kidney Transplant Rejection

Vladimir Savransky, Mark Haas, and Hamid Rabb

Johns Hopkins University School of Medicine

Approximately 6 million people in the United States have some degree of impaired kidney function, and more than 400,000 individuals suffer from end-stage renal disease (ESRD). Renal transplantation is the treatment of choice for ESRD, with approximately 14,000 kidney transplants performed in the United States and 50,000 worldwide each year. Except in the case of identical twins, the foreign graft antigens will activate the humoral and cellular arms of the immune response, which will in turn lead to allograft rejection. The events leading to allograft rejection are a consequence of the donor antigen recognition by the recipient's T cells. This is based on the recipient's recognition of foreign peptides associated with the major histocompatibility complex (MHC) class 1 and 2 molecules on the cell surface of transplanted organs and the graft-derived peptides bound to self MHC. The transplantation immune response comprises virtually all aspects of immune function.

Basic Immunobiology of Kidney Allograft Rejection

Donor's MHC Molecules as a Main Target of Recipient Immune System

Histocompatibility molecules are encoded by an array of genes on chromosome 6 and now completely sequenced. A consortium of laboratories has determined the sequence of 3,673,800 nucleotides on chromosome 6 that encode the genes of the MHC [1]. MHC class 1 molecules consist of a transmembrane protein that is noncovalently attached to a molecule of β-2 microglobulin and a short peptide. MHC class 1 molecules are encoded by three loci: HLA-A, HLA-B, and HLA-C. They are expressed at the surface of almost all the cells of the body (except for red blood cells and the cells of the central nervous system). Class 2 molecules consist of two transmembrane polypeptides. They are represented by alpha (α) chain and beta (β) chain. The α and β chains are encoded by clusters of loci in the region of chromosome 6 designated HLA-D. In contrast to MHC class 1 molecules, MHC class 2 molecules are not as widely expressed in the body. However, cells at the site of inflammation strongly express MHC class 2 molecules and provide a powerful stimulus to the immune system. The core function of MHC molecules is to "present" antigenic peptides to the T cells of the immune system. The peptides—usually about nine amino acids long—are bound by noncovalent forces in the α groove at the surface of the MHC molecule.

Alloantigen Recognition

Transplanted organs express donor MHC molecules, resulting in two pathways of antigen recognition (allorecognition) by T cells: direct and indirect. Allorecognition refers to T cell recognition of genetically encoded polymorphisms between members of the same species. The primary targets of the immune response to allogeneic tissues are MHC molecules on donor cells. Direct and indirect pathways of T-cell allorecognition are mediated by different antigen-presenting cells (APCs), and their cellular mechanisms are distinct. The direct pathway consists of recipient T cells recognizing intact donor MHC molecules expressed on donor cells. Allorecognition via the indirect pathway requires that recipient APCs process the donor-MHC antigens before presenting them to recipient T cells. The direct pathway is more closely associated with acute allograft rejection, and the indirect pathway with chronic rejection. Transplantation of a vascular organ induces MHC sensitization by direct stimulation of circulating host immune cells that encounter donor MHC antigens on allograft cell surfaces.

T-Cell Activation

T-cell activation is essential for allograft rejection. T-cell activation is associated with nuclear translocation of specific transcription factors that regulate expression of genes critical for T-cell function. NF-κB plays the central role in this process. The activation of T cells is a key start mechanism of immune response and requires two distinct, but synergistic, signals. The first signal is provided by a specific antigen and is delivered via the T-cell receptor. The second signal (costimulatory signal) is not antigen specific. Indeed, many T-cell molecules may serve as receptors for costimulation. The most well characterized costimulatory molecule is CD28, which has two ligands (B7-1 [CD80] and B7-2 [CD86]) that are expressed primarily on APCs. Another molecule, CTLA-4, is similar to CD28 and is also expressed on T cells. Although CTLA-4 binds B7-1 and B7-2, it transmits an inhibitory signal that serves to terminate the immune response.

Cellular and Humoral Mechanisms of Allograft Rejection

One or more attacks of acute cellular (Figure 1) or humoral (antibody-mediated) rejection (Figure 2) usually occur in almost half of the kidney transplant recipients despite active immunosuppressive strategies. Until recently, most studies on the mechanisms of renal allograft rejection have focused on the central role of T cells and of other cellular mechanisms of tissue injury. It has been established that CD4 T cells are crucial in initiating most acute rejection episodes, and that alloactivated CD4 T cells, cytotoxic CD8 T cells, monocytes/macrophages, and NK (natural killer) cells play a major role in cell-mediated mechanisms that eventually result in allograft destruction. Perforin and granzyme B are two proteins that are present in the cytoplasmic granules of cytotoxic T cells and NK, cells which are an integral part of the effector mechanisms of cell-mediated allograft rejection. In recent years, it has also become increasingly appreciated that detection of anti-MHC donor specific antibodies (DSA) de novo after transplantation is associated with rejection due to antibody-mediated effector mechanisms of tissue injury. The identification of the complement fragment C4d as a specific marker for humoral rejection in peritubular capillaries (PTCs) of renal allograft biopsies has helped to define and characterize these syndromes.

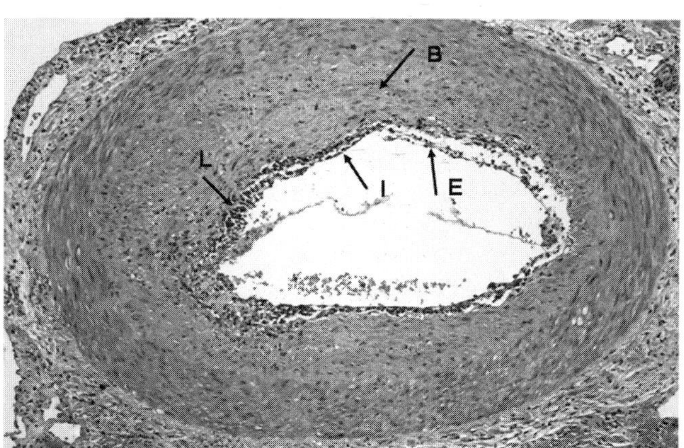

Figure 1 Acute cellular rejection. Lymphocytes (L) infiltrate beneath endothelium (E) with edema of intima (I) and subintimal layers (B). (H × E, original magnification 100×.) (see color insert)

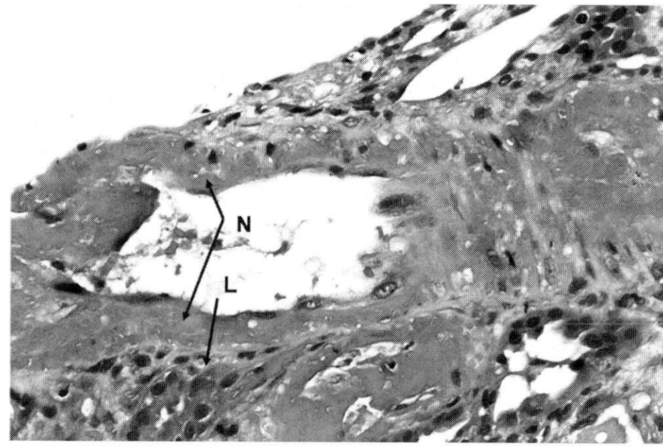

Figure 2 Acute humoral rejection. Fibrinoid necrosis (N) of the thickened arterial wall surrounded by polymorphonuclear and lymphoid cell infiltrates (L). (H × E, original magnification 400×.) (see color insert)

Cytokines and Cell Adhesion Molecules in Transplant Immunity

Cytokines are any of numerous low-molecular-weight proteins that regulate the intensity and duration of the immune response by exerting a variety of effects on lymphocytes and other immune cells. A diversity of cytokines, each with many functions, is involved in an immune response. While IL-12 facilitates differentiation towards the Th1 phenotype and IL-4 towards the Th2 phenotype, other cytokines such as INF-8 and IL-2 secreted by the Th1 cells promote cell-mediated immune responses, and IL-4, IL-6, and IL-7 released by Th2 cells are important in B cell maturation. IL-2 and IFN-γ play crucial roles in graft rejection. They are important for recruitment, activation, and proliferation of various leukocytes, for the induction or upregulation of cell adhesion molecules and MHC molecules, and for mediating communication between leukocytes and parenchymal cells. Therefore, IL-2 has been studied as a potential target for suppression of graft rejection. The complexity of the cytokine network, particularly the plethora of cytokines and their overlapping functions, is a major obstacle in achieving this objective.

Cell adhesion molecules, particularly intercellular adhesion molecule-1 (ICAM-1) and vascular cell adhesion

Table I The Role of Some Cytokines and Chemoattractive Molecules in Renal Transplant Rejection.

Cytokine	Function in allograft rejection
IL-1	Causes neointimal formation and pathogenesis of chronic rejection
IL-2	Enhances all types of allograft rejection
IL-4	Promotes a delay in vasculopathy in the graft
IL-5	Mediates transplant vasculopathy
IL-10	Prevents ischemia–reperfusion injury and decreases acute rejection
IL-15	Activates allospecific CD8 T cells during acute rejection
IL-16	Plays an activation role rather than an inhibition of anti-graft reaction
IL-17	Stimulates early alloimmune responses
IL-18	Plays the activation role in acute rejection
IFN-γ	Promotes acute rejection of kidney allografts
TNF-α	Participates in pathogenesis of acute and chronic rejection
TGF-β1	Expression is linked with chronic vasculopathy
VEGF	Influences adhesion and migration of leukocytes across the endothelium
MCP-1	Associated with premature kidney graft failure
ICAM-1 and VCAM-1	Early leukocyte and lymphocyte recruitment in the microvasculature of rejecting allograft, costimulation T cell activation
PDGF	Mediates mesenchymal cell proliferation in chronic rejection
M-CSF	Promotes macrophage recruitment and proliferation

molecule-1 (VCAM-1), are also key regulatory molecules in immune responses. They are important in migration and localization of leukocytes into tissues as well as in a variety of cell-to-cell interactions that include signaling between cells and even cell-mediated cytotoxicity. The expression of these molecules, which occurs in a sequential fashion, is important for orchestrating the various steps in graft rejection, though the actual sequence of events and their underlying mechanism is not yet clear. A brief summary of the role of some cytokines involved in allograft rejection is given in Table I.

Microvasculature and Renal Transplant Rejection

Endothelium of the allograft vasculature is the interface between an allograft and the recipient's immune system. In this boundary position, endothelial cells may play important roles in the afferent and efferent phases of allograft rejection. The expression by endothelial cells of granule membrane protein-140 (GMP-140/P-selectin) and endothelial leukocyte adhesion molecule-1 (ELAM-1/E-Selectin) increases tissue factor activity, augments secretion of plasminogen activator inhibitor, and decreases thrombomodulin, contributing to hyperacute rejection. Similarly, endothelial cells may actively participate in acute cellular rejection and in the development of transplant-associated arteriopathy as a result of induction of antigen-presenting function (i.e., MHC class 2 expression), upregulation of adhesion molecules for lymphocytes and monocytes, and release of platelet-derived growth factors. Therefore, endothelial cell functions, which are important for normal inflammatory responses and vessel behavior, may be pathogenic in the allograft [2].

Microvascular Injury in Hyperacute and Accelerated Kidney Transplant Rejection

Hyperacute rejection occurs within minutes to hours after the vascular clamps to the transplanted organ are released. This dramatic event is caused by preexisting cytotoxic, anti-HLA class 1 (IgG) or anti-ABO blood group antibodies (IgM) in the recipient. The antibodies bind to the endothelial surface of the arterioles on the graft, activate complement, and lead to severe microvascular injury including thrombosis and obliteration of the graft vasculature. The endothelial cells are stimulated to secrete von Willebrand factor (vWF), which mediates platelet adhesion and aggregation. Complement activation initiates coagulation cascade and the generation of multiple inflammatory mediators. Eventually, transplanted tissue suffers irreversible ischemic damage. Hyperacute rejection is mediated by antibodies against alloantigens that have appeared in response to previous exposure to these antigens through blood transfusion, prior transplantation, or multiple pregnancies. Pathologic findings show fibrin thrombus formation, margination of neutrophils, and ultimately fibrinoid necrosis of the vessel walls. The transplant may become flaccid or cyanotic and hard and may rupture.

Accelerated acute rejection taking place within 1 to 4 days after transplantation occurs when the recipient has been sensitized by prior interaction with graft antigen, generally by prior transplantations but also by transfusions, and is thought to represent an immunologic memory response to prior sensitization. This type of rejection may represent a combination of cellular and antibody-mediated injury, but the cellular infiltration may not be as intensive as with acute rejection.

Renal Microvasculature and Acute Transplant Rejection

Acute renal graft rejection is able to activate human endothelial cells leading to upregulation of mRNAs coding for VCAM-1 and ICAM-1 and plays a direct role in the pathogenesis of acute rejection [3]. Endothelial deposition of the complement split product C4d is an established marker of antibody-mediated acute renal allograft rejection. Cells of the monocyte/macrophage system have active contribution to acute allograft destruction. Monocytes are recovered from both the central and the marginal blood pool by perfusing either the recipient's circulation or the allograft vasculature. During allograft rejection MHC class 2 molecules, CD161 (NKR-P1A), CD62L, and CD8, are upregulated, while CD4 and CD43 are down-modulated. Activated monocytes participate in the kidney allograft destruction by directly damaging endothelial cells and by promoting intravascular coagulation.

Recruitment of leukocytes during immune responses requires the coordinate expression of adhesion molecules in concert with chemokines and their receptors. The Duffy antigen receptor for chemokines (DARC) binds multiple chemokines and is expressed on postcapillary venules in the normal kidney. The chemokine receptor CCR5, which shares the ligand regulated upon activation, normal T-cell expressed and secreted (RANTES) with DARC, is expressed by infiltrating T cells in the renal interstitium. DARC is involved in the attraction of CCR5-positive cells. Therefore, the increased number of DARC-positive venules in areas of interstitial injury and the co-localization with CCR5-positive infiltrating leukocytes indicate a role for endothelial DARC expression during leukocyte adhesion and interstitial infiltration [4]. Histopathology of the allografts reveals edema and interstitial cellular infiltration as well as tubulitis, and necrosis and hemorrhage in severe cases. Destruction PTCs and tubules accompanied by disruption of basement membrane (BM) occurred with capillaritis or tubulitis in areas with a severe cellular infiltrate. Glomerular changes notably included swelling of the tufts due to hypercellularity, which is consistent with transplant glomerulitis. The intrarenal arteries exhibit intimal or in severe cases transmural mononuclear cell inflammation with or without fibrinoid necrosis of vessel wall (the latter often with antibody mediated rejection) [5].

Microvasculopathy during Chronic Kidney Transplant Rejection

Chronic rejection is associated with the development of interstitial fibrosis and PTC, endothelial cell, and tubular

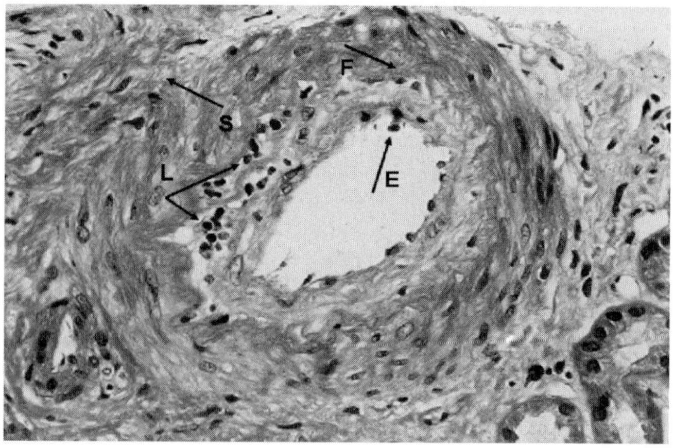

Figure 3 Chronic rejection. Lymphocyte (L) infiltration and fibrosis of intima (F). There is perivascular sclerosis and edema (S). Endothelial cells are swollen and some are pyknotic (E) (Mallory, original magnification 400×). (see color insert)

epithelial cell death associated with CD3+ cell infiltration (Figure 3). During the development of chronic rejection, capillaritis of PTCs and tubulitis are maintained by persistent T-cell infiltration, and the remaining PTCs and tubules exhibit progressive atrophy with thickening and/or lamination of BM. Then identifiable PTCs and tubules are lost in areas of interstitial fibrosis. Proliferating myofibroblasts accumulate around PTCs and tubules and in interstitium, and there is widespread interstitial fibrosis. The contribution of alloantibody-dependent immune reactions to chronic rejection is being increasingly appreciated.

Influence of Ischemic Reperfusion Injury on Renal Graft Function and Kidney Microvasculature

All allografts undergo some degree of ischemic reperfusion injury (IRI) during transplantation. IRI causes renal vascular endothelial damage and plays an important role in kidney transplant pathophysiology. IRI induces allograft endothelial cell swelling, alters endothelial cell–cell connection, and alters endothelial cell–basement membrane attachment. Functional consequences of these morphological changes include altered vascular reactivity, increased leukocyte adherence and extravasation, altered coagulation due to loss of normal endothelial function and/or barrier, and increased interstitial edema. Increased levels of gene transcripts involved in cellular adhesion, chemotaxis, apoptosis, and monocyte recruitment and activation dominate the immediate postreperfusion state. T cells are a fundamental link between IRI injury and alloimmunity. This phenomenon is highlighted by data demonstrating that T-cell depletion can improve the course of experimental renal IRI [6]. Damage during IRI predisposes to acute and chronic rejection.

Microvascular Injury Caused by Immunosuppressive Therapy

Calcineurin inhibitors such as cyclosporine A (CyA) and tacrolimus (FK506) drastically enhance the survival of organ transplants and recipients. But they themselves can affect endothelial function in renal transplant patients, as administration of these immunosuppressants is correlated with a high incidence of transplant arteriolopathy. The endothelium-dependent and -independent vasodilation of the patients on FK506 is better preserved than in patients on CsA therapy. Vascular endothelial cells naturally express a death factor, Fas ligand, that inhibits detrimental leukocyte infiltration. It has been shown that CyA and FK506 down-regulate Fas ligand expression on endothelial cells with accompanying decrease in the cytotoxicity toward Fas-bearing cells. These data suggest a mechanism by which immunosuppressive treatment contributes to atherogenesis [7]. Development of thrombotic microangiopathy has been associated with CyA and FK506 toxicity.

Glossary

Allograft: A transplant of an organ or tissue that is donated either by a genetically matched relative of the patient or by an unrelated (but genetically similar) donor.

Cytokines: Nonantibody proteins secreted by inflammatory leukocytes, and some nonleukocytic cells, that act as intercellular mediators. They differ from classical hormones in that they are produced by a number of tissue or cell types rather than by specialized glands. They generally act locally in a paracrine or autocrine rather than endocrine manner.

Endothelium: The layer of epithelial cells that lines the cavities of the heart and of the blood and lymph vessels, originating from the mesoderm.

Lymphokines: Soluble protein factors generated by activated lymphocytes that affect other cells, primarily those involved in cellular immunity.

Rejection: Any immune process leading to the destruction or detachment of a graft or other specified structure.

References

1. Complete sequence and gene map of a human major histocompatibility complex The MHC sequencing consortium. *Nature* (1999). **28**, 921–923.
2. Sedmak, D. D., and Orosz, C. G. (1991). The role of vascular endothelial cells in transplantation. *Arch. Pathol. Lab. Med.* **115**(3), 260–265.
3. Lucchiari, N., Panajotopoulos, N., Xu, C., Rodrigues, H., Ianhez, L. E., Kalil, J., and Glotz, D. (2000). Antibodies eluted from acutely rejected renal allografts bind to and activate human endothelial cells. *Hum. Immunol.* **61**(5), 518–527.
4. Segerer, S., Regele, H., MacK, M., Kain, R., Cartron, J. P., Colin, Y., Kerjaschki, D., and Schlondorff, D. (2000). The Duffy antigen receptor for chemokines is up-regulated during acute renal transplant rejection and crescentic glomerulonephritis. *Kidney Int.* **58**(4), 1546–1556.
5. Haishima, A., Kawakami, Y., Mizuno, S., Kageyama, T., Muto, M., Suzuki, T., Inoue, K., and Shirota, K. (2002) Acute vascular and interstitial rejection following renal allograft transplantation in dogs. *J. Vet. Med. Sci.* **64**(12), 1137–1140.
6. Yokota, N., Daniels, F., Crosson, J., and Rabb, H. (2002). Protective effect of T cell depletion in murine renal ischemia-reperfusion injury. *Transplantation* **74**(6), 759–763.
7. Sata, M., and Walsh, K. (1999). Cyclosporine downregulates Fas ligand expression on vascular endothelial cells: implication for accelerated vasculopathy by immunosuppressive therapy. *Biochem. Biophys. Res. Commun.* **263**(2), 430–432.

Further Reading

Inston, N., and Cockwell, P. (2002). The evolving role of chemokines and their receptors in acute allograft rejection. *Nephrol. Dial. Transplantation* **17**, 1374–1379. *The article is concentrated on the expression of chemokine receptors that direct the trafficking of alloactivated T cells into the graft in response to local production of chemokines, initially by resident cells. There is now deep interest in this area that reflects the recent identification of restricted chemokine–receptor interactions as key functional events in T-cell recruitment and potential therapeutic targets for the prophylaxis of acute allograft rejection.*

Shimizu, A., Colvin, R. B., and Yamanaka, N. (2000). Rejection of peritubular capillaries in renal allo- and xeno-graft. *Clin. Transplantation* **14**, 6–14. *The review shows that microvasculature plays an important role in the pathogenesis of humoral- and cell-mediated renal allograft rejection. PTC endothelium expresses the MHC antigens in the resting phase, as does the glomerular capillary endothelium.*

Thiru, S., and Waldmann, H. (eds.) (2001). *Pathology and Immunology of Transplantation and Rejection.* Boston: Blackwell Science. *This textbook offers a good insight into various aspects of organ transplantation.*

Capsule Biography

Vladimir Savransky, M.D., Ph.D., received his medical school and surgical training in St. Petersburg, Russia. He is currently a Postdoctoral Research Fellow in the Division of Nephrology, Johns Hopkins University School of Medicine.

Mark Haas, M.D., Ph.D., received his medical school training at Duke University and specialization in kidney pathology at Yale. He is currently Professor in the Department of Pathology, Johns Hopkins University School of Medicine.

Hamid Rabb, M.D., received his medical school training at McGill University and specialization in kidney diseases at Harvard. He is currently Director, Transplant Nephrology and Associate Professor of Medicine, Johns Hopkins University School of Medicine.

SECTION M

Leak Syndrome

CHAPTER 123

Sepsis-Induced Microvascular Leak Syndrome

Gernot Marx and Martin Leuwer
University Department of Anaesthesia, University of Liverpool, Liverpool, United Kingdom

Introduction

Despite major advances in critical care and surgery, the incidence of sepsis and the number of sepsis-related deaths are high and increasing. A recent epidemiology survey of severe sepsis in the United States reported 751,000 cases per year, with a mortality of 28.6 percent [1]. Sepsis is associated with prolonged stay both in the intensive care unit and in the hospital. Over the first year following an episode of sepsis, mortality remains high and an increased sepsis-associated risk of dying may persist for up to 5 years after hospitalization. Additionally, the long-term quality of life of sepsis survivors is reduced compared to age- and sex-matched controls. The economic burden associated with sepsis has recently been estimated at US$17 billion (€16.8 billion) each year in the United States alone [1].

Inflammatory cascading reactions that occur in sepsis induce increased microvascular permeability and microvascular leakage resulting in interstitial fluid accumulation, loss of protein, and tissue edema [2]. In this situation hypoalbuminemia frequently occurs as a result of transcapillary loss and impaired hepatic synthesis of albumin leading to reduced intravascular colloid osmotic pressure, which, in turn, compromises the ability to preserve intravascular volume. Consequently, sepsis and septic shock are characterized by a relative as well as an absolute intravascular volume deficit. The absolute volume deficit is caused by fever, increased perspiration and increased insensible loss, vomiting, diarrhea, and volume loss by drains or sequestration. The relative volume deficit is due to vasodilatation, venous pooling, and alterations in the endothelial barrier. The functional disturbances during sepsis-induced microvascular leak are associated with impaired tissue perfusion and organ oxygenation causing organ dysfunction, which is reflected by increased blood lactate concentrations, oliguria, coagulation abnormalities, and altered mental state.

Pathogenesis

Sepsis and Systemic Inflammatory Response

During sepsis and systemic inflammatory response the endothelium is modulated by a number of different mechanisms [3]. Usually, components of the bacterial wall (e.g., lipopolysaccharide [LPS]) activate pattern recognition receptors on the surface of the endothelium. Activation of endothelial cells results from the release and activation of inflammatory mediators such as cytokines, activation of neutrophils, reactive oxygen species (ROS), histamine, bradykinin, platelet-activating factor, serine proteases, complement, coagulation, fibrinolysis, and/or changes in oxygenation or blood flow. Changes occur through direct toxic damage to the endothelial barrier, or through functional alterations involving the cytoskeleton. The mediators can interact with the microvascular endothelium, eliciting a series of intracellular signaling reactions that compromise the barrier structure and enhance transendothelial flux of fluid and macromolecules. In response to this interaction the microvascular endothelial cells may undergo structural changes (e.g., vacuolization, cytoplasmic swelling) as well as functional changes, such as increased cell adhesion and leukocyte trafficking and increased microvascular permeability.

Microvascular Permeability and Edema

Increased microvascular leakage and edema are the main characteristics of inflammation-induced organ injury.

Because of the increased microvascular permeability, the transcapillary loss of macromolecules leads to intravascular fluid loss, lower intravascular colloid osmotic pressure, and hypovolemia. Several phases of endothelial barrier dysfunction contributing to an increased microvascular permeability can be identified. In sepsis, inflammatory mediators and altered balance of vasodilators (i.e., nitric oxide [NO] and prostacyclin) and vasoconstrictors (i.e., endothelin, thromboxane A_2, and platelet activating factor) induce loss of junctional integrity, a process that involves actin–myosin interaction. Subsequently, the interaction of leukocytes amplifies leakage by the leukocyte-derived mediators.

The physiological concept is based on Starling's hypothesis about filtration and reabsorption of water in capillaries and the formation of lymph [4]. Starling's hypothesis was that the difference in concentration of plasma proteins between the plasma and tissue is the main determinant of oncotic pressure, which opposes hydrostatic filtration. Thus, the driving force for fluid filtration rate across the vessel wall is determined by four pressures: the hydraulic and colloid osmotic pressures in the vessel and in the tissue space:

$$J_V/A = L_P[P_c - P_i - \sigma(\pi_c - \pi_i)]$$

Here, J_V/A is the fluid filtration flux across the capillary wall per unit area; L_P is the hydraulic permeability of the capillary wall; σ is the oncotic reflection coefficient; and P_c, P_i, π_c, π_i are global values for the hydrostatic and colloid osmotic pressures in the capillary and interstitial compartments. Thus, edema formation in any tissue may be the result of increased hydrostatic driving pressures, or altered integrity of the microvascular membrane. Starling's equation has been applied across the entire transendothelial barrier. However, there is growing recognition that the application of the Starling equation is much subtler than has been previously realized.

Whereas specialized pathways between and through endothelial cells enable water and small solutes such as ions, lactate, urea, and glucose to pass, the passage of macromolecules (i.e., proteins) is restricted [5]. Transvascular macromolecular transport involves convective (i.e., by large pores) and diffusive (i.e., paracellular transport through intercellular junctional pathways or via small pores) forces [5]. Inflammatory mediators can interact with the microvascular endothelium, eliciting a series of intracellular signaling reactions that compromise the barrier structure, and allowing transendothelial flux of fluid and macromolecules. It has been suggested that endothelial hyperpermeability is related to alterations of the cellular cytoskeleton. Regulation of paracellular transport is associated with actin-based systems linking cells by cadherins, proteins that are crucial for tight junction formation. Activation of cell contraction and disturbance of junctional organization subsequently result in the induction of interendothelial gaps followed by enhanced paracellular endothelial permeability. Major initiators of this process are polymorphonuclear leukocyte–derived oxygen metabolites, pore-forming bacterial exotoxins, and endogenous proinflammatory mediators. Furthermore vascular cellular adhesion molecule-1 (VCAM-1), upregulated during sepsis, induces an increase of permeability by modulating cadherin function through the production of ROC. The transport of solutes across the microvascular walls depends mainly on mechanical pressure or shear stress forces, plasma and interstitial protein concentration, wall thickness, and perivascular barriers to albumin diffusion, but intrinsic properties of the endothelium such as the presence of surface binding proteins, the charge of subendothelial matrix proteins, and the surface charge are important as well.

At a later stage, the entire microvasculature undergoes dramatic remodeling initiated by angiogenic factors. Not only do these angiogenic growth factors affect the integrity of the cell junctions by induction of endothelial migration; some of them, in particular vascular endothelial growth factor (VEGF), directly induce a hyperpermeable status of the vasculature. VEGF has been shown to induce transendothelial pathways by formation of so-called vesiculo-vacuolar organelles, which are interconnected chains of vesicles forming a kind of a pore through endothelial cells [6].

Assessment of Sepsis-Induced Microvascular Leak Syndrome

EXPERIMENTAL

Plasma and lung lymphatic flow and protein concentration can be used to measure microvascular permeability. Using this approach increased microvascular permeability caused by *Pseudomonas* bacteremia could be demonstrated in sheep lungs. Measurement of wet weight to dry ratios is another frequently used method for experimental assessment of pulmonary edema. Furthermore, radioactive tracers can be used to measure increased microvascular permeability accurately. In sheep pulmonary transvascular protein flux was measured using ^{113m}In-labeled transferrin and ^{99m}Tc-labeled erythrocytes using a gamma camera. After the initial evaluation of a sepsis-induced increased microvascular permeability in the pulmonary circulation, histological and ultrastructural changes in nonpulmonary organs during early hyperdynamic sepsis were demonstrated. Increased microvascular permeability could be shown in the lung and abdomen using a dual radionuclide method (^{99m}Tc and ^{131}I serum albumin) in septic pigs. A tissue/organ dependent and insult-dependent alteration in radioiodine-labeled albumin flux was demonstrated in septic rats. Following an abdominal bacteremia or endotoxic challenge, microvascular permeability increased mainly in the liver, heart, colon, and kidneys. There were also regional time-dependent differences in permeability. As a measure of systemic increase of sepsis-induced microvascular permeability to albumin the albumin escape rate (^{51}Cr-tagged erythrocytes and ^{125}I albumin) is well established in rodents and pigs. Another measurement of microvascular permeability is based on monitoring (video images) the leakage of fluorescein (FITC)-labeled albumin and rhodamine dye from the pulmonary capillaries into the alveoli. This latter method was used to evaluate pulmonary microvascular changes during sepsis or to assess the modulation of coronary venular permeability to albumin by different flow rates.

Clinical

Unfortunately, to date there are no standardized criteria for the clinical diagnosis of increased microvascular permeability, because the assessment of fluid distribution and fluid balance in septic patients is very difficult. Clearly, early diagnosis of increased microvascular permeability would be valuable, because it would allow early and specified treatment as well as the evaluation of the efficacy of therapeutic efforts. It is important to distinguish between increased microvascular permeability and other hypo-oncotic conditions leading to fluid retention, that is, caused by renal or hepatic failure. Proposed criteria for clinical assessment of the microvascular leak syndrome so far either are nonspecific or have limited bedside applicability: Microvascular leak was defined as noncardiogenic generalized edema and hemodynamic instability or more than 3 percent increase of body weight within 24 hours, combined with generalized edema. In this context, it is worth noting that calculated fluid balances are not predictive for actual weight changes in critically ill patients. A dual-radionuclide method using ^{67}Ga transferrin and ^{99m}Tc-labeled erythrocytes was used to measure pulmonary edema. This is certainly an interesting approach, but the required sophisticated technique is not widely available. The transcapillary escape rate of ^{125}I-labeled albumin was suggested as a surrogate measurement of microvascular leak syndrome. This method is limited by the fact that the recirculation of radiolabeled albumin from the tissues via the lymphatic system cannot be quantified. Moreover, the clinical applicability of radioactive tracers is limited due to radioactive contamination and dye accumulation. The disadvantages, especially in repeated measurements, are obvious.

Increased extravascular lung water has been suggested as a morphological correlate of pulmonary edema and can be used for analysis of pulmonary microvascular leakage. Currently available systems measure extravascular lung water by a double-indicator (indocyanin green and heat) or a single-indicator thermodilution (heat) technique. The accuracy of both thermodilution techniques was demonstrated gravimetrically. In animals, the sensitivity was 81 percent, the specificity was 97 percent. The coefficient of variation for repeated measurements of extravascular lung water was 8 to 9 percent in human beings and 6 to 7 percent in animals, respectively. The limitations of the thermal-dye method include overestimation of extravascular lung water at normal levels of water content and perfusion dependence. Severe alterations of lung perfusion may lead to an underestimation of the water content using the thermal dye dilution method. When significant proportions of pulmonary tissue are excluded from the pulmonary circulation, the indicators do not reach the nonperfused areas and, therefore, both intravascular and extravascular fluid pools can go undetected.

A modification of venous congestion plethysmography allows noninvasive assessment of the filtration capacity as a measure of microvascular permeability and isovolumetric venous pressure, a value that is related to the balance of filtration forces across the microvasculature. Using this method an increased microvascular permeability was demonstrated in septic shock patients. This technique requires a sedated or a cooperative patient, because movements cause artifacts. Furthermore, if the isovolumetric venous pressure is high and the diastolic blood pressure is low (as in some septic patients) the measurement of the filtration capacity is not accurate because fluid filtration is only observed at high cuff pressures that are close to the diastolic arterial pressure. In nonseptic blunt trauma patients renal albumin excretion can be used as a surrogate for microvascular leakage. However, this approach is not recommendable in septic patients because of the frequent occurrence of renal dysfunction/failure. Our own group has recently suggested a set of noninvasive diagnostic determinants for the microvascular leak syndrome [7]. Initially reliability of a noninvasive measurement of extracellular fluid volume using bioelectrical impedance analysis compared with inulin was assessed. On that basis measurement of an increased extracellular fluid volume using bioelectrical impedance analysis combined with the response of colloid osmotic pressure to albumin infusion in septic shock patients was demonstrated as a noninvasive method for the diagnosis of sepsis-induced microvascular leak syndrome applicable at the bedside.

Drug Therapy

Fluids

Sepsis is associated with a profound intravascular fluid deficit due to microvascular leak syndrome. In this clinical situation, fluid therapy is essential for restoration and maintenance of an adequate intravascular volume in order to improve tissue perfusion and nutritive microcirculatory flow. Circulatory stability following fluid resuscitation in the septic patient is usually achieved at the expense of tissue edema formation that may significantly influence vital organ function. The risk of edema has been used to discredit each type of fluid. Because crystalloid fluid distributes primarily in the interstitial space, edema is an expected feature of crystalloid fluid resuscitation. However, edema is also a risk with colloid fluid resuscitation, especially in the presence of increased microvascular permeability, as colloids do not remain in the intravascular compartment and the leakage of macromolecules might result in an increase of interstitial oncotic pressure and the expansion of the interstitial compartment. On the other hand, the advocates of colloid therapy in sepsis argue that by maintenance of an increased colloid osmotic pressure, fluid is retained in the intravascular space, even in the presence of increased permeability.

Until today, there has been no definitive answer to this question. In four meta-analyses comparing the effects of crystalloids and colloids on patient outcome, either no clear difference between crystalloids and colloids, or a slight benefit related to crystalloids, has been found. In view of the clinical relevance and the fact that this has been an ongoing discussion for decades, there is a striking lack of contemporary studies including sufficient numbers of patients aiming at the investigation of the optimal fluid strategy. One reason is the lack of appropriate clinical study end points for fluid resuscitation. Although mortality is an obvious end point, fluid therapy is only one factor within a very complex situation, which may influence the outcome. Considering these problems it is not surprising that there are more data from animal models available in comparison to clinical studies.

In a hyperdynamic porcine septic shock small-volume model, hypertonic saline–dextran or 6 percent dextran 60 was superior in restoring intravascular volume at constant plasma COP in comparison to Ringers' lactate, resulting in higher cardiac output and intestinal blood flow. In septic hamsters hypertonic saline attenuated plasma volume loss with or without dextran. A relationship could be confirmed between reduced colloid osmotic pressure and intestinal hypoxia and development of edema. In an in vivo model, using cat skeletal muscle, synthetic colloids such as HES, gelatin, and dextran had no direct effect on albumin microvascular permeability. A particular HES solution called pentafraction, containing a selected category of medium-weight molecules, might reduce microvascular leakage by a direct sealing effect. This hypothesis, which implies that appropriately sized HES molecules might act as plugs and seal or even restore microvascular integrity at capillary–endothelial junctions, is mainly based on laboratory investigations using ischemia–reperfusion models. In septic pigs less pentafraction in comparison to pentastarch was required to prevent hemoconcentration. Pentafraction was associated with less hepatic and pulmonary structural damage. In septic sheep pentafraction diminished tissue injury, but did not show an advantage in comparison to pentastarch. However, very few investigators used accurate methods such as radionuclide tracers in order to evaluate the interaction between fluid therapy and microvascular leak syndrome in sepsis. In endotoxemic rodents, plasma volume decreased after infusion of a crystalloid solution and increased after the administration of gelatin. Interestingly, no difference in the degree of microvascular leakage between septic rats treated with normal saline or gelatin could be demonstrated. In septic sheep similar amounts of crystalloid or colloid solutions were required to maintain plasma volume. Despite similar circulatory response and increased organ blood flows, colloid infusion for 48 hours preserved microvascular integrity and cellular structures in the left ventricle and gastrocnemius muscle. In accordance there are more data indicating beneficial effects of colloid solutions in sepsis under well-defined experimental conditions. Recently, our group demonstrated, in a porcine model of septic shock with concomitant microvascular leak syndrome, that it is possible to maintain plasma volume by the artificial colloids modified fluid gelatin 4 percent and 8 percent (MFG4%, MFG8%), and 6 percent HES 200/0.5, but not with Ringer's solution despite increased microvascular permeability [8].

Theoretically, in the presence of a microvascular leak—which allows the escape of albumin—one would have expected the escape of the smaller gelatin molecules, as well as the escape of HES 200/0.5, which contains a substantial fraction of molecules smaller than albumin. Some experimental work suggest that the presence of surface binding proteins, the charge of subendothelial matrix proteins, and the surface charge are important. The loss of negative endothelial charge in sepsis due to an increased protein extravasation was demonstrated in a hyperdynamic septic rats. Although this may have contributed to the retention of colloids, the explanation seems to be speculative at the moment and further studies are needed to elucidate the exact mechanism involved in the intravascular retention of colloids in microvascular leak syndrome.

Septic patients receiving albumin 5 percent had an expansion of the extracellular volume twice the infused volume compared to those receiving normal saline. This suggests that infusing excessive amount of colloid can cause interstitial fluid overload. Indeed, in rats expansion of plasma volume with colloids enhanced transport of plasma protein from the vascular to the interstitial compartment because of dissipative transport of albumin [9].

In the porcine septic shock model just mentioned, animals receiving Ringer's solution demonstrated impaired systemic oxygenation compared to the colloid solutions [8]. The underlying mechanism may be that Ringer's solution increases tissue edema compared to hyperoncotic colloid solutions. One effect of such an edema would be to retard oxygen uptake by increasing distances from the blood vessel to the mitochondria. Furthermore, in the colloid groups HES-infused animals showed a significantly higher cardiac output, systemic oxygen delivery, and lower oxygen extraction ratio compared to those receiving MFG4% and MFG8%. Recent experimental work suggested that HES improves rheology by decreasing blood viscosity. Additionally, HES may improve rheology by removing plasma proteins from the endothelial glycocalix. Impaired systemic hemodynamics in the MFG4 and MFG8% groups might indicate an influence on rheology and impaired tissue oxygenation due to an increase in plasma viscosity in porcine sepsis. In septic patients, there were differences in hemodynamics after receiving albumin 5 percent or HES 10 percent 260/0.5. The administration of HES 10 percent 200/0.5 compared to lactated Ringer's solution improved cardiac index and oxygen transport variables in septic patients, which could not be achieved by the lactated Ringer's solution. In septic patients HES 10 percent 200/0.5 preserved splanchnic perfusion assessed by pHi measurements for more than 5 days, whereas the pHi decreased in patients receiving albumin 20 percent, indicating deteriorated splanchnic perfusion. The administration of HES 10 percent

200/0.5 compared to 20 percent albumin in sepsis resulted in a lower plasma concentration of adhesion molecules. These results suggesting effects of specific fluid therapy on the immune function during inflammation. In vitro HES, compared to albumin, inhibited lipopolysaccharide-stimulated vWF release but not endothelial E-selectin and neutrophil CD11bCD18 expression in a dose-dependent manner, thus suggesting an inhibition of endothelial cell activation by HES. Hence, there is some evidence that HES solutions may have some beneficial effect on the inflammatory process, which might in turn explain beneficial effects on systemic hemodynamics and oxygenation.

Anti-inflammatory Agents

Despite identification of several targets for pharmacological manipulations, at present there are no clinically effective specific therapies available to counteract sepsis-induced microvascular leak. In vitro and experimental data suggest that anti-inflammatory drugs may have the potential to attenuate or even to prevent microvascular leak. A variety of drugs interfering with the proinflammatory response including blockade of TNF-α, blockade platelet activating factor, or β-adrenergic-agent have been shown to effectively attenuate endothelial cell dysfunction in animal models of inflammation, but they generally failed when tested in large trials in septic patients. Recently, more specific targets have been identified [6]. First, the blockade of endothelial receptor is a potential target. Antihistamines and endothelin antagonists, as well as blockers of adhesion molecules and integrin CD11/CD18, are currently under clinical investigation. Furthermore, the administration of angiopoietin-1 has been shown to counteract VEGF and inflammatory-induced microvascular leak.

Modulation of signal transduction has been identified as another promising target to modulate endothelial function. The elevation of cytoplasmic calcium ion concentration of the endothelial cell is central for the induction of the endothelial hyperpermeability. The role of several protein kinases in changing intracellular calcium ion concentration and regulating endothelial permeability has been recognized as well. Another therapeutic strategy could be the enforcement of the barrier function by inhibiting the decrease in cAMP. Suitable agents include β-adrenergic-agents or phosphodiesterase inhibitors. As complement activation contributes significantly to the inflammatory reaction in sepsis, inhibition of complement activation has become a target for pharmacological research [10]. Endogenous soluble complement inhibitors, such as C1 esterase inhibitor, recombinant complement receptor 1, or blocking C5a have been developed to attenuate complement cascade reaction. An example of a successful application of the principle of complement inhibition in treatment of microvascular leak is the clinical use of C1 esterase inhibitor concentrate purified from human plasma. C1 esterase inhibitor concentrate is primarily indicated for the treatment of angioedema, but experimental and clinical evidence suggest that complement inhibition might also be an effective tool in the treatment of sepsis-induced microvascular leak.

Conclusion

Sepsis is associated with a marked intravascular fluid deficit caused by vasodilatation, venous pooling, and increased microvascular leak. The most striking clinical feature is fluid imbalance and the associated hemodynamic instability despite extensive fluid therapy. So far, the treatment options are fairly unspecific. It seems to be most important to guarantee adequate volume replacement in septic patients aiming at restoration and maintenance of intravascular volume in order to improve organ perfusion and nutritive microcirculatory flow. The type of fluid used is apparently less important. Despite new information, the current understanding of the complex, dynamic pathophysiology of sepsis-induced microvascular leak is still incomplete. Therefore, at present, neither standardized criteria for assessment of sepsis-induced microvascular leak, nor any *specific* therapeutic interventions are available. Promising agents for the treatment of microvascular leak have been developed in animal models. It remains to be seen whether they can be incorporated clinically.

Glossary

Edema: Excessive amount of watery fluid accumulated in the intercellular spaces, most commonly present in subcutaneous tissue.

Microvasculature leakage: Leakage of intravascular fluids into the extravascular space. This syndrome is observed in patients who demonstrate a state of generalized leaky capillaries following shock syndromes, low-flow states, ischemia–reperfusion injuries, toxemias, or poisoning. It can lead to generalized edema and multiple organ failure.

Multiple organ failure: Progressive condition usually characterized by combined failure of several organs such as the lungs, liver, and kidney, along with some clotting mechanisms.

Sepsis: Presence of pathogenic microorganisms or their toxins in tissues or in the blood.

Septic shock: Infection-induced hypotension not reversed with fluid resuscitation and associated with organ dysfunction or hypoperfusion abnormalities.

Severe sepsis: Infection-induced organ dysfunction or hypoperfusion abnormalities.

References

1. Angus, D. C., Linde-Zwirble, W. T., Lidicker, J., Clermont, G., Carcillo, J., and Pinsky, M. R. (2001). Epidemiology of severe sepsis in the United States: Analysis of incidence, outcome, and associated costs of care. *Crit. Care Med.* **29**, 1303–1310.
2. Groeneveld, A. B., Bronsveld, W., and Thijs, L. G. (1986). Hemodynamic determinants of mortality in human septic shock. *Surgery* **99**, 140–153.
3. Aird, W. C. (2003). The role of the endothelium in severe sepsis and multiple organ dysfunction syndrome. *Blood* **101**, 3765–3777. *A review article with a special emphasis on the role of the endothelium in the host response in sepsis.*
4. Starling, E. (1896). On the absorption of fluids from the connective tissue spaces. *J. Physiol. London* **19**, 312–326.

5. Michel, C. C., and Curry, F. E. (1999). Microvascular permeability. *Physiol. Rev.* **79**, 703–761. *A detailed review of the mechanisms involved in the pathogenesis of microvascular permeability.*
6. van Nieuw Amerongen, G. P., and van Hinsbergh, V. W. (2002). Targets for pharmacological intervention of endothelial hyperpermeability and barrier function. *Vasc. Pharmacol.* **39**, 257–272.
7. Marx. G., Vangerow. B., Burczyk. C., Gratz. K. F., Maassen. N., Cobas Meyer. M., Leuwer. M., Kuse. E., and Rueckoldt. H. (2000). Evaluation of noninvasive determinants for capillary leakage syndrome in septic shock patients. *Intensive Care Med.* **26**, 1252–1258.
8. Marx, G., Cobas Meyer, M., Schuerholz, T., Vangerow, B., Gratz, K. F., Hecker, H., Sumpelmann, R., Rueckoldt, H., and Leuwer, M. (2002). Hydroxyethyl starch and modified fluid gelatin maintain plasma volume in a porcine model of septic shock with capillary leakage. *Intensive Care Med.* **28**, 629–635. *This study demonstrates that plasma volume can be maintained in septic pigs despite increased microvascular permeability by artificial colloids but not by crystalloids.*
9. Renkin, E. M., Tucker, V., Rew, K., O Loughlin, D., Wong, M., and Sibley, L. (1992). Plasma volume expansion with colloids increases blood-tissue albumin transport. *Am. J. Physiol.* **262**, H1054–H1067.
10. Kirschfink, M. (2001). Targeting complement in therapy. *Immunol. Rev.* **180**, 177–189.

Capsule Biography

Dr. G. Marx has been Senior Lecturer at the University Department of Anaesthesia in Liverpool and Honorary Consultant in Intensive Care Medicine at the Royal Liverpool University Hospital since 2000. His key research topics are clinical and preclinical aspects of sepsis and increased microvascular leak syndrome. He has a particular interest in the development of new colloids and their potential therapeutic implications.

M. Leuwer has been Professor of Anaesthesia and Head of the University Department of Anaesthesia in Liverpool since 2000. His research interests are the interactions of phenol derivates with ion channels in view of their potential as a new class of local anesthetic/antidysrhythmic and clinical and preclinical aspects of sepsis and increased microvascular leak syndrome.

SECTION N

Lung Injury

CHAPTER 124

Microvascular Responses to Acute Lung Injury

Kazunori Murakami and Daniel L. Traber
Department of Anesthesiology, University of Texas Medical Branch and Shriners Hospital for Children, Galveston, Texas

Introduction

Acute lung injury (ALI) is observed in various serious conditions such as sepsis, multiple trauma, pneumonia, and smoke inhalation. ALI is defined as a Pa_{O_2} (partial arterial oxygen pressure)/Fi_{O_2} (fractional concentration of inspired oxygen) ratio of less than 300. The decrease in Pa_{O_2}/Fi_{O_2} ratio is observed because of the alveolar wall thickness (pulmonary edema) and ventilation–perfusion mismatching. The microvascular responses involve the major part of the pathophysiology [1]. In this review, we focus on the various factors that play a role in the microvascular responses associated with ALI.

Pulmonary Circulation Is Unique

There are several points that show the unique character of pulmonary circulation:

- All the blood in the body passes via the pulmonary artery to the pulmonary capillary bed
- The chest cavity is usually at a negative pressure
- The pulmonary capillary is very small (smaller than erythrocytes: diameter is less than 8 μ
- The pulmonary circulation is at a lower pressure than systemic circulation (normal mean pulmonary artery pressure is 9 to 16 mmHg)
- The alveolar macrophage stimulates pulmonary endothelial cells
- The pulmonary endothelial cells are easily activated because of endotoxin or bacterial stimulation from the airways
- Pulmonary vascular blood is provided from two sources: the pulmonary artery and bronchial arteries, a branch of the systemic arterial system

Because of these unique circulation systems, the lung easily becomes a target organ for systemic inflammation. Endotoxins or bacteria stimulate pulmonary endothelial cells to express adhesion molecules such as P-and E-selectin and intercellular adhesion molecule-1 (ICAM-1). Since the neutrophils are also activated to express those ligands such as P-selectin glycoprotein ligand-1 (PSGL-1), CD11b/ CD18, or L-selectin, activated neutrophils accumulate in the lung. Activated neutrophils starts rolling and sticking on the surfaces of endothelial cells. While the neutrophils are rolling, they release neutrophil elastase and reactive oxygen species (ROS). Those inflammatory mediators injure the endothelial cells and the vascular permeability increases. The changes in pulmonary microvascular permeability may result in interstitial edema as well as alveolar flooding, leading to serious problems with gas exchange.

There is another mechanism of neutrophil accumulation to the lung: the adhesion molecule–independent accumulation. When the neutrophils are activated by cytokines or platelet activating factor, intracellular calcium levels markedly increase via G protein–coupled receptors, serine/threonine kinase, Akt, and phosphatidylinositol 3-kinases (PI_3), which results in F-actin activation. As a consequence, cell deformability decreases significantly. Since the diameter of the neutrophil (approximately 12 μ) is greater than that of the pulmonary capillary, the stiffened neutrophils are stopped in the pulmonary microcirculation. This is an adhesion molecule–independent white cell accumulation in the lung, and it also causes lung injury because of perfusion disturbances.

Selectins may play a role in this process of neutrophil activation as well as cellular adhesion. We have reported that some activated neutrophils escape from the lung into aortic blood following acute lung injury induced by smoke inhalation. Activation of these cells was prevented by treatment with an antibody to L-selectin.

Activation of Akt/PI3 also inhibits neutrophil apoptosis by increasing transcription of the anti-apoptotic proteins Mcl-1 and Bcl-2, and also inhibiting caspase-9, a cell death protein. Without apoptosis the life of the neutrophils is prolonged, thus increasing the duration of the inflammatory process.

In terms of bronchial circulation, the adherence process may play a more important role since these systemic capillaries are larger and there is a better defined venular area. During the passage of neutrophils through this systemic circuit, they may become activated and the cells then pass on into bronchial venous drainage to lodge in the pulmonary microvasculature.

Nitric Oxide and Superoxide Play a Role

Nitric oxide (NO) is formed from three known isoforms of nitric oxide synthase (NOS). Endothelial NOS (eNOS) is constitutively expressed in endothelial cells and produces basic amounts of NO, which regulates microcirculation by dilating vascular smooth muscle cells and preventing platelet aggregation. NO also prevents neutrophil and macrophage activation and inhibits cytokine production. Neuronal NOS (nNOS) is another constitutive NOS in brain and nerve endings and is also reported to be present in neutrophils. We have determined that the administration of an inhibitor of nNOS would reduce the acute lung injury noted with ALI induced both by a combination of pneumonia and smoke inhalation. Similar beneficial effect was observed in ALI induced and by 40 percent burns combined with smoke inhalation. There is also an inducible NOS (iNOS) in macrophages, monocytes, lung epithelial cells, and so on. iNOS is expressed as a result of cytokine or endotoxin stimulation. New evidence also suggests that iNOS may be induced directly or indirectly by poly(ADP-ribose) polymerase (PARP). This latter enzyme is formed as a result of oxidative damage to DNA. PARP is present after injury. When iNOS is induced during acute lung injury, enormous amounts of NO are formed. This NO has been shown to form reactive nitrogen species (RNS) that result in leakage of the pulmonary microcirculation, since activated neutrophils release reactive oxygen species that can also combine with NO to contribute to the microvascular injury.

Normally the microcirculation to nonventilated lung is vasoconstricted, a process called hypoxic pulmonary vasoconstriction (HPV). Excess NO formed in acute lung injury abolishes HPV consequently; some of the blood going through the lung is not oxygenated. Thus with acute lung injury arterial blood is poorly saturated with oxygen. Administration of NOS inhibitors has been shown to restore HPV and arterial oxygen saturation.

RNS, especially peroxynitrite, injures DNA and induces single-strand breaks. As a consequence, PARP is activated. The activation of PARP has been shown in various ALI models. PARP inhibition has been shown to reduce pulmonary transvascular fluid flux and bronchial blood flow after injury by bacteria or burn/smoke [2]. PARP has been shown to play a role in activation of nuclear regulatory factors, NOS, and IL-8. A multicenter clinical trial using PARP inhibitors in ALI began in early 2004.

Neutrophil Elastase Injures Pulmonary Endothelial Cells

A protease called neutrophil elastase exists in the neutrophil granule. This protease effectively digests foreign bodies that come into the neutrophils by phagocytosis. In acute lung injury, activated neutrophils leak some elastase into the microcirculation. Under normal conditions, released neutrophil elastase in the blood is immediately inactivated by α1-antitrypsin (or α1 protease inhibitor). However, α1-antitrypsin is easily inactivated by both ROS and RNS; thus neutrophil elastase can damage the microvasculature, contributing to edema formation (Figure 1). Recently, the specific neutrophil elastase inhibitor ONO-5046 was shown to be effective in acute lung injury and is now on the market for clinical use in several countries.

What Factors Increase the Pulmonary Transvascular Fluid Flux?

Pulmonary vascular permeability is regulated by several factors. Starling's equation tells us what is involved:

$$Q = K[(P_c - P_i) - \sigma(\pi_c - \pi_i)]$$

where Q is the net flux of fluid out of the microvasculature, K is the filtration coefficient (an index of microvascular permeability to small molecules), P_c and P_i are the hydrostatic pressures in the capillary and interstitial spaces, respectively, π_c and π_i are the corresponding colloid osmotic pressures, and σ is the reflection coefficient; an index of the microvascular permeability to protein. The reflection coefficient is reduced when endothelial cells are injured and the permeability increases. The reflection coefficient is reduced and the pulmonary microvascular pressure is increased with ALI [3].

As we mentioned earlier, the blood flow to the lung is supplied by the pulmonary and bronchial arteries. In the normal condition, the bronchial arterial blood flow is approximately 1 percent of cardiac output. However, in the pathologic condition, it increases tenfold to fifteenfold. The marked reduction in bronchial vascular resistance results in an increase in microvascular pressure (P_c) and airway edema formation [3]. Some of the edema formed also leaks into the airway lumen. The airway fluid formed has a high protein content, suggesting that the bronchial microvasculature has also become leaky to colloid. In our ovine ALI model with burn and smoke inhalation injury, the increase in bronchial blood flow was significantly inhibited by specific

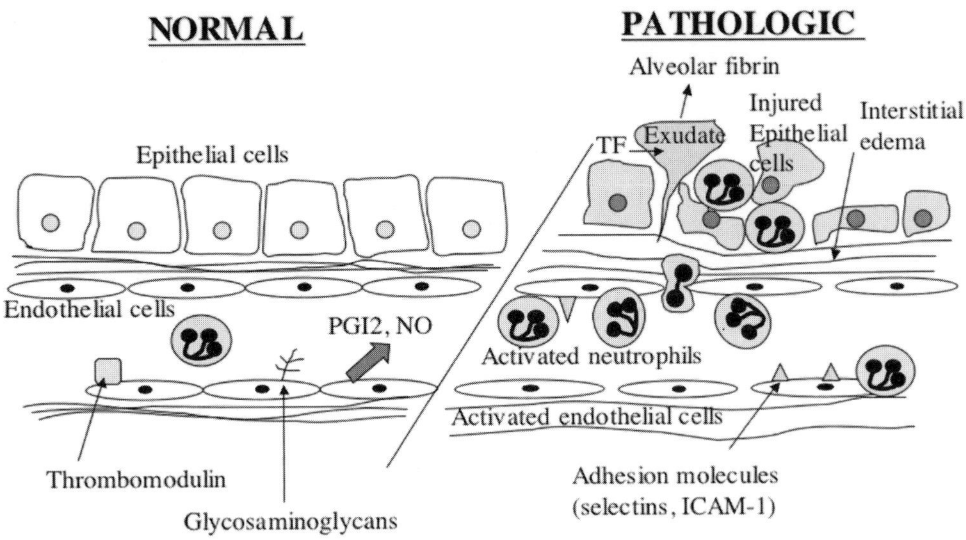

Figure 1 Under normal conditions (*left*), microvascular blood flow is regulated by endothelial cells. Several anticoagulant mechanisms as well as PGI_2 and NO production are involved in regulating microvascular circulation. When the endothelial cells are activated by inflammatory cytokines or endotoxin (*right*), adhesion molecules are expressed and neutrophils accumulate in the lung. Also, the activated neutrophil-induced endothelial injury causes an increase in vascular permeability, which results in edema formation. Exudates in the airways make clots and cause ventilation–perfusion mismatching. (see color insert)

iNOS and nNOS inhibitors, suggesting that the bronchial blood flow is regulated by both iNOS and nNOS [4, 5].

Airway Obstruction Plays Role in ALI

In addition to the transudate formed as a result of the leaky bronchial microvasculature, there is also a marked secretion from the airway mucous glands. The airway secretions contain mucins and blood coagulation factors such as fibrinogen. Broncho-pulmonary epithelial cells and alveolar macrophages express tissue factor when they are activated, which contributes to fibrin clots (Figure 1).

There are several reasons that the airway cast formation should be avoided. First of all, airway obstruction causes ventilation–perfusion mismatching. As a consequence of airway obstruction, some parts of the lung are ventilated but some are not. The blood flow to the nonventilated part will be a shunt flow. Second, if the patients are under mechanical ventilation, the tidal volume of air goes only into the ventilated part of lung. When the ventilator is volume controlled, it causes a significant increase in airway pressure. From our experience, histological investigation showed less than 10 percent obstruction in noninjured sheep bronchi or bronchioles, but after smoke inhalation with burn injury or smoke inhalation with pneumonia, the obstructed area significantly increased to 30 to 40 percent of the area [6]. The sheep with elevation in airway pressure showed marked increases in mRNA for the chemokine IL-8 in the pulmonary microcirculatory area, along with evidence of neutrophil infiltration. In order to prevent barotrauma to the lung, we put the animals with ALI on a CO_2 removal device, which allows a reduction in tidal volume and ventilatory pressures. Those animals showed significantly lower expression of IL-8, suggesting that the overextension of the airways is involved in chemokine induction and further inflammatory reactions. Perhaps this was the result of reduced cast formation. On the other hand, more than 90 percent of the bronchial venous drainage flows directly into the pulmonary microvasculature. Leukocytes activated in the airway as well as other materials such as ROS, RNS, or proteases would be directly transported to the pulmonary microvasculature.

Another strategy to reduce airway cast formation is to prevent an increase in bronchial blood flow. As described earlier, the bronchial blood flow increases tenfold to fifteenfold after injury; the increase of blood flow causes the increase in vascular permeability, which results in cast formation. Ablation of the bronchial circulation is reported to prevent the ovine model of ALI [7], suggesting that the blood flow through bronchial artery plays a role in the pathophysiology. Since upregulation of nitric oxide causes an increase in blood flow, the inhibition of NOS is effective to prevent cast formation. From our study in sheep, both the inhibition of iNOS and that of nNOS was effective in reducing cast formation [4, 5]. However, these NOS inhibitors did not completely block cast formation. We have not investigated the effect of dual inhibition of iNOS and nNOS. The effect might be additive. This hypothesis should be confirmed. There is much debate over whether or not mice have a blood supply from the bronchial artery, but still we do not have the exact answer. Thus gene-modified mice such as iNOS and/or nNOS knockout are not suitable for this type of study.

Neutrophils and Pulmonary Endothelial Cells Are Targets to Treat ALI

Activated neutrophils are obviously playing role in the lung microvascular pathophysiology. The removal of neutrophils by filtration, by anti-neutrophil antibody, or by chemical agents such as nitrogen mustard is beneficial for the experimental ALI. Antiadhesion molecule antibodies are also beneficial [8]. We have reported that the inhibition of L-selectin attenuated ovine ALI [9], but anti-P-selectin antibody did not. Since P-selectin is induced at very early time points and decreases soon after the peak expression, the inhibition of P-selectin could not be effective. In rat or mouse models, there are several papers showing the beneficial effects of E-selectin or ICAM-1 inhibition. However, in clinical trials in sepsis, none of these antiadhesion molecule studies were shown to be effective. The reason for this is not known. Perhaps the adhesion molecule–independent accumulation of neutrophils, such as increased cell stiffness, may be involving in sepsis-related ALI.

To inhibit neutrophil activation, there are several potent agents. Cepharanthine, an alkaloid extracted from an herbal plant, inhibits neutrophil activation and has been shown effective in ALI models in rats and sheep. It inhibits neutrophil elastase release and oxygen radical production. Intracellular cyclic AMP (cAMP) is known to inhibit neutrophil activation. The increase in neutrophil cAMP level inhibits F-actin activation, which results in the attenuation of deformability. Also, it inhibits neutrophil elastase release and superoxide production. Since cAMP is metabolized into 5'AMP by phosphodiesterase (PDE), PDE inhibitors increase the intracellular cAMP levels. Caffeine or aminophylline are well-known PDE inhibitors and are effective in asthma attack because increased cAMP levels in bronchial smooth muscle cells attenuate the bronchial constriction and spasm. In addition to that, those PDE inhibitors may have a beneficial effect through their anti-inflammatory properties. The increase in cAMP levels in the microvascular endothelial cells attenuates the vascular leakage, so there is a possibility that PDE inhibitors also inhibit cast formation. Adenosine administration shows similar effects in various animal models, but again, we need to confirm this efficacy in clinical trials.

Endothelial Cell Injury Induces Microcirculation Disturbances

On the surface of endothelial cells, there are three kinds of anticoagulant systems (Figure 1). First, a heparin-like substance, glycosaminoglycan (GAG), covers the endothelial cell surface. GAG is a receptor for antithrombin and promotes its antithrombotic property. Thrombin and antithrombin both have an affinity for GAG and the inhibitory reaction is accelerated by GAG. In ALI, inflammatory mediators such as neutrophil elastase cleave GAG and the endothelial GAG content decreases severely. Second, thrombomodulin is another anticoagulant protein on the endothelial cells. When thrombin is formed in the blood circulation, it binds to thrombomodulin. Once thrombin binds to thrombomodulin, its procoagulant potential is converted into anticoagulant and it activates a physiologic anticoagulant called protein C. Activated protein C inhibits the active forms of coagulation factors V and VIII so that thrombomodulin–thrombin complexes inhibit further thrombus formation. In the case of ALI, cytokines such as TNF or IL-1 activate endothelial cells and downregulate the thrombomodulin expression (Figure 1). Neutrophil elastase also damages thrombomodulin. Third, prostaglandin I_2 (PGI_2) and NO are produced from the endothelial cells and regulate platelet aggregation and vascular smooth muscle cell contraction (Figure 1). NO, which regulates proper vascular tone, is mainly produced by eNOS. In ALI or severe inflammatory diseases, endothelial production of PGI_2 and the activity of eNOS decrease. In addition to these losses of anticoagulant potential in endothelial cells, cytokine-activated endothelial cells express tissue factor, which stimulates the coagulation cascade. Taken together, microthrombi are easily formed in the pulmonary microcirculation and interrupt smooth blood flow.

Conclusion

Recent progress in molecular biology and various gene-modified animals shows numerous interesting results. However, we need to be careful about the anatomical differences in those small animals. In addition, animal models should mimic the clinical situation so that the obtained results are useful in the future clinical trials. In fact, various compounds targeting the microvascular response in ALI have been investigated in both animals and humans. Several clinical trials also begin in the near future.

References

References 2, 4, and 5 are our recent original work. The findings are novel. Reference 3 is our previous work, which clearly showed the pathophysiologically important phenomenon.

1. Traber, D. L., Herndon, D. N., et al. (2002). The pathophysiology of inhalation injury. In *Total Burn Care*. (D. N. Herndon ed.). pp. 221–232. London, Edinburgh, New York, Philadelphia, St. Louis, Sydney, Toronto: W.B. Saunders.
2. Shimoda, K., Murakami, K., Enkhbaatar, P., Traber, L. D., Cox, R. A., Hawkins, H. K., Schmalstieg, F. C., Komjati, K., Mabley, J. G., Szabo, C., Salzman, A. L., and Traber, D. L. (2003). Effect of poly(ADP ribose) synthetase inhibition on burn and smoke inhalation injury in sheep. *Am. J. Physiol. Lung Cell. Mol. Physiol.* **285**, L240–L249.
3. Isago, T., Noshima, S., Traber, L. D., Herndon, D. N., and Traber, D. L. (1991). Analysis of pulmonary microvascular permeability after smoke inhalation. *J. Appl. Physiol.* **71**, 1403–1408.
4. Enkhbaatar, P., Murakami, K., Shimoda, K., Mizutani, A., McGuire, R., Schmalstieg, F., Cox, R., Hawkins, H., Jodoin, J., Lee, S., Traber, L., Herndon, D., and Traber, D. (2003). Inhibition of neuronal nitric oxide synthase by 7-nitroindazole attenuates acute lung injury in an

ovine model. *Am. J. Physiol. Regul. Integr. Comp. Physiol.* **285**, R366–R372.

5. Enkhbaatar, P., Murakami, K., Shimoda, K., Mizutani, A., Traber, L., Phillips, G. B., Parkinson, J. F., Cox, R., Hawkins, H., Herndon, D., and Traber, D. (2003). The inducible nitric oxide synthase inhibitor BBS-2 prevents acute lung injury in sheep after burn and smoke inhalation injury. *Am. J. Respir. Crit, Care Med.* **167**, 1021–1026.
6. Murakami, K., Bjertnaes, L. J., Schmalstieg, F. C., McGuire, R., Cox, R. A., Hawkins, H. K., Herndon, D. N., Traber, L. D., and Traber, D. L. (2002). A novel animal model of sepsis after acute lung injury in sheep. *Crit. Care Med.* **30**, 2083–2090.
7. Efimova, O., Volokhov, A. B., Iliaifar, S., and Hales, C. A. (2001). Ligation of the bronchial artery in sheep attenuates early pulmonary changes following exposure to smoke. *J. Appl. Physiol.* **88**, 888–893.
8. Tasaki, O., Mozingo, D. W., Ishihara, S., Brinkley, W. W., Johnson, A. A., Smith, R. H., Srivastava, O., Mason, A. D., Jr., Pruitt, B. A., Jr., and Cioffi, W. G., Jr. (1998). Effect of Sulfo Lewis C on smoke inhalation injury in an ovine model. *Crit. Care Med.* **26**, 1238–1243.
9. Katahira, J., Murakami, K., Schmalstieg, F. C., Cox, R., Hawkins, H., Traber, L. D., and Traber, D. L. (2002). Role of anti-L-selectin antibody in burn and smoke inhalation injury in sheep. *Am. J. Physiol. Lung Cell. Mol. Physiol.* **283**, L1043–L1050.

Capsule Biography

Dr. Murakami has been working for Dr. Traber since 1999. He was winner of the Respiratory Specialty Award, Critical Care Medicine, in 2004. His work is supported by the Shriners North America.

Dr. Traber is the Charles Robert Allen Professor of Physiology. A long-term investigator in the area of ARDS, he has more than 300 publication in this area. Dr. Traber has received support for his work from the NIH, The Shriners of North America, and many pharmaceutical companies. He is a holder of the both the Distinguished Alumni and Distinguished Teaching (Inaugural holder) Awards from The Graduate School of Biomedical Science at the University of Texas Medical Branch.

SECTION O

Organ Transplantation

CHAPTER 125

Microvascular Responses to Organ Transplantation

[1]Michael D. Menger and [2]Brigitte Vollmar

[1]*Institute for Clinical and Experimental Surgery, University of Saarland, Homburg-Saar, Germany*
[2]*Department of Experimental Surgery, University of Rostock, Rostock, Germany*

Introduction

Although widespread interest in tissue engineering for organ replacement has forced major developments during the past few years, the most successful clinical approach to counteract whole organ failure still remains allograft transplantation. Refinements of surgical techniques but in particular the increasing knowledge of the pathophysiology and immunology of graft destruction have additionally improved both early engraftment and long-term outcome in transplantation medicine, including kidney, pancreas, and liver grafting. Of interest, the transplantation procedure is able to substantially restore quality of life of these patients.

Nonetheless, the shortage of donor organs together with the increasing numbers of patients with need for a transplant turned transplant surgeons toward the use of organs of critical quality. Those organs, however, are more susceptible to developing graft failure and, thus, require additional improvements of organ procurement and post-transplant treatment.

The majority of primary and delayed loss of graft function in clinical transplantation is caused by cold preservation–reperfusion-mediated graft injury and rejection-associated immunologic tissue destruction. Whereas reperfusion injury is considered to be a nonspecific, granulocyte-mediated inflammatory process, rejection is considered a highly specific, alloantigen-triggered disease. Although these two events may differ distinctly in nature, recent studies indicate a linkage in that postischemic reperfusion injury sensitizes whole-organ transplants for the development of graft rejection [1].

Because the interaction of inflammatory and immunocompetent blood cells with the grafts' microvascular endothelial lining is thought to be the first step in the cascade of events involved in ischemia–reperfusion injury and graft rejection, the microcirculation of the transplanted organs represents the central target in post-transplant graft failure [2]. In the development of microcirculatory dysfunction, a variety of parameters affect the manifestation of ischemia–reperfusion injury, including organ donor conditions, donor organ quality, organ preservation and storage, the technical surgical procedure, and, finally, reperfusion-associated inflammation (Figure 1).

Donor Organ Conditions

The quality of post-transplantation microvascular perfusion depends on both the general condition of the organ donor and the individual condition of the donor organ. Steatosis and fibrosis of the donor organ alter the quality of microcirculation under normal conditions but in particular after an ischemia–reperfusion insult, including not only warm ischemia but also cold ischemia, preservation, and transplantation. After transplantation, steatotic livers show narrow and irregular sinusoids, associated with altered perfusion and blood cell adhesions to endothelial cells. In those fatty livers the increased leukocyte accumulation and adhesion within the microvasculature is associated with a more pronounced generation of oxygen radicals as well as with a reduction of mitochondrial membrane potential, resulting in marked hepatocellular disintegration. As a consequence, the transplantation of steatotic livers is associated with an increased rate of primary dysfunction or nonfunction of the grafts, which, in clinical practice, requires emergency retransplantation.

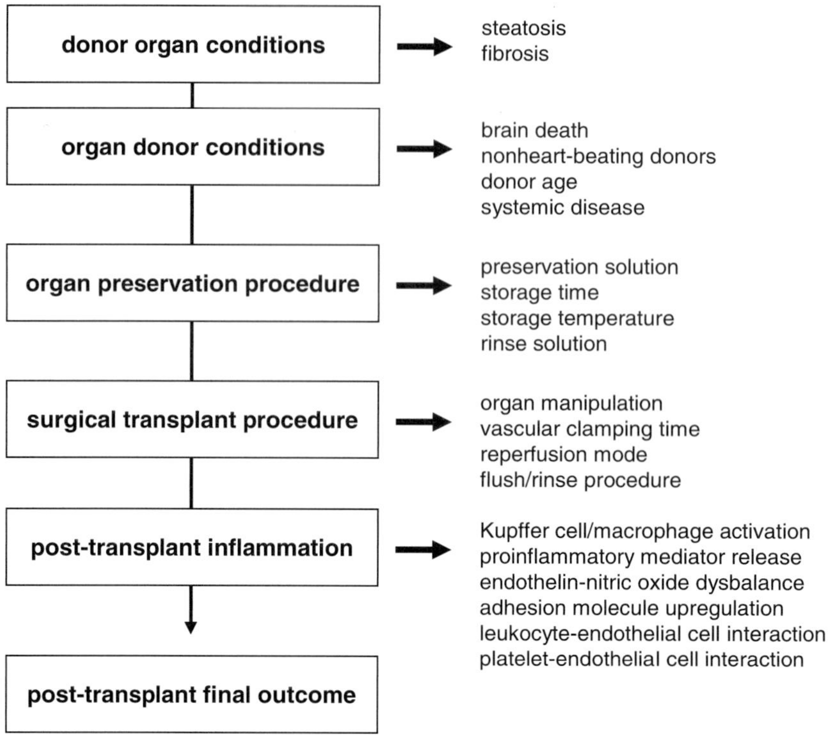

Figure 1 Factors influencing microcirculatory reperfusion injury after whole-organ cold storage and transplantation.

Organ Donor Conditions

Besides donor organ dysfunction, brain death of the organ donor per se induces arterial hypotension, which is associated with liver sinusoidal perfusion failure and leukocyte–endothelial cell interaction as well as hepatocellular disintegration and dysfunction, as indicated by hepatocyte vacuolization and decreased bile production. The microcirculatory failure, the inflammatory response, and the hepatocellular dysfunction seem not to be due to the arterial hypotension, because a comparable extent of hypotension without brain death does not affect hepatic microcirculation and function, and induction of normotension during brain death is not capable of preventing deterioration of the microcirculation and hepatocellular injury.

With the growing disparity between the number of patients awaiting organ transplantation and the limited supply of cadaveric organs, there is renewed interest in the use of organs retrieved from nonheart-beating donors. However, it should be taken into account that the addition of a warm ischemic period to the cold preservation period may dramatically aggravate the post-transplant ischemia–reperfusion injury, including the deterioration of the microcirculation. In both kidneys and livers from nonheart-beating donors microvascular perfusion with histidine–tryptophan–ketoglutarate (HTK) solution has been shown to be slightly deteriorated after 30 minutes and markedly deteriorated after 60 minutes of cardiac arrest. Perfusion solutions with higher viscosity than that of HTK, as represented by the University of Wisconsin solution, show even more pronounced microvascular perfusion deficits in both kidneys and livers. The microcirculatory preservation quality of those organs from nonheart-beating donors may be improved by dual perfusion (in livers) and increasing the perfusion pressure, by pretreatment of the donor with heparin and phentolamine, or by a warm preflush of the organs with streptokinase [3]. In addition to an appropriate equilibration of the microvasculature with preservation solution, the oxygenation of nonheart-beating donor organs during cold storage may be a promising approach to avoid parenchymal cell injury and to improve post-transplant organ function. This is achieved by retrograde venous oxygen persufflation. In fact, in long-term preserved livers harvested from nonheart-beating donors mitochondrial redox state and thus ATP content were found to be significantly improved when pure oxygen was applied with a persufflation pressure of 9 mmHg or 18 mmHg.

Donor Organ Conditioning

Preservation of organ grafts may induce hyperinflammation after transplantation. In rat livers, 24 hours of storage in cold Euro-Collins solution has been shown to double Kupffer cell activation and phagocytosis. Accordingly, blockade of Kupffer cell function of organ donors by methyl palmitate is not only effective in reducing Kupffer cell activation, but also in improving post-transplant survival. In line with these results, organ donor pretreatment with epoprostenol, the anti-inflammatory prostaglandin-I_2, results in amelioration of post-transplant sinusoidal perfusion failure and microvascular leukocyte adherence, which is associated with an overall improvement of parenchymal graft function. The efficacy of anti-inflammatory conditioning of the organ donor has recently been confirmed in an experimental model of porcine liver transplantation, demonstrating a reduction of ischemia–reperfusion injury by elimination of Kupffer cell function through gadolinium chloride donor pretreatment. Gadolinium chloride given to donor livers can reduce post-transplant AST-levels, improve hepatic macro- and microcirculation, and increase survival. Although those studies showed that blockade of Kupffer cells is not associated with the risk of increased infection, but, in contrast, improved hepatic endotoxin clearance, the approach to attenuate post-transplant reperfusion injury by targeting Kupffer cell function of the organ donor has not yet been introduced in clinical practice yet.

Other donor conditioning programs, such as induction of heat shock proteins, short-term immunosuppression by mycophenolate mophetil, or pretreatment with interleukin-10, prostacyclin, and pentoxifylline, have also been shown to attenuate post-transplant reperfusion injury. However, the role of the microcirculation in this protection has not been analyzed yet. Again, those programs have also not been introduced to patients until now.

The only donor organ conditioning program that is presently being tested in clinical practice is the ischemic preconditioning procedure. Ischemic preconditioning refers to a phenomenon in which brief episodes of vascular occlusion render a tissue resistant to the deleterious effects of subsequent prolonged ischemia. Recent experimental and clinical studies have demonstrated that ischemic preconditioning is able to protect from injury after both warm and cold ischemia–reperfusion by improving the microvascular perfusion and reducing leukocyte- and platelet–endothelial cell interaction in the transplanted organs. The mechanisms of action of ischemic preconditioning are not fully understood yet; however, it seems to involve the action of adenosine, nitric oxide, and heat shock proteins, including downregulation of nuclear factor kappa B and inhibition of caspase activation [4].

Organ Preservation Procedures

Organ preservation is a major determinant for post-transplant reperfusion injury, including preservation time, temperature, and solution. Experiments with liver, pancreas, and intestinal transplantation have demonstrated that the cold ischemia time determines post-transplant microcirculatory derangements, inasmuch as an increase of duration of preservation is associated with aggravation of both microvascular perfusion failure (no-reflow) and leukocytic response (reflow-paradox). In clinical practice, a temperature of 4°C for preservation is chosen because of the practicability of cooling the organ with ice and is based on the idea that the temperature should be as low as possible to achieve the utmost protection. However, recent experimental studies have demonstrated that for a 90-minute ischemia period, preservation at 15°C and 26°C is as effective as cooling to 4°C in preventing postischemic microcirculatory failure, leukocytic inflammatory response, and hepatocellular disintegration. The protection achieved with only moderate cooling is probably the result of attenuation of intratissue reactive oxygen species production, associated with a reduced adenosine monophosphate hydrolysis during the ischemic time period. In fact, there is some evidence that a mild hypothermia of about 20°C may protect the grafts' microvascular endothelium even more efficiently than a storage temperature of 4°C.

Daily clinical practice has revealed that post-transplant graft function is strongly dependent on the preservation solution used for organ cold storage. Today, University of Wisconsin (UW) solution and HTK solution are the most applied introduced preservation solutions in parenchymal organ transplantation, including the liver, the kidney, and the pancreas. Although it has not been finally clarified which ingredients within the solutions are protective to prevent the manifestation of post-transplant microcirculatory dysfunction, there is evidence that preservation solutions that contain antioxidative compounds, such as UW or HTK solution, are more effective in reducing microvascular cold ischemia–reperfusion injury when compared with others lacking antioxidative components, such as Euro-Collins solution. The efficacy of UW and HTK solution to improve post-transplant microvascular perfusion after cold storage may, however, be due not only to their antioxidative action, but also to the solution-associated capacity to inhibit cell swelling, as shown directly by electron microscopy, and, indirectly, by the prevention of cold ischemia–reperfusion-induced sinusoidal diameter reduction.

Surgical Transplant Procedures

Surgical manipulation during organ harvest, that is, touching, retracting and moving of the organ, induces marked deterioration of the microcirculation, resulting in tissue hypoxia, parenchymal cell death, and, finally, a decrease of transplant survival [5]. The manipulation-induced microcirculatory derangements seem to involve Kupffer cell activation and oxygen free radical release, because inactivation or deletion of Kupffer cells can significantly reduce microcirculatory disturbances and parenchymal tissue hypoxia, thereby improving recipient survival. In experimental liver transplantation, it has been further demonstrated that the innervation of the organ also contributes to the manipulation-induced deterioration of the microcirculation, and that denervation before organ harvesting prevents circulatory deterioration and primary nonfunction of the transplant.

The manifestation of post-transplant microcirculatory dysfunction depends additionally on the surgical procedure applied, in particular in liver transplants. Although it was found that the portal vein clamping time does not strongly determine the quality of graft microcirculation after transplantation, deterioration of hepatic arterial or portal venous inflow during reperfusion is well known to dramatically alter the function of the transplanted organ. A morphometric analysis in experimental rat liver transplantation revealed that liver samples from rats that underwent hepatic artery reconstruction had preserved lobular architecture and estimated bile duct and connective tissue volumes, whereas liver samples from rats that did not undergo hepatic artery reconstruction showed bile duct proliferation and an increase in connective tissue volume associated with a decrease in hepatocyte volume. In vivo experiments have further confirmed the importance of the hepatic dual blood supply, inasmuch as reconstruction of the hepatic artery with restitution of hepatic arterial blood flow has been shown to significantly reduce reperfusion-associated microcirculatory perfusion failure compared to liver grafts in which solely the portal vein was anastomosed. In addition, the simultaneous onset of hepatic arterial and portal venous blood flow after the transplantation procedure compared to the "classical" delayed onset of hepatic arterial reperfusion is considered effective in reducing microvascular perfusion failure, leukocyte adhesion, and Kupffer cell activation, which finally results in an improved primary graft function. These results have been confirmed in that in nonarterialized liver transplants a persistence of impaired hepatic microcirculation could be detected, which is associated with hepatocellular and endothelial cell swelling as well as bile duct injury. Apart from this, the prevention of intestinal congestion by insertion of a porto-jugular shunt during the surgical transplantation procedure also significantly reduces post-transplant reperfusion injury, including sinusoidal perfusion failure and microvascular leukocyte accumulation and adhesion.

In clinical practice, the preservation solution is flushed out by a rinse solution just prior to reperfusion of the transplant. Thus, the compounds of the preservation solution are removed before they can act during early reperfusion. Therefore, a rinse solution was designed to counteract early manifestation of reperfusion injury. This solution, named Carolina rinse, is composed of a variety of components potentially capable of counteracting post-transplant organ injury, including nicardipine, hydroxyethyl starch, fructose, glucose, and insulin, as well as adenosine and the three antioxidants desferrioxamine, glutathione, and allopurinol [6]. The first in vivo studies clearly indicated that this solution is a superior alternative to Ringer's solution to protect liver grafts from reperfusion injury. Further studies then could demonstrate that Carolina rinse prevents cold ischemia–reperfusion-induced endothelial cell killing and detachment, and improves the quality of early microvascular reperfusion. The effectiveness of the rinse solution may additionally be increased by prewarming the solution before infusion into the donor organ. As demonstrated in rat liver and intestine, prewarming of either Carolina rinse or Ringer's lactate indeed reduces post-transplant no-reflow, ameliorates microvascular leukocyte adhesion, improves lymphatic capillary perfusion, and prevents graft failure.

Transplant Reperfusion-Associated Inflammation

Reperfusion-associated microcirculatory dysfunction involves the activation of distinct humoral, cellular, and molecular systems. In contrast to warm ischemia–reperfusion, after cold ischemia (preservation) and transplantation (reperfusion) the nonparenchymal cell population, including the microvascular endothelial cells, has to be considered as a major target for the manifestation of injury and thus graft failure [7]. Nonetheless, the nature of the microcirculatory dysfunction is similar when compared with that observed after warm ischemia. In vivo studies with orthotopic rat liver transplantation showed both sinusoidal no-reflow- (perfusion failure) and reflow-paradox-associated events (leukocyte and platelet accumulation and adherence) as determinants in the development of primary graft dysfunction. Leukocyte–endothelial cell interaction seems to be mediated classically by involvement of P-selectin in leukocyte rolling and Mac-1 and ICAM-1 in leukocyte firm adhesion, because blockade of selectins, β2-integrins, and ICAM-1 has been shown effective in reducing post-transplant leukocyte adhesion. In parallel, platelet recruitment within the microvasculature seems to strongly involve P-selectin, because mice lacking P-selectin showed a reduction of platelet–endothelial cell interaction and improvement of survival after warm ischemia–reperfusion. Accordingly, the application of a soluble P-selectin glycoprotein ligand in a cold ischemia–reperfusion kidney transplantation model has been proven able to protect against blood flow perturbations and parenchymal injury.

A large number of recent experimental studies brought evidence that an imbalance of the nitric oxide–endothelin system contributes to the manifestation of both no-reflow and reflow-paradox in post-transplant reperfusion injury. Endothelin-1 seems to be the "bad guy" in this scenario. This view is supported by the fact that endothelin-1 and big-endothelin concentrations are elevated in liver graft tissue during cold storage and reperfusion, and that the cold storage–induced endothelin-1 release is associated with an increase in portal pressure at the time of reperfusion. In parallel, a considerable number of studies could demonstrate that the blockade of the endothelin A receptor results in an attenuation of post-transplant microcirculatory perfusion failure and leukocyte adhesion in liver, pancreas, and small intestine [8]. In line with this, inhibition of nitric oxide by using the nitric oxide synthase inhibitor L-NAME aggravates hepatic microcirculatory ischemia–reperfusion injury, while nitric oxide donors, such as L-arginine or FK409, improve post-transplant microcirculation and reduce leukocyte adhesion and microvascular permeability in pancreas and small-bowel transplants.

Apart from the imbalance of the nitric oxide–endothelin system, a considerable number of other inflammatory mediators contribute to the intravascular leukocyte accumulation in cold ischemia reperfusion, including oxygen radicals, nuclear factor kappa B-related cytokines, complement, phospholipase A2, leukotrienes, thromboxane, and platelet-activating factor. Kupffer cells have to be considered as the source of these inflammatory mediators. In fact, there is substantial evidence from liver transplantation experiments that cold ischemia–reperfusion markedly activates Kupffer cells and that this activation is associated with depressed hepatocellular function post-transplantation. As a consequence, blockade or elimination of Kupffer cells by methyl palmitate or liposome-encapsulated dichloromethylene diphosphonate attenuates sinusoidal endothelial cell damage and improves post-transplant survival. Whether oxygen radicals produced after onset of reperfusion trigger the activation of Kupffer cells is not yet clear. Treatment with the antioxidant *N*-acetylcysteine did not affect Kupffer cell phagocytotic activity, although it reduced the leukocytic response after liver transplantation. In contrast, UW solution and in particular Carolina rinse, both containing a whole armamentarium of antioxidants, have been shown to be effective in reducing Kupffer cell activation. In addition, the source of postischemic radical production is still a matter of controversy. Whereas some studies with cold ischemia–reperfusion indicate that hepatocytes may serve as a major site of reactive oxygen species generation, triggered through Kupffer cell–independent mechanisms, others have shown that microcirculatory failure after rat liver transplantation is related to Kupffer cell–derived oxidant stress.

Little is known as to whether pericytes and hepatic stellate cells are also involved in mediating microcirculatory dysfunction after cold preservation and transplantation. Stellate cells regulate sinusoidal vascular tone via endothelin and nitric oxide action. In conditions associated with ischemia–reperfusion an imbalance toward endothelin may result in sinusoidal narrowing and thus microvascular perfusion failure. Accordingly, a recent study has demonstrated in warm ischemia–reperfusion as well as after orthotopic liver transplantation that the inhibition of stellate cell contraction by a rock inhibitor can prevent postischemic hepatic microcirculatory disruption and, by this means, can improve post-transplant survival.

Conclusions and Clinical Perspectives

A considerable number of experimental studies have demonstrated that the microcirculation represents a primary target in the manifestation of transplantation-associated graft dysfunction. Accordingly, analysis of the microcirculation in transplantation surgery has been introduced into clinical practice during the last few years. This should aid in both achieving early diagnosis of injury and monitoring the appropriateness of therapy, which should target the compromised organ donor and donor organ conditions, the organ preservation procedure, the surgical transplant procedure, and the reperfusion-associated post-transplantation inflammation (Figure 2). The techniques introduced in clinics for the analysis of the microcirculation of organ transplants are laser Doppler flowmetry, polarographic oximetry,

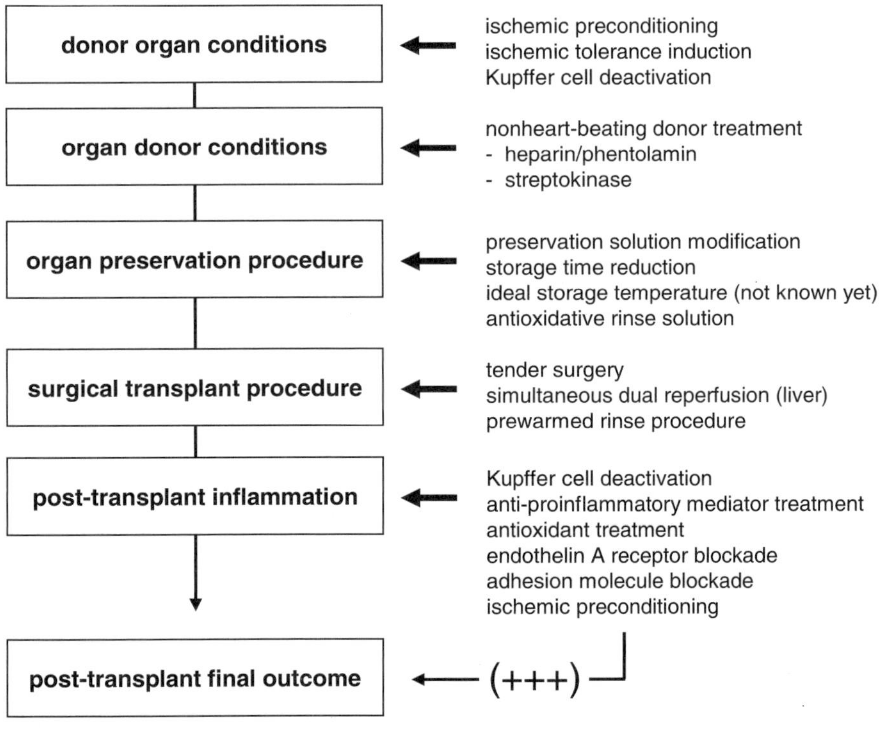

Figure 2 Targets and potential novel treatment strategies to attenuate microcirculatory reperfusion injury after whole-organ cold storage and transplantation.

thermodiffusion analysis [9], near-infrared spectrometry, and orthogonal polarization spectral (OPS) imaging [10]. The use of these techniques may further open the door for a more detailed understanding of the mechanisms of microcirculatory dysfunction in transplant patients, which may help in developing novel therapeutic strategies to ameliorate reperfusion injury, and, thus, improve outcome in transplantation surgery.

Glossary

Inflammation: Reaction of the body to defend the tissue invasion of micro-organisms and foreign materials by activation of the pro-inflammatory mediator cascade and recruitment of immunocompetent cells, in particular polymorphonuclear leukocytes.

Microcirculation: The microcirculation is the blood flow through the finest vascular structures in the body, including terminal arterioles, nutritive capillaries and draining venules, which guarantees the oxygen supply to the tissue and the drainage of accumulated toxic substances.

Organ transplantation: Organ transplantation represents the transfer of functional tissue from a donor to a recipient, which may be done between species as a xenograft or within the species as an allograft.

Reperfusion injury: Although indispensably needed after complete arrest of blood flow, reperfusion exerts additional damage to tissue, termed reperfusion injury, involving both prolongation of hypoxia and induction of hyperinflammation.

References

1. Land, W. (2002). Postischemic reperfusion injury to allografts—a case for innate immunity? *Eur. Surg. Res.* **34**, 160–169.
2. Menger, M. D., and Vollmar, B. (2000). Role of microcirculation in transplantation. *Microcirculation* **7**, 291–306.
3. Yamauchi, J. I., Richter, S., Vollmar, B., Menger, M. D., and Minor, T. (2000). Warm preflush with streptokinase improves microvascular procurement and tissue integrity in liver graft retrieval from nonheart-beating donors. *Transplantation* **69**, 1780–1784. *This study showed for the first time in an experimental setup with in situ fluorescence microscopy that microvascular procurement of organs from nonheart-beating donors with UW solution is associated with marked perfusion deficits resulting in aggravation of organ injury. The novelty of the study is the use of streptokinase in a Ringer's lactate flush solution, which was effective in significantly reducing microvascular perfusion deficits and thus attenuating organ injury. Based on these original studies, the application of streptokinase in nonheart-beating donors was recently introduced in clinical practice.*
4. Selzner, N., Rüdiger, H., Graf, R., and Clavien, P. A. (2003). Protective strategies against ischemic injury of the liver. *Gastroenterology* **125**, 917–936.
5. Schemmer, P., Schoonhoven, R., Swenberg, J. A., Bunzendahl, H., and Thurman, R. G. (1998). Gentle in situ liver manipulation during organ harvest decreases survival after rat liver transplantation: Role of Kupffer cells. *Transplantation* **65**, 1015–1020. *This study demonstrated for the first time that manipulation of the donor organ markedly aggravates microcirculatory and parenchymal reperfusion injury after transplantation, and that this manipulation-mediated aggravation of reperfusion injury is mediated by Kupffer cells, because blockade of Kupffer cells blunted manifestation of this additional injury. For the clinical transplant surgeon these results are highly important, because it indicates the necessity of careful operation without manipulation of the tissue, in particular during the explant procedure.*
6. Gao, W. S., Takei, Y., Marzi, I., Lindert, K. A., Caldwell-Kenkel, J. C., Currin, R. T., Tanaka, Y., Lemasters, J. J., and Thurman, R. G. (1991). Carolina rinse solution—a new strategy to increase survival time after orthotopic liver transplantation in the rat. *Transplantation* **52**, 417–424.
7. Caldwell-Kenkel, J. C., Currin, R. T., Tanaka, Y., Thurman, R. G., and Lemasters, J. J. (1989). Reperfusion injury to endothelial cells following cold ischemic storage of rat livers. *Hepatology* **10**, 292–299. *This study demonstrated for the first time that in contrast to warm ischemia–reperfusion, the endothelial cell is a primary target of injury in cold ischemia–reperfusion. After 24 hours and 48 hours of cold storage virtually all parenchymal cells (liver) remained viable, while about 40 percent of the nonparenchymal cells were trypan positive, exhibiting cellular rounding, nuclear pyknosis, and protrusion into the sinusoidal lumen. Thus, this early study indicated a major role of microvascular endothelial cell injury in cold storage–transplantation-associated ischemia–reperfusion.*
8. Witzigmann, H., Ludwig, S., Armann, B., Gabel, G., Teupser, D., Kratzsch, J., Pietsch, U. C., Tannapfel, A., Geissler, F., Hauss, J., and Uhlmann, D. (2003). Endothelin(A) receptor blockade reduces ischemia/reperfusion injury in pig pancreas transplantation. *Ann. Surg.* **238**, 264–274. *This study confirmed results of a number of other previous studies in that the blockade of the endothelin A receptor is effective in attenuating microcirculatory dysfunction after cold storage and transplantation. The study may be considered important because it nicely shows in a large-animal (porcine) model of pancreas transplantation that the attenuation of microcirculatory dysfunction of the transplant is associated with improvement of tissue oxygenation, reduction of parenchymal injury, and, as a consequence, prevention of transplant pancreatitis (which is a major cause of loss of pancreas grafts).*
9. Klar, E., Kraus, T., Bredt, M., Osswald, B., Senninger, N., Herfarth, C., and Otto, G. (1996). First clinical realization of continuous monitoring of liver microcirculation after transplantation by thermodiffusion. *Transplantation Int.* **9**(Suppl 1), S140–S143.
10. Puhl, G., Schaser, K. D., Vollmar, B., Menger, M. D., and Settmacher, U. (2003). Noninvasive in vivo analysis of the human hepatic microcirculation using orthogonal polarization spectral imaging. *Transplantation* **75**, 756–761.

Capsule Biography

Dr. Menger is professor for experimental surgery at the University of Saarland and chairs the Institute for Clinical and Experimental Surgery. He was winner of the Bernhard von Langenbeck Award of the German Society for Surgery in 1993 and is an honorary member of the Society of University Surgeons, USA. His research focus is microcirculatory dysfunction in shock, sepsis, ischemia–reperfusion, and transplantation. His work is supported by grants from the Deutsche Forschungsgemeinschaft.

Dr. Vollmar is professor for experimental surgery at the University of Rostock and chairs the Department of Experimental Surgery. Winner of the Otto-Goetze-Award of the Bavarian Society of Surgery in 1994 and the Calogero-Pagliarello-Award of the University of Saarland in 2000. Her research focus is microcirculatory dysfunction and mechanisms in shock, sepsis, ischemia–reperfusion, and microvascular thrombosis. Her work is supported by grants from the Deutsche Forschungsgemeinschaft.

SECTION P

Periodontal Disease

CHAPTER 126

The Pathobiology of the Periodontium: Insights into Vascular Dysfunction and Acceleration of Atherosclerosis

[1]Evanthia Lalla, [1]Panos N. Papapanou, [1]Ira B. Lamster, and [2]Ann Marie Schmidt

[1]*School of Dental and Oral Surgery, Columbia University, New York, New York*
[2]*College of Physicians and Surgeons, Columbia University, New York, New York*

Atherosclerosis is a chronic disease of the vasculature in which structural lesions in the form of atherosclerotic plaques may progress to highly unstable lesions susceptible to rupture, clinically manifested as events such as heart attacks and strokes. Hyperlipidemia is classically implicated as the key risk factor leading to generation of fatty streaks, followed often by progression to more complex and unstable plaques. A variety of inflammatory, metabolic, infectious, and hemodynamic states accelerate the development and progression of atherosclerosis. In this review, we will focus on the role of oral—particularly periodontal—infection in modulation of local *and* systemic inflammation, linked epidemiologically and experimentally to acceleration of vascular dysfunction and atherosclerosis. Underlying mechanisms and the role of modifying conditions such as diabetes will be discussed.

Periodontal Disease, Vascular Inflammation and Dysfunction

Periodontal diseases are chronic infections affecting the supporting structures of teeth. Several cross-sectional, case-control, and longitudinal studies have reported an increased risk for vascular disease and related clinical events in individuals with periodontal disease. In these studies the risk was independent of other known risk factors, including smoking, diabetes, and the blood lipid profile [1].

Multiple pathogenic mechanisms likely underlie these observations, including, at least in part, direct infection of the vasculature by oral pathogenic bacteria, and/or generalized inflammation secondary to lipopolysaccharide or other bacterial products that engage host defense mechanisms. Indeed, more recent evidence has indicated that patients with severe periodontal disease have increased serum levels of CRP, IL-1, and IL-6, hyperfibrinogenemia, and moderate leukocytosis compared to unaffected controls [2].

The findings above have been extended, as a relationship between periodontal disease, tooth loss, and *subclinical* vascular disease (carotid intima-media thickening or carotid plaque presence) has also been recently reported [3, 4]. In the Oral Infections and Vascular Disease Epidemiology Study (INVEST), the investigators identified that tooth loss, a measure of cumulative periodontal disease, was associated with the prevalence of carotid artery plaque. Among subjects with no to nine missing teeth, 46 percent displayed carotid artery plaque, whereas individuals with 10 or more missing teeth displayed a prevalence of carotid plaque of

60 percent. These statistically significant findings suggested that measures of subclinical atherosclerosis paralleled the severity of periodontal disease in these subjects.

Although periodontal infection is not a cause of atherosclerosis, it may act as a potent modifier of disease, synergizing with classical risk factors such as hyperlipidemia, diabetes, or elevated blood pressure. However, some published evidence is not supportive and data from a randomized controlled clinical trial evaluating the effect of periodontal therapy on the risk for vascular disease are not available to date.

The state of microvascular function has been suggested to correlate with frank macrovascular disease. Thus, assessment of flow-mediated dilation may provide a window into vascular inflammation and atherosclerosis. These correlations have been found in subjects with periodontal disease as well; subjects with advanced periodontal disease displayed significantly lower flow-mediated dilation of the brachial artery on ultrasonography compared with age-matched control subjects. These considerations have led to the suggested use of flow-mediated dilation as a surrogate marker for the presence/extent of atherosclerosis. However, the link between such surrogates and definitive clinical events such as myocardial infarctions/death and the response to therapeutic intervention is not yet fully clarified, and conflicting evidence exists.

The development of suitable animal models provides a way to assess these concepts experimentally and dissect the pathways by which periodontal infections are linked to vascular inflammation and atherosclerosis.

A Murine Model of Accelerated Atherosclerosis in Mice Infected with a Periodontal Pathogen

Toward that end, our laboratory employed mice deficient in apolipoprotein E (apo E null). In these animals, spontaneous atherosclerosis develops on a normal chow diet driven primarily by hypercholesterolemia. We, and others, have shown that induction of superimposed stresses, such as hyperglycemia or hyperhomocysteinemia, accelerates atherosclerosis in this model. To test the premise that oral infection may accelerate vascular inflammation and macrovascular disease, male apo E null mice were inoculated with the human periodontal pathogen *Porphyromonas gingivalis*, strain 381, by oral gavage and oral/anal topical application, beginning at age 6 weeks. Control animals received vehicle (phosphate buffered saline; PBS). In parallel with accelerated alveolar bone loss in infected mice versus those animals receiving PBS, *P. gingivalis*-infected mice displayed a significant 40 percent increase in atherosclerotic lesion area at the aortic root versus vehicle at age 17 weeks (Figure 1). Although the majority of lesions in both groups of animals were limited to fatty streaks, it is possible that upon longer evaluation periods, acceleration of lesion area and progression of fatty streaks to more unstable plaques would occur.

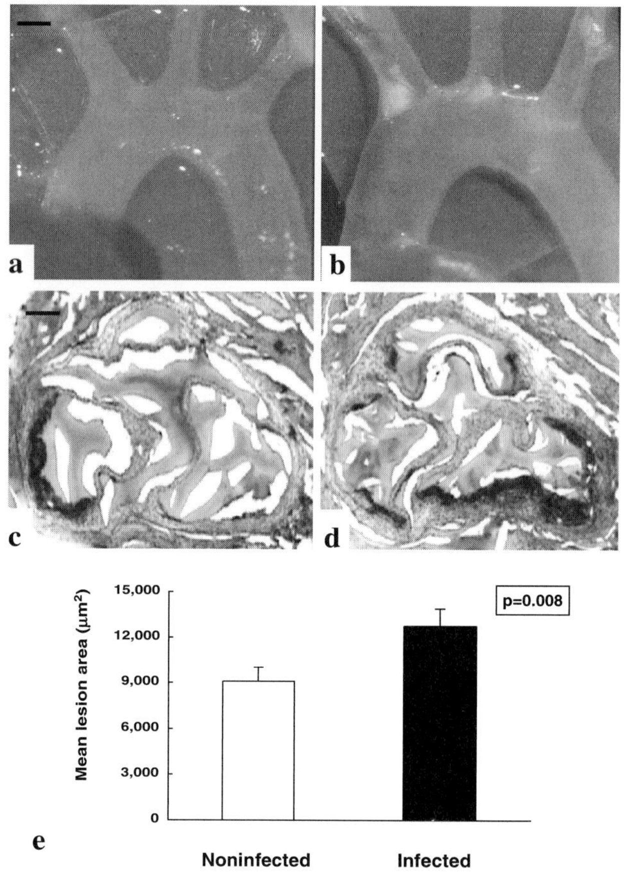

Figure 1 Oral infection with *P. gingivalis* accelerates early atherosclerosis in apoE-null mice. Representative photographs of the proximal aorta demonstrate increased atherosclerosis in a *P. gingivalis*-infected (**b**) versus a noninfected (**a**) apo E-null mouse at 17 weeks of age. Scale bar, 1 mm. Oil red O staining of cryosections at the aortic sinus from control mice revealed few small fatty streaks (**c**), whereas atherosclerotic lesions were greater in number and size in infected animals (**d**), scale bar 50 μm. Quantitation of atherosclerotic lesion area at the aortic sinus (**e**) revealed that mean lesion area was significantly increased in infected versus noninfected mice (12,753 ± 1,128 versus 9,080 ± 951 mm^2, $n = 25$/group, $p = 0.008$). Republished with permission of *Arteriosclerosis Thrombosis and Vascular Biology* from: Lalla, E., et al. (2003). Oral infection with a periodontal pathogen accelerates early atherosclerosis in apolipoprotein E-null mice. *Arterioscler. Thromb. Vasc. Biol.* **23**(8), 1405–1411. (see color insert)

The key question to address is: what are the mechanisms underlying these observations? We postulated that direct invasion of the vasculature by pathogenic bacteria accounted, at least in part, for these findings. Consistent with this premise, PCR on DNA prepared from aortic tissue at age 17 weeks (8 weeks after infection) revealed transcripts encoding *P. gingivalis* in two of nine infected mice; in contrast, none of the PBS-treated mice demonstrated transcripts for *P. gingivalis* in the aorta. Direct evidence for enhanced vascular inflammation was obtained by assessment of proinflammatory/prothrombotic markers in the vasculature. By Western blotting, levels of vascular cell adhesion molecule-1 (VCAM-1) and tissue factor antigen were significantly increased in infected versus noninfected aortic lysates. In parallel, levels of matrix metalloproteinase-2 antigen and activity were increased in aortas retrieved from *P. gingivalis*-infected mice versus controls. These findings suggested that either by direct invasion of the vessel wall, or by exposure to inflammation-evoking bacterial products, or both, the vasculature of infected mice expressed increased inflammatory molecule profile and activity.

Strongly suggestive of activation of the systemic host response by infection with *P. gingivalis* was the finding that statistically significant increases in levels of plasma IL-6, an acute phase reactant, were evident in *P. gingivalis*-infected mice versus PBS-treated controls (Figure 2a). A significant correlation between the levels of IL-6 and extent of atherosclerotic lesion area was observed in these animals (Figure 2b; $r = 0.37$, $p = 0.007$). Further, a trend in plasma levels of another acute phase reactant, serum amyloid A, was observed in infected versus noninfected apo E null mice at age 17 weeks.

Critical to the premise that oral infection accelerated atherosclerosis was the observation that infection with *P. gingivalis* did not modify established risk factors for atherogenesis

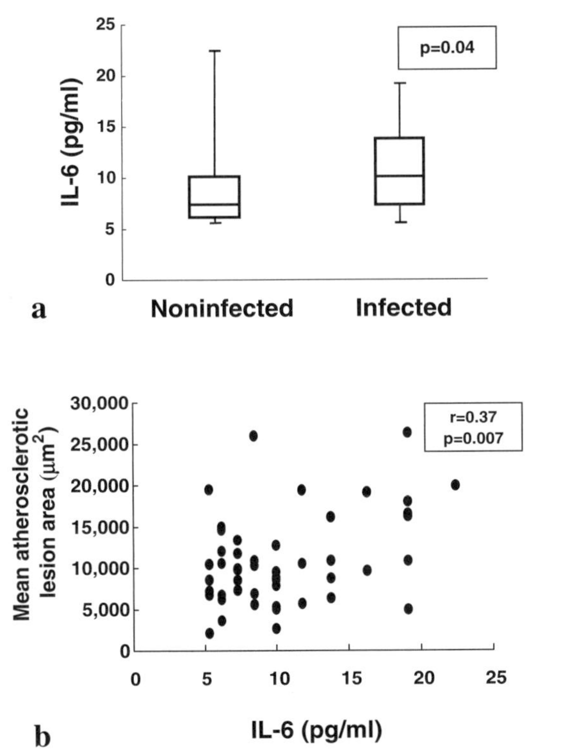

Figure 2 Oral infection and the systemic host response. **(a)** Box-and-whisker plot. At sacrifice, levels of plasma IL-6 by ELISA were significantly increased in infected apo E-null mice versus controls. Horizontal lines represent median values (noninfected: 7.3 pg/mL versus infected: 10 pg/mL). Top and bottom of each box mark the first and third quartile of the values and lines extending from the end of each box mark the minimum and maximum for the group. Nonparametric analysis, $p = 0.04$, $n = 25$/group. **(b)** Pearson's correlation analysis revealed a statistically significant positive correlation between levels of plasma IL-6 and extent of atherosclerotic lesion area ($r = 0.37$, $p = 0.007$; $n = 50$). Republished with permission of *Arteriosclerosis Thrombosis and Vascular Biology* from: Lalla, E., et al. (2003). Oral infection with a periodontal pathogen accelerates early atherosclerosis in apolipoprotein E-null mice. *Arterioscler. Thromb. Vasc. Biol.* **23**(8), 1405–1411.

in this model. Thus, levels and profile of total plasma cholesterol and triglyceride did not differ between infected and noninfected mice. Similarly, no differences in glucose, insulin, or creatinine were observed between the two groups.

Where do we go from here? The next series of studies must examine the contribution of each cellular component implicated in atherogenesis and progression of atherosclerosis to the acceleration of vascular disease triggered by infection with *P. gingivalis*. Specifically, the impact of circulating monocytes and lymphocytes in diverse processes such as engulfment/phagocytosis of bacteria and antigen presentation/processing must be studied. Resident cells of the vessel wall must also be examined, such as endothelial cells and smooth muscle cells. Interestingly, it is also possible that platelets interface with bacterial pathogens such as *P. gingivalis* to provide a hospitable environment for vascular activation, perhaps by increasing expression of chemokines such as MCP-1. In addition, the potential impact of distinct oral pathogens on atherosclerosis should be addressed experimentally.

The importance of such endeavors cannot be overstated. Only by a thorough understanding of the molecular and biochemical pathways by which bacterial infection activates signaling pathways linked to cellular migration/activation can logical therapeutic targets be identified. In this context, recent studies have shown that immunization against *P. gingivalis* may reduce acceleration of atherosclerosis in a murine model [5].

Other Systemic Diseases and Acceleration of Microvascular Dysfunction and Atherosclerosis

A common systemic disease linked to acceleration of vascular pathology and exaggerated periodontal disease is diabetes. Diabetes is a heterogeneous group of disorders affecting millions of people in the United States. Worldwide, the incidence of diabetes is burgeoning. Multiple sequelae of this disorder lead to a wide range of complications in target tissues such as the retina, kidney, peripheral nerves, and micro- and macrovasculature. Multiple epidemiologic studies have definitively linked the incidence and severity of atherosclerosis to diabetes.

A key defining feature of diabetes is the presence of microvascular dysfunction. Earlier studies showed that enhanced vascular permeability, as measured by tissue–blood isotope ratios in human diabetic subjects, was associated with increased risk of cardiovascular events. Flow-mediated dilation is abnormal in both types 1 and 2 diabetes and often has been found to be reflective of diffuse atherosclerosis. In addition to heightened vascular inflammation, enhanced oxidant stress in diabetes has been ascribed a central role in the development of vascular pathology in diabetes. Increased generation of reactive oxygen species may lead to excessive generation of superoxide; in certain settings such as diabetes, this may lead to enhanced generation of peroxynitrite and concomitant tissue injury. In this context, flow-mediated dilation has suggested that treatment of type 1 diabetic subjects with vitamin E resulted in enhanced vascular function by this measure. How such improvement relates to long-term improvement in atherosclerosis and the risk of coronary/cerebral events is not well delineated, especially since clinical trials of antioxidant use have failed to provide objective evidence of improved survival or decreased morbid events.

What is the link to oral pathology? Epidemiologic studies have strongly suggested that diabetes increases the prevalence and severity of periodontitis. Our laboratory developed a murine model with which to test these concepts. Oral/anal inoculation with *P. gingivalis* of mice rendered diabetic with streptozotocin enhanced alveolar bone loss in hyperglycemic animals, in parallel with increased expression of inflammatory and gelatinolytic mediators in the gingival tissue. Multiple pathogenic mechanisms underlie diabetes-accelerated periodontal disease, including the chronic accumulation of advanced glycation end products (AGEs), the products of nonenzymatic glycation/oxidation of proteins/lipids that may form in settings of hyperglycemia, oxidant stress, and inflammation. AGEs engage a central receptor in the vasculature, receptor for AGE (RAGE). Our previous studies have shown that pharmacological blockade of this receptor, using the extracellular soluble form of RAGE, suppresses accelerated alveolar bone loss in streptozotocin-induced diabetic mice infected with *P. gingivalis* compared to vehicle-treated infected mice. In parallel, gingival levels of inflammatory mediators such as IL-6 and TNF-α were significantly reduced by RAGE blockade.

These considerations underscore the concept that RAGE is a central mechanism amplifying the inflammatory response. RAGE is expressed in a wide array of cell types, thereby potentially implicating this receptor in a broad range of immune/inflammatory disorders. In this context, we and others have shown that in addition to AGEs, proinflammatory S100/calgranulins are signal transduction ligands of RAGE. S100/calgranulins trigger activation of MAP kinases and activation of NF-κB via RAGE. In atherosclerotic lesions, AGEs and S100/calgranulins accumulate to enhanced degrees. When tested in rodent models of vascular dysfunction in diabetes, we demonstrated that pharmacological blockade of RAGE was associated with decreased tissue blood isotope ratio in diabetic rats. In apo E-null mice rendered diabetic with streptozotocin, blockade of RAGE attenuated early acceleration and late progression of atherosclerosis [6, 7].

"Closing the Loop" on RAGE, Periodontal Disease, and Vascular Inflammation

These data highlight the concept that a key common denominator in infected periodontium and diabetic blood vessels is enhanced inflammation. Our findings on the role of RAGE blockade in attenuating inflammation and tissue injury in both settings place RAGE at the center of a cascade of systemic/local inflammatory conditions that may accelerate microvascular dysfunction. We thus propose that periodontal disease presents a sufficient stimulus to augment both local and systemic vascular activation—processes that, if left unchecked, may predispose to acceleration of atherosclerosis. In the setting of diabetes, however, inflammation in local and vascular tissues is further fueled, at least in part via ligation of RAGE, by accumulation of AGEs, proinflammatory molecules such as S100/calgranulins, and oxidant species.

Taken together, these considerations underscore the premise that control of the exaggerated inflammatory response in bacterial infection of the periodontium and in the diabetic vessel wall may provide therapeutic benefit in vascular disease and atherosclerosis.

Glossary

Advanced glycation end products: Products of nonenzymatic glycation and oxidation of proteins and lipids; these species may be formed by such stimuli as hyperglycemia, inflammation, and oxidant stress.

Flow-mediated dilation: Measure of endothelial function and the bioavailability/activity of nitric oxide; flow-mediated dilation is generally measured by assessment of blood flow in the brachial artery by ultrasonography.

Periodontium: The tissues that invest and support the teeth including the gingiva, cementum (covering the roots), alveolar bone, and periodontal ligament (the ligament that attaches the tooth to the bone).

References

1. Beck, J. D., and Offenbacher, S. (2001). The association between periodontal diseases and cardiovascular diseases: A state-of-the-science review. *Ann. Periodontol.* **6**, 9–15. *A review of the evidence for an association between cardiovascular and periodontal diseases.*
2. De Nardin, E. (2001). The role of inflammatory and immunological mediators in periodontitis and cardiovascular disease. *Ann. Periodontol.* **6**, 30–40. *A comprehensive review of host-related mechanisms that may explain the association between cardiovascular and periodontal diseases.*
3. Beck, J. D., Elter, J. R., Heiss, G., Couper, D., Mauriello, S. M., and Offenbacher, S. (2001). Relationship of periodontal disease to carotid artery intima-media wall thickness: The atherosclerosis risk in communities (ARIC) study. Arterioscler. *Thromb. Vasc. Biol.* **21**, 1816–1822.
4. Desvarieux, M., Demmer, R. T., Rundek, T., Boden-Albala, B., Jacobs, D.R., Jr., Papapanou, P. N., and Sacco, R. L. (2003). Relationship between periodontal disease, tooth loss, and carotid artery plaque: The Oral Infections and Vascular Disease Epidemiology Study (INVEST). *Stroke* **34**, 2120–2125.
5. Li, L., Messas, E., Batista, E. L., Jr., Levine, R. A., and Amar, S. (2002). Porphyromonas gingivalis infection accelerates the progression of atherosclerosis in a heterozygous apolipoprotein E-deficient murine model. *Circulation* **105**, 861–867.
6. Park, L., Raman, K. G., Lee, K. J., Lu, Y., Ferran, L. J., Jr., Chow, W.S., Stern, D., and Schmidt, A. M. (1998). Suppression of accelerated diabetic atherosclerosis by the soluble receptor for advanced glycation endproducts. *Nat. Med.* **4**, 1025–1031.
7. Bucciarelli, L. G., Wendt, T., Qu, W., Lu, Y., Lalla, E., Rong, L. L., Goova, M. T., Moser, B., Kislinger, T., Lee, D. C., Kashyap, Y., Stern, D. M., and Schmidt, A. M. (2002). RAGE blockade stabilizes estab-

lished atherosclerosis in diabetic apolipoprotein E-null mice. *Circulation* **106**, 2827–2835.

Further Reading

Amar, S., Gokce, N., Morgan, S., Loukideli, M., Van Dyke, T. E., and Vita, J. A. (2003). Periodontal disease is associated with brachial artery endothelial dysfunction and systemic inflammation. *Arterioscler. Thromb. Vasc. Biol.* **23**, 1245–1249. *The authors report that human subjects with severe periodontal disease exhibit endothelial dysfunction (decreased flow-mediated dilation) and increased serum CRP.*

Brevetti, G., Silvestro, A., Di Giacomo, S., Bucur, R., Di Donato, A., Schiano, V., and Scopacasa, F. (2003). Endothelial dysfunction in peripheral arterial disease is related to increase in plasma markers of inflammation and severity of peripheral circulatory impairment but not to classic risk factors and atherosclerotic burden. *J. Vasc. Surg.* **38**, 374–379.

Lalla, E., Lamster, I. B., Feit, M., Huang, L., and Schmidt, A. M. (1998). A murine model of accelerated periodontal disease in diabetes. *J. Periodont. Res.* **33**, 387–399.

Lalla, E., Lamster, I. B., Feit, M., Huang, L., Spessot, A., Qu, W., Kislinger, T., Lu, Y., Stern, D. M., and Schmidt, A. M. (2000). Blockade of RAGE suppresses periodontitis-associated bone loss in diabetic mice. *J. Clin. Invest.* **105**, 1117–1124.

Lalla, E., Lamster, I. B., Hofmann, M. A., Bucciarelli, L., Jerud, A. P., Tucker, S., Lu, Y., Papapanou, P. N., and Schmidt, A. M. (2003). Oral infection with a periodontal pathogen accelerates early atherosclerosis in apolipoprotein e-null mice. *Arterioscler. Thromb. Vasc. Biol.* **23**, 1405–1411.

Manson, J. E., Hsia, J., Johnson, K. C., Rossouw, J. E., Assaf, A. R., Lasser, N. L., Trevisan, M., Black, H. R., Heckbert, S. R., Detrano, R., Strickland, O. L., Wong, N. D., Crouse, J. R., Stein, E., and Cushman, M. (2003). Estrogen plus progestin and the risk of coronary heart disease. *N. Engl. J. Med.* **349**, 523–534.

Sitges, M., Heras, M., Roig, E., Duran, M., Masotti, M., Zurbano, M. J., Roque, M., and Sanz, G. (2001). Acute and mid-term combined hormone replacement therapy improves endothelial function in post-menopausal women with angina and angiographically normal coronary arteries. *Eur. Heart J.* **22**, 2116–2124.

Taylor, G. (2001). Bi-directional interrelationships between diabetes and periodontal diseases: An epidemiologic perspective. *Ann. Periodontol.* **6**, 99–112. *A comprehensive review of the evidence for the association between diabetes mellitus and periodontal diseases.*

Yan, S. F., Ramasamy, R., Naka, Y., and Schmidt, A. M. (2003). Glycation, inflammation, and RAGE: a scaffold for the macrovascular complications of diabetes and beyond. *Circ. Res.* **93**, 1159–1169. *An overview of the role of the receptor for AGE in diabetes, atherosclerosis, and vascular dysfunction.*

Capsule Biography

Evanthia Lalla, an Associate Professor of Dentistry at Columbia University, has been studying mechanisms underlying the association between diabetes, atherosclerosis, and periodontal infections.

Panos N. Papapanou and Ira B. Lamster are Professors of Dentistry at Columbia University and established investigators in the fields of epidemiology and pathogenesis of periodontal diseases.

Ann Marie Schmidt is a Professor of Surgery. Her laboratory primarily focuses on vascular biology and the receptor for advanced glycation end products.

Their work is supported by grants from the NIH, JDRFI, AHA, and Burroughs Wellcome Fund.

CHAPTER 127

The Possible Roles of Vascular Endothelial Growth Factor in Periodontal Disease

[1]Kanyarat Suthin, [2]Miho Machigashira, and [2]Yuichi Izumi

[1]*Prince of Songkla University, Thailand*

[2]*Kagoshima University, Japan*

Definition

Vascular Endothelial Growth Factor (VEGF) is considered to be a potent inducer of vascular permeability and angiogenesis and has an effect on a variety of physiological and pathological processes including embryogenesis, wound healing, tumor growth, ocular neovascular disease, myocardial ischemia, and chronic inflammatory diseases, such as rheumatoid arthritis and periodontal disease. During the progression of periodontal disease inflamed tissue enhances the expression of inflammatory mediators, which in their turn promote angiogenesis, possibly due to the role of VEGF. Since the increases of vascular permeability and angiogenesis are characteristic features of wound healing, VEGF has been suggested to have a role in the wound-healing process.

Introduction

Vascular endothelial growth factor is a glycoprotein, which shares homology with platelet-derived growth factor and placenta growth factor (PlGF). It is a macromolecule that has recently attracted attention as a potent inducer of vascular permeability and angiogenesis. VEGF has also been known as vascular permeability factor. Other possible, but less-studied, roles include lymphangiogenesis, hypotensive effects in vivo, and stimulation of tissue factor production and monocyte chemotaxis.

VEGF has been found to stimulate angiogenesis and has an important role in the development of embryonic blood vessels. This has been confirmed by findings that the inactivation of a VEGF gene in mice can lead to impairment of angiogenesis and vasculogenesis. While angiogenesis is regulated by many cytokines and growth factors such as transforming growth factor-α (TGF-α), transforming growth factor-β (TGF-β), hepatocyte growth factor, tumor necrosis factor-α (TNF-α), prostaglandin E_2 (PGE_2), angiogenin, and interleukin-8 (IL-8), VEGF is considered to be among its most potent inducers.

Periodontal disease is an infection of highly vascularized supporting tissues of the teeth usually leading to loss of bone and periodontal ligament. Persistent granulation and aberrant angiogenesis in gingiva have often been observed in periodontal disease, and these phenomena may contribute significantly to its pathogenesis. There is probably an association of VEGF with the pathogenesis of periodontal disease, particularly disorder in infected gingiva, which is accompanied by heavy neovascularization. Furthermore, in the recovery phase of inflamed periodontal tissue, sufficient supply of blood and angiogenesis is required. Therefore, VEGF might be useful for the cure of that tissue.

VEGF Family

Several members of the VEGF family have been described during the past few years, including VEGF-A, PlGF, VEGF-B, VEGF-C, VEGF-D, and VEGF-E. VEGF-

A is the original human VEGF and has multiple roles in the human body. All members of the family share the ability to enhance endothelial cell proliferation but are thought to have different roles in vascularization; VEGF-B is related to vascularization of skeletal muscle, and PlGF to that of the placenta. VEGF-E is the viral homolog of VEGF. In addition, VEGF-B and VEGF-D are associated with bone and tooth development, VEGF-C with lymphangiogenesis, and VEGF-A, VEGF-B, and VEGF-C have a role in tumor angiogenesis.

Alternative splicing from a single gene containing eight exons gives rise to at least five different isoforms of VEGF-A, containing 121, 145, 165, 189, and 206 amino acid residues. The isoforms are differentiated by the presence of peptides encoded by exons 6 and 7 of the VEGF gene. $VEGF_{206}$ and $VEGF_{189}$ contain both exons 6 and 7 (but differ in the number of amino acids), $VEGF_{165}$ contains exon 7 only, $VEGF_{145}$ contains exon 6 only, and $VEGF_{121}$ lacks both exons. The best studied VEGF isoforms are $VEGF_{121}$, $VEGF_{165}$, and $VEGF_{189}$, because of their abundance and the fact that they are usually produced simultaneously by VEGF-producing cells. $VEGF_{145}$ and $VEGF_{206}$ are rarer and less studied forms.

The VEGF isoforms differ in their heparin binding affinities. $VEGF_{121}$ does not bind to heparin, whereas both $VEGF_{165}$ and $VEGF_{189}$ do, because of the presence of exons 6 and 7. The binding abilities of $VEGF_{189}$ are higher, which has been explained by the presence of both exons.

VEGF Receptors

VEGF receptors have been classified in three groups; VEGFR-1 (flt-1), VEGFR-2 (KDR/flk-1), and VEGFR-3 (flt-4) (see Figure 1). VEGFR-1 and VEGFR-2 are predominantly expressed on vascular endothelial cells. VEGFR-3 is expressed on vascular and lymphatic endothelial cells during embryogenesis, but at later stages is expressed only on lymphatic endothelium. The receptors possess intracellular domains containing ligand-stimulatable tyrosine kinase, a single hydrophobic transmembrane-spanning membrane, and extracellular domains comprising seven immunoglobulin-like loops. Extracellular domains of VEGFR-3 contain a disulfide bridge, whereas VEGFR-1 and VEGFR-2 do not, which may explain their different binding abilities.

VEGF-A reveals high-affinity binding to both VEGFR-1 and VEGFR-2. VEGF-B and PlGF bind with high affinity to VEGFR-1 only, whereas VEGF-E binds to VEGFR-2 only. VEGF-C and VEGF-D bind to VEGFR-2 and VEGFR-3.

In addition, some types of VEGF bind to co-receptors; PlGF binds to neuropilin-1, and $VEGF_{165}$ to neuropilin-1 and neuropilin-2. The association of $VEGF_{165}$ with neuropilin-1 has been reported to increase the affinity of $VEGF_{165}$ with VEGFR-2 about tenfold, resulting in $VEGF_{165}$ being the strongest signal transducer among the VEGF subtypes.

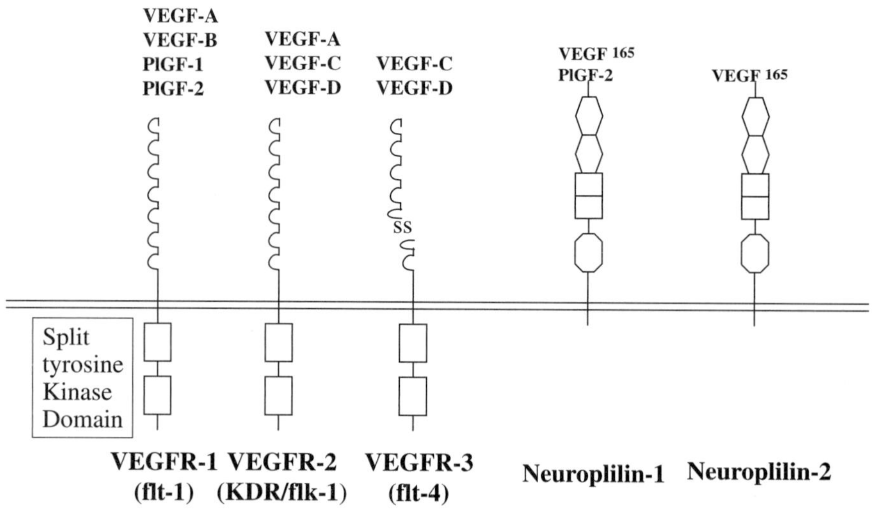

Figure 1 Family of VEGF receptors. The diagram illustrates the interaction of VEGF with the family of VEGF receptors, VEGFR-1 (flt-1), VEGFR-2 (KDR/flk-1), and VEGFR-3 (flt-4). The receptors possess intracellular domains containing ligand-stimulatable tyrosine kinase, a single hydrophobic transmembrane-spanning membrane, and extracellular domains comprising seven immunoglobulin-like loops. Extracellular domains of VEGFR-3 contain a disulfide bridge, whereas VEGFR-1 and VEGFR-2 do not. VEGF-A binds to both VEGFR-1 and VEGFR-2. VEGF-B and PlGF bind with high affinity to VEGFR-1 only, whereas VEGF-E binds to VEGFR-2 only. VEGF-C and VEGF-D bind to VEGFR-2 and VEGFR-3. Neuropilin is a co-receptor for some types of VEGF: PlGF binds with neuropilin-1, and $VEGF_{165}$ with neuropilin-1 and neuropilin-2. (G. Neufeld et al. FASEB J. 1999)

Synthesis and Regulation of VEGF

VEGF is produced by various cells, such as hepatocytes, tumor cells, smooth-muscle cells, osteoblasts, and pericytes around vascular endothelium, keratinocytes, and human dental pulp cells. It has been reported that VEGF is produced in response to various components of pathogenic bacteria in periodontal tissues within vascular endothelial cells, plasma cells, and macrophages and in junctional, sulcular, and gingival epithelium. VEGF is also detectable in human gingival fibroblasts and human periodontal ligament fibroblasts.

Several mechanisms are known to have a role in the regulation of VEGF gene expression. Hypoxia is the main stimulus for increased VEGF expression both *in vitro* and *in vivo*. VEGF mRNA expression has also been shown to be upregulated by some types of cytokines and growth factors. Thus IL-6 significantly induces VEGF in several cell lines. In fibroblastic and epithelial cells VEGF can be upregulated in response to TGF-α. In vascular smooth muscle cells basic fibroblast growth factor induces the expression of VEGF. Other types of cytokines upregulating VEGF expression include IL-1β, PGE_2, and TGF-α.

In addition, alterations in cellular regulatory pathways may result in VEGF upregulation. For example, VEGF mRNA is upregulated during the myogenic differentiation of C2C12 cells and during the conversion of 3T3 preadipocytes into adipocytes. Specific transformation is another event that can result in VEGF gene expression induction. This could include oncogenic mutations of *ras*, mutation of the murine p53 tumor suppressor gene, and overexpression of *v-raf*.

Role in Disease Processes

VEGF has an effect on a variety of physiological and pathological biological processes including embryogenesis, normal growth and differentiation, wound healing, tumor growth, ocular neovascular disease, myocardial ischemia, and chronic inflammatory diseases such as rheumatoid arthritis and periodontal disease.

VEGF mRNA is significantly upregulated in the majority of human tumors such as lung carcinomas, thyroid carcinomas, breast carcinomas, gastrointestinal tract tumors, urinary tract tumors, female reproductive tract tumors, germ cell tumors, angiosarcoma, and some intracranial tumors. Postoperative surgeries have indicated that the relapse-free survival rate of patients with VEGF-rich tumors is significantly lower than that of VEGF-poor tumors. Patients with VEGF-positive tumors have a worse prognosis than those with VEGF-negative tumors.

Diabetes mellitus or occlusion of central retinal vein can be associated with intraocular neovascularization. The new blood vessels may lead to vitreous hemorrhage, retinal detachment, neovascular glaucoma, and eventual blindness. VEGF is a possible candidate for a mediator of intraocular neovascularization. Increases in VEGF levels in the aqueous and vitreous of eyes with proliferative retinopathy have been observed. VEGF is also involved in the pathogenesis of rheumatoid arthritis (RA). The RA synovium is characterized by the formation of pannus, an extensively vascularized tissue that invades and destroys the articular cartilage.

Role in Periodontal Disease

During the progression of periodontal disease, the periodontal vasculature is profoundly affected and there is evidence that inflamed tissue enhances the expression of inflammatory mediators, which in their turn may promote angiogenesis. It has been found that there are greater amounts of VEGF in gingival crevicular fluid collected from clinically diseased sites than in fluid from healthy sites. VEGF production in gingival fibroblasts and periodontal ligament fibroblasts may enhance vascular permeability and the accumulation of inflammatory cells, which is similar to the inflammatory phase of periodontal disease. Therefore, VEGF may be associated with the etiology of periodontal disease, particularly specific to the gingiva, which is characterized by swelling, edema, gingival exudation, and heavy neovascularization.

Porphyromonas gingivalis and *Actinobacillus actinomycetemcomitans* are major etiologic agents of periodontal disease. They contain a number of cell surface bioactive components such as fimbriae, extracellular vesicles, lipopolysaccharide, and outer membrane protein. These components induce numerous inflammatory mediators, such as prostaglandins, IL-8, and TNF-α, from human peripheral mononuclear cells and fibroblasts in periodontal tissue and are involved in the progression of periodontal disease. Recent research shows that gingival fibroblasts and periodontal ligament fibroblasts express VEGF in response to periodontopathic bacteria.

The possibility that VEGF is produced by gingival fibroblast and periodontal ligament fibroblasts has important implications for the study of periodontal disease and its progression, as well as the healing process that follows periodontal treatment.

Therapeutic Applications

VEGF is known to be expressed by some types of tumor cells. VEGF production stimulates the growth of endothelial cells in the tumor and thus enhances its further growth. Consequently it has been shown that the suppression of VEGF formation could suppress the tumor growth through containing endothelial cells development. Different approaches to suppressing tumor angiogenesis include the use of neutralizing antibodies to VEGF and the development of drugs that inhibit VEGF receptor kinase activity.

The promotion of new vessel growth by the agent could be useful in the treatment of disorders characterized by inadequate tissue perfusion and might give an alternative to surgical reconstruction procedures. Chronic limb ischemia, one of the diseases caused by obstructive atherosclerosis affecting the superficial femoral artery, is related to a high rate of morbidity and mortality and is limited to surgical revascularization procedures. The development of collateral vessels has been observed after a single intra-arterial administration of $rhVEGF_{165}$ in rabbits with chronic hind limb ischemia. There is a possibility that the recombinant VEGF or gene therapy with VEGF genes may be used to promote endothelial growth and collateral vessel formation, as well as to prevent occlusive artery and ischemic heart and limb disease.

VEGF also has an effect on periodontal healing. There are several kinds of cells related to the healing process of periodontal disease, such as periodontal ligament fibroblasts and osteoblasts. Periodontal ligament fibroblast originates in part from the ectomesenchyme of the cranial neural crest, and this developmental origin may give these cells specialized properties. In particular, pluripotential mesenchymal stem cells can differentiate into mineralized tissue-forming cells such as osteoblasts. VEGF induces periodontal ligament fibroblast to express osteogenic markers such as *c-fos*, osteopontin, and osteocalcin and stimulate bone formation by turning into osteoblast-like cells.

One of the early events of periodontal tissue healing and regeneration is angiogenesis, in which neovascularization prompts delivery of inflammatory cells and fibroblasts to the wound site. Angiogenesis, the mechanism responsible for the development of blood vessels, is an essential step in the process of healing, in particular for periodontal tissue healing and bone healing. Numerous studies have demonstrated the critical role of angiogenesis for successful osteogenesis during endochondral ossification and bone repair. It has been reported that VEGF is important for the regulation of bone remodeling by recruiting endothelial cells and osteoclasts and by stimulating osteoblast differentiation. VEGF has not only an angiogenic but also an osteogenic function by the direct stimulation of the osteoblasts through the enhanced expression of a potent osteogenic factor.

Future Perspectives

Research has established the important role VEGF plays in the development of the embryonic vascular system, in angiogenesis, and in maintaining or augmenting endothelial functions in the adult vasculature. Both positive and negative regulation of endothelial cells by VEGF can be critical for the maintenance of the vascular system.

Future research needs to concentrate on the mechanisms through which VEGF induces its effects. Research has to clarify the effect of VEGF on gene expression, the interactions between different signaling pathways in the regulation of endothelial cells, and their integration within the cell. Firmly establishing the functions of VEGF could help advance research on the use of VEGF for endothelial and periodontal regeneration. Identifying VEGF inhibitors inducing apoptosis of endothelial cells could be crucial for advances in the treatment of angiogenesis-dependent diseases such as cancer and rheumatoid arthritis.

Glossary

Angiogenesis: The formation of capillaries from preexisting vessels.

Growth factors: A diverse group of polypeptides that play important roles in regulation and development of a variety of organs.

Periodontal disease: Inflammation of highly vascularized supporting tissues of the teeth usually leading to loss of bone and periodontal ligament.

Vasculogenesis: The development of blood islands in primary vascular plexus, which is further developed and processed to form the mature vascular system.

Bibliography

Breier, G. (2000). Angiogenesis in embryonic development—a review. *Placenta* **21**(14), s11–s15.

Cochran, D. L., and Wozney, J. M. (1999). Biological mediators for periodontal regeneration. *Periodontol. 2000* **19**, 40–58. *This article summarized biological mediators related to periodontal and alveolar bone regeneration in periodontal disease.*

Deckers, M. M., Karperien, M., Van der Bent, C., Yamashita, T., Papapoulos, S. E., and Lowik, C. W. (2000). Expression of vascular endothelial growth factors and their receptors during osteoblast differentiation. *Endocrinology* **141**, 1667–1674.

Ferrara, N. (2003). The biology of VEGF and its receptors. *Nat. Med.* **9**(6), 669–676. *This article is an excellent review of the role of vascular endothelial growth factor and its receptors in angiogenesis.*

Gerwins, P., Sköldenberg, E., and Claesson-Welsh, L. (2000). Function of fibroblast growth factors and vascular endothelial growth factors and their receptors in angiogenesis. *Crit. Rev. Oncol. Hematol.* **34**, 185–194.

Johnson, R. B., Serio, F. G., and Dai, X. (1998). Vascular endothelial growth factor and progression of periodontal disease. *J. Periodontol.* **70**(8), 848–852.

Kinane, D. F., and Lindhe, J. (1998). Pathogenesis of periodontitis. In: *Clinical Periodontology and Implant Dentistry*, Lindhe, J., Karring, T., and Lang, N. P., eds., 3rd ed., pp. 189–225. Copenhagen: Munkgaard. *An excellent summary of etiology and pathogenic processes of periodontal disease.*

Zachary, I. (2001). Signaling mechanisms mediating vascular protective actions of vascular endothelial growth factor. *Am. J. Physiol. Cell Physiol.* **280**(6), C1375–C1386.

Capsule Biography

Dr. Kanyarat Suthin is head of Periodontology at the Faculty of Dentistry, Prince of Songkla University, Thailand, and her research focus is on pathogenesis and tissue regeneration. Her work is supported by grants from the Ministry of Education, Science, Sports and Culture of Japan.

Dr. Miho Machigashira is a assistant professor of the Department of Periodontology, Kagoshima University Graduate School of Medical and Dental Sciences, Kagoshima, Japan, and her research focus is on periodontal pathogens and the role and mechanism of hepatocyte growth factor (HGF) in periodontal regeneration. Her work is supported by grants from the Japan Society for the Promotion of Science.

Prof. Yuichi Izumi has headed the Department of Periodontology, Kagoshima University Graduate School of Medical and Dental Sciences, Kagoshima, Japan, since 1999. He has numerous publications in the area of pathogenesis and regeneration in periodontology, and his work is supported by numerous grants from the Ministry of Education, Science, Sports and Culture of Japan.

SECTION Q

Pollutants

CHAPTER 128

Microvascular Responses to Cigarette Smoke

Hans-Anton Lehr
Institute of Pathology, University of Mainz, Medical Center, Mainz, Germany

Cigarette smoke contains large amounts of carcinogenic and cocarcinogenic substances that explain the increased burden of malignant tumors in smokers, particularly in the lungs, the upper gastrointestinal system, and the urinary bladder. Nonneoplastic diseases associated with cigarette smoking include chronic bronchitis and pulmonary emphysema, gastrointestinal ulcers, and a significantly increased risk of cardiovascular diseases, ranging from peripheral artery disease to myocardial infarction and cerebrovascular incidences. Although the exact pathophysiological link between cigarette smoking and cardiopulmonary disease progression remains incompletely understood, it has been recognized that a key role is played by effects on diverse aspects of the microcirculation both in terms of morphological (i.e., vessel wall injury, capillary loss) and functional aspects. The latter concerns predominantly changes in tissue perfusion and its regulatory mechanisms (i.e., reactive hyperemia, sequestration of blood cells in the microcirculation). The mechanisms of action of cigarette smoking on the microcirculation include compromised endothelium-dependent vasorelaxation, platelet aggregation, endothelial cell dysfunction, the activation of circulating leukocytes, and eventually the aggregation and adhesion of leukocytes and/or platelets to the microvascular endothelium. In the following, an effort is made to review the existing knowledge on microcirculatory dysfunction(s) induced by cigarette smoking. For references, the reader is referred to a recent review on this subject [1].

Morphological Aspects of Cigarette Smoke–Induced Microvascular Dysfunction

Histomorphological studies have demonstrated thickening of the walls of pulmonary arterioles, as well as arterioles in trachea, esophagus, stomach, pancreas, and kidneys in smokers, depending on the length and severity of the smoking habit. Subsequent investigations have shown fibrous thickening in coronary and intramyocardial arterioles in smokers and in beagle dogs trained to smoke. Another organ that has been subject to intensive interest in this context is the placenta: Attempting to explain the reduced birth weight of children born to smoking mothers and the increased perinatal morbidity and mortality associated with maternal smoking, several authors have studied the placentas of smoking versus nonsmoking mothers. Smoking-related findings included a significantly reduced surface area of the fetal microvasculature, degeneration of villous capillary endothelial cells, subendothelial edema formation, and—eventually—increased syncytiotrophoblastic necrosis in placentas of smoking mothers. More recently, Lova and coworkers have demonstrated microvascular remodeling in the labial mucosa of smokers, revealing increased numbers of dystrophic capillary loops, exhibiting smaller calibers, increased tortuosity, and increased frequencies of microaneurysms [2].

Functional Aspects of Cigarette Smoke–Induced Microvascular Dysfunction

In terms of functional effects of the cigarette smoke impaired microvasculature, the accessibility of the skin microcirculation to noninvasive diagnostic techniques (i.e., plethysmography, nailfold microscopy, laser Doppler fluxmetry) has resulted in an abundant literature on the effect of cigarette smoking on skin perfusion in humans and experimental animals. These studies have found that baseline blood flow in the skin of smokers is reduced either permanently and/or after smoking one or more cigarettes. Also, reactive hyperemia is compromised in the skin of neonatal infants born to smoking mothers and in the skin of adult smokers—in terms of both peak blood flow and recovery time. The study by Asano and Branemark is of particular interest to microcirculation researchers because of its unique experimental approach: These authors implanted titanium observation chambers into the upper arm skin of human volunteers and recorded by intravital microscopy the microcirculatory flow velocity during cigarette smoke exposure [3]. Similar results have later been reproduced using intravital microscopy on the nailfold capillary bed, the labial mucosa, and the retina of smoking human subjects. Intravital microscopic studies have found that chronic exposure of rats and pigs to tobacco smoke or to oral nicotine significantly reduces red blood cell velocity and blood flow through capillaries of the mesentery, the intervertebral disc, and the testis. The situation has been found more complex for cigarette smoke effects on the cerebrovascular microcirculation. Whereas several studies have demonstrated reductions in cerebral blood flow in chronic smokers, acute cigarette smoke exposure has resulted in contradictory results in both animals and humans, with some authors reporting increases and some decreases in cerebral blood flow.

Cellular Targets of Cigarette Smoke–Induced Microvascular Dysfunction

Endothelial Cells

In the mechanistic insight into cigarette smoke–induced microcirculatory dysfunction, the endothelium appears to play a central role and has thus been called the cornerstone of smoking pathophysiology. Direct toxic effects of cigarette smoking on endothelial cells have been demonstrated in rat aortas by electron microscopy, by the demonstration of anuclear endothelial cell carcasses in the circulating blood of smokers, and by the demonstration of reduced prostacyclin production by cigarette smoke–exposed endothelial cells. Since the endothelium is involved in the regulation of vasomotor tone, intensive research endeavors have focused on the impact of cigarette smoking on endothelial function. Coronary angiographic studies have shown a significant vasoconstriction after cigarette smoking, and an increase in coronary resistance despite an increase in oxygen demand. In line with these observations, several animal experiments and clinical studies have demonstrated a significant reduction of reactive hyperemia in otherwise healthy smokers. Indeed, the attenuation of endothelium-dependent vasodilation was inversely correlated with the number of pack years smoked by the subjects rather than with blood cotinine levels. Similar findings have been demonstrated in veins and large-caliber arteries. A plausible explanation for these findings has been derived from studies on porcine coronary artery rings, where cigarette smoke extracts inactivate endothelium-derived nitric oxide (NO) through reactive oxygen–dependent mechanisms and where NO inactivation prevented subsequent acetylcholine-induced vasodilation. A similar reduction of basal NO activity has later been reported in coronary and brachial arteries of chronic smokers, and recent evidence suggests that cigarette smoke extract decreases the inducible form of NO synthase in an endothelial cell culture and that NO synthase is reduced in saphenous vein grafts explanted from chronic smokers. Several experimental and clinical observations have focused on the oxidative stress hypothesis, suggesting that reactive oxygen species either contained in cigarette smoke or released from smoke-activated neutrophils may contribute the endothelial dysfunction.

Intravital microscopy on the hamster cheek pouch model found that topical superfusion with cigarette smoke extract attenuates acetylcholine-induced dilation of resistance arterioles and that application of superoxide dismutase effectively restored endothelium-dependent arteriolar dilation after nicotine superfusion, suggesting a key role of reactive oxygen species in the induction of endothelial dysfunction. Similar results have been obtained with aqueous extracts of smokeless tobacco and the studies were later extended to the effect of cigarette smoke extract and/or nicotine on other important cellular vasodilator pathways, including adenylate cyclase and ATP-sensitive potassium channels.

Platelets

For more than 25 years it has been known that cigarette smoking induces platelet aggregation (closely linked to the serum levels of nicotine) and increases urinary excretion of thromboxane. At the same time, smoker's platelets (i) exhibit a significantly increased surface density of receptors for thromboxane and serotonin, (ii) produce increased amounts of eicosanoids, and (iii) show reduced formation of nitric oxide. Of interest, an essential role seems to be played by the lipid profile of the patients, since the most pronounced abnormalities in platelet aggregation have been described in hypercholesterolemic smokers rather than in normocholesterolemic smokers. Platelet aggregation in the bloodstream is an ubiquitous phenomenon that is not confined to either large-caliber conductance vessels or the microcirculation. In intravital microscopic studies from our laboratory, we found that exposure of hamsters to the smoke of one cigarette induced the rapid formation in the bloodstream of platelet/leukocyte aggregates and scanning electron microscopic studies demonstrated the dendritic nature of activated platelets involved in these aggregates. Platelet aggregates were also seen on the endothelial surface of the aorta studied *en face* in smoke-exposed hamsters [4]. In collaboration with Tom McIntyre and his coworkers at the University of Utah at Salt Lake City, we were able to demonstrate in the blood of smoke-exposed hamster the formation of platelet-activating factor–like lipids. In a series of ex vivo experiments, we could demonstrate that these PAF-like lipids were responsible for the induction of platelet aggregates in the hamster bloodstream [5]. Pharmacological blockade of the PAF receptor (through injection of WEB2170) or inhibition of the formation of PAF-like lipids (through the administration of the water-soluble antioxidant vitamin C) effectively prevented platelet aggregation after cigarette smoke challenge in the hamsters, emphasizing the relevance of these smoke–induced lipid mediators in the intact organism. Since similar lipids have been described to occur in human smokers, it is likely that similar pathomechanisms are operative during the induction of platelet activation in human smokers.

Leukocytes

The organism undergoes an inflammatory-type response with every cigarette smoked. This is caused either by combustion products, reactive oxygen species, and other compounds that are inhaled with the cigarette smoke, or by reactive oxygen species and inflammatory mediators released from the organism in response to cigarette smoking (particularly from phagocytic cells that are sequestered in the pulmonary microcirculation in response to cigarette smoke). This is most obviously reflected in an activation of white blood cells. Although leukocyte adhesion and emigration are involved in host defense and phagocytosis and thus serve a beneficial role during a well-contained inflammatory response, leukocytes may also turn against the host and contribute to tissue damage, characterized by the breakdown of capillary perfusion, the loss of endothelial integrity, and the extravasation of fluid and macromolecules into the interstitial space [6]. A feature common to the pathomechanisms of most cigarette smoke–associated diseases (such as atherosclerosis, chronic bronchitis, pulmonary emphysema, and periodontitis) is the activation and adhesion of circulating leukocytes to micro- and macrovascular endothelium, followed by acute or chronic leukocyte-mediated tissue damage. Peripheral blood monocytes isolated from habitual smokers demonstrate significantly increased adherence to cultured endothelial cells, positively correlated with the duration of the smoking habit. Likewise, exposure of peripheral blood mononuclear cells to cigarette smoke condensate in vitro increased cell surface expression of adhesion molecules CD11b and subsequent adhesion to cultured bovine and human endothelial cells through a protein kinase C–dependent mechanism. Direct chemotactic effects of cigarette smoke for leukocytes have been described, as well as enhanced generation of leukocyte chemotaxins or cochemotaxins including leukotrienes, complement, and Gc globulin. Also, cigarette smoke stimulates the generation of oxidatively modified low-density lipoproteins that exert powerful chemotactic and adhesion-promoting effects on. With the demonstration of all these proinflammatory changes in leukocytes, it is not surprising that leukocytes have been found sequestered in different organs, particularly in the pulmonary and tracheal microcirculation, but also in pancreatic microvessels of cigarette smoke–exposed experimental animals. Likewise, leukocyte recruitment was found in the pulmonary microcirculation of cigarette smoke–exposed experimental animals and habitual smokers. An alternative explanation for the enhanced sequestration of neutrophils in the pulmonary microcirculation has been proposed by Drost and coworkers, who demonstrated significantly reduced deformability of peripheral blood neutrophils from habitual smokers and of neutrophils exposed to cigarette smoke in vitro [7]. Cigarette smoke condensate induces functional changes in endothelial cells, leading to the increased expression of adhesion molecules for leukocyte–endothelial cell interaction. Increased levels of soluble intercellular adhesion molecule-1 (ICAM-1) have recently been found in the serum of habitual smokers. Through the inhibition of nitric oxide generation and/or its inactivation through radical-dependent mechanisms, cigarette smoking might also contribute to the stimulation of leukocyte and platelet interaction with endothelial cells. Along the same lines, the well-recognized inhibition of prostacyclin generation by cigarette smoking could predispose to increased leukocyte and platelet adhesion. In a recent study, Sikora and her colleagues have demonstrated in an elegant experimental approach using intravital microscopy on transplanted lung tissue in the dorsal skinfold chamber model in nude mice that nicotine-induced inflammation of the airways involves selectin- and MAP kinase-dependent rolling and adhesion of leukocytes [8].

In an effort to further characterize the mechanism of cigarette smoke–induced leukocyte adhesion to the microvascular endothelium, we have exposed hamsters to the smoke generated by one cigarette and could demonstrate rolling and subsequent adhesion of fluorescently labeled leukocytes to the endothelium of microvascular and macrovascular endothelium [4]. Cigarette smoke–induced leukocyte adhesion was significantly blunted in animals in which leukotriene generation was blocked pharmacologically by an inhibitor of 5-lipoxygenase, suggesting a mediator role of leukotrienes in this event. Likewise, administration of superoxide dismutase almost entirely prevented cigarette smoke–induced leukocyte adhesion. Of particular interest was the observation not only that leukocyte adhesion was confined to the venular segment of the microcirculation, but that leukocytes were found to adhere to the endothelium in arterioles, as well as on the aortic endothelium. In both microvascular segments, as well as the aortic endothelium, leukocytes tended not to interact with the endothelium as single cells, but mostly in the form of aggregates of two or more cells, loosely held together by activated platelets. Indeed, any intervention that prevented the formation of these aggregates also prevented the interaction of leukocytes in the arteriolar branch of the microcirculation. In analogy to the experiments described earlier in the section on platelets, we found that PAF-like lipids elicited the interaction of leukocytes with endothelial cells in vitro in the intact hamster organism: Pharmacological blockade of the PAF receptor (through injection of WEB2170) or inhibition of the formation of PAF-like lipids (through the administration of the water-soluble antioxidant vitamin C) effectively prevented leukocyte adhesion to venular or arteriolar endothelium in cigarette smoke-exposed hamsters [5].

Beside the stimulation of leukocyte adhesion to endothelial cells, several authors have demonstrated a breakdown of endothelial barrier function in response to cigarette smoke and/or nicotine. Ultrastructural studies have demonstrated modifications in intercellular cleft morphology, the opening of interendothelial junctions, and the formation of subendothelial edema in response to cigarette smoke. Likewise, a compromised endothelial barrier function of microvessels in the oral mucosa has been found in response to smokeless tobacco exposure. Similar observations have been made by intravital microscopy in the hamster cheek pouch and in the gastric microcirculation in rats. The latter findings are of particular importance because they provide an adequate pathophysiological explanation for the clinical observation of delayed gastric ulcer healing and increased frequency of ulcer recurrence in smokers. Indeed, laser Doppler flowmetric studies in heavy smokers have demonstrated significantly reduced mucosal blood flow in the gastric antral region, and the reduced mucosal perfusion was associated with reduced basal bicarbonate secretion.

Consequences of Cigarette Smoke–Induced Microvascular Dysfunction

Most smoke-related diseases point toward effects on the microcirculation. For one, the development of pulmonary emphysema occurs at the microcirculatory level of the delicate alveolar membranes, presumably due to a dysbalance between aggressive factors such as tissue degrading mediators and reactive oxygen species released from smoke-activated neutrophils (i.e., elastase, hydrolytic enzymes) and protective factors such as α-1 antiprotease, which are inactivated by cigarette smoke. Also, the association of cigarette smoking and chronic inflammatory conditions such as periodontal disease is a largely microcirculatory phenomenon that involves phagocyte dysfunction and other pathomechanisms. Another important problem related to cigarette smoking–induced microcirculatory dysfunction is of major concern to plastic surgeons. There is an impressive literature on the adverse effects of cigarette smoking on the outcome after surgical interventions, such as (i) a significantly higher incidence of skin necrosis after face lifts, (ii) reduced viability of transplanted skin flaps, (iii) increased graft failure after microvascular toe or finger replantation, (iv) deleterious effects on bone grafting for fracture stabilization, and (v) an increased incidence of anastomotic dehiscence after colonic resection. The authors even suggested that surgeons should have a lower threshold for using a proximal diversion in smokers in order to lessen mortality and morbidity after colonic surgery. The fact that smoking has been linked to erectile dysfunction may appear of secondary importance—at least to those who are not affected by this microcirculatory dysfunction.

Prevention of Cigarette Smoke–Induced Microvascular Dysfunction

Based on the knowledge that we have gained of the pathomechanism of cigarette smoke–induced microvascular dysfunction, we must acknowledge that while some effects may be due to nicotine, many other effects are due to the reactive oxygen species that are either introduced into the organism by the smoke per se, or else released from activated neutrophils, primarily in the lungs of smokers—in response to cigarette smoke inhalation. Based on our initial finding that cigarette smoke–induced microcirculatory dysfunction was effectively counteracted by superoxide dismutase, we tried to inhibit the generation of reactive oxygen species in the organism of a smoke-exposed animal with the antioxidants vitamin C (which acts as water-soluble antioxidant to prevent the initiation of lipid peroxidation) or vitamin E (which is lipid-soluble and acts as a chain-breaking antioxidant). When the animals were then exposed to cigarette smoke, we found that microcirculatory dysfunction was almost entirely prevented in vitamin C–treated animals, whereas no such inhibitory effect was seen with vitamin E [4]. In further studies we could demonstrate that vitamin C acted by preventing the formation of platelet activating factor–like lipids in the hamster blood and that these lipids accounted for most of the microcirculatory dysfunction secondary to cigarette smoke exposure [5]. In the past few years, our studies on the inhibitory action of vitamin C have been reproduced and translated into the human situation: Several authors have independently demonstrated that oral or parenteral application of vitamin C to human subjects restores cigarette smoke–induced endothelial dependent vasoregulatory dysfunction in the calf, the forearm, and the nailfold microcirculation. It remains to be seen whether these interventions will eventually reduce cigarette smoke–associated pathology.

Glossary

Cigarette smoke: Cigarette smoke can be divided into two phases, gas-phase smoke and tar (which gets trapped in filter pores). Tar contains high concentrations of radicals, which are sufficiently stable to be detectable by electron spin resonance. Many of the components of cigarette tar are water soluble; hence, aqueous extracts of cigarette tar reduce oxygen to form activated oxygen species that can cause biological damage. In contrast, the radicals in the gas phase are very short-lived, yet entertain a remarkable steady state, yielding high concentrations even in "aged" smoke. Many of these radicals enter the lungs of smokers and avidly interact with biological matter, as becomes manifest in the rapid inactivation of α-1 proteinase inhibitor—a key mechanism of cigarette smoke–induced emphysema formation.

PAF-like lipids: Platelet-activating factor (PAF) is a potent proinflammatory phospholipid with diverse pathological and physiological effects. This bioactive phospholipid mediates processes as diverse as wound healing, inflammation, ischemia–reperfusion injury, angiogenesis, and reproduction. As first described by a team of researchers at the University of Utah, the unregulated production of structural analogs of PAF by nonspecific oxidative reactions has expanded the superfamily of signaling molecules to include "PAF-like" lipids whose mode of action is identical to that of authentic PAF. Inappropriate activation of this signaling pathway is associated with many diseases in which inflammation is thought to be one of the underlying features.

Vitamin C: Vitamin C is a water-soluble vitamin that is necessary for normal growth and development. It is required in the synthesis of collagen in connective tissue, neurotransmitters, steroid hormones, carnitine, and the conversion of cholesterol to bile acids, and it enhances iron bioavailability. Vitamin C plays a paramount role as an antioxidant and free radical scavenger and helps protect the body against the aggressive action of reactive oxygen species.

References

1. Lehr, H. A. (2000). Microcirculatory dysfunction induced by cigarette smoking. *Microcirculation* **7**, 367–384.
2. Lova et al. (2002).
3. Asano, M., and Branemark, P. I. (1970). Cardiovascular and microvascular responses to smoking in man. *Adv. Microcirc.* **3**, 152–158. *This article is outstanding in the field of microcirculation research, being the first study of the human microcirculation by intravital microscopy using an implanted observation chamber. Implantation of the observation chamber was performed in the delicate, rather loose skin in the thoracic aspect of the upper arm in volunteers. Using this model, Asano could demonstrate for the first time significant changes in blood flow through microvessels in human volunteers during cigarette smoke exposure.*
4. Lehr, H. A., Frei, B., and Arfors, K. E. (1994). Vitamin C prevents cigarette smoke–induced leukocyte aggregation and adhesion to endothelium *in vivo*. *Proc. Natl. Acad. Sci. USA* **91**, 7688–7692. *Lehr and coworkers described in a paper in Free Radical Biology & Medicine for the first time the effect of cigarette smoke on the microcirculation, involving the generation and action of superoxide dismutase. In the PNAS paper from 1994, feeding hamsters for 1 week with a vitamin C–supplemented diet entirely prevented cigarette smoke–induced leukocyte adhesion in postcapillary venules. No such beneficial effect was seen for the lipid-soluble antioxidant vitamin E. As explanation for these observations, the authors proposed that vitamin C scavenges oxidant species that enter the bloodstream through the cigarette smoke and hence prevents their action in the microcirculation (see Ref. [5]).*
5. Lehr, H. A., Weyrich, A. S., Jurek, A., Saetzler, R. K., Arfors, K. E., Zimmermann, G. A., Prescott, S. M., McIntyre, T. M. (1997). Vitamin C blocks antiinflammatory platelet-activating factor like lipids created by cigarette smoking. *J. Clin. Invest.* **99**, 2358–2364. *Based on the landmark observation of PAF-like lipids (PAF-LL) by the group of Tom McIntyre and coworkers that are generated by the attack of reactive oxygen species on phospholipids, Lehr and coworkers went on to study whether the inhibitory effects on vitamin C in their smoking hamster model was linked to the formation of PAF-LL in the bloodstream and their inhibition by vitamin C. The two groups teamed up to demonstrate that indeed cigarette smoke exposure generates in the bloodstream of smoke-exposed hamsters the formation of a factor that elicits leukocyte adhesion and leukocyte–platelet aggregation through a mechanism that is both dependent on the PAF receptor and inhibited by vitamin C. The demonstration that the smoke induced adhesive action was blocked by PAF receptors in vitro and in vivo alike suggests that indeed PAF-LL are formed in an intact animal organism in response to free radical attack.*
6. Lehr, H. A., and Arfors, K. E. (1994). Mechanisms of tissue damage by leukocytes. *Curr. Opin. Hematol.* **1**, 92–99.
7. Drost, E. M., Selby, C., Bridgeman, M. M., and MacNee, W. (1993). Decreased leukocyte deformability after acute cigarette smoking in humans. *Am. Rev. Respir. Dis.* **148**, 1277–1283.
8. Sikora et al. (2003).

Capsule Biography

Dr. Hans-Anton Lehr has been involved for almost 15 years in microcirculation research, predominantly by intravital fluorescence microscopy using the dorsal skinfold chamber model in hamsters and mice. His studies have focused on questions of ischemia–reperfusion injury, atherogenesis, cigarette smoke–induced microcirculatory dysfunction, and, lately, tumor angiogenesis. Dr. Lehr is a diagnostic pathologist currently acting as Professor of Pathology at the University of Mainz Medical Center, Germany.

CHAPTER 129

Effects of Cigarette Smoke on Rheological Properties of Leukocytes Influencing Their Passage through the Microvasculature

[1]G. B. Nash and [2]W. MacNee
[1]*The Medical School, The University of Birmingham, Birmingham, England*
[2]*The Medical School, University of Edinburgh, Edinburgh, Scotland*

Introduction

This article discusses how exposure to cigarette smoke or substances derived from it may change the properties of leukocytes, or other cells of the vasculature, in such a way that circulation of leukocytes is modified. These changes are thought to contribute to smoking-related pathology, especially in the lungs. Although emphasis is placed on modifications affecting the microcirculation, the processes discussed may also influence arterial disease.

Cigarette Smoking and Microvascular Pathology

Cigarette smoking is recognized as an etiological factor in arterial diseases, possibly through the ability of components of the smoke to cause damage to the vessel wall and to modify functions of the vascular endothelial cells. The ability of cigarette smoke (CS) to promote the inflammatory response causing atheroma of large arteries is also manifest in the link between cigarette smoking and microvascular pathology, in organs such as the lungs, kidney, retina, and gastrointestinal tract. In the lungs, cells of the pulmonary microvasculature and of the blood may be functionally modified by direct exposure to components of cigarette smoke. In remote organs, dissolved components of the smoke, modified plasma constituents, or substances released from the pulmonary cells may be responsible for changes in the local functions of microvascular endothelium. However, it is also possible that leukocytes exposed to products from CS in the blood not only cause local problems in the lungs, but are themselves carried to other sites and affect the microvasculature there. In principle, this could occur if the rheological properties (either mechanical or adhesive) of leukocytes are modified, so that microvascular transit is impaired.

Rheological Properties of Leukocytes Potentially Sensitive to Cigarette Smoking

MECHANICAL RESISTANCE TO DEFORMATION

In general, resistance of both red and white blood cells to deformation depends on the cell geometry (volume and surface-area-to-volume ratio) and on the separate viscoelastic properties of the membrane and of the cytoplasmic structures. Cells must have an excess of membrane surface area over the minimum required to enclose the cell volume; otherwise they cannot adapt their shape without expansion of surface area or decrease in volume. Neither change is possible on the time scales of microvessel transit. Leukocytes have a large excess surface area in the form of microvilli and surface folds. The limiting factor in their rate of passive deformation arises from resistance elements in their cytoplasm, especially the cytoskeleton, which is rich in polymerized actin filaments (F-actin). The nucleus is also relatively rigid, but this appears to be a more important element in mononuclear cells than in polymorphonuclear granulocytes. The internal structures of leukocytes make them much more resistant to deformation than red cells, which lack a nucleus and true cytoskeleton.

In general, passage of leukocytes through capillaries is achieved by passive deformation under the action of the local blood pressure gradient. The leukocytes enter small capillaries much more slowly than red blood cells and may cause intermittent temporary occlusions of capillaries or hold up trains of following red cells [1]. In the case of the pulmonary microcirculation, all of the blood leukocytes pass through repeated capillary segments at relatively low pressure. The average pulmonary capillary diameter is 5 μm, compared to the average diameter of a polymorphonuclear granulocyte of about 8 μm. Thus a proportion of neutrophils (the most numerous granulocytes) are delayed in transit in the pulmonary capillary bed (illustrated in Figure 1), which provides a noncirculating or slowly circulating pool of neutrophils that can be mobilized in times of stress.

The slow transit through microvessels may be hindered if local pressure gradients are decreased (e.g., in shock), or if the leukocytes, especially neutrophilic granulocytes, become "activated" by chemical stimuli that increase their rigidity [1]. On neutrophil activation, a rapid increase in F-actin content followed by rearrangement of F-actin into pseudopodia is associated with greatly increased rigidity. Shape changes induced by activation are illustrated in Figure 1. In this connection, one must be clear about the difference between passive resistance to deformation, and active deformation required for migration. Structural changes in the cytoskeleton allow leukocytes to change shape actively and move slowly over and through endothelial cell monolayers (see the following subsection). These changes will typically occur in response to chemotactic agents such as chemokines or bacterial products. The structural changes, however, actually make the cell more rigid in the sense that

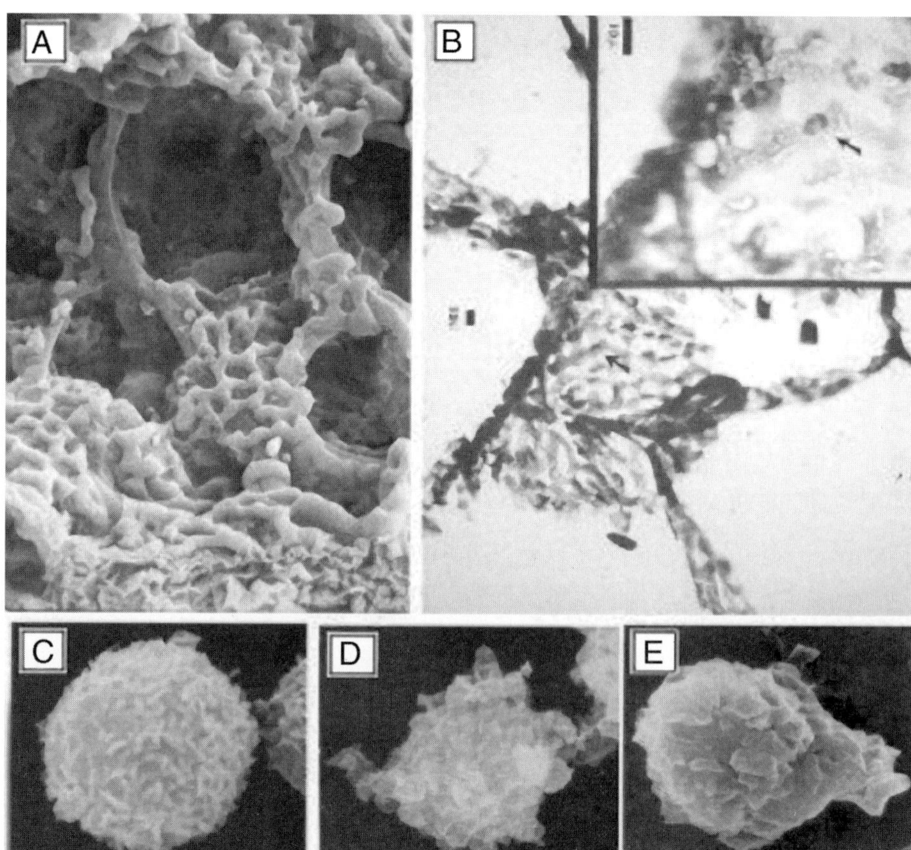

Figure 1 Morphology of the pulmonary microcirculation and of resting and stimulated neutrophils. **(A)** Scanning electron micrograph of alveoli, showing walls lined with many capillary segments. **(B)** Cross section of alveoli and interstitium, showing capillaries and (*arrow*) a neutrophil (with segmented nucleus) in transit. **(C, D, E)** Scanning electron micrographs of neutrophils resting (unstimulated), exposed to cigarette smoke, or treated with bacterial peptide (fMLP), respectively. (see color insert)

larger external forces will be required to make them flow through a narrow capillary, on a much faster time scale. Thus active cytoskeletal rearrangements are appropriate for leukocyte migration out of vessels, but impair passive flow of cells through capillary lumens.

LEUKOCYTE ADHESION AND MIGRATION

In addition to mechanical resistance to deformation, transit of leukocytes depends on any tendency to adhere to the endothelium lining microvessels. Adhesion and migration into tissue typically occur in the postcapillary venules, as part of the protective inflammatory response to infection and trauma. However, uncontrolled adhesion has the potential to occlude microvessels and to cause damage to the vessel wall and surrounding tissue. In the pulmonary microcirculation, adhesion and migration of leukocytes, particularly neutrophils, tends to occur in the pulmonary capillaries rather than in venules [2].

The process whereby leukocytes move from the bloodstream into inflamed tissue is shown schematically in Figure 2 and can be divided into distinct steps [2]. In the case of neutrophils, flowing cells are captured through fast-forming bonds between selectin receptors (E- and P-selectin on endothelial cells and L-selectin on neutrophils) and protein-borne carbohydrate ligands. These bonds support a rolling form of adhesion, slowing the neutrophils down. Neutrophils can then interact with chemotactic agents such as platelet-activating factor (PAF) and the CXC-chemokine interleukin-8 (IL-8) presented on the surface of endothelial cells. Engagement of these agents by specific receptors causes activation of the neutrophil β2-integrins (CD11a/CD18, CD11b/CD18). The integrins immobilize the neutrophils by binding to intercellular adhesion molecule-1 (ICAM-1) and, in the case of CD11b/CD18, possibly other structural proteins. The neutrophils next spread and migrate first over and then through the endothelial surface, in a few minutes. Migration requires polymerization of cytoskeletal actin, and cyclical regulation of integrin binding. Lymphocytes and monocytes follow a similar sequence. They have the added ability to use their α4β1-integrin to bind vascular cell adhesion molecules-1 (VCAM-1) to support capture and immobilization, and they utilize a largely separate group of CC- and CXC-chemokines.

The foregoing paradigm has largely been developed with reference to leukocyte recruitment to endothelium in postcapillary venules of the systemic circulation during inflammation. In the lungs, inflammatory recruitment in microvessels supplying the large airways appears to conform to this pattern, but adhesion in the pulmonary circulation may follow a different pathway (see Figure 2 for comparison of pathways). Here, capillaries are the main sites of leukocyte adhesion, and because the cells travel slowly through these segments, the requirement for selectin-mediated adhesion is reduced. Thus, chemokines can activate the slow-flowing cells that can go directly to stable integrin-mediated adhesion and then migration. In all cases, endothelial cells control the stages of recruitment, through their own response to inflammatory mediators and cytokines, such as tumor necrosis factor-α (TNF) and interleukin-1 (IL-1). These agents regulate endothelial expression of the selectins, chemotactic agents, or integrin ligands that are required in the different regions.

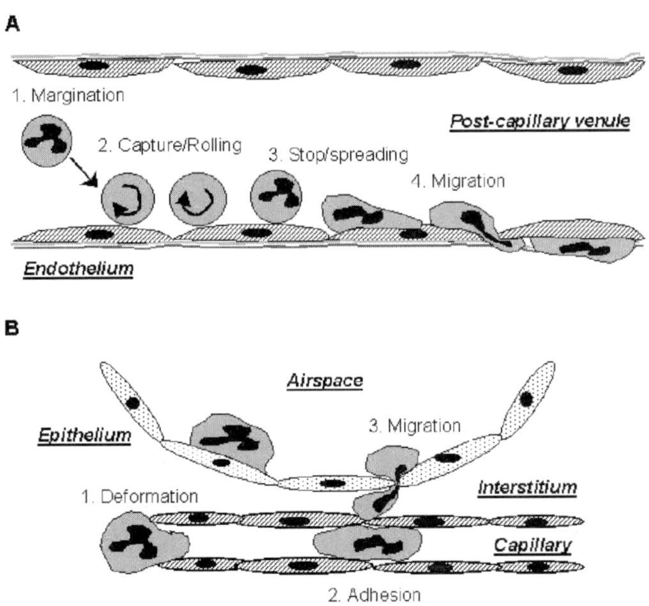

Figure 2 Comparison of the steps involved in traffic of neutrophils from the blood into tissue in **(A)** postcapillary venules and **(B)** the pulmonary microcirculation. In postcapillary venules, rapidly moving leukocytes are marginated by centrally flowing red blood cells, and initial attachment is supported by selectin receptors on the endothelium. An unstable rolling form of adhesion is established, which is converted to stable adhesion in response to a chemotactic agent such as interleukin-8. This stable, integrin-mediated adhesion also supports movement over and through the endothelial monolayer, into tissue. In the pulmonary capillaries, neutrophils must deform to enter and are in close contact with the endothelium. Selectin-mediated adhesion appears redundant for these slow-moving cells, and stable adhesion occurs if integrins become activated, again by local chemotactic agents. Then the cells will migrate not only into the interstitial space, but also, potentially, across alveolar epithelium into the airspace.

Effects of Cigarette Smoking on Behavior of Leukocytes in the Microcirculation

In assessing the effects of cigarette smoke on leukocyte rheology and passage through the microcirculation, one must consider potential modifications in their mechanical and adhesive properties, and also possible effects on the endothelial cells that can promote adhesion. In general, studies of the effects of cigarette smoke or its constituents on leukocyte behavior fall into one of the following categories: (i) studies in which humans or animals are acutely exposed to cigarette smoke and observations are made on changes in circulatory behavior in vivo; (ii) studies in which blood is withdrawn from human cigarette smokers or non-smokers (with or without additional acute smoking), and properties of leukocytes are tested ex vivo; (iii) studies in which blood or isolated leukocytes from humans are exposed to cigarette smoke or components of cigarette smoke in vitro; (iv) occasional studies in which endothelial cells are cultured and exposed to cigarette smoke or components of cigarette smoke in vitro, and ability to bind leukocytes is tested. The direct in vivo observations will be described first, before considering the more reductive analyses related to cell mechanics and adhesion that may explain the phenomena seen in vivo and predict behavior in the microcirculation.

In Vivo Studies of Effects of Cigarette Smoke on Leukocyte Behavior in Humans

It is possible to extract leukocytes (usually neutrophils) from blood, to radiolabel them, and then to reinject them, in order to examine their fate in the circulation. Such an approach can be used to locate sites of inflammation where cells accumulate, but can also be used to study the kinetics of passage of leukocytes through the human lung. It has been shown, for instance, that a large pool of neutrophils is usually moving more slowly through the lung than the bulk of red blood cells, so that about 50 percent of all circulating neutrophils are located there at any time [2]. This is believed to arise from the restricted rate of passage of the neutrophils through the many capillary segments in the pulmonary circulation. Of particular interest here are studies made in 1989, which showed that when humans smoked a cigarette there was an immediate sequestration of neutrophils in the lungs that took longer than usual to wash out of the lungs [3]. Linked studies indicated that in vitro measurements of the resistance to flow of neutrophils through capillary-sized pores correlated, on a person-by-person basis, with the number of cells sequestered in the pulmonary circulation. Such changes have been associated with evidence of systemic oxidative stress [4]. The implication is that upon cigarette smoking, neutrophils are exposed to agents in smoke (such as oxidants) that delay their passage through the lungs, possibly by causing an increase in their rigidity. However, from these studies, it cannot be ruled out that passage through the microcirculation is also impaired through changes in neutrophil or endothelial adhesive properties. Indeed, acute cigarette smoking has been associated with increased migration of neutrophils into the lung airspaces.

In Vivo Studies of Effects of Cigarette Smoke in Experimental Animals

Direct visual observations of the systemic microcirculation have been made before and after exposure to cigarette smoke in animals, and larger vessels have been removed and examined [5]. In these studies, inhalation of cigarette smoke was associated with an increase in rolling and stationary adhesion of leukocytes in arterioles and venules within minutes. Postmortem examination also revealed deposition of leukocytes in large arteries, typically associated with deposition of platelets. In general, induction of rolling adhesion requires no change in the properties of leukocytes, but a stimulus to endothelial cells is needed to express selectins on their surface. This can occur within minutes after stimulation with agonists, such as thrombin and histamine, within approximately hours of cytokine stimulation, and after an intermediate period upon stimulation by oxidants such as oxidized LDL.

Transformation to stationary adhesion requires either that the endothelial cells generate and present chemotactic agents, or that the leukocytes are directly activated by soluble agents. It is also worth noting that activated platelets attached to the vessel wall can support capture, rolling, and, under some conditions, immobilization of flowing leukocytes. It is instructive that oxidative mechanisms were strongly implicated in the response to CS because treatment with superoxide dismutase or the water-soluble antioxidant vitamin C reduced the adhesive response to smoke inhalation. Thus, dramatic changes in leukocyte–vessel wall interactions occurred in tissues remote from the lungs in animals inhaling cigarette smoke, and it is possible that these arose from responses in leukocytes, endothelial cells, and/or platelets.

Exposure to cigarette smoke also leads to sequestration of neutrophils in the pulmonary microcirculation in animals. This sequestration has been linked to changes in expression of adhesion molecules suggesting local activation, and also to release of immature and less deformable neutrophils from the bone marrow [6].

Effects of Cigarette Smoke on Leukocytes: Changes in Resistance to Flow

Detailed analysis has been made of the changes in resistance to flow of isolated neutrophils upon exposure to components of CS transferred across a gas–liquid interface (e.g., Ref. 7). Summarizing a number of studies (involving the authors and coworkers), there was a marked increase in the resistance of suspensions of neutrophils to flow through filters with 5-μm pores soon after exposure to CS. Changes in cell shape occurred in many cells on the same time scale, with marked formation of blebs (see Figure 1). Studies of mechanical properties of individual cells indicated that these distorted cells had the greatest increase in flow resistance. Although the morphological changes were not reminiscent of those induced by chemotactic agents, they were associated with formation of polymerized F-actin, which also occurs and leads to rigidification with these agents. The changes in cell mechanics were also linked to oxidative processes, since antioxidants were protective against the smoke-induced deterioration in flow properties. These studies imply that substances transferred from cigarette smoke into the blood may directly cause changes in the structure, shape, and rigidity of neutrophils, apparently through a form of oxidative damage. This response might occur specifically as neutrophils transit the lung but might also be disseminated in a dilute form throughout the vasculature, so that microvascular impairment might be systemic as well as local. Systemic oxidative stress has been linked with cigarette smoking [4].

Effects of Cigarette Smoke on Leukocyte Adhesion

Direct Effects on Leukocytes

The effects of cigarette smoke directly on adhesive behavior of leukocytes (as opposed to effects via endothelial cells or platelets) are not so clear. In the case of monocytes, in vitro exposure to condensate derived from cigarette smoke caused upregulation of β2-integrin expression and activation of integrin function over tens of minutes, so that adhesion to endothelial cells was enhanced [8]. Ex vivo studies of monocytes from smokers have also indicated that there is an upregulation of β2-integrin expression and adhesiveness associated with cigarette smoking [9]. However, early studies of neutrophils indicated that adhesion of these cells to various surfaces, supported by β2-integrin receptors, was inhibited if the cells had been treated with CS [10]. Neutrophil spreading was also impaired. In vitro exposure, or ex vivo studies after cigarette smoking, did not show changes in level of β2-integrin expression by neutrophils. All of these studies were in static systems where selectin-mediated capture was not tested. In our own unpublished observation, we found that in a flow system, neutrophils that had been exposed to water-soluble components of CS could adhere effectively to P-selectin. Interestingly, soluble components of CS could cause transformation of rolling to stationary adhesion (suggesting integrin activation), but this stabilization of adhesion lasted only a few minutes.

In summary, exposure to components of CS may be able to activate integrins of either monocytes or neutrophils and upregulate surface expression of integrins in the former. Time courses may differ, in that changes in neutrophils appear to reverse rapidly so that there is no stable enhancement of adhesion, whereas monocytes exposed to CS appear to retain increased integrin expression and adhesiveness. Morphological changes may also hinder stable attachment of neutrophils. Uncertainty arises when comparing the in vitro data for the different types of leukocytes, because monocytes have typically been exposed to condensate from CS, whereas neutrophils have been exposed to nonparticulate components that exchange into an aqueous phase. The responsiveness of neutrophils and monocytes to cigarette smoke is also manifest in changes in other forms of functional response, which may be linked to adhesion or circulation. For instance, monocytes and neutrophils exposed to elements of CS or derived from cigarette smokers show increased secretion of cytokines and granule enzymes that might promote inflammation, although ability of neutrophils to release oxidants may be impaired rather than enhanced by exposure to CS.

Effects on Endothelial Cells or Platelets

Adhesion of leukocytes may also be promoted if endothelial cells become activated to present adhesion receptors of the selectin family, and this attachment will be stabilized if chemokines or modified lipids such as PAF are presented. Condensate of CS has been shown to increase endothelial expression of ICAM-1, VCAM-1, and E-selectin over a period of approximately hours [10]. This, along with the direct effects noted earlier, increased adhesion of monocytes. Plasma levels of soluble ICAM-1 are also elevated in smokers, presumably as a result of shedding from "activated" endothelial cells. We recently investigated effects of water-soluble components of CS partitioned into flowing culture medium, perfused over endothelial cells downstream in an in vitro vascular model [11]. We found that over a period of about 60 to 150 minutes EC responded by upregulating expression of P- and E-selectin and becoming able to efficiently capture flowing neutrophils. Some of these neutrophils became stationarily attached, implying that the EC also generated activators, as neutrophils were not themselves exposed to CS. Thus direct-contact models using CS condensate, and perfusion models of effects of soluble mediators, indicate that endothelial cells will take on a "proinflammatory" or proadhesive phenotype as a result of cigarette smoking. The different models may be relevant to effects that might occur in the lungs themselves and in tissue remote from lungs, respectively. The response of the endothelial cells develops over hours and might become chronic in a regular cigarette smoker. This contrasts with the rapid responses of leukocytes, which might exacerbate the underlying "mild" inflammation sporadically.

In addition to endothelial cells, it is increasingly recognized that in pathological conditions at least, platelets may act as promoters of leukocyte attachment to the vessel wall. The platelets themselves may bind to intact endothelium in ischemic or atherosclerotic vessels, or to exposed collagen in damaged vessels, and capture flowing leukocytes through presentation of P-selectin. As noted earlier, circumstantial evidence indicates that platelets may promote leukocyte adhesion in arteries of animals inhaling CS, but we are not aware of any studies clarifying the steps by which this might occur. It is known, however, that cigarette smoking can modify the adhesive properties of platelets and their release of vasoactive compounds. The ability of cigarette smoke to cause platelets to bind in the microcirculation and then cause thrombotic or inflammatory responses appears worthy of further investigation.

Table I Changes in the Properties of the Cells of the Vascular System Induced by Exposure to Cigarette Smoke, with Consequences for Circulation of Leukocytes.

Type of leukocyte	Form of exposure	Modification induced
Neutrophil	Human—smoking	Delay in transit through lungs
Neutrophils	Isolated cells exposed to water-soluble components of CS	Shape change, increased rigidity, and impaired migration
Monocyte	Human smokers—ex vivo	Increased integrin expression and adhesiveness
Monocytes	Isolated cells exposed to CS condensate	Increased integrin expression and adhesiveness
Leukocytes (type undefined)	Hamster breathing CS	Adhesion to wall of large and small blood vessels
Endothelial cells	Isolated cells exposed to CS condensate	Upregulation of adhesion receptors for leukocytes
Endothelial cells	Isolated cells exposed to water-soluble component(s) of CS	Upregulation of adhesion receptors and activating agents for leukocytes
Platelets	Human smokers—ex vivo	Activation, increased adhesion and secretion

Conclusions

Table I summarizes changes in leukocyte rheology, and consequent changes in their circulation, that are caused by cigarette smoke. Direct exposure to water soluble components of CS cause neutrophils to become rigid, but the pattern of shape change and F-action formation does not closely resemble that seen with classical chemotactic agents such as formyl peptides. Monocytes show upregulation of integrins and signs of increased adhesion when exposed to CS condensate. Both types of cells show some signs of activation when taken from smokers. The cells harvested from the blood may have experienced direct effects of CS, but may also be responding indirectly to CD through agents released into the circulation from the lungs. In endothelial cells, direct contact with condensate and perfusion of soluble agents from smoke both upregulate adhesion molecules, and in the latter case neutrophil-activating agent(s) are also produced. In all cases, oxidizing components of CS or oxidized components of plasma exposed to CS may play causative roles in cellular responses.

Changes in rigidity or in integrin adhesion might hold up leukocytes in the lung or in microvasculature of other organs if changes are maintained over minutes when cells are released from the lungs. Local changes in endothelial adhesiveness may contribute to holdup of leukocytes in the lungs, although selectin receptors at least do not appear to be important in that organ. In remote tissue, endothelial activation by transmitted CS components could initiate an inflammatory response or add impetus to other inflammatory agonists. Whereas mechanical changes are only likely to affect circulation in microvessels, adhesive responses are likely to be relevant to arterial as well as small-vessel disease.

Glossary

F-actin: Filamentous or polymerized actin forming relatively rigid structures within cells (cytoskeleton), in dynamic equilibrium with cytoplasmic pool of monomeric globular G-actin.

Leukocyte migration: Process whereby leukocytes move from blood into tissue, by first adhering to the vascular endothelium and then migrating actively through it.

Oxidants: Reactive forms of oxygen and their derivatives (e.g., found in cigarette smoke) that can oxidize and modify cellular structures (i.e., cause oxidative damage).

Rheology: Field of physics dealing with deformation and flow of materials. Applied here to physical properties of leukocytes controlling their circulation.

References

1. Adams, D. H., and Nash, G. B. (1996). Disturbance of leucocyte circulation and adhesion to the endothelium as factors in circulatory pathology. *Br. J. Anaesth.* **77**, 17–31.
2. Wagner, J. G., and Roth, R. A. (2000) Neutrophil migration mechanisms, with an emphasis on the pulmonary vasculature. *Pharmacol. Rev.* **52**, 349–372. *Reviews the steps by which neutrophils move from blood into tissue and describes how this process may differ between microvessels in the lungs and other parts of the body.*
3. MacNee, W., Wiggs, B., Belzberg, A. S., and Hogg, J. C. (1989) The effect of cigarette smoking on neutrophil kinetics in human lungs. *N. Engl. J. Med.* **321**, 924–928. *Demonstrates directly in humans that neutrophils transit slowly through the lungs and are further delayed when the subject smokes a cigarette.*
4. MacNee, W. (2000). Oxidants/antioxidants and COPD. *Chest* **117**(Suppl 1), 303S–317S.
5. Lehr, H. A. (2000). Microcirculatory dysfunction induced by cigarette smoking. *Microcirculation* **7**, 367–384. *Reviews several aspects of the effects of cigarette smoke on the microcirculation, especially studies showing that exposure to cigarette smoke in animals rapidly causes adhesion of leukocytes to walls of various blood vessels, and that antioxidants inhibit these responses.*
6. Terashima, T., Klut, M. E., English, D., Hards, J., Hogg, J. C., and van Eeden, S. F. (1999). Cigarette smoking causes sequestration of polymorphonuclear leukocytes released from the bone marrow in lung microvessels. *Am. J. Respir. Cell. Mol. Biol.* **20**, 171–177.
7. Drost, E. M., Selby, C., Lannan, S., Lowe, G. D., and MacNee, W. (1992). Changes in neutrophil deformability following in vitro smoke exposure: Mechanism and protection. *Am. J. Respir. Cell. Mol. Biol.* **6**, 287–295.
8. Shen, Y., Rattan, V., Sultana, C., and Kalra, V. K. (1996). Cigarette smoke condensate-induced adhesion molecule expression and transendothelial migration of monocytes. *Am. J. Physiol.* **270**, H1624–H1633.

9. Weber, C., Erl, W., Weber, K., and Weber, P. C. (1996). Increased adhesiveness of isolated monocytes to endothelium is prevented by vitamin C intake in smokers. *Circulation* **93**, 1488–1492.
10. Selby, C., Drost, E., Brown, D., Howie, S., and MacNee, W. (1992). Inhibition of neutrophil adherence and movement by acute cigarette smoke exposure. *Exp. Lung Res.* **18**, 813–827.
11. Stone, P. C., Fisher, A. C., Rainger, G. E., and Nash, G. B. (2002). Neutrophil capture by selectins on endothelial cells exposed to cigarette smoke. Biochem. Biophys. Res. Commun. **295**, 1150–1155.

Capsule Biography

Professor Nash heads the Cardiovascular Rheology group in the Medical School of Birmingham University. The group focuses on the physicochemical factors regulating the recruitment of leukocytes to tissue, in relation to vascular diseases. The work is mainly supported by the British Heart Foundation, BBSRC, and Arthritis Research Campaign.

Professor MacNee is Professor of Respiratory and Environmental Medicine, University of Edinburgh Medical School. His interest is in lung inflammation and specifically in the role of oxidative stress and lung inflammation in airway disease. His work is sponsored by the Medical Research Council, National Institute for Health, British Lung Foundation, and the Colt Research Foundation.

CHAPTER 130

Pulmonary Response to Environmental and Occupational Pollutants

[1]Nicola Verna and [2]Mario Di Gioacchino

[1]*Department of Medicine and Science of Aging, Section of Allergy, Clinical Immunology, and Occupational Medicine, "G. D'Annunzio" University School of Medicine, Chieti, Italy*

[2]*Department of Medicine and Science of Aging, Section of Allergy, Clinical Immunology, and Occupational Medicine, "G. D'Annunzio" University School of Medicine, Chieti, Italy*

Introduction

The lung is a vulnerable target for volatile substances because of its large alveolar surface area, estimated to be about 100 square meters, leading to exposure to large air volumes. Indeed, humans inhale a volume of about 0.5 liters per breath or 8.640 liters per day. Moreover, it is estimated that the cardiac output is about 7.0 L/day. This flow traverses a rich pulmonary capillary bed that provides ease of absorption and transport of inhaled toxic materials throughout the body. Thus volatile agents can directly act in the lung and/or in some remote target organ.

The effects of air pollution on health have been recognized from long time. In the past, the majority of air pollution was derived from domestic and industrial burning of fossil fuels. The introduction of legislation to reduce the use of such fuels has changed the composition of air pollutants. At present, fossil fuel–associated pollutants have been replaced by a steady increase in traffic-associated pollutants such as finer, respirable particulates, oxides of nitrogen (NO_x), ozone (O_3), and volatile organic compounds.

Usually, all inhaled toxins are associated with environmental exposure. This is largely related to an occupation. The toxins may be responsible for causing chronic inflammation, granulomatous lung diseases, immunosuppression leading to a heightened incidence of respiratory infections, as well as cancer (Table I) [1].

In terms of incidence, the agents producing immune-inflammatory responses (*hypersensitivities*) are more common than those responsible for irritant reactions or fibrogenesis.

The immune toxins can be classified on the basis of the hypersensitivity mechanism (Table II). *Immediate hypersensitivity* is associated with IgE-mediated mast cell activation, resulting in the accumulation of eosinophils and Th2 lymphocytes in the airways. *Immune-complex mediated hypersensitivity* is related to the immune-complex deposition due to excessive antibody production or deficient clearance. *T-cell mediated hypersensitivity* is elicited by T cells secreting cytokines that activate macrophages.

Additionally, some substances are able to elicit lung injury involving different cells from those of the immune system such as fibroblasts (silicosis and asbestosis). The last substances to be mentioned are those able to enter the body by the lung but causing their effects in different organs (carbon monoxide).

Irritant Reactions

Respiratory tract irritants may damage different lung districts on the basis of their hydro-solubility. Therefore, highly soluble gases—such as ammonia, aldehyde gases, hydrogen chloride, and other acids—are dissolved in watery mucus of

Table I Toxin Classification Based on Site of Action in the Lungs.

Upper airways and bronchi	Terminal bronchioli, alveoli	Systemic absorption (with only slight or no pulmonary toxicity)
Chlorine	Nitrogen dioxide	Toluene
Ammonia	Nitrogen oxide	Xylene
Aldehydes of lower molecular weight (formaldehyde, acetaldehyde)	Phosgene	Carbon dioxide
	Ozone	Carbon monoxide
	Dusts from hemp, flax, and cotton processing	Hydrogen cyanide
Acrolein		Propane
Sulfur dioxide		Carbon tetrachloride
Hydrogen chloride		
Hydrogen fluoride		
Acetic acid		

Table II Examples of Agents Causing Immune-Toxic Reactions.

Immunologic mechanism	Tissue reaction	Example
Immediate hypersensitivity IgE-mediated	Bronchial edema, plugging	*Low molecular weight* Nickel Isocyanates Aldehydes (formaldehyde, acetaldehyde) *High molecular weight* Detergent enzymes House dust mite Pollens Molds Latex
Immune-complex-mediated hypersensitivity	Interstitial infiltrate, granuloma	Farmer's lung
T-cell mediated hypersensitivity	Noncaseating granuloma	Berylliosis
Local immunosuppression	Higher incidence of infections	Benzo[*a*]pyrene Phosgene Nitrogen dioxide

the nose, throat, and trachea and have their primary effects there. They can provoke acute painful irritation with sneezing, cough, sore throat, and acute bronchitis. Asphyxiation is possible due to laryngeal spasm and bronchoconstriction. Generally, it is easy to detect the presence of these agents in the environment by their odor and the discomfort evoked.

Gases with lower water solubility—such as phosgene, nitrogen dioxide, chlorine, and isocyanates—evade the first line of defense in the upper airways and have their primary effect on the lower respiratory tract. Their irritative actions appear hours or days after the exposure and may evoke severe inflammatory responses. In most cases it is quite difficult to detect the presence of these gases in the environment.

In addition to their chemical proprieties, the toxicity of irritant agents is related to the intensity of exposure. Therefore, air concentration and duration of exposure can be combined into a cumulative exposure value (*concentration x time*). These are important parameters to be evaluated in cases of intoxication. Indeed, any irritant gas, at sufficiently high concentration, may be a factor in producing pulmonary edema. The following examples of irritant gases are not exhaustive. We list only the most common irritant agents and their mechanisms of action.

Phosgene

Phosgene gas (carbonyl chloride; carbon oxychloride; chloroformyl chloride) is a toxic inhalant that directly damages the lungs. Although phosgene has not been deployed in warfare since the Geneva Protocol prohibited chemical agent use in 1925, its continued application in common industrial processes, such as dye or plastic manufacturing, makes it an ongoing potential industrial hazard. Phosgene exposure also can occur in fires associated with organochlorine compounds (e.g., vinyl chloride), the use of carbon tetrachloride fire extinguishers, and during arc welding procedures.

Phosgene gas has the appearance of a white cloud and the characteristic odor of newly mown hay. Odor alone is insufficient for the detection of phosgene, since toxic exposures may occur at concentrations below the olfactory threshold. Phosgene is one of the most volatile chemical warfare agents, its density is greater than air, and it tends to accumulate in low areas.

Phosgene exerts a direct effect on the respiratory tract, causing extensive cellular damage to the alveolar–capillary membrane. Phosgene reacts with intraalveolar water to form hydrochloric acid. When hydrochloric acid itself is aspirated experimentally there is activation of the alternative pathway of complement leading to deposition of the membrane attack complex and cell injury. It is appealing to believe that phosgene injury follows the same mechanism.

The clinical effects of phosgene are dose dependent. At low concentrations, victims may complain only of a mild cough, dyspnea, and chest tightness. At moderate concentrations, they may also complain of tearing. At high concentrations, victims develop noncardiogenic pulmonary edema within 2 to 6 hours of exposure, producing a clinical picture similar to the acute respiratory distress syndrome (ARDS). Laryngospasm also occurs at higher concentrations, which in turn may cause sudden death. Physical exertion within 72 hours of exposure can trigger dyspnea and pulmonary edema in otherwise asymptomatic patients. Toxic manifestations often are clinically silent at rest, because patients are able to compensate for pulmonary damage in the absence of stress. Death from phosgene inhalation often is caused by latent noncardiogenic pulmonary edema.

Nitrogen Dioxide ($\cdot NO_2$) and Reactive Nitrogen Species (RNS)

The lung can be exposed to a variety of reactive nitrogen intermediates through the inhalation of environmental oxidants and those produced during inflammation. Reactive nitrogen species (RNS) include nitrogen dioxide ($\cdot NO_2$), and peroxynitrite ($ONOO^-$). Classically known as a major component of both indoor and outdoor air pollution, $\cdot NO_2$ is a toxic free radical gas. The gas is formed primarily from burning fuel in motor vehicles, electric power plants, and other industrial, commercial, and residential sources that burn fuel. $\cdot NO_2$ can also be formed during inflammation by the decomposition of $ONOO^-$ or through peroxidase-catalyzed reactions. Since nitrogen dioxide is a traffic-related pollutant, emissions are generally highest in urban rather than rural areas.

Although $\cdot NO_2$ found in outdoor air is a significant source of exposure, the concentration present in indoor air often exceeds that from the outside. Indoor sources of $\cdot NO_2$ include kerosene heaters, gas cooking stoves, gas-powered ice scrapers used in hockey rinks, and tobacco smoke. Nitrogen oxides are an important precursor to the formation of ground-level ozone and acid rain and may affect both terrestrial and aquatic systems.

Because of their reactive nature, RNS may play an important role in disease pathology. Depending on the dose and the duration of administration, $\cdot NO_2$ has been documented to cause pulmonary injury in both animal and human studies.

The main site of $\cdot NO_2$ deposition and injury is at the distal conducting airways at the level of the terminal bronchioles. Injury to the lung epithelial cells following exposure to $\cdot NO_2$ is characterized by airway denudation followed by compensatory proliferation. Loss of epithelial cells due to injury causes the proliferation of alveolar type II cells, Clara cells, and/or other bronchiolar cells, which repopulate the injured areas of airway epithelium. The persistent injury and repair process may contribute to airway remodeling, including the development of fibrosis. Furthermore, inhaled $\cdot NO_2$ can exacerbate asthma and is associated with an increased susceptibility to respiratory infections.

Ozone

Ozone (O_3) is a secondary pollutant formed in the troposphere through a series of sunlight-driven reactions of atmospheric oxygen with volatile organic compounds and nitrogen oxides, which are produced through combustion. Levels of ozone tend to be higher during hot summer days and in cities with large amounts of traffic and with temperature inversions[1] (Los Angeles, Mexico City), but may also be very high in workplaces such as welding plants and paper mills. Ozone is virtually insoluble in water and, therefore, it may be deposited anywhere along the airway, but as it is highly reactive it is thought that very little of the inspired ozone reaches the alveolar epithelium. The majority of the effects of ozone are mediated by a cascade of secondary products derived from free radicals that cause cellular damage when antioxidant defenses have been overwhelmed. Experimental exposure of humans to high concentrations of ozone causes irritation and cough, with decrements in forced vital capacity and forced expiratory volume at first second (FEV_1). In healthy subjects ozone induces neutrophilic inflammation of the airway mucosa accompanied by increased levels of inflammatory mediators and proteins in bronchoalveolar lavage (BAL). Albumin, IgG, and α_1-antitrypsin are increased in the epithelial lining fluid, indicating increased vascular permeability. Exposure to ozone also induces release of the neutrophil chemoattractant IL-8, as well as GM-CSF, and increases expression of the endothelial adhesion molecule ICAM-1, suggesting a likely mechanism for neutrophil recruitment. Some studies have reported recruitment of eosinophils and mast cells after ozone exposure. In vitro, cultured human epithelial cells exposed to ozone release increased amounts of lipid mediators PGE_2, LTC_4, LTD_4, and LTE_4. Fibronectin and the proinflammatory cytokines IL-6 and IL-8 are also released

[1] Temperature inversions are defined as a layer of air with negative lapse rate or simply as a region within the troposphere where temperature increases with height.

by epithelial cells in response to ozone exposure. Other in vitro studies carried out using human epithelial cells have shown that both NFκB and AP-1, which regulate the IL-8 gene, increase in response to ozone exposure in respiratory epithelial cells.

Immunologic Reactions

The exposure of lung to proteic molecules or molecules that can be carried by endogenous proteins (aptens) may elicit an activation of the immune system (hypersensitivity). The immune response to inhaled antigens is potentially a two-edged sword that may either protect (for example, against infections) or, via a hypersensitivity reaction, harm the host (Table III).

Immediate Hypersensitivity and Asthma

Immediate hypersensitivity is an allergic reaction induced by a specific antigen (allergen), provoked by reexposure to the same antigen, and is mediated by specific IgE antibodies in genetically susceptible individuals (atopics). In the most extreme systemic form of the reaction, called *anaphylaxis*, mast cell- and/or basophil-derived mediators can restrict airways leading to asphyxiation and producing potential fatal cardiovascular collapse.

IgE antibodies are cytophilic and bind on the surface of circulating basophils and on mast cells in various tissues. These cells constitutively express high-affinity surface receptors for the Fc component of IgE (FcεRI). Binding and cross-linking of the allergen to surface-receptor bound IgE triggers the immediate release from cytoplasmic granules of mast cells and basophils of preformed (primary) vasoactive mediators of immediate hypersensitivity and also initiates de novo synthesis and release of other (secondary) mediators of hypersensitivity. These last-mentioned mediators are mostly responsible for the *late phase reaction* characterized by an inflammatory infiltrate of eosinophils, basophils, neutrophils, and lymphocytes, ensuing 4 to 6 hours after mast cell and basophil degranulation.

Allergies may be thought of as Th2-dependent disease because Th2 cells produce IL-4, which is required for IgE production, and IL-5, which stimulates eosinophilic inflammation, a characteristic of many allergic diseases. This reaction differs from *delayed-type hypersensitivity*, which is mediated by CD4+ Th1 and CD8+ T cells and macrophages with no central role of antibodies.

The inhaled allergens (aeroallergens) are usually proteins associated with biogenic particles sized 2 to 60 μm (Table IV). Such a size allows them to be easily carried, suspended in the atmosphere. The most common and practical classification of the aeroallergens is based on their source. It is possible to distinguish aeroallergens spontaneously derived from sources in nature and aeroallergens generated through domestic or occupational practices.

The natural resource–related aeroallergens (outdoor allergens) show a seasonal fluctuation of their environ-

Table III Immunological Mechanisms of Tissue Injury and Inflammation.

Hypersensitivity reaction	Immediate hypersensitivity IgE-mediated (anaphylaxis)	Immune-complex-mediated hypersensitivity	T-cell mediated hypersensitivity (delayed hypersensitivity)
Time elapsed	Seconds to minutes	Hours to days	2–3 days
Specific immune agent	IgE	IgG, IgM	T-cell reactant
Chemical mediators of tissue injury and inflammation	Vasoactive products of mast cells/basophils (histamine, arachidonate derivatives)	Complement and Fc receptor (cytolytic, chemotactic, vasoactive components)	Lymphokines and monokines
Cell pathology and pathophysiology	Accumulation of neutrophils and eosinophils. Smooth muscle contraction	Accumulation of neutrophils, macrophages. Release of lytic lysosomal enzymes	Lymphocytes and macrophages; granulomas
Clinical examples	Anaphylaxis Atopic disorders (allergic rhinitis, hay fever, bronchial asthma)	Extrinsic allergic alveolitis, serum sickness	Berylliosis, tuberculosis

Table IV Characteristics of Aeroallergens.

	Natural resource-related (outdoor aeroallergens)	Human activity-related (indoor and occupational aeroallergens)
Patterns of prevalence	Regionally with flora and fauna	Determined by domestic and occupational practice
Period of prevalence	Usually seasonal	Often perennial or prolonged
Avoidance	Difficult or unfeasible	Usually simple and effective
Microscopic appearance	Often recognizable distinctive units	Often amorphous or nondiagnostic
Examples	Pollens, fungus spores of *Alternaria*	House dust, animal allergens, vegetable dusts, fungus spores of *Penicillium*

mental concentration related to the local flora and fauna. Despite aerobiological techniques that are generally able to detect their presence in the atmosphere, it is quite difficult to avoid exposure to them. Classical examples of this group of aeroallergens are pollens (Table V) and fungus spores of *Alternaria* species. On the other hand, the indoor aeroallergen concentrations show fewer fluctuations and preventive exposure strategies are usually effective. In this group are included house dust mites, animal skin allergens, and spores of *Penicillium* species (Table IV). All aeroallergens listed in Tables IV and V are potentially able to provoke asthma.

Asthma is an inflammatory disease caused by repeated immediate hypersensitivity reactions in the lung leading to a variable degree of airflow obstruction, bronchial hyperresponsiveness, and airway inflammation.

Although the fraction of patients suffering from allergic asthma is increasing for various and still unknown reasons, there is a subpopulation of asthmatic patients with no personal or family history of allergy, with negative skin tests, and with normal serum levels of IgE. These patients are said to have *idiosyncratic asthma*.

For many patients, the allergic asthma has its roots in infancy, and both genetic factors (atopy) and environmental factors (viruses, allergens, and occupational exposures) contribute to its inception and evolution.

The avoidance of allergens—when it is possible—should be the first recommendation in the treatment of allergic asthma. The diagnosis of asthma should be clearly established and the baseline severity of the disease classified to help establish the recommended course of pharmacological therapy (Tables VI and VII). An early diagnosis and treatment can reduce the decline in lung function and airway remodeling. Remodeling entails thickening of the airway walls, with increases in submucosal tissue, the adventitia, and smooth muscle. These features differ in asthma and chronic obstructive pulmonary diseases, in allergic and nonallergic asthma, and with the severity of asthma. The precise

Table V Pollen Allergens from Various Sources for Which the Entire Coding Sequences Have Been Cloned.

	Common name	Scientific name	Pollen allergen
Weeds	Short ragweed	*Ambrosia artemisiifolia*	Amb a 1, Amb a 2, Amb a 5, Amb a 6
	Western ragweed	*Ambrosia psilostachya*	Amb p 5
	Giant ragweed	*Ambrosia trifida*	Amb t 5
		Parietaria judaica	Par j 1, Par j 2
Grasses	Bermuda grass	*Cynodon dactylon*	Cyn d 1, Cyn d 7, Cyn d 12
	Orchard grass	*Dactylis glomerata*	Dac g 2, Dac g 3
	Velvet grass	*Holcus lanatus*	Hol l 1, Hol l 5
	Perennial ryegrass	*Lolium perenne*	Lol p 1, Lol p 2, Lol p 3, Lol p 5
		Lolium italicum	Lol i 1
	Canary grass	*Phalaris aquatica*	Pha a 1, Pha a 5
	Timothy grass	*Phleum pretense*	Phl p 1, Phl p 2, Phl p 5, Phl p 6, Phl p 7, Phl p 11, Phl p 13
	Kentucky blue grass	*Poa pratensis*	Poa p 2, Poa p 9
	Johnson grass	*Sorghum halepence*	Sor h 1
	Wheat	*Triticum aestivum*	Tri a 2
	Maize	*Zea mays*	Zea m 1, Zea m 11
	Barley	*Hordeum vulgare*	Hor v 9
	Rice	*Oryza sativa*	Ory s 1
Trees	Alder	*Alnus glutinosa*	Aln g 1, Aln g 4
	White birch	*Betula verrucosa*	Bet v 1, Bet v 2, Bet v 3, Bet v 4, Bet v 5
	Hornbeam	*Carpinus betulus*	Car b 1
	Japanese cypress	*Chamaecyparis obtuse*	Cha o 1, Cha o 2
	Hazel	*Corylus avellana*	Cor a 1
	Japanese cedar	*Cryptomeria japonica*	Cry j 1, Cry j 2
		Cupressus arizonica	Cup a 1
		Cupressus sempervirens	Cup s 1
		Juniperus ashei	Jun a 1, Jun a 2, Jun a 3
		Juniperus oxycedrus	Jun o 2
		Juniperus oxycedrus	Jun o 2
	Eastern red cedar	*Juniperus virginiana*	Jun v 1, Jun v 3
	Privet	*Ligustrum vulgare*	Lig v 1
	Lilac	*Syringea vulgaris*	Syr v 1
	Annual dog's mercury	*Mercurialis annua*	Mer a 1
	Olive	*Olea europea*	Ole e 1, Ole e 2, Ole e 3, Ole e 6, Ole e 8
	Turnip	*Brassica rapa*	Bra r 1, Bra r 2
Field crops	Oilseed rape	*Brassica napus*	Bra n 1, Bra n 2
	Sunflower	*Helianthus annuus*	Hel a 2

Table VI Classification of Asthma Severity [2].

Classification	Step	Days with symptoms	Nights with symptoms	For adults and children aged > 5 years who can use a spirometer or peak flow meter	
				FEV_1 or PEF % predicted of normal	PEF variability (%)
Severe Persistent	4	Continual	Frequent	≤ 60	> 30
Moderate Persistent	3	Daily	> 1/week	> 60 – < 80	> 30
Mild persistent	2	> 2/week, but < 1 time/day	> 2/month	≥ 80	20–30
Mild intermittent	1	≤ 2/week	< 2/month	≥ 80	< 20

Table VII Medications Used in Different Levels of Asthma Severity [4].

Classification	Step	Daily medication	Quick relief medication
Severe persistent	4	High-dose inhaled steroids (ICS) and long-acting inhaled β_2-agonist If needed, add oral steroids!	Short-acting inhaled β_2-agonist, as needed; oral steroids may be required
Moderate persistent	3	Low-to-medium-dose ICS and long-acting β_2-agonist (preferred) *or* Medium-dose ICS (another preferred option for children aged < 5 years) *or* Low-to-medium-dose ICS and either leukotriene modifier or theophylline	Short-acting inhaled β_2-agonist, as needed; oral steroids may be required
Mild persistent	2	Low-dose inhaled steroids (preferred) *or* Cromolyn, leukotriene modifier, or (except for children aged < 5 years) nedocromil or sustained release theophylline to serum concentration of 5–15 μg/mL	Short-acting inhaled β_2-agonist, as needed; oral steroids may be required
Mild intermittent	1	No daily medicine needed	Short-acting inhaled β_2-agonist, as needed; oral steroids may be required

mechanisms underlying the remodeling process are under intense study.

The diagnosis of asthma is established by demonstrating reversible airway obstruction. Reversibility is traditionally defined as a 15 to 20 percent or greater increase in FEV_1 after β-adrenergic agonist administration. When the spirometry results are normal at presentation, the diagnosis can be made by showing heightened airway responsiveness to challenge with methacoline.

In the past decade, the treatment of asthma has emphasized long-term suppression of airway inflammation plus relief of symptoms with quick-acting bronchodilators (primarily aerosolized β-agonists). Inhaled corticosteroids are the most effective agents available for the symptomatic control of asthma and improvement in pulmonary function, but their potential side effects when used in escalating doses have led to the use of adjunctive therapies. Concomitant treatments with long-acting β-agonists, theophylline, and leukotriene antagonists have all been shown to help control asthma while minimizing the doses of inhaled corticosteroids that are needed.

Nevertheless, whether used alone or in combination with other therapies, corticosteroids do not consistently abrogate airway inflammation in patients with asthma. For this reason, other approaches that modulate IgE-associated immunologically mediated inflammatory responses are in use or under development. Allergen immunotherapy can be effective in many, but not all, patients [3]. Immunotherapy can be administered by subcutaneous injection, topically to the nasal mucosa, or sublingually. When given by injection, exceptionally severe, potentially fatal, anaphylactic reactions have been reported, and hence it is best if a specialist prescribes immunotherapy.

DNA vaccines and other molecular methods of down-regulating antigen-specific Th2-mediated responses are currently being studied. Agents directed at diminishing the production of IgE through effects on interleukin-4 or on IgE itself have also been evaluated. One such compound is a soluble recombinant interleukin-4 receptor that can be delivered in nebulized form.

Another compound, a recombinant humanized monoclonal antibody that forms complexes with free IgE (rhuMAB-E25, or omalizumab), blocks the interaction of IgE with mast cells and basophils. Several clinical studies have been performed in adults and children with moderate-to-severe allergic asthma to evaluate the efficacy and safety of this agent. Treatment with omalizumab was well tolerated and showed clinical benefit in terms of a reduction in the frequency and number of asthma exacerbation episodes and lower usage of corticosteroids and other medications to control disease, along with improved quality of life.

The efficacy of these therapies emphasizes the important contribution of allergic inflammatory mechanisms in the pathophysiology of asthma in many patients [5].

Immune-Complex-Mediated Hypersensitivity and Extrinsic Allergic Alveolitis (EAA)

When large amounts of antigen enter the bloodstream and bind to antibody, circulating immune complexes are noted. If antigen is in excess, small complexes form; because these are not easily cleared by the phagocytic cells, they can cause complement activation and tissue damage.

Hypersensitivity pneumonitis (HP), also called *extrinsic allergic alveolitis* (EAA), is a complex syndrome of varying intensity, clinical presentation, and natural history, rather than a single uniform disease. It can progress to disabling or even fatal end-stage lung disease. The only truly effective treatment is early recognition and control of exposure. Although patients produce antibody exuberantly, the immunopathogenesis involves also cellular immunity—notably, $CD8^+$ cytotoxic lymphocytes, multinucleate giant-cell granulomas, and ultimately interstitial fibrosis. Many causative agents have been recognized in occupational dusts or mists, but most current new cases arise from residential exposure to pet birds (pigeons and parakeets), contaminated humidifiers, and indoor molds. Pathologically, acute HP is characterized by poorly formed noncaseating interstitial granulomas and mononuclear cell infiltration in a peribronchial distribution with prominent giant cells.

The subacute, or intermittent, form produces more well-formed noncaseating granulomas. There is bronchiolitis with or without organizing pneumonia, and interstitial fibrosis. Chronic forms reveal the additional findings of chronic interstitial inflammation and alveolar destruction (honeycombing) associated with dense fibrosis. Cholesterol clefts or asteroid bodies are present within or outside granulomas.

Most patients have circulating immunoglobulin G (IgG) antibodies that are specific for the offending antigen. The antibody will react with a specific antigen to form a precipitate. Surprisingly, approximately 50 percent of asymptomatic people exposed to the sensitizing antigen also will develop these precipitating antibodies.

The late onset of symptoms, about 4 to 8 hours after antigen inhalation, suggests that the disease mechanism may resemble an Arthus-type skin reaction following intradermal skin tests that has the same kinetics. An antibody-dependent inflammatory process has been suggested for both. The Arthus skin histology demonstrates deposition of immune complexes and complement with an influx of neutrophils, similar to lung histology of biopsies taken during the acute phase. Generally biopsies are taken from patients with more long-standing lung disease and show a lymphocytic infiltrate, with minimal evidence for immune complexes or complement. Therefore the pathogenesis of EAA is thought to be a hypersensitivity reaction against inhaled antigens with involvement of both humoral and cellular immune responses.

The symptoms and physical findings are nonspecific. Serum IgG contains high titers of specific antibody to the offending antigen. Pulmonary function tests show restrictive and diffusion defects with hypoxemia, especially after exercise. Occasionally, small-airway disease causes obstruction. Radiographic changes vary according to the stage of the disease and are best evaluated by means of high-resolution computed tomography. In typical cases, the history of a known exposure and the presence of a characteristic interstitial lung infiltrate with serologic confirmation of IgG antibody to the offending antigen suffice for diagnosis. In more obscure cases, observation of changes after a natural environmental exposure, along with BAL and lung biopsy, might be indicated.

Early response to the antigen is characterized by an increase in neutrophils in the alveoli and small airways followed by an influx of mononuclear cells. These cells release proteolytic enzymes, prostaglandins, and leukotrienes. In addition the production and release of interleukins, cytokines, growth factors, and various other mediators from T lymphocytes and macrophages play an important role in HP pathogenesis.

Most syndromes are occupation-related and symptom prevalence can be up to 16 percent of exposed subjects, with significant morbidity. In addition to its clinical and economic importance, the disease has considerable research potential as a model of inflammatory lung disease in which the populations at risk can be identified, the relevant antigens purified, and the immunological and clinical consequences of antigen exposure monitored. Many of the pathological characteristics of HP overlap with other lung diseases and may therefore provide a means to identify common mechanisms of pathogenesis.

Prevalence varies by region, climate, and farming practices. Most patients recover completely after the inciting exposure ceases. The sources of antigen causing HP are diverse even though the clinical presentation of the different allergens is similar. This suggests that although the antigen is specific for each syndrome, the subsequent hypersensitivity pathway reactions are common. For this simple reason it could be argued that there can be no disease-specific antigen. There are however, some nonspecific agents common to the alveolitis-associated antigens. These are generally derived from organic material, or from organisms growing on organic waste. The agents are about one micron in diameter, which is the correct aerodynamic size to reach and deposit soluble antigen in the alveoli, where the disease process is most evident. These particles generally have adjuvant activity.

On the basis of the antigens we can distinguish several syndromes. *Farmer's lung*, first described by Campbell in 1932, is a hypersensitivity pneumonitis caused by the inhalation of thermophilic actinomycetes that grow in moldy hay or straw (*Saccharopolyspora rectivirgula*, *Micropolyspora faeni*, and *Aspergillus umbrosis*). *Bird fancier's (breeder's) lung* is caused by inhaling transuded serum and secretory proteins on feather dust (bloom) and in bird droppings. There is also described a *humidifier fever* caused by antigens dispersed by contaminated water from recirculating air-conditioning systems that harbor the protozoan *Naegleria gruberi*. *Isocyanate alveolitis* is an example of occupational exposure to a bioactive inorganic molecule that can cause HP [6].

T-Cell-Mediated Reaction or Delayed-Type Hypersensitivity (DTH) and Berylliosis

The delayed-type hypersensitivity reactions develop when antigen activates sensitized CD4+ T cells of the Th1 subset and CD8+ cells, both of which secrete cytokines that activate macrophages and induce inflammation. Berylliosis is a classic example of T-cell-mediated immune lung injury.

Beryllium (Be) is a lightweight metal with excellent thermal and electrical conductivity. Initially used in the manufacture of fluorescent lamps, beryllium is now widely used in the ceramics, nuclear weapons, computer, and aerospace industries. It is also a common component of many household appliances. Inhalation of beryllium has been associated with two pulmonary syndromes, which are an acute chemical pneumonitis and a granulomatous lung disease known as *chronic beryllium disease* (CBD), or *berylliosis*.

In the acute beryllium disease, the metal acts as a direct chemical irritant causing a nonspecific inflammatory reaction (acute chemical pneumonitis). Because of strict regulation of acceptable exposure levels, acute berylliosis rarely occurs.

CBD continues to be observed in industries where beryllium is manufactured and processed and where workers are exposed to beryllium fumes or dust. Its clinical and histopathological features are similar to those of other granulomatous diseases such as sarcoidosis.

The histopathologic findings in this disease are primarily diffuse alveolar damage. Although CBD may affect multiple organs, the lung usually is most severely involved. Exertional dyspnea is the most common presenting complaint, followed by cough and chest pain; less common symptoms include weakness, weight loss, fevers, and arthralgias. The diagnosis of CBD may be suggested by a history of significant exposure to beryllium, consistent radiographic findings, and abnormal pulmonary function tests. Confident diagnosis, however, requires demonstration of granulomas in tissue, immunologic evidence that granuloma formation is caused by beryllium hypersensitivity, and a proliferative response of lymphocytes obtained by bronchoalveolar lavage to beryllium. This last test, known as the *bronchoalveolar lymphocyte transformation test*, is highly accurate and can diagnose CBD before onset of clinical symptoms or pulmonary function abnormalities. Most beryllium is excreted in the urine, and the pulmonary half-life ranges from several weeks to 6 months. Relatively insoluble chemical forms of beryllium may be retained for years.

The key to the pathogenesis of CBD is a delayed-type hypersensitivity reaction in which beryllium most likely functions as a hapten and acts as a class II restricted antigen, stimulating local proliferation and accumulation in the lung of beryllium-specific T cells. Following inhalation of beryllium, large numbers of $CD4^+$ lymphocytes accumulate in the lungs. These helper T cells demonstrate a marked proliferative response as a result of exposure to beryllium. Beryllium not only has antigen-specific effects but also acts in a nonspecific inflammatory manner to promote the cellular events leading to granuloma formation. It may induce changes in lung permeability and the production of proinflammatory cytokines and growth factors that lead to granuloma formation and maintenance. As the disease progresses, the granulomas become organized and eventually form small, fibrous nodules; progressive impairment of pulmonary function occurs [7].

Fibrogenic Reactions

Fibrotic processes affecting the lung and other tissues are characterized by stimulation of fibroblast proliferation and connective tissue deposition. Conventional therapy, consisting of glucocorticoids or cytotoxic agents, is usually ineffective in blocking progression of disease.

Pneumoconiosis (Silicosis)

Pneumoconiosis is the general term for lung disease caused by inhalation of mineral dust. Thus, silicosis is a fibronodular lung disease caused by inhalation of dust containing crystalline silica (α-quartz or silicon dioxide), which is distributed widely, or its polymorphs (tridymite or cristobalite), which are distributed less widely. Workers with potential for exposure are miners, sandblasters, foundry workers, tunnel drillers, quarry workers, stone carvers, ceramic workers, and silica flour production workers.

Although silicosis has been recognized for many centuries, its prevalence increased markedly with the introduction of mechanized mining. In recent decades the high prevalence has declined in developed countries because of effective industrial hygiene measures. The disease is caused by small particles (< 1 mm) deposited distally in the respiratory bronchioles, alveolar ducts, and alveoli. The surface of these particles generates silicon-based radicals that lead to the production of hydroxyl and other oxygen radicals as well as hydrogen peroxide. These oxygen free radicals damage cell membranes by lipid peroxidation and inactivate essential cell proteins.

Alveolar macrophages ingest the silica particles, become activated, and release cytokines, including tumor necrosis factor, interleukin-1, and leukotriene B_4, as well as chemotactic factors that recruit other inflammatory cells. The ensuing inflammation damages resident cells and the extracellular matrix. Transforming growth factor–α induces proliferation of type 2 pneumocytes and stimulates synthesis of other cytokines (e.g., platelet-derived growth factor, insulin-like growth factor) and activates fibroblasts to proliferate and produce collagen; fibrosis results. Silica particles outlive the alveolar macrophages that ingest them, thereby continuing the cycle of injury.

Exposure to silica can lead to one of three disease patterns: (1) chronic simple silicosis, which usually follows more than 10 years of exposure to respirable dust with less than 30 percent quartz; (2) subacute or accelerated silicosis, which generally follows shorter, heavier exposures (i.e., 2 to 5 years); and (3) acute silicosis, which is often seen following intense exposure to fine dust of high silica content over a several-month period.

There are few symptoms and signs attending chronic simple silicosis. The diagnosis is usually made by chest radiographs, which reveal small round opacities (less than 10 mm in diameter) in both lungs, with a predilection for the upper lung zones. If an adequate occupational history is obtained from the patient along with a thorough review of the chest radiographs, the diagnosis of silicosis should not present any great difficulty. Pulmonary function tests in patients with simple silicosis usually are normal, but a mild obstructive impairment is occasionally found because of chronic dust-induced bronchitis. With complicated silicosis involving progressive fibrosis (nodules more than 10 mm in diameter), increasing dyspnea is noted, initially with exertion and progressing to dyspnea at rest. Complicated chronic silicosis is associated with reduced lung volumes, decreased diffusing capacity, and hypoxemia with exercise.

There is an increased incidence of mycobacterial disease, both typical and atypical, in silicosis. Fungal diseases (especially cryptococcosis, blastomycosis, and coccidioidomycosis) are also seen with greater frequency. The mechanism by which immune–inflammatory responses to inhaled silica lead to the increased incidence of mycobacterial and fungal infections is not clearly understood.

No treatment for silicosis is known, so management is directed toward the prevention of progression and complications. Continued exposure should be avoided and surveillance for tuberculosis should be instituted. In tuberculin skin-test positive (more than 10 mm induration) silicotic workers, treatment of latent tuberculosis infection with isoniazid for 9 months is recommended. In acute silicosis, therapeutic whole-lung lavage has been employed to physically remove silica from the alveoli.

The prognosis for patients with chronic silicosis is good, especially if they are removed from exposure. Mortality remains high, however, in those who develop progressive massive fibrosis (PMF) [8].

Asbestosis

Pulmonary fibrosis caused by asbestos inhalation is called asbestosis. Asbestos is the name for the fibrous forms of a group of mineral silicates including chrysolite, amosite, anthophyllite, and crocidolite. The word *asbestos* is derived from Greek and means "inextinguishable," and asbestos is a group of naturally occurring, heat-resistant fibrous silicates. Asbestosis is another example of a pneumoconiosis caused by inhalation and deposition of mineral dust.

Major occupational exposures occurred with asbestos mining and milling, manufacture or installation of insulation for ships or buildings, manufacture of friction materials for brake linings and clutch facings, asbestos cement manufacture, asbestos textile manufacture, and asbestos-containing spray products for decorative, acoustical, and fireproofing purposes.

Asbestosis refers to the diffuse interstitial pulmonary fibrosis caused by inhalation of asbestos fibers. Many factors are believed to play a role in disease initiation and progression, including the type and size of fiber, the intensity and duration of exposure, history of cigarette smoking, and individual susceptibility. A dose–response relationship exists such that asbestosis is more common in workers with a higher exposure level. Once asbestosis has begun, it may progress irrespective of removal from continued exposure. Finally, there is a considerable latency period (at least 10 years) between exposure and development of clinically apparent disease.

The diagnosis of asbestosis is made by a thorough exposure history, clinical examination, appropriate imaging studies and pulmonary function testing. The symptoms of asbestosis are indistinguishable from any other gradually progressive interstitial pulmonary fibrosing disorder, with progressive dyspnea and nonproductive cough being the most prominent. Bibasilar crackles with a "Velcro" quality can be auscultated over the posterolateral chest in the mid to late phase of inspiration. The crackles of asbestosis are unaffected by coughing.

Imaging studies that are helpful in the evaluation of asbestos-exposed patients are the chest radiograph and high-resolution CT scanning. The chest radiograph shows small, irregular or linear opacities distributed throughout the lung fields but more prominent in the lower zones, a pattern similar to many other forms of interstitial lung disease. The most useful radiographic finding is the presence of bilateral pleural thickening, which does not commonly occur with other diseases causing interstitial pulmonary fibrosis. Diaphragmatic or pericardial calcification is almost a pathognomonic sign of asbestos exposure. High-resolution CT scanning is the most sensitive imaging method for detecting early asbestosis.

Depending on the severity of disease, pulmonary function testing will show varying degrees of restrictive impairment. Because asbestosis begins as a peribronchiolar process, reduced flow rates at low lung volumes, indicative of small airways obstruction, may be seen.

There is no known treatment for asbestosis. Workers with the disease should be removed from further exposure, because the risk that parenchymal scarring will progress appears to increase with cumulative asbestos exposure. Any other factors that may contribute to respiratory disease should be reduced or eliminated. This is especially true of cigarette smoking because there is some evidence that it may contribute to the initiation and progression of asbestosis.

Lung cancer, either *squamous cell carcinoma* or *adenocarcinoma*, is the most frequent cancer associated with asbestos exposure. Epidemiological studies describe a significant multiplicative effect that leads to a far greater risk of lung cancer in persons who are cigarette smokers and have asbestos exposure than would be expected from the additive risk of each factor. *Mesotheliomas*, both pleural and peritoneal, are also associated with asbestos exposure. These tumors do not appear associated with smoking.

The substitution of other fibrous materials for asbestos and the institution of strict environmental controls where it is still present have led to a dramatic reduction of occupational exposures to asbestos. Medical surveillance of currently exposed workers in the United States is required by OSHA regulation [9].

Particle Problem and Its Multiple Mechanisms

Finally, there are the air fine particles often monitored in many countries as PM_{10} (less than 10 μ in aerodynamic diameter). These consist of a mixture of particle components, including traffic- and combustion-derived carbon-centered ultrafine particles (less than 100 nm in diameter), secondary particles (nitrates and sulfates, wind-blown dust of geological origin, potentially containing endotoxin), and biological particles (e.g., spores, pollens) with their associated allergens.

Because of the complex nature of PM_{10}, many studies have used model components or surrogates such as ultrafine carbon black, solutions of metals, or direct treatment with endotoxins.

There is epidemiological evidence of an association of PM_{10} levels with cardiovascular and respiratory mortality and morbidity. The precise mechanisms underlying the cardiovascular effects of PM_{10} remain uncertain. Current attention is particularly focused on ultrafine particles (diameter larger than 0.05 to 0.10 μm), which are highly reactive and are present in large numbers in the urban environment. Ultrafine particles can penetrate the epithelium and vascular walls and enter the bloodstream and, in animal models, have been reported to produce alterations in blood coagulability and increased rates of cardiovascular disorders, as well as increased carcinogenicity and potentiation of autoimmune disorders.

Some other epidemiological studies have reported that patients with asthma are adversely affected by PM pollution. As with the cardiovascular effects, there seems to be no threshold below which PM effects disappear.

There is also a considerable support for the importance of metals in the proinflammatory effects of PM_{10}. Transition metals are known to redox-cycle by Fenton chemistry, generating hydroxyl radicals capable of inducing oxidative stress and damage within biological systems.

Further studies in humans will no doubt help to elucidate the relevant mechanisms of air fine particles, and experimental use of defined mixtures will allow investigation of their potential additive or synergistic properties [10].

Abbreviation

BAL: Bronchoalveolar lavage
FEV_1: Forced expiratory volume at first second
HP: Hypersensitivity
PEF: Peak expiratory flow

References

1. Rom, W. N. (1998). *Environmental and Occupational Medicine*, 3rd ed. Philadelphia: Lippincott Williams & Wilkins. *The book provides encyclopedic coverage of the history, causes, prevention, and treatment of environmental and occupational diseases. The third edition features new chapters on molecular carcinogenesis, biological markers, genetic susceptibility to environmentally induced diseases, and molecular mechanisms of asthma and particle-induced lung disease.*
2. Bousquet, J. (2000). Global initiative for asthma (GINA) and its objectives. *Clin. Exp. Allergy* **30**(Suppl 1), 2–5. *This is the asthma classification most accepted worldwide, and it clearly recognizes inflammation as an important feature of asthma and as a target of therapy. More information for doctors and patients can be found at http://www.ginasthma.com/.*
3. Nelson, H. S. (2003). Advances in upper airway diseases and allergen immunotherapy. *J. Allergy Clin. Immunol.* **111**(3 Suppl), S793–S798.
4. Williams, S. G., et al. (2003). Key clinical activities for quality asthma care. Recommendations of the National Asthma Education and Prevention Program. *MMWR Recomm. Rep.* **52**(RR-6), 1–8.
5. Adkinson, N. F., Jr., Yunginger, J. W., Busse, W. W., Bochner, B. S., Holgate, S. T., and Simons, F. E. R. (2003). *Middleton's Allergy: Principles and Practice*, 6th ed. St. Louis: Mosby. *Focusing on human immunology and biology, while also reporting on relevant animal experimentation, this book provides comprehensive coverage of state-of-the-art basic science as well as authoritative guidance on the practical aspects of day-to-day diagnosis and management.*
6. Greenberger, P. A. (2002). Allergic bronchopulmonary aspergillosis, allergic fungal sinusitis, and hypersensitivity pneumonitis. *Clin. Allergy Immunol.* **16**, 449–468.
7. Rossman, M. D. (2001). Chronic beryllium disease: A hypersensitivity disorder. *Appl. Occup. Environ. Hyg.* **16**(5), 615–618.
8. Kuschner, W. G., and Stark, P. (2003). Occupational lung disease. Part 2. Discovering the cause of diffuse parenchymal lung disease. *Postgrad. Med.* **113**(4), 81–88.
9. Bourdes, V., Boffetta, P., and Pisani, P. (2000). Environmental exposure to asbestos and risk of pleural mesothelioma: review and meta-analysis. *Eur. J. Epidemiol.* **16**(5), 411–417.
10. Wilson, M. R., et al. (2002). Interactions between ultrafine particles and transition metals in vivo and in vitro. *Toxicol. Appl. Pharmacol.* **184**(3), 172–179.

Capsule Biography

Dr. Nicola Verna is an allergist and clinical immunologist with a Ph.D. in metabolic diseases. He worked as research fellow at the Trauma Center of Brigham and Women's Hospital in Boston, Massachusetts. He collaborated on the organization of international meetings focused on occupational medicine and environmental medicine.

SECTION R

Pre-eclampsia

CHAPTER 131

Microvascular Permeability in Pre-eclampsia

Nick Anim-Nyame, John Gamble, and Philip J. Steer

Academic Department of Obstetrics and Gynaecology, Faculty of Medicine, Imperial College London, Chelsea and Westminster Hospital, London, United Kingdom

Introduction

Profound maternal cardiovascular changes occur during pregnancy in response to altered nutritional requirements, added thermoregulatory needs and changes in the hormonal environment. Since the essential functions of the cardiovascular system occur at the level of the microcirculation, changes in microvascular parameters, such as vascular remodeling in response to altered local demand, are likely to occur in normal pregnancy. Pathophysiological changes are also likely to occur at the microvascular level in pre-eclampsia. The primary objective of this chapter is to consider how changes in the forces that govern both microvascular exchange and the microvascular endothelial barrier contribute to the pathophysiology of the multisystem disorder of pre-eclampsia. In the first part of the chapter, the cardiovascular changes in normal pregnancy that influence forces governing transvascular fluid movement and their effects on microvascular exchange are discussed. Since changes in microvascular permeability are likely to affect plasma volume, a brief discussion of volume homeostasis in pregnancy is included. Primary source references have been provided.

Volume Homeostasis in Normal Pregnancy

During normal pregnancy, blood volume increases by 40 percent, and this increase correlates positively with infant birth weight. The expansion starts in the first trimester, reaches a peak at about 30 weeks gestation, and remains elevated to term. Low maternal blood volume is associated with complications such as pre-eclampsia and intrauterine growth restriction (IUGR). Total body water (intracellular and extracellular) also increases significantly during normal pregnancy, accounting for 60 to 70 percent of the weight gain after the first 20 weeks of pregnancy. The expansion in plasma volume is part of the generalized increase in all compartments of total body water and results from fluid retention rather than a shift within the extracellular fluid compartments.

Mechanisms for Plasma Volume Changes in Normal Pregnancy

The precise mechanism(s) for plasma volume expansion during pregnancy remain unexplained. Although 950 mmol of sodium is accumulated and distributed between the products of conception and maternal extracellular volumes in pregnancy, neither changes in the maternal renal handling of sodium nor changes in sodium intake during normal pregnancy correlate with changes in plasma volume. Other suggested mechanisms for the increased plasma volume include increased levels of antinatriuretic factors, such as mineralocorticoids (aldosterone and deoxycorticosterone), and elevated natriuretic factors, such as oxytocin, prostacyclin, and melanocyte stimulating hormone (MSH). Increased progesterone levels may contribute to antinatriuresis by increasing deoxycorticosterone production via 21-hydroxylation.

Three theories, comprising vascular "underfill," "normalfill," and "overfill," have also been postulated to explain these observations. The "underfill" theory proposes that the hypervolemia of normal pregnancy is a suboptimal response to the generalized vasodilatation of pregnancy. Evidence in support of this theory includes activation of the renin–angiotensin system and data that adrenectomy and/or sodium restriction are tolerated less well by pregnant rats than by nonpregnant controls. These observations are supported by serial hemodynamic studies of both baboons and women in early pregnancy. The normal-fill theory proposes a continual resetting of volume sensing mechanisms as pregnancy advances, so that the increases in volume are sensed as normal. This theory is supported by evidence that sodium and water excretion by the proximal nephron is unaltered and in pregnancy, urine dilution appears to be normal during fluid overload. Furthermore, the sodium excretory responses to saline infusion are similar in the normal pregnant and nonpregnant states. The "overfill" theory, on the other hand, considers the plasma volume expansion as an epiphenomenon, secondary to factors such as the marked increased in circulating mineralocorticoids. This explanation is supported by the absolute increases in extracellular water volumes, high levels of circulating natriuretic factors, and/or inhibitors of the membrane pump increases in renal hemodynamics and also by increased sodium excretion in response to saline infusions.

Transcapillary Forces and Microvascular Permeability in Normal Pregnancy

Edema is one of the features of normal pregnancy, affecting up to 80 percent of all pregnant women. Whereas edema formation could result from a disturbance of the Starling forces that govern fluid exchange between vascular and interstitial spaces, the lymphatic drainage, or changes in permeability of the microvascular endothelial barrier itself, the precise mechanisms contributing to edema formation during normal pregnancy remain unexplained. The higher incidence of peripheral and pulmonary edema is suggestive that microvascular permeability is increased in normal pregnancy. Venous occlusion plethysmography studies support this concept, as do disappearance studies using Evan's Blue (a nontoxic dye that is firmly bound to plasma albumin). The edema of pregnancy appears to be due to increases in all extracellular fluid compartments, since the ratio of intravascular to interstitial compartment remains unchanged.

Transcapillary Colloid Osmotic Gradient in Normal Pregnancy

In normal pregnancy, plasma albumin concentration decreases from nonpregnant values to about 10 g/L by 30 weeks of gestation, and then decreases more slowly toward term. The reduced plasma albumin concentration (resulting in a fall in plasma oncotic pressure) would be expected to result in a loss of fluid from the intravascular to the interstitial compartment. Such a change would be expected to result in a fall in plasma volume, a rise in interstitial fluid volume and a reduction in the interstitial oncotic pressure. In fact, plasma volume increases during normal pregnancy and interstitial fluid volume remains fairly constant until late pregnancy, when edema usually becomes clinically evident. Examination of transcapillary fluid balance during normal pregnancy has shown evidence of a fall in plasma colloid osmotic pressure from 23.2 mm Hg during the first trimester to 21.2 mm Hg in the third trimester (a gradient of 2 mm Hg). In these studies however, the interstitial oncotic pressure has been shown to fall even more (4 to 5 mm Hg). The transcapillary colloid osmotic pressure difference was therefore increased in spite of reduced plasma oncotic pressure. Thus, it would seem that the reduced interstitial oncotic pressure maintains the normal transcapillary oncotic pressure gradient, thus preserving the circulating blood volume and protecting against edema. Although the mechanism(s) for the fall in interstitial oncotic pressure during normal pregnancy remain unexplained, it could be due to upregulation of lymphatic flow stimulated by increased capillary pressure or increased rate of transcapillary fluid flux. Increased lymph flow removes filtered proteins from the interstitium, thereby restoring a normal plasma/interstitial fluid protein concentration gradient.

Plasma oncotic pressure has been shown to fall from 22 mm Hg at the onset of labor to a nadir of 16 mm Hg at 6 hours after delivery. This is irrespective of the mode of delivery or method of analgesia, although it does appear to be exacerbated by use of excessive amounts of intravenous crystalloids during labor and delivery. The value of plasma oncotic pressure at the time of delivery is close to values at which noncardiogenic pulmonary edema occurs, which is 13 to 16 mm Hg.

Hydrostatic Pressure Gradient in Normal Pregnancy

Capillary hydrostatic pressure, calculated from Starling's equation using values of plasma oncotic pressure, interstitial oncotic pressure (wick method), and interstitial fluid hydrostatic pressure (wick-in needle technique), increases by about 30 percent between the first and third trimesters. However, in these studies, no significant difference in interstitial hydrostatic pressure was observed. According to Starling's equation, the combined effects of increased hydrostatic and oncotic pressure gradients would increase net fluid filtration into the interstitial compartment, thereby reducing intravascular volume. In fact, plasma volume actually increases during pregnancy, while the interstitial fluid volume remains constant, until late pregnancy when edema is usually observed.

The pulmonary vascular resistance, as well as the systemic vascular resistance, also falls during normal pregnancy. Pulmonary wedge pressure is slightly reduced from nonpregnant values, in spite of an increased cardiac output. Pulsed Doppler pulmonary blood velocities have been used to calculate pulmonary artery pressure. The results indicated that pulmonary pressure falls from 2.8 resistance units before pregnancy to 2.7 resistance units at 8 weeks of gestation. It then remains unchanged for the rest of the pregnancy, returning to nonpregnant values by 6 months after delivery.

Given the fact that plasma oncotic pressure at the time of delivery falls close to values at which noncardiogenic pulmonary edema occurs (i.e., 13 to 16 mm Hg), it is perhaps surprising that the modern management of labor, with liberal use of crystalloid intravenous fluids, does not result in pulmonary edema more frequently.

Microvascular Permeability in Pre-eclampsia

Pre-eclampsia is a multisystem disorder of the second half of pregnancy, characterized by hypertension and proteinuria. Although the primary pathology remains unexplained, it appears that abnormal implantation and placental ischemia may be the basis of the disease. The mechanisms by which reduced placental perfusion translates into profound alterations of maternal physiology are still unclear. There is functional, morphological, and biochemical evidence that generalized endothelial dysfunction underlies the clinical manifestations of the disease. Clinical features of the disease, such as edema, proteinuria, and reduced plasma volume, suggest that increased microvascular permeability may play a role in the pathophysiology and the resulting complications of the disease. For example, plasma volume, which is less than normal in pre-eclampsia, correlates positively with birth weight and inversely with disease severity. Edema affecting organs such as the brain, eyes, and lungs increases the morbidity and mortality associated with the disease. The risk of pulmonary edema is significantly greater in pre-eclampsia compared to normal pregnancy and is a major cause of death from the disease.

The reduced plasma volume seen in pre-eclampsia is due to an altered distribution of total extracellular fluid volume (ECFV). Although total ECFV remains unaltered in pregnancies complicated by the disease, redistribution of plasma volume into the interstitial space as a result of increased microvascular permeability occurs, as evidenced by the reduction in the ratio of intravascular to interstitial fluid volumes. Studies of Evans Blue dye disappearance have shown increased microvascular permeability to proteins in pregnancies complicated by pre-eclampsia. The albumin-bound Evans Blue dye disappears faster from the intravascular compartment of women with pre-eclampsia compared to normal pregnant controls. Moreover, the change correlates significantly with the degree of proteinuria and with the level of edema. Since the dye is firmly bound to albumin, the increased permeability implies a reduction in the value of the osmotic reflection coefficient (σ), thereby reducing the effective oncotic pressure generated by the plasma proteins.

However, it should be noted that the results of different ways of examining this phenomenon do not give consistent results. For example, although increased microvascular permeability in pre-eclampsia is observed using the Evans Blue dye redistribution technique (confirmed by various investigators), other workers failed to demonstrate any correlation between permeability measured with different methodologies and the degree of proteinuria. Furthermore, there is a poor correlation between plasma volume and microvascular permeability. These results suggest that other components of the Starling equation that govern the movement of fluid across the microvascular interface may also play a role in the observed plasma volume reduction in pre-eclampsia. It should be noted that the oncotic pressure measured in vitro, where the value of (σ) for the examining membrane is 1.0, might differ from the effective (in vivo) oncotic pressure at the microvascular interface, which is influenced by pathophysiological and physicochemical changes at the endothelial interface.

Oncotic Pressure Gradient in Pre-eclampsia

It has been shown that compared to normal pregnancy, plasma albumin concentration and oncotic pressures are reduced by 6 percent and 27 percent in moderate and severe pre-eclampsia, respectively. In moderate pre-eclampsia the reduction in plasma oncotic pressure is compensated for by a similar reduction in interstitial oncotic pressure, so that oncotic pressure gradient and therefore plasma volume remain unchanged. The reduction in interstitial fluid oncotic pressure may be due interstitial protein washout, attributable to increased lymphatic protein transport and/or dilution of preexisting interstitial protein by the increased microvascular filtration of protein-poor fluid. In contrast, in severe pre-eclampsia, the interstitial oncotic pressure increases so that the transcapillary oncotic pressure gradient falls. This increase in interstitial oncotic pressure could be due to reduced capillary pressure (secondary to raised precapillary resistance, for instance), increased microvascular permeability to plasma proteins, or reduced lymphatic flow. It has been noted that interstitial oncotic pressure appears to correlate with the degree of hypertension and the level of proteinuria, as well as with peripheral edema, suggesting that increased microvascular permeability to proteins may be a major contributor to the increase. This may explain the reduced plasma volume and hemoconcentration often observed in severe pre-eclampsia.

Hydrostatic Pressure Gradient in Pre-eclampsia

Interstitial hydrostatic pressure increases in pre-eclampsia, and this may be due to raised interstitial fluid volume as a result of increased microvascular filtration, which is not adequately compensated by increased lymph flow. However, the reported increase in hydrostatic pressure is small and may be attributable to high interstitial tissue compliance. Capillary hydrostatic pressure calculated from other components of the Starling equation increases by 30 percent between the first and third trimester in normal pregnancy, but decreases by 40 percent during the same period in pre-eclampsia. Since capillary pressure is dependent on the values of pre- and postcapillary resistance, the low values may be due to increased precapillary tone from the generalized vasospasm in pregnancies complicated by the disease. However, intravascular pressures at the microvascular interface may be influenced by postcapillary pathophysiological changes.

Endothelial Barrier Function in Pre-eclampsia

Although changes in the oncotic and hydrostatic pressure gradients do occur in pre-eclampsia compared to normal pregnancy, they do not appear adequately to explain the increased fluid flux observed in pregnancies complicated by the disease. Whereas increased microvascular filtration capacity resulting from the generalized endothelial dysfunction of pre-eclampsia may account for the enhanced microvascular permeability, the precise local mechanism(s) remain unexplained. Such mechanisms may include redistribution of endothelial junctional proteins, as well as the effects of circulating permeability factor(s), several of which are elevated in pregnancies complicated by the condition.

Endothelial Junctional Proteins in Pre-eclampsia

There is evidence that the endothelial cell dysfunction known to occur in pre-eclampsia results in redistribution of the endothelial cell junctional proteins: VE-cadherin and occludin. VE-cadherin is an endothelium-specific cadherin that regulates junction organization in endothelial cells and is selectively expressed in all types of endothelial cells. It plays an important role in the organization of lateral endothelial junctions, and its expression is required for the maintenance of normal endothelial barrier function. Occludin is a transmembrane glycoprotein that is found in tight junctions and has a dual role in terms of barrier function. Tight junctions create the primary barriers to solute diffusion through paracellular pathways. They also serve as a barrier between apicolateral and basolateral plasma membrane domains, maintaining cell polarity. VE-cadherin and occluding junctional protein organizations and the permeability of endothelial cell monolayers isolated from human umbilical vein endothelial cells (HUVECs) from both normal pregnant and pre-eclamptic pregnancies have been compared using electron microscopy.

In pre-eclampsia, junctional proteins are disorganized, and in addition, the permeability of the endothelial monolayer is significantly increased. Furthermore, significant correlation has been observed between the monolayer permeability and junctional protein redistribution. Although the data provided an insight into the molecular basis for increased microvascular permeability, the HUVEC samples used in this study were of fetal origin and may differ functionally from maternal endothelial cells. Thus, although interesting, the observations may not reflect maternal endothelial barrier function.

Circulating Permeability Factors in Pre-eclampsia

The altered junctional protein expression associated with increased microvascular permeability appears to be a consequence of some pathophysiological events that affect endothelial function rather than an inherent endothelial cell defect. Although the main reason for this is given as the reversibility of the adverse changes, it should be remembered that endothelial cells have a slow but significant rate of turnover, so that the "recovery" might reflect cell replacement. Pathological features of pre-eclampsia such as oxidative stress, neutrophil activation, and the release of inflammatory cytokines could be responsible for initiating the endothelial barrier dysfunction associated with pregnancies complicated by the disease.

There is evidence that sera from women with pre-eclampsia rather that normal pregnancy increase the permeability of HUVEC monolayers. This suggests that there may be a factor or factors present in the maternal circulation that are responsible for the increased microvascular permeability in pregnancies complicated by the disease. Maternal circulating factors that are elevated in pre-eclampsia and may affect microvascular permeability include vascular endothelial growth factor (VEGF), angiopoietins, leptin, proinflammatory cytokines such as tumor necrosis factor-α (TNF-α), and, more recently, neurokinin B. The possible mechanism(s) of action of these circulating factors on microvascular permeability are discussed next.

VEGF and Angiopoietin-1

VEGF is a 43- to 46-kDa glycoprotein that serves as a key factor in the maintenance of a confluent, normally functioning vascular endothelium. Physiological levels of VEGF in the serum are pivotal for maintaining vascular endothelial cell homeostasis. VEGF is one of the most potent microvascular permeability inducing agents known. Circulating levels are elevated in pre-eclampsia, and this appears to be related to tissue hypoxia associated with the disease. Angiopoietins are vascular endothelial cell-specific growth factors that play an important role, principally during the last stages of the angiogenesis occurring after the induction of new capillaries by VEGF. There is evidence that overexpression of angiopoietin-1 (Ang-1) results in nonleaky vessels by inhibiting the effects of VEGF. Although combined overexpression of VEGF and Ang-1 has been shown to have an additive effect on new vessel formation, this combination results in leak-resistant vessels typical of Ang-1. Furthermore, Ang-1 suppresses VEGF-induced expression of cell adhesion.

Although the role of VEGF in microvascular permeability in the nonpregnant state is well established, its role in microvascular permeability in pregnancies complicated by pre-eclampsia remains to be established. Using scanning electron microscopy, it has been shown that incubation of subcutaneous arteries from women with pre-eclampsia with VEGF increases vascular permeability with disorganization of the vascular endothelium and development of intercellular gaps. In the same study, Ang-1 was shown to reverse the VEGF-induced increase in permeability as well as the development of gap junctions. These observations certainly support the notion that Ang-1 can act as antipermeability factor. However, in a clinical study using venous occlusion plethysmography to measure microvascular filtration capacity as an index of permeability, no correlation was observed between microvascular permeability and plasma concentrations of VEGF. It is therefore possible that the effect of increased VEGF on vascular permeability is suppressed by the presence of other circulating factor(s), such as Ang-1, in pregnancies complicated by pre-eclampsia.

TUMOR NECROSIS FACTOR-α (TNF-α)

TNF-α is a proinflammatory cytokine and is associated with increased microvascular permeability, for example in sepsis. It appears to do this in studies of nonpregnant women by increasing the permeability coefficient (Lp) of venular endothelial cells. Plasma levels of TNF-α are significantly increased in pre-eclampsia and may play a role in the pathophysiology of the disease. TNF-α released by the ischemic placenta may provide the link between abnormal placentation and maternal endothelial dysfunction. It has also been postulated that the effects of TNF-α may be mediated by oxidative stress. Thus, it would appear that upregulation of TNF-α in pre-eclampsia might provide a mechanism for the increased microvascular permeability seen in this disease. There is indeed evidence of a positive correlation between plasma TNF-α concentrations and microvascular permeability in pre-eclamptic patients, although not in pregnant controls.

LEPTIN

Leptin, a protein product of the obesity gene, plays an important role in regulation of body weight in the nonpregnant state through its receptors in the satiety center of the hypothalamus. Endothelial cells also express leptin receptors. Their activation induces angiogenesis and increases microvascular permeability. It has been suggested that the increased circulating leptin found during pregnancy is of placental origin. Although the functions of leptin in the placenta are not known, in addition to its angiogenic effect it has been suggested that may help to increase the exchange of small molecules between the maternal circulation and the fetus by enhancing convective mechanisms and by the induction and maintenance of increased vascular permeability. Placental expression and circulating levels of leptin are significantly increased in pregnancies complicated by pre-eclampsia. Structurally, leptin has been shown to resemble class I cytokines. Moreover, it may regulate placental cytokine production during pregnancy itself.

NEUROKININ B

Neurokinin B is a decapeptide of the tachykinin family, a group of neuropeptides that include substance P, a well-established proinflammatory neuropeptide and a potent mediator of increased microvascular permeability. There is evidence that neurokinin B is a potent stimulator of plasma extravasation through two distinct pathways: via activation of NK_1 receptors, and via a neurokinin-independent pathway. The human placenta expresses neurokinin B, and plasma concentrations are significantly elevated in pregnancies complicated by pre-eclampsia. It has been suggested that elevated levels may explain edema formation in pre-eclampsia in a manner similar to its effect in causing plasma extravasation in animal studies. Although neurokinin B may provide the candidate permeability factor in pre-eclampsia, edema is not specific to pre-eclampsia and occurs in up to 80 percent of pregnant women.

Future Research

Microvascular endothelial cells form the semipermeable wall of capillaries and venules that are essential for regulating blood–tissue exchange. Although clinical studies suggest that increased transvascular flux of fluid and proteins in pre-eclampsia might be due to endothelial barrier failure in pre-eclampsia, the molecular mechanisms underlying the pathological regulation of microvascular endothelium remain to be investigated. Transendothelial movement of fluid and solutes is a dynamic process regulated by a complex interaction between signaling molecules and structural elements comprising the cell cytoskeleton, cell–cell adherence, and cell matrix attachment. Further work is required to unravel the intracellular signaling pathways that mediate increased permeability and the molecular events that alter the endothelial barrier structure. There is convincing evidence that pre-eclampsia is an exaggerated inflammatory state, and, as indicated earlier in the chapter, clinical studies have shown a significant correlation between microvascular permeability and levels of TNF-α. Further research is required to investigate the specific intracellular pathways by which inflammatory mediators increase microvascular permeability, particularly the roles of phospholipase C, cytosolic calcium, protein kinase C, nitric oxide synthase, and guanylate cyclase. More work is also required on how the permeability-inducing signals result in the formation of intercellular gaps through disorganization of junctional proteins and focal adhesion phosphorylation and redistribution, if we are to fully understand the mechanism of enhanced microvascular permeability in pre-eclampsia.

Summary

Microvascular permeability is increased in pre-eclampsia and may well play an important role in the pathophysiology of the disease. Abnormal transvascular flux of fluid and proteins resulting in tissue edema and dysfunction accounts for some of the clinical manifestations and complications of the disease. Although changes in oncotic pressure gradient resulting from a failure of the interstitial edema preventing mechanisms may contribute to the tissue edema in pre-eclampsia, it is likely that loss of endothelial barrier function due to generalized endothelial dysfunction plays a role in the significantly increased transvascular fluid flux. It is apparent that factor(s) present in the maternal circulation, such as cytokines and growth factors, as well as oxidants released from activated neutrophils, may play a role in the loss of endothelial barrier function in pregnancies complicated by the disease. However, the mechanism(s) of increased microvascular permeability remain unexplained.

Glossary

Filtration capacity: An index of microvascular permeability; reflects the product of the area available for fluid filtration and the permeability per unit surface area.

Osmotic reflection coefficient (σ): An index of the microvascular permeability to plasma proteins. $\sigma = 1.0$ means the microvascular surface interface is wholly impermeable to the plasma protein under consideration, a situation rarely encountered in vivo. Values less than 1 imply less complete rejection of the solute

Pre-eclampsia: A hypertensive disorder of pregnancy diagnosed as hypertension (BP of 140/90 or higher) and proteinuria (24 hours urine protein of 0.3 g/L or higher) occurring for the first time after 20 weeks gestation, and reversal of both after delivery.

Bibliography

Anim-Nyame, N., Gamble, J., Sooranna, S. R., Johnson, M. R., and Steer, P. J. (2003). Microvascular permeability is related to circulating levels of TNF-α in pre-eclampsia. *Cardiovasc. Res.* **58**(1), 162–169. *This article provides evidence that microvascular permeability in pre-eclampsia correlates with circulating levels of TNF-α but not VEGF.*

Bates, D. O., Lodwick, D., and Williams, B. (1999). Vascular endothelial growth factor and microvascular permeability. *Microcirculation* **6**(2), 83–96. *A review of the mechanisms by which VEGF increases microvascular permeability.*

Haller, H., Hempel, A., Homuth, V., Mandelkow, A., Busjahn, A., Maasch, C., Drab, M., Lindschau, C., Jupner, A., Vetter K., Dudenhausen, J., and Luft, F. C. (1998). Endothelial-cell permeability and protein kinase C in pre-eclampsia. *Lancet* 3 51, 945–949. *This article provided evidence for the first time that a factor(s) present in the sera of women with pre-eclampsia increased microvascular permeability.*

Redman, C. W., Sacks, G. P., and Sargent, I. L. (1999). Pre-eclampsia: An excessive maternal inflammatory response to pregnancy. *Am. J. Obstet. Gynecol.* **180**, 499–506. *A detailed review of the evidence that the endothelial dysfunction of pre-eclampsia is part of a generalized inflammatory response that is common to pregnancy but exaggerated in pregnancies complicated by the disease.*

Yuan, S. Y. (2003). Protein kinase signalling in the modulation of microvascular permeability. *Vasc. Pharmacol.* **39**, 213–223. *A review of the complex interactions between signaling molecules and structural proteins that influence endothelial barrier integrity.*

Capsule Biography

Dr. Nick Anim-Nyame is an obstetrician with a research interest in microvascular physiology. He was winner of the Burroughs Wellcome prize at the summer 2000 FASEB Conference and of the European Microcirculation Society travel award (2001). His current research interests are regulation of microvascular blood flow, particularly mechanisms for sensing tissue hypoxia, and microvascular changes in complicated pregnancies.

Dr. John Gamble pioneered the plethysmographic system for noninvasive assessment of microvascular permeability and other microvascular parameters in humans. He is currently honorary Senior Research Fellow at the School of Sports and Exercise Sciences. University of Birmingham, United Kingdom.

Professor Philip J. Steer is Academic Head of Obstetrics and Gynaecology at Chelsea and Westminster Hospital, London. He is a member of the editorial board of *The British Journal of Obstetrics & Gynaecology* and was president of the British Association of Perinatal Medicine. His current research interest is maternal factors affecting fetal growth.

SECTION S

Wound Healing

CHAPTER 132

Wound Healing: An Orchestration of Humoral and Cellular Forces

Rodney K. Chan, Perry H. Liu, and Dennis P. Orgill
Department of Plastic Surgery, Brigham and Women's Hospital, Harvard Medical School, Boston, Massachusetts

Introduction

The healing of chronic wounds is a major health problem. Through studying the normal healing of acute wounds and the use of impaired wound healing models such as the genetically diabetic mouse, we have dissected the cell types and growth factors that are involved in this complex and collaborative process. This review discusses the classical phases of wound healing and the signals important for cell recruitment and proliferation, epithelialization, and angiogenesis. The complexity of this process is highlighted by numerous clinical trials performed to translate these growth factors into clinical use but only one currently approved drug recombinant human platelet derived growth factor-ββ. There is a desperate need for therapies that are both efficacious and cost-effective.

Unlike amphibians, humans lack tissue regenerative abilities and generally respond to injury through scarring and contraction. Some exceptions, such as regeneration of liver tissue after partial hepatectomy, suggest that the capacity for regeneration exists. The skin also has the potential for repair and regeneration. In fact, wounds in the second trimester not only heal in utero with minimal scarring, but also achieve the strength of unwounded skin. Yet, in adults, chronic nonhealing wounds are becoming a major problem, one made worse by our growing elderly population and a dramatic rise in the incidence of diabetes.

A wide variety of treatments for chronic wounds exist. These include surgical and enzymatic debridements, dressings, topical and systemic antibiotics, moist environment, surgical revascularization, nutritional supplementation, skin grafts, tissue flaps, and tissue engineering products. These options have allowed many wounds to heal without the need to resort to amputations or radical reconstructions. Yet many chronic wounds persist, especially among diabetics, in areas of prior irradiation, as well as in areas under constant pressure or venous congestion.

In the past two decades, the mechanisms of normal wound healing have slowly been elucidated. The purpose of this chapter is to review our current understanding of wound healing biology and to provide a basis for devising new treatments, especially those involving the application of growth factors to poorly healing wounds.

Wound Healing Cascade

The creation of an acute wound initiates an overlapping sequence of events involving specific cells, chemokines, and cytokines directed at limiting the initial damage and restoring tissue integrity. The four classical phases of this process are coagulation, inflammation, proliferation, and maturation [1].

In the coagulation phase, traumatically exposed collagen IV and V initiate the coagulation cascade, which begins with platelet aggregation and plugging. Within seconds to minutes, the formation of a fibrin scaffold traps red blood cells and provides a lattice framework for endothelial cells, inflammatory cells, and fibroblasts to operate during wound healing.

This is followed by the inflammatory phase when activated platelets and endothelial cells release mediators that increase vascular permeability and recruit inflammatory cells such as neutrophils and, subsequently, macrophages and fibroblasts. The resulting inflammatory infiltrate clears the wound of contaminating debris and bacteria. Over the following 2 to 3 days, neutrophils are replaced by macrophages and mononuclear phagocytic cells. Aside from their phagocytic abilities, macrophages provide more than 30 different growth factors and cytokines that stimulate angiogenesis and extracellular matrix production.

The proliferative phase of wound healing is characterized by the replacement of the provisional fibrin matrix, established during the initial clot formation, with collagen and proteoglycans. Activated monocytes and platelets synthesize platelet-derived growth factor (PDGF) and transforming growth factor-β (TGF-β) to stimulate fibroblast proliferation and collagen synthesis. Type III collagen initially predominates but is gradually replaced by type I collagen as the wound matures. Endothelial cells also proliferate to form new capillaries to deliver the necessary oxygen and nutrients for wound repair.

During wound maturation, the final phase of wound healing, random collagen fibrils are replaced by organized fibrils with increasing numbers of intermolecular bonds. During the first 6 weeks, wound strength increases rapidly but then gradually levels off in an asymptotic fashion. Although healed wounds never achieve the strength or degree of collagen organization found in normal tissue, they do eventually reach 70 percent of the strength of nonwounded skin.

Growth Factors in Wound Healing

Cutaneous wound healing is an orchestra of cell division, cell migration, and the production and degradation of extracellular matrix. For each stage of the process, large number of growth factors act as triggers, modulators, and terminators. Our basic understanding of wound healing comes from pathological descriptions of cellular migration and the appearance and disappearance of various types of extracellular matrices. More recently, the increasing availability of knockout animals and specific protein antagonists has allowed the elucidation of critical elements in the healing of cutaneous wounds.

PDGF was the first factor described to have an effect on wound healing. It induces neutrophil chemotaxis, fibroblast proliferation, and extracellular matrix production. Released in large quantities by platelets within the wound, it has been shown to be absent during the dysfunctional healing of diabetic wounds. Preclinical studies using PDGF on wounds have been mixed, but on average encouraging [2]. PDGF-treated wounds have shown a consistent increase in granulation tissue formation. Its effects on epithelialization and wound contraction have, however, been variable and sometimes nonreproducible. In a guinea pig partial thickness skin wound model, the application of PDGF showed no effects on either epithelialization or contraction. However, there was a greater amount of granulation tissue compared to vehicle alone. Similar results were obtained using a pig full-thickness skin wound model. Only in rabbit ear and diabetic mice wound models did PDGF increase the amount of granulation tissue as well as the extent of epithelialization. PDGF was found to have a dramatic effect in enhancing wound closure as well as the amount of fibroblasts and capillaries in the wound. The combination of these studies led to the testing of PDGF in a series of clinical trials, which ultimately resulted in its approval by the U.S. Food and Drug Administration (FDA) for the treatment of chronic neuropathic ulcers in diabetic patients. In addition to PDGF, platelets also produce many other growth factors (Table I). Many of them have also been tested in the context of wound healing.

Table I Table of Platelet-Derived Factors and Their Presumed Role in Wound Healing.

Growth factor	Effects
PDGF	Neutrophil chemotaxis, fibroblast proliferation, extracellular matrix production
EGF	Keratinocyte proliferation and migration
TGF-β	Fibroblast chemotaxis, angiogenesis, extracellular matrix production
Basic FGF	Angiogenesis
Acidic FGF	Angiogenesis
PF-4	Antiangiogenic

Despite the seemingly central role that platelets and their reservoir of growth factors play in wound healing, the addition of a neutralizing anti-platelet antibody to mice has surprisingly led to no change in the wound healing rates, but significantly increased the number of macrophages and T cells present in the wound [3]. This is just a reminder that our understanding of wound healing is still rudimentary and much work still needs to be done.

Wound Reepithelialization

Rapid wound epithelialization, the migration of keratinocytes to seal an open wound, is considered by many to be a good predictor of successful wound healing. A fully epithelialized wound is more likely to heal. Keratinocytes begin this process by invading the provisional clot made of fibrin and fibronectin. Plasmin is upregulated soon after wounding, presumably for breaking down the fibrin clot. Indeed, plasminogen knockout mice demonstrate significantly diminished wound reepithelialization [4]. Other degradation enzymes, such as matrix metalloproteinase (MMP)-1, -9, and -10, are also upregulated around wound edges. Analysis of chronic wound fluid revealed high levels of proteolytic activity from MMPs, suggesting that their overexpression retards wound healing. In normal wound healing, epithelial cells stretch and divide to cover a denuded wound surface with a single layer of cells. Once a single epithelial cell layer has been established, in a process inhibited through integrin signaling and contact inhibition, keratinocyte proliferation ceases.

Fibronectin, which binds fibrin, is a potent stimulus of keratinocyte migration. The α-3, β-1 integrin fibronectin receptor is upregulated in keratinocytes after wounding, suggesting that it is important in keratinocyte migration. The fragment of fibronectin that binds to keratinocytes has been found in large quantities in wounds and has been shown to stimulate monocyte and fibroblast chemotaxis through the dermis [5]. Indeed, it appears that the invasive/migratory phenotype expressed by epithelial cells and fibroblasts after wounding may be the result of interactions between this soluble fibronectin fragment and α-3, β-1 integrin signaling. Topical application to this fragment has been shown to accelerate wound healing in diabetic obese mice.

Epidermal growth factor (EGF), transforming growth factor-α, and heparin binding-EGF are all part of the EGF superfamily and have been considered key regulators of keratinocyte proliferation. All three of these factors are released in abundance at the site of injury. Application of the first two to burnt pig skin hastens the reepithelization process. Other stimulants of epithelialization include keratinocyte growth factor (KGF), which is upregulated one-hundredfold in normal wound margins but significantly less in those of diabetic wounds and steroid-treated wounds. Surprisingly, however, KGF knockout mice do not exhibit delayed reepithelialization. This is thought to be secondary to redundancy since mice with the dominant mutant KGF receptor demonstrate significantly delayed wound healing. Application of exogenous KGF also has rate-enhancing effects with increased expression of tissue type plasminogen activator and MMP-10. Furthermore, transplantation of both normal and transgenic keratinocytes into full-thickness wounds has improved wound healing.

Surgeons have long known that a key factor in wound healing is excellent tissue perfusion. Vascular endothelial growth factor (VEGF), fibroblast growth factor (FGF), and TGF-β are the critical factors involved in wound angiogenesis. VEGF is expressed at low levels in normal skin but highly upregulated in keratinocytes during wound healing. Angiogenesis appears to be tied to matrix deposition and granulation tissue formation. Application of a neutralizing anti-VEGF antibody to a wound is associated with decreased wound angiogenesis. Conversely, application of topical VEGF increases wound angiogenesis, granulation tissue formation, and wound closure rates [6]. Diabetic wounds are found to have decreased levels of VEGF expression, but demonstrate very high levels of angiopoetin-2, an antiangiogenic protein. Although angiopoetin-2 production in wounds is not unusual during specific phases of wound healing when vessel growth requires destabilizing preexisting vessels, such high levels are not usually seen. The impaired angiogenesis observed in diabetic mice is thought mainly to be due to free radical mediated lipid peroxidation, dysfunctional keratinocytes, and a resultant decreased endothelial cell VEGF production.

TGF-β is a multipurpose growth factor expressed ubiquitously during wound healing, but its major roles are the stimulation of endothelial cell migration and matrix deposition. There are several different TGF-β isoforms, but they interestingly all bind to the same receptor. TGF-β1 is expressed early on in wound healing, whereas TGF-β3 is expressed late. TGF-β3 has also been shown to decrease scarring. Despite the expectedly decreased matrix deposition in TGF-β-receptor 1 deficient animals, increased wound angiogenesis is observed. This apparent paradox highlights the apparently contradictory effects of TGF-β in a wound and in culture. TGF-β inhibits angiogenesis in vitro, whereas the opposite occurs in vivo.

Wound Healing and the Immune System

Thought initially to be the hallmark of the inflammatory phase, a large cellular infiltrate is now known not to be beneficial to the rate of wound closure. Neutrophil-depleted mice demonstrate markedly faster wound closures than their wild-type counterparts. Also, mice deficient in the tumor necrosis factor receptor p55 have a reduced leukocyte infiltration and improved wound closure. Mice deficient in TGF-β signaling protein SMAD3 have reduced monocytic infiltration, reduced matrix, and an increased rate of wound closure [7]. Similarly, estrogen has shown positive effects on cutaneous wound healing by downregulating the macrophage migration inhibitory factor (MIF) and decreasing inflammation. MIF knockout animals also have fast wound closure, decreased inflammatory cell infiltrate, and increased matrix deposition. All these studies indicate that reduced cellular infiltration is associated with a faster wound healing phenotype. It is, however, important to determine whether faster healing is necessarily better healing, as the hastily healed wound of SMAD3-deficient animals might lack resilience to breaking forces.

Whereas a large neutrophilic infiltrate adversely affects wound closure rates, other cell types are also important in positively or negatively modifying the wound-healing process. γ-δ T cells are a subset of the T-cell population. Unlike the majority of the T-cells, which are of an α-β subtype and involved in cellular and humoral immunity, these cells reside mostly in epithelial tissue. Previously thought to be unimportant, they are now known to proliferate in response to injury. When stimulated, they produce FGF-7 and KGF-2 (FGF-10). Animals deficient in γ-δ T-cells exhibit delayed wound closure, which can be rescued with the application of either γ-δ T cells or KGF [8]. Thus, this T-cell subtype and its products serve to enhance wound closure. The results of similar studies in T-cell deficient nude mice are less clear. Wound strength was increased in the first 2 weeks post injury, but decreased below that of wild-type

Table II Summary of All Clinical Trials Pertaining to the Use of Growth Factors in Wound Healing.

Drug[a]	Year	Type	No.	Outcome	
For chronic diabetic neuropathic ulcers					
PDGF-ββ	—	III	250	No difference between good ulcer care alone and 100 μg/g PDGF	
PDGF-ββ	—	III	172	Worse healing in PDGF group in comparison to good ulcer care alone	
PDGF-ββ	—	III	382	Improved healing only in high-dose 100 μg/g group	
PDGF-ββ	1996	III	118	Improved healing in PDGF 30 μg/g group	
FGF	1995	II	17	No difference	
For chronic nonhealing venous ulcers					
rhKGF-2	2001	II	94	More patients achieve 75% wound closure in KGF-2 treated group	
Epithelial cultures	2002	I	11	7 patients healed, 4 did not	
PR	2000	II	86	No difference after biweekly prescription for 9 months	
PR	1999	II	15	No significant difference	
GM-CSF	1999	I	38	Complete healing seen in 90.4% of patients without recurrence	
GM-CSF	1999	II	60	Improvement of number of animals with complete healing by 13 weeks	
EGF	1992	II	35	Improvement in rate of complete healing	
For chronic pressure ulcers					
TGF-β3	2001	II	270	High dose (2.5 μg/cm^2) TGF-β3 healed faster on fourth visit but no difference at the end	
GM-CSF and/or FGF	2000	II	—	FGF alone achieved the fast rate of wound closure	
PDGF-ββ	1999	II	124	Significant increase in wounds achieving > 90% healing	
PDGF-ββ	1994	II	41	Effective in decreasing wound area after 28 days	
PDGF-ββ	1992	I/II	20	Significant improvement in closure	
For miscellaneous wounds					
PDGF-ββ	2002	II	21	Abdominal wound separation	Decreased in time to complete healing
GM-CSF	2002	I	5	Chronic leg ulcers	Appears effective
GH	2002	II	24	Donor site healing in burned adults	No improvement in donor site healing time
Bovine basic FGF	2000	II	1024	Burn wounds, donor sites, and chronic dermal ulcers	Treatment arm (bFGF) improved wound closure with superficial second best
PDGF-ββ	2001	Ia	7	4-mm full-thickness punch biopsy	Daily PDGF treatment significantly improved closure rate
PDGF-ββ	2000	I	134	Chronic lower extremity diabetic ulcers	57.5% achieved complete healing in 20 weeks with 21% recurrence
GM-CSF	1998	I	29	Chronic refractory wounds	One third healed completely in 6 weeks and another 11 decreased in size by 50%
GM-CSF	1998	Ia	10	5-mm punch wound	Did not show improvement in healing time
GM-CSF	1997	II	16	Chronic leg ulcers	Nonsignificant difference
EGF	1995	Ia	17	Bilateral split-thickness skin graft donor site	No difference
PR	1993	I	—	Chronic nonhealing wounds	Wounds of > 75 weeks old reepithelialized in 10 weeks posttreatment
PR	1992	II	13	Chronic diabetic foot ulcers > 8 weeks	Results in much faster healing than control
GM-CSF	1992	II	35	Leprosy wounds	Rapid filling of treated wounds
PR	1991	II	18	Nonhealing lower extremity wounds	No improvement in healing rate
EGF	1991	II	9	Chronic wounds > 12 m	All wounds heal in 34 days
PR	1990	II	32	Chronic nonhealing wounds	Significantly more wounds in intervention wound achieving epithelialization
EGF	1989	Ia	12	Donor sites	Improvement in epithelialization by 1–1.5 days on average

[a] Abbreviations: PDGF-ββ, platelet-derived growth factor-ββ; FGF, fibroblast growth factor; rhKGF-2, recombinant human keratinocyte growth factor; PR, platelet releasates; GM-CSF, granulocyte/macrophage colony stimulating factor; EGF, epidermal growth factor; TGF-β3, transforming growth factor β3; GH, growth hormone.

mice after 6 weeks, indicating that T cells can have both a stimulatory and inhibitory effect.

Clinical Trials in Wound Healing Using Growth Factors

Currently, PDGF is the only growth factor approved by the FDA for clinical use in the treatment of chronic wounds. This undertaking, however, took pooled data from four Phase III clinical studies treating neuropathic ulcers in almost 1,000 diabetic patients [9]. The variability seen in the four studies underlines the difficulty in performing wound healing clinical trials. With all the trials taken together, PDGF gave a modest 10 percent increase in the overall rate of complete wound healing. Other ulcers that could potentially benefit from its use include chronic pressure ulcers that showed promise among three Phase II trials totaling 175 patients. However, Phase III studies led to no improvement.

Even though only one growth factor has been approved clinically for chronic wounds, many studies on growth factors have been done. A summary of clinical trials performed categorized by indication is given in Table II. Many drugs show early promise in Phase I studies but fail to show efficacy in blinded studies performed in multiple institutions. Alternatively, the improvement in healing is so slight that it does not achieve clinical or statistical significance. Such small differences in healing rates would not justify the expense needed for the production of the drug. Difficulty in these trials also can be attributed to confounders such as small size (many trials contain fewer than 20 patients), patient noncompliance (i.e., pressure sores), and rapid degradation of the growth factors.

Granulocyte/macrophage-colony stimulating factor (GM-CSF) is a growth factor that has shown early clinical promise in the treatment of chronic nonhealing venous ulcers in numerous phase I and phase II studies. Similarly, recombinant keratinocyte growth factor (KGF) has shown success in a Phase II study involving 94 patients with chronic venous ulcers. In the treated group, more patients achieved 75 percent wound closure than in the placebo group. The next few years will tell whether their efficacy persists in Phase III studies.

Future Directions: Involvement of the Nervous System

Whereas significant wound healing research in the past has focused on the role of circulating humoral growth factors, evidence now points to the nervous system as being central in cutaneous healing. It first came from observations that poorly innervated areas of the body are more resistant to wound healing. Diabetic patients, with associated peripheral neuropathies, suffer from poorly healing extremities. They are 15 to 20 times more likely to undergo amputations than their nonneuropathic diabetic counterparts. Although there could be many contributory factors, impaired healing in diabetic wounds has been postulated to be a result of their reduced levels of substance P, a proinflammatory cytokine released by nerve endings after wounding to recruit cells and promote vasodilatation. Because of the diabetic patients' increased levels of neutral endopeptidases, substance P is degraded more rapidly. Indeed, topical application of substance P or endopeptidase inhibitors have resulted in better wound healing. Similarly, mice lacking the low-affinity NGF receptor p75 demonstrate impaired healing while the topical application of a nerve growth factor to their wounds improved healing [10].

Vacuum-Assisted Wound Closure

Although significant progress in wound healing has focused on the role of growth factors, many studies have highlighted the success of an emerging modality that applies subatmospheric pressures to a wound. The VAC (Kinetic Concepts Inc.) is one such device that utilizes a modified dressing consisting of a porous sponge with an inlaid suction tubing that is secured to the wound using an occlusive dressing [10]. An attached pump maintains a continuous negatively pressured wound environment while removing edematous fluid. In addition to decreasing tissue edema and bacterial load, it has been hypothesized that this device also applies micromechanical forces to the wound stretching individual cells, thereby promoting proliferation in the wound micro-environment. The VAC has been associated with accelerated development of granulation tissue, early reepithelialization, and faster healing of both burn and complex wounds. Although many studies are underway to understand the physiological effects of this device, several clinical studies have clearly found that this treatment facilitates wound healing.

Conclusion

Cutaneous wound healing is a potpourri of humoral and cellular factors. Although great strides have been made in the treatment of chronic wounds through the control of infection as well as nutritional supplementation, this chapter is clearly incomplete. The next stride will hinge on finding an optimal combination of growth factors and defining other factors that lead to effective wound healing such as micromechanical forces. Any newly developed treatment options have to be both efficacious and cost-effective. Only when such options are available will the surging tide of poorly healed chronic wounds ebb.

Glossary

Chronic wounds: Open skin surfaces that fail to close after normal expected time for wound closure. Common causes include metabolic disorders such as malnutrition, diabetes, and steroid dependence; or mechanical causes such as pressure, venous congestion, arterial occlusion, and after radiation or burns.

Micromechanical forces: Tension, surface tension, shear, compression, and gravity are all examples of micromechanical forces. It is known that cellular morphology can directly influence cellular function, growth, and differentiation. Lack of micromechanical forces can result in apoptosis. Conversely, application of these micromechanical forces to wounds has been associated with improved cell growth and improved wound healing.

Reepithelialization: The migration of keratinocytes to seal an open wound, considered by many to be a good predictor of successful wound healing.

References

1. Werner, S., and Grose, R. (2003). Regulation of wound healing by growth factors and cytokines. *Physiol. Rev.* **83**(3), 835–870. *This review summarizes the results of expression studies that have been performed in rodents, pigs, and humans to localize growth factors and their receptors in skin wounds.*
2. LeGrand, E. K. (1998). Preclinical promise of becaplermin (rhPDGF-ββ) in wound healing. *Am. J. Surg.* **176**(2A Suppl), 48S–54S. *This article reviews how the study of rhPDGF-ββ in animal wound healing models has assisted in addressing the potential clinical utility of rhPDGF-ββ.*
3. Joris, I., Braunstein, P. W., Jr., Pechet, L., and Majno, G. (1980). Effect of thrombocytopenia on wound healing. A study in the rat. *Exp. Mol. Pathol.* **33**(3), 283–291. *The addition of a neutralizing anti-platelet antibody to mice has surprisingly led to no change in the wound healing rates, but significantly increased the number of macrophages and T-cells present in the wound.*
4. Martin, P. (1997). Wound healing—aiming for perfect skin regeneration. *Science* **276**(5309), 75–81. *A review on wound healing focusing on the contributing cell types during the phases of proliferation, migration, matrix synthesis, and contraction, as well as the growth factor and matrix signals present at a wound site.*
5. Livant, D. L., Brabec, R. K., Kurachi, K., Allen, D. L., Wu, Y., Hoaseth, R., Andrews, P., Ethier, S. P., and Markwart, S. (2000). The PHSRN sequence induces extracellular matrix invasion and accelerates wound healing in obese diabetic mice. *J. Clin. Invest.* **105**(11), 1537–1545. *The PHSRN sequence of the plasma fibronectin (pFn) cell-binding domain induces human keratinocytes and fibroblasts to migrate and also stimulates reepithelialization and contraction of dermal wounds in healing-impaired, obese diabetic mice.*
6. Kirchner, L. M., Meerbaum, S. O., Gruber, B. S., Knoll, A. K., Bulgrin, J., Taylor, R. A., and Schmidt, S. P. (2003). Effects of vascular endothelial growth factor on wound closure rates in the genetically diabetic mouse model. *Wound Repair Regen.* **11**(2), 127–131. *Topically applied vascular endothelial growth factor improves wound closure rates in diabetic animals in a full-thickness wound model in genetically diabetic mice.*
7. Ashcroft, G. S., Yang, X., Anzano, M., Green-Wild, T., Wahl, S. M., Glick, A. B., Weinstein, M., Letterio, J. L., Mizel, D. E., Deng, C., and Roberts, A. B. (1999). Mice lacking Smad3 show accelerated wound healing and an impaired local inflammatory response. *Nat. Cell Biol.* **1**(5), 260–266. *In contrast to predictions made on the basis of the ability of exogenous TGF-β to improve wound healing, Smad3-null (Smad3ex8/ex8) mice paradoxically show accelerated cutaneous wound healing compared with wild-type mice, characterized by increased reepithelialization and reduced local infiltration of monocytes.*
8. Jameson, J., Ugarte, K., Chen, N., Yachi, P., Fuchs, E., Biosmenn, R., and Hauran, W. L. (2002). A role for skin γδ T cells in wound repair. *Science* **296**(5568), 747–749. *Activated DETCs produce keratinocyte growth factors (KGFs) and chemokines, raising the possibility that DETCs play a role in tissue repair. We performed wound healing studies and found defects in keratinocyte proliferation and tissue reepithelialization in the absence of wild-type DETCs.*
9. Wieman, T. J. (1998). Clinical efficacy of becaplermin (rhPDGF-ββ) gel. Becaplermin Gel Studies Group. *Am. J. Surg.* **176**(2A Suppl), 74S–79S. *The results of four multicenter, randomized, placebo-controlled, parallel group studies of the efficacy of becaplermin gel are reviewed here.*
10. Gibran, N. S., Jang, Y. C., Isik, F. F., Greenhalgh, D. G., Muffley, L. A., Underwood, R. A., Uswi, M. L., Larsen, J., Smith, D. G., Bunett, N., Ansel, J. C., and Olarnd, J. E. (2002). Diminished neuropeptide levels contribute to the impaired cutaneous healing response associated with diabetes mellitus. *J. Surg. Res.* **108**(1), 122–128. *We demonstrated fewer nerves in the epidermis and papillary dermis of skin from human subjects with diabetes. Likewise, db/db murine skin had significantly fewer epidermal nerves than nondiabetic littermates. The use of substance P in improving wound healing is substantiated by a murine model of impaired healing.*

Capsule Biography

Dr. Chan is a surgical resident at the Brigham and Women's Hospital and a research fellow in the Tissue Engineering and Wound Healing Laboratory in the Department of Plastic Surgery at Harvard Medical School.

Dr. Liu is a surgical resident at Our Lady of Mercy Medical Center in The Bronx, New York. He is taking time off from residency to work as a research fellow at the Tissue Engineering and Wound Healing Laboratory of Brigham and Women's Hospital/Harvard Medical School. His current research, supported in part by a grant from the Center for Integration of Medicine & Innovative Technology, focuses on the use of micromechanical forces, platelet releasates, and stem cells to enhance wound healing.

Dr. Orgill is an Associate Professor of Surgery at Brigham & Women's Hospital/Harvard Medical School, Associate Chief of the Division of Plastic Surgery, and Associate Chief of the Burn Unit. His Tissue Engineering and Wound Healing Laboratory is currently investigating the use of micromechanical forces in wound healing, the use of melanocytes to correct skin and hair pigment disorders, and the engineering of durable tissue substitutes for repairing tissue defects. His research is supported by BWH and the Center for Integration of Medicine & Innovative Technology.

Chapter 133

Microvascular Repair

Jamie Y. Jeremy and Michael R. Dashwood

Bristol Heart Institute, University of Bristol, United Kingdom, and The Royal Free and University College Medical School, London, United Kingdom

Introduction

Microvessels (MVs) control oxygenation of tissues, the removal of metabolic waste products (including carbon dioxide), blood pressure, inflammatory responses, permeability to macromolecules, and hemostasis. Since damage to the MV system can have serious consequences, it is essential that the repair of the MV system following injury be rapid and complete. MVs therefore possess the capacity to rapidly regenerate and reestablish an integrated microcirculation by means of angiogenesis. This process is mediated principally by the proliferation, migration, and organization of endothelial cells into new microvessels. Infection or surgery can result in loss of the endothelium without breaching the MV wall. Thus, endothelial regrowth is another critical aspect of MV repair. In this chapter, we will consider the different types of injury that MVs can sustain, the exogenous triggers that promote repair, the role of hypoxia, oxygen tension, and reactive oxygen species, and the expression of factors that perpetuate repair.

The Diversity of Microvascular Damage

MV repair (endothelial regrowth and angiogenesis) is a rapid response to physical trauma that ranges from minor cuts and abrasions to gunshot wounds and burns elicited by both fire and chemicals. In this context, MV repair occurs as a component of "wound healing."

Surgery also disrupts the integrity of the MV system. For example, in organ transplantation or vein graft bypass surgery, whole organs or tissues are excised and reimplanted. Since the microvessels that supply these tissues are severed, it is implicit that the tissue is no longer supplied with blood and therefore is deprived of oxygen and nutrients. This applies to all forms of surgery, and therefore rapid MV repair is crucial for the success of surgical procedures.

The infection of wounds or systemic infection (septicemia) is known to damage the MVs. An example of this scenario is acute respiratory distress syndrome, in which endotoxins derived from bacteria precipitate extensive adhesion of leukocytes and platelets to the pulmonary vascular endothelium. In turn, these cells release a battery of cytokines and factors that are cytotoxic or apoptotic and that eventually destroy the endothelium.

Cardiovascular risk factors and metabolic status can compromise MV repair. For example, diabetes mellitus is associated with impaired angiogenesis and endothelial regrowth, as are oxidized lipids and homocysteine.

Angiogenesis

Angiogenesis (the formation of new vessels from resident capillaries) involves distinct stages: matrix degradation, migration, proliferation, and organization of endothelial cells (ECs) into tubes (Figure 1). An increase in vascular permeability is one of the earliest steps in the process. ECs dissociate themselves from surrounding tissue by secrcting enzymes that degrade extracellular matrix (basement membrane). Simultaneously, ECs replicate and align themselves into "tubules," a process described as sprouting (Figure 1). EC proliferation and migration will then continue and the sprout continues to forge its way through matrix proteins ever elongating and increasing in lumenal size. This it achieves by dissolving matrix components. The cells ultimately anastomose to existing vessels, which reestablishes blood flow through the newly formed vessel and therefore restores oxygenation of the compromised tissue. Maturation and stabilization of the new vessels occur through the recruitment of pericytes and vascular smooth muscle cells (VSMC).

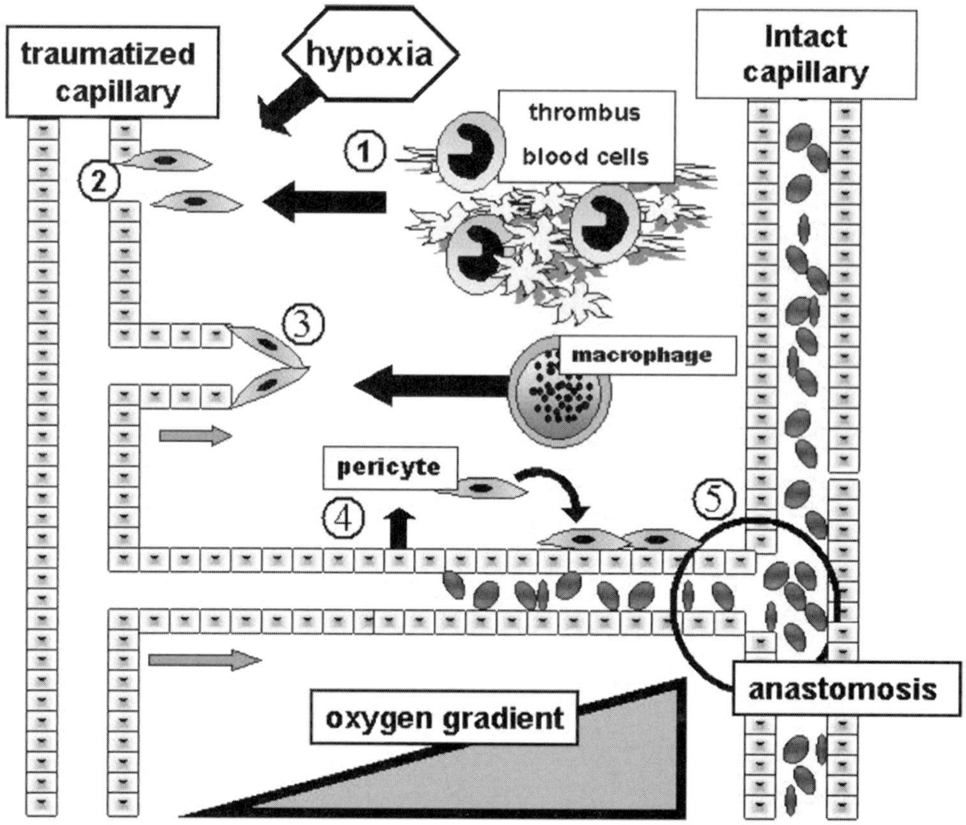

Figure 1 Principal steps in microvascular repair: angiogenesis. (1) Trauma following wounding or surgery disrupts normal blood flow, elicits the formation of a thrombus and the accumulation of platelet neutrophils, monocytes, and immune cells, and renders tissue hypoxic. Release substances from these bloods cells and thrombus components, coupled with hypoxia, trigger the expression of angiogenic factors including VEGF, angiopoietins, and HIF-1α. (2) Endothelial cells (ECs) take on a migratory and replicative phenotype and detach from the matrix by secretion of matrix proteases. (3) ECs continue to replicate and migrate simultaneously aligning themselves to form tubules (sprouting). (4) The new microvessel (MV) grows in the direction of an oxygen gradient toward a capillary that carries red blood cells. MVs can release chemotactic substances that promote pericytes to migrate and associate with the MV. (5) The new MV (bio)anastomoses with an intact neighboring capillary restoring blood supply to the tissue. The subsequent intrinsic flow imparts longevity to the MVs and adapts functionally according to the needs of the tissue it supplies. (see color insert)

Angiogenesis under normal conditions is self-limiting so that when sufficient vessel formation has occurred to satisfy the oxygen demands of the tissue, the endothelial cells become quiescent and the vessels either remain or regress if they are not needed. Apoptosis plays a central role in regression or cessation of angiogenesis. Once a microvessel has formed it becomes stable in that ECs become resistant to exogenous factors (i.e., quiescent) and survive for years. Newly formed microvessels, of course, adopt the prerequisite properties required for the normal function of any given tissue.

The mechanisms underlying this process are expanded upon in the following.

The Cell Cycle and Proliferation

In angiogenesis, ECs undergo division through a highly controlled and orchestrated process known as the cell cycle. On binding to their receptors, growth promoters trigger an alteration of ECs from a quiescent or resting state (G_0) into a replicative phase (G_1); G represents "gap." During this time window (the G_0/G_1 phase), extracellular signals are transmitted through cytoplasmic signal cascades (see later discussion), which in turn induce nuclear responses that control progression of the cell through mitosis. Once a cell has passed a specific point in the G_1 phase (after about 11 hours), it is committed to divide and enters the S phase. Once past this point the cells cycle is irreversible. The S phase is complete within 8 hours and the cell enters the G_2 phase (2 hours). Finally the cell enters the M phase (1 hour) where cell replication takes place. The whole cell cycle is therefore complete within 24 hours after initial stimulation with growth promoters.

The early signal transduction pathways that trigger VSMC replication can be described as a series of protein phosphorylations that are mediated both sequentially and sometimes in parallel by protein kinases (PKs) (Figure 2). A phosphorylated protein kinase will then activate another PK through phosphorylation of that PK. It is by this means that

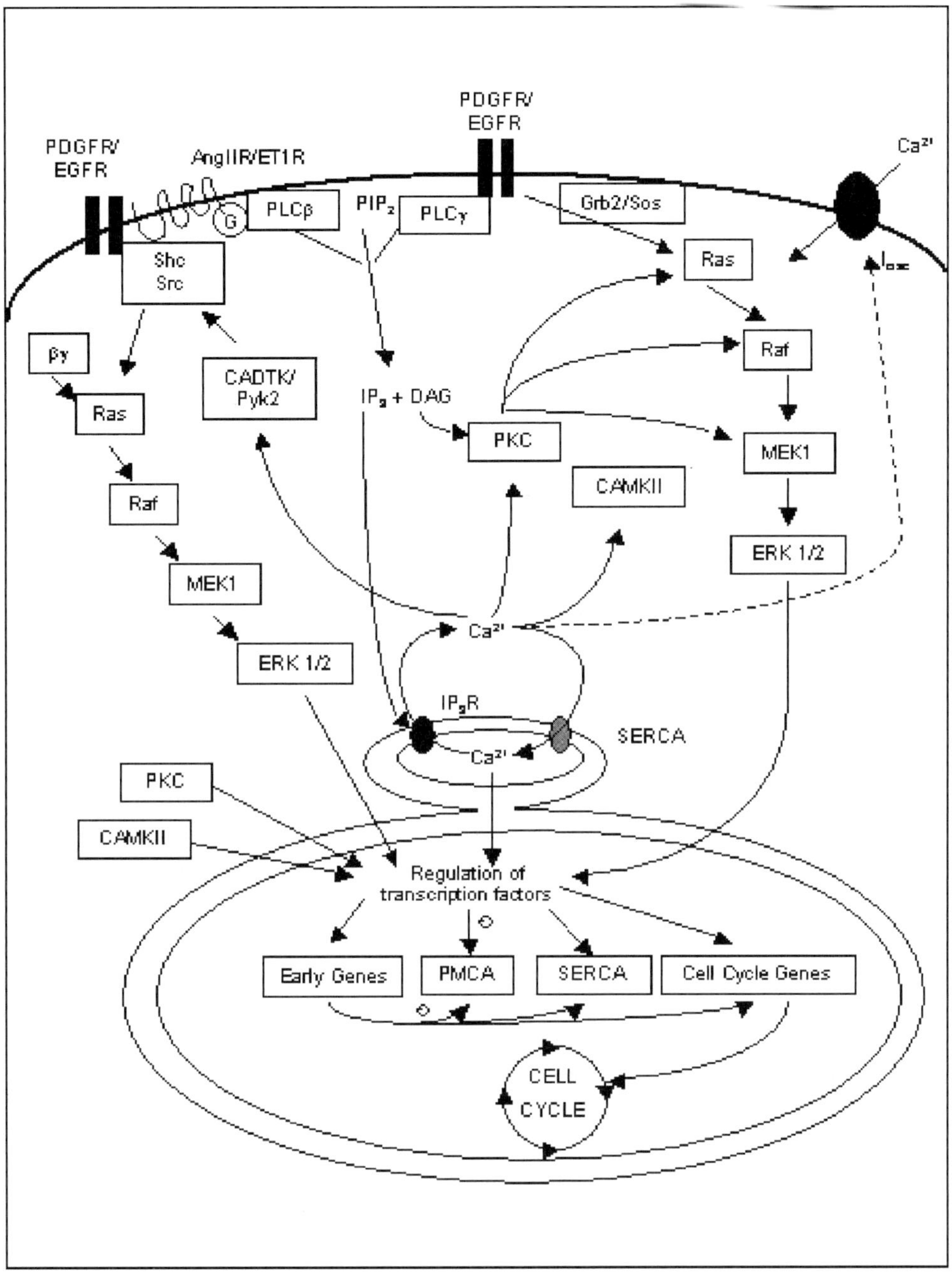

Figure 2 Proliferative signaling in endothelial cells originating from RTK and G-protein linked receptors and involving Ca^{2+} and ERK1/2. Detailed discussion are presented in the text. Abbreviations: PDGFR, platelet-derived growth factor receptor; EGFR, epidermal growth factor receptor; AngIIR, angiotensin II receptor; ET1R, endothelin 1 receptor; PLC, phospholipase; PIP_2, phosphoinositol bisphosphate; IP_3, inositol trisphosphate; IP_3R, inositol trisphosphate receptor; DAG, diacylgylcerol; PKC, protein kinase C; CAMKII, calcium/calmodulin dependent kinase II; MEK, mitogen activated/extracellular regulated kinase; ERK, extracellular signal regulated kinase; G, heterotrimeric G-protein; CADTK (also Pyk2), calcium dependent tyrosine kinase; PMCA, plasma membrane calcium ATPase; SERCA, sarco(endo)plasmic calcium ATPase; βγ, βγ subunits of heterotrimeric G-proteins; I_{CRAC}, putative capacitative entry channel.

the G_1 phase triggering replication can be tightly controlled. The central components of signal transduction in ECs are (1) receptor linked tyrosine kinases (TKs), (2) phospholipase C (PLC), (3) the Ras-Raf-1 proteins, (4) phosphoinositide hydrolysis: diacylglycerol and inositol trisphosphate generation, (5) the MAP kinase system, and (6) calcium mobilization. How these are interrelated is described in Figure 2.

Ultimately, these signal transduction cascades converge on the nucleus. The ordered sequence of events of the cell cycle is controlled by protein complexes composed of cyclin-dependent kinases (CDKs) and their catalytic part-

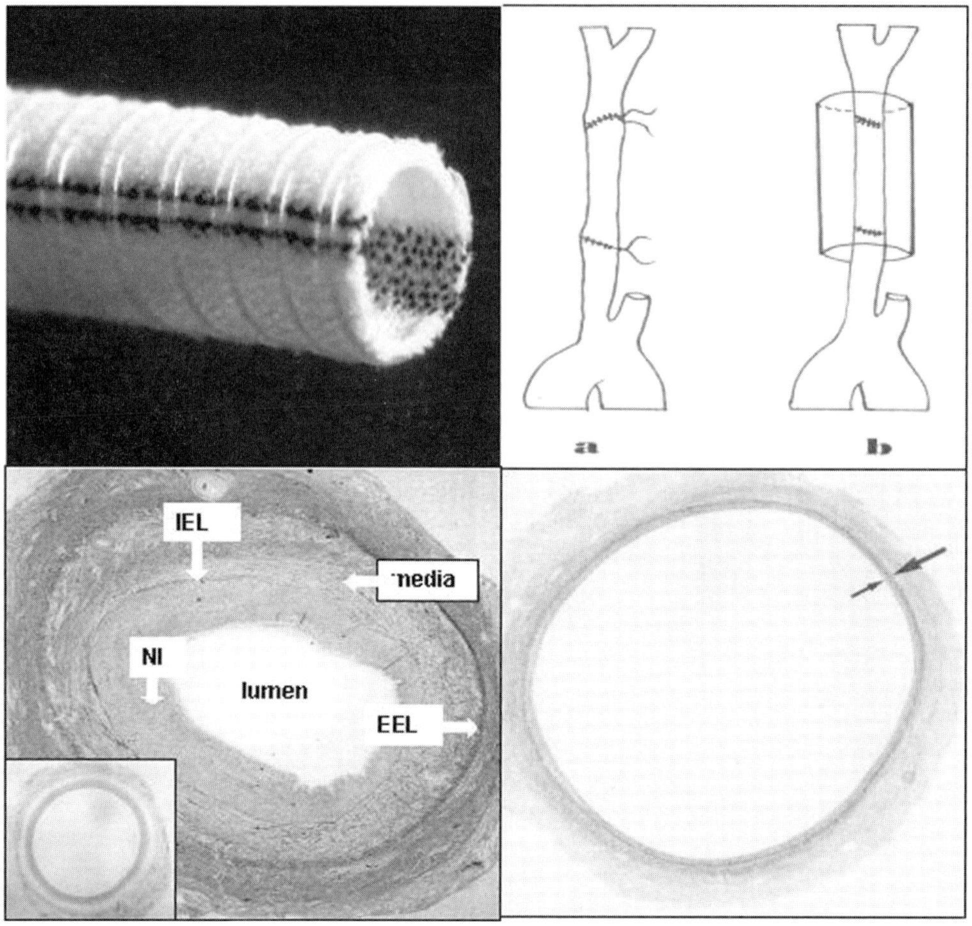

Figure 3 Effect of an external sheath (stent) on neointima formation and vein graft thickening in a pig model. In these studies, a loose-fitting, that is, nonrestrictive, polyester sheath or stent (*upper left panel*) was placed around a saphenous vein into carotid artery interposition graft (*upper right panel*). After 1 month, the graft was excised and studied histologically. As can be seen (*lower panel*), there is a marked increase in graft size and neointima (NI) formation (the layer between the internal elastic lamina, IEL, and the lumen) compared to the original ungrafted saphenous vein (*inset*). The graft fitted with the external stent, however, shows a profound reduction of graft thickening (small arrow IEL and large arrow external elastic lamina, EEL) and a complete inhibition of neointima formation (see Figure 4 for higher magnification). This stented graft was characterized by profound microvessel growth (Figure 4). (see color insert)

ners, the cyclins. Cyclins bind to CDKs (designated cdc2 for the cell cycle), which then phosphorylate selected proteins. The G_1-phase specific D cyclins (D_1,D_2,D_3) are key regulators in the transition of cells from quiescence to proliferation and progression through the G_1 phase and the G_1/S transition. Cyclin D mRNAs are induced by growth factors, including vascular endothelial growth factor (VEGF), and are suppressed by antiproliferative agents. Induced overexpression of the D cyclins accelerates cell cycle progression and shortens the length of the cycle in many cell types.

A principal target for the cyclins in the control of cell cycle progression is the retinoblastoma (Rb) tumor suppressor gene Rb protein (Rbp), which suppresses mitosis. Phosphorylation of Rbp by the cyclin–CDK system negates its action, allowing replication to progress. Growth factors also elicit the rapid expression of proto-oncogenes, in particular c-fos and c-myc, which in turn are associated with the G_0/G_1 phase of the cell cycle. Another important intracellular mediator of VSMC proliferation is NFκB, which is classically activated in tissues subjected to inflammation.

Migration

The migration of ECs requires the cells to "crawl" and "burrow" through the surrounding tissue matrix. Motility is dependent on the intracellular cytoskeleton, which is composed principally of actin and myosin. As in VSMCs, contraction/relaxation is mediated principally by Ca^{2+} mobilization, phosphorylation of myosin light-chain kinase (MLCK) and the rho kinases. Motility and migration are orchestrated by the focal adhesion complexes that link sites of adhesion with intracellular activity of actin. Components of a focal adhesion, which are complex dynamic structures, include the β1 and β3 integrins, vinculin, tensin, talin, zyxin, α-actinin, and focal adhesion kinase (FAK).

In order for motility to take place, migrating ECs need a fulcrum of attachment, which is achieved through adhesion of ECs to the matrix proteins via specialized adhesion molecules: (1) the integrins, including the $\beta 1$ series and the vitronectin receptor; the integrins consist of heterodimeric transmembrane proteins with an extracellular domain and an intracellular domain that is linked to the cytoskeleton; (2) the immunoglobulin superfamily, including intracellular adhesion molecule (ICAM), vascular cell adhesion molecule (VCAM), and platelet endothelial cell adhesion molecule-1 (PECAM-1); (3) the selectins (e.g., E- and P-selectin); and (4) the cadherins that form complexes with catenins, which are components of focal adhesions. The matrix proteins bound by EC integrins are principally vitronectin, laminin, fibrinogen, collagen, thrombospondin, von Willebrand factor, and fibronectin.

Matrix and Matrix Metalloproteinases

In order for ECs to migrate, proliferate, and form tubes within the confines of any tissue, they must first forge a path through that tissue. This is achieved by the activation and secretion of matrix metalloproteinases (MMPs) that dissolve the extracellular matrix (ECM). The ECM comprises collagen, fibronectin, elastin, vitronectin, and proteoglycans. MMPs include interstitial collagenase (MMP-1), gelatinase A (MMP-2), stromelysin-1 (MMP-3), matrilysin (MMP-7), and gelatinase B (MMP-9). MMP activity, in turn, is regulated by tissue inhibitors of MMPs (TIMPs). Other proteases that have been implicated in mediating angiogenesis include urokinase-type plasminogen activator (u-PA).

In this context, constituents of the matrix exert a strong influence on angiogenesis. Collagen and procollagen, for example, are required physically for migration to occur. Proteoglycans (PGs), too, are axiomatic in mediating the proliferation and migration of ECs. ECs synthesize heparan sulfate, including the major endothelial extracellular matrix PG, perlecan, members of the syndecan family of transmembrane PGs, and the cell surface-associated PG, glypican-1. Vascular endothelial cells also constitutively synthesize and secrete the large aggregating chondroitin sulfate PG, versican, and the small leucine-rich chondroitin/dermatan sulfate PG, biglycan. In addition, the expression of another small leucine-rich PG, decorin, is induced during formation of neovessels both in vitro and in vivo. Both high-molecular-weight hyaluronate and chondroitin sulfate elicit a dose-dependent stimulation of tube formation. Heparanase promotes angiogenesis. Chondroitin sulfate is capable of binding fibrinogen/fibrin, thereby mediating endothelial cell migration and invasion into the fibrin provisional matrix during wound repair. Decorin expression seems to be of special importance for the survival of endothelial cells as well as for cord and tube formation. Heparan sulfate proteoglycans act as docking molecules for MMP-7 in the basement membrane.

Triggers for Vascular Repair in Wound Healing

The primary triggers for MV repair associated with wounds and trauma are those derived from blood cells and thrombi. Immediately after severance of microvessels, platelets, neutrophils, monocytes, and T cells accumulate at the sites of injury. Platelets are programmed to recognize vascular severance and rapidly change shape and adhere to block the breach in the vessel wall by forming a platelet plug. They simultaneously release a battery of prothrombotic factors to promote thrombus formation to consolidate the plug, as well as releasing vasoconstrictors (thromboxane A_2 and serotonin) that further restrict blood loss. Components of thrombogenesis, such as fibrin, fibrinogen, tissue factor, and factor Xa, are among the most potent angiogenic factors known. Other platelet release substances are also proangiogenic.

Once blood loss has been checked, neutrophils and monocytes rapidly accumulate in a second wave that has a twofold purpose: (1) to destroy bacteria and prevent wound infection and (2) to engender tissue repair including the reestablishment of the local MVs. Thus, neutrophils and monocytes release yet another battery of factors that promote angiogenesis. Monocytes then become resident macrophages, which again release factors that modulate tissue repair and angiogenesis. Indeed, macrophages are deemed the principal orchestrators of inflammatory angiogenesis. Giant cells, which are amalgamations of many macrophages, are coincidentally associated with microvessel proliferation in arteritis.

Hypoxia is a major initiating factor for and indeed the raison d'être for angiogenesis. Stimulated by the barrage of triggers mentioned earlier, as well as hypoxia, ECs are activated and respond by immediately expressing genes that will result in EC proliferation, migration, and angiogenesis, principal among which are the hypoxia inducible factors (HIFs), peptide growth factors, activator protein-1 (AP-1), early growth response 1 (EGR-1), nuclear factor interleukin-6 (NF-IL6), and nuclear factor kappa B (NFκB). HIF is an $\alpha\beta$ heterodimer. HIF-1α and HIF-1β are constitutively expressed, whereas HIF-1α subunits are inducible by hypoxia. HIF-1 binds to a consensus region of the gene termed the hypoxia response element (HRE). HIF-1α is subjected to rapid ubiquination and proteasomal degradation, a process that is inhibited under normoxic conditions. Apart from hypoxia, cytokines, peptide growth factors, insulin and basic fibroblast growth factor (bFGF) all upregulate hypoxic inhibitory factor 1α (HIF1α) protein expression as well as HIF-1 DNA binding activity and HIF-1 target gene expression. Phosphorylation plays an important role in mediating HIF activity. Intracellular signal transduction systems include Ras/Raf/MAP kinase and receptor tyrosine kinase, PI3 kinase, and PTEN Akt kinase pathways. There are a large number of target genes for HIF-1, which include VEGF, VEGF receptor FLT-1, PAI1, inducible nitric oxide synthase (iNOS), and ceruloplasmin.

Factors Influencing Angiogenesis

Vascular Endothelial Growth Factor

One of the earliest responses to hypoxia/MV repair is the expression of vascular endothelial growth factor, the tie-1 and tie-2 receptors, and angiopoietin (Ang), which are all obligatory for angiogenesis. The VEGF family comprises at least five isoproteins, whose effects are mediated by three VEGF receptors (VGR), which in turn communicate with ECs via RTKs. Regulation of the VEGF gene is controlled by the stabilization of hypoxia. VEGF mRNA levels dramatically increase when endothelial cell cultures are exposed to hypoxia and return to normal when oxygen supply is restored. VEGF isoforms vary in their actions. For example, $VEGF_{121}$ and $VEGF_{165}$ and their receptors increase and $VEGF_{189}$ decreases MV lumen size.

Other Peptide Growth Factors

Other peptide growth factors that promote MV repair are basic fibroblast growth factor (bFGF), platelet-derived growth factor (PDGF), transforming growth factor β (TGFβ), and insulin-like growth factor (IGF-1). However, unlike VEGF, these peptides tend not to be specific for ECs and also promote the proliferation of fibroblasts, VSMCs, and other cells. One facet of the angiogenic process is that MMPs liberate these growth factors from tight association with the matrix. Apart from augmenting angiogenesis directly, bFGF augments the accumulation of inflammatory cells and PDGF recruits pericytes and VSMCs around nascent vessel sprouts.

Angiopoietins

Angiopoietins are axiomatic in mediating angiogenesis and interact at many levels with VEGF. Angiopoietin 1 (Ang 1), a ligand for the endothelial receptor, Tie2, inhibits permeability, and overexpressing Ang1 promotes an increase in lumen diameter. Ang 1 via phosphorylation of Tie 2 is chemotactic for ECs and induces sprouting and stimulates EC-pericyte interactions. Angiopoietin 2 (Ang 2) is a natural antagonist of Ang1 and is involved in detaching VSMCs and loosening underlying matrix, thereby allowing ECs to migrate. Ang 2 in concert with VEGF promotes angiogenesis, but alone it can be an inhibitor.

Endothelin, Nitric Oxide, Prostacyclin, and Other Endothelial Factors

Perhaps not surprisingly, angiogenesis is controlled and modulated by those factors manufactured by ECs. These include nitric oxide (NO), endothelin-1 (ET-1), tissue plasminogen activator (tPA), thrombospondin, von Willebrand factor, and prostaglandins.

NO plays an axiomatic role in mediating angiogenesis. VEGF augments NO release from ECs and NO mediates VEGF-stimulated EC proliferation. VEGF-stimulated NO release from simian virus-40 immortalized microvascular ECs and induced cell migration, whereas the nitric oxide synthase (NOS) inhibitor L-NAME or antisense oligonucleotides to endothelial NOS (eNOS) suppressed this effect. In an in vivo model of angiogenesis in the rabbit cornea, NO potentiated the proangiogenic effect of substance P that was prevented by the absence of L-arginine, the substrate for NO formation.

Endothelin-1

ET-1 is a potent mitogen for both ECs and VSMCs but appears to act more as a promoter of other angiogenic factors, rather than being a primary stimulator. For example, when combined with VEGF, ET-1 induces angiogenesis in subcutaneously implanted matrigel plugs in mice. ET-1 may induce angiogenesis via the upregulation of VEGF and/or iNOS. There are conflicting reports as to which receptor subtype mediates angiogenesis, with both ET_A and ET_B receptors having been implicated. There may also be considerable species-dependent variations and differences between different vascular beds.

Prostaglandins, Thromboxanes, and Leukotrienes (Eicosanoids)

Thromboxane A_2, a major release factor of platelets, appears to be the most potent in promoting angiogenesis. In a corneal model, activated human microvascular cells generate TXA_2, and selective cyclooxygenase-2 (COX-2) antagonists inhibited TXA_2 production along with endothelial cell migration and corneal angiogenesis. bFGF and VEGF increase TXA_2 synthesis. The TXA_2-mimetic U46619 stimulates endothelial cell migration, whereas a TXA_2 antagonist (SQ29548) inhibited bFGF- and VEGF-stimulated endothelial cell migration.

With regard to the prostaglandins (PGs), which are produced principally by blood vessels, PGE_1 and PGE_2 promote angiogenesis in a rat femoral artery model. PGE_2 induces expression of VEGF in fibroblasts. Indomethacin (a nonselective cyclooxygenase inhibitor) reduces EC migration.

Lipoxygenase derived 12-*S*-hydroxyeicosatetraenoic acid (12-*S*-HETE) and cytochrome P450 derived 12-*R*-HETE (derived principally from leukocytes) are pro-angiogenic factors. The angiogenic potency of 12-*R*-HETE is equal to that of bFGF. VEGF-induced endothelial cell migration can be partially reversed by adding 12-*S*-HETE. Leukotrienes (LTs), in particular LTC_4, stimulate endothelial cell migration and tube formation. In an in vivo chick chorioallantoic membrane system LTC_4 and LTD_4 promoted angiogenesis in a dose-dependent manner that was abolished by selective LT antagonists.

Reactive Oxygen Species and Angiogenesis

Reactive oxygen species (ROS) play a role in modulating angiogenesis. Principal among ROS are superoxide (O_2^-), hydrogen peroxide (H_2O_2), and products of the reactions between ROS and NO. Potential sources of ROS include NADPH oxidase, the mitochondrial electron transport chain, xanthine oxidase, NOS, cyclooxygenase, and lipoxygenase. Recent studies have indicated that NADPH oxidase may be axiomatic in mediating angiogenesis. In keeping with the wound healing mechanisms described earlier, virtually all the factors that promote angiogenesis also promote the upregulation of vascular NADPH oxidase, including cytokines, peptide growth factors, angiotensin II, and TXA_2. Hypoxia itself has also been shown to induce the expression of NADPH oxidase. It is of interest that copper is both a potent angiogenic factor as well as a catalyst for the generation of ROS. The mechanisms involved, however, are still obscure.

Antiangiogenic Factors

Angiogenesis is under tight negative control and factors are expressed that limit neovascularization in a wound healing or hypoxic response. First, oxygenation status is a powerful determinant. Hyperoxia is a potent inhibitor of EC proliferation, angiogenesis, and the expression of key growth factors including VEGF. Inhibitors of angiogenesis include angiostatin, endostatin, interferon-$\alpha/\beta/\gamma$, interleukins, platelet factor 4, thrombospondin-1, TIMP, and TNF-α. Some MMPs also act as inhibitors. MMP-1 and MMP-3 interfere with integrin binding and MMP-7 and -9 generate angiostatin from plasminogen, thereby inhibiting EC proliferation. Apoptosis of ECs appears to be a key event that mediates the cessation of angiogenesis.

Angiogenesis in Vein Grafts: Impact of the External Stent

Surgery, ipso facto, elicits MV damage. In turn, the efficacy of revascularization may be axiomatic in determining the outcome of surgical procedures. One clinical scenario in which many of these factors come into play is in coronary artery bypass graft surgery (CABG) using autologous saphenous vein. CABG involves the removal of the vein from the patient, which is then implanted by anastomosis into the aortic arch and then below an atherosclerotic lesion in the coronary artery, effectively "bypassing" the occlusion and restoring normal blood flow. However, as many as 50 percent of grafts fail within 10 years because of reocclusion. CABG disrupts the integrity of the microvascular supply of the saphenous vein (the vasa vasorum). The vasa vasorum supplies larger blood vessels with oxygen and nutrients and removes metabolites and carbon dioxide since large vessels cannot derive their oxygen and nutrient requirements from blood flowing through their own lumens. The loss of vasa vasorum coupled with a rapidly thickening vein graft creates an imbalance between oxygen supply and demand that results in graft hypoxia. Hypoxia promotes VSMC proliferation, procoagulant pathways, and the expression of adhesion molecules in vascular tissue, all components of vein graft thickening and atherogenesis. However, as was mentioned earlier, the formation of a neo–vasa vasorum is designed to counteract the pathological impact of hypoxia.

The importance of angiogenesis in vein graft disease is highlighted by the effect of an external cuff or stent on vein graft morphology and angiogenesis. In the pig, placement of a nonrestrictive, highly porous external stent around implanted saphenous vein grafts markedly inhibits graft thickening and neointima formation (Figure 3). These stented vein grafts were characterized by a distinct "neoadventitia" and a microvasculature that extended into the media of the graft. In contrast, the adventitia of unstented vein grafts was dispersed and poorly organized.

It was noted that within 1 week there was an exudate in the gap between the stent and the graft that contained components of fibrinolysis and leukocytes (Figure 4). At 2 weeks after implantation, this graft–stent interface was completely filled with a semiopaque "gel" or exudate that was fibrin-rich. At 1 month this had organized into a well-defined structure comprising a dense population of fibroblasts and microvessels (or neo–vasa vasorum). It was proposed, therefore, that the central mechanism underlying the effect of the external stent was the promotion of angiogenesis by this exudate. Once angiogenesis has been triggered, other endogenous growth factors come into play in the process. These adventitial microvessels contain a high density of B subtypes of the endothelin-1 receptor. This is of particular interest since ET_B receptor agonists have been reported to promote angiogenesis. There are high levels of NOS in the endothelium of neoadventitial microvessels of stented porcine vein graft. Since NO promotes the proliferation and migration of endothelial cells, it was suggested that these observations pointed to a role for NO in stent-induced angiogenesis.

Risk Factors for Impairment of Microvascular Repair

Risk factors for vascular disease, which include diabetes mellitus, hypercholesterolemia, and hyperhomocysteinemia, all impair angiogenesis. In turn, this can result in impaired wound healing and complications following surgery. This has been exemplified in bypass graft surgery in the preceding section. Metabolic and dietary factors, can also influence the process. For example, vitamin B_6 has been shown to

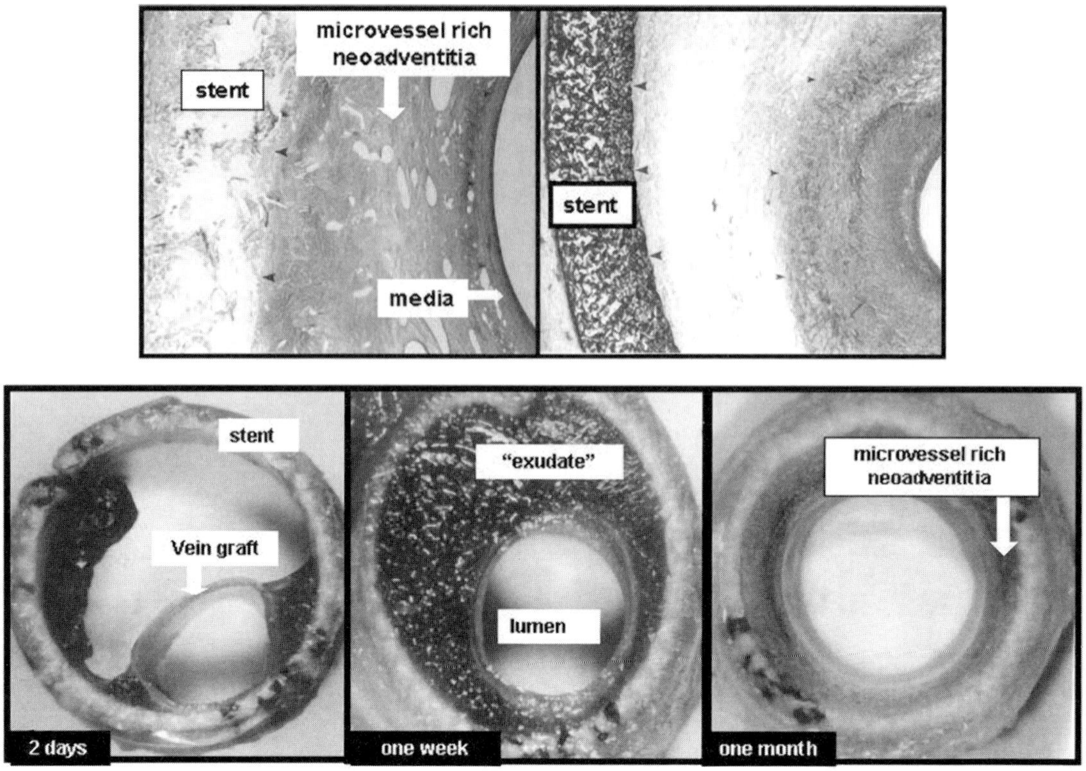

Figure 4 At a higher magnification of the graft fitted with the polyester external stent (*upper left panel*), the presence of a highly developed microvascular system can be seen in the space between the stent and the media. It was suggested that the stent promotes this microvessel (MV) growth, which in turn provides oxygenation of the vein graft, thereby obviating hypoxia and graft thickening. The importance of microvessel formation can be seen with another kind of stent made of PTFE (*upper right panel*), which clearly does not promote MV formation but also does not prevent graft thickening. It has been suggested that the principal mechanism underlying this effect is that the stent concentrates an exudate (derived from the anastomotic wound) in the space between the graft and the stent (*lower three panels*). This exudate is rich in fibrin and leukocytes, which promotes the growth of microvessels via the classic wound-healing pathways. (see color insert)

inhibit angiogenesis. Interestingly, vitamin B_6 controls the levels of homocysteine, an amino acid that also inhibits angiogenesis.

Concluding Remarks

Microvascular repair (angiogenesis) following trauma or surgery is a complex process that is triggered initially by hypoxia and inflammation. Angiogenesis is controlled by an integrated but extremely complex interaction between exogenous and intracellular factors that ultimately result in the reestablishment of a fully functioning microvascular system that services any given organ or tissue. Since an impairment of this process, particularly after surgery, may compromise the outcome of the procedure, further understanding of microvascular repair may be of great clinical importance. How metabolic factors and risk factors for vascular disease influence microvascular repair and how this can be obviated is therefore an area that needs to be addressed in further research.

Glossary

Angiogenesis: The formation of new microvessels from existing endothelial cells. The process involves the migration and proliferation of endothelial cells, which form tubes. These elongate until they reach another microvessel or capillary with which they connect to restore blood flow.

Endothelial cell: Specialized cells that line all blood vessels. Their principal functions are the prevention of thrombosis, the control of vascular tone, and microvessel formation. Endothelial cells possess the capacity to generate modulators of these functions, which include nitric oxide, prostacyclin, and peptide growth factors.

Migration: A process by which cells, including endothelial cells, move from one site to another. Migration is dependent on the intracellular cytoskeleton, which is formed principally of actin and myosin. Migration is orchestrated by the focal adhesion complexes that link sites of adhesion with intracellular actin.

Proliferation: A process by which cells, including endothelial cells, divide (replicate) and multiply; axiomatic in angiogenesis. Cell replication is a highly controlled and orchestrated process known as the cell cycle. Peptide growth factors, including vascular endothelial growth factor, are key initiators of cell proliferation.

Bibliography

Abo-Auda, W., and Benza, R. L. (2003). Therapeutic angiogenesis: Review of current concepts and future directions. *J. Heart Lung Transplant.* **22**, 370–382. *This review covers the current therapeutic applications of angiogenesis, as does the review by Carmeliet (2003). Both are indispensable reading.*

Conway, E. M., Collen, D., and Carmeliet, P. (2001). Molecular mechanisms of blood vessel growth. *Cardiovasc. Res.* **49**, 507–521; Carmeliet, P. (2003). Angiogenesis in health and disease. *Nat. Med.* **9**, 653–660. *Both these reviews are "must" reading. They provide the reader not only with up-to-date concepts and directions but with a superb basic grounding even for those not versed in the intricacies of angiogenesis.*

Folkman, J. (2003). Angiogenesis and apoptosis. *Semin. Cancer Biol.* **13**, 159–167. *Apoptosis is a key factor in the stabilization and regression of newly formed microvessels that is likely to be a key line of research in the coming years. This review provides an excellent grounding for those interested in pursuing this area.*

Li, J., Zhang, Y. P., and Kirsner, R. S. (2003). Angiogenesis in wound repair: Angiogenic growth factors and the extracellular matrix. *Microsc. Res. Tech.* **60**, 107–114. *Matrix proteins play an axiomatic role in mediating angiogenesis. This review provides an excellent and accessible overview of the topic.*

Martin, A., Komada, M. R., and Sane, D. C. (2003). Abnormal angiogenesis in diabetes mellitus. *Med. Res. Rev.* **23**, 117–145. *Although this review focuses on diabetes, it highlights how disorders of metabolism and risk factors for cardiovascular disease can influence angiogenesis, which in turn can influence the progression of vascular disease.*

Pugh, C. W., and Ratcliffe, P. J. (2003). Regulation of angiogenesis by hypoxia: Role of the HIF system. *Nat. Med.* **9**, 677–684; Wenger, R. H. (2002). Cellular adaptation to hypoxia: O_2-sensing protein hydroxylases, hypoxia-inducible transcription factors, and O_2-regulated gene expression. *FASEB J.* **16**, 1151–1162; De Boer, R. A., Pinto, Y. M., and Van Veldhuisen, D. J. (2003). The imbalance between oxygen demand and supply as a potential mechanism in the pathophysiology of heart failure: The role of microvascular growth and abnormalities. *Microcirculation* **10**, 113–126. *These three excellent reviews cover the emerging and complex relationship between hypoxia and angiogenesis. Hypoxia and its role in new microvessel growth is likely to become a prominent area of research in the coming years.*

Ware, J. A. (1999). Cellular mechanisms of repair. In: *Angiogenesis and Cardiovascular Disease* Ware, J. A., and Simons, J. A., eds., pp. 30–78. Oxford: Oxford University Press. *This chapter gives a comprehensive overview of all the key components of angiogenesis, ranging from definitions to intracellular signal transduction. Indeed, this book as a whole is highly recommended as an ideal reference source and for all interested in angiogenesis and microvascular repair.*

Capsule Biography

Dr. Jamie Jeremy is a Reader in Vascular Biology at The Bristol Heart Institute, UK. For the past 10 years Dr. Jeremy has focused his research on the pathophysiology of vein graft failure with a particular emphasis on the role of microvessel regrowth.

Dr. Michael Dashwood is a Senior Research Fellow in the Department of Clinical Biochemistry, Royal Free Hospital. His main research interests are the mechanisms underlying vein graft failure in patients undergoing coronary artery bypass surgery. Another major focus of research is the role of microvessels in various aspects of vascular disease.

SECTION T

Sclerosis

CHAPTER 134

The Role of Microvascular Pericytes in Systemic Sclerosis

Vineeth Rajkumar, David Abraham, and Carol Black
Royal Free and University College Medical School

Introduction

Systemic sclerosis (SSc) encompasses a spectrum of connective tissue disorders of unknown etiology. Whereas the pathological hallmark of SSc is increased synthesis and deposition of extracellular matrix within the skin and internal organs resulting in clinical fibrosis and scarring [1], the origins of the disease are believed to lie within the microvasculature. More than 90 percent of SSc patients initially present with Raynaud's phenomenon (RP), a disorder defined as episodic vasoconstriction of the microvasculature. Although microvascular injury is believed to be one of the key initiating events in SSc pathophysiology, both the root cause of the underlying microvascular injury and the precise cellular and molecular mechanisms by which it gives rise to fibrosis are unknown. Microvascular pericytes reside at the interface between the microvasculature and interstitium and are thus ideally situated to mediate interactions between microvessels and extravascular tissue. Furthermore, pericytes have been identified as mesenchymal precursor cells with the ability to transdifferentiate into other mesenchymal cell types including collagen-synthesizing fibroblasts, potentially making them key players in the link between microvascular damage and fibrosis (Figure 1).

Microvascular Abnormalities in SSc

The prevalence of RP in the general UK population is estimated to be approximately 14 percent. The majority of these patients (95 percent) are classified as having primary RP, a benign though painful condition with no associated structural damage and abnormalities to the microvessels and that will not develop any associated connective tissue disease.

However, in more than 90 percent of SSc patients, RP is frequently the first reported symptom of underlying connective tissue disease and appears to be a unifying characteristic across a heterogeneous disease spectrum; it is therefore likely to be of etiological significance. The presence of RP in SSc patients is associated with irreversible tissue damage, capillary dilation and dropout, and structural damage to the digital microvasculature. In SSc, microvascular damage appears to be a systemic phenomenon, as affected internal organs also show significantly perturbed blood flow. Ultrastructurally, endothelial cells (ECs) show evidence of vacuolization, nuclear degeneration, loss of cell–cell contacts, and multilayering of the basement membrane prior to the onset of fibrosis [2]. Serum and plasma levels of factors considered to be markers of endothelial activation such as von Willebrand factor, adhesion molecules, and endothelin-1 (ET-1) are all increased in SSc patients. However, it can be argued that soluble levels of these factors may reflect activation of cell types other than endothelial cells. More relevant are in situ analyses of SSc skin and other affected organs that have shown increased expression of adhesion molecules, endothelin (ET)-1, platelet-derived growth factor-B (PDGF-B), and transforming growth factor (TGF)-β by endothelial cells. A stimulus of early endothelial cell activation in SSc is likely to be ischemic associated hypoxia, which is known to have profound effects on endothelial cell metabolism. Hypoxia has been shown to rapidly induce the synthesis by cultured endothelial cells of adhesion molecules and key inflammatory cytokines and growth factors.

Anti-endothelial cell autoantibodies (AECAs) may also be responsible for regulating endothelial cell phenotype in SSc. AECAs have been detected in the sera of SSc patients

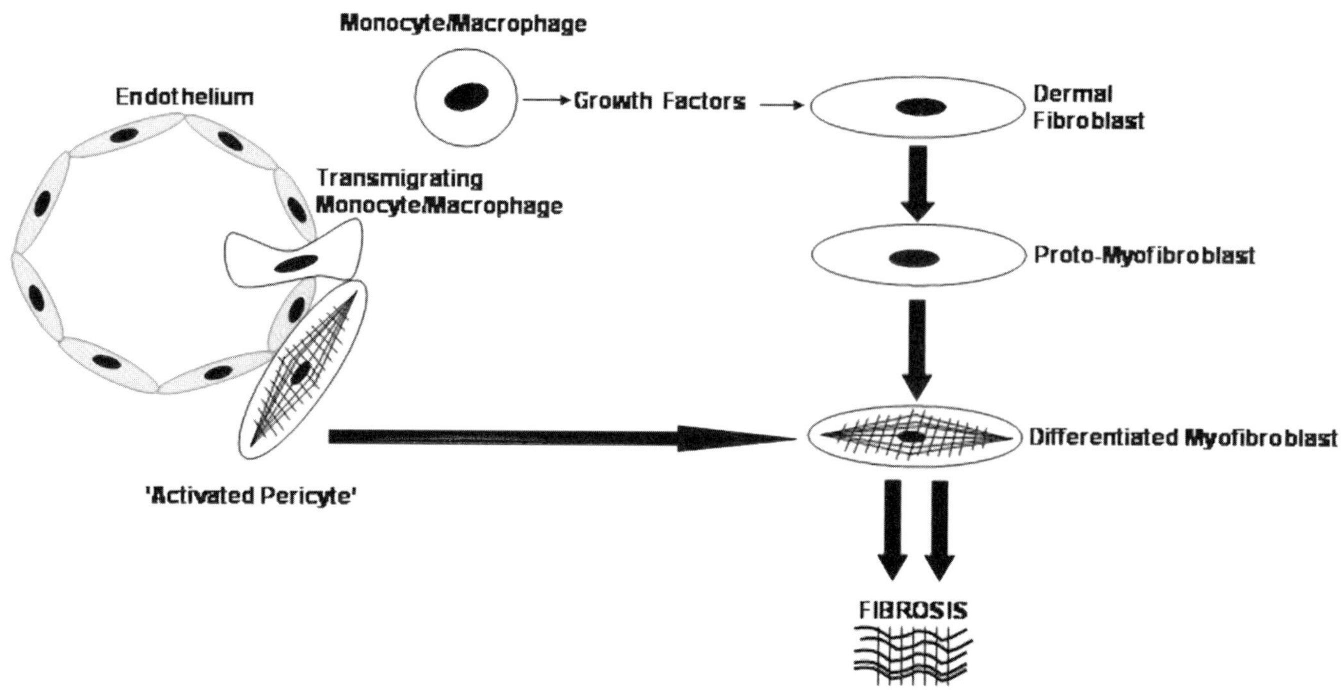

Figure 1 Fibrosis and scar formation is a multistage process involving complex interactions between numerous participating cell types. Pericytes are intimately associated with this process. Preceding transmigration of monocytes is an increase in vascular permeability, brought about by a change in the interactions between endothelial cells and pericytes. Pericytes are also known to produce profibrotic growth factors that can directly modulate fibroblast function and biosynthetic profile. Finally, pericytes are considered mesenchymal precursor cells and as such can transdifferentiate directly into matrix-synthesizing myofibroblasts. (see color insert)

with a prevalence of up to 50 percent, suggesting that the microvasculature maybe the target of an autoimmune response. Although the antigens to which they bind appear to be heterogeneous within and between sera, it is known that AECA can induce pathogenic changes in endothelial cells in vitro, such as increased leukocyte adhesion, and secretion of inflammatory cytokines, such as interleukin (IL)-1. More recently, autoimmune sera from SSc patients have been shown to cross-react with both the human cytomegalovirus late protein UL94 and SSc associated autoantigens, such as fibrillarin and heterogeneous nuclear ribonucleoprotein. Therefore molecular mimicry of cytomegalovirus may trigger an SSc associated autoimmune response and result in endothelial cell activation and apoptosis. Increased endothelial cell apoptosis has been shown to occur in vivo, prior to the onset of fibrosis in both human SSc tissue and in the UCD 2000 avian model of SSc, an animal model of SSc that shows similar vascular abnormalities.

Microvascular Pericytes

Over the past 20 years, the function and role of the endothelium in SSc has been extensively studied; however, almost nothing is known about the role of microvascular pericytes. Developmentally derived from the mesenchyme, pericytes are located outside the vascular endothelium in continuous and fenestrated capillaries, venules, and arterioles less than 30 μm in diameter. They envelop the endothelium and are embedded within a basement membrane, which is synthesized by both endothelial cells and pericytes. During blood vessel formation in the developing embryo, recruitment of pericytes and smooth muscle cells occurs via the bidirectional release of soluble mediators between endothelial cells and smooth muscle cell/pericyte precursor cells. A number of mouse models such as the PDGF-β receptor, PDGF-BB, and angiopoietin-1 knockout mice have demonstrated that failure to recruit smooth muscle cells/pericytes to the nascent vascular compartments during vasculogenesis results in a dysfunctional vascular network and embryonic lethality [3]. In the adult, angiogenesis occurs during certain normal processes such as wound healing and a number of pathological conditions such as tumors and rheumatoid arthritis. Studies have shown that disrupting the relationship between pericytes and endothelial cells leads to an ablation of angiogenesis. These studies have highlighted that the physical and molecular interactions between endothelial cells and pericytes are involved not only in maintaining normal vascular homeostasis but also during pathological activation [4] and that phenotypic alterations in one cell type will almost certainly result in phenotypic alterations in the other cell type. Therefore, in pathological conditions characterized by endothelial cell activation, the contribution of pericytes will be as critical as that of endothelial cells.

Role of Pericytes in Fibrosis

It is becoming increasingly clear that microvascular pericytes contribute to the pathogenesis of fibrosis in a number of diverse tissues by both direct and indirect mechanisms. Microvascular pericytes have been proposed as mesenchymal precursor cells, and it has been argued that pericytes play a direct role in the pathophysiology of fibrosis by their transdifferentiation into collagen-synthesizing cells. Evidence from morphological studies has shown that during angiogenesis, pericytes migrate from the microvascular wall into the interstitium and acquire a fibroblast-like morphology. In hypertrophic scars and SSc skin, collagen-synthesizing cells are predominantly located adjacent to microvessels. More recently, studies have shown that in a number of diverse dermal conditions associated with increased synthesis of extracellular matrix such as SSc, wound healing, and excessive dermal scarring, pericytes have been found to have a common activated phenotype. Specifically they express PDGF-β receptors and the high-molecular-weight melanoma-associated antigen (HMW-MAA), a known activation marker [5, 6]. Concurrent in vitro studies showed that explanted pericytes with the same activated phenotype underwent a spontaneous transdifferentiation to collagen-synthesizing fibroblasts. A similar phenomenon has also been identified in the development of liver fibrosis where liver pericytes or Ito cells are the principal collagen-producing cells in the fibrotic lesion. Central to this fibrogenic potential is a phenotypic transdifferentiation from liver pericyte to an activated myfibroblastic stellate cell. Like pericytes, myofibroblasts express α-SMA and share phenotypic traits of both fibroblasts and smooth muscle cells. Myofibroblasts were initially described in wound healing where their principal function is the contraction of the provisional granulation tissue prior to their removal by apoptosis. They have subsequently been identified in a number of different conditions associated with fibrosis and scarring, and it has been postulated that a failure in the clearance of myofibroblasts may be important in distinguishing an acute wound response from a chronic fibrotic disorder. Therefore, targeted prevention of a pericyte to myofibroblast transdifferentiation may be of therapeutic value in fibrosis.

Pericytes can also modulate the activity and function of other cells involved in the fibrotic response. In wound healing and SSc, fibrosis is preceded by an infiltration of inflammatory mononuclear cells. Recently knockout mouse models have shown that pericytes play a key role in determining transendothelial cell permeability. Furthermore, liver pericytes are known to express adhesion molecules during liver injury and modulate the recruitment and migration of mononuclear cells within the perisinusoidal space of diseased livers. Pericytes are also able to synthesize a number of growth factors, notably TGF-β, connective tissue growth factor, and ET-1, all of which have been established as potent profibrotic modulators of fibroblast function.

Perhaps a key pathway in the pathological activation of microvascular pericytes during fibrosis is the PDGF-BB/β receptor pathway. In normal tissue, PDGF-β receptors are not constitutively expressed; however, in a number of diverse fibrotic conditions in skin, lung, and kidney, pericytes have been shown to overexpress PDGF-β receptors. The PDGF-BB/β receptor axis is a key developmental pathway, central to the recruitment of pericytes to developing vessels. Whereas PDGF-BB has been shown to induce both pericyte proliferation and migration in vitro, its role in vivo is currently less well understood.

Pericytes as Tissue-Specific Mesenchymal Stem Cells

The fibrogenic potential of pericytes is in part dependent upon their ability to transdifferentiate into other functional mesenchymal cell types. Over the past few years several studies have highlighted the pluripotent nature of pericytes. They have been shown both in vitro and in vivo to differentiate into osteoblasts, adipocytes, chondrocytes, and smooth muscle cells. Interestingly, pericytes also share a strong phenotypic similarity to a population of mesenchymal stem cells found in the adult bone marrow. The bone marrow is composed of two distinct cell lineages, the hematopoietic component and the supporting stroma, each with their own distinct population of stem cells. Bone marrow–derived mesenchymal stem cells can be identified by their expression of STRO-1, a cell surface protein whose function is as yet unknown. Interestingly, the STRO-1 antigen has also been shown to be expressed by both endothelial cells and microvascular pericytes in a number of different tissues. Furthermore, bone marrow stem cells also commonly express α-smooth muscle actin (α-SMA) in culture and have a phenotype more like that of pericytes than smooth muscle cells. Although a strong phenotypic similarity between pericytes and bone marrow–derived stem cells exists, further studies are needed before pericytes can classified as in situ stem cells.

Concluding Remarks

Pericytes were discovered a century ago, yet their roles as indispensable partners to endothelial cells and putative mesenchymal stem cells have only recently been identified. Although we have acquired a basic understanding of how endothelial cells and pericytes communicate, the details are still poorly understood. However, even a basic understanding has dispelled the idea that pericytes are mere support cells; rather, they are involved in both the normal homeostatic maintenance of microvessels and most likely in pathologies associated with microvascular activation and angiogenesis.

Glossary

Fibrosis: Thickening and scarring of connective tissue leading to impaired organ function. Most often a consequence of inflammation or injury.

Mesenchymal stem cell: A cell that can give rise to a number of different mesenchymal cell lineages.

Vasoconstriction: A reduction in the diameter of blood vessels especially arterioles and capillaries leading to reduced blood flow to that part.

References

1. Black, C. M. (1995). The aetiopathogenesis of systemic sclerosis: Thick skin-thin hypotheses. The Parkes Weber Lecture 1994. J. R. Coll. Physicians London **29**, 119–130. *A thorough review of the pathophysiology and epidemiology of systemic sclerosis.*
2. Prescott, R. J., et al. (1992). Sequential dermal microvascular and perivascular changes in the development of scleroderma. *J. Pathol.* **166**, 255–263.
3. Carmeliet, P., Moons, L., and Collen, D. (1998). Mouse models of angiogenesis, arterial stenosis, atherosclerosis and hemostasis. *Cardiovasc. Res.* **39**, 8–33. *A very good review summarizing the murine models that have been generated for studying and manipulating angiogenesis and other related vascular pathologies.*
4. Gerhardt, H., and Betsholtz, C. (2003). Endothelial–pericyte interactions in angiogenesis. *Cell Tissue Res.* **314**, 15–23.
5. Rajkumar, V. S., et al. (1999). Activation of microvascular pericytes in autoimmune Raynaud's phenomenon and systemic sclerosis. *Arthritis Rheum.* **42**, 930–941.
6. Sundberg, C., et al. (1996). Pericytes as collagen-producing cells in excessive dermal scarring. *Lab. Invest.* **74**, 452–466.

Further Reading

Bianco, P., et al. (2001). Bone marrow stromal stem cells: Nature, biology, and potential applications. *Stem Cells* **19**, 180–192.

Herrick, A. L., and Clark, S. (1998). Quantifying digital vascular disease in patients with primary Raynaud's phenomenon and systemic sclerosis. *Ann. Rheum. Dis.* **57**, 70–78.

Schor, A. M., et al. (1995). Pericyte differentiation. *Clin. Orthop.* **313**, 81–91.

Tomasek, J. J., et al. (2002). Myofibroblasts and mechano-regulation of connective tissue remodelling. *Nat. Rev. Mol. Cell Biol.* **3**, 349–363. *A comprehensive review of myofibroblast differentiation and function in fibrosis.*

Capsule Biography

Vineeth Rajkumar is a research scientist based in the Centre for Rheumatology and Connective Tissue Disease. The primary focus of his research is the role of the microvasculature in the development of connective tissue fibrosis.

David Abraham is the head of research at the Centre for Rheumatology and Connective Tissue Disease. His laboratory focuses primarily on the cellular and molecular mechanisms of connective tissue fibrosis.

Carol Black is the director of the Centre for Rheumatology and Connective Tissue Disease. She is an internationally recognized expert in the treatment and management of systemic sclerosis and oversees an extensive program of scientific and clinical research into systemic sclerosis. She was elected President of the Royal College of Physicians (London) in 2002.

SECTION U

Skin

CHAPTER 135

Impaired Skin Microcirculatory Function in Human Obesity

Alexei V. Agapitov[1] and William G. Haynes[2]
[1]*SUNY at Buffalo*
[2]*University of Iowa*

Obesity is associated with hypertension, left ventricular hypertrophy, atherosclerosis, and cardiovascular morbidity and mortality. Patients with obesity have an exaggerated pressor response to mental stress and diminished forearm and systemic vasodilatation to mental stress. Acute mental stress also leads to a greater pressor response in hypertensive individuals. Chronic psychosocial stress is associated with the development of hypertension. An exaggerated blood pressure elevation to external stressors is associated with higher subsequent resting diastolic blood pressure 6.5 years later. Like hypertension, obesity is associated with a greater blood pressure increase to acute and chronic psychosocial stress. The skin microcirculation plays an important role in systemic blood pressure regulation. Dilatation of skin microvessels to mental stress is blunted in obese normotensive subjects, which may account for the increased propensity of obese individuals to develop hypertension under conditions of chronic psychosocial stress.

Skin Microvasculature and Techniques for Measuring Skin Blood Flow

Skin microvasculature is organized as two horizontal plexuses, with one located 1 to 1.5 mm below the skin surface and the other at the dermal–subcutaneous junction. The lower plexus is formed by perforating vessels from the underlying muscles and subcutaneous fat. The arterioles and venules in the fat lobules are identical in size and structure to those of the lower horizontal plexus. Descending venules and ascending arterioles connect the two plexuses and are arranged in pairs and supply the hair bulbs and sweat glands. The nutritive capillary loops of the dermal papillae arise from the upper plexus. The arterioles in the papillary dermis are terminal arterioles and function as part of the resistance vessels in the skin. The other vessels of the papillary dermis are capillaries and postcapillary venules. Postcapillary venules present the majority of vessels and are associated with a variety of functions—migration of inflammatory cells from vessels into the tissue and increasing vascular permeability in response to acute inflammation. Postcapillary venules join larger valve-containing vessels that pass from the deep dermis into the superficial layer of the adipose tissue. These valves in collecting veins at dermal–fat interface ensure the forward propulsion of blood [1]. Direct cannulation of human finger nailfold capillaries has demonstrated that the blood pressure is pulsatile in both arterioles and venules with systolic pressures fluctuating between 11 and 75 mmHg.

Skin blood flow is regulated by the nervous system and local factors. Several reflex arcs are involved, including central reflexes, short reflexes through the spinal cord, and local reflex arcs within the skin. Sympathetic nervous system tone usually keeps arteriovenous anastomoses in a constricted state. Loss of sympathetic tone results in opening of these shunts and decreases the blood flow through the skin capillaries [2]. Spectral analysis of vasomotor waves measured by laser-Doppler flowmetry in skin demonstrated that the beat-to-beat rate of the cardiac cycle varies around two means: 0.1 and 0.25 Hz. The 0.1-Hz frequency likely represents sympathetic activity, as it becomes accentuated with position change from supine to erect. These postural responses can be reduced significantly by sympathetic blockade of the limb space [3].

Laser-Doppler flowmetry involves applying a beam of laser light to the skin. Blood cells moving within the area will alter the wavelength of the light (Doppler shift). The scattered light is converted to an electronic signal and analyzed. The greater the number and velocity of moving cells, the higher the Doppler shift. The laser-Doppler technique gives an estimation of skin perfusion, which is expressed as arbitrary units. It does not measure exact skin perfusion in mL/min/100 g of tissue. Nevertheless, the laser-Doppler technique is useful in measuring changes in skin blood flow in response to various stimuli and interventions, such as local administration of vasoactive medications by iontophoresis [4]. The laser-Doppler flowmetry technique gathers signals from 1 to 1.5 mm below the epidermis; therefore, it does not measure blood flow to the structures located 3 to 5 mm below the skin surface, such as sweat glands and hair follicles. However, some researchers have used the laser-Doppler flux technique to measure adipose tissue blood flow by introducing the Doppler probe directly into adipose tissue through a needle.

Iontophoresis and microdialysis are techniques allowing drug delivery through the skin and could be used for extraction of substances from the skin and measuring concentrations of these substances. These techniques are described in detail elsewhere [5]. In addition, intradermal electrode techniques can be used to directly measure nitric oxide in skin. Nitric oxide is involved in skin vasodilatation in response to heat and ischemia–reperfusion.

Dynamic capillaroscopy provides the best information about nutritional status of a circumscribed skin area. This technique involves using a microscope, television camera, monitor, and software to analyze the nutritional blood flow. Several observations can be made with this technique, such as number and type of capillaries in the nailfold, intercapillary distance, caliber, tortuosity, and capillary blood cell velocity. Transcapillary diffusion of sodium fluorescein through the wall of a single capillary can be measured by this technique to assess capillary permeability.

Skin adipose tissue blood flow can vary greatly if measured in milliliters per unit of tissue. However, blood flow is relatively constant if expressed per adipocyte. Adipose tissue blood flow expressed in milliliters per amount of tissue decreases with increasing adipocyte mass in lipid accumulation [6].

A useful technique of local catheterization involves insertion of a catheter into a vein draining anterior abdominal wall, as this adipose depot has separate venous outflow from muscle. Adipose tissue blood flow in humans is determined by the clearance of previously injected radiolabeled molecules. The xenon washout technique involves administration of ^{133}Xe into adipose tissue and serial sampling of venous radioactivity for several hours. Adipose tissue blood flow is thought to be equivalent to the washout of xenon into the venous blood.

An additional methodology is a microdialysis technique that allows measurement of blood flow by introducing ethanol into human adipose tissue through a dialysis probe, and then measuring its escape into the dialysate spectrophotometrically.

Importance of Skin Microvasculature and Alterations of Skin Microcirculation in Obesity

Adipose tissue receives a rich capillary blood supply. Capillary rarefaction and increased vasoconstrictor response to local cooling in the human skin has been demonstrated in individuals with essential hypertension [7, 8]. Alterations in skeletal muscle microvascular function are paralleled by alterations in skin microvascular function in hypertensive humans [9]. Sympathetic control of skin blood flow is reduced in obesity [10]. Even young children with obesity have morphologic alterations in skin microvasculature, such as increased number of tortuous loops in finger nailfold capillaries. Obese children have also been shown to have decreased capillary blood cell velocity after a 1-minute arterial occlusion [11]. These children's blood flow response to 1-minute arterial occlusion, as measured by laser-Doppler flowmetry, was not altered. This could represent shunting of blood from capillaries through arteriovenous anastomoses. Increased arteriovenous shunting of blood in human skin has also been demonstrated in diabetes [2]. Obesity precedes insulin resistance and diabetes, and these early changes in skin microvascular morphology and reactivity demonstrated by Chin et al. could be precursors of microvascular changes leading to the development of skin ulcers with the development of diabetes.

Individuals with impaired glucose tolerance have been shown to have blunted vasodilatory response to local skin heating. This abnormality was found to be associated with elevated fasting plasma insulin concentration, but not with fasting plasma glucose [12]. The mechanisms responsible for the association between defective skin microvascular hyperemic response and insulin resistance remain to be elucidated. Body mass index has a significant inverse correlation with endothelium-dependent skin vasodilatation [13]. Normotensive obese subjects exhibit deficient skin microcirculatory vasodilatation to mental stress. There is evidence that subcutaneous adipose tissue blood flow contributes significantly to total forearm blood flow. Mental stress causes an increase in skin sympathetic nerve activity and cutaneous vasodilatation in the normal range of skin temperatures. This is consistent with a predominant β-adrenoreceptor-mediated vasodilatation. Cardiopulmonary baroreceptor unloading leads to skin vasoconstriction, which demonstrates that skin microvasculature plays a role in regulation of systemic vascular resistance and blood pressure. Recently, our group demonstrated that deficient skin as well as whole-forearm vasodilatation occurs after mental stress in obese normotensive subjects (Figure 1). Deficient skin vasodilatation in conjunction with increased cardiac output in response to psychosocial stress may account for the development of hypertension and cardiovascular disease in obesity.

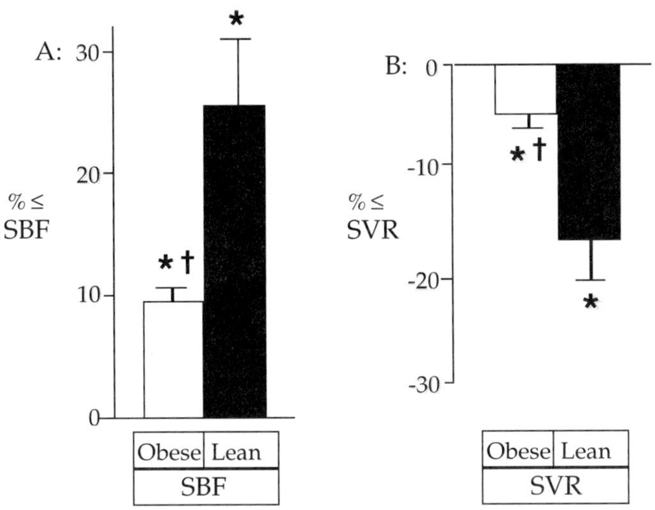

Figure 1 **(A)** Skin blood flow and **(B)** skin vascular resistance percent changes from baseline during mental stress in obese and lean subjects. SBF, skin blood flow; SVR, skin vascular resistance. *$P < 0.05$ versus baseline. †$P < 0.05$ versus lean subjects. $n = 11$.

Potential Mechanisms Responsible for Deficient Skin Vasodilatation to Mental Stress in Obesity

Several mechanisms could possibly lead to the observed impairment in skin vasodilatation to mental stress in obesity. Skin vasodilatation in response to mental stress in humans occurs due to both local sympathetic nerve traffic withdrawal and β2-adrenergic vasodilatation from circulating epinephrine and norepinephrine. It is possible that circulating norepinephrine causes vasodilatation by stimulating β-adrenoreceptors located predominantly on the inner layer of smooth muscle. Given that nerve terminals are located in the adventitia of the blood vessels, neurally released norepinephrine causes vasoconstriction by stimulation of α-adrenoreceptors, located predominantly in the outer layer of vascular smooth muscle [14]. The sympathetic withdrawal during mental stress might be attenuated, because it is known that baroreflex control of muscle sympathetic nerve activity is impaired in obesity and improves with weight loss in obese individuals. There is an endothelium-dependent component of β2-adrenergic vasodilatation in the human forearm microvasculature. There is also a cholinergic system in human endothelium. Endothelium-dependent forearm vasodilatation to mental stress is associated with cholinergic stimulation of the vascular endothelium. Obesity without complications leads to endothelial dysfunction in forearm resistance vessels, and skin microvasculature constitutes a significant component of the forearm vascular bed. Forearm vasodilatation to mental stress is endothelium dependent. Obesity is associated with defective β2-adrenergic vasodilatation in the human forearm. Obesity-associated endothelial dysfunction may lead to deficient skin vasodilatory response to mental stress and, therefore, may account for the development of hypertension and cardiovascular disease in obese individuals.

In addition, insulin resistance might play a role in abnormal microcirculatory responses to mental stress in obesity. It is known that obesity leads to insulin resistance. Many obese individuals have insulin resistance but normal fasting glucose values. Skin capillary recruitment and acetylcholine-mediated vasodilatation have a strong and positive correlation with insulin sensitivity. A reduction in skin microvascular vasodilator response to ischemia and acetylcholine is associated with insulin resistance [9]. Deficiency of skin microvascular dilatation has also been associated with higher 24-hour systolic blood pressures [9]. Exaggerated blood pressure increase to mental and physical stress has been observed in normotensive obese women with insulin resistance and abnormal glucose tolerance tests. Endothelial dysfunction appears to contribute to the absence of insulin-induced vasodilatation in obesity [15].

Adipose tissue has a local renin–angiotensin system. Angiotensinogen, angiotensin-converting enzyme, and AT1 and AT2 receptors are all present in human adipose tissue. Angiotensinogen expression in adipose tissue correlates with waist-to-hip ratio in obese humans. Central fat distribution is associated with increased systemic vascular resistance and hypertension. Increased activity of this local renin–angiotensin system may impair skin vasodilatation in obesity. Also, increased levels of proinflammatory cytokines might play a role, because chronic subclinical inflammation was shown to be a part of the insulin resistance syndrome. C-reactive protein has been shown to positively correlate with measures of obesity, such as body mass index and waist circumference. Tumor necrosis factor-α and interleukin-6 concentrations are higher in obese individuals. Weight loss leads to a decrease in the levels of these proinflammatory cytokines, along with a decrease in adhesion molecules and improvement in endothelial function.

Elevated free fatty acid secretion by adipose tissue in obesity may adversely affect skin microvascular function, because these lipid metabolites have been shown to impair endothelium-dependent vasodilatation at the microcirculatory level [15] and to increase vasoconstrictor responses in dorsal hand veins [16].

Obese individuals are known to have increased levels of low-density lipoprotein (LDL) cholesterol and triglycerides and decreased high-density lipoprotein (HDL) cholesterol. In healthy women, skin vasodilatation in response to iontophoresis of acetylcholine, isoprenaline, and nitroprusside has been shown to be positively correlated to HDL cholesterol, and negatively correlated to the ratio of total cholesterol/HDL and triglycerides [4]. Skin postischemic microcirculatory dilatation is blunted in patients with hypertriglyceridemia [17].

Further studies are necessary to evaluate the significance of all the potential mechanisms just described in relation to skin microcirculatory impairments in obesity.

Conclusions

Skin microcirculatory dilatation to mental stress in obese human subjects is impaired. This may be implicated as a pathophysiologic mechanism for the known greater pressor response to chronic psychosocial stress in obesity and contribute to the cardiovascular complications of obesity.

Glossary

Microcirculation: The smallest blood vessels, including arterioles, capillaries and venules, serving to control peripheral resistance to blood flow, tissue perfusion, and blood–tissue exchange.

Obesity: Excessive subcutaneous and visceral fat deposition, associated with insulin resistance, type II diabetes, and cardiovascular disease.

Vasodilatation: Increase in diameter of blood vessel as a result of vascular smooth muscle relaxation, leading to an increase in blood flow through the vessel.

References

1. Mahler, F., Muheim, M., Intaglietta, M., Bollinger, A., and Anliker, M. (1979). Blood pressure fluctuations in human nailfold capillaries. *Am. J. Physiol.* **236**, H888–H893.
2. Fagrell, B., Jorneskog, G., and Intaglietta, M. (1999). Disturbed microvascular reactivity and shunting: a major cause for diabetic complications. *Vasc. Med.* **4**, 125–127.
3. Bernardi, L., Rossi, M., Fratino, P., Finardi, G., Mevio, E., and Orlandi, C. (1989). Relationship between phasic changes in human skin blood flow and autonomic tone. *Microvasc. Res.* **37**, 16–27.
4. Algotsson, A. (1996). Serum lipids and lipoproteins are correlated to skin vessel reactivity in healthy women. *Journal of Internal Medicine* **239**, 147–152.
5. Vinik, A., Erbas, T., Park, T., Pierce, K., and Stansberry, K. (2001). Methods for evaluation of peripheral neurovascular dysfunction. *Diabetes Technol. Ther.* **3**, 29–50.
6. Blaak, E., van Baak, M., Kemerink, G., Pakbiers, M., Heidendal, G., and Saris, W. (1995). Beta-adrenergic stimulation and abdominal subcutaneous fat blood flow in lean, obese, and reduced-obese subjects. *Metabolism.* **44**, 183–187.
7. Prasad, A., Dunhill, G., Mortimer, P., and MacGregor, G. (1995). Capillary rarefaction in the forearm skin in essential hypertension. *Journal of Hypertension* **13**, 265–268.
8. Gasser, P., and Buhler, F. R. (1992). Nailfold microcirculation in normotensive and essential hypertensive subjects, as assessed by videomicroscopy. *Journal of Hypertension* **10**, 83–86.
9. Serne, E., Stehouwer, C., Maaten, J., Wee, P., Rauwerda, J., Donker, A., and Gans, R. (1999). Microvascular function relates to insulin sensitivity and blood pressure in normal subjects. *Circulation* **99**, 896–902.
10. Smagghue, O., Velayoudon, P., Paries, J., Attali, J., and Valensi, P. (1996). The Laser-Doppler microcirculation technique in the study of the orthosympathetic system in obese and non-insulin-dependent diabetic patients. *Journal des Maladies Vasculaires* **21**, 299–302.
11. Chin, L., Huang, T., Yu, C., Wu, C., Hsu, C., and Yu, H. (1999). Increased cutaneous blood flow but impaired post-ischemic response of nutritional flow in obese children. *Atherosclerosis* **146**, 179–185.
12. Jaap, A., Shore, A., and Tooke, J. (1997). Relationship of insulin resistance to microvascular dysfunction in subjects with fasting hyperglycaemia. *Diabetologia* **40**, 238–243.
13. Caballero, A., Arora, S., Saouaf , R., Lim, S., Smakowski, P., Park, J., King, J., LoGerfo, F., Horton, E., and Veves, A. (1999). Microvascular and macrovascular reactivity is reduced in subjects at risk for type 2 diabetes. *Diabetes* **48**, 1856–1862.
14. Ballard , K. (1977). Functional characteristics of the microcirculation in white adipose tissue. *Microvasc. Res.* **16**, 1–18.
15. Steinberg, H., Tarshoby, M., Monestel, R., Hook, G., Cronin, J., Johnson, A., Bayazeed, B., and Baron, A. (1997). Elevated circulating free fatty acid levels impair endothelium-dependent vasodilation. *Journal of Clinical Investigation* **100**, 1230–1239.
16. Stepniakowski, K., Goodfriend, T., and Egan, B. (1995). Fatty acids enhance vascular alpha-adrenergic sensitivity. *Hypertension* **25**, 774–778.
17. Tur, E., Politi, Y., and Rubinstein, A. (1994). Cutaneous blood flow abnormalities in hypertriglyceridemia. *J. Invest. Dermatol.* **103**, 597–600.

Further Reading

Agapitov, A., Correia, M., Sinkey, C., Dopp, J., and Haynes, W. (2002). Impaired skeletal muscle and skin microcirculatory function in human obesity. *J. Hypertens.* **20**, 1401–1405. *This recent article provides insights into the role of the skin microvasculature in the pathophysiology of obesity-induced hypertension.*

Braverman, I. (2000). The cutaneous microcirculation. *J Invest Dermatol. Symp. Proc.* **5**, 3–9. *A recent review on anatomy and physiology of skin microcirculation.*

Grandall, D., Hausman, G., and Kral, J. (1997). A review of the microcirculation of adipose tissue: Anatomic, metabolic, and angiogenic perspectives. *Microcirculation* **4**, 211–232. *A recent comprehensive review of adipose tissue microcirculation.*

Narkiewicz, K. (2002). Obesity-related hypertension: Relevance of vascular responses to mental stress. *J. Hypertens.* **20**, 1277–1278. *A recent article discussing the relevance of stress-induced microvascular responses to obesity-related hypertension.*

Rosell, S., and Belfrage, E. (1979). Blood circulation in adipose tissue. *Physiol. Rev.* **59**, 1078–1104.

Capsule Biography

Alexei Agapitov, M. D., received his research training at the University of Iowa. His research interests include the sympathetic nervous system, obesity, and hypertension.

Dr. William Haynes received his medical education at the University of Sheffield, England. In 1995, Dr. Haynes moved to the University of Iowa, where he now holds a position as Professor of Internal Medicine in the Divisions of Cardiovascular Diseases and Clinical Pharmacology. He is Program Director of the NIH-funded Clinical Research Center at the University. Dr. Haynes is also Associate Editor of the journal *Arteriosclerosis, Thrombosis and Vascular Biology*. His research is focused on mechanisms of vascular dysfunction in obesity, hypertension, and atherosclerosis.

CHAPTER 136

Cutaneous Microcirculation in Patients with Scleroderma

[1]Michael Junger and [2]Anja Schlez

[1]*Department of Dermatology of the Ernst-Moritz-Arndt-University, Greifswald, Germany*
[2]*Department of Dermatology of the Eberhard-Karls-University, Tübingen, Germany*

Circulatory disorders of the extremities in the form of Raynaud's phenomenon belong to the earliest clinical manifestations of systemic scleroderma. The patients complain of cold hands and feet, stiff joints, and loss of mobility in the extremities. Later cutaneous manifestations of the disease include edematous swelling in the extremities and, in more extreme cases, often very painful, refractory acral necroses.

Nutritive Nailfold Capillaries

Characteristic changes in microcirculation can affect function or morphology. Morphological changes of the nailfold capillaries in systemic sclerosis (SSc) have been studied extensively. The illness is characterized from the very onset by cutaneous microangiopathy in the form of aneurysms, capillary rarefaction and microthromboses, and the development of megacapillaries [1]. Morphological changes of nailfold capillaries are detected in more than 90 percent of patients [2].

For quantitative evaluation of human TV nailfold capillary microscopy a new semiautomated image-analyzing system called "CapiShape" was presented. The outer limits or boundaries are recognized by a special image processing algorithm to describe the shape of capillaries [3].

Enhanced Transcapillary and Interstitial Diffusion

By means of video microscopy after intravenous injection of sodium fluorescein (NaF) and indocyanin green as tracers, a more accurate morphological description became possible, providing measurements of the diameters of the capillary lumen, red blood cell column, plasma layer, and pericapillary halo. Halo diameter was significantly increased at the apex only and not at the two capillary limbs. This finding confirms the visual impression of "apical capping" after NaF injection as a important sign for microangiopathy due to collagen vascular disease [4].

Other morphological changes observed with the disease include vascular obliteration and excessive collagen production in the pericapillary area [5]. In many patients, these changes preceded detection of autoantibodies [6].

Reduced Capillary Flow and Pressure

The capillaries of the skin not only undergo morphological changes, but also display functional deficits. Both blood flow velocity and capillary pressure in the nailfold capillaries of SSc patients were significantly lower than in healthy volunteers [7]. Cold exposure leads to a marked slowing of blood flow velocity in the capillaries of the patients, in extreme cases to the point of stasis [8].

These pathological dysfunctions are documented by means of intravital capillaroscopy combined with a cold provocation challenge. In contrast to healthy persons, patients with SSc usually react to cold with prolonged stasis. Vasodilators or the application of heat increases blood cell velocity. Capillaroscopy in combination with a special video image analysis method allowed us to quantify changes in blood cell velocity.

In Vivo Proof of Endothelial Dysfunction

Recently endothelial function, a factor in the regulation of perfusion, was examined in detail in SSc patients and healthy volunteers by administering intracapillary microinjections of vasoactive substances and monitoring the resulting endothelium-dependent (acetylcholine) or endothelium-independent (sodium nitroprusside) vasodilatation, increase in vascular diameter, and decrease in blood flow velocity. The microinjection method used for the first time in this study made it possible to establish that endothelium-independent vasomotion is substantially disturbed in patients with scleroderma [9].

Intimal damage to the blood vessels of SSc patients was already demonstrated serologically in studies showing elevated levels of anti-bleeding factor, endothelin-1 and ACE [10–12]. The trigger for the endothelial damage is still unknown. One potential cause presently under discussion is a vasculopathy caused by cytomegalovirus or parvovirus B19 [13]. Recently, it was possible to show that autoantibodies to human cytomegalovirus cause the apoptosis of endothelial cells in systemic scleroderma patients [14].

Therapeutical Effects to the Microcirculation of the Skin

The treatment of cutaneous circulatory problems can range from general measures such as protecting the hands from cold, exercise therapy, and lymph drainage to the use of vasoactive medication such as the prostanoids.

Numerous studies have demonstrated the efficacy of prostaglandin E_1 in the treatment of vasospastic and obliterative vascular diseases [15, 16]. In almost all studies to date on the use of prostaglandin E_1 to treat collagen diseases, medication was given intravenously. The first clinical studies on iloprost for secondary Raynaud's phenomenon were carried out in Europe. They included open [17], crossover, placebo-controlled [18], controlled single-blind [19], and iloprost/nifedipine comparison studies [20]. Watson summarized the European studies in a meta-analysis. Overall, iloprost led to a 20 to 30 percent reduction in the number of Raynaud's episodes. Iloprost was also more effective than placebo in the treatment of ischemic ulcerations. In an American double-blind study on iloprost, of 35 patients with secondary Raynaud's phenomenon, seven patients in the iloprost group and four patients in the placebo group had ischemic lesions. After a treatment period of 10 weeks, six of the seven patients in the iloprost group experienced complete healing of the lesions on their fingers, whereas the ulcerations failed to heal in any of the patients of the placebo group [16]. The transdermal application of prostaglandin E_1 ethyl ester was also shown to have a favorable effect on nutritive blood flow in the capillaries of the skin in systemic scleroderma patients. The patients experienced clinical benefits: The number of Raynaud's episodes dropped and the trophic skin lesions began to heal [21, 22].

Clinical improvement, that is, abating Raynaud's attacks, is accompanied by an improvement of nutritive cutaneous blood flow shown in decreased blood flow stasis times. Thus the assessment of capillary blood flow in response to cold challenge has prognostic value.

Glossary

Halo: The bright yellow fluorescence of the pericapillary interstitial space after intravenous injection of sodium fluorescein.

Stasis: Blood flow stops in the nailfold capillaries, particularly after cold exposure. It can be observed by means of capillaroscopy.

Systemic scleroderma: A rare, chronic autoimmune disease. The systemic forms can affect any part of the body (skin, blood vessels, and internal organs). The systemic forms of scleroderma cause fibrosis (scar tissue) to be formed in the skin and/or internal organs. The function of the diseased organs is impaired.

References

1. Maricq, H. R., LeRoy, E. C., D'Angelo, W. A., Medsger, T. A. Jr, Rodnan, G. P., Sharp, G. C., and Wolfe, J. F. (1980). Diagnostic potential of in vivo capillary microscopy in scleroderma and related disorders. *Arthritis. Rheum.* **23**, 183–189.
2. Grassi, W., Medico, P. D., Izzo, F., and Cervini, C. (2001). Microvascular involvement in systemic sclerosis: capillaroscopic findings. *Semin. Arthritis. Rheum.* **30**, 397–402.
3. Hu, Q., and Mahler, F. (1999). New system for image analysis in nailfold capillaroscopy. *Microcirculation* **6**, 227–235.
4. Brulisauer, M., and Bollinger, A. (1991). Measurement of different human microvascular dimensions by combination of videomicroscopy with Na-fluorescein (NaF) and indocyanin green (ICG) in normals and patients with systemic sclerosis. *Int. J. Microcirc. Clin. Exp.* **10**, 21–31.
5. LeRoy, E. C. (1996). Systemic Sclerosis. A vascular perspective. *Rheum. Dis. Clin. North. Am.* **22**, 675–694.
6. Jünger, M., Klyscz, T., and Hahn, M. (1998). Clinical relevance of cutaneous microangiopathy in systemic sclerosis (SSc). *JEADV* 1 (Suppl. 2), **202**. *Morphological changes of capillaries in scleroderma patients preceded detection of autoantibodies.*
7. Hahn, M., Heubach, T., Steins, A., and Jünger, M. (1998). Hemodynamics in nailfold capillaries of patients with systemic scleroderma: Synchronous measurements of capillary blood pressure and red blood cell velocity. *J. Invest. Dermatol.* **110**, 982–985. *This study show that blood flow velocity and capillary pressure in nailfold capillaries of SSc patients were significantly lower than in healthy volunteers.*
8. Hahn, M., Klyscz , T., Jünger, M., and Rassner, G. (1995). Local cold exposure test as therapy control in patients with Raynaud's phenomenon: Comparison between laser Doppler fluxmetry and simultaneous red blood cell velocity measurements in nailfold capillaries. *Br. J. Dermatol.* **133**, 704–709.
9. Schlez, A., Kittelm, M., Braun, S., Häfner, H. M., and Jünger, M. (2003). Endothelium-dependent regulation of cutaneous microcirculation in patients with systemic scleroderma. *J. Invest. Dermatol.* **120**, 332–334.
10. Maeda, M., Kachi, H., Takagi, H., and Kitajima, Y. (1997). Is there circadian variation of plasma endothelin (ET-1) in patients with systemic scleroderma (SSc)? *J. Dermatol. Sci.* **16**, 38–44.
11. Gordon, M. M., and Madhok, R. (1996). Circulating endothelin-1 levels in systemic sclerosis subsets—a marker of fibrosis or vascular dysfunction? *J. Rheumatol.* **23**, 788–789.
12. Herrick, A. L., Illingworth, K., Blann, A., Hay, C. R., Hollis, S., and Jayson, M. I. (1996). Von Willebrand factor, thrombomodulin, thromboxane, beta-thromboglobulin and markers of fibrinolysis in primary Raynaud's phenomenon and systemic sclerosis. *Ann. Rheum. Dis.* **55**, 122–127.
13. Ferri, C., Zakrzewska, K., Longombardo, G., Giuggiols, D., Storino, F., Pasero, G., and Azzi, A. (1999). Parovirus B19 infection of bone marrow in systemic sclerosis patients. *Clin. Exp. Rheumatol.* **17**, 718–720.
14. Lunardi, C., Bason, C., Navone, R., Millo, E., Damonte, G., Corrocher, R., and Puccetti, A. (2000). Systemic sclerosis immunoglobulin G autoantibodies bind the human cytomegalovirus late protein UL94 and induce apoptosis in human endothelial cells. *Nat. Med.* **6**, 1183–1186.

This study shows that autoantibodies to human cytomegalovirus cause the apoptosis of endothelial cells in systemic scleroderma patients.

15. Clifford et al. (1980).
16. Fagrell, B., Lundberg, G., Ollson, A., and Östergen, J. (1986). PgE1 treatment of severe skin ischaemia in patients with peripheral arterial insufficiency—the effect of skin microcirculation. *Vasa* **15**, 56–60.
17. Rademaker, M., Thomas, R. H., Provost, G., Beacham, J. A., Cooke, E. D., and Kirby, J. D. (1987). Prolonged increase in digital blood flow following iloprost infusion in patients with systemic sclerosis. *Postgraduate Medical Journal* **63**, 617–620.
18. McHugh, N. J., Csuka, M., Watson, H., Belcher, G., Amadi, A., Ring, E. F., Black, C. M., and Maddison, P. J. (1988). Infusion of iloprost, a prostacyclin analogue, for treatment of Raynaud´s phenomenon in systemic Sclerosis. *Ann. Rheum. Dis.* **47**, 43–47.
19. Kyle, V., Parr, G., Salisbury, R., Thomas, P. P., and Hazleman, B. (1985). Prostaglandin-E1 vasospastic disease and thermography. *Ann. Rheum. Dis.* **44**, 73–78.
20. Rademaker, M., Cooke, E. D., Almond, N. E., Beacham, J. A., Smith, R. E., Mant, T. G., and Kirby, J. D. (1989). Comparision of intravenous infusion of iloprost and oral nifedipine in treatment of Raynaud´s phenomenon in patients with systemic sclerosis. *BMJ* **298**, 561–564.
21. Schlez, A., Kittel, M., Scheuerle, B., Diehm, C., and Jünger, M. (2003). Transdermal application of prostaglandin E1 ethyl ester for treatment of trophic acral skin lesions in a patient with systemic scleroderma. *JEADV* **16**, 526–528.
22. Schlez, A., Häfner, H. M., Kittel, M., Braun, S., Diehm, C., and Jünger, M. (2003). Systemic scleroderma patients have improved skin perfusion after the transdermal application of PGE1 ethyl ester. *Vasa* **32**, 83–86.

Capsule Biography

Dr. Jünger was responsible for angiology and phlebology in the Department of Dermatology in Greifswald. He has headed the Department of Dermatology at the University of Greifswald since 2001. Systematically he and his coworkers examined the microangiopathy of the skin in patients with chronic venous disease. His work was supported by grants from the Deutsche Forschungsgemeinschaft and the German Society of Phlebology.

Dr. Schlez has worked in the Department of Dermatology of the University in Tübingen, Germany, since 1997. Systematically she and her coworkers examined the microangiopathy of the skin in patients with scleroderma and chronic venous disease. Her work was supported by a grant of the German Scleroderma Foundation.

CHAPTER 137

Angiogenic Diseases of the Skin

[1]Rainer Kunstfeld and [2]Michael Detmar

[1]*Department of Dermatology, Massachusetts General Hospital, Harvard Medical School, Boston, Massachusetts*
[2]*Institute of Pharmaceutical Sciences, Swiss Federal Institute of Technology (ETH) Zurich*

Structure and Function of Cutaneous Blood Vessels

The cutaneous blood vessels are involved in the control of body temperature and provide a conduit for the supply of nutrients and oxygen to the skin and for the rapid disposal of metabolic waste products. The vascular system of the skin consists of a superficial and a deep vascular plexus with additional vascular networks surrounding sweat glands and hair follicles [1]. Consistently, blood vessels are found immediately below the nonvascularized epidermis and surrounding anagen (growth phase) hair follicles. Whereas the richness of the follicular blood supply depends on the size of the hair follicles, recent studies suggest that, conversely, the extent of perifollicular vascularization may directly influence the size of hair follicles [2].

All cutaneous blood vessels contain a continuous inner monolayer of flat endothelial cells that are surrounded by a continuous basement membrane. Smaller vessels contain a second, sometimes discontinuous layer of pericytes that can be visualized by staining for alpha-smooth muscle actin and desmin. The walls of most cutaneous arterioles and larger venules and veins contain contractile smooth muscle cells that are surrounded by a basement membrane. Vascular basement membranes contain collagen type IV, collagen type XVIII, laminin, fibronectin, and other extracellular matrix proteins.

Molecular Control of Skin Angiogenesis

Vascular Endothelial Growth Factor

Whereas angiogenesis is a common feature in developing skin during embryogenesis, blood vessels in healthy adult skin are quiescent with the exception of the cyclic expansion of perifollicular vessels during the hair cycle. In normal skin, vascular quiescence is maintained by the dominant influence of angiogenesis inhibitors over angiogenic stimuli. The major proangiogenic factor in the skin, vascular endothelial growth factor (VEGF), occurs in at least four isoforms of 121, 165, 189, and 201 amino acid residues. VEGF121 and VEGF165 are the predominant isoforms found in human skin. VEGF mediates its activity mainly through interaction with two type III tyrosine kinase receptors, VEGF receptor-1 (flt-1) and VEGF receptor 2 (KDR, flk-1), which are selectively expressed on the cutaneous endothelium. VEGF-165 also binds the neuropilin receptors on endothelial and other cells. VEGF is expressed at low levels in normal epidermis, whereas keratinocyte VEGF expression is rapidly upregulated by hypoxia and is also induced by several growth factors that mediate epidermal hyperplasia (Figure 1). In psoriasis, healing wounds, and squamous cell carcinomas, transforming growth factor-α and other ligands of the epidermal growth factor receptor are released by suprabasal keratinocytes. In an autocrine loop, these growth factors induce hyperplasia of the epidermis. Simultaneously, they induce VEGF expression and secretion by keratinocytes, leading to paracrine induction of angiogenesis through interaction with VEGF receptors on cutaneous microvessels (Figure 1). This mechanism links epidermal hyperplasia with increased vascularization, thereby providing enhanced vascular support to meet the enhanced nutritional needs of proliferating keratinocytes. Several other growth factors that stimulate keratinocyte proliferation, including keratinocyte growth factor and hepatocyte growth factor/scatter factor, have recently joined the group of molecules that induce keratinocyte VEGF expression.

The biological importance of epidermis-derived VEGF for cutaneous angiogenesis in vivo has been confirmed in transgenic mouse models, using the keratin 14 promoter to selectively target expression of murine VEGF164 to basal epidermal keratinocytes and to follicular keratinocytes of the outer root sheath of the hair follicle. VEGF transgenic

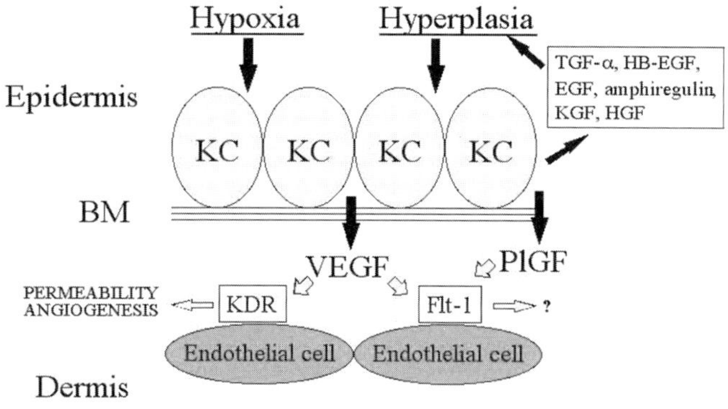

Figure 1 Molecular regulation of skin angiogenesis by VEGF and PlGF. BM, Basement membrane.

mice are characterized by elongated and tortuous dermal blood vessels that are hyperpermeable for circulating plasma proteins. VEGF overexpressing mice also show enhanced rolling and adhesion of peripheral blood mononuclear cells in cutaneous postcapillary venules, suggesting that VEGF also contributes to the recruitment of leukocytes to inflamed skin [3].

Placental Growth Factor and Other Angiogenic Factors

Placental growth factor (PlGF) is a recently identified new member of the VEGF family that occurs in at least three different isoforms of 149 (PlGF-1), 170 (PlGF-2), and 221 (PlGF-3) amino acids. Both PlGF-1 and PlGF-2 are expressed in human skin. Similar to VEGF, the expression of PlGF is upregulated in cutaneous squamous cell carcinomas, in epidermal keratinocytes at the advancing wound edge, and in the hyperplastic psoriatic epidermis. In contrast to VEGF, PlGF does not activate VEGFR-2 (KDR), but selectively binds to VEGFR-1. In addition, the heparin-binding isoform PlGF-2, but not PlGF-1, binds to the neuropilin receptor. Recent evidence from genetic mouse models indicates that PlGF and VEGF act in synergy to induce skin angiogenesis and vascular leakage, and that the presence of PlGF is essential for distinct VEGF effects to occur [4]. A number of additional proangiogenic molecules are upregulated in angiogenic skin conditions, including interleukin-8, platelet-derived growth factors, and fibroblast growth factors.

Endogenous Inhibitors of Skin Angiogenesis

Thrombospondin (TSP)-1 and TSP-2 are major endogenous inhibitors of skin angiogenesis and both are expressed in normal human skin. TSP-1 is expressed by dermal cells and by epidermal keratinocytes, and it is deposited in the dermo-epidermal basement membrane zone, contributing to the barrier that prevents ingrowth of blood vessels into the dermis. In contrast, TSP-1 expression is downregulated in squamous cell carcinomas of the skin. Reintroduction of the TSP-1 gene into squamous cell carcinomas inhibits tumor growth in mice, associated with inhibition of tumor angiogenesis and enhanced tumor cell necrosis. The potential mechanisms by which TSP-1 inhibits skin angiogenesis include the induction of endothelial cell apoptosis through specific interactions with the endothelial CD36 receptor, the activation of latent TGF-beta, and the inhibition of matrix metalloproteinase activity. Overexpression of TSP-1 in the skin of transgenic mice results in impaired granulation tissue formation and wound vascularization [5] in delayed and reduced skin carcinogenesis, and in diminished photodamage induced by chronic ultraviolet-B irradiation of the skin.

TSP-2 expression during embryonic development and in adult tissues is spatially and temporally different from TSP-1. The expression of TSP-2 is downregulated in epithelial squamous cell carcinoma cells, whereas a strong upregulation of TSP-2 is found in the mesenchymal stroma during skin carcinogenesis, representing a natural tumor defense mechanism [6]. Mice that are deficient in TSP-2 show increased skin vascularization, enhanced and prolonged inflammatory reactions, and enhanced skin carcinogenesis, confirming the important role of endogenous TSP-2 in the control of skin angiogenesis. Accordingly, genetic overexpression of TSP-2 protects from skin cancer development and inhibits the growth of established skin cancers. A number of additional endogenous inhibitors of angiogenesis are likely involved in the maintenance of normal vascular quiescence in the skin, including interferons and fragments of collagens type IV, XV, and XVIII. Their potential contribution to the antiangiogenic environment in normal skin remains to be established.

Angiogenesis in Skin Diseases

Epithelial Skin Cancers

Squamous cell carcinoma (SCC), a malignant tumor of keratinocytes with destructive growth pattern and the capacity to metastasize, arises as a result of exogenous carcinogens such as chronic exposure to sunlight, ionizing radiation, or chemicals. The stroma of SCC is richly vascularized and SCC tumor cells strongly express VEGF. Whereas the inhibition of VEGFR-2 (flk-1) prevents SCC growth and invasion, experimental overexpression of VEGF in highly differentiated SCC cell lines promotes invasiveness, tumor growth, and angiogenesis. These experimental data provide evidence for a critical function of VEGF in SCC progression, and they indicate that blockade of VEGF in patients might reduce the malignant progression of SCC. In contrast, only low levels of VEGF expression have been detected in basal cell carcinomas of the skin that are also richly vascularized. Preliminary evidence suggests that other angiogenic factors such as fibroblast growth factors and platelet-derived growth factors might play a role in these tumors.

Malignant Melanoma

The observation that cutaneous melanoma cells acquire the capacity to actively induce the growth of new blood vessels dates back to the earliest days of tumor angiogenesis research. The clinical and prognostic significance of tumor angiogenesis for melanoma progression and metastasis has, however, remained controversial. Several studies have reported an inverse correlation between tumor microvessel density and disease-free and overall survival of melanoma patients. Other studies, in contrast, failed to detect any correlation between melanoma vascularization and prognosis. Thus, the potential prognostic value of tumor vascularization in human cutaneous melanomas remains unresolved [7].

Although VEGF expression is not as prominent in melanomas as it is in most epithelial cancers and, therefore, might not represent the major angiogenic activity in these tumors, the expression of functional VEGF receptors on human melanoma cells suggests the intriguing possibility that VEGF might also exert autocrine effects on the tumor cells themselves. Several other angiogenic factors have also been implicated in the pathology of human melanomas. Expression of basic fibroblast growth factor has been detected in metastatic and primary invasive melanomas, whereas melanocytes in benign nevi did not express this factor. Interleukin-8 was found to be absent from normal epidermis and benign melanocytic lesions but was expressed at high levels in the majority of cutaneous melanomas examined. Increased expression levels of PlGF, platelet-derived growth factor (PDGF)-AA and -BB, and angiogenin have also been found in human melanomas; however, their relative contribution to melanoma progression remains at present unclear.

Psoriasis

Psoriasis is associated with chronic inflammatory skin lesions that are characterized by epidermal hyperplasia, impaired epidermal differentiation, and accumulation of distinct leukocyte subpopulations. Cutaneous blood vessels show major abnormalities in psoriatic lesions and are found to be enlarged, tortuous, and hyperpermeable. In 1994, VEGF was identified as a major epidermis-derived vessel-specific growth factor that was strongly upregulated in psoriatic skin lesions [8]. Since then, several studies have demonstrated that VEGF expression is increased in lesional psoriatic skin, that the serum levels of circulating VEGF protein are significantly elevated in patients with severe disease, and that VEGF serum levels were directly correlated with disease activity. A major role of VEGF in the pathogenesis of psoriasis was further corroborated by the phenotype of transgenic mice with epidermis-specific overexpression of VEGF. At about 6 months of age, these mice spontaneously develop chronic inflammatory skin lesions that histologically closely resemble human psoriasis [9]. It is of interest that selective targeting of skin vessels via epidermal overexpression of an angiogenesis factor was able to reproduce the complete psoriatic phenotype, including epidermal hyperplasia and altered epidermal differentiation, upregulation of adhesion molecules, and accumulation of CD4-positive T-lymphocytes within the dermis and of CD8-positive cells within the epidermis. Moreover, VEGF transgenic mice show the characteristic Koebner phenomenon, with induction of chronic psoriasis-like lesions by unspecific skin irritation. Additional genetic evidence supports a role of VEGF in psoriasis. The +405C/C genotype of the VEGF gene, and the +405C allele, were found significantly more often in patients with severe psoriasis, and this genotype was also significantly more frequent in psoriasis patients with disease onset between the 20th and 40th years. Based on previous reports that the +405C allele is associated with elevated serum levels of VEGF in healthy individuals, these findings indicate that distinct genetic polymorphisms might contribute to enhanced VEGF production and to individually increased psoriasis susceptibility, and they suggest that therapeutic blockade of the VEGF/VEGF receptor system might represent a novel, pharmacogenomic approach for the future treatment of psoriasis.

Other Skin Diseases and Antiangiogenic Therapy

Proliferative hemangiomas of infancy represent benign vascular hyperproliferations and have been found to respond, at least in part, to treatment with interferon-alpha [10]. Other vascular lesions that might respond to antiangiogenic treatment include kaposiform hemangioendotheliomas (Table I). Similarly, teleangiectasias, in particular in rosacea, appear to represent prime targets for antiangiogenic

Table I Skin Conditions and Diseases Associated with Angiogenesis.

1. Hair growth and cycling
2. Wound healing
3. Skin neoplasias
   — squamous cell carcinoma
   — basal cell carcinoma
   — malignant melanoma
   — malignant cutaneous lymphomas
4. Vascular tumors
   — angiosarcoma
   — Kaposi's sarcoma
   — infantile hemangiomas
   — hemangioendothelioma
5. Inflammatory dermatoses
   — psoriasis
   — dermatitis (atopic, contact)
6. Bullous diseases
   — bullous pemphigoid
   — erythema multiforme
7. Other diseases
   — viral warts
   — rosacea
8. UV irradiation

therapy. Mutations of the vascular tie-2 receptor have been associated with vascular malformations, and mutations in the genes of the low-affinity TGF-β receptor endoglin and of activin receptor-like kinase have been found to be associated with the autosomal dominant vascular malformations of hereditary hemorrhagic telangiectasia type I and II. Angiogenesis and vascular activation also play a major role in mediating ultraviolet-B induced skin damage, indicating the potential use of angiogenesis inhibitors for chemoprevention. The challenge and opportunity for dermatology will be to develop topical angiogenesis inhibitors that will be able to penetrate the skin but that will not reach potentially toxic systemic levels.

Glossary

Melanoma: Malignant tumor of melanocytes (pigment cells).

Proliferative hemangioma of infancy: Most frequent benign tumor of infancy, due to proliferation of vascular endothelial cells.

Psoriasis: Chronic inflammatory skin disease of unknown etiology.

Squamous cell carcinoma: Epithelial skin cancer derived from epidermal keratinocytes.

References

1. Detmar, M. (2003). Vascular biology. In: *Textbook of Dermatology*, Bologna, J., Jorizzo, J., and Rapini, R., eds., Harcourt Health Sciences. *This chapter provides a comprehensive review of the development, physiology, and pathology of the cutaneous vasculature.*
2. Yano, K., Brown, L. F., and Detmar, M. (2001). Control of hair growth and follicle size by VEGF-mediated angiogenesis. *J. Clin. Invest.* **107**, 409–417.
3. Detmar, M., Brown, L. F., Schön, M. P., Elicker, B. M., Velasco, P., Richard, L., Fukumura, D., Monsky, W., Claffey, K. P., and Jain, R. K. (1998). Increased microvascular density and enhanced leukocyte rolling and adhesion in the skin of VEGF transgenic mice. *J. Invest. Dermatol.* **111**, 1–6.
4. Carmeliet, P., Moons, L., Luttun, A., Vincenti, V., Compernolle, V., De Mol, M., De Mol, M., Wu, Y., Bono, F., Devy, L., Beck, H., Scholz, D., Acker, T., DiPalma, T., Dewerchin, M., Noel, A., Stalmans, I., Barra, A., Blacher, S., Vandendriessche, T., Ponten, A., Eriksson, U., Plate, K. H., Foidart, J. M., Schaper, W., Charnock-Jones, D. S., Hicklin, D. J., Herbert, J. M., Collen, D., and Persico, M. G. (2001). Synergism between vascular endothelial growth factor and placental growth factor contributes to angiogenesis and plasma extravasation in pathological conditions. *Nat. Med.* **7**, 575–583. *This article identified an essential role of PlGF in pathological skin angiogenesis and revealed that PlGF is needed for VEGF's biological activity.*
5. Streit, M., Velasco, P., Riccardi, L., Spencer, L., Brown, L. F., Janes, L., Lange-Asschenfeldt, B., Yano, K., Hawighorst, T., Iruela-Arispe, L., and Detmar, M. (2000). Thrombospondin-1 suppresses wound healing and granulation tissue formation in the skin of transgenic mice. *EMBO J.* **19**, 3272–3282.
6. Hawighorst, T., Velasco, P., Streit, M., Kyriakides, T. R., Brown, L. F., Bornstein, P., and Detmar, M. (2001). Thrombospondin-2 plays a protective role in multistep carcinogenesis: A novel host anti-tumor defense mechanism. *EMBO J.* **20**, 2631–2640.
7. Streit, M., and Detmar, M. (2003). Angiogenesis, lymphangiogenesis and melanoma metastasis. *Oncogene* **22**, 3172–3179.
8. Detmar, M., Brown, L. F., Claffey, K. P., Yeo, K. T., Kocher, O., Jackman, R. W., et al. (1994). Overexpression of vascular permeability factor/vascular endothelial growth factor and its receptors in psoriasis. *J. Exp. Med.* **180**, 1141–1146. *This paper, for the first time, identified a specific angiogenesis factor to be upregulated in psoriasis.*
9. Xia, Y. P., Li, B., Hylton, D., Detmar, M., Yancopoulos, G. D., and Rudge, J. S. (2003). Transgenic delivery of VEGF to mouse skin leads to an inflammatory condition resembling human psoriasis. *Blood* **102**, 161–168. *This article describes the spontaneous development of inflammatory, psoriasis-like skin lesions in transgenic mice selectively overexpressing VEGF in the skin.*
10. Arbiser, J. L. (1996). Angiogenesis and the skin: A primer. *J. Am. Acad. Dermatol.* **34**, 486–497.

Capsule Biography

Dr. Rainer Kunstfeld obtained his M.D. degree from the University of Vienna, Austria, where he also trained in dermatology. He currently is a research fellow at the Cutaneous Biology Research Center in Boston.

Dr. Michael Detmar obtained his M.D. degree from the University of Freiburg, Germany, and trained in dermatology at the Free University of Berlin. At present, he is an Associate Professor of Dermatology at the Massachusetts General Hospital and Harvard Medical School, Boston, Massachusetts. He is also a Principal Investigator at the Cutaneous Biology Research Center and Associate Chief of Research, Department of Dermatology, Massachusetts General Hospital.

CHAPTER 138

Skin Microcirculation in the Elderly

Marco Rossi and Carlo Palombo
Department of Internal Medicine, University of Pisa, Medical School, Pisa, Italy

Introduction

Aging is one of the major challenges facing developed societies in the current century. In the United States today there are 35 million people 65 years of age or older, and their number will double by the year 2030 [1]. Thus, age-related problems will increasingly engage all levels of social organization: individual citizens, health care providers, and government authorities fixing priorities for funding allocation are called to tackle different facets of the aging mosaic. Aging classically reminds us of the Janus myths, with a favorable side represented by the gain in individual life expectancy, and a second one with its weight of increased susceptibility to disease and disability. Widening knowledge about age-related disease mechanisms and clinical consequences may help improve both quality of life for elder subjects and global community health.

Aging and Cardiovascular Disease

Aging is a major risk factor for congestive heart failure, stroke, coronary events, and peripheral arterial obstructive disease. A number of age-associated changes in cardiovascular structure and function are implicated in the increased risk for cardiovascular disease in older persons. An increased intimal–medial thickness (IMT) and wall stiffness of large arteries, as well as endothelial dysfunction, are recognized so far as the best established fingerprints of vascular aging in apparently otherwise healthy older persons. Recent advances in vascular biology have pointed out age-associated cellular, enzymatic, and molecular mechanisms that underlie vascular remodeling [1]. The activation of stress-responsive genes is thought yield elevated levels or activity of growth factors, such as platelet-derived growth factor (PDGF), transforming growth factor-β (TGF-β), and angiotensin II, proteolytic enzymes (metalloproteases, MMPs), adhesion molecules, and inflammatory cytokines (ICAM-1 and interleukin-6, IL-6). Such a "molecular remodeling" provides the metabolic milieu for smooth muscle cell (SMC) proliferation, migration, and apoptosis, as well as for synthesis of extracellular matrix proteins (i.e., fibrosis), disruption of basal membrane and elastin, and increased endothelial permeability.

The concept of endothelial senescence represents a recent breakthrough for understanding age-related and atherosclerotic vascular diseases and also exhibits great potential from a diagnostic point of view, for the follow-up of disease progression/regression, and possibly for therapeutic interventions [1]. Endothelial senescence can be identified in cell cultures or in bioptic samples by histochemistry (β-galactosidase staining) and is characterized by the suppression of telomerase activity. Telomere length is a marker of cellular turnover and is inversely associated with age, atherosclerotic grade, and pulse pressure [2]. Loss of telomere function was recently shown to induce endothelial dysfunction in vascular endothelial cells, whereas inhibition of telomere shortening suppresses the age-associated dysfunction in these cells [3]. Very recently, endothelial progenitor cells (PECs) deriving from bone marrow have been identified in peripheral blood, and it has been proposed that circulating PECs normally repair and rejuvenate vascular endothelium; thus, their exhaustion due to a persisting challenge such as atherosclerosis, chronic inflammatory processes, or aging could hamper vascular repair and facilitate progression of vascular disease independent of the major risk factors [4].

Two key theoretical questions are open, with significant implications for vascular aging. The first one confronts the programmatic theory of aging, stating that aging is an inherent genetic process, with the stochastic theory, suggesting that it represents the result of random environmental damage [5]; the latter relates to the overlap between aging and atherosclerosis. Age-associated changes in vascular structure and function and atherosclerosis are intertwined and interdependent. In fact, the prevalence of risk factors for atherosclerotic disease such as high blood pressure, obesity, insulin resistance, and physical inactivity increases with age, and vascular inflammation, endothelial dysfunction, and oxidative stress are the common end pathways underlying vascular disease. To date, evidence is mounting that subclinical vascular disease in older persons represents specific aspects of vascular aging and is not synonymous with low-grade atherosclerosis [1]. Rather, aging increases susceptibility to atherosclerosis, facilitating an imbalance between protective and insulting vascular factors for a given atherogenic load, whereas a different burden from modifiable risk factors could accelerate or delay the impact of age, resulting in the well-known possible discrepancy between biological and chronological age.

The question also arises as to what extent different vascular beds may share similar mechanisms of disease, and knowledge gained from studying large arteries and cell cultures can be extrapolated to human microvasculature in vivo. Despite deep differences between regional circulations depending on the specific organ function, the basic pattern of response of SMCs and endothelial cells to inflammatory and oxidative injury seems to be quite uniform throughout the vascular tree, which allows us to foresee an extension to peripheral microvasculature of modern discoveries about vascular senescence.

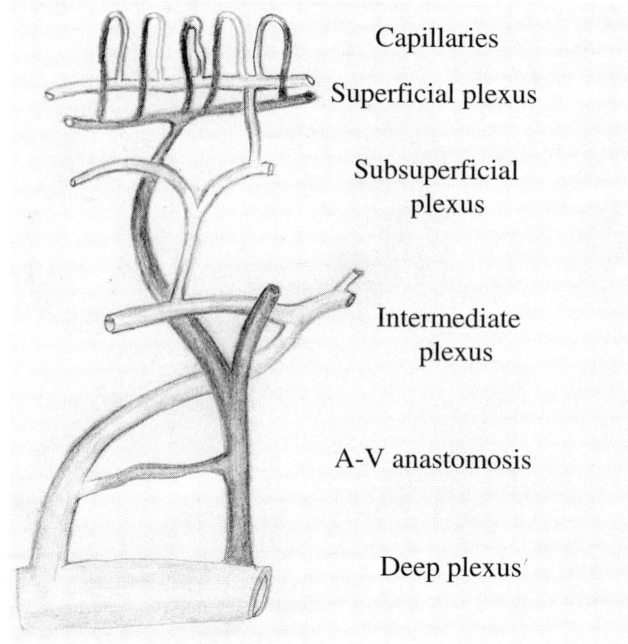

Figure 1 Diagram of vascular anatomy in plantar skin. From Conrad, M. C. *Functional Anatomy of the Circulation of Lower Extremities*. Copyright © 1971 by Year Book Medical Publishers, Inc., Chicago.

Functional Anatomy of the Skin Microvasculature

Skin is a complex organ, with a wide surface area of about 2 square meters, which consists of two layers. The outer one, the epidermis, is a waterproof layer of keratinizing stratified squamous epithelium. The inner layer, dermis or corium, is connective tissue that supports the epidermis and hosts the microvasculature.

Skin microcirculation guarantees three main functions [6]: skin tissue nutrition, heat exchange for thermoregulation, and blood flow redistribution during stress. Skin blood flow spans throughout an extremely wide range (in nonacral skin, from 1.0 mL/min per 100 g, needed to meet the relatively low intrinsic tissue metabolic demand, to 8.0 mL/min per 100 g, and in acral regions from 0.2 to 50 mL/min per 100 g). Under extreme thermal stress, skin blood flow can account for up to 50 percent of total cardiac output. Such a functional reserve is made possible by a fine control of cutaneous vascular resistance, capable of large dynamic changes in response to local, mechanical, and humoral factors, as well as to the autonomic nervous signaling. A peculiar microvascular architecture provides the anatomic substrate for this functional dynamics.

Microcirculatory bed in human dermis is organized in four plexuses (Figure 1): deep, intermediate, subsuperficial, and superficial (the so-called papillary plexus). These vessels tend to be arranged parallel to the skin surface, with the exception of, first, the capillary loops that are arranged perpendicular to the cutaneous surface and, second, the arterioles and venules which come from the deep to the superficial plexus. Skin microvessels have relatively thick walls that protect them against the shearing stresses to which the skin is exposed. The greater thickness of their walls is due to a membrane basement layer in which smooth muscle cells, pericytes, collagen, and elastin are embedded. Three to four branching orders of arterioles (from order 0, corresponding to capillaries, to order 4, assigned to the largest arterioles, with a diameter of 100 to 150 μm), are reported in the skin microvascular bed. Arterioles show spontaneous diameter oscillations, referred to as vasomotion, with a specific range of frequencies that vary according to the vessel order, as demonstrated by the work of Colantuoni, Bertuglia, and others. As shown by Griffith and others, arteriolar diameter oscillations may modify local blood flow distribution. In particular, they yield a reduced resistance in microvascular networks compared with those with steady conditions, and ensure an intermittent but adequate flow to the tissue in presence of a decreased blood flow supply [7].

A special feature of cutaneous microvasculature is the presence of numerous arteriovenous anastomoses (AV), which are coiled vessels with an average lumen of 35 μm, connecting arterioles and venules in the acral skin. They are present mainly in the hands, feet, ears, and nail beds, but have not been found in the skin of the forearm or the calf. Their function is to allow the blood flow directly from the arterioles to the venules of the deep plexus, bypassing the high-resistance arterioles and capillaries of the more superficial plexus. Having a dense innervation and a thick layer of smooth muscle cells in their walls, they play a major role in determining the neurally mediated changes in arteriolar tone and microvascular resistance that occur in response to thermal stimuli.

Control Mechanisms of the Skin Microcirculatory Perfusion

The fine tuning of skin microvascular resistance and blood flow distribution is obtained through the dynamic interaction of sympathetic vasoconstriction, pressure-dependent vasoconstriction, flow-dependent endothelium-mediated vasodilation, metabolic vasodilation, and spontaneous myogenic activity [6]. At a given time, not all skin microvascular units are open and perfused. The functional recruitment of previously inactive units represents a further mechanism for increasing capillary perfusion during exercise or passive thermal stress. The different control mechanisms modulate diameter oscillations of the arterioles, that is, vasomotion, which results, as shown by Bertuglia and Colantuoni, in blood flow oscillations, the so-called flowmotion. By power spectral analysis of Laser Doppler perfusion monitoring signals, a low-frequency component of flowmotion oscillating at 0.1 Hz has been unequivocally associated to the sympathetic activity [8]. Thus, the well-known age-related increase in basal sympathetic activity and reduced tonic baroreflex sympathoinhibition may have an impact on skin microcirculatory control and be revealed by the analysis of variability component on laser-Doppler recordings.

Skin Microcirculation in the Elderly

Cutaneous Tissue Modifications with Aging

Age-related structural alterations in the skin have been well documented by histologic and ultrastructural examination of skin biopsy specimens from younger compared to older people [9]. The fine, regular epidermal surface pattern changes to coarser and less regular ridges with aging. Epidermal projections into the dermis are retracted and the dermal–epidermal junction is flattened. The dermis becomes thinner, and there are fewer fibrillar collagen and elastic fibers in older skin. A progressive reduction with age in the dermis hyaluronic acid content has been also demonstrated, which could account for some of the most striking alterations of the aged skin, including decreased turgidity, altered elasticity, and less support for microvessels. Basic mechanisms underlying skin aging can be represented by a reduced cellular turnover and a prolonged time to recover after an injury [10]. According to the majority of the experts in this field, the morphofunctional state of skin, to a large extent, depends on the state of microcirculation. As the skin nutrition is provided by the microvessels, an altered microcirculatory function is thought to affect the trophic state of the skin. A number of age-associated functional and structural alterations are reported in the skin microcirculation and discussed in the following section (Table I).

Table I Age-Associated Changes in Skin Microvasculature.

A. Skin microvascular remodeling
— Arteriolar wall hypertrophy
— Reduced capillary density ("capillary rarefaction")
— Venular dilation and congestion
B. Hemorheological changes
— Raised plasma fibrinogen
— Increased plasma viscosity
— Increased red-cell aggregability
— Decreased red-cell deformability
C. Skin perfusion
Baseline perfusion
— Reduced capillary blood flow velocity
— Reduced blood flow through A-V anastomoses
— Reduced flow motion
Stimulated perfusion
— Reduced hyperemic response to heat
— Reduced vasodilation during exercise
— Reduced sympathetic vasoconstriction to cold
— Reduced reactive hyperemic response to ischemia and pressure
— Reduced endothelium-dependent vasodilation

Skin Microvascular Remodeling with Aging

With aging, density and size of blood vessels are reduced and, in particular, dermal papillary loops are significantly reduced in old skin compared with young skin, as described by Tenland and others since the early 1980s, and confirmed later on by Kelly and others. Skin capillary rarefaction has been recently reported by MacGregor and Serné in middle-aged patients with essential hypertension. The observation that the same abnormality is detectable in normotensive offspring of hypertensive subjects pointed out the hypothesis that skin capillary rarefaction, possibly dependent on a defective neoangionetic capacity, could represent a pathophysiologic mechanism underlying the development of high blood pressure [11, 12]. Whether or not a reduction in skin capillary density may contribute to the hypertension in the elderly is still unknown.

Capillaroscopic observations in old subjects documented a permanent dilatation and congestion of venules and capillaries in nailfold microvasculature (Figure 2), whereas other studies showed a decreased blood flow through AV anastomoses.

Age-Associated Hemorheological Changes

Besides remodeling of skin microvasculature, aging is associated with hemorheological changes, which, in turn, have negative influence on the microcirculation of various

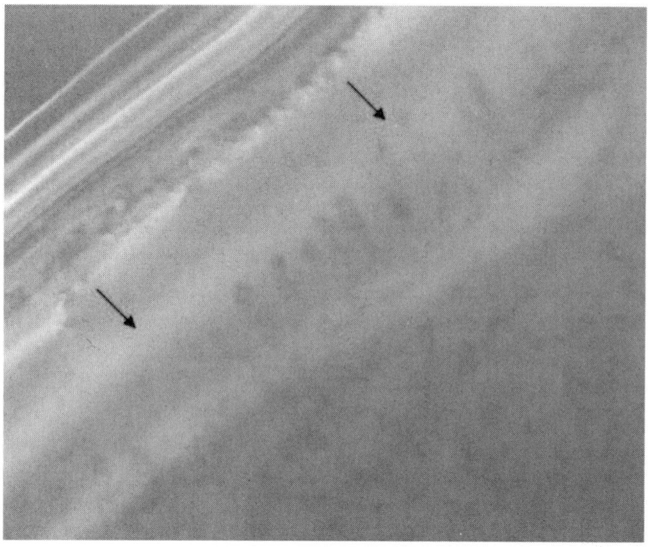

Figure 2 Nailfold capillaroscopy from a 75-year-old man. Areas of focal capillary rarefaction are detectable (*arrows*). (see color insert)

tissues [13]. In fact, a number of clinical observations show a rise of fibrinogen with advancing age, which contributes to a rise in plasma viscosity and red cells aggregation. Rheological properties of red cells also change, resulting in a decreased cell deformability and sluggish blood flow. All these hemorheological changes affect blood flow in large vessels, but play an even more important role in the microcirculation, where the inertial forces, the corpuscular behavior of blood, and the aggregation phenomenon have a critical influence on blood flow dynamics. Although some observations show a relationship between the rise in blood viscosity with age and a decreased cerebral blood flow, there is no conclusive evidence about a negative influence of these age-dependent hemorheological changes on skin blood flow.

Age-Associated Changes in Skin Perfusion

Baseline Perfusion

Differences in skin perfusion between young and old people have been observed by various methods. More than 60 years ago, Brown and Roth showed a lower blood cell velocity in finger nailfold capillaries of old versus young subjects. These findings were confirmed subsequently by modern dynamic video microscopy: Quantitative capillary blood flow velocity measurement showed a significantly lower velocity in the toe and the finger nailfold of elderly compared to young subjects. However, this phenomenon was not demonstrated in other skin regions, such as the ventral surface of the forearm, where average capillary blood flow velocity was not significantly different between young and old subjects.

A marked heterogeneity of cutaneous microcirculation in resting conditions has been demonstrated by laser Doppler flowmetry, which allows continuous, noninvasive, real-time assessment of skin perfusion in a hemispheric illuminated tissue volume of 1 to 1.5 mL under a measuring probe [14]. To understand the pathophysiologic meaning of the measurements, one must remember that the laser Doppler signal is generated by the movement of blood cells in both skin microvascular networks, that is, the subpapillary thermoregulatory bed and the nutritive capillaries. Therefore this procedure yields information about nutritional and nonnutritional perfusion. Skin blood flow is significantly higher in the face than in the trunk and extremities, with the highest perfusion values detected at lip level, followed by the chin, forehead, and cheek. The lowest perfusion is found in the gluteal region: on the back of the foot and on the soles. Men exhibit a higher cutaneous perfusion than women, whereas age-related differences in skin perfusion at rest have not been demonstrated by most studies. In contrast, a significant age-related reduction in resting blood flow was observed at the dorsum of the hand and of the foot using a xenon-133 clearance method.

More evident are the differences in skin microcirculatory pefusion between young and old subjects when the responses to physiologic or pharmacologic stress are considered.

Stimulated Perfusion

Skin microvasculature shows a decreased efficiency of vasodilator and vasoconstrictor adaptive responses to both neurally mediated and humoral stimuli. A reduced capacity of skin microcirculation in the elderly to adapt to a thermal stress is well demonstrated. Many recent studies have examined this aspect, and the overall results is that aging decreases the hyperemic response of cutaneous blood flow to direct heat. This lowered capacity is particularly evident during dynamic exercise. Other observations, however, suggest that the age-associated decline in heat loss during exercise may be masked by repeated exercise training, and that when the effects of chronic diseases and sedentary lifestyle keep physical activity to a minimum, heat tolerance appears to be minimally compromised by age. An inadequate vasodilation during exercise can hamper thermoregulatory capacity in hypertensive patients [15]. On the other side, sympathetic vasoconstriction to cold appears to be reduced with aging, with a decreased capacity for thermoconservation, which may explain the intolerance to cold of which the elderly often complains.

The hyperemic response to prolonged pressure is another function of skin microcirculation that is blunted by aging. The cutaneous reactive hyperemic response to ischemia and pressure over the sacrum and the gluteal region has been found to be significantly reduced in geriatric patients compared to younger subjects. This could easily induce development of decubitus ulcers in the elderly during long periods of immobility.

Besides the capillary rarefaction and arteriolar wall hypertrophy responsible for the structural remodeling of skin microcirculation, defective functional mechanisms also contribute to an impaired capillary recruitment. A reduction of microcirculatory skin flow-motion in elderly people may play a role in the attenuated cutaneous vasoreactivity in response to heat and ischemic stress.

Endothelial senescence is supposed to represent a critical factor underlying microvascular aging. The molecular mechanism that could be mainly involved by aging is the activation of telomerase in human dermal microvascular endothelial cells, which controls their durability both in vitro and in vivo. Aging per se is associated with a mild impaired endothelium-dependent vasodilation of skeletal muscle resistance vessels, and brachial and coronary arteries. However, the study of endothelial function in the skin microvasculature demonstrated only a mild impairment of the endothelium-dependent vasoreactivity in elderly subjects with low risk profile of atherosclerosis [16]. An aging-related decrease in the bioavailability of nitric oxide (NO), due to either reduced biosynthesis or increased breakdown triggered by enhanced oxidative stress, could account for this impairment. However, it is worth noting that, as suggested by studies of Noon and Khan, the skin microcirculatory response to acetylcholine (ACh) is mainly mediated by endothelium-derived prostanoids rather than NO, which is the mediator of ACh-induced vasodilation in brachial and coronary arteries. In fact, the skin response to ACh is 50 percent reduced by pretreatment with ASA (prostanoid synthesis inhibitor), whereas the forearm response to ACh is reduced by L-NMMA. By contrast, in older subjects a NO-dependent mechanism accounts for approximately 60 percent of the active skin vasodilation induced by hyperthermia, against the 20 percent representing the contribution of NO-dependent vasodilation in the skin of young subjects. This suggests that attenuated cutaneous vasodilation with age may be due to a reduction in, or decreased vascular responsiveness to, other mediators than NO.

Clinical Implications and Perspectives

In conclusion, aging is associated with many skin microvasculature changes, all with a negative influence on tissue nutrition, which can explain the increased fragility and vulnerability of elderly skin to the different noxae and to the environment stress. Thermoregulatory adaptive mechanisms are also affected, resulting in a reduced capability to cope with both heat and cold stress.

Even without endeavoring to analyze the intricacy of the interactions between aging and modifiable risk factors for atherosclerosis, it is worth remembering that an age-altered vascular substrate increases susceptibility to the injury from modifiable risk factors. The presence of risk factors, in turn, may accelerate vascular aging.

The skin microcirculatory bed represents a window for estimating "biological age" as opposed to chronological age and a useful clinical model for noninvasive investigation of mechanisms involved in vascular senescence.

Glossary

Aging: The natural and continuous process of becoming old. For medical purposes, at present the age at which a subject is elderly is set at 65 years.

Endothelium: Cell monolayer aligned along all the vascular tree, at the interface between bloodstream and vascular wall.

Flowmotion: Rhythmic oscillations of the blood flow in microvessels.

Skin perfusion: Blood flow supply to the skin.

Vasomotion: Rhythmic oscillations of the diameter of microvessels.

References

1. Lakatta, E. G. (2003). Arterial and cardiac aging: Major shareholders in cardiovascular disease enterprises. Part III: Cellular and molecular clues to heart and arterial aging. *Circulation* **107**, 490–497. *An updated overview of vascular biology of aging.*
2. Benetos, A., Okuda, K., Khan, M. Y., Skurnick, J., Labat, C., Bean, K., and Aviv, A. (2001). Telomere length as an indicator of biological aging: The gender effect and relation with pulse pressure and pulse wave velocity. *Hypertension* **37**, 381–385.
3. Minamino, T., Miyauchi, H., Yoshida, T., Ishida, Y., Yoshida, H., and Komuro, I. (2002). Endothelial cell senescence in human atherosclerosis: Role of telomere in endothelial dysfunction. *Circulation* **105**, 1541–1544.
4. Rauscher, F. M., Goldschmidt-Clermont, P. J., Davis, B. H., Wang, T., Gregg, D., Ramaswami, P., Pippen, A. M., Annex, B. H., Dong, C., and Taylor, D. A. (2003). Aging, progenitor cell exhaustion, and atherosclerosis. *Circulation* **108**, 457–463.
5. Sherr, C., and DePinho, R. (2001). Cellular senescence: Mitotic clock or culture shock? *Cell Biol.* **102**, 407–410.
6. Holtz, J. (1996). Hemodynamics in regional circulatory beds and local vascular reactivity. In *Comprehensive Human Physiology*, R. Greger and U. Windhorst, eds., Vol. 2, pp. 1917–1939. Berlin: Springer-Verlag. *Reference text for detailed information about physiology of regional microcirculatory beds.*
7. Bertuglia, S., Colantuoni, A., Coppini, G., and Intaglietta, M. (1991). Hypoxia- or hyperoxia-induced changes in arteriolar vasomotion in skeletal muscle microcirculation. *Am. J. Physiol.* **260**, H362–H372.
8. Bernardi, L., Hayoz, D., Wenzel, R., Passino, C., Calciati, A., Weber, R., and Noll, G. (1997). Synchronous and baroceptor-sensitive oscillations in skin microcirculation: Evidence for central autonomic control. *Am. J. Physiol.* **273**, H1867–H1878.
9. Smith, L. (1989). Histopathologic characteristics and ultrastructure of aging skin. *Cutis* **43**, 414–424.
10. Grove, G. (1989). L. Physiologic changes in older skin. *Clin. Geriatr. Med.* **5**, 115–125.
11. Noon, J. P., Walker, B. R., Webb, D. J., Shore, A. C., Holton, D. W., Edwards, H. V., and Watt, G. C. (1997). Impaired microvascular dilatation and capillary rarefaction in young adults with a predisposition to high blood pressure. *J. Clin. Invest.* **99**, 1873–1879.
12. Antonios, T. F., Rattray, F. M., Singer, D. R., Markandu, N. D., Mortimer, P. S., and MacGregor, G. A. (2003). Rarefaction of skin capillaries in normotensive offspring of individuals with essential hypertension. *Heart* **89**, 175–178.
13. Ajami, R. S., and Rifkind, J. M. (1998). Hemorheological changes during human aging. *Gerontology* **44**, 111–120.
14. Van den Brande, P., von Kemp, K., De Coninck, A., and Debing, E. (1997). Laser Doppler flux characteristics at the skin of the dorsum of the foot in young and in elderly healthy human subjects. *Microvasc. Res.* **53**, 156–162. *Clear summary of the differences in skin microvascular perfusion between elderly and young normal subjects.*
15. Kellogg, D. L., Jr., Morris, S. R., Rodriguez, S. B., Liu, Y., Grossmann, M., Stagni, G., and Shepherd, A. M. (1998). Thermoregulatory reflexes and cutaneous active vasodilation during heat stress in hypertensive humans. *J. Appl. Physiol.* **85**, 175–180.

16. Rossi, M., Cupisti, A., Mariani, S., Santoro, G., and Pentimone, F. (2002). Endothelium-dependent and endothelium-independent skin vasoreactivity in the elderly. *Aging Clin. Exp. Res.* **14**, 343–346.

Further Reading

Chang, E., Yang, J., Nagavarapu, U., and Herron, G. S. (2002). Aging and survival of cutaneous microvasculature. *J. Invest. Dermatol.* **118**, 752–758.

Pierzga, J. M., Frymoyer, A., and Kenney, W. L. (2003). Delayed distribution of active vasodilation and altered vascular conductance in aged skin. *J. Appl. Physiol.* **94**, 1045–1053.

Capsule Biography

Marco Rossi, M.D., graduated in Medicine from the Medical School of Pisa University. His is a specialist in cardiovascular diseases and in internal medicine and a researcher at the University of Pisa, Department of Internal Medicine, with clinical and teaching activity at the same department. A Fellow of European Society of Microcirculation, he is a member of the National Committee of Italian Society of Clinical Hemorheology and Microcirculation and of Tuscan-Umbrian Section of the Italian Society of Angiology and Vascular Pathology, as well as of Working Groups on diabetic arteriopathy of the Italian Society of Angiology of Tuscan-Umbrian section and Vascular Pathology.

He is author of numerous full papers in peer-reviewed international Journals. His main research fields are represented by the study of peripheral microcirculation in aging and in subjects with peripheral arterial diseases and other pathological conditions and by the study of vascular disease and pathophysiology by ultrasound. He is author of numerous full papers in peer-reviewed international journals.

Carlo Palombo, M.D., graduated in Medicine from the Medical School of Pisa University. He is a specialist in cardiology and clinical pharmacology and an Associate Professor of Medicine at the University of Pisa, Department of Internal Medicine.

He is a Fellow of the European Society of Cardiology, and a member of the Italian Societies of Cardiology, Hypertension, Cardiovascular Research; European and International Societies of Hypertension; Working Group on Echocardiography of the European Society of Cardiology; and the Study Group on Arterial Mechanics of the Italian Society of Hypertension.

His clinical and teaching activity is at the Department of Internal Medicine, University of Pisa, and at the Institute of Clinical Physiology of the National Research Council, Pisa. He is author of more than 150 full papers, most of them in peer-reviewed international journals.

His main research fields are represented by the study of hypertensive heart disease; coronary and peripheral microcirculation in subjects with risk factors for atherosclerosis; arterial mechanics; vascular ultrasound; and cardiovascular aspects of insulin resistance.

SECTION V

Tumors

CHAPTER 139

Tumor Growth Patterns and Angiogenesis

Cecile G. Colpaert and Peter B. Vermeulen
Translational Cancer Research Group, University Hospital Antwerp and General Hospital Sint-Augustinus, University of Antwerp, Antwerp, Belgium

The cancer cells of a neoplasm invade preexisting host tissues and can thereby elicit the formation of reactive vascular tumor stroma, composed of extracellular matrix and host cells including fibroblasts, endothelial cells, and inflammatory cells. This newly generated stroma shows many similarities with granulation tissue that is formed in wound healing. Although cancer research in past decades has mainly focused on the cancer cell itself, there is growing awareness that the surrounding vascular stroma conditions include both local tumor progression and the metastatic process.

Routine pathological examination of human carcinomas reveals striking differences in the way cancer cells interact with host tissues, that is, the growth pattern of the cancer cells. In several primary and metastatic carcinomas, different growth patterns reflect different degrees of reactive vascular tumor stroma formation. Some growth patterns imply destruction of the preexisting tissues and intense remodeling of the stroma, whereas in other growth patterns there is remarkable preservation of the preexisting architecture and no or minimal new stroma formation. This has important prognostic and predictive implications.

Growth Patterns and Angiogenesis in Primary Breast Cancer

It has become evident that tumor growth cannot be understood without considering the interactions between the tumor cells and the components of the stromal microenvironment, such as the vasculature. Much research has been devoted to determining the impact of neovascularization—that is, angiogenesis—on tumor progression. Many authors have investigated the prognostic value of quantifying tumor vascularity. Although some contradictory results have been published, mainly due to differences in methodology, almost all studies with multivariate analysis that used either CD31 or CD34 for immunostaining the vessels showed a significant association between high tumor vascularity and unfavorable clinical outcome.

The recommended methodology of angiogenesis quantification has been described in an international consensus report [1]. It is unlikely, however, that this methodology will ever be applied to every breast carcinoma in routine pathology practice if more practical surrogate markers can be used to estimate tumor vascularity. Histological surrogate markers should give a reliable, fast, and easy estimate of the amount of angiogenesis in a tumor. Ideally, a standard histochemical technique, such as a hematoxylin–eosin stain, should suffice. Interpretation of the slides should not rely on extensive training and should not be time-consuming. Yet, a clear pathophysiological mechanism should corroborate the association of these markers with parameters directly reflecting angiogenesis, such as microvessel density and the fraction of proliferating endothelial cells.

Two candidate surrogate markers of angiogenesis are the fibrotic focus and the growth pattern.

A fibrotic focus (Figure 1) was proposed in 1996 [2] as an indicator of tumor aggressiveness in invasive ductal carcinoma of the breast. It appears as a central, radially expanding fibrosclerotic core and consists of loose, dense, or hyalinized collagen bundles, a variable number of fibroblasts, blood vessels, and inflammatory cells. It shows many similarities with granulation tissue and subsequent scar formation in wound repair and can be regarded as a focus of exaggerated reactive tumor stroma formation.

Different growth patterns have been shown to reflect differences in angiogenesis, both in lung tumors and in liver metastases [3, 4]. In invasive breast carcinoma, we defined two different growth patterns: An expansive carcinoma is a well-circumscribed tumor consisting of carcinoma cells and reactive tumor stroma without intervening preexisting structures, whereas an infiltrative carcinoma consists of carcinoma cells infiltrating between preserved normal structures. Tumors consisting of a central expansive nodule with infiltrative tumor at the periphery are included in the expansive category. The association of these growth patterns with other prognostic features was studied in 104 $T_{1\text{-}2}N_0M_0$ breast carcinoma patients (Table I). An expansive growth pattern was most often observed in ductal carcinomas with poor histological differentiation and brisk mitotic activity. Necrosis was rarely present in carcinomas with an infiltrative growth pattern, and the presence of a fibrotic focus was also significantly less frequent in these tumors than in the expansive ones. Necrosis and fibrotic foci are routine histological markers of intratumoral hypoxia. This has been confirmed by studying the expression of the endogenous hypoxia marker carbonic anhydrase IX (CA IX) in a group of 184 breast cancer patients [5]. CA IX expression in carcinoma cells and in intratumoral fibroblasts was significantly associated with the presence of a fibrotic focus. Intratumoral hypoxia is assuming growing importance in the pathogenesis of neoplastic progression. In addition to being a driver of angiogenesis, some genes controlling this process being

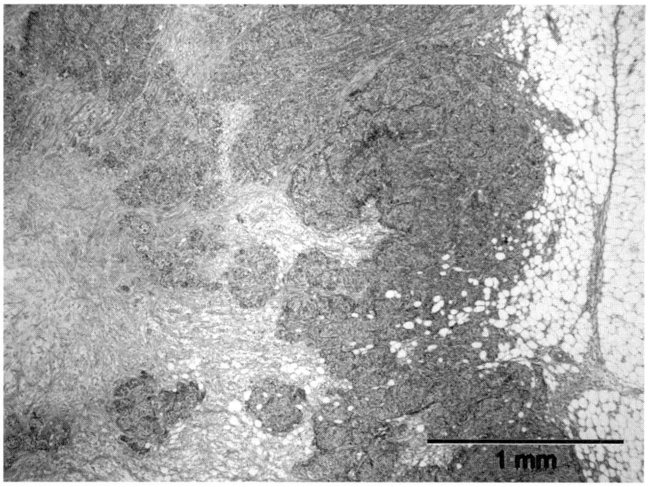

Figure 1 Expansively growing carcinoma of the breast containing a large fibrotic focus. The tumor shows a pushing margin (*right*) and a central area of fibrous scar-like tissue, the fibrotic focus (*left*).

Table I Correlation of Growth Pattern with Other Variables in 104 $T_{1\text{-}2}N_0M_0$ Breast Cancer Patients.

Variable	Expansive growth	Infiltrative growth	*p*-value
Histological type			
Lobular ($n = 12$)	2	10	
Ductal ($n = 92$)	62	30	0.0007
Size (cm)	1.9 ± 0.8 (1.7)	1.7 ± 0.8 (1.6)	> 0.1
Histological grade			
Well differentiated ($n = 31$)	18	13	
Moderately differentiated ($n = 33$)	19	14	
Poorly differentiated ($n = 28$)	25	3	0.007
Mitotic activity index (MAI)	15.9 ± 16.4 (10.0)	5.1 ± 6.0 (2.5)	< 0.0001
Necrosis			
Present ($n = 22$)	19	3	
Absent ($n = 82$)	45	37	0.004
Fibrotic focus			
Present ($n = 55$)	45	10	
Absent ($n = 49$)	19	30	< 0.0001
CA IX score in carcinoma cells	14.3 ± 27.6 (2.0)	1.0 ± 2.9 (0.0)	0.0001
CA IX score in intratumoral fibroblasts	23.5 ± 43.5 (2.0)	1.7 ± 9.7 (0.0)	< 0.0001
Tumor vascularity			
Chalkley max	9.5 ± 2.7 (9.0)	7.5 ± 2.1 (8.0)	0.0007
Chalkley mean	8.5 ± 2.2 (8.0)	6.6 ± 1.6 (7.0)	< 0.0001

Adapted from Colpaert, C. G., et al. (2001). *J. Pathol.* **193**, 442–449.

For continuous variables, figures are mean ± standard deviation (median).

MAI: mitotic activity index; the number of mitotic figures per 10 high-power fields of the microscope (magnification 400×).

CA IX: Carbonic anhydrase IX; an endogenous marker of hypoxia. Its expression is semiquantitatively scored as the product of the percentage of immunostained cells with an immunostaining intensity score ranging from 0 (no staining) to 3 (strong staining).

Chalkley count: Point overlap morphometric technique to quantify the relative area occupied by vessels in a vascular "hot spot," that is, an area of high vascular density. Chalkley max is the count obtained in the most vascular hot spot; Chalkley mean is the mean of the three highest counts.

oxygen regulated, hypoxia induces proteomic and genetic changes that promote an aggressive tumor phenotype.

Expression of the hypoxia marker CA IX was significantly more pronounced in expansively growing tumors than in infiltrative ones (Table I). The assumption is that the latter are less hypoxic because the tumor cells grow along preexisting well-formed blood vessels that are probably more efficient in providing oxygen and nutrients to the tumor than newly formed vessels. Therefore, the necessity to initiate angiogenesis is less pronounced in infiltrative tumors than in expansive ones, which is reflected by the significantly lower vascular density (estimated by Chalkley counting; see Table I) in the infiltrative tumors. The presence of a fibrotic focus is also significantly associated with vascular density estimated by Chalkley counting [6] and with high endothelial cell proliferation fractions [5]. Both growth pattern and fibrotic focus can therefore be used as routine histopathological surrogate markers of hypoxia-driven angiogenesis.

Growth Patterns and Angiogenesis in Cutaneous Breast Cancer Deposits

The most common tumor to metastasize to the skin is breast cancer. These skin deposits show remarkably different growth patterns with distinct angiogenic profiles. In a study of 51 surgically resected cutaneous deposits [7], 26 cases had an infiltrative growth pattern: The carcinoma cells infiltrated between preexisting dermal structures without significant disturbance of the dermal architecture. In nine cases, the growth pattern was expansive, the dermal deposit forming a well-circumscribed nodule consisting of carcinoma cells and reactive vascular tumor stroma. Preexisting dermal structures were pushed aside by the expansively growing nodule. The growth pattern was mixed infiltrative–expansive in the remaining 16 cases, meaning that these deposits consisted of a central expansive nodule surrounded by carcinoma cells showing an infiltrative growth pattern. The interobserver consistency of the assignment of the cutaneous deposits to the different growth pattern categories was 92 percent. The growth pattern of the skin deposit was positively correlated with the growth pattern of the respective primary tumor.

Cutaneous deposits with an infiltrative growth pattern only rarely contained necrosis or a fibrotic focus and infrequently showed expression of CA IX in a minority of tumor cells, whereas these histological markers of intratumoral hypoxia were significantly more frequent in deposits containing an expansive nodule (expansive and mixed growth patterns). These differences in hypoxia markers between the growth patterns were reflected in differences in angiogenesis. Microvessel density, quantified by the Chalkley method, was significantly higher in the expansive and mixed growth patterns than in the infiltrative growth pattern. Mean Chalkley count was higher when a fibrotic focus was present and when CA IX was expressed. The fraction of proliferating endothelial cells was highest in the expansive growth pattern, intermediate in the mixed pattern, and low in the infiltrative growth pattern.

One major mechanism by which oxygen deficiency stimulates angiogenesis is hypoxia-modulated gene expression, including activation of vascular endothelial growth factor (VEGF) gene transcription. VEGF is also known as the vascular permeability factor and induces the extravasation of plasma proteins, including fibrinogen and prothrombin. VEGF also induces the expression of tissue factor on the endothelial cells, as does hypoxia. Tissue factor, which is also present in the subendothelial matrix and on many tumor cells, triggers the formation of fibrin by activation of thrombin and subsequent polymerization of thrombin-cleaved fibrinogen. Fibrin is an essential component of the provisional matrix that supports tissue remodeling, angiogenesis, and tumor growth. In the cutaneous breast cancer deposits, fibrin was present in all deposits with an expansive growth pattern, but in fewer than half of the deposits with an infiltrative pattern. The deposition of fibrin was positively correlated with microvessel density and with endothelial cell proliferation fractions.

The distinct angiogenic profiles in the different growth pattern categories of cutaneous breast cancer deposits may have important consequences for therapy. Different degrees of angiogenesis may predict a different response to anti-angiogenic therapy.

Growth Patterns and Angiogenesis in Liver Metastases

The liver is a densely vascularized organ that frequently hosts metastases of colorectal and breast adenocarcinomas. Hypothetically, tumors that would be able to preserve this vasculature, well adapted to the high metabolic needs of physiological liver function, would not necessarily induce hypoxia-driven angiogenesis. Also, the gradient of angiogenic factors necessary for endothelial cell migration and proliferation might be diluted by the high vascular flow. Comparison of the histology of liver metastases in breast cancer and colorectal cancer patients has revealed a growth pattern that preserves the liver architecture and thus co-opts the preexisting sinusoidal blood vessels ([4]; Stessels F. et al., 2004). This growth pattern was expressed in nearly all breast cancer liver metastases (43/45) and in only a minority of the colorectal cancer liver metastases (9/28), and was characterized by tumor cells replacing the hepatocytes in the liver plates without inducing inflammation or fibrosis (Figure 2). In this "replacement" growth pattern, the endothelial cells of the coopted sinusoidal blood vessels lost their constitutive LYVE-1 expression and started to express CD34, suggesting paracrine interactions between coopted endothelial cells and tumor cells. The endothelial cell proliferation fraction in this growth pattern, as a measure of ongoing angiogenesis, was low (about 3%) and only a small minority of the breast cancer liver

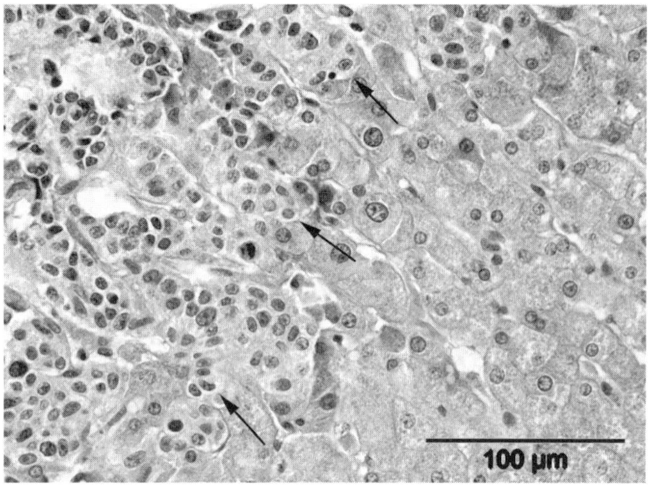

Figure 2 Replacement growth pattern in a liver metastasis of a breast adenocarcinoma. The tumor cells (*left*) are replacing the hepatocytes in the liver plates (*right*), thereby coopting the sinusoidal blood vessels. There is close apposition of tumor cells and hepatocytes at the tumor–liver interface (*arrows*) without induction of inflammation or fibrosis.

metastases with a replacement growth pattern expressed the hypoxia marker CA IX or had fibrin depositions at the tumor–liver interface (Table II). Probably, the well-described mechanisms of invasive tumor growth, such as fibroblast–myofibroblast transdifferentiation, TGFβ pathways, proinflammatory signaling, hyaluronic acid action, and hypoxia-responsive gene activation, are not involved. The search for gene sets that are responsible for the phenotype of nonangiogenesis-dependent colonization of a distant site is ongoing. Selective induction of apoptosis in hepatocytes at the interface by tumor cells might be one of the mechanisms of growth of blood-vessel-coopting metastases. The other growth patterns in the liver were characterized by destruction of the architecture of the liver parenchyma and were associated with desmoplasia and new blood vessel formation. The metastases were (desmoplastic growth pattern) or were not (pushing growth pattern) surrounded by a fibrotic capsule.

The consequences of this heterogeneity of human liver metastases are the limited value of model systems that selectively reproduce the well-studied angiogenesis-dependent growth of metastases and the difficulties in analyzing the results of clinical trials applying biomodulatory drugs. Imaging of the vascular flow and leakage by contrast-enhanced CT or MR might be helpful in selecting patients with angiogenic versus nonangiogenic liver metastases. The existence of different growth patterns also stresses the superior value of the endothelial cell proliferation (ECP) fraction for angiogenesis quantification compared to microvessel density [1]. Liver metastases with a replacement growth pattern indeed have a high microvessel density as a result of cooption of the liver vasculature but have a low ECP due to lack of ongoing angiogenesis.

Growth Patterns and Angiogenesis in Primary Lung Cancer

Bronchiolo-alveolar lung adenocarcinomas have a typical "lepidic" growth pattern: Tumor cells replace the normal pneumocytes, thereby preserving the stromal component of the alveolar wall and coopting the capillary blood vessels. The structure of the lung parenchyma remains intact, which probably explains the growth of satellite lesions in the lung due to transportation of tumor cells by airflow. This nonangiogenic growth has been shown to be present in more common types of nonsmall-cell lung carcinomas (NSCLC) and in lung metastases [3]. In 16 percent of 500 NSCLC, the tumors were growing without parenchymal destruction and without the formation of desmoplastic stroma. In this "alveolar" growth pattern, tumor cells were filling the alveoli as solid nests. The only blood vessels present were those in the preserved alveolar septa. The coopted blood vessels did not express the integrin alpha-v-beta-3 necessary for angiogenesis. In another study, outcome of 283 patients with operable NSCLC was studied and linked to the growth pattern of the lung tumors (Sardari Nia P. et al., 2004). Whereas the majority of the patients had a tumor with associated desmoplasia and angiogenesis in which the alveolar architecture of the lung was not preserved, 18 percent of the patients had a NSCLC with an alveolar, nonangiogenic growth at the tumor–lung interface. The alveolar growth pattern was not associated with a specific histiotype (30% adenocarcinoma, 40% squamous cell carcinoma, and 30% large cell carcinoma and other types). In univariate analysis, T-stage, N-stage, and growth pattern predicted overall and disease-free survival. Multiple logistic regression showed that TN-stage and growth pattern were independent prognostic factors. Hazard ratios for the alveolar growth pattern were 2.0 (95% confidence interval: 1.3 to 3.2) for overall survival and 2.4 (95% confidence interval: 1.5 to 3.8) for disease-free survival, if compared to NSCLC with associated desmoplasia and angiogenesis. When stage I tumors were analyzed separately (174 patients), growth pattern retained its independent prognostic value. This confirms the study of Pastorino et al. [8], which described 137 pT1N0 patients. Both a nonangiogenic type of vascular pattern and epidermal growth factor receptor expression were associated with a poorer survival rate.

Assessing the growth pattern in primary NSCLC is potentially important since it can predict prognosis, but probably also the response to different treatment modalities. The growth pattern is indeed an integrative parameter containing information of the relationship between tumor cells and stromal cells. Surgical pathologists can easily determine the growth pattern on a standard hematoxylin–eosin stained tissue section and integrate it in the pathology report. Another consequence of the growth patterns of lung carcinomas is that the prognostic value of microvessel density in NSCLC can only be investigated within the subgroup of angiogenic tumors.

Table II Comparison of Glandular Differentiation, Fibrin Deposition, CAIX Expression, and the Macrophage Content of Breast Cancer and Colorectal Cancer Liver Metastases.

	Breast	Colorectal	
Glandular differentiation:	($n = 45$)	($n = 28$)	
1	1 (3%)	21 (75%)	
2	8 (17%)	5 (18%)	
3	36 (80%)	2 (7%)	$p < 0.0001$
Fibrin, central:	($n = 37$)	($n = 24$)	
0	16 (43%)	6 (25%)	
1	10 (27%)	4 (17%)	
2	2 (6%)	4 (17%)	
3	9 (24%)	10 (41%)	$p = 0.15$
Fibrin, interface:	($n = 38$)	($n = 25$)	
0	30 (79%)	11 (44%)	
1	6 (15%)	10 (40%)	
2	1 (3%)	1 (4%)	
3	1 (3%)	3 (12%)	$p = 0.037$
CA IX, central:	($n = 44$)	($n = 24$)	
Absent	32 (73%)	1 (4%)	
Present	12 (27%)	23 (96%)	$p < 0.0001$
CA IX, interface:	($n = 45$)	($n = 24$)	
Absent	38 (84%)	11 (46%)	
Present	7 (16%)	13 (54%)	$p = 0.002$
Global CA IX score:	14.4 ± 8.6 (0)	74.5 ± 14.9 (52.5)	$p < 0.0001$
Macrophage count:	4.57 ± 0.28 (4.25)	8.25 ± 0.60 (7.50)	$p < 0.0001$

Differences in categorical variables are validated by two-tail Fisher's exact testing. Continuous variables are expressed as mean ± standard error (median). Differences are validated by Wilcoxon testing.

Glandular differentiation score: 1, > 75% tubule formation; 2, 10–75% tubule formation; 3, < 10% tubule formation.

Fibrin deposition: Detected immunohistochemically with NYB.T2G1monoclonal antibody. 0, No staining; 1, minimal staining; 2, moderate staining; 3, extensive staining.

Global CA IX score: Carbonic anhydrase IX is an endogenous marker of hypoxia. Its expression is semiquantitatively scored as the product of the percentage of immunostained cells with an immunostaining intensity score ranging from 0 (no staining) to 3 (strong staining).

CA IX: absent, no immunostaining; present, immunostaining in any percentage of tumor cells.

Macrophage count: the relative area occupied by CD68 immunostained macrophages, quantified with the Chalkley morphometric point counting method.

Conclusion

Different growth patterns of primary and metastatic tumors are a reflection of different interactions of the cancer cells with the surrounding tissue structures. The observation of nonangiogenic growth patterns in human carcinomas challenges the hypothesis that tumor growth is always dependent on angiogenesis. It would be more correct to say that neoplastic growth depends on an adequate blood supply. If this can be obtained from a vascular bed that already exists, the tumor can grow without the formation of new blood vessels.

Glossary

Alveolar growth pattern: Nonangiogenic growth pattern described in primary nonsmall-cell lung cancer and in pulmonary metastases; tumor cells fill the alveoli and exploit the interalveolar capillaries for their blood supply.

Fibrotic focus: Focus of exaggerated reactive tumor stroma formation in the center of a carcinoma consisting of collagen, a variable number of fibroblasts, blood vessels, and inflammatory cells; practical histopathological surrogate marker of hypoxia-driven angiogenesis in breast cancer.

Nonangiogenic growth: Tumor growth without induction of angiogenesis in which tumor cells obtain adequate blood supply by exploiting a preexisting vascular bed.

Replacement growth pattern: Nonangiogenic growth pattern described in liver metastases; tumor cells replace the hepatocytes in the liver plates and exploit the sinusoidal blood vessels for their blood supply.

References

1. Vermeulen, P. B., Gasparini, G., Fox, S. B., Colpaert, C. G., Marson, L., Gion, M., Beliën, J. A. M., de Waal, R. M. W., Van Marck, E. A., Magnani, E., Weidner, N., Harris, A. L., and Dirix, L. Y. (2002). Second international consensus on the methodology and criteria of evaluation of angiogenesis quantification in solid human tumours. *Eur. J. Cancer* **38**, 1564–1579. *Guidelines for the estimation of ongoing angiogenesis and the amount of blood vessels in a solid tumor, integrating new concepts and mechanisms of tumor vascularization.*

2. Hasebe, T., Tsuda, H., Hirohashi, S., Shimosato, Y., Iwai, M., Imoto, S., and Mukai, K. (1996). Fibrotic focus in invasive ductal carcinoma: An indicator of high tumor aggressiveness. *Jpn. J. Cancer Res.* **87**, 385–394. *First report presenting the fibrotic focus as a prognostic factor in invasive breast carcinoma.*
3. Pezzella, F., Pastorino, U., Tagliabue, E., Andreola, S., Sozzi, G., Gasparini, G., Menard, S., Gatter, K. C., Harris, A. L., Fox, S., Buyse, M., Pilotti, S., Pierotti, M., and Rilke, F. (1997). Non-small-cell lung carcinoma tumor growth without morphological evidence of neo-angiogenesis. *Am. J. Pathol.* **151**, 1417–1423. *Investigation of the pattern of vascularization in a series of 500 lung carcinomas with the description of an alveolar growth pattern characterized by the lack of parenchymal destruction and absence of both tumor associated stroma and new vessels.*
4. Vermeulen, P. B., Colpaert, C. G., Salgado, R., Royers, R., Hellemans, H., Van de Heuvel, E., Goovaerts, G., Dirix, L. Y., and Van Marck, E. A. (2001). Liver metastases from colorectal adenocarcinomas grow in three patterns with different angiogenesis and desmoplasia. *J. Pathol.* **195**, 336–342. *Description of different patterns of vascularization in liver metastases with identification of a growth pattern characterized by tumor cells replacing the hepatocytes in the liver plates and exploiting the preexisting sinusoidal blood vessels.*
5. Colpaert, C. G., Vermeulen, P. B., Fox, S. B., Harris, A. L., Dirix, L. Y., and Van Marck, E. A. (2003). The presence of a fibrotic focus in invasive breast carcinoma correlates with the expression of carbonic anhydrase IX and is a marker of hypoxia and poor prognosis. *Breast Cancer Res. Treat.* **81**, 137–147.
6. Colpaert, C. G., Vermeulen, P. B., van Beest, P., Goovaerts, G., Weyler, J., Van Dam, P., Dirix, L. Y., and Van Marck, E. A. (2001). Intratumoral hypoxia resulting in the presence of a fibrotic focus is an independent predictor of early distant relapse in lymph node-negative breast cancer patients. *Histopathology* **39**, 416–426.
7. Colpaert, C. G., Vermeulen, P. B., van Beest, P., Goovaerts, G., Dirix, L. Y., Harris, A. L., and Van Marck, E. A. (2003). Cutaneous breast cancer deposits show distinct growth patterns with different degrees of angiogenesis, hypoxia and fibrin deposition. *Histopathology* **42**, 530–540.
8. Pastorino, U., Andreola, S., Tagliabue, E., Pezzella, F., Incarbone, M., Sozzi, G., Buyse, M., Menard, S., Pierotti, M., and Rilke, F. (1997). Immunocytochemical markers in Stage I lung cancer: Relevance to prognosis. *J. Clin. Oncol.* **15**, 2858–2865.

Capsule Biography

Cecile G. Colpaert is a pathologist working in a teaching hospital. Her main interest is breast cancer and the application of research findings from the field of tumor biology in diagnostic pathology practice.

Peter B. Vermeulen is a diagnostic pathologist doing translational breast cancer research mainly focused on angiogenesis and tumor–stroma interactions.

CHAPTER 140

Breast Cancer Resistance Protein in Microvessel Endothelium

Hiran C. Cooray and Margery A. Barrand
University of Cambridge, Cambridge, United Kingdom

Introduction

Breast Cancer Resistance Protein (BCRP) is a transporter recently identified as a member of the ATP Binding Cassette (ABC) superfamily of transmembrane proteins (discussed in the following section). It was recognized initially in several drug-resistant tumor cell lines (as discussed later) and was subsequently identified in certain tumor tissues where its presence has been putatively associated with poor clinical response to chemotherapy. However, BCRP is found on many normal, that is, nonmalignant, cells at a number of different sites in the body (see later discussion). These include endothelial cells lining various vasculature. Clues to the possible physiological role or roles of BCRP in these locations may be derived from the many studies that are now being conducted, which examine its mechanisms of action and its substrate (discussed later) and inhibitor profiles in different cell systems or that analyze its influence on the distribution of drugs in whole animals.

ABC Transporters

Transporters belonging to the ATP-Binding Cassette (or ABC) family of proteins have the ability to transport substrates across cellular membranes against a concentration gradient using the energy of ATP hydrolysis. They are so called because of their distinctive ATP-binding domains, which contain highly conserved sequences, Walker A and Walker B and an additional ABC "signature" sequence. ABC proteins constitute the largest subclass of transmembrane proteins and are expressed ubiquitously in all living organisms, including bacteria, plants, and animals. Collectively, they are responsible for transporting a multitude of diverse substrates including sugars, ions, lipids and phospholipids, peptides, bile acids, sterols, pigments, and xenobiotics across membranes, thus affecting the distribution of molecules at subcellular, cellular, and tissue levels. There are many different ABC transporters. The mammalian ones have now been classified into subfamilies, termed ABCA through ABCG (http://www.humanabc.org/). In this scheme, BCRP is named as ABCG2. The reader is directed to http://www.ncbi.nlm.nih.gov/books/bookres.fcgi/mono_001/mono_001.pdf and http://arjournals.annualreviews.org/doi/pdf/10.1146/annurev.biochem.71.102301.093055 for two comprehensive reviews on ABC transporters.

Structural Organization of ABC Transporters

Many of the ABC transporters are constructed as a tandem repeat of a basic unit containing two domains: an N-terminal transmembrane domain (TD) of 6 to 11 α-helices that provide substrate specificity to the protein, and a C-terminal nucleotide-binding domain (NBD) located in the cytoplasm that binds and cleaves ATP in order to generate the energy for substrate transport. Binding and subsequent hydrolysis of ATP result in a conformational change that causes the substrate to be translocated across the membrane. This general structure is typified by the mammalian multidrug transporter, P-glycoprotein, otherwise named ABCB1 (Figure 1). Some ABC transporters including the Multidrug Resistance-associated Protein, MRP1 (ABCC1), have an additional N-terminal TD. However others contain only one NBD and one TD. This "half transporter" structure is

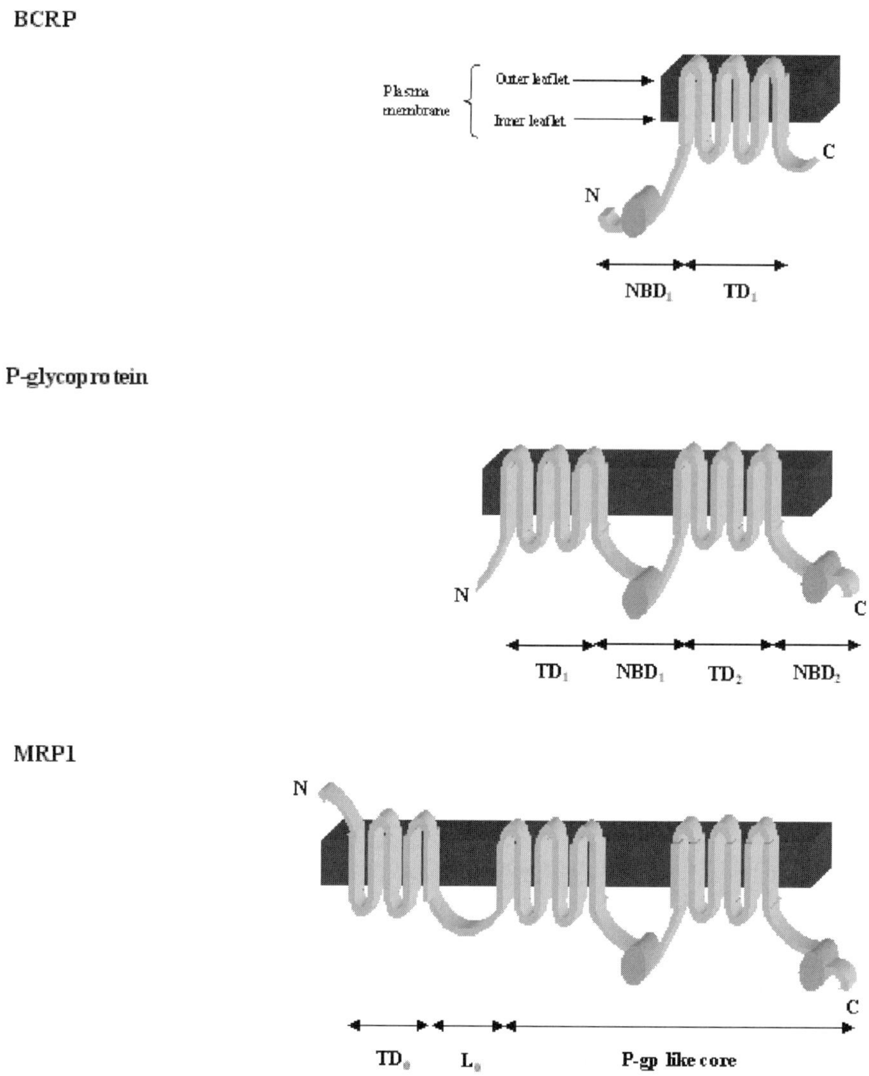

Figure 1 Putative membrane topology of the three main multidrug transporters. BCRP consists of one transmembrane domain (TD) and one nucleotide binding domain (NBD) and is termed a *half-transporter*. P-glycoprotein, like many ABC proteins, is a full transporter with a TD_1-NBD_1-TD_2-NBD_2 structure. Several of the MRP family members, including MRP1, have an additional N-terminal TD linked to the P-gp-like core by a linker region (L_0). Note that the TD and NBD in BCRP are in reverse orientation to those of the other two ABC proteins. (*N* and *C* refer to the N and C termini of the transporter, respectively.) (see color insert)

exemplified by members of several ABC subfamilies including the ABCG subfamily of which BCRP is a member. In addition, members of this subfamily have their TD and NBD in reverse orientation (see Figure 1); hence, BCRP is referred to as a "reverse" half-transporter. Half transporters have to dimerize with a partner protein (either with itself to form a homodimer or with another protein to form a heterodimer) in order to be functional. Currently, the bulk of the evidence points to BCRP being a homodimer. Interestingly however, BCRP shows closest homology with the *white*, *brown*, and *scarlet* proteins that heterodimerize (*white/brown* and *white/scarlet*) to transport pigment precursors (guanine and tryptophan) in the *Drosophila* eye.

ABC Transporters Associated with Multidrug Resistance

Drug resistance can be a serious obstacle to successful anticancer treatment with tumors often failing to respond either to initial chemotherapy (intrinsic resistance) or to subsequent rounds of treatment (acquired resistance). Studies conducted in the laboratory on tumor cell lines cultured in the presence of cytotoxic drugs reveal that resistance can develop, not only to the selecting drug, but to a number of structurally and functionally dissimilar drugs as well, hence providing multidrug resistance (MDR). This MDR phenomenon may involve several different types of mechanisms, but a common cause is the presence of multidrug ABC transporters, which prevent access of the drugs to their intracellular targets sites. They accomplish this task by either effluxing the drug out of the cell via the plasma membrane, or by sequestering the drug within intracellular organelles such as the endoplasmic reticulum, lysosome, or peroxisome.

Identification of BCRP as a Multidrug Transporter

P-gp was the first ABC transporter to be associated with MDR (around the mid-1970s) with the MRPs being later discoveries (MRP1 sequenced in 1992). It was not until 1998 that BCRP was revealed as another MDR-associated ABC transporter. It was becoming apparent that there were certain tumor cell lines that showed resistance to several drugs (mainly mitoxantrone, bisantrene, topotecan, and doxorubicin) yet did not overexpress P-glycoprotein or MRPs. Recognition of this atypical, non-P-gp, non-MRP resistance phenotype initiated a search for another multidrug transporter, leading to the ultimate discovery of BCRP, otherwise termed Mitoxantrone Resistance Protein (MXR) or ABC Transporter in Placenta (ABCP). The different names of this protein derive from the fact that it was characterized and cloned from three different sources in independent laboratories at about the same time: from a multidrug-resistant breast cancer cell line selected in doxorubicin (MCF7/AdrVp3000, hence BCRP), from a colon cancer line selected in mitoxantrone (S1-M1-80, hence MXR), and from human placenta (hence ABCP) [1].

Though commonly called BCRP by virtue of its initial discovery in a drug-selected breast cancer cell line, it is not clear that the protein is actually often overexpressed in breast cancers in vivo. Indeed only weak BCRP expression has been found in breast tumors. There are, however, many different human tumors in which BCRP expression has been clearly demonstrated, including tumors of the kidney, ovary, stomach, colon, thyroid, brain, endometrium, and testis; squamous tumors (lung, head and neck, and esophagus); soft tissue sarcomas; pheochromocytomas; and hepatocarcinomas. This may reflect the distribution of BCRP in normal tissues (see later discussion).

Substrate and Inhibitor Profiles of BCRP Compared to Other Multidrug Transporters

The ABC transporters associated with MDR vary somewhat in their mechanisms of action, substrate and inhibitor profiles, and in their tissue locations. Nevertheless there is a significant overlap in these characteristics between the transporters (Figure 2).

Functional studies on both BCRP overexpressing drug-selected and BCRP-transfected cell lines show that the transporter can confer resistance to anthracyclines (doxorubicin, daunorubicin), anthracenediones (mitoxantrone), camptothecins (topotecan, irinotecan, and its active metabolite, SN-38), and etoposide, but not to vincristine, taxol, or colchicine, which are classical P-gp substrates. Substrates currently recognized to be transported by BCRP are shown in Table I.

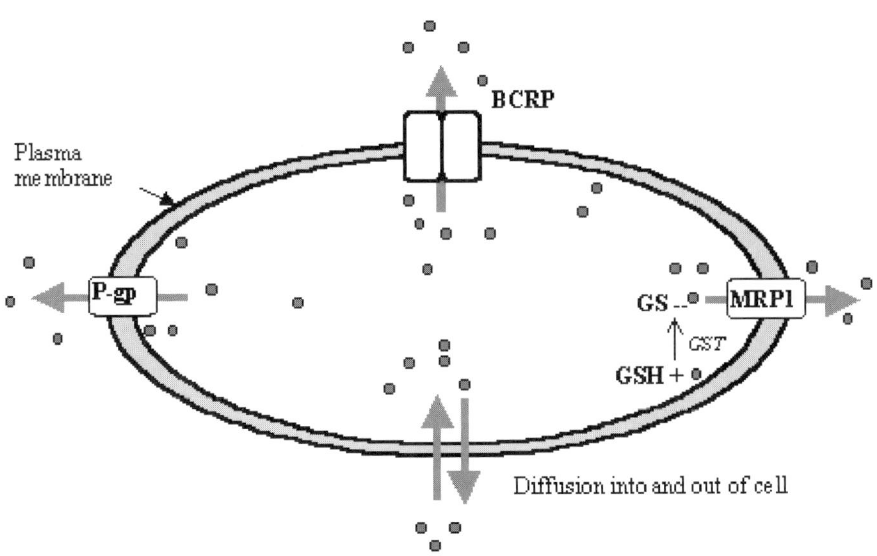

Figure 2 Actions of multidrug transporters BCRP, P-gp, and MRP1 in efflux of substances from cells. Drugs can enter and leave cells by passive diffusion along a concentration gradient. Multidrug transporters provide a second route for drug exit and can drive drugs out of the cell against a concentration gradient by exploiting the energy of ATP hydrolysis. This additional efflux reduces drug concentrations inside cells to sublethal levels. P-gp and BCRP (in the form of a dimer) can efflux drugs unmodified (mainly hydrophobic, amphipathic compounds). The MRPs require the presence of reduced glutathione (GSH) to transport unmodified drugs (predominantly organic anions) but can transport drugs following their conjugation (GST, Glutathione transferase). (see color insert)

Table I Substrate Profile of BCRP Compared with P-gp and MRP1.

Substrate	BCRP	P-gp	MRP1
Paclitaxel		✓	
Verapamil		✓	
Colchicine		✓	
Vinblastine		✓	✓
Etoposide (VP-16)	✓	✓	✓
Daunorubicin	✓	✓	✓
Doxorubicin	✓	✓	✓
Epirubicin	✓	✓	✓
Mitoxantrone	✓	✓	✓
Methotrexate	✓[a]	✓	✓
Prazosin	✓	✓	
Topotecan	✓	✓	
Bisantrene	✓	✓	
Rhodamine-123	✓	✓	
Flavopiridol	✓		
Lysotracker Green	✓		
SN-38	✓		
E_2-GLU	✓		✓
Estrone 3-sulfate	✓		✓
LTC_4			✓
GSH			✓

E_2-GLU, 17β-Estradiol 17-(β-D-glucuronide); GSH, reduced glutathione; LTC_4, leukotriene C_4. Rhodamine 123 and Lysotracker Green are fluorescent dyes.

[a] Methotrexate is effluxed only by the wild-type BCRP.

A point mutation affecting substrate specificity has recently been reported in BCRP. Though wild-type BCRP has an arginine at the 482nd position (at the start of the third transmembrane segment), BCRP overexpressed in certain drug-selected cell lines was shown to contain either a glycine or threonine at this site. This point mutation causes a paradigm shift in the protein's substrate specificity—wild-type BCRP is incapable of effluxing the fluorescent dye rhodamine 123 or anthracycline drugs such as doxorubicin, but the mutants can handle both these substrates. Conversely, only wild-type BCRP transports the antifolate cytotoxic methotrexate [1]. Polymorphisms of the BCRP gene have also been described in human populations.

Subcellular Location of BCRP

BCRP seems to be predominantly localized to the plasma membrane of both drug-selected and transfected cell lines. This sets it apart from other half-transporters that are localized mainly to intracellular membranes such as the mitochondrion (ABCB7), the peroxisome (ALD subfamily), and the endoplasmic reticulum (Tap1/Tap2). Such a location for BCRP is consistent with a putative role in the efflux of substrates from the cell. Furthermore, it is apparent in polarized cell lines such as BCRP-transfected MDCK-II Madine-Darby canine kidney cells that the transporter becomes localized primarily to the apical aspect of the plasma membrane, where it mediates the translocation of substrates from basal to apical side. However, a role for BCRP in the intracellular trafficking of molecules cannot be ruled out as some immunocytochemical studies have reported perinuclear staining for BCRP in several topotecan- and mitoxantrone-resistant cell lines [1].

Table II Tissue Distribution and Putative Functions of BCRP at These Locations.

Localization	Putative function
Placenta—syncytiotrophoblast	Protection of fetus, excretion of substrates
Apical membrane of epithelium of small intestine and colon	Reduced uptake/excretion of substrates into maternal circulation
Liver canalicular membrane	Excretion of substrates by the liver into bile
Apical membrane of lobules and lactiferous ducts of breast	Unknown
Endothelium of capillaries and veins	Unknown
"Side" population of hematopoietic stem cells	Unknown

Tissue Distribution of BCRP Compared to Other Multidrug Transporters

The apical siting of BCRP on polarized cells is of particular relevance to the possible role or roles of BCRP in normal tissues. The protein is found in many tissues, including barrier sites, as outlined in Table II.

The highest BCRP expression is found in the placenta on the syncytiotrophoblast facing the maternal circulation. This suggests a role for the protein in the elimination of substrates from the fetus. This has been established for mouse Bcrp1; in both wild-type and P-gp knockout mice, inhibition of Bcrp1 by GF120918 (a common inhibitor of human and mouse P-gp and BCRP) resulted in at least a twofold increase in the fetal uptake of orally administered topotecan, a BCRP substrate. BCRP is also expressed at more modest levels in the colon, small intestine, liver, ovary, and breast, where it may be concerned with elimination of material from these tissues. In a recent clinical study utilizing GF120918, it was shown that the oral bioavailability of topotecan more than doubled (from 40% to 97%) when the drug was coadministered with the inhibitor, thus underlining the functional significance of BCRP expression in the intestine.

This distribution of BCRP shows similarities to that of the multidrug transporter P-glycoprotein, which is also expressed in various epithelia, particularly in organs associated with drug absorption and disposition, such as hepatocyte canalicular membrane and the intestinal mucosa. P-gp is thought to provide a first line of defense against the entry of many types of xenobiotics into the body. Knockout mice deficient in functional P-gp, although viable, fertile, and without obvious histological or developmental abnormalities, show significantly altered pharmacokinetics (and toxicity of several drugs) [2]. The third subfamily of multidrug transporters, the MRPs, are widely distributed throughout the body in tissues including the choroid plexus, oral mucosa, small intestine, testis, and respiratory tract. Because there are several MRP homologs with overlapping substrate specificities, the importance of each for the elimination of particular substances is difficult to assess.

Both BCRP and P-gp are to be found on the endothelium lining the blood–brain barrier (see later discussion). The presence of multidrug transporters at such barrier sites creates "pharmacological sanctuaries" within the body, permitting certain organs and tissues to function in relative isolation from the rest of the body. Indeed, in P-gp knockout mice, the integrity of the blood–brain barrier is shown to be significantly compromised, with much higher brain penetration of P-gp substrates such as vinblastine and ivermectin being demonstrated. The relevance of BCRP at these sites is still under investigation (see later discussion).

The generation of the Bcrp1 knockout mouse [3] has thrown new light on the putative physiological function of this transporter. Though these mice were anatomically normal and fertile, a defect was seen in their ability to handle a metabolite of chlorophyll, pheophorbide *a*, resulting in severe phototoxicity in mice exposed to light. They also exhibited a previously uncharacterized form of porphyria. Thus it became known that BCRP performs an essential function at the gut epithelium in effluxing toxic products of chlorophyll metabolism. BCRP knockout mice generated independently by Zhou et al. [4] were used to demonstrate that this transporter, rather than P-gp, is responsible for the dye efflux in the cells. This allows analysis of the "side-population," enriched in murine hematopoietic stem cells, which have high bone-marrow repopulating activity. The role BCRP plays at this location is still to be elucidated.

Expression of BCRP in Endothelia of Normal Tissues and of Tumors

BCRP differs from P-gp in being expressed on the endothelial lining of vascular beds in many tissues, not just at the blood-brain barrier. Interestingly, BCRP is evident in venules and capillaries (see Table II) but not in arterioles [5]. Hence the transporter is distributed in the regions of the vasculature where the bulk of the exchange of materials between blood and tissues occurs. On endothelial cells of vasculature-supplying tumors (for example, testicular germ-cell tumors, endometrial, ovarian and colon carcinomas, and brain tumors), antibody staining for BCRP has been described as moderate to strong, stronger indeed than on the vascular endothelium in the surrounding normal regions [6]. This raises the interesting possibility that BCRP expression is perhaps upregulated in the endothelium of blood vessels during neoplastic vasculogenesis.

Localization of BCRP in the Specialized Endothelium of the Blood–Brain Barrier

The presence of multidrug transporters is of particular importance in vascular endothelial cells at special barrier sites such as the blood–brain and blood–testis barriers. Here the vessels possess tight junctions that place severe restrictions on the free paracellular diffusion of many substances seen in peripheral endothelia.

Recent studies have explored the localization of BCRP in human brain material using fresh-frozen samples of both normal and tumor brain (meningiomas and gliomas) [7]. Western blot results show a higher degree of expression of BCRP protein in the gliomas over the normal and meningioma samples. It could be seen by immunostaining that BCRP is primarily localized to blood vessels within the brain. In the case of two meningioma samples, notable heterogeneous staining for BCRP was seen in brain parenchymal cells in addition to endothelial cells. Diestra et al. [6] also reported a higher expression of BCRP in several unspecified brain tumors over normal brain parenchyma using immunohistochemical staining with a well-characterized anti-BCRP antibody.

By exploiting the powerful resolving capabilities of the confocal microscope, it has been possible to gain some understanding of the subcellular distribution of BCRP within brain microvessels. Utilizing the fact that the brain endothelial glucose transporter GLUT-1 is localized on both sides of brain endothelial cells (both luminal and abluminal membranes), dual-staining with antibodies for GLUT-1 and for BCRP revealed the main sites of BCRP expression in microvessels in both normal and tumor brain sections. The distribution of BCRP staining was seen to be inner to that of GLUT-1 in all microvessels viewed, which suggests that BCRP is localized toward the luminal membrane of human brain endothelial cells in the in vivo blood–brain barrier [7]. It is probable therefore that BCRP, localized strategically at the luminal membrane of endothelial cells, has a protective function at the blood–brain barrier in limiting entry of substrates into the brain.

P-gp also has been localized to the luminal aspect of the brain capillary endothelium. It is already well documented that in this situation it performs what has been described as a "gatekeeper" role at the blood–brain barrier, pumping out a variety of xenobiotics that would otherwise gain access to the brain via the transcellular pathway due to their lipophilicity [2]. The number of drugs known to be excluded from the brain by P-gp is large, ranging from nonsedating antihistamines, antiepileptics, and beta-blockers to anti-HIV reverse transcriptase inhibitors. What additional protection BCRP may bring to bear is as yet not well defined and would require the advent of knockout mice lacking several of the MDR transporters or use of combinations of specific inhibitors so that the influence of BCRP can be distinguished from that of P-gp in vivo.

The presence and importance of MRPs at the blood–brain barrier is even less clear. This is due both to the multiplicity of transporters in this family and to existing controversies in the literature. In contrast to BCRP, MRP1 is known to be functionally active in vivo at the *epithelium* of the choroid plexus, regulating the distribution of several xenobiotics into the CSF. But it has not been definitively localized in vivo at the blood–brain barrier. MRP1 does, however, become upregulated in cultured brain endothelial cells [8]. This has allowed its functionality to be explored in vitro in brain endothelial cells cultured from several sources including human brain. In the case of MRP2, results of recent studies using in vivo microdialysis hint at a functional role for this protein at the rat blood–brain barrier, limiting the brain uptake of the anticonvulsant phenytoin. However, these observations need further investigation.

A homolog of BCRP has been described in porcine brain endothelial cells by Eisenblatter et al. [9]. This protein, named Brain Multidrug-Resistance Protein (BMDP), shows 86 percent amino acid identity with human BCRP and is predicted to have the typical architecture of a "reverse" half-transporter. BMDP was shown, using immunohistochemistry, to be localized to the cell membrane of cultured porcine brain endothelial cells. A blood vessel location in vivo for the message was inferred from RNA isolation experiments in which the mRNA of BMDP appeared to be concentrated in the brain microvessels, with the levels of transcript higher in isolated capillaries than in homogenized brain tissue.

High levels of BMDP transcript were also detectable in cultured porcine brain endothelial cells. These appeared approximately 30 times higher than equivalent P-gp expression, suggesting that at least in the porcine endothelium, BMDP plays a more prominent role than P-gp. This was corroborated via functional studies using the radiolabeled substrate ^{3}H-daunorubicin (a substrate common to both P-gp and BCRP) performed on cultured porcine brain endothelial cells grown as monolayers—GF120918 (which inhibits both P-gp and BCRP) abrogated almost completely the transport of daunorubicin from the basolateral to the apical side of the porcine brain endothelial cell monolayer. However, specific P-gp inhibitors gave only moderate inhibition. This strongly suggests that the contribution by BMDP to transport of substrates across the porcine blood–brain barrier may be greater than by P-gp. These studies are the first to report functional BCRP in cells derived from the blood–brain barrier.

Many questions still remain regarding the in vivo functionality of BCRP. Its ubiquitous expression in the endothelium of veins and capillaries of every tissue so far examined suggest that it might efflux substrates that are potentially toxic to many tissues but are incapable of passing between endothelial cells. In particular, its expression at the blood–brain barrier may also be of paramount significance to limiting the brain penetration of substrates. The vast body of research available on P-gp and members of the MRP family has pointed to a number of specific–roles performed by these transporters at various sites in the body. There is still much to be learned about BCRP and the function it may perform in the microvessel endothelium.

Acknowledgments

The authors thank the Cancer Research Campaign for their contributions to the authors' own research work and the Cambridge Commonwealth Trust for assistance toward a studentship for HCC, who also holds an award from Universities UK.

References

1. Allen, J. D., and Schinkel, A. H. (2002). Multidrug resistance and pharmacological protection mediated by the Breast Cancer Resistance Protein (BCRP/ABCG2). *Mol. Can. Ther.* **1**, 427–434. *A very useful and comprehensive recent review on all aspects of BCRP.*
2. Schinkel, A. H. (1999). P-Glycoprotein, a gatekeeper at the blood–brain barrier. *Adv. Drug Deliv. Rev.* **36**, 179–194.
3. Jonker, J. W., Buitelaar, M., Wagenaar, E., van der Valk, M. A., Scheffer, G. L., Scheper, R. J., Plosch, T., Kuipers, F., Oude Elferink, R. P. J., Rosing, H., Beijnen, J. H., and Schinkel, A. H. (2002). The breast cancer resistance protein protects against a major chlorophyll-derived dietary phototoxin and protoporphyria. *Proc. Natl. Acad. Sci. USA* **99**, 15649–15654.
4. Zhou, S., Morris, J. J., Barnes, Y., Lan, L., Schuetz, J. D., and Sorrentino, B. P. (2002). BCRP1 gene expression is required for normal numbers of side-population stem cells in mice, and confers relative protection to mitoxantrone in hematopoietic cells *in vivo*. *Proc. Natl. Acad. Sci. USA* **99**, 12339–12344.
5. Maliepaard, M., Scheffer, G. L., Faneyte, I. F., van Gastelen M. A., Pijnenborg, A. C. L. M., Schinkel, A. H., van de Vijver, M. J., Scheper, R. J., and Schellens, J. H. M. (2001). Subcellular localization and distribution of the breast cancer resistance protein transporter in normal human tissues. *Cancer Res.* **61**, 3458–3464.
6. Diestra, J. E., Scheffer, G. L., Catal, I., Maliepaard, M., Schellens, J. H. M., and Scheper, R. J. (2002). Frequent expression of the multidrug resistance associated protein BCRP/MXR/ABCP/ABCG2 in human tumors detected by the BXP-21 monoclonal antibody in paraffin-embedded material. *J. Pathol.* **198**, 213–219.
7. Cooray, H. C., Blackmore, C. G., Maskell, L., and Barrand, M. A. (2002). Localization of Breast Cancer Resistance Protein in microvessel endothelium of human brain. *Neuroreport* **13**, 2059–2063. *Establishes a luminal localization for BCRP at the human blood–brain barrier.*
8. Seetharaman, S., Barrand, M. A., Maskell, L., and Scheper, R. J. (1998). Multidrug resistance–related transport proteins in isolated human brain microvessels and in cells cultured from these isolates. *J. Neurochem.* **70**, 1151–1159.
9. Eisenblatter, T., Huwel, S., and Galla, H. J. (2003). Characterisation of the brain multidrug resistance protein (BMDP/ABCG2/BCRP) expressed at the blood–brain barrier. *Brain Res.* **971**, 221–231. *The first paper to report a porcine homolog of BCRP highly expressed in cultured endothelial cells derived from the blood–brain barrier.*

Further Reading

Litman, T., Brangi, M., Hudson, E., Fetsch, P., Abati, A., Ross, D. D., Miyake, K., Resau, J. H., and Bates, S. E. (2000). The multidrug-resistant phenotype associated with overexpression of the new ABC half-transporter, MXR (ABCG2). *J. Cell. Sci.* **113**, 2011–2021.

Capsule Biography

Hiran C. Cooray is in the final year of his doctorate studying the expression and putative roles of BCRP in human brain material and in cultured endothelial cells.

Dr. Margery Barrand is a Senior Lecturer in the Department of Pharmacology in the University of Cambridge. Her group has strong research interests in multidrug transporters and, in particular, transport systems at the blood–brain barrier.

CHAPTER 141

Tumor versus Normal Microvasculature

Moritz A. Konerding
Johannes Gutenberg-Universität Mainz, Germany

Introduction and Definitions

Therapeutic concepts in oncology such as antiangiogenic or antivascular approaches have reawakened the scientific community's interest in comparative microvascular anatomy and biology. Microvascularity and angiogenesis play a significant role during normal growth, in physiological conditions, and in a variety of pathological conditions such as inflammation, diabetes, macular degeneration, and wound healing. Angiogenesis is thus, not a specific phenomenon in tumors, but instead an integral element of numerous different normal and pathological conditions.

After the short early embryonic phase of primary angiogenesis, which comprises the formation of poorly differentiated vessels composed of endothelial cells derived from angioblasts of blood islands in the extraembryonal mesoderm of the yolk sac, all further vascular growth in physiological and pathological conditions is summarized as secondary angiogenesis. In secondary angiogenesis formation of new vessel segments can be accomplished either by vessel sprouting, which requires endothelial cell mitoses, or by intussusceptive growth without need for endothelial cell mitoses. Primary angiogenesis never works without mitoses.

When describing microvessel morphology, we have to discern between structure and architecture, both of which determine functional properties in terms of blood flow. *Structure* in morphological terms means vessel wall building, cellular differentiation, cell shape, cell contact structure and cell surface differentiation, and organelle and cytoskeleton contents. Microvessel *architecture* encompasses pattern formation, vessel course, density, and all parameters defining the 3D arrangement of the microvascular unit.

Microvessel architecture can be studied best by means of scanning electron microscopy of corrosion casts, which allows also for 3D imaging, reconstruction, and quantitative analysis. For microvascular corrosion casting a resin is injected into the vascularity either systemically or locally. After polymerization of the casting medium the tissue is corroded in an alkaline solution. After drying and mounting the vessel system replicas can be studied in detail. Intravital microscopy, especially after injection of fluorescent dyes, is effective in the observation of two-dimensional vascular networks as given in most angiogenesis assay systems. The possibility of examining the vasculature in the time course is the biggest advantage; the limited resolution the major shortcoming. Classical injection methods with light microscopic evaluation are time consuming, require laborious reconstruction work, and are still of limited value, unless sophisticated computerized techniques are used. The variety of morphological techniques for assessing microvascular structure from light to confocal scanning and transmission microscopy was increased significantly by immunostaining, allowing for further structure–function correlations.

Structural Differences between Tumor and Normal Microvasculature

Normal microvasculature is hierarchically organized. The capillary network branches from arterioles with continuous or metarterioles with discontinuous layers of smooth muscle cells. In both cases precapillary sphincters may regulate blood flow. Blood flow regulation is also facilitated by arteriovenous anastomoses. The capillaries are usually made up by a continuous layer of endothelial cells that are joined together by tight junctions. In some organs, such as in parts of the gut and in endocrine glands, fenestrated endothelium prevails. Pericytes originating from mesenchymal cells and a basement membrane regularly embrace continuous endothelium-type capillaries, whereas pericytes are rarely found in fenestrated capillaries. Gaps between endothelial cells are seen only in discontinuous, sinusoidal vessels in the spleen, liver, and bone marrow. In these vessels basement membranes are missing.

Tumor microvasculature is characterized by a lack of hierarchy and differentiation. Even large caliber vessels are mainly composed of no more components than an endothelium and a basement membrane (Figure 1). Because of the usually higher interstitial pressure the endothelium may decrease in height and appear flattened. Only sprout formations and early vessel forms, such as in the tumor invasion front, reveal an organelle-rich and comparatively high endothelium (Figure 1b,c). Villus-like protrusions of the luminal surfaces indicate the young age of these vessels. Abluminal protrusions of cell ramifications, again, indicate active migration and new sprouting. The thinning of the endothelium sets in with further stabilization and increase in the luminal diameter. However, only a part of the early forms persists: Many early forms undergo degeneration and destruction just like those sprouts that did not fuse with others. Until now no reliable morphometric data have been available on the fate of sprouts and early forms. We have the impression that at least two thirds of all formed sprouts will not differentiate into early forms or established vessels.

Fenestrations shut by diaphragms are rarely seen (Figure 1d), whereas discontinuities or gaps that allow for hemorrhage and facilitate permeability are common features irrespective of the origin of the tumor (Figure 1e). Cell contacts are usually poorly differentiated, and no complex contact structures exist even in well-established vessels and late forms. Pericytes are frequently missing.

Comparisons of the structural features of different human sarcomas, melanomas, and carcinomas xenografted onto nude mice as well as a variety of human primary tumors (colorectal, renal, and larynx carcinomas) did not show tumor cell line or tumor entity specific cell structure pattern. This might well have consequences for targeting approaches. In general, the number of intracytoplasmic contractile filaments is reduced, which contradicts any possible blood flow regulation by pericyte or endothelial cell contractility. Blood flow is regulated primarily on the level of preexisting arteries and arterioles nourishing the tumor, resulting in heterogeneous blood flow. Apart from incorporated vessels no clearly demarcated arterial sphincters are visible within the newly formed tumor vascular network. Autonomous nerves are also detectable only in the tumor periphery as incorporated nerve fibers. Thus, low structural stability is accompanied by lacking regulation.

The difficulties frequently seen when using panendothelial cell markers such as anti-CD31 or anti-FVIII for labeling tumor microvessels indicate a functional and structural heterogeneity of lumen-confining cells in tumors. This is at least in part due to the participation of nonendothelial cells in the vessel wall formation. Newly formed vessels composed of cells with extremely different cell organelle contents indicate the involvement of different cell types (Figure 1d). Sometimes "endothelial cells" could be observed that engulfed bundles of collagen fibers.

It has been well known for 30 years that pericytes may be formed not only by the angioblastic pathway, but also from mesenchymal cells. Likewise, since then evidence has accumulated that tumor cells themselves may be involved in vessel wall building. Interstitial plasma flow may take place in channels completely lined by nonendothelial cells (Figure 1f), for which the term *vascular mimicry* was recently coined.

Architectural Differences between Normal and Tumor Microvasculature

All tissues and organs develop a very specific vascular architecture reflecting the specific demands. The large intestine microvasculature, for example, is characterized by a flatly extended mucosal capillary plexus with a hexagonal, honeycomb pattern around the tissue of the mucosal glands (Figure 2a). This subepithelial plexus is supplied by arteries that divide within the submucosa and is drained by venules originating immediately under the mucosal surface. Elongated capillaries in the renal medulla are densely packed in vascular bundles permitting the essential countercurrent exchange mechanism, whereas the capillaries derived from the afferent arterioles branch within the renal corpuscle with its unique architecture (Figure 2b). Microvessel plexuses within loose connective tissue such as the subcutis are flatly organized with undulating capillary courses that enable compliance with shear stress exerted by the skin (Figure 2c). Capillaries coursing into dermal papillae show typical hairpin loops that are involved in temperature regulation (Figure 2d).

In tumors, the organ- and tissue specific vascular architecture is not retained but is replaced by newly formed vessels without significant hierarchy (Figure 2e–h). Vessel densities may vary considerably; the highest vessel densities are usually found in the periphery within the invasion front. Vessel density within "hot spots," that is, densely vascularized areas, may be even higher than in the autochthonous tissue. In desmoplasias next to the neoplastic tissue, sprouting, dilatation, and structural adaptation may change the original architecture of the preexisting vessels as result of the release of proangiogenic factors.

Common features of human primary and of experimental tumor vascularities—irrespective of origin, size, and growth behavior—are missing hierarchy, the formation of large-caliber sinusoidal vessels (Figure 2e, f), and markedly expressed vessel density heterogeneity (Figure 2e, g). The diameters within individual tumor vessels vary significantly. Sinusoids originating from and draining into venous vessels increase vessel densities, but do not contribute to nutritive blood flow. Capillary elongations by more than tenfold explain the low intratumoral po_2. Vessel compressions and blind ends are found close together.

Tortuous courses and elongated sinusoidal vessels may occur also in other forms of secondary angiogenesis; however, in wound healing and chronic inflammation a real remodeling of the newly formed vasculature takes place with differentiation into arteries and veins and elimination of ineffective vessel segments.

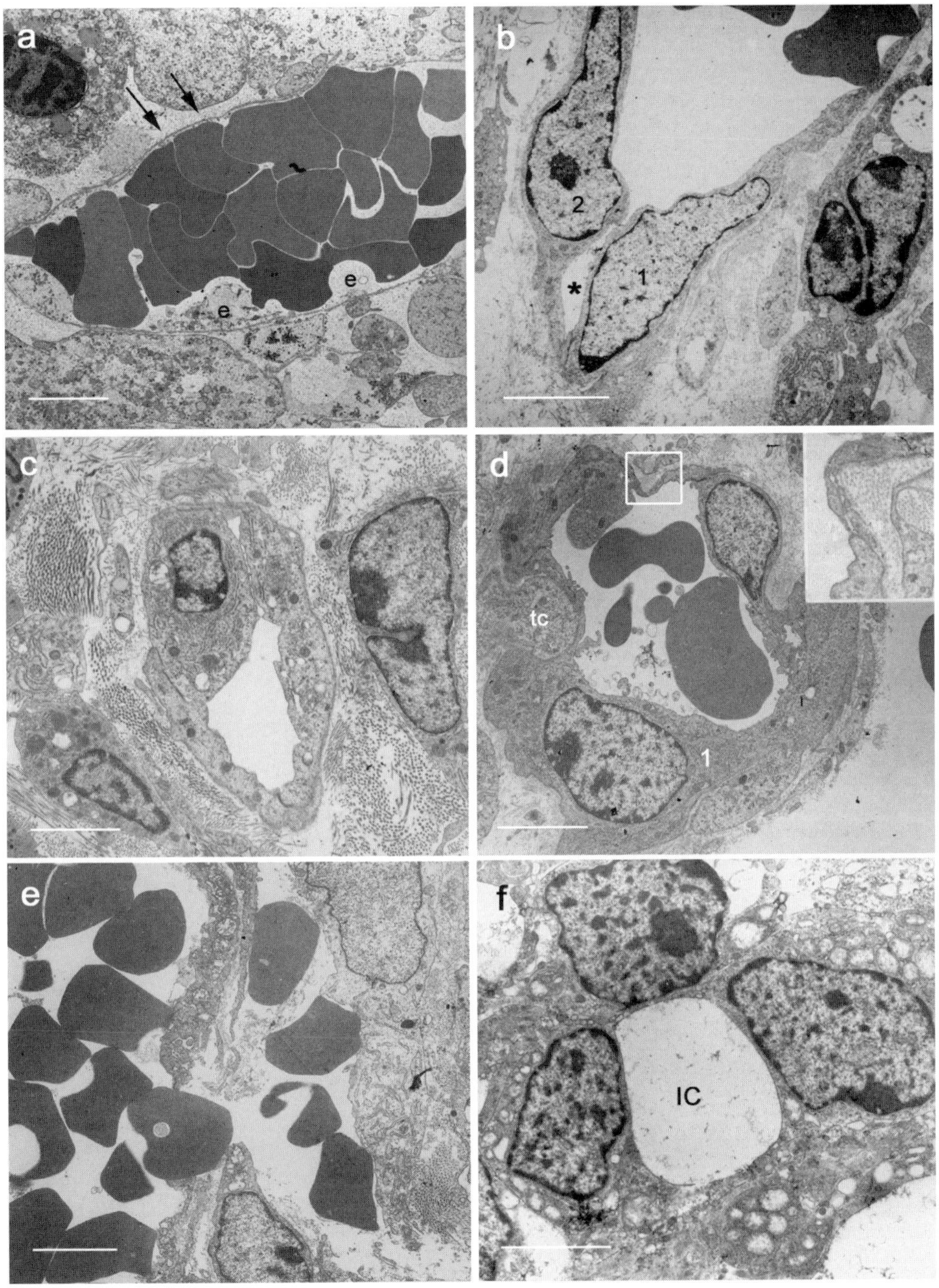

Figure 1 Transmission electron microscopy of tumor vascularity. **(a)** Established sinusoidal vessel (late form) of a renal cell carcinoma transplanted under murine renal capsule with flattened endothelium (*arrows*) without medial layer. Note the edema (e) within some endothelial cell protrusions. **(b)** Intercellular bud formation through endothelial cell migration in a human melanoma xenografted onto nude mice. The eccentrically located perikaryon of cell 2 and a pseudopodium of cell 1 veer out from the endothelial structure, forming a new lumen, which is not fully connected to the primary lumen. **(c)** Early form in a melanoma. Note the height of the endothelium and the unusual overlapping of the three lumen-confining cells. **(d)** Partial compression of an early form by extravascular tumor cells (tc) next to a lumen-confining cell (1), which results in a thinning of the endothelium. (*Inset*) Fenestrated endothelium is superimposed by continuous endothelium. **(e)** Structural defects in sinusoidal vessel of a xenografted squamous cell carcinoma with evasions. The endothelial cells (*left*) show some bleb formation indicative of cytoskeleton damage, whereas the pericytes appear normal. **(f)** Interstitial channel (ic) lined completely by hypoxic tumor cells in a xenografted spindle cell sarcoma. All bars = 5 μm.

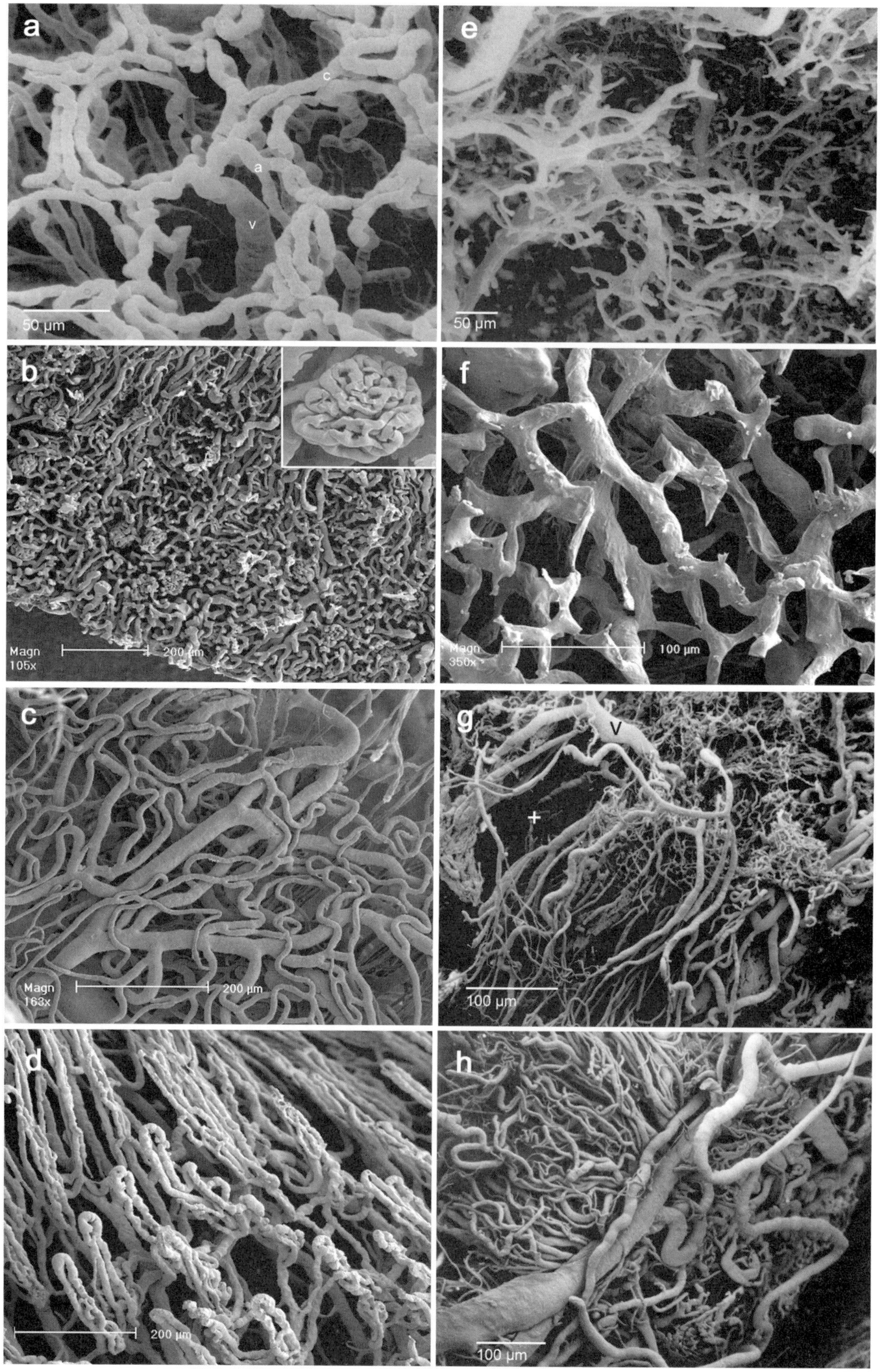
a
c
a
v
50 µm
e
50 µm
b
Magn
105x
200 µm
f
Magn
350x
100 µm
c
Magn
163x
200 µm
g
V
+
100 µm
d
200 µm
h
100 µm

Figure 2 Scanning electron microscopy of normal (a–d) and tumor microvascular corrosion casts (e–h). **(a)** The human colonic mucosal capillary plexus (*c*) is arranged in a honeycomb pattern and shows numerous intercapillary connections. The supplying arterioles (*a*) and draining veins (*v*) take a straight course from the underlying submucosal vessels. **(b)** Mouse kidney vessels with cortical vessels and glomeruli (*lower left and inset*) and parallel-oriented medullary vessels (*upper part*). **(c)** Subcutaneous vessel plexus on the fascia with undulating vessel courses. **(d)** Hairpin loops of skin capillaries oriented along the dermal papillae (sheep ear skin). **(e)** Loss of original vessel architecture in human colonic carcinoma (pT3, pNo, pMx; G2). Note the heterogeneous distribution of the sinusoids and the loss of hierarchy. **(f)** Murine RenCa-tumor sinusoids with varying diameters and numerous blind ends. **(g)** Xenografted sarcoma with compressed main veins (*v*) and avascular areas (+) next to hot spots. **(h)** Tumor vascular envelope in a xenografted squamous cell carcinoma.

Frequently xenografted tumors induce the formation of a vascular envelope of sprouting and preexistent host vessels embedded in connective tissue, in which a certain hierarchy is retained (Figure 2h). This is not true for human primary tumors since the invasive growth prevents the formation of such a vascular and/or fibrous tissue capsule.

Despite the common features, which are expressed to different extents, the architecture of tumor vasculature is tumor-type specific. This is in contrast to the structure of the tumor vascular cells. Morphometry of parameters determining the architecture of the microvascular unit such as intervessel and interbranch distances as well as diameter and variability of diameter prove that individual tumor entities express characteristic vascular patterns. The inherent architecture of the tumor seems to be primarily determined by the tumor cells themselves; experiments involving transfected tumor cells with different capacity to produce and release FGFII showed significant differences in tumor growth, but not in microvascular architecture.

Last, it should be pointed that the features seen in primary tumor vascularity can be seen qualitatively as well in precancerous lesions—although to a lesser extent—long before the transformation to the malignant phenotype.

Glossary

Microvascular corrosion casting: Replication of vessel systems by injection of low-viscosity casting media and digestion of all surrounding tissues, allowing for scanning electron microscopic examination.

Microvascular unit: Capillary bed segment fed by an individual metarteriole.

Tumor sinusoids: Relatively undifferentiated vessels with capillary wall building and increased diameter.

Vascular mimicry: Nonendothelial cells lining vessels or perfused channels.

Further Reading

Burri, P. H., and Tarek, M. R. (1990). A novel mechanism of capillary growth in the rat pulmonary microcirculation. *Anat. Rec.* **228**, 35–45. *The first paper describing intussusceptive angiogenesis.*

Cliff, W. J. (1981). Endothelial structure and ultrastructure during growth, development and aging. In *Structure and Function of the Circulation*, C. J. Schwartz and N. T. Werthessen, eds., Vol. 2, pp. 695–718. New York: Plenum Press.

Hammersen, F., Endrich, B., and Messmer, K. (1985). The fine structure of tumor blood vessels. I. Participation of nonendothelial cells in tumor angiogenesis. *Int. J. Microcirc. Clin. Exp.* **4**, 31–43. *The first article proving the involvement of tumor cells in angiogenesis.*

Konerding, M. A., Fait, E., and Gaumann, A. (2001). 3D microvascular architecture of pre-cancerous lesions and invasive carcinomas of the colon. *Br. J. Cancer* **84**, 1354–1362.

McDonald, D. M., and Choyke, P. L. (2003). Imaging of angiogenesis: From microscope to clinic. *Nat. Med.* **9**, 713–725. *Comprehensive review.*

Capsule Biography

Dr. Konerding has been Professor of Anatomy at the Institute of Anatomy in Mainz, Germany, since 1993. His group focuses on secondary angiogenesis and antiangiogenesis as well as on clinical anatomy. His work is mainly supported by grants of the DFG and EC.

CHAPTER 142

Endothelial Cell Heterogeneity Targeting the Tumor Vasculature

Robert R. Langley

Department of Cancer Biology, The University of Texas M.D. Anderson Cancer Center, Houston, Texas

The growth and dissemination of malignant tumors are dependent on a blood supply. In response to increasing metabolic pressures, tumors enhance their vascular supply by inducing resident endothelial cells to form new vascular networks (angiogenesis), by mobilizing populations of endothelial precursor cells to the tumor site (vasculogenesis), or by modifying the structural architecture of preexisting blood vessels (vascular remodeling). Understanding the molecular mechanisms that regulate these complex processes in different organs is essential for developing antivascular therapy.

Introduction

In his 1945 report examining the vascular response to tumor implantation, Algire concluded "an outstanding characteristic of the tumor cell is its capacity to elicit continuously the growth of new capillary endothelium from the host" [1]. Some 25 years later, enough supportive evidence had accumulated to advance the hypothesis that the progressive growth of tumors is, indeed, dependent upon the induction of angiogenesis. Since that time, considerable effort has been extended toward defining the molecular mechanisms that regulate tumor neovascularization and characterizing the phenotype of tumor blood vessels. Targeting of tumor-associated endothelium is an attractive approach to control tumor growth in that the endothelial component of the tumor is considered to be genetically stable and, therefore, not prone to develop resistance to therapy. Moreover, the tumor endothelium provides an accessible target as compared to malignant cell populations embedded within the underlying tissue parenchyma. However, although a number of angiogenic inhibitors have advanced from the laboratory into the clinical setting, their success in the patient population has, to date, been less than remarkable. One potential explanation for the limited therapeutic response is that an extensive body of evidence clearly indicates that the microcirculation of different tissues exhibits diversity at the structural, molecular, and functional levels. Determining which angiogenic factors are specific for the vessels associated with a given tumor is central for designing appropriate therapeutic regimes.

Emerging evidence indicates that hematopoietic stem cells (HSCs) and endothelial progenitor cells (EPCs) that originate from either the bone marrow or other tissues may contribute to the neovascularization of tumors. HSCs and EPCs have also been found in newly developed blood vessels associated with wound healing, limb ischemia, atherosclerosis, and post–myocardial infarction. Preliminary experimental studies indicate that the homing properties of precursor cells may be exploited to produce therapy of tumors. Considerable effort is now directed toward determining the factors that mobilize and direct these cells to sites of neovascularization.

The extent to which tumors are dependent on the process of angiogenesis varies among different types of tumors. Some neoplasms, such as certain nonsmall-cell lung tumors, have been shown to meet their metabolic requirements by proliferating along the surface of preexisting blood vessels and, thus, are independent of angiogenesis. Therefore, this form of tumor growth could be impervious to pharmacologic interventions that are directed against dividing endothelial cell populations. Perhaps a more effective approach for the management of this type of tumor growth would be to deliver therapy in a regional manner by targeting tissue-specific receptors that are present on the vascular endothelium and, thereby, alleviate much of the systemic toxicity associated with chemotherapy.

Angiogenesis

Angiogenesis refers to the development of new blood vessels from the existing vascular bed. During embryonic development, angiogenesis provides ectoderm- and mesoderm-derived tissues, including the brain and kidney, with a vascular supply. In the adult, angiogenic events occur less frequently and are primarily restricted to tightly regulated processes that occur during the female reproductive cycle (formation of corpus luteum and placenta) and in response to wound healing. However, angiogenesis does play a prominent role in the progression of several pathologic conditions including arthritis, diabetic retinopathy, psoriasis, and tumor growth. In contrast to the transient highly regulated vascular development that occurs during normal tissue growth, the angiogenesis associated with the aforementioned disease states is persistent and unrelenting.

The onset of angiogenesis is presumed to be the result of an imbalance between inhibitor and stimulator molecules. Normal tissues are exposed to an excess of inhibitor molecules and, consequently, the vascular endothelium remains in a quiescent state. Studies examining the kinetics of endothelial cell proliferation in several normal tissues estimate that the turnover time of endothelial cells can be measured in years. Tumors may disrupt the equilibrium between inhibitor and stimulator molecules, however, by inducing alterations in the local microenvironment, by downregulating tumor-suppressor genes, or through the activation of oncogenes. Microenvironmental changes appear to result from increasing metabolic demands by an expanding tumor cell population that are not satisfactorily met by available blood delivery. Insufficient blood flow, in turn, leaves resident tumor cells exposed to an inhospitable milieu that is characteristically devoid of oxygen and nutrients and rich in metabolic waste products. For some tumors, such as carcinomas of the head and neck, the decline in oxygen tension functions as a selection pressure that leads to the proliferation of cells with enhanced metastatic potential. However, for tumor cells subjected to severe hypoxia, the anaerobic environment may have a detrimental effect on tumor cell survival. Indeed, Fidler and colleagues have demonstrated that if a tumor cell resides more than 170 μm from a vascular supply, there is a strong probability that this cell's destiny is one of death (Figure 1).

Tumor Blood Flow

Studies conducted in experimental animal tumors indicate that insufficient blood delivery is a common feature of a number of tumor types. For some, the reduction in blood flow can be profound when compared to measurements obtained in adjacent, nonaffected tissue. For example, measurements of tumor blood flow in ovarian implants show that blood flow rates in these tumors are some 50 times less than that observed in normal ovarian tissue. In rodent models, blood flow to tumors continues to decline as the size of the tumor increases, irrespective of the intensity of the angiogenic response. Many tumors display a progressive rarefaction of the vascular bed (*i.e.*, a decrease in the number of patent vessels per gram of tissue). Because the rate of neovascularization is considerably reduced in comparison to tumor cell proliferation, the vascular space becomes a progressively smaller component of the total mass. To determine how the vascular surface area of experimental tumors compared to that of normal tissues, we injected a radiolabeled monoclonal antibody directed against CD31 into the systemic circulation of mice harboring subcutaneous tumors and compared antibody binding on tumor vessels to uptake on vessels supplying other tissues (Figure 2). The results show that the tumor vascular surface area is a fraction of that found in normal tissues.

Hypoxia Inducible Factor

Recent studies indicate that many tumors rely on the same hypoxia-sensing mechanism utilized by normal cells during conditions of low oxygen availability to activate genes associated with angiogenesis. Specifically, hypoxia inducible factor-1 (HIF-1) has been shown by *in situ* hybridization to be expressed in hypoxic regions of tumors, and overexpression has been documented in several primary human cancers and their metastases. HIF is composed of an $\alpha\beta$ heterodimer in which the α subunits are inducible by hypoxia, whereas the β subunits are constitutively expressed nuclear proteins. Declining oxygen tension stabilizes HIF-1α and promotes its accumulation in the cell nucleus, where it heterodimerizes with HIF-1β and initiates the transcription of a number of genes implicated in angiogenesis, invasion, and metastasis.

One of the most intensely studied genes targeted by HIF activation is vascular endothelial cell growth factor/vascular permeability factor (VEGF/VPF). Although originally characterized based on its ability to induce protein extravasation from tumor vessels, VEGF has since been determined to be essential for the establishment of new blood vessels that are associated with embryonic development and maturation of the ovarian follicle, as well as a number of pathologic processes. At present, five VEGF ligands (VEGFs A–E) have been identified that possess binding potential for three distinct VEGF receptors: VEGFR1 (Flt-1), VEGFR2 (KDR/Flk-1), and VEGFR3 (Flt-4). Of the VEGF ligands, the various VEGFA isoforms appear to be key mediators of new blood vessel growth, whereas VEGFC and VEGFD have been implicated in lymphatic angiogenesis.

VEGF can elicit cell migration, protease production, and proliferation necessary for angiogenesis. VEGF also

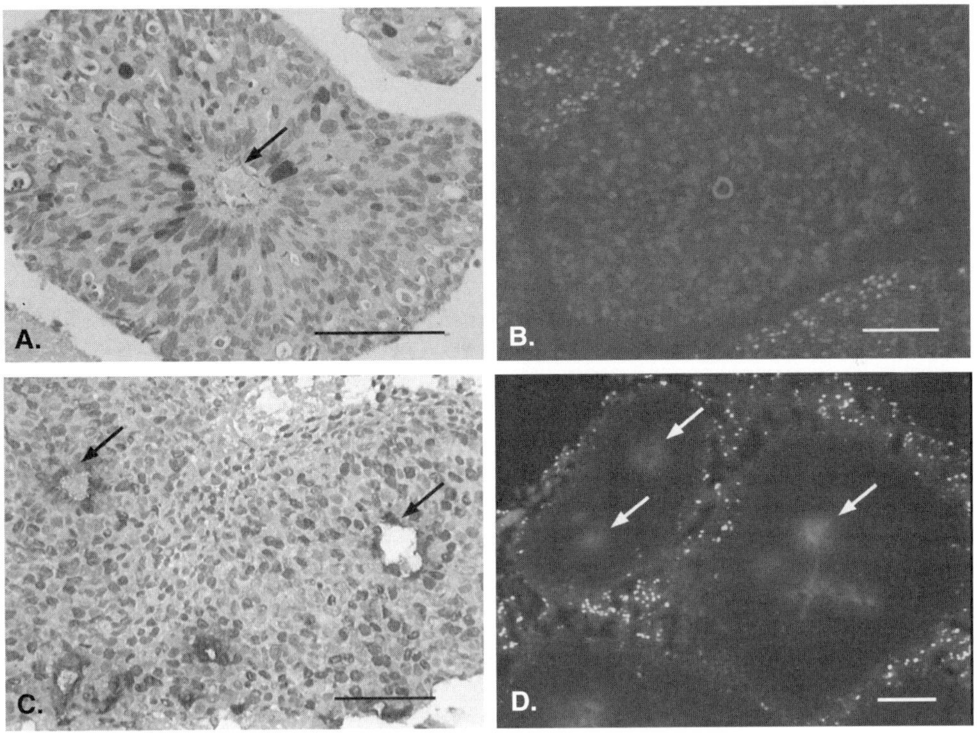

Figure 1 Location of dividing and apoptotic tumor cells in relation to blood vessels in brain metastases. **(a)** Autochthonous human lung-cancer brain metastases. BrdU-positive cells (cells engaged in cell division) are stained red and are located within 75 μm of the nearest blood vessel (*arrow*). **(b)** Autochthonous human lung-cancer brain metastasis. TUNEL-positive (apoptotic) cells stained bright green and are positioned 160–170 μm from the nearest blood vessel. The vessel is marked in red. **(c)** Human KM12C metastasis in the brain of nude mice. BrdU-positive nuclei stained red. Arrow points to blood vessel stained for CD31 (brown). **(d)** Human KM12C metastasis in the brain of nude mice. CD31 staining (red) identified blood vessels and bright green nuclei were TUNEL-positive. Bar = 100 μm. Reproduced with permission from Fidler et al. (2002). *Lancet Oncology* **3**, 53–57. (see color insert)

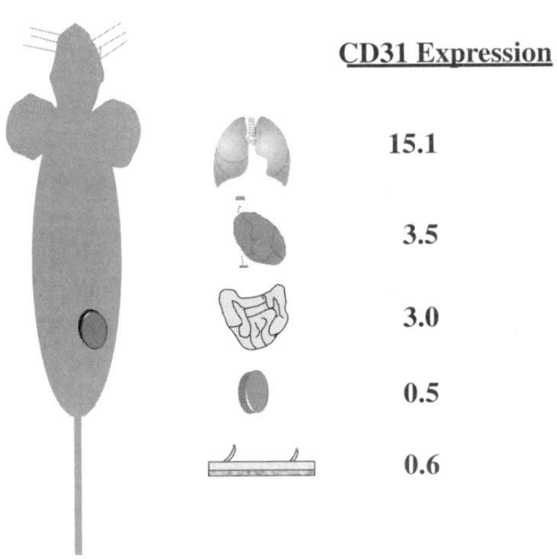

Figure 2 CD31 expression levels in lung, heart, small intestine, tumor, and skin of C57Bl/6 mice. Radiolabeled monoclonal antibody (mAb) directed against CD31 was injected intravenously into mice harboring subcutaneous murine RM1 tumors, and binding on tumor blood vessels was compared to values obtained from blood vessels supplying normal tissues. Values are expressed as μg mAb/g tissue. This technique has been shown to provide a reliable index of microvascular density in different tissues. Modified from Langley et al. (1999). *Am. J. Physiol.* **277**, H1156–H1166.

functions as a survival factor for endothelial cells by upregulating the phosphatidylinositol-3 OH kinase/Akt signal transduction pathway and enhancing the expression of the antiapoptotic proteins Bcl-2 and A1. Overexpression of VEGF is a hallmark of a variety of human tumors, and inhibition of Flk-1 signaling using several different approaches can significantly impair tumor growth in experimental models. Considering the important role that VEGF and its receptors play in the neovascularization process of malignant tumors, it is not surprising that this signaling cascade is at the forefront of antiangiogenic research efforts.

Also included among the genes induced by HIF-1α activation are those that encode for the polypeptide chains of platelet-derived growth factor (PDGF). To date, four PDGF polypeptide chains have been identified that combine to form five PDGF isoforms: PDGF-AA, -AB, -BB, -CC, and -DD. These isoforms exert their effects by activating two protein tyrosine kinase receptors, denoted the α-receptor and β-receptor. Although each of the different PDGF isoforms has been shown to be capable of contributing to tumor neovascularization, their specific role in this process appears to be dependent upon the location of the tumor. Indeed, the results obtained from studies examining PDGF-R signaling cascades in different experimental tumors underscore the diverse roles that tyrosine kinase receptors may assume in different vascular beds. For example, in pancreatic tumors, PDGF signaling has been shown to provide integrity to developing vascular networks by recruiting mural cells to support the immature vessel wall. In gliomas, PDGF BB promotes the neovascularization of cerebral lesions, in part, by stimulating the release of VEGF from tumor-associated endothelium. Tumors growing in the subcutaneous space, however, rely on PDGF BB to regulate the level of interstitial fluid pressure within the tumor. In the bone microenvironment, activation of PDGF-R on tumor-associated vessels has been shown to be essential for progression of androgen-independent prostate tumors. With the advent of small-molecule inhibitors that selectively target PDGF-R, such as STI571 (Glivec), much attention has been directed toward determining whether other malignancies are critically dependent on this tyrosine kinase.

Role of the Microenvironment and Angiogenesis

The anatomical location of a tumor may be an important determinant in regulating the phenotype of its vasculature. Unfortunately, many preclinical studies neglect to account for influence of the microenvironment on tumor growth and vascularization by utilizing models in which different types of tumors are all implanted into the subcutaneous space. One may argue that, with the exception of those tumors originating from or metastasizing to skin, subcutaneously implanted tumors provide relatively little useful information regarding antiangiogenic intervention. This is perhaps true in that this model evaluates only one type of endothelial cell (dermal) and, equally important, because frequently the tumor is located in an ectopic site. Recently, Blouw and coworkers [2] utilized an elegant strategy involving HIF-1α-deficient transformed astrocytes to illustrate how HIF-1α can act as either a negative or a positive regulator of astrocytoma progression, depending on the microenvironment in which the tumor is located. Specifically, the loss of HIF-1α impaired the growth of astrocytomas in the subcutis, but not in the brain. This finding was attributed to the inability of HIF-1α-deficient cells to mobilize VEGF that, consequently, blocked their capacity to recruit new vessels in the inherently vessel-poor subcutaneous compartment (see Figure 2), whereas the same HIF-1α-deficient cells exploited the preexisting rich vascular networks of the brain. The results of this study are noteworthy in several respects. First, had the examination been conducted solely within the confines of the subcutaneous space, one may have reached the conclusion, albeit an incorrect one, that targeting HIF-1α may be an effective approach to control the growth of glioblastomas. In addition, recently there has been an appreciable interest in the development of pharmaceuticals capable of targeting HIF-1α in tumors. If this interest is extended to the production phase, it would be very informative to determine whether agents directed against HIF-1α possess efficacy in the treatment of cancer metastases located in lung, brain, bone, or liver tissue, which are well-vascularized tissues.

Studies evaluating angiogenesis in human tumors also conclude that there is considerable heterogeneity regarding the intensity of endothelial cell proliferation among different types of tumors. Eberhard and colleagues utilized a double-labeling immunohistochemical approach to measure proliferating endothelial cells in a broad panel of human tumors. The data from this study indicate that glioblastomas and renal cell carcinomas possess the highest number of capillary beds with evidence of endothelial cell proliferation (9.6% and 9.4%, respectively), whereas lung and prostate tumors possess the fewest (2.6% and 2.0%, respectively). These results caution against accepting rapidly dividing murine tumors that grow exponentially over a period of a few weeks as the model for the more measured expansion that takes place in human malignancies.

Recent improvements in the ability to selectively isolate and propagate endothelial cells from different organs can now permit more detailed examinations into the tissue-specific factors that regulate angiogenesis. Indeed, our laboratory has applied a series of multicolor flow cytometry selection strategies that target inducible endothelial cell adhesion molecules to *H-2K^b-tsA58* murine tissues in order to generate microvascular endothelial cell lines from a number of different organs [3]. Cells derived from *H-2K^b-tsA58* mice all harbor a temperature-sensitive SV40 large T antigen and, thus, permit detailed molecular examinations on both activated and differentiated phenotypes of endothelial cells. Preliminary studies conducted on several endothelial cell lines reveal that each possesses distinct growth factor profiles. For example, we have noted that for uterine-derived endothelial cells, the most potent endothelial cell mitogen is epidermal growth factor (EGF). However, for endothelial cells originating from bone tissue, basic fibroblast growth factor (bFGF) elicits the most robust increase in proliferation (unpublished data). Similarly, a recent study conducted on microvascular endothelial cells from the human intestine has determined that interleukin-8 can promote both cell division and migration. cDNA expression arrays have also been utilized to construct gene expression profiles on human endothelial cells obtained from brain, lung, and lymphatic tissue. Additional reports have provided evidence that some tissues elaborate organ-restricted endothelial cell mitogens and that, moreover, these growth factors can be detected in tumors arising from these anatomic regions.

Targeting Angiogenesis in Tumors

Kerbel and Folkman have proposed a classification scheme that places agents that target tumor neovascularization into two classes. Direct inhibitors, such as endostatin, angiostatin, and vitaxin, function by preventing endothelial cell proliferation, migration, and initiation of antiapoptotic programs in response to various angiogenic proteins. Indirect inhibitors, on the other hand, block production of protein products (bFGF, VEGF) from tumor cells or interfere with endothelial cell receptor tyrosine kinase signaling cascades. Perhaps one of the most valuable lessons learned from the preclinical evaluations of these agents is that their effectiveness at limiting tumor growth is much more pronounced if they are administered in combination with chemotherapy or radiotherapy.

One of the most active areas of antiangiogenic investigation has centered on the VEGF system. Although several of the small-molecule inhibitors of VEGFR2, such as SU5416 and SU6668, demonstrated efficacy in limiting human tumor xenograft growth in mice, these agents were found to produce unacceptable toxicity profiles in humans that has halted their development. Adverse side effects observed in patients receiving these agents have included pulmonary emboli, myocardial infarction, and cerebrovascular events. rhuMAB VEGF (avastin) is a recombinant humanized monoclonal antibody to VEGF that has undergone initial evaluation in the treatment in two different types of human tumors. In a Phase II study conducted with patients harboring stage IIIB/IV nonsmall-cell lung cancer, individuals were randomized to standard chemotherapy with carboplatin and paclitaxel alone or in combination with either 7.5 or 15 mg/kg rhuMAB. Although the response rates were slightly increased in the group receiving chemotherapy and high-dose rhuMAB, median survival was only improved by 3 months in this group. A randomized trial employing rhuMAB has also been completed in a group of 116 patients with metastatic clear-cell renal carcinoma [4]. Interestingly, in this study design, patients were assigned to groups receiving placebo or 3 or 10 mg/kg rhuMAB that was administered every 2 weeks. Clear-cell renal carcinoma is perhaps an ideal candidate for VEGF therapy in that the von Hippel–Lindau tumor suppressor gene is usually mutated in this cancer and, as a result, VEGF is overproduced because of a mechanism involving HIF1-α. Unfortunately, monotherapy with rhuMAB failed to demonstrate an increase in survival in this patient population.

Many preclinical studies are based on rapidly growing subcutaneous neoplasms. Moreover, treatment of human tumor xenograft models is usually initiated at early time points. This is in sharp contrast to the clinical setting, where patients usually present with advanced disease and are eligible to receive antiangiogenic therapies only after failing one or more conventional treatment modalities. In addition, because tumors are made up of heterogeneous populations of cells, treatment with only one angiogenic agent may result in the selective expansion of cells expressing different proangiogenic proteins.

Although there remain many unanswered questions regarding which angiogenic inhibitor is appropriate for a given tumor, there have been some recent advances in determining which signaling cascades are operative in some of the preferred sites of metastasis. The bone is the fourth most common site of tumor metastasis, and recent estimates predict that approximately 350,000 cancer patients die each year with evidence of tumor in the skeleton. Consequently, much effort has been extended toward identifying those components in the bone microenvironment that support tumor growth. Recently, members of our laboratory evaluated the efficacy of STI571, a small-molecule inhibitor of the PDGF-R tyrosine kinase, in a human tumor xenograft model of androgen-independent prostate cancer implanted into the bone [5]. In that model, inhibition of PDGF-R activation with STI571, when administered in combination with taxotere, produced a profound reduction in both tumor mass and lymphatic metastases. It was determined that the effectiveness of the combination therapy was directly related to apoptosis of tumor-associated endothelial cells. The promising results obtained in this study have prompted the initiation of a clinical trial to evaluate the efficacy of this treatment regime in men with refractory prostate cancer bone metastasis. In addition, these findings may have important implications for other cancers that metastasize to bone such as renal, breast, and lung tumors. However, it will be necessary to determine whether these tumors are operating through a PDGF-dependent pathway. Indeed, a recent report has shown that renal cell carcinoma may preferentially exploit EGF signaling cascades in the bone microenvironment in order to promote their vascularization and ensure their growth [6].

Paracrine EGF signaling from tumor cells to microvascular endothelial cells has also been shown to be an important component for the growth of pancreatic metastases in the liver. Using either a monoclonal antibody blocking strategy (C225) or small molecule inhibitor of EGFR signaling in combination with gemcitabine, Bruns and coworkers were able to dramatically reduce both primary pancreatic tumor growth and liver metastatic burden. In both tumor models, combination therapy was associated with downregulation of VEGF in the tumor region and apoptosis of tumor-associated endothelial cells. Independent studies have also identified VEGF as an important proangiogenic cytokine in the liver. Reports examining liver regeneration in rats following partial hepatectomy have shown that VEGF plays a critical role in the revascularization process by stimulating the proliferation of hepatic sinusoidal endothelial cells. VEGF also appears to play a role in colon cancer liver metastasis, as blockade of either VEGF or VEGFR2 in a human tumor xenograft model results in the induction of both tumor cell and endothelial cell apoptosis [7].

One of the most elusive organs for targeted treatment of metastasis is the brain. Cerebral blood vessels possess highly resistant tight junctions that are further reinforced by the end feet processes of astrocytes. In addition, this highly specialized vascular network is also enriched with a number of transport proteins, such as the multidrug-resistant protein P-glycoprotein, which restrict the passage of several chemotherapeutic agents into the tissue parenchyma. In fact, it has been shown that P-glycoprotein can also mediate the efflux of small-molecule inhibitors such as Glivec. Adding to the complexity of the cerebral vasculature, Fidler and colleagues demonstrated that neovascularization of metastases in the brain occurs by a mechanism that is distinct from traditional sprouting angiogenesis. In brain metastasis, blood vessel expansion occurs by the insertion of dividing endothelial cells into the preexisting blood vessel, a process that is referred to as angioectasia. In addition, because of the detrimental consequences that ensue upon cessation of cerebral blood flow, it is likely that there is considerable redundancy in angiogenic proteins in this anatomic compartment.

Tumor Vasculogenesis

Although most of the examinations into the neovascularization of tumors have focused on angiogenesis, new evidence indicates that hematopoietic stem cells (HSCs) and endothelial precursor cells (EPCs) may also play an important role in this process. Asahara et al. [8] were the first to demonstrate the EPCs could be isolated from human peripheral blood and incorporate themselves into active areas of angiogenesis in animal models of ischemia. EPCs may be differentiated from mature circulating endothelial cells that have been shed from the vascular wall by their significantly enhanced capacity to proliferate, and also by their unique expression of cell-surface markers that include VEGFR2, AC133, CXCR4, and CD146. Recent studies are beginning to provide insight into the mechanisms that facilitate the mobilization of HSCs and EPCs from the bone marrow and to sites of neovascularization. The recruitment process appears to be initiated by angiogenic products that are released from tumor cells and lead to the activation and secretion of matrix metalloproteinase-9 (MMP-9) by hematopoietic and stromal elements in the bone marrow. MMP-9 activation, in turn, leads to liberation of soluble KIT ligand that promotes cell proliferation and also directs the transfer of these cells into the peripheral circulation.

Direct evidence that HSCs and EPCs contribute to tumor neovascularization has come from examinations conducted in mice that are deficient in Id proteins. Mutant mice with the $Id1^{+/-}Id3^{-/-}$ phenotype possess defective angiogenic responses and, thus, are unable to support tumor growth. However, when these mice are transplanted with wild-type bone marrow or HSCs, tumor growth in the subcutaneous compartment can be restored. This process appears to require cooperation of both VEGFR1 and VEGFR2 in that treatment of $Id1^{+/-}Id3^{-/-}$ mice with neutralizing antibodies against both of these receptors results in vascular disruption and tumor cell death. The contribution of HSCs and EPCs to the tumor vasculature appears to be influenced by the tumor type, but most reports suggest that these cells represent a small fraction (6 to 10%) of tumor-associated blood vessels, and that they arrive at tumor vessels during the initial phase of malignant growth. Despite the relatively low number of progenitor cells that lend support to tumor vessels, evidence suggest that their ability to traffic to tumor sites may be exploited for therapeutic targeting. Genetically modified endothelial progenitors that were stably transfected with thymidine kinase or the soluble truncated form of VEGFR2 have been shown to home to subcutaneous tumors and significantly impair tumor growth. Although these initial reports are promising, more studies are needed to determine which types of malignant lesions are critically dependent upon progenitor cells.

Conclusion

Over the past decade, the understanding of the cellular and molecular mechanisms that tumors utilize to promote their blood delivery and facilitate their growth has advanced. However, therapies that are directed toward the vascular compartment of malignant lesions are still in the early developmental phase. Continued investigations into the specific factors that regulate angiogenesis in different organs, coupled with detailed characterization of the products released by tumors in different tissues, should improve targeted therapy of angiogenesis. Similarly, an enhanced understanding of the biology of HSCs and EPCs could provide clinicians with a superior method of delivering therapy to tumors. The identification of tissue-specific receptors, on both normal and tumor-associated endothelial cells, will also enhance the development of vascular-based therapies. Employing orthotopic tumor models that more closely approximate the clinical reality could well facilitate translation of laboratory findings to the clinic.

Glossary

Angioectasia: Blood vessel dilation that is due to endothelial cell division occurring within the wall of the blood vessel. This process is distinct from angiogenesis and has been reported in brain metastasis.

Angiogenesis: Refers to the generation of new vascular networks from preexisting blood vessels. During this process, endothelial cells elaborate proteolytic enzymes, exhibit migratory behavior, and undergo cell division. Ultimately, developing capillary sprouts will coalesce to form new vascular structures.

Endothelium: Specialized simple squamous epithelium that lines the intimal surface of all blood vessels.

Acknowledgment

The author thanks Dr. I. J. Fidler for helpful discussions and editorial assistance.

References

1. Algire, G. H., Chalkley, H. W., Legallais, F. Y., and Park, H. D. (1945). Vascular reactions of normal and malignant tumors in vivo. I. Vascular reactions of mice to wounds and to normal and neoplastic transplants. *J. Natl. Cancer Inst.* **6**, 73–85.
2. Blouw, B., Song, H., Tihan, T., Bosze, J., Ferrara, N., Gerber, H. P., Johnson, R. S., and Bergers, G. (2003). The hypoxic response of tumors is dependent on their microenvironment. *Cancer Cell* **4**, 133–146.

3. Langley, R. R., Ramirez, K. M., Tsan, R. Z., Van Arsdall, M., Nilsson, M. B., and Fidler, I. J. (2003). Tissue-specific microvascular endothelial cell lines from H-2K(b)-tsA58 mice for studies of angiogenesis and metastasis. *Cancer Res.* **63**, 2971–2976.
4. Yang, J. C., Haworth, L., Sherry, R. M., Hwu, P., Schwartzentrube, D. J., Topalian, S. L., Steinberg, S. M., Chen, H.X., and Rosenberg, S. A. (2003). A randomized trial of bevacizumab, an anti-vascular endothelial growth factor antibody, for metastatic renal cancer. *N. Engl. J. Med.* **349**, 427–434.
5. Uehara, H., Kim, S. J., Karashima, T., Shepherd, D. L., Fan, D., Tsan, R., Killion, J. J., Logothetis, C., Mathew, P., and Fidler, I. J. (2003). Effects of blocking platelet-derived growth factor-receptor signaling in a mouse model of experimental prostate cancer bone metastases. *J. Natl. Cancer Inst.* **95**, 458–470.
6. Weber, K. L., Doucet, M., Price, J. E., Baker, C., Kim, S. J., and Fidler, I. J. (2003). Blockade of epidermal growth factor receptor signaling leads to inhibition of renal cell carcinoma growth in the bone of nude mice. *Cancer Res.* **63**, 2940–2947.
7. Shaheen, R. M., Davis, D. W., Liu, W., Zebrowski, B. K., Wilson, M. R., Bucana, C. D., McConkey, D. J., McMahon, G., and Ellis, L. M. (1999). Antiangiogenic therapy targeting the tyrosine kinase receptor for vascular endothelial growth factor receptor inhibits the growth of colon cancer liver metastasis and induces tumor and endothelial cell apoptosis. *Cancer Res.* **59**, 5412–5416.
8. Asahara, T., Murohara, T., Sullivan, A., Silver, M., van der Zee, R., Li, T., Witzenbichler, B., Schatteman, G., and Isner, J. M. (1997). Isolation of putative progenitor endothelial cells for angiogenesis. *Science* **275**, 964–967.

Further Reading

Arap, W., Pasqualini, R., and Ruoslahti, E. (1998). Cancer treatment by targeted drug delivery to tumor vasculature in a mouse model. *Science* **279**, 377–380.

Dvorak, H. F. (2003). How tumors make bad blood vessels and stroma. *Am. J. Pathol.* **162**, 1747–1757.

Ferrara, N., Gerber, H. P., and LeCouter, J. (2003). The biology of VEGF and its receptors. *Nat. Med.* **9**, 669–676.

Fidler, I. J., Uano, S., Zhang, R., Fujimaki, T., and Bucana, C. D. (2002). The seed and soil hypothesis: Vascularisation and brain metastases. *Lancet Oncol.* **3**, 53–57.

Hood, J. C., Bednarski, M., Frausto, R., Guccione, S., Reisfeld, R. A., Xiang, R., and Cherish, D. A. (2002). Tumor regression by targeted gene delivery to the neovasculature. *Science* **296**, 2404–2407. *The authors utilize nanosphere-assisted targeting of the neovasculature with mutant Raf-1 to induce apoptosis of tumor-associated endothelium and promote regression of established primary tumors and metastatic lesions.*

McIntosh, D. P., Tan, X. Y., Oh, P., and Schnitzer, J. E. (2002). Targeting endothelium and its dynamic caveolae for tissue-specific transcytosis in vivo: A pathway to overcome cell barriers to drug and gene delivery. *Proc. Natl. Acad. Sci. USA* **99**, 1996–2001. *Study demonstrates that the molecular composition of lung microvascular caveolae is distinct from caveolae found in other regional circulations and that, moreover, this discriminating feature can be exploited to selectively transport immunotoxin to the lung.*

Pugh, C. W., and Ratcliffe, P. J. (2003). Regulation of angiogenesis by hypoxia: Role of the HIF system. *Nat. Med.* **9**, 677–684.

Raffi, S., Lyden, D., Benezra, R., Hattori, K., and Heissig, B. (2002). Vascular and haematopoietic stem cells: Novel targets for anti-angiogenesis therapy? *Nat. Rev. Cancer* **2**, 826–835. *This important review discusses contribution of HSCs to tumor vasculature and the therapeutic potential of precursor cells for targeting tumor-associated blood vessels.*

Capsule Biography

Robert R. Langley received his Ph.D. from the Department of Molecular and Cellular Physiology, Louisiana Health Sciences Center, Shreveport, Louisiana, in 1999. He received postdoctoral training in the Department of Cancer Biology at the University of Texas M.D. Anderson Cancer Center in Houston, Texas, from 2000 to 2003. He is currently serving as an Instructor in the Department of Cancer Biology and examining the contribution of endothelial cells to the metastatic process.

A Metronomic Approach to Antiangiogenesis

Raffaele Longo and Giampietro Gasparini
Division of Medical Oncology, "S. Filippo Neri" Hospital, Rome, Italy

Introduction

Chemotherapy has been the mainstay of medical approaches to the treatment of solid neoplasia. It is usually given in the form of bolus infusion at maximum tolerated doses (MTDs) with the goal of complete tumor kill. With the exception of a few tumors of the adult, such as lymphoid, germ cell, and some pediatric cancers, however, eradication of advanced cancer has been elusive, even with high doses and autologous bone-marrow rescue [1]. It was suggested that the rest periods between each cycle of therapy provide the tumor endothelium an opportunity to repair the chemotherapy-induced damages. The harsh side effects and the ultimate failures of conventional chemotherapy fueled broad investigation of therapeutic alternatives, including drugs targeting not only tumor cells, but also genetically stable cells of tumor stroma, such as endothelial cells (ECs). The emerging paradigm is that patient survival is not incompatible with tumor persistence. A therapy aimed at making the cancer a chronic disease, with tumor burden held at the lowest achievable volume, may prove to be a more appropriate therapeutic strategy for human solid tumors. (Figure 1)

Metronomic Chemotherapy

Based on the results of experimental studies [2, 3], Hanahan et al. [4] proposed the term *metronomic chemotherapy* for schedules of cytotoxic agents given regularly at subcytotoxic doses and with suppression of the "activated" endothelium as the principal target (i.e., the antiangiogenesis chemotherapy paradigm).

Angiogenesis is necessary to sustain the growth of primary tumor and metastases. Metronomic schedules are more effective in targeting tumor "activated" ECs than large single high-dose bolus doses followed by long rest periods [2], because intratumor ECs, in contrast to quiescent mature ECs of normal adult tissues, proliferate rapidly and are more vulnerable to cytotoxic agents [5]. However, the rest period between cycles of conventional chemotherapy permits the survival and regrowth of a fraction of ECs, allowing tumor angiogenesis to persist [5]. Indeed, other functions such as EC motility, invasion, and vessel remodeling may be blocked or altered by metronomic chemotherapy [5]. Suppression of mobilization of bone marrow–derived EC progenitors to sites of angiogenesis is another possibility [2, 3]. Tumor-cell heterogeneity may allow the coexistence of cells having different sensitivities to the same therapeutic agent, and the ability of those cells to shift among the different sensitivity compartments over time, with a "resensitization" effect that may best be exploited using metronomic chemotherapy.

A further theoretical advantage of metronomic chemotherapy is that it minimizes the toxic effects, allowing combinations of potentially synergistic selective inhibitors of angiogenesis [3, 5].

However, a phenomenon that may limit the advantages of low-dose metronomic or continuous-dose delivery is a threshold effect for drug activity. Specifically, there may exist a minimum concentration of drug below which no tumor inhibition takes place [1].

Recently, Miller et al. defined the criteria for antiangiogenic activity of cytotoxic agents: (a) camptothecin analogs, vinca alkaloids, and taxanes are active against ECs at doses lower than those required for tumor-cell cytotoxicity; (b) cisplatin, cyclophosphamide, methotrexate, and doxorubicin interfere with EC function without cytotoxic effects; and (c) purine analogues, cisplatin, and anthracyclines block specific steps of the angiogenic cascade [5].

Experimental Studies

Recent studies have demonstrated that the antitumor activity of metronomic schedules of chemotherapy is mediated through an antiangiogenic effect [5].

Browder et al. compared cyclophosphamide given by a metronomic schedule with conventional single, high-dose chemotherapy [2]. The metronomic schedule was more effective in eradicating Lewis lung carcinoma and L1210 leukemia. When the drug was given at the MTD, which requires long periods of rest to allow bone marrow recovery, apoptosis of ECs in the tumor microvasculature was observed, but this damage was largely repaired during the rest periods between the cycles [2].

However, if the same drug was administered chronically on a once a week schedule, without long breaks, at a lower dose, the repair process was compromised and the antiangiogenic effects of the drug were enhanced. It was also observed that tumors resistant to conventional dosage of cyclophosphamide responded to the metronomic schedule. Reversal of acquired resistance was attributed to the shifting of the target from the cancer cell to the genetically stable and sensitive ECs.

Similarly, in human orthotopic breast or ectopic colon cancer xenografts in nude or SCID mice, cyclophosphamide given PO at low doses through drinking water showed a relevant antitumor effect, particularly when used in combination with an anti-VEGF-R_2 antibody [6].

In an in vivo mouse corneal model, O'Leary et al. showed a significant antiangiogenic activity of camp-

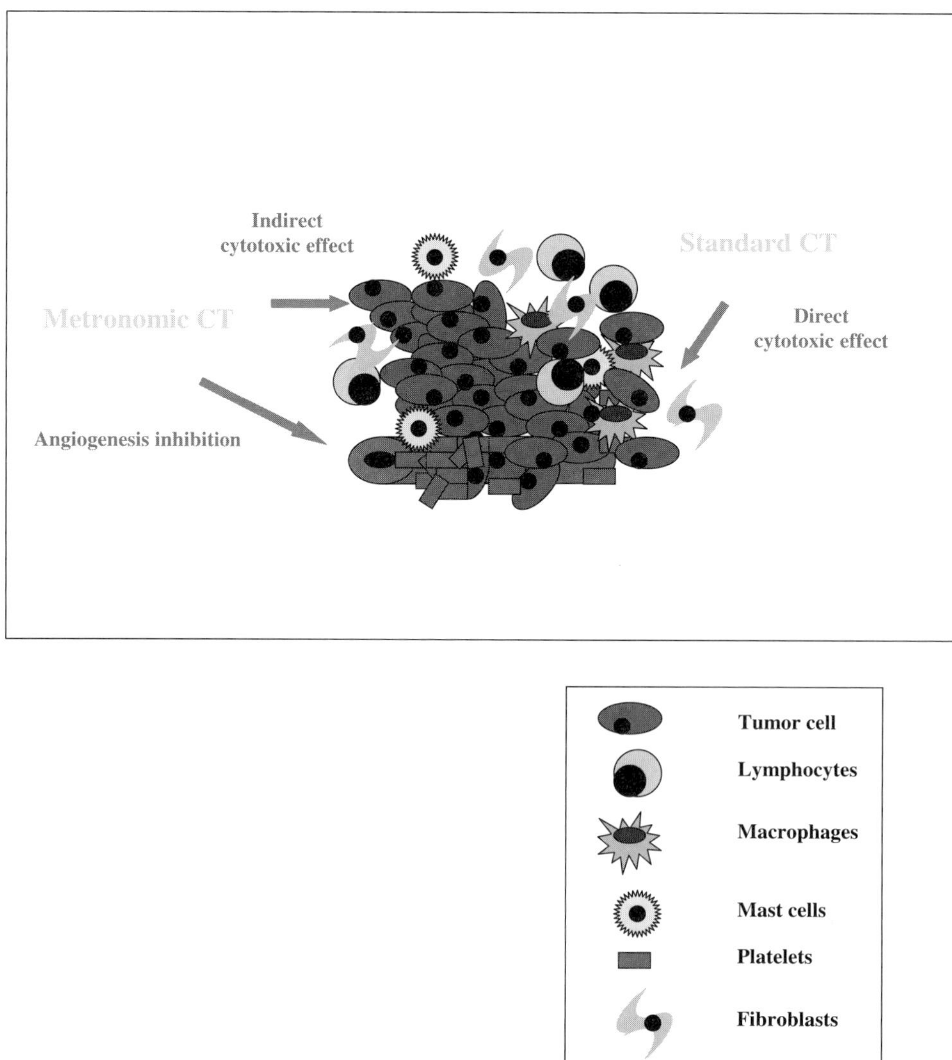

Figure 1 Metronomic chemotherapy interferes with tumor microvasculature and the stromal compartment, in contrast to standard chemotherapy that presents a direct cytotoxic effect. (see color insert)

tothecin analogs at a dose of 359 nmol/kg (210 mg/kg) delivered over 6 days [5].

Presta et al. reported that 6-methylmercaptopurine riboside, a purine analog, inhibits angiogenesis induced by FGF-2 in in vitro and in vivo rabbit cornea and chorioallantoic membrane models at topical doses of 100 nmol twice a day for 10 days, or at the single dose of 25 μmol onto the implants, respectively [5].

Low doses of vinblastine (0.1–1 pmol/L in vitro, and 0.5–1 pmol/L in vivo) had reversible antiangiogenic activity, without cytotoxicity [5]. The highest antiangiogenic dose was 1 pmol/L, which is equivalent to a dose of 16 μg in a 70-kg adult, a much lower dose than currently used in the clinic.

More recently, Klement et al. demonstrated that vinblastine at one tenth to one twentieth of the MTD caused a significant inhibition of angiogenesis with partial tumor regression in subcutaneous implanted tumors. This effect was significantly enhanced by combination with an anti-VEGF R_2 (flk-1) antibody [3].

In Swiss male nude mice implanted intracranially with human glioblastoma cells, Bello et al. showed that low and semicontinuous chemotherapy in combination with the recombinant human PEX (a fragment of matrix metalloproteinase-2) was associated with longer survival, marked decreases in tumor volume, vascularity, and proliferative index, and increased apoptosis, without major side effects [7].

Paclitaxel selectively inhibits the proliferation of human ECs at noncytotoxic concentrations (0.1 to 100 pM) by blocking the formation of sprouts and tubes in the three-dimensional fibrin matrix. This activity does not affect the cellular microtubule structure, and the treated cells do not show G_2/M cell cycle arrest and apoptosis, suggesing a novel, but as yet unknown, mechanism of action [8].

Available data obtained from studies in tumor-bearing animals specifically aimed to investigate the antiangiogenic effect of cytotoxic agents have found two patterns of anti-angiogenic effect [9]. The first "type" is defined by an angiogenesis inhibition due to a direct action on ECs regardless of the tumor cell line used and observed at dose levels lower than that required for the cytotoxic effect (e.g., bleomycin, vinca alkaloids, paclitaxel). The type-2 antiangiogenic effect is not consistently present within a variety of tumor cell lines resulting from an antiproliferative effect on tumor cells or non-ECs of the tumor host. In the alginate tumor angiogenesis model, the type-2 reaction pattern has been shown for doxorubicin, epirubicin, etoposide, and 5-fluorouracil. However, these effects may be dependent on the scheduling of drug administration [9].

All these preclinical studies present certain important problems: (i) the experimental model used: The studies with subcutaneous transplantation of rapidly proliferating tumors do not reflect the slow growth of spontaneous tumors and the characteristics of the stromal component of the organ of origin [5]; (ii) the lack of systematic comparative studies with conventional schedules; and (iii) the possible mechanisms of acquired resistance have not been adequately investigated [5].

In addition, the best results were obtained using combinations with selective antiangiogenic inhibitors, leading to additive and/or synergistic effects.

In conclusion, before proceeding to clinical trials, more compelling and appropriate experimental studies on metronomic chemotherapy are needed, and, although several cytotoxic agents affect ECs in vitro, only a few have been shown to produce substantial antiangiogenic activity in vivo, namely, cyclophosphamide, vinblastine, and paclitaxel [2, 3, 5].

Clinical Studies

There is currently no published clinical study in which metronomic chemotherapy is prospectively compared with conventional schedules, or that describes appropriate in vitro or in vivo methods of monitoring the antiangiogenic activity [5].

In a Phase I study, Retain et al. explored the continuous infusion of low doses of vinblastine up to 36 weeks and found that 0.7 mg/mq daily was the most effective dose, associated with a good tolerability and some clinical benefit [5].

Colleoni et al. demonstrated the activity of the combination of low oral doses of cyclophosphamide and methotrexate in patients with metastatic breast cancer, without relevant side effects. Serum VEGF levels measured at 2 months were lower than at baseline, with statistically significant reduction only in the subgroup of responding patients [10]. The major weakness of the study are the dosages administered, being in the range of cytotoxic effects; the low response rate in previously untreated patients; and, finally, the fact that the method of determination of VEGF is not standardized.

Other Phase I–II studies are investigating low, continuous oral doses of cytotoxic agents (uracil/tegafur and leucovorin, etoposide, 5-fluorouracil) or continuous intravenous infusion (5-fluorouracil, idarubicin, methotrexate, irinotecan) resembling an antiangiogenic schedule [5]. Forty percent of the patients with nonsmall-cell lung cancer not responsive to standard doses of etoposide responded to the same drug given orally at a much lower single dose, with only a 1-week break every month [5].

Promising results have been reported with weekly taxane treatment in breast and ovarian cancer, even in patients with progressive disease after the same drug given at the MTD every 3 weeks [11,12]. Thus, reversal of an apparent state of clinical drug resistance could be achieved by altering the dosing and frequency of the drug. However, at present, it is unknown if this effect is really related to an antiangiogenic activity.

At our center, we are evaluating the combination of weekly paclitaxel, at 80 mg/mq, and celecoxib, at 400 mg bid, in patients with nonsmall-cell lung cancer, as second-line chemotherapy. The preliminary data regarding both the tolerability and efficacy are encouraging and in accordance with the results of other similar ongoing Phase II studies.

In a Phase I dose-finding study, we evaluated the combination of rofecoxib with intravenous weekly irinotecan on days 1, 8, 15, 22 and infusional 5-fluorouracil at the fixed dosage of 200 mg/mq/day, as second-line therapy of metastatic colorectal cancer. The dose levels of irinotecan explored were from 75 to 125 mg/mp. Seven of 15 patients assessable for response obtained a partial response (46%) with a median duration of 5 months and another six had a stable disease with a median duration of 5 months. The acute and subacute hematological toxicity was moderate, and mucosal side effects were less than those expected with the same regimen without rofecoxib (submitted). A Phase II study testing the activity of such a schedule is ongoing.

Future Directions and Open Questions

There are several theoretical advantages and opportunities for metronomic chemotherapy (Table I). However, there are also potential problems and challenges in terms of appropriate experimental study design and clinical testing [5].

First, combined metronomic chemotherapy should be tested by adequate experimental models, such as orthotopic and metastatic tumors. The EC heterogeneity, which also extends to morphology, function, and response to antiangiogenic molecules, suggests that agents that affect angiogenesis in one organ may not be effective in other sites. Tumor cells implanted into mice usually produce rapidly growing lesions, which can double in size every few days and contain a high proportion of dividing ECs.

Second, human solid tumors are heterogeneous, with different molecular abnormalities, even in the same tumor histotype [5]. Gene expression profiling and cDNA microarrays may categorize tumors into biologically homogeneous subgroups and may be of help to design individually tailored treatments [5].

Third, the identification of specific surrogate biomarkers can allow the selection of patients as well as the possibility of monitoring tumor response. "Biological" response criteria should replace the currently used the clinical end point based on the objective response.

Table I Potential Advantages of Metronomic over Conventional Schedules of Chemotherapy.

- The targets are genetically stable ECs
- Activity against both the parenchymal and stromal tumor components
- Enhanced antiangiogenic and proapoptotic activity
- Reduced likelihood of emergence of acquired resistance
- Fewer systemic side effects
- Feasibility of long-term administration
- Possibility of combination with other cytostatic, molecular-targeted treatments

Fourth, experimental results have emphasized the critical need for combining metronomic regimens with selective antiangiogenic agents. The intrinsic elevated sensitivity of activated ECs to metronomic chemotherapy, compared with that of other cells, may not be related to the presence of high concentrations of EC-specific survival factors, such as VEGF [12]. Such combinations may be particularly effective in inducing higher levels of apoptosis in activated ECs coupled with the inhibition of cell proliferation. It is possible that the inhibition of ECs proliferation or induction of apoptosis may not be a direct effect of the drug, but rather an induced secondary event: e.g., a change in expression of genes or proteins that mediate the antiangiogenic effects.

In conclusion, certain cytotoxic drugs with antiangiogenic properties retain a potent antitumor activity when used in a protracted manner at very low concentrations, being able to affect both the tumoral parenchyma and vascular compartments. In contrast to the concept of "more is better," it appears that survival depends more on a cytostatic effect of chemotherapy than on rapid tumor shrinkage [5].

It is unlikely that metronomic chemotherapy will lead to significant clinical benefit if given alone. We suggest that it should be considered a novel approach making feasible for long periods the administration of cytotoxic agents in combination with selective molecular-targeting compounds directed against specific growth factors or blocking angiogenesis.

Even though the initial development of these novel combined treatments is in the context of advanced disease, the major therapeutic benefits are expected in the adjuvant setting or as maintenance therapy.

Glossary

Angiogenesis: Intratumor formation of new blood vessels from a preexisting vascular network.

Anticancer therapy: Conventional cytotoxic chemotherapy.

Metronomic chemotherapy: Chemotherapy given regularly at subcytotoxic soles with the "activated" endothelium as principal target.

References

1. Hahnfeldt, P., Folkman, J., and Hlatky, L. (2003). Minimizing long-term tumor burden: The logic for metronomic chemotherapeutic dosing and its antiangiogenic basis. *J. Theor. Biol.* **220**, 545–554.
2. Browder, T., Butterfield, C. E., Kraling, B. M., Shi, B., Marshall, B., O'Reilly, M. S., and Folkman, J. (2000). Antiangiogenic scheduling of chemotherapy improves efficacy against experimental drug-resistant cancer. *Cancer Res.* **60**, 1878–1886. *The first experimental study testing an antiangiogenic scheduling of chemotherapy. The study also demonstrated that metronomic chemotherapy reverses drug resistance to cyclophosphamide.*
3. Klement, G., Baruchel, S., Rak, J. R., Man, S., Clark, K., Hiclin, D. J., Bohlen, P., and Kerbel, R. S. (2000). Continuous low-dose therapy with vinblastine and VEGF receptor-2 antibody induces sustained tumor regression without over toxicity. *J. Clin. Invest.* **105**, 15–24. *The first study suggesting that the combined therapy of metronomic*

chemotherapy with a selective inhibitor of angiogenesis has a supra-additive antitumoral effect.
4. Hanahan, D., Bergers, G., and Bergsland, E. (2000). Less is more, regularly: Metronomic dosing of cytotoxic drugs can target tumor angiogenesis in mice. *J. Clin. Invest.* **105**, 1045–1047.
5. Gasparini, G. (2001). Metronomic scheduling: The future of chemotherapy? *Lancet Oncol.* **2**, 733–740. *A review paper detailing the rationale, the results of experimental and clinical studies, and challenges of metronomic chemotherapy.*
6. Mann, S., Bocci, G., Francia, G., Green, S. K., Jothy, S., Hanahan, D., Bohlen, P., Hicklin, D. J., Bergers, O., and Kerbel, R. S. (2002). Antitumor effects in mice of low-dose (metronomic) cyclophosphamide administered continuously through the drinking water. *Cancer Res.* **62**, 2731–2735.
7. Bello, L., Carrabba, G., Giussani, C., Iucini, V., Cerntti, F., Scaglione, F., Landre, J., Pluderi, M., Tomei, G., Villani, R., Carrull, R. S., Black, P. M., and Bikfalvi, A. (2001). Low-dose chemotherapy combined with an antiangiogenic drug reduces human glioma growth in vivo. *Cancer Res.* **61**, 7501–7506.
8. Wang, J., Lou, P., Lesniewski, R., and Henkin, J. (2003). Paclitaxel at ultra low concentrations inhibits angiogenesis without affecting cellular microtubule assembly. *Anticancer Drugs.* **14**, 13–19.
9. Schirner, M. (2003). Antiangiogenic chemotherapeutic agents. *Cancer Metastasis Rev.* **19**, 67–73.
10. Colleoni, M., Rocca, A., Sandri, M. T., Zorzino, L., Masci, G., Nole F., Perruzzotti, G., Robertson, C., Orlando, L., Cinieri, S., de B., F., Viale G, and Goldhirsch, A. (2002). A. Low-dose oral methotrexate and cyclophosphamide in metastatic breast cancer: Antitumor activity and correlation with vascular endothelial growth factor levels. *Ann Oncol.* **13**, 73–80.
11. Burstein, H. J., Manola, J., Younger, J., Parker, L. M., Bunnel, C. A., Scheib, R., Matrilonis, U. A., Garber, J. E., Clarke, K. D., Shulman, L. N., and Winer, E. P. (2000). Docetaxel administered on a weekly basis for metastatic breast cancer. *J. Clin. Oncol.* **18**, 1212–1219.
12. Sweeney, C. J., Miller, K. D., Sissons, S. E., Nazaki, S., Heilman, D. K., Shen, J., and Sledge, GWjr. (2001). The antiangiogenic property of docetaxel is synergistic with a recombinant humanized monoclonal antibody against vascular endothelial growth factor or 2-methoxyestradiol but antagonized by endothelial growth factors. *Cancer Res.* **61**, 3369–3372.
13. Klement, G., Mayer, B.Huang, P., Green, S. K., Man, S. Bohlen, P., Hicklin, D., and Kerbel, R. S. (2002). Differences in therapeutic indexes of combination metronomic chemotherapy and an anti-VEGFR-2 antibody in multidrug resistant human breast cancer xenograft. *Clin. Cancer Res.* **8**, 221–232.

Capsule Biography

Raffaele Longo is medical doctor and physician researcher at the Division of Medical Oncology, "S. Filippo Neri" Hospital, Rone (Italy), since March 2002. Ongoing research topics are tumor angiogenesis, Cox-2 inhibitors in solid cancers, and clinical development of new targeted molecular drugs.

Professor Gasparini has been Director of the Medical Oncology Unit at the San Filippo Neri Hospital in Rome since 2000. His scientific interests primarily focus on translational research on angiogenesis and molecular-targeted anticancer treatments. He is author of 250 publications and a member of the editorial boards of 15 international oncological journals.

CHAPTER 144

Platelet–Endothelial Interaction in Tumor Angiogenesis

[1,2]Philipp C. Manegold and [1,3]Marc Dellian

[1]*Institute for Surgical Research, Klinikum Grosshadern, University of Munich, Germany*

[2]*Department of Surgery, Klinikum Grosshadern, University of Munich, Germany*

[3]*Department of Otorhinolaryngology, Head and Neck Surgery, Klinikum Grosshadern, University of Munich, Germany*

Tumor angiogenesis, the growth of new blood vessels from preexisting blood vessels, is indispensable for tumor growth and tumor metastasis. Activated blood coagulation has been linked to tumor angiogenesis since clotting factors were found to promote endothelial cell proliferation. Beside plasmatic clotting factors, platelets may be involved in angiogenic processes: they release pro- and anti-angiogenic growth factors upon activation and aggregation and promote formation of capillary-like structures in vitro [1, 2].

Tumor angiogenesis is a complex process regulated by a versatile number of mediators [3]. Focusing on general mechanisms, tumor angiogenesis is induced by endothelium-specific growth factors VEGF and FGF, but also by cytokines (TNF, IL-1) and clotting factors that are released from tumor cells as well as from macrophages and fibroblasts. On preexisting venules growth factors and cytokines stimulate endothelial cells and induce loosening of intercellular junctions of endothelial cells (PECAM, VE-cadherin) and smooth muscle cells. Therefore, angiogenic endothelium is highly permeable, leading to extravasation of plasma proteins such as fibrinogen and clotting factors. Extravascular fibrin meshes provide a provisional tumor matrix. Proteases, especially urokinase plasminogen activator (uPA) and matrix metalloproteinases (MMPs, e.g., progelatinase A) expressed on proliferating endothelial cells, facilitate sprouting of blood vessels into the fibrin matrix toward the stimuli.

The established tumor microcirculation is characterized by its high vascular density, chaotic vascular branching, and aneurysmatic sacculations. High vascular permeability is a consequence of sustained influence of growth factors and cytokines and is a result of intercellular gaps in tumor microvessels. Mosaic blood vessels composed of endothelial cells and nonendothelial cells forming vascular lumen are frequently found [4]. High vascular resistance and high blood viscosity result in sluggish blood flow within the tumor microcirculation. Anoxic, acidotic areas within the tumor center appear in consequence of redistribution of blood perfusion to peripheral areas [5].

Therefore, tumor angiogenesis might provide a prothrombotic environment that favors interactions of platelets with the angiogenic endothelium. Platelet–endothelial interactions might affect tumor angiogenesis by release of angiogenic growth factors from activated platelets. Spontaneous thrombosis, however, might result in tumor necrosis.

Physiology of Platelet–Endothelial Interaction

In differentiated tissues, activation and aggregation of platelets on the endothelial surface is abrogated by endothelial antithrombotic mediators such as nitric oxide (NO) and prostacyclin (PGI). The small amount of continuously expressed adhesion molecules on the endothelial surface does not sufficiently provoke platelet–endothelial interactions.

As a result of disruption of the endothelial surface and exposure of subendothelial matrix as well as in response to inflammatory cytokines (TNF, IL-1, IFN) and angiogenic growth factors (VEGF), platelet–endothelial interactions are mediated by endothelial and platelet-specific adhesion molecules. Platelet interaction with the endothelial surface and subendothelial matrix occurs in a sequence of initial short-timed contact (rolling) followed by adhesion and aggregation of activated platelets. Initiation of platelet rolling appears to be independent of platelet activation.

Platelet rolling on stimulated endothelium is mediated by endothelial P- and E-selectin. P-selectin and von Willebrand factor are released from Weibel–Palade bodies of endothelial cells and are presented on the endothelial surface within minutes following stimulation [6]. Expression of E-selectin and a second phase of P-selectin expression depend on de novo protein synthesis and occur hours after the primary stimulus. The main ligand for endothelial P-selectin on platelets has been identified to be P-selectin binding ligand-1. Binding of endothelial P-selectin by glycoprotein (GP) Ib/IX/V as a second ligand depends on previous platelet activation. Platelets store P-selectin along with adhesion proteins fibronectin, fibrinogen, and von Willebrand factor (vWF) in their α-granules and release these factors within minutes following stimulation. P-selectin expression on activated platelets may also facilitate platelet rolling. The endothelial ligand for platelet P-selectin has not been identified.

Slowing down platelet trafficking along the endothelial surface ensures stable platelet adhesion by a fibrin-dependent bridging mechanism between GPIIb/IIIa on platelets and endothelial ICAM-1 or $\alpha_v\beta_3$-integrin on activated and proliferating endothelial cells. Adhesion proteins vWF, fibrin, and fibronectin are required for further platelet aggregation by GPIIb/IIIa binding. Receptors involved in platelet interaction with intact endothelial surface are depicted in Figure 1.

At sites of endothelial damage and exposure of subendothelial matrix platelet rolling and aggregation depend mainly on vWF [7]. Plasmatic and endothelial vWF become immobilized on subendothelial matrix. Platelet rolling and adhesion are initiated by the binding of GPIb/IX/V to immobilized vWF that goes along with platelet activation. ATP and serotonin released from dense granules of activated platelets stimulate the aggregation of further platelets. This is accompanied by secretion of vWF, P-selectin, fibrinogen, and fibronectin from α-granules of platelets and transformation of GPIIb/IIIa into its active configuration. Thus, platelet adhesion to the subendothelial matrix is enhanced by fibrin-mediated GPIIb/IIIa binding to immobilized vWF. Furthermore, different types of collagen receptors on platelets can cause a fast platelet activation on contact to the subendothelial matrix. Depending on the type of vessel wall, collagen receptors contribute differently to platelet–vessel wall interactions. Receptors involved in platelet rolling, adhesion, and aggregation on the subendothelial matrix are summarized in Figure 2.

Platelet–Endothelial Interaction in Tumor Angiogenesis

The microvascular architecture of tumors has been studied in different experimental tumors. However, platelets in tumor microcirculation have rarely been described. Ultrastructural studies on different melanomas of the hamster revealed that platelets were attached to the microvascular endothelium only occasionally at sites of tumor invasion into the microvasculature [8]. In vivo observations of platelets in tumor microvessels were initially described in the amelanotic melanoma (A-Mel-3) implanted into the dorsal skinfold chamber of hamsters using intravital microscopy and intravenous application of fluorescence marker for platelets. When tumor necrosis became obvious, masses slowing down blood flow and partly occluding tumor blood vessels were detected. This phenomenon was interpreted as platelet aggregation and vascular thrombosis [5]. However, extensive ultrastructural studies on the microvasculature of the amelanotic melanoma revealed infrequent contact of platelets to the microvascular endothelium even at sites of endothelial leakage and erythrocyte

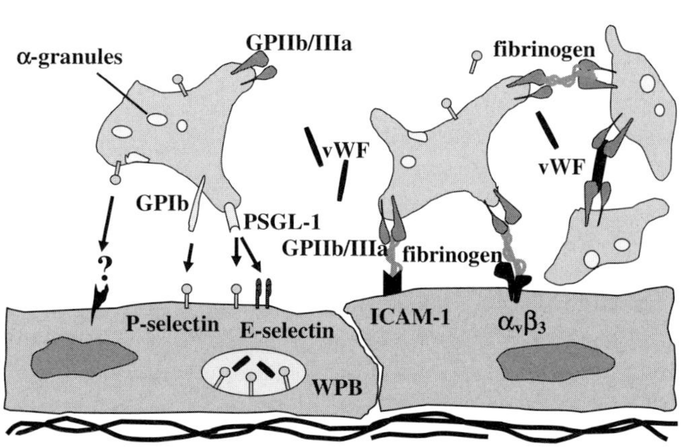

Figure 1 Platelet adherence to the intact endothelial surface. (*Left*) Mechanisms of platelet rolling. (*Right*) Platelet adherence and aggregation on the endothelial surface. GP, glycoprotein; PSGL-1, P-selectin glycoprotein ligand-1; vWF, von Willebrand factor; WPB, Weibel–Palade body.

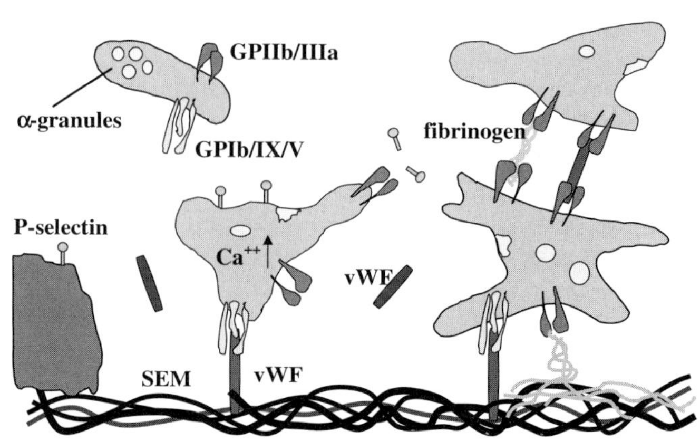

Figure 2 Platelet interactions on the subendothelial matrix. (*Left*) Initial platelet rolling on immobilized vWF induces activation of glycoprotein (GP) Ib/IX/V and GP IIb/IIIa. (*Right*) Adherence on the subendothelial matrix and aggregation of platelets is mediated by glycoproteins and adhesion proteins. SEM, subendothelial matrix.

extravasation. Tumor microvessels were regularly composed of endothelial and tumor cells forming the vessel lumen [4].

We have studied platelet–endothelial interactions in angiogenesis and growth of Lewis lung carcinoma (LLC-1) and methylcholanthrene-induced fibrosarcoma (BFS-1). The study was carried out using the dorsal skinfold chamber on mice and intravital fluorescence microscopy. Ex vivo fluorescently labeled syngenic platelets were transfused into experimental animals. Both tumors developed a typical heterogeneous tumor microcirculation. In contrast to the amelanotic melanoma of hamsters, tissue necrosis did not appear in LLC-1 carcinoma and BFS-1 fibrosarcoma within 14 days after tumor cell implantation into the dorsal skinfold chamber. Platelet–endothelial interaction was rarely observed in tumor microvessels. No differences in platelet–endothelial interactions were seen in tumor microvessels compared to subcutaneous venules in tumor-free tissue (Figure 3).

Under physiologic conditions, low interactions between platelets and endothelial cells have been described in different organ systems that might be due to weak platelet agonist (ADP, epinephrine, serotonin). We found equivalent baseline platelet–endothelial interactions in subcutaneous venules of tumor-free tissue. In the early phase of tumor growth, platelet rolling was slightly enhanced in angiogenic microvessels in close vicinity to implanted tumor cells. The angiogenic microvessels appeared as dilated and highly permeable microvessels due to growth factors and inflammatory cytokines.

Since tumor microcirculation was supposed to present a highly prothrombotic environment, we would have expected more significant platelet–endothelial interactions in tumor angiogenesis and tumor microcirculation. However, the compromised hemodynamics within the tumor microvasculature was not associated with activation or aggregation of platelets. Our results were confirmed by a study indicating that radioactively labeled fibrinogen accumulates in Lewis-lung carcinoma; however, the study failed to detect accumulation of radioactively labeled platelets within the same tumor. Furthermore, electron microscopy showed single platelets adhering to endothelial gaps in tumor microvessels only occasionally.

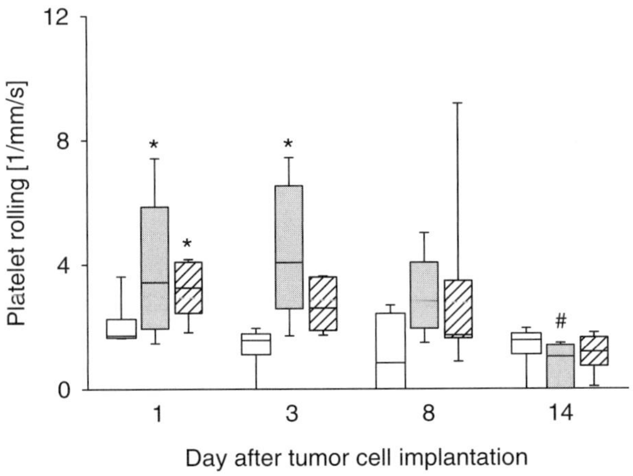

Figure 3 Platelet–endothelial interaction is not an intense phenomenon during tumor angiogenesis. Platelet rolling was increased on day 1 in pre-existing microvessels of the host in response to both LLC-1 (gray bars) and BFS-1 (hatched bars) and also on day 3 in LLC-1 tumor microvessels, but was only slightly above the low baseline level quantified in postcapillary venules of controls (open bars). $n = 6$ experimental animals per group, $^{*}p < 0.05$ versus controls; $^{\#}p < 0.05$ versus day 1 and day 3 (Kruskal–Wallis test).

Platelets Are Potentially Involved in Tumor Angiogenesis

Platelets are potentially involved in tumor angiogenesis because they contain numerous angiogenic factors (e.g., VEGF, bFGF, PDGF, PAI-1). Clinical studies revealed that platelets are the main source of the serum VEGF concentration typically correlated with poor prognosis in cancer patients.

Although the mechanisms of growth factor release from platelets have not been entirely resolved, in vitro experiments have demonstrated that growth factor release depends on platelet activation and aggregation. Although platelet secretion occurs independently of platelet aggregation, release of growth factors becomes detectable only above threshold doses of thrombin and other platelet agonists at which platelet aggregation is inevitable. A further prerequisite for the promoting effect of platelets on the formation of microvascular tubuli seems to be direct cell-to-cell interactions between platelets and the proliferating endothelium.

Growth factors, especially VEGF, are secreted by a high percentage of malignant animal and human tumors, and also by stromal cells adjacent to the tumors. Only a few exceptional tumors are known that express little or no VEGF mRNA and protein. These tumors, renal papillary carcinoma and lobular breast cancer, are found to be almost avascular and to have better prognosis because of their minor invasiveness. A positive correlation between platelets and disease progression has only been reported for advanced cancer diseases characterized by dense vascularization, invasiveness, and metastatic diseases.

VEGF may be taken up by platelets from blood plasma since there is a strong correlation between plasma VEGF levels and serum VEGF load per platelet in tumor patients. Therefore, VEGF content in platelets might be not only a result of increased protein syntheses in megakaryocytes in response to tumor released cytokines and growth factors, but also a consequence of endocytosis of tumor-derived VEGF by circulating platelets [9].

In conclusion, tumor microvessels do not express a highly prothrombotic environment favoring platelet aggregation. Hence, platelets do not become sufficiently activated to release angiogenic growth factors at sites of tumor angiogenesis. It is more probable that platelets take up and preserve angiogenic growth factors released by tumor cells. The role of platelets as a reservoir for angiogenic growth factors in tumor growth has not yet been identified.

Platelet–Endothelial Interaction Following Endothelial Stimulation

In a second set of experiments on the microcirculation of LLC-1 carcinoma and BFS-1 fibrosarcoma, platelet–endothelial interactions were assessed in response to endothelial stimulation by calcium ionophore A23187. Calcium ionophore A23187 increases intracellular calcium concentration and thereby induces the release of adhesion molecules vWF and P-selectin from endothelial cell–specific Weibel–Palade bodies sparing endothelial integrity. The secretion of Weibel–Palade bodies from endothelial cells plays a central role in wound healing, coagulation, inflammation, and ischemia–reperfusion injury. In addition, angiogenic growth factor VEGF induces the secretion of Weibel–Palade bodies and adhesion molecule presentation. Hence, tumor cells might influence adhesion molecule expression in microvessels [10].

The application of calcium ionophore A23187 did not affect platelet–endothelial interaction in BFS-1 fibrosarcoma, and only a slight increase in platelet rolling was detected in microvessels of the LLC-1 carcinoma. However, in subcutaneous venules of tumor-free tissue, platelet rolling increased significantly (Figure 4). Platelet adherence was only occasionally seen following superfusion with calcium ionophore. The results indicate that tumor microvessels do not express sufficient amounts of vWF and P-selectin in response to stimulation with calcium ionophore A23187. In respect to the stimulus used, this phenomenon can be attributed to the reduced or missing secretion of adhesion molecules from endothelial Weibel–Palade bodies [10].

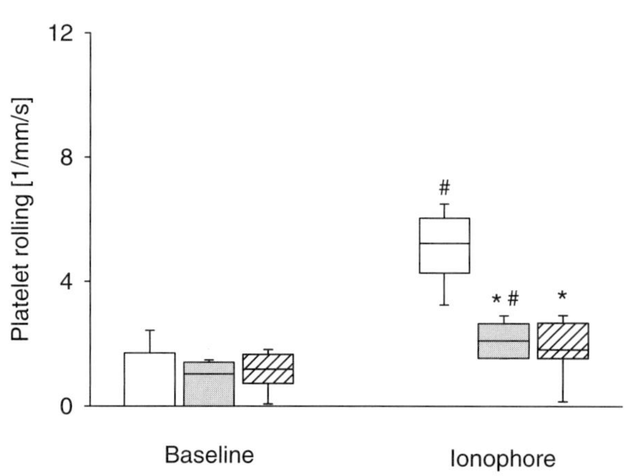

Figure 4 Calcium ionophore A23187 induces platelet rolling favorably in normal and less in tumor microvessel. Superfusion of tumor tissues with calcium ionophore A23187 (20 µM) over 15 minutes causes a twofold increase of rolling platelets in LLC-1 (gray bars) and exerts no effects on BFS-1 (hatched bars) on day 14 after tumor cell implantation. In contrast, the treatment of tumor-free preparations resulted in a threefold increase of rolling platelets in postcapillary venules of controls (open bars) on corresponding days to tumor groups after chamber preparation. $n = 6$ experimental animals per group, $^{*}p < 0.05$ versus controls, $^{\#}p < 0.05$ versus baseline (Kruskal–Wallis test and Wilcoxon test).

Immunohistological studies on vWF and P-selectin stored in Weibel–Palade bodies of endothelial cells describe a granular staining pattern in the microcirculation of differentiated tissues [11]. In tumor microvessels, however, vWF expression appears to be extremely heterogeneous. Immunohistological techniques showed reduced or even absent staining for vWF alternating with areas highly positive for vWF. In follicular thyroid cancer, vWF was not detectable, whereas it was highly significant in normal thyroid tissue. A similar immunohistological pattern has been found for P-selectin expression within the tumor microcirculation. The distribution pattern of vWF and P-selectin was never associated with angiogenic processes.

Hence, immunohistological studies disclose that vWF and P-selectin storage in Weibel–Palade bodies of endothelial cells is reduced in tumor microvessels. This might be a consequence of Weibel–Palade body exocytosis due to sustained stimulation by angiogenic growth factors and cytokines preventing redistribution and regeneration of vWF and P-selectin into their intracellular storage granules. Furthermore, angiogenic growth factors prevent protein synthesis of adhesion molecules that are required for interactions of platelets as well as of leukocytes with the microvascular endothelium. Distribution of Weibel–Palade bodies within the tumor microcirculation seems to depend on tumor type and species. Weibel–Palade bodies may be completely missing in the microvasculature of some tumors.

Conclusions

Tumor angiogenesis and tumor growth are accompanied by only a transient increase in platelet rolling in angiogenic microvessels in close vicinity to tumor cells. This might be due to endothelial stimulation by growth factors and cytokines released partly from tumor cells. Within tumor microcirculation, adhesion molecules mediating platelet–endothelial interactions appeared to be downregulated or even absent. Platelet–endothelial interaction in response to endothelial stimulation was significantly reduced in tumor microcirculation.

Future studies are needed to identify mechanisms regulating adhesion molecule expression in tumor microcirculation. Further attempts should focus on induction of platelet aggregation by selective transport of prothrombotic agents into the tumor microcirculation. The results might provide further therapeutic aspects in targeting tumor angiogenesis.

Bibliography

1. Möhle, R., Green, D., Moore, M. A., Nachman, R. L., and Rafii, S. (1997). Constitutive production and thrombin-induced release of vascular endothelial growth factor by human megakaryocytes and platelets. *Proc. Natl. Acad. Sci. USA* **94**, 663–668.
2. Pinedo, H. M., Verheul, H. M., D'Amato, R. J., and Folkman, J. (1998). Involvement of platelets in tumour angiogenesis? *Lancet* **352**, 1775–1777. *This article proposes a possible role for platelets in tumor*

angiogenesis based on a review of the literature, which was the starting point of our experimental work.

3. Carmeliet, P. and Jain, R. K. (2000). Angiogenesis in cancer and other diseases. *Nature* **407**, 249–257.
4. Hammersen, F., Osterkamp-Baust, U., and Endrich, B. (1983). Ein Beitrag zum Feinbau terminaler Strombahnen und ihrer Entstehung in bösartigen Tumoren. In *Mikrozirkulation in malignen Tumoren*, P. Vaupel and F. Hammersen, eds., pp. 15–51. Basel, CH: Karger. *An early publication that already described the absence of platelets in tumor microvasculature.*
5. Asaishi, K., Endrich, B., Goetz, A., and Messmer, K. (1981). Quantitative analysis of microvascular structure and function in the amelanotic melanoma A-Mel-3. *Cancer Res.* **41**, 1898–1904. *A detailed, quantitative analysis of tumor microcirculation based on intravital microscopy of a newly developed transparent chamber preparation.*
6. Frenette, P. S., Johnson, R. C., Hynes, R. O., and Wagner, D. D. (1995). Platelets roll on stimulated endothelium in vivo: An interaction mediated by endothelial P-selectin. *Proc. Natl. Acad. Sci. USA* **92**, 7450–7454.
7. Konstantopoulos, K., Kukreti, S., and McIntire, L. V. (1998). Biomechanics of cell interactions in shear fields. *Adv. Drug. Deliv. Rev.* **33**, 141–164.
8. Warren, B. A., Shubik, P., and Feldman, R. (1978). Metastasis via the blood stream: The method of intravasation of tumor cells in a transplantable melanoma of the hamster. *Cancer Lett.* **4**, 245–251.
9. Salgado, R., Benoy, I., Bogers, J., Weytjens, R., Vermeulen, P., Dirix, L., and van Marck, E. (2001). Platelets and vascular endothelial growth factor (VEGF): A morphological and functional study. *Angiogenesis* **4**, 37–43.
10. Manegold, P. C., Hutter, J., Pahernik, S. A., Messmer, K., and Dellian, M. (2003). Platelet–endothelial interaction in tumor angiogenesis and microcirculation. *Blood* **101**, 1970–1976.
11. Schlingemann, R. O., Rietveld, F. J., Kwaspen, F., van de Kerkhof, P. C., De Waal, R. M., and Ruiter, D. J. (1991). Differential expression of markers for endothelial cells, pericytes, and basal lamina in the microvasculature of tumors and granulation tissue. *Am. J. Pathol.* **138**, 1335–1347.

Capsule Biography

Dr. Dellian has been leading a research laboratory on tumor microcirculation and angiogenesis at the Institute for Surgical Research, University of Munich, since 1995. He was introduced to his research by the pioneers in tumor microcirculation, Prof. Konrad Messmer, Prof. Alwin Goetz, and Prof. Rakesh Jain.

CHAPTER 145

Von Hippel–Lindau Disease: Role of Microvasculature in Development of the Syndrome

David R. Mole and Peter J. Ratcliffe
University of Oxford, Oxford, United Kingdom

Positional cloning of defective genes that underlie rare hereditary cancer syndromes, such as von Hippel–Lindau disease, has provided many important insights into more common nonhereditary malignancies and the fundamental cellular processes that are involved in oncogenesis. In general, the tumor suppressor functions defined by analysis of such genes have involved a direct role in the regulation of cell proliferation (such as the retinoblastoma gene product, Rb), a direct role in DNA repair (such as *MSH2* and *MLH1*), or a role in linking DNA damage to the arrest of cellular proliferation and apoptosis (such as p53). Such functions fit readily into a simple genetic model of cancer whereby the accrual of mutations allows uncontrolled growth, and genomic instability reinforces the generation of mutant clones with an increasingly aggressive growth advantage.

Nevertheless tumor development involves many other processes, such as the maintenance of adequate blood oxygen delivery by induction of angiogenesis (indeed the Greek name *cancer* derives from the crab-like pattern of tumor neovasculature). Von Hippel–Lindau (VHL) disease is one of several hereditary cancer syndrome syndromes associated with excessive angiogenesis, and recent insights into the role of the VHL protein (pVHL) in the development of the abnormal microvasculature seen in this syndrome have led to a greater mechanistic understanding of tumor angiogenesis and cellular oxygen sensing.

The von Hippel–Lindau Hereditary Cancer Syndrome

VHL disease is inherited as an autosomal dominant trait affecting 1 in 36,000 of the population. Affected individuals bear a germ-line mutation in the VHL tumor suppressor gene. However, it is the predisposition to cancer rather than the cancer itself that is inherited, and the presence of a single mutation at the susceptibility locus is insufficient for tumor formation. Tumors are associated with somatic loss or inactivation of the remaining wild-type allele, in accordance with the Knudson "two-hit" hypothesis. The high likelihood of somatic mutation affecting the single remaining wild-type allele over the lifetime of an individual with the inherited syndrome explains the dominant inheritance pattern of tumors, the greater than 90 percent penetrance by 65 of years age, and the multifocal nature of associated tumors. Sporadic tumors can arise in nonaffected individuals, by the occurrence of somatic loss of both alleles within the same cell. However, on average, it takes longer to accrue two "hits" within the same cell, accounting for the lower prevalence (within the much larger at-risk population) and older age distribution of sporadic tumors, compared to those associated with the hereditary VHL syndrome.

The first report of patients with what was probably VHL syndrome was by Treacher Collins in 1894, when he reported two siblings with retinal angiomas. In 1904 Eugene von Hippel, who gives his name to the syndrome, reported families with blood-vessel tumors (angiomas) of the retina.

The correlation with microscopically indistinguishable tumors (hemangioblastomas) of the central nervous system (especially cerebellum and spinal cord) was first described, in 1926, by Arvind Lindau. This description also included cysts in the kidney, pancreas, and epididymis. However, the first diagnostic criteria to include renal cell carcinoma as part of the syndrome was that of Melmon and Rosen in 1964. Other tumors now known also to be associated with the syndrome include pheochromocytoma (a hormone-secreting tumor of the adrenal medulla), endolymphatic-sac tumors of the inner ear, and islet-cell tumors of the pancreas.

Hemangioblastomas and renal-cell carcinomas are both highly vascular tumors and demonstrate disordered growth of the microvasculature. Hemangioblastomas are the commonest tumors in the VHL syndrome, affecting 60 to 80 percent of individuals, and have a predilection for the posterior fossa and spinal cord. They are well-defined thinly encapsulated benign tumors formed from a mixture of "stromal" cells and a rich plexus of sometimes telangiectatic capillary blood vessels, and are frequently associated with edema and cysts. It is the stromal cells that are the tumor cells. These cells present a finely vacuolated or foamy cytoplasm, and sometimes evoke an epitheloid appearance of the tumor mimicking that of metastatic renal cell carcinoma. They lack functional pVHL and overproduce angiogenic growth factors such as vascular endothelial growth factor (VEGF) and platelet-derived growth factor B chain (PDGF-β). These in turn drive proliferation of endothelial cells and pericytes, respectively, leading to the angiogenic phenotype. Furthermore, VEGF, also known as vascular permeability factor (VPF), increases capillary leakiness, leading to edema.

Renal cell carcinomas, also known as clear-cell carcinomas, arise from the renal tubular epithelium and again are highly vascular, giving them their characteristic red color at operation. The tumors develop from preneoplastic cysts lined with VHL–/– epithelial cells. Importantly, the very large number of associated benign lesions compared to malignant lesions suggests a complex oncogenic process, most probably involving several additional genetic alterations following VHL inactivation. However, restoration of pVHL function in fully transformed VHL–/– renal carcinoma cells suppresses their ability to form tumors in nude mice, indicating an ongoing requirement for VHL inactivation. Interestingly, restoration of pVHL does not affect the ability of tumor cells to grow in adherent tissue culture, suggesting that the tumor suppressor action is in some way environmentally specific. Like the stromal cells in hemangioblastomas, renal-cell carcinoma cells also overexpress VEGF and PDGF-β, leading to their highly vascular appearance.

Although the angiogenic tumor phenotype is most apparent in VHL-associated tumors, this increased angiogenesis is not unique to them. A considerable body of evidence spanning more than three decades has documented that tumor growth and metastasis of many cancers requires persistent new blood-vessel growth. When tumor cells were transplanted into avascular sites, such as the cornea, the implants attracted new capillary growth. In the absence of access to an adequate vasculature, tumor cells became necrotic and/or apoptotic. Furthermore, ingress of new blood vessels in these tumor transplantation models suggested that tumors released diffusible activators of angiogenesis that could stimulate a quiescent vasculature to begin capillary sprouting. In a variety of transgenic mouse tumor models, an "angiogenic switch" could be detected during the early stages of tumorigenesis, preceding the development of solid tumors. Recent analyses of von Hippel–Lindau disease have provided the first direct link between a tumor suppressor gene function and the molecular mechanisms of angiogenesis.

The VHL Protein (pVHL)

The VHL gene was first cloned from chromosome 3p25-26 by a large consortium in 1993, and comprises just three exons coding for a protein of 213 amino acids. A second isoform of 150 residues is produced from an in frame ATG at codon 54. Notably all disease-causing mutations affect sequences C-terminus to this second start site, so that tumorigenesis is associated with inactivation of both forms.

The majority of familial point mutations lie within two regions of the protein, suggesting two major functional domains. Although the primary VHL sequence did not immediately suggest a function, protein association experiments defined a series of pVHL interacting proteins, including elongins B and C, CUL2, and Rbx1, which bound to the frequently mutated α-domain. Cul2, a member of the cullin family, resembles yeast Cdc53, and elongin C resembles yeast Skp1. In yeast, Cdc53 and Skp1 bind to one another to form ubiquitin ligases referred to as SCF complexes (Skp1/Cdc53/F-box protein), which covalently attach a ubiquitin polymer to cell-cycle control proteins targeting them for destruction by the proteasome. In such complexes, the target protein destined for polyubiquitination, and hence destruction by the proteasome, is recognized or bound by the F-box protein (so named because of a collinear Skp1-binding motif present in cyclin F), suggesting an analogous role for pVHL. This putative role was strengthened by recognition of a second frequently mutated subdomain of pVHL, the β-domain, which has features of a substrate-docking site. Experimental evidence confirmed that anti-pVHL immunoprecipitates exhibit ubiquitin ligase activity in vitro.

The discovery that pVHL was acting as a ubiquitin ligase substrate recognition module raised the question of what was its target protein, and how was this related to the angiogenic phenotype of the tumors? Repetition of the protein association studies in the presence of proteasomal blockade demonstrated two novel pVHL binding partners, the α-subunits of the transcription factors hypoxia inducible factor 1 and 2. Furthermore, in vitro ubiquitylation studies showed that these two proteins were targeted for ubiquitylation by pVHL (see Figure 1).

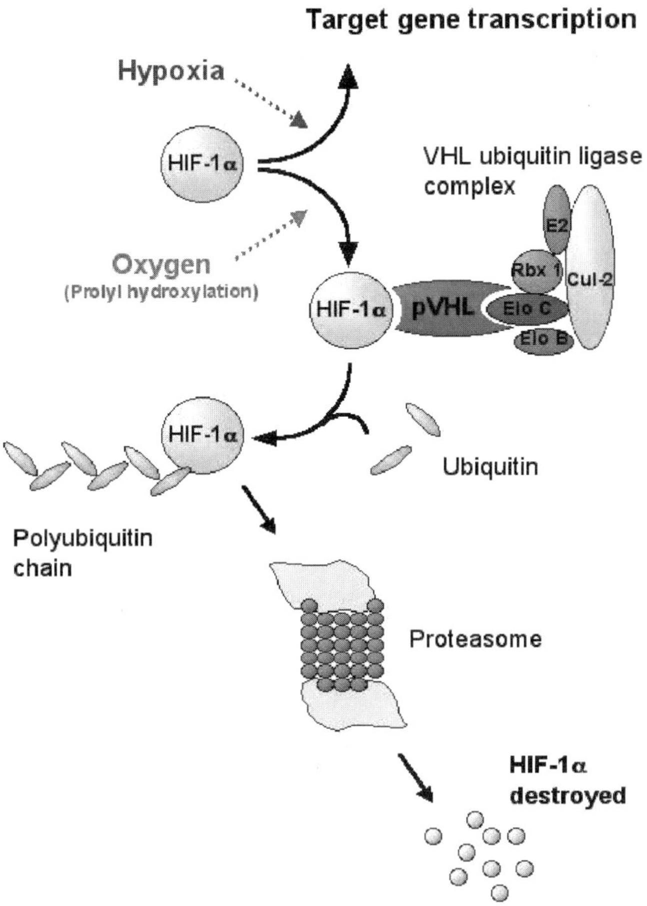

Figure 1 Regulation of HIF-α by pVHL. Under oxygen-replete conditions HIF-α is enzymatically modified by prolyl hydroxylase activity to a form that is recognized by pVHL. The pVHL ubiquitin ligase complex covalently attaches a polyubiquitin tag, which then targets HIF-α for rapid destruction in by the proteasome, so that it is no longer available to effect target gene transcription. (see color insert)

pVHL and Oxygen Sensing

Hypoxia inducible factor (HIF) is an αβ-heterodimer that was first recognized as a DNA-binding protein responsible for mediating the hypoxia-inducible activity of the gene encoding the hematopoietic growth factor erythropoietin. Both subunits contain basic helix–loop–helix (bHLH)-PAS domains. Whereas the bHLH domain defines a superfamily of eukaryotic transcription factors, the PAS domain was first defined in the PER, ARNT, and SIM proteins and defines a subset of the bHLH family. HIF-β subunits are constitutively expressed nuclear proteins that are involved via other dimerization partners in a variety of transcriptional responses. The HIF-α subunits exist as three isoforms, all of which are inducible by lack of oxygen (hypoxia). However, in oxygenated cells these α-subunits are rapidly degraded, with HIF-1α and HIF-2α having an exceptionally short half-life of just a few minutes. When the α-subunits are stabilized, HIF dimerizes and interacts with *cis*-acting hypoxia response elements (HREs) to induce transcriptional activity. A large and rapidly expanding array of genes have now been shown to contain HREs and to be transcription targets of HIF. Examples of these responses include not only erythropoietin, but also endothelial nitric oxide synthase and endothelin, which control vascular tone; glucose transporters and many key enzymes involved in the metabolism of glucose; enzymes involved in catecholamine synthesis; proteins concerned with iron transport and handling; and checkpoints in cell proliferation and quiescence. However, most importantly with regard to the angiogenic phenotype of VHL disease, many genes that play an essential role in angiogenesis are HIF target genes.

HIF-α subunits contain two domains that regulate their activity in response to oxygen availability: an internal oxygen-dependent degradation domain (ODDD), which regulates protein destruction, and the C-terminal transactivation domain (CAD), which regulates transactivating ability through recruitment of the coactivator p300/CBP. In normoxia, enzymatic hydroxylation of two prolyl residues, within the HIF-α ODDD, by a newly defined prolyl hydroxylase activity facilitates recognition of HIF-α by the VHL E3 ubiquitin ligase complex that targets HIF for destruction by the proteasome. In mammalian cells three closely related enzymes, each the product of a different gene, have been shown to have HIF prolyl hydroxylase activity. These proteins are all nonheme Fe (II) enzymes that are members of the superfamily of 2-oxoglutarate dependent dioxygenases. An absolute requirement for molecular oxygen renders their function oxygen sensitive. In the absence of pVHL, hydroxylated HIF-α cannot be targeted for ubiquitination and subsequent proteasomal degradation, leading to constitutive upregulation of HIF-mediated gene transcription even in the presence of adequate oxygen for hydroxylase function.

Further oxygen-dependent control is regulated by factor inhibiting HIF (FIH) through hydroxylation of an asparaginyl residue in the C-terminal transactivation domain. Hydroxylation at this residue interferes with HIF transactivating ability by blocking interaction with the transcriptional coactivator CBP/p300. Interestingly, the fact that pVHL loss results in full, rather than partial, activation suggests that pVHL could be involved in aspects of the HIF response to oxygen beyond ODDD-mediated proteolysis. It has been suggested that pVHL might be involved directly in the process of transcriptional inactivation both by promoting the function of FIH and by recruitment of transcriptional repressors.

Hypoxia-Inducible Factor and Angiogenesis

It is now clear that vascular architecture in tissues is largely controlled by angiogenic signals from the constituent cells. Local oxygen tension appears to be a critical signal in detecting and responding to local vascular insufficiency; for example, in the retina local ischemia almost always precedes new vessel growth.

HIF and pVHL play a critical role in these responses and many points of interaction with molecular processes

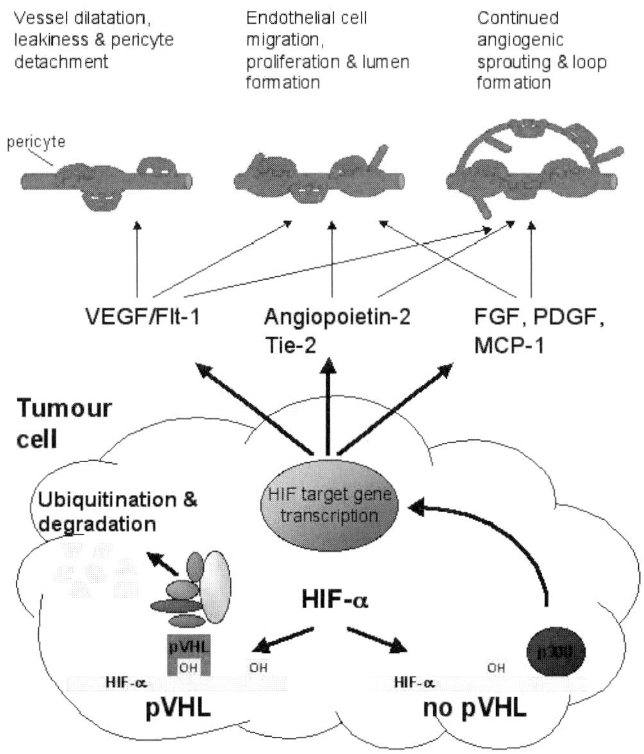

Figure 2 The mechanism of angiogenesis in VHL disease. pVHL is required for the normoxic degradation of HIF-α subunits. In the pVHL-deficient tumor cells, HIF accumulates and effects transcription of its target genes. These include secreted factors such as VEGF, Flt-1, angiopoietins-2, Tie-2, FGF, PDGF, and MCP-1, which act on different stages of angiogenesis, leading to localized new vessel growth. (see color insert)

regulating angiogenesis have been defined (see Figure 2). The vasculature is usually quiescent in the adult, and endothelial cells are among the longest lived outside the nervous system. High levels of VEGF alone are capable of initiating angiogenesis in these quiescent vessels, and several isoforms of VEGF and its receptors are hypoxia inducible via HIF-dependent processes. In the earliest stages of angiogenesis, vasodilatation and increased vascular permeability, potentially mediated through HIF-regulated genes such as VEGF-A, Flt-1 and nitric oxide synthase, allow extravasation of plasma proteins, which lay down a provisional matrix on which activated endothelial cells migrate. This is accompanied by a loosening of the surrounding pericyte cells that is thought to be mediated by angiopoietin-2 and its tyrosine kinase receptor, Tie-2, both of which again demonstrate induction by hypoxia. The vascular basement membrane and extracellular matrix are then degraded under the balanced control of matrix metalloproteinases, and the HIF-regulated tissue inhibitor of metalloproteinase (TIMP-1). Endothelial cells proliferate and migrate into the perivascular space under the control of VEGF, angiopoietins, fibroblast growth factor (FGF), monocyte chemoattractant protein (MCP-1), and platelet-derived growth factor (PDGF), all of which are hypoxically regulated. The endothelial cells then change shape, adhere to one another, and form a lumen. Continued proliferation within the vascular wall allows for enlargement of the vessel lumen. Perivascular cells are recruited and a basal lamina is formed around the new vessel, and finally the vascular sprouts fuse to form loops that circulate the blood.

The involvement of hypoxia and in particular the HIF system at so many points in this complex process suggests that HIF directs a coordinated transcriptional response that is itself sufficient to activate productive angiogenesis. The tight association between VHL mutations that dysregulate HIF and susceptibility to retina and CNS angiomata (see later discussion) strongly suggests that dysregulated HIF activity is directly responsible for this manifestation of the disease. The central role of HIF in the regulation of angiogenesis is also supported by genetic manipulation of the system in rodents. The abnormal vascular development and embryonic lethality seen in the HIF-1α–/– knockout mouse clearly demonstrates the importance of HIF-1α to developmental angiogenesis. Further evidence of the importance of HIF to angiogenesis has come from transgenic expression of constitutively stable HIF-1α mutants or chimeric genes, which result in increased vascularity without the permeability and edema associated with VEGF overexpression alone.

HIF and Tumorigenesis

The importance of HIF in the regulation of angiogenesis, together with the striking upregulation of this system following inactivation of the VHL tumor suppressor, raises the question as to the role played by HIF activation in the oncogenic process. Currently this important question is not fully resolved. However, recent advances in the characterization of the HIF/VHL system have enabled the question to be considered in VHL-associated malignancy and in cancer in general. In the following, we have summarized current knowledge as to the role of HIF in both non-VHL and VHL-associated malignancy.

Non-VHL Associated Cancer

Upregulation of the HIF system is observed in many common cancers and occurs by a multiplicity of genetic and environmental factors. Microenvironmental activation of HIF in cancer occurs at the simplest level by physiological activation of the oxygen-sensitive pathways by hypoxia within the growing mass of cells. However, in addition to VHL inactivation, a wide range of tumor suppressor and oncogenic mutations have been reported to activate the HIF system by a variety of mechanisms. These include the association of p53 loss-of-function with decreased ubiquitylation of HIF-1α; PTEN loss-of-function, PI3K/AKT/FRAP signaling, and SRC gain-of-function with increased HIF-1α synthesis; and $p14^{ARF}$ loss-of-function with decreased nucleolar sequestration of HIF-1α.

In addition to promoting angiogenesis, activation of the HIF system may also mediate metabolic abnormalities associated with cancer. Tumor cells are characterized by a marked increase in glycolytic metabolism, even when cultured in the presence of high oxygen concentrations. First described by Otto Warburg more than 70 years ago, these abnormalities are, like angiogenesis, particularly associated with aggressive rapid growth tumors. HIF transcriptional targets include the glucose transporter isoforms, glycolytic isoenzymes and regulatory enzymes [e.g., 6-phosphofructo-2-kinase/fructose-2,6-bisphosphatase-3 (PFKFB3)] whose upregulation is classically associated with the Warburg effect. Thus HIF activation mediates coordinate transcriptional activation of the entire pathway, from glucose uptake to lactate production, providing a molecular explanation for this "metabolic switch." HIF transcriptional targets include other molecules that have potential protumorigenic actions, such as transforming growth factor-α and the c-Met protooncogene that mediates hepatocyte growth factor signaling.

These observations, together with clinico-pathological studies that have associated high levels of HIF-α subunits in human tumors with an aggressive phenotype, have led to the proposal that HIF activation contributes causally to the oncogenic processes. However, not all data support this conclusion. First, the correlation with tumor aggression is not universal. Second, some HIF-inducible genes are apparently anti-tumorigenic. For instance, in mouse embryonic stem cells HIF-1α can have proapoptotic effects, possibly through induction of target genes such as NIP3, which is involved in apoptotic pathways and binds Bcl-2. Third, though most studies of experimental tumor growth by cells bearing genetic modifications of the HIF system have demonstrated pro-tumorigenic effects, this finding has again not been universal, with some studies reporting the opposite finding. One straightforward explanation for these apparently conflicting results is that the protumorigenic effects of HIF activation are cell type specific—such a phenomenon being commonly observed with tumor suppressor inactivation. Potentially in keeping with this, VHL-associated tumor formation is highly tissue specific. Nevertheless even in VHL disease the causative role of HIF activation in the oncogenic process as opposed to the development of benign angiomata remains unclear.

VHL-Associated Cancer

VHL disease shows clear genotype–phenotype associations so that effects of particular mutations on HIF activation can be correlated with particular patterns of tumor predisposition. Families have been divided into type 1 or type 2A, 2B, or 2C, according to the risk of renal cell carcinoma, pheochromocytoma, and hemangioblastoma (Table I). The tight correlation between hemangioblastoma susceptibility and HIF dysregulation strongly supports a causal role of HIF activation in these benign vascular tumors consistent with the mechanistic findings linking the VHL/HIF pathway to the regulation of angiogenic growth factors. The associations are also compatible with a role for HIF activation in the development of renal cell carcinoma, but they indicate that HIF activation is unlikely to underlie pheochromocytoma development, strongly suggesting the existence of other VHL tumor suppressor mechanisms.

In kidneys removed from patients with VHL disease a very large number of foci of HIF activation can be observed and enhanced vascular density can be observed adjacent to these microscopic lesions. This suggests that HIF-mediated enhancement of angiogenesis is an early event following VHL inactivation and that other events (most probably several other events) are required for tumor formation. Though indices of cell proliferation are low in these early lesions, they must in some way provide an advantageous background on which subsequent oncogenic mutations can progress cancer development. Whether this is related to HIF activation and the early enhancement of angiogenesis is unclear.

Some direct support for a role of HIF in renal cell carcinoma development has been provided by studies of the effect of transfected HIF-α genes on experimental tumors derived from renal cell carcinoma cell lines and grown in

Table I Genotype–Phenotype Correlations in VHL Disease.

Type of VHL disease	Most prevalent type of VHL mutation	Effect on pVHL function in the HIF pathway	Tumor risk		
			Pheochromocytoma	Hemangioblastoma	Renal cell carcinoma
Type 1	Loss or protein misfolding	Impaired binding to HIF-α. Upregulation of HIF and its target genes	Low	High	High
Type 2a	Missense	Impaired binding to HIF-α. Upregulation of HIF and its target genes	High	High	Low
Type 2b	Missense	Impaired binding to HIF-α. Upregulation of HIF and its target genes	High	High	High
Type 2c	Missense	Retains ability to bind and degrade HIF-α	Yes	No	No

nude mice. Transfection of VHL-defective renal cell carcinoma lines with a wild-type VHL gene suppresses tumor formation, but this tumor suppression can be relieved by supertransfection with a HIF-α gene that has been mutated so as to retain transcriptional activity but escape proteolytic destruction by VHL. Interestingly reversal of VHL tumor suppression was observed for HIF-2α but not HIF-1α, suggesting that HIF-2α but not HIF-1α upregulation contributes to the oncogenic process in these tumors. Despite these findings it is notable that direct (as opposed to indirect by VHL) mutational activation of HIF has not yet been observed in this tumor type—suggesting that mutational selection driving the tumor development may be affecting other VHL-dependent pathways.

A variety of VHL interacting proteins and VHL functions have been described that are apparently unrelated to effects on the HIF system. These include effects on other transcription factors such as Sp1, effects on the general transcriptional apparatus, effects on signaling molecules such as protein kinase C, and effects on matrix metabolism. Perhaps of greatest interest is evidence for a role of VHL in fibronectin assembly, given associations between abnormal fibronectin function and the cancer phenotype in other settings.

Interaction assays have demonstrated that pVHL can bind to fibronectin, and it is notable that type 2C, pheochromocytoma-causing pVHL mutants, which retain normal HIF-α regulation, are defective in fibronectin binding, suggesting a possible causative role in formation of these tumors. How intracellular pVHL interacts with the secreted fibronectin protein is uncertain. There is no evidence that pVHL is secreted into the lumen of the endoplasmic reticulum, although some fibronectin (possibly malfolded or misprocessed) may undergo retrograde transport to the cytosolic surface and be accessible to pVHL. What is clear, however, is that although cells lacking pVHL secrete fibronectin, they are defective with respect to fibronectin-matrix assembly. This phenotype is potentially complicated by the HIF dependence of several other genes involved in extracellular matrix turnover, such as tissue inhibitors of metalloproteinases (TIMPs), matrix metalloproteinases (MMPs), and collagen prolyl hydroxylase. However, VHL-deficient cells are unable to process exogenous fibronectin, suggesting a problem with pVHL handling rather than synthesis, and have been shown to have a defect in β1-integrin function, which is involved in assembly of fibrillary adhesions.

Thus in summary HIF-mediated angiogenesis most likely accounts for VHL-associated hemangioblastoma, but its relation to other aspects of the cancer predisposition is less clear. If HIF activation is not directly implicated in causing tumor development, it is necessary to consider other explanations for the common association of HIF activation with aggressive malignancy, and the paradox that although the HIF pathway is commonly activated by oncogenic and tumor suppressor mutations, it is not itself subject to oncogenic mutation.

A potential explanation exists in the function of HIF as a coordinator of extensive physiological pathways that link the metabolic demands of a proliferating cell mass to the need for development of an adequate oxygen supply. The multiple links between the stimuli that promote cell proliferation and the activation of HIF suggest that such pathways are "hard-wired" into development as part of the fundamental physiological need to preserve oxygen homeostasis. In this case, the classical genetic model of cancer involving the progressive selection of mutant cells that manifest a cell autonomous survival advantage would be expected to evolve other new properties, such as activation of HIF and associated angiogenic or metabolic phenotypes through the coselection of these intrinsically linked pathways. Such a coselection process would be compatible with the observation that hypoxic and angiogenic pathways are indirectly rather than directly activated by oncogenic mutations, and could also explain how properties such as angiogenesis, which confer a common advantage to a mass of cells rather than a cell autonomous advantage, can be selected. Since the concept of clonal selection in cancer requires that a cell gain predominance over its neighbors by individual rather than group advantage, it is otherwise difficult to understand how selection of an angiogenic phenotype occurs. The operation of coselection also predicts that angiogenesis might be activated in a disorganized way since activation of the relevant pathways reflects links to stochastic events driving clonal selection rather than a coordinated physiological response. If the links between a tumor suppressor gene or oncogene and proangiogenic pathways are strong, then early and excessive angiogenesis might be expected, as is the case for VHL disease. If the links were less powerful, then angiogenesis might lag behind tumor growth, and then develop more rapidly when further mutations are selected with more robust links to proangiogenic pathways. Given the enormous complexity of the pathways that are being revealed by molecular analysis of cell physiology, it is likely that the role played by coselection in the generation of cancer phenotypes has been significantly underestimated. Further analysis of the HIF system in cancer biology should be illuminating in this respect.

Glossary

Angiogenic switch: A discrete step in the development of tumors in which the balance between pro- and anti-angiogenic factors changes to favor constant growth of new tumor blood vessels. It may occur at different stages of the tumor progression pathway, depending on the type of tumor and its microenvironment, but generally occurs early. Prior to the switch tumor growth is limited to a few cubic millimeters.

Hypoxia-inducible factor: A heterodimeric transcription factor that senses and responds to ambient oxygen tension, through oxygen-dependent regulation of its abundance and transactivating ability. This in turn controls the transcription of whole families of hypoxia-inducible genes, including many involved in the regulation of angiogenesis.

Ubiquitin ligase: A protein complex that recognizes its substrate and covalently attaches a series of ubiquitin molecules. This polyubiquitin chain then tags the protein for destruction by a cellular structure known as the proteasome. pVHL acts as the recognition component for a ubiquitin ligase complex that targets HIF-α.

Bibliography

Bergers, G., and Benjamin, L. E. (2003). Tumorigenesis and the angiogenic switch. *Nat. Rev. Cancer* **3**(6), 401–410. *Using a multistep model of tumorigenesis, this review examines the impact of angiogenesis on tumor development and progression.*

Clifford, S. C., Cockman, M. E., Smallwood, A. C., Mole, D. R., Woodward, E. R., Maxwell, P. H., Ratcliffe, P. J., and Maher, E. R. (2001). Contrasting effects on HIF-1alpha regulation by disease-causing pVHL mutations correlate with patterns of tumourigenesis in von Hippel–Lindau disease. *Hum. Mol. Genet.* **10**, 1029–1038. *This paper characterizes the phenotype–genotype correlations in VHL syndrome.*

Hanahan, D., and Folkman J. (1996). Patterns and emerging mechanisms of the angiogenic switch during tumorigenesis. *Cell* **86**(3), 353–364.

Kaelin, W. G., Jr. (2002). Molecular basis of the VHL hereditary cancer syndrome. *Nat. Rev. Cancer* **2**(9), 673–682. *This is a clear and concise review of the VHL syndrome, and the known functions of the VHL protein, with respect to both HIF and other interacting proteins.*

Maxwell, P. H., Pugh, C. W., and Ratcliffe, P. J. (2001). Activation of the HIF pathway in cancer. *Curr. Opin. Genet. Dev.* **11**(3), 293–299.

Maxwell, P. H., Wiesener, M. S., Chang, G. W., Clifford, S. C., Vaux, E. C., Cockman, M. E., Wykoff, C. C., Pugh, C. W., Maher, E. R., and Ratcliffe, P. J. (1999). The tumour suppressor protein VHL targets hypoxia-inducible factors for oxygen-dependent proteolysis. *Nature* **399**(6733), 271–275. *This is the first demonstration that pVHL is involved in the proteolytic regulation of HIF-α subunits.*

Pugh, C. W., and Ratcliffe, P. J. (2003). Regulation of angiogenesis by hypoxia: Role of the HIF system. *Nat. Med.* **9**(6), 677–684. *This is a clear review of the regulation of HIF by oxygen, and a summary of the action of hypoxia on molecular signals involved in different stages of angiogenesis. It forms part of a special issue of reviews on angiogenesis.*

Semenza, G. L. (2000). Hypoxia, clonal selection, and the role of HIF-1 in tumor progression. *Crit. Rev. Biochem. Mol. Biol.* **35**(2), 71–103.

Capsule Biographies

Dr. David R. Mole is Clinical Lecturer in Nephrology and General Medicine at Oxford University. He has worked on the oxygen-dependent interaction between pVHL and HIF-α, defining post-translational modification by prolyl hydroxylation as a key regulatory event leading to recognition of HIF-α by the VHL ligase complex in normoxia. By demonstrating a requirement for iron, oxygen, and 2-oxoglutarate in this hydroxylation of prolyl residues, he has helped to define the involvement of a new family of 2-oxoglutarate-dependent dioxygenases (PHD1, 2, and 3) in the regulation of HIF-α abundance.

Peter J. Ratcliffe is Professor of Renal Medicine and Head of the Nephrology and Cell Physiology Group at the Henry Wellcome Building of Genomic Medicine, University of Oxford. Initially studying the regulation of erythropoietin by the kidney, his group was the first to recognize the general role of this pathway in directing cellular responses to hypoxia. He has published more than 100 papers in the field of cellular oxygen sensing and the HIF/VHL/hydroxylase pathway, and in 2002 was elected to the Fellowship of the Royal Society.

CHAPTER 146

Angiostatin Inhibits Signaling in Endothelial and Smooth-Muscle Cells

David C. Sane
Section of Cardiology, Department of Internal Medicine, Wake Forest University School of Medicine, Winston-Salem, North Carolina

Overview

Angiogenesis is the process of generating new capillaries from preexisting blood vessels. Excessive angiogenesis contributes to the pathophysiology of a variety of diseases including tumor growth, metastasis, and proliferative diabetic retinopathy. Inadequate angiogenesis, on the other hand, is a factor in delayed wound healing and ischemic vascular disease. It follows, that angiogenesis is a tightly regulated process, controlled by a variety of growth factors and inhibitors. In this review, we consider the molecular mechanisms of action of angiostatin, one of the known inhibitors of angiogenesis. Although the major interest in angiostatin has surrounded its use in inhibiting tumor angiogenesis, its biological effects are not limited to this arena. Angiostatin could affect virtually any vascular function in which a healthy, proliferating endothelium is required. Furthermore, there is evidence that angiostatin also targets other cells, including circulating endothelial progenitor cells, smooth muscle cells, and even neutrophils. In this review, we consider the multiple mechanisms by which angiostatin could affect microvascular structure and function.

Angiostatin: An Inhibitor of Angiogenesis

Angiostatin is a 38-kDa protein, derived from an internal fragment of plasminogen spanning the first four kringle domains that. Kringles 1–3 of plasminogen also exhibit antiangiogenic activity. The conversion of plasminogen to angiostatin appears to require reduction of disulfide bonds followed by proteolysis with one or more proteinases. Angiostatin was originally purified from the urine of mice with Lewis lung carcinoma and is generated endogenously by a variety of human tumors as well. Angiostatin inhibits neovascularization in the mouse corneal angiogenesis assay and in the chick chorioallantoic membrane assay. Angiostatin inhibits the growth of primary tumors and metastases in mice and induces complete regression of human tumors implants in nude mice.

Cellular Mechanisms for the Antiangiogenic Activity of Angiostatin

Angiostatin inhibits proliferation, migration, and tube formation by endothelial cells. Angiostatin induces the apoptosis of endothelial cells in culture and in mouse models. Although most studies have focused on the endothelial effects of angiostatin, other cells may also be targets for angiostatin. Angiostatin may have a more potent effect on circulating, bone marrow–derived endothelial progenitor cells than it has on endothelial cells. We have shown that smooth muscle cell (SMC) migration and proliferation are inhibited by angiostatin. Furthermore, hepatocyte growth factor (HGF) signaling pathways were inhibited by angiostatin in SMCs. Benelli et al. [1] have demonstrated that angiostatin inhibits monocyte migration to MCP-1 and fMLP. An even more potent effect of angiostatin was exhibited in inhibiting neutrophil migration to CXCR1 and CXCR2 agonists (IL-8 and MIP-2). These nonendothelial cellular effects of angiostatin may be important because leukocytes contribute to angiogenesis and smooth muscle cells are essential for arteriogenesis.

Proposed Molecular Targets of Angiostatin

One of the first known binding sites for angiostatin on endothelial cells was the F_1–F_0 ATP synthase (Figure 1). The ATP synthase has historically been considered to be an exclusively mitochondrial enzyme. Moser et al. not only localized ATP synthase on the endothelial cell (EC) surface, but demonstrated that it is active in extracellular ATP synthesis. Antibodies to ATP synthase prevented the antiproliferative effect of angiostatin on endothelial cells. The proton pumping function of ATP synthase is also blocked by angiostatin, endangering the ability of ECs to survive in hypoxic, acidic microenvironments.

Tarui and colleagues reported that bovine aortic endothelial cells adhered to angiostatin via the integrin receptor $\alpha_v\beta_3$. Since angiostatin did not induce stress fiber formation upon binding to the integrin, these investigators hypothesized that angiostatin competes with the natural ligands for $\alpha_v\beta_3$, producing an antiangiogenic effect. Blockade of $\alpha_v\beta_3$ by angiostatin could inhibit EC migration. It is also possible that this effect reduces MMP-2 activation, which is enhanced by the ligation of pro-MMP-2 with $\alpha_v\beta_3$.

The competitive inhibition of $\alpha_v\beta_3$ by angiostatin could alter focal adhesion kinase (FAK) activity. Integrins do not possess intrinsic kinase activity, but instead form complexes with signaling molecules at focal adhesion contact sites. FAK is a tyrosine kinase that is activated by integrin clustering, although there are also other mechanisms for activating FAK independent of cell–ECM interactions. FAK is also activated by ligation of some growth factor receptors. Claesson-Welch and colleagues [2] have shown that angiostatin upregulates FAK activity in endothelial cells in an RGD-independent and therefore probably integrin-independent fashion. These investigations suggested that the aberrant upregulation of FAK could contribute to inhibition of EC migration and to the induction of apoptosis.

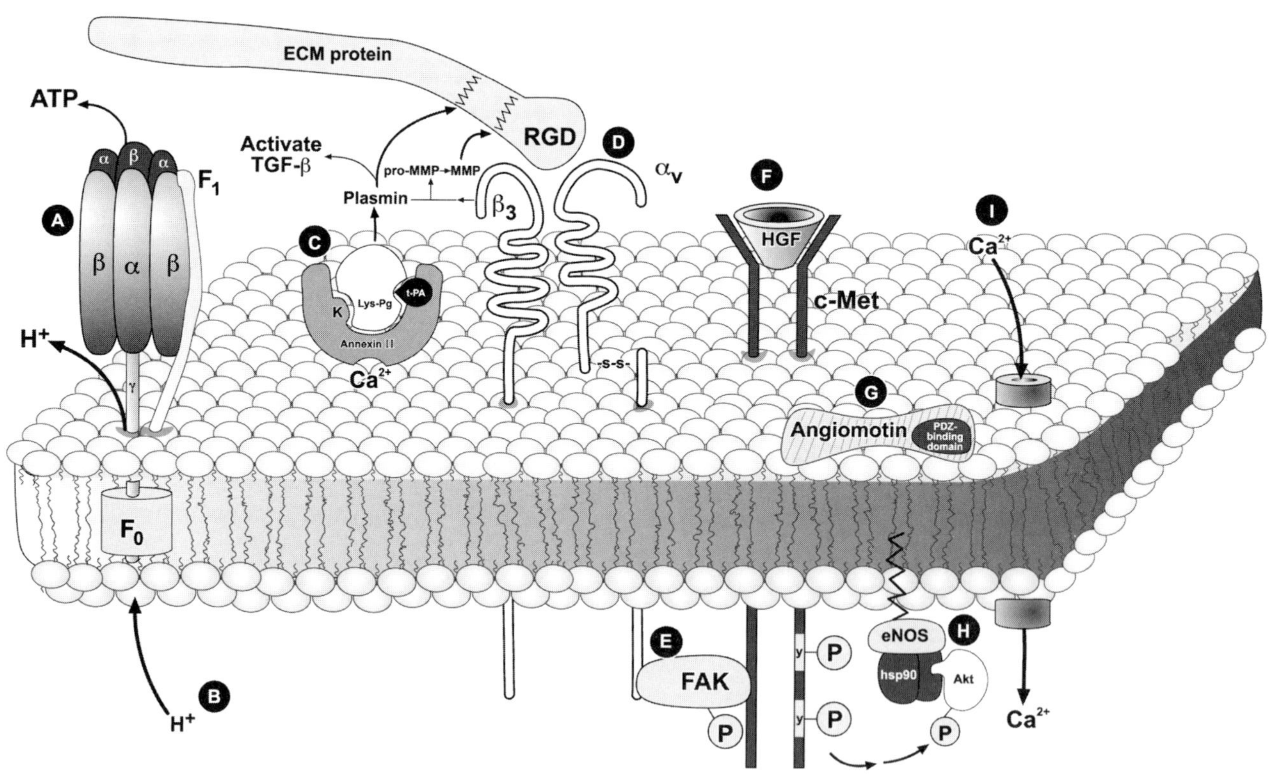

Figure 1 Reported molecular targets of angiostatin. A section of a cell membrane is diagrammed showing the molecular targets of angiostatin. Angiostatin binds to the α/β subunits of ATP synthase **(A)**, limiting extracellular ATP synthesis and increasing the intracellular $[H^+]$ **(B)**. Angiostatin also binds to annexin II **(C)**, displacing lys-plasminogen, thereby preventing plasmin generation. Plasmin has several important pericellular activities, including activation of TGF-β, degrading ECM proteins, and activating MMPs. Integrin receptors may also contribute to MMP activation. Angiostatin has also been reported to bind to the vitronectin receptor ($\alpha_v\beta_3$), probably inhibiting the binding of ECM protein ligands **(D)**. Focal adhesion kinase (FAK) is coupled to the intracellular domain of integrin receptors and also to some growth factor receptors **(E)**. Angiostatin has been reported to activate FAK phosphorylation. Angiostatin also prevents signaling through the hepatocyte growth factor (HGF) c-Met pathway **(F)**, by binding to c-Met. One of the actions of HGF c-Met that is blocked by Angiostatin is phosphorylation and activation of Akt (protein kinase B). Angiostatin inhibits migration and tube formation that is promoted by the cell-surface associated protein angiomotin **(G)**. Angiostatin uncouples eNOS from hsp90 **(H)**, leading to reduced production of NO, with increased production of $O_2^{\bullet -}$. Finally, angiostatin promotes Ca^{2+} influx by an unknown mechanism **(I)**. Several molecular targets of angiostatin, including ATP synthase, eNOS, integrins, and annexin II, may associate with caveolae, representing a potential mechanism by which an inhibitory effect on a variety of targets could be enhanced.

We have demonstrated that angiostatin inhibits HGF-induced signaling in endothelial and smooth muscle cells. We hypothesized that angiostatin, which has significant homology to HGF, would competitively inhibit HGF from binding to its receptor (c-Met). Angiostatin inhibited the HGF-induced tyrosine phosphorylation of c-Met, as well as the phosphorylation of Akt and ERK1/2.

Angiostatin-mediated EC apoptosis has been reported to involve calcium signaling. Jiang and colleagues [3] have shown that angiostatin induces an acute rise in $[Ca^{2+}]_i$ of human and bovine endothelial cells. Treatment with endostatin evoked a similar response and attenuated the acute elevation in $[Ca^{2+}]_i$ with subsequent administration of VEGF and FGF-2. Angiostatin is able to deplete thapsigargin-sensitive intracellular Ca^{2+} stores that are normally released by muscarinic agonists. The cellular target responsible for the effects on $[Ca^{2+}]_i$ in response to angiostatin are currently unknown. Kringle 5 of plasminogen, also an inhibitor of angiogenesis, has been reported to induce a rise in $[Ca^{2+}]_i$ in HUVEC by binding to a cell-surface voltage-dependent anion channel [4]. It is unknown whether angiostatin affects the same or similar ion channels.

Angiostatin may inhibit pericellular plasmin generation, limiting the ability of the EC to dissolve its basement membrane, and preventing migration in the early phases of angiogenesis. There are at least two mechanisms for this effect. Stack and colleagues [5] demonstrated that angiostatin binds to t-PA and inhibits plasminogen activation. Another mechanism could be through the interaction of angiostatin with annexin II. Annexin II is an endothelial cell surface protein that binds to both lys-plasminogen and t-PA, enhancing plasmin generation. Tuszynski et al. reported that angiostatin binds to annexin II on the EC surface, providing another potential mechanism by which angiostatin could inhibit plasmin production at the EC surface. Plasmin has important roles in angiogenesis by activating MMPs and TGF-β1 and promoting cell migration.

Angiomotin is a 72-kDa cell surface associated protein localized to the lamellipodia of the leading edge of migrating capillary endothelial cells and is colocalized with FAK. Angiomotin-transfected cells also have increased FAK activity in response to angiostatin. When angiomotin-transfected cells are treated with angiostatin, migration, and tube formation are significantly inhibited. Although a detailed mechanism of action of angiomotin has not been determined, a PDZ-binding domain appears to have a critical role.

Nontumor Effects of Angiostatin

Angiostatin could have important effects on a variety of tissue targets, its benefit not being exclusively limited to the tumor microvasculature. Advanced atherosclerosis and neointimal hyperplasia are associated with extensive neovascularization. Angiostatin inhibits atherosclerosis development in apolipoprotein E null mice and reduces neointimal hyperplasia in response to balloon injury in rabbits. Adipose tissue is highly vascularized and exhibits angiogenic properties. Weight reduction in response to angiogenesis inhibitors is accompanied by evidence of adipose endothelial cell apoptosis, suggesting an antiangiogenic mechanism of action. Angiostatin is effective in reducing weight gain in obese (ob/ob) mice.

Angiostatin could be generated in virtually any tissue with the appropriate environmental conditions: ischemia, high oxidative stress, reduced NO production, and elevated MMP-2, -7, and -9 activity. Angiostatin has been demonstrated in tear fluid, where it has been proposed to protect against corneal neovascularization. In a canine model of repetitive coronary ischemia, treatment with L-NAME was associated with the accumulation of angiostatin in the myocardial interstitial fluid. It was postulated that angiostatin production may limit the development of coronary artery collaterals in response to ischemia.

An endothelium with intact angiogenesis capability may be important in the health of many tissues. Therefore angiostatin's actions on the microvasculature, especially its effects on the endothelium, could produce effects that extend well beyond an exclusive role in inhibiting angiogenesis. An example is the reported ability of angiostatin to impair endothelium-dependent vasodilation. By reducing the interaction between eNOS and hsp90, angiostatin enhances $O_2^{\cdot-}$ production, while decreasing NO production. This effect could regulate vasomotor tone and also have antiangiogenic effects, since many angiogenic growth factors including VEGF and FGF induce NO production. Furthermore, increased NO production with vasodilation is an early step in capillary beds undergoing angiogenesis. It is also possible that angiostatin could also inhibit vasodilation by blocking $\alpha_v\beta_3$. Mogford et al. [6] have demonstrated that RGD peptides applied to rat arterioles produced concentration-dependent, sustained endothelium-independent vasodilation. Vasodilatory RGD-containing peptides can be produced endogenously during collagenolysis. These fragments may be produced during arterial remodeling and may contribute to maintaining luminal expansion.

The production of angiostatin during tissue hypoxia could exacerbate the development of pulmonary hypertension. The expression of angiogenic growth factors may be one protective mechanism to reduce the severity of pulmonary arterial hypertension (PAH). When angiostatin is overexpressed in the lung of chronically hypoxic mice, they developed more severe pulmonary artery hypertension, right ventricular hypertrophy, and muscularization of pulmonary artery vessels [7]. It is possible that a strategy aimed at inhibiting endogenous angiostatin production could slow the progression of PAH.

Summary

Angiostatin is an interesting inhibitor of angiogenesis with several reported molecular targets. The principal targets for its effect on endothelial and other cells needs to be clarified. Angiostatin and other angiogenesis inhibitors may have a variety of effects on normal vascular function both in modulating vascular anatomy and in producing transient effects on vasomotor tone.

Glossary

Angiogenesis: The process of sprouting new capillaries from existing blood vessels. Although classically distinguished from vasculogenesis, the de novo formation of blood vessels during embryogenesis, recent studies have demonstrated overlap in these processes. Circulating endothelial progenitor cells may engraft in the vessel wall and contribute to neovascularization in the adult.

Angiostatin: An anti-angiogenic 38-kDa protein derived by proteolysis of plasminogen, consisting of the first 3 or 4 kringle domains.

Hepatocyte growth factor: An angiogenic growth factor also known as "scatter factor" that stimulates mitogenesis, motogenesis, and morphogenesis in cells that express its receptor, c-met.

References

1. Benelli, R., Morini, M., Carrozzino, F., Ferrari, N., Minghelli, S., Santi, L., Cassatella, M., Noonan, D. M., and Albini, A. (2002). Neutrophils as a key cellular target for angiostatin: implications for regulation of angiogenesis and inflammation. *FASEB J* **16**, 267–269.
2. Claesson-Welsh, L., Welsh, M., Ito, N., Anand-Apte, B., Soker, S., Zetter, B., O'Reilly, M., and Folkman, J. (1998). Angiostatin induces endothelial cell apoptosis and activation of focal adhesion kinase independently of the integrin-binding motif RGD. *Proc. Natl. Acad. Sci. USA* **95**, 5579–5583.
3. Jiang, L., Jha, V., Dhanabal, M., Sukhatma, V. P., and Alper, S. I. (2001). Intracellular Ca^{2+} signaling in endothelial cells by the angiogenesis inhibitors endostatin and angiostatin. *Am. J. Physiol. Cell Physiol.* **280**, C1140–C1150.
4. Gonzalez-Gronow, M., Kalfa, T., Johnson, C. E., Gawdi, G., and Pizzo, S. V. (2003). The voltage-dependent anion channel is a receptor for plasminogen kringle 5 on human endothelial cells. *J. Biol. Chem.* **278**, 27312–27318.
5. Stack, M. S., Gately, S., Bafetti, L. M., Enghild, J. J., and Soff, G. A. (1999). Angiostatin inhibits endothelial and melanoma cellular invasion by blocking matrix-enhanced plasminogen activation. *Biochem. J.* **340**, 77–84.
6. Mogford, J. E., Davis, G. E., Platts, S. H., and Meininger, G. A. (1996). Vascular smooth muscle alpha sub v beta sub 3 integrin mediates arteriolar vasodilation in response to RGD peptides. *Circ. Res.* **79**, 821–826.
7. Pascaud, M-A., Griscelli, F., Raoul, W., Marcos, E., Opolon, P., Raffestin, B., Perricaudet, M., Adnot, S., and Eddahibi, S. (2003). Lung overexpression of angiostatin aggravates pulmonary hypertension in chronically hypoxic mice. *Am. J. Respir. Cell Miol. Biol.* **29**, 449–457.

Further Reading

Koshida, R., Ou, J., Matsunaga, T., Chilian, W. M., Oldham, K. T., Ackerman, A. W., and Pritchard, K. A., Jr. (2003). Angiostatin. A negative regulator of endothelial-dependent vasodilation. *Circulation* **107**, 803–806. *This study demonstrates a novel activity of angiostatin as a modulator of vascular tone. Angiostatin uncoupled eNOS from hsp90, decreasing NO production, while increasing $O_2^{\bullet}$.*

Matsunaga, T., Weihrauch, D. W., Moniz, M.C., Tessmer, J., Warltier, D. C., and Chilian, W. M. (2002). Angiostatin inhibits coronary angiogenesis during impaired production of nitric oxide. *Circulation* **105**, 2185–2191.

Moser, T. L., Stack, M. S., Wahl, M., and Pizzo, S. V. (2002). The mechanism of action of angiostatin: Can you teach an old dog new tricks? *Thromb. Haemost.* **87**, 394–401. *An excellent review of angiostatin with a focus on one of its first known cellular targets, ATP synthase.*

Moulton, K. S., Vakili, K., Zurakowski, D., Soliman, M., Butterfield, C., Sylvin, E., Lo, K.-M., Gillies, S., Javaherian, K., and Folkman, J. (2003). Inhibition of plaque neovascularization reduces macrophage accumulation and progression of advanced atherosclerosis. *Proc. Natl. Acad. Sci. USA* **100**, 4736–4741.

O'Reilly, M. S., Homgren, L. L., Shing, Y., Chen, C., Rosenthal, R. A., Moses, M., Lane, W. S., Cao, Y., Sage, E. H., and Folkman, J. (1994). Angiostatin: A novel angiogenesis inhibitor that mediates the suppression of metastases by a Lewis lung carcinoma. *Cell* **79**, 315–328. *The original description of angiostatin with a particularly interesting discussion on the clinical observations suggesting that some primary tumors elaborate angiogenesis inhibitors. The discovery of a novel angiogenesis inhibitor derived from a known protein had major implications for the search for other novel bioactive protein derivatives and served as an excellent example for the value of proteomic-based solutions to diseases.*

Rupnick, M. A., Panigraphy, D., Zhang, C.-Y., Dallabrida, S. M., Lowell, B. B., Langer, R., and Folkman, M. J. (2002). Adipose tissue mass can be regulated through the vasculature. *Proc. Natl. Acad. Sci. USA* **99**, 10730–10735. *This study demonstrates that obesity can be inhibited in mouse models by treatment with angiogenesis inhibitors including angiostatin. The study emphasizes the potential roles that angiogenesis inhibitors could have in a variety of diseases.*

Tarui, T., Miles, L. A., and Takada, Y. (2001). Specific interaction of angiostatin with integrin $\alpha_v\beta_3$ in endothelial cells. *J. Biol. Chem.* **276**, 39562–39568.

Troyanovsky, B., Levchenko, T., Månsson, G., Matvijenko, O., and Holmgren, L. (2001). Angiomotin: An angiostatin binding protein that regulates endothelial cell migration and tube formation. *J. Cell. Biol.* **152**, 1247–1254.

Tuszynski, G. P., Sharma, M. R., Rothman, V. L., and Sharma, M. C. (2002). Angiostatin binds to tyrosine kinase substrate annexin II through the lysine-binding domain in endothelial cells. *Microvasc. Res.* **64**, 448–462.

Wajih, N., and Sane, D. C. (2003). Angiostatin selectively inhibits signaling by hepatocyte growth factor in endothelial and smooth muscle cells. *Blood* **101**, 1857–1863.

Capsule Biography

David C. Sane, M.D., is an Associate Professor of Internal Medicine in the Section of Cardiology, Wake Forest University School of Medicine, Winston-Salem, North Carolina. Dr. Sane has studied structure–function relationships of the extracellular matrix protein vitronectin and its interaction with tranglutaminases. He is also interested in thrombosis and vascular biology and has examined the effect of angiostatin on HGF/C-met signaling.

Endothelial Cell Transformation by the Kaposi's Sarcoma–Associated Herpesvirus/Human Herpesvirus-8

Akrit Sodhi and J. Silvio Gutkind

Cell Growth Regulation Section, Oral and Pharyngeal Cancer Branch, National Institute of Dental Research, National Institutes of Health, Bethesda, Maryland

Summary

In 1872 dermatopathologist Moritz Kaposi first described Kaposi's sarcoma (KS) as a rare "idiopathic multiple pigmented sarcoma of the skin" found primarily in the lower extremities of elderly Mediterranean men. For more than a century since this initial description, despite extensive investigation by clinicians and basic science researchers, the cause of KS remained a mystery. In 1994, the Kaposi's sarcoma–associated herpesvirus/human herpesvirus-8 (KSHV/HHV-8) was identified as the etiologic agent for KS. KSHV is believed to infect and immortalize endothelial cells, thereby generating the KS tumor (or spindle) cell. Molecular characterization of the KSHV genome by numerous laboratories has since revealed several genes that harbor potential for KSHV pathogenesis. Examination of candidate KSHV oncogenes is shedding new light onto the cause of this enigmatic tumor and has further provided considerable insight into the complex interplay among angiogenic growth factors, endothelial cells, and cancer.

Introduction

Kaposi's sarcoma (KS) is the most common cancer arising in patients with acquired immunodeficiency syndrome (AIDS). KS is a multifocal neovascular tumor that most often manifests with skin lesions, although involvement of the oral mucosa, lymph nodes, and visceral organs is not uncommon. The recent introduction of highly active antiretroviral therapy (HAART) has witnessed a dramatic decrease in the proportion of new AIDS-defining KS cases as well as a regression in the size of existing KS lesions. However, in parts of the developing world, KS has tragically emerged as one of the most frequent cancers among children and adult men, and it can be an aggressive disease that remains a significant cause of morbidity and mortality among the AIDS population [1].

The clinical course of AIDS-related KS is highly variable, ranging from minimal stable disease to explosive growth. In early-stage KS (also known as patch stage), dermal lesions are composed of small, irregular, endothelium-lined spaces that surround normal blood vessels, with perivascular infiltrates of lymphocytes and plasma cells. As the disease progresses into the plaque stage, the tumor (or spindle) cells form slit-like, erythrocyte-replete vascular channels that expand throughout the dermis. In late (or nodular) stage lesions, the dermis is replaced by sheets of spindle cells with interspersed vascular spaces that give the lesion the appearance of a honeycomb. More aggressive lesions are characterized by mitotic events and cellular atypia [2].

The dominant cell of KS lesions, the spindle cell, is believed to be of endothelial origin, although the precise histogenesis of these cells remains in debate. While not fully transformed, the KS spindle cell elaborates a variety of proinflammatory and angiogenic factors believed to be essential for the development of KS. Recent data support the

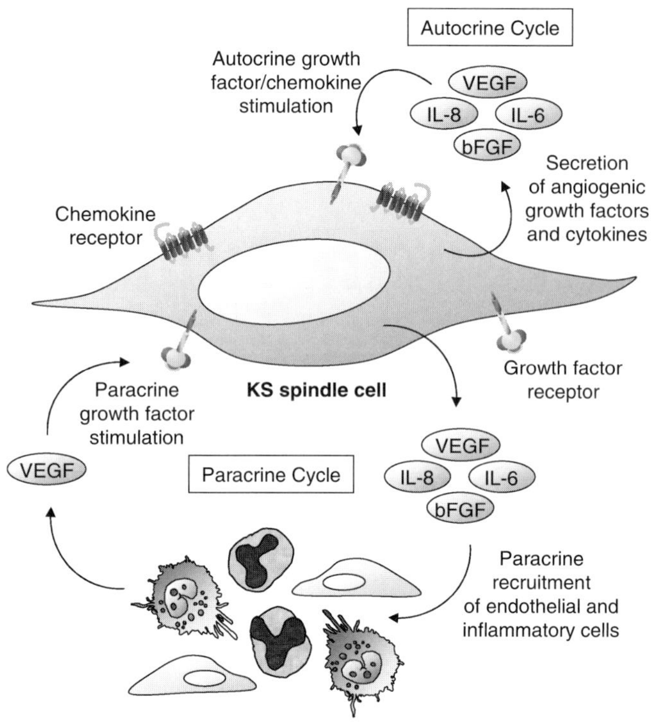

Figure 1 Autocrine and paracrine growth factor cycles promoted by the Kaposi's sarcoma spindle cell. The KS spindle cell, believed to be of endothelial origin, is the driving force of KS with its secreted growth factors and cytokines sustaining its own proliferation while recruiting the remaining cell types in a paracrine fashion. (see color insert)

role of the spindle cell as the driving force of KS pathogenesis (Figure 1), with its secreted products responsible for the recruitment of the remaining cell types in a paracrine fashion [3]. However, the mechanism as to what initiates spindle cell formation and how secreted growth factors and cytokines sustain spindle cell growth is still poorly understood.

Kaposi's Sarcoma: A Disease in Search of a Cause

The epidemiology of KS suggested an infectious etiology for this disease. As a consequence of its close association with the AIDS epidemic, initial research focus was concentrated on a newly identified virus, HIV, as the KS etiologic agent. Early work suggested that the potent transactivator HIV-1 protein, Tat, plays a major role in the pathogenesis of AIDS-related KS. Among many of its KS-promoting activities, the Tat protein augments the angiogenic activities of basic fibroblast growth factor (bFGF), interferon gamma, and vascular endothelial growth factor (VEGF); mimics the effects of the extramedullary matrix proteins fibronectin and vitronectin; and increases the expression of matrix metalloproteinases [4].

However, as KS is found in only a subpopulation of HIV-infected individuals, HIV did not appear to be sufficient for KS development. Moreover, the cause of KS in HIV-negative individuals remained unexplained. This suggested that a second etiologic agent or cofactor might be involved in KS pathogenesis.

Kaposi's Sarcoma–Associated Herpesvirus (KSHV or HHV8)

In 1994, Chang et al. used representational difference analysis (RDA) to search for DNA sequences present in AIDS-associated KS but not in adjacent normal skin. Using this approach, they were able to identify a then unknown human γ-herpesvirus in KS tissue, which they called the Kaposi's sarcoma–associated herpesvirus (KSHV) or human herpesvirus 8 (HHV-8) [5]. KSHV belongs to the genus *Rhadinovirus*, and its genomic structure is most similar to the closely related lymphotropic γ-herpesviruses Epstein-Barr virus (EBV), the only other known human herpesvirus associated with human cancers. Compelling evidence points to KSHV as the infectious etiologic agent for KS: (i) Epidemiological studies have established a pattern of KSHV infection that precedes KS development and overlaps with KS risks. (ii) Examination of KS biopsies revealed that KSHV DNA is present in KS tumor (spindle) cells and in endothelial cells lining vascular spaces of KS lesion. (iii) In vitro studies have demonstrated that KSHV can infect and transform endothelial cells (thought to be premalignant for KS). Of note, these studies further revealed that KSHV was present in only a subset of cultured immortalized endothelial cells, suggestive of a paracrine mechanism in the immortalization of uninfected cells [6]. (iv) Unlike HIV, KSHV has been found to be associated with all four forms of KS (classic, iatrogenic, endemic, and AIDS-related), in addition to two other neoplastic disorders: primary effusion lymphoma (PEL) and multicentric Castleman's disease (MCD) [7]. However, while PEL and MCD primarily involve KSHV-infected lymphocytes, KS involves KSHV infection of endothelial cells. As KSHV is a lymphotropic herpesvirus, it remains to be determined how and why KSHV makes the transition from lymphocyte to endothelial cell.

Following endothelial infection, KSHV can enter one of two pathways: the latent cycle or the lytic cycle. In the latent cycle, the KSHV genome is tethered to the host cell chromosome as an independent episome. Replication of the host cell results in replication of the viral genome, which is passed on to each daughter cell. The few genes that are expressed during latency function primarily to maintain the episomal viral DNA and to promote host-cell proliferation, thereby promoting replication of the viral genetic material within the latently infected host cell. Reentry into the lytic cycle (viral reactivation) is usually triggered in latently infected cells under special circumstances. Upon reactivation, viral early genes—required for the transcription and translation of the viral genome—are first expressed, which then facilitate the expression of viral structural proteins (late genes). Viral genes that play a role in sustaining host-cell survival or in modulating the host immune response are also expressed during the lytic cycle, promoting efficient production of viral progeny and ensuring successful viral propagation.

Current efforts are now focused on defining the mechanism whereby this virus causes KS. KSHV contains more than 90 open reading frames (ORFs), including several homologs to mammalian proteins likely pirated by KSHV from its cellular host. Further investigation has revealed that several of these genes bear potential for KS pathogenesis [8]. These include both latent and lytic genes that may be directly transforming or could contribute to KS indirectly by promoting the production of angiogenesis-activating growth factors and inflammatory cytokines.

KSHV Latent Genes and the Promotion of Cell Proliferation and Survival

Efforts to identify the KSHV gene(s) responsible for the genesis of KS have focused primarily on the latent genes (LANA-1, LANA-2, vCyclin, vFlip, and Kaposin). These genes are expressed in almost all spindle cells in late KS lesions and are therefore expected to play a critical role in the progression of KS. Several recent studies have suggested that many of the KSHV latent genes bear potential for promoting endothelial cell transformation (Table I).

LANA-1 and 2

In addition to maintaining the KSHV episome and ensuring its efficient segregation to progeny cells, LANA-1 has been shown to interact with and inhibit the tumor suppressor proteins p53 and pRB. Similarly, LANA-2 also inhibits p53 and may cooperate with LANA-1 to repress the function of this tumor suppressor protein. More recently, LANA-1 has also been shown to stabilize β-catenin by binding to its negative regulator GSK-3β, causing a cell cycle–dependent nuclear accumulation of GSK-3β. Although neither LANA protein is directly transforming, LANA-1 can prolong the life span of primary human endothelial cells, suggestive of a role for this latent protein in endothelial cell immortalization.

Table I Role for KSHV Latent Genes in Endothelial Cell Immortalization.

Latent gene	Mechanism(s) for enhancing cell proliferation
LANA-1	Interacts with and inhibits the two tumor suppressor proteins p53 and Rb. Stabilizes β-catenin by binding to its negative regulator GSK-3β, causing a cell cycle–dependent nuclear accumulation of GSK-3β
LANA-2	Interacts with p53, antagonizing p53-mediated apoptosis
vCyclin	Complexes with cellular cdk6 to phosphorylate and inactivate the tumor suppressor Rb and the CDK inhibitor p27, thereby promoting cell cycle progression
vFlip	Binds to procaspase-8, inhibiting its recruitment and activation at the death-induced signaling complex, resulting in diminished Fas-mediated apoptotic cascade initiation
Kaposin (K12)	Interacts with cytohesin-1, an ARF guanine nucleotide exchange factor, leading to the activation of the mitogenic kinase ERK1/2.

vCyclin and vFLIP

vCyclin, which shares significant homology to cellular cyclin D2, complexes with CDK6 to phosphorylate and inhibit pRB and other cell cycle regulators. vFlip, encoded by a bicistronic mRNA species also encoding vCyclin, is a death effector domain (DED)-containing protein that interferes with apoptosis signaled through death receptors and stimulates the transcription factor, nuclear factor kappa B (NFκB).

Kaposin

Kaposin (or K12) is a small highly polymorphic protein that activates the mitogenic activated protein kinase (MAPK) signaling pathway. Genetic mutations leading to overactivation of MAPK is a common theme in many human cancers, suggesting that Kaposin may play an analogous role in KS.

Although the latent genes are individually not transforming, endothelial cells immortalized by the SV40 Large T Ag (which mimics the inhibition of p53 by LANA-1 and -2) also stably expressing both vCyclin and vFlip formed small tumors in nude mice [9]. However, the full tumorigenic potential of this cell line was only realized in the presence of the paracrine secretions of a KSHV lytic gene, vGPCR, suggestive of a cooperative role for KSHV genes in Kaposi's sarcomagenesis.

KSHV Lytic Genes and Paracrine Neoplasia

Lytic genes are expressed during the phase of the viral life cycle when viral progeny are produced. Consequently, these viral genes are expressed in cells destined to die (lyse). It is therefore generally assumed that lytic genes are not likely to play a significant role in tumorigenesis. However, emerging evidence supports a role for paracrine secretions released by cells expressing KSHV lytic genes in the promotion of KS (Table II).

vGPCR

vGPCR is a G protein-linked receptor (GPCR) that exhibits ligand-independent activities, enabling constitutive signaling through multiple mitogenic and survival pathways. Cells expressing vGPCR also constitutively secrete

Table II Role of KSHV Lytic Genes in Paracrine Neoplasia.

Lytic gene	Paracrine mechanism(s) for enhancing cell proliferation
vGPCR	Induces VEGF transcription and secretion by activating the transcription factor HIF-1α; induces the transcription and secretion of NFκB-dependent chemokines (e.g., IL-8, Groα, IL-6); activates the transcription factor NFAT leading to upregulation of NFAT-dependent cytokines
vIL-6	Binds to gp130 signal transducing subunit of the IL-6R; promotes VEGF secretion
K1	Induces NFAT activity, leading to expression of several NFAT-dependent cytokines
vMIPs	Binds cellular chemokine receptors; proangiogenic

angiogenic growth factors and chemokines, including VEGF, interleukin-6 (IL-6), interleukin-8 (IL-8), and growth-related oncogene protein alpha (Groα). Therefore, it can be argued that in KS lesions, the expression of this viral receptor by a subset of infected cells could make a major contribution to KSHV-induced angiogenesis, functioning primarily by paracrine mechanisms.

vIL-6

The KSHV IL-6 homolog (vIL-6) has sequence similarity (25% amino acid identity) to human IL-6. Like its cellular homolog, vIL-6 activates specific JAK/STAT mitogenic signaling pathways by acting on the gp130 signal transducing subunit of the IL-6 receptor. vIL-6 promotes hematopoiesis and acts as an angiogenic factor through the induction of VEGF, suggesting that vIL-6 could play an essential role in the pathogenesis of KS as an autocrine or paracrine factor.

K1

KSHV K1 encodes a transmembrane glycoprotein that interacts with several cellular signal transduction proteins including Vav, p85, and Syk kinase. K1 also induces nuclear factor of activated T cells (NFAT) activity, leading to expression of several cytokines, supportive of a role for this lytic gene in paracrine neoplasia.

vMIPs

Three KSHV genes vMIP-I, vMIP-II, and vMIP-III, are viral chemokines (virokines) that have significant protein sequence similarity to cellular CC chemokines, a superfamily of proinflammatory cytokines. Both vMIP-I and vMIP-II are proangiogenic, and all three are agonists for cellular chemokine receptors, consistent with a paracrine role for these virokines in KS progression.

Other lytic genes (e.g., vBcl-2, vIRF-1, Rta, K8) may further contribute to Kaposi's sarcomagenesis by promoting the survival of lytically infected cells. Ongoing efforts examining the roles of these genes in KSHV pathogenesis may provide additional insight into the development of KS.

Triggering Kaposi's Sarcomagenesis

Thus, although alone not transforming, latent genes may still enhance tumor cell proliferation and survival, likely contributing to the more aggressive nature of late-stage or "nodular" KS tumors in which most tumor cells express KSHV latent genes. Similarly, lytic genes may also contribute to KS tumor progression by activating mitogenic signaling cascades by their paracrine secretions. Nonetheless, the question as to what initially triggers this unusual cancer remains unclear, as expression of KSHV latent and lytic genes does not normally result in the development of KS. In this regard, recent work suggests that dysregulated expression of a single KSHV lytic gene, *vGPCR*, may be responsible for initiating KS.

Initiation of Kaposi's Sarcomagenesis by vGPCR

Compelling evidence implicates vGPCR in the initiation of Kaposi's sarcomagenesis. In particular, when a KS model in which endothelial-specific retroviral transduction was utilized to target candidate KSHV oncogenes to the vascular endothelium of mice, it was observed that only vGPCR produced vascular tumors in mice that were strikingly similar to human KS lesions [9]. vGPCR experimental tumors expressed key histopathological and molecular hallmarks for KS, suggestive of an important role for the vGPCR in the initiation of this enigmatic disease.

Dysregulated Expression of vGPCR Triggers KS

However, as vGPCR is a lytic gene, normally expressed only in cells destined for lysis, it was unclear how it could be the gene responsible for initiating KS. One explanation could be that dysregulation of the normal viral gene program, leading to expression of vGPCR in *nonlytic* cells, may trigger KS. Under these circumstances, unregulated vGPCR signaling to mitogenic and survival pathways may promote direct cell transformation while the continuous secretion of cytokines and growth factors may induce the recruitment and subsequent indirect (paracrine) cell transformation of neighboring endothelial cells (Figure 2). vGPCR oncogenesis may thereby provide a "double-hit" that requires susceptibility to both direct and paracrine cell transformation [10], consistent with the profound vulnerability of endothelial cells to vGPCR oncogenesis. Of note, the paracrine nature of vGPCR experimental tumors is remarkably similar to paracrine transformation of KSHV-infected endothelial cells in vitro [6]. However, the circumstances leading to vGPCR dysregulation have yet to be defined. Similarly, the contribution of other KSHV latent and lytic genes to the initiation and progression of KS remains poorly understood.

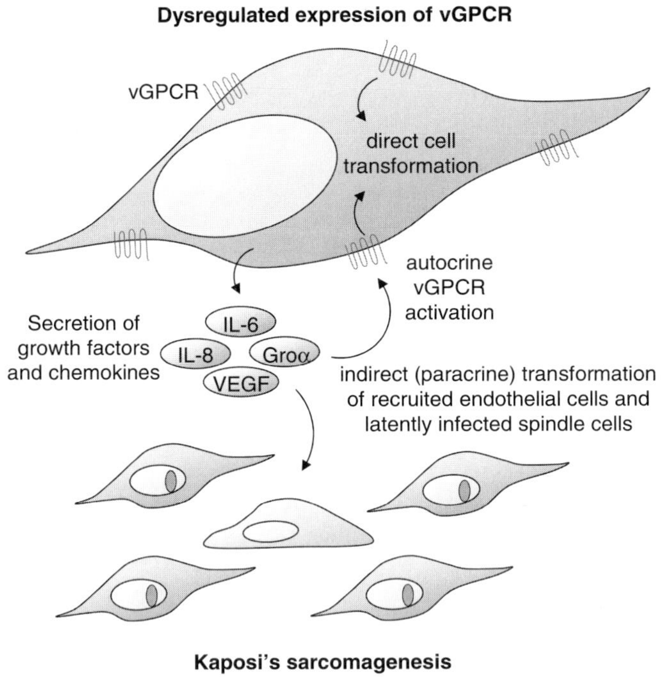

Figure 2 Dysregulated (nonlytic) expression of vGPCR. Unregulated vGPCR signaling to survival pathways may promote direct cell transformation while continuous secretion of chemokines and growth factors may induce the recruitment and subsequent indirect (paracrine) cell transformation of neighboring "bystander" endothelial cells. (see color insert)

Conclusion

The recent discovery of KSHV, the virus responsible for causing the neovascular tumor, KS, has prompted renewed interest into the genesis of this enigmatic neoplasm. Nonetheless, despite a greater understanding of its pathogenesis, KS remains an incurable disease. Examination of the molecular mechanisms whereby KSHV latent and lytic genes contribute to Kaposi's sarcomagenesis may ultimately expose novel therapeutic targets for the treatment of KS and may further provide insight into the complex relationship between tumorigenesis and angiogenesis.

Glossary

Neoplasia: The pathological process that results in the formation and growth of a tumor.

Sarcoma: A usually malignant tumor arising from connective tissue (bone, cartilage, fat, muscle, blood vessels, or other connective or supportive tissue); one of the four major types of cancer.

Transformation: Conversion of normal eukaryotic cells to a cancer-like state of uncontrolled division.

Tumor suppressor gene: Gene that suppresses the formation of tumors. Deletion or inactivation of this gene is believed to be a necessary prerequisite for tumor development.

Viral oncogene: A viral gene that contributes to cancer development in vertebrate hosts. Otherwise normal cells expressing a viral oncogene change into cancerous tumor cells.

References

1. Dezube, B. J. (2000). Acquired immunodeficiency syndrome-related Kaposi's sarcoma: clinical features, staging, and treatment. *Semin. Oncol.* **27**, 424–430.
2. Hengge, U. R., Ruzicka, T., Tyring, S. K., Stuschke, M., Roggendorf, M., Schwartz, R. A., and Seeber, S. (2002). Update on Kaposi's sarcoma and other HHV8 associated diseases. Part 1: epidemiology, environmental predispositions, clinical manifestations, and therapy. *Lancet Infect. Dis.* **2**, 281–292.
3. Ensoli, B., Barillari, G., and Gallo, R. C. (1992). Cytokines and growth factors in the pathogenesis of AIDS-associated Kaposi's sarcoma. *Immunol. Rev.* **127**, 147–155.
4. Noonan, D., and Albini, A. (2000). From the outside in: extracellular activities of HIV Tat. *Adv. Pharmacol.* **48**, 229–250.
5. Chang, Y., Cesarman, E., Pessin, M. S., Lee, F., Culpepper, J., Knowles, D. M., and Moore, P. S. (1994). Identification of herpesvirus-like DNA sequences in AIDS-associated Kaposi's sarcoma. *Science* **266**, 1865–1869. *Using representational difference analysis (RDA), the authors identified DNA sequences from an unknown human g-herpesvirus present in AIDS-associated KS but not in adjacent normal skin. They called the virus the Kaposi's sarcoma-associated herpesvirus (KSHV), which was later confirmed to be the viral etiologic agent for KS.*
6. Flore, O., Rafii, S., Ely, S., O'Leary, J. J., Hyjek, E. M., and Cesarman, E. (1998). Transformation of primary human endothelial cells by Kaposi's sarcoma-associated herpesvirus. *Nature* **394**, 588–592. *The authors demonstrate that infection of human primary endothelial cells with purified KSHV particles causes long-term proliferation and survival of these cells. However, KSHV was present in only a subset of cells, and paracrine mechanisms were found to be responsible for the survival of uninfected cells, suggesting that transformation of endothelial cells by KSHV may be mediated by paracrine mechanisms.*
7. Antman, K., and Chang, Y. (2000). Kaposi's sarcoma. *N. Engl. J. Med.* **342**, 1027–1038.
8. Boshoff, C., and Chang, Y. (2001). Kaposi's sarcoma-associated herpesvirus: a new DNA tumor virus. *Annu. Rev. Med.* **52**, 453–470.
9. Montaner, S., Sodhi, A., Molinolo, A., Bugge, T. H., Sawai, E. T., He, Y., Li, Y., Ray, P. E., and Gutkind, J. S. (2003). Endothelial infection with KSHV genes in vivo reveals that vGPCR initiates Kaposi's sarcomagenesis and can promote the tumorigenic potential of viral latent genes. *Cancer Cell* **3**, 23–36. *The authors engineered a novel transgenic mouse line expressing the avian retroviral receptor, TVA, under the endothelial cell–specific, TIE2, promoter, enabling endothelial-specific infection with avian leucosis virus-derived vectors expressing candidate KSHV oncogenes in vivo. They found that a single KSHV lytic gene, vGPCR, was able to induce angioproliferative tumors in mice that were strikingly similar to human KS. However, expression of vGPCR was limited to only a subset of cells, and paracrine mechanisms were found to be responsible for the survival of neighboring endothelial cells, suggesting that vGPCR transformation of endothelial cells, similar to KSHV, may be mediated by paracrine mechanisms. Using a KS allograft model, the authors further demonstrate that endothelial cells expressing vGPCR were able to promote tumor formation by cells expressing KSHV latent genes, suggestive of a cooperative role among latent and lytic genes in the promotion of Kaposi's sarcomagenesis.*
10. Sodhi, A., Montaner, S., Patel, V., Gomez-Roman, J. J., Li, Y., Sausville, E. A., Sawai, E. T., and Gutkind, J. S. (2004). Akt plays a central role in sarcomagenesis induced by Kaposi's sarcoma herpesvirus-encoded G protein-coupled receptor. Akt plays a central role in vGPCR-induced sarcomagenesis: a novel therapeutic target for Kaposi's sarcoma. *Proc. Natl. Acad. Sci. USA* **14**, 4821–4826.

Capsule Biography

Dr. Akrit Sodhi is a research fellow in the Oral and Pharyngeal Cancer Branch, National Institute of Dental and Craniofacial Research, at the National Institutes of Health. Dr. Sodhi helped develop a novel high-throughput vascular endothelial-specific, retroviral gene transfer system. He is currently conducting research examining the molecular pathogenesis of the Kaposi's sarcoma–associated herpesvirus *vGPCR* in the laboratory of Dr. J. Silvio Gutkind.

Dr. J. Silvio Gutkind has been the Branch Chief in the Oral and Pharyngeal Cancer Branch, National Institute of Dental and Craniofacial Research, at the National Institutes of Health since 1997. His lab has elucidated key signaling pathways by which cellular and viral receptors, coupled to heterotrimeric G proteins, can regulate gene expression. His pioneering discoveries provided novel insights into the biochemical routes by which these membrane receptors control normal and aberrant cell growth.

CHAPTER 148

Endothelial Cells in Kawasaki Disease

[1]Seiichiro Takeshita and [2]Keigo Nakatani
[1]*Ibaraki University, Faculty of Education*
[2]*Kobayashi City Hospital*

Kawasaki disease (KD) is an acute febrile illness that affects predominantly infants and children. Although its etiology remains unknown, KD is characterized as one type of multisystemic vasculitis. This disease leads to the activation and injury of endothelial cells (ECs) in the acute phase. The most critical complication of KD is the coronary artery aneurysm that may lead to myocardial infarction. KD patients have structural, immunological, and functional abnormalities of ECs (Table I).

Pathological Findings

Histopathological studies of vascular changes in autopsied patients with KD revealed the EC degeneration and the increased vascular permeability with edema and degeneration of the media. In the acute phase, vasculitis in subcutaneous regions begins with EC necrosis and subcutaneous edema, followed by degenerative changes in the muscle cells. Active remodeling of the coronary artery lesions (CAL) such as intimal proliferation and neoangiogenesis continues for several years after the onset of KD. Furthermore, Takahashi et al. suggest that CAL in this disease might become a risk factor for atherosclerosis of coronary arteries later in life.

Immunological Abnormalities

Immunological abnormalities during the acute phase of KD are characterized by a marked activation of the immune system: functional activation of monocytes and neutrophils and an excessive production of such inflammatory mediators as cytokines (IL-1, IFN-γ, TNF-α), proteases (neutrophil elastase and myeloperoxidase), and toxic oxygen radicals. They can induce EC activation and/or injury. The activated EC increase the expressions of endothelium–leukocyte adhesion molecule-1 (ELAM-1), intercellular adhesion molecule-1 (ICAM-1), matrix metalloproteinases-2 and -9 (MMP-2 and -9), vascular endothelial growth factor (VEGF), and its receptor (fms-like tyrosine kinase-1 receptor, flt-1). Furthermore, ECs secrete such soluble molecules as von Willebrand factor, thrombomodulin, intercellular adhesion molecule-1 (ICAM-1), and E-selectin (CD62E), which are secreted from EC into the circulation. However, no specific serological markers of EC damage in KD have been identified so far. Although the presence and titer of anti-endothelial cell antibodies (AECA) has been reported to be correlated with disease activity in systemic vasculitis, pathogenic role of AECA in the pathogenesis of KD vasculitis remains controversial.

Impaired EC Function

EC dysfunction is a key event of the process of atherogenesis. EC can release nitric oxide (NO) in response to shear stress, and the loss of NO production leads to abnormal control of vascular tone. The EC-dependent dilation of vascular diameter is reduced in the coronary artery of KD patients with aneurysms. Furthermore, Dhillon et al. demonstrate that flow-mediated dilation of brachial artery is markedly reduced in KD patients without CAL compared with control subjects, and that abnormalities of systemic EC function are present many years after resolution of acute KD.

Table I Structural, Immunological and Functional Abnormalities of EC in KD Vasculitis.

I. Pathological findings
- EC degeneration and necrosis
- Intimal proliferation and neoangiogenesis

II. Immunological findings
1) Increased expression of adhesion molecules on EC
   - Endothelial-leukocyte adhesion molecule-1 (ELAM-1)
   - Intercellular adhesion molecule-1 (ICAM-1)
   - Matrix metalloproteinases 2 and 9 (MMP-2 and -9)
   - Vascular endothelial growth factor (VEGF) and its receptor (fms-like tyrosine kinase-1 receptor, flt-1)
2) Increased production of serum soluble molecules derived from EC
   - von Willbrand factor
   - Thrombomodulin
   - Intercellular adhesion molecule-1 (ICAM-1)
   - E-selectin (CD62E)

III. Functional findings
- Impaired EC-dependent dilation in systemic vessels including the coronary artery

Other Findings

Recently, circulating endothelial cells (CECs) are reported to be observed in several diseases with vascular injury. Nakatani et al. reveal that the number of CEC also increases during the acute phase of KD. Furthermore, the number of endothelial progenitor cells (EPCs), which were derived from the bone marrow, increases during the subacute phase in KD patients with CAL. Therefore, the increased numbers of CECs and EPCs are suggested to reflect the EC damage of this disease. Although the main origin of CECs may be mature ECs that have detached from the vessel walls, CECs also contain a small population of EPCs that might be involved in both the repair of EC damage and the potential vasculogenesis.

EC injury mediated by activated neutrophils has been seen in systemic inflammatory response syndrome, acute respiratory distress syndrome, and multiple organ failure. In the acute phase of KD, circulating leukocytes, especially neutrophils, increase in number, and the neutrophil function is also activated. Pathological studies reveal that neutrophils infiltrate the CAL in the early phase of KD. The activated neutrophils also secrete a large amount of such autotoxic mediators as proteases, toxic oxygen radicals, and arachidonic metabolites, which are believed to induce EC injury. It is thus suggested that activated neutrophil-mediated EC injury may be involved in the pathogenesis of KD vasculitis.

Therapy for KD Vasculitis

Although treatment with a combination of aspirin and intravenous immunoglobulin (IVIG) is generally effective for KD patients, 5 to 15 percent of the patients develop CAL. Aspirin has possible antiplatelet activity and antithrombotic effect. Although mechanisms of action for IVIG have still not been elucidated fully, this preparation exhibits a broad spectrum of immunomodulatory activities. However, the therapeutic efficacy of both aspirin and IVIG is not specific for EC injury in KD vasculitis. Additional therapy focused on EC for KD vasculitis will be needed in future.

Glossary

Circulating endothelial cells (CECs): Endotheliums mobilized in the circulation.

Coronary artery lesions (CAL): Dilatation, aneurysm and stenosis of coronary artery.

Endothelial cells (ECs): Please erase this term in Glossary, because this definition was provided in "CECs".

Kawasaki disease (KD): Mucocutaneous Lymph Node Syndrome.

References

Dhillon, R., Clarkson, P., Donald, A. E., Powe, A. J., Nash, M., Novelli, V., Dillon, M. J., and Deanfield, J. E. (1996). Endothelial dysfunction late after Kawasaki disease. *Circulation* **94**, 2103–2106. *This article provides an important evidence that abnormalities of systemic endothelial function are present many years after resolution of acute KD.*

Gavin, P. J., Crawford, S. E., Shulman, S. T., Garcia, F. L., and Rowley, A. H. (2003). Systemic arterial expression of matrix metalloproteinases 2 and 9 in acute Kawasaki disease. *Arterioscler. Thromb. Vasc. Biol.* **23**, 576–581.

Leung, D. Y., Cotran, R. S., Kurt-Jones, E., Burns, J. C., Newburger, J. W., and Pober, J. S. (1989). Endothelial cell activation and high interleukin-1 secretion in the pathogenesis of acute Kawasaki disease. *Lancet* **2**, 1298–1302. *This article represents convincing evidence that endothelial cell activation plays an important role in the pathogenesis of acute KD.*

Nakatani, K., Takeshita, S., Tsujimoto, T., Kawamura, Y., Tokutomi, T., and Sekine, I. (2003). Circulating endothelial cells in Kawasaki disease. *Clin. Exp. Immunol.* **131**, 536–540. *This article represents conclusive evidence that the number of CECs increases during the acute phase of KD.*

Praprotnik, S., Rozman, B., Blank, M., and Shoenfeld, Y. (2000). Pathogenic role of anti-endothelial cell antibodies in systemic vasculitis. *Wien. Klin. Wochenschr.* **112**, 660–664. *This article provides a detailed review of pathogenic role of anti-endothelial cell antibodies in systemic vasculitis including Kawasaki disease.*

Takahashi, K., Oharaski, T., and Naoe, S. (2001). Pathological study of postcoronary arteritis in adolescents and young adults: With reference to the relationship between sequelae of Kawasaki disease and atherosclerosis. *Pediatr. Cardiol.* **22**, 138–142.

Takeshita, S., Dobashi, H., Nakatani, K., Koike, Y., Tsujimoto, H., Hirayama, K., Kawamura, Y., Mori, K., Sekine, I., and Yoshioka, S. (1997). Circulating soluble selectins in Kawasaki disease. *Clin. Exp. Immunol.* **108**, 446–450.

Takeshita, S., Tokutomi, T., Kawase, H., Nakatani, K., Tsujimoto, H., Kawamura, Y., and Sekine, I. (2001). Elevated serum levels of matrix metalloproteinase-9 (MMP-9) in Kawasaki disease. *Clin. Exp. Immunol.* **125**, 340–344.

Yasukawa, K., Terai, M., Shulman, S. T., Toyozaki, T., Yajima, S., Kohno, Y., and Rowley, A. H. (2002). Systemic production of vascular endothelial growth factor and fms-like tyrosine kinase-1 receptor in acute Kawasaki disease. *Circulation* **105**, 766–769.

Capsule Biography

Dr. Takeshita worked at the National Defense Medical College from 1993 to 2002, and he became a professor of Ibaraki University, Faculty of Education in 2005. He focuses on the pathogenesis and etiology of Kawasaki disease. His work is supported by a grant for Research on Specific Diseases from the Ministry of Health, Labor and Welfare, Japan.

Dr. Nakatani worked at the National Defense Medical College from 1996 to 2000 and at Kobayashi city hospital in 2003–2004. He focuses on the pathogenesis and etiology of Kawasaki disease.

Response of Colon Carcinoma Cells to Proinflammatory Cytokines

M. ten Kate and C.H.J. Van Eijck
Erasmus Medical Center Rotterdam, The Netherlands

Introduction

For colon carcinoma, local as well as distant recurrences remain a major problem, even with the introduction of new treatment modalities. Dissemination of cancer cells in the peritoneal cavity as well as in primary tumor drainage and peripheral blood vessels during operation plays an important role in the metastasis process.

The process of hematogenous metastasis is a multistep event in which tumor cells first have to detach from the primary tumor, then invade the blood circulation where they must escape from immunological control and mechanical disruption, and subsequently adhere to the endothelium within a distant organ and finally invade and multiply to form a metastasis (Figure 1).

Metastasizing tumor cells have to proceed through all of these sequential steps to form metastases.

During this process, tumor cells circulate in the vasculature where they encounter host defense mechanisms. Furthermore, the life span of circulating tumor cells is limited. If tumor cells do not adhere to and migrate through the endothelium within a certain time, they will not survive. Taken together, hematogenous metastasis appears to be highly inefficient.

Although under normal conditions tumor cell arrest in distant organs is negligible, clinical and experimental observations indicate preferential tumor cell arrest at sites of injury, healing, and inflammation [1]. Further clinical observations gave the impression of accelerated metastasis to distant organs after surgical trauma.

Abdominal surgery associated with major tissue injury provokes an inflammatory reaction with release of these proinflammatory cytokines, not only locally, but also systemically.

The systemically presence of the proinflammatory cytokines after surgery may be related with apparent accelerated metastasis to distant organs after surgical trauma. Intentionally curative resection of gastrointestinal carcinomas may therefore involve a serious risk of developing metastases. Several studies on this topic have been performed, some regarding in vitro studies and others in vivo studies. In the following section we will give an overview of these studies.

Proinflammatory Cytokines

During inflammation a plethora of mediators are released. These mediators attract neutrophils, macrophages and other leukocytes to the site of inflammation. Proinflammatory cytokines like TNF-α, IL-1β, and IL-6 are the key mediators of the inflammatory reaction. Tissue injury triggers the synthesis and release of these cytokines. The proinflammatory cytokines may act locally, but in higher concentrations these factors also act systemically. The cytokine response to surgical trauma is correlated to the degree and duration of surgical trauma. Normal inflammatory response will enable the host to recover from trauma, but will become detrimental if the balance between mediators of the inflammatory response is lost.

Endothelial Cells

The vascular endothelium comprises of a layer of thin, flattened cells lining the inside surfaces of blood vessels and is a major regulator in many homeostatic and pathophysiologic mechanisms. In inflammation and injury

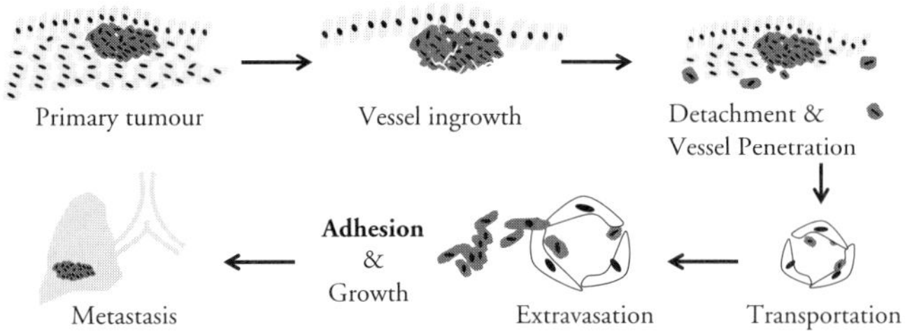

Figure 1 A schematic reproduction of the metastasis process. Tumor cells first have to detach from the primary tumor, invade the blood circulation, subsequently adhere to the endothelium within a distant organ and finally invade and multiply to form a metastasis. (see color insert)

the endothelium is activated. The activation of endothelial cells leads to induction of adhesion molecules and the release of mediators, such as proinflammatory cytokines, chemokines, and reactive oxygen species. These processes lead to migration of inflammatory cells to the wounded or inflamed site. The vasculature constitutes the primary route by which tumor cells disseminate to distant sites. Interactions between circulating tumor cells and the endothelium significantly influence the outcome of the metastatic process.

Selectins, integrins, cadherins, immunoglobulins, and other unclassified molecules have been demonstrated to govern the adhesive interactions between tumor cells and the endothelium, analogous to their ability to effect localization of leukocytes at sites of inflammation. Some of these adhesion molecules are expressed constitutively and appear to have organ specificity in their distribution. Others are inducible and under the influence of environmental mediators such as the proinflammatory cytokines.

There are macrovascular endothelial cells surrounding large blood vessels, such as arteries and veins, and microvascular endothelial cells surrounding small vessels, the arterioles, venules, and capillaries. Microvascular and macrovascular cells are heterogeneous as microvascular cells react differently to stimuli compared to macrovascular endothelial cells. Differences exist not only between macrovascular and microvascular endothelial cells, but also between endothelial cells from different vascular beds. Each organ has its own microenvironment with different characteristics, such as a different pattern of endothelial adhesion molecules, reacting particularly to stimuli.

In Vitro Studies

Macrovascular Endothelium

Human umbilical vein endothelial cells (HUVECs) are macrovascular endothelial cells of embryonic origin. Because HUVECs are relatively easy to obtain and culture, they are used in many models studying the pathophysiology of the endothelium.

There are several reports on the influence of proinflammatory cytokines on tumor cell adhesion to HUVECs. Activation of HUVECs by TNF-α or IL-1β promotes the adhesion of many tumor cell lines, including colon, pancreas, gastric, breast, lung, melanoma, cervix, osteosarcoma, and kidney carcinoma cell lines. All studies show that the increased adhesion is due to induction of adhesion molecules on HUVECs, although there is heterogeneity between tumor cells in the adhesion to HUVECs. For instance, melanoma cells express enhanced adhesion to cytokine-stimulated HUVECs that can be blocked by VCAM-1 antibody. Colon carcinoma cell lines express enhanced adhesion to stimulated HUVECs, which is not VCAM-1 dependent, but E-selectin dependent. In these studies, HUVECs were preincubated with cytokines for different periods, with peak adhesion for colon carcinoma cells after 4 hours of preincubation, after which adhesion decreased. This peak adhesion correlates with the E-selectin expression on stimulated HUVECs, and indeed adhesion was inhibited with an E-selectin antibody.

The differences between particular tumor cell types indicate specific adhesion molecule patterns on tumor cells that can interfere with a specific adhesion molecule or complex of adhesion molecules on endothelial cells. Indeed, melanoma cells express high levels of the VLA-4 integrin, a counterpart of VCAM-1. On the other hand, most gastrointestinal carcinoma cells, such as colon carcinoma cells, do express relatively high levels of sialyl Lewis ^{a} and/or ^{x} and low levels of VLA-4. Sialyl Lewis ^{a} and ^{x} are able to bind to endothelial E-selectin. The heterogeneity between tumor–endothelial interactions can therefore be explained on the basis of particular adhesion molecule patterns of different tumor cells. In summary, the adhesion of tumor cells to activated HUVECs seems to be mediated by an upregulation of adhesion molecules and displays heterogeneity between different tumor cells.

Microvascular Endothelium

Although several studies of the influence of cytokines on tumor cell adhesion to HUVECs have been performed, studies using microvascular endothelium are limited. However, tumor cell adhesion does not occur in the macrovasculature; it occurs in the microvasculature. Because of the heterogeneity between microvascular and macrovascular cells, it is a prerequisite to use the correct endothelial cell type in studies investigating endothelial pathophysiology. Furthermore, HUVECs are fetal cells, which may involve other characteristics compared to normal cells.

Therefore, we studied the influence of proinflammatory cytokines on tumor cell adhesion to microvascular endothelial cells. In this model, we used human microvascular endothelial cells of the lung (HMVEC-L). We investigated the adhesion of two human colon carcinoma cell lines, HT29 and Caco2. Stimulation of HMVEC-L with the proinflammatory cytokines TNF-α and IL-1β, but not IL-6, increased adhesion of both tumor cell lines to HMVEC-L

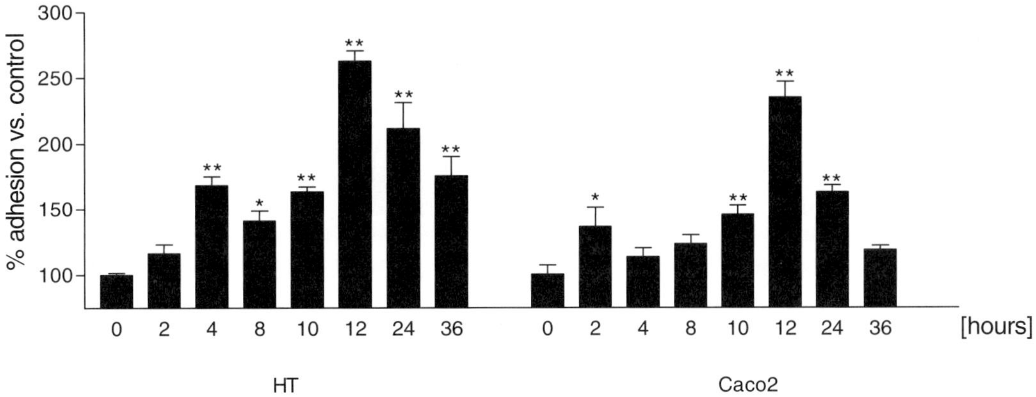

Figure 2 Tumor cell adhesion to HMVEC-L. HMVEC-L were pre-incubated for varying times (in hours) with TNF-α (HT29) or IL-1β (Caco2). Means (n = 6, % vs. control) + SEM are shown. * = $p < 0.05$; ** = $p < 0.01$.

significantly. The enhancement is concentration dependent, with maximal increase at 1 ng/mL TNF-α or 10 ng/mL IL-1β (Figure 2). Furthermore, tumor cell adhesion shows time dependence. Enhancement started after 2 hours of preincubation of the HMVEC-L monolayer with the cytokines. Maximal adhesion was achieved after 12 hours, and adhesion declined thereafter (Figure 2). This outcome differs from results obtained with HUVEC studies, in which adhesion had already peaked after 4 hours of preincubation, followed by a decline.

Another difference between HUVECs and HMVEC-L is E-selectin dependence in tumor cell adhesion. Cytokine-stimulated HUVECs demonstrate E-selectin dependent adhesion of colon carcinoma cells. However, an E-selectin antibody did not inhibit the adhesion of both colon carcinoma cells to stimulated HMVEC-L.

Unfortunately, as far as we know, there are no other reports on the adhesion of colon carcinomas to microvascular endothelium in vitro. However, there are a few reports regarding other tumor cell types. Kwang et al. [2] found differences between macrovascular and microvascular endothelial cells in melanoma cell–endothelial cell adhesion. The binding of melanoma cells to HUVECs occurs via VLA-4 on melanoma cells and VCAM-1 on cytokine-stimulated HUVECs. However, the binding of melanoma cells to cytokine-stimulated human dermal microvascular endothelium (HDMEC) could not be inhibited by a VCAM-antibody. Therefore, another mechanism must be responsible for the binding to HDMEC.

Endothelial cells and tumor cells are both exposed to proinflammatory cytokines during surgical trauma. However, stimulation of colon carcinoma cells does not enhance the adhesion to endothelial cells. Therefore, the mechanism of enhanced tumor cell endothelial cell interactions under influence of proinflammatory cytokines seems to be dependent on endothelial changes. Although information about tumor cell interactions with the microvasculature is scarce, it is obvious that using the correct endothelial cell type studying the metastasis process is a prerequisite.

In Vivo Studies

To validate the hypothesis that surgically derived cytokines accelerate metastasis of colon carcinomas, in vivo experiments were carried out. These experiments were carried out in different tumor systems: human tumors in nude mice, murine or rat tumors in syngeneic mice or rats, respectively, and tumors that formed metastases to liver, lung, or intraperitoneal dissemination.

Injection of the proinflammatory cytokines TNF-α or IL-1β before tumor cell injection in experimental animals promoted formation of metastases. This effect was also found when animals received surgical stress instead of proinflammatory cytokines prior to tumor cell injection. In these animals serum levels of proinflammatory cytokines were elevated after surgery. Administering antibodies against the cytokines inhibited the enhanced metastasis after surgery. For example, we performed a study in which rats underwent a hemihepatectomy, ileum resection, or sham operation after injection of CC531 rat colon carcinoma cells. After 3 weeks, the animals were killed. Tumor recurrence in the lung was significantly higher in hemihepatectomy and ileum resection groups compared to sham groups, indicating that surgical trauma influences the metastasis process. Studies comparing laparotomy versus laparoscopy show the relation between the degree of surgical trauma and metastasis, since less tumor recurrence was found after laparoscopy than after laparotomy.

From these in vivo studies we can conclude that surgery-derived cytokines play a major role in tumor recurrence. However, cytokine pretreatment did not induce metastasis by tumor cell lines that were not metastatic, nor did it change their metastatic homing, indicating that cytokines quantitatively modulate the metastasis process without changing the intrinsic metastatic properties of tumor cells.

Because tumor cell adhesion to the endothelium seems to be an important step in metastasis, experimental studies have been performed to investigate the role of endothelial adhesion molecules in enhanced metastasis. In these studies, antibodies against adhesion molecules on tumor cells or on endothelial cells were administered to animals prior to surgery and indeed inhibited the enhanced metastasis.

Another mechanism of enhanced tumor recurrence after surgery may be the accelerated growth of tumor cells under the influence of surgery-derived mediators, such as cytokines and growth factors. However, several studies contradict the hypothesis that enhanced growth of tumor cells is responsible for accelerated metastasis formation after surgery. Therefore, acceleration of metastasis appeared to be independent of the effect of the cytokine on the primary tumor growth.

These in vivo studies demonstrate a relation between surgery and accelerated metastasis via the release of proinflammatory cytokines. These cytokines induce adhesion molecules on the vascular endothelium, causing increased tumor cell adhesion.

Adhesion Molecules

Because of the observed time dependence in tumor–endothelial cell interactions, a role of protein synthesis leading to the induction or upregulation of adhesion molecules is suggested in the enhanced tumor cell adhesion.

In vitro studies using HUVECs revealed the importance of the adhesion molecules ICAM-1, VCAM-1, and E-selectin in tumor cell–endothelial cell interactions. For gastrointestinal carcinomas such as colon, gastric, and pancreas carcinoma, E-selectin seems to be the major adhesion molecule in cytokine-stimulated endothelial interactions.

For HUVECs it is known that under the influence of proinflammatory cytokines, adhesion molecules are inducted or upregulated. In resting conditions HUVECs lack E-selectin and VCAM-1 expression. By proinflammatory cytokines such as TNF-α, E-selectin is rapidly synthesized and expressed on HUVECs after 4 hours of stimulation, with peak values between 4 and 6 hours. VCAM-1 expression reaches peak values after 6 to 12 hours of stimulation, and expression declines thereafter. ICAM-1 is constitutively expressed on HUVECs but is upregulated by TNF-α after 8 hours, remaining high for approximately 24 hours.

Only a few reports regarding the influence of cytokines on adhesion molecule expression on microvascular endothelium are available.

On HMVEC-L, E-selectin is rapidly expressed after TNF-α and IL-1β stimulation, with peak expression after 8 hours, followed by a decline (Figure 3). ICAM-1 expression slowly increases, with enhancement still present 24 hours after starting stimulation with TNF-α and IL-1β. The same kinetics, but less pronounced, is observed for VCAM-1 expression. IL-6 does not increase E-selectin expression or ICAM-1 or VCAM-1 expression. Both E-selectin and ICAM-1 expression are correlated with cytokine concentration; the expression increases with increasing concentration.

Adhesion molecule expression in relation to cytokine stimulation is also described for human intestinal, dermal, and CNS microvascular endothelial cells. These studies show induction of the adhesion molecules ICAM-1, VCAM-1, and E-selectin largely similar to HUVECs, although there are discrepancies in time and extent of expression. Other studies show the existence of organ specific adhesion molecules. For example, the group of Pauli [3] reported a new lung-specific adhesion molecule that mediated lung metastasis of murine melanomas.

The influence of proinflammatory cytokines on the expression of adhesion molecules on tumor cells is described only in few reports. These reports found only minor changes that were not impressive enough to cover the enhancement in tumor cell endothelial cell interactions under influence of cytokines.

The kinetics of E-selectin expression on HMVEC-L are conflicting with the adhesion assays, because maximal E-selectin expression is seen after 8 hours of stimulation, whereas maximal adhesion of colon and pancreas carcinoma cells occurs after 12 hours of stimulation. These results exclude E-selectin as the principal adhesion molecule responsible for the enhanced adhesion. And since adhesion of both colon and pancreas carcinoma cells to HMVEC-L was not inhibitable by an E-selectin antibody, it is more

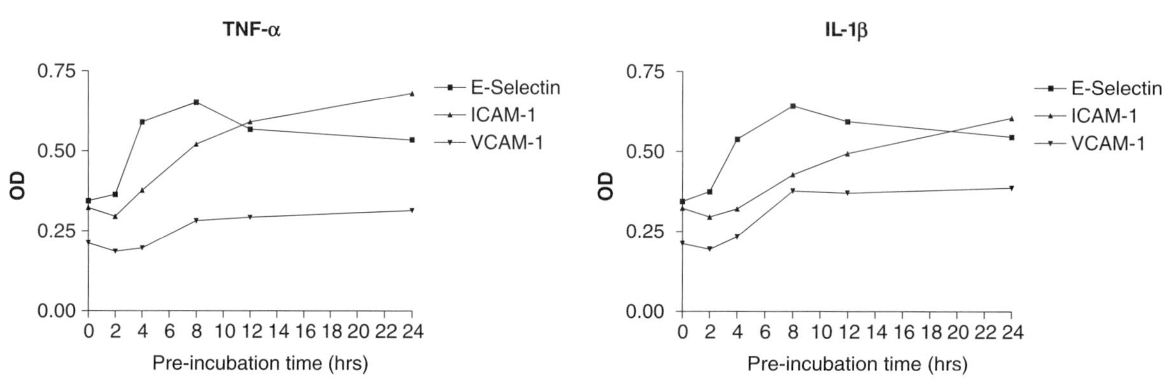

Figure 3 Kinetics of adhesion molecule expression on HMVEC-L assessed by EIA. HMVEC-L monolayers were pre-incubated with TNF-α or IL-1β for varying times. Data represent mean absorbance values (OD 405 nm) ± SD.

likely that another adhesion molecule or complex of adhesion molecules is responsible for the enhanced adhesion.

Discussion

The studies just mentioned give compelling evidence that interactions between tumor cells and the endothelium play a pivotal role in metastasis formation to distant sites, involving a multiplicity of adhesion molecule complexes. These adhesive interactions are susceptible to regulation by proinflammatory cytokines, which are capable of causing tumor cell arrest in the microvasculature and strongly influence the outcome of the metastatic process.

Comparing the results obtained with macrovascular cells with those obtained with microvascular cells, there are discrepancies. According to Paget's "seed and soil" hypothesis, certain tumor cells have their preferential organs to metastasize to, which cannot be explained by simple anatomical or mechanical hypotheses. Each organ microenvironment has different characteristics, such as a different pattern of endothelial adhesion molecules, reacting in particular ways to stimuli. This can explain typical metastasis patterns for specific tumors. Therefore, it is a prerequisite to study endothelial cells derived from the vascular bed of interest rather than extrapolate from results obtained with HUVECs.

Most studies are performed under static conditions. However, the metastatic process is a dynamic one. Tumor cells circulating in the vasculature encounter shear forces that are associated with the blood flow. These shear forces may prevent tumor cell and endothelial cell interactions that could occur under static conditions. The studies mentioned earlier use culture systems lacking shear forces. Nevertheless, useful data have been acquired with static assays, providing evidence that adhesive interactions are susceptible to regulation by environmental stimuli such as cytokines. Furthermore, tumor cell endothelial cell interactions take place in the microvasculature where the role of shear forces is limited, since blood flow in the microvasculature is low.

Dynamics are not only related to blood flow. The inflammatory response is a dynamic process involving a whole range of mediators. Those mediators can affect each other, resulting in a complicated network. Stimulation of endothelial cells with proinflammatory cytokines not only bring about the induction of adhesion molecules; endothelial cells are also triggered to release mediators themselves. The production of endothelial proinflammatory cytokines therefore will enhance the inflammatory state, leading to a vicious circle. Not only endothelial cells can produce proinflammatory cytokines; tumor cells can as well. Research regarding the issue whether tumor cells produce enough cytokines to stimulate endothelial cells is contradictory. Some reports claim that tumor cell–derived cytokines indeed stimulate endothelial cells, resulting in increased tumor cell adhesion, whereas others could not detect such a stimulative effect.

Other mediators are released during the inflammatory response and tissue injury, such as reactive oxygen species, chemokines, prostaglandins, and growth factors. Moreover, circulating cells, such as monocytes, neutrophils, platelets, and erythrocytes, may prevent or support tumor cell–endothelial cell interactions. All these factors produce dynamic interactions that may influence the metastasis process.

The apparent relationship between the inflammatory response after surgical trauma and distant tumor recurrence necessitates further unraveling of the mechanisms involved. This may bring about a treatment modality to reduce distant recurrences and benefit a significant proportion of patients in terms of survival or quality of life.

Glossary

Cell adhesion molecule: Although this could mean any molecule involved in cellular adhesive phenomena, it has acquired a more restricted sense, namely a molecule on the surface of animal tissue cells, antibodies against which specifically inhibit some form of intercellular adhesion.

Colon cancer: Disease in which malignant (cancer) cells are found in the tissues of the colon.

Microvascular endothelium: Comprises of a layer of thin flattened cells lining the inside surfaces of small blood vessels, like arterioles, venules and capillaries.

Proinflammatory cytokines: Primarily produced by white blood cells (monocytes and macrophages); has an antineoplastic effect but causes inflammation.

References

1. Sugarbaker, E. V. (1981). Patterns of metastasis in human malignancies. *Cancer Biol. Rev.* **2**, 235–278.
2. Lee, K. H., Lawley, T. J., Xu, Y. L., and Swerlick, R. A. (1992). VCAM-1-, ELAM-1-, and ICAM-1-independent adhesion of melanoma cells to cultured human dermal microvascular endothelial cells. *J. Invest. Dermatol.* **98**(1), 79–85.
3. Pauli, B. U., Augustin-Voss, H. G., el-Sabban, M. E., Johnson, R. C., and Hammer, D. A. (1990). Organ-preference of metastasis. The role of endothelial cell adhesion molecules. *Cancer Metastasis Rev.* **9**(3), 175–189.

Further Reading

Aosasa, S., Ono, S., Mochizuki, H., Tsujimoto, H., Osada, S., Takayama, E., Seki, S., and Hiraide, H. (2000). Activation of monocytes and endothelial cells depends on the severity of surgical stress. *World. J. Surg.* **24**, 10–16.

Auerbach, R., Alby, L., Morrissey, L. W., Tu, M., and Joseph, J. (1985). Expression of organ-specific antigens on capillary endothelial cells. *Microvasc. Res.* **29**, 401–411. *Auerbach et al. showed that capillary endothelial cells express an array of antigens on their cell surface that manifest organ selectivity.*

Chambers, A. F., Groom, A. C., and MacDonald, I. C. (2002). Dissemination and growth of cancer cells in metastatic sites. *Nature Reviews* **2**, 563–572.

Giavazzi, R. (1986). Cytokine-mediated tumor–endothelial cell interaction in metastasis. *Curr. Top. Microbiol. Immunol.* **213**(Pt 2), 13–30. *This review gives an excellent overview of the current knowledge about tumor cell–endothelial cell interactions in metastasis under influence of cytokines.*

Haier, J., and Nicolson, G. L. (2001). The role of tumor cell adhesion as an important factor in formation of distant colorectal metastasis. *Dis. Colon Rectum* **44**, 876–884. *Haier and Nicolson summarized and dis-*

cussed different aspects of colorectal carcinoma cell adhesion during hematogenous formation of distant metastases.

Orr, F. W., and Wang, H. H. (2001). Tumor cell interactions with the microvasculature. A rate-limiting step in metastasis. *Surg. Oncol. Clin. North Am.* **10**, 357–381. *This comprehensive review describes the vascular phase of cancer metastasis.*

Weiss, L. (1990). Metastatic inefficiency. *Adv. Cancer Res.* **54**, 159–211.

Capsule Biography

As a Ph.D. researcher, M. ten Kate, M.D., is investigating tumor cell endothelial interactions under influence of surgical trauma. This research is carried out under supervision of C. H. J. van Eijck, M.D., Ph.D., an oncologic surgeon at the Erasmus Medical Center in Rotterdam.

SECTION W

Vasculitis

Chapter 150

Vasculitis

Ross Milner and Robert B. Smith, III
Emory University School of Medicine, Department of Surgery

Introduction

Vasculitis is an inflammatory disorder that affects blood vessels in characteristic locations throughout the body. More than 30 types of vasculitis have been described by separate names. Although vasculitis has a characteristic distribution based on the affected vascular structures, it remains a systemic disease with systemic manifestations. Patients present with fever, weight loss, fatigue, tachycardia, and diffuse "aches and pains." In addition to these constitutional symptoms, almost every organ system can be affected.

Unfortunately, the cause of many of the disorders discussed in this chapter is unknown. It is clear, though, that the immune system plays a large role in the tissue damage caused by vasculitis. Inflammation in the affected blood vessels leads to critical stenosis creating an end-organ destructive process.

This chapter provides an overview of some of the most common vasculitides. In addition, a review of the mechanism of the immunologic development of vasculitis will be discussed.

Common Vasculitides

Behçet's Disease

This triad of aphthous oral ulcers, genital lesions, and recurrent eye inflammation was first described by a Turkish dermatologist, Hulusi Behçet, in the 1930s. Patients are most commonly affected in the Far East and Mediterranean regions. In fact, it is the leading cause of blindness in Japan. This disease occurs infrequently in the United States.

The vascular involvement is unusual in that it includes arterial structures of all sizes and venous structures as well. The variety of vascular beds affected allows for a diversity of involved organs, but the eyes, mouth, skin, lungs, joints, brain, genitals, and gastrointestinal tract are the most commonly reported systems.

Behçet's is one of only a few of the vasculitides in which a genetic predisposition has been described. The presence of the gene, HLA-B51, is a risk factor for development of the disease. Importantly, many people possess this gene and do not develop the disease. In fact, familial cases are not the rule and represent only about 5 percent of the total cases. Therefore, additional factors, such as environmental exposures, must play a large role. In fact, the mechanism for development of the disease is not understood.

Buerger's Disease

Also known as thromboangiitis obliterans, Buerger's disease was initially described in 1908. It is characterized by acute inflammation and thrombosis of arteries and veins, usually affecting the hands and the feet. The classic patient is a young male who is a heavy smoker. It most commonly occurs in Southeast Asia, Orient, India, and the Middle East; Buerger's disease does not commonly occur in African-Americans.

The disorder causes manifestations of peripheral vascular occlusive disease including claudication, rest pain, and the development of ulcerations. Despite the severity of the extremity manifestations that can develop, Buerger's usually spares other organ systems. The pathophysiology of this organ distribution is not known.

The association with tobacco use, including smokeless tobacco, is crucial for the development of this disease. It is postulated that the mechanism is autoimmune in nature and triggered by a factor in the tobacco. Interestingly, anti-inflammatory medications and anticoagulation lack efficacy in treating Buerger's disease; abstinence from tobacco use is the only effective therapy.

Central Nervous System Vasculitis

Many forms of vasculitis can involve the central nervous system. Central nervous system vasculitis (CNSV) is a specific form of vasculitis in which the disease is confined to the central nervous system and no infectious etiology can be identified. CNSV can be difficult to diagnose because the symptoms of the disease are subtle and a brain biopsy is required for accurate diagnosis. The diagnosis can sometime be suggested findings on magnetic resonance angiography (MRA).

CNSV is also known as *isolated angiitis of the central nervous system* (IACNS) and *granulomatous angiitis of the nervous system* (GANS). The current preferred name of the disorder is *primary angiitis of the central nervous system* (PACNS). This disease process affects the small and medium blood vessels of the brain and spinal cord. Some patients appear to have a milder form of the disorder that is described as an angiopathy rather than an arteritis, in other words, *benign angiopathy of the central nervous system* (BACNS).

PACNS occurs in an equal distribution of men and women and can occur in any age group. It is most common, though, in the fifth decade. BACNS occurs in young women with a history of migraines. Although the exact mechanism is unclear, these patients tend to have exposure to caffeine, nicotine, and oral contraceptive medications.

The pathophysiology is not understood. It is presumed that infectious agents, such as a viral infection, may stimulate the inflammatory process. This initial process then somehow becomes self-sustaining. It appears that genetics do not play a role in the development of CNSV.

Churg-Strauss Syndrome

Churg-Strauss is a systemic vasculitis described in 1951 by Jacob Churg and Lotte Strauss. They reported a syndrome combining asthma, eosinophilia, fever, and accompanying vasculitis of various organ systems. This disease shares features with polyarteritis nodosa (PAN) but can be differentiated by the presence of granulomas and the abundance of eosinophils. It affects men and women equally and begins in middle age with development of a new-onset diagnosis of asthma.

The cause is unknown, but appears to be multifactorial. Although genetics may play a small role in the disease, it appears that two individuals within a family unit are never affected. Environmental factors and infectious etiologies may contribute, but have not been definitively implicated in the causation of the disease.

Cryoglobulinemia

This name literally means cold antibody in the blood. Cryoglobulinemia is a disease state in which antibodies precipitate in the blood under cold conditions. These immunoglobulins resorb upon rewarming. Hepatitis C is a likely initiating event for the development of these antibodies.

Three types of antibodies exist in cryoglobulinemia. Type I is a monoclonal antibody that does not reflect rheumatoid factor activity. Because type I antibodies do not commonly activate complement, it takes a high level of antibody to develop a hyperviscosity syndrome. Types II and III antibodies are rheumatoid factors, meaning that they bind to the Fc fragment of IgG. In type II, the rheumatoid factor is monoclonal. In type III, it is polyclonal. It is understood that most patients with type II or III antibodies have an underlying diagnosis of hepatitis C.

Treatment depends upon the type of antibody present. Severe hyperviscosity requires plasmapheresis and the treatment of any underlying malignancy. Types II and III cryoglobulinemia usually responds to corticosteroid therapy, cyclophosphamide, or both.

Giant-Cell Arteritis

Giant-cell arteritis, also known as one of its variants, temporal arteritis, is the most common form of vasculitis in adults. This disease is a panarteritis of medium- and large-sized arterial structures, especially the extracranial branches of the carotid artery. This form of vasculitis causes facial pain, headaches, joint pain, fever, difficulties with vision, and the possibility of permanent blindness in one or both eyes. Women are affected three times more commonly than men. Most patients are in the eighth decade of life with onset of the disease.

Although the underlying reason for developing giant-cell arteritis is unclear, immunologic mechanisms have been well defined and will be discussed in more detail later in this chapter. Briefly, the immune system inappropriately attacks the target arteries and leads to significant inflammation with the accompanying symptoms. Laboratory blood tests are nonspecific, but almost every individual afflicted with temporal arteritis has an elevated erythrocyte sedimentation rate (ESR). A definitive diagnosis is made by temporal artery biopsy. If clinical suspicion is high and the initial biopsy is negative, the contralateral vessel can be biopsied for diagnosis. Prompt initiation of corticosteroid therapy is mandatory. Biopsies will likely remain positive for up to 2 weeks after initiation of therapy, and corticosteroid treatment should not be withheld while awaiting biopsy.

Polyarteritis Nodosa

Polyarteritis nodosa (PAN) was initially described in 1866 as periarteritis nodosa. The name was changed to recognize the inflammation that occurred throughout the entire arterial wall. PAN is also known as "systemic necrotizing vasculitis." It is a disease of small and medium sized arteries that commonly develops in the fourth and fifth decades. It is twice as common in men as compared to women. Once again, the cause is unknown. Hepatitis B and C are likely underlying causes of PAN, explaining the high incidence of the condition in substance abusers.

PAN can affect nearly every organ, but has a predilection for the nerves, skin, kidney, and gastrointestinal tract. A large number of patients present with hypertension and an elevated ESR. Once again, there is no specific laboratory test to confirm the diagnosis of PAN. Urinalysis may reveal proteinuria in patients with affected kidneys.

The American College of Rheumatology (ACR) has designated 10 criteria that are diagnostic of PAN and assist with distinguishing it from other forms of vasculitis. Patients must have at least three of the following criteria to be given the diagnosis:

1. Weight loss of more than 4 kg since beginning of illness
2. Livedo reticularis
3. Testicular pain or tenderness
4. Myalgias, weakness, or leg tenderness
5. Mononeuropathy or polyneuropathy
6. Development of hypertension
7. Elevated blood urea nitrogen or creatinine unrelated to dehydration or obstruction
8. Presence of hepatitis B surface antigen or antibody in serum
9. Arteriogram demonstrating aneurysms or occlusions of the visceral arteries
10. Biopsy of small or medium-sized artery containing granulocytes

Confirmation of the diagnosis requires a biopsy specimen that includes small or medium-sized arteries. Also, a mesenteric arteriogram with microaneurysms may be diagnostic, but arteriography is usually reserved for situations in which there is not a symptomatic site from which to obtain a biopsy. Without treatment, most patients die within 2 to 5 years of the initial diagnosis. A combination of cyclophosphamide and prednisone has altered the course to a survival of 70 percent over 10 years.

Takayasu's Arteritis

Mikito Takayasu first described the disease in 1908 as an abnormal appearance of the blood vessels on retinal examination. Takayasu's arteritis is commonly known as "pulseless disease" because of the difficulty in palpating peripheral pulses secondary to the multiple arterial stenoses. The usual patient is a woman under the age of 40. In fact, there is a 9 : 1 distribution of women to men affected by Takayasu's. In addition, it appears to be more common in Asian women. Overall, it is a very rare disease with only two to three cases per 1 million people.

Takayasu's involves the larger arteries of the body. In fact, the aorta and its major branches are the most commonly afflicted vessels. The syndrome is clinically divided into two phases: (1) a systemic phase and (2) an occlusive phase. In the systemic phase, patients have classic constitutional symptoms associated with inflammatory diseases. Conversion to the occlusive phase occurs as patients proceed to chronic arterial stenoses leading to abnormal perfusion. It can be exceptionally difficult to obtain an accurate blood pressure in these patients using the standard cuff technique in the arms. Pulmonary arteries may also be affected by Takayasu's, and this arterial bed can develop beading of the vessels, occlusions, and aneurysms.

The exact cause of this arteritis is unknown. A hypothesis has been proposed that patients with a certain genetic predisposition develop the disease after a specific exposure occurs, such as a viral or bacterial infection. Anemia and an elevated erythrocyte sedimentation rate are markers of active disease. The diagnosis can be made by standard contrast arteriography or magnetic resonance angiography. The vast majority of patients respond in the systemic phase to corticosteroid therapy. If surgical intervention is required for complications of the disease and tissue is sent for pathologic analysis, the artery wall appears identical to that in giant-cell arteritis. Surgical interventions are usually reserved as a treatment modality until the disease is in an inactive state.

Wegener's Granulomatosis

A German medical student, Heinz Klinger, was the first to actually describe this disease in 1931. A few years later, Friederich Wegener, a German pathologist, reported three additional cases and recognized the disorder as a form of vasculitis. It occurs in an even distribution between sexes and in people of all ages, but is more common in the 40-year-old age group. As with the other vasculitides, Wegener's can affect any organ in the body. The most common sites of involvement are the upper respiratory tract, lungs, and kidneys.

The diagnosis of Wegener's can be made by a blood test or direct biopsy of the affected organ. Anemia, mild leukocytosis, and an elevated erythrocyte sedimentation rate are nonspecific markers of the disease. Anti-neutrophil cytoplasmic antibody (ANCA) is the blood test used to diagnose Wegener's, specifically C-ANCA. This agent is directed against serine proteinase-3 and is relatively sensitive and highly specific for Wegener's. If positive, the test is useful to diagnose the disease early and to allow rapid institution of therapy, but it can be inaccurately interpreted and lead to misdiagnosis and mistreatment. Even with positive results from an ANCA, a biopsy proven diagnosis is still prudent. Lung biopsy and kidney biopsy are commonly performed in an attempt to confirm the diagnosis. The pathologic specimen is considered positive if a triad of granulomata, vasculitis, and tissue necrosis are observed.

Wegener's was almost uniformly fatal until the 1970s. Now, it is well treated with a combination of cyclophosphamide and prednisone. Even when successfully treated initially, patients can develop "flares" of the disease again in the future. With the described regimen, more than 90 percent of patients are successfully managed.

Immunologic Mechanisms of Disease

Giant-cell, or temporal, arteritis has provided the basis for a majority of the understanding of the immunologic mechanisms in the destructive processes of vasculitis. Three key observations have recently been made in defining this process. The first is that this vasculitis is a T-cell-dependent process. The second is that T-cell activation in the nonlymphoid environment of the arterial wall requires the activation of antigen-presenting cells, otherwise known as dendritic cells. And the third observation is that the blood vessel determines the site specificity of giant-cell arteritis.

The CD4+ T cell is the predominant cell in the vasculitic lesion. These data arise from animal models using mouse chimeras. It is also supported by observations in humans that clonally expanded populations of CD4+ T cells from isolated vascular lesions have identical receptors.

This concept of a T-cell response leading to arterial injury and the initiation of giant-cell arteritis is dependent on three crucial events: (1) T cells arrive and gain access to a site in which they do not normally reside; (2) there is an inciting antigen that is accessible to these T cells; and (3) antigen-presenting cells that are capable of T-cell stimulation go through a differentiation process. The site of entry for T cells in giant-cell arteritis is thought to be the vasa vasorum, but, before T-cell entry, dendritic cells play a large role. In fact, endothelial cells and dendritic cells together lead to recruitment of T cells into the vessel wall.

Dendritic cells were only recently recognized to reside in medium-sized vessels, and they appear to be isolated to the adventitia of these vessels. In healthy arteries, these cells remain immature and help maintain T-cell unresponsiveness. In fact, if T cells encounter antigen on an immature dendritic cell, an inhibitory signal is provided to the T cell. Therefore, one of the postulated roles of dendritic cells in the adventitial location is to prevent the activation of T cells. This finding of an immature dendritic cell is not seen in arteries of patients with giant-cell arteritis. Such lesions have mature, activated dendritic cells. These mature dendritic cells now foster and activate T cells as opposed to the unresponsiveness in their immature state. Therefore, it is reasoned that the activation of dendritic cells is an initiating step in the development of giant-cell arteritis.

A difference in the activated dendritic cells seen in temporal arteritis versus other inflammatory disorders also explains the pathophysiology of the disease. Usually, activated dendritic cells migrate to the regional lymph nodes and stimulate other cells by the production of chemokines when they reach that location. This function does not occur in temporal arteritis in which the dendritic cells stay localized to the arterial wall and elaborate chemokines activating migration of cells to the affected vessel.

The formation of granulomas is dependent on T cells and is a result of their response to indigestible antigens. It also appears that interferon-γ is required for the development of a granuloma. The media of the arterial wall is the location of the formation of granulomas in temporal arteritis. This development is dependent on the production of interferon-γ by T cells located in the adventitia of the arterial wall. The granulomatous reaction can wall off immunogens, but leads to damage of the surrounding tissue. Tissue necrosis that occurs from the release of lytic enzymes is seen in other vasculitides, such as Wegener's granulomatosis and Churg-Strauss syndrome. Interestingly, necrosis is not a feature of giant-cell arteritis, and its presence excludes this diagnosis. Thinning of the arterial wall can occur and has been implicated in the development of aneurysmal formation in these patients.

The chief target of oxidative attack in patients with giant-cell arteritis is the medial smooth muscle cell. In normal blood vessels, the media is inaccessible to inflammatory cells because of the lack of capillaries in this location. The formation of neocapillaries in an inflamed artery leads to the recruitment of cellular activity causing arterial damage. The cell responsible for this response is the macrophage.

Macrophages in the media of blood vessels produce reactive oxygen intermediates resulting in the oxidative attack just mentioned. Oxygen-derived free radicals injure tissue through several mechanisms, the most important of which is the oxidation of membrane lipids. This results in structural disintegration of the cell and cellular death. Another mechanism is the interruption of intracellular signaling pathways. This interruption can affect the balance between the destructive and protective mechanisms characteristic of the pattern of response to injury in giant-cell arteritis. The usual course of the injury process is to cause intimal hyperplasia with resultant ischemic complications (Figure 1). The most feared complication of temporal arteritis, blindness, is produced by severe vascular stenoses caused by the oxidative process of injury to nutrient arteries. The variable response seen among patients is not well understood.

Conclusions

Vasculitic disease remains poorly understood for the majority of the subtypes. Interestingly, therapeutic regimens have been developed that can significantly alter the course of the diseases despite a lack of advancement in full understanding of the pathophysiology of the vasculitis process. It is hoped that the growing knowledge of the inflammatory mechanisms seen in giant-cell arteritis will further assist with the management of all types of vasculitis.

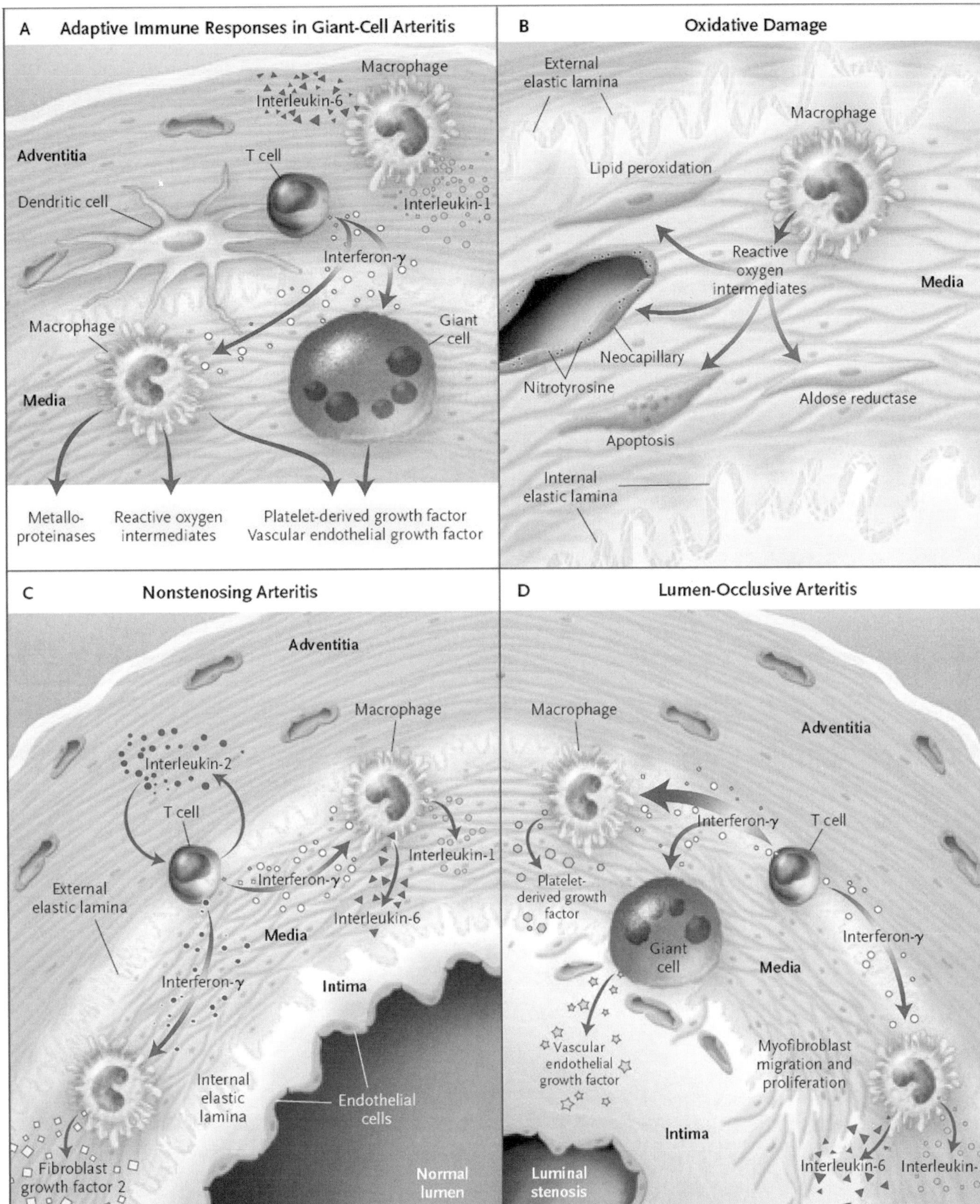

Figure 1 Adaptive immune responses in vasculitis and the consequences of arterial-wall injury. **(A)** Activation and trapping of dendritic cells in the arterial adventitia generate the conditions required for the recruitment and stimulation of antigen-specific T cells. CD4+ T cells that enter the microenvironment of the arterial wall interact with dendritic cells and begin secreting cytokines. Interferon-γ is a critical cytokine that regulates the differentiation and function of macrophages. The functional commitment of the macrophages in the vascular infiltrates is closely linked to their location in the arterial wall. Macrophages in the adventitial layer supply the inflammatory cytokines interleukin-1 and interleukin-6. Macrophages in the media secrete metalloproteinases and play a critical part in oxidative injury through the production of reactive oxygen intermediates. **(B)** Three aspects of oxidative damage in the media. Protein nitration occurs in endothelial cells lining neocapillaries. Toxic aldehydes are formed in the process of lipid peroxidation, and smooth-muscle cells undergo apoptosis. In parallel, reactive oxygen intermediates also trigger cellular activation, as exemplified by the induction of aldose reductase. The response of the artery to injury is shown in panels C and D. Arteritis does not necessarily result in luminal stenosis and may proceed without compromising blood flow **(C)**. In patients with ample production of platelet-derived growth factor and vascular endothelial growth factor, rapid and exuberant intimal hyperplasia ensues, causing lumen-occlusive arteritis **(D)**. Accordingly, the clinical presentation of arteritis may or may not include ischemic complications. From Weyand, C. M., and Goronzy, J. J. (2003). Medium- and large-vessel vasculitis. *N. Engl. J. Med.* **349**(2), 160–169.

References

Abril, A., Calamia, K. T., and Cohen, M. D. (2003). The Churg Strauss syndrome (allergic granulomatous angiitis): Review and update. *Semin. Arthritis Rheum.* **33**(2), 106–114.

Frankel, S. K., Sullivan, E. J., and Brown, K. K. (2002). Vasculitis: Wegener granulomatosis, Churg-Strauss syndrome, microscopic polyangiitis, polyarteritis nodosa, and Takayasu arteritis. *Crit. Care Clin.* **18**(4), 855–879.

Manna, R., Miele, L., La Regina, M., Grieco, A., and Gasbarrini, G. Cryoglobulinemia: A true internistic disease? *Int. J. Immunopathol. Pharmacol.* **16**(1), 33–41.

Mills, J. L., Sr. (2003). Buerger's disease in the 21st century: Diagnosis, clinical features, and therapy. *Semin. Vasc. Surg.* **16**(3), 179–189.

Nordborg, E., and Nordborg, C. (2004). Giant cell arteritis: Strategies in diagnosis and treatment. *Curr. Opin. Rheumatol.* **16**(1), 25–30.

Parra, J. R., and Perler, B. A. (2003). Takayasu's disease. *Semin. Vasc. Surg.* **16**(3), 200–208.

Shepherd, R. F., and Rooke, T. (2003). Uncommon arteriopathies: What the vascular surgeon needs to know. *Semin. Vasc. Surg.* **16**(3), 240–251. *An excellent review of the manifestations of the vasculitides as they affect the vascular surgeon. Surgical treatment is reviewed as well as the expected outcomes based on interesting case presentations.*

Stone, J. H. (2002). Polyarteritis nodosa. *JAMA* **288**(13), 1632–1639.

Tamargo, R. J., Connolly, E. S., Jr., McKhann, G. M., Khandji, A., Chang, Y., Libien, J., and Adams, D. (2003). Clinicopathological review: Primary angiitis of the central nervous system in association with cerebral amyloid angiopathy. *Neurosurgery* **53**(1), 136–143; discussion 143.

Weyand, C. M., and Goronzy, J. J. (2003). Medium- and large-vessel vasculitis. *N. Engl. J. Med.* **349**(2), 160–169. *The authors thoroughly review the most advanced understanding of the pathologic mechanisms of medium- and large-vessel vasculitis. This article is clearly written with excellent illustrations.*

Yazici, H. (2003). Behçet's syndrome: An update. *Curr. Rheumatol. Rep.* **5**(3), 195–199.

Capsule Biography

Dr. Milner is an Assistant Professor of Surgery in the Division of Vascular Surgery at Emory University School of Medicine. He has a strong interest in the endovascular management of peripheral vascular disease.

Dr. Smith is John E. Skandalakis Professor of Surgery Emeritus and former Head of the Vascular Surgery Division, Emory University School of Medicine, where he has been a member of the faculty for more than 38 years. He is currently Associate Chairman of the Department of Surgery and Medical Director of Emory University Hospital. In addition, he serves on the Executive Committee of the Board of Commissioners, Joint Commission on Accreditation of Healthcare Organizations.

PART IV

Therapy

SECTION A

Hemorrhagic Shock

CHAPTER 151

The Effect of Fluid Treatment on the Microvasculature in Uncontrolled Hemorrhagic Shock

Michael M. Krausz

Department of Surgery A and the Laboratory for Shock and Trauma Research, Rambam Medical Center, and The Bruce Rappaport Faculty of Medicine, Technion-Israel Institute of Technology, Haifa, Israel

Clinical as well as laboratory studies on hemorrhagic shock have questioned the initial therapeutic goal of normalizing systemic blood pressure by isotonic or hypertonic crystalloid solutions.

The current regimen of prehospital fluid administration consists of using massive volumes of isotonic crystalloid solutions such as Ringer's lactate administered as rapidly as possible. These recommendations for the management of trauma victims who are hypotensive from acute hemorrhage are based on the interpretation of data from studies conducted beginning in the 1940s. These classic studies utilized a Wiggers model of hemorrhage in which animals were bled from an intravascular catheter into a reservoir and maintained at a prescribed level of hypotension prolonged periods of time prior to initiating resuscitation. When the animals were subjected to several hours of hypotension that included a short period of severe hypotension (mean arterial pressure of 30 to 40 mmHg) they developed microvascular damage, termed "irreversible shock," which ultimately led to death of the animals despite what was considered appropriate resuscitation. This prolonged period of hemorrhagic hypotension was associated with the development of microvascular injury with marked extracellular fluid deficit that could be corrected only by the administration of isotonic crystalloids in volumes 2 to 3 times the estimated blood loss to achieve survival. This was the basis of the current well known dogma "3 to 1 rule" for the treatment of hemorrhagic hypotension, which was adopted by the ATLS for the treatment of trauma casualties.

Based on these data, advocates of early aggressive resuscitation argued that the need for increasing cardiac output and oxygen delivery, to maintain microvascular perfusion and oxygenation, exceeds any risk of accentuating hemorrhage, and therefore trauma victims in hypotensive hemorrhage should receive large volumes of fluids as early as possible.

Data published by Stern et al. support Wiggers's observations that severe hemorrhagic hypotension left untreated for several hours will progress to end-organ dysfunction and ultimately death. However, substantial periods of moderate hypotension (MAP 50 to 60 mmHg), such as the interval from time of injury to the time of operative intervention, can be tolerated effectively without resulting in significant permanent microvascular damage and end-organ failure.

These fixed-pressure (Wiggers) and fixed-volume models of hemorrhagic shock were developed and refined over the past century and have served as a basis for developing the standard therapeutic approaches of aggressive fluid therapy, whereas studies using uncontrolled hemorrhage models that have emerged more recently have been used increasingly over the past decade.

The model of uncontrolled continuing hemorrhage was assumed to be more relevant to the clinical scenario of trauma casualties with internal bleeding that cannot be effectively controlled in the prehospital environment.

The fundamental problem in patients with exsanguinating hemorrhage is a bleeding source that cannot be readily controlled. In this situation the common practice of "raising blood pressure" makes no clinical sense because this action is likely to increase blood pressure, reduce vasoconstriction, and accelerate the rate of bleeding.

The goal of therapy in traumatic internal hemorrhage leading to severe hemorrhagic shock was therefore altered to rapid evacuation to a trauma center, termed "scoop and run," and development of therapies aimed at buying time until surgical control of bleeding is possible.

The goal of deliberate underresuscitation was recognized as early as World War I by Cannon, who wrote in 1918:

Injection of fluid that will increase blood pressure has dangers in itself. Hemorrhage in a case of shock may not have occurred to a marked degree because blood pressure has been too low and the flow too scant to overcome the obstacle offered by the clot. If the pressure is raised before the surgeon is ready to check the bleeding that may take place, blood that is sorely needed, will be lost.

More than 30 years ago, Milles et al. and Shaftan et al. observed that spontaneous control of arterial hemorrhage depends on clot formation and a fall in blood pressure. Animals subjected to arterial injury and subsequently made normotensive with either fluid resuscitation or vasopressor infusion experienced larger hemorrhage volumes, longer hemorrhage durations, and a higher incidence of rebleeding compared to animals subjected to a similar injury but left untreated and hypotensive.

In 1988 two types of experimental hemorrhagic shock were defined: controlled hemorrhagic shock, in which bleeding was induced through an arterial catheter that was occluded immediately after hemorrhage (as in Wiggers-type models); and uncontrolled hemorrhagic shock, in which internal abdominal blood loss was induced by injury to major intra-abdominal blood vessels (ileocolic artery). In this model blood loss stopped only after blood pressure was markedly reduced with vasoconstriction and temporary clot formation. Bolus small-volume hypertonic saline infusion in this rat model of uncontrolled hemorrhagic shock or in uncontrolled hemorrhage induced by rat tail resection led to resumed blood loss from the injured blood vessels, hemodynamic deterioration, and increased mortality compared to no resuscitation.

These results were later confirmed by Bickell et al. in 1992 in other animal models such as a pig model with an acute aortotomy.

The mechanisms suggested for increased bleeding following bolus fluid resuscitation in uncontrolled hemorrhagic shock include increased blood pressure, vasodilatation, and disruption of an early unstable thrombus followed by a fatal secondary hemorrhage. In addition, rapid intravenous infusion of crystalloid may promote hemorrhage by diluting coagulating factors and by lowering blood viscosity in the microvasculature, thereby decreasing the resistance to flow around an incomplete thrombus. Shaftan et al. described direct observations regarding the effect of blood pressure elevation on clot formation. They noted that following blood vessel injury and hypotension a large but "soft and jelly like" extramural clot was formed that surrounded the vascular defect. As time elapsed the clot was observed to become more palpably firm. With fluid resuscitation and blood pressure elevation, the margins of the clot began to leak, and rebleeding subsequently occurred. This gross and visual description is consistent with light and electron microscopic observations that transformation of an initially friable unstable clot to a more rigid hemostatic plug requires the formation and deposition of a significant amount of fibrin, which occurs in an intact coagulation system after at least 20–30 minutes. Therefore, it is not surprising that resuscitation strategies that result in early increase in blood flow and blood pressure would present a higher risk of accentuating ongoing or initiating new hemorrhage.

Following these animal studies of uncontrolled hemorrhage, Bickell et al. published in 1994 a clinical study in patients with penetrating torso injury, comparing those who received delayed fluid resuscitation only after the source of hemorrhage was controlled to patients receiving immediate fluid resuscitation according to the ATLS guidelines. It was found that delayed resuscitation was followed by improved survival of 70 percent compared to 62 percent in the early treatment group.

These results went against the conventional dogma that early prompt administration of large volumes of intravenous crystalloids improves outcome in trauma patients.

Subsequently more sophisticated animal studies in models of uncontrolled hemorrhage by Kowalenko et al. in 1992 suggested that fluid resuscitation to a mean arterial pressure of 40 to 60 mmHg improved microvascular perfusion optimally to a sufficient level to improve survival compared to over- or underresuscitation and gained sufficient time until the bleeding can be surgically controlled. This was termed *hypotensive resuscitation*.

Most laboratory models and clinical observations of uncontrolled hemorrhagic shock in trauma casualties were studied in large-vessel injury models and did not incorporate solid-organ tissue injury that may be a critical variable in blood loss and survival.

In an effort to more accurately simulate the tissue damage and the rapid blood loss clinically seen in severe blunt abdominal trauma, animal models of solid organ injury, with a standardized liver injury in rats, lethal hemorrhagic shock in a swine liver injury model, and massive splenic injury in rats have been developed. In these models, animal survival was critically determined by the severity of injury, rate of blood loss, and the type of fluid that was used for resuscitation.

Matsuoka et al., using a standardized model of liver injury of uncontrolled hemorrhagic shock, claimed that large-vessel-injury animal models are relevant only to a small segment of penetrating injuries but not to most patients with blunt abdominal trauma with solid organ injury. In this liver-injury model the infusion of hypertonic saline or large volumes of isotonic solutions maintained circulatory stability in spite of increased bleeding with no increased mortality.

Further studies in uncontrolled hemorrhage following solid-organ injury were performed in a rat model of massive or moderate splenic injury. In this model bolus infusion of large volumes of normal saline, Ringer's lactate, or hydroxyethyl starch was followed by increased bleeding and increased mortality similar to large-vessel injury, whereas hypertonic saline (HS) infusion in this model of solid organ injury did not increase blood loss or mortality.

The clinical use of HS solutions for the resuscitation of trauma patients has been debated for many years. Although HS has been approved for use throughout Europe, North America, and South America, a clear endorsement for its use has not emerged. The few large studies on prehospital administration of HS and dextran (HSD) have not demonstrated a significant survival advantage. Proponents of using these solutions cite beneficial physiologic responses seen with HSD. Those opposed to using HSD express concern over its application in settings of uncontrolled hemorrhage because of a propensity to increase bleeding.

HS rapidly improves blood pressure and cardiac output in the severely injured patient. This is generally attributed to its ability to draw water into the vascular space, leading to increase in plasma volume. The beneficial effects of HS are, however, transitory because it increases microvascular permeability by shrinkage of endothelial cells leading to opening gap junctions. Dextran was therefore added to prolong its ability to hold water in the vascular space. The reflection coefficient of dextran (0.8), which is similar to that of albumin (0.9), provides the oncotic driving force for water flow through the newly expanded gap junctions into the vascular space.

The difference in the response to hypertonic saline in uncontrolled hemorrhagic shock, between large-vessel injury and solid-organ injury, is not readily explained. HS infusion in large-vessel injury leads to increased blood loss, probably by dislodgement of the temporary clot formed in the injured large vessel induced by the temporary increase in blood pressure. It is hypothesized that in solid organ (hepatic or splenic) injury the multiple small clots that were formed in the small-caliber vessels throughout the parenchyma of the injured organ cannot presumably be dislodged by the increase in blood pressure, and thus rebleeding is not initiated, and the net effect of the increase in plasma volume by the fluid shift into the circulation caused by HS is manifested.

In a further study aimed at simulating more closely the clinical scenario of uncontrolled hemorrhagic shock following solid-organ injury treated by isotonic solutions, slow continuous volume infusion of Ringer's lactate combined with splenectomy was used to treat massive splenic injury. In this model of uncontrolled hemorrhagic shock, continuous infusion of moderate volumes of crystalloid or colloid solutions resulted in less blood loss and improved hemodynamics and survival when compared to no fluid infusion.

These results were later confirmed by direct comparison of rapid bolus infusion of large volumes of Ringer's lactate (RL) with splenectomy to slow continuous infusion and splenectomy in the rat model of uncontrolled hemorrhagic shock after massive splenic injury. It was found that continuous infusion of RL combined with splenectomy resulted in significantly less blood loss and improved survival compared to bolus infusion of RL or untreated animals. Another significant finding of this study was that bolus as well as continuous infusion of small-volume hypertonic saline (5 mL/kg NaCl 7.5%) did not increase blood loss in this model of solid-organ injury and resulted in improved survival compared to untreated animals.

Similar results were also observed by Stern et al. in a swine model of uncontrolled hemorrhagic shock induced by large-vessel injury (4-mm aortic tear). In this model of near-lethal uncontrolled hemorrhage slow infusion of HSD (7.5% NaCl/6% dextran-70) restored cardiodynamics while minimizing hemorrhage volume and mortality. Rapid infusion of HSD in this model led to an immediate rise in aortic pressure and flow within 1 minute leading immediately to rebleeding from the aortic tear.

Conclusion

Resuscitation strategies of bolus large-volume resuscitation with isotonic solutions in uncontrolled hemorrhagic shock induced by large-vessel injury, as well as solid-organ injury, result in early abrupt increase in blood pressure and microvascular blood flow, which lead to clot dislodgement and increased rebleeding from large and small injured blood vessels, hemodynamic deterioration, and increased mortality. Limited continuous volume resuscitation titrated to achieve optimal microvascular perfusion and pressure that is sufficient to maintain organ viability until the bleeding source can be surgically controlled (termed hypotensive resuscitation) results in less blood loss and improved survival.

Bolus hypertonic saline resuscitation in uncontrolled hemorrhagic shock following large-vessel injury also results in increased blood loss, hemodynamic instability, and increased mortality. Bolus infusion of HS in this setting of acute hemorrhagic shock should therefore be avoided. Bolus as well as continuous HS resuscitation following solid-organ injury does not increase blood loss from the injured organ and leads to improves hemodynamics and survival.

Glossary

Controlled hemorrhage: hemorrhage that was topped
Fluid resuscitation: infusion of fluids to improve hemodynamics
Hypertonic saline: NaCl—7.5%
Large-vessel injury: damage to blood vessels larger than 3mm
Solid-organ injury: injury to intra-abdominal organs
Uncontrolled hemorrhage: hemorrhage that was not occluded

Bibliography

American College of Surgeons Committee on Trauma (1997). *Advanced Trauma Life Support Course for Physicians*, pp. 87–107. American College of Surgeons.

Bickell, W. H., Bruttig, S. P., Millnamow, G. A., O'Benar, J., and Wade, C. E. (1991). The detrimental effects of intravenous crystalloid after aortotomy in swine. *Surgery* **110**, 529–536.

Bickell, W. H., Wall, M. J., Pepe, P. E., Martin, R. R., Ginger, W. F., Allen, M. K., and Mattox, K. L. (1994). Immediate versus delayed fluid resuscitation for hypotensive patients with penetrating torso injuries. *N. Engl. J. Med.* **33**, 1105–1109.

Cannon, W. B., Fraser, J., and Cowell, E. M. (1918). The preventive treatment of wound shock. *JAMA* **70**, 618–621.

Gross, D., Landau, E. H., Assalia, A., and Krausz, M. M. (1988). Is hypertonic saline resuscitation safe in "uncontrolled" hemorrhagic shock? *J. Trauma* **28**, 751–756. *This article is the first study that demonstrates the danger of rebleeding induced by hypertonic saline treatment of uncontrolled hemorrhagic shock.*

Kowalenko, T., Stern, S. A., Dronen, S. C., and Wang, X. (1992). Improved outcome with hypotensive resuscitation of uncontrolled hemorrhagic shock in a swine model. *J. Trauma* **33**, 349–353.

Krausz, M. M. (1995). Controversies in shock research: Hypertonic resuscitation pros and cons. *Shock* **3**, 69–72. *This work stresses the advantages and disadvantages of hypertonic saline resuscitation of uncontrolled hemorrhagic shock.*

Manning, J. E., Katz, L. M., Brownstein, M. R., Pearce, L. B., Gawryl, M. S., Baker, C. C., and the Carolina Resuscitation Research Group (2000). Bovine hemoglobin-based oxygen carrier (HBOC-201) for resuscitation of uncontrolled exsanguinating liver injury in swine. *Shock* **13**, 152–159.

Mattox, K. L., Maningas, P. A., Moore, E. E., Mateer, J. R., Marx, J. A., Aprahamian, C., Burch, J. M., and Pepe, P. E. (1991). Prehospital hypertonic saline/Dextran infusion for post-traumatic hypotension. *Ann. Surg.* **1213**, 482–491.

Mazzoni, M. C., Borgstrom, P., Arfors, K. E., and Intaglietta, M. (1988). Dynamic fluid redistribution in hyperosmotic resuscitation of hypovolemic hemorrhage. *Am. J. Physiol.* **255**, H629–H637.

Sixma, J. J., and Webster, J. (1977). The hemostatic plug. *Semin. Hematol.* **14**, 265–299.

Solomonov, E., Hirsh, M., Yahiya, A., and Krausz, M. M. (1977). The effect of vigorous fluid resuscitation in uncontrolled hemorrhagic shock following massive splenic injury. *Crit. Care Med.* **28**, 749–754. *This article is the first study that demonstrates that massive fluid resuscitation induces rebleeding and increased mortality following solid organ injury.*

Stern, S. A., Kowalenko, T., Younger, J., Wang, X., and Dronen, S. C. (2000). Comparison of the effects of bolus vs. slow infusion of 7.5% Na Cl/6% dextran-70 in a model of near lethal uncontrolled hemorrhage. *Shock* **14**, 616–622.

Victorino, G. P., Newton, C. R., and Curran, B. (2002). The effect of tonicity and hypertonic solutions on microvascular permeability. *J. Surg. Res.* **104**, 101–105.

Wiggers, C. J. (1950). Experimental hemorrhagic shock. In: *Physiology of Shock*, pp. 121–146. New York: Commonwealth Fund.

Capsule Biography

Dr. Krausz has headed the Department of Surgery A at the Rambam Medical center in Haifa, Israel, since 1996. His main research interests are fluid treatment of trauma casualties in civilian as well as military scenarios; lately his research has also focused on gender differences in hemorrhagic as well as septic shock, and the role of leukocytes in pathophysiology of these phenomena. He has received grants from the Israel Academy of Sciences and the Israeli Ministry of Health.

SECTION B

Preconditioning

Chapter 152

Mechanisms Involved in Delayed Preconditioning

Initiating, Signaling, and Effector Components

Tao Rui, Gediminas Cepinskas, and Peter R. Kvietys
Vascular Biology/Inflammation Program, Lawson Health Research Institute, London, Ontario, Canada

Introduction

Reperfusion of ischemic myocardium results in pathological changes in the heart, such as myocardial stunning, an acute inflammatory response, and infarction. Paradoxically, pretreatment of the heart with an ischemia–reperfusion (I/R) challenge confers protection against a subsequent I/R insult. This phenomenon, referred to as *ischemic preconditioning* (PC), has two distinct phases based on the time course and mechanisms involved. An early phase of PC occurs within minutes after the initial I/R challenge and persists for only a few hours. This early PC is independent of protein synthesis and, thus, relies on activation of existing effector molecules. There is also a later phase of PC, which becomes apparent 24 hours or later after the initial I/R challenge and persists for up to several days (delayed PC). Delayed PC is dependent on protein synthesis and, thus, represents a genetically mediated adaptive response. Herein, we will focus on some of the proposed mechanisms involved in the development of delayed PC with respect to the initiating (reactive oxygen metabolites), signaling (nuclear transcription factors), and effector (nitric oxide synthase and superoxide dismutase) components.

The heart is a complex organ consisting of different cell types with specialized functions (myocytes, endothelial cells, and so on). These individual cells may respond differently to I/R and use distinctly different mechanisms to contribute to delayed PC in the myocardium. Thus, both in vivo and in vitro approaches have been used to address the mechanisms involved in the development of delayed PC. In vivo studies generally involve an initial series of short I/R challenges to the heart followed 24 hours (or later) by a more prolonged I/R challenge (Figure 1A). The in vitro approaches simulate I/R by targeting the oxygenation aspects of I/R. Typically isolated cardiac myocytes are exposed to anoxia (or hypoxia) and, subsequently, reoxygenated (anoxia/reoxygenation; A/R) (Figure 1B).

Initiating Molecules

Reactive oxygen metabolites (ROM) are generated within cells as a normal consequence of aerobic metabolism. Mammalian cells contain antioxidant enzymes (catalase, SOD, and so on) to scavenge and neutralize these oxidants. It is generally believed that the oxidant production after an I/R challenge overwhelms the detoxifying capacity of the endogenous antioxidants and initiates tissue injury.

The following lines of evidence support a role for ROM in the initiation of delayed PC in vivo. An I/R challenge to the heart in situ results in the local generation of oxidants. Administration of antioxidants during the initial I/R

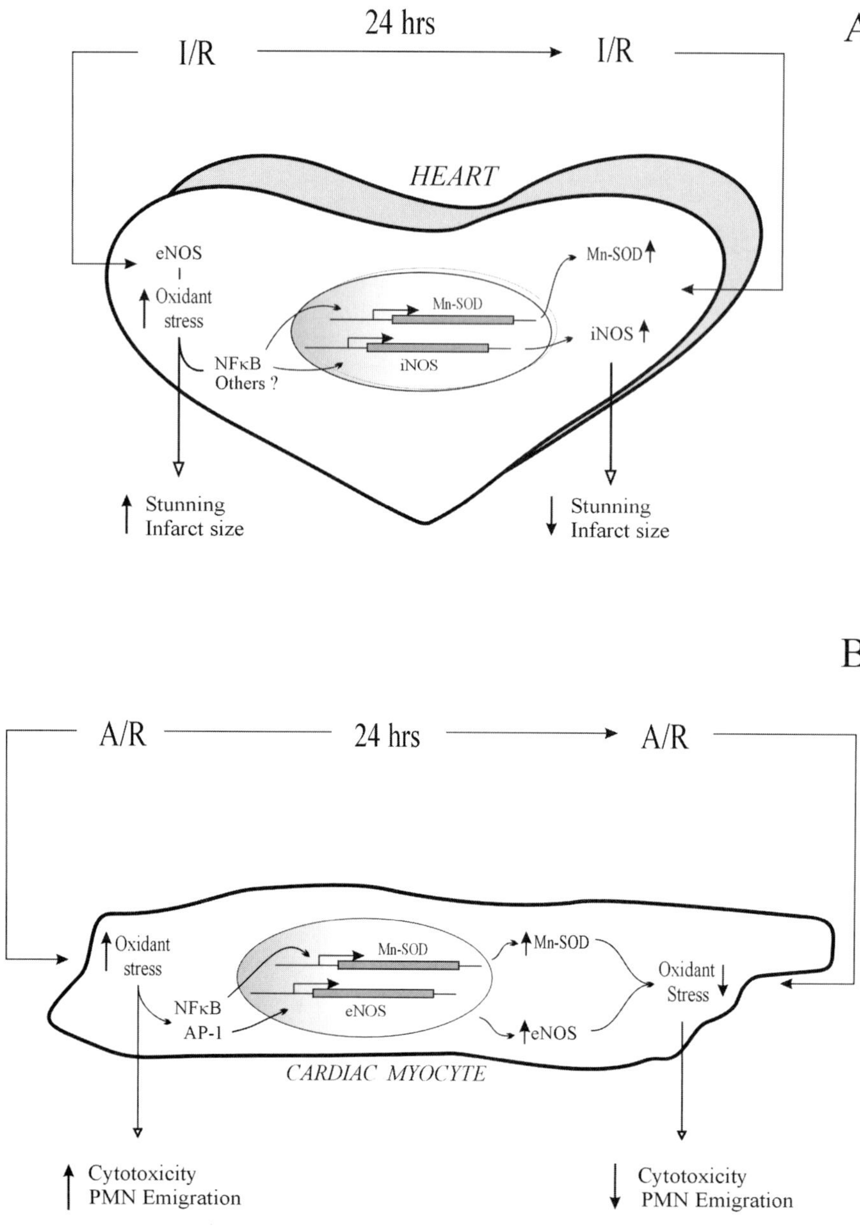

Figure 1 Current hypotheses based on in vivo (**A**) and in vitro (**B**) studies on the mechanisms involved in the development of delayed preconditioning. **(A)** In vivo studies involve an initial series of short ischemia–reperfusion (I/R) challenges (e.g., six cycles of 4-minute coronary artery occlusion followed by 4-minute reperfusion) and 24 hours later challenged with a more prolonged period of I/R (e.g., 30-minute period of occlusion and one 24-hour period of reperfusion). **(B)** In vitro studies involve exposing isolated cells to a 30-minute period of anoxia, reoxygenating them (A/R), and 24 hours later subjecting these cells to the same A/R protocol. See text for the lines of evidence supporting the proposed initiating (e.g., oxidants), signaling (nuclear transcription factors), and effector (NOS and SOD) components of the development of delayed preconditioning.

challenge prevents the development of delayed preconditioning with respect to myocardial stunning and infarct size. Local administration of oxidant-generating chemicals (rather than an initial I/R challenge) can induce the development of delayed PC.

In vitro studies indicate that an oxidant stress is also important in initiating the development of delayed PC in isolated cardiac myocytes. Exposure of myocytes to a mild A/R challenge (30/30 minutes) induces (1) an oxidant stress and (2) the rapid development of a proinflammatory phenotype (these myocytes can promote PMN transendothelial migration). Interestingly, the in vitro studies have uncovered an important communication link between myocytes and endothelial cells, that is, the A/R-induced increase in myocardial oxidant stress is transferred to the adjacent endothelial cells and contributes to the PMN transendothelial migration [1]. When cardiac endothelial cells derived from mice overexpressing Mn-SOD are used in assays, the A/R-challenged myocytes no longer induce an oxidant stress in the adjacent endothelial cells and the myocytes no longer promote PMN transendothelial migration (Figure 2). A more severe A/R challenge (1 hour of anoxia) can also induce cell death in some of the myocytes. If the myocytes are pretreated with a mild A/R challenge 24 hours earlier, (1) the subsequent A/R challenge does not result in an oxidant stress, (2) a proinflammatory phenotype does not develop, and (3) there is less cytotoxicity (delayed PC). Delayed PC can also be induced by pretreatment of the myocytes with H_2O_2 rather than an A/R challenge. Finally, delayed PC can be prevented by pretreating the myocytes during the initial A/R challenge with an antioxidant.

Taken together, both in vivo and in vitro studies indicate that induction of an oxidant stress during the initial I/R or A/R challenge plays an important role in initiating the development of delayed PC (Figure 1).

The source of the ROM within the myocytes is not clear and appears to be controversial. There are three isoforms of NOS in the myocardium: endothelial NOS (eNOS), neuronal NOS (nNOS), and inducible NOS (iNOS). Of these three isoforms, eNOS appears to be involved in the initiation

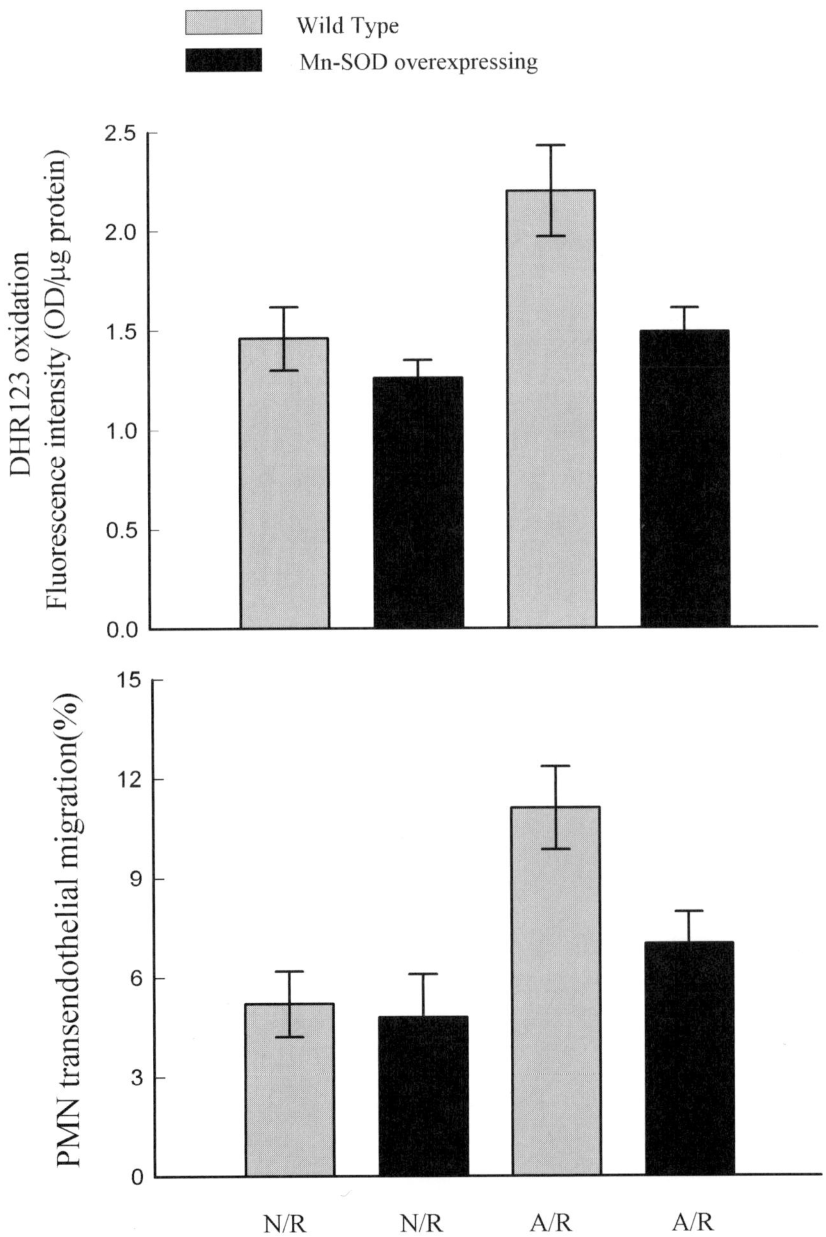

Figure 2 Endothelial cell oxidant stress (*top*) and PMN transendothelial migration (*bottom*) induced by supernatants from A/R-conditioned myocytes was diminished when endothelial cells derived from hearts of Mn-SOD overexpressing mice were used. From Rui et al. (2001). *Am. J. Physiol. Heart Circ. Physiol.* **281**, H440–H447.

of delayed PC in the heart. Brief periods of I/R challenge are associated with NO biosynthesis, the most likely source being eNOS. Based on pharmacological approaches, NO generated by eNOS appears to play a role in initiating delayed PC in the heart with respect to stunning and infarction. Administration of nonspecific NOS inhibitors (but not specific iNOS inhibitors) prior to the initial I/R challenge prevents delayed PC. Conversely, local administration of NO donors (rather than an initial I/R challenge) can induce delayed PC. Taken together, these observations suggest that NO derived from eNOS initiates the development of delayed PC (presumably via the generation of ROM). In vitro studies using isolated cardiac myocytes are not in agreement with the in vivo studies. The initial A/R challenge does not result in the production of NO, and pharmacological blockade of NO synthase does not prevent the oxidant stress induced by the initial A/R challenge. Furthermore, the initial A/R challenge can still induce an oxidant stress in cardiac myocytes isolated from eNOS deficient mice. Similar results have been noted with cardiac endothelial cells, that is, the initial A/R challenge induces an oxidant stress in endothelial cells harvested from eNOS-deficient mice.

Signaling Molecules

Of the potential nuclear transcription factors that could be involved in the development of delayed PC, NFκB has received the most attention. ROM can activate NFκB, thereby allowing it to translocate to the nucleus. Increased levels of NFκB have been noted in nuclear extracts obtained from hearts subjected to I/R, an effect prevented by the administration of an antioxidant during the I/R challenge. Furthermore, pharmacological prevention of NFκB activation and translocation to the nucleus abrogates the development of delayed PC with respect to myocardial stunning and infarction. In vitro studies are in general agreement, that is, an initial A/R challenge imposed on cardiac myocytes results in the mobilization of NFκB to the nucleus. A similar observation was noted using isolated human umbilical vein endothelial cells (HUVECs). Pharmacological prevention of NFκB activation or translocation to the nucleus after the initial A/R challenge to cardiac myocytes prevented the development of delayed PC with respect to (1) an oxidant stress and (2) the development of a proinflammatory phenotype. That is, under these conditions, the oxidant stress and myocyte-mediated PMN transendothelial migration was the same after the initial and second A/R challenges. In HUVECs, introduction of an oligonucleotide with consensus binding sites for NFκB during the first A/R challenge prevented the development of delayed PC with respect to PMN adhesion to HUVECs. Thus, both in vivo studies and in vitro studies support a role for NFκB as an important signaling transcription factor in the development of delayed PC (Figure 1).

In vitro studies have also uncovered a potential role for another nuclear transcription factor in delayed PC in cardiac myocytes. After the initial challenge there was an increased accumulation of AP-1 in the nucleus. Furthermore, introduction of an oligonucleotide containing binding sites for AP-1 during the first A/R challenge prevented the development of delayed PC with respect to (1) myocyte oxidative stress and (2) myocyte-mediated PMN transendothelial migration (unpublished observation).

Effector Molecules

Superoxide Dismutase (SOD)

Administration of antioxidant enzymes (e.g., SOD, catalase) can ameliorate I/R-induced myocardial injury. Thus, it has been proposed that an increase in the antioxidant status of the heart can contribute to the development of delayed PC. Of the endogenous antioxidants that are upregulated after an I/R challenge, SOD (particularly Mn-SOD) has received the most attention. Overexpression of Mn-SOD (transgenic mice) results in a reduction in the I/R-induced myocardial infarct size. Furthermore, endogenous Mn-SOD protein and activity are increased in the myocardium during the development of delayed PC. There appears to be a direct correlation between the increase in myocardial Mn-SOD activity and the reduction in infarct size. Finally, inhibition of Mn-SOD protein synthesis and activity by the use of an antisense approach prevents the development of delayed PC with respect to infarct size. In isolated cardiac myocytes, Mn-SOD, but not Cu/Zn-SOD, activity is increased 24 hours after the initial A/R challenge. Pretreatment of the myocytes with a Mn-SOD antisense oligonucleotide prior to the initial A/R challenge (to prevent Mn-SOD synthesis) prevents the increase in SOD activity typically observed 24 hours after the initial A/R challenge. In addition, this procedure also prevents the development of delayed PC with respect to (1) PMN transendothelial migration and (2) myocyte death. Collectively, these in vivo and in vitro observations provide strong support for the contention that the development of delayed PC is dependent on the induction of Mn-SOD (Figure 1).

The in vitro studies have also provided some strong evidence favoring a role for NFκB as the nuclear transcription factor involved in upregulating Mn-SOD. Pharmacological inhibition of NFκB activation or translocation during the initial A/R challenge prevents the induction of Mn-SOD. In addition, this maneuver prevents the development of delayed PC with respect to (1) myocyte oxidative stress and (2) myocyte-mediated PMN transendothelial migration.

Nitric Oxide Synthase (NOS)

Of the three isoforms of NOS (eNOS, nNOS, and iNOS), eNOS and iNOS appear to be involved in the development of delayed PC in the heart.

The iNOS isoform appears to play an important role as an effector molecule in the development of delayed PC at the whole organ level (Figure 1A). NO production is increased in the myocardium at 24 hours after the initial I/R challenge. Message levels of iNOS are increased within 3 hours after the initial I/R challenge, and iNOS protein and activity are increased at 24 hours. By contrast, eNOS protein and activity are unaltered at 24 hours after the initial I/R insult. Administration of selective iNOS inhibitors during the second I/R challenge prevents delayed PC with respect to both myocardial stunning and infarction. Finally, delayed PC (infarct size) is abrogated in iNOS-deficient mice [2].

In vitro studies have yielded strikingly different conclusions regarding the role of eNOS and iNOS as effector enzymes in the development of delayed PC [3]. The initial challenge of myocytes derived from iNOS-deficient mice resulted in an increase in myocyte oxidative stress. If these myocytes were pretreated with an A/R challenge 24 hours earlier, the second A/R challenge did not induce an increase in myocyte oxidant stress (delayed PC). In a similar fashion, the initial challenge of myocytes derived from eNOS-deficient animals resulted in an increase in myocyte oxidant stress. However, an A/R challenge imposed 24 hours after an initial A/R challenge to myocytes from eNOS-deficient mice still resulted in an increase in myocyte oxidant stress (no delayed PC). Identical results were obtained using isolated cardiac endothelial cells, rather than myocytes in the in vitro assays, that is, delayed PC (with respect to an oxidant stress) could be demonstrated in endothelial cells derived from iNOS-deficient mice, but not in those derived from eNOS-deficient mice. In myocytes derived from wild-type mice, mRNA for eNOS, but not iNOS, increased after the initial A/R challenge. Similarly, mRNA for eNOS increased after an A/R challenge imposed on myocytes derived from iNOS-deficient mice. Taken together these findings indicate that eNOS-derived NO is a prerequisite for the development of delayed PC in isolated cardiac myocytes (Figure 1B).

The reason for the disparate results between in vivo and in vitro studies regarding the isoform of NOS (iNOS versus eNOS) serving as an effector enzyme in delayed PC is not entirely clear. However, one possible explanation deserves comment. The in vivo studies targeted the heart in situ with an intact systemic circulation, while the in vitro studies used cardiac myocytes isolated from their natural external milieu. Based on studies of the systemic inflammatory response syndrome, it is quite possible that multiple I/R challenges to the heart in vivo would result in the systemic accumulation of cytokines, which in turn would impinge upon the heart during the 24-hour hiatus before the second I/R challenge. Although I/R, per se, is a weak stimulus for induction of iNOS, cytokines are potent stimuli for induction of iNOS. Thus, in the in vivo studies it is quite possible that cytokines contributed to the I/R-induced iNOS upregulation. Such an effect would be precluded in the in vitro studies. Further, studies are warranted to directly assess this possibility.

The in vitro studies have also provided evidence supporting a role for AP-1 in upregulating eNOS protein and activity. Pretreatment of cardiac myocytes with an oligonucleotide that contains binding sites for AP-1 (prevention of AP-1 translocation to the nucleus) during the initial A/R challenge abrogates the A/R-induced upregulation of eNOS. In addition, this maneuver prevents the development of delayed PC with respect to (1) myocyte oxidative stress and (2) myocyte-induced PMN transendothelial migration (unpublished observation).

Interaction between NOS and SOD

The in vitro studies indicate that both eNOS and Mn-SOD appear to be critical effector enzymes in the development of delayed PC. These two enzyme systems may be acting in concert. In myocytes isolated from either iNOS-deficient mice (which developed delayed PC) or eNOS-deficient mice (which did not develop delayed PC), Mn-SOD protein increased to the same extent 24 hours after the initial A/R insult. However, in myocytes isolated from iNOS-deficient mice Mn-SOD activity was increased, whereas in myocytes derived from eNOS-deficient mice Mn-SOD activity was not affected. Collectively, these observations indicate that NO derived from eNOS is modulating the activity of Mn-SOD, but not affecting its transcription [3]. The exact mechanism(s) by which eNOS-derived NO is modulating Mn-SOD activity under these conditions is not entirely clear and warrants further attention.

Summary

A significant amount of progress has been made in understanding the mechanisms involved (initiating, signaling, and effector components) in the development of delayed PC in the heart. In general, the results of in vivo and in vitro studies tend to be congruent. However, there are some important differences worth noting. In terms of the initiating molecules, ROM generated during the initial stress are critical to the development of delayed PC. However, the source of the ROM after an I/R (or A/R) challenge is not clear, that is, in vivo studies suggest that NO derived from eNOS generates the ROM, while the in vitro studies do not agree with this notion. In terms of the effector molecules, both NOS and Mn-SOD have been implicated as playing important roles. Indeed, there is some indication that these two enzyme systems may act in concert in the development of delayed PC. The only serious controversial issue is whether iNOS or eNOS is the isoform involved, that is, in vivo studies favor a role for iNOS, whereas in vitro studies favor a role for eNOS. In terms of the signaling molecules that link the initiating factors (ROM) and the effector molecules (NOS and SOD), both in vivo and in vitro studies agree that NFκB is pivotal. However, the in vivo studies indicate that NFκB is important in the upregulation of iNOS, whereas the in vitro studies suggest that this transcription factor upregulates Mn-SOD. In vitro studies have also implicated a role for AP-1 as a potentially important nuclear transcription factor in delayed PC. Further complementary in vivo and in vitro studies are warranted to resolve these issues in order for clinical applications of these bench studies to be fully realized.

References

1. Rui, T., Cepinskas, G., Feng, Q., Ho, Y.-S., and Kvietys, P. R. (2001). Cardiac myocytes exposed to anoxia–reoxygenation promote neutrophil transendothelial migration. *Am. J. Physiol. Heart Circ Physiol.* **281**, H440–H447. *This is the first in vitro study to show that there is a communication between cardiac myocytes and endothelial cells during an A/R challenge.*
2. Guo, Y., Jones, W. K.. Xuan, Y.-T., Tang, X.-L., Bao, W., Wu, W.-J., Han, H., Laubach, V. E., Ping, P., Yang, Z., Qiu, Y., and Bolli, R. (1999). The late phase of preconditioning is abrogated by targeted disruption of the inducible NO synthase gene. *Proc. Natl. Acad. Sci. USA* 9 R. 6, 1507–1512. *This is the first in vivo study to show that iNOS is the effector component of delayed preconditioning using genetically altered mice.*
3. Rui, T., Cepinskas, G., Feng, Q., and Kvietys, P. R. (2003). Delayed preconditioning in cardiac myocytes with respect to development of a proinflammatory phenotype: Role of SOD and NOS. *Cardiovasc. Res.* **59**, 901–911. *This is the first in vitro study to demonstrate that eNOS is the effector component of delayed preconditioning using cardiac myocytes and endothelial cells isolated from genetically altered mice. In addition, evidence was provided for an interaction between Mn-SOD and eNOS in the development of delayed preconditioning.*

Further Reading

Cepinskas G., Rui, T., and Kvietys, P. R. (2001). Neutrophil–endothelial cell interactions during the development of tolerance to ischemia/reperfusion in isolated cells. *Acta Physiol. Scand.* **173**, 23–33.

Dawn, B., and Bolli, R. (2002). Role of nitric oxide in myocardial preconditioning. *Ann. N. Y. Acad. Sci.* **962**, 18–41.

Hoshida, S., Yamashita, N., Otsu, K., and Hori, M. (2002). The importance of manganese superoxide dismutase in delayed preconditioning: Involvement of reactive oxygen species and cytokines. *Cardiovasc. Res.* **55**, 495–505.

Marber, M. S. (2000). Ischemic preconditioning in isolated cells. *Circ. Res.* **86**, 926–931.

Capsule Biography

Dr. Tao Rui received his M.D. in 1986 from Nanjing Medical University and his M.Sc. in 1992 from Shanghai Medical University, P.R. China. Currently, he is a postdoctoral researcher at Lawson Health Research Institute, London, Ontario, Canada. His research interests include myocyte oxidative stress, ischemia–reperfusion, and delayed preconditioning.

Dr. Gediminas Cepinskas received his D.V.M. in 1984 from the Lithuanian Academy of Veterinary Sciences. He is a Scientist at Vascular Cell Biology/Inflammation Program at Lawson Health Research Institute, London, Ontario, Canada. His research is focused on the molecular mechanisms of inflammation. In particular, leukocyte–endothelial cell interactions, mechanisms of ischemia–reperfusion tolerance, and cell signaling are the prime subjects of his research.

Dr. Peter R. Kvietys received his Ph.D. in physiology in 1979 from Michigan State University. His research interests include inflammation and ischemia–reperfusion. Dr. Kvietys is a member of the Vascular Cell Biology/Inflammation Program at Lawson Health Research Institute, London, Ontario, Canada, and a Fellow of the Cardiovascular Section of American Physiological Society.

SECTION C

Statins

CHAPTER 153

Statins and the Microvasculature

Reda E. Girgis and Jeffrey R. Jacobson
Division of Pulmonary and Critical Care Medicine, Johns Hopkins University

Introduction

The statin group of drugs has revolutionized the treatment of coronary atherosclerosis since their introduction nearly two decades ago. By competitively inhibiting the rate-limiting enzyme in cholesterol biosynthesis, 3-hydroxy-3-methylglutaryl (HMG)-CoA reductase, intrahepatic cholesterol is reduced, thereby leading to an increase in low-density lipoprotein (LDL) receptor expression and consequent enhanced endocytosis of plasma LDL cholesterol. Older therapies for hyperlipidemia, although capable of lowering cholesterol, did not have a significant impact on clinical outcomes in cardiovascular disease. The beneficial effects of statin therapy correlate poorly with reductions in serum cholesterol. In addition, improvements in clinical outcomes have been noted well before any changes in lipid levels. These observations led to the hypothesis that statins have vasculoprotective effects independent of cholesterol lowering [1]. In 1996, the first such mechanism to be elucidated was the inhibition of vascular smooth muscle cell (VSMC) proliferation, followed in 1998 by the discovery that statins increase the expression and activity of endothelial nitric oxide synthase (eNOS) in vascular endothelial cells (EC). Since that time a host of other, so-called pleiotropic actions have been described, making this pharmacologic class of agents the target of intensive research efforts in a broad variety of disease states.

The microvascular endothelium plays a critical role in the regulation of regional blood flow, and endothelial dysfunction is central to many cardiovascular diseases. Several beneficial effects on the endothelium are induced by statins that maintain adequate tissue perfusion, such as enhanced eNOS and fibrinolytic activity. The microvascular endothelium is a major target, as well as a source of oxidant and immune/inflammatory-induced injury. HMG-CoA reductase inhibitors possess potent antioxidant and anti-inflammatory properties that may prove beneficial in diverse pathologic processes. Capillary angiogenesis is crucial to increasing blood flow in chronically ischemic tissues, and statin effects on the endothelium have been shown to alter this process as well. Thus, the vasculoprotective effects of statins serve to maintain the integrity of the microvasculature, particularly its endothelial lining, allowing it to perform its central function of controlling local blood flow and exchange of nutrients with the interstitial space, while preventing edema formation and the efflux of inflammatory cells.

This review will begin with a discussion of the molecular mechanisms mediating the pleiotropic actions of statins. Then, those effects with particular relevance to the microcirculation will be summarized, followed by an overview of the potential clinical applications.

Mechanisms of Action

HMG-CoA Reductase Inhibition and Protein Prenylation

The molecular mechanism(s) involved in many of the cholesterol-independent effects of statins have not been completely characterized. Most of these can be reversed by provision of supplemental mevalonate, indicating that inhibition of HMG-CoA reductase is responsible. The generation of mevalonate is the rate-limiting step in cholesterol synthesis (Figure 1). The next step is isopentenyl pyrophosphate (IPPP), which has an unknown function in mammalian cells but can modify tRNA. IPPP is the precursor of the isoprenoid compounds, farnesylpyrophosphate (FPP; C-15) and geranylgeranylpyrophosphate (GGPP; C-20). These lipid intermediates are required for post-translational modification of certain proteins by forming cysteine thioethers at their carboxy termini. Prenylated proteins include the Ras and Rho families of small GTP-binding signaling molecules. Prenylation allows localization to the cytoplasmic

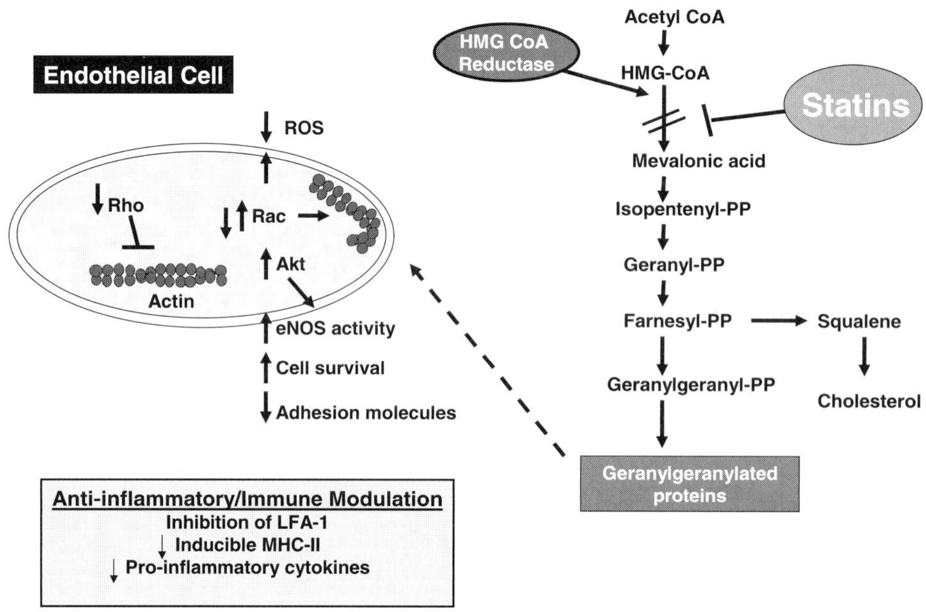

Figure 1 Statin effects on the microvasculature. Effects of statins that target the endothelium and inflammation are critical to their vasculoprotective action. Although specific mechanisms remain to be fully elucidated, inhibition of geranylgeranylation is thought to play an important role. Rho is inhibited by statins, attenuating actin stress fiber formation. Although inhibition of Rac membrane translocation accounts for decreased ROS production, increased Rac activation by statins is associated with enhanced cortical actin and barrier protection (see text). (see color insert)

surface of cell membranes where they can respond to upstream signals by converting from the inactive GDP-bound state to the active GTP-bound state, thereby transmitting a downstream signal (Figure 1). Intrinsic GTPase activity converts it back to the inactive GDP-bound state, releasing the bound downstream effector and allowing another cycle of signal transduction [2]. The Rho family (Rho, Rac, Cdc42, and others) is preferentially geranylgeranylated by the enzyme geranylygeranyl transferase (GGTase), whereas Ras is selectively farnesylated by farnesyl transferase (FTase). Many of the vascular actions of statins are antagonized by GGPP, but not FPP and mimicked by GGTase inhibitors, suggesting that inhibition of GGPP synthesis and Rho prenylation underlies these effects.

Functions of Rho-GTPases

The Rho family of small G-proteins are best known for their ability to mediate actin cytoskeletal reorganization. Rho activation promotes actin polymerization and myosin light chain phosphorylation, which lead to the formation of contractile actomyosin stress fibers. Rho-kinase (ROCK) is an important downstream effector of this process, which leads to intercellular gap formations between EC with disruption of barrier function and enhanced tone in VSMC. Rac activation produces lamellipodia and membrane ruffles and Cdc42 induces filopodia formation. Rho proteins are thus critical for many actin cytoskeleton-dependent processes, such as cell migration, adhesion, phagocytosis, and cytokinesis. Rho proteins can regulate gene expression via activation of c-jun N-terminal kinase (JNK), p38 mitogen—activated kinase and the transcription factors, serum response factor (SRF), and nuclear factor κB (NFκB). The Rho-GTPases are also required for G_1 cell cycle progression, either through direct effects on gene transcription or due to their effects on the cytoskeleton.

Effects of Statins on the Microvasculature

Enhanced eNOS Activity

Nitric oxide (NO) generated by the enzyme NO synthase III (eNOS) constitutively expressed in EC plays a major role in the maintenance of adequate perfusion within the microcirculation. eNOS activity is enhanced in response to shear stress in the precapillary arterioles, mediating VSMC relaxation and increased blood flow. NO also has potent antiplatelet properties and inhibits leukocyte–endothelial cell adhesion, particularly in postcapillary venules, thereby maintaining blood flow.

Statins increase EC eNOS protein levels in a time- and concentration-dependent manner. As little as 0.1 μmol/L of simvastatin induces a nearly twofold increase in eNOS expression after 48 hours. A post-transcriptional mechanism is involved since eNOS mRNA stability increases without changes in gene transcription. Statins prevent the downregulation of eNOS induced by hypoxia and oxidized LDL. These effects are reversed by provision of either mevalonate or GGPP. Statin treatment also rapidly (within 1 hour) enhances NO production by endothelial cells without necessarily affecting eNOS protein levels [3]. This increase in eNOS activity is mediated by phosphorylation of eNOS at Ser1177, which in turn is dependent on statin-induced eNOS association with the chaperone Hsp 90 and activation of the protein kinase Akt/PKB. Statin treatment also reduces caveolin abundance and the consequent inhibitory effect on eNOS.

Effects on iNOS

The role of inducible NOS II (iNOS) in the microcirculation is controversial. Whereas large quantities of NO contribute to hypotension in septic shock and can induce nitrosative tissue damage, there is evidence that iNOS-derived NO attenuates leukocyte–endothelial adhesion. Statins (1–10 μM) enhance cytokine-mediated induction of iNOS in VSMC and perhaps other cell types. In contrast, reduced iNOS induction by cytokines is noted in EC and macrophages.

Statins as Antioxidants

A critical balance exists between NO and superoxide (O_2^-). Under physiologic conditions, NO levels are 1,000 times those of O_2^-, and the normal antioxidant mechanisms are able to detoxify small amounts of reactive oxygen species (ROS). During ischemia–reperfusion and inflammatory disorders, these defenses are overwhelmed and oxidant-mediated injury ensues, including peroxidation of cell membranes, impaired NO bioactivity, induction of leukocyte–endothelial adhesion, and thrombosis. Oxidative stress also contributes to the pathogenesis of chronic vascular diseases such as atherosclerosis, diabetes, and hypertension.

The NADPH oxidase is a major source of O_2^- within the vascular wall, and Rac-GTPase is an important regulator of this enzyme in EC, VSMC, and phagocytes. Statins block the translocation of GTP-bound Rac to the membrane, consequently reducing transcription of critical NADPH subunits and vascular wall production of ROS in response to various stimuli, such as angiotensin II and PKC agonists. An increase in catalase expression has also been observed in aortic walls of spontaneously hypertensive rats treated with statins [4]. A direct scavenging effect of fluvastatin on hydroxyl and superoxide radicals has been described, providing another antioxidant mechanism of statins.

Effects on Other Vasoactive Mediators

Statins inhibit the production of endothelin-1 and the expression of angiotensin receptor I. Beyond their effects on vasomotor tone, these mediators are involved in leukocyte adhesion, thrombosis, and oxidant generation. Adenosine synthesis by EC is enhanced by statins through inhibition of ecto-5′-nucleotidase endocytosis, resulting in increased activity. The activity of this enzyme is normally increased in response to hypoxia and has been implicated in myocardial ischemic preconditioning.

Regulation of the Endothelial Cytoskeleton by Statins

The primary function of the Rho signaling proteins is regulation of the cytoskeleton. In the endothelium, cytoskeletal filament rearrangement determines barrier function. In response to certain stimuli, such as thrombin, RhoA is activated and localized to the membrane. Rho-kinase is consequently activated, leading to the formation of actin stress fibers, focal adhesions, intercellular gap formations, and disruption of EC barrier function. Simvastatin (5 μM for 16 to 24 hours) attenuates thrombin-induced disruption of barrier function by roughly two thirds in both systemic and pulmonary EC and reduces vascular leakage in hyperlipidemic animals. Morphologically, ECs demonstrate reduced stress fibers at baseline and in response to thrombin (Figure 2).

In addition, we have observed cortical actin ring enhancement associated with increased translocation of cortactin to the cell periphery in statin-treated pulmonary EC that persists after thrombin stimulation. This effect has previously been shown to be mediated by Rac activation through Edg receptor agonists, such as sphingosine 1-phosphate and shear stress. Indeed, statins increase GTP-bound (activated) Rac in whole-cell lysates, whereas membrane-bound Rac is reduced. Prenylation of Rac is required for its attachment to membranes and activation of certain downstream effectors such as NADPH oxidase, but also reduces its binding to Rho-GDI (guanine nucleotide dissociation inhibitor), which normally maintains Rac in its inactive GDP-bound state [5]. Statins may also alter various guanine nucleotide exchange factor proteins (GEPs) or GTPase-activating proteins (GAPs) that positively regulate Rho family activity. Thus, some functions of Rac, such as its effects on cytoskeletal rearrangement, may not require membrane localization and these may be increased in response to statin treatment.

Statin Effects on Vascular Wall Cell Proliferation and Survival

As a result of inhibition of GGPP synthesis, statins inhibit cell proliferation and induce apoptosis. Simvastatin and other lipophilic statins, at concentrations of 1 to 10 μM, suppress proliferation and migration of both systemic and pulmonary VSMC in the presence of serum or growth factors such as PDGF. A direct effect on the cytoskeleton or changes in gene expression could underlie the antiproliferative effects of statins, such as an upregulation of the cyclin-dependent kinase inhibitor $p27^{kip1}$. In serum-free conditions, statins induce apoptosis in a concentration

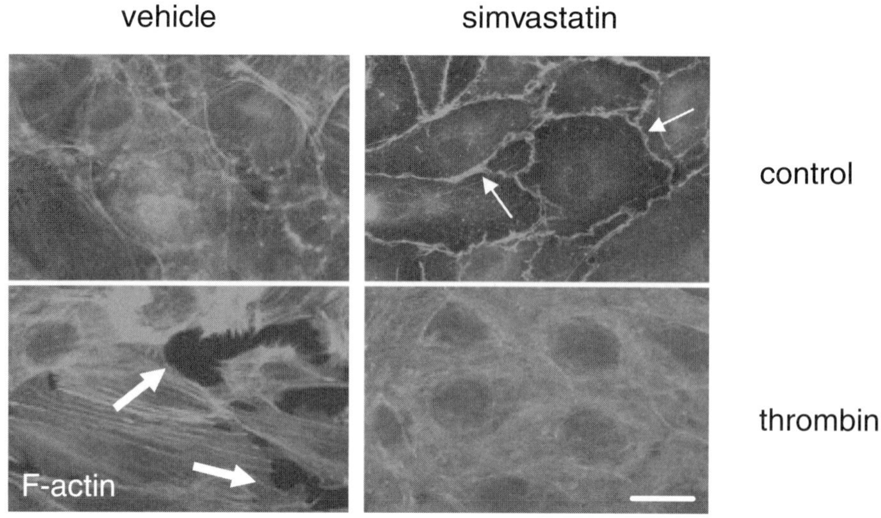

Figure 2 Effect of simvastatin on endothelial cell (EC) cytoskeletal rearrangement. Relative to vehicle controls, ECs treated with simvastatin (5 μM, 16 hours) demonstrate enhanced cortical actin (*small arrows*) and reduced central stress fibers. Additionally, upon thrombin stimulation (1 U, 5 minutes) few paracellular gaps are observed in simvastatin-treated cells compared to vehicle controls (*large arrows*). Bar = 10 μm. (see color insert)

dependent manner with as little as 0.5 μM having an effect. Downregulation of Bcl-2 and consequent caspase activation may mediate apoptosis.

Moderate to high concentrations of statins (> 0.1 μM atorvastatin) similarly inhibit EC proliferation and migration and induce apoptosis, thereby impairing angiogenesis. In contrast, low concentrations (0.01 to 0.05 μM) enhance VEGF release, upregulate VEGF receptor 2, induce proliferation, inhibit hypoxia-induced apoptosis, and increase EC migration by up to 75 percent. These actions promote capillary tube formation in vitro, that is, angiogenesis, similar to VEGF stimulation. In mice, inflammation induced angiogenesis is enhanced by low dose (0.5 mg/kg/day) atorvastatin, whereas high-dose therapy (2.5 mg/kg/day) is inhibitory and also reduces tumor angiogenesis [6]. The high-dose antiangiogenic properties of statins are reversed by GGPP, indicating that inhibition of Rho prenylation is responsible. The mechanisms mediating EC survival in response to low concentrations of statins are not completely understood. They are reversed by mevalonate, but not by Rho-kinase inhibitors. The proangiogenic effects of statins are linked to phosphorylation and activation of Akt at serine 473 and subsequent eNOS Ser1177 phosphorylation and activation, which are observed at very low concentrations (0.0001 μM atorvastatin) and remain evident with higher concentrations. Phosphatidylinositol-3-kinase inhibition reverses low-dose statin effects on EC indicating activation of the PI3K/Akt pathway. Provision of mevalonate also blocks statin-inducted Akt activation. Whether GGPP also blocks low-dose statin effects on ECs is unknown. Activation of eNOS may be ultimately responsible since NOS inhibition by LNMA prevents the EC stimulating effects of statins in some studies. *Importantly, Akt activation is not observed in VSMCs.*

In addition to angiogenesis (sprouting of EC from preexisting capillaries), the mobilization and incorporation of bone-marrow derived endothelial progenitor cells (EPCs) is a mechanism of postnatal neovascularization. Statins, in a fashion similar to VEGF, increase circulating EPCs in animals and humans and activates EPCs (increased adhesion, proliferation, survival). These actions are also mediated by activation of the PI3K/Akt pathway, but are not reversed by LNMA, GGPP, or Rho-kinase inhibitors. Similar to the effects on mature ECs, however, mevalonate prevents EPC differentiation [7].

Investigators have recently begun to dissect the pathway(s) involved in statin-mediated activation of Akt/PKB in endothelial cells [8]. Within 10 minutes of application of simvastatin, membrane protrusions (lamellipodia and filopodia) and ruffles are formed and Akt is translocated from the nucleus and cytoplasm to these sites, accompanied by increased phosphorylation of Akt. Mevalonate or cholesterol repletion prevents these changes, as does PI3K inhibition. This is in contrast to VEGF-mediated Akt activation, which is not affected by cholesterol loading. Depletion of cholesterol from lipid rafts in the membrane may thus underlie statin-induced Akt activation. Alternatively, the membrane changes observed are reminiscent of Rac and Cdc42 activation. As mentioned previously, there is evidence to suggest that statins increase active GTP-bound Rac, while reducing membrane translocation. Moreover, Ras, which acts upstream of PI3K, has been found to be activated by statins, and statin-mediated Akt phosphorylation in EC is inhibited by overexpression of a dominant-negative

Ras mutant. The PI3K/Akt signaling pathway is highly complex, and the exact mechanism of activation by statins is not clear; however, inhibition of Rho prenylatin may not be primarily responsible.

Anti-inflammatory Effects of Statins

Statins have been shown to possess multiple anti-inflammatory actions that play an important role in mediating their vasculoprotective properties. Conditions associated with impairment of NO availability and increased oxidant stress, such as hypoxia and ischemia–reperfusion, induce an inflammatory endothelial phenotype characterized by enhanced expression of the surface adhesion molecules, P-selectin, ICAM-1, and VCAM-1. By enhancing eNOS function and reducing ROS, statins decrease expression of adhesion molecules and inhibit leukocyte–EC binding and transmigration in vitro at concentrations as low as 0.01 μM. In animal models, statins decrease leukocyte adhesion within the microvasculature in response to thrombin, bacterial toxins, and low-dose LNMA. High-dose LNMA reverses these effects, and they are not observed in eNOS knockout mice. Statins may directly inhibit the transcription factors NF-κB and activator protein-1 (AP-1), and induce the nuclear receptors PPARα and γ which regulate inflammatory responses in EC. Anti-inflammatory effects on leukocytes have also been demonstrated, consisting of decreased expression of integrins, which mediate binding to EC, and proinflammatory cytokines.

T-lymphocyte-mediated immune responses are primarily responsible for allograft rejection after organ transplantation. An increasingly important role for T cells is now being recognized in mediating the inflammatory phenotype of the microvasculature in response to injurious stimuli such as ischemia–reperfusion and hypercholesterolemia. High concentrations (1 to 10 μM) of simvastatin completely block interferon-γ induction of MHC class II antigen on microvascular ECs. A similar effect is noted in monocytes–macrophages, but not in highly specialized antigen-presenting dendritic cells and B-lymphocytes. This effect is mediated through transcriptional inhibition of promoter IV of the MHC II transactivator, CIITA, and is completely reversed by mevalonate or GGPP. The consequences repressing MHC II induction are an inhibition of T-lymphocyte proliferation and IL-2 release upon exposure to allogeneic EC or macrophages.

All the anti-inflammatory effects of statins just described are dependent on HMG CoA reductase inhibition, in that they can be blocked by mevalonate. A novel HMG CoA reductase–independent mechanism has been demonstrated consisting of allosteric inhibition of lymphocyte function antigen-1 (LFA-1) by binding to a novel L-(lovastatin) site within the molecule, locking it in an inactive state. As a result, LFA-1 mediated adhesion of leukocytes with the corresponding ligand on EC (ICAM-1) is inhibited, as is lymphocyte costimulation. Based on this discovery, researchers designed a novel lovastatin-based LFA-1 inhibitor that inhibits T-cell costimulation at concentrations of around 0.01 μM compared with 1 μM for the parent compound [9].

Antithrombotic Properties of Statins

Activation of thrombotic processes contributes to microvascular dysfunction in a variety of disease states such as sepsis and ischemia–reperfusion. Statins inhibit platelet activity by several mechanisms, including increased endothelial NO and adenosine release and decreased platelet thromboxane production and cholesterol content. Tissue factor expression and activity in response to thrombin is reduced in EC by simvastatin at concentrations as low as 100 nM. The fibrinolytic system may also be favorably affected by statins. An increase in tissue plasminogen activator and a decrease in plasminogen activator inhibitor type I in EC in response to statins have been demonstrated, but these effects require fairly high concentrations and human studies have been inconsistent.

Potential Clinical Applications of Statins Targeting the Microvasculature

Cardiovascular Disease

The pleiotropic actions of statins have been postulated to account for the impressive clinical benefits observed in atherosclerotic cardiovascular disease independent of cholesterol lowering. Improved microvascular function likely contributes to these benefits. Hypercholesterolemia, diabetes, and hypertension are all associated with impaired endothelium-dependent vasodilation, which is improved within days by statin therapy in humans. The stimulation of angiogenesis by these agents would be expected to promote the normal adaptive response of the microcirculation to chronic ischemia. In a murine model of hind limb ischemia, simvastatin at a dose of 0.1 mg/kg promoted new capillary formation and collateral flow. The microcirculation also plays an important role in the progression of large vessel plaques. The vasa vasorum of these arteries is the source of inflammatory cells that can promote plaque rupture. Statins may help to stabilize such lesions. Some investigators have speculated that the anti-angiogenic effects of high-dose statin therapy, by reducing plaque neovascularization, contributes to their clinical efficacy. The role of the anti-inflammatory effects of statins in atherosclerosis is suggested by the observation that these agents reduce C-reactive protein, an important risk factor for cardiovascular events.

Vascular smooth muscle cell hyperplasia and endothelial dysfunction are felt to underlie the pathogenesis of pulmonary hypertension. Statins attenuate and reverse experimental models of pulmonary hypertension.

Ischemia–Reperfusion Injury (IRI)

An impairment of eNOS activity and increased ROS generation, followed by adherence of leukocytes and thrombosis within the microvasculature, are key events in the development of IRI. In multiple studies of cardiac IRI, pretreatment with statins reduces infarct size and left ventricular dysfunction. Significant protection has also been demonstrated in animal models of IRI in lung, brain, and kidney at physiologically relevant doses (e.g., 0.5 mg/kg of simvastatin). In most studies, an important mechanism appears to be enhancement of NO activity, as NOS antagonists reversed the statin effect.

IRI is thought to underlie the pathogenesis of the multiple organ dysfunction syndrome complicating sepsis. A provocative retrospective study reported a mortality of 6 percent in bacteremic patients taking statins compared with 28 percent in nonusers of statins.

Neurologic Disease [10]

Statins inhibit the migration of lymphocytes across the blood–brain barrier and attenuate the severity of experimental autoimmune encephalomyelitis, a model of multiple sclerosis. A small clinical trial demonstrated a 45 percent reduction in the number of gadolinium-enhancing lesions on MRI. Alzheimer's dementia is becoming increasingly recognized as a disorder of the cerebral microcirculation. Epidemiologic data suggest a 70 percent reduction in the prevalence of dementia in statin users, and large trials are currently underway.

Transplant Rejection

The use of pravastatin in heart transplant recipients is associated with a significant reduction in acute rejection, as well as transplant coronary vasculopathy. A recent retrospective study reported a surprisingly lower incidence of acute rejection and obliterative bronchiolitis after lung transplantation. No reduction in renal allograft rejection has been observed with statin therapy, although statins may improve survival in these patients by reducing cardiovascular events.

Pharmacokinetic Considerations

Individual drugs differ in their ability to mediate the various pleiotropic effects of statins. The lipophilic agents (simvastatin, atorvastatin, cerivastatin, fluvastatin, and lovastatin) have prominent effects on VSMC proliferation and contractility, whereas the more hydrophilic pravastatin has minimal effects. A similar difference has been found for certain actions on leukocytes, such as ROS production and inhibition of LFA-1. On the other hand, effects targeting EC have generally been equivalent, when compared. Most studies have employed one of the lipophilic statins, and thus it is not known whether pravastatin would have similar actions. Because of its hydrophilic nature, pravastatin requires a specific organic anion transporter, LST-1, for its uptake into hepatocytes. The distribution of pravastatin to extrahepatic sites is significantly less than the lipophilic compounds. For those actions dependent upon HMG Co-A reductase inhibition, the potency of the various agents will also depend on their IC_{50} for this enzyme in a given cell type. In hepatocytes, the IC_{50} of simvastatin is 8 nM compared with 4.1 and 2.0 nM for lovastatin and pravastatin, respectively. In contrast, the IC_{50} in myoblasts is 4 nM for simvastatin, compared with 19 and 110 nM for lovastatin and pravastatin, respectively.

The range of serum levels achievable with standard clinical doses of atorvastatin or simvastatin is 0.002 to 0.2 μmol/L. Based on in vitro studies, the main non-cholesterol-lowering effects that would be expected with these concentrations are those targeting endothelial cells. These consist of Akt activation, with resultant phosphorylation and stimulation of eNOS activity and promotion of EC survival. A reduction in EC adhesion molecule expression would also be anticipated, an action that largely appears to be mediated by enhanced eNOS activity. An antiangiogenic effect may occur with concentrations toward the upper end of this range.

Conclusion

Statins have proved to be a crucial component of cardiovascular disease treatment. Much of their benefit may be unrelated to their cholesterol-lowering properties. A wide array of vasculoprotective actions has been described, the most potent targeting endothelial cells. The relatively strong safety profile of HMG Co-A reductase inhibitors in humans makes them an attractive pharmacologic class to study in other disease processes besides atherosclerosis. Moreover, delineation of the mechanisms involved in mediating their pleiotropic effects generates opportunities to develop novel compounds that can target a more specific pathway, e.g., LFA-1 inhibitors, Rho-kinase inhibitors, GGTase inhibitors, and potentially Akt activators.

Glossary

Actin stress fibers: Intracellular filaments in vascular endothelial cells formed by polymerization of actin and myosin leading to cell contracture and disruption of barrier function.

Angiogenesis: The process whereby endothelial cells of existing capillaries proliferate to form new vessels.

eNOS (endothelial nitric oxide synthase): The constitutively expressed isoform of the enzyme present in vascular endothelial cells and an important regulator of basal microvascular function.

Pleiotropic effects: A term used to describe the multiple cholesterol-independent actions of statins on vascular function and structure.

Prenylation or isoprenylation: Binding of lipid intermediates to protein signaling molecules.

Statins: A pharmacological class of drugs that competitively inhibit the rate rate-limiting enzyme in cholesterol biosynthesis, 3-hydroxy-3-methylglutaryl (HMG)-CoA reductase.

References

1. Mason, J. C. (2003). Statins and their role in vascular protection. *Clin. Sci. (Lond.)* **105**(3), 251–266. *A comprehensive review of the non-cholesterol-lowering effects of statins, including enhanced eNOS activity and the antithrombotic, antioxidant, and anti-inflammatory actions. The implications of the biphasic effects on angiogenesis are discussed. Cites 200 references.*
2. Takai, Y., Sasaki, T., and Matozaki, T. (2001). Small GTP-binding proteins. *Physiol. Rev.* 81(1), 153–208.
3. Kureishi, Y., Luo, Z., Shiojima, I., Bialik, A., Fulton, D., Lefer, D. J., Sessa, W. C., and Walsh, K. (2000). The HMG-CoA reductase inhibitor simvastatin activates the protein kinase Akt and promotes angiogenesis in normocholesterolemic animals. *Nat. Med.* **6**(9), 1004–1010. *This is the first study linking the enhanced eNOS activity to Akt stimulation and angiogenesis, both in vitro and in vivo in a mouse model of hind limb ischemia.*
4. Wassmann, S., Laufs, U., Muller, K., Konkol, C., Ahlbory, K., Baumer, A. T., Linz, W., Bohm, M., and Nickenig, G. (2002). Cellular antioxidant effects of atorvastatin in vitro and in vivo. *Arterioscler. Thromb. Vasc. Biol.* **22**(2), 300–305.
5. Vecchione, C., and Brandes, R. P. (2002). Withdrawal of 3-hydroxy-3-methylglutaryl coenzyme A reductase inhibitors elicits oxidative stress and induces endothelial dysfunction in mice. *Circ. Res.* **91**(2), 173–179. *This study demonstrates attenuation of endothelium-dependent vasodilatation upon statin withdrawal in mice, due to increased vascular superoxide generation as a consequence of Rac-1 translocation, which was activated during statin treatment.*
6. Weis, M., Heeschen, C., Glassford, A. J., and Cooke, J. P. (2002). Statins have biphasic effects on angiogenesis. *Circulation* **105**(6), 739–745.
7. Dimmeler, S., Aicher, A., Vasa, M., Mildner-Rihm, C., Adler, K., Tiemann M., Rutten, H., Fichtlscherer, S., Martin, H., and Zeiher, A. M. (2001). HMG-CoA reductase inhibitors (statins) increase endothelial progenitor cells via the PI 3-kinase/Akt pathway. *J. Clin. Invest.* **108**(3), 391–397.
8. Skaletz-Rorowski, A., Lutchman, M., Kureishi, Y., Lefer, D. J., Faust, J. R., and Walsh, K. (2003). HMG-CoA reductase inhibitors promote cholesterol-dependent Akt/PKB translocation to membrane domains in endothelial cells. *Cardiovasc. Res.* **57**(1), 253–264.
9. Weitz-Schmidt, G., Welzenbach, K., Brinkmann, V., Kamata, T., Kallen, J., Bruns, C., Cottens, S., Takada, Y., and Hommel, U. (2001). Statins selectively inhibit leukocyte function antigen-1 by binding to a novel regulatory integrin site. *Nat. Med.* **7**(6), 687–692.
10. Stuve, O., Youssef, S., Steinman, L., Zamvil, S. S. (2003). Statins as potential therapeutic agents in neuroinflammatory disorders. *Curr. Opin. Neurol.* **16**(3), 393–401.

Capsule Biography

Dr. Girgis is Associate Director of the Pulmonary Hypertension and Lung Transplant Programs at Johns Hopkins University. His research focuses on the effects of statins on vascular remodeling in pulmonary hypertension and is supported by the NIH and Pulmonary Hypertension Association.

Dr. Jacobson is a research fellow in the division of Pulmonary and Critical Care Medicine at Johns Hopkins. He is investigating the effects of statins on cytoskeletal rearrangement in endothelial cells.

SECTION D

Therapeutic Angiogenesis

CHAPTER 154

Therapeutic Angiogenesis

[1]Michael Simons and [2]Mark J. Post

[1]*The Angiogenesis Research Center, Section of Cardiology, Departments of Medicine and Pharmacology and Toxicology, Dartmouth Medical School, Dartmouth-Hitchcock Medical Center, Lebanon, New Hampshire*

[2]*Department of Physiology and Biomedical Technology, University of Maastricht, Maastricht, The Netherlands*

Introduction

Disruption of blood flow to tissues leads to reduction in oxygenation and deterioration of function. In the myocardium this leads to reduction in myocardial contractility with subsequent development of heart failure. Similar functional consequences can be observed in other organs. Although the majority of tissues have the ability to partially compensate for the reduction in blood supply by stimulation of growth of new blood vessels, in most cases this response does not result in full compensation. Thus, in coronary arteries, chronic occlusions typically result in development of anginal symptoms and, frequently, in chronic myocardial stunning. To date, treatment of such an obstructive coronary disease relies either on the use of drugs designed to reduce, by various means, the oxygen requirement of the myocardium or on mechanical revascularization procedures including coronary bypass and various forms of percutaneous catheter-based approaches. Recent advances in our understanding of vessel growth have opened the possibility of therapeutic angiogenesis, that is, induction of neovascularization in the desired locations designed to restore blood flow to ischemic tissues. In this chapter we consider biological foundations of therapeutic angiogenesis and the current state of clinical research in this field.

Biology of Angiogenesis

The formation of mature vasculature in the course of embryonic development includes three distinct sequential steps—formation of primary capillary plexus from embryonic stem cells (vasculogenesis), sprouting of endothelium-lined vascular structures (angiogenesis), and remodeling of these structures into fully fledged vessels (arteriogenesis). These series of events involve numerous growth factors and regulatory molecules that drive various parts of the pathway while preventing excessive vascular growth or formation of defective vasculature.

During vasculogenesis, primitive blood vessels originate from embryonic endothelial stem cells called angioblasts while the blood cells develop from hematopoietic cells. To date, little information is available with regard to the nature of these cells and the factors regulating their appearance and survival. They are thought to develop from a common blood cells/endothelial cell precursor referred to as hemangioblast and express VEGF receptor flk-1 as well as fibroblast growth factor (FGF) receptors. The key growth factors involved in vasculogenesis include VEGF and TGF-β1. Disruptions of VEGF, flk-1, or TGF-β1 genes result in embryonic lethality with a failure of endothelial precursors to differentiate and to form the primary vascular plexus. In addition, FGF2, IGF-1, and GM-CSF have the ability to stimulate differentiation and mobilization of angioblasts from the bone marrow. Whereas vasculogenesis clearly plays a key role during embryogenesis, the occurrence of this process in adult tissues is fairly controversial. There appears to be a bone-marrow-derived population of endothelial precursor cells (EPCs) that under some circumstances can contribute to neovascularization.

Once formed, the primary plexus is then transformed by processes of branching angiogenesis into a primitive vascular system. The deletion of a single VEGF allele leads to a profound failure of branching, suggesting that this process is heavily VEGF-dependent. In addition, VEGF receptor flt-1 and the angiopoietin-1/tie-2 system are also critical for this series of events. At the same time, the extent of sprouting, that is, the total number of vascular branches and generations of branches, is likely regulated by FGFs. This is suggested by a marked reduction in branching of the trachea and bronchial tree in *Drosophila* following disruption of its single FGF gene and by increased arterial branching in mice overexpressing FGF1.

The differentiation of the primary vascular network into adult vasculature involves the envelopment of these structures by vascular smooth muscle and other mural cells, including pericytes, forming, in larger vessels, tunica media and tunica adventitia, a process in which platelet-derived growth factor (PDGF)-BB, TGF-β, Ang-1, and Ang-2 appear to play a role. The differentiation of the forming vasculature into arterial and venous systems and the formation of connections between the two involves the ephrin family of proteins, whereas development of the lymphatic system may depend on a specific subset of VEGFs.

Whereas the process of vascular growth in the embryo is fairly well understood, the process of neovascularization in the adult is much less clear. There appear to be at least two and possibly three distinct processes—capillary growth in ischemic tissues (angiogenesis); growth and remodeling of preexisting vessels into arterioles and arteries; or de novo formation of such vessels (arteriogenesis) and formation of vessels, capillaries, or larger vascular structures from circulating EPCs (vasculogenesis). Although it is superficially similar to embryonic development, neovascularization in mature tissues is not well understood. Furthermore, the relative physiological importance of various forms of arterialization under various pathological conditions is not established.

The primary stimulus driving angiogenesis is tissue ischemia, which results in activation of hypoxia-inducible factor (HIF)-1α that in turn stimulates expression of VEGF, VEGF receptor flt-1, PDGF-BB, Ang-2, iNOS, and a number of other genes. Although it is certainly effective in increasing capillary numbers in ischemic tissues, the contribution of this process to the restoration of perfusion is not clear.

In contrast to angiogenesis taking place in the ischemic myocardium (in the case of myocardial ischemia), the formation of epicardial collaterals by the process of arteriogenesis occurs in nonischemic areas. Furthermore, this process has the potential to fully restore the blood flow to the distal coronary bed, resulting in an effective "biological bypass." Therefore, arteriogenesis, and not angiogenesis, is the most appealing target for a therapeutic intervention. Remarkably, there is very little understanding of the arteriogenesis process itself or of the factors that regulate it. For example, it is not clear whether arteriogenesis proceeds by the enlargement of preexisting vessels that remodel into arterioles large enough to carry significant volume of blood flow or by the de novo growth of these vessels. This is a fundamental distinction since agents that might affect vascular remodeling are likely to be very different from agents that primarily stimulate endothelial cell proliferation and migration, although the latter is definitely required for remodeling as well. The distinction will also have consequences for the preferred site and route of administration of any such agent. Furthermore, we have only a vague notion of the stimuli that initiate arteriogenesis and of the growth factors involved.

Accumulation of blood-derived monocytes/macrophages appears to be critical to this process, as these cells secrete a number of growth factors and cytokines involved in endothelial, pericyte, and smooth muscle cell growth and differentiation, including VEGF, FGF2, TGF-β, Il-8, and MCP-1. Animal studies have demonstrated that activation of monocyte/macrophage accumulation at the arterial occlusion sites when stimulated by MCP-1 leads to a robust arteriogenic response that in turn can lead to significant restoration of tissue perfusion. At the same time, deficiency of the VEGF-related growth factor PLGF prevents collateral growth by impairing monocyte recruitment. Furthermore, monocyte ability to respond to hypoxic stress by increasing their HIF-1α protein levels appears to correlate with the extent of coronary collateral development in patients with advanced coronary artery disease (CAD). The factors responsible for monocyte recruitment to sites of arterial narrowing have not been defined, although shear stress has been suggested as a potential contributor.

Angiogenic Growth Factors

A large number of genes can influence a process as complex as angiogenesis. At the same time, a smaller number of proteins have the ability to directly stimulate endothelial cell proliferation. Traditionally, such proteins have been referred to as angiogenic growth factors. Among these, VEGFs, PDGFs, and FGFs have received the greatest attention and are discussed briefly here.

The VEGF family includes five VEGFs (A–E) and a closely related protein PlGF (placental growth factor). VEGF-A (commonly referred to as simply VEGF) was first isolated as a vascular permeability factor (VPF) and was subsequently shown to have endothelial cell growth stimulatory properties. Four different isoforms are generated by alternative splicing from a single VEGF-A gene: $VEGF_{206}$, $VEGF_{189}$, $VEGF_{165}$, and $VEGF_{121}$. Other rare splice variants such as $VEGF_{145}$ probably also exist. $VEGF_{189}$ and $VEGF_{205}$ isoforms possess high avidity for heparan sulfates and as a result are tightly bound to the cell surface. This property likely severely limits their therapeutic utility. $VEGF_{165}$ and $VEGF_{145}$ also demonstrate heparan sulfate binding (significantly less than the other two), whereas $VEGF_{121}$ completely lacks the heparan sulfate binding site.

Angiogenic activity of VEGF in large measure is dependent on release of nitric oxide (NO) as blockade of NO generation markedly reduces VEGF activity. Whereas VEGF plays a crucial role in embryonic vascular development, its role in adult tissues is less clear, as deactivation of the VEGF-A gene in mature animals does not lead to significant vascular defects. This may not be the case, however, for newly forming vasculature in adult tissues where the loss of VEGF leads to a rapid involution of newly formed vasculature.

Other VEGF genes include VEGF-C (also known as VEGF-2), VEGF-B (also known as VEGF-3), VEGF-D, and VEGF-E. VEGF-C is predominantly involved in lymphangiogenesis, whereas VEGF-B may play a role in development of coronary capillaries. The functions of VEGF-D and E isoforms have not been fully established.

VEGF levels are very low in normal tissues and are exquisitely sensitive to hypoxia and ischemia with even a transient episode resulting in rapid and substantial increases in expression. This sensitivity is driven by the hypoxia-inducible factor (HIF)-1α, a transcription factor with very short half-life in normal cells that gets rapidly and profoundly extended by the onset of cellular hypoxia. In addition, VEGF expression in the myocardium is regulated by tissue stretch, with higher intraventricular pressures inducing significant increases in VEGF expression in a TNFα-dependent manner.

Although all VEGFs are able to stimulate endothelial cell proliferation in vitro and in vivo, they are, by themselves, very weak mitogens. This suggests that biological effects of these molecules may have a lot more to do with activities other than direct stimulation of endothelial cell growth. One such activity is the ability to induce a local inflammatory response by increasing vascular permeability that leads to chemoattraction and accumulation of blood-derived monocytes. Other biological activities include stimulation of local production of nitric oxide, enhancement of monocyte/leukocyte adhesion to the endothelium, activation of tissue digesting enzymes such as matrix metalloproteinases (MMPs), and stimulation of expression of growth factors such as FGF2. Finally, VEGF may also play an important role in stimulation of bone marrow release of endothelial precursor cells.

Platelet-derived growth factors constitute a four-gene family. Whereas PDGF A and B have been extensively studied, comparatively little is known about PDGF C and D. The A and B chains can form homodimers (AA or BB) or a heterodimer (AB), which have distinctly different properties and biological activities. PDGFs act by binding to high-affinity tyrosine kinase receptors (PDGF-R) that are composed of two chains, PDGF-Rα and PDGF-Rβ, which can also form either homodimers ($\alpha\alpha$ or $\beta\beta$) or a heterodimer ($\alpha\beta$). PDGF-AA binds only to PDGF-R$\alpha\alpha$ and PDGF-BB binds to all three PDGF receptors, whereas PDGF-AB is limited to $\alpha\alpha$ or $\alpha\beta$. Homozygous disruptions of either PDGF-B or PDGF-β receptor have shown that PDGF-B is responsible for vascular maturation. Respective knockout mice die late in gestation with abnormal kidney glomeruli and vascular wall abnormalities that are particularly prominent in the brain and the heart and are related to insufficient recruitment and organization of pericytes and smooth muscle cells. The PDGF-A and PDGF-α knockouts show less overlapping phenotypes with less specific vascular pathology.

PDGF-BB appears to be a potent angiogenic growth factor with the ability to induce formation of both capillaries and larger vessels. However, in addition to its angiogenic activity, PDGF-BB has also been implicated in the formation of intimal hyperplasia following arterial injury and in progression of atherosclerosis. These aspects of PDGF-BB's activity profile raise concerns that therapeutic benefits of application of this growth factor in the setting of atherosclerotic cardiovascular disease might be offset by stimulation of restenosis and atherosclerosis.

Fibroblast growth factors are a 23-member family of closely related proteins. One of the key differences between various FGFs is the presence or absence of the leader sequence required for conventional peptide secretion (absent in FGF 1 and 2 but present in FGF 4 and 5 and most other FGFs) and the difference in affinity for various isoforms of FGF receptors. All FGFs demonstrate high affinity for heparan sulfates recognizing specific sequences in glycosaminoglycan chains. For most FGFs, their ability to bind cell surface and matrix heparan sulfates serves both to prolong effective tissue half-life and to facilitate binding to corresponding high-affinity receptors. A pattern of specific FGF-FGF receptor and FGF–heparan sulfate interactions probably accounts for differences in activity of various FGFs.

In cell culture as well as in vivo studies, FGF-1,2,4 and 5 are potent mitogens for cells of mesenchymal, neural, and epithelial origin, including all cell types found in the vascular wall (endothelial cells, smooth muscle cells, and pericytes). FGF2, and probably other FGFs, also is able to activate nitric oxide release, to induce synthesis of plasminogen activator and matrix metalloproteinases, and to stimulate smooth muscle cell and monocyte chemotaxis. One interesting aspect of FGF2 biology is the synergy of its biological activity with VEGF. A combination of FGF2 and VEGF is far more potent in inducing angiogenesis in vitro and in vivo than either growth factor alone. Furthermore, FGF2 induces VEGF expression in smooth muscle and endothelial cells.

Both FGF1 and FGF2 are present in the normal myocardium (as well as other tissues) where their expression is not significantly affected by hypoxia, tissue ischemia, or hemodynamic stress. However, despite significant levels of FGFs in normal tissues, the growth factors do not appear to be biologically active. In part this lack of activity may be due to their sequestration in the extracellular matrix by virtue of binding to the heparan sulfate–carrying proteoglycan perlecan that would make them unavailable to cell surface receptors. In addition, normal tissues demonstrate low levels of expression of FGF receptors FGF-R1, R2, and syndecan-4. Thus, unlike VEGF, where biological activity appears to be driven by the amount of the ligand present, FGF activity is likely controlled by the level of expression of FGF receptors and their ability to bind the ligand.

Among other angiogenic growth factors, angiopoietins and hepatocyte growth factor (HGF) may also play particularly important biologic and therapeutic roles. The angiopoietin family consists of four members. Overexpression of Ang-1 in the skin leads to a striking increase in vascularization that is characterized by a pronounced increase in the vessel size with only a modest increase in the vessel number. A combined VEGF/Ang-1 overexpression leads to a very pronounced increase in vascularity that, in contrast to overexpression of VEGF alone, does not exhibit increased vascular permeability. These findings suggest that Ang-1 plays a role in stabilizing the existing vessel in a yet undefined manner.

The role played by Ang-2 in angiogenesis in adult tissues is not clear. Whereas its overexpression during embryonic development leads to early mortality with morphological defects resembling those of Ang-1 and Tie-2 knockouts, induction of ischemia in adult hearts led to a rapid initial rise in Ang-2 expression that equally rapidly declines while Ang-1 expression shows a gradual increase. These observations suggest that Ang-2 might provide a destabilizing signal necessary for the initiation of angiogenic response.

HGF, also known as the scatter factor, is a powerful endothelial cell mitogen. HGF and its receptor c-met are essential for the normal embryonic development, but are not primarily involved in the development of cardiovascular system. However, the HGF/c-met expression is markedly upregulated in ischemic hearts, and the lack of this increase in expression correlates with poor native collateral development. HGF has a robust angiogenic effect in vivo, which may at least in part be mediated by stimulation of VEGF expression. For use as a therapeutic agent, the definite pro-oncogenic properties of HGF and c-met are a major concern. Transgenic overexpression of HGF leads to a variety of tumors of mesenchymal and epithelial origin, and mutations in the tyrosine kinase domain of MET that lead to its constitutive activation produce papillary renal carcinomas.

Therapeutic Applications of Angiogenesis Concepts

The concept of using angiogenic growth factors as a treatment for occlusive coronary artery and peripheral vascular disease has been extensively tested in both animal and clinical studies. Although in principle the concept appears to be correct, a number of significant challenges prevented an easy transition from successful animal to human usage. In chronic-ischemia large-animal models of myocardial angiogenesis, single-bolus intracoronary, periadventitial, or intrapericardial administration of several growth factors including FGF1, FGF2, $VEGF_{165}$, and PDGF-BB enhanced neovascularization and restored blood flow in the ischemic territory to essentially normal levels. Likewise, gene therapy approaches have been used with equal success including intracoronary injections of FGF-4, FGF-5, $VEGF_{121}$, and $VEGF_{165}$ adenoviruses and $VEGF_{165}$ plasmid.

Similar success has been seen in studies that utilized a hind limb ischemia model. In these studies, an occlusion of the femoral artery is typically associated with a severe reduction in blood flow to the ipsilateral limb. Treatment with angiogenic growth factors accelerates recovery and, in some cases, preserves tissue loss in the foot and the distal ankle.

At the same time, clinical trials have been less successful in demonstrating benefits of therapeutic angiogenesis. Whereas the initial series of open label studies testing $VEGF_{165}$, FGF1, and FGF2 all demonstrated significant improvement in every parameter measured, including measures of myocardial perfusion and ventricular function, larger randomized double-blind trials have shown similarly significant improvements in both placebo and treatment groups. These results suggest the presence of a large placebo effect as well as a potential for a significant improvement in the control group.

The efficacy of $VEGF_{165}$ treatment was tested in the VIVA trial in which the protein was given by a combination of an intracoronary and intravenous infusion. The trial enrolled 178 "no-option" patients with stable exertional angina, who were randomized to receive placebo, low-dose recombinant human VEGF (17 ng/kg/min), or high-dose rhVEGF (50 ng/kg/min) by intracoronary infusion on day 0, followed by intravenous infusions on days 3, 6, and 9. Assessment of exercise capacity, angina class, quality of life, and myocardial perfusion were performed 60 and 120 days later. The primary end point, change in exercise treadmill test (ETT) time from baseline to day 60, showed equally significant improvements in all three. Angina class and quality of life were also significantly improved within each group, with no difference among groups.

The therapeutic efficacy of FGF2 was studied in several trials. A single bolus protein intracoronary administration of FGF2 protein was tested in the double-blind, randomized, placebo-controlled trial (FIRST). The 337 patients were randomized to an intracoronary injection of rFGF2 (0.3, 3, or 30 μg/kg) or placebo. Efficacy was evaluated at 90 and 180 days by exercise tolerance test, SPECT imaging, and quality-of-life questionnaires. Similar to the VIVA finding, exercise tolerance was increased at 90 days in all groups but was not significantly different between the placebo and FGF2-treated groups. rFGF2 treatment reduced angina symptoms with the differences more pronounced in highly symptomatic patients. However, none of the differences were significant at 180 days because of continued improvement in the placebo group.

The concept that the prolonged presence of FGF2 in tissues would be superior to a single bolus therapy was tested in a small trial of a heparin alginate sustained-release FGF2 formulation implanted into the unrevascularizable but viable myocardium at the time of coronary artery bypass grafting. Twenty-four patients were randomized to implantations of 10 heparin-alginate capsules containing 0, 10, or 100 μg of FGF2. At 3-month follow-up, there was a significant reduction in the occurrence of angina in the high-dose rFGF2-treated patients compared to other groups. Remarkably, this benefit was preserved after 3 years of follow-up.

An alternative to achieving sustained local levels of a growth factor involves a gene transfer. This approach was used in the AGENT trial that examined the therapeutic efficacy of adenoviral FGF-4 therapy. Seventy-nine patients were randomized to a single IC infusion of placebo ($n = 19$) or Ad5-FGF4 ($n = 60$). Overall, patients receiving Ad5-FGF4 tended to have greater improvements in exercise time at 4 weeks. Improvement was the greatest in patients with more severe baseline impairment in exercise capacity. Interestingly, patients with low baseline anti-adenoviral antibody titers showed a significantly better response than patients with high levels of antibodies. The safety and feasibility of catheter-based local intracoronary $VEGF_{165}$ gene transfer in the prevention of in-stent restenosis and in the treatment of chronic myocardial ischemia were recently also studied. Although this procedure had no effect on the extent of restenosis, adenoviral but not plasmid-based gene transfer resulted in increased myocardial perfusion as determined by SPECT imaging 6 months later, after percutaneous interventions. The significance of this finding is not clear, as the patients were not randomized to assess this specific end point. Furthermore, there was no difference in nitrates requirement and exercise time in all study groups. Two other gene transfer trials, one involving adenoviral $VEGF_{121}$ (REVASC trial) and one plasmid-based $VEGF_{165}$ (Euroinject One trial), reported essentially negative results. Yet another trial examined the utility of plasmid-based gene transfer by direct intramyocardial injections of VEGF2 plasmid in 27 inoperable patients with class III or IV angina. Twelve weeks after injection, there was a significant improvement in the angina class in the VEGF2 versus placebo-treated patients. However, the small size of the trial with a particularly small control group and the absence of the usual placebo effect (likely another reflection of the small trial size) make efficacy conclusions tenuous.

Assessment of Current Experience

Although one may conclude from this description of clinical trials that therapeutic angiogenesis has been pretty much a dismal failure, it can and should be argued that such a conclusion is premature. Indeed, a number of factors likely have influenced these outcomes and the lessons learned should facilitate future studies.

Perhaps one of the biggest reasons for failure is that none of these trials has explicitly addressed the arteriogenesis versus angiogenesis issues in trial design, drug delivery, or evaluation. Indeed, VEGFs are not primarily arteriogenic and, given the typically short duration of tissue residence following protein injection or expression after adenoviral or plasmid-based transfer in the myocardium, the negative results are not surprising. Similarly, although FGFs are arteriogenic, the short duration of tissue exposure following protein or adenoviral-based transfer likely limited their efficacy. No trial to date has evaluated the efficacy of what currently appears to be the most potent arteriogenic growth factor, PDGF-BB, whether given in combination with FGF2 or not.

Indeed, it could be argued that augmentation of monocyte accumulation at the site of desired arteriogenesis could be the most effective therapeutic strategy. In that regard it should be noted that a small clinical trial that evaluated GM-CSF infusions in patients with CAD reported positive results. Yet another approach may involve administration of MCP-1, a potent monocyte chemoattractant. The side effects of such strategies, however, have not been evaluated.

Another very important issue is the length of exposure of tissues to the therapeutic agent, such as a growth factor. This has been brought into a stark focus by the study of Dor et al. showing that 32-day duration of expression of $VEGF_{165}$ in mouse myocardium is needed for induction of stable vessel growth, whereas a shorter time course (12 days) had no lasting effect. These numbers likely translate into much longer time intervals in patients. In light of these considerations it is interesting that the trial demonstrating the most benefit involved prolonged sustained-release delivery of FGF2.

An issue closely related to the length of tissue exposure is the means of delivery and the formulation used for the growth factor administration. Preclinical studies have clearly demonstrated minimal tissue deposition and the fast washout of growth factors given by an intracoronary route. The efficiency of adenoviral transfer by this method appears to be equally dismal. In contrast, intramyocardial injections result in much higher local concentrations and increased duration of tissue exposure to the agent. However, it is unlikely that intramyocardial protein therapy in the absence of sustained-release polymer formulations will be effective. Furthermore, it is unclear if the adenoviral or the plasmid-based gene transfer produces a length of expression long enough for the therapeutic effect to take place.

Finally, patient selection also presents an important challenge. It has become abundantly clear that populations under study demonstrate a very significant placebo effect as well as spontaneous improvement even in such "hard" end points as myocardial blood flow and function. In addition, the choice of "no-option" patients in clinical trails to date may have inadvertently selected a particularly resistant population with a failure of the native angiogenic response due to any number of defects in the angiogenic cascade. This premise is illustrated by the differences in response to hypoxia of monocytes selected from patients with CAD that correlated with the prevalence of coronary collaterals.

Summary

Therapeutic angiogenesis has evolved markedly over the past decade. Although the initial clinical experience has been rather disappointing, the wealth of basic and preclinical data strongly suggests that this ultimately will be an effective therapeutic strategy. One novel approach to this therapy that has not yet received vigorous clinical testing involves the use of EPC. This can be achieved either by stimulation of their release (for example, with the use of agents such as GM-CSF) or by physical extraction of such cells from the bone marrow followed by their injection into the target tissues. This and other strategies will undoubtedly be tested in the coming years.

Glossary

Angiogenesis: Growth of capillaries or capillary-like structures (blood vessels with a single endothelial wall layer).

Arteriogenesis: Growth of arterioles or arteries (blood vessels with more than one wall layer).

Gene therapy: Means of transfer of gene encoding for a desired growth factor to a host organism.

Growth factor: A protein capable of induction blood vessel growth.

Vasculogenesis: Growth of blood vessels from circulating precursor cells.

Bibliography

Arras, M., Ito, W. D., Scholz, D., Winkler, B., Schaper, J., and Schaper, W. (1998). Monocyte activation in angiogenesis and collateral growth in the rabbit hindlimb. *J. Clin. Invest.* **101**, 40–50.

Asahara, T., Murohara, T., Sullivam, A., Silver, M., van der Zee, R., Li, T., Witzenbichler, B., Schatterman, G., and Isner, J. M. (1997). Isolation of putative progenitor endothelial cells for angiogenesis. *Science* **275**, 964–967.

Carmeliet, P. (2003). Angiogenesis in health and disease. *Nat. Med.* **9**, 653–660.

Dor, Y., Djonov, V., Abramovitch, R., Itin, A., Fishman, G. I., Carmeliet, P., Goelman, G., and Keshet, E. (2002). Conditional switching of VEGF provides new insights into adult neovascularization and pro-angiogenic therapy. *EMBO J.* **21**, 1939–1947. *First demonstration of the importance of prolonged growth factor expression in maintenance of newly developed vasculature.*

Helisch, A., and Schaper, W. (2003). Arteriogenesis: The development and growth of collateral arteries. *Microcirculation* **10**, 83–97.

Li, J., Hampton, T., Morgan, J. P., and Simons, M. (1997). Stretch-induced VEGF expression in the heart. *J. Clin. Invest.* **100**, 18–24.

Rafii, S., and Lyden, D. (2003). Therapeutic stem and progenitor cell transplantation for organ vascularization and regeneration. *Nat. Med.* **9**, 702–712.

Schultz, A., Lavie, L., Hochberg, I., Beyar, R., Stone, T., Skorecki, K., Lavie, P., Roguin, A., and Levy, A. P. (1999). Interindividual heterogeneity in the hypoxic regulation of VEGF: Significance for the development of the coronary artery collateral circulation. *Circulation* **100**, 547–552. *Significance of clinical phenotypes in determining native arteriogenic response to myocardial ischemia.*

Simons, M., Bonow, R. O., Chronos, N. A., Cohen, D. J., Giordano, F. J., Hammond, H. K., Laham, R. J., Li, W., Pike, M., Sellke, F. W., Stegmann, T. J., Udelson, J. E., and Rosengart, T. K. (2000). Clinical trials in coronary angiogenesis: Issues, problems, consensus: An expert panel summary. *Circulation* **102**, E73–E86. *A comprehensive review of issues involving design and execution of clinical trials of therapeutic angiogenesis.*

Yla-Herttuala, S., and Alitalo, K. (2003). Gene transfer as a tool to induce therapeutic vascular growth. *Nat. Med.* **9**, 694–701.

Capsule Biography

Dr. Simons is Chief of Cardiology and Director of the Angiogenesis Research Center at Dartmouth Medical School. His research activities are centered on endothelial cell matrix interactions and therapeutic angiogenesis therapies in advanced coronary artery disease.

Dr. Post is Chair of Physiology at the University of Maastricht. His research focuses on ischemic models of angiogenesis and development of growth factor protein and gene delivery tools.

SECTION E

Vascular Endothelium

CHAPTER 155

Delivery of Therapeutic and Genetic Materials to Vascular Endothelium

[1]Vladimir R. Muzykantov and [2]Sergei M. Danilov

[1]*Department of Pharmacology, University of Pennsylvania, Philadelphia, Pennsylvania*

[2]*Department of Anesthesiology, University of Illinois in Chicago*

Introduction

Endothelial cells control vascular tone, permeability, angiogenesis, and interactions of blood with tissues and thus represent a key therapeutic target in treatment disease conditions including inflammation, thrombosis, oxidant stress, angiogenesis, acute lung injury, tumor growth, immune disorders, diabetes, hypertension, and atherosclerosis. Agents useful for diagnostic, prophylactic, and therapeutic interventions in endothelial cells in these conditions include label compounds (isotopes), antioxidants, fibrinolytics, anticoagulants, anti-inflammatory agents, pro- and anti-angiogenic mediators, NO donors, and other drugs. In theory, proteins performing some of these functions and genes encoding these proteins can be employed as highly potent and specific "natural" therapeutic agents.

Alas, most exogenous proteins, genetic materials, imaging agents, and drugs are rapidly cleared from blood and lack specific affinity to endothelium. Thus, an inadequately low percent of active agents is delivered to the target cells, while a major fraction of injected drugs is a potentially dangerous waste. Multiple injections of high doses to achieve and sustain a therapy cause side effects. Therapeutic proteins and genetic materials have specific shortcomings—a high propensity for immune reactions, inactivation by inhibitors and enzymes (proteases, nucleases), and the necessity of specific subcellular localization. This chapter briefly describes strategies for optimization of endothelial delivery of drugs. It focuses on targeting to endothelial surface determinants, promising specific delivery of protein and genetic therapeutic agents to selected subcellular compartments of endothelial cells localized in resting or altered vasculature.

Means for Improvement of Pharmacokinetics and Vascular Delivery of Drugs

Design of most drug or gene delivery strategies pursues protection of therapeutic cargoes from inactivation en route, reduction of interaction with nontarget cells, and providing a sufficient circulation time to permit accumulation in the target [1]. These goals can be achieved by loading of therapeutic cargoes into vehicles, either natural (blood cells, albumin, lipoproteins, viruses) or artificial (polymers, liposomes, nanoparticles). Vehicles separating cargoes from blood (polymer shells, liposomes) help to minimize their systemic adverse effects. In general, the diameter of vehicles and carriers for drug and gene delivery in microvasculature is less than 300 to 500 nm, that is, a size that provides free circulation through capillaries in the vascular system and, in some cases, permits intracellular delivery into endothelial cells.

Liposomes, bilayer vesicles formed from phospholipids, represent versatile vehicles (size from 50 nm to microns) and are good carriers for small hydrophobic drugs [1, 2]. Liposomes containing positively charged phospholipids (cationic liposomes) are potentially useful for gene delivery, since they form condensed complexes with DNA, thus protecting it from inactivation and facilitating intracellular delivery [1, 3]. Micro- (0.5- to 50-μm diameter) and nano- (100 to 500 nm diameter) vehicles and carriers consisting of biocompatible, biodegradable polymers [e.g., poly(lactic acid)] are being designed for drug delivery [1]. Delivery of anticancer agents due to enhanced permeation of tumor vasculature and prolonged retention of extravasated materials in the tumor due to poor lymphatic drainage is an active area of exploration for polymer nanocarriers loaded with antibiotics. Loading of large doses of active large hydrophilic drugs (e.g., proteins) into liposomal or polymer vehicles is a more challenging and not fully accomplished goal.

Unless specific countermeasures are undertaken, some drugs and most drug vehicles and carriers are rapidly cleared from the bloodstream and activate host defense systems. For example, complement activation may cause liposome destruction, uptake by phagocytes, and generation of proinflammatory mediators (e.g., cytokines). These concerns are especially acute when antibodies are used for vascular targeting, since their Fc fragments initiate defense reactions including activation of complement, blood cells, and phagocytes. Macrophages have receptors for Fc fragments of immunoglobulins and components of complement that serve for docking and endocytosis of objects coated (opsonized) with antibodies or complement. Macrophages in the liver, bone, and spleen have good access to circulation and effectively eliminate liposomes, viral particles, and even larger drug vehicles (several micrometers in diameter).

Elimination of interaction with Fc receptors, attainable by utilization of antigen-binding domains of antibodies (Fab and single-chain, scFv fragments) helps to reduce some of these untoward effects and prolong circulation time. On the other hand, attachment of poly(ethylene glycol) (PEG, a hydrophilic linear polymer) to proteins, liposomes, polymers, and viral particles creates a "PEG brush" on the surface of these carriers. PEG attracts water, which prevents aggregation of particles and reduces their accessibility for antibodies, complement, phagocytes, and immune cells. PEG-ylation markedly prolongs circulation time of drugs and drug–carrier complexes and minimizes immune reactions and other side effects associated with recognition by host defense—"stealth technologies" [1]. A few types of FDA-approved stealth PEG liposomes are currently in clinical studies. The stealth effect increases proportionally to length and number of PEG molecules attached to a protein, DNA, or liposome; the more dense the PEG brush, the more profound the stealth effect. However, an excessive PEG coupling inactivates protein and genetic therapeutic agents, whereas incorporation of PEG moieties into more than 10 percent of liposomal phospholipids destabilizes lipid bilayer and destroys liposomes. Design of synthetic, totally PEG-coated vesicles consisting of di-block PEG copolymers with structural biocompatible polymers such as poly(lactic acid) is an attractive novel technological avenue to produce inert, long-circulating vehicles for drug delivery.

Passive Drug Delivery to Microvasculature

Circulation of a drug encapsulated in a vehicle or conjugated with a carrier creates a relatively homogenous level in the blood. Only a small fraction accumulates in the area of therapeutic interest. One approach to improve delivery of a drug to vascular areas of interest is to load it into large vehicles (e.g., microspheres with diameters 20 to 50 μm) that mechanically lodge in the precapillary arterioles, creating an elevated local level of a drug or imaging agent. For example, clinically used albumin microspheres labeled with technetium-99, indium-111, or other isotopes can be visualized in a gamma-camera and thus help to image perfusion of blood vessels downstream from the injection site: pulmonary circulation after injection via intravenous route, coronary or cerebral vasculature after injection via the coronary or carotid artery routes, or other microvascular beds after infusion into afferent arteries.

A similar principle, lodging in the precapillary and capillary vascular bed in an organ of interest, is being also explored in the context of gene therapy. Thus, cationic liposomes complexed with negatively charged plasmid DNA rapidly form multimicrometer aggregates in blood, which lodge downstream the injection site. This technique has been employed for a transgene expression in the lungs and in angiogenic endothelial cells in tumors [3]. Further, infusion via vascular catheters inserted in the arteries supplying a pathologically altered organ creates a high local level of therapeutics or genetic materials, promoting their delivery to or beyond endothelium in the area of interest.

However, mechanical or electrostatic retention of small aggregates affords no selective delivery of cargoes into certain cell types (e.g., endothelial cells), nor any control over subcellular localization of the delivered cargoes. In addition, perfusion rapidly removes drugs released from a lodged vehicle or carrier, reducing the time of contact with target cells that is necessary for binding and uptake. In some cases, use of the catheters permitting temporarily cessation of perfusion in the site of delivery helps to circumvent this problem. However, interruption of blood flow creates ischemia and activates flow-sensitive endothelium, causing potentially harmful side effects.

Targeted Drug Delivery to Endothelial Surface Determinants

Stealth liposomes, protein carriers, and other delivery systems prolong the circulation time of drugs and may enhance their cellular uptake, but do not confer an affinity to vascular endothelium. Thus, the major fraction of injected materials ends up in the liver, not in the endothelial cells. In order to facilitate targeting, cargoes or their carriers can be conjugated with affinity moieties that bind to endothelial cells. Immunostaining of tissues, in vivo selection of peptide ligands using phage display, and tracing labeled antibodies in animals helped to identify endothelial determinants potentially useful as targets [4–6].

No universal or ideal affinity carrier, however, suits all therapeutic needs. Specific therapeutic goals require different secondary effects mediated by binding to the endothelium, drug targeting to different subpopulations of endothelial cells (e.g., resting versus inflammation-engaged endothelium), and to diverse cellular compartments. Targeted delivery of antioxidants, antithrombotics, or NO donors to normal or resting endothelial cells can be useful for either prophylaxis or therapies. On the other hand, specific recognition and drug delivery to abnormally activated or pathologically altered endothelium might permit more specific means for treatment of such maladies as localized tumor growth and inflammation. Inhibition of some endothelial antigens leads to harmful adverse effects. For example, blocking a constitutive endothelial antigen thrombomodulin may aggravate thrombosis. In addition, pathological conditions suppress expression of some endothelial antigens.

The pulmonary vasculature comprises a preferred target for vascular drug delivery to endothelium, because (i) it is the first major capillary network encountered by intravenously injected drugs; (ii) it contains roughly one-third of the endothelium in the body; (iii) it is exposed to the entire cardiac output of venous blood; and (iv) a relatively slow blood perfusion rate through high-capacity, low-resistance pulmonary vessels kinetically favors binding of ligands to endothelium. Therefore, affinity carriers recognizing panendothelial determinants (e.g., antibodies directed against constitutive cell adhesion molecules) tend to accumulate in the lungs after systemic administration. Because of its accessibility to external insults from the airways and its function as a filter for blood clots, debris, activated white blood cells, drugs, and toxins, pulmonary vasculature represents an important therapeutic target [5].

Panendothelial Surface Determinants Candidates for Targeting

Constitutively expressed determinants are good candidates for targeting either normal or pathologically altered endothelium. For example, platelet-endothelial adhesion molecule-1 (PECAM, CD31) is a panendothelial transmembrane Ig superfamily glycoprotein, predominantly localized in the sites of cellular contacts in the endothelial monolayer. Platelets and white blood cells also express PECAM, but at levels that are orders of magnitude lower than endothelial cells. PECAM is involved in the cellular recognition, adhesion, signaling, and transendothelial migration of leukocytes, key in pathogenesis of many disease conditions. Blocking PECAM by anti-PECAM suppresses inflammation and protects organs against leukocyte-mediated oxidative stress. Therefore, anti-PECAM targeting may provide secondary benefits for management of inflammation, perhaps by attenuation of leukocyte transmigration. PECAM is abundant in endothelial cells, which express millions of anti-PECAM binding sites. In addition, PECAM is a stable endothelial cell antigen: cytokines and ROS do not downregulate its expression and surface density on the endothelium. This permits a robust PECAM-targeted drug delivery to either normal or pathologically altered vasculature, for either prophylaxis or therapies. Endothelial cells bind anti-PECAM without internalization, but readily internalize multimeric anti-PECAM complexes within 100- to 350-nm diameter via a novel internalization pathway different from clathrin- and caveoli-mediated endocytosis, namely CAM-mediated endocytosis (common for PECAM-1 and ICAM-1). Thus, by rational design of anti-PECAM conjugates of a proper size, one can achieve either intracellular or surface targeting to endothelium [5]. An active reporter enzyme, β-galactosidase, conjugated to anti-PECAM has been shown to accumulate intracellularly in the pulmonary endothelium as soon as 10 minutes after IV injection in mice and pigs. Targeting of anti-PECAM/catalase conjugate protects pulmonary vasculature against acute oxidant stress in mice and rats [7].

Intercellular adhesion molecule-1 (ICAM-1, CD54) is another Ig superfamily surface glycoprotein with a short cytoplasmic domain, a transmembrane domain, and a large extracellular domain. It is normally expressed by endothelial cell at relatively high surface density (2×10^4 to 2×10^5 surface copies per cell). Some other cell types also express ICAM, yet the major fraction of blood-accessible ICAM is exposed on the luminal surface of endothelium. Robust and specific binding of ICAM antibodies and anti-ICAM conjugates to vascular endothelium after intravenous administration has been documented in animals. Pathological stimuli, such oxidants, cytokines, abnormal shear stress, and hypoxia, stimulate surface expression of ICAM by endothelial cells. Thus, inflammation and other pathological conditions do not suppress, but rather markedly facilitate ICAM targeting. ICAM, a counter-receptor for integrins on leukocytes, supports their adhesion to endothelial cells and thus contributes to inflammation. In addition, ICAM serves as a natural ligand for certain viruses. ICAM may also serve as a signaling molecule, yet the significance of this signaling remains to be more fully elucidated. ICAM antibodies suppress leukocytes adhesion, thus producing anti-inflammatory effects in animal models and clinical pathological settings associated with vascular injury, such as acute inflammation, ischemia–reperfusion, and oxidative stress. The anti-inflammatory effect of multivalent anti-ICAM conjugates may be even more potent because of the higher affinity/valency of the conjugate binding and down-regulation of surface ICAM due to internalization via CAM-mediated endocytosis (which was described earlier for PECAM). Thus, subcellular localization of anti-ICAM conjugates can be controlled by their size: Anti-ICAM carriers permit targeted delivery of antithrombotic agents to the endothelial surface and antioxidant enzymes to the endothelial interior [5].

Angiotensin-converting enzyme (ACE) is a transmembrane glycoprotein expressed on the endothelial luminal surface, which converts Ang I into Ang II to induce vasoconstricting, pro-oxidant, and proinflammatory activities. Pulmonary vasculature is enriched in ACE: Nearly 100 percent of endothelial cells in the alveolar capillaries are ACE positive, whereas 10 to 15 percent in the extrapulmonary capillaries are ACE-positive. Radiolabeled anti-ACE and reporter compounds conjugated with anti-ACE accumulate in the lungs after injections via diverse routes in animals and humans. Anti-ACE does not cause acute harmful reactions in animals and humans. ACE antibodies suppress its activity; this might enhance therapeutic utility of the targeting and provide beneficial effects in conditions associated with vascular oxidative stress, ischemia, and inflammation. Endothelial cells internalize anti-ACE that may deliver drugs intracellularly. Anti-ACE-conjugated antioxidant enzymes, such as catalase, accumulate in rat lungs in vivo and protect perfused rat lungs against H_2O_2. Proinflammatory agents suppress ACE expression and may inhibit therapeutic targeting directed to ACE, yet anti-ACE is a good candidate for targeting to the pulmonary endothelium for a prophylactic use and for gene delivery [8, 9].

Targeting Specific Vascular Areas

Injecting of anti-CAM conjugates via catheters inserted in a conduit artery facilitates local delivery in the downstream vascular area. Local delivery may also be enhanced by surface endothelial determinants enriched in particular vascular areas or in focal pathological processes.

For example, endothelium contains surface determinants localized to cholesterol-enriched plasma membrane microdomains, including caveoli. Ligands of these determinants accumulate in the pulmonary vasculature after intravenous injection in rats, then enter and traverse endothelial cells [4]. The functions and human counterparts of these caveoli-localized determinants, interesting candidates for drug delivery, are not known.

Cerebral endothelium is an especially important and difficult target. Carrier antibodies and peptides directed to surface determinants relatively enriched in cerebral endothelium, including receptors for transferrin, insulin, and some growth factors, permit delivery of reporter compounds and genes into the brain in animals. Some of these receptors apparently permit transendothelial delivery into the brain tissue and neurons [10].

Endothelial cells exposed to inflammatory mediators and abnormal shear stress show cell surface expression of P-selectin, normally stored intracellularly and mobilized rapidly to the surface, and E-selectin, which is newly synthesized by activated endothelium. Thus, selectins are transiently exposed on the surface of stressed endothelial cells. Experiments in cell cultures and limited animal studies show that selectins may permit targeting of drugs to cytokine-activated endothelium [2]. However, kinetics and persistence of selectin exposure upon endothelial activation are difficult to follow even in animal studies, where exact initiation of cytokine activation could be easily controlled; thus the transient character of surface exposure hinders targeting of selectins. Further, even at the activation peak, selectins are exposed at relatively low surface densities; hence, robustness of the targeting may be suboptimal for therapeutic interventions requiring delivery of large doses. However, it may suffice for diagnostic visualization of inflammation foci [11] and, perhaps, for gene delivery [12].

Endothelial cells in solid tumors also represent an especially important and challenging target for delivery of agents designed to visualize tumors, inhibit angiogenesis, or eradicate malignant cells. Tumor vasculature is characterized by numerous morphological abnormalities. Endothelial cells in tumor vessels expose abnormal determinants including VCAM, selectins, integrins, apoptosis markers, and receptors for growth factors [13]. Targeting these determinants in tumors might be useful to accomplish two goals: (i) delivery of antitumor agents to the proper malignant cells using, for example, PEG-immunoliposomes or polymers loaded with taxol or doxorubicin; and, (ii) inflicting damage in the tumor vasculature leading to thrombosis, infarction and starvation of malignant cells [14].

Targeting of Genetic Materials and Viruses to Endothelial Cells

Delivery of protein therapeutics affords very fast but transient effects due to inevitable degradation. Selective and safe transfection of endothelial cells with genes encoding therapeutic proteins may help to overcome this problem and bring about new means for treatment of diverse disease conditions, acute (thrombosis, respiratory distress syndrome, stroke, inflammation) and chronic (tumor growth, systemic and pulmonary hypertension, atherosclerosis, diabetes). However, endothelial cells are difficult to transfect even in cell cultures; the challenges of implementation of this goal in vivo are formidable [3, 10, 12, 15].

Nonviral vehicles for gene delivery (e.g., cationic liposomes) are generally viewed as safer than viral means, yet the latter offer more effective and prolonged transgene expression. However, both nonviral and viral vehicles have no tropism to endothelium, whereas nontarget cells may take up circulating liposomes and/or express high levels of receptors for viruses (e.g., hepatic cells). Thus most transgene proteins are produced predominantly in liver and other tissues, which are at risk of toxicity due to inadvertent transduction. Targeting of either nonviral or viral carriers of genetic materials to endothelial determinants, for example using monoclonal antibodies to ACE, PECAM-1, or selectins conjugated to the liposomes or viral particles, might enhance effectiveness of transfection of endothelial cells and help to reduce side effects [9, 12]. Coupling of artificial affinity moieties (antibodies or peptides) to rationally selected viral determinants also blocks their function of receptor-mediated docking to nontarget cells, thus altering viral tropism (viral retargeting) and permitting partially reduced side effects [9].

Identification and utilization of promoters for PECAM-1, Flt-1 (a receptor for vascular endothelial growth factor), and other endothelium-specific proteins also enhances specificity of transfection of vascular endothelium [15]. Loading genetic materials driven by endothelial-specific promoters into vehicles utilizing affinity carriers directed to endothelial surface determinants synergistically combines advantages of transductional and transcriptional targeting [9]. For example, successful specific gene expression in the pulmonary endothelium was achieved using ACE antibody–conjugated adenoviruses driven by an endothelium-specific Flt-1 promoter (Figure 1).

Viral retargeting can be attained either by chemical conjugation of targeting antibodies or peptides to the surface of viral capsid or by genetic incorporation of these affinity moieties into the capsid structure. Genetic retargeting approaches have to a large extent been limited by the structural constraints of viral capsid proteins: Only a few relatively small ligands have been inserted without disruption of the viability of the virus. Efforts are currently underway to incorporate bigger single-chain fragments of antibodies (scFv) directed against identified vascular targets into viral capsid and/or to use phage panning in vivo to define unique vascular addresses for specific vascular beds [6].

Conclusion

Diagnostic and therapeutic cargoes including enzymes and genetic materials can be delivered to endothelial cells using diverse vehicles, natural (viruses, blood components) and synthetic (liposomes, nanoparticles). Affinity carriers that bind to specific endothelial determinants selectively and avidly permit targeting of reporter and therapeutic cargoes to endothelial cells in experimental studies in cell cultures and laboratory animals. A growing number of surface determinants, expressed predominantly or selectively in diverse normal and pathological vascular areas and types of vessels, are being identified. Affinity carriers for endothelial targeting include monoclonal antibodies and their fragments, peptides, growth factors, hormones, and nutrients. In addition to function of targeted delivery, some carriers provide additional means to control rate of internalization and subcellular or transcellular traffic, which is especially important in

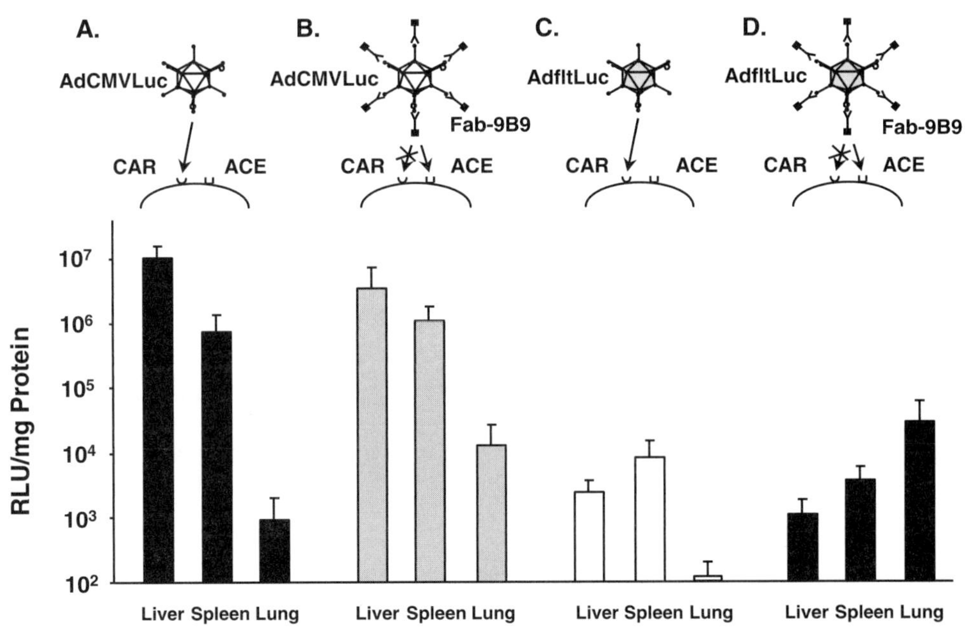

Figure 1 Enhancement of specificity and effectiveness of a model transgene expression in pulmonary endothelium in rats using combined transductional and transcriptional targeting of adenoviral gene delivery. Rats were injected intravenously with adenoviral vectors carrying the luciferase reporter gene, and 3 days later organs were harvested and luciferase activity determined. **(A)** Untargeted vector, AdCMVLuc, showing dominant transgene expression in liver and spleen due to biological tropism of the virus to cells in these organs expressing receptors for viral capsid proteins. **(B)** Conjugation of ACE antibody Fab fragments to these virus surface proteins blocks their binding to hepatic and splenic receptors and endows the virus with affinity to endothelial cells. This provides targeting to endothelial cells and enhanced pulmonary and reduced hepatic gene expression (note log scale). **(C)** AdfltLuc virus, containing an endothelium-specific promoter Flt-1, but lacking specific affinity to endothelial cells, provides ineffective gene expression in all organs. **(D)** Conjugation of AdfltLuc with anti-ACE restores transgene expression levels in lungs, but still further reduces expression in nontarget liver and spleen, thereby improving selectivity lung/liver ratio net 300,000-fold. RLU, Relative light units. (Adapted from Ref. [9].) (see color insert)

case of gene and enzyme replacement therapies. Several promoters specific for resting or pathologically altered endothelial cells are being identified; utilization of these promoters in gene delivery means greatly enhanced effectiveness and specificity of endothelial transfection of reporter and potentially therapeutic proteins. Ongoing investigations will characterize the practical feasibility of translation of these promising concepts to the clinical domain.

Glossary

Angiotensin-converting enzyme (ACE): A transmembrane endothelial glycoprotein, an ectopeptidase that controls activity of the vasoactive peptides angiotensin and bradykinin, a determinant for targeting pulmonary endothelium.

Caveoli: Specialized cholesterol-rich domains in the endothelial plasma membrane that form characteristic flask-shaped invaginations and serve as sensing, signaling, and endocytotic compartments. Caveoli represent a unique target for delivery of drugs and genes into and through endothelial cells.

Flt-1: Type 1 receptor for vascular endothelial growth factor (VEGFR-1), an endothelial cell–specific receptor. Flt-1 promoter is useful for driving endothelium-specific gene transfection.

Intercellular adhesion molecule-1 (ICAM-1): A transmembrane endothelial glycoprotein that facilitates leukocytes adhesion, a determinant for targeting pulmonary and systemic endothelium, either resting or pathologically altered.

Liposome: An artificial vehicle suitable for drug or gene delivery, a phospholipid vesicle (100 to 1,000 nm in size) that can be loaded with various hydrophilic and hydrophobic drugs into the aqueous interior or phospholipid membrane of the liposome, respectively.

Luciferase gene: The gene for an enzyme that catalyzes a reaction that emits light (e.g., as in firefly luciferase) that can be detected in a luminometer. It is a common, sensitive reporter gene assay used in gene delivery studies.

Platelet-endothelial adhesion molecule-1 (PECAM-1): A transmembrane endothelial glycoprotein that facilitates leukocyte transmigration from blood to tissues, a determinant for targeting resting endothelium.

Poly(ethylene glycol) (PEG): A linear organic polymer. PEG conjugation to proteins, liposomes, and cells forms a hydrophilic shell around these objects that repels opsonins and defense cells. PEG-ylation reduces immune reactions and elimination of the injected materials ("stealth technology").

Selectins: Selectins P and E, surface adhesion molecules involved in leukocyte rolling on endothelium, transiently expressed on the surface of pathologically altered endothelial cells, for example, in inflammation foci; candidates for selective targeting of pathological vasculature.

Targeting moiety: A molecule (usually, antibody) that specifically binds to characteristic surface determinants in cells, organs, or areas of interest. Being attached to a drug molecule or drug vehicle, such a molecule enhances their accumulation in therapeutic targets.

Vascular endothelial growth factor (VEGF): A multifunctional protein that facilitates vascular permeability and stimulates vascular proliferation and angiogenesis.

VCAM-1: Vascular cell adhesion molecule-1 is a proinflammatory adhesion molecule expressed on endothelial and other cells altered by cytokines, oxidants, and other pathological mediators.

Vector: A vehicle to achieve gene delivery.

References

1. Torchilin, V. (2003). Strategies and means for drug targeting. In: *Biomedical Aspects of Drug Targeting*, V. Muzykantov and V. Torchilin, eds., pp. 3–26. Dordrecht: Kluwer Academic.
2. Spragg, D. D., Alford, D. R., Greferath, R., Larsen, C. E., Lee, K. D., Gurtner, G. C., Cybulsky, M. I., Tosi, P. F., Nicolau, C., and Gimbrone, M. A., Jr. (1997). Immunotargeting of liposomes to activated vascular endothelial cells: A strategy for site-selective delivery in the cardiovascular system. *Proc. Natl. Acad. Sci. USA* **94**, 8795–8800.
3. Thurston, G., McLean, J. W., Rizen, M., Baluk, P., Haskell, A., Murphy, T. J., Hanahan, D., and McDonald, D. M. (1998). Cationic liposomes target angiogenic endothelial cells in tumors and chronic inflammation in mice. *J. Clin. Invest.* **101**(7), 1401–1413.
4. McIntosh, D. P., Tan, X. Y., Oh, P., and Schnitzer, J. E. (2002). Targeting endothelium and its dynamic caveolae for tissue-specific transcytosis in vivo: A pathway to overcome cell barriers to drug and gene delivery. *Proc. Natl. Acad. Sci. USA* **99**, 1996–2001. *This experimental paper describes a strategy for targeting specific surface determinants localized in rat vasculature in endothelial caveoli. In addition to enhanced effectiveness of drug delivery to endothelial cells, targeting caveoli facilitates internalization of endothelium-bound materials and their transcellular traffic via a unique endocytotic pathway.*
5. Muzykantov, V. R. (2002). Targeting pulmonary endothelium. In: *Biomedical aspects of drug targeting*, V. Muzykantov and V. Torchilin, eds., pp. 129–148. Dordrecht: Kluwer Academic.
6. Pasqualini, R., Arap, W., and McDonald, D. M. (2002). Probing the structural and molecular diversity of tumor vasculature. *Trends Mol. Med.* **8**(12), 563–571. *An authorized review of an original strategy for identification of determinants expressed on normal and pathologically altered endothelial cells in diverse organs and vascular areas. It employs in vivo administration of phage libraries displaying high numbers of diverse specific peptides and selecting peptides providing specific targeting of phages in the area of interest.*
7. Kozower, B., Christofidou-Solomidou, M., Sweitzer, T., Muro, S., Buerk, D., Solomides, C., Albelda, S., Patterson, G., and Muzykantov, V. (2003). Immunotargeting of catalase to the pulmonary endothelium alleviates oxidative stress and reduces acute transplantation lung injury. *Nat. Biotechnol.* **21**, 392–398.
8. Muzykantov, V. R., and Danilov, S. M. (1995). Targeting of radiolabeled monoclonal antibody against ACE to the pulmonary endothelium. In: *Targeted Delivery of Imaging Agents*, V. Torchilin, ed., pp. 465–485. Boca Raton, FL: CRC Press.
9. Reynolds, P. N., Nicklin, S. A., Kaliberova, L., Boatman, B. G., Grizzle, W. E., Balyasnikova, I. V., Baker, A. H., Danilov, S. M., and Curiel, D. T. (2001). Combined transductional and transcriptional targeting improves the specificity of transgene expression in vivo. *Nat. Biotechnol.* **19**, 838–842.
10. Pardridge, W. M. (2002). Drug and gene targeting to the brain with molecular Trojan horses. *Nat. Rev. Drug. Discov.* **1**(2), 131–139.
11. Lindner, J. R., Klibanov, A. L., and Ley, K. (2003). Targeting inflammation. In: *Biomedical Aspects of Drug Targeting*, V. Muzykantov and V. Torchilin, eds., pp. 149–172. Dordrecht: Kluwer Academic Publishers.
12. Harari, O. A., Wickham, T. J., Stocker, C. J., Kovesdi, I., Segal, D. M., Huehns, T. Y., Sarraf, C., and Haskard, D. O. (1999). Targeting an adenoviral gene vector to cytokine-activated vascular endothelium via E-selectin. *Gene Ther.* **6**, 801–807.
13. Oh, P., Li, Y., Yu, J., Durr, E., Krasinska, K., Carver, L., Testa, J., and Schnitzer, J. E. (2004) Subtractive proteomic mapping of the endothelial surface in lung and solid tumours for tissue-specific therapy. *Nature* **429** (6992), 629–635.
14. Huang, X., Molema, G., King, S., Watkins, L., Edgington, T. S., and Thorpe, P. E. (1997). Tumor infarction in mice by antibody-directed targeting of tissue factor to tumor vasculature. *Science* **275**(5299), 547–550. *This experimental paper gives an example of tumor vasculature targeting strategy, which is potentially useful for destruction of solid tumors by cutting off their blood supply. This strategy is pursued by chemical or genetic conjugation of agents producing blood clots or endothelial damage are conjugated with monoclonal antibodies or peptides that specifically bind to surface determinants selectively expressed in tumor vasculature.*

15. Nicklin, S. A., Reynolds, P. N., Brosnan, J. M., White, S. J., Curiel, D. T., Dominiczak, A. F., and Baker, A. H. (2001). Analysis of cell-specific promoters for viral gene therapy targeted at the vascular endothelium. *Hypertension* **38**, 65–70.

Capsule Biography

Vladimir Muzykantov obtained an M.D. at First Moscow Medical School, Russia, in 1980, and a Ph.D. at the Russian Cardiology Research Center, Moscow, in 1985. He is an Associate Professor of Pharmacology and Medicine at the University of Pennsylvania. Muzykantov's research is focused on design of strategies for targeted delivery of therapeutic agents in cardiovascular system, in particular targeting of enzymes to endothelial cells.

Sergei Danilov obtained an M.D. from the Second Moscow Medical School, Russia, in 1975, and a Ph.D. in 1980 and Dr.Sci. in 1994 from the Russian Cardiology Research Center, Moscow. He was a Visiting Professor at INSERM in Paris, France, from 1993 to 1995. He is now an Assistant Professor of Anesthesiology at the University of Illinois in Chicago. For 20 years, Danilov's research interests have been focused on structural–functional studies of angiotensin-converting enzymes using monoclonal antibodies and on the molecular design of affinity carriers to ACE for targeting of genetic materials to endothelial cells. Dr. Danilov has published more than 80 peer-reviewed papers on these subjects.

This work was supported by NIH (RO1 HL71175, HL78785 and HL73940) and Department of Defense Grant (PR 012262).

PART V

New Research Modes and Procedures

SECTION A

DNA Microarray

CHAPTER 156

The Identification of Cell Type–Specific Transcripts by DNA Microarray Analysis

Andrew Leask

Centre for Rheumatology, Royal Free and University College Medical School, Hampstead, London, United Kingdom

Introduction

The constellation of mRNAs that are selectively and differentially expressed in a cell or tissue is a critical determinant of its specific structure or function. Microarray technology has transformed the study of biology by permitting the analysis of the global pattern of mRNA expressed in cells and tissues of interest. This technology has resulted in novel insights into cell and tissue function. mRNAs expressed are reverse transcribed into labeled cDNA probes, which are then hybridized onto glass slides, more commonly known as gene chips, and analyzed in chip readers. This genome-wide transcriptional profiling of individual cell types allows the identification of genes that are expressed preferentially in one cell type as compared to another. In addition, genes that are responsive to external stimuli in a cell type–specific response can also be identified. This article summarizes DNA microarray technology and its applications and includes a summary of recent uses of DNA microarray technology to study the microvasculature.

Background

The recent sequencing of the human genome has opened a new era in biomedical research. However, the creative energy expended on the completion of this task has now been necessarily refocused on a central, essential question, namely how to properly convert the dry, linear sequence data generated by the Human Genome Project into practical use—for example, into insights into cellular function such as the characterization of the role of gene products in normal physiology and pathological disease.

Initially developed through data obtained through the Human Genome Project, DNA microarrays have emerged as a key, powerful technology that has related gene sequence information to functional insights into biological mechanisms. High-density DNA probe arrays are fast transforming gene-based biomedical research and are becoming routine tools for the high-throughput analysis of gene expression in a wide range of biological systems. Although a number of approaches can be taken when implementing microarray-based studies, all are capable of providing important insights into biologic function. Microarrays will continue to have a significant impact on research, especially in terms of identifying transcripts whose expression is enriched in one cell type as compared to another, either in the resting state of the cell, or in the response to external stimuli.

The Gene Chip

A so-called gene chip is actually a silicon or glass slide to which DNA is bound. The glass slide has an inherent adhesiveness for DNA that is enhanced with a coating of polylysine or saline. DNA is "spotted" on the glass under precise positional control using specialized equipment. Photolithographic DNA synthesis has enabled the large-scale production of GeneChip probe arrays containing hundreds of thousands of oligonucleotide sequences on a glass "chip" about 1.5 cm^2 in size. Up to 400,000 distinct sequences can be placed on one slide. The manufacturing process integrates solid-phase photochemical oligonucleotide synthesis with lithographic techniques similar to those used in the microelectronics industry. Because of their very high information content, GeneChip probe arrays are finding widespread use in the hybridization-based detection and analysis of mutations and polymorphisms ("genotyping"), and in a wide range of gene expression studies. Commercial chips can be purchased; alternatively, with the appropriate equipment, the laboratory can manufacture chips to its own specifications. Some of the commercial chips contain multiple copies of the same oligonucleotide, and this arrangement can enhance the validity of the obtained result.

Detection of Genes on the Chip

To estimate the relative expression levels of each gene within a particular cell type or tissue, total cellular RNA is extracted by standard techniques and reverse transcribed into radiolabeled (for example, with ^{32}P or ^{33}P-UTP) or fluorescently labeled (for example, with Cy-3 or Cy-5-UTP) cDNA probes. These probes are then hybridized to the chips. Fluorescently labeled probes are more commonly used as they are convenient and quantitative and can be measured with more precision using commercially available equipment.

After the hybridization of the probes, slides are scanned by laser. The laser causes the excitation of fluorescently labeled cDNA probes. Only the spots representing mRNAs in the sample give emission signals. The emission is measured using a scanning confocal laser microscope, and the fluorescence data at each and every spot are analyzed by appropriate software. The absence of fluorescence in a specific spot means that complementary mRNA is not present in the sample. If the fluorescence is present, the intensity of the signal obtained is a measure of the level of particular mRNA in the examined cell population.

Usually an investigator wants to compare mRNA abundance between two different samples (or a sample and a control). In this case, cDNAs from the sample and a control are labeled with two different fluorescent dyes (e.g., red label for the cDNAs from the sample and a green label for the control), and these two cDNA populations are allowed to hybridize to the same microarray slide. If a particular mRNA from the sample is in abundance, the spot with a complementary probe will be red; if the concentration of the particular mRNA is higher in the control, the spot will be green. If both samples contain the same amount of a given mRNA, the spot will be yellow. Thus one can draw conclusions about the relative expression levels of the genes based on the colors and fluorescence intensities of the microarray spots. However, some commercially available systems permit the hybridization of only one sample per chip. In this case, data from one chip are compared with data from other different chips using appropriate software. Thus, the relative expression levels of genes in one cell type and another cell type can be compared directly.

Data Analysis

Often the amount of data generated by gene chip analysis is overwhelming. Conventional genomic analysis involves the description of the position of genes on chromosomes and does not per se lead to functional insights into cell or tissue function. However, gene arrays allow for the simultaneous measurement of all transcribed RNA, and hence of all active genes, in any one state of activity or differentiation. This pattern of mRNA expression is termed an expression profile. The collective constellation of proteins that are selectively and differentially expressed in a cell or tissue is a critical determinant of its structure or function. There are many possible applications for the analysis of this profile; for example, transcripts that are cell-type specific or possess a cell type–specific response to a particular stimulus can be identified.

DNA array analysis permits the organization of expressed genes into clusters based on their differential expression patterns. For example, a large number of whole-genome observations in the yeast cell have been compared using several different experimental conditions and a template in which many thousands of measurements can be conveniently displayed. This model takes advantage of the fact that color-coded probes were hybridized to the array, and that the intensity of the color is directly correlated with expression levels of the different transcripts. Based on the relative intensity of expression of each gene, genes are clustered in hierarchies based on their similarities and differences in expression across the various experiments, for example in relation to cell cycle regulation. Such tabulations allow unforeseen correlations to take place. Grouping genes according to known function may also help the functional interpretation of the results, especially when a gene identified has been little studied, and hence the total functions of the gene in question may not be fully appreciated. These analyses may be facilitated by computer programs and algorithms especially designed for these applications.

Limitations of the Technology

DNA microarrays are a powerful tool for monitoring thousands of transcript levels simultaneously. However, many transcripts important to function have low expression levels or are expressed in relatively few cells, making them difficult to detect in the complex background of whole tissue. Thus, more reliable results may be obtained using cultured cells than tissue samples. In addition, gene chips generally only reliably detect medium- to high-abundance transcripts because detection of low-abundance transcripts, many of which have great relevance to biological function, is inconsistent. Thus, it may still be necessary to identify low-abundance transcripts important for cell-type or tissue-type function by alternative methods, such as by subtractive hybridization or differential display approaches. Finally, it is essential to confirm experimental conclusions derived from data generated by array analysis using additional, more sensitive methods, such as real-time polymerase chain reaction technology.

The Use of Arrays to Study the Vasculature

Microarray technology has been successfully used in several studies regarding the vasculature, including the identification of cell-type enriched transcripts, the basis of pathological disease, and the external response to stimuli in a cell type–specific fashion.

For example, microarray technology has been successfully used to investigate the mechanism of cell-type resistance to apoptotic effects of the fatal disease idiopathic thrombotic thrombocytopenic purpura (TTP) [1]. TTP is a disease characterized by the apoptotic injury of all microvascular endothelial cells (MVECs) except those of pulmonary origin. Oligonucleotide gene chip technology identified genes differentially overexpressed in TTP-resistant (lung) MVECs relative to TTP-sensitive (skin) MVECs and hence that may contribute to the resistance of lung MVECs to be resistant to apoptosis induced by TTP plasma. This study showed that exposure of cells to TTP plasma yielded 157 genes that were overexpressed in primary human lung MVEC, including specific pro-survival signals such as the TRAIL antagonist, osteoprotegerin, and the vascular endothelial growth factors, VEGF/VPF and VEGF-C, and their receptors, VEGFR-2 (KDR) and VEGFR-3 (Flt4) [1].

Similar studies identified genes overexpressed in angiogenesis. Angiogenesis is a key process in a variety of human diseases, including cancer. The ability to target selectively the tumor vasculature is potentially useful for the diagnosis and treatment of cancer. Tumor endothelial cells (TECs), in contrast to normal endothelial cells, are resistant to apoptosis, are proadhesive for renal carcinoma cells, and are able to grow and organize in the absence of serum in persistent capillary-like structures [2]. cDNA array technology was recently used to identify markers specific for TECs [2, 3]. In one study, IGFBP-3 was found to be restricted exclusively to ECs, indicating that IGFBP-3 is a potential novel marker of angiogenesis. In another study, expression of angiopoietin-1 and vascular endothelial growth factor (VEGF)-D and the Akt survival pathway were markedly upregulated in TECs. These results suggest the possibility of using IGFBP-3, VEGF-D, or the Akt pathway as therapeutic targets for antiangiogenic strategies.

DNA array technology has been recently used to establish a molecular link between hyperhomocystinemia (HHCy) and atherosclerosis. Microarray analysis showed that homocysteine (Hcy) was linked to cholesterol dysregulation by showing that Hcy treatment of human umbilical vein endothelial cells (HUVECs) resulted in the upregulation of 3-hydroxy-3-methylglutaryl coenzyme A reductase (HMGCR), an enzyme involved with cholesterol synthesis [4]. Further biochemical analysis performed in this study confirmed that Hcy resulted in an increase in cellular cholesterol levels, suggesting a novel solid explanation for the observed proatherogenic effect of Hcy.

In addition, gene array systems were used to study the cellular response of endothelial cells to external stimuli. In one study, the response of human dermal microvascular endothelial cell (DMVEC) cultures to infection by Kaposi's sarcoma (KS)-associated herpesvirus (KSHV) was analyzed [5]. Significant dysregulation was shown of several cytokine-related genes or receptors, as well as endothelial cell and macrophage markers and other genes associated with angiogenesis or transformation. Microarrays have also been used to analyze changes in gene expression following stimulation of myometrial microvascular endothelial cells (MMECs) with vascular endothelial growth factor (VEGF). In this study, a total of 110 genes were identified as upregulated by VEGF, the vast majority of which (81%) were not previously reported as upregulated by VEGF or by angiogenesis [6].

Functionality

Ultimately, of course, it is essential to determine functionality of the overexpressed transcript. These analyses can take the form of overexpressing the encoded protein of interest, either transiently or stably in cell lines or by the generation of transgenic mice. The advent of siRNA makes it substantially more accessible, without necessarily having to generate knockout mice in which the gene of interest has been removed, to assess the effect of eliminating the expression of the transcript in question. Whether overexpression or knockout technology is used, it is necessary to possess a sufficiently robust assay to assess the effect of modifying expression of the protein of interest. Such assays might range from a global mRNA expression array analysis of mutant and wild-type cells to a promoter- or polymerase chain reaction–based assay to specifically assess the role of a protein in a specific cellular outcome of interest.

Conclusion

In summary, microarray technology creates an exciting new challenge in experimental design and control and in data handling and interpretation. However, the power of this technology is its ability to assign previously unknown genes to functions and to discover new functions for known genes or in clinical medicine to identify targets around which to develop cell-type specific therapies. The place of microarray technology in the whole lexicon of molecular and genetic approaches is not yet fully developed or defined. However, microarray systems can be expected to invigorate research into the microvasculature for many years to come.

Glossary

Expression profiling: Using gene microarray or gene chip technology to assess the overall transcriptional state of a cell or tissue.

Gene chip: A silicon or glass slide upon which cDNA or oligonucleotides have been placed, also known as a microarray.

References

1. Kim et al. (2001). *Microvasc. Res.* **62**, 83–93.
2. Bussolati et al. (2003). *FASEB J.* **17**, 1159–1161.
3. Schmid et al. (2003). *Int. J. Cancer* **103**, 577–586.
4. Li et al. (2002). *Circulation* **105**, 1037–1043.
5. Poole et al. (2002). *J. Virol.* **76**, 3395–3420.
6. Weston et al. (2002). *Mol. Hum. Reprod.* **8**, 855–863.

Bibliography

Cheung, V. G., Morley, M., Aguilar, F., Massimi, A., Kucherlapati, R., and Childs, G. (1999). Making and reading microarrays. *Nat. Genet.* **21S**, 15–19.

Duggan, D. J., Bittner, M., Chen, Y., Meltzer, P., and Trent, J. M. (1999). Expression profiling using cDNA microarrays. *Nat. Genet.* **21S**, 10–14. *A good review focusing on the general principles of cDNA microarray technology.*

Evans, S. J., Datson, N. A., Kabbaj, M., Thompson, R. C., Vreugdenhil, E., De Kloet, E. R., Watson, S. J., and Akil, H. (2002). Evaluation of Affymetrix Gene Chip sensitivity in rat hippocampal tissue using SAGE analysis. Serial Analysis of Gene Expression. *Eur. J. Neurosci.* **16**, 409–413.

Ferea, T. L., Botstein, D., Brown, P. O., and Rosenzweig, R. F. (1999). Systematic changes in gene expression patterns following adaptive evolution in yeast. *Proc. Natl. Acad. Sci. USA* **96**, 9721–9726.

Gaasterlaand, T., and Baikaranov, S. (2000). Making the most of microarray data. *Nat. Genet.* **24**, 204–206. *A concise review highlighting the latest advances in computational analysis of microarray data.*

Lander, E. S. (1999). Array of hope. *Nat. Genet.* **21S**, 3–4. *A well written review on the strengths and limitations of microarray technology from one of the pioneers of the human genome project.*

Capsule Biography

Dr. Leask is currently Visiting Lecturer in the Centre of Rheumatology, Royal Free and University College Medical School, London. Prior to that, he was Senior Scientist at FibroGen, Inc., San Francisco, CA. His research interest is the molecular control of fibrosis, including the roles of connective tissue growth factor, TGFβ and endothelin in this process. Dr. Leask was born in Vancouver, B.C., Canada, and received his undergraduate degree (First Class) at the University of British Columbia and his Ph.D. at the University of Chicago. He was a Medical Research Council of Canada Postdoctoral Fellow at Stanford University.

CHAPTER 157

Gene Expression Profiling of Human Endothelial Cells by DNA Microarray and SAGE

Shui Qing Ye, Shwu-Fan Ma, Dmitry Grigoryev, Mohan Parigi, and William L. Shao

Division of Pulmonary and Critical Care Medicine, Department of Medicine, Johns Hopkins University School of Medicine, Baltimore, Maryland

Introduction

The endothelium is the largest organ in the body, consisting of endothelial cells (ECs) lining every blood vessel. Historically ECs were considered to be merely an inert structural component of vasculature to form a semipermeable barrier between the blood and the interstitium of each organ. Currently it is widely appreciated that ECs function in a multitude of physiologic processes, including the control of cellular trafficking, the regulation of vasomotor tone, the maintenance of blood fluidity, and the growth of new blood vessels [1, 2]. Endothelial cell heterogeneity has been recognized at the level of morphology, function, antigen composition, and signaling network [3]. Genetic determinants and environmental cues may critically regulate endothelial cell heterogeneity and contribute to its normal function and its response to pathophysiological stimuli. With the full human genome sequence in hand [4] and the advent of new powerful DNA microarray technology [5, 6] and serial analysis of gene expression (SAGE [7]), the research paradigm has been shifted from the traditional search for a single or a few specific genes to the current understanding of the biochemical and molecular functioning of hundreds and thousands of genes and how complicated networks of interaction determine the normal function of ECs and lead to the dysfunction of ECs and pathogenesis of various human diseases. In this chapter, the first part synopsizes DNA microarray and SAGE technologies and the second part highlights new biological insights derived from the application of DNA microarray and SAGE in the gene expression profiling of EC. The epilogue contains perspectives.

DNA Microarray and SAGE Technologies

A. DNA Microarray Technology

DNA microarray was derived from an initial report in the mid-1970s (Burgess, 2001) that gene expression could be monitored by nucleic acid molecules attached to a solid support. There are two common types of DNA microarrays: oligonucleotide microarrays and cDNA microarray. Oligonucleotide microarrays [5, 8] use photolithography or ink-jet technology to synthesize oligonucleotides in situ, on silicon wafers. High-density-oligonucleotide microarrays from Affymetrix, http://www.affymetrix.com, are examples of the use of photolithography technology. Similarly, procedures developed by Rosetta Inpharmatics, http://www.rii.com, and licensed to Agilent Technologies represents an example of the use of ink-jet technology. Alternatively, presynthesized oligonucleotides can be printed onto glass slides. Methods based on synthetic oligonucleotides offer the advantage that sequence information alone is sufficient to generate the DNA to be arrayed; cDNAs do not have to be produced. Affymetrix (Santa

Clara, CA, USA) has pioneered the use of this form of array production with the development of the GeneChip. Their newest GeneChip, Human Genome U133 Plus 2.0 Array, contains more than 54,000 probe sets representing approximately 39,000 well-characterized human genes on a single microarray. The GeneChip Mouse Genome 430 2.0 Array covers the mouse genome representing more than 34,000 well-substantiated mouse genes, and Rat Genome 230 2.0 Array can analyze the expression level of more than 28,000 well-substantiated rat genes. These newly empowered GeneChips will facilitate the gene expression profiling of human diseases in animal models. A typical oligonucleotide microarray operational schema is illustrated in Figure 1A.

The cDNA microarray [6], another type of DNA microarray, is created using a precise *xyz* robot that is programmed to spot cDNA samples onto a solid substrate, usually a glass microscope slide, in a high-density pattern. cDNA arrays are also produced on nylon membranes but with spots of larger size in a lower-density pattern. Patrick Brown's laboratory at Stanford University created the first *xyz* arrayer, and instructions on how to build an arrayer can be found on their Web site (http://cmgm.stanford.edu/pbrown/, Table I). Many companies now produce cDNA microarrayers commercially. These machines differ primarily in the way the spot is placed on the substrate. Spotted arrays allow a greater degree of flexibility in the choice of arrayed elements, particularly for the preparation of smaller, customized samples. Accordingly, cDNA gridded arrays have been the technique most frequently used. With prices for oligonucleotide synthesis becoming more reasonable for large-scale studies, spotted long-oligonucleotide arrays could be a viable alternative to full-length cDNA array. A typical cDNA microarray operational schema is illustrated in Figure 1B.

Data analysis is the most demanding and challenging part in the use of microarray tools. This stems from at least three reasons: (1) an unprecedented volume of data; (2) lack of complete functional annotations of all genes in humans, animals, and other organisms; and (3) the fact that no method, strategy, or software program is universally suited to reveal the same biologically meaningful knowledge in all data sets, although increasing number of software tools are available. Different biological insights may be uncovered by different approaches to analyze the same data set. Within the scope of this chapter, we only briefly describe basic steps and principles in microarray data analysis: data normalization, comparison analysis, and clustering analysis. The reader can refer to several Web sites listed in Table I to obtain in-depth information on the microarray technology and data analysis. Data normalization attempts to identify the biological information by removing the impact of non-biological influence on the data, and by correcting for systematic bias in expression data. Systematic bias can be caused by differences in mRNA amounts and labeling efficiencies, uneven hybridization, scanner malfunction, printing and tip problems, different concentrations of DNA on the arrays (reporter bias), and other microarray batch bias as well as experimenter-related issues. There are several techniques that widely used to normalize gene expression data such as total intensity normalization, linear regression, ratio statistics, and locally weighted scatter-plot smoothing (LOWESS) or locally weighted regression (LOESS) correction [9]. Two common approaches have been used for the comparison analysis to select the differentially expressed genes. Earlier studies used arbitrary cutoff values such as twofold increase or decrease after the data normalization without the justification of theoretical background. Newer approaches apply parametric statistical tests such as Student's *t*-test and ANOVA or nonparametric tests such as the Mann–Whitney *U* test or the Kruskal–Wallis test for every individual gene. One caveat for statistical analysis of microarray data is that the number of replicates is usually low because of the costs of microarray experiments, thereby

Table I A Selection of DNA Microarray, SAGE Technology, and Data Information on the Internet.

Web site	URL	Feature
DNA Microarray		
Stanford University	cmgm.stanford.edu/pbrown	The Brown lab's complete guide to microarraying for the molecular biologist
TREX	pga.tigr.org/software/	Microarray software, freely downloadable
Affymetrix	www.affymetrix.com	Oligonucleotide microarray technology, product, and data analysis guides
HOPGENE	www.hopkins-genomics.org/	Gene expression data on 10 different cardiopulmonary diseases and some software
PGA	pga.lbl.gov/resources/PGA/	NHLBI Program for Genomic Applications resources include microarray software and data collections
NCBI	www.ncbi.nlm.nih.gov/geo/	The Gene Expression Omnibus (GEO) is a public gene expression data repository
EMBI-EBI	www.ebi.ac.uk/ArrayExpress	Another public gene expression data repository
SAGE		
SAGEnet	www.sagenet.org	SAGE introduction, software, protocols, references, and links to other SAGE resources
SAGEmap	www.ncbi.nlm.nih.gov/SAGE	Online SAGE data analysis, tag to gene mapping, download CGAP SAGE data, submitting SAGE data to GEO
Genzyme Molecular	www.genzyme.com/sage/	SAGE technology and oncology applications information for commercial users

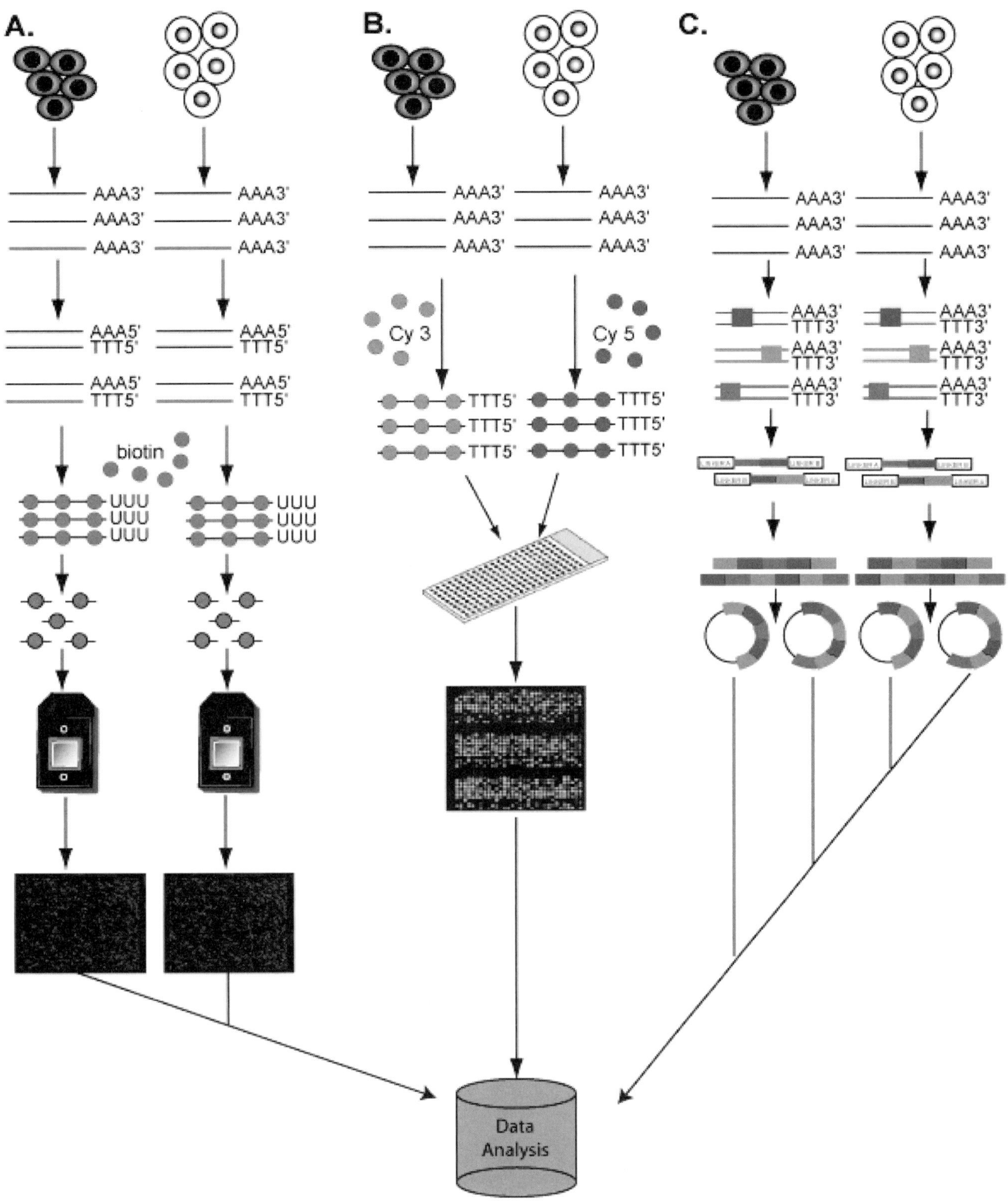

Figure 1 Schematics of DNA microarrays and SAGE. **(A)** Schematic of oligonucleotide microarray. RNA from different tissues or cell populations is used to generate double-stranded cDNA carrying a transcriptional start site for T7 DNA polymerase. It is followed by in vitro transcription, during which biotin-labeled nucleotides are incorporated into cRNA molecules. Each target sample is hybridized to a separate probe array, and target binding is detected by staining with a fluorescent dye coupled to streptavidin. Signal intensities of probe array element sets on different arrays are used to calculate relative mRNA abundance for the genes represented on the array. **(B)** Schematic of cDNA microarray. RNA from different tissues or cell populations is used to synthesize single-stranded cDNA in the presence of nucleotides labeled with two different fluorescent dyes (for example, Cy-3 and Cy-5). Both samples are mixed in a small volume of hybridization buffer and hybridized to the array surface, usually by stationary hybridization under a coverslip, resulting in competitive binding of differentially labeled cDNAs to the corresponding array elements. High-resolution confocal fluorescence scanning of the array with two different wavelengths corresponding to the dyes used provides relative signal intensities and ratios of mRNA abundance for the genes represented on the array. **(C)** Schematic of SAGE. Poly(A)$^{+}$RNA is isolated by an oligo-dT column and double-stranded cDNA is synthesized. It is followed by an anchoring enzyme digestion (e.g., *Nla*III), ligated to linkers, PCR amplification, concatenation, cloning into a sequencing vector, and DNA sequencing. The sequencing files are used to extract the numbers of SAGE tags in a digital format, which can reveal the difference in expression copies of interested genes between the diseased sample and a normal control. A relatively detailed description of the SAGE procedure is provided in the text. (see color insert)

leading to inaccurate estimate of variance. It is highly recommended that at minimum triplicate independent biological samples be analyzed for each condition. Clustering analysis [10] can identify the patterns of gene expression and grouping genes into expression classes, thus providing much greater insight into their potential biological relevance than simple lists of up- and downregulated genes. Clustering approaches can be divided into unsupervised learning and supervised learning. Unsupervised learning analyzes data without a priori input on genes or cases. It can be used for class discovery. The technique includes feature determination using principal component analysis, cluster determination using nearest neighbor clustering, self-organizing map, *k*-mean clustering, or one- and two-dimensional dendrograms, and network determination using Boolean, Bayesian, or relevance networks. Supervised learning analyzes data based on existing biological information about specific genes that are functionally related to "guide" the clustering algorithm. It can be used for class prediction. The technique includes single feature determination using nearest-neighbor or *t*-test, and multiple feature determination using decision trees, neural networks, or support vector machines. Clustering can also be classified into hierarchical clustering and nonhierarchical clustering. The former is used to visualize a set of samples or genes by organizing them into a phylogenetic tree, often referred as a dendrogram. The latter, such as *k*-mean clustering, is used to partition objects into different clusters without trying to specify the relationship between individual elements. The computer tools necessary to analyze the data are rapidly evolving and no clear consensus exists as to the best method for revealing patterns of gene expression, so complementary methods may be helpful in analyzing a single data set. Many studies have demonstrated that analysis of microarray data holds great promise for identifying early disease markers, classifying disease groups, revealing signal transduction pathways, and providing therapeutic insights. Several examples are provided in a later section.

SAGE Technology

SAGE was originally developed by Velculescu et al. [7]. Two basic principles underlie the SAGE methodology: (1) a short sequence tag (10 bp) in a defined position in the cDNA that contains sufficient information to uniquely identify a transcript; and (2) the concatenation of tags that allows for efficient sequence-based analysis of transcription. The schematic of the SAGE procedure is depicted in Figure 1C. Briefly, poly(A)$^+$ RNA is isolated by oligo-dT column chromatography. cDNA is synthesized from poly(A)$^+$ RNA using a primer of biotin-5′-T_{18}-3′. The cDNA is cleaved with an anchoring enzyme (e.g., *Nla*III), and the 3′-terminal cDNA fragments are bound to streptavidin-coated beads. An oligonucleotide linker containing recognition sites for a tagging enzyme (e.g., *BsmF*I) is linked to the bound cDNA. The tagging enzyme is a class II restriction endonuclease that cleaves the DNA at a constant number of bases 3′ to the recognition site. This results in the release of a short tag plus the linker from the beads after digestion with *BsmF*I. The 3′-ends of the released tags plus linkers are then blunted and ligated to one another to form 102 bp linked ditags. After PCR amplification of the 102-bp ditags, the linkers and tags are released by digestion with the anchoring enzyme. The 26-28mer tags are then gel purified, concatenated, and cloned into a sequence vector. Sequencing the concatemers enables individual tags to be identified and the abundance of the transcripts for a given cell line or tissue to be determined.

Although SAGE has become an extremely powerful technique for global analysis of gene expression, its requirement for a large amount of input mRNA (2.5 to 5.0 μg, which is equivalent to 250 to 500 μg total RNA) limits its utility. Several laboratories have attempted SAGE gene profiling using smaller amounts of RNA, but these attempts have all involved either PCR amplification of starting cDNA materials, such as SAGE-Lite and PCR-SAGE, or PCR reamplification of SAGE ditags generated by a first round of PCR amplification such as microSAGE and SAGE adaptation for downsized extracts. These additional PCR amplifications potentially introduce bias and compromise the quantitative aspects of the SAGE method. A recently modified method, miniSAGE, was successfully applied to profile gene expression of human fibroblasts from 1 μg total RNA without extra PCR amplification [11]. Invitrogen (www.invitrogen.com) developed a convenient I-SAGE kit, which packages all necessary quality controlled reagents together, enough for 5 SAGE library syntheses. These modifications have contributed to the broad application of SAGE. Saha et al. [12] developed a longSAGE method that generates 21-bp tags, instead of 14-bp, from the 3′ ends of transcripts. This method is similar to the original SAGE approach, but uses a different type IIS restriction endonuclease (*Mme*I) and incorporates other modifications to produce longer transcript tags. More than 75 percent of 21-bp tags, but not 14-bp tags, can be uniquely assigned to the human genome based on actual sequence information from approximately 16,000 known genes, although 14-bp tags do allow such assignment to ESTs and previously characterized mRNAs.

Analyses of SAGE data such as quality control, normalization, and cross-library comparison are relatively easier than those of DNA microarray data, because the DNA sequence file is the only common form of SAGE data output in a digital format. Because of the space limitation, no detailed description on SAGE data analysis is presented. Three major Web sites for SAGE technology, software, and data information are listed in Table I. SAGE 2000 software (Dr. K. Kinzler, Johns Hopkins University, Baltimore, MD, http://www.sage.org) can be used to extract SAGE tags, tabulate tag counts, and remove duplicate ditags, linker sequences and 1-bp variation sequences. It is freely available to academic users. SAGE commercial inquiries should be directed to Genzyme (Genzyme Corporate, Cambridge, MA, http://www.genzyme.com/sage/). Several other SAGE software products are also available, such as the Expression profile viewer (ExProView, Dr. M. Larsson, Royal Institute of Technology, Stockholm, Sweden, http://www.biochem.kth. se/exproview), eSAGE (Dr. E. H. Margulies, University of Michigan, Ann Arbor, MI), USAGE (Dr. A. H. van Kampen, Academic Medical Center, Amsterdam, The Netherlands, http://www.cmbi.kun.nl/usage/), POWER_SAGE (Dr. M. Man, PGRD, Ann Arbor, MI), and SAGE Genie (Dr. G. J. Riggins, Duke University, Durham, NC). Similar analysis strategies on gene expression patterns to pathways and networks of microarray data have been also applied to the analysis of SAGE data. The latest example illustrated the power of high-throughput gene expression profiling SAGE data. Buckhaults et al. [13] combined supervised and unsupervised clustering methods to generate a gene expression-based classification map of five genes identified from the training set of 21,321 unique SAGE transcript tags derived from 11 libraries. This map correctly categorized adenocarcinoma of the ovary, breast, colon, and pancreas. It may provide a practical approach to determine tumor type in cases of metastatic carcinoma of clinically unknown origin.

Comparison between DNA Microarray and SAGE Technologies

DNA microarray and SAGE are the two most popular platforms for the gene expression profiling; however, each technology has its pros and cons. Table II lists several major differences between microarray and SAGE technology. Because the preparation of microarrays requires prior knowledge of the sequence of the gene transcripts to be analyzed, an advantage of SAGE is that it can identify novel genes and can be used to analyze gene expression in organisms whose genomes are largely uncharacterized. This is a serious limitation for microarrays, even for organisms with completely sequenced genomes such as humans, because genome annotation and gene prediction remains technically challenging. There are additional advantages to SAGE. SAGE provides data in the digital format, whereas microarray data are analog. Although SAGE data are directly comparable, the differences in microarray formats and normalization methodologies make direct comparison of data sets between microarray platforms somewhat difficult. SAGE not only can accurately determine the absolute abundance of mRNAs but also can detect even slight differences in expression levels between samples. Microarray is only reliable in detecting genes whose expression differences are relatively large. Setup cost for SAGE is low. It does not require expensive instruments other than a DNA sequencer, which is available at most institutions. Microarrays need expensive robotic arrayers and scanners, presently available only in core facilities of major institutions. Microarrays, on the other hand, are relatively easy to use and more suitable for high-throughput applications. Also, SAGE requires several enzymatic manipulation steps, which are relatively taxing, especially for the novice—although Invitrogen has recently marketed a kit, the I-SAGE kit, which helps the new SAGE users. Finally, although the cost of DNA sequencing keeps falling, the relative high cost for sequencing a SAGE library is still a concern for potential SAGE users. Despite the great power of SAGE using 10-bp tags as a transcript identifier, a small percentage of tags are ambiguous. A single tag may identify multiple genes, or multiple tags may identify a single gene. This remains a challenge for identifying genes. A technique involving generation of longer cDNA fragments from SAGE tags for gene identification (GLGI), developed by Dr. S. M. Wang at the University of Chicago, has alleviated this problem by the simultaneous conversion of a large number of SAGE tags into corresponding 3′ cDNAs and the efficient identification of the correct genes for SAGE tags with multiple matches. Since microarrays use well-characterized immobilized sequences (cDNAs or oligonucleotides), identification of expressed genes is much easier.

The choice of gene expression technique is determined by the question being asked. Expression profiling of hundreds of disease samples is certainly more efficient using microarrays. SAGE, however, seems to be a better choice for the identification of new genes, especially in the analysis of previously uncharacterized organisms. SAGE also provides for more sensitive quantification of gene expression. Regular 14-bp tag SAGE is still useful for the quantification of mRNA level, whereas 21-bp tag SAGE (LongSAGE) is more suitable for the identification of new genes.

Table II Differences between SAGE and Microarray Technology.

Parameters	SAGE	Microarray
New gene identification	Yes	No
Data format	Digital	Analog
Sensitivity	Higher	Lower
Setup cost	Low	High
Operation procedure	Relatively cumbersome	Simpler
Application to multiple samples	Higher sequencing cost involved	Economical

Application of DNA Microarray and SAGE to the Gene Expression Profiling of Endothelial Cells

DNA microarray and SAGE have been employed to profile gene expression patterns in endothelial cells of different types with or without a perturbation. Selected publications on these applications are listed in Table III. Here we do not intend to review the subject in a comprehensive and exhaustive way; rather we highlight some successful examples that have emerged from their application into the gene expression profiling of endothelial cells.

Identifying Endothelial Cell–Specific Genes and EC Phenotypically Specific Genes

The endothelium plays a pivotal role in many physiological and pathological processes and is known to be an

Table III Publications on Human EC Using DNA Microarray (a) and SAGE Technologies (b).

Cell Source	Model	Major Findings	Ref.
a. Using Microarray Technology			
HAEC	+ Laminar shear stress for 24 hours	Genes for inflammation and proliferation↓ Tie2 and Flk-1, MMP-1	[14]
HCAEC	+ Cigarette smoke condensate	Matrix degradation enzyme↑ Proinflammatory cytokines↑	[15]
	+ Nicotine	Expression of PIP kinase, DG kinase NF-κB- and cAMP-responsive genes changed	[16]
	+ IL-1β + TNF-α	209 genes' expression changed	[17]
HMVEC	+ Hypoxia	Adenosine RA2B↑	[18]
HUVEC	+ LPS	Pro- and anti-inflammatory genes↑	[19]
	+ IL-1β	33 genes changed	[20]
	+ TNF-α	58 genes changed	
	+ TNF-α	ICAM-1, TRAF1↑	[21]
	+ Demethoxycurcumin	MMP-9↓	[22]
	+ VEGF	Many genes regulated	[23]
	+ Laminar shear stress or turbulent SS	Distinctive gene expression patterns associated with phenotypes	[24]
	+ Shear stress	KFL2↑	[25]
	+ Shear stress	CytoP450 1A1 &1B1↑ Endothelin-1, MCP-1↓	[26]
	+ Homocysteine	HMG-COA reductase↑	[27]
	+ E coli Braun Lipoprotein	Gene expression pattern like that induced by LPS	[28]
	+ LPS	MCP-1, IL-8↑	[29]
	+ TF Ets-1	Neurophilin-1↑	[30]
HEC	+ Native HDL3 Oxidized HDL3	MMP1, MMP14 ↑ PAI-1 and MMP-1↑	[31]
EPC	+ Atorvastatin	Cyclin↑, p27Kip1↓	[32]
Lymphatic EC	Compared to MVEC-S	Identification of lymphatic EC specific genes	[33]
Myometrial MVEC	+ VEGF	BDGF, oxytocin R, estrogen sulfotransferase↑	[34]
b. Using SAGE Technologies			
HUVEC	+ Transfected c-MYC	216 tags up, 260 tags down	[35]
HUVEC	+ Oxidized LDL	56 genes up or down ≥ 5-fold	[36]
HMVEC	+ VEGF	52,579 tags (−VEGF), 58,004 tags (+VEGF)	SAGEmap, NCBI
Brain AEC	Aneurysm	Expression of many genes changes	[37]
EC from colorectum	Tumor	46 genes up, 39 genes down in tumor	[38]

exceptionally active transcriptional site. Databases with the depository of gene expression data such as Gene Expression Omnibus (GEO, NCBI), SAGEmap (NCBI), and ArrayExpress (EMBI-EBI) have made possible the in silico cloning of novel endothelium-specific genes. Huminiecki and Bicknell [39] combined SAGE library subtraction with the UniGene/EST approach and reverse transcription polymerase chain reaction to examine endothelial cell–specific genes. Four novel genes were identified and labeled: endothelial cell–specific molecules (ECSM) 1 to 3 and magic roundabout (similar to the axon guidance protein roundabout). Magic roundabout is a new member of the roundabout receptor family that is endothelial specific and expressed at sites of active angiogenesis. A novel endothelial protein disulfide isomerase (EndoPDI) is highly expressed in ECs, and it has a protective effect only in endothelial cells exposed to hypoxia. The loss of EndoPDI expression under hypoxia caused a significant decrease in the secretion of adrenomedullin, endothelin-1, and CD105, molecules that protect endothelial cells from hypoxia-initiated apoptosis. The identification of an endothelial PDI further extends this increasing multigene family, and EndoPDI, unlike archetypal PDI, may be a molecule with which to target tumor endothelium. Ho et al. [40] also identified 64 pan-endothelial markers by combined database mining and microarray analysis. Similarly, St. Croix et al. [38] identified 93 panendothelial markers (PEMs) that were expressed at levels at least twentyfold higher in ECs in vivo compared with nonendothelial cells in culture by filtering approximately 1.8 million SAGE tags from a variety of cell lines. Identification of PEMs by this database search and analysis paved the way for their subsequent identification of tumor endothelial markers (see later section). Identification of endothelial cell–specific genes will contribute to advance our understanding of EC biology and to elucidate potential pharmaceutical targets.

There is an increasing appreciation that endothelial cells in different vascular beds differ from one another in structure and function. To examine whether differential gene expression may underlie the phenotypic heterogeneity among different vascular beds will provide a powerful foundation for designing vascular bed-specific therapies. Transcriptional profiling studies by Hirakawa et al. [41] revealed increased expression of several extracellular matrix and adhesion molecules including versican, collagens, laminin, and N-cadherin, and of the growth factor receptors endoglin and vascular endothelial growth factor receptor-1/Flt-1 in blood vascular ECs but not in lymphatic ECs, whereas cultured lymphatic ECs but not blood vascular ECs expressed the lymphatic markers Prox1 and LYVE-1 and formed LYVE-1-positive vascular tubes after implantation in vivo. These results demonstrate that some lineage-specific genes are only expressed during distinct developmental stages. This study identified new molecular markers for blood vascular and lymphatic endothelium with important implications for future studies of vascular development and function. In humans, endothelial barrier properties are not uniform throughout the pulmonary vasculature, with greater macromolecule diffusion in postcapillary venules compared to pulmonary arterioles in whole-lung models, whereas cultured microvascular endothelial cell exhibit tenfold higher barrier properties than macrovascular ECs measured by electrical resistance across monolayers. Other differences between them also exist. Gene expression profiling of ECs from 28 large vessels and 25 microvascular ECs by Chi et al. [42] found that pervasive differences in gene expression patterns distinguish the ECs of large vessels from microvascular ECs. The distinct gene expression patterns are likely to be related to the characteristic differences in physiological functions of these vascular channels. Large-vessel ECs differentially express several genes involved in the biosynthesis and remodeling of ECM, such as fibronectin, collagen 5α1, collagen 5α2, and osteonectin. These genes of high-level expression may be responsible for the relatively thick vascular wall of the large vessels. Microvascular ECs express higher levels of basement membrane protein genes, such as laminin, collagen 4α1, collagen 4α2, and collagen 4α-binding protein and ECM-interacting proteins, such as CD36, α1 integrin, α4 integrin, α9 integrin, and β4 integrin, perhaps related to the intimate association of microvascular ECs with the basement membrane and ECM. These may account for the higher barrier properties of the microvasculature. Chi et al. [42] also reported that ECs from different blood vessels and microvascular ECs from different tissues have distinct and characteristic gene expression profiles. Hey2, the human homolog of the zebrafish gene *gridlock*, was selectively expressed in arterial ECs, whereas several genes critical in the establishment of left/right asymmetry were expressed preferentially in venous ECs, suggesting coordination between vascular differentiation and body plan development. Tissue-specific expression patterns in different tissue microvascular ECs suggest they are distinct differentiated cell types that play roles in the local physiology of their respective organs and tissues.

Deciphering New Molecular Mechanisms from the "Signature" Gene Expression Profile

DNA microarray and SAGE have been employed in ECs to reveal previously unrecognized molecular mechanisms of pharmaceutical actions, disease risk factors, and biomechanical stimuli with physiological and pathophysiological relevance. Joyce et al. [43] defined new mechanisms of human protein C in modulating inflammation and apoptosis in HUVECs treated with the recombinant activated protein C based on the gene expression pattern using Hu6800 oligonucleotide arrays of the Affymetrix (Santa Clara, CA). Human protein C is a natural anticoagulant factor, and the recombinant activated form of the molecule, recombinant human activated protein C (rhAPC, Xigris), was approved by FDA for treatment of severe sepsis. Because of the pathophysiologic role of endothelial dysfunction in severe inflammatory disease and sepsis, the authors explored the possibility that rhAPC might directly modulate endothelial function, independent of its anticoagulant activity. Using broad transcriptional profiling, they demonstrated that rhAPC directly modulates patterns of endothelial cell gene expression clustering into anti-inflammatory and cell survival pathways. rhAPC directly suppressed expression of p50 and p52 NF-κB subunits, resulting in a functional decrease in NF-κB binding at target sites. Further, rhAPC blocked expression of downstream NF-κB regulated genes following tumor necrosis factor-α (TNF-α) induction, including dose-dependent suppression of cell adhesion expression and functional binding of intracellular adhesion molecule-1, vascular cell adhesion molecule-1, and E-selectin. Further, rhAPC modulated several genes in the endothelial apoptosis pathway, including the Bcl-2 homolog protein and inhibitor of apoptosis protein. These pathway changes resulted in the ability of rhAPC to inhibit the induction of apoptosis by the potent inducer staurosporine. This new mechanistic understanding of endothelial regulation and the modulation of TNF-α induced endothelial dysfunction creates a novel link between coagulation, inflammation, and cell death and provides insight into the molecular basis for the efficacy of APC in systemic inflammation and sepsis. This may also shed some light on the ever-growing list of APC's pleiotropic pharmacological effects and broader clinical applications.

A second example is the discovery of a new atherogenic mechanism of homocysteine by Li et al. [27]. Using a cDNA microarray, they uncovered an unexpected link between homocysteine and cholesterol dysregulation based on the finding of increased mRNA abundance of HMGCoA reductase and other genes involved in cholesterol synthesis such as caveolin-1 in HUVECs and renal microvascular endothelial cells treated with homocysteine. This effect was confirmed using quantitative reverse transcriptase–polymerase chain reaction. Actinomycin D studies revealed that homocysteine stabilized HMGCoA reductase. Expression of immunodetectable HMGCoA reductase and caveolin-1 in both HUVECs and renal microvascular endothelial cells was increased in homocysteine-treated cells. Additional biochemical analysis also confirmed the increased amount of total cellular cholesterol by homocysteine. It has been established that hyperhomocyst(e)inemia is an independent and graded risk factor for atherosclerosis, but the molecular link to the atherosclerotic process remains obscure. The new observation in this study by Li et al. [27] using cDNA microarray may provide a solid explanation for the observed proatherogenic effect of homocysteine.

The third example was to discover the distinctive gene expression pattern underlying the EC phenotypic plasticity in response to the different type of biomechanical stimuli. One of the striking features of vascular endothelium is its phenotypic plasticity in response to biomechanical stimuli. To explore the functional implications of this biomechanical paradigm of EC activation, Garcia-Cardena et al. [24] performed a high-throughput gene expression analysis of cultured human endothelial cells exposed to two well-defined biomechanical stimuli—a steady laminar shear stress (LSS) and a turbulent shear stress (TSS) of equivalent spatial and temporal average intensity. Comparison of the transcriptional activity of 11,397 unique genes revealed distinctive patterns of up- and downregulation associated with each type of stimulus. Cluster analyses of transcriptional profiling data were coupled with other molecular and cell-biological techniques to examine whether these global patterns of biomechanical activation are translated into distinct functional phenotypes. Confocal immunofluorescence microscopy of structural and contractile proteins confirmed and revealed the formation of a complex apical cytoskeleton in response to LSS but not TSS. Cell cycle analysis documented and related the different effects of laminar and turbulent shear stresses on cell proliferation to their differential regulation on some genes such as cyclin D1, cyclin B1, cullin 3, and retinoblastoma binding protein-2. The demonstration that different gene expression patterns or different molecular mechanisms are responsive to different hemodynamic stimuli has important implications for our understanding of the mechanisms of vascular homeostasis and pathogenesis. In our lab, we demonstrated that excessive cyclic stretch (CS, 18% elongation relevant levels of strain) applied to human pulmonary artery endothelial cells (HPAECs) directly causes vascular barrier disruption or enhances endothelial cell sensitivity to edemagenic agents compared to physiologically (5% elongation CS) relevant levels of strain [45]. We compared the gene expression profiling between 18 percent CS treated HPAECs and 5 percent CS treated HPAECs and found that a number of genes including a TGF-β superfamily, ras related rho, proteinase-activated receptor-2 with the graded expressions are compatible with the magnitude-dependent regulation of pulmonary endothelial cell barrier function by cyclic stretch. Comprehensive comparative analysis of these gene expression profiling data (Ye et al., manuscript in preparation) may reveal the distinct gene expression patterns, and thus the distinct molecular mechanisms, which may explain the magnitude-dependent regulation of pulmonary endothelial cell barrier function by cyclic stretch and contribute the mechanistic insight into ventilator-induced lung injury, an important clinical condition.

Discovery of Candidate Disease Markers

One of the goals of a large scale of gene expression profiling by DNA microarray and SAGE has been to identify disease markers. St. Croix et al. [38] reported in *Science* that they had obtained the most detailed sketch yet of how cancerous tumors secure the blood supplies that nourish their growth by comparing gene expression patterns of endothelial cells derived from blood vessels of normal and malignant colorectal tissues using SAGE. Of more than 170 transcripts predominantly expressed in the endothelium, 49 transcripts were expressed at substantially higher levels (more than tenfold) in tumor endothelium than in normal endothelium, and 33 transcripts were expressed at substantially lower levels in tumor than in normal endothelium. Among the top 25 transcripts with the highest expression in tumor, only seven genes (MMP-2, collagen type I α1, collagen type I α2 A, collagen type I α2 B, collagen type VI α3, nidogen, and Thy-1 cell surface antigen) were previously shown to be upregulated in angiogenic vessels, and the other 18 genes have not been previously characterized. These genes are characteristically expressed in tumors derived from several different tissue types including lung, brain, and metastatic liver cancers, demonstrating that tumor endothelium, in general, is different from the endothelium in surrounding normal tissue. Although most of the genes expressed differentially in tumor endothelium are also expressed during angiogenesis of corpus luteum formation and wound healing, suggesting that tumors may recruit vasculature by means of the same signals elaborated during other physiological or pathological processes, the fact that the expression of a tumor endothelial marker 8 was not detectable in developing corpus luteum suggests that there may be discrete differences between tumor angiogenesis and normal angiogenesis. Thus, these differentially expressed genes may represent potential tumor endothelial markers. Researchers had previously identified a few "markers" that differ between normal blood vessels and the presumably newly formed vessels of tumors, but nothing on the scale revealed by this Hopkins group using SAGE, one of the powerful and efficient genomics techniques. Even though researchers have a great deal of work to do to find out whether all the differentially expressed genes identified by St. Croix et al. [38] turn out to be true tumor endothelial markers and whether any of them will be suitable targets for drug therapy, this landmark discovery has already set off a flurry of work by other investigators with the eventual goal of identifying the tumor-specific markers and therapeutic targets.

Another example was to identify the candidate markers for intracranial aneurysm using SAGE. Approximately 6 percent of human beings harbor an unruptured intracranial aneurysm. Each year in the United States, more than 30,000 people suffer a ruptured intracranial aneurysm, resulting in subarachnoid hemorrhage. Despite the high incidence and catastrophic consequences of a ruptured intracranial aneurysm and the fact that there is considerable evidence that predisposition to intracranial aneurysm has a strong genetic component, very little is known with regard to the pathology and pathogenesis of this disease. To begin characterizing the molecular pathology of intracranial aneurysm, Peters et al. [37] used a global gene expression analysis approach (SAGE-Lite) in combination with a novel data-mining approach to perform a high-resolution transcript analysis of a single intracranial aneurysm, obtained from a 3-year-old girl. SAGE-Lite provided a detailed molecular snapshot of a single intracranial aneurysm. The study indicated that the aneurysmal dilation resulted in a highly dynamic cellular environment in which extensive wound healing and tissue/extracellular matrix remodeling were taking place. Specifically, the authors observed significant overexpression of genes encoding extracellular matrix components (e.g., collagen type III $\alpha1$, collagen type I $\alpha1$, collagen type I $\alpha2$, collagen type VI $\alpha1$, collagen type VI $\alpha2$, elastin) and genes involved in extracellular matrix turnover (TIMP-3, OSF-2), cell adhesion and antiadhesion (SPARC, hevin), cytokinesis (PNUTL2), and cell migration (tetraspanin-5). Interestingly, the investigators did the gene expression profiling from the whole lump of the aneurysmal tissue, but the list of most upregulated genes including collagen type III $\alpha1$, collagen type I $\alpha1$, collagen type I $\alpha2$, SPARC, and hevin jibed well with that of the report by St Croix et al. [38] that they were predominantly expressed in endothelial cells and upregulated in tumor endothelium. Although this study represents analysis of only one individual, it did present a unique first insight into the molecular basis of aneurysmal disease and defined numerous candidate markers for future biochemical, physiological, and genetic studies of intracranial aneurysm.

Future Perspectives

Recent advances in genomic technology such as DNA microarray and SAGE and their applications into endothelial cells have begun to unravel some new mechanisms that underpin the endothelial basic biology and clinical phenotype. When considering the central role of endothelial cells in control of normal vascular function and in progression of vasculopathic disease states, endothelial cell researchers still face daunting challenges to leverage the complete knowledge of the normal development program, vascular bed–specific heterogeneity, and genetic and environmental cues in physiology and pathophysiology of endothelial cells. Although the focus of high-throughput discovery has progressed through the genome and the transcriptome and is now moving toward assessing the proteome, glycome, and metabolome, the last are more difficult problems and they are at a much earlier stage of development. Practically, DNA microarray or SAGE are the real high-throughput functional genomic technology available at the moment, which can be extensively used to examine vascular bed–specific transcriptomes of endothelial cells in normal and disease states to gain global perspective. The combination of gene profiling by DNA microarray or SAGE with genetic strategies will facilitate the intelligent and productive use of gene expression data in a genetic context. It may be possible to extract more information than that gained from conventional studies of a more descriptive or correlational nature. It is hoped that the combination of genome-wide expression profiling and genetic strategies will allow the eventual molecular delineation of site-specific anatomy of endothelial cells, leading to the site- and organ-specific delivery of therapeutics to endothelial cells.

Glossary

Bioinformatics: The merger of biotechnology and information technology with the goal of revealing new insights and principles in biology.

cDNA microarray: Probe cDNA (500 to 5,000 bases long) is immobilized to a solid surface such as glass using robot spotting and exposed to a set of targets either separately or in a mixture to allow massively parallel gene expression and gene discovery studies.

Gene expression: The process by which a gene's coded information is converted into the structures present and operating in the cell. Expressed genes include those that are transcribed into mRNA and then translated into protein and those that are transcribed into RNA but not translated into protein (e.g., transfer and ribosomal RNAs).

Genome: All the genetic material in the chromosomes of a particular organism; its size is generally given as its total number of base pairs.

Genomics: The study of genes and their function.

Oligonucleotide microarray: An array of oligonucleotide (20- to 80-mer oligos) or peptide nucleic acid (PNA) probes is synthesized either in situ (on-chip) or by conventional synthesis followed by on-chip immobilization. The array is exposed to labeled probe, hybridized, and scanned to allow the identity/abundance of thousands of gene sequences to be simultaneously determined.

Polymerase chain reaction (PCR): A method for amplifying a DNA base sequence using a heat-stable polymerase and two 20-base primers, one complementary to the (+) strand at one end of the sequence to be amplified and one complementary to the (−) strand at the other end. Because the newly synthesized DNA strands can subsequently serve as additional templates for the same primer sequences, successive rounds of primer annealing, strand elongation, and dissociation produce rapid and highly specific amplification of the desired sequence. PCR also can be used to detect the existence of the defined sequence in a DNA sample.

Probe: The known nucleic acid sequence tethered on a solid surface.

SAGE: Serial Analysis of Gene Expression. It is a powerful tool that allows the analysis of overall gene expression patterns with digital analysis. Because SAGE does not require a preexisting clone, it can be used to identify and quantitate new genes as well as known genes.

Target: The free nucleic acid sample whose identity/abundance is being detected.

Transcriptome: The full complement of activated genes, mRNAs, or transcripts in a particular tissue at a particular time.

References

1. Dudek, S. M., and Garcia, J. G. (2001). Cytoskeletal regulation of pulmonary vascular permeability. *J. Appl. Physiol.* **91**(4), 1487–1500. *This paper was a comprehensive review of how the balance between competing EC contractile forces and adhesive cell–cell and cell–matrix tethering forces regulate pulmonary vascular barrier function. It focused on the critical role of the endothelial cytoskeleton in*

integrating multiple aspects of pulmonary vascular permeability, which pointed to a fertile area for the development of clinically important barrier-modulating therapies.

2. Stevens, T., Rosenberg, R., Aird, W., Quertermous, T., Johnson, F. L., Garcia, J. G., Hebbel, R. P., Tuder, R. M., and Garfinkel, S. (2001). NHLBI workshop report: Endothelial cell phenotypes in heart, lung, and blood diseases. *Am. J. Physiol. Cell Physiol.* **281**(5), C1422–C1433. *This paper was an NHLBI workshop report. It included experts' bird's-eye views on the advance of endothelial cell physiology and pathophysiology as well as future goals and research opportunities in the field of endothelial cell research.*
3. Aird, W. C. (2003). Endothelial cell heterogeneity. *Crit. Care Med.* **31**(4 Suppl), S221–S230. *This paper extensively reviewed recent advances in the field of endothelial cell heterogeneity and applied this knowledge to understand site-specific vasculopathy, including acute lung injury.*
4. Guttmacher, A. E., and Collins, F. S. (2003). Welcome to the genomic era. *N. Engl. J. Med.* **349**(10), 996–998.
5. Lockhart, D. J., Dong, H., Byrne, M. C., Follettie, M. T., Gallo, M. V., Chee, M. S., Mittmann, M., Wang, C., Kobayashi, M., Horton, H., and Brown, E. L. (1996). Expression monitoring by hybridization to high-density oligonucleotide arrays. *Nat. Biotechnol.* **14**, 1675–1680.
6. Schena, M., Shalon, D., Davis, R. W., and Brown, P. O. (1995). Quantitative monitoring of gene expression patterns with a complementary DNA microarray. *Science* **270**, 467–470.
7. Velculescu, V. E., Zhang, L., Vogelstein, B., and Kinzler, K. W. (1995). Serial analysis of gene expression. *Science* **270**, 484–487.
8. Schulze, A., and Downward, J. (2001). Navigating gene expression using microarrays—a technology review. *Nat. Cell Biol.* **3**(8), E190–E195.
9. Quackenbush, J. (2001). Computational analysis of microarray data. *Nat. Rev. Genet.* **2**(6), 418–427.
10. Eisen, M. B., Spellman, P. T., Brown, P. O., and Botstein, D. (1998). Cluster analysis and display of genome-wide expression patterns. *Proc. Natl. Acad. Sci. U.S.A.* **95**(25), 14863–14868.
11. Ye, S. Q., Zhang, L. Q., Zheng, F., Virgil, D., and Kwiterovich, P. O. (2000). miniSAGE: Gene expression profiling using serial analysis of gene expression from 1 microgram total RNA. *Anal. Biochem.* **287**, 144–152.
12. Saha, S., Sparks, A. B., Rago, C., Akmaev, V., Wang, C. J., Vogelstein, B., Kinzler, K. W., and Velculescu, V. E. (2002). Using the transcriptome to annotate the genome. *Nat. Biotechnol.* **20**(5), 508–512.
13. Buckhaults, P., Zhang, Z., Chen, Y. C., Wang, T. L., St. Croix, B., Saha, S., Bardelli, A., Morin, P. J., Polyak, K., Hruban, R. H., Velculescu, V. E., and Shih, IeM. (2003). Identifying tumor origin using a gene expression-based classification map. *Cancer Res.* **63**(14), 4144–4149.
14. Chen, B. P., Li, Y. S., Zhao, Y., Chen, K. D., Li, S., Lao, J., Yuan, S., Shyy, J. Y., and Chien, S. (2001). DNA microarray analysis of gene expression in endothelial cells in response to 24-h shear stress. *Physiol. Genomics.* **7**(1), 55–63.
15. Nordskog, B. K., Blixt, A. D., Morgan, W. T., Fields, W. R., and Hellmann, G. M. (2003). Matrix-degrading and pro-inflammatory changes in human vascular endothelial cells exposed to cigarette smoke condensate. *Cardiovasc. Toxicol.* **3**(2), 101–117.
16. Zhang, S., Day, I. N., and Ye, S. (2001). Microarray analysis of nicotine-induced changes in gene expression in endothelial cells. *Physiol Genomics.* **5**(4), 187–192.
17. Bandman, O., Coleman, R. T., Loring, J. F., Seilhamer, J. J., and Cocks, B. G. (2002). Complexity of inflammatory responses in endothelial cells and vascular smooth muscle cells determined by microarray analysis. *Ann. NY Acad. Sci.* **975**, 77–90.
18. Eltzschig, H. K., Ibla, J. C., Furuta, G. T., Leonard, M. O., Jacobson, K. A., Enjyoji, K., Robson, S. C., and Colgan, S. P. (2003). Coordinated adenine nucleotide phosphohydrolysis and nucleoside signaling in posthypoxic endothelium: role of ectonucleotidases and adenosine A2B receptors. *J. Exp. Med.* **198**(5), 783–796.
19. Ahn, S. K., Choe, T. B., and Kwon, T. J. (2003). The gene expression profile of human umbilical vein endothelial cells stimulated with lipopolysaccharide using cDNA microarray analysis. *Int. J. Mol. Med.* **12**(2), 231–236.
20. Zhao, B., Stavchansky, S. A., Bowden, R. A., and Bowman, P. D. (2003). Effect of interleukin-1beta and tumor necrosis factor-alpha on gene expression in human endothelial cells. *Am. J. Physiol. Cell Physiol.* **284**(6), C1577–C1583.
21. Murakami, T., Mataki, C., Nagao, C., Umetani, M., Wada, Y., Ishii, M., Tsutsumi, S., Kohro, T., Saiura, A., Aburatani, H., Hamakubo, T., and Kodama, T. (2000). The gene expression profile of human umbilical vein endothelial cells stimulated by tumor necrosis factor alpha using DNA microarray analysis. *J. Atheroscler. Thromb.* **7**(1), 39–44.
22. Kim, J. H., Shim, J. S., Lee, S. K., Kim, K. W., Rha, S. Y., Chung, H. C., and Kwon, H. J. (2002). Microarray-based analysis of anti-angiogenic activity of demethoxycurcumin on human umbilical vein endothelial cells: crucial involvement of the down-regulation of matrix metalloproteinase. *Jpn. J. Cancer Res.* **93**(12), 1378–1385.
23. Abe, M., and Sato, Y. (2001). cDNA microarray analysis of the gene expression profile of VEGF-activated human umbilical vein endothelial cells. *Angiogenesis.* **4**(4), 289–298.
24. Garcia-Cardena, G., Comander, J., Anderson, K. R., Blackman, B. R., Gimbrone, M. A., Jr. (2001). Biomechanical activation of vascular endothelium as a determinant of its functional phenotype. *Proc. Natl. Acad. Sci. U.S.A.* **98**(8), 4478–4485.
25. Dekker, R. J., van Soest, S., Fontijn, R. D., Salamanca, S., de Groot, P. G., VanBavel, E., Pannekoek, H., and Horrevoets, A. J. (2002). Prolonged fluid shear stress induces a distinct set of endothelial cell genes, most specifically lung Kruppel-like factor (KLF2). *Blood.* **100**(5), 1689–1698.
26. McCormick, S. M., Eskin, S. G., McIntire, L. V., Teng, C. L., Lu, C. M., Russell, C. G., and Chittur, K. K. (2001). DNA microarray reveals changes in gene expression of shear stressed human umbilical vein endothelial cells. *Proc. Natl. Acad. Sci. U.S.A.* **98**(16), 8955–8960.
27. Li, H., Lewis, A., Brodsky, S., Rieger, R., Iden, C., and Goligorsky, M. S. Homocysteine induces 3-hydroxy-3-methylglutaryl coenzyme a reductase in vascular endothelial cells: a mechanism for development of atherosclerosis? *Circulation.* **105**(9), 1037–1043.
28. Neilsen, P. O., Zimmerman, G. A., and McIntyre, T. M. (2001). Escherichia coli Braun lipoprotein induces a lipopolysaccharide-like endotoxic response from primary human endothelial cells. *J. Immunol.* **167**(9), 5231–5239.
29. Zhao, B., Bowden, R. A., Stavchansky, S. A., and Bowman, P. D. (2001). Human endothelial cell response to gram-negative lipopolysaccharide assessed with cDNA microarrays. *Am. J. Physiol. Cell Physiol.* **281**(5), C1587–C1595.
30. Teruyama, K., Abe, M., Nakano, T., Takahashi, S., Yamada, S., Sato, Y. (2001). Neurophilin-1 is a downstream target of transcription factor Ets-1 in human umbilical vein endothelial cells. *FEBS Lett.* **504**(1–2), 1–4.
31. Norata, G. D., Pellegatta, F., Hamsten, A., Catapano, A. L., and Eriksson, P. (2003). Effects of HDL3 on the expression of matrix-degrading proteases in human endothelial cells. *Int. J. Mol. Med.* **12**(1), 73–78.
32. Assmus, B., Urbich, C., Aicher, A., Hofmann, W. K., Haendeler, J., Rossig, L., Spyridopoulos, I., Zeiher, A. M., and Dimmeler, S. (2003). HMG-CoA reductase inhibitors reduce senescence and increase proliferation of endothelial progenitor cells via regulation of cell cycle regulatory genes. *Circ. Res.* **92**(9), 1049–1055.
33. Podgrabinska, S., Braun, P., Velasco, P., Kloos, B., Pepper, M. S., Skobe, M. (2003). Molecular characterization of lymphatic endothelial cells. *Proc. Natl. Acad. Sci. U.S.A.* **99**(25), 16069–16074.
34. Weston, G. C., Haviv, I., and Rogers, P. A. (2002). Microarray analysis of VEGF-responsive genes in myometrial endothelial cells. *Mol. Hum. Reprod.* **8**(9), 855–863.
35. Menssen, A., and Hermeking, H. (2002). Characterization of the c-MYC-regulated transcriptome by SAGE: identification and analysis of c-MYC target genes. *Proc. Natl. Acad. Sci. U.S.A.* **99**(9), 6274–6279.
36. de Waard, V., van den Berg, B. M., Veken, J., Schultz-Heienbrok, R., Pannekoek, H., and van Zonneveld, A. J. (1999). Serial analysis of

gene expression to assess the endothelial cell response to an atherogenic stimulus. *Gene.* **226**(1), 1–8.
37. Peters, D. G., Kassam, A. B., Feingold, E., Heidrich-O'Hare, E., Yonas, H., Ferrell, R. E., and Brufsky, A. (2001). Molecular anatomy of an intracranial aneurysm: coordinated expression of genes involved in wound healing and tissue remodeling. *Stroke.* **32**(4), 1036–1042.
38. St Croix, B., Rago, C., Velculescu, V., Traverso, G., Romans, K. E., Montgomery, E., Lal, A., Riggins, G. J., Lengauer, C., Vogelstein, B., and Kinzler, K. W. (2000). Genes expressed in human tumor endothelium. *Science.* **289**(5482), 1197–1202.
39. Huminiecki, L., and Bicknell, R. (2000). In silico cloning of novel endothelial-specific genes. *Genome Res.* **10**(11), 1796–1806.
40. Ho, M., Yang, E., Matcuk, G., Deng, D., Sampas, N., Tsalenko, A., Tabibiazar, R., Zhang, Y., Chen, M., Talbi, S., Ho, Y. D., Wang, J., Tsao, P. S., Ben-Dor, A., Yakhini, Z., Bruhn, L., and Quertermous, T. (2003). Identification of endothelial cell genes by combined database mining and microarray analysis. *Physiol. Genomics.* **13**(3), 249–262.
41. Hirakawa, S., Hong, Y. K., Harvey, N., Schacht, V., Matsuda, K., Libermann, T., and Detmar, M. (2003). Identification of vascular lineage-specific genes by transcriptional profiling of isolated blood vascular and lymphatic endothelial cells. *Am. J. Pathol.* **162**(2), 575–586.
42. Chi, J. T., Chang, H. Y., Haraldsen, G., Jahnsen, F. L., Troyanskaya, O. G., Chang, D. S., Wang, Z., Rockson, S. G., van de Rijn, M., Botstein, D., and Brown, P. O. (2003). Endothelial cell diversity revealed by global expression profiling. *Proc. Natl. Acad. Sci. USA* **100**(19), 10623–10628. *The authors determined the expression profile of 53 cultured ECs using DNA microarrays to explore EC specialization on a global scale. They found that ECs from different blood vessels and microvascular ECs from different tissues have distinct and characteristic gene expression profiles.*
43. Joyce, D. E., Gelbert, L., Ciaccia, A., DeHoff, B., and Grinnell, B. W. (2001). Gene expression profile of antithrombotic protein c defines new mechanisms modulating inflammation and apoptosis. *J. Biol. Chem.* **276**(14), 11199–11203.
44. Birukov, K. G., Jacobson, J. R., Flores, A. A., Ye, S. Q., Birukova, A. A., Verin, A. D., and Garcia, J. G. (2003). Magnitude-dependent regulation of pulmonary endothelial cell barrier function by cyclic stretch. *Am. J. Physiol. Lung Cell Mol. Physiol.* **285**(4), L785–L797.

Capsule Biography

Dr. Shui Qing Ye is a faculty member and the Director of the Gene Expression Profiling Core (http://162.129.178.254/ctrm/index.php) at the Center for Translational Respiratory Medicine, Johns Hopkins University School of Medicine. His current research focuses on the biomarkers of acute lung injury using genomic, biochemical, cell, and transgenic technologies.

Dr. Shwu-Fan Ma is a faculty member in the Center for Translational Respiratory Medicine, Johns Hopkins University School of Medicine. She leads the data analysis unit in the Gene Expression Profiling Core.

Dr. Dmitry Grigoryev is a research fellow in the Center for Translational Respiratory Medicine, Johns Hopkins University School of Medicine. He is an innovative data analyst in the Gene Expression Profiling Core.

Mr. Mohan Parigi is currently Manager of Information Technology, Division of Pulmonary and Critical Care Medicine, Johns Hopkins University. He leads initiatives in the division relating to infrastructure development, centralization of divisional databases, and integration of clinical databases with microarray expression data repositories.

Mr. William L. Shao is an experienced data analyst in the Division of Pulmonary and Critical Care Medicine, Johns Hopkins University. He developed the Web-based MIAME database system to store and retrieve microarray data.

SECTION B

Engineered Tissue

CHAPTER 158

Angiogenesis of Engineered Tissue

Craig Neville and Joseph P. Vacanti
Massachusetts General Hospital and Harvard Medical School

Tissue engineering is a multidisciplinary field that utilizes principles of engineering and biological sciences for the development of biomedical structures that restore or augment tissue function. Tissue and organ transplantation is the current therapy for the end-stage failure or loss of many types of organs and tissues. Significant improvements in recent years in patient selection, surgical technique, histocompatibility typing, and immunological suppression have combined to greatly improve both patient and graft survival. However, a severe shortage of available organs limits the number of transplants performed, resulting in a patient waiting list that continues to grow. Scientists are striving to fabricate biosynthetic devices for transplant that are not dependent upon organ donation. A very significant aspect of this endeavor is to provide an adequate vascular system to provide nourishment and oxygen for these tissues.

Introduction

The major approach to tissue engineering utilizes cells, either from the patient or a donor, and combines and organizes them with a three-dimensional synthesized matrix or support composed of biodegradable polymers. The entire structure is implanted in the recipient, where the cells continue to proliferate and remodel to form new tissue. Simultaneously, the artificial structural components are gradually broken down, leaving a natural, functional organ.

A significant challenge in the biosynthesis of tissues and organs is the difficulty in supplying a vascular network to provide oxygen and nutrients, and remove metabolic byproducts. Survival of tissue engineered constructs generally rely upon diffusion from the media while growing in vitro, and the ingrowth of microvessels from the host after grafting. As it can take several days to weeks after implantation for neovascularization to occur by the host, detrimental necrosis often results in all but the smallest of engineered tissues. A variety of approaches are presently being evaluated in order to solve the problem of how to furnish an optimal vascular system to biosynthetic tissues and organs. The continued development of techniques to create functional small-diameter blood vessels within tissue-engineered organs is a formidable problem that requires further integration of the fields of biomaterial development and design, molecular and cellular physiology of the microvasculature, and clinical procedure.

Bioengineered Tissue

Patients with severe burns face life-threatening complications such as bodily fluid loss and infection, resulting in shock or sepsis. Treatment consists of removing burned skin and replacing it with temporary grafts consisting of natural or synthetic polymers, or skin grafts from cadavers (allografts). As the immune system of the patient frequently rejects these grafts, normally they are subsequently replaced with permanent autograft or split-thickness (containing the epidermis and a fraction of the dermal layer) skin autograft. Full-thickness autografts tend to give better results with less scarring and wound contractures than split-thickness, but can have a higher engraftment rejection rate. Nascent grafts are not initially vascularized and are therefore dependent upon diffusion for nutrients. Increasing the thickness of the graft increases the distance of the outermost cell layer from the remaining vasculature at the burn site, making it more prone to rejection.

If the patient does not have sufficient undamaged skin to serve as donor, autologously derived composite cultured skin substitute (CSS) is often the next best option. Current bioengineered skin substitutes are also not vascularized at the time of engraftment and so present difficulties similar to those with full- and split-thickness skin grafts. Small biopsies taken from the patient provide an autologous source of fibroblasts and keratinocytes that are readily expanded in vitro to large quantities. When seeded on biopolymer scaffolds, the cells provide permanent replacement of both dermal and epidermal layers of the skin. The two cell types can provide connective tissue and a barrier function; however, the absence of a vascular plexus leads to even further delayed vascularization relative to natural skin autograft. A major complication is that the cells frequently do not survive long enough for neovascularization of the dermis to occur, resulting in loss of the epidermal CSS graft (which is dependent upon nutrient diffusion from dermal capillaries) and increased clinical morbidity and mortality.

Significant improvements in care are resulting in patients with far more extensive burns surviving and requiring grafts. As the burned surface area of a patient increases, the available skin available for harvesting and grafting decreases. As a result, CSS has been playing an increasingly greater role in burn treatment. Several courses of investigation are presently being vigorously pursued for the further enhancement of CSS graft vascularization.

- *Reduction or delayed onset of apoptosis.* Much of the loss of grafted tissue has been attributed to programmed cell death induced by nutritional deprivation and ischemic stress. The forced expression of antiapoptotic factors such as Bcl2 in keratinocytes of CSS significantly increases successful engraftment rates in experimental systems. Temporarily blocking apoptosis for a few days is often sufficient to allow the neovasculature to form and begin supplying the graft.
- *Expression of proangiogenic factors.* Genetically modified keratinocytes engineered to express angiogenic cytokines such as vascular endothelial growth factor (VEGF) or basic fibroblast growth factor (bFGF) incorporated into CSS demonstrate enhanced vascularization in animal models. Overexpression of VEGF appears to accelerate early graft vascularization, enhancing engraftment and tissue development. Long-term grafts displayed similar densities of microcapillaries for both VEGF-expressing and control tissues.
- *In vitro construction of a capillary-like network.* Successful inosculation of the capillary plexus already present within allo- and autografts with the remaining underlying microvascular network of a burn patient's wound can appreciably decrease the time required for the dermis to acquire functional circulation. CSS contains no microvasculature, necessitating a more prolonged period for capillary formation in the dermal component. Present efforts of several laboratories are focused upon directing the in vitro formation of a capillary-like network within the CSS that can anastomose with existing subdermal vasculature after engraftment.

Seminal work by Black et al. showed that the inclusion of specific components of the extracellular matrix (ECM, composed of chondroitin sulfates, chitosan, and collagen biopolymers) instructed introduced human endothelial cells to form capillary-like structures [1]. When the ECM was introduced with human umbilical vein endothelial cells to the fibroblasts and keratinocytes normally used for the in vitro phase of CSS formation, a morphologically complex network of tubular endothelial cells containing basement membrane and intercellular junction formation was observed. Others have subsequently shown in animal models that these structures can differentiate into functional microvasculature within several days of transplantation.

The difficulties encountered in creating vascularized tissue-engineered skin are greatly magnified when dealing with thicker, more complex tissues and organs. As a result, progress in bioengineering organs such as liver, kidney, and even glands is more preliminary. Neovascularization arising from the host is protracted for such tissues and so is unlikely to be able to serve the immediate needs of a new graft. One promising method developed by Nor et al. [2] seeks to initially provide a vascular network within a highly porous matrix, and then allow infiltration of parenchymal cells. When human endothelial cells seeded onto a three-dimensional biopolymer scaffolding were implanted subcutaneously in an immunodeficient mouse, they spontaneously self-organized into tubular structures. Although initially the structures were nonfunctional, within a week they began to anastomose with host vessels and transport murine blood cells. Within the areas of anastomosis, chimeric blood vessels were discernible, composed of both human and murine cells. Initial invasion of host smooth muscle cells became apparent within 2 weeks of implantation, leading to rapid formation of a perivascular layer. This model system is likely to prove valuable in evaluating potential strategies for enhancement of anastomosis between engineered tissues and host.

Delivery of Angiogenic Factors

Many investigators are developing methods to induce neovascularization of tissue-engineered structures with growth factors. Bolus injection of angiogenesis regulatory proteins is not practical in clinical applications as they can have deleterious systemic effects, and dosage is difficult to control. Furthermore, the proteins have short half-lives in vivo, often on the order of minutes. Instead, a variety of mechanisms are being developed to allow targeted sustained release of angiogenic factors.

A variety of matrices and polymers have been used to encapsulate synthetic or recombinant-produced polypeptides. Inert materials such as expanded polytetrafluoroethylene (ePTFE, Gore-Tex) were originally used as implants to provide growth factors in vivo, although efficacy was low because of the short period of time that the protein could be delivered. Inclusion of the proteins within a relatively long-lasting biodegradable matrix was found to protect the growth factor from proteases and binding proteins, allowing it to leach out more gradually over a period spanning days or weeks. A derivative of chitin (hydroxypropylchitosan acetate), an oligosaccharide derived from the exoskeletons of marine crustaceans, can be incorporated into Gore-Tex disks as a delivery vehicle of growth factors such as bFGF. The negatively charged sulfate groups of hydroxypropylchitosan acetate bind to the basic region of growth factors and serves as a biodegradable scaffolding to provide more prolonged sustained release of proteins relative to the original Gore-Tex implants. Other alternatives that are completely resorbable have been developed as well. For example, beads of heparin-Sepharose have been used for well over a decade to control release of bFGF to support angiogenesis. Another biodegradable material, poly(lactide-*co*-glycolide) acid (PLGA), has also been used to synthesize microspheres to control release of angiogenic factors. More recently, PLGA has begun to be used as part of a porous scaffold in engineered tissue to circumvent point-source administration of growth factors provided by disks and beads. Pharmacological levels can be maintained by this method for more than 2 weeks by incorporation of VEGF in a porous biodegradable scaffold of PLGA. Synergism between multiple growth factors both in the formation and maturation of nascent microvessels is just beginning to be explored.

An alternative to promoting angiogenesis in engineered tissue by providing growth factors in the form of polypeptides is to introduce genetically modified cells to the constructs. Tissues arising from these cells synthesize and secrete the angiogenic factors. Genes encoding the factors are commonly introduced by recombinant viruses, owing to their much greater efficiency of entering primary cells than naked plasmid DNA. Prolonged expression (longer than a few weeks) of angiogenic promoting genes is not normally desirable; although retrovirus is most commonly used as a vector, several other viruses that provide high initial expression levels and are subsequently attenuated (e.g., adenovirus, herpes simplex virus amplicon) have proven useful. Alternatively, recombinant genes whose expression can be regulated by systemic application of a synthetic compound such as doxycycline have also been used. Cells without proliferation potential have also been used as growth factor delivery vehicles, for they are not replaced and eventually senesce.

Fabrication of Microvascular Structures

A. Microelectromechanical Systems

Relying upon ingrowth of capillaries into large implanted tissue-engineered grafts, particularly vital organs with thick or complex structures such as the liver, kidney, and heart, is at present impractical because of the extensive time periods required. Other techniques are clearly needed to augment de novo vascularization from the host. Several highly novel strategies are being developed to preform a permanent microvasculature to sustain the parenchymal cells of neo-organs and tissues.

Microelectromechanical systems (MEMS) is a rapidly emerging field derived from fabrication technologies used for semiconductor devices. MEMS is currently being used to address problems in biomedical engineering. Standard MEMS technology allows synthesis of mechanical structures at scales ranging from greater than a centimeter to smaller than a micrometer. One application has been to create patterns that direct the assembly of cells into engineered tissue at the micrometer scale. A branching vascular system was modeled, based on normal blood flow rheology, accounting for relative hematocrit, flow rate, and shear stress. Micromachining allows designated patterns to be imprinted into silicon wafers using a series of subtractive etching methods. The microfluidics design was done in two dimensions to accommodate MEMS constraints, with a strategy to create three-dimensional tissue structures by folding, stacking, or rolling layers. The vascular channels begin with a single inlet with a diameter of 500 μm, and repeatedly bifurcates to generate capillary-scale channels of 10 μm in diameter. This pattern was then etched upon silicon and Pyrex wafers utilizing MEMS technology (Figure 1). Although endothelial cells were originally directly cultured upon the wafers to line the channels and lifted off as sheets, current fabrication processes use patterned silicon wafers to replica mold polymers, such as polydimethylsiloxane (PDMS), which are then used to coculture endothelial lined channels and parenchymal cells such as hepatocytes.

Nylon Strand Supports

Another innovative technique being developed is the creation of perfused microvessels in vitro that can then be incorporated into artificial tissue [3]. Laminin-coated nylon strands suspended between two pieces of polyester tubing

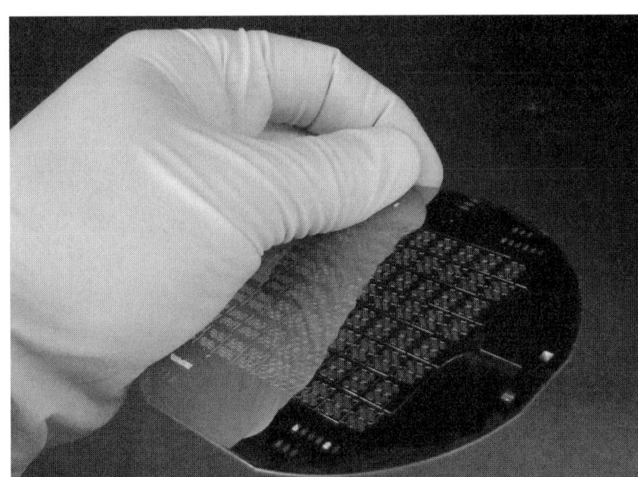

Figure 1 The branching vascular network pattern was modeled to reproduce microvascular blood flow rates and shear stress. The 2D pattern was micromachined on a 100-mm diameter semiconductor-grade silicon wafer using standard MEMS photoresist lithography technology. The patterned surfaces of such wafers are transferred to elastomeric polymers such as PDMS to provide guidance for endothelial cell growth and differentiation. When stacked, the layers serve as a 3D scaffold for growing engineered tissue.

were seeded with vascular smooth muscle cells and cultured. Uninterrupted monolayers formed within 24 hours, and further proliferation created multilayered sleeves. When the nylon lines were extracted from the sleeves, the resulting artificial microvessels were able to be perfused for extended periods of time (at least several days) without any measurable loss of fluid from within the system. Present investigations focus upon the inclusion of an endothelial layer to modulate the effects of shear stress, and eventual incorporation of the vessels into engineered tissue.

At present, it remains difficult to engineer large three-dimensional tissues with an adequate vascular supply. To date, all engineered tissue used for clinical purposes has relied upon neovascularization from the recipient to provide mass transfer of oxygen and nutrients. However, emerging technologies, combined with advances in the fields of biomaterials and cell biology, hold great promise to aid in the fabrication of tissue and organs specifically designed for human therapy.

Glossary

Allogenic: Materials originate from a genetically different individual of the same species.

Anastomosis: Connection of separate or severed tubular hollow structures to form a continuous channel.

Apoptosis: Programmed cell death initiated by the nuclei of normally functioning cells in response to age or stress.

Autogenic: Materials originate with the individual to which graft is applied.

Biomaterial: A biocompatible material used to construct artificial organs or prostheses and replace natural body tissues.

Inosculate: To unite (blood vessels) by generating small openings.

Scaffold: The complex mixture of proteins comprising the extracellular matrix that both directs cellular morphogenesis and serves as a supporting structure of the tissue or organ.

References

1. Black et al. (1998). *FASEB J.* **12**, 1331–1340.
2. Nor, J. E., Peters, M. C., Christensen, J. B., Sutorik, M. M., Linn, S., Khan, M. K., Addison, C. L., Mooney, D. J., and Polverini, P. J. (2001). Engineering and characterization of functional human microvessels in immunodeficient mice. *Lab. Invest.* **81**, 453–463. *Elegant demonstration of the innate ability of endothelial cells to form 3D microvascular structures in vivo. Provides model system to examine the molecular and physiologic regulators of angiogenesis.*
3. Neumann et al. (2003). *Microvasc. Res.* **66**, 59–67.

Further Reading

Germain, L., Remy-Zolghadri, M., and Auger, F. (2000). Tissue engineering of the vascular system: From capillaries to larger blood vessels. *Med. Biol. Eng. Comput.* **38**, 232–240. *Demonstrated endothelialized human tissue-engineered skin with a network of capillary-like tubes.*

Kaihara, S., Borenstein, J., Koka, R., Lalan, S., Ochoa, E. R., Ravens, M., Pien, H., Cunningham, B., and Vacanti, J. P. (2000). Silicon micromachining to tissue engineer branched vascular channels for liver fabrication. *Tissue Eng.* **6**, 105–117.

Langer, R. S., and Vacanti, J. P. (1999). Tissue engineering: The challenges ahead. *Sci. Am.* **280**, 86–89. *Describes present challenges and future benefits of tissue engineering. Issue focuses on various aspects of bioengineering.*

Moldovan, N. I., and Ferrari, M. (2002). Prospects for microtechnology and nanotechnology in bioengineering of replacement microvessels. *Arch. Pathol. Lab. Med.* **126**, 320–324.

Nomi, M., Atala, A., Coppi, P. D., and Soker, S. (2002). Principals of neovascularization for tissue engineering. *Mol. Aspects Med.* **23**, 463–483.

Sipe, J. D., Kelley, C. A., and McNichol, L. A., eds. (2002). Reparative medicine: Growing tissues and organs. *Ann. N. Y. Acad. Sci.* **261**. *Volume contains the proceedings of a symposium sponsored by the National Institutes of Health Bioengineering Consortium on June 25–26, 2001. Several contributions focus upon principles guiding vascular assembly in engineered tissue.*

Supp, D. M., and Boyce, S. T. (2002). Overexpression of vascular endothelial growth factor accelerates early vascularization and improves healing of genetically modified cultured skin substitutes. *J. Burn Care Rehabil.* **23**, 10–20.

Capsule Biography

Dr. Craig Neville is an Instructor at Harvard Medical School and Massachusetts General Hospital.

Dr. Joseph P. Vacanti is John Homans Professor of Surgery at Massachusetts General Hospital and Harvard Medical School and Chief of Pediatric Surgery. He is cofounder of the Tissue Engineering Society and was elected to the Institute of Medicine in 2001.

CHAPTER 159

Engineering of Blood Vessels

Joe Tien, Andrew P. Golden, and Min D. Tang
Department of Biomedical Engineering, Boston University, Boston, Massachusetts

Introduction

Are human blood vessels replaceable? If so, what types of materials—synthetic or biological, living or nonliving—are best suited for these replacements? The field of vascular tissue engineering has emerged to answer these questions, with the goal of providing constructs that can repair or replace damaged vessels in vivo. This review summarizes the key concepts, requirements, and designs for the synthesis of artificial blood vessels.

In many vascular pathologies, the simplest replacement for a diseased vessel is an autologous transplant. For instance, in atherosclerosis, a narrowed coronary artery may be readily replaced by vessels of similar diameters, such as the internal mammary artery, radial artery, or saphenous vein; "bypass" surgeries of this sort are now routine, with half a million procedures performed annually. Healthy autologous vessels are not always available for transplantation, however, especially in patients whose vessels have already been harvested for previous operations. The clinical promise of tissue engineering lies in its potential to produce implantable vessels—either as single vessels or as entire networks—for the treatment of vascular obstruction or disease.

Whether an engineered vessel is suitable as a replacement depends on several factors: (1) *Strength:* Is the vessel sufficiently robust to withstand cyclic mechanical stresses generated by a pulsatile flow of blood through it? (2) *Biochemistry:* Does the vessel provide an antithrombotic, anti-inflammatory surface? Does the vessel respond appropriately to vasoactive compounds? (3) *Adaptability:* Does the vessel change its structure when faced with chronic changes in pressures or flows? Do the mechanical properties of the vessel adjust to approximate those of adjacent vessels? (4) *Organization:* Is the vessel histologically similar to a native one? The relative importance of these factors depends on the specific application in mind. Thus, it is not necessary to synthesize vessels that are histologically indistinguishable from native ones (although this goal is desirable) before using these vessels in the clinical setting.

Engineered Arteries

Synthetic, Large-Diameter Arteries (Greater Than 5 mm)

Modern efforts to synthesize artificial blood vessels began with the seminal work of the French surgeon Alexis Carrel in the early 1900s. In experiments with transplantation in small animals, Carrel demonstrated that many synthetic materials—rubber, glass, metals—could be used to form large-diameter arteries that were stable for several months in vivo. Successful incorporation of these foreign substances into the vascular tree required precise surgical techniques that Carrel pioneered; these techniques minimized trauma of adjacent vascular segments during suture and attempted to preserve laminar flow of blood downstream from the synthetic graft. With the development of advanced polymers in the 1930s and 1940s and the subsequent refinement of techniques to mold, weave, or extrude these polymers into three-dimensional structures, it became possible to create grafts whose mechanical properties resembled those of native vessels (in contrast to the rigid tubes used by Carrel). These flexible grafts reduced radial mismatches between grafted and host vessels during expansion and contraction in vivo. Large-diameter vascular grafts are now commercially available for clinical use in a variety of different chemistries, including polytetrafluoroethylene (Teflon), polyester (Dacron), polyurethane, and polyacrylate (Figure 1A).

The primary limitation of these grafts is their poor long-term patency (20 to 70 percent remain open in 2 years, depending on anatomical location) due to gradual intimal

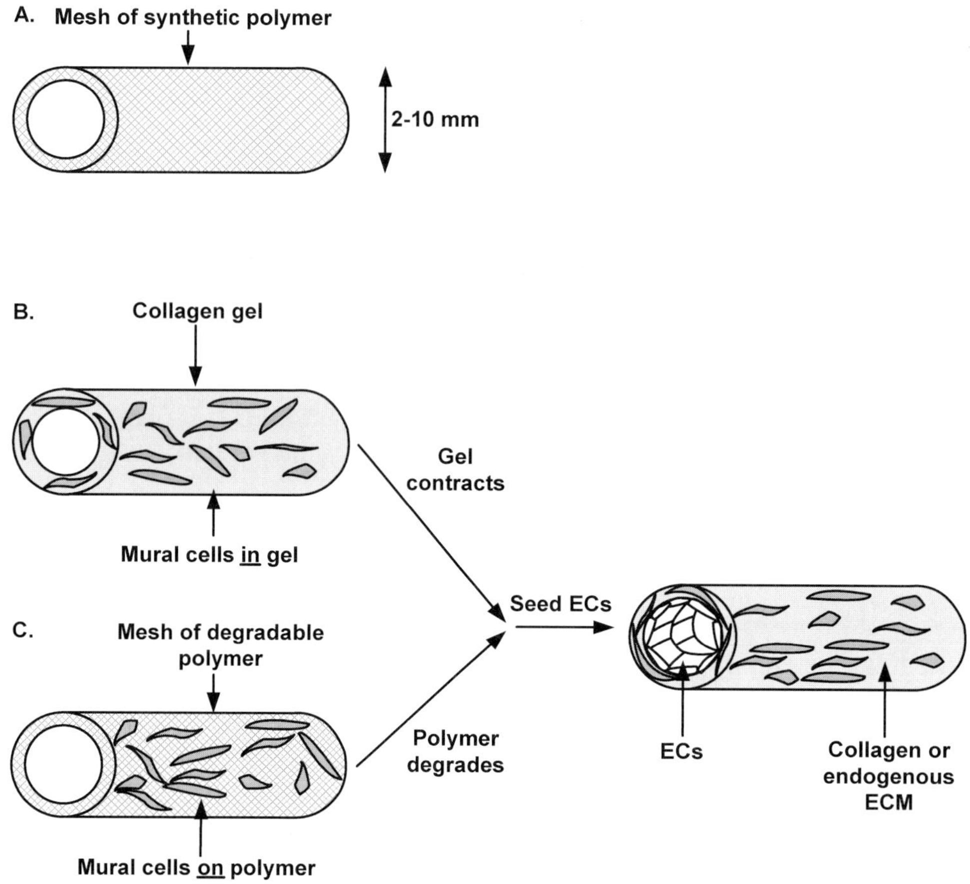

Figure 1 Processes for engineering arteries. **(A)** Commercially available large-diameter grafts consist of a woven polymer tube. **(B)** Collagen-based small-diameter grafts consist of a contracted sheath of collagen and mural cells that is lined by a monolayer of ECs. **(C)** Small-diameter grafts made from degradable polymers consist of tubes of mural cells and endogenously produced ECM, with a lining of ECs. Collagen-based and degradable polymer-based approaches may also be used to synthesize large-diameter arteries.

growth and occlusion. Several strategies to address this issue have been proposed; a common theme is the use of an endothelial lining to prevent the adherence of blood cells, adsorption of plasma proteins, and migration of mesenchymal cells to the surface of the synthetic polymers. Initial attempts to implement this idea relied on injection of a suspension of autologous endothelial cells (ECs) into the lumen of a graft during or before implantation of the vessel. Seeded ECs often detached from the graft, however, upon exposure to hemodynamic stresses mediated by the flow of blood. Recent approaches to enhance the adhesion of ECs to a graft have focused on derivatization of its surface with adhesive peptides (either in the form of adsorbed extracellular matrix proteins, or as small adhesive motifs such as Arg-Gly-Asp). Clinical trials of these EC-lined synthetic vessels in humans have demonstrated that the presence of an internal lining of ECs reduces the extent of intimal formation in implanted grafts.

A complementary strategy to enhance patency relies on the stimulation of ECs to migrate over the surface of a graft from junctions between host vessels and the graft, and thereby to form a complete monolayer. Postsurgical injection of chemotactic and chemokinetic agents such as vascular endothelial growth factor (VEGF) and acidic and basic fibroblast growth factors (aFGF and bFGF) appears to promote migration of ECs over some synthetic materials in animal models. It remains to be seen how much the incorporation of ECs—whether injected into lumens of grafts or attracted from host vessels—can improve the patency rates of synthetic large-diameter vessels in humans, and whether there are inherent limitations to the long-term performance of synthetic materials as blood vessels in vivo.

Biological, Small-Diameter Arteries (Less Than 5 mm)

In contrast to the clinically successful application of synthetic vessels for replacement of large arteries, the use of synthetic vessels as small-diameter arteries has not yet proven to be feasible. The intimal growth that slowly leads to possible occlusion in large-diameter grafts rapidly induces failure of small-diameter ones: Occlusion of small, synthetic vessels by thrombus occurs readily, and the mismatch in mechanical compliances between synthetic polymers and native tissues often results in intimal formation. These deficiencies have led to the idea that engineered vessels should ultimately be made of living, biologically active materials, which may provide signals to reduce the occurrence of thrombosis and intimal hyperplasia, and which may resemble native arteries in mechanical properties.

Collagen-Based Arteries

Some groups have attempted to synthesize purely biological vessels from gels, particularly collagen, with ECs and mural cells cultured on and in the gel, respectively (Figure 1B). In 1986, Weinberg and Bell demonstrated this design by gelling collagen, in which smooth muscle cells (SMCs) were suspended, around a tubular mesh [1]. Subsequent gelling of collagen and adventitial fibroblasts around the muscular layer and injection of ECs into the lumen of the tube generated a construct that segregated ECs, SMCs, and fibroblasts into compartments that grossly resembled the intima, media, and adventitia of arteries. The endothelial layer served as a selective barrier to large proteins and produced measurable amounts of prostacyclin, a vasodilator and potent inhibitor of platelet aggregation in vivo.

Engineered vessels made from gelled collagen primarily suffer from a lack of sufficient mechanical strength, due to abnormal cellular organization and low densities of collagen and cells. Although contraction of collagen by SMCs and fibroblasts helps initially to increase the burst strengths of engineered arteries, these strengths are still much lower than those of native vessels. Moreover, a gradual decrease in strength often follows the initial increase, because the collagen used in these vessels has a low density of cross-links, and thus is particularly sensitive to digestion by collagenases expressed by SMCs and fibroblasts. Improvements on the model of Weinberg and Bell have therefore focused on inactivation of collagenases (by transfection, or by addition of protease inhibitors), and on increasing the strength of the vessel (by cross-linking collagen with aldehydes or radiation). Although promising, improvements based on transfection or addition of small molecules run counter to the idea that engineered vessels should be composed solely of cells and proteins found in vivo.

To address these concerns, L'Heureux and others developed a process in 1998 to engineer vessels with high burst strength solely from human vascular cells and the extracellular matrix (ECM) produced by them [2]. This process coaxed mural cells to produce thick sheets of ECM, by supplementing culture media with antioxidants and metal salts that promoted cross-linking of deposited collagen. Winding of these sheets of cells and matrix around a cylindrical mandrel, removal of the mandrel, and seeding of ECs into the lumen of the resulting tube generated a construct with burst strengths greater than that of human saphenous vein. The superior performance of these vessels, compared to that of vessels synthesized from gelled collagen, may result from their higher cellular densities and greater numbers of endogenous cross-links in the ECM.

Arteries Made from Degradable Polymers

A second strategy to form biological vessels uses biodegradable polymers, as proposed by Langer and Vacanti in 1993 (Figure 1C) [3]. Here, the principle is to use a degradable polymer as a mechanical support for vascular cells. While the polymer slowly degrades, the seeded cells will produce ECM to replace the lost volume of the polymer; if carefully controlled, the rates of degradation and synthesis balance to eventually yield a construct composed solely of cells and endogenous ECM. Appropriate selection and placement of hydrolyzable bonds in these polymers enable the tuning of degradation times on the order of days to years. Because these scaffolds may be molded or woven into a variety of shapes, it may be possible to use degradable polymers to produce noncylindrical arteries (in contrast to the method of L'Heureux, which appears limited to cylindrical geometries).

In 1999, Niklason and others demonstrated this strategy by seeding SMCs onto a degradable tubular mesh in a bioreactor [4]. As in the work of L'Heureux, the culture media was supplemented to enhance the secretion of collagen by SMCs and the cross-linking of ECM. Upon degradation of the mesh, seeding of ECs into the lumen of the tubes formed a biological vessel. This construct contracted when exposed to the vasoconstrictor $PGF_{2\alpha}$, displayed a histologically appropriate organization, and remained patent 1 month after implantation in small animals. Efforts to extend this work to the synthesis of human vessels are currently focused on methods to obtain sufficient populations of autologous human vascular cells.

Maturation of Engineered Arteries

Whether synthesized from collagen or degradable scaffolds, engineered vessels rarely possess sufficient mechanical strength as is to withstand in vivo stresses. Many groups have examined the use of shear and transmural stresses transduced by lumenal flow to enhance the strength of these constructs. Shear stress acts on the endothelial lining to induce the release of growth factors that may promote the proliferation of SMCs, while transmural stretch (especially as a result of pulsatile flow) increases the synthesis of ECM by mural cells and encourages the circumferential alignment of SMCs. The strengths of matured, engineered vessels made of human cells can exceed 2,000 mmHg; by comparison, the burst strength of a human saphenous vein is approximately 1,500 mmHg, whereas that of immature constructs is on the order of 200 mmHg. To date, useful schedules of maturation have arisen mostly by trial and error. As the effects of mechanical stresses on vascular cells become better studied, rational design of processes to control the mechanical properties of engineered arteries will become possible.

Engineered Microvascular Networks

Whereas artificial arteries are used to treat dysfunction of single vessels, engineered microvessels are envisioned for the treatment of ischemia mediated by the degeneration or absence of entire microvascular networks. As with small-diameter arteries, engineered microvessels will require the synthesis of completely biological tissues to remain patent in vivo. In contrast to the manual preparation of constructs used for macroscale arteries, it is neither practical nor necessary to assemble microvessels by hand into an interconnected network. Two complementary strategies have been proposed for vascularization of a tissue (Figure 2): the addition of angiogenic growth factors to attract capillaries from a host into the desired tissue, and the organization of ECs in an implanted tissue into networks that anastomose with microvascular beds from the host.

Microvascular Networks Induced by Growth Factors

Many growth factors, such as VEGF, aFGF, and bFGF, induce angiogenesis in vivo; presentation of these factors in a concentration gradient results in directional growth of capillary sprouts toward the regions of high concentration. To take advantage of this natural biological process for the vascularization of tissues, many groups have tried to infuse an ischemic tissue or transplant with angiogenic factors (Figure 2A). The factors may be slowly released from micropellets of degradable polymers impregnated with the desired proteins, or locally produced by transfected cells (whether exogenously added or generated in situ from the host). Initial results have indicated that strategies for vascularization that rely on release of a single growth factor, such as VEGF, often do not yield fully functional microvascular beds. Instead, the vessels that form are usually highly permeable, poorly invested by mural cells, and prone to regression over time.

Because signals provided by mural cells play a crucial role in the maturation of vascular beds in vivo, current efforts have focused on the release of a combination of growth factors to promote angiogenesis and to recruit mural cells. In 2001, Mooney and others demonstrated that implantation of composites that sequentially release an angiogenic factor (VEGF) and a factor chemotactic and

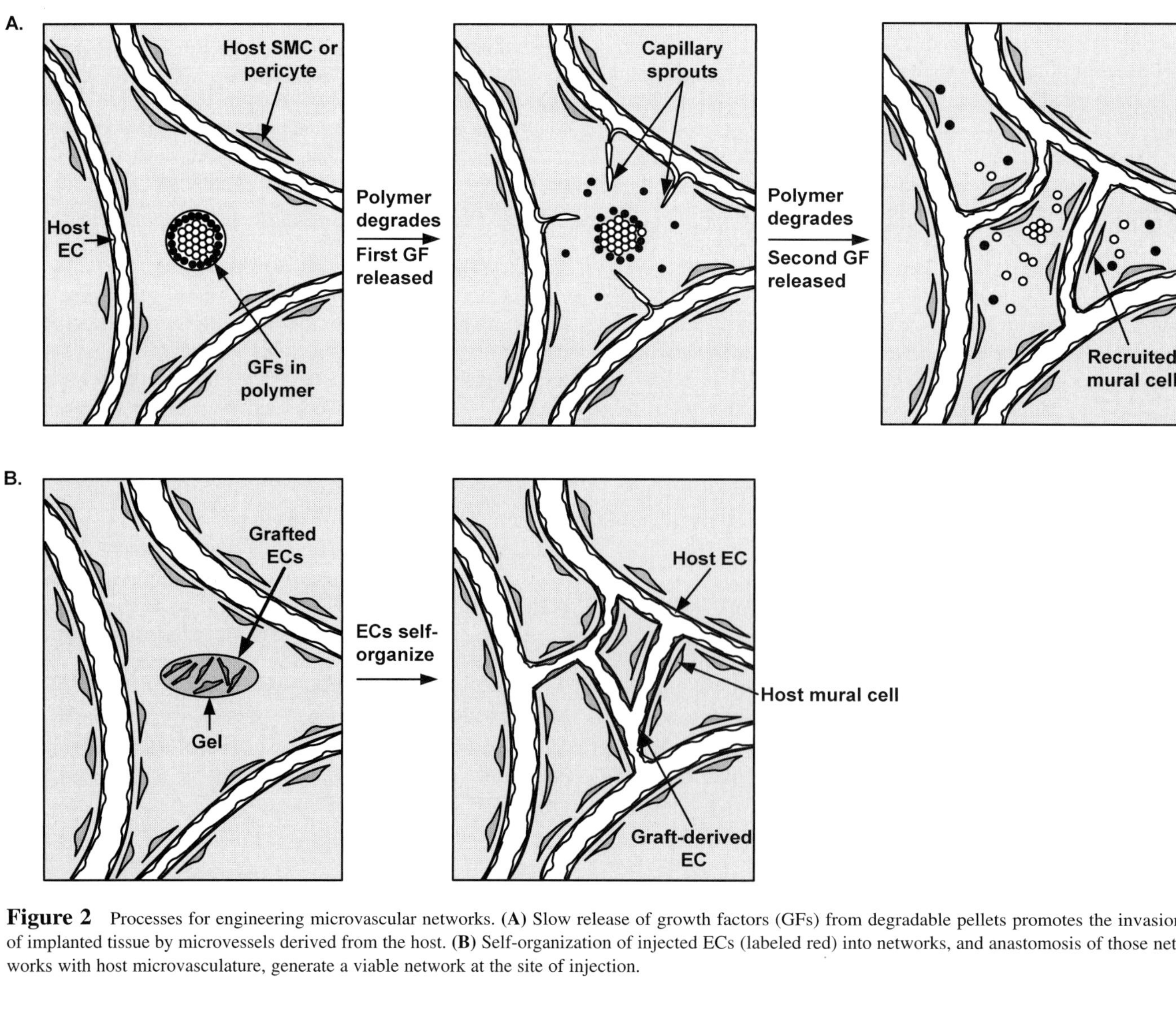

Figure 2 Processes for engineering microvascular networks. **(A)** Slow release of growth factors (GFs) from degradable pellets promotes the invasion of implanted tissue by microvessels derived from the host. **(B)** Self-organization of injected ECs (labeled red) into networks, and anastomosis of those networks with host microvasculature, generate a viable network at the site of injection.

mitogenic for mural precursors (platelet-derived growth factor; PDGF) induced the sprouting and maturation of capillary networks from the host into the grafted area [5]. These composites consisted of layers of a degradable polymer, one of which degraded more quickly than the other and hence allowed a gradual, two-stage release of growth factors. The final microvascular networks induced by these composites were well-invested by mural cells, exhibited appropriate permeability, and regressed less than vessels induced by a single growth factor. The idea of using the release of two or more factors to promote angiogenesis and mural recruitment has expanded to encompass many permutations (VEGF/PDGF, bFGF/PDGF, VEGF/bFGF/PDGF, and so on), each of which may display different potencies for inducing vascularization in different organs. Exhaustive searches for an optimal combination of growth factors and an effective schedule of release will undoubtedly yield improvements in the densities and stabilities of engineered vascular networks.

Microvascular Networks Formed from Grafted Endothelial Cells

It has been known since the early 1980s that cultures of ECs embedded in a collagen gel may organize to form short "cords" comprised of several cells linked in a chain. Some of these cords have open lumens and interconnect to form rudimentary networks. The application of these self-organized structures to microvascular engineering lies in their potential to anastomose with preexisting microvascular networks. In theory, implantation of exogenous ECs into an ischemic tissue may allow the organization of these ECs into

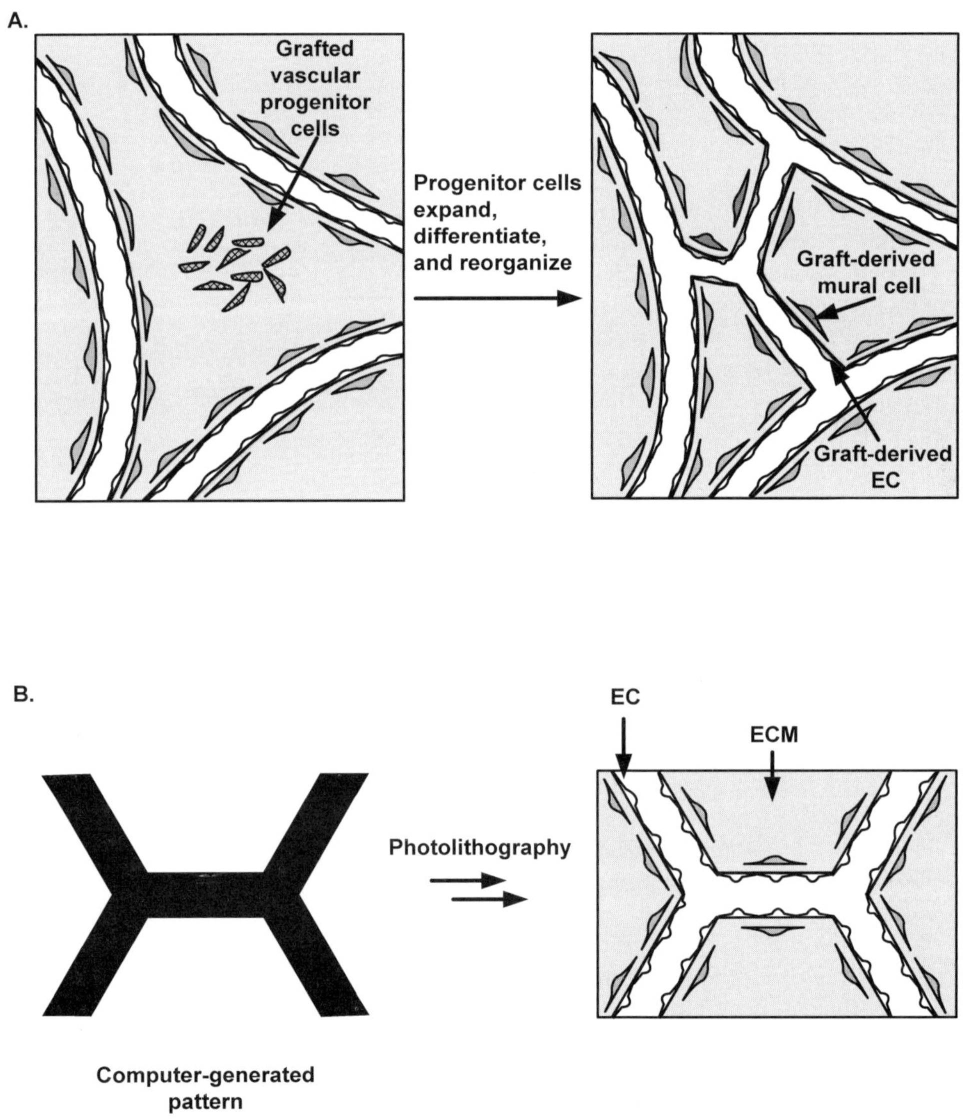

Figure 3 New strategies in vascular tissue engineering. **(A)** Vascular progenitor cells may differentiate in situ to form both ECs and mural cells, which subsequently organize into networks. Graft-derived cells are labeled red. **(B)** Lithographically formed patterns may guide the placement of vascular cells and enable the engineering of microvessels that have predefined shapes.

networks that connect to host vessels (Figure 2B). Because cellular self-organization occurs in parallel throughout the volume of the grafted tissue, this approach appears particularly well suited for the vascularization of thick tissues.

Initial attempts to engineer microvascular networks by implanting gels in which ECs were embedded did not succeed, primarily because the ECs did not survive long enough to anastomose with vessels from the host. Subsequent improvements have focused on the enhancement of cellular survival: For example, Schechner and others used ECs transfected with the antiapoptotic gene Bcl-2 to prolong survival of ECs [7], whereas Herron and others used immortalized telomerase-active ECs [7]. In both of these studies, grafts of transfected ECs survived for a longer period of time than those of untransfected cells did, and eventually organized into networks that anastomosed with host vasculature. Portions of the engineered networks were well invested by layers of mural cells and were stable for about 2 months.

As in the engineering of arteries, the use of transfected cells for microvascular engineering may compromise the biological activity of the formed tissue in unpredictable ways (e.g., whether telomerized ECs exhibit the same functional responses as normal ECs is debatable). Some groups have attempted to extend the viability of grafted ECs without the use of transfection by coinjecting mural precursors with ECs. In vitro, coculture of mural cells and ECs in collagen gels results in the formation of invested vascular networks; in 2004, Jain and others demonstrated that co-injection of mural cells and ECs in vivo results in networks that anastomose with host microvasculature and that are stable for longer than 1 year [8].

Future Directions

Vascular tissue engineering has evolved from the substitution of macroscale tubes for portions of the arterial tree, to the combined use of growth factors, woven polymers, purified populations of cells, and genetic manipulation to construct macro- and microvessels. It has resulted in the clinical application of large-diameter tubular grafts for repair of aortae and peripheral vessels, and in the development of small-diameter vessels that may someday be implanted in humans. Clinical trials are underway to test the effectiveness of angiogenic growth factors in inducing the formation of microvascular networks in ischemic tissues. Strategies based on the use of seeded cells for the engineering of arteries and microvessels appear promising and await the development of methods to culture sufficient numbers of autologous vascular cells before these therapies can be widely applied in humans.

Two recent developments may potentially revolutionize the engineering of blood vessels (Figure 3): First, in 1997, Isner and others discovered that human peripheral blood contains multipotent progenitor cells that can differentiate into cells that express EC- or SMC-specific markers [9]. This work has spawned the idea of using these progenitors to produce large numbers of autologous vascular cells in vitro for cell-based engineering, or perhaps, to allow injected progenitor cells to expand in vivo into invested networks. Second, in 2000, Vacanti demonstrated that photolithography, a technique widely used in the microelectronics industry to pattern metals, ceramics, and semiconductors, may be applied to construct polymeric structures that resemble microvascular networks in shape [10]. Lithographic tools may be able to provide micrometer-scale control over the placement of vascular cells in three dimensions, and thus enable the synthesis of microvessels of a selected topology. Further advances in these areas will expand the toolkit available to tissue engineers and should lead to the synthesis of custom-made vascular tissues for the treatment of human disease.

Glossary

Autologous: Derived from the same organism (e.g., an autologous transplant).

Biodegradable polymer: A polymer that degrades in physiological conditions into small molecules that are readily absorbed or secreted.

Patency: The ability of a vessel to support the flow of blood through it (usually in reference to large vessels).

Photolithography: A process that uses irradiation of light-sensitive materials through a selective mask to generate patterns with micrometer-scale resolution.

Tissue engineering: The synthesis of living structures that can repair or replace defective tissues.

References

1. Weinberg, C. B., and Bell, E. (1986). A blood vessel model constructed from collagen and cultured vascular cells. *Science* **231**, 397–400.
2. L'Heureux, N., Paquet, S., Labbe, R., Germain, L., and Auger, F. A. (1998). A completely biological tissue-engineered human blood vessel. *FASEB J.* **12**, 47–56.
3. Langer, R., and Vacanti, J. P. (1993). Tissue engineering. *Science* **260**, 920–926.
4. Niklason, L. E., Gao, J., Abbott, W. M., Hirschi, K. K., Houser, S., Marini, R., and Langer, R. (1999). Functional arteries grown in vitro. *Science* **284**, 489–493. *This study was the first to demonstrate the use of degradable polymers in vascular tissue engineering, as originally envisioned in the 1993 review by Langer and Vacanti.*
5. Richardson, T. P., Peters, M. C., Ennett, A. B., and Mooney, D. J. (2001). Polymeric system for dual growth factor delivery. *Nat. Biotechnol.* **19**, 1029–1034.
6. Schechner, J. S., Nath, A. K., Zheng, L., Kluger, M. S., Hughes, C. C., Sierra-Honigmann, M. R., Lorber, M. I., Tellides, G., Kashgarian, M., Bothwell, A. I., and Pober, J. S. (2000). In vivo formation of complex microvessels lined by human endothelial cells in an immunodeficient mouse. *Proc. Natl. Acad. Sci. USA* **97**, 9191–9196.
7. Yang, J., Nagavarapu, U., Relloma, K., Sjaastad, M. D., Moss, W. C., Passaniti, A., and Herron, G. S. (2001). Telomerized human microvasculature is functional in vivo. *Nat. Biotechnol.* **19**, 219–224.
8. Koike, N., Fukumura, D., Gralla, O., Au, P., Schechner, J. S., and Jain, R. K. (2004). Creation of long-lasting blood vessels. *Nature* **428**, 138–139.
9. Asahara, T., Murohara, T., Sullivan, A., Silver, M., van der Zee, R., Li, T., Witzenbichler, B., Schatteman, G., and Isner, J. M. (1997). Isolation

of putative progenitor endothelial cells for angiogenesis. *Science* **275**, 964–967.

10. Kaihara, S., Borenstein, J., Koka, R., Lalan, S., Ochoa, E. R., Ravens, M., Pien, H., Cunningham, B., and Vacanti, J. P. Silicon micromachining to tissue engineer branched vascular channels for liver fabrication. *Tissue Eng.* **6**, 105–117.

Further Reading

Nerem, R. M., and Seliktar, D. (2001). Vascular tissue engineering. *Annu. Rev. Biomed. Eng.* **3**, 225–243. *In this review, Nerem and Seliktar describe many important considerations in the construction and maturation of engineered arteries, with a particular emphasis on the use of lumenal flow to strengthen the walls of as-synthesized constructs.*

Yannas, I. V. (2000). Synthesis of organs: In vitro or in vivo? *Proc. Natl. Acad. Sci. USA* **97**, 9354–9356. *This review by Yannas outlines two complementary approaches for generation of functional microvascular networks in a tissue: recruitment of microvessels from a host, and anastomosis of graft-derived microvessels with those of the host.*

Capsule Biography

Dr. Tien, Mr. Golden, and Ms. Tang are members of the Department of Biomedical Engineering at Boston University. Their research focuses on the engineering of living tissues that have complex three-dimensional architectures. Their work is supported by the NIH and the Whitaker Foundation.

Chapter 160

Blood Substitutes' Efficacy Microvascular and Rheological Determinants

[1]Amy G. Tsai, [1]Pedro Cabrales, [2]Hiromi Sakai, and [1]Marcos Intaglietta
[1]*Department of Bioengineering, University of California, San Diego, La Jolla, California*
[2]*Department of Polymer Chemistry, Waseda University, Tokyo, Japan*

The development of a blood substitute, also called "artificial" blood, or more exactly an oxygen-carrying plasma expander (OCPE), is still a major goal of transfusion medicine, driven by blood shortages, problems associated with the transmission of disease by available blood, and the complex logistics of acquiring, analyzing, storing, distributing, and delivering the needed blood. To date blood is unequaled (or appears to be) in its capacity to restore circulatory volume; however, it is often remarked that if blood were proposed today as an oxygen-carrying volume restoration fluid, it would not be approved by regulatory agencies.

Blood is needed in the presence of blood losses; however, the initial anemia is usually inconsequential to the organism's survival, whereas the associated hypovolemia is tolerated only within a narrow margin. As a result, a blood substitute should ideally target both events, possibly sequentially. Therefore, the development of an effective blood substitute is also in part related to availability and detailed understanding of an effective plasma expander, which is the term used to describe a volume restoration fluid used prior to reaching the transfusion trigger, or the point in volume restitution at which the introduction of an oxygen carrier, blood, is determined to be essential.

Blood exerts its principal functions in the microcirculation, where the materials it transports are exchanged. Thus any attempt to introduce a blood surrogate must ensure efficacy at this level. In contrast, most efforts aimed at developing artificial blood were made in the absence of detailed information and analysis of how oxygen is managed at the level of the microcirculation. The cornerstones in the development of artificial blood up to the present are that this fluid should restore most of the oxygen-carrying capacity of the shed blood, that it is beneficial for the resulting mixture of remaining blood and resuscitation fluid (e.g., OCPE) to have a viscosity lower than that of natural blood, and that the material should have very low oxygen affinity so that oxygen would be readily released when blood arrives to the microcirculation.

These principles have guided the development of products that are now undergoing clinical trials. It is apparent that the initial impetus for the development of artificial blood was based on the restoration of systemic functions after acute blood losses with little or no emphasis on understanding the behavior of the resuscitation fluids in the microcirculation, which is the organ system where blood performs its functions. This was in part due to the imperfect understanding of how oxygen is managed in the microcirculation in both normal and pathophysiological conditions, the role and regulation of shear stress–dependant mediators produced by the endothelium, and the lack of techniques for measuring the key transport parameters that determine efficacy in maintaining microvascular function upon the introduction of a blood surrogate.

Oxygen-carrying capacity and oncotic pressure were prescribed to be similar to that of blood, while the experience with hemodilution suggested that improvements in transport would be obtained by lowering blood viscosity to values significantly below those of whole blood. An additional presumed beneficial modification was the use of oxygen carriers based on modified hemoglobins with high p50s, believed to facilitate oxygen unloading and tissue oxygenation. As acellular modified molecular hemoglobin became the oxygen carrier of choice, it was found this material was vasoactive, causing hypertension, which is deleterious in resuscitation. Vasoactivity was attributed to hemoglobin scavenging NO, leading to vasoconstriction, which gave rise to a significant effort aimed at modifying the hemoglobin molecule so that its affinity for NO was reduced.

At present an optimal OCPE is still perceived to have the following properties: oxygen-carrying capacity equivalent to 10 to 14 g/dL hemoglobin, p50 greater than 30 mmHg, viscosity around 1 cP, oncotic pressure approximately 25 mmHg, and low NO binding. However, studies in the microcirculation show that a fluid configured according these concepts yields problematic outcomes in terms of resuscitation from anemic hypovolemia. Furthermore, since the most favored source of hemoglobin is human, even if it were possible to obtain a one to one conversion from blood to blood substitute, the problem of blood shortages is not solved. In an effort to circumvent the human hemoglobin source there have been various attempts to obtain hemoglobin by recombinant technology. Biopure Inc. (Boston, MA) has progressed to Phase III clinical trials with a molecular hemoglobin-based fluid derived from bovine blood. However, recombinant technology has not progressed to the development of an efficacious product to date, and the bovine-based product was not developed on the basis of microcirculatory data.

The Design of an Efficacious Oxygen-Carrying Plasma Expander

To date there are virtually no rivals to hemoglobin as a transporter of oxygen from the lung to the tissue because of its ability to bind a large amount of oxygen through a chemical reaction. The discovery that fluorocarbons could dissolve a comparatively large amount of oxygen, albeit at high oxygen partial pressures, suggested using this vehicle as the oxygen transporter. However, the use of fluorocarbon-based blood replacement fluids has not materialized, in part because of the lack of definitive experimental studies on the physiology related to altered blood physical properties and changes in the distribution of oxygen partial pressure in the circulation.

Various modifications of hemoglobin have optimized its performance and mostly eliminated the vasoactivity of this molecular species. Human hemoglobin remains the most favored source because of the well-defined methodology for obtaining blood from donors, which in advanced medical systems is virtually free of parasitic, bacterial, or viral contamination. The present formulations of bovine hemoglobin appear to be vasoactive and in the long term could present unknown risks of introducing diseases that may have extraordinarily long incubation times.

If the perceived and now frequently reported blood shortage is the driving force behind the development of hemoglobin-based oxygen-carrying blood substitutes, then the use of human hemoglobin is problematic since the processing technology and formulation would require that a unit of original blood yield at least an equivalent unit of "artificial blood," a zero-sum result that does not relieve shortages. A realistic process should produce several units of equivalent hemoglobin-based oxygen-carrying blood substitute from a unit of natural blood.

The present goal in devising a human hemoglobin-based blood substitute is to circumvent the inherent toxicity of the hemoglobin molecule and to be as efficacious as an equivalent unit of natural blood but at a lower hemoglobin concentration than blood, thus introducing a multiplying factor between the original source of human hemoglobin and the final product. Advances in microvascular technology allow us to critically analyze each of the "cornerstone" precepts that have guided the development of blood substitutes to date, namely the viscosity of the material, the affinity for oxygen, the effective concentration, and the resulting colloidal osmotic pressure when the material is present in the circulation. In the following these parameters will be analyzed from a microcirculatory perspective.

The Role of Viscosity in Oxygen Transport

Blood viscosity depends on red blood cell concentration (hematocrit) and on plasma viscosity. The manipulation of these two viscosities is the basis of the clinical practice of hemodilution. Accordingly, the restitution of blood losses with conventional plasma expanders can be effectively and safely accomplished up to a 50 percent loss of the red blood cell mass, using fluids with plasma-like viscosity. The decrease in viscosity due to hemodilution causes a compensatory increase in cardiac output due to the lowered flow resistance, thus maintaining oxygen delivery.

A specific decrease in oxygen-carrying capacity is one of the parameters that defines the so-called transfusion trigger. Microcirculatory experimental studies do not support the contention that lack of oxygen-carrying capacity is the actual determining factor in the decision of transfusing blood. In the awake hamster window chamber model, neither oxygen-carrying capacity nor tissue oxygenation is in jeopardy with red blood cell losses of two-thirds of the original mass (Figure 1).

A factor that is significantly changed upon reaching the transfusion trigger is blood viscosity, which is approximately half of normal, because of the loss of red blood cells. Thus, additional losses of red blood cells will further reduce blood viscosity, which is strongly dependent on hematocrit. The reduction of blood viscosity is initially compensated by increased cardiac output. However, cardiac output seldom doubles, and blood viscosity has the potential of decreasing

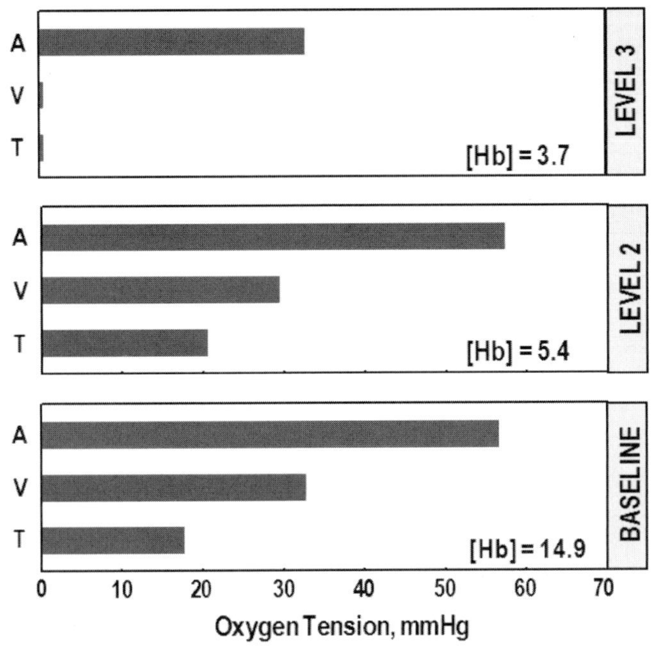

Figure 1 Distribution of oxygen in arterioles (A) and venules (V) in the tissue (T) in the awake hamster window model as a function of total hemoglobin (Hb, grams per deciliter). Normovolemic hemodilution was achieved using dextran 70 kDa. It is apparent that a loss of the order of two thirds of red blood cells has no influence on microcirculatory po_2 of this model. (see color insert)

to 25 percent of normal upon further reductions of hematocrit, leading to the significant lowering of functional capillary density (FCD, capillaries with red blood cell transit), a condition detrimental to survival in hemorrhagic shock.

Although capillaries do not appear to be the determinant structure for the supply of oxygen in some tissues, the maintenance of FCD in shock is a critical parameter in determining outcome independently of tissue oxygen tension (po_2), suggesting that extraction of metabolic by-products may be as critical to a capillary function as oxygenation. The relationship between FCD and survival has been demonstrated in experimental analysis of conditions during prolonged hemorrhagic shock, where the principal microvascular functional difference between survivors and nonsurvivors was that survivors maintained FCD above a threshold of about 40 percent of that present in the normal organism; there were no other significant differences between groups [1].

The blood viscosity threshold that causes the decrease in FCD appears to coincide with the decision to transfuse blood. In other words, the transfusion trigger may also be a viscosity trigger; therefore results obtained with blood transfusions may also be achieved by increasing plasma viscosity. Thus use of red blood cells solely for the purpose of increasing blood viscosity may be avoided by introducing a material that increases plasma viscosity in the circulation. In this scenario blood viscosity resulting from the balance between the diminished red blood cell concentration and the increased plasma viscosity leads to the maintenance of vascular resistance. Tsai et al. [2] explored this phenomenon by inducing extreme hemodilution with low-viscosity dextran 70 kDa. At 11 percent hematocrit, FCD and microvascular flow were significantly reduced from control. However, when plasma viscosity was maintained above 2 cP by the introduction of high-viscosity dextran 500 kDa, FCD was maintained near to control values, and microvascular flow increased significantly above control, though hematocrit was 11 percent. This effect was not found if extreme hemodilution was performed with the Biopure product Oxyglobin, even at a total blood hemoglobin content of 6.7 g Hb/dL. High-viscosity plasma also caused blood flow to increase significantly above nonhemodiluted values because of the release of shear-dependent generated endothelial relaxing factors.

Counterintuitive Rheological Findings

High-viscosity plasma restores mean arterial blood pressure (MAP) in hypotension without vasoconstriction. Systemic blood viscosity depends on hematocrit squared; thus viscous losses in major vessels of the circulation can be minimized even when plasma viscosity is increased, shifting pressure and pressure gradients from the systemic to the peripheral circulation. Reduced blood viscosity decreases shear stress and the release of vasodilators, causing vasoconstriction, which negates the benefit of reducing the rheological component of vascular resistance. Vascular resistance depends on the first power of blood viscosity and the 4th power of vessel radius. Therefore reducing blood viscosity with low viscosity plasma might decrease flow and oxygen delivery to the tissue if there is an associated vasoconstrictor stimulus. However, increased flow and/or increased viscosity augments shear stress on the endothelium since the elevation of plasma viscosity causes sustained NO-mediated dilatation in the hamster muscle microcirculation, supporting this interpretation. Enhancement of shear stress on the vessel wall results in the release of prostacyclin and NO from the vascular endothelium. Endogenous NO release reduces total peripheral resistance during moderate hemodilution. Increased wall shear stress following increased blood flow induces vasodilation [3] showing that the link between shear stress and vasodilatation is well established.

Experimental verification of the beneficial effects of high-viscosity plasma during hemodilution in the microcirculation is evidenced by effects on FCD, perfusion and vasodilatation. More recently it was demonstrated by Cabrales et al. [4] that an increase in capillary pressure is the principal mechanical event that governs the effects due to perfusion with high-viscosity plasma.

Hemodilution, Blood Viscosity, and Vasoactivity

It is now apparent that low viscosity molecular hemoglobin solutions lower FCD independently of the intrinsic

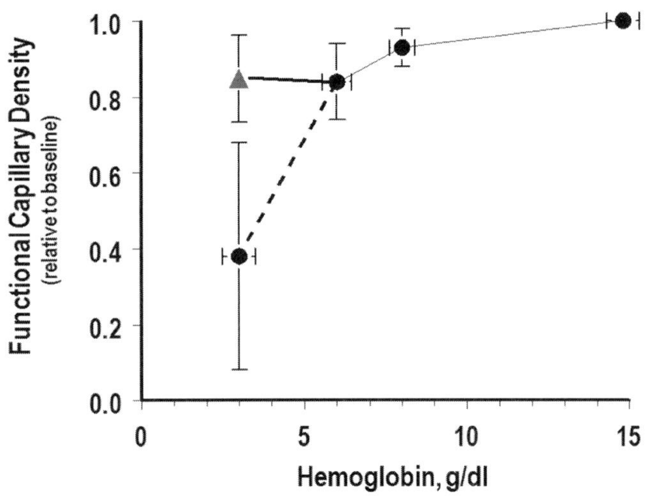

Figure 2 Changes of FCD following isovolemic hemodilution with low- and high-viscosity plasma expanders. ● Hemodilution with dextran 70 kDa maintains FCD up to a hematocrit (hemoglobin) that is 40 percent of normal. Further hemodilution with the same low-viscosity diluent causes the fall of FCD to pathologically low levels. ▲ Continuation of hemodilution with dextran 500 kDa after reaching 40 percent of normal hemoglobin using dextran 70. FCD is maintained to normal levels by the increased plasma viscosity. Redrawn from Tsai et al. [2].

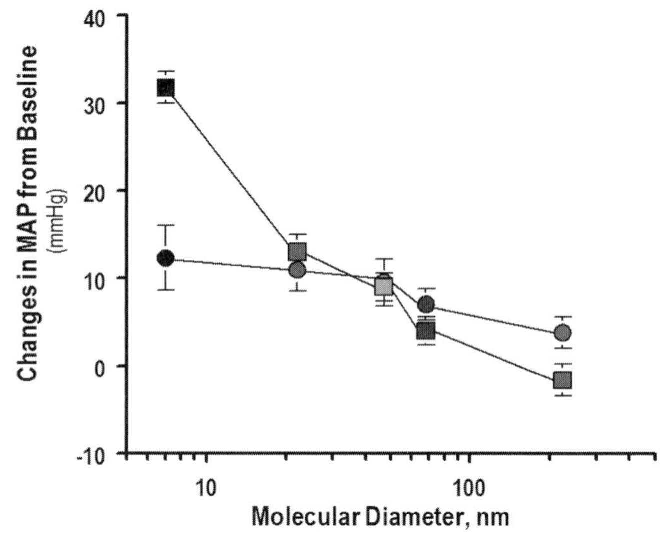

Figure 3 Changes in mean arterial blood pressure after a 5 percent by volume (5 Hb g/dL) topload infusion of free hemoglobin solutions of different molecular diameters and vesicle encapsulated hemoglobin. ■ Pressor effect after infusion. ● Pressor effect 3 hours after infusion. Redrawn from Sakai et al. [5]. (see color insert)

vasoactive properties of the hemoglobin molecule because they cause a significant decrease in blood viscosity after reaching the transfusion trigger. An additional factor attendant to the restoration of blood volume upon reaching the transfusion trigger with a plasma-like viscosity fluid is that this process brings the organism to near extreme hemodilution conditions, characterized by decreased shear stress on the endothelium, lowering the production of endothelial-derived vasodilators. Increasing plasma viscosity to about 2.0 to 2.5 cP increases shear stress and the production of vasodilators, which breaks up the vicious circle caused by extreme hemodilution, compensatory vasoconstriction and low viscosity.

Experimental results in Figure 2 show that the maintenance of FCD is not directly linked to oxygen delivery, but to mechanical factors related to the viscosity of the perfusion fluid and the production of vasodilators by mechanotransduction in the endothelium. Therefore an acellular oxygen carrier should maintain plasma viscosity above a specific threshold, while ensuring that overall blood viscosity does not exceed normal values.

Low blood viscosity can be compensated for by hemoglobin solutions with high viscosity. This can be achieved by mixing the hemoglobin molecule with a viscogenic material such as hydroxyethyl starch (HES) at suitable concentrations, or by modifying the hemoglobin molecule to produce an intrinsically viscous solution by increasing its molecular dimension. This latter process can be implemented by polymerization or conjugation with various molecules such as starch and polyethylene glycol, as described by Sakai et al. [5], who showed that the pressor effect is inversely related to molecular size (Figure 3). Hemoglobin molecules with these and several other beneficial features are polyethylene glycol conjugated hemoglobin molecules [6].

The Vasoconstrictive Effect of Hemoglobin

Natural hemoglobin molecules are presumed to be vasoconstrictive because of their ability to scavenge NO. However, recent experimental evidence shows that whereas NO binding is virtually identical for most hemoglobin molecules [7], the vasoconstrictive effect is not, being essentially absent in polyethylene glycol modified hemoglobins and in some very large hemoglobin polymers.

NO is produced by the endothelium as a result of shear stress and other processes. The affinity of the hemoglobin molecule for NO is due to the physical similarity between NO and O_2. Thus in general, hemoglobins with high affinity for O_2 generally also have a high affinity for NO, and vice versa. The production of genetically modified hemoglobins that appear to have little affinity for NO, while maintaining a normal affinity for O_2, may challenge this generalization; however, the fact remains that interfering with NO production with administration of L-arginine methyl ester hydrochloride (L-NAME) or scavenging NO with cell free unmodified hemoglobin causes the constriction of aortic rings, and an increase in blood pressure in experimental subjects.

The concept that hemoglobin extravasation and its location between the endothelium and smooth muscle is the principal factor causing hypertension and vasoconstriction is also questionable because the extravasated molecule will eventually saturate. In fact the presence of a NO-avid molecule in plasma is sufficient to distort the diffusion field of NO from the endothelium, whereby hemoglobin does not need to extravasate to be vasoconstrictive.

NO scavenging does not provide a consistent explanation for the pressor effect of free hemoglobin in the circulation that is applicable to the different hemoglobin modifications. The lack of correlation between pressor responses and NO scavenging characteristic of hemoglobin molecules led McCarthy et al. [8] to propose that hypertension following the introduction of molecular hemoglobin in the circulation is caused by a mechanism related to the process of facilitated diffusion of oxyhemoglobin. According to this hypothesis the presence of molecular hemoglobin causes an additional flux of oxygen in the plasma layer due to the diffusion of oxyhemoglobin. Although the diffusion constant of hemoglobin is low, the amount of oxygen carried is large because hemoglobin binds a large amount of oxygen. The net result of this process is that a comparatively small concentration of molecular hemoglobin augments oxygen transfer to the vessel wall, leading to a hyperoxia signal, and consequently a vasoconstrictive response.

In vivo, peripheral vascular resistance is autoregulated at the level of the arterioles by a mechanism that senses oxygen tension, producing vasodilatory signals when blood and tissue po_2 is low, and vice versa. This conceptualization is supported by the finding that large hemoglobin molecules are not vasoactive, although they carry oxygen. As an example, polyethylene glycol (PEG) surface decorated hemoglobins (PEG-Hb) have consistently been shown to be vasoinactive. These molecules have a large volume because of the water bound by PEG. Since the diffusion constant is inversely proportional to molecular radius, it can be shown that PEG-Hb has a diffusion constant that is about one fifth that of the native hemoglobin.

Experimentation with different levels of hemoglobin surface decoration show that vasoinactivity may be partially related to the degree to which the surface of the hemoglobin molecule is shielded by the water–PEG combination [6]. This phenomenon suggests that free hemoglobin may also cause a pharmacological effect mediated at the surface of the endothelium, and that conjugation of hemoglobin with PEG may produce a shield that prevents this process.

The vasoconstrictive effects of molecular hemoglobin may have several components that sometimes reinforce each other. When blood viscosity becomes too low, there is a reflex vasoconstriction that attempts to maintain perfusion pressure, a phenomenon independent of blood oxygen-carrying capacity. Oxygen regulation plays a crucial role since the arteriolar walls and the tissue sense both the rate of oxygen delivery from the red blood cell column and local po_2. When molecular hemoglobin is present in plasma, there is a significant additional flux of oxygen to the arteriolar wall by facilitated diffusion, a process enhanced with right-shifted oxygen dissociation hemoglobin molecules. NO scavenging can also be a factor that may be balanced by increased NO (and/or prostacyclin) production resulting from elevated shear stress caused by high-viscosity hemoglobin molecules. Furthermore, considering that modest amounts of small molecular hemoglobin can elicit a pressor response, a pharmacological effect due to "naked" small hemoglobin molecules in the circulation may also be present.

Vasoconstriction limits perfusion and decreases FCD. Although healthy organisms could probably compensate for moderate hypertensive episodes leading to corresponding decreases in FCD, these same episodes may place the organism in jeopardy if they are superposed to other vasoconstrictive stimuli, such as those inherent to hemorrhagic shock. Conversely, high plasma viscosity is critical in resuscitation, as an OCPE is administered in conditions of extreme hemodilution because there is no need for using these products prior to reaching the transfusion trigger.

Optimal Oxygen Disassociation Properties

The development of oxygen carriers has implicitly assumed that the oxygen dissociation curve should be right shifted, thus facilitating the release of oxygen. This approach does not consider the longitudinal gradient of oxygen tension in the circulation, whereby a right-shifted dissociation curve favors oxygen unloading from small arteries and arterioles. Hemodilution with hemoglobin-filled vesicles of different p50s in the hamster window chamber model has shown that improved tissue oxygenation is obtained when this parameter is 16 mmHg, instead of 34 mmHg (Department of Polymer Chemistry, Waseda University, Tokyo, Japan). PEG-conjugated hemoglobin (Hemospan, 4% Mal-PEG hemoglobin) produced by Sangart (San Diego, CA), with a p50 of 5 mmHg, used at low concentration in hemodilution maintains FCD and positive acid–base balance.

This apparent paradox may be understood by analyzing the distribution of oxygen in the microcirculation as shown in Figure 4, where oxygen tension in the microcirculation in normal conditions has a baseline tissue po_2 level of 22 to 24 mmHg (which appears to be common for most tissues). It is notable that although oxygen is regulated to achieve this partial pressure in the tissue, anaerobic metabolism occurs when tissue po_2 is below 2.4 to 2.9 mmHg.

A possible rationale for the high po_2 tissue regulation is that the organism has excess oxygen-carrying capacity, not only as a requirement for extreme efforts, but also for compensation of oxygen delivery inhomogeneity in the microcirculation. The effect of this inhomogeneity becomes apparent in considering the variability of oxygen partial pressure distribution in the hamster window chamber, which is of the order of $\pm$ 4 mmHg. This variability is a consequence of the quasirandom distribution of the transport properties of the microcirculation, and therefore intrinsic to any level of tissue oxygenation. In conditions of extreme hemodilution tissue po_2 decreases to 3 to 5 mmHg; thus, if

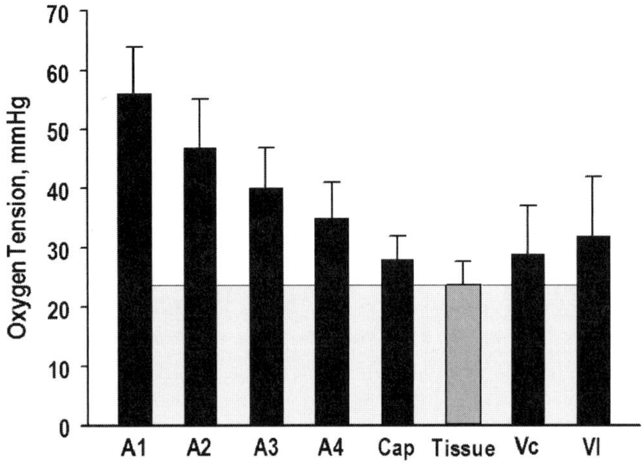

Figure 4 Distribution of po_2 in the microvessel of the microcirculation of the hamster window preparation as a function of the arteriolar order of branching (As) and the venular order of branching (Vs). It is apparent that tissue po_2 is narrowly regulated to a value in the range of 22 to 24 mmHg, which is significantly higher than the level associated with anaerobic metabolism.

the same variability is present, there is a significant amount of tissue that is anoxic. In the presence of a fraction of the oxygen-carrying capacity that can only be released at very low po_2s, a form of targeted oxygen delivery, the effects of this variability will be nullified, ensuring that all the tissue, even at low po_2 is oxygenated above the anaerobic threshold.

Tissue po_2 levels that may be considered harmful could, in fact, be quite safe if it were possible to eliminate the inherent variability of oxygen delivery shown by the variability of tissue po_2. A small quantity of a low-p50 hemoglobin oxygen carrier in the circulation accomplishes this because it delivers oxygen only to portions of the tissue where the anoxic threshold is passed, while the presence of even significant amounts of right-shifted hemoglobin would have no effect since most of the bound oxygen would be unloaded in oxygenated regions.

Cross-linked or polymerized hemoglobins developed so far have a high p50, presumed to be beneficial since it facilitates oxygen unloading. However, po_2 in the microcirculation is regulated so that there is a significant decrease in oxygen tension from the systemic circulation to the capillaries, which typically have a po_2 of about 30 mmHg. At this p50 half of the blood oxygen is delivered by arterioles in normal conditions; however, if the p50 of the OCPEs is above this value, as in the case of Oxyglobin (p50 = 54 mmHg), most of the oxygen in the blood should be delivered by the arterioles if this material were to replace blood. These vessels extract a significant amount of oxygen from the circulation while consuming a major portion of this oxygen flux, thus increasing their oxygen supply increases tissue oxygen inhomogeneity, which is further aggravated by the vasoconstrictor autoregulatory response already discussed.

Oxygen-Carrying Capacity

Measurements of po_2 in the microcirculation utilizing the technique of phosphorescence oxygen quenching show that when hemodilution carried out to a total hemoglobin content in red blood cells of 5.6 g/dL, then tissue oxygen is somewhat higher than normal but not statistically significant. The required oxygen-carrying capacity can also be obtained by a simple calculation that relates the whole-body oxygen consumption and cardiac output, which yields a nearly identical number for the organism at rest. Therefore, in principle, the oxygen-carrying capacity of an OCPE does not need to reproduce the value for normal blood and can be significantly lower.

Colloid Osmotic Pressure

It is generally assumed that a blood substitute should have a colloid osmotic pressure similar to that of blood and in the range of 20 to 25 mmHg; however, several plasma expanders have zero colloid osmotic pressure (saline, Ringer's lactate) and small-volume resuscitation utilizes fluids with very high osmotic properties. To date there is no definitive answer on what is the osmotic and/or oncotic property that is most appropriate, and in all probability this is a variable that depends on the type of blood loss to be corrected. Resuscitation with noncolloidal fluids leads to tissue edema. Conversely fluids with high colloidal and osmotic pressures cause tissue fluid to come into the vascular compartment, thus decreasing the amount of fluid to be administered. Most conditions of hemorrhage are associated with endothelial edema, which has been demonstrated to be rapidly reversed upon the introduction of hyperosmotic and hyperoncotic fluids. Volume expansion fluids such as hydroxyethyl starch have relative high colloid osmotic pressures, typically in the range of 30 to 50 mmHg depending on formulation. Small molecule hemoglobin–based OCPEs have their oncotic pressure adjusted to be that of plasma, but PEG-hemoglobin modified OCPEs tend to have higher oncotic pressures.

Synthesis of an Effective Oxygen-Carrying Plasma Expander

An OCPE based on the preceding concepts is a fluid with properties fundamentally different from those of blood, since it has low oxygen-carrying capacity, p50 is low and in the neighborhood of 5 mmHg, viscogenic properties are such that when introduced into the circulation plasma viscosity should be of the order of 2.0 to 2.5 cP, and colloidal osmotic pressure can be high. A fluid with these properties can be obtained by conjugating hemoglobin with PEG, and various formulations have been tested in both animal experiments and human trials with excellent results. Notably this formulation is vasoinactive, and its NO-scavenging characteristics do not appear to be relevant since these fluids have the same NO binding constant as other vasoactive formulations [7].

These fluids are in some cases more effective than blood because they are designed to maintain FCD, which is as necessary as restoring tissue oxygenation for recovery from blood losses. Because in the foreseeable future OCPEs will use human hemoglobin, these fluids are practical: Their hemoglobin content is low, and more than two units of blood equivalent unit of resuscitation fluid can be obtained from one unit of blood. Finally, this low oxygen-carrying capacity is practical and safe because it yields a significant improvement of microvascular function.

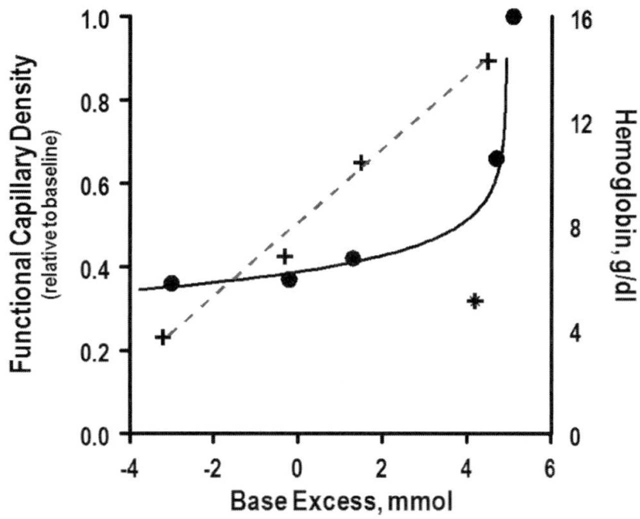

Figure 5 Relationship between total circulating hemoglobin and base excess, and FCD and base excess, for different hemoglobin modifications and concentrations, including hemoglobin vesicles, in normovolemic hemodilution experiments. The data marked ● shows the relationship between FCD and base excess, showing that MAL-Peg-Hb (▼) yields high FCD and base excess at low hemoglobin concentrations. It is apparent that base excess is a direct function of hemoglobin concentration (+) with the exception of MAL-Peg-Hb (✱), which presents normal base excess at a very low total hemoglobin content. (see color insert)

Experimental Evidence

The effectiveness of different resuscitation modalities was tested experimentally in studies of extreme hemodilution and hemorrhagic shock in the microcirculation of the hamster chamber window model, which allows microcirculatory monitoring in the awake condition for a period up to 1 week, after the effects of the surgical intervention have subsided. Extreme hemodilution was chosen because in most instances, lowering systemic hematocrit to 50 percent of baseline with a suitable plasma expander does not alter microvascular hemodynamics and transport in our experimental model. Animals were hemodiluted to 60 percent of normal with dextran 70 kDa, and further hemodiluted to a final hematocrit of 11 percent using the different products simulating blood losses initially remedied with conventional plasma expanders, which upon passing the transfusion trigger are corrected with an oxygen-carrying blood substitute.

A compendium of findings in extreme hemodilution to 50 percent of normal with dextran 70 kDa and further hemodilution to a final hematocrit of 11 percent with the different products is shown in Figure 5, including results obtained with PEG-Hb vesicles developed at Waseda University, Tokyo, using a somewhat different protocol where extreme hemodilution was achieved with a continuous exchange of a hemoglobin vesicle suspension. FCD is shown as a function of blood base excess, which represents systemic conditions and suggests the definition of *critical functional capillary density* as the value for this parameter at which base excess is no longer sustained and drops following modest reductions of total blood hemoglobin, that is, in the neighborhood of a 50 percent FCD reduction. The most important result is that normal base excess is obtained with total blood hemoglobin of 5 percent, if 1 percent of this is Mal-PEG-Hb—a result not found with other OCPEs.

Extreme hemodilution is not a clinically relevant procedure and serves only to study basic mechanisms. A clinically relevant test is to rescue a subject in hemorrhagic shock. Studies were therefore conducted to determine the effects of resuscitation with blood, starch, and Mal-PEG-Hb in a conventional 50 percent bleed shock protocol. The animals were resuscitated after 1 hour without any additional volume manipulation using shed blood, HES, and Mal-PEG-Hb with 25 percent of the blood volume. The results, shown in Figure 6, indicate that Mal-PEG-Hb is superior to both HES and blood in reestablishing microvascular function. Concurrently it was found that base excess was higher in the Mal-PEG hemoglobin-resuscitated animals than in the blood-resuscitated animals. An explanation for these findings is that low p50 hemoglobin targets oxygen delivery of oxygen to only the anoxic tissue.

An extreme hemorrhage study was performed with Mal-PEG-Hb in which rats were 50 percent exchange transfused before hemorrhage with either $\alpha\alpha$-cross-linked hemoglobin, or 4 percent Mal-PEG hemoglobin (Figure 7). These animals were then subjected to a continuous exponential bleed (1 hour, 60 percent of blood volume) whereby at the end of the second hour after the start of bleeding 50 percent of the control animals succumbed. In these experiments it was found that at the end of one hour all animals that received Mal-PEG hemoglobin before hemorrhage survived, while all of those receiving $\alpha\alpha$-cross-linked hemoglobin did not survive.

Summary and Conclusions

The revision of microvascular physiology related to modifying basic transport properties of blood such as plasma viscosity, p50, and hemoglobin concentration shows

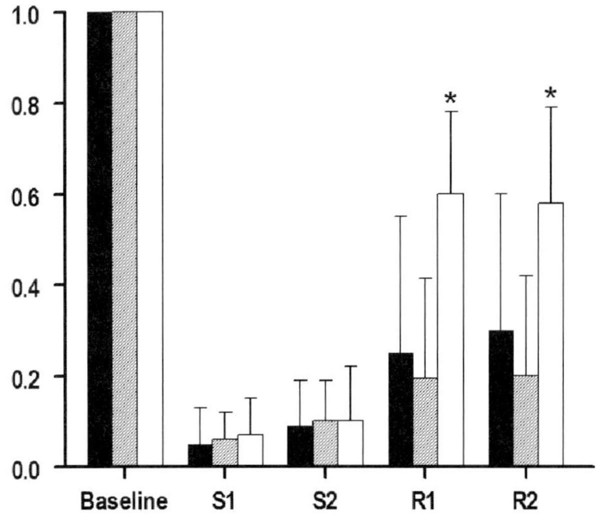

Figure 6 Recovery of FCD during resuscitation from 1 hour hemorrhagic shock with identical volumes of shed blood (*black bars*), 5 percent HES (*shaded bars*), and 4 percent MAL-Peg-Hb (*white bars*). S1 and S2 initial and final conditions during the shock period. R1, Recovery immediately after resuscitation; R2, 1 hour after resuscitation. $p < 0.05$ relative to shed blood and HES.

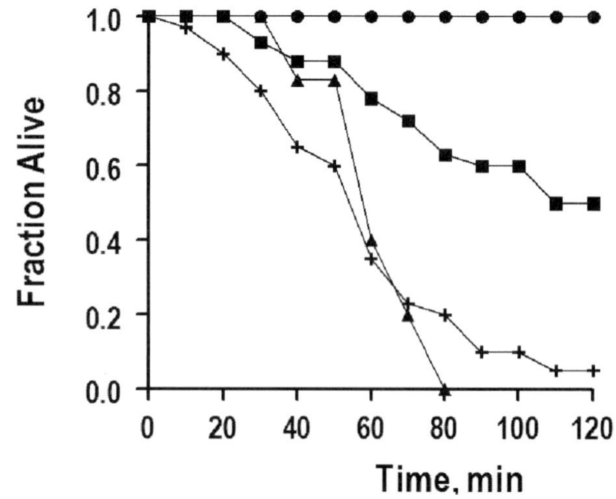

Figure 7 Controlled bleeding in rats that are 50 percent exchange transfused with MAL-Peg-Hb (●), αα-cross-linked hemoglobin (▲), and a polymerized hemoglobin (+), versus controls (■) with no treatment. The study was designed so that 50 percent of the untreated (not transfused controls) would survive 120 minutes.

that blood or a bloodlike fluid may not be the optimal oxygen-carrying volume-restoring fluid. A critical parameter for either oxygen-carrying or noncarrying blood replacements is their viscosity, which is a factor in maintaining capillary flow.

Analysis of the microvascular consequences of changing blood rheological conditions and particularly plasma shows that low plasma viscosity is not of universal benefit. Patients following trauma, peripheral arterial occlusive disease, and acute myocardial infarction have elevated plasma viscosity, a condition presumed to be pathological. However, there are situations where increased viscosity may be a protective or beneficial mechanism.

Plasma expanders are not used after reaching the transfusion trigger because the reduction of blood oxygen-carrying capacity beyond this point is assumed to jeopardize tissue oxygenation, according to the systemic evaluation of the organism portrayed by blood gases. Conditions in the microcirculation and local microscopic tissue environment when the reduction of red blood cells is extended beyond the transfusion trigger have not been consistently explored and presently show that oxygen-carrying capacity is not the major factor in determining tissue survival.

Studies show that the transfusion trigger is also the limit for the organism to adapt to low blood viscosity in acute conditions; thus *the conventional transfusion trigger is also a viscosity trigger*. Since the administration of a molecular oxygen carrier is physically similar to continuing fluid therapy after reaching the transfusion trigger, the maintenance of FCD requires the increase of plasma viscosity which through shear stress–dependent mechanisms operating in the endothelium ensures the maintenance of optimal microvascular function. Oxygen-carrying capacity is exhausted upon red blood cell (or hemoglobin) losses that are significantly greater than those represented by the transfusion trigger. However, these losses of oxygen-carrying capacity do not need to be compensated on a one-to-one basis, if microvascular function (i.e., FCD) is maintained and an oxygen carrier is introduced only to deliver oxygen to anoxic tissue regions. This approach ensures a uniform maintenance of the whole organism above the anaerobic threshold, while limiting the amount of oxygen carrier needed to maintain metabolism. Thus the combination of maintenance of microvascular function and targeted oxygen delivery is the primary determinant of an efficacious human hemoglobin-based blood substitute that is more effective than blood in acute conditions and that also expands the available blood supply, since a unit of blood yields more than two units of surrogate blood.

Glossary

Functional capillary density: Number of capillaries in a unit volume of tissue that presents the passage of red blood cells. This parameter is experimentally determined by measuring the length of red blood cell–perfused capillaries in a microscopic field of view.

Microvascular function: A combination of parameters including flow, number of open capillaries, intact vascular permeability, and level of vessels tone that allows for the proper interaction between blood and tissue at the microscopic level.

Oxygen-carrying capacity: The amount of oxygen in milliliters at standard atmospheric conditions and temperature contained in a fluid.

p50: Partial pressure of oxygen at which hemoglobin is 50 percent saturated with oxygen.

Plasma expander: A fluid used to restore circulatory volume when oxygen-carrying capacity is adequate.

Transfusion trigger: Level of blood hemoglobin at which the decision is made to introduce red blood cells into the circulation in order to restore oxygen-carrying capacity.

Vasoactivity: Inherent property of compounds that cause vasoconstriction and the elevation of systemic blood pressure.

Acknowledgments

This work was supported by Bioengineering Research Partnership grant R24-HL64395 and grants R01-HL62354 and R01-HL62318 to M.I.

References

1. Kerger, H., Saltzman, D. J., Menger, M. D., Messmer, K., and Intaglietta, M. (1996). Systemic and subcutaneous microvascular po_2 dissociation during 4-h hemorrhagic shock in conscious hamsters. *Am. J. Physiol.* **270**, H827–H836. *Reports the basic data showing the direct correlation between survival and maintenance of functional capillary density.*
2. Tsai, A. G., Friesenecker, B., McCarthy, M., Sakai, H., and Intaglietta, M. (1998). Plasma viscosity regulates capillary perfusion during extreme hemodilution in hamster skin fold model. *Am. J. Physiol.* **275**, H2170–H2180. *Experimental demonstration that high viscosity plasma in extreme hemodilution maintains microvascular function and systemic conditions, an effect that is not present at an identical reduction of hematocrit (oxygen-carrying capacity) when plasma viscosity is normal.*
3. Smiesko, V., and Johnson, P. C. (1993). The arterial lumen is controlled by flow-related shear stress. *News Physiol. Sci.*, 34–38.
4. Cabrales, P., Tsai, A. G., and Intaglietta, M. (2004). Microvascular pressure and functional capillary density in extreme hemodilution with low and high plasma viscosity expanders. *Am. J. Physiol. Heart Circ. Physiol.* **287**, H363–H373. *Experimental demonstration that capillary blood pressure and functional capillary density are directly related. In extreme hemodilution increased plasma viscosity elevates capillary pressure at reduced overall oxygen delivery. By comparison, hemodilution to the same hematocrit, but increased blood hemoglobin following the introduction of a low-viscosity vasoactive hemoglobin solution, lowers capillary pressure and FCD.*
5. Sakai, H., Hara, H., Yuasa, M., Tsai, A. G., Takeoka, S., Tsuchida E., and Intaglietta, M. (2000). Molecular dimensions of Hb-based O_2 carriers determine constriction of resistance arteries and hypertension. *Am. J. Physiol.* **279**, H908–H915.
6. Manjula, B. N., Tsai, A. G., Intaglietta, M., Tsai, C.-H., Ho, C., Smith, P. K., Perumalsamy, K., Kanika, N. D., Friedman, J. M., and Acharya, S. A. (2005). Conjugation of multiple copies of polyethylene glycol to hemoglobin facilitated through thiolation: Influence on hemoglobin structure and function. *Protein J.* (in press).
7. Rohlfs, R. J., Brunner, E., Chiu, A., Gonzales, A., Gonzales, M. L., Magde, D., Magde, M. D., Jr., Vandegriff, K. D., and Winslow, R. M. (1998). Arterial blood pressure responses to cell-free hemoglobin solutions and the reaction with nitric oxide. *J. Biol. Chem.* **273**, 12128–12134. *Direct demonstration that vasoactivity and the induction of hypertension are not dependent on NO scavenging, since hemoglobin solutions with the same NO binding capacity cause a range of responses, varying from no effect on blood pressure found with pegylated hemoglobin to a maximal increase in blood pressure found with $\alpha\alpha$-hemoglobin.*
8. McCarthy, M. R., Vandegriff, K. D., and Winslow, R. M. (2001). The role of facilitated diffusion in oxygen transport by cell-free hemoglobins: Implications for the design of hemoglobin-based oxygen carriers. *Biophys. Chem.* **92**, 103–117.

Further Reading

Kobayashi, K., Tsuchida, E., and Horinouchi, H. (2005), *Artificial Oxygen Carrier: Its Frontline*, Vol. 12. Tokyo: Springer-Verlag.

Rudolph, A. S., Rabinovich, R., and Feurestein, G. Z. (1998). *Red Blood Cell Substitutes. Basic Principles and Clinical Applications.* New York: Marcel Dekker.

Tsuchida, E. (1998). *Blood Substitutes: Present and Future Perspectives.* Amsterdam: Elsevier Science.

Winslow, R. M., Vandegriff, K. D., and Intaglietta, M. (1995). *Blood Substitutes. Physiological Basis of Efficacy.* Boston: Birkhäuser.

Winslow, R. M., Vandegriff, K. D., and Intaglietta, M. (1996). *Blood Substitutes. New Challenges.* Boston: Birkhäuser.

Winslow, R. M., Vandegriff, K. D., and Intaglietta, M. (1997). *Advances in Blood Substitutes. Industrial Opportunities and Medical Challenges.* Boston: Birkhäuser.

Biographies

Dr. Amy G. Tsai obtained her Ph.D. in bioengineering at the University of California, San Diego, where she is currently a Senior Research Scientist. She is widely recognized for her findings on oxygen consumption by the microvasculature and the development of high-viscosity plasma expanders. She is an expert in mathematical modeling, measuring methods for the in vivo study of the microcirculation and small animal experimentation.

Dr. Pedro Cabrales received his Ph.D. from the Universidad de los Andes, Bogotá, Colombia, studying the microvascular effects of extreme hemodilution with perfluorocarbons. He specializes in hemodynamic transport phenomena, having developed techniques for the analysis of tissue oxygenation at the microscopic level. He is presently at the Laboratory of Microhemodynamics of the University if California, San Diego.

Dr. Hiromi Sakai received his Ph.D. in polymer chemistry from Waseda University, Tokyo, Japan, where he is now Associate Professor. He specialized in the synthesis and characterization of oxygen carriers from the viewpoint of molecular assembly. For several years he was a visiting scholar at the University of California, San Diego, where he developed expertise in microhemodynamics. He is currently working on the optimization of oxygen carriers using in vivo methods in order to determine their safety and efficacy.

Prof. Marcos Intaglietta received his Ph.D. in applied mechanics from the California Institute of Technology in Pasadena and developed his academic career at the University of California, San Diego, where he is one of the founders of the bioengineering program and department. His specialty is the study of transport phenomena in the microcirculation and the development of blood substitutes. He has developed and implemented most of the methods presently used for the study of the microcirculation.

SECTION C

Endothelial Biomarker

CHAPTER 161

N(Carboxymethyl)lysine as a Biomarker for Microvascular Complications

Jean-Luc Wautier and Marie-Paule Wautier

Cellular Biology, Université Denis Diderot, Paris, Institut National de la Transfusion Sanguine (INTS), Institut National de la Santé et Recherche Médicale U.76

Introduction

In recent decades evidence has been presented in favor of a potential role for advanced glycation end products (AGEs) in the development of vascular dysfunction in diabetes mellitus, renal failure, and aging. Since the first description of AGEs by L. C. Maillard, they have appeared to be a complex class of molecules including different products of glycoxidation but also of lipoxidation [1]. Their chemical detection is relatively complicated, and high-performance liquid chromatography and mass spectrometry are frequently used. Two well-defined molecules, carboxymethyllysine (CML) and pentosidine, are antigenic and can be measured by immunological techniques.

Biochemistry of Carboxymethyllysine

AGEs result from a reaction between free amino groups of lysine or arginine of proteins and carbohydrates. CML is formed by the cleavage of a glycated protein (fructose-lysine) into CML and erythronic acid (Figure 1). Carboxyethyllysine and CML belong to the carboxyalkyllysine group of molecules. CML was shown to accumulate in different organs, and its potential role in the genesis of nephropathy or retinopathy was suggested. CML adducts demonstrated a high-affinity binding capacity to the receptor for AGE (RAGE). RAGE is a molecule that belongs to the immunoglobulin superfamily of membrane receptors and is present on different cell types, endothelial cells, macrophages, lymphocytes, granulocytes, and neurons [2].

CML and Vascular Cell Pathology

CML proteins bind to endothelial cells and macrophages and activate these cells. CML binding to RAGE is followed by a series of reactions including NADPH-oxidase activation. NADPH-oxidase produces reactive oxygen intermediates that induce NF-κB activation and subsequently gene transcription. As a consequence, vascular cell adhesion molecule (VCAM-1) is induced, synthesis of cytokines such as interleukin-6 or monocyte-chemotactic protein 1 is potentiated, and tissue factor is expressed. These alterations of endothelial cells favored leukocyte adhesion and migration and the development of an inflammatory reaction. Vascular permeability is increased in diabetic animals [3]. Using radiolabeled albumin or fluorescent probes, it was possible to demonstrate that the increase in vascular permeability may be induced by AGE infusion. The increase in vascular permeability observed in diabetic rats as well as when induced by AGE infusion can be blocked when AGE–RAGE interactions were inhibited using anti-RAGE antibodies or soluble RAGE [4].

These results demonstrated that, at least, vascular hyperpermeability observed in diabetic rats was related to AGE-RAGE interaction (Figure 2).

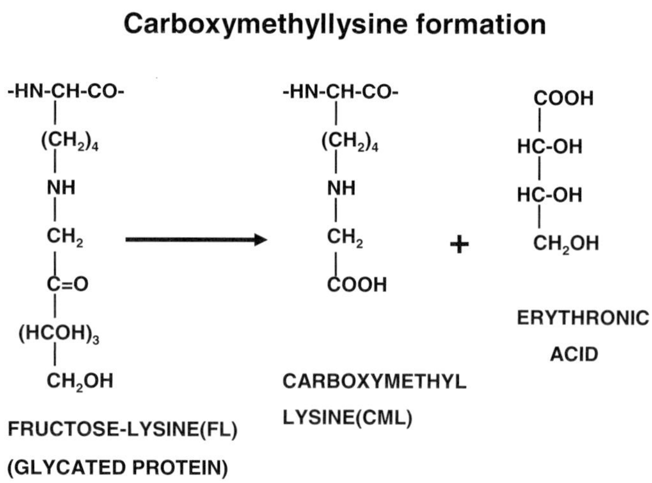

Figure 1 Formation of carboxymethyllysine, an advanced glycation end product (AGE).

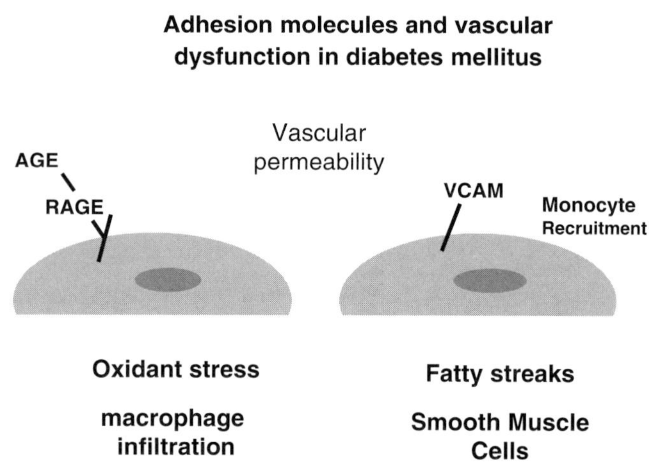

Figure 2 Vascular dysfunction in diabetes mellitus: AGE binding to RAGE activates endothelial cells producing an oxidant stress, inducing vascular cell adhesion molecules, and increasing vascular permeability.

CML in Diabetes Mellitus, Aging, and Renal Failure

Using selective electrospray ionization mass spectrometry, different AGEs can be measured simultaneously, but in most of the clinical investigations that have been conducted in patients with diabetes or renal failure, only one or two AGEs have been measured. In several studies, since CML adducts can be measured by immunological methods, CML was determined by ELISA assay. In patients with type II diabetes mellitus CML–human serum albumin (CML-HSA) blood levels were found to be significantly augmented compared to normal subjects. CML proteins are also increased in patients with chronic renal failure [5].

During aging CML proteins accumulate in different organs. Concentrations of CML in collagen of glomerular membrane increased with age in rats. CML-AGEs were detected in retina of patients with Eale's syndrome, but also

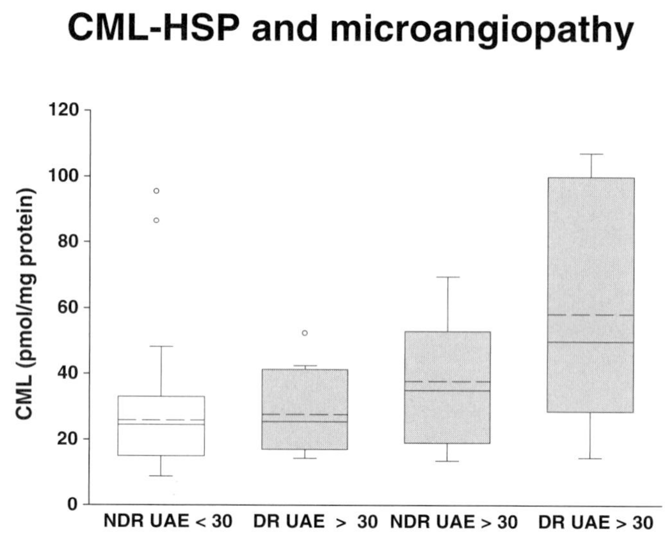

Figure 3 Carboxymethyllysine–human serum protein (CML-HSP) blood level was measured by ELISA assay. CML-HSP was significantly higher in diabetic patients with diabetic retinopathy (DR) or microalbuminuria (UAE greater than 30) or when having these two microvascular complications ($p < 0.01$).

in diabetic retinopathy. CML also deposits in skin collagen of patients with chronic renal failure.

CML and Microvascular Complications

CML-HSA blood levels are higher in type II diabetic patients with microangiopathy compared to those without microvascular complications. CML-HSA was more augmented in diabetic patients who have microalbuminuria and retinopathy than in those who have only one complication (Figure 3).

The statistically significant correlation between CML-HSA blood level and the extent of microvascular lesions suggested that CML may participate in the vascular dysfunction. This hypothesis fits with the deleterious effect of AGE on the vascular functions observed in animal and prevented by AGE-lowering agents or by blocking AGE receptors.

CML adducts are present in excess in eye or kidney of patients with retinopathy or nephropathy. Oxidatively formed CML is present in nondiabetic retinas as a regular constituent, but increases in diabetics both in neurological and vascular components.

The prevalence of the allele *249C* of the RAGE gene is significantly increased in patients with type II diabetes and retinopathy, further indicating a crucial role for CML, the high-affinity ligand to RAGE, in the development of retinopathy. Genetically modified mice that overexpressed RAGE and were diabetic developed an accelerated glomerulosclerosis that is prevented by AGE-limiting compounds.

In the presence of impaired renal function serum CML levels showed a significant inverse relation with creatinine clearance. CML accumulation may be due to both a low serum clearance and/or increased in situ CML formation in chronic renal failure.

CML modification of proteins engages RAGE, thereby activating key cell signaling pathways such as NF-κB and modulating gene expression. Thus, CML–RAGE interactions lead to an accelerated deleterious process affecting vascular functions. In addition to cytokine production, leukocyte adhesion molecule induction favor an inflammatory reaction. CML adducts can trigger vascular inflammation but can also result from oxidative reaction.

Clinical investigation demonstrated that CML proteins are increased in the blood of patients with microvascular complications when suffering from diabetes mellitus. Since the CML modifications in proteins are irreversible, it has been suggested that CML may be an integrative biomarker for the accumulative oxidative stress of proteins that had been exposed.

CML-HSA is an interesting marker for glycoxidation and is related to microangiopathy genesis via a receptor-mediated mechanism.

Glossary

Advanced glycation end products: The result of the nonenzymatic reaction between a sugar and the free amino group(s) of proteins.

Microangiopathy: Disease of the capillaries, often applied to vascular changes in diabetes mellitus. Synonym: capillaropathy.

Vascular permeability: The fact that blood vessels permit the passage of substances, as through a membrane or other structure.

References

1. Brownlee, M. (2000). Negative consequences of glycation. *Metabolism* **49**, 9–13.
2. Schmidt, A. M., Yan, S. D., Wautier, J. L., and Stern, D. (1999). Activation of receptor for advanced glycation endproducts: A mechanism for chronic vascular dysfunction in diabetic vasculopathy and atherosclerosis. *Circ. Res.* **84**, 489–497.
3. Wautier, J. L., Zoukourian, C., Chappey, O., Wauticr, M. P., Guillausseau, P. J., Cao, R., Hori, O., Stern, D., and Schmidt, A. M. (1996). Receptor-mediated endothelial cell dysfunction in diabetic vasculopathy. Soluble receptor for advanced glycation end products hyperpermeability in diabetic rats. *J. Clin. Invest.* **97**, 238–243.
4. Wautier, J. L., and Guillausseau, P. J. (2001). Advanced glycation end products, their receptors, and diabetic angiopathy. *Diabetes Metab.* **27**, 535–542.
5. Wautier, M. P., Massin, P., Guillausseau, P. J., Huijberts M., Levy, B. I., Boulanger, E., Laloi-Michelin, M., and Wautier, J. L. (2003). *N*-(Carboxymethyl)lysine as a biomarker for microvascular complications in type II diabetic patients. *Diabetes Metab.* **29**, 44–52.

Further Reading

Raj, D., Choudhury, D., Welbourne, T., and Levi, M. (2000). Advanced Glycation end products: A nephrologist's prospective. *Am. J. Kidney Dis.* **35**, 365–380. *A well-documented review on renal complications (145 references).*

Schmidt, A. M., Yan, S. D., Yan, S. F., and Stern, D. (2001). The multiligand receptor RAGE as a progression amplifying immune and inflammatory responses. *J. Clin. Invest.* **108**, 949–955. *A comprehensive, concise article on RAGE.*

Thornalley, P. (1998). Cell activation by glycated proteins AGE receptors, receptor recognition factors and functional classification of AGEs. *Cell. Mol. Biol.* **44**, 1013–1023. *A didactic approach.*

Wautier, M. P., Chappey, O., Corda, S., Stern, D., Schmidt, A. M., and Wautier, J. L. (2001). Activation of NADPH oxidase by AGE links oxidant stress to altered gene expression via RAGE. *Am. J. Physiol. Endocrinol. Metab.* **280**, 685–694. *A convincing demonstration of the relationship between AGEs and reactive oxygen intermediates.*

Yamamoto, Y., Kato, I., Doi, T., Yorekura, H., Ohaski, S., Takeuchi, M., Watanabe, T., Yamagishi, S., Sakurai, S., Takasawa, S., Okamoto, H., and Yamamoto, H. (2001). Development and prevention of advanced diabetic nephropathy in RAGE overexpressing mice. *J. Clin. Invest.* **108**, 261–268.

Capsule Biography

Jean Luc Wautier is Professor of Hematology at the University of Paris and is the director of the research laboratory on cellular and vascular biology. He has published articles that are now milestones in the pathophysiology of diabetic angiopathy (*N. Engl. J. Med.* 1981, *Proc. Natl. Acad Sci USA* 1994, *J. Clin. Invest.* 1996, *Am. J. Physiol.* 2001, *Kidney Int.* 2002).

SECTION D

Knockout/transgenic Models

CHAPTER 162

Knockout Mouse Models Have Provided Invaluable Insights into Microvascular Development and Functions

[1]Kristofer Rubin, [1]Alexei V. Salnikov, and [1,2]Christian Sundberg

[1]*Department of Medical Biochemistry and Microbiology, Uppsala University, Uppsala, Sweden*
[2]*Department of Pediatrics, Uppsala Academic Hospital, Uppsala, Sweden*

Introduction

The primary function of microvessels forming the capillary bed is to enable exchange of nutrients, waste products, gases, and fluid between the blood and the parenchyma of a tissue. Microvessels can thus be regarded as the effector unit of the vascular system by fulfilling its primary task of transport. The microvascular units consist of the blood vessel (capillaries and venules) and surrounding loose interstitial connective tissues in which most, if not all, microvessels outside of the central nervous system are embedded. Comparisons of the genome of flies, worms, and humans reveal that many genes encoding extracellular matrix (ECM) proteins characteristic of loose interstitial connective tissues, such as fibronectin, elastin, interstitial fibrillar collagens, fibulin, and fibrillin, evolved in the chordate lineage in parallel with the vascular system. A functional role for fluid transport from the bloodstream to the interstitium by the loose connective tissues has been demonstrated. It has been proposed that the cells of the loose connective tissue control fluid fluxes by regulating the tension they exert on the interstitial ECM. The "microvascular unit" encompasses both the vessel proper and its surrounding loose interstitial connective tissue (Figure 1).

Null Mutants That Affect the Development of the Microvascular Unit

The use of null mutant mice has helped to define molecular mechanisms that are involved in the development of the embryonic vascular system. The knockout strategy has, however, some shortcomings that make the analyses of phenotypic changes complex and general conclusions about the proper function of a particular gene ambivalent. Many of the phenotypic changes that result from ablation of specific genes have been shown to depend on the genetic background (mouse strains) in which the ablations exist. Furthermore, many gene products are nonredundant; mice deficient in such genes are often not viable and die at various stages of development. To enable the study of functions of nonredundant genes, chimeric mice have been used. Here, the fates of null-mutant cells can be followed in a viable mouse. Furthermore, conditional knockouts in which the gene deletion only occurs in desired cell types and/or after induction also enable phenotype analyses of genetic ablations that result in embryonic lethality. Targeted gene deletions, however, often result in no or very subtle phenotypic changes, requiring extensive analyses to detect any phenotypic change that could be attributed to the genetic ablation introduced. Occasionally the phenotype becomes

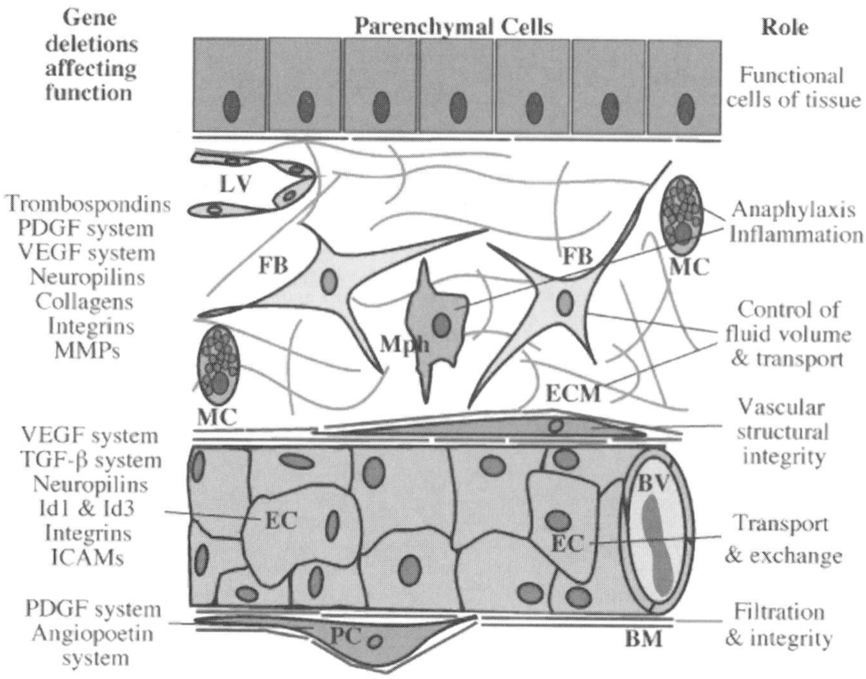

Figure 1 The structure of the microvascular unit and gene deletions that affect its function. LV, lymphatic vessel; FB, fibroblast; MC, mast cell; Mph, macrophage; EC, endothelial cell; PC, pericyte; BV, blood vessel; BM, basement membrane; ECM, extracellular matrix. (see color insert)

evident only during pathological processes such as tumor growth or inflammation.

Angiogenesis and vasculogenesis are commonly considered to be regulated by a balance of stimulatory and inhibitory factors. In the embryo, the loss of a particular gene may be compensated by alternative genes that, in the adult animal, may have become redundant. Similarly, overexpression of certain gene products in adult animals, such as by adenoviral transfer of genes into a tissue, may shift the natural balance between stimulatory and inhibitory factors, resulting in excessive or disturbed vessel formation or function. The latter case, however, does not necessarily imply that the particular gene has the same function during normal development of the vascular system. A parallel can also be drawn with neovascularization during malignant growth in which angiogenic factors are overproduced, causing a distorted and dysfunctional vascularity in the tumor.

The Endothelial Lining

Several families of growth factors and their receptors are expressed by endothelial cells during embryonic development. These growth factors orchestrate distinct phases of vasculogenesis and angiogenesis. Targeted deletion of genes encoding molecules governing endothelial cell proliferation, differentiation, migration, and survival emphasize the importance of a coordinated sequence of events for the development of a functional vascular system.

Vascular Endothelial Growth Factors (VEGFs)

VEGF was first described as a vascular permeability factor (VPF) and was later shown to be a specific and potent mitogen for endothelial cells. The VEGF family of endothelial cell growth factors includes several members (VEGF A–E) and splice variants that signal through cell surface receptors containing intracellular tyrosine kinase domains (VEGFR1–3). *Vegf* receptors are largely confined to endothelial cells. Mice lacking the *Vegf-A*, *VegfR-1*, *VegfR-2*, or *VegfR-3* genes die in utero during developmental stages E8.5 to E11.5. These deletions result in severe malformations of the cardiovascular system and in defective vascular remodeling. Both homozygous and heterozygous loss of the *Vegf-A* gene are lethal, with a severe vascular phenotype. In contrast, ablation of the *Vegf-B* gene in mice generates viable animals, although the heart is underdeveloped and small. Embryos deficient in *VegfR-1* fail to assemble endothelial cells into functional vessels. *VegfR-2* null embryos represent an early defect in the development of hematopoietic and endothelial cells. Analysis of *VegfR-3* null embryos revealed that although vasculogenesis and angiogenesis occurred, VEGFR-3 is needed for the remodeling and maturation of the primary vascular network into functional larger blood vessels. The fact that deletion of any one *Vegf* gene results in embryonic lethality demonstrates that the *Vegf* system is not functionally redundant during embryonal development.

FIBROBLAST GROWTH FACTORS (FGFS)

In contrast to the VEGF system, single- and double-knockout mice for *Fgf1* and *Fgf2* genes display no developmental vascular phenotype, suggesting functional redundancy among related molecules. Data on the role of FGFR-1 in vascular development are limited because of embryonal death before gastrulation after targeted inactivation of the *FGFR-1* gene.

TRANSFORMING GROWTH FACTOR-β (TGF-β)

Members of the TGF-β family of growth factors participate in the regulation of endothelial cell function during embryonal vascular development. Around half of mouse embryos lacking the *TGF-β1* gene die at E10.5, with embryos defective in vasculogenesis in a process that appears to involve endothelial cell differentiation, ultimately leading to inadequate capillary tube formation and integrity. Since there is no obvious phenotypic overlap between *TGF-β2-* and *TGF-β3*-null mice, TGF-β1 seems to have isoform-specific functions during vasculogenesis. TGF-β receptor (*TβRI-II*)- or *Alk1*-deficient mice die at mid-gestation, exhibiting severe vascular abnormalities reminiscent of animals lacking the *TGF-β1* gene. The blood vessel defects in *Alk1* null mice are characterized by excessive fusion of capillary plexuses and hyperdilation of large vessels. Deletion of the endoglin gene encoding an auxiliary component of the TGF-β receptor complex has led to defective vascular development due to arrested endothelial remodeling and poor vascular smooth muscle development. Similar vascular defects were observed in mice lacking SMAD5 protein.

NEUROPILINS

Neuropilin (Nrp)1 and Nrp2, nonkinase receptors for class 3 semaphorins and co-receptors for VEGF165 and VEGF145 isoforms, are expressed on endothelial cells and other cell types. Nrp1 has been found to potentiate VEGF-A binding to VEGFR-2. *Nrp1* knockout animals have defects in the yolk sac, in embryonal vascularization, and in the development of large vessels in the heart. Homozygous *Nrp2*-null mutants are partially or completely deficient in small lymphatic vessels and capillaries during development. Nrp2, therefore, is selectively required for the formation of small lymphatic vessels and capillaries. Double mutant embryos for *Nrp1* and *Nrp2* die at E8.5 and have an abnormal vascular phenotype resembling that of *Vegf-A*, and *VegfR-2* knockout animals.

ID1 AND ID3 TRANSCRIPTION FACTORS

Id1 and Id3 proteins are expressed in endothelial cells and are required for angiogenesis in the brain. *Id1–Id3* double knockout mice display vascular malformations in the forebrain and an absence of branching and sprouting of blood vessels into the neuroectoderm.

THROMBOSPONDINS

The thrombospondins are large, secreted oligomeric multidomain glycoproteins that exert their functions at the cell surface. The family comprises five members of which thrombospondin-1 and -2 (TSP-1 and -2) have antiangiogenic effects. Mice deficient in TSP-2 have increased vascular density and increased angiogenesis in subcutaneously implanted sponges. These results support the concept of a balance between angiogenic and antiangiogenic factors for neovascularization and identify thrombospondins as a modulator of angiogenesis in vivo.

INTEGRINS

Integrins are a large family of heterodimeric cell surface glycoproteins comprising noncovalently associated α- and β-chains that mediate intercellular and cell–ECM adhesion. Several mouse lines deficient in specific integrins have been generated. Whereas genetic ablation of some integrin subunits, notably the integrin β1 subunit, is lethal in the early embryo, genetic ablation of other integrin subunits causes no, or very subtle, phenotypes. Loss of the genes encoding the integrin α5 subunit results in severe vascular defects and early death of the embryo. The integrin α5 gene negative embryos have grossly distended and leaky blood vessels both in the embryo itself and in the vitellin yolk sac. Approximately 80 percent of mouse embryos lacking the integrin αv subunit gene die at mid-gestation, and about 20 percent are born at term. Embryos deficient in the genes encoding integrin β8 and αv subunits have a similar phenotype characterized by brain vessel abnormalities and underdeveloped placental labyrinth layers. It therefore seems likely that αvβ8 integrin functions to anchor brain vessels to the parenchyma and in the development of placental blood vessels. The fact that αv integrin gene null mice have extensive angiogenesis and vasculogenesis demonstrates that the αv family of integrins is not essential for blood-vessel development. This is further supported by the fact that mice deficient in the integrin β3 and/or β5 deficient are fertile, and, apart from a Glanzmann's thrombasthenia–like condition in integrin β3 gene null mice, their vasculature is normal. β3 and/or β5 deficient mice also support tumor angiogenesis and growth. This contrasts with results showing that selective inhibitors of the αvβ3 and αvβ5 integrins inhibit angiogenesis in tumors and in chick allantoic membranes. It has been proposed that this apparent contradiction can be explained by the fact that αvβ3 and αvβ5 integrins regulate endothelial cell apoptosis when bound to the proper ligand. This αv integrin-mediated control of apoptosis and angiogenesis may be controlled by specific inhibitors. Mouse embryos lacking the integrin α4 subunit die during E11 to E14 as a result of defects in the heart and placenta.

INTERCELLULAR CELL ADHESION RECEPTORS

Endothelial adhesion receptors mediating cell–cell adhesion, such as vascular endothelial cadherin (VE-cadherin), connexins (Cx), and platelet endothelial cell adhesion molecule (PECAM-1/CD31), are essential for various aspects of endothelial cell structure, morphogenesis, and integrated vessel functions. VE-cadherin mediates adhesion between endothelial cells, and targeted genetic inactivation of this cadherin induces apoptosis and impairs remodeling and maturation of vascular plexuses. Mice lacking the genes encoding either Cx37 or Cx40, the predominant gap junction proteins present in the vascular endothelium, are viable and exhibit largely nonblood-vessel-related phenotypes. However, double knockout mice for Cx37 and Cx40 display severe vascular abnormalities and die during the perinatal period. PECAM-1/CD31 has also been linked to vasculogenesis and angiogenesis and has been implicated as a key mediator of the transendothelial migration of leukocytes. Mice deficient in PECAM-1/CD31 are viable and show no apparent vascular phenotype. However, in response to inflammatory stimuli, leukocytes are arrested between the vascular endothelium and the basement membrane of mesenteric microvessels at the inflammatory site in PECAM-1/CD31 deficient mice. The later verifies a role for PECAM-1/CD31 in the migration of leukocytes through the subendothelial ECM.

The Pericytes

Pericytes are juxtapositioned to the abluminal surface of endothelial cells, continuous with the basement membrane in venules, capillaries, and small arterioles. They are structurally and functionally similar to smooth muscle cells. Intermediate cells, possessing characteristics of both cell types, exist in the transitional portion between micro- and macrovessels, and evidence exists that pericytes in fact differentiate into smooth muscle cells. Furthermore, close similarities also exist with fibroblasts/myofibroblasts, and a growing body of evidence suggests that pericytes act as progenitors of these cells in adult reactive tissues. Pericytes have also been suggested to control endothelial cell behavior, vascular structural integrity, and remodeling of the microvascular bed.

PLATELET-DERIVED GROWTH FACTOR (PDGF)

PDGF β-receptors are expressed on microvascular pericytes in human pathologic conditions, therefore suggesting that pericytes play a central role in PDGF β-receptor signaling in adult tissues in vivo. Gene deletion studies in mice have shown that PDGF β-receptor signaling is important in pericyte behavior during embryogenesis. Disruption of the PDGF β-receptor system results in a lack of recruitment and vessel coverage of pericytes, poor development of the muscular coat in larger vessels, edema formation, distension of the heart, placental defects, and deficient mesangial cell development in the kidney. The collective consequence of these disturbances leads to vascular microaneurysm formation and extensive hemorrhaging, resulting in near-term lethality of the embryo. The results suggest that PDGF β-receptor expressing embryonic vascular cells act as the progenitors of pericytes and vascular smooth muscle cells. Data supporting these findings are found in diseases in human adults, for instance, as a part of the pathophysiology of pulmonary hypertension.

Although pericytes act as progenitors for vascular smooth muscle cells during embryogenesis, PDGF β-receptor chimeric mice are deficient in fibroblast development during reactive conditions, such as wound healing in adult animals. This suggests that pericytes act as progenitor cells for fibroblasts during reactive conditions in the adult animal, a notion that is supported by studies of PDGF β-receptors in many human pathologies. This illustrates the important point that events occurring during embryonic development may not necessarily be used to predict events in adult reactive tissues.

Mice lacking genes encoding the PDGF α- or β-receptors, as well as PDGF A- or B-chains, have been generated. Although different organs are affected in different animals, one common denominator for the function of the PDGF system has been identified. PDGF receptors are expressed on the pericytes and myofibroblasts that are involved in the folding of endothelial and epithelial sheaths, maximizing the functional surface area (and thus the interexchange of solutes) at the interface between two tissue compartments. This is important in the placenta, microvascular bed, kidney glomeruli, small intestine, and lung, organs that are all affected in mice lacking various components of the PDGF system.

PDGF-BB participates in the control of capillary-to-interstitium transport of fluid by stimulating contractility of fibrocytes or pericytes in the loose interstitial connective tissue surrounding peripheral blood vessels. Evidence for this physiological function of PDGF-BB includes data obtained from "knock-in" mice harboring mutated PDGF β-receptors that are unable to activate phosphatidylinositide-3-kinase after ligand stimulation. PDGF-BB normalizes the lowered dermal interstitial fluid pressure following anaphylaxis in wild-type mice but not in those with mutated PDGF β-receptors.

ANGIOPOIETIN/TIE RECEPTOR

The angiopoietin (Ang)/Tie receptor system is important in vascular integrity, maturation, and maintenance. Mice deficient in the *Ang-1* gene exhibit diminished pericyte coverage, hemorrhages, reduced vessel branching, distended structurally aberrant vessels, and increased vessel permeability, suggesting a defect in vessel maturation. In adult reactive tissues, Ang-1 is expressed in pericytes, whereas the Tie receptors are expressed on endothelium during the latter stages of angiogenesis. This suggests that rather than acting directly on pericytes, Ang-1 produced by pericytes stimulates Tie-2 receptor-bearing endothelial cells. They in turn produce a yet unknown factor creating a paracrine reciprocal stimulation resulting in pericyte recruitment and vessel maturation. The Tie-1 and Tie-2 receptors are tyrosine kinases expressed in both mature and developing endothelial cells and are important in angiogenesis. Analysis of *Tie-1* null mice identified the requirement of Tie-1 for the establishment of the structural integrity of vascular endothelial cells and their survival. Mouse embryos with a deleted *Tie-1* gene develop severe edema and hemorrhages. However, the ligand for the Tie-1 receptor has not been identified yet. In contrast, the Tie-2 receptor binds Ang-1, Ang-2, and Ang-4. Tie-2 has a significant role in angiogenesis, especially for vascular network formation. Consequently, although vasculogenesis in mice proceeds in the absence of members of the Ang/Tie system, vessel maturation and functional properties of the vessels are severely compromised.

The Extracellular Matrix

BASEMENT MEMBRANES

The major components of basement membranes are type IV collagens, laminins, nidogens, perlecan, and fibronectins. Extensive studies on the in vivo role of specific basement membrane proteins have been performed using gene targeting. The importance of basement membranes for normal development is evident from that mice carrying a null mutation in the laminin γ1 chain gene die as early as around E5.5. Ablation of some other genes encoding basement membrane components give less severe phenotypes, presumably due to their less widespread tissue distribution. This can be exemplified by that null mutant mice lacking β2 chains of a laminin variant present in the glomerulus basement membranes die early from renal malfunction. Collagen types XV and XVIII are present in the endothelial basement membrane zones in several tissues. Mice lacking type XV collagen are viable and fertile but suffer from muscle and cardiac defects. Type XV collagen appears to be a structural component necessary to stabilize microvessels in skeletal muscle and in the heart.

INTERSTITIAL COLLAGENS

Collagen type I is the most abundant protein in the body. Insertional mutation by Mov 13 retrovirus in the procollagen α1(I) gene has been used to generate mouse embryos lacking collagen type I. The embryos die due to rupture of major blood vessels between days 12 and 14 of gestation. Genetic ablation of the procollagen α1(III) gene in mice results in a more subtle phenotype in which vessel rupture occurs in adult animals. These studies show the role of the interstitial collagens in providing tensile strength to vessels.

FIBULIN

Inactivation of fibulin-1, another ECM protein of the vessel wall, has been shown to lead to dilation and rupture in the endothelial lining of microvessels and massive hemorrhages in tissues starting at mid-gestation.

FIBRONECTINS

Genetic ablation of fibronectin (FN), the only known ligand for the α5β1 integrin, results in a phenotype similar to that of integrin α5 subunit–deficient mice. FN null mice die at E8.5 with severe defects in tissues derived from the mesoderm. Embryonic vessels and the heart are deformed. The FN-null phenotype is more severe than the phenotypes arising from ablation of any of the integrin α-subunits and, therefore, it is likely that the integrins, notably the α5β1 and αv integrins, can partly compensate for each other during early embryonic development.

PROTEINASES

Mice deficient in various MMPs show certain defects in angiogenesis. For example, genetic ablation of MMP-9 results in a delay in endochondral bone formation, an effect that is reportedly due to delayed neovascularization. Transgenic insulinomas growing in MMP-9-deficient mice fail to undergo the angiogenic switch, indicating a role for MMP-9 in tumor angiogenesis. Mice deficient in genes encoding MMP-2, MMP-7, or MMP-11 display reduced carcinogenesis in various models, possibly dependent partly on defective or disturbed tumor angiogenesis. Genetic ablation of the membrane-anchored MT1-MMP (MMP-14) gives a phenotype with a defective angiogenesis during development.

Summary

The mechanisms behind the formation of functional microvascular units are now being elucidated at the cell-biological and biochemical levels. Our understanding of vasculogenesis and angiogenesis during embryogenesis has increased dramatically by studying murine null mutants, deficient in expression of specific proteins. However, caution in extrapolating these findings to adult tissues and pathological processes is warranted. The field of angiogenesis research must utilize a broad repertoire of models in order to identify differences and similarities between embryonic events and pathological conditions in the adult. Thus, a combination of these models will provide information regarding pathological angiogenesis, but also, and just as important, mechanisms by which a hierarchically functional vascular bed is formed. Furthermore, previously unrecognized functions of certain molecules in vasculogenesis and angiogenesis have been identified by experiments using murine null mutants.

Acknowledgements

The authors are supported by grants from the Swedish Cancer Foundation (to K.R. and C.S.), the Swedish Science Council (to C.S.), Konung Gustaf V:s 80-årsfond (to K.R. and C.S.), and Clas Groschinsky foundation (to C.S.). Editorial assistance by Dr. Alan McWhirter is gratefully acknowledged.

Bibliography

Aukland, K., and Reed, R. K. (1993). Interstitial-lymphatic mechanisms in the control of extracellular fluid volume. *Physiol. Rev.* **73**, 1–78. *An important review of the physiology of microcirculation.*

Carmeliet, P. (2000). Mechanisms of angiogenesis and arteriogenesis. *Nat. Med.* **6**, 389–395. *An excellent review of mechanisms for angiogenesis and vasculogenesis.*

Folkman, J. (1995). Angiogenesis in cancer, vascular, rheumatoid and other disease. *Nat. Med.* **1**, 27–31. *An excellent summary of the concept of antiangiogenic therapy written by a leader in the field.*

Gustafsson, E., and Fässler, R. (2000). Insights into extracellular matrix functions from mutant mouse models. *Exp. Cell Res.* **261**, 52–68. *An excellent summary of knockout mice lacking extracellular matrix proteins.*

Hynes, R. O. (2002). A reevaluation of integrins as regulators of angiogenesis [Review]. *Nat. Med.* **8**, 918–921. *An interesting discussion of the importance of specific integrins in angiogenesis.*

Hynes, R. O., and Zhao, Q. (2000). The evolution of cell adhesion [Review]. *J. Cell Biol.* **150**, F89–96.

Jain, R. K. (2003). Molecular regulation of vessel maturation. *Nat. Med.* **9**, 685–693. *An excellent review of angiogenesis and vasculogenesis.*

Sundberg, C., Ljungström, M., Lindmark, G., Gerdin, B., and Rubin, K. (1993). Microvascular pericytes express platelet-derived growth factor β-receptors in human healing wounds and colorectal adenocarcinoma. *Am. J. Pathol.* **143**, 1377–1188.

Veikkola, T., Karkkainen, M, Claesson-Welsh, L., and Alitalo, K. (2000). Regulation of angiogenesis via vascular endothelial growth factor receptors. *Cancer Res.* **60**, 203–212. *An excellent review of growth factors involved in angiogenesis.*

Wiig, H., Rubin, K., and Reed, R. K. (2003). A new and active role of the interstitium in control of interstitial fluid pressure—potential therapeutic consequences. *Acta Anaesthesiol. Scand.* 47:111–121. *This review article summarizes evidence implying that connective tissue cells control capillary-to-interstitium transport of fluid in the microcirculation.*

Capsule Biography

Kristofer Rubin, Ph.D., has been a professor of connective tissue biochemistry at Uppsala University, Sweden, since 1995. His laboratory focuses on the role of growth factors and adhesion receptors in tumor stroma physiology. His work is supported by The Swedish Cancer Foundation.

Alexei V. Salnikov earned his M.D., Ph.D. in histology and physiology from the Siberian Medical University in Russia in 1999. He held joint appointments at the Department of Anatomy and General Pathology at the same institution and at Children's Hospital. His current work focuses on angiogenesis and pathophysiology of tumor stroma.

Christian Sundberg, M.D., Ph.D., early on studied the role of integrins and growth factors in inflammation and tumors at Uppsala University, Sweden, and experimental pathology at Harvard Medical School. He is now heading a research group focused on mechanisms underlying tissue fibrosis and optimal tissue regeneration.

SECTION E

Mathematical Models

Chapter 163

Mathematical Models of Angiogenesis

James W. Baish

Department of Mechanical Engineering, Bucknell University, Lewisburg, Pennsylvania

Definition

Mathematical models of angiogenesis are symbolic representations of the physical principles and physiological processes by which new blood vessels arise from existing vessels. They are used to predict vascular density, vascular architecture, kinetics of vascular growth, physiological function, and response to treatment. Models have been developed for angiogenesis during primary development of the organism, wound healing, and growth of tumors in experimental assays or that arise spontaneously in situ.

Introduction

Mathematics can be a powerful tool to aid in the study of angiogenesis. Whereas much is known about the molecular mechanisms of angiogenesis, considerably less is understood about how genetic and molecular processes translate into the variety of vascular architectures that are observed in health and disease. This article focuses on some of the general principles of mathematical modeling as it is applied to angiogenesis and presents examples from several types of models that illustrate some widely used techniques.

A mathematical perspective is especially valuable for integrating existing and new experimental observations, testing and generating hypotheses, planning new experiments, and guiding the scale-up of animal experiments to clinical trials. Although seldom able to predict the full spectrum of complex behavior observed in biological systems, mathematical models aid in establishing the limits imposed by physical principles and yield insight into the sensitivity of the organism to changes in parameters.

Conceptual Framework

Mathematical models of angiogenesis typically consider the tissue as consisting of two subvolumes: the vascular space and the extravascular space. Separating these subvolumes is the blood vessel wall consisting of endothelial cells and perivascular cells, such as smooth muscle cells, along with a matrix component. The surrounding extravascular space is a porous solid consisting of extracellular matrix, tumor cells, or mesenchymal cells specific to the host tissue. The vascular space is the site of the most vigorous fluid movement, but fluid may also slowly flow through the interstices of the extravascular space. The primary goal of most mathematical models of angiogenesis is to predict the changes in the vascular architecture that occur over time by tracking either the density of blood vessels or the location and diameter of each vessel segment.

Angiogenesis involves a series of steps. First, the existing vessel walls respond to angiogenic stimuli such as diffusing cytokines through vasodilation and increased permeability. Second, as the basement membrane supporting the endothelial cells and components of the extracellular matrix are degraded, migration and proliferation of the endothelial cells can begin. Third, the newly formed sprouts interconnect, form lumens, and conduct blood flow. Finally, the provisional vasculature stabilizes and remodels to meet the needs of the specific site in a series of steps. These maturation steps include tightening of intercellular junctions to regulate vascular permeability, regression of redundant vessels, and recruitment of perivascular cells. The connectivity of the vascular network is regulated to meet the metabolic needs of the tissue while the structure of the vessel walls responds to the pressure and shear stress of the flowing blood. In healthy tissues, the maturation processes lead to a tree structure of arteries and veins that serve as highly efficient conduits for transporting blood over large distances to and from an orderly capillary bed, which efficiently brings blood close enough to each cell so that diffusion can complete the delivery of nutrients. Under pathological conditions such as those found in solid tumors or diabetic retinopathy, one or more of the maturation processes goes awry, leading to chaotic patterns of vessels that show little evidence of a hierarchical artery and vein structure or an orderly capillary bed.

Physical Principles

A handful of physical principles underlie most mathematical models of angiogenesis. Typically, these include conservation principles that account for the concentration of specific chemical species and the number of cells of various types. The rate of change of each conserved quantity is equal to the rate of spatial movement plus the net rate of production balanced against net losses. Each conservation law must be coupled to a set of rules that govern the rates of movement, production, and loss of each conserved quantity. Typical models of chemical species and cell populations are described in the following sections.

Conservation of Chemical Species

In the case of chemical species we track the concentration c of a substance such as VEGF, bFGF, or O_2 by

$$\underbrace{\frac{\partial c}{\partial t}}_{\text{Change with time}} = \underbrace{-\nabla \cdot \mathbf{J}_c}_{\text{Movement}} + \underbrace{g}_{\text{Net Generation}}$$

where $\mathbf{J}_c$ is the flux of the species and g is the net generation of the chemical species. Movement may occur by two mechanisms, diffusion and convection, such that $\mathbf{J}_c = \mathbf{J}_{diffusion} + \mathbf{J}_{convection}$. Diffusion arises from the net movement of molecules from regions of high concentration to those of low concentration as modeled by $\mathbf{J}_{diffusion} = -D_c\nabla c$, where the diffusion coefficient D_c represents the molecular diffusivity. Diffusion coefficients for relevant substances are primarily a function of their molecular weights, with small molecules such as O_2 and NO diffusing more readily than ones with large molecular weights. Convection is the bulk movement of material such that $J_{convection} = \mathbf{v}c$ where $\mathbf{v}$ is the local average velocity of the fluid (or in exceptional cases, solid undergoing large deformations). The convective velocity arises from fluid pressure differences in the tissue that can be modeled by standard tools from fluid mechanics. Convection is vigorous within blood vessels and in tumors, and it may also be significant in the extravascular space because of pressure gradients in the interstitium that arise as a result of leakage through the relatively permeable vessels found there. When the effects of many blood vessels are averaged, an alternative formulation may be used in which $\nabla \cdot \mathbf{J}_{convection} = Q(c_{out} - c_{in})$, where Q is the blood flow rate per unit volume and c_{out} and c_{in} are the concentrations coming from and going to the organ via the bloodstream, respectively. The generation and loss terms can arise from secretion or absorption by cells or reaction with other substances. The kinetics of the reactions are specific to particular models.

Conservation of Cell Populations

The principles governing cell populations are close analogs to those for chemical species, but represent processes on a cellular rather than molecular scale. The conservation equation for a population of cells such as endothelial cells is given by

$$\underbrace{\frac{\partial n}{\partial t}}_{\text{Change with time}} = \underbrace{-\nabla \cdot \mathbf{J}_n}_{\text{Movement}} + \underbrace{M}_{\text{Net Generation}}$$

where the first term represents the rate at which the cell density changes within a small unit of tissue due to the flux of cells and the net increase in cell numbers M that includes gains due to mitosis and differentiation and losses due to cell death. The flux of cells may have four contributions: $\mathbf{J}_n = \mathbf{J}_{random} + \mathbf{J}_{chemo} + \mathbf{J}_{hapto} + \mathbf{J}_{convection}$, where $\mathbf{J}_{random}$ is the random motility of the cells, $\mathbf{J}_{chemo}$ is the chemotactic flux, $\mathbf{J}_{hapto}$ is the haptotactic flux, and $\mathbf{J}_{convection}$ is the convective flux of cells. The random motility is similar in principle to the random diffusion of molecules. Cells have a net, random migration from regions of high cell density to regions of lower density as modeled by $\mathbf{J}_{random} = -D_n\nabla n$, where D_n is the cell random motility coefficient, analogous to the diffusion coefficient for molecules, and ∇n is the gradient in the density of cells.

The chemotactic flux arises from the ability of some cells to move selectively toward regions having a higher concentration of specific chemical species. The chemotactic flux may be represented by $\mathbf{J}_{chemo} = \chi(c)n\nabla c$, where the flux is proportional to the density of cells and the gradient of the chemical species as well as a chemotactic function $\chi(c)$ that depends on the kinetics of the receptors on the cells. A suitable form of the chemotactic function that reflects a progressively decreasing sensitivity to the chemical species at higher concentrations is $\chi(c) = \chi_0 k_1/(k_1 + c)$, where the chemotactic coefficient χ_0 and k_1 are positive constants.

The haptotactic flux arises from the cellular interaction with components of the extracellular matrix such as fibronectin. The haptotactic flux is modeled by $\mathbf{J}_{hapto} = \rho_0 n\nabla f$, where ρ_0 is the haptotactic coefficient, and ∇f is the gradient in the concentration of an extracellular matrix component.

And as with chemical species, the convective flux can be represented by $J_{convection} = n\mathbf{v}$, where $\mathbf{v}$ is the rate of displacement of the cells and matrix from their original configuration. Cells in the extravascular space are not generally carried along by moving fluid, so the only bulk motion that can arise is from growth of the tissue, an effect seldom considered. In contrast, cells inside the vessels can be carried by the blood. A recent model by Stoll et al. shows how circulating progenitor cells can add significantly to the population of endothelial cells that are available to line blood vessels.

The models just described are best suited when a high degree of homogeneity in the cell phenotypes and extracellular matrix may be assumed. Heterogeneity in the underlying substrate requires some different perspectives, such as those from percolation theory, which is discussed later.

Examples

A Model of Proangiogenic and Antiangiogenic Factors without Explicit Vascular Architecture

The preceding principles may be utilized to test the hypothesis that a local imbalance between pro- and antiangiogenic factors in regions of growing solid tumors may lead to focal suppression of angiogenesis. This model divides the diffusing cytokines into two categories: those stimulating angiogenesis and those inhibiting angiogenesis. The steady state conservation of diffusing factors is modeled by

$$0 = \underbrace{D\nabla^2 c}_{Diffusion} + \underbrace{g}_{Generation} - \underbrace{kc}_{Degradation}$$

where c is the concentration of either a pro- or an antiangiogenic factor, D is the diffusivity, k is the degradation rate constant, and g is the generation rate of the factor. Although the specific molecular mechanisms of their actions are not modeled, the pro- and antiangiogenic substances are distinguished by their diffusivity, activity, and generation and degradation rates—all of which can take on different values in the tumor and host tissues.

Analytical solutions are obtained for a spherically symmetric tumor in the midst of tumor-free host tissue. The concentrations are normalized relative to the concentrations far from the tumor in the host tissue (proangiogenic: $\theta^+ = c^+/c^+_\infty$, antiangiogenic: $\theta^- = c^-/c^-_\infty$) where there exists a stable vascular network. A net antiangiogenic effect occurs if $\theta^+/\theta^- < 1$, leading to a local suppression of angiogenesis, and a net angiogenic effect occurs if $\theta^+/\theta^- > 1$, leading to growth of new vessels.

Figure 1 shows profiles of the dimensionless concentration ratio for angiogenesis stimulators and inhibitors for the baseline set of parameter values given in Table I. For the baseline case, only a weak imbalance between stimulation and inhibition is present, which produces regions of vascular growth as well as focal suppression of angiogenesis. Greater imbalances in parameter values can induce either suppression of angiogenesis everywhere or global stimulation of vessel growth. Small tumors tend to have an excess of proangiogenic factors promoting angiogenesis throughout. In contrast, angiogenic suppression arises at the centers of larger tumors with stimulation persisting only near the tumor periphery, as shown in Figure 1.

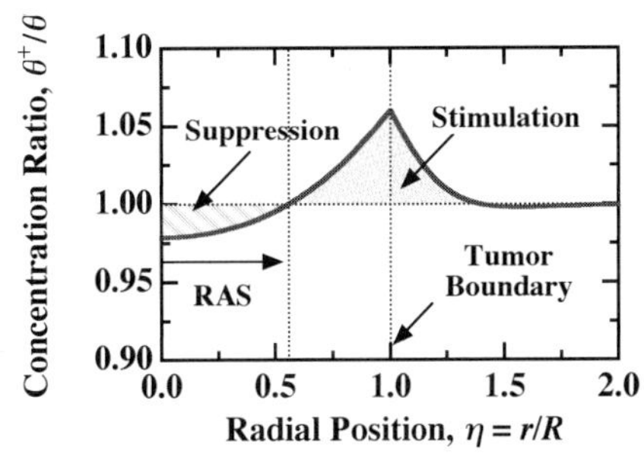

Figure 1 Concentration ratio profile in tumor ($\eta < 1$) and host tissues ($\eta > 1$) for a typical parameter set with crossover from net angiogenesis inhibition to stimulation at $r = 0.56R$ and peak angiogenesis stimulation at the tumor surface. RAS denotes the region of angiogenic suppression. Reprinted from Ramanujan et al. (2000). *Cancer Res.* **60**, 1442–1448, with permission.

Table I Parameter Values Used to Produce the Results in Figure 1.

	Proangiogenic factors	Antiangiogenic factors
D_{tumor}/D_{host}	1.4	1.2
Dimensionless production rate	Tumor: 136 Host: 31	Tumor: 109 Host: 29
Dimensionless degradation rate constants	Tumor: 4.8 Host: 5.6	Tumor: 4.1 Host: 5.4

This model points to the need for precise knowledge of chemical kinetic constants and diffusion constants for proangiogenic and antiangiogenic factors in tumor and host tissue. The model illustrated in Figure 1 has been extended to assess the impact of primary tumors on angiogenesis in tumors at a secondary site as well as the impact of circulating endothelial progenitor cells on tumor growth and angiogenesis.

A Model of Explicit Vascular Architecture Induced by Diffusing Angiogenic Factors

The previous model was useful for identifying local tissue conditions that would be conducive to angiogenesis. In the second model we consider how diffusing angiogenic factors can affect endothelial cell migration and vascular network production. This model is from a class of models used to investigate the corneal angiogenesis assay in which a few tumor cells or other angiogenic stimulus is implanted into a previously avascular tissue, the cornea. Sprouting and migration of the vascular network may then occur from the existing vessels at the periphery toward the implant. In this case, conservation of an angiogenic factor is modeled by

$$0 = \underbrace{D\nabla^2 c}_{Diffusion} - \underbrace{kc}_{inactivation} - \underbrace{uLc}_{uptake}$$

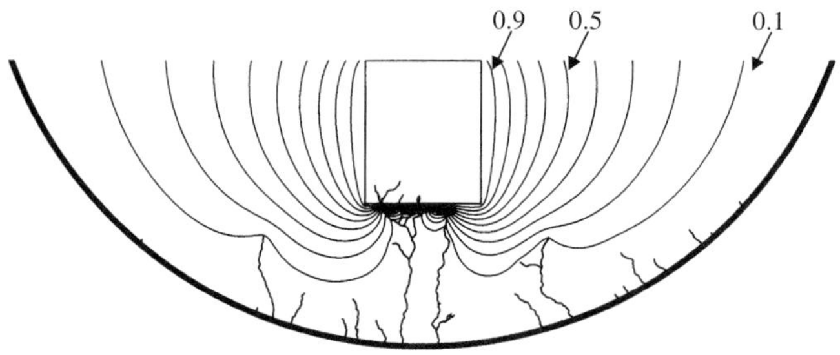

Figure 2 Vascular network at a relatively high rate of bFGF uptake (u = 10,000 μm/hour, at 120 hours). The figure shows only the lower part of the cornea. The box at the center represents an implant of tumor cells. The angiogenic stimulus was too weak to produce sprouting elsewhere. Contours of relative concentration for bFGF are shown to illustrate how the uptake by the vessels affects the concentration field. Reprinted from Tong and Yuan (2001). *Microvasc. Res.* **61**, 14–27, with permission from Elsevier.

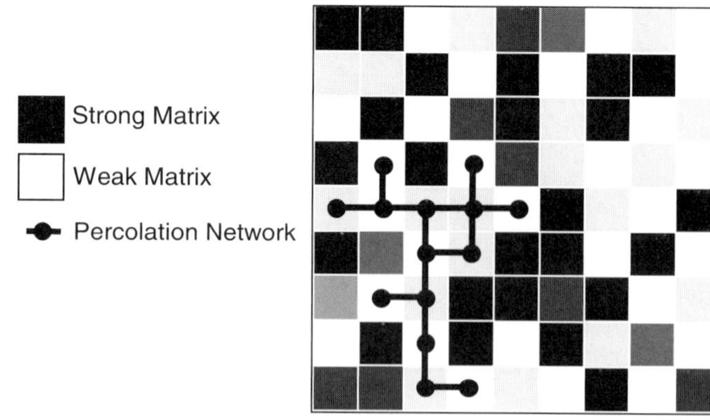

Figure 3 Small sample of a percolation network formed by invading the weakest sites in the underlying substrate. The network is shown after 13 steps of growth beginning at the center of the array. The relative strength of the substrate is assigned by a random number generator with no spatial correlation. Reprinted from Baish and Jain (2000). *Cancer Res.* **60,** 3683–3688, with permission.

where k is the inactivation rate constant that applies throughout the cornea, u is the rate constant for uptake by the vessels when they are present, and L is vessel density, defined as the length of vessel per unit area. Angiogenesis is assumed to occur only when the concentration of an angiogenic factor exceeds a fixed threshold. This is consistent with qualitative observations, but is not yet based quantitatively on experimental evidence. Sprout formation is modeled as a stochastic process. The probability of adding a new sprout is assumed to be proportional to the length of the existing segment and the local concentration of angiogenic factors. Once a sprout appears, it tends to grow in the direction of the concentration gradient in angiogenic factors and also tends to persist in its growth in a consistent direction. The persistence of direction for the chemotactic response is governed by a user-selected parameter. The rate of growth is proportional to the concentration of angiogenic factors. Figure 2 shows how vessel growth responds to angiogenic factors and how dynamic changes in the vascular architecture can modify the distribution of the angiogenic factors themselves.

Models of Heterogeneity of the Extracellular Matrix: Percolation Models

The preceding models reflect the response of endothelial cells in a homogeneous substrate in which they may move freely to any location. Whereas normal capillaries eventually form relatively regular patterns, the chaotic architecture of tumor vasculature suggests that a model reflecting some underlying heterogeneity in the substrate may be useful.

Percolation is such a model. Widely used in other settings, such as the movement of oil through randomly fractured rock, it provides a rich set of simulation tools and analytical results for transport in random media. A primary motivation for developing percolation models of tumor vasculature is that vascular patterns in animal models show avascular voids of many sizes that closely match the fractal scaling characteristics known to arise from percolation processes. Percolation networks also mimic the vascular tortuosity observed in tumors.

Figure 3 illustrates how percolation can be used to model the expansion of a network of blood vessels. Beginning from an initial network, the blood vessels invade the readily accessible, "weaker" material while growing around randomly placed, "stronger" obstacles. Local heterogeneity may occur as a result of variations in the actual mechanical strength of the extracellular matrix as it is degraded and secreted, or alternatively can reflect heterogeneity in the phenotype of the cells themselves with some cells having greater angiogenic activity than others. The relative degree of substrate heterogeneity determines whether its effects are more significant than gradients established by diffusing factors. Flow through the resulting network of connected tubes can be readily modeled by standard fluid mechanical methods so that the transport of bloodborne diffusible substances such as nutrients, drugs, and imaging tracers may be studied.

In a real tumor, the length scales within which the percolation model may be applied are bounded. At the absolute upper limit is the size of the largest avascular space. Above this so-called correlation length, the blood vessels fill the tissue more or less homogeneously. The diameter of the blood vessels themselves puts an absolute lower limit on self-similarity. The measurements available to date suggest that percolation-like scaling exists in many tumors between the limits of about 50 μm to 5 mm.

Gamba et al. recently showed that percolation-like, fractal scaling also occurs in experiments of in vitro formation of blood vessels by cells grown in a gel matrix. A critical density of cells exists, below which cells form disconnected clusters and above which a single connected network forms.

Models of Vessel Maturation: Branching Models with Vascular Diameter

In healthy tissues, blood vessels undergo a maturation process that prunes and remodels the vasculature into a morphology that is nearly optimal from an engineering perspective. The vascular network delivers nutrients near to every cell in a three-dimensional volume while consuming a near minimum of energy for transport. Although the vasculature apparently has no direct means of seeking such a global optimum, considerable evidence suggests that mechanochemical transduction permits regulation of both vessel diameter and connectivity in response to local conditions of shear stress and pressure. A classic model of branching for mature arteries was developed by Murray in 1926. He found that when vessel *i* divides into vessels *j* and *k*, the diameters of the vessels that minimize the pumping and blood volume costs to the organism satisfy $d_i^3 = d_j^3 + d_k^3$. A consequence of what is now known as Murray's law is that the shear stress on the endothelial cells lining all vessels of all sizes should be constant. Empirically, $d_i^{2.7} = d_j^{2.7} + d_k^{2.7}$ has been found to hold remarkably well over many generations of branching. Recent extensions of this style of modeling yield the spatial distribution of the vessels and incorporate other aspects of the physiology, including diffusing factors and nutrients. West et al. have recently applied scaling arguments to the branching of larger blood vessels to explain why the metabolic rates of entire organism scale with mass as $m^{3/4}$. These models give insights into the number and size of the vessels, but not their actual locations.

Future Directions

Many recent advances in the specific molecular and genetic mechanisms underlying angiogenesis have yet to be translated into a mathematical framework. Mathematics should be particularly useful for elucidating why different tumor cell lines and different angiogenic and antiangiogenic stimuli produce grossly different vascular morphologies. Valuable insights may also be obtained into how the rates of vascular remodeling depend on the dose and means of administering angiogenic inhibitors. Specifically, work is needed on the kinetics of specific molecular and cellular responses. In addition, much remains to be learned about the dynamics of the extracellular matrix. There are also gaps in our ability to integrate models of processes on the endothelial-cell and capillary level with models of the arteries and veins that connect to these smaller vessels. A key here appears to be improved models of perivascular cells. Overall, the value of mathematical modeling for angiogenesis will be enhanced as more of the models fully incorporate physiological details as well as physical principles based on independently measured parameter values.

Glossary

Chemotaxis: The ability of cells to move preferentially toward regions of higher concentration of a diffusing factor.

Conservation: The principle that the rate of change of a quantity such as mass or cell number depends on the net balance between sources and sinks for the quantity.

Haptotaxis: The ability of cells to move preferentially toward regions of higher concentration of specific components of the extracellular matrix.

Murray's law: A rule that relates the diameter of the daughter vessels at a bifurcation to the diameter of the parent vessel in such a way that total pumping and blood volume costs are minimized.

Percolation: A theory of networks formed by randomly connected neighboring sites.

Bibliography

A. Models of Cell Migration, Angiogenesis in the Corneal Angiogenesis Assay and Wound Healing

Chaplain, M. A. J., Singh, G. D., and McLachlan, J. C. eds. (1999). *On Growth and Form: Spatio-temporal Pattern Formation in Biology.* New York: Wiley.

Stokes, C. L., Lauffenburger, D. A., and Williams, S. K. (1991). Migration of individual microvessel endothelial cells: Stochastic model and parameter measurement. *J. Cell Sci.* **99**, 419–430.

Stoll, Brian R., Migliorini, Cristiano, Kadambi, Ananth, Munn, Lance L., Jain, Rakesh K., Gamba, A., Ambrosi, D., Coniglio, A., de Candia, A., Di Talia, S., Giraudo, E., Serini, G., Preziosi, L., Bussolini, F.

Tong, S., and Yuan, F. (2001). Numerical simulations of angiogenesis in the cornea. *Microvasc. Res.* **61**, 14–27.

Zawicki, D. F., Jain, R. K., Schmid-Schoenbein, G. W., and Chien, S. (1981). Dynamics of neovascularization in normal tissue. *Microvasc. Res.* **21**, 27–47.

B. Population Effects of Circulating Progenitor Cells

Stoll, B. R., Migliorini, C., Kadambi, A., Munn, L. L., and Jain, R. K. (2003). A mathematical model of the contribution of endothelial progenitor cells to early angiogenesis in solid tumors. *Blood* **102**(7), 2555–2561.

C. Use of Fractals and Percolation in Angiogenesis

Baish, J. W., and Jain, R. K. (2000). Fractals and cancer. *Cancer Res.* **60**, 3683–3688.

Craciunescu, O. I., Das, S. K., and Clegg, S. T. (1999). Dynamic contrast-enhanced MRI and fractal characteristics of percolation clusters in two-dimensional tumor blood perfusion. *J. Biomech. Eng.* **121**, 480–486.

Gamba, A., Ambrosi, D., Coniglio, A., de Candia, A., Di Tali, S., Giraudo, E., Serini, G., Preziosi, L., and Bussolini, F. (2003). Percolation, morphogenesis, and Burgers dynamics in blood vessel formation. *Phys. Rev. Lett.* **90**(11), 118101–1-4.

D. Branching Models

Murray, C. D. (1926). The physiological principle of minimum work. I. The vascular system and the cost of blood volume. *Proc. Natl. Acad. Sci. USA* **12**, 207–214.

West, G. B., Brown, J. H., and Enquist, B. J. (1997). A general model for the origin of allometric scaling laws in biology. *Science* **276**, 122–126.

Biography

James W. Baish is Professor of Mechanical and Biomedical Engineering at Bucknell University. He is chair of the Mechanical Engineering Department. His primary interests are the application of mathematical and engineering methods to the study of transport processes in tumors.

CHAPTER 164

Mathematical and Computational Models of the Microcirculation

Aleksander S. Popel
Department of Biomedical Engineering, Whitaker Biomedical Engineering Institute, Johns Hopkins University, Baltimore, Maryland

Introduction

Significant progress has been achieved in quantitative understanding of the microcirculation in recent decades, owing to new experimental methodologies and development of mathematical and computational models. The microcirculation is the site of the major part of the hemodynamic resistance to blood flow and of molecular exchange between the blood and surrounding tissue. Interactions between the endothelial cells that line the vasculature and leukocytes involved in the immune response to inflammation occur primarily in the venules. Endothelial cell permeability, leukocyte–endothelial interactions, and vascular tone are exquisitely regulated by both systemic and local mechanisms. The complexity of the system makes it a natural area for application of mathematical and computational ("in silico") modeling. There is a growing realization that modeling is necessary for quantitative understanding of complex systems, which contain a large number of interacting subsystems. Quantitation of microcirculatory phenomena has a rich history, beginning with Poiseuille's studies of blood flow in small tubes leading to formulation of the pressure–flow relationship, Krogh's model for capillary oxygen transport, Starling's relationship for water permeability of blood capillaries, and Murray's law for vessel diameter change along the vascular bed. Development of realistic and accurate models requires experimental data; model predictions, in turn, lead to formulation of new biological hypotheses and stimulate new experiments. Experimental description of the microcirculation has progressed greatly in large part because of the developments in intravital microscopy and image analysis, development of fluorescent probes for in vivo measurements, and techniques for measuring concentrations with high spatial and temporal resolution. Experimental methods developed in cell and molecular biology are being extended to microvascular research; these include fluorescent imaging of intracellular Ca^{2+}, modification of cell surface receptors and ion channels, introduction of recombinant DNA and RNA and their fragments into cells, and designing transgenic animals. In parallel with the experimental developments in cell and molecular biology, including the emergence of genomic and proteomic data, modeling studies are under way of cell genetic, signaling, and metabolic networks, cell mechanics that explicitly treat the cell's molecular components, cell electrophysiology, and cell–cell interactions. This rapid progress in modeling at all levels of biological organization has led to an integrative program, call the Physiome Project, whose mission is to develop mathematical and computational models and archive and disseminate quantitative data and models of the functional behavior of biological molecules, cells, tissues, organs, and eventually organisms [1]; it envisions creation of a hierarchy of models at the different spatial and temporal scales for which parameters of a model at one scale can be interpreted in terms of the models at the finer scale. Applications of this concept to the microcirculation are referred to as the Microcirculation Physiome. The concept provides an appropriate framework to review the state of the art in modeling of the microcirculation.

Models of Microvascular Blood Flow

Models of blood flow in microvascular networks must include as their elements flow in individual arterioles, capillaries, venules, and vascular bifurcations, and quantitative geometric and structural information about the network. Although the definition of macro-versus microvessels is not commonly accepted, in this review we will refer to vessels with internal diameter smaller than 200 μm as microvessels. In models of arteries and veins the blood is usually treated as a homogeneous Newtonian fluid with constant viscosity; in contrast, for flow in microvessels blood cannot be considered as a homogeneous fluid because the dimensions of its formed elements (red cells, white cells or leukocytes, and platelets) are comparable to those of the vessels. In fact, capillary diameter is typically smaller than the dimensions of undeformed red cells and leukocytes.

Flow in Capillaries

The first mathematical models of flow of membrane-bound red blood cells (RBCs) in a capillary were formulated by Richard Skalak and coworkers around 1980 and subsequently extended by many investigators. The models included a continuum rheological model of the red cell membrane, subsequently called the Evans–Skalak model, allowing for large deformations. This rheological model has been extended to include membrane bending stiffness and membrane viscosity (viscoelastic behavior). Parameters of the model have been measured in vitro by several laboratories. Recently, detailed molecular models of the red cell membrane have been constructed based on properties of two major cytoskeletal proteins, spectrin and actin. The original models assumed a cylindrical shape of the capillary; subsequent models described a "corrugated" shape attributed to the endothelial cell nuclei bulging into the lumen and a layer of endothelial glycocalyx with a thickness on the order of a micron. The experimental finding by Duling and coworkers that endothelial glycocalyx could be as thick as a micrometer (and therefore occupy up to 40 percent of the cross-sectional area of a 5-μm capillary) is having a significant impact on our understanding of capillary hemodynamics, leukocyte–endothelial interactions, microvascular permeability. and endothelial mechanotransduction (i.e., transduction of external mechanical forces into cellular response). Theoretical models of these phenomena based on the notion of proteoglycans with elastic protein cores and side chains of glycosaminoglycans (GAGs) have been summarized by Weinbaum et al. [2]. They have shown that the foregoing phenomena can be explained in terms of a matrix composed of the core proteins and unsteady movement of the blood plasma through this layer. There are alternative models that also consider electrokinetic and osmotic effects of the glycocalyx as well as dynamic adsorption of plasma proteins forming so-called endothelial surface layer (ESL). Future experiments should elucidate the dynamic physicochemical properties of the endothelial glycocalyx and the contribution of different factors under different physiological conditions.

Under normal conditions leukocytes in blood occupy a small fraction of the total blood cell volume, about 1/600; thus they do not significantly affect the bulk rheological properties of blood. However, their undeformed diameter (about 8 μm) is larger than capillary diameter and they are much stiffer than the red cells; thus the hydrodynamic resistance per cell and residence time in capillaries for leukocytes are as much as three orders of magnitude greater than that of the red cells. When a leukocyte traverses a capillary, the capillary glycocalyx is deformed or stripped away, and it takes about a second for it to be restored. Leukocytes have been modeled as viscoelastic bodies or as a volume of viscous fluid representing the nucleus surrounded by a layer of viscous fluid with a different viscosity representing the cytosol.

Flow in Arterioles and Venules

RBCs flowing in arterioles tend to migrate away from the wall as a result of hydrodynamic forces resulting in a marginal cell-free or cell-depleted layer several micrometers thick next to the endothelium. The hematocrit and velocity profiles in arterioles tend to be approximately axisymmetric, except near the bifurcations. The result of these nonuniform hematocrit profiles is the Fahraeus effect (vessel hematocrit is smaller than the discharge hematocrit; discharge hematocrit is defined as the hematocrit of the effluent blood if it were collected into a reservoir), and the Fahraeus–Lindqvist effect (apparent viscosity of blood in microvessels is smaller than the bulk viscosity of blood at the same discharge hematocrit). These effects are qualitatively similar in arterioles and in artificial tubes of same diameter. For venules, the cell-free layer is typically smaller and hematocrit distribution is often nonaxisymmetric, possibly resulting from converging flows with unequal hematocrits from two venules; this results in nonaxisymmetric velocity profiles that persist through the venular segment. The glycocalyx may also play a role in small arterioles and venules; hence the different relationships between apparent viscosity and diameter under in vivo and in vitro conditions.

In many mammalian species, RBCs have a tendency to aggregate at low shear rates in the presence of specific macromolecules, primarily fibrinogen in vivo; aggregation in vitro can be induced by other macromolecules, such as high-molecular-weight dextran. The process of aggregation is reversible and the aggregates disperse at higher shear rates due to hydrodynamic forces. In a number of tissues shear rates are low enough on the venous side, so aggregates are observed; under ischemic, low-flow pathological conditions aggregation could take place on both venous and arterial sides, although direct microvascular observations are scarce. Potentially, red cell aggregation could play a significant role under in vivo conditions, for example, by increasing vascular resistance. Phenomenological models of blood flow in arterioles and venules are plentiful, mostly representing flow as two fluids: a low-viscosity concentric region near the wall and a higher viscosity core region containing red cells. Although qualitatively the models agree with the experimental data in narrow tubes, quantitative agreement cannot be achieved without introducing features that reflect the presence of discrete red cells, even in continuum models. Computational modeling of flow in arterioles and venules with discrete red cells, leukocytes, and platelets remains a challenging unresolved problem. Progress has been made with models of flow of suspensions of rigid particles and fluid droplets, with only limited success dealing with deformable RBC-like particles. Developments in algorithms of computational fluid dynamics (CFD) and availability of high-performance computers give promise that these problems will be solved by direct numerical simulation in the foreseeable future.

Flow in Microvascular Networks

Microvascular networks in an organ comprise millions of microvessels. Arteriolar networks are mostly diverging and venular networks are mostly converging tree-like structures, with anastomotic connections present in many tissues. Capillaries are found in the immediate vicinity of every cell, with the exception of a few avascular tissues such as cartilage and early stages of tumors. The topology and geometry of microvascular networks are organ specific. The relationship between their structure and function is not fully understood. To aid in quantitative analysis of microvascular topology and geometry, vascular classification schemes have been used, such as the Strahler scheme, adapted from geophysics. Anastomoses are not included in these schemes. Blood flow in microvascular networks is heterogeneous, in part because complex network geometry results in nonuniform pressure in vessels of similar size. Experimental studies and theoretical estimates show a tendency for higher flow daughter branches at arteriolar bifurcations to draw a disproportionately higher number of RBCs, thus causing a heterogeneous hematocrit distribution in microvascular networks. The flow and hematocrit heterogeneity results in a lower average network hematocrit, an effect called the network Fahraeus effect. An empirical relationship between RBC flows and volumetric blood flows in the arteriolar branches is referred to as a bifurcation law. Microvascular networks have been mapped experimentally in several tissues, with nearly complete measurements of vessel interconnections, vascular segment diameters, and lengths, as well as hemodynamics measurements at selected sites. However, such measurements in three-dimensional tissues are very difficult, because it is difficult to follow vessels not lying in a plane perpendicular to the microscope axis or lying deeper than 100 μm from the tissue surface. Progress in multiphoton microscopy should make measurements in deep tissue layers possible. The most detailed measurements have been done in thin "two-dimensional" preparation of rat mesentery where the complete networks comprise several hundred to a thousand vascular segments. Poor agreement between experimental and predicted flow and hematocrit distributions was obtained from the model that incorporates the following elements: experimentally measured network geometry; diameter- and hematocrit-dependent apparent viscosity based on in vitro measurements; and the bifurcation law [3]. However, when a glycocalyx-type layer was assumed that completely retarded the flow of plasma, the agreement was significantly improved. These calculations described a network-scale effect of the glycocalyx that has been extended in subsequent models. The calculations also resulted in an "in vivo" relationship for apparent viscosity in microvessels, different from that in vitro. To date, predictions of the microvascular network flow model have only been compared with experiment on a vessel-by-vessel basis for rat mesentery. Much remains to be accomplished in extending such models to other tissues, although progress has been made in modeling blood flow in skeletal muscle, heart, brain, kidney, and tumors in some level of detail.

Since the early 1980s when the Strahler vessel classification scheme was applied to microvascular networks, there was a realization that vessel diameters, lengths, and number of branches of the same order follow power laws, characteristic of scale-free fractal structures. These properties have been explored in different organs. Not only the structure, but also blood flow distribution follows fractal relationships; for example, blood flow per unit volume versus volume size follows a power law with a noninteger exponent. The values of the structural or flow exponents may change as the tissue undergoes physiological or pathological changes, such as aging, hypertension, or cancer. Modeling studies show that fractal algorithms lead to creation of realistic microvascular networks with just a few parameters, due to the intrinsic similarity of the network at different scales. Thus, the two approaches, vessel-by-vessel mapping and fractal description of vascular structure and hemodynamics, are complementary and both have advantages and disadvantages: Although complete mapping provides most detailed description of individual networks and allows validation of blood flow models on a vessel-by-vessel basis, the collection of data is very labor intensive and in many cases not practical; the fractal approach allows simple characterization of network properties and may be appropriate for exploring general structure–function relationships under physiological and pathological conditions.

Understanding the relations between form/structure and function has been in the center of biological science; the monograph by D'Arcy Thompson "On Growth and Form," published in 1917, is an example of conceptual thinking on the subject. In application to vascular networks, Murray in 1926 raised the question whether vascular networks grow to optimize a certain quantity. He showed that for a vascular bifurcation, minimization of a "cost function" equal to the sum of the viscous dissipation and the energy required to maintain the vascular volume yields the ratio of the parent to daughter branches of $2^{1/3}$; this relationship is referred to as Murray's law. One of the consequences of Murray's law is a cubic relationship between blood flow and vessel diameter in a network, that is, when blood flow, Q, is plotted on logarithmic scale against vessel diameter, D, for different vascular segments in a network, the slope of the relationship is approximately equal to 3 (not to be confused with the fourth-power relationship in Poiseuille's law). Numerous studies explored the question of optimality and showed agreement with the cubic law in some cases and disagreement in others. Analysis of microvascular data on vessel diameters at bifurcations showed that whereas the cost function considered in Murray's law reaches a minimum when the exponent is equal to 3, the deviation from this minimum value is small, less than 5 percent, within a wide range of the exponent values, between 2 and 10, and it increases outside this range. Thus, accurate adherence to Murray's law (or similar minimization principles) could not be expected since moderate deviations from the minimum are not costly and other factors might affect the structure of the network. The search for universal principles of vascular network design continues; this theme also arises in vascular development, angiogenesis, and remodeling.

Models of Microvascular Transport of Oxygen and Nitric Oxide

Delivery of oxygen (O_2) and removal of waste products, such as carbon dioxide (CO_2), is one of the main functions of the cardiovascular system. In most organs and tissues, the limitation of tissue metabolism is based on O_2 delivery. Therefore, we review the current understanding of the processes of oxygen transport and the corresponding mathematical models. In arterial blood under normal conditions, more than 98 percent of O_2 is bound to hemoglobin (Hb) inside RBCs while the rest is dissolved in the plasma. Oxygen is transported convectively through the macrocirculation with negligible losses. Research in many laboratories in the past 30 years has shown that a gradual decrease of blood O_2 content starts in arterioles; for example, in resting striated muscle, precapillary losses amount to about two-thirds of the total O_2 exchanged between blood and tissue, with approximately one-third exchanged in the capillaries. The site of utilization of the O_2 exchanged in the arterioles is a subject of controversy: Experimental evidence exists that the O_2 lost by arterioles is transported to nearby capillaries; other evidence suggests that the arteriolar wall consumes a significant part of the oxygen. Models of O_2 transport in the microcirculation began with the famous Krogh model formulated in 1918 comprising a single cylindrical capillary surrounded by a concentric tissue cylinder. A simple solution to the diffusion equation was obtained expressing the concentration of oxygen in the tissue cylinder as a function of the radial position. The model has played a major role in conceptual thinking about tissue gas exchange. However, in view of the more recent experimental findings, significant changes in modeling gas exchange became necessary and numerous models have emerged. Here are some of the important elements that the models consider. Pre- and post-capillary transport of oxygen through the vascular wall and into the surrounding tissue has been considered for individual microvessels and microvascular networks. Intravascular resistance to O_2 transport has been shown to be important; for example, in skeletal muscle approximately half of the partial pressure of oxygen (po_2) drop between RBCs and distal mitochondria occurs inside the capillary. Intravascular resistance is characterized by the vessel mass transfer coefficient (defined as the ratio of the O_2 flux and the transmural po_2 difference), which is a function of vessel diameter and hematocrit and, under certain conditions, of hemoglobin saturation with O_2. A consequence of this resistance, in addition to the tissue transport resistance, is that under conditions of maximal consumption, such as in skeletal muscle and heart, it is necessary to maintain a po_2 above 20 mmHg at the venous end of the capillaries to maintain adequate O_2 flux to the distal mitochondria. Heterogeneity of blood flow (both RBC velocity and hematocrit), intrinsic in the microvascular structure as discussed earlier, also plays a significant role in tissue oxygenation. Accounting for these factors has resulted in much more complex models, compared to the Krogh model. Owing to this complexity, no "standard" model has emerged that has been quantitatively validated against experimental data. Development of such models, specialized to individual organs and tissues, would be important for understanding physiology and pathophysiology, as well as in such applications as designing therapeutic interventions and tissue engineering. Recent models also describe O_2 transport in the presence of hemoglobin-based blood substitutes.

The role of nitric oxide (NO) as a major signaling molecule with an impact on a myriad of biological processes is being actively investigated. Nitric oxide is produced endogenously in the body through the enzymatic degradation of L-arginine by several isoforms of the enzyme nitric oxide synthase (NOS). Endothelial cells express a constitutive form of the enzyme (eNOS). NO can diffuse freely, which enables it to act both in an autocrine and in a paracrine fashion. It is a reactive species with a short half-life in vivo. It can be degraded by a number of reactions, but under physiological conditions NO concentrations are submicromolar and it is the fast reactions with superoxide and heme-containing proteins such as hemoglobin (Hb), myoglobin (Mb), guanylate cyclase (GC), and cytochrome *c* oxidase that dominate its chemistry. In response to hemodynamic or agonist stimuli, vascular endothelial cells alter production of NO, which can diffuse across cell membranes to the adjacent smooth muscle where it activates soluble guanylate cyclase (sGC), leading to an increase in the intracellular cyclic guanosine monophosphate (cGMP) concentration and to smooth muscle relaxation. Thus, NO is a major factor in maintaining vascular tone. Models of NO transport in arterioles show that even though Hb inside RBCs is the major scavenger of freely diffusible NO, under physiological conditions the concentration of endothelium-derived NO in the smooth muscle can be maintained at a level of several hundred nM. Neuronal NOS (nNOS) of extramural origin could further contribute to maintaining NO bioavailability. Important issues in modeling microvascular NO distribution remain to be resolved: NO release rates via eNOS and nNOS pathways in vivo have not been reported and their values obtained in vitro are used in the simulations; release rates in arterioles, capillaries, and venules may be different; release could be in the form of time-dependent bursts rather than sustained. Since the reaction rate of NO with RBC-contained Hb is known to be two to three orders of magnitude lower than that with free Hb, the presence of free Hb in the vascular lumen drastically reduces smooth muscle NO concentration. Free Hb is present in the lumen in sickle cell anemia and during administration of Hb solutions as blood substitutes. Free Hb can also extravasate into the interstitial space and act as an NO scavenger; this effect can be reduced by Hb polymerization in the case of administration of Hb-based O_2 carrier. NO belongs to a large family of free radicals that play important roles under physiological and pathophysiological conditions, such as inflammation and cancer. Diffusion-reaction models have been formulated that consider simultaneous transport of NO, superoxide, peroxynitrite and other free radicals with the goal of achieving a quantitative understanding of spatiotemporal distribution of these species in the microcirculation [4].

Models of Regulation of Microvascular Blood Flow

Short-Term Regulation: Metabolic, Myogenic, and Flow-Dependent

Studies of local regulation of vascular tone are central to microvascular physiology. First we describe models of short-term regulation of vascular tone, typically on the time scale of minutes, leading to changes in blood flow and pressure distribution within the network. Blood flow rate is known to be locally regulated such that tissue demand for oxygen and nutrients is matched to supply; in addition, capillary blood pressure is regulated such that water balance between the blood and tissue compartments is maintained. Metabolites are released by parenchymal and stromal cells in response to changes in local po_2. Molecular mechanisms of oxygen sensing have been identified, but the molecular nature of the metabolic signal reaching the arteriolar wall and causing vasodilation or vasoconstriction is not completely understood; the nature of this signal is likely to be multifactorial, involving multiple molecular species, and tissue dependent. Typically, arterioles constrict in response to an increase in transmural pressure and dilate in response to a decrease in transmural pressure (the myogenic response). In addition, arterioles dilate in response to an increased shear stress acting on the endothelium, mediated by NO (flow-dependent mechanism). Much has been done to elucidate cellular and molecular mechanisms of metabolic, myogenic, and flow-dependent modes of vascular regulation and of the mechanisms of communication between vascular segments, but only a few quantitative models are available. Most models consider simple series–parallel vascular network topology where vessels of the same diameter are connected in parallel and vessels of different diameter in series. The models postulate relationships between arteriolar diameter and average tissue po_2 or concentration of a metabolic factor (metabolic mechanism), transmural pressure (myogenic mechanism), or shear stress (flow-dependent mechanism). Predictions include the presence of complex oscillations (including regimes of "deterministic chaos") of arteriolar diameter in qualitative agreement with experimentally observed rhythmic variations of arteriolar lumen diameter (vasomotion). A step in detailed molecular modeling of the vascular wall has been made with developing a model of a vascular smooth muscle cell, with an explicit representation of ionic channels (Ca^{2+}, K^+, Cl^-), Ca^{2+}-ATPase pumps, Na^+-K^+-ATPase, and other cellular transport entities [5]. Solution of the detailed kinetic equations demonstrates the existence of complex oscillations in intracellular Ca^{2+} and membrane potential that would lead to smooth muscle contractions. This modeling approach should eventually allow the response to pharmacological interventions to be understood at the cellular and molecular levels. A similar approach can be developed for endothelial cells, and these models could then be combined and extended to result in a detailed molecular-level model of the vascular wall.

Long-Term Regulation: Vascular Remodeling, Angiogenesis

Long-term changes of vascular network structure occur on the time scale of days or weeks and longer; vascular networks may adapt in response to such factors as prolonged changes of arterial pressure, ischemia, hypoxia, glucose level, and tumor growth. Adaptation includes structural changes of the vascular wall as well as changes in the network topology, either growth of new vessels (angiogenesis) or retraction of the existing vessels (vascular rarefaction). Pries, Secomb, and coworkers have conducted systematic modeling studies of microvascular network adaptation based on the geometric and hemodynamic measurements for rat mesenteric networks comprising hundreds of segments. They hypothesized rules for diameter adaptation in response to changes in metabolic stimuli, intravascular pressure, and shear stress, and information transfer from segment to segment (conducted response) to match the computer-simulated and experimentally observed network structure. They obtained good quantitative agreement under specific adaptation rules. These studies integrate complex relationships between different mechanisms of local control of the microvasculature and could serve as a guide for formulating quantitative hypotheses and designing further experiments. At the cellular and molecular level, processes of vascular remodeling, growth, and retraction involve myriad of molecular species, including growth factors (e.g., vascular endothelial growth factor, VEGF; fibroblast growth factor, FGF; platelet-derived growth factor, PDGF; tissue transforming factor, TGF, and their receptors), heparan sulfate proteoglycans, integrins, components of the extracellular matrix, and intracellular signaling molecules. During angiogenesis by capillary sprouting, endothelial cells undergo proliferation, migration, differentiation, and apoptosis, and the new sprouts are stabilized by recruiting stromal cells that differentiate into pericytes or smooth muscle cells. During vascular remodeling, microvessels recruit progenitor cells or fibroblasts residing in the surrounding tissue that are transformed into smooth muscle cells. The complexity of the relationships between these molecular and cellular species, many of which are poorly understood, preclude formulating global molecular-level models at the present time. However, general principles for in silico simulations of microvascular remodeling have been put forward [6].

Summary and Future Studies

This brief review of mathematical and computational models of the microcirculation aims at outlining major approaches to modeling of diverse processes, from mechanics of blood flow, to intravascular and tissue transport of molecules, to regulation of vascular tone. The brevity of the review precludes comprehensive coverage; for example, such important issues as transendothelial transport of water and proteins are not included. The biological level of detail in the models vary: In some models individual molecular species are explicitly represented, whereas in others phenomenological, mostly empirical, relationships between some of the microcirculatory parameters are used. Models at these different scales are complementary and it is possible to structure them so that the molecular-level models are used to predict certain relationships that are part of the larger scale models. Such hierarchical modeling systematically covering the scales from the microcirculatory down to the molecular, accompanied by databases of the necessary cell- and tissue-dependent parameters represents the emerging concept of the Microcirculation Physiome. Clearly, specific models will continue to be developed by independent researchers; however, integration of these models will require a modeling infrastructure—for example, standardized languages for representation and exchange of molecular-, cellular-, and tissue-level models such as CellML or the systems biology markup language, SBML, as well as standards for biological physiome-type databases. These developments would not only lead to a deeper quantitative understanding of physiology and pathophysiology of the microcirculation, but would also facilitate design of drugs and drug delivery vehicles, and design of microvascular networks for tissue-engineered constructs.

Glossary

Computational model: Computer simulation (numerical algorithm and implementation) that solves a mathematical model.

Mathematical model of a biological system: The use of mathematical language, such as algebraic or differential equations, to describe the behavior of a biological system.

Myogenic mechanism: Vasoconstriction in response to an increase in transmural pressure or vasodilation in response to a decrease in transmural pressure.

NOS: The enzyme nitric oxide synthase that participates in synthesizing nitric oxide from arginine and oxygen.

Physiome: The quantitative description of the functional state of an organism or its part.

Wall shear stress: Force per unit area acting tangentially on the vessel wall and arising from flow of viscous fluid.

References

1. Hunter, P. J., and Borg, T. K. (2003). *Nat. Rev. Mol. Cell. Biol.* **4**, 237–243.
2. Weinbaum, S., Zhang, X., Han, Y., Vink, H., and Cowin, S. C. (2003). Mechanotransduction and flow across the endothelial glycocalyx. *Proc. Natl. Acad. Sci. USA* **100**(13), 7988–7995. *Description of several models involving endothelial glycocalyx for water permeability, RBC– and leukocyte–endothelial interactions, and mechanotransduction.*
3. Pries, A. R., Secomb, T. W., Gessner, T., Sperandio, M. B., Gross, J. F., and Gaehtgens, P. (1994). Resistance to blood flow in microvessels in vivo. *Circ. Res.* **75**(5), 904–915. *This study describes formulation of a model for blood flow in a microvascular network and simulates flow distribution in the rat mesentery.*
4. Buerk, D. G., Lamkin-Kennard, K., and Jaron, D. (2003). Modeling the influence of superoxide dismutase on superoxide and nitric oxide interactions, including reversible inhibition of oxygen consumption. *Free Radic. Biol. Med.* **34**(11),1488–1503.
5. Parthimos, D., Edwards, D. H., and Griffith, T. M. (1999). Minimal model of arterial chaos generated by coupled intracellular and membrane Ca^{2+} oscillators. *Am. J. Physiol.* **277**(3 Part 2), H1119–H1144.
6. Peirce, S. M., Van Gieson, E. J., and Skalak, T. C. (2004). Multicellular simulation predicts microvascular patterning and in silico tissue assembly. *FASEB J.* **18**(6), 731–733.

Further Reading

Beard, D. A., Schenkman, K. A., and Feigl, E. O. (2003). Myocardial oxygenation in isolated hearts predicted by an anatomically realistic microvascular transport model. *Am. J. Physiol.* **285**(5), H1826–H1836, in press.

Glenny, R. W., Robertson, H. T., Yamashiro, S., and Bassingthwaighte, J. B. (1991). Applications of fractal analysis to physiology. *J. Appl. Physiol.* **70**(6), 2351–2367.

Mac Gabhann, F., and Popel, A. S. (2004). Model of competitive binding of vascular endothelial growth factor and placental growth factor to VEGF receptors on endothelial cells. *Am. J. Physiol.* **286**(1), H153–H1564.

Popel, A. S., and Johnson, P. C. (2005). Microcirculation and hemorheology. *Am. Rev. Fluid Mech.* **37**, 43–69.

Pries, A. R., and Secomb, T. W. (2005). Control of blood vessel structure: insights from theoretical models. *Am. J. Physiol.* **288**(3), H1010–H1015.

Secomb, T. W., Hsu, R., and Pries, A. R. (2001). Motion of red blood cells in a capillary with an endothelial surface layer: Effect of flow velocity. *Am. J. Physiol. Heart Circ. Physiol.* **281**(2), H629–H636.

Sun, C., Migliorini, C., and Munn, L. L. (2003). Red blood cells initiate leukocyte rolling in postcapillary expansions: a lattice Boltzmann analysis. *Biophys. J.* **85**(1), 208–222.

Tsoukias, N. M., Kavdia, M., and Popel, A. S. (2004). A theoretical model of nitric oxide transport in arterioles: frequency- vs. amplitude-dependent control of cGMP formation. *Am. J. Physiol.* **286**(3), H1043–H1056.

West, G. B., Brown, J. H., and Enquist, B J. (1997). A general model for the origin of allometric scaling laws in biology. *Science* **276**(5309), 122–126.

Capsule Biography

Dr. Aleksander S. Popel is a professor of Biomedical Engineering at Johns Hopkins University. His main interests are in mathematical and computational modeling at the cellular and microvascular levels, including angiogenesis, cell mechanics, transport of nitric oxide and oxygen, and regulation of blood flow.

SECTION F

Proteomics

CHAPTER 165

Proteomics of the Microvasculature

[1]Emma McGregor and [2]Michael J. Dunn
[1]Proteome Sciences plc, London
[2]Conway Institute of Biomolecular and Biomedical Research, University College Dublin

Introduction

The concept of mapping the human complement of protein expression was first proposed more than 25 years ago with the development of a technique where proteins are separated by two-dimensional polyacrylamide gel electrophoresis (2DE) [1]. The term *proteome*, the protein complement of a genome, was coined in the mid-1990s. Proteomics is defined as the study of proteomes but also refers to the plethora of complementary techniques that are used. Despite its longevity, 2DE remains the core technology of choice for the majority of applied proteomic projects. It has the ability to separate simultaneously thousands of proteins and to indicate post-translational modifications that result in alterations in protein p*I* and/or M_r. High-sensitivity visualization of 2D protein separations, compatibility with quantitative computer analysis to detect differentially regulated proteins, and the relative ease with which proteins from 2D gels can be identified and characterized by mass spectrometry are the main advantages of using 2DE in proteomics research.

Two-Dimensional Gel Electrophoresis

2DE involves the separation of solubilized proteins in the first dimension according to their charge properties (isoelectric point, p*I*) by isoelectric focusing (IEF) under denaturing conditions, followed by their separation in the second dimension according to their relative molecular mass (M_r) by sodium dodecyl sulfate polyacrylamide gel electrophoresis (SDS-PAGE). As the charge and mass properties of proteins are essentially independent parameters, this orthogonal combination of charge (p*I*) and size (M_r) separations results in the sample proteins being distributed across the two-dimensional gel profile (Figure 1). Developments over the past few years that have resulted in the current 2DE method, which combines increased resolving power and high reproducibility with relative simplicity of use, can be found in recent reviews [2].

Protein Detection

Following separation by 2DE, proteins must be visualized at high sensitivity using a gel staining method compatible with subsequent protein identification by mass spectrometry. Staining with Coomassie Brilliant Blue (CBB) has for many years been a standard method for protein detection following gel electrophoresis, but its limited sensitivity (around 100 ng protein) stimulated the development of a more sensitive (around 10 ng protein) method utilizing CBB in a colloidal form. Presently, the most common staining methods employ silver because of its high sensitivity (around 0.1 ng protein). More recently there has been an increase in the popularity of fluorescent gel stains as these have the potential of increasing sensitivity even further than silver and are also combined with an extended dynamic range for improved quantitation.

Protein Identification

A major breakthrough in the analysis of proteins and peptides came about with the development of sensitive methods based on the use of mass spectrometry (MS). The importance of these developments was recognized in 2002 with the awarding of the Nobel Prize in Chemistry to John Fenn (electrospray ionization, ESI) and Koichi Tanaka (matrix-assisted laser desorption/ionization ionization, MALDI), pioneers in the development of methods of ionization that made protein and peptide MS a practicable procedure. These

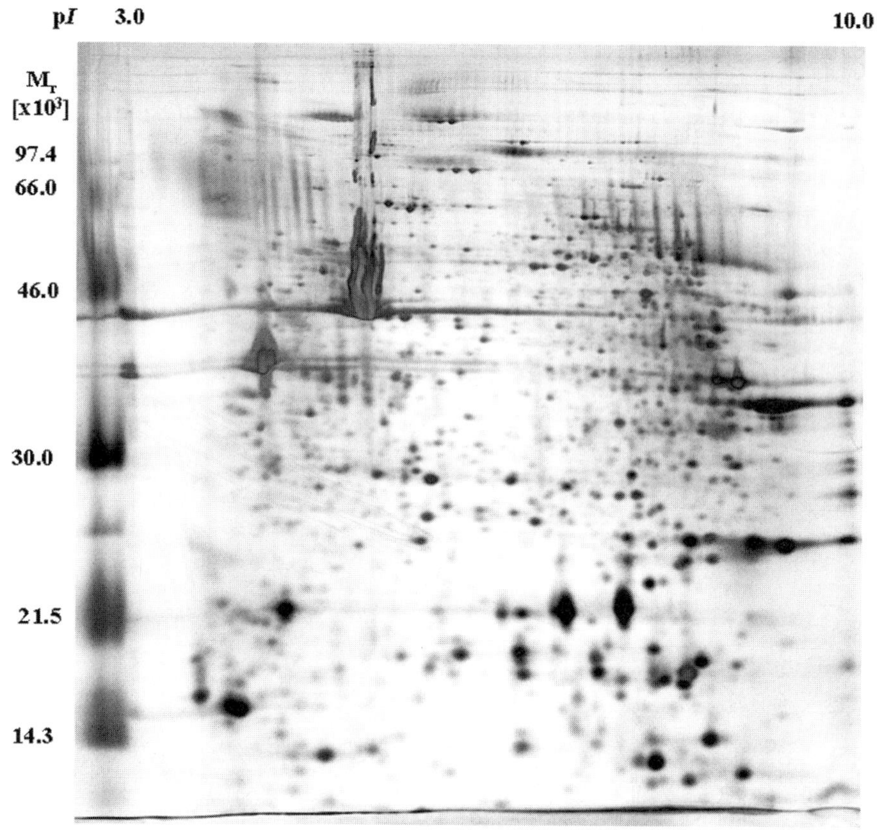

Figure 1 2D electrophoretic separation of intact contractile human saphenous vein smooth muscle. Proteins (400 μg load) were separated by isoelectric focusing in the first dimension using a pH 3–10 NL, 18 cm, IPG DryStrip (Amersham Biosciences). The pH gradient of the strip is illustrated across the top of the 2D gel with the most acidic pH to the left and the most basic pH to the right of the gel image. Proteins were then separated in the second dimension by SDS-PAGE through a 12 percent acrylamide gel. Standard molecular weight markers (Amersham Biosciences), ranging from 14.3 to 97.4 kDa, were run on the same gel and are annotated on the far left of the gel image. Visualization of proteins was achieved using silver staining.

developments have led to MS being the method of choice for protein identification and characterization of protein and peptides. Peptide mass fingerprinting (PMF) is the primary tool for MS identification of proteins in proteomic studies. In this method protein spots of interest are excised and the gel plug containing the spot is digested with a protease, typically trypsin, which cleaves polypeptide chains at basic amino acids (arginine/lysine residues). In this way the intact protein is broken down into a mixture of peptides. The masses of the resulting peptides, or more strictly their mass-to-charge ratios (m/z), are then measured by mass spectrometry to produce a characteristic mass profile or "fingerprint" of that protein. The mass profile is then compared with peptide masses predicted by theoretical digestion of known protein sequences contained within current protein databases or predicted from nucleotide sequence databases. This approach is usually very effective when trying to identify proteins from species whose genomes are completely sequenced, but is not reliable for organisms with incomplete genomic information.

In cases where it proves impossible to assign an unequivocal identity to a protein based on PMF alone, peptide sequence information is then required to confirm an identity. This can be generated by conventional automated chemical Edman microsequencing, but is most readily accomplished using tandem mass spectrometry (MS/MS). MS/MS is a two-stage process, by means of MALDI-MS with post-source decay (PSD), MALDI-TOF-TOF-MS/MS, or ESI-MS/MS triple-quadrupole, ion-trap, or Q-TOF machines, to induce fragmentation of peptide bonds. One approach, termed *peptide sequence tagging*, is based on the interpretation of a portion of the MS/MS or PSD fragmentation data to generate a short partial sequence or "tag." Using the tag in combination with the mass of the intact parent peptide ion provides significant additional information for the homology search [3]. A second approach uses a database searching algorithm, SEQUEST. This matches uninterpreted experimental MS/MS spectra with predicted fragment patterns generated in silico from sequences in protein and nucleotide databases. These techniques are highly sensitive, with the current generation of instruments capable of identifying proteins in the femtomole range, representing a 2D gel spot containing only a few nanograms of protein. This approach can also be combined with single- or multidimensional LC coupled either online (ESI-MS/MS) or offline (MALDI-TOF/TOF) to MS to fractionate the peptide mixture, thereby facilitating the identification of multiple protein species that can co-migrate in a single 2D gel spot.

Vessel Proteomics

Investigating protein expression of different cell types making up a blood vessel can be extremely challenging. One approach is to isolate primary cells from vessels, such as endothelial cells, smooth muscle cells, and fibroblasts, and solubilize protein directly from these cells for proteomic analysis. This can be achieved easily using larger blood vessels; for example, there are numerous studies where human umbilical vein endothelial cells (HUVECs) have been successfully isolated and analyzed using a 2DE approach [4, 5], as well as fibroblasts [6]. In addition, recent studies have illustrated the relative ease of dissecting intact contractile smooth muscle from human saphenous vein (HSV), and analyzing quantitative protein expression of this tissue using 2DE [7, 8] (Figure 1). These approaches can be applied to microvessels, but because of the very small size and the difficulty in isolating these vessels it becomes much more difficult. However, there are examples where primary cells (endothelial cells) have been successfully isolated from microvessels from within the heart.

The protein yield of potential 2DE samples, following protein extraction, via standard 2DE methods, must be sufficient for running the different types of 2D gels: analytical gels (100 μg protein) used for quantifying protein expression, via the analysis of gel images by means of specialized computer software packages, and preparative gels (400 μg protein), where protein spots are excised from the 2D gel for identification by mass spectrometry techniques. The approaches highlighted earlier, where primary cells and intact tissue are isolated from vessels, provide a limited amount of material that can be subjected to 2DE, and this becomes much further reduced when this material is harvested from the microvasculature. One way to counter act this is to separate limited samples out on small format (7 × 7 cm) mini-2DE gels rather than large format (20 × 20 cm) 2DE gels. However, for protein identification, sample protein yield should ideally be concentrated enough to be separated on a large format 2DE preparative gel intended for subsequent analysis by mass spectrometry. An alternative method for increasing protein yield from cell types associated with the (micro)vasculature is to grow specific cell lines in culture.

There are many proteomics studies where cells associated with blood vessels have been cultured and then subjected to protein analysis via 2DE. The advantage of using cultured cell lines is that there is always a ready supply of purified material, which can be concentrated by harvesting increased numbers of cells as required. Investigations of this type have used cultured HUVECs and cultures of immortalized HSV endothelial cells (Figure 2), as well as explanted HSV smooth muscle cells, where sections of dissected HSV medial smooth muscle are placed in culture conditions, thus stimulating the growth of smooth muscle cells outward from the original explanted intact tissue. A recent example of this type of study, using primary human endothelial cells cultured from HUVECs and a human endothelial cell line (HUVEC-derived, immortalized EC-RF24), is that by Sprenger et al. [5]. Using a classical 2DE approach, coupled with optimized fractionation techniques, two subdomains of similar lipid composition, caveolae and rafts, representing 0.5 percent of total cellular protein and less than 2 percent of total plasma membrane protein, were analyzed and compared [5]. This reference study was able to demonstrate the power of subproteomics, allowing enhanced separation and identification of membrane proteins in particular.

In addition to the foregoing, there are a limited number of studies that have attempted to analyze cells derived from vessels associated with the human brain and kidney. One such study, performed using cultured human cerebral endothelial cells (HCECs), was designed to identify potential pharmaceutical targets for the treatment of multiple sclerosis. Gene expression measurements (Affymetrix Hu6800 oligonucleotide arrays) and analysis of protein via 2DE with mass spectrometry, were used to establish early alterations in HCECs activated by TNFα [9], where HUVECs were used as the reference system. Results indicated the involvement of the urokinase plasminogen activator system and cytoskeletal rearrangements unique to TNFα activation of cerebral endothelial cells [9].

Although using cultured cells provides abundant protein yields for many parallel experiments to be undertaken, as described by Franzen et al. [9], the main disadvantage of using cultured cells is that one is using an in vitro system, where once cells are cultured through many passages they can be subjected to changes in their gene expression and sometimes morphology (e.g., fibroblasts), which can influence gene expression measurements and protein expression analysis by 2DE. This is a particular problem when analyzing cultured smooth muscle cells, which lose their contractile properties in culture, and is why it is preferential to use intact smooth muscle [7, 8] (Figure 1).

Recently we have started to apply the technique of laser capture microdissection (LCM) in which a laser beam is used to isolate specific regions of interest from microscope sections of tissue. Although this technique generally results in the isolation of relatively small amounts of material, it has been shown to be possible to perform proteomic studies of the resulting protein samples [10]. The heart is an example of an organ dominated by a particular cell type, that is, cardiac myocytes, but which also contains lower amounts of proteins derived from other cell types including fibroblasts, smooth muscle, and endothelial cells found in the microvessels within the heart. In preliminary studies we have been able to generate sufficient material by LCM of human cardiac tissue sections to produce large-format 2D gels of proteins from isolated cardiac myocytes and microvessels to

Figure 2 2D electrophoretic separation of cultured human saphenous vein endothelial cell proteins. Proteins were separated by isoelectric focusing in the first dimension using a 24 cm, pH 3–10 IPG DryStrip (Amersham Biosciences). The pH gradient of the strip is illustrated across the top of the 2D gel with the most acidic pH to the left and the most basic pH to the right of the gel image. Proteins were then subjected to SDS-PAGE in the second dimension through a 12 percent acrylamide gel. Standard molecular weight markers (Amersham Biosciences), ranging from 14.3 to 97.4 kDa, were run on the gel and are annotated on the far left of the gel image. Visualization of proteins was achieved using silver staining.

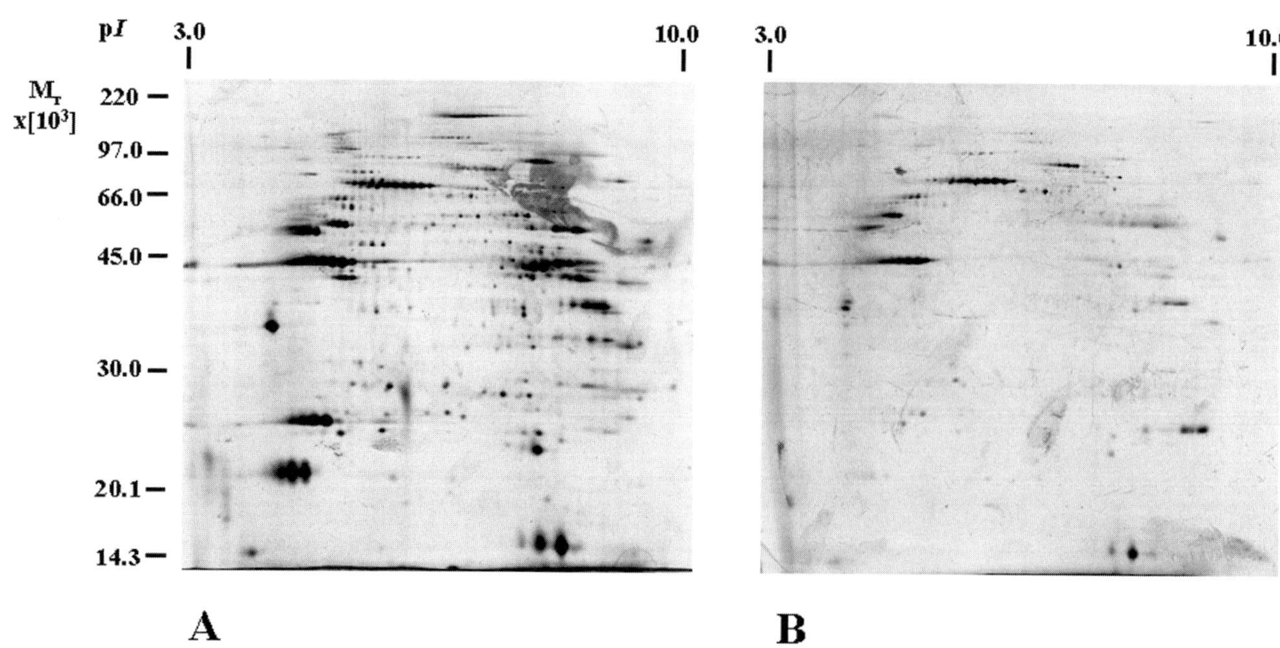

Figure 3 2D electrophoretic separations of laser capture microdissected human cardiac tissue. **(A)** Cardiac myocytes. **(B)** Blood vessels. Proteins were separated by isoelectric focusing in the first dimension using a pH 3–10 NL, 18 cm, IPG DryStrip (Amersham Biosciences). The pH gradient of the strip is illustrated across the top of the 2D gel with the most acidic pH to the left and the most basic pH to the right of the gel image. Proteins were then separated in the second dimension by SDS-PAGE through a 12 percent acrylamide gel. Standard molecular weight markers (Amersham Biosciences), ranging from 14.3 to 97.4 kDa, were run on the same gel and are annotated on the far left of the gel image. Visualization of proteins was achieved using silver staining.

investigate (i) the effects of fixation and staining on cardiac proteins separated by two-dimensional gel electrophoresis (2DE) and (ii) feasibility of using laser microdissection to separately prepare myocytes and blood vessels for 2DE gel analysis (Figure 3) [11].

Thus, proteomic analysis of blood vessels is in its infancy and has been largely confined to the study of larger vessels. These investigations promise to yield important new information on protein expression in vessels in different biological systems. However, a major problem, if the studies are to be extended to analysis of the microvasculature, will be obtaining sufficient purified starting material.

Glossary

Proteome: The protein complement of a genome.

Proteomics: The study of proteomes by 2DE, mass spectrometry and additional complementary techniques.

Two-dimensional electrophoresis (2DE): The separation of solubilized proteins in the first dimension according to their charge properties (isoelectric point, p*I*) by isoelectric focusing (IEF) under denaturing conditions, followed by their separation in the second dimension according to their relative molecular mass (M_r) by sodium dodecyl sulfate polyacrylamide gel electrophoresis (SDS-PAGE).

Acknowledgment

We are grateful to Professor Janet Powell and Dr. Lee Kempster (Imperial College London, UK) for providing us with the 2D separation of cultured human saphenous vein endothelial cells used in Figure 2.

References

1. O'Farrell, P. H. (1975). High resolution two-dimensional electrophoresis of proteins. *J. Biol. Chem.* **250**, 4007–4021.
2. Rabilloud, T. (2002). Two-dimensional gel electrophoresis in proteomics: Old, old fashioned, but it still climbs up the mountains. *Proteomics* **2**, 3–10. *Provides the reader with a general background in the techniques of two-dimensional polyacrylamide gel electrophoresis.*
3. Mann, M., and Wilm, M. (1994). Error-tolerant identification of peptides in sequence databases by peptide sequence tags. *Anal. Chem.* **66**(24), 4390–4399.
4. Bruneel, A., Labas, V., Mailloux, A., Sharma, S., Vinh, J., Vaubourdolle, M., and Baudin B. (2003). Proteomic study of human umbilical vein endothelial cells in culture. *Proteomics* **3**(5), 714–723.
5. Sprenger, R. R., Speijer, D., Back, J. W., De Koster, C. G., Pannekoek, H., and Horrevoets, A. J. (2004). Comparative proteomics of human endothelial cell caveolae and rafts using two-dimensional gel electrophoresis and mass spectrometry. *Electrophoresis* **25**(1), 156–172. *This recent reference illustrates how proteomics can be effectively used to analyze protein expression in cultured cells associated with vessels.*
6. Bumke, M. A., Neri, D., and Elia, G. (2003). Modulation of gene expression by extracellular pH variations in human fibroblasts: A transcriptomic and proteomic study. *Proteomics* **3**(5), 675–688.
7. McGregor, E., Kempster, L., Wait, R., Welson, S. Y., Gosling, M., Dunn, M. J., and Powell, J. T. (2001). Identification and mapping of human saphenous vein medial smooth muscle proteins by two-dimensional polyacrylamide gel electrophoresis. *Proteomics* **1**(11),1405–1414.
8. McGregor, E., Kempster, L., Wait, R., Gosling, M., Dunn, M. J., and Powell, J. T. (2004). F-actin capping (CapZ) and other contractile saphenous vein smooth muscle proteins are altered by hemodynamic stress: A proteomic approach. *Mol. Cell. Proteomics* **3**, 115–124.
9. Franzen, B., Duvefelt, K., Jonsson, C., Engelhardt, B., Ottervald, J., Wickman, M., Yang, Y., and Schuppe-Koistinen, I. (2003). Gene and protein expression profiling of human cerebral endothelial cells activated with tumor necrosis factor-alpha. *Brain Res. Mol. Brain Res.* **115**(2), 130–146.
10. Banks, R. E., Dunn, M. J., Forbes, M. A., Stanley, A., Pappin, D., Naven, T., Gough, M., Harnden, P., and Selby, P. J. (1999). The potential use of laser capture microdissection to selectively obtain distinct populations of cells for proteomic analysis—preliminary findings. *Electrophoresis* **20**(4–5), 689–700.
11. De Souza, A. I., McGregor, E., Dunn, M. J., and Rose, M. L. (2004). Preparation of human heart for laser microdissection and proteomics. *Proteomics* **4**, 578–586.

Further Reading

Dunn, M. J., and Gorg, A. (2001). Two-dimensional polyacrylamide gel electrophoresis for proteome analysis. In *Proteomics, from Protein Sequence to Function*, S. R. Pennington and M. J. Dunn, eds., pp. 43–63. BIOS Scientific. *A comprehensive explanation of how to perform two-dimensional polyacrylamide gel electrophoresis, the theory behind it, and problems that may be encountered.*

Capsule Biography

Dr. Emma McGregor works for a leading proteomics company, Proteome Sciences plc, in London. Her research focuses on the proteomic analysis of human disease.

Mike Dunn is Professor of Biomedical Proteomics at the Conway Institute of Biomolecular and Biomedical Research, University College Dublin. His research focuses on the proteomic analysis of human disease. He is Editor in Chief of the monthly journal *Proteomics*.

CHAPTER 166

Proteomics and the Microvasculature

[1]John H. Stone and [2]Emanuel F. Petricoin

[1]*Division of Rheumatology and The Johns Hopkins Vasculitis Center, Johns Hopkins University School of Medicine*
[2]*Division of Therapeutic Products, Center for Biologics Evaluation and Research, Food and Drug Administration and National Cancer Institute/Food and Drug Administration Clinical Proteomics Program*

Abstract

The human proteome is made up of the sum of the body's proteins, their numerous isoforms, and untold numbers of post-translational modifications (PTMs) that are still unclassified. Because of its proximity to physiology and to the development of disease, the proteome is linked inextricably with the microvasculature, which is the interface between individual cells and the circulatory system at large. The vocabulary of proteomics and the swiftly developing, technological nature of the field constitute substantial barriers to investigators. In this chapter, we provide an overview of the proteome and its field of investigation, proteomics. We review both classic proteomic techniques (two-dimensional gel electrophoresis) and newer approaches to studying the proteome: isotope-coded affinity tagging (ICAT), multidimensional protein identification technology (MudPIT), mass spectrometry (MS, including tandem mass spectrometry, MS/MS), and protein microarrays.

Introduction

Proteins and the Microvasculature

Long before the term *proteomics* was coined in the mid- to late 1990s, clinicians recognized the connections among the microvasculature, proteins, and normal physiologic function. Writing in 1912, Sir William Osler noted:

> In the capillary lake into which the arterial stream widens, the current slows and the pressure lessens. . . . In the brief fraction of a second . . . the business of life is transacted, for here is the mart or exchange in which the raw and the manufactured articles from the intestinal and hepatic shops are spread out for sale.

In this graphic but enlightened observation, Osler acknowledged that the critical mechanisms of both health and disease are elaborated at the level of the smallest blood vessels in the body. The essential "manufactured articles" to which Osler referred were recognized—even in his day—as *proteins*. As the effectors of the genetic code, proteins determine the phenotype not only of each cell, but also of every tissue and organ, and ultimately the entire organism.

Proteins play a role in nearly all aspects of cell function (structure, signal transduction, DNA transcription, macromolecular synthesis, energy production). Their involvement in these processes is highly regulated by post-translational modifications (PTMs). PTMs, which include phosphorylation, glycosylation, lipidation, sulfation, and proteolytic modifications, are critical determinants of protein trafficking, cell distribution, activation, and function. The number of documented protein co- and PTMs has now reached several hundred. PTMs either increase or decrease the molecular mass of affected amino acid residues. Consequently, detailed understandings of PTMs are often possible now through such newer proteomic techniques such as tandem mass spectrometry (MS/MS), a method by which analytes can be weighed and sequenced with accuracy.

The Proteome versus the Genome

A *proteome* is the complement to a certain set or subset of genes, that is, the total of all proteins linked to a given

Proteome

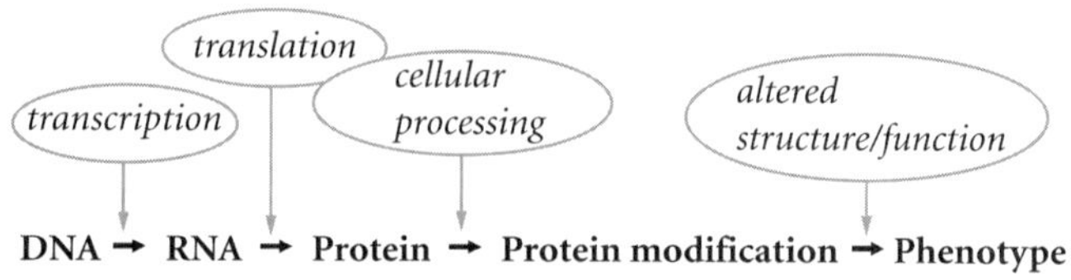

Figure 1 Diagram of assorted cellular processes leading to phenotype and the relationship of those processes to proteomics.

genome (Figure 1). Each type of nucleated cell possesses all of the genetic information necessary to make and maintain the entire organism that contains it. Judging from the wide range in function of individual cell types (compare, for example, the neutrophil and the neuron), it is obvious that not all genes are expressed in each cell type. In general, certain fundamental genes, the protein products of which are necessary for the physiological processes such as glycolysis, are expressed by all cells. Different cell types are distinguished by expression of genes whose protein products provide structure or function that is cell specific. Whereas the genome is generic for all cells in an organism, the proteome, tailored to the environment of each cell, is characterized by a diversity that is staggering by comparison. Investigators also refer to *subproteomes*, which may be restricted to specific biological compartments, such as the inner mitochondrial membrane. Heightening the complexity of the proteome further is the fact that although a given protein is the product of a single gene, multiple variants or isotypes exist among different cell types and even within the same cell. PTMs determine protein trafficking, tertiary structure, function, and turnover. Levels of mRNA do not predict corresponding cellular protein content. Protein composition and levels are determined by message stability, by specific translational efficiencies, and by turnover of the protein.

The proteome is not static, but rather changes with time in response to any biological stimulus the organism may encounter. A proteome is the result not only of genetic expression, but also of ribosomal synthesis, PTMs, proteolytic degradation, and other alterations that appear to be "unprogrammed" by the genetic code. In contrast to the 30,000 to 50,000 genes that constitute the human genome, the exact number of proteins contained within the human body is not known with any degree of certainty; the range of estimates varies enormously—from a mere 1 million to perhaps 1 billion. A humbling fact today is that even with the most robust MS techniques, far fewer than 1,000 proteins have been identified to date. (These include many of the proteins used today in clinical evaluations, e.g., creatine kinase, troponin, and aspartate and alanine aminotransferase.) Contrary to the case of the human genome, a full description of the human proteome is a task for which completion is not even nearly in sight, nor for which are all of the essential tools yet available.

A major challenge to investigating and describing the full proteome is its *dynamic range*. This term refers to the proteins and peptides within the proteome that are defined by a set of certain characteristics, such as molecular weight (MW), charge, or abundance. Proteins in the serum array themselves across a dynamic range spanning 10 orders of magnitude. The proteins of lowest abundance include cytokines (e.g., interleukin-6), with concentrations in some cases on the order of between 1 and 5 pg/mL. In contrast, high-abundance proteins such as albumin achieve concentrations as high as 3.5 to 5.0 mg/dL, accounting for 55 to 60 percent of all proteins in the serum. Comparing the concentrations of analytes of the lowest abundance to those with the highest abundance is analogous to comparing the mass of one human being to the combined mass of all 6 billion people on earth! Designing an instrument that can measure both ends of this spectrum remains an unmet challenge in proteomics. All currently available proteomics tools evaluate only a narrow spectrum of the proteome's dynamic range. Approaches to proteomic analyses are evaluated in part by the dynamic range of the proteome to which they provide access. With regard to MW, proteomic techniques that are highly effective in analyses of ions and peptides in the range of 700 to 12,000 daltons (Da), for example, have little utility at molecular weights beyond this range.

Proteomics

Proteomics is the application of tools from fields as diverse as clinical medicine, molecular biology, mass spectrometry, and bioinformatics to explore the separation, identification, and characterization of proteins, and to shape the wealth of information offered by the proteome into new knowledge. The three major steps involved in a proteomic analysis are sample preparation, protein (or peptide) separation, and protein characterization/identification. Many different technologies may be brought to bear on each of these steps, particularly the tasks of protein separation and identification. The ever-changing technologies can be mixed and matched to exploit the strengths of each and address the biological question at hand. Thus, proteomics is not a single discipline but rather a collection of highly specialized forms of expertise, all of which may be applied to many types of clinical problems.

The field of proteomics is applicable to a broad spectrum of diseases because the microvasculature is continuously bathed in the proteome. Through modification of study designs and the adaptation of techniques, tissues ranging from blood samples to tissue biopsies may be studied by proteomics. Samples may be obtained by methods as simple as phlebotomy or as complex as laser capture microdissection. One principal advantage of utilizing serum or plasma in proteomic studies is the comparative ease of acquisition for such specimens. Conversely, one major disadvantage is that analyses of blood are more difficult than those of most other tissue samples because of the presence of highly abundant proteins that interfere with the detection of proteins of lower abundance.

"Classic" Proteomics

Classic proteomics studies are multistep processes that involve the separation and simplification of complex protein mixtures; proteolytic digestion of individual proteins; peptide analysis by MS; and finally, protein identification by matching the mass data to a corresponding peptide sequence library. Proteomic techniques such as two-dimensional gel electrophoresis (2DE) and MS have been available for decades, yet the study of proteinaceous substances in the blood has much earlier origins. Examinations of a blood substance called "albumin" began as early as the 1830s. Thus, in some ways, only the name "proteomics" is new. Steady progress with 2DE and MS over the past 25 years, coupled with developments in such related fields as bioinformatics and genomics, has now made detailed studies of the proteome possible and relevant.

Preparation of the Sample

"Fractionation" is a step in the preparation of samples for some analyses. Fractionation involves the removal of certain proteins from a sample (e.g., high-abundance proteins). In some cases, the removal of "interference" from such proteins facilitates the analysis of other proteins of interest (e.g., those of lower abundance). One problem with this approach, however, is that the preponderance of low-abundance proteins bind avidly to high-abundance "carrier" proteins. Removing the carrier proteins results inevitably in the loss of many analytes of potential interest. *Fractionation* must be distinguished from *separation*, which is the differentiation of proteins and peptides from each other that usually occurs (by MS or another technique) *after* fractionation has been performed. Identification of proteins in a complex mixture generally proceeds via one of two approaches: separation of proteins and subsequent digestion into respective peptides, or digestion of the entire protein mixture followed by separation of the resulting peptides.

To Fractionate or Not?

Simple stoichiometry dictates that most small, low-abundance peptides within the serum will be bound to larger, charged species that are far more numerous. This point has profound implications for any attempts to fractionate serum specimens, simply because the removal of high-abundance proteins almost certainly means that lower abundance proteins (peptides) are removed, as well. The fraction of low-abundance proteins that exist typically in a bound state is inherently interesting and is now the focus of explorations of the albuminome—the proteins bound to carrier molecules.

For proteomic approaches predicated on the recognition of patterns of analytes by mass:charge (m/z) value (as opposed to identification of individual peptides), fractionation may be unnecessary [1]. The retention of albumin in unfractionated serum samples may be essential to the diagnostic power of information gleaned from serum through this method. Low-molecular-weight molecules bound to albumin, immunoglobulins, and other carrier proteins are the underpinning of the diagnostic patterns identified by this approach. The sequestration of low-molecular-weight, low-abundance molecules by proteins such as albumin serves two important functions: (1) amplification of these molecules to concentrations readily detectable by high-resolution spectrometry; and, (2) prevention of loss of the molecules through renal excretion. Finally, the ease of use of this technique, which requires minimal processing, increases its potential for clinical applications.

Two-Dimensional Gel Electrophoresis

For three decades, the mainstay of proteomic analysis has been 2DE [2]. The basic separation principles behind 2DE are essentially unchanged since the 1970s. Resolution of proteins by charge in the first dimension (isoelectric focusing) is followed by separation by relative mass (SDS-PAGE) in the second. Proteins separated by 2DE can be visualized by conventional staining methods including silver, Coomassie and amido black stains, and fluorescent stains. Fluorescence-based protein detection methods have recently surpassed conventional technologies such as colloidal Coomassie blue and silver staining in terms of quantitative accuracy, detection sensitivity, and compatibility with modern downstream protein characterization procedures such as MS.

Peptides derived from separation techniques (2DE, SDS-PAGE, chromatography) or entire digests of protein mixtures can be analyzed by many types of MS coupled to one of two ionization techniques: matrix-assisted laser desorption ionization (MALDI), or electrospray ionization (ESI). Spots of interest are selected, digested, and then analyzed by MS. Even so, this application has a limited dynamic range and is generally effective at the identification of only high-abundance proteins. Two-dimensional gels have two major limitations as a tool for proteomics: (1) general ineffectiveness at distinguishing low-abundance proteins; and (2) underrepresentation of basic and membrane proteins.

Newer Approaches to Differential Display

Three approaches to differential display proteomics are now commonly employed: difference gel electrophoresis ("dual color labeling"), isotope-coded affinity tagging (ICAT), and multidimensional protein identification technology (MudPIT). The fundamental principle behind dual color labeling is the detection of protein differences within samples of interest by the ratios of differently colored fluorescent signals [3]. ICAT and MudPIT are discussed in additional detail later.

Isotope-Coded Affinity Tags (ICAT)

ICAT peptide labeling is used to distinguish between two populations of proteins using isotope ratios [4]. ICAT reagents consist of a protein-reactive group, an ethylene glycol linker region, and a biotin tag. The commercialized ICAT reagent employs a reactive functionality specific for the thiol group of cysteine residues. Using linkers, two different isotope tags are generated containing either eight hydrogen atoms (the "light reagent") or eight deuterium atoms (the "heavy reagent"). A reduced protein mixture from one protein specimen is derivatized with the isotopically light version of the ICAT reagent, while the other reduced protein specimen is derivatized with the isotopically heavy version of the ICAT reagent. Following combination of the two samples, the mixture is digested with trypsin or Lys-C and then fractionated by avidin affinity chromatography, a step that retrieves only peptides containing cysteine. The purified mixture of peptides contains roughly tenfold fewer peptides than the original sample, thus simplifying the analysis.

MudPIT ("Multiplex Proteomics")

MudPIT [5, 6] involves proteolytic digestion of the entire protein mixture in solution followed by two-dimensional chromatographic separation of peptides. A key advantage of MudPIT is its ability to determine amino acid sites of PTMs. This approach, sometimes referred to as "top-down" proteomics, assesses all proteins within a mixture and does not select for certain classes of proteins. After collection of the peptides using a variety of techniques, 2-D chromatography is followed by ESI and tandem MS.

Mass Spectrometry

An MS (Figures 2A, B) is an instrument designed to measure the masses of individual molecules that have been converted to ions, that is, that are electrically charged. Compared with 2DE, MS offers dramatic improvements in throughput, the quantity of biological specimen required for analysis, and sensitivity to low-molecular-weight and low-abundance peptides and proteins. MS can also detect tens of thousands of molecules concomitantly, without the need for antibody development and validation in advance, a major shortcoming of protein arrays in their current state (see later discussion).

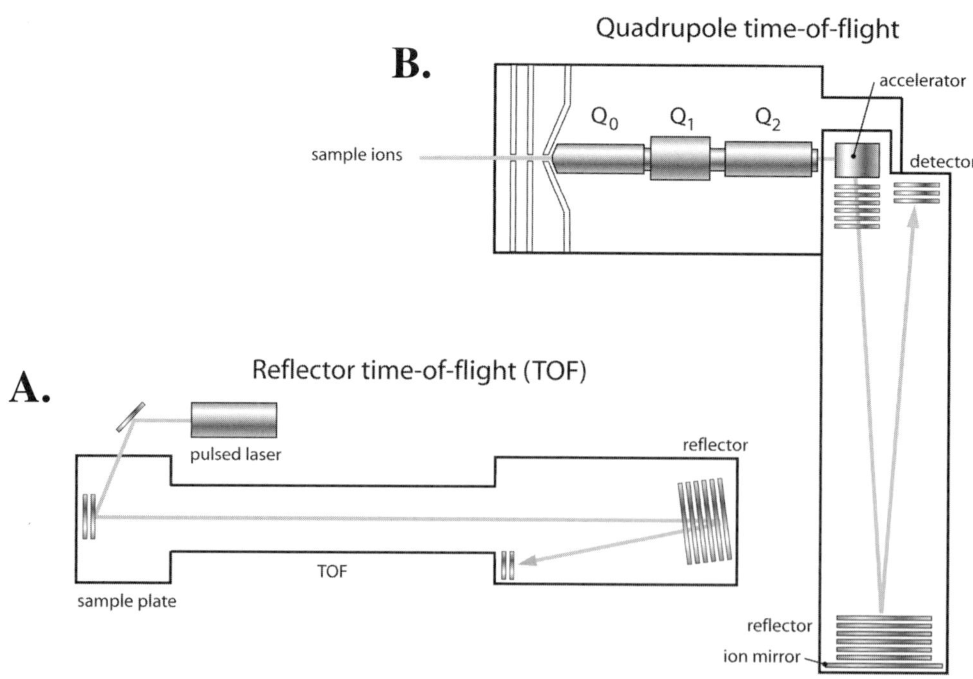

Figure 2 Mass spectrometry. **(A)** Standard low-resolution mass spectrometer, typical of those used in MALDI and SELDI analyses. **(B)** Tandem mass spectrometer. The principal distinguishing feature from low-resolution instruments is the presence of a collision cell in which peptides are broken apart by collision with an inert gas, permitting in many cases sequence analysis and parent protein identification. **(C)** Positions on protein chip spots. Each spot is divided into tiny coordinates (the 20, 50, and 80 marks shown in the figure) known as positions that are used to direct the laser to strike at precisely the same point on each spot. Smaller ions fly more swiftly to the detector plate, permitting their differentiation by mass : charge value. The right side of the figure shows the proteomic profile generated by the sample on the spot. (see color insert) **(D)** Peak map (lower part of the figure) showing how numerous analytes may be found around individual *m/z* values.

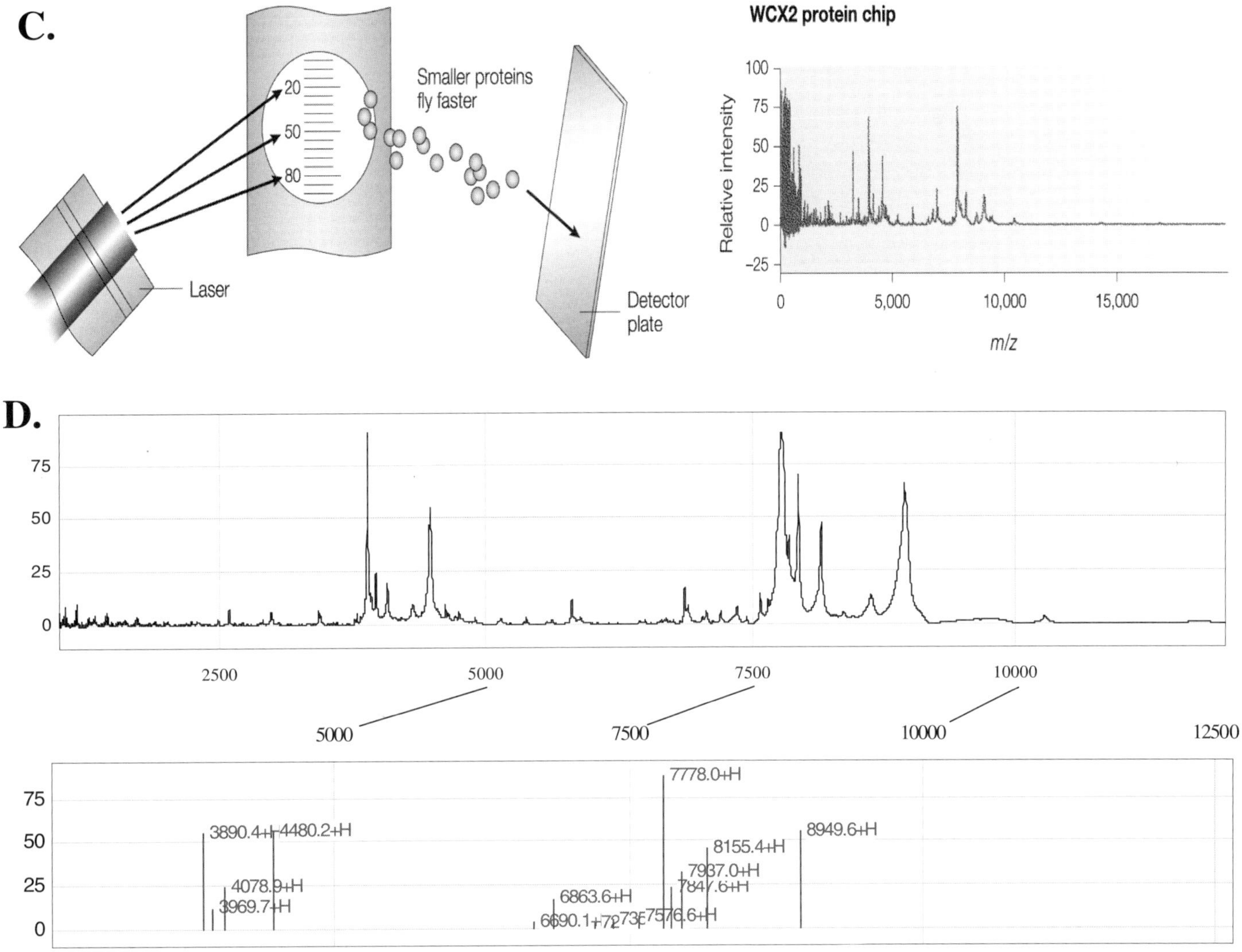

Figure 2 *Continued.*

In general, an MS has three components: (1) a chamber that holds the ion source (the clinical sample from which analytes are ionized via laser); (2) a detector that registers the number of ions at each *m/z* value; and (3) a mass analyzer that measures the *m/z* ratio of the ionized analytes. A mass spectrometer has the ability to analyze samples processed on a variety of platforms, including MALDI, ESI, and surface-enhanced laser desorption ionization (SELDI).

Ionization Techniques

Two concepts central to MS are (1) the ionization of analytes out of solution, which may be accomplished by a variety of methods, and (2) the time-of-flight differences determined by analytes' *m/z* values. It is this latter concept that permits distinction of the ions based on their size and charge. MALDI and ESI are "soft" ionization methods that produce ions with low internal energy, thus preserving analyte structure and allowing the transfer of larger biomolecules (e.g., peptides, proteins, oligosaccharides, and oligonucleotides) into the gas phase. The peptide data provided by these mass spectrometers can then be used to identify the proteins of origin by searching genomic databases for matches among the theoretical masses of all peptides. Traditional MALDI- or ES-MS instruments measure peptide masses. In contrast, tandem MS instruments (which also typically achieve ionization of analytes through either MALDI or ESI) permit the determination of peptide sequences. The various approaches to ionization in MS are discussed separately later.

Time of Flight

Time of flight refers to the length of time required for proteins and peptides ionized from the surface of a protein chip to travel through the MS chamber to its detector plate (Figure 2C). "Time-of-flight" is abbreviated "TOF," as in "SELDI-TOF" or "MALDI-TOF" or "quadrupole-TOF (Q-TOF)." The fundamental principle that permits MS to separate analytes is the fact that small ions fly faster than large ones.

MALDI

MALDI consists of a stainless steel plate onto which the sample is spotted directly. With the MALDI technique, analytes are sublimated (i.e., taken directly from the solid to the gaseous phase) and ionized out of a dry, crystalline matrix by laser pulses. Protein separation using affinity columns or other fractionation techniques is usually performed *before* the application of mass spectrometry by MALDI.

ESI

ESI ionizes analytes out of solution. It is readily coupled, therefore, to liquid-based protein separation tools such as liquid chromatography (LC). Integrated systems of LC and mass spectrometry (LC-MS), now based on ESI, are the preferred technique for the analyzing complex samples.

SELDI

SELDI involves the retention of proteins on a solid-phase chromatographic surface (ProteinChip Array) and direct detection of the retained proteins by time-of-flight mass spectrometry [Tang]. The SELDI technique performs protein separation on the protein chip, based on analytes' surface charge. Samples can be preprocessed on the basis of size exclusion, pH, pI, and other features to isolate proteins of interest further. First applied to clinical medicine in the late 1990s, SELDI represents a breakthrough in protein separation techniques because of its superiority (compared to two-dimensional gel electrophoresis, 2DE) in the detection of low MW ions and ions of basic charge. SELDI offers on-chip preparation and capture of proteins based on a variety of chemical and biological matrices, coupling direct mass spectral analysis to chemical and affinity-based separations. SELDI profiles a subset of proteins distinct from those separated by 2DE and is most appropriate for low molecular weight components. The various chemistries (hydrophobic, cationic, or anionic) of these affinity surfaces capture the proteins with complementary chemical characteristics. Chips are available that focus on a particular class of proteins (e.g., SNARE proteins) or proteins in a specific pathway (e.g., signal transduction). Some substrates capture proteins with weakly positive charges, whereas others have affinities for metal ions such as nickel or copper. Because of overlapping dynamic ranges that they target, different chip surfaces may be complementary. In general, the same protein chips used for SELDI analyses may also be used with tandem MS platforms.

A significant disadvantage of SELDI is that the technique does not provide a sequence-based identification, because there are many proteins close to a given *m/z* ratio (Figure 2D). The resolution of SELDI is inferior to that of the most current MS instruments, particularly tandem MS. Moreover, with SELDI, the protein peaks representing potential markers cannot be identified without conventional peptide mapping and significant additional effort.

Tandem Mass Spectrometry

Tandem MS measurements now provide the means to characterize specific PTMs and to identify structural differences between related proteins, differentially modified proteins, and protein isoforms [7]. Individual proteins can be identified through the analysis of collision-induced spectra, which provide information about peptide sequences.

For proteomic studies, ESI sources are most commonly paired with tandem mass analyzers. Tandem MS instruments are also referred to as MS/MS (and, when coupled to liquid chromatography, as LC-MS/MS). MS/MS, an analysis that occurs in two stages, is generally used to derive structural information as well as data related to the masses of analytes. The instruments for tandem MS fall into two categories: those with two analyzers in series (tandem in space; e.g., triple quadrupole and hybrid quadrupole-TOF configurations) and those instruments with sequential analyzers [tandem in time; e.g., quadrupole ion trap and Fourier transform–ion cyclotron resonance spectrometers (FT-ICR)].

The operational principles behind all tandem analyzers are similar: A single peptide ion species, selected on the basis of its *m/z* value (first mass selection stage), is fragmented by a process known as collision-induced dissociation (CID). CID involves collision of the ion with an inert gas (e.g., argon or nitrogen). In this second stage of mass selection, the resulting "daughter" ion fragments are then detected (second stage of mass selection) and an *m/z* spectrum is produced. Mass differences between daughter fragments, which correspond to successive losses of individual amino acid fragments, can be used to construct a partial amino acid sequence. If this amino acid sequence is sufficiently long—perhaps as few as five residues—the parent protein can be identified if the sequence is not derived from a highly conserved motif. This approach to protein identification is termed *product ion searching* (see later discussion).

An important but counterintuitive point is that even tandem mass spectrometers are, at best, only semiquantitative instruments. For both MALDI and ESI platforms, the relationship between the amount of a given analyte present and the measured signal intensity is complex and nonlinear. The reasons for this phenomenon remain poorly understood. A technique known as stable isotope labeling now permits quantification of peptide levels by MS/MS [8, 9].

Approaches to Protein Identification

Two approaches are used for protein identification: peptide mass fingerprinting (PMF) and product ion searching. Both methods use software-assisted database searching, and both are continuously evolving as the process of informatics improves. For PMF, the predominant MS approach is MALDI-based. PMF uses an algorithm to search a database and match peptide masses from known proteins with peptide masses of unknown proteins determined by MS. These databases contain theoretical masses developed from genomic information. PMF is therefore most effective when the genome of the organism has been sequenced entirely and an extensive peptide mass database corresponding to the genome exists. PMF is least effective for expressed sequence tags (ESTs), in which only short stretches of sequence are present in the database, or for unusual organisms, for which the database is limited.

For product ion searching (peptide sequence analysis), ESI-tandem MS is the most commonly employed method. Peptide fragmentation and analysis of corresponding product ions by tandem MS can yield sequence information. This strategy for protein identification is more definitive than ones relying merely on molecular mass determination. In some cases, a protein from a well-characterized genome can be identified from a single peptide.

Protein Microarrays

Protein arrays contain collections of immobilized protein spots, each of which can contain a homogeneous (or heterogeneous) set of "bait" molecules (Figure 3) [10]. A spot on the array may display an antibody, a cell or phage lysate, a

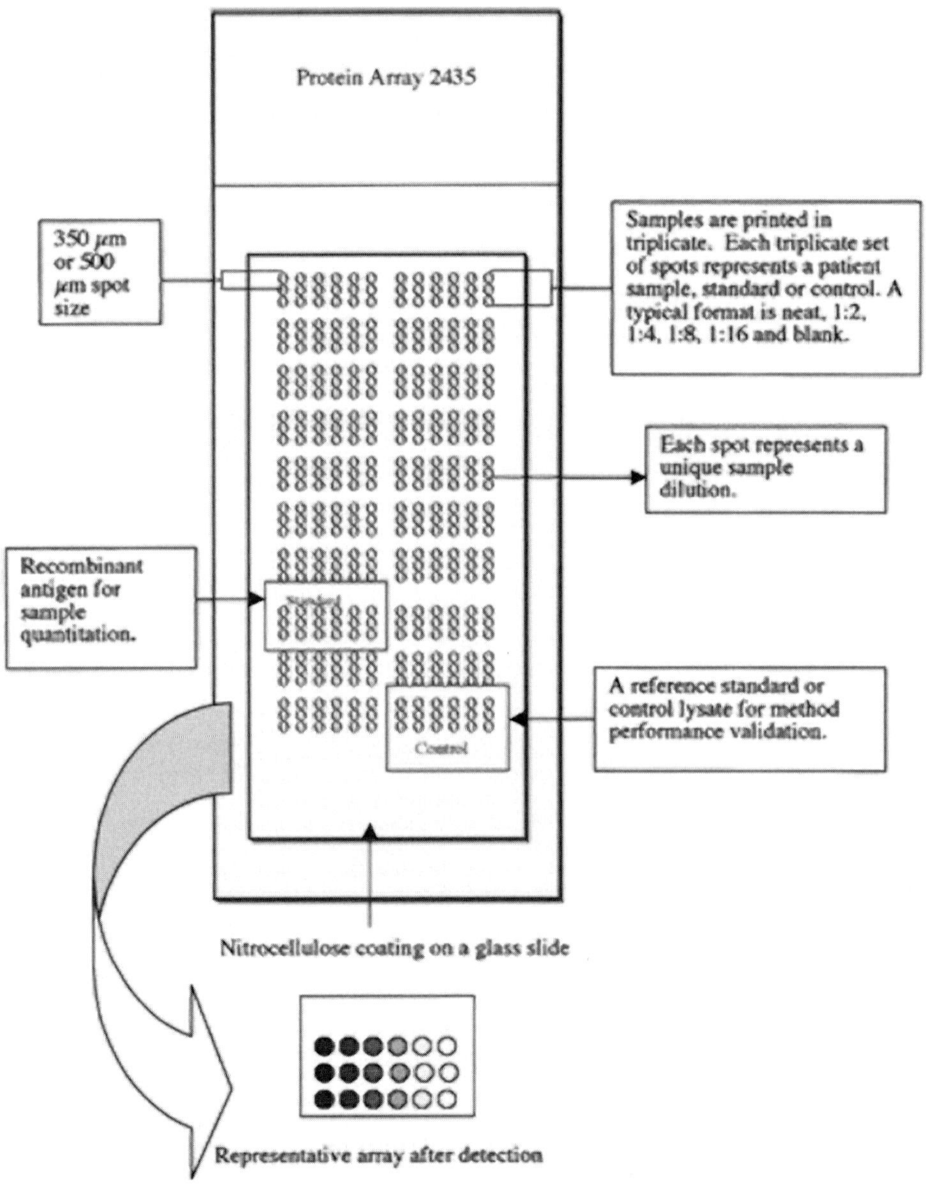

Figure 3 An idealized reverse phase array format. Triplicate samples are printed in dilution curves representing undiluted, 1 : 2, 1 : 4, 1 : 8, and 1 : 16 dilutions. The sixth spot represents a negative control, consisting of extraction buffer without sample. Each set of triplicate spots represents a patient sample before or after treatment. A reference sample or control sample is printed on each array for the purpose of monitoring assay performance. A recombinant antigen is printed on each array for comparative quantitation of patient samples.

recombinant protein or peptide, a drug, or a nucleic acid. The array is queried with (1) a probe (labeled antibody or ligand), or (2) an unknown biologic sample (e.g., cell lysate or serum sample) containing analytes of potential interest. By tagging the query molecules with a signal-generating moiety, a pattern of positive and negative spots is generated. For each spot, the intensity of the signal is proportional to the quantity of applied query molecules bound to the bait molecules. An image of the spot pattern is captured, analyzed, and interpreted.

Protein microarrays may be used to (1) discover novel ligands or drugs that bind to specific bait molecules on the array (protein arrays offer one means of conducting massive screening of drugs); (2) develop miniature panels of serum biomarkers or cytokines through multiplexing immunoassays; and (3) profile the state of specific members of known signal pathways and protein networks.

As illustrated in Figure 4, protein microarray formats fall into two major classes: forward phase arrays (FPA) and reverse phase arrays (RPA), depending on whether the analyte is captured from the solution phase or bound to the solid phase. In the FPA format, each array is incubated with one test sample, and multiple analytes are measured at once. In contrast, the RPA format immobilizes an individual test sample in each array spot, such that an array is composed of hundreds of different patient samples. In the RPA format, each array is incubated with one detection protein (e.g., antibody), and a single analyte end point is measured and directly compared across multiple samples. RPAs have achieved the sensitivity and precision acceptable for the analysis of clinical tissue biopsy specimens.

The most important starting point for the development of any protein array method is the selection of antibodies with high specificity and adequate affinity. If the experimental focus is cellular signaling analysis, the platform will require at least two different kinds of antibodies for each protein, one chosen to recognize the phosphorylated form of the protein, and the second to recognize the protein regardless of its phosphorylation status. One challenge in protein microarray technology is the fact that high-quality antibodies are currently available for only a small percentage of the known proteins involved in signal networks and gene regulation.

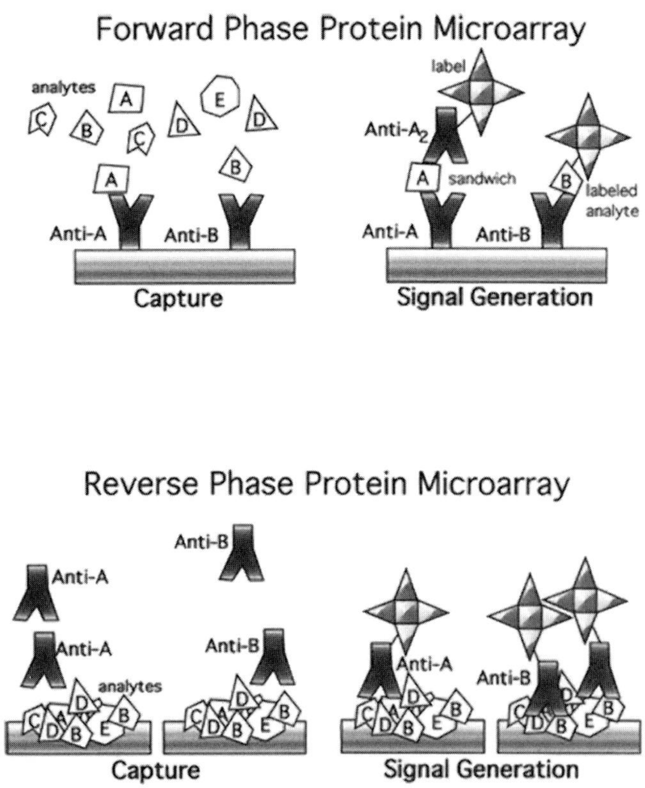

Figure 4 Forward and reverse phase arrays: schematic. Forward phase arrays (*top*) immobilize a bait molecule such as an antibody designed to capture specific analytes with a mixture of test sample proteins. The bound analytes are detected by a second sandwich antibody, or by labeling the analyte directly (*upper right*). Reverse phase arrays immobilize the test sample analytes on the solid phase. An analyte-specific ligand (e.g., antibody, *lower left*) is applied in the solution phase. Bound antibodies are detected by secondary tagging and signal amplification (*lower right*). (see color insert)

Glossary

Abundance: The proteomics literature refers to "low-abundance" proteins (e.g., cytokines) and "high-abundance" proteins (e.g., albumin or immunoglobulins). The term *abundance* simply means "concentration." The development of methods by which low-abundance proteins may be studied in the setting of high-abundance proteins that dwarf them by many orders of magnitude is one of the greatest conundrums confronting proteomics today.

Dynamic range: Refers to the proteins and peptides within the proteome that are defined by a set of certain characteristics, such as molecular weight (MW), charge, abundance, or other features. Approaches to proteomic analyses may be evaluated in part by the dynamic range of the proteome to which they provide access. With regard to MW, for example, some proteomic techniques (e.g., SELDI) that are highly effective in analyses of ions and peptides in the range of 700 to 12,000 daltons (Da) may have little utility at MW beyond this range.

Fractionation: A step in the preparation of samples for some types of proteomic analysis. Fractionation involves the use of a variety of techniques to remove certain proteins from a sample (e.g., high-abundance proteins). In some cases, the removal of "interference" by such proteins facilitates the analysis of other proteins of interest (e.g., those of lower abundance) that are theoretically more pertinent to the disease of interest. *Fractionation* must be distinguished from *separation*, which is the differentiation of proteins and peptides from each other that usually occurs (by mass spectrometry or another technique) *after* fractionation has been performed.

Resolution: Refers to the ability of a mass spectrometer (or, more specifically, of its mass analyzer) to distinguish between discrete analytes with similar characteristics (see, for example, the peak map in Figure 2). In general, tandem MS techniques have greater resolution than their predecessors, albeit their dynamic ranges may be considerably narrower.

Time-of-flight: Refers to the length of time required for proteins and peptides ionized from the surface of a protein chip to travel through the MS chamber to the detector plate. "Time of flight" is abbreviated "TOF," as in "SELDI-TOF" or "MALDI-TOF" or "quadrupole-TOF (Q-TOF)." The fundamental principle that permits MS to separate analytes is the fact that small ions fly faster than large ones. The ions' m/z ratios may be calculated from the time that each requires to reach the detector plate. Differences in TOF permit the distinction and, in many cases, the identification (by tandem MS) of different peptides.

References

1. Petricoin, E. F., Ardekani, A. M., Hitt, B. A., Levine, P. J., Fusaro, V. A., Steinberg, S. M., Mills, G. B., Simone, C., Fishman, D. A., Kohn, E. C., and Liotta, L. A. (2002). Use of proteomic patterns in serum to identify ovarian cancer. *Lancet* **359**(9306), 572–577.
2. O'Farrell, P. H. (1975). High resolution two-dimensional electrophoresis of proteins. *J. Biol. Chem.* **250**, 4007–4021.
3. Patton, W. F. (2002). Detection technologies in proteome analysis. *J. Chromatogr. B Analyt. Technol. Biomed. Life Sci.* **771**(1–2), 3–31.
4. Gygi, S. P., Rist, B., Gerber, S. A., Turecek, F., Gelb, M. H., and Aebersold, R. (1999). Quantitative analysis of complex protein mixtures using isotope-coded affinity tags. *Nat. Biotechnol.* **17**, 994–999.
5. Washburn, M. P., Wolters, D., and Yates, J. R. (2001). Large-scale analysis of the yeast proteome by multi-dimensional protein identification technology. *Nat. Biotechno.* **19**, 242–247.
6. Reid, G. A., and McLuckey, S. A. (2002). "Top-down" protein characterization via tandem mass spectrometry of biomolecules. *J. Mass. Spectrom.* **37**, 663–675.
7. Aebersold, R., and Mann, M. (2003). Mass spectrometry-based proteomics. *Nature* **422**, 198–207.
8. Conrads, T. P., Issaq, H. J., and Veenstra, T. D. (2002). New tools for quantitative phosphoproteome analysis. *Biochem. Biophys. Res. Commun.* **290**, 885–890.
9. Ong, S. E., Blagoev, B., Kratchmarova, I., Kristensen, D. B., Steen, H., Pandey, A., and Mann, M. (2002). Stable isotope labeling by amino acids in cell culture, SILAC, as a simple and accurate approach to expression proteomics. *Mol. Cell Proteomics* **1**, 376–386.
10. Liotta, L. A., Espina, V., Mehta, A. I., Calvert, V., Rosenblatt, K., Geho, D., Munson, P. J., Young, L., Wulfkuhle, J., and Petricoin, E. F. 3rd. (2003). Protein microarrays: Meeting analytical challenges for clinical applications. *Cancer Cell* **3**, 317–325.

Capsule Biography

Dr. Stone is the Director of the Johns Hopkins Vasculitis Center and the Deputy Director for Clinical Research at the Johns Hopkins Bayview Medical Center. His work, funded by the National Institute of Arthritis, Musculoskeletal, and Skin Diseases and the Immune Tolerance Network (National Institute of Allergy and Infectious Disease), focuses on translational research in proteomics and the conduct of clinical trials of novel therapeutic agents.

Dr. Petricoin is the Co-Director of the NCI/FDA Clinical Proteomics Program. Dr. Petricoin and his coworkers were among the first to identify the JAK-STAT pathway as critical to global cell growth and cytokine signaling. In his current role as Chief of the Tissue Proteomics Unit in the Division of Cytokine Biology at the FDA, Dr. Petricoin directs studies of laser capture microdissection and proteomics, with a focus on human carcinogenesis. He has performed groundbreaking work in the area of serum proteomic studies that employ pattern-recognition approaches.

SECTION G

Stem Cells

CHAPTER 167

The Therapeutic Promise of Endothelial Stem Cells

[1,2]Denis English, [1]Kevin Harvey, and [3]Ursula Gehling
[1]Department of Neurosurgery, University of South Florida College of Medicine, Tampa, FL
[2]Methodist Hospital of Indiana, Indianapolis, Indiana
[3]University Hospital Hamburg-Eppendorf, Hamburg, Germany

Definition of the Subject

The vascular system consists of the heart and lungs, arteries, veins, and capillaries as well as the cells and plasma they transport. It provides an essential mechanism of communication between all organs and tissues while delivering cells indispensable for host defense, oxygenation, wound healing, and hemostasis. The vascular system is complex, and defects within it lie at the very foundation of many chronic and untreatable diseases. Although chronic diseases associated with vascular dysfunction resist therapy, transplantation with vascular progenitor cells offers therapeutic hope. To reach this goal, the identity, evolution, differentiation, and development of stem cells involved in forming the vasculature must be understood in depth. This chapter will review the impact of results of recent stem cell research upon our understanding of the mechanisms involved in the development of the vascular system.

The Essence of the Vascular System

The development of the vascular system during embryogenesis has fascinated investigators for more than a century. This fascination has sparked research that now promises to provide new and effective treatment for diseases ranging from diabetic retinopathy to Alzheimer's disease. The vascular system is indeed unique. It is a multicompartmental system in both its cellular composition and its varied critical function. Angiogenesis, the formation of new blood vessels from existing capillaries, facilitates vascular repair in adults, whereas vasculogenesis, the process of blood vessel formation from stem cells, establishes the vascular system in embryonic development. Recent studies indicate that vascular stem cells residing in the hematopoietic system of adults contribute to neovascularization, suggesting that vasculogenesis occurs in adults. This concept has important implications for the biology of development of the vascular system as well as for the potential therapeutic utility of vascular and endothelial stem cells.

Vascular Development and Divergent Endothelial Cell Populations

As we gain a better understanding of the angiogenic response, we also gain a better understanding of the development of the vasculature from stem cells, including the widely heralded embryonic stem cells of the blastocyst, as well as the adult vascular stem cell, the putative hemangioblast.

Most tissues and organs possess the capability to mount a vigorous angiogenic response. Thus, therapy to promote angiogenesis may be most useful in tissues in which angiogenesis is sluggish or defective. Our studies indicate that these tissues include the lung and the brain, which often are the first organs to show pathology associated with defective angiogenesis in otherwise healthy individuals. To reach this goal, it is essential to identify the characteristics of the cells we hope to use therapeutically. Many subpopulations of endothelial progenitor cells may be involved in vascular development. Thus, if the angiogenic potential of endothelial cells of differing organs differs because the cells derive initially from divergent tissue-specific endothelial stem cells, then tissue specific endothelial progenitor cells will be required for effective therapy. However, if local factors are the cause of divergent endothelial responses of different tissues, a common primitive vascular stem cell may hold therapeutic promise. The answers to these questions are not known. Clinical trials are underway to assess the utility of putative endothelial progenitor cells for various diseases.

The Diverse Endothelium

During development, the vasculature is dynamic and heterogeneic. Chronic vascular defects are extremely devastating and resistant to treatment in organs that are well protected from injury throughout life, in particular the lung and the brain. Endothelial cells derived from the capillaries of these organs display sluggish angiogenic responses in culture systems. What could explain the development of such a diverse vascular system, where endothelial cells and their progenitors of some organ systems, such as that of the oral mucosa and the derma, display vigorous angiogenic responses, while others do not?

Throughout development, the endothelium is dynamic as it assumes crucial and diverse functions. In early development, the endothelium is one of the first structures formed. Its cells line developing blood vessels and exert essential roles in homeostasis and development. Many of these functions are lost as development progresses. Thus, early vascular cells can display phagocytic properties, providing essential host defense. Indeed, the Kupffer cells that line the liver sinusoids are endothelial cells that originate from the yolk sac splanchnic mesoderm and are introduced into the liver by sprouting of the vitellin vein.

Primitive Hematopoiesis

The primitive hematopoietic system derives from these yolk sac vessels, which possess the ability to generate both primitive erythrocytes and new endothelium. These derive from progeny of embryonic stem (ES) cells, which can be produced in culture for therapy (Figure 1). Later, in the embryo, the primitive yolk-sac-derived "hemogenic" endothelium, as well as embryonic blood islands, generate other blood cells while providing new vessels to developing organs. Thus the hematopoietic system derives from specialized stem cells called *hemangioblasts*, which produce both hematopoietic cells and vascular endothelial cells upon a single cell division. The ability of hemangioblasts to provide both hematopoietic cells and endothelium decreases as the embryo matures as the "definitive" hematopoietic system develops (Figure 2). Recent reports suggest that the adult equivalent of the hemangioblast also derives from these primitive stem cells and ultimately takes up residence in hematopoietic tissue. If it exists, the adult hemangioblast holds tremendous therapeutic potential.

The Diverse Endothelium

The endothelium indeed is a functionally diverse tissue. During development, the skeletal system protected highly specialized organs in a manner that allowed them to develop in an environment that posed little traumatic threat. Other tissues, continuously subject to trauma and injury, needed to develop a vascular system that healed faster than that of the former. Specialized, protected organs used this opportunity to develop a specialized endothelium, which optimizes the function of the organ they protect, and as a result did not develop the vigorous angiogenic capabilities less specialized tissues critically required, tissues that were constantly under attack, such as the skin, the mucous membranes, the intestines, and the oral cavity. Further, as skeletal protection developed late in embryogenesis, the endothelium of these tissues also did not acquire other functions associated with fetal endothelial cells, such as the ability to generate stem cells. In evolution, one only has to survive until the age of reproduction to successfully compete. Thus, early man, evolving with a well-protected ectoderm on the pristine environment of the earth, required only 20 to 30 years of life to bear children and raise them to reproduce against evolutionary forces. However, humans evolved so successfully that they then changed the equation dramatically by extending their life expectancy. Further complicating development was humans' unparalleled alteration of the environment

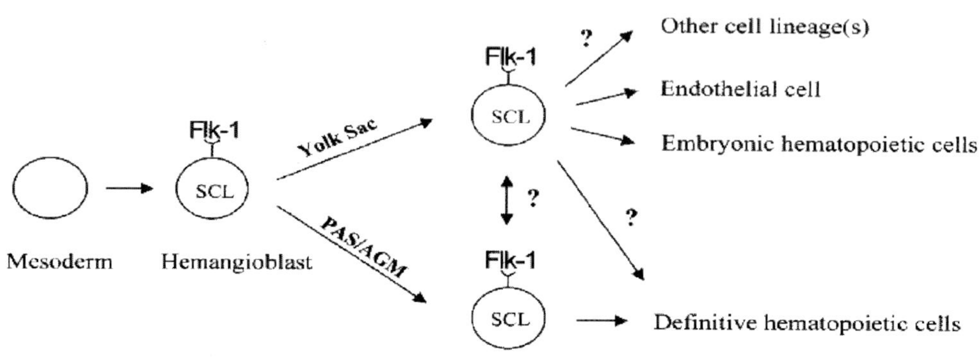

Figure 1 Isolation of embryonic stem cells for investigation and therapy. (see color insert)

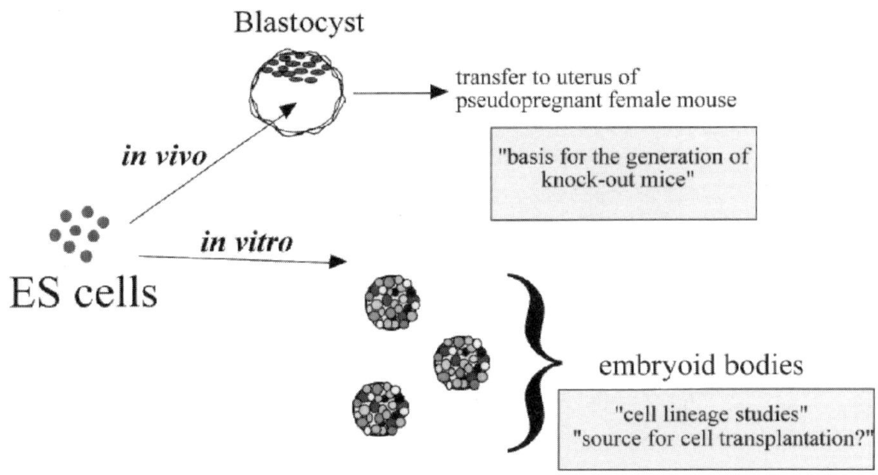

Figure 2 Schematic view of primitive and definitive hematopoietic development in the developing embryo.

with new toxins and chemicals, the likes of which the forces of evolution never previously encountered and were not able to cope with. Thus, much of the pathology and the therapeutic resistance associated with many diseases observed in aging individuals, such as Alzheimer's and Parkinson's disease, as well as chronic lung diseases, are a consequence of the diminished angiogenic capabilities of vascular endothelial cells in these tissues. The chronic nature and resistance to treatment of these diseases reflects ineffective inability to promote angiogenesis in these organs, thereby rendering them attractive candidates for therapeutic angiogenesis using vascular stem cells.

Adult Endothelial Progenitor Cells

Several methods are being evaluated for using adult stem cells to restore vascular function. Within the population of adult stem cells, there is thought to exist a precursor of the endothelium, or endothelial progenitor cell (EPC), which holds great therapeutic promise. As stated earlier, it is of paramount importance to define the basis of the differences in function of endothelial cells of diverse organs, and how these differences may impinge upon therapy. Recent results render this consideration even more important, since adult EPCs have not yet lived up to their promise as predicted from animal experiments, and because adult EPCs isolated by various groups show differences in phenotypes. Initial studies by Isner and Asahara claimed that endothelial progenitors circulated in blood and expressed CD34 and Flk-1. Later studies indicate that $CD14^+$ monocytes were the source of these cells, shedding some question on their identity as stem or progenitor cells. However, many studies referred to $CD14^+$ monocytes as "endothelial progenitor cells," although $CD14^+$ monocytes are considered to be mature, end stage blood cells. Eventually, most groups working with *bloodborne* putative endothelial progenitor cells in fact used $CD14^+$ monocytes. The blood monocyte is a remarkably adaptable cell and can express antigens that endothelial cells express, incorporate into the developing vasculature and change shape in response to local needs and clues. However, it serves this function temporarily, until actual endothelial cells are able to take their place.

The blood cells discussed previously do not express markers associated today by most groups with the characteristics of a progenitor cell, such as CD133 or P1H12 (CD146). In addition, many reports showed that their presence in the new capillary decreased within a few weeks. At this time, studies from Raffii's laboratory as well as studies by Gehling and associates reported the identification and characterization of specialized stem cells that appear more likely to be authentic endothelial progenitor cells. Thus, Gehling and associates showed that the $CD133^+$ population from mobilized blood contained progenitor cells that could differentiate into functional endothelial cells that form blood vessels in vivo (Figure 3). At present, the term *endothelial progenitor cells* should be reserved to denote a specialized stem cell with the ability to generate cells that not only form functional vessels, but also stably incorporate into the developing vasculature during angiogenesis. Current dogma characterizes these cells as $CD34^+/CD133^+/VEGFR2^+$ stem cells. These are localized in hematopoietic tissue but mobilize in response to certain cytokines to the circulation.

An additional subset of physiological endothelial progenitor cells has been described that is defined by expression of CD34, CD133, and VEGFR-3. This cell represents lymphatic vascular progenitor cells. Observations in Dr. Gehling's laboratory show that another subset of endothelial cell exists that does not express CD34, CD133, CD14, or VEGF R-2. Identification of vascular or endothelial progenitors is in its infancy, and ongoing studies should clarify this area in the near future.

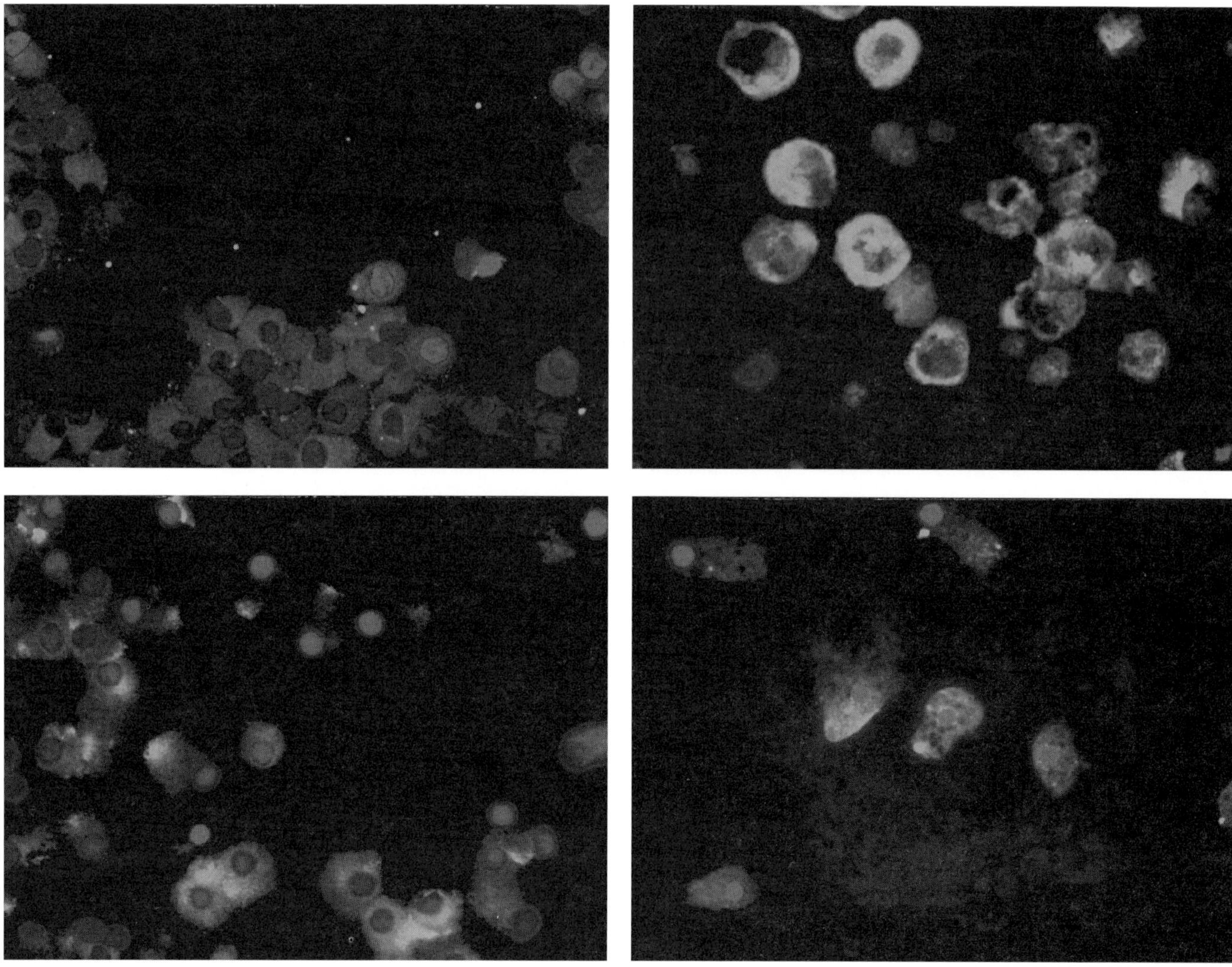

Figure 3 Phenotypic characteristics of the putative adult hemangioblast. CD133 enriched cells were obtained after 14 days of expansion and differentiated for an additional 14 days in defined media. All cells are positive for VE-cadherin (*upper left panel*, weak staining), KDR (*upper right panel*), CD31 (*lower left panel*), and vWF (*lower right panel*), respectively. Addition of VEGF caused these cells to generate endothelial cells, whereas hematopoietic growth factors induced hematopoiesis. (See Lodges et al., *Stem Cells and Development*, April 2004.)

Embryonic Vascular Progenitor Cells

Vasculogenesis precedes angiogenesis, but both processes are required for development of the vascular system. Study of vascular development during embryogenesis and the stem cells involved shed light on the characteristics of adult vascular stem cells.

The vascular system consists of several components, the origins and functions of which may well be interconnected in a unique manner: a manner that renders the vascular and circulatory system as a single system, at least in its development. The two most important components, the vascular endothelium and the blood that percolates through it, may, at first glance, seem to be entirely independent. But this raises serious questions in development, since these two components are useless alone. Blood needs vessels, and vessels have no use without blood.

The answer to this problem was intuitive to the pioneering and insightful investigator Florence Sabin, who in 1917 identified a common stem cell that connects the two systems, which she termed the angioblast. Her tedious and elegantly detailed rendition of the development of a vascular system, replete with endothelium, blood, and plasma, in a 2-day-old chick blastocyte observed in a hanging drop culture is stunning:

The question of the origin of the vascular system can be solved by the method of studying the living blastoderm of the chick in hanging-drop suspensions; by watching vascular development of chicks in their second day of incubation in these preparations, it is possible to see all the processes by which blood vessels and their later blood cells form." (From *The Anatomical Record*, **13**, 199–204; an exact duplicate of this classic article can be found in *J. Hematother. Stem Cell Res.* **11**, 5–7, 2002.)

Dr. Sabin's observations indicate that in the early stages of embryogenesis the vasculature and the cells that it carries develop simultaneously in an interdependent manner, thereby solving the problem of how two completely different tissues that depend on each other arose simultaneously during development. In early development, cells termed *angioblasts* differentiate throughout the wall of the yolk sac and embryo from mesodermal derivatives. These cells, observed in the blastoderm of a chick 2 days after fertilization, produce blood plasma, endothelial cells, and primitive red blood cells. Confined to blood islands throughout the mesoderm, the angioblasts increase in both size and number by cell division and growth as well as by the constant addition of new angioblasts that differentiate from mesodermal parenchyma. These masses join and project cytoplasmic processes from which blood vessels are derived. In the middle of this dense mass, as cells within the mass liquefy, vacuoles appear, leaving endothelial cells along the border. Primitive red blood cells are generated from the angioblastic endothelium, as cytoplasmic projectiles of nearby similar structures seek and enjoin one another, developing a primitive vascular bed into which the young erythrocytes are released. In the developing chick, this process was completed in some cases in as little as 2 hours, and was effected in some instances by a single cell, allowing the conclusion that the endothelium lumen may initially consist of a single, vacuolated cell. The process appeared to continue well into gestation to supply oxygen to organs of the developing embryo. Dr. Sabin never reported the generation of cells other than fetal erythroblasts by angioblasts, and their relation to the putative and somewhat elusive adult hemangioblasts discussed earlier is unclear. However, the angioblast of the early yolk sac does add conceptual support for the concept that a single endothelial cell can indeed produce both hematopoietic and endothelial progeny.

The Hemangioblast

Indeed, the very concept of the hemangioblast in both embryonic and adult vascular development affords a degree of simplicity to the processes involved, prompting many to search tediously for the cell in adult tissue. Because recent studies clearly show the involvement of vasculogenesis in adult wound healing and neoplastic development (i.e., vasculogenesis is not restricted to embryonic vascular development), many have turned their attention to the adult human hemangioblast. Recent reports describe the putative hemangioblast as a derivative of a primitive population of adult cells that express the developmental antigen CD133. These cells may be extremely useful therapeutically, if they are expanded and infused or implanted in patients with torn or diseased vessels.

In this respect, two aspects of endothelial stem cell research require more attention to determine the therapeutic promise of these cells. First, the tissue specificity of the progeny of these cells must be assessed. If, as discussed previously, differences in endothelial cell function of different tissues are a result of local factors, then the EPCs that are isolated from hematopoietic tissue may in fact be useful therapeutically for a wide number of tissues. However, if the functional capabilities of endothelial cells of divergent organs reflect fundamental differences in cellular biochemistry as a result of their derivation from specific endothelial stem cells that derive from different precursors during embryogenesis, then optimal therapy may depend on isolation and expansion of tissue-specific EPCs.

Undifferentiated Embryonic Stem (ES) Cells

Although the role of the hemangioblast in embryonic development as well in adult homeostasis is debated, it is clear that harnessing the multifaceted embryonic stem cell of the blastocyte could lead to immediate therapeutic goals and avoid some of the potential problems associated with the specificity and identity of the adult endothelial progenitors (Figure 4). During embryogenesis, derivatives of these cells provide progenitors for all tissues and organs. The steps involved are complex, but can be recapitulated in vitro. However, it should be noted that in vitro differentiation of ES cells dose not correlate temporally with that of cells in the developing embryo, and experimental artifacts may cloud recent studies showing vascular development from ES cells.

Precursors of these hematopoietic clusters are thought—by some—to be endothelial cells, as they express CD31 as well as CD34, bind *Ulex europus* lecithin, and take up acetylated low-density lipoprotein. The endothelial basal lamina is intact where it tightly enjoins these hematopoietic clusters of the "hemogenic endothelium," which appear to arise from the combined tight connections of endothelial cells and hematopoietic cells. The former mesodermal precursors—the endothelial progenitors—are thought to express KDR or VEGF R2, but not CD45, the leukocyte common antigen, or CD34. CD34 is expressed by hematopoietic stem cells and, to a lesser extent, most mature endothelial cells. In addition to CD45, hematopoietic progenitors strongly express CD34 and c-Kit. Immature hematopoietic progenitors express CD133, and despite reports to the contrary, we observe slight but consistent expression of CD133 by mature endothelial cells (English and Harvey, unpublished). However, there is no clear consensus on development of the definitive hemato-vascular system, or the distinct and independent development of the hematopoietic system and the prehemogenic endothelium. Indeed, recent studies have implicated both $CD34^+$ and $CD34^-$ cells as hematopoietic stem cells, and very recent reports show that hematopoiesis can be completely restored in ablated hosts by infusion of $CD45^-$/ flt 1^+/$CD31^+$ endothelial-like fibroblasts.

Embryonic Vascular Development

As evidenced microscopically and by the effects of loss of function mutations, blood and blood vessels develop in close association early in development, perhaps by more

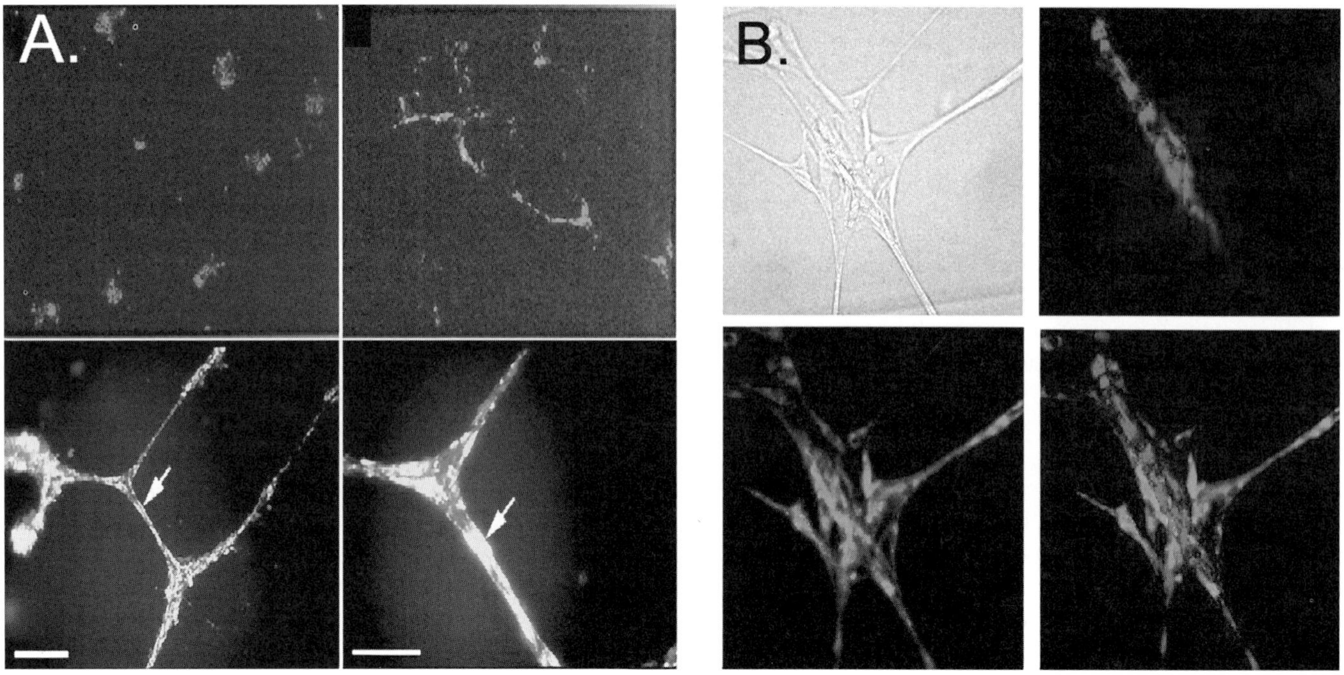

Figure 4 Vascular development from adult differentiated endothelial cells and mural pericytes in coculture. **(A)** Endothelial cells are stained green and the pericytes cells red; overlap is evident by yellow areas, expanded in the lower right panel. Endothelial cells and pericytes within the developed three-dimensional structure are shown isolated in the upper left and right panels, respectively. Lower panels show how the two cell types interact during coculture. From Darland, D. C., and D'Amore, P. A. (2001). TGFβ is required for the formation of capillary-like structures in three-dimensional cocultures of 10T1/2 and endothelial cells. *Angiogenesis* **4**, 11–20. **(B)** Coculture of pulmonary microvascular endothelial cells (red) and smooth muscle cells (green) on a defined angiogenic support. Note that under the conditions used, the red endothelial cells accumulate firmly within the mural covering. The photograph at the upper right shows a phase contrast image of the structure. This image from our laboratory depicts the model we currently employ to assess vascular maturation.

than one pathway. The first population to develop is thought by some to be posterior to the cement glands and gives rise to embryonic blood and vitellin veins in the anterior blood islands, as well as the endocardium of the heart. Later, from the dorsal plate mesoderm, precursors of adult blood cells and major vessels begin to develop. Both populations migrate from their point of origin to differentiate in a permissive environment, and both give rise to endothelial cells as well as hematopoietic cells. Some think they are seeded from the yolk sac to the dorsal plate of the embryo. Others dismiss that notion as inconsistent with the fundamentals of embryology, and strongly feel that the vascular system of the embryo develops in the embryo.

Vascular Maturation

Endothelial progenitors also can differentiate into mural cells, including pericytes and smooth muscle cells, which migrate to the developing endothelial lumen in response to platelet derived growth factor-BB and other cytokines (Figure 2). This process provides both strength and functionality to the developing vessel. Hematopoietic development appears to become definitive when the vascular endothelium expresses vascular endothelial (VE)-cadherin. Unlike other embryonic murine hematopoietic precursors, those produced by ES cells that express VE–cadherin effected long-term hematopoietic reconstitution when transplanted into SCID mice. Again, the precursors of these definitive hematopoietic cells expressed endothelial markers but not the leukocyte common antigen, CD45. This activity may initially arise within a few cells of the endothelium of the aorta. Later, definitive blood islands found both in the yolk sac and the embryo proper possess hemangioblasts, angioblasts, and hematopoietic stem cells. As observed by Sabin, projectiles emanate from these blood islands that coalesce into a vascular network.

Stem Cell Therapy

Despite questions of development, it would appear that the vascular progenitor cells of the embryo, like embryonic stem cells of the blastoderm, are neither committed nor tissue restricted, and therefore may provide a means to treat vascular disease in a variety of adult tissues. Although it is difficult to extrapolate the findings of experiments with cultured cells to physiological development, several groups have now demonstrated the generation of an intact endothelial lumen from mouse, monkey, and human embryonic stem (ES) cells. Others have shown that murine stem cells can, upon careful culture and amplification, produce structures that resemble bilayered blood vessels. Recently, these results were reported using human embryonic stem cells as the starting material; again, caution is urged in evaluating these results, as the structures produced may arise from cellular contaminants, in vitro artifacts in gene expression, and other factors, as evidenced by the extremely rapid maturation of these putative ES cells to vascular cells. In both approaches, embryonic stem cells were cultured and a subpopulation selected and exposed to collagen to induce vascular differentiation. After further selection and culture on specific extracellular matrices in the presence of both mural and endothelial growth factors, bilayered tubular structures somewhat resembling adult vessels were formed.

Recent Developments

We shall not understand vascular development in the embryo until we understand the process of angiogenesis. Dr. Patricia D'Amore has developed refined, three-dimensional vessels in cocultures of adult ECs and mural cells in vitro (Figure 4A). Her work has identified cytokines involved in vascular maturation as mural pericytes descended on the developing endothelium. Recently, Dr. D'Amore has investigated the role of transforming growth factor in vascular formation from preexisting, disrupted cells, and she has generated a stunning replication of intact vessels using a refined coculture system in which this unique cytokine plays a key role in development.

We have generated three-dimensional bilayer tubes of endothelial and smooth muscle cells by plating these cells under defined conditions using a defined extracellular support designed to mimic that which facilitates the development of highly invasive neoplastic cells (Figure 4B). In these experiments, we found that the bioactive lipid we previously implicated as the physiological regulator of angiogenesis, sphingosine 1-phosphate, slowed mural cell association with the developing endothelial lumen as it enhanced lumen development. However, once formed, the lumen effectively metabolized the platelet-derived lipid, allowing the mural cells eagerly waiting nearby to quickly descend upon the developing endothelium. Current investigations being carried out by Dr. English with David N. Brindley at the University of Alberta have defined the processes involved in this agonist-dependent retardation of migration, an event first associated with developmental arrest and an event that appears necessary for orchestrated vascular development. These results provide another rather simple explanation for what once appeared to be a hopelessly complex process leading to the orchestration of development of multicellular structures. In this case, the agonist that promotes cells of the inner core of the vascular structure to develop simultaneously prevents outer cells from interfering with lumen assembly until the lumen is intact. At this time, the intact lumen degrades the regulatory small molecule, releasing its stranglehold on mural cells, and development proceeds. This simple scheme ensures that outer cells will not enter or interfere with development of the lumen until the lumen is developed.

One of the most interesting properties of EPCs is their ability to mobilize from the circulation when needed and their ability to specifically home to sites of angiogenesis. Factors involved in mobilization and homing are not well defined, but the cells mobilize in response to trauma, cardiac assault, neoplastic growth, and ischemia, as well as to infusion of vascular and hematopoietic growth factors. Curiously, administration of cholesterol-lowering statins results in vigorous mobilization, a phenomenon that has been exploited for collection of these cells from blood rather than marrow, and to define mechanisms involved in cellular recruitment in vivo, a process that seems to depend on the activity of small GTPases of the Rho family. Once in the circulation, the EPCs efficiently home to areas where they are needed for angiogenesis. As a result, several investigators have suggested using these cells to target gene therapy. This result has been achieved in mice and provides a rationale for using EPCs to target vectors to reduce inflammation, kill cancer cells, or expedite healing at sites of angiogenesis.

Summary and Outlook

Many studies as well as hypotheses support the concept of therapy for chronic irreversible disease using endothelial or vascular stem cells. However, we have a long way to go before this goal is reached. First, we must clarify the differences in the endothelium of various organs of the adult to determine if a single source of vascular cells is appropriate or if multiple sources or tissue specific sources will be necessary. Methods are required for optimal expansion of these cells in culture, and further investigation is needed to characterize the adult and embryonic cell of choice for therapy. Although the problems are formidable, the outcome may border on a medical miracle. Vascular transplants using isolated stem cells may provide relief for countless individuals with chronic diseases of the lung and brain, and of the circulatory as well as the hormonal system; with cancer and arthritis; or with chronic infection and severe trauma. Yes, the problems ahead are formidable, but the goal is admirable. Stem cell therapy offers a chance, but must be implemented rationally and cautiously, or unfounded conclusions will cloud their utility, distract investigators, and limit the progress they may otherwise make.

Glossary

Embryonic stem (ES) cells: Recovered from fertilized growing blastocytes early in embryonic development. ES cells can be induced to differentiate into cells of all germ layers and are thus pluripotential. These cells hold tremendous therapeutic promise.

Hemangioblast: A cell with the potential to develop either hematopoietic cells or endothelial cells upon stimulation. The existence of the hemangioblast during early embryonic development is clear. Here, these vascular cells provide stem cells for other organs and systems that are targeted by the vasculature to their destination. However, the physiological role of the adult hemangioblast remains to be defined. If it indeed exists, the cell could hold valuable therapeutic potential.

MSC: Mesenchymal or mesodermal stem cells, often referred to as "adult" stem cells, which are thought to be pluripotential equivalents of ES cells and to hold similar therapeutic potential. However, the identity, characteristics, and derivation of this cell, if it indeed exists, are not universally accepted.

Stem cell: Any cell that is capable of giving rise to a copy of itself, differentiated progeny, or, through asymmetric cell division, both types of daughter cells. Only the embryonic stem cell has been defined with certainty.

Vasculogenesis: The process that results in the development of blood vessels from stem cells. This process involves angiogenesis, the formation of vessels from preexisting differentiated cells.

Bibliography

Chung, Y. S., Zhang, W. J., Arentson, E., Kingsley, P. D., Palis, J., and Choi, K. (2002). Lineage analysis of the hemangioblast as defined by FLK1 and SCL expression. *Development* **129**(23), 5511–5520. *Dr. Choi and his team generated knockin embryonic stem (ES) cells carrying a nonfunctional human CD4 at the Scl locus. By using in vitro differentiated ES cells, they demonstrated that FLK1 and SCL are molecular determinants of the hemangioblast. Their studies demonstrated that hematopoietic and endothelial cells develop via distinct, sequential generation of FLK1 and SCL-expressing cells. FLK1(+)CD4(−) cells first arise in developing embryoid bodies. The Scl gene is turned on within FLK1(+)CD4(−) cells to give rise to FLK1(+)CD4(+) cells. Replating studies demonstrated that hematopoietic progenitors are enriched within FLK1(+)CD4(+) and FLK1(−)CD4(+) cells, whereas endothelial cells develop from FLK1(+)CD4(+) and FLK1(+)CD4(−) cell populations.*

Darland, D. C., and D'Amore, P. A. (2001). TGFB is required for the formation of capillary-like structures in three-dimensional cocultures of 10T1/2 and endothelial cells. *Angiogenesis* **4**, 11–20. *In prior reports, Dr. D'Amore has defined the role of cytokines in vascular maturation using three-dimensional culture systems, such as that shown in Figure 4A. Here, she adds to the results of Gualandris et al. by showing that TGF-B is required for structural assembly in a coculture system. The article provides the most recent culture techniques for assessment of vascular development in a multicomponent system.*

Dzierzak, E. (2003). Ontogenic emergence of definitive hematopoietic stem cells. *Adv. Exp. Med. Biol.* **522**, 59–74.

Gualandris, A., Annes, J. P., Arese, M., Noguera, I., Jurukovski, V., and Rifkin, D. B. (2000). The latent transforming growth factor-beta-binding protein-1 promotes in vitro differentiation of embryonic stem cells into endothelium. *Mol. Biol. Cell* **11**(12), 4295–4308.

Honig, G. R., Li, F., Lu, S. J., and Vida, L. (2004). Hematopoietic differentiation of rhesus monkey embryonic stem cells. *Blood Cells Mol. Dis.* **32**, 5–10.

Levenberg, S., Golub, J. S., Amit, M., Itskovitz-Eldor, J., and Langer, R. (2002). Endothelial cells derived from human embryonic stem cells. *Proc. Natl. Acad. Sci. USA* **99**(7), 4391–4396. *The authors describe the differentiation of human ES cells to endothelial cells in a culture system that may be useful as a source of cells for transplantation or tissue engineering. The differentiation steps of human embryonic stem cells into endothelial cells forming vascular-like structures are documented as was the behavior and potential in tissue engineering of the cells generated. The authors conclude that endothelial cells generated in culture from ES cells could provide a source of human endothelial cells that could be beneficial for potential applications, such as engineering new blood vessels, endothelial cell transplantation into the heart for myocardial regeneration, and induction of angiogenesis for treatment of regional ischemia.*

Mikkola, H. K., Fujiwara, Y., Schlaeger, T. M., Traver, D., and Orkin, S. H. (2003). Expression of CD41 marks the initiation of definitive hematopoiesis in the mouse embryo. *Blood* **101**(2), 508–516.

Milner, R., and Campbell, I. L. (2002). Developmental regulation of B1 integrins during angiogenesis in the central nervous system. *Mol. Cell. Neurosci.* **20**, 616–626.

Peault, B., Oberlin, E., and Tavian, M. (2002). Emergence of hematopoietic stem cells in the human embryo. *C. R. Biol.* **325**(10), 1021–1026.

Schuh, A. C., Faloon, P., Hu, Q. L., Bhimani, M., and Choi, K. (1999). In vitro hematopoietic and endothelial potential of flk-1(−/−) embryonic stem cells and embryos. *Proc. Natl. Acad. Sci. USA* **96**(5), 2159–2164. *Mice deficient in the Flk-1 receptor tyrosine kinase are known to die in utero because of defective vascular and hematopoietic development. However, the authors show that flk-1(−/−) embryonic stem cells are able to differentiate into hematopoietic and endothelial cells in vitro, although they give rise to a greatly reduced number of blast colonies, a measure of hemangioblast potential. Furthermore, normal numbers of hematopoietic progenitors are found in 7.5-day postcoitum flk-1(−/−) embryos, even though 8.5-day postcoitum flk-1(−/−) embryos are known to be deficient in such cells.*

Shattil, S. J., and Ginsberg, M. H. (1997). Integrin signaling in vascular biology. *J. Clin. Invest.* **100**, 1–5. *This article and others in the series Cell Adhesion in Vascular Biology present in a clearly written format the important role of integrins and cell adherence in vascular on vascular development.*

Sonja, L., Boris, F., Marc, A. B., Katrin, L., Martin, B., Markus, G., Gunter, S., Süleyman, E., Uta, F., Axel R. Z., Dieter, K. H., Walter, F., and Ursula, M. G. (2004). Identification of the Adult Human Hemangioblast. *Stem. Cells and Development* **13**(3), 229–242.

Stogard, C. M., Stupack, D. G., Jonczyk, A., Goodman, S. L., Fox, R. I., and Cheresh, D. A. (1999). Decreased angiogenesis and arthritic disease in rabbits treated with an $\alpha_v B_3$ antagonist. *J. Clin. Invest.* **103**, 47–54 *An excellent example of the potential therapeutic relevance of disruption of integrin signaling is given in this article, whose authors were among the first to recognize the importance of integrin and cell adherence in vascular maturation.*

Walmsley, M., Ciau-Uitz, A., and Patient, R. (2002). Adult and embryonic blood and endothelium derive from distinct precursor populations which are differentially programmed by BMP in *Xenopus*. *Development* **129**(24), 5683–5695. *Blood and blood vessels develop in close association in vertebrate embryos, and loss-of-function mutations suggest common genetic regulation. Here, the investigators locate two distinct populations of progenitors for blood and endothelial cells in developing Xenopus embryos. The first population is located immediately posterior to the cement gland during neurula stages and gives rise to embryonic blood and vitellin veins in the anterior ventral blood island (aVBI), and to the endocardium of the heart. The second population resides in the dorsal lateral plate mesoderm and contains precursors of adult blood stem cells and the major vessels. Both populations differentiate into endothelial cells in situ, but migrate to new locations to differentiate into blood, suggesting that their microenvironments are unsuitable for hematopoietic differentiation. Both require BMP for their formation. The results indicate that both adult and the anterior component of embryonic blood in Xenopus embryos derive from populations of progenitors that also give rise to endothelial cells.*

Yamashita, J., Itoh, H., Hirashima, M., Ogawa, M., Nishikawa, S., Yurugi, T., Naito, M., Nakao, K., and Nishikawa, S. (2000). Flk1-positive cells derived from embryonic stem cells serve as vascular progenitors. *Nature* **408**(6808), 43, 45.

An issue of Journal of Hematotherapy & Stem Cell Research: Focus on Hematopoietic Angiogenesis and Vascular Stem Cells (Vol. 11, No. 1) presents several state-of-the-art reviews on vascular development and the use of vascular stem cells, as well as the classic reprint of Dr. Sabin, referred to earlier. The authors used a novel system to analyze the differentiation of single cells. Adult human CD133$^+$ *cells incubated with either granulocyte colony-stimulating factor or with stem cell growth factor in the presence of and VEGF robustly generated either neutrophils or endothelial cells, respectively. Conditions were defined to promote the simultaneous, dual differentiation of a single cell to both lineages showing for the first time the presence of a definite population of hemangioblasts.*

Capsule Biography

Dr. English of the Methodist Hospital in Indianapolis is editor-in-chief of the journal *Stem Cells and Development*. His research has defined important links between blood clotting and angiogenesis and more recently identified processes that regulate vascular maturation and stem cell differentiation. He is funded by the National Institutes of Health and several local agencies. As a research scientist of the Methodist Research Institute, Mr. Harvey assists Dr. English. Dr. Gehling's research focuses on the development of the human vasculature, and the role of the hemangioblast in adults.

Index

A

B

C

D

E

F

G

H

I

J

K

L

M

N

O

P

Q

R

S

W

X

Z

Color Insert

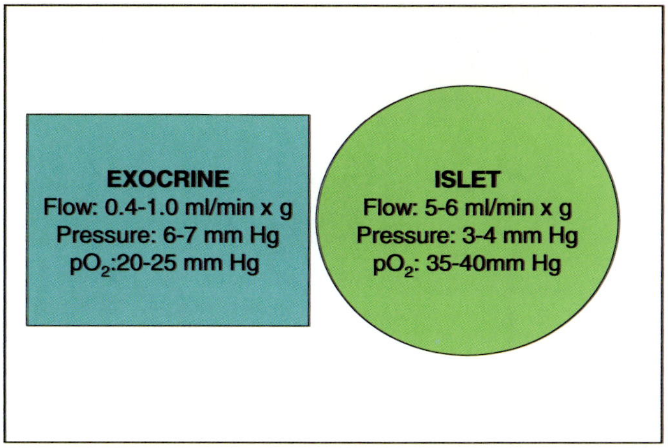

Chapter 84 Figure 1 Summary of microvascular parameters in the endocrine and exocrine pancreas of rodents.

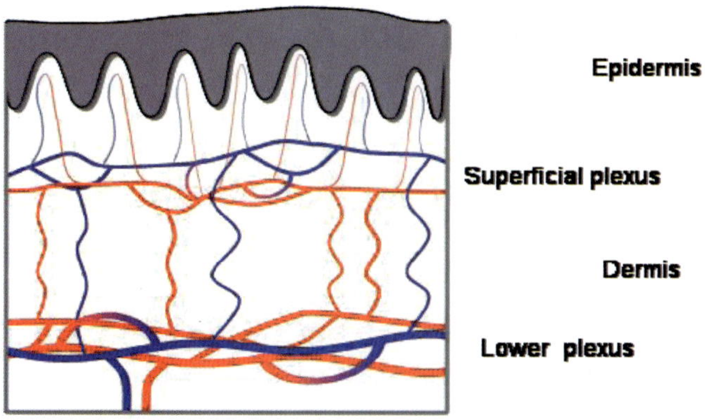

Chapter 86 Figure 1 Diagram of the organization of the cutaneous microvascular bed.

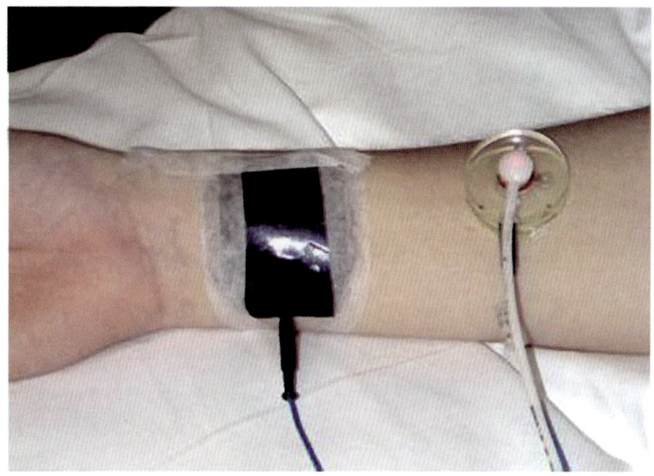

Chapter 86 Figure 2 Ions of soluble salt are introduced into the tissue by iontophoretic current. The amount of drug delivered to the skin is proportional to the magnitude and duration of the applied current.

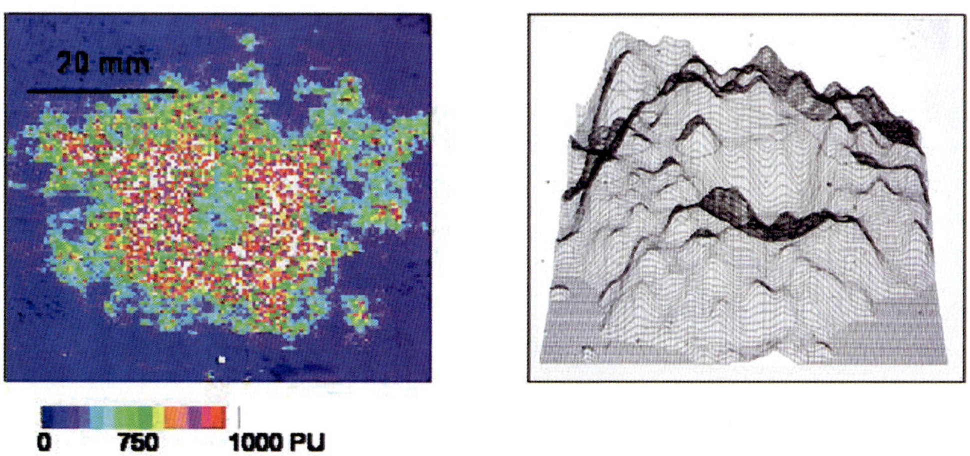

Chapter 86 Figure 3 High-resolution scanning laser Doppler imaging uses a low-power (~2 mW), 633-nm red He–Ne laser to scan the skin in a raster pattern and build up a two-dimensional image of red blood cell flux up to a depth of approximately 0.6 to 1 mm.

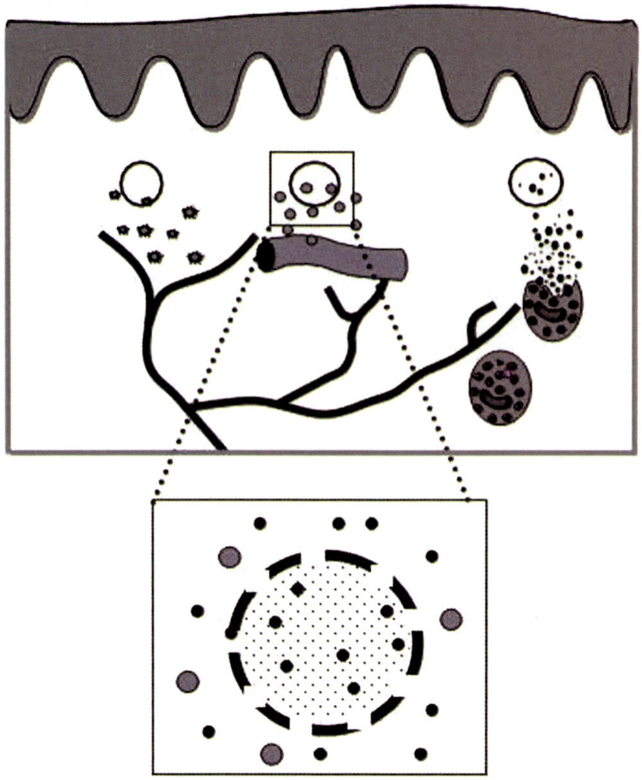

Chapter 86 Figure 4 Principles of microdialysis. The recovery of a substance by microdialysis is governed by the same principles as that of exchange of small hydrophilic molecules across the microvascular wall, that is, passive diffusion down a concentration gradient as described by the Fick principle. It is influenced by a number of factors including the characteristics of the dialysis membrane and the recovered solute (its charge, size, solubility) and the concentration of the solute outside the membrane (which in turn may be influenced by tissue metabolism, hydration, and local blood flow).

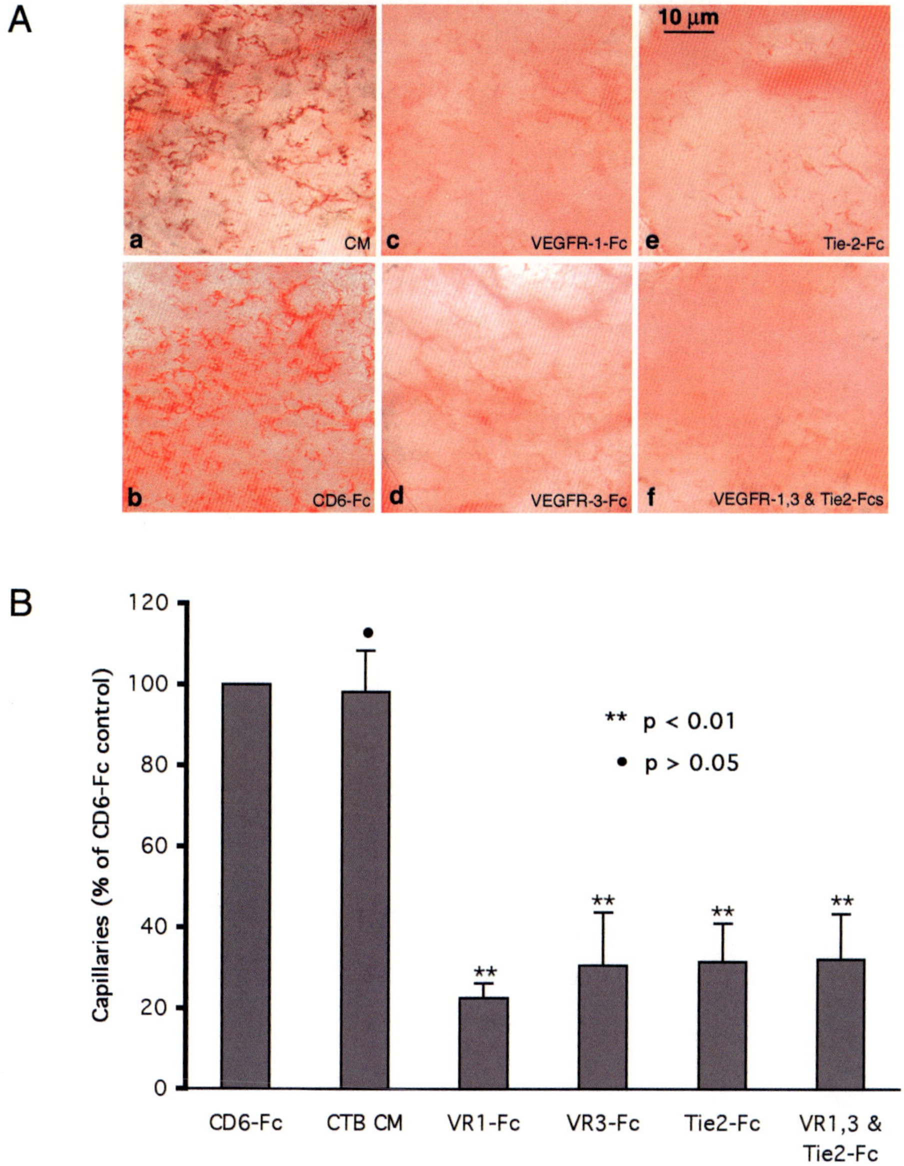

Chapter 87 Figure 2 PlGF, VEGF-C, and Ang2 in cytotrophoblast-conditioned medium stimulate angiogenesis in vivo. The angiogenic effects of factors in cytotrophoblast-conditioned medium were assessed by using the chick chorioallantoic membrane assay. (**A**) The membranes were photographed after exposure to the samples for 48 hours. Cytotrophoblast-conditioned medium (a) and medium treated with a control CD6-Fc protein (b) induced a robust angiogenic response that was comparable to the response induced by basic FGF (data not shown). Removal of PlGF (c), VEGF-C (d), or Ang2 (e) diminished this effect by 70 to 80 percent. (f) Depletion of all three factors produced no further decrease in the angiogenic capacity of cytotrophoblast-conditioned medium. (**B**) The data were quantified by counting the number of new capillary sprouts. Data are expressed as means and standard deviations of seven experiments. The significance of the data from four separate experiments was determined by analysis of variance (ANOVA). (Reproduced with permission from *Developmental Biology*.)

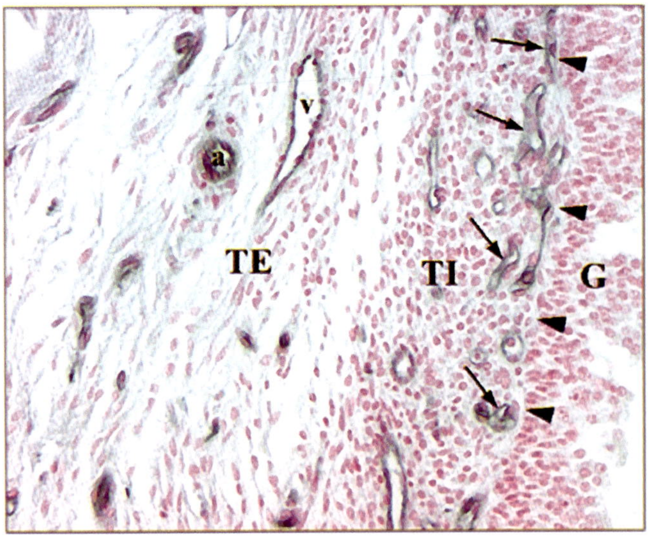

Chapter 88 Figure 2 Histochemical staining of a preovulatory sheep follicle for BS-1 lectin, which stains microvessels. Note the dense "capillary wreath" (arrows) that is located in the theca interna (TI) adjacent to the basement membrane (delineated approximately by the arrowheads) surrounding the granulosa layer (G). Also note that the granulosa layer is avascular. An arteriole (a) and venule (v) are indicated within the theca externa (TE). The follicular fluid-filled antrum is to the right of the granulosa layer (a portion of the antrum is visible in the lower right corner of the micrograph), and the oocyte, which would be located antrally and surrounded by granulosa cells, is not visible. The section has been counterstained with nuclear fast red, and the micrograph was taken with a 20× objective.

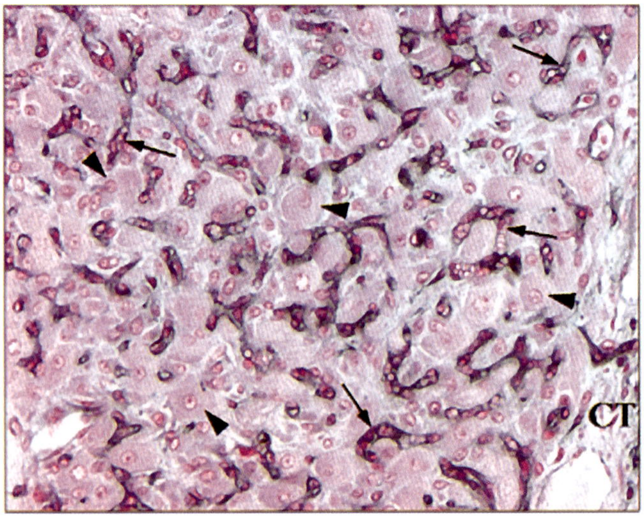

Chapter 88 Figure 3 Histochemical staining of a mature sheep corpus luteum for BS-1 lectin, which stains microvessels. Note that nearly every steroidogenic epithelial cell (a few are indicated by arrowheads) in the parenchymal area is in contact with a capillary (a few are indicated by the arrows; some are filled with red blood cells). Also note a few larger lymphatic capillaries located in the parenchymal area (upper right and lower left of the micrograph). A portion of a connective tissue tract (CT), which is where the vast majority of the luteal arterioles and venules are located, is visible in the lower right corner. The section has been counterstained with nuclear fast red, and the micrograph was taken with a 20× objective.

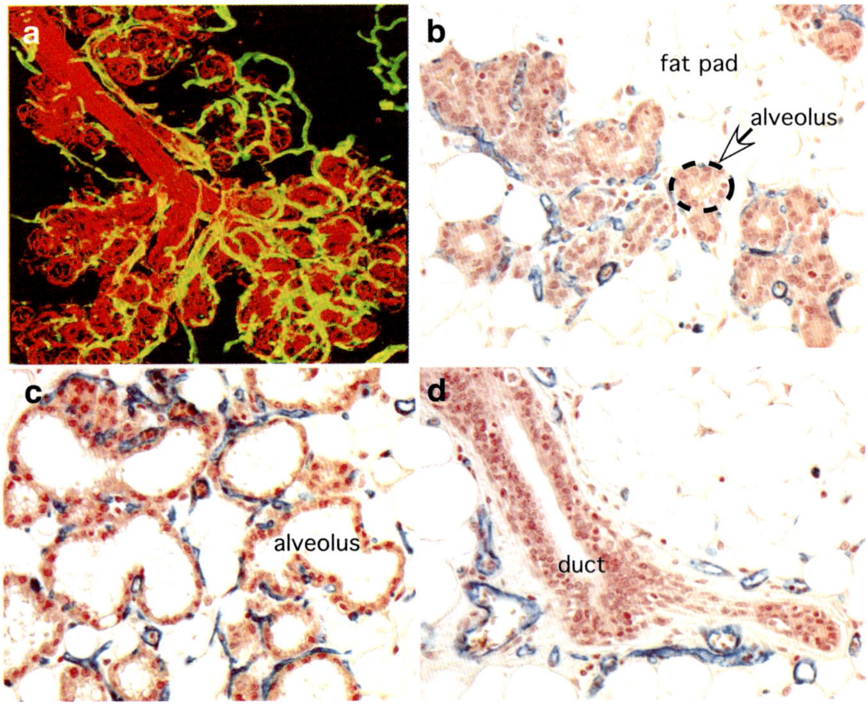

Chapter 90 Figure 2 Tools to visualize the mammary-associated vasculature. **(a)** A three-dimensional composite image of a lectin and phalloidin-stained mammary gland harvested in the last third of pregnancy. The pregnant host was injected intravenously with lectin-FITC (green), the inguinal mammary gland was embedded in OCT and sectioned at 50 μm, and then poststained with phalloidin–Alexa Red 455 (image courtesy of Bryan Welm and Jeffrey Rosen, Baylor College of Medicine). Note the network of vessels enveloping the alveoli and the vessels that run in parallel with and encircle the ducts. To observe the fine microvessel structure, anti-CD31 immunohistochemistry (blue color) was performed on zinc-fixed paraffin sections counterstained with nuclear fast red **(b–d)** isolated from mice at day 15 of pregnancy **(b)**, day 2 of lactation **(c)**, or from nulliparous mice **(d)** (staining courtesy of Debbie Liao, UCSD).

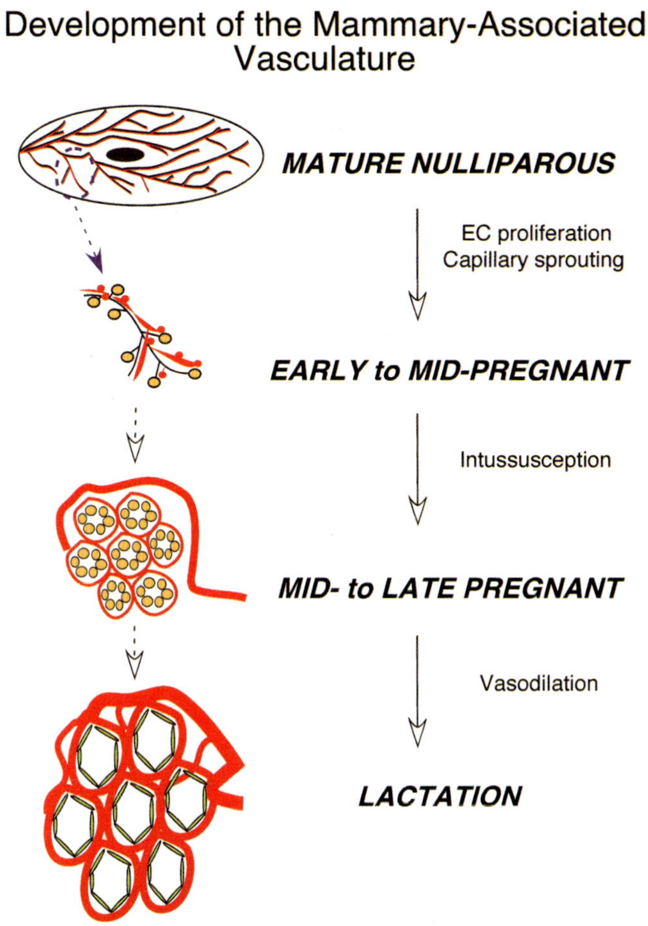

Chapter 90 Figure 3 Cartoon representation of the various stages of mammary-associated vasculature development in conjunction with epithelial cell development. In nulliparous mice, the ducts (black lines) do not contain any alveolar structures (indicated in yellow), and the vasculature (indicated in red) runs in parallel with the ductal tree. In the first half of pregnancy, the vasculature expands by rapid endothelial cell (EC) proliferation and capillary sprouting, doubling from day 5 to day 10 of pregnancy in the rat mammary gland. In the latter half of pregnancy, when the alveoli have organized into lobuloalveolar structures, the vasculature expands by nonproliferative intussusceptive angiogenesis. At lactation, the epithelial cells become flattened since the alveoli are expanded and engorged with milk and the vessels dilate and become more tortuous. Finally, the vasculature regresses along with the epithelium during involution. The entire process will repeat over the course of each successive pregnancy.

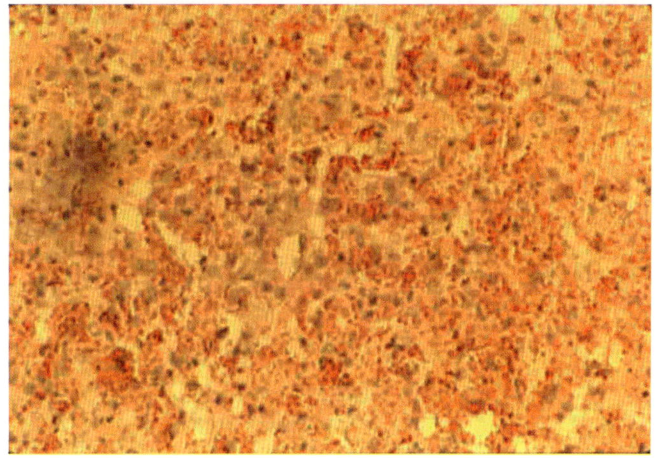

Chapter 92 Figure 1 ARDS: Acute phase. Intense lung congestion and inflammation is evident in the interstitial and intra-alveolar space.

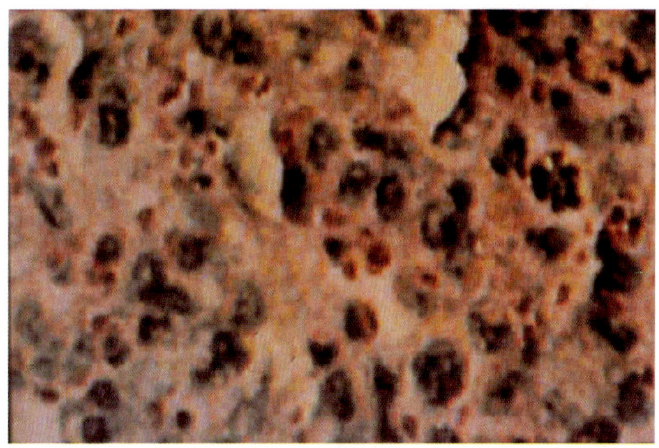

Chapter 92 Figure 2 ARDS: Late phase. Macrophage infiltration and increased collagen (pink) is evident. Macrophages release inflammatory cytokines that increase the inflammatory response. These cytokines can produce both lung and systemic organ injury. Increased fibroblasts are also present as precursors to later lung fibrosis.

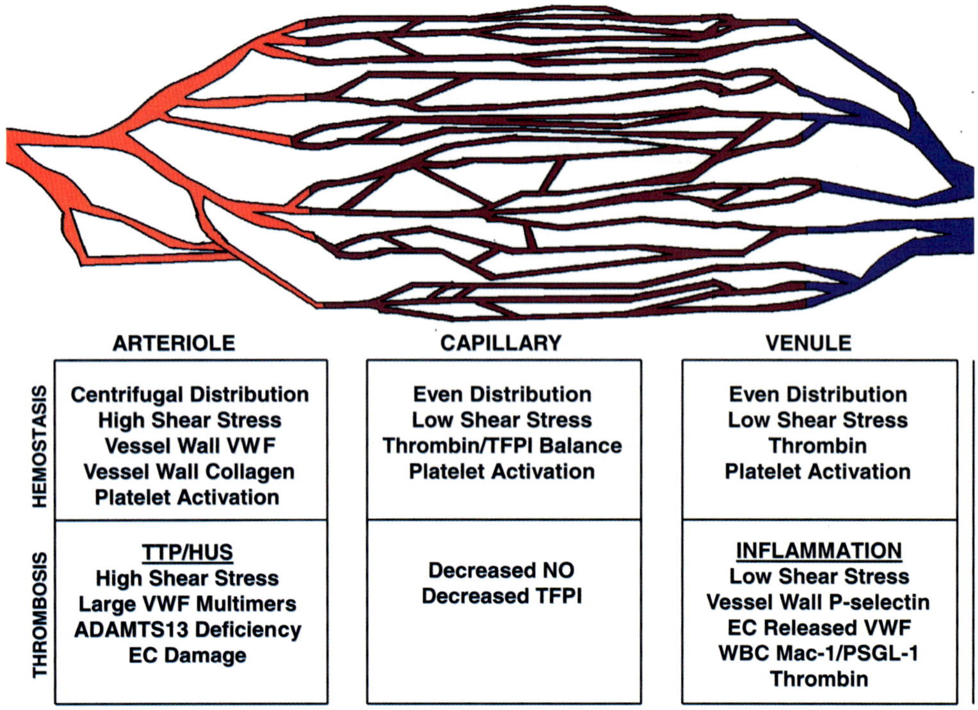

Chapter 95 Figure 1 The elements within Virchow's triad that direct hemostasis or thrombosis in the microcirculation. The microcirculation—which is generally defined by a vascular lumen diameter less than 100 μm—encompasses three vessel beds of strikingly different morphology and rheology: arterioles, capillaries, and venules. Each possesses a unique repertoire of physiological and pathological attributes and responses. Distribution refers to where the majority of blood platelets are found within a cross-sectional vessel lumen. VWF, von Willebrand factor; TTP, thrombotic thrombocytopenic purpura; HUS, haemolytic–uremic syndrome; ADAMTS13, a disintegrin and metalloproteinase with thrombospondin type-1 motif 13; EC, endothelial cell; TFPI, tissue factor pathway inhibitor; WBC, white blood cell; PSGL-1, P-selectin glycoprotein ligand-1.

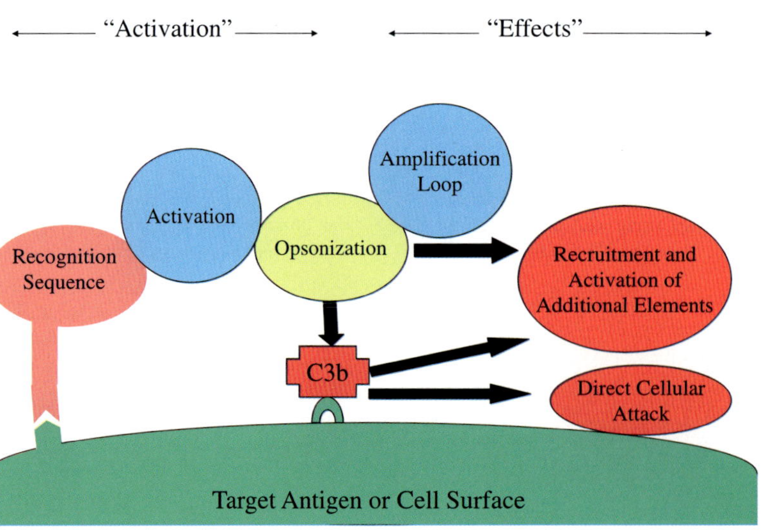

Chapter 96 Figure 1 Schematic diagram of complement functional units.

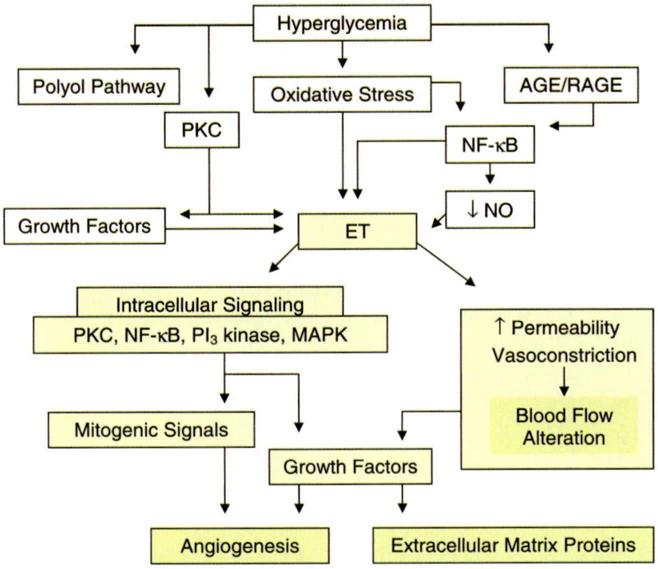

Chapter 99 Figure 1 Putative mechanisms and consequences of ET alteration in diabetes. A schematic outlining various hyperglycemia-induced pathways leading to upregulation of endothelin levels is shown in the upper panel. Some of the major effects of increased ET levels are also presented.

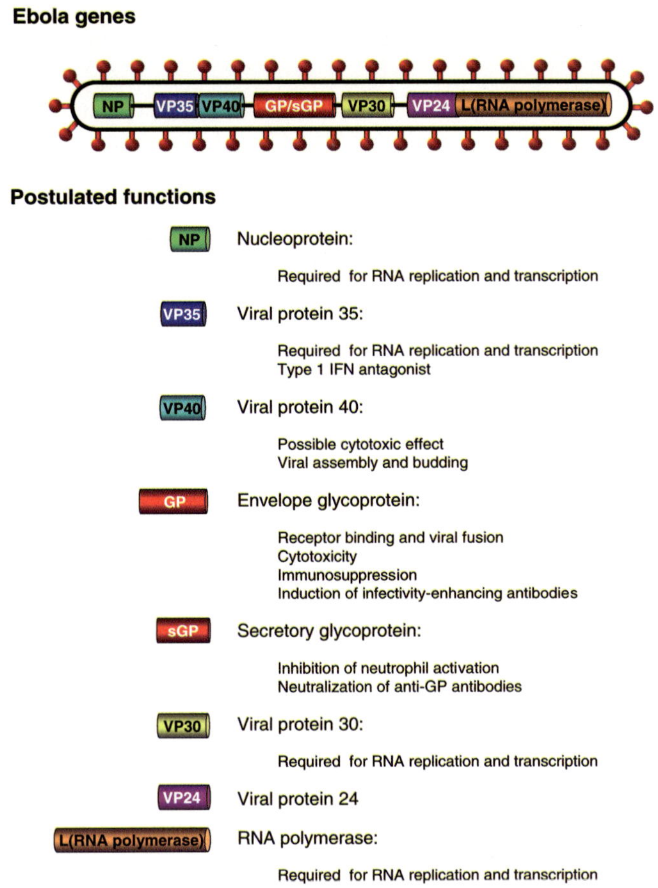

Chapter 102 Figure 2 The Ebola genome.

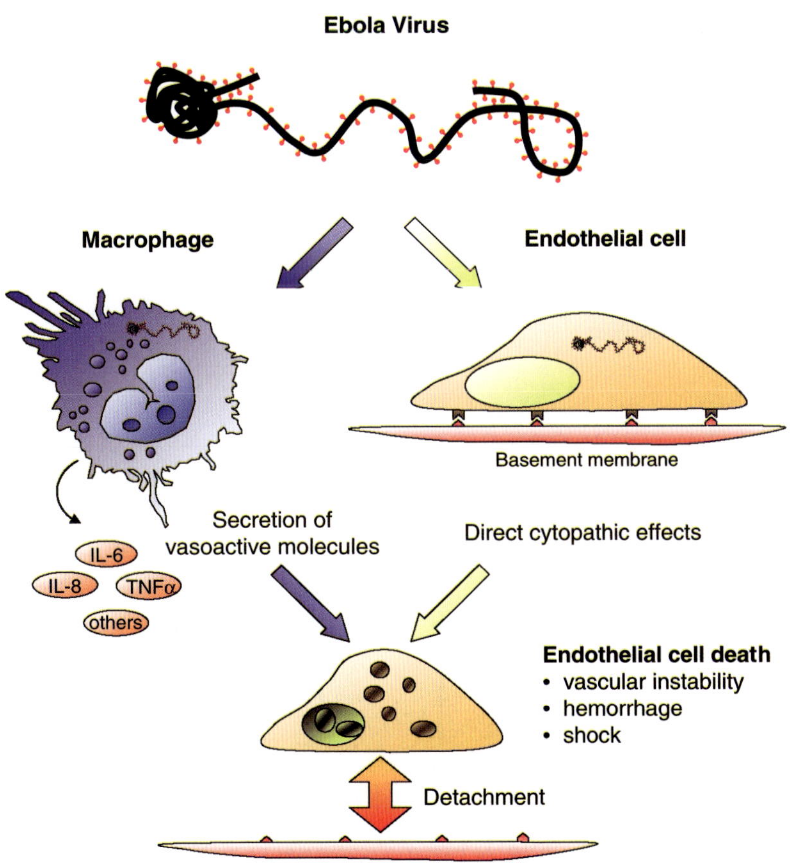

Chapter 102 Figure 1 Effects of Ebola virus infection on the vascular endothelium.

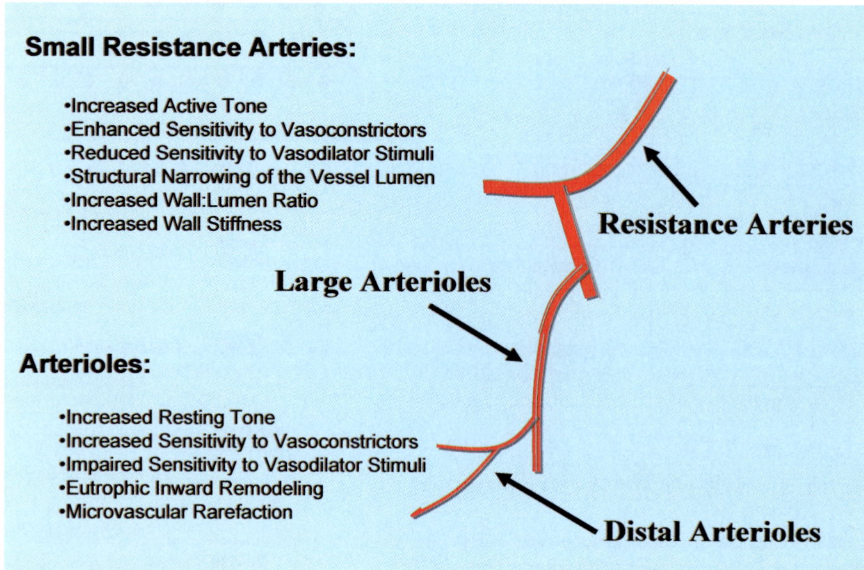

Chapter 104 Figure 1 Summary diagram of the mechanisms by which changes in arteriolar structure and function may contribute to an elevated vascular resistance in hypertension. (Adapted with thanks from an original drawing by Dr. Francis A. Sylvester.)

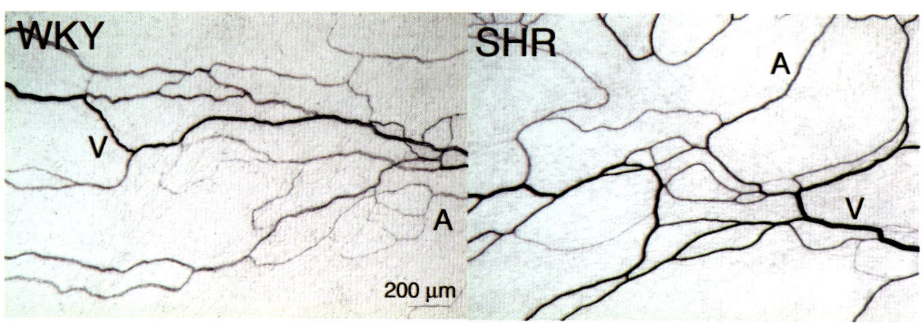

Chapter 105 Figure 2 Light microscopic cross-section of initial lymphatic in rat skeletal muscle. The vessel has a single endothelial lining and is positioned in the vicinity of adipose cells (A) and a terminal arteriole (TA), embedded in collagen fibers (C), and positioned between two arcade arterioles (AA) with vascular smooth muscle media (VSM).

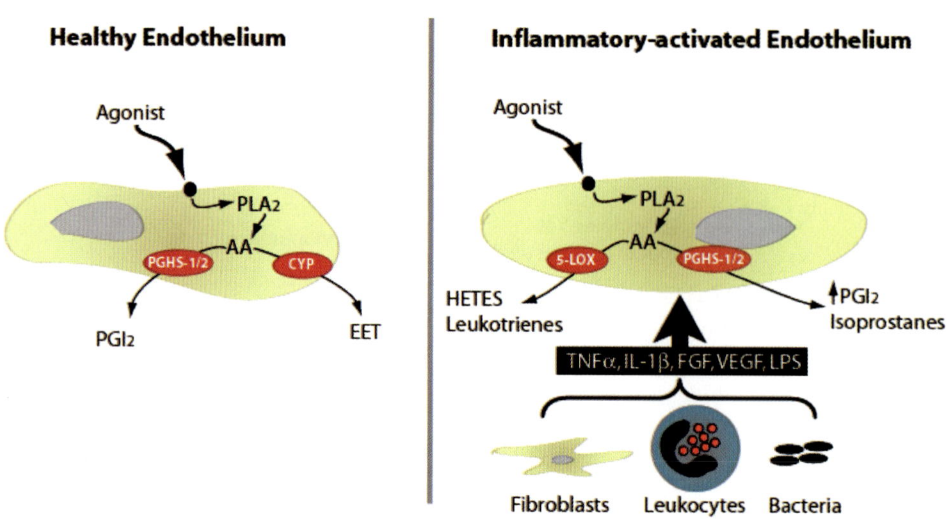

Chapter 106 Figure 1 Generation of bioactive eicosanoids by healthy and inflammatory-activated endothelium. Following its hydrolysis from the membrane by phospholipase A_2 (PLA_2), arachidonate is oxidized to prostacyclin (PGI_2) or epoxyeicosatetraenoic acid (EET) by prostaglandin H synthase (PGHS).

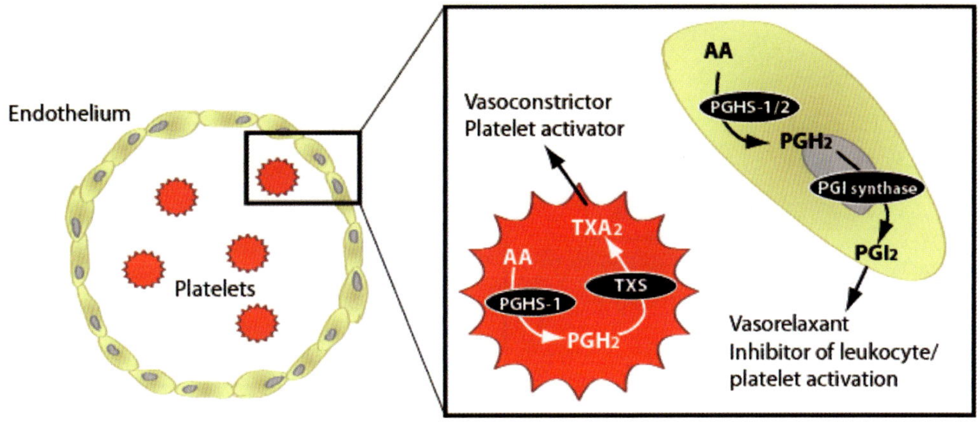

Chapter 106 Figure 2 Localization of PGHS isoforms in vascular cells. Platelets contain PGHS-1, which forms prostaglandin H_2 (PGH_2) and is subsequently metabolized by thromboxane synthase (TXS) to thromboxane A_2 (TXA_2). Endothelial cells contain PGHS-1 and -2, both of which are responsible for providing PGH_2 for PGI_2 synthesis by PGI synthase.

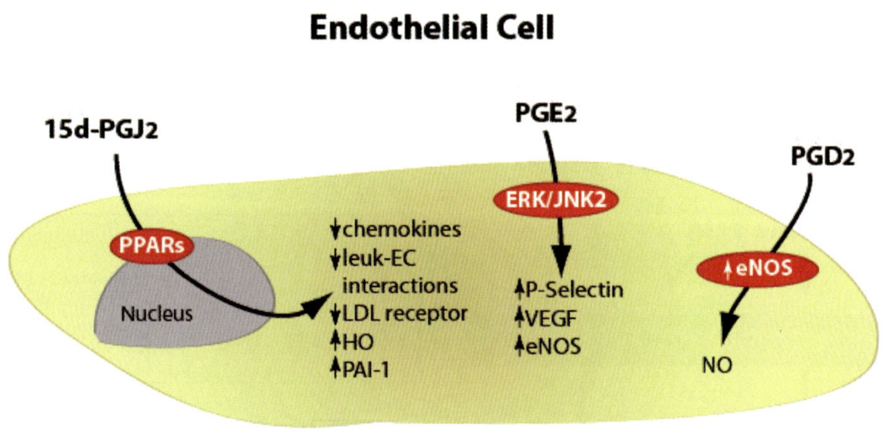

Chapter 106 Figure 3 Effects of PGHS products on endothelial cell function. A series of PGHS metabolites have potent biological effects on endothelial cell function. PPAR, peroxisomal proliferator activating receptor; HO, heme oxygenase; VEGF, vascular endothelial growth factor; 15δ-PGJ_2, 15-deoxy-δ (12,14)-prostaglandin J_2; PAI-1, plasminogen activator inhibitor type-1.

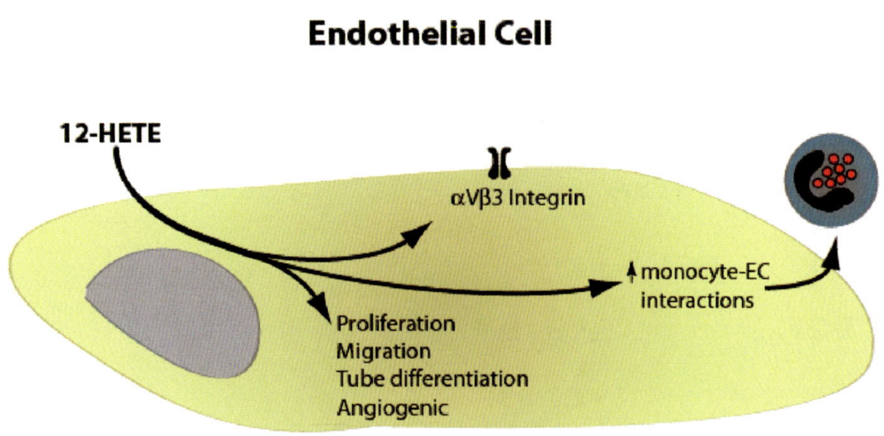

Chapter 106 Figure 4 Signaling properties of 12-HETE in endothelial cells. The LOX product 12-HETE induces multiple biological effects in endothelium.

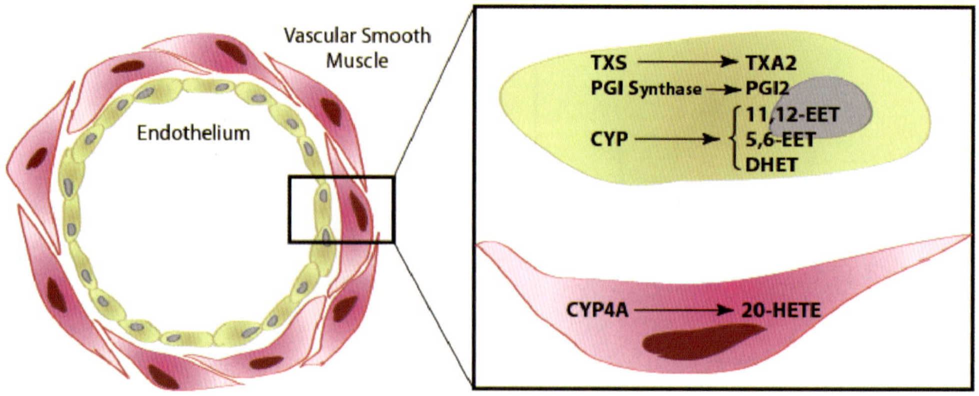

Chapter 106 Figure 5 Localization of CYP isoforms in vascular tissue. Endothelium contains a number of CYP enzymes that generate bioactive lipid products. An additional important isoform is CYP4A, in smooth muscle that generates 20-HETE. TXS, thromboxane synthase; EET, epoxyeicosatetraenoic acid; DHET, dihydroxyeicosatetraenoic acid.

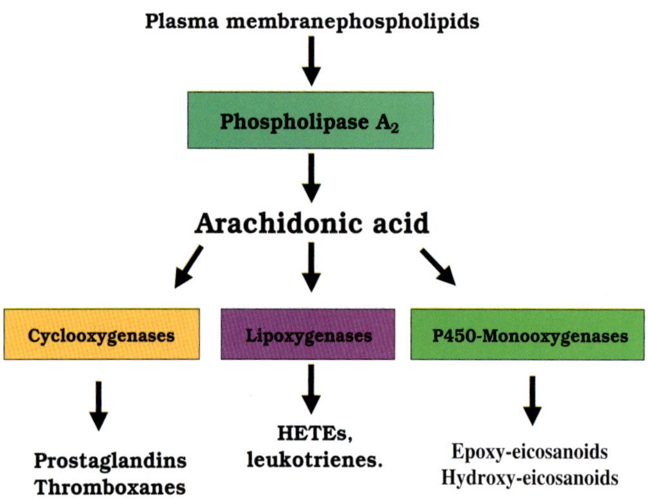

Chapter 107 Figure 1 Arachidonic acid metabolic pathways.

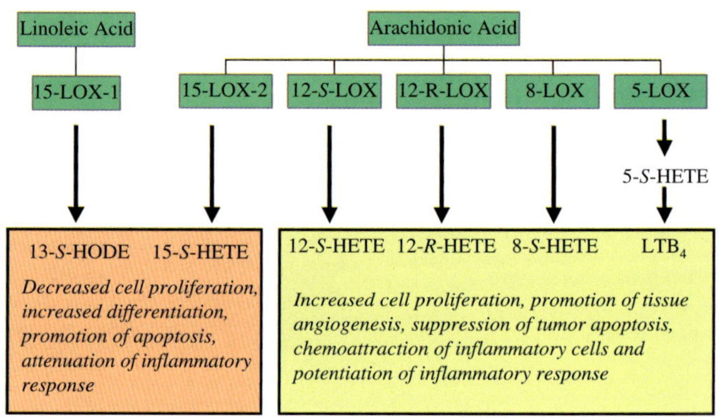

Chapter 107 Figure 3 Products of lipoxygenase metabolism.

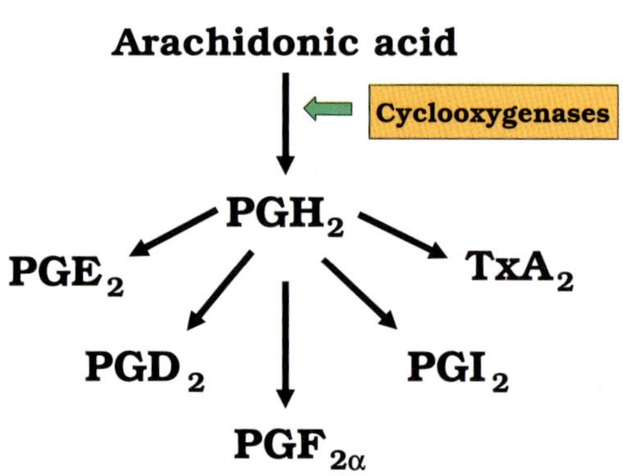

Chapter 107 Figure 2 Tissue-specific products of cyclooxygenase activity.

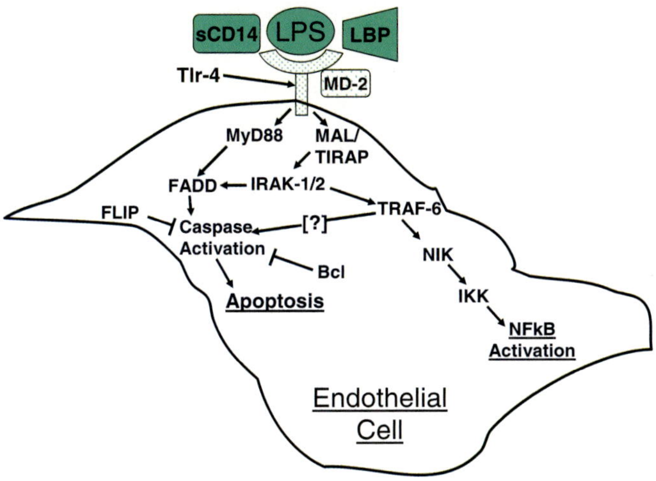

Chapter 108 Figure 1 Schematic diagram of LPS proapoptotic signaling in endothelial cells. After LPS binding to TLR-4, EC signaling pathways leading to nuclear translocation of NFκB and apoptosis are activated. These pathways share some of the same signaling molecules, including MyD88, IRAK-1, and TRAF-6. A redundant signaling pathway involving MAL/TIRAP similarly mediates both NFκB activation and apoptosis. Downstream of TRAF-6, the signaling pathways that activate these two processes diverge. The molecules that link the upstream NFκB signaling molecules to the recruitment and activation of caspases remains unknown. LBP, LPS-binding protein; NIK, NFκB-inducing kinase; IKK, IκB kinase; sCD14, soluble CD14; FADD, Fas-associated death domain; FLIP, FADD-like interleukin converting enzyme–like inhibitor protein; TRAF-6, TNF receptor-associated factor-6; IRAK-1/2, IL-1 receptor-associated kinase-1; MyD88, myeloid differentiation factor 88.

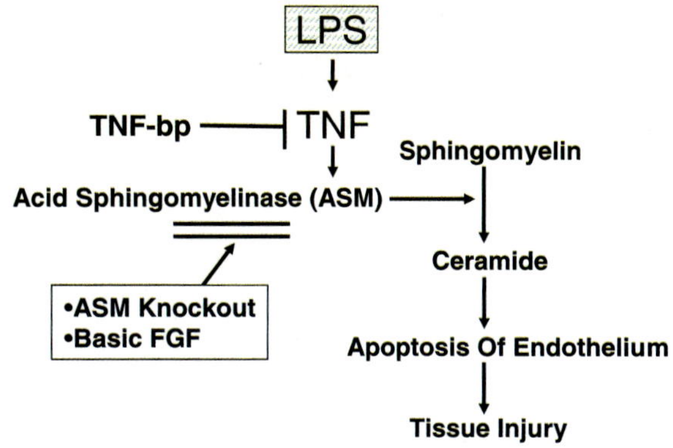

Chapter 108 Figure 2 Entry of LPS into the circulation promotes TNFα release from monocytes and macrophages. TNFα promotes hydrolysis of sphingomyelin producing ceramide through the action of acid sphingomyelinase. Endothelial cell apoptosis results. Basic fibroblast growth factor and the use of acid sphingomyelinase knockout mice attenuate the generation of ceramide and the production of apoptosis.

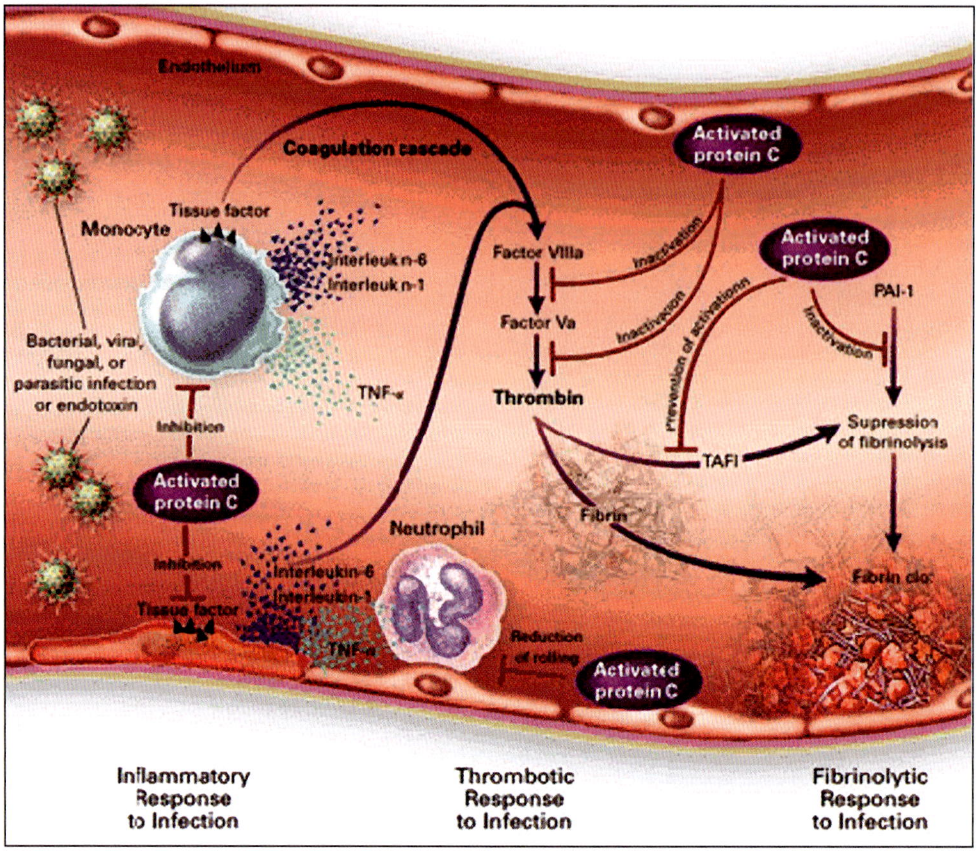

Chapter 108 Figure 3 Proinflammatory cytokines such as TNFα and IL-1 shift the microvascular milieu to a procoagulant environment through generation of tissue factor from monocytes and the endothelium. Tissue factor generates thrombin and a fibrin deposition. Cytokines and thrombin inhibit fibrinolytic activity by promoting the release of plasminogen-activator inhibitor 1 (PAI-1). PAI-1 inhibits tissue plasminogen activator activity. Thrombin activity further suppresses fibrinolytic system by activating thrombin-activatable fibrinolysis inhibitor (TAFI). Protein C activation is secondarily dramatically reduced. Microvascular injury decreases thrombomodulin, an essential factor producing protein C activation. Widespread microvascular thrombosis results, leading to multiorgan failure and enhanced mortality. Infusion of activated protein C produces a potent antithrombotic effect by inhibiting formation of factors Va and VIIIa and the subsequent generation of thrombin. Thrombin levels in the circulation fall and anti-inflammatory effects occur by reduction of TNFα and IL-1 transcription in monocytes. Further, activated protein C infusion enhances fibrinolytic activity through its ability to inhibit PAI-1.

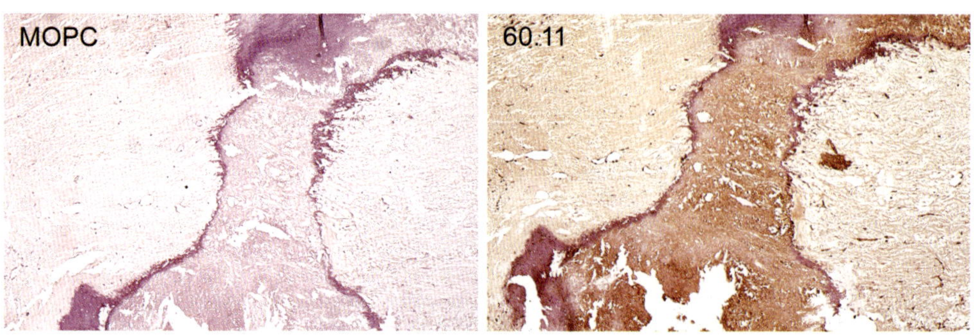

Chapter 109 Figure 1 Immunohistochemical staining of serial sections of the necrotic core of an atherosclerotic lesion of human coronary artery stained with anti-gC1q-R mAb 60.11 (*right panel*) or with isotype- and species-matched antibody (MOPC-21). Upon closer microscopic examination, foam cells in necrotic areas in the deeper intima and the vessel walls in the adventitia showed the strongest gC1q-R staining. Moreover, the necrotic center of calcified atherosclerotic plaques stained strongly for gC1q-R. Staining of endothelial cells was also noted.

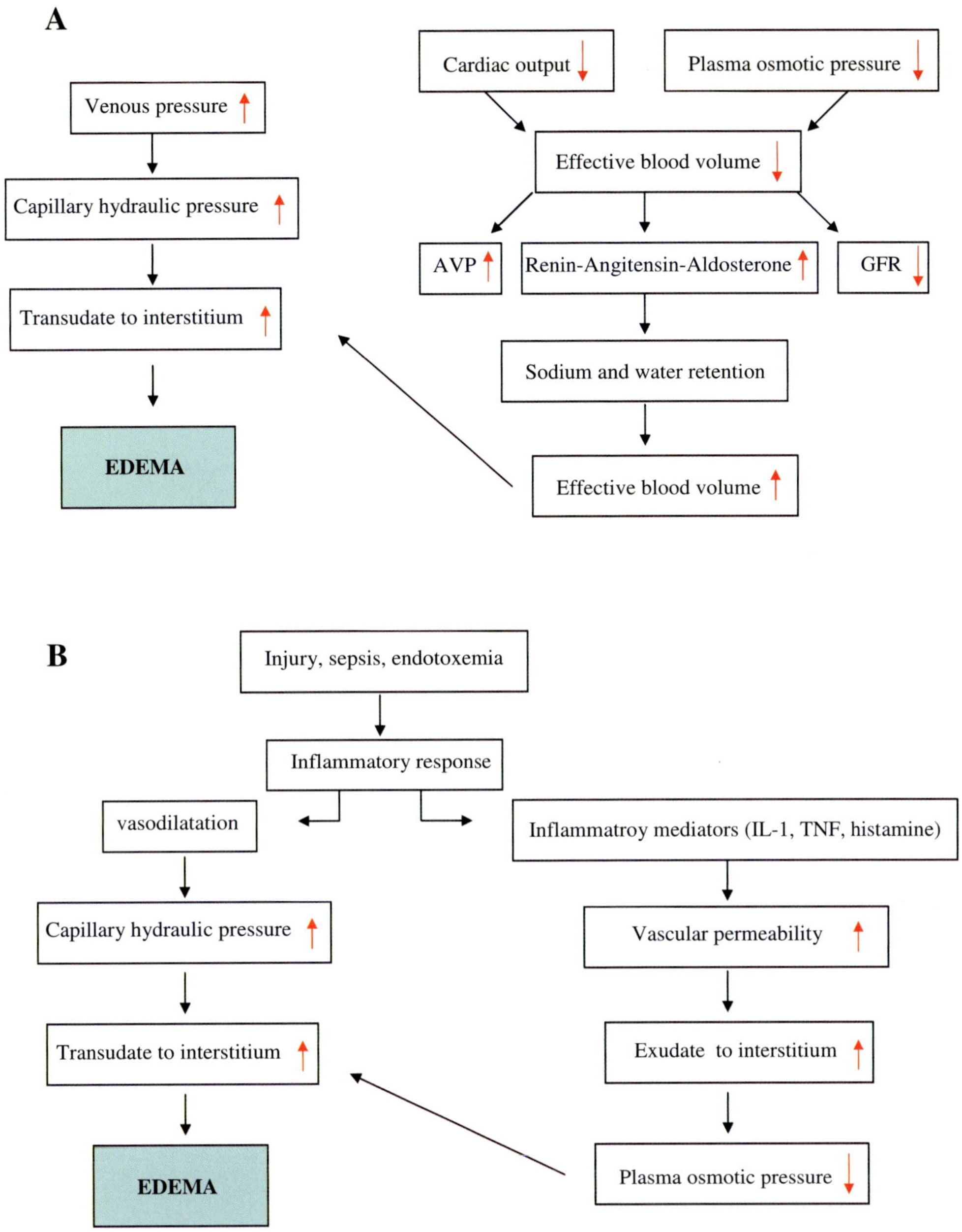

Chapter 111 Figure 1 The hemodynamic **(A)** and inflammatory **(B)** events leading to edema formation.

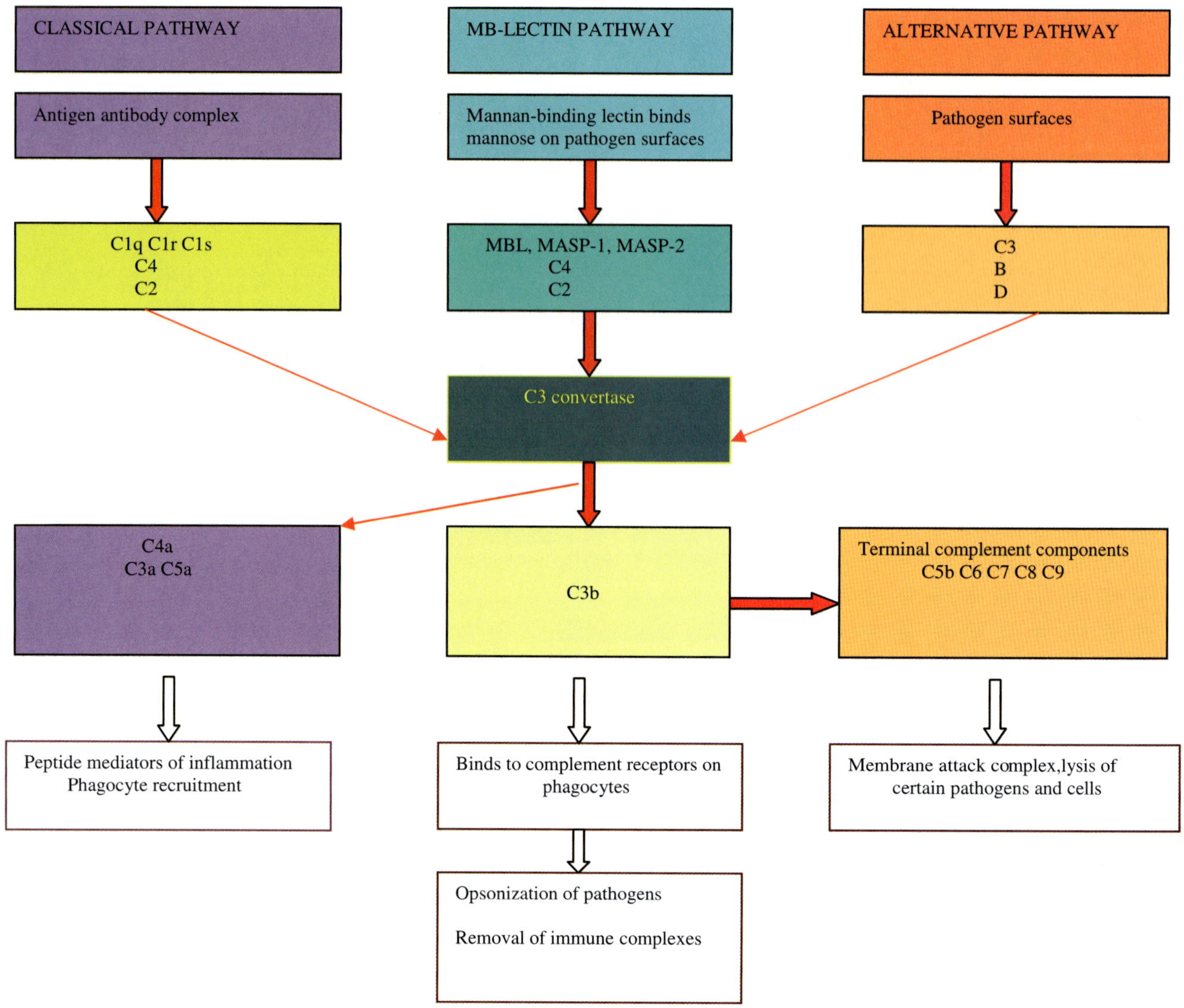

Chapter 113 Figure 1 The three complement pathways.

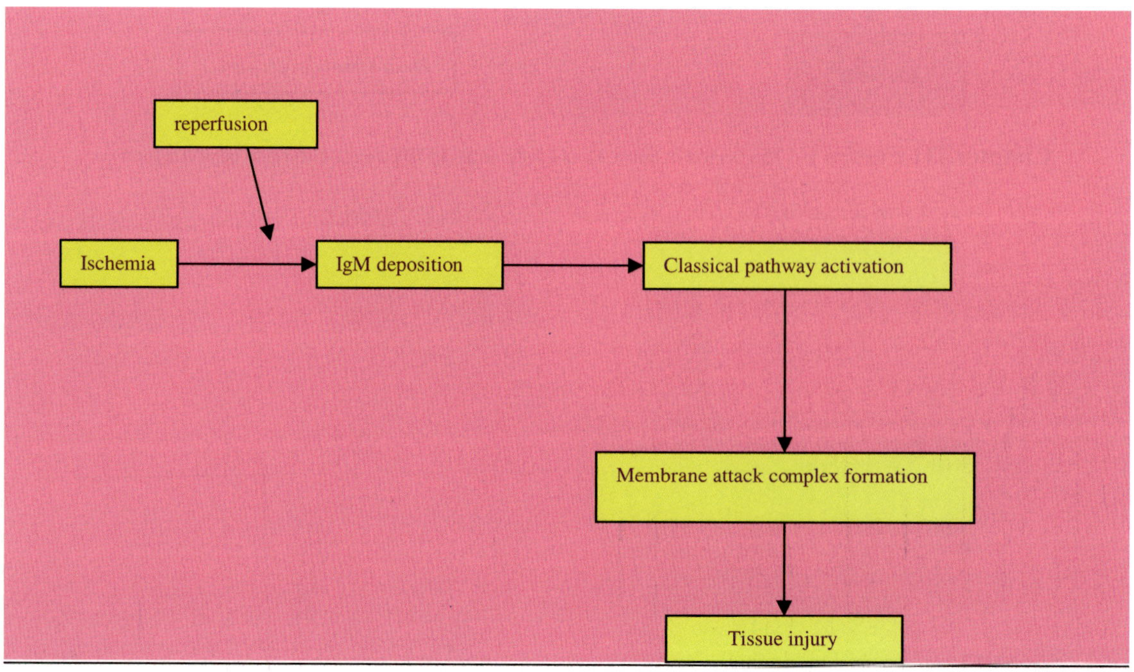

Chapter 113 Figure 2 IR activates the classical complement pathway with formation of the membrane attack complex.

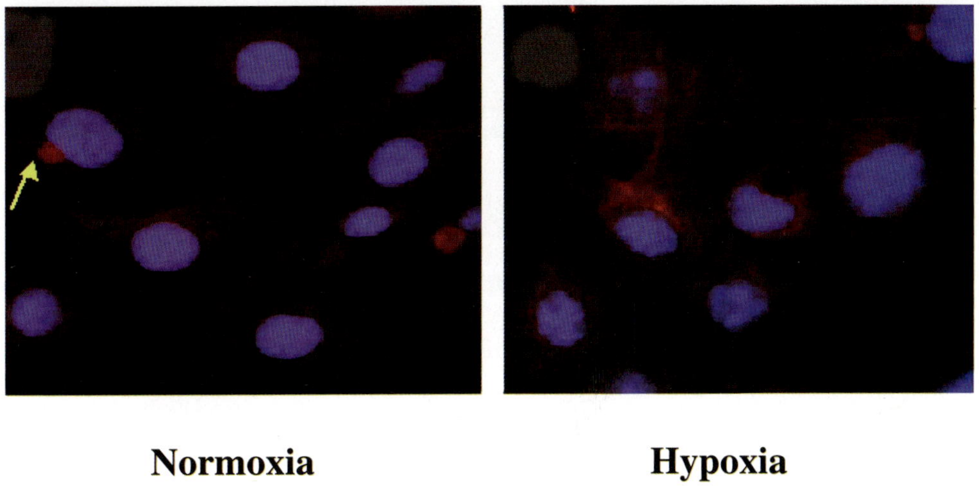

Chapter 114 Figure 1 XOR is localized in discrete vesicles in pulmonary microvascular endothelial cells (arrow) and undergoes redistribution in response to hypoxia.

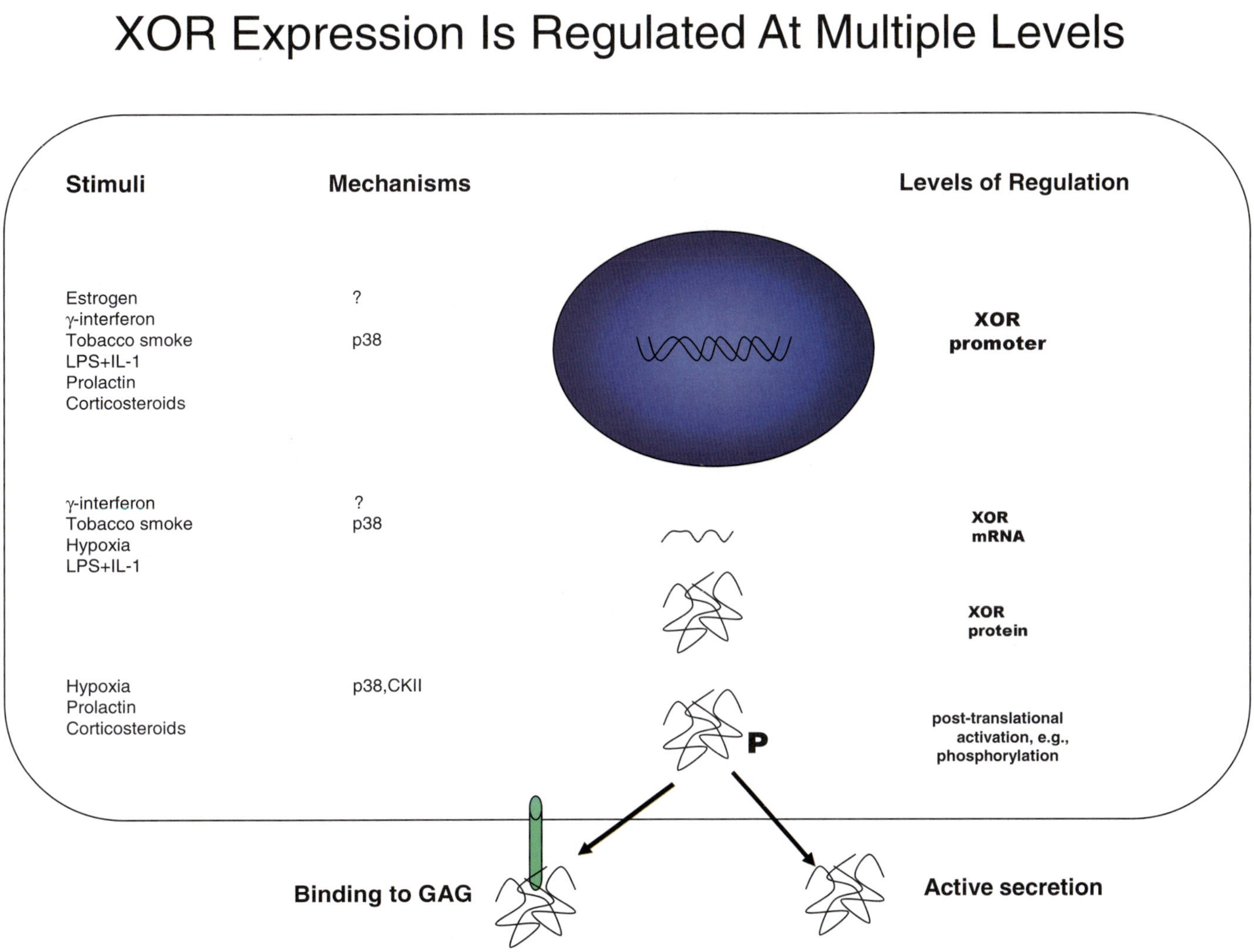

Chapter 114 Figure 2 XOR expression is regulated at multiple levels.

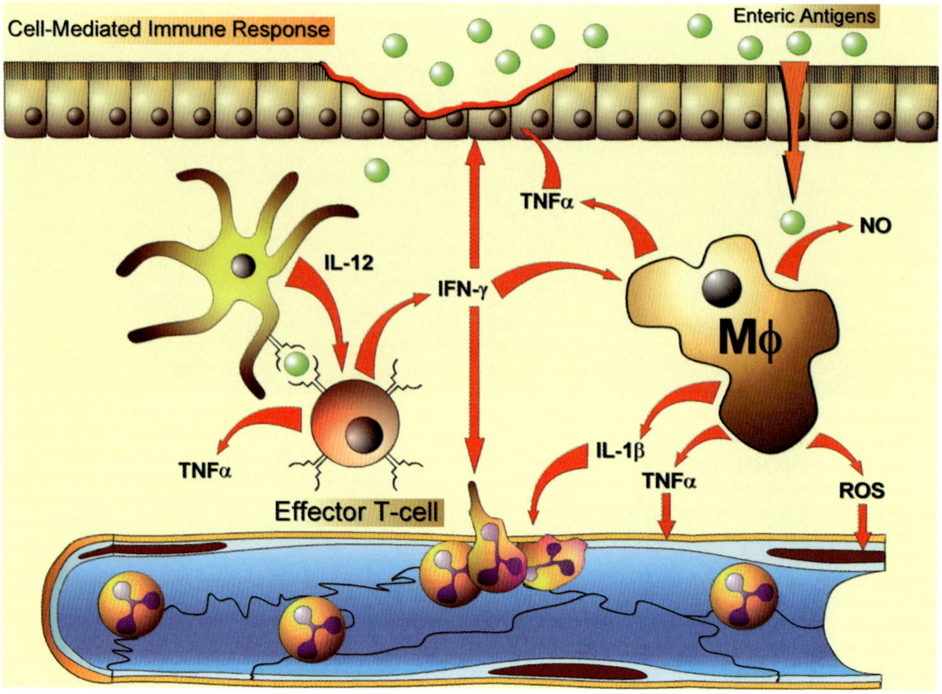

Chapter 116 Figure 1 Intestinal immune response to enteric antigens in the absence of regulatory cells. Effector $CD4^+$ T-cells produce Th1-type cytokines in response to T-cell receptor engagement of antigens processed and presented by APCs. These cytokines (such as IFN-γ) may affect the gut epithelium directly and/or activate resident Mϕ to release large amounts of proinflammatory mediators and cytokines as well as reactive metabolites of oxygen (ROS) and NO. The net result is the recruitment of additional leukocytes and subsequent inflammation and injury. APC, antigen presenting cell; Mϕ, macrophage; NO, nitric oxide; IFN-γ, interferon-γ.

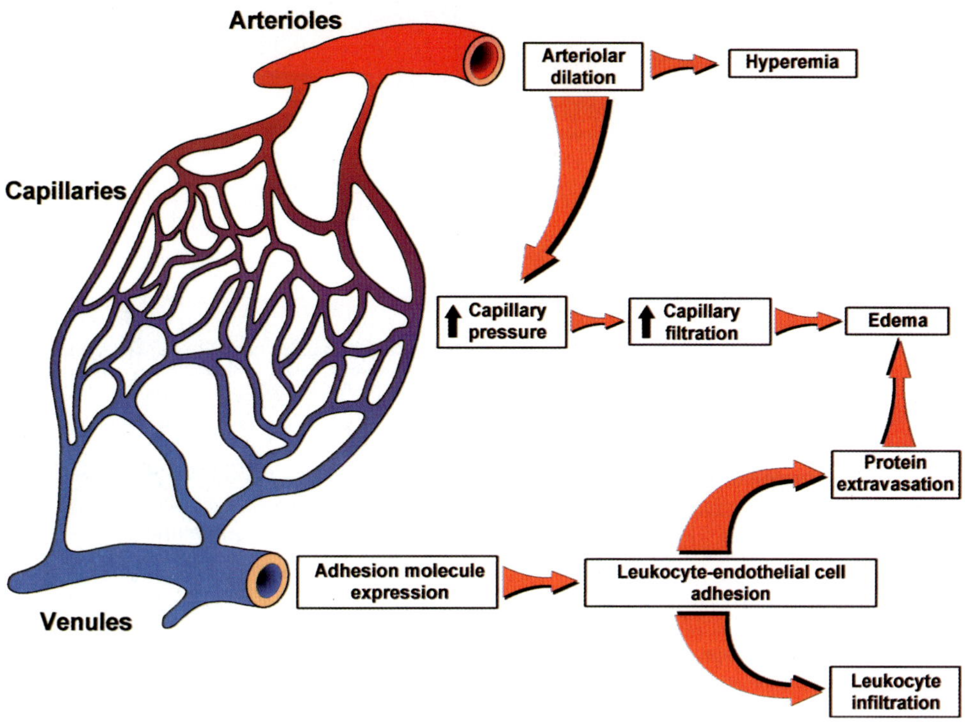

Chapter 116 Figure 2 All segments of the microcirculation contribute to the pathophysiology of chronic gut inflammation. A variety of inflammatory mediators (e.g., histamine, bradykinin, nitric oxide, prostaglandins) produced by the affected tissue relax the vascular smooth muscle surrounding arterioles. The consequent dilation of arterioles leads to an increased blood flow (hyperemia), thereby producing erythema. Another consequence of arteriolar dilation is an increased hydrostatic pressure in the downstream capillaries. The increased capillary hydrostatic pressure alters the balance of forces that govern fluid movement across intestinal capillaries to favor net fluid filtration. The increased capillary filtration rate contributes to the interstitial edema associated with gut inflammation. Proinflammatory cytokines (TNF, IFN-γ, IL-12) released by activated mast cells, macrophages, and lymphocytes activate venular endothelial cells and increase expression of endothelial cell adhesion molecules that mediate leukocyte–endothelial cell adhesion and eventual emigration (tissue infiltration) of leukocytes. Leukocyte emigration is often associated with vascular protein leak (extravasation). Consequently, inflammation generally results in a diminished endothelial barrier function in venules that promotes the accumulation of albumin and other plasma proteins in the interstitium. The resultant increase in interstitial oncotic pressure further promotes fluid filtration across capillaries and accelerates the development of interstitial edema.

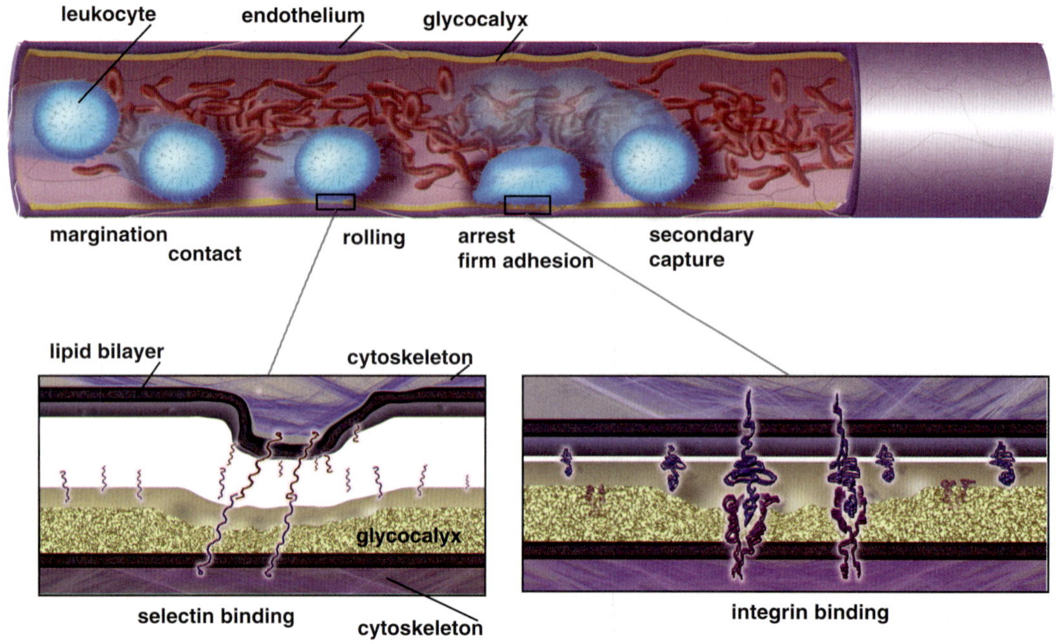

Chapter 117 Figure 1 Leukocyte adhesion in a postcapillary venule. Leukocytes flowing in a dense suspension of RBCs first marginate toward the vessel wall, influenced by vessel geometry and the forces exerted by passing RBCs. Upon contact with the wall, selectin binding can occur, resulting in slowing of the cell velocity and commencement of rolling along the endothelium. If activation of the leukocyte and endothelium are sufficient, firm integrin binding can take over, and the cell stops rolling. Eventually, this cell will extravasate into the tissue to carry out its immune function. Secondary capture can occur downstream of the adherent leukocyte to amplify the recruitment of cells.

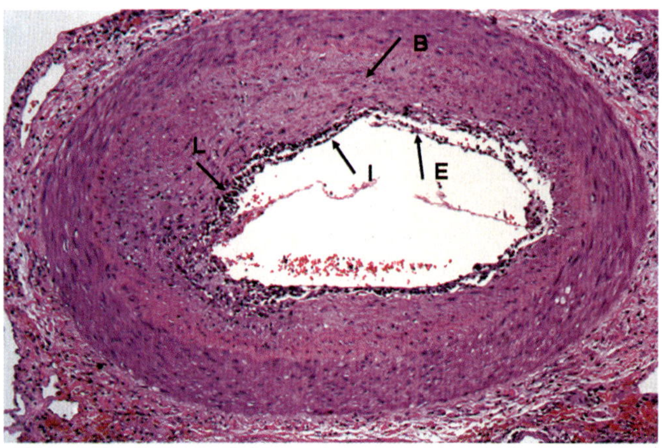

Chapter 122 Figure 1 Acute cellular rejection. Lymphocytes (L) infiltrate beneath endothelium (E) with edema of intima (I) and subintimal layers (B). (H × E, original magnification 100×).

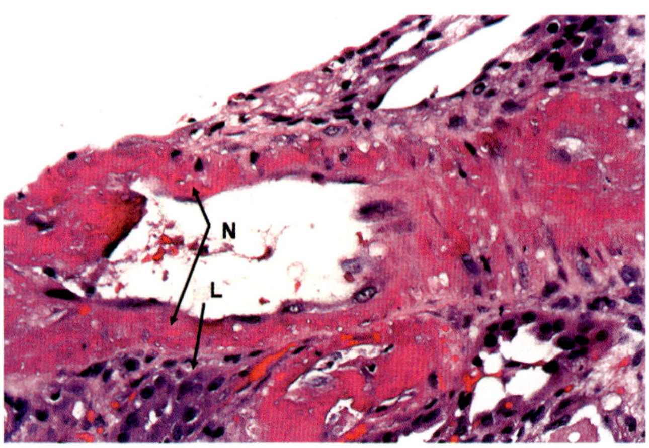

Chapter 122 Figure 2 Acute humoral rejection. Fibrinoid necrosis (N) of the thickened arterial wall surrounded by polymorphonuclear and lymphoid cell infiltrates (L). (H × E, original magnification 400×).

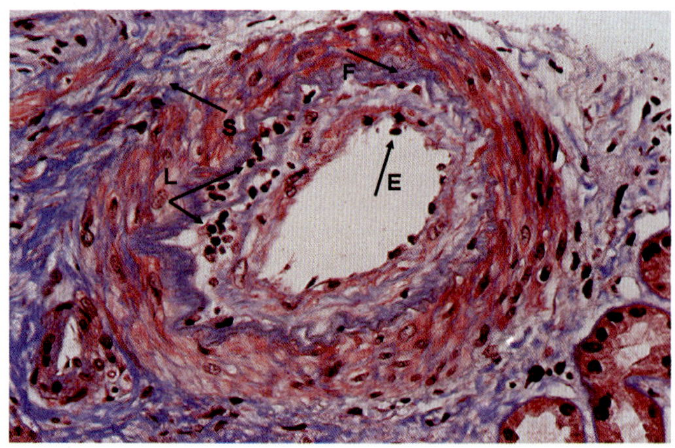

Chapter 122 Figure 3 Chronic rejection. Lymphocyte (L) infiltration and fibrosis of intima (F). There is perivascular sclerosis and edema (S). Endothelial cells are swollen and some are pyknotic (E) (Mallory, original magnification 400×).

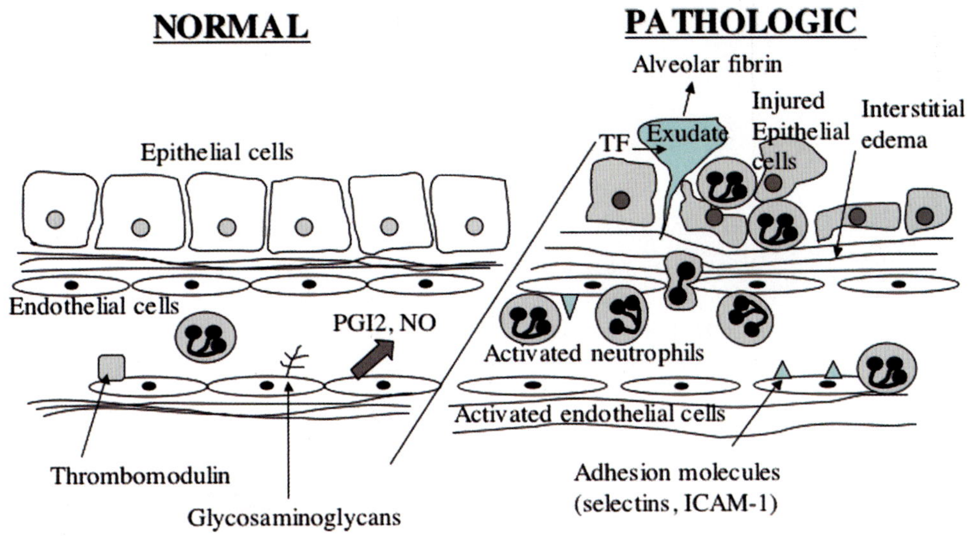

Chapter 124 Figure 1 Under normal conditions (*left*), microvascular blood flow is regulated by endothelial cells. Several anticoagulant mechanisms as well as PGI_2 and NO production are involved in regulating microvascular circulation. When the endothelial cells are activated by inflammatory cytokines or endotoxin (*right*), adhesion molecules are expressed and neutrophils accumulate in the lung. Also, the activated neutrophil-induced endothelial injury causes an increase in vascular permeability, which results in edema formation. Exudates in the airways make clots and cause ventilation–perfusion mismatching.

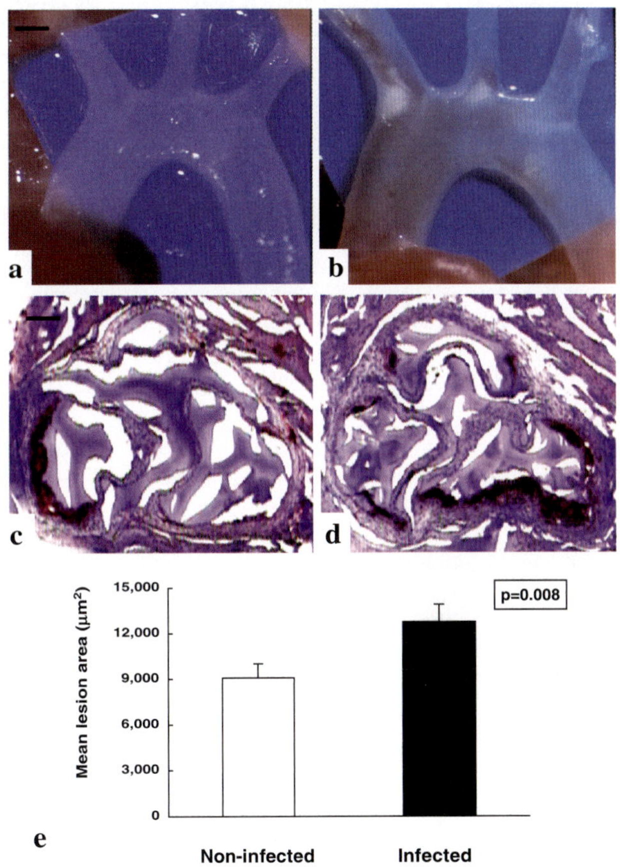

Chapter 126 Figure 1 Oral infection with *P. gingivalis* accelerates early atherosclerosis in apoE-null mice. Representative photographs of the proximal aorta demonstrate increased atherosclerosis in a *P. gingivalis*-infected **(b)** versus a noninfected **(a)** apoE-null mouse at 17 weeks of age. Scale bar, 1 mm. Oil red O staining of cryosections at the aortic sinus from control mice revealed few small fatty streaks **(c)**, whereas atherosclerotic lesions were greater in number and size in infected animals **(d)**, scale bar 50 μm. Quantitation of atherosclerotic lesion area at the aortic sinus **(e)** revealed that mean lesion area was significantly increased in infected versus noninfected mice ($12{,}753 \pm 1{,}128$ versus $9{,}080 \pm 951$ mm^2, $n = 25$/group, $p = 0.008$). Republished with permission of *Arteriosclerosis Thrombosis and Vascular Biology* from: Lalla, E., et al. (2003). Oral infection with a periodontal pathogen accelerates early atherosclerosis in apolipoprotein e-null mice. *Arterioscler. Thromb. Vasc. Biol.* **23**(8), 1405–1411.

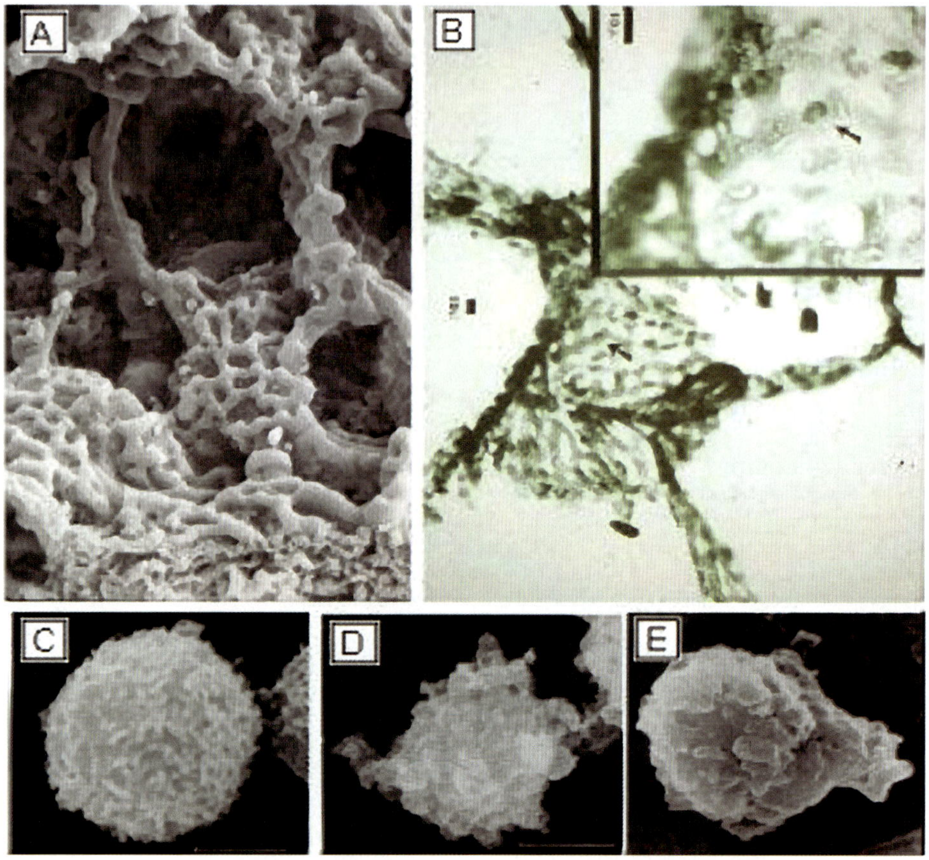

Chapter 129 Figure 1 Morphology of the pulmonary microcirculation and of resting and stimulated neutrophils. **(A)** Scanning electron micrograph of alveoli, showing walls lined with many capillary segments. **(B)** Cross section of alveoli and interstitium, showing capillaries and (*arrow*) a neutrophil (with segmented nucleus) in transit. **(C, D, E)** Scanning electron micrographs of neutrophils resting (unstimulated), exposed to cigarette smoke, or treated with bacterial peptide (fMLP), respectively.

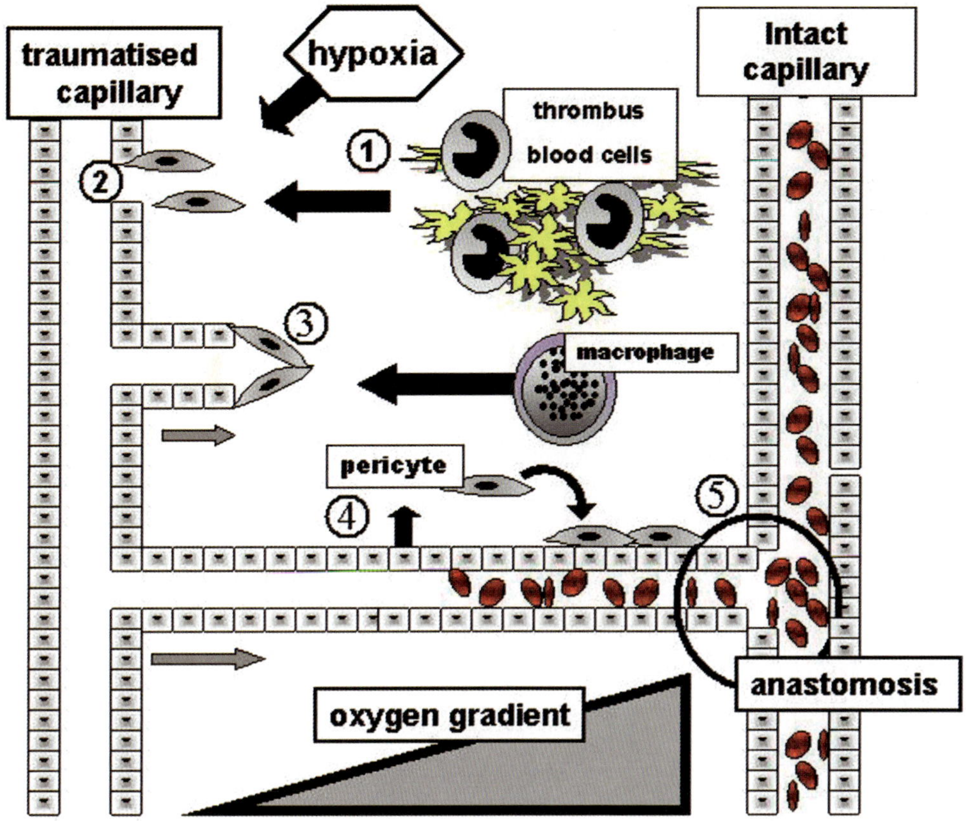

Chapter 133 Figure 1 Principal steps in microvascular repair: angiogenesis. (1) Trauma following wounding or surgery disrupts normal blood flow, elicits the formation of a thrombus and the accumulation of platelet neutrophils, monocytes, and immune cells, and renders tissue hypoxic. Release substances from these bloods cells and thrombus components, coupled with hypoxia, trigger the expression of angiogenic factors including VEGF, angiopoietins, and HIF-1α. (2) Endothelial cells (ECs) take on a migratory and replicative phenotype and detach from the matrix by secretion of matrix proteases. (3) ECs continue to replicate and migrate simultaneously aligning themselves to form tubules (sprouting). (4) The new microvessel (MV) grows in the direction of an oxygen gradient toward a capillary that carries red blood cells. MVs can release chemotactic substances that promote pericytes to migrate and associate with the MV. (5) The new MV (bio)anastomoses with an intact neighboring capillary restoring blood supply to the tissue. The subsequent intrinsic flow imparts longevity to the MVs and adapts functionally according to the needs of the tissue it supplies.

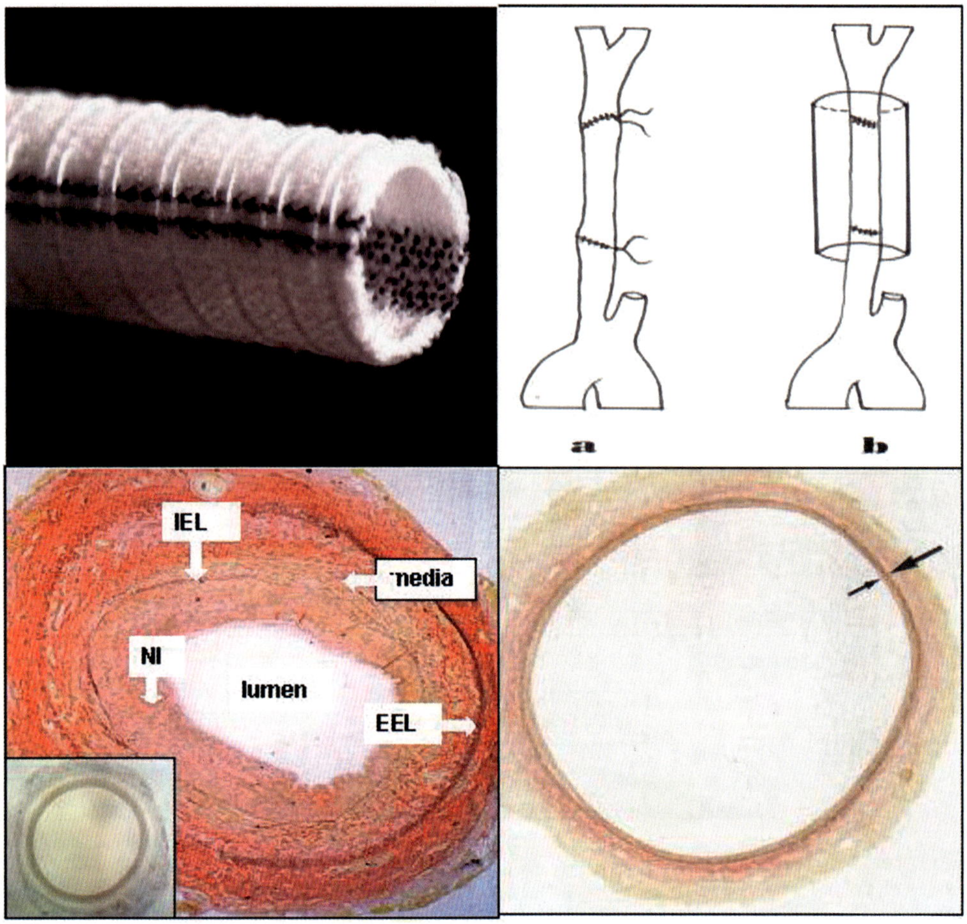

Chapter 133 Figure 3 Effect of an external sheath (stent) on neointima formation and vein graft thickening in a pig model. In these studies, a loose-fitting, that is, nonrestrictive, polyester sheath or stent (*upper left panel*) was placed around a saphenous vein into carotid artery interposition graft (*upper right panel*). After 1 month, the graft was excised and studied histologically. As can be seen (*lower panel*), there is a marked increase in graft size and neointima (NI) formation (the layer between the internal elastic lamina, IEL, and the lumen) compared to the original ungrafted saphenous vein (*inset*). The graft fitted with the external stent, however, shows a profound reduction of graft thickening (small arrow IEL and large arrow external elastic lamina, EEL) and a complete inhibition of neointima formation (see Figure 4 for higher magnification). This stented graft was characterized by profound microvessel growth (Figure 4).

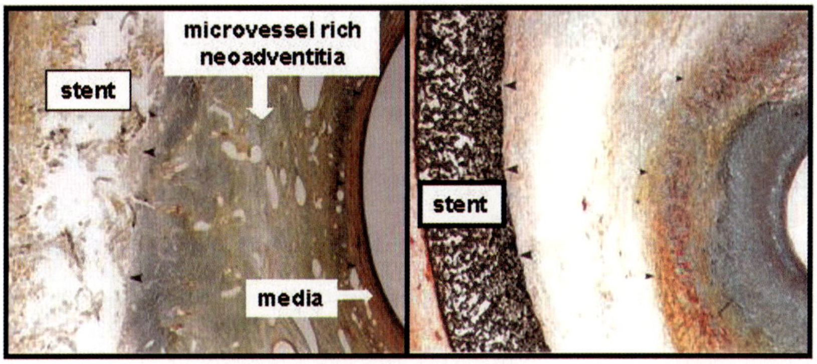

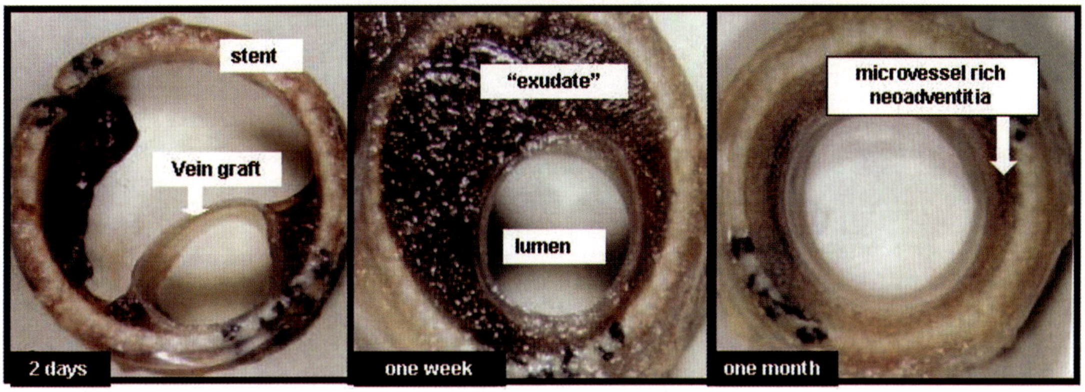

Chapter 133 Figure 4 At a higher magnification of the graft fitted with the polyester external stent (*upper left panel*), the presence of a highly developed microvascular system can be seen in the space between the stent and the media. It was suggested that the stent promotes this microvessel (MV) growth, which in turn provides oxygenation of the vein graft, thereby obviating hypoxia and graft thickening. The importance of microvessel formation can be seen with another kind of stent made of PTFE (*upper right panel*), which clearly does not promote MV formation but also does not prevent graft thickening. It has been suggested that the principal mechanism underlying this effect is that the stent concentrates an exudate (derived from the anastomotic wound) in the space between the graft and the stent (*lower three panels*). This exudate is rich in fibrin and leukocytes, which promotes the growth of microvessels via the classic wound-healing pathways.

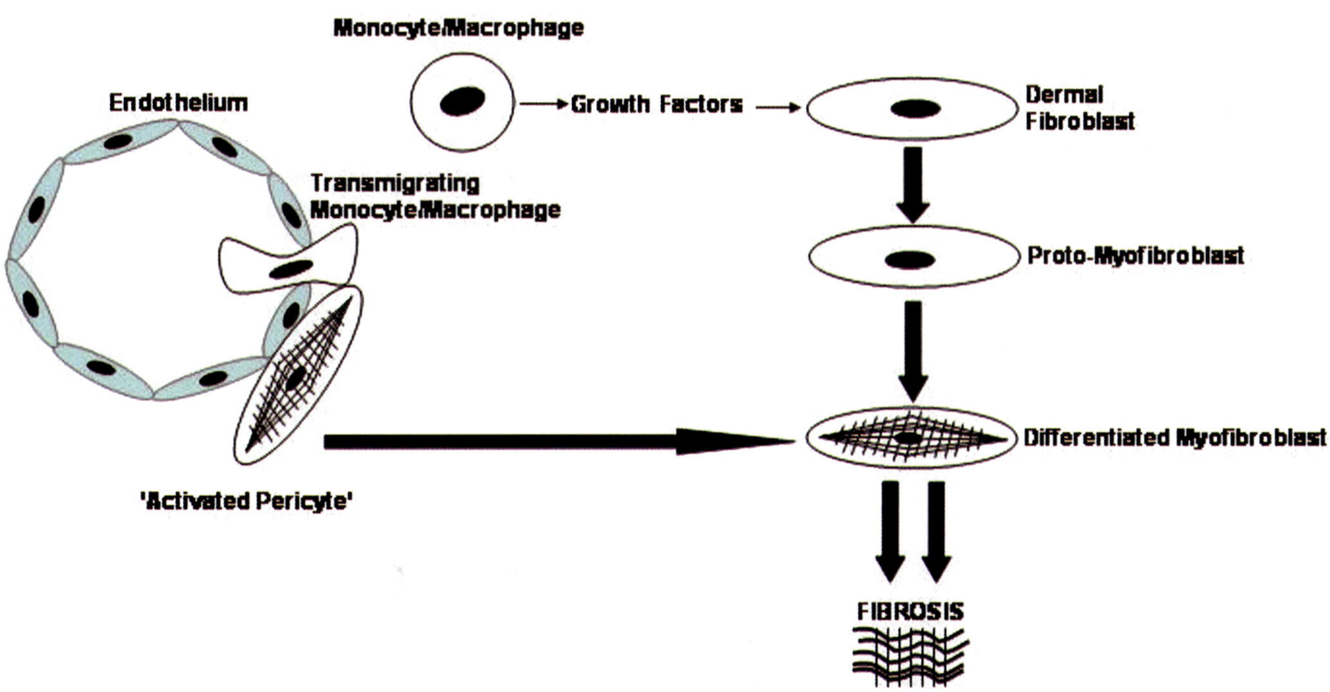

Chapter 134 Figure 1 Fibrosis and scar formation is a multistage process involving complex interactions between numerous participating cell types. Pericytes are intimately associated with this process. Preceding transmigration of monocytes is an increase in vascular permeability, brought about by a change in the interactions between endothelial cells and pericytes. Pericytes are also known to produce profibrotic growth factors that can directly modulate fibroblast function and biosynthetic profile. Finally, pericytes are considered mesenchymal precursor cells and as such can transdifferentiate directly into matrix-synthesizing myofibroblasts.

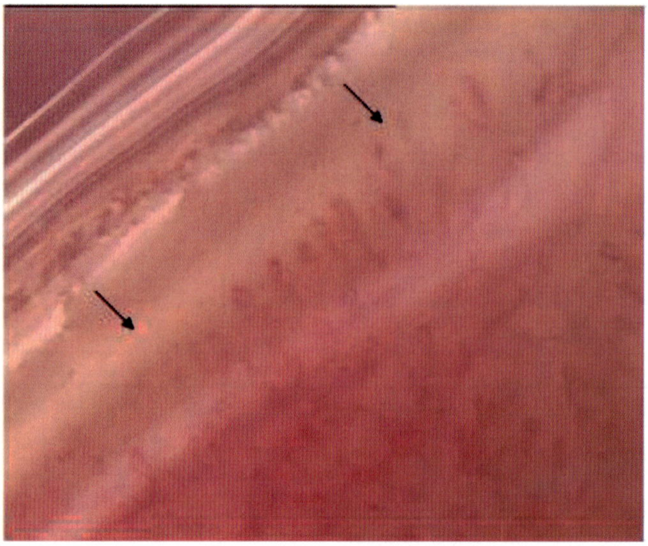

Chapter 138 Figure 2 Nailfold capillaroscopy from a 75-year-old man. Areas of focal capillary rarefaction are detectable (*arrows*).

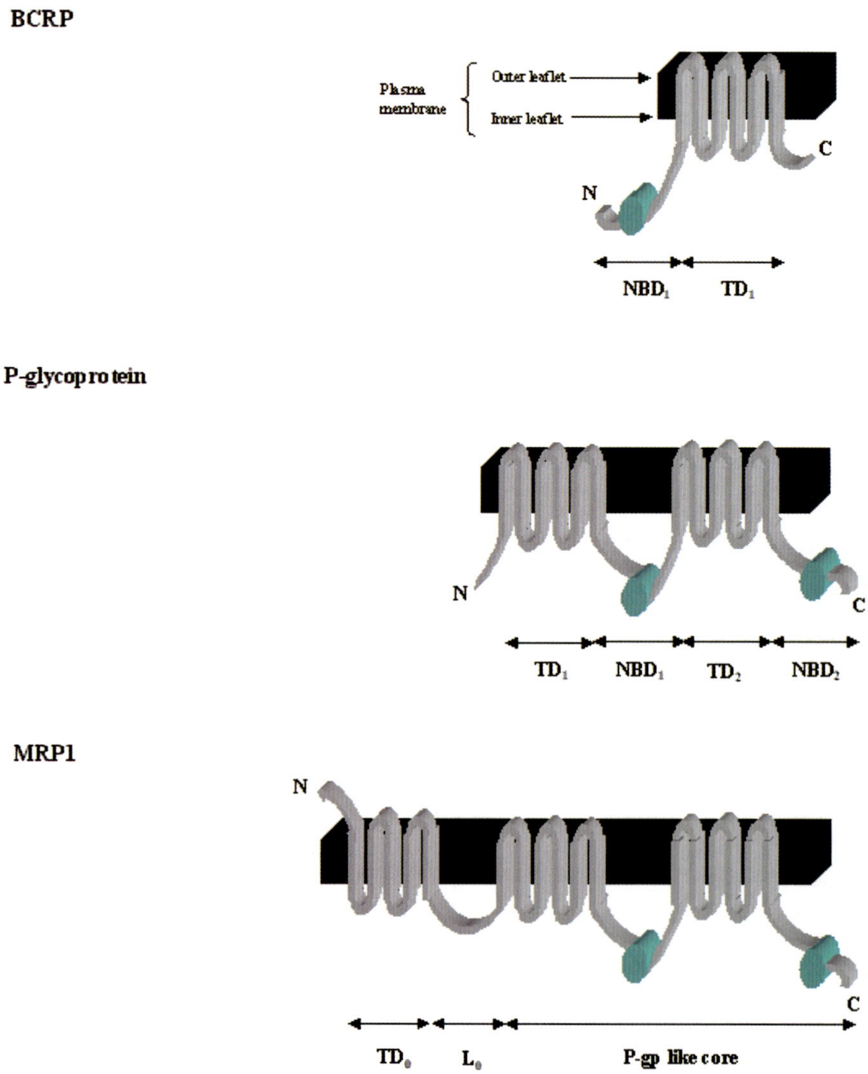

Chapter 140 Figure 1 Putative membrane topology of the three main multidrug transporters. BCRP consists of one transmembrane domain (TD) and one nucleotide binding domain (NBD) and is termed a *half-transporter*. P-glycoprotein, like many ABC proteins, is a full transporter with a TD_1-NBD_1-TD_2-NBD_2 structure. Several of the MRP family members, including MRP1, have an additional N-terminal TD linked to the P-gp-like core by a linker region (L_0). Note that the TD and NBD in BCRP are in reverse orientation to those of the other two ABC proteins. (*N* and *C* refer to the N and C termini of the transporter, respectively.)

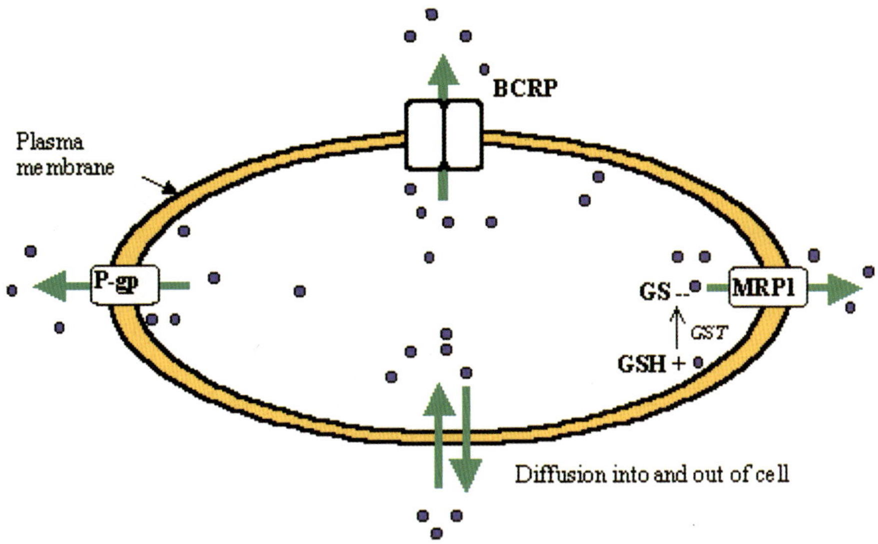

Chapter 140 Figure 2 Actions of multidrug transporters BCRP, P-gp and MRP1 in efflux of substances from cells. Drugs can enter and leave cells by passive diffusion along a concentration gradient. Multidrug transporters provide a second route for drug exit and can drive drugs out of the cell against a concentration gradient by exploiting the energy of ATP hydrolysis. This additional efflux reduces drug concentrations inside cells to sublethal levels. P-gp and BCRP (in the form of a dimer) can efflux drugs unmodified (mainly hydrophobic, amphipathic compounds). The MRPs require the presence of reduced glutathione (GSH) to transport unmodified drugs (predominantly organic anions) but can transport drugs following their conjugation (GST, Glutathione transferase).

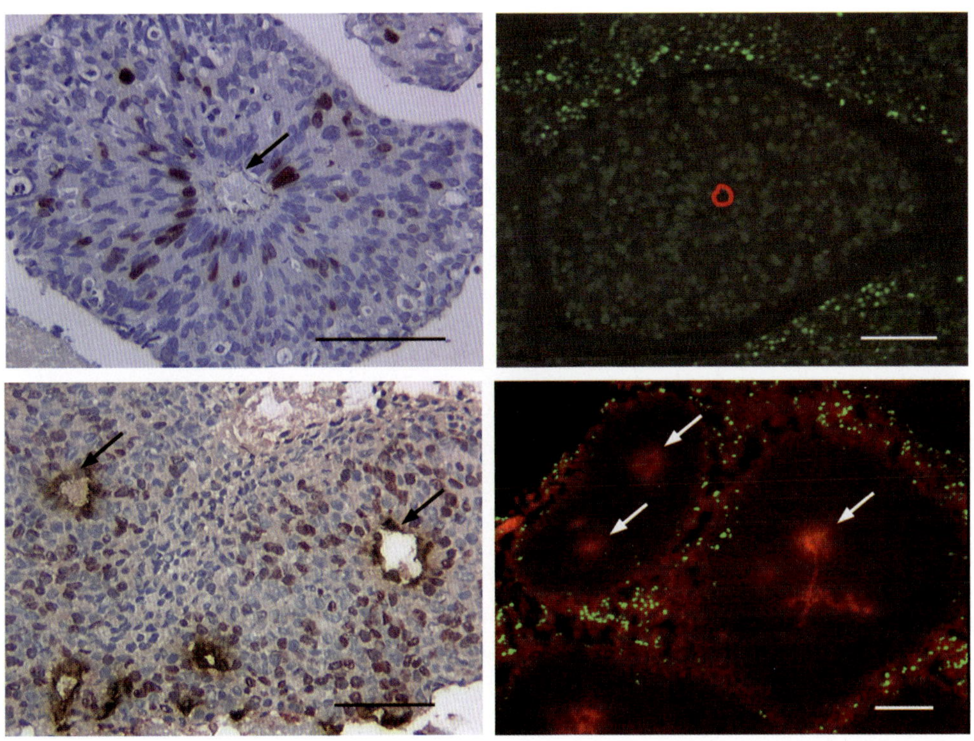

Chapter 142 Figure 1 Location of dividing and apoptotic tumor cells in relation to blood vessels in brain metastases. **(a)** Autochthonous human lung-cancer brain metastases. BrdU-positive cells (cells engaged in cell division) are stained red and are located within 75 μm of the nearest blood vessel (*arrow*). **(b)** Autochthonous human lung-cancer brain metastasis. TUNEL-positive (apoptotic) cells stained bright green and are positioned 160–170 μm from the nearest blood vessel. The vessel is marked in red. **(c)** Human KM12C metastasis in the brain of nude mice. BrdU-positive nuclei stained red. Arrow points to blood vessel stained for CD31 (brown). **(d)** Human KM12C metastasis in the brain of nude mice. CD31 staining (red) identified blood vessels and bright green nuclei were TUNEL-positive. Bar = 100 μm. Reproduced with permission from Fidler et al. (2002). *Lancet Oncology* **3**, 53–57.

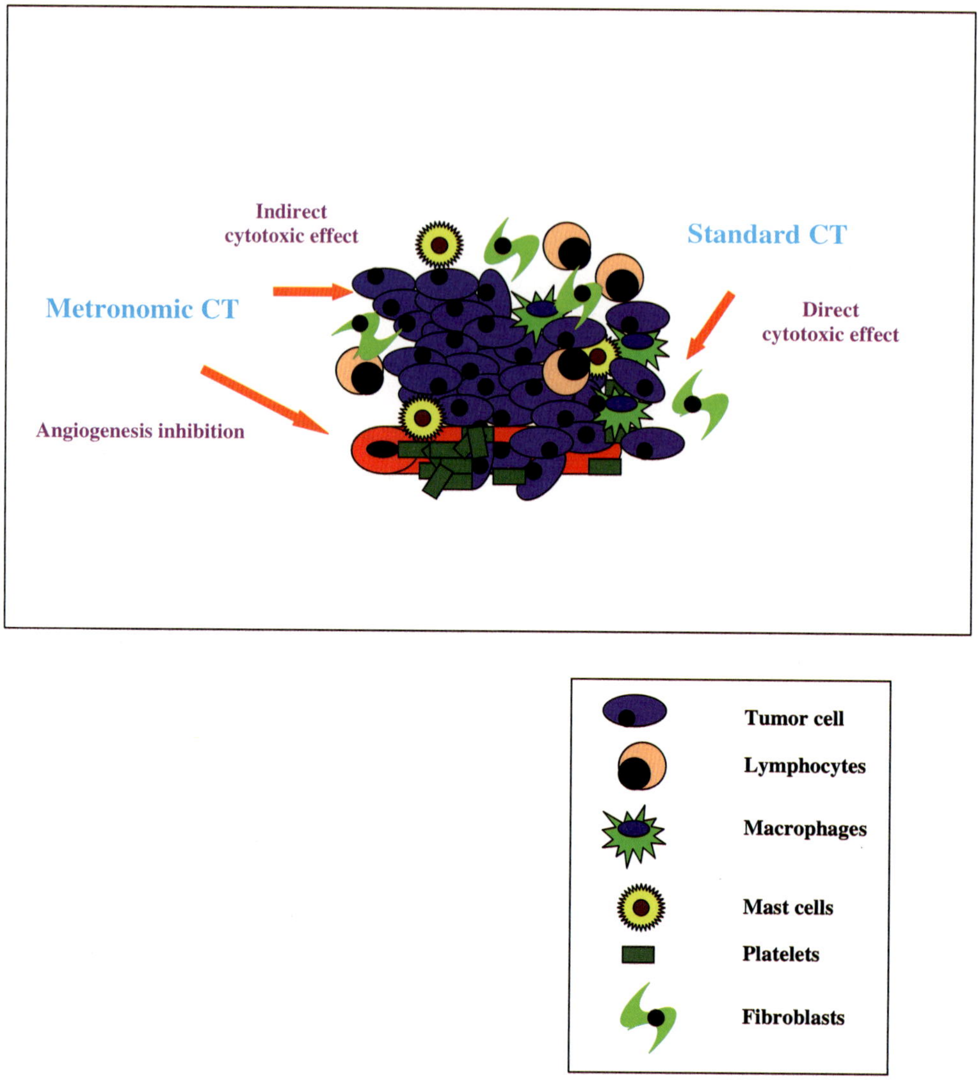

Chapter 143 Figure 1 Metronomic chemotherapy interferes with tumor microvasculature and the stromal compartment, in contrast to standard chemotherapy that presents a direct cytotoxic effect.

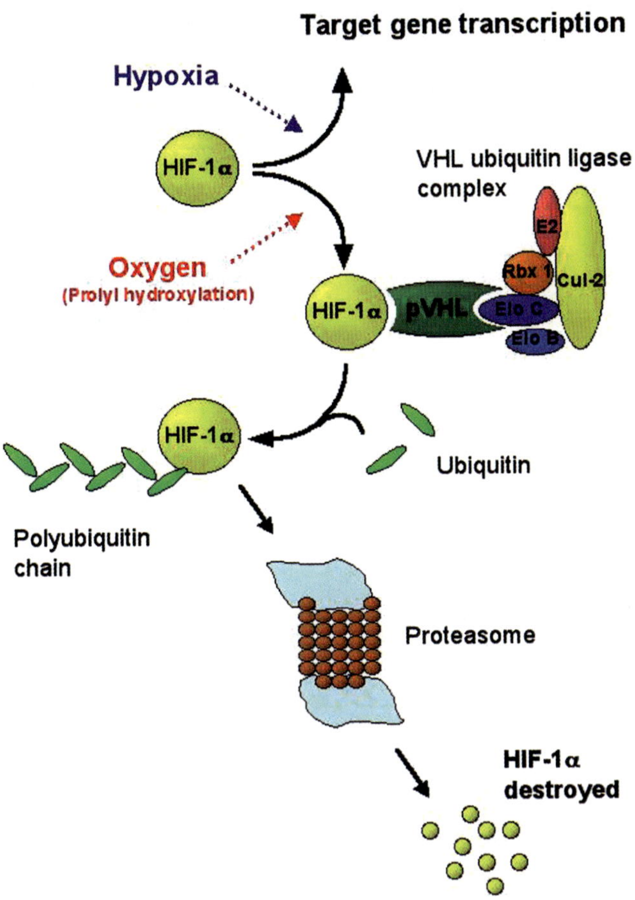

Chapter 145 Figure 1 Regulation of HIF-α by pVHL. Under oxygen-replete conditions HIF-α is enzymatically modified by prolyl hydroxylase activity to a form that is recognized by pVHL. The pVHL ubiquitin ligase complex covalently attaches a polyubiquitin tag, which then targets HIF-α for rapid destruction in by the proteasome, so that it is no longer available to effect target gene transcription.

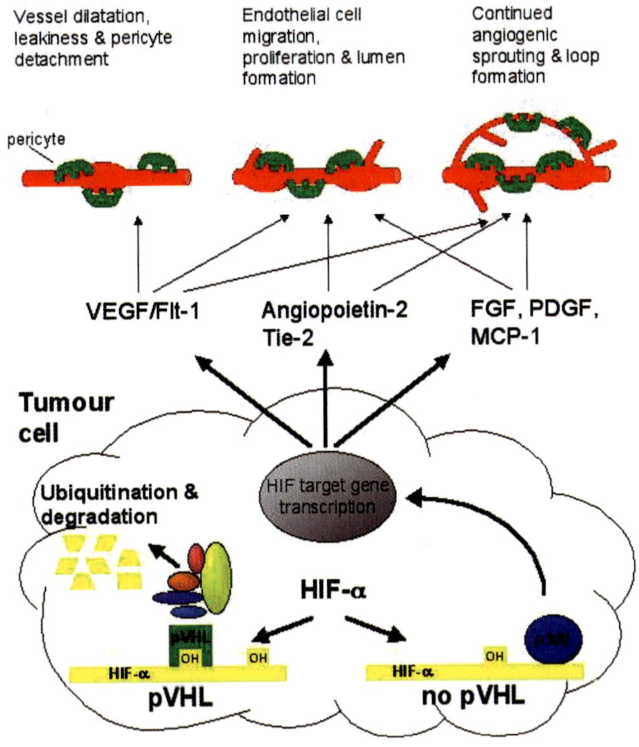

Chapter 145 Figure 2 The mechanism of angiogenesis in VHL disease. pVHL is required for the normoxic degradation of HIF-α subunits. In the pVHL-deficient tumor cells, HIF accumulates and effects transcription of its target genes. These include secreted factors such as VEGF, Flt-1, angiopoietins-2, Tie-2, FGF, PDGF, and MCP-1, which act on different stages of angiogenesis, leading to localized new vessel growth.

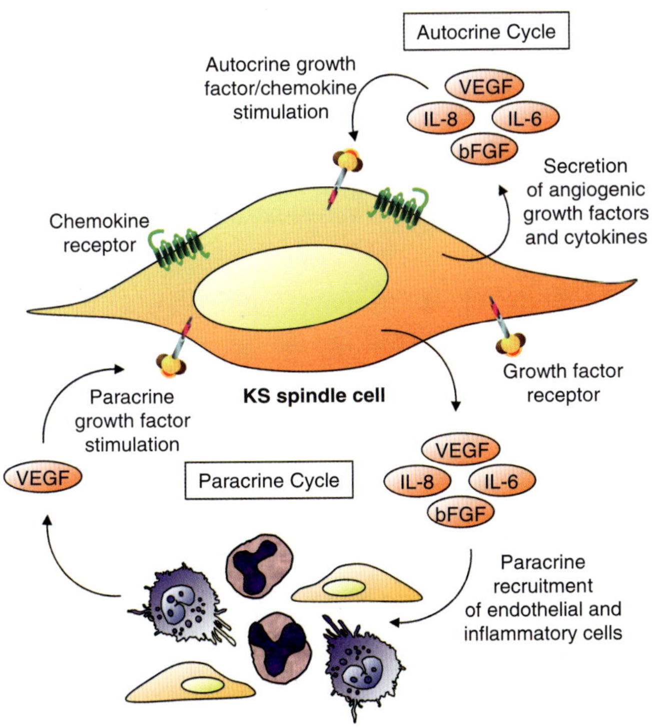

Chapter 147 Figure 1 Autocrine and paracrine growth factor cycles promoted by the Kaposi's sarcoma spindle cell. The KS spindle cell, believed to be of endothelial origin, is the driving force of KS with its secreted growth factors and cytokines sustaining its own proliferation while recruiting the remaining cell types in a paracrine fashion.

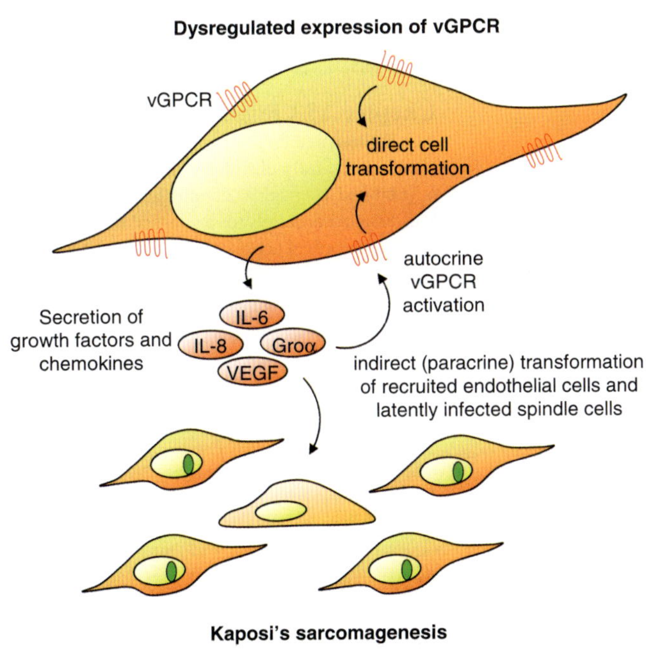

Chapter 147 Figure 2 Dysregulated (nonlytic) expression of vGPCR. Unregulated vGPCR signaling to survival pathways may promote direct cell transformation while continuous secretion of chemokines and growth factors may induce the recruitment and subsequent indirect (paracrine) cell transformation of neighboring "bystander" endothelial cells.

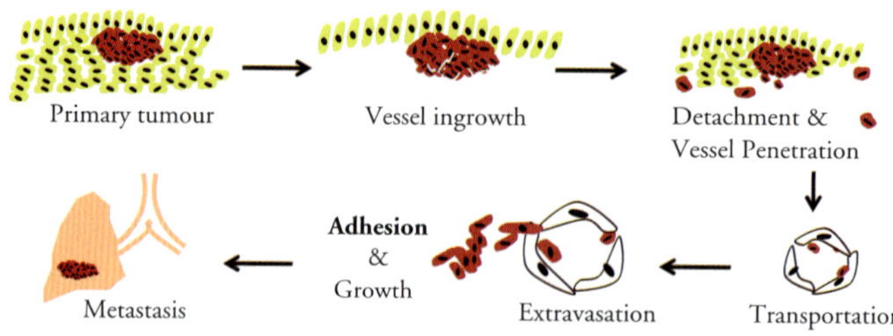

Chapter 149 Figure 1 A schematic reproduction of the metastasis process. Tumor cells first have to detach from the primary tumor, invade the blood circulation, subsequently adhere to the endothelium within a distant organ and finally invade and multiply to form a metastasis.

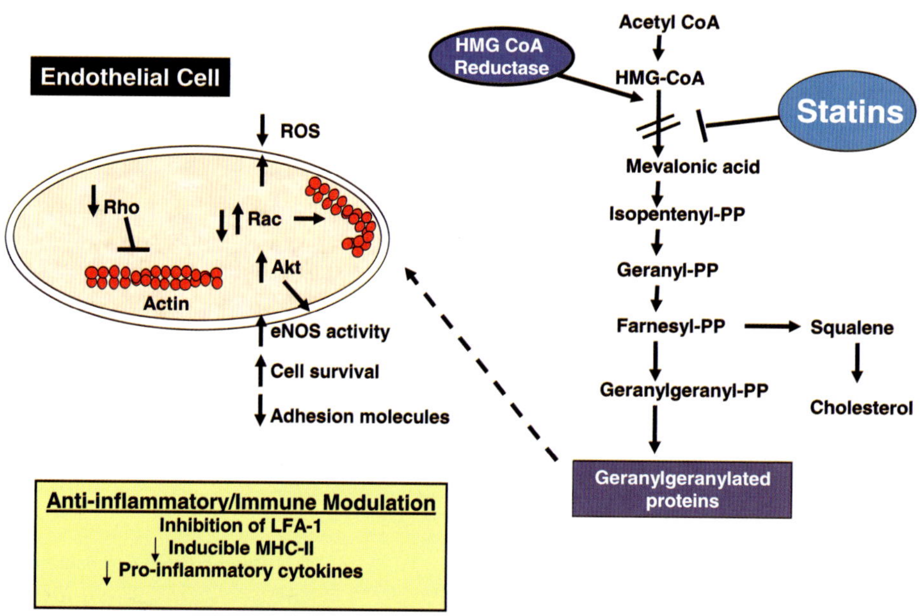

Chapter 153 Figure 1 Statin effects on the microvasculature. Effects of statins that target the endothelium and inflammation are critical to their vasculoprotective action. Although specific mechanisms remain to be fully elucidated, inhibition of geranylgeranylation is thought to play an important role. Rho is inhibited by statins, attenuating actin stress fiber formation. Although inhibition of Rac membrane translocation accounts for decreased ROS production, increased Rac activation by statins is associated with enhanced cortical actin and barrier protection (see text).

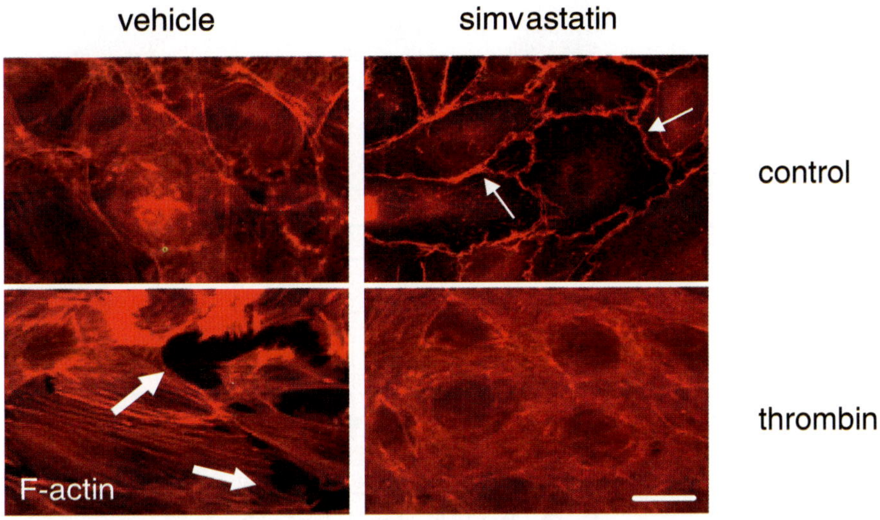

Chapter 153 Figure 2 Effect of simvastatin on endothelial cell (EC) cytoskeletal rearrangement. Relative to vehicle controls, ECs treated with simvastatin (5 μM, 16 hours) demonstrate enhanced cortical actin (*small arrows*) and reduced central stress fibers. Additionally, upon thrombin stimulation (1 U, 5 minutes) few paracellular gaps are observed in simvastatin-treated cells compared to vehicle controls (*large arrows*). Bar = 10 μm.

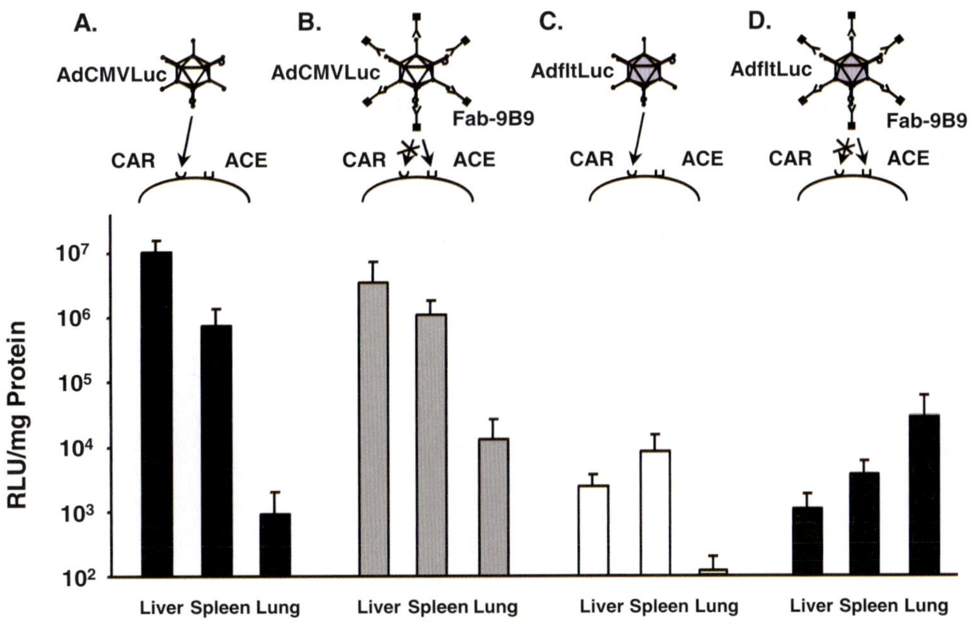

Chapter 155 Figure 1 Enhancement of specificity and effectiveness of a model transgene expression in pulmonary endothelium in rats using combined transductional and transcriptional targeting of adenoviral gene delivery. Rats were injected intravenously with adenoviral vectors carrying the luciferase reporter gene, and 3 days later organs were harvested and luciferase activity determined. **(A)** Untargeted vector, AdCMVLuc, showing dominant transgene expression in liver and spleen due to biological tropism of the virus to cells in these organs expressing receptors for viral capsid proteins. **(B)** Conjugation of ACE antibody Fab fragments to these virus surface proteins blocks their binding to hepatic and splenic receptors and endows the virus with affinity to endothelial cells. This provides targeting to endothelial cells and enhanced pulmonary and reduced hepatic gene expression (note log scale). **(C)** AdfltLuc virus, containing an endothelium-specific promoter Flt-1, but lacking specific affinity to endothelial cells, provides ineffective gene expression in all organs. **(D)** Conjugation of AdfltLuc with anti-ACE restores transgene expression levels in lungs, but still further reduces expression in non-target liver and spleen, thereby improving selectivity lung/liver ratio net 300,000-fold. RLU, Relative light units. (Adapted from [9].)

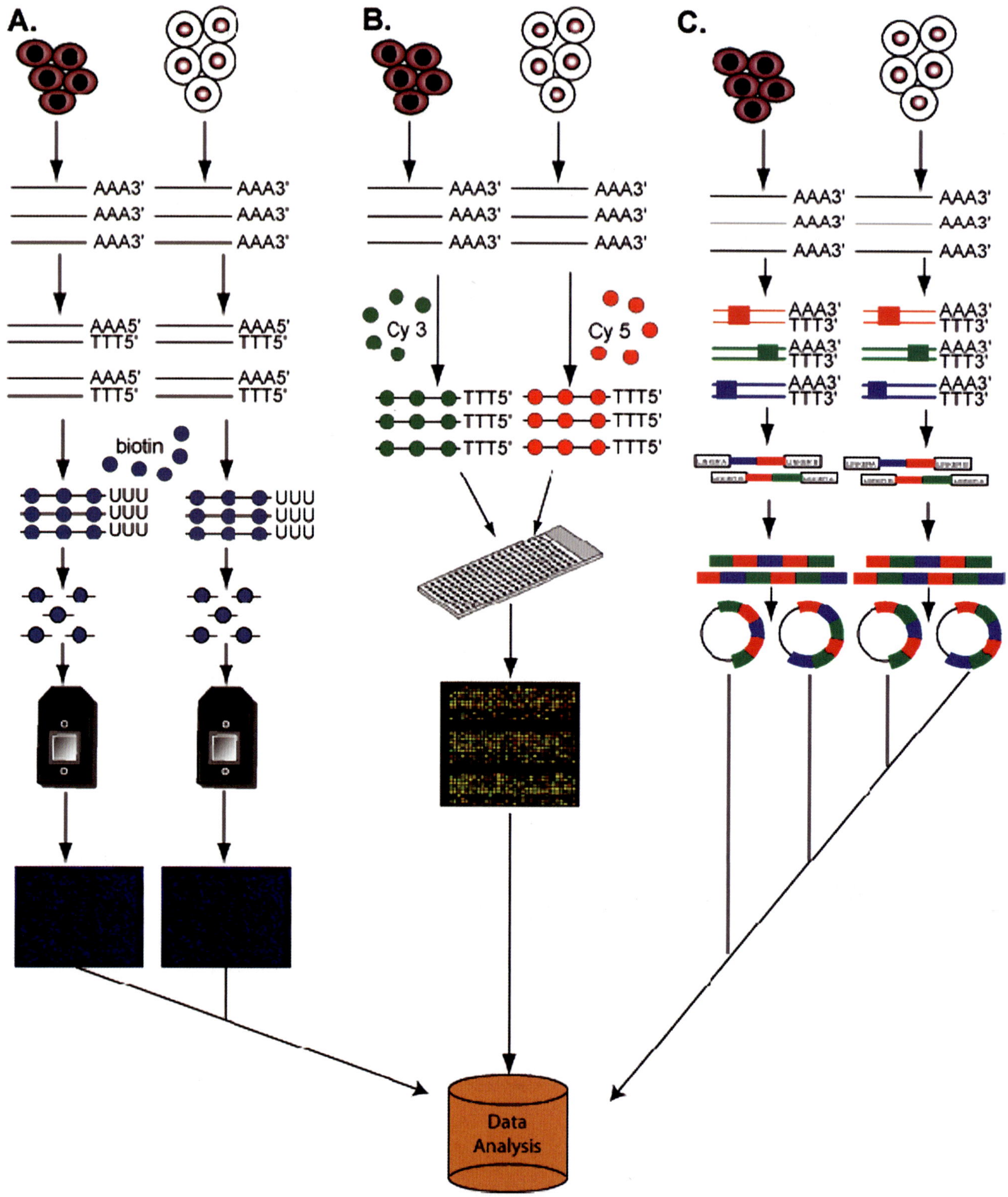

Chapter 157 Figure 1 Schematics of DNA microarrays and SAGE. **(A)** Schematic of oligonucleotide microarray. RNA from different tissues or cell populations is used to generate double-stranded cDNA carrying a transcriptional start site for T7 DNA polymerase. It is followed by in vitro transcription, during which biotin-labeled nucleotides are incorporated into cRNA molecules. Each target sample is hybridized to a separate probe array, and target binding is detected by staining with a fluorescent dye coupled to streptavidin. Signal intensities of probe array element sets on different arrays are used to calculate relative mRNA abundance for the genes represented on the array. **(B)** Schematic of cDNA microarray. RNA from different tissues or cell populations is used to synthesize single-stranded cDNA in the presence of nucleotides labeled with two different fluorescent dyes (for example, Cy-3 and Cy-5). Both samples are mixed in a small volume of hybridization buffer and hybridized to the array surface, usually by stationary hybridization under a coverslip, resulting in competitive binding of differentially labeled cDNAs to the corresponding array elements. High-resolution confocal fluorescence scanning of the array with two different wavelengths corresponding to the dyes used provides relative signal intensities and ratios of mRNA abundance for the genes represented on the array. **(C)** Schematic of SAGE. Poly(A)$^{+}$RNA is isolated by an oligo-dT column and double-stranded cDNA is synthesized. It is followed by an anchoring enzyme digestion (e.g., *Nla*III), ligated to linkers, PCR amplification, concatenation, cloning into a sequencing vector, and DNA sequencing. The sequencing files are used to extract the numbers of SAGE tags in a digital format, which can reveal the difference in expression copies of interested genes between the diseased sample and a normal control. A relatively detailed description of the SAGE procedure is provided in the text.

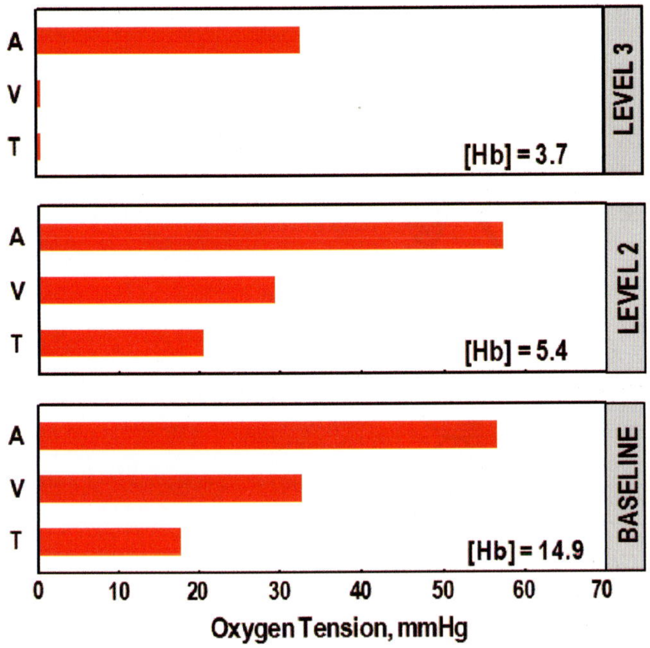

Chapter 160 Figure 1 Distribution of oxygen in arterioles (A) and venules (V) in the tissue (T) in the awake hamster window model as a function of total hemoglobin (Hb, grams per deciliter). Normovolemic hemodilution was achieved using dextran 70 kDa. It is apparent that a loss of the order of two thirds of red blood cells has no influence on microcirculatory po_2 of this model.

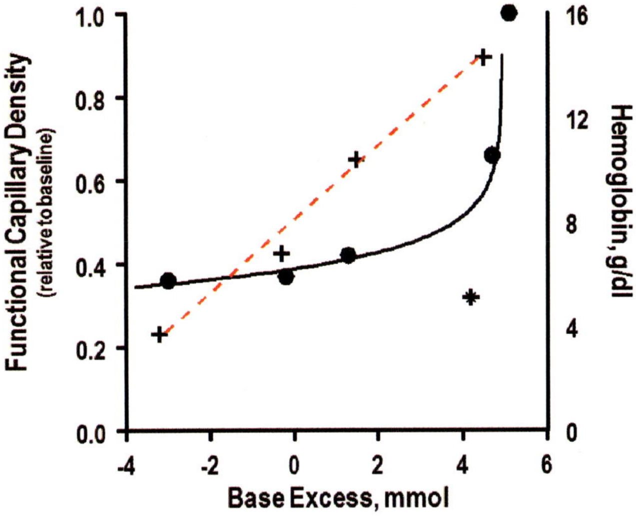

Chapter 160 Figure 5 Relationship between total circulating hemoglobin and base excess, and FCD and base excess, for different hemoglobin modifications and concentrations, including hemoglobin vesicles, in normovolemic hemodilution experiments. The data marked ● shows the relationship between FCD and base excess, showing that MAL-Peg-Hb (▼) yields high FCD and base excess at low hemoglobin concentrations. It is apparent that base excess is a direct function of hemoglobin concentration (+) with the exception of MAL-Peg-Hb (✱), which presents normal base excess at a very low total hemoglobin content.

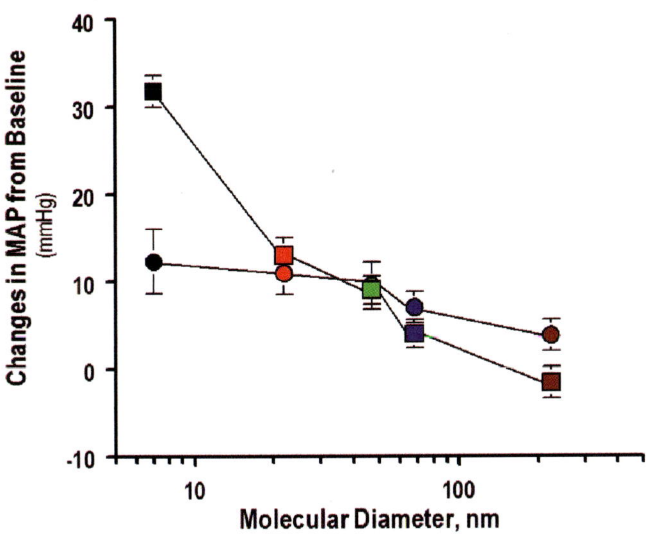

Chapter 160 Figure 3 Changes in mean arterial blood pressure after a 5 percent by volume (5 Hb g/dL) topload infusion of free hemoglobin solutions of different molecular diameters and vesicle encapsulated hemoglobin. ■ Pressor effect after infusion. ● Pressor effect 3 hours after infusion. Redrawn from Sakai et al. [5].

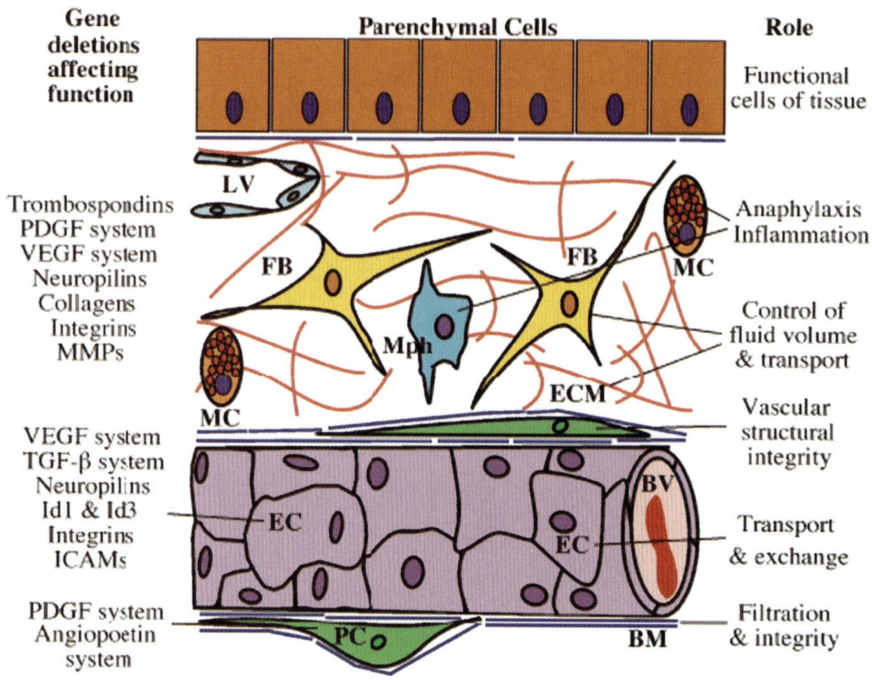

Chapter 162 Figure 1 The structure of the microvascular unit and gene deletions that affect its function. LV, lymphatic vessel; FB, fibroblast; MC, mast cell; Mph, macrophage; EC, endothelial cell; PC, pericyte; BV, blood vessel; BM, basement membrane; ECM, extracellular matrix.

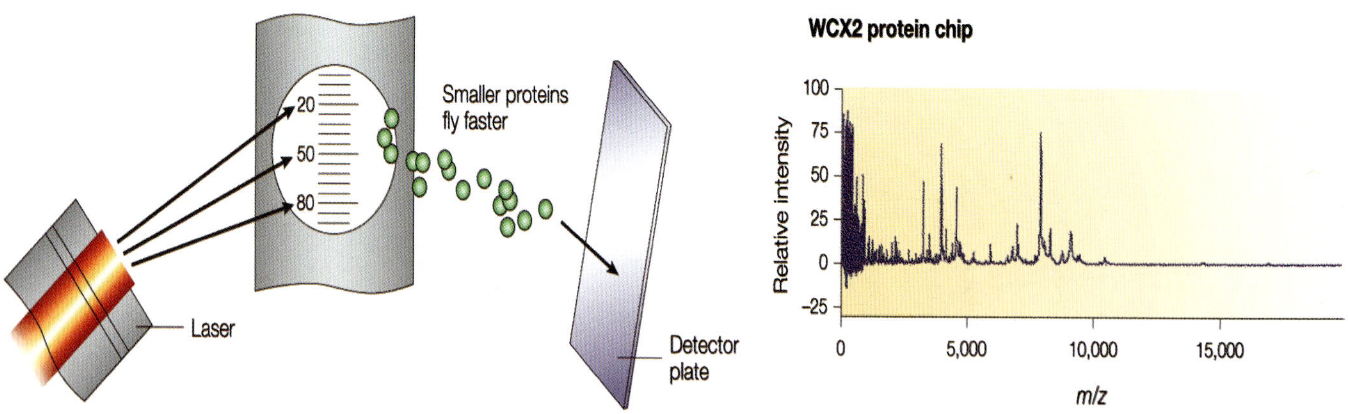

Chapter 166 Figure 2 Mass spectrometry. **(A)** Standard low-resolution mass spectrometer, typical of those used in MALDI and SELDI analyses. **(B)** Tandem mass spectrometer. The principal distinguishing feature from low-resolution instruments is the presence of a collision cell in which peptides are broken apart by collision with an inert gas, permitting in many cases sequence analysis and parent protein identification. **(C)** Positions on protein chip spots. Each spot is divided into tiny coordinates (the 20, 50, and 80 marks shown in the figure) known as positions that are used to direct the laser to strike at precisely the same point on each spot. Smaller ions fly more swiftly to the detector plate, permitting their differentiation by mass : charge value. The right side of the figure shows the proteomic profile generated by the sample on the spot. **(D)** Peak map (lower part of the figure, in red) showing how numerous analytes may be found around individual *m/z* values.

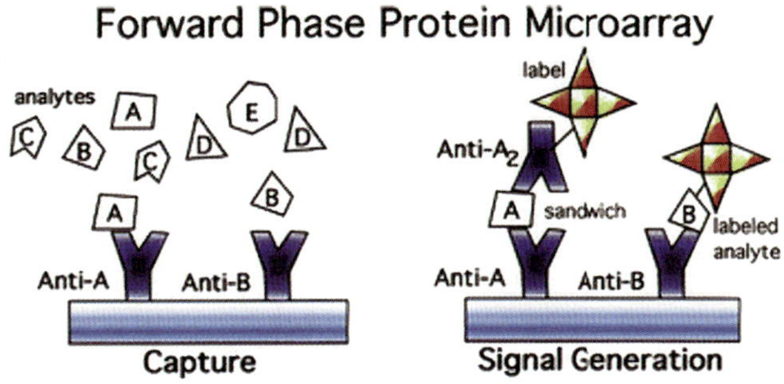

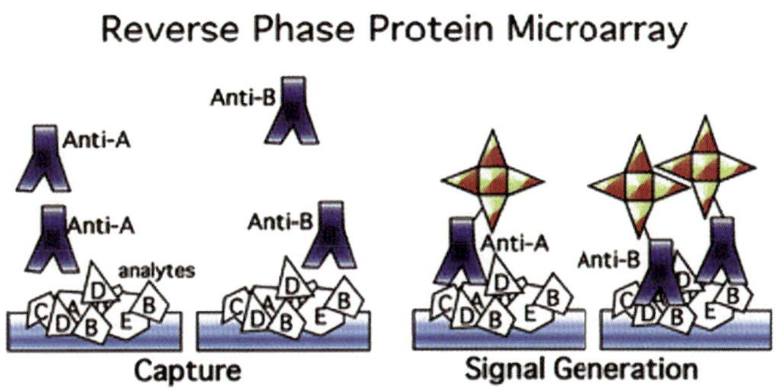

Chapter 166 Figure 4 Forward and reverse phase arrays: schematic. Forward phase arrays (*top*) immobilize a bait molecule such as an antibody designed to capture specific analytes with a mixture of test sample proteins. The bound analytes are detected by a second sandwich antibody, or by labeling the analyte directly (*upper right*). Reverse phase arrays immobilize the test sample analytes on the solid phase. An analyte-specific ligand (e.g., antibody, *lower left*) is applied in the solution phase. Bound antibodies are detected by secondary tagging and signal amplification (*lower right*).

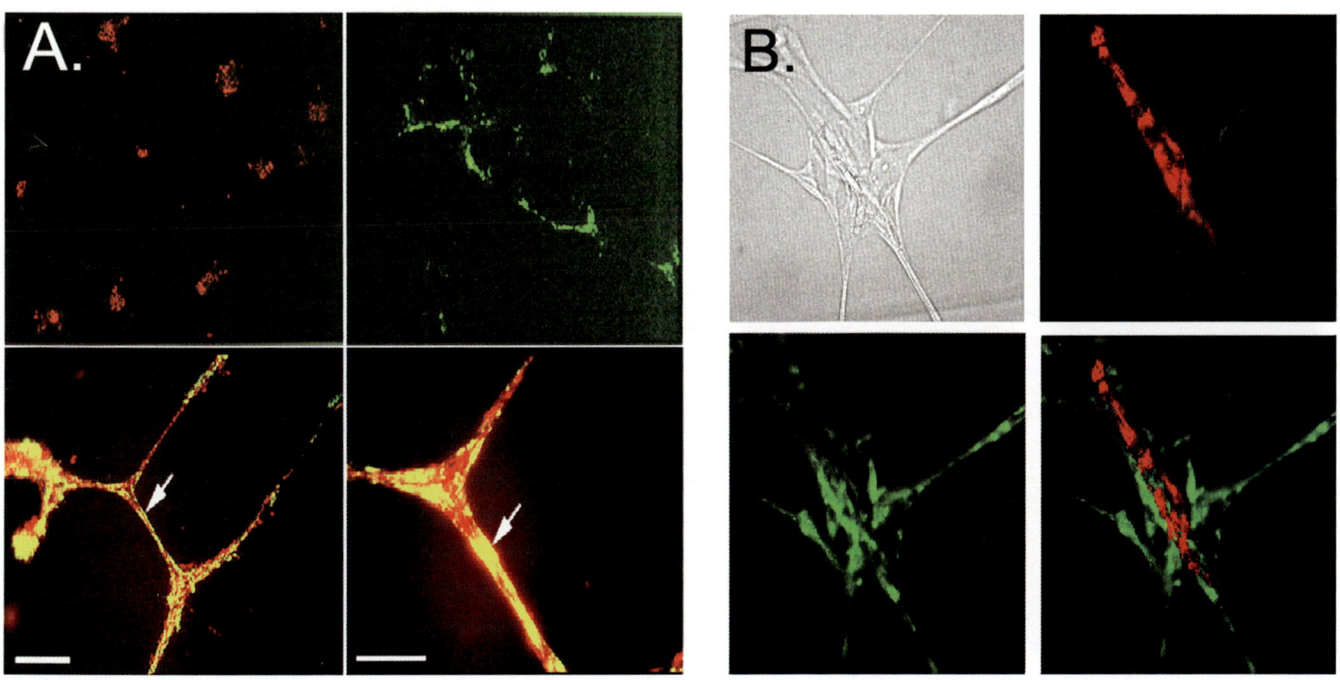

Chapter 167 Figure 4 Vascular development from adult differentiated endothelial cells and mural pericytes in coculture. (**A**) Endothelial cells are stained green and the pericytes cells red; overlap is evident by yellow areas, expanded in the lower right panel. Endothelial cells and pericytes within the developed three-dimensional structure are shown isolated in the upper left and right panels, respectively. Lower panels show how the two cell types interact during coculture. From Darland, D. C., and D'Amore, P. A. (2001). TGFβ is required for the formation of capillary = like structures in three-dimensional cocultures of 10T1/2 and endothelial cells. *Angiogenesis* **4**, 11–20. (**B**) Coculture of pulmonary microvascular endothelial cells (red) and smooth muscle cells (green) on a defined angiogenic support. Note that under the conditions used, the red endothelial cells accumulate firmly within the mural covering. The photograph at the upper right shows a phase contrast image of the structure. This image from our laboratory depicts the model we currently employ to assess vascular maturation.